THE COLLECTED WORKS OF ALONZO CHURCH

Alonzo Church on August 29, 1934,
at about the time of his work on
the decision problem and Church's Thesis.

THE COLLECTED WORKS OF ALONZO CHURCH

Tyler Burge
Chief Editor

Herbert B. Enderton
Editor

The MIT Press
Cambridge, Massachusetts London, England

ASL

Association for Symbolic Logic

This book was typeset in the ASL proprietary Monotype Times Roman fonts using a version of AMS-LaTeX by Cathy Dohm Young. Printed and bound in the United States of America.

Library of Congress Cataloging-in-Publication Data is available.

ISBN: 978-0-262-02564-5

10 9 8 7 6 5 4 3 2 1

CONTENTS

PREFACE AND EDITOR'S NOTE

This collection presents all of Alonzo Church's published articles, his monograph ***The Calculi of Lambda-Conversion***, and the Introduction to his famous textbook ***Introduction to Mathematical Logic***, Volume I. The collection omits some short published comments in conference proceedings that either were judged of slight general interest or could not be located. It omits some lecture notes that were mimeographed or otherwise "informally" published, but were judged to be superceded by later work. It does not reprint all of Church's textbook, which is of course available elsewhere. Otherwise the collection is complete as regards Church's self-standing published work. The collection includes a large selection of Church's reviews—essentially those that were judged to be of greatest independent intellectual value. It includes a substantial amount of previously unpublished work—some short papers, some lectures, and two chapters from the projected second volume of ***Introduction to Mathematical Logic***. Some of the other chapters that were projected to be in this never-completed work were published by Church as articles—and are reprinted here. There is also a selection from Church's correspondence. Letters that formed coherent and interesting groups were chosen for this collection. Some of the correspondence with Gödel is published in ***Kurt Gödel, Collected Works***, Volume IV: ***Correspondence A–G***, Solomon Feferman et al. editors. That correspondence is not reprinted here. No attempt has been made to be complete in the selection of unpublished work or in the selection of correspondence. Unpublished writing of obvious interest that was found in Church's *Nachlass*, or elsewhere, was chosen for inclusion. A complete investigation of the *Nachlass* is a separate project. Further delay of the present publication to allow for such investigation was deemed unwise.

Originals of previously unpublished materials by Church that are collected in this volume, for whose publication the Church estate gave permission, and further unpublished material in Church's *Nachlass*, are part of the Alonzo Church Papers, Manuscripts Division, Department of Rare Books and Special Collections, Princeton University Library. Material from the Church archives that is deemed of sufficient interest will be posted, in its original format, on the Web site of the MIT Press, http://mitpress.mit.edu. Permission to reprint the photographs was granted by the Church family, with one exception. The rights holders for the photograph of the 1946 Mathematics Conference at Princeton University could not be reached. More generally on the individual papers, every effort has been made to contact those who hold the rights. Any rights holders not credited should contact the publisher so that a correction can be made in the next printing.

Church's work has been edited here on the principle that it is to speak for itself. Commentary on the work, except for the brief Introduction, is left for other venues. Editorial notes have been kept to a minimum—mainly explaining the provenance of the papers. These notes and the picture captions are the responsibility of the Chief Editor. He benefited from technical advice on the former from Tony Anderson, Herbert Enderton, David Kaplan, Tony Martin, and Charles Parsons, and historical advice on the latter from John Addison, Mary Ann Addison, Alonzo Church Jr., and Mildred Church Dandridge.

The presentation is historical in two senses. First, works are presented chronologically. The reviews in ***The Journal of Symbolic Logic*** (as distinguished from invited review articles), the letters, and unpublished work that is incomplete or not definitively dated by Church himself, or that stems from a variety of periods, are exceptions. All of these are placed after the other material. Second, the articles have been left in their original form, with relatively little attempt to regiment their format. The idea has been to retain some of the "look" of the originals.

There are several departures from this latter practice, which will now be noted. Obvious typographical errors in the originals, of which there were very few—a tribute to Church's meticulous proofreading and to his editorial skills—are corrected without comment. In some cases, typographical errors that Church himself later caught, and entered into his own reprints of his articles, have been entered into the texts, without editorial comment. Occasionally, putative but not obvious typographical errors have been corrected, but noted. In several cases, Church wanted to make major additions or changes from the original publications, for this collection. Usually these changes are modernizations of notation, but sometimes they involve retrospective comment. These have been entered and noted.

In the *texts* of the articles, the method of citation of titles—whether this be by font style or by use of quotation marks—has been left unaltered. In all *footnotes* of the articles, fonts in citations of titles have been regimented, according to the following practice: titles of articles are cited in regular italics; titles of books, monographs, proceedings, and journals are cited in bold italics. Capitalization and abbreviation practice, and other practices of citation—in both texts and footnotes in the original articles—have been allowed to vary with the style of the original article. Where bibliographies occur at the ends of articles as originally printed, the bibliographies are left in the same form as in the originals.

The editorial notes and the two bibliographies at the end of this collection have, of course, a regimented style: Again, titles of articles are cited in regular italics; titles of books, monographs, proceedings, and journals are cited in bold italics. In the editorial notes and the two bibliographies, major words in titles are capitalized, with one exception: In editorial notes in the chapter Reviews from ***The Journal of Symbolic Logic***, the practice of minimal capitalization has been retained, as a respectful and whimsical nod to Church's practice in the reviews section of ***The Journal of Symbolic Logic***. The bibliography, at the end of the volume, of items cited by Church follows Church's most common citation style—omitting names of publishers of books.

Fonts of titles of the individual articles at the beginnings of "chapters" in this collection were chosen to conform to the practice of ***The Journal of Symbolic Logic***—except that font size is here slightly larger. Church surely formulated the ***Journal's*** practice.

In printing symbols, variations in the way a given symbol was printed—for example, variations in the way an epsilon was printed—have not been retained. Here it was thought that the variations depended on printer variations or technological variations. Where there are differences in the symbols themselves—for example, between epsilon and varepsilon—differences have, of course, been retained. Another example of the principle that differences in symbols themselves are to be retained is the following:

Church used both centered dots and base-aligned dots for ellipsis. Although base-aligned dots are much more common nowadays, it was decided that since Church used centered dots sometimes in his own handwriting, these two forms of ellipsis should be considered different symbols; and differences in the original publications are respected, even though the variations may seem odd to the modern eye.

This project began in the late 1970s. Church participated in it sporadically until his death in 1995. It was originally envisioned as issuing in a collection that would be published during his lifetime. Some of the delay in publication is attributable to the fact that Church wished to enter corrections and comments on the original articles before they were republished, but he did not want to take time from his original research activity to carry out this wish. Sometime in the late 1980s, Church acknowledged the pragmatically self-defeating character of his position, and said with his deep chuckle that carrying through the project would have to be left until after his death. Even so, Church continued until his death to make contributions specifically to the project. Further delay well beyond 1995 is attributable to the immense complexity of the typesetting, proofreading, entering of corrections, implementation of editorial decisions, unification of styles, and so on—and to human weakness of various sorts. No fewer than nine typesetters contributed to the end product, which was presented in completely edited, camera-ready form to the publisher. The excellent, final, unifying typesetting was carried through by Cathy Dohm Young, whose skill and commitment were exemplary, and beyond adequate praise.

The funding for the project came from two sources, beyond personal funds. A large grant from the Association for Symbolic Logic and a smaller grant from the Flint Fund at UCLA, both in the early 1990s, were indispensable, and are hereby gratefully acknowledged. Many people contributed without being financially remunerated, and no one who was thus remunerated was compensated adequately. Church's example of meticulous work, motivated by love of the subject, served as a source of inspiration to all who participated. In his case and ours, love's labour will not be lost.

Acknowledgment with gratitude goes to the following individuals for their contributions: Alonzo Addison, John Addison, Mary Ann Addison, Tony Anderson, Paul Benacerraf, Steve Blackman, Amy Brand, Daniel Burge, Dorli Burge, Johannes Burge, Lyman Chaffee, Alonzo Church, Alonzo Church Jr., Joshua Cohen, Brian Copenhaver, Mildred Church Dandridge, Martin Davis, Katherine Dunlop, Ellen Faran, Solomon Feferman, Julian Fischer, Valerie Geary, Warren Goldfarb, James Hardy, Gilbert Harman, Patricia Jasper, Chuck Hurwitz, David Kaplan, Jonathan Lear, Mary MacDonald, Joe Mandel, Tony Martin, Colin McLarty, Yiannis Moschovakis, Kaoru Mulvihill, Neil Nelson, Leo Novel, Charles Parsons, Thomas Polacik, Robert Prior, Hartley Rogers, Nathan Salmon, Wilfried Sieg, Robert Stalnaker, Brian Walter, Diane Wells, Betty Wilson, Cathy Young, Michael Zelëny, and Erin Zhou. I am grateful to the following publishers for waiving or significantly reducing their fee for reprinting articles by Church for which they hold copyright: The American Mathematical Society, The Association for Symbolic Logic, The Encyclopedia Brittanica, The Journal of Philosophy, Stanford University Press, University of Notre Dame. They did so in honor of Church and his contributions to logic and philosophy.

—T.B.

INTRODUCTION

Alonzo Church: Life and Work

Life is not the same as *work*, but in the case of Alonzo Church the connection is tight. His name is preserved in "Church's Thesis," referring to the identification of effective calculability with a precisely defined notion, and in "Church's Theorem," referring to the undecidability of the concepts of validity and of satisfiability in first-order logic. And ω_1^{CK} is an ordinal number (the Church–Kleene ordinal) associated with his name.

Church's work in symbolic logic spans a wide range, both in time and in subject matter. His first published paper, *Uniqueness of the Lorentz Transformation*, appeared in 1924. The following year *On Irredundant Sets of Postulates* appeared. Seven decades later, his paper *A Theory of the Meaning of Names* was published in 1995. This amazing span of seventy-two years embraces a remarkable collection of publications.

The nature of Church's contributions to logic will be discussed below, arranged under the three headings (i) calculability, (ii) set theory and foundations, and (iii) philosophy and intensional logic. The subjects Church chose to work on were not selected at random. (Indeed, one suspects that very few of Church's decisions involved any elements of randomness.) He was guided by a sense of what the field needed. Being a person with great determination—even ambition—he then resolved to get at the heart of the important problems. For example, his extensive work (especially from 1950 on) in intensional logic seems to have stemmed from a feeling that following symbolic logic's success in explicating extensional logic, the field of philosophy stood in need of an analogous analysis of intensional logic. As he wrote in 1951, "intensional logic also must ultimately receive treatment by the logistic method."

Church's commitment to the needs of the young field of symbolic logic is also evident in his deep involvement with ***The Journal of Symbolic Logic***. Church was influential in the founding in 1935 of the Association for Symbolic Logic, the publisher of JSL. From the first volume (1936), he was an editor and also edited the Reviews Section of JSL.

The field needed a precise and sufficiently comprehensive textbook for undergraduate students. By 1936, mimeographed notes, *Mathematical Logic*, had been prepared by Church and his students. In 1944, ***Introduction to Mathematical Logic***, Part I, was published in the Annals of Mathematical Studies series, based in part on notes by C. A. Truesdell on Church's 1943 lectures. Finally in 1956, the greatly revised and enlarged textbook, ***Introduction to Mathematical Logic***, Volume I, was published by Princeton University Press. This was the book that defined the subject for a generation of logicians. It remains in print after more than fifty years, and has set a high standard of rigor and precision.

Page x of this textbook gives a "tentative table of contents for volume two," listing Chapters VI–XII (to follow the Chapters I–V in Volume I). While Volume II was never published, material for Chapter VII (*Second Order Arithmetic*) was mimeographed for class use, and is printed in these ***Collected Works*** (see *Chapter VII. The Logistic System A^2*). Material for Chapter VIII (*Gödel's Incompleteness Theorems*) is also included in these ***Collected Works*** (see *Partial Outline of Chapter VIII*), based in part on notes written by Church and in part on student notes.

The topics for some of the other chapters were addressed in subsequent research papers. In connection with Chapter VI (*Functional Calculi of Higher Order*) see

Church's 1972 paper, *Axioms for Functional Calculi of Higher Order*. In connection with Chapter IX (*Recursive Arithmetic*) see his 1957 paper, *Binary Recursive Arithmetic* (and his 1965 paper, *An Independence Question in Recursive Arithmetic*).

Chapter X was to be *An Alternative Formulation of the Simple Theory of Types*. See his 1974 paper, *Russellian Simple Type Theory*. Chapter XI was to be *Axiomatic Set Theory*. While it is uncertain what was to be included in this chapter, certainly the topic was represented in such papers as his 1974 article *Set Theory with a Universal Set*. Finally, Chapter XII was to be *Mathematical Intuitionism*. Here it is even less certain what was to be included.

Church's influence on the field of symbolic logic remains strong today. Thirty-one students received doctorates under his supervision, from 1931 to 1985. In addition, he taught many more students, either in person or through his textbook. And his papers continue to form the basis for further research. The purpose of these ***Collected Works*** is to facilitate the continued study of Church's writings.

Life. Alonzo Church was born on June 14 (Flag Day), 1903, in Washington, D.C. His great-grandfather (also named Alonzo Church) had moved from Vermont to Georgia, where he was a professor of mathematics and astronomy—and then president —of Franklin College, which later became the University of Georgia. His grandfather, Alonzo Webster Church, was at one time Librarian of the U.S. Senate. His father, Samuel Robbins Church, was a Justice of the Municipal Court of the District of Columbia, until failing vision and hearing compelled him to give up that post. The family then moved to rural Virginia, where Alonzo Church and his younger brother grew up.

Alonzo Church had an uncle (also named Alonzo Church) living in Newark, New Jersey, who was financially helpful to the family, taking an interest in the children's education. An airgun incident in high school left Church blind in one eye. He attended the Ridgefield School, a preparatory school in Connecticut, graduating in 1920.

After graduating from Ridgefield School, Church enrolled at Princeton, where his uncles had attended college. For a while he worked part-time in the dining hall to help pay his way. He was an exceptional student; his first published paper, *Uniqueness of the Lorentz Transformation*, was written while he was an undergraduate. Its object was to obtain a set of logically independent postulates that uniquely determine the Lorentz transformation in one dimension. He graduated with an A.B. in mathematics, 1924. He then continued graduate work at Princeton, completing a Ph.D. in three years (in 1927) under Oswald Veblen. The title of his dissertation was ***Alternatives to Zermelo's Assumption***.

While a graduate student, he married (1925) Mary Julia Kuczinski, who was training to be a nurse. (This in spite of the fact that his senior class had voted him the "most likely to remain a bachelor"—his handsome features notwithstanding. In the summer of 1924, Church stepped off a curb and was hit by a trolley car coming from his blind side; Mary was a nurse-trainee at the hospital.) They were inseparable for the next 51 years, until Mary's death in 1976. Mary was an excellent cook; over the years many a mathematician enjoyed dining at the Church home.

On receiving his degree, he was awarded a two-year National Research Fellowship. So after serving briefly as an Instructor at the University of Chicago in the summer of 1927, he spent two years visiting first Harvard (1927–28) and then Göttingen and

Amsterdam (1928–29). In particular, he met with Bernays and with Heyting on this trip.

At the end of his term as a National Research Fellow, Church was invited to return to the Princeton Mathematics Department, to begin his academic career there. He was an Assistant Professor 1929–39, an Associate Professor 1939–47, and a Professor 1947–67 (of Mathematics and Philosophy, 1961–67).

Meanwhile, he and Mary had three children. Alonzo Church, Jr., was born in 1929 (in Amsterdam), Mary Ann in 1933, and Mildred in 1938. (Mary Ann later married the logician John Addison.) While Church was developing what came to be known as Church's Thesis, there were two small children in the house. Somehow he balanced Sunday afternoon family outings with the demands of his work.

Early in his career, Church put together *A Bibliography of Symbolic Logic*—nothing less than a complete annotated bibliography of every publication in symbolic logic up to that time. Thus it would be an understatement to say that he had a complete familiarity with the literature: he had read, organized, and indexed (by authors and subject) that literature. Of course, the literature was in many languages. Church not only had a wide knowledge of modern languages; he also had studied Latin and Greek.

Princeton in the 1930s was an exciting place for logic. There was Church together with his students Stephen Kleene and Barkley Rosser. John von Neumann was there. Alan Turing, who had been thinking about the concept of effective calculability, came as a visiting graduate student in 1936. He ended up writing his dissertation under Church. Kurt Gödel visited the Institute for Advanced Study in 1933–34 (when he lectured on his then-recent incompleteness theorem) and 1935, before moving there permanently. Since Church had recently been to Göttingen and Amsterdam, he knew personally almost every logician.

As mentioned above, Church was one of the principal founders of the Association for Symbolic Logic, in 1935. He made certain that ***The Journal of Symbolic Logic*** got off to a good start; he was an editor for contributed papers for its first 15 volumes (1936–50). (The other editor of the early volumes was C. H. Langford.) More importantly, for most of his career Church was its editor for reviews, a task he performed for its first 44 volumes (1936–79).

Church had two goals for the Reviews Section. First, it continued his *A Bibliography of Symbolic Logic* (which was published in the first volume of JSL) in providing a complete and accurate bibliographical record of publications in symbolic logic. And secondly, it provided critical analytical commentary on these publications. In a 1950 letter to Rudolf Carnap, Church explained the importance he attached to this commentary: "The situation in mathematical logic has always been much worse than in other branches of mathematics, in the matter of the frequency of publication of matter containing errors and even absurdities, with the effect that there is serious danger of the field itself falling into disrepute. ... I believe that the Journal has made already considerable progress, not only in reducing somewhat the proportion of erroneous or purposeless publications, but more important, in making it possible for the general body of mathematicians and philosophers to have at least some approximate idea as to how the publications in the field are to be sorted out and which ones are worth giving attention to." The reviews were needed to defend the field and to separate wheat from chaff.

And a motivation for Church's 1967 departure from Princeton after so many years was that after his retirement (which would have been mandatory in 1971 at age 68),

Princeton would be unwilling to accommodate the small staff working on the JSL reviews, while UCLA promised to support the reviews office as long as Church was its editor.[1]

Thus Church left Princeton in 1967, and moved to UCLA, where he was Flint Professor of Philosophy and Mathematics, 1967–90.

Finally in 1979, Church retired from editing JSL reviews, after 44 years. He retired from teaching at UCLA in 1990, at age 87, ending a teaching career that had started 63 years before. But his research continued until he died in 1995, at the age of 92.

Throughout his career, Church supervised research of graduate students. Altogether, he had 31 doctoral students. The first 28 were at Princeton, 1931–67; the last three were at UCLA, 1976–85. They form a distinguished list (see *Doctoral Dissertation Students of Alonzo Church* in these ***Collected Works***).

Alonzo Church had the polite manners of a gentleman who had grown up in Virginia. He was never known to be rude, even with people with whom he had strong disagreements. A deeply religious person, he was a lifelong member of the Presbyterian church.

In his habits he was careful and deliberate—very careful and very deliberate. The students in his classes would discover this on the first day, when they saw how he would erase a blackboard. The material he wrote out on paper (he did not type) was often done in several colors of ink—sometimes colors made by mixing bottles together—and always done in his distinctive unslanted handwriting. He was a master at using white-out fluid to eliminate imperfections. Finally, an important piece of writing could be made permanent by coating it with a thin layer of Duco cement. (Duco was the one brand of household cement that did not shrink the paper.)

Church enjoyed reading science fiction magazines. He did not like to see writers get their facts wrong; he was known to have written several letters to the editors.

He preferred a nocturnal schedule, working late at night, when it was quiet and he would not be disturbed. His staff at the JSL reviews office would leave material on his desk in the afternoon. On arrival in the morning, they would find the replies from Church.

He never drove a car. But he would walk substantial distances, in varying weather and at all hours. Many a student crossing the campus at night would see a well-dressed portly gentleman with white hair, carrying a briefcase and humming softly to himself.

While living in Princeton in the 1960s, Church and his family liked to visit the Bahamas. Eventually Church bought property there and built two duplexes. Even after moving to Los Angeles, Church would spend summers at his place in the Bahamas. He did not spend his time lying on the beach, but it was both a place to escape to, and a place where the children and grandchildren could gather.

Although Church had solitary work habits, that is not to say that he lived in isolation from his colleagues. He attended and spoke at professional meetings. As time allowed, he answered his mail. He maintained an active correspondence both with those whose viewpoints were close to his (such as Kleene and Rosser) and with those whose viewpoints were not (such as Carnap and Quine).[2]

---

[1]See H. B. Enderton, *Alonzo Church and the Reviews*, ***The Bulletin of Symbolic Logic***, vol. 4 (1998), pp. 172–180.

[2]See also María Manzano, *Alonzo Church: His Life, His Work and Some of His Miracles*, ***History and Philosophy of Logic***, vol. 18 (1997), pp. 211–232.

Church belonged to a large number of professional organizations, including the American Academy of Arts and Sciences, the American Association for the Advancement of Science, the American Mathematical Society, the American Philosophical Association, the Association for Symbolic Logic, Circolo Matematico di Palermo, and Académie Internationale de Philosophie des Sciences.

In 1966 he was elected a Corresponding Fellow of the British Academy. And in 1978 he was elected to the National Academy of Sciences. He received honorary D.Sc. degrees from Case Western Reserve (1969), Princeton (1985), and SUNY at Buffalo (1990).

Church died in Hudson, Ohio (where his son lived), on August 11, 1995. He was buried in the Princeton Cemetery, where his wife and his parents had been buried. The papers he left behind were donated to the Princeton University Library.

Work in calculability. Church focused on the concept of effective calculability at the latest by 1934, the year in which both Stephen Kleene and Barkley Rosser received their Ph.D.s under Church's direction. The concept of effective calculability arises in problems throughout mathematics (e.g., given two simplicial complexes, can we effectively decide whether or not they are homeomorphic?). But it is central to the logical notion of an *acceptable proof*. A proof must be *verifiable*, that is, there must in principle be some effective procedure that, given an alleged proof, will verify its syntactic correctness. But exactly what is an effective procedure?

In 1936 a pair of papers by Church changed the course of logic. *An Unsolvable Problem of Elementary Number Theory* presents a definition and a theorem: "The purpose of the present paper is to propose a definition of effective calculability which is thought to correspond satisfactorily to the somewhat vague intuitive notion . . . , and to show, by means of an example, that not every problem of this class is solvable." The "definition" now goes by the name Church's Thesis: "We now define the notion . . . of an *effectively calculable* function of positive integers by identifying it with the notion of a recursive function of positive integers (or of a λ-definable function of positive integers)." (The name, "Church's Thesis," was introduced by Kleene.)

The theorem in the paper is that there is a set that can be defined in the language of elementary number theory (viz., the set of Gödel numbers of formulas of the λ-calculus having a normal form) that is not recursive—although it is recursively enumerable. Thus truth in elementary number theory is an effectively unsolvable problem.

A sentence at the end of the paper adds the consequence that if the system of ***Principia Mathematica*** is ω-consistent, then its decision problem is unsolvable. It also follows that the system (if ω-consistent) is incomplete, but of course Gödel had shown that in 1931.

As indicated above, the paper identifies the effective calculability of a function with two equivalent precisely defined concepts: One is *recursiveness*, which here means that a set of recursion equations exists from which can be derived exactly the correct values of the function (a concept Gödel formulated in his 1934 lectures at Princeton, crediting in part a suggestion by Jacques Herbrand). The other is *λ-definability*, meaning that for a suitable formula of the λ-calculus, exactly the correct values of the function are derivable. (In a footnote, Church writes, "The question of the relationship between effective calculability and recursiveness . . . was raised by Gödel in conversation with the author. The corresponding question of the relationship between effective calculability and λ-definability had previously been proposed by the author independently.")[3]

Church had been working on the λ-calculus in connection with his two-part *A Set of Postulates for the Foundations of Logic* (1932 and 1933), where the intent was to develop axioms that "would lead to a system of mathematical logic free of some of the complications entailed by Bertrand Russell's theory of types, and would at the same time avoid the well known paradoxes, in particular the Russell paradox." As it turned out, the future of the λ-calculus lay not in that direction (it turned out that contradictions had not been avoided), but in computer science. Church's 1941 monograph ***The Calculi of Lambda-Conversion*** was later useful to others in the development of semantics for programming languages.[4] Today the λ-calculus is a major research topic in theoretical computer science.

One person who read Church's paper with great interest—not to say dismay—was Alan Turing. Turing at that time was a student at Cambridge, working with Max Newman. Turing had independently formulated an exact mathematical concept to make rigorous the informal concept of effective calculability. Turing's concept was equivalent to Church's, as Turing showed in an appendix to his paper. But it was formulated in very different terms, involving a step-by-step simulation—by an imaginary machine—of a calculational procedure a person might carry out. (The phrase "Turing machine" first appeared in Church's review of Turing's paper in ***The Journal of Symbolic Logic***.) Thus Turing's definition stood in contrast to the formulations in terms of formal logical calculi.

Turing wrote to Church, and he came to Princeton as a visiting graduate student. At the encouragement of John von Neumann, Turing stayed a second year, eventually completing his Ph.D. degree under Church. Turing's dissertation involved transfinite extensions of logical systems, a topic taken up much later by Solomon Feferman.

It is an interesting fact that the idea of formalizing the concept of effective calculability occurred at roughly the same time to several people: Gödel (who mentions the problem in his 1934 Princeton lectures), Church, Turing, and Emil Post (who independently developed a formulation somewhat along the lines Turing used).

The name for the class of effectively computable functions (as defined by any one of these equivalent definitions) came to be "recursive functions." The name made sense from the viewpoint of the Gödel–Herbrand formulation. (Gödel had used the term "rekursiv" in his 1931 paper for what are now called the primitive recursive functions. Kleene applied the name "general recursive functions" to the functions obtained from the primitive recursive functions by the least-zero search operator.) But the other equivalent formulations did not involve *recursion* in an essential way. So the name of the class of functions (and of the subject of "recursive function theory" or "recursion theory") was something of a misnomer. The adjective Church used was "calculable"; the adjective Turing used was "computable." Recently there has been a movement to change the name of the subject to "computability theory."

Another 1936 paper that changed the course of logic was *A Note on the Entschiedungsproblem*. This short paper (two pages, followed by a two-page correction) presents what is now called Church's Theorem: The problem of deciding validity of formulas

---

[3]See Wilfried Sieg, *Step by Recursive Step: Church's Analysis of Effective Calculability*, ***The Bulletin of Symbolic Logic***, vol. 3 (1997), pp. 154–180.

[4]See Henk Barendregt, *The Impact of the Lambda Calculus in Logic and Computer Science*, ***The Bulletin of Symbolic Logic***, vol. 3 (1997), pp. 181–215.

in first-order logic is unsolvable. The method of proof is first to make a finitely axiomatizable theory (in an expanded language) in which a certain function enumerating a recursively enumerable, but non-recursive, set can be represented. And then one can eliminate the added function symbols.

Moreover, Church published two other papers in 1936, together with his former students. *Some Properties of Conversion*, with Barkley Rosser, dealt with the λ-calculus. *Formal Definitions in the Theory of Ordinal Number*, with Stephen Kleene, was followed by Church's 1938 paper, *The Constructive Second Number Class*. It is here that the least non-constructive ordinal, now called ω_1^{CK}, arises.

Work in set theory and foundations. Church's dissertation, *Alternatives to Zermelo's Assumption* (published in 1927), already displayed a broad-minded (and even skeptical) attitude toward set theory. "Zermelo's assumption" is of course the axiom of choice, and Church's attitude was not to regard the axiom of choice as received doctrine, but instead was to examine what array of other set theories might serve as alternatives for the foundations of mathematics.

A similar spirit can be seen in his much later paper *Set Theory with a Universal Set* (presented at a symposium honoring Alfred Tarski in 1971 and published in 1974). In this instance it was the so-called principle of limitation of size that was the issue. This paper presents a system of set theory in which every set has a complement, and proves its consistency relative to ZF. As the paper says, "there is room for exploration of the axiomatic possibilities." And it concludes with the intriguing statement: "Indeed an interesting possibility which must not at this stage be excluded is a synthesis or partial synthesis of ZF set theory and Quine set theory."

It is not surprising that Church found Quine's system New Foundations (NF) interesting. While many mathematicians—through fondness for the principle of limitation of size or for universal choice—found NF unappealing, Church was more than willing to take it seriously. The key idea of avoiding the paradoxes by restricting comprehension to stratified formulas fit well into the Frege–Russell tradition. What was lacking was a relative consistency proof.

Church devoted a substantial effort, especially in the 1970s, toward supplying a consistency result for NF. As he said, his "purpose is not specially to advocate the Quine set theory but to explore the possibilities in regard to non-Neumannian set theories." In the end, the effort was unsuccessful. He left in his *Nachlass* several three-ring binders presenting his work—describing what he had tried, what worked, what did not work. The work was left in well-organized form, written out in several colors of ink. These notebooks, with his other papers, went to the Princeton University Library. Future graduate students may yet mine this material for dissertations.

In connection with set theory, it might also be mentioned that in his 1953 paper, *Non-normal Truth-tables for the Propositional Calculus*, where Church considers replacing the set $\{0, 1\}$ of two truth values by larger Boolean algebras, he wrote: "Another motive is the suggestion, which was made to me by Paco Lagerström ten years or more ago, that use may be made [of the method] in order to extend to the functional calculi of first and higher order, and other related systems, the method of proving independence of axioms." A decade later, Boolean-valued models of set theory were used to present Paul Cohen's results on the relative independence of the axiom of choice in Zermelo–Fraenkel set theory.

—H.B.E.

Philosophy and intensional logic.[5] In a letter to the UCLA Philosophy Department, December 17, 1969, Alfred Tarski cites some of Church's articles from the 1930s, his paper *The Need for Abstract Entities in Semantic Analysis*, and his textbook, as important published contributions. Tarski goes on to mention some of Church's eminent students and his editorship of ***The Journal of Symbolic Logic***. He concludes with the memorable declaration, "Hardly any other cases are known in which a branch of science is so much indebted for its powerful expansion and splendid development to a single man."

It is noteworthy that Tarski makes his case for Church's heroic effect on the science of mathematical logic in a letter to a philosophy department. Following the examples set by Frege and Russell, Church took developments in mathematical logic as a basis for philosophical reflection, and approached philosophy as a subject continuous with parts of mathematical science.

Church's contributions to philosophy, though extremely influential, have not been *as* influential—nor are they as wide-ranging—as those of his contemporaries, Quine and Carnap. However, Church's work has a depth and integrity that promise long life. One sees in it, and in his correspondence, a clarity and power of mind that are certainly not surpassed, and possibly not equaled, by any of his contemporaries, with the lone exception of Gödel.

Fundamental to Church's work in philosophy is a pragmatic methodology grounded in his practice as a logician: formulate logistic systems and reflect on the insights they yield into the inferential structures governing the subject matter at hand. Church's method presupposes a rationalism that downplays the centrality of immediate insight. He allows that some axioms are self-evident, but holds that many axioms have to be accepted provisionally (*Mathematics and Logic*, 1962). Warrant for a system lies partly in the hypothetical-deductive testing of the system against a wide variety of intuitive and systemic considerations. It is this method that led Church not to support any of the standard approaches to the philosophy of mathematics in his day—formalism, intuitionism, logicism. In conversation, he said, with a grin, that all the approaches had made contributions. Church's insistence on this method is constant throughout his career. It is commonly directed against philosophers working in the philosophy of language or philosophy of logic whose claims require evaluation against formalizations that those philosophers never present. Church's insistence is not on its face very different from similar enunciations of method by Carnap and Quine. What distinguishes Church's methodology is the purity of his pragmatism in adhering to it. The example that he sets in the use of this methodology is one of his finest contributions to philosophy. There is in his work a refreshing, steady resistance to many of the main philosophical trends that marked the period of his career and that have now subsided into period pieces.

Church never advocated logical positivism or any other sort of empiricism. He praises attempts to formulate the Verificationist Theory of Meaning in a way clear enough to evaluate it, but presents important criticisms of such attempts. (See the Review of Ayer, 1949.)

---

[5] For a fuller account of Church's philosophical contributions, see C. Anthony Anderson, *Alonzo Church's Contributions to Philosophy and Intensional Logic*, ***The Bulletin of Symbolic Logic***, vol. 4 (1998), pp. 129–171. The present survey is indebted to this article in various ways.

Early in his career (see Review of ***Principia***, 1928; *A Set of Postulates for the Foundation of Logic*, 1932; and Review of Chwistek, 1937) Church believed that mathematical entities are fictions. This view was replaced in his mature philosophical work by a steady ontological realism, very probably encouraged by his exposure to Frege. This is not a realism based on Platonic intuition. It is rather the relaxed realism about abstract entities, backed by a pragmatic epistemology, that is an articulation of the natural realist position in mathematics and logic. It is closely associated with reflection on the semantical commitments of successful formalizations (see *Intensional Semantics* (*The Need for Abstract Entities in Semantic Analysis*), 1951).

Church provides powerful criticism of nominalism in the philosophy of logic. He criticizes the extreme nominalism, proposed by Goodman and Quine, pursued by Scheffler, and later abandoned by Quine, that attempts to do without abstract entities altogether. (See the sober *Ontological Commitment*, 1958; the witty *Misogyny and Ontological Commitment*, 1958; and the formidable *On Scheffler's Approach to Indirect Quotation*, 1973.)

Church is famous for the series of papers that attack a more moderate nominalism, defended notably by Carnap. This form of nominalism attempts, in formalizations of reasoning about belief and knowledge, to dispense with "intensional entities," such as propositions, in favor of linguistic entities, such as sentences. Church's papers on this issue include *On Carnap's Analysis of Statements of Assertion and Belief*, 1950; *Intensional Isomorphism and Identity of Belief*, 1954; and *Propositions and Sentences*, 1956. The discussion of this issue occupies much of the correspondence with Carnap, which forms a background for Church's 1950 paper. Church's main arguments turn on attempts to formulate principles that enable one to treat reasoning about belief, in ordinary human interaction or in cognitive psychology, in a systematic and general way. This motive—not some intuitive commitment to abstractions about meaning—lies behind Church's famous translation test. Interestingly, the idea behind this test goes back at least to 1943. (See the correspondence with Quine, July 26, 1943.)

The criticism of nominalism is closely associated with Church's resistance to Quine's ontological extensionalism—the view that formalizations should quantify only over individuals, and extensions or sets. Quine held that although commitment to set-theoretic abstract entities is acceptable in mathematical or natural science, commitment to "intensional" abstract entities like propositions or meanings is not acceptable in any science. His position began with the view that postulation of "intensional entities" is unacceptably unclear and requires analysis or replacement in other terms. Ultimately, it is this view that expanded into an elaborate theory about the indeterminacy and linguistic relativity of any notion of meaning.

Church acknowledged that clarification of identity conditions for "intensional entities" is more difficult than for sets. In his characteristic way, he set out several logistic systems each of which incorporates a different alternative. He argued that a wholesale attempt to avoid postulating such entities was misguided. He found formalizations of reasoning about belief, meaning, and modality unacceptably barren in the absence of commitments to "intensional entities," such as propositions. He was unpersuaded by Quine's arguments that there is a general reason why propositions or meanings are inherently unacceptable. He argues that suitably generalized, Quine's criterion of ontological commitment itself has intensional implications. (See *Ontological*

Commitment and *Misogyny and Ontological Commitment*, both 1958, and the correspondence with Quine.) Church points out the inadequacies of attempts to provide logistic systems for semantics, psychology, and modality that do without "intensional entities"—inadequacies in representing intuitively faultless inferences. (See the papers on ontological commitment, on belief sentences, and on intensional semantics mentioned earlier; also the correspondence with Quine; the Review of Quine, 1943 and Postscript 1968; *A Remark Concerning Quine's Paradox about Modality*, 1982; and *Intensionality and the Paradox of the Name Relation*, 1989.) From the present vantage point, it is plausible that Church's pragmatic position has been supported by ongoing work in semantics, cognitive psychology, and modal logic—fields that have been unmoved by Quine's scepticism about such enterprises and unmoved by his urging the adoption of extensionalist constraints.

Sometime in the late 1930s or early 1940s Church was converted to a Fregean approach to semantics and intensional logic. In his Review of Carnap's ***Introduction to Semantics***, 1943, Church notes his conversion to a Fregean point of view. (He had already discussed the conversion in a talk in 1942.) Preparatory to offering a powerful, and later influential, reconstruction of Frege's implicit argument that all true sentences designate the truth value, truth, Church writes laconically, "On this point the reviewer confesses to have changed his own former opinion, but not without compelling reason." This argument initiated a nearly career-long development of Fregean systems. Church's presentation of the argument was accompanied, from the outset, by a recognition that a Russellian system had a resource for blocking the argument. As applied to languages or theories that deal with their subject matters in a context-free way, a Russellian system could circumvent any formal analogy between truth values and referents of singular terms by invoking Russell's theory of descriptions. Church regarded the Russellian alternative as tenable but inferior, on systematic grounds. It is typical of Church's pragmatism that he retained an interest in Russellian semantics and made several significant contributions to its development. (See, for example, *Comparison of Russell's Resolution of the Semantical Antinomies with that of Tarski*, 1976; *How Far Can Frege's Intensional Logic be Reproduced in Russell's Theory of Descriptions?*, 1979; *Russell's Theory of Identity of Propositions*, 1984.)

The main line of Church's contributions to semantics and intensional logic, on which he grounded much of his philosophical reflection, runs through a Fregean point of view. One can see these contributions as following two primary branches. One concerns a Fregean semantics. The other concerns the logic of the relations directly between sense and denotation, without treating the semantical relations among linguistic terms, senses, and denotations.

The first branch is initiated in *Intensional Semantics* (*The Need for Abstract Entities in Semantical Analysis*), 1951, and in ***Introduction to Mathematical Logic***, 1956. Church develops, with great precision and philosophical sophistication, the details of a Fregean semantical theory. The footnotes in the latter work are widely admired for the richness of their insights. Church modifies Frege's theory by allowing conversions of functional designations of functions into singular designations of the same functions. This innovation, already elaborated logistically in ***The Calculi of Lambda-Conversion***, 1941, in effect corrects Frege's error of insisting that the concept, or function, *horse* is not a concept, or a function. This innovation deepens Frege's very important contributions to the understanding of predication and functional application. Further, Church makes explicit the semantical implications of Frege's notions of sense and denotation; and he elaborates the formal analogy between a sentence's truth value and a term's denotation.

Church aimed at a general semantical theory that is, at its most abstract levels, invariant over languages. To this end, he eschewed, correctly, the then-widespread interpretation of Tarski's work as having defined truth. (Tarski's work presents a method for defining truth—in the sense of giving a mathematically equivalent characterization in non-semantical terms—for given languages, one by one: "truth-in-L," not "truth in any language L." It used to be common to acknowledge this fact, but still to think, mistakenly, that Tarski had defined or eliminated the concept of truth quite generally in favor of non-semantic terms.) Church took semantical notions to be primitive. Church was, of course, concerned with the obvious threat to any such general semantic theory—paradox. His various papers on the Liar, Grelling, and Berry paradoxes—and on more intricate paradoxes deriving from Russell's paradox modified to apply to a totality of propositions—should be seen in the light of his interest in a completely general semantic theory. Although he did not propose a final solution, it is clear that in most of his work Church favored some sort of hierarchical approach.

The second branch of Church's Fregean work concerns intensional logic. It centers more on the ontological than on the specifically semantical aspects of the Fregean approach that he outlined in the 1951 paper and in his textbook. Church develops axioms for the non-linguistic relation *concept of* between the senses of linguistic expressions and the denotations of those expressions. He concentrates on principles for the identity of senses (and of propositions, which are senses of sentences) and on the logic of the *concept-of* relation. This work is presented in a series of papers on the Logic of Sense and Denotation. (See *A Formulation of the Logic of Sense and Denotation*, 1951; *Outline of a Revised Formulation of the Logic of Sense and Denotation*, 1973, 1974; *Revised Formulation of the Logic of Sense and Denotation, Alternative (1)*, 1993.) These papers have their roots in Church's reflection on Fregean theory and on Quine's early treatment of modality. See the Review of Quine, 1943, and Postscript 1968, which produce examples of valid inferences in intensional contexts, and develop dissatisfaction with the standard box-diamond *operator* formalization of sentences, or propositions, about modality.

In these papers on the Logic of Sense and Denotation, Church worked on three logical systems corresponding to three identity criteria for "propositions." Of course, each criterion yields a different conception or abstraction associated with the word 'proposition'. One criterion is that "propositions" are identical if and only if their material equivalence is true on logical grounds alone. This criterion is known as Alternative (2). Church regarded a logic for propositions thus individuated as suitable, or at least convenient, in the study of modality.[6]

Church regarded a second criterion, Alternative (1), not so much as a formulation of an intuitive notion as a preliminary, simplified model for the full-scale development of a logic for propositions, individuated according to the third criterion, Alternative (0).

---

[6]For refinement and further development, see David Kaplan, *How to Russell a Frege-Church*, ***Journal of Philosophy***, vol. 72 (1975), pp. 716–729; Charles Parsons, *Intensional Logic in Extensional Language*, ***The Journal of Symbolic Logic***, vol. 47 (1982), pp. 289–328; and Terry Parsons, *The Logic of Sense and Denotation: Extensions and Applications* in ***Logic, Meaning and Computation: Essays in Memory of Alonzo Church***, C. A. Anderson and M. Zelëny, eds. (Dordrecht, The Netherlands: Kluwer, 2001).

Alternative (0) is aimed at capturing the fundamental conception of proposition, a conception fine-grained enough to allow plausible formalization of all the main types of reasoning that depend on distinguishing propositions. Alternative (0) is, roughly, that propositions expressed by sentences A and B are identical if and only if either sentence can be obtained from the other by structure-preserving exchange of synonyms among non-complex expressions. Roughly, such synonymies are stipulated synonymies of sense given in the construction of a language. (See *Intensional Isomorphism and Identity of Belief*, 1954, for motivation of this criterion.) Stipulated synonymies—abbreviative definitions—are common in formal languages. They may be seen as implicit, at least occasionally, in the practice of natural languages. Church emphasized that applying an idealized notion such as synonymy in concrete cases is subject to vagueness and different standards for application with different natural languages (see the correspondence with Quine). He maintained, however, that notions of sense and proposition are necessary for a general logic capable of formalizing reasoning about belief in ordinary discourse and in cognitive psychology, reasoning about knowledge in epistemology, and reasoning involving discovery through proof in the mathematical sciences.

Unfortunately, initial formulations of both Alternative (1) and Alternative (0) lead to paradox.[7] A logic for Alternative (0) has not yet been successfully formulated and is still being studied along lines set by Church.

It remains to comment on the superb Runes dictionary articles (1959) and ***Encyclopædia Britannica*** articles (1956–1972). These articles offer characterizations of important notions in logic and philosophy of language. Church's articles on logic are classics of exposition. Nearly all these articles seem now as fresh and undated as they were when they were written, and constitute an important pedagogical contribution. It is rare for such a major thinker—for purely pedagogical purposes—to undertake to explain basic notions in such a systematic way. This work is continuous with Church's commitment—also evident in his reviews, his editing work, and his textbook—to help to set standards in mathematical logic and philosophy.

Church's primary original contributions to philosophy lie in his insistence on, and exemplification of, applying the logistic method within philosophy; in his formulation and adherence to pragmatic rationalist epistemology; in his development of a general conception of semantics through a Fregean point of view; in his clarification of functional application and predication through correcting Frege's error about the concept *horse*; in his work on intensional logic; and in his criticism of various forms of nominalism. Much of his work in pure logic or mathematics is of direct philosophical import—most notably his analysis in Church's Thesis of the nature of effective calculability and (implicitly) the nature of human reasoning. The philosophical significance of this work is abundantly evident in the correspondence with Post. Here as elsewhere it is pointless to try to draw a sharp boundary between Church's mathematical and philosophical work. Both are of fundamental importance, and continuing interest.

—T.B.

---

[7]See C. Anthony Anderson, *Some New Axioms for the Logic of Sense and Denotation: Alternative (0)*, ***Noûs*** vol. 14 (1980), pp. 217–234; *Alonzo Church's Contributions to Philosophy and Intentional Logic*, op. cit.; and *Alternative (1*): A Criterion of Identity for Intensional Entities*, in ***Logic, Meaning and Computation: Essays in Memory of Alonzo Church***, op. cit.

Church with his mother, Mildred, and his father, Samuel, 1906.

Three Alonzo Churches, 1910: Alonzo Church, an uncle, Chancellor of New Jersey, who helped put Church through Princeton; Alonzo Church at age seven; and Alonzo Church, a grandfather, last librarian of the U.S. Senate before it became the Library of Congress.

Church's last year in high school.
The handwriting is his.

Church in his early college
years, approximately 1922.

The year of Church's graduation
from college, 1924.

H. H. Mitchell, Church, E. T. Bell, Warren Weaver, Chicago, 1927.
Mitchell was apparently the author of books on biochemistry and nutrition;
Bell wrote ***Men of Mathematics*** and other popular historical books;
Weaver, an Association for Symbolic Logic member,
wrote an article on Lewis Carroll as mathematician, ***Scientific American***, 1956.

Passport photos, Landwoort, Holland, August 14, 1929.
An unconfirmed family story indicates that Church was approached by
Hollywood and firmly rejected the offer. These and other pictures from this
period lend some credence to the story.

Church with his son Alonzo Jr., Princeton 1933 or 1934.

Asbury Park, New Jersey, 1944,
with daughter Mildred.

Tower of London, 1954, with daughter Mary Ann, wife Mary, and guard.

August 27, 1955, at the wedding of daughter Mary Ann.

Lecturing at Princeton, 1955.

International Congress of Logic, Methodology and Philosophy of Science,
Stanford, California, September 1, 1960.
Front row: Adrianne Rogers, Hartley Rogers,
daughter Mildred Church Dandridge, Church, Mary Church.
Back row: Kai Lai Chung, Leon Henkin, John Addison.

Church with grandchildren, Tom and West Addison,
September 4, 1965.

Oriskany, New York, 1965.

Stephen Kleene, Church, Anatoly Ivanovich Malcev,
Moscow, 1966.

J. B. Rosser and Church, Association for Computing Machinery,
August 17, 1982, Pittsburgh, Pennsylvania.

Church receiving from William G. Bowen an honorary degree,
Princeton, 1985.

Niagara Falls, New York, 1990.

December 17–19, 1946, Conference on Problems of Mathematics
celebrating the Bicentenniel of Princeton University.
Church is fifth from the left on the fourth row.

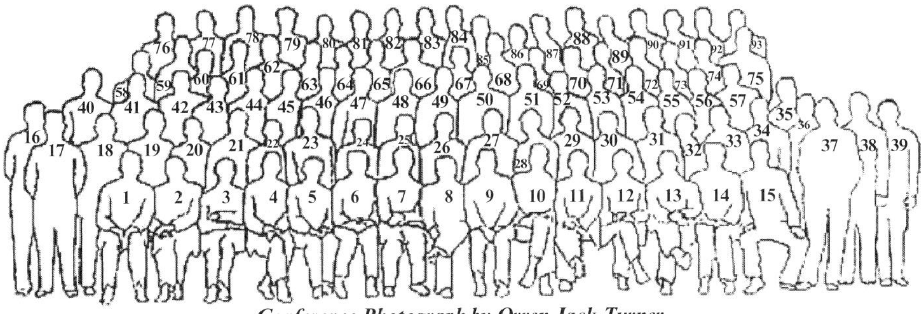

Conference Photograph by Orren Jack Turner

THE PROBLEMS OF MATHEMATICS

1. Morse, M., Institute for Advanced Study
2. Ancochea, G., University of Salamanaca, Spain
3. Borsuk, K., University of Warsaw, Poland
4. Cramér, H. University of Stockholm, Sweden
5. Hlavaty, V., University of Prague, Czechoslovakia
6. Whitehead, J. H. C., University of Oxford, England
7. Garding, L. J., Princeton
8. Riesz, M., University of Lund, Sweden
9. Lefschetz, S., Princeton
10. Veblen, O., Institute for Advanced Study
11. Hopf, H., Federal Technical School, Switzerland
12. Newman, M. H. A., University of Manchester, England
13. Hodge, W. V. D., Cambridge University, England
14. Dirac, P. A. M., Cambridge University, England
15. Hua, L. K., Tsing Hua University, China
16. Tukey, J. W., Princeton
17. Harrold, O. G., Princeton
18. Mayer, W., Institute for Advanced Study
19. Mautner, F. I., Institute for Advanced Study
20. Gödel, K., Institute for Advanced Study
21. Levinson, N., Massachusetts Institute of Technology
22. Cohen, I. S., University of Pennsylvania
23. Seidenberg, A., University of California
24. Kline, J. R., University of Pennsylvania
25. Eilenberg, S., Indiana University
26. Fox, R. H., Princeton
27. Wiener, N., Massachusetts Institute of Technology
28. Rademacher, H., University of Pennsylvania
29. Salem, R., Massachusetts Institute of Technology
30. Tarski, A., University of California
31. Bargmann, V., Princeton
32. Jacobson, N., The Johns Hopkins University
33. Kac, M., Cornell University
34. Stone, M. H., University of Chicago
35. Von Neumann, J., Institute for Advanced Study
36. Hedlund, G. A., University of Virginia
37. Zariski, O., University of Illinois
38. Whyburn, G. T., University of Virginia
39. McShane, E. J., University of Virginia
40. Quine, W. V., Harvard
41. Wilder, R. L., University of Michigan
42. Kaplansky, I., Institute for Advanced Study
43. Bochner, S., Princeton
44. Leibler, R. A., Institute for Advanced Study
45. Hildebrandt, T. H., University of Michigan
46. Evans, G. C., University of California
47. Widder, D. V., Harvard
48. Hotelling, H., University of North Carolina
49. Peck, L. G., Institute for Advanced Study
50. Synge, J. L., Carnegie Institute of Technology
51. Rosser, J. B., Cornell
52. Murnaghan, F. D., The Johns Hopkins University
53. Mac Lane, S., Harvard
54. Cairns, S. S., Syracuse University
55. Brauer, R., University of Toronto, Canada
56. Schoenberg, I. J., University of Pennsylvania
57. Shiffman, M., New York University
58. Milgram, A. N., Institute for Advanced Study
59. Walker, R. J., Cornell
60. Hurewicz, W., Massachusetts Institute of Technology
61. McKinsey, J. C. C., Oklahoma Agricultural and Mechanical
62. Church, A., Princeton
63. Robertson, H. D., Princeton
64. Bullitt, W. M., Bullitt and Middleton, Louisville, Ky.
65. Hille, E., Yale University
66. Albert, A. A., University of Chicago
67. Rado, T., The Ohio State University
68. Whitney, H., Harvard
69. Ahlfors, L. V., Harvard
70. Thomas, T. Y., Indiana University
71. Crosby, D. R., Princeton
72. Weyl, H., Institute for Advanced Study
73. Walsh, J. L., Harvard
74. Dunford, N., Yale
75. Spencer, D. C., Stanford University
76. Montgomery, D., Yale
77. Birkhoff, B., Harvard
78. Kleene, S. C., University of Wisconsin
79. Smith, P. A., Columbia University
80. Youngs, J. W. T., Indiana University
81. Steenrod, N. E., University of Michigan
82. Wilks, S. S., Princeton
83. Boas, R. P., Mathematical Reviews, Brown University
84. Doob, J. L., University of Illinois
85. Feller, W., Cornell University
86. Zygmund, A., University of Pennsylvania
87. Artin, E., Princeton
88. Bohneblust, H. F., California Institute of Technology
89. Allendoerfer, C. B., Haverford College
90. Robinson, R. M., Princeton
91. Bellman, R., Princeton
92. Begle, E. G., Yale
93. Tucker, A. W., Princeton

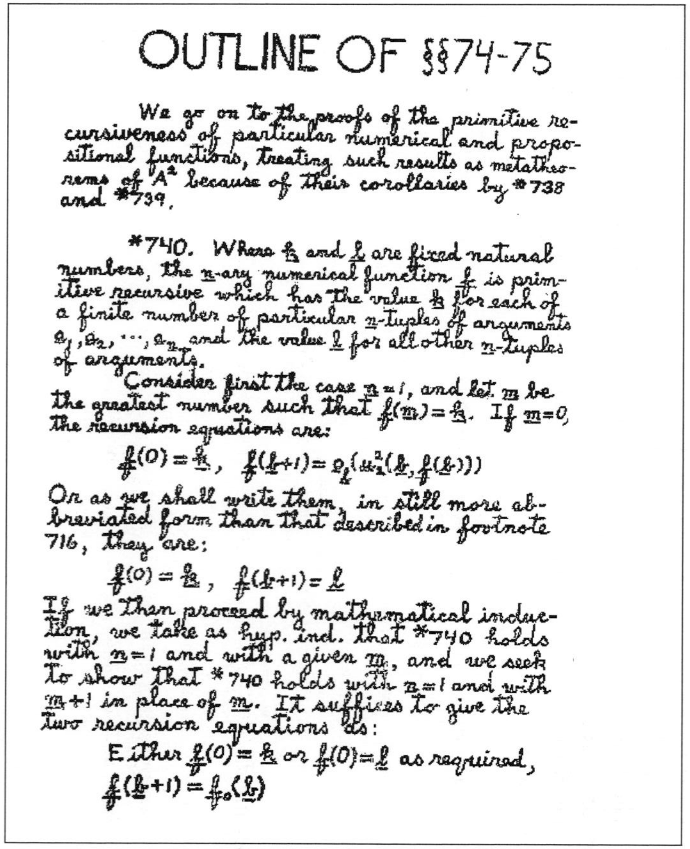

OUTLINE OF §§74-75

We go on to the proofs of the primitive recursiveness of particular numerical and propositional functions, treating such results as metatheorems of A^2 because of their corollaries by *738 and *739.

*740. Where k and l are fixed natural numbers, the n-ary numerical function f is primitive recursive which has the value k for each of a finite number of particular n-tuples of arguments $a_1, a_2, \cdots, a_n$ and the value l for all other n-tuples of arguments.

Consider first the case $n=1$, and let m be the greatest number such that $f(m)=k$. If $m=0$, the recursion equations are:

$$f(0)=k, \quad f(b+1)=c_l(u^2_1(b, f(b)))$$

Or as we shall write them, in still more abbreviated form than that described in footnote 716, they are:

$$f(0)=k, \quad f(b+1)=l$$

If we then proceed by mathematical induction, we take as hyp. ind. that *740 holds with $n=1$ and with a given m, and we seek to show that *740 holds with $n=1$ and with $m+1$ in place of m. It suffices to give the two recursion equations as:

Either $f(0)=k$ or $f(0)=l$ as required,

$f(b+1)=f_0(b)$

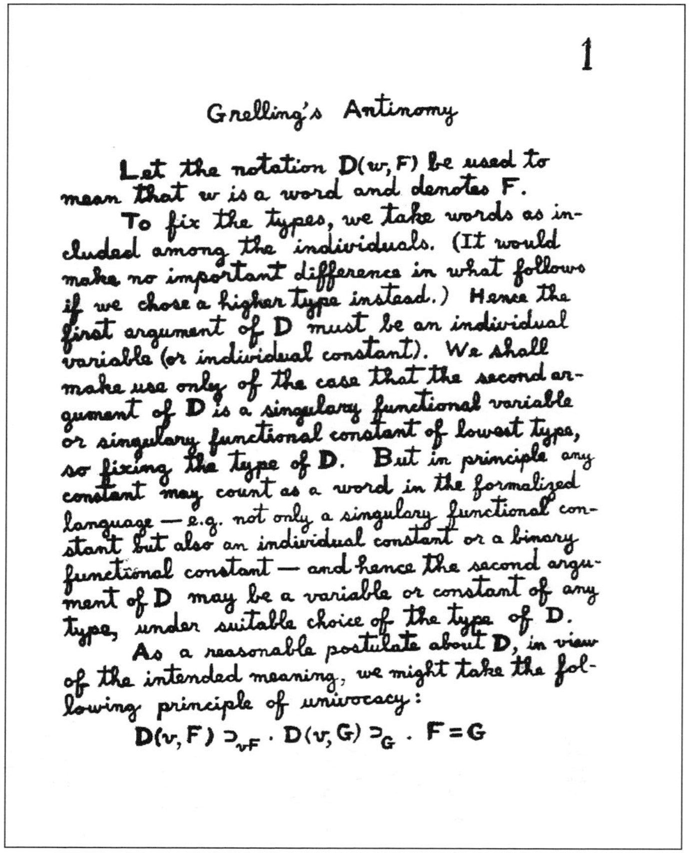

1

Grelling's Antinomy

Let the notation $\mathbf{D}(w, F)$ be used to mean that w is a word and denotes F.

To fix the types, we take words as included among the individuals. (It would make no important difference in what follows if we chose a higher type instead.) Hence the first argument of $\mathbf{D}$ must be an individual variable (or individual constant). We shall make use only of the case that the second argument of $\mathbf{D}$ is a singulary functional variable or singulary functional constant of lowest type, so fixing the type of $\mathbf{D}$. But in principle any constant may count as a word in the formalized language — e.g. not only a singulary functional constant but also an individual constant or a binary functional constant — and hence the second argument of $\mathbf{D}$ may be a variable or constant of any type, under suitable choice of the type of $\mathbf{D}$.

As a reasonable postulate about $\mathbf{D}$, in view of the intended meaning, we might take the following principle of univocacy:

$$\mathbf{D}(v, F) \supset_{vF} \cdot \mathbf{D}(v, G) \supset_G \cdot F = G$$

Samples of Alonzo Church's handwriting:
(left) see page 882 of this work, and (right) see page 929. The latter piece is one of the last things he wrote for publication.

UNIQUENESS OF THE LORENTZ TRANSFORMATION (1924)

1. The object of this inquiry is to obtain a set of logically independent postulates which uniquely determine the Lorentz transformation for one dimension. For this purpose we propose the following set:

1. The required transformation expresses $\bar{x}$ and $\bar{t}$ as functions of x, t, and v, which have continuous partial derivatives with respect to x and t, where $\bar{x}$, $\bar{t}$, x, and t are real variables, and v is a parameter which may have any value less than 1 and greater than -1.
2. For no value of v is the partial derivative of $\bar{t}$ with respect to t negative for every value of x and t.
3. $dx/dt = 1$ implies that $d\bar{x}/d\bar{t} = 1$.
4. $dx/dt = v$ implies that $d\bar{x}/d\bar{t} = 0$.
5. The inverse transformation, which expresses x and t in terms of $\bar{x}$ and $\bar{t}$, is obtained from the direct transformation by replacing $\bar{x}$ by x, and $\bar{t}$ by t, and x by $\bar{x}$, and t by $\bar{t}$, and v by $-v$.
6. The transformation is unchanged if $\bar{x}$ be replaced by $-\bar{x}$, and x by $-x$, and v by $-v$.

If we suppose our units of measurement so chosen that the velocity of light is 1, the physical meaning of the last five of these postulates is as follows:

2. The transformed time, $\bar{t}$, does not flow backwards with respect to t.
3. The velocity of light is invariant.
4. The origin of the transformed coördinate system is moving along the x-axis, with velocity v in the original coördinate system.
5. The resultant of two velocities with the same numerical value and with opposite directions is zero.
6. The form of the transformation is independent of our choice of a positive direction on the x-axis.

2. Let the required transformation have the form

$$\bar{x} = \varphi(x, t, v), \quad \bar{t} = \psi(x, t, v).$$

Then

$$\left(\psi_x \frac{dx}{dt} + \psi_t\right) \frac{d\bar{x}}{d\bar{t}} = \varphi_x \frac{dx}{dt} + \varphi_t. \tag{1}$$

Setting $dx/dt = v$ in (1), it follows from postulate 4 that

$$v\varphi_x + \varphi_t = 0. \tag{2}$$

It is a consequence of postulates 4 and 5 that $dx/dt = 0$ implies that $d\bar{x}/d\bar{t} = -v$. Therefore, setting $dx/dt = 0$ in (1), it follows that

$$\varphi_t + v\psi_t = 0. \tag{3}$$

Originally published in ***The American Mathematical Monthly***, vol. 31 (1924), pp. 376–382.

Integrating this with respect to t, we obtain

$$\varphi + v\psi = X, \tag{4}$$

where X is independent of t.

Setting $dx/dt = 1$ in (1), it follows from postulate 3 that

$$\psi_x + \psi_t = \varphi_x + \varphi_t. \tag{5}$$

From (2) and (3) it follows that

$$\varphi_x = \psi_t \tag{6}$$

and therefore by (5)

$$\psi_x = \varphi_t. \tag{7}$$

Substituting in (2) this gives

$$v\varphi_x + \psi_x = 0$$

and therefore integrating with respect to x,

$$v\varphi + \psi = T, \tag{8}$$

where T is independent of x.

For values of v between $+1$ and -1 equations (4) and (8) can be solved simultaneously. In this way we obtain:

$$\varphi = X_1 + T_1, \quad \psi = X_2 + T_2,$$

where X_1 and X_2 are independent of t, and T_1 and T_2 are independent of x.

From equations (6) and (7) it now follows that

$$\frac{dX_1}{dx} = \frac{dT_2}{dt}, \quad \frac{dX_2}{dx} = \frac{dT_1}{dt}.$$

These conditions, however, can be satisfied only if dX_1/dx, dX_2/dx, dT_1/dt and dT_2/dt are all independent of both x and t. Therefore X_1, X_2, T_1, and T_2 are of the first degree in x and t. Therefore φ and ψ are of the first degree in x and t. Let us accordingly write the transformation in the form,

$$\bar{x} = p_1x + q_1t + r_1, \quad \bar{t} = p_2x + q_2t + r_2,$$

where p_1, p_2, q_1, q_2, r_1, and r_2 are functions of v.

Then, by (6),

$$p_1 = q_2,$$

by (7)

$$p_2 = q_1,$$

and by (3)

$$q_1 = -vq_2.$$

Therefore

$$p_2 = -vq_2.$$

The required transformation can therefore be written in the form

$$\bar{x} = \beta(x - vt) + r_1, \quad \bar{t} = \beta(t - vx) + r_2, \tag{9}$$

where β, r_1, and r_2 are functions of v. In order that postulate 6 be satisfied it is necessary that β and r_2 be even functions of v, and that r_1 be an odd function.

By solving (9) for x and t we obtain

$$x = \frac{\bar{x} + v\bar{t}}{(1 - v^2)\beta} - \frac{r_1 + vr_2}{(1 - v^2)\beta}, \quad t = \frac{\bar{t} + v\bar{x}}{(1 - v^2)\beta} - \frac{r_2 + vr_1}{(1 - v^2)\beta}.$$

Consequently postulate 5 requires that

$$r_1 = \frac{r_1 + vr_2}{(1 - v^2)\beta}, \quad r_2 = -\frac{r_2 + vr_1}{(1 - v^2)\beta}, \quad \beta = \frac{1}{(1 - v^2)\beta}.$$

Solving these equations simultaneously for β, r_1, and r_2, we find that

$$r_1 = 0, \quad r_2 = 0, \quad \beta = \pm\frac{1}{\sqrt{1 - v^2}}.$$

The expression obtained for β represents, not two but infinitely many determinations of β as a function of v, since it is possible to take the upper sign for some values of v and the lower sign for the remaining values, in any arbitrary manner which makes β an even function of v. It is for this reason that postulate 2 is required.

In accordance with postulate 2, there is only one possible determination of β, namely,

$$\beta = \frac{1}{\sqrt{1 - v^2}}.$$

The required transformation, therefore, has the form:

$$\bar{x} = \frac{x - vt}{\sqrt{1 - v^2}}, \quad \bar{t} = \frac{t - vx}{\sqrt{1 - v^2}},$$

which is the Lorentz transformation for one dimension.

3. We shall give independence proofs for the last five of our postulates. There is no immediately evident independence proof for the first postulate. It is quite possible that it is not necessary to assume the existence of the partial derivatives. $\varphi_x, \psi_x, \varphi_t, \psi_t$.

As an independence proof for postulate 2 we may cite the transformation:

$$\bar{x} = f(v)\frac{x - vt}{\sqrt{1 - v^2}}, \quad \bar{t} = f(v)\frac{t - vx}{\sqrt{1 - v^2}},$$

where $f(v)$ is a function of v which is restricted to have always one of the two values $+1$ and -1, in such a way that $f(v) = f(-v)$ for all values of v, and $f(v)$ is not equal to $+1$ for every value of v.

An independence proof for postulate 3 is the transformation:

$$\bar{x} = x - vt, \quad \bar{t} = t.$$

The simplest independence proof for postulate 4 is:

$$\bar{x} = x, \quad \bar{t} = t,$$

but we may also take the transformation:

$$\bar{x} = \frac{x - kvt}{\sqrt{1 - k^2v^2}}, \quad \bar{t} = \frac{t - kvx}{\sqrt{1 - k^2v^2}},$$

where k is any constant which is real and less than 1 in numerical value.

An independence proof for postulate 5 is:

$$\bar{x} = x - vt, \quad \bar{t} = t - vx.$$

And, finally, for postulate 6 we have:

$$\bar{x} = \frac{(1-v)^n}{(1+v)^{n+1}}(x - vt), \quad t = \frac{(1-v)^n}{(1+v)^{n+1}}(t - vx), \tag{10}$$

where n is any real number different from $-\frac{1}{2}$.

4. The transformations (10) have all the essential properties of the Lorentz transformation except the property of symmetry expressed by postulate 6.

The transformation corresponding to $n = -\frac{1}{2}$ is the Lorentz transformation. The transformations corresponding to the other values of n are arranged symmetrically about the value $-\frac{1}{2}$ in the sense that the transformation corresponding to $n = a$ and that corresponding to $n = -(a-1)$ apply to the same space with a different choice of the positive direction on the x-axis.

In a space in which one of these transformations is valid there is an effect on the apparent length of a moving body analogous to that given by the Lorentz transformation, but this effect may be either a lengthening or a shortening, and, in accordance with the asymmetric character of the transformations, it depends not only on the speed of the motion but also on the direction.

The invariant of the transformation is:

$$\frac{(t-x)^{n+1}}{(t+x)^n}.$$

The proper time is given by

$$dr = \frac{(dt - dx)^{n+1}}{(dt + dx)^n}.$$

The transformation obtained by setting $n = 0$ in (10) is:

$$\bar{x} = \frac{x - vt}{1+v}, \quad \bar{t} = \frac{t - vx}{1+v}.$$

This transformation has the properties just enumerated. The proper time given by it is $dr = dt - dx$. Since $\int(dt - dx)$ is independent of the path of integration, the absolute character of time is preserved in a way in which it is not by the Lorentz transformation. This means that there are no geodesics in the two-dimensional space-time to which this transformation applies; and that a unique proper time can be attached to every event, independently of a moving particle to which the time is referred.

5. We shall prove that the transformations (10) are the only transformations which satisfy postulates 1 to 5 above and the following three in addition:

7. $x = 0, t = 0$ implies that $\bar{x} = 0, \bar{t} = 0$.

8. The transformations corresponding to the various possible values of v constitute a group.

9. $\bar{x}$ and $\bar{t}$ are continuous as functions of v.

A transformation which satisfies these postulates must be of the form, (9), above. By postulate 7 we have $r_1 = r_2 = 0$. Therefore the transformation has the form:

$$\bar{x} = \beta(v)(x - vt), \quad \bar{t} = \beta(v)(t - vx).$$

A second transformation of the set may have the form:

$$\bar{\bar{x}} = \beta(v')(\bar{x} - v'\bar{t}), \quad \bar{\bar{t}} = \beta(v')(\bar{t} - v'\bar{x}).$$

Combining these two transformations, we obtain:

$$\bar{\bar{x}} = \beta(v)\beta(v')(1 - vv')\left(x - \frac{v + v'}{1 + vv'}t\right),$$

$$\bar{\bar{t}} = \beta(v)\beta(v')(1 + vv')\left(t - \frac{v + v'}{1 + vv'}x\right).$$

Therefore, in order that postulate 8 be satisfied, it is necessary that

$$\beta(v)\beta(v')(1 + vv') = \beta\left(\frac{v + v'}{1 + vv'}\right).$$

Let

$$\beta(v) = \frac{\varphi(x)}{1 + v}.$$

Then

$$\varphi(v)\varphi(v') = \varphi\left(\frac{v + v'}{1 + vv'}\right). \tag{11}$$

Now we know, by postulate 2, that $\varphi(v)$ is positive for some value of v between $+1$ and -1. Therefore, since it follows from postulate 9 that $\varphi(v)$ is continuous, there is some interval between $+1$ and -1 throughout which $\varphi(v)$ is positive. In order, therefore, to find the form of $\varphi(v)$ in this interval, we are justified in taking the logarithm of both sides of (11). It will turn out that the form of $\varphi(v)$ so obtained cannot vanish for any value of v between $+1$ and -1, from which it will follow that $\varphi(v)$ has this form throughout the interval from -1 to $+1$.

Taking the logarithm of each side of (11), we obtain:

$$\log\varphi(v) + \log\varphi(v') = \log\varphi\left(\frac{v + v'}{1 + vv'}\right).$$

Since v and v' lie between $+1$ and -1, we may make the following substitutions without going outside the real number system:

$$v = \tanh u$$
$$v' = \tanh w$$
$$\log\varphi(\tanh u) = F(u).$$

Making these substitutions, we obtain

$$F(u) + F(w) = F(u + w). \tag{12}$$

Since u and w are independent variables, we may assign any relationship between them that we please without affecting the validity of (12). Setting $u = w$, we have:

$$F(2w) = 2F(w).$$

Similarly, setting $u = 2w$,

$$F(3w) = 3F(w)$$

and so on. It is clear that we may prove by induction that

$$F(aw) = aF(w), \tag{13}$$

for every positive integer a.

Setting $u = 0$ in (12), we obtain:

$$F(w) + F(0) = F(w).$$

Since $F(w)$ is not infinite for all values of w, it follows that

$$F(0) = 0.$$

Therefore, setting $u = -w$ in (12),

$$F(-w) = -F(w),$$

from which it follows that (13) is also true when a is a negative integer.

Setting $w = w'/a$ in (13), we obtain:

$$F\left(\frac{w'}{a}\right) = \frac{1}{a}F(w').$$

Therefore if p/q is any rational number,

$$F\left(\frac{p}{q}w\right) = \frac{p}{q}F(w),$$

and therefore, since $F(w)$ is continuous,

$$F(aw) = aF(w)$$

for every real number, a. And, dividing by aw,

$$\frac{F(w)}{w} = \frac{F(aw)}{aw}.$$

Therefore

$$\frac{F(w)}{w} = b,$$

or

$$F(w) = bw,$$

where b is an arbitrary constant. In order to make sure that b is actually arbitrary, we must, of course, substitute this expression for $F(w)$ in equation (12).

We now have

$$\varphi(v) = e^{b \tanh^{-1} v}$$

and therefore

$$\varphi(v) = \left(\frac{1-v}{1+v}\right)^n,$$

where n is an arbitrary constant. This gives

$$\beta(v) = \frac{(1-v)^n}{(1+v)^{n+1}},$$

from which it follows that the required transformation is in the form (10).

ON IRREDUNDANT SETS OF POSTULATES[1]

(1925)

1. Definitions. We shall say that the postulate A is weaker than the postulate B if B implies A and A does not imply B. This is a definition of the relative strength of two postulates in an absolute sense, as distinguished from their relative strength in the presence of other postulates.

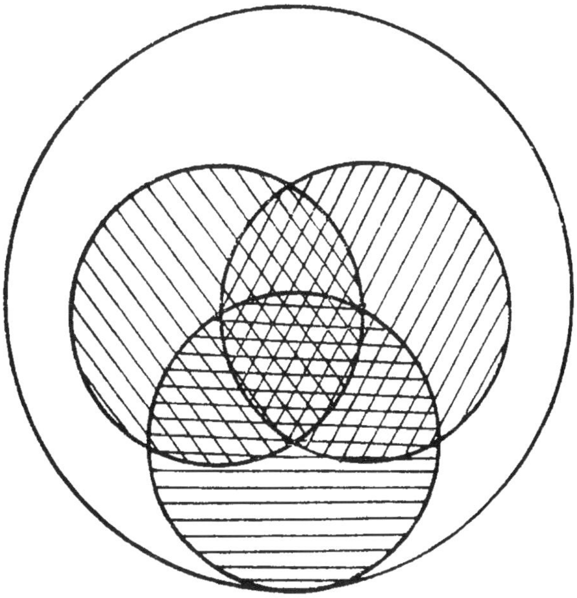

FIGURE 1

The postulate A_1 can be weakened with respect to the set $A_1, A_2, A_3, \cdots, A_n$ if there exists a postulate, A_1', weaker than A_1, such that the logical product $A_1' A_2 A_3 \cdots A_n$ implies A_1. Under this definition we say that a postulate can be weakened with respect to a set of postulates, of which it is one, only when it can be replaced in the set by a weaker postulate in such a way that none of the implications of the set as a whole is lost.

A set of postulates is *irredundant* if the postulates are independent and no one of them can be weakened with respect to the set.

Originally published in ***Transactions of the American Mathematical Society***, vol. 27 (1925), pp. 318–328.

[1]Presented to the Society, December 30, 1924.

2. The relation of irredundance to complete independence.[2] If we represent the set of all possible mathematical systems by the set of all points in the interior of a circle, we may represent a postulate by a curve which divides the interior of the circle into two regions, which represent, one the set of systems which satisfy the postulate, and one the set of those which do not. We may distinguish between these two regions by shading the one within which the postulate is false and leaving the other unshaded. It is then clear that, if the postulate A is weaker than the postulate B, the shaded area corresponding to A in the diagram will lie entirely within that corresponding to B.

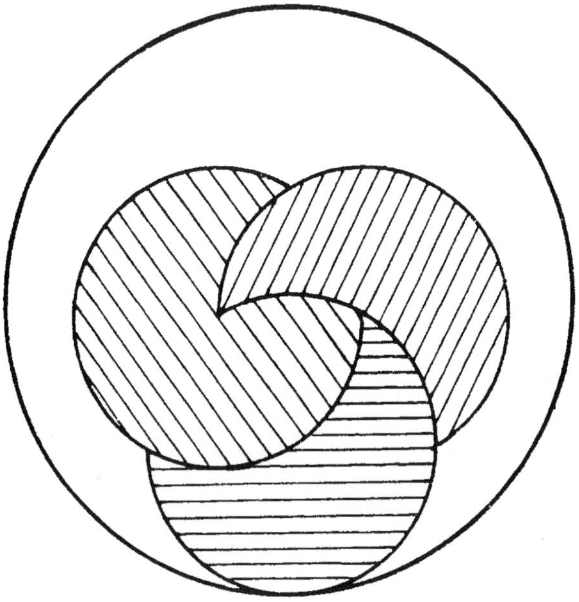

FIGURE 2

In this way we may represent a set of three completely independent postulates as shown in Figure 1. The postulates are completely independent because, as is easily verified from the figure, systems exist which satisfy some of the postulates and fail to satisfy the others in any arbitrary assigned manner. This corresponds to the fact that the three small circles which represent the postulates divide the interior of the large circle into 2^3, or 8, regions.

If this set is to be made irredundant, the postulates must be weakened until no two of the shaded regions overlap; for so long as two of the shaded regions do overlap, it is possible to shrink one of them without reducing the total shaded area. The weakened set of postulates is represented in Figure 2. On the other hand, if a set of postulates which are not completely independent are to be made so, it may be necessary to

[2]The notion of complete independence is due to E. H. Moore. See ***Introduction to a form of general analysis***, New Haven Colloquium, 1906, p. 82. For a discussion of the significance of this notion, see E. V. Huntington, *A new set of postulates for betweenness, with proof of complete independence*, these ***Transactions***, vol. 26 (1924), p. 277.

strengthen one or more of them.[3] In particular, if the set is irredundant, all but one of the postulates, at least, must be strengthened.

A set of n completely independent postulates divides the set of all possible systems into 2^n parts, represented in a diagram by regions, the greatest number possible. The division effected by an irredundant set of n postulates, if the set is consistent, is a division into $n + 1$ parts, the least number which is possible if the postulates are to be consistent and independent.

In general, an irredundant set of postulates is as far removed as possible from a completely independent set.[4]

3. A criterion for irredundance. In the diagram just described the relation of implication between postulates is represented by the relation of inclusion between regions. Thus, if A implies B, the shaded area corresponding to A contains that corresponding to B, and if the negative of A implies B, the unshaded area corresponding to A contains the shaded one corresponding to B. The observations made in the preceding section, therefore, suggests the following theorem:

A necessary and sufficient condition that the postulate A_1 cannot be weakened with respect to the set $A_1, A_2, A_3, \cdots, A_n$ is that the negative of A_1 imply each of the postulates $A_2, A_3, \cdots, A_n$.

The condition is necessary. For let B stand for the logical product $A_2A_3 \cdots A_n$, and suppose that *not*-A_1 does not imply B. Then *not*-B does not imply A_1, and, therefore, A_1 *or not*-B does not imply A_1. The postulate A_1 *or not*-B is, therefore, weaker than A_1, and can be substituted for A_1 in the set $A_1, A_2, A_3, \cdots, A_n$, without loss of any of the implications of the set.

The condition is sufficient. For suppose that *not*-A_1 implies B, but that there exists a postulate A_1', weaker than A_1, such that the logical product $A_1'B$ implies B. Then since *not*-A_1 implies B, it follows that *not*-B implies A_1; but $A_1'B$ implies A_1, and, by the law of excluded middle, B or *not*-B; therefore A_1' implies A_1, contrary to the hypothesis that A_1' was weaker than A_1.

The validity of this theorem depends on the validity of certain postulates of logic on which it is based, among which, as we have just seen, is the law of excluded middle. The conception with which we began, that a postulate divides the set of all possible mathematical systems into two parts, also depends, of course, on this law.

From the theorem which has just been proved follows as a corollary

A necessary and sufficient condition that a set of postulates be irredundant is that the postulates be independent and that the negatives of every two be contradictory.

Thus if the independent postulates A, B, C are to be irredundant, we must have *not*-A incompatible with *not*-B, *not*-A incompatible with *not*-C, and *not*-B incompatible with *not*-C.

---

[3]For an interesting example of this, see J. S. Taylor, *Complete existential theory of Bernstein's set of four postulates for Boolean algebras*, ***Annals of Mathematics***, vol. 19 (1917), p. 68, and the footnote on that page.

[4]The idea of a set of mutually prime postulates as defined by H. M. Sheffer is similar to that of an irredundant set as here defined. The two ideas are not, however, the same, because mutual primeness implies complete independence in the sense of E. H. Moore. See an abstract of a paper by H. M. Sheffer, *Mutually prime postulates*, ***Bulletin of the American Mathematical Society***, vol. 22 (1916), p. 287. See also *The general theory of notational relativity*, a privately circulated paper.

There is a mechanical method by which any set of independent postulates can be made irredundant. If A, B, C, D are independent postulates, the set A, B', C', D' is an equivalent set which is irredundant, if B' is the proposition *if* A *then* B, and C' is the proposition *if* A *and* B *then* C, and D' the proposition *if* A *and* B *and* C *then* D. This method is open to the objection that it is likely to lead to postulates very complicated in form, and that, in general, the hypothesis of a postulate obtained in this way will be concerned with elements which are not mentioned in the conclusion. For example, if we applied this method to Peano's postulates for the system of positive integers,[5] we should obtain as our second postulate the following: "If 1 is a number, then every number has a unique successor."

The question arises to what extent it is possible to obtain irredundant sets of postulates which are not open to this objection. As a partial answer to this question, we shall give two examples of such sets of postulates.

4. A simple example of an irredundant set. The following set of postulates describes a system of a finite number of elements, which we call numbers, arranged in cyclic order. The undefined terms are *number* and *successor*. The postulates are the following:

1. There is a set of numbers, not null, such that every number in the set has a successor which is in the set.

2. In every proper part P, not null, of the set of all numbers there is a number none of whose successors is in P.

The negatives of these two postulates are

Not-1. There is no set of numbers, other than the null set, such that every number in the set has a successor which is in the set.

Not-2. In some proper part P, not null, of the set of all numbers every number has a successor which is in P.

Since these two statements are plainly contradictory, it will follow, as soon as we give independence examples, that the set of postulates 1 and 2 is irredundant.

An independence example for the first postulate is a system of a finite number of elements arranged in linear order. An independence example for the second postulate is the system of positive integers.

In the case of each example, when we have verified that it fails to satisfy the postulate for which it is given as independence proof, we need not afterwards verify that it satisfies the other postulates, because we know that the negatives of the two postulates are contradictory. This corresponds to the fact that if a set of postulates satisfy the condition of the preceding section, that is, if the negatives of every two be contradictory, then a postulate of the set can fail to be independent only if its negative be self-contradictory in the system of logic on which the postulates are based.

From postulate 1 above follows the existence of at least one number. From postulate 2 it follows that the set whose existence is required by postulate 1 is the set of all numbers, and, therefore, that every number has a successor.

If a is any number, every number can be reached from a by proceeding from number to successor a finite number of times, for if A is the set of all numbers which can be so reached from a, it is true that every number in A has a successor in A, and therefore, by postulate 2, that A is the set of all numbers. If any number is successor of itself,

[5]See Peano, *Sul concetto di numero*, ***Rivista di Matematica***, vol. 1 (1891), pp. 87–102. See also below.

it follows from postulate 2 that it is the only number, and in this case we have a cycle which contains just one element. In any other case we can find distinct numbers, a and b, such that b is a successor of a. Then there are one or more routes by which a can be reached from b in a finite number of steps from number to successor. And it follows from postulate 2 that any arbitrarily chosen one of these routes must contain every number. The set of all numbers is, therefore, finite, and can be arranged in a cycle in such a way that every number is a successor of the number which next precedes it in the cycle.

Finally, there is no number a which has two successors, b and c. For, if there were, we could find a finite route from b to c, and this route would not contain a, because every finite route from b to a contains all numbers. Similarly there would be a finite route from c to b which did not contain a, and the set P of all numbers which were in either of these routes would have the property that every number in it has a successor in it. But P would not contain a, and would, therefore, violate postulate 2.

We have, therefore, shown that postulates 1 and 2 uniquely determine a system of a finite number of elements arranged in cyclic order.

5. An irredundant set of postulates for the system of positive and negative integers.[6]

The following postulates, which deal with a set of undefined elements, *numbers*, and an undefined relation, *successor* among them, describe the system of positive and negative integers.

1. There is a set of numbers, S, not null, such that every number in S has a successor which is in S.

2. If the set of all numbers is in such a set S, it is not the only such set.

3. If such sets S exist, one of them has the property that every number in the set is successor of some number in the set.

4. In every proper part P, not null, of the set of all numbers, either some number has no successor which is in P, or some number is successor of no number in P.

The negatives of these postulates are:

Not-1. There is no set of numbers S, other than the null set, such that every number in S has a successor which is in S.

Not-2. The set of all numbers is an S and the only S.

Not-3. Sets S exist, but no one of them has the property that every number in the set is successor of some number in the set.

Not-4. There is a proper part P, not null, of the set of all numbers, such that every number in P has a successor in P and is successor of some number in P.

Of these four statements every two are contradictory. This is evident at once, except in the case of *not*-2 and *not*-3. In order to see that *not*-2 and *not*-3 are contradictory, we observe that if the set of all numbers, N, were an S and the only S, then every number in N would be successor of some number in N, because if the number a in N were successor of no number in N we could, by omitting a from N, obtain an S not the set of all numbers.

[6]For a set of independent postulates for the system of positive and negative integers, see A. Padoa, *Un nouveau système irréductible de postulats pour l'algèbre*, ***Deuxième Congrès International des Mathématiciens***, Paris, 1900, pp. 249–256, and Padoa, *Numeri interi relativi*, ***Rivista di Matematica***, vol. 7 (1901), pp. 73–84. This set is, however, not categorical, because it is satisfied by the integers reduced modulo m, where m is any odd integer.

Independence examples are as follows. For postulate 1, a system of a finite number of elements arranged in linear order; for postulate 2, a finite cycle; for postulate 3, the system of positive integers; for postulate 4, a system of two sets of a finite number of elements each, no element common to the two sets, and each set arranged separately in cyclic order. As pointed out in the preceding section, it is sufficient to see, in the case of the first example, that it satisfies *not*-1, because every system which satisfies *not*-1 must also satisfy postulates 2, 3, and 4; and similarly in the case of each of the other examples.

The set of postulates 1, 2, 3, 4 is, therefore, irredundant.

6. A proof that the foregoing set of postulates is categorical. We shall prove that the postulates given above uniquely determine the system of positive and negative integers. In order to do this, we shall show that the set of all numbers, as defined by these postulates, can be divided into two parts which have in common one and only one number, a, of which one satisfies Peano's postulates for the system of positive integers[7] when "1" is replaced in them by "a", and the other satisfies Peano's postulates when "1" is replaced in them by "a" and "successor" by "predecessor".

Peano's postulates are as follows:

1. 1 is a number.
2. Every number has a unique successor.
3. If a and b are numbers, and if their successors are equal, then a and b are equal.
4. 1 is not the successor of any number.
5. If S is a class of numbers which contains 1, and the class of all successors of numbers in S is contained in S, then every number is contained in S.

If not imperative, it is at least desirable that the proposed proof should be made without the use of the notion of finite cardinal number, although this necessitates an argument somewhat longer than might otherwise be given.

THEOREM 1. *There exists a number.* (Postulate 1 of §5.)

THEOREM 2. *Every number has a successor.*

For it follows from postulate 4 that the set whose existence is required by postulate 3 is the set of all numbers.

DEFINITION. If the number a is successor of the number b, then b is *predecessor* of a.

THEOREM 3. *Every number has a predecessor.* (Postulates 1, 3, 4.)

THEOREM 4. *No number is successor of itself.*

For if a were successor of itself, then a would be the only number, by postulate 4. But in that case postulate 2 would be violated.

THEOREM 5. *If the number b is successor of the number a, a is not successor of b.*

For if a were successor of b, a and b would be the only numbers, by postulate 4. But in that case postulate 2 would be violated.

DEFINITION. The set of numbers Q is an *interval* if it contains numbers, a and b, which have, respectively, no predecessor and no successor in Q, and every number other than a has a unique predecessor in Q, and every number other than b has a unique successor in Q, and finally, there is no part of Q, not null, in which every

[7] Loc. cit. See also Peano, ***Formulaire de Mathématiques***, vol. 3 (1901), pp. 39–44.

number has a successor. The numbers a and b are the *end numbers* of Q, and Q is *an interval ab*.

Theorem 6. *The set of numbers obtained by omitting b from an interval ab, if* $b \neq a$, *is an interval ac, where c is the predecessor of b in ab. The set of numbers obtained by omitting a from an interval ab, if* $b \neq a$, *is an interval db, where d is the successor of a in ab.*

Theorem 7. *If intervals ab and bc have the property that no number in either, other than b, has a successor in the other, other than b, then the sum, P, of ab and bc is an interval ac.*

Suppose that in some part T of P, not null, every number had a successor. Then the set R of numbers common to T and bc would have the property that every number in it had a successor in it, for otherwise some number in ab would be successor of some number in bc. But R cannot be null, because, if it were, I would be a part of ab. Therefore there is no part of P in which every number has a successor. Therefore, since it is easily seen that P has the other properties required by the definition, it follows that P is an interval ac.

Corollary 1. *If c is a successor of b, and, in the interval ab, c has no successor and no predecessor other than b, then the set of numbers obtained by adding c to ab is an interval ac.*

Corollary 2. *If c is a predecessor of a, and, in an interval ab, c has no predecessor and no successor other than a, then the set of numbers obtained by adding c to ab is an interval cb.*

Theorem 8. *If c is a number of an interval ab, this interval can be divided into intervals, ac and cb, whose sum is ab, and which have only c in common.*

Let P be the set of all numbers, c, in ab of which this is not true. If the theorem is true of any number, d, in ab, it is true also of the predecessor of d in ab (Theorem 6 and Theorem 7, Corollary 1). Therefore, if any number c is contained in P, the successor of c in ab is also contained in P. This is true without exception, since b is not contained in P, because ab can be divided into ab and bb, where bb consists of b alone. Therefore, by the definition of interval, the set P is null.

Theorem 9. *This division is unique.*

For suppose there were another such division of ab, say a division into $(ac)'$ and $(cb)'$. Let P be the set of all numbers in ac and not in $(ac)'$. Since every number in ac, other than c, has a successor in ac, and every number in $(ac)'$, other than a, has a predecessor in $(ac)'$, it follows that if P contains d it contains also the successor of d in ab. Therefore P is null. And, therefore, every number in ac is contained also in $(ac)'$. By the same reasoning, every number in $(ac)'$ is contained also in ac. Therefore ac and $(ac)'$ are identical. Therefore cb and $(cb)'$ are identical.

Definition. The number d, in ab, different from c, in ab, *precedes* or *follows c in ab* according as d is in ac or cb.

Theorem 10. *If d precedes c in ab, then c follows d in ab; and if c follows d in ab, then d precedes c in ab.*

For by Theorems 7, 8, 9, we have

$$ab = ac + cb = ad + dc + cb = ad + db,$$

where db contains c.

THEOREM 11. *If c_1 precedes c_2 in ab and c_2 precedes c_3 in ab, then c_1 precedes c_3 in ab.*

For we have

$$ab = ac_1 + c_1b = ac_1 + c_1c_2 + c_2b = ac_1 + c_1c_2 + c_2c_3 + c_3b = ac_1 + c_1c_3 + c_3b,$$

and since $c_1c_3 + c_3b = c_1b$, by Theorem 7, it follows that c_1b contains c_3.

THEOREM 12. *In any subset P, not null, of an interval ab there is a first number and a last number.*

Let p stand for any number in P. Then in the set R of all numbers which occur in any interval ap, there is a number, c, which has no successor in the set. This number c is a member of P. For, if it were not, it would be a number, not an end number, in some interval ap, and its successor in ap would be a member of R. And there is no number p, in P, which follows c in ab. For, if there were, c would be contained in ap, and its successor in ap would be a member of R. Finally, if d is any number in P, different from c, d must precede c in ab, for c does not precede d, and it follows at once from the definition that one of the numbers c and d precedes the other in ab. Therefore c is the last number in P.

In the set T of all numbers which occur in every interval ap is a number, e, which has no successor in the set. This number e is a member of P. For, if it were not, it would be a number, not an end number, in every interval ap, and its successor in ab would be in every ap, and would, therefore, be a member of T. And there is no number p, in P, which precedes e in ab. For, if there were, e would not be contained in ap, and would, therefore, not be a member of T. And if d is any number in P, different from e, d must follow e in ab. Therefore e is the first number in P.

THEOREM 13. *If a and b are distinct numbers, there do not exist both an interval ab and an interval ba.*

Suppose both do exist. Then, by postulate 4, $ab + ba$ is the set of all numbers.

Suppose that, apart from a and b, ab and ba have no numbers in common, and no number in either is successor of any number in the other. Then, by postulate 2, there is some proper part, T, of $ab + ba$ in which every number has a successor. There is a number, c, in $ab + ba$ which is not contained in T. It is possible to choose c distinct from a and b. For, if a is not in T, the predecessor of a in ba is also not in T, and if b is not in T, the predecessor of b in ab is also not in T; and since, by Theorem 5, either ab or ba must contain numbers other than a and b, we can always find a number distinct from a and b and not in T, as required. Let c be so chosen, and suppose that c is in ab. Let d and e be the predecessor and the successor, respectively, of c in ab. Then $eb + ba + ad$ is an interval ed by Theorem 7. But this interval contains a part, T, in which every number has a successor, contrary to the definition of interval.

Suppose that, apart from a and b, ab and ba have no number in common, and that some number in one of them, different from a and b, say c in ab, has a successor in the other different from a and b, say d in ba. Then $ac + da$ is a proper part of the set of all numbers in which every number has a successor and a predecessor, contrary to postulate 4.

Suppose that *ab* and *ba* have a number, *c*, in common, other than *a* and *b*. Then $ac + ca$ is a proper part of the set of all numbers in which every number has a successor and a predecessor, contrary to postulate 4.

Therefore, in every case, the supposition that both *ab* and *ba* exist leads to a contradiction.

THEOREM 14. *If c is a successor of b, the existence of an interval ab implies the existence of an interval ac.*

The number *c* has no successor, *d*, in *ab*. For, if it had, either *d* would be the predecessor of *b* in *ab*, in which case *b*, *c*, and *d* would be the only numbers, by postulate 4, and postulate 2 would be violated, or, if *d* were not the predecessor of *b*, the numbers *c* and *d* would constitute an interval *cd* and *c* together with *db* would constitute an interval *dc*, contrary to the preceding theorem. Now there are numbers in *ab* of which *c* is successor, for *b* is such a number. Let *d* be the first such number (Theorem 12). Then *ad* and *c* together constitute an interval *ac*, by Theorem 7, Corollary 1.

THEOREM 15. *If c is a predecessor of a, the existence of an interval ab implies the existence of an interval cb.*

This follows by reasoning similar to the preceding, using Theorem 7, Corollary 2, in place of Theorem 7, Corollary 1.

DEFINITION. If *a* is any number, the *ray A* is the set of all numbers *b* such that an interval *ab* exists, and the *ray A′* is the set of all numbers *c* such that an interval *ca* exists.

From the theorems which precede, we can infer at once that the ray *A* contains *a*, and contains no predecessor of *a*, that every number in *A* has a successor in *A*, and that every number but *a* in *A* has a predecessor in *A*. Similarly, the ray A' contains *a*, and contains no successor of *a*, and every number in A' has a predecessor in A', and every number but *a* in A' has a successor in A'.

THEOREM 16. *The rays A and A' have no number other than a in common.*

For, if they had *b*, different from *a*, in common, intervals *ab* and *ba* would both exist, contrary to Theorem 13.

THEOREM 17. *If the numbers b and c are successors of the number a, then b and c are the same number.*

For suppose that *b* and *c* were distinct. Then, if the ray *B* did not contain *c*, $A' + B$ would be a proper part of the set of all numbers of which postulate 4 would not be true. Therefore *B* contains *c*, and, for the same reason, *C* contains *b*. But neither *B* nor *C* contains *a*. Therefore $B + C$ is a proper part of the set of all numbers in which every number has a successor and a predecessor, contrary to postulate 4.

THEOREM 18. *If the number a is a successor of the numbers b and c, then b and c are the same number.*

THEOREM 19. *$A + A'$ is the set of all numbers.* (Postulate 4.)

THEOREM 20. *If S is a part of the ray A, such that a is a member of S, and the class of all successors of numbers in S is contained in S, then S contains every number of A.*

For otherwise $S + A'$ would be a proper part of the set of all numbers in which every number had a successor and a predecessor.

THEOREM 21. *If S is a part of the ray A', such that a is a member of S, and the class of all predecessors of numbers in S is contained in S, then S contains every number of A'.*

For otherwise $S + A$ would violate postulate 4.

We have now shown that the set of all numbers can be divided into two parts, A and A', which have a only in common and which satisfy Peano's postulates in the manner described above. Therefore the set of postulates 1, 2, 3, 4 with which we began is categorical.

PRINCETON UNIVERSITY,
PRINCETON, N.J.

ON THE FORM OF DIFFERENTIAL EQUATIONS OF A SYSTEM OF PATHS[1] (1927)

In a recent paper[2] Veblen and Thomas have proposed, as a generalization of the affine geometry of paths studied by them, the study of paths determined by a system of differential equations of the form

$$\delta^{ij}_{ab}\frac{d^2x^a}{ds^2}\frac{dx^b}{ds}+\Pi^{ij}_{pqr}\frac{dx^p}{ds}\frac{dx^q}{ds}\frac{dx^r}{ds}=0 \tag{1}$$

in which the Π's are functions of the x's, symmetric in the indices p, q, r and alternating in the indices i and j, and subject to an appropriate law of transformation.

It is the purpose of this paper to show that, if equations (1) are not the differential equations of a system of paths which has an affine connection, then the curves which satisfy (1) fail to constitute a system of paths at all, in that they do not have the property that, locally, two paths are joined by one and only one curve.[3]

Since equations (1) are invariant under all analytic transformations of the parameter s, we are at liberty to choose the parameter as we please. Let us, therefore, fix a particular integer k such that $1 \leqq k \leqq n$, and choose $s = \alpha x^k + \beta$ where α and β are constants. Then $d^2x^k/ds^2 = 0$. Of the equations (1) choose those in which $i = k$ and solve them for d^2x^j/ds^2, so obtaining

$$\frac{d^2x^j}{ds^2}=\Pi^{kj}_{pqr}\frac{dx^p}{ds}\frac{dx^q}{ds}\frac{dx^x}{ds}\left(\frac{dx^k}{ds}\right)^{-1} \tag{2}$$

where the index k has the fixed value just chosen. Since the right hand side is homogeneous of the second degree in the first derivatives, these equations determine a system of paths[4] (that is, a system of curves which has the property that, locally, two points are joined by one and only one curve). Consequently if equations (1) are to determine a system of paths it must be that the remaining equations of (1), in which $i \neq k$, are none of them independent of equations (2).

If we substitute in (1) the values of the second derivatives given by (2), we obtain a set of equations which, after multiplication by dx^k/ds, can be written in the form

$$\left(\delta^j_b\Pi^{ki}_{pqr}+\delta^i_b\Pi^{jk}_{pqr}+\delta^k_b\Pi^{ij}_{pqr}\right)\frac{dx^b}{ds}\frac{dx^p}{ds}\frac{dx^q}{ds}\frac{dx^r}{ds}=0. \tag{3}$$

Since these equations, unless the left hand side vanishes identically, are clearly independent of (2), a necessary condition that (1) determine a system of paths is that the

---

Originally published in ***Annals of Mathematics***, ser. 2 vol. 28 (1927), pp. 629–630.

[1] Received May 23, 1927.

[2] O. Veblen and J. M. Thomas, *Projective invariants of affine geometry of paths*, ***Annals of Math.*** (2), vol. 27 (1926), p. 295.

[3] O. Veblen, *Remarks of the foundations of geometry*, ***Bull. Amer. Math. Soc.***, vol. 31 (1925), p. 128. O. Veblen and J. M. Thomas, loc. cit.

[4] J. Douglas, *The most general geometry of paths*, read before the American Mathematical Society May 7, 1927.

left hand side of (3) vanish identically; that is

$$\delta^j_b \Pi^{ki}_{pqr} + \delta^i_b \Pi^{jk}_{pqr} + \delta^k_b \Pi^{ij}_{pqr} + \delta^j_r \Pi^{ki}_{bpq} + \delta^i_r \Pi^{jk}_{bpq} + \delta^k_r \Pi^{ij}_{bpq} + \delta^j_q \Pi^{ki}_{brp} + \delta^i_q \Pi^{jk}_{brp} + \delta^k_q \Pi^{ij}_{brp} + \delta^j_p \Pi^{ki}_{bqr} + \delta^i_p \Pi^{jk}_{bqr} + \delta^k_p \Pi^{ij}_{bqr} = 0.$$

Contracting this for k and b and setting

(4) $$\Pi^{ki}_{kpq} = -\frac{n+1}{3}\Pi^i_{pq},$$

an equation which implies that the quantities Π^i_{pq} transform like a projective connection, we obtain, after dividing through by $n+1$, the condition

(5) $$\Pi^{ij}_{pqr} - \frac{1}{3}\left(\delta^{ij}_{kp}\Pi^k_{qr} + \delta^{ij}_{kq}\Pi^k_{rp} + \delta^{ij}_{kr}\Pi^k_{pq}\right) = 0.$$

Now this is the condition obtained by Veblen and Thomas that the paths determined by (1) have an affine connection. In fact, if we substitute for Π^{ij}_{pqr} in (1) the expression $\frac{1}{3}(\delta^{ij}_{kp}\Pi^k_{qr} + \delta^{ij}_{kq}\Pi^k_{rp} + \delta^{ij}_{kr}\Pi^k_{pq})$, we obtain the equations

$$\delta^{ij}_{ab}\frac{d^2x^a}{ds^2}\frac{dx^b}{ds} + \delta^{ij}_{kp}\Pi^k_{qr}\frac{dx^p}{ds}\frac{dx^q}{ds}\frac{dx^r}{ds} = 0$$

which are known to represent a system of paths which has an affine connection.

Hence we infer that a system of differential equations of the form (1) cannot be used to determine a system of paths which does not have an affine connection.

It is pointed out by Veblen and Thomas that the left hand side of (5) is a tensor. From this it follows that Π^{ij}_{pqr} can be expressed uniquely in the form

$$\Pi^{ij}_{pqr} = T^{ij}_{pqr} + \frac{1}{3}\left(\delta^{ij}_{kp}\Pi^k_{qr} + \delta^{ij}_{kq}\Pi^k_{rp} + \delta^{ij}_{kr}\Pi^k_{pq}\right)$$

where T^{ij}_{pqr} is a tensor and Π^k_{pq} is the projective connection determined by (4). Hence the theory of the invariants of Π^{ij}_{pqr} is equivalent to the theory of the simultaneous invariants of a projective connection and a tensor.

PRINCETON UNIVERSITY

ALTERNATIVES TO ZERMELO'S ASSUMPTION[1]
(1927)

1. The axiom of choice. The object of this paper is to consider the possibility of setting up a logic in which the axiom of choice is false. The way of approach is through the second ordinal class, in connection with which there appear certain alternatives to the axiom of choice. But these alternatives have consequences not only with regard to the second ordinal class but also with regard to other classes, whose definitions do not involve the second ordinal class, in particular with regard to the continuum. And therefore it is possible to consider these alternatives as, in some sense, postulates of logic. In what follows we proceed, after certain introductory considerations, to state these postulates, to inquire into their character, and to derive as many as possible of their consequences.

The axiom of choice, which is also known as Zermelo's assumption,[2] and, in a weakened form, as the multiplication axiom,[3] is a postulate of logic which may be stated in the following way:

Given any set X of classes which does not contain the null class, there exists a one-valued function, F, such that if x is any class of the set X then $F(x)$ is a member of the class x.

An equivalent statement is that there exists an assignment to every class x belonging to the set X of a unique element p such that p is contained in x.

The important case is that in which the set X contains an infinite number of classes, because the assertion of the postulate is obviously capable of proof when the number of classes is finite. Accordingly a convenient, although not quite precise, characterization of the axiom of choice is obtained by saying that it is a postulate which justifies the employment of an infinite number of acts of arbitrary choice.

Instead of assuming that the function F exists in the case of every set X of classes, it is possible to assume only that the function F exists if the set X contains a denumerable infinity of classes.[4] Or we may assume that the function F exists if the set X contains either $\aleph_1$ classes or some less number of classes. In this way we obtain a sequence of postulates, each stronger than those which precede it, all of them weakened forms of Zermelo's assumption, which we may call, respectively, the axiom of choice for sets of $\aleph_0$ classes, the axiom of choice for sets of $\aleph_1$ classes, and so on.

Originally published in the ***Transactions of the American Mathematical Society***, vol. 29 (1927), pp. 178–208. This was Church's doctoral dissertation, Princeton University, 1926.

[1]Presented to the Society, May 2, 1925; received by the editors in June, 1926.

[2]E. Zermelo, *Beweis dass jede Menge wohlgeordnet werden kann*, ***Mathematische Annalen***, vol. 59 (1904), p. 514. See also Zermelo, *Neuer Beweis für die Möglichkeit einer Wohlordnung*, ***Mathematische Annalen***, vol. 65 (1908), p. 110, and Zermelo, *Untersuchungen über die Grundlagen der Mengenlehre*, ***Mathematische Annalen***, vol. 65 (1908), p. 266, where Zermelo states the weaker form of the axiom of choice which Russell has called the multiplication axiom.

[3]B. Russell, *On some difficulties in the theory of transfinite numbers and order types*, ***Proceedings of the London Mathematical Society***, vol. 4 (1907), p. 48; Whitehead and Russell, ***Principia Mathematica***, vol. I, 1910, p. 561.

[4]Cf. B. Russell, ***Introduction to Mathematical Philosophy***, 1919, p. 129.

For our present purpose we wish to exclude all forms of the axiom of choice from among the postulates of logic, so that in what follows no appeal to the axiom of choice is to be allowed.

2. The second ordinal class. As defined by Cantor,[5] the second ordinal class consists of all those ordinals α such that a well-ordered sequence of ordinal number α has $\aleph_0$ as its cardinal number. Instead of this definition we prefer a definition in terms of order alone such as that given in the next paragraph. This definition probably cannot be proved equivalent to Cantor's except with the aid of the axiom of choice for sets of $\aleph_0$ classes. The relation between the two definitions will appear more clearly in §§8, 9, and 10 below.

We shall, therefore, define the second ordinal class by means of the following set of postulates:[6]

1. *The second ordinal class is a simply ordered aggregate.*
2. *There is a first ordinal ω in the second ordinal class.*
3. *If α is any ordinal of the second ordinal class, there is a first ordinal, $\alpha + 1$, of the set of ordinals of the second ordinal dass which follow α.*
4. *If the ordinals $\beta_0, \beta_1, \beta_2, \cdots$ of the second ordinal class are all distinct and form, in their natural order, an ordered sequence ordinally similar to the sequence* $0,1,2,3,\cdots$ *of positive integers, there is an ordinal β of the second ordinal class, the upper limit of the sequence $\beta_0, \beta_1, \beta_2, \cdots$, which is the first ordinal in the set of ordinals which follow every ordinal β_i of this sequence.*
5. *There is no proper subset of the second ordinal class which contains the ordinal ω and which has the property that if it contains the ordinal α it contains also $\alpha + 1$, and if it contains a sequence $\beta_0, \beta_1, \beta_2, \cdots$ of the kind described in Postulate 4 it contains also the upper limit β.*

The fifth postulate makes possible the process of transfinite induction.

The positive integers, including 0, are thought of as forming the first ordinal class and as preceding the ordinal ω in their natural order, so that ω is the upper limit of the sequence $0, 1, 2, 3, \cdots$.

The ordinal Ω is the first ordinal which follows all the ordinals of the second ordinal class. It belongs to the third ordinal class, or would belong to this class if we chose to construct it, as could be done by means of a set of postulates analogous to Postulates 1–5.

If an ordinal α precedes an ordinal β in the arrangement of the ordinals just described (which we shall call the natural order of the ordinals), we say that α is *less than* β and β is *greater than* α.

An ordinal β, other than 0, of the first or the second ordinal class is of the first kind or of the second kind according as there is or is not a greatest ordinal less than β.

The *upper limit* α of a sequence s of ordinals in their natural order such that s contains no greatest ordinal is the least ordinal greater than all the ordinals in s. This

[5] G. Cantor, *Beiträge zur Begründung der transfiniten Mengenlehre*, zweiter Artikel, ***Mathematische Annalen***, vol. 49 (1897), p. 227.

[6] A closely similar definition of the second ordinal class has been given by O. Veblen, *Definition in terms of order alone in the linear continuum and in well-ordered sets*, these ***Transactions***, vol. 6 (1905), p. 170.

ordinal α always exists if s contains no greatest ordinal, but α may sometimes belong to a higher ordinal class than any ordinal in s.

A sequence of distinct ordinals of the first and second ordinal classes, in their natural order, $\beta_0, \beta_1, \beta_2, \cdots$, ordinally similar to the sequence $0, 1, 2, 3, \cdots$ of positive integers, is said to be a *fundamental sequence* of its upper limit β.

A sequence t of ordinals in their natural order is *internally closed* if it contains the upper limits of all its sub-sequences which have an upper limit different from the upper limit of t.

We shall not prove explicitly as consequences of Postulates 1–5 all the theorems about the second ordinal class which we shall need, but we shall make use freely of known theorems whenever these theorems do not depend on the axiom of choice.

The set of all ordinals which are less than a given ordinal α forms, when these ordinals are arranged in their natural order, a well-ordered sequence, which is called the *segment* determined by α, and α is said to be the *ordinal number* of this sequence and of all well-ordered sequences ordinally similar to it. In particular, ω is the ordinal number of the sequence of positive integers in their natural order.

The notions of addition, multiplication, and exponentiation of ordinals, of which we shall need to make some use, either may be defined[7] in terms of the notion of the ordinal number of a well-ordered sequence or may be defined more directly from the postulates by means of a process of induction.

3. Notations for cardinal numbers. The cardinal number of the segment determined by an ordinal α is called the cardinal number corresponding to α. Those cardinals which correspond to some ordinal greater than or equal to ω are called aleph cardinals. When arranged in order of magnitude they form a well-ordered sequence. The first of them is the cardinal number corresponding to ω, which we call[8] $\aleph_0$. The remainder of them are denoted by the letter $\aleph$ (aleph) with an ordinal as a subscript, this ordinal indicating the position of the number in the well-ordered sequence of aleph cardinals.

The cardinal number corresponding to Ω is different[9] from $\aleph_0$. The question whether or not it is $\aleph_1$, the first aleph cardinal after $\aleph_0$ is left open for discussion below.

If the cardinal number of an aggregate S is $\aleph_\alpha$, the cardinal number $\beth_\alpha$ of the aggregate of subsets of S is greater than $\aleph_\alpha$. These cardinal numbers we call beth cardinals and denote them by the letter $\beth$ (beth) with the same subscript as the corresponding aleph. Besides these we may define a set of beth cardinals with two subscripts as follows. If the cardinal number of an aggregate S is $\aleph_\alpha$, and $\aleph_\beta$ is an aleph cardinal less than $\aleph_\alpha$ then $\beth_{\alpha,\beta}$ is the cardinal number of the aggregate of those subsets of S which have a cardinal number not greater than[10] $\aleph_\beta$.

[7] G. Cantor, *Beiträge zur Begründung der transfiniten Mengenlehre*, zweiter Artikel, ***Mathematische Annalen***, vol. 49 (1897), pp. 207–218, and pp. 231–235.

[8] G. Cantor, *Beiträge zur Begründung der transfiniten Mengenlehre*, erster Artikel, ***Mathematische Annalen***, vol. 46 (1895), p. 492.

[9] G. Cantor, *Beiträge zur Begründung der transfiniten Mengenlehre*, zweiter Artikel, ***Mathematische Annalen***, vol. 49 (1897), p. 227.

[10] The cardinal numbers defined in this paragraph are all infinite both in the sense of being non-inductive and in the sense of being reflexive. We shall not be concerned with non-inductive non-reflexive cardinals, although the existence of these cardinals seems to be possible if we deny the axiom of choice. On this class of cardinals see Whitehead and Russell, ***Principia Mathematica***, vol. II, 1912, p. 278 and p. 288.

The question of the distinctness of these cardinals from the aleph cardinals and from one another must be left open.

The cardinal number $\beth_0$ is, by definition, the cardinal number of the class of all classes of positive integers. It is the cardinal number of the continuum of real numbers.[11] It is also the cardinal number of the set of real numbers on a segment of the continuum, the cardinal number of the set of irrational numbers of the continuum, and the cardinal number of the set of irrational numbers on a segment of the continuum. And by means of the expansion of an irrational number as a continued fraction it can be shown that $\beth_0$ is the cardinal number of the class of sequences of positive integers of ordinal number ω.

4. The cardinal number of the class of all well-ordered rearrangements of the positive integers. By a well-ordered rearrangement of the positive integers is to be understood a well-ordered sequence of positive integers such that in it every positive integer occurs once and but once. The well-ordered sequence may be of any possible ordinal number, but no positive integer may be repeated in the sequence and none may be omitted from it. With this understanding we shall prove the following theorem:

THEOREM 1. *The class of all well-ordered rearrangements of the positive integers has the cardinal number* $\beth_0$ *of the continuum.*

For the class of all ordered pairs of positive integers is of cardinal number[12] $\aleph_0$. Therefore the set P of all classes of ordered pairs of positive integers is of cardinal number $\beth_0$.

Now to every well-ordered rearrangement W of the positive integers corresponds a class Q of ordered pairs of positive integers such that the ordered pair (a, b) is contained in Q if and only if a precedes b in W. And no such class Q of ordered pairs of positive integers corresponds in this way to more than one well-ordered rearrangement of the positive integers. Therefore the class of all well-ordered rearrangements of the positive integers can be put into one-to-one correspondence with a part of the set P and therefore with a part of the continuum (because P can be put into one-to-one correspondence with the continuum).

But the class of all well-ordered rearrangements of the positive integers contains a part which can be put into one-to-one correspondence with the continuum, namely the class O of those well-ordered rearrangements of ordinal number ω which have the property that in them the set of odd positive integers and the set of even positive integers occur each in its natural order (so that O contains, for example, the sequence $0, 2, 1, 4, 6, 3, 8, 10, 5, 12, 14, 7, \cdots$). For to the well-ordered rearrangement $a_O, a_l, a_2, \cdots$ contained in O can be correlated the irrational number $(b_0/2)+(b_l/2^2)+(b_2/2^3)+\cdots$ where b_i is equal to O or to 1 according as a_i, is even or odd. In this way can be set up a one-to-one correspondence between O and the set of irrational numbers between O and 1 and therefore between O and the continuum.

The theorem to be proved, therefore, follows by an appeal to the theorem of Schröder and Bernstein[13] which states that if each of two classes can be put into one-to-one

[11] G. Cantor, loc. cit., erster Artikel, p. 488. The class of subsets of a class of cardinal number b evidently has the cardinal number 2^b as defined by Cantor.

[12] G. Cantor, loc. cit., erster Artikel, p. 494.

[13] E. Schröder, *Über zwei Definitionen der Endlichkeit und G. Cantor'sche Satze*, ***Nova Acta Academiae Caesareae Leopoldino-Carolinae Germanicae Naturae Curiosorum***, vol. 71 (1898), pp. 336–340; E. Borel,

correspondence with a part of the other, then the two classes can be put into one-to-one correspondence.[14]

COROLLARY 1. *The continuum can be divided into* $\aleph_1$ *mutually exclusive subsets, each of cardinal number*[15] $\beth_0$.

For the class of well-ordered rearrangements of the positive integers can be so divided by classifying the well-ordered rearrangements of the positive integers according to their ordinal number.

COROLLARY 2. *The class of all well-ordered sequences of positive integers which contain no repetition of any integer is of cardinal number* $\beth_0$.

This corollary and the two following can be proved by the same argument as that used in proving Theorem 1.

COROLLARY 3. *The class of all permutations of the positive integers is of cardinal number*[16] $\beth_0$, *where a permutation of the positive integers is restricted to be of ordinal number* ω.

COROLLARY 4. *The class of all simply ordered sequences of positive integers which contain no repetition of any integer is of cardinal number*[17] $\beth_0$.

The preceding theorem and corollaries are independent of the axiom of choice.

5. The categorical character of the set of postulates 1–5. Between any two aggregates J and J' both of which satisfy Postulates 1–5 of §2 it is possible to set up in the following way a one-to-one correspondence which preserves order. Let the first element ω of J correspond to the first element ω' of J'. Then make the requirement that if the element α of J correspond to the element α' of J' then the element $\alpha + 1$ of J shall correspond

Leçons sur la Theorie des Fonctions, pp. 102–107; A. Korselt, *Uber einen Beweis des Äquivalenzsatzes*, ***Mathematische Annalen***, vol. 70 (1911), pp. 294–296.

A very neat proof is given by J. König, *Sur la théorie des ensembles*, ***Comptes Rendus***, vol. 143 (1906), pp. 110–112.

[14]The Schröder-Bernstein theorem can be proved in such a way that an explicit one-to-one correspondence between the classes in question is set up. Therefore we are able here actually to set up an explicit one-to-one correspondence between the continuum and the class of well-ordered rearrangements of the positive integers. And this means, of course, that the sets of Corollary I are explicitly defined subsets of the continuum.

When we have made explicit in this way the one-to-one correspondence between the continuum and the class of well-ordered rearrangements of the positive integers, it is not difficult to say in simple cases what well-ordered rearrangement of the positive integers corresponds to a given number a of the continuum. But in certain cases the answer to this question involves the solving of difficult (conceivably unsolvable) problems about the dual fractional expansion of a similar to that proposed by L. E. J. Brouwer, ***Mathematische Annalen***, vol. 83 (1921), pp. 209–210, for the decimal expansion π, but more complicated in character. But the correspondence which we have set up is none the less explicit.

[15]W. Sierpinski, in ***Bulletin International de l'Académie des Sciences de Cracovie***, 1918, p. 110, gives another proof, independent of the axiom of choice, that the continuum can be divided into $\aleph_1$ mutually exclusive subsets. The division of the continuum which he effects is actually a division into subsets each of cardinal number $\beth_0$, although he does not prove this.

[16]F. Bernstein, *Untersuchungen aus der Mengenlehre*, ***Mathematische Annalen***, vol. 61 (1905), p. 142.

[17]Cf. F. Bernstein, *Untersuchungen aus der Mengenlehre*, ***Mathematische Annalen***, vol. 61 (1905), pp. 140–145, where it is proved that $\beth_0$ is the cardinal number of the class of all order types in which the set of positive integers can be arranged. Two proofs of this theorem are given, but in both of them it is necessary to use the axiom of choice. Nevertheless an obvious modification of the first of these proofs suffices to prove without the aid of the axiom of choice, not the theorem stated by Bernstein, but the related theorem stated in Corollary 4 above.

to the element $\alpha' + 1$ of J' and to no other element of J', and that if the elements of a fundamental sequence $\beta_0, \beta_1, \beta_2, \cdots$ in J correspond respectively to the elements of a fundamental sequence $\beta'_0, \beta'_1, \beta'_2, \cdots$ in J' then the upper limit β of the first sequence shall correspond to the upper limit β' of the second sequence and to no other element of J'.

Now no element of either aggregate follows next after more than one element of the aggregate. And in either aggregate two fundamental sequences have the same upper limit if and only if it is true that any given ordinal of either sequence precedes some ordinal of the other sequence. From this it follows that if a correspondence between J and J' which satisfies the requirement just stated is one-to-one and preserves order in the case of every element which precedes a given element α of J, then it is one-to-one and preserves order in the case of α also. Therefore we can establish by transfinite induction the existence of a one-to-one correspondence between J and J' which preserves order, the correspondence being constructed in a step by step fashion in accordance with the requirement of the preceding paragraph.

If α is an element of the second kind in J, the corresponding element α' in J' is obtained by choosing a fundamental sequence $\alpha_0, \alpha_1, \alpha_2, \cdots$ for α. Then if $\alpha'_0, \alpha'_1, \alpha'_2, \cdots$ in J' correspond respectively to $\alpha_0, \alpha_1, \alpha_2, \cdots$ in J, α' is the upper limit of the fundamental sequence $\alpha'_0, \alpha'_1, \alpha'_2, \cdots$ in J'. It is true, however, that no matter what fundamental sequence is chosen for α the same corresponding element α' in J' is obtained. The construction of the one-to-one correspondence between J and J' involves, accordingly, no appeal to the axiom of choice.

It will be convenient in this connection to use the following definition of categorical character. A set of postulates is categorical if, given any two systems both of which satisfy all the postulates of the set, there exists a one-to-one correspondence between these two systems which preserves all the relations among their elements which appear as undefined terms in the postulates.

The only such relation which appears in the set of postulates 1–5 under discussion is the relation of order among the ordinals. Therefore we have proved that this set is categorical in the sense just defined.

6. Nature of a categorical set. Suppose that a categorical set S of postulates is given, and two contradictory statements K and L in the form of theorems about the system described by S. Then we are not at liberty to suppose that there exist two systems s_1 and s_2 both of which satisfy S, in one of which K is true, and in the other of which L is true, because, if this were the case, we could obtain a contradiction at once by means of the one-to-one correspondence between s_1 and s_2. In a certain sense, therefore, a categorical set is a complete set, because it is impossible to employ simultaneously two distinct systems which satisfy the same categorical set of postulates.

It does not, however, follow that one of the statements K or L must be inconsistent with the set of postulates[18] S. It is quite conceivable that, although the coexistence of s_1 and s_2 lead to contradiction, nevertheless neither the existence of s_1 alone nor that of s_2 alone should lead to contradiction.

---

[18] E. B. Wilson, *Logic and the continuum*, ***Bulletin of the American Mathematical Society***, vol. 14 (1908), pp. 432–443. See also E. V. Huntington, *A set of postulates for ordinary complex algebra*, these ***Transactions***, vol. 6 (1905), p. 210 and the second footnote on that page.

It is clear that the completeness of the set of postulates at the basis of our logic is involved.[19] We might, not unnaturally, make it one of the requirements for completeness of a set of postulates for logic that, in all such cases as that described above, one of the statements K or L should lead to a contradiction when taken in conjunction with the set S. In the absence, however, of any demonstration that the set of postulates on which our logic is based satisfies this requirement, we must not infer from the fact that a set of postulates S is categorical that there do not exist one or more independent postulates which can be added to the set.

It is not improbable that the set of postulates at the basis of our logic is not complete, even if the axiom of choice is included in the set, because if it were complete it ought to be possible, in the case of every set X of classes which does not contain the null class, to construct a particular function F of the kind whose existence is required by the axiom of choice, a construction the possibility of which is, in many cases, doubtful. If the axiom of choice is excluded the probability of completeness is even more remote.

The question whether the set of postulates at the basis of our logic is or is not complete is evidently equivalent to the question whether or not every mathematical problem can be solved.[20] On the other hand, since each of these questions is in the form of a theorem about the postulates of logic, into the truth of which theorem it is proposed to inquire, neither has a direct connection with the law of excluded middle,[21] which is itself a postulate of logic. Suppose, for example, that we have before us a certain consistent set W of postulates for logic among which is the law of excluded middle. There may be, if W is not complete, a postulate p such that either p or *not-p* can be added to the set W without destroying the consistent character of the set. In this case there may be a universe of discourse U_1 in which p and the postulates of W are satisfied and also a universe of discourse U_2 in which *not-p* and the postulates of W are satisfied. Then p would satisfy the law of excluded middle both in U_1 and in U_2, in U_1 by being true, and in U_2 by being false. Accordingly our inability to conclude on the basis of W whether p is true or false does not prevent our concluding on the basis of W that p is either true or false.

7. Alternatives to Zermelo's assumption. The foregoing discussion is intended to prepare the way for the suggestion that there may be one or more additional independent postulates which can be added to the set of postulates 1–5 and to forestall the objection that this set is already categorical.

[19] It is possible to demonstrate the completeness of a certain portion of the postulates of logic in the sense that no new independent and consistent postulate can be added to this portion without introducing a new undefined term. The portion in question consists of the postulates given by Whitehead and Russell in part I, section A, of the ***Principia mathematica***. The proof of the completeness of these postulates has been given by E. L. Post, *Introduction to a general theory of elementary propositions*, ***American Journal of Mathematics***, vol. 43 (1921), pp. 163–185. But this does not imply the completeness in any sense of the full set of postulates for logic (as at present known), because this full set involves additional undefined terms.

[20] The latter question is proposed by D. Hilbert in *Mathematische Probleme*, ***Göttinger Nachrichten***, 1900, pp. 261–262, and ***Archiv der Mathematik und Physik***, vol. 1 (1901), p. 52, and ***Archiv der Mathematik und Physik***, (3), vol. 1 (1901), p. 52 , and *Mathematical problems*, ***Bulletin of the American Mathematical Society***, vol. 8 (1902), pp. 444–445.

[21] The opposite view is maintained by L. E. J. Brouwer. See for example, *Intuitionistische Mengenlehre*, ***Jahresbericht der Deutschen Mathematiker-Vereinigung***, vol. 28 (1920), pp. 203–208.

With this possibility in mind we propose to examine the consequences of each of the following postulates when it is taken in conjunction with Postulates 1–5:

A. *There exists an assignment of a unique fundamental sequence to every ordinal of the second kind in the second ordinal class.*

B. *There exists no assignment of a unique fundamental sequence to every ordinal of the second kind in the second ordinal class; but given any ordinal α of the second ordinal class, there exists an assignment of a unique fundamental sequence to every ordinal of the second kind less than α.*

C. *There is an ordinal θ of the second ordinal class such that there exists no assignment of a unique fundamental sequence to every ordinal of the second kind less than θ.*

It is an immediate consequence of Postulates 1–5 that there exist fundamental sequences for any particular ordinal of the second kind in the second ordinal class. The preceding postulates A, B, and C are concerned with the possibility of assigning a particular fundamental sequence to every such ordinal in a simultaneous manner.[22]

Postulate A can be derived as a consequence of the axiom of choice for sets of $\aleph_1$ classes. Postulate B implies a denial of the axiom of choice for sets of $\aleph_1$ classes but seems to be consistent with this axiom for sets of $\aleph_0$ classes. Postulate C implies a denial of the axiom of choice for sets of $\aleph_0$ classes. There is no reason, however, to suppose that postulate B implies the axiom of choice for sets of $\aleph_0$ classes or that Postulate A implies this axiom for sets of $\aleph_1$ classes.

Postulates A, B, and C are mutually exclusive and it is clear that, together, they exhaust the conceivable alternatives. There are, therefore, three conceivable kinds of second ordinal classes, one corresponding to each of these postulates. If any one of these involve a contradiction it is reasonable to expect that a systematic examination of its properties will ultimately reveal this contradiction. But if a considerable body of theory can be developed on the basis of one of these postulates without obtaining inconsistent results, then this body of theory, when developed, could be used as presumptive evidence that no contradiction existed.

If there be two of these postulates neither of which leads to contradiction, then there are corresponding to them two distinct self-consistent second ordinal classes, just as euclidean geometry and Lobachevskian geometry are distinct self-consistent geometries, with, however, this difference, that the two second ordinal classes are incapable of existing together in the same universe of discourse.

It is not unlikely that no one of the three postulates A, B, C leads to any contradiction.

8. Consequences of Postulate A.

THEOREM A_1. *If α is any ordinal of the second ordinal class, the cardinal number corresponding to α is*[23] *is $\aleph_0$.*

---

[22]For a discussion of the problem of carrying out such an assignment of fundamental sequences see O. Veblen, *Continuous increasing functions of finite and transfinite ordinals*, these ***Transactions***, vol. 9 (1908), pp. 280–292.

[23]See G. Cantor, loc. cit., zweiter Artikel, p. 221. As already explained, Cantor uses the property described in Theorem A_1 in defining the second ordinal class. He then proves, with the aid of the axiom of choice, that the second ordinal class has the properties expressed by the postulates of §2 above.

The way in which the axiom of choice is involved is pointed out by Whitehead and Russell, ***Principia Mathematica***, vol. III, 1913, p. 170.

For assign to every ordinal β of the second kind which is less than or equal to α fundamental sequence u_β (Postulate A).

The ordinals which precede ω form, when arranged in their natural order, a sequence of ordinal number ω. With this as a starting point, we assign to the ordinals which follow ω, one by one in order, an arrangement of all preceding ordinals in a sequence of ordinal number ω, in the following way.

When we have assigned an arrangement in a sequence t of ordinal number ω of all ordinals which are less than an ordinal γ, an arrangement in a sequence of ordinal number ω of all ordinals which are less than $\gamma + 1$ is obtained by placing γ before t.

When, to every ordinal ζ which is less than an ordinal β of the second kind, we have assigned an arrangement in a sequence t_ζ of ordinal number ω of all ordinals which are less than ζ, the sequences $t_{\beta_0}, t_{\beta_1}, t_{\beta_2}, \cdots$, where $\beta_0, \beta_1, \beta_2, \cdots$ is the fundamental sequence u_β, may be written one below the other so as to obtain the following array:

$$\begin{array}{cccc} \delta_{00}, & \delta_{01}, & \delta_{02}, & \cdots \\ \delta_{10}, & \delta_{11}, & \delta_{12}, & \cdots \\ \delta_{20}, & \delta_{21}, & \delta_{22}, & \cdots \\ . & . & . & . \end{array}$$

where $\delta_{i0}, \delta_{i1}, \delta_{i2}, \cdots$ is the sequence t_{β_i} and contains all the ordinals which are less than β_i. Then the ordinals δ_{ij} may be arranged in a sequence of ordinal number ω as follows:

$$\delta_{00}, \delta_{01}, \delta_{10}, \delta_{02}, \delta_{11}, \delta_{20}, \delta_{03}, \delta_{12}, \delta_{21}, \delta_{30}, \cdots .$$

By omitting from this sequence all occurrences of any ordinal after the first occurrence, so that a sequence without repetitions results, we obtain an arrangement in a sequence of ordinal number ω, of all ordinals less than β.

We may prove by induction that this process continues until we obtain an arrangement in a sequence of ordinal number ω of all ordinals less than α. Therefore the cardinal number corresponding to α is $\aleph_0$.

COROLLARY 1. *There exists an assignment to every ordinal β in the second ordinal class of an arrangement in a sequence of ordinal number ω of all ordinals less than β.*

For, under Postulate A, we may assign a fundamental sequence u_β to every ordinal of the second kind in the second ordinal class. The process just described then continues until we have assigned to every ordinal β of the second ordinal class an arrangement in a sequence of ordinal number ω of all ordinals less than β.

COROLLARY 2. *There exists an assignment to every ordinal β in the second ordinal class of a well-ordered rearrangement of the positive integers of ordinal number β.*

Because an arrangement in a sequence of ordinal number ω of all ordinals less than β determines a one-to-one correspondence between the ordinals less than β and the positive integers, and this one-to-one correspondence determines in turn a well-ordered rearrangement of the positive integers of ordinal number β.

THEOREM A_2. *If the class R consists of all well-ordered sequences of positive integers (allowing any number of repetitions of the same integer) whose ordinal numbers belong to the second ordinal class, the cardinal number of R is $\beth_{1,0}$.*

Let O be the class of ordinals less than Ω. Then the cardinal number of O is $\aleph_1$. Let Q be the class of the subsets of O which have a cardinal number not greater than $\aleph_0$. Then the cardinal number of Q is $\beth_{1,0}$.

The class Q can be put into one-to-one correspondence with a part of R as follows. To the subset S of O contained in Q correlate the sequence $a_0, a_1, a_2, \cdots, a_\omega, \cdots$ of R, where a_μ is 1 or 0 according as μ is or is not contained in S, and the sequence consists of those a's whose subscripts are less then β, where β is the least ordinal greater than every ordinal in S and not less than ω.

The class R can be put into one-to-one correspondence with a part of Q as follows. To the sequence $b_0, b_1, b_2, \cdots, b_\omega, \cdots$ of R, of ordinal number γ, correlate the subset of O consisting of the following ordinals: $0, 1, 2, \cdots, b_0; \omega, \omega + 1, \cdots, \omega + b_1; 2\omega, 2\omega + 1, \cdots, 2\omega + b_2; \cdots, \gamma\omega$.[24]

Therefore, by the Schröder-Bernstein theorem,[25] Q and R can be put into one-to-one correspondence. Therefore the cardinal number of R is $\beth_{1,0}$.

THEOREM A_3. *The cardinal numbers* $\beth_0$ *and* $\beth_{1,0}$ *are identical.*

In accordance with Corollary 1 of Theorem A_1, assign to every ordinal β of the second ordinal class an arrangement in a sequence t_β of ordinal number ω of all ordinals less than β.

Let b be one of the sequences belonging to the class R of the preceding theorem, and let β be the ordinal number of b. Then, corresponding to b, we can determine a sequence c of positive integers of ordinal number ω by the rule that if κ_i is the ith ordinal of t_β then the ith positive integer in c shall be the κ_ith positive integer of b. In this way we can set up a one-to-one correspondence between all the sequences b of R which have a fixed ordinal number β and the class of sequences of positive integers of ordinal number ω. And in exactly the same way we can set up a one-to-one correspondence between those well-ordered rearrangements of the positive integers which have a fixed ordinal number β and the class of permutations of the positive integers, where a permutation of the positive integers is restricted to be of ordinal number ω. But the class of sequences of positive integers of ordinal number ω and the class of permutations of the positive integers can be put into one-to-one correspondence, since each is of cardinal number $\beth_0$. Therefore, choosing a particular such one-to-one correspondence C, we can set up a one-to-one correspondence K_β between the sequences of R which have a fixed ordinal number β and the well-ordered rearrangements of the positive integers which have the ordinal number β. Moreover, since C can be chosen once for all, we have a uniform method of setting up the one-to-one correspondences K_β, and therefore, without appeal to the axiom of choice, we can suppose them all set up, for every ordinal β of the second ordinal class. But as soon as this is done we have a one-to-one correspondence between R and the class of well-ordered rearrangements of the positive integers. And, in Theorems A_2 and 1, we have shown that these two classes have, respectively, the cardinal number $\beth_{1,0}$ and the cardinal number $\beth_0$. Therefore these two cardinal numbers are identical.

COROLLARY. *The class R of Theorem A_2 can be put into one-to-one correspondence with the continuum.*

THEOREM A_4. *The continuum contains a subset of cardinal number*[26] $\aleph_1$.

---

[24]We adopt Cantor's earlier notation, placing the multiplier before the multiplicand.

[25]Loc. cit.

[26]A proof of this theorem has been given by G. H. Hardy, *A theorem concerning the infinite cardinal numbers*, ***Quarterly Journal of Pure and Applied Mathematics***, vol. 35 (1903), pp. 87–94. The use of a

For the class R, which has just been shown to have the cardinal number of the continuum, contains such a subset, namely the set of all those sequences belonging to R which consist entirely of 2's.

The same theorem can be proved by means of Corollary 2 of Theorem A_1, because, in accordance with that corollary, the class of well-ordered rearrangements of the positive integers contains a subset of cardinal number $\aleph_1$, and, by Theorem 1, this class can be put into one-to-one correspondence with the continuum.

The fact that the set of postulates 1–5 and A, which are all statements about the second ordinal class, has consequences about an entirely different aggregate, namely the continuum, is evidently connected with the fact that these postulates contain more than is necessary to render them categorical.

We have already observed that if we are to think of the three second ordinal classes, that corresponding to Postulate A, that corresponding to B, and that corresponding to C, as all existing we must think of them as each existing in a different universe of discourse. Therefore when we single out one of these second ordinal classes for consideration we thereby restrict the character of the universe of discourse within which we are working. In this way we may think of Postulate A as being indirectly a postulate of logic, although it is in form a statement about the second ordinal class. And the same remark applies to Postulates B and C.

9. Consequences of Postulate B.

THEOREM B_1. *If α is any ordinal of the second ordinal class, the cardinal number corresponding to α is $\aleph_0$.*

The proof of Theorem A_1 applies without change.

THEOREM B_2. *There exists no assignment to every ordinal β in the second ordinal class of an arrangement in a sequence of ordinal number ω of all ordinals less than β.*

For, given an arrangement in a sequence t, of ordinal number ω, of all ordinals less than β, we could obtain a fundamental sequence for β by omitting from t all the ordinals α which did not have the property that every ordinal which preceded α in t also preceded α in the natural order of the ordinals. Therefore if there existed an assignment to every ordinal β in the second ordinal class of an arrangement in a sequence of ordinal number ω of all ordinals less than β, there would exist an assignment of a fundamental sequence to every ordinal β in the second ordinal class, contrary to Postulate B.

COROLLARY. *There exists no assignment to every ordinal β in the second ordinal class of a well-ordered rearrangement of the positive integers of ordinal number β.*

THEOREM B_3. *The continuum cannot be well-ordered.*

For it follows from Theorem 1 that, if the continuum could be well-ordered, the set of all well-ordered rearrangements of the positive integers could be arranged in a well-ordered sequence u. And it would then be possible to assign to every ordinal β in the second ordinal class a well-ordered rearrangement of the positive integers of ordinal number β, because we could choose, for every β, the first rearrangement of

simultaneous assignment of a fundamental sequence to every ordinal of the second kind in the second ordinal class is essential to Hardy's proof, as it is to the proof given above.

ordinal number β that occurred in u. This, however, is contrary to the corollary of the preceding theorem.

THEOREM B4. *If the class R consists of all well-ordered sequences of positive integers (allowing any number of repetitions of the same integer) whose ordinal numbers belong to the second ordinal class, the cardinal number of R is* $\beth_{1,0}$.

The proof of Theorem A_2 applies without change.

DEFINITION. Let t be an internally closed sequence of ordinals of the first and second ordinal classes, all distinct, and arranged in their natural order. Then there is one and only one way in which t can be put into one-to-one correspondence (preserving order) with the sequence of ordinals less than Ω or a segment of this sequence. The ordinals of the first and second ordinal classes which are correlated to themselves in this correspondence form, when arranged in their natural order, an internally closed sequence t', the *first derived sequence*[27] of t. The first derived sequence of t' is the *second derived sequence* of t and so on. If ν is an ordinal of the second kind in the second ordinal class, the *νth derived sequence* t^{ν} of t consists of those ordinals which are contained in all second ordinal class the $(\mu+1)$*th derived sequence* $t^{\mu+1}$ of t is the first derived sequence of t^{μ}.

If μ is an ordinal of the first or second class, the μth derived sequence t^{μ} of t is, like t, an internally closed sequence of ordinals of the first and second ordinal classes, all distinct, and arranged in their natural order. And if the ordinal number of t is Ω that of t^{μ} is also[28] Ω.

If the first term of t is greater than 0 the sequence formed by taking the first term of each of the derived sequences of t in order is an internally closed sequence of distinct ordinals in their natural order.[29]

DEFINITION. An internally closed sequence r of ordinals of the second kind of the second ordinal class, all distinct and arranged in their natural order, is a *reduction sequence* if there exists an assignment to every ordinal κ of the second kind in the second ordinal class of a sequence υ_{κ} of distinct ordinals arranged in their natural order, such that the upper limit of υ_{κ} is κ and the ordinal number of υ_{κ} is either ω or one of the ordinals of r.

It follows at once from Postulate B that the ordinal number of a reduction sequence must be Ω.

The sequence of self-residual[30] ordinals of the second ordinal class in their natural order is a reduction sequence. Its first derived sequence, the sequence of ϵ-numbers,[30] is also a reduction sequence.

THEOREM B5. *If the first derived sequence of a reduction sequence r is a reduction sequence, then all the successive derived sequences of r are reduction sequences.*

---

[27]The sequence t determines by its correspondence with the sequence of ordinals less than Ω or a segment thereof a continuous increasing function. The sequence t' consists of the set of values of the first derived function. See O. Veblen, *Continuous increasing functions of finite and transfinite ordinals*, these ***Transactions***, vol. 9 (1908), p. 281.

[28]O. Veblen, loc. cit., pp. 283–285.

[29]O. Veblen, loc. cit., p. 285.

[30]For definitions see G. Cantor, loc. cit., zweiter Artikel.

For let r^ν be the νth derived sequence of r. Assign a fundamental sequence to every ordinal of the second kind less than ν (Postulate B). And to every ordinal of the second kind in the second ordinal class assign a sequence ν_κ of distinct ordinals arranged in their natural order such that the upper limit of ν_κ is κ and the ordinal number of ν_κ is either ω or one of the ordinals of r'. This is possible since, by hypothesis, r' is a reduction sequence.

Let ρ be an ordinal of the second kind which occurs in r' but not in r^ν. Then there is a last ordinal μ, which is less than ν, such that ρ occurs in r^μ. Let α be the ordinal which corresponds to ρ in a one-to-one correspondence (preserving order) between r^μ and the sequence of all ordinals less than Ω.

If μ is of the second kind, let $\mu_0, \mu_1, \mu_2, \cdots$ be the fundamental sequence which we have assigned to μ.

If μ is of the second kind and α is equal to 0, so that ρ stands in the first place in r^μ, the ordinals $\rho_0, \rho_1, \rho_2, \cdots$ which stand in the first place in $r^{\mu_0}, r^{\mu_1}, r^{\mu_2}, \cdots$, respectively, constitute a fundamental sequence for ρ.

If μ is of the second kind and α is of the first kind, let σ be the ordinal which next precedes ρ in r^μ, and let ρ_i be the ordinal which next follows α in r^{μ_i}. The ordinals $\rho_0, \rho_1, \rho_2, \cdots$ are in their natural order, because if ρ_{i+1} were less than ρ_i it would follow that ρ_{i+1} did not occur in r^{μ_i}, contrary to the definition of derived sequences, which requires that all the terms of r^{μ_i+1} shall be ordinals of r^{μ_i}. And $\rho_0, \rho_1, \rho_2, \cdots$ are all distinct, because if ρ_{i+1} were equal to ρ_i it would follow that both were equal to $\beta + 1$, where β was the ordinal which corresponded to σ in a one-to-one correspondence (preserving order) between r^{μ_i} and the sequence of all ordinals less than Ω, contrary to the requirement that all the ordinals of r, and therefore all those of r^{μ_i}, shall be ordinals of the second kind. The upper limit of the sequence $\rho_0, \rho_1, \rho_2, \cdots$, since it is also the upper limit of the sequence $\rho_i, \rho_{i+1}, \rho_{i+2}, \cdots$, necessarily occurs in r^{μ_i}, and, since it occurs in every r^{μ_i}, it occurs also in r^μ. This upper limit cannot be greater than ρ, because, if it were, some ρ_i, say ρ_n, would be greater than ρ, and it would follow that ρ did not occur in r^{μ_n}. But the upper limit of $\rho_0, \rho_1, \rho_2, \cdots$ is greater than σ, because each term is greater than σ. Therefore the upper limit is ρ, so that $\rho_0, \rho_1, \rho_2, \cdots$ is a fundamental sequence for ρ.

If μ is of the first kind (so that $\mu = \lambda + 1$) and α is equal to 0, let ρ_0 be the first ordinal of r^λ, ρ_1 the ρ_0th ordinal of r^λ, ρ_2 the ρ_1th ordinal of r^λ, and so on. Then the sequence $\rho_0, \rho_1, \rho_2, \cdots$ is an increasing sequence, because $\rho_{i+1} < \rho_i$ is impossible on account of the increasing character of r^λ, and $\rho_{i+1} = \rho_i$ would imply $\rho_1 = \rho_0$, a situation which is impossible because ρ_0 cannot be equal to 0 and ρ_1 is the ρ_0th ordinal of r^λ. And the upper limit of $\rho_0, \rho_1, \rho_2, \cdots$ is ρ.

If μ is of the first kind (so that $\mu = \lambda + 1$) and α is also of the first kind, let σ be the ordinal which next precedes ρ in r^μ. Let ρ_0 be the $(\sigma + 1)$th ordinal of r^λ, ρ_1 the ρ_0th ordinal of r^λ, ρ_2 the ρ_1th ordinal of r^λ, and so on. Then the sequence $\rho_0, \rho_1, \rho_2, \cdots$ is an increasing sequence, because $\rho_{i+1} < \rho_i$ is impossible on account of the increasing character of r^λ, and $\rho_{i+1} = \rho_i$ would imply $\rho_0 = \sigma + 1$, whereas ρ_0 must be an ordinal of the second kind. And the upper limit of $\rho_0, \rho_1, \rho_2, \cdots$ is ρ.

The case $\mu = 1$ is taken account of in each of the two preceding paragraphs, for in that case $\lambda = 0$ and r^λ is the sequence r itself.

If α is of the second kind, we have assigned to α a sequence υ_α whose upper limit is α and whose ordinal number is an ordinal τ of r'. If γ is any ordinal less than τ, let α_γ be the γth term of υ_α, and let ρ_γ be the α_γth term of r^μ. Then the sequence $\rho_2, \rho_3, \rho_4, \cdots, \rho_\omega, \rho_{\omega+1}, \cdots$ is a sequence of ordinal number τ whose upper limit is ρ. And α, and therefore τ, is less than ρ. The problem of finding an increasing sequence whose upper limit is ρ and whose ordinal number is either ω or an ordinal of r^ν is, therefore, reduced to the corresponding problem for the smaller ordinal τ, and this reduction continues until we obtain such a sequence or until one of the cases already considered arises.

We can now conclude that it is possible to assign to every ordinal ρ in r' an increasing sequence of ordinals whose upper limit is ρ and whose ordinal number is either ω or an ordinal of r^ν, because we have just described a systematic method of making such an assignment. But we have assigned to every ordinal κ of the second kind in the second ordinal class an increasing sequence υ_κ of ordinals whose upper limit is κ and whose ordinal number is either ω or an ordinal of r'. Therefore we can assign to every ordinal κ of the second kind in the second ordinal class an increasing sequence of ordinals whose upper limit is κ and whose ordinal number is either ω or an ordinal of r^ν. Therefore r^ν is a reduction sequence.

COROLLARY 1. *The sequence $\tilde{r}$ of those ordinals which occur in the first place in the sequences r^θ, where θ takes on all values which make it an ordinal of the second kind, is a reduction sequence.*

For, by the method just given, we can assign to every ordinal ρ of r' an increasing sequence of ordinals whose upper limit is ρ and whose ordinal number is an ordinal of $\tilde{r}$.

If ρ occurs in one of the sequences r^θ in such a way that it has an immediate predecessor σ in r^θ, let ρ_α be the ordinal which next follows σ in r^α, where α is any ordinal less than θ. Then the ordinals ρ_α form an increasing sequence of ordinal number θ whose upper limit is ρ. From this, by means of the sequence υ_θ, we obtain an increasing sequence whose upper limit is ρ and whose ordinal number is an ordinal ρ' of r' less than ρ. The problem of finding an increasing sequence of ordinals whose upper limit is ρ and whose ordinal number is an ordinal of $\tilde{r}$ therefore reduces to the corresponding problem for the smaller ordinal ρ'. And this reduction continues until we obtain such a sequence or one of the other possible cases arises.

In all other cases we proceed exactly as we did in proving Theorem B_5.

COROLLARY 2. *The sequence $\bar{r}$ of those ordinals which occur in the first place in the successive derived sequences of r is a reduction sequence.*

For $\bar{r}$ contains $\tilde{r}$.

COROLLARY 3. *The first derived sequence $\bar{r}'$ of $\bar{r}$, and therefore all the derived sequences of $\bar{r}$, are reduction sequences.*

Since $\tilde{r}$ is a reduction sequence, we can assign to every ordinal κ of the second kind in the second ordinal class an increasing sequence υ_κ of ordinals whose upper limit is κ and whose ordinal number α is either ω or an ordinal of $\tilde{r}$. If α is not an ordinal of $\bar{r}'$, it occurs in the κ'th place in $\bar{r}$, where κ' is an ordinal of the second kind less than κ. The problem of finding an increasing sequence of ordinals whose upper limit is κ and whose ordinal number is either ω or an ordinal of $\bar{r}'$ therefore reduces to the corresponding problem for the smaller ordinal κ', and this reduction continues until an ordinal $\kappa^{(k)}$ is obtained which is either ω or an ordinal of $\bar{r}'$.

COROLLARY 4. *The first derived sequence $\tilde{r}'$ of $\tilde{r}$, and therefore all the derived sequences of $\tilde{r}$, are reduction sequences.*

For $\tilde{r}$ contains $\tilde{r}'$, and therefore $\tilde{r}'$ contains $\tilde{r}''$.

It is possible that it can be proved that every internally closed sequence of ordinal number Ω and consisting of ordinals of the second kind is a reduction sequence. At any rate, we are not able to show the contrary.

In particular, the question naturally arises in this connection whether there exists a reduction sequence whose first derived sequence is not a reduction sequence. It is clear, in view of Theorem B_5, that if such a reduction sequence existed, it could not be the first derived sequence of any sequence. But it does not follow from this alone that such reduction sequences do not exist. In fact it is possible to find internally closed sequences of ordinal number Ω in the second ordinal class which are not first derived sequences of other sequences. In constructing an example we have only to choose 2ω as the first ordinal of the sequence, because, in any order preserving one-to-one correspondence between the set of ordinals less than 2ω and a subset of them, the ordinal ω necessarily corresponds to itself. As an example of a sequence which not only is not the first derived sequence of any sequence but retains that property no matter how many ordinals are omitted from the beginning of it, we may take the sequence s of the ordinals $(2\omega)^{\alpha}$ arranged in order of magnitude, where α takes on all values less than Ω except the value 0. The sequence s and its first derived sequence (namely the sequence of ϵ-numbers) are, however, both reduction sequences.

10. Consequences of Postulate C.

DEFINITION. Under Postulate C there is an ordinal ϕ of the second ordinal class such that there exists no assignment of a unique fundamental sequence to every ordinal of the second kind less than ϕ. The ordinal υ_1 is the least such ordinal ϕ in the second ordinal class.

THEOREM C_1. *The ordinal υ_1 is an ordinal of the second kind.*

For if υ_1 were an ordinal of the first kind, there would be an ordinal α which next preceded υ_1, and there would exist an assignment of a unique fundamental sequence to every ordinal of the second kind less than α. Then, in order to obtain an assignment of a unique fundamental sequence to every ordinal of the second kind less than υ_1, we would have to make at most one arbitrary choice, namely a choice of a fundamental sequence for α if α were of the second kind. Since a single arbitrary choice is always permissible, this would lead to a contradiction with the definition of υ_1. Therefore υ_1 is an ordinal of the second kind.

The way in which Postulate C involves a denial of the axiom of choice for sets of $\aleph_0$ classes can be made clear in this connection by proposing the following argument which purports to show that Postulate C leads to a contradiction. Let $\alpha_0, \alpha_1, \alpha_2, \cdots$ be a fundamental sequence for υ_1. Since α_0 is less than υ_1 it follows at once from the definition of υ_1 that there exists an assignment A_0 of a unique fundamental sequence to every ordinal of the second kind less than α_0, and similarly that there exists an assignment A_1 of a unique fundamental sequence to every ordinal of the second kind less than α_1, an assignment A_2 of a unique fundamental sequence to every ordinal of the second kind less than α_2, and so on. Then, in order to obtain an assignment of a unique fundamental sequence to every ordinal of the second kind less than υ_1, we may use A_0 for ordinals less than α_0, A_1 for ordinals less than α_1 and not less than α_0, A_2 for ordinals less than α_2 and not less than α_1, and so on.

This argument fails to obtain a contradiction from Postulate C because the assignments $A_0, A_1, A_2, \cdots$ are each of them only one of the many existing assignments of the required character, so that there is an element of arbitrary choice involved in fixing upon the particular assignment A_0, another in fixing upon A_1, and so on. The use of all the assignments, $A_0, A_1, A_2, \cdots$, simultaneously, therefore, involves an appeal to Zermelo's assumption, and this is, of course inadmissible in this connection.

THEOREM C_2. *If α is an ordinal of the second ordinal class less than υ_1, the cardinal number corresponding to α is $\aleph_0$.*

Since α is less than υ_1 it is possible to assign to every ordinal β of the second kind less than or equal to α a fundamental sequence u_β. The proof of Theorem A_1, can, therefore, be used here without change.

THEOREM C_3. *The cardinal number corresponding to υ_1 is not $\aleph_0$.*

For suppose that the set of ordinals less than υ_1 could be arranged in a sequence t of ordinal number ω. Then for any ordinal κ less than υ_1, we could obtain a fundamental sequence by omitting from t, first all ordinals which were not less than κ, then all ordinals which did not have the property of being greater than every ordinal less than κ which preceded them in t. This would enable us to assign a fundamental sequence to every ordinal κ of the second kind less than υ_1, contrary to the definition of υ_1.

COROLLARY. *The cardinal number corresponding to υ_1 is $\aleph_1$.*

We have pointed out in §2 above that the definition of the second ordinal class which we are using differs from Cantor's definition. Under the latter the second ordinal class consists of all those ordinals to which the corresponding cardinal number is $\aleph_0$. In connection with Postulate C, these would be the ordinals less than υ_1 and not less than ω, whereas, under the definition which we are using, the second ordinal class contains ordinals greater than υ_1.

The convenience of a definition in terms of order, such as the one which we are using, lies in the fact that it enables us to use unchanged many known theorems about the second ordinal class which have been proved by means of order properties and therefore apply to the more extensive class of ordinals rather than to the ordinals between ω and υ_1. An example is the theorem that if f is a continuous increasing function[31] defined for the set of ordinals less than Ω and its value is always an ordinal less than Ω then the first derived function[31] of f exists.[32] The truth of this theorem is not affected by our choice among the postulates A, B, C, if we adhere to the definition of Ω which we have given. But, as we shall prove in Theorem C_8 below, Postulate C implies the falsehood of the theorem just stated if we take Ω to be the ordinal which we have called υ_1, as we should have to if we used Cantor's definition of the second ordinal class. Another example is the theorem, to which we shall refer below, that every ordinal of the second ordinal class can be expressed in Cantor's normal form. This theorem is true not only of ordinals of the second ordinal class less than υ_1 but of all the ordinals of the second ordinal class as we have defined it, because Cantor's

[31] For definitions, see below.

[32] O. Veblen, loc. cit., p. 283.

proof is equally applicable to ordinals less than υ_1 and to ordinals of the second ordinal class greater than υ_1.

Nevertheless it is true that, in many ways, the ordinal υ_1 plays, in connection with Postulate C, a role similar to that played by Ω in connection with Postulates A and B.

THEOREM C_4. *There exists a denumerable set of subsets $S_0, S_1, S_2, \cdots$ of the continuum such that there is no choice of one element out of each of the sets $S_0, S_1, S_2, \cdots$.*

By Theorem 1, there exists a one-to-one correspondence between the continuum and the set of well-ordered rearrangements of the positive integers. Let K be such a one-to-one correspondence. Let $\alpha_0, \alpha_1, \alpha_2, \cdots$ be a fundamental sequence for υ_1. And let S_i be the set of those numbers of the continuum which correspond under K to well-ordered rearrangements of the positive integers of ordinal number α_i. Then there is no choice of one element out of each of the sets S_i, because, if there were, this would lead to a choice, for each ordinal α_i, of a well-ordered rearrangement of the positive integers of ordinal number α_i, and this would lead in turn to a choice, for each ordinal α_i, of an arrangement of the ordinals less than α_i in a sequence of ordinal number ω, and then, by the method of Theorem A_1, we could obtain an arrangement of the ordinals less than υ_1 in a sequence of ordinal number ω, contrary to Theorem C_3.

COROLLARY. *The continuum cannot be well-ordered.*

THEOREM C_5. *There exists a set T of points of the continuum and a set I of intervals which covers T such that no denumerable subset of I covers[33] T.*

On account of the possibility of setting up a one-to-one correspondence between the continuum and the segment $(0, 1/2)$ of the continuum, the sets S_i of the preceding theorem can be so chosen that all their elements lie between 0 and 1/2. Let the sets S_i be chosen in this way, and let T consist of the points $0, 1, 2, 3, \cdots$. Let I_i consist of all intervals $(i - s_i, i + s_i)$, where s_i is an element of S_i. Then there is a one-to-one correspondence between I_i and S_i. Let I be the sum of all the sets I_i, so that I contains every interval which occurs in any one of the sets I_i.

It is clear that I covers T. Suppose that some denumerable subset J of I covered T. Then J would contain at least one interval in common with each of the sets I_i. And J could be arranged in a sequence of ordinal number ω, and for each set I_i could be chosen the first interval of J which belonged also to I_i. And this would effect a choice of one interval out of each of the sets I_i, corresponding to which there would be a choice of one element out of each of the sets S_i, contrary to the preceding theorem.

COROLLARY. *No subset of I which can be well-ordered covers T.*

---

[33]The contrary of this theorem has been proved by W. H. Young with the aid of the axiom of choice. See ***Proceedings of the London Mathematical Society***, vol. 35 (1902), pp. 384–388, and also W. H. Young and G. C. Young, ***The Theory of Sets of Points***, 1906, pp. 38–40.

See also E. Lindelöf, *Sur quelques points de la théorie des ensembles* ***Comptes Rendus***, vol. 137 (1903), pp. 697–700, and Lindelöf, *Remarques sur un théorème fondamental de la théorie des ensembles*, ***Acta Mathematica***, vol. 29 (1905), pp. 187–189. Lindelöf's theorem is stronger than that given by Young in that it applies to space of any finite number of dimensions and weaker in that there must be a one-to-one correspondence between the points of I and the intervals (or n-spheres) of T, each point being at the center of the corresponding interval (or n-sphere). The axiom of choice is necessary to the proof.

The falsity of Lindelöf's theorem for one dimension can be proved as a consequence of Postulate C by means of the following modification of the proof of Theorem C_5. Let T consist of all points $i + (1/2)s_i$, $i = 0, 1, 2, 3, \cdots$, and to the point $i + (1/2)s_i$, let correspond the interval $(i, i + s_i)$, I_i consisting of all intervals $(i, i + s_i)$ for a fixed value of i, and I being the sum of the sets I_i.

DEFINITION. An internally closed sequence r of ordinals of the second kind less than υ_1 all distinct and arranged in their natural order is a *reduction sequence* if there exists an assignment to every ordinal κ of the second kind less than υ_1 of a sequence υ_κ of distinct ordinals arranged in their natural order, such that the upper limit of υ_κ is κ and the ordinal number υ_κ is either ω or one of the ordinals of r.

We shall use the same definition of a derived sequence as that given in the preceding section.

DEFINITION. A function f defined for all ordinals less than a certain ordinal Ξ, the value of f being always an ordinal, is a *continuous increasing function*,[34] if, for every pair of ordinals ξ_1 and ξ_2, both less than Ξ, such that ξ_1 is less than ξ_2, it is true that $f(\xi_1)$ is less than $f(\xi_2)$, and the set t of all ordinals $f(\xi)$, where ξ takes on all values less than Ξ, forms an internally closed sequence when the ordinals are arranged in their natural order. The *αth derived function*[34] of f is the continuous increasing function of ξ, $f(\xi,\alpha)$, determined by the one-to-one correspondence between the αth derived sequence t^α of t and the whole or a segment of the sequence of all ordinals less than Ξ.

THEOREM C_6. *The ordinal υ_1 is an ϵ-number and occupies the υ_1th place in the sequence of ϵ-numbers arranged in their natural order.*

Since ω^ξ is a continuous increasing function of ξ, we know that ω^{υ_1} is greater than or equal to υ_1.

Suppose that ω^{υ_1} is greater than υ_1. Then, since the sequence of ordinals in their natural order is internally closed, there is a greatest ordinal α such that ω^α is not greater than υ_1. And α is less than υ_1. Assign to every ordinal β of the second kind less than or equal to α a fundamental sequence $\beta_0, \beta_1, \beta_2, \cdots$. Since the ordinals ω^ξ are the self-residual ordinals, every ordinal κ of the second kind less than υ_1 can be written in a unique way in the form $\gamma + \omega^\beta$, where γ has its least possible value, which may be 0, and β is not greater than α. If β is of the first kind it has an immediate predecessor ζ, and a fundamental sequence for κ is $\gamma + \omega^\zeta, \gamma + 2\omega^\zeta, \gamma + 3\omega^\zeta, \cdots$ (if ζ is equal to 0, ω^ζ is to be taken equal to 1). If β is of the second kind, the sequence $\gamma + \omega^{\beta_0}, \gamma + \omega^{\beta_1}, \gamma + \omega^{\beta_2}, \cdots$ is a fundamental sequence for κ. In this way we are able to assign a fundamental sequence to every ordinal κ of the second kind less than υ_1, contrary to the definition of υ_1.

Therefore ω^{υ_1} is equal to υ_1. Therefore υ_1 is an ϵ-number.

In the sequence of ϵ-numbers in their natural order let the αth place be the place occupied by υ_1. Then α is less than or equal to υ_1.

Suppose that α is less than υ_1. To every ordinal θ of the second kind less than α assign α a fundamental sequence $\theta_0, \theta_1, \theta_2, \cdots$. Every ordinal κ of the second kind less than υ_1 can be written in a unique way in the form $\gamma + \omega^{\kappa'}$ where γ has its least possible value, which may be 0, and κ' is not greater than κ.

If κ' is an ϵ-number, so that $\kappa' = \omega^{\kappa'} = \epsilon_\theta$, then θ is less than α. If θ is of the second kind, a fundamental sequence for κ is $\gamma + \epsilon_{\theta_0}, \gamma + \epsilon_{\theta_1}, \gamma + \epsilon_{\theta_2}, \cdots$. If θ is of the first kind it has an immediate predecessor δ, and a fundamental sequence for κ is[35]

$$\gamma + \epsilon_\delta + 1, \gamma + \omega^{\epsilon_\delta + 1}, \gamma + \omega^{\omega^{\epsilon_\delta + 1}}, \cdots.$$

[34]O. Veblen, loc. cit.

[35]G. Cantor, loc. cit., zweiter Artikel, p. 243.

If κ' is of the first kind it has an immediate predecessor ζ, and a fundamental sequence for κ is $\gamma + \omega^\zeta, \gamma + 2\omega^\zeta, \gamma + 3\omega^\zeta, \cdots$.

If κ' is of the second kind but is not an ϵ-number, then the problem of finding a fundamental sequence for κ reduces to that of finding a fundamental sequence for the ordinal κ', less than κ, and this reduction continues in the same way until an ordinal $\kappa^{(k)}$ is obtained which either is of the first kind or is an ϵ-number.

In this way we are able to assign a fundamental sequence to every ordinal κ of the second kind less than υ_1, contrary to the definition of υ_1.

Therefore α is equal to υ_1.

COROLLARY 1. *The ordinal υ_1 is a self-residual ordinal.*

COROLLARY 2. *The sequence of self-residual ordinals less than υ_1 arranged in their natural order, and its first derived sequence, the sequence of ϵ-numbers less than υ_1 arranged in their natural order, are reduction sequences.*

THEOREM C_7. *If the first derived sequence of a reduction sequence r is a reduction sequence, then the ordinal number of r is υ_1 and the first υ_1 derived sequences of r exist and are reduction sequences of ordinal number υ_1.*

If ν is an ordinal less than υ_1 we can prove by the same argument as that used in proving Theorem B_5 that r^ν is a reduction sequence, and this argument applies even if we suppose r^ν empty. And then, as soon as we have proved that r^ν is a reduction sequence, it follows from Postulate C that r^ν is not empty. Therefore the first υ_1 derived sequences of r exist and are reduction sequences.

It remains to prove that r and its first υ_1 derived sequences are all of ordinal number υ_1.

The ordinal number α of r^ν cannot be greater than υ_1 because r^ν consists entirely of ordinals less than υ_1. Suppose that α is less than υ_1. Then $r^{\nu+1}$ contains no ordinal greater than or equal to α. Therefore we can assign a fundamental sequence $\rho_0, \rho_1, \rho_2, \cdots$ to every ordinal ρ in $r^{\nu+1}$. But to every ordinal κ of the second kind less than υ_1 we can assign an increasing sequence $\kappa_0, \kappa_1, \kappa_2, \cdots, \kappa_\omega, \cdots$ of ordinals whose upper limit is κ and whose ordinal number is either ω or an ordinal ρ of $r^{\nu+1}$. If the ordinal number of this increasing sequence is not ω, a fundamental sequence for κ is $\kappa_{\rho_0}, \kappa_{\rho_1}, \kappa_{\rho_2}, \cdots$. Therefore we can assign a fundamental sequence to every ordinal κ of the second kind less than υ_1, contrary to the definition of υ_1.

Therefore the ordinal number α of r^ν is υ_1. And in the same way we can prove that the ordinal number of r is υ_1.

COROLLARY 1. *Let $f(\xi) = \omega^\xi$. Then, if α is less than υ_1, $f(\upsilon_1, \alpha) = \upsilon_1$.*

COROLLARY 2. *If $f(\xi) = \omega^\xi$, then $f(0, \upsilon_1) = \upsilon_1$.*

COROLLARY 3. *The sequence $\bar{r}$ of those ordinals which occur in the first place in the successive derived sequences of r is a reduction sequence of ordinal number υ_1, and the first υ_1 derived sequences of $\bar{r}$ are reduction sequences of ordinal number υ_1.*

The proof of this is the same as that of the corollaries to Theorem B_5.

COROLLARY 4. *Let $f(\xi) = \omega^\xi$ and $\phi(\xi) = f(0, \xi)$. Then if α is less than υ_1, $\phi(\upsilon_1, \alpha) = \upsilon_1$ and $\phi(0, \upsilon_1) = \upsilon_1$.*

THEOREM C_8. *There exists a continuous increasing function defined for the set of ordinals less than υ_1, the value of which is always an ordinal less than υ_1 and the first derived function of which does not exist.*

Since υ_1 is of the second kind, there exists a fundamental sequence $\alpha_0, \alpha_1, \alpha_2, \cdots$ for υ_1. Let $f(\xi) = \omega^\xi$, and let $\beta_i = f(0, \alpha_i + 1)$ for $i = 0, 1, 2, \cdots$. Then the upper limit of the sequence $\beta_0, \beta_1, \beta_2, \cdots$ is υ_1. And β_i is the least value of ξ such that $f(\xi, \alpha_i) = \xi$.

Let a function F be defined as follows. If $0 \leqq \xi \leqq \beta_0$, $F(\xi) = f(\xi, \alpha_1)$, and if $\beta_i < \xi \leqq \beta_{i+1}$, $F(\xi) = f(\xi, \alpha_{i+2})$. Then F is a continuous increasing function defined for the set of ordinals less than υ_1, and its value is always an ordinal less than υ_1, but there is no ordinal ξ less than υ_1 such that $F(\xi) = \xi$, and therefore the first derived function of F does not exist.

11. Properties of υ-numbers.

DEFINITION. The υ-numbers are those ordinals κ of the second ordinal class, greater than ω, which have the property that the cardinal number corresponding to κ is greater than that corresponding to any ordinal less than κ.

If Postulate C is denied, it follows that υ-numbers do not exist (Theorems A_1 and B_1).

If Postulate C is accepted, we are assured of the existence of at least one υ-number, namely υ_1. Postponing the question how many υ-numbers the second ordinal class contains, we are able to say that, in any case, the numbers arranged in order of magnitude form a well-ordered sequence, so that the αth ordinal of this sequence, counting from υ_1 as the first ordinal of the sequence, may be indicated by the symbol υ_α.

The following theorems are consequences of Postulates 1–5 alone, but since they become vacuous if Postulate A or Postulate B is accepted, we think of them as belonging with Postulate C.

THEOREM C_9. *The υ-numbers are ordinals of the second kind.*

For let $\alpha + 1$ be any ordinal of the first kind in the second ordinal class. Then the set of ordinals of the second ordinal class less than $\alpha + 1$ can be arranged in a sequence of ordinal number α by placing α first and letting the remaining ordinals follow in their natural order, thus, $\alpha, 0, 1, 2, 3, \cdots, \omega, \omega + 1, \cdots$. But if $\alpha + 1$ were a υ-number, the set of ordinals of the second ordinal class less than $\alpha + 1$ could not be arranged in a sequence of ordinal number less than $\alpha + 1$. Therefore $\alpha + 1$ is not a υ-number.

THEOREM C_{10}. *Given any increasing sequence s of υ-numbers, $\upsilon_{\alpha_0}, \upsilon_{\alpha_1}, \upsilon_{\alpha_2}, \cdots$ of ordinal number ω, the upper limit υ_α of s is a υ-number.*

The cardinal number corresponding to υ_α cannot be equal to that corresponding to any ordinal υ_{α_i} of s, because, if it were, it would be less than that corresponding to $\upsilon_{\alpha_{i+1}}$. Therefore the cardinal number corresponding to υ_α is greater than that corresponding to any ordinal of s. Therefore the cardinal number corresponding to υ_α is greater than that corresponding to any less ordinal. Therefore υ_α is a υ-number.

THEOREM C_{11}. *The υ-numbers are ϵ-numbers.*

We have already shown that υ_1 is an ϵ-number. We shall show by transfinite induction that the remaining υ-numbers (if any other υ-numbers exist) are also ϵ-numbers.

If υ_α is a υ-number which is also an ϵ-number, then the next following υ-number $\upsilon_{\alpha+1}$ (if it exist) is also an ϵ-number. For suppose $\upsilon_{\alpha+1}$ is not an ϵ-number. Then it can be written in Cantor's normal form:[36]

$$a_0\omega^{\upsilon_0} + a_1\omega^{\upsilon_1} + \cdots + a_n\omega^{\upsilon_n}$$

where $\upsilon_{\alpha+1} > \upsilon_0 > \upsilon_1 > \cdots > \upsilon_n$, the coefficients a_i are finite ordinals, and the sum contains a finite number of terms in all. Since υ_α is an ϵ-number, $\omega^{\upsilon_\alpha} = \upsilon_\alpha$. Therefore υ_0 is greater than υ_α. Therefore $\upsilon_0 + 1$ is greater than υ_α. And, since υ_0 is less than $\upsilon_{\alpha+1}$, $\upsilon_0 + 1$ is also less than $\upsilon_{\alpha+1}$, by Theorem C_9. Consequently, since $\upsilon_{\alpha+1}$ is the next υ-number after υ_α the set of ordinals less than $\upsilon_0 + 1$ can be put into one-to-one correspondence with the set of ordinals less than υ_α. Let such a one-to-one correspondence be set up, and if κ is any ordinal of the set of ordinals less than $\upsilon_0 + 1$, let κ' be the corresponding ordinal of the set of ordinals less than υ_α. Now every ordinal less than than $\upsilon_{\alpha+1}$ can be written in Cantor's normal form, $\sum_i b_i\omega^{\mu_i}$, where $\upsilon_0 + 1 > \mu_0 > \mu_1 > \cdots$, the coefficients b_i are finite ordinals, and the sum contains a finite number of terms in all. The understanding is that one of the exponents μ_i may have the value 0, and that ω^0 is to be taken equal to 1. To the ordinal $\sum_i b_i\omega^{\mu_i}$, less than $\upsilon_{\alpha+1}$, let correspond the ordinal $\sum_i b_i\omega^{\mu_i'}$, where the terms of the sum are to be arranged in order of magnitude, the greatest first. Then $\sum_i b_i\omega^{\mu_i'}$ is less than υ_α, because $\upsilon_\alpha = \omega^{\upsilon_\alpha}$, and all the exponents μ_i' are less than υ_α. We have, accordingly, set up in this way a one-to-one correspondence between the set of all ordinals less than $\upsilon_{\alpha+1}$ and a certain set of ordinals less than υ_α. But this is impossible, because the cardinal number corresponding to $\upsilon_{\alpha+1}$ is greater than that corresponding to υ_α. Therefore the supposition that $\upsilon_{\alpha+1}$ was not an ϵ-number was incorrect.

If every ordinal of the increasing sequence of υ-numbers $\upsilon_{\beta_0}, \upsilon_{\beta_1}, \upsilon_{\beta_2}, \cdots$ is an ϵ-number, the upper limit υ_β of the sequence is an ϵ-number, because the ϵ-numbers form in order of magnitude an internally closed sequence.

Therefore, by transfinite induction, every υ-number is an ϵ-number.

COROLLARY. *If α is any ordinal of the second ordinal class, the cardinal number corresponding to α^ω is the same as that corresponding to α. And, therefore, if β is any ordinal greater than α and less than α^ω, the cardinal number corresponding to β is the same as that corresponding to α.*

For the least ϵ-number ϵ_β greater than α is the upper limit of the sequence[37] $\alpha + 1, \omega^{\alpha+1}, \omega^{\omega^{\alpha+1}}, \cdots$ and is, therefore, greater than $\omega^{\omega^{\alpha+1}}$, or $(\omega^{\omega^\alpha})^\omega$. And ω^{ω^α} is greater than or equal to α. Therefore ϵ_β is greater than α^ω. But the least υ-number greater than α is greater than or equal to ϵ_β and therefore greater than α^ω. Therefore the cardinal number corresponding to α^ω is the same as that corresponding to α.

12. Postulates F and G. In connection with Postulate C there appear two possibilities, which we shall state as Postulates F and G, inconsistent with each other, but each apparently consistent with Postulates 1–5 and C. These possibilities are the following:

F. *If Ψ is any ordinal of the second ordinal class, there is some ordinal α of the second ordinal class, such that there exists no assignment to every ordinal κ of the second kind less than α of an increasing sequence υ_κ of ordinals such that the upper limit of υ_κ is κ and the ordinal number of υ_κ is less than Ψ.*

[36] G. Cantor, loc. cit., zweiter Artikel, p. 237.

[37] G. Cantor, loc. cit., zweiter Artikel, p. 243.

G. *There is an ordinal Ψ of the second ordinal class such that, given any ordinal α of the second ordinal class, there exists an assignment to every ordinal κ of the second kind less than α of an increasing sequence υ_κ of ordinals such that the upper limit of υ_κ is κ and the ordinal number of υ_κ is less than Ψ.*

Postulate F is stated in such a way that it implies Postulate C, but Postulate G does not.

We shall examine briefly the consequences of each of the postulates just stated when taken in conjunction with Postulates 1–5 and C, taking the same experimental attitude as that which we took in the case of Postulates A, B, and C.

13. Consequences of Postulate F.

THEOREM F_1. *If υ_η is any υ-number, there exists a υ-number greater than υ_η.*

For suppose the contrary. Then there exists a greatest υ-number, υ_β. Let ψ be the ordinal $\upsilon_\beta + 1$ and let α be any ordinal of the second ordinal class. Then the set of all ordinals less than α can be arranged in a sequence t_α of ordinal number less than ψ. Then there exists an assignment to every ordinal κ of the second kind less than α of an increasing sequence υ_κ of ordinals such that the upper limit of υ_κ is κ and the ordinal number of υ_κ is less than ψ. For we could choose υ_κ to be the sequence obtained by omitting from t_α, first all ordinals not less than κ, and than all ordinals which do not have the property of being greater than every ordinal less than κ which precedes them in t_α.

This, however, is contrary to Postulate F. Therefore if υ_η is any υ-number, there exists a υ-number greater than υ_η.

THEOREM F_2. *The sequence of the υ-numbers of the second ordinal class arranged in order of magnitude is an internally closed sequence of ordinal number Ω.*

This follows at once from the preceding theorem and Theorem C_{10}.

COROLLARY. *The cardinal number corresponding to Ω is $\aleph_\Omega$.*

14. Consequences of Postulate G. Turning now to the consequences of Postulates C and G taken together, we recall that, in accordance with Postulate G, there exist ordinals ψ in the second ordinal class such that, given any ordinal α of the second ordinal class, there exists an assignment to every ordinal κ of the second kind less than α of an increasing sequence υ_κ of ordinals such that the upper limit of υ_κ is κ and the ordinal number of υ_κ is less than ψ. Let Υ be the least such ordinal ψ. Then

THEOREM CG_1. *The ordinal Υ is an ordinal of the second kind.*

For suppose that Υ is an ordinal of the first kind. Then there exists an ordinal β such that Υ is equal to $\beta + 1$.

In accordance with the definition of Υ, given any ordinal α of the second ordinal class, there exists an assignment to every ordinal κ of the second kind less than α of an increasing sequence υ_κ of ordinals such that the upper limit of υ_κ is κ and the ordinal number of υ_κ is less than Υ and therefore less than or equal to β.

If β is an ordinal of the first kind the ordinal number υ_κ cannot be equal to β, because only those sequences in which there is no greatest ordinal have an upper limit. Therefore in this case the ordinal number of υ_κ is always less than β, contrary to the definition of Υ.

If β is an ordinal of the second kind it has a fundamental sequence $\beta_0, \beta_1, \beta_2, \cdots$. Those sequences υ_κ which are of ordinal number β can then be replaced by sequences υ'_κ of ordinal number ω obtained by omitting from υ_κ all ordinals except those in the positions $\beta_0, \beta_1, \beta_2, \cdots$. And in this way we obtain again a contradiction of the definition of Υ.

Therefore Υ is an ordinal of the second kind.

THEOREM CG_2. *The ordinal Υ is a υ-number.*

Suppose that to some ordinal β less than Υ corresponds the same cardinal number as to Υ. Then the set of ordinals less than Υ can be rearranged in a sequence of ordinal number β. Choosing a particular such rearrangement t of the set of ordinals less than Υ, we have a uniform method of rearranging any given well-ordered sequence s of ordinal number greater than β but not greater than Υ in a well-ordered sequence of ordinal number less than or equal to β, because there is a one-to-one correspondence between s and the whole or a segment of the sequence u of ordinals less than Υ arranged in order of magnitude, so that the rearrangement t of u determines the desired rearrangement of s.

Now in accordance with the definition of Υ, given any ordinal α of the second ordinal class, there exists an assignment to every ordinal κ of the second kind less than α of an increasing sequence υ_κ of ordinals such that the upper limit of υ_κ is κ and the ordinal number of υ_κ is less than Υ. In accordance with the preceding paragraph we can rearrange υ_κ as a well-ordered sequence of ordinal number less than or equal to β, and by omitting from the rearranged sequence all ordinals which do not have the property of being greater than every ordinal which precedes them in the sequence we obtain an increasing sequence w_κ of ordinals such that the upper limit of w_κ is κ and the ordinal number of w_κ is less than or equal to β and therefore less than $\beta + 1$.

Since, by hypothesis, β is less than Υ so that Υ cannot be less than $\beta + 1$, it follows, in view of the definition of Υ, that Υ is equal to $\beta + 1$. This, however, is contrary to Theorem CG_1.

Therefore there is no ordinal β less than Υ such that the same cardinal number corresponds to β as to Υ.

But it follows at once from Postulate C that Υ is greater than ω. Therefore Υ is a υ-number.

THEOREM CG_3. *The ordinal Υ is the greatest υ-number.*

Let α be an ordinal of the second ordinal class, greater than Υ. Assign to every ordinal κ of the second kind which is less than or equal to α an increasing sequence υ_κ of ordinals such that the upper limit of υ_κ is κ and the ordinal number of υ_κ is less than Υ.

The set of ordinals which precede Υ form, when arranged in their natural order, a sequence of ordinal number Υ. With this as a starting point assign to the ordinals which follow Υ, one by one in order, an arrangement of all preceding ordinals in a sequence of ordinal number Υ, in the following way.

When we have assigned an arrangement in a sequence t of ordinal number Υ of all ordinals which are less than an ordinal γ, an arrangement in a sequence of ordinal number Υ of all ordinals which are less than $\gamma + 1$ is obtained by placing γ before t.

When we have assigned to every ordinal ζ which is less than an ordinal β of the second kind an arrangement in a sequence t_ζ of ordinal number Υ of all ordinals which are less than ζ, the sequences $t_{\beta_0}, t_{\beta_1}, t_{\beta_2}, \cdots, t_{\beta_\omega}, \cdots$, where $\beta_0, \beta_1, \beta_2, \cdots, \beta_\omega, \cdots$ is the sequence υ_β, may be written one after the other so as to obtain a sequence u_β of ordinal number not greater than Υ^2. In accordance with the corollary of Theorem C_{11} the set of ordinals less than Υ^2 can be rearranged in a sequence of ordinal number Υ. Choosing a particular such rearrangement s of the set of ordinals less than Υ^2, we have a uniform method of rearranging any sequence u_β in a well-ordered sequence w_β of ordinal number Υ, because there is a one-to-one correspondence between u_β and the whole or a segment of the sequence υ of ordinals less than Υ^2 arranged in their natural order, so that the rearrangement s of υ determines the desired rearrangement w_β of u_β (the ordinal number of w_β cannot be less than Υ on account of the fact that Υ is a υ-number). By omitting from w_β all occurrences of any ordinal after the first occurrence, so that a sequence without repetition results, we obtain an arrangement in a sequence of ordinal number Υ of all ordinals less than β.

We may prove by induction that this process continues until we obtain an arrangement in a sequence of ordinal number Υ of all ordinals less than α. Therefore the cardinal number corresponding to α is the same as that corresponding to Υ. But α was any ordinal of the second ordinal class greater than Υ. Therefore Υ is the greatest υ-number.

COROLLARY. *The sequence of the υ-numbers of the second ordinal class arranged in order of magnitude is an internally closed sequence whose ordinal number is an ordinal of the first kind less than Ω.*

It should be noted that there is nothing in the preceding to preclude the possibility that υ_1 and Υ are the same ordinal, in which case υ_1 would be the only υ-number.

PRINCETON UNIVERSITY,
PRINCETON, N.J.

ON THE LAW OF EXCLUDED MIDDLE (1928)

1. Introduction. The purpose of this paper is to discuss the possibility of a system of logic in which the law of excluded middle is not assumed, and also to point out what seem to be errors in a recent paper[1] in which the conclusion is reached that such a system of logic is self-contradictory.

The law of excluded middle is the logical principle in accordance with which every proposition is either true or false. This principle is used, in particular, whenever a proof is made by the method of *reductio ad absurdum*. And it is this principle, also, which enables us to say that the denial of the denial of a proposition is equivalent to the assertion of the proposition.

The simplest alternative to the inclusion of the law of excluded middle among the principles of logic is its bare omission without assertion of any contrary principle. The effect of such an omission is, of course, to reduce the number of theorems which can be proved and also to render of interest certain theorems otherwise vacuous. We cannot derive theorems which contradict theorems obtained with the aid of the law of excluded middle unless we make some assertion of a principle which contradicts the law of excluded middle.

It is not possible, as an alternative to the law of excluded middle, to assert that some proposition is neither true nor false, because by so doing not only the law of excluded middle would be denied but also the law of contradiction. In fact, to assert that a proposition p is not true and is also not false is to assert at once *not*-p and *not*-(*not*-p) and consequently to assert that *not*-p is both true and false.

It may be, however, that, by introducing the middle ground between true and false as an undefined term, let us say "tiers" (adopting from the French the word used by Barzin and Errera), and making an appropriate set of assumptions about the existence and properties of tiers propositions, we can produce a system of logic which is consistent with itself but which becomes inconsistent if the law of excluded middle be added.

2. The position of L. E. J. Brouwer. L. E. J. Brouwer proposes that the law of excluded middle should not be regarded as an admissible logical principle,[2] and expresses, as a basis for his proposal, doubts concerning the truth of this law. He says, for example, that the law of excluded middle has been extended to the mathematics of infinite classes by an unjustified analogy with that of finite classes. He says also that to assert the law of excluded middle is equivalent to asserting the doubtful proposition that every proposed theorem can be either proved or disproved if the proper method be found. The latter point depends, of course, on identifying the truth of a proposition with the possibility of proving the proposition. But it seems more in accord with our usual ideas to think of truth as a property of a proposition independent of our ability

---

Originally published in the ***Bulletin of the American Mathematical Society***, vol. 34 (1928), pp. 75–78.

[*Editor's note*: In the original Church was marked as National Research Fellow in Mathematics.]

[1] M. Barzin and A. Errera, *Sur la logique de M. Brouwer*, Académie Royale de Belgique, ***Bulletins de la Classe des Sciences***, (5), vol. 13 (1927), pp. 56–71.

[2] See, for example, *Intuitionistische Mengenlehre*, ***Jahresbericht der Deutschen Mathematiker-Vereinigung***, vol. 28 (1919), pp. 203–208.

to prove it. Consequently we prefer to take the truth of a proposition merely as an undefined term subject to certain postulates, among them, if we choose to include it, the law of excluded middle.

In connection with geometry and other branches of mathematics it is commonly recognized that it is meaningless to ask about the absolute truth of a postulate and that the choice between one of two contrary postulates must be made on the basis of simplicity and serviceability. It seems reasonable to recognize the same thing with regard to the postulates of logic, in particular the law of excluded middle, and to say on this basis that it is meaningless to ask about the truth of the law of excluded middle.

Taking this point of view, we may accept a system of logic in which the law of excluded middle is assumed, a system in which the law of excluded middle is omitted without making a contrary assumption, and a system which contains assumptions not in accord with the law of excluded middle as all three equally admissible, unless one of them can be shown to lead to a contradiction. If we had to choose among these systems of logic, we could choose the one most serviceable for our purpose, and we might conceivably make different choices for different purposes.

3. Barzin and Errera's Paper. Barzin and Errera, however, reach the conclusion (loc. cit.) that the system of logic proposed by Brouwer leads to contradiction. This conclusion we believe to be erroneous, for the following reasons.

The method of the argument is the method of *reductio ad absurdum*. It is assumed that if the law of excluded middle is not accepted then it must be explicitly denied by asserting the existence of tiers propositions, and on this basis contradictory results are obtained. This argument is clearly not effective against one who merely omits the law of excluded middle from his system of logic without assuming any contrary principle, because the insistence that one who refuses to accept a proposition must deny it can be justified only by an appeal to the law of excluded middle, the very principle in doubt. The method of *reductio ad absurdum*, in fact, necessarily employs the law of excluded middle and cannot be used against one who does not admit this law.

The argument of Barzin and Errera would not be effective, however, even against one who asserted the existence of tiers propositions, because in the course of the argument it is necessary to assume what the authors name the principle of excluded fourth, that every proposition is either true, or false, or tiers, a principle which seems to be a restricted form of the law of excluded middle. This assumption is defended on the basis of the definition of a tiers proposition as one which is neither true nor false, but this definition, as pointed out above, is not consistent with itself and consequently cannot be used. If the concept *tiers* is introduced at all, it must be as an undefined term.

If we admit the concept tiers and the principle of excluded fourth, the argument of Barzin and Errera is then quite correct. They show that "p is tiers" together with "p obeys the principle of excluded fourth" implies a contradiction; hence that p is not tiers; and hence, by an appeal to the principle of excluded fourth, that p is either true or false. But one who asserts the existence of tiers propositions need not assert the principle of excluded fourth; and one who merely omits the law of excluded middle without introducing the concept *tiers* certainly could not assert the principle of excluded fourth.

It would not be permissible to say that the word *tiers* is to cover all possibilities other than true and false, however many such possibilities there may be. This language is, in fact, equivalent to defining tiers to mean neither true nor false, and is open to the same objections. But the following objection could also be raised. Using p' to stand for "p is tiers," it seems likely that $(p')'$, or p'', would constitute a fourth possibility, $(p'')'$ a fifth possibility, and so on. Uniting these possibilities into one under a new name, let us say p^*, we then have to reckon with the likelihood that $(p^*)'$ gives still another possibility which we may designate by the transfinite ordinal ω. And after this we have to reckon with an $(\omega + 1)$th possibility, and so on, so that ultimately we reach any given transfinite ordinal whatever. Any attempt to unite all these possibilities into one must be regarded as extremely doubtful, because of its connection with Burali-Forti's[3] paradox concerning the ordinal number of the sequence of all transfinite ordinals. This paradox, in fact, compels us to regard as illegitimate the consideration of this sequence as a whole.

Harvard University

[3]C. Burali-Forti, *Una questione sui numeri transfiniti*, ***Rediconti del Circolo Matematico di Palermo***, vol. 11 (1897), pp. 154–164. Whitehead and Russell, ***Principia mathematica***, vol. III, pp. 73–75.

REVIEW: *PRINCIPIA MATHEMATICA*, VOLUMES II AND III (1928)

Principia Mathematica. By Alfred North Whitehead and Bertrand Russell. Volumes II and III. Second Edition. Cambridge University Press, 1927. xxxi+742 pp., and viii+491 pp.

The second and third volumes of the second edition of Whitehead and Russell's *Principia Mathematica* do not differ from the corresponding volumes of the first edition. An account of the changes which the authors think desirable is contained in the introduction and appendices to the first volume of the second edition, of which a review has previously appeared,[1] but the text of all three volumes has been left unchanged.

The second and third volumes of the *Principia Mathematica* are devoted to building up, on the basis of the system of logic developed in the first volume, the theories of cardinal numbers, relations and relation-numbers, series, well-ordered series and ordinal numbers, and finally of the continuum and of real numbers. The task of developing these theories on the basis of the theorems and processes of logic only, as well as that, undertaken in the first volume of investigating logic itself by mathematical methods, are both of the highest importance to our understanding of the foundations of mathematics. The magnitude of the undertaking thoroughly justifies the formidable appearance of the three volumes, which contain what is, so far, the most nearly successful attempt to accomplish these tasks. In spite of serious difficulties which remain unsolved, the work has established its claim to be ranked as a notable achievement.

Volume II begins with an account of cardinal numbers, based on the definition, given originally by Frege, that the cardinal number of a class α is the class of all classes similar to α. This definition has a pragmatic justification, in that it leads to cardinal numbers which have the properties we require. But it seems worth while to notice the possibility of another definition[2] more in accord with our intuitive idea of number, namely that the cardinal number of a class α is the abstraction from α with respect to the propositional function *x is similar to y*. In order to make this definition, we require the following postulate:[3] If ϕ is any propositional function of two variables which is transitive and symmetric, and if A is any term such that $\phi(A, x)$ is true for some x, then A_ϕ, the abstraction from A with respect to ϕ, exists and has the property that $A_\phi = B_\phi$ if and only if $\phi(A, B)$ holds. The obvious objection is that this is an unnecessary additional postulate which ought to be avoided, but this objection, we believe, can be removed by a consideration of the treatment of classes in the first volume of the *Principia Mathematica.*

According to this treatment the existence of classes is not assumed and the word class itself is not defined, but a determinate meaning is given to propositions about classes. On this basis, classes appear as shadowy things without actual existence. In a certain sense this is the correct view, for classes are, essentially, not a part of reality but fictions devised for their usefulness as instruments in understanding reality. But

---

Originally published in ***Bulletin of the American Mathematical Society***, vol. 34 (1928), pp. 237–240.

[1] B. A. Bernstein, this ***Bulletin***, vol. 32 (1926), pp. 711–713.

[2] G. Peano, ***Formulaire de Mathématiques***, vol. 3 (1901), p. 70.

[3] See B. Russell, ***The Principles of Mathematics***, 1903, p. 166, where a similar postulate is proposed.

the same statement is true of propositions and propositional functions and, in fact, all the terms of logic and of mathematics. A postulate asserting the existence of a logical or mathematical term ought, indeed, to be understood merely as making this term, by fiat, a part of a certain abstract structure which is being built. The treatment of classes given by Whitehead and Russell seems not well balanced in that it ascribes a greater degree of reality to propositions and propositional functions than to classes. A postulate asserting the existence of classes probably ought to be included in their treatment.

If, however, we accept the postulate about abstractions proposed above, it is not necessary to postulate the existence of classes, because they can be introduced as follows. A class is determined by a propositional function, but classes differ from propositional functions in that two equivalent but distinct propositional functions determine the same class. By analogy with Whitehead and Russell's definition of cardinal number, we would, except for the obvious circle, define the class determined by a propositional function ϕ as the class of propositional functions equivalent to ϕ. On the basis of the postulate which asserts the existence of abstractions, we could define the class determined by a propositional function ϕ as the abstraction from ϕ with respect to equivalence, and so avoid the necessity of another postulate for the introduction of classes.

These considerations would be changed, of course, if we accepted the postulate proposed in the introduction to the second edition of the *Principia Mathematica*, that functions of propositions are always truth-functions, and that a function can occur in a proposition only through its values, for this implies that equivalent propositional functions are identical, and hence that classes and propositional functions may be regarded as the same. But the authors themselves are doubtful about this proposal and their arguments in support of it are not fully convincing. In fact "*A* asserts *p*," which they take as a crucial instance, and which is certainly not a truth-function, seems to be a function of the proposition *p*. The statement that *A* asserts *p* is correctly analyzed not as meaning that *A* utters certain sounds (the analysis proposed by Whitehead and Russell) but that *A* utters sounds which have a certain content of meaning, and it is this content of meaning which constitutes the proposition *p*, in the usual sense of the word proposition. The proposal under discussion can mean only that the word proposition is to be used in some quite different sense, in which case the *Principia Mathematica* would fail to give any account of propositions in the usual sense of the word.

Continuing our consideration of the contents of volumes II and III, we find of especial interest the construction of the system of inductive cardinals (positive integers) on the basis of logical postulates only, because it is well known how the real number system, the space of euclidean or projective geometry, and other important mathematical systems, can be constructed on the basis of the system of positive integers, apart from difficulties which arise in connection with the theory of logical types.

The success of this construction of the system of inductive cardinals depends largely on the adoption of the following definition: Consider all those classes of cardinal numbers which contain 0 and which have the property that if they contain ν then they contain $\nu + 1$; the common part of all these classes is the class of inductive cardinals. The numbers 0 and 1 and the operation of addition have, of course, been previously defined. Apart from difficulties connected with the theory of types (which theory we hope to see supplanted or greatly modified) this definition gives us the complete system of positive integers with all its usual properties, something certainly not present in the original assumptions (except in the sense that these assumptions enable us to construct the system of positive integers in the way just described).

This definition seems to be the only one that will lead to the desired result. It might suggest itself to consider the cardinal numbers of those classes which are similar to no proper part of themselves, but, without the aid of the multiplicative axiom or a weakened form of it, it is not known how to prove that this class of cardinals is not more extensive than that of the inductive cardinals defined as above.

The axiom of infinity, which it is found necessary to assume in proving important familiar properties of the finite cardinals, is of interest because, as one of the authors has pointed out elsewhere,[4] the existence of infinite classes can be proved if we disregard restrictions imposed by the theory of types.

Of the remaining subjects treated, that of real numbers and the treatment of Dedekindian and continuous series are of greatest interest.

It is well known how to construct the system of real numbers once the series of positive integers is given, but the usual method takes no account of difficulties raised by the theory of types. According to their original scheme, the authors proposed to overcome these difficulties by means of the axiom of reducibility, but in the introduction to the second edition they abandon this axiom in favor of the postulate mentioned above, that functions of propositions are always truth-functions, and that a function can occur in a proposition only through its values. As a result of this substitution the system of real numbers can no longer be adequately dealt with by any method known to the authors.

If for no other reason than that it leads to important results to which the other does not, the axiom of reducibility seems to be distinctly preferable to the postulate which Whitehead and Russell now propose as a substitute for it.

It is to be hoped, however, that the question will ultimately be settled by the complete abandonment of the theory of logical types or by an alteration in it more radical than any yet proposed. For the theory of types, although adequate to obtain the results we require, provided that the axiom of infinity and the axiom of reducibility are admitted, introduces many complications and creates some awkward situations, one of them the following, referred to at the beginning of volume II of the *Principia Mathematica*. Having proved the theorems that we require about functions of the first $n-1$ types, then in order to obtain the same theorems about functions of the nth type we must make a new assumption of all our postulates, applying them to functions of the nth type instead of functions of some lower type, and must then prove all our theorems anew. We "see," by symbolic analogy, that this can always be done. But the statement that this is so is impossible under the theory of types.

The well known contradictions, of which a list is given in the first volume of the *Principia Mathematica*, must, of course, be dealt with, and the theory of types at present affords the best known method of doing this. Our hope is that another method can be found which will entail fewer complications.

ALONZO CHURCH[5]

---

[4]B. Russell, ***The Principles of Mathematics***, 1903, p. 357.

[5]National Research Fellow in Mathematics.

REVIEW OF RAMSEY'S *THE FOUNDATIONS OF MATHEMATICS* (1932)

The Foundations of Mathematics. By F. P. Ramsey. New York, Harcourt, Brace, & Co., 1931. xviii + 292 pages.

This is a collection of essays in mathematics, mathematical logic, and philosophy, including the most important of the author's previously published works, and a number of papers, some of them fragmentary, which were still unpublished at the time of the author's death in 1930.

The first of these essays, which was originally published in the *Proceedings of The London Mathematical Society* in 1926, sets forth the author's proposal for revision of the system of logic contained in Whitehead and Russell's *Principia Mathematica*. Three important changes are advocated, namely, abandonment of the principle that x and y are identical (or equal) when every propositional function satisfied by x is also satisfied by y, abandonment of the principle that every class is determined by a propositional function, and a modification of the theory of types designed to avoid the much discussed axiom of reducibility.

Of these proposals, the first seems to be open to serious objection. For if x and y are any two things which have all their properties in common, and if we allow that x has the property of being identical with x, then we must allow that y also has the property of being identical with x, that is, that $y = x$. The possibility that x may not be a definable object, and therefore that the property of being identical with x may not be a definable property, is clearly irrelevant.

And the second proposal is open to an entirely similar objection, for if α is a class, then the propositional function $\hat{x} \in \alpha$ determines α, and also the propositional function, "$\hat{x}$ has all the elementary properties which are common to all the members of α," determines α. Of course, there may be such things as undefinable classes, and if α is an undefinable class then there is no definable propositional function which determines α. But this, as we understand it, is not denied by the authors of *Principia Mathematica*.

The third proposal, however, is worthy of more serious consideration. Mr. Ramsey observes that the contradictions with which the theory of types is intended to deal fall into two classes, firstly what we may call the mathematical paradoxes (for example the paradox of Burali-Forti) which can be expressed entirely by means of the symbols of formal logic, and secondly what we may call the epistemological paradoxes (for example Richard's paradox) which require the use of the verb "means." Two hierarchies of types, of rather different sorts, are required, one of them for the sake of avoiding the mathematical paradoxes, and the other to avoid the epistemological paradoxes. And it is the second of these two hierarchies which renders the axiom of reducibility necessary. Mr. Ramsey's proposal is to replace this second hierarchy of types by another hierarchy of types, which he defines, and under which an axiom of reducibility is not necessary. And he then gives, in terms of this new hierarchy, a solution of the epistemological paradoxes which depends on an ambiguity in the meaning of "means."

Originally published in ***American Mathematical Monthly***, vol. 39 (1932), pp. 355–357.

Distrust of the axiom of reducibility is, of course, widespread, being shared even by the authors of *Principia Mathematica*, and there seems to be no doubt of the desirability of a theory which avoids this axiom. But we cannot agree with Mr. Ramsey, that the reason for the desirability of avoiding it is that the axiom is not a tautology in the sense of Wittgenstein, or that it is desirable or necessary that all the axioms of logic should be tautologies. For the notion of a tautology in this sense depends for its intuitive significance on the identification of $(x).\varphi x$ with the logical product of all possible values of φx, and this identification seems to be doubtful, because the nature of the logical product is such that its meaning cannot be understood without first understanding the meaning of each proposition which enters into the logical product.

And certainly the notion of a tautology loses much of its connotation of "necessary" when we discover that the axiom of infinity is a tautology if it be true, but a contradiction if it be false.

It is worth remarking that the proof that the axiom of infinity is a tautology (if true) depends, not, as the author seems to imply, on his proposed revision of the notion of identity, but wholly on the introduction of propositional functions in extension. This notion of a propositional function in extension is certainly legitimate, but it seems doubtful whether the distinction can successfully be maintained between ordinary propositional functions and propositional functions in extension.

In the second essay of the book, the author defends his treatment of mathematical logic against the formalism of Hilbert and the intuitionism of Brouwer and Weyl. The third essay, which is devoted to the solution of a particular case of the Entscheidungsproblem, contains also a theorem of the theory of classes which is not without interest on its own account. And the remaining papers are, for the most part, on subjects which are philosophical, rather than mathematical, in character. All of them are well worth reading, for the sake of the author's insight into the questions of which he treats, and for their power of stimulating thought on the part of the reader.

ALONZO CHURCH

A SET OF POSTULATES FOR THE FOUNDATION OF LOGIC[1,2] (1932)

1. Introduction. In this paper we present a set of postulates for the foundation of formal logic, in which we avoid use of the free, or real, variable, and in which we introduce a certain restriction on the law of excluded middle as a means of avoiding the paradoxes connected with the mathematics of the transfinite.

Our reason for avoiding use of the free variable is that we require that every combination of symbols belonging to our system, if it represents a proposition at all, shall represent a particular proposition, unambiguously, and without the addition of verbal explanations. That the use of the free variable involves violation of this requirement, we believe is readily seen. For example, the identity

$$a(b+c) = ab + ac \tag{1}$$

in which a, b, and c are used as free variables, does not state a definite proposition unless it is known what values may be taken on by these variables, and this information, if not implied in the context, must be given by a verbal addition. The range allowed to the variables a, b, and c might consist of all real numbers, or of all complex numbers, or of some other set, or the ranges allowed to the variables might differ, and for each possibility equation (1) has a different meaning. Clearly, when this equation is written alone, the proposition intended has not been completely translated into symbolic language, and, in order to make the translation complete, the necessary verbal addition must be expressed by means of the symbols of formal logic and included, with the equation, in the formula used to represent the proposition. When this is done we obtain, say,

$$R(a)R(b)R(c) \supset_{abc} . \, a(b+c) = ab + ac \tag{2}$$

where $R(x)$ has the meaning "x is a real number," and the symbol $\supset_{abc}$ has the meaning described in §§5 and 6 below. And in this expression there are no free variables.

A further objection to the use of the free variable is contained in the duplication of symbolism which arises when the free, or real, variable and the bound, or apparent variable are used side by side.[3] Corresponding to the proposition, represented by equation (1) when a, b, and c stand for any three real numbers, there is also a proposition expressed without the use of free variables, namely (2), and between these two propositions we know of no convincing distinction. An attempt to identify the two propositions is, however, unsatisfactory, because substitution of (1) for (2), when the latter occurs as a part of a more complicated expression, cannot always be allowed without producing confusion. In fact, the only feasible solution seems to be the complete abandonment of the free variable as a part of the symbolism of formal logic.[4]

Originally published in ***Annals of Mathematics***, vol. 33 (1932), pp. 346–366.

[1]Received October 5, 1931.

[2]This paper contains, in revised form, the work of the author while a National Research Fellow in 1928–29.

[3]Cf. the introduction to the second edition of Whitehead and Russell's ***Principia Mathematica***.

[4]Unless it is retained as a mere abbreviation of notation.

Rather than adopt the method of Russell for avoiding the familiar paradoxes of mathematical logic,[5] or that of Zermelo,[6] both of which appear somewhat artificial, we introduce for this purpose, as we have said, a certain restriction on the law of excluded middle.

This restriction consists in leaving open the possibility that a propositional function **F** may, for some values **X** of the independent variable, represent neither a true proposition nor a false proposition. For such a value **X** of the independent variable we suppose that **F**(**X**) is undefined and represents nothing, and we use a system of logical symbols capable of dealing with propositional functions whose ranges of definition are limited.

In the case of the Russell paradox the relevance of this proposed restriction on the law of excluded middle is evident. The formula **P** which leads to this paradox may be written, in the notation explained below, $\{\lambda\varphi \,.\, \sim\varphi(\varphi)\}(\lambda\varphi \,.\, \sim\varphi(\varphi))$. It has the property that if we assume $\sim$ **P** then we can infer **P** and if we assume **P** then we can infer $\sim$ **P**. On ordinary assumptions both the truth and the falsehood of **P** can be deduced in consequence of this property, but the system of this paper, while it provides for the existence of a propositional function $\lambda\varphi \,.\, \sim\varphi(\varphi)$ does not provide either that this propositional function shall be true or that it shall be false, for the value $\lambda\varphi \,.\, \sim\varphi(\varphi)$ of the independent variable.

Other paradoxes either disappear in the same way, or else, as in the case of the Epimenides or the paradox of the least undefinable ordinal, they contain words which are not definable in terms of the undefined symbols of our system, and hence need not concern us.

The paradox of Burali-Forti is not, however, so readily disposed of. The question whether this paradox is a consequence of our postulates, or what modification of them will enable us to avoid it, probably must be left open until the theory of ordinal numbers which results from the postulates has been developed.

Whether the system of logic which results from our postulates is adequate for the development of mathematics, and whether it is wholly free from contradiction, are questions which we cannot now answer except by conjecture. Our proposal is to seek at least an empirical answer to these questions by carrying out in some detail a derivation of the consequences of our postulates, and it is hoped either that the system will turn out to satisfy the conditions of adequacy and freedom from contradiction or that it can be made to do so by modifications or additions.

2. Relation to intuitionism. Since, in the postulate set which is given below, the law of the excluded middle is replaced by weaker assumptions, the question arises what the relation is between the system of logic which results from this set, and the intuitionism of L. E. J. Brouwer.[7]

---

[5]B. Russell, *Mathematical logic as based on the theory of types*, ***Amer. Jour. Math.***, vol. 30 (1908), pp. 222–262. A list of some of these paradoxes, with a reference to the source of each, will be found in this article, or in Whitehead and Russell, ***Principia Mathematica***, vol. 1, pp. 63–64.

[6]E. Zermelo, *Untersuchungen über die Grundlagen der Mengenlehre*, ***Math. Annalen***, vol. 65 (1908), pp. 261–281.

[7]See L. E. J. Brouwer, *Intuitionistische Mengenlehre*, ***Jahresbericht der D. Math. Ver.***, vol. 28 (1919), pp. 203–208, and Brouwer, *Mathematik, Wissenschaft und Sprache*, ***Monatshefte für Math. u. Phys.***, vol. 36 (1929), pp. 153–164, and many other papers.

The two systems are not the same, because, although both reject a certain part of the principle of the excluded middle, the parts rejected are different. The law of double negation, denied by Brouwer, is preserved in the system of this paper, and the principle, which Brouwer accepts, that a statement[8] from which a contradiction can be inferred is false, we find it necessary to abandon in certain cases.

Our system appears, however, to have the property, which relates it to intuitionism, that a statement of the form $\Sigma\mathbf{x} \,.\, \mathbf{F}(\mathbf{x})$ (read, "there exists **x** such that $\mathbf{F}(\mathbf{x})$") is never provable unless there exists a formula **M** such that $\mathbf{F}(\mathbf{M})$ is provable.

3. The abstract character of formal logic. We do not attach any character of uniqueness or absolute truth to any particular system of logic. The entities of formal logic are abstractions, invented because of their use in describing and systematizing facts of experience or observation, and their properties, determined in rough outline by this intended use, depend for their exact character on the arbitrary choice of the inventor.

We may draw the analogy of a three dimensional geometry used in describing physical space, a case for which, we believe, the presence of such a situation is more commonly recognized. The entities of the geometry are clearly of abstract character, numbering as they do planes without thickness and points which cover no area in the plane, point sets containing an infinitude of points, lines of infinite length, and other things which cannot be reproduced in any physical experiment. Nevertheless the geometry can be applied to physical space in such a way that an extremely useful correspondence is set up between the theorems of the geometry and observable facts about material bodies in space. In building the geometry, the proposed application to physical space serves as a rough guide in determining what properties the abstract entities shall have, but does not assign these properties completely. Consequently there may be, and actually are, more than one geometry whose use is feasible in describing physical space. Similarly, there exist, undoubtedly, more than one formal system whose use as a logic is feasible, and of these systems one may be more pleasing or more convenient than another, but it cannot be said that one is right and the other wrong.

In consequence of this abstract character of the system which we are about to formulate, it is not admissible, in proving theorems of the system, to make use of the meaning of any of the symbols, although in the application which is intended the symbols do acquire meanings. The initial set of postulates must of themselves define the system as a formal structure, and in developing this formal structure reference to the proposed application must be held irrelevant. There may, indeed, be other applications of the system than its use as a logic.

4. Intuitive logic. It is clear, however, that formulas composed of symbols to which no meaning is attached cannot define a procedure of proof or justify an inference from one formula to another. If our postulates were expressed wholly by means of the symbols of formal logic without use of any words or symbols having a meaning, there would be no theorems except the postulates themselves. We are therefore obliged to use in some at least of our postulates other symbols than the undefined terms of the formal system, and to presuppose a knowledge of the meaning of these symbols, as

[8] We purposely use the word, "statement", because we wish to preserve the word, "proposition", for something either true or false. A statement, in the form of a proposition, which fails to be either true or false, we regard as a mere group of symbols, without significance.

well as to assume an understanding of a certain body of principles which these symbols are used to express, and which belong to what we shall call intuitive logic.[9] It seems desirable to make these presuppositions as few and as simple as we can, but there is no possibility of doing without them.

Before proceeding to the statement of our postulates, we shall attempt to make a list of these principles of intuitive logic which we find it necessary to assume and of the symbols a knowledge of whose meanings we presuppose. The latter belong to what we shall call the language of intuitive logic, as distinguished from the language of formal logic which is made up of the undefined terms of our abstract system.

We assume that we know the meaning of the words *symbol* and *formula* (by the word formula we mean a set of symbols arranged in an order of succession, one after the other). We assume the ability to write symbols and to arrange them in a certain order on a page, and the ability to recognize different occurrences of *the same* symbol and to distinguish between such a double occurrence of a symbol and the occurrence of *distinct* symbols. And we assume the possibility of dealing with a formula as a unit, of copying it at any desired point, and of recognizing other formulas as being *the same* or *distinct*.

We assume that we know what it means to say that a certain symbol or formula *occurs* in a given formula, and also that we are able to pick out and discuss a particular *occurrence* of one formula in another.

We assume an understanding of the operation of *substituting* a given symbol or formula *for a particular occurrence* of a given symbol or formula.

And we assume also an understanding of the operation of substitution throughout a given formula, and this operation we indicate by an S, $S_{\mathbf{Y}}^{\mathbf{X}}\mathbf{U}|$ representing the formula which results when we operate on the formula **U** by replacing **X** by **Y** throughout, where **Y** may be any symbol or formula but **X** must be a single symbol, not a combination of several symbols.

We assume that we know how to recognize a given formula as *being obtainable* from the formulas of a certain set by *repeated combinations* of the latter according to a given law. This assumption is used below in defining the term "well-formed." It may be described as an assumption of the ability to make a definition by induction, when dealing with groups of symbols.

We assume the ability to make the assertion that a given formula is one of those belonging to the abstract system which we are constructing, and this assertion we indicate by the words *is true*. As an abbreviation, however, we shall usually omit the words *is true*, the mere placing of the formula in an isolated position being taken as a sufficient indication of them.

We assume further the meaning and use of the word *every* as part of the language of intuitive logic, and the use in connection with it of variable letters, which we write in bold face type to distinguish them from variable letters used in the language of formal logic. These variable letters, written in bold face type stand always for a variable (or undetermined) symbol or formula.

We assume the meaning and use of the following words from the language of intuitive logic: *there is, and, or, if ··· then, not*, and *is* in the sense of identity. That is, we assume

[9]The principles of intuitive logic which we assume initially form, of course, a part of the body of facts to which the formal system, when completed, is to be applied. We should not, however, allow this to confuse us as to the clear cut distinction between intuitive logic and formal logic.

that we know what combinations of these words with themselves and our other symbols constitute propositions, and, in a simple sense, what such propositions mean.

We assume that we know how to distinguish between the words and symbols we have been enumerating, which we shall describe as *symbols of intuitive logic*, and other symbols, which are mere symbols, without meaning, and which we shall describe as *formal symbols*. We assume that we know what it is to be a proposition of intuitive logic, and that we are able to assert such propositions, not merely one proposition, but various propositions in succession. And, finally, we assume the permanency of a proposition once asserted, so that it may at any later stage be reverted to and used as if just asserted.

In making the preceding statements it becomes clear that a certain circle is unavoidable in that we are unable to make our explanations of the ideas in question intelligible to any but those who already understand at least a part of these ideas. For this reason we are compelled to assume them as known in the beginning independently of our statement of them. Our purpose has been, not to explain or convey these ideas, but to point out to those who already understand them what the ideas are to which we are referring and to explain our symbolism for them.

5. Undefined terms. We are now ready to set down a list of the undefined terms of our formal logic. They are as follows:

$$\{\ \}(\ ),\lambda[\ ],\Pi,\Sigma,\&,\sim,\iota,A.$$

The expressions $\{\ \}(\ )$ and $\lambda[\ ]$ are not, of course, single symbols, but sets of several symbols, which, however, in every formula which will be provable as a consequence of our postulates, always occur in groups in the order here given, with other symbols or formulas between as indicated by the blank spaces. In addition to the undefined terms just set down, we allow the use, in the formulas belonging to the system which we are constructing, of any other formal symbol, and these additional symbols used in our formulas we call *variables*.

An occurrence of a variable x in a given formula is called an occurrence of x as a *bound variable* in the given formula if it is an occurrence of x in a part of the formula of the form $\lambda\mathsf{x}[\mathsf{M}]$; that is, if there is a formula M such that $\lambda\mathsf{x}[\mathsf{M}]$ occurs in the given formula and the occurrence of x in question is an occurrence in $\lambda\mathsf{x}[\mathsf{M}]$. All other occurrences of a variable in a formula are called occurrences as a *free variable*.

A formula is said to be *well-formed* if it is a variable, or if it is one of the symbols $\Pi,\Sigma,\&,\sim,\iota,A$, or if it is obtainable from these symbols by repeated combinations of them of one of the forms $\{\mathsf{M}\}(\mathsf{N})$ and $\lambda\mathsf{x}[\mathsf{M}]$, where x is any variable and M and N are symbols or formulas which are being combined. This is a definition by induction. It implies the following rules (1) a variable is well-formed (2) $\Pi,\Sigma,\&,\sim,\iota$ and A are well-formed (3) if M and N are well-formed then $\{\mathsf{M}\}(\mathsf{N})$ is well-formed (4) if x is a variable and M is well-formed then $\lambda\mathsf{x}[\mathsf{M}]$ is well-formed.

All the formulas which will be provable as consequences of our postulates will be well-formed and will contain no free variables.

The undefined terms of a formal system have, as we have explained, no meaning except in connection with a particular application of the system. But for the formal system which we are engaged in constructing we have in mind a particular application, which constitutes, in fact, the motive for constructing it, and we give here the meanings which our undefined terms are to have in this intended application.

If **F** is a function and **A** is a value of the independent variable for which the function is defined, then $\{\mathbf{F}\}(\mathbf{A})$ represents the value taken on by the function **F** when the independent variable takes on the value **A**. The usual notation is $\mathbf{F}(\mathbf{A})$. We introduce the braces on account of the possibility that **F** might be a combination of several symbols, but, in the case that **F** is a single symbol, we shall often use the notation $\mathbf{F}(\mathbf{A})$ as an abbreviation for the fuller expression.[10]

Adopting a device due to Schönfinkel,[11] we treat a function of two variables as a function of one variable whose values are functions of one variable, and a function of three or more variables similarly. Thus, what is usually written $\mathbf{F}(\mathbf{A}, \mathbf{B})$ we write $\{\{\mathbf{F}\}(\mathbf{A})\}(\mathbf{B})$, and what is usually written $\mathbf{F}(\mathbf{A}, \mathbf{B}, \mathbf{C})$ we write $\{\{\{\mathbf{F}\}(\mathbf{A})\}(\mathbf{B})\}(\mathbf{C})$, and so on. But again we frequently find it convenient to employ the more usual notations as abbreviations.

If **M** is any formula containing the variable **x**, then $\lambda\mathbf{x}[\mathbf{M}]$ is a symbol for the function whose values are those given by the formula. That is, $\lambda\mathbf{x}[\mathbf{M}]$ represents a function, whose value for a value **L** of the independent variable is equal to the result $\mathrm{S}^{\mathbf{x}}_{\mathbf{L}}\mathbf{M}|$ of substituting **L** for **x** throughout **M**, whenever $\mathrm{S}^{\mathbf{x}}_{\mathbf{L}}\mathbf{M}|$ turns out to have a meaning, and whose value is in any other case undefined.

The symbol Π stands for a certain propositional function of two independent variables, such that $\Pi(\mathbf{F}, \mathbf{G})$ denotes, "$\mathbf{G}(x)$ is a true proposition for all values of x for which $\mathbf{F}(x)$ is a true proposition." It is necessary to distinguish between the proposition $\Pi(\mathbf{F}, \mathbf{G})$ and the proposition 'x . $\mathbf{F}(x) \supset \mathbf{G}(x)$ (read, "For every x, $\mathbf{F}(x)$ implies $\mathbf{G}(x)$"). The latter proposition justifies, for any value **M** of x, the inference $\mathbf{F}(\mathbf{M}) \supset \mathbf{G}(\mathbf{M})$, and hence can be used only in the case that the functions **F** and **G** are defined for all values of their respective independent variables. The proposition $\Pi(\mathbf{F}, \mathbf{G})$ does not, on the other hand, justify this inference, although, when $\{\mathbf{F}\}(\mathbf{M})$ is known to be true, it does justify the inference $\{\mathbf{G}\}(\mathbf{M})$. And the proposition $\Pi(\mathbf{F}, \mathbf{G})$ is, therefore, suitable for use in the case that the ranges of definition of the functions **F** and **G** are limited.

The symbol Σ stands for a certain propositional function of one independent variable, such that $\Sigma(\mathbf{F})$ denotes, "There exists at least one value of x for which $\mathbf{F}(x)$ is true."

The symbol & stands for a certain propositional function of two independent variables, such that, if **P** and **Q** are propositions, $\&(\mathbf{P}, \mathbf{Q})$ is the logical product **P**-and-**Q**.

The symbol $\sim$ stands for a certain propositional function of one independent variable, such that, if **P** is a proposition, then $\sim(\mathbf{P})$ is the negation of **P** and may be read, "Not-**P**".

The symbol ι stands for a certain function of one independent variable, such that, if **F** is a propositional function of one independent variable, then $\iota(\mathbf{F})$ denotes, "The object x such that $\{\mathbf{F}\}(\mathbf{x})$ is true."

The symbol A stands for a certain function of two independent variables, the formula $A(\mathbf{F}, \mathbf{M})$ being read, "The abstraction from **M** with respect to **F**."

---

[10]The braces { } are, as a matter of fact, superfluous and might have been omitted from our list of undefined terms, but their inclusion makes for readability of formulas.

[11]M. Schönfinkel, *Über die Bausteine der mathematischen Logik*, ***Math. Annalen***, vol. 92 (1924), pp. 305–316.

6. Abbreviations and definitions. In practice we do not use actually the notation just described, but introduce various abbreviations and substituted notations, partly for the purpose of shortening our formulas and partly in order to render them more readable. We do not, however, regard these abbreviations as an essential part of our theory but rather as extraneous. When we use them we do not literally carry out the development of our system, but we do indicate in full detail how this development can be carried out, and this is for our purpose sufficient.

As has been said above, we use $\{\mathbf{F}\}(\mathbf{A}, \mathbf{B})$ as an abbreviation for $\{\{\mathbf{F}\}(\mathbf{A})\}(\mathbf{B})$ and similarly in the case of functions of larger numbers of variables. Moreover, alike in the case of functions of one variable and in the case of functions of two or more variables, we omit the braces { } whenever the function is represented by a single symbol rather than by an expression consisting of several symbols. Thus, if $\mathbf{F}$ is a single symbol, we write $\mathbf{F}(\mathbf{A})$ instead of $\{\mathbf{F}\}$ $(\mathbf{A})$ and $\mathbf{F}(\mathbf{A}, \mathbf{B})$ instead of $\{\mathbf{F}\}(\mathbf{A}, \mathbf{B})$ or $\{\{\mathbf{F}\}(\mathbf{A})\}(\mathbf{B})$.

We shall usually write $[\mathbf{M}][\mathbf{N}]$ instead of $\&(\mathbf{M}, \mathbf{N})$ or $\{\{\&\}(\mathbf{M})\}(\mathbf{N})$, and $\sim[\mathbf{M}]$ instead of $\sim(\mathbf{M})$ or $\{\sim\}(\mathbf{M})$.

When $\mathbf{F}$ is a single symbol, we shall often write $\mathbf{Fx}[\mathbf{M}]$ instead of $\mathbf{F}(\lambda\mathbf{x}\,[\mathbf{M}])$.

Instead of $\Pi(\lambda\mathbf{x}[\mathbf{M}], \lambda\mathbf{x}[\mathbf{N}])$, we shall write $[\mathbf{M}] \supset_{\mathbf{x}} [\mathbf{N}]$. And $[\Sigma\mathbf{y}[\mathbf{M}]] \supset_{\mathbf{x}} [[\mathbf{M}] \supset_{\mathbf{y}} [\mathbf{N}]]$ we abbreviate further to $[\mathbf{M}] \supset_{\mathbf{xy}} [\mathbf{N}]$, and $[\Sigma\mathbf{y}[\Sigma\mathbf{z}[\mathbf{M}]]] \supset_{\mathbf{x}} [[\Sigma\mathbf{z}[\mathbf{M}]] \supset_{\mathbf{y}} [[\mathbf{M}] \supset_{\mathbf{z}} [\mathbf{N}]]]$ we abbreviate to $[\mathbf{M}] \supset_{\mathbf{xyz}} [\mathbf{N}]$ and so on.

Moreover, whenever possible without ambiguity, we omit square brackets [ ], whether the brackets belong to the undefined term $\lambda[\ ]$ or whether they appear as a part of one of our abbreviations. In order to allow the omission of square brackets as often as possible, we adopt the convention that, whenever there is more than one possibility, the extent of the omitted square brackets shall be taken as the shortest. And when the omission of the square brackets is not possible without ambiguity, we can sometimes substitute for them a dot, or period. This dot, when it occurs within a parenthesis, enclosed by either square brackets [ ], round parentheses ( ), or braces { }, stands for square brackets extending from the place where the dot occurs and up to the end of the parenthesis, or, if the parenthesis is divided into sections by commas as in the case of functions of two or more variables, extending from the place where the dot occurs and up to the first of these commas or to the end of the parenthesis, whichever is first reached. And when the dot is not within any parenthesis, it stands for square brackets extending from the place where it occurs and up to the end of the entire formula. In other words, a dot represents square brackets extending the greatest possible distance forward from the point where it occurs.

In addition to these abbreviations, we allow freely the introduction of abbreviations of a simpler sort, which we call definitions,[12] and which consist in the substitution of a particular single symbol for a particular well-formed formula containing no free variables.

We introduce at once the following definitions, using an arrow $\longrightarrow$ to mean "Stands for", or, "Is an abbreviation for":

$V \longrightarrow \lambda\mu\lambda\nu\,.\sim.\sim\mu\,.\sim\nu$

$U \longrightarrow \lambda\mu\lambda\nu\,.\sim.\,\mu\,.\sim\nu$

$Q \longrightarrow \lambda\mu\lambda\nu\,.\,\Pi(\mu, \nu)\,.\,\Pi(\nu, \mu)$

[12] There seems to be, as a matter of fact, no serious objections to treating definitions as an essential part of the system rather than as extraneous, but we believe it more consistent to class them with our other and more complicated abbreviations.

$E \longrightarrow \lambda\pi\Sigma\varphi \,.\, \varphi(\pi)$

$V(\mathsf{P},\mathsf{Q})$ is to be read, "**P** or **Q**", $U(\mathsf{P},\mathsf{Q})$ is to be read, "**P** implies **Q**", $Q(\mathsf{F},\mathsf{G})$ is to be read "**F** and **G** are equivalent" and $E(\mathsf{M})$ is to be read "**M** exists".

And in connection with these symbols just defined we introduce some further abbreviations. For $V(\mathsf{P},\mathsf{Q})$ we write $[\mathsf{P}] \vee [\mathsf{Q}]$, and for $U(\mathsf{P},\mathsf{Q})$ we write $[\mathsf{P}] \supset [\mathsf{Q}]$. We abbreviate $Q(\lambda\mathsf{x} \,.\, \mathsf{M}, \lambda\mathsf{x} \,.\, \mathsf{N})$ to $[\mathsf{M}] \equiv_{\mathsf{x}} [\mathsf{N}]$. And we abbreviate $E(\mathsf{x}) \supset_{\mathsf{x}} [\mathsf{M}]$ to '$\mathsf{x}[\mathsf{M}]$, which may be read, "For every **x**, **M**". The notion of a class may be introduced by means of the definition:

$$K \longrightarrow A(Q).$$

The formula $K(\mathsf{F})$ is then to be read, "the class of x's such that $\{\mathsf{F}\}(x)$ is true."

7. Postulates. We divide our postulates into two groups, of which the first consists of what we shall call rules of procedure and the second of what we shall call formal postulates. The latter assert that a given formula is true, and contain nothing from the language of intuitive logic other than the words *is true* (and even these words, as already explained, we leave unexpressed when we write the postulates). And the former, the rules of procedure, contain other words from the language of intuitive logic.

The theorems which are proved as consequences of these postulates are of the same form as the postulates of the first group, namely, that a certain formula is true. And the proof of a theorem consists of a series of steps which, from a set of one or more postulates of the first group as a starting point, leads us to the theorem, each step being justified by an appeal to one of the rules of procedure.

The postulates of our first group, the rules of procedure, are five in number:

I. *If* **J** *is true, if* **L** *is well-formed, if all the occurrences of the variable* **x** *in* **L** *are occurrences as a bound variable, and if the variable* **y** *does not occur in* **L**, *then* **K**, *the result of substituting* $\mathrm{S}^{\mathsf{x}}_{\mathsf{y}}\mathsf{L}|$ *for a particular occurrence of* **L** *in* **J**, *is also true.*

II. *If* **J** *is true, if* **M** *and* **N** *are well-formed, if the variable* **x** *occurs in* **M**, *and if the bound variables in* **M** *are distinct both from the variable* **x** *and from the free variables in* **N**, *then* **K**, *the result of substituting* $\mathrm{S}^{\mathsf{x}}_{\mathsf{N}}\mathsf{M}|$ *for a particular occurrence of* $\{\lambda\mathsf{x} \,.\, \mathsf{M}\}(\mathsf{N})$ *in* **J**, *is also true.*

III. *If* **J** *is true, if* **M** *and* **N** *are well-formed, if the variable* **x** *occurs in* **M**, *and if the bound variables in* **M** *are distinct both from* **x** *and from the free variables in* **N**, *then* **K**, *the result of substituting* $\{\lambda\mathsf{x} \,.\, \mathsf{M}\}(\mathsf{N})$ *for a particular occurrence of* $\mathrm{S}^{\mathsf{x}}_{\mathsf{N}}\mathsf{M}|$ *in* **J**, *is also true.*

IV. *If* $\{\mathsf{F}\}(\mathsf{A})$ *is true and* **F** *and* **A** *are well-formed, then* $\Sigma(\mathsf{F})$ *is true.*

V. *If* $\Pi(\mathsf{F},\mathsf{G})$ *and* $\{\mathsf{F}\}(\mathsf{A})$ *are true, and* **F**, **G**, *and* **A** *are well-formed, then* $\{\mathsf{G}\}(\mathsf{A})$ *is true.*

And our formal postulates are the thirty-seven following:

1. $\Sigma(\varphi) \supset_{\varphi} \Pi(\varphi,\varphi)$.
2. $`x \,.\, \varphi(x) \supset_{\varphi} . \, \Pi(\varphi,\psi) \supset_{\psi} \psi(x)$.
3. $\Sigma(\sigma) \supset_{\sigma} .[\sigma(x) \supset_{x} \varphi(x)] \supset_{\varphi} . \, \Pi(\varphi,\psi) \supset_{\psi} . \, \sigma(x) \supset_{x} \psi(x)$.
4. $\Sigma(\varrho) \supset_{\varrho} . \, \Sigma y[\varrho(x) \supset_{x} \varphi(x,y)] \supset_{\varphi} .[\varrho(x) \supset_{x} \Pi(\varphi(x),\psi(x))] \supset_{\psi} .$ $[\varrho(x) \supset_{x} \varphi(x,y)] \supset_{y} . \, \varrho(x) \supset_{x} \psi(x,y)$.
5. $\Sigma(\varphi) \supset_{\varphi} . \, \Pi(\varphi,\psi) \supset_{\psi} . \, \varphi(f(x)) \supset_{fx} \psi(f(x))$.

6. $`x \,.\, \varphi(x) \supset_{\varphi} .\, \Pi(\varphi, \psi(x)) \supset_{\psi} \psi(x, x)$.
7. $\varphi(x, f(x)) \supset_{\varphi f x} .\, \Pi(\varphi(x), \psi(x)) \supset_{\psi} \psi(x, f(x))$.
8. $\Sigma(\varrho) \supset_{\varrho} .\, \Sigma y[\varrho(x) \supset_{x} \varphi(x, y)] \supset_{\varphi} .[\varrho(x) \supset_{x} \Pi(\varphi(x), \psi)] \supset_{\psi} .$
$[\varrho(x) \supset_{x} \varphi(x, y)] \supset_{y} \psi(y)$.
9. $`x \,.\, \varphi(x) \supset_{\varphi} \Sigma(\varphi)$.
10. $\Sigma x \varphi(f(x)) \supset_{f\varphi} \Sigma(\varphi)$.
11. $\varphi(x, x) \supset_{\varphi x} \Sigma(\varphi(x))$.
12. $\Sigma(\varphi) \supset_{\varphi} \Sigma x \varphi(x)$.
13. $\Sigma(\varphi) \supset_{\varphi} .[\varphi(x) \supset_{x} \psi(x)] \supset_{\psi} \Pi(\varphi, \psi)$.
14. $p \supset_{p} .\, q \supset_{q} pq$.
15. $pq \supset_{pq} p$.
16. $pq \supset_{pq} q$.
17. $\Sigma x \Sigma \theta[\varphi(x) \,.\, \sim \theta(x) \,.\, \Pi(\psi, \theta)] \supset_{\varphi\psi} \sim \Pi(\varphi, \psi)$.
18. $\sim \Pi(\varphi, \psi) \supset_{\varphi\psi} \Sigma x \Sigma \theta \,.\, \varphi(x) \,.\, \sim \theta(x) \,.\, \Pi(\psi, \theta)$.
19. $\Sigma x \Sigma \theta[\sim \varphi(u, x) \,.\, \sim \theta(u) \,.\, \Sigma(\varphi(y)) \supset_{y} \varphi(y)] \supset_{\varphi u} \sim \Sigma(\varphi(u))$.
20. $\sim \Sigma(\varphi) \supset_{\varphi} \Sigma x \,.\, \sim \varphi(x)$.
21. $p \supset_{p} .\, \sim q \supset_{q} \sim .\, pq$.
22. $\sim p \supset_{p} .\, q \supset_{q} \sim .\, pq$.
23. $\sim p \supset_{p} .\, \sim q \supset_{q} \sim .\, pq$.
24. $p \supset_{p} .[\sim .\, pq] \supset_{q} \sim q$.
25. $[\sim .\, \varphi(u) \,.\, \psi(u)] \supset_{\varphi\psi u} .[[[\varphi(x) \,.\, \sim \psi(x)] \supset_{x} \varrho(x)] \,.$
$[[\sim \varphi(x) \,.\, \psi(x)] \supset_{x} \varrho(x)] \,.[\sim \varphi(x) \,.\, \sim \psi(x)] \supset_{x} \varrho(x)] \supset_{\varrho} \varrho(u)$.
26. $p \supset_{p} \sim\sim p$.
27. $\sim\sim p \supset_{p} p$.
28. $\sim \Sigma(\varphi) \supset_{\varphi} .\, \Sigma(\psi) \supset_{\psi} \Pi(\varphi, \psi)$.
29. $\sim \Sigma(\varphi) \supset_{\varphi} .\, \sim \Sigma(\psi) \supset_{\psi} \Pi(\varphi, \psi)$.
30. $`x \,.\, \varphi(x) \supset_{\varphi} .[\theta(x) \,.\, \psi(x) \supset_{\psi} \Pi(\theta, \psi)] \supset_{\theta} \varphi(\iota(\theta))$.
31. $\varphi(\iota(\theta)) \supset_{\theta\varphi} \Pi(\theta, \varphi)$.
32. $E(\iota(\theta)) \supset_{\theta} \Sigma(\theta)$.
33. $[\varphi(x, y) \supset_{xy} .\, \varphi(y, z) \supset_{z} \varphi(x, z)][\varphi(x, y) \supset_{xy} \varphi(y, x)] \supset_{\varphi} .\, \varphi(u, v) \supset_{uv} .$
$\psi(A(\varphi, u)) \supset_{\psi} \psi(A(\varphi, v))$.
34. $[\varphi(x, y) \supset_{xy} .\, \varphi(y, z) \supset_{z} \varphi(x, z)][\varphi(x, y) \supset_{xy} \varphi(y, z)] \supset_{\varphi} .$
$[\varphi(x, y) \supset_{xy} \theta(x, y)] \supset_{\theta} .\, \sim \theta(u, v) \supset_{uv} \sim .\, \psi(A(\varphi, u)) \supset_{\psi} \psi(A(\varphi, v))$.
35. $[\psi(A(\varphi, u)) \supset_{\psi} \psi(A(\varphi, v))] \supset_{\varphi uv} \varphi(u, v)$.
36. $E(A(\varphi)) \supset_{\varphi} .\, \varphi(x, y) \supset_{xy} .\, \varphi(y, z) \supset_{z} \varphi(x, z)$.
37. $E(A(\varphi)) \supset_{\varphi} .\, \varphi(x, y) \supset_{xy} \varphi(y, x)$.

8. The relation between free and bound variables. By a *step* in a proof we mean an application of one of the rules of procedure IV or V, occurring in the course of the proof. And in counting the *number of steps* in a proof, each step is to be counted with its proper multiplicity. That is, if a formula **M** is proved and then used r times as a premise for subsequent steps of the proof, then each step in the proof of **M** is to be counted r times.

If **M** and **N** are well-formed and if **N** can be derived from **M** by successive applications of the rules of procedure I, II, and III, then **M** is said to be *convertible* into **N**, and the process is spoken of as a *conversion* of **M** into **N**.

The formula **N** is said to be *provable as a consequence of* the formula **M**, if **M** is well-formed, and $\mathbf{N}$ could be made a provable formula by adding $\mathbf{M}$ to our list of formal postulates as a thirty-eighth postulate. Either of the formulas, $\mathbf{M}$ or $\mathbf{N}$, or both, may contain free variables, since although none of our formal postulates contains free variables, there is, formally, nothing to prevent our adding a thirty-eighth postulate which does contain free variables.

We conclude by proving about our system of postulates the three following theorems:

THEOREM I. *Suppose that* $\mathbf{M}$ *contains* $\mathbf{x}$ *as a free variable, and that* $\Sigma\mathbf{x}\,.\,\mathbf{M}$ *is provable, and that* $\mathbf{N}$ *is provable as a consequence of* $\mathbf{M}$. *Then, if* $\mathbf{N}$ *contains* $\mathbf{x}$ *as a free variable, the formula* $[\mathbf{M}]\supset_{\mathbf{x}}.\,\mathbf{N}$ *is provable. And if* $\mathbf{N}$ *does not contain* $\mathbf{x}$ *as a free variable, the formula* $\mathbf{N}$ *itself is provable.*

We shall prove this theorem by induction with respect to the number of steps in the proof of $\mathbf{N}$ as a consequence of $\mathbf{M}$.

If this number of steps is zero, then $\mathbf{M}$ is convertible into $\mathbf{N}$, and, since $\mathbf{M}$ contains $\mathbf{x}$ as a free variable, so does $\mathbf{N}$. Hence in this case $[\mathbf{M}]\supset_{\mathbf{x}}.\,\mathbf{N}$ may be proved as follows.

Before each formula we give a symbol which will subsequently be used in referring to it. And after each formula we give the means by which it is inferred, that is, the number of the rule of procedure and a reference to the premise or premises, or, if the formula is one of our formal postulates, the number of that postulate.

$\mathbf{B}_1$: $\Sigma\mathbf{x}\,.\,\mathbf{M}$ —provable by hypothesis.
$\mathbf{A}_1$: $\{\lambda\varphi\Sigma(\varphi)\}(\lambda\mathbf{x}\,.\,\mathbf{M})$ —III, $\mathbf{B}_1$.
$\mathbf{A}_2$: $\Sigma(\varphi)\supset_\varphi \Pi(\varphi,\varphi)$ —1.
$\mathbf{A}_3$: $\{\lambda\varphi\Pi(\varphi,\varphi)\}(\lambda\mathbf{x}\,.\,\mathbf{M})$ — V, $\mathbf{A}_2$, $\mathbf{A}_1$.
$\mathbf{A}_4$: $[\mathbf{M}]\supset_{\mathbf{x}}.\,\mathbf{M}$ — II, $\mathbf{A}_3$.
$\mathbf{A}_5$: $[\mathbf{M}]\supset_{\mathbf{x}}.\,\mathbf{N}$ — by conversion, from $\mathbf{A}_4$.

Suppose now, as the hypothesis of our induction, that our theorem is true whenever the number of steps in the proof of $\mathbf{N}$ as a consequence of $\mathbf{M}$ is not greater than n and consider a case in which this number of steps is $n+1$.

The last step in the proof of $\mathbf{N}$ as a consequence of $\mathbf{M}$ might be an application of Rule IV, with premise $\{\mathbf{F}\}(\mathbf{A})$ and conclusion $\Sigma(\mathbf{F})$ or it might be an application of Rule V, with premises $\Pi(\mathbf{F},\mathbf{G})$ and $\{\mathbf{F}\}(\mathbf{A})$ and conclusion $\{\mathbf{G}\}(\mathbf{A})$. In the former case, $\mathbf{F}$ might or might not contain $\mathbf{x}$ as a free variable, and $\mathbf{A}$ might or might not contain $\mathbf{x}$ as a free variable. In the latter case, $\mathbf{F}$, $\mathbf{G}$, and $\mathbf{A}$ each might or might not contain $\mathbf{x}$ as a free variable. All these possibilities must be considered separately.

Case 1, the last step in the proof of $\mathbf{N}$ as a consequence of $\mathbf{M}$ is an application of Rule V, $\mathbf{A}$ is identical with $\mathbf{x}$, and neither $\mathbf{F}$ nor $\mathbf{G}$ contains $\mathbf{x}$ as a free variable. We can prove $[\mathbf{M}]\supset_{\mathbf{x}}.\,\mathbf{N}$ as follows:

$\mathbf{B}_1$: $\Sigma\mathbf{x}\,.\,\mathbf{M}$ —provable, by hypothesis.
$\mathbf{A}_1$: $\{\lambda(\sigma)\Sigma(\sigma)\}(\lambda\mathbf{x}\,.\,\mathbf{M})$ —III, $\mathbf{B}_1$.
$\mathbf{A}_2$: $\Sigma(\sigma)\supset_\sigma.[\sigma(x)\supset_x\varphi(x)]\supset_\varphi.\,\Pi(\varphi,\psi)\supset_\psi.\,\sigma(x)\supset_x\psi(x)$ —3.
$\mathbf{A}_3$: $\{\lambda\sigma\,.[\sigma(x)\supset_x\varphi(x)]\supset_\varphi.\,\Pi(\varphi,\psi)\supset_\psi.\,\sigma(x)\supset_x\psi(x)\}(\lambda\mathbf{x}\,.\,\mathbf{M})$ —V, $\mathbf{A}_2$, $\mathbf{A}_1$.
$\mathbf{A}_4$: $\{\lambda\sigma\,.[\sigma(\mathbf{x})\supset_{\mathbf{x}}\varphi(\mathbf{x})]\supset_\varphi.\,\Pi(\varphi,\psi)\supset_\psi.\,\sigma(\mathbf{x})\supset_{\mathbf{x}}\psi(\mathbf{x})\}(\lambda\mathbf{x}\,.\,\mathbf{M})$ —I, $\mathbf{A}_3$.
$\mathbf{A}_5$: $[\{\lambda\mathbf{x}\,.\,\mathbf{M}\}(\mathbf{x})\supset_{\mathbf{x}}\varphi(\mathbf{x})]\supset_\varphi.\,\Pi(\varphi,\psi)\supset_\psi.\{\lambda\mathbf{x}\,.\,\mathbf{M}\}(\mathbf{x})\supset_{\mathbf{x}}\psi(\mathbf{x})$ —II, $\mathbf{A}_4$.
$\mathbf{A}_6$: $[[\mathbf{M}]\supset_{\mathbf{x}}\varphi(\mathbf{x})]\supset_\varphi.\,\Pi(\varphi,\psi)\supset_\psi.\{\lambda\mathbf{x}\,.\,\mathbf{M}\}(\mathbf{x})\supset_{\mathbf{x}}\psi(\mathbf{x})$ —II, $\mathbf{A}_5$.[13]

[13] Evidently, we may suppose, without loss of generality, that $\mathbf{M}$ does not contain $\mathbf{x}$ as a bound variable, because, if the formula $\mathbf{M}$ did contain $\mathbf{x}$ as a bound variable, it would be convertible into a formula $\mathbf{M}'$ which did not contain $\mathbf{x}$ as a bound variable. A similar remark applies frequently below.

$\mathbf{A}_7$: $[[\mathbf{M}] \supset_{\mathbf{x}} \varphi(\mathbf{x})] \supset_{\varphi} . \Pi(\varphi, \psi) \supset_{\psi} .[\mathbf{M}] \supset_{\mathbf{x}} \psi(\mathbf{x})$ —II, $\mathbf{A}_6$.
$\mathbf{B}_2$: $[\mathbf{M}] \supset_{\mathbf{x}} \{\mathbf{F}\}(\mathbf{x})$ —provable, by hypothesis of induction.
$\mathbf{A}_8$: $\{\lambda\varphi .[\mathbf{M}] \supset_{\mathbf{x}} \varphi(\mathbf{x})\}(\mathbf{F})$ —III, $\mathbf{B}_2$.
$\mathbf{A}_9$: $\{\lambda\varphi . \Pi(\varphi, \psi) \supset_{\psi} .[\mathbf{M}] \supset_{\mathbf{x}} \psi(\mathbf{x})\}(\mathbf{F})$ —V, $\mathbf{A}_7$, $\mathbf{A}_8$.
$\mathbf{A}_{10}$: $\Pi(\mathbf{F}, \psi) \supset_{\psi} .[\mathbf{M}] \supset_{\mathbf{x}} \psi(\mathbf{x})$ —II, $\mathbf{A}_9$.
$\mathbf{B}_3$: $\Pi(\mathbf{F}, \mathbf{G})$ —provable, by hypothesis of induction.
$\mathbf{A}_{11}$: $\{\lambda\psi\Pi(\mathbf{F}, \psi)\}(\mathbf{G})$ —III, $\mathbf{B}_3$.
$\mathbf{A}_{12}$: $\{\lambda\psi .[\mathbf{M}] \supset_{\mathbf{x}} \psi(\mathbf{x})\}(\mathbf{G})$ —V, $\mathbf{A}_{10}$, $\mathbf{A}_{11}$.
$\mathbf{A}_{13}$: $[\mathbf{M}] \supset_{\mathbf{x}} \{\mathbf{G}\}(\mathbf{x})$ —II, $\mathbf{A}_{12}$.
$\mathbf{A}_{14}$: $[\mathbf{M}] \supset_{\mathbf{x}} . \mathbf{N}$ —by conversion, from $\mathbf{A}_{13}$, since $\{\mathbf{G}\}(\mathbf{x})$ is convertible into $\mathbf{N}$.

Case 2, the last step in the proof of $\mathbf{N}$ as a consequence of $\mathbf{M}$ is an application of Rule V, and $\mathbf{F}$ contains $\mathbf{x}$ as a free variable, and $\mathbf{G}$ and $\mathbf{A}$ do not. We can prove $\mathbf{N}$ as follows:

$\mathbf{B}_1$: $\Sigma\mathbf{x} . \mathbf{M}$ —provable by hypothesis.
$\mathbf{A}_1$: $\{\lambda\varrho\Sigma(\varrho)\}(\lambda\mathbf{x} . \mathbf{M})$ —III, $\mathbf{B}_1$.
$\mathbf{A}_2$: $\Sigma(\varrho) \supset_{\varrho} . \Sigma y[\varrho(x) \supset_x \varphi(x, y)] \supset_{\varphi} .[\varrho(x) \supset_x \Pi(\varphi(x), \psi)] \supset_{\psi} .$
$[\varrho(x) \supset_x \varphi(x, y)] \supset_y \psi(y)$ —8.
$\mathbf{A}_3$: $\{\lambda\varrho . \Sigma y[\varrho(x) \supset_x \varphi(x, y)] \supset_{\varphi} .[\varrho(x) \supset_x \Pi(\varphi(x), \psi)] \supset_{\psi} .$
$[\varrho(x) \supset_x \varphi(x, y)] \supset_y \psi(y)\}(\lambda\mathbf{x} . \mathbf{M})$ —V, $\mathbf{A}_2$, $\mathbf{A}_1$.
$\mathbf{A}_4$: $\{\lambda\varrho . \Sigma y[\varrho(\mathbf{x}) \supset_{\mathbf{x}} \varphi(\mathbf{x}, y)] \supset_{\varphi} .[\varrho(\mathbf{x}) \supset_{\mathbf{x}} \Pi(\varphi(\mathbf{x}), \psi)] \supset_{\psi} .$
$[\varrho(\mathbf{x}) \supset_{\mathbf{x}} \varphi(\mathbf{x}, y)] \supset_y \psi(y)\}(\lambda\mathbf{x} . \mathbf{M})$ —I, $\mathbf{A}_3$.
$\mathbf{A}_5$: $\Sigma y[\{\lambda\mathbf{x} . \mathbf{M}\}(\mathbf{x}) \supset_{\mathbf{x}} \varphi(\mathbf{x}, y)] \supset_{\varphi} .[\{\lambda\mathbf{x} . \mathbf{M}\}(\mathbf{x}) \supset_{\mathbf{x}} \Pi(\varphi(\mathbf{x}), \psi)] \supset_{\psi} .$
$[\{\lambda\mathbf{x} . \mathbf{M}\}(\mathbf{x}) \supset_{\mathbf{x}} \varphi(\mathbf{x}, y)] \supset_y \psi(y)$ —II, $\mathbf{A}_4$.
$\mathbf{A}_6$: $\Sigma y[[\mathbf{M}] \supset_{\mathbf{x}} \varphi(\mathbf{x}, y)] \supset_{\varphi} .[\{\lambda\mathbf{x} . \mathbf{M}\}(\mathbf{x}) \supset_{\mathbf{x}} \Pi(\varphi(\mathbf{x}, \psi)] \supset_{\psi} .$
$[\{\lambda\mathbf{x} . \mathbf{M}\}(\mathbf{x}) \supset_{\mathbf{x}} \varphi(\mathbf{x}, y)] \supset_y \psi(y)$ —II, $\mathbf{A}_5$.
$\mathbf{A}_7$: $\Sigma y[[\mathbf{M}] \supset_{\mathbf{x}} \varphi(\mathbf{x}, y)] \supset_{\varphi} .[[\mathbf{M} \supset_{\mathbf{x}} \Pi(\varphi(\mathbf{x}), \psi)] \supset_{\psi} .$
$[\{\lambda\mathbf{x} . \mathbf{M}\}(\mathbf{x}) \supset_{\mathbf{x}} \varphi(\mathbf{x}, y)] \supset_y \psi(y)$ —II, $\mathbf{A}_6$.
$\mathbf{A}_8$: $\Sigma y[[\mathbf{M}] \supset_{\mathbf{x}} \varphi(\mathbf{x}, y)] \supset_{\varphi} .[[\mathbf{M} \supset_{\mathbf{x}} \Pi(\varphi(\mathbf{x}), \psi)] \supset_{\psi} .$
$[[\mathbf{M}] \supset_{\mathbf{x}} \psi(\mathbf{x}, y)] \supset_y \psi(y)$ —II, $\mathbf{A}_7$.
$\mathbf{B}_2$: $[\mathbf{M}] \supset_{\mathbf{x}} \{\mathbf{F}\}(\mathbf{A})$ —provable, by hypothesis of induction.
$\mathbf{A}_9$: $[\mathbf{M}] \supset_{\mathbf{x}} \{\lambda\mathbf{x} . \mathbf{F}\}(\mathbf{x}, \mathbf{A})$ —III, $\mathbf{B}_2$.
$\mathbf{A}_{10}$: $\{\lambda y .[\mathbf{M}] \supset_{\mathbf{x}} \{\lambda\mathbf{x} . \mathbf{F}\}(\mathbf{x}, y)\}(\mathbf{A})$ —III, $\mathbf{A}_9$.
$\mathbf{A}_{11}$: $\Sigma y .[\mathbf{M}] \supset_{\mathbf{x}} \{\lambda\mathbf{x} . \mathbf{F}\}(\mathbf{x}, y)$ —IV, $\mathbf{A}_{10}$.
$\mathbf{A}_{12}$: $\{\lambda\varphi\Sigma y .[\mathbf{M}] \supset_x \varphi(\mathbf{x}, y)\}(\lambda\mathbf{x} . \mathbf{F})$ —III, $\mathbf{A}_{11}$.
$\mathbf{A}_{13}$: $\{\lambda\varphi .[[\mathbf{M}] \supset_{\mathbf{x}} \Pi(\varphi(\mathbf{x}), \psi)] \supset_{\psi} .$
$[[\mathbf{M}] \supset_{\mathbf{x}} \varphi(\mathbf{x}, y)] \supset_y \psi(y)\}(\lambda\mathbf{x} . \mathbf{F})$ —V, $\mathbf{A}_8$, $\mathbf{A}_{12}$.
$\mathbf{A}_{14}$: $[[\mathbf{M}] \supset_{\mathbf{x}} \Pi(\{\lambda\mathbf{x} . \mathbf{F}\}(\mathbf{x}), \psi)] \supset_{\psi} .[[\mathbf{M}] \supset_{\mathbf{x}} \{\lambda\mathbf{x} . \mathbf{F}\}(\mathbf{x}, y)] \supset_y \psi(y)$ —II, $\mathbf{A}_{13}$.
$\mathbf{A}_{15}$: $[[\mathbf{M}] \supset_{\mathbf{x}} \Pi(\mathbf{F}, \psi)] \supset_{\psi} .[[\mathbf{M}] \supset_{\mathbf{x}} .\{\lambda\mathbf{x} . \mathbf{F}\}(\mathbf{x}, y)] \supset_y \psi(y)$ —II, $\mathbf{A}_{14}$.
$\mathbf{A}_{16}$: $[[\mathbf{M}] \supset_{\mathbf{x}} \Pi(\mathbf{F}, \psi)] \supset_{\psi} .[[\mathbf{M}] \supset_{\mathbf{x}} \{\mathbf{F}\}(y)] \supset_y \psi(y)$ —II, $\mathbf{A}_{15}$.
$\mathbf{B}_3$: $[\mathbf{M}] \supset_{\mathbf{x}} \Pi(\mathbf{F}, \mathbf{G})$ —provable, by hypothesis of induction.
$\mathbf{A}_{17}$: $\{\lambda\psi .[\mathbf{M}] \supset_{\mathbf{x}} \Pi(\mathbf{F}, \psi)\}(\mathbf{G})$ —III, $\mathbf{B}_3$.
$\mathbf{A}_{18}$: $\{\lambda\psi .[[\mathbf{M}] \supset_{\mathbf{x}} \{\mathbf{F}\}(y)] \supset_y \psi(y)\}(\mathbf{G})$ —V, $\mathbf{A}_{16}$, $\mathbf{A}_{17}$.
$\mathbf{A}_{19}$: $[[\mathbf{M}] \supset_{\mathbf{x}} \{\mathbf{F}\}(y)] \supset_{\mathbf{x}} \{\mathbf{G}\}(y)$ —II, $\mathbf{A}_{18}$.
$\mathbf{A}_{20}$: $\{\lambda y .[\mathbf{M}] \supset_{\mathbf{x}} \{\mathbf{F}\}(y)\}(\mathbf{A})$ —III, $\mathbf{B}_2$.
$\mathbf{A}_{21}$: $\{\lambda y\{\mathbf{G}\}(y)\}(\mathbf{A})$ —V, $\mathbf{A}_{19}$, $\mathbf{A}_{20}$.
$\mathbf{A}_{22}$: $\{\mathbf{G}\}(\mathbf{A})$ —II, $\mathbf{A}_{21}$.

$\mathbf{N}$ —by conversion, from $\mathbf{A}_{22}$.

Case 3, the last step in the proof of $\mathbf{N}$ as a consequence of $\mathbf{M}$ is an application of Rule IV, $\mathbf{A}$ is identical with $\mathbf{x}$, and $\mathbf{F}$ does not contain $\mathbf{x}$ as a free variable. By hypothesis, $\Sigma\mathbf{x}.\ \mathbf{M}$ is provable, and from it, by conversion and an application of Rule IV, we can obtain $\Sigma\varphi\ .\ \varphi(\lambda\mathbf{x}\ .\ \mathbf{M})$. Therefore, using Postulate 9, we can prove $\varphi(\lambda\mathbf{x}\ .\ \mathbf{M}) \supset_\varphi \Sigma(\varphi)$. And hence, using Postulate 5, we can prove $f(\mathbf{x}, \lambda\mathbf{x}\ .\ \mathbf{M}) \supset_{f\mathbf{x}} \Sigma(f(\mathbf{x}))$. Moreover $\Sigma\mathbf{x}\ .\ \mathbf{M}$ becomes by conversion $\Sigma\mathbf{x}\ .\{\lambda y\lambda\varphi\ .\ \varphi(y)\}(\mathbf{x}, \lambda\mathbf{x}\ .\ \mathbf{M})$, and this combined with the preceding yields $\{\lambda y\lambda\varphi\ .\ \varphi(y)\}(\mathbf{x}, \lambda\mathbf{x}\ .\ \mathbf{M}) \supset_\mathbf{x} \Sigma(\{\lambda y\lambda\varphi\ .\ \varphi(y)\}(\mathbf{x}))$. And this last formula becomes by conversion $[\mathbf{M}] \supset_\mathbf{x} \{\lambda\mathbf{x}E(\mathbf{x})\}(\mathbf{x})$.

Now using Postulate 3, since $\Sigma\mathbf{x}\ .\ \mathbf{M}$ is provable, we can obtain

$$[[\mathbf{M}] \supset_\mathbf{x} \varphi(\mathbf{x})] \supset_\varphi .\ \Pi(\varphi, \psi) \supset_\psi .[\mathbf{M}] \supset_\mathbf{x} \psi(\mathbf{x}).$$

Hence, using the last formula in the preceding paragraph, we can prove

$$\Pi(\lambda\mathbf{x}E(\mathbf{x}), \psi) \supset_\psi .[\mathbf{M}] \supset_\mathbf{x} \psi(\mathbf{x}).$$

And hence, using Postulate 9, we can prove $[\mathbf{M}] \supset_\mathbf{x} .\ \varphi(\mathbf{x}) \supset_\varphi \Sigma(\varphi)$.

Thus we are able to prove $[\mathbf{M}] \supset_\mathbf{x} \Pi(\mathbf{F}', \mathbf{G}')$, where $\mathbf{F}'$ stands for $\lambda\varphi\ .\ \varphi(x)$ and $\mathbf{G}'$ for $\lambda\varphi\Sigma(\varphi)$. And, by the hypothesis of our induction, we are able to prove $[\mathbf{M}] \supset_\mathbf{x} \{\mathbf{F}'\}(\mathbf{A}')$, where $\mathbf{A}'$ stands for $\mathbf{F}$. Therefore, by the method of Case 2, we can prove $\{\mathbf{G}'\}(\mathbf{A}')$, and from it by conversion, $\mathbf{N}$.

Case 4, the last step in the proof of $\mathbf{N}$ as a consequence of $\mathbf{M}$ is an application of Rule IV, and $\mathbf{A}$ contains $\mathbf{x}$ as a free variable, and $\mathbf{F}$ does not. Since $\{\mathbf{F}\}(\mathbf{A})$ is convertible into $\{\lambda\mathbf{x}\{\mathbf{F}\}(\mathbf{A})\}(\mathbf{x})$, we are able, by the method of Case 3, to prove $\Sigma\mathbf{x}\ .\{\mathbf{F}\}(\mathbf{A})$. This last formula is convertible into $\Sigma\mathbf{x}\ .\{\mathbf{F}\}(\{\lambda\mathbf{x}\ .\ \mathbf{A}\}(\mathbf{x}))$. Hence, using Postulate 10, we can prove $\Sigma(\mathbf{F})$ and from it by conversion, $\mathbf{N}$.

Case 5, the last step in the proof of $\mathbf{N}$ as a consequence of $\mathbf{M}$ is an application of Rule IV, and $\mathbf{F}$ contains $\mathbf{x}$ as a free variable, and $\mathbf{A}$ does not. By the hypothesis of our induction, we can prove $[\mathbf{M}] \supset_\mathbf{x} \{\mathbf{F}\}(\mathbf{A})$. Hence we can prove $E(\mathbf{A})$, and hence, using Postulate 9, $\varphi(\mathbf{A}) \supset_\varphi \Sigma(\varphi)$. And combining this result with Postulate 5, we can obtain $f(x, \mathbf{A}) \supset_{fx} \Sigma(f(x))$.

Now, by the method of Case 3, we can prove $\Sigma\mathbf{x}\ .\{\mathbf{F}\}(\mathbf{A})$ which is convertible into $\Sigma\mathbf{x}\ .\{\lambda\mathbf{x}\ .\ \mathbf{F}\}(\mathbf{x}, \mathbf{A})$. Combining this with the preceding, we can obtain

$$\{\lambda\mathbf{x}\ .\ \mathbf{F}\}(\mathbf{x}, \mathbf{A}) \supset_x \Sigma(\{\lambda\mathbf{x}\ .\ \mathbf{F}\}(\mathbf{x})).$$

And this is convertible into $\{\mathbf{F}\}(\mathbf{A}) \supset_x \Sigma(\mathbf{F})$.

Thus we are able to prove $\Pi(\mathbf{F}', \mathbf{G}')$ where $\mathbf{F}'$ stands for $\lambda x\ .\{\mathbf{F}\}(\mathbf{A})$ and $\mathbf{G}'$ stands for $\lambda x\Sigma(\mathbf{F})$. And, by the hypothesis of our induction, we are able to prove $[\mathbf{M}] \supset_\mathbf{x} \{\mathbf{F}'\}(\mathbf{x})$. Therefore, by the method of Case 1, we can prove $[\mathbf{M}] \supset_\mathbf{x} \{\mathbf{G}'\}(\mathbf{x})$, and from it by conversion, $[\mathbf{M}] \supset_\mathbf{x} .\ \mathbf{N}$.

Case 6, the last step in the proof of $\mathbf{N}$ as a consequence of $\mathbf{M}$ is an application of Rule IV, and $\mathbf{F}$ contains $\mathbf{x}$ as a free variable, and $\mathbf{A}$ is identical with $\mathbf{x}$. By hypothesis, $\Sigma\mathbf{x}\ .\ \mathbf{M}$ is provable, and, by the hypothesis of our induction, $[\mathbf{M}] \supset_\mathbf{x} \{\mathbf{F}\}(\mathbf{x})$ is provable. Therefore, by the method of Case 3, we can prove $\Sigma\mathbf{x}\ .\{\mathbf{F}\}(\mathbf{x})$, and this is convertible into $\Sigma\mathbf{x}\ .\{\lambda\mathbf{x}\ .\ \mathbf{F}\}(\mathbf{x}, \mathbf{x})$. Hence, using Postulate 11, we can obtain $\{\mathbf{F}\}(\mathbf{x}) \supset_\mathbf{x} .\ \Sigma(\mathbf{F})$.

Thus we are able to prove $\Pi(\mathbf{F}', \mathbf{G}')$ where $\mathbf{F}'$ stands for $\lambda\mathbf{x}\ .\{\mathbf{F}\}(\mathbf{x})$ and $\mathbf{G}'$ stands for $\lambda\mathbf{x}\Sigma(\mathbf{F})$. And, by the hypothesis of our induction, we are able to prove $[\mathbf{M}] \supset_\mathbf{x} \{\mathbf{F}'\}(\mathbf{x})$.

Therefore, by the method of Case 1, we can prove $[\mathbf{M}] \supset_{\mathbf{x}} \{\mathbf{G}'\}(\mathbf{x})$, and from it by conversion, $[\mathbf{M}] \supset_{\mathbf{x}} . \mathbf{N}$.

Case 7, the last step in the proof of **N** as a consequence of **M** is an application of Rule IV, and both **F** and **A** contain **x** as a free variable. By the hypothesis of our induction,

$$[\mathbf{M}] \supset_{\mathbf{x}} \{\mathbf{F}\}(\mathbf{A})$$

is provable, and this is convertible into $[\mathbf{M}] \supset_{\mathbf{x}} \{\lambda y \{\mathbf{F}\}(\{\lambda \mathbf{x} . \mathbf{A}\}(y))\}(\mathbf{x})$. Therefore, by the method of Case 6, we can prove $[\mathbf{M}] \supset_{\mathbf{x}} \Sigma y . \{\mathbf{F}\}(\{\lambda \mathbf{x} . \mathbf{A}\}(y))$.

This last formula is convertible into $[\mathbf{M}] \supset_{\mathbf{x}} \{\lambda \varphi \Sigma y . \varphi(\{\lambda \mathbf{x} . \mathbf{A}\}(y))\}(\mathbf{F})$. Therefore, by the method of Case 4, we can prove $\Sigma \varphi \Sigma y . \varphi(\{\lambda \mathbf{x} . \mathbf{A}\}(y))$. And, combining this with Postulate 10, we can obtain $\Sigma y[\varphi(\{\lambda \mathbf{x} . \mathbf{A}\}(y))] \supset_{\varphi} \Sigma(\varphi)$. And hence, combining with Postulate 5, we can obtain

$$\Sigma y . [f(x, \{\lambda \mathbf{x} . \mathbf{A}\}(y))] \supset_{fx} \Sigma(f(x)).$$

By the method of Case 3, using $[\mathbf{M}] \supset_{\mathbf{x}} \Sigma y . \{\mathbf{F}\}(\{\lambda \mathbf{x} . \mathbf{A}\}(y))$, we can prove

$$\Sigma \mathbf{x} \Sigma y . \{\mathbf{F}\}(\{\lambda \mathbf{x} . \mathbf{A}\}(y))$$

and this is convertible into $\Sigma x \Sigma y . \{\lambda \mathbf{x} . \mathbf{F}\}(x, \{\lambda \mathbf{x} . \mathbf{A}\}(y))$. Combining this last formula with the formula at the end of the preceding paragraph, we can obtain $\Sigma y[\{\lambda \mathbf{x} . \mathbf{F}\}(x, \{\lambda \mathbf{x} . \mathbf{A}\}(y))] \supset_{x} \Sigma(\{\lambda \mathbf{x} . \mathbf{F}\}(x))$. And this is convertible into

$$\Sigma y[\{\mathbf{F}\}(\{\lambda \mathbf{x} . \mathbf{A}\}(y))] \supset_{\mathbf{x}} \Sigma(\mathbf{F}).$$

Thus we are able to prove $\Pi(\mathbf{F}', \mathbf{G}')$ and $[\mathbf{M}] \supset_{\mathbf{x}} \{\mathbf{F}'\}(\mathbf{x})$ where $\mathbf{F}'$ stands for

$$\lambda \mathbf{x} \Sigma y . \{\mathbf{F}\}(\{\lambda \mathbf{x} . \mathbf{A}\}(y))$$

and $\mathbf{G}'$ stands for $\lambda \mathbf{x} \Sigma(\mathbf{F})$. Therefore, by the method of Case 1, we can prove $[\mathbf{M}] \supset_{\mathbf{x}} \{\mathbf{G}'\}(\mathbf{x})$, and from it by conversion, $[\mathbf{M}] \supset_{\mathbf{x}} . \mathbf{N}$.

Case 8, the last step in the proof of **N** as a consequence of **M** is an application of Rule V, and **A** contains **x** as a free variable, and **F** and **G** do not. By the hypothesis of our induction, $[\mathbf{M}] \supset_{\mathbf{x}} \{\mathbf{F}\}(\mathbf{A})$ is provable. Hence, by the method of Case 4, we can prove $\Sigma(\mathbf{F})$. And, by the method of Case 3, we can prove $\Sigma \mathbf{x} . \{\mathbf{F}\}(\mathbf{A})$, which is convertible into $\Sigma \mathbf{x} . \{\mathbf{F}\}(\{\lambda \mathbf{x} . \mathbf{A}\}(\mathbf{x}))$.

Also, by the hypothesis of our induction, $\Pi(\mathbf{F}, \mathbf{G})$ is provable. Hence, using Postulate 5, we can obtain $\{\mathbf{F}\}(f(x)) \supset_{fx} \{\mathbf{G}\}(f(x))$ and from this, using

$$\Sigma \mathbf{x} . \{\mathbf{F}\}(\{\lambda \mathbf{x} . \mathbf{A}\}(\mathbf{x})),$$

we can obtain $\{\mathbf{F}\}(\{\lambda \mathbf{x} . \mathbf{A}\}(\mathbf{x})) \supset_{\mathbf{x}} \{\mathbf{G}\}(\{\lambda \mathbf{x} . \mathbf{A}\}(\mathbf{x}))$.

From this last formula, using Rule II, we can obtain $\Pi(\mathbf{F}', \mathbf{G}')$, where $\mathbf{F}'$ and $\mathbf{G}'$ stand respectively for $\lambda \mathbf{x} . \{\mathbf{F}\}(\mathbf{A})$ and $\lambda \mathbf{x} . \{\mathbf{G}\}(\mathbf{A})$. And, applying Rule III to $[\mathbf{M}] \supset_{\mathbf{x}} \{\mathbf{F}\}(\mathbf{A})$, we can prove $[\mathbf{M}] \supset_{\mathbf{x}} \{\mathbf{F}'\}(\mathbf{x})$. Therefore, by the method of Case 1, we can prove $[\mathbf{M}] \supset_{\mathbf{x}} \{\mathbf{G}'\}(\mathbf{x})$ and from it by conversion, $[\mathbf{M}] \supset_{\mathbf{x}} . \mathbf{N}$.

Case 9, the last step in the proof of **N** as a consequence of **M** is an application of Rule V, and **G** contains **x** as a free variable, and **F** and **A** do not. By the hypothesis of our induction, $\{\mathbf{F}\}(\mathbf{A})$ is provable, and hence, using Postulate 2, $\Pi(\mathbf{F}, \psi) \supset_{\psi} \psi(\mathbf{A})$ is provable.

Also, by the hypothesis of our induction, $[\mathbf{M}] \supset_{\mathbf{x}} \Pi(\mathbf{F}, \mathbf{G})$ is provable, and from this by conversion we can obtain $[\mathbf{M}] \supset_{\mathbf{x}} \{\lambda \psi \Pi(\mathbf{F}, \psi)\}(\mathbf{G})$.

Thus we can prove $\Pi(\mathbf{F}', \mathbf{G}')$ and $[\mathbf{M}] \supset_{\mathbf{x}} \{\mathbf{F}'\}(\mathbf{A}')$ where $\mathbf{F}'$ stands for $\lambda\psi\Pi(\mathbf{F}, \psi)$ and $\mathbf{G}'$ stands for $\lambda\psi . \psi(\mathbf{A})$, and $\mathbf{A}'$ stands for $\mathbf{G}$. Therefore, by the method of Case 8, we can prove $[\mathbf{M}] \supset_{\mathbf{x}} \{\mathbf{G}'\}(\mathbf{A}')$ and from it by conversion, $[\mathbf{M}] \supset_{\mathbf{x}} . \mathbf{N}$.

Case 10, the last step in the proof of $\mathbf{N}$ as a consequence of $\mathbf{M}$ is an application of Rule V, and $\mathbf{F}$ and $\mathbf{G}$ contain $\mathbf{x}$ as a free variable, and $\mathbf{A}$ does not. We can prove $[\mathbf{M}] \supset_{\mathbf{x}} . \mathbf{N}$ as follows:

$\mathbf{B}_1$: $\Sigma\mathbf{x} . \mathbf{M}$ —provable, by hypothesis.

$\mathbf{A}_1$: $\{\lambda\varrho\Sigma(\varrho)\}(\lambda\mathbf{x} . \mathbf{M})$ —III, $\mathbf{B}_1$.

$\mathbf{A}_2$: $\Sigma(\varrho) \supset_\varrho . \Sigma y[\varrho(x) \supset_x \varphi(x, y)] \supset_\varphi . [\varrho(x) \supset_x \Pi(\varphi(x), \psi(x))] \supset_\psi .$
$[\varrho(x) \supset_x \varphi(x, y)] \supset_y . \varrho(x) \supset_x \psi(x, y)$ —4.

$\mathbf{A}_3$: $\{\lambda\varrho . \Sigma y[\varrho(x) \supset_x \varphi(x, y)] \supset_\varphi . [\varrho(x) \supset_x \Pi(\varphi(x), \psi(x))] \supset_\psi .$
$[\varrho(x) \supset_x \varphi(x, y)] \supset_y . \varrho(x) \supset_x \psi(x, y)\}(\lambda\mathbf{x} . \mathbf{M})$ —V, $\mathbf{A}_2$, $\mathbf{A}_1$.

$\mathbf{A}_4$: $\{\lambda\varrho . \Sigma y[\varrho(\mathbf{x}) \supset_{\mathbf{x}} \varphi(\mathbf{x}, y)] \supset_\varphi . [\varrho(\mathbf{x}) \supset_{\mathbf{x}} \Pi(\varphi(\mathbf{x}), \psi(\mathbf{x}))] \supset_\psi .$
$[\varrho(\mathbf{x}) \supset_{\mathbf{x}} \varphi(\mathbf{x}, y)] \supset_y . \varrho(\mathbf{x}) \supset_{\mathbf{x}} \psi(\mathbf{x}, y)\}(\lambda\mathbf{x} . \mathbf{M})$ —I, $\mathbf{A}_3$.

$\mathbf{A}_5$: $\Sigma y[\{\lambda\mathbf{x} . \mathbf{M}\}(\mathbf{x}) \supset_{\mathbf{x}} \varphi(\mathbf{x}, y)] \supset_\varphi . [\{\lambda\mathbf{x} . \mathbf{M}\}(\mathbf{x}) \supset_{\mathbf{x}} \Pi(\varphi(\mathbf{x}), \psi(\mathbf{x}))] \supset_\psi .$
$[\{\lambda\mathbf{x} . \mathbf{M}\}(\mathbf{x}) \supset_{\mathbf{x}} \varphi(\mathbf{x}, y)] \supset_y . \{\lambda\mathbf{x} . \mathbf{M}\}(\mathbf{x}) \supset_{\mathbf{x}} \psi(\mathbf{x}, y)$ —II, $\mathbf{A}_4$.

$\mathbf{A}_6$: $\Sigma y[[\mathbf{M}] \supset_{\mathbf{x}} \varphi(\mathbf{x}, y)] \supset_\varphi . [\{\lambda\mathbf{x} . \mathbf{M}\}(\mathbf{x}) \supset_{\mathbf{x}} \Pi(\varphi(\mathbf{x}), \psi(\mathbf{x}))] \supset_\psi .$
$[\{\lambda\mathbf{x} . \mathbf{M}\}(\mathbf{x}) \supset_{\mathbf{x}} \varphi(\mathbf{x}, y)] \supset_y . \{\lambda\mathbf{x} . \mathbf{M}\}(\mathbf{x}) \supset_{\mathbf{x}} \psi(\mathbf{x}, y)$ —II, $\mathbf{A}_5$.

$\mathbf{A}_7$: $\Sigma y[[\mathbf{M}] \supset_{\mathbf{x}} \varphi(\mathbf{x}, y)] \supset_\varphi . [[\mathbf{M}] \supset_{\mathbf{x}} \Pi(\varphi(\mathbf{x}), \psi(\mathbf{x}))] \supset_\psi .$
$[\{\lambda\mathbf{x} . \mathbf{M}\}(\mathbf{x}) \supset_{\mathbf{x}} \varphi(\mathbf{x}, y)] \supset_y . \{\lambda\mathbf{x} . \mathbf{M}\}(\mathbf{x}) \supset_{\mathbf{x}} \psi(\mathbf{x}, y)$ —II, $\mathbf{A}_6$.

$\mathbf{A}_8$: $\Sigma y[[\mathbf{M}] \supset_{\mathbf{x}} \varphi(\mathbf{x}, y)] \supset_\varphi . [[\mathbf{M}] \supset_{\mathbf{x}} \Pi(\varphi(\mathbf{x}), \psi(\mathbf{x}))] \supset_\psi .$
$[[\mathbf{M}] \supset_{\mathbf{x}} \varphi(\mathbf{x}, y)] \supset_y . \{\lambda\mathbf{x} . \mathbf{M}\}(\mathbf{x}) \supset_{\mathbf{x}} \psi(\mathbf{x}, y)$ —II, $\mathbf{A}_7$.

$\mathbf{A}_9$: $\Sigma y[[\mathbf{M}] \supset_{\mathbf{x}} \varphi(\mathbf{x}, y)] \supset_\varphi . [[\mathbf{M}] \supset_{\mathbf{x}} \Pi(\varphi(\mathbf{x}), \psi(\mathbf{x}))] \supset_\psi .$
$[[\mathbf{M}] \supset_{\mathbf{x}} \varphi(\mathbf{x}, y)] \supset_y . [\mathbf{M}] \supset_{\mathbf{x}} \psi(\mathbf{x}, y)$ —II, $\mathbf{A}_8$.

$\mathbf{B}_2$: $[\mathbf{M}] \supset_{\mathbf{x}} \{\mathbf{F}\}(\mathbf{A})$ —provable by hypothesis of induction.

$\mathbf{A}_{10}$: $[\mathbf{M}] \supset_{\mathbf{x}} \{\lambda\mathbf{x} . \mathbf{F}\}(\mathbf{x}, \mathbf{A})$ —III, $\mathbf{B}_2$.

$\mathbf{A}_{11}$: $\{\lambda y . [\mathbf{M}] \supset_{\mathbf{x}} \{\lambda\mathbf{x} . \mathbf{F}\}(\mathbf{x}, y)\}(\mathbf{A})$ —III, $\mathbf{A}_{10}$.

$\mathbf{A}_{12}$: $\Sigma y . [\mathbf{M}] \supset_{\mathbf{x}} \{\lambda\mathbf{x} . \mathbf{F}\}(\mathbf{x}, y)$ —IV, $\mathbf{A}_{11}$.

$\mathbf{A}_{13}$: $\{\lambda\varphi\Sigma y . [\mathbf{M}] \supset_{\mathbf{x}} \varphi(\mathbf{x}, y)\}(\lambda\mathbf{x} . \mathbf{F})$ —III, $\mathbf{A}_{12}$.

$\mathbf{A}_{14}$: $\{\lambda\varphi . [[\mathbf{M}] \supset_{\mathbf{x}} \Pi(\varphi(\mathbf{x}), \psi(\mathbf{x}))] \supset_\psi . [[\mathbf{M}] \supset_{\mathbf{x}} \varphi(\mathbf{x}, y)] \supset_y .$
$[\mathbf{M}] \supset_{\mathbf{x}} \psi(\mathbf{x}, y)\}(\lambda\mathbf{x} . \mathbf{F})$ —V, $\mathbf{A}_9$, $\mathbf{A}_{13}$.

$\mathbf{A}_{15}$: $[[\mathbf{M}] \supset_{\mathbf{x}} \Pi(\{\lambda\mathbf{x} . \mathbf{F}\}(\mathbf{x}), \psi(\mathbf{x}))] \supset_\psi . [[\mathbf{M}] \supset_{\mathbf{x}} \{\lambda\mathbf{x} . \mathbf{F}\}(\mathbf{x}, y)] \supset_y .$
$[\mathbf{M}] \supset_{\mathbf{x}} \psi(\mathbf{x}, y)$ —II, $\mathbf{A}_{14}$.

$\mathbf{A}_{16}$: $[[\mathbf{M}] \supset_{\mathbf{x}} \Pi(\mathbf{F}, \psi(\mathbf{x}))] \supset_\psi . [[\mathbf{M}] \supset_{\mathbf{x}} \{\lambda\mathbf{x} . \mathbf{F}\}(\mathbf{x}, y)] \supset_y .$
$[\mathbf{M}] \supset_{\mathbf{x}} \psi(\mathbf{x}, y)$ —II, $\mathbf{A}_{15}$.

$\mathbf{A}_{17}$: $[[\mathbf{M}] \supset_{\mathbf{x}} \Pi(\mathbf{F}, \psi(\mathbf{x}))] \supset_\psi . [[\mathbf{M}] \supset_{\mathbf{x}} \{\mathbf{F}\}(y)] \supset_y . [\mathbf{M}] \supset_{\mathbf{x}} \psi(\mathbf{x}, y)$ —II, $\mathbf{A}_{16}$.

$\mathbf{B}_3$: $[\mathbf{M}] \supset_{\mathbf{x}} \Pi(\mathbf{F}, \mathbf{G})$ —provable, by hypothesis of induction.

$\mathbf{A}_{18}$: $[\mathbf{M}] \supset_{\mathbf{x}} \Pi(\mathbf{F}, \{\lambda\mathbf{x} . \mathbf{G}\}(\mathbf{x}))$ —III, $\mathbf{B}_3$.

$\mathbf{A}_{19}$: $\{\lambda\psi . [\mathbf{M}] \supset_{\mathbf{x}} \Pi(\mathbf{F}, \psi(\mathbf{x}))\}(\lambda\mathbf{x} . \mathbf{G})$ —III, $\mathbf{A}_{18}$.

$\mathbf{A}_{20}$: $\{\lambda\psi . [[\mathbf{M}] \supset_{\mathbf{x}} \{\mathbf{F}\}(y)] \supset_y . [\mathbf{M}] \supset_{\mathbf{x}} \psi(\mathbf{x}, y)\}(\lambda\mathbf{x} . \mathbf{G})$ —V, $\mathbf{A}_{17}$, $\mathbf{A}_{19}$.

$\mathbf{A}_{21}$: $[[\mathbf{M}] \supset_{\mathbf{x}} \{\mathbf{F}\}(y)] \supset_y . [\mathbf{M}] \supset_{\mathbf{x}} \{\lambda\mathbf{x} . \mathbf{G}\}(\mathbf{x}, y)$ —II, $\mathbf{A}_{20}$.

$\mathbf{A}_{22}$: $[[\mathbf{M}] \supset_{\mathbf{x}} \{\mathbf{F}\}(y)] \supset_y . [\mathbf{M}] \supset_{\mathbf{x}} \{\mathbf{G}\}(y)$ —II, $\mathbf{A}_{21}$.

$\mathbf{A}_{23}$: $\{\lambda y . [\mathbf{M}] \supset_{\mathbf{x}} \{\mathbf{F}\}(y)\}(\mathbf{A})$ —III, $\mathbf{B}_2$.

$\mathbf{A}_{24}$: $\{\lambda y . [\mathbf{M}] \supset_{\mathbf{x}} \{\mathbf{G}\}(y)\}(\mathbf{A})$ —V, $\mathbf{A}_{22}$, $\mathbf{A}_{23}$.

$\mathbf{A}_{25}$: $[\mathbf{M}] \supset_{\mathbf{x}} \{\mathbf{G}\}(\mathbf{A})$ —II, $\mathbf{A}_{24}$.

$\mathbf{A}_{26}$: $[\mathbf{M}] \supset_{\mathbf{x}} . \mathbf{N}$ —by conversion, from $\mathbf{A}_{25}$.

Case 11, the last step in the proof of $\mathbf{N}$ as a consequence of $\mathbf{M}$ is an application of Rule V, and $\mathbf{F}$ and $\mathbf{G}$ and $\mathbf{A}$ all three contain $\mathbf{x}$ as a free variable. We include, in particular, the case that $\mathbf{F}$ and $\mathbf{G}$ both contain $\mathbf{x}$ as a free variable and $\mathbf{A}$ is identical with $\mathbf{x}$, since there is nothing to be gained by treating this case separately. By the hypothesis of our induction, $[\mathbf{M}] \supset_{\mathbf{x}} \{\mathbf{F}\}(\mathbf{A})$ is provable, and this is convertible into

$$[\mathbf{M}] \supset_{\mathbf{x}} \{\lambda x \{\lambda y \,.\, \mathrm{S}^{\mathbf{x}}_{y}\mathbf{F}|\}(x, \{\lambda y \,.\, \mathrm{S}^{\mathbf{x}}_{y}\mathbf{A}|\}(x))\}(\mathbf{x}).$$

Therefore, by the method of Case 3, we can prove $\Sigma x\{\lambda y \,.\, \mathrm{S}^{\mathbf{x}}_{y}\mathbf{F}|\}(x, \{\lambda y \,.\, \mathrm{S}^{\mathbf{x}}_{y}\mathbf{A}|\}(x))$. Hence, using Postulate 7, we can prove

$$\{\lambda y \,.\, \mathrm{S}^{\mathbf{x}}_{y}\mathbf{F}|\}(x, \{\lambda y \,.\, \mathrm{S}^{\mathbf{x}}_{y}\mathbf{A}|\}(x)) \supset_{x} \,.\, \Pi(\{\lambda y \,.\, \mathrm{S}^{\mathbf{x}}_{y}\mathbf{F}|\}(x), \psi(x)) \supset_{\psi} \psi(x, \{\lambda y \,.\, \mathrm{S}^{\mathbf{x}}_{y}\mathbf{A}|\}(x)),$$

and this is convertible into

$$\{\mathbf{F}\}(\mathbf{A}) \supset_{\mathbf{x}} \,.\, \Pi(\mathbf{F}, \psi(\mathbf{x})) \supset_{\psi} \psi(\mathbf{x}, \mathbf{A}).$$

And hence, by the method of Case 1, since $[\mathbf{M}] \supset_{\mathbf{x}} \{\mathbf{F}\}(\mathbf{A})$ is provable, we can prove $[\mathbf{M}] \supset_{\mathbf{x}} \,.\, \Pi(\mathbf{F}, \psi(\mathbf{x})) \supset_{\psi} \psi(\mathbf{x}, \mathbf{A})$.

Thus we are able to prove $[\mathbf{M}] \supset_{\mathbf{x}} \Pi(\mathbf{F}', \mathbf{G}')$ where $\mathbf{F}'$ stands for $\lambda\psi\Pi(\mathbf{F}, \psi(\mathbf{x}))$ and $\mathbf{G}'$ stands for $\lambda\psi \,.\, \psi(\mathbf{x}, \mathbf{A})$. And, by the hypothesis of our induction, we are able to prove $[\mathbf{M}] \supset_{\mathbf{x}} \{\mathbf{F}'\}(\mathbf{A}')$, where $\mathbf{A}'$ stands for $\lambda x \,.\, \mathbf{G}$. Therefore, by the method of Case 10, we can prove $[\mathbf{M}] \supset_{\mathbf{x}} \{\mathbf{G}'\}(\mathbf{A}')$ and from it by conversion, $[\mathbf{M}] \supset_{\mathbf{x}} \,.\, \mathbf{N}$.

Case 12, the last step in the proof of $\mathbf{N}$ as a consequence of $\mathbf{M}$ is an application of Rule V, and $\mathbf{F}$ and $\mathbf{A}$ both contain $\mathbf{x}$ as a free variable, and $\mathbf{G}$ does not contain $\mathbf{x}$ as a free variable. As before, $\mathbf{A}$ may, in particular, be identical with $\mathbf{x}$. By the hypothesis of our induction, $[\mathbf{M}] \supset_{\mathbf{x}} \{\mathbf{F}\}(\mathbf{A})$ is provable, and this is convertible into $[\mathbf{M}] \supset_{\mathbf{x}} \{\lambda\varphi \,.\, \varphi(\mathbf{A})\}(\mathbf{F})$. Therefore, by the method of Case 7, we can prove $[\mathbf{M}] \supset_{\mathbf{x}} \Sigma\varphi \,.\, \varphi(\mathbf{A})$, and this is convertible into $[\mathbf{M}] \supset_{\mathbf{x}} \{\lambda x E(x)\}(\mathbf{A})$. Hence, using Postulate 2, and the method of Case 8, we can prove $[\mathbf{M}] \supset_{\mathbf{x}} \,.\, \varphi(\mathbf{A}) \supset_{\varphi} \,.\, \Pi(\varphi, \psi) \supset_{\psi} \psi(\mathbf{A})$. And hence, by the method of Case 11, since $[\mathbf{M}] \supset_{\mathbf{x}} \{\mathbf{F}\}(\mathbf{A})$ is provable, we can prove $[\mathbf{M}] \supset_{\mathbf{x}} \,.\, \Pi(\mathbf{F}, \psi) \supset_{\psi} \psi(\mathbf{A})$.

Thus we are able to prove $[\mathbf{M}] \supset_{\mathbf{x}} \Pi(\mathbf{F}', \mathbf{G}')$ where $\mathbf{F}'$ stands for $\lambda\psi\Pi(\mathbf{F}, \psi)$ and $\mathbf{G}'$ stands for $\lambda\psi \,.\, \psi(\mathbf{A})$. And, by the hypothesis of our induction, we are able to prove $[\mathbf{M}] \supset_{\mathbf{x}} \{\mathbf{F}'\}(\mathbf{A}')$ where $\mathbf{A}'$ stands for $\mathbf{G}$. Therefore, by the method of Case 10, we can prove $[\mathbf{M}] \supset_{\mathbf{x}} \{\mathbf{G}'\}(\mathbf{A}')$, and from it by conversion, $[\mathbf{M}] \supset_{\mathbf{x}} \,.\, \mathbf{N}$.

Case 13, the last step in the proof of $\mathbf{N}$ as a consequence of $\mathbf{M}$ is an application of Rule V, and $\mathbf{A}$ is identical with $\mathbf{x}$, $\mathbf{G}$ contains $\mathbf{x}$ as a free variable and $\mathbf{F}$ does not. By the hypothesis of our induction, $[\mathbf{M}] \supset_{\mathbf{x}} \{\mathbf{F}\}(\mathbf{x})$ is provable, and this is convertible into $[\mathbf{M}] \supset_{\mathbf{x}} \{\lambda\varphi \,.\, \varphi(\mathbf{x})\}(\mathbf{F})$. Therefore, by the method of Case 5, we can prove $[\mathbf{M}] \supset_{\mathbf{x}} \Sigma\varphi \,.\, \varphi(\mathbf{x})$ and this is convertible into $[\mathbf{M}] \supset_{\mathbf{x}} E(\mathbf{x})$. Hence, using Postulate 6, and the method of Case 1, we can prove $[\mathbf{M}] \supset_{\mathbf{x}} \,.\, \varphi(\mathbf{x}) \supset_{\varphi} \,.\, \Pi(\varphi, \psi(\mathbf{x})) \supset_{\psi} \psi(\mathbf{x}, \mathbf{x})$. And hence, by the method of Case 10, since $[\mathbf{M}] \supset_{\mathbf{x}} \{\mathbf{F}\}(\mathbf{x})$ is provable, we can prove $[\mathbf{M}] \supset_{\mathbf{x}} \,.\, \Pi(\mathbf{F}, \psi(\mathbf{x})) \supset_{\psi} \psi(\mathbf{x}, \mathbf{x})$.

Thus we are able to prove $[\mathbf{M}] \supset_{\mathbf{x}} \Pi(\mathbf{F}', \mathbf{G}')$ where $\mathbf{F}'$ stands for $\lambda\psi\Pi(\mathbf{F}, \psi(\mathbf{x}))$, and $\mathbf{G}'$ stands for $\lambda\psi \,.\, \psi(\mathbf{x}, \mathbf{x})$. And, by the hypothesis of our induction, we are able to prove $[\mathbf{M}] \supset_{\mathbf{x}} \{\mathbf{F}'\}(\mathbf{A}')$ where $\mathbf{A}'$ stands for $\lambda\mathbf{x} \,.\, \mathbf{G}$. Therefore, by the method of Case 10, we can prove $[\mathbf{M}] \supset_{\mathbf{x}} \{\mathbf{G}'\}(\mathbf{A}')$ and from it by conversion, $[\mathbf{M}] \supset_{\mathbf{x}} \,.\, \mathbf{N}$.

Case 14, the last step in the proof of $\mathbf{N}$ as a consequence of $\mathbf{M}$ is an application of Rule V, and $\mathbf{G}$ and $\mathbf{A}$ both contain $\mathbf{x}$ as a free variable, and $\mathbf{F}$ does not contain $\mathbf{x}$ as a free variable. By the hypothesis of our induction, we can prove $[\mathbf{M}] \supset_{\mathbf{x}} \{\mathbf{F}\}(\mathbf{A})$ and hence, by the method of Case 4, we can prove $\Sigma(\mathbf{F})$. Hence, using Postulate 5, we can prove $\Pi(\mathbf{F}, \psi) \supset_{\psi} . \{\mathbf{F}\}(f(x)) \supset_{fx} \psi(f(x))$. And, by the hypothesis of our induction, we can prove $[\mathbf{M}] \supset_{\mathbf{x}} \Pi(\mathbf{F}, \mathbf{G})$, which is convertible into $[\mathbf{M}] \supset_{\mathbf{x}} \{\lambda \psi \Pi(\mathbf{F}, \psi)\}(\mathbf{G})$. Therefore, by the method of Case 8, we can prove

$$[\mathbf{M}] \supset_{\mathbf{x}} \{\lambda \psi . \{\mathbf{F}\}(f(x)) \supset_{fx} \psi(f(x))\}(\mathbf{G}),$$

which is convertible into $[\mathbf{M}] \supset_{\mathbf{x}} \{\mathbf{F}\}(f(x)) \supset_{fx} \{\mathbf{G}\}(f(x))$.

Moreover, by conversion from $[\mathbf{M}] \supset_{\mathbf{x}} \{\mathbf{F}\}(\mathbf{A})$ we can prove

$$[\mathbf{M}] \supset_{\mathbf{x}} \{\lambda x . \{\mathbf{F}\}(\{\lambda \mathbf{x} . \mathbf{A}\}(x))\}(\mathbf{x}),$$

and therefore, by the method of Case 3, we can prove $\Sigma x . \{\mathbf{F}\}(\{\lambda \mathbf{x} . \mathbf{A}\}(x))$. Hence, by the method of Case 9, using the formula at the end of the preceding paragraph, we can prove

$$[\mathbf{M}] \supset_{\mathbf{x}} . \{\mathbf{F}\}(\{\lambda \mathbf{x} . \mathbf{A}\}(x)) \supset_{x} \{\mathbf{G}\}(\{\lambda \mathbf{x} . \mathbf{A}\}(x)).$$

Hence, using $[\mathbf{M}] \supset_{\mathbf{x}} \{\lambda x . \{\mathbf{F}\}(\{\lambda x . \mathbf{A}\}(x))\}(\mathbf{x})$ and the method of Case 13, we can prove

$$[\mathbf{M}] \supset_{\mathbf{x}} \{\lambda x . \{\mathbf{G}\}(\{\lambda \mathbf{x} . \mathbf{A}\}(x))\}(\mathbf{x}).$$

And hence finally, by conversion, we can prove $[\mathbf{M}] \supset_{\mathbf{x}} . \mathbf{N}$.

Case 15, the last step in the proof of $\mathbf{N}$ as a consequence of $\mathbf{M}$ is an application of Rule V, and neither $\mathbf{F}$ nor $\mathbf{G}$ nor $\mathbf{A}$ contains $\mathbf{x}$ as a free variable. By the hypothesis of our induction, $\Pi(\mathbf{F}, \mathbf{G})$ and $\{\mathbf{F}\}(\mathbf{A})$ are both provable. Hence, using Rule V, $\{\mathbf{G}\}(\mathbf{A})$ is provable, and from it by conversion, $\mathbf{N}$.

Case 16, the last step in the proof of $\mathbf{N}$ as a consequence of $\mathbf{M}$ is an application of Rule IV, and neither $\mathbf{F}$ nor $\mathbf{A}$ contains $\mathbf{x}$ as a free variable. By the hypothesis of our induction, $\{\mathbf{F}\}(\mathbf{A})$ is provable. Hence, using Rule IV, $\Sigma(\mathbf{F})$ is provable, and from it by conversion, $\mathbf{N}$.

COROLLARY. *Suppose that* $\mathbf{M}$ *contains* $\mathbf{x}$ *and* $\mathbf{y}$ *as free variables and that* $\Sigma \mathbf{x} \Sigma \mathbf{y} . \mathbf{M}$ *is provable, and that* $\mathbf{N}$ *is provable as a consequence of* $\mathbf{M}$. *Then if* $\mathbf{N}$ *contains neither* $\mathbf{x}$ *nor* $\mathbf{y}$ *as a free variable, the formula* $\mathbf{N}$ *is a provable formula. If* $\mathbf{N}$ *contains* $\mathbf{y}$ *as a free variable, the formula* $[\mathbf{M}] \supset_{\mathbf{xy}} . \mathbf{N}$ *is a provable formula. And if* $\mathbf{N}$ *contains* $\mathbf{x}$ *as a free variable, the formula* $[\mathbf{M}] \supset_{\mathbf{yx}} . \mathbf{N}$ *is a provable formula.*

And similarly for cases where $\mathbf{M}$ contains a greater number of free variables.

THEOREM II. *If* $\Sigma(\mathbf{F})$ *is provable, and* $\mathbf{G}$ *does not contain* $\mathbf{x}$ *as a free variable, and* $\{\mathbf{G}\}(\mathbf{x})$ *is provable as a consequence of* $\{\mathbf{F}\}(\mathbf{x})$, *then* $\Pi(\mathbf{F}, \mathbf{G})$ *is provable.*

For, using Postulate 12, we can prove $\Sigma \mathbf{x} . \{\mathbf{F}\}(\mathbf{x})$. Therefore, by Theorem I, $\{\mathbf{F}\}(\mathbf{x}) \supset_{\mathbf{x}} \{\mathbf{G}\}(\mathbf{x})$ is provable. And hence, using Postulate 13, we can prove $\Pi(\mathbf{F}, \mathbf{G})$.

THEOREM III. *If* $\Sigma(\mathbf{F})$ *is provable and* $\mathbf{N}$ *does not contain* $\mathbf{x}$ *as a free variable, and* $\mathbf{N}$ *is provable as a consequence of* $\{\mathbf{F}\}(\mathbf{x})$, *then* $\mathbf{N}$ *is provable.*

For, using Postulate 12, we can prove $\Sigma \mathbf{x} . \{\mathbf{F}\}(\mathbf{x})$. Therefore, by Theorem I, $\mathbf{N}$ is provable.

Moreover it is readily seen that if our rules of procedure are to be Rules I–V and if Theorems II and III are to be true of our system then Postulates 1–13 must be, if not postulates (as we have taken them to be), at least provable formulas.

PRINCETON UNIVERSITY, PRINCETON, N.J.

A SET OF POSTULATES FOR THE FOUNDATION OF LOGIC (SECOND PAPER)[1] (1933)

1. Revision of the list of formal postulates. In a recent paper[2] the author proposed a set of postulates which, it was believed, would lead to a system of mathematical logic free of some of the complications entailed by Bertrand Russell's theory of types, and would at the same time avoid the well known paradoxes, in particular the Russell paradox, by weakening the classical principle of reductio ad absurdum. But, in the course of working with these postulates, it has since become clear that the set as originally given requires some modification in order to render it free from contradiction.

Postulate 19 makes it possible to prove $\sim\Sigma|(\{\mathbf{F}\}(\mathbf{U}))$ if there can be found some consequence of $\Sigma(\{\mathbf{F}\}(y))$ which is false of $\mathbf{U}$, subject only to the restriction that $\{\mathbf{F}\}(\mathbf{U}, x)$ shall be known to be false for some x. This is not far from assuming the principle of reductio ad absurdum in its full strength as applied to propositions of the form $\Sigma(\mathbf{A})$. But, given a proposition which is not of the form $\Sigma(\mathbf{A})$, it is ordinarily possible in various ways, to write another proposition of much the same import, which is of the form $\Sigma(\mathbf{A})$. And in this way we find that Postulate 19 contains enough of the principle of reductio ad absurdum to lead to a modified form of the Russell paradox, in outline as follows.

We take $\mathfrak{Y}$ to be $\lambda u \lambda v E(u)E(v)$ and prove, in a straightforward manner,

$$\sim\sim\{\mathfrak{Y}\}\left(\Lambda\varphi\,.\sim\varphi(\{\mathfrak{Y}\}(\varphi)), \{\mathfrak{Y}\}\left(\{\mathfrak{Y}\}(\Lambda\varphi\,.\sim\varphi(\{\mathfrak{Y}\}(\varphi)))\right)\right),$$

where Λ is defined as in §5 below. Hence, by Rule III,

$$\sim\{\lambda\varphi\,.\sim\varphi(\{\mathfrak{Y}\}(\varphi))\}\left(\{\mathfrak{Y}\}(\Lambda\varphi\,.\sim\varphi(\{\mathfrak{Y}\}(\varphi)))\right).$$

Hence, by Theorem 8 below, $\sim\{\Lambda\varphi\,.\sim\varphi(\{\mathfrak{Y}\}(\varphi))\}\left(\{\mathfrak{Y}\}(\Lambda\varphi\,.\sim\varphi(\{\mathfrak{Y}\}(\varphi)))\right)$. And hence we prove $\Sigma y \Sigma x\,.\sim x(y(x))\,.\,x = \Lambda\varphi\,.\sim\varphi(y(\varphi))$. But, assuming

$$\Sigma x\,.\sim x(y(x))\,.\,x = \Lambda\varphi\,.\sim\varphi(y(\varphi)),$$

we can prove $\Sigma r\,.\,r(r)\,.\sim r(y(r))$, because

$$\sim\{\Lambda\varphi\,.\sim\varphi(y(\varphi))\}\left(y(\Lambda\varphi\,.\sim\varphi(y(\varphi)))\right)$$

is convertible into $\{\lambda\varphi\,.\sim\varphi(y(\varphi))\}(\Lambda\varphi\,.\sim\varphi(y(\varphi)))$, which, by Theorem 6 below, yields in turn $\{\Lambda\varphi\,.\sim\varphi(y(\varphi))\}(\Lambda\varphi\,.\sim\varphi(y(\varphi)))$. Therefore, by Theorem I, $\Sigma(\{\mathfrak{F}\}(y)) \supset_y \{\mathfrak{G}\}(y)$, where $\{\mathfrak{F}\}(y)$ stands for $\lambda x\,.\sim x(y(x))\,.\,x = \Lambda\varphi\,.\sim\varphi(y(\varphi))$ and $\{\mathfrak{G}\}(y)$ stands for $\Sigma r\,.\,r(r)\,.\sim r(y(r))$. Now we take $\mathfrak{U}$ to be λzz and, with the aid of Postulate 19, we prove $\sim\{\mathfrak{G}\}(\mathfrak{U})$. Therefore, by Postulate 19, $\sim\Sigma(\{\mathfrak{F}\}(\mathfrak{U}))$, that is, $\sim\Sigma x\,.\sim x(x)\,.\,x = \Lambda\varphi\,.\sim\varphi(\varphi)$. Then, by means of Theorem I, we prove $\sim\varphi(\varphi) \supset_\varphi \Sigma x\,.\sim x(x)\,.\,x = \varphi$. Therefore, by Theorem 9 below,

$$\sim\{\Lambda\varphi\,.\sim\varphi(\varphi)\}(\Lambda\varphi\,.\sim\varphi(\varphi)).$$

Originally published in ***Annals of Mathematics***, vol. 34 (1933), pp. 839–864. © Princeton University. Reprinted by permission.

[1]Received January 3, 1933.

[2]Church, *A set of postulates for the foundation of logic*, ***Annals of Mathematics***, vol. 33 (1932), pp. 346–366.

Therefore, by Rule III, $\{\lambda\varphi \,.\sim \varphi(\varphi)\}(\Lambda\varphi \,.\sim \varphi(\varphi))$. Therefore, by Theorem 6 below, $\{\lambda\varphi \,.\sim \varphi(\varphi)\}(\Lambda\varphi \,.\sim \varphi(\varphi))$, a contradiction.

For this reason we shall omit Postulate 19 from our list. And with it we shall omit Postulates 20, 28, and 29, because these three postulates entirely lose their point after the omission of Postulate 19.

The chief effect of the omission of Postulate 19 is that we are no longer able to prove formulas of the form $\sim \Sigma(\mathbf{A})$.[3] We believe, however, that this restriction will not render our system inadequate, because the formula $\sim\{\Lambda y \Sigma(\{\mathbf{F}(y)\})\}(\mathbf{U})$ can often be proved and used instead of $\sim \Sigma(\{\mathbf{F}\}(\mathbf{U}))$.

Let us call a formula *vacuous* when it has, or is convertible into, the form $\Pi(\mathbf{F}, \mathbf{G})$, and there is no formula $\mathbf{M}$ such that $\{\mathbf{F}\}(\mathbf{M})$ is provable. The omission of Postulates 28 and 29 does away with our chief means of proving vacuous formulas. It remains possible to prove vacuous formulas of certain particular kinds by means of Postulate 25 or of Postulate 34, but this property of these two postulates is largely accidental, and we propose to eliminate it by means of appropriate slight changes in them. It then becomes impossible, we think, to prove any vacuous formula. And we recognize this property of our system by adding to it (tentatively) the postulate, $\Sigma\psi\Pi(\varphi, \psi) \supset_\varphi \Sigma(\varphi)$. This postulate is, probably, not indispensable, but it is sometimes convenient. Its addition to the list renders Postulate 32 non-independent, and the latter postulate may, therefore, be omitted.

And it is now also necessary to modify Postulates 17 and 18 to read, respectively, $\Sigma x[\varphi(x) \,.\sim \psi(x)] \supset_{\varphi\psi} \sim \Pi(\varphi, \psi)$ and $\sim \Pi(\varphi, \psi) \supset_{\varphi\psi} \Sigma x \,.\, \varphi(x) \,.\sim \psi(x)$. For otherwise it would be impossible to prove any formula $\sim \Pi(\mathbf{F}, \mathbf{G})$ unless $\Sigma(\mathbf{G})$ were provable.

And finally, Postulate 2 fails to be independent, and may, therefore, be omitted from the list. For the proofs of cases 9 and 12 under Theorem I can be modified so that Postulate 7 takes the place of Postulate 2, and Theorem I can therefore be proved from Postulates 1, 3–11.[4]

The question whether our other postulates are all independent remains at present open to doubt, but there appears to be no reason why this question cannot be investigated by the usual method of constructing independence examples.

Our revised list of formal postulates is now as follows:

1. $\Sigma(\varphi) \supset_\varphi \Pi(\varphi, \varphi)$.
3. $\Sigma(\sigma) \supset_\sigma .[\sigma(x) \supset_x \varphi(x)] \supset_\varphi .\, \Pi(\varphi, \psi) \supset_\psi .\, \sigma(x) \supset_x \psi(x)$.
4. $\Sigma(\varrho) \supset_\varrho .\, \Sigma y[\varrho(x) \supset_x \varphi(x, y)] \supset_\varphi .[\varrho(x) \supset_x \Pi(\varphi(x), \psi(x))] \supset_\psi .$ $[\varrho(x) \supset_x \varphi(x, y)] \supset_y .\, \varrho(x) \supset_x \psi(x, y)$.
5. $\Sigma(\varphi) \supset_\varphi .\, \Pi(\varphi, \psi) \supset_\psi .\, \varphi(f(x)) \supset_{fx} \psi(f(x))$.
6. $`x \,.\, \varphi(x) \supset_\varphi .\, \Pi(\varphi, \psi(x)) \supset_\psi \psi(x, x)$.
7. $\varphi(x, f(x)) \supset_{\varphi f x} .\, \Pi(\varphi(x), \psi(x)) \supset_\psi \psi(x, f(x))$.

---

[3]It is to be observed, however, that another symbol, meaning, "There exists", can be defined as follows:

$$\exists \longrightarrow \lambda\varphi\lambda\psi \sim .\, \varphi(x) \supset_x \sim \psi(x).$$

It will then be found that $\exists(\mathbf{A}, \mathbf{B})$ is provable whenever $\Sigma x \,.\{\mathbf{A}\}(x)\{\mathbf{B}\}(x)$ is provable and conversely. And, for particular formulas $\mathbf{A}$ and $\mathbf{B}$, it frequently is possible to prove $\sim \exists(\mathbf{A}, \mathbf{B})$.

The expression $\exists(\mathbf{A}, \mathbf{B})$ is to be read, "Some $\mathbf{A}$ is $\mathbf{B}$".

[4]This observation is due to Mr. J. B. Rosser.

8. $\Sigma(\varrho) \supset_{\varrho} . \Sigma y[\varrho(x) \supset_{x} \varphi(x, y)] \supset_{\varphi} .[\varrho(x) \supset_{x} \Pi(\varphi(x), \psi)] \supset_{\psi} .$ $[\varrho(x) \supset_{x} \varphi(x, y)] \supset_{y} \psi(y).$
9. $`x . \varphi(x) \supset_{\varphi} \Sigma(\varphi).$
10. $\Sigma x\varphi(f(x)) \supset_{f\varphi} \Sigma(\varphi).$
11. $\varphi(x, x) \supset_{\varphi x} \Sigma(\varphi(x)).$
12. $\Sigma(\varphi) \supset_{\varphi} \Sigma x\varphi(x).$
13. $\Sigma(\varphi) \supset_{\varphi} .[\varphi(x) \supset_{x} \psi(x)] \supset_{\psi} \Pi(\varphi, \psi).$
14. $p \supset_{p} . q \supset_{q} pq.$
15. $pq \supset_{qp} p.$
16. $pq \supset_{pq} q.$
17. $\Sigma x[\varphi(x) . \sim \psi(x)] \supset_{\varphi\psi} \sim \Pi(\varphi, \psi).$
18. $\sim \Pi(\varphi, \psi) \supset_{\varphi\psi} \Sigma x . \varphi(x) . \sim \psi(x).$
21. $p \supset_{p} . \sim q \supset_{q} \sim . pq.$
22. $\sim p \supset_{p} . q \supset_{q} \sim . pq.$
23. $\sim p \supset_{p} . \sim q \supset_{q} \sim . pq.$
24. $p \supset_{p} .[\sim . pq] \supset_{q} \sim q.$
25. $[\sim[\varphi(u)\psi(u)] .[[\varphi(x) . \sim \psi(x)] \supset_{x} \varrho(x)] .[[\sim \varphi(x) . \psi(x)] \supset_{x} \varrho(x)] .$ $[\sim \varphi(x) . \sim \psi(x)] \supset_{x} \varrho(x)] \supset_{\varphi\psi\varrho u} \varrho(u).$
26. $p \supset_{p} \sim \sim p.$
27. $\sim \sim p \supset_{p} p.$
30. $`x . \varphi(x) \supset_{\varphi} .[\theta(x) . \psi(x) \supset_{\psi} \Pi(\theta, \psi)] \supset_{\theta} \varphi(\iota(\theta)).$
31. $\varphi(\iota(\theta)) \supset_{\theta\varphi} \Pi(\theta, \varphi).$
33. $[\varphi(x, y) \supset_{xy} . \varphi(y, z) \supset_{z} \varphi(x, z)][\varphi(x, y) \supset_{xy} \varphi(y, x)] \supset_{\varphi} . \varphi(u, v)$ $\supset_{uv} . \psi(A(\varphi, u)) \supset_{\psi} \psi(A(\varphi, v)).$
34. $[\varphi(x, y) \supset_{xy} . \varphi(y, z) \supset_{z} \varphi(x, z)][\varphi(x, y) \supset_{xy} \varphi(y, x)][\varphi(x, y) \supset_{xy} \theta(x, y)]$ $[\sim \theta(u, v)] \supset_{\varphi\theta uv} \sim . \psi(A(\varphi, u)) \supset_{\psi} \psi(A(\varphi, v)).$
35. $[\psi(A(\varphi, u)) \supset_{\psi} \psi(A(\varphi, v))] \supset_{\varphi uv} \varphi(u, v).$
36. $E(A(\varphi)) \supset_{\varphi} . \varphi(x, y) \supset_{xy} . \varphi(y, z) \supset_{z} \varphi(x, z).$
37. $E(A(\varphi)) \supset_{\varphi} . \varphi(x, y) \supset_{xy} \varphi(y, x).$
38. $\Sigma\psi\Pi(\varphi, \psi) \supset_{\varphi} \Sigma(\varphi).$

2. Possibility of a proof of freedom from contradiction. Our present project is to develop the consequences of the foregoing set of postulates, until a contradiction is obtained from them, or until the development has been carried so far consistently as to make it empirically probable that no contradiction can be obtained from them. And in this connection it is to be remembered that just such empirical evidence, although admittedly inconclusive, is the only existing evidence of the freedom from contradiction of any system of mathematical logic which has a claim to adequacy.

It is worth observing, however, that there may be a possibility of proving that there is no formula **A** such that both **A** and $\sim$ **. A** are consequences of our postulates, or, failing this, of proving the same theorem about some related set of postulates. This is conceivable on account of the entirely formal character of the system which makes it possible to abstract from the meaning of the symbols and to regard the proving of theorems (of formal logic) as a game played with marks on paper according to a certain arbitrary set of rules. It is then proposed to use our intuitive logic (some portion of which we have already had to presuppose) to prove about this game, regarded objectively, that if it be played according to the rules it cannot lead to combinations of marks of certain particular kinds. Indeed the whole set of undefined terms of the formal logic, including enumerably many symbols available for use as variables, is enumerable, and hence the whole set of formulas and of possible proofs is also enumerable. And it is therefore even to be hoped that the portion of intuitive logic necessary to the proof of freedom from contradiction should not go beyond the logic of enumerable classes or employ any number system more elaborate than the system of positive integers.[5]

The impossibility of such a proof of freedom from contradiction for the system of *Principia Mathematica* has recently been established by Kurt Gödel.[6] His argument, however, makes use of the relation of implication U between propositions in a way which would not be permissible under the system of this paper, and there is no obvious way of modifying the argument so as to make it apply to the system of this paper. It therefore remains, at least for the present, conceivable that there should be found a proof of freedom from contradiction for our system.

3. Abbreviations of the statement of proofs. In the proofs of theorems which follow, we indicate after each formula in the proof the means by which it is inferred, that is, we give the number of the rule of procedure and a reference to the premise or premises. Formal postulates are referred to by the number written alone, and theorems are referred to by the abbreviation "Thm" followed by the number.

We use Theorems I, II, and III, and the corollary of Theorem I as if they were rules of procedure. This is justified because our proof of these theorems was constructive in character, so that their use in this way can be regarded as a mere device for abbreviating the statement of a proof. For instance, whenever we have two particular formulas M and N, each of which contains x as a free variable, and a particular proof of $\Sigma\mathsf{x} \,.\, \mathsf{M}$, and a particular scheme for proving N as a consequence of M, then, by the method used in the proof of Theorem I, we can obtain a particular proof of $[\mathsf{M}] \supset_{\mathsf{x}} . \,\mathsf{N}$.

Moreover we allow uses of Theorems I, II, and III, and the corollary of Theorem I, in the role of rules of procedure, even in the course of proving N as a consequence of M preparatory to another application of one of the Theorems I, II, III, or Theorem I Corollary. For our proof of these theorems remains valid even after the formula M is added to our list as an additional formal postulate, provided only that the variable x, which occurs in M as a free variable, is not used elsewhere as a bound variable.

In the proof of a theorem we give before each formula a symbol which will be used in referring to it, namely a letter A, C, or D, with a subscript. The formulas marked A_i have been proved outright. Those marked D_i are assumed, in order to derive other formulas from them, preparatory to an application of Theorem I, or II, or III, or Theorem I Corollary. And those marked C_i are proved on the basis of the assumption of certain of the formulas D_i, as indicated in each case.

We abbreviate further by omitting steps which involve only an application of one of the rules of procedure I, II, III, or by condensing several such steps into one, using

[5]The conception of a consistency proof of this kind is due to David Hilbert. See *Die logischen Grundlagen der Mathematik*, ***Math. Annalen***, vol. 88 (1923), pp. 151–165, and *Die Grundlagen der Mathematik*, ***Abh. a. d. Math. Sem. d. Hamb. Univ.***, vol. 6 (1928), pp. 65–92.

[6]Kurt Gödel, *Über formal unentscheidbare Sätze der Principia Mathematica und verwandter Systeme I*, ***Monatshefte für Mathematik und Physik***, vol. 38 (1931), pp. 173–198.

"conv" to mean, "By conversion", or, "By application of Rules I, II, and III".

We use IV^n to stand for n successive applications of Rule IV. And we use V^n to stand for n successive applications of Rule V, the result of each application of Rule V being used in turn as the major premise[7] for the next application of Rule V. After V^n, so used, we give a reference, first to the major premise of the first application of Rule V, and then to the successive minor premises[7] in order. But in passing from $\{\mathbf{F}\}(\mathbf{x}, \mathbf{y}) \supset_{\mathbf{xy}} \{\mathbf{G}\}(\mathbf{x}, \mathbf{y})$ and $\{\mathbf{F}\}(\mathbf{A}, \mathbf{B})$ to $\{\mathbf{G}\}(\mathbf{A}, \mathbf{B})$ or in passing from $\{\mathbf{F}\}(\mathbf{x}, \mathbf{y}) \supset_{\mathbf{xy}} \{\mathbf{G}\}(\mathbf{y})$ and $\{\mathbf{F}\}(\mathbf{A}, \mathbf{B})$ to $\{\mathbf{G}\}(\mathbf{B})$, we refer to $\{\mathbf{F}\}(\mathbf{A}, \mathbf{B})$ as the only minor premise, omitting reference to the other minor premise $\Sigma \mathbf{y}\{\mathbf{F}\}(\mathbf{A}, \mathbf{y})$. Similarly we shall sometimes refer to a formula $\mathbf{M}$ as minor premise when the true minor premise is not $\mathbf{M}$ but some formula obtained from $\mathbf{M}$ by use of Rule IV or conversion or both. And we shall sometimes refer to two minor premises, $\mathbf{P}$ and $\mathbf{Q}$, when the true minor premise is the logical product $[\mathbf{P}][\mathbf{Q}]$, obtained from the $\mathbf{P}$ and $\mathbf{Q}$ by use of Postulate 14.

4. Identity. We adopt the following definitions of identity, or equality,[8] and of distinctness, or inequality:

$$= \longrightarrow \lambda\mu\lambda\nu \,.\, \psi(\mu) \supset_\psi \psi(\nu).$$

$$\neq \longrightarrow \lambda\mu\lambda\nu \sim .\, \psi(\mu) \supset_\psi \psi(\nu).$$

We shall abbreviate $\{=\}(\mathbf{A}, \mathbf{B})$ as $[\mathbf{A}] = [\mathbf{B}]$, and $\{\neq\}(\mathbf{A}, \mathbf{B})$ as $[\mathbf{A}] \neq [\mathbf{B}]$.

In terms of these definitions, Postulates 33, 34, 35 can be rewritten as follows:

33′. $[\varphi(x, y) \supset_{xy} .\, \varphi(y, z) \supset_z \varphi(x, z)][\varphi(x, y) \supset_{xy} \varphi(y, x)] \supset_\varphi .\, \varphi(u, v) \supset_{uv} .\, A(\varphi, u) = A(\varphi, v)$.

34′. $[\varphi(x, y) \supset_{xy} .\, \varphi(y, z) \supset_z \varphi(x, z)][\varphi(x, y) \supset_{xy} \varphi(y, x)][\varphi(x, y) \supset_{xy} \theta(x, y)] [\sim \theta(u, v)] \supset_{\varphi\theta uv} .\, A(\varphi, u) \neq A(\varphi, v)$.

35′. $[A(\varphi, u) = A(\varphi, v)] \supset_{\varphi uv} \varphi(u, v)$.

It is seen that the propositions 33′, 34′, 35′ are derivable by conversion from Postulates 33, 34, 35, respectively. And Postulate 30 is seen to be essentially the same as Theorem 4 below.

THEOREM 1. $`x \,.\, x = x$.

$\mathbf{D}_1$: $E(x)$.
$\mathbf{C}_1$: $\Sigma\psi \,.\, \psi(x)$ —conv, $\mathbf{D}_1$.
$\mathbf{C}_2$: $\psi(x) \supset_\psi \psi(x)$ —V, 1, $\mathbf{C}_1$; $\mathbf{D}_1$.
$\mathbf{C}_3$: $x = x$ —conv, $\mathbf{C}_2$; $\mathbf{D}_1$.
$\mathbf{A}_1$: $\Sigma\psi \,.\, \psi(\lambda\varphi\Pi(\varphi, \varphi))$ —IV, 1.
$\mathbf{A}_2$: $E\varphi\Pi(\varphi, \varphi)$ —conv, $\mathbf{A}_1$.
$\mathbf{A}_3$: $\Sigma x E(x)$ —IV, $\mathbf{A}_2$.
$\mathbf{A}_4$: $`x \,.\, x = x$ —Thm I, $\mathbf{C}_3$, $\mathbf{A}_3$.

THEOREM 2. $x = y \supset_{xy} .\, y = x$.

$\mathbf{D}_1$: $x = y$.
$\mathbf{C}_1$: $\Sigma\varphi \,.\, \varphi(x)$ —IV, $\mathbf{D}_1$.

---

[7]In an application of Rule V, we call the formula $\Pi(\mathbf{F}, \mathbf{G})$ the *major premise* and the formula $\{\mathbf{F}\}(\mathbf{A})$ the *minor premise*.

[8]This definition is similar to the one used in ***Principia Mathematica***. It is a translation into symbolic notation of the definition originally given by Leibniz. Cf. C. I. Lewis, ***A Survey of Symbolic Logic***, 1918, p. 373.

C$_2$: $E(x)$ —III, **C**$_1$; **D**$_1$.
C$_3$: $x = x$ —V, Thm 1, **C**$_2$; **D**$_1$.
C$_4$: $\{\lambda z \,.\, z = x\}(x)$ —III, **C**$_3$; **D**$_1$.
C$_5$: $\{\lambda z \,.\, z = x\}(y)$ —V, **D**$_1$, **C**$_4$; **D**$_1$.
C$_6$: $y = x$ —II, **C**$_5$; **D**$_1$.
A$_1$: $Ex \,.\, x = x$ —IV, Thm 1.
A$_2$: $[\lambda x \,.\, x = x] = \,.\, \lambda x \,.\, x = x$ —V, Thm 1, **A**$_1$.
A$_3$: $\Sigma x \Sigma y \,.\, x = y$ —IV^2, **A**$_2$.
A$_4$: $x = y \supset_{xy} . \, y = x$ —Thm I Cor, **C**$_6$, **A**$_3$.

THEOREM 3. $x = y \supset_{xy} . \, y = z \supset_z . \, x = z$.

D$_1$: $x = y$.
D$_2$: $y = z$.
C$_1$: $x = z$ —V, **D**$_2$, **D**$_1$.
C$_2$: $E(y)$ —IV, **D**$_1$.
C$_3$: $y = y$ —V, Thm 1, **C**$_2$; **D**$_1$.
C$_4$: $\Sigma z \,.\, y = z$ —IV, **C**$_3$; **D**$_1$.
C$_5$: $y = z \supset_z . \, x = z$ —Thm I, **C**$_1$, **C**$_4$; **D**$_1$.
A$_1$: $x = y \supset_{xy} . \, y = z \supset_z . \, x = z$ —Thm I Cor, **C**$_5$, **A**$_3$ under Thm 2.

THEOREM 4. $[\theta(x) \,.\, \theta(y) \supset_y . \, x = y] \supset_{\theta x} . \, x = \iota(\theta)$.

D$_1$: $\theta(x) \,.\, \theta(y) \supset_y . \, x = y$.
C$_1$: $\theta(x)$ —V^2, 15, **D**$_1$.
C$_2$: $E(x)$ —IV, **C**$_1$; **D**$_1$.
C$_3$: $x = x$ —V, Thm 1, **C**$_2$; **D**$_1$.
C$_4$: $\theta(y) \supset_y . \, x = y$ —V^2, 16, **D**$_1$.
D$_2$: $\theta(y)$.
C$_5$: $x = y$ —V, **C**$_4$, **D**$_2$; **D**$_1$, **D**$_2$.
D$_3$: $\psi(x)$.
C$_6$: $\psi(y)$ —V, **C**$_5$, **D**$_3$; **D**$_1$, **D**$_2$, **D**$_3$.
C$_7$: $\Sigma(\theta)$ —IV, **C**$_1$; **D**$_1$.
C$_8$: $\pi(\theta, \psi)$ —Thm II, **C**$_6$, **C**$_7$; **D**$_1$, **D**$_3$.
C$_9$: $\Sigma\psi \,.\, \psi(x)$ —IV, **C**$_1$; **D**$_1$.
C$_{10}$: $\psi(x) \supset_\psi \pi(\theta, \psi)$ —Thm I, **C**$_8$, **C**$_9$; **D**$_1$.
C$_{11}$: $x = \iota(\theta)$ —V^3, 30, **C**$_2$, **C**$_3$, **C**$_1$, **C**$_{10}$; **D**$_1$.
A$_1$: $\Sigma y \,.[\lambda x \,.\, x = x] = y$ —IV, **A**$_2$ under Thm 2.
A$_2$: $[[\lambda x \,.\, x = x] = y] \supset_y .[\lambda x \,.\, x = x] = y$ —V, 1, **A**$_1$.
A$_3$: $[[\lambda x \,.\, x = x] = \,.\, \lambda x \,.\, x = x] \,.[[\lambda x \,.\, x = x] = y] \supset_y .[\lambda x \,.\, x = x] = y$
—V^2, 14, **A**$_2$ under Thm 2, **A**$_2$.
A$_4$: $\Sigma\theta\Sigma x \,.\, \theta(x) \,.\, \theta(y) \supset_y . \, x = y$ —IV^2, **A**$_3$.
A$_5$: $[\theta(x) \,.\, \theta(y) \supset_y . \, x = y] \supset_{\theta x} . \, x = \iota(\theta)$ —Thm I Cor, **C**$_{11}$, **A**$_4$.

THEOREM 5. ‘$x \,.\, x = \,.\, \iota y \,.\, x = y$.

D$_1$: $E(x)$.
C$_1$: $x = x$ —V, Thm 1, **D**$_1$.
C$_2$: $\Sigma y \,.\, x = y$ —IV, **C**$_1$; **D**$_1$.
C$_3$: $[x = y] \supset_y . \, x = y$ —V, 1, **C**$_2$; **D**$_1$.
C$_4$: $x = \,.\, \iota y \,.\, x = y$ —V, Thm 4, **C**$_1$, **C**$_3$; **D**$_1$.

$\mathbf{A}_1$: $`x \,.\, x = \,.\, \iota y \,.\, x = y$ —Thm I, $\mathbf{C}_4$, $\mathbf{A}_3$ under Thm 1.

5. The completion of a propositional function. We call a propositional function **F** *significant* for the value **A** of the independent variable if $\{\mathbf{F}\}(\mathbf{A})$ is either a true proposition or a false proposition.

We define the symbol Λ as follows:

$$\Lambda \longrightarrow \lambda\mu\lambda\pi \,.\, \Pi(\mu, \gamma) \supset_\gamma \gamma(\pi).$$

If **F** is any propositional function, then $\Lambda(\mathbf{F})$ is a propositional function which we call the *completion* of **F**. We shall prove (in Theorems 6 and 7) that if **F** is at least once true then $\Lambda(\mathbf{F})$ is equivalent to **F**, that is, that for every value of the independent variable for which either is true the other is also true. We shall prove (in Theorem 9 and its corollary) that, of all propositional functions which are equivalent to **F**, $\Lambda(\mathbf{F})$ is significant for the greatest range of values of the independent variable. And we shall prove (in Theorems 14 and 15) that there exist functions **F** for which the range of significance of $\Lambda(\mathbf{F})$ is greater than that of **F**.

We define the symbol G as follows:

$$G \longrightarrow \lambda\nu \,.\, \sim \Lambda(\nu, x) \supset_x \sim \nu(x).$$

And $G(\mathbf{F})$ is to be read, "**F** is complete". In Theorem 16 we prove that if **F** is at least once true and at least once false, then $\Lambda(\mathbf{F})$ is complete. And Theorem 20 can be regarded as asserting a certain form of the principle of reductio ad absurdum, as applied to complete propositional functions.

THEOREM 6. $\Sigma(\varphi) \supset_\varphi \Pi(\varphi, \Lambda(\varphi))$.

$\mathbf{D}_1$: $\Sigma(\varphi)$.
$\mathbf{C}_1$: $\Pi(\varphi, \varphi)$ —V, 1, $\mathbf{D}_1$.
$\mathbf{C}_2$: $\Sigma\gamma\Pi(\varphi, \gamma)$ —IV, $\mathbf{C}_1$; $\mathbf{D}_1$.
$\mathbf{D}_2$: $\varphi(x)$.
$\mathbf{D}_3$: $\Pi(\varphi, \gamma)$.
$\mathbf{C}_3$: $\gamma(x)$ —V, $\mathbf{D}_3$, $\mathbf{D}_2$.
$\mathbf{C}_4$: $\Pi(\varphi, \gamma) \supset_\gamma \gamma(x)$ —Thm I, $\mathbf{C}_3$, $\mathbf{C}_2$; $\mathbf{D}_1$, $\mathbf{D}_2$.
$\mathbf{C}_5$: $\Lambda(\varphi, x)$ —conv, $\mathbf{C}_4$; $\mathbf{D}_1$, $\mathbf{D}_2$.
$\mathbf{C}_6$: $\Pi(\varphi, \Lambda(\varphi))$ —Thm II, $\mathbf{C}_5$, $\mathbf{D}_1$; $\mathbf{D}_1$.
$\mathbf{A}_1$: $\Sigma(\Pi(\lambda\varphi\Sigma(\varphi)))$ —IV, 1.
$\mathbf{A}_2$: $\Sigma\varphi\Sigma(\varphi)$ —IV, $\mathbf{A}_1$.
$\mathbf{A}_3$: $\Sigma(\varphi) \supset_\varphi \Pi(\varphi, \Lambda(\varphi))$ —Thm I, $\mathbf{C}_6$, $\mathbf{A}_2$.

COROLLARY. $\varphi(x) \supset_{\varphi x} \Lambda(\varphi, x)$.

THEOREM 7. $\Sigma(\varphi) \supset_\varphi \Pi(\Lambda(\varphi), \varphi)$.

$\mathbf{D}_1$: $\Sigma(\varphi)$.
$\mathbf{C}_1$: $\Pi(\varphi, \varphi)$ —V, 1, $\mathbf{D}_1$.
$\mathbf{D}_2$: $\Lambda(\varphi, x)$.
$\mathbf{C}_2$: $\Pi(\varphi, \gamma) \supset_\gamma \gamma(x)$ —conv, $\mathbf{D}_2$.
$\mathbf{C}_3$: $\varphi(x)$ —V, $\mathbf{C}_2$, $\mathbf{C}_1$; $\mathbf{D}_1$, $\mathbf{D}_2$.
$\mathbf{C}_4$: $\Pi(\varphi, \Lambda(\varphi))$ —V, Thm 6, $\mathbf{D}_1$.
$\mathbf{D}_3$: $\varphi(y)$.
$\mathbf{C}_5$: $\Lambda(\varphi, y)$ —V, $\mathbf{C}_4$, $\mathbf{D}_3$; $\mathbf{D}_1$, $\mathbf{D}_3$.

$\mathbf{C}_6$: $\Sigma(\Lambda(\varphi))$ —IV, $\mathbf{C}_5$; $\mathbf{D}_1$, $\mathbf{D}_3$.
$\mathbf{C}_7$: $\Lambda(\varphi)$ —Thm III, $\mathbf{C}_6$, $\mathbf{D}_1$; $\mathbf{D}_1$.
$\mathbf{C}_8$: $\Pi(\Lambda(\varphi), \varphi)$ —Thm II, $\mathbf{C}_3$, $\mathbf{C}_7$; $\mathbf{D}_1$.
$\mathbf{A}_1$: $\Sigma(\varphi) \supset_\varphi \Pi(\Lambda(\varphi), \varphi)$ —Thm I, $\mathbf{C}_8$, $\mathbf{A}_2$ under Thm 6.

Corollary. $\Lambda(\varphi, x) \supset_{\varphi x} \varphi(x)$.

Postulate 38 is necessary to the proof of the corollary.

Theorem 8. $[\Sigma(\varphi) \centerdot \sim \varphi(x)] \supset_{\varphi x} \sim \Lambda(\varphi, x)$.

$\mathbf{D}_1$: $\Sigma(\varphi) \centerdot \sim \varphi(x)$.
$\mathbf{C}_1$: $\Sigma(\varphi)$ —V^2, 15, $\mathbf{D}_1$.
$\mathbf{C}_2$: $\Pi(\varphi, \varphi)$ —V, 1, $\mathbf{C}_1$; $\mathbf{D}_1$.
$\mathbf{C}_3$: $\sim \varphi(x)$ —V^2, 16, $\mathbf{D}_1$.
$\mathbf{C}_4$: $\Pi(\varphi, \varphi) \centerdot \sim \varphi(x)$ —V^2, 14, $\mathbf{C}_2$, $\mathbf{C}_3$; $\mathbf{D}_1$.
$\mathbf{C}_5$: $\Sigma\gamma \centerdot \Pi(\varphi, \gamma) \centerdot \sim \gamma(x)$ —IV, $\mathbf{C}_4$; $\mathbf{D}_1$.
$\mathbf{C}_6$: $\sim \Pi(\varphi, \gamma) \supset_\gamma \gamma(x)$ —V^2, 17, $\mathbf{C}_5$; $\mathbf{D}_1$.
$\mathbf{C}_7$: $\sim \Lambda(\varphi, x)$ —conv, $\mathbf{C}_6$.
$\mathbf{A}_1$: $\sim \sim \centerdot \Sigma(\varphi) \supset_\varphi \Pi(\varphi, \varphi)$ —V, 26, 1.
$\mathbf{A}_2$: $\Sigma(\sim)$ —IV, $\mathbf{A}_1$.
$\mathbf{A}_3$: $\Sigma(\sim) \centerdot \sim \sim \centerdot \Sigma(\varphi) \supset_\varphi \Pi(\varphi, \varphi)$ —V, 14, $\mathbf{A}_2$, $\mathbf{A}_1$.
$\mathbf{A}_4$: $\Sigma\varphi\Sigma x \centerdot \Sigma(\varphi) \centerdot \sim \varphi(x)$ —IV^2, $\mathbf{A}_3$.
$\mathbf{A}_5$: $[\Sigma(\varphi) \centerdot \sim \varphi(x)] \supset_{\varphi x} \sim \Lambda(\varphi, x)$ —Thm I Cor, $\mathbf{C}_7$, $\mathbf{A}_4$.

Theorem 9. $[\Pi(\varphi, \psi) \centerdot \sim \psi(x)] \supset_{\varphi\psi x} \sim \Lambda(\varphi, x)$.

$\mathbf{D}_1$: $\Pi(\varphi, \psi) \centerdot \sim \psi(x)$.
$\mathbf{C}_1$: $\sim \centerdot \Pi(\varphi, \gamma) \supset_\gamma \gamma(x)$ —V^2, 17, $\mathbf{D}_1$.
$\mathbf{C}_2$: $\sim \Lambda(\varphi, x)$ —conv, $\mathbf{C}_1$; $\mathbf{D}_1$.
$\mathbf{A}_1$: $\sim \sim \centerdot \Sigma(\varphi) \supset_\varphi \Pi(\varphi, \varphi)$ —V, 26, 1.
$\mathbf{A}_2$: $\Sigma(\sim)$ —IV, $\mathbf{A}_1$.
$\mathbf{A}_3$: $\Pi(\sim, \sim)$ —V, 1, $\mathbf{A}_2$.
$\mathbf{A}_4$: $\Pi(\sim, \sim) \centerdot \sim \sim \centerdot \Sigma(\varphi) \supset_\varphi \Pi(\varphi, \varphi)$ —V^2, 14, $\mathbf{A}_3$, $\mathbf{A}_1$.
$\mathbf{A}_5$: $\Sigma\varphi\Sigma\psi\Sigma x \centerdot \Pi(\varphi, \psi) \centerdot \sim \psi(x)$ —IV^3, $\mathbf{A}_4$.
$\mathbf{A}_6$: $[\Pi(\varphi, \psi) \centerdot \sim \psi(x)] \supset_{\varphi\psi x} \sim \Lambda(\varphi, x)$ —Thm I Cor, $\mathbf{C}_2$, $\mathbf{A}_5$.

Corollary. $[Q(\varphi, \psi) \centerdot \sim \psi(x)] \supset_{\varphi\psi x} \sim \Lambda(\varphi, x)$.

Theorem 10. $\sim \Pi(E, \sim)$.

$\mathbf{A}_1$: $\sim \sim \centerdot p \supset_p \sim \sim p$ —V, 26, 26.
$\mathbf{A}_2$: $E(p \supset_p \sim \sim p)$ —IV, $\mathbf{A}_1$.
$\mathbf{A}_3$: $E(p \supset_p \sim \sim p) \centerdot \sim \sim \centerdot p \supset_p \sim \sim p$ —V^2, 14, $\mathbf{A}_2$, $\mathbf{A}_1$.
$\mathbf{A}_4$: $\sim \Pi(E, \sim)$ —V^2, 17, $\mathbf{A}_3$.

Theorem 11. $\sim \Pi(E, \lambda zz)$.

$\mathbf{A}_1$: $\sim \sim \centerdot p \supset_p \sim \sim p$ —V, 26, 26.
$\mathbf{A}_2$: $E(\sim \centerdot p \supset_p \sim \sim p)$ —IV, $\mathbf{A}_1$.
$\mathbf{A}_3$: $E(\sim \centerdot p \supset_p \sim \sim p) \centerdot \sim \{\lambda zz\}(\sim \centerdot p \supset_p \sim \sim p)$ —V^2, 14, $\mathbf{A}_2$, $\mathbf{A}_1$.
$\mathbf{A}_4$: $\sim \Pi(E, \lambda zz)$ —V^2, 17, $\mathbf{A}_3$.

Theorem 12. $\Sigma(\iota)$.

$\mathbf{A}_1$: $`x \centerdot x = \centerdot \iota y \centerdot x = y$ —Thm 5.
$\mathbf{A}_2$: $E(`x \centerdot x = \centerdot \iota y \centerdot x = y)$ —IV, $\mathbf{A}_1$.

$\mathbf{A}_3$: $[{}^{\iota}x \centerdot x = \centerdot \iota y \centerdot x = y] = \centerdot \iota y \centerdot [{}^{\iota}x \centerdot x = \centerdot \iota y \centerdot x = y] = y$ —V, $\mathbf{A}_1$, $\mathbf{A}_2$.
$\mathbf{A}_4$: $\iota y \centerdot [{}^{\iota}x \centerdot x = \centerdot \iota y \centerdot x = y] = y$ —V, $\mathbf{A}_3$, $\mathbf{A}_1$.
$\mathbf{A}_5$: $\Sigma(\iota)$ —IV, $\mathbf{A}_4$.

THEOREM 13. $\Sigma x \sim \iota(x)$.

$\mathbf{A}_1$: $\sim\sim \centerdot p \supset_p \sim\sim p$ —V, 26, 26.
$\mathbf{A}_2$: $E(\sim \centerdot p \supset_p \sim\sim p)$ —IV, $\mathbf{A}_1$.
$\mathbf{A}_3$: $[\sim \centerdot p \supset_p \sim\sim p] = \centerdot \iota y \centerdot [\sim \centerdot p \supset_p \sim\sim p] = y$ —V, Thm 5, $\mathbf{A}_2$.
$\mathbf{A}_4$: $\sim \iota y \centerdot [\sim \centerdot p \supset_p \sim\sim p] = y$ —V, $\mathbf{A}_3$, $\mathbf{A}_1$.
$\mathbf{A}_5$: $\Sigma x \sim \iota(x)$ —IV, $\mathbf{A}_4$.

THEOREM 14. $\sim \Lambda(\iota, E)$.

$\mathbf{D}_1$: $\iota(x)$.
$\mathbf{C}_1$: $\Pi(x, \lambda zz)$ —V^2, 31, $\mathbf{D}_1$.
$\mathbf{A}_1$: $\Pi(\iota, \lambda x \Pi(x, \lambda zz))$ —Thm II, $\mathbf{C}_1$, Thm 12.
$\mathbf{A}_2$: $\Pi(\iota, \lambda x \Pi(x, \lambda zz)) \centerdot \sim \Pi(E, \lambda zz)$ —V^2, 14, $\mathbf{A}_1$, Thm 11.
$\mathbf{A}_3$: $\sim \centerdot \Pi(\iota, \gamma) \supset_\gamma \gamma(E)$ —V^2, 17, $\mathbf{A}_2$.
$\mathbf{A}_4$: $\sim \Lambda(\iota, E)$ —conv, $\mathbf{A}_3$.

THEOREM 15. $\sim\{\Lambda x \sim \iota(x)\}(E)$.

$\mathbf{D}_1$: $\sim \iota(x)$.
$\mathbf{C}_1$: $\Pi(x, \sim)$ —V^2, 31, $\mathbf{D}_1$.
$\mathbf{A}_1$: $\sim \iota(x) \supset_x \Pi(x, \sim)$ —Thm I, $\mathbf{C}_1$, Thm 13.
$\mathbf{A}_2$: $[\sim \iota(x) \supset_x \Pi(x, \sim)] \sim \Pi(E, \sim)$ —V^2, 14, $\mathbf{A}_1$, Thm 10.
$\mathbf{A}_3$: $\sim \centerdot \Pi(\lambda x \sim \iota x, \gamma) \supset_\gamma \gamma(E)$ —V^2, 17, $\mathbf{A}_2$.
$\mathbf{A}_4$: $\sim\{\Lambda x \sim \iota(x)\}(E)$ —conv, $\mathbf{A}_3$.

THEOREM 16. $\Sigma x \Sigma y[\varphi(x) \centerdot \sim \varphi(y)] \supset_\varphi G(\Lambda(\varphi))$.

$\mathbf{D}_1$: $\varphi(z)$.
$\mathbf{C}_1$: $\Lambda(\varphi, z)$ —V^2, Thm 6 Cor, $\mathbf{D}_1$.
$\mathbf{C}_2$: $\Lambda(\Lambda(\varphi), z)$ —V^2, Thm 6 Cor, $\mathbf{C}_1$; $\mathbf{D}_1$.
$\mathbf{D}_2$: $\varphi(x) \centerdot \sim \varphi(y)$.
$\mathbf{C}_3$: $\varphi(x)$ —V^2, 15, $\mathbf{D}_2$.
$\mathbf{C}_4$: $\Sigma(\varphi)$ —IV, $\mathbf{C}_3$; $\mathbf{D}_2$.
$\mathbf{C}_5$: $\Pi(\varphi, \Lambda(\Lambda(\varphi)))$, —Thm II, $\mathbf{C}_2$, $\mathbf{C}_4$; $\mathbf{D}_2$.
$\mathbf{D}_3$: $\sim \Lambda(\Lambda(\varphi), t)$.
$\mathbf{C}_6$: $\sim \Lambda(\varphi, t)$ —V^3, Thm 9, $\mathbf{C}_5$, $\mathbf{D}_3$; $\mathbf{D}_2$, $\mathbf{D}_3$.
$\mathbf{C}_7$: $\Lambda(\varphi, x)$ —V^2, Thm 6 Cor, $\mathbf{C}_3$; $\mathbf{D}_2$.
$\mathbf{C}_8$: $\Sigma(\Lambda(\varphi))$ —IV, $\mathbf{C}_7$; $\mathbf{D}_2$.
$\mathbf{C}_9$: $\sim \varphi(y)$ —V^2, 16, $\mathbf{D}_2$.
$\mathbf{C}_{10}$: $\sim \Lambda(\varphi, y)$ —V^2, Thm 8, $\mathbf{C}_4$, $\mathbf{C}_9$; $\mathbf{D}_2$.
$\mathbf{C}_{11}$: $\sim \Lambda(\Lambda(\varphi), y)$ —V^2, Thm 8, $\mathbf{C}_8$, $\mathbf{C}_{10}$; $\mathbf{D}_2$.
$\mathbf{C}_{12}$: $\Sigma t \sim \Lambda(\Lambda(\varphi), t)$ —IV, $\mathbf{C}_{11}$; $\mathbf{D}_2$.
$\mathbf{C}_{13}$: $\sim \Lambda(\Lambda(\varphi), t) \supset_t \sim \Lambda(\varphi, t)$ —Thm I, $\mathbf{C}_6$, $\mathbf{C}_{12}$; $\mathbf{D}_2$.
$\mathbf{C}_{14}$: $G(\Lambda(\varphi))$ —conv, $\mathbf{C}_{13}$; $\mathbf{D}_2$.
$\mathbf{D}_4$: $\Sigma x \Sigma y \centerdot \varphi(x) \centerdot \sim \varphi(y)$.
$\mathbf{C}_{15}$: $G(\Lambda(\varphi))$ —Thm I Cor, $\mathbf{C}_{14}$, $\mathbf{D}_4$; $\mathbf{D}_4$.
$\mathbf{A}_1$: $E(\Pi(E, \sim))$ —IV, Thm 10.
$\mathbf{A}_2$: $\Sigma(E)$ —IV, $\mathbf{A}_1$.

A$_3$: $\Pi(E, E)$ — V, 1, **A**$_2$.
A$_4$: $\Pi(E, E) \,.\, \sim \Pi(E, \sim)$ — V^2, 14, **A**$_3$, Thm 10.
A$_5$: $\Sigma\varphi\Sigma x\Sigma y \,.\, \varphi(x) \,.\, \sim \varphi(y)$ — IV^3, **A**$_4$.
A$_6$: $\Sigma x\Sigma y[\varphi(x) \,.\, \sim \varphi(y)] \supset_\varphi G(\Lambda(\varphi))$ — Thm I, **C**$_{15}$, **A**$_5$.

Corollary. $\Sigma x\Sigma y[\varphi(x) \,.\, \sim \Lambda(\varphi, y)] \supset_\varphi G(\Lambda(\varphi))$.

Theorem 17. $\Sigma(G)$.

A$_1$: $\Sigma x\Sigma y \,.\, \Pi(E, x) \,.\, \sim \Pi(E, y)$ — IV^2, **A**$_4$ under Thm 16.
A$_2$: $G(\Lambda(\Pi(E)))$ — V, Thm 16, **A**$_1$.
A$_3$: $\Sigma(G)$ IV, **A**$_2$.

Corollary. $\Sigma\varphi G(\varphi)$.

Theorem 18. $G(\varphi) \supset_\varphi \Sigma(\varphi)$.

D$_1$: $G(\varphi)$.
C$_1$: $\sim \Lambda(\varphi, x) \supset_x \sim \varphi(x)$ — II, **D**$_1$.
C$_2$: $\Sigma x \sim \Lambda(\varphi, x)$ — V, 38, **C**$_1$; **D**$_1$.
D$_2$: $\sim \Lambda(\varphi, x)$.
C$_3$: $\sim . \,\Pi(\varphi, \gamma) \supset_\gamma \gamma(x)$ — conv, **D**$_2$.
C$_4$: $\Sigma\gamma \,.\, \Pi(\varphi, \gamma) \,.\, \sim \gamma(x)$ — V^2, 18, **C**$_3$; **D**$_2$.
D$_3$: $\Pi(\varphi, \gamma) \,.\, \sim \gamma(x)$.
C$_5$: $\Pi(\varphi, \gamma)$ — V^2, 15, **D**$_3$.
C$_6$: $\Sigma(\varphi)$ V, 38, **C**$_5$; **D**$_3$.
C$_7$: $\Sigma(\varphi)$ — Thm I, **C**$_6$, **C**$_4$; **D**$_2$.
C$_8$: $\Sigma(\varphi)$ — Thm I, **C**$_7$, **C**$_2$; **D**$_1$.
A$_1$: $G(\varphi) \supset_\varphi \Sigma(\varphi)$ — Thm I, **C**$_8$, Thm 17 Cor.

Theorem 19. $G(\varphi) \supset_\varphi \Sigma x \sim \varphi(x)$.

D$_1$: $G(\varphi)$.
C$_1$: $\sim \Lambda(\varphi, x) \supset_x \sim \varphi(x)$ — II, **D**$_1$.
C$_2$: $\Sigma x \sim \Lambda(\varphi, x)$ — V, 38, **C**$_1$; **D**$_1$.
D$_2$: $\sim \Lambda(\varphi, x)$.
C$_3$: $\sim \varphi(x)$ — V, **C**$_1$, **D**$_2$; **D**$_1$, **D**$_2$.
C$_4$: $\Sigma x \sim \varphi(x)$ — IV, **C**$_3$; **D**$_1$, **D**$_2$.
C$_5$: $\Sigma x \sim \varphi(x)$ — Thm I, **C**$_4$, **C**$_2$; **D**$_1$.
A$_1$: $G(\varphi) \supset_\varphi \Sigma x \sim \varphi(x)$ — Thm I, **C**$_5$, Thm 17 Cor.

Theorem 20. $G(\varphi) \supset_\varphi .[\Pi(\varphi, \psi) \,.\, \sim \psi(x)] \supset_{\psi x} \sim \varphi(x)$.

D$_1$: $\Pi(\varphi, \psi) \,.\, \sim \psi(x)$.
C$_1$: $\sim \Lambda(\varphi, x)$ — V, Thm 9, **D**$_1$.
D$_2$: $G(\varphi)$.
C$_2$: $\sim \Lambda(\varphi, x) \supset_x \sim \varphi(x)$ — II, **D**$_2$.
C$_3$: $\sim \varphi(x)$ — V, **C**$_2$, **C**$_1$; **D**$_1$, **D**$_2$.
C$_4$: $\Sigma(\varphi)$ V, Thm 18, **D**$_2$.
C$_5$: $\Pi(\varphi, \varphi)$ — V, 1, **C**$_4$; **D**$_2$.
C$_6$: $\Sigma y \sim \varphi(y)$ — V, Thm 19, **D**$_2$.
D$_3$: $\sim \varphi(y)$.
C$_7$: $\Pi(\varphi, \varphi) \,.\, \sim \varphi(y)$ — V^2, 14, **C**$_5$, **D**$_3$; **D**$_2$, **D**$_3$.
C$_8$: $\Sigma\psi\Sigma x \,.\, \Pi(\varphi, \psi) \,.\, \sim \psi(x)$ — IV^2, **C**$_7$; **D**$_2$, **D**$_3$.
C$_9$: $\Sigma\psi\Sigma x \,.\, \Pi(\varphi, \psi) \,.\, \sim \psi(x)$ — Thm I, **C**$_8$, **C**$_6$; **D**$_2$.

$\mathbf{C}_{10}$: $[\Pi(\varphi,\psi) \,.\, \sim\psi(x)] \supset_{\psi x} \sim\varphi(x)$ —Thm I Cor, $\mathbf{C}_3$, $\mathbf{C}_9$; $\mathbf{D}_2$.
$\mathbf{A}_1$: $G(\varphi) \supset_\varphi .[\Pi(\varphi,\psi) \,.\, \sim\psi(x)] \supset_{\psi x} \sim\varphi(x)$ —Thm I, $\mathbf{C}_{10}$, Thm 17 Cor.

THEOREM 21. $G(\varphi) \supset_\varphi Gx \sim\{\Lambda y \sim\varphi(y)\}(x)$.

$\mathbf{D}_1$: $G(\varphi)$.
$\mathbf{C}_1$: $\Sigma y \sim\varphi(y)$ —V, Thm 19, $\mathbf{D}_1$.
$\mathbf{D}_2$: $\varphi(u)$.
$\mathbf{C}_2$: $\sim\sim\varphi(u)$ —V, 26, $\mathbf{D}_2$.
$\mathbf{C}_3$: $\sim\{\Lambda y \sim\varphi(y)\}(u)$ —V^2, Thm 8, $\mathbf{C}_1$, $\mathbf{C}_2$; $\mathbf{D}_1$, $\mathbf{D}_2$.
$\mathbf{D}_3$: $\Pi(\lambda x \sim\{\Lambda y \sim\varphi(y)\}(x), \gamma) \,.\, \sim\gamma(z)$.
$\mathbf{C}_4$: $\Pi(\lambda x \sim\{\Lambda y \sim\varphi(y)\}(x), \gamma)$ —V^2, 15, $\mathbf{D}_3$.
$\mathbf{C}_5$: $\gamma(u)$ —V, $\mathbf{C}_4$, $\mathbf{C}_3$; $\mathbf{D}_1$, $\mathbf{D}_2$, $\mathbf{D}_3$.
$\mathbf{C}_6$: $\Sigma(\varphi)$ —V, Thm 18, $\mathbf{D}_1$.
$\mathbf{C}_7$: $\Pi(\varphi,\gamma)$ —Thm II, $\mathbf{C}_5$, $\mathbf{C}_6$; $\mathbf{D}_1$, $\mathbf{D}_3$.
$\mathbf{C}_8$: $\sim\gamma(z)$ —V, 16, $\mathbf{D}_3$.
$\mathbf{C}_9$: $\Pi(\varphi,\gamma) \,.\, \sim\gamma(z)$ —V^2, 14, $\mathbf{C}_7$, $\mathbf{C}_8$; $\mathbf{D}_1$, $\mathbf{D}_3$.
$\mathbf{C}_{10}$: $\Sigma\gamma \,.\, \Pi(\varphi,\gamma) \,.\, \sim\gamma(z)$ —IV, $\mathbf{C}_9$; $\mathbf{D}_1$, $\mathbf{D}_3$.
$\mathbf{C}_{11}$: $\sim\Lambda(\varphi,z)$ —V^2, 17, $\mathbf{C}_{10}$; $\mathbf{D}_1$, $\mathbf{D}_3$.
$\mathbf{D}_4$: $\sim\{\Lambda x \sim\{\Lambda y \sim\varphi(y)\}(x)\}(z)$.
$\mathbf{C}_{12}$: $\Sigma\gamma \,.\, \Pi(\lambda x \sim\{\Lambda y \sim\varphi(y)\}(x), \gamma) \,.\, \sim\gamma(z)$ —V^2, 18, $\mathbf{D}_4$.
$\mathbf{C}_{13}$: $\sim\Lambda(\varphi,z)$ —Thm I, $\mathbf{C}_{11}$, $\mathbf{C}_{12}$; $\mathbf{D}_1$, $\mathbf{D}_4$.
$\mathbf{C}_{14}$: $\sim\Lambda(\varphi,x) \supset_x \sim\varphi(x)$ —II, $\mathbf{D}_1$.
$\mathbf{C}_{15}$: $\sim\varphi(z)$ —V, $\mathbf{C}_{14}$, $\mathbf{C}_{13}$; $\mathbf{D}_1$, $\mathbf{D}_4$.
$\mathbf{C}_{16}$: $\{\Lambda y \sim\varphi(y)\}(z)$ —V^2, Thm 6 Cor, $\mathbf{C}_{15}$; $\mathbf{D}_1$, $\mathbf{D}_4$.
$\mathbf{C}_{17}$: $\sim\sim\{\Lambda y \sim\varphi(y)\}(z)$ —V, 26, $\mathbf{C}_{16}$; $\mathbf{D}_1$, $\mathbf{D}_4$.
$\mathbf{D}_5$: $\sim\varphi(v)$.
$\mathbf{C}_{18}$: $\{\Lambda y \sim\varphi(y)\}(v)$ —V^2, Thm 6 Cor, $\mathbf{D}_5$.
$\mathbf{C}_{19}$: $\sim\sim\{\Lambda y \sim\varphi(y)\}(v)$ —V, 26, $\mathbf{C}_{18}$; $\mathbf{D}_5$.
$\mathbf{C}_{20}$: $\Sigma x \sim\{\Lambda y \sim\varphi(y)\}(x)$ —IV, $\mathbf{C}_3$; $\mathbf{D}_1$, $\mathbf{D}_2$.
$\mathbf{C}_{21}$: $\Sigma x \sim\{\Lambda y \sim\varphi(y)\}(x)$ —Thm III, $\mathbf{C}_{20}$, $\mathbf{C}_6$; $\mathbf{D}_1$.
$\mathbf{C}_{22}$: $\sim\{\Lambda x \sim\{\Lambda y \sim\varphi(y)\}(x)\}(v)$ —V^2, Thm 8, $\mathbf{C}_{21}$, $\mathbf{C}_{19}$; $\mathbf{D}_1$, $\mathbf{D}_5$.
$\mathbf{C}_{23}$: $\Sigma z \sim\{\Lambda x \sim\{\Lambda y \sim\varphi(y)\}(x)\}(z)$ —IV, $\mathbf{C}_{22}$; $\mathbf{D}_1$, $\mathbf{D}_5$.
$\mathbf{C}_{24}$: $\Sigma z \sim\{\Lambda x \sim\{\Lambda y \sim\varphi(y)\}(x)\}(z)$ —Thm I, $\mathbf{C}_{23}$, $\mathbf{C}_1$; $\mathbf{D}_1$.
$\mathbf{C}_{25}$: $\sim\{\Lambda x \sim\{\Lambda y \sim\varphi(y)\}(x)\}(z) \supset_z \sim\sim\{\Lambda y \sim\varphi(y)\}(z)$ —Thm I, $\mathbf{C}_{17}$, $\mathbf{C}_{24}$; $\mathbf{D}_1$.
$\mathbf{C}_{26}$: $Gx \sim\{\Lambda y \sim\varphi(y)\}(x)$ —conv, $\mathbf{C}_{25}$; $\mathbf{D}_1$.
$\mathbf{A}_1$: $G(\varphi) \supset_\varphi Gx \sim\{\Lambda y \sim\varphi(y)\}(x)$ —Thm I, $\mathbf{C}_{26}$, Thm 17 Cor.

COROLLARY. $Gx \sim\varphi(x) \supset_\varphi Gx \sim\Lambda(\varphi,x)$.

THEOREM 22. $[\Sigma z \,.\, x \neq z] \supset_x Gy \,.\, x = y$.

$\mathbf{D}_1$: $\Sigma z \,.\, x \neq z$.
$\mathbf{C}_1$: $E(x)$ —IV, $\mathbf{D}_1$.
$\mathbf{C}_2$: $x = x$ —V, Thm 1, $\mathbf{C}_1$; $\mathbf{D}_1$.
$\mathbf{C}_3$: $\{\lambda u \,.\, x = u\}(x)$ —III, $\mathbf{C}_2$; $\mathbf{D}_1$.
$\mathbf{C}_4$: $\{\Lambda u \,.\, x = u\}(x)$ —V^2, Thm 6 Cor, $\mathbf{C}_3$; $\mathbf{D}_1$.
$\mathbf{D}_2$: $\sim\{\Lambda u \,.\, x = u\}(y)$.
$\mathbf{C}_5$: $\{\Lambda u \,.\, x = u\}(x) \,.\, \sim\{\Lambda u \,.\, x = u\}(y)$ —V^2, 14, $\mathbf{C}_4$, $\mathbf{D}_2$; $\mathbf{D}_1$, $\mathbf{D}_2$.
$\mathbf{C}_6$: $\Sigma\varrho \,.\, \varrho(x) \,.\, \sim\varrho(y)$ —IV, $\mathbf{C}_5$; $\mathbf{D}_1$, $\mathbf{D}_2$.
$\mathbf{C}_7$: $x \neq y$ —V^2, 17, $\mathbf{C}_6$; $\mathbf{D}_1$, $\mathbf{D}_2$.

$\mathbf{C}_8$: $\Sigma u \,.\, x = u$ —IV, $\mathbf{C}_2$; $\mathbf{D}_1$.
$\mathbf{D}_3$: $x \neq z$.
$\mathbf{C}_9$: $\sim\{\lambda u \,.\, x = u\}(z)$ —III, $\mathbf{D}_3$.
$\mathbf{C}_{10}$: $\sim\{\Lambda u \,.\, x = u\}(z)$ —V^2, Thm 8, $\mathbf{C}_8$, $\mathbf{C}_9$; $\mathbf{D}_1$, $\mathbf{D}_3$.
$\mathbf{C}_{11}$: $\Sigma y \sim\{\Lambda u \,.\, x = u\}(y)$ —IV, $\mathbf{C}_{10}$; $\mathbf{D}_1$, $\mathbf{D}_3$.
$\mathbf{C}_{12}$: $\Sigma y \sim\{\Lambda u \,.\, x = u\}(y)$ —Thm I, $\mathbf{C}_{11}$, $\mathbf{D}_1$; $\mathbf{D}_1$.
$\mathbf{C}_{13}$: $\sim\{\Lambda u \,.\, x = u\}(y) \supset_y x \neq y$ —Thm I, $\mathbf{C}_7$, $\mathbf{C}_{12}$; $\mathbf{D}_1$.
$\mathbf{C}_{14}$: $Gy \,.\, x = y$ —conv, $\mathbf{C}_{13}$; $\mathbf{D}_1$.
$\mathbf{A}_1$: $\Sigma\varrho \,.\, \varrho(E) \,.\, \sim\varrho(\sim)$ —IV, $\mathbf{A}_4$ under Thm 16.
$\mathbf{A}_2$: $E \neq \sim$ —V^2, 17, $\mathbf{A}_1$.
$\mathbf{A}_3$: $\Sigma x \Sigma z \,.\, x \neq z$ —IV^2, $\mathbf{A}_2$.
$\mathbf{A}_4$: $[\Sigma z \,.\, x \neq z] \supset_x Gy \,.\, x = y$ —Thm I, $\mathbf{C}_{14}$, $\mathbf{A}_3$.

6. Classes. We recall the following definitions:

$$Q \longrightarrow \lambda\mu\lambda\nu \,.\, \Pi(\mu, \nu) \,.\, \Pi(\nu, \mu).$$
$$K \longrightarrow A(Q).$$

And to these we add two new definitions:

$$\varepsilon \longrightarrow \lambda\alpha\lambda\beta \,.[\beta = K(\varphi)] \supset_\varphi \varphi(\alpha).$$
$$O \longrightarrow \lambda\gamma K\pi \sim\{\varepsilon\}(\pi, \gamma).$$

We shall abbreviate $\varepsilon(\mathbf{A}, \mathbf{B})$ as $[\mathbf{A}]\,\varepsilon[\mathbf{B}]$, and hence we may rewrite the definition of O as follows:

$$O \longrightarrow \lambda\gamma K\pi \sim .\, \pi\,\varepsilon\,\gamma.$$

If $\mathbf{F}$ and $\mathbf{G}$ are propositional functions, $Q(\mathbf{F}, \mathbf{G})$ is to be read, "$\mathbf{F}$ and $\mathbf{G}$ are equivalent", and $K(\mathbf{F})$ is to be read, "The class of x's such that $\{\mathbf{F}\}(x)$ is true". And $[\mathbf{A}]\,\varepsilon[\mathbf{B}]$ is to be read, "$\mathbf{A}$ is a member of the class $\mathbf{B}$". And $O(\mathbf{B})$ is to be read, "The class of things which are not members of the class $\mathbf{B}$", or, "The complement of the class $\mathbf{B}$".

Theorem 23. $Q(x, y) \supset_{xy} .\, Q(y, z) \supset_z Q(x, z)$.

$\mathbf{D}_1$: $Q(x, y)$.
$\mathbf{C}_1$: $\Pi(x, y)$ —V^2, 15, $\mathbf{D}_1$.
$\mathbf{D}_2$: $x(r)$.
$\mathbf{C}_2$: $y(r)$ —V, $\mathbf{C}_1$, $\mathbf{D}_2$; $\mathbf{D}_1$, $\mathbf{D}_2$.
$\mathbf{D}_3$: $Q(y, z)$.
$\mathbf{C}_3$: $\Pi(y, z)$ —V^2, 15, $\mathbf{D}_3$.
$\mathbf{C}_4$: $z(r)$ —V, $\mathbf{C}_3$, $\mathbf{C}_2$; $\mathbf{D}_1$, $\mathbf{D}_2$, $\mathbf{D}_3$.
$\mathbf{C}_5$: $\Sigma(x)$ —V, 38, $\mathbf{C}_1$; $\mathbf{D}_1$.
$\mathbf{C}_6$: $\Pi(x, z)$ —Thm II, $\mathbf{C}_4$, $\mathbf{C}_5$; $\mathbf{D}_1$, $\mathbf{D}_3$.
$\mathbf{C}_7$: $\Pi(z, y)$ —V, 16, $\mathbf{D}_3$.
$\mathbf{D}_4$: $z(s)$.
$\mathbf{C}_8$: $y(s)$ —V, $\mathbf{C}_7$, $\mathbf{D}_4$; $\mathbf{D}_3$, $\mathbf{D}_4$.
$\mathbf{C}_9$: $\Pi(y, x)$ —V, 16, $\mathbf{D}_1$.
$\mathbf{C}_{10}$: $x(s)$ —V, $\mathbf{C}_9$, $\mathbf{C}_8$; $\mathbf{D}_1$, $\mathbf{D}_3$, $\mathbf{D}_4$.
$\mathbf{C}_{11}$: $\Sigma(z)$ —V, 38, $\mathbf{C}_7$; $\mathbf{D}_3$.
$\mathbf{C}_{12}$: $\Pi(z, x)$ —Thm II, $\mathbf{C}_{10}$, $\mathbf{C}_{11}$; $\mathbf{D}_1$, $\mathbf{D}_3$.
$\mathbf{C}_{13}$: $Q(x, z)$ —V^2, 14, $\mathbf{C}_6$, $\mathbf{C}_{12}$; $\mathbf{D}_1$, $\mathbf{D}_3$.

$\mathbf{C}_{14}$: $\Sigma(y)$ —V, 38, $\mathbf{C}_9$; $\mathbf{D}_1$.
$\mathbf{C}_{15}$: $\Pi(y, y)$ —V, 1, $\mathbf{C}_{14}$; $\mathbf{D}_1$.
$\mathbf{C}_{16}$: $Q(y, y)$ —V^2, 14, $\mathbf{C}_{15}$, $\mathbf{C}_{15}$; $\mathbf{D}_1$.
$\mathbf{C}_{17}$: $\Sigma z Q(y, z)$ —IV, $\mathbf{C}_{16}$; $\mathbf{D}_1$.
$\mathbf{C}_{18}$: $Q(y, z) \supset_z Q(x, z)$ —Thm I, $\mathbf{C}_{13}$, $\mathbf{C}_{17}$; $\mathbf{D}_1$.
$\mathbf{A}_1$: $Q(E, E)$ —V^2, 14, $\mathbf{A}_3$ under Thm 16, $\mathbf{A}_3$ under Thm 16.
$\mathbf{A}_2$: $\Sigma x \Sigma y Q(x, y)$ —IV^2, $\mathbf{A}_1$.
$\mathbf{A}_3$: $Q(x, y) \supset_{xy} . Q(y, z) \supset_z Q(x, z)$ —Thm I Cor, $\mathbf{C}_{18}$, $\mathbf{A}_2$.

THEOREM 24. $Q(x, y) \supset_{xy} Q(y, x)$.

$\mathbf{D}_1$: $Q(x, y)$.
$\mathbf{C}_1$: $\Pi(x, y)$ —V^2, 15, $\mathbf{D}_1$.
$\mathbf{C}_2$: $\Pi(y, x)$ —V^2, 16, $\mathbf{D}_1$.
$\mathbf{C}_3$: $Q(y, x)$ —V^2, 14, $\mathbf{C}_2$, $\mathbf{C}_1$; $\mathbf{D}_1$.
$\mathbf{A}_1$: $Q(x, y) \supset_{xy} Q(y, x)$ —Thm I Cor, $\mathbf{C}_3$, $\mathbf{A}_2$ under Thm 23.

THEOREM 25. $Q(\varphi, \psi) \supset_{\varphi\psi} . K(\varphi) = K(\psi)$ —V, 33, Thm 23, Thm 24.

THEOREM 26. $[K(\varphi) = K(\psi)] \supset_{\varphi\psi} Q(\varphi, \psi)$.

$\mathbf{A}_1$: $K(E) = K(E)$ —V^2, Thm 25, $\mathbf{A}_1$ under Thm 23.
$\mathbf{A}_2$: $\Sigma u \Sigma v . K(u) = K(v)$ —IV^2, $\mathbf{A}_1$.
$\mathbf{A}_3$: $[K(u) = K(v)] \supset_{uv} Q(u, v)$ —V, 35, $\mathbf{A}_2$.
$\mathbf{A}_4$: $[K(\varphi) = K(\psi)] \supset_{\varphi\psi} Q(\varphi, \psi)$ —conv, $\mathbf{A}_3$.

THEOREM 27. $\sim Q(\varphi, \psi) \supset_{\varphi\psi} . K(\varphi) \neq K(\psi)$.

$\mathbf{D}_1$: $\Sigma y Q(x, y)$.
$\mathbf{C}_1$: $Q(x, y) \supset_y Q(x, y)$ —V, 1, $\mathbf{D}_1$.
$\mathbf{A}_1$: $Q(x, y) \supset_{xy} Q(x, y)$ —Thm I, $\mathbf{C}_1$, $\mathbf{A}_2$ under Thm 23.
$\mathbf{D}_2$: $\sim Q(\varphi, \psi)$.
$\mathbf{C}_2$: $K(\varphi) \neq K(\psi)$ —V^4, 34, Thm 23, Thm 24, $\mathbf{A}_1$, $\mathbf{D}_2$.
$\mathbf{D}_3$: $\sim p$.
$\mathbf{C}_3$: $E(p)$ —IV, $\mathbf{D}_3$.
$\mathbf{A}_2$: $\Pi(\sim, E)$ —Thm II, $\mathbf{C}_3$, $\mathbf{A}_2$ under Thm 9.
$\mathbf{A}_3$: $\sim Q(E, \sim)$ —V^2, 22, Thm 10, $\mathbf{A}_2$.
$\mathbf{A}_4$: $\Sigma\varphi\Sigma\psi \sim Q(\varphi, \psi)$ —IV^2, $\mathbf{A}_3$.
$\mathbf{A}_5$: $\sim Q(\varphi, \psi) \supset_{\varphi\psi} . K(\varphi) \neq K(\psi)$ —Thm I Cor, $\mathbf{C}_2$, $\mathbf{A}_4$.

THEOREM 28. $\Sigma x \Sigma y[\varphi(x) . \sim \varphi(y)] \supset_\varphi G\psi . K(\varphi) = K(\psi)$.

$\mathbf{D}_1$: $Q(x, y)$.
$\mathbf{C}_1$: $K(x) = K(y)$ —V^2, Thm 25, $\mathbf{D}_1$.
$\mathbf{C}_2$: $\{\Lambda\psi . K(x) = K(\psi)\}(y)$ —V^2, Thm 6 Cor, $\mathbf{C}_1$; $\mathbf{D}_1$.
$\mathbf{A}_1$: $Q(x, y) \supset_{xy} \{\Lambda\psi . K(x) = K(\psi)\}(y)$ —Thm I Cor, $\mathbf{C}_2$, $\mathbf{A}_2$ under Thm 23.
$\mathbf{D}_2$: $\sim\{\Lambda\psi . K(u) = K(\psi)\}(v)$.
$\mathbf{C}_3$: $K(u) \neq K(v)$ —V^4, 34, Thm 23, Thm 24, $\mathbf{A}_1$, $\mathbf{D}_2$.
$\mathbf{D}_3$: $u(x) . \sim u(y)$.
$\mathbf{C}_4$: $u(x)$ —V^2, 15, $\mathbf{D}_3$.
$\mathbf{C}_5$: $\sim\sim u(x)$ —V, 26, $\mathbf{C}_4$; $\mathbf{D}_3$.
$\mathbf{C}_6$: $\sim \Pi(u, \lambda x \sim u(x))$ —V^2, 17, $\mathbf{C}_4$, $\mathbf{C}_5$; $\mathbf{D}_3$.
$\mathbf{C}_7$: $\sim u(y)$ —V^2, 16, $\mathbf{D}_3$.
$\mathbf{C}_8$: $\sim \Pi(\lambda x \sim u(x), u)$ —V^2, 17, $\mathbf{C}_7$, $\mathbf{C}_7$; $\mathbf{D}_3$.

C$_9$: $\sim Q(u, \lambda x \sim u(x))$ —V^2, 23, **C**$_6$, **C**$_8$; **D**$_3$.
C$_{10}$: $K(u) \neq Kx \sim u(x)$ —V^2, Thm 27, **C**$_9$; **D**$_3$.
C$_{11}$: $\Sigma(u)$ —IV, **C**$_4$; **D**$_3$.
C$_{12}$: $\Pi(u, u)$ —V, 1, **C**$_{11}$; **D**$_3$.
C$_{13}$: $Q(u, u)$ —V^2, 14, **C**$_{12}$, **C**$_{12}$; **D**$_3$.
C$_{14}$: $K(u) = K(u)$ —V^2, Thm 25, **C**$_{13}$; **D**$_3$.
C$_{15}$: $\Sigma\psi \,.\, K(u) = K(\psi)$ —IV, **C**$_{14}$; **D**$_3$.
C$_{16}$: $\sim\{\Lambda\psi \,.\, K(u) = K(\psi)\}(\lambda x \sim u(x))$ —V^2, Thm 8, **C**$_{15}$, **C**$_{10}$; **D**$_3$.
C$_{17}$: $\Sigma v \sim\{\Lambda\psi \,.\, K(u) = K(\psi)\}(v)$ —IV, **C**$_{16}$; **D**$_3$.
C$_{18}$: $\sim\{\Lambda\psi \,.\, K(u) = K(\psi)\}(v) \supset_v . \, K(u) \neq K(v)$ —Thm I, **C**$_3$, **C**$_{17}$; **D**$_3$.
C$_{19}$: $G\psi \,.\, K(u) = K(\psi)$ —conv, **C**$_{18}$; **D**$_3$.
D$_4$: $\Sigma x \Sigma y \,.\, u(x) \,.\sim u(y)$.
C$_{20}$: $G\psi \,.\, K(u) = K(\psi)$ —Thm I Cor, **C**$_{19}$, **D**$_4$; **D**$_4$.
A$_2$: $\Sigma u \Sigma x \Sigma y \,.\, u(x) \,.\sim u(y)$ —I, **A**$_5$ under Thm 16.
A$_3$: $\Sigma x \Sigma y[u(x) \,.\sim u(y)] \supset_u G\psi \,.\, K(u) = K(\psi)$ —Thm I, **C**$_{20}$, **A**$_2$.
A$_4$: $\Sigma x \Sigma y[\varphi(x) \,.\sim \varphi(y)] \supset_\varphi G\psi \,.\, K(\varphi) = K(\psi)$ —I, **A**$_3$.

THEOREM 29. $\varphi(x) \supset_{\varphi x} . \, x \,\varepsilon\, K(\varphi)$.

D$_1$: $K(\psi) = K(\varphi)$.
C$_1$: $Q(\psi, \varphi)$ —V^2, Thm 26, **D**$_1$.
C$_2$: $\Pi(\psi, \varphi)$ —V^2, 15, **C**$_1$; **D**$_1$.
D$_2$: $\psi(x)$.
C$_3$: $\varphi(x)$ —V, **C**$_2$, **D**$_2$; **D**$_1$, **D**$_2$.
C$_4$: $\Sigma(\psi)$ —IV, **D**$_2$.
C$_5$: $\Pi(\psi, \psi)$ —V, 1, **C**$_4$; **D**$_2$.
C$_6$: $Q(\psi, \psi)$ —V^2, 14, **C**$_3$, **C**$_5$; **D**$_2$.
C$_7$: $K(\psi) = K(\psi)$ —V^2, Thm 25, **C**$_6$; **D**$_2$.
C$_8$: $\Sigma\varphi \,.\, K(\psi) = K(\varphi)$ —IV, **C**$_7$; **D**$_2$.
C$_9$: $[K(\psi) = K(\varphi)] \supset_\varphi \varphi(x)$ —Thm I, **C**$_3$, **C**$_8$; **D**$_2$.
C$_{10}$: $x \,\varepsilon\, K(\psi)$ —conv, **C**$_9$; **D**$_2$.
A$_1$: $\Sigma\psi\Sigma x \,.\, \psi(x)$ —IV^2, 1.
A$_2$: $\psi(x) \supset_{\psi x} . \, x \,\varepsilon\, K(\psi)$ —Thm I Cor, **C**$_{10}$, **A**$_1$.
A$_3$: $\varphi(x) \supset_{\varphi x} . \, x \,\varepsilon\, K(\varphi)$ —I, **A**$_2$.

THEOREM 30. $[x \,\varepsilon\, K(\varphi)] \supset_{\varphi x} \varphi(x)$.

D$_1$: $x \,\varepsilon\, K(\psi)$.
C$_1$: $E(K(\psi))$ —IV, **D**$_1$.
C$_2$: $K(\psi) = K(\psi)$ —V, Thm 1, **C**$_1$; **D**$_1$.
C$_3$: $[K(\psi) = K(\varphi)] \supset_\varphi \varphi(x)$ —conv, **D**$_1$.
C$_4$: $\psi(x)$ —V, **C**$_3$, **C**$_2$; **D**$_1$.
A$_1$: $[\lambda\varphi\Pi(\varphi, \varphi)] \,\varepsilon\, K(\Pi(\lambda\varphi\Sigma(\varphi)))$ —V^2, Thm 29, 1.
A$_2$: $\Sigma\psi\Sigma x \,.\, x \,\varepsilon\, K(\psi)$ —IV^2, **A**$_1$.
A$_3$: $[x \,\varepsilon\, K(\psi)] \supset_{\psi x} \psi(x)$ —Thm I Cor, **C**$_4$, **A**$_2$.
A$_4$: $[x \,\varepsilon\, K(\varphi)] \supset_{\varphi x} \varphi(x)$ —I, **A**$_3$.

COROLLARY 1. $\Sigma(\varphi) \supset_\varphi Q(\varphi, \lambda x \,.\, x \,\varepsilon\, K(\varphi))$.

COROLLARY 2. $\Sigma(\varphi) \supset_\varphi . \, K(\varphi) = Kx \,.\, x \,\varepsilon\, K(\varphi)$.

THEOREM 31. $E(K(\varphi)) \supset_\varphi \Sigma(\varphi)$.

$\mathbf{D}_1$: $E(K(\varphi))$.
$\mathbf{C}_1$: $K(\varphi) = K(\varphi)$ —V, Thm 1, $\mathbf{D}_1$.
$\mathbf{C}_2$: $Q(\varphi, \varphi)$ —V^2, Thm 26, $\mathbf{C}_1$; $\mathbf{D}_1$.
$\mathbf{C}_3$: $\Pi(\varphi, \varphi)$ —V^2, 15, $\mathbf{C}_2$; $\mathbf{D}_1$.
$\mathbf{C}_4$: $\Sigma(\varphi)$ —V, 38, $\mathbf{C}_3$; $\mathbf{D}_1$.
$\mathbf{A}_1$: $E\left(K(\Pi(\lambda\varphi\Sigma(\varphi)))\right)$ —IV, $\mathbf{A}_1$ under Thm 30.
$\mathbf{A}_2$: $\Sigma\varphi E(K(\varphi))$ —IV, $\mathbf{A}_1$.
$\mathbf{A}_3$: $E(K(\varphi)) \supset_\varphi \Sigma(\varphi)$ —Thm I, $\mathbf{C}_4$, $\mathbf{A}_2$.

COROLLARY 1. $E(K(\varphi)) \supset_\varphi \Sigma x \,.\, x \,\varepsilon\, K(\varphi)$.

COROLLARY 2. $\Sigma x[\sim .\, x \,\varepsilon\, \alpha] \supset_\alpha \Sigma x \,.\, x \,\varepsilon\, \alpha$.

THEOREM 32. $\Sigma x[x \,\varepsilon\, \alpha] \supset_\alpha .\, \alpha = Kx \,.\, x \,\varepsilon\, \alpha$.

$\mathbf{D}_1$: $\alpha = K(\varphi)$.
$\mathbf{C}_1$: $E(K(\varphi))$ —IV, $\mathbf{D}_1$.
$\mathbf{C}_2$: $\Sigma(\varphi)$ —V, Thm 31, $\mathbf{C}_1$; $\mathbf{D}_1$.
$\mathbf{C}_3$: $K(\varphi) = Kx \,.\, x \,\varepsilon\, K(\varphi)$ —V, Thm 30 Cor 2, $\mathbf{C}_2$; $\mathbf{D}_1$.
$\mathbf{C}_4$: $\alpha = Kx \,.\, x \,\varepsilon\, K(\varphi)$ —V^3, Thm 3, $\mathbf{D}_1$, $\mathbf{C}_3$; $\mathbf{D}_1$.
$\mathbf{C}_5$: $K(\varphi) = \alpha$ —V^2, Thm 2, $\mathbf{D}_1$.
$\mathbf{C}_6$: $\alpha = Kx \,.\, x \,\varepsilon\, \alpha$ —V, $\mathbf{C}_5$, $\mathbf{C}_4$; $\mathbf{D}_1$.
$\mathbf{D}_2$: $x \,\varepsilon\, \alpha$.
$\mathbf{C}_7$: $[\alpha = K(\varphi)] \supset_\varphi \varphi(x)$ —conv, $\mathbf{D}_2$.
$\mathbf{C}_8$: $\Sigma\varphi \,.\, \alpha = K(\varphi)$ —V, 38, $\mathbf{C}_7$; $\mathbf{D}_2$.
$\mathbf{C}_9$: $\alpha = Kx \,.\, x \,\varepsilon\, \alpha$ —Thm I, $\mathbf{C}_6$, $\mathbf{C}_8$; $\mathbf{D}_2$.
$\mathbf{D}_3$: $\Sigma x \,.\, x \,\varepsilon\, \alpha$.
$\mathbf{C}_{10}$: $\alpha = Kx \,.\, x \,\varepsilon\, \alpha$ —Thm I, $\mathbf{C}_9$, $\mathbf{D}_3$; $\mathbf{D}_3$.
$\mathbf{A}_1$: $\Sigma\alpha\Sigma x \,.\, x \,\varepsilon\, \alpha$ —IV^2, $\mathbf{A}_1$ under Thm 30.
$\mathbf{A}_2$: $\Sigma x[x \,\varepsilon\, \alpha] \supset_\alpha .\, \alpha = Kx \,.\, x \,\varepsilon\, \alpha$ —Thm I, $\mathbf{C}_{10}$, $\mathbf{A}_1$.

COROLLARY. $[[x \,\varepsilon\, \alpha] \equiv_x .\, x \,\varepsilon\, \beta] \supset_{\alpha\beta} .\, \alpha = \beta$.

THEOREM 33. $[\Sigma(\varphi) \,.\, \sim \varphi(x)] \supset_{\varphi x} \sim .\, x \,\varepsilon\, K(\varphi)$.

$\mathbf{D}_1$: $\psi(y)$.
$\mathbf{C}_1$: $y \,\varepsilon\, K(\psi)$ —V^2, Thm 29, $\mathbf{D}_1$.
$\mathbf{C}_2$: $E(K(\psi))$ —IV, $\mathbf{C}_1$; $\mathbf{D}_1$.
$\mathbf{C}_3$: $K(\psi) = K(\psi)$ —V, Thm 1, $\mathbf{C}_2$; $\mathbf{D}_1$.
$\mathbf{D}_2$: $\Sigma(\psi) \,.\, \sim \psi(x)$.
$\mathbf{C}_4$: $\sim \psi(x)$ —V^2, 16, $\mathbf{D}_2$.
$\mathbf{C}_5$: $K(\psi) = K(\psi) \,.\, \sim \psi(x)$ —V^2, 14, $\mathbf{C}_3$, $\mathbf{C}_4$; $\mathbf{D}_1$, $\mathbf{D}_2$.
$\mathbf{C}_6$: $\Sigma\varphi \,.\, K(\psi) = K(\varphi) \,.\, \sim \varphi(x)$ —IV, $\mathbf{C}_5$; $\mathbf{D}_1$, $\mathbf{D}_2$.
$\mathbf{C}_7$: $\sim .[K(\psi) = K(\varphi)] \supset_\varphi \varphi(x)$ —V^2, 17, $\mathbf{C}_6$; $\mathbf{D}_1$, $\mathbf{D}_2$.
$\mathbf{C}_8$: $\sim .\, x \,\varepsilon\, K(\psi)$ —conv, $\mathbf{C}_7$; $\mathbf{D}_1$, $\mathbf{D}_2$.
$\mathbf{C}_9$: $\Sigma(\psi)$ —V^2, 15, $\mathbf{D}_2$.
$\mathbf{C}_{10}$: $\sim .\, x \,\varepsilon\, K(\psi)$ —Thm III, $\mathbf{C}_8$, $\mathbf{C}_9$; $\mathbf{D}_2$.
$\mathbf{A}_1$: $\Sigma\psi\Sigma x \,.\, \Sigma(\psi) \,.\, \sim \psi(x)$ —I, $\mathbf{A}_4$ under Thm 8.
$\mathbf{A}_2$: $[\Sigma(\psi) \,.\, \sim \psi(x)] \supset_{\psi x} \sim .\, x \,\varepsilon\, K(\psi)$ —Thm I Cor, $\mathbf{C}_{10}$, $\mathbf{A}_1$.
$\mathbf{A}_3$: $[\Sigma(\varphi) \,.\, \sim \varphi(x)] \supset_{\varphi x} \sim .\, x \,\varepsilon\, K(\varphi)$ —I, $\mathbf{A}_2$.

THEOREM 34. $\Sigma y[\sim .\, y \,\varepsilon\, \alpha] \supset_\alpha Gx \,.\, x \,\varepsilon\, \alpha$.

$\mathbf{D}_1$: $x \,\varepsilon\, \alpha$.

C$_1$: $\Sigma x \,.\, x \varepsilon \alpha$ —IV, **D**$_1$.
C$_2$: $\Pi(\Lambda x \,.\, x \varepsilon \alpha, \lambda x \,.\, x \varepsilon \alpha)$ —V, Thm 7, **C**$_1$; **D**$_1$.
C$_3$: $\Pi(\lambda x \,.\, x \varepsilon \alpha, \Lambda x \,.\, x \varepsilon \alpha)$ —V, Thm 6, **C**$_1$; **D**$_1$.
C$_4$: $Q(\Lambda x \,.\, x \varepsilon \alpha, \lambda x \,.\, x \varepsilon \alpha)$ —V^2, 14, **C**$_2$, **C**$_3$; **D**$_1$.
C$_5$: $K(\lambda x \,.\, x \varepsilon \alpha) = Kx \,.\, x \varepsilon \alpha$ —V^2, Thm 25, **C**$_4$; **D**$_1$.
C$_6$: $\alpha = Kx \,.\, x \varepsilon \alpha$ —V, Thm 32, **C**$_1$; **D**$_1$.
C$_7$: $Kx[x \varepsilon \alpha] = \alpha$ —V^2, Thm 2, **C**$_6$; **D**$_1$.
C$_8$: $K(\Lambda x \,.\, x \varepsilon \alpha) = \alpha$ —V^3, Thm 3, **C**$_5$, **C**$_7$; **D**$_1$.
C$_9$: $\{\Lambda x \,.\, x \varepsilon \alpha\}(x)$ —V, **C**$_3$, **D**$_1$; **D**$_1$.
C$_{10}$: $\Sigma(\Lambda x \,.\, x \varepsilon \alpha)$ —IV, **C**$_9$; **D**$_1$.
D$_2$: $\sim\{\Lambda x \,.\, x \varepsilon \alpha\}(z)$.
C$_{11}$: $\sim . \, z \varepsilon K(\Lambda x \,.\, x \varepsilon \alpha)$ —V^2, Thm 33, **C**$_{10}$, **D**$_2$; **D**$_1$, **D**$_2$.
C$_{12}$: $\sim . \, z \varepsilon \alpha$ —V, **C**$_8$, **C**$_{11}$; **D**$_1$, **D**$_2$.
D$_3$: $\sim . \, y \varepsilon \alpha$.
C$_{13}$: $\sim\{\Lambda x \,.\, x \varepsilon \alpha\}(y)$ —V^2, Thm 8, **C**$_1$, **D**$_3$; **D**$_1$, **D**$_3$.
C$_{14}$: $\Sigma x \sim\{\Lambda x \,.\, x \varepsilon \alpha\}(z)$ —IV, **C**$_{13}$; **D**$_1$, **D**$_3$.
C$_{15}$: $\sim\{\Lambda x \,.\, x \varepsilon \alpha\}(z) \supset_z \sim . \, z \varepsilon \alpha$ —Thm I, **C**$_{12}$, **C**$_{14}$; **D**$_1$, **D**$_3$.
C$_{16}$: $Gx \,.\, x \varepsilon \alpha$. —conv, **C**$_{15}$; **D**$_1$, **D**$_3$.
D$_4$: $\Sigma y \sim . \, y \varepsilon \alpha$.
C$_{17}$: $\Sigma x \,.\, x \varepsilon \alpha$ —V, Thm 31 Cor 2, **D**$_4$.
C$_{18}$: $Gx \,.\, x \varepsilon \alpha$ —Thm I, **C**$_{16}$, **C**$_{17}$; **D**$_3$, **D**$_4$.
C$_{19}$: $Gx \,.\, x \varepsilon \alpha$ —Thm I, **C**$_{18}$, **D**$_4$; **D**$_4$.
A$_1$: $\Sigma(\Pi(E))$ —IV, **A**$_3$ under Thm 16.
A$_2$: $\sim . [\sim] \varepsilon K(\Pi(E))$ —V^2, Thm 33, **A**$_1$, Thm 10.
A$_3$: $\Sigma\alpha\Sigma y \sim . \, y \varepsilon \alpha$ —IV^2, **A**$_2$.
A$_4$: $\Sigma y[\sim . \, y \varepsilon \alpha] \supset_\alpha Gx \,.\, x \varepsilon \alpha$ —Thm I, **C**$_{19}$, **A**$_3$.

THEOREM 35. $[\sim . \, x \varepsilon \alpha] \supset_{x\alpha} . \, x \varepsilon O(\alpha)$.

D$_1$: $\sim . \, x \varepsilon \alpha$.
C$_1$: $x \varepsilon K\pi \sim . \, \pi \varepsilon \alpha$ —V^2, Thm 29, **D**$_1$.
C$_2$: $x \varepsilon O(\alpha)$ —III, **C**$_1$; **D**$_1$.
A$_1$: $\Sigma x \Sigma\alpha \sim . \, x \varepsilon \alpha$ —IV^2, **A**$_2$ under Thm 34.
A$_2$: $[\sim . \, x \varepsilon \alpha] \supset_{x\alpha} . \, x \varepsilon O(\alpha)$ —Thm I Cor, **C**$_2$, **A**$_1$.

THEOREM 36. $[x \varepsilon O(\alpha)] \supset_{x\alpha} \sim . \, x \varepsilon \alpha$.

D$_1$: $x \varepsilon O(\alpha)$.
C$_1$: $x \varepsilon K\pi \sim . \, \pi \varepsilon \alpha$ —II, **D**$_1$.
C$_2$: $\sim . \, x \varepsilon \alpha$ —V^2, Thm 30, **C**$_1$; **D**$_1$.
A$_1$: $[\sim] \varepsilon O(K(\Pi(E)))$ —V^2, Thm 35, **A**$_2$ under Thm 34.
A$_2$: $\Sigma x \Sigma\alpha \,.\, x \varepsilon O(\alpha)$ —IV^2, **A**$_1$.
A$_3$: $[x \varepsilon O(\alpha)] \supset_{x\alpha} \sim . \, x \varepsilon \alpha$ —Thm I Cor, **C**$_2$, **A**$_2$.

THEOREM 37. $\Sigma y[\sim . \, y \varepsilon \alpha] \supset_\alpha . [x \varepsilon \alpha] \supset_x \sim . \, x \varepsilon O(\alpha)$.

D$_1$: $x \varepsilon \alpha$.
C$_1$: $\sim\sim . \, x \varepsilon \alpha$ —V, 26, **D**$_1$.
D$_2$: $\Sigma y \sim . \, y \varepsilon \alpha$.
C$_2$: $\sim . \, x \varepsilon Ky \sim . \, y \varepsilon \alpha$ —V^2, Thm 33, **D**$_2$, **C**$_1$; **D**$_1$, **D**$_2$.
C$_3$: $\sim . \, x \varepsilon O(\alpha)$ —conv, **C**$_2$; **D**$_1$, **D**$_2$.
C$_4$: $\Sigma x \,.\, x \varepsilon \alpha$ —V, Thm 31 Cor 2, **D**$_2$.

$\mathbf{C}_5$: $[x \varepsilon \alpha] \supset_x \sim . x \varepsilon O(\alpha)$ —Thm I, $\mathbf{C}_3$, $\mathbf{C}_4$; $\mathbf{D}_2$.
$\mathbf{A}_1$: $\Sigma y[\sim . y \varepsilon \alpha] \supset_\alpha .[x \varepsilon \alpha] \supset_x \sim . x \varepsilon O(\alpha)$ —Thm I, $\mathbf{C}_5$, $\mathbf{A}_3$ under Thm 34.

COROLLARY 1. $\Sigma y[\sim . y \varepsilon \alpha] \supset_\alpha .[x \varepsilon \alpha] \supset_x x \varepsilon O(O(\alpha))$.

COROLLARY 2. $[\sim . x \varepsilon \alpha] \supset_{x\alpha} \sim . x \varepsilon O(O(\alpha))$.

COROLLARY 3. $\Sigma y[\sim . y \varepsilon \alpha] \supset_\alpha .[\sim . x \varepsilon O(O(\alpha))] \supset_x \sim . x \varepsilon \alpha$.

COROLLARY 4. $\Sigma y[\sim . y \varepsilon \alpha] \supset_\alpha .[\sim . x \varepsilon O(O(\alpha))] \supset_x . x \varepsilon O(\alpha)$.

THEOREM 38. $\Sigma y[\sim . y \varepsilon \alpha] \supset_\alpha Gx \sim . x \varepsilon O(\alpha)$.

$\mathbf{D}_1$: $\Sigma y \sim . y \varepsilon \alpha$.
$\mathbf{D}_2$: $x \varepsilon \alpha$.
$\mathbf{C}_1$: $\sim . x \varepsilon O(\alpha)$ —V^2, Thm 37, $\mathbf{D}_1$, $\mathbf{D}_2$.
$\mathbf{C}_2$: $\{\Lambda y \sim . y \varepsilon O(\alpha)\}(x)$ —V^2, Thm 6 Cor, $\mathbf{C}_1$; $\mathbf{D}_1$, $\mathbf{D}_2$.
$\mathbf{C}_3$: $\Sigma y . y \varepsilon \alpha$ —V, Thm 31 Cor 2, $\mathbf{D}_1$.
$\mathbf{C}_4$: $\Pi(\lambda y . y \varepsilon \alpha, \Lambda y \sim . y \varepsilon O(\alpha))$ —Thm II, $\mathbf{C}_2$, $\mathbf{C}_3$; $\mathbf{D}_1$.
$\mathbf{C}_5$: $Gy . y \varepsilon \alpha$ —V, Thm 34, $\mathbf{D}_1$.
$\mathbf{D}_3$: $\sim\{\Lambda y \sim . y \varepsilon O(\alpha)\}(z)$.
$\mathbf{C}_6$: $\sim . z \varepsilon \alpha$ —V^3, Thm 20, $\mathbf{C}_5$, $\mathbf{C}_4$, $\mathbf{D}_3$; $\mathbf{D}_1$, $\mathbf{D}_3$.
$\mathbf{C}_7$: $z \varepsilon O(\alpha)$ —V^2, Thm 35, $\mathbf{C}_6$; $\mathbf{D}_1$, $\mathbf{D}_3$.
$\mathbf{C}_8$: $\sim\sim . z \varepsilon O(\alpha)$ —V, 26, $\mathbf{C}_7$; $\mathbf{D}_1$, $\mathbf{D}_3$.
$\mathbf{D}_4$: $\sim . u \varepsilon \alpha$.
$\mathbf{C}_9$: $u \varepsilon O(\alpha)$ —V^2, Thm 35, $\mathbf{D}_4$.
$\mathbf{C}_{10}$: $\sim\sim . u \varepsilon O(\alpha)$ —V, 26, $\mathbf{C}_9$; $\mathbf{D}_4$.
$\mathbf{C}_{11}$: $\Sigma y \sim . y \varepsilon O(\alpha)$ —IV, $\mathbf{C}_1$; $\mathbf{D}_1$, $\mathbf{D}_2$.
$\mathbf{C}_{12}$: $\sim\{\Lambda y \sim . y \varepsilon O(\alpha)\}(u)$ —V^2, Thm 8, $\mathbf{C}_{11}$, $\mathbf{C}_{10}$; $\mathbf{D}_1$, $\mathbf{D}_2$, $\mathbf{D}_4$.
$\mathbf{C}_{13}$: $\Sigma z \sim\{\Lambda y \sim . y \varepsilon O(\alpha)\}(z)$ —IV, $\mathbf{C}_{12}$; $\mathbf{D}_1$, $\mathbf{D}_2$, $\mathbf{D}_4$.
$\mathbf{C}_{14}$: $\Sigma z \sim\{\Lambda y \sim . y \varepsilon O(\alpha)\}(z)$ —Thm I, $\mathbf{C}_{13}$, $\mathbf{D}_1$; $\mathbf{D}_1$, $\mathbf{D}_2$.
$\mathbf{C}_{15}$: $\Sigma z \sim\{\Lambda y \sim . y \varepsilon O(\alpha)\}(z)$ —Thm I, $\mathbf{C}_{14}$, $\mathbf{C}_3$; $\mathbf{D}_1$.
$\mathbf{C}_{16}$: $\sim\{\Lambda y \sim . y \varepsilon O(\alpha)\}(z) \supset_z \sim\sim . z \varepsilon O(\alpha)$ —Thm I, $\mathbf{C}_8$, $\mathbf{C}_{15}$; $\mathbf{D}_1$.
$\mathbf{C}_{17}$: $Gx \sim . x \varepsilon O(\alpha)$ —conv, $\mathbf{C}_{16}$; $\mathbf{D}_1$.
$\mathbf{A}_1$: $\Sigma y[\sim . y \varepsilon \alpha] \supset_\alpha Gx \sim . x \varepsilon O(\alpha)$ —Thm I, $\mathbf{C}_{17}$, $\mathbf{A}_3$ under Thm 34.

7. The Russell paradox. The existence of the propositional function $\lambda\varphi . \sim\varphi(\varphi)$ follows readily from our assumptions. In fact, it can be shown that the value of this function is a true proposition for the value $\lambda x . \sim E(x)$ of the independent variable (among others), and that the value of the function is a false proposition for each of the values E and Σ of the independent variable (among others). If, however, we inquire whether the value of the function $\lambda\varphi . \sim\varphi(\varphi)$ is a true proposition or a false proposition for the value $\lambda\varphi . \sim\varphi(\varphi)$ of the independent variable, we are led at once into difficulty. For let $\mathfrak{B}$ stand for the formula $\{\lambda\varphi . \sim\varphi(\varphi)\}(\lambda\varphi . \sim\varphi(\varphi))$. Then $\mathfrak{B}$ is convertible into $\sim . \mathfrak{B}$, and $\sim . \mathfrak{B}$ is convertible into $\mathfrak{B}$. And hence if $\mathfrak{B}$ be a true proposition it must also be a false proposition, and if $\mathfrak{B}$ be a false proposition it must also be a true proposition.

We are not, however, forced to the conclusion that our system is self-contradictory. For no way appears by which we can prove either of the formulas $\mathfrak{B}$ or $\sim . \mathfrak{B}$, and in the absence of such a proof there is no obvious method of obtaining a contradiction out of the situation just described. There is, in fact, no reason to suppose that the propositional function $\lambda\varphi \,.\, \sim\varphi(\varphi)$ is significant for all values of the independent variable, or that the formula $\mathfrak{B}$ is anything but a meaningless aggregate of symbols.

Of course, if there were any way of constructing a propositional function **F**, equivalent to $\lambda\varphi \,.\, \sim\varphi(\varphi)$ and significant for all values of the independent variable, then a contradiction could be obtained by considering the formula $\{\mathbf{F}\}(\mathbf{F})$. But there is no reason to suppose that such a function **F** can be constructed. In particular, it does not appear that $\Lambda\varphi \,.\, \sim\varphi(\varphi)$ could be taken as **F**.

The property of the formula $\mathfrak{B}$, that its falsehood is provable as a consequence of its truth, and its truth as a consequence of its falsehood, is shared by many other formulas. We cite as examples the formulas

$$\{\lambda\varphi\lambda x \,.\, \sim\varphi(x,\varphi)\}(\lambda\varphi\lambda x \,.\, \sim\varphi(x,\varphi), \lambda\varphi\lambda x \,.\, \sim\varphi(x,\varphi)),$$

and

$$\{\Lambda\varphi \,.\, \sim\varphi(\varphi)\}(\Lambda\varphi \,.\, \sim\varphi(\varphi)),$$

and

$$Kx[\sim . \, x \,\varepsilon\, x] \,\varepsilon\, Kx \sim . \, x \,\varepsilon\, x.$$

Let $\mathfrak{D}$ stand for the formula $Kx[\sim . \, x \,\varepsilon\, x] \,\varepsilon\, Kx \sim . \, x \,\varepsilon\, x$. Then if $\mathfrak{D}$ were true, Theorem 30 would enable us to prove $\sim . \, \mathfrak{D}$. And if $\sim . \, \mathfrak{D}$ were true, Theorem 29 would enable us to prove $\mathfrak{D}$. This means that if we assume $\mathfrak{D}$ we can deduce the contradiction $\mathfrak{D} \,.\, \sim . \, \mathfrak{D}$, and hence, under the classical principle of reductio ad absurdum, we should be justified in inferring that $\mathfrak{D}$ is false, that is, $\sim . \, \mathfrak{D}$. And once we had $\sim . \, \mathfrak{D}$, we could infer $\mathfrak{D}$ by Theorem 29.

In this way we see that the classical principle of reductio ad absurdum, and the assumption that for every propositional function there is an equivalent function significant for all values of the independent variable, are from a certain point of view the same.

And our omission of the principle of reductio ad absurdum from our list of assumptions is justified by reference to cases, like that just discussed, to which the principle appears to be inapplicable.

8. Properties of the expression $\lambda\varphi \,.\, \sim\varphi(x(\varphi))$. We now proceed to the proof of a number of theorems which bear a close relation to the Russell paradox.

Theorem 39. $\Sigma x\{\lambda\varphi \,.\, \sim\varphi(x(\varphi))\}\,(\lambda\varphi \,.\, \sim\varphi(x(\varphi)))$.

$\mathbf{A}_1$: $Ep \,.\, q \supset_q pq$ —IV, 14.
$\mathbf{A}_2$: $E(E)$ —IV, $\mathbf{A}_1$.
$\mathbf{A}_3$: $E(E)E(E)$ —V^2, 14, $\mathbf{A}_2$, $\mathbf{A}_2$.
$\mathbf{A}_4$: $Ev \,.\, E(E)E(v)$ —IV, $\mathbf{A}_3$.
$\mathbf{A}_5$: $\sim\sim Ev \,.\, E(E)E(v)$ —V, 26, $\mathbf{A}_4$.
$\mathbf{A}_6$: $E\varphi \,.\, \sim\varphi(\lambda v \,.\, E(\varphi)E(v))$ —IV, $\mathbf{A}_5$.
$\mathbf{A}_7$: $E\varphi[\sim\varphi(\lambda v \,.\, E(\varphi)E(v))]E(E)$ —V^2, 14, $\mathbf{A}_6$, $\mathbf{A}_2$.
$\mathbf{A}_8$: $Ew \,.\, E\varphi[\sim\varphi(\lambda v \,.\, E(\varphi)E(v))]E(w)$ —IV, $\mathbf{A}_7$.
$\mathbf{A}_9$: $Ew[E\varphi[\sim\varphi(\lambda v \,.\, E(\varphi)E(v))]E(w)]E(E)$ —V^2, 14, $\mathbf{A}_8$, $\mathbf{A}_2$.
$\mathbf{A}_{10}$: $Ey \,.\, Ew[E\varphi[\sim\varphi(\lambda v \,.\, E(\varphi)E(v))]E(w)]E(y)$ —IV, $\mathbf{A}_9$.
$\mathbf{A}_{11}$: $E\varphi[\sim\varphi(\lambda v \,.\, E(\varphi)E(v))] \,.\, Ey \,.\, Ew[E\varphi[\sim\varphi(\lambda v \,.\, E(\varphi)E(v))]E(w)]E(y)$ —V^2, 14, $\mathbf{A}_6$, $\mathbf{A}_{10}$.

$\mathbf{A}_{12}$: $\sim\sim . E\varphi[\sim\varphi(\lambda v . E(\varphi)E(v))] . Ey . Ew[E\varphi[\sim\varphi(\lambda v . E(\varphi)E(v))]E(w)]E(y)$ —V, 26, $\mathbf{A}_{11}$.

$\mathbf{A}_{13}$: $\{\lambda\varphi . \sim\varphi(\lambda v . E(\varphi)E(v))\}(\lambda\varphi . \sim\varphi(\lambda v . E(\varphi)E(v)))$ —conv, $\mathbf{A}_{12}$.

$\mathbf{A}_{14}$: $\Sigma x\{\lambda\varphi . \sim\varphi(x(\varphi))\}(\lambda\varphi . \sim\varphi(x(\varphi)))$ —IV, $\mathbf{A}_{13}$.

THEOREM 40. $\Sigma x \sim . \{\lambda\varphi . \sim\varphi(x(\varphi))\}(\lambda\varphi . \sim\varphi(x(\varphi)))$.

$\mathbf{A}_1$: $Ep . \sim q \supset_q \sim . pq$ —IV, 21.

$\mathbf{A}_2$: $E(E)$ —IV, $\mathbf{A}_1$.

$\mathbf{A}_3$: $\sim\sim E(E)$ —IV, 26, $\mathbf{A}_2$.

$\mathbf{A}_4$: $\sim . E(E) . \sim E(E)$ —V^2, 21, $\mathbf{A}_2$, $\mathbf{A}_3$.

$\mathbf{A}_5$: $Ev . E(E) . \sim E(v)$ —IV, $\mathbf{A}_4$.

$\mathbf{A}_6$: $\sim\sim Ev . E(E) . \sim E(v)$ —V, 26, $\mathbf{A}_5$.

$\mathbf{A}_7$: $E\varphi . \sim\varphi(\lambda v . E(\varphi) . \sim E(v))$ —IV, $\mathbf{A}_6$.

$\mathbf{A}_8$: $\sim . E\varphi[\sim\varphi(\lambda v . E(\varphi) . \sim E(v))] . \sim E(E)$ —V^2, 21, $\mathbf{A}_7$, $\mathbf{A}_3$.

$\mathbf{A}_9$: $Ew . E\varphi[\sim\varphi(\lambda v . E(\varphi) . \sim E(v))] . \sim E(w)$ —IV, $\mathbf{A}_8$.

$\mathbf{A}_{10}$: $\sim . Ew[E\varphi[\sim\varphi(\lambda v . E(\varphi) . \sim E(v))] . \sim E(w)] . \sim E(E)$ —V^2, 21, $\mathbf{A}_9$, $\mathbf{A}_3$.

$\mathbf{A}_{11}$: $Ey . Ew[E\varphi[\sim\varphi(\lambda v . E(\varphi) . \sim E(v))] . \sim E(w)] . \sim E(y)$ —IV, $\mathbf{A}_9$.

$\mathbf{A}_{12}$: $\sim\sim Ey . Ew[E\varphi[\sim\varphi(\lambda v . E(\varphi) . \sim E(v))] . \sim E(w)] . \sim E(y)$ —V, 26, $\mathbf{A}_{11}$.

$\mathbf{A}_{13}$: $\sim . E(\varphi)[\sim\varphi(\lambda v . E(\varphi) . \sim E(v))] . \sim Ey . Ew[E\varphi[\sim\varphi(\lambda v . E(\varphi) . \sim E(v))] . \sim E(w)] . \sim E(y)$ —V^2, 21, $\mathbf{A}_7$, $\mathbf{A}_{12}$.

$\mathbf{A}_{14}$: $\sim\sim\sim . E(\varphi)[\sim\varphi(\lambda v . E(\varphi) . \sim E(v))] . \sim Ey . Ew[E\varphi[\sim\varphi(\lambda v . E(\varphi) . \sim E(v))] . \sim E(w)] . \sim E(y)$ —V, 26, $\mathbf{A}_{13}$.

$\mathbf{A}_{15}$: $\sim . \{\lambda\varphi . \sim\varphi(\lambda v . E(\varphi) . \sim E(v))\}(\lambda\varphi . \sim\varphi(\lambda v . E(\varphi) . \sim E(v)))$ —conv, $\mathbf{A}_{14}$.

$\mathbf{A}_{16}$: $\Sigma x \sim . \{\lambda\varphi . \sim\varphi(x(\varphi))\}(\lambda\varphi . \sim\varphi(x(\varphi)))$ —IV, $\mathbf{A}_{15}$.

THEOREM 41. $\{\lambda\varphi . \sim\varphi(x(\varphi))\}(\lambda\varphi . \sim\varphi(x(\varphi))) \supset_x . x \neq \lambda zz$.

$\mathbf{D}_1$: $\{\lambda\varphi . \sim\varphi(x(\varphi))\}(\lambda\varphi . \sim\varphi(x(\varphi)))$.

$\mathbf{C}_1$: $\sim\{\lambda\varphi . \sim\varphi(x(\varphi))\}\left(x(\lambda\varphi . \sim\varphi(x(\varphi)))\right)$ —II, $\mathbf{D}_1$.

$\mathbf{C}_2$: $\sim\sim\{\lambda\varphi . \sim\varphi(x(\varphi))\}(\lambda\varphi . \sim(x(\varphi)))$ —V, 26, $\mathbf{D}_1$.

$\mathbf{C}_3$: $\sim\sim\{\lambda\varphi . \sim\varphi(x(\varphi))\}\left(\{\lambda zz\}(\lambda\varphi . \sim\varphi(x(\varphi)))\right)$ —III, $\mathbf{C}_2$; $\mathbf{D}_1$.

$\mathbf{C}_4$: $\sim\{\lambda\varphi . \sim\varphi(x(\varphi))\}\left(x(\lambda\varphi . \sim\varphi(x(\varphi)))\right) . \sim\sim\{\lambda\varphi . \sim\varphi(x(\varphi))\} \left(\{\lambda zz\}(\lambda\varphi . \sim\varphi(x(\varphi)))\right)$ —V^2, 14, $\mathbf{C}_1$, $\mathbf{C}_3$; $\mathbf{D}_1$.

$\mathbf{C}_5$: $\Sigma y . y(x) . \sim y(\lambda zz)$ —IV, $\mathbf{C}_4$; $\mathbf{D}_1$.

$\mathbf{C}_6$: $x \neq \lambda zz$ —V^2, 17, $\mathbf{C}_5$; $\mathbf{D}_1$.

$\mathbf{A}_1$: $\{\lambda\varphi . \sim\varphi(x(\varphi))\}(\lambda\varphi . \sim\varphi(x(\varphi))) \supset_x . x \neq \lambda zz$ —Thm I, $\mathbf{C}_6$, Thm 39.

COROLLARY. $\sim . \{\Lambda x\{\lambda\varphi . \sim\varphi(x(\varphi))\}(\lambda\varphi . \sim\varphi(x(\varphi)))\}(\lambda zz)$.

THEOREM 42. $\sim\{\lambda\varphi . \sim\varphi(x(\varphi))\}(\lambda\varphi . \sim\varphi(x(\varphi))) \supset_x . x \neq \lambda zz$.

$\mathbf{D}_1$: $\sim\{\lambda\varphi . \sim\varphi(x(\varphi))\}(\lambda\varphi . \sim\varphi(x(\varphi)))$.

$\mathbf{C}_1$: $\sim\sim\{\lambda\varphi . \sim\varphi(x(\varphi))\}\left(x(\lambda\varphi . \sim\varphi(x(\varphi)))\right)$ —II, $\mathbf{D}_1$.

$\mathbf{C}_2$: $\sim\sim\sim\{\lambda\varphi . \sim\varphi(x(\varphi))\}(\lambda\varphi . \sim\varphi(x(\varphi)))$ —V, 26, $\mathbf{D}_1$.

$\mathbf{C}_3$: $\sim\sim\sim\{\lambda\varphi . \sim\varphi(x(\varphi))\}\left(\{\lambda zz\}(\lambda\varphi . \sim\varphi(x(\varphi)))\right)$ —III, $\mathbf{C}_2$; $\mathbf{D}_1$.

$\mathbf{C}_4$: $\sim\sim\{\lambda\varphi . \sim\varphi(x(\varphi))\}\left(x(\lambda\varphi . \sim\varphi(x(\varphi)))\right) . \sim\sim\sim\{\lambda\varphi . \sim\varphi(x(\varphi))\} \left(\{\lambda zz\}(\lambda\varphi . \sim\varphi(x(\varphi)))\right)$ —V^2, 14, $\mathbf{C}_1$, $\mathbf{C}_3$; $\mathbf{D}_1$.

$\mathbf{C}_5$: $\Sigma y . y(x) . \sim y(\lambda zz)$ —IV, $\mathbf{C}_4$; $\mathbf{D}_1$.

$\mathbf{C}_6$: $x \neq \lambda zz$ — V^2, 17, $\mathbf{C}_5$; $\mathbf{D}_1$.
$\mathbf{A}_1$: $\sim\{\lambda\varphi \,.\, \sim\varphi(x(\varphi))\}(\lambda\varphi \,.\, \sim\varphi(x(\varphi))) \supset_x \,.\, x \neq \lambda zz$ — Thm I, $\mathbf{C}_6$, Thm 40.
COROLLARY. $\sim \,.\, \{\Lambda x \sim\{\lambda\varphi \,.\, \sim\varphi(x(\varphi))\}(\lambda\varphi \,.\, \sim\varphi(x(\varphi)))\}(\lambda zz)$.

9. Positive integers. We define

$$1 \longrightarrow \lambda f \lambda x \,.\, f(x).$$
$$S \longrightarrow \lambda \varrho \lambda f \lambda x \,.\, f(\varrho(f, x)).$$
$$N \longrightarrow \lambda y \,.[\varphi(1) \,.\, \varphi(x) \supset_x \varphi(S(x))] \supset_\varphi \varphi(y).$$

The symbol 1 is to be read, "One", and $S(\mathbf{A})$ is to be read, "The successor of **A**", and $N(\mathbf{A})$ is to be read, "**A** is a positive integer".

If we define $2 \longrightarrow S(1)$, and $3 \longrightarrow S(2)$, and so on, we find that 2 is convertible into $\lambda f \lambda x \,.\, f(f(x))$, and 3 is convertible into $\lambda f \lambda x \,.\, f(f(f(x)))$, and so on. The form of these expressions provides a convenient method of writing definitions by induction, as we may illustrate in the case of the definition of the sum, $m + n$, of two positive integers m and n. The equivalent of the recursion formulas, $m + 1 = S(m)$, and $m + (k + 1) = S(m + k)$, is, in fact, obtained by defining:

$$+ \longrightarrow \lambda m \lambda n \,.\, n(S, m).$$

And the operations of subtraction and multiplication may then be defined as follows:

$$- \longrightarrow \lambda r \lambda s \iota x \,.\{+\}(x, s) = r$$
$$\times \longrightarrow \lambda m \lambda n \,.\{-\}(m(n(S), 1), 1).$$

Peano's axioms for the positive integers[9] may be expressed in our notation as follows:

1. $N(1)$.
2. $N(x) \supset_x N(S(x))$.
3. $[N(x) \,.\, N(y) \,.\, S(x) = S(y)] \supset_{xy} \,.\, x = y$.
4. $N(x) \supset_x \,.\, S(x) \neq 1$.
5. $[\varphi(1) \,.\, \varphi(x) \supset_x \varphi(S(x))] \supset_\varphi \,.\, N(y) \supset_y \varphi(y)$.

It is believed that each of these five propositions will turn out to be a theorem which can be proved as a consequence of our postulates. And indeed proofs of 1,2, and 5 are immediately evident.

Our program is to develop the theory of positive integers on the basis which we have just been describing, and then, by known methods or appropriate modifications of them, to proceed to a theory of rational numbers and a theory of real numbers.

PRINCETON UNIVERSITY, PRINCETON, N.J.

[9]G. Peano, *Sul concetto di numero*, ***Rivista di Mathematica***, vol. 1 (1891), pp. 87–102.

FERMAT'S LAST THEOREM
(1934)

Our topic forms a chapter in the theory of Diophantine equations. A treatment of it will be found in any standard text on number theory (e.g. Dickson's or Carmichael's). Our present purpose is simply to make a discussion of the one topic in isolation, without presupposing any acquaintance with number theory beyond what is contained in arithmetic and elementary algebra. This is as a matter of fact one of the few cases in which such an elementary presentation is possible in brief space of a question which is an important object of current mathematical research.

A Diophantine equation is an algebraic equation in two or more variagles $x, y, \ldots$ about which the problem is proposed, to find its solutions in integers (or, it may be, in positive integers). If there are only a finite number of ways of giving integral (positive integral) values to $x, y, \ldots$ so as to satisfy the equation, the solution will consist in writing all these out; if there is no way of doing this, the solution will consist in proving that fact; if there are an infinite number of ways, the solution will consist in a formula which generates all of them (in a sense which will be illustrated below in connection with a particular example), this formula being the nearest possible equivalent to writing all the particular solutions out.

In the case of Diophantine equations of the first degree it is possible to give a general method of solution—understanding "solution" in the sense just explained. This will be found in any text of advanced algebra; there is an excellent treatment of it, for instance, in Fine's College Algebra.

On the other hand, even in the case of Diophantine equations of the second degree no general method of solution is known, and it is even probable that no such method exists. Solutions are known, however, for many particular equations, or equations of particular forms.

Before taking up our main topic, we set down here, for convenience of reference, a number of theorems about integers which are either easily proved or may be regarded as already familiar.

I. The sum of two even integers, or of two odd integers, is even. The sum of an even integer and an odd integer is odd.

II. If the integers a and b are each divisible by the integer c, then $a + b$ is divisible by c and $a - b$ is divisible by c. (For if $a = cA$ and $b = cB$, then $a + b = c(A + B)$ and $a - b = c(A - B)$.)

III. The square of an even integer is divisible by 4.

IV. If the square of an odd integer is divided by 4, the remainder is 1. (For $(2a + 1)^2 = 4(a^2 + a) + 1$.)

V. Hence if upon dividing a number by 4 it is found that the remainder is either 2 or 3, it may be inferred that the number is not a perfect square.

VI. A positive integer can be factored into prime factors in one and only one way.

VII. If the positive integers a and b have no common factor, and ab is a perfect square, then a is a perfect square and b is a perfect square.

A Lecture given at the Galois Institute of Mathematics at Long Island University, 300 Pearl Street, Brooklyn, N.Y., 1934.

Of the above theorems, VI is familiar in a sense to every one with a knowledge of arithmetic. Nevertheless to give an explicit proof of it is not entirely trivial.

The proof of VII is immediate from VI. For consider a and b, and hence their product ab, factored into prime factors. Each prime factor of ab appears an even number of times. But a and b have no prime factor in common. Hence each prime factor of a appears an even number of times, and each prime factor of b appears an even number of times.

We turn now to consideration of the Diophantine equation $x^2 + y^2 = z^2$. There is clearly no essential difference between the problem to solve this equation in integers and the problem to solve it in positive integers. As a matter of convenience we base our discussion on the problem to solve it in positive integers.

This problem can be regarded geometrically, as the problem to find a right triangle the lengths of all three of whose sides are whole numbers. Certain particular solutions, such as $3^2 + 4^2 = 5^2$, are familiar to every one who has worked textbook problems in algebra and geometry; but we wish to find all solutions.

Now it is easily seen that, having one solution, other solutions may be obtained by multiplying the values of x, y, z all three by the same number. For instance, since 3, 4, 5 is a solution, we have 6, 8, 10 as another solution, 9, 12, 15 as another solution, and so on. Moreover, if we knew that 6, 8, 10 was a solution, we could infer from that alone that 3, 4, 5 was a solution; and the like generally.

Hence it will be sufficient if we find all solutions in which x, y, x have no common factor.

Suppose then that $x^2 + y^2 = z^2$ and that x, y, z have no common factor.

x and y cannot both be even. For then, by I, z^2 and therefore z would be even, and x, y, z would have the common factor 2.

x and y cannot both be odd. For if they were we would have by IV that x^2 and y^2 when divided by 4 would each leave the remainder 1. That is, $x^2 = 4m+1$, $y^2 = 4n+1$. Then we would have $z^2 = x^2 + y^2 = 4m + 1 + 4n + 1 = 4(m + n) + 2$. That is z^2, when divided by 4, would leave the remainder 2—contrary to V.

Hence one of the numbers x, y is even and the other odd. Let us say that x is even and y is odd. Then, by I, z^2 is odd, and therefore z is odd. Therefore by I, $z + y$ and $z - y$ are both even numbers.

Transposing terms in the equation $x^2 + y^2 = z^2$ and factoring, we get

$$(z + y)(z - y) = x^2,$$

and therefore, dividing by 4,

$$\left(\frac{z+y}{2}\right)\left(\frac{z-y}{2}\right) = \left(\frac{x}{2}\right)^2$$

Here $\frac{z+y}{2}$, $\frac{z-y}{2}$, and $\frac{x}{2}$ are integers.

Now $\frac{z+y}{2}$ and $\frac{z-y}{2}$ cannot have a common factor. For suppose that $\frac{z+y}{2}$ and $\frac{z-y}{2}$ had the common prime factor d. Then, by II, z and y would both have the factor d (because $z = \frac{z+y}{2} + \frac{z-y}{2}$ and $y = \frac{z+y}{2} - \frac{z-y}{2}$). Therefore $z^2 - y^2$ would have the factor d^2. That is x^2 would have the factor d^2, and therefore x would have the factor d. Thus x, y, z would all three have the factor d.

Consequently it follows from VII that $\frac{z+y}{2}$ and $\frac{z-y}{2}$ are both perfect squares. Let

$$\frac{z+y}{2} = u^2$$
$$\frac{z-y}{2} = v^2$$

Solving these equations, and the equation $x^2 + y^2 = z^2$, as simultaneous equations, for x, y, z, we obtain

$$\begin{aligned} x &= 2uv \\ y &= u^2 - v^2 \\ z &= u^2 + v^2 \end{aligned} \tag{1}$$

Since x, y, z have no common factor, we can infer from these equations that u and v have no common factor. Since y is odd, u and v are not both odd. Since y is positive, u is greater than v.

Finally, if we substitute in $x^2 + y^2 = z^2$ the values of x, y, z given by (1), we find the equation satisfied.

If in (1) we take u and v to be any integers we have a solution of the Diophantine equation $x^2 + y^2 = z^2$.

All the solutions in which x, y, z have no common factor may be obtained from (1) by taking u and v to be positive integers, with u greater than v, u and v having no common factor, and either u or v even.

The general solution of the Diophantine equation $x^2 + y^2 = z^2$ may be written

$$\begin{aligned} x^2 &= r(2uv) \\ y^2 &= r(u^2 - v^2) \\ z^2 &= r(u^2 + v^2) \end{aligned} \tag{2}$$

where r, u, v are any integers.

The number of solutions is infinite. Hence they cannot all be written down. But there is a sense in which formulas (1) and (2) can be thought of as the equivalent of writing them all down.

If it is required to write down all solutions in which z is less than some fixed number, say 100, this can be done immediately by means of (1) and (2).

Thus we consider these formulas as being the solution of the Diophantine equation.

It is suggested to the reader as an exercise that he undertake to solve by similar means the Diophantine equations,

$$\begin{aligned} 2x^2 + y^2 &= z^2, \\ 3x^2 + y^2 &= z^2 \\ 6x^2 + y^2 &= z^2 \\ x^2 + y^2 &= z^4, \end{aligned}$$

finding in each case also the least solution in positive integers.

The solution given above of the equation $x^2 + y^2 = z^2$ in integers goes back to Diophantus (third century A.D.), from whom Diophantine equations take their name.

The French mathematician Pierre de Fermat (1601–1665) was the discoverer of many theorems in number theory, including at least one of central importance to the whole subject, and is regarded by many as the founder of modern number theory. Some of his results were published in his life time, but many more were not. A part of his work in number theory has been preserved through marginal notes made by him in a Latin translation of Diophantus which he owned. These notes, which give original solutions of problems, and theorems, usually, but not always with proofs, were published in 1670 by his son Samuel.

In one of these notes Fermat states that a perfect cube cannot be the sum of two cubes, or a perfect fourth power of two fourth powers, and so on to infinity; that is that, if n is a positive integer greater than 2, the Diophantine equation $x^n + y^n = z^n$ has no solution (except the trivial solution in which x or y or z is equal to 0). To this he adds: "Of which thing I have found a remarkable proof. This the narrowness of the margin would not receive." ("Cujus rei demonstrationem mirabilem sane detexi. Hanc marginis exiguitas non caperet.")

This is the proposition now known as Fermat's Last Theorem—not altogether correctly, since a proposition not known to be true cannot properly be called a theorem. The combined efforts of generations of mathemeticians since Fermat's day have not sufficed either to prove this proposition or to prove it false, so that it stands as one of the famous unsolved problems of mathematics.

Probably most number-theorists of today believe that Fermat's Last Theorem is true and will some day be proved. But if so this will almost certainly not be by such elementary means as those employed in the present exposition, or even by means which could conceivably have been known to Fermat in the seventeenth century.

Nevertheless it may be worth while, as an illustration of what a proof of Fermat's last theorem might be like, to give a proof for the particular case $n = 4$ (proofs for many other particular cases are also known).

That is, we shall prove that the Diophantine equation $x^4 + y^4 = z^4$ has no solution in which x, y, z are all three different from 0.

In order to do this we shall prove the stronger theorem that the Diophantine equation $x^4 + y^4 = z^2$ has no solution in positive integers; and from this the other theorem will follow immediately.

Suppose that perfect squares do exist which can be expressed as the sum of two fourth powers. Then let z^2 be the smallest such perfect square. Then

$$x^4 + y^4 = z^2$$

x and y have no common factor. For if d were a common prime factor of x and y it would follow by II that z^2 was divisible by d^4 and therefore that z was divisible by d^2; thus we would have $\left(\frac{x}{d}\right)^4 + \left(\frac{y}{d}\right)^4 = \left(\frac{z}{d^2}\right)^2$, and z^2 would thus not be the smallest perfect square expressible as the sum of two fourth powers.

x and y cannot both be odd. For then we would have, by IV, $x^4 = 4b+1$, $y^4 = 4c+1$; hence $z^2 = x^4 + y^4 = 4(b + c) + 2$, contrary to V.

Hence of the numbers x and y one is even and the other odd. Say x is even.

Then, by our solution of $x^2 + y^2 = z^2$, we have in the present case

$$x^2 = 2uv$$
$$y^2 = u^2 - v^2$$
$$z = u^2 + v^2$$

where u and v have no common factor and cannot both be odd.

v cannot be odd. For if it were u would be even, and by III and IV we would have,

$$y^2 = u^2 - v^2 = 4b - (4c + 1) = 4(b - c - 1) + 3,$$

and hence y^2, when divided by 4 would leave a remainder 3, contrary to V.

Therefore,

$$v = 2k$$

where k is a positive integer.

Substituting this in $x^2 = 2uv$ and then dividing by 4, we get

$$\left(\frac{x}{2}\right)^2 = ku$$

Here $\frac{x}{2}$ is a positive integer (since x is even), and k and u have no common factor (since v and u do not).

Therefore, by VII,

$$k = m^2$$
$$u = n^2$$

where m and n are positive integers.

Now from $y^2 = u^2 - v^2$ we get $v^2 + y^2 = u^2$. Hence, again by our earlier solution of $x^2 + y^2 = z^2$, we get,

$$v = 2rs,$$
$$y = r^2 - s^2,$$
$$u = r^2 + s^2,$$

where r and s have no common factor.

Substituting $2k$ for v,

$$2k = 2rs,$$
$$k = rs.$$

But we had above $k = m^2$. Therefore,

$$m^2 = rs.$$

Therefore by VII,

$$r = p^2,$$
$$s = q^2$$

where p and q are positive integers.

But we had also $u = n^2$, and $u = r^2 + s^2$. Therefore

$$n^2 = p^4 + q^4.$$

Now n is less than or equal to u (since $u = n^2$), and u is less than or equal to u^2, and u^2 is less than z (since $z = u^2 + v^2$). Therefore n is less than z, and n^2 is less than z^2.

We started by supposing that z^2 was the smallest perfect square expressible as the sum of two fourth powers. But we now have $n^2 = p^4 + q^4$ and n^2 is less than z^2. This is a contradiction. So our conclusion must be that no perfect square can be expressed as the sum of two fourth powers.

From this follows as already pointed out the truth of Fermat's Last Theorem for the case $n = 4$; also, as the reader will see, for the cases $n = 8, 16, 32, 64, \ldots$.

Hence as a corollary we have the well known remark that if Fermat's Last Theorem could be proved for all prime values of n greater than 2, its truth would follow for all values of n.

Finally we suggest a further exercise to the reader: Prove that the Diophantine equation $x^4 - y^4 = z^2$ has no solution in which x, y, z are all different from 0.

THE RICHARD PARADOX[1]
(1934)

A system of symbolic logic must begin with a list of undefined symbols, a list of formal axioms, and a list of rules of inference.

Let us call any finite sequence of the undefined symbols of the system a *formula*. Then each of the formal axioms is a formula. And each of the rules of inference states an operation which enables us, out of given formulas, to obtain new ones. The theorems of the system are the formulas which can be obtained from the formal axioms by a finite number of applications of the rules of inference.

It would seem natural to require that the list of undefined symbols and the list of formal axioms should each be finite, but, as a matter of fact, in most of the systems of symbolic logic which have actually been proposed, one or both of these lists is enumerably infinite.

Now it is well known that in any particular system the set of all formulas is enumerable. For we may arrange the list of undefined symbols in a fixed order, and then define the *grade* of a formula to be the least positive integer n, such that the formula does not contain more than n symbols and does not contain any symbol beyond the nth in the list of undefined symbols. The set of all formulas of a particular grade is then always finite. And therefore, after fixing upon a rule for ordering formulas of the same grade (as is easily done), we can enumerate the set of all formulas by arranging them in the order of their grades.

For convenience, let us use the phrase *function of positive integers* to mean a single-valued function of one variable, $f(x)$, which takes on a value which is a positive integer, whenever x takes on a value which is a positive integer. We can prove, by a familiar argument, that the set of all functions of positive integers is not enumerable. For, if $f_1(x), f_2(x), f_3(x), \cdots$ is any enumeration of functions of positive integers, then $1 + f_x(x)$ is a function of positive integers not included in the enumeration.

Since, in any system of symbolic logic, the set of all formulas is enumerable, whereas the set of all functions of positive integers is not enumerable, it seems to follow that, in the case of any system of symbolic logic, there exists a function of positive integers such that there is no formula which stands for it. And surely the existence of a function of positive integers which has no representation as a formula in the system means that the system is inadequate even for elementary number theory.

The Richard paradox can be said to consist in the following problem. How is it possible that a system of symbolic logic, in which the set of all formulas is enumerable, should be adequate for any branch of mathematics which deals with the members of a non-enumerable set (in particular for elementary number theory)?

Given a system of symbolic logic, let us try to construct the function of positive integers such that there is no formula in the system that stands for it. What we must do is first to enumerate all formulas, by the method which we have described, and then, going through this enumeration, to pick out in order those formulas which stand

---

Originally published in ***American Mathematical Monthly***, vol. 41 (1934), pp. 356–361.

[1] An address delivered at the meeting of the Mathematical Association in Cambridge, Mass., Dec. 30, 1933.

for functions of positive integers. The result is an enumeration of all formulas which stand for functions of positive integers. And if we let $f_n(x)$ be the function of positive integers represented by the nth formula in this enumeration, then $1 + f_x(x)$ is the function of positive integers such that there is no formula in the system that stands for it.

But this function $1 + f_x(x)$ is not, in general, defined in such a way that it is always possible to calculate its value for a given positive integer x. For, in the process of going through the list of all formulas and picking out those which stand for functions of positive integers, we may at some stage find a formula about which we do not know whether or not it stands for a function of positive integers. For example, we may find a formula whose intuitive meaning is, "The least positive integer n, greater than x, such that the equation $u^n + v^n = w^n$ has a solution in positive integral values of u, v, w." And we could not determine whether this formula stood for a function of positive integers without first proving, or disproving, Fermat's last theorem. Indeed, to be sure of always being able to determine whether a given formula stands for a function of positive integers, we must have discovered a method of procedure which would enable us to solve any problem of number theory whatever. Therefore the infinite sequence (about which we have been talking) of all formulas which stand for functions of positive integers almost certainly is not such an infinite sequence that it is possible to calculate as many terms of it as we please. And therefore the function $1 + f_x(x)$ has not been defined in a way which could be called constructive, but has merely been proved by an indirect argument to exist.

Now a particular system of symbolic logic could be adjudged inadequate only in the presence of a particular function of positive integers, which could be defined intuitively, but has no formula in the system to stand for it. We require of our formal system merely that it shall be adequate to define any function of positive integers which can be defined intuitively. And hence an existence proof which cannot be supported with an effective construction has no significance for our present problem.

Hence it appears to be possible that there should be a system of symbolic logic containing a formula to stand for every definable function of positive integers, and I fully believe that such systems exist.

But surely a function of positive integers which cannot be defined by any means whatever is no function of positive integers at all. If you agree, then it seems to follow that the non-enumerable set of all functions of positive integers can be put into one-to-one correspondence with a subset of the enumerable set of all formulas of an adequate symbolic logic. In fact, we are presented with the alternative of supposing that there is no adequate system of symbolic logic or of supposing that an enumerable set can contain a non-enumerable subset. Of the two alternative suppositions the latter is clearly preferable and (as we have seen) by no means untenable.

Let us turn, however, to another aspect of the problem of the possible adequacy of a system of symbolic logic. In order that a system be adequate it is necessary not only that it contain a formula to stand for every function of positive integers, but also that, in the case of every formula f which stands for a function of positive integers, the formal theorem $N(x) \supset_x N(f(x))$ shall be provable; where $N(x)$ is the formula whose intuitive meaning is "x is a positive integer," and $\supset_x$ is the formula which stands for the relation of implication between propositional functions, so that $N(x) \supset_x N(f(x))$ is to be read, "x is a positive integer implies that $f(x)$ is a positive integer."

But in the case of any system of symbolic logic the set of all provable theorems is enumerable.

This is most easily seen in the case of a system of the simplest sort, for which the number of formal axioms and the number of rules of inference are alike finite.[2] For in such a case the number of theorems provable in n steps is finite, for any fixed value of n. By inspection of the rules of inference and the formal axioms we can obtain a complete list of the theorems provable in one step. And, in general, by inspection of the rules of inference, the formal axioms, and the finite list of theorems provable in no more than $n-1$ steps, we can obtain a complete list of the theorems provable in n steps. Hence we may enumerate all the theorems of the system provable in n steps. Hence we may enumerate all the theorems of the system by enumerating first the formal axioms, then the theorems provable in one step, then the theorems provable in two steps, and so on.

In the case of a system of a more complicated sort, for which the number of rules of inference, or the number of formal axioms, or both, are infinite, an evident modification of the foregoing method will still enable us to obtain an enumeration of all the formal theorems.

Out of this enumeration of all theorems select those which have the form $N(x) \supset_x N(f(x))$. This gives an enumeration $N(x) \supset_x N(f_1(x)), N(x) \supset_x N(f_2(x)), N(x) \supset_x N(f_3(x)), \cdots$. And hence we obtain an enumeration $f_1(x), f_2(x), f_3(x), \cdots$ of all formulas about which we can prove the formal theorem that they are functions of positive integers; whereas the set of all (intuitively definable) functions of positive integers is not enumerable.

Since the enumeration of all formal theorems is effective and since there is a uniform procedure by which we can recognize whether any given formula has the form $N(x) \supset_x N(f(x))$, it follows that in the case of any particular system $f_1(x), f_2(x), f_3(x), \cdots$ to as many terms as we care to. Hence we cannot escape from the present dilemma in the same way that we did before.

Since the set of all formulas about which we can prove the formal theorem that they are functions of positive integers is (effectively) enumerable, while the set of all functions of positive integers is not enumerable, we conclude, either that there is some function of positive integers which is definable intuitively but about which we cannot prove the formal theorem that it is a function of positive integers, or else that there is some formula about which we can prove the formal theorem that it is a function of positive integers but which on the basis of the intuitive meanings given to our undefined terms does not stand for a function of positive integers. That is, briefly, every system of symbolic logic either is inadequate to prove all theorems which are intuitively true or else suffices to prove theorems which are intuitively false.

Or, if we prefer not to assume the existence of an intuitive logic which is right in an absolute sense, then it is sufficient to observe that in any system of symbolic logic not hopelessly inadequate there would be a formal equivalent of the intuitive argument which we have just set forth, and hence that this system of symbolic logic would contain the formal theorem that this same system regarded objectively, was either insufficient

[2]We assume that with a fixed premise or premises and a fixed rule of inference the formula obtained as conclusion is unique (a suggestion due to J. B. Rosser). This property by no means holds of all systems which have actually been proposed. But at the possible cost of increasing the number of rules of inference, it can always be made to hold, without essentially altering the system.

or over-sufficient. This means that the assumption that this system of symbolic logic was a true and complete representation of what is logically correct would defeat itself.

This, of course, is a deplorable state of affairs. It plainly implies that the whole program of the mathematical logician is futile.

For in the presence of such a situation not only is it impossible to obtain a single set of postulates which would lead to all mathematics (as, for example the authors of *Principia Mathematica* would do), but it is even impossible to obtain a set of postulates adequate to a particular branch of mathematics, such as number theory, analysis, or Euclidean plane geometry. That is, provided we assume, as I think we must, that a satisfactory formalization of, say, Euclidean plane geometry, means formalizing, not only the geometric terms, such as "point" and "line," which appear in its propositions, but also the logical terms, such as "if," "and," "is."

Indeed, if there is no formalization of logic as a whole, then there is no exact description of what logic is, for it is in the very nature of an exact description that it implies a formalization. And if there is no exact description of logic, then there is no sound basis for supposing that there is such a thing as logic.

Under these circumstances, a definition, for instance of Euclidean plane geometry, as that body of propositions which follows logically from a certain set of axioms, is altogether vague and unsatisfactory, because any attempted definition of the adverb "logically" is necessarily incomplete. We are led to despair of the currently accepted search for mathematical rigor, which amounts essentially to an appeal from the realm of spatial and other intuitions to the realm of logic.

Fortunately, however, there is a way out of this condition of nihilism. The theorem which led us to such pessimistic conclusions does not really apply to all systems of symbolic logic but only to systems which satisfy certain conditions. And one of these conditions is, either that there shall be a unique symbol for implication between propositional functions, or that there shall be a set of symbols for implication and an effective way by which we can always determine whether a given formula is one of the symbols for implication. For, in the contrary case, there would be no effective way of picking out from a list of theorems those which had the form $N(x) \supset_x N(f(x))$, and hence we could escape from our second dilemma in the same way that we did from our first one.

Therefore we seek a system of symbolic logic in which the notion of implication between propositional functions is obtained by definition, and in which there are a variety of notions of implication, obtainable by different definitions. In the case of each definition we desire that it shall be possible by an intuitive argument to prove the character of the defined symbol as an implication symbol. But there shall be no uniform means of determining whether a given formula is an implication symbol.

A system of this sort not only escapes our unpleasant theorem that it must be either insufficient or oversufficient, but I believe that it escapes the equally unpleasant theorem of Kurt Gödel to the effect that, in the case of any system of symbolic logic which has a claim to adequacy, it is impossible to prove its freedom from contradiction in the way projected in the Hilbert program. This theorem of Gödel is, in fact, more closely related to the foregoing considerations than appears from what has been said.

As I speak, I have in mind a particular set of postulates for symbolic logic, whose freedom from contradiction can be proved, and which lead to a non-enumerable multiplicity of definitions of implication, in the manner we desire.

It seems probable that the system of logic which results from these postulates is adequate at least for elementary number theory, but how far it is adequate for analysis there is at present no safe basis for conjecture.

Apparently, however, in view of the theorem of Gödel, and of the difficulties arising in connection with the Richard paradox, a system of symbolic logic of this kind is the most general which can be regarded as satisfactory from our present point of view. If it be true that no system of this kind can lead to analysis, then it seems to follow that the indictment against the soundness of analysis which is contained in the Richard paradox must be allowed to stand.

A PROOF OF FREEDOM FROM CONTRADICTION (1935)

1. *A System of Logic.*—We take as undefined the symbols {, }, (,), λ, [,], δ and an infinite list of variables $a, b, c, \ldots$. Following previous usage,[1] with minor changes, we define a *formula* to be a finite sequence of undefined symbols, and define the terms *well formed, free, bound*, by induction, as follows. The symbol δ standing alone is a well-formed formula; any variable $\mathbf{x}$ standing alone is a well-formed formula and the occurrence of $\mathbf{x}$ in it is an occurrence as a free variable in it; if the formulas $\mathbf{M}$ and $\mathbf{N}$ are well formed, $\{\mathbf{M}\}(\mathbf{N})$ is well formed, and an occurrence of a variable $\mathbf{x}$ as a free (bound) variable in $\mathbf{M}$ or in $\mathbf{N}$ is an occurrence of $\mathbf{x}$ as a free (bound) variable in $\{\mathbf{M}\}(\mathbf{N})$; if the formula $\mathbf{R}$ is well formed and contains an occurrence of $\mathbf{x}$ as a free variable in $\mathbf{R}$, then $\lambda\mathbf{x}[\mathbf{R}]$ is well formed, any occurrence of $\mathbf{x}$ in $\lambda\mathbf{x}[\mathbf{R}]$ is an occurrence of $\mathbf{x}$ as a bound variable in $\lambda\mathbf{x}[\mathbf{R}]$, and an occurrence of a variable $\mathbf{y}$, other than $\mathbf{x}$, as a free (bound) variable in $\mathbf{R}$ is an occurrence of $\mathbf{y}$ as a free (bound) variable in $\lambda\mathbf{x}[\mathbf{R}]$.

A formula $\{\mathbf{F}\}(\mathbf{X})$ is abbreviated as $\mathbf{F}(\mathbf{X})$ in all cases where $\mathbf{F}$ is or is represented by a single symbol. A formula $\{\{\mathbf{F}\}(\mathbf{X})\}(\mathbf{Y})$ is abbreviated as $\{\mathbf{F}\}(\mathbf{X}, \mathbf{Y})$, or, if $\mathbf{F}$ is or is represented by a single symbol, as $\mathbf{F}(\mathbf{X}, \mathbf{Y})$. And $\{\{\{\mathbf{F}\}(\mathbf{X})\}(\mathbf{Y})\}(\mathbf{Z})$ is abbreviated as $\{\mathbf{F}\}(\mathbf{X}, \mathbf{Y}, \mathbf{Z})$, or as $\mathbf{F}(\mathbf{X}, \mathbf{Y}, \mathbf{Z})$, and so on. A formula $\lambda\mathbf{x}_1[\lambda\mathbf{x}_2[\ldots\lambda\mathbf{x}_n[\mathbf{R}]\ldots]]$ is abbreviated as $\lambda\mathbf{x}_1\mathbf{x}_2\ldots\mathbf{x}_n[\mathbf{R}]$. If $\mathbf{F}$ is or is represented by a single symbol, $\{\mathbf{F}\}(\lambda\mathbf{x}[\mathbf{M}])$ is abbreviated as $\mathbf{F}\mathbf{x}[\mathbf{M}]$. Whenever possible without ambiguity, brackets [ ] are omitted, whether the brackets are the undefined symbols [ ], or whether they are themselves part of an abbreviation. A dot occurring in a formula stands for omitted brackets extending from the position of the dot forward the maximum distance that is consistent with the formula's being well formed. And when omitted brackets are not replaced by a dot, they are understood to extend forward the minimum distance.

Following Kleene (1934), we reserve heavy type letters to represent undetermined formulas in metamathematical discussions, and adopt the conventions that each heavy type letter shall represent a well-formed formula and each set of symbols standing apart which contains a heavy type letter shall represent a well-formed formula. A formula is said to be in *λ-normal form* if it is well formed and has no part of the form $\{\lambda\mathbf{x}\mathbf{M}\}(\mathbf{N})$. A formula is said to be in *normal form* if it is in λ-normal form and has no part which has the form $\delta(\mathbf{M}, \mathbf{N})$ and which contains no occurrence of any variable as a free variable in $\delta(\mathbf{M}, \mathbf{N})$.

The expression $S^{\mathbf{x}}_{\mathbf{N}}\mathbf{M}|$ is used to stand for the result of substituting $\mathbf{N}$ for $\mathbf{x}$ throughout $\mathbf{M}$.

We adopt a single formal postulate, $\lambda f x \,.\, f(f(x))$, and seven rules of procedure:

I. To replace any part $\lambda\mathbf{x}\mathbf{R}$ of a formula by $\lambda\mathbf{y}S^{\mathbf{x}}_{\mathbf{y}}\mathbf{R}|$, where $\mathbf{y}$ is any variable which does not occur in $\mathbf{R}$.

Originally published in the ***Proceedings of the National Academy of Sciences***, vol. 21 (1935), no. 5, pp. 275–281.

[1] Alonzo Church, *A Set of Postulates for the Foundation of Logic*, ***Ann. Math.***, 2nd ser., vol. 33 (1932), pp. 346–366, and *A Set of Postulates for the Foundation of Logic* (Second Paper), ***Ann. Math.***, vol. 34 (1933), pp. 839–864. S. C. Kleene, *Proof by Cases in Formal Logic*, ***Ann. Math.***, 2nd ser., vol. 35, (1934), pp. 529–544. We shall refer to the latter paper as Kleene 1934.

II. To replace any part $\{\lambda \mathbf{x}\mathbf{M}\}(\mathbf{N})$ of a formula by $S^{\mathbf{x}}_{\mathbf{N}}\mathbf{M}|$, provided that the bound variables in **M** are distinct both from **x** and from the free variables in **N**.

III. To replace any part $S^{\mathbf{x}}_{\mathbf{N}}\mathbf{M}|$ (not immediately following λ) of a formula by $\{\lambda \mathbf{x}\mathbf{M}\}(\mathbf{N})$, provided that the bound variables in **M** are distinct both from **x** and from the free variables in **N**.

IV. To replace any part $\delta(\mathbf{M}, \mathbf{N})$ of a formula by $\lambda fx \,.\, f(f(x))$, where **M** and **N** are in normal form and contain no free variables, and **M** conv-I **N**.

V. To replace any part $\delta(\mathbf{M}, \mathbf{N})$ of a formula by $\lambda fx \,.\, f(x)$, where **M** and **N** are in normal form and contain no free variables, and it is not true that **M** conv-I **N**.

VI. To replace any part $\lambda fx \,.\, f(f(x))$ of a formula by $\delta(\mathbf{M}, \mathbf{N})$, where **M** and **N** are in normal form and contain no free variables, and **M** conv-I **N**.

VII. To replace any part $\lambda fx \,.\, f(x)$ of a formula by $\delta(\mathbf{M}, \mathbf{N})$, where **M** and **N** are in normal form and contain no free variables, and it is not true that **M** conv-I **N**.

We are here using the notation **M** *conv-I* **N** to mean that **N** is obtainable from **M** by a sequence of applications of Rule I. It should be observed that, given any two formulas, **M** and **N**, it is constructively possible to determine whether **M** conv-I **N**.

A sequence of applications of Rules I–VII will be called a *conversion*, a sequence of applications of Rules I–III a *λ-conversion*, and a sequence of applications of Rules IV–VII a *δ-conversion*. It will be said that **M** is *convertible* (*λ-convertible, δ-convertible*) into **N** if **N** is obtainable from **M** by a conversion (λ-conversion, δ-conversion). And the phrase *is convertible into* will be abbreviated as *conv*. A *reduction* is a conversion, each step of which is an application of one of the rules I, II, IV, V, and in which one and only one step is an application of a rule other than Rule I. The formula **Y** is said to be *a normal form* (*a λ-normal form*) *of* the formula **X** if **Y** is in normal form (λ-normal form) and **X** conv **Y** (**X** λ-conv **Y**).

The following theorem may be proved by induction on the length of the formula:

THEOREM I. *If a formula is in normal form no reduction of it is possible.*

The following will be proved in a forthcoming paper:[2]

THEOREM II. *If a formula has a normal form, this normal form is unique to within applications of Rule I, and any sequence of reductions of the formula must (if continued) terminate in the normal form.*

And from this follows readily:

THEOREM III. *If a well-formed part of a formula has no normal form, the whole formula likewise has no normal form.*

2. *Equality, Negation, Logical Product.*—We introduce the following definitions and abbreviations, where the arrow is to be read, "stands for," or, "is an abbreviation for."

$1 \rightarrow \lambda fx \,.\, f(x)$,
$2 \rightarrow \lambda fx \,.\, f(f(x))$, and so on, for all the positive integers.
$\sim \rightarrow \lambda x \,.\, 6 - [\delta(x, 1) + 2\delta(x, 2)]$.
$\sim[\mathbf{P}] \rightarrow \sim(\mathbf{P})$.
$\& \rightarrow \lambda xy \,.\, \sim \,.\, 6 - \min(\delta(x, 1) + 2\delta(x, 2), \delta(y, 1) + 2\delta(y, 2))$.
$[\mathbf{P}][\mathbf{Q}] \rightarrow \&(\mathbf{P}, \mathbf{Q})$.

---

[2]Alonzo Church and J. B. Rosser, *Some Properties of Conversion*. [*Editor's note*: This paper was published in ***Transactions of the American Mathematical Society***, vol. 39 (1936), pp. 472–482.]

Here the functions denoted by $+$, $-$, min and multiplication, denoted by juxtaposition, as in $2\delta(x, 2)$, are supposed defined as by Kleene. Where necessary to avoid confusion with the abbreviation for $\&(\mathbf{P}, \mathbf{Q})$, multiplication will be denoted by $\times$. If $\mathbf{m}$ and $\mathbf{n}$ are positive integers, $\mathbf{m} + \mathbf{n}$ and $\mathbf{mn}$ are respectively the sum and the product of $\mathbf{m}$ and $\mathbf{n}$ in the ordinary sense, $\min(\mathbf{m}, \mathbf{n})$ is the lesser of $\mathbf{m}$ and $\mathbf{n}$, and $\mathbf{m} - \mathbf{n}$ is the difference of $\mathbf{m}$ and $\mathbf{n}$ in the ordinary sense if $\mathbf{m} > \mathbf{n}$ and is 1 if $\mathbf{m} \leqq \mathbf{n}$.[3]

Using the preceding definitions, and Theorem II, we obtain the following:

THEOREM IV. *If* $\mathbf{P}$ *conv* 2 *then* $\sim\mathbf{P}$ *conv* 1. *If* $\mathbf{P}$ *conv* 1, *then* $\sim\mathbf{P}$ *conv* 2. *If* $\mathbf{P}$ *has a normal form different from* 1 *and* 2, *then* $\sim\mathbf{P}$ *conv* 3.

THEOREM V. *If* $\mathbf{P}$ *conv* 2 *and* $\mathbf{Q}$ *conv* 2, *then* $\&(\mathbf{P}, \mathbf{Q})$ *conv* 2. *If* $\mathbf{P}$ *conv* 1 *and* $\mathbf{Q}$ *conv* 2, *or if* $\mathbf{P}$ *conv* 2 *and* $\mathbf{Q}$ *conv* 1, *or if* $\mathbf{P}$ *conv* 1 *and* $\mathbf{Q}$ *conv* 1, *then* $\&(\mathbf{P}, \mathbf{Q})$ *conv* 1. *If* $\mathbf{P}$ *and* $\mathbf{Q}$ *have normal forms, and one of the two normal forms is different from* 1 *and* 2, *then* $\&(\mathbf{P}, \mathbf{Q})$ *conv* 3.

In the intuitive interpretation of our formal system, we identify the formula 1 with the truth value *false*, and the formula 2 with the truth value *true*. This identification is artificial, especially since the same two formulas are to be identified with the positive integers *one* and *two*, but it is apparently harmless. The viewpoint taken is that formal logic requires nothing of the ideas *true* and *false* except that they be distinct, which property the formulas 2 and 1 do possess. On this basis, the formulas δ, $\sim$ and $\&$ are identified respectively as equality, negation and logical product.

The theorem that our system is free of contradiction now follows immediately.

THEOREM VI. *There is no formula* $\mathbf{P}$ *such that both* $\mathbf{P}$ *and* $\sim\mathbf{P}$ *are provable.*

For suppose that $\mathbf{P}$ is provable. Then, since the process of conversion is reversible, $\mathbf{P}$ conv 2. Therefore, by Theorem IV, $\sim\mathbf{P}$ conv 1. Therefore, by Theorem II, it is not true that $\sim\mathbf{P}$ conv 2. Therefore $\sim\mathbf{P}$ is not provable.

3. *Combinations.*—We define:[4]

$I \rightarrow \lambda x \,.\, x.$
$J \rightarrow \lambda f x y z \,.\, f(x, f(z, y)).$
$T \rightarrow J(I, I).$

The term *combination* is defined by induction as follows. The formulas I, J, δ and any variable standing alone are combinations. If $\mathbf{M}$ and $\mathbf{N}$ are combinations, $\{\mathbf{M}\}(\mathbf{N})$ is a combination.

We define an intuitive operator $\lambda_{\mathbf{x}}|$ such that, if $\mathbf{M}$ is a combination containing $\mathbf{x}$ as a free variable, $\lambda_{\mathbf{x}}\mathbf{M}|$ is another combination, which does not contain $\mathbf{x}$ as a free variable. The definition of this operator is by induction, as follows. $\lambda_{\mathbf{x}}\mathbf{x}|$ is I. If $\mathbf{A}$ contains $\mathbf{x}$ as a free variable and $\mathbf{F}$ does not, $\lambda_{\mathbf{x}}\{\mathbf{F}\}(\mathbf{A})|$ is $J(T, \lambda_{\mathbf{x}}\mathbf{A}|, J(I, \mathbf{F}))$. If $\mathbf{F}$ contains $\mathbf{x}$ as a free variable and $\mathbf{A}$ does not, $\lambda_{\mathbf{x}}\{\mathbf{F}\}(\mathbf{A})|$ is $J(T, \mathbf{A}, \lambda_{\mathbf{x}}\mathbf{F}|)$. If both $\mathbf{F}$ and $\mathbf{A}$ contain $\mathbf{x}$ as a free variable, $\lambda_{\mathbf{x}}\{\mathbf{F}\}(\mathbf{A})|$ is $J(T, T, J(I, J(T, T, J(T, \lambda_{\mathbf{x}}\mathbf{A}|, J(T, \lambda_{\mathbf{x}}\mathbf{F}|, J)))))$.

---

[3]Explicit formal definitions are given by S. C. Kleene, *A Theory of Positive Integers in Formal Logic*, ***Amer. Jour. Math.***, vol. 57 (1935), pp. 153–173. This paper, and a second part of it, forthcoming in the ***Amer. Jour. Math.***, will be referred to as Kleene 1935.

[4]Cf. J. B. Rosser, *A Mathematical Logic without Variables*, ***Ann. Math.***, 2nd ser., vol. 36 (1935), pp. 127–150, and a second part of the same paper, forthcoming in the [***Duke Math. J.***, vol. 1 (1935), pp. 328–355]. Cf. also Kleene 1934, p. 536.

The *combination belonging* to a formula is defined by induction as follows. The combination belonging to a formula which consists of δ or a variable standing alone is the same as the formula itself. The combination belonging to $\{\mathbf{M}\}(\mathbf{N})$ is $\{\mathbf{M}'\}(\mathbf{N}')$ where $\mathbf{M}'$ and $\mathbf{N}'$ are the combinations belonging to **M** and **N** respectively. The combination belonging to $\lambda\mathbf{x}[\mathbf{R}]$ is $\lambda_{\mathbf{x}}\mathbf{R}'|$, where $\mathbf{R}'$ is the combination belonging to **R**.

Obviously, every well-formed formula has a unique combination belonging to it. And it is readily proved that the combination belonging to a formula is λ-convertible into the formula. Also that the same combination belongs to two formulas **A** and **B** if and only if **A** conv-I **B**. Given a combination, it is constructively possible to determine whether it belongs to a formula and, if so, to obtain the formula.

4. *Metads.*—We introduce the abbreviation, $[\mathbf{a}, \mathbf{b}] \rightarrow \lambda f \mathrel{.} f(\mathbf{a}, \mathbf{b})$. And we modify the notion of *metad* employed by Kleene (1935 Part II), defining this term, and the term *rank* of a metad, by induction as follows. The formulas 1,2,3, ... (the positive integers) are metads of rank 1, and the formulas obtainable from them by conversion are metads of rank 1. If **a** and **b** are metads, $[\mathbf{a}, \mathbf{b}]$, as well as any formula obtainable from $[\mathbf{a}, \mathbf{b}]$ by conversion, is a metad whose rank is greater by one than the greater of the ranks of the two metads **a** and **b**. The *metad of* a combination is defined by induction as follows. The metad of I is 1, the metad of J is 2, the metad of δ is 3 and the metads of the variables $a, b, c, \ldots$ are 4,5,6, ... respectively. The metad of $\{\mathbf{A}\}(\mathbf{B})$ is $[\mathbf{a}, \mathbf{b}]$, where **a** is the metad of **A** and **b** is the metad of **B**. The *metad belonging* to a formula is defined to be the metad of the combination belonging to the formula.

Using methods of formal definition which are due to Kleene (1935 Part II) and, in the case of thm, the combinatory analysis of λ-conversion which is due to Rosser (loc. cit.), we are able to define formulas having the following properties.

$\mathfrak{M}_1$ and $\mathfrak{M}_2$ such that, if $[\mathbf{a}, \mathbf{b}]$ is a metad, $\mathfrak{M}_1([\mathbf{a}, \mathbf{b}])$ conv **a** and $\mathfrak{M}_2([\mathbf{a}, \mathbf{b}])$ conv **b**, and if **a** is a metad of rank 1, $\mathfrak{M}_1(\mathbf{a})$ conv 1 and $\mathfrak{M}_2(\mathbf{a})$ conv **a**.

$\mathfrak{G}$ such that, if **a** is a metad belonging to a formula **A** which contains no free variables, $\mathfrak{G}(\mathbf{a})$ conv **A**.

met such that, if **A** is a formula which contains no free variables and has a normal form, met(**A**) is convertible into the metad belonging to the normal form of **A**.

$\mathfrak{n}$ such that, if **a** is the metad belonging to one of the formulas $\lambda fx \mathrel{.} f(x)$, $\lambda fx \mathrel{.} f(f(x))$, $\lambda fx \mathrel{.} f(f(f(x)))$, ..., then $\mathfrak{n}(\mathbf{a})$ conv 2, and if **a** is any other metad, then $\mathfrak{n}(\mathbf{a})$ conv 1.

thm, a formula which enumerates the metads belonging to provable formulas, in the sense that in the infinite sequence, thm(1), thm(2), thm(3), ... , every term is a metad belonging to a provable formula, and every metad which belongs to a provable formula occurs at least once.

5. *Quantifiers.*—We define:

$N \rightarrow \lambda x \mathrel{.} \mathfrak{n}(\mathrm{met}(x))$.

$\iota \rightarrow \lambda f \mathfrak{G}(\mathfrak{M}_2(\mathrm{thm}(\mathfrak{p}(\lambda n \mathrel{.} \delta(f, \mathfrak{G}(\mathfrak{M}_1(\mathrm{thm}(n)))), 1))))$.

$\Sigma \rightarrow \lambda f \mathrel{.} f(\iota(f))$.

$\Xi \rightarrow \lambda fgx \mathrel{.} \delta(f, \mathfrak{G}(\mathfrak{M}_1(\mathrm{thm}(\mathfrak{p}(\lambda n \mathrel{.} \sim[\sim\delta(f, \mathfrak{G}(\mathfrak{M}_1(\mathrm{thm}(n)))) \mathrel{.}$
$\sim\delta(g, \mathfrak{G}(\mathfrak{M}_1(\mathrm{thm}(n))))] \mathrel{.} \delta(x, \mathfrak{G}(\mathfrak{M}_2(\mathrm{thm}(n)))), 1)))))$.

Here $\mathfrak{p}$ is the Kleene $\mathfrak{p}$-function (defined in Kleene 1935 Part II), which has the property that, if, for every positive integer **m**, **F**(**m**) conv 1 or 2, and **r** is the least positive integer **m**, $\geq$ **n**, such that **F**(**m**) conv 2, then $\mathfrak{p}$(**F**, **n**) conv **r**.

The formula N has the property that, if **X** has a normal form, N(**x**) conv 2 or 1 according to whether **X** is or is not a positive integer. The formula ι has the property that, if there is any **X** such that **F**(**X**) conv 2, then ι(**F**) is such an **X**, and in the contrary case ι(**F**) has no normal form. Consequently Σ(**F**) conv 2 if there is any **X** such that **F**(**X**) conv 2, and in the contrary case Σ(**F**) has no normal form. The formula Ξ has the property that, if **F**(**X**) conv 2 and **G** has a normal form but **G**(**X**) is not convertible into 2 then Ξ(**F**, **G**, **X**) conv 2, and if **G**(**X**) conv 2 and **F** has a normal form but **F**(**X**) is not convertible into 2 then Ξ(**F**, **G**, **X**) conv 1, and if neither **F**(**X**) nor **G**(**X**) is convertible into 2 then Ξ(**F**, **G**, **X**) has no normal form.

Making use of metads and of the $\mathfrak{p}$-function, it is possible to define a set of formulas $\Pi_1, \Pi_2, \Pi_3, \ldots, \Pi_\omega, \Pi_{\omega+1}, \ldots$ related to the universal quantifier as follows. If **G**(**x**) is found among the consequences of **F**(**x**) where **x** is a variable, then Π_α(**F**, **G**) conv 2, and in any other case Π_α(**F**, **G**) has no normal form.

For this purpose, of course, the notion of the consequences of a formula must be defined formally. It turns out, as might be expected in connection with the theorem of Gödel,[5] that there is no all inclusive definition of the notion of the consequences of a formula, but merely a sequence, extending into the second number class, of more and more inclusive definitions of the notion. The subscript α, in the symbol Π_α, refers to the different definitions of the notion of the consequences of a formula.

If α is of the second kind, the definition of Π_α has a special form, as may be illustrated in connection with the definition of Π_ω. It is possible to define a formula π such that, for every positive integer **n**, π(**n**) conv Π_n. Then, $\Pi_\omega \rightarrow \lambda f g \Sigma x \,.\, N(x) \,.\, \pi(x, f, g)$.

We also define:

$$\Pi^\alpha \rightarrow \Xi(\Pi_\alpha(N), \lambda f \Sigma x \,.\, N(x) \,.\, \sim f(x)).$$
$$\Sigma^\alpha \rightarrow \lambda f \sim . \,\Pi^\alpha x \,.\, \sim f(x).$$

If now we select out of the second number class a particular constructively defined ordinal α of the second kind, and identify Π^α and Σ^α respectively with the intuitive notions of the universal quantifier over the class of positive integers and the existential quantifier over the class of positive integers, it is believed that the system of logic with which we are dealing will then be found to be adequate to a portion of elementary number theory which can be described as the portion which does not require proofs by induction of higher than a certain order. In particular, if α is ω, proofs by induction of any finite order are possible. And it is believed that if α increases indefinitely, the order of induction possible increases indefinitely, so that we obtain, in a certain sense, a metamathematical proof of the freedom from contradiction of elementary number theory.

Of course, the considerations adduced by Gödel (loc. cit.) make it certain that the intuitive proof that Π^α has the properties of a universal quantifier cannot be formalized in terms of Π^α as the universal quantifier (although the intuitive proof probably can be formalized in terms of a quantifier of higher order). The proof of freedom from contradiction which we have outlined is believed, nevertheless, to have an interest on

[5]Kurt Gödel, *Über formal unentscheidbare Sätze der Principia Mathematica und verwandter Systeme I*, ***Monatsh. für Math. u. Phys.***, vol. 38 (1931), pp. 173–198.

the ground that it is easier to recognize a particular intuitive proof as valid than it is to recognize intuitively the general proposition that every proof formalizable in a given system is valid.

There is also, perhaps, the possibility that the methods here used might lead to a proof of the consistency of some portion of analysis.

QUINE ON LOGISTIC
(1935)

A System of Logistic. By Willard Van Orman Quine. Harvard University Press, 1934. x + 204 pp.

In this book is presented a system of symbolic logic based on that of Whitehead and Russell's *Principia Mathematica*, but involving a number of fundamental changes. The most important of these changes are: (1) the representation of functions of two or more variables as functions of one variable through the introduction, as an undefined term, of the operation of ordination, that is, the operation of combining two elements a and b into the ordered pair a, b; (2) the use of this same notion of ordination to replace the notion of predication, the proposition φa, obtained by predicating the propositional function φ of the argument a, being identified with the ordered pair φ, a; (3) the introduction in connection with the operation of abstraction, $\hat{}$, of a rule of inference, the rule of concretion, which takes the place of that tacit rule of *Principia* which, to speak somewhat inexactly, allows the substitution for φx, in any proved expression in which φ is a free variable, of any appropriate expression containing x; (4) a liberalization of the theory of types, by which the axiom of reducibility is rendered unnecessary; (5) the use of the notion of classial referent, introduced by an actual nominal definition, to replace almost entirely the clumsy descriptions introduced in *Principia* as incomplete symbols; (6) the introduction, under the name of congeneration, of the relation of implication between propositional functions, as an undefined term, out of which both the relation of implication between propositions and the universal and existential quantifiers are obtained by definition.

Quine's propositional functions have the property that equivalence implies equality, and for this reason he speaks of them as classes rather than as propositional functions. Nevertheless he uses them for the purposes for which propositional functions are used in *Principia* and in other systems, and hence, for the sake of comparison, we continue to call them propositional functions.

In regard to Quine's use of ordination, it is, of course, clear, as he points out, that the introduction as primitive ideas of an infinite number of different notions of predication, one for functions of one variable, another for functions of two variables, another for functions of three variables, and so on, is awkward and that it is therefore desirable to find some device by which functions of two or more variables can be regarded as special cases of functions of one variable. It is not so clear, however, that the introduction of the ordered pair as an undefined term is the best method of doing this. From some points of view the more natural and more elegant method is that of Schönfinkel,[1] under which a function of $n + 1$ variables is regarded as a function of one variable whose values are functions of n variables. For example, instead of what is ordinarily written $\varphi(a, b)$, Schönfinkel writes $(\varphi a)b$, where φa is regarded as a function which, when taken of the argument b, yields the proposition $(\varphi a)b$ and φ is regarded as a function which, when taken of the argument a, yields the function $\varphi(a)$. Of course, if the Schönfinkel device be adopted, it is necessary to introduce the

Originally published in ***Bulletin of the American Mathematical Society***, vol. 41 (1935), pp. 598–603.

[1] ***Mathematische Annalen***, vol. 92 (1924), pp. 305–316.

notion of predication for functions of one variable as a primitive idea, but the number of primitive ideas is not thereby increased, because the ordered pair a, b can then be defined, in terms of predication and abstraction, as $\hat{\varphi}((\varphi a)b)$, that is, as the class of relations which hold between a and b.

On the other hand, Quine's identification of the notion of predication with that of ordination not only raises the difficult philosophical problem of justifying the assumption that, if x is of one type higher than y, then any assertion about x and y can be construed as an assertion about the proposition x, y, but also introduces unnecessary formal complications, for example, in the rule of concretion. The use of predication as a primitive idea has the advantage that it neither accepts nor denies Quine's special philosophy concerning the nature of predication.

The use of the device of Schönfinkel just referred to is not incompatible with the theory of types, but it does require modification of the usage of *Principia* by which propositional functions are regarded as entities of an entirely different sort from other functions, modification at least to the extent of allowing that functions of one variable whose values are propositions, and functions of one variable whose values are propositional functions, are concepts sufficiently similar so that one notion of application, or predication, and one method of symbolizing this notion, are sufficient for both.

As a matter of fact, it is the contention of the present reviewer that the distinction between propositional functions and functions of other sorts is no more fundamental than, say, the distinction between functions of a real variable and functions of a complex variable, and that the one notion of application, or predication, should suffice for all functions of one variable. Quine, however, following *Principia*, has two notations for the application of a function, one for propositional functions and one for descriptive functions, and in the same way two notations for the operation of abstraction. Thus if M is an expression which contains x as a free variable and which takes on propositions as values when x takes on particular values, he uses $\hat{x}$M to denote the corresponding propositional function, and $\hat{x}$M, a to denote the result of application of this propositional function to the argument a. But if M takes on classes as values when x takes on particular values, then he uses

$$\hat{w}(\text{Ǝ}, \hat{\alpha}(\text{Ǝ}, \hat{x}(w = \alpha,\ x \cdot \alpha = \mathrm{M})))$$

to denote the corresponding function (his usage amounts to that), and

$$\hat{w}(\text{Ǝ}, \hat{\alpha}(\text{Ǝ}, \hat{x}(w = \alpha,\ x \cdot \alpha = \mathrm{M})))\text{‘}a$$

to denote the result of application of this function to the argument a.

The rule of inference of *Principia* referred to under (3) in the first paragraph above is (like the simple rule of substitution) entirely suppressed by the authors of that work, who use it repeatedly but make no mention of it. Hilbert and Ackermann[2] make this rule explicit, but their statement of it is inadequate. Quine's revised statement of the rule, on page 187 of his book, is perhaps adequate as applied to the system of *Principia*, but his statement of what the analogous rule would have to be for his own system is again inadequate, as can be seen, for example, by considering the proposition

$$\alpha, \hat{v}(\alpha, \iota\text{‘}v) \cdot \supset : \cdot \sim \cdot \alpha, \wedge : \supset \cdot \text{Ǝ}, \hat{w}(\alpha, \iota\text{‘}w)$$

[2] ***Grundzüge der theoretischen Logik***, p. 53.

and raising the question of substituting for the free variable α on the basis that α, x shall mean $x, \hat{y}(x, [y])$. The following is proposed as a correct statement of what this rule of inference should be for use in Quine's system (in order to make possible in Quine's system the equivalent of what Quine's (26) on page 187 makes possible in the system of *Principia*).

Let A be a significant expression, and let α be a variable whose occurrences as a free variable in A are occurrences as the first symbol of parts of A of the form α, U. We may list these parts as $\alpha, \mathrm{U}_1, \cdots, \alpha, \mathrm{U}_n$ where $\mathrm{U}_1, \cdots, \mathrm{U}_n$ are significant, and where the listing is in such an order that, if U_j contains α, U_i, then $i < j$. Let M be a significant expression in which x occurs as a free variable, and let M_1 stand for the result of substituting (in the sense of Quine, page 42) U_1 for x in M. Let A_1 be the expression obtained from A by substituting M_1 for the part α, U_1 and let $\alpha, \mathrm{U}_{12}, \cdots, \alpha, \mathrm{U}_{1n}$ be the parts of A_1 into which the parts $\alpha, \mathrm{U}_2, \cdots, \alpha, \mathrm{U}_n$ of A are transformed by the substitution. Let M_2 stand for the result of substituting U_{12} for x in M. Let A_2 be the expression obtained from A_1 by substituting M_2 for the part α, U_{12}, and let $\alpha, \mathrm{U}_{23}, \cdots, \alpha, \mathrm{U}_{2n}$ be the parts of A_2 into which the parts $\alpha, \mathrm{U}_{13}, \cdots, \alpha, \mathrm{U}_{1n}$ of A_1 are transformed by the substitution, and so on, until M_n and A_n are defined. If A is a proved expression, the rule which we are stating allows us to infer A_n, provided that A_n is a propositional expression (as defined below).[3]

The superior simplicity of the rule of concretion, and the advantage of avoiding the foregoing complicated rule by introducing the simpler one (as Quine does), is obvious. In fact, the effect is to analyze a complicated inference into a series of simpler inferences by substitution and concretion, as can be illustrated in connection with the example just mentioned, by substituting $\hat{x}(x, \hat{y}(x, [y]))$ for α in the proposition in question and then making a number of successive applications of the rule of concretion (four or five according to the order in which they are made). Unfortunately, as already remarked, the essential simplicity of the rule of concretion is partially obscured by the peculiar use of ordination as a substitute for predication.

Nowhere in Quine's book is there a definition of the word *significant*, which is used in his statement of the rule of substitution, but from scattered remarks about types and by observation of how the rule of substitution is actually used, it is possible to surmise what probably is meant by the word. Since this is a matter of some importance, especially in view of the fact that it is only through this word (or the related term *propositional expression*) that the theory of types enters the formal system at all, an explicit definition of *significant* is attempted here.

The four metamathematical (or "prosystematic") terms, *significant, classial expression, propositional expression, type*, must be defined simultaneously by induction, as follows. A variable standing alone is significant and may be assigned any type out of the scheme of types explained in Quine's second chapter; if assigned a type of the form $a!$, the variable is a classial expression, and if assigned a type of the form $a! \uparrow a$, it is a propositional expression. If M and N are significant and are assigned the types m and n, respectively, and if it is true of every variable x which occurs in both M and N that the same type was assigned to x in assigning the type m to M that was assigned to x in assigning the type n to N, then (M, N) is significant and must be assigned the type

[3]With appropriate modifications to adapt it to the notation of Quine, the statement of this rule is taken from a set of notes by S. C. Kleene on lectures of Kurt Gödel, which the reviewer has before him.

$m \uparrow n$; moreover, if m is $n!$, then (M, N) is a propositional expression. If M is assigned the type m and is a classial expression, then [M] is significant, is a classial expression, and must be assigned the type $m!$. If M is assigned the type m and is a propositional expression, and x is any variable, then $\hat{x}$M is significant, is a classial expression, *and must be assigned the type* $r!$ *where* r *is the type that was assigned to* x *in assigning the type* m *to* M, *or, if* x *does not occur in* M, *where* r *is any type whatever*. When no particular assignment of types to the parts of an expression is in question, the expression shall be called significant (a classial expression, a propositional expression) if types can be assigned to its parts so as to make it significant (a classial expression, a propositional expression).

Quine further requires that an expression set down as a postulate or theorem shall not be considered significant unless it is a propositional expression. But this seems to be an unnecessary complication of terminology, which could be avoided by no greater change than replacing the word "significant" by "a propositional expression" in the statement of the rule of substitution.

The italicized clause in the foregoing definition marks a sharp divergence of Quine's theory of types from that of *Principia Mathematica*; for if the analogy with the theory of types of *Principia* were preserved, $\hat{x}$M could not be of lower type than M.[4] It is true, of course, in *Principia*, that if α is a class then a proposition of the form $x \epsilon \alpha$ must be of type just one higher than the type of x, but it is to be remembered that this situation is brought about only with the aid of the axiom of reducibility, and that, in any case, the classes of *Principia* are incomplete symbols defined only contextually. Since Quine's $\hat{x}$M is not an incomplete symbol, a truer comparison of the two systems appears to be obtained if we compare expressions in Quine of the form $\hat{x}$M with the propositional functions of *Principia* rather than with the classes of *Principia*. And from this point of view it is seen that, without claiming to do so, Quine has really made an important modification in the theory of types, in a direction which seems to have been first suggested by F. P. Ramsey.[5]

This modification in the theory of types renders the axiom of reducibility unnecessary in the system of Quine. In particular, the difficulty in regard to the least upper bound of a bounded set of real numbers[6] disappears. For let real numbers be segments of rational numbers, and let λ be a bounded set of real numbers. Then $\epsilon`\lambda$ (replacing the $s`\lambda$ of *Principia*) is the least upper bound of λ, and is of the same type as the real numbers in the set λ.

Quine's statements of his rules of inference require a number of corrections. In his definition, in connection with the rule of substitution, of what he means by the process of substitution, it should be provided that when the bound variable is rewritten in E′ it should be made alphabetically distinct, not only from all variables in E, but also from all variables previously occurring in E′. In the rule of subsumption the proviso should be added that $[\alpha], \hat{x}(---)$ be a propositional expression; otherwise, since $x, \alpha \cdot \supset \cdot x, \alpha$ is provable, the rule could be used to infer $[\alpha], \hat{x}(x, \alpha \cdot \supset \cdot x, \alpha)$. In the rule of concretion it should be stipulated that $---$ and $\cdots$ be significant; otherwise the rule could be applied to $\hat{z}(\hat{x}\, x, y), z), \hat{t}(\exists, \hat{u}(u, t)$, taking $---$ to be $x, y), z$, and $\cdots$ to be $\hat{t}(\exists, \hat{u}(u, t)$.

[4]See ***Principia Mathematica***, 2d ed., vol. 1, p. 48 et seq., and introduction to the second edition, p. xxxix.

[5]***Proceedings of the London Mathematical Society***, (2), vol. 25 (1926), p. 362 et seq.

[6]See ***Principia Mathematica***, introduction to the second edition, pp. xliv–xlv.

Apparently a rule of inference allowing an alphabetical change of a bound variable in any theorem or postulate should be added to Quine's four rules, since such a rule of inference is used in the proof of 3.8 on page 83. Of course, it may be that such a rule of inference is unnecessary on the ground that the effect of it can be obtained by some succession of applications of the four rules of inference and the postulates. But if so, this should be explained.

The contention on page 51 that the use of z, x, y to replace $z, (x, y)$ is not to be construed as a definition, or abbreviation, is definitely untenable. For if z, x, y is to be regarded otherwise than as an abbreviated notation for $z, (x, y)$, the system of Quine is open to the same charge as that which he brings against the system of *Principia*, namely, that of being incompletely formalized and leaving lacunae to be bridged by the common sense of the reader. If z, x, y is not an abbreviation, and if the rule of substitution is to be taken literally, then we may substitute z, x for t in the proposition $t, y = u, \hat{p}(\sim p), (\sim \cdot v = v) :\supset y$ and so obtain $z, x, y = u, \hat{p}(\sim p), (\sim \cdot v = v) :\supset y$. In the opinion of the reviewer the remedy for this situation is to introduce the notation for an ordered pair as (x, y) instead of x, y then to use x, y and z, x, y as abbreviations for (x, y) and $(z, (x, y))$, respectively, whenever no ambiguity is thereby created, of course with the understanding that the rules of inference are applicable only to the unabbreviated form of an expression. In this way the parentheses in (x, y) would become as much a part of the formal system as the brackets in $[\alpha]$, and the notion of parentheses as an extra-formal convention would disappear.

The classial referent of x with respect to the relation α, denoted by $\alpha\text{‘}x$, is defined in such a way that its intuitive meaning is, "the class of all members of classes bearing the relation α to x." Thus, if there is one and only one class bearing the relation α to x, then $\alpha\text{‘}x$ denotes that class. Consequently the classial referent can be used as a description in any case where the thing described is a class, and it happens in the system of Quine that nearly everything worth describing is a class. The superiority of a nominal definition over the use of descriptions as incomplete symbols (as in *Principia*) requires no elaboration.

A considerable economy in the number of primitive ideas is effected by introducing the notion of congeneration, denoted by [ ], as a primitive idea. This notion is explained by Quine on the basis that $[\alpha]$ means the class of classes containing α. But it may also be thought of as implication between propositional functions, because, if α and β are propositional functions (classes), then $[\alpha], \beta$ is the proposition, "α implies β," expressed in *Principia* as $\alpha x \supset_x \beta x$. It is perhaps worth while to observe that [ ] is a propositional function of two variables, not in the sense of Quine, but in the sense of Schönfinkel, since, if α is a propositional function of one variable, $[\alpha]$ is a propositional function of one variable.

There is no slur on the invaluable pioneer work of Whitehead and Russell when it is said that their system is unsatisfactory from the viewpoints of formal definiteness and of mathematical elegance. The work of Quine is in both respects an important improvement over the system of *Principia*, and, although open to criticism in certain directions, is probably not too highly praised by Whitehead when he calls it, "A landmark in the history of the subject."

ALONZO CHURCH

AN UNSOLVABLE PROBLEM OF ELEMENTARY NUMBER THEORY

Preliminary Report

(1935)

205. Professor Alonzo Church: *An unsolvable problem of elementary number theory*. Preliminary report.

Following a suggestion of Herbrand, but modifying it in an important respect, Gödel has proposed (in a set of lectures at Princeton, N.J., 1934) a definition of the term *recursive function*, in a very general sense. In this paper a definition of *recursive function of positive integers* which is essentially Gödel's is adopted. And it is maintained that the notion of an effectively calculable function of positive integers should be identified with that of a recursive function, since other plausible definitions of effective calculability turn out to yield notions which are either equivalent to or weaker than recursiveness. There are many problems of elementary number theory in which it is required to find an effectively calculable function of positive integers satisfying certain conditions, as well as a large number of problems in other fields which are known to be reducible to problems in number theory of this type. A problem of this class is the problem to find a complete set of invariants of formulas under the operation of conversion (see abstract 41-5-204). It is proved that this problem is unsolvable, in the sense that there is no complete set of effectively calculable invariants. (Received March 22, 1935.)

Originally published in ***Bulletin of the American Mathematical Society***, vol. 41 (1935), pp. 332–333.

AN UNSOLVABLE PROBLEM OF ELEMENTARY NUMBER THEORY[1] (1936)

1. Introduction. There is a class of problems of elementary number theory which can be stated in the form that it is required to find an effectively calculable function f of n positive integers, such that $f(x_1, x_2, \cdots, x_n) = 2$[2] is a necessary and sufficient condition for the truth of a certain proposition of elementary number theory involving $x_1, x_2, \cdots, x_n$ as free variables.

An example of such a problem is the problem to find a means of determining of any given positive integer n whether or not there exists positive integers x, y, z, such that $x^n + y^n = z^n$. For this may be interpreted, required to find an effectively calculable function f, such that $f(n)$ is equal to 2 if and only if there exist positive integers x, y, z, such that $x^n + y^n = z^n$. Clearly the condition that the function f be effectively calculable is an essential part of the problem, since without it the problem becomes trivial.

Another example of a problem of this class is, for instance, the problem of topology, to find a complete set of effectively calculable invariants of closed three-dimensional simplicial manifolds under homeomorphisms. This problem can be interpreted as a problem of elementary number theory in view of the fact that topological complexes are representable by matrices of incidence. In fact, as is well known, the property of a set of incidence matrices that it represent a closed three-dimensional manifold, and the property of two sets of incidence matrices that they represent homeomorphic complexes, can both be described in purely number-theoretic terms. If we enumerate, in a straightforward way, the sets of incidence matrices which represent closed three-dimensional manifolds, it will then be immediately provable that the problem under consideration (to find a complete set of effectively calculable invariants of closed three-dimensional manifolds) is equivalent to the problem, to find an effectively calculable function f of positive integers, such that $f(m, n)$ is equal to 2 if and only if the m-th set of incidence matrices and the n-th set of incidence matrices in the enumeration represent homeomorphic complexes.

Other examples will readily occur to the reader.

The purpose of the present paper is to propose a definition of effective calculability[3] which is thought to correspond satisfactorily to the somewhat vague intuitive notion in terms of which problems of this class are often stated, and to show, by means of an example, that not every problem of this class is solvable.

Originally published in ***American Journal of Mathematics***, vol. 58 (1936), pp. 345–363. © The Johns Hopkins University Press. Reprinted with permission of The Johns Hopkins University Press. Certain changes, made by Church in 1971, especially some changes of terminology that Church believed were required in order to conform to the terminology of later papers, have been incorporated here. The original form of the article may be found on the MIT Press Web site for this book, http://mitpress.mit.edu.

[1] Presented to the American Mathematical Society, April 19, 1935.

[2] The selection of the particular positive integer 2 instead of some other is, of course, accidental and non-essential.

[3] As will appear, this definition of effective calculability can be stated in either of two equivalent forms,

2. Conversion and λ-definability. We select a particular list of symbols, consisting of the symbols {, }, (,), λ, [,], and an enumerably infinite set of symbols $a, b, c, \cdots$ to be called *variables*. And we define the word *formula* to mean any finite sequence of symbols out of this list. The terms *well-formed formula*, *free variable*, and *bound variable* are then defined by induction as follows. A variable $\boldsymbol{x}$ standing alone is a well-formed formula and the occurrence of $\boldsymbol{x}$ in it is an occurrence of $\boldsymbol{x}$ as a free variable in it; if the formulas $\boldsymbol{F}$ and $\boldsymbol{X}$ are well-formed, $\{\boldsymbol{F}\}(\boldsymbol{X})$ is well-formed, and an occurrence of $\boldsymbol{x}$ as a free (bound) variable in $\boldsymbol{F}$ or $\boldsymbol{X}$ is an occurrence of $\boldsymbol{x}$ as a free (bound) variable in $\{\boldsymbol{F}\}(\boldsymbol{X})$; if the formula $\boldsymbol{M}$ is well-formed and contains an occurrence of $\boldsymbol{x}$ as a free variable in $\boldsymbol{M}$, then $\lambda\boldsymbol{x}[\boldsymbol{M}]$ is well-formed, any occurrence of $\boldsymbol{x}$ in $\lambda\boldsymbol{x}[\boldsymbol{M}]$ is an occurrence of $\boldsymbol{x}$ as a bound variable in $\lambda\boldsymbol{x}[\boldsymbol{M}]$, and an occurrence of a variable $\boldsymbol{y}$, other than $\boldsymbol{x}$, as a free (bound) variable in $\boldsymbol{M}$ is an occurrence of $\boldsymbol{y}$ as a free (bound) variable in $\lambda\boldsymbol{x}[\boldsymbol{M}]$.

We shall use heavy type letters to stand for variable or undetermined formulas. And we adopt the convention that, unless otherwise stated, each heavy type letter shall represent a well-formed formula and each set of symbols standing apart which contains a heavy type letter shall represent a well-formed formula.

When writing particular well-formed formulas, we adopt the following abbreviations. A formula $\{\boldsymbol{F}\}(\boldsymbol{X})$ may be abbreviated as $\boldsymbol{F}(\boldsymbol{X})$ in any case where $\boldsymbol{F}$ is or is represented by a single symbol. A formula $\{\{\boldsymbol{F}\}(\boldsymbol{X})\}(\boldsymbol{Y})$ may be abbreviated as $\{\boldsymbol{F}\}(\boldsymbol{X}, \boldsymbol{Y})$, or, if $\boldsymbol{F}$ is or is represented by a single symbol, as $\boldsymbol{F}(\boldsymbol{X}, \boldsymbol{Y})$. And $\{\{\{\boldsymbol{F}\}(\boldsymbol{X})\}(\boldsymbol{Y})\}(\boldsymbol{Z})$ may be abbreviated as $\{\boldsymbol{F}\}(\boldsymbol{X}, \boldsymbol{Y}, \boldsymbol{Z})$, or as $\boldsymbol{F}(\boldsymbol{X}, \boldsymbol{Y}, \boldsymbol{Z})$, and so on. A formula $\lambda\boldsymbol{x}_1[\lambda\boldsymbol{x}_2[\cdots\lambda\boldsymbol{x}_n[\boldsymbol{M}]\cdots]]$ may be abbreviated as $\lambda\boldsymbol{x}_1\boldsymbol{x}_2\cdots\boldsymbol{x}_n\,.\,\boldsymbol{M}$ or as $\lambda\boldsymbol{x}_1\boldsymbol{x}_2\cdots\boldsymbol{x}_n\boldsymbol{M}$.

We also allow ourselves at any time to introduce abbreviations of the form that a particular symbol α shall stand for a particular sequence of symbols $\boldsymbol{A}$, and indicate the introduction of such an abbreviation by the notation $\alpha \longrightarrow \boldsymbol{A}$, to be read, "$\alpha$ stands for $\boldsymbol{A}$."

---

(1) that a function of positive integers shall be called effectively calculable if it is λ-definable in the sense of §2 below, (2) that a function of positive integers shall be called effectively calculable if it is recursive in the sense of §4 below. The notion of λ-definability is due jointly to the present author and S. C. Kleene, successive steps toward it having been taken by the present author in the ***Annals of Mathematics***, vol. 34 (1933), p. 863, and by Kleene in the ***American Journal of Mathematics***, vol. 57 (1935), p. 219. The notion of recursiveness in the sense of §4 below is due jointly to Jacques Herbrand and Kurt Gödel, as is there explained. And the proof of equivalence of the two notions is due chiefly to Kleene, but also partly to the present author and to J. B. Rosser, as explained below. The proposal to identify these notions with the intuitive notion of effective calculability is first made in the present paper (but see the first footnote to §7 below).

With the aid of the methods of Kleene (***American Journal of Mathematics***, 1935), the considerations of the present paper could, with comparatively slight modification, be carried through entirely in terms of λ-definability, without making use of the notion of recursiveness. On the other hand, since the results of the present paper were obtained, it has been shown by Kleene (see his forthcoming paper, *General recursive functions of natural numbers*) that analogous results can be obtained entirely in terms of recursiveness, without making use of λ-definability. The fact, however, that two such widely different and (in the opinion of the author) equally natural definitions of effective calculability turn out to be equivalent adds to the strength of the reasons adduced below for believing that they constitute as general a characterization of this notion as is consistent with the usual intuitive understanding of it.

We introduce at once the following infinite list of abbreviations,

$$\mathbf{1} \longrightarrow \lambda ab \,.\, a(b),$$
$$\mathbf{2} \longrightarrow \lambda ab \,.\, a(a(b)),$$
$$\mathbf{3} \longrightarrow \lambda ab \,.\, a(a(a(b))),$$

and so on, each positive integer in Arabic notation standing for a formula of the form $\lambda ab \,.\, a(a(\cdots a(b)\cdots))$.

The expression $S^{\boldsymbol{x}}_{\boldsymbol{N}}\boldsymbol{M}|$ is used to stand for the result of substituting $\boldsymbol{N}$ for $\boldsymbol{x}$ throughout $\boldsymbol{M}$.

We consider the three following operations on well-formed formulas:

I. *To replace any part* $\lambda\boldsymbol{x}[\boldsymbol{M}]$ *of a formula by* $\lambda\boldsymbol{y}[S^{\boldsymbol{x}}_{\boldsymbol{y}}\boldsymbol{M}|]$, *where* $\boldsymbol{y}$ *is a variable which does not occur in* $\boldsymbol{M}$.

II. *To replace any part* $\{\lambda\boldsymbol{x}[\boldsymbol{M}]\}(\boldsymbol{N})$ *of a formula by* $S^{\boldsymbol{x}}_{\boldsymbol{N}}\boldsymbol{M}|$, *provided that the bound variables in* $\boldsymbol{M}$ *are distinct both from* $\boldsymbol{x}$ *and from the free variables in* $\boldsymbol{N}$.

III. *To replace any part* $S^{\boldsymbol{x}}_{\boldsymbol{N}}\boldsymbol{M}|$ *(not immediately following* λ*) of a formula by* $\{\lambda\boldsymbol{x}[\boldsymbol{M}]\}(\boldsymbol{N})$, *provided that the bound variables in* $\boldsymbol{M}$ *are distinct both from* $\boldsymbol{x}$ *and from the free variables in* $\boldsymbol{N}$.

Any finite sequence of these operations is called a *conversion*, and if $\boldsymbol{B}$ is obtainable from $\boldsymbol{A}$ by a conversion we say that $\boldsymbol{A}$ is *convertible* into $\boldsymbol{B}$, or "$\boldsymbol{A}$ conv $\boldsymbol{B}$." If $\boldsymbol{B}$ is identical with $\boldsymbol{A}$ or is obtainable from $\boldsymbol{A}$ by a single application of one of the operations I, II, III, we say that $\boldsymbol{A}$ is *immediately convertible* into $\boldsymbol{B}$.

A conversion which contains exactly one application of Operation II, and no application of Operation III, is called a *reduction*.

A formula is said to be *in normal form* if it is well-formed and contains no part of the form $\{\lambda\boldsymbol{x}[\boldsymbol{M}]\}(\boldsymbol{N})$. And $\boldsymbol{B}$ is said to be a *normal form of* $\boldsymbol{A}$ if $\boldsymbol{B}$ is in normal form and $\boldsymbol{A}$ conv $\boldsymbol{B}$.

The original given order $a, b, c, \cdots$ of the variables is called their *natural order*. And a formula is said to be *in principal normal form* if it is in normal form, and no variable occurs in it both as a free variable and as a bound variable, and the variables which occur in it immediately following the symbol λ are, when taken in the order in which they occur in the formula, in natural order without repetitions, beginning with a and omitting only such variables as occur in the formula as free variables.[4] The formula $\boldsymbol{B}$ is said to be the *principal normal form of* $\boldsymbol{A}$ if $\boldsymbol{B}$ is in principal normal form and $\boldsymbol{A}$ conv $\boldsymbol{B}$.

Of the three following theorems, proof of the first is immediate, and the second and third have been proved by the present author and J. B. Rosser:[5]

THEOREM I. *If a formula is in normal form, no reduction of it is possible.*

---

[4]For example, the formulas $\lambda ab \,.\, b(a)$ and $\lambda a \,.\, a(\lambda c \,.\, b(c))$ are in principal normal form, and $\lambda ac \,.\, c(a)$, and $\lambda bc \,.\, c(b)$, and $\lambda a \,.\, a(\lambda a \,.\, b(a))$ are in normal form but not in principal normal form. Use of the principal normal form was suggested by S. C. Kleene as a means of avoiding the ambiguity of determination of the normal form of a formula, which is troublesome in certain connections.

Observe that the formulas 1, 2, 3, $\cdots$ are all in principal normal form.

[5]Alonzo Church and J. B. Rosser, *Some properties of conversion*, forthcoming (abstract in ***Bulletin of the American Mathematical Society***, vol. 41, p. 332) [*Editor's note*: later published, ***Transactions of the American Mathematical Society***, vol. 39 (1936), pp. 472–482].

THEOREM II. *If a formula has a normal form, this normal form is unique to within applications of Operation I, and any sequence of reductions of the formula must (if continued) terminate in the normal form.*

THEOREM III. *If a formula has a normal form, every well-formed part of it has a normal form.*

We shall call a function a *function of positive integers* if the range of each independent variable is the class of positive integers and the range of the dependent variable is contained in the class of positive integers. And when it is desired to indicate the number of independent variables we shall speak of a function of one positive integer, a function of two positive integers, and so on. Thus if F is a function of n positive integers, and $a_1, a_2, \cdots, a_n$ are positive integers, then $F(a_1, a_2, \cdots, a_n)$ must be a positive integer.

A function F of one positive integer is said to be *λ-definable* if it is possible to find a formula $\boldsymbol{F}$ such that, if $F(m) = r$ and $\boldsymbol{m}$ and $\boldsymbol{r}$ are the formulas for which the positive integers m and r (written in Arabic notation) stand according to our abbreviations introduced above, then $\{\boldsymbol{F}\}(\boldsymbol{m})$ conv $\boldsymbol{r}$.

Similarly, a function F of two positive integers is said to be λ-definable if it is possible to find a formula $\boldsymbol{F}$ such that, whenever $F(m, n) = r$, the formula $\{\boldsymbol{F}\}(\boldsymbol{m}, \boldsymbol{n})$ is convertible into $\boldsymbol{r}$ (m, n, r being positive integers and $\boldsymbol{m}, \boldsymbol{n}, \boldsymbol{r}$ the corresponding formulas). And so on for functions of three or more positive integers.[6]

It is clear that, in the case of any λ-definable function of positive integers, the process of reduction of formulas to normal form provides an algorithm for the effective calculation of particular values of the function.

3. The Gödel number of a formula. Adapting to the formal notation just described a device which is due to Gödel,[7] we associate with every formula a positive integer to represent it, as follows. To each of the symbols {, (, [we let correspond the number 11, to each of the symbols },),] the number 13, to the symbol λ the number 1, and to the variables $a, b, c, \cdots$ the prime numbers 17, 19, 23, $\cdots$ respectively. And with a formula which is composed of the n symbols $\tau_1, \tau_2, \cdots, \tau_n$ in order we associate the number $2^{t_1} 3^{t_2} \cdots p_n^{t_n}$, where t_i is the number corresponding to the symbol τ_i, and where p_n stands for the n-th prime number.

This number $2^{t_1} 3^{t_2} \cdots p_n^{t_n}$ will be called the *Gödel number* of the formula $\tau_1 \tau_2 \cdots \tau_n$.

Two distinct formulas may sometimes have the same Gödel number, because the numbers 11 and 13 each correspond to three different symbols, but it is readily proved that *no two distinct well-formed formulas can have the same Gödel number*. It is clear, moreover, that there is an effective method by which, given any formula, its Gödel number can be calculated; and likewise that there is an effective method by which, given any positive integer, it is possible to determine whether it is the Gödel number of a well-formed formula and, if it is, to obtain that formula.

---

[6]Cf. S. C. Kleene, *A theory of positive integers in formal logic*, ***American Journal of Mathematics***, vol. 57 (1935), pp. 153–173 and 219–244, where the λ-definability of a number of familiar functions of positive integers, and of a number of important general classes of functions, is established. Kleene uses the term *definable*, or *formally definable*, in the sense in which we are here using *λ-definable*.

[7]Kurt Gödel, *Über formal unentscheidbare Sätze der Principia Mathematica und verwandter Systeme I*, ***Monatshefte für Mathematik und Physik***, vol. 38 (1931), pp. 173–198.

In this connection the Gödel number plays a rôle similar to that of the matrix of incidence in combinatorial topology (cf. §1 above). For there is, in the theory of well-formed formulas, an important class of problems, each of which is equivalent to a problem of elementary number theory obtainable by means of the Gödel number.[8]

4. Recursive functions. We define a class of expressions, which we shall call *elementary expressions*, and which involve, besides parentheses and commas, the symbols 1, s, an infinite set of numerical variables $x, y, z, \cdots$, and, for each positive integer n, an infinite set $f_n, g_n, h_n, \cdots$ of functional variables with subscript n. This definition is by induction as follows. The symbol 1 or any numerical variable, standing alone, is an elementary expression. If A is an elementary expression, then $s(A)$ is an elementary expression. If $A_1, A_2, \cdots, A_n$ are elementary expressions and f_n is any functional variable with subscript n, then $f_n(A_1, A_2, \cdots, A_n)$ is an elementary expression.

The particular elementary expressions $1, s(1), s(s(1)), \cdots$ are called *numerals*. And the positive integers 1, 2, 3, $\cdots$ are said to correspond to $1, s(1), s(s(1)), \cdots$.

An expression of the form $A = B$, where A and B are elementary expressions, is called an *elementary equation*.

The *derived equations* of a set E of elementary equations are defined by induction as follows. The equations of E themselves are derived equations. If $A = B$ is a derived equation containing a numerical variable x, then the result of substituting a particular numeral for all the occurrences of x in $A = B$ is a derived equation. If $A = B$ is a derived equation containing an elementary expression C (as part of either A or B), and if either $C = D$ or $D = C$ is a derived equation, then the result of substituting D for a particular occurrence of C in $A = B$ is a derived equation.

Suppose that no derived equation of a certain finite set E of elementary equations has the form $k = l$ where k and l are different numerals, that the functional variables which occur in E are $f^1_{n_1}, f^2_{n_2}, \cdots, f^r_{n_r}$ with subscripts $n_1, n_2, \cdots, n_r$ respectively, and that, for every value of i from 1 to r inclusive, and for every set of numerals $k^i_1, k^i_2, \cdots, k^i_{n_i}$, there exists a unique numeral k^i such that $f^i_{n_i}(k^i_1, k^i_2, \cdots, k^i_{n_i}) = k^i$ is a derived equation of E. And let $F^1, F^2, \cdots, F^r$ be the functions of positive integers defined by the condition that, in all cases, $F^i(m^i_1, m^i_2, \cdots, m^i_{n_i})$ shall be equal to m^i, where $m^i_1, m^i_2, \cdots, m^i_{n_i}$, and m^i are the positive integers which correspond to the numerals $k^i_1, k^i_2, \cdots, k^i_{n_i}$, and k^i respectively. Then the set of equations E is said to *define*, or to be a set of *recursion equations* for, any one of the functions F^i, and the functional variable $f^i_{n_i}$ is said to *denote* the function F^i.

A function of positive integers for which a set of recursion equations can be given is said to be *recursive*.[9]

---

[8]This is merely a special case of the now familiar remark that, in view of the Gödel number and the ideas associated with it, symbolic logic in general can be regarded, mathematically, as a branch of elementary number theory. This remark is essentially due to Hilbert (cf. for example, ***Verhandlungen des dritten internationalen Mathematiker-Kongresses in Heidelberg***, 1904, p. 185; also Paul Bernays in ***Die Naturwissenschaften***, vol. 10 (1922), pp. 97 and 98) but is most clearly formulated in terms of the Gödel number.

[9]This definition is closely related to, and was suggested by, a definition of recursive functions which was proposed by Kurt Gödel, in lectures at a Princeton, N.J., 1934, and credited by him in part to an unpublished suggestion of Jacques Herbrand. The principal features in which the present definition of recursiveness differs from Gödel's are due to S. C. Kleene.

In a forthcoming paper by Kleene to be entitled, *General recursive functions of natural numbers* (abstract in

It is clear that for any recursive function of positive integers there exists an algorithm using which any required particular value of the function can be effectively calculated. For the derived equations of the set of recursion equations E are effectively enumerable, and the algorithm for the calculation of particular values of a function F^i, denoted by a functional variable $f^i_{n_i}$, consists in carrying out the enumeration of the derived equations of E until the required particular equation of the form $f^i_{n_i}(k^i_1, k^i_2, \cdots, k^i_{n_i}) = k^i$ is found.[10]

We call an infinite sequence of positive integers recursive if the function F such that $F(n)$ is the n-th term of the sequence is recursive.

We call a propositional function of positive integers recursive if the function whose value is 2 or 1, according to whether the propositional function is true or false, is recursive. By a recursive property of positive integers we shall mean a recursive propositional function of one positive integer, and by a recursive relation between positive integers we shall mean a recursive propositional function of two or more positive integers.

A function F, for which the range of the dependent variable is contained in the class of positive integers and the range of the independent variable, or of each independent variable, is a subset (not necessarily the whole) of the class of positive integers, will be called *potentially recursive*, if it is possible to find a recursive function F' of positive integers (for which the range of the independent variable, or of each independent variable, is the whole of the class of positive integers), such that the value of F' agrees with the value of F in all cases when the latter is defined.

By an *operation on* well-formed formulas we shall mean a function for which the range of the dependent variable is contained in the class of well-formed formulas and the range of the independent variable, or of each independent variable, is the whole class of well-formed formulas. And we call such an operation recursive if the corresponding function obtained by replacing all formulas by their Gödel number is potentially recursive.

Similarly any function for which the range of the dependent variable is contained either in the class of positive integers or in the class of well-formed formulas, and for which the range of each independent variable is identical either with the class of positive integers or with the class of well-formed formulas (allowing the case that some of the ranges are identical with one class and some with the other), will be said to be recursive if the corresponding function obtained by replacing all formulas by their Gödel numbers is potentially recursive. We call an infinite sequence of well-formed formulas recursive if the corresponding infinite sequence of Gödel numbers is recursive. And we call a property of, or relation between, well-formed formulas recursive if the

---

Bulletin of the American Mathematical Society, vol. 41), several definitions of recursiveness will be discussed and equivalences among them obtained. In particular, it follows readily from Kleene's results in that paper that every function recursive in the present sense is also recursive in the sense of Gödel (1934) and conversely.

Added 1971: this paper afterwards appeared in ***Mathematische Annalen***, vol. 112 (1936), pp. 727–742.

[10]The reader may object that this algorithm cannot be held to provide an effective calculation of the required particular value of F^i unless the proof is constructive that the required equation $f^i_{n_i}(k^i_1, k^i_2, \cdots, k^i_{n_i}) = k^i$ will ultimately be found. But if so this merely means that he should take the existential quantifier which appears in our definition of a set of recursion equations in a constructive sense. What the criterion of constructiveness shall be is left to the reader.

The same remark applies in connection with the existence of an algorithm for calculating the values of a λ-definable function of positive integers.

corresponding property of, or relation between, their Gödel numbers is potentially recursive. A set of well-formed formulas is said to be recursively enumerable if there exists a recursive infinite sequence which consists entirely of formulas of the set and contains every formula of the set at least once.[11]

In terms of the notion of recursiveness we may also define a *proposition of elementary number theory*, by induction as follows. If ϕ is a recursive propositional function of n positive integers (defined by giving a particular set of recursion equations for the corresponding function whose values are 2 and 1) and if $x_1, x_2, \cdots, x_n$ are variables which take on positive integers as values, then $\phi(x_1, x_2, \cdots, x_n)$ is a proposition of elementary number theory. If P is proposition of elementary number theory involving x as a free variable, then the result of substituting a particular positive integer for all occurrences of x as a free variable in P is a proposition of elementary number theory, and $(x)P$ and $(\exists x)P$ are propositions of elementary number theory, where (x) and $(\exists x)$ are respectively the universal and existential quantifiers of x over the class of positive integers.

It is then readily seen that the negation of a proposition of elementary number theory or the logical product or the logical sum of two propositions of elementary number theory is equivalent, in a simple way, to another proposition of elementary number theory.

5. Recursiveness of the Kleene p-function. We prove two theorems which establish the recursiveness of certain functions which are definable in words by means of the phrase, "The least positive integer such that," or, "The n-th positive integer such that."

THEOREM IV. *If F is a recursive function of two positive integers, and if for every positive integer x there exists a positive integer y such that $F(x, y) > 1$, then the function F^*, such that, for every positive integer x, $F^*(x)$ is equal to the least positive integer y for which $F(x, y) > 1$, is recursive.*

For a set of recursion equations for F^* consists of the recursion equations for F together with the equations,

$$i_2(1, 2) = 2,$$
$$i_2(s(x), 2) = 1,$$
$$i_2(x, 1) = 3,$$
$$i_2(x, s(s(y))) = 3,$$
$$j_2(1, y) = y,$$
$$j_2(s(x), y) = x,$$
$$g_2(x, 1) = i_2(f_2(x, 1), 2),$$
$$g_2(x, s(y)) = i_2(f_2(x, s(y)), g_2(x, y)),$$
$$h_2(s(x), y) = x,$$
$$h_2(g_2(x, y), x) = j_2(g_2(x, y), y),$$
$$f_1(x) = h_2(1, x),$$

where the functional variables f_2 and f_1 denote the functions F and F^* respectively, and 2 and 3 are abbreviations for $s(1)$ and $s(s(1))$, respectively.[12]

[11] It can be shown, in view of Theorem V below, that, if an infinite set of formulas is recursively enumerable in this sense, it is also recursively enumerable in the sense that there exists a recursive infinite sequence which consists entirely of formulas of the set and contains every formula of the set exactly once.

[12] Since this result was obtained, it has been pointed out to the author by S. C. Kleene that it can be proved more simply by using the methods of the latter in ***American Journal of Mathematics***, vol. 57 (1935), p. 231 et seq. His proof will be given in his forthcoming paper already referred to.

THEOREM V. *If F is a recursive function of one positive integer, and if there exist an infinite number of positive integers x for which $F(x) > 1$, then the function F^0, such that, for every positive integer n, $F^0(n)$ is equal to the n-th positive integer x (in order of increasing magnitude) for which $F(x) > 1$, is recursive.*

For a set of recursion equations for F^0 consists of the recursion equations for F together with the equations,

$$g_2(1, y) = g_2(f_1(s(y)), s(y)),$$
$$g_2(s(x), y) = y,$$
$$g_1(1) = k,$$
$$g_1(s(y)) = g_2(1, g_1(y)),$$

where the functional variables g_1 and f_1 denote the functions F^0 and F respectively, and where k is the numeral to which corresponds the least positive integer x for which $F(x) > 1$.[13]

6. Recursiveness of certain functions of formulas. We list now a number of theorems which will be proved in detail in a forthcoming paper by S. C. Kleene[14] or follow immediately from considerations there given. We omit proofs here, except for brief indications in some instances.

Our statement of the theorems and our notation differ from Kleene's in that we employ the set of positive integers (1, 2, 3, $\cdots$) in the rôle in which he employs the set of natural numbers (0, 1, 2, $\cdots$). This difference is, of course, unessential. We have selected what is, from some points of view, the less natural alternative, in order to preserve the convenience and naturalness of the identification of the formula $\lambda ab \,.\, a(b)$ with 1 rather than with 0.

THEOREM VI. *The property of a positive integer, that there exists a well-formed formula of which it is the Gödel number is recursive.*

THEOREM VII. *The set of well-formed formulas is recursively enumerable.*

This follows from Theorems V and VI.

THEOREM VIII. *The function of two variables, whose value, when taken of the well-formed formulas **F** and **X**, is the formula $\{\boldsymbol{F}\}(\boldsymbol{X})$, is recursive.*

THEOREM IX. *The function, whose value for each of the positive integers $1, 2, 3, \cdots$ is the corresponding formula $\boldsymbol{1}, \boldsymbol{2}, \boldsymbol{3}, \cdots$, is recursive.*

THEOREM X. *A function, whose value for each of the formulas $\boldsymbol{1}, \boldsymbol{2}, \boldsymbol{3}, \cdots$ is the corresponding positive integer, and whose value for other well-formed formulas is a fixed positive integer, is recursive. Likewise the function, whose value for each of the formulas $\boldsymbol{1}, \boldsymbol{2}, \boldsymbol{3}, \cdots$ is the corresponding positive integer plus one, and whose value for other well-formed formulas is the positive integer 1, is recursive.*

---

[13]This proof is due to Kleene.

[14]S. C. Kleene, *λ-definability and recursiveness*, forthcoming (abstract in ***Bulletin of the American Mathematical Society***, vol. 41). In connection with many of the theorems listed, see also Kurt Gödel, ***Monatshefte für Mathematik und Physik***, vol. 38 (1931), p. 181 et seq., observing that every function which is recursive in the sense in which the word is there used by Gödel is also recursive in the present more general sense.

Added 1971: this paper afterwards appeared in ***Duke Mathematical Journal***, vol. 2 (1936), pp. 340–353.

THEOREM XI. *The relation of immediate convertibility, between well-formed formulas, is recursive.*

THEOREM XII. *It is possible to associate simultaneously with every well-formed formula an enumeration of the formulas obtainable from it by conversion, in such a way that the function of two variables, whose value, when taken of a well-formed formula **A** and a positive integer n, is the n-th formula in the enumeration of the formulas obtainable from **A** by conversion, is recursive.*

THEOREM XIII. *The property of a well-formed formula, that it is in principal normal form, is recursive.*

THEOREM XIV. *The set of well-formed formulas which are in principal normal form is recursively enumerable.*

This follows from Theorems V, VII, XIII.

THEOREM XV. *The set of well-formed formulas which have a normal form is recursively enumerable.*[15]

For by Theorems XII and XIV this set can be arranged in an infinite square array which is recursively defined (i.e., defined by a recursive function of two variables). And the familiar process by which this square array is reduced to a single infinite sequence is recursive (i.e., can be expressed by means of recursive functions).

THEOREM XVI. *Every recursive function of positive integers is λ-definable.*[16]

THEOREM XVII. *Every λ-definable function of positive integers is recursive.*[17]

For functions of one positive integer this follows from Theorems IX, VIII, XII, XIII, IV, X. For functions of more than one positive integer it follows by the same method, using a generalization of Theorem IV to functions of more than two positive integers.

7. The notion of effective calculability. We now define the notion, already discussed, of an *effectively calculable* function of positive integers by identifying it with the notion of a recursive function of positive integers[18] (or of a λ-definable function of positive integers). This definition is thought to be justified by the considerations which follow, so far as positive justification can ever be obtained for the selection of a formal definition to correspond to an intuitive notion.

It has already been pointed out that, for every function of positive integers which is effectively calculable in the sense just defined, there exists an algorithm for the calculation of its values.

---

[15]This theorem was first proposed by the present author, with the outline of proof here indicated. Details of its proof are due to Kleene and will be given by him in his forthcoming paper, *λ-definability and recursiveness* [see the preceding footnote].

[16]This theorem can be proved as a straightforward application of the methods introduced by Kleene in the ***American Journal of Mathematics*** (loc. cit.). In the form here given it was first obtained by Kleene. The related result had previously been obtained by J. B. Rosser that, if we modify the definition of *well-formed* by omitting the requirement that $\boldsymbol{M}$ contain $\boldsymbol{x}$ as a free variable in order that $\lambda\boldsymbol{x}[\boldsymbol{M}]$ be well-formed, then every recursive function of positive integers is λ-definable in the resulting modified sense.

[17]This result was obtained independently by the present author and S. C. Kleene at about the same time.

[18]The question of the relationship between effective calculability and recursiveness (which it is here proposed to answer by identifying the two notions) was raised by Gödel in conversation with the author. The corresponding question of the relationship between effective calculability and λ-definability had previously been proposed by the author independently.

Conversely it is true, under the same definition of effective calculability, that for every function, an algorithm for the calculation of the values of which exists, is effectively calculable. For example, in the case of a function F of one positive integer, an algorithm consists in a method by which, given any positive integer n, a sequence of expressions (in some notation) $E_{n1}, E_{n2}, \cdots, E_{nr_n}$, can be obtained; where E_{n1} is effectively calculable when n is given; where E_{ni} is effectively calculable when n and the expressions E_{nj}, $j < i$, are given; and where, when n and all the expressions E_{ni} up to and including E_{nr_n} are given, the fact that the algorithm has terminated becomes effectively known and the value of $F(n)$ is effectively calculable. Suppose that we set up a system of Gödel numbers for the notation employed in the expressions E_{ni}, and that we then further adopt the method of Gödel of representing a finite sequence of expressions $E_{n1}, E_{n1}, \cdots, E_{ni}$ by the single positive integer $2^{e_{n1}} 3^{e_{n2}} \cdots p_i^{e_{ni}}$ where $e_{n1}, e_{n2}, \cdots, e_{ni}$ are respectively the Gödel numbers of $E_{n1}, E_{n2}, \cdots, E_{ni}$ (in particular representing a vacuous sequence of expressions by the positive integer 1). Then we may define a function G of two positive integers such that, if x represents the finite sequence $E_{n1}, E_{n2}, \cdots, E_{nk}$, then $G(n, x)$ is equal to the Gödel number of E_{ni}, where $i = k + 1$, or is equal to 10 if $k = r_n$ (that is if the algorithm has terminated with E_{nk}), and in any other case $G(n, x)$ is equal to 1. And we may define a function H of two positive integers, such that the value of $H(n, x)$ is the same as that of $G(n, x)$, except in the case that $G(n, x) = 10$, in which case $H(n, x) = F(n)$. If the interpretation is allowed that the requirement of effective calculability which appears in our description of an algorithm means the effective calculability of the functions G and H,[19] and if we take the effective calculability of G and H to mean recursiveness (λ-definability), then the recursiveness (λ-definability) of F follows by a straightforward argument.

Suppose that we are dealing with some particular system of symbolic logic, which contains a symbol, $=$, for equality of positive integers, a symbol $\{\ \}(\ )$ for the application of a function of one positive integer to its argument, and expressions $\mathbf{1}, \mathbf{2}, \mathbf{3}, \cdots$ to stand for the positive integers. The theorems of the system consist of a finite, or enumerably infinite, list of expressions, the *axioms*, together with all the expressions obtainable from them by a finite succession of applications of operations chosen out of a given finite, or enumerably infinite, list of operations, the *rules of inference*. If the system is to serve at all the purposes for which a system of symbolic logic is usually intended, it is necessary that each rule of inference be an effectively calculable operation, that the complete set of rules of inference (if infinite) be effectively enumerable, that the complete set of axioms (if infinite) be effectively enumerable, and that the relation between a positive integer and the expression which stands for it be effectively determinable. Suppose that we interpret this to mean that, in terms of a system of Gödel numbers for the expressions of the logic, each rule of inference must be a recursive operation,[20] the complete set of rules of inference must be recursively enumerable (in the sense that there exists a recursive function Φ such that $\Phi(n, x)$ is the

[19] If this interpretation or some similar one is not allowed, it is difficult to see how the notion of an algorithm can be given any exact meaning at all.

[20] As a matter of fact, in known systems of symbolic logic, e.g., in that of ***Principia Mathematica***, the stronger statement holds, that the relation of *immediate consequence* (*unmittelbare Folge*) is recursive. Cf. Gödel, loc. cit., p. 185. In any case where the relation of immediate consequence is recursive it is possible to find a set of rules of inference, equivalent to the original ones, such that each rule is a (one-valued) recursive operation, and the complete set of rules is recursively enumerable.

number of the result of applying the n-th rule of inference to the ordered finite set of formulas represented by x), the complete set of axioms must be recursively enumerable, and the relation between a positive integer and the expression which stands for it must be recursive.[21] And let us call a function F of one positive integer[22] *calculable within the logic* if there exists an expression f in the logic such that $\{f\}(\mu) = \nu$ is a theorem when and only when $F(m) = n$ is true, μ and ν being the expressions which stand for the positive integers m and n. Then, since the complete set of theorems of the logic is recursively enumerable, it follows by Theorem IV above that every function of one positive integer which is calculable within the logic is also effectively calculable (in the sense of our definition).

Thus it is shown that no more general definition of effective calculability than that proposed above can be obtained by either of two methods which naturally suggest themselves (1) by defining a function to be effectively calculable if there exists an algorithm for the calculation of its values (2) by defining a function F (of one positive integer) to be effectively calculable if, for every positive integer m, there exists a positive integer n such that $F(m) = n$ is a provable theorem.

8. Invariants of conversion. The problem naturally suggests itself to find invariants of that transformation of formulas which we have called conversion. The only effectively calculable invariants at present known are the immediately obvious ones (e.g., the set of free variables contained in a formula). Others of importance very probably exist. But we shall prove (in Theorem XIX) that, under the definition of effective calculability proposed in §7, *no complete set of effective calculable invariants of conversion exists* (cf. §1).

The results of Kleene (***American Journal of Mathematics***, 1935) make it clear that, if the problem of finding a complete set of effectively calculable invariants of conversion were solved, most of the familiar unsolved problems of elementary number theory would thereby also be solved. And from Theorem XVI above it follows further that to find a complete set of effectively calculable invariants of conversion would imply the solution of the decision problem (Entscheidungsproblem) for any system of symbolic logic whatever (subject to the very general restrictions of §7). In the light of this it is hardly surprising that the problem to find such a set of invariants should be unsolvable.

It is to be remembered, however, that, if we consider only the statement of the problem (and ignore things which can be proved about it by more or less lengthy arguments), it appears to be a problem of the same class as the problems of number theory and topology to which it was compared in §1, having no striking characteristic by which it can be distinguished from them. The temptation is strong to reason by analogy that other important problems of this class may also be unsolvable.

LEMMA. *The problem, to find a recursive function of two formulas* ***A*** *and* ***B*** *whose value is* 2 *or* 1 *according as* ***A*** *conv* ***B*** *or not, is equivalent to the problem, to find a recursive*

[21] The author is here indebted to Gödel, who, in his 1934 lectures already referred to, proposed substantially these conditions, but in terms of the more restricted notion of recursiveness which he had employed in 1931, and using the condition that the relation of immediate consequence be recursive instead of the present conditions on the rules of inference.

[22] We confine ourselves for convenience to the case of functions of one positive integer. The extension to functions of several positive integers is immediate.

function of one formula **C** *whose value is* 2 *or* 1 *according as* **C** *has a normal form or not*.[23]

For, by Theorem X, the formula $\boldsymbol{a}$ (the formula $\boldsymbol{b}$), which stands for the positive integer which is the Gödel number of the formula $\boldsymbol{A}$ (the formula $\boldsymbol{B}$), can be expressed as a recursive function of the formula $\boldsymbol{A}$ (the formula $\boldsymbol{B}$). Moreover, by Theorems VI and XII, there exists a recursive function F of two positive integers such that, if m is the Gödel number of a well-formed formula $\boldsymbol{M}$, then $F(m, n)$ is the Gödel number of the n-th formula in an enumeration of the formulas obtainable from $\boldsymbol{M}$ by conversion. And, by Theorem XVI, F is λ-definable, by a formula $\mathfrak{f}$. If we define,

$$Z_1 \longrightarrow \mathcal{Q}(\lambda x \, . \, x(I), I),$$
$$Z_2 \longrightarrow \mathcal{Q}(\lambda xy \, . \, s(x) - y, I),$$

where $\mathcal{Q}$ is the formula defined by Kleene (***American Journal of Mathematics***, vol. 57 (1935), p. 226), then Z_1 and Z_2 λ-define the functions of one positive integer whose values, for a positive integer n, are the n-th terms respectively of the infinite sequences $1, 1, 2, 1, 2, 3, \cdots$ and $1, 2, 1, 3, 2, 1, \cdots$. By Theorem VIII the formula,

$$\{\lambda xy \, . \, \mathfrak{p}(\lambda n \, . \, \delta(\mathfrak{f}(x, Z_1(n)), \mathfrak{f}(y, Z_2(n))), 1)\}(\boldsymbol{a}, \boldsymbol{b}),$$

where $\mathfrak{p}$ and δ are defined as by Kleene (loc. cit., p. 173 and p. 231), is a recursive function of $\boldsymbol{A}$ and $\boldsymbol{B}$, and this formula has a normal form if and only if $\boldsymbol{A}$ conv $\boldsymbol{B}$.

Again, by Theorem X, the formula $\boldsymbol{c}$, which stands for the positive integer which is the Gödel number of the formula $\boldsymbol{C}$, can be expressed as a recursive function of the formula $\boldsymbol{C}$. By Theorems VI and XIII, there exists a recursive function G of one positive integer such that $G(m) = 2$ if m is the Gödel number of a formula in principal normal form, and $G(m) = 1$ in any other case. And, by Theorem XVI, G is λ-definable, by a formula $\mathfrak{g}$. By Theorem VIII the formula,

$$\{\lambda x \, . \, \mathfrak{p}(\lambda n \, . \, \mathfrak{g}(\mathfrak{f}(x, n), 1, 1))\}(\boldsymbol{c})$$

where $\mathfrak{f}$ is the formula $\mathfrak{f}$ used in the preceding paragraph, is a recursive function of $\boldsymbol{C}$, and this formula is convertible into the formula $\mathbf{1}$ if and only if $\boldsymbol{C}$ has a normal form.

Thus we have proved that a formula $\boldsymbol{C}$ can be found as a recursive function of formulas $\boldsymbol{A}$ and $\boldsymbol{B}$, such that $\boldsymbol{C}$ has a normal form if and only if $\boldsymbol{A}$ conv $\boldsymbol{B}$; and that a formula $\boldsymbol{A}$ can be found as a recursive function of a formula $\boldsymbol{C}$, such that $\boldsymbol{A}$ conv $\mathbf{1}$ if and only if $\boldsymbol{C}$ has a normal form. From this the lemma follows.

THEOREM XVIII. *There is no recursive function of a formula* **C**, *whose value is* 2 *or* 1 *according as* **C** *has a normal form or not*.

That is, the property of a well-formed formula, that it has a normal form, is not recursive.

For assume the contrary.

Then there exists a recursive function H of one positive integer such that $H(m) = 2$ if m is the Gödel number of a formula which has a normal form, and $H(m) = 1$ in any other case. And, by Theorem XVI, H is λ-definable by a formula $\mathfrak{h}$.

[23] These two problems, in the forms, (1) to find an effective method of determining of any two formulas $\boldsymbol{A}$ and $\boldsymbol{B}$ whether $\boldsymbol{A}$ conv $\boldsymbol{B}$, (2) to find an effective method of determining of any formula $\boldsymbol{C}$ whether it has a normal form, were both proposed by Kleene to the author, in the course of a discussion of the properties of the $\mathfrak{p}$-function, about 1932. Some attempts toward solution of (1) by means of numerical invariants were actually made by Kleene at about that time.

By Theorem XV, there exists an enumeration of the well-formed formulas which have a normal form, and a recursive function A of one positive integer such that $A(n)$ is the Gödel number of the n-th formula in this enumeration. And, by Theorem XVI, A is λ-definable, by a formula $\mathfrak{a}$.

By Theorems VI and VIII, there exists a recursive function B of two positive integers such that, if m and n are Gödel numbers of well-formed formulas $\boldsymbol{M}$ and $\boldsymbol{N}$, then $B(m, n)$ is the Gödel number of $\{\boldsymbol{M}\}(\boldsymbol{N})$. And, by Theorem XVI, B is λ-definable, by a formula $\mathfrak{b}$.

By Theorems VI and X, there exists a recursive function C of one positive integer such that, if m is the Gödel number of one of the formulas $\mathbf{1}, \mathbf{2}, \mathbf{3}, \cdots$, then $C(m)$ is the corresponding positive integer plus one, and in any other case $C(m) = 1$. And, by Theorem XVI, C is λ-definable, by a formula $\mathfrak{c}$.

By Theorem IX there exists a recursive function Z^{-1} of one positive integer, whose value for each of the positive integers $1, 2, 3, \cdots$ is the Gödel number of the corresponding formula $\mathbf{1}, \mathbf{2}, \mathbf{3}, \cdots$. And, by Theorem XVI, Z^{-1} is λ-definable, by a formula $\mathfrak{z}$.

Let $\mathfrak{f}$ and $\mathfrak{g}$ be the formulas $\mathfrak{f}$ and $\mathfrak{g}$ used in the proof of the Lemma. By Kleene 15III Cor. (loc. cit., p. 220), a formula $\mathfrak{d}$ can be found such that:

$$\mathfrak{d}(\mathbf{1}) \text{ conv } \lambda x \,.\, x(\mathbf{1})$$

$$\mathfrak{d}(\mathbf{2}) \text{ conv } \lambda u \,.\, \mathfrak{c}(\mathfrak{f}(u, \mathfrak{p}(\lambda m \,.\, \mathfrak{g}(\mathfrak{f}(u, m)), \mathbf{1}))).$$

We define,

$$\mathfrak{e} \longrightarrow \lambda n \,.\, \mathfrak{d}(\mathfrak{h}(\mathfrak{b}(\mathfrak{a}(n), \mathfrak{z}(n))), \mathfrak{b}(\mathfrak{a}(n), \mathfrak{z}(n))).$$

Then if $\boldsymbol{n}$ is one of the formulas $\mathbf{1}, \mathbf{2}, \mathbf{3}, \cdots$, $\mathfrak{e}(\boldsymbol{n})$ is convertible into one of the formulas $\mathbf{1}, \mathbf{2}, \mathbf{3}, \cdots$ in accordance with the following rules: (1) if $\mathfrak{b}(\mathfrak{a}(\boldsymbol{n}), \mathfrak{z}(\boldsymbol{n}))$ conv a formula which stands for the Gödel number of a formula which has no normal form, $\mathfrak{e}(\boldsymbol{n})$ conv $\mathbf{1}$, (2) if $\mathfrak{b}(\mathfrak{a}(\boldsymbol{n}), \mathfrak{z}(\boldsymbol{n}))$ conv a formula which stands for the Gödel number of a formula which has a principal normal form which is not one of the formulas $\mathbf{1}, \mathbf{2}, \mathbf{3}, \cdots$, $\mathfrak{e}(\boldsymbol{n})$ conv $\mathbf{1}$, (3) if $\mathfrak{b}(\mathfrak{a}(\boldsymbol{n}), \mathfrak{z}(\boldsymbol{n}))$ conv a formula which stands for the Gödel number of a formula which has a principal normal form which is one of the formulas $\mathbf{1}, \mathbf{2}, \mathbf{3}, \cdots$, $\mathfrak{e}(\boldsymbol{n})$ conv the next following formula in the list $\mathbf{1}, \mathbf{2}, \mathbf{3}, \cdots$.

By Theorem III, since $\mathfrak{e}(\mathbf{1})$ has a normal form, the formula $\mathfrak{e}$ has a normal form. Let $\mathfrak{G}$ be the formula which stands for the Gödel number of $\mathfrak{e}$. Then, if $\boldsymbol{n}$ is any one of the formulas $\mathbf{1}, \mathbf{2}, \mathbf{3}, \cdots$, $\mathfrak{G}$ is not convertible into the formula $\mathfrak{a}(\boldsymbol{n})$, because $\mathfrak{b}(\mathfrak{G}, \mathfrak{z}(\boldsymbol{n}))$ is, by the definition of $\mathfrak{b}$, convertible into the formula which stands for the Gödel number of $\mathfrak{e}(\boldsymbol{n})$, while $\mathfrak{b}(\mathfrak{a}(\boldsymbol{n}), \mathfrak{z}(\boldsymbol{n}))$ is, by the preceding paragraph, convertible into the formula which stands for the Gödel number of a formula definitely not convertible into $\mathfrak{e}(\boldsymbol{n})$ (Theorem II). But, by our definition of $\mathfrak{a}$, it must be true of one of the formulas $\boldsymbol{n}$ in the list $\mathbf{1}, \mathbf{2}, \mathbf{3}, \cdots$ that $\mathfrak{a}(\boldsymbol{n})$ conv $\mathfrak{G}$.

Thus, since our assumption to the contrary has led to a contradiction, the theorem must be true.

In order to present the essential ideas without any attempt at exact statement, the preceding proof may be outlined as follows. We are to deduce a contradiction from the assumption that it is effectively determinable of every well-formed formula whether or not it has a normal form. If this assumption holds, it is effectively determinable of every well-formed formula whether or not it is convertible into one of the formulas $\mathbf{1}, \mathbf{2}, \mathbf{3}, \cdots$; for, given a well-formed formula $\boldsymbol{R}$, we can first determine whether or not it has a normal form, and if it has we can obtain the principal normal form by enumerating the formulas into which $\boldsymbol{R}$ is convertible (Theorem XII) and picking out the first formula in principal normal form which occurs in the enumeration, and we can then determine whether the principal normal form is one of the formulas $\mathbf{1}, \mathbf{2}, \mathbf{3}, \cdots$. Let $A_1, A_2, A_3, \cdots$ be an effective enumeration of the well-formed formulas which have a normal form (Theorem XV). Let E be a function of one positive integer, defined by the rule that, where $\boldsymbol{m}$ and $\boldsymbol{n}$ are the formulas which stand for the positive integers m and n respectively, $E(n) = 1$ if $\{A_n\}(\boldsymbol{n})$ is not convertible into one of the formulas $\mathbf{1}, \mathbf{2}, \mathbf{3}, \cdots$, and $E(n) = m+1$ if $\{A_n\}(\boldsymbol{n})$ conv $\boldsymbol{m}$ and $\boldsymbol{m}$ is one of the formulas $\mathbf{1}, \mathbf{2}, \mathbf{3}, \cdots$. The function E is effectively calculable and is therefore λ-definable, by a formula $\mathfrak{e}$. The formula $\mathfrak{e}$ has a normal form, since $\mathfrak{e}(\mathbf{1})$ has a normal form. But $\mathfrak{e}$ is not any one of the formulas $A_1, A_2, A_3, \cdots$, because, for every n, $\mathfrak{e}(\boldsymbol{n})$ is a formula which is not convertible into $\{A_n\}(\boldsymbol{n})$. And this contradicts the property of the enumeration $A_1, A_2, A_3, \cdots$ that it contains all well-formed formulas which have a normal form.

COROLLARY 1. *The set of well-formed formulas which have no normal form is not recursively enumerable.*[24]

For, to outline the argument, the set of well-formed formulas which have a normal form is recursively enumerable, by Theorem XV. If the set of those which do not have a normal form were also recursively enumerable, it would be possible to tell effectively of any well-formed formula whether it had a normal form, by the process of searching through the two enumerations until it was found in one or the other. This, however, is contrary to Theorem XVIII.

This corollary gives us an example of an effectively enumerable set (the set of well-formed formulas) which is divided into two non-overlapping subsets of which one is effectively enumerable and the other not. Indeed, in view of the difficulty of attaching any reasonable meaning to the assertion that a set is enumerable but not effectively enumerable, it may even be permissible to go a step further and say that here is an example of an enumerable set which is divided into two non-overlapping subsets of which one is enumerable and the other non-enumerable.[25]

COROLLARY 2. *Let a function F of one positive integer be defined by the rule that $F(n)$ shall equal 2 or 1 according as n is or is not the Gödel number of a formula which has a normal form. Then F (if its definition be admitted as valid at all) is an example of a non-recursive function of positive integers.*[26]

This follows at once from Theorem XVIII.

Consider the infinite sequence of positive integers, $F(1), F(2), F(3), \cdots$. It is impossible to specify effectively a method by which, given any n, the n-th term of this sequence could be calculated. But it is also impossible ever to select a particular term of

[24] This corollary was proposed by J. B. Rosser.

The outline of proof here given for it is open to the objection, recently called to the author's attention by Paul Bernays, that it ostensibly requires a non-constructive use of the principle of excluded middle. This objection is met by a revision of the proof, the revised proof to consist in taking any recursive enumeration of formulas which have no normal form and showing that this enumeration is not a complete enumeration of such formulas, by constructing a formula $\mathfrak{e}(\boldsymbol{n})$ such that (1) the supposition that $\mathfrak{e}(\boldsymbol{n})$ occurs in the enumeration leads to contradiction (2) the supposition that $\mathfrak{e}(\boldsymbol{n})$ has a normal form leads to contradiction.

[25] Cf. the remarks of the author in ***The American Mathematical Monthly***, vol. 41 (1934), pp. 356–361.

[26] Other examples of non-recursive functions have since been obtained by S. C. Kleene in a different connection. See his forthcoming paper, *General recursive functions of natural numbers* [citation in footnote 9].

this sequence and prove about that term that its value cannot be calculated (because of the obvious theorem that if this sequence has terms whose values cannot be calculated then the value of each of those terms is 1). Therefore it is natural to raise the question whether, in spite of the fact that there is no systematic method of effectively calculating the terms of this sequence, it might not be true of each term individually that there existed a method of calculating its value. To this question perhaps the best answer is that the question itself has no meaning, on the ground that the universal quantifier which it contains is intended to express a mere infinite succession of accidents rather than anything systematic.

There is in consequence some room for doubt whether the assertion that the function F exists can be given a reasonable meaning.

THEOREM XIX. *There is no recursive function of two formulas* ***A*** *and* ***B***, *whose value is* 2 *or* 1 *according as* ***A*** *conv* ***B*** *or not.*

This follows at once from Theorem XVIII and the Lemma preceding it.

As a corollary of Theorem XIX, it follows that the decision problem is unsolvable in the case of any system of symbolic logic which is ω-consistent (ω-widerspruchsfrei) in the sense of Gödel (loc. cit., p. 187) and is strong enough to allow certain comparatively simple methods of definition and proof. For in any such system the proposition will be expressible about two positive integers a and b that they are Gödel numbers of formulas ***A*** and ***B***, such that ***A*** is immediately convertible into ***B***. Hence, utilizing the fact that a conversion is a finite sequence of immediate conversions, the proposition $\Psi(a, b)$ will be expressible that a and b are Gödel numbers of formulas ***A*** and ***B*** such that ***A*** conv ***B***. Moreover if ***A*** conv ***B***, and a and b are the Gödel numbers of ***A*** and ***B*** respectively, the proposition $\Psi(a, b)$ will be provable in the system, by a proof which amounts to exhibiting, in terms of Gödel numbers, a particular finite sequence of immediate conversions, leading from ***A*** to ***B***; and if ***A*** is not convertible into ***B***, the ω-consistency of the system means that $\Psi(a, b)$ will not be provable. If the decision problem for the system were solved, there would be a means of determining effectively of every proposition $\Psi(a, b)$ whether it was provable, and hence a means of determining effectively of every pair of formulas ***A*** and ***B*** whether ***A*** conv ***B***, contrary to Theorem XIX.

In particular, if the system of *Principia Mathematica* be ω-consistent, its decision problem is unsolvable.

PRINCETON UNIVERSITY,
PRINCETON, N.J.

A NOTE ON THE ENTSCHEIDUNGSPROBLEM (1936)

In *An unsolvable problem of elementary number theory*[1] the author proposed a definition of the commonly used term "effectively calculable" and showed on the basis of this definition that the general case of the decision problem is unsolvable in any system of symbolic logic which is adequate to a certain portion of arithmetic and is ω-consistent. The purpose of the present note is to outline an extension of this result to the *engere Funktionenkalkül* of Hilbert and Ackermann.[2]

In the cited paper it is pointed out that there can be associated recursively with every well-formed formula[3] a recursive enumeration of the formulas into which it is convertible.[3] This means the existence of a recursively defined function a of two positive integers such that, if y is the Gödel number of a well-formed formula $\boldsymbol{Y}$ then $a(x, y)$ is the Gödel number of the xth formula in the enumeration of the formulas into which $\boldsymbol{Y}$ is convertible.

Consider the system L of symbolic logic which arises from the *engere Funktionenkalkül* by adding to it: as additional undefined symbols, a symbol 1 for the number 1 (regarded as an individual), a symbol = for the propositional function = (equality of individuals), a symbol s for the arithmetic function $x+1$, a symbol a for the arithmetic function a described in the preceding paragraph, and symbols $b_1, b_2, \cdots, b_k$ for the auxiliary arithmetic functions which are employed in the recursive definition of a; and as additional axioms, the recursion equations for the functions $a, b_1, b_2, \cdots, b_k$ (expressed with free individual variables, the class of individuals being taken as identical with the class of positive integers), and a finite list of axioms of equality which may be described as follows. First we take the two axioms of equality,

$$(x)[x = x],$$

$$(x)(y)(z)[x = y \supset [x = z \supset y = z]],$$

then for s the axiom,

$$(x)(y)[x = y \supset s(x) = s(y)],$$

[*Editor's note*: First published, ***The Journal of Symbolic Logic***, vol. 1, no. 1 (1936), pp. 40–41. © Association for Symbolic Logic. Reprinted, with the changes noted in the following note, by permission. The following footnote was written by Church in 1971.]

This paper was first published in ***The Journal of Symbolic Logic***, vol. 1 (1936), pp. 40–41, with a correction appearing in the same volume, pp. 101–102. The text of the paper has here been rewritten by the author [in 1971] to incorporate the correction. At the same time some changes in terminology have been made, to conform to later usage, and the logical notation of Hilbert and Ackermann — which was used in the original paper — has been changed to the Peano–Russell notation, for the sake of agreement with the notation used in later papers. There has been an effort not to depart otherwise from the point of view from which the paper was written in 1936 — except in passages which are explicitly marked as "added in 1971" and in footnotes to such passages.

The author is indebted to Paul Bernays for pointing out the error in the first version of the proof and suggesting the method by which it was corrected in 1936, and also for calling attention to the desirability of distinguishing in this connection between proofs which are constructive (*finit*) and those which are not.

[1] ***American Journal of Mathematics***, vol. 58 (1936), pp. 345–353.

[2] Hilbert and Ackermann, ***Grundzüge der theoretischen Logik***, Berlin, 1928. (Added 1971: For a sketch of the pure functional calculus of first order and an explanation of its decision problem, see *Special cases of the decision problem*, ***Revue philosophique de Louvain***, vol. 49 (1951), pp. 203–221, or for a more detailed account see the author's ***Introduction to mathematical logic***, Vol. I, Princeton 1956, sixth printing 1970.)

[3] As these terms are defined in *An unsolvable problem of elementary number theory*.

then for a the two axioms,

$(x)(y)(z)[x = y \supset a(x, z) = a(y, z)]$,

$(x)(y)(z)[x = y \supset a(z, x) = a(z, y)]$,

and then similar axioms for each of the functions $b_1, b_2, \cdots, b_k$. In the presence of the axioms and rules of the pure functional calculus of first order, there then follow as theorems of L all well-formed formulas $\mathbf{c} = \mathbf{d} \supset [\mathbf{M} \supset \mathbf{N}]$ of L, where **c** and **d** are any individual variables, and **N** is obtained from **M** by substituting **d** for all free occurrences of **c** throughout **M** (subject to the restriction that no bound occurrences of **d** result by the substitution).[4]

The consistency of the system L follows by the methods of existing proofs.[5] The ω-consistency of L is a matter of more difficulty, but for our present purpose the following weaker property of L is sufficient: *If* **P** *contains no quantifiers and* $(\exists x)\mathbf{P}$ *is provable in* L, *then some one of* $\mathbf{P}_1, \mathbf{P}_2, \mathbf{P}_3, \cdots$ *is provable in* L (*where* $\mathbf{P}_1, \mathbf{P}_2, \mathbf{P}_3, \cdots$ *are respectively the results of substituting* $1, s(1), s(s(1)), \cdots$ *for* x *throughout* **P**). This property has been proved by Bernays[6] for any one of a class of systems of which L is one. Hence, by the argument of the author's cited paper follows:

The general case of the decision problem of the system L *is unsolvable.*

Now by a device which is well known, it is possible to replace the system L by an equivalent system L′ which has no symbols for arithmetic functions. This is done by replacing $s, a, b_1, b_2, \cdots, b_k$ by the symbols $S, A, B_1, B_2, \cdots, B_k$ for the propositional functions $x = s(y), x = a(y, z)$, etc., and making corresponding alterations in the axioms of L as follows.

The recursion equations for $a, b_1, b_2, \cdots, b_k$ are re-expressed in the new notation, all the individual variables being again bound by initially placed universal quantifiers. The first two axioms of equality remain the same. In place of the equality axiom for s, we have four axioms for S,

$(y)(\exists x)S(x, y)$,

$(x)(y)(z)[S(x, z)S(y, z) \supset x = y]$,

$(x)(y)(z)[x = y \supset [S(x, z) \supset S(y, z)]]$,

$(x)(y)(z)[x = y \supset [S(z, x) \supset S(z, y)]]$.

In place of the two equality axioms for a, we have five axioms in which A is involved,

$(y)(z)(\exists x)A(x, y, z)$,

$(x_1)(x_2)(y)(z)[A(x_1, y, z)A(x_2, y, z) \supset x_1 = x_2]$,

$(x_1)(x_2)(y)(z)[x_1 = x_2 \supset [A(x_1, y, z) \supset A(x_2, y, z)]]$,

$(x)(y_1)(y_2)(z)[y_1 = y_2 \supset [A(x, y_1, z) \supset A(x, y_2, z)]]$,

$(x)(y)(z_1)(z_2)[z_1 = z_2 \supset [A(x, y, z_1) \supset A(x, y, z_2)]]$.

And there are similar sets of axioms for each of $B_1, B_2, \cdots, B_k$.

The system L′ differs from the pure functional calculus of first order by the additional primitive symbols $1, =, S, A, B_1, B_2, \cdots, B_k$ and by a finite number of well-formed

---

[4]In other words, all results of substituting for F in $\mathbf{c} = \mathbf{d} \supset [F(\mathbf{c}) \supset F(\mathbf{d})]$, such that no functional variables remain after the substitution. For proof, see Hilbert and Bernays, ***Grundlagen der Mathematik***, Vol. I, Berlin 1934, pp. 373–375.

[5]Cf. Wilhelm Ackermann, *Begründung des "tertium non datur" mittels der Hilbertschen Theorie der Widerspruchsfreiheit*, ***Mathematische Annalen***, vol. 93 (1924–5), pp. 1–136; J. v. Neumann, *Zur Hilbertschen Beweistheorie*, ***Mathematische Zeitschrift***, vol. 26 (1927), pp. 1–46; Jacques Herbrand, *Sur la non-contradiction de l'arithmétique*, ***Journal für die reine und angewandte Mathematik***, vol. 166 (1931–2), pp. 1–8.

[6]In lectures at Princeton, N.J., 1936. The methods employed are those of existing consistency proofs.

formulas which are introduced as additional axioms. Let T be the conjunction of these additional axioms, and say that u is an individual variable which does not occur in T. Choose functional (i.e., propositional-function) variables $F_1, F_2, \cdots, F_{k+3}$, each one being of the appropriate number of arguments — viz. F_1 is to be binary, F_2 binary, F_3 ternary, and so on. And let W be the result of substituting throughout T the variables $u, F_1, F_2, \cdots, F_{k+3}$ for the symbols $1, =, S, A, B_1, B_2, \cdots, B_k$ respectively.

Let **Q** be a well-formed formula of L′. We may suppose without loss of generality that **Q** contains none of the variables $u, F_1, F_2, \cdots, F_{k+3}$. Let **R** be the result of substituting throughout **Q** the variables $u, F_1, F_2, \cdots, F_{k+3}$ for the symbols $1, =, S, A, B_1, B_2, \cdots, B_k$ respectively. Then **Q** is provable in L′ if and only if $W \supset \mathbf{R}$ is provable in the pure functional calculus of first order.

Thus a solution of the general case of the decision problem of the pure first-order functional calculus would lead to a solution of the general case of the decision problem of L′ and hence of L. Therefore:

The general case of the decision problem of the pure functional calculus of first order is unsolvable.[7]

The foregoing argument provides a constructive proof of the unsolvability of what we may call the *deducibility problem*[8] of the pure functional calculus of first order, that is, the problem to find an effective procedure which is capable of determining, about any given well-formed formula, whether there exists a proof of it. The inference, however, to the unsolvability of the other form of the decision problem, which concerns a procedure for determining universal validity, depends on the non-constructively proved theorem of Gödel that every universally valid formula in the pure functional calculus of first order has a proof,[9] as well as on the converse of this, that every provable formula is universally valid. The unsolvability of this second form of the decision problem of the pure functional calculus of first order cannot, therefore, be regarded as established beyond question.[10]

(Added 1971.) But the pessimism of this last sentence is unnecessary, as it is possible to avoid the non-constructiveness by the following modification of the argument, in which no use is made of the completeness theorem of Gödel. In the system L we have in fact the unsolvability of that special case of the decision problem which concerns

[7]From this follows further the unsolvability of the particular case of the decision problem of the pure first-order functional calculus which concerns the provability of well-formed formulas of the form $(\exists \mathbf{b}_1)(\exists \mathbf{b}_2)(\exists \mathbf{b}_3)(\mathbf{c}_1)(\mathbf{c}_2)\cdots(\mathbf{c}_n)\mathbf{M}$, where **M** contains no quantifiers and no individual variables except $\mathbf{b}_1, \mathbf{b}_2, \mathbf{b}_3, \mathbf{c}_1, \mathbf{c}_2, \cdots, \mathbf{c}_n$. Cf. Kurt Gödel, *Zum Entscheidungsproblem des logischen Funktionenkalküls*, ***Monatshefte für Mathematik und Physik***, vol. 40 (1933), pp. 433–443.

[8](Added 1971.) The writer now prefers the terms *decision problem for provability* and *decision problem for validity*.

[9]*Die Vollständigkeit der Axiome des logischen Funktionenkalküls*, ***Monatshefte für Mathematik und Physik***, vol. 37 (1930), pp. 349–360.

[10](Added 1971.) This is the original wording of this sentence. The writer professes to have vacillated in his estimate of the value of certain non-constructive methods of proof. But it is certainly possible to see an important distinction between, e.g., existence proofs which do not even in principle provide an example of anything of the sort which is said to exist and others which proceed more directly or constructively — without therefore necessarily adopting either Brouwer's mathematical subjectivism or such Brouwerian doctrines as that "mathematics is prior to logic." And indeed there now seems to be a substantial consensus that it is desirable at least to make and maintain the distinction between constructive proofs and others, and even (following Hilbert) that it is especially important to do this in connection with metamathematical proofs.

formulas of the form $(\exists x)[\mathbf{m} = a(x, \mathbf{n})]$, where **m** and **n** are numerals,[11] and this follows constructively for both forms of the decision problem. For any particular numerals **m** and **n**, let **Q** be the well-formed formula of L′ which corresponds to $(\exists x)[\mathbf{m} = a(x, \mathbf{n})]$, re-expressing it in the new notation.[12] Then **Q** is valid in L′ if and only if $W \supset \mathbf{R}$ is valid in the pure functional calculus of first order.[13] So we may argue that solution of the general case of the decision problem for validity in the pure first-order functional calculus would lead to the solution of *a special case of the decision problem for validity in* L *which is known to be unsolvable.*

[11]The numerals are the formulas 1, $s(1), s(s(1)), \cdots$ already mentioned in stating Bernays's theorem. And the reason for the unsolvability of this special case of the decision problem of L is that it corresponds (by way of Gödel numbers) to the known unsolvable decision problem in the calculus of λ-conversion, to determine of well-formed formulas **M** and **N** whether **M** conv **N** (*An unsolvable problem of elementary number theory*, op. cit., Theorem XIX.)

[12]In a fuller treatment this heuristic indication must be replaced by a precise syntactical statement specifying uniquely how **Q** is obtained from $(\exists x)[\mathbf{m} = a(x, \mathbf{n})]$.

[13]Briefly, if $(\exists x)[\mathbf{m} = a(x, \mathbf{n})]$ is valid in L, there must be some numeral **x** such that $\mathbf{m} = a(\mathbf{x}, \mathbf{n})$ is valid. Then this last equation must be derivable by the standard calculation process from the recursion equations for $a, b_1, b_2, \cdots, b_k$ (since all the recursion equations are valid, the calculation process preserves validity, and it is known that some equation,

$$\text{numeral} = a(\mathbf{x}, \mathbf{n}),$$

must be derivable by the calculation process). The calculation process may be parallel for any choice of a non-empty domain of individuals and any system of values of the variables $u, F_1, F_2, \cdots, F_{k+3}$ that gives to W the value truth, to show that **R** then also has the value truth; that is, $W \supset \mathbf{R}$ is valid. —On the other hand if we are given that $W \supset \mathbf{R}$ is valid, we consider the system of values $1, =, S, A, B_1, B_2, \cdots, B_k$ of the variables $u, F_1, F_2, \cdots, F_{k+3}$; this system of values of the variables give to W the value truth; hence it must give to **R** the value truth; hence **Q** is valid.

SOME PROPERTIES OF CONVERSION[1]

(1936)

Alonzo Church and J. B. Rosser

Our purpose is to establish the properties of conversion which are expressed in Theorems 1 and 2 below. We shall consider first conversion defined by Church's Rules I, II, III[2] and shall then extend our results to several other kinds of conversion.[3]

1. Conversion defined by Church's Rules I, II, III. In our study of conversion we are particularly interested in the effects of Rules II and III and consider that applications of Rule I, though often necessary to prevent confusion of free and bound variables, do not essentially change the structure of a formula. Hence we shall omit mention of applications of Rule I whenever it seems that no essential ambiguity will result. Thus when we speak of replacing $\{\lambda \boldsymbol{x} \,.\, \boldsymbol{M}\}(\boldsymbol{N})$[4] by $\mathrm{S}^{\boldsymbol{x}}_{\boldsymbol{N}}\boldsymbol{M}|$ it shall be understood that any applications of I are made which are needed to make this substitution an application of II. Also we may write bound variables as unchanged throughout discussions even though tacit applications of I in the discussion may have changed them.

A conversion in which III is not used and II is used exactly once will be called a *reduction*. If II is not used and III is used exactly once, the conversion will be called an *expansion*. "***A*** imr ***B***," read "***A*** is immediately reducible to ***B***," shall mean that it is possible to go from ***A*** to ***B*** by a single reduction. "***A*** red ***B***," read "***A*** is reducible to ***B***," shall mean that it is possible to go from ***A*** to ***B*** by one or more reductions.[5] "***A*** conv-I ***B***," read "***A*** conv ***B*** by applications of I only," shall mean just that (including the case of a zero number of applications). "***A*** conv-I–II ***B***," read "***A*** conv ***B*** by applications of I and II only," shall mean just that (including the case of a zero number of applications).

We shall say that we *contract* or *perform a contraction on* $\{\lambda \boldsymbol{x} \,.\, \boldsymbol{M}\}(\boldsymbol{N})$ if we replace it by $\mathrm{S}^{\boldsymbol{x}}_{\boldsymbol{N}}\boldsymbol{M}|$.

It is possible to visualize the process of conversion by drawing a broken line in which the segments correspond to successive steps of the conversion, horizontal segments indicating applications of I, segments of negative slope applications of II, and segments of positive slope applications of III. Thus, in the figure A_1 conv A_2 and B_1 conv B_2 each by a single use of I, A_2 conv A_3 and B_2 conv B_3 each by a single use of III, and A_7

Originally published in ***Transactions of the American Mathematical Society***, vol. 39 (1936), pp. 472–482.

[1]Presented to the Society, April 20, 1935; received by the editors June 4, 1935.

[2]By Church's rules we shall mean the rules of procedure given in A. Church, *A set of postulates for the foundation of logic*, ***Annals of Mathematics***, (2), vol. 33 (1932), pp. 346–366 (see pp. 355–356), as modified by S. C. Kleene, *Proof by cases in formal logic*, ***Annals of Mathematics***, (2), vol. 35 (1934), pp. 529–544 (see p. 530). We assume familiarity with the material on pp. 349–355 of Church's paper and in §§1, 2, 3, 5 of Kleene's paper. We shall refer to the latter paper as "Kleene."

[3]The authors are indebted to Dr. S. C. Kleene for assistance in the preparation of this paper, in particular for the detection of an error in the first draft of it and for the suggestion of an improvement in the proof of Theorem 2.

[4]Note carefully the convention at the beginning of §3, Kleene, which we shall constantly use.

[5]Our use of "conv" allows us to write "***A*** conv ***B***" even in the case that no applications of I, II, or III are made in going from ***A*** to ***B*** and ***A*** is the same as ***B***. But we write "***A*** red ***B***" only if there is at least one reduction in the process of going from ***A*** to ***B*** by applications of I and II, and use the notation "***A*** conv-I–II ***B***" if we wish to allow the possibility of no reductions.

conv A_8 and C_1 conv C_2 each by a single use of II. The dotted lines represent various alternative conversions to the conversion given by the solid line.

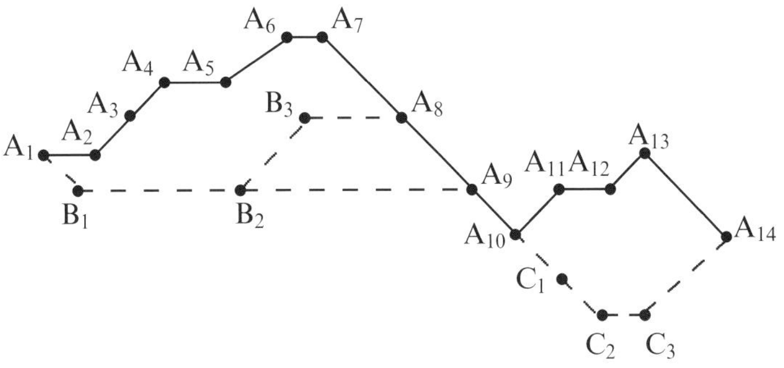

FIGURE 1

A conversion in which no expansions follow any reductions will be called a *peak* and one in which no reductions follow any expansions will be called a *valley*. The central theorem of this paper states that if $\boldsymbol{A}$ conv $\boldsymbol{B}$, there is a conversion from $\boldsymbol{A}$ to $\boldsymbol{B}$ which is a valley. We prove it by means of a lemma which states that a peak in which there is a single reduction can always be replaced by a valley. Then the theorem becomes obvious. For example, in the conversion pictured by the solid line in the figure we replace the peak $A_1A_2 \cdots A_8$ by the valley $A_1B_1B_2B_3A_8$, then the peak $B_2B_3A_8A_9$ by the valley B_2A_9, then the peak $A_{10}A_{11} \cdots A_{14}$ by the valley $A_{10}C_1C_2C_3A_{14}$, getting the valley $A_1B_1B_2A_9A_{10}C_1C_2C_3A_{14}$.

Suppose that a formula $\boldsymbol{A}$ has parts $\{\lambda \boldsymbol{x}_j \,.\, \boldsymbol{M}_j\}(\boldsymbol{N}_j)$ which may or may not be parts of each other (cf. Kleene 2VIII (p. 532)). We suppose that, if $p \neq q$, $\{\lambda \boldsymbol{x}_p \,.\, \boldsymbol{M}_p\}(\boldsymbol{N}_p)$ is not the same part as $\{\lambda \boldsymbol{x}_q \,.\, \boldsymbol{M}_q\}(\boldsymbol{N}_q)$, though it may be the same formula. The $\{\lambda \boldsymbol{x}_j \,.\, \boldsymbol{M}_j\}(\boldsymbol{N}_j)$ need not be all of the parts of $\boldsymbol{A}$ which have the form $\{\lambda \boldsymbol{y} \,.\, \boldsymbol{P}\}(\boldsymbol{Q})$. We define the *residuals* of the $\{\lambda \boldsymbol{x}_j \,.\, \boldsymbol{M}_j\}(\boldsymbol{N}_j)$ after a sequence of applications of I and II (these residuals being certain well-formed parts of the formula which results from the sequence of applications of I and II). If no applications of I or II occur, each part $\{\lambda \boldsymbol{x}_j \,.\, \boldsymbol{M}_j\}(\boldsymbol{N}_j)$ is its own residual. If a series of applications of I occur, then each part $\{\lambda \boldsymbol{x}_j \,.\, \boldsymbol{M}_j\}(\boldsymbol{N}_j)$ is changed into a part $\{\lambda \boldsymbol{y}_j \,.\, \boldsymbol{M}'_j\}(\boldsymbol{N}'_j)$ of the resulting formula and this part $\{\lambda \boldsymbol{y}_j \,.\, \boldsymbol{M}'_j\}(\boldsymbol{N}'_j)$ is the residual of $\{\lambda \boldsymbol{x}_j \,.\, \boldsymbol{M}_j\}(\boldsymbol{N}_j)$ in the resulting formula. Clearly the residuals of two parts coincide only if the parts coincide. Next consider the case where a single reduction occurs. Let $\{\lambda \boldsymbol{x} \,.\, \boldsymbol{M}\}(\boldsymbol{N})$ be the part of $\boldsymbol{A}$ which is contracted in performing the reduction and $\boldsymbol{A}'$ be the formula resulting from $\boldsymbol{A}$ by the reduction. Let $\{\lambda \boldsymbol{x}_p \,.\, \boldsymbol{M}_p\}(\boldsymbol{N}_p)$ be an arbitrary one of the $\{\lambda \boldsymbol{x}_j \,.\, \boldsymbol{M}_j\}(\boldsymbol{N}_j)$.

Case 1. Let $\{\lambda \boldsymbol{x}_p \,.\, \boldsymbol{M}_p\}(\boldsymbol{N}_p)$ not be part of $\{\lambda \boldsymbol{x} \,.\, \boldsymbol{M}\}(\boldsymbol{N})$. Then (a) $\{\lambda \boldsymbol{x} \,.\, \boldsymbol{M}\}(\boldsymbol{N})$ has no part in common with $\{\lambda \boldsymbol{x}_p \,.\, \boldsymbol{M}_p\}(\boldsymbol{N}_p)$ or else (b) $\{\lambda \boldsymbol{x} \,.\, \boldsymbol{M}\}(\boldsymbol{N})$ is part of $\{\lambda \boldsymbol{x}_p \,.\, \boldsymbol{M}_p\}(\boldsymbol{N}_p)$ (by Kleene 2VIII). If (a) holds, then under the reduction from $\boldsymbol{A}$ to $\boldsymbol{A}'$, $\{\lambda \boldsymbol{x}_p \,.\, \boldsymbol{M}_p\}(\boldsymbol{N}_p)$ goes into a definite part of $\boldsymbol{A}'$ which we shall call the residual in $\boldsymbol{A}'$ of $\{\lambda \boldsymbol{x}_p \,.\, \boldsymbol{M}_p\}(\boldsymbol{N}_p)$. In this case the residual of $\{\lambda \boldsymbol{x}_p \,.\, \boldsymbol{M}_p\}(\boldsymbol{N}_p)$ is the same formula as $\{\lambda \boldsymbol{x}_p \,.\, \boldsymbol{M}_p\}(\boldsymbol{N}_p)$. If (b) holds then $\{\lambda \boldsymbol{x} \,.\, \boldsymbol{M}\}(\boldsymbol{N})$ is part either of $\boldsymbol{M}_p$ or of $\boldsymbol{N}_p$ since $\{\lambda \boldsymbol{x}_p \,.\, \boldsymbol{M}_p\}(\boldsymbol{N}_p)$ is not part of $\{\lambda \boldsymbol{x} \,.\, \boldsymbol{M}\}(\boldsymbol{N})$ (see Kleene, 2X and 2XII). Hence if we contract $\{\lambda \boldsymbol{x} \,.\, \boldsymbol{M}\}(\boldsymbol{N})$ we perform a reduction on the formula $\{\lambda \boldsymbol{x}_p \,.\, \boldsymbol{M}_p\}(\boldsymbol{N}_p)$

which carries it into a formula $\{\lambda \boldsymbol{x}'_p \,.\, \boldsymbol{M}'_p\}(\boldsymbol{N}'_p)$. Then the reduction from $\boldsymbol{A}$ to $\boldsymbol{A}'$ can be considered as consisting of the replacement of the part $\{\lambda \boldsymbol{x}_p \,.\, \boldsymbol{M}_p\}(\boldsymbol{N}_p)$ of $\boldsymbol{A}$ by $\{\lambda \boldsymbol{x}'_p \,.\, \boldsymbol{M}'_p\}(\boldsymbol{N}'_p)$ and this particular occurrence of $\{\lambda \boldsymbol{x}'_p \,.\, \boldsymbol{M}'_p\}(\boldsymbol{N}'_p)$ in $\boldsymbol{A}'$ is called the residual in $\boldsymbol{A}'$ of $\{\lambda \boldsymbol{x}_p \,.\, \boldsymbol{M}_p\}(\boldsymbol{N}_p)$.

Case 2. Let $\{\lambda \boldsymbol{x}_p \,.\, \boldsymbol{M}_p\}(\boldsymbol{N}_p)$ be part of $\{\lambda \boldsymbol{x} \,.\, \boldsymbol{M}\}(\boldsymbol{N})$. By Kleene 2X this case breaks up into three subcases.

(a) Let $\{\lambda \boldsymbol{x}_p \,.\, \boldsymbol{M}_p\}(\boldsymbol{N}_p)$ be $\{\lambda \boldsymbol{x} \,.\, \boldsymbol{M}\}(\boldsymbol{N})$. Then we say that $\{\lambda \boldsymbol{x}_p \,.\, \boldsymbol{M}_p\}(\boldsymbol{N}_p)$ has no residual in $\boldsymbol{A}'$.

(b) Let $\{\lambda \boldsymbol{x}_p \,.\, \boldsymbol{M}_p\}(\boldsymbol{N}_p)$ be part of $\lambda \boldsymbol{x} \,.\, \boldsymbol{M}$ and hence part of $\boldsymbol{M}$ (by Kleene 2XII). Let $\boldsymbol{M}'$ be the result of replacing all free $\boldsymbol{x}$'s of $\boldsymbol{M}$ except those occurring in $\{\lambda \boldsymbol{x}_p \,.\, \boldsymbol{M}_p\}(\boldsymbol{N}_p)$ by $\boldsymbol{N}$. Under these changes the part $\{\lambda \boldsymbol{x}_p \,.\, \boldsymbol{M}_p\}(\boldsymbol{N}_p)$ of $\boldsymbol{M}$ goes into a definite part of $\boldsymbol{M}'$ which we shall denote also by $\{\lambda \boldsymbol{x}_p \,.\, \boldsymbol{M}_p\}(\boldsymbol{N}_p)$, since it is the same formula. If now we replace $\{\lambda \boldsymbol{x}_p \,.\, \boldsymbol{M}_p\}(\boldsymbol{N}_p)$ in $\boldsymbol{M}'$ by $\mathrm{S}^{\boldsymbol{x}}_{\boldsymbol{N}}\{\lambda \boldsymbol{x}_p \,.\, \boldsymbol{M}_p\}(\boldsymbol{N}_p)|$, $\boldsymbol{M}'$ becomes $\mathrm{S}^{\boldsymbol{x}}_{\boldsymbol{N}}\boldsymbol{M}|$ and we denote by $\mathrm{S}^{\boldsymbol{x}}_{\boldsymbol{N}}\{\lambda \boldsymbol{x}_p \,.\, \boldsymbol{M}_p\}(\boldsymbol{N}_p)|$ the particular occurrence of $\mathrm{S}^{\boldsymbol{x}}_{\boldsymbol{N}}\{\lambda \boldsymbol{x}_p \,.\, \boldsymbol{M}_p\}(\boldsymbol{N}_p)|$ in $\mathrm{S}^{\boldsymbol{x}}_{\boldsymbol{N}}\boldsymbol{M}|$ that resulted from replacing $\{\lambda \boldsymbol{x}_p \,.\, \boldsymbol{M}_p\}(\boldsymbol{N}_p)$ in $\boldsymbol{M}'$ by the formula $\mathrm{S}^{\boldsymbol{x}}_{\boldsymbol{N}}\{\lambda \boldsymbol{x}_p \,.\, \boldsymbol{M}_p\}(\boldsymbol{N}_p)|$. Now the residual in $\boldsymbol{A}'$ of $\{\lambda \boldsymbol{x}_p \,.\, \boldsymbol{M}_p\}(\boldsymbol{N}_p)$ in $\boldsymbol{A}$ is defined to be the part $\mathrm{S}^{\boldsymbol{x}}_{\boldsymbol{N}}\{\lambda \boldsymbol{x}_p \,.\, \boldsymbol{M}_p\}(\boldsymbol{N}_p)|$ in the particular occurrence of $\mathrm{S}^{\boldsymbol{x}}_{\boldsymbol{N}}\boldsymbol{M}|$ in $\boldsymbol{A}'$ that resulted from replacing $\{\lambda \boldsymbol{x} \,.\, \boldsymbol{M}\}(\boldsymbol{N})$ in $\boldsymbol{A}$ by $\mathrm{S}^{\boldsymbol{x}}_{\boldsymbol{N}}\boldsymbol{M}|$.

(c) Let $\{\lambda \boldsymbol{x}_p \,.\, \boldsymbol{M}_p\}(\boldsymbol{N}_p)$ be part of $\boldsymbol{N}$. And let $\{\lambda \boldsymbol{y}_i \,.\, \boldsymbol{P}_i\}(\boldsymbol{Q}_i)$ respectively stand for the particular occurrences of the formula $\{\lambda \boldsymbol{x}_p \,.\, \boldsymbol{M}_p\}(\boldsymbol{N}_p)$ in $\mathrm{S}^{\boldsymbol{x}}_{\boldsymbol{N}}\boldsymbol{M}|$ which are the part $\{\lambda \boldsymbol{x}_p \,.\, \boldsymbol{M}_p\}(\boldsymbol{N}_p)$ in each of the particular occurrences of the formula $\boldsymbol{N}$ in $\mathrm{S}^{\boldsymbol{x}}_{\boldsymbol{N}}\boldsymbol{M}|$ that resulted from replacing the free $\boldsymbol{x}$'s of $\boldsymbol{M}$ by $\boldsymbol{N}$. Now the residuals in $\boldsymbol{A}'$ of $\{\lambda \boldsymbol{x}_p \,.\, \boldsymbol{M}_p\}(\boldsymbol{N}_p)$ in $\boldsymbol{A}$ are the parts $\{\lambda \boldsymbol{y}_i \,.\, \boldsymbol{P}_i\}(\boldsymbol{Q}_i)$ in the particular occurrence of the formula $\mathrm{S}^{\boldsymbol{x}}_{\boldsymbol{N}}\boldsymbol{M}|$ in $\boldsymbol{A}'$ that resulted from replacing $\{\lambda \boldsymbol{x} \,.\, \boldsymbol{M}\}(\boldsymbol{N})$ in $\boldsymbol{A}$ by $\mathrm{S}^{\boldsymbol{x}}_{\boldsymbol{N}}\boldsymbol{M}|$.

This completes the definition of the residuals in $\boldsymbol{A}'$ of $\{\lambda \boldsymbol{x}_p \,.\, \boldsymbol{M}_p\}(\boldsymbol{N}_p)$ in $\boldsymbol{A}$ in the case that $\boldsymbol{A}'$ is obtained from $\boldsymbol{A}$ by a single reduction. Clearly the residuals of $\boldsymbol{A}'$ of $\{\lambda \boldsymbol{x}_p \,.\, \boldsymbol{M}_p\}(\boldsymbol{N}_p)$ in $\boldsymbol{A}$ (if any) have the form $\{\lambda \boldsymbol{x} \,.\, \boldsymbol{M}\}(\boldsymbol{N})$. Call them $\{\lambda \boldsymbol{y}_i \,.\, \boldsymbol{P}_i\}(\boldsymbol{Q}_i)$. Then if $\boldsymbol{A}'$ imr $\boldsymbol{A}''$, the residuals in $\boldsymbol{A}''$ of $\{\lambda \boldsymbol{x}_p \,.\, \boldsymbol{M}_p\}(\boldsymbol{N}_p)$ in $\boldsymbol{A}$ are defined to be the residuals in $\boldsymbol{A}''$ of the $\{\lambda \boldsymbol{y}_i \,.\, \boldsymbol{P}_i\}(\boldsymbol{Q}_i)$ in $\boldsymbol{A}'$. We continue in this way, defining the residuals after each successive reduction as the residuals of the formulas that were residuals before the reduction, and noting that the residuals always have the form $\{\lambda \boldsymbol{x} \,.\, \boldsymbol{M}\}(\boldsymbol{N})$. We also note that a residual in $\boldsymbol{B}$ of the part $\{\lambda \boldsymbol{x} \,.\, \boldsymbol{M}\}(\boldsymbol{N})$ in $\boldsymbol{A}$ cannot coincide with a residual in $\boldsymbol{B}$ of the part $\{\lambda \boldsymbol{x}' \,.\, \boldsymbol{M}'\}(\boldsymbol{N}')$ in $\boldsymbol{A}$ unless $\{\lambda \boldsymbol{x} \,.\, \boldsymbol{M}\}(\boldsymbol{N})$ coincides with $\{\lambda \boldsymbol{x}' \,.\, \boldsymbol{M}'\}(\boldsymbol{N}')$.

We say that a sequence of reductions on $\boldsymbol{A}$, say $\boldsymbol{A}$ imr $\boldsymbol{A}_1$ imr $\boldsymbol{A}_2$ imr $\cdots$ imr $\boldsymbol{A}_{n+1}$, is a *sequence of contractions on the parts* $\{\lambda \boldsymbol{x}_j \,.\, \boldsymbol{M}_j\}(\boldsymbol{N}_j)$ *of* $\boldsymbol{A}$ if the reduction from $\boldsymbol{A}_i$ to $\boldsymbol{A}_{i+1}$ ($i = 0, \cdots, n$; $\boldsymbol{A}_0$ the same as $\boldsymbol{A}$) is a contraction on one of the residuals in $\boldsymbol{A}_i$ of the $\{\lambda \boldsymbol{x}_j \,.\, \boldsymbol{M}_j\}(\boldsymbol{N}_j)$. Moreover, if no residuals of the $\{\lambda \boldsymbol{x}_j \,.\, \boldsymbol{M}_j\}(\boldsymbol{N}_j)$ occur in $\boldsymbol{A}_{n+1}$ we say that the sequence of contractions on the $\{\lambda \boldsymbol{x}_j \,.\, \boldsymbol{M}_j\}(\boldsymbol{N}_j)$ terminates and that $\boldsymbol{A}_{n+1}$ is the result.

In some cases we wish to speak of a sequence of contractions on the parts $\{\lambda \boldsymbol{x}_j \,.\, \boldsymbol{M}_j\}(\boldsymbol{N}_j)$ of $\boldsymbol{A}$ where the set $\{\lambda \boldsymbol{x}_j \,.\, \boldsymbol{M}_j\}(\boldsymbol{N}_j)$ may be vacuous. To handle this we shall agree that if the set $\{\lambda \boldsymbol{x}_j \,.\, \boldsymbol{M}_j\}(\boldsymbol{N}_j)$ is vacuous, the sequence of contractions shall be a vacuous sequence of reductions.

LEMMA 1. *If* $\{\lambda \boldsymbol{x}_j \,.\, \boldsymbol{M}_j\}(\boldsymbol{N}_j)$ *are parts of* $\boldsymbol{A}$, *then a number* m *can be found such that any sequence of contractions on the* $\{\lambda \boldsymbol{x}_j \,.\, \boldsymbol{M}_j\}(\boldsymbol{N}_j)$ *will terminate after at most* m

contractions, and if A' *and* A'' *are two results of terminating sequences of contractions on the* $\{\lambda x_j \,.\, M_j\}(N_j)$, *then* A' *conv-I* A''.

Proof by induction on the number of proper symbols of A.

The lemma is true if A is a proper symbol, the number m being 0.

Assume the lemma true for formulas with n or less proper symbols. Let A have $n+1$ proper symbols.

Case 1. A is $\lambda x \,.\, M$. Then all the parts $\{\lambda x_j \,.\, M_j\}(N_j)$ of A must be parts of M. However M has only n proper symbols and so we use the hypothesis of the induction.

Case 2. A is $\{F\}(X)$.

(a) $\{F\}(X)$ is not one of the $\{\lambda x_j \,.\, M_j\}(N_j)$. Then any sequence of contractions on the parts $\{\lambda x_j \,.\, M_j\}(N_j)$ of A can be replaced[6] by two sequences of contractions performed successively, one on those of the $\{\lambda x_j \,.\, M_j\}(N_j)$ which are parts of F and one on those of the $\{\lambda x_j \,.\, M_j\}(N_j)$ which are parts of X, in such a way that if the original sequence of contractions carried $\{F\}(X)$ into $\{F'\}(X')$ then the two sequences of contractions by which it is replaced carry F into F' and X into X' respectively and the total number of contractions on residuals of parts of F or on residuals of parts of X is the same as before.[7] Then we use the hypothesis of the induction, since F and X each have n or less proper symbols.

(b) $\{F\}(X)$ is one of the $\{\lambda x_j \,.\, M_j\}(N_j)$, say $\{\lambda x_p \,.\, M_p\}(N_p)$. As long as the residual of $\{\lambda x_p \,.\, M_p\}(N_p)$ has not been contracted, the argument of (a) applies. Hence if any sequence of contractions on the $\{\lambda x_j \,.\, M_j\}(N_j)$ of A is continued long enough, the residual of $\{\lambda x_p \,.\, M_p\}(N_p)$ must be contracted (since we prove readily that one and only one residual of $\{\lambda x_p \,.\, M_p\}(N_p)$ occurs, until a contraction on the residual of $\{\lambda x_p \,.\, M_p\}(N_p)$). Let a sequence of contractions, μ, on $\{\lambda x_j \,.\, M_j\}(N_j)$ consist of a sequence, ϕ, of contractions on the $\{\lambda x_j \,.\, M_j\}(N_j)$ which are different from $\{\lambda x_p \,.\, M_p\}(N_p)$, a contraction, β, on the residual of $\{\lambda x_p \,.\, M_p\}(N_p)$, and a sequence, θ, consisting of the remaining contractions of μ. Now, as in (a), we can replace ϕ by a sequence, α, of contractions on the $\{\lambda x_j \,.\, M_j\}(N_j)$ which are parts of $\lambda x_p \,.\, M_p$ (and therefore parts of M_p), followed by a sequence, η, of contractions on the $\{\lambda x_j \,.\, M_j\}(N_j)$ which are parts of N_p, and this without changing the total number of contractions on residuals of parts of M_p or on residuals of parts of N_p. Then η followed by β can be replaced by β', a contraction on the residual, $\{\lambda y \,.\, P\}(N_p)$, of $\{\lambda x_p \,.\, M_p\}(N_p)$, followed by a set of applications of η on each of the occurrences of N_p in $S^{y}_{N_p} P|$ that arose by substituting N_p for y in P. Our sequence of contractions now has a special form, namely a sequence of contractions, α, on parts of M_p, followed by a contraction, β', on the residual of $\{\lambda x_p \,.\, M_p\}(N_p)$, followed by other contractions. We will now indicate a process whereby this sequence of contractions can be replaced by another having the same special form but having the property that after the contraction on the residual of $\{\lambda x_p \,.\, M_p\}(N_p)$ one less contraction on the residuals of parts of M_p occurs. This process can then be successively applied until no

[6]We say that a sequence, μ, of reductions on A can be replaced by a sequence, ν, of reductions on A, if both μ and ν give the same end formula B and residuals in B of any part $\{\lambda x_j \,.\, M_j\}(N_j)$ are the same under both μ and ν.

[7]The reader will easily understand the convention which we use when we say that the same sequence of contractions which carries F into F' will carry $\{F\}(X)$ into $\{F'\}(X)$ and that the same sequence of contractions which carries X into X' will carry $\{F'\}(X)$ into $\{F'\}(X')$. This convention will also be used in (b).

contraction on a residual of a part of $\boldsymbol{M}_p$ occurs after the contraction on the residual of $\{\lambda \boldsymbol{x}_p \,.\, \boldsymbol{M}_p\}(\boldsymbol{N}_p)$. Moreover this process increases by one the number of contractions which precede the contraction on the residual of $\{\lambda \boldsymbol{x}_p \,.\, \boldsymbol{M}_p\}(\boldsymbol{N}_p)$, so that the total number of contractions on the residuals of parts of $\boldsymbol{M}_p$ is the same after the process as before, and hence the same as in μ.

Let us consider that sequence of reductions ζ which is composed of β' and the contractions that follow it, up to and including the first contraction on a residual of a part of $\boldsymbol{M}_p$. Denoting the formula on which ζ acts by $\{\lambda \boldsymbol{y} \,.\, \boldsymbol{P}\}(\boldsymbol{N}_p)$, we see that ζ can be considered as the act of first replacing the free $\boldsymbol{y}$'s of $\boldsymbol{P}$ by various formulas, $\boldsymbol{N}_{pk}$, got from $\boldsymbol{N}_p$ by sets of reductions, and then contracting on a residual, $\{\lambda \boldsymbol{z} \,.\, \boldsymbol{R}\}(\boldsymbol{S})$, of one of the $\{\lambda \boldsymbol{x}_j \,.\, \boldsymbol{M}_j\}(\boldsymbol{N}_j)$ which are parts of $\boldsymbol{M}_p$, say $\{\lambda \boldsymbol{x}_q \,.\, \boldsymbol{M}_q\}(\boldsymbol{N}_q)$. From this point of view, we see that none of the free $\boldsymbol{z}$'s of $\boldsymbol{R}$ are parts of any $\boldsymbol{N}_{pk}$, and hence ζ can be replaced by a contraction on the residual in $\lambda \boldsymbol{y} \,.\, \boldsymbol{P}$ of $\{\lambda \boldsymbol{x}_q \,.\, \boldsymbol{M}_q\}(\boldsymbol{N}_q)$ of which $\{\lambda \boldsymbol{z} \,.\, \boldsymbol{R}\}(\boldsymbol{S})$ is a residual, followed by a contraction on the residual of $\{\lambda \boldsymbol{x}_p \,.\, \boldsymbol{M}_p\}(\boldsymbol{N}_p)$, followed by contractions on residuals of parts of $\boldsymbol{N}_p$. This completes the indication of what the process is.

Hence μ can be replaced by a sequence of contractions, α, on the $\{\lambda \boldsymbol{x}_j \,.\, \boldsymbol{M}_j\}(\boldsymbol{N}_j)$ which are parts of $\boldsymbol{M}_p$ (such that α contains as many contractions as there are contractions in μ on residuals of parts of $\boldsymbol{M}_p$), followed by a contraction, β, on the residual of $\{\lambda \boldsymbol{x}_p \,.\, \boldsymbol{M}_p\}(\boldsymbol{N}_p)$, followed by a sequence of contractions, γ, on residuals of parts of $\boldsymbol{N}_p$. Moreover, after α and β, $\{\lambda \boldsymbol{x}_p \,.\, \boldsymbol{M}_p\}(\boldsymbol{N}_p)$ has become a formula containing several occurrences of $\boldsymbol{N}_p$ and γ is a sequence of contractions on parts of these occurrences of $\boldsymbol{N}_p$. Hence, since $\lambda \boldsymbol{x}_p \,.\, \boldsymbol{M}_p$ and $\boldsymbol{N}_p$ each contain n or less proper symbols, we can use the hypothesis of the induction in connection with α and γ to show that if α followed by β followed by γ terminates the result is unique to within applications of I. But if μ terminates, so does α followed by β followed by γ. Hence the results of any two terminating sequences are unique to within applications of I.

It remains to be shown that a number m can be found such that each sequence of contractions terminates after at most m contractions.

By the hypothesis of induction, a number a can be found such that any sequence of contractions on those $\{\lambda \boldsymbol{x}_j \,.\, \boldsymbol{M}_j\}(\boldsymbol{N}_j)$ which are parts of $\boldsymbol{M}_p$ terminates after at most a contractions, and a number b can be found such that any sequence of contractions on those $\{\lambda \boldsymbol{x}_j \,.\, \boldsymbol{M}_j\}(\boldsymbol{N}_j)$ which are parts of $\boldsymbol{N}_p$ terminates after at most b contractions. Then any sequence of contractions on the $\{\lambda \boldsymbol{x}_j \,.\, \boldsymbol{M}_j\}(\boldsymbol{N}_j)$ of $\boldsymbol{A}$ must, if continued, include a contraction on the residual of $\{\lambda \boldsymbol{x}_p \,.\, \boldsymbol{M}_p\}(\boldsymbol{N}_p)$, after at most $a + b + 1$ contractions. Hence we may confine our attention to sequences of contractions μ on the $\{\lambda \boldsymbol{x}_j \,.\, \boldsymbol{M}_j\}(\boldsymbol{N}_j)$ of $\boldsymbol{A}$ which include a contraction on the residual of $\{\lambda \boldsymbol{x}_p \,.\, \boldsymbol{M}_p\}(\boldsymbol{N}_p)$ and which can therefore be replaced by a sequence of contractions of the form, α followed by β followed by γ. Moreover it will be clear, on examination of our preceding argument, that, when the sequence μ is replaced by the sequence, α followed by β followed by γ, the total number of contractions in the sequence is either increased or left unchanged (because each step in the process of transforming μ into α followed by β followed by γ has the property that it cannot decrease the total number of contractions). Therefore it is sufficient to find a number m such that any sequence of contractions on the $\{\lambda \boldsymbol{x}_j \,.\, \boldsymbol{M}_j\}(\boldsymbol{N}_j)$ of $\boldsymbol{A}$ which has the form, α followed by β followed by γ, must terminate after at most m contractions.

If we start with the formula $\boldsymbol{M}_p$ and perform a terminating sequence of contractions on those $\{\lambda \boldsymbol{x}_j \,.\, \boldsymbol{M}_j\}(\boldsymbol{N}_j)$ which are parts of $\boldsymbol{M}_p$, the result is a formula $\boldsymbol{M}'_p$, which is unique to within applications of I, and which contains a certain number, c, $\geqq 1$, of occurrences of $\boldsymbol{x}_p$ as a free symbol. And in the case of any sequence of contractions on those $\{\lambda \boldsymbol{x}_j \,.\, \boldsymbol{M}_j\}(\boldsymbol{N}_j)$ which are parts of $\boldsymbol{M}_p$, whether terminating or not, the result (that is, the formula into which $\boldsymbol{M}_p$ is transformed) contains at most c occurrences of $\boldsymbol{x}_p$ as a free symbol.

Hence the required number m is $a + 1 + cb$.

LEMMA 2. *If* $\boldsymbol{A}$ *imr* $\boldsymbol{B}$ *by a contraction on the part* $\{\lambda \boldsymbol{x} \,.\, \boldsymbol{M}\}(\boldsymbol{N})$ *of* $\boldsymbol{A}$, *and* $\boldsymbol{A}$ *is* $\boldsymbol{A}_1$, *and* $\boldsymbol{A}_1$ *imr* $\boldsymbol{A}_2$, $\boldsymbol{A}_2$ *imr* $\boldsymbol{A}_3$, $\cdots$, *and, for all* k, $\boldsymbol{B}_k$ *is the result of a terminating sequence of contractions on the residuals in* $\boldsymbol{A}_k$ *of* $\{\lambda \boldsymbol{x} \,.\, \boldsymbol{M}\}(\boldsymbol{N})$, *then*:

I. $\boldsymbol{B}_1$ *is* $\boldsymbol{B}$.

II. *For all* k, $\boldsymbol{B}_k$ *conv-I–II* $\boldsymbol{B}_{k+1}$.

III. *Even if the sequence* $\boldsymbol{A}_1, \boldsymbol{A}_2, \cdots$ *can be continued to infinity, there is a number* ϕ_m, *depending on the formula* $\boldsymbol{A}$, *the part* $\{\lambda \boldsymbol{x} \,.\, \boldsymbol{M}\}(\boldsymbol{N})$ *of* $\boldsymbol{A}$, *and the number* m, *such that, starting with* $\boldsymbol{B}_m$, *at most* ϕ_m *consecutive* $\boldsymbol{B}_k$*'s occur for which it is not true that* $\boldsymbol{B}_k$ *red* $\boldsymbol{B}_{k+1}$.

Part I is obvious.

We prove Part II readily as follows. Let $\{\lambda \boldsymbol{y}_i \,.\, \boldsymbol{P}_i\}(\boldsymbol{Q}_i)$ be the residuals in $\boldsymbol{A}_k$ of $\{\lambda \boldsymbol{x} \,.\, \boldsymbol{M}\}(\boldsymbol{N})$ and let $\boldsymbol{A}_k$ imr $\boldsymbol{A}_{k+1}$ be a contraction on the part $\{\lambda \boldsymbol{z} \,.\, \boldsymbol{R}\}(\boldsymbol{S})$ of $\boldsymbol{A}_k$. Then $\boldsymbol{B}_{k+1}$ is the result of a terminating sequence of contractions on $\{\lambda \boldsymbol{z} \,.\, \boldsymbol{R}\}(\boldsymbol{S})$ and the parts $\{\lambda \boldsymbol{y}_i \,.\, \boldsymbol{P}_i\}(\boldsymbol{Q}_i)$ of $\boldsymbol{A}_k$. Now if $\{\lambda \boldsymbol{z} \,.\, \boldsymbol{R}\}(\boldsymbol{S})$ is one of the $\{\lambda \boldsymbol{y}_i \,.\, \boldsymbol{P}_i\}(\boldsymbol{Q}_i)$, then no residuals of $\{\lambda \boldsymbol{z} \,.\, \boldsymbol{R}\}(\boldsymbol{S})$ occur in $\boldsymbol{B}_k$, and $\boldsymbol{B}_k$ conv-I $\boldsymbol{B}_{k+1}$. If however $\{\lambda \boldsymbol{z} \,.\, \boldsymbol{R}\}(\boldsymbol{S})$ is not one of the $\{\lambda \boldsymbol{y}_i \,.\, \boldsymbol{P}_i\}(\boldsymbol{Q}_i)$, then a set of residuals of $\{\lambda \boldsymbol{z} \,.\, \boldsymbol{R}\}(\boldsymbol{S})$ does occur in $\boldsymbol{B}_k$ and a terminating sequence of contractions on these residuals in $\boldsymbol{B}_k$ gives $\boldsymbol{B}_{k+1}$ by Lemma 1.

In order to prove Part III we note that $\boldsymbol{B}_k$ red $\boldsymbol{B}_{k+1}$ unless the reduction from $\boldsymbol{A}_k$ to $\boldsymbol{A}_{k+1}$ consists of a contraction on a residual in $\boldsymbol{A}_k$ of $\{\lambda \boldsymbol{x} \,.\, \boldsymbol{M}\}(\boldsymbol{N})$ (see preceding paragraph). But if we start with any particular $\boldsymbol{A}_k$ this can only be the case a finite number of successive times, by Lemma 1. Hence we define ϕ_m as follows. Perform m successive reductions on $\boldsymbol{A}$ in all possible ways. This gives a finite set of formulas (except for applications of I). In each formula find the largest number of reductions that can occur in a terminating sequence of contractions on the residuals of $\{\lambda \boldsymbol{x} \,.\, \boldsymbol{M}\}(\boldsymbol{N})$. Then let ϕ_m be the largest of these.

THEOREM 1. *If* $\boldsymbol{A}$ *conv* $\boldsymbol{B}$, *there is a conversion from* $\boldsymbol{A}$ *to* $\boldsymbol{B}$ *in which no expansion precedes any reduction.*

That is, any conversion can be replaced by a conversion which is a valley. This follows from Lemma 2 by the process already indicated.

COROLLARY 1. *If* $\boldsymbol{B}$ *is a normal form*[8] *of* $\boldsymbol{A}$, *then* $\boldsymbol{A}$ *conv-I–II* $\boldsymbol{B}$.

For no reductions are possible on a normal form.

COROLLARY 2. *If* $\boldsymbol{A}$ *has a normal form, its formal form is unique* (*to within applications of Rule I*).

[8] Kleene §5, p. 535.

For if $\boldsymbol{B}$ and $\boldsymbol{B}'$ are both normal forms of $\boldsymbol{A}$, then $\boldsymbol{B}'$ is a normal form of $\boldsymbol{B}$. Hence $\boldsymbol{B}$ conv-I–II $\boldsymbol{B}'$. Hence $\boldsymbol{B}$ conv-I $\boldsymbol{B}'$, since no reductions are possible on the normal form $\boldsymbol{B}$.

Note that only parts I and II of Lemma 2 are needed for Theorem 1 and its corollaries.

THEOREM 2. *If $\boldsymbol{B}$ is a normal form of $\boldsymbol{A}$, then there is a number m such that any sequence of reductions starting from $\boldsymbol{A}$ will lead to $\boldsymbol{B}$ (to within applications of Rule I) after at most m reductions.*

We prove by induction on n that, if a formula $\boldsymbol{B}$ is a normal form of some formula $\boldsymbol{A}$, and there is a sequence of n reductions leading from $\boldsymbol{A}$ to $\boldsymbol{B}$, then there is a number $\psi_{\boldsymbol{A},n}$ depending on the formula $\boldsymbol{A}$ and the number n such that any sequence of reductions starting from $\boldsymbol{A}$ will lead to a normal form of $\boldsymbol{A}$ (which will be $\boldsymbol{B}$ to within applications of I by Theorem 1, Corollary 2) in at most $\psi_{\boldsymbol{A},n}$ reductions.

If $n = 0$, we take $\psi_{\boldsymbol{A},0}$ to be 0.

Assume our statement for $n = k$. Let $\boldsymbol{A}$ imr $\boldsymbol{C}$, $\boldsymbol{C}$ imr $\boldsymbol{C}_1$, $\boldsymbol{C}_1$ imr $\boldsymbol{C}_2, \cdots, \boldsymbol{C}_{k-1}$ imr $\boldsymbol{B}$. Let $\boldsymbol{A}$ be the same as $\boldsymbol{A}_1$, $\boldsymbol{A}_1$ imr $\boldsymbol{A}_2$, $\boldsymbol{A}_2$ imr $\boldsymbol{A}_3, \cdots$. By Lemma 2 there is a sequence ($\boldsymbol{D}_1$ the same as $\boldsymbol{C}$) $\boldsymbol{D}_1$ conv-I–II $\boldsymbol{D}_2$, $\boldsymbol{D}_2$ conv-I–II $\boldsymbol{D}_3, \cdots$, such that $\boldsymbol{A}_j$ conv-I–II $\boldsymbol{D}_j$ for all j's for which $\boldsymbol{A}_j$ exists, and also, if the reduction from $\boldsymbol{A}$ to $\boldsymbol{D}_1$ (or $\boldsymbol{C}$) is a contraction on $\{\lambda \boldsymbol{x} \,.\, \boldsymbol{M}\}(\boldsymbol{N})$, such that, starting with $\boldsymbol{D}_m$, at most ϕ_m consecutive $\boldsymbol{D}_j$'s occur for which it is not true that $\boldsymbol{D}_j$ red $\boldsymbol{D}_{j+1}$. The sequence $\boldsymbol{C}$ imr $\boldsymbol{C}_1$, $\boldsymbol{C}_1$ imr $\boldsymbol{C}_2, \cdots$ leads to $\boldsymbol{B}$ in k reductions and so by the hypothesis of the induction there is a number $\psi_{\boldsymbol{C},k}$ such that any sequence of reductions on $\boldsymbol{C}$ leads to a normal form (i.e., terminates) after at most $\psi_{\boldsymbol{C},k}$ reductions. Hence there are at most $\psi_{\boldsymbol{C},k}$ reductions in the sequence $\boldsymbol{D}_1$ conv-I–II $\boldsymbol{D}_2$, $\boldsymbol{D}_2$ conv-I–II $\boldsymbol{D}_3, \cdots$, and hence this terminates after at most $f(\psi_{\boldsymbol{C},k})$ steps where $f(p)$ is defined as follows:

$$f(0) = \phi_1,$$
$$f(p+1) = f(p) + M + 1,$$

where M is the greatest of the numbers $\phi_1, \phi_2, \cdots, \phi_{f(p)+1}$.[9]

Then, since the sequence of $\boldsymbol{D}_j$'s continues as long as there are $\boldsymbol{A}_j$'s on which reductions can be performed, it follows that after at most $f(\psi_{\boldsymbol{C},k})$ reductions we come to an $\boldsymbol{A}_j$ on which no reductions are possible. But this is equivalent to saying that this $\boldsymbol{A}_j$ is a normal form. Hence any sequence of $k + 1$ reductions from $\boldsymbol{A}$ to $\boldsymbol{B}$ determines an upper bound which holds for all sequences of reductions starting from $\boldsymbol{A}$. We then take all sequences of $k + 1$ reductions starting from $\boldsymbol{A}$ (this is a finite set of sequences, since we reckon two sequences as the same if they differ only by applications of I) and find the upper bounds determined by each of them that leads to a normal form. Then we define $\psi_{\boldsymbol{A},k+1}$ to be the least of these upper bounds.

This completes our induction. But, by Theorem 1, Corollary 1, if $\boldsymbol{B}$ is a normal form of $\boldsymbol{A}$, there is a sequence of some finite number of reductions leading from $\boldsymbol{A}$ to $\boldsymbol{B}$. Hence Theorem 2 follows.

COROLLARY. *If a formula has a normal form, every well-formed part of it has a normal form.*

---

[9] $f(p)$ depends, of course, on the formula $\boldsymbol{A}$ and the part $\{\lambda \boldsymbol{x} \,.\, \boldsymbol{M}\}(\boldsymbol{N})$ of $\boldsymbol{A}$, as well as on p, because ϕ_m depends on $\boldsymbol{A}$ and $\{\lambda \boldsymbol{x} \,.\, \boldsymbol{M}\}(\boldsymbol{N})$.

2. Other kinds of conversion. There are also other systems of operations on formulas, similar to the system which we have been discussing and in which there can be distinguished reductions and expansions and possibly neutral operations (such as applications of Rule I). For convenience, we speak of the operations of any such system as conversions, and we define a normal form to be a formula on which no reductions are possible.

A kind of conversion which appears to be useful in certain connections is obtained by taking a new undefined term δ (restricting ourselves by never using δ as a bound symbol) and add to Church's Rules I, II, III the following rules:

IV. *Suppose that* $\boldsymbol{M}$ *and* $\boldsymbol{N}$ *contain no free symbols other than* δ, *that there is no part* $\{\lambda \boldsymbol{z} \,.\, \boldsymbol{P}\}(\boldsymbol{Q})$ *of either* $\boldsymbol{M}$ *or* $\boldsymbol{N}$, *and there is no part* $\delta(\boldsymbol{R}, \boldsymbol{S})$ *of either* $\boldsymbol{M}$ *or* $\boldsymbol{N}$ *in which* $\boldsymbol{R}$ *and* $\boldsymbol{S}$ *contain no free symbols other than* δ. *Then we may pass from a formula* $\boldsymbol{J}$ *to a formula* $\boldsymbol{K}$ *obtained from* $\boldsymbol{J}$ *by substituting for a particular occurrence of* $\delta(\boldsymbol{M}, \boldsymbol{N})$ *in* $\boldsymbol{J}$ *either* $\lambda f x \,.\, f(f(x))$ *or* $\lambda f x \,.\, f(x)$ *according to whether it is or is not true that* $\boldsymbol{M}$ *conv-I* $\boldsymbol{N}$.

V. *The inverse operation of that described in Rule IV is allowable. That is, we may pass from* $\boldsymbol{K}$ *to* $\boldsymbol{J}$ *under the same circumstances.*

We call an application of Rule II together with applications of Rule I, or an application of Rule IV together with applications of Rule I, a reduction, and the reverse operations (involving Rule III or Rule V) expansions. We call any sequence of applications of various ones of the five rules a conversion. Also we say that we contract $\{\lambda \boldsymbol{x} \,.\, \boldsymbol{M}\}(\boldsymbol{N})$ if we replace it by $\mathrm{S}^{\boldsymbol{x}}_{\boldsymbol{N}}\boldsymbol{M}|$, and that we contract $\delta(\boldsymbol{M}, \boldsymbol{N})$ if we replace it by $\lambda f x \,.\, f(f(x))$ or $\lambda f x \,.\, f(x)$ in accordance with Rule IV.

We define the residuals of $\{\lambda \boldsymbol{x} \,.\, \boldsymbol{M}\}(\boldsymbol{N})$ after an application of I or II in the same way as before, and after an application of IV as what $\{\lambda \boldsymbol{x} \,.\, \boldsymbol{M}\}(\boldsymbol{N})$ becomes (the restrictions in IV ensure that it becomes something of the form $\{\lambda \boldsymbol{y} \,.\, \boldsymbol{P}\}(\boldsymbol{Q})$). The residuals of $\delta(\boldsymbol{M}, \boldsymbol{N})$ after an application of I, II, or IV are defined only in the case that $\boldsymbol{M}$ and $\boldsymbol{N}$ are in normal form and contain no free symbols other than δ. In that case the residuals of $\delta(\boldsymbol{M}, \boldsymbol{N})$ are whatever part or parts of the entire resulting formula $\delta(\boldsymbol{M}, \boldsymbol{N})$ becomes, except that after an application of IV which is a contraction of $\delta(\boldsymbol{M}, \boldsymbol{N})$ itself, $\delta(\boldsymbol{M}, \boldsymbol{N})$ has no residual. Thus residuals of $\delta(\boldsymbol{M}, \boldsymbol{N})$ are always of the form $\delta(\boldsymbol{P}, \boldsymbol{Q})$, where $\boldsymbol{P}$ and $\boldsymbol{Q}$ are in normal form and contain no free symbols other than δ. We define a sequence of contractions on the parts $\{\lambda \boldsymbol{x}_j \,.\, \boldsymbol{M}_j\}(\boldsymbol{N}_j)$ and $\delta(\boldsymbol{P}_i, \boldsymbol{Q}_i)$ of $\boldsymbol{A}$, where $\boldsymbol{P}_i$ and $\boldsymbol{Q}_i$ are in normal form and contain no free symbols other than δ, by analogy with our former definition. Similarly for a terminating sequence of such contractions. Then we prove Lemma 1 by an obvious extension of our former argument. Lemma 2 and Theorems 1 and 2 then follow as before. Of course we replace "conv-I–II" by "conv-I–II–IV" in Lemma 2 and in general throughout the proofs of Theorems 1 and 2. In Lemma 1 we allow that the set of parts of $\boldsymbol{A}$ on which a sequence of contractions is taken should include not only parts of the form $\{\lambda \boldsymbol{x}_j \,.\, \boldsymbol{M}_j\}(\boldsymbol{N}_j)$ but also parts of the form $\delta(\boldsymbol{P}_k, \boldsymbol{Q}_k)$ in which $\boldsymbol{P}_k$ and $\boldsymbol{Q}_k$ are in normal form and contain no free symbols other than δ. And in Lemma 2 we consider also the case that $\boldsymbol{A}$ imr $\boldsymbol{B}$ by a contraction on the part $\delta(\boldsymbol{P}, \boldsymbol{Q})$ of $\boldsymbol{A}$.

We may also consider a third kind of conversion, namely the conversion that results if we modify Kleene's definitions of *well-formed*, *free*, and *bound* by omitting the requirement that $\boldsymbol{x}$ be a free symbol of $\boldsymbol{R}$ from (3) of the definition (see Kleene, top of p. 530) and then modify Church's rules I, II, III by using the new meanings of *well-formed*, *free* and *bound* and by omitting "if the proper symbol $\boldsymbol{x}$ occurs in $\boldsymbol{M}$" from both II and III.[10] Then we call an application of the modified II together with applications of the modified I a reduction, and the reverse operation an expansion, and applications of all three rules conversion. Also we say that we contract $\{\lambda \boldsymbol{x} \,.\, \boldsymbol{M}\}(\boldsymbol{N})$ if we replace it by $\mathrm{S}^{\boldsymbol{x}}_{\boldsymbol{N}}\boldsymbol{M}|$. If $\boldsymbol{A}$ has the form $\{\lambda \boldsymbol{x} \,.\, \boldsymbol{M}\}(\boldsymbol{N}_1, \boldsymbol{N}_2, \cdots, \boldsymbol{N}_r)$, then $\{\lambda \boldsymbol{x} \,.\, \boldsymbol{M}\}(\boldsymbol{N}_1)$ is said to be of order one in $\boldsymbol{A}$ and a contraction of $\{\lambda \boldsymbol{x} \,.\, \boldsymbol{M}\}(\boldsymbol{N}_1)$ is said to be a reduction of order one of $\boldsymbol{A}$. Then Lemma 1 and parts I and II of Lemma 2 hold (although some modification is required in the proofs). Hence Theorem 1 and its corollaries hold. But Theorem 2 is false. Instead a weaker form of Theorem 2 can be proved, namely:

THEOREM 3. *If* $\boldsymbol{A}$ *has a normal form, then there is a number* m *such that at most* m *reductions of order one can occur in a sequence of reductions on* $\boldsymbol{A}$.

PRINCETON UNIVERSITY,
PRINCETON, N.J.

[10]The first kind of conversion which we have considered is essentially equivalent to a certain portion of the combinatory axioms and rules of H. B. Curry (***American Journal of Mathematics***, vol. 52 (1930), pp. 509–536, 789–834), as has been proved by J. B. Rosser (see ***Annals of Mathematics***, (2), vol. 36 (1935), p. 127). In terms of Curry's notation, our third kind of conversion can be thought of as differing from the first kind by the addition of the constancy function K.

In dealing with properties of conversion, use of the Schönfinkel-Curry combinatory analysis appears in certain connections to be an important, even indispensable, device. But to recast the present discussion and results entirely into a combinatory notation would, it is thought, be awkward or impossible, because of the difficulty in finding a satisfactory equivalent, for combinations, of the notions of *reduction* and *normal form* as employed in this paper.

FORMAL DEFINITIONS IN THE THEORY OF ORDINAL NUMBERS[1]

(1936)

Alonzo Church and S. C. Kleene

1. The purpose of this paper is to extend to the transfinite ordinals, and functions of transfinite ordinals, the theory of formal definition which has been developed by one of the present authors[2] for functions of positive integers. The notation and terminology are those previously employed by the authors.[3]

For the convenience of the reader we give here a brief explanation of this notation and terminology.

Consider the infinite list of symbols: {, }, (,), λ, [,], $a, b, c, \ldots$ of which $a, b, c, \ldots$ are to be called *proper symbols*. A formula is any finite sequence of symbols from this list.

A formula is called *well-formed*, and an occurrence of a proper symbol in a formula is said to be an occurrence as a *free* symbol in accordance with the following rules, and those only: (1) a formula consisting of a single proper symbol $\boldsymbol{x}$ is well-formed, and the occurrence of $\boldsymbol{x}$ in the formula $\boldsymbol{x}$ is an occurrence as a free symbol, (2) if $\boldsymbol{F}$ and $\boldsymbol{X}$ are well-formed, $\{\boldsymbol{F}\}(\boldsymbol{X})$ is well-formed, and the occurrences of proper symbols as free symbols in $\boldsymbol{F}$ and in $\boldsymbol{X}$ are occurrences as free symbols in $\{\boldsymbol{F}\}(\boldsymbol{X})$, (3) if $\boldsymbol{M}$ is well-formed, and $\boldsymbol{x}$ is a proper symbol which occurs as a free symbol in $\boldsymbol{M}$, $\lambda\boldsymbol{x}[\boldsymbol{M}]$ is well-formed, and the occurrences of proper symbols other than $\boldsymbol{x}$ as free symbols in $\boldsymbol{M}$ are occurrences as free symbols in $\lambda\boldsymbol{x}[\boldsymbol{M}]$. An occurrence of a proper symbol in a formula which is not an occurrence as a free symbol is called an occurrence as a *bound*

Originally published in ***Fundamenta Mathematica***, vol. 28 (1936), pp. 11–21.

[1]Presented to the American Mathematical Society, September 1935.

[2]S. C. Kleene, *A theory of positive integers in formal logic*, ***Amer. Jour. Math.***, vol. 57 (1935), pp. 153–173, 219–244. It should be observed that the part of this paper of Kleene which is concerned with problems of formal definition depends only on Church's rule of procedure I–III and is therefore unaffected by the fact that the complete system of Church, making use of the properties of Π and &, is now known to lead to contradiction (see Kleene and Rosser, *The inconsistency of certain formal logics*, ***Ann. Math.***, vol. 36 (1935), pp. 630–636). Indeed these problems of formal definition have an interest which is independent even of the question whether there exists a consistent, adequate system of symbolic logic which embodies the rules of procedure I–III. The fact of this independent interest is made especially clear by the results of S. C. Kleene, loc. cit., §18, and of Alonzo Church, *An unsolvable problem in elementary number theory*, forthcoming (abstract in ***Bull. Amer. Math. Soc.***, vol. 41 (1935), p. 332). Also it is thought that the considerations of the present paper have a bearing on the distinction between constructive and non-constructive ordinals, and on questions of enumerability and effective enumerability.

[3]See Alonzo Church, *A set of postulates for the foundation of logic*, ***Ann. Math.***, vol. 41, 1932; S. C. Kleene, loc. cit., and *Proof by cases in formal logic*, ***Ann. Math.***, vol. 35 (1934), pp. 529–544; Alonzo Church and J. B. Rosser, *Some properties of conversion*, forthcoming (abstract in ***Bull. Amer. Math. Soc.***, vol. 41 (1935), p. 332). These papers will hereafter be referred to by author or by author and date.

Note particularly the definition of *well-formed* (Kleene 1934, §1), the abbreviations of well-formed formulas (Church 1932, §6 and Kleene 1934, §3), the use of heavy-type letters (Kleene 1934, §3, (1) and (2), and 1935, footnotes on pp. 219–220), the rules of procedure I–III (Church 1932, pp. 350, 355–356, and Kleene 1934, §1) and the definition of *conversion* (Church 1932, p. 357, and Kleene 1934, p. 535), the definition of *normal form* (Kleene 1934, p. 535), the definitions of I (Kleene 1934, p. 536) and of $1, 2, 3, 4, \ldots, S$ (Kleene 1935, p. 156), and definition of *formal definability* (Kleene 1935, p. 219).

symbol. By a free (bound) symbol of $\boldsymbol{K}$, is meant a proper symbol which occurs as a free (bound) symbol in $\boldsymbol{K}$.

Heavy type letters are used to represent undetermined well-formed formulas. The expression $\mathrm{S}_{\boldsymbol{N}}^{\boldsymbol{x}}\boldsymbol{M}|$ is used to stand for the result of substituting $\boldsymbol{N}$ for $\boldsymbol{x}$ throughout $\boldsymbol{M}$.

An *immediate conversion* is any one of the following operations on well-formed formulas:

I. To replace any part $\lambda\boldsymbol{x}[\boldsymbol{M}]$ by $\lambda\boldsymbol{y}[\mathrm{S}_{\boldsymbol{y}}^{\boldsymbol{x}}\boldsymbol{M}|]$, where $\boldsymbol{y}$ is any proper symbol which does not occur in $\boldsymbol{M}$.

II. To replace any part $\{\lambda\boldsymbol{x}[\boldsymbol{M}]\}(\boldsymbol{N})$ by $\mathrm{S}_{\boldsymbol{N}}^{\boldsymbol{x}}\boldsymbol{M}|$, provided that the bound symbols of $\boldsymbol{M}$ are distinct both from $\boldsymbol{x}$ and from the free symbols of $\boldsymbol{N}$.

III. To replace any part $\mathrm{S}_{\boldsymbol{N}}^{\boldsymbol{x}}\boldsymbol{M}|$ (other than a proper symbol which immediately follows a λ) by $\{\lambda\boldsymbol{x}[\boldsymbol{M}]\}(\boldsymbol{N})$, provided that the bound symbols of $\boldsymbol{M}$ are distinct both from $\boldsymbol{x}$ and from the free symbols of $\boldsymbol{N}$.

A finite sequence of immediate conversions is called a *conversion*; and it is said that $\boldsymbol{A}$ is convertible into $\boldsymbol{B}$, or $\boldsymbol{A}$ conv $\boldsymbol{B}$, if $\boldsymbol{B}$ is obtainable from $\boldsymbol{A}$ by a conversion.

A formula is said to be in *normal form* if it is well-formed and has no part of the form $\{\lambda\boldsymbol{x}[\boldsymbol{M}]\}(\boldsymbol{N})$. A formula $\boldsymbol{B}$ is said to be a normal form of a formula $\boldsymbol{A}$ if $\boldsymbol{B}$ is in normal form and $\boldsymbol{A}$ conv $\boldsymbol{B}$.

As a matter of convenience in the actual writing down of formulas, various abbreviations are employed. $\{\ldots\{\{\boldsymbol{F}\}(\boldsymbol{X}_1)\}(\boldsymbol{X}_2)\ldots\}(\boldsymbol{X}_n)$ is written as

$$\{\boldsymbol{F}\}(\boldsymbol{X}_1,\ldots,\boldsymbol{X}_n) \text{ or } \boldsymbol{F}(\boldsymbol{X}_1,\ldots,\boldsymbol{X}_n);$$

and

$$\lambda\boldsymbol{x}_1[\lambda\boldsymbol{x}_2[\ldots\lambda\boldsymbol{x}_n[\boldsymbol{M}]\ldots]] \text{ as } \lambda\boldsymbol{x}_1\boldsymbol{x}_2\ldots\boldsymbol{x}_n\,.\,\boldsymbol{M}.$$

We use $I, S, 1, 2, 3, \ldots$ as abbreviations respectively for

$$\lambda x\,.\,x,\ \ \lambda nfx\,.\,f(n(f,x)),\ \ \lambda fx\,.\,f(x),\ \ \lambda fx\,.\,f(f(x)),\ \ \lambda fx\,.\,f(f(f(x))),\ldots.$$

The formulas $1, 2, 3, \ldots$ are taken to represent the positive integers.

2. Using an arrow to mean "stands for" or "is an abbreviation for", we let:

$$0_o \rightarrow \lambda m\,.\,m(1).$$

$$S_o \rightarrow \lambda am\,.\,m(2,a).$$

$$L \rightarrow \lambda arm\,.\,m(3,a,r).$$

$$1_o \rightarrow S_o(0_o),\quad 2_o \rightarrow S_o(1_o),\quad etc.$$

The formulas $0_o, 1_o, 2_o, \ldots$ are taken to represent the finite ordinals.

Here we are using the subscript o to distinguish notations used in connection with the present theory from like notations used in other connections. When the context precludes ambiguity, we may omit this subscript o, writing for instance $\omega + 2$ instead of $\omega +_o 2_o$ (cf. §4 below).

We adopt the following rules for the assignment of formulas to represent particular ordinals of the first and second number classes:

i. If $\boldsymbol{a}$ represents the ordinal a, and $\boldsymbol{b}$ conv $\boldsymbol{a}$, then $\boldsymbol{b}$ also represents a.
ii. 0_o represents the ordinal zero.

iii. If $\boldsymbol{a}$ represents the ordinal a, then $S_o(\boldsymbol{a})$ represents the successor of a.

iv. If b is the limit of an increasing sequence (series) of ordinals, $b_0, b_1, b_2, \ldots$, of order type ω, and if $\boldsymbol{r}$ is a formula such that the formulas $\boldsymbol{r}(0_o), \boldsymbol{r}(1_o), \boldsymbol{r}(2_o), \ldots$ represents the ordinals $b_0, b_1, b_2, \ldots$, respectively, then $L(0_o, \boldsymbol{r})$ represents b.

Since one of the formulas assigned by these rules to represent the least ordinal ω of the second number class is $L(0_o, I)$, we let,

$$\omega \to L(0_o, I).$$

We shall call an ordinal of the first or second number class *formally definable*, or *λ-definable*, if there is, under Rules i–iv, at least one formula assigned to represent it.

From the possibility under iv of using different sequences for any particular b (or of using different $\boldsymbol{r}$'s for any particular sequence) it follows that every formally definable ordinal of the second number class has an infinite number of non-interconvertible formulas assigned to represent it.

We shall say that a sequence of ordinals of order type ω is *formally defined as a function of ordinals* by $\boldsymbol{r}$ if, for every formula $\boldsymbol{a}$ which represents a finite ordinal a, $\boldsymbol{r}(\boldsymbol{a})$ represents the $(1 + a)$th ordinal of the sequence.

It is clear that the question what ordinals of the second number class are formally definable is tied up with the question what sequences of ordinals are formally definable as functions of ordinals, in such a way that neither question can be regarded as prior to the other.

From Theorem 2 below and the enumerability of the set of all well-formed formulas, it follows (by a non-constructive argument) that the set of all formally definable ordinals is enumerable, and hence that there is a least ordinal ξ in the second number class which is not formally definable. It is then readily proved that no ordinal in the second class greater than ξ is formally definable.

It is to be emphasized, however, that the definition of ξ which has just been given is not constructive, in any usual sense of that term, and that no means is known to the authors of obtaining constructively an ordinal of the second number class which is not formally definable. In particular, it will be shown below that, if a and b are formally definable ordinals out of the first and second number classes, then the sum of a and b, the product of a and b, and the result of raising a to the power b are formally definable, and hence all ordinals less than ε_o are formally definable. Moreover, if a is formally definable, then ε_a is formally definable, and hence it follows that all ordinals less than the least solution of $\varepsilon_x = x$ are formally definable. Similar theorems can be established in which various generalizations of the ε-numbers[4] appear. Thus the boundary of ordinals in the second number class known to be formally definable is continually pushed upwards, and no constructively obtainable limit to this process is seen.

We prove the two following theorems by transfinite induction.

THEOREM 1. *Every formula which represents an ordinal number under Rules* i–iv *has a normal form.*

---

[4]Cf. O. Veblen, *Continuous increasing functions of finite and transfinite ordinals*, ***Trans. Amer. Math. Soc.***, vol. 9 (1908), pp. 280–292. All the particular ordinals defined by Veblen in this paper are formally definable in our present sense, including the ordinals $E(1), E(2), \ldots$ and even the first (second, and so on) solutions of the equation $f(1_1, \ldots, 1_\alpha) = \alpha$, page 292.

Any formula which represents the ordinal zero must be convertible into 0_o and hence has the normal form 0_o. Suppose that every formula which represents an ordinal less than b has a normal form. If b is the successor of an ordinal a, any formula which represents b must be convertible into $S_0(\boldsymbol{a})$, where $\boldsymbol{a}$ represents a, and hence must have the normal form $\lambda m \,.\, m(2, \boldsymbol{A})$, where $\boldsymbol{A}$ is the normal form of $\boldsymbol{a}$. If b is the limit of an increasing sequence of ordinals of order type ω, any formula which represents b must be convertible into $L(0_o, \boldsymbol{r})$, where there is some increasing sequence of ordinals of order type ω which has b as its limit and which is formally defined as a function of ordinals by $\boldsymbol{r}$; moreover $\boldsymbol{r}$ must have a normal form $\boldsymbol{R}$, because otherwise $\boldsymbol{r}(0_o)$ would lack a normal form,[5] and hence $L(0_o, \boldsymbol{r})$ must have the normal form $\lambda m \,.\, m(3, 0_o, \boldsymbol{R})$.

THEOREM 2. *If, under Rules* i–iv, *a formula* $\boldsymbol{b}$ *represents an ordinal* b, *then* $\boldsymbol{b}$ *cannot represent an ordinal distinct from* b.

The theorem is true if b is zero, because $\boldsymbol{b}$ then has $\lambda m \,.\, m(1)$ as a normal form, and formulas which represent ordinals distinct from zero have a normal form of one of the forms, $\lambda m \,.\, m(2, \boldsymbol{A})$, or $\lambda m \,.\, m(3, 0_0, \boldsymbol{R})$, and hence cannot have $\lambda m \,.\, m(1)$ as a normal form.[6] We proceed by induction with respect to b.

If b is the successor of an ordinal a, then b is not the limit of any increasing sequence of ordinals and is not the successor of any ordinal other than a, and consequently $\boldsymbol{b}$ must have a normal form $\lambda m \,.\, m(2, \boldsymbol{A})$, where $\boldsymbol{A}$ represents a. Therefore[6] every normal form of b must be of the form $\lambda \boldsymbol{n} \,.\, \boldsymbol{n}(\boldsymbol{2}, \boldsymbol{A}')$, where $\boldsymbol{2}$ conv 2 and $\boldsymbol{A}'$ conv $\boldsymbol{A}$ (being the same formulas except for possible alphabetical differences of bound symbols). Therefore $\boldsymbol{b}$ can represent only a successor of an ordinal represented by $\boldsymbol{A}$. By hypothesis of induction, $\boldsymbol{A}$ cannot represent an ordinal distinct from a. Therefore $\boldsymbol{b}$ cannot represent an ordinal distinct from b.

If b is the limit of an increasing sequence of ordinals of order type ω, then b is not the successor of any ordinal, and consequently $\boldsymbol{b}$ must have a normal form $\lambda m \,.\, m(3, 0_o, \boldsymbol{R})$, where $\boldsymbol{R}$ formally defines as a function of ordinals an increasing sequence of ordinals of order type ω which has b as its limit. Therefore[6] every normal form of $\boldsymbol{b}$ must be of the form $\lambda \boldsymbol{n} \,.\, \boldsymbol{n}(\boldsymbol{3}, \boldsymbol{O}, \boldsymbol{R}')$, where $\boldsymbol{3}$ conv 3, $\boldsymbol{O}$ conv 0_o, $\boldsymbol{R}'$ conv $\boldsymbol{R}$. Therefore $\boldsymbol{b}$ can represent only a limit of an increasing sequence of ordinals which is formally defined as a function of ordinals by $\boldsymbol{R}$. But it follows from the hypothesis of induction that $\boldsymbol{R}$ can define only one increasing sequence of ordinals of order type ω. Therefore $\boldsymbol{b}$ can represent only b.

3. We shall call a function a *function in the first and second number classes* if the range of the independent variable (or of each independent variable) consists of the first and second number classes and the range of the dependent variable is contained in the first and second number classes.

We shall say that a function f, of n variables, in the first and second number classes, is *formally defined* by a formula $\boldsymbol{f}$ if, for every set of n formulas $\boldsymbol{x}_1, \boldsymbol{x}_2, \ldots, \boldsymbol{x}_n$ which represent ordinals $x_1, x_2, \ldots, x_n$, respectively, in the first and second number classes, $\boldsymbol{f}(\boldsymbol{x}_1, \boldsymbol{x}_2, \ldots, \boldsymbol{x}_n)$ is a formula which represents the ordinal $f(x_1, x_2, \ldots, x_n)$. We shall call a function f in the first and second number classes *formally definable*, or *λ-definable*, if there is at least one formula $\boldsymbol{f}$ by which it is formally defined.

[5] Church and Rosser, Thm. 2 Cor.

[6] Church and Rosser, Thm. 1 Cor. 2.

We remark that if a formula $\boldsymbol{f}$ is to define (formally) a function, of n variables, in the first and second number classes, not only must $\boldsymbol{f}(\boldsymbol{x}_1, \boldsymbol{x}_2, \ldots, \boldsymbol{x}_n)$ represent an ordinal in the first and second number classes for every set of n formulas $\boldsymbol{x}_1, \boldsymbol{x}_2, \ldots, \boldsymbol{x}_n$ which represent ordinals in the first and second number classes, but also, if $\boldsymbol{y}_1, \boldsymbol{y}_2, \ldots, \boldsymbol{y}_n$ are formulas which represent ordinals equal respectively to the ordinals represented by $\boldsymbol{x}_1, \boldsymbol{x}_2, \ldots, \boldsymbol{x}_n$, then $\boldsymbol{f}(\boldsymbol{y}_1, \boldsymbol{y}_2, \ldots, \boldsymbol{y}_n)$ must represent an ordinal equal to the ordinal represented by $\boldsymbol{f}(\boldsymbol{x}_1, \boldsymbol{x}_2, \ldots, \boldsymbol{x}_n)$.

It is clear that any function in the first and second number classes which is formally definable must be in a certain sense constructive, because given a formal definition of the function, the process of reduction to normal form provides an algorithm for calculating the values of the function (provided that we consider an ordinal to have been calculated if a formula in the normal form which represents it has been obtained). Therefore certain functions which are not constructive in this sense are clearly not formally definable; in particular, the function f, such that $f(x, y)$ is equal to the ordinal zero or the ordinal one according to whether the ordinals x and y are equal or distinct, is not formally definable.[7]

It is conjectured, however, that every constructive function in the first and second number class is formally definable. This conjecture is vague because of the lack of a satisfactory definition of the notion of a constructive function of ordinals, but it is supported by analogy with the case of functions of positive integers, where it is possible to give a plausible definition of constructivity and to prove the equivalence of constructivity and λ-definability.[8]

4. We proceed now to the proof of a theorem which is useful in many particular cases where it is required to obtain a formal definition of some given function in the first and second number classes.

THEOREM 3. *If* $\boldsymbol{A}$, $\boldsymbol{G}$ *and* $\boldsymbol{H}$ *are given formulas having no free symbols, it is possible to find a set of eight formulas* $\boldsymbol{f}_{ijk}$ *(where the subscripts* i, j, k *take the values* 1 *and* 2*) satisfying the following conditions*:

$$\begin{array}{ll}
\boldsymbol{f}_{1jk}(0_o) \text{ conv } \boldsymbol{A} & \boldsymbol{f}_{2jk}(0_o) \text{ conv } \boldsymbol{A}(\boldsymbol{f}_{2jk}) \\
\boldsymbol{f}_{i1k}(S_0(\boldsymbol{a})) \text{ conv } \boldsymbol{G}(\boldsymbol{a}) & \boldsymbol{f}_{i2k}(S_o(\boldsymbol{a})) \text{ conv } \boldsymbol{G}(\boldsymbol{f}_{i2k}, \boldsymbol{a}) \\
\boldsymbol{f}_{ij1}(L(\boldsymbol{a}, \boldsymbol{r})) \text{ conv } \boldsymbol{H}(\boldsymbol{a}, \boldsymbol{r}) & \boldsymbol{f}_{ij2}(L(\boldsymbol{a}, \boldsymbol{r})) \text{ conv } \boldsymbol{H}(\boldsymbol{f}_{ij2}, \boldsymbol{a}, \boldsymbol{r}).
\end{array}$$

This theorem is to be understood to mean, not merely that the formulas $\boldsymbol{f}_{ijk}$ "exist", but that means are at hand by which they can be effectively obtained in any given case.

In order to construct the formulas $\boldsymbol{f}_{ijk}$ we proceed as follows. We let,

$$\mathfrak{G}_1 \to \lambda f b \,.\, b(4, f),$$
$$\mathfrak{G}_2 \to \lambda f b \,.\, f(\lambda x \,.\, x(\lambda n \,.\, b(n, b))).$$

Using Kleene 1935, 15 III, we construct eight formulas $\boldsymbol{B}_{ijk}$ such that

$$\boldsymbol{B}_{ijk}(1) \text{ conv } \mathfrak{G}_i(\boldsymbol{A}), \quad \boldsymbol{B}_{ijk}(2) \text{ conv } \mathfrak{G}_j(\boldsymbol{G}), \quad \boldsymbol{B}_{ijk}(3) \text{ conv } \mathfrak{G}_k(\boldsymbol{H}), \quad \text{and} \quad \boldsymbol{B}_{ijk}(4) \text{ conv } I.$$

[7]The full proof of this assertion makes use of the results of Church, *An unsolvable problem of elementary number theory*.

[8]See Church, *An unsolvable problem of elementary number theory*, and S. C. Kleene, *λ-definability and recursiveness*, forthcoming (abstract in ***Bull. Amer. Math. Soc.***, vol. 41 (1935), no. 7). [*Editor's note*: S. C. Kleene, *λ-Definability and Recursiveness*, ***Duke Mathematical Journal***, vol. 2 (1936), pp. 340–353.]

Then we let,

$$\boldsymbol{f}_{ijk} \to \lambda x \,.\, x(\lambda n \,.\, \boldsymbol{B}_{ijk}(n, \boldsymbol{B}_{ijk})).$$

The proof that the formulas $\boldsymbol{f}_{ijk}$ have the required conversion properties is then straightforward.

In particular, we can, according to Theorem 3, obtain a formula $\mathfrak{f}$ such that $\mathfrak{f}(0_o)$ conv I, $\mathfrak{f}(S_o(\boldsymbol{a}))$ conv $\lambda x \,.\, S_o(\mathfrak{f}(\boldsymbol{a}, x))$, and $\mathfrak{f}(L(\boldsymbol{a}, \boldsymbol{r}))$ conv $\lambda x \,.\, L(\boldsymbol{a}, \lambda m \,.\, \mathfrak{f}(\boldsymbol{r}(m), x))$. Then it can be proved, by transfinite induction, that, if $\boldsymbol{a}$ and $\boldsymbol{b}$ represent ordinals a and b, respectively, in the first and second number classes, $\mathfrak{f}(\boldsymbol{a}, \boldsymbol{b})$ represents the sum of b and a. Hence we let,

$$[\boldsymbol{b}] +_o [\boldsymbol{a}] \to \mathfrak{f}(\boldsymbol{a}, \boldsymbol{b}).$$

Likewise we may obtain a formula $\mathfrak{f}$ such that $\mathfrak{f}(0_o)$ conv I, $\mathfrak{f}(S_o(\boldsymbol{a}))$ conv $\mathfrak{f}(\boldsymbol{a})$, and $\mathfrak{f}(L(\boldsymbol{a}, \boldsymbol{r}))$ conv $\mathfrak{f}(\boldsymbol{a}, \mathfrak{f}(\boldsymbol{r}(0_o)))$. Then, by transfinite induction, if $\boldsymbol{a}$ represents an ordinal, $\mathfrak{f}(\boldsymbol{a})$ conv I. Hence a constancy function, K_o, of ordinals may be defined by letting,

$$K_o \to \lambda xy \,.\, \mathfrak{f}(y, x).$$

If $\boldsymbol{b}$ represents an ordinal, $K_o(\boldsymbol{a}, \boldsymbol{b})$ conv $\boldsymbol{a}$.

Likewise we may obtain a formula $\mathfrak{f}$ such that

$$\mathfrak{f}(0_o) \text{ conv } K_o(0_o), \qquad \mathfrak{f}(S_o(\boldsymbol{a})) \text{ conv } \lambda x \,.\, \mathfrak{f}(\boldsymbol{a}, x) +_o x,$$

and

$$\mathfrak{f}(L(\boldsymbol{a}, \boldsymbol{r})) \text{ conv } \lambda x \,.\, L(\boldsymbol{a}, \lambda m \,.\, \mathfrak{f}(\boldsymbol{r}(m), x)).$$

Then, if $\boldsymbol{a}$ and $\boldsymbol{b}$ represent ordinals a and b respectively, $\mathfrak{f}(\boldsymbol{a}, \boldsymbol{b})$ represents the product of a and b (with a as multiplier and b as multiplicand). Hence we let,

$$[\boldsymbol{a}] \times_o [\boldsymbol{b}] \to \mathfrak{f}(\boldsymbol{a}, \boldsymbol{b}).$$

The expression $[\boldsymbol{a}] \times_o [\boldsymbol{b}]$ may be abbreviated as $[\boldsymbol{a}][\boldsymbol{b}]$, when no ambiguity arises thereby.

Similarly, we may obtain a formula $\mathfrak{f}$ such that

$$\mathfrak{f}(0_o) \text{ conv } K_o(1_o), \qquad \mathfrak{f}(S_o(\boldsymbol{a})) \text{ conv } \lambda x \,.\, x \times_o \mathfrak{f}(\boldsymbol{a}, x),$$

and

$$\mathfrak{f}(L(\boldsymbol{a}, \boldsymbol{r})) \text{ conv } \lambda x \,.\, L(\boldsymbol{a}, \lambda m \,.\, \mathfrak{f}(\boldsymbol{r}(m), x)).$$

Then, if $\boldsymbol{a}$ and $\boldsymbol{b}$ represent ordinals a and b, $\mathfrak{f}(\boldsymbol{a}, \boldsymbol{b})$ represents the result of raising b to the power a. Hence we let,

$$\exp_o \to \lambda xy \,.\, \mathfrak{f}(y, x)$$

and in any case where no confusion with exponentiation of positive integers, or other forms of exponentiation, is produced, we abbreviate $\exp_o(\boldsymbol{b}, \boldsymbol{a})$ as $[\boldsymbol{b}]^{[\boldsymbol{a}]}$.

A predecessor function of ordinals is formally defined by a formula P_o chosen to satisfy the relations[9] $P_o(0_o)$ conv 0_o, $P_o(S_o(\boldsymbol{a}))$ conv $\boldsymbol{a}$, and $P_o(L(\boldsymbol{a}, \boldsymbol{r}))$ conv $L(\boldsymbol{a}, \boldsymbol{r})$.

Let $\mathfrak{F}$ be so chosen that

$$\mathfrak{F}(0_o) \text{ conv } 0_o, \quad \mathfrak{F}(S_o(\boldsymbol{a})) \text{ conv } \mathfrak{F}(\boldsymbol{a}), \quad \text{and} \quad \mathfrak{F}(L(\boldsymbol{a}, \boldsymbol{r})) \text{ conv } K_o(1_o, L(\boldsymbol{a}, \boldsymbol{r})).$$

[9] The choice of the third relation was made somewhat arbitrarily. Any other of the form required by Theorem 3 might have been used instead.

Then, if $\boldsymbol{a}$ represents an ordinal a, $\mathfrak{F}(\boldsymbol{a})$ is convertible into 0_o or 1_o according as a is finite or infinite.

Choose $\mathfrak{R}$ so that

$$\mathfrak{R}(0_o) \text{ conv } 0_o, \quad \mathfrak{R}(S_o(\boldsymbol{a})) \text{ conv } K_o(1_o, \boldsymbol{a}), \quad \text{and} \quad \mathfrak{R}(L(\boldsymbol{a}, \boldsymbol{r})) \text{ conv } K_o(2_o, L(\boldsymbol{a}, \boldsymbol{r})).$$

Then $\mathfrak{R}$ formally defines the *kind* of an ordinal in the following fashion: If $\boldsymbol{a}$ represents an ordinal a, $\mathfrak{R}(\boldsymbol{a})$ is convertible into 0_o or 1_o or 2_o according as a is zero, the successor of some ordinal, or the limit of some increasing sequence of ordinals.

The formula $\lambda n \,.\, P_o(n(S_o, 0_o))$ defines the nth (finite) ordinal number as a function of the nth positive integer. Inversely, if $\mathfrak{T}$ is so chosen that $\mathfrak{T}(0_o) \text{ conv } 1$, $\mathfrak{T}(S_o(\boldsymbol{a})) \text{ conv } S(\mathfrak{T}(\boldsymbol{a}))$, and $\mathfrak{T}(L(\boldsymbol{a}, \boldsymbol{r})) \text{ conv } K_o(1, L(\boldsymbol{a}, \boldsymbol{r}))$, then $\mathfrak{T}$ defines the nth positive integer as a function of the nth (finite) ordinal number. Also, if a is any ordinal number of the second kind, $\mathfrak{T}$ defines the nth positive integer as a function of the nth ordinal number in the set of ordinal numbers from a on. With the aid of these transformations between ordinal numbers and positive integers, much of the theory of formal definition of functions of positive integers (Kleene 1935) can be carried over at once to the ordinal numbers.

Choose $\mathfrak{e}$ so that

$$\mathfrak{e}(0_o) \text{ conv } L(0_o, \lambda n \,.\, \mathfrak{T}(n, \exp_o(\omega), 0_o)),$$

$$\mathfrak{e}(S_o(\mathfrak{a})) \text{ conv } L(0_o, \lambda n \,.\, \mathfrak{T}(n, \exp_o(\omega), S_o(\mathfrak{e}(\boldsymbol{a})))),$$

and

$$\mathfrak{e}(L(\boldsymbol{a}, \boldsymbol{r})) \text{ conv } L(\boldsymbol{a}, \lambda n \,.\, \mathfrak{e}(\boldsymbol{r}(n))).$$

Then if $\boldsymbol{a}$ represents an ordinal number a, the formula $\mathfrak{e}(\boldsymbol{a})$ (which we abbreviate as $\varepsilon_{\boldsymbol{a}}$) represents the $(1 + a)$th epsilon number.

Similarly formal definitions may be obtained for various generalizations of the ε-numbers, using Theorem 3 and generalizations of Theorem 3.

5. We let

$$Z_1 \to Q(\lambda x \,.\, x(I), I),$$
$$Z_2 \to Q(\lambda xy \,.\, S(x) - y, I),$$

where Q is the formula defined in the first paragraph of §17, Kleene 1935. Then Z_1 and Z_2 define the sequences 1,1,2,1,2,3,1,2,3,4, ... and 1,2,1,3,2,1,4,3,2,1, ... respectively, as functions of positive integers (cf. Kleene 1935, 17 I).

Using Kleene 1935, 15 III, we obtain a formula $\mathfrak{B}$ such that

$$\mathfrak{B}(1) \text{ conv } \lambda f x \,.\, K_o(x, f(0_o, 1))$$

and, for all formulas $\boldsymbol{n}$ which represent positive integers,

$$\mathfrak{B}(S(\boldsymbol{n})) \text{ conv } \lambda f x \,.\, f(x, \boldsymbol{n}).$$

Then using Theorem 3 we obtain a formula enm such that

$$\text{enm}(0_o) \text{ conv } \lambda n \,.\, n(I, 0_o),$$
$$\text{enm}(S_o(\boldsymbol{a})) \text{ conv } \lambda n \,.\, \mathfrak{B}(n, \text{enm}, \boldsymbol{a}),$$

and

$$\mathrm{enm}(L(\boldsymbol{a},\boldsymbol{r})) \text{ conv } \lambda n \,.\, K_o(\mathrm{enm}(\boldsymbol{r}(P_o(Z_1(n, S_o, 0_o)))),\ Z_2(n)),\ \boldsymbol{a}).$$

The formula enm has the property that, if $\boldsymbol{a}$ represents an ordinal a of the first or the second number class, other than zero, the infinite sequence

$$\mathrm{enm}(\boldsymbol{a}, 1),\ \mathrm{enm}(\boldsymbol{a}, 2),\ \mathrm{enm}(\boldsymbol{a}, 3),\ \ldots$$

is an enumeration, with repetitions, of the ordinals less than a (i.e., every term of the sequence represents such an ordinal, and every such ordinal is represented by at least one term of the sequence).

It is not, however, true, if $\boldsymbol{b}$ represents an ordinal equal to a, and $\boldsymbol{n}$ represents a positive integer, that $\mathrm{enm}(\boldsymbol{b}, \boldsymbol{n})$ must represent an ordinal equal to $\mathrm{enm}(\boldsymbol{a}, \boldsymbol{n})$. There is no formula which has this property in addition to the property of enm just described.

6. The assignment of formulas to represent ordinals which was set up in §2 can be extended to larger ordinals (not, however, to all ordinals) by replacing Rule iv by the following transfinite set of rules, iv_a, where the subscript a may take on as value any ordinal a formula to represent which has been previously assigned:

iv_a. If ι_a is the least ordinal a formula to represent which is not assigned by the rules i, ii, iii, iv_i $(0 \leqq i < a)$, if b is the limit of an increasing sequence of ordinals of order type ι_a, which is formally defined as a function of ordinals by $\boldsymbol{r}$, and if $\boldsymbol{a}$ represents a, then $L(\boldsymbol{a}, \boldsymbol{r})$ represents b.

For the rule iv_0 we take ι_0 to be ω, so that Rule iv_0 is identical with Rule iv.

It can be proved as before that a formula which represents an ordinal must have a normal form, and that distinct ordinals cannot be represented by the same formula; and it will be found that the formal definitions of particular functions of ordinals which are obtained in §4 remain valid under the extended assignment of formulas to represent ordinals.

The ordinals ι_a are defined formally by

$$\iota_{\boldsymbol{a}} \rightarrow L(\boldsymbol{a}, I).$$

Moreover, using Theorem 3, we may find a formula $\mathfrak{o}$ such that $\mathfrak{o}(0_o)$ conv $L(0_o, I)$, $\mathfrak{o}(S_o(\boldsymbol{a}))$ conv $L(1_o +_o \boldsymbol{a}, I)$, and $\mathfrak{o}(L(\boldsymbol{a}, \boldsymbol{r}))$ conv $L(\boldsymbol{a}, \lambda n \,.\, L(\boldsymbol{r}(n), I))$, and so obtain a set of ordinals ω_a defined formally by

$$\omega_{\boldsymbol{a}} \rightarrow \mathfrak{o}(\boldsymbol{a}).$$

It can be proved, by a non-constructive argument, that all the ordinals ω_a are contained within the second number class. Nevertheless there is an analogy between the ordinals ω_a and the first ordinals of the successive number classes, in which the notion of λ-definability corresponds, in general, to the notion of existence. In particular, Ω, the first ordinal of the third number class, can be described as the least ordinal of the second kind which is not the limit of an increasing sequence of ordinals of order type ω. Correspondingly, ω_1 is the least ordinal of the second kind which is not the limit of a λ-definable increasing sequence of ordinals of order type ω.

In fact, if we are willing to take the position that an ordinal less than ω_1 is not constructively calculated unless a formula representing it is constructively calculated, then we may regard an infinite sequence of ordinals (of order type ω) as constructive if some function f such that $f(n)$ is the Gödel representation of a formula representing the nth ordinal of the sequence is effectively calculable in the sense of Church, *An unsolvable problem of elementary number theory*, and hence conclude, by use of $\mathfrak{T}$ and of Kleene, *λ-definability and recursiveness*, (25), that ω_1 is not the limit of any constructive increasing sequence of ordinals of order type ω.

THE CONSTRUCTIVE SECOND NUMBER CLASS*
(1938)

The existence of at least a vague distinction between what I shall call the constructive and the non-constructive ordinals of the second number class, that is, between ordinals which can in some sense be effectively built up to step by step from below and those for which this cannot be done (although there may be existence proofs), is, I believe, somewhat generally recognized. My purpose here is to propose an exact definition of this distinction and of the related distinction between constructive and non-constructive functions of ordinals in the second number class; where, again to speak vaguely, a function is constructive if there is a rule by which, whenever a value of the independent variable (or a set of values of the independent variables) is effectively given, the corresponding value of the dependent variable can be effectively obtained, effectiveness in the case of ordinals of the second number class being understood to refer to a step by step process of building up to the ordinal from below.

Much of the interest of the proposed definition lies, of course, in its absoluteness, and would be lost if it could be shown that it was in any essential sense relative to a particular scheme of notation or a particular formal system of logic. It is my present belief that the definition is absolute in this way — towards those who do not find this convincing the definition may perhaps be allowed to stand as a challenge, to find either a less inclusive definition which cannot be shown to exclude some ordinal which ought reasonably to be allowed as constructive, or a more inclusive definition which cannot be shown to include some ordinal of the second class which cannot be seen to be constructive.

It is believed that the distinction which is proposed to develop between constructive and non-constructive ordinals (and functions of ordinals) should be of interest generally in connection with applications of the transfinite ordinals to mathematical problems. The relevance of the distinction is especially clear, however, in the case of applications of ordinals to certain questions of symbolic logic (for example, various questions more or less closely related to the well known theorem of Gödel on undecidable propositions)[1] — this is because of the criterion of effectiveness or "definiteness" which necessarily applies to the rules of procedure of a formal system of logic.[2] The distinction is also presumably relevant to proposals to classify recursive functions of

---

Originally published in ***Bulletin of the American Mathematical Society***, vol. 44 (1938), pp. 224–232.

*An address delivered by invitation of the Program Committee at the Indianapolis meeting of the American Mathematical Society, December 29, 1937.

[1] Kurt Gödel, *Über formal unentscheidbare Sätze der Principia Mathematica und verwandter Systeme I*, ***Monatshefte für Mathematik und Physik***, vol. 38 (1931), pp. 173–198. See also ***On Undecidable Propositions of Formal Mathematical Systems***, mimeographed lecture notes, Princeton, N.J., 1934.

[2] There is some current tendency to apply the name "logic" to schemes which are similar to accepted systems of symbolic logic but involve one or more rules of procedure which lack this characteristic of effectiveness. Such schemes may perhaps be of interest as abstract definitions of classes of formulas, but they cannot in my opinion be called "logics" except by a drastic (and possibly misleading) change in the usual meaning of that word. For they do not provide an *applicable* criterion as to what constitutes a valid proof.

natural numbers according to ordinals of the second number class[3] — because the possibility of an effective step by step calculation of the values of the function may reasonably be taken as an essential part of the notion of a recursive function.

For present purposes it will be convenient to make a minor departure from the established terminology, using "second number class" in such a sense that the first number class is included as a part of the second number class (thus avoiding the sometimes awkward phrase, "first and second number classes"). On this basis, the *second number class* may be described as the simply ordered set which results when we take 0 as the first (or least) element of the set and allow the two following processes of generation: (1) given any element of the set, to generate the element which next follows it (the least element greater than it); (2) given any infinite increasing sequence of elements, of the order type of the natural numbers, to generate the element which next follows the sequence (the least element greater than every element of the sequence).[4] The elements of the set are *ordinals*. The ordinal which next follows a given ordinal is the *successor* of that ordinal. The ordinal which next follows a given infinite increasing sequence of ordinals, of the order type of the natural numbers, is the *upper limit* of that sequence. The ordinal which has a given ordinal as a successor is the *predecessor* of that ordinal. An infinite increasing sequence of ordinals, of the order type of the natural numbers, which has a given ordinal as upper limit is a *fundamental sequence* of that ordinal. Every ordinal except 0 has either a predecessor or a fundamental sequence but not both; in the first case the ordinal is of the *first kind*, in the second case of the *second kind*.

As a definition of the distinction between constructive and non-constructive ordinals in the second number class might be proposed the following:

An ordinal ξ is constructive if it is possible to devise a system of notation which assigns a unique notation to every ordinal less than or equal to ξ and, associated with the system of notation, three effective processes by which, respectively, (1) given the notation for any ordinal it can be determined whether the ordinal is of the first or the second kind, (2) given the notation for any ordinal of the first kind the notation for the predecessor of the ordinal can be obtained, (3) given the notation for any ordinal of the second kind a fundamental sequence of that ordinal can be obtained, in the sense of an effective process for calculating the notations for the successive terms of the sequence.[5]

It will be seen that the contemplated system of notation for ordinals less than or equal to ξ will then also admit an effective process by which, given the notations for any two ordinals, it can be determined which ordinal is greater.

Moreover, of course this contemplated system of notation is required (by definition) to yield an effective simultaneous selection, for every ordinal μ of the second kind less than or equal to ξ, of one out of the various fundamental sequences of μ. In fact,

---

[3]Cf. David Hilbert, *Über das Unendliche*, ***Mathematische Annalen***, vol. 95 (1926), pp. 161–190; Wilhelm Ackermann, *Zum Hilbertschen Aufbau der reellen Zahlen*, ***Mathematische Annalen***, vol. 99 (1928), pp. 118–133.

[4]This definition of the second number class is selected as fundamental because it represents the way in which, ordinarily, the ordinals actually arise in applications to mathematical problems — in particular the way in which they arise in connection with those questions of symbolic logic to which reference was just made.

[5]A similar but essentially different property of an ordinal ξ is employed by S. C. Kleene; see *On notation for ordinal numbers*, abstract in the ***Bulletin of the American Mathematical Society***, vol. 43 (1937), p. 41.

such an assignment of a unique fundamental sequence to every ordinal of the second kind less than or equal to ξ is held evidently to be a necessary consequence of the step by step process of building up to ξ from below which was first taken as characterizing constructiveness.

The present definition of constructiveness of ξ (that there exists a system of notation, of the kind described, for the ordinals less than or equal to ξ) is thought to correspond satisfactorily to the vaguer notion with which we began, and also to be satisfactorily exact, except for one thing, the vagueness of the notion of an effective process.

This notion of an effective process occurs frequently in connection with mathematical problems, where it is apparently felt to have a clear meaning, but this meaning is commonly taken for granted without explanation. For our present purpose it is desirable to give an explicit definition.

Perhaps the most convincing form in which this definition of an effective process can be put results from the adoption of an idea of Turing.[6] A process is effective if it is possible to devise a calculating machine, with a finite number of parts of finite size, which, with the aid of an endless tape, running through the machine, on which symbols are printed, will carry out the process in any particular case—of course only a finite portion of the tape being used in any particular case. (It will be seen that a human calculator, provided with pencil and paper and explicit instructions, can be thought of as a machine of this kind, the paper taking the place of the tape.)

In the case of a process which, applied to a natural number, yields a natural number, another form of the definition of effectiveness is that the process is effective if it corresponds to an arithmetic function which is recursive in the most general sense.[7] This definition can be extended to processes upon the formulas of an arbitrary system of notation by employing the now familiar device of representing the formulas by Gödel numbers.

Still another form of the definition of an effective process is obtained by replacing the condition of general recursiveness by the condition of λ-definability.[8]

The equivalence of these three definitions is established in papers by Kleene and Turing.[9]

This completes the explanation of the proposed definition of the distinction between constructive and non-constructive ordinals in the second number class. In order, however, to obtain a definition of the related distinction between constructive and non-constructive functions of ordinals, and in order to establish that not every ordinal in the second number class is constructive, it is desirable to extend the notion of λ-definition, which was first introduced for positive integers, to the transfinite ordinals by introducing appropriate formulas (of the λ-formalism) to represent the ordinals. For the constructive ordinals of the second number class this is done as follows.

---

[6]A. M. Turing, *On computable numbers, with an application to the Entscheidungsproblem*, ***Proceedings of the London Mathematical Society***, (2), vol. 42 (1936–1937), pp. 230–265.

[7]I.e., general recursive in the sense of Herbrand and Gödel. Cf. Kurt Gödel, ***On Undecidable Propositions of Formal Mathematical Systems***, pp. 26–27; S. C. Kleene, *General recursive functions of natural numbers*, ***Mathematische Annalen***, vol. 112 (1935–1936), pp. 727–742.

[8]Cf. Alonzo Church, *An unsolvable problem of elementary number theory*, ***American Journal of Mathematics***, vol. 58 (1936), pp. 345–363.

[9]S. C. Kleene, *λ-definability and recursiveness*, ***Duke Mathematical Journal***, vol. 2 (1936), pp. 340–353; A. M. Turing, *Computability and λ-definability*, ***The Journal of Symbolic Logic***, vol. 2 (1937), pp. 153–163.

Using an arrow to mean "stands for" or "is an abbreviation for," let

$$0_o \rightarrow \lambda m \,.\, m(1),$$
$$S_o \rightarrow \lambda am \,.\, m(2, a),$$
$$L \rightarrow \lambda arm \,.\, m(3, a, r).$$

where,

$$1 \rightarrow \lambda fx \,.\, f(x),$$
$$2 \rightarrow \lambda fx \,.\, f(f(x)),$$
$$3 \rightarrow \lambda fx \,.\, f(f(f(x))).$$

Then let a class of formulas, to be called *ordinal-formulas*, and a relation $<$ between ordinal-formulas be defined simultaneously by induction as follows (1–6):

1. If $\boldsymbol{a}$ is an ordinal-formula and $\boldsymbol{b}$ conv $\boldsymbol{a}$, then $\boldsymbol{b}$ is an ordinal-formula; further any ordinal-formula which bears the relation $<$ to $\boldsymbol{a}$ bears that relation also to $\boldsymbol{b}$; further $\boldsymbol{b}$ bears the relation $<$ to any ordinal-formula to which $\boldsymbol{a}$ bears this relation.

2. 0_o is an ordinal-formula.

3. If $\boldsymbol{a}$ is an ordinal-formula, $S_o(\boldsymbol{a})$ is an ordinal formula, and $\boldsymbol{a} < S_o(\boldsymbol{a})$.

4. If $\boldsymbol{r}$ is a well-formed formula and each of the formulas in the infinite list, $\boldsymbol{r}(0_o)$, $\boldsymbol{r}(S_o(0_o))$, $\boldsymbol{r}(S_o(S_o(0_o)))$, $\cdots$, is an ordinal-formula and bears the relation $<$ to the formula which follows it in the list, then $L(0_o, \boldsymbol{r})$ is an ordinal-formula, and each of the formulas in the infinite list bears the relation $<$ to $L(0_o, \boldsymbol{r})$.

5. If $\boldsymbol{a}, \boldsymbol{b}, \boldsymbol{c}$ are ordinal-formulas and $\boldsymbol{a} < \boldsymbol{b}$ and $\boldsymbol{b} < \boldsymbol{c}$, then $\boldsymbol{a} < \boldsymbol{c}$.

6. The ordinal-formulas are the smallest class of formulas possible consistently with 1–5, and the relation $<$ subsists between two ordinal formulas only when compelled to do so by 1–5.

It will seen that under the relation $<$ the ordinal-formulas form, not a simply ordered, but a partially ordered set.

Then an assignment of formulas to represent ordinals in the second number class is defined by induction as follows (i–v):

i. If $\boldsymbol{a}$ represents an ordinal a, and $\boldsymbol{b}$ conv $\boldsymbol{a}$, then $\boldsymbol{b}$ also represents a.

ii. 0_o represents the ordinal 0.

iii. If $\boldsymbol{a}$ represents an ordinal a, then $S_o(\boldsymbol{a})$ represents the successor of a.

iv. If $\boldsymbol{r}$ is a well-formed formula and each of the formulas in the infinite list, $\boldsymbol{r}(0_o)$, $\boldsymbol{r}(S_o(0_o))$, $\boldsymbol{r}(S_o(S_o(0_o)))$, $\cdots$, is an ordinal-formula and bears the relation $<$ to the formula which follows it in the list, and if the formulas in the infinite list represent respectively the ordinals $b_0, b_1, b_2, \cdots$, then $L(0_o, \boldsymbol{r})$ represents the upper limit of the sequence $b_0, b_1, b_2, \cdots$.[10]

v. A formula represents an ordinal only if compelled to do so by i–iv.

Evidently, every formula which represents an ordinal is an ordinal-formula (as previously defined) and every ordinal-formula represents an ordinal. If the relation $<$

---

[10]This constitutes a minor but essential correction to the assignment of formulas to represent ordinals which is proposed by Alonzo Church and S. C. Kleene, *Formal definitions in the theory of ordinal numbers*, ***Fundamenta Mathematicae***, vol. 28 (1937), pp. 11–21. The correction is minor in the sense that no further changes in the cited paper are necessitated by it.

The correction is regarded as essential from the point of view of intuitive justification. Since this address was delivered, however, I have seen a proof by Kleene that the two definitions of λ-definability of ordinals in the second number class (the definition given here and our previous definition) are equivalent.

(between ordinal-formulas) holds between two given ordinal-formulas, the relation *less than* (between ordinals) must hold between the ordinals which they represent—but not conversely.

Moreover, under this assignment of formulas to represent ordinals in the second number class, every formula which represents an ordinal has a normal form, and no formula represents more than one ordinal, as has been proved by Church and Kleene.[11] In general, however, the same ordinal is represented by an infinite number of non-interconvertible formulas.

Let us call an ordinal in the second number class *λ-definable* if there is at least one formula which represents it under the foregoing scheme. Then the class of λ-definable ordinals coincides with the class of constructive ordinals previously defined.

For the ordinal-formulas in principal normal form[12] which bear the relation $<$ to a given ordinal-formula form a simply ordered set under the relation $<$, as can be proved by transfinite induction. If ξ is a λ-definable ordinal, and $\boldsymbol{\xi}$ is a formula in principal normal form which represents ξ, then $\boldsymbol{\xi}$ together with the ordinal-formulas in principal normal form which bear the relation $<$ to $\boldsymbol{\xi}$ constitute a system of notation for the ordinals less than or equal to ξ which is of the kind required by the definition of constructiveness. Hence every λ-definable ordinal is constructive.

Moreover, if ξ is a constructive ordinal, the system of notation for ordinals less than or equal to ξ which is referred to in the definition of constructiveness can be replaced by an assignment of a positive integer to each ordinal less than or equal to ξ, the notation for each ordinal being replaced by the Gödel number of the notation. And then the positive integer assigned to each ordinal may in turn be replaced by the formula (of the λ-formalism) which represents that positive integer.[13] Thus there is correlated to each ordinal less than or equal to ξ a formula (of the λ-formalism), and in fact one of the formulas which represent the positive integers. In view of the definition of the constructiveness of ξ, and since every effective function of positive integers is λ-definable, there will be three formulas $\boldsymbol{K}$, $\boldsymbol{P}$, $\boldsymbol{f}$, having the following properties: if $\boldsymbol{m}$ is any formula (of the λ-formalism) which is correlated (as just described) to an ordinal μ less than or equal to ξ, then $\boldsymbol{K}(\boldsymbol{m})$ conv 1 if μ is of the first kind, $\boldsymbol{K}(\boldsymbol{m})$ conv 2 if μ is of the second kind, $\boldsymbol{K}(\boldsymbol{m})$ conv 3 if μ is 0_o; also if μ is of the first kind, $\boldsymbol{P}(\boldsymbol{m})$ conv the formula correlated to the predecessor of μ; and if μ is of the second kind and $\boldsymbol{n}$ is a formula which represents a finite ordinal n,[14] then $\boldsymbol{f}(\boldsymbol{m}, \boldsymbol{n})$ conv the formula correlated to the $(1 + n)$th term of that fundamental sequence of μ which is given by the effective process (3) referred to in the definition of the constructiveness of ξ. Using a theorem of Kleene,[15] one may obtain a formula $\boldsymbol{T}$ having the following conversion properties,

---

[11]In the paper cited in the preceding footnote.

[12]For definition of the terms *normal form* and *principal normal form* reference may be made to *An unsolvable problem of elementary number theory*. The normal form of a formula is ambiguous only to the extent of possible alphabetical changes of the bound variables which appear, but it is sometimes convenient to remove even this ambiguity by adopting a device due to Kleene by which a particular one of the normal forms of a formula is designated as the principal normal form.

[13]For the representation of positive integers by formulas (of the λ-formalism) see, for example, *An unsolvable problem of elementary number theory*.

[14]The finite ordinals are here taken as distinct from the corresponding non-negative integers, as is done in the cited paper of Church and Kleene. In view of the formula $\mathfrak{T}$ derived in that paper (page 19), there is no difficulty caused here by taking n to be a finite ordinal rather than a positive integer.

[15]*λ-definability and recursiveness*, Theorem 19 and footnote 17.

where $\boldsymbol{m}$ is any formula which represents a positive integer:

$$\boldsymbol{T}(\boldsymbol{m}) \text{ conv } S_o(\boldsymbol{T}(\boldsymbol{P}(\boldsymbol{m}))), \text{ if } \boldsymbol{K}(\boldsymbol{m}) \text{ conv } 1;$$
$$\boldsymbol{T}(\boldsymbol{m}) \text{ conv } L(0_o, \lambda n\, \boldsymbol{T}(\boldsymbol{f}(\boldsymbol{m}, n))), \text{ if } \boldsymbol{K}(\boldsymbol{m}) \text{ conv } 2;$$
$$\boldsymbol{T}(\boldsymbol{m}) \text{ conv } 0_o, \text{ if } \boldsymbol{K}(\boldsymbol{m}) \text{ conv } 3.$$

Then if $\boldsymbol{m}$ is the formula which is correlated to an ordinal μ less than or equal to ξ, the formula $\boldsymbol{T}(\boldsymbol{m})$ will represent the ordinal μ (proof by transfinite induction). In particular if $\boldsymbol{x}$ is the formula which is correlated to ξ, then $\boldsymbol{T}(\boldsymbol{x})$ will represent ξ. Hence every constructive ordinal is λ-definable.

Now let a function $F(x_1, x_2, \cdots, x_r)$ of ordinals in the second number class be called *λ-definable* if there is a formula $\boldsymbol{F}$ such that whenever the formulas $\boldsymbol{x}_1, \boldsymbol{x}_2, \cdots, \boldsymbol{x}_r$ represent the ordinals $x_1, x_2, \cdots, x_r$ respectively, the formula $\boldsymbol{F}(\boldsymbol{x}_1, \boldsymbol{x}_2, \cdots, \boldsymbol{x}_r)$ will represent the ordinal $F(x_1, x_2, \cdots, x_r)$. As a definition of the notion of a constructive function of ordinals in the second number class, it is proposed simply to identify this notion with that of a λ-definable function of ordinals in the second number class. This is rendered plausible by the known properties of the λ-formalism, and no definition with a more direct appeal suggests itself.

It has been proved by Church and Kleene[16] that a large class of functions of ordinals are λ-definable, including addition, multiplication, exponentiation, the function ϵ_x, a predecessor function of ordinals, and others. The function $\phi(x, y)$ of ordinals in the second number class whose value is the ordinal 0 when x is less than y, and 1 when x is equal to y, and 2 when x is greater than y is , however, demonstrably not λ-definable.[17] This is taken to mean that from a strictly finitary point of view the second number class as a simply ordered set is inadmissible and must be replaced by a partially ordered set which has the structure of the set of ordinal-formulas under the relation $<$. Many of the classical theorems about the second number class may, however, be represented by corresponding finitary theorems about this partially ordered set.

In order to establish the existence of non-constructive ordinals in the second number class it is sufficient to observe that the class of all formulas (of the λ-formalism) is enumerable, and hence (by a non-constructive argument) that the class of all λ-definable ordinals is enumerable, whereas the class of all ordinals in the second number class is non-enumerable.

Let ω_1 be the least non-constructive ordinal in the second number class. Then evidently every ordinal greater than ω_1 in the second number class is also non-constructive.

It is readily proved that ω_1 cannot be of the first kind. Thus by an indirect argument the existence of a fundamental sequence of ω_1 is established. But demonstrably there cannot be an effective process of calculating the successive terms of a fundamental

[16] Loc. cit.

[17] This is not surprising. It is, for instance, not difficult to give examples of pairs of constructive definitions of ordinals such that the question whether the ordinals defined are equal, or which of the two is greater, depends on this or that unsolved problem of number theory; and indeed this may be done without employing any ordinal greater than ω^2.

sequence of ω_1.[18]

From a finitary point of view, ω_1 belongs to the third number class but not the second.

PRINCETON UNIVERSITY

[18] If it be required that the process of calculating the successive terms of the sequence shall not only provide a step by step process of building up to each term from below, but shall also effectively exhibit the increasing character of the sequence by including the step by step process of building up to any term as a part of the corresponding process for each subsequent term, the impossibility is an immediate consequence of the definition of ω_1. If, however, the condition be omitted that the process effectively exhibit the increasing character of the sequence in this way, the proof of impossibility depends on the theorem of Kleene previously referred to concerning the equivalence of the two definitions of λ-definability of ordinals in the second number class.

THE PRESENT SITUATION IN THE FOUNDATION OF MATHEMATICS (1939)

1. I think it may be accepted as a starting point that the central problem in the foundation of mathematics is the construction of a symbolic system within which the body of extant mathematics may be derived in accordance with sharply stated and immediately applicable formal rules, and that other questions must be regarded as secondary on the ground that they cannot be given a definite meaning, in fact do not have a definite *subject*, until such a construction is accomplished. Indeed it is perhaps not going too far to say that this view of the central role of a formalized system is emerging as a point of at least approximate agreement even among those whose views on foundational questions are otherwise widely divergent. Not only is it shared, for instance, by the so-called logistic school (WHITEHEAD and RUSSELL) and the school of HILBERT—to whose programs it is, of course, alike fundamental—but such a formalization has been adopted by HEYTING [**1**] as a means of rendering precise, and precisely communicable, the ideas of BROUWER which must otherwise have remained in large part vague.[1]

The question is non-essential whether we contemplate a single formalized system embracing the whole of extant mathematics or separate systems one for each branch of extant mathematics. The systems will in any case be finite in number and therefore capable of synthesis into a single system—and as a matter of fact this synthesis tends to be possible in a way to yield an organized whole.

On the other hand it is a consequence of GÖDEL's theorem [**2**] that the insistence on *extant* mathematics is unavoidable, on the ground that a single system adequate to "all conceivable mathematics" is impossible (in fact the GÖDEL result must be taken to mean that there is no definition of the phrase in quotations which provides a genuinely applicable criterion). In this connection, however, the word *extant* obviously should not be construed in so narrow a sense that the contemplated formal system might consist of expressions for all mathematical theorems actually in the present literature (a finite set), together with the null set of rules of inference, or a set of entirely trivial rules. In practice the search is for a formalized basis which is simple and brief beside the entire body of concretely existing mathematics but capable, under its rules of inference, of yielding this entire body and (infinite) further material; this further material is then taken as also belonging to extant mathematics (as here intended) and the degree of simplicity of the formalized basis is in a sense a measure of the legitimacy of the construction. Thus the notion of "extant mathematics" is itself indefinite, but it is capable of being made definite in a way in which that of "all mathematics" is not. As we have phrased the matter, it turns out that the problem of making the notion of

Originally published in ***Philosophie mathématique***, by F. Gonseth. Actualités scientifiques et industrielles 837, Hermann et Cie, Paris 1939, pp. 67–72. Church's contribution was part of a larger article.

[1] It does not affect the present point that HEYTING regards his use of a formalized system, or language, as only a *Hilfsmittel*. The indispensability of the *Hilfsmittel* is evident (and apparently admitted) for communication not only between HEYTING and others but also between HEYTING himself at one moment of his life and at a later moment.

extant mathematics definite is the same as the problem of formalization which we are urging is central.

Among other problems of current interest concerning the foundations of mathematics may be mentioned the question of the validity of the logistic thesis, that mathematics is a branch of formal logic in the sense that it may be developed by purely logical methods out of postulates purely logical in character; also the question whether mathematics—and, for those who hold the logistic thesis, with it also logic—consists merely in a symbolic formalism and its manipulation according to rule, or whether the formalism is to be regarded as the visible sign, and means of communication, of something less tangible which lies behind it, whether an activity of the human intellect (the intuitionistic *Denktätigkeit*) or a self-existent realm of abstractions which the symbols mean. Both these questions, in my opinion, can be satisfactorily discussed and answered only in connection with the construction of a formalized system of extant mathematics, either in consequence of such a construction or simultaneously with it. Thus the question concerning the validity of the logistic thesis requires an exact definition of mathematics and of logic, or at least of extant mathematics (and perhaps of extant logic), and the symbolic method here advocated offers the only means of accomplishing this—in the very nature of the subject. The second question, concerning the nature of mathematics, evidently is in some danger of degenerating into a debate over a difference merely in fashion of speaking; I think we may feel assured of the genuineness of the question only in the light of a formalized development of mathematics and of the effect of the answer upon the character and content of this development.

2. There are at the present time two widely accepted symbolic systems which offer a formalized basis, and thus a proposed definition of extant mathematics. These are the axiomatic set theory of ZERMELO **[3]** and the system of *Principia Mathematica* **[4]**.

The Zermelo set theory may be said to have been in its inception opposed to the symbolic method, and in accordance rather with the common method under which only the specially mathematical concepts are treated symbolically and the underlying logical concepts by means of which they are organized into a theory are left at the unformalized level.[2] The step taken by ZERMELO lay in transferring the concepts of class and class-membership from the (unformalized) logical level to the (formalized) mathematical level. However, subsequent development and improvement of the ZERMELO theory—by FRAENKEL, VON NEUMANN, SKOLEM, and others—has been bringing it continually closer to the symbolic method, until it is now possible to formulate the theory as a system of symbolic logic in the same sense as the system of *Principia* **[5–6]**.

The system of *Principia* has also undergone modifications since its proposal—of which the most important is the now generally accepted replacement of the original ramified theory of types by the so-called simple theory of types, with consequent elimination of the necessity for the much disputed (and clearly inelegant) axiom of reducibility[3] It is also sometimes convenient to allow what may be called the *extended simple theory of types*, under which a propositional function (say of one variable) is

---

[2]The difference between the two methods consists not in the use of symbols—since words are also symbols, not in principle different from other kinds—but in the exactness with which the laws governing the logical symbols are formulated.

[3]The simple theory of types which had its origin in suggestions of CHWISTEK and RAMSEY is set forth in CARNAP [7].

not restricted to arguments of the *next* lower type but may take an argument of *any lower type*. From this extended theory it is but a step to a theory allowing, in addition to the finite types a certain number of transfinite types.[4]

In spite of superficial differences, the two systems, ZERMELO's and RUSSELL's, are in their currently accepted forms essentially similar, both as regards the character and extent of the mathematics to which they lead and as regards the nature of the restrictions which they impose upon naive logic in order to avoid the paradoxes of which RUSSELL's and BURALI-FORTI's are typical. In what follows the two will be treated together.

There seems to be a sense, however, in which the ZERMELO theory is the stronger, since QUINE **[5]** has shown how to derive the simple theory of types within the ZERMELO theory, whereas the inverse derivation is probably impossible, unless by employing transfinite types, or by allowing a much more radical reinterpretation of *class* and *class-membership* than Quine found necessary.

3. The widely held opinion that the RUSSELL system, or that of ZERMELO, constitutes a final solution of the problem of a rigorously formalized basis for extant mathematics seems to me unjustified. In particular it may be urged that there is no convincing basis for a belief in the consistency[5] of either system, even as probable.

Apparently the only arguments which can at present be brought in support of the consistency of either system are: I) that much formal work has so far been done in the system without discovery of an inconsistency; II) that the primitive propositions and rules are self-evident logical truths and therefore necessarily consistent.

As against I) it is perhaps sufficient to point out that the length and complication of formal derivations which have so far been carried out are necessarily small by comparison with the length and complication of those which the system in principle allows. The most that can be inferred from an empirical argument is that, if the system is inconsistent, the inconsistency must arise by way of a formal derivation of more than such and such a length and degree of complication.

As against II) it may be urged that the self-evident character of these propositions and rules is by no means universally accepted[6] Moreover, as a matter of history, it has happened that systems which might have been, and in effect were, put forward on such a basis of evidence in view of their content, later were found to be inconsistent. The conspicuous example is, of course, the system of FREGE **[10]** but the same thing should perhaps be said of the (never explicitly formulated) "classical" logic which underlay the mathematics of the past until the crisis produced by the discovery of the paradoxes; and perhaps also of the system of PEANO, although PEANO's vagueness in the statement of some of his rules, especially in connection with his abstraction operation, makes a positive determination difficult.

---

[4]See GÖDEL **[2]** footnote 48a; also TARSKI **[8]** p. 395 et seq.

[5]A formalized system is called consistent, relatively to a certain symbol interpreted as negation, if there is no expression such that both this expression and its negation can be derived (proved) under the rules of the system. Evidently no system can be regarded as satisfactory unless consistent.

[6]Cf. the remark of WEYL **[9]** writing in the connection with the system of HILBERT (not essentially different for present purpose from that of ***Principia Mathematica***): "Erst die Durchführung des Widerspruchslosigkeitsbeweises oder die Bemühungen darum decken uns die höchste verzwickte logische Struktur der Mathematik auf, ein Gewirr von zirkelhaften Rückverknüpfungen, von denen sich gar nicht übersehen lässt, ob sei nicht zu eklatanten Widersprüchen führen."

A cogent basis for acceptance of the ZERMELO system or that of *Principia* might be found in an objective proof, finitary in the sense of HILBERT, of the consistency of the system. At the present time, however, it seems at least doubtful that such a proof is possible—although it is noteworthy that GENTZEN **[11]**[7] has accomplished a finitary consistency proof of what may be roughly described as a truncated portion of the system of *Principia*, adequate to (elementary) number theory.

The significance of such a finitary consistency proof—when possible—is twofold. A finitary proof of a general proposition is convincing in a sense in which many mathematical proofs (formalizable, e.g., in *Principia*) are not, because by the method of the proof itself it is possible to verify the proposition in any particular case, not through another proof involving generalities, but through a visible construction. Further in the presence of a consistency proof of a system, the acceptance of the system as consistent involves only the recognition of the validity of that particular proof, whereas to ask that a system be accepted as consistent on the basis of an intuition that its primitive propositions and rules are valid is to ask that the intuition somehow obtain a comprehensive survey of all that is possible within the system by means of those propositions and rules.

The latter point is less emphasized in the literature than it deserves. From this point of view it is seen that even a system so framed that all proofs possible within it are finitary nevertheless requires its Hilbertian consistency proof, because the difficulty of recognizing its consistency intuitively may turn precisely on the difficulty of recognizing in advance that all proofs within it will be finitary. Thus (perhaps *a fortiori*) such a system as HEYTING's formalization of intuitionism is open to doubts of the same order as those which affect *Principia*, and has the same need of a proof of consistency.

4. The system of *Principia* and the ZERMELO axiomatic set theory are also open to attack from the other side as too restrictive of the generality which we should otherwise wish to attach to the laws of logic.[8] Indeed both theories have an element of artificiality in the imposition of *ad hoc* restrictions in order to avoid the known paradoxes.

On the other hand it must be said that proposals to resolve the paradoxes in other ways for the most part either have never made definite incorporation into a specific formal system, and thus cannot be satisfactorily evaluated **[14]**, or else have led to systems which were actually inconsistent.[9]

BIBLIOGRAPHIE

Editor's note: The asterisks on the numeration of items have to do with cross-referencing within the larger work of which Church's contribution is a part.

1. HEYTING (A.), *Die formalen Regeln der intuitionistischen Logik*, Sitzungsberichte der Preussischen Akademie der Wissenschaften, Phy. Math. Klasse, 1930, pp. 42–71, 158–169.

[7]To describe GENTZEN's proof as transfinite, as is sometimes done, because it involves complete induction up to the ordinal ε_0 seems to me misleading. If this method be transfinite, what shall be said of induction up to ω^ω? Of induction up to ω^2?

[8]Cf. FREGE's comment **[12]** on a proposal of SCHRÖDER **[13]** which, except as regards motivation by the paradoxes, precisely anticipates the simple theory of types.

[9]This is the case with a system proposed by me and a related system due to CURRY, the inconsistency of both being demonstrated by KLEENE and ROSSER **[15]** (Both systems, however, contain elements which can be separated out and have proved to be of value in other connections).

Editor's note: Church asked that the last paragraph of this paper be omitted, which we have done.

2. GÖDEL (K.). *Ueber formal unentscheidbare Sätze der Principia Mathematica und verwandter Systeme*, Monatsh. Math. Phys., 1931, **38**, pp. 173–198.

3. ZERMELO (E.). *Untersuchungen über die Grundlagen der Mengenlehre*, Mathematischen Annalen, 1908, **65**, pp. 261–281.

4. RUSSELL (B.). *Mathematical logic as based on the theory of types*, Amer. Jour. Math., 1908, **30**, pp. 222–262.

5. QUINE (W. V.). *Set-theoretic foundations for logic*, The Journal of Symbolic Logic, 1936, **1**, pp. 47–57.

*6. ACKERMANN (W.). *Mengentheoretische Begründung der Logik*, Mathematische Annalen, 1937, **115**, pp. 1–22.

7. CARNAP (R.). *Abriss der Logistik*, Julius Springer, Vienne, 1929.

8. TARSKI (A.). *Der Wahrheitsbegriff in den formalisierten Sprachen*, Studia Philosophica, 1936, **1**, pp. 261–405.

9. WEYL (H.). Philosophie der Mathematik und Naturwissenschaften, *Handbuch der Philosophie*, 1926, **2**, Beitrag A., pp. 1–162.

10. FREGE (G.). *Grundgesetze der Arithmetik*, Jena, 1893, **1**; 1903, **2**.

11. GENTZEN (G.). *Die Widerspruchsfreiheit der reinen Zahlentheorie*, Mathematische Annalen, 1936, **112**, pp. 493–565.

12. FREGE (G.). *Kritische Beleuchtung einiger Punkte in E. Schröders Vorlesungen über die Algebra der Logik*, Archiv für Systematische Philosophie, 1895, **1**, pp. 439–440.

13. SCHRÖDER (E.). *Vorlesungen über die Algebra der Logik*, **1**, Leipzig, 1890.

14. BEHMANN (H.). *Zu den Widersprüchen der Logik und der Mengenlehre*, Jahresbericht der deutschen Mathematiker Vereinigung, 1931, **40**, pp. 37–48.

15. KLEENE (S. C.) and ROSSER (J. B.). *The inconsistency of certain formal logics*, Annals of mathematics, 1935, 2ᵉ s., **36**, pp. 630–636.

*16. QUINE (W. V.). *New foundations for mathematical logic*, The American Mathematical Monthly, 1937, **44**, pp. 70–80.

*17. QUINE (W. V.). *On Cantor's theorem*, The Journal of Symbolic Logic, 1937, **2**, pp. 120–124.

18. CHURCH (A.). *A proof of freedom from contradiction*, Proceedings of the National Academy of Sciences of the U. S. A., 1935, **21**, p. 275–281.[10]

19. CHURCH (A.). *Mathematical logic*, Mimeographed, Princeton, 1936.

20. CHURCH (A.) and ROSSER (J. B.). *Some properties of conversion*, Transactions of the American Mathematical Society, 1936, **39**, pp. 472–482.

21. LÖWENHEIM (L.). *Ueber Möglichkeiten im Relativkalkül*, Math. Annalen, 1915, **76**, pp. 447–470.

22. SKOLEM (Th.). *Logisch-kombinaritorische Untersuchungen über die Erfüllbarkeit oder Beweisbarkeit mathematischer Sätze nebst ein Theorem über dichte Mengen*, Vid Skrifter Oslo, Math., nat. Klasse, 1920, **4**.

23. CHURCH (A.). *An unsolvable problem of elementary number theory*, Amer. Journal of Mathematics, 1936, **58**.

24. CHURCH (A.). *A note on the Entscheidungsproblem ; Correction to a note on the Entscheidungsproblem*, The Journal of Symbolic Logic, 1936, **1**.

*25. HILBERT (D.) und ACKERMANN (W.). *Grundzüge der theoretischen Logik*, Julius Springer, Berlin, 1938.

26. GÖDEL (K.). *Zur intuitionistischen Logik*, Ergebnisse eines mathematischen Kolloquiums, Wien, 1932.

27. WEYL (H.). *Die Stufen des Unendlichen*, Vortrag, Jena, 1931.

---

[10] In this paper the following typographical corrections are necessary: p. 277, l. 11, insert before the 6 a negation sign followed by a dot; p. 277, l. 22–23, P should be in bold type (six times); p. 279, l. 29, N should be italic (twice).

28. CHWISTEK (L.), *Neue Grundlagen der Logik und Mathematik*, Mathematische Zeitschrift, 1929, **30**, pp. 704–724.

29. CHWISTEK (L.), HETPER (W.) et HERZBERG (J.), *Fondements de la métamathématique rationnelle*, Bulletin International de l'Académie Polonaise des Sciences et des Lettres, 1933, série A, pp. 253–264.

*30. CHWISTEK (L.) et HETPER (W.), *New foundation of formal metamathematics*, The Journal of Symbolic Logic, 1938, **3**.

*31. HETPER (W.), *Simple systems and metasystems and generalized metasystems* (à paraître).

*32. HERZBERG (J.), *Sur la notion de collectif*, Annales de la Société Polonaise de Mathématiques, 1939.

REVIEW OF CARNAP'S *FOUNDATIONS OF LOGIC AND MATHEMATICS* (1939)

Foundations of Logic and Mathematics. By Rudolf Carnap. (International Encyclopedia of Unified Science, vol. 1, no. 3.) Chicago, University Press, 1939. 8+71 pp.

This monograph presents in condensed form and with a minimum of formal detail the author's views concerning the relation of logistically formalized calculi to language in the ordinary sense, and concerning the application of such calculi in empirical science. It is a noteworthy contribution to philosophy of science and in particular to analysis of the relationship between pure and applied mathematics, the questions involved being made much more precise and intelligible than would otherwise be possible, through use of the methods of modern symbolic logic.

In many respects the author's views are here modified or clarified in such a way as to remove serious objections previously urged against them.

The "principle of tolerance" is explicitly restricted to uninterpreted logistic calculi and it is said that "a system of logic is not a matter of choice, but either right or wrong, if an interpretation of the logical signs is given in advance." The quoted sentence—by failing to take account of the fact that an interpretation, in advance of *some* formalization, must have a considerable element of vagueness—may even admit too much to the anti-conventionalists. The author adds, however: "It is important to be aware of the conventional components in the construction of a language system."

The purely syntactical method of the author's previous publications is here supplemented by an account of semantics. Designata are admitted not only for concrete terms but also, in some cases at least, for abstract symbols and expressions. Thus predicates are said to designate properties of things (p. 9), (declarative) sentences are allowed to designate "states of affairs" (p. 11), and "functors" are said to be signs for functions (p. 57). (The more usual terminology is "proposition" instead of "state of affairs" and "function symbol" instead of "functor.") The reviewer would prefer a still more liberal admission of abstract designata, not on any realistic ground, but on the basis that this is the most intelligible and useful way of arranging the matter—it would apparently be meaningless to ask whether abstract terms *really* have designata, but it is rather a matter of taste or convenience whether abstract designata shall be postulated.

The point brought out in §16, that a postulate set in the usual mathematical sense must be regarded as added to an underlying system of logic—which, for exactness, must be logistically formalized—is, of course, not new. But it deserves attention, because neglect of just this point has resulted in much misunderstanding concerning the significance of a set of postulates for a particular mathematical discipline.

On page 23, instead of distinguishing between finite and transfinite rules, it would seem to be better to distinguish between effective and non-effective rules. The matter is complicated by the fact that "finite" is often used in this connection substantially as a synonym of "effective." But a rule might well be non-effective without being transfinite in Carnap's sense.

Originally published in ***Bulletin of the American Mathematical Society***, vol. 45 (1939), pp. 821–822.

In §14 there appears to be an oversimplification of the relation between logic and arithmetic, partly through failure to make explicit mention of the axiom of infinity, and partly through an unsound use of recursive definition. An example of the latter is Definition 14, which is in effect a schema providing separate definitions for $m + 0, m + 1, m + 2, \cdots$. That this is no definition of the function $+$ may be seen by considering that the sentence, "For all natural numbers x and y, $x + y = y + x$," for example, remains undefined. This section (like most of the monograph) undertakes only to provide an outline statement with omission of formal detail; nevertheless it seems to the reviewer that an unfortunately misleading impression is given.

SCHRÖDER'S ANTICIPATION OF THE SIMPLE THEORY OF TYPES (1939)

The attempt to exclude the well-known paradoxes of set theory from a system of formal logic, without at the same time rendering the system so weak as to be inadequate for the purposes of mathematics, may be made in various ways. Currently the most favored methods are the Zermelo axiomatic set theory and the simple theory of types.

As is well known, the simple theory of types may be described in the following terms. A particular domain, to be called the domain of *individuals* is selected; this may be any domain within very wide limits, but in any particular system must be treated as fixed. Classes and relations (and functions, if the system provides for such) are then classified into a hierarchy of types. For simplicity in the description of this hierarchy it is convenient to think of relations (and functions) as defined in terms of classes—following a suggestion first made by Wiener in 1914. On this basis, the first type is composed of the individuals, the second type of classes of individuals, the third type of classes of classes of individuals, and so on. The restriction imposed by the simple theory of types is that the members of a class which belongs to the type $n + 1$ must all belong to the type n. There is no provision for classes which do not belong to a type: classes not obeying the restriction just stated are regarded as non-existent, and names for such classes are excluded from the system. And even to say or write $a \varepsilon b$ (that a is a member of the class b) is excluded as not well-formed unless the variables or constants a, b show by their syntactical form that, for some n, they refer to types n and $n + 1$ respectively.

(Of course, the foregoing statement cannot be made within the system—based on the simple theory of types—which is being described, but only within some stronger system, e.g., one employing transfinite types. For the purpose of the formal construction of the system it is not necessary to make this statement at all, but only to state the corresponding syntactical formation rules.)

What is now known as the ramified theory of types was proposed by Russell in 1908 as a means of avoiding the paradoxes, and was afterwards incorporated into *Principia mathematica* by Whitehead and Russell. The simple theory of types was proposed as a modification of this theory by Chwistek (1921) and Ramsey (1926), and was adopted by Carnap (1929), Gödel, Tarski, Quine and others.

My present purpose is to point out a striking anticipation of the simple theory of types which appears in the first volume of Schröder's *Algebra der Logik* (1890). Schröder regards the universal class 1 which appears in his algebra, not as an absolute universal class, but as composed of all the elements of a certain domain fixed in advance. The domain is arbitrary except that it must be a *reine Mannigfaltigkeit*, i.e., it

---

Originally published in ***Erkenntnis***, vol. 10 (1976), pp. 407–411. *Editor's note*: In Church's words: "This paper was presented at the Fifth International Congress for the Unity of Science in Cambridge, Massachusetts, in 1939. Preprints of the paper were distributed to the members of the Congress, and the paper was to have been published in ***The Journal of Unified Science*** (***Erkenntnis***), vol. 9 (1939), pp. 149–152. But this volume never appeared and the paper has not otherwise had publication. The present version follows the original preprints except that purely typographical errors have been corrected and an additional sentence has been inserted at the end of the second paragraph to remedy a defect which was pointed out by W. V. Quine."

must have the property "*dass unter ihren als 'Individuen' gegebenen Elementen sich keine Klassen befinden, welche ihrerseits Elemente derselben Mannigfaltigkeit als Individuen unter sich begreifen*" (p. 248). Given one such *reine Mannigfaltigkeit* a second one may be obtained, to which the algebra is equally applicable, by taking the subsets of the first to be the individuals of the second. But, Schröder emphasizes, "*man darf die Betrachtungen innerhalb der ersten mit denjenigen innerhalb der zweiten Mannigfaltigkeit nich vermengen*" (p. 249); in particular, the null classes associated with the two domains must be kept distinct. Finally, this hierarchy of *reine Mannigfaltigkeiten* may be extended to infinity (top of p. 248).

The exactness of the anticipation of the simple hierarchy of types fails only in that Schröder regards a unit class as identical with the single element which it contains. If 0 is the null class associated with the first domain, $\mathbf{O}$ the null class associated with the second ("derived") domain, and A the class composed of the two elements 0, 1, where 1 is the universal class associated with the first domain, then Schröder would regard $0 \subseteqq \mathbf{O}$ as false (not meaningless) and $0 \subseteqq A$ as true. Actually, however, this divergence from an exact anticipation of the simple theory of types is apparent rather than real; it means that we must interpret Schröder's symbol 0 within the algebra of the first domain as meaning Λ, but within the algebra of the second domain as meaning $\iota`\Lambda$, and likewise in other cases.

The paradoxes of set theory are unknown to Schröder at that time. His reason for introducing (in effect) a hierarchy of types is that he has no symbol for the relation of class-membership (ε) and for the most part confuses or identifies this relation entirely with that of class-inclusion ($\subseteqq$)—cf. p. 245 in particular. The distinction of types serves Schröder essentially as a substitute for the distinction between ε and $\subseteqq$ (cf. e.g., p. 482).

Speaking in more general terms, we may say that Schröder regarded a class as somehow built of its members, as a wall is built of bricks or (to adopt Frege's simile) as a wood is composed of trees. From such a point of view, according to which the members of a class are prior to the class, something like a theory of types does appear natural and inevitable.

Account must here be taken of the claim sometimes made on behalf of Frege that his *Stufen* (cf. his *Grundgesetze der Arithmetik*, vol. 1, 1893) constitute an anticipation of the simple theory of types. I believe this claim to be based on a misunderstanding.

With Frege a function is not properly an (abstract) object at all, but is a sort of incompleted abstraction. A function is *ungesättigt*—is like an incomplete symbol in that it requires something additional, argument or quantifier, to complete its meaning—but nevertheless partakes sufficiently of the nature of an object to be represented by a variable. The notion of a function as an actual object, a completed abstraction in intension, is not contemplated by Frege; such a notion, as a matter of fact, raises difficult questions which it is unnecessary to go into here. On the other hand, the corresponding completed abstraction in extension plays an important role in Frege's work under the name of *Werthverlauf* of a function (or *Begriffsumfang* in the special case of a propositional function).—Parenthetically, I add that, while there may be some uncertainty or conflict of usage, on the whole it is the *Werthverlauf* which corresponds to the notion of a function as used in mathematics; $\sin 2x$ and $2 \sin x \cos x$, for instance, are regarded as the same function of x, two projective collineations in a plane are the same if they have the same effect on every point on the plane, etc.

The peculiar character of Frege's functions, which are not purely syntactical entities (incomplete symbols) and not quite actual abstract objects, automatically necessitates their division into *Stufen*. The function $(\ )^2 + 1$ may take a number as argument, but there is no significant way for it to take itself as argument; to be sure, there is a function $((\ )^2 + 1)^2 + 1$, but to regard this as a *value* of the function $(\ )^2 + 1$ involves an obvious confusion; and no other possibility offers itself.

Thus the division of Frege's functions into *Stufen* is not a theory of types. It might become so if it were denied that besides functions there were also the corresponding completed abstractions, or if a similar restriction were imposed upon the completed abstractions. Actually, Frege introduces the corresponding completed abstractions as *Werthverläufe*, and explicitly denies that they are subject to the restriction associated with division into *Stufen*.

The characteristic features of the simple theory of types — that a domain of individuals is fixed upon, and the laws of logic stated first for (classes or functions over) this domain and then restated successively for other domains derived one by one from the original domain and from one another — is not only not adopted by Frege but is vigorously rejected by him.

In the *Archiv für systematische Philosophie*, vol. 1 (1895), pp. 433–456, Frege published a violent criticism of certain aspects of Schröder's algebra, especially Schröder's notion of a class. The following quotation from this article indicates what Frege would think (or, indeed, did think) of the simple theory of types (pp. 439–440):

> "Herr Schröder zieht hieraus den Schluss, dass die ursprüngliche Mannigfaltigkeit 1 so beschaffen sein müsse, dass unter ihren als Individuen gegebenen Elementen sich keine Klassen befinden, welche ihrerseits Elemente derselben Mannigfaltigkeit als Individuen unter sich begreifen. Dieses Auskunftsmittel erscheint wie ein nachträgliches Abbringen des Schiffes von einer Sandbank, die bei guter Führung ganz hätte vermieden werden können. Jetzt wird es klar, weshalb in kluger Voraussicht der drohenden Gefahr gleich anfangs eine gewisse Mannigfaltigkeit als Schauplatz der Vorgänge eingeführt wurde, wofür im reinen Gebietekalkül kein Grund war. Schön ist die nachträgliche Einschränkung dieses Feldes unserer logischen Bethätigung in keinem Falle. Während sonst die Logik den Anspruch erheben darf, dass ihre Gesetze unumschränkte Geltung haben, wird uns hier zugemutet, eine Mannigfaltigkeit sorgsam prüfend vorher abzugrenzen, innerhalb deren wir uns dann nur bewegen dürfen."

The modern reader is ironically reminded that Frege's own ship in 1902 ran upon another and very different sandbank, the Russell paradox.

In spite of this afterknowledge, Frege's criticism seems to me still to retain much of its force, and to deserve serious consideration by those who hold that the simple theory of types is the final answer to the riddle of the paradoxes.

Princeton University

ON THE CONCEPT OF A RANDOM SEQUENCE[1] (1940)

Von Mises has based his frequency theory of probability on the notion of a *Kollektiv*,[2] that is, of an infinite sequence of trials of an event whose possible outcomes have each a definite probability but otherwise appear entirely at random. (Convenient illustrative examples are an infinite sequence of tosses of a coin, an infinite sequence of rolls of a die, and the like.)[3]

Abstractly the *Kollektiv* may be represented by an infinite sequence of points of an appropriate space, the *Merkmalraum*. Or if the number of possible outcomes of a trial is finite (and it may well be argued that this is always the case for any actual physical observation[4]), it is sufficient to employ an infinite sequence of natural numbers which are less than a fixed natural number. This infinite sequence—of points or of natural numbers—satisfies certain conditions which correspond to those appearing in the description of a *Kollektiv* as just given, and which we shall express by saying that it is a random sequence (*regellose Folge*).

For the present purpose it is largely sufficient to confine attention to the case that each trial has only two possible outcomes, as with the toss of a coin adjudged as falling heads or tails, or the roll of a die adjudged as showing or not showing an ace. The *Kollektiv* may then be represented abstractly by a random sequence of 0's and 1's: in the case of the coin, for instance, we may let 1 correspond to the fall of heads and 0 to tails.

The definition of a random sequence of 0's and 1's as given by von Mises may perhaps be put in the following form:

An infinite sequence $a_1, a_2, \cdots$ of 0's and 1's is a random sequence if the two following conditions are satisfied:

(1) If $f(r)$ is the number of 1's among the first r terms of $a_1, a_2, \cdots$, then $f(r)/r$ approaches a limit p as r approaches infinity.

(2) If $a_{n_1}, a_{n_2}, \cdots$ is any infinite sub-sequence of $a_1, a_2, \cdots$, formed by deleting some of the terms of the latter sequence according to a rule which makes the deletion or retention of a_n depend only on n and $a_1, a_2, \cdots, a_{n-1}$, and if $g(r)$ is the number of 1's among the first r terms of $a_{n_1}, a_{n_2}, \cdots$, then $g(r)/r$ approaches the same limit p as r approaches infinity.

Originally published in ***Bulletin of the American Mathematical Society***, vol. 46 (1940), pp. 130–135.

[1]Presented to the Society, April 8, 1939.

[2]Richard von Mises, *Grundlagen der Wahrscheinlichkeitsrechnung*, ***Mathematische Zeitschrift***, vol. 5 (1919), pp. 52–99; ***Wahrscheinlichkeit, Statistik und Wahrheit***, Vienna, 1928; ***Wahrscheinlichkeitsrechnung***, Leipzig and Vienna, 1931; and see especially the second edition of ***Wahrscheinlichkeit, Statistik und Wahrheit***, Vienna, 1936, for its account of the objections which have been raised to von Mises's theory and the alternatives which have been proposed.

[3]The introduction of an *infinite* sequence of trials (tosses of a coin, and so on) is, of course, an abstraction from the realities of the situation, made for the sake of the mathematical theory. It is an instance of the familiar device of employing the infinite as being, for certain purposes, a convenient and useful approximation to the large finite.

[4]In cases where the number of possible outcomes of a trial is taken as infinite, either there is a further element of abstraction involved (the infinite again replacing the large finite), or else the problem considered has no direct physical application.

The inclusion of condition (2) corresponds to the *Prinzip vom ausgeschlossenen Spielsystem* of von Mises.[5] If a fixed number of wagers of "heads" are to be made, at fixed odds and in fixed amount, on the tosses of a coin, no advantage is gained in the long run if the player, instead of betting at random, follows some system, such as betting on every seventh toss, or (more plausibly) betting on the next toss after the appearance of four tails in succession, or (still more plausibly) making his nth bet after the appearance of $n+4$ tails in succession. This is accepted by von Mises as a sufficiently familiar and uncontroverted empirical generalization to be made fundamental to his theory in this way.

However, this definition, as given by von Mises or as rephrased above, while clear as to general intent, is too inexact in form to serve satisfactorily as the basis of a mathematical theory.

A plausible attempt to state the definition more exactly is the following (the numbers b_i serve as a convenient device to represent a function of a variable number of variables as a function of one variable):

An infinite sequence $a_1, a_2, \cdots$ of 0's and 1's is a random sequence if the two following conditions are satisfied:

(1) If $f(r)$ is the number of 1's among the first r terms of $a_1, a_2, \cdots$, then $f(r)/r$ approaches a limit p as r approaches infinity.

(2) If ϕ is any function of positive integers, if $b_1 = 1$, $b_{n+1} = 2b_n + a_n$, $c_n = \phi(b_n)$, and the integers n such that $c_n = 1$ form in order of magnitude an infinite sequence $n_1, n_2, \cdots$, and if $g(r)$ is the number of 1's among the first r terms of $a_{n_1}, a_{n_2}, \cdots$, then $g(r)/r$ approaches the same limit p as r approaches infinity.

However it has been pointed out by various authors[6] that the definition in this form is self-contradictory, in the sense that it makes the class of random sequences associated[7] with any probability other than 0 or 1 an empty class. For the failure of (2) may always be shown by taking $\phi(x) = a_{\mu(x)}$, where $\mu(x)$ is the least positive integer m such that $2^m > x$: the sequence $a_{n_1}, a_{n_2}, \cdots$ will then consist of those and only those terms of $a_1, a_2, \cdots$ which are 1's. This means that the definition in this form does not satisfactorily represent the requirement that the deletion or retention of a_n shall not depend on a_n or on the sequence $a_1, a_2, \cdots$ as a whole, but only on n and $a_1, a_2, \cdots, a_{n-1}$. Grave question is raised whether this requirement, made in vague terms by von Mises, can be satisfactorily represented in an exact definition at all.

This difficulty may be avoided by abandoning the attempt to define a random sequence and substituting some less restricted class of sequences, such as the admissible

---

[5] Loc. cit.

[6] Erhard Tornier, *Wahrscheinlichkeitsrechnung und Zahlentheorie*, ***Journal für die reine und angewandte Mathematik***, vol. 160 (1929), pp. 177–198; Hans Reichenbach, *Axiomatik der Wahrscheinlichkeitsrechnung*, ***Mathematisch Zeitschrift***, vol. 34 (1931–1932), pp. 568–619; E. Kamke, *Über neuere Begründungen der Wahrscheinlichkeitsrechnung*, ***Jahresbericht der deutschen Mathematiker-Vereinigung***, vol. 42 (1932–1933), pp. 14–27; Arthur H. Copeland, *Point set theory applied to the random selection of the digits of an admissible number*, ***American Journal of Mathematics***, vol. 58 (1936), pp. 181–192.

[7] An infinite sequence of 0's and 1's will be said to be associated with probability p if $f(r)/r$ approaches p as r approaches infinity, where $f(r)$ is the number of 1's among the first r terms of the sequence.

numbers of Copeland[8] or the equivalent normal sequences of Reichenbach.[9]

These admissible numbers (to adopt Copeland's term) are closely related to the normal members of Borel[10]—indeed an admissible number associated with the probability 1/2 is the same as a number *entièrement normal* to the base 2. The definition may be stated as follows: An infinite sequence $a_1, a_2, \cdots$ of 0's and 1's is an admissible number if it is associated with a probability p and if, for every positive integer m and every set of distinct positive integers $r_1, r_2, \cdots, r_k$ which are all less than or equal to m, the sequence whose nth term is the product $a_{nm+r_1} a_{nm+r_2} \cdots a_{nm+r_k}$ is associated with the probability[11] p^k.

The admissible numbers have properties which are sufficient to form a basis for a large part of the theory of probability, and they have the important advantage that their existence, for any assigned probability p, can be proved.[12] Their use for this purpose, however, is open to certain objections from the point of view of completeness of the theory, as has been forcibly urged by von Mises,[13] and it is therefore desirable to consider further the question of finding a satisfactory form for the definition of a random sequence.

The purpose of the present note is to call attention to the following possibility in this connection.

It may be held that the representation of a *Spielsystem* by an arbitrary function ϕ is too broad. To a player who would beat the wheel at roulette a system is unusable which corresponds to a mathematical function known to exist but not given by explicit definition; and even the explicit definition is of no use unless it provides a means of calculating the particular values of the function. As a less frivolous example, the scientist concerned with making predictions or probable predictions of some phenomenon must employ an effectively calculable function: if the law of the phenomenon is not approximable by such a function, prediction is impossible. Thus a *Spielsystem* should be represented mathematically, not as a function, or even as a definition of a function, but as an effective algorithm for the calculation of the values of a function.

Now a formal definition of effective calculability, for functions of positive integers, has been proposed by the author,[14] and the adequacy of this definition to represent the empirical notion of an effective calculation finds strong support in a recent result

---

[8]Arthur H. Copeland, *Admissible numbers in the theory of probability*, ***American Journal of Mathematics***, vol. 50 (1928), pp. 535–552. The infinite sequences of 0's and 1's are there taken as binary fractional expansions of real numbers between zero and one.

[9]Loc. cit. (1931–1932). See also Hans Reichenbach, *Les fondements logiques du calcul des probabilités*, ***Annales de l'Institut Henri Poincaré***, vol. 7 (1937), pp. 267–348.

[10]Émile Borel, *Les probabilités dénombrables et leurs applications arithmétiques*, ***Rendiconti del Circolo Matematico di Palermo***, vol. 27 (1909), pp. 247–271, reprinted as Note V to Borel's ***Leçons sur la Théorie des Fonctions***, 2nd edition, 1914, and 3rd edition, 1928.

[11]Copeland imposes the further condition $p \neq 0$, $p \neq 1$.

[12]Copeland, loc. cit. (1928). See also Borel, loc. cit. (1909).

[13]***Mathematische Annalen***, vol. 108 (1933), pp. 771–772; ***Wahrscheinlichkeit, Statistik und Wahrheit***, 2d edition, pp. 116–117.

[14]Alonzo Church, *An unsolvable problem of elementary number theory*, ***American Journal of Mathematics***, vol. 58 (1936) pp. 345–363. A function of positive integers is defined to be effectively calculable if it has either of the two equivalent properties of λ-definability (in the sense of Church-Kleene) or general recursiveness (in the sense of Herbrand-Gödel). Cf. S. C. Kleene, *λ-definability and recursiveness*, ***Duke Mathematical Journal***, vol. 2 (1936), pp. 340–353, and *General recursive functions of natural numbers*, ***Mathematische Annalen***, vol. 112 (1936), pp. 727–742.

of Turing.[15] It is therefore suggested that this definition of effective calculability be employed in order to define a random sequence as follows:

An infinite sequence $a_1, a_2, \cdots$ of 0's and 1's is a random sequence if the two following conditions are satisfied:

(1) If $f(r)$ is the number of 1's among the first r terms of $a_1, a_2, \cdots$, then $f(r)/r$ approaches a limit p as r approaches infinity.

(2) If ϕ is any effectively calculable function of positive integers, if $b_1 = 1$, $b_{n+1} = 2b_n + a_n$, $c_n = \phi(b_n)$, and the integers n such that $c_n = 1$ form in order of magnitude an infinite sequence $n_1, n_2, \cdots$, and if $g(r)$ is the number of 1's among the first r terms of $a_{n_1}, a_{n_2}, \cdots$, then $g(r)/r$ approaches the same limit p as r approaches infinity.

The existence of random sequences in this sense is an immediate consequence of a result of Doob,[16] or alternatively of a theorem of Wald,[17] if use is made of the fact that the set of effectively calculable functions can be represented as a subset of an effectively enumerable set and is therefore itself (noneffectively) enumerable. From Doob's theorem, taken in conjunction with Borel's result[18] that the infinite sequences of 0's and 1's associated with the probability 1/2 (regarded as binary fractional expansions of real numbers between zero and one) form a set of measure one, it follows that the random sequences associated with the probability 1/2 (similar regarded) form a set of measure one; and from the existence of random sequences associated with the probability 1/2 the existence of random sequences associated with other probabilities is readily derived. From Wald's Theorem I it follows as a corollary that the set of random sequences associated with a fixed probability has the power of the continuum.

That every random sequence is an admissible number is also easily demonstrated. On the other hand, the set of random sequences is more restricted than the set of admissible numbers; this follows, for example, from the existence of admissible numbers $a_1, a_2, \cdots$ such that a_n is an effectively calculable function of n, a property[19] which clearly cannot be possessed by any random sequence.

Thus an existence proof for random sequences is necessarily non-constructive.

---

[15]A. M. Turing, *On computable numbers, with an application to the Entscheidungsproblem*, ***Proceedings of the London Mathematical Society***, (2), vol. 42 (1936–1937), pp. 230–265; *A correction*, ibid., vol. 43 (1937), pp. 544–546; and *Computability and λ-definability*, ***Journal of Symbolic Logic***, vol. 2 (1937), pp. 153–163. Turing proves the equivalence of λ-definability and general recursiveness to a notion of a computability whose definition, briefly stated, is as follows: A function ϕ is computable if it is possible to make a computing machine, with a finite number of parts of finite size, which will calculate $\phi(n)$ for any assigned n, printing intermediate calculations and the final result on a tape with which the machine must be supplied (no upper limit is place on the time or on the length of tape required for a particular calculation). Actually, Turing imposes several further conditions on the computing machine, but these are more or less clearly nonessential.

[16]J. L. Doob, *Note on probability*, Annals of Mathematics, (2), vol. 37 (1936), pp. 363–367. The author is indebted to A. H. Copeland for calling his attention to the significance of Doob's theorem in this connection, as well as to the matter of effective constructibility of admissible numbers (footnote 19), and for other suggestions.

[17]Abraham Wald, *Sur la notion de collectif dans le calcul des probabilités*, ***Comptes Rendus des Séances de l'Académie des Sciences***, vol. 202 (1936), pp. 180–183, and *Die Widerspruchsfreiheit des Kollektivbegriffes der Wahrscheinlichkeitsrechnung*, ***Ergebnisse eines mathematischen Kolloquiums***, vol. 8 (1937), pp. 38–72.

[18]Loc. cit. (1909).

[19]The example of von Mises, ***Mathematische Annalen***, vol. 108 (1933), p. 769, is readily specialized so as to have this property. A very neat effective construction of an admissible number is implied in a paper of D. G. Champernowne, *The construction of decimals normal in the scale of ten*, ***Journal of the London Mathematical Society***, vol. 8 (1933), pp. 254–260.

Use of the above proposed definition of a random sequence as fundamental to the theory of probability is consequently open to the objection that by its means such otherwise apparently combinatorial matters as elementary questions of probability in connection with the tossing of a coin are made to depend on the powerful (and dubious) non-constructive methods of analysis. It is clear, however, that any definition of a random sequence more stringent than this one would have the same disadvantage, and on the other hand that no definition in any respect less stringent could be regarded as even approximately representing von Mises's intention or as being free from such objections as those brought by him against the use of admissible numbers or normal sequences.

Nevertheless it would seem to be of interest to investigate criteria of randomness of intermediate strength, in particular the definition of a random sequence which results if the condition that ϕ be effectively calculable is replaced by the condition that ϕ be primitive recursive. Since the primitive recursive functions are effectively enumerable, sequences satisfying this criterion can be effectively constructed in accordance with Wald's Theorem V.[20]

PRINCETON UNIVERSITY

[20] Loc. cit. (1937). Wald relies on the common notion of effectiveness and has no exact definition. His proof is entirely applicable here. Wald also remarks on a criterion of randomness—in general more stringent than that proposed in the present paper—which consists, in effect, in replacing the condition that ϕ be effectively calculable by the condition that ϕ be definable within a fixed system of symbolic logic L. There are, however, several objections to this criterion. It is unavoidably relative to the choice of the particular system L, and thus has an element of arbitrariness which is artificial. If used within the system L, it requires the presence in L of the semantical relation of *denotation* (known to be problematical on account of the Richard paradox). It it is used outside of L, it becomes necessary to say more exactly what is meant by "definable in L," and the questions of consistency and completeness of L are likely to be raised in a peculiarly uncomfortable way.

A FORMULATION OF THE SIMPLE THEORY OF TYPES (1940)

The purpose of the present paper is to give a formulation of simple theory of types[1] which incorporates certain features of the calculus of λ-conversion.[2] A complete incorporation of the calculus of λ-conversion into the theory of types is impossible if we require that λx and juxtaposition shall retain their respective meanings as an abstraction operator and as denoting the application of function to argument. But the present partial incorporation has certain advantages from the point of view of type theory and is offered as being of interest on this basis (whatever may be thought of the finally satisfactory character of the theory of types as a foundation for logic and mathematics).

For features of the formulation which are not immediately connected with the incorporation of λ-conversion, we are heavily indebted to Whitehead and Russell,[3] Hilbert and Ackermann,[4] Hilbert and Bernays,[5] and to forerunners of these, as the reader familiar with the works in question will recognize.

1. The hierarchy of types. The class of *type symbols* is described by the rules that ι and o are each type symbols and that if α and β are type symbols then $(\alpha\beta)$ is a type symbol; it is the least class of symbols which contain the symbols ι and o and is closed under the operation of forming the symbol $(\alpha\beta)$ from the symbols α and β.

As exemplified in the statement just made, we shall use the Greek letters α, β, γ to represent variables or undetcrmined type symbols. We shall abbreviate type symbols by omission of parentheses with the convention that association is to the left — so that, for instance, $o\iota$ will be an abbreviation for $(o\iota)$, $\iota\iota\iota$ for $((\iota\iota)\iota)$, $\iota\iota(\iota\iota)$ for $((\iota\iota)(\iota\iota))$, etc. Moreover, we shall use α' as an abbreviation for $((\alpha\alpha)(\alpha\alpha))$, a'' as an abbreviation for $((\alpha'\alpha')(\alpha'\alpha'))$, etc.

The type symbols enter our formal theory only as subscripts upon variables and constants. In the interpretation of the theory it is intended that the subscript shall indicate the type of the variable or constant, o being the type of propositions, ι the type of individuals, and $(\alpha\beta)$ the type of functions of one variable for which the range of the independent variable comprises the type β and the range of the dependent variable is contained in the type α. Functions of several variables are explained, after

---

Originally published in ***The Journal of Symbolic Logic***, vol. 5 (1940), pp. 56–68.

[1] See Rudolf Carnap, ***Abriss der Logik***, Vienna 1929, §9. (The simple theory of types was suggested as a modification of Russell's ramified theory of types by Leon Chwistek in 1921 and 1922 and by F. P. Ramsey in 1926.)

[2] See, for example, Alonzo Church, ***Mathematical Logic***, mimeographed, Princeton, New Jersey, 1936 and ***The calculi of lambda-conversion***, forthcoming monograph.

[3] Bertrand Russell, *Mathematical logic as based on the theory of types*, ***American Journal of Mathematics***, vol. 30 (1908), pp. 222–262; Alfred North Whitehead and Bertrand Russell, ***Principia Mathematica***, vol. 1, Cambridge, England, 1910 (second edition, 1925), vol. 2, Cambridge, England, 1912 (second edition, 1927), and vol. 3, Cambridge, England, 1913 (second edition, 1927).

[4] D. Hilbert and W. Ackermann, ***Grundzüge der theoretischen Logik***, Berlin, 1928 (second edition, 1938).

[5] D. Hilbert and P. Bernays, ***Grundlagen der Mathematik***, vol. 1, Berlin 1934, and vol. 2, Berlin, 1939.

Schönfinkel,[6] as functions of one variable whose values are functions, and propositional functions are regarded simply as functions whose values are propositions. Thus, e.g., $o\iota\iota$ is the type of propositional functions of two individual variables.

We purposely refrain from making more definite the nature of the types o and ι, the formal theory admitting of a variety of interpretations in this regard. Of course the matter of interpretation is in any case irrelevant to the abstract construction of the theory, and indeed other and quite different interpretations are possible (formal consistency assumed).

2. Well-formed formulas. The *primitive symbols* are given in the following infinite list:

$$\lambda, (,), N_{oo}, A_{ooo}, \Pi_{o(o\alpha)}, \iota_{\alpha(o\alpha)}, a_\alpha, b_\alpha, \cdots, z_\alpha, \bar{a}_\alpha, \bar{b}_\alpha, \cdots.$$

Of these, the first three are *improper symbols*, and the others are *proper symbols*. Of the proper symbols, N_{oo}, A_{ooo}, $\Pi_{o(o\alpha)}$, and $\iota_{\alpha(o\alpha)}$ are *constants*, and the remainder are *variables*.

(The inclusion of $\Pi_{o(o\alpha)}$ in this list of primitive symbols is meant in this sense, that, if α is any type symbol, $\Pi_{o(o\alpha)}$ is a primitive symbol, a proper symbol, and a constant; similarly in the case of $\iota_{\alpha(o\alpha)}$, a_α, etc.)

Any finite sequence of primitive symbols is a *formula*. Certain formulas are distinguished as being *well-formed* and as having a certain *type*, in accordance with the following rules: (1) a formula consisting of single proper symbol is well-formed and has the type indicated by the subscript; (2) if $\boldsymbol{x}_\beta$ is a variable with subscript β and $\boldsymbol{M}_\alpha$ is a well-formed formula of type α, then $(\lambda \boldsymbol{x}_\beta \boldsymbol{M}_\alpha)$ is a well-formed formula having the type $\alpha\beta$; (3) if $\boldsymbol{F}_{\alpha\beta}$ and $\boldsymbol{A}_\beta$ are well-formed formulas of types $\alpha\beta$ and β respectively, then $(\boldsymbol{F}_{\alpha\beta}\boldsymbol{A}_\beta$ is a well-formed formula having the type α. The well-formed formulas are the least class of formulas which these rules allow, and the type of a well-formed formula is that determined (uniquely) by these rules. An occurrence of a variable $\boldsymbol{x}_\beta$ in a well-formed formula is *bound* or *free* according as it is or is not an occurrence in a well-formed part of the formula having the form $(\lambda \boldsymbol{x}_\beta \boldsymbol{M}_\alpha)$. The bound variables of a well-formed formula are those which have bound occurrences in the formula, and the free variables are those which have free occurrences.

In making metamathematical (syntactical) statements, we shall use bold capital letters as variables for well-formed formulas, and bold small letters as variables for variables, employing subscripts to denote the type—as in the preceding paragraph. Moreover we shall adopt the customary, self-explanatory usage, according to which symbols belonging to the formal language serve in the syntax language (English) as names for themselves, and juxtaposition serves to denote juxtaposition.

In writing well-formed formulas we shall often employ various conventions of abbreviation. In particular, we may omit parentheses ( ) when possible without ambiguity, using the convention in restoring omitted parentheses that the formula must be well-formed and that otherwise association is to the left. Thus, for instance, $a_{\iota'}b_{\iota\iota}\ (c_{\iota\iota}d_\iota)$ is an abbreviation for $((a_{((\iota\iota)(\iota\iota))}b_{(\iota\iota)})(c_{\iota\iota}d_\iota))$, and $\lambda b_{\iota\iota}\lambda c_{\iota\iota}(a_{\iota'}b_{\iota\iota}(c_{\iota\iota}d_\iota))$ is an abbreviation for $(\lambda b_{\iota\iota}(\lambda c_{\iota\iota}((a_{((\iota\iota)(\iota\iota))}b_{(\iota\iota)})(c_{\iota\iota}d_\iota))))$.

[6]M. Schönfinkel, *Über die Bausteine der Mathematischen Logik*, ***Mathematische Annalen***, vol. 92 (1924), pp. 305–316.

As indicated in the examples just given, type-symbol subscripts may be abbreviated in the way described in §1. When the subscript is o it may be omitted altogether: thus a small italic letter without subscript is to be read as having the subscript o.

We introduce further the following conventions of abbreviation (reading the arrow as "stands for," or "is an abbreviation for"):

$$[\sim \boldsymbol{A}_o] \rightarrow N_{oo}\boldsymbol{A}_o.$$
$$[\boldsymbol{A}_o \vee \boldsymbol{B}_o] \rightarrow A_{ooo}\boldsymbol{A}_o\boldsymbol{B}_o.$$
$$[\boldsymbol{A}_o\boldsymbol{B}_o] \rightarrow [\sim[[\sim \boldsymbol{A}_o] \vee [\sim \boldsymbol{B}_o]]].$$
$$[\boldsymbol{A}_o \supset \boldsymbol{B}_o] \rightarrow [[\sim \boldsymbol{A}_o] \vee \boldsymbol{B}_o].$$
$$[\boldsymbol{A}_o \equiv \boldsymbol{B}_o] \rightarrow [[\boldsymbol{A}_o \supset \boldsymbol{B}_o][\boldsymbol{B}_o \supset \boldsymbol{A}_o]].$$
$$[(\boldsymbol{x}_\alpha)\boldsymbol{A}_o] \rightarrow \Pi_{o(o\alpha)}(\lambda \boldsymbol{x}_\alpha \boldsymbol{A}_o).$$
$$[(\exists \boldsymbol{x}_\alpha)\boldsymbol{A}_o] \rightarrow [\sim[(\boldsymbol{x}_\alpha)[\sim \boldsymbol{A}_o]]].$$
$$[(\imath \boldsymbol{x}_\alpha)\boldsymbol{A}_o] \rightarrow \imath_{\alpha(o\alpha)}(\lambda \boldsymbol{x}_\alpha \boldsymbol{A}_o).$$
$$Q_{o\alpha\alpha} \rightarrow \lambda x_\alpha \lambda y_\alpha[(f_{o\alpha})[f_{o\alpha}x_\alpha \supset f_{o\alpha}y_\alpha]].$$
$$[\boldsymbol{A}_\alpha = \boldsymbol{B}_\alpha] \rightarrow Q_{o\alpha\alpha}\boldsymbol{A}_\alpha\boldsymbol{B}_\alpha.$$
$$[\boldsymbol{A}_\alpha \neq \boldsymbol{B}_\alpha] \rightarrow [\sim[\boldsymbol{A}_\alpha = \boldsymbol{B}_\alpha]].$$
$$I_{\alpha\alpha} \rightarrow \lambda x_\alpha x_\alpha.$$
$$K_{\alpha\beta\alpha} \rightarrow \lambda x_\alpha \lambda y_\beta x_\alpha.$$
$$0_{\alpha'} \rightarrow \lambda f_{\alpha\alpha}\lambda x_\alpha x_\alpha,$$
$$1_{\alpha'} \rightarrow \lambda f_{\alpha\alpha}\lambda x_\alpha(f_{\alpha\alpha}x_\alpha),$$
$$2_{\alpha'} \rightarrow \lambda f_{\alpha\alpha}\lambda x_\alpha(f_{\alpha\alpha}(f_{\alpha\alpha}x_\alpha)),$$
$$3_{\alpha'} \rightarrow \lambda f_{\alpha\alpha}\lambda x_\alpha(f_{\alpha\alpha}(f_{\alpha\alpha}(f_{\alpha\alpha}x_\alpha))), \text{ etc.}$$
$$S_{\alpha'\alpha'} \rightarrow \lambda n_{\alpha'}\lambda f_{\alpha\alpha}\lambda x_\alpha(f_{\alpha\alpha}(n_{\alpha'}f_{\alpha\alpha}x_\alpha)).$$
$$N_{o\alpha'} \rightarrow \lambda n_{\alpha'}[(f_{o\alpha'})[f_{o\alpha'}0_{\alpha'} \supset [[(x_{\alpha'}[f_{o\alpha'}x_{\alpha'} \supset f_{o\alpha'}(S_{\alpha'\alpha'}x_{\alpha'})]] \supset f_{o\alpha'}n_{\alpha'}]]].$$
$$\omega_{\alpha''\alpha'\alpha'} \rightarrow \lambda y_{\alpha'}\lambda z_{\alpha'}\lambda f_{\alpha'\alpha'}\lambda g_{\alpha'}\lambda h_{\alpha\alpha}\lambda x_\alpha(y_{\alpha'}(f_{\alpha'\alpha'}g_{\alpha'}h_{\alpha\alpha})(z_{\alpha'}(g_{\alpha'}h_{\alpha\alpha})x_\alpha)).$$
$$\langle \boldsymbol{A}_{\alpha'}, \boldsymbol{B}_{\alpha'}\rangle \rightarrow \omega_{\alpha''\alpha'\alpha'}\boldsymbol{A}_{\alpha'}\boldsymbol{B}_{\alpha'}.$$
$$P_{\alpha'\alpha'''} \rightarrow \lambda n_{\alpha'''}(n_{\alpha'''}(\lambda p_{\alpha''}\langle S_{\alpha'\alpha'}(p_{\alpha''}(K_{\alpha'\alpha'\alpha'}I_{\alpha'})0_{\alpha'}), p_{\alpha''}(K_{\alpha'\alpha'\alpha'}I_{\alpha'})0_{\alpha'}\rangle)\langle 0_{\alpha'}, 0_{\alpha'}\rangle(K_{\alpha'\alpha'\alpha'}0_{\alpha'})I_{\alpha'}.$$
$$T_{\alpha''\alpha'} \rightarrow \lambda x_{\alpha'}[(\imath x_{\alpha''}[(N_{o\alpha''}x_{\alpha''})[x_{\alpha''}S_{\alpha'\alpha'}0_{\alpha'} = x_{\alpha'}]]].$$
$$P_{\alpha'\alpha'} \rightarrow \lambda x_{\alpha'}(P_{\alpha'\alpha'''}(T_{\alpha'''\alpha''}(T_{\alpha''\alpha'}x_{\alpha'}))).$$

As a further abbreviation, we omit square brackets [ ] introduced by the above abbreviations, when possible without ambiguity. When, in omitting square brackets, the initial bracket is replaced by a bold dot **.**, it is to be understood that the scope of the omitted pair of brackets is from the dot forward the maximum distance which is consistent with the whole expression's being well-formed or interpretable as an abbreviation of a well-formed formula. When omitted brackets are not thus replaced by a dot, the convention in restoring omitted brackets is association to the left, except as modified by the understanding that the abbreviated formulas are well-formed and by the following relation of precedence among the different kinds of brackets. The brackets in $[\sim \boldsymbol{A}_o]$ and $[\boldsymbol{A}_o\boldsymbol{B}_o]$ are of the lowest rank, those in $[(\boldsymbol{x}_\alpha)\boldsymbol{A}_o]$ and $[(\exists \boldsymbol{x}_\alpha)\boldsymbol{A}_o]$ and $[(\imath \boldsymbol{x}_o)\boldsymbol{A}_o]$ and $[\boldsymbol{A}_\alpha = \boldsymbol{B}_\alpha]$ and $[\boldsymbol{A}_\alpha \neq \boldsymbol{B}_\alpha]$ are of next higher rank, those in $[\boldsymbol{A}_o \vee \boldsymbol{B}_o]$ are of next higher rank, and those in $[\boldsymbol{A}_o \supset \boldsymbol{B}_o]$ and $[\boldsymbol{A}_o \equiv \boldsymbol{B}_o]$ are of highest rank; in restoring omitted brackets(not represented by a dot), those of lower rank are to be put in before those of higher rank, so that the smaller scope is allotted to those of lower rank. For example,

$$\sim p \supset q \supset . \, pq \vee rs \supset \sim . \, q \vee s \supset \sim p \sim r$$

is an abbreviation for

$$[[[\sim p] \supset q] \supset [[[pq] \vee [rs]] \supset [\sim[[q \vee s] \supset [[\sim p][\sim r]]]]]],$$

which is in turn an abbreviation for

$$((A_{ooo}(N_{oo}((A_{ooo}(N_{oo}(N_{oo}p_o)))q_o)))((A_{ooo}(N_{oo}((A_{ooo}(N_{oo}((A_{ooo}(N_{oo}p_o))(N_{oo}q_o))))\\ (N_{oo}((A_{ooo}(N_{oo}r_o))(N_{oo}s_o))))))\\ (N_{oo}((A_{ooo}(N_{oo}((A_{ooo}q_o)s_o)))(N_{oo}((A_{ooo}(N_{oo}(N_{oo}p_o)))(N_{oo}(N_{oo}r_o)))))))).$$

In the intended interpretation of the formal system λ will have the rôle of an abstraction operator, N_{oo} will denote negation, A_{ooo} will denote disjunction, $\Pi_{o(o\alpha)}$ will denote the universal quantifier (as a propositional function of propositional functions), $\iota_{\alpha(o\alpha)}$ will denote a selection operator (as a function of propositional functions), and juxtaposition, between parentheses, will denote application of a function to its argument. Such a logical construction of the natural numbers in each type α' is intended that $0_{\alpha'}$ will denote the natural number 0, $1_{\alpha'}$ will denote 1, $2_{\alpha'}$ will denote 2, etc. Then $S_{\alpha'\alpha'}$ will denote the successor function of natural numbers; or, more exactly, it will denote a function which has the entire type α' as the range of its arguments and which operates as a successor function in the case that the argument is a natural number. Moreover, $N_{o\alpha'}$ will denote the propositional function "to be a natural number (of type α')." If $N_{\alpha''}$ denotes a natural number of type α'', then $N_{\alpha''}S_{\alpha'\alpha'}0_{\alpha'}$ denotes the same (more exactly, the corresponding) natural number in the type α'. Hence if $N_{\alpha'}$ denotes a natural number of type α', the same natural number in the type α'' will be denoted by $T_{\alpha''\alpha'}N_{\alpha'}$. The formula $P_{\alpha'\alpha'''}$ is adapted from Kleene's formula P employed in the calculus of λ-conversion[7] and has the property that if $n_{\alpha'''}$ denotes a natural number of type α''' then $P_{\alpha'\alpha'''}N_{\alpha'''}$ denotes the predecessor of that natural number in type α'. The true predecessor function, which gives the predecessor in the same type, is denoted by $P_{\alpha'\alpha'}$; it follows from the independence of the axiom of infinity (§4) that this predecessor function cannot be defined without using descriptions (i.e., the selection operator $\iota_{\beta(o\beta)}$).

3. Rules of inference. The rules of inference (or rules of procedure) are the six following:

I. *To replace any part* $\boldsymbol{M}_\alpha$ *of a formula by the result of substituting* $\boldsymbol{y}_\beta$ *for* $\boldsymbol{x}_\beta$ *throughout* $\boldsymbol{M}_\alpha$, *provided that* $\boldsymbol{x}_\beta$ *is not a free variable of* $\boldsymbol{M}_\alpha$ *and* $\boldsymbol{y}_\beta$ *does not occur in* $\boldsymbol{M}_\alpha$. (I.e., to infer from a given formula the formula obtained by this replacement.)

II. *To replace any part* $((\lambda \boldsymbol{x}_\beta \boldsymbol{M}_\alpha)\boldsymbol{N}_\beta)$ *of a formula by the result of substituting* $\boldsymbol{N}_\beta$ *for* $\boldsymbol{x}_\beta$ *throughout* $\boldsymbol{M}_\alpha$, *provided that the bound variables of* $\boldsymbol{M}_\alpha$ *are distinct both from* $\boldsymbol{x}_\beta$ *and from the free variables of* $\boldsymbol{N}_\beta$.

III. *Where* $\boldsymbol{A}_\alpha$ *is the result of substituting* $\boldsymbol{N}_\beta$ *for* $\boldsymbol{x}_\beta$ *throughout* $\boldsymbol{M}_\alpha$, *to replace any part* $\boldsymbol{A}_\alpha$ *of a formula by* $((\lambda \boldsymbol{x}_\beta \boldsymbol{M}_\alpha)\boldsymbol{N}_\beta)$, *provided that the bound variables of* $\boldsymbol{M}_\alpha$ *are distinct both from* $\boldsymbol{x}_\beta$ *and from the free variables of* $\boldsymbol{N}_\beta$.

IV. *From* $\boldsymbol{F}_{o\alpha}\boldsymbol{x}_\alpha$ *to infer* $\boldsymbol{F}_{o\alpha}\boldsymbol{A}_\alpha$, *provided that* $\boldsymbol{x}_\alpha$ *is not a free variable of* $\boldsymbol{F}_{o\alpha}$.

---

[7] S. C. Kleene, *A theory of positive integers in formal logic*, ***American Journal of Mathematics***, vol. 57 (1935), pp. 153–173, 219–244.

V. *From* $\boldsymbol{A}_o \supset \boldsymbol{B}_o$ *and* $\boldsymbol{A}_o$, *to infer* $\boldsymbol{B}_o$.
VI. *From* $\boldsymbol{F}_{o\alpha}\boldsymbol{x}_\alpha$ *to infer* $\Pi_{o(o\alpha)}\boldsymbol{F}_{o\alpha}$, *provided that* $\boldsymbol{x}_\alpha$ *is not a free variable of* $\boldsymbol{F}_{o\alpha}$.

The word *part* of a formula is to be understood here as meaning *consecutive well-formed part* other than a variable immediately following the occurrence of λ. Moreover, as already explained, bold capital letters represent *well-formed* formulas and bold small letters represent variables, the subscript in each case showing the type. When (as in the rules I, II, III) we speak of replacing a part $\boldsymbol{M}_\alpha$ of a formula by something else, it is to be understood that, if there are several occurrences of $\boldsymbol{M}_\alpha$ as a part of the formula, any *one* of them may be so replaced. When we speak of the result of substituting $\boldsymbol{N}_\beta$ for $\boldsymbol{x}_\beta$ throughout $\boldsymbol{M}_\alpha$, the case is not excluded that $\boldsymbol{x}_\beta$ fails to occur in $\boldsymbol{M}_\alpha$, the result of the substitution in that case being $\boldsymbol{M}_\alpha$.

The rules I–III are called rules of *λ-conversion*, and any chain of applications of these rules is called a λ-conversion, or briefly, a conversion. Rule IV is the rule of *substitution*, Rule V is the rule of *modus ponens*, and Rule VI is the rule of *generalization*. In an application of Rule IV, we say that the variable $\boldsymbol{x}_\alpha$ is *substituted for*; and in an application of Rule VI, we say that the variable $\boldsymbol{x}_\alpha$ is *generalized upon*.

The two following rules of inference are derived rules, in the sense that the indicated inference can be accomplished in each case by a chain of applications of I–VI (the effect of IV′ can be obtained by means of λ-conversion and Rule IV, the effect of VI′ can be obtained by means of λ-conversion and Rule VI):

IV′. *From* $\boldsymbol{M}_o$ *to infer the result of substituting* $\boldsymbol{A}_\alpha$ *for the free occurrences of* $\boldsymbol{x}_\alpha$ *throughout* $\boldsymbol{M}_o$, *provided that the bound variables of* $\boldsymbol{M}_o$ *other than* $\boldsymbol{x}_\alpha$ *are distinct from the free variables of* $\boldsymbol{A}_\alpha$.
VI′. *From* $\boldsymbol{M}_o$ *to infer* $(\lambda \boldsymbol{x}_\alpha)\boldsymbol{M}_o$.

4. Formal axioms. The formal axioms are the formulas in the following infinite list:

1. $p \vee p \supset p$.
2. $p \supset p \vee q$.
3. $p \vee q \supset q \vee p$.
4. $p \supset q \supset . r \vee p \supset r \vee q$.

5^α. $\Pi_{o(o\alpha)} f_{o\alpha} \supset f_{o\alpha} x_\alpha$.
6^α. $(x_\alpha)[p \vee f_{o\alpha} x_\alpha] \supset p \vee \Pi_{o(o\alpha)} f_{o\alpha}$.
7. $(\exists x_\iota)(\exists y_\iota) . x_\iota \neq y_\iota$.
8. $N_{o\iota'} x_{\iota'} \supset . N_{o\iota'} y_{\iota'} \supset . S_{\iota'\iota'} x_{\iota'} = S_{\iota'\iota'} y_{\iota'} \supset x_{\iota'} = y_{\iota'}$.
9^α. $f_{o\alpha} x_\alpha \supset . (y_\alpha)[f_{o\alpha} y_\alpha \supset x_\alpha = y_\alpha] \supset f_{o\alpha}(\iota_{\alpha(o\alpha)} f_{o\alpha})$.
$10^{\alpha\beta}$. $(x_\beta)[f_{\alpha\beta} x_\beta = g_{\alpha\beta} x_\beta] \supset f_{\alpha\beta} = g_{\alpha\beta}$.
11^α. $f_{o\alpha} x_\alpha \supset f_{o\alpha}(\iota_{\alpha(o\alpha)} f_{o\alpha})$.

The *theorems* of the system are the formulas obtainable from the formal axioms by a succession of applications of the rules of inference. A *proof* of a theorem of the system is a finite sequence of formulas, the last of which is the theorem, and each of which is either a formal axiom or obtainable from previous formulas in the sequence by an application of a rule of inference.

We must, of course, distinguish between *formal theorems*, or theorems of the system, and *syntactical theorems*, or theorems about the system, this and related distinctions being a necessary part of the process of using a known language (English) to set up another (more exact) language. (We deliberately use the word "theorem" ambiguously, sometimes for a proposition and sometimes for a sentence or formula meaning the proposition in some language.)

Axioms 1–4 suffice for the propositional calculus and Axioms 1–6^α for the logical functional calculus.

In order to obtain elementary number theory it is necessary to add (to 1–6^α) Axioms 7, 8, and 9^α. Of these, 9^α are *axioms of descriptions*, and 7 and 8 taken together have the effect of an *axiom of infinity*. The independence of Axiom 7 may be established by considering an interpretation of the primitive symbols according to which there is exactly one individual, and that of Axiom 8 by considering an interpretation according to which there are a finite number, more than one, of individuals.

In order to obtain classical real number theory (analysis) it is necessary[8] to add also Axioms $10^{\alpha\beta}$ and 11^α. Of these, $10^{\alpha\beta}$ are *axioms of extensionality* for functions, and 11^α are *axioms of choice*.

Axioms $10^{\alpha\beta}$, although weaker in some directions than axioms of extensionality which are sometimes employed, are nevertheless adequate. For classes may be introduced in such a way that the class associated with the propositional function denoted by $\boldsymbol{F}_{o\alpha}$ is denoted by $\lambda x_\alpha(\imath y_{\iota'})\,.\,(F_{o\alpha}x_\alpha)[y_{\iota'} = 0_{\iota'}] \vee (\sim F_{o\alpha}x_\alpha)[y_{\iota'} = 1_{\iota'}]$. We remark, however, on the possibility of introducing the additional axiom of extensionality, $p \equiv q \supset p = q$, which has the effect of imposing so broad a criterion of identity between propositions that there are in consequence only two propositions, and which, in conjunction with $10^{\alpha\beta}$, makes possible the identification of classes with propositional functions.[8½]

Axioms 9^α obviously fail to be independent of 1–4 and 11^α. We have nevertheless included the axioms 9^α because of the desirability of considering the consequences of Axioms 1–9^α without $10^{\alpha\beta}$, 11^α.

If 1–9^α are the only formal axioms, each of the axioms 9^α is then independent, but if 10^α is added there is a sense in which those other than 9° and 9^ι, although independent, are superfluous. For, of the symbols $\imath_{\alpha(o\alpha)}$, we may introduce only $\imath_{o(oo)}$ and $\imath_{\iota(o\iota)}$ as primitive symbols and then introduce the remainder by definition (i.e., by conventions of abbreviation) in such a way that the formulas 9^α, read in accordance with these definitions (conventions of abbreviation), become theorems provable from the formal axioms 1–8, 9°, 9^ι, $10^{\alpha\beta}$. The required definitions are summarized in the following schema, which states the definition of $\imath_{\alpha\beta(o(\alpha\beta))}$ in terms of $\imath_{\alpha(o\alpha)}$:

$$\imath_{\alpha\beta(o(\alpha\beta))} \to \lambda h_{o(\alpha\beta)}\lambda x_\beta(\imath y_\alpha)(\exists f_{\alpha\beta})\,.\,h_{o(\alpha\beta)}f_{\alpha\beta}\,.\,y_\alpha = f_{\alpha\beta}x_\beta.$$

5. The deduction theorem. Derivation of the formal theorems of the propositional calculus from Axioms 1–4 by means of Rules IV′ and V is well known and need not be

[8]Devices of contextual definition, such as Russell's methods of introducing classes and descriptions (loc. cit.), are here avoided, and assertions concerning the necessity of axioms and the like are to be understood in the sense of this avoidance.

[8½](Added 1970) By taking o to be the type of propositions and excluding the axiom $p \equiv q \supset p = q$, Russell's theory of meaning (as it appears in ***Principia mathematica***) is here being followed. The writer's preference would now be to follow Frege's theory of meaning in order to attain full extensionality; this means that the axioms $p \equiv q \supset p = q$ would be added, and the type o would then preferably be thought of as the type of truth-values.

repeated here.[9] In what follows we shall employ theorems of the propositional calculus as needed, assuming the proof as known.

It is also clear that, by means of Rules I and IV′, alphabetical changes of the variables (free and bound) may be made in any formal axiom, provided that the types of the variables are not altered, that variables originally the same remain the same, and that variables originally different remain different. Formal theorems obtained in this way (including the formal axioms themselves) will be called *variants* of the axioms and will be employed as needed without explicit statement of the proof.

By a *proof* of a formula $\boldsymbol{B}_o$ *on the assumption of* the formulas $\boldsymbol{A}_o^1, \boldsymbol{A}_o^2, \cdots, \boldsymbol{A}_o^n$, we shall mean a finite sequence of formulas, the last of which is $\boldsymbol{B}_o$, and each of which is either one of the formulas $\boldsymbol{A}_o^1, \boldsymbol{A}_o^2, \cdots, \boldsymbol{A}_o^n$, or a variant of a formal axiom, or obtainable from preceding formulas in the sequence by an application of a rule of inference subject to the condition that no variable shall be substituted for or generalized upon which appears as a free variable in any of the formulas $\boldsymbol{A}_o^1, \boldsymbol{A}_o^2, \cdots, \boldsymbol{A}_o^n$. In order to express that there is a proof of $\boldsymbol{B}_o$ on the assumption of $\boldsymbol{A}_o^1, \boldsymbol{A}_o^2, \cdots, \boldsymbol{A}_o^n$, we shall employ the (syntactical) notation:

$$\boldsymbol{A}_o^1, \boldsymbol{A}_o^2, \cdots, \boldsymbol{A}_o^n \vdash \boldsymbol{B}_o.$$

In the use of this notation, it is not excluded that n should be 0 and the set of formulas $\boldsymbol{A}_o^i$ vacuous; i.e., the notation $\vdash \boldsymbol{B}_o$ will be used to mean that $\boldsymbol{B}_o$ is a (formal) theorem. (This use of the sign $\vdash$ must be distinguished from the entirely different use of the assertion sign by Russell and earlier by Frege.)

The following syntactical theorem is known as the *deduction theorem*:

VII. *If* $\boldsymbol{A}_o^1, \boldsymbol{A}_o^2, \cdots, \boldsymbol{A}_o^n \vdash \boldsymbol{B}_o$, *then* $\boldsymbol{A}_o^1, \boldsymbol{A}_o^2, \cdots, \boldsymbol{A}_o^{n-1} \vdash \boldsymbol{A}_o^n \supset \boldsymbol{B}_o$, $(n = 1, 2, 3, \cdots)$.

In order to prove this, we suppose that the finite sequence of formulas $\boldsymbol{B}_o^1, \boldsymbol{B}_o^2, \cdots, \boldsymbol{B}_o^m$ is a proof of $\boldsymbol{B}_o$ on the assumption of $\boldsymbol{A}_o^1, \boldsymbol{A}_o^2, \cdots, \boldsymbol{A}_o^n$, the formula $\boldsymbol{B}_o^m$ being the same as $\boldsymbol{B}_o$, and we show in succession, for each value of i from 1 to m, that

$$\boldsymbol{A}_o^1, \boldsymbol{A}_o^2, \cdots, \boldsymbol{A}_o^{n-1} \vdash \boldsymbol{A}_o^n \supset \boldsymbol{B}_o^i.$$

This is done by cases, according as $\boldsymbol{B}_o^i$ is $\boldsymbol{A}_o^n$, is one of $\boldsymbol{A}_o^1, \boldsymbol{A}_o^2, \cdots, \boldsymbol{A}_o^{n-1}$, is a variant of an axiom, or is obtained from a preceding formula or pair of formulas by one of the rules I–VI. If $\boldsymbol{B}_o^i$ is $\boldsymbol{A}_o^n$, we may obtain $\boldsymbol{A}_o^n \supset \boldsymbol{B}_o^i$ from $p \supset p$ by IV′. If $\boldsymbol{B}_o^i$ is one of $\boldsymbol{A}_o^1, \boldsymbol{A}_o^2, \cdots, \boldsymbol{A}_o^{n-1}$ or is a variant of an axiom, we may obtain $\boldsymbol{A}_o^n \supset \boldsymbol{B}_o^i$ from $p \supset . q \supset p$ by a succession of applications of IV′ and V. If $\boldsymbol{B}_o^i$ is obtained from $\boldsymbol{B}_o^a$ $(a < i)$ by one of the rules I, II, III, we may obtain $\boldsymbol{A}_o^n \supset \boldsymbol{B}_o^i$ from $\boldsymbol{A}_o^n \supset \boldsymbol{B}_o^n$ by the same rule. If $\boldsymbol{B}_o^i$ is obtained from $\boldsymbol{B}_o^a$ $(a < i)$ by Rule IV, we may obtain $\boldsymbol{A}_o^n \supset \boldsymbol{B}_o^i$ from $\boldsymbol{A}_o^n \supset \boldsymbol{B}_o^a$ by IV′. If $\boldsymbol{B}_o^i$ is obtained from $\boldsymbol{B}_o^a$ and $\boldsymbol{B}_o^b$ $(a < i, b < i)$ by Rule V, we may obtain $\boldsymbol{A}_o^n \supset \boldsymbol{B}_o^i$ from $\boldsymbol{A}_o^n \supset \boldsymbol{B}_o^a$ and $\boldsymbol{A}_o^n \supset \boldsymbol{B}_o^b$, and $p \supset [q \supset r] \supset . p \supset q \supset . p \supset r$ by a succession of applications of IV′ and V. If $\boldsymbol{B}_o^i$ is obtained from $\boldsymbol{B}_o^a$ $(a < i)$ by Rule VI, we may obtain $\boldsymbol{A}_o^n \supset \boldsymbol{B}_o^i$ from $\boldsymbol{A}_o^n \supset \boldsymbol{B}_n^a$ and 6^α by a succession of applications of IV′, V, and VI′. Proof of the following theorems,[10] which are consequences of the

[9] Cf. Hilbert and Ackermann, loc. cit.; P. Bernays, *Axiomatische Untersuchungen des Aussagen-Kalkuls der "Principia Mathematica"*, ***Mathematische Zeitschrift***, vol. 25 (1926), pp. 305–320.

[10] The same device of typical ambiguity which was employed in stating the rules of inference and formal axioms now serves us, not only to condense the statement of an infinite number of theorems (differing only in the type subscripts of the proper symbols which appear) into a single schema of theorems, but also to condense the proof of the infinite number of theorems into a single schema of proof. Of course, in the explicit formal development of the system, a stage would never be reached at which all of the theorems 12°, 12^ι,

formal axioms 1–6^α, is left to the reader (it will be found convenient in most cases to abbreviate the proof by employing the deduction theorem in the role of a derived rule):

12^α. $(x_\alpha)f_{o\alpha}x_\alpha \supset f_{o\alpha}y_\alpha$.
13^α. $f_{o\alpha}y_\alpha \supset (\exists x_\alpha)f_{o\alpha}x_\alpha$.
14^α. $(x_\alpha)[p \supset f_{o\alpha}x_\alpha] \supset . p \supset (x_\alpha)f_{o\alpha}x_\alpha$.
15^α. $(x_\alpha)[f_{o\alpha}x_\alpha \supset p] \supset .(\exists x_\alpha)f_{o\alpha}x_\alpha \supset p$.
16^α. $x_\alpha = x_\alpha$.
17^α. $x_\alpha = y_\alpha \supset . f_{o\alpha}x_\alpha \supset f_{o\alpha}y_\alpha$.
$18^{\beta\alpha}$. $x_\alpha = y_\alpha \supset f_{\beta\alpha}x_\alpha = f_{\beta\alpha}y_\alpha$.
19^α. $x_\alpha = y_\alpha \supset y_\alpha = x_\alpha$.
20^α. $x_\alpha = y_\alpha \supset . y_\alpha = z_\alpha \supset x_\alpha = z_\alpha$.

The following theorems are consequences of the formal axioms 1–4 and $10^{\alpha\beta}$ (no use will be made of them below because we shall be concerned entirely with consequences of 1–9^α):

$21^{\alpha\beta}$. $f_{\alpha\beta} = \lambda x_\beta(f_{\alpha\beta}x_\beta)$.

6. Peano's postulates for arithmetic. Three of the five Peano postulates for arithmetic[11] are represented by the following formal theorems:

22^α. $N_{o\alpha'}0_{\alpha'}$.
23^α. $N_{o\alpha'}x_{\alpha'} \supset N_{o\alpha'}(S_{\alpha'\alpha'}x_{\alpha'})$.
24^α. $f_{o\alpha'}0_{\alpha'} \supset .(x_{\alpha'})[N_{o\alpha'}x_{\alpha'} \supset . f_{o\alpha'}x_{\alpha'} \supset f_{o\alpha'}(S_{\alpha'\alpha'}x_{\alpha'})] \supset .$
$N_{o\alpha'}x_{\alpha'} \supset f_{o\alpha'}x_{\alpha'}$.

These theorems are consequences of 1–6^α; proofs are left to the reader.

From 24^α and the deduction theorem we obtain the following syntactical theorem which we shall call the *induction theorem*:[11½]

VIII. *If* $x_{\alpha'}$ *is not a free variable of* $\mathbf{A}_o^1, \mathbf{A}_o^2, \cdots, \mathbf{A}_o^n$, $\mathbf{F}_{o\alpha'}$, *if* $\mathbf{A}_o^1, \mathbf{A}_o^2, \cdots, \mathbf{A}_o^n \vdash \mathbf{F}_{o\alpha'}0_{\alpha'}$, *and if* $\mathbf{A}_o^1, \mathbf{A}_o^2, \cdots, \mathbf{A}_o^n, N_{o\alpha'}\mathbf{x}_{\alpha'}, \mathbf{F}_{o\alpha'}\mathbf{x}_{\alpha'} \vdash \mathbf{F}_{o\alpha'}(S_{\alpha'\alpha'}\mathbf{x}_{\alpha'})$, *then* $\mathbf{A}_o^1, \mathbf{A}_o^2, \cdots, \mathbf{A}_o^n \vdash N_{o\alpha'}x_{\alpha'} \supset \mathbf{F}_{o\alpha'}\mathbf{x}_{\alpha'}$.

A proof which is or can be abbreviated by employing the induction theorem in the role of a derived rule will be called a proof by (*mathematical*, or *complete*) *induction* on the variable $\mathbf{x}_{\alpha'}$.

Another of the Peano postulates is represented by the following formal theorems:

25^α. $N_{o\alpha'}x_{\alpha'} \supset S_{\alpha'\alpha'}x_{\alpha'} \neq 0_{\alpha'}$.

$12^{\prime\prime}, \cdots$, (for example) had been proved, but by the device of a schema of proof with typical ambiguity we obtain metamathematical assurance that any required *one* of the theorems in the infinite list can be proved. Cf. the Prefatory Statement to the second volume of ***Principia Mathematica***.

[11] G. Peano, *Sul concetto di numero*, ***Revista di Matematica***, vol. 1 (1891), pp. 87–102, 256–267.

[11½] (Added 1970) Both the deduction theorem and the induction theorem, though they have a more complex form than, e.g., IV′ and VI′ above, nevertheless qualify as derived rules in the following sense. There is an effective method by which, whenever the deduction theorem or the induction theorem or both have been used in order to abbreviate the proof of a formal theorem, it is possible to find and actually write out the full proof in the object language in the literal sense of §4.

These theorems are consequences of 1–6^α and 7, as we shall show (for certain types α they are consequences of 1–6^α only).

The remaining Peano postulate would correspond to the following:

$$26^\alpha. \quad N_{o\alpha'}x_{\alpha'} \supset . N_{o\alpha'}y_{\alpha'} \supset . S_{\alpha'\alpha'}x_{\alpha'} = S_{\alpha'\alpha'}y_{\alpha'} \supset x_{\alpha'} = y_{\alpha'}.$$

These formulas are demonstrably not theorems (consistency assumed) in the case of type symbols α consisting entirely of o's with no ι's. We shall show that the formulas $26^\iota, 26^{\iota'}, 26^{\iota''}, \cdots$ are theorems—in fact they are consequences of 1–6^α and 8, the formula 26^ι being the same as 8.

A proof of the theorem

$$27^o. \quad (\exists x_o)(\exists y_o) . x_o \neq y_o,$$

may be made as follows. In 17° substitute $p \vee \sim p$ for x_o, and $\sim . p \vee \sim p$ for y_o, and $\lambda r \sim . p \vee \sim p \supset \sim r$ for f_{oo}, by successive applications of IV$'$, and then apply Rule II twice, so obtaining

$$[p \vee \sim p] = [\sim . p \vee \sim p] \supset .[\sim . p \vee \sim p \supset \sim . p \vee \sim p] \supset \sim . p \vee \sim p \supset \sim\sim . p \vee \sim p.$$

Hence using the theorems of the propositional calculus,

$$\sim . p \vee \sim p \supset \sim . p \vee \sim p,$$

$$p \supset [r \supset s] \supset . r \supset . q \supset s,$$

and the rules IV$'$ and V, obtain

$$[p \vee \sim p] = [\sim . p \vee \sim p] \supset \sim . p \vee \sim p \supset \sim\sim . p \vee \sim p.$$

Hence, using the theorems of the propositional calculus,

$$p \vee \sim p \supset \sim\sim . p \vee \sim p,$$

$$q \supset . r \supset \sim q \supset \sim r,$$

and IV$'$ and V (method of *reductio ad absurdum*), obtain

$$[p \vee \sim p] \neq [\sim . p \vee \sim p].$$

Hence by two successive uses of 13°, with I, II, III, IV$'$, V, obtain 27°.

In regard to proof of the theorems,

$$27^\alpha. \quad (\exists x_\alpha)(\exists y_\alpha) . x_\alpha \neq y_\alpha,$$

since we have a proof of 27°, and 27^ι is Axiom 7, it is sufficient to show how to obtain a proof of $27^{\alpha\beta}$ if a proof of 27^α is given.

By conversion $z_\alpha \neq t_\alpha \vdash K_{\alpha\beta\alpha}z_\alpha x_\beta \neq K_{\alpha\beta\alpha}t_\alpha x_\beta$.

Hence by 17^α (using II, IV$'$, V), $z_\alpha \neq t_\alpha$,

$$K_{\alpha\beta\alpha}z_\alpha = K_{\alpha\beta\alpha}t_\alpha \vdash K_{\alpha\beta\alpha}t_\alpha x_\beta \neq K_{\alpha\beta\alpha}t_\alpha x_\beta.$$

Hence by the deduction theorem,

$$z_\alpha \neq t_\alpha \vdash K_{\alpha\beta\alpha}z_\alpha = K_{\alpha\beta\alpha}t_\alpha \supset K_{\alpha\beta\alpha}t_\alpha x_\beta \neq K_{\alpha\beta\alpha}t_\alpha x_\beta.$$

By 16^α (using IV$'$), $K_{\alpha\beta\alpha}t_\alpha x_\beta = K_{\alpha\beta\alpha}t_\alpha x_\beta$.

Hence by *reductio ad absurdum*, as above, $z_\alpha \neq t_\alpha \vdash K_{\alpha\beta\alpha}z_\alpha \neq K_{\alpha\beta\alpha}t_\alpha$.

Hence by two successive uses of $13^{\alpha\beta}$ (with I, II, III, IV′, V),

$$z_\alpha \neq t_\alpha \vdash (\exists x_{\alpha\beta})(\exists y_{\alpha\beta}) \centerdot x_{\alpha\beta} \neq y_{\alpha\beta}.$$

Hence by the deduction theorem, $\vdash z_\alpha \neq t_\alpha \supset (\exists x_{\alpha\beta})(\exists y_{\alpha\beta}) \centerdot x_{\alpha\beta} \neq y_{\alpha\beta}$.
Hence using VI′, $\vdash (t_\alpha) \centerdot z_\alpha \neq t_\alpha \supset (\exists x_{\alpha\beta})(\exists y_{\alpha\beta}) \centerdot x_{\alpha\beta} \neq y_{\alpha\beta}$.
Hence by 15^α (using I, II, III, IV′, V),

$$\vdash (\exists t_\alpha)[z_\alpha \neq t_\alpha] \supset (\exists x_{\alpha\beta})(\exists y_{\alpha\beta}) \centerdot x_{\alpha\beta} \neq y_{\alpha\beta}.$$

Hence by using VI′, $\vdash (z_\alpha) \centerdot (\exists t_\alpha)[z_\alpha \neq t_\alpha] \supset (\exists x_{\alpha\beta})\, (\exists y_{\alpha\beta}) \centerdot x_{\alpha\beta} \neq y_{\alpha\beta}$.
Hence by 15^α (using I, II, III, IV′, V),

$$\vdash (\exists z_\alpha)(\exists t_\alpha)[z_\alpha \neq t_\alpha] \supset (\exists x_{\alpha\beta})(\exists y_{\alpha\beta}) \centerdot x_{\alpha\beta} \neq y_{\alpha\beta}.$$

Hence if $\vdash 27^\alpha$ then, using I and V, $\vdash 27^{\alpha\beta}$.
Thus for every type α we have a proof of 27^α. Using this, we proceed to the proof of

28^α. $\quad S_{\alpha'\alpha'}x_{\alpha'} \neq 0_{\alpha'}$.

By conversion, $z_\alpha \neq t_\alpha \vdash S_{\alpha'\alpha'}x_{\alpha'}(K_{\alpha\alpha\alpha}z_\alpha)t_\alpha \neq 0_{\alpha'}(K_{\alpha\alpha\alpha}z_\alpha)t_\alpha$. Hence by the method illustrated in the preceding proof, using in order 17^α, the deduction theorem, 16^α, and *reductio ad absurdum*, $z_\alpha \neq t_\alpha \vdash S_{\alpha'\alpha'}x_{\alpha'} \neq 0_{\alpha'}$. Eliminating the assumption $z_\alpha \neq t_\alpha$ by the method of the preceding proof, using in order the deduction theorem, VI′, 15^α, VI′, 15^α, 27^α, we have $\vdash 28^\alpha$.

Having 28^α, we prove 25^α by using $p \supset \centerdot q \supset p$.
We need also the theorems:

29^α. $\quad N_{o\alpha''}n_{\alpha''} \supset N_{o\alpha'}(n_{\alpha''}S_{\alpha'\alpha'}0_{\alpha'})$.

The (schema of) proof of these theorems is a simple example of proof by induction.
From 22^α by conversion, $\vdash N_{o\alpha'}(0_{\alpha''}S_{\alpha'\alpha'}0_{\alpha'})$.
By 23^α, $N_{o\alpha'}(n_{\alpha''}S_{\alpha'\alpha'}0_{\alpha'}) \vdash N_{o\alpha'}(S_{\alpha'\alpha'}(n_{\alpha''}S_{\alpha'\alpha'}0_{\alpha'}))$.
Hence by conversion, $N_{o\alpha'}(n_{\alpha''}S_{\alpha'\alpha'}0_{\alpha'}) \vdash N_{o\alpha'}(S_{\alpha''\alpha''}n_{\alpha''}S_{\alpha'\alpha'}0_{\alpha'})$.
Hence by the induction theorem, taking $F_{o\alpha''}$ to be $\lambda x_{\alpha''}(N_{o\alpha'}(x_{\alpha''}S_{\alpha'\alpha'}0_{\alpha'}))$ and $x_{\alpha''}$ to be $n_{\alpha''}$, and employing conversion as required, we have $\vdash 29^\alpha$.
Returning now to 26^α, we consider in connection with it:

30^α. $\quad N_{o\alpha''}m_{\alpha''} \supset \centerdot N_{o\alpha''}n_{\alpha''} \supset \centerdot m_{\alpha''}S_{\alpha'\alpha'}0_{\alpha'} = n_{\alpha''}S_{\alpha'\alpha'}0_{\alpha'} \supset m_{\alpha''} = n_{\alpha''}$.

As in the case of 26^α, not all the formulas 30^α are theorems. We shall show that 26^α and 30^α are theorems if α is one of the types $\iota, \iota', \iota'', \cdots$. Since 26^ι is Axiom 8, we may do this by showing that (1) if $\vdash 26^\alpha$ then $\vdash 30^\alpha$, and (2) if $\vdash 26^\alpha$ and $\vdash 30^\alpha$ then $\vdash 26^{\alpha'}$.[12]

By $18^{\alpha'\alpha''}$, $S_{\alpha''\alpha''}x_{\alpha''} = S_{\alpha''\alpha''}y_{\alpha''} \vdash S_{\alpha''\alpha''}x_{\alpha''}S_{\alpha'\alpha'}0_{\alpha'} = S_{\alpha''\alpha''}y_{\alpha''}S_{\alpha'\alpha'}0_{\alpha'}$.
Hence by conversion, we have

$$S_{\alpha''\alpha''}x_{\alpha''} = S_{\alpha''\alpha''}y_{\alpha''} \vdash S_{\alpha'\alpha'}(x_{\alpha''}S_{\alpha'\alpha'}0_{\alpha'}) = S_{\alpha'\alpha'}(y_{\alpha''}S_{\alpha'\alpha'}0_{\alpha'}).$$

[12]The question suggests itself whether 30^ι could be used in place of Axiom 8 as the second part of the axiom of infinity. The writer has a proof (depending on the properties of $P_{\iota'\iota'''}$) that 30^ι and $30^{\iota'}$ are together sufficient, in the presence of 1–6^α, to replace Axiom 8. A proof has also been carried out by A. M. Turing that, in the presence of 1–7 and 9^α, 30^ι is sufficient alone to replace Axiom 8. Whether 8 is independent of 1–7 and 30^ι remains an open problem (familiar methods of eliminating descriptions do not apply here).

Hence if $\vdash 26^\alpha$, we have by 29^α,

$$N_{o\alpha''}x_{\alpha''},\ N_{o\alpha''}y_{\alpha''},\ S_{\alpha''\alpha''}x_{\alpha''} = S_{\alpha''\alpha''}y_{\alpha''} \vdash x_{\alpha''}S_{\alpha'\alpha'}0_{\alpha'} = y_{\alpha''}S_{\alpha'\alpha'}0_{\alpha'}.$$

Hence if $\vdash 26^\alpha$ and $\vdash 30^\alpha$, we have

$$N_{o\alpha''}x_{\alpha''},\ N_{o\alpha''}y_{\alpha''},\ S_{\alpha''\alpha''}x_{\alpha''} = S_{\alpha''\alpha''}y_{\alpha''} \vdash x_{\alpha''} = y_{\alpha''}.$$

Hence by three applications of the deduction theorem, if $\vdash 26^\alpha$ and $\vdash 30^\alpha$, then $\vdash 26^{\alpha'}$. This is (2) above.

Now by conversion, $0_{\alpha''}S_{\alpha'\alpha'}0_{\alpha'} = 0_{\alpha''}S_{\alpha'\alpha'}0_{\alpha'} \vdash 0_{\alpha'} = 0_{\alpha'}$. Hence by the deduction theorem, $\vdash 0_{\alpha''}S_{\alpha'\alpha'}0_{\alpha'} = 0_{\alpha''}S_{\alpha'\alpha'}0_{\alpha'} \supset 0_{\alpha'} = 0_{\alpha'}$.

By conversion, $0_{\alpha''}S_{\alpha'\alpha'}0_{\alpha'} = S_{\alpha''\alpha''}n_{\alpha''}S_{\alpha'\alpha'}0_{\alpha'} \vdash 0_{\alpha'} = S_{\alpha'\alpha'}(n_{\alpha''}S_{\alpha'\alpha'}0_{\alpha'})$. Hence by $19^{\alpha'}$, $0_{\alpha''}S_{\alpha'\alpha'}0_{\alpha'} = S_{\alpha''\alpha''}n_{\alpha''}S_{\alpha'\alpha'}0_{\alpha'} \vdash S_{\alpha'\alpha'}(n_{\alpha''}S_{\alpha'\alpha'}0_{\alpha'}) = 0_{\alpha'}$. By 28^α, $\vdash S_{\alpha'\alpha'}(n_{\alpha''}S_{\alpha'\alpha'}0_{\alpha'}) \neq 0_{\alpha'}$. Hence using $p \sim p \supset q$, we have

$$0_{\alpha''}S_{\alpha'\alpha'}0_{\alpha'} = S_{\alpha''\alpha''}n_{\alpha''}S_{\alpha'\alpha'}0_{\alpha'} \vdash 0_{\alpha''} = S_{\alpha''\alpha''}n_{\alpha''}.$$

Hence by the deduction theorem,

$$\vdash 0_{\alpha''}S_{\alpha'\alpha'}0_{\alpha'} = S_{\alpha''\alpha''}n_{\alpha''}S_{\alpha'\alpha'}0_{\alpha'} \supset 0_{\alpha''} = S_{\alpha''\alpha''}n_{\alpha}.$$

Hence by the induction theorem, followed by VI′,

$$\vdash (n_{\alpha''}) \centerdot N_{o\alpha''}n_{\alpha''} \supset \centerdot 0_{\alpha''}S_{\alpha'\alpha'}0_{\alpha'} = n_{\alpha''}S_{\alpha'\alpha'}0_{\alpha'} \supset 0_{\alpha''} = n_{\alpha''}.$$

By conversion, $S_{\alpha''\alpha''}m_{\alpha''}S_{\alpha'\alpha'}0_{\alpha'} = 0_{\alpha''}S_{\alpha'\alpha'}0_{\alpha'} \vdash S_{\alpha'\alpha'}(m_{\alpha''}S_{\alpha'\alpha'}0_{\alpha'}) = 0_{\alpha'}$. By 28^α, $\vdash S_{\alpha'\alpha'}(m_{\alpha''}S_{\alpha'\alpha'}0_{\alpha'}) \neq 0_{\alpha'}$. Hence, using $p \sim p \supset q$, we have

$$S_{\alpha''\alpha''}m_{\alpha''}S_{\alpha'\alpha'}0_{\alpha'} = 0_{\alpha''}S_{\alpha'\alpha'}0_{\alpha'} \vdash S_{\alpha''\alpha''}m_{\alpha''} = 0_{\alpha''}.$$

Hence by the deduction theorem,

$$\vdash S_{\alpha''\alpha''}m_{\alpha''}S_{\alpha'\alpha'}0_{\alpha'} = 0_{\alpha''}S_{\alpha'\alpha'}0_{\alpha'} \supset S_{\alpha''\alpha''}m_{\alpha''} = 0_{\alpha''}.$$

By conversion,

$$S_{\alpha''\alpha''}m_{\alpha''}S_{\alpha'\alpha'}0_{\alpha'} = S_{\alpha''\alpha''}n_{\alpha''}S_{\alpha'\alpha'}0_{\alpha'} \vdash S_{\alpha'\alpha'}(m_{\alpha''}S_{\alpha'\alpha'}0_{\alpha'}) = S_{\alpha'\alpha'}(n_{\alpha''}S_{\alpha'\alpha'}0_{\alpha'}).$$

Hence if $\vdash 26^\alpha$, we have by 29^α, $N_{o\alpha''}m_{\alpha''}$, $N_{o\alpha''}n_{\alpha''}$,

$$S_{\alpha''\alpha''}m_{\alpha''}S_{\alpha'\alpha'}0_{\alpha'} = S_{\alpha''\alpha''}n_{\alpha''}S_{\alpha'\alpha'}0_{\alpha'} \vdash m_{\alpha''}S_{\alpha'\alpha'}0_{\alpha'} = n_{\alpha''}S_{\alpha'\alpha'}0_{\alpha'}.$$

Hence if $\vdash 26^\alpha$, we have (using $12^{\alpha''}$) $N_{o\alpha''}m_{\alpha''}$, $(n_{\alpha''}) \centerdot N_{o\alpha''}n_{\alpha''} \supset \centerdot m_{\alpha''}S_{\alpha'\alpha'}0_{\alpha'} = n_{\alpha''}S_{\alpha'\alpha'}0_{\alpha'} \supset m_{\alpha''} = n_{\alpha''}$, $N_{o\alpha''}n_{\alpha''}$, $S_{\alpha''\alpha''}m_{\alpha''}S_{\alpha'\alpha'}0_{\alpha'} = S_{\alpha''\alpha''}n_{\alpha''}S_{\alpha'\alpha'}0_{\alpha'} \vdash m_{\alpha''} = n_{\alpha''}$. Hence, using $18^{\alpha''\alpha''}$, to obtain $S_{\alpha''\alpha''}m_{\alpha''} = S_{\alpha''\alpha''}n_{\alpha''}$ and then applying the deduction theorem, we have (if $\vdash 26^\alpha$), $N_{o\alpha''}m_{\alpha''}$, $(n_{\alpha''}) \centerdot N_{o\alpha''}n_{\alpha''} \supset \centerdot m_{\alpha''}S_{\alpha'\alpha'}0_{\alpha'} = n_{\alpha''}S_{\alpha'\alpha'}0_{\alpha'} \supset m_{\alpha''} = n_{\alpha''}$, $N_{o\alpha''}n_{\alpha''} \vdash S_{\alpha''\alpha''}m_{\alpha''}S_{\alpha'\alpha'}0_{\alpha'} = S_{\alpha''\alpha''}n_{\alpha''}S_{\alpha'\alpha'}0_{\alpha'} \supset S_{\alpha''\alpha''}m_{\alpha''} = S_{\alpha''\alpha''}n_{\alpha''}$.

Hence by the induction theorem, followed by VI′, we have (if $\vdash 26^\alpha$) $N_{o\alpha''}m_{\alpha''}$, $(n_{\alpha''}) \centerdot N_{o\alpha''}n_{\alpha''} \supset \centerdot m_{\alpha''}S_{\alpha'\alpha'}0_{\alpha'} = n_{\alpha''}S_{\alpha'\alpha'}0_{\alpha'} \supset m_{\alpha''} = n_{\alpha''} \vdash (n_{\alpha''}) \centerdot N_{o\alpha''}n_{\alpha''} \supset \centerdot S_{\alpha''\alpha''}m_{\alpha''}S_{\alpha'\alpha'}0_{\alpha'} = n_{\alpha''}S_{\alpha'\alpha'}0_{\alpha'} \supset S_{\alpha''\alpha''}m_{\alpha''} = n_{\alpha''}$.

Hence again, applying the induction theorem to preceding results, we have (if $\vdash 26^\alpha$), $\vdash N_{o\alpha''}m_{\alpha''} \supset \centerdot (n_{\alpha''}) \centerdot N_{o\alpha''}n_{\alpha''} \supset \centerdot m_{\alpha''}S_{\alpha'\alpha'}0_{\alpha'} = n_{\alpha''}S_{\alpha'\alpha'}0_{\alpha'} \supset m_{\alpha''} = n_{\alpha''}$.

Hence using V and $12^{\alpha''}$, we have (if $\vdash 26^\alpha$), $N_{o\alpha''}m_{\alpha''}$, $N_{o\alpha''}n_{\alpha''} \vdash m_{\alpha''}S_{\alpha'\alpha'}0_{\alpha'} = n_{\alpha''}S_{\alpha'\alpha'}0_{\alpha'} \supset m_{\alpha''} = n_{\alpha''}$.

Hence by two applications of the deduction theorem, if $\vdash 26^\alpha$ then $\vdash 30^\alpha$. This is (1) above.

7. Properties of $T_{\alpha''\alpha'}$. We proceed now to proofs of the following theorems:

31^α. $N_{o\alpha'}x_{\alpha'} \supset N_{o\alpha''}(T_{\alpha''\alpha'}x_{\alpha'})$.
32^α. $N_{o\alpha'}x_{\alpha'} \supset T_{\alpha''\alpha'}x_{\alpha'}S_{\alpha'\alpha'}0_{\alpha'} = x_{\alpha'}$.

The proofs require $9^{\alpha''}$ and are possible only for types α for which there is a proof of 30^α.

We begin by proving as a lemma:

33^α. $N_{o\alpha'}x_{\alpha'} \supset (\exists x_{\alpha''}) \centerdot N_{o\alpha''}x_{\alpha''} \centerdot x_{\alpha''}S_{\alpha'\alpha'}0_{\alpha'} = x_{\alpha'}$.

Proof of this requires only the axioms 1–6^α and is possible for an arbitrary type α.

By 16^α, using IV′ and conversion, we have $\vdash 0_{\alpha''}S_{\alpha'\alpha'}0_{\alpha'} = 0_{\alpha'}$. Hence using $22^{\alpha'}$ and $p \supset \centerdot q \supset pq$ and $13^{\alpha''}$, we have $\vdash (\exists x_{\alpha''}) \centerdot N_{o\alpha''}x_{\alpha''} \centerdot x_{\alpha''}S_{\alpha'\alpha'}\, 0_{\alpha'} = 0_{\alpha'}$.

By $18^{\alpha'\alpha'}$, $x_{\alpha''}S_{\alpha'\alpha'}0_{\alpha'} = x_{\alpha'} \vdash S_{\alpha'\alpha'}(x_{\alpha''}S_{\alpha'\alpha'}0_{\alpha'}) = S_{\alpha'\alpha'}x_{\alpha'}$. Hence by conversion, $x_{\alpha''}S_{\alpha'\alpha'}0_{\alpha'} = x_{\alpha'} \vdash S_{\alpha''\alpha''}x_{\alpha''}S_{\alpha'\alpha'}0_{\alpha'} = S_{\alpha'\alpha'}x_{\alpha'}$. Also, by $23^{\alpha'}$, $N_{o\alpha''}x_{\alpha''} \vdash N_{o\alpha''}(S_{\alpha''\alpha''}x_{\alpha''})$. Hence using $pq \supset p$ and $pq \supset q$ and $p \supset \centerdot q \supset pq$, we have $N_{o\alpha''}x_{\alpha''} \centerdot x_{\alpha''}S_{\alpha'\alpha'}0_{\alpha'} = x_{\alpha'} \vdash N_{o\alpha''}(S_{\alpha''\alpha''}x_{\alpha''}) \centerdot S_{\alpha''\alpha''}x_{\alpha''}\, S_{\alpha'\alpha'}0_{\alpha'} = S_{\alpha'\alpha'}x_{\alpha'}$. Hence employing in order $13^{\alpha''}$, the deduction theorem, and VI′, we have

$$\vdash (x_{\alpha''}) \centerdot [N_{o\alpha''}x_{\alpha''} \centerdot x_{\alpha''}S_{\alpha'\alpha'}0_{\alpha'} = x_{\alpha'}] \supset (\exists x_{\alpha''}) \centerdot N_{o\alpha''}x_{\alpha''} \centerdot x_{\alpha''}S_{\alpha'\alpha'}0_{\alpha'} = S_{\alpha'\alpha'}x_{\alpha'}.$$

Hence by $15^{\alpha''}$,

$$\vdash (\exists x_{\alpha''})[N_{o\alpha''}x_{\alpha''} \centerdot x_{\alpha''}S_{\alpha'\alpha'}0_{\alpha'} = x_{\alpha'}] \supset (\exists x_{\alpha''}) \centerdot N_{o\alpha''}x_{\alpha''} \centerdot x_{\alpha''}S_{\alpha'\alpha'}0_{\alpha'} = S_{\alpha'\alpha'}x_{\alpha'}.$$

Hence by the induction theorem, $\vdash 33^\alpha$.

Now proceeding with the proof of 31^α and 32^α (for types α for which $\vdash 30^\alpha$), we may—with the aid of 30^α—show that

$$N_{o\alpha''}x_{\alpha''} \centerdot x_{\alpha''}S_{\alpha'\alpha'}0_{\alpha'} = x_{\alpha'}, \quad N_{o\alpha''}y_{\alpha''} \centerdot y_{\alpha''}S_{\alpha'\alpha'}0_{\alpha'} = x_{\alpha'} \vdash x_{\alpha''} = y_{\alpha''}.$$

Hence by the deduction theorem and VI′,

$$N_{o\alpha''}x_{\alpha''} \centerdot x_{\alpha''}S_{\alpha'\alpha'}0_{\alpha'} = x_{\alpha'} \vdash (y_{\alpha''}) \centerdot [N_{o\alpha''}y_{\alpha''} \centerdot y_{\alpha''}S_{\alpha'\alpha'}0_{\alpha'} = x_{\alpha'}] \supset x_{\alpha''} = y_{\alpha''}.$$

Hence by $9^{\alpha''}$ (refer to the definition of $T_{\alpha''\alpha'}$, §2),

$$N_{o\alpha''}x_{\alpha''} \centerdot x_{\alpha''}S_{\alpha'\alpha'}0_{\alpha'} = x_{\alpha'} \vdash N_{o\alpha''}(T_{\alpha''\alpha'}x_{\alpha'}) \centerdot T_{\alpha''\alpha'}x_{\alpha'}S_{\alpha'\alpha'}0_{\alpha'} = x_{\alpha'}.$$

Hence employing in order the deduction theorem, VI′, and $15^{\alpha''}$, we have

$$\vdash (\exists x_{\alpha''})[N_{o\alpha''}x_{\alpha''} \centerdot x_{\alpha''}S_{\alpha'\alpha'}0_{\alpha'} = x_{\alpha'}] \supset \centerdot N_{o\alpha''}(T_{\alpha''\alpha'}x_{\alpha'}) \centerdot T_{\alpha''\alpha'}x_{\alpha'}S_{\alpha'\alpha'}0_{\alpha'} = x_{\alpha'}.$$

Hence using 33^α and $p \supset q \supset \centerdot [q \supset rs] \supset \centerdot p \supset r$ and $p \supset q \supset \centerdot [q \supset rs] \supset \centerdot p \supset s$, we have $\vdash 31^\alpha$ and $\vdash 32^\alpha$.

A further property of $T_{\alpha''\alpha'}$ is contained in the following theorem (if α is a type for which there is a proof of 30^α):

34^α. $N_{o\alpha'}x_{\alpha'} \supset T_{\alpha''\alpha'}(S_{\alpha'\alpha'}x_{\alpha'}) = S_{\alpha''\alpha''}(T_{\alpha''\alpha'}x_{\alpha'})$.

Proof of this depends on using $23^{\alpha'}$ to prove $N_{o\alpha''}(S_{\alpha''\alpha''}(T_{\alpha''\alpha'}x_{\alpha'}))$ on the assumption of $N_{o\alpha'}x_{\alpha'}$, and using $16^{\alpha'}$ and conversion to prove $S_{\alpha''\alpha''}(T_{\alpha''\alpha'}\, x_{\alpha'})\, S_{\alpha'\alpha'}\, 0_{\alpha'} = S_{\alpha'\alpha'}(T_{\alpha''\alpha'}x_{\alpha'}S_{\alpha'\alpha'}0_{\alpha'})$ and hence $S_{\alpha''\alpha''}(T_{\alpha''\alpha'}x_{\alpha'})S_{\alpha'\alpha'}0_{\alpha'} = S_{\alpha'\alpha'}x_{\alpha'}$ on the assumption of $N_{o\alpha'}x_{\alpha'}$—then using 30^α (with 23^α, 32^α, 31^α).

A similar use of 30^{α} leads to a proof of the following (where α is a type for which there is a proof of 30^{α}):

35^{α}. $T_{\alpha''\alpha'}0_{\alpha'} = 0_{\alpha''}$.

8. Definition by primitive recursion. The formalization of definition by primitive recursion requires that, given formulas $\boldsymbol{A}_{\alpha'}$ and $\boldsymbol{B}_{\alpha'\alpha'\alpha'}$, we find a formula $\boldsymbol{F}_{\alpha'\alpha'}$ such that the following are theorems (where $\boldsymbol{x}_{\alpha'}$ is not a free variable of $\boldsymbol{A}_{\alpha'}$, $\boldsymbol{B}_{\alpha'\alpha'\alpha'}$ or $\boldsymbol{F}_{\alpha'\alpha'}$):

$$\boldsymbol{F}_{\alpha'\alpha'}0_{\alpha'} = \boldsymbol{A}_{\alpha'}.$$

$$N_{o\alpha'}\boldsymbol{x}_{\alpha'} \supset \boldsymbol{F}_{\alpha'\alpha'}(S_{\alpha'\alpha'}\boldsymbol{x}_{\alpha'}) = \boldsymbol{B}_{\alpha'\alpha'\alpha'}\boldsymbol{x}_{\alpha'}(\boldsymbol{F}_{\alpha'\alpha'}\boldsymbol{x}_{\alpha'}).$$

This may be done by taking $\boldsymbol{F}_{\alpha'\alpha'}$ to be the following formula (where $\boldsymbol{x}_{\alpha'}, \boldsymbol{y}_{\alpha''}$ are not free variables of $\boldsymbol{A}_{\alpha'}$ or $\boldsymbol{B}_{\alpha'\alpha'\alpha'}$):[13]

$$\lambda \boldsymbol{x}_{\alpha'} \,\boldsymbol{.}\, T_{\alpha'''\alpha''}(T_{\alpha''\alpha'}\boldsymbol{x}_{\alpha'})(\lambda \boldsymbol{y}_{\alpha''}\langle S_{\alpha'\alpha'}(\boldsymbol{y}_{\alpha''}(K_{\alpha'\alpha'\alpha'}I_{\alpha'})0_{\alpha'}), \boldsymbol{B}_{\alpha'\alpha'\alpha'}(\boldsymbol{y}_{\alpha''}(K_{\alpha'\alpha'\alpha'}I_{\alpha'})0_{\alpha'})(\boldsymbol{y}_{\alpha''}(K_{\alpha'\alpha'\alpha'}0_{\alpha'})I_{\alpha'})\rangle)\langle 0_{\alpha'}, \boldsymbol{A}_{\alpha'}\rangle(K_{\alpha'\alpha'\alpha'}0_{\alpha'})I_{\alpha'}.$$

The definition of $P_{\alpha'\alpha'}$ already given is a particular case and may be used as an illustration. The following theorems may be proved in order:

36^{α}. $N_{o\alpha'}n_{\alpha'} \supset \lambda f_{\alpha\alpha}\lambda x_{\alpha}(n_{\alpha'}f_{\alpha\alpha};x_{\alpha}) = n_{\alpha'}$. (By induction, using $16^{\alpha'}$, $18^{\alpha'\alpha'}$, and conversion.)

37^{α}. $N_{o\alpha'}m_{\alpha'} \supset \boldsymbol{.}\, N_{o\alpha'}n_{\alpha'} \supset \langle m_{\alpha'}, n_{\alpha'}\rangle(K_{\alpha'\alpha'\alpha'}I_{\alpha'})0_{\alpha'} = m_{\alpha'}$. (By induction on $n_{\alpha'}$, using 36^{α}.)

38^{α}. $N_{o\alpha'}m_{\alpha'} \supset \boldsymbol{.}\, N_{o\alpha'}n_{\alpha'} \supset \langle m_{\alpha'}, n_{\alpha'}\rangle(K_{\alpha'\alpha'\alpha'}0_{\alpha'})I_{\alpha'} = n_{\alpha'}$. (By induction on $m_{\alpha'}$, using 36^{α}.)

39^{α}. $N_{o\alpha'''}n_{\alpha'''} \supset n_{\alpha'''}(\lambda p_{\alpha''}\langle S_{\alpha'\alpha'}(p_{\alpha''}(K_{\alpha'\alpha'\alpha'}I_{\alpha'})0_{\alpha'}), p_{\alpha''}(K_{\alpha'\alpha'\alpha'}I_{\alpha'})0_{\alpha'}\rangle)\langle 0_{\alpha'}, 0_{\alpha'}\rangle(K_{\alpha'\alpha'\alpha'}I_{\alpha'})0_{\alpha'} = n_{\alpha'''}S_{\alpha''\alpha''}0_{\alpha''}S_{\alpha'\alpha'}0_{\alpha'}$. (By induction, using 29^{α}, 37^{α}.)

40^{α}. $N_{o\alpha'''}n_{\alpha'''} \supset P_{\alpha'\alpha'''}(S_{\alpha'''\alpha'''}n_{\alpha'''}) = n_{\alpha'''}S_{\alpha''\alpha''}0_{\alpha''}S_{\alpha'\alpha'}0_{\alpha'}$. (By 39^{α}, 38^{α}, using 29^{α}.)

41^{α}. $P_{\alpha'\alpha'''}0_{\alpha'''} = 0_{\alpha'}$. (By $16^{\alpha'}$ and conversion.)

42^{α}. $P_{\alpha'\alpha'}0_{\alpha'} = 0_{\alpha'}$. (By 41^{α}, $35^{\alpha'}$, 35^{α}.)

43^{α}. $N_{o\alpha'}n_{\alpha'} \supset P_{\alpha'\alpha'}(S_{\alpha'\alpha'}n_{\alpha'}) = n_{\alpha'}$. (By 40^{α}, $31^{\alpha'}$, 31^{α}, $34^{\alpha'}$, 34^{α}, $32^{\alpha'}$, 32^{α}.)

PRINCETON UNIVERSITY

---

[13]This schema employs descriptions, through the appearance in it of $T_{\alpha'''\alpha''}$ and $T_{\alpha''\alpha'}$. In certain cases a formula $\boldsymbol{F}_{\alpha'\alpha'}$ may be obtained which does not involve descriptions. In particular, for addition and multiplication of non-negative integers we may use the definitions due to J. B. Rosser:

$\tau_{\alpha'\alpha'\alpha'} \to \lambda m_{\alpha'}\lambda n_{\alpha'}\lambda f_{\alpha\alpha}\lambda x_{\alpha}(m_{\alpha'}f_{\alpha\alpha}(n_{\alpha'}f_{\alpha\alpha}x_{\alpha}))$.

$B_{\alpha'\alpha'\alpha'} \to \lambda m_{\alpha'}\lambda n_{\alpha'}\lambda f_{\alpha\alpha}(m_{\alpha'}(n_{\alpha'}f_{\alpha\alpha}))$.

$[\boldsymbol{A}_{\alpha'} + \boldsymbol{B}_{\alpha'}] \to \tau_{\alpha'\alpha'\alpha'}\boldsymbol{A}_{\alpha'}\boldsymbol{B}_{\alpha'}$.

$[\boldsymbol{A}_{\alpha'} \times \boldsymbol{B}_{\alpha'}] \to B_{\alpha'\alpha'\alpha'}\boldsymbol{A}_{\alpha'}\boldsymbol{B}_{\alpha'}$.

ELEMENTARY TOPICS IN MATHEMATICAL LOGIC
I. THE ALGEBRA OF CLASSES
(1941)

1. The purpose of this first number is to give an expository account of the algebra of subclasses of a (fixed) domain, not in axiomatic form, but with reliance on the reader's intuition in order to introduce him as rapidly as possible to the actual use of the algebra.

A *fundamental domain* (or "universe of discourse") is determined upon, and letters, a, b, c, . . . , are used to stand for classes of things in this domain. If a is a class (of things in the fundamental domain), the *complement* of a, denoted by a′, is the class consisting of all of the fundamental domain except a. If a and b are classes (of things in the fundamental domain), the *logical product* of a and b, denoted by ab, is the common part of the classes a and b; and the *logical sum* of a and b, denoted by a + b, consists of a and b taken together (including, of course, the common part of a and b).

The ideas involved are most conveniently explained by means of some simple examples.

First example

As our first example, suppose that we determine the fundamental domain to consist of all inhabitants of the United States, and let b stand for the class of Bostonians, f for farmers (in the United States), n for New Englanders, r for Republicans, y for New Yorkers.

In order that the algebra apply to this case, we must agree upon meanings of the words, "farmer," "inhabitant of the United States," etc., more exact than those in everyday use, so that it can in principle be said of every one either that he or she is not a farmer, etc. Moreover we must agree upon a fixed date and understand that the reference is to inhabitants of the United States at the date and to those who are farmers, Republicans, etc. at that date.

On this basis, with the above meanings for the letters b, f, n, r, y, we have that b′ is the class of inhabitants of the United States outside of Boston — f′ consists of non-farmers in the United States — r′ of non-Republicans (whether Democrats, members of minor parties, Independents, or simply political non-entities) — etc.

As illustration of the logical product we may take the class ry, consisting of those who are both Republicans and New Yorkers. Evidently ry and yr are the same class; we express it by the equation

$$ry = yr$$

(in words, Republican New Yorkers are the same as New York Republicans). Other examples are nf (New England farmers) and n(rf) (New England Republican farmers). We have:

$$n(rf) = (nr)f$$

Originally published in mimeograph, The Galois Institute of Mathematics and Art, Brooklyn, New York, 1940, 1941.

(New England Republican farmers are the same as New England Republicans who are farmers). Hence it is unnecessary to distinguish between n(rf) and (nr)f, and we may write simply nrf. Of course, we also have the equations,

$$\text{nrf} = \text{nfr} = \text{rnf} = etc.$$

Another logical product is bn. Since, however, all Bostonians are New Englanders, we have he equation,

$$\text{bn} = \text{b}.$$

Null class

If we suppose that no Bostonian can be also a New Yorker (this will depend, perhaps, on what we choose as the exact definition of the words), then the logical product, by, will be the *empty class* or *null class*. We denote this by the small letter o and write the equation:

$$\text{by} = \text{o}$$

(in words, Bostonian New Yorkers are nil).

Universal Class

The complement of the null class is the *universal class*, which comprises the entire fundamental domain, and which we denote by u. Thus, $o' = u$, and $u' = o$.

Extensionality

The reader will observe that we call two classes the same, and write the sign = between the letters or expressions denoting the two classes, if they have the same content, or comprise the same portion of the fundamental domain, regardless of whether the definitions of the two classes are at all the same in meaning. Thus if it should be found that in fact all New Englanders are Republicans and all Republicans New Englanders, we would then write $n = r$, although what it is to be a Republican is something very different from what it is to be a New Englander. From this rule of *extensionality*, as it is called, it follows in particular that there is only one universal class and only one null class, so that we may correctly speak of *the* universal class and *the* null class (once the fundamental domain is fixed).

Frequently the numerals 0 and 1 are used as symbols for the null class and the universal class respectively. We are here substituting the small letters o and u in order to avoid any danger of confusion between the *classes* 0 and 1 and the *numbers* 0 and 1. (It is true that we are using the same notation for logical sum and logical product as is commonly used for the numerical or arithmetical sum and product, but there is no danger of confusion here if we keep in mind what each letter being used stands for: where the sign + stands between two letters or expressions denoting classes it is to be understood as meaning the logical sum, and where it stands between two letters denoting numbers it is to be understood as a numerical sum.)

As illustrations of the logical sum we may take $b + y$ (Bostonians and New Yorkers), $y + f$ (the class composed of those who are either New Yorkers or farmers including those, if any, who are both), $b + y'$ (this is the same as y'), $n + b'$ (this is the universal class, $n + b' = u$), $y + f + r$, etc. Evidently we have:

$$y + f = f + y;$$
$$y + (f + r) = (y + f) + r$$

(hence we need not distinguish between $y+(f+r)$ and $(y+f)+r$, but may write simply $y+f+r$);

$$y+f+r = y+r+f = f+y+r = etc.$$

More complex expressions which may be used as illustrations are: $br+r'y$ (Bostonian Republicans and non-Republicans New Yorkers); $(r+f)'$ (those not included in the class Republicans-and-farmers, or in other words, those who are not either-Republicans-or-farmers); $n(r+f)'$ (New Englanders and those who are not either-Republicans-or-farmers); $(n(r+f))'$ (those who are not both New Englanders and either Republicans or farmers).

The reader should see, on reflection, the truth of the following equations:

$$(r+f)' = r'f',$$
$$n(r+f)' = nr'f',$$
$$n(r+f) = nr + nf,$$
$$(n(r+f))' = n' + r'f'.$$

Second example

As a second example, suppose that we determine the fundamental domain to consist of all positive whole numbers (i.e., of the numbers 1, 2, 3, 4, ...), and let e stand for the class of even numbers, k for the class of (positive whole) numbers less than 6, p for the class of prime numbers, c for the class of composite numbers, s for the class of square numbers.

(A prime number is a whole number greater than 1 which is not exactly divisible by any positive whole number except itself and 1. A composite number is a number which can be expressed as the product of two or more whole numbers greater than 1. A square number is a number which is the exact square of a positive whole number.)

For brevity in the paragraphs immediately following we shall use the word *numbers* with the understanding of a restriction to positive whole numbers.

Then e' is the class of odd numbers, k' is the class of number greater than 5, p' is the class of non-prime numbers (i.e., 1 and composite numbers), s' is the class of non-square numbers.

Moreover, $e's$ is the class of odd square numbers, $e'p$ is the class of odd prime numbers (i.e., prime numbers except 2), ep' is the class of even non-prime numbers (i.e., even numbers except 2). The class $p+s+k'$ consists of all prime numbers together with all square numbers and all numbers greater than 5, in other words, all numbers $(p+s+k' = u)$. The class psk' consists of prime square numbers greater than 5, in other words, the null class, since no number is both prime and square ($ps = o$, therefore $psk' = o$).—We must, of course, distinguish between psk' and $(psk)'$; in fact, while one of these classes is the null class, the other one is the universal class.

The class sk consists of the two numbers, 1 and 4.

The class ep consists of just one number, namely the number 2. Likewise $es'k$ consists of the single number 2. Hence we write:

$$ep = es'k$$

(even prime numbers are the same as even non-square numbers less than 6). It would not, however, be correct to write ep = 2, because 2 is a *number*, whereas ep is a *class of numbers*, and they are not to be identified.

Unit class

A class which consists of just one element is called a *unit class*. It is often convenient to use the Greek letter iota as a notation for a unit class, writing, for example, $\iota 2$ to mean the unit class whose single element is the number 2 (thus ep = es′k = $\iota 2$).

As an exercise it is suggested to the reader to translate into words the following symbolic expressions for classes, to find a simpler symbolic expression for the same class where possible (without introducing new letters beyond those used in the text), and, where finite, to make a complete list of the numbers composing the class: ep + k, ps + e′, c′s + es′k, (cs + ek′)′, (c + s)k, (p + s)(e + k).

Fundamental laws

We are now ready to set down the principal formal laws of the algebra of classes. We make no attempt to give proofs, or to designate some of the laws as axioms from which the others follow; but the reader should verify on an intuitive basis, in view of the meanings of the operations of the algebra as already explained, that each of the equations given is an *identity*, i.e., that it is true whatever the fundamental domain is and no matter what particular classes the letters (other than o and u) are taken to stand for.

The two commutative laws:

$$\mathrm{ab} = \mathrm{ba}, \qquad \mathrm{a} + \mathrm{b} = \mathrm{b} + \mathrm{a}.$$

The two associative laws:

$$\mathrm{a}(\mathrm{bc}) = (\mathrm{ab})\mathrm{c}, \qquad \mathrm{a} + (\mathrm{b} + \mathrm{c}) = (\mathrm{a} + \mathrm{b}) + \mathrm{c}.$$

The two distributive laws:

$$\mathrm{a}(\mathrm{b} + \mathrm{c}) = \mathrm{ab} + \mathrm{ac}, \qquad \mathrm{a} + \mathrm{bc} = (\mathrm{a} + \mathrm{b})(\mathrm{a} + \mathrm{c}).$$

The two laws of tautology:

$$\mathrm{aa} = \mathrm{a}, \qquad \mathrm{a} + \mathrm{a} = \mathrm{a}.$$

The two laws of absorption:

$$\mathrm{a}(\mathrm{a} + \mathrm{b}) = \mathrm{a}, \qquad \mathrm{a} + \mathrm{ab} = \mathrm{a}.$$

The two laws of complementation:

$$\mathrm{aa}' = \mathrm{o}, \qquad \mathrm{a} + \mathrm{a}' = \mathrm{u}.$$

The law of double complementation:

$$(\mathrm{a}')' = \mathrm{a}.$$

The two laws of De Morgan:

$$(\mathrm{ab})' = \mathrm{a}' + \mathrm{b}', \qquad (\mathrm{a} + \mathrm{b})' = \mathrm{a}'\mathrm{b}'.$$

The laws of operations with o and u:

$$\mathrm{oa} = \mathrm{o}, \qquad \mathrm{u} + \mathrm{a} = \mathrm{u},$$

$$\text{ua} = \text{a}, \qquad\qquad \text{o} + \text{a} = \text{a},$$

$$\text{o}' = \text{u}, \qquad\qquad \text{u}' = \text{o}.$$

The analogies with ordinary numerical algebra should be noticed, and also the differences. The o is closely analogous to the 0 of numerical algebra, and the u is analogous, but less closely, to 1. From the first law of tautology it follows that there are no exponents in the algebra, since a^2, a^3, a^4, etc. reduce simply to a. Similarly, in view of the second law of tautology, there are no numerical coefficients, since a + a, a + a + a, etc., do not give 2a, 3a, etc., but each reduces simply to a.

Duality

It is seen that the equations above are arranged in pairs, one equation of the pair being obtained from the other by interchanging logical sum and logical product and at the same time interchanging o and u. This arrangement in pairs holds generally for all identities of the algebra of classes. *Given any identity, the equation which results from it by replacing logical sum by logical product, logical product by logical sum,* o *by* u, *and* u *by* o *is also identity.* An equation, or an expression obtained from another by thus interchanging logical sum and logical product, o and u, is called its *dual*; and the underscored principle just stated is called the *principle of duality*. It is necessary to remember that the principle of duality applies only to identities, not to equations which are true by virtue of special meanings attached to letters (other than o and u) which appear.

It can happen in special cases that the dual of an identity is the same identity, and when this happens the identity is called *self-dual.* The law of double complementation is thus self-dual. A less trivial example of a self-dual identity is the following:

$$\text{abc} + \text{a}'\text{bc} + \text{ab}'\text{c} + \text{abc}' = (\text{a} + \text{b} + \text{c})(\text{a}' + \text{b} + \text{c})(\text{a} + \text{b}' + \text{c})(\text{a} + \text{b} + \text{c}').$$

(The self-duality is immediately evident; it is left to the reader to verify that the equation actually is an identity, by multiplying out and simplifying the right-hand side in accordance with instructions below.)

For convenience in the description of formal manipulations, we adopt the words "monomial" and "polynomial" from numerical algebra. We shall also use the word *variables* to mean letters other than o and u (this terminology is justified because we shall be dealing with processes which are valid independently of the particular meanings of the letters other than o and u).

Simplification of monomials

A *monomial* is either: (1) a single letter (a variable or o or u) with or without an accent ′, denoting complementation, after it; or (2) a product of two or more factors of the kind (1). Thus, some examples of monomials are: o, u, u′, a, x′, ab, x′y, ab′x′y′c, aaaa, uo′ax, aa′a′ba′.

Monomials may be simplified by using the first law of tautology, the first law of complementation, and the laws of operation with o and u. When a monomial is reduced to simplest form, it may happen that the result is o, or that it is u. Otherwise, (i.e., except these two cases) the simplest form of a monomial will not have either o, u, o′, u′ appearing as a factor; and no letter will be repeated as a factor, whether both times without accent after it, or both times with accent, or once with and once without accent.—Examples: $aa'x$ simplifies to ox and hence to o; $o'uu$ simplifies to uuu and hence to u; $aaab'ab'$ simplifies to ab'; $o'o'x'x'x'$ simplifies to x'; $abbccab'abbc'$ simplifies to o.

Simplification of polynomials

A *polynomial* is either a monomial or a sum of two or more monomials. The separate monomials are called the *terms* of the polynomial.

In simplifying a polynomial, each term may, of course, be simplified by the method of simplifying monomials just described. However, even after each separate term is in simplest form, the polynomial may often still be simplified by means of the second law of tautology, the second law of absorption, the second law of complementation, and the laws of operations with o and u. Frequently other laws are used in conjunction with these. We give some examples:

The polynomial $abx + abcxy + axa'$ simplifies, by the second law of absorption, to $abx + axa'$; this simplifies further to $abx + o$; and hence finally to abx. The polynomial $abc + a'bc + b' + c'$, by using the first distributive law, may be changed to $(a + a')bc + b' + c'$; hence, by the second law of complementation, it becomes $ubc + b' + c'$; this simplifies to $bc + b' + c'$; by the first law of De Morgan, this is the same as $bc + (bc)'$; and this finally, by the second law of complementation, reduces to u. The polynomial $abcx' + a'c + b'c + cx$, by using the first distributive law, may be changed to $abcx' + (a' + b' + x)c$; by using the first law of De Morgan and the law of double complementation, this becomes $abcx' + (abx')'c$; by the first distributive law, this becomes $c(abx' + (abx')')$; by the second law of complementation, this reduces to cu, and hence finally to c.

For many purposes, however, it is preferable, instead of reducing a polynomial to simplest form, to reduce it to what is called normal form, as will be explained below.

Multiplication of polynomials

From the first distributive law, in conjunction with the commutative and associative laws, it follows that two polynomials may be multiplied together (multiplication in the sense of logical product) by a process exactly similar to that used in ordinary numerical algebra for multiplication of polynomials.

For example, if it is required to multiply out $(a + b + c' + a'x)(a' + c + bx)$, we may arrange the work as follows:

$$
\begin{array}{r}
a + b + c' + a'x \\
\underline{a' + c + bx\quad\;\;} \\
0 + a'b + a'c' + a'x \\
+\, ac + bc + o + a'cx \quad\;\; \\
\underline{+\, abx + bx + bc'x + a'bx} \\
a'b + a'c' + a'x + ac + bc + bx
\end{array}
$$

Thus $(a+b+c'+a'x)(a'+c+bx) = a'b+a'c'+a'x+ac+bc+bx$. (The second law of absorption was used to reduce $a'x+a'cx+a'bx$ to $a'x$, and to reduce $abx+bx+bc'x$ to bx.)

Use of De Morgan's laws

To simplify such an expression as $(ab+ac+a'x'y)(ab'c+a'x'y'+a'by)'$, where not all the multipliers are polynomials, it is necessary first to reduce the expression to a product of polynomials by using De Morgan's laws and the law of double complementation, and then to multiply out as above. For the particular example given, the work is as follows:

$$
\begin{aligned}
&(ab + ac + a'x'y)(ab'c + a'x'y' + a'by)' \\
&= (ab + ac + a'x'y)(ab'c)'(a'x'y')'(a'by)' \\
&= (ab + ac + a'x'y)(a' + b + c')(a + x + y)(a + b' + y')
\end{aligned}
$$

$$
\begin{array}{l}
\quad ab + ac + a'x'y \\
\quad a' + b + c' \\
\hline
\quad o + o + a'x'y \\
+ ab + abc + a'bx'y \\
+ abc' + o + a'c'x'y \\
\hline
\quad ab + a'x'y \\
\quad a + x + y \\
\hline
\quad ab + o \\
+ abx + o \\
+ aby + a'x'y \\
\hline
\quad ab + a'x'y \\
\quad a + b' + y' \\
\hline
\quad ab + o \\
+ o + a'b'x'y \\
+ aby' + o \\
\hline
\quad ab + a'b'x'y
\end{array}
$$

Thus $(ab + ac + a'x'y)(ab'c + a'x'y' + a'by)'$ simplifies to $ab + a'b'x'y$.

Normal form of a polynomial

We return now to consideration of polynomials in *normal form*.

A polynomial is said to be in normal form in the four variables a, b, c, d if (1) every term has exactly four factors appearing, the first factor being either a or a', the second factor either b or b', the third factor either c or c', the fourth factor either d or d', and (2) no two terms are the same (exactly).

For example, $abc'd + abcd' + abcd + a'b'c'd'$ is a polynomial in normal form in the four variables a, b, c, d. But $ab'cd + abcx$ is not in normal form in four variables a, b, c, d, because the fourth factor in the second term is neither d nor d′ but x; and ab′c′d + a′b′c′d + ab′c′d is not in the normal form in the four variables a, b, c, d because the first and the third terms are the same.

The definition of being in the normal form in any particular set of variables is exactly similar. For example, a polynomial is in normal form in the two variables x, y if (1) every term has exactly two factors appearing, the first factor being either x or x′, the second factor being either y or y′, and (2) no two terms are the same. There are thus altogether just fifteen different polynomials which are in normal form in the two variables x, y. The complete list is: xy, xy′, x′y, x′y′, xy + xy′, xy + x′y, xy + x′y′, xy′ + x′y, xy′ + x′y′, x′y + x′y′, xy + xy′ + x′y, xy + xy′ + x′y′, xy + x′y + x′y′, xy′ + x′y + x′y′, xy + xy′ + x′y + x′y′. (Of course this is on the basis that two polynomials differing only as regards the order of their terms are counted the same, so that, for instance, the polynomial x′y′ + xy′ + x′y is the same as the next to last polynomial in the list.) Of the fifteen polynomials listed, the last one has the greater number of terms, four, and is called the complete normal form in the two variables x, y; it can be simplified to u (the reader should verify this).

There are just three polynomials in normal form in *one* variable, say x. These are x, x′, and x + x′. The last one, x + x′, is the complete normal form in the one variable x.

The monomials o and u are counted as polynomials in normal form in *no variables*. Of these, u is the complete normal form.

In general, in the complete list of polynomials in normal form in a particular set of n variables (n greater than 0), there is one which as the greatest number of terms, namely 2^n terms. This one is called the *complete normal form* in that set of variables. It is always reducible to u and is the only polynomial in the list which is reducible to u.

Moreover the polynomials in the list—other than the complete normal form—have some but not all of the terms which appear in the complete normal form. The complement of any such polynomial can be expressed by simply writing the sum of the remaining terms which appear in the complete normal form but not in it. The reader should verify this in some particular cases, showing, for example, that the following equations are identities:

$$(xy + x'y')' = xy' + x'y.$$
$$(xy' + x'y + x'y')' = xy.$$
$$(x'y)' = xy + xy' + x'y'.$$

It is also of importance in applications that the separate terms of any polynomial in normal form must stand for classes which have no elements in common, i.e., for classes the logical product of any two of which is o.

Reduction to normal form

Given any polynomial, it is possible to reduce it to normal form. This is a systematic method for doing this, which we explain by means of an example.

Required to reduce to normal form the polynomial:

$$a'b + a'c' + a'x + ac + bc + bx.$$

The variables which appear in the polynomial are a, b, c, x. Hence we seek to reduce it to a polynomial in normal form in the four variables a, b, c, x. The first term fails to contain the variables c, x (with or without accents). Hence we multiply this term by (c + c′)(x + x′)—this is permissible, i.e., the new expression obtained stands for the same class that the original polynomial does, because, by the second law of complementation, $c + c'$ and $x + x'$ are both equal to u. Similarly, the second term fails to contain b, x, and hence we multiply it by $(b + b')(x + x')$. Treating all six terms in this way, we obtain:

$$a'b(c + c')(x + x') + a'c'(b + b')(x + x') + a'x(b + b')(c + c')$$
$$+ ac(b + b')(x + x') + bc(a + a')(x + x') + bx(a + a')(c + c').$$

Multiplying this out, we obtain:

$$a'bcx + a'bc'x + a'bcx' + a'bc'x' + a'bc'x + a'b'c'x$$
$$+ a'bc'x' + a'b'c'x' + a'bcx + a'b'cx + a'bc'x + a'b'c'x$$
$$+ abcx + ab'cx + abcx' + ab'cx' + abcx + a'bcx$$
$$+ abcx' + a'bcx' + abcx + a'bcx + abc'x + a'bc'x.$$

Canceling duplicate terms (law of tautology) and rearranging the order of terms (commutative law), we obtain the required normal form:

$$abcx + abcx' + abc'x + ab'cx + a'bcx + ab'cx' + a'bcx'$$
$$+ a'bc'x + a'b'cx + a'bc'x' + a'b'c'x + a'b'c'x'.$$

Since we have already explained methods by which any expression in the algebra of classes can be reduced to a polynomial, it now follows that *any expression in the algebra can be reduced to a polynomial in normal form.*

After an expression has been reduced to a polynomial in normal form, it is always possible to increase the number of variables in the normal form if desired. For example, if it is desired to reduce the polynomial

$$a'b + a'c' + a'x + ac + bc + bx,$$

to a polynomial in normal form in the five variables a, b, c, x, y, we may take the normal form in the four variables a, b, c, x, as just obtained, and multiply it by $y + y'$. On multiplying out, the desired result is obtained.

Sometimes it is possible to reverse this process, and reduce the number of variables in the normal form. For example, take the polynomial,

$$abcx' + a'c + b'c + cx,$$

and reduce it to a polynomial in normal form in the four variables a, b, c, x. The first step is:

$$abcx' + a'c(b + b')(x + x') + b'c(a + a')(x + x') + cx(a + a')(b + b').$$

Multiplying this out, cancelling duplicate terms, and rearranging gives the normal form:

$$abcx + abcx' + ab'cx + a'bcx + ab'cx' + a'bcx' + a'b'cx + a'b'cx'.$$

In order to see whether the variable x can be eliminated, we compare the terms in which x appears as a factor to those in which x' appears as a factor. This leads to:

$$(abc + ab'c + a'bc + a'b'c)x + (abc + ab'c + a'bc + a'b'c)x'.$$

By the first distributive law, this leads to the factorization:

$$(abc + ab'c + a'bc + a'b'c)(x + x').$$

(This step is possible, of course, only because it happens that the expression in parentheses multiplying x is the same as the expression in parentheses multiplying x′.) Since $x + x' = u$, this gives:

$$abc + ab'c + a'bc + a'b'c.$$

This is a polynomial in normal form in the three variables a, b, c. — In this case, as it happens, the process can be continued, and the variables a and b also eliminated. The successive steps are:

$$(ac + a'c)b + (ac + a'c)b'.$$
$$(ac + a'c)(b + b').$$
$$ac + a'c.$$
$$c(a + a').$$
$$c.$$

As a means of acquiring some facility in the technique of the algebra of classes, the following exercises are suggested to the reader. It is asked in each case to reduce the given expression to a polynomial in normal form having the smallest possible number of variables.

$$xy + x'y + x'y'z + xy'z.$$
$$a' + c'e' + bcd' + b'cd.$$
$$(a + b + c)(ab + a'c)'.$$
$$(ax + a'x')(x'y + xy')(b + xy).$$
$$(x'y + xyz' + xy'z + x'y'z't + t')'.$$
$$a'bc' + a + b' + c + d.$$
$$[(ab + a'b')' + (a'x + ax')' + (bx + b'x + b'x')']'.$$
$$(a'bcd' + ab'c'd + a'bcd + a'b'c'd' + a'bc'd' + a'b'cd)'.$$
$$(a'b + bc + c'd + a'd' + ac' + b'd)'.$$
$$(x' + y + z' + a + v' + w')'(x' + y + z + a' + v')'.$$

Equations of condition

We turn now to consideration of equations which are not (necessarily) identities, but which are true in view of special meanings attached to the letters which they contain, or which may be considered as imposing conditions on the letters which they contain.

In manipulating equations, it is of course permissible to perform on either side of an equation the operations which we have been discussing; in particular, either side may be simplified, or be reduced to a polynomial in normal form. Moreover the following operations are permissible, in the sense that when applied to a true equation they must yield a true equation: (1) to take the complement of both sides of an equation; (2) to multiply both sides of an equation by the same thing, or by equal things; (3) to add the same thing (equal things) to both sides of an equation.

Use may also be made of the principle that if a logical product is equal to u each separate factor must be equal to u; and of the dual principle to this, that if a logical sum is equal to o each separate term of the sum must be equal to o. For example, from the equation $a(x'y + xy') = u$ we may infer both of the equations $a = u$ and $x'y + xy' = u$. From the latter equation, taking complements of both sides, we may obtain $xy + x'y' = o$, and hence we may infer both of the equations $xy = o$ and $x'y' = o$. Taking complements again, we may infer $x' + y' = u$ and $x + y = u$.

One equation is said to be derivable from another if it can be obtained from it by a succession of the permissible operations listed in the two preceding paragraphs. For example, $x + y = u$ is derivable from $a(x'y + xy') = u$; $ax = ay$ is derivable from $x = y$; etc. Whenever one equation is derivable from another, the derived equation must be true if the original one is, *but not necessarily conversely*; for example, if it is true that $x = y$ then it must be true that $ax = ay$, but if it is known only that $ax = ay$, it may not be inferred that necessarily $x = y$.

The reader should be warned against certain operations on equations which are familiar in connection with ordinary numerical algebra but are not valid in the algebra of classes. It is not permissible in the algebra of classes to divide out a factor from both sides of an equation—as e.g., the inference from $ax = ay$ to $x = y$ would require. It is not permissible to cancel a term which appears on both sides of an equation, or to transpose a term from one side of an equation to the other. The method of solving an equation by factoring, which is used in ordinary algebra, is not valid in the algebra of classes, because this depends on the proposition that a product cannot be equal to 0 unless one of the factors is equal to 0, and the corresponding proposition, with o instead of 0, does not hold in the algebra of classes.

Equivalence of equations

One equation is said to be *equivalent* to another if each equation is derivable from the other. The two notions of equivalence and derivability should be carefully distinguished. For many purposes, what is needed is to obtain equations equivalent to a given equation rather than merely derivable from it.

Reduction of an equation to form with o *on right*

Given any equation in the algebra of classes, it is always possible to obtain an equivalent equation whose right-hand side is o. In fact, if A and B are any two expressions, the equation,

$$A = B,$$

is equivalent to the equation,

$$A'B + AB' = o.$$

To prove this, we must show how to derive each of these equations from the other. Starting with the equation $A = B$, we may multiply both sides of the equation by B', so obtaining $AB' = o$; also we may multiply both sides of the equation by A' and then change the order of the two sides, so obtaining $A'B = o$; adding the equations $AB' = o$ and $A'B = o$, we obtain $A'B + AB' = o$. Starting with the equation $A'B + AB' = o$, we may take the complement of both sides, so obtaining $AB + A'B' = u$; then multiplying both sides of this equation by B, we obtain $AB = B$; also, multiplying both sides of $A'B + AB' = o$ by B', we obtain $AB' = o$; adding the equations $AB = B$ and $AB' = o$, we obtain $AB + AB' = B$, an equation which, on simplifying the left-hand side, reduces to $A = B$.

(The operation of adding two equations together, which was twice used in the preceding paragraph, of course comes under the head of "adding the same thing (equal things) to both sides of an equation." For example, having the equation AB = B, we add AB′ to one side of the equation and o to the other side, since AB′ and o are known to be equal things.)

Normal form of an equation

An equation is said to be *in normal form* if its right-hand side is o and its left-hand side is a polynomial in normal form. *Every equation in the algebra of classes is equivalent to an equation in normal form.* For we may first use the above method to put the equation in such a form that the right-hand side is o, and then use the previous methods to reduce the left-hand side of the resulting equation to a polynomial in normal form.

Example: to reduce to normal form the equation,

$$x + y = a + b.$$

By the theorem proved above, this equation is equivalent to

$$(x + y)'(a + b) + (x + y)(a + b)' = o.$$

By the second law of De Morgan, this is equivalent to

$$x'y'(a + b) + a'b'(x + y) = o.$$

By the first distributive law, this is equivalent to

$$ax'y' + bx'y' + a'b'x + a'b'y = o.$$

By the second law of complementation (etc.), this is equivalent to

$$ax'y'(b + b') + bx'y'(a + a') + a'b'x(y + y') + a'b'y(x + x') = o.$$

By the first distributive law, this is equivalent to

$$abx'y' + ab'x'y' + abx'y' + a'bx'y' + a'b'xy + a'b'xy' + a'b'xy + a'b'x'y = o.$$

Cancelling duplicate terms, by the second law of tautology, we obtain the desired equation in normal form:

$$abx'y' + ab'x'y' + a'bx'y' + a'b'xy + a'b'xy' + a'b'x'y = o$$

Simultaneous equations equivalent to a single equation

Given any set of simultaneous equations, there is always a single equation to which it is equivalent, in the sense that the single equation is derivable from the set of simultaneous equations and each of the simultaneous equations is derivable from the single equation.

For example, suppose that three simultaneous equations are given. By the methods which have just been explained, each of the three equations can be replaced by an equivalent equation with o as the right-hand side. This gives three equations,

$$A = o,$$
$$B = o,$$
$$C = o$$

(where A, B, C are expressions of algebra, involving some or all of the letters which appeared in the original equations). Then this set of three equations is equivalent to the single equation:

$$A + B + C = o.$$

For this reason it is unnecessary to make any separate treatment of the subject of simultaneous equations in the algebra of classes. Given a set of simultaneous equations, we may simply replace it by a single equation to which it is equivalent and thereafter deal with this single equation.

(There is, of course, nothing analogous to this in ordinary numerical algebra.)

Relation $\subset$

We shall use the notation $a \subset b$ to mean that all elements of the class a are also elements of b, in other words that a is either a part of b or the same as b. This notation is really superfluous, since $a \subset b$ thus means the same thing as $ab' = o$, but it is nevertheless often convenient. The notation may be read "a is contained in b" or "a is a subclass of b" (the subclasses of a class b are the classes which are parts of b, not excluding the class b itself as one of the subclasses of b).

Relation ϵ

The relation between a and b expressed by $a \subset b$ must be sharply distinguished from the relation that a is *an element of* the class b. The latter relation is usually expressed by the Greek letter epsilon: $a \epsilon b$, to be read "a epsilon b" or "a is an element of b."

For illustration, we may return to our example that the fundamental domain is determined to consist of all positive whole numbers, s being the class of square numbers, and p the class of prime numbers. Then we have $s \subset p'$ (the square numbers are a subclass of the non-prime numbers); likewise we have $4 \epsilon p'$ (4 is one the non-prime numbers). It would be wrong, however, to write $s \epsilon p'$; it is not true that the *class* of square numbers *is one* of the non-prime numbers. It would equally be wrong to write $4 \subset p'$. But $\iota 4 \subset p'$ is correct. And $\iota 4 + \iota 9 \subset p'$ is correct (the class which consists of the two numbers 4 and 9 and no others is a subclass of p').

The relation $\subset$ has the important property that it is transitive; i.e., if $a \subset b$ and $b \subset c$ it necessarily follows that $a \subset c$.—For example, since the class of positive integral powers of 4 (i.e., the class consisting of the numbers $4, 4^2, 4^3, \ldots$) is a subclass of the square numbers, and the square numbers are a subclass of the non-prime numbers, it follows that the class of positive integral powers of 4 is a subclass of the non-prime numbers.

The relation ϵ, on the other hand, is *not* transitive. Thus if John Doe is a Republican and the Republicans are a political party it does not follow that John Doe is a political party. In symbols, let J stand for John Doe, r for the Republican party, and q for the class of political parties in the United States; from $J \epsilon r$ and $r \epsilon q$ it does not follow that $J \epsilon q$.

Solution of an equation

Returning now to the subject of equations, suppose that in a given equation the meaning of all the letters but one, say x, are known, while x is unknown. If it is required to solve the equation for x, we may proceed as follows.

First we replace the given equation by an equivalent equation of the form,

$$Ax + Bx' = o,$$

where A and B are expressions which do not contain the letter x (this can be done, for example, by reducing the given equation to normal form, grouping together the terms containing x as a factor, and those containing x′ as a factor, and applying the first distributive law). Then we reason that this equation is equivalent to the pair of simultaneous equations,

$$Ax = o, \qquad Bx' = o,$$

and hence, using the relation $\subset$, that the solution is

$$B \subset x \subset A'$$

(in words, "x contains B and is contained in A′").

Thus the solution is to be understood in the sense that the equation is satisfied by any value of x (any class x) which contains the class denoted by B and is contained in the class denoted by A′, and further that these are all the values of x which satisfy the equation.

It follows that there are *no* solutions (i.e., the given equation is impossible) *unless* the condition is satisfied,

$$B \subset A',$$

or, equivalently,

$$AB = o.$$

When this condition is satisfied, there is a least solution, $x = B$, and a greatest solution $x = A'$, and any value of x intermediate between these two (in the sense $B \subset x \subset A'$) is also a solution.

Evidently the given equation has a *unique* solution if and only if $A' = B$, and in this case the solution is $x = B$.

Elimination of a letter

The same method may be used to eliminate a particular letter, say x, from a given equation. Reduce the given equation to the form,

$$Ax + Bx' = o,$$

where A and B are expressions which do not contain the letter x. By the preceding reasoning, this equation has a solution for x (is possible) if and only if the condition is satisfied,

$$AB = o.$$

The letters other than x must have such values that this last equation is satisfied, and the original equation is possible if and only if it is true.

Solution for two or more unknowns

The solution of an equation for two or more unknowns is best explained by means of an example.

In the equation,

$$ab'c'x + ab'xy = bc'x'y',$$

suppose that a, b, c are known and it is required to solve for x and y.

The given equation is equivalent to

$$(ab'c'x + ab'xy)'bc'x'y' + (ab'c'x + ab'xy)(bc'x'y')' = o.$$

This simplifies to

$$bc'x'y' + ab'c'x + ab'xy = o.$$

This is equivalent to

$$bc'x'y' + ab'c'x(y + y') + ab'xy = o.$$

Multiplying out and using the second law of absorption, we obtain

$$bc'x'y' + ab'c'xy' + ab'xy = o.$$

Using the first distributive law (etc.), we obtain

$$(ab'x)y + (bc'x' + ab'c'x)y' = o. \tag{1}$$

Eliminating y by the method explained above, we get

$$(ab'x)(bc'x' + ab'c'x) = o,$$

which reduces to

$$ab'c'x = o,$$

or

$$ab'c'x + ox' = o.$$

Solving this for x by the method explained above, we get

$$o \subset x \subset a' + b + c. \tag{2}$$

Solving the equation (1) for y by the same method, we get

$$bc'x' + ab'c'x \subset y \subset a' + b + x'.$$

Because $ab'c'x = o$, this reduces to

$$bc'x' \subset y \subset a' + b + x'. \tag{3}$$

Conditions (2) and (3) together constitute the solution of the given equation.

This solution means that x may have any value between the extremes o and $a' + b + c$, inclusive, and that y may then have any value between certain extremes which depend on the particular value of x (as given in condition (3)). For example, x may have the value o, and y may then have any value between bc' and u inclusive; x may have the value b, and y may then have any value between o and u inclusive (i.e., any value at all); x may have the value c, and y may then have any value between bc' and $a' + b + c'$ inclusive, x may have the value $a' + b + c$, and y may then have any value between o and $a' + b + c'$ inclusive; etc.

Of course, instead of solving as above, we could also solve the given equation first for y and then for x in terms of y. This would give the solution:

$$o \subset y \subset u,$$
$$bc'y' \subset x \subset a' + b + cy'.$$

The result of *eliminating* x and y from the given equation (by, say, first eliminating y and then eliminating x from the result) is the equation o = o. This means that the given equation has solutions for x and y no matter what particular classes the letters a, b, c stand for.

Elimination of two and more letters

The elimination of two and more letters from a given equation can be accomplished by a standard method, the same for any number of letters.

For example, suppose that it is required to eliminate x, y, z, from a given equation. By first reducing it to normal form (or otherwise), the given equation can be transformed into an equivalent equation of the form,

$$Axyz + Bxyz' + Cxy'z + Dx'yz + Exy'z' + Fx'yz' + Gx'y'z + Hx'y'z' = o.$$

The result of the elimination is then the equation,

$$ABCDEFGH = o.$$

(The reader should verify this by carrying out the successive eliminations, say first for x and then for y and then for z.)

Illustrations of the application of the algebra of classes to concrete problems will be taken up in the next number. We conclude this number with a list of formal exercises in solving equations and in carrying out eliminations.

Each of the following equations or sets of simultaneous equations is to be solved for x, and the condition is to be determined which the letters other than x must satisfy in order that there shall be a solution for x:

$ax + bx' = b'x + ax'.$ (Answer: $x = a'b + ab'$, no condition on a and b.)

$x + ax' = a'x'.$

$ax + bx' = cx + b'x'.$

$ax + bx' = cx + dx'.$

$ax + bx' = ax + cx'$ and $dx + bx' = d'x + cx'.$
(Answer: $x = o$, condition $b = c$.)

$ax + bx' = cx + dx'$ and $ax + dx' = c'x + b'x'.$

$a'b'x + ax' + bx' = o.$

Each of the following equations or sets of simultaneous equations is to be solved for x and y, and the condition is to be determined which the letters other that x and y must satisfy in order that there shall be a solution for x and y:

$$x'y + xy' + x'y' = o.$$

$$xy + x'y' = o.$$

$$axy + bxy' + cx'y + dx'y' = o.$$

$$a'xy + ax'y' = x'y + xy'.$$

$$ax + by' = bx' + a'y.$$

$$axy + x'y' = o \text{ and } a'xy' + x'y = o.$$

$$ax = b'y' \text{ and } by = a'x'.$$

$$a'x = ab' + x'y \text{ and } by' = b'x' + ay.$$

The letters x. y, z are to be eliminated from each of the following equations or sets of simultaneous equations:

$$axy + bxz + cyz = a'x' + b'y' + c'z'. \qquad \text{(Answer: } a'bc' = o.\text{)}$$

$$ax + by = z' \text{ and } cy + az = x' \text{ and } bz + cx = y'. \qquad \text{(Answer: } o = o.\text{)}$$

$$ax + by = cz \text{ and } mx + ny = pz'.$$

THE CALCULI OF LAMBDA-CONVERSION (1941)

Chapter I

INTRODUCTORY

1. THE CONCEPT OF A FUNCTION. Underlying the formal calculi which we shall develop is the concept of a function, as it appears in various branches of mathematics, either under that name or under one of the synonymous names, "operation" or "transformation." The study of the general properties of functions, independently of their appearance in any particular mathematical (or other) domain, belongs to formal logic or lies on the boundary line between logic and mathematics. This study is the original motivation for the calculi — but they are so formulated that it is possible to abstract from the intended meaning and regard them merely as formal systems.

A *function* is a rule of correspondence by which when anything is given (as *argument*) another thing (the *value* of the function for that argument) may be obtained. That is, a function is an operation which may be applied on one thing (the argument) to yield another thing (the value of the function). It is not, however, required that the operation shall necessarily be applicable to everything whatsoever; but for each function there is a class, or range, of possible arguments — the class of things to which the operation is significantly applicable — and this we shall call the *range of arguments*, or *range of the independent variable*, for that function. The class of all values of the function, obtained by taking all possible arguments, will be called the *range of values*, or *range of the dependent variable*.

If f denotes a particular function, we shall use the notation (fa) for the value of the function f for the argument a. If a does not belong to the range of arguments of f, the notation (fa) shall be meaningless.

It is, of course, not excluded that the range of arguments or range of values of a function should consist wholly or partly of functions. The derivative, as this notation appears in the elementary differential calculus, is a familiar mathematical example of a function for which both ranges consist of functions. Or, turning to the integral calculus, if in the expression $\int_0^1 (fx)dx$ we take the function f as the independent variable, we are led to a function for which the range of arguments consists of functions and the range of values, of numbers. Formal logic provides other examples; thus the existential quantifier, according to the present account, is a function for which the range of arguments consists of propositional functions, and the range of values consists of truth-values.

In particular it is not excluded that one of the elements of the range of arguments of a function f should be the function f itself. This possibility has frequently been denied, and indeed, if a function is defined as a correspondence between two previously given ranges, the reason for the denial is clear. Here, however, we regard the operation or rule of correspondence, which constitutes the function, as being first given, and the range of arguments then determined as consisting of the things to which the operation

---

Originally published in ***Annals of Mathematics Studies***, no. 6, Princeton University Press, 1941.

is applicable. This is a departure from the point of view usual in mathematics, but it is a departure which is natural in passing from consideration of functions in a special domain to the consideration of function in general, and it finds support in consistency theorems which will be proved below.

The *identity function* I is defined by the rule that (Ix) is x, whatever x may be; then in particular (II) is I. If a function H is defined by the rule that (Hx) is I, whatever x may be, then in particular (HH) is I. If Σ is the existential quantifier, then $(\Sigma\Sigma)$ is the truth-value *truth*.

The functions I and H may also be cited as examples of functions for which the range of arguments consists of all things whatsoever.

2. EXTENSION AND INTENSION. The foregoing discussion leaves it undetermined under what circumstances two functions shall be considered the same.

The most immediate and, from some points of view, the best way to settle this question is to specify that two functions f and g are the same if they have the same range of arguments and, for every element a that belongs to this range, (fa) is the same as (ga). When this is done we shall say that we are dealing with *functions in extension*.

It is possible, however, to allow two functions to be different on the ground that the rule of correspondence is different in meaning in the two cases although always yielding the same result when applied to any particular argument. When this is done we shall say that we are dealing with *functions in intension*. The notion of difference in meaning between two rules of correspondence is a vague one, but, in terms of some system of notation, it can be made exact in various ways. We shall not attempt to decide what is the true notion of difference in meaning but shall speak of functions in intension in any case where a more severe criterion of identity is adopted than for functions in extension. There is thus not one notion of function in intension, but many notions, involving various degrees of intensionality.

In the calculus of λ-conversion and the calculus of restricted λ-K-conversion, as developed below, it is possible, if desired, to interpret the expressions of the calculus as denoting functions in extension. However, in the calculus of λ-δ-conversion, where the notion of identity of functions is introduced to the system by the symbol δ, it is necessary, in order to preserve the finitary character of the transformation rules, so to formulate these rules that an interpretation by functions in extension becomes impossible. The expressions which appear in the calculus of λ-δ-conversion are interpretable as denoting functions in intension of an appropriate kind.

3. FUNCTIONS OF SEVERAL VARIABLES. So far we have tacitly restricted the term "function" to functions of one variable (or, of one argument). It is desirable, however, for each positive integer n, to have the notion of a function of n variables. And, in order to avoid the introduction of a separate primitive idea for each n, it is desirable to find a means of explaining functions of n variables as particular cases of functions of one variable. For our present purpose, the most convenient and natural method of doing this is to adopt an idea of Schönfinkel [49], according to which a function of two variables is regarded as a function of one variable whose values are functions of one variable, a function of three variables as a function of one variable whose values are functions of two variables, and so on.

Thus if f denotes a particular function of two variables, the notation $((fa)b)$—which we shall frequently abbreviate as (fab) or fab—represents the value of f for the arguments a, b. The notation (fa)—which we shall frequently abbreviate as fa—represents the function of one variable, whose value for any argument x is fax. The function f has a range of arguments, and the notation fa is meaningful only when a belongs to that range; the function fa again has a range of arguments, which is, in general, different for different elements a, and the notation fab is meaningful only when b belongs to the range of arguments of fa.

Similarly, if f denotes a function of three variables, $(((fa)b)c)$ or $fabc$ denotes the value of f for the arguments a, b, c, fa denoting a certain function of two variables, and $((fa)b)$ or fab denoting a certain function of one variable—and so on.

(According to another scheme, which is the better one for certain purposes, a function of two variables is regarded as a function (of one variable) whose arguments are ordered pairs, a function of three variables as a function whose arguments are ordered triads, and so on. This other concept of a function of several variables is not, however, excluded here. For, as will appear below, the notions of ordered pair, ordered triad, etc., are definable by means of abstraction (§4) and the Schönfinkel concept of a function of several variables; and thus functions of several variables in the other sense are also provided for.)

An example of a function of two variables (in the sense of Schönfinkel) is the *constancy function* K, defined by the rule that Kxy is x, whatever x and y may be. We have, for instance that KII is I, KHI is H, and so on. Also KI is H (where H is the function defined above in §1). Similarly KK is a function whose value is constant and equal to K.

Another example of a function of two variables is the function whose value for the arguments f, x is (fx); for reasons which will appear later we designate this function by the symbol 1. The function 1, regarded as a function of one variable, is a kind of identity function, since the notation $(1f)$ whenever significant, denotes the same function as f; the functions I and 1 are not, however, the same function, since the range of arguments consists in one case of all things whatever, in the other case merely of all functions.

Other examples of functions of two or more variables are the function H, already defined, and the functions T, J, B, C, W, S, defined respectively by the rules that Txf is (fx), $Jfxyz$ is $fx(fzy)$, $Bfgx$ is $f(gx)$, $Cfxy$ is (fyx), Wfx is (fxx), $Snfx$ is $f(nfx)$.

Of these, B and C may be more familiar to the reader under other names, as the *product* or *resultant* of two transformations f and g, and as the *converse* of a function of two variables f. To say that BII is I is to say that the product of the identity transformation by the identity transformation is the identity transformation, whatever the domain within which the transformations are being considered; to say that $B11$ is 1 is to say that within any domain consisting entirely of functions the product of the identity transformation by itself is the identity transformation. BI is 1, since it is the operation of composition with the identity transformation, and thus an identity operation, but one applicable only to transformations.

The reader may further verify that CK is H, CT is 1, $C1$ is T, CI is T—that 1 and I have the same converse is explained by the fact that, while not the same function, they have the same effect in all cases where they can significantly be applied to two arguments. The function BCC, the converse of the converse, has the effect of an identity when applied to a function of two variables, but when applied to a function of one variable it has the effect of so restricting the range of arguments as to transform the function into a function of two variables (if possible); thus $BCCI$ is 1.

There are many similar relations between these functions, some of them quite complicated.

4. ABSTRACTION. For our present purpose it is necessary to distinguish carefully between a symbol or expression which denotes a function and an expression which contains a variable and denotes ambiguously some value of the function — a distinction which is more or less obscured in the usual language of mathematical function theory.

To take an example from the theory of functions of natural numbers, consider the expression $(x^2+x)^2$. If we say, "$(x^2+x)^2$ is greater than 1,000," we make a statement which depends on x and actually has no meaning unless x is determined as some particular natural number. On the other hand, if we say, "$(x^2+x)^2$ is a primitive recursive function," we make a definite statement whose meaning in no way depends on a determination of the variable x (so that in this case x plays the rôle of an apparent, or bound, variable). The difference between the two cases is that in the first expression $(x^2+x)^2$ serves as an ambiguous, or variable, denotation of a natural number, while in the second case it serves as the denotation of a particular function. We shall hereafter distinguish by using $(x^2+x)^2$ when we intend an ambiguous denotation of a natural number, but $(\lambda x(x^2+x)^2)$ as the denotation of the corresponding function — and likewise in other cases.

(It is, of course, irrelevant here that the notation $(x^2+x)^2$ is commonly used also for a certain function of real numbers, a certain function of complex numbers, etc. In a logically exact notation the functions, addition of natural numbers, addition of real number, addition of complex numbers, would be denoted by different symbols, say $+_{\mathrm{n}}, +_{\mathrm{r}}, +_{\mathrm{c}}$; and the three functions, square of a natural number, square of a real number, square of a complex number, would be similarly distinguished. The uncertainty as to the exact meaning of the notation $(x^2+x)^2$, and the consequent uncertainty as to the range of arguments of the function $(\lambda x(x^2+x)^2)$, would then disappear.)

In general, if M is an expression containing a variable x (as a free variable, i.e., in such a way that the meaning of M depends on a determination of x), then $(\lambda x\mathsf{M})$ denotes a function whose values, for an argument a, is denoted by the result of substituting (a symbol denoting) a for x in M. The range of arguments of the function $(\lambda x\mathsf{M})$ consists of all objects a such that the expression M has a meaning when (a symbol denoting) a is substituted for x.

If M does not contain the variable x (as a free variable), then $(\lambda x\mathsf{M})$ might be used to denote a function whose value is constant and equal to (the thing denoted by) M, and whose range of arguments consists of all things. This usage is contemplated below in connection with the calculi of λ-K-conversion, but is excluded from the calculi of λ-conversion and λ-δ-conversion — for technical reasons which will appear.

Notice that, although x occurs as a free variable in M, nevertheless, in the expression $(\lambda x\mathsf{M})$, x is a bound, or apparent, variable. Example: the equation $(x^2+x)^2 = (y^2+y)^2$ expresses a relation between the natural numbers denoted by x and y and its truth depends on a determination of x and of y (in fact, it is true if and only if x and y are determined as denoting the same natural number); but the equation $(\lambda x(x^2 + x)^2) = (\lambda y(y^2 + y)^2)$ expresses a particular proposition—namely that $(\lambda x(x^2 + x)^2)$ is the same function as $(\lambda y(y^2 + y)^2)$—and it is true (there is no question of a determination of x and y).

Notice also that λ, or λx, is not the name of any function or other abstract object, but is an *incomplete symbol*—i.e., the symbol has no meaning alone, but appropriately formed expressions containing the symbol have a meaning. We call the symbol λx an *abstraction operator*, and speak of the function which is denoted by $(\lambda x\mathsf{M})$ as obtained from the expression M by *abstraction*.

The expression $(\lambda x(\lambda y\mathsf{M}))$, which we shall often abbreviate as $(\lambda xy\,.\,\mathsf{M})$, denotes a function whose value, for any argument denoted by x, is denoted by $(\lambda y\mathsf{M})$—thus a function whose values are functions, or a function of two variables. The expression $(\lambda y(\lambda x\mathsf{M}))$, abbreviated as $(\lambda yx\,.\,\mathsf{M})$, denotes the converse function to that denoted by $(\lambda xy\,.\,\mathsf{M})$. Similarly $(\lambda x(\lambda y(\lambda z\mathsf{M})))$, abbreviated as $(\lambda xyz\,.\,\mathsf{M})$, denotes a function of three variables, and so on.

Functions introduced in previous sections as examples can now be expressed, if desired, by means of abstraction operators. For instance, I is (λxx); J is $(\lambda fxyz\,.\,fx(fzy))$; S is $(\lambda nfx\,.\,f(nfx))$; H is (λxI), or $(\lambda x(\lambda yy))$, or $(\lambda xy\,.\,y)$; K is $(\lambda xy\,.\,x)$; 1 is $(\lambda fx\,.\,fx)$.

Chapter II

LAMBDA-CONVERSION

5. PRIMITIVE SYMBOLS, AND FORMULAS. We turn now to the development of a formal system, which we shall call *the calculus of λ-conversion*, and which shall have as a possible interpretation or application the system of ideas about functions described in Chapter I.

The primitive symbols of this calculus are three symbols,

$$\lambda, \quad (, \quad),$$

which we shall call *improper symbols*, and an infinite list of symbols

$$a, b, c, \ldots, x, y, z, \bar{a}, \bar{b}, \ldots, \bar{z}, \bar{\bar{a}}, \ldots,$$

which we shall call *variables*. The order in which the variables appear in this originally given infinite list shall be called their *alphabetical order*.

A *formula* is any finite sequence of primitive symbols. Certain formulas are distinguished as *well-formed formulas*, and each occurrence of a variable in a well-formed formula is distinguished as *free* or *bound*, in accordance with the following rules (1–4), which constitutes a definition of these terms by recursion:

1. A variable $\boldsymbol{x}$ is a well-formed formula, and the occurrence of the variable $\boldsymbol{x}$ in this formula is free.
2. If $\mathbf{F}$ and $\mathbf{A}$ are well-formed, $(\mathbf{FA})$ is well-formed, and an occurrence of a variable $\boldsymbol{y}$ in $\mathbf{F}$ is free or bound in $(\mathbf{FA})$ according as it is free or bound in $\mathbf{F}$, and an occurrence of a variable $\boldsymbol{y}$ in $\mathbf{A}$ is free or bound in $(\mathbf{FA})$ according as it is free or bound in $\mathbf{A}$.
3. If $\mathbf{M}$ is well-founded and contains at least one free occurrence of $\boldsymbol{x}$, then $(\lambda\boldsymbol{x}\mathbf{M})$ is well-formed, and an occurrence of a variable $\boldsymbol{y}$, other than $\boldsymbol{x}$, in $(\lambda\boldsymbol{x}\mathbf{M})$ is free or bound in $(\lambda\boldsymbol{x}\mathbf{M})$ according as it is free or bound in $\mathbf{M}$. All occurrences of $\boldsymbol{x}$ in $(\lambda\boldsymbol{x}\mathbf{M})$ are bound.
4. A formula is well-formed, and an occurrence of a variable in it is free, or is bound, only when this follows from 1–3.

The *free variables* of a formula are the variables which have at least one free occurrence in the formula. The bound variables of a formula are the variables which have at least one bound occurrence in the formula.

Hereafter (as was just done in the statement of the rules 1–4) we shall use bold capital letters to stand for variable or undetermined *formulas*, and bold small letters to stand for variable or undetermined *variables*. Unless otherwise indicated in a particular case, it is to be understood that formulas represented by bold capital letters are well-formed formulas. Bold letters are thus not part of the calculus which we are developing but a device for use in *talking about* the calculus: they belong, not to the system itself, but to the *metamathematics* or *syntax* of the system.

Another syntactical notation which we shall use is the notation,

$$S^{\boldsymbol{x}}_{\mathbf{N}}\mathbf{M}|$$

which shall stand for the formula which results by substitution of $\mathbf{N}$ for $\boldsymbol{x}$ throughout $\mathbf{M}$. This formula is well-formed, except in the case that $\boldsymbol{x}$ is a bound variable of $\mathbf{M}$ and

$\mathbf{N}$ is other than a single variable — see §7. (In the special case that $\boldsymbol{x}$ does not occur in $\mathbf{M}$, it is the same formula as $\mathbf{M}$.)

For brevity and perspicuity in dealing with particular well-formed formulas, we often do not write them in full but employ various abbreviations.

One method of abbreviation is by means of a *nominal definition*, which introduces a particular new symbol to replace or stand for a particular well-formed formula. We indicate such a nominal definition by an arrow, pointing from the new symbol which is being introduced to the well-formed formula which it is to replace (the arrow may be read "stands for"). As an example we make at once the nominal definition:

$$I \to (\lambda aa).$$

This means that I will be used as an abbreviation for (λaa) — and consequently that (II) will be used as an abbreviation for $((\lambda aa)(\lambda aa))$, $(\lambda a(aI))$ as an abbreviation for $(\lambda a(a(\lambda aa)))$, etc.

Another method of abbreviation is by means of a *schematic definition*, which introduces a class of new expressions of a certain form, specifying a scheme according to which each of the new expressions stands for a corresponding well-formed formula. Such a schematic definition is indicated in a similar fashion by an arrow, but the expressions on each side of the arrow contain bold letters. When a bold small letter — one or several — occurs in the expression following the arrow (the definiens) but not in the expression preceding the arrow (the definiendum), the following convention is to be understood:

> $\boldsymbol{a}$ stands for the first variable in alphabetical order not otherwise appearing in the definiens, $\boldsymbol{b}$ stands for the second such variable in alphabetical order, $\boldsymbol{c}$ the third, and so on.

As examples, we make at once the following schematic definitions:

$$[\mathbf{M}+\mathbf{N}] \to (\lambda \boldsymbol{a}(\lambda \boldsymbol{b}((\mathbf{M}\boldsymbol{a})((\mathbf{N}\boldsymbol{a})\boldsymbol{b})))).$$

$$[\mathbf{M}\times\mathbf{N}] \to (\lambda \boldsymbol{a}(\mathbf{M}(\mathbf{N}\boldsymbol{a}))).$$

$$[\mathbf{M}^{\mathbf{N}}] \quad \to (\mathbf{NM}).$$

The first of these definitions means that, for instance $[x+y]$ will be used as an abbreviation for $(\lambda a(\lambda b((xa)((ya)b))))$, and $[a+c]$ will be used as an abbreviation for $(\lambda b(\lambda d((ab)((cb)d))))$, and $[I+I]$ as an abbreviation for $(\lambda b(\lambda c((Ib)((Ib)c))))$, etc.

As a further device of abbreviation, we shall allow the omission of the parentheses ( ) in $(\mathbf{FA})$ when this may be done without ambiguity, whether $(\mathbf{FA})$ is the entire formula being written or merely some part of it. In restoring such omitted parentheses, the convention is to be followed that association is to the left (cf. Schönfinkel [49], Curry [17]). For example, fxy is an abbreviation of $((fx)y)$, $f(xy)$ is an abbreviation of $(f(xy))$, $fxyz$ is an abbreviation of $(((fx)y)z)$, $f(xy)z$ is an abbreviation of $((f(xy))z)$, $f(\lambda xx)y$ is an abbreviation of $((f(\lambda xx))y)$, etc.

In expressions which (in consequence of schematic definitions) contain brackets [ ], we allow a similar omission of brackets, subject to a similar convention of association to the left; thus $x+y+z$ is an abbreviation for $[[x+y]+z]$, which expression is in turn an abbreviation for a certain well-formed formula in accordance with the schematic definition already introduced. Moreover we allow, as an abbreviation, omitting a pair of brackets and at the same time putting a dot or period in the place of the original bracket [; in this case the convention, instead of association to the left, is that the omitted bracket extends from the bold period as far to the right as possible, consistently with the formula's being well-formed — so that, for instance, $x + \mathbf{.}\, y + z$ is an abbreviation for $[x + [y + z]]$, and $x + \mathbf{.}\, y + \mathbf{.}\, z + t$ is an abbreviation for $[x + [y + [z + t]]]$, and $(\lambda x \,\mathbf{.}\, x + x)$ is an abbreviation for $(\lambda x[x + x])$.

We also introduce the following schematic definitions:

$$(\lambda \boldsymbol{x} \,\mathbf{.}\, \mathsf{FA}) \rightarrow (\lambda \boldsymbol{x}(\mathsf{FA})),$$
$$(\lambda \boldsymbol{xy} \,\mathbf{.}\, \mathsf{FA}) \rightarrow (\lambda \boldsymbol{x}(\lambda \boldsymbol{y}(\mathsf{FA}))),$$
$$(\lambda \boldsymbol{xyz} \,\mathbf{.}\, \mathsf{FA}) \rightarrow (\lambda \boldsymbol{x}(\lambda \boldsymbol{y}(\lambda \boldsymbol{z}(\mathsf{FA})))),$$

and so on for any number of variables $\boldsymbol{x}, \boldsymbol{y}, \boldsymbol{z}, \ldots$ (which must be all different). And we allow similar omission of λ's, preceding a bold period which represents an omitted bracket in the way described in the previous paragraph — using, e.g., $\lambda xyz \,\mathbf{.}\, x + y + z$ as an abbreviation for $(\lambda x(\lambda y(\lambda z[[x + y] + z])))$.

Finally, we allow omission of the outside parentheses in $(\lambda \boldsymbol{x}\mathsf{M})$, or in $(\lambda \boldsymbol{x} \,\mathbf{.}\, \mathsf{FA})$, or $(\lambda \boldsymbol{xy} \,\mathbf{.}\, \mathsf{FA})$, or $(\lambda \boldsymbol{xyz} \,\mathbf{.}\, \mathsf{FA})$, etc., when this is the entire formula being written — but not when one of these expressions appears as a proper part of a formula.

Hereafter, in writing definitions, we shall abbreviate the definiens in accordance with previously introduced abbreviations and definitions. Thus the definition of $[\mathsf{M} + \mathsf{N}]$ would now be written:

$$[\mathsf{M} + \mathsf{N}] \rightarrow \lambda \boldsymbol{ab} \,\mathbf{.}\, \mathsf{M}\boldsymbol{a}(\mathsf{N}\boldsymbol{ab}).$$

> Definitions and other abbreviations are introduced merely as matters of convenience and are not properly part of the formal system at all. When we speak of the free variables of a formula, the bound variables of a formula, the length (number of symbols) of a formula, the occurrence of one formula as a part of another, etc., the reference is always to the unabbreviated form of the formulas in question.

The introduction and use of definitions and other abbreviations is, of course, subject to the restriction that there shall never be any ambiguity as to what formula a given abbreviated form stands for. In practice certain further restrictions are also desirable, e.g., that all free variables of the definiens be represented explicitly in the definiendum. Exact formulation of these restrictions is unnecessary for our present purpose, since all definitions and abbreviations are extraneous to the formal system, as just explained, and in principle dispensable.

6. CONVERSION. We introduce now the three following operations, or transformation rules, on well-formed formulas:

I. To replace any part M of a formula by $S^{\boldsymbol{x}}_{\boldsymbol{y}}\mathsf{M}|$, provided that $\boldsymbol{x}$ is not a free variable of M and $\boldsymbol{y}$ does not occur in M.

II. To replace any part $((\lambda \boldsymbol{x}\mathsf{M})\mathsf{N})$ of a formula by $S^{\boldsymbol{x}}_{\mathsf{N}}\mathsf{M}|$, provided that the bound variables of M are distinct both from $\boldsymbol{x}$ and from the free variables of N.

III. To replace any part $S^x_{\mathsf{N}}\mathsf{M}|$ of a formula by $((\lambda x\mathsf{M})\mathsf{N})$, provided that $((\lambda x\mathsf{M})\mathsf{N})$ is well-formed and the bound variables of M are distinct both from x and from the free variables of N.

In the statement of these rules—and hereafter generally—it is to be understood that the word *part* (of a formula) means *consecutive well-formed part not immediately following an occurrence of the symbol* λ.

When the same formula occurs several times as such a part of another formula, each occurrence is to be counted as a different part. Thus, for instance, Rule I may be used to transform $ab(\lambda aa)(\lambda aa)$ into $ab(\lambda bb)(\lambda aa)$. Rule III may be used to transform λaa into $\lambda a\, .(\lambda aa)a$. But Rule III may not be used to transform (λaa) into $(\lambda((\lambda aa)a)a)$—the latter formula is, in fact, not even well-formed.

Rules I–III have the important property that they are effective or "definite," i.e., there is a means of always determining of any two formulas **A** and **B** whether **A** can be transformed into **B** by application of one of the rules (and, if so, of which one).

If **A** can be transformed into **B** by an application of one of the Rules I–III, we shall say that **A** is *immediately convertible* into **B** (abbreviation, "**A** imc **B**"). If there is a finite sequence of formulas, in which **A** is the first formula and **B** the last, and in which each formula except the last is immediately convertible into the next one, we shall say that **A** is *convertible* into **B** (abbreviation, "**A** conv **B**"); and the process of obtaining **B** from **A** by a particular finite sequence of applications of Rules I–III will be called a *conversion* of **A** into **B** (no reference is intended to conversion in the sense of forming the converse—but for the corresponding noun we use, not "converse," but "convert"). It is not excluded that the number of applications of Rules I–III in a conversion of **A** into **B** should be zero, **B** being then the same formula as **A**.

The relation which holds between **A** and **B** when **A** conv **B** will be called *interconvertibility*, and we shall use the expression "**A** and **B** are interconvertible" as synonymous with "**A** conv **B**." The relation of interconvertibility is transitive, symmetric, and reflexive—symmetric because Rules II and III are inverses of each other and Rule I is its own inverse.

If there is a conversion of **A** into **B** which contains no application of Rule II or Rule III, we shall say that **A** is *convertible-I* into **B** (**A** conv-I **B**). Similarly we define "**A** conv-I-II **B**" and "**A** conv-I-III **B**."

A conversion which contains no application of Rule II and exactly one application of Rule III will be called an *expansion*. A conversion which contains no application of Rule III and exactly one application of Rule II will be called a *reduction*. If there is a reduction of **A** into **B**, we shall say that **A** is *immediately reducible* to **B** (**A** imr **B**). If there is a conversion of **A** into **B** which consists of one or more successive reductions, we shall say that **A** is *reducible* to **B** (**A** red **B**). (The meaning of "**A** red **B**" thus differs from that of "**A** conv-I-II **B**" only in that the former implies the presence of at least one application of Rule II in the conversion of **A** into **B**.)

An application of Rule II to a formula will be called a *contraction* of the part $((\lambda x\mathsf{M})\mathsf{N})$ which is affected.

A well-formed formula will be said to be in *normal form* if it contains no part of the form $((\lambda x\mathsf{M})\mathsf{N})$. We shall call **B** a *normal form of* **A** if **B** is in normal form and **A** conv **B**. We shall say that **A** *has a normal form* if there is a formula **B** which is a normal form of **A**.

A well-formed formula will be said to be in *principal normal form* if it is in normal form, and no variable is both a bound variable and free variable of it, and the first bound variable occurring in it (in the left-to-right order of the symbols which compose the formula) is the same as the first variable in the alphabetical order which is not a free variable of it, and the symbols which occur in it immediately following the symbol λ are, when taken in the order in which they occupy in the formula, in alphabetical order, without repetitions, and without omissions except of variables which are free variables of the formula. For example, $\lambda ab \, . \, ba$, and $\lambda a \, . \, a(\lambda c \, . \, bc)$, and $\lambda b \, . \, ba$ are in principal normal form; and $\lambda ac \, . \, ca$, and $\lambda bc \, . \, cb$, and $\lambda a \, . \, a(\lambda a \, . \, ba)$ are in normal form but not in principal normal form.

We shall call **B** a *principal normal form* of **A** if **B** is in principal normal form and **A** conv **B**. A formula in normal form is always convertible-I into a corresponding formula in principal normal form, and hence every formula which has a normal form has a principal normal form. We shall show in the next section that the principal normal form of a formula, if it exists, is unique.

An example of a formula which has no normal form (and therefore no principal normal form) is $(\lambda x \, . \, xxx)(\lambda x \, . \, xxx)$.

It is intended that, in any interpretation of the formal calculus, only those well-formed formulas which have a normal form shall be meaningful, and, among these, interconvertible formulas shall have the same meaning. The condition of being well-formed is thus a necessary condition for meaningfulness but not a sufficient condition.

It is important that the condition of being well-formed is *effective* in the sense explained at the beginning of this section, whereas the condition of being well-formed and having a normal form is not effective.

7. FUNDAMENTAL THEOREMS ON WELL-FORMED FORMULAS AND ON THE NORMAL FORM. The following theorems are taken from Kleene [34] (with non-essential changes to adapt them to the present modified notation). Their proof is left to the reader; or an outline of the proof may be found in Kleene, loc. cit.

7 I. In a well-formed formula **K** there exists a unique pairing of the occurrences of the symbol (, each with a corresponding occurrence of the symbol), in such a way that two portions of **K**, each lying between an occurrence of (and the corresponding occurrence of) inclusively, either are non-overlapping or else are contained one entirely within the other. Moreover, if such a pairing exists in the portions of **K** lying between the nth and the $(n+r)$th symbol of **K** inclusively, it is a part of the pairing in **K**.

7 II. A necessary and sufficient condition that the portion **N** of a well-formed formula **K** which lies between a given occurrence of (in **K** and a given occurrence of) in **K** inclusively be well-formed is that the given occurrence of (and the given occurrence of) correspond.

7 III. Every well-formed formula has one of the three forms, $\boldsymbol{x}$, where $\boldsymbol{x}$ is a variable, or (**FA**), where **F** and **A** are well-formed, or ($\lambda\boldsymbol{x}$**M**), where **M** is well-formed and $\boldsymbol{x}$ is a free variable of **M**.

7 IV. If (**FA**) and either **F** or **A** is well-formed, then both **F** and **A** are well-formed.

7 V. If ($\lambda\boldsymbol{x}$**M**) is well-formed, $\boldsymbol{x}$ being a variable, then **M** is well-formed and $\boldsymbol{x}$

is a free variable of **M**.

7 VI. A well-formed formula can be of the form (**FA**), where **F** (or **A**) is well-formed, in only one way.

7 VII. A well-formed formula can be of the form (λx**M**), where x is a variable, in only one way.

7 VIII. If **P** and **Q** are well-formed parts of a well-formed formula **K**, then either **P** is a part of **Q**, or **Q** is a part of **P**, or **P** and **Q** are non-overlapping.

7 IX. Two distinct occurrences of the same well-formed formula **P** as part of a well-formed formula **K** must be non-overlapping.

7 X. If **P**, **F**, and **A** are well-formed and **P** is a part of (**FA**), then **P** is (**FA**) or **P** is a part of **F** or **P** is a part of **A**.

7 XI. If **P** and **M** are well-formed and x is a variable and **P** is a part of (λx**M**), then **P** is (λx**M**) [or **P** is x] or **P** is a part of **M**. (The clause in brackets is superfluous because of the meaning we give to the word *part* of a formula—see §6.)

7 XII. An occurrence of a variable x in a well-formed formula is bound or free according as it is or is not an occurrence in a well-formed part of **K** of the form (λx**M**). (Hence, in particular, no occurrence of a variable in a well-formed formula is both bound and free.)

7 XIII. If **M** is well-formed and the variable x is not a free variable of **M** and the variable y does not occur in **M**, then S^x_y**M**| is well-formed and has the same free variables as **M**.

7 XIV. If **M** and **N** are well-formed and the variable x occurs in **M** and the bound variables of **M** are distinct both from x and from the free variables of **N**, then $S^x_{\mathbf{N}}$**M**| and ((λx**M**)**N**) are well-formed and have the same free variables.

7 XV. If **K**, **P**, **Q** are well-formed and all free variables of **P** are also free variables of **Q**, the formula obtained by substituting **Q** for a particular occurrence of **P** in **K**, not immediately following an occurrence of λ, is well-formed.

7 XVI. If **A** is well-formed and **A** conv **B**, then **B** is well-formed.

7 XVII. If **A** is well-formed and **A** conv **B**, then **A** and **B** have the same free variables.

7 XVIII. If **K**, **P** and **Q** are well-formed, and **P** conv **Q**, and **L** is obtained by substituting **Q** for a particular occurrence of **P** in **K**, not immediately following an occurrence of λ, then **K** conv **L**.

We shall call a well-formed part **P** of a well-formed formula **K** a *free* occurrence of **P** in **K** if every free occurrence of a variable in **P** is also a free occurrence of that variable in **K**; in the contrary case (if some free occurrence of a variable in **P** is at the same time a bound occurrence of that variable in **K**) we shall call the part **P** of **K** a *bound* occurrence of **P** in **K**. If **P** is an occurrence of a variable in **K**, not immediately following an occurrence of λ, this definition is in agreement with our previous definition of free and bound occurrence of variables.

Moreover we shall extend the notation $S^x_{\mathbf{N}}$**M**| introduced in §5 by allowing $S^{\mathbf{P}}_{\mathbf{N}}$**M**|

to stand for the result of substituting $\mathbf{N}$ for $\mathbf{P}$ throughout $\mathbf{M}$, where $\mathbf{N}, \mathbf{P}, \mathbf{M}$ are any well-formed formulas. This is possible without ambiguity, by 7 IX.

7 XIX. A well-formed part $\mathbf{P}$ of a well-formed formula $\mathbf{K}$ is a bound or free occurrence of $\mathbf{P}$ in $\mathbf{K}$ according as it is or is not an occurrence in a well-formed part of $\mathbf{K}$ of the form $(\lambda x\mathbf{M})$ where x is a free variable of $\mathbf{P}$.

7 XX. If $\mathbf{K}, \mathbf{P}, \mathbf{Q}$ are well-formed, the formula obtained by substituting $\mathbf{Q}$ for a particular free occurrence of $\mathbf{P}$ in $\mathbf{K}$ is well-formed.

7 XXI. If $\mathbf{K}, \mathbf{P}, \mathbf{Q}$ are well-formed and there is no bound occurrence of $\mathbf{P}$ in $\mathbf{K}$, then $S^{\mathbf{P}}_{\mathbf{Q}}\mathbf{K}|$ is well-formed.

7 XXII. Let x be a free variable of the well-formed formula $\mathbf{N}$ and let $\mathbf{P}$ be the formula obtained by substituting $\mathbf{N}$ for the free occurrences of x in $\mathbf{M}$. If the resulting occurrences of $\mathbf{N}$ in $\mathbf{P}$ are free, $((\lambda x\mathbf{M})\mathbf{N})$ conv $\mathbf{P}$.

In what follows we shall frequently make tacit assumption of these theorems.

In stating these theorems, it has been necessary to hold in abeyance the convention that formulas represented by bold capital letters are well-formed. Hereafter this convention will be restored, and formulas so represented are to be taken always as well-formed.

We turn now to a group of theorems on conversion taken from Church and Rosser [16]. In order to state these, it is necessary first to define the notion of the *residuals* of a set of parts $((\lambda x_j\mathbf{M}_j)\mathbf{N}_j)$ of a formula $\mathbf{A}$ after a sequence of applications of Rules I and II to $\mathbf{A}$ (§6).

We assume that, if $p \neq q$, then $((\lambda x_p\mathbf{M}_p)\mathbf{N}_p)$ is not the same part of $\mathbf{A}$ as $((\lambda x_q\mathbf{M}_q)\mathbf{N}_q)$—though it may be the same formula. The parts $((\lambda x_j\mathbf{M}_j)\mathbf{N}_j)$ of $\mathbf{A}$ need not be all the parts of $\mathbf{A}$ which have the form $((\lambda y\mathbf{P})\mathbf{Q})$. The *residuals* of the $((\lambda x_j\mathbf{M}_j)\mathbf{N}_j)$ after a particular sequence of applications of Rules I and II to $\mathbf{A}$ are then certain parts, of the form $((\lambda y\mathbf{P})\mathbf{Q})$, of the formula into which $\mathbf{A}$ is converted by this sequence of applications of Rules I and II. They are defined as follows:

If the sequence of applications of Rules I and II in question is vacuous, each part $((\lambda x_j\mathbf{M}_j)\mathbf{N}_j)$ is its own residual.

If the sequence consists of a single application of Rule I, each part $((\lambda x_j\mathbf{M}_j)\mathbf{N}_j)$ is changed into a part $((\lambda y_j\mathbf{M}'_j)\mathbf{N}'_j)$ of the resulting formula, and this part $((\lambda y_j\mathbf{M}'_j)\mathbf{N}'_j)$ is the residual of $((\lambda x_j\mathbf{M}_j)\mathbf{N}_j)$.

If the sequence consists of a single application of Rule II, let $((\lambda x\mathbf{M})\mathbf{N})$ be the part of $\mathbf{A}$ which is contracted (§6), and let $\mathbf{A}'$ be the resulting formula into which $\mathbf{A}$ is converted. Let $((\lambda x_p\mathbf{M}_p)\mathbf{N}_p)$ be a particular one of the $((\lambda x_j\mathbf{M}_j)\mathbf{N}_j)$, and distinguish the six following cases.

Case 1: $((\lambda x\mathbf{M})\mathbf{N})$ and $((\lambda x_p\mathbf{M}_p)\mathbf{N}_p)$ do not overlap. Under the reduction of $\mathbf{A}$ to $\mathbf{A}'$, $((\lambda x_p\mathbf{M}_p)\mathbf{N}_p)$ goes into a definite part of $\mathbf{A}'$, which is the same formula as $((\lambda x_p\mathbf{M}_p)\mathbf{N}_p)$. This part of $\mathbf{A}'$ is the residual of $((\lambda x_p\mathbf{M}_p)\mathbf{N}_p)$.

Case 2: $((\lambda x\mathbf{M})\mathbf{N})$ is a part of $\mathbf{M}_p$. Under the reduction of $\mathbf{A}$ to $\mathbf{A}'$, $\mathbf{M}_p$ goes into a definite part $\mathbf{M}'_p$ of $\mathbf{A}$, which arises from $\mathbf{M}_p$ by contraction of $((\lambda x\mathbf{M})\mathbf{N})$, and $((\lambda x_p\mathbf{M}_p)\mathbf{N}_p)$ goes into the part $((\lambda x_p\mathbf{M}'_p)\mathbf{N}_p)$ of $\mathbf{A}'$. This part $((\lambda x_p\mathbf{M}'_p)\mathbf{N}_p)$ of $\mathbf{A}'$ is the residual of $((\lambda x_p \mathsf{M}_p)\mathsf{N}_p)$.

Case 3: $((\lambda x\mathsf{M})\mathsf{N})$ is a part of N_p. Under the reduction of A to A', N_p goes into a definite part N'_p of A, which arises from N_p by contraction of $((\lambda x\mathsf{M})\mathsf{N})$, and $((\lambda x_p \mathsf{M}_p)\mathsf{N}_p)$ goes into the part $((\lambda x_p \mathsf{M}_p)\mathsf{N}'_p)$ of A'. This part $((\lambda x_p \mathsf{M}_p)\mathsf{N}'_p)$ of A' is the residual of $((\lambda x_p \mathsf{M}_p)\mathsf{N}_p)$.

Case 4: $((\lambda x\mathsf{M})\mathsf{N})$ is $((\lambda x_p \mathsf{M}_p)\mathsf{N}_p)$. In this case $((\lambda x_p \mathsf{M}_p)\mathsf{N}_p)$ has no residual in A'.

Case 5: $((\lambda x_p \mathsf{M}_p)\mathsf{N}_p)$ is a part of M. Let M' be the result of replacing all x's of M except those occurring in $((\lambda x_p \mathsf{M}_p)\mathsf{N}_p)$ by N. Under these changes the part $((\lambda x_p \mathsf{M}_p)\mathsf{N}_p)$ of M goes into a definite part of M' which we shall denote also by $((\lambda x_p \mathsf{M}_p)\mathsf{N}_p)$, since it is the same formula. If now we replace $((\lambda x_p \mathsf{M}_p)\mathsf{N}_p)$ in M' by $S^x_{\mathsf{N}}((\lambda x_p \mathsf{M}_p)\mathsf{N}_p)|$, M' becomes $S^x_{\mathsf{N}}\mathsf{M}|$ and we denote by $S^x_{\mathsf{N}}((\lambda x_p \mathsf{M}_p)\mathsf{N}_p)|$ the particular occurrence of $S^x_{\mathsf{N}}((\lambda x_p \mathsf{M}_p)\mathsf{N}_p)|$ in $S^x_{\mathsf{N}}\mathsf{M}|$ that resulted from replacing $((\lambda x_p \mathsf{M}_p)\mathsf{N}_p)$ in M' by the formula $S^x_{\mathsf{N}}((\lambda x_p \mathsf{M}_p)\mathsf{N}_p)|$. Then the residual in A' of $((\lambda x_p \mathsf{M}_p)\mathsf{N}_p)$ in A is defined to be the part $S^x_{\mathsf{N}}((\lambda x_p \mathsf{M}_p)\mathsf{N}_p)|$ in the particular occurrence of $S^x_{\mathsf{N}}\mathsf{M}|$ in A' that resulted from replacing $((\lambda x\mathsf{M})\mathsf{N})$ in A by $S^x_{\mathsf{N}}\mathsf{M}|$.

Case 6: $((\lambda x_p \mathsf{M}_p)\mathsf{N}_p)$ is a part of N. Let $((\lambda y_i \mathsf{P}_i)\mathsf{Q}_i)$ respectively stand for the particular occurrence of the formula $((\lambda x_p \mathsf{M}_p)\mathsf{N}_p)$ in $S^x_{\mathsf{N}}\mathsf{M}|$ which are the part $((\lambda x_p \mathsf{M}_p)\mathsf{N}_p)$ in each of those particular occurrences of the formula N in $S^x_{\mathsf{N}}\mathsf{M}|$ that resulted from replacing the x's of M by N. Then the residuals in A' of $((\lambda x_p \mathsf{M}_p)\mathsf{N}_p)$ in A are the parts $((\lambda y_i \mathsf{P}_i)\mathsf{Q}_i)$ in the particular occurrence of the formula $S^x_{\mathsf{N}}\mathsf{M}|$ in A' that resulted from replacing $((\lambda x\mathsf{M})\mathsf{N})$ in A by $S^x_{\mathsf{N}}\mathsf{M}|$.

Finally, in the case of a sequence of two or more successive applications of Rules I, II to A, say A imc A' imc A'' imc . . . , we define residuals in A' of the parts $((\lambda x_j \mathsf{M}_j)\mathsf{N}_j)$ of A in the way just described, and we define the residuals in A'' of $((\lambda x_j \mathsf{M}_j)\mathsf{N}_j)$ of A to be the residuals of the residuals in A', and so on.

7 XXIII. After a sequence of applications of Rules I and II to A, under which A is converted into B, the residuals of the parts $((\lambda x_j \mathsf{M}_j)\mathsf{N}_j)$ of A are a set (possibly vacuous) of parts of B which each have the form $((\lambda y\mathsf{P})\mathsf{Q})$.

7 XXIV. After a sequence of applications of Rules I and II to A, no residual of the part $((\lambda x\mathsf{M})\mathsf{N})$ of A can coincide with a residual of the part $((\lambda x'\mathsf{M}')\mathsf{N}')$ of A unless $((\lambda x\mathsf{M})\mathsf{N})$ coincides with $((\lambda x'\mathsf{M}')\mathsf{N}')$.

We say that a sequence of reductions on A_1, say A_1 imr A_2 imr A_3 . . . imr A_{n+1}, is a *sequence of contractions on the parts* $((\lambda x_j \mathsf{M}_j)\mathsf{N}_j)$ *of* A_1 if the reduction from A_i to A_{i+1} $(i = 1, \ldots, n)$ involves a contraction of a residual of the $((\lambda x_j \mathsf{M}_j)\mathsf{N}_j)$. Moreover, if no residuals of the $((\lambda x_j \mathsf{M}_j)\mathsf{N}_j)$ occur in A_{n+1} we say that the sequence of contractions on the $((\lambda x_j \mathsf{M}_j)\mathsf{N}_j)$ *terminates* and that A_{n+1} is the result.

In some cases we wish to speak of a sequence of contractions on the parts $((\lambda x_j \mathsf{M}_j)\mathsf{N}_j)$ of A where the set $((\lambda x_j \mathsf{M}_j)\mathsf{N}_j)$ may be vacuous. To handle this we agree that, if the set $((\lambda x_j \mathsf{M}_j)\mathsf{N}_j)$ is vacuous, the sequence of contractions shall be a vacuous sequence of reductions.

7 XXV. If $((\lambda x_j \mathsf{M}_j)\mathsf{N}_j)$ are parts of A, then a number m can be found such that any sequence of contractions on the $((\lambda x_j \mathsf{M}_j)\mathsf{N}_j)$ will terminate after at most m contractions, and if A' and A'' are two results of terminating sequences

of contractions on the $((\lambda x_j \mathsf{M}_j)\mathsf{N}_j)$, then A' conv-I A''.

This is proved by induction on the length of A. It is trivially true if the length of A is 1 (i.e., if A consists of a single symbol), the number m being then 0. As hypothesis of induction, assume that the proposition is true of every formula A of length less than n. On this hypothesis we have to prove that the proposition is true of an arbitrary given formula A of length n. This we proceed to do, by means of a proof involving three cases.

Case 1: A has the form $\lambda x\mathsf{M}$. All the parts $((\lambda x_j \mathsf{M}_j)\mathsf{N}_j)$ of A must be parts of M. Since M is of length less than n, we apply the hypothesis of induction to M.

Case 2: A has the form FX, where FX is not one of the $((\lambda x_j \mathsf{M}_j)\mathsf{N}_j)$. All the parts $((\lambda x_j \mathsf{M}_j)\mathsf{N}_j)$ of A must be parts either of F or of X. Since F and X are each of length less than n, we apply the hypothesis of induction.

Case 3: A is $((\lambda x_p \mathsf{M}_p)\mathsf{N}_p)$, where $((\lambda x_p \mathsf{M}_p)\mathsf{N}_p)$ is one of the $((\lambda x_j \mathsf{M}_j)\mathsf{N}_j)$. By the hypothesis of induction, there is a number a such that any sequence of contractions on those $((\lambda x_j \mathsf{M}_j)\mathsf{N}_j)$ which are parts of M_p terminate after at most a contractions, and there is a number b such that any sequence of contractions on those $((\lambda x_j \mathsf{M}_j)\mathsf{N}_j)$ which are parts of N_p terminates after at most b contractions; moreover, if we start with the formula M_p and perform a terminating sequence of contractions on those $((\lambda x_j \mathsf{M}_j)\mathsf{N}_j)$ which are parts of M_p, the result is a formula M, which is unique to within application of Rule I, and which contains a certain number c, $\geqq 1$, of free occurrences of the variable x_p.

Now one way of performing a terminating sequence of contractions on the parts $((\lambda x_j \mathsf{M}_j)\mathsf{N}_j)$ of A is as follows. First perform a terminating sequence of contractions on those $((\lambda x_j \mathsf{M}_j)\mathsf{N}_j)$ which are parts of M_p, so converting A into $((\lambda t\mathsf{M})\mathsf{N}_p)$. Then there is one and only one residual of $((\lambda x_p \mathsf{M}_p)\mathsf{N}_p)$, namely the entire formula $((\lambda x\mathsf{M})\mathsf{N}_p)$. Perform a contraction of this, so obtaining

$$S^{t}_{\mathsf{N}_p}\mathsf{M}'|$$

where M' differs from M at most by applications of Rule I. Then in this formula there are c occurrences of N_p resulting from the substitution of N_p for t. Take each of these occurrences of N_p in order and perform a terminating sequence of contractions on the residuals of the $((\lambda x_j \mathsf{M}_j)\mathsf{N}_j)$ occurring in it.

Let us call such a terminating sequence of contractions on the parts $((\lambda x_j \mathsf{M}_j)\mathsf{N}_j)$ of A a *special* terminating sequence of contractions on the parts $((\lambda x_j \mathsf{M}_j)\mathsf{N}_j)$ of A. Clearly such a special sequence of contractions contains at most $a+1+cb$ contractions.

Consider now any sequence of contractions, μ, on the parts $((\lambda x_j \mathsf{M}_j)\mathsf{N}_j)$ of A. The part $((\lambda x_p \mathsf{M}_p)\mathsf{N}_p)$ of A will have just one residual (which will always be the entire formula) up to the point that a contraction of its residual occurs, and thereafter will have no residual; moreover, if the sequence of contractions is continued, a contraction of the residual of $((\lambda x_p \mathsf{M}_p)\mathsf{N}_p)$ must occur within at most $a+b+1$ contractions. Hence we may suppose, without loss of generality, that μ consists of a sequence of contractions, ϕ, on the $((\lambda x_j \mathsf{M}_j)\mathsf{N}_j)$ which are different from $((\lambda x_p \mathsf{M}_p)\mathsf{N}_p)$, followed by a contraction β_0 of the residuals of $((\lambda x_p \mathsf{M}_p)\mathsf{N}_p)$, followed by a sequence of contractions, θ, on the then remaining residuals of the $((\lambda x_j \mathsf{M}_j)\mathsf{N}_j)$. Clearly, ϕ can be replaced by a sequence of contractions, α_0, on the $((\lambda x_j \mathsf{M}_j)\mathsf{N}_j)$ which are parts of

$\mathbf{M}_p$, followed by a sequence of contractions, η, on the $((\lambda x_j \mathbf{M}_j)\mathbf{N}_j)$ which are parts of $\mathbf{N}_p$ — in the sense that α_0 followed by η gives the same end formulas as ϕ and the same set of residuals for each of the $((\lambda x_j \mathbf{M}_j)\mathbf{N}_j)$. Moreover, replacing ϕ by α_0 followed by η does not change the total number of contractions of residuals of parts $\mathbf{M}_p$ or of residuals of parts of $\mathbf{N}_p$. Next, η followed by β_0 can be replaced by a contraction β' of the residual $((\lambda y\mathbf{P})\mathbf{N}_p)$ of $((\lambda x_p\mathbf{M}_p)\mathbf{N}_p)$ followed by a set of applications of η on each of those occurrences of $\mathbf{N}_p$ in the resulting formula

$$S^{y}_{\mathbf{N}_p}\mathbf{P}'|$$

which arose by substituting $\mathbf{N}_p$ for y in $\mathbf{P}'$. (Here $\mathbf{P}'$ differs from $\mathbf{P}$ at most by applications of Rule I. Since η may be thought of as a transformation of the formula $\mathbf{N}_p$, the convention will be understood which we use when we speak of the sequence of reductions of a given formula which results from applying η to a particular occurrence of $\mathbf{N}_p$ in that formula.)

By this means the sequence of contractions, μ, is replaced by a sequence of contractions, μ', which consists of a sequence of contractions, α_0, on the $((\lambda x_j \mathbf{M}_j)\mathbf{N}_j)$ which are parts of $\mathbf{M}_p$, followed by a contraction β' of the residual of $((\lambda x_p\mathbf{M}_p)\mathbf{N}_p)$, followed by further contractions on the then remaining residuals of the $((\lambda x_j \mathbf{M}_j)\mathbf{N}_j)$.

Consider now the part ζ of μ', consisting of β' and the contractions that follow it, up to and including the first contraction of a residual of a part of $\mathbf{M}_p$. Denoting the formula on which ζ acts by $((\lambda y\mathbf{P})\mathbf{N}_p)$, we see that ζ can be considered as the act of first replacing the free y's of $\mathbf{P}'$ by various formulas $\mathbf{N}_{pk}$, got from $\mathbf{N}_p$ by various sequences of reductions (which may be vacuous), and then (possibly after some applications of Rule I) contracting a residual $((\lambda z\mathbf{R})\mathbf{S})$ of one of the $((\lambda x_j \mathbf{M}_j)\mathbf{N}_j)$ which are parts of $\mathbf{M}_p$, say $((\lambda x_q\mathbf{M}_q)\mathbf{N}_q)$. From this point of view, we see that none of the free x's of $\mathbf{R}$ are parts of any $\mathbf{N}_{pk}$, and hence ζ can be replaced by a contraction (possibly after some application of Rule I) of that residual in $\mathbf{P}$ of $((\lambda x_q\mathbf{M}_q)\mathbf{N}_q)$ of which $((\lambda z\mathbf{R})\mathbf{S})$ is a residual, followed by a contraction (possibly after some applications of Rule I) of the residuals of $((\lambda x_p\mathbf{M}_p)\mathbf{N}_p)$, followed by a sequence of contractions on residuals of parts of $\mathbf{N}_p$.

If μ' is altered by replacing ζ in this way, the result is a sequence of contractions, μ'', having the same form as μ', but having the property that after the contraction of the residual of $((\lambda x_p\mathbf{M}_p)\mathbf{N}_p)$ one less contraction of residuals of parts of $\mathbf{M}_p$ occurs.

By repetition of this process, μ is finally replaced by a sequence of contractions ν, which consists of a sequence of contractions, α, on the $((\lambda x_j \mathbf{M}_j)\mathbf{N}_j)$ which are parts of $\mathbf{M}_p$, followed by a contraction β of the residual of $((\lambda x_p\mathbf{M}_p)\mathbf{N}_p)$, followed by a sequence of contractions γ on residuals of the $((\lambda x_j \mathbf{M}_j)\mathbf{N}_j)$ which are parts of $\mathbf{N}_p$. Moreover, ν contains at least as many contractions as μ — for in the process of obtaining ν from μ there is no step which can decrease the number of contractions. The sequence of contractions, α, contains at most a contractions, and γ contains at most cb contractions. Thus ν, and consequently μ, contains at most $a+1+cb$ contractions.

Thus we have proved that any sequence of contractions on the parts $((\lambda x_j \mathbf{M}_j)\mathbf{N}_j)$ of $\mathbf{A}$ will terminate after at most $a+1+cb$ contractions.

Now suppose that μ is a *terminating* sequence of contractions. Then ν either is a *special* terminating sequence of contractions (see above) or can be made so by some evident changes in the order in which the contractions in γ are performed. By the hypothesis of induction, applied to $\mathbf{M}_p$ and $\mathbf{N}_p$, the result of a special terminating sequence of contractions is unique to within possible application of Rule I. Therefore the result of any terminating sequence of contractions, μ, is unique to within possible applications of Rule I.

7 XXVI. If $\mathbf{A}$ imr $\mathbf{B}$ by a contraction of the part $((\lambda x\mathbf{M})\mathbf{N})$ of $\mathbf{A}$, and $\mathbf{A}_1$ is $\mathbf{A}$, and $\mathbf{A}_1$ imr $\mathbf{A}_2$, $\mathbf{A}_2$ imr $\mathbf{A}_3, \ldots$, and, for all k, $\mathbf{B}_k$ is the result of a terminating sequence of contractions on the residuals in $\mathbf{A}_k$ of $((\lambda x\mathbf{M})\mathbf{N})$, then:

(1) $\mathbf{B}_1$ is $\mathbf{B}$.

(2) For all k, $\mathbf{B}_k$ conv-I-II $\mathbf{B}_{k+1}$.

(3) Even if the sequence $\mathbf{A}_1, \mathbf{A}_2, \ldots$ can be continued to infinity, there is a number u_m, depending on the formula $\mathbf{A}$, the part $((\lambda x\mathbf{M})\mathbf{N})$ of $\mathbf{A}$, and the number m, such that, starting with $\mathbf{B}_m$, at most u_m consecutive $\mathbf{B}_k$'s occur for which it is not true that $\mathbf{B}_k$ red $\mathbf{B}_{k+1}$.

(1) is obvious.

To prove (2), let $((\lambda y_i \mathbf{P}_i)\mathbf{Q}_i)$ be the residuals in $\mathbf{A}_k$ of $((\lambda x\mathbf{M})\mathbf{N})$ and let the reduction of $\mathbf{A}_k$ into $\mathbf{A}_{k+1}$ involve a contraction of (a residual of) the part $((\lambda z\mathbf{R})\mathbf{S})$ of $\mathbf{A}_k$. Then $\mathbf{B}_{k+1}$ is the result of a terminating sequence of contractions on $((\lambda z\mathbf{R})\mathbf{S})$ and the parts $((\lambda y_i \mathbf{P}_i)\mathbf{Q}_i)$ of $\mathbf{A}_k$. If $((\lambda z\mathbf{R})\mathbf{S})$ is one of the $((\lambda y_i \mathbf{P}_i)\mathbf{Q}_i)$, no residuals of $((\lambda z\mathbf{R})\mathbf{S})$ occur in $\mathbf{B}_k$, and $\mathbf{B}_k$ conv-I $\mathbf{B}_{k+1}$ by 7 XXV. If, however, $((\lambda z\mathbf{R})\mathbf{S})$ is not one of the $((\lambda y_i \mathbf{P}_i)\mathbf{Q}_i)$, a set of residuals of $((\lambda z\mathbf{R})\mathbf{S})$ does occur in $\mathbf{B}_k$ and a terminating sequence of contractions on these residuals in $\mathbf{B}_k$ gives $\mathbf{B}_{k+1}$ by 7 XXV.

Thus $\mathbf{B}_k$ red $\mathbf{B}_{k+1}$ unless the reduction of $\mathbf{A}_k$ into $\mathbf{A}_{k+1}$ involves a contraction of a residual of $((\lambda x\mathbf{M})\mathbf{N})$; but if we start with any particular $\mathbf{A}_k$ this can be the case only a finite number of successive times by 7 XXV. Hence (3) is proved, u_m being defined as follows:

Perform m successive reductions on $\mathbf{A}$ in all possible ways. This gives a finite set of formulas (since, for this purpose, we need not distinguish formulas differing only by applications of Rule I). In each formula find the largest number of reductions that can occur in a terminating sequence of contractions on the residuals of $((\lambda x\mathbf{M})\mathbf{N})$. Then let u_m be the largest of these.

7 XXVII. If $\mathbf{A}$ conv $\mathbf{B}$, there is a conversion of $\mathbf{A}$ into $\mathbf{B}$ in which no expansion precedes any reduction.

In the given conversion of $\mathbf{A}$ into $\mathbf{B}$, let the last expansion which precedes any reduction be an expansion of $\mathbf{B}_1$ into $\mathbf{A}_1$. This expansion is followed by a sequence of one or more reductions, say $\mathbf{A}_1$ imr $\mathbf{A}_2$, $\mathbf{A}_2$ imr $\mathbf{A}_3, \ldots, \mathbf{A}_{n-1}$ imr $\mathbf{A}_n$, and $\mathbf{A}_n$ conv-I-III $\mathbf{B}$. The inverse of the expansion of $\mathbf{B}_1$ into $\mathbf{A}_1$ is a reduction of $\mathbf{A}_1$ into $\mathbf{B}_1$; let $((\lambda x\mathbf{M})\mathbf{N})$ be the part of $\mathbf{A}$ which is contracted in this reduction, and let $\mathbf{B}_k$ $(k = 2, 3, \ldots, n)$ be the result of a terminating sequence of contractions on the residuals in $\mathbf{A}_k$ of $((\lambda x\mathbf{M})\mathbf{N})$. By 7 XXVI, $\mathbf{B}_1$ conv-I-II $\mathbf{B}_2$, $\mathbf{B}_2$ conv-I-II $\mathbf{B}_3, \ldots, \mathbf{B}_{n-1}$ conv-I-II $\mathbf{B}_n$, $\mathbf{B}_n$ conv-I-III $\mathbf{A}_n$, $\mathbf{A}_n$ conv-I-III $\mathbf{B}$. This provides an alternative conversion of $\mathbf{B}_1$ into $\mathbf{B}$ in which no expansion precedes any reduction. The given conversion of $\mathbf{A}$ into $\mathbf{B}$ may be altered by employing this alternative conversion of $\mathbf{B}_1$ into $\mathbf{B}$ instead of the one originally involved, with the result that the number of expansions which are out of place (precede reductions) in the conversion of $\mathbf{A}$ into $\mathbf{B}$ is decreased by one. Repetition of this process lead to a conversion of $\mathbf{A}$ into $\mathbf{B}$ in which no expansion precedes reductions.

7 XXVIII. If $\mathbf{B}$ is a normal form of $\mathbf{A}$, then $\mathbf{A}$ conv-I-II $\mathbf{B}$.

This is a corollary of 7 XXVII, since no reductions are possible of a formula in normal form.

7 XXIX. If **A** has a normal form, its normal form is unique to within application of Rule I.

For if **B** and **B**$'$ are both normal forms of **A**, then **B**$'$ is a normal form of **B**. Hence **B** conv-I-II **B**$'$. Hence **B** conv-I **B**$'$, since no reductions are possible of the normal form **B**.

Note that 7 XXIX ensures a kind of consistency of the calculus of λ-conversion, in that certain formulas for which different interpretations are intended are shown not to be interconvertible.

7 XXX. If **A** has a normal form, it has a unique principal normal form.

7 XXXI. If **B** is a normal form of **A**, then there is a number m such that any sequence of reductions starting from **A** will lead to **B** (to within applications of Rule I) after at most m reductions.

In order to prove 7 XXXI, we first prove the following lemma by induction on n:

> If **B** is a normal form of **A** and there is a sequence of n reductions leading from **A** to **B**, then there is a number $v_{\mathbf{A},n}$ such that any sequence of reductions starting from **A** will lead to a normal form of **A** in at most $v_{\mathbf{A},n}$ steps.

If $n = 0$, we take $v_{\mathbf{A},n}$ to be 0.

Assume, as hypothesis of induction, that the lemma is true when $n = k$. Suppose **A** imr **C**, **C** imr $\mathbf{C}_1$, $\mathbf{C}_1$ imr $\mathbf{C}_2$, $\mathbf{C}_2$ imr $\mathbf{C}_3$, ... , $\mathbf{C}_{k-1}$ imr **B**. Also, where $\mathbf{A}_1$ is the same as **A**, suppose $\mathbf{A}_1$ imr $\mathbf{A}_2$, $\mathbf{A}_2$ imr $\mathbf{A}_3$, ... By 7 XXVI there is a sequence ($\mathbf{D}_1$ the same as **C**), $\mathbf{D}_1$ conv-I-II $\mathbf{D}_2$, $\mathbf{D}_2$ conv-I-II $\mathbf{D}_3$, ... , such that $\mathbf{A}_j$ conv-I-II $\mathbf{D}_j$ for all j's for which $\mathbf{A}_j$ exists; and, if the reduction from **A** to **C** involves a contraction of $((\lambda x\mathbf{M})\mathbf{N})$, then, starting with $\mathbf{D}_m$, at most u_m consecutive $\mathbf{D}_j$'s occur for which it is not true that $\mathbf{D}_j$ red $\mathbf{D}_{j+1}$.

Since the sequence **C** imr $\mathbf{C}_1$, $\mathbf{C}_1$ imr $\mathbf{C}_2$, ... , leads to **B** in k reductions, there is, by hypothesis of induction, a number $v_{\mathbf{C},k}$ such that any sequence of reductions starting from **C** leads to a normal form (and thus termination) after at most $v_{\mathbf{C},k}$ reductions. Hence there are at most $v_{\mathbf{C},k}$ reductions in the sequence $\mathbf{D}_1$ conv-I-II $\mathbf{D}_2$, $\mathbf{D}_2$ conv-I-II $\mathbf{D}_3$, ... , and this sequence must terminate after at most $f(v_{\mathbf{C},k})$ steps, $f(x)$ being defined as follows:

$$f(0) = u_1,$$
$$f(x+1) = f(x) + M + 1,$$

where M is the greatest of the numbers $u_1, u_2, \ldots, u_{f(x)+1}$. (Of course $f(x)$ depends on the formula **A** and the part $((\lambda x\mathbf{M})\mathbf{N})$ of **A**, as well as on x, because u_m depends on **A** and $((\lambda x\mathbf{M})\mathbf{N})$.)

Since the sequence of $\mathbf{D}_j$'s continues as long as there are $\mathbf{A}_j$'s on which reduction can be performed, it follows that after at most $f(v_{\mathbf{C},k})$ reductions an $\mathbf{A}_j$ is reached on which no reductions are possible. But this is equivalent to saying that this $\mathbf{A}_j$ is in normal form. Thus any reductions of **A** to a formula **C**, such that there is a sequence of k reductions leading from **C** to a normal form of **A**, determines an upper bound, $f(v_{\mathbf{C},k})$, which holds for all sequences of reductions starting from **A**. Since the number of possible reductions of **A** to such formulas **C** is finite (reductions, or formulas **C**, which differ only by application of Rule I need not be distinguished as different), we take $v_{\mathbf{A},k+1}$ to be the least of the numbers $f(v_{\mathbf{C},k})$.

This completes the proof of the lemma. Hence 7 XXXI follows by 7 XXVIII.

7 XXXII. If **A** has a normal form, every [well-formed] part of **A** has a normal form.

This follows from 7 XXXI, since any sequence of reductions on a part of **A** implies a sequence of reductions on **A** and therefore must terminate.

Chapter III

LAMBDA-DEFINABILITY

8. LAMBDA-DEFINABILITY OF FUNCTIONS OF POSITIVE INTEGERS.

We define,

$$1 \to \lambda ab \,.\, ab,$$
$$2 \to \lambda ab \,.\, a(ab),$$
$$3 \to \lambda ab \,.\, a(a(ab)),$$

and so on, each numeral (in the Arabic decimal notation) being introduced as an abbreviation for a corresponding formula of the indicated form. But where a numeral consists of more than one digit, a bar is used over it, in order to avoid confusion with other notations; thus,

$$\overline{11} \to \lambda ab \,.\, a(a(a(a(a(a(a(a(a(a(ab)))))))))),$$

but 11, without the bar, is an abbreviation for

$$(\lambda ab \,.\, ab)(\lambda ab \,.\, ab).$$

In connection with these definitions an interpretation of the calculus of λ-conversion is contemplated under which each of the formulas abbreviated as a numeral is interpreted as denoting the corresponding positive integer. Since it is intended at the same time to retain the interpretation of the formulas of the calculus (which have a normal form) as denoting certain functions in accordance with the ideas of Chapter I, this means that the positive integers are identified with certain functions. For example, the number 2 is identified with the function which, when applied to the function f as argument, yields the product of f by itself (product in the sense of the product, or resultant, of two transformation); similarly the number 14 is identified with the function, when applied to the function f as argument, yields the fourteenth power of f (power in the sense of power of a transformation). This is allowable on the ground that abstract number theory requires of the positive integers only that they form a progression and, subject to this condition, the integers may be identified with any entities whatever; as a matter of fact, logical construction of the positive integers by identifying them with entities thought to be logically more fundamental are possible in many ways (the present method should be compared with that familiar in the works of Frege and Russell, according to which the non-negative integers are identified with classes of similar finite classes).

A function F of positive integers—i.e., a function of one variable for which the range of arguments and the range of values each consist of positive integers—is said to be *λ-definable* if there is a formula **F** such that (1) whenever m and n are positive integers, and $Fm = n$, and **M** and **N** are the formulas which represent (denote) the integers m and n respectively, then **FM** conv **N**, and (2) whenever the function F has no value for the positive integer m as argument, and **M** represents m, then **FM** has no normal form. Similarly the function F of two integer variables is said to be *λ-definable* if there is a formula **F** such that (1) if l, m, n are positive integers, and $Flm = n$, and **L**, **M**, **N** represent the integers l, m, n respectively, then **FLM** conv **N**, and (2) if the function F has no value for the positive integers l, m as arguments, and **L**, **M** represent

l, m respectively, then **FLM** has no normal form. And so on, for functions of any number of variables.

We shall say also, under the circumstances described, that the formula **F** *λ-defines* the function F (we use the word "λ-defines" rather than "denotes" or "represents" only because the function which **F** denotes, in general has other elements than positive integers in its range — or ranges — of arguments).

The successor function of the positive integers (i.e., the function $x + 1$) is λ-defined by the formula S, where

$$S \rightarrow \lambda abc \,.\, b(abc).$$

It is left to the reader to verify this, and also to verify that addition, and multiplication, and exponentiation of positive integers are λ-defined by the formulas $\lambda mn \,.\, m+n$, and $\lambda mn \,.\, m \times n$, and $\lambda mn \,.\, m^n$ respectively (see definition in §5).

These λ-definitions of addition, multiplication, and exponentiation are due to Rosser (see Kleene [35]). The definability of multiplication depends on the observation that the product of two positive integers in the sense of the product of transformations is the same as their product in the arithmetic sense, and the definition of exponentiation then follows because, when the positive integer n is taken of any function f as argument, there results the nth power of f in the sense of the product of transformation.

The reader may also verify that, for any formulas **L**, **M**, **N** (whether representing positive integers or not):

$$[\mathbf{L} + \mathbf{M}] + \mathbf{N} \text{ conv } \mathbf{L} + [\mathbf{M} + \mathbf{N}],$$
$$[\mathbf{L} \times \mathbf{M}] \times \mathbf{N} \text{ conv } \mathbf{L} \times [\mathbf{M} \times \mathbf{N}],$$
$$[\mathbf{L} + \mathbf{M}] \times \mathbf{N} \text{ conv } [\mathbf{L} \times \mathbf{N}] + [\mathbf{M} \times \mathbf{N}],$$
$$\mathbf{L}^{\mathbf{M}+\mathbf{N}} \text{ conv } \mathbf{L}^{\mathbf{M}} \times \mathbf{L}^{\mathbf{N}},$$
$$\mathbf{L}^{\mathbf{M}\times\mathbf{N}} \text{ conv } [\mathbf{L}^{\mathbf{N}}]^{\mathbf{M}},$$
$$S\mathbf{M} \text{ conv } 1 + \mathbf{M}.$$

9. ORDERED PAIRS AND TRIADS, THE PREDECESSOR FUNCTION. We now introduce formulas which may be thought of as representing ordered pairs and ordered triads, as follows:

$$[\mathbf{M}, \mathbf{N}] \rightarrow \lambda \boldsymbol{a} \,.\, \boldsymbol{a}\mathbf{MN},$$
$$[\mathbf{L}, \mathbf{M}, \mathbf{N}] \rightarrow \lambda \boldsymbol{a} \,.\, \boldsymbol{a}\mathbf{LMN},$$
$$2_1 \rightarrow \lambda a \,.\, a(\lambda bc \,.\, cIb),$$
$$2_2 \rightarrow \lambda a \,.\, a(\lambda bc \,.\, bIc),$$
$$3_1 \rightarrow \lambda a \,.\, a(\lambda bcd \,.\, cIdIb),$$
$$3_2 \rightarrow \lambda a \,.\, a(\lambda bcd \,.\, bIdIc),$$
$$3_3 \rightarrow \lambda a \,.\, a(\lambda bcd \,.\, bIcId).$$

If **L**, **M**, **N** are formulas representing positive integers, then $2_1(\mathbf{M}, \mathbf{N})$ conv **M**, $2_2(\mathbf{M}, \mathbf{N})$ conv **N**, $3_1(\mathbf{L}, \mathbf{M}, \mathbf{N})$ conv **L**, $3_2(\mathbf{L}, \mathbf{M}, \mathbf{N})$ conv **M**, and $3_3(\mathbf{L}, \mathbf{M}, \mathbf{N})$ conv **N**.

Verification of this depends on the observation that, if **M** is a formula representing a positive integer, **M**I conv I (the mth power of the identity is the identity).

By the predecessor function of positive integers we mean the function whose value for the argument 1 is 1 and whose value for any other positive integer argument x is $x-1$. This function is λ-defined by

$$P \rightarrow \lambda a \,.\, 3_3(a(\lambda b[S(3_1 b), 3_1 b, 3_2 b])[1, 1, 1]).$$

For if **K**, **L**, **M** represent positive integers,

$$(\lambda b[S(3_1 b), 3_1 b, 3_2 b])[\mathbf{K}, \mathbf{L}, \mathbf{M}] \operatorname{conv}[S\mathbf{K}, \mathbf{K}, \mathbf{L}],$$

and hence if **A** represents a positive integer,

$$\mathbf{A}(\lambda b[S(3_1 b), 3_1 b, 3_2 b])[1, 1, 1] \operatorname{conv} [S\mathbf{A}, \mathbf{A}, \mathbf{B}],$$

where **B** represents the predecessor of the positive integer represented by **A**. (The method of λ-definition of the predecessor due to Kleene [35] is here modified by employment of a different formal representation of ordered triads.)

A kind of subtraction of positive integers, which we distinguish by placing a dot above the sign of subtraction, and which differs from the usual kind in that $x \dot{-} y = 1$ if $x \leqq y$, may now be shown to be λ-definable:

$$[\mathbf{M} \dot{-} \mathbf{N}] \rightarrow \mathbf{N}P\mathbf{M}.$$

The functions *the lesser of the two positive integers x and y and the greater of the two positive integers x and y* are λ-definable respectively by

$$\min \rightarrow \lambda ab \,.\, Sb \dot{-} \,.\, Sb \dot{-} a,$$
$$\max \rightarrow \lambda ab \,.[a+b] \dot{-} \min ab.$$

The parity of a positive integer, i.e., the function whose value is 1 for an odd positive integer and 2 for an even positive integer, is λ-defined by

$$\operatorname{par} \rightarrow \lambda a \,.\, a(\lambda b \,.\, 3 \dot{-} b)2.$$

Using ordered pairs in a way similar to that in which ordered triads were used to obtain a λ-definition of the predecessor function, we give a λ-definition of the function *the least integer not less than half of x* — or, in other words, the quotient upon dividing $x+1$ by 2, in the sense of division with a remainder:

$$H \rightarrow \lambda a \,.\, P(2_1(a(\lambda b[P[2_1 b + 2_2 b], 3 \dot{-} 2_2 b])[1, 2])).$$

Of course this H is unrelated to the — entirely different — function H which was introduced for illustration in §1.

If we let

$$\mathfrak{L} \rightarrow \lambda b \,.\, b(\lambda c \lambda d[dPc(\lambda e \,.\, e1I)(\lambda fg \,.\, fgS)c, \\ dPc(\lambda h \,.\, h1IS)(\lambda ijk \,.\, kij(\lambda l \,.\, l1))d])$$

$$\mathfrak{A} \rightarrow \lambda a \,.\, a\mathfrak{L}[1, 1],$$
$$Z \rightarrow \lambda a \,.\, 2_2(\mathfrak{A}a),$$
$$Z' \rightarrow \lambda a \,.\, \mathfrak{A}a(\lambda bc \,.\, b \dot{-} c),$$

then, if **M**, **N** represent the positive integers m, n respectively, $\mathfrak{L}(\mathbf{M}, \mathbf{N}]$ conv $[S\mathbf{M}, 1]$ if $m \dot{-} n = 1$ and conv $[\mathbf{M}, S\mathbf{N}]$ if $m \dot{-} n > 1$; hence $\mathfrak{A}1, \mathfrak{A}2, \ldots$ are convertible respectively into

$$[2, 1],\ [3, 1],\ [3, 2],\ [4, 1],\ [4, 2],\ [4, 3],\ [5, 1],\ \ldots;$$

hence $Z1, Z2, \ldots$ are convertible respectively into

$$1,\ 1,\ 2,\ 1,\ 2,\ 3,\ 1,\ 2,\ 3,\ 4,\ 1,\ 2,\ 3,\ 4,\ 5,\ \ldots,$$

and $Z'1, Z'2, \ldots$ are convertible respectively into

$$1,\ 2,\ 1,\ 3,\ 2,\ 1,\ 4,\ 3,\ 2,\ 1,\ 5,\ 4,\ 3,\ 2,\ 1,\ \ldots.$$

Thus the infinite sequence of ordered pairs,

$$[Z1, Z'1],\ [Z2, Z'2],\ [Z3, Z'3],\ \ldots,$$

contains all ordered pairs of positive integers, with no repetition. The function whose value for the arguments x, y is the number of the ordered pair $[x, y]$ in this enumeration is λ-defined by

$$\text{nr} \rightarrow \lambda ab \,.\, S(H[[a+b] \times P[a+b]]) \dot{-} b.$$

10. PROPOSITIONAL FUNCTIONS, THE KLEENE $\mathfrak{p}$-FUNCTION. By a *propositional function* we shall mean a function (of one or more variables) whose values are *truth values*—i.e., truth and falsehood. A *property* is a propositional function of one variable; a *relation* is a propositional function of two variables. The *characteristic function* associated with a propositional function is the function whose value is 2 when (i.e., for an argument or arguments for which) the value of the propositional function is truth, whose value is 1 when the value of the propositional function is falsehood, and which has no value otherwise.

A propositional function of positive integers will be said to be λ-definable if the associated characteristic function is a λ-definable function. (It can readily be shown that the choice of the particular integers 2 and 1 in the definition of *characteristic function* is here non-essential; the class of λ-definable propositional functions of positive integers remain unaltered if any other pair of distinct positive integers is substituted.)

In particular, the relation $>$ and $=$ between positive integers are λ-definable, as is shown by giving λ-definitions of the associated characteristic functions:

$$\text{exc} \rightarrow \lambda ab \,.\, \min 2[Sa \dot{-} b],$$

$$\text{eq} \rightarrow \lambda ab \,.\, 4 \dot{-} \,.\, \text{exc}\, ab + \text{exc}\, ba.$$

From this follows the λ-definability of a great variety of properties and relations of positive integers which are expressible by means of equations and inequalities; conjunction, disjunction, and negation of equations and inequalities can be provided for by using min, max, and $\lambda a \,.\, 3 \dot{-} a$ respectively.

We prove also the two following theorems from Kleene [35], and a third closely related theorem.

10 I. If R is a λ-definable propositional function of $n+1$ positive integer arguments, then the function F is λ-definable (1) whose value for the positive integer arguments $x_1, x_2, \ldots, x_n$ is the least positive integer y such that $Rx_1x_2 \ldots x_my$ holds (i.e., has the value truth), provided that there is such a least positive

integer y and that, for every positive integer z less than this y, $Rx_1x_2\dots x_mz$ has a value, truth or falsehood, and (2) which has no value otherwise.

In the case that R has a value for every set of $n+1$ positive integer arguments, F may be described simply by saying that $Fx_1x_2\dots x_n$ is the least positive integer y such that $Rx_1x_2\dots x_ny$ holds.

Let

$$\mathfrak{G} \to \lambda n \,.\, n(\lambda r \,.\, r(\lambda s \,.\, s1II(\lambda xgt \,.\, g1(tx)Ix)))$$
$$(\lambda f \,.\, fI1II)(\lambda xgt \,.\, g(t(Sx))(Sx)gt).$$

Then

$$\mathfrak{G}1 \text{ red } \lambda xgt \,.\, g(t(Sx))(Sx)gt,$$
$$\mathfrak{G}2 \text{ red } \lambda xgt \,.\, g1(tx)Ix.$$

Hence if $\mathbf{N}$ represents a positive integer and $\mathbf{TN}$ conv either 1 or 2, we have (using 7 XXVIII to show that $\mathbf{TN}$ red 1 or 2),

$$\mathfrak{G}1\mathbf{N}\mathfrak{G}\mathbf{T} \text{ red } \mathfrak{G}(\mathbf{T}(S\mathbf{N}))(S\mathbf{N})\mathfrak{G}\mathbf{T},$$

$$\mathfrak{G}2\mathbf{N}\mathfrak{G}\mathbf{T} \text{ red } \mathbf{N}.$$

Hence if we let

$$\mathfrak{p} \to \lambda tx \,.\, \mathfrak{G}(tx)x\mathfrak{G}t,$$

we have $\mathfrak{p}\mathbf{TN}$ red $\mathbf{N}$ if $\mathbf{TN}$ conv 2, and $\mathfrak{p}\mathbf{TN}$ conv $\mathfrak{p}\mathbf{T}(S\mathbf{N})$ if $\mathbf{TN}$ conv 1, and (by 7 XXXI, 7 XXXII) $\mathfrak{p}\mathbf{TN}$ has no normal form if $\mathbf{TN}$ has no normal form.

If $\mathbf{N}$ represents the positive integer n and $\mathbf{T}$ λ-defines the characteristic function associated with the property T of positive integers, it follows that $\mathfrak{p}\mathbf{TN}$ is convertible into the formula which represents the least positive integer y, not less than n, for which Ty holds, provided that there is such a least positive integer y and that, for every positive integer x less than this y and not less than n, Tx has a value, truth or falsehood; and that in any other case $\mathfrak{p}\mathbf{TN}$ has no normal form (in the case that Ty has the value falsehood for all positive integers y not less than n, we have

$$\mathfrak{p}\mathbf{TN} \text{ red } \mathfrak{G}(\mathbf{TN})\mathbf{N}\mathfrak{G}\mathbf{T} \text{ red } \mathfrak{G}(\mathbf{T}(S\mathbf{N}))(S\mathbf{N})\mathfrak{G}\mathbf{T} \text{ red } \mathfrak{G}(\mathbf{T}(S(S\mathbf{N})))(S(S\mathbf{N})\mathfrak{G}\mathbf{T}$$
$$\text{red} \dots$$

to infinity, and hence no normal form by 7 XXXI).

Let $\mathbf{R}$ be a formula which λ-defines the characteristic function associated with the propositional function R referred to in 10 I. Then F is λ-defined by

$$\lambda \boldsymbol{x}_1\boldsymbol{x}_2\dots\boldsymbol{x}_n \,.\, \mathfrak{p}(\mathbf{R}\boldsymbol{x}_1\boldsymbol{x}_2\dots\boldsymbol{x}_n)1.$$

10 II. If T is a λ-definable property of positive integers, the function F is λ-definable (1) whose value for the positive integer argument x is the xth positive integer y (in the order of magnitude of the positive integers) such that Ty holds, provided that there is such a positive integer y and that, for every positive integer z less than y, Tz has a value, truth or falsehood, and (2) which has no value otherwise.

For let $\mathbf{T}$ be a formula which λ-defines the characteristic function associated with T, and let

$$\mathscr{P} \rightarrow \lambda tx \,.\, P(x(\lambda n \,.\, S(\mathfrak{p}tn))1).$$

Then $\mathscr{P}\mathbf{T}$ λ-defines F.

10 III. If R_1 and R_2 are λ-definable propositional functions each of $n+1$ positive integer arguments, then the propositional function R is λ-definable

(1) whose value for the positive integer arguments $x_1, x_2, \dots, x_n$ is falsehood if there is positive integer y such that $R_1x_1x_2\dots x_ny$ holds and $R_1x_1x_2\dots x_nz$ and $R_2x_1x_2\dots x_nz$ both have the value falsehood for every positive integer z less than y, and

(2) whose value for the positive integer arguments $x_1, x_2, \dots, x_n$ is truth if there is a positive integer y such that $R_2x_1x_2\dots x_ny$ holds and $R_1x_1x_2\dots x_ny$ has the value falsehood and $R_1x_1x_2\dots x_nz$ and $R_2x_1x_2\dots x_nz$ both have the value falsehood for every positive integer z less than y, and

(3) which has no value otherwise.

Let

$$\text{alt} \rightarrow \lambda xyn \,.\, \text{par}\, n(\lambda a \,.\, a(\lambda b \,.\, b1Iy))(\lambda c \,.\, c(\lambda def \,.\, fde))x(Hn).$$

$$\pi \rightarrow \lambda xy \,.\, \text{par}(\mathfrak{p}(\text{alt}\, xy)1).$$

If F and G are functions of positive integers, each being a function of one argument and including the integer 1 in the range of arguments, and if $\mathbf{F}$ and $\mathbf{G}$ λ-define F and G respectively, then alt $\mathbf{FG}$ λ-defines the function whose value for the odd integer $2x-1$ is $\mathbf{F}x$ and whose value for the even integer $2x$ is $\mathbf{G}x$.

If $\mathbf{R}_1$ and $\mathbf{R}_2$ λ-defines the characteristic functions associated with R_1 and R_2 respectively, then the characteristic function associated with R is λ-defined by

$$\lambda \boldsymbol{x}_1\boldsymbol{x}_2\dots\boldsymbol{x}_n \,.\, \pi(\mathbf{R}_1\boldsymbol{x}_1\boldsymbol{x}_2\dots\boldsymbol{x}_n)(\mathbf{R}_2\boldsymbol{x}_1\boldsymbol{x}_2\dots\boldsymbol{x}_n)$$

—this completes the proof of 10 III.

Formulas having the essential properties of $\mathfrak{p}$ and $\mathscr{P}$ were first obtained by Kleene. These formulas λ-define (in a sense which will be readily understood without explicit definition) certain functions of functions of positive integers, as already indicated.

As a further application of the formula $\mathfrak{p}$, we give λ-definition of subtraction of positive integers in the ordinary sense (so that $x-y$ has no value if $x \leqq y$) and exact division (so that $x \div y$ has no value unless x is a multiple of y):

$$[\mathbf{M} - \mathbf{N}] \rightarrow \mathfrak{p}(\lambda \boldsymbol{a} \,.\, \text{eq}\ \mathbf{M}[\mathbf{N} + \boldsymbol{a}])1.$$

$$[\mathbf{M} \div \mathbf{N}] \rightarrow \mathfrak{p}(\lambda \boldsymbol{a} \,.\, \text{eq}\ \mathbf{M}[\mathbf{N} \times \boldsymbol{a}])1.$$

11. DEFINITION BY RECURSION. A function F of n positive integers is said to be defined by *composition* in terms of functions G and $H_1, H_2, \dots, H_m$ of positive integers (of the indicated number of arguments) by the equation,

$$Fx_1x_2\dots x_n = G(H_1x_1x_2\dots x_n)(H_2x_1x_2\dots x_n)\dots(H_mx_1x_2\dots x_n).$$

(The case is not excluded that m or n or both are 1.)

A function F of $n+1$ positive integer arguments is said to be defined by *primitive recursion* in terms of the functions G_1 and G_2 of positive integers (of the indicated number of arguments) by the pair of equations:

$$Fx_1x_2\dots x_n1 = G_1x_1x_2\dots x_n,$$

$$Fx_1x_2\dots x_n(y+1) = G_2x_1x_2\dots x_ny(Fx_1x_2\dots x_ny).$$

(The case is not excluded that $n=0$, the function G_1 being replaced in that case by a given positive integer a.)

The class of *primitive recursive functions* of positive integers is defined by the three following rules, a function being primitive recursive if and only if it is determined as such by these rules:

(1) The function C such that $Cx=1$ for every possible integer x, the successor function of positive integers, and the functions U_i^n (where n is any positive integer and i is any positive integer not greater than n) such that $U_i^nx_1x_2\dots x_n = x_i$, are primitive recursive.

(2) If the function F of n arguments is defined by composition in terms of the functions G and $H_1, H_2, \dots, H_m$ and if $G, H_1, H_2, \dots, H_m$ are primitive recursive, then F is primitive recursive.

(3) If the function F of $n+1$ arguments is defined by primitive recursion in terms of functions G_1 and G_2 and if G_1 and G_2 are primitive recursive, then F is primitive recursive; or in the case that $n=0$, if F is defined by primitive recursion in terms of the integer a and the function G_2 and if G_2 is primitive recursive, then F is primitive recursive.

In order to show that every primitive recursive function of positive integers is λ-definable, we must show that all the functions mentioned in (1) are λ-definable; that if F is defined by composition in terms of G and $H_1, H_2, \dots, H_m$ and if $G, H_1, H_2, \dots, H_m$ are λ-definable, then F is λ-definable; and that if F is defined by primitive recursion in terms of G_1 and G_2 (or, in the case $n=0$, in terms of a and G_2) and if G_1 and G_2 are λ-definable (or, in the case $n=0$, if G_2 is λ-definable), then F is λ-definable.

Only the last of these three things makes any difficulty. Suppose that F is defined by primitive recursion in terms of G_1 and G_2, and that G_1 and G_2 are λ-defined respectively by G_1 and G_2. Then in order to obtain a formula F which λ-defines F we employ ordered triads:

$$\mathsf{F} \rightarrow \lambda \boldsymbol{x}_1\boldsymbol{x}_2\dots\boldsymbol{x}_n\boldsymbol{y} \,.\, 3_3(\boldsymbol{y}(\lambda \boldsymbol{z}[S(3_1\boldsymbol{z}),$$
$$\mathsf{G}_2\boldsymbol{x}_1\boldsymbol{x}_2\dots\boldsymbol{x}_n(3_1\boldsymbol{z})(3_2\boldsymbol{z}), (3_2\boldsymbol{z})])[1, \mathsf{G}_1\boldsymbol{x}_1\boldsymbol{x}_2\dots\boldsymbol{x}_n, 1])$$

($\boldsymbol{x}_1\boldsymbol{x}_2\dots\boldsymbol{x}_n, \boldsymbol{y}, \boldsymbol{z}$ being any $n+2$ distinct variables). In the case $n=0$, this reduces to:

$$\mathsf{F} \rightarrow \lambda \boldsymbol{y} \,.\, 3_3(\boldsymbol{y}(\lambda \boldsymbol{z}[S(3_1\boldsymbol{z}), \mathsf{G}_2(3_1\boldsymbol{x})(3_2\boldsymbol{x}), 3_2\boldsymbol{z}])[1, \mathsf{A}, 1]),$$

where A represents the positive integer a.

(The λ-definition of the predecessor function given in §9 may be regarded as a special case of the foregoing in which a is 1 and G_2 is U_1^2. The extension of the method used for the predecessor function to the general case of definition by primitive recursion is due to Paul Bernays, in a letter of May 27th, 1935—where, however, the matter is stated within the context of the calculus of λ-K-conversion and ordered pairs are consequently used instead of ordered triads. As remarked by Bernays, this method of dealing with definition by primitive recursion has the advantage that it shows also, for each n, the λ-definability of the function ρ of functions of positive integers whose value for the arguments G_1 and G_2 is the function F defined by primitive recursion in terms of G_1 and G_2 — i.e., essentially, the function ρ of Hilbert [31].)

Thus we have:

11 I. Every primitive recursive function of positive integers is λ-definable.

The class of primitive recursive functions is known to include substantially all the ordinarily used numerical functions — of, e.g., Skolem [50], Gödel [27], Péter [41] (it is readily seen to be a non-essential difference that some of these authors deal with primitive recursive functions of non-negative integers rather than of positive integers). Primitive recursive, in particular, are functions corresponding to the quotient and remainder in division, the greatest common divisor, the xth prime number, and many related functions; λ-definitions of these functions can consequently be obtained by the method just given.

The two schemata, of definition by composition and by primitive recursion, have this property in common, that — on the hypothesis that all particular values are known of the functions in terms of which F is defined — the given equations make possible the calculation of any required particular value of F by a series of steps each consisting of a substitution, either of a (symbol for a) particular number for (all occurrences of) a variable, or of one thing for another known to be equal to it. By allowing additional, or more general, schemata having this property, various more extensive notions of recursiveness are obtainable (cf. Hilbert [31], Ackermann [1], Péter [41, 42, 43, 44]). If the definition of primitive recursiveness is modified by allowing, in place of (2) and (3), *any* definition by a set of equations having this property, the functions obtained are called *general recursive* — if it is required of all functions defined that they have a value for every set of the relevant number of positive integer arguments — or *partial recursive* if this is not required. For a more exact statement (which may be made in any one of several equivalent ways), the reader is referred to Gödel [28], Church [9], Kleene [36, 39], Hilbert and Bernays [33].

That every general recursive function of positive integers is λ-definable can be proved in consequence of 10 I and 11 I by using the result of Kleene [36], that every general recursive function of n positive integer arguments $x_1, x_2, \ldots, x_n$ can be expressed in the form $F(\epsilon y(Rx_1x_2\ldots x_ny))$, where F is a primitive recursive function of positive integers, R is a propositional function of positive integers whose associated characteristic function is primitive recursive, and "ϵy" is to be read "the least positive integer y such that." (Cf. Kleene [37].) The converse proposition, that every λ-definable function of positive integers, having a value for every set of the relevant number of positive integer arguments, is general recursive, is proved by the method of Church [9] or Kleene [37] (the proof makes use of the fact that, by 7 XXXI, the process of reduction to normal form provides a method of calculating explicitly any required particular value of a function whose λ-definition is given, and proceeds by setting up a set of recursion equations which in effect describe this process of calculation).

These proofs may be extended to the case of partial recursive functions without major modification (cf. Kleene [39]). Hence are obtained the following theorems (proofs omitted here):

11 II. Every partial recursive function of positive integers is λ-definable.

11 III. Every λ-definable function of positive integers is partial recursive.

The notion of a method of effective calculation of the values of a function, or the notion of a function for which such a method of calculation exists, is of not uncommon occurrence in connection with mathematical questions, but it is ordinarily left on the intuitive level, without attempt at explicit definition. The known theorems concerning λ-definability, or recursiveness, strongly suggest that the notion of an *effectively calculable function of positive integers* be given an exact definition by identifying it with that of a λ-definable function, or equivalently of a partial recursive function. As in all cases where a formal definition is offered of what was previously an intuitive or empirical idea, no complete proof is possible; but the writer has little doubt of the finality of identification. (Concerning the origin of this proposal, see Church [9], footnotes 3, 18.)

An equivalent definition of effective calculability is to identify it with *calculability within* a formalized system of logic whose postulates and rules have appropriate properties of recursiveness—cf. Church [9], §7, Hilbert and Bernays [33], Supplement II.

Another equivalent definition, having a more immediate intuitive appeal is that of Turing [55], who calls a function computable if (roughly speaking) it is possible to make a finite calculating machine capable of computing any required value of the function. The machine is supplied with a tape on which computations are printed (the analogue of the paper used by a human calculator), and no upper limit is placed on the length of tape or on the time required for computation of a particular value of the function, except that it be finite in each case. Further restrictions imposed on the character of the machine are more or less clearly either non-essential or necessarily contained in the requirement of finiteness. The equivalency of computability to λ-definability and general recursiveness (attention being confined to functions of one argument for which the range of arguments consists of all positive integers) is proved in Turing [57].

Mention should also be made of the notion of a *finite combinatory process* introduced by Post [46]. This again is equivalent to the other concepts of effective calculability.

Examples of functions which are not effectively calculable can now be given in various ways. In particular, it is proved in Church [9] that if the set of well-formed formulas of the calculus of λ-conversion be enumerated in a straightforward way (any one of the particular enumerations which immediately suggest themselves may be employed), and if F is the function such that F is 2 or 1 according as the xth formula in this enumeration has or has not a normal form, then F is not λ-definable. This may be taken as the exact meaning of the somewhat vague statement made at the end of §6, that the condition of having a normal form is not effective.

In the explicit proofs of many of the theorems which have been stated without proof in this section, use is made of the notion of the Gödel number of a formula or formal expression. In the published papers referred to, this notion is introduced by a method closely similar to that employed by Gödel [27]. In the case of well-formed formulas of the calculus of λ-conversion, however, it would be equally possible to use the somewhat different method of our next chapter.

Chapter IV

COMBINATIONS, GÖDEL NUMBERS

12. COMBINATIONS. If s is any set of well-formed formulas, the class of *s-combinations* is defined by the two following rules, a formula being an s-combination if and only if it is determined as such by these rules:

(1) Any formula of the set s, and any variable standing alone, is an s-combination.

(2) If A and B are s-combinations, AB is an s-combination.

In the cases in which we shall be interested the formulas of s will contain no free variables and will none of them be of the form AB. In such a case it is possible to distinguish the *terms* of an s-combination, each occurrence of a free variable or one of the formulas of s being a term.

If s is the null set, the s-combinations will be called *combinations of variables*.

If s consists of the two formulas I, J, where

$$I \to \lambda aa,$$
$$J \to \lambda abcd \,.\, ab(adc),$$

the s-combinations will be called simply *combinations*.

We shall prove that *every well-formed formula is convertible into a combination*. This theorem is taken from Rosser [47], the present proof of it from Church [8]; the ideas involved go back to Schönfinkel [49] and Curry [18, 21].

Let:

$$\tau \to JII.$$

Then τ conv $\lambda ab \,.\, ba$, and hence $\tau\mathsf{AB}$ conv BA.

If M is any combination containing x as a free variable, we define an associated combination $\lambda_{\mathsf{x}}\mathsf{M}|$, which does not contain x as a free variable but otherwise contains the same free variables as M. This definition is by recursion, according to the following rules:

(1) $\lambda_x x|$ is I.

(2) If B contains x as a free variable and A does not, $\lambda_x\mathsf{AB}|$ is $J\tau\lambda_x\mathsf{B}|(JI\mathsf{A})$.

(3) If A contains x as a free variable and B does not, $\lambda_x\mathsf{AB}|$ is $J\tau\mathsf{B}\lambda_x\mathsf{A}|$.

(4) If both A and B contain x as a free variable,
$\lambda_x\mathsf{AB}|$ is $J\tau\tau(JI(J\tau\tau(J\tau\lambda_x\mathsf{B}|(J\tau\lambda_x\mathsf{A}|J))))$.

12 I. If M is a combination containing x as a free variable, $\lambda_x\mathsf{M}|$ conv $\lambda x\mathsf{M}$.

We prove this by induction with respect to the number of terms of M.

If M has one term, then M is x, and $\lambda_x\mathsf{M}|$ is I, which is convertible into λxx.

If M is AB and B contains x as a free variable and A does not, then $\lambda_x\mathsf{M}|$ is $J\tau\lambda_x\mathsf{B}|(JI\mathsf{A})$, which (see definitions of I, J, τ) is convertible into $\lambda d \,.\, \mathsf{A}(\lambda_x\mathsf{B}|d)$, which, by hypothesis of induction, is convertible into $\lambda d \,.\, \mathsf{A}((\lambda x\mathsf{B})d)$ which finally is convertible into $\lambda x \,.\, \mathsf{AB}$.

If M is AB and A contains x as a free variable and B does not, then $\lambda_x\mathsf{M}|$ is $J\tau\mathsf{B}\lambda_x\mathsf{A}|$, which is convertible into $\lambda d \,.\, \lambda_x\mathsf{A}|d\mathsf{B}$, which, by hypothesis of induction is convertible into $\lambda d \,.\, (\lambda x\mathsf{A})d\mathsf{B}$, which finally is convertible into $\lambda x \,.\, \mathsf{AB}$.

If $\mathbf{M}$ is $\mathbf{AB}$ and both $\mathbf{A}$ and $\mathbf{B}$ contain x as a free variable, then $\lambda_x\mathbf{M}|$ is

$$J\tau\tau(JI(J\tau\tau(J\tau\lambda_x\mathbf{B}|(J\tau\lambda_x\mathbf{A}|J)))),$$

which is convertible into $\lambda d \,.\, \lambda_x\mathbf{A}|d(\lambda_x\mathbf{B}|d)$, which, by hypothesis of induction, is convertible into $\lambda d \,.(\lambda x\mathbf{A})d((\lambda x\mathbf{B})d)$, which finally is convertible into $\lambda x \,.\, \mathbf{AB}$.

The foregoing tacitly assumes that $\mathbf{A}$ and $\mathbf{B}$ do not contain d as a free variable. The modification necessary for the contrary case is, however, obvious.

This completes the proof of 12 I. We define the combination belonging to a well-formed formula, by recursion as follows:

(1) The combination belonging to x is x (where x is any variable).

(2) The combination belonging to $\mathbf{FA}$ is $\mathbf{F}'\mathbf{A}'$, where $\mathbf{F}'$ and $\mathbf{A}'$ are the combinations belonging to $\mathbf{F}$ and $\mathbf{A}$ respectively.

(3) The combination belonging to $\lambda x\mathbf{M}$ is $\lambda_x\mathbf{M}'|$, where $\mathbf{M}'$ is the combination belonging to $\mathbf{M}$.

12 II. Every well-formed formula is convertible into the combination belonging to it.

Using 12 I, this is proved by induction with respect to the length of the formula. The proof is straightforward and details are left to the reader.

12 III. The combination belonging to $\mathbf{X}$ and the combination belonging to $\mathbf{Y}$ are identical if and only if $\mathbf{X}$ conv-I $\mathbf{Y}$.

13. PRIMITIVE SETS OF FORMULAS. A set s of well-formed formulas is called a *primitive set*, if the formulas of s contain no free variables and are none of them of the form $\mathbf{AB}$, and every well-formed formula is convertible into an s-combination. (When necessary to distinguish this idea from the analogous idea in the calculus of λ-K-conversion, the calculus of λ-δ-conversion, etc. — see Chapter V — we may speak of primitive sets of λ-formulas, primitive sets of λ-K-formulas, primitive sets of λ-δ-formulas, etc.)

It was proved in §12 that the formulas I, J are a primitive set. Another primitive set of formulas, suggested by the work of Curry, consists of the four formulas B, C, W, I, where:

$$B \rightarrow \lambda abc \,.\, a(bc).$$
$$C \rightarrow \lambda abc \,.\, acb.$$
$$W \rightarrow \lambda ab \,.\, abb.$$

In order to prove this it is sufficient to express J as a $\{B, C, W, I\}$-combination, as follows:

$$J \text{ conv } B(BC(BC))(B(W(BBB))C).$$

Still another primitive set of formulas consists of the four formulas B, T, D, I, where:

$$T \rightarrow \lambda ab \,.\, ba.$$
$$D \rightarrow \lambda a \,.\, aa.$$

In order to prove this it is sufficient to express C and W as $\{B, T, D, I\}$-combinations, as follows:

$$C \text{ conv } B(T(BBT))(BBT).$$

$$W \text{ conv } B(B(T(BD(B(TT)(B(BBB)T))))(BBT))(B(T(B(TI)(TI)))B).$$

[*Added by Church in the second printing, 1951*: " ... the amendment should also be taken into account which is suggested by Rosser [109]. The following simpler expression for W is available:

$$W \text{ conv } B(T(B(BDB)T))(BBT).$$

Hence replace [the preceding formula] by this."]

A primitive set of formulas is said to be *independent* if it ceases to be a primitive set upon omission of any one of the formulas. It seems plausible that each of the three primitive sets which have been named is independent.—In the case of the set $\{I, J\}$, the independence of J follows (using 7 XVII) from the fact that any combination all of whose terms are I is convertible into I; and the independence of I follows (using 7 XXVIII) from the fact that if $\mathbf{A}$ imr $\mathbf{B}$ and $\mathbf{B}$ contains a (well-formed) part convertible-I into I then $\mathbf{A}$ must contain a (well-formed) part convertible-I into I.

14. AN APPLICATION OF THE THEORY OF COMBINATIONS. We prove now the following theorems, due to Kleene [34, 35, 37].

14 I. If $\mathbf{A}_1$ and $\mathbf{A}_2$ contain no free variables, a formula $\mathbf{L}$ can be found such that $\mathbf{L}1 \text{ conv } \mathbf{A}_1$ and $\mathbf{L}2 \text{ conv } \mathbf{A}_2$.

For, by 12 II, $\mathbf{A}_1$ and $\mathbf{A}_2$ are convertible into combinations $\mathbf{A}_1'$ and $\mathbf{A}_2'$ respectively. We take $\mathbf{A}_1'$ to be the combination belonging to $\mathbf{A}_1$, unless that combination fails to contain an occurrence of J, in which case we take $\mathbf{A}_1'$ to be $JIIII$; and $\mathbf{A}_2'$ is similarly determined relative to $\mathbf{A}_2$. Let $\mathbf{A}_1''$ and $\mathbf{A}_2''$ be the result of replacing all occurrences of J by the variable j in $\mathbf{A}_1'$ and $\mathbf{A}_2'$ respectively, and let $\mathbf{B}_1$ and $\mathbf{B}_2$ be $\lambda j \mathbf{A}_1''$ and $\lambda j \mathbf{A}_2''$ respectively. Then $\mathbf{B}_1 J \text{ conv } \mathbf{A}_1$, and $\mathbf{B}_2 J \text{ conv } \mathbf{A}_2$, and $\mathbf{B}_1 I \text{ conv } I$, and $\mathbf{B}_2 I \text{ conv } I$. Consequently a formula $\mathbf{L}$ having the required property is:

$$\lambda n \,.\, n(\lambda x \,.\, x(\lambda y \,.\, yI\mathbf{B}_2))(\lambda z \,.\, zII)\mathbf{B}_1 J.$$

14 II. If $\mathbf{A}_1, \mathbf{A}_2, \ldots, \mathbf{A}_n$ contain no free variables, a formula $\mathbf{L}$ can be found such that $\mathbf{L}1 \text{ conv } \mathbf{A}_1$, $\mathbf{L}2 \text{ conv } \mathbf{A}_2$, $\ldots$, $\mathbf{LN} \text{ conv } \mathbf{A}_n$ ($\mathbf{N}$ being the formula which represents n).

For the case that n is 1 or 2, this follows from 14 I. For larger values of n, we prove it by induction.

Let $\mathbf{L}_2$ be a formula such that $\mathbf{L}_2 1 \text{ conv } \mathbf{A}_1$, and let $\mathbf{L}_1$ be a formula such that $\mathbf{L}_1 1 \text{ conv } \mathbf{A}_2$, $\mathbf{L}_2 \text{ conv } \mathbf{A}_3$, $\ldots$, $\mathbf{L}_1 \mathbf{M} \text{ conv } \mathbf{A}_n$ (where $\mathbf{M}$ represents $n-1$). Also let $\mathbf{G}$ be a formula such that $\mathbf{G}1 \text{ conv } \mathbf{L}_1$ and $\mathbf{G}2 \text{ conv } \mathbf{L}_2$. Then a formula $\mathbf{L}$ having the required property is:

$$\lambda i \,.\, \mathbf{G}[3 \dot{-} i](Pi).$$

14 III. If $\mathbf{A}_1, \mathbf{A}_2, \ldots, \mathbf{A}_n, \mathbf{F}_1, \mathbf{F}_2, \ldots, \mathbf{F}_m$ contain no free variables, a formula $\mathbf{E}$ can be found which represents an enumeration of the least set of formulas which contains $\mathbf{A}_1, \mathbf{A}_2, \ldots, \mathbf{A}_n$ and is closed under each of the operations of forming $\mathbf{F}_\alpha \mathbf{XY}$ from the formulas $\mathbf{X}, \mathbf{Y}$ ($\alpha = 1, 2, \ldots, n$), in the sense that every

formula of this set is convertible into one of the formulas in the infinite sequence

$$\mathbf{E}_1, \mathbf{E}_2, \ldots,$$

and every formula in this infinite sequence is convertible into one of the formulas of the set.

We prove this first for the case $m = 1$, using a device due to Kleene for obtaining formulas satisfying arbitrary conversion conditions of the general kind illustrated in (1) below.

Using 14 II, let $\mathbf{U}$ be a formula such that

$$\mathbf{U}1 \operatorname{conv} I,$$
$$\mathbf{U}2 \operatorname{conv} \lambda xy \,.\, \mathbf{F}_1(y(S[\mathbf{N}' \dot{-} Zx])[Zx \dot{-} \mathbf{N}]y)(y(S[\mathbf{N}' \dot{-} Z'x])[Z'x \dot{-} \mathbf{N}]y),^*$$
$$\mathbf{U}3 \operatorname{conv} \lambda xy \,.\, yx\mathbf{A}_1,$$
$$\mathbf{U}4 \operatorname{conv} \lambda xy \,.\, yx\mathbf{A}_2,$$
$$\cdots\cdots\cdots\cdots$$
$$\mathbf{UN}' \operatorname{conv} \lambda xy \,.\, yx\mathbf{A}_n,$$

where $\mathbf{N}$ represents n and $\mathbf{N}'$ represents $n+2$, and Z and Z' are the formulas introduced in §9. Let $\mathbf{E}$ be the formula,

$$\lambda i \,.\, \mathbf{U}(S[\mathbf{N}' \dot{-} i])[i \dot{-} \mathbf{N}]\mathbf{U}.$$

Then we have

$$\begin{aligned} &\mathbf{E}1 \operatorname{conv} \mathbf{A}_n,\\ &\mathbf{E}2 \operatorname{conv} \mathbf{A}_{n-1},\\ &\cdots\cdots\cdots\cdots\\ &\mathbf{EN} \operatorname{conv} \mathbf{A}_1,\\ &\mathbf{EK} \operatorname{conv} \mathbf{F}_1(\mathbf{E}(Z[\mathbf{K} \dot{-} \mathbf{N}]))(\mathbf{E}(Z'[\mathbf{K} \dot{-} \mathbf{N}])),\end{aligned} \tag{1}$$

$\mathbf{K}$ being any formula which represents an integer greater than n. From this it follows that $\mathbf{E}$ is a formula of the kind required.

Consider now the case $m > 1$. Let $\mathbf{M}$ represent m and let $\mathbf{F}$ be a formula such that $\mathbf{F}1 \operatorname{conv} \mathbf{F}_1$, $\mathbf{F}2 \operatorname{conv} \mathbf{F}_2, \ldots, \mathbf{FM} \operatorname{conv} \mathbf{F}_m$. By the preceding proof for the case $m = 1$, a formula $\mathbf{E}'$ can be found which represents an enumeration of the least set of formulas which contains $[1, \mathbf{A}_1], [2, \mathbf{A}_1], \ldots, [\mathbf{M}, \mathbf{A}_1], [1, \mathbf{A}_2], [2, \mathbf{A}_2], \ldots, [\mathbf{M}, \mathbf{A}_2], \ldots, [1, \mathbf{A}_n], [2, \mathbf{A}_n], \ldots, [\mathbf{M}, \mathbf{A}_n]$ and is closed under the operation of forming $\mathbf{Y}(\lambda xy[x, \mathbf{XF}y])$ from the formulas $\mathbf{X}, \mathbf{Y}$. Then a formula $\mathbf{E}$ of the kind required is:

$$\lambda i \,.\, 2_2(\mathbf{E}'i).$$

It is immaterial that the enumeration so obtained contains repetitions. (Notice that $2_2(\mathbf{B}, \mathbf{C}) \operatorname{conv} \mathbf{C}$ if $\mathbf{B}$ is any formula such that $\mathbf{B}I \operatorname{conv} I$, in particular if $\mathbf{B}$ is any formula representing a positive integer; the case considered in §9 that $\mathbf{B}$ and $\mathbf{C}$ *both* represent positive integers is thus only a special case.)

---

* *Editor's note*: The original text reads $[Zx - \mathbf{N}]$ with a minus sign in the formula instead of $\dot{-}$. The latter is probably right.

14 IV. If $\mathbf{A}_1, \mathbf{A}_2, \ldots, \mathbf{A}_n, \mathbf{F}_1, \mathbf{F}_2, \ldots, \mathbf{F}_m, \mathbf{F}_{m+1}, \mathbf{F}_{m+2}, \ldots, \mathbf{F}_{m+r}$ contain no free variables, a formula $\mathbf{E}$ can be found which represents an enumeration of the least set of formulas which contain $\mathbf{A}_1, \mathbf{A}_2, \ldots, \mathbf{A}_n$ and is closed under each of the operations of forming $\mathbf{F}_\alpha \mathbf{X}\mathbf{Y}$ from the formulas $\mathbf{X}, \mathbf{Y}$ $(\alpha = 1, 2, \ldots, m)$ and of forming $\mathbf{F}_{m+\beta}\mathbf{X}$ from the formula $\mathbf{X}$ $(\beta = 1, 2, \ldots, r)$—in the sense that every formula of this set is convertible into one of the formulas in the infinite sequence

$$\mathbf{E}_1, \mathbf{E}_2, \ldots,$$

and every formula in this infinite sequence is convertible into one of the formulas of the set.

(The case is not excluded that $m = 0$ or that $r = 0$, provided that m and r are not both 0.)

By the method used in the proof of 14 I, find formulas $\mathbf{B}_1, \mathbf{B}_2, \ldots, \mathbf{B}_n, \mathbf{G}_1, \mathbf{G}_2, \ldots, \mathbf{G}_{m+r}$ such that $\mathbf{B}_1 J$ conv $\mathbf{A}_1$, $\mathbf{B}_2 J$ conv $\mathbf{A}_2, \ldots, \mathbf{B}_n J$ conv $\mathbf{A}_n$, $\mathbf{G}_1 J$ conv $\mathbf{F}_1$, $\mathbf{G}_2 J$ conv $\mathbf{F}_2, \ldots, \mathbf{G}_{m+r} J$ conv $\mathbf{F}_{m+r}$, and $\mathbf{B}_1 I$ conv I, $\mathbf{B}_2 I$ conv $I, \ldots, \mathbf{B}_n I$ conv I, $\mathbf{G}_1 I$ conv I, $\mathbf{G}_2 I$ conv $I, \ldots, \mathbf{G}_{m+r} I$ conv I. By 14 III, a formula $\mathbf{E}'$ can be found which represents an enumeration of the least set of formulas which contain $\mathbf{B}_1, \mathbf{B}_2, \ldots, \mathbf{B}_n$ and is closed under each of the operations of forming $\lambda x \,.\, \mathbf{G}_\alpha x(\mathbf{X}x)(\mathbf{Y}x)$ from the formulas $\mathbf{X}, \mathbf{Y}$ $(\alpha = 1, 2, \ldots, m)$ and of forming $\lambda x \,.\, \mathbf{Y} I \mathbf{G}_{m+\beta} x(\mathbf{X}x)$ from the formulas $\mathbf{X}, \mathbf{Y}$ $(\beta = 1, 2, \ldots, r)$. Then a formula $\mathbf{E}$ of the kind required is:

$$\lambda i \,.\, \mathbf{E}' i J.$$

15. A COMBINATORY EQUIVALENT OF CONVERSION. [*Added by Church in the second printing, 1951*: "[In this section], the combinatory equivalent of conversion which is given can be simplified by the method of Rosser [110], and in particular the proof of the equivalence to conversion can be greatly shortened. Details of this, including the proof of equivalence, may be obtained from Rosser's paper; and the formula o of §16, and the formula do of §20, may then be modified correspondingly.

For a combinatory equivalent of λ-K-conversion, and also of λ-K-conversion with the addition of a rule by which BI and I are interchangeable, see [70] – where Curry employs Rosser's method in order to simplify his earlier treatments of the theory of combinators (which are referred to at the end of §15)."]

It is desirable to have a set of operations (upon combination) which have the property that they always change a combination into a combination and which constitute an equivalent of conversion in the sense that a combination $\mathbf{X}$ can be changed into a combination $\mathbf{Y}$ by a sequence of (0 or more of) these operations if and only if $\mathbf{X}$ conv $\mathbf{Y}$. Such a set of operations is the following (0I–0XXXVIII)—where $\mathbf{F}, \mathbf{A}, \mathbf{B}, \mathbf{C}, \mathbf{D}$ are arbitrary combinations, β, γ, ω are defined as indicated blow, and the sign $\vdash$ is used to mean that the combination which precedes $\vdash$ is changed by the operation into the combination which follows:

0I. $I\mathbf{A} \vdash \mathbf{A}$.

0II. $\mathbf{A} \vdash I\mathbf{A}$.

0III. $\mathbf{F}(I\mathbf{A}) \vdash \mathbf{F}\mathbf{A}$.

0IV. $\mathbf{F}\mathbf{A} \vdash \mathbf{F}(I\mathbf{A})$.

0V. $\mathbf{F}(I\mathbf{AB}) \vdash \mathbf{F}(\mathbf{AB})$.

0VI. $\mathbf{F}(\mathbf{AB}) \vdash \mathbf{F}(I\mathbf{AB})$.

0VII. $\mathbf{F}(J\mathbf{ABCD}) \vdash \mathbf{F}(\mathbf{AB}(\mathbf{ADC}))$.

0VIII. $\mathbf{F}(\mathbf{AB}(\mathbf{ADC})) \vdash \mathbf{F}(J\mathbf{ABCD})$.

0IX. $\mathbf{F}J \vdash \mathbf{F}(\omega(\beta\gamma(\beta(\beta(\beta\gamma))(\beta(\beta(\beta\beta\beta))I))))$.

0X. $\mathbf{F}(\omega(\beta\gamma(\beta(\beta(\beta\gamma))(\beta(\beta(\beta\beta\beta))I)))) \vdash \mathbf{F}J$.

0XI. $\mathbf{F}\beta \vdash \mathbf{F}(\beta(\beta(\beta I))\beta)$.

0XII. $\mathbf{F}(\beta(\beta(\beta I))\beta) \vdash \mathbf{F}\beta$.

0XIII. $\mathbf{F}\gamma \vdash \mathbf{F}(\beta(\beta(\beta I))\gamma)$.

0XIV. $\mathbf{F}(\beta(\beta(\beta I))\gamma) \vdash \mathbf{F}\gamma$.

0XV. $\mathbf{F}I \vdash \mathbf{F}(\beta II)$.

0XVI. $\mathbf{F}(\beta II) \vdash \mathbf{F}I$.

0XVII. $\mathbf{F}(\gamma(\beta\beta(\beta\beta\beta))\beta) \vdash \mathbf{F}(\beta(\beta\beta)\beta)$.

0XVIII. $\mathbf{F}(\beta(\beta\beta)\beta) \vdash \mathbf{F}(\gamma(\beta\beta(\beta\beta\beta))\beta)$.

0XIX. $\mathbf{F}(\gamma(\beta\beta(\beta\beta\beta))\gamma) \vdash \mathbf{F}(\beta(\beta\gamma)(\beta\beta\beta))$.

0XX. $\mathbf{F}(\beta(\beta\gamma)(\beta\beta\beta)) \vdash \mathbf{F}(\gamma(\beta\beta(\beta\beta\beta))\gamma)$.

0XXI. $\mathbf{F}(\gamma(\beta\beta\beta)\omega) \vdash \mathbf{F}(\beta(\beta\omega)(\beta\beta\beta))$.

0XXII. $\mathbf{F}(\beta(\beta\omega)(\beta\beta\beta)) \vdash \mathbf{F}(\gamma(\beta\beta\beta)\omega)$.

0XXIII. $\mathbf{F}(\gamma\beta I) \vdash \mathbf{F}(\beta(\beta I)I)$.

0XXIV. $\mathbf{F}(\beta(\beta I)I) \vdash \mathbf{F}(\gamma\beta I)$.

0XXV. $\mathbf{F}(\beta\beta\gamma) \vdash \mathbf{F}(\beta(\beta(\beta\gamma)\gamma)(\beta\beta))$.

0XXVI. $\mathbf{F}(\beta(\beta(\beta\gamma)\gamma)(\beta\beta)) \vdash \mathbf{F}(\beta\beta\gamma)$.

0XXVII. $\mathbf{F}(\beta\beta\omega) \vdash \mathbf{F}(\beta(\beta(\beta(\beta(\beta\omega)\omega)(\beta\gamma))(\beta(\beta\beta)))\beta)$.

0XXVIII. $\mathbf{F}(\beta(\beta(\beta(\beta(\beta\omega)\omega)(\beta\gamma))(\beta(\beta\beta)))\beta) \vdash \mathbf{F}(\beta\beta\omega)$.

0XXIX. $\mathbf{F}(\beta\gamma\gamma) \vdash \mathbf{F}(\beta(\beta I))$.

0XXX. $\mathbf{F}(\beta(\beta I)) \vdash \mathbf{F}(\beta\gamma\gamma)$.

0XXXI. $\mathbf{F}(\beta(\beta(\beta\gamma)\gamma)(\beta\gamma)) \vdash \mathbf{F}(\beta(\beta\gamma(\beta\gamma))\gamma)$.

0XXXII. $\mathbf{F}(\beta(\beta\gamma(\beta\gamma))\gamma) \vdash \mathbf{F}(\beta(\beta(\beta\gamma)\gamma)(\beta\gamma))$.

0XXXIII. $\mathbf{F}(\beta\gamma\omega) \vdash \mathbf{F}(\beta(\beta(\beta\omega)\gamma)(\beta\gamma))$.

0XXXIV. $\mathbf{F}(\beta(\beta(\beta\omega)\gamma)(\beta\gamma)) \vdash \mathbf{F}(\beta\gamma\omega)$.

0XXXV. $\mathbf{F}(\beta\omega\gamma) \vdash \mathbf{F}\omega$.

0XXXVI. $\mathbf{F}\omega \vdash \mathbf{F}(\beta\omega\gamma)$.

0XXXVII. $\mathbf{F}(\beta\omega\omega) \vdash \mathbf{F}(\beta\omega(\beta\omega))$.

0XXXVIII. $\mathbf{F}(\beta\omega(\beta\omega)) \vdash \mathbf{F}(\beta\omega\omega)$.

$$\gamma \to J\tau(J\tau)(J\tau).$$
$$\beta \to \gamma(JI\gamma)(JI).$$
$$\omega \to \gamma(\gamma(\beta\gamma(\gamma(\beta J\tau)\tau))\tau).$$

(Note that $\tau, \gamma, \beta, \omega$ are convertible respectively into T, C, B, W.)

These thirty-eight operations have characteristics of simplicity not possessed by the operations I, II, III of §6, namely: (1) they are one-valued, i.e., given the combination operated on and the particular one of the thirty-eight operations which is applied, the combination resulting is uniquely determined; (2) they do not involve the idea of substitution at an arbitrary place, but only that of substitution at a specified place. This has the effect of rendering some of the developments in §16 much simpler than they otherwise might be.

The proof of the equivalence of 0I–0XXXVIII to conversion is too long to be included here. It may be found in Rosser's dissertation [47] (cf. Section H therein). Many of the important ideas and methods involved derive from Curry [17, 18, 20, 21]; in fact, Curry has results which may be thought of as constituting an approximate equivalent to the one in question here but which are nevertheless sufficiently different so that we are unable to use them directly.

16. GÖDEL NUMBERS. The *Gödel number of* a combination is defined by induction as follows:

(1) The Gödel number of I is 1.

(2) The Gödel number of J is 3.

(3) The Gödel number of the nth variable in alphabetical order (see §5) is $2n + 5$.

(4) If m and n are the Gödel numbers of **A** and **B** respectively, the Gödel number of **AB** is $(m + n)(m + n - 1) - 2n + 2$.

The *Gödel number belonging to* a formula is defined to be the Gödel number of the combination belonging to the formula. (Notice that the Gödel number belonging to a combination is thus in general not the same as the Gödel number of the combination.)

It is left to the reader to verify that the Gödel numbers of two combinations **A** and **B** are the same if and only if **A** and **B** are the same; and that the Gödel numbers belonging to two formulas **A** and **B** are the same if and only if **A** conv-I **B** (cf. 12 III). (Notice that the Gödel number of **AB**, according to (4), is twice the number of the order pair $[m, n]$ in the enumeration of ordered pairs described at the end of §9.)

The usefulness of Gödel numbers arises from the fact that our formalism contains no notations for formulas—i.e., for sequences of symbols. (It is not possible to use formulas as notations for themselves, because interconvertible formulas must denote the same thing although they are not the same formula, and because formulas containing free variables cannot denote any [fixed] thing.) The Gödel number belonging to a formula serves in many situations as a substitute for a notation for the formula and often enables us to accomplish things which might have been thought to be impossible without a formal notation for formulas.

This use of Gödel numbers is facilitated by the existence of a formula, form, such that, if **N** represents the Gödel number belonging to **A**, and **A** contains no free variables, then, form **N** conv **A**. In order to obtain this formula, first notice that par **N** conv 2 if $\mathbf{N}$ represents the Gödel number of a combination having more than one term, and par $\mathbf{N}$ conv 1 if $\mathbf{N}$ represents the Gödel number of a combination having only one term; also that if $\mathbf{N}$ represents the Gödel number of a combination $\mathbf{AB}$, then $Z(H\mathbf{N})$ is convertible into the formula representing the Gödel number of $\mathbf{A}$, and $Z'(H\mathbf{N})$ is convertible into the formula representing the Gödel number of $\mathbf{B}$ (see §9). We introduce the abbreviations:

$$\mathbf{N}_1 \to Z(H\mathbf{N}).$$
$$\mathbf{N}_2 \to Z'(H\mathbf{N}).$$

Subscripts used in this way may be iterated, so that, for instance

$$\mathbf{N}_{122} \to Z'(H(Z'(H(Z(H\mathbf{N}))))).$$

By the method of §14, find a formula $\mathfrak{B}$ such that

$$\mathfrak{B}1 \text{ conv } \lambda x \,.\, x12.$$
$$\mathfrak{B}2 \text{ conv } I.$$
$$\mathfrak{B}3 \text{ conv } \lambda x \,.\, x12J,$$

and a formula $\mathfrak{U}$ such that

$$\mathfrak{U}1 \text{ conv } \mathfrak{B}.$$
$$\mathfrak{U}2 \text{ conv } \lambda xy \,.\, y(\text{par } x_1)x_1y(y(\text{par } x_2)x_2y),$$

(these formulas $\mathfrak{B}$ and $\mathfrak{U}$ can be explicitly written down by referring to the proofs of 14 I and 14 II).

Let

$$\text{form} \to \lambda n \,.\, \mathfrak{U}(\text{par } n)n\mathfrak{U}.$$

Then

$$\text{form } 1 \text{ conv } I,$$
$$\text{form } 3 \text{ conv } J, \quad \text{and}$$
$$\text{form } \mathbf{N} \text{ conv form } \mathbf{N}_1(\text{form } \mathbf{N}_2)$$

if $\mathbf{N}$ represents an even positive integer. From this it follows that form has the property ascribed to it above; for if $\mathbf{N}$ represents the Gödel number of a combination $\mathbf{A}'$ belonging to a formula $\mathbf{A}$, containing no free variables, then form $\mathbf{N}$ conv $\mathbf{A}'$, and $\mathbf{A}'$ conv $\mathbf{A}$.

Let:

$$\begin{aligned}\sigma \to \lambda n \,.\, & [\text{par} n + \text{par } n_1 + \text{eq } \overline{24812}n_{11} + [3 \dot{-} \text{eq } \overline{156}n_{12}] + \text{par } n_2 + \text{eq } \overline{12}n_{21} \dot{-} \overline{10}] \\ & + [2\times [\text{par } n + \text{par } n_1 + \text{eq } \overline{24812}n_{11} + [3 \dot{-} \min(\text{par } n_2)(\text{eq } \overline{12}n_{21})] \dot{-} 6]] \\ & + [3\times [\text{par } n + \text{eq } \overline{623375746}n_1 + \text{par } n_2 + \text{eq } \overline{12}n_{21} + \text{par } n_{22} \\ & \quad + \text{eq } \overline{623375746}n_{221} + \text{par } n_{222} + \text{par } n_{2221} + \text{eq } \overline{24812}n_{22211} \\ & \quad + \text{par } n_{2222} + \text{par } n_{22221} + \text{eq } \overline{24812}n_{222211} + \text{eq } 3n_{22222} \dot{-} \overline{24}]] \\ & \dot{-} \; 5.\end{aligned}$$

Noting that the Gödel numbers of $JI, \tau, J\tau, J\tau\tau$ are respectively 12, 156, 24812, 623375746, the reader may verify that:

$\sigma\mathbf{N}$ conv 1, 2, 3, or 4 if $\mathbf{N}$ represents a positive integer;

$\sigma\mathbf{N}$ conv 2 if $\mathbf{N}$ represents the Gödel number of a combination of the form $J\tau\mathbf{B}(JI\mathbf{A})$, with $\mathbf{B}$ different from τ;

$\sigma\mathbf{N}$ conv 3 if $\mathbf{N}$ represents the Gödel number of a combination of the form $J\tau\mathbf{BA}$ but not of the form $J\tau\mathbf{B}(JI\mathbf{A})$;

$\sigma\mathbf{N}$ conv 4 if $\mathbf{N}$ represents the Gödel number of a combination of the form $J\tau\tau(JI(J\tau\tau(J\tau\mathbf{B}(J\tau\mathbf{A}J))))$;

$\sigma\mathbf{N}$ conv 1 if $\mathbf{N}$ represents the Gödel number of a combination not of these three forms.

Again using §14, we find a formula u such that

$u1$ conv $\lambda xy \,.\, y5x$,
$u2$ conv $\lambda xy \,.\, y(\sigma x_{12})x_{12}y$,
$u3$ conv $\lambda xy \,.\, y(\sigma x_2)x_2y$,
$u4$ conv $\lambda xy \,.\, \min(y(\sigma x_{22212})x_{22212}y)(y(\sigma x_{222212})x_{222212}y)$,
$u5$ conv $\lambda x \,.\, 3 \dot{-} x$,

and we let

$$\mathrm{o} \to \lambda n \,.\, u(\sigma n)nu.$$

Then o λ-defines a function of positive integers whose value is 2 for an argument which is the Gödel number of a combination of the form $\lambda_x\mathbf{M}|$, and 1 for an argument which is the Gödel number of a combination not of this form—or, as we shall say briefly, o λ-defines the property of a combination of being of the form $\lambda_x\mathbf{M}|$.

By similar construction, involving lengthy detail but nothing new in principle, the following formulas may be obtained:

1) A formula, occ; such that, if $\mathbf{N}$ represents a positive integer n, we have that occ $\mathbf{N}$ λ-defines the property of a combination of containing the nth variable in alphabetical order, as a free variable (i.e., as a term).

2) A formula $\mathfrak{e}$, such that, $\mathbf{N}$ representing a positive integer n, if $\mathbf{G}$ represents the Gödel number of a combination not of the form $\lambda_x\mathbf{M}|$, then $\mathfrak{e}\mathbf{NG}$ conv $\mathbf{G}$, and if $\mathbf{G}$ represents the Gödel number of a combination $\lambda_x\mathbf{M}|$, then $\mathfrak{e}\mathbf{NG}$ is convertible into the formula representing the Gödel number of the combination obtained from $\mathbf{M}$ by substituting for all free occurrences of $\boldsymbol{x}$ in $\mathbf{M}$ the nth variable in alphabetical order.

3) A formula $\mathfrak{G}$, such that, if $\mathbf{G}$ represents the Gödel number of a combination not of the form $\lambda_x\mathbf{M}|$, then $\mathfrak{G}\mathbf{G}$ conv $\mathbf{G}$, and if $\mathbf{G}$ represents the Gödel number of a combination $\lambda_x\mathbf{M}|$, then $\mathfrak{G}\mathbf{G}$ is convertible into the formula representing the Gödel number of the combination obtained from $\mathbf{M}$ by substituting for all free occurrences of $\boldsymbol{x}$ in $\mathbf{M}$ the first variable in alphabetical order which does not occur in $\mathbf{M}$ as a free variable.

4) A formula $\mathfrak{r}$ which λ-defines the property of a combination, that there is a formula to which it belongs.

5) A formula $\wedge$ which λ-defines the property of a combination of belonging to a formula of the form $\lambda\boldsymbol{x}\mathbf{M}$.

6) A formula, prim, which λ-defines the property of a combination of containing no free variables.

7) A formula, norm, which λ-defines the property of a combination of belonging to a formula which is in normal form.

8) A formula 0_1 which corresponds to the operation 0I of §15, in the sense that, if $\mathbf{G}$ represents the Gödel number of a combination of such a form that 0I is not applicable to it, then $0_1\mathbf{G}$ conv $\mathbf{G}$, and if $\mathbf{G}$ represents the Gödel number of a combination $\mathbf{M}$ to which 0I is applicable, then $0_1\mathbf{G}$ is convertible into the formula representing the Gödel number of the combination obtained from $\mathbf{M}$ by applying 0I.

9) Formulas $0_2, 0_3, \dots, 0_{38}$ which correspond respectively to the operations 0II, 0III, ..., 0XXXVIII of §15, in the same sense.

By 14 III, a formula, cb, can be found which represents an enumeration of the least set of formulas which contains 1 and 3 and is closed under the operation of forming $(\lambda ab \,.\, 2 \times \text{nr}\ ab)\mathbf{XY}$ from the formulas $\mathbf{X}, \mathbf{Y}$. But if $\mathbf{X}, \mathbf{Y}$ represent the Gödel numbers of combinations $\mathbf{A}, \mathbf{B}$ respectively, then $(\lambda ab \,.\, 2 \times \text{nr}\ ab)\mathbf{XY})$ is convertible into the formula which represents the Gödel number of $\mathbf{AB}$. Hence the formula, cb, enumerates the Gödel number of combinations containing no free variables, in the sense that every formula representing such a Gödel number is convertible into one of the formulas in the infinite sequence

$$\text{cb}\ 1,\ \text{cb}\ 2,\ \dots,$$

and every formula in this infinite sequence is convertible into a formula representing such a Gödel number.

If now we let

$$\text{ncb} \to \lambda n \,.\, \text{cb}\ (\mathscr{P}(\lambda x \,.\, \text{norm}(\text{cb}\ x))n),$$

then ncb enumerates, in the same sense, the Gödel numbers of combinations which belong to formulas in normal form and contain no free variables (cf. 10 II).

By 14 IV, a formula 0 can be found which represents an enumeration of the least set of formulas which contains I and is closed under each of the thirty-eight operations of forming $(\lambda ab \,.\, 0_\beta(ab))\mathbf{X}$ from the formula $\mathbf{X}$ ($\beta = 1, 2, \dots, 38$). Let

$$\text{cnvt} \to \lambda ab \,.\, 0ba.$$

Then if $\mathbf{G}$ represents the Gödel number of a combination $\mathbf{M}$, the formula, cnvt $\mathbf{G}$, enumerates (again in the same sense as in the two preceding paragraphs) the Gödel numbers of combinations obtainable from $\mathbf{M}$ by conversion—cf. §15.

Let

$$\text{nf} \to \lambda n \,.\, \text{cnvt}\ n(\mathfrak{p}(\lambda x \,.\, \text{norm}\ (\text{cnvt}\ nx))1).$$

Then nf λ-defines the operation *normal form of* a formula, in the sense that (1) if $\mathbf{G}$ represents the Gödel number of a combination $\mathbf{M}$, then nf $\mathbf{G}$ is convertible into the formula representing the Gödel number belonging to the normal form of $\mathbf{M}$; and hence (2) if $\mathbf{G}$ represents the Gödel number belonging to a formula $\mathbf{M}$, then nf $\mathbf{G}$ is convertible into the formula representing the Gödel number belonging to the normal form of $\mathbf{M}$. If $\mathbf{G}$ represents the Gödel number of a combination (or belonging to a formula) which has no normal form, then nf $\mathbf{G}$ has no normal form (cf. 10 I).

Let i and s be the formulas representing the Gödel numbers belonging to 1 and S respectively. Then the formulas

$$Z'(H(1(\lambda x \,.\, 2 \times \text{nr}\ sx)i)),\quad Z'(H(2(\lambda x \,.\, 2 \times \text{nr}\ sx)i)),$$
$$Z'(H(3(\lambda x \,.\, 2 \times \text{nr}\ sx)i)), \dots,$$

are convertible respectively into formulas representing Gödel numbers belonging to

$$1,\ S1,\ S(S1),\ \ldots .$$

Hence a formula v which λ-defines the property of a combination of belonging to a formula in normal form which represents a positive integer, may be obtained by defining

$$v \rightarrow \lambda n \,.\, \pi(\mathrm{eq}\ n)(\lambda m \,.\, \mathrm{eq}\ n(\mathrm{nf}(Z'(H(m(\lambda x \,.\, 2 \times \mathrm{nr}\ sx)i))))).$$

(It is necessary, in order to see this, to refer to 10 III, and to observe that the Gödel number belonging to a formula in normal form representing a positive integer is always greater than that integer.)

Chapter V

THE CALCULI OF λ-K-CONVERSION AND λ-δ-CONVERSION

17. THE CALCULUS OF λ-K-CONVERSION. The calculus of *λ-K-conversion* is obtained if a single change is made in the construction of the calculus of λ-conversion which appears in §§5, 6: namely, in the definition of *well-formed formula* (§5) deleting the words "and contains at least one free occurrence of $\boldsymbol{x}$" from the rule 3. The rules of conversion, I, II, III, in §6 remain unchanged, except that *well-formed* is understood in the new sense.

Typical of the difference between the calculi of λ-conversion and λ-K-conversion is the possibility of defining in the latter the constancy function,

$$K \to \lambda a(\lambda ba),$$

and the integer zero, by analogy with definition of the positive integers in §8,

$$0 \to \lambda a(\lambda bb).$$

Many of the theorems of §7 hold also in the calculus of λ-K-conversion. But obvious minor modifications must be made in 7 III and 7V, and the following theorems fail: 7 XVII, clause (3) of 7 XXVI, and 7 XXXI, and 7 XXXII. Instead of 7 XXXI, the following weaker theorem can be proved, which is sufficient for certain purposes, in particular for the definition of $\mathfrak{p}$ (see §10):

17 I. Let a reduction be called of order one if the application of Rule II involved is a contraction of the initial $(\lambda\boldsymbol{x}\mathbf{M})\mathbf{N}_1$ in a formula of the form

$$(\lambda\boldsymbol{x}\mathbf{M})\mathbf{N}_1\mathbf{N}_2\dots\mathbf{N}_r, \qquad (r = 1, 2, \dots).$$

Then if $\mathbf{A}$ has a normal form, there is a number m such that at most m reductions of order one can occur in a sequence of reductions on $\mathbf{A}$.

A notion of *λ-K-definability* of functions of non-negative integers may be introduced, analogous to that of λ-definability of functions of positive integers, and the development of Chapter III may then be completely paralleled in the calculus of λ-K-conversion. The same definitions may be employed for the successor function and for addition and multiplication as in Chapter III. Many of the developments are simplified by the presence of the zero: in particular, ordered pairs may be employed instead of ordered triads in the definition of the predecessor function, and the definition of $\mathfrak{p}$ may be simplified as in Turing [58].

It can be proved (see Kleene [37], Turing [57]) that a function F of one non-negative integer argument is λ-K-definable if and only if $\lambda x \,.\, F(x-1)+1$ is λ-definable—and similarly for functions of more than one argument.

The calculus of λ-K-conversion has obvious advantages over the calculus of λ-conversion, including the possibility of defining the constancy function and of introducing the integer zero in a simpler and more natural way. However, for many purposes—in particular for the development of a system of symbolic logic such as that sketched in §21 below—these advantages are more than offset by the failure of 7 XXXII. Indeed if we regard those and only those formulas as meaningful which have a normal form, it becomes clearly unreasonable that $\mathbf{FN}$ should have a normal form and $\mathbf{N}$ have no normal form (as may happen in the calculus of λ-K-conversion); or even if we impose a more stringent condition of meaningfulness, Rule III of the calculus of λ-K-conversion can be objected to on the ground that if **M** is a meaningful formula containing no free variables, the substitution of $(\lambda x\mathbf{M})\mathbf{N}$ for **M** ought not to be possible unless **N** is meaningful. This way of putting the matter involves the meanings of the formulas, and thus an appeal to intuition, but corresponding difficulties do appear in the formal developments in certain directions.

18. THE CALCULUS OF RESTRICTED λ-K-CONVERSION. In order to avoid the difficulty just described, Bernays [4] has proposed a modification of the calculus of λ-K-conversion which consists in adding to Rules II and III the proviso that **N** shall be in normal form (notice that the condition of being in normal form is effective, although that of having a normal form is not). We shall call the calculus so obtained the *calculus of restricted λ-K-conversion*. In it, as follows by the methods of §7, a formula which in the calculus of λ-K-conversion has a normal form and has no parts without normal form will continue to have the same normal form; in particular, no possibility of conversion into a normal form is lost which existed in the calculus of λ-conversion. On the other hand, all of the theorems 7 XXVIII–7 XXXII remain valid in the calculus of restricted λ-K-conversion—and are much more simply proved than in the calculus of λ-conversion. (It should be added that the content of the theorems 7 XXVIII–7 XXXII for the calculus of restricted λ-K-conversion is in a certain sense much less than the content of these theorems for the calculus of λ-conversion, and in fact cannot be regarded as sufficient to establish the satisfactoriness of the calculus of restricted λ-K-conversion from an intuitive viewpoint without addition of such a theorem as that asserting the equivalence to the calculus of (unrestricted) λ-K-conversion in the case of formulas all of whose parts have normal forms.)

The development of the calculus of restricted λ-K-conversion may follow closely that of the calculus of λ-conversion (as in Chapter II–IV), with such modifications as are indicated in §17 for the calculus of λ-K-conversion. Many of the theorems must have added hypotheses asserting that certain of the formulas involved have normal forms.

19. TRANSFINITE ORDINALS. Church and Kleene [15] have extended the concept of λ-definability to ordinal numbers of the second number class and functions of such ordinal numbers. There results from this on the one hand an extension of the notion of effective calculability to the second number class (cf. Church [13], Kleene [39], Turing [59]), and on the other hand a method of introducing some theory of ordinal numbers into the system of symbolic logic of §21 below.

Instead of reproducing here this development within the calculus of λ-conversion, we sketch briefly an analogous development within the calculus of restricted λ-K-conversion.

According to the idea underlying the definition of §8, the positive integers (or the non-negative integers) are certain functions of functions, namely the finite powers of a function in the sense of iteration. This idea might be extended to the ordinal numbers of the second number class by allowing them to correspond in the same way to the transfinite powers of a function, provided that we first fixed upon a limited process relative to which the transfinite powers should be taken. Thus the ordinal ω could be taken as the function whose value for a function f as an argument is the function g such that gx is the limit of the sequence $x, fx, f(fx), \ldots$. Then $\omega + 1$ would be

$\lambda x \centerdot f(\omega f x)$, and so on.

Or, instead of fixing upon a limiting process, we may introduce the limiting process as an additional argument a (for instance taking the ordinal ω to be the function whose value for a and f as arguments is the function g such that gx is the limit of the sequence $x, fx, f(fx), \ldots$, relative to the limiting process a). This leads to the following definition in the calculus of restricted λ-K-conversion, the subscript o being used to distinguish these notations from similar notations used in other connections:

$$0_o \rightarrow \lambda a(\lambda b(\lambda cc)),$$
$$1_o \rightarrow \lambda abc \centerdot bc,$$
$$2_o \rightarrow \lambda abc \centerdot b(bc), \qquad \text{and so on.}$$
$$S_o \rightarrow \lambda dabc \centerdot b(dabc).$$
$$L_o \rightarrow \lambda rabc \centerdot a(\lambda d \centerdot rdabc).$$
$$\omega_o \rightarrow \lambda abc \centerdot a(\lambda d \centerdot dabc).$$

We prescribe that 0_o shall represent the ordinal 0; if $\mathbf{N}$ represents the ordinal n, the principal normal form of $S_o\mathbf{N}$ shall represent the ordinal $n+1$; if $\mathbf{R}$ represents the monotone increasing infinite sequence of ordinals $n_0, n_1, n_2, \ldots$, in the sense that $\mathbf{R}_o^0, \mathbf{R}_o^1, \mathbf{R}_o^2, \ldots$ are convertible into formulas representing $n_0, n_1, n_2, \ldots$, respectively, then the principal normal form of $L_o\mathbf{R}$ shall represent the upper limit of this infinite sequence of ordinals. The transfinite ordinals which are represented by formulas then turn out to constitute a certain segment of the second number class, which may be described as consisting of those ordinals which can be effectively built up from below (in a sense which we do not make explicit here).

The formula representing a given ordinal of the second number class is not unique: for example, the ordinal ω is represented not only by ω_o but also by the principle normal form L_oS_o, and by many other formulas. Hence the formula representing ordinals are not to be taken as denoting ordinals but rather as denoting certain things which are in many-one correspondence with ordinals.

A function F of ordinal members is said to be *λ-K-defined* by a formula $\mathbf{F}$ if (1) whenever $Fm = n$ and $\mathbf{M}$ represents m, the formula $\mathbf{FM}$ is convertible into a formula representing n, and (2) whenever an ordinal m is not in the range of $\mathbf{F}$ and $\mathbf{M}$ represents m, the formula $\mathbf{FM}$ has no normal form.

The foregoing account presupposes the classical second number class. By suitable modifications (cf. Church [13]), this presupposition may be eliminated, with the result that the calculus of restricted λ-K-conversion is used to obtain a definition of a (non-classical) *constructive* second number class, in which each classical ordinal is represented, if at all, by an infinity of elements.

20. THE CALCULUS OF λ-δ-CONVERSION. The *calculus of λ-δ-conversion* is obtained by making the following changes in the construction of the calculus of λ-conversion which appears in §§5,6: adding to the list of primitive symbols a symbol δ, which is neither an improper symbol nor a variable, but is classed with the variables as a *proper symbol*; adding to the rule 1 in the definition of *well-formed formulas* that the symbol δ is a well-formed formula; and adding to the rules of conversion in §6 four additional rules, as follows:

IV. To replace any part $\delta\mathsf{MN}$ of a formula by 1, provided that M and N are in δ-normal form and contain no free variables and M is not convertible-I into N.
V. To replace any part 1 of a formula by $\delta\mathsf{MN}$, provided that M and N are in δ-normal form and contain no free variables and M is not convertible-I into N.
VI. To replace any part $\delta\mathsf{MM}$ of a formula by 2, provided that M is in δ-normal form and contains no free variables.
VII. To replace any part 2 of a formula by $\delta\mathsf{MM}$, provided that M is in δ-normal form and contains no free variables.

Here a formula is said to be in *δ-normal* form if it contains no part of the form $(\lambda x\mathsf{P})\mathsf{Q}$ and contains no part of the form $\delta\mathsf{RS}$ with R and S containing no free variables. It is necessary to observe that both the condition of being in δ-normal form and the condition that M is not convertible-I into N are effective.

A *conversion* (or a *λ-δ-conversion*) is a finite sequence of applications of Rules I–VII. A λ-δ-conversion is called a *reduction* (or a *λ-δ-reduction*) if it contains no application of Rules III, V, VII and exactly one application of one of the Rules II, IV, VI. A is said to be *immediately reducible to* B if there is a reduction of A into B, and A is said to be *reducible* to B if there is a conversion of A into B which consists of one or more successive reductions.

All the theorems of §7 hold also in the calculus of λ-δ-conversion, if some appropriate modifications are made (see Church and Rosser [16]). The *residuals* of $(\lambda x_p\mathsf{M}_p)\mathsf{N}_p$ after an application of Rule I or II are defined in the same way as before, and after an application of IV or VI they are defined as what $(\lambda x_p\mathsf{M}_p)\mathsf{N}_p$ becomes (this is always something of the form $(\lambda x_p\mathsf{M}'_p)\mathsf{N}'_p$). The residuals of $\delta\mathsf{M}_p\mathsf{N}_p$ after an application of I, II, IV, or VI are defined only in the case that M_p and N_p are in δ-normal form and contain no free variables. In that case the residuals of $\delta\mathsf{M}_p\mathsf{N}_p$ are whatever part or parts of the entire resulting formula $\delta\mathsf{M}_p\mathsf{N}_p$ becomes, except that after an application of IV or VI in which $\delta\mathsf{M}_p\mathsf{N}_p$ itself is contracted (i.e., replaced by 1 or 2), $\delta\mathsf{M}_p\mathsf{N}_p$ has no residual. Thus residuals of $\delta\mathsf{M}_p\mathsf{N}_p$ are always of the form $\delta\mathsf{MN}$, where M and N are in δ-normal form and contain no free variables. A *sequence of contractions on a set of parts* $(\lambda x_j\mathsf{M}_j)\mathsf{N}_j$ and $\delta\mathsf{R}_i\mathsf{S}_i$ of A_1, where R_i and S_i are in δ-normal form and contain no free variables, is defined by analogy with the definition in §7. Similarly a *terminating* sequence of such contractions. In 7 XXV, the set of parts of A on which a sequence of contractions is taken is allowed to include not only parts of the form $(\lambda x_j\mathsf{M}_j)\mathsf{N}_j$, but also parts of the form $\delta\mathsf{R}_i\mathsf{S}_i$ in which R_i and S_i are in δ-normal form and contain no free variables. The modified 7 XXV may then be proved by an obvious extension of the proof given in §7, and thereupon 7 XXVI–7 XXXII follow as before. In 7 XXVI–7 XXXII "conv-I-II" must be replaced throughout by "conv-I-II-IV-VI" and in 7 XXVI the case must also be considered that A imr B by a contraction of the part $\delta\mathsf{MN}$ of A. For 7 XXX, there must be supplied a definition of *principle δ-normal form of* a formula, analogous to the definition in §6 of the principle (λ-)normal form.

In connection with the calculus of λ-δ-conversion we shall use both of the terms *λ-conversion* and *λ-δ-conversion*, the former meaning a finite sequence of applications of Rules I–III, the latter a finite sequence of applications of Rules I–VII. The term *conversion* will be used to mean a λ-δ-conversion, as already explained.

Similarly we shall use both of the terms *λ-normal form of* a formula and *δ-normal form of* a formula. A formula will be called a λ-normal form of another if it is in

λ-normal form and can be obtained from the other by λ-conversion. A formula will be called a δ-normal form of another if it is in δ-normal form and can be obtained from the other by λ-δ-conversion. By 7 XXIX applied to the calculus of λ-conversion, the λ-normal form of a formula (in the calculus of λ-δ-conversion), if it exists, is unique to within application of Rule I. By the analogue of 7 XXIX for the calculus of λ-δ-conversion, the δ-normal form of a formula, if it exists, is unique to within applications of Rule I.

In order to see that the calculus of λ-δ-conversion requires an intensional interpretation (cf. §2), it is sufficient to observe that, for example, although 1 and $\lambda ab \,.\, \delta ab1ab$ correspond to the same function in extension, they are nevertheless not interchangeable, since $\delta 11$ conv 2 but $\delta 1(\lambda ab \,.\, \delta ab1ab)$ conv 1.

A constancy function κ may be defined:

$$\kappa \rightarrow \lambda ab \,.\, \delta bbIa.$$

Then $\kappa\mathbf{AB}$ conv $\mathbf{A}$, if $\mathbf{B}$ has a δ-normal form and contains no free variables, and in that case only (the conversion properties of κ are thus weaker than those of the formula K in either of the calculi of λ-K-conversion).

The entire theory of λ-definability of functions of positive integers carries over into the calculus of λ-δ-conversion, since the calculus of λ-conversion is contained in that of λ-δ-conversion as a part. It only requires proof that the notion of λ-δ-definability of functions of positive integers is not more general than that of λ-definability, and this can be supplied by known methods (e.g., those of Kleene [37]).

The theory of combinations carries over into the calculus of λ-δ-conversion, provided that we redefine a *combination* to mean an $[I, J, \delta]$-combination. In defining the *combination belonging to* a formula, it is necessary to add the provision that the combination belonging to δ is δ.

If $\mathbf{A}_1$ is a well-formed formula of the calculus of λ-δ-conversion and contains no free variables, a formula $\mathbf{B}_1$ can be found such that $\mathbf{B}_1 J$ conv $\mathbf{A}_1$ and $\mathbf{B}_1 I$ conv I. For let $\mathbf{A}_1''$ be the combination belonging to $\mathbf{A}_1$, unless that combination fails to contain an occurrence of either J or δ, in which case let $\mathbf{A}_1'$ be $JIIII$. Let $\mathbf{A}_1''$ be obtained from $\mathbf{A}_1'$ by replacing J and δ throughout by j and $\delta Ij(\lambda x \,.\, x(\lambda y \,.\, yIII))(\lambda z \,.\, zI)\delta$ respectively. Then $\mathbf{B}_1$ may be taken as $\lambda j \mathbf{A}_1''$.

Hence 14 I, and the remaining theorems of §14, may be proved for the calculus of λ-δ-conversion in the same way as for the calculus of λ-conversion.

In order to obtain a combinatory equivalent of λ-δ-conversion, analogous to the combinatory equivalent of λ-conversion given in §15, it is necessary to add to 0I–0XXXVIII the following four additional operations — where $\mathbf{F}$, $\mathbf{A}$, $\mathbf{B}$, $\mathbf{C}$ are combinations, and $\mathbf{A}$ and $\mathbf{B}$ belong to formulas in δ-normal form, contain no free variables, and are not the same, and $\mathbf{C}$ belongs to the formula which represents the Gödel number of $\mathbf{A}$:

0XXXIX. $\mathbf{F}(\delta\mathbf{AB}) \vdash \mathbf{F}(\beta I)$.

0XL. $\mathbf{FAB}(\beta I) \vdash \mathbf{FAB}(\delta\mathbf{AB})$.

0XLI. $\mathbf{F}(\delta\mathbf{AA}) \vdash \mathbf{F}(\omega\beta)$.

0XLII. $\mathbf{FC}(\omega\beta) \vdash \mathbf{FC}(\delta\mathbf{AA})$.

The reader should verify that the conditions on **A**, **B**, **C** — although complex in character — are effective (§6).

In order to see that these four operations are equivalent, in the presence of 0I–0XXXVIII, to the rules of conversion IV–VII, it is necessary to observe that βI and $\omega\beta$ are λ-convertible into 1 and 2 respectively.

To show that 0XLII provides an equivalent to Rule VII, we must show that it enables us to change $\mathbf{G}(\omega\beta)$ into $\mathbf{G}(\delta\mathbf{AA})$. Since 0I–0XXXVIII are equivalent to λ-conversion, this can be done as follows: $\mathbf{G}(\omega\beta)$ is λ-convertible into $\gamma(\tau I)\mathbf{GC}(\omega\beta)$, and this becomes, by 0XLII, $\gamma(\tau I)\mathbf{GC}(\delta\mathbf{AA})$, and this in turn is λ-convertible into $\mathbf{G}(\delta\mathbf{AA})$.

Similarly, to show that 0XL provides an equivalent of Rule V, we must show that it enables us to change $\mathbf{G}(\beta I)$ into $\mathbf{G}(\delta\mathbf{AB})$. This can be done as follows: $\mathbf{G}(\beta I)$ is λ-convertible into $\gamma(\gamma(\tau I)\mathbf{G})(\beta I)(\omega\beta)$; and this can be changed by the method of the preceding paragraph into $\gamma(\gamma(\tau I)\mathbf{G})(\beta I)(\delta\mathbf{BB})$; and this is λ-convertible into $\gamma(\gamma(\tau I)(\gamma(\gamma(\tau I)\mathbf{G})(\beta I)))(\delta\mathbf{BB})(\omega\beta)$; and this can be changed by the method of the preceding paragraph into $\gamma(\gamma(\tau I)(\gamma(\gamma(\tau I)\mathbf{G})(\beta I)))(\delta\mathbf{BB})(\delta\mathbf{AA})$; and this is λ-convertible into $\gamma(\beta(\gamma(\beta\gamma(\gamma(\gamma(\beta(\beta\beta)(\omega\delta))I)(\gamma(\gamma(\tau I)\mathbf{G})))))(\omega\delta))\mathbf{AB}(\beta I)$; and this becomes, by 0XL, $\gamma(\beta(\gamma(\beta\gamma(\gamma(\gamma(\beta(\beta\beta)(\omega\delta))I)(\gamma(\gamma(\tau I)\mathbf{G})))))(\omega\delta))\mathbf{AB}(\delta\mathbf{AB})$; and this is λ-convertible into $\gamma(\gamma(\tau I)(\gamma(\gamma(\tau I)\mathbf{G})(\delta\mathbf{AB})))(\delta\mathbf{BB})(\delta\mathbf{AA})$; and this becomes, by 0XLI, $\gamma(\gamma(\tau I)(\gamma(\gamma(\tau I)\mathbf{G})(\delta\mathbf{AB})))(\delta\mathbf{BB})(\omega\beta)$; and this is λ-convertible into $\gamma(\gamma(\tau I)\mathbf{G})(\delta\mathbf{AB}(\delta\mathbf{BB})$; and this becomes, by 0XLI, $\gamma(\gamma(\tau I)\mathbf{G})(\delta\mathbf{AB})(\omega\beta)$; and this, finally, is λ-convertible into $\mathbf{G}(\delta\mathbf{AB})$.

Only minor modifications are necessary in §16 in order to carry over its results to the calculus of λ-δ-conversion. In the definition of the Gödel number of a combination the clause must be added: (2a) The Gödel number of δ is 5. In the construction of the formula, form, it is only necessary to impose on $\mathfrak{B}$ the further condition that $\mathfrak{B}5 \operatorname{conv} \lambda x \,.\, x12\delta$, so insuring that form 5 conv δ. The construction of o remains unchanged. The formulas occ, $\mathfrak{e}$, $\mathfrak{G}$, $\mathfrak{r}$, $\wedge$, prim, norm, and 0_1–0_{38} may then be obtained, having the properties described in §16 (norm λ-defines the property of a combination of belonging to a formula which is in λ-normal form). The formulas cb, ncb, 0, cnvt, nf (*the λ-normal form of*), and v may then also be obtained as before. The formula, cb, represents an enumeration of the least set of formulas which contains 1, 3, and 5 and is closed under the operation of forming $(\lambda ab \,.\, 2 \times \text{nr}\ ab)\mathbf{XY}$ from the formulas **X**, **Y**.

Besides norm it is also possible to obtain a formula, dnorm, which λ-defines the property of a combination of belonging to a formula in δ-normal form. Details of this are left to the reader.

Formulas 0_{39}–0_{42} may be obtained, related to the operations 0XXXIX–0XLII in the same way that 0_1–0_{38} are related to 0I–0XXXVIII. We give details in the case of 0_{40} and 0_{42}. Let $\mathfrak{F}_{40}$ be a formula such that $\mathfrak{F}_{40}1 \operatorname{conv} I$ and $\mathfrak{F}_{40}2 \operatorname{conv} \lambda x \,.\, 2 \times \text{nr}\ x_1[2 \times \text{nr}\ 5x_{112}]x_{12}]$; then let

$$0_{40} \to \lambda x \,.\, \mathfrak{F}_{40}[\text{par}\, x + \text{par}\, x_1 + \text{par}\, x_{11} + \text{prim} x_{112} + \text{dnorm} x_{112} + \text{prim} x_{12} + \text{dnorm} x_{12} + \text{eq}\ \eta x_2 \dot{-} \text{eq}\ x_{112}x_{12} \dot{-} \overline{13}]x,$$

η being the formula representing the Gödel number of βI. Let $\mathfrak{F}_{42}$ be a formula such that $\mathfrak{F}_{42}1$ conv I and $\mathfrak{F}_{42}2$ conv $\lambda x \,.\, 2 \times \text{nr}\ x_1[2 \times \text{nr}[2 \times \text{nr}\ 5(\text{form}\ x_{12})](\text{form}\ x_{12})]$; then let

$$0_{42} \rightarrow \lambda x \,.\, \mathfrak{F}_{42}[\text{par}\, x + \text{par}\, x_1 + h(\nu x_{12})x_{12} + \text{eq}\ \zeta x_2 \dot{-} 6]x,$$

where ζ is the formula representing the Gödel number of $\omega\beta$, and h is such a formula that $h1$ conv $\lambda x \,.\, x1$ and $h2$ conv $\lambda x \,.\, \text{min}(\text{prim}(\text{form}\ x))(\text{dnorm}(\text{form}\ x))$.

Then a formula, do, may be obtained, analogous to o but involving all of 0_1–0_{42} instead of only 0_1–0_{38}. Let

$$\text{dcnvt} \rightarrow \lambda ab \,.\, \text{do}\ ba.$$

Then, if **G** represents the Gödel number of a combination **M**, the formula, dcnvt **G**, enumerates the Gödel numbers of combinations obtainable from **M** by λ-δ-conversion (whereas cnvt **G** enumerates merely the Gödel numbers of combinations obtainable from **M** by λ-conversion).

It is also possible, by using the formula, dnorm, to obtain a formula, dnf, which λ-defines the operation *δ-normal form of* a formula, and a formula, dncb, which enumerates the Gödel numbers of combinations which belong to formulas in δ-normal form and contain no free variables. The definitions parallel those of nf and ncb.

Finally, in the calculus of λ-δ-conversion, a formula, met, may be obtained which provides a kind of inverse of the function, form: if **M** is a formula which contains no free variables and has a δ-normal form, then met **M** is convertible into the formula representing the Gödel number belonging to the δ-normal form of **M**. The definition is as follows:

$$\text{met} \rightarrow \lambda \,.\, \text{dncb}\ (\mathfrak{p}(\lambda n \,.\, \delta(\text{form}\ (\text{dncb}\ n))x)1).$$

21. A SYSTEM OF SYMBOLIC LOGIC. If we identify the truth values, truth and falsehood, with the positive integers 2 and 1 respectively, we may base a system of symbolic logic on the calculus of λ-δ-conversion. This system has one primitive formula or axiom, namely the formula 2, and seven rules of inference, namely the rules I–VII of λ-δ-conversion; the provable formulas, or theses, of the system are the formulas which can be derived from the formula 2 by sequences of applications of the rules of inference. (As a matter of fact, the rules of inference II, IV, VI are superfluous, in the sense that their omission would not decrease the class of provable formulas, as follows from 7 XXVII, or rather from the analogue of this theorem for the calculus of λ-δ-conversion.)

The identification of the truth values, truth and falsehood, with the positive integers 2 and 1 is, of course artificial, but apparently it gives rise to no actual formal difficulty. If it be thought objectionable, the artificiality may be avoided by a minor modification in the system, which consists in introducing a symbol $\vdash$ and writing $\vdash 2$, instead of 2, as the primitive formula; all the theses of the system will then be preceded by the sign $\vdash$, which may be interpreted as asserting that that which follows is equal to 2.

In this system of symbolic logic the fundamental operations of the propositional calculus — negation, conjunction, disjunction — may be introduced by the following definitions:

$$[\sim \mathsf{A}] \to \pi(\lambda a \,.\, aI(\delta 2\mathsf{A}))(\lambda a \,.\, aI(\delta 1\mathsf{A})).$$
$$[\mathsf{A} \,\&\, \mathsf{B}] \to 4 \dot{-} \,.[\sim \mathsf{A}] \div [\sim \mathsf{B}].$$
$$[\mathsf{A} \vee \mathsf{B}] \to \sim .[\sim \mathsf{A}] \&[\sim \mathsf{B}].$$

It follows from these definitions that $\mathsf{A} \vee \mathsf{B}$ cannot be a thesis unless either **A** or **B** is a thesis — and this situation apparently cannot be altered by any suitable change in the definitions. Since this property is known to fail for classical systems of logic, e.g., that of Whitehead and Russell's *Principia Mathematica*, it is clear that the present system therefore differs from the classical systems in a direction which may be regarded as finitistic in character.

Functions of positive integers are of course represented in the system by the formulas λ-defining these functions, and properties of and relations between positive integers are represented by the formulas λ-defining the corresponding characteristic functions. The propositional function *to be a positive integer* is represented in the system as a formula N, defined as follows (referring to §§16, 20):

$$N \to \lambda x \,.\, v(\text{met } x).$$

The general relation of equality or identity (in intension) is represented by δ.

An existential quantifier Σ may be introduced:

$$\iota \to \lambda f \,.\, \text{form } (Z'(H(\text{dcnvt } \alpha(\mathfrak{p}(\lambda n \,.\, \delta f$$
$$(\text{form } (Z(H(\text{dcnvt } \alpha n)))))1)))),$$

where α is the formula representing the Gödel number belonging to the formula 2;

$$\Sigma \to \lambda f \,.\, f(\iota f).$$

Here ι represents a general selection operator. Given a formula **F**; if there is any formula **A** such that **FA** conv 2, then ι**F** is one of the formulas **A** having this property; and in the contrary case ι**F** has no normal form. Consequently Σ represents an existential quantifier without a negation: Σ**F** conv 2 if there is a formula **A** such that **FA** conv 2, and in the contrary case Σ**F** has no normal form.

The operator ι should be compared with Hilbert's operator ε [31 and elsewhere], or, perhaps better, the η-operator of Hilbert and Bernays [33]. The ι should be used with the caution that the equivalence of propositional functions represented in the system by **F** and **G** need not imply the equality of ι**F** and ι**G**.

The interpretation of ι as a selection operator and of Σ as an existential quantifier depends on an identification of formal provability in the system with truth. But this is justified by a completeness property which the system possesses: a formula which is not provable, unless it is convertible into a principle normal form other than 2 and hence is disprovable, must have no normal form, and hence be meaningless.

For convenience in the further development of the system, or for the sake of comparison with more usual notations, we may introduce the abbreviations:

$$[\iota x\mathsf{M}] \to \iota(\lambda x\mathsf{M}).$$

$$[\exists x\mathsf{M}] \to \Sigma(\lambda x\mathsf{M}).$$

The problem of introducing universal quantifiers into the system, or, equivalently, of introducing existential quantifiers having a negation, is beyond the scope of the present treatise. It follows by the methods of Gödel [27] that any universal quantifier introduced by definition will have a certain character of incompleteness; this is in effect the same incompleteness property which, in accordance with the results of Gödel, almost any consistent and satisfactorily adequate system of formal logic must have, except that it here appears transferred from the realm of provability to the realm of meaning of the quantifiers.

The *consistency* of the system of symbolic logic just outlined is a corollary of 7 XXX, or rather of the analogue of this theorem for the calculus of λ-δ-conversion. This consistency proof is of a strictly constructive or finitary nature.

(The failure in this system of the known paradoxes of set theory depends, in some of the simpler cases, merely on the fact that the formula which would otherwise lead to the paradox fails to have a normal form. Thus, in the case of Russell's paradox, we find that $(\lambda x \,.\, \sim(xx))(\lambda x \,.\, \sim(xx))$ has no normal form; and in the case of Grelling's paradox concerning heterological words, or, as we shall put it, concerning heterological Gödel numbers, we find that $(\lambda x \,.\, \sim(\text{form}\, xx))(\text{met}\, (\lambda x \,.\, \sim(\text{form}\, xx)))$ has no normal form. In more complicated cases, where the expression of the paradox requires a universal quantifier, the failures may depend on the above indicated incompleteness property of the quantifier.)

INDEX OF THE PRINCIPAL FORMULAS INTRODUCED BY DEFINITION

§5. I, $[\mathsf{M}+\mathsf{N}]$, $[\mathsf{M}\times\mathsf{N}]$, $[\mathsf{M}^{\mathsf{N}}]$.

§8. 1, 2, 3, ... , S, $[\mathsf{M}+\mathsf{N}]$, $[\mathsf{M}\times\mathsf{N}]$, $[\mathsf{M}^{\mathsf{N}}]$.

§9. $[\mathsf{M}, \mathsf{N}]$, $[\mathsf{L}, \mathsf{M}, \mathsf{N}]$, 2_1, 2_2, 3_1, 3_2, 3_3, P, $[\mathsf{M} \dot{-} \mathsf{N}]$, min, max, par, H, Z, Z', nr.

§10. exc, eq, $\mathfrak{p}$, $\mathscr{P}$, π.

§12. I, J, τ.

§13. B, C, W, T, D.

§15. γ, β, ω.

§16. form, o, occ, $\mathfrak{e}$, $\mathfrak{G}$, $\mathfrak{r}$, $\wedge$, prim, norm, 0_1–0_{38}, cb, ncb, 0, cnvt, nf, v.

§17. K, 0.

§19. 0_o, 1_o, 2_o, ... , S_o, L_o, ω_o.

§20. k, form, o, occ, $\mathfrak{e}$, $\mathfrak{G}$, $\mathfrak{r}$, $\wedge$, prim, norm, 0_1–0_{38}, cb, ncb, 0, cnvt, nf, v, dnorm, 0_{39}–0_{42}, do, dcnvt, dnf, dncb, met.

§21. $[\sim \mathsf{A}]$, $[\mathsf{A} \& \mathsf{B}]$, $[\mathsf{A} \vee \mathsf{B}]$, N, ι, Σ, $[\iota x\mathsf{M}]$, $[\exists x\mathsf{M}]$.

BIBLIOGRAPHY

1. Wilhelm Ackermann, *Zum Hilbertschen Aufbau der reelen Zahlen*, Mathematische Annalen, vol. 99 (1928), pp. 118–133.
2. Paul Bernays, *Sur le platonisme dans les mathématiques*, l'Enseignement mathématique, vol. 34 (1935), pp. 52–69.
3. Paul Bernays, *Quelques points essentiels de la métamathématique*, ibid., pp. 70-95.
4. Paul Bernays, Review of Church and Rosser [16], The journal of symbolic logic, vol. 1 (1936), pp. 74–75.
5. Alonzo Church, *A set of postulates for the foundation of logic*, Annals of mathematics, ser. 2, vol. 33 (1932), pp. 346–366.
6. Alonzo Church, *A set of postulates for the foundation of logic (second paper)*, ibid., ser. 2, vol. 34 (1933), pp. 839–864.
7. Alonzo Church, *The Richard paradox*, The American mathematical monthly, vol. 41 (1934), pp. 356–361.
8. Alonzo Church, *A proof of freedom from contradiction*, Proceedings of the National Academy of Sciences of the United States of America, vol. 21 (1935), pp. 275–281.
9. Alonzo Church, *An unsolvable problem of elementary number theory*, American journal of mathematics, vol. 58, (1936), pp. 345–363.
10. Alonzo Church, *Mathematical logic*, mimeographed lecture notes, Princeton University, 1936.
11. Alonzo Church, *A note on the Entscheidungsproblem*, The journal of symbolic logic, vol. 1 (1936), pp. 40–41.
12. Alonzo Church, *Correction to A note on the Entscheidungsproblem*, ibid., pp. 101–102.
13. Alonzo Church, *The constructive second number class*, Bulletin of the American Mathematical Society, vol. 44 (1938), pp. 224–232.
14. Alonzo Church, *On the concept of a random sequence*, ibid., vol. 46 (1940), pp. 130–135.
15. Alonzo Church and S. C. Kleene, *Formal definitions in the theory of ordinal numbers*, Fundamenta mathematicae, vol. 28 (1937), pp. 11–21.
16. Alonzo Church and J. B. Rosser, *Some properties of conversion*, Transactions of the American Mathematical Society, vol. 39 (1936), pp. 472–482.
17. H. B. Curry, *An analysis of logical substitution*, American journal of mathematics, vol. 51 (1929), pp. 365–384.
18. H. B. Curry, *Grundlagen der kombinatorischen Logik*, ibid., vol. 52 (1930), pp. 509–536, 789–834.
19. H. B. Curry, *The universal quantifier in combinatory logic*, Annals of mathematics, ser. 2, vol. 32 (1931), pp. 154–180.

20. H. B. Curry, *Some additions to the theory of combinators*, American journal of mathematics, vol. 54 (1932), pp. 551–558.

21. H. B. Curry, *Apparent variables from the standpoint of combinatory logic*, Annals of mathematics, ser. 2, vol. 34 (1933), pp. 381–404.

22. H. B. Curry, *Some properties of equality and implication in combinatory logic*, ibid., ser. 2, vol. 35 (1934), pp. 849–860.

23. H. B. Curry, *Functionality in combinatory logic*, Proceedings of the National Academy of Sciences of the United States of America, vol. 20 (1934), pp. 584–590.

24. H. B. Curry, *First properties of functionality in combinatory logic*, The Tôhoku mathematical journal, vol. 41 (1936), pp. 371–401.

25. H. B. Curry, Review of Church [10], The journal of symbolic logic, vol. 2 (1937), pp. 39–40.

26. Frederic B. Fitch, *A system of formal logic without an analogue to the Curry ω operator*, The journal of symbolic logic, vol. 1 (1936), pp. 92–100.

27. Kurt Gödel, *Über formal unentscheidare Sätze der Principia Mathematica und verwandter Systeme I*, Monatshefte für Mathematik und Physik, vol. 38 (1931), pp. 175–198.

28. Kurt Gödel, *On undecidable propositions of formal mathematical systems*, mimeographed lecture notes, The Institute for Advanced Study, 1934.

29. Kurt Gödel, *Über die Länge von Beweisen*, Ergebnisse eines mathematischen Kolloquiums, no. 7 (1936), pp. 23–24.

30. Jacques Herbrand, *Sur la non-contradiction de l'arithmétique*, Journal für die reine und angewandte Mathematik, vol. 166 (1931), pp. 1–8.

31. David Hilbert, *Über das Unendliche*, Mathematische Annalen, vol. 95 (1926), pp. 161–190.

32. David Hilbert and Paul Bernays, *Grundlagen der Mathematik*, vol. 1, Julius Springer, Berlin, 1934.

33. David Hilbert and Paul Bernays, *Grundlagen der Mathematik*, vol. 2, Julius Springer, Berlin, 1939.

34. S. C. Kleene, *Proof by cases in formal logic*, Annals of mathematics, ser. 2, vol. 35 (1934), pp. 529–544.

35. S. C. Kleene, *A theory of positive integers in formal logic*, American journal of mathematics, vol. 57 (1935), pp. 153–173, 219–244.

36. S. C. Kleene, *General recursive functions of natural numbers*, Mathematische Annalen, vol. 112 (1936), pp. 727–742; see [45], and [39] footnote 4.

37. S. C. Kleene, *λ-definability and recursiveness*, Duke mathematical journal, vol. 2 (1936), pp. 340-353.

38. S. C. Kleene, *A note on recursive functions*, Bulletin of the American Mathematical

Society, vol. 42 (1936), pp. 544–546.

39. S. C. Kleene, *On notation for ordinal numbers*, The journal of symbolic logic, vol. 3 (1938), pp. 150-155.

40. S. C. Kleene and J. B. Rosser, *The inconsistency of certain formal logics*, Annals of mathematics, ser. 2, vol. 36 (1935), pp. 630-636.

41. Rózsa Péter, *Über den Zusammenhang der verschiedenen Begriffe der rekursiven Funktion*, Mathematische Annalen, vol. 110 (1934), pp. 612–632.

42. Rózsa Péter, *Konstruktion nichtrekursiver Funktionen*, ibid., vol. 11 (1935), pp. 42–60.

43. Rózsa Péter, *A rekurzív függvények elméletéhez (Zur Theorie der rekursiven Funktionen)*, Matematikai és fizikai lapok, vol. 42 (1935), pp. 25–44.

44. Rózsa Péter, *Über die mehrfache Rekursion*, Mathematische Annalen, vol. 113 (1936), pp. 489–527.

45. Rózsa Péter, Review of Kleene [36], The journal of symbolic logic, vol. 2 (1937), p. 38; see *Errata*, ibid., vol. 4 (1939), p. iv.

46. Emil L. Post, *Finite combinatory processes - formulation 1*, ibid., vol. 1 (1936), pp. 103–105.

47. J. B. Rosser, *A mathematical logic without variables*, Annals of mathematics, ser. 2, vol. 36 (1935), pp. 127–150, and Duke mathematical journal, vol. 1 (1935), pp. 328–355.

48. J. B. Rosser, *Extensions of some theorems of Gödel and Church*, The journal of symbolic logic, vol. 1 (1936), pp. 87–91.

49. Moses Schönfinkel, *Über die Bausteine der mathematischen Logik*, Mathematische Annalen, vol. 92 (1924), pp. 305–316.

50. Thoralf Skolem, *Begründung der elementaren Arithmetik durch die rekkurrierende Denkweise ohne Anwendung scheinbarer Veränderlichen mit unendlichem Ausdehnungsbereich*, Skrifter utgit av Videnskapsselskapet i Kristiania, I. Matematisk-naturvidenskabelig klasse 1923, no. 6.

51. Thoralf Skolem, *Über die Zurückführbarkeit einiger durch Rekursionen definierter Relationen auf "arithmetische"*, Acta scientiarum mathematicarum, vol. 8 (1937), pp. 73–88.

52. G. Sudan, *Sur le nombre transfini* ω^ω, Bulletin mathématique de la Société Roumaine des Sciences, vol. 30 (1927), pp. 11–30.

53. Alfred Tarski, *Pojęcie prawdy w językach nauk dedukcyjnych*, Travaux de la Société des Sciences et des Lettres de Varsovie, Classe III, Sciences mathématiques et physiques, no. 34, Warsaw 1933.

54. Alfred Tarski, *Der Wahrheitsbegriff in den formalisierten Sprachen*, German translation of [53] with added *Nachwort*, Studia philosophica, vol. 1 (1935), pp. 261–405.

55. A. M. Turing, *On computable numbers, with an application to the Entscheidungsproblem*, Proceedings of the London Mathematical Society, ser. 2, vol. 42 (1936), pp. 230-265.

56. A. M. Turing, *On computable numbers, with an application to the Entscheidungsproblem, A correction*, ibid., ser. 2, vol. 43 (1937), pp. 544–546.

57. A. M. Turing, *Computability and λ-definability*, The journal of symbolic logic, vol. 2 (1937), pp. 153–163.

58. A. M. Turing, *The* 𝔭*-function in λ-K-conversion*, ibid., p. 164.

59. A. M. Turing, *Systems of logic based on ordinals*, Proceedings of the London Mathematical Society, ser. 2, vol. 45 (1939), pp. 161–228.

Addenda

60. Alonzo Church, *A formulation of the simple theory of types*, The journal of symbolic logic, vol. 5 (1940), pp. 56–68.

61. H. B. Curry, *A formalization of recursive arithmetic*, American journal of mathematics, vol. 63 (1941), pp. 263–282.

62. H. B. Curry, *A revision of the fundamental rules of combinatory logic*, The journal of symbolic logic, vol. 6 (1941), pp. 41–53.

63. H. B. Curry, *Consistency and completeness of the theory of combinators*, ibid., pp. 54–61.

Further addenda (1951)

64. Alonzo Church, Review of Post [101], The journal of symbolic logic, vol. 8 (1943), pp. 50-52; see erratum, ibid., p. iv. See note thereon by Post, Bulletin of the American Mathematical Society, vol. 52 (1946), p. 264.

65. Paul Csillig, *Eine Bemerkung zur Auflösung der eingeschachtelten Rekursion*, Acta scientiarum mathematicarum, vol. 11 (1947), pp. 169–173.

66. H. B. Curry, *The paradox of Kleene and Rosser*, Transactions of the American Mathematical Society, vol. 50 (1941), pp. 454–516.

67. H. B. Curry, *The combinatory foundations of mathematical logic*, The journal of symbolic logic, vol. 7 (1942), pp. 49–64; see erratum, ibid., vol. 8, p. iv.

68. H. B. Curry, *The inconsistency of certain formal logics*, ibid., vol. 7 (1942), pp. 115–117; see erratum ibid., p. iv.

69. H. B. Curry, *Some advances in the combinatory theory of quantification*, Proceedings of the National Academy of Sciences of the United States of America, vol. 28 (1942), pp. 564–569.

70. H. B. Curry, *A simplification of the theory of combinators*, Synthese, vol. 7 no. 6A (1949), pp. 391–399.

71. Martin Davis, *On the theory of recursive unsolvability*, dissertation, Princeton 1950.

72. Robert Feys, *La technique de la logique combinatoire*, Revue philosophique de

Louvain, vol. 44 (1946), pp. 74–103, 237–270.

73. Frederic B. Fitch, *A basic logic*, The journal of symbolic logic, vol. 7 (1942), pp. 105–114; see erratum, ibid., p. iv.

74. Frederic B. Fitch, *Representations of calculi*, ibid., vol. 9 (1944), pp. 57–62; see errata, ibid., p. iv.

75. Frederic B. Fitch, *A minimum calculus for logic*, ibid., pp. 89–94; see erratum, ibid., vol. 10, p. iv.

76. Frederic B. Fitch, *An extension of basic logic*, ibid., vol. 13 (1948), pp. 95–106.

77. Frederic B. Fitch, *The Heine-Borel theorem in extended basic logic*, ibid., vol. 14 (1949), pp. 9–15.

78. Frederic B. Fitch, *On natural numbers, integers, and rationals*, ibid., pp. 81–84.

79. Frederic B. Fitch, *A further consistent extension of basic logic*, ibid., pp. 209–218.

80. Frederic B. Fitch, *A demonstrably consistent mathematics – Part I*, ibid., vol. 15 (1950), pp. 17–24.

81. Hans Hermes, *Definite Begriffe und berechenbare Zahlen*, Semester-Berichte (Münster i. W.), summer 1937, pp. 110-123.

82. S. C. Kleene, *Recursive predicates and quantifiers*, Transactions of the American Mathematical Society, vol. 53 (1943), pp. 41–73.

83. S. C. Kleene, *On the forms of the predicates in the theory of constructive ordinals*, American journal of mathematics, vol. 66 (1944), pp. 41–58.

84. S. C. Kleene, *On the interpretation of intuitionistic number theory*, The journal of symbolic logic, vol. 10 (1945), pp. 109–124.

85. S. C. Kleene, *On the intuitionistic logic*, Proceedings of the Tenth International Congress of Philosophy, North-Holland Publishing Company, Amsterdam 1949, pp. 741–743.

86. A. Markoff, *On the impossibility of certain algorithms in the theory of associative systems*, Comptes rendus (Doklady) de l'Académie des Sciences de l'URSS, n. s. vol. 55 no. 7 (1947), pp. 583–586.

87. A. Markoff, *Névozmožnost' nékotoryh algorifmov v téorii associativnyh sistém*, Doklady Akadémii Nauk SSSR, vol. 55 (1947), pp. 587–590, vol. 58 (1947), pp. 353–356, and vol. 77 (1951), pp. 19–20.

88. A. Markoff, *O nékotoryh nérazréšimyh problémah kasaúščihsá matric*, ibid., vol. 57 (1947), pp. 539–542.

89. A. Markoff, *O prédstavlénii rékursivnyh funkcij*, ibid., vol. 58 (1947), pp. 1891–1892.

90. Andrzej Mostowski, *On definable sets of positive integers*, Fundamenta mathematicae, vol. 34 (1946), pp. 81–112.

91. Andrzej Mostowski, *On a set of integers not definable by means of one-quantifier predicates*, Annales de la Société Polonaise de Mathématique, vol. 21 (1948), pp.

114–119.

92. Andrzej Mostowski, *Sur l'interprétation géométrique et topologique des notions logiques*, Proceedings of the Tenth International Congress of Philosophy, North-Holland Publishing Company, Amsterdam 1949, pp. 767–769.

93. John R. Myhill, *Note on an idea of Fitch*, The journal of symbolic logic, vol. 14 (1949), pp. 175–176.

94. David Nelson, *Recursive functions and intuitionistic number theory*, Transactions of the American Mathematical Society, vol. 61 (1947), pp. 307–368; see errata, ibid., p. 556.

95. David Nelson, *Constructible falsity*, The journal of symbolic logic, vol. 14 (1949), pp. 16–26.

96. M. H. A. Newman, *On theories with a combinatorial definition of "equivalence,"*, Annals of mathematics, ser. 2, vol. 43, (1942), pp. 223–243.

97. M. H. A. Newman, *Stratified systems of logic*, Proceedings of the Cambridge Philosophical Society, vol. 39 (1943), pp. 69–83.

98. Rózsa Péter, *Zusammenhang der mehrfachen und transfiniten Rekursionen*, The journal of symbolic logic, vol. 15 (1950), pp. 248–272.

99. Rózsa Péter, *Zum Begriff der rekursiven reellen Zahl*, Acta scientiarum mathematicarum, vol. 12 part A (1950), pp. 239–245.

100. Rózsa Péter, *Rekursive Funktionen*, Akademischer Verlag, Budapest, 1951.

101. Emil L. Post, *Formal reductions of the general combinatorial decision problem*, American journal of mathematics, vol. 65 (1943), pp. 197–215.

102. Emil L. Post, *Recursively enumerable sets of positive integers and their decision problems*, Bulletin of the American Mathematical Society, vol. 50 (1944), pp. 284–316.

103. Emil L. Post, *Recursive unsolvability of a problem of Thue*, The journal of symbolic logic, vol. 11 (1946), pp. 1–11.

104. Emil L. Post, *Note on a conjecture of Skolem*, ibid., pp. 73–74.

105. Julia Robinson, *Definability and decision problems in arithmetic*, The journal of symbolic logic, vol. 14 (1949), pp. 98–114.

106. Julia Robinson, *General recursive functions*, Proceedings of the American Mathematical Society, vol. 1 (1950), pp. 703–718.

107. Raphael Robinson, *Primitive recursive functions*, Bulletin of the American Mathematical Society, vol. 53 (1947), pp. 925–942.

108. Raphael Robinson, *Recursion and double recursion*, ibid., vol. 54 (1948), pp. 987–993.

109. J. B. Rosser, Review of this monograph, The journal of symbolic logic, vol. 6 (1941), p. 171.

110. J. B. Rosser, *New sets of postulates for combinatory logics*, ibid., vol. 7 (1942), pp.

18–27; see errata, ibid., vol. 7, p. iv, and vol. 8, p. iv.

111. Thoralf Skolem, *Einfacher Beweis der Unmöglichkeit eines allgemeinen Lösungsverfahrens für arithmetische Probleme*, De Kongelige Norske Videnskabers Selskab, Forhandlinger, vol. 13 (1940), pp. 1–4.

112. Thoralf Skolem, *Remarks on recursive functions and relations*, ibid., vol. 17 (1944), pp. 89–92.

113. Thoralf Skolem, *Some remarks on recursive arithmetic*, ibid., pp. 103–106.

114. Thoralf Skolem, *A note on recursive arithmetic*, ibid., pp. 107–109.

115. Thoralf Skolem, *Some remarks on the comparison between recursive functions*, ibid., pp. 126–129.

116. Thoralf Skolem, *Den rekursive aritmetikk*, Norsk matematisk tidsskrift, vol. 28 (1946), pp. 1–12.

117. Thoralf Skolem, *The development of recursive arithmetic*, Den 10. Skandinaviske Matematiker, Kongres, Jul. Gjellerups Forlag, Copenhagen 1947, pp. 1–16.

118. Ernst Specker, *Nicht konstruktiv beweisbare Sätze der Analysis*, The journal of symbolic logic, vol. 14 (1949), pp. 145–158.

119. Alfred Tarski, Andrzej Mostowski and Alfred Tarski, Julia Robinson, abstracts in The journal of symbolic logic, vol. 14 (1949), pp. 75–78.

DIFFERENTIALS
(1942)

1. I am interested in the note of Kac and Randolph in the February number of the MONTHLY (pp. 110–112), because I agree with them that the usual definition of the differential which they criticize is unsound, and that this unsoundness is within the understanding of the more intelligent beginning student in the calculus—or at least that it is sufficiently near the threshold of his understanding so that the definition causes him difficulty (even if he cannot make explicit the reason for his difficulty).

I would urge, however, that the objection which they make to the usual definition is not sufficient to reveal an unsoundness in it. In effect their objection is that, in the equation

$$\frac{dy}{dx} = \lim_{\Delta x \to 0} \frac{\Delta y}{\Delta x},$$

if dx is identified with Δx, the same variable Δx appears in the equation in two different rôles, on the left as a free variable and on the right as a bound variable.[1] But this must not be considered an error. For example, in a context where π has been defined as

$$\int_{-1}^{1} \frac{dx}{\sqrt{1-x^2}},$$

no one would think of objecting to the equation,

$$\tan(x+\pi) = \tan x,$$

on the ground that π stands for an expression containing x as a bound variable. Likewise, in a context where $D_x y$ has been defined as

$$\lim_{\Delta x \to 0} \frac{\Delta y}{\Delta x},$$

there should be no objection to writing an equation which contains $D_x y$ and at the same time contains a separate occurrence of Δx as a free variable.

The modification of Kac and Randolph in the definition of the differential must therefore be considered as designed to remove a difficulty for the student, rather than as correcting an actual error. Its advantage is that it avoids the possible necessity for an added explanation, which would, at least in effect, have to reproduce the distinction between free and bound variables.

2. On the other hand, there is a more serious objection which applies alike to the definition of Kac and Randolph and to the more usual definition which they wish to replace.

This objection may be formulated in the following terms. Both definitions agree in defining the differential of the independent variable, say x, by taking dx to be a

Originally published in ***American Mathematical Monthly***, vol. 49 (1942), pp. 389–392.

[1]A variable is *free* in a given expression (in which it occurs) if the meaning or value of the expression depends upon determination of the value of the variable; in other words, if the expression can be considered as representing a function with that variable as argument. In the contrary case the variable is called a *bound* (or *apparent*, or *dummy*) variable.

new independent variable — hence it should with equal correctness be possible to use a single letter, say z, to represent this new variable. Then the differential of a dependent variable, say y, is defined by taking dy to be $(D_x y)dx$ — *i.e.*, $(D_x y)z$. But a survey of the more usual purposes for which differentials are employed in the calculus will show that not all of these are adequately served by taking dx and dy to be z and $(D_x y)z$ respectively.

In particular, the use of differentials in connection with integration fails to be provided for. Thus $\int x\,dx$ becomes simply $\int xz$, and the whole significance of the notion is lost. (What plausibly is the operation $\int$ which, applied to the product of two independent variables x and z, yields $\frac{1}{2}x^2 + C$?) — It should be emphasized that the facility which comes with the use of differentials is more marked in the integral calculus than in the differential calculus, and that any definition of differentials which fails to account for their use with the sign $\int$ has therefore lost more than half their value. Compare, *e.g.*, any one of the following processes in terms of differentials with the clumsier parallel process which regards the notation $\int \cdots dx$ as indivisible and employs only derivatives without differentials: (1) the integration $\int x\sqrt{4x+3}\,dx$ by the substitution $t^2 = 4x + 3$, using the equation between differentials, $dx = \frac{1}{2}t\,dt$; (2) the solution of the differential equation $dy/dx = xy$ by multiplying both sides by dx/y and then applying the operation $\int$ to both sides; (3) the discovery by inspection of an integrating factor for a differential equation of the first order and first degree in two (or more) variables.

Another aspect of the foregoing objection lies in the point that the same notation, dx or dy, is given different meanings according as x is independent or dependent variable. This is especially unfortunate because it is often precisely one of the advantages in the use of differentials that various variables are symmetrically treated, without arbitrarily singling out one or more of them as the independent variable or variables (compare, *e.g.*, the equation $ds^2 = dx^2 + dy^2$ with the corresponding equation for $(ds/dx)^2$, or for $(ds/dy)^2$).

Sometimes a student will ask why the result,

$$\frac{dy}{dx} = \frac{dy}{du}\frac{du}{dx},$$

cannot be obtained by simple cancellation of du against du, and the more difficult argument which employs properties of limits thus avoided. On the basis of the usual definition of a differential, the reply is to point out that du has different meanings in its two occurrences; but this immediately reveals the weakness of this usual definition.

If desired, the objection that the usual definition of a differential does not treat the variables symmetrically can be regarded as the fundamental objection. The operation $\int$ must, of course, be the inverse of the operation d, however the latter operation is defined; and if the operation d fails to treat the variables symmetrically, the same lack of symmetry must affect the inverse operation. The difficulty in connection with integration reduces in part to the point that the embarrassment occasioned by the lack of symmetry is more acute in the case of the inverse operation. But the matter is further complicated by the way in which the usual definition of dy introduces a new independent variable z.

Unless some solution of these difficulties can be found, it seems that it would be preferable to introduce differentials in a frankly inaccurate and heuristic manner as small values of the increment or "little bits" of the "variable quantity" involved, rather than to clothe the idea with the deceptive appearance of logical accuracy.

3. There is a method of introducing differentials which suggests itself as a possible remedy, but unfortunately it may not be suitable for use in an elementary course except by devoting a disproportionate amount of time to the study of parametric equations. The statement of it which follows is at all events not intended to be in form for presentation to the student.

This method is simply to define dx and dy to be $D_\tau x$ and $D_\tau y$ respectively, where τ is an arbitrary parameter.[2]

This does not contradict the usual statement that differentials are direction numbers of the tangent line. On the contrary it implies that statement. But it also adds a supplement to it which is needed to provide for certain ordinary uses of differentials. In particular the sign $\int$ taken by itself is then naturally understood to mean integration with respect to τ.

The extension of this idea to the differential du of a function u of two independent variables x and y is possible but somewhat cumbrous. It would perhaps be preferable to interpret equations involving du as relative to an arbitrary functional relationship between x and y (*i.e.*, as holding for every such functional relationship which satisfies appropriate conditions).

No very convenient method is provided of introducing second differentials, or of associating differentials with small values of the increment. In fact it seems that, if this definition of differentials is adopted, the notion of a differential should be kept entirely separate from considerations connected with small values of the increment. A special notation to represent $f'(x)\Delta x$ may be unnecessary; if such a notation is introduced, it should be something like $\delta_x f(x)$ or $\delta_x y$ (the subscript being dropped only in cases where it is irrelevant which variable is independent).

[2]There should be no objection from the point of view of rigor to the introduction of an arbitrary parameter, as opposed to a particular parameter. It means that, in a more explicit formulation, function variables (or relation variables) would appear, corresponding to the fixed functions (or relations) which would be used in introducing a particular parameter.

CARNAP'S *INTRODUCTION TO SEMANTICS*[1] (1943)

The author understands the word *language* in a wide sense, so as to cover both word-languages, spoken and written, and sign-languages of various kinds, and both natural and artificial languages, the latter including what we shall here call *formalized languages*, which have arisen by explicit construction of calculi for special purposes (*e.g.*, for the systematization of various branches of logic, of mathematics, or of natural science). Following C. W. Morris, he applies the name *semiotic* to the general theory of languages, or of a language, and divides this theory into three parts — namely *pragmatics*, in which there is explicit reference to the speakers or users of a language and their behavior in using it, *semantics*, which analyses the relation between the expressions of a language and their meaning, but without reference to the users, and *syntax*, which analyses formal relations between the expressions of a language in abstraction both from their meaning and from the users of the language. The syntax and semantics of a language thus have an abstract character, like that of mathematics: they do not relate how in historical fact the language was used by so-and-so and so-and-so, but they propose a definition of what shall be considered the correct usage of the language, and base an abstract theory on this definition. The distinction between *pure syntax* and *descriptive syntax*, and that between *pure semantics* and *descriptive semantics*, are analogous to the distinction between pure and applied mathematics; or in fact it would be better to say that they are special cases of the latter distinction, pure syntax and pure semantics being branches of pure mathematics (this is not in contradiction with the fact that, from another point of view, any branch of mathematics requires to be formalized by means of a calculus and thus to have its own syntax and semantics).

In constructing a syntax and semantics for a natural language, say English, it becomes necessary to resolve certain uncertainties, vaguenesses, and inconsistencies, which are found in the existing (pragmatical) usage, and thus to replace the natural language by a formalized counterpart. It may indeed be required that the formalized counterpart of English shall have the minimum divergence from existing English — even at the cost of much complication in the construction. Nevertheless it appears that, strictly speaking, only a formalized language may have a syntax and semantics.

It should be added further that Carnap uses the word *language* in such a sense that the relation of logical deducibility between sentences of a language is a part of the semantics of the language, and any change in the relation of logical deducibility, even if accompanied by no other change, would mean that the language itself was no longer the same language.

The present book is the first of an intended series of "Studies in Semantics", and is designed to give a general introduction to the fields of syntax and semantics and to explain the principal concepts of these fields. It deals primarily, not with any particular language, but with things common to the syntax and semantics of all or of many

Originally published in ***The Philosophical Review***, vol. 52, no. 3 (1943), pp. 298–304.

[1]A review of Rudolf Carnap's ***Introduction to Semantics*** (Studies in Semantics, Volume I), Cambridge, Mass., Harvard University Press, 1942. Pp. xii, 263.

different languages; several simple formalized languages are, however, introduced as illustrations, and the natural languages English and German are also drawn upon for this purpose. Because of the newness of the subject of semantics, the book has at the same time the character of an introductory exposition and of an original contribution. In both respects it is an important pioneering work.

It is to be expected that semantics, as Carnap here develops it, may not prove to be in its best or its final form, and that fundamental changes may become necessary. In fact he calls his development "a first attempt" and explicitly leaves the way open for such changes.

The reviewer would take the opportunity to urge two such changes which seem to him immediately necessary.

The first of these has to do with the question of designata of sentences. Carnap takes it as an assumption that the designata of sentences are propositions, and makes this his primary usage (although he does also mention the possibility of truth-values as designata of sentences). However, if a language, in addition to certain other common properties, contains an abstraction operator '(λx)' such that '$(\lambda x)(\ldots)$' means 'the class of all x such that …', then—independently of the question whether the language is intensional or extensional—it is possible to prove that the designata of sentences of the language must be truth-values rather than propositions. For the sake of uniformity, it therefore seems desirable to take the designatum of a sentence always to be a truth-value. On this point the reviewer confesses to have changed his own former opinion, but not without compelling reason.

In fact let '…' be a sentence, in the language S, which is true but not L-true (true but not analytic), and let $\mathfrak{U}$ be the expression '$(\lambda x)(x = x \,.\, \sim \ldots)$'. Further let $\mathfrak{S}_1$ and $\mathfrak{S}_2$ be the respective sentences '$(\lambda x)(x = x \,.\, \sim \ldots) = \Lambda$' and '$\Lambda = \Lambda$' in S, where 'Λ' is a symbol of S meaning simply 'the null class' (or instead of 'Λ' we could use '$(\lambda x)(\sim x = x)$'). We consider a metalanguage S' of S, which we may suppose to contain the whole of S (it must at least contain expressions synonymous with those in S, and they may as well be taken to be the same expressions), and in addition to contain semantical terms appropriate to S, in particular the predicate 'Des' ('designates'). Then the following are true sentences of S':

$$\text{`Des}(\mathfrak{S}_1, (\lambda x)(x = x \,.\, \sim \ldots) = \Lambda)\text{'},$$
$$\text{`Des}(\mathfrak{S}_2, \Lambda = \Lambda)\text{'}.$$

Moreover (page 55), since they have the same designatum, namely the null class, $\mathfrak{U}$ and 'Λ' are synonymous, whether in S or in S'. Also, synonymous expressions are interchangeable (Carnap seems to assert this on page 75—or in any case it can be proved by means of what seem to be the inevitable semantical and syntactical rules for '='). Hence, using the interchangeability of $\mathfrak{U}$ and 'Λ' in S', we obtain a third true sentence of S':

$$\text{`Des}(\mathfrak{S}_1, \Lambda = \Lambda)\text{'}.$$

Hence, again using the definition of synonymy (page 55), but this time within S' and in the sense of synonymy in S, we obtain, as a true sentence of S',

$$\text{`Syn}(\mathfrak{S}_1, \mathfrak{S}_2)\text{'},$$

where 'Syn' is the predicate 'are synonymous'. This is already sufficient to show that the designata of $\mathfrak{S}_1$ and $\mathfrak{S}_2$ cannot be propositions, since the corresponding propositions are certainly not the same for any ordinary meaning of the word 'proposition' (one sentence is L-true and the other not!). However, Carnap assumes (page 92) that L-equivalent sentences are synonymous — and, of various ways which suggest themselves of settling the synonymy of sentences, this seems indeed one very natural choice. On this basis we can go further. For '...' and $\mathfrak{S}_1$ are L-equivalent and therefore synonymous. Hence, using the obvious transitivity of synonymy, we obtain as a true sentence of S':

$$\text{'Syn('...', } \mathfrak{S}_2\text{)'.}$$

Thus we have a means by which any true sentence (as illustrated in the case of '...') can be shown to be synonymous with $\mathfrak{S}_2$. Hence we have a means of showing any two true sentences to be synonymous. By a similar method any two false sentences can be shown to be synonymous. Therefore finally no possibility remains for the designata of sentences except that they be truth-values.

(In the preceding paragraph, of course the notation ''...'' does not mean, as it normally should, 'the symbol or expression composed of three dots in a horizontal line'. Rather the three dots function as a blank; the reader is to select some suitable language S and a sentence which is true but not L-true in S; and then this sentence is to be filled into the blank everywhere throughout the paragraph, without disturbing quotation-marks which appear.)

According to Frege ("Über Sinn und Bedeutung", *Zeitschrift für Philosophie und philosophische Kritik*, C (1892) 25–50) a *sentence* (Behauptungssatz) *expresses a proposition* (drückt aus einen Gedanken) but denotes or *designates a truth-value* (bedeutet einen Wahrheitswerth). His argument in support of this distinction lends itself to reproduction in more exact form by means of Carnap's semantical terminology, and this is what we have just done.

Frege makes this same distinction between the intensional meaning, the *sense* (Sinn) which a name expresses, and the extensional meaning, the *designatum* (Bedeutung) which the name denotes or designates, in the case of all *names* (*i.e.*, expressions which designate). Similar distinctions had been made before, *e.g.* between "comprehension" and "extension" by the authors of the Port-Royal Logic, and between "connotation" and "denotation" by J. S. Mill, but these earlier distinctions were associated with traditional logic and shared with it certain serious shortcomings (especially the confusion between class-membership and class-inclusion). Frege's distinction between two kinds of meaning is related especially to Mill's but is in more satisfactory form and is applied to better purpose.

Briefly, the sense of an expression is in its linguistic meaning, the meaning which is known to any one familiar with the language and for which no knowledge of extralinguistic fact is required; the sense is what we have grasped when we are said to *understand* the expression. On the other hand the designatum of an expression often requires to be discovered by an empirical investigation or other considerations in addition to the knowledge of the language; it may well be possible to understand an expression yet not know its designatum. If a name forming part of a longer name is replaced by another having the same sense, the sense of the whole is not altered; if replaced by another having the same designatum, the designatum of the whole is not altered but the sense may be. *Examples*: The designatum of 'unicorn' is the null class; the sense is the quality or property of unicornhood (here we make a minor departure from the account of Frege, who does not identify "Sinn eines Begriffsumfangsnamens" with any more familiar concept). The designatum of 'the author of Waverley' is the man Sir Walter Scott; the sense is the description of a man by his being the author of Waverley (not, of course, a property in this case, but another sort of an abstract object, which let us call a description). The designatum of 'Sir Walter Scott' is again the man Sir Walter Scott; the sense is again a description, but of a different sort, involving British customs as to titles and personal names. The designatum of 'All men are mortal' is the truth-value truth; the sense is the proposition which the sentence expresses. The designatum of 'that all men are mortal', as it occurs, *e.g.*, in 'I believe that all men are mortal', is the proposition, while the sense is a certain description of a proposition by its structure and constituents. The last example illustrates the English idiom sometimes employs a word or expression to designate what would normally be its sense, although this usage followed systematically would be ambiguous; in a formalized counterpart of English, such a name of the proposition that all men are mortal would contain, not 'man' and 'mortal', but names of the corresponding senses, to wit, 'humanity' and 'mortality'.

The reader may think this artificial or unnecessarily complicated. But the difficulties faced by the most obvious alternative account, allowing only one kind of meaning for names, are conclusive against it. To see this it is sufficient to consider Russell's comparison of the sentences 'Scott is Scott' and 'Scott is the author of Waverley'. If 'Scott' and 'the author of Waverley' are synonymous names, why do not the two sentences convey the same information? Or a similar question may be asked about the sentences 'The null class is the null class' and 'The class of unicorns is the null class' (here the synonymous names have the form of class abstracts, '$(\lambda x)(\dots)$').

Apparently the only tenable alternative to Frege's account is that proposed by Russell ("On denoting", *Mind*, XIV (1905) 479–493).

Russell's reasons for rejecting Frege's notion of sense are, in the reviewer's opinion, without force. The point that some expressions, *e.g.*, 'the king of France in 1905', have a sense but no designatum simply does not constitute a difficulty, except in the sense of a complicating factor in the construction of a formalized language. And Russell's other objections, it would seem, are traceable merely to confusion between use and mention of expressions, of a sort which Frege is careful to avoid by the employment of quotation-marks. Russell applies quotation-marks to distinguish the sense of an expression from its denotation, but leaves himself without any notation for the expression itself; upon introduction of (say) a second kind of quotation-marks to signalize names of expressions, Russell's objections to Frege completely vanish.

If, however, we do reject the notion of sense—perhaps in an attempt to secure economy of semantical ideas—then Russell's well known conclusion seems inescapable that descriptions (in the syntactical sense of names of the form 'the x such that ... ') may not be admitted except as introduced by contextual definition, *i.e.*, that a sentence ostensibly containing a description must be construed as a mere abbreviation for a sentence of another form in which neither the description nor any name synonymous to it appears. Russell confines his discussion to descriptions, but the same thing applies to class abstracts and other sorts of names. As a result, the notion of designation largely disappears along with that of sense—except perhaps in the case of sentences, which may then be taken, if desired, as designating propositions. Predicates must be construed as having no meaning in isolation, and must be distinguished from class names, which do not appear (their place being taken by a contextual definition of class abstracts). The principal remaining semantical concept is that of truth.

Now Carnap's semantics, in view of its heavy reliance on the notion of designation, clearly is not intended to follow Russell's account. Moreover, in the opinion of the reviewer, the initial economy in the construction of a semantics which is attained by rejecting languages which contain names and so eliminating both sense and designation is more than offset by the artificiality of this course and by the probability that it will be a complicating factor at a later stage, in the application of semantics to specific philosophical or linguistic questions. This brings us therefore to the second change which is here urged in Carnap's development of semantics, namely that the notion of sense not only should receive treatment but should be taken into account throughout as of equal importance with that of designation. In particular, the notion of sense should be prominent in connection with propositions, truth-conditions, absolute L-concepts, extensionality, L-synonymy. The statement about "meaning" in connection with L-synonymy on page 75 probably should be replaced by an explicit postulate about sense, that two expressions have the same sense if and only if they are L-synonymous. The assumption on page 92 that L-equivalent sentences have the same designatum, and are therefore synonymous, should then be replaced by the assumption that L-equivalent sentences have the same sense, and are therefore L-synonymous (unless indeed this can be proved).

In a concluding section of the appendix, Carnap indicates briefly how some of the principal views exhibited in his *Logical Syntax of Language* have become modified since the first publication of that book. The underlying change is that he now takes into account semantical as well as syntactical concepts, a point in which, as he explains, he has been greatly influenced by Tarski. "Many of the earlier discussions and analyses", he writes, "are now seen to be incomplete, although correct; they have to be supplemented by corresponding semantical analyses."

The *principle of tolerance* is still maintained in its original form, but its effect seems to be greatly modified. The abstract construction of a calculus is said to be a matter of convention. But the definitions for the L-concepts within a given semantical system and the construction of a calculus in accordance with a given semantical system are not purely conventional, though they are partly so. It is not clear to what extent the author's previous criticism of mathematical intuitionism, for example, would continue to hold on this basis. For the intuitionistic rejection of certain principles of classical mathematics (*e.g.*, the principle of excluded middle) refers, not to the abstract construction of a calculus, but to the construction of a calculus whose interpretation is at least partly given. Some statement by Carnap regarding the application of the principle of tolerance to this and like cases would seem to be desirable in order to clarify his revised position.

Various syntactical definitions are now replaced by semantical definitions which are expected better to serve the intended purpose. *E.g.*, 'analytic' is abandoned in favor of 'L-true'.

The *thesis of extensionality* is said to be still held as a supposition, but on the basis of a semantical concept of extensionality.—However, if the designatum of a sentence is always a truth-value, then Carnap's definition of 'extensional' fails in that under it every language (every semantical system) is extensional, even those which contain names of propositions and modal operators, or which contain names of properties as opposed to class names. Apparently a more satisfactory definition of extensionality of a language, or of a semantical system, must be found before the thesis of extensionality can be considered.

The *thesis that philosophy is the syntax of the language of science* is now changed to the following: *The task of philosophy is semiotical analysis.* It is still maintained that theoretical philosophy is the *logic of science*, by which is meant the syntax and semantics of the language (or a language?) of science; but the possibility is considered of using the term 'philosophy' to include also certain empirical problems, belonging mostly to pragmatics.—Of course Carnap intends here a thesis about the nature of philosophy, not merely a definition. It is meant that problems which have traditionally been called philosophical should be transformed into problems about the semiotical structure of the language of science, and that all those which cannot be so transformed (especially much or all of metaphysics) are nonsense. Thus the thesis, even as now revised, represents an extreme view. It is not necessary to share it in order to appreciate the importance of semantical investigations to philosophy, or in order to agree with Carnap in his expectation of fruitfulness for this new line of approach.

ALONZO CHURCH

PRINCETON UNIVERSITY

CARNAP'S *FORMALIZATION OF LOGIC* (1944)

Formalization of Logic. By RUDOLF CARNAP. (Studies in Semantics, Volume II.) Cambridge, Mass., Harvard University Press, 1943. Pp. xviii, 159.

In this work the author applies the concepts and methods of his *Introduction to Semantics*[1] to a particular problem, namely the development of the semantics of the propositional calculus and (more briefly) the functional calculus of first order. After some preliminary developments—which are in part routine translation, into more exact terminology, of things which are otherwise well known—we come to the author's main results, concerning the existence of "non-normal" true interpretations of the propositional calculus and functional calculus of first order, and the matter of what he calls "full formalizations."

A form of the propositional calculus is used (primarily) in which the primitive connectives are $\sim$ and $\vee$ and in which the rule of substitution and the necessity for propositional variables are avoided by the now familiar device of introducing schemata of primitive sentences instead of single primitive sentences. As atomic sentences there may be either propositional constants or propositional variables or both, and the molecular sentences are constructed in the usual fashion from the atomic sentences by means of the two primitive connectives. The discussion of normal and non-normal interpretations is confined to the case that all the atomic sentences are propositional constants.

A non-normal interpretation of the propositional calculus is one which in some way contravenes the usual interpretation by means of truth-tables with two truth-values. It is shown that the non-normal true interpretations are of two mutually exclusive kinds: (1) those which violate the semantical rule that the negation of a true sentence must be false, and (2) those which violate the rule that the negation of a false sentence must be true.

For Carnap an interpretation of a calculus is merely an assignment of truth-conditions for its sentences, but for purposes of exposition it will be convenient here to speak of the more familiar idea of interpreting a calculus by assigning to each sentence a meaning as expressing a certain proposition. Then, for the propositional calculus, non-normal true interpretations of the first kind are easily obtained by interpreting *every* sentence to express a true proposition—*e.g.*, the various propositional constants could be interpreted as expressing various true propositions, $\vee$ (disjunction) could be interpreted as disjunction in the usual sense, and $\sim$ (negation) could be so interpreted that $\sim \mathfrak{S}$ expresses in every case the same proposition as $\mathfrak{S}$.

Examples of non-normal interpretations of the propositional calculus which are of the second kind are somewhat less immediately obvious. In place of Carnap's abstractly formulated examples we here offer the following more concrete one. Suppose that there are just two propositional constants, A_1 and A_2. Let A_1 be interpreted to mean "There are 14 days in a week", and let A_2 be interpreted to mean "There are 21 days in a week". Moreover let every molecular sentence be interpreted to express a proposition "There

Originally published in ***The Philosophical Review***, vol. 53 (1944), pp. 493–498.

[1] See this Review [***The Philosophical Review***], LII (1943), 298–304 [in this volume, the preceding review].

are n in a week", where n is some number, and let this be done according to the following rules: (a) if $\mathfrak{S}$ expresses a proposition "There are x days in a week", then $\sim\mathfrak{S}$ shall express "There are z days in a week", where z is obtained as the quotient $294 \div x$; (b) if $\mathfrak{S}_1$ and $\mathfrak{S}_2$ express respectively "There are x days in a week", and "There are y days in a week", then $\mathfrak{S}_1 \vee \mathfrak{S}_2$ shall express "There are z days in a week", where z is obtained as the greatest common divisor of x and y. Then every sentence is assigned a unique meaning as expressing a certain proposition. The interpretation of the calculus so obtained is a *true interpretation* in the sense that every sentence which is true according to the syntactical rules of the propositional calculus (*i.e.*, every theorem of the propositional calculus) is interpreted as expressing a true proposition; and that further, whenever every sentence of a certain class $\mathfrak{K}$ of sentences is interpreted as expressing a true proposition, and the sentence $\mathfrak{S}$ is derivable from $\mathfrak{K}$ according to the syntactical rules of the propositional calculus, then $\mathfrak{S}$ is interpreted as expressing a true proposition. Nevertheless the interpretation is non-normal, since A_1 and $\sim A_1$ both express false propositions; and also since $A_1 \vee A_2$ expresses the true proposition "There are 7 days in a week", although A_1 and A_2 both express false propositions.

The existence of non-normal true interpretations of the propositional calculus is related to the known fact that there exist other truth-tables which yield as true (as tautologies) the same sentences as do the usual two-valued tables. In particular, the elements of any Boolean algebra may be taken as truth-values, with the universal class as the truth-value truth and other elements of the Boolean algebra as kinds of falsehood, the Boolean sum corresponding to disjunction and the Boolean complement to negation. The above example of a non-normal interpretation of the second kind is an elaborated version of the use of a four-element Boolean algebra in this way. The observation that there exist non-normal interpretations of the propositional calculus was made in particular by B. A. Bernstein.[2] His statement is less exact than Carnap's, but it is evidently non-essential that he uses[3] plane regions rather than propositions as interpretations of the sentences of the propositional calculus—for each plane region could be replaced by the proposition that it is identical with U, the universal or maximum region. Bernstein also anticipates in effect[4] Carnap's conclusion that not every true interpretation of the propositional calculus obeys the laws of contradiction and excluded middle in their semantical versions—the laws, namely, that for every sentence $\mathfrak{S}$ not containing (free) variables not both $\mathfrak{S}$ and $\sim\mathfrak{S}$ are true, and either $\mathfrak{S}$ or $\sim\mathfrak{S}$ is true. The reviewer cannot agree with the opinion of Bernstein and Carnap that this constitutes a shortcoming of the usual logistic formulation of the propositional calculus, nor with Bernstein's proposed remedy, to replace the propositional calculus by "the Boolean logic of propositions", which is not a logistic system but rather a "mathematical science" in his sense.[5] The defect of a Bernsteinian "mathematical science" is that it leaves the underlying logic undetermined, or at best only vaguely determined by a reference to "ideas of general language"; since the consequences of a set of postulates depend as much on the underlying logic as on the specifically

[2] ***Bulletin of the American Mathematical Society***, 38 (1932), 390, and 592.

[3] *Loc. cit.*, 390.

[4] *Loc. cit.*, 592. Also in his review of ***Principia Mathematica*** in the ***Bulletin of the American Mathematical Society***, 32 (1926), see 712–713.

[5] In addition to papers already cited, see a paper by Bernstein in the ***Bulletin of the American Mathematical Society***, 37 (1931), 480–488, and a criticism of it by E. J. Nelson, *ibid.*, 40 (1934), 478–486.

mathematical postulates, this does not provide an ultimately satisfactory foundation for any branch of mathematics, but the difficulties are especially acute when the method is applied to logic itself.

Carnap's proposed remedy, although different from Bernstein's, is in the opinion of the reviewer open to objections of somewhat the same order. The importance remains, however, of the striking facts about the propositional calculus which are brought out by Bernstein and by Carnap, and of the contributions to theoretical syntax which are contained in Carnap's procedure.

Carnap first introduces the notions of *conjunctives* and *disjunctives*. There are both thought of as classes of sentences, but a conjunctive and a disjunctive which contain exactly the same sentences are nevertheless not the same; technically, as Carnap points out, we might define conjunctives and disjunctives as ordered pairs, the first member of the pair being a class of sentences in every case, but the second member being say the universal class in the case of a conjunctive, the null class in the case of a disjunctive. A *junctive* is either a single sentence or a conjunctive or a disjunctive. Then the usual syntactical rules for the propositional calculus are modified as follows: the primitive sentences are stated as consequences of the null conjunctive; in the rule of *modus ponens* the two premisses $\mathfrak{S}_1$ and $\sim\mathfrak{S}_1 \vee \mathfrak{S}_2$ are replaced by a single premiss, the conjunctive containing these two sentences; and additional rule of inference makes the null disjunctive a consequence of the universal conjunctive; and another additional rule makes the disjunctive containing $\mathfrak{S}_1$ and $\mathfrak{S}_2$ a consequence of $\mathfrak{S}_1 \vee \mathfrak{S}_2$, in case $\mathfrak{S}_1$ and $\mathfrak{S}_2$ contain no (free) variables. These rules define the relation of *direct derivability* for junctives. A junctive $\mathfrak{T}_2$ is *derivable* from a junctive $\mathfrak{T}_1$ if $\mathfrak{T}_2$ belongs to every class of junctives which (1) contains $\mathfrak{T}_1$, (2) whenever it contains a junctive $\mathfrak{T}$ contains all junctives directly derivable from $\mathfrak{T}$, (3) contains a conjunctive $\mathfrak{K}$ if and only if it contains every sentence in $\mathfrak{K}$, and (4) contains a disjunctive $\mathfrak{K}$ if and only if it contains at least one sentence in $\mathfrak{K}$. This yields a "full formalization of propositional logic", for which non-normal true interpretations are impossible, although the theorems, in the sense of sentences derivable from the null conjunctive, are the same as in usual formations of the propositional calculus.

The similarity should be noted between Carnap's use of disjunctives and the proposal of E. V. Huntington[6] to include it as a postulate, or as a theorem, of the propositional calculus that if *a + b is in T then at least one of a and b is in T*—where it is intended that the sign + shall be interpretable as representing disjunction, and *T* as the class of true propositions.

The most immediate objection to Carnap's procedure is that it is necessary to make it a part of the definition of a true interpretation of the propositional calculus—when the formalization involving junctives is used—that a disjunctive shall be true if and only if at least one sentence in it is true. *I.e.*, the definition requires that the initially purely syntactical notion of a disjunctive shall be given this particular semantical significance in the interpretation. It would seem to be quite as reasonable to make it a part of the definition of a true interpretation that $\mathfrak{S}_1 \vee \mathfrak{S}_2$ shall be true if and only if at least one of $\mathfrak{S}_1$ and $\mathfrak{S}_2$ is true.

In view of his requirement that disjunctives be interpreted in a particular way,

[6] ***Transactions of the American Mathematical Society***, 35 (1933), 301; also ***Bulletin of the American Mathematical Society***, 40 (1934), 129.

Carnap's use of them is a concealed use of semantics; and in fact, if this arbitrary requirement is dropped, non-normal interpretations of his "full formalization" become possible. The reviewer suggests that non-normal interpretations of the propositional calculus can be excluded only by semantical (as opposed to purely syntactical) rules.

As an alternative to the use of conjunctives and disjunctives, Carnap outlines a method of obtaining a "full formalization" by the introduction of a relation of "involving" (or "involution") between classes of sentences. Syntactically, involution is a generalization of the notion of derivability, or of the related notion of C-implication. Semantically, $\mathfrak{K}_1$ involves $\mathfrak{K}_2$ if the existence of a true sentence in $\mathfrak{K}_2$ is a consequence of the simultaneous truth of all the sentences of $\mathfrak{K}_1$. This method is open to the same objection as the other, namely that the requirement is arbitrary that the syntactical notion of involution shall receive a particular kind of interpretation. (As a matter of fact Carnap's requirement, in connection with usual formulations of the propositional calculus, that the syntactical relation of derivability receive a particular kind of interpretation, is also objectionable as being arbitrary; but this point happens not to be crucial for the present purpose.)

The reviewer believes, however, that there is a more fundamental criticism to be made of Carnap's procedure. For the purpose of stating this, let us distinguish between *elementary syntax* and *theoretical syntax* on the following basis.

The elementary syntax of a language comprises what syntax must be known (at least implicitly) in order to use the language correctly at all. It is concerned solely with stating the formally correct usage of a language, or calculus—or equivalently, with teaching this correct usage to a learner who understands only the syntax language. As a consequence elementary syntax should require of the learner only that he understand and be able to follow general directions, stated in the syntax language for the concrete manipulation of physical objects (symbols).—This is a minimum presupposition for the construction of a calculus. It is important that this presupposition is much less, and of a different kind, than the intended content of logical and mathematical calculi which are ultimately constructed by this method, dealing, as the latter do, with abstract objects, infinite classes, relations which may be non-effective (or in Carnap's terminology, indefinite).

Theoretical syntax comprises the remainder of syntax. It deals with questions of what can and cannot be done in the object language, and generally with the mathematical theory of the object language as a formal structure, in abstraction from its meaning. The syntax language used must normally be logically stronger than for elementary syntax, and it may often be as strong or stronger than the object language in its intended meaning.

In appraising the distinction between the two kinds of syntax it is to be remembered that the object language is intended, at least in theory, for use as a language in its own right (however exclusively logicians may be concerned with talking *about* the language). It is true indeed, as Carnap says (113, 114), that the syntax language is necessary in order to state that a particular sequence of sentences is a proof and in order to state derivations from premisses (unless the object language is capable of expressing its own syntax). But the users of the object language are not in general concerned with stating that a sequence of sentences is a proof, but rather with making the successive assertions which constitute the proof. And (again unless the object language is capable of expressing its own syntax) a derivation from premisses should not be expected in the object language except when the user is willing to assert the premisses outright; in the contrary case its place is to be taken by a proof of a corresponding implication sentence.

Now the users of a language assert single sentences, one at a time. They deal in classes of sentences only indirectly, to the extent that a class of sentences can be represented in this way. *E.g.*, a finite disjunctive may be representable by a disjunction sentence, or an infinite disjunctive may be representable by a sentence containing an existential quantifier. Likewise an involution between two finite classes of sentences may be representable by a sequence of implication sentences.

Carnap's notions of conjunctives, disjunctives, and involution therefore belong to theoretical syntax. They are foreign to elementary syntax, and may not be used in the construction of a calculus.

Finally, in Carnap's treatment of the functional calculus of first order, non-normal interpretations are again obtained on the basis of a formalization of the usual kind, and a similar proposal for "full formalization" is made. The reviewer has the same criticisms as before. There is, however, the additional point that Carnap now makes it explicit that his method employs non-effective (indefinite) syntactical rules. It is clearly not possible for the users of a language systematically to follow a non-effective rule in practice, and such rules must therefore be excluded from elementary syntax. (They have a place, and perhaps an important one, in theoretical syntax.)

As a matter of fact non-effective rules occur already in the "full formalization" of propositional logic, *e.g.*, the rule which makes each sentence in a conjunctive derivable from the conjunctive—since the definition of a particular class of sentence, even of a finite class of sentences, may well be non-effective.

ALONZO CHURCH

PRINCETON UNIVERSITY

ABSTRACT OF *A FORMULATION OF THE LOGIC OF SENSE AND DENOTATION*[1] (1946)

The distinction of *Sinn und Bedeutung* (see abstract of a paper by the author in this JOURNAL, vol. 7, p. 47) is incorporated into a logistic system. Types: ι_0, names of individuals; ι_{i+1}, names of senses of names of type ι_i; o_0, names of truth-values; o_{i+1}, names of senses of names of type o_i; for any types α, β, a type $(\alpha\beta)$ of names of functions, such that if $\boldsymbol{F}_{(\alpha\beta)}$ and $\boldsymbol{A}_\beta$ are names, of types indicated by the subscripts, then $(\boldsymbol{F}_{(\alpha\beta)}\boldsymbol{A}_\beta)$ is a name of type α. Subscripts upon constants and variables of the system, and upon syntactical variables (bold letters) indicate the type. To represent variable or undetermined type symbols are used Greek letters, α, β, etc.; and subscript i is used upon such Greek letters to indicate the result of increasing all subscripts in the type symbol by i. As an abbreviation, parentheses are omitted under the convention of association to the left. Primitive symbols: constants f_{o_i}, $C_{o_io_io_i}$, $\Pi_{o_i(o_i\alpha_i)}$, $\iota_{\alpha_i(o_i\alpha_i)}$, $\Delta_{o_i\iota_{j+i+1}\iota_{j+i}}$, $\Delta_{o_io_{j+i+1}o_{j+i}}$; an infinite list of variables of each type; the abstraction operator λ; parentheses. Definitions: $[\boldsymbol{A}_{o_i} \supset \boldsymbol{B}_{o_i}] \rightarrow C_{o_io_io_i}\boldsymbol{B}_{o_i}\boldsymbol{A}_{o_i}$, $[(x_{\alpha_i})\boldsymbol{A}_{o_i}] \rightarrow \Pi_{o_i(o_i\alpha_i)}(\lambda x_{\alpha_i}\boldsymbol{A}_{o_i})$, $Q_{o_i\alpha\alpha} \rightarrow \lambda x_\alpha \lambda y_\alpha (f_{o_i\alpha}) \mathbin{.} f_{o_i\alpha}y_\alpha \supset f_{o_i\alpha}x_\alpha$, $[\boldsymbol{A}_\alpha = \boldsymbol{B}_\alpha] \rightarrow Q_{o_0\alpha\alpha}\boldsymbol{B}_\alpha\boldsymbol{A}_\alpha$, $\Delta_{o_i(\alpha_{i+1}\beta_{i+1})(\alpha_i\beta_i)} \rightarrow \lambda f_{\alpha_i\beta_i}\lambda f_{\alpha_{i+1}\beta_{i+1}}(x_{\beta_i})(x_{\beta_{i+1}}) \mathbin{.} \Delta_{o_i\beta_{i+1}\beta_i}x_{\beta_i}x_{\beta_{i+1}} \supset \Delta_{o_i\alpha_{i+1}\alpha_i}(f_{\alpha_i\beta_i}x_{\beta_i})(f_{\alpha_{i+1}\beta_{i+1}}x_{\beta_{i+1}})$. Conventions for omission of brackets and punctuation by dots are the same as in the author's *Formulation of the simple theory of types*, this JOURNAL, vol. 5 (1940), pp. 56–68). Notation is such that a (constant) name becomes a name of its sense by increasing subscripts in all type symbols by 1; and $\Delta_{o_0\alpha_1\alpha}$ is interpreted as the relation between the sense and the denotation of a name of type α. Rules of inference and axioms are based on those of *Formulation of the simple theory of types*, but with modifications such that no asserted formula contains free variables, with the Wajsberg-Quine axioms for the propositional calculus (see a paper by W. V. Quine in this JOURNAL, vol. 3 (1938), pp. 37–40), and with the added axiom $(p)(q) \mathbin{.} p \supset q \supset \mathbin{.} q \supset p \supset \mathbin{.} p = q$ (where the omitted subscripts are o_0). To these are added the axioms $\Delta_{o_0o_{i+1}o_i}f_{o_i}f_{o_{i+1}}$, $\Delta_{o_0(o_{i+1}o_{i+1}o_{i+1})(o_io_io_i)}C_{o_io_io_i}C_{o_{i+1}o_{i+1}o_{i+1}}$, etc. Then further axioms about the notion of sense are adjoined according to either of two alternative heuristic principles: (1) two names are assumed to have different senses in all cases where it is not already a consequence that the senses are the same; (2) $\boldsymbol{A}_\alpha$ and $\boldsymbol{B}_\alpha$ are assumed to have the same sense if and only if $\boldsymbol{A}_\alpha = \boldsymbol{B}_\alpha$ is logically valid. Alternative (1) seems to be that intended by Frege. (2) leads to notions of necessity and strict implication akin to those of Lewis.

Added April 29, 1946. From the foregoing it follows that, if $\boldsymbol{A}_\alpha$ conv $\boldsymbol{B}_\alpha$ (i.e., if $\boldsymbol{A}_\alpha$ can be changed to $\boldsymbol{B}_\alpha$ by a chain of applications of Rules I–III), and if there are no free variables, then $\boldsymbol{A}_\alpha$ and $\boldsymbol{B}_\alpha$ must have the same sense. This accords well with the

---

Originally published in ***The Journal of Symbolic Logic***, vol. 11 (1946), p. 31.

[1]The abstract is of *A Formulation of the Logic of Sense and Denotation*, ***Structure, Method, and Meaning: Essays in Honor of Henry M. Sheffer***, New York, 1951. [Added by Church, 1971:] In 1951 the modification (or amendment) of Alternative (1) which had been described in the addendum to the abstract was called Alternative (0). Compare footnote 10 of *Intensional isomorphism and identity of belief*, ***Philosophical studies***, vol. 5 (1954), pp. 65–73.

tendency of Alternative (2) just mentioned, but not with that of (1). Therefore, (only) for the case that the direction of Alternative (1) is followed, the author now wishes to make the following amendments.

The primitive symbol λ is replaced by an infinite list of primitive symbols λ_i, of which λ_0 is abbreviated as λ and interpreted as an abstraction operator. If $\boldsymbol{M}_{\alpha_i}$ is well-formed (is a constant or variable name) and is of type α_i, if $\boldsymbol{x}_{\beta_i}$ is a variable of type β_i, then $(\lambda_i \boldsymbol{x}_{\beta_i} \boldsymbol{M}_{\alpha_i})$ is well-formed of type $\alpha_i\beta_i$, and in it $\boldsymbol{x}_{\beta_i}$ is a bound variable. Notation is such that a (constant) name becomes a name of its sense by increasing subscripts in every type symbol and after every λ by 1. In the definitions given above λ is replaced throughout by λ_i, and in the third definition α is replaced by α_i. In Rules II, III, λ is restricted to be λ_0. Then (1) involves the following principle, capable of expression as an axiom of the system: If the (constant) names $\boldsymbol{F}_{\alpha\beta}\boldsymbol{A}_\beta$ and $\boldsymbol{G}_{\alpha\beta}\boldsymbol{B}_\beta$ have the same sense, then $\boldsymbol{F}_{\alpha\beta}$ and $\boldsymbol{G}_{\alpha\beta}$ have the same sense, and $\boldsymbol{A}_\beta$ and $\boldsymbol{B}_\beta$ have the same sense.

By the argument of the Richard paradox, not every nameable infinite sequence of natural numbers can have a name in the logistic system. *A fortiori* it is false that for every sense there is a name in the system having that sense. Nor can this situation be remedied by any extension of the system. At appropriate points the preceding discussion must therefore be applied to names (and their senses) which might be added to the system, as well as to names in it.

Added 1971. This paper was not successful — not even as nearly successful as I had hoped at the time — in obtaining a satisfactory logic of the *concept relation for type* α, i.e., the relation $\Delta_{o_0\alpha_1\alpha}$ between a thing of type α and a concept of it.[2] Its value is only as an indication of direction, of what, or what kind of thing, I believe has to be sought in formulating a logic of the concept relation.

Difficulties in the formulation of 1951 were pointed out to me originally by A. F. Bausch, with regard especially to Alternative (2), and somewhat later by John Myhill with regard to Alternative (1). In lectures at the University of California — at Berkeley in the spring of 1960 and at Los Angeles in the spring of 1961 — I attempted a set-theoretic model of concepts (or sense) and the concept relation by supposing that there are one or more contingent propositions $P_1, P_2, \ldots$ such that any given concept c_{α_1} will generally be a concept of various different things $c_\alpha, d_\alpha, \ldots$ according to which of the contingent propositions P_i are true and which false — not excluding that in certain cases, determined by the truth and falsity of the P_i, it may happen that c_α is an empty concept in the sense that there is nothing of which it is a concept. And while in the real situation which the model is intended to represent there will no doubt be a very large number of the P_i, it would seem that most (and perhaps all) of the logical points that are critical will arise already if we suppose that there is only one contingent proposition P_1. Thus a concept c_{α_1} will appear in the model as an ordered pair of things of type α — with appropriate provision for the possibility that one of the two terms of the ordered pair may in special cases appear only as a blank. David Kaplan pointed out to me the resemblance of these ideas to ideas regarding an extensional model of modal logic which had been presented in lectures by Carnap.[3]

---

[2]This terminology, by which the sense of a name is called a *concept of* its denotation, was introduced in the paper of 1951. This use of the word *concept* must be distinguished from Frege's use of *Begriff*, in German, to which it is unrelated.

[3]This is described in Carnap's *Replies and systematic expositions*, in ***The philosophy of Rudolf Carnap***, Paul Arthur Schilpp ed., La Salle, Illinois, 1963, see pages 889–900. As Carnap explains (see also page

My attempt was to use this model to find a correction of the axioms for the concept relation. And while the attempt of 1960 and 1961 was again unsuccessful, it is certainly in this direction in which clarification of the difficulties is to be sought and in which hope of success therefore lies.

In fact the model suggests that the replacement of the primitive symbol λ by an infinite list of primitive symbols λ_i will be desirable for Alternative (2) as well as for Alternative (0). And the first four of the definitions (more correctly, definition schemata) in the abstract are then to be modified in the same way for Alternative (2) as was already done for Alternative (0) in the addendum to the abstract. But the fifth definition (schema) must probably be abandoned in favor of taking all of the symbols $\Delta_{o_i\alpha_{i+1}\alpha_i}$ as primitive, and then making more cautious assumptions connecting $\Delta_{o_i(\alpha_{i+1}\beta_{i+1})(\alpha_i\beta_i)}$ with $\Delta_{o_i\alpha_{i+1}\alpha_i}$ and $\Delta_{o_i\beta_{i+1}\beta_i}$.

The model, as described, is applicable to the case of Alternative (2), but success in this case would be a first step towards formulating a logic of sense and denotation in the important and more difficult case of Alternative (0).

This question of a logic of the concept relation — which is to be formulated in an object language — must of course be distinguished from the question of an intensional semantics — which is to be formulated in a meta-language of that object language, and for which there is a descriptive proposal or sketch in the paper *Intensional Semantics* (1951, 1985).

1045 of the same book), these ideas were treated by him in a seminar in the spring of 1955, and afterwards included in a privately circulated paper of June 1955. But for the earlier publication of at least related ideas about the semantics of modal logic the credit must go to others — see a review by David Kaplan of Saul Kripke's *Semantical analysis of modal logic I*, in ***The journal of symbolic logic***, vol. 31 (1966), pp. 120–122.

CONDITIONED DISJUNCTION AS A PRIMITIVE CONNECTIVE FOR THE PROPOSITIONAL CALCULUS (1948)

In the (classical, two-valued) propositional calculus, consider the ternary connective which is characterized by the following truth-table, 0 standing for truth and 1 for falsehood:

p	q	r	$[p, q, r]$
0	0	0	0
0	0	1	0
0	1	0	0
0	1	1	1
1	0	0	1
1	0	1	1
1	1	0	0
1	1	1	1

We shall call this connective *conditioned disjunction* and, as indicated in the table, we shall use the notation $[p, q, r]$—or, more generally, $[\mathbf{A}, \mathbf{B}, \mathbf{C}]$, where $\mathbf{A}$, $\mathbf{B}$, and $\mathbf{C}$ are arbitrary expressions of the propositional calculus. In words, $[p, q, r]$ may be read, "p or r according as q or not q."

In writing expressions of the propositional calculus we shall make use also of the letter t, for truth, and f, for falsehood. These are propositional constants; or, as we may also think of them, they are 0-ary connectives.

Conditioned disjunction, t, and f, are a complete set of independent primitive connectives for the propositional calculus. This has been proved by E. L. Post as a part of his comprehensive study of primitive connectives for the propositional calculus.[1] But it seems worth while to give a separate proof here.[2]

By saying that these are a complete set of primitive connectives we mean that, given an arbitrary truth-table (in the two truth-values, truth and falsehood) involving any number of propositional variables, it is always possible to find an expression of the propositional calculus which contains only the given propositional variables and the three connectives, conditioned disjunction, t, and f, and which has the given truth-table as its truth-table.[3] Or, as we may say more suggestively but less accurately, every possible connective is definable by means of the three primitive connectives,

---

Originally published in ***Portugaliae Mathematica***, vol. 6 (1948), pp. 87–90.

Recebido em 1948, Agosto, 19.

Research supported by the United States Office of Naval Research.

[1] E. L. Post, ***The Two-Valued Iterative Systems of Mathematical Logic*** (Annals of Mathematics Studies, no. 5), Princeton, N.J., 1941.

[2] The method of the proof of completeness is that used by E. L. Post in 1921 in connection with a different set of primitive connectives. See the ***American Journal of Mathematics***, vol. 43, pp. 167–168.

[3] By the truth-table of an expression of the propositional calculus is meant a table, arranged like that given above for the expression $[p, q, r]$, which shows the truth-value of the expression for every set of truth-values of its variables.

conditioned disjunction, t, and f. The proof is by mathematical induction with respect to the number of different variables involved. If the number of variables involved is 0, there are just two possible truth-tables; the expression consisting of the letter t alone has one of these truth-tables as its truth-table, and the expression consisting of the letter f alone has the other truth-table. Suppose now that every truth-table involving n different (propositional) variables can be represented by an expression which contains only those n variables and the three connectives, conditioned disjunction, t, and f, and which has that truth-table as its truth-table. And consider a truth-table T involving $n+1$ different variables. Suppose (as we may without loss of generality) that the first column of T is for the variable p. From T form the truth table T′ by deleting all the rows which have 1 in the first column and then deleting the first column; and form the truth table T″ by deleting all the rows which have 0 in the first column and then deleting the first column. Let **A** and **C** be the expressions (in the three connectives, conditioned disjunction, t, and f) which, by hypothesis of induction, have the truth tables T′ and T″ respectively. Then [**A**, p, **C**] has the truth-table T.

By saying that conditioned disjunction, t, and f are independent connectives we mean no two of the three connectives, taken alone, are a complete set of primitive connectives for the propositional calculus. In the case of t and f taken alone, this is immediately obvious. In the case of conditioned disjunction and t, it follows from the first row in the truth-table of conditioned disjunction that no expression in these two connectives only can have the value 1 (falsehood) for all values of its variables. In the case of conditioned disjunction and f it follows similarly from the last row in the truth-table of conditioned disjunction that no expression in these two connectives only can have the value 0 (truth) for all values of its variables.

In the usual sense of duality in the propositional calculus, the dual of $[p, q, r]$ is $[r, q, p]$ as may be seen from the truth-table of conditioned disjunction. In order, therefore, to dualize an expression of the propositional calculus in which the only connectives occurring are conditioned disjunction, t, and f, it is sufficient to write the expression backwards and at the same time to interchange the letters t and f.[4]

The purpose of this note is to point out the advantages in connection with duality in the propositional calculus which result from using conditioned disjunction, t, and f as primitive connectives. Indeed in any standard treatment of duality in the propositional calculus[5] it may be seen that matters would be substantially simplified by using a self-dual set of primitive connectives.

Negation, conjunction, and (the ordinary, inclusive) disjunction are a natural set of primitive connectives which is self-dual and complete. But these primitive connectives have the disadvantage that they are not independent.

If we are not to go beyond binary connectives, there is available essentially only one self-dual complete set of independent primitive connectives.[6] These are (material) implication and its dual, converse non-implication. But this set has the disadvantage that a self-dual definition of negation is impossible — so that it becomes necessary, in

[4]The writer is indebted to Mr. A. F. Bausch for suggesting the notation which is used for conditioned disjunction, and for pointing out its superior convenience over other notations in connection with dualization.

[5]E.g., that the writer in *Introduction to Mathematical Logic*, Volume I (***Annals of Mathematics Studies***, no. 13), Princeton, N.J., 1944.

[6]As shown by Post, loc. cit. (1941), and by William Wernick in ***The Transactions of the American Mathematical Society***, vol. 51 (1942), pp. 117–132.

the treatment of duality, to introduce two negations that are dual to each other.

On the other hand the primitive connectives, conditioned disjunction, t, and f, not only are independent and form a self-dual complete set (as we have shown), but also admit the following self-dual definition of negation:

$$\overline{\mathbf{A}} \quad \text{stands for} \quad [f, \mathbf{A}, t]$$

By using a bar over an expression as the notation for its negation, we retain the convenient rule given above for dualizing—i.e., this rule is applicable also to expressions containing negations as introduced by the foregoing definition.

The notation for conditioned disjunction might be simplified by omitting the commas, and abbreviation would also be possible by omitting the outer brackets in [**A**, **B**, **C**] or [**A B C**] whenever covered by a bar.

We add the conjecture that the idea of a self-dual complete set of independent primitives can be extended (1) to the discipline which Russell calls theory of implication[7] and Łukasiewicz and Tarski call *erweiterter Aussagenkalkül*[8] and (2) to the functional calculus of second order[9] by employing conditioned disjunction and the universal and existential quantifiers as primitive. The sufficiency of these primitives is shown by the fact that (following an idea of Russell) we can define t and f as $(\mathrm{E}p)p$ and $(p)p$ respectively, or as $(\mathrm{EF})(\mathrm{E}x)\mathrm{F}(x)$ and $(\mathrm{F})(x)\mathrm{F}(x)$ respectively.

[7] ***American Journal of Mathematics***, vol. 28 (1906), pp. 159–202.

[8] See ***Comptes Rendus des Séances de la Société des Sciences et des Lettres de Varsovie***, Classe III, vol. 23 (1930), pp. 44–50.

[9] In the terminology of the monograph by the writer, cited in footnote 5. Or *Prädikatenkalkül der zweiten Stufe* in the terminology of Hilbert and Ackermann, ***Grundzüge der theoretischen Logik***, second edition, 1938.

GRELLING'S ANTINOMY FOR THE SYSTEM OF *A FORMULATION OF THE SIMPLE THEORY OF TYPES* (1949)

In order to obtain Grelling's antinomy in connection with the system of *A formulation of the simple theory of types* it is of course necessary to make certain additions to the system.

First of all we introduce an infinite list of additional primitive symbols $D_{o\iota'\alpha}$, which are to be *proper symbols*, and which in the intended interpretation of the system are to correspond to the relation of *designation* or *denotation* in a way which we shall explain more precisely below. As a matter of fact only one of the additional primitive symbols, namely $D_{o\iota'(o\iota')}$, is needed for our immediate purpose — but it seems preferable to introduce the entire infinite list at once, for the sake of background, of possible discussions of side issues, and of possible formulations of some of the other semantical antinomies (such as Berry's, Richard's, or König's).

Since in the verbal statement of Grelling's antinomy it is necessary to *speak about* words and expressions (or wffs) as well as to *use* them, it would seem at first sight that we must also add to the system primitive notations which are designed to provide, for every wff, another wff which denotes it. This complication way, however, be avoided by adopting a device due to Gödel. Namely we introduce a system of numbering the wffs — so that with each wff is associated a natural number, to be called *the number of* that wff, and no two wffs have the same number. Then instead speaking about a wff directly, we may speak merely about its number, thus using the numbers of the wffs as proxies for the wffs themselves (in much the same way that the houses on a street or the inmates of a prison may be numbered and thereafter referred to only by number).

Therefore we let $D_{o\iota'\alpha}$ be the only additional primitive symbols. The definition of "wff" remains the same as in the original paper, except for the addition of the primitive symbols $D_{o\iota'\alpha}$ and their inclusion among the proper symbols (each one having the type indicated by its subscript). Details of the method by which numbers are assigned to wffs are not important for our present purpose, and we therefore skip the lengthy but mechanical task of stating such details: the method of Gödel's paper of 1931 may be followed with obvious necessary modifications, or any one of many other methods may be used. We resolve the ambiguity of type involved in speaking of "the number of a wff" by specifying that the type shall be ι' — i.e., we use natural numbers of the lowest type.

Now consider the relation holding between the number of a wff of type α without free variables, and that which the wff denotes. In other words a natural number has this relation to an entity (function, individual, etc.) of type α if and only if the given natural number is the number of a wff of type α without free variables and this wff denotes the given entity of type α. In the formal system let us take the primitive symbol $D_{o\iota'\alpha}$ to denote this relation. It would seem plausible that this interpretation of the symbol $D_{o\iota'\alpha}$ is admissible, since this symbol has the right type subscript, and its denotation has not yet otherwise been assigned.

This previously unpublished paper was found in type-script, dated February 14, 1949 in Church's hand. The title refers to Church's *A Formulation of the Simple Theory of Types*, ***The Journal of Symbolic Logic***, vol. 5 (1940), pp. 56–68. Printed by permission of the Alonzo Church estate.

In the natural languages there often occur expressions which are equivocal, i.e., which have more than one denotation. But in any formalized language, i.e., in any admissible interpretation of a logistic system, presumably such equivocacy is avoided. The meaning which we have assigned to the symbols $D_{o\iota'\alpha}$ therefore leads us to set down the following *axioms of non-equivocacy* (and to add them as axioms to the system):

$$0^{\alpha}\ .\ D_{o\iota'\alpha}g_{\alpha}x_{\iota'} \supset .\ D_{o\iota'\alpha}f_{\alpha}x_{\iota'} \supset .\ f_{\alpha} = g_{\alpha}.$$

With these additions to the system we go on to exhibit Grelling's antinomy.

The intended interpretation of the system is that described in the original paper, with the foregoing addition regarding the denotation assigned to $D_{o\iota'\alpha}$. Under this interpretation each natural number of type ι' is denoted by many different wffs. But for each such natural number let us select one particular wff denoting it, and call this wff its *standard name*. Namely let the standard names of the natural numbers of type ι' be the wffs which are abbreviated in the original paper as $0_{\iota'}, 1_{\iota'}, 2_{\iota'}, 3_{\iota'}, \ldots$

Now let the wff

$$\lambda x_{\iota'}(f_{o\iota'}) \sim [(f_{o\iota'}x_{\iota'})(D_{o\iota'(o\iota')}f_{o\iota'}x_{\iota'})]$$

(which contains no free variables) be abbreviated as $H_{o\iota'}$.

Here the letter H is intended to suggest the word "heterological." It is important to notice, however, that the introduction of this abbreviation is merely a matter of convenience, and that we could carry out without use of it the entire argument which follows, always writing out the wff $H_{o\iota'}$ in full.

Let the standard name of the number of the wff $H_{o\iota'}$ be the wff $\eta_{\iota'}$.

Here again the use of the Greek letter η is merely a matter of convenience. Having once decided upon and stated explicitly the details of the method by which we assign numbers to wffs, we could actually compute the number of the particular wff $H_{o\iota'}$, and could then write out in full the standard name of that number (instead of abbreviating it as $\eta_{\iota'}$).

Now by $13^{o\iota'}$ (see page 63 of the original paper):

$$D_{o\iota'(o\iota')}H_{o\iota'}\eta_{\iota'},\ H_{o\iota'}\eta_{\iota'} \vdash (\exists f_{o\iota'})\ [(f_{o\iota'}\eta_{\iota'})(D_{o\iota'(o\iota')}f_{o\iota'}\eta_{\iota'})].$$

I.e.,

$$D_{o\iota'(o\iota')}H_{o\iota'}\eta_{\iota'},\ H_{o\iota'}\eta_{\iota'} \vdash \sim(f_{o\iota'}) \sim [(f_{o\iota'}\eta_{\iota'})(D_{o\iota'(o\iota')}f_{o\iota'}\eta_{\iota'})].$$

Hence by conversion:

$$D_{o\iota'(o\iota')}H_{o\iota'}\eta_{\iota'},\ H_{o\iota'}\eta_{\iota'} \vdash \sim H_{o\iota'}\eta_{\iota'}.$$

Hence by the deduction theorem:

$$D_{o\iota'(o\iota')}H_{o\iota'}\eta_{\iota'} \vdash H_{o\iota'}\eta_{\iota'} \supset \sim H_{o\iota'}\eta_{\iota'}.$$

Hence by propositional calculus:

$$D_{o\iota'(o\iota')}H_{o\iota'}\eta_{\iota'} \vdash \sim H_{o\iota'}\eta_{\iota'}. \qquad \text{(I)}$$

Now by the axiom of non-equivocacy, $0^{o\iota'}$, substitution, and *modus ponens*:

$$D_{o\iota'(o\iota')}H_{o\iota'}\eta_{\iota'} \vdash D_{o\iota'(o\iota')}f_{o\iota'}\eta_{\iota'} \supset .\ f_{o\iota'} = H_{o\iota'}.$$

Hence by propositional calculus, *modus ponens,* and $17^{o\iota'}$:

$$D_{o\iota'(o\iota')}H_{o\iota'}\eta_{\iota'},\ [(f_{o\iota'}\eta_{\iota'})(D_{o\iota'(o\iota')}f_{o\iota'}\eta_{\iota'})] \vdash H_{o\iota'}\eta_{\iota'}.$$

Hence by the deduction theorem:

$$D_{o\iota'(o\iota')}H_{o\iota'}\eta_{\iota'} \vdash [(f_{o\iota'}\eta_{\iota'})(D_{o\iota'(o\iota')}f_{o\iota'}\eta_{\iota'})] \supset H_{o\iota'}\eta_{\iota'}.$$

Hence by propositional calculus:

$$D_{o\iota'(o\iota')}H_{o\iota'}\eta_{\iota'} \vdash \sim H_{o\iota'}\eta_{\iota'} \supset \sim [(f_{o\iota'}\eta_{\iota'})(D_{o\iota'(o\iota')}f_{o\iota'}\eta_{\iota'})].$$

Hence by (I) above, and *modus ponens*:

$$D_{o\iota'(o\iota')}H_{o\iota'}\eta_{\iota'} \vdash \sim [(f_{o\iota'}\eta_{\iota'})(D_{o\iota'(o\iota')}f_{o\iota'}\eta_{\iota'})].$$

Hence by generalization:

$$D_{o\iota'(o\iota')}H_{o\iota'}\eta_{\iota'} \vdash (f_{o\iota'})\sim [(f_{o\iota'}\eta_{\iota'})(D_{o\iota'(o\iota')}f_{o\iota'}\eta_{\iota'})].$$

Hence by conversion:

$$D_{o\iota'(o\iota')}H_{o\iota'}\eta_{\iota'} \vdash H_{o\iota'}\eta_{\iota'}.$$

Hence by (I) above, and propositional calculus:

$$D_{o\iota'(o\iota')}H_{o\iota'}\eta_{\iota'} \vdash H_{o\iota'}\eta_{\iota'} \,.\, \sim H_{o\iota'}\eta_{\iota'}.$$

Thus although the meaning we have assigned to the symbol $D_{o\iota'(o\iota')}$ seems to compel us to make the assumption $D_{o\iota'(o\iota')}H_{o\iota'}\eta_{\iota'}$, it now appears that this assumption leads to a contradiction. This is Grelling's antinomy in its usual form. It is better, however, to continue the formal argument a little further, as follows.

By the deduction theorem:

$$\vdash D_{o\iota'(o\iota')}H_{o\iota'}\eta_{\iota'} \supset . \, H_{o\iota'}\eta_{\iota'} \,.\, \sim H_{o\iota'}\eta_{\iota'}.$$

Hence by propositional calculus:

$$\vdash \sim D_{o\iota'(o\iota')}H_{o\iota'}\eta_{\iota'}.$$

Thus we have proved a theorem which contradicts the meaning which we attempted to assign to the symbol $D_{o\iota'(o\iota')}$. The upshot is that it is demonstrably impossible to adjoin to the original formal system a symbol with the meaning which we described above for $D_{o\iota'(o\iota')}$.

On the other hand no objection appears to giving the following meaning to the symbols $D_{o\iota'\alpha}$. Consider the relation which holds between a natural number and an entity of type α if and only if the given natural number is the number of a wff of type α which contains no free variables and no occurrences of any of the symbols D (with any subscript), and this wff denotes the given entity of type α. We may then assign this relation to the symbol $D_{o\iota'\alpha}$ as its denotation.

These conclusions about the relation of denotation (or designation) are just the analogues of those which are obtained by Tarski about the property of truth. We have reached conclusions about the relation of denotation rather than about the property of truth, partly because of the different character of the logistic system we have employed, partly because we started from Grelling's paradox (which, in the form here taken, concerns denotation) whereas Tarski started from the paradox of the liar (which concerns truth).

Alonzo Church
February 14, 1949

ON CARNAP'S ANALYSIS OF STATEMENTS OF ASSERTION AND BELIEF[1] (1950)

1. FOR statements such as (1) *Seneca said that man is a rational animal* and (A) *Columbus believed the world to be round*, the most obvious analysis makes them statements about certain abstract entities which we shall call 'propositions' (though this is not the same as Carnap's use of the term), namely the proposition that man is a rational animal and the proposition that the world is round; and these propositions are taken as having been respectively the object of an assertion by Seneca and the object of a belief by Columbus. We shall not discuss this obvious analysis here except to admit that it threatens difficulties and complications of its own, which appear as soon as the attempt is made to formulate systematically the syntax of a language in which statements like (1) and (A) are possible. But our purpose is to point out what we believe may be an insuperable objection against alternative analyses that undertake to do away with propositions in favor of such more concrete things as sentences.

As attempts which have been or might be made to analyze (1) in terms of sentences we cite: (2) *Seneca wrote the words 'Man is a rational animal'*; (3) *Seneca wrote the words 'Rationale enim animal est homo'*; (4) *Seneca wrote words whose translation from Latin into English is 'Man is a rational animal'*; (5) *Seneca wrote words whose translation from some Language S′ into English is 'Man is a rational animal'*; (6) *There is a language S′ such that Seneca wrote as sentence of S′ words whose translation from S′ into English is 'Man is a rational animal.'* In each case, 'wrote' is to be understood in the sense, "wrote with assertive intent." And to simplify the discussion, we ignore the existence of spoken languages, and treat all languages as written.

Of those proposed analyses of (1), we must reject (2) on the ground that it is no doubt false although (1) is true. And each of (3)–(6), though having the same truth-value as (1), must be rejected on the ground that it does not convey the same information as (1). Thus (1) conveys the content of what Seneca said without revealing his actual words, while (3) reproduces Seneca's words without saying what meaning was attached to them. In (4) the crucial information is omitted (without which (1) is not even a consequence) that Seneca intended his words as a Latin sentence, rather than as a sentence of some other language in which conceivably the identical words 'Rationale enim animal est homo' might have some quite different meaning. To (5) the objection is the same as to (4), and indeed if we take 'language' in the abstract sense of Carnap's 'semantical system' (so that it is not part of the concept of a language that a language must have been used in historical fact by some human kindred or tribe), then (5) is *L*-equivalent merely to the statement that Seneca wrote something.

(5) and (6) are closely similar to the analysis of belief statements which is offered by Carnap in "Meaning and Necessity", and although he does not say so explicitly it seems clear that Carnap must have intended also such an analysis as this for statements of assertion. However, (6) is likewise unacceptable as analysis of (1). For it is not even possible to infer (1) as a consequence of (6), on logical grounds alone—but only by

Originally published in ***Analysis***, vol. 10 (1950), pp. 97–99.

[1] Presented to the Association for Symbolic Logic, December 28, 1949.

making use of the item of factual information, not contained in (6), that 'Man is a rational animal' means in English that man is a rational animal.

Following a suggestion of Langford[2] we may bring out more sharply the inadequacy of (6) as an analysis of (1) by translating into another language, say German, and observing that the two translated statements would obviously convey different meanings to a German (whom we may suppose to have no knowledge of English). The German translation of (1) is (1′) *Seneca hat gesagt, dass der Mensch ein vernünftiges Tier sei.* In translating (6), of course 'English' must be translated as 'Englisch' (not as 'Deutsch') and ''Man is a rational animal'' must be translated as ''Man is a rational animal'' (not as ''Der Mensch ist ein vernünftiges Tier'').

Replacing the use of translation (as it appears in (6)) by the stronger requirement of intensional isomorphism, Carnap would analyze the belief statement (A) as follows: (B) *There is a sentence* $\mathfrak{S}_i$ *in a semantical system* S' *such that* (*a*) $\mathfrak{S}_i$ *is intensionally isomorphic to 'The world is round' and* (*b*) *Columbus was disposed to an affirmative response to* $\mathfrak{S}_i$. However, intensional isomorphism, as appears from Carnap's definition of it, is a relation between ordered pairs consisting each of a sentence and a semantical system. Hence (B) must be rewritten as: (C) *There is a sentence* $\mathfrak{S}_i$ *in a semantical system* S' *such that* (*a*) $\mathfrak{S}_i$ *as sentence of* S' *is intensionally isomorphic to 'The world is round' as English sentence and* (*b*) *Columbus was disposed to an affirmative response to* $\mathfrak{S}_i$ *as sentence of* S'.

For the analysis of (1), the analogue of (C) would seem to be: (7) *There is a sentence* $\mathfrak{S}_i$ *in a semantical system* S' *such that* (*a*) $\mathfrak{S}_i$ *as sentence of* S' *is intensionally isomorphic to 'Man is a rational animal' as English sentence and* (*b*) *Seneca wrote* $\mathfrak{S}_i$ *as sentence of* S'.

Again Langford's device of translation makes evident the untenability of (C) as an analysis of (A), and of (7) as an analysis of (1).

2. The foregoing assumes that the word 'English' in English and the word 'Englisch' in German have a sense which includes a reference to matters of pragmatics (in the sense of Morris and Carnap)—something like, e.g., "the language which was current in Great Britain and the United States in 1949 A.D."

As an alternative we might consider taking the sense of these words to be something like "the language for which such and such semantical rules hold," a sufficient list of rules being given to ensure that there is only one language satisfying the description. The objection would then be less immediate that (1) is not a logical consequence of (6) or (7), and it is possible that it would disappear.

In order to meet this latter alternative without discussing in detail the list of semantical rules which would be required, we modify as follows the objection to (7) as an analysis of (1). Analogous to the proposal, for English, to analyze (1) as (7), we have, for German, the proposal to analyze (1′) as (7″) *Es gibt einen Satz* $\mathfrak{S}_i$ *auf einem semantischen system* S', *so dass* (*a*) $\mathfrak{S}_i$ *als Satz von* S' *intensional isomorph zu 'Der Mensch ist ein vernünftiges Tier' als deutscher Satz ist, und* (*b*) *Seneca* $\mathfrak{S}_i$ *als Satz von* S' *geschrieben hat*. Because of the exact parallelism between them, the two proposals stand or fall together. Yet (7″) in German and (7) in English are not in any acceptable sense translations of each other. In particular, they are not intensionally isomorphic.

[2] In ***The Journal of Symbolic Logic***, vol. 2 (1937), p. 53.

And if we consider the English sentence (α) *John believes that Seneca said that man is a rational animal* and its German translation (α'), we see that the sentences to which we are led as supposed analyses of (α) and (α') may even have opposite truth-values in their respective languages; for John, though knowing the semantical rules of both English and German, may nevertheless fail to draw certain of their logical (or other) consequences.

Princeton University

INTENSIONAL SEMANTICS (THE NEED FOR ABSTRACT ENTITIES IN SEMANTIC ANALYSIS) (1951)

We distinguish between a *logistic system* and a *formalized language* on the basis that the former is an abstractly formulated calculus for which no interpretation is fixed, and thus has a syntax but no semantics; but the latter is a logistic system together with an assignment of meanings to its expressions.

As primitive basis of a logistic system it suffices to give, in familiar fashion: (1) The list of primitive symbols, or *vocabulary* of the system (together usually with a classification of the primitive symbols into categories, which will be used in stating the formation rules and rules of inference). (2) The *formation rules*, determining which finite sequences of primitive symbols are to be *well-formed* expressions, determining certain categories of well-formed expressions, among which we shall assume that at least the category of *sentence* is included, and determining (in case *variables* are included among the primitive symbols) which occurrences of variables in a well-formed expression are *free* occurrences and which are *bound* occurrences.[1] (3) The transformation rules or *rules of inference*, by which from the *assertion* of certain sentences (the *premisses*, finite in number) a certain sentence (the *conclusion*) may be *inferred*. (4) Certain asserted sentences, the *axioms*.

In order to obtain a formalized language it is necessary to add to these *syntactical rules* of the logistic system, *semantical rules* assigning meanings (in some sense) to the well-formed expressions of the system.[2] The character of the semantical rules will depend on the theory of meaning adopted, and this in turn must be justified by the purpose which it is to serve.

Let us take it as our purpose to provide an abstract theory of the actual use of language for human communication—not a factual or historical report of what has been observed to take place, but a norm to which we may regard everyday linguistic behavior as an imprecise approximation, in the same way that e.g. elementary (applied) geometry is a norm to which we may regard as imprecise approximations the practical activity of the land-surveyor in laying out a plot of ground, or of the construction foreman in seeing that building plans are followed. We must demand of such a theory that it have a place for all observably informative kinds of communication—including such notoriously troublesome cases as belief statements, modal statements, conditions contrary to fact—or at least that it provides a (theoretically) workable substitute for them. And solutions must be available for puzzles about meaning which may arise, such as the so-called "paradox of analysis."

Originally published under the title *The Need for Abstract Entities in Semantic Analysis*, in the ***Proceedings of the American Academy of Arts and Sciences***, vol. 80, no. 1, July 1951, pp. 100–112. Reprinted in ***The Structure of Language***, edited by J. Fodor and J. Katz, Prentice-Hall, Englewood Cliffs 1964, pp. 437–445. Reprinted in ***The Philosophy of Language***, edited by A. P. Martinich, Oxford University Press, New York 1985, pp. 40–47.

[1] For convenience of the present brief exposition we make the simplifying assumption that sentences are without free variables, and that only sentences are asserted.

[2] The possibility that the meaningful expressions may be a proper subclass of the well-formed expressions must not ultimately be excluded. But again for the present sketch it will be convenient to treat the two classes as identical — the simplest and most usual case. Compare, however, footnote 13.

There exist more than one theory of meaning showing some promise of fulfilling these requirements, at least so far as the formulation and development have presently been carried. But the theory of Frege seems to recommend itself above others for its relative simplicity, naturalness, and explanatory power—or, as I would advocate, Frege's theory as modified by elimination of his somewhat problematical notion of a function (and in particular of a *Begriff*) as *ungesättigt*, and by some other changes which bring it closer to present logistic practice without loss of such essentials as the distinction of sense and denotation.

This modified Fregean theory may be roughly characterized by the tendency to minimize the category of *syncategorematic* notations—i.e., notations to which no meaning at all is ascribed in isolation but which may combine with one or more meaningful expressions to form a meaningful expression[3]—and to reduce the categories of meaningful expressions to two, (proper) *names* and *forms*, for each of which two kinds of meaning are distinguished in a parallel way.

A name, or a *constant* (as we shall also say, imitating mathematical terminology), has first its *denotation*, or that of which it is a name.[4] And each name has also a *sense*—which is perhaps more properly to be called its meaning, since it is held that complete understanding of a language involves the ability to recognize the sense of any name in the language, but does not demand any knowledge beyond this of the denotations of names. (Declarative) *sentences*, in particular, are taken as a kind of names, the denotation being the *truth-value* of the sentence, *truth* or *falsehood*, and the sense being the *proposition* which the sentence expresses.

A name is said to *denote* its denotation and to *express* its sense, and the sense is said to be *a concept of* the denotation. The abstract entities which serve as senses of names let us call *concepts*—although this use of the word 'concept' has no analogue in the writings of Frege, and must be carefully distinguished from Frege's use of 'Begriff'. Thus anything which is or is capable of being the sense of some name in some language, actual or possible, is a concept.[5] The terms *individual concept*, *function concept*, and the like are then to mean a concept which is a concept of an individual, of a function, etc. A *class concept* may be identified with a *property*, and a *truth-value concept* (as already indicated) with a proposition.

Names are to be meaningful expressions without free variables, and expressions which are analogous to names except that they contain free variables, we call forms (a rather wide extension of the ordinary mathematical usage, here adopted for lack of a better term).[6] Each variable has a *range*, which is the class of admissible *values* of the

---

[3]Such notations can be reduced to at most two, namely the notation (consisting, say, of juxtaposition between parentheses) which is used in application of a singulary function to its argument, and the abstraction operator λ. By the methods of the Schönfinkel–Curry combinatory logic it may even be possible further to eliminate the abstraction operator, and along with it the use of variables altogether. But this final reduction is not contemplated here—nor even necessarily the simpler reduction to two syncategorematic notions.

[4]The complicating possibility is here ignored of *denotationless names*, or names which have a sense but no denotation. For though it may be held that these do occur in the natural language, it is possible, as Frege showed, to construct a formalized language in such a way as to avoid them.

[5]This is meant only as a preliminary rough description. In logical order, the notion of a concept must be postulated and that of a possible language defined by means of it.

[6]Frege's term in German is *Marke*.—The form or *Marke* must of course not be confused with its associated abstract entity, the *function*. The function differs from the form in that it is not a linguistic entity, and belongs to no particular language. Indeed the same function may be associated with different forms;

variable.[7] And analogous to the denotation of a name, a form has a *value* for every system of admissible values of its free variables.[8]

The assignment of a value to a variable, though it is not a syntactical operation, corresponds in a certain way to the syntactical operation of substituting a constant for the variable. The denotation of the substituted constant represents the value of the variable.[9] And the sense of the substituted constant may be taken as representing a *sense-value* of the variable. Thus every variable has, beside its range, also a *sense-range*, which is the class of admissible sense-values of the variable. And analogous to the sense of a name, a form has a *sense-value* for every system of admissible sense-values of its free variables.[10]

The following principles are assumed:[11] (i) Every concept is a concept of at most one thing. (ii) Every constant has a unique concept as its sense. (iii) Every variable has a non-empty class of concepts as its sense-range. (iv) For any assignment of sense-values, one to each of the free variables of a given form, if each sense-value is admissible in the sense that it belongs to the sense-range of the corresponding variable, the form has a unique concept as its sense-value. (v) The denotation of a constant is that of which its sense is a concept. (vi) The range of a variable is the class of those things of which the members of the sense-range are concepts. (vii) If $S, s_1, s_2, \ldots, s_m$

---

and if there is more than one free variable the same form may have several associated functions. But in some languages it is possible from the form to construct a name (or names) of the associated function (or functions) by means of an abstraction operator.

[7]The idea of allowing variables of different ranges is not Fregean, except in the case of functions in Frege's sense (i.e., as *ungesättigt*), the different categories of which appear as ranges for different variables. The introduction of *Gegenstandsbuchstaben* with restricted ranges is one of the modifications here advocated in Frege's theory.

[8]Exceptions to this are familiar in common mathematical notation. E.g. the form x/y has no value for the system of values $0, 0$ of x, y. However, the semantics of a language is much simplified if a value is assigned to a form for every system of values of the free variable which are admissible in the sense that each value belongs to the range of the corresponding variable. And for purposes of the present exposition we assume that this has been done. (Compare footnote 4.)

[9]Even if the language contains no constant denoting the value in question, it is possible to consider an extension of the language obtained by adjoining such a constant.

[10]The notion of a sense-value of a form is not introduced by Frege, at least not explicitly, but it can be argued that it is necessarily implicit in his theory. For Frege's question, "How can $\mathbf{a} = \mathbf{b}$ if true ever differ in meaning from $\mathbf{a} = \mathbf{a}$?" can be asked as well for forms **a** and **b** as for constants, and leads to the distinction of value and sense-value of a form just as it does to the distinction of denotation and sense of a constant. Even in a language like that of ***Principia Mathematica***, having no forms other than propositional forms, a parallel argument can be used to show that from the equivalence of two propositional forms **A** and **B** the identity in meaning of **A** and **B** in all respects is not to be inferred. For otherwise how could $\mathbf{A} \equiv \mathbf{B}$ if true (i.e., true for all values of the variables) ever differ in meaning from $\mathbf{A} \equiv \mathbf{A}$?

[11]For purposes of the preliminary sketch, the metalanguage is left unformalized, and such questions are ignored as whether the metalanguage shall conform to the theory of types or to some alternative such as transfinite type theory or axiomatic set theory. Because of the extreme generality which is attempted in laying down these principles, it is clear that there may be some difficulty in rendering them precise (in their full attempted generality) by restatement in a formalized metalanguage. But it should be possible to state the semantical rules of a particular object language so as to conform, so that the principles are clarified to this extent by illustration.

It is not meant that the list of principles is necessarily complete or in final form, but rather a tentative list is here proposed for study and possible amendment. Moreover it is not meant that it may not be possible to formulate a language not conforming to the principles, but only that a satisfactory general theory may result by making conformity to these principles a part of the definition of a formalized language (compare footnote 12).

are concepts of $A, a_1, a_2, \ldots, a_m$ respectively, and if S is the sense-value of a form **F** for the system of sense-values $s_1, s_2, \ldots, s_m$ of its free variables $\mathbf{x}_1, \mathbf{x}_2, \ldots, \mathbf{x}_m$, then the value of **F** for the system of values $a_1, a_2, \ldots, a_m$ of $\mathbf{x}_1, \mathbf{x}_2, \ldots, \mathbf{x}_m$ is A. (viii) If $\mathbf{C}'$ is obtained from a constant **C** by replacing a particular occurrence of a constant **c** by a constant $\mathbf{c}'$ that has the same sense as **c**, then $\mathbf{C}'$ is a constant having the same sense as **C**.[12] (ix) If $\mathbf{C}'$ is obtained from a constant **C** by replacing a particular occurrence of a constant **c** by a constant $\mathbf{c}'$ that has the same denotation as **c**, then $\mathbf{C}'$ is a constant having the same denotation as **C**.[13] (x) If $\mathbf{C}'$ is obtained from a constant **C** by replacing a particular occurrence of a form **f** by a form $\mathbf{f}'$ that has the same free variables as **f**, and if, for every admissible system of sense-values of their free variables, **f** and $\mathbf{f}'$ have the same sense-value, then $\mathbf{C}'$ is a constant having the same sense as **C**.[12] (xi) If $\mathbf{C}'$ is obtained from a constant **C** by replacing a particular occurrence of a form **f** by a form $\mathbf{f}'$ that has the same free variables as **f**, and if, for every system of values of their free variables which are admissible in the sense that each value belongs to the range of the corresponding variable, **f** and $\mathbf{f}'$ have the same value, then $\mathbf{C}'$ is a constant having the same denotation as **C**.[13] (xii) If $\mathbf{x}_1, \mathbf{x}_2, \ldots, \mathbf{x}_m$ are all the distinct variables occurring (necessarily as bound variables) in a constant **C**, if $\mathbf{y}_1, \mathbf{y}_2, \ldots, \mathbf{y}_m$ are distinct variables having the same sense-range as $\mathbf{x}_1, \mathbf{x}_2, \ldots, \mathbf{x}_m$ respectively, and if $\mathbf{C}'$ is obtained from **C** by substituting $\mathbf{y}_1, \mathbf{y}_2, \ldots, \mathbf{y}_m$ throughout for $\mathbf{x}_1, \mathbf{x}_2, \ldots, \mathbf{x}_m$ respectively, then $\mathbf{C}'$ is a constant having the same sense as **C**. (xiii) If $\mathbf{x}_1, \mathbf{x}_2, \ldots, \mathbf{x}_m$ are the distinct variables occurring in a constant **C**, if $\mathbf{y}_1, \mathbf{y}_2, \ldots, \mathbf{y}_m$ are distinct variables having the same ranges as $\mathbf{x}_1, \mathbf{x}_2, \ldots, \mathbf{x}_m$ respectively, and if $\mathbf{C}'$ is obtained from **C** by substituting $\mathbf{y}_1, \mathbf{y}_2, \ldots, \mathbf{y}_m$ throughout for $\mathbf{x}_1, \mathbf{x}_2, \ldots, \mathbf{x}_m$ respectively, then $\mathbf{C}'$ is a constant having the same denotation as **C**. (xiv) The result of substituting constants for all free variables of a form is a constant, if the sense of each substituted constant belongs to the sense-range of the corresponding variable.[12] (xv) The sense of a constant **C** thus obtained by substituting constants $\mathbf{c}_1, \mathbf{c}_2, \ldots, \mathbf{c}_m$ for the free variables $\mathbf{x}_1, \mathbf{x}_2, \ldots, \mathbf{x}_m$ of a form **F**

[12]In the case of some logistic systems which have been proposed (e.g., by Hilbert and Bernays), if semantical rules are to be added, in conformity with the theory here described and with the informally intended interpretation of the system, it is found to be impossible to satisfy (viii), (x), and (xiv), because of restriction imposed on the bound variables which may appear in a constant or form used in a particular context. But it would seem that modifications in the logistic system necessary to remove the restriction may be considered nonessential, and that in this sense (viii), (x), (xiv) may still be maintained.

In regard to all of the principles it should be understood that nonessential modifications in existing logistic systems may be required to make them conform. In particular the principles have been formulated in a way which does not contemplate the distinction in typographical style between free and bound variables that appears in systems of Frege and of Hilbert–Bernays.

In (x) and (xi), the condition that f' have the same free variables as f can in many cases be weakened to the condition that every free variable of f' occur also as a free variable of f.

[13]Possibly (ix) and (xi) should be weakened to require only that if $\mathbf{C}'$ is well-formed then it is a constant having the same denotation as **C**. Since there is in general no syntactical criterion by which to ascertain whether two constants **c** and $\mathbf{c}'$ have the same denotation, or whether two forms have always the same values, there is the possibility that the stronger forms of (ix) and (xi) might lead to difficulty in some cases. However, (ix) as here stated has the effect of preserving fully the rule of substitutivity of equality—where the equality sign is so interpreted that $[\mathbf{c}_1 = \mathbf{c}_2]$ is a sentence denoting truth if and only if $\mathbf{c}_1$ and $\mathbf{c}_2$ are constants having the same denotation—and if in some formalized languages, (ix) and (xi) should prove to be inconsistent with the requirement that every well-formed expression be meaningful (footnote 2), it may be preferable to abandon the latter. Indeed the preservation of the rule of substitutivity of equality may be regarded as an important advantage of a Fregean theory of meaning over some of the alternatives that suggest themselves.

is the same as the sense-value of **F** when the senses of $\mathbf{c}_1, \mathbf{c}_2, \ldots, \mathbf{c}_m$ are assigned as the sense-values of $\mathbf{x}_1, \mathbf{x}_2, \ldots, \mathbf{x}_m$.

To these must still be added principles which are similar to (viii)–(xv), except that substitution is made in forms instead of constants, or that forms and variables as well as constants are substituted for the free variables of a form. Instead of stating these here, it may be sufficient to remark that they follow if arbitrary extensions of the language are allowed by adjoining (as primitive symbols) constants which have as their senses any concepts that belong to sense-ranges of variables in the language, if the foregoing principles are assumed to hold also for such extensions of the language, and if there is assumed further: (xvi) Let an expression **F** contain the variables $\mathbf{x}_1, \mathbf{x}_2, \ldots, \mathbf{x}_m$; and suppose that in every extension of the language of the kind just described and for every substitution of the constants $\mathbf{c}_1, \mathbf{c}_2, \ldots, \mathbf{c}_m$ for the variables $\mathbf{x}_1, \mathbf{x}_2, \ldots, \mathbf{x}_m$ respectively, if the sense of each constant belongs to the sense-range of the corresponding variable, **F** becomes a constant; then **F** is a form having $\mathbf{x}_1, \mathbf{x}_2, \ldots, \mathbf{x}_m$ as its free variables.

To those who find forbidding the array of abstract entities and principles concerning them which is here proposed, I would say that the problems which give rise to the proposal are difficult and a simpler theory is not known to be possible.[14]

To those who object to the introduction of abstract entities at all I would say that I believe there are more important criteria by which a theory should be judged. The extreme demand for a simple prohibition of abstract entities under all circumstances perhaps arises from a desire to maintain the connection between theory and observa-

---

[14] At the present stage it cannot be said with assurance that a modification of Frege's theory will ultimately prove to be the best or the simplest. Alternative theories demanding study are: the theory of Russell, which relies on the elimination of names by contextual definition to an extent sufficient to render the distinction of sense and denotation unnecessary; the modifications of Russell's theory, briefly suggested by Smullyan [*Editor's note*: ***The Journal of Symbolic Logic***, 13 (1948), pp. 31–37], according to which descriptive phrases are to be considered as actually contained in the logistic system rather than being (in the phrase of Whitehead and Russell) "mere typographical conveniences," but are to differ from names in that they retain their need for scope indicators; and finally the theory of Carnap's ***Meaning and Necessity***.

Though the Russell theory has an element of simplicity in avoiding the distinction of two kinds of meaning, it leads to complications of its own of a different sort, in connection with the matter of scope of descriptions. The same should be said of Smullyan's proposed modification of the theory. And the distinctions of scope become especially important in modal statements, where they cannot be eliminated by the convention of always taking the minimum scope, as Smullyan has shown (loc. cit.).

Moreover, in its present form it would seem that the Russell theory requires some supplementation. For example, "I am thinking of Pegasus," "Ponce de Leon searched for the fountain of youth," "Barbara Villiers was less chaste than Diana" cannot be analyzed as "$(\exists c)[x$ is a Pegasus $\equiv_x x = c]$[I am thinking of c]," "$(\exists c)$ [x is a fountain of youth $\equiv_x x = c$] [Ponce de Leon searched for c]," "$(\exists c)[x$ is a Diana $\equiv_x x = c]$ [Barbara Villiers was less chaste than c]" respectively—if only because of the (probable or possible) difference of truth-value between the given statements and their proposed analyses. On a Fregean theory of meaning the given statements might be analyzed as being about the individual concepts of Pegasus, of the fountain of youth, and of Diana rather than about some certain winged horse, some certain fountain, and some certain goddess. For the Russell theory it might be suggested to analyze them as being about the property of being a Pegasus, the property of being a fountain of youth, and the property of being a Diana. This analysis in terms of properties would also be possible on a Fregean theory, though perhaps slightly less natural. On a theory of the Russell type the difficulty arises that names of properties seem to be required, and on pain of readmitting Frege's puzzle about equality (which leads to the distinction of sense and denotation in connection with names of any kind), such names of properties either must be analyzed away by contextual definition—it is not clear how—or must be so severely restricted that two names of the same property cannot occur unless trivially synonymous.

tion. But the preference of (say) *seeing* over *understanding* as a method of observation seems to me capricious. For just as an opaque body may be seen, so a concept may be understood or grasped. And the parallel between the two cases is indeed rather close. In both cases the observation is not direct but through intermediaries—light, lens of eye or optical instrument, and retina in the case of visible body, linguistic expressions in the case of the concept. And in both cases there are or may be tenable theories according to which the entity in question, opaque body or concept, is not assumed, but only those things which would otherwise be called its effects.[14½]

[14½] (Added 1971.) In spite of its Platonic origin this remark is not intended to support Platonism, or even the extreme realism of Frege and Gödel, but rather only in opposition to the claim of superior reality for physical or spatio-temporal entities.

There is an interesting comment on this passage by Carnap in his *Replies and Systematic Expositions* (***The Philosophy of Rudolf Carnap***, edited by Paul Arthur Schilpp, La Salle, Illinois, 1963, see pages 924–925): "Relations of the casual type can indeed hold only among physical objects (or states or processes), not between a physical object and an abstract entity. It seems typical of Platonism, which both Sellars and I reject, that it speaks of relations of this causal type (called 'commerce' or 'intercourse' or the like) as holding between physical objects (or persons or minds) and abstract entities. My reason for regarding the two sentences 'John observes the table' and 'John observes (is aware of) the number 13' as not being analogous is just this: the first sentence states a causal relation between the table and John (mediated by light rays, the retina, etc., as Church indicates) but the second does not. Only spatio-temporal objects, not numbers, can have a causal effect on John. On the other hand, it seems to me that some psychological concepts may be regarded or reconstructed as relations (in the wide sense of the logical terminology, not in the causal sense) between a person and an abstract entity: e.g., believing may be taken as a relation between a person and a proposition (as is done by Church, compare §9 VII), and thinking-of as a relation between a person and a concept (intension or sense) and the like. In particular, there seems to be no objection to the use of relations of this kind *in a theoretical language* (compare my remarks on semantical concepts in a theoretical language in §10 V)."

The notion of a *spatio-temporal* entity—i.e., one for which it is appropriate to give space-time coordinates—and the distinction between spatio-temporal entities and others must no doubt be conceded to Carnap as familiar and intelligible. But it is argued in a later paper (*Ontological Commitment*, ***The Journal of Philosophy***, vol. 55 (1958), pp. 1008–1014) that the notion of a *temporal* entity, as one that changes in time, is not so clear—except in the case that the entity is in fact spatio-temporal.

However, Carnap's comment just quoted depends also on his notion of a theory, and on his distinction between the theoretical terms, which are introduced only in a theoretical language, and the vocabulary used to state the observational data. Here also the writer is largely in agreement with Carnap's approach, provided that (as Carnap would perhaps agree) the term "theory" is not restricted to what is scientific in the narrow sense but is allowed to cover also, as being at least inchoate theories, various systems of notions which arose long before scientific method came to be consciously used and which have therefore become embedded as matters of "common sense" in the natural languages. The difference with Carnap lies rather in his tacit assumption, in the quoted passage, that if entities are spatio-temporal (as physical objects, states, processes, persons, events), then names of and terminology for them must belong to the vocabulary in which the observational data are stated. The last sentence of the paragraph in the text above was intended to suggest that physical objects in particular are better understood as theoretical constructs, so that the terminology of physical objects belongs to a theoretical language.

The issue as to what entities may enter into a causality relation is of lesser importance. But it should be said that the cause of an event is never a physical object or even another event but a total situation, in the description of which many entities, both abstract and concrete, must in general be mentioned.

For example the cause of John's suicide is neither the bell tower from which he leaped nor the hard pavement below, though both of these enter into the causal situation. To describe a sufficient cause we must explain, among other things, that John believed that Mary did not love him and that he knew that the bell tower was nearby, and unguarded, and of such a height above the pavement that his leap would be fatal. Granted that the object of knowledge or belief must be a proposition, the sense in which various propositions enter into the cause thus seems to be not significantly different from the sense in which various physical

The variety of entities (whether abstract or concrete) which a theory assumes is indeed one among other criteria by which it may be judged. If multiplication of entities is found beyond the needs of workability, simplicity, and generality of the theory, then the razor shall be applied.[15] The theory of meaning here outlined I hold

---

objects enter into the cause. Indeed Carnap would perhaps express John's belief by a disposition predicate of John. But it remains essential that John's belief was a belief in the specific proposition that Mary did not love him—not e.g. a belief that Anna did not love him, which could not (in the existing circumstances) have had so drastic an effect.

I anticipate the question, if the terminology of physical objects does not belong to the vocabulary in which the observational data are to be stated, then in what vocabulary can they be stated.

The answer must take into account that in order to state anything at all it is necessary to have an organized language and hence a theory, or at least what was called above an inchoate theory. And once an accepted theory of physical objects is available, one may use the vocabulary it provides to state observational data, and such data may then be used to support or refute some further theory (or even perhaps this same theory).

Before a theory of enduring physical objects is available, perhaps the observational data may be stated in terms of instantaneous physical objects, or perhaps a vocabulary of sense data may be used. But this is not to say that either instantaneous physical objects or sense data are ultimate in the way that Carnap takes spatio-temporal entities generally as ultimates. Rather at some not very definite stage the regress of theories must cease to be a regress of sharply formulated theories and become a regress to cruder and cruder theories until we end in mankind's earliest language of cries, grunts and gestures.

Indeed, we may choose to end the regress of *sharply* organized languages and sharply formulated theories at a point at which we still have vocabulary for all presently accepted spatio-temporal entities (including such sophisticated ones as e.g. magnetic fields). But the choice seems injudicious, and I suppose that Carnap does not intend this. Rather Carnap lays down (see page 959 of ***The philosophy of Rudolf Carnap***) that the vocabulary of the observation language is to cover directly observable properties and relations, of *what* is not said, but presumably of directly observable things. It seems that one may choose what are the directly observable things; and it is not immediately clear whether tables and bell towers shall be among them, and if so, whether as enduring physical objects or as instantaneous physical objects.

Curiously, Carnap himself says (p. 868 ff in the same book) that the statement asserting the reality of the external world and statements of the reality or irreality of abstract entities are pseudo-statements "i.e., devoid of cognitive content" [when taken as "external existential statements" laid down prior to the decision as to acceptance of some particular language]. Yet the passage quoted from pages 924–925 does seem to ascribe some sort of superior reality to "physical objects (or states or processes)"—or at least to those which are observable (perhaps I should not say "directly observable," as the passage seems to allow that observation may be mediated).

[15]Here a warning is necessary against spurious economies, since not every subtraction from the entities which a theory assumes is a reduction in the variety of entities.

For example, in the simple theory of types it is well-known that the individuals may be dispensed with if classes and relations of all types are retained; or one may abandon also classes and relations of the lowest type, retaining only those of higher type. In fact any finite number of levels at the bottom of the hierarchy may be deleted. But this is no reduction in the variety of entities, because the truncated hierarchy of types, by appropriate deletions of entities in each type, can be made isomorphic to the original hierarchy—and indeed the continued adequacy of the truncated hierarchy to the original purposes depends on this isomorphism.

Similarly the idea may suggest itself to admit the distinction of sense and denotation at the nth level and above in the hierarchy of types, but below the nth level to deny this distinction and to adopt instead Russell's device of contextual elimination of names. The entities assumed would thus include only the usual extensional entities below the nth level, but at the nth level and above they would include also concepts, concepts of concepts, and so on. However, this is no reduction in the variety of entities assumed, as compared to the theory which assumes at all levels in the hierarchy of types not only the extensional entities but also the concepts of them, concepts of concepts of them, and so on. For the entities assumed by the former theory are reduced again to isomorphism with those assumed by the latter, if all entities below the nth level are deleted and appropriate deletions are made in every type at the nth level and above.

Some one may object that the notion of isomorphism is irrelevant which is here introduced, and insist that any subtraction from the entities assumed by a theory must be considered a simplification. But to such objector I would reply that his proposal leads (in the cases just named, and others) to perpetual oscillation

exempt from such treatment no more than any other, but I do advocate its study.

Let us return now to our initial question, as to the character of the semantical rules which are to be added to the syntactical rules of a logistic system in order to define a particular formalized language.

On the foregoing theory of meaning the semantical rules must include at least the following: (5) *Rules of sense*, by which a sense is determined for each well-formed expression without free variables (all such expressions thus becoming names). (6) *Rules of sense-range*, assigning to each variable a sense-range. (7) *Rules of sense-value*, by which a sense-value is determined for every well-formed expression containing free variables and every admissible system of sense-values of its free variables (all such expressions thus becoming forms).

In the case of both syntactical and semantical rules there is a distinction to be drawn between *primitive* and *derived* rules, the primitive rules being those which are stated in giving the primitive basis of the formalized language, and the derived rules being rules of similar kind which follow as consequences of the primitive rules. Thus besides primitive rules of inference there are also derived rules of inference, besides primitive rules of sense also derived rules of sense, and so on. (But instead of "derived axioms" it is usual to say *theorems*.)

A statement of the denotation of a name, the range of a variable, or the value of a form does not necessarily belong to the semantics of a language. For example, that 'the number of planets' denote the number nine is a fact as much of astronomy as it is of the semantics of the English language, and can be described only as belonging to a discipline broad enough to include both semantics and astronomy. On the other hand, a statement that 'the number of planets' denotes the number of planets is a purely semantical statement about the English language. And indeed it would seem that a statement of this kind may be considered as purely semantical only if it is a consequence of the rules of sense, sense-range, and sense-value, together with the syntactical rules and the general principles of meaning (i)–(xvi).

Thus as derived semantical rules rather than primitive, there will be also: (8) *Rules of denotation*, by which a denotation is determined for each name. (9) *Rules of range*, assigning to each variable a range. (10) *Rules of value*, by which a value is determined for every form and for every admissible system of values of its free variables.

By stating (8), (9), and (10) as primitive rules, without (5), (6), and (7) there results what may be called the *extensional part* of the semantics of a language. The remaining *intensional part* of the semantics does not follow from the extensional part. For the sense of a name is not uniquely determined by its denotation, and thus a particular rule of denotation does not of itself have as a consequence the corresponding rule of sense.

On the other hand, because the metalinguistic phrase which is used in the rule of denotation must itself have a sense, there is a certain sense (though not that of logical consequence) in which the rule of denotation, by being given as a primitive rule of denotation, uniquely indicates the corresponding rule of sense. Since the like is true of the rules of range and rules of value, it is permissible to say that we have fixed an *interpretation* of a given logistic system, and thus a formalized language, if we have

between two theories T_1 and T_2, T_1 being reduced to T_2, and T_2 to T_1 by successive "simplifications" *ad infinitum*.

stated only the extensional part of the semantics.[16]

Although all the foregoing account has been concerned with the case of a formalized language, I would go on to say that in my opinion there is no difference in principle between this case and that of one of the natural languages. In particular, it must not be thought that a formalized language depends for its meaning or its justification (in any sense in which a natural language does not) upon some prior natural language, say English, through some system of translation of its sentences into English—or, more plausibly, through the statement of its syntactical and semantical rules in English. For speaking in principle, and leaving all questions of practicality aside, the logician must declare it a mere historical accident that you and I learned from birth to speak English rather than a language with less irregular, and logically simpler, syntactical rules, similar to those of one of the familiar logistic systems in use today—or that we learned in school the content of conventional English grammars and dictionaries rather than a more precise statement of a system of syntactical and semantical rules of the kind which has been described in this present sketch. The difference of a formalized language from a natural language lies not in any matter of principle, but in the degree of completeness that has been attained in the laying down of explicit syntactical and semantical rules and the extent to which vagueness and uncertainties have been removed from them.

For this reason the English language itself may be used as a convenient though makeshift illustration of a language for which syntactical and semantical rules are to be given. Of course only a few illustrative examples of such rules can be given in brief space. And even for this it is necessary to avoid carefully the use of examples involving English constructions that raise special difficulties or show too great logical irregularities, and to evade the manifold equivocacy of English words by selecting and giving attention to just one meaning of each word mentioned. It must also not be asked whether the rules given as examples are among the "true" rules of the English language or are "really" a part of what is implied in an understanding of English; for the laying down of rules for a natural language, because of need to fill gaps and to decide doubtful points, is as much a process of legislation as of reporting.

With these understandings, and with no attempt made to distinguish between primitive and derived rules, following are some examples of syntactical and semantical rules of English according to the program which has been outlined.[17]

(1) Vocabulary: 'equals' 'five' 'four' 'if' 'is' 'nine' 'number' 'of' 'planet' 'planets' 'plus' 'round' 'the' 'then' 'the world'—besides the bare list of primitive symbols (words) there must be statements regarding their classification into categories and systematic relations among them, e.g., that 'planet' is a common noun,[18] that 'planets' is the

---

[16]As is done in the revised edition of my ***Introduction to Mathematical Logic***, Volume I.

[17]For convenience, English is used also as the metalanguage, although this gives a false appearance of triviality or obviousness to some of the semantical rules. Since the purpose is only illustrative, the danger of semantical antinomies is ignored.

[18]For present illustrative purposes the question may be avoided whether common nouns in English, in the singular, shall be considered to be variables (e.g., 'planet' or 'a planet' as a variable having planets as its range), or to be class names (e.g., 'planet' as a proper name of the class of planets), or to have "no status at all in a logical grammar" (see Quine's ***Methods of Logic***, p. 207), or perhaps to vary from one of these uses to another according to context.

plural of 'planet',[19] that 'the world' is a proper noun, that 'round' is an adjective.

(2) Formation Rules: If **A** is the plural of a common noun, then 'the'⌒'number'⌒'of'⌒**A** is a singular term. A proper noun standing alone is a singular term. If **A** and **B** are singular terms, then **A**⌒'equals'⌒**B** is a sentence. If **A** is a singular term and **B** is an adjective, then **A**⌒'is'⌒**B** is a sentence.[20] If **A** and **B** are sentences, then 'if'⌒**A**⌒'then'⌒**B** is a sentence.—Here singular terms and sentences are to be understood as categories of well-formed expressions; a more complex list of formation rules would no doubt introduce many more such.

(3) Rules of Inference: Where **A** and **B** are sentences, from 'if'⌒**A**⌒'then'⌒**B** and **A** to infer **B**. When **A** and **B** are singular terms and **C** is an adjective, from **A**⌒'equals'⌒**B** and **B**⌒'is'⌒**C** to infer **A**⌒'is'⌒**C**.

(4) Axioms-Theorems: 'if the world is round, then the world is round'; 'four plus five equals nine.'

(5) Rules of Sense: 'round' expresses the property of roundness. 'the world' expresses the (individual) concept of the world. 'the world is round' expresses the proposition that the world is round.

(6) Rules of Denotation: 'round' denotes the class of round things. 'the world' denotes the world. 'the world is round' denotes the truth-value thereof that the world is round.[21]

On a Fregean theory of meaning, rules of truth in Tarski's form—e.g., "'the world is round' is true if and only if the world is round"—follow from the rules of denotation for sentences. For that a sentence is true is taken to be the same as that it denotes truth.

---

[19]Or possibly 'planet' and 's' could be regarded as two primitive symbols, by making a minor change in existing English so that all common nouns form the plural by adding 's'.

[20]If any of you finds unacceptable the conclusion that therefore 'the number of planets is round' is a sentence, he may try to alter the rules to suit, perhaps by distinguishing different types of terms. This is an example of a doubtful point, on the decision of which there may well be differences of opinion. The advocate of a set-theoretic language may decide one way and the advocate of type theory another, but it is hard to say that either decision is the "true" decision for the English language as it is.

[21]But of course it would be wrong to include as a rule of denotation: 'the world is round' denotes truth. For this depends on a fact of geography extraneous to semantics (namely that the world is round).

A FORMULATION OF THE LOGIC OF SENSE AND DENOTATION (1951)

The intensional aspects of Frege's logical doctrine, and his distinction between the sense (*Sinn*) and the denotation (*Bedeutung*) of a name, were explained by him informally in his paper, *Über Sinn und Bedeutung*,[1] and in incidental passages in a number of his other publications, including the first volume of his book, *Grundgesetze der Arithmetik* (Jena, 1893). In his more formal work, Frege's formalized language (*Begriffsschrift*, or *Formelsprache*) has an entirely extensional interpretation, and it may even be that his interest in intensional logic was primarily to clear up certain difficulties regarding its relationship to extensional logic,[2] so as to be able to proceed with development of the latter unhampered. Nevertheless, it seems that Frege would agree that intensional logic also must ultimately receive treatment by the logistic method. And it is the purpose of this paper to make a tentative beginning toward such a treatment, along the lines of Frege's doctrine.

While we preserve what we believe to be the important features of the theory of Frege, we do make certain changes to which he would probably not agree. One of these is the introduction of the simple theory of types as a means of avoiding the logical antinomies. Another is the abandonment of Frege's notion of a function (including propositional functions) as something *ungesättigt*, in favor of a notion according to which the name of a function may be treated in the same manner as any other name, provided that distinctions of type are observed. (But it is even possible that Frege might accept this latter change, on the basis of an understanding that what we call a function is the same thing which he calls *Werthverlauf einer Funktion*.)

It should be added that, while Frege makes no complete discussion of the circumstances under which two names shall be considered to have the same sense, it is clear that he intends that two names, **A** and **B**, may have different senses even when the equation **A** = **B** is logically valid. Nevertheless, the criterion that **A** and **B** have the same sense if and only if **A** = **B** is logically valid is a natural one to consider. It is adopted by Carnap in his book, *Meaning and Necessity*,[3] for what he calls the intension

---

Originally published in ***Structure, Method and Meaning: Essays in Honor of Henry M. Sheffer***, edited by Paul Henley, H. M. Kallen, and S. K. Langer, The Liberal Arts Press, New York, 1951, pp. 3–24.

[1]In ***Zeitschrift für Philosophie und philosophische Kritik***, C (1892), 25–50. See English translations of this paper by Black, in ***The Philosophical Review***, LVII (1948), 207–230, and by Feigl, in ***Readings in Philosophical Analysis*** (New York, 1949); and also a discussion of Frege's doctrines by Russell, in Appendix A of ***The Principles of Mathematics***. In reading these, it is necessary to make allowance for differences in the translations that are adopted of some of Frege's terms. We shall here translate Frege's *ausdrücken* as "express" and Frege's *bedeuten* or *bezeichnen* as "denote" or "be a name of," so that a name is said to express its sense and to denote or to be a name of its denotation.

[2]We mention the doctrine of Frege's ***Begriffsschrift*** of 1879, according to which the relation of identity or equality is a relation between names rather than between the things named, apparently on the ground that identity construed in the latter sense would be too trivial a relation to serve its intended purpose. If use and mention are not to be confused, the idea of identity as a relation between names renders a formal treatment of the logic of identity all but impossible. Solution of this difficulty is made the central theme of *Über Sinn und Bedeutung* and is actually a prerequisite to Frege's treatment of identity in ***Grundgesetze der Arithmetik***.

[3]Chicago, 1947.

of a designator. And it is one of (in effect) three alternatives regarding a criterion for identity of sense which were considered by the present writer in an abstract of the same title as this paper.[4]

In the last-named abstract, as originally written, two different criteria of sense were proposed, which, following the original abstract, we shall here call Alternative (1) and Alternative (2). Both of these have the consequence that **A** and **B** have the same sense if **A** conv **B** (that is, if **B** can be obtained from **A** by a succession of the three operations which are given in Rules I–III below). But Alternative (1) is in the direction of making the senses of **A** and **B** different, whenever this is possible under the assumptions which are outlined in the abstract (and which will be repeated in this paper). And Alternative (2), on the other hand, makes the senses of **A** and **B** the same whenever $\mathbf{A} = \mathbf{B}$ is logically valid.

In the addendum to the abstract, dated April 29, 1946, an amended form of Alternative (1) is introduced, which — to distinguish it from the original Alternative (1) — we shall now call Alternative (0). This alternative may be described roughly by saying that it makes the notion of sense correspond to Carnap's notion of intensional structure, with the one difference that it retains the notion of sense as something to be dealt with in the object language, whereas Carnap's intensional structure is a metatheoretic notion and is dealt with in his meta-language.[5]

---

[4]In ***The Journal of Symbolic Logic***, XI (1946), 31. In the system of the present paper, as compared to that of the abstract, there are some changes of detail. These include the introduction of all the symbols Δ as primitive symbols, instead of using the scheme given in the abstract by which some of the symbols Δ are defined in terms of the others. Then the place of these definitions is taken by Axioms $15^{\alpha\beta}$ and $16^{\alpha\beta}$ (given below).

As a matter of fact, if the direction of Alternative (0) is followed, Axioms $16^{\alpha\beta}$ must simply be dropped, and it is an oversight to the addendum to the abstract that this situation is not pointed out. And for Alternatives (1) and (2), there is moreover the possibility that Axioms $16^{\alpha\beta}$ may have to be modified in connection with considerations concerning vacuous concepts (in the sense of footnote 18).

[5]The notion of intensional structure is applied by Carnap to the analysis of such statements as "Seneca said that man is a rational animal" or "Columbus believed that the world is round," understood in such a way that they do not reveal what was the actual succession of letters that Seneca wrote down or what languages may have been known to Columbus. But in the writer's opinion, Carnap's analysis of these statements is open to a fatal objection, which is connected with the fact that intensional structure has an essentially metatheoretic character, and which seems likely to hold against any analysis according to which assertion or belief is a relation between a person and a sentence (rather than between a person and a proposition or the like).

In order to explain this objection, we shall make use of the device of translation from one language into another, following a suggestion of Langford in ***The Journal of Symbolic Logic***, II (1937), 53–54. This device is not essential to the explanation, but is helpful in order to dispel any remnants of an illusion that there is something in some way necessary or transparent about the connection between a word or a sentence and its meaning, whereas, of course, this connection is entirely artificial and arbitrary.

For convenience of explanation, let us suppose that Carnap's semantical system S is English (although ultimately Carnap's analysis is intended to apply to a formalized language or semantical system, rather than to so loose and uncertain a structure as the English language). The statement that Columbus believed that the world is round is analyzed by Carnap as follows: There is a sentence $\mathfrak{S}_i$ in a semantical system S' such that (*a*) $\mathfrak{S}_i$ is intensionally isomorphic to "The world is round," and (*b*) Columbus was disposed to an affirmative response to $\mathfrak{S}_i$. Before discussing this, we must supply an omission in it, the importance of which depends on the fact that the same sentence $\mathfrak{S}$ may have entirely different meanings in different semantical systems. (This is obvious if "semantical system" is taken in an abstract sense, but in view of the uncounted billions of human languages, past, present, and future, it is no doubt also true of the natural languages in a strictly historical sense.) Namely, intensional isomorphism, as Carnap's definition of it makes clear, is a relation not between two sentences, but between two ordered pairs consisting each of a sentence

Because of the need to use an infinite hierarchy of operators λ, distinguished from one another by subscripts, as explained in the addendum to the abstract, it is clear that Alternative (0) is likely to lead to a more complicated system than either of the other two alternatives. Moreover, a preliminary survey suggests that additional complications, not contemplated when the abstract and the addendum were written, may be necessary in the case of Alternative (0) or Alternative (1), or both, in order to avoid admitting into the system an analogue of Richard's antinomy.[6] In the present paper, attention will be given primarily to the (seemingly) less difficult Alternative (2). Alternative (0) will be left aside entirely, and only some brief indications regarding Alternative (1) will be given at the end.[7]

The system of the writer's *A Formulation of the Simple Theory of Types*[8] is taken,

---

and a semantical system. And a like amendment must be made in regard to the relation of being disposed to an affirmative response, as this relation also involves a particular system or a language. (For example, if I enter a crowded restaurant with a party, my response to the syllable "nine" uttered by the head waiter will be quite different in New York from what it would be in Vienna.) Therefore Carnap's analysis must be reworded as follows: There is a sentence $\mathfrak{S}_i$ in a semantical system S' such that (*a*) $\mathfrak{S}_i$ as a sentence of S' is intensionally isomorphic to "The world is round" as an English sentence, and (*b*) Columbus was disposed to an affirmative response to $\mathfrak{S}_i$ as a sentence of S'.

Now this last is supposed to be in some sense synonymous with "Columbus believed that the world is round" (so that it is permissible to construe the shorter sentence as being, by contextual definition, a mere abbreviation of the longer one). If the two English sentences are synonymous, then of course their German translations must be synonymous; let us therefore test the synonymy by translating both into German. It is left to the reader to carry out the actual translation, remembering that the German translation of "English" is "*englisch*" (not "*deutsch*"!) and that, although the German translation of "that the world is round" is "*dass die Welt rund sei*," the German translation of " "The world is round" " is simply " "The world is round." " The synonymy of the German sentences is then to be tested by considering the information which each will convey to a German who does not understand English but does fully understand German, and who in particular is acquainted with the notions of semantical system, intensional isomorphism, and disposal to an affirmative response, as explained to him under their German names and in his own language.

[6]The writer is indebted to Leon Henkin for raising the question of the cardinal number of the concepts (senses) of a given type, in connection with the answer to which an antinomy may easily appear, at least unless appropriate caution has been exercised in regard to the assumptions which are made (in the form of axioms and rules). Because of this and other possibilities of self-contradiction, no logistic treatment of sense and denotation can be accepted as more than provisional until its consistency has been thoroughly studied.

[7]Alternative (1) is selected for this concluding sketch only because its lesser divergence from Alternative (2) makes it possible to discuss it in briefer space. The indications regarding Alternative (1) are given partly for the sake of their bearing on Alternative (0). And the writer attaches the greater importance to Alternative (0) because it would seem that it is in this direction (provided it can be followed through in spite of threatening difficulties) that a satisfactory analysis is to be sought of statements regarding assertion and belief.

Moreover, the suggestion is obvious that if a sound logistic treatment of Alternative (0) could once be obtained, it might be possible to dispense with separate treatment of Alternatives (1) and (2) by taking senses in the sense of Alternative (1) and of Alternative (2) as certain kinds of classes of senses in the sense of Alternative (0). But whether this can actually be done must await the determination of just what is possible in the treatment of Alternative (0).

(There is also the possibility that the treatment of Alternative (0) and (1) might be combined in a single system by using both the primitive symbols Δ which are introduced in this paper and the symbols Δ which are introduced by definition in the abstract of the same title—see footnote 4—of course introducing some difference of notation to distinguish the latter symbols from the former. The latter symbols would then be thought of as denoting the relation of being a concept of, in the sense of Alternative (1), and the former as denoting the relation of being a concept of, in the sense of Alternative (0).)

[8]***The Journal of Symbolic Logic***, V (1940), 56–58.

with some modifications, as a basis for those proposed here, and familiarity with that paper will be assumed in what follows. One of the modifications is for the purpose of avoiding the assertion of formulas containing free variables,[9] as is stated in the abstract. But the primitive symbols adopted here are in part different from those of the abstract, as well as from those of *Formulation of the Simple Theory of Types*, and other modifications correspond to this change.

The *type symbols* include the two infinite lists of symbols, $o_0, o_1, o_2, \cdots$ and $\iota_0, \iota_1, \iota_2, \cdots$, of which o_0 and ι_0 are also written as o and ι respectively, and are identified with the type symbols o and ι of *Formulation of the Simple Theory of Types*. All the remaining type symbols are obtained by the rule that, if α and β are any type symbols, then $(\alpha\beta)$ is a type symbol. For practical convenience in writing type symbols, we shall abbreviate them by omission of parentheses, with the convention that association is to the left; also by omission of the subscript 0 after o and ι, as already stated.

As illustrated in the preceding paragraph, we use Greek letters $\alpha, \beta, \gamma, \delta$ (also the same letters with any number of bars over them) as syntactical variables whose values are type symbols. Moreover, when such a Greek letter — say, for example, α — is being used for a type symbol, then we use α_n for the type symbol obtained from it by increasing every subscript by n. (Here n is any of the natural numbers $0, 1, 2, \cdots$ or the corresponding numeral denoting the natural numbers.)[10]

Among the type symbols we distinguish certain ones as *preferred type symbols*, according to the following rules: o_n is a preferred type symbol; $\alpha\iota$ is a preferred type symbol; if α is a preferred type symbol, then $\alpha\beta$ is a preferred type symbol.

Our primitive symbols include, for each type symbol α, an infinite list of *variables* which have the type symbol α as subscript and are said to be of type α. Namely any small italic letter, with or without one or more bars over it and with any type symbol as subscript, is used as a variable.

Besides variables, the primitive symbols include the following *constants* (infinite in number):

$$C_{o_n o_n o_n}, \quad \Pi_{o_n(o_n\alpha_n)}, \quad \iota_{\beta_n(o_n\beta_n)}, \quad \Delta_{o_n\alpha_{n+1}\alpha_n},$$

for all type symbols α, all preferred type symbols β, and all natural numbers n.

Then, to complete the list of primitive symbols, we add the three symbols λ, (,). These are called *improper symbols*, to distinguish them from primitive constants and variables, which are called *proper symbols*.

Any finite sequence of primitive symbols is a *formula*. Certain formulas are distinguished as being *well-formed* and as having a certain *type*, in accordance with the following rules: (1) a formula consisting of a single proper symbol is well-formed and has the type indicated by its subscript; (2) if $\mathbf{x}_\beta$ is a variable of type β and $\mathbf{M}_\alpha$ is a well-formed formula of type α, then $(\lambda\mathbf{x}_\beta\mathbf{M}_\alpha)$ is a well-formed formula having the type $\alpha\beta$; (3) if $\mathbf{F}_{\alpha\beta}$ and $\mathbf{A}_\beta$ are well-formed formulas of types $\alpha\beta$ and β, respectively, then $(\mathbf{F}_{\alpha\beta}\mathbf{A}_\beta)$ is a well-formed formula having the type α. The well-formed formulas are the least class of formulas which these rules allow, and the type of a well-formed

[9]Other formulations of the logic of quantifiers that avoid the assertion of formulas containing free variables are due to Quine, Fitch, and Berry — see W. V. Quine's ***Mathematical Logic***, edition of 1947, §16. However, these are of a rather different character from that presented here.

[10]For example, if α is $o(o_1\iota_1)(o\iota)$, then α_1 is $o_1(o_2\iota_2)(o_1\iota_1)$ and α_2 is $o_2(o_3\iota_3)(o_2\iota_2)$. When $n = 0$, of course α_n is the same as α.

formula is that determined (uniquely) by these rules. And occurrence of a variable $\mathbf{x}_\beta$ in a well-formed formula is bound or free, according as it is or is not an occurrence in a well-formed part of the formula having the form $(\lambda \mathbf{x}_\beta \mathbf{M}_\alpha)$. The bound variables of a well-formed formulas are those which have bound occurrences in the formula, and the free variables are those which have free occurrences.

As illustrated in the foregoing statement, we use bold letters as syntactical variables, bold capital letters for well-formed formulas, and bold small letters for variables, a subscript being added in either case to show the type of the well-formed formula or the variable which the bold letter is to have as value. Moreover, in making metatheoretic statements, we adopt the customary, self-explanatory usage, according to which symbols belonging to the object language serve in the meta-language (English) as names of themselves, and juxtaposition is used for juxtaposition. (Thus we avoid the rather cumbersome apparatus of quotation and quasi-quotation, without—it is hoped—any real loss of clearness.)

The same conventions will be adopted for abbreviating well-formed formulas by omitting parentheses and brackets, and by using bold dots to replace brackets, as explained in *Formulation of the Simple Theory of Types*. And we introduce also the following conventions of abbreviation (reading the arrow as "stands for" or "is an abbreviation for"):[11]

$$
\begin{array}{l}
[\mathbf{A}_{o_n} \supset \mathbf{B}_{o_n}] \to C_{o_n o_n o_n} \mathbf{A}_{o_n} \mathbf{B}_{o_n} \\
(\mathbf{x}_{\alpha_n}) \mathbf{A}_{o_n} \to \Pi_{o_n (o_n \alpha_n)} (\lambda \mathbf{x}_{\alpha_n} \mathbf{A}_{o_n}) \\
T \to (a_o) \centerdot a_o \supset a_o \\
T_{o_n} \to (a_{o_n}) \centerdot a_{o_n} \supset a_{o_n} \\
F \to (a_o) a_o \\
F_{o_n} \to (a_{o_n}) a_{o_n} \\
\sim \mathbf{A}_{o_n} \to \mathbf{A}_{o_n} \supset F_{o_n} \\
(\boldsymbol{E} \mathbf{x}_{\alpha_n}) \mathbf{A}_{o_n} \to \sim (\mathbf{x}_{\alpha_n}) \sim \mathbf{A}_{o_n} \\
(\imath \mathbf{x}_{\beta_n}) \mathbf{A}_{o_n} \to \imath_{\beta_n (o_n \beta_n)} (\lambda \mathbf{x}_{\beta_n} \mathbf{A}_{o_n}) \\
Q_{o_n \alpha\alpha} \to \lambda a_\alpha \lambda b_\alpha (f_{o_n \alpha}) \centerdot f_{o_n \alpha} b_\alpha \supset f_{o_n \alpha} a_\alpha \\
[\mathbf{A}_\alpha = \mathbf{B}_\alpha] \to Q_{o\alpha\alpha} \mathbf{B}_\alpha \mathbf{A}_\alpha \\
[\mathbf{A}_\alpha =_n \mathbf{B}_\alpha] \to Q_{o_n \alpha\alpha} \mathbf{B}_\alpha \mathbf{A}_\alpha \\
[\mathbf{A}_\alpha \neq \mathbf{B}_\alpha] \to \sim \centerdot \mathbf{A}_\alpha = \mathbf{B}_\alpha \\
[\mathbf{A}_\alpha \neq_n \mathbf{B}_\alpha] \to \sim \centerdot \mathbf{A}_\alpha =_n \mathbf{B}_\alpha \\
N_{o_m o_n} \to Q_{o_m o_n o_n} T_{o_n}
\end{array}
$$

As indicated above, we may omit the subscript after F or T whenever the subscript is o, and we may omit the subscript after either of the signs $=$ or $\neq$ whenever the subscript is 0. As further abbreviations of the same kind we add the following:

We may omit the subscript of $N_{o_m o_n}$ if m is 0 and if the omitted subscript can be restored in a unique manner from the information that m is 0 and that the formula being written is well-formed. We may omit the subscript of any of the primitive constants,

$$C_{o_n o_n o_n}, \qquad \Pi_{o_n (o_n \alpha_n)}, \qquad \imath_{\beta_n (o_n \beta_n)}, \qquad \Delta_{o_n \alpha_{n+1} \alpha_n},$$

if n is 0 and if the omitted subscript can be restored in a unique manner from the information that n is 0 and that the formula being written is well-formed. We may

[11] Such conventions of abbreviation we call also "definitions." The m and n occurring in these definitions are arbitrary numerals, and, in particular, m or n or both may be 0.

omit the subscript after one of the letters p, q, r, s (with or without bars over them) with the understanding that the omitted subscript is o. We may omit the subscript after one of the letters f, g, h (with or without bars over them) with the understanding that the omitted subscript has one of the forms $o\alpha$, $o\alpha\beta$, $o\alpha\beta\gamma$, and so on, if the omitted subscript can be restored in a unique manner on the basis of this understanding and the information that the formula being written is well-formed.

We add also the following convention of abbreviation—which, however, is not actually used in this paper. The subscript of a variable may be omitted at all of its occurrences in a formula except the first one, the understanding being then that the same subscript is to be supplied after later occurrences of the letter as after its first occurrence. (Of course, this abbreviation is never to be used in any case where the same small italic letter occurs with different subscripts in one formula.)

In the intended interpretation, there is contemplated for each type symbol α a corresponding domain which we call the *type* α. This domain is to constitute the range of possible values of a variable of type α. And not only the variable but also its values, the members of the domain, will be said to be *of type* α. (Generally, this duplication of terminology causes no confusion; and, where necessary to avoid misunderstanding, it is always possible to substitute a longer phrase or to add an explanation.)

In order to describe what the members of each type are to be, it will be convenient to introduce the term *concept* in a sense which is entirely different from that of Frege's *Begriff*, but which corresponds approximately to the use of the word by Russell and others in the phrase "class concept" and rather closely to the recent use of the word by Carnap, in *Meaning and Necessity*. Namely anything which is capable of being the sense of a name of x is called a *concept of* x.

A concept in this sense is not to be thought of as associated with any particular language or system of notation, since names in different languages may express the same sense (or concept). We suppose that a concept may in some sense exist even if there is no language in actual use that contains a name expressing this concept. And we may even wish to admit a non-enumerable infinity of concepts—thus more concepts than there can be names to express in any one actual language.[12]

The type o is to consist of the truth-values, truth and falsehood. That is, it is to have just these two members.

For the type ι any well-defined domain may be selected, which may be infinite or finite or even empty. That is, many different interpretations are admitted, one for each possible determination of the type ι. But we suppose that a particular determination of the type ι has been fixed upon, and we call the members of the type ι *individuals*.

The type o_1 is to consist of concepts of truth-values—which, following Frege, we identify as *propositions* (*Gedanken*). The type o_2 is to consist of concepts of propositions, or, as we shall say, *propositional concepts*. And, generally, the type o_{n+1} is to consist of concepts of the members of the type o_n.

Similarly, the type ι_1 is to consist of concepts of individuals, or *individual concepts*. And, generally, the type ι_{n+1} is to consist of concepts of members of the type ι_n.[13]

---

[12]It is entirely usual to assume the existence of more things—e.g., real numbers—than there can be names to denote in any one actual language. The assumption of more concepts than there can be names to express would seem to be at least no more objectionable.

[13]The hierarchy of concepts of successively higher order arises as soon as we suppose that a concept, like

All the remaining types are to consist of functions in the way which is described in *Formulation of the Simple Theory of Types*. Namely, given any two types, corresponding to type symbols α and β, we may consider the functions such that the range of the argument (or of the independent variable) comprises the second of the two types, and the values of the function are members of the first type. All these functions together are to constitute a new type, corresponding to the type symbol $(\alpha\beta)$.

For example, the members of the type $\iota\iota$ are functions from individuals to individuals—that is, functions which take individuals as arguments and which, when applied to an individual as argument, yield an individual as the value of the function. Similarly, members of the type $o\iota$ are functions from individuals to truth values. Members of the type $o\iota\iota$ are functions from individuals to things of type $o\iota$; and, since in this case the values of the function are themselves functions, which may be applied to individuals as arguments and will then yield truth-values as values, we may also regard members of the type $o\iota\iota$ as binary functions of individuals having truth-values as their values.

Similarly, the members of the type ooo may be regarded as binary functions of truth-values having truth-values as their values—or, as we shall say, following the terminology of *Principia Mathematica*, as binary *truth-functions*. One such binary truth-function is the familiar material implication, which, when applied to two truth-values as arguments, has the value truth if either the first argument is falsehood or the second argument is truth, and has the value falsehood otherwise.[14] We interpret the primitive constant C_{ooo} as denoting material implication[15]—and the convention by which $[\mathbf{A}_o \supset \mathbf{B}_o]$ stands for $C_{ooo}\mathbf{A}_o\mathbf{B}_o$ therefore simply reintroduces the more usual notation for material implication.

In general, the primitive constants shall denote each a particular member of the type that is indicated by the subscript (just as, for example, C_{ooo} denotes a particular member of the type ooo).

For each type symbol α, the primitive constant $\Pi_{o(o\alpha)}$ denotes a function of the indicated type which, when applied to a function of the right type as argument will yield as value either truth, in case the latter function has the value truth for all arguments of the

anything else which can be discussed at all, is capable of having a name given to it. For a sense of a name of a concept is a concept of the next higher order, and so on.

Attempts might be made to avoid the infinite regress by supposing that a name of a concept differs from other names in some way as to the analysis of its meaning—e.g., that a name of a concept has a denotation but no sense, or that the sense and denotation coincide (contrary to the intent of Axioms 17^{α} below), or that the relation between the name and the concept is some third kind of meaning, neither that of expressing nor of denoting. But these meet with the difficulty that the paradox about identity statements which Frege discusses in *Über Sinn und Bedeutung* ("How can $\mathbf{A} = \mathbf{B}$, if true, ever differ in meaning from $\mathbf{A} = \mathbf{A}$?") can be reproduced at every level in the hierarchy and always leads by the same argument to the introduction of the next higher level—and this will be true independently of which of the three alternatives (0), (1), (2) we adopt. See a discussion of the "paradox of analysis" by Black and White in ***Mind***, LIII and LIV, and a review of it by the present writer in ***The Journal of Symbolic Logic***, XI (1946), 132–133.

[14]This is Frege's conditional as it must be modified to conform to the theory of types which is here adopted, or Russell's material implication as it must be modified to conform to a Fregean theory of meaning. There is, of course, also a corresponding binary function of propositions, having propositions as values, which might be called "material implication"; this is denoted by C with the subscript $o_1o_1o_1$, in our present system of notation, as will appear below.

[15]Although this statement mentions explicitly only the denotation of C_{ooo}, it is to be understood as giving implicitly also the sense. In a more formal treatment of the semantics of the system, the two statements—one about the denotation of C_{ooo} and one about the sense of C_{ooo}—should be made separately and explicitly. A like remark applies to other semantical statements made below.

right type, or otherwise falsehood. The relationship of this to the universal quantifier is such that the convention by which $(\mathbf{x}_\alpha)\mathbf{A}_o$ stands for $\Pi_{o(o\alpha)}(\lambda\mathbf{x}_\alpha\mathbf{A}_o)$ amounts to introducing the usual notation for the universal quantifier as an abbreviation.

As already presupposed in some of the foregoing explanations, we use juxtaposition between parentheses as a notation for application of function to argument (the parentheses being often omitted as an abbreviation), and we interpret λ as an abstraction operator, in the same way as explained in *Formulation of the Simple Theory of Types*.

It follows that the formulas T and F denote the truth-values truth and falsehood, respectively. And it follows that the notations $\sim, =, \neq$ which are introduced above by conventions of abbreviation, may be read in the usual way as "negation," "equality," and "non-equality," respectively — but provided, in the case of $\sim$, that the formula to which it is prefixed is of type o and, in the case of $=$ and $\neq$, that the subscript is 0 (or that there is no subscript after the sign). Similarly, $(\boldsymbol{E}\mathbf{x}_\alpha)$ may be read as the existential quantifier.

For the interpretation of the primitive constants $\iota_{\beta(o\beta)}$, we suppose that, in some way (depending on the determination of the domain of individuals), there has been selected in each preferred type a particular member of the type which will be called the *designated* member of that type. The preferred types are simply those which are necessarily non-empty, even if the domain of individuals is empty; and the reason for restricting the primitive constants $\iota_{\beta(o\beta)}$ to those for which β is a preferred type symbol is in order to be able to make use of designated members of the corresponding type in providing an interpretation. (If desired, of course, primitive constants $\iota_{\beta(o\beta)}$ may be added also for non-preferred types β; but as soon as this is done, the interpretation of the system is restricted to domains of individuals which are non-empty, or perhaps, at least in the case of Alternative (2), to those which are necessarily non-empty.)

For each preferred type symbol β, the primitive constant $\iota_{\beta(o\beta)}$ denotes a function of the indicated type, which, when applied to a function of the right type as argument, will yield as value either the argument for which the latter function has the value truth, in case there is one and only one such argument, or otherwise the designated member of the type corresponding to the type symbol β.[16]

---

[16]The constant $\iota_{\beta(o\beta)}$ is thus given a rather artificially complicated sense, in order to assure that names of the form $(\iota_{\beta(o\beta)}\mathbf{A}_{o\beta})$ shall never lack a denotation. This device is due to Frege and serves the purpose of greatly simplifying the construction of a formalized language that contains a symbol or notation, such as $\iota_{\beta(o\beta)}$, which can be used to form descriptions.

It would also be possible, by recognizing functions having less than an entire type as the range of the argument, to fix the sense of $\iota_{\beta(o\beta)}$ in such a way that names of the form $(\iota_{\beta(o\beta)}\mathbf{A}_{o\beta})$ would occur that have a sense but no denotation. Frege held that such names do exist in the natural languages, but avoided them in constructing a formalized language. The writer believes that the construction of a formalized language containing denotationless names should also be possible. And it might well be worth while to carry out the construction of such a language in spite of probable complications — if only as a museum piece, to show that the avoidance of denotationless names in a formalized language is a matter of option rather than theoretical necessity. Some suggestions toward such a language may perhaps be taken from two papers by the writer in ***Annals of Mathematics***, XXXIII (1932), 346–366, and Vol. XXXIV (1933), 839–864; for although the formalized language introduced in these papers turned out to involve an inconsistency, it may be that the inconsistency should be traced only to the lack of anything in the nature of a theory of types and that the feature of allowing denotationless names can be preserved.

There are sound reasons for the opinion generally held that the meaningfulness of any expression of a formalized language must not be allowed to depend on any question of extralinguistic fact. But these reasons refer to meaningfulness in the sense of having a sense, rather than in the sense of having a denota-

Thus the notation $(\imath \mathbf{x}_\beta)$, which was introduced above by a convention of abbreviation, may be read as a description operator. Use of the ι as upright instead of inverted may serve to distinguish this description operator from the contextually defined description operator of Russell, as employed in *Principia Mathematica*.[17]

Now consider a function ϕ of type corresponding to the type symbol $\alpha\beta$ and a function ϕ' of type corresponding to the type symbol $\alpha_1\beta_1$. We shall say that the function ϕ' *characterizes* the function ϕ if the following condition is satisfied, that the value of the function ϕ for an argument ξ is η if, and only if, the value of the function ϕ' for any concept of ξ as argument is always a concept of η.

For every concept σ of the function ϕ, there is a unique corresponding function ϕ' that characterizes ϕ, determined by the rule that the value of ϕ' for an argument ξ' is the sense of $\mathbf{F}_{\alpha\beta}\mathbf{X}_\beta$, where $\mathbf{F}_{\alpha\beta}$ is a name having the concept σ as its sense and $\mathbf{X}_\beta$ is a name having the concept ξ' as its sense. (If such names are not available in the language which is being treated, then, for this purpose, we must consider an enlarged language obtained by adding such names as primitive constants.) The characterizing function ϕ' being thus uniquely determined by the concept σ, we introduce the assumption that they are identical. That is, *we identify each concept σ of a function ϕ with the characterizing function ϕ' of ϕ that is determined by σ*—and indeed, if the type symbols α and β consist of more than one letter, this identification was already anticipated in ascribing to ϕ' the type symbol $\alpha_1\beta_1$ when $\alpha\beta$ is the type symbol for ϕ. Moreover, *we assume that every characterizing function ϕ' of ϕ is to be regarded as a concept of ϕ*.[18]

The foregoing assumptions are not found in Frege's writings, but at least the first of the two seems to be not inconsistent with anything he says, and they do greatly simplify the logistic treatment of his doctrine without interfering with its serviceability

tion. Therefore, tentatively, we shall allow that in some language, although not in this present one (the formalized language here being discussed), there may be names which have a sense but no denotation. Hence we also admit concepts that are not concepts of anything; and although no name in the present language has such a concept as its sense, we may wish in the construction of the language to allow for existence of such concepts.

[17]Strictly speaking, there is no description operator and there are no descriptions in the formalized language of ***Principia Mathematica***. For the authors of ***Principia*** state explicitly that they regard definitions generally as "mere typographical conveniences"—thus not a part of their formalized language, but only a means for its easier presentation. In the case of an expression—say, a description—introduced by contextual definition, this means that when the longer expression in which it occurs is rewritten in full there is found to be no well-formed part which can be identified with the description. Hence the Fregean analysis of meaning must not be applied to Russell's descriptions. And, in particular, such a description must not be said to denote, except as a manner of speaking, introduced by a contextual definition applied to the meta-language.

It is even possible that not only the "denotation" but also the "sense" of a Russellian description might be introduced by contextual definition into the metatheory. But until this has actually been done for a particular formalized meta-language, these expressions—especially the "sense"—must be used with a great deal of caution, if at all. (Moreover, it seems unlikely that by such a device the need can be done away with for a direct treatment of intensional notions.)

[18]Of course a function ϕ', belonging to a type having a type symbol of the form $\alpha_1\beta_1$, may be such that its values for two concepts of the same thing as arguments are sometimes concepts of different things. It may be tentatively suggested to think of such a function ϕ' as being also a concept, but one which is, as we shall say, vacuous—i.e., not a concept of anything actual (cf. footnote 16). On the other hand, if ϕ' has the property that its values for two concepts of the same thing as arguments always either are themselves concepts of the same thing or are both vacuous, it seems preferable to regard ϕ' as a concept of a function which may have less than an entire type as its range of arguments.

or explanatory power.[19]

For each type symbol α, the primitive constant $\Delta_{o\alpha_1\alpha}$ denotes a binary function whose value, for a pair of arguments of the indicated types, is truth in case the second argument is a concept of the first argument and is falsehood in the contrary case. For example,

$$\Delta_{o(o_1o_1o_1)(ooo)} C_{ooo} C_{o_1o_1o_1}$$

denotes truth, because we construe the primitive constant $C_{o_1o_1o_1}$ as denoting the sense of the primitive constant C_{ooo}.

We may now conclude our account of the intended meanings of well-formed formulas summarily, without separate mention of the remaining primitive constants, by saying that if, in a well-formed formula without free variables, all the subscripts in all the type symbols appearing are increased by 1, the resulting well-formed formula denotes the sense of the first one. Moreover, our conventions of abbreviation have been arranged so as to preserve this when the formulas are written in abbreviated form, provided that omitted subscripts 0 and omitted type-symbol subscripts are taken into account, and provided that the subscripts on the signs $=$ and $\neq$ are increased by 1 also.

Returning now to the formal development of our system, we introduce the five following rules of inference:

I. To replace any part $\mathbf{M}_\alpha$ of a formula by the result of substituting $\mathbf{y}_\beta$ for $\mathbf{x}_\beta$ throughout $\mathbf{M}_\alpha$, provided that $\mathbf{x}_\beta$ is not a free variable of $\mathbf{M}_\alpha$ and $\mathbf{y}_\beta$ does not occur in $\mathbf{M}_\alpha$. (That is, to infer from a given formula the formula obtained by this replacement.)

II. To replace any part $((\lambda \mathbf{x}_\beta \mathbf{M}_\alpha)\mathbf{N}_\beta)$ of a formula by the result of substituting $\mathbf{N}_\beta$ for $\mathbf{x}_\beta$ throughout $\mathbf{M}_\alpha$, provided that the bound variables of $\mathbf{M}_\alpha$ are distinct both from $\mathbf{x}_\beta$ and from the free variables of $\mathbf{N}_\beta$.

III. Where $\mathbf{A}_\alpha$ is the result of substituting $\mathbf{N}_\beta$ for $\mathbf{x}_\beta$ throughout $\mathbf{M}_\alpha$, to replace any part $\mathbf{A}_\alpha$ of a formula by $((\lambda \mathbf{x}_\beta \mathbf{M}_\alpha)\mathbf{N}_\beta)$, provided that the bound variables of $\mathbf{M}_\alpha$ are distinct both from $\mathbf{x}_\beta$ and from the free variables of $\mathbf{N}_\beta$.

IV. From $\Pi_{o(o\alpha)}\mathbf{F}_{o\alpha}$ to infer $\mathbf{F}_{o\alpha}\mathbf{A}_\alpha$, provided that $\mathbf{A}_\alpha$ has no free variables.

V. From $C_{ooo}\mathbf{A}_o\mathbf{B}_o$ and $\mathbf{A}_o$ to infer $\mathbf{B}_o$.

Then we require axioms which will provide, in effect, the usual laws of quantifiers and of the propositional calculus. For this purpose, we put down the following axioms:[20]

$1^{\alpha\beta}$. $(f) \centerdot (x_\alpha)(y_\beta) f x_\alpha y_\beta \supset (y_\beta)(x_\alpha) f x_\alpha y_\beta$

$2^{\alpha\beta}$. $(f) \centerdot \Pi f \supset (g_{\alpha\beta})(x_\beta) f(g_{\alpha\beta} x_\beta)$

3^{α}. $(f)(g) \centerdot (x_\alpha)[f x_\alpha \supset g x_\alpha] \supset \centerdot \Pi f \supset \Pi g$

4^{α}. $(f) \centerdot (x_\alpha)(y_\alpha) f x_\alpha y_\alpha \supset (x_\alpha) f x_\alpha x_\alpha$

5^{α}. $(p) \centerdot p \supset (x_\alpha) p$

6^{α}. $(f)(x_\alpha) \centerdot \Pi f \supset f x_\alpha$

[19]This should be qualified by the remark that the second of the two italicized assumptions has to be dropped for Alternative (0). The first assumption is represented formally by Axioms $15^{\alpha\beta}$ and the second one by Axioms $16^{\alpha\beta}$; hence compare also footnote 4.

[20]Where a superscript appears after the number, an infinite number of axioms are being condensed into a single axiom schema, by the device of typical ambiguity in the sense of ***Principia Mathematica***. See the explanation which is given in footnote 10 of *Formulation of the Simple Theory of Types*, loc. cit.

7. $(p)(q)(r)(s) \centerdot p \supset q \supset r \supset \centerdot r \supset p \supset \centerdot s \supset p$

Here Axiom 7 is borrowed from Łukasiewicz.[21] It is intended to give those laws of the propositional calculus which involve material implication only (of course after some of the more fundamental laws of quantifiers have been developed from Axioms 1–6). No special axioms about negation are necessary because, in view of the definition of negation, its properties all follow from the laws of quantifiers.

To these we add the following *axioms of extensionality*:

8. $(p)(q) \centerdot p \supset q \supset \centerdot q \supset p \supset \centerdot p = q$

$9^{\alpha\beta}$. $(f_{\alpha\beta})(g_{\alpha\beta}) \centerdot (x_\beta)[f_{\alpha\beta}x_\beta = g_{\alpha\beta}x_\beta] \supset \centerdot f_{\alpha\beta} = g_{\alpha\beta}$

Also the following *axioms of descriptions*, in which β is restricted to be a preferred type symbol:

10^{β}. $(f)(x_\beta) \centerdot fx_\beta \supset \centerdot (y_\beta)[fy_\beta \supset \centerdot y_\beta = x_\beta] \supset f(\iota f)$

(If it is desired to add axioms to assume that no type is empty, this might be done by removing the restriction in the axiom schema 10^{β} that β must be a preferred type symbol. Or, for the purpose of dealing with a fragment of the system not involving descriptions, it might be done by adding the axioms $(p) \centerdot (x_\alpha)p \supset p$ for non-preferred type-symbols α. Also an axiom of infinity may be added if desired—for example, the axiom of infinity, which is used in *A Formulation of the Simple Theory of Types*.)

This brings us to the axioms which specially concern the notion of a concept (or sense). In the first place, we assume the following axioms of this kind, restricting β to be a preferred type symbol in the case of $13^{n\beta}$:

11^{n}. $\Delta C_{o_n o_n o_n} C_{o_{n+1} o_{n+1} o_{n+1}}$

$12^{n\alpha}$. $\Delta \Pi_{o_n(o_n\alpha_n)} \Pi_{o_{n+1}(o_{n+1}\alpha_{n+1})}$

$13^{n\beta}$. $\Delta \iota_{\beta_n(o_n\beta_n)} \iota_{\beta_{n+1}(o_{n+1}\beta_{n+1})}$

$14^{n\alpha}$. $\Delta \Delta_{o_n\alpha_{n+1}\alpha_n} \Delta_{o_{n+1}\alpha_{n+2}\alpha_{n+1}}$

$15^{\alpha\beta}$. $(f_{\alpha\beta})(f_{\alpha_1\beta_1})(x_\beta)(x_{\beta_1}) \centerdot \Delta_{o(\alpha_1\beta_1)(\alpha\beta)} f_{\alpha\beta} f_{\alpha_1\beta_1} \supset \centerdot$
$\Delta_{o\beta_1\beta} x_\beta x_{\beta_1} \supset \Delta_{o\alpha_1\alpha}(f_{\alpha\beta}x_\beta)(f_{\alpha_1\beta_1}x_{\beta_1})$

$16^{\alpha\beta}$. $(f_{\alpha\beta})(f_{\alpha_1\beta_1}) \centerdot (x_\beta)(x_{\beta_1})[\Delta_{o\beta_1\beta} x_\beta x_{\beta_1} \supset$
$\Delta_{o\alpha_1\alpha}(f_{\alpha\beta}x_\beta)(f_{\alpha_1\beta_1}x_{\beta_1})] \supset \Delta_{o(\alpha_1\beta_1)(\alpha\beta)} f_{\alpha\beta} f_{\alpha_1\beta_1}$

17^{α}. $(x_\alpha)(y_\alpha)(x_{\alpha_1}) \centerdot \Delta_{o\alpha_1\alpha} x_\alpha x_{\alpha_1} \supset \centerdot \Delta_{o\alpha_1\alpha} y_\alpha y_{\alpha_1} \supset \centerdot x_\alpha = y_\alpha$

So far these are axioms which we wish to adopt regardless of whether we take the direction of Alternative (1) or Alternative (2). The remaining axioms will depend on which of these two directions we follow.

Consider first the case of Alternative (2). On the basis of this alternative we may interpret the formula N_{oo_1} as meaning logical validity, or necessity. Therefore, on the basis of heuristic principle that whatever is a theorem of the system is also necessary (since the system is itself intended to constitute a proposal, at least in part, as to what shall be considered logically valid, or necessary), we are led to put down the following axioms, restricting β to be a preferred type symbol in the case of 27^{β} and $30^{n\beta}$:

[21] See ***Proceedings of the Royal Irish Academy***, Vol. LVII, Section A, No. 3 (1948), 25–33.

$18^{\alpha\beta}$. $N(f_{o_1\beta_1\alpha_1}) \centerdot (x_{\alpha_1})(y_{\beta_1})f_{o_1\beta_1\alpha_1}x_{\alpha_1}y_{\beta_1} \supset (y_{\beta_1})(x_{\alpha_1})f_{o_1\beta_1\alpha_1}x_{\alpha_1}y_{\beta_1}$

$19^{\alpha\beta}$. $N(f_{o_1\alpha_1}) \centerdot \Pi_{o_1(o_1\alpha_1)}f_{o_1\alpha_1} \supset (g_{\alpha_1\beta_1})(x_{\beta_1})f_{o_1\alpha_1}(g_{\alpha_1\beta_1}x_{\beta_1})$

20^{α}. $N(f_{o_1\alpha_1})(g_{o_1\alpha_1}) \centerdot (x_{\alpha_1})[f_{o_1\alpha_1}x_{\alpha_1} \supset g_{o_1\alpha_1}x_{\alpha_1}] \supset \centerdot \Pi_{o_1(o_1\alpha_1)}f_{o_1\alpha_1} \supset \Pi_{o_1(o_1\alpha_1)}g_{o_1\alpha_1}$

21^{α}. $N(f_{o_1\alpha_1\alpha_1}) \centerdot (x_{\alpha_1})(y_{\alpha_1})f_{o_1\alpha_1\alpha_1}x_{\alpha_1}y_{\alpha_1} \supset (x_{\alpha_1})f_{o_1\alpha_1\alpha_1}x_{\alpha_1}x_{\alpha_1}$

22^{α}. $N(p_{o_1}) \centerdot p_{o_1} \supset (x_{\alpha_1})p_{o_1}$

23^{α}. $N(f_{o_1\alpha_1})(x_{\alpha_1}) \centerdot \Pi_{o_1(o_1\alpha_1)}f_{o_1\alpha_1} \supset f_{o_1\alpha_1}x_{\alpha_1}$

24. $N(p_{o_1})(q_{o_1})(r_{o_1})(s_{o_1}) \centerdot p_{o_1} \supset q_{o_1} \supset r_{o_1} \supset \centerdot r_{o_1} \supset p_{o_1} \supset \centerdot s_{o_1} \supset p_{o_1}$

25. $N(p_{o_1})(q_{o_1}) \centerdot p_{o_1} \supset q_{o_1} \supset \centerdot q_{o_1} \supset p_{o_1} \supset \centerdot p_{o_1} =_1 q_{o_1}$

$26^{\alpha\beta}$. $N(f_{\alpha_1\beta_1})(g_{\alpha_1\beta_1}) \centerdot (x_{\beta_1})[f_{\alpha_1\beta_1}x_{\beta_1} =_1 g_{\alpha_1\beta_1}x_{\beta_1}] \supset \centerdot f_{\alpha_1\beta_1} =_1 g_{\alpha_1\beta_1}$

27^{β}. $N(f_{o_1\beta_1})(x_{\beta_1}) \centerdot f_{o_1\beta_1}x_{\beta_1} \supset \centerdot (y_{\beta_1})[f_{o_1\beta_1}y_{\beta_1} \supset \centerdot y_{\beta_1} =_1 x_{\beta_1}] \supset f_{o_1\beta_1}(\iota_{\beta_1(o_1\beta_1)}f_{o_1\beta_1})$

28^{n}. $N(\Delta_{o_1(o_{n+2}o_{n+2}o_{n+2})(o_{n+1}o_{n+1}o_{n+1})}C_{o_{n+1}o_{n+1}o_{n+1}}C_{o_{n+2}o_{n+2}o_{n+2}})$

$29^{n\alpha}$. $N(\Delta_{o_1(o_{n+2}(o_{n+2}\alpha_{n+2}))(o_{n+1}(o_{n+1}\alpha_{n+1}))}\Pi_{o_{n+1}(o_{n+1}\alpha_{n+1})}\Pi_{o_{n+2}(o_{n+2}\alpha_{n+2})})$

$30^{n\beta}$. $N(\Delta_{o_1(\beta_{n+2}(o_{n+2}\beta_{n+2}))(\beta_{n+1}(o_{n+1}\beta_{n+1}))}\iota_{\beta_{n+1}(o_{n+1}\beta_{n+1})}\iota_{\beta_{n+2}(o_{n+2}\beta_{n+2})})$

$31^{n\alpha}$. $N(\Delta_{o_1(o_{n+2}\alpha_{n+3}\alpha_{n+2})(o_{n+1}\alpha_{n+2}\alpha_{n+1})}\Delta_{o_{n+1}\alpha_{n+2}\alpha_{n+1}}\Delta_{o_{n+2}\alpha_{n+3}\alpha_{n+2}})$

$32^{\alpha\beta}$. $N(f_{\alpha_1\beta_1})(f_{\alpha_2\beta_2})(x_{\beta_1})(x_{\beta_2}) \centerdot \Delta_{o_1(\alpha_2\beta_2)(\alpha_1\beta_1)}f_{\alpha_1\beta_1}f_{\alpha_2\beta_2} \supset \centerdot$
$\Delta_{o_1\beta_2\beta_1}x_{\beta_1}x_{\beta_2} \supset \Delta_{o_1\alpha_2\alpha_1}(f_{\alpha_1\beta_1}x_{\beta_1})(f_{\alpha_2\beta_2}x_{\beta_2})$

$33^{\alpha\beta}$. $N(f_{\alpha_1\beta_1})(f_{\alpha_2\beta_2}) \centerdot (x_{\beta_1})(x_{\beta_2})[\Delta_{o_1\beta_2\beta_1}x_{\beta_1}x_{\beta_2} \supset$
$\Delta_{o_1\alpha_2\alpha_1}(f_{\alpha_1\beta_1}x_{\beta_1})(f_{\alpha_2\beta_2}x_{\beta_2})] \supset \Delta_{o_1(\alpha_2\beta_2)(\alpha_1\beta_1)}f_{\alpha_1\beta_1}f_{\alpha_2\beta_2}$

$33^{*\alpha\beta}$. $(f_{\alpha_1\beta_1})(f_{\alpha_2\beta_2})N \centerdot (x_{\beta_1})(x_{\beta_2})[\Delta_{o_1\beta_2\beta_1}x_{\beta_1}x_{\beta_2} \supset$
$\Delta_{o_1\alpha_2\alpha_1}(f_{\alpha_1\beta_1}x_{\beta_1})(f_{\alpha_2\beta_2}x_{\beta_2})] \supset \Delta_{o_1(\alpha_2\beta_2)(\alpha_1\beta_1)}f_{\alpha_1\beta_1}f_{\alpha_2\beta_2}$

34^{α}. $N(x_{\alpha_1})(y_{\alpha_1})(x_{\alpha_2}) \centerdot \Delta_{o_1\alpha_2\alpha_1}x_{\alpha_1}x_{\alpha_2} \supset \centerdot \Delta_{o_1\alpha_2\alpha_1}y_{\alpha_1}x_{\alpha_2} \supset \centerdot x_{\alpha_1} =_1 y_{\alpha_1}$

35^{α}. $(f_{o_1\alpha_1})(x_{\alpha})(x_{\alpha_1})(x_{\alpha_2}) \centerdot \Delta x_{\alpha}x_{\alpha_1} \supset \centerdot \Delta x_{\alpha_1}x_{\alpha_2} \supset \centerdot$
$N(\Delta_{o_1\alpha_2\alpha_1}x_{\alpha_1}x_{\alpha_2}) \supset \centerdot N(\Pi_{o_1(o_1\alpha_1)}f_{o_1\alpha_1}) \supset N(f_{o_1\alpha_1}x_{\alpha_1})$

36. $(p_{o_1})(q_{o_1}) \centerdot N[p_{o_1} \supset q_{o_1}] \supset \centerdot Np_{o_1} \supset Nq_{o_1}$

37^{α}. $N(f_{o_2\alpha_2})(x_{\alpha_1})(x_{\alpha_2})(x_{\alpha_3}) \centerdot \Delta_{o_1\alpha_2\alpha_1}x_{\alpha_1}x_{\alpha_2} \supset \centerdot \Delta_{o_1\alpha_3\alpha_2}x_{\alpha_2}x_{\alpha_3} \supset \centerdot$
$N_{o_1o_2}(\Delta_{o_2\alpha_3\alpha_2}x_{\alpha_2}x_{\alpha_3}) \supset \centerdot N_{o_1o_2}(\Pi_{o_2(o_2\alpha_2)}f_{o_2\alpha_2}) \supset N_{o_1o_2}(f_{o_2\alpha_2}x_{\alpha_2})$

38. $N(p_{o_2})(q_{o_2}) \centerdot N_{o_1o_2}[p_{o_2} \supset q_{o_2}] \supset \centerdot N_{o_1o_2}p_{o_2} \supset N_{o_1o_2}q_{o_2}$

The relationship should be noted of Axioms $18^{\alpha\beta}$–$33^{\alpha\beta}$ and 34^{α} to Axioms $1^{\alpha\beta}$–17^{α} (in order), of Axioms 35^{α}–36 to Rules IV–V, and of Axioms 37^{α}–38 to Axioms 35^{α}–36.

The heuristic principle involved would lead us to introduce also axioms having a similar relationship to Axioms $18^{\alpha\beta}$–38, and so on *ad infinitum*. For example, corresponding to $18^{\alpha\beta}$ there would be:

$$N_{oo_1}\,(N_{o_1o_2}(f_{o_2\beta_2\alpha_2}) \centerdot (x_{\alpha_2})(y_{\beta_2})f_{o_2\beta_2\alpha_2}x_{\alpha_2}y_{\beta_2} \supset (y_{\beta_2})(x_{\alpha_2})f_{o_2\beta_2\alpha_2}x_{\alpha_2}y_{\beta_2})$$

However, since the principles of double necessity,

$$(p_{o_2}) \centerdot N_{oo_2}p_{o_2} \supset N_{oo_1}(N_{o_1o_2}p_{o_2}),$$
$$(p_{o_2}) \centerdot N_{oo_1}(N_{o_1o_2}p_{o_2}) \supset N_{oo_2}p_{o_2},$$

are theorems (it is believed), we prefer to introduce instead the axioms:

$18^{2\alpha\beta}$. $N_{oo_2}(f_{o_2\beta_2\alpha_2}) \centerdot (x_{\alpha_2})(y_{\beta_2}) f_{o_2\beta_2\alpha_2} x_{\alpha_2} y_{\beta_2} \supset (y_{\beta_2})(x_{\alpha_2}) f_{o_2\beta_2\alpha_2} x_{\alpha_2} y_{\beta_2}$

Similar considerations then lead us to introduce Axioms $18^{3\alpha\beta}$, and so on. Thus we have for every positive integer m the axioms:

$18^{m\alpha\beta}$. $N_{oo_m}(f_{o_m\beta_m\alpha_m}) \centerdot (x_{\alpha_m})(y_{\beta_m}) f_{o_m\beta_m\alpha_m} x_{\alpha_m} y_{\beta_m} \supset (y_{\beta_m})(x_{\alpha_m}) f_{o_m\beta_m\alpha_m} x_{\alpha_m} y_{\beta_m}$

(Of these, $18^{1\alpha\beta}$ are the same as $18^{\alpha\beta}$. Axioms $18^{0\alpha\beta}$ could also be included if desired, but they are rather trivially related to $1^{\alpha\beta}$ and hence are perhaps better omitted.)

Similar considerations lead us to introduce Axioms $19^{m\alpha\beta}$ in the same way, and so on. We need not take the space to write these out here, as they are easily supplied by analogy.

Our complete list of axioms, for the case of Alternative (2), is now the following, where β is restricted to be a preferred type symbol in every case in which one or more of the primitive symbols ι appear: $1^{\alpha\beta}$, $2^{\alpha\beta}$, 3^{α}, 4^{α}, 5^{α}, 6^{α}, 7, 8, $9^{\alpha\beta}$, 10^{β}, 11^{n}, $12^{n\alpha}$, $13^{n\beta}$, $14^{n\alpha}$, $15^{\alpha\beta}$, $16^{\alpha\beta}$, 17^{α}, $18^{m\alpha\beta}$, $19^{m\alpha\beta}$, $20^{m\alpha}$, $21^{m\alpha}$, $22^{m\alpha}$, $23^{m\alpha}$, 24^{m}, 25^{m}, $26^{m\alpha\beta}$, $27^{m\beta}$, 28^{mn}, $29^{mn\alpha}$, $30^{mn\beta}$, $31^{mn\alpha}$, $32^{m\alpha\beta}$, $33^{m\alpha\beta}$, $33^{*m\alpha\beta}$, $34^{m\alpha}$, 35^{α}, 36, $37^{m\alpha}$, 38^{m}.

Certainly not all of these axioms are independent. Indeed, in infinitely many cases, the non-independence is immediately seen in consequence of the principle that whatever is necessary is true, which is expected to be a theorem when expressed in the notation of the systems as:

$$(p_o)(p_{o_1}) \centerdot \Delta p_o p_{o_1} \supset \centerdot N p_{o_1} \supset p_o$$

However, we shall not here attempt to treat questions of independence or to examine how the system of axioms might be reduced or simplified.

Alternative (2) itself, if restricted to some fixed type (corresponding, say, to the type symbol α), can now be expressed in the notation of the system by means of the two following formulas:

$$(x_\alpha)(y_\alpha)(x_{\alpha_1})(y_{\alpha_1}) \centerdot \Delta x_\alpha x_{\alpha_1} \supset \centerdot \Delta y_\alpha y_{\alpha_1} \supset \centerdot N[x_{\alpha_1} =_1 y_{\alpha_1}] \supset \centerdot x_{\alpha_1} = y_{\alpha_1}$$

$$(x_\alpha)(y_\alpha)(x_{\alpha_1})(y_{\alpha_1}) \centerdot \Delta x_\alpha x_{\alpha_1} \supset \centerdot \Delta y_\alpha y_{\alpha_1} \supset \centerdot x_{\alpha_1} = y_{\alpha_1} \supset N \centerdot x_{\alpha_1} =_1 y_{\alpha_1}$$

It is believed that these will be theorems, for every α, else they also would have been assumed as axioms.

Similarly, we do not put down as an axiom that propositions which strictly imply each other are identical, because this is believed to be a theorem:[22]

$$(p_{o_1})(q_{o_1}) \centerdot N[p_{o_1} \supset q_{o_1}] \supset \centerdot N[q_{o_1} \supset p_{o_1}] \supset \centerdot p_{o_1} = q_{o_1}$$

It will be seen that our axioms and rules are all consistent with the interpretation that a concept of anything is always the same as the thing itself—so that propositions coincide with truth-values, individual concepts with individuals, and so on. This observation can no doubt be made the heuristic basis of a relative consistency proof, to the effect, roughly, that the present system with axiom of infinity added is consistent if the simple theory of types with axiom of infinity is consistent.

[22]This must be distinguished from the principle expressed in Axiom 8, that truth-values which materially imply each other are identical. And, of course, we do not assume that propositions which materially imply each other are identical, although this also can be expressed in our notation.

On the other hand, it may be thought a defect in the system that it admits this other and comparatively trivial interpretation besides the one actually intended. If so, it is not clear what addition to the system is desirable in order to remedy matters. For though it is possible to express in the notation of the system that there exists a proposition which is true but not necessary (or even, in each type, that everything which has a concept has at least two different concepts), we have refrained from adding this as an axiom, not because of any doubt that it is true, but because of a doubt whether it is necessary and therefore appropriate to be assumed as an axiom of logic. This question — whether it is a necessary truth that some truths are not necessary — might indeed be dismissed on the ground that the familiar but vague notion of logical necessity with which we began, and which the system itself is intended to make precise, is not sufficiently definite to provide an answer. But the writer would be inclined on the whole to give a negative answer.[23]

Returning now to Alternative (1), we retain in this case all of the Axioms $1^{\alpha\beta}$ to 17^{α}, at least as a basis for discussion, and for possible amendment after examining their consequences. In addition to these, we shall want axioms to the effect that concepts are under certain circumstances not the same; and indeed we are to be guided by the idea that such assumptions of the inequality of concepts are to be carried as far as is possible consistently with the other axioms (especially with Axioms $9^{\alpha_1\beta_1}$). The following axioms suggest themselves (where β is restricted to be a preferred type symbol in every case in which one or more of the primitive symbols $\iota_{\beta_n(o_n\beta_n)}$ appear):

$39^{n\alpha}$. $(p_{o_{n+1}})(q_{o_{n+1}})(f_{o_{n+1}\alpha_{n+1}}) \centerdot C_{o_{n+1}o_{n+1}o_{n+1}} p_{o_{n+1}} q_{o_{n+1}} \neq \Pi_{o_{n+1}(o_{n+1}\alpha_{n+1})} f_{o_{n+1}\alpha_{n+1}}$

40^{n}. $(p_{o_{n+1}})(q_{o_{n+1}})(g_{o_{n+1}o_{n+1}}) \centerdot C_{o_{n+1}o_{n+1}o_{n+1}} p_{o_{n+1}} q_{o_{n+1}} \neq \iota_{o_{n+1}(o_{n+1}o_{n+1})} g_{o_{n+1}o_{n+1}}$

$41^{n\alpha}$. $(p_{o_{n+1}})(q_{o_{n+1}})(x_{\alpha_{n+1}})(x_{\alpha_{n+2}}) \centerdot C_{o_{n+1}o_{n+1}o_{n+1}} p_{o_{n+1}} q_{o_{n+1}} \neq \Delta_{o_{n+1}\alpha_{n+2}\alpha_{n+1}} x_{\alpha_{n+1}} x_{\alpha_{n+2}}$

$42^{n\alpha}$. $(f_{o_{n+1}\alpha_{n+1}})(g_{o_{n+1}o_{n+1}}) \centerdot \Pi_{o_{n+1}(o_{n+1}\alpha_{n+1})} f_{o_{n+1}\alpha_{n+1}} \neq \iota_{o_{n+1}(o_{n+1}o_{n+1})} g_{o_{n+1}o_{n+1}}$

$43^{n\alpha\beta}$. $(f_{o_{n+1}\alpha_{n+1}})(x_{\beta_{n+1}})(x_{\beta_{n+2}}) \centerdot \Pi_{o_{n+1}(o_{n+1}\alpha_{n+1})} f_{o_{n+1}\alpha_{n+1}} \neq \Delta_{o_{n+1}\beta_{n+2}\beta_{n+1}} x_{\beta_{n+1}} x_{\beta_{n+2}}$

$44^{n\alpha}$. $(g_{o_{n+1}o_{n+1}})(x_{\alpha_{n+1}})(x_{\alpha_{n+2}}) \centerdot \iota_{o_{n+1}(o_{n+1}o_{n+1})} g_{o_{n+1}o_{n+1}} \neq \Delta_{o_{n+1}\alpha_{n+2}\alpha_{n+1}} x_{\alpha_{n+1}} x_{\alpha_{n+2}}.$

45^{n}. $(p_{o_{n+1}})(q_{o_{n+1}})(r_{o_{n+1}})(s_{o_{n+1}}) \centerdot C_{o_{n+1}o_{n+1}o_{n+1}} p_{o_{n+1}} q_{o_{n+1}} = C_{o_{n+1}o_{n+1}o_{n+1}} r_{o_{n+1}} s_{o_{n+1}} \supset \centerdot p_{o_{n+1}} = r_{o_{n+1}}$

---

[23] Other treatments of modality in connection with quantifiers — differing from that which is tentatively sketched here in that they are not based on Frege's theory of meaning — are found in Lewis and Langford's ***Symbolic logic***; a series of papers by Ruth Barcan in ***The Journal of Symbolic Logic***, XI and XII; a paper by Carnap, in ***The Journal of Symbolic Logic***, XI; and a paper by Fitch in ***Portugaliae Mathematica***, VII, No. 2. The first systematic treatments are those of Barcan and Carnap (independent, and nearly simultaneous).

The difference between the theory of modality which we have here outlined and other theories of modality may be seen especially clearly in connection with Becker's axiom of double necessity — i.e., with the question whether, if a proposition is necessary, it is therefore also necessary that it is necessary. According to the present theory, the answer to this question depends on what concept of the proposition is employed. For example, *that it is necessary that everything has some property or other* is no doubt itself necessary; but *that the proposition mentioned on lines 27–28 of page 272 of Lewis and Langford's* ***Symbolic Logic*** *is necessary* is true but not necessary. For although the proposition mentioned on lines 27–28 of page 272 of Lewis and Langford's ***Symbolic Logic*** in fact *is* that everything has some property or other, this identity is presumably not a necessary one. That is, in "It is necessary that p is necessary," we take the type of p to be o_2; the quoted expression may have the value truth for one propositional concept as value of p and have the value falsehood for another propositional concept, although the two propositional concepts are concepts of the same proposition and although the proposition in question is necessary. (There is also a similar difference between our symbolic expression of the proposition that everything — i.e., every individual — has some property or other and Lewis and Langford's, since we take properties of individuals to be of type $o_1\iota_1$; but at least for purposes of the present illustration, we assume that the propositions are nevertheless the same.)

46^n. $(p_{o_{n+1}})(q_{o_{n+1}})(r_{o_{n+1}})(s_{o_{n+1}}) \centerdot C_{o_{n+1}o_{n+1}o_{n+1}} p_{o_{n+1}} q_{o_{n+1}} = C_{o_{n+1}o_{n+1}o_{n+1}} r_{o_{n+1}} s_{o_{n+1}} \supset \centerdot q_{o_{n+1}} = s_{o_{n+1}}$

$47^{n\alpha}$. $(f_{o_{n+1}\alpha_{n+1}})(g_{o_{n+1}\alpha_{n+1}}) \centerdot \Pi_{o_{n+1}(o_{n+1}\alpha_{n+1})} f_{o_{n+1}\alpha_{n+1}} = \Pi_{o_{n+1}(o_{n+1}\alpha_{n+1})} g_{o_{n+1}\alpha_{n+1}} \supset \centerdot f_{o_{n+1}\alpha_{n+1}} = g_{o_{n+1}\alpha_{n+1}}$

$48^{n\beta}$. $(f_{o_{n+1}\beta_{n+1}})(g_{o_{n+1}\beta_{n+1}}) \centerdot \iota_{\beta_{n+1}(o_{n+1}\beta_{n+1})} f_{o_{n+1}\beta_{n+1}} = \iota_{\beta_{n+1}(o_{n+1}\beta_{n+1})} g_{o_{n+1}\beta_{n+1}} \supset \centerdot f_{o_{n+1}\beta_{n+1}} = g_{o_{n+1}\beta_{n+1}}$

$49^{n\alpha}$. $(x_{\alpha_{n+1}})(x_{\alpha_{n+2}})(y_{\alpha_{n+1}})(y_{\alpha_{n+2}}) \centerdot \Delta_{o_{n+1}\alpha_{n+2}\alpha_{n+1}} x_{\alpha_{n+1}} x_{\alpha_{n+2}} = \Delta_{o_{n+1}\alpha_{n+2}\alpha_{n+1}} y_{\alpha_{n+1}} y_{\alpha_{n+2}} \supset \centerdot x_{\alpha_{n+1}} = y_{\alpha_{n+1}}$

$50^{n\alpha}$. $(x_{\alpha_{n+1}})(x_{\alpha_{n+2}})(y_{\alpha_{n+1}})(y_{\alpha_{n+2}}) \centerdot \Delta_{o_{n+1}\alpha_{n+2}\alpha_{n+1}} x_{\alpha_{n+1}} x_{\alpha_{n+2}} = \Delta_{o_{n+1}\alpha_{n+2}\alpha_{n+1}} y_{\alpha_{n+1}} y_{\alpha_{n+2}} \supset \centerdot x_{\alpha_{n+2}} = y_{\alpha_{n+2}}$

$51^{mn\beta}$. $(f_{o_{n+1}\beta_{m+n+2}})(f_{o_{m+n+2}\beta_{m+n+2}}) \centerdot \iota_{\beta_{m+n+2}(o_{n+1}\beta_{m+n+2})} f_{o_{n+1}\beta_{m+n+2}} \neq \iota_{\beta_{m+n+2}(o_{m+n+2}\beta_{m+n+2})} f_{o_{m+n+2}\beta_{m+n+2}}$

Finally also the following axioms, in which α and β are restricted to be different type symbols:[24]

$52^{n\alpha\beta}$. $(f_{o_{n+1}\alpha_{n+1}})(f_{o_{n+1}\beta_{n+1}}) \centerdot \Pi_{o_{n+1}(o_{n+1}\alpha_{n+1})} f_{o_{n+1}\alpha_{n+1}} \neq \Pi_{o_{n+1}(o_{n+1}\beta_{n+1})} f_{o_{n+1}\beta_{n+1}}$

$53^{n\alpha\beta}$. $(x_{\alpha_{n+1}})(x_{\alpha_{n+2}})(y_{\beta_{n+1}})(y_{\beta_{n+2}}) \centerdot \Delta_{o_{n+1}\alpha_{n+2}\alpha_{n+1}} x_{\alpha_{n+1}} x_{\alpha_{n+2}} \neq \Delta_{o_{n+1}\beta_{n+2}\beta_{n+1}} y_{\beta_{n+1}} y_{\beta_{n+2}}$

PRINCETON UNIVERSITY

[24]Apparently, these axioms impose an additional restriction on the interpretation of the system, to the effect that the members of two types must not coincide or must not coincide necessarily. Without them, for example, it would be a possible interpretation that the domain of individuals coincides with that of truth-values, whereupon the domain of individual concepts would coincide with that of propositions, and so on.

THE WEAK THEORY OF IMPLICATION (1951)

1. *The weak positive implicational propositional calculus.*[1]

The positive implicational propositional calculus of Hilbert[2] is determined by the four following axioms,

$$s \supset [s \supset q] \centerdot s \supset q,$$
$$p \supset q \supset \centerdot s \supset p \supset \centerdot s \supset q,$$
$$s \supset [p \supset q] \supset \centerdot p \supset \centerdot s \supset q,$$
$$p \supset \centerdot q \supset p,$$

where the sign of implication, $\supset$, is the single primitive connective, and the rules of inference are, as usual, substitution and *modus ponens*. From this is obtained what we shall call the weak positive implicational propositional calculus, or briefly, the *weak implicational calculus*, if the last axiom is weakened to $p \supset p$, so that the four axioms become:[3]

1. $s \supset [s \supset q] \supset \centerdot s \supset q$
2. $p \supset q \supset \centerdot s \supset p \supset \centerdot s \supset q$
3. $s \supset [p \supset q] \supset \centerdot p \supset \centerdot s \supset q$
4. $p \supset p$

By a *proof of* $\mathbf{B}$ *from hypotheses* $\mathbf{A}_1, \mathbf{A}_2, \ldots, \mathbf{A}_n$ is meant a finite sequence of well-formed formulas $\mathbf{B}_1, \mathbf{B}_2, \ldots, \mathbf{B}_m$ in which $\mathbf{B}_m$ is the same as $\mathbf{B}$ and in which each $\mathbf{B}_i$, in order, either is a variant of an axiom (i.e., is obtained from one of the axioms by alphabetic changes of the variables), or is one of $\mathbf{A}_1, \mathbf{A}_2, \ldots, \mathbf{A}_n$, or is inferred by *modus ponens* from two earlier well-formed formulas in the sequence, or finally is inferred by substitution from an earlier well-formed formula in the sequence, subject to the condition that the substitution must not be for a variable occurring as a free variable in any of $\mathbf{A}_1, \mathbf{A}_2, \ldots, \mathbf{A}_n$. As a notation to express that we have a proof of $\mathbf{B}$ from the hypotheses $\mathbf{A}_1, \mathbf{A}_2, \ldots, \mathbf{A}_n$ we write:

$$\mathbf{A}_1, \mathbf{A}_2, \ldots, \mathbf{A}_n \vdash \mathbf{B}$$

Originally published in ***Kontrolliertes Denken***, edited by A. Menne, A. Wilhelmy, and H. Angstl, Kommissions-Verlag Karl Alber, Freiburg and Munich 1951, pp. 22–37.

[1]This paper was presented to the Association for Symbolic Logic on April 28, 1951, under the title *The weak positive implicational propositional calculus*, and an abstract will appear in volume 16 of ***The Journal of Symbolic Logic***. However, since its presentation to the Association a number of additions to the paper have been made, and its title has therefore been changed. Concerning the reasons for attaching some special significance to the weak positive implicational propositional calculus, as compared to other fragments of the propositional calculus which might be studied, see an abstract of a paper entitled *Minimal logic*, presented to the same meeting of the Association for Symbolic Logic, which will also appear in volume 16 of ***The Journal of Symbolic Logic***.

[2]David Hilbert in ***Abhandlungen aus dem Mathematischen Seminar der Hamburgischen Universität***, vol. 6 (1928), pp. 65–85; reprinted in ***Grundlagen der Geometrie***, 7th edn., pp. 289–312. David Hilbert and Paul Bernays, ***Grundlagen der Mathematik***, vol. 1 (1934), §3, and vol. 2 (1939), Supplement III A.

[3]The idea of a propositional calculus in which the law $p \supset \centerdot q \supset p$ fails is not new. See for example Paulette Destouches-Février in ***Comptes Rendus des Séances de l'Académie des Sciences*** (Paris), vol. 226 (1948), pp. 38–39, and references there given.

This notation is used also in the case $n = 0$. I.e., we write $\vdash$ **B** to mean that we have a proof of **B**. The distinction is not important here that in a proof from hypotheses variants of the axioms may be used, whereas a proof (proper) only the axioms themselves may be used—because of course for any variant of an axiom a proof may easily be supplied.

We proceed to sketch the proofs of a number of theorems and metatheorems of the weak implicational calculus.

5. $\vdash r \supset [p \supset q] \supset . r \supset . s \supset p \supset . s \supset q$

From axiom 2 by substitution and *modus ponens*.

6. $\vdash r \supset [p \supset q] \supset . s \supset p \supset . r \supset . s \supset q$

From 5 by axiom 3 and axiom 2.

7. $\vdash s \supset [p \supset q] \supset . s \supset p \supset . s \supset . s \supset q$

By substitution in 6.

8. $\vdash s \supset [p \supset q] \supset . s \supset p \supset . s \supset q$

From 7, by using axiom 1, and axiom 2 (twice).

9. $\vdash p \supset q \supset . r \supset p \supset [s \supset p] \supset . r \supset p \supset . s \supset q$

By substituting in axiom 2, simultaneously,[4] $s \supset p$ for p, $s \supset q$ for q, and $r \supset p$ for s; and then using axiom 2 again (twice).

10. $\vdash s \supset p \supset . p \supset q \supset . s \supset q$

By axiom 2 and axiom 3.

11. $\vdash s \supset r \supset . p \supset q \supset . r \supset p \supset . s \supset q$

From 9, by axiom 3, and 10.

12. $\vdash p \supset . p \supset q \supset q$

By substituting $p \supset q$ for p in axiom 4, and then using axiom 3.

13. $\vdash r \supset [p \supset q] \supset . r_1 \supset [s \supset p] \supset . r_1 \supset . r \supset . s \supset q$

By 6, and axiom 2.

14. $\vdash r \supset [p \supset q] \supset . s \supset p \supset . s_1 \supset r \supset . s_1 \supset . s \supset q$

By 6 and axiom 2.

15. $\vdash p \supset q \supset p \supset . p \supset q \supset q$

By substituting $p \supset q$ for p in axiom 4, and then using 8.

16. $\vdash p \supset p \supset [p \supset p \supset p] \supset p$

By substituting in axiom 1, simultaneously, $p \supset p$ for s and p for q, and then using axioms 3 and 4.

17. $\vdash p \supset q \supset . q \supset p \supset . p \supset q$

By substituting q and p for r and s respectively in 11 and then using axiom 1.

18. $\vdash p \supset . q \supset p \supset . p \supset q \supset q$

By 17, axiom 3, and axiom 2.

19. *The weak deduction theorem.*

[4]The indicated simultaneous substitution can of course be accomplished by an appropriate succession of single substitutions, so that ultimately the rule of single substitution suffices as rule of inference for this purpose.

If $\mathbf{A}_1, \mathbf{A}_2, \ldots, \mathbf{A}_n \vdash \mathbf{B}$, then either $\mathbf{A}_1, \mathbf{A}_2, \ldots, \mathbf{A}_{n-1} \vdash \mathbf{A}_n \supset \mathbf{B}$ or $\mathbf{A}_1, \mathbf{A}_2, \ldots, \mathbf{A}_{n-1} \vdash \mathbf{B}$.

Given $\mathbf{B}_1, \mathbf{B}_2, \ldots, \mathbf{B}_m$, a proof of $\mathbf{B}$ from the hypotheses $\mathbf{A}_1, \mathbf{A}_2, \ldots, \mathbf{A}_n$, we prove by mathematical induction that, for each $\mathbf{B}_i$, either $\mathbf{A}_1, \mathbf{A}_2, \ldots, \mathbf{A}_{n-1} \vdash \mathbf{A}_n \supset \mathbf{B}_i$ or $\mathbf{A}_1, \mathbf{A}_2, \ldots, \mathbf{A}_{n-1} \vdash \mathbf{B}_i$. Consider then a particular $\mathbf{B}_i$, and (if $i > 1$) suppose, as hypothesis of induction, that, for every $j < i$, either $\mathbf{A}_1, \mathbf{A}_2, \ldots, \mathbf{A}_{n-1} \vdash \mathbf{A}_n \supset \mathbf{B}_j$ or $\mathbf{A}_1, \mathbf{A}_2, \ldots, \mathbf{A}_{n-1} \vdash \mathbf{B}_j$. The following cases arise.

Case 1: $\mathbf{B}_i$ is either a variant of an axiom or one of $\mathbf{A}_1, \mathbf{A}_2, \ldots, \mathbf{A}_{n-1}$. In this case we have immediately $\mathbf{A}_1, \mathbf{A}_2, \ldots, \mathbf{A}_{n-1} \vdash \mathbf{B}_i$, the proof of $\mathbf{B}_i$ from the hypotheses $\mathbf{A}_1, \mathbf{A}_2, \ldots, \mathbf{A}_{n-1}$ consisting of the single well-formed formula $\mathbf{B}_i$.

Case 2: $\mathbf{B}_i$ is $\mathbf{A}_n$. In this case $\mathbf{A}_1, \mathbf{A}_2, \ldots, \mathbf{A}_{n-1} \vdash \mathbf{A}_n \supset \mathbf{B}_i$ because $\mathbf{A}_n \supset \mathbf{B}_i$ is inferred by substitution from an appropriate variant of axiom 4.

Case 3a: $\mathbf{B}_i$ is inferred by substitution from $\mathbf{B}_j$, where $j < i$, and $\mathbf{A}_1, \mathbf{A}_2, \ldots, \mathbf{A}_{n-1} \vdash \mathbf{A}_n \supset \mathbf{B}_j$. Then $\mathbf{A}_1, \mathbf{A}_2, \ldots, \mathbf{A}_{n-1} \vdash \mathbf{A}_n \supset \mathbf{B}_i$, since the same substitution by which $\mathbf{B}_i$ is inferred from $\mathbf{B}_j$ may be used to infer $\mathbf{A}_n \supset \mathbf{B}_i$ from $\mathbf{A}_n \supset \mathbf{B}_j$. (Here the condition is essential that the substitution is for a variable not occurring as a free variable in $\mathbf{A}_n$.)

Case 3b: $\mathbf{B}_i$ is inferred by substitution from $\mathbf{B}_j$, where $j < i$, and $\mathbf{A}_1, \mathbf{A}_2, \ldots, \mathbf{A}_{n-1} \vdash \mathbf{B}_j$. Then $\mathbf{A}_1, \mathbf{A}_2, \ldots, \mathbf{A}_{n-1} \vdash \mathbf{B}_i$, since the same substitution may be used to infer $\mathbf{B}_i$ from $\mathbf{B}_j$.

Case 4a: $\mathbf{B}_i$ is inferred by *modus ponens* from $\mathbf{B}_j$ and $\mathbf{B}_k$, where $j < i$, $k < i$, $\mathbf{B}_j$ is $\mathbf{B}_k \supset \mathbf{B}_i$, and $\mathbf{A}_1, \mathbf{A}_2, \ldots, \mathbf{A}_{n-1} \vdash \mathbf{A}_n \supset \mathbf{B}_j$, and $\mathbf{A}_1, \mathbf{A}_2, \ldots, \mathbf{A}_{n-1} \vdash \mathbf{A}_n \supset \mathbf{B}_k$. Then $\mathbf{A}_1, \mathbf{A}_2, \ldots, \mathbf{A}_{n-1} \vdash \mathbf{A}_n \supset \mathbf{B}_i$ by 8.

Case 4b: $\mathbf{B}_i$ is inferred by *modus ponens* from $\mathbf{B}_j$ and $\mathbf{B}_k$, where $j < i$, $k < i$, $\mathbf{B}_j$ is $\mathbf{B}_k \supset \mathbf{B}_i$, and $\mathbf{A}_1, \mathbf{A}_2, \ldots, \mathbf{A}_{n-1} \vdash \mathbf{A}_n \supset \mathbf{B}_j$, and $\mathbf{A}_1, \mathbf{A}_2, \ldots, \mathbf{A}_{n-1} \vdash \mathbf{B}_k$. Then $\mathbf{A}_1, \mathbf{A}_2, \ldots, \mathbf{A}_{n-1} \vdash \mathbf{A}_n \supset \mathbf{B}_i$ by axiom 3.

Case 4c: $\mathbf{B}_i$ is inferred by *modus ponens* from $\mathbf{B}_j$ and $\mathbf{B}_k$, where $j < i$, $k < i$, $\mathbf{B}_j$ is $\mathbf{B}_k \supset \mathbf{B}_i$, and $\mathbf{A}_1, \mathbf{A}_2, \ldots, \mathbf{A}_{n-1} \vdash \mathbf{B}_j$, and $\mathbf{A}_1, \mathbf{A}_2, \ldots, \mathbf{A}_{n-1} \vdash \mathbf{A}_n \supset \mathbf{B}_k$. Then $\mathbf{A}_1, \mathbf{A}_2, \ldots, \mathbf{A}_{n-1} \vdash \mathbf{A}_n \supset \mathbf{B}_i$ by axiom 2.

Case 4d: $\mathbf{B}_i$ is inferred by *modus ponens* from $\mathbf{B}_j$ and $\mathbf{B}_k$, where $j < i$, $k < i$, $\mathbf{B}_j$ is $\mathbf{B}_k \supset \mathbf{B}_i$, and $\mathbf{A}_1, \mathbf{A}_2, \ldots, \mathbf{A}_{n-1} \vdash \mathbf{B}_j$, and $\mathbf{A}_1, \mathbf{A}_2, \ldots, \mathbf{A}_{n-1} \vdash \mathbf{B}_k$. Then $\mathbf{A}_1, \mathbf{A}_2, \ldots, \mathbf{A}_{n-1} \vdash \mathbf{B}_i$ by *modus ponens*.

Since it now follows by mathematical induction for all $\mathbf{B}_i$ — and thus in particular for $\mathbf{B}_m$ — that either $\mathbf{A}_1, \mathbf{A}_2, \ldots, \mathbf{A}_{n-1} \vdash \mathbf{A}_n \supset \mathbf{B}_i$ or $\mathbf{A}_1, \mathbf{A}_2, \ldots, \mathbf{A}_{n-1} \vdash\supset \mathbf{B}_i$, the proof of the weak deduction theorem is complete.

It should be observed that, although the statement of the proof does not make this quite explicit, the proof is as a matter of fact *effective*, in the sense that it provides a mechanical procedure by which whenever a proof of a well-formed formula $\mathbf{B}$ from hypotheses $\mathbf{A}_1, \mathbf{A}_2, \ldots, \mathbf{A}_n$ is given, a proof can be constructed of either $\mathbf{A}_n \supset \mathbf{B}$ or $\mathbf{B}$ from the hypotheses $\mathbf{A}_1, \mathbf{A}_2, \ldots, \mathbf{A}_{n-1}$.

Moreover the proof from the hypotheses $\mathbf{A}_1, \mathbf{A}_2, \ldots, \mathbf{A}_{n-1}$ so constructed is of $\mathbf{A}_n \supset \mathbf{B}$ or of $\mathbf{B}$ according as $\mathbf{A}_n$ is or not *actually used* in the given proof of $\mathbf{B}$ from the hypotheses $\mathbf{A}_1, \mathbf{A}_2, \ldots, \mathbf{A}_n$, in the sense that $\mathbf{A}_n$ occurs in this proof from hypotheses in such a way that there is a chain of inferences connecting $\mathbf{A}_n$ with the final well-formed formula $\mathbf{B}$. Thus we have the following corollary (also effective):

20. If $\mathbf{A}_1, \mathbf{A}_2, \ldots, \mathbf{A}_n \vdash \mathbf{B}$, and $\mathbf{A}_{n-r}, \mathbf{A}_{n-r+1}, \ldots, \mathbf{A}_n$ are actually used in the proof

of $\mathbf{B}$ from the hypotheses $\mathbf{A}_1, \mathbf{A}_2, \ldots, \mathbf{A}_n$, then $\mathbf{A}_1, \mathbf{A}_2, \ldots, \mathbf{A}_{n-r-1} \vdash \mathbf{A}_{n-r} \supset . \mathbf{A}_{n-r+1} \supset . \ldots \mathbf{A}_{n-1} \supset . \mathbf{A}_n \supset \mathbf{B}$.

We prove also the following metatheorem (again effectively):

21. *Rule of substitutivity of mutual implication.*[5]

If $\mathbf{B}$ is obtained from $\mathbf{A}$ by substitution of $\mathbf{N}$ for $\mathbf{M}$ at some (or all or none) of the places at which $\mathbf{M}$ occurs in $\mathbf{A}$, then $\mathbf{M} \supset \mathbf{N}, \mathbf{N} \supset \mathbf{M} \vdash \mathbf{A} \supset \mathbf{B}$ and $\mathbf{M} \supset \mathbf{N}, \mathbf{N} \supset \mathbf{M} \vdash \mathbf{B} \supset \mathbf{A}$.

In the special case (i) that $\mathbf{M}$ coincides with $\mathbf{A}$ and the substitution of $\mathbf{N}$ for $\mathbf{M}$ is at this one place in $\mathbf{A}$, the result of the metatheorem is immediate, because $\mathbf{M} \supset \mathbf{N}$ and $\mathbf{N} \supset \mathbf{M}$ coincide with $\mathbf{A} \supset \mathbf{B}$ and $\mathbf{B} \supset \mathbf{A}$ respectively. In the special case (ii) that the substitution of $\mathbf{N}$ for $\mathbf{M}$ is at no places in $\mathbf{A}$, so that $\mathbf{A}$ and $\mathbf{B}$ coincide, the result of the metatheorem follows by substitution in axiom 4.

In order to prove the metatheorem generally, we proceed by mathematical induction with respect to the length of $\mathbf{A}$. The shortest length occurs when $\mathbf{A}$ consists of a variable standing alone, and the metatheorem then reduces to one of the two special cases (i) and (ii) already considered. Therefore let $\mathbf{A}$ be a well-formed formula of other than the shortest length, and suppose as hypothesis of induction that the metatheorem holds for any shorter well-formed formula in place of $\mathbf{A}$. Necessarily $\mathbf{A}$ has the form $\mathbf{A}_1 \supset \mathbf{A}_2$; and, except in case (i) already considered, $\mathbf{B}$ will then have the form $\mathbf{B}_1 \supset \mathbf{B}_2$, where $\mathbf{B}_1$ is obtained from $\mathbf{A}_1$ and $\mathbf{B}_2$ from $\mathbf{A}_2$ by substitution of $\mathbf{N}$ for $\mathbf{M}$ at some (or all or none) of its occurrences. By hypothesis of induction we have:

$$\mathbf{M} \supset \mathbf{N}, \mathbf{N} \supset \mathbf{M} \vdash \mathbf{A}_1 \supset \mathbf{B}_1$$
$$\mathbf{M} \supset \mathbf{N}, \mathbf{N} \supset \mathbf{M} \vdash \mathbf{B}_1 \supset \mathbf{A}_1$$
$$\mathbf{M} \supset \mathbf{N}, \mathbf{N} \supset \mathbf{M} \vdash \mathbf{A}_2 \supset \mathbf{B}_2$$
$$\mathbf{M} \supset \mathbf{N}, \mathbf{N} \supset \mathbf{M} \vdash \mathbf{B}_2 \supset \mathbf{A}_2$$

Hence by substitution in 11 and *modus ponens*, we get:

$$\mathbf{M} \supset \mathbf{N}, \mathbf{N} \supset \mathbf{M} \vdash \mathbf{A} \supset \mathbf{B}$$
$$\mathbf{M} \supset \mathbf{N}, \mathbf{N} \supset \mathbf{M} \vdash \mathbf{B} \supset \mathbf{A}$$

2. *The weak implicational calculus with negation.*

Following an idea of Johansson,[6] let us adjoin to the weak implicational calculus a propositional constant f, which we think of as denoting falsehood but for which no new axioms are introduced. And let us then define the negation of $\mathbf{A}$ as:[7]

$$\sim \mathbf{A} \rightarrow \mathbf{A} \supset f$$

Among others the following theorems involving negation may be provided:

22. $\vdash p \supset \sim q \supset . q \supset \sim p$

By substitution in axiom 3.

23. $\vdash p \supset q \supset . \sim q \supset \sim p$

[5]Compare M. Wajsberg in ***Monatshefte für Mathematik und Physik***, vol. 42 (1935), pp. 221–242, and in ***Wiadomości Matematyczne***, vol. 43 (1936), pp. 131–168; W. V. Quine in ***The Journal of Symbolic Logic***, vol. 3 (1938), pp. 37–40.

[6]Ingebrigt Johansson in ***Compositio Mathematica***, vol. 4 (1936), pp. 119–136. The *Minimalkalkül* of Johansson must of course not be confused with what is called *minimal logic* in the abstract referred to in footnote 1.

[7]The arrow is to be read, "stands for".

By substitution in 10.

24. $\vdash\ p \supset \sim\sim p$

By substitution in 12.

25. $\vdash\ \sim\sim\sim p \supset \sim p$

By 24 and 23.

26. $\vdash\ p \supset \sim p \supset \sim p$

By substitution in axiom 1.

27. $\vdash\ p \supset q \supset . p \supset \sim q \supset \sim p$

By 8 and axiom 3.

28. $\vdash\ p \supset . \sim q \supset \sim . p \supset q$

By 12 and 23 and axiom 2.

29. $\vdash\ \sim\sim[p \supset q] \supset . p \supset \sim\sim q$

By 28, 23, and axioms 2 and 3.

30. $\vdash\ \sim\sim[\sim\sim p \supset \sim\sim q] \supset . \sim\sim p \supset \sim\sim q$

By 29, 25, and axiom 2.

31. $\vdash\ \sim\sim p \supset \sim\sim q \supset \sim\sim . \sim\sim p \supset \sim\sim q$

By substitution in 24.

32. $\vdash\ \sim\sim . \sim\sim[\sim\sim s \supset \sim\sim . \sim\sim p \supset \sim\sim q] \supset \sim\sim .$
$\sim\sim[\sim\sim s \supset \sim\sim p] \supset \sim\sim . \sim\sim s \supset \sim\sim q$

By 8, 30, 31, and 21.

33. $\vdash\ \sim\sim p \supset \sim\sim q \supset . \sim q \supset \sim p$

By 22, 25, and axiom 2.

34. $\vdash\ \sim\sim . \sim\sim[\sim\sim\sim\sim\sim p \supset \sim\sim\sim\sim\sim q] \supset \sim\sim . \sim\sim q \supset \sim\sim p$

By 33, 24, 25, 30, 31, and 21.

Suppose now that we add the axiom:

100. $f \supset . p \supset f$

Then we have also:

101. $\vdash\ \sim p \supset . p \supset \sim q$

By axiom 100 and axiom 2.

102. $\vdash\ \sim p \supset . q \supset \sim p$

By 101, 22, and axiom 2.

103. $\vdash\ p \supset . q \supset \sim\sim p$

By 101, axiom 3, 22, and axiom 2.

104. $\vdash\ \sim\sim . \sim\sim p \supset \sim\sim . \sim\sim q \supset \sim\sim p$

By 102, 31, and axiom 2.

By the *double-negation analogue* of a well-formed formula we mean the expression obtained from it by prefixing a double negation, $\sim\sim$, to every well-formed part of it—so that, for example, 32 is the double-negation analogue of 8. As a notation for the double-negation analogue of a well-formed formula **A**, we use **A***. And we prove the following metatheorem (in which the word "*tautology*" means *tautology according to the usual two-valued truth-tables for the propositional calculus*):[8]

105. If **A** is a well-formed formula of the propositional calculus with $\supset$ and $\sim$ as primitive connectives, and if in **A*** the negation sign, $\sim$, is understood according to the definition given at the beginning of this section, then **A*** is a theorem of the weak implicational calculus, with axiom 100 added, if and only if **A** is a tautology.

That **A** is a tautology if $\vdash$ **A***, follows immediately because axioms 1–4, 100 are tautologies and because of the known properties of double negation in the two valued propositional calculus. In order to prove the converse, we make use of the fact that the three well-formed formulas,

$$p \supset . q \supset p$$

$$s \supset [p \supset q] \supset . s \supset p \supset . s \supset q$$

$$\sim p \supset \sim q \supset . q \supset p$$

are a sufficient system of axioms for the two-valued propositional calculus, all tautologies following from them by substitution and *modus ponens*.[9] Since the double-negation analogues of these three well-formed formulas have already been shown to be theorems (104, 32, 34), it is sufficient to show that the double-negation analogues of the rules of substitution and *modus ponens* hold in the following senses:

If **B** is inferred from **A** by the rule of substitution, then **B*** can be inferred from **A*** be the rule of substitution. Namely if **B** is obtained from **A** by substituting **C** for the variable **c**, and if **C*** is $\sim\sim$ **D**, then **B*** is obtained from **A*** by substituting **D** for **c**.

If **B** is inferred from **C** and **A** by *modus ponens*, then **C** is **A** $\supset$ **B**, and **C*** is $\sim\sim .$ **A*** $\supset$ **B***. Therefore **B*** can be inferred from **C*** and **A*** by 30, substitution, and *modus ponens*.

This completes the proof of 105. From it by means of 24, 25, 30, 31, and 21 we infer immediately the following simpler metatheorem:[8]

106. Let **A** be a well-formed formula of the propositional calculus with $\supset$ and $\sim$ as primitive connectives, and let **A**** be obtained from **A** by prefixing a double negation, $\sim\sim$, to every variable. And in **A**** let the negation sign, $\sim$, be understood according to the definition given at the beginning of this section. Then **A**** is a theorem of the weak implicational calculus, with axiom 100 added, if and only if **A** is a tautology.

3. *Independence examples.*

The five following truth tables serve to show, in order, that axioms 1–4 of the weak implicational calculus are independent, and that $p \supset . q \supset p$ is not a theorem of the weak implicational calculus. In the first four tables, 0 is the designated truth-value, and in the fifth table, 0 and 1 are designated truth-values.

---

[8]This metatheorem, for a related but stronger propositional calculus, having $p \supset . q \supset p$ as an axiom in place of 100, and a similar metatheorem for a corresponding functional calculus of first order, are due to A. Kolmogoroff, in ***Recueil Mathématique de la Société Mathématique de Moscou***, vol. 32 (1925), pp. 646–667.

[9]The result is due to Jan Łukasiewicz. As a convenient place in which to find a proof of it, the reader may be referred to my ***Introduction to Mathematical Logic***, Volume I, either the edition of 1944 or the forthcoming revised and enlarged edition.

p	q	$p \supset q$	$p \supset q$	$p \supset q$	$p \supset q$	$p \supset q$
0	0	0	0	0	0	0
0	1	1	0	1	1	1
0	2	1	2	1	1	2
1	0	1	0	0	0	2
1	1	0	0	0	0	1
1	2	0	0	0	0	2
2	0	1	0	0	0	1
2	1	0	0	1	1	1
2	2	0	0	0	1	1

By taking the value of the constant f to be 0, we may use the same five truth-tables to show that the axioms 1–4, 100 are independent of one another. And by taking the value of the constant f to be 1, the fifth truth-table may also be used to show that $p \supset . q \supset p$ is not a theorem of the weak implicational calculus with axiom 100 added.

In all cases it is meant that the rules of inference are substitution and *modus ponens*. The detailed verification of the independence examples is left to the reader.

4. *The weak theory of implication.*

The system obtained from the implicational propositional calculus by allowing universal quantification with respect to propositional variables, and adding appropriate axioms involving it, was called by Russell *the theory of implication.*[10]

To the system similarly obtained from the weak implicational calculus let us give the name *weak theory of implication.*[11]

In this system the rule of substitution will be modified to read as follows: from **A** to infer the result of substituting **C** for all free occurrences of the variable **c** throughout **A**, provided that no free occurrences of **c** in **A** is in a well-formed part of **A** of the form (**d**)**D** where **d** is a free variable of **C**. The rule of *modus ponens* remains unchanged. And one further rule of inference, the rule of generalization, is added: from **A** to infer (**a**)**A**, where **a** is any variable. The axioms are axiom 1–4 or §1, plus the two following axiom schemata (in which **a** and **b** are any two different variables, and **A** is any well-formed formula):

200. $(\mathbf{a})[\mathbf{b} \supset \mathbf{A}] \supset . \mathbf{b} \supset (\mathbf{a})\mathbf{A}$
201. $(\mathbf{a})\mathbf{A} \supset \mathbf{A}$

To show that $p \supset . q \supset p$ is not a theorem of the weak theory of implication, we may use the fifth truth-table of §3, evaluating the universal quantifier as follows: For a given system of values of the free variables (**a**)**A**, the value of (**a**)**A** is 2 if the value of **A** is 2 for one or more (at least one) value of **a**; the value of (**a**)**A** is 1 if the value of **A** is 1 for every value of **a**; and the value of (**a**)**A** is 0 in the remaining cases. Details of the verification are left to the reader.

---

[10]Bertrand Russell in ***American Journal of Mathematics***, vol. 28 (1906), pp. 159–202. Russell uses negation as an additional primitive in his formulation of the system, but points out the possibility of eliminating it by means of a suitable definition.

[11]The weak theory of implication, like the weak implicational calculus, is a part of minimal logic in the sense of the abstract referred to in footnote 1.

In connection with the weak theory of implication we shall mean by a *proof of* **B** *from the hypotheses* $\mathbf{A}_1, \mathbf{A}_2, \ldots, \mathbf{A}_n$, a finite sequence of well-formed formulas $\mathbf{B}_1, \mathbf{B}_2, \ldots, \mathbf{B}_m$ in which $\mathbf{B}_m$ is the same as **B** and in which each $\mathbf{B}_i$, in order, either is a variant of an axiom (i.e., is an instance of one of the axiom schemata 200, 201, or is obtained from one of the axioms 1–4 by alphabetic changes of the variables), or is one of $\mathbf{A}_1, \mathbf{A}_2, \ldots, \mathbf{A}_n$, or is inferred by *modus ponens* from two earlier well-formed formulas in the sequence, or is inferred by substitution or by generalization from an earlier well-formed formula in the sequence, subject to the condition that no variable which occurs as a free variable in any of $\mathbf{A}_1, \mathbf{A}_2, \ldots, \mathbf{A}_n$ may be either substituted for or generalized upon.

The theorems 5–18 of the weak implicational calculus are of course also theorems of the weak theory of implication.

The weak deduction theorem may also be extended, without change, to the weak theory of implication. The proof is the same except that there are the two following additional cases:

Case 5a: $\mathbf{B}_i$ is inferred by generalization from $\mathbf{B}_j$, where $j < i$, and $\mathbf{A}_1, \mathbf{A}_2, \ldots, \mathbf{A}_{n-1} \vdash \mathbf{A}_n \supset \mathbf{B}_j$. Say $\mathbf{B}_i$ is $(\mathbf{a})\mathbf{B}_j$. Then by generalization, $\mathbf{A}_1, \mathbf{A}_2, \ldots, \mathbf{A}_{n-1} \vdash (\mathbf{a}) \,.\, \mathbf{A}_n \supset \mathbf{B}_j$. Hence by an appropriate instance of axiom schema 200, and *modus ponens*, $\mathbf{A}_1, \mathbf{A}_2, \ldots, \mathbf{A}_{n-1} \vdash \mathbf{A}_n \supset \mathbf{B}_i$. (Here the condition is essential that the variable **a** which is generalized upon does not occur as a free variable in $\mathbf{A}_n$.)

Case 5b: $\mathbf{B}_i$ is inferred by generalization from $\mathbf{B}_j$, where $j < i$, and $\mathbf{A}_1, \mathbf{A}_2, \ldots, \mathbf{A}_{n-1} \vdash \mathbf{B}_j$. Then by generalization, $\mathbf{A}_1, \mathbf{A}_2, \ldots, \mathbf{A}_{n-1} \vdash \mathbf{B}_i$.

The corollary of the weak deduction theorem, 20, then follows also for the weak theory of implication.

In order to extend the rule of substitutivity of mutual implication to the weak theory of implication, we first establish the following theorem schema of the weak theory of implication:

202. $\vdash (\mathbf{a})[\mathbf{A} \supset \mathbf{B}] \supset .(\mathbf{a})\mathbf{A} \supset (\mathbf{a})\mathbf{B}$

For by 201 we have:

$\vdash (\mathbf{a})[\mathbf{A} \supset \mathbf{B}] \supset . \mathbf{A} \supset \mathbf{B}$
$\vdash (\mathbf{A})\mathbf{A} \supset \mathbf{A}$

Hence by 6, using substitution and *modus ponens*, we have:

$\vdash (\mathbf{a})[\mathbf{A} \supset \mathbf{B}] \supset .(\mathbf{a})\mathbf{A} \supset \mathbf{B}$

From this, by generalizing upon **a** and using 200, we get:

$\vdash (\mathbf{a})[\mathbf{A} \supset \mathbf{B}] \supset .(\mathbf{a}) .(\mathbf{a})\mathbf{A} \supset \mathbf{B}$

But by substitution in an appropriate instance of 200, we have also:

$\vdash (\mathbf{a})[(\mathbf{a})\mathbf{A} \supset \mathbf{B}] \supset .(\mathbf{a})\mathbf{A} \supset (\mathbf{a})\mathbf{B}$

Hence 202 follows by axiom 2.

For the weak theory of implication, the rule of substitutivity of mutual implication then takes the following form:

203. If **B** is obtained from **A** by substitution of **N** for **M** at some (or all or none) of the places at which **M** occurs in **A**, and if $\mathbf{c}_1, \mathbf{c}_2, \ldots, \mathbf{c}_n$ is a complete list of the free variables

in **M** and **N**, then $(\mathbf{c}_1)(\mathbf{c}_2)\ldots(\mathbf{c}_n)\,.\,\mathbf{M}\supset\mathbf{N}, (\mathbf{c}_1)(\mathbf{c}_2)\ldots(\mathbf{c}_n)\,.\,\mathbf{N}\supset\mathbf{M}\vdash\mathbf{A}\supset\mathbf{B}$, and $(\mathbf{c}_1)(\mathbf{c}_2)\ldots(\mathbf{c}_n)\,.\,\mathbf{M}\supset\mathbf{N}, (\mathbf{c}_1)(\mathbf{c}_2)\ldots(\mathbf{c}_n)\,.\,\mathbf{N}\supset\mathbf{M}\vdash\mathbf{B}\supset\mathbf{A}$.

The proof parallels the proof of 21. But use must be made of the fact that, by 201:

$$(\mathbf{c}_1)(\mathbf{c}_2)\ldots(\mathbf{c}_n)\,.\,\mathbf{M}\supset\mathbf{N}\vdash\mathbf{M}\supset\mathbf{N}$$
$$(\mathbf{c}_1)(\mathbf{c}_2)\ldots(\mathbf{c}_n)\,.\,\mathbf{N}\supset\mathbf{M}\vdash\mathbf{N}\supset\mathbf{M}$$

And in the final part of the proof, account must be taken of the possibility, not only that **A** may have the form $\mathbf{A}_1\supset\mathbf{A}_2$, but also that **A** may have the form $(\mathbf{c})\mathbf{A}_1$. If **A** has the latter form, and excepting case (i), we find that **B** must have the form $(\mathbf{c})\mathbf{B}_1$, where $\mathbf{B}_1$ is obtained from $\mathbf{A}_1$ by substitution of **N** for **M** at some (or all or none) of its occurrences. Then by hypothesis of induction we have:

$$(\mathbf{c}_1)(\mathbf{c}_2)\ldots(\mathbf{c}_n)\,.\,\mathbf{M}\supset\mathbf{N}, (\mathbf{c}_1)(\mathbf{c}_2)\ldots(\mathbf{c}_n)\,.\,\mathbf{N}\supset\mathbf{M}\vdash\mathbf{A}_1\supset\mathbf{B}_1$$
$$(\mathbf{c}_1)(\mathbf{c}_2)\ldots(\mathbf{c}_n)\,.\,\mathbf{M}\supset\mathbf{N}, (\mathbf{c}_1)(\mathbf{c}_2)\ldots(\mathbf{c}_n)\,.\,\mathbf{N}\supset\mathbf{M}\vdash\mathbf{B}_1\supset\mathbf{A}_1$$

Hence by generalizing upon **c** and using 202, we get:

$$(\mathbf{c}_1)(\mathbf{c}_2)\ldots(\mathbf{c}_n)\,.\,\mathbf{M}\supset\mathbf{N}, (\mathbf{c}_1)(\mathbf{c}_2)\ldots(\mathbf{c}_n)\,.\,\mathbf{N}\supset\mathbf{M}\vdash\mathbf{A}\supset\mathbf{B}$$
$$(\mathbf{c}_1)(\mathbf{c}_2)\ldots(\mathbf{c}_n)\,.\,\mathbf{M}\supset\mathbf{N}, (\mathbf{c}_1)(\mathbf{c}_2)\ldots(\mathbf{c}_n)\,.\,\mathbf{N}\supset\mathbf{M}\vdash\mathbf{B}\supset\mathbf{A}$$

In order to introduce negation into the weak theory of implication, we may use the definition at the beginning of §2, taking f to be some particular well-formed formula without free variables which is not a theorem. We consider here the two following possibilities

$$f \rightarrow (p)(q)\,.\,p\supset\,.\,q\supset p$$
$$f \rightarrow (p)p$$

We call these respectively *the weak definition of* f and *the strong definition of* f. And the corresponding definitions of negation we speak of as *the weak definition of negation* and *the strong definition of negation* respectively.[12]

We leave it to the reader to show that, under either of the two definitions of f and of negation, axiom 100 becomes a theorem, and hence 22–34, 101–104 are theorems of the weak theory of implication.

Under the strong definition of f and of negation, such additional theorems are obtained as $f\supset p$ and $\sim p\supset\,.\,p\supset q$. On the other hand, a larger class of theorems of the form $\sim\mathbf{A}$ is obtained under the weak definition.

In order to extend 105 to the weak theory of implication, we need a sufficient system of axioms and rules for the full theory of implication (with $\supset$ and the universal quantifier as primitive, and negation introduced by either the weak or the strong definition). As such a system we may take the three axioms,

$$p\supset\,.\,q\supset p,$$
$$s\supset[p\supset q]\supset\,.\,s\supset p\supset\,.\,s\supset q,$$
$$\sim p\supset\sim q\supset\,.\,q\supset p,$$

together with the axiom schemata 200 and 201 and the same three rules of inference as for the weak theory of implication.[13]

---

[12]The strong definition of negation (as we here call it) was suggested by Russell in his paper of 1906, and—perhaps independently—by Jan Łukasiewicz and Alfred Tarski in a paper in ***Comptes Rendus des Séances de la Société des Sciences et des Lettres de Varsovie***, Classe III, vol. 23 (1930), pp. 30–50.

[13]The sufficiency of this system of axioms and rules—we are not here concerned with questions of independence or with simplifications of the axioms—is a consequence of theorem 34 of the paper of Łukasiewicz and Tarski cited in the preceding footnote. Or since Łukasiewicz and Tarski do not provide

We must then show that the double-negation analogues of all of these hold in the weak theory of implication. For the first three axioms and the rules of substitution and *modus ponens*, this has already been done (substantially) in §2. For the rule of generalization, the double-negation analogue consists in first generalizing and then prefixing a double negation to the result; and this can be accomplished in the weak theory of implication by means of the rule of generalization, 24, substitution, and *modus ponens*. For 200 and 201, the double-negation analogues are:

$$\sim\sim . \sim\sim(\mathbf{a})\sim\sim[\sim\sim\mathbf{b}\supset\mathbf{A}^*]\supset\sim\sim . \sim\sim\mathbf{b}\supset\sim\sim(\mathbf{a})\mathbf{A}^*$$

$$\sim\sim . \sim\sim(\mathbf{a})\mathbf{A}^*\supset\mathbf{A}^*$$

And these are instances of the following theorem schemata of the weak theory of implication:

204. $\vdash\sim\sim . \sim\sim(\mathbf{a})\sim\sim[\sim\sim\mathbf{b}\supset\sim\sim\mathbf{A}]\supset\sim\sim . \sim\sim\mathbf{b}\supset\sim\sim(\mathbf{a})\sim\sim\mathbf{A}$

where **a** and **b** are any two distinct variables.

205. $\vdash\sim\sim\ \sim\sim(\mathbf{a})\sim\sim\mathbf{A}\supset\sim\sim\mathbf{A}$

For the proof of 204 and 205 it is convenient to have also the following theorem schemata:

206. $\vdash\sim\sim\mathbf{a})\sim\sim\mathbf{A}\supset(\mathbf{a})\sim\sim\mathbf{A}$

207. $\vdash(\mathbf{a})\sim\sim\mathbf{A}\supset\sim\sim(\mathbf{a})\sim\sim\mathbf{A}$

The theorem schema 207 follows by substitution in 24.

To prove 206, we take the following instance of 201:

$$\vdash(\mathbf{a})\sim\sim\mathbf{A}\supset\sim\sim\mathbf{A}$$

From this, by using 23 twice and then 25, we get:

$$\vdash\sim\sim(\mathbf{a})\sim\sim\mathbf{A}\supset\sim\sim\mathbf{A}$$

Hence, by generalization upon **a** and then using 200, we obtain 206.

Also, using again that

$$\vdash\sim\sim(\mathbf{a})\sim\sim\mathbf{A}\supset\sim\sim\mathbf{A}$$

(as just proved), and 24, we obtain 205.

To prove 204, we use an instance of 200 in which **A** has been taken to be $\sim\sim\mathbf{A}$, and in it substitute $\sim\sim\mathbf{b}$ for **b**. From this 204 follows by 24, 30, 31, 206, 207, 203.

Hence results the following extension of 105 to the weak theory of implication:

208. If **A** is a well-formed formula of the weak theory of implication, and if in $\mathbf{A}^*$ the negation sign is understood according to either the weak or the strong definition, then $\mathbf{A}^*$ is a theorem of the weak theory of implication if and only if **A** is a theorem of the full (i.e., two-valued) theory of implication.

From this by 206, 207, 30, 31, 203 there follows as a corollary the following extension of 106 to the weak theory of implication:

209. Let **A** be a well-formed formula of the weak theory of implication, and let $\mathbf{A}^{**}$ be obtained from **A** by prefixing a double negation, $\sim\sim$, to every variable, at all of its occurrences, both free and bound, except (of course) occurrences of the variable

---

a proof of their result, the reader may prefer to undertake a direct proof of the sufficiency of our present system of axioms and rules using the obvious definition of validity by means of two-valued truth-tables for the full theory of implication.

between the parentheses of the universal quantifier. Then $\mathbf{A}^{**}$ is a theorem of the weak theory of implication if and only if $\mathbf{A}$ is a theorem of the full (two-valued) theory of implication.

5. *Conjunction, equivalence, disjunction*

It is clear that there cannot be a definition of conjunction such that both $p \supset . q \supset pq$ and $pq \supset p$ are theorems—because from these two together, by axiom 2, we would get $p \supset . q \supset p$ as a theorem. Similarly, there cannot be a definition of disjunction such that $p \supset p \vee q$ and $q \supset p \vee q$ and $p \supset r \supset . q \supset r \supset . p \vee q \supset r$ are theorems—because from these three theorems, by a slightly longer argument, we would again get $p \supset . q \supset p$ as a theorem.

In order to reproduce as nearly as possible the essential properties of conjunction, equivalence, and disjunction, the following definitions suggest themselves:[14]

$$[\mathbf{AB}] \to (\mathbf{c}) . \mathbf{A} \supset [\mathbf{B} \supset \mathbf{c}] \supset \mathbf{c}$$
$$[\mathbf{A} \equiv \mathbf{B}] \to [\mathbf{A} \supset \mathbf{B}][\mathbf{B} \supset \mathbf{A}]$$
$$[\mathbf{A} \vee \mathbf{B}] \to \mathbf{B} \supset \mathbf{A} \supset . \mathbf{A} \supset \mathbf{B} \supset \mathbf{B}$$

Here $\mathbf{c}$ must be chosen (according to some specific rule, given as part of the definition) to be a variable which is not free in either $\mathbf{A}$ or $\mathbf{B}$.

We leave to the reader the proof of the following theorems:[15]

210. $\vdash p \supset [q \supset r] \supset . pq \supset r$
211. $\vdash p \supset . q \supset pq$
212. $\vdash pq \supset r \supset . p \supset . q \supset r .$
213. $\vdash pq \supset qp$
214. $\vdash [pq]r \supset p[qr]$
215. $\vdash p[qr] \supset [pq]r$
216. $\vdash pq \supset \sim\sim p$
217. $\vdash pq \supset \sim\sim q$
218. $\vdash p \supset q \supset . q \supset p \supset . p \equiv q$
219. $\vdash p \equiv q \supset . p \supset q$
220. $\vdash p \equiv q \supset . q \supset p$
221. $\vdash p \supset . p \vee q$
222. $\vdash q \supset . p \vee q$
223. $\vdash p \vee q \supset . q \vee p$
224. $\vdash p \vee p \supset p$
225. $\vdash [p \vee q] \vee r \supset . p \vee [q \vee r]$
226. $\vdash p \vee [q \vee r] \supset . [p \vee q] \vee r$
227. $\vdash p \supset . \sim p \vee \sim q \supset \sim q$
228. $\vdash q \supset . \sim p \vee \sim q \supset \sim p$
229. $\vdash \sim p \supset \sim r \supset . \sim q \supset \sim r \supset . \sim p \vee \sim q \supset \sim r$

---

[14]The definition of conjunction is due to Bertrand Russell in ***The Principles of Mathematics*** (1903). And the definition of disjunction is a modification of Russell's definition (ibid.) of $p \vee q$ as $p \supset q \supset q$.

[15]Theorems 221 and 224 are in fact the same as theorems 18 and 16 respectively. Theorem 219 follows from 17 by means of 210.

The following additional theorems require the strong definition of negation:

$$\sim p \supset . p \vee q \supset \sim\sim q$$

$$\sim q \supset . p \vee q \supset \sim\sim p$$

$$p \supset \sim r \supset . q \supset \sim r \supset . p \vee q \supset \sim r$$

SPECIAL CASES OF THE DECISION PROBLEM (1951)

Functional calculus. To facilitate the statement of results about the decision problem, we shall adopt a particular formulation of the pure functional calculus of first order, as follows. (All of the results treated, however, may easily be restated so as to apply to other formulations.)

There is one primitive propositional constant, f (falsehood), one primitive sentence connective $\supset$ (material implication), and two primitive quantifiers, the universal and the existential quantifier.[1]

As individual variables we use $x, y, z, x_0 \ldots$; as propositional variables, $p, q, r, p_0, \ldots$; and as n-ary functional variables, $F^n, G^n, H^n, F_0^n, \ldots$. The superscript n after the functional variable may, however, usually be omitted without any danger of misunderstanding.

Where $\mathbf{a}$ is any individual variable, we use $(\mathbf{a})$ as universal quantifier and $(E\mathbf{a})$ as existential quantifier.

An *elementary well-formed formula* either consists of the propositional constant f standing alone, or consists of a propositional variable alone, or has the form $\mathbf{f}(\mathbf{a}_1, \mathbf{a}_2, \ldots, \mathbf{a}_n)$, where $\mathbf{f}$ is an n-ary functional variable and $\mathbf{a}_1, \mathbf{a}_2, \ldots, \mathbf{a}_n$ are individual variables (not necessarily all distinct). A *well-formed formula* is an expression built up in the usual way by means of the connective $\supset$ and the quantifiers from elementary well-formed formulas. And the elementary well-formed formulas thus used in building up a well-formed formula are called its *elementary parts*.

Hereafter we shall use the abbreviations, "wff" for "well-formed formula," and "wf" for "well-formed."

As already illustrated, we use bold letters in the rôle of syntactical variables, bold small letters (as **a**, **b**) standing for variables of the object language, and bold capital letters (as **A**, **B**) for wffs.

If $(\mathbf{a})\mathbf{A}$ or $(E\mathbf{a})\mathbf{A}$ occurs as a wf part of a wff **L**, then that particular occurrence of the wff **A** in **L** is called the *scope* of that particular occurrence of the quantifier $(\mathbf{a})$ or $(E\mathbf{a})$. All occurrences of an individual variable **a** in a wf part of $(\mathbf{a})\mathbf{A}$ or $(E\mathbf{a})\mathbf{A}$ of a wff **L** are called *bound occurrences* of **a** in **L**; and all other occurrences of **a** in **L** are called *free occurrences*. All occurrences of propositional and functional variables are free occurrences. The *bound variables* of **L** are those (individual) variables which have bound occurrences in **L**, and the *free variables* of **L** are those variables which have free occurrences in **L**.

A particular occurrence of a wff **P** as a wf part of a wff **L** is called an *occurrence as a P-constituent* in **L** if it is not within the scope of any quantifier in **L**, and **P** does not have the form, either of an implication $[\mathbf{A} \supset \mathbf{B}]$ or of the constant f standing alone. And the *P-constituents* of **L** are those wffs which have occurrences as P-constituents in **L**.[2]

Originally published in ***Revue Philosophique de Louvain***, vol. 49 (1951), pp. 203–221.

[1]For our present purpose it is more convenient to treat both quantifiers as primitive.

[2]Thus every P-constituent of a wff either is an elementary part (other than f) or else begins with a quantifier and consists of the quantifier together with its scope. And every wff can be thought of as obtained from a wff of the propositional calculus by substituting various P-constituents for the propositional variables.

We shall speak in the usual sense of validity and satisfiability of wffs. These notions are capable of a precise syntactical definition by a method due to Tarski.[3] But we content ourselves here with the following brief statement. If a particular non-empty domain D is taken as the range of the individual variables, then the range of the n-ary functional variables is to consist of all n-ary propositional functions over this domain, and the range of the propositional variables is to consist of the two truth-values, truth and falsehood. (The value of the constant f is always to be the truth-value falsehood.) Upon taking the ranges of the variables in this way, a wff is said to be *valid in the domain* D if it has the value truth for every system of values of its free variables, *satisfiable in the domain* D if it has the value truth for at least one system of values of its free variables. A wff is said to be *valid* if it is valid in every non-empty domain, *satisfiable* if it is satisfiable in some non-empty domain.

We require axioms and rules of inference for the functional calculus of first order satisfying the two conditions, that every theorem of the calculus is a valid wff, and every valid wff is a theorem of the calculus. As is well known, this may be accomplished in various ways. And since the particular choice of axioms and rules is not important for our present purpose, we shall not take the space to list them, but shall simply assume the proofs of familiar theorems of the calculus as these are needed.

Merely as a matter of convenience in the metatheoretic presentation, we shall use the following abbreviations of wffs:

$\sim \mathbf{A}$ as an abbreviation of $[\mathbf{A} \supset f]$.

$[\mathbf{A} \vee \mathbf{B}]$ as an abbreviation of $[\sim \mathbf{B} \supset \mathbf{A}]$.

$[\mathbf{AB}]$ as an abbreviation of $\sim [\mathbf{B} \supset \sim \mathbf{A}]$.

$[\mathbf{A} \equiv \mathbf{B}]$ as an abbreviation of $[[\mathbf{A} \supset \mathbf{B}][\mathbf{B} \supset \mathbf{A}]]$.

We shall also abbreviate by omitting brackets under the convention of association to the left—so that, e.g., $F(x) \supset G(y) \supset H(x, y)$ is an abbreviation of $[[F(x) \supset G(y)] \supset H(x, y)]$, and $p \vee q \vee [r_1 \supset r_2]$ is an abbreviation of $[[p \vee q] \vee [r_1 \supset r_2]]$. And a bold dot will be used to stand for a pair of brackets, one at the position of the dot, and the other, either at the end of the wff or otherwise at the latest position which is consistent with the formula's being wf—so that, e.g., $F(x) \supset G(y) \supset \centerdot\, H(x, y) \supset H(x, y)$ is an abbreviation of $[[F(x) \supset G(y)] \supset [H(x, y) \supset H(x, y)]]$, and $[p \supset \centerdot\, q \supset \centerdot\, r_1 \supset r_2] \supset r$ is an abbreviation of $[[p \supset [q \supset [r_1 \supset r_2]]] \supset r]$.

A wff $\mathbf{A}_1 \vee \mathbf{A}_2 \vee \ldots \vee \mathbf{A}_\mu$ will be called a *disjunction*, and $\mathbf{A}_1, \mathbf{A}_2, \ldots, \mathbf{A}_\mu$ will be called its terms.[4] Similarly, $\mathbf{A}_1\mathbf{A}_2 \ldots \mathbf{A}_\mu$ is a *conjunction* and $\mathbf{A}_1, \mathbf{A}_2, \ldots, \mathbf{A}_\mu$ are its terms; $\mathbf{A} \supset \mathbf{B}$ is an *implication*, $\mathbf{A}$ is its *antecedent*, and $\mathbf{B}$ is its *consequent*; $\mathbf{A} \equiv \mathbf{B}$ is an *equivalence*, and $\mathbf{A}$ and $\mathbf{B}$ are its *sides*. Also a wff $\sim \mathbf{A}$ will be called the *negation* of $\mathbf{A}$.

We shall say that two wffs $\mathbf{A}$ and $\mathbf{B}$ are *equivalent* if the wff $\mathbf{A} \equiv \mathbf{B}$ is a theorem (of the pure functional calculus of first order). And we shall assume the following metatheorems: If in any theorem a wf part is replaced by an equivalent wf part, the wff so obtained is also a theorem. Hence, in particular, if two wffs are equivalent, and if either is a theorem, then the other is a theorem.

In order to provide a convenient syntactical notation for the operation of substitution, we shall often write $\mathbf{M}(\mathbf{a}_1\mathbf{a}_2 \ldots \mathbf{a}_\nu)$ instead of merely $\mathbf{M}$ as a notation for a wff having no other free individual variables than $\mathbf{a}_1, \mathbf{a}_2, \ldots, \mathbf{a}_\nu$. Then $\mathbf{M}(\mathbf{b}_1\mathbf{b}_2 \ldots \mathbf{b}_\nu)$ will

---

[3] Alfred TARSKI, *Der Wahrheitsbegriff in den formalisierten Sprachen*, ***Studia Philosophica***, 1 (1936).

[4] As a particular case, n may be 1; i.e., an arbitrary wff $\mathbf{A}_1$ may be considered as a disjunction having a single term; or also as a conjunction having a single term.

be used as a notation for the result of substituting $\mathbf{b}_1, \mathbf{b}_2, \ldots, \mathbf{b}_\nu$ for free occurrences of $\mathbf{a}_1, \mathbf{a}_2, \ldots, \mathbf{a}_\nu$ respectively, simultaneously throughout $\mathbf{M}(\mathbf{a}_1\mathbf{a}_2 \ldots \mathbf{a}_\nu)$.

We shall also sometimes make use of the notation $(\mathbf{M})^{\mathbf{a}}_{\mathbf{b}}$ for the result of substituting **b** for all free occurrences of **a** in **M**. (Notice that the use of this notation for substitution does not imply but what **M** may contain other free individual variables than **a**.)

Tautology. We shall call a wff *tautologous* if it has the value truth for every system of truth-values of its P-constituents. Assuming this notion to be familiar, we recall only that there is an effective test, the truth-table decision procedure, by which to recognize whether a given wff is tautologous. Evidently, every tautologous wff is valid, but not conversely.

If a wff **L** is not tautologous, there exist one or more *falsifying* systems of truth-values of its P-constituents—i.e., systems of truth-values of the P-constituents for which the value of **L** is falsehood. Let the (distinct) P-constituents of **L** be $\mathbf{P}_1, \mathbf{P}_2, \ldots, \mathbf{P}_\mu$; and let the complete list of falsifying systems of truth values of these P-constituents consist in the systems of values $\tau^i_1, \tau^i_2, \ldots, \tau^i_\mu$, where $i = 1, 2, \ldots, \nu$. Let $\mathbf{P}^i_j$ be $\mathbf{P}_j$ or $\sim \mathbf{P}_j$ according as the corresponding truth-value τ^i_j is truth or falsehood. Then the wff

$$\mathbf{L} \equiv [\mathbf{P}^1_1 \supset . \mathbf{P}^1_2 \supset . \ldots \mathbf{P}^1_\mu \supset f][\mathbf{P}^2_1 \supset . \mathbf{P}^2_2 \supset . \ldots \mathbf{P}^2_\mu \supset f] \ldots [\mathbf{P}^\nu_1 \supset . \mathbf{P}^\nu_2 \supset . \ldots \mathbf{P}^\nu_\mu \supset f]$$

is tautologous, and therefore a theorem. The right hand side of this equivalence we shall here call the *conjunctive normal form* of **L** (as is only a minor deviation from standard terminology). The separate terms $\mathbf{P}^i_1 \supset . \mathbf{P}^i_2 \supset . \ldots \mathbf{P}^i_\mu \supset f$ we call the *terms of* the conjunctive normal form.

If **L** is tautologous, then $\nu = 0$, and there is no conjunctive normal form of **L**. In this case, the terms of the conjunctive normal form of **L** shall be understood to be the null set.

Prenex normal form. An occurrence of a quantifier, (**a**) or $(E\mathbf{a})$, in a wff **L** is said to be *vacuous* if **a** is not a free variable of **A**, where **A** is the scope.

An occurrence of a quantifier in a wff **L** is said to be *initially placed* if it is at the beginning of **L** or is preceded in **L** only by other quantifiers, and if at the same time its scope extends to the end of **L**. Thus, e.g., in $(x)(y) . (z)F(x, z) \supset F(x, y)$, the quantifiers (x) and (y) are initially placed, but not the quantifier (z).

A wff is said to be in *prenex normal form* if all its quantifiers are initially placed and none of them are vacuous.

Thus a wff in prenex normal form may be thought of as consisting of two parts: the *matrix*, which is a quantifier-free wff, preceded by the *prefix*, which is composed entirely of quantifiers (0 or more) each with its own variable. The variables in the prefix must be all different and must all occur in the matrix.

According to a familiar metatheorem, *every wff can be reduced to prenex normal form.* More explicitly, given any wff **L**, and equivalent wff $\mathbf{L}^*$ in prenex normal form can be obtained from **L** by applying the following reduction steps (1)–(4) a finite number of times:

(1) To delete a vacuous occurrence of a quantifier.

(2) **b** being and individual variable which does not otherwise occur, to replace a wf part $\mathbf{C}_1 \supset . \mathbf{C}_2 \supset . \ldots \mathbf{C}_\mu \supset \mathbf{D}$ by $(E\mathbf{b}) . \mathbf{C}'_1 \supset . \mathbf{C}'_2 \supset . \ldots \mathbf{C}'_\mu \supset \mathbf{D}'$; where in case **D**

has the form $(E\mathbf{a})\mathbf{B}$, $\mathbf{D}'$ is $(\mathbf{B})^{\mathbf{a}}_{\mathbf{b}}$, and otherwise $\mathbf{D}'$ is $\mathbf{D}$; and where (for $i = 1, 2, \ldots, \mu$), in case $\mathbf{C}_i$ has the form $(\mathbf{a}_i)\mathbf{A}_i$, $\mathbf{C}_i$ is $(\mathbf{A}_i)^{\mathbf{a}_i}_{\mathbf{b}}$, and otherwise $\mathbf{C}_i'$ is $\mathbf{C}_i$; but provided, finally, that $\mu > 0$ and that the quantifier $(E\mathbf{b})$ so introduced is not vacuous.

(3) **b** being an individual variable which does not otherwise occur, to replace a wf part $\mathbf{C} \supset (\mathbf{a})\mathbf{B}$ by $(\mathbf{b}) \mathbin{.} \mathbf{C} \supset (\mathbf{B})^{\mathbf{a}}_{\mathbf{b}}$.

(4) **b** being an individual variable which does not otherwise occur, to replace a wf part $(E\mathbf{a})\mathbf{A} \supset \mathbf{D}$ by $(\mathbf{b}) \mathbin{.} (\mathbf{A})^{\mathbf{a}}_{\mathbf{b}} \supset \mathbf{D}$.

The observation is due to Behmann (1922) and more recently to Quine that, in connection with certain cases of the decision problem, it may be useful to apply also what we shall call the *inverse of reduction to prenex normal form*. This process consists in the following reduction steps (i)–(vii) to be applied any number of times in succession:

(i) To delete a vacuous occurrence of a quantifier.

(ii) To replace a wf part $(E\mathbf{a}) \mathbin{.} \mathbf{A}_1 \supset \mathbin{.} \mathbf{A}_2 \supset \mathbin{.} \ldots \mathbf{A}_\mu \supset \mathbf{B}$ by $(\mathbf{a})\mathbf{A}_1 \supset \mathbin{.} (\mathbf{a})\mathbf{A}_2 \supset \mathbin{.} \ldots (\mathbf{a})\mathbf{A}_\mu \supset (E\mathbf{a})\mathbf{B}$, if $\mu > 0$.

(iii) To replace a wf part $(\mathbf{a})\mathbf{B}$ by $[(\mathbf{a})\mathbf{B}_1 \supset \mathbin{.} (\mathbf{a})\mathbf{B}_2 \supset \mathbin{.} \ldots (\mathbf{a})\mathbf{B}_\mu \supset f] \supset f$, where $\mathbf{B}_1, \mathbf{B}_2, \ldots, \mathbf{B}_\mu$ are the terms of the conjunctive normal form of **B**, and where $\mu > 1$.

(iv) To replace a wf part $(\mathbf{a})\mathbf{B}$ by $(\mathbf{a})\mathbf{B}_1$, if **B** is not in conjunctive normal form, and the conjunctive normal form of **B** has a single term $\mathbf{B}_1$.

(v) To replace a wf part $(\mathbf{a})\mathbf{B}$ by $f \supset f$ if **B** is tautologous.

(vi) To replace a wf part $(\mathbf{a}) \mathbin{.} \mathbf{A}_1 \supset \mathbin{.} \mathbf{A}_2 \supset \mathbin{.} \ldots \mathbf{A}_\mu \supset \mathbin{.} \mathbf{C} \supset \mathbf{B}$ by $\mathbf{C} \supset (\mathbf{a}) \mathbin{.} \mathbf{A}_1 \supset \mathbin{.} \mathbf{A}_2 \supset \mathbin{.} \ldots \mathbf{A}_\mu \supset \mathbin{.} \mathbf{B}$, if **a** is not a free variable of **C**, and $\mu \geqq 0$.

(vii) To replace a wf part $(\mathbf{a}) \mathbin{.} \mathbf{A} \supset \mathbf{D}$ by $(E\mathbf{a})\mathbf{A} \supset \mathbf{D}$, if **a** is not a free variable of **D**.

Starting with a given wff **L** and applying steps (i)–(vii) repeatedly until the process is blocked, we obtain an equivalent wff $\mathbf{L}_*$ which is in a certain sense as far removed as possible from the prenex normal form of **L**.

In connection with the process of reduction of **L** to prenex normal form and the inverse process, it is possible to give effective instructions for building up proofs of the equivalences $\mathbf{L} \equiv \mathbf{L}^*$ and $\mathbf{L} \equiv \mathbf{L}_*$ as theorems of the functional calculus. We here leave it to the reader to supply these.

Decision problem. By a solution of a special case of *the decision problem of the pure functional calculus of first order* (or briefly, of *the decision problem*) we mean that there shall be given a special class of wffs, an effective test for recognizing whether a wff belongs to this class, an effective procedure (the *decision procedure*) for determining whether a wff **L** of this class is a theorem, and finally an effective method of finding a proof of a wff thus determined to be a theorem.

In calling a procedure or a method *effective* we mean, roughly speaking, that there is assurance of reaching the desired result in every particular case by a faithful and purely mechanical following of fixed instruction, without the need for exercise of ingenuity. We forgo a more exact definition here, because it is believed that in the cases treated below there can hardly be doubt of the effectiveness of the particular procedures introduced.[5]

The purpose of this note is to outline briefly the important special cases of the decision problem for which solutions either exist explicitly or are easily obtained as

[5]Representing formulas by Gödel numbers in a straightforward manner, we find in fact that all the procedures introduced are represented by primitive recursive functions — which is surely sufficient.

corollaries of methods found in the literature. The actual decision procedures will be given only in the simpler cases (the reader being referred to the original papers for the rest), and their proofs will be at best roughly indicated.

For purposes of exposition it will be convenient to divide existing solutions of special cases of the decision problem into two types, namely (I) those in which the special class of wffs is characterized by conditions on the prefix and the matrix of a wff **L** in prenex normal form and the decision procedure is applied to **L** directly in this prenex normal form, and (II) those in which the decision procedure is indirect, the decision problem for a given wff **L** being reduced to the decision problem for one or more wffs which fall under special cases of type I.

In stating special cases of type I, we shall assume that the given wff **L** is in prenex normal form and without free individual variables. The latter condition involves no loss of generality, since $\mathbf{L}(\mathbf{a}_1\mathbf{a}_2\ldots\mathbf{a}_\nu)$ is a theorem if and only if $(\mathbf{a}_1)(\mathbf{a}_2)\ldots(\mathbf{a}_\nu)\mathbf{L}(\mathbf{a}_1\mathbf{a}_2\ldots\mathbf{a}_\nu)$ is a theorem.

For special cases of type II we shall assume, for convenience, that every P-constituent is in prenex normal form.

Type I. Solutions exist of the following special cases:[6]

CASE 1.[7] Prefix $(\mathbf{a}_1)(\mathbf{a}_2)\ldots(\mathbf{a}_l)(E\mathbf{b}_1)(E\mathbf{b}_2)\ldots(E\mathbf{b}_m)$.

CASE 2.[8] Prefix $(\mathbf{a}_1)(\mathbf{a}_2)\ldots(\mathbf{a}_l)(E\mathbf{b})(\mathbf{c}_1)(\mathbf{c}_2)\ldots(\mathbf{c}_n)$.

CASE 3.[9] Prefix $(\mathbf{a}_1)(\mathbf{a}_2)\ldots(\mathbf{a}_l)(E\mathbf{b}_1)(E\mathbf{b}_2)(\mathbf{c}_1)(\mathbf{c}_2)\ldots(\mathbf{c}_n)$.

CASE 4.[10] Prefix $(\mathbf{a}_1)(\mathbf{a}_2)\ldots(\mathbf{a}_l)(E\mathbf{b}_1)(E\mathbf{b}_2)\ldots(E\mathbf{b}_m)(\mathbf{c}_1)(\mathbf{c}_2)\ldots(\mathbf{c}_n)$, and a matrix in which every elementary part that contains individual variables other than $\mathbf{a}_1, \mathbf{a}_2, \ldots, \mathbf{a}_l$ contains either all of the variables $\mathbf{b}_1, \mathbf{b}_2, \ldots, \mathbf{b}_m$ or at least one of the variables $\mathbf{c}_1, \mathbf{c}_2, \ldots, \mathbf{c}_n$.

CASE 5.[10] Prefix beginning with $(\mathbf{a}_1)(\mathbf{a}_2)\ldots(\mathbf{a}_l)$ and ending with $(\mathbf{c}_1)(\mathbf{c}_2)\ldots(\mathbf{c}_n)$, and a matrix in which every elementary part that contains individual variables other than $\mathbf{a}_1, \mathbf{a}_2, \ldots, \mathbf{a}_l$ contains at least one of the variables $\mathbf{c}_1, \mathbf{c}_2, \ldots, \mathbf{c}_n$.

CASE 6.[11] There is at most one falsifying system of truth-values of the P-constituents

[6]In each case, it is not excluded that l, m, n may some or all of them be 0.

[7]Paul BERNAYS and Moses SCHÖNFINKEL, *Zum Entscheidungsproblem der Mathematischen Logik*, ***Mathematische Annalen***, 99 (1928). The subcase $l = 0, n = 1$ of Case 2 is also treated by Bernays and Schönfinkel in this paper.

[8]Wilhelm ACKERMANN, *Ueber die Erfüllbarkeit gewisser Zählausdrücke*, ***Mathematische Annalen***, 100 (1928); Thoralf SKOLEM, *Ueber die Mathematische Logik*, ***Norsk Matematisk Tidsskrift***, 10 (1928); Jacques HERBRAND, *Sur le Problème Fondamental de la Logique Mathématique*, ***Comptes Rendus des séances de la Société des Sciences et des Lettres de Varsovie***, Classe III, 24 (1931). In some of these papers it is the decision problem for satisfiability which is treated directly, i.e., the problem to find an effective procedure to determine whether wffs are satisfiable. But we have here modified and restated their results so as to apply to the decision problem in the sense in which we are using the term. (Compare the discussion of Bernays and Schönfinkel regarding the relationship of the two decision problems.)

[9]Kurt GÖDEL, *Ein Spezialfall des Entscheidungsproblems der Theoretischen Logik*, ***Ergebnisse eines Mathematischen Kolloquiums***, 2 (1932); László KALMÁR, *Ueber die Erfüllbarkeit derjenigen Zählausdrücke, welche in der Normalform zwei Benachbarte Allzeichen Enthalten*, ***Mathematische Annalen***, 108 (1933); Kurt GÖDEL, *Zum Entscheidungsproblem de Logischen Funktionenkalküls*, ***Monatshefte für Mathematik und Physik***, 40 (1933); Kurt SCHÜTTE, *Untersuchungen zum Entscheidungsproblem der Mathematischen Logik*, ***Mathematische Annalen*** 109 (1934).

[10]Cf. SKOLEM, *loc. cit.*

[11]HERBRAND, *loc. cit.* Another way of stating the condition which characterizes this case is that the matrix is a disjunction of elementary parts and negations of elementary parts, or equivalent to such a disjunction.

of the matrix.

CASE 7.[12] Prefix $(E\mathbf{b}_1)(E\mathbf{b}_2)\dots(E\mathbf{b}_m)(\mathbf{c})$, where $m \leqq 4$, and a matrix of the form $\mathbf{N} \supset \mathbf{f}(\mathbf{b}_1, \mathbf{c})$, where $\mathbf{c}$ does not occur in $\mathbf{N}$.

The treatment of some of these cases may be unified by means of the following metatheorem. Let the ordered m-tuples of natural numbers be enumerated in such a way that one m-tuple comes later in the enumeration than another if it contains as member a natural number greater than the greatest natural number contained in the other, and the m-tuples having the same greatest natural number are arranged among themselves in lexicographic order. For the ith member of the kth m-tuple in this enumeration, use the notation $[mki]$. Let $\mathbf{M}(\mathbf{a}_1\mathbf{a}_2\dots\mathbf{a}_l\mathbf{b}_1\mathbf{b}_2\dots\mathbf{b}_m\mathbf{c}_1\mathbf{c}_2\dots\mathbf{c}_n)$ be any quantifier-free wff, let $\mathbf{M}_k$ be

$$\mathbf{M}(x_0x_1\dots x_{l-1}x_{[mk1]}x_{[mk2]}\dots x_{[mkm]}x_{l+(k-1)n}x_{l+(k-1)n+1}\dots x_{l+kn-1}),$$

let $\mathbf{D}_k$ be the disjunction $\mathbf{M}_1 \vee \mathbf{M}_2 \vee \dots \vee \mathbf{M}_k$, and let $\mathbf{L}$ be $(\mathbf{a}_1)(\mathbf{a}_2)\dots(\mathbf{a}_l)(E\mathbf{b}_1)(E\mathbf{b}_2)\dots(E\mathbf{b}_m)(\mathbf{c}_1)(\mathbf{c}_2)\dots(\mathbf{c}_n)\mathbf{M}(\mathbf{a}_1\mathbf{a}_2\dots\mathbf{a}_l\mathbf{b}_1\mathbf{b}_2\dots\mathbf{b}_m\mathbf{c}_1\mathbf{c}_2\dots\mathbf{c}_n)$. *Then* $\mathbf{L}$ *is a theorem if and only if* $\mathbf{D}_k$ *is tautologous for some* k.

This metatheorem is, in fact, a corollary of Gödel's proof[13] of the completeness of the functional calculus of first order.

Moreover, if a particular $\mathbf{D}_k$ has been found which is tautologous, it is well known how (by methods belonging essentially to propositional calculus) to find a proof of $\mathbf{D}_k$. Then to go on from a proof of $\mathbf{D}_k$ to a proof of $\mathbf{L}$ is a matter of using familiar laws of quantifiers (it is suggested that the reader carry this out in a representative particular case — say the case that $l = 1, m = 2, n = 1, k = 7$). Thus an effective method may be provided, to find a proof of $\mathbf{L}$ when a number k is given such that $\mathbf{D}_k$ is tautologous.

For a class of wffs in prenex normal form, if all the prefixes are of the form given above (i.e., if no universal quantifiers occur between existential quantifiers) it will therefore constitute a solution of the decision problem to find a particular value K of k, depending in an effective way on the given wff $\mathbf{L}$, such that either $\mathbf{D}_K$ is tautologous or none of the wffs $\mathbf{D}_k$ is tautologous. And we may state the solution just by giving this number K.

To illustrate the method, consider a wff $\mathbf{L}$ falling under Case 2 with $l = 0$ and $n = 1$, and suppose that the matrix $\mathbf{M}(\mathbf{bc})$ of $\mathbf{L}$ contains no propositional variables, and no functional variables but one binary functional variable $\mathbf{f}$. Taking the domain of natural numbers as the range of the individual variables, we attempt as follows to find

---

But if the prefix begins with the universal quantifiers $(\mathbf{a}_1)(\mathbf{a}_2)\dots(\mathbf{a}_l)$ $(l \geqq 0)$, a somewhat weaker condition is actually sufficient, namely that the matrix shall be or be equivalent to a disjunction in which each term is a conjunction of elementary parts and negations of elementary parts, at most one of the terms of each such conjunction containing individual variables other than $\mathbf{a}_1, \mathbf{a}_2, \dots, \mathbf{a}_l$. For the method of solution of Case 6 which is indicated below may be readily extended to solve this more general case.

[12]Wilhelm ACKERMANN, *Beiträge zum Entscheidungsproblem der Mathematischen Logik*, ***Mathematische Annalen***, 112 (1936); I. GÉGALKINE, *Sur l'Entscheidungsproblem*, ***Recueil Mathématique***, 6 (1939). In spite of the rather special appearance of the condition which characterizes it, this case has some considerable interest because it is here the only case that includes examples of wffs which though valid in every finite domain, are not valid in an infinite domain. Moreover, as Ackermann points out, there follows as an immediate corollary a solution of the decision problem for wffs with prefix $(E\mathbf{a})(\mathbf{b})(E\mathbf{c})$ or $(E\mathbf{a})(\mathbf{b})(E\mathbf{c})(E\mathbf{d})$; for a wff with either of these prefixes is reduced to case 7 by the usual process for reduction to Skolem normal form.

[13]Kürt GÖDEL, *Die Vollständigkeit der Axiome des Logischen Funktionenkalküls*, ***Monatshefte für Mathematik unk Physik***, 37 (1930).

a propositional function φ as value of the variable **f** so that the corresponding value of **L** shall be falsehood. For the value 0 of **b** we must find a corresponding value of **c** for which the value of **M**(**bc**) is falsehood, and we may suppose without loss of generality that this value of **c** is 1. The elementary parts of **M**(**bc**), other than f, are some or all of **f**(**b**, **b**), **f**(**b**, **c**), **f**(**c**, **b**), **f**(**c**, **c**), and by giving suitable truth-values to these we may, in 0 or more ways, give to **M**(**bc**) the value falsehood. Thus we determine the possibilities as to what the truth-values $\varphi(0,0)$, $\varphi(0,1)$, $\varphi(1,0)$, $\varphi(1,1)$ may be. Next, for the value 1 of **b** we must find a corresponding value of **c** for which the value of **M**(**bc**) is falsehood, and we suppose this value of **c** to be 2. And again in the same way, by considering the truth-values to be assigned to the elementary parts of **M**(**bc**) so as to give to **M**(**bc**) the value falsehood, we determine the possibilities as to what the truth-values $\varphi(1,1)$, $\varphi(1,2)$, $\varphi(2,1)$, $\varphi(2,2)$ may be. At this stage there are the two following alternatives: (a) It may happen that the two determinations of the truth-value $\varphi(1,1)$ cannot be reconciled with each other by using any of the possible assignments of truth-values to **f**(**b**,**b**), **f**(**b**,**c**), **f**(**c**,**b**), **f**(**c**,**c**) that give to **M**(**bc**) the value falsehood (either by using the same assignment of truth-values to **f**(**b**,**b**), **f**(**b**,**c**), **f**(**c**,**b**), **f**(**c**,**c**) both times or by using two different assignments); then **L** is valid in the domain of natural numbers, and moreover, by the metatheorem quoted above, **L** is a theorem.[14] (b) It may happen that the two determinations of $\varphi(1,1)$ can be reconciled with each other; then we may go on to find corresponding to the value 2 of **b** a value 3 of **c** for which the value of **M**(**bc**) is falsehood, and so on; since no further hindrance can be encountered, it follows that **L** is not valid in the domain of natural numbers and therefore is not a theorem. Consequently, for the present subcase of Case 2, the solution of the decision problem is given by $K = 2$.

If there are two binary functional variables instead of one, and the other conditions remain the same, then the same procedure may be followed as just described, except that there are eight elementary parts **f**(**b**,**b**), **g**(**b**,**b**), **f**(**b**,**c**), **g**(**b**,**c**), **f**(**c**,**b**), **g**(**c**,**b**), **f**(**c**,**c**), **g**(**c**,**c**) to which truth-values are to be assigned. It is convenient to think of pairing these elementary parts together in the way indicated — **f**(**b**,**b**) with **g**(**b**,**b**), **f**(**b**,**c**) with **g**(**b**,**c**), and so on. For the pair of elementary parts, **f**(**b**,**b**) and **g**(**b**,**b**), there are four available truth-value pairs, namely truth-truth, truth-falsehood, falsehood-truth, falsehood-falsehood. And in this way the solution of the decision problem is found to be $K = 4$. The procedure of successive assignments of truth-values to $\varphi(0,0)$, $\psi(0,0)$, $\varphi(0,1)$, etc., is summarized in Table 1, and the meaning of this table should now be clear from the foregoing explanations. (The same table may be used also for the case of a single binary functional variable by deleting the columns with **g** in the heading.)

The reader should now apply the same method to solve in general the subcase of Case 2 in which $l = 0$, $n = 1$, i.e., in which the prefix is $(E\mathbf{b})(\mathbf{c})$, showing in this case that $K = 2^N$, where N is the number of different functional variables appearing. As an illustration or exercise, the solution may be applied to Skolem's example $(Ex)(y) \mathbin{.} F(x,x) \supset F(y,y) \supset F(x,y)G(x) \supset G(y)$.

It is left to the reader further to supply the general solution of Case 2 by the same method. For example, if the prefix is $(E\mathbf{b})(\mathbf{c}_1)(\mathbf{c}_2)$ we find $K = 2^{2^N} - 1$ (the procedure

[14]According to the metatheorem of Löwenheim–Skolem, if a wff is valid in the domain of natural numbers, it is valid in every non-empty domain. However, no actual use is made here of this metatheorem, but rather only of certain methods which may also be used in a proof of the metatheorem.

of successive assignments of truth-values is summarized in Table 2 for the case that $N = 1$). For the prefix $(\mathbf{a})(E\mathbf{b})(\mathbf{c})$ we find $K = 2^W + 1$, where W is the sum of the weights of the different functional variables that appear, the *weight* of a functional variable **f** being the number of different elementary parts that contain **f** in the wff $\mathbf{M}(\mathbf{abb})$, with the exception of the elementary part $\mathbf{f}(\mathbf{a}, \mathbf{a}, \ldots, \mathbf{a})$, which is not to be counted (see Table 3, where $W = 3$ if all the different elementary parts shown in the table actually occur in **L**).

The same method may be used to solve Case 1, as is indicated in Table 4 for the subcase $l = 3$, $m = 2$. The table shows two singulary functional variables, but actually the number and kinds of the functional variables is irrelevant. After the rows of the table that correspond to those ordered m-tuples of natural numbers which contain natural numbers $< l$ only, the remaining rows of the table impose no further restrictions on the assignment of truth-values to elementary parts of the matrix. E.g., in Table 4, we may use for the tenth row of the table the same assignment of truth-values that was used for the first row (thus making $\varphi(3)$, $\psi(3)$ the same truth-values as $\varphi(0)$, $\psi(0)$), and for the eleventh row of the table the same assignment of truth-values as for the third row, and so on. Thus the solution of Case 1 is given by $K = l^m$; or if $l = 0$, then $K = 1$.

Extension of the same method to cases in which $m > 1$, $n > 0$ is not successful in general, but it may be used to solve the decision problem when the matrix satisfies one or another of various special conditions, as stated above for Cases 4–6.[15]

Table 5 is for the prefix $(E\mathbf{b}_1)(E\mathbf{b}_2)(\mathbf{c})$, and with the aid of this table we may solve the subcases of Cases 5 and 6 in which the prefix is of this form. In fact, for this subcase of Case 5, after deleting those columns of the table that correspond to elementary parts not containing **c**, we find that there are no two rows of the table which are *related* (i.e., which have at least one entry in common); therefore $K = 1$. And for this subcase of Case 6, since the matrix has the value falsehood for at most one assignment of truth-values to its elementary parts other than f, we need follow the table only far enough to exhaust every possible pattern of common entries between two rows; and even this may be reduced since it is not necessary to consider a pattern of common entries between two rows which is merely a part of a pattern already considered (as, e.g., the pattern of common entries between the second and fifth rows is a part of that between the first and second rows); therefore we find $K = 4$.

We may also use Table 5 to illustrate the solution of Case 4 taking the subcase in which the prefix is $(E\mathbf{b}_1)(E\mathbf{b}_2)(\mathbf{c})$, and deleting the first and fifth columns from the table. Then let two rows of the table be associated together as a pair if the value of $\mathbf{b}_1$ is in each row the same as the value of $\mathbf{b}_2$ in the other row (e.g., the eighth row is thus paired with the sixth row, and the ninth row is paired with itself). It will be seen that every such pair of rows is related to (has entries in common with) at most one earlier row in the table — and in fact that the situation is sufficiently analogous to that in Tables 1 and 2 so that it leads to a termination of the testing process. If, as shown in the table there is only a single binary functional variable appearing, then $K = 81$; or, instead of stating the solution in this way, we may obtain a shorter decision procedure by remarking that, after the first eleven rows in Table 5, only the thirteenth, fourteenth, sixteenth, eighteenth, twenty-second, twenty-fifth, thirty-sixth,

[15]We do not discuss here the more difficult Case 3.

forty-ninth, sixty-fourth, and eighty-first rows need be considered.

As an example, the reader may take:
$(Ex)(Ey)(z) \centerdot F(x, y) \equiv F(y, x) \equiv F(z, y) \supset \centerdot F(x, z) \equiv F(z, x) \supset \centerdot F(x, z) \equiv F(y, z) \supset \centerdot F(x, y) \supset F(y, x) \equiv \sim F(z, z) \supset \centerdot F(x, y) \vee F(y, x) \equiv F(x, z)$.

The general solution of Case 4 may now also be found by the same method.

Cases 5 and 6 may be solved by an analogous method. But because more general prefixes are allowed in these cases, it will be necessary to allow a more general form for the terms of the disjunction $\mathbf{M}_1 \vee \mathbf{M}_2 \vee \ldots \vee \mathbf{M}_k$. E.g., if $\mathbf{L}$ is $(E\mathbf{b})(\mathbf{c})(E\mathbf{d})\mathbf{M}(\mathbf{bcd})$, where $\mathbf{M}(\mathbf{bcd})$ is the matrix, then $\mathbf{M}_k$ is

$$\mathbf{M}(x_{[2k1]}x_{[2k1]+1}x_{[2k2]}),$$

as may be seen in Table 6. And if $\mathbf{L}$ is $(\mathbf{a})(E\mathbf{b})(\mathbf{c})(E\mathbf{d})(\mathbf{e})\mathbf{M}(\mathbf{abcde})$, where $\mathbf{M}(\mathbf{abcde})$ is the matrix, then $\mathbf{M}_k$ is

$$\mathbf{M}(x_0 x_{[2k1]} x_{([2k1]+1)^2} x_{[2k2]} x_{k+h}),$$

where h is the greatest integer such that $h^2 - h < k$.

Type II. For cases of type II, the principal method is to use the conjunctive normal form of the given wff $\mathbf{L}$, and the obvious metatheorem that $\mathbf{L}$ *is a theorem if and only if every term of the conjunctive normal form of* $\mathbf{L}$ *is a theorem.*

As each term $\mathbf{P}_1^i \supset \centerdot \mathbf{P}_2^i \supset \centerdot \ldots \mathbf{P}_\mu^i \supset f$ of the conjunctive normal form is reduced to prenex normal form, advantage is taken of available liberty as to the order in which steps are performed in order to obtain if possible a form for which the decision problem is solved. Evidently it will be desirable, in general, first to bring out as many universal quantifiers at the beginning of the prefix as possible, then to bring out the existential quantifiers into the prefix in such a way, if possible, as to avoid separating existential quantifiers by universal quantifiers, and then to bring the remaining universal quantifiers into the prefix. The form of the terms of the conjunctive normal form is well adapted to control of this process, since each wff $\mathbf{P}_j^i$ is in either prenex normal form or the negation of prenex normal form; in the former case the quantifiers in $\mathbf{P}_j^i$ (universal and existential) are changed into quantifiers of the opposite kind (existential and universal) when brought into the prefix, while in the latter case the quantifiers in $\mathbf{P}_j^i$ remain of the same kind when brought into the prefix.

Of the most general case in which the decision problem can be solved in this way, no simpler characterization can be given than the instructions to reduce each term of the conjunctive normal form to prenex normal form in the fashion just described, then, wherever free individual variables appear, to bind them by universal quantifiers added at the beginning of the prefix, then finally to examine each of the resulting wffs separately to determine whether it falls under one of the cases 1–7 of type I. However, the four following special cases are, in particular, always solvable in this way, and seem to be worthy of special mention because of the relative simplicity of their characterization.

CASE 8. In the prefix of each P-constituent, the quantifiers are all of the same kind.

CASE 9. The P-constituents have prefixes of the following forms only: null, $(\mathbf{a})$, $(E\mathbf{b})$, $(\mathbf{a}_1)(\mathbf{a}_2)$, $(\mathbf{a})(E\mathbf{b})$, $(E\mathbf{b}_1)(E\mathbf{b}_2)$, $(E\mathbf{b})(\mathbf{c})$, $(\mathbf{a}_1)(\mathbf{a}_2)(E\mathbf{b})$, $(\mathbf{a})(E\mathbf{b}_1)(E\mathbf{b}_2)$, $(E\mathbf{b}_1)(E\mathbf{b}_2)(\mathbf{c})$, $(E\mathbf{b})(\mathbf{c}_1)(\mathbf{c}_2)$, $(\mathbf{a}_1)(\mathbf{a}_2)(E\mathbf{b}_1)(E\mathbf{b}_2)$, $(E\mathbf{b}_1)(E\mathbf{b}_2)(\mathbf{c}_1)(\mathbf{c}_2)$.

CASE 10. Each P-constituent $\mathbf{P}_i$ has a prefix of one of the forms $(\mathbf{a}_1)(\mathbf{a}_2)\ldots(\mathbf{a}_{l_i})(E\mathbf{b}_1)(E\mathbf{b}_2)\ldots(E\mathbf{b}_m)$ or $(E\mathbf{a}_1)(E\mathbf{a}_2)\ldots(E\mathbf{a}_{l_i})(\mathbf{b}_1)(\mathbf{b}_2)\ldots(\mathbf{b}_m)$ and a matrix in which every elementary part containing individual variables contains all of the variables $\mathbf{b}_1, \mathbf{b}_2, \ldots, \mathbf{b}_m$ — where m is fixed and $0 \leqq l_i \leqq m$.

CASE 11. The matrix of every P-constituent is an elementary part or the negation of an elementary part.

Evidently the chance of success in solving the decision problem by the process under consideration will be increased, and the length of the process itself will be decreased if, before starting, we replace equivalent P-constituents as far as possible by identical P-constituents. In particular, this should always be done in the following cases, in which the equivalence of two P-constituents is effectively recognizable: (1) two P-constituents differ only by alphabetic changes of bound variable; (2) the prefixes of two P-constituents can be rendered identical by alphabetic changes of bound variables, and after this is done it is found that the equivalence whose sides are the matrices of the two P-constituents is tautologous.

It may also happen that one P-constituent is equivalent to the negation of another, and in this case again advantage should be taken of the equivalence, whenever it is effectively recognizable, in order to reduce the number of distinct P-constituents. To this end, each P-constituent should be compared with the prenex normal form of the negation of each other P-constituent; if the two prefixes can be rendered identical by alphabetic changes of bound variables, this should be done, and the equivalence whose sides are the two matrices should then be tested to determine whether it is tautologous.

More important than the foregoing remarks, however, is the following. Roughly speaking, the finer is the division of **L** which consists in dividing **L** into its P-constituents, the greater is the chance of success in solving the decision problem for **L** by the process under consideration. Therefore it may often be desirable, before starting the process, first to apply to **L** the inverse of reduction to prenex normal form, so putting **L** into the form $\mathbf{L}_*$.

This leads to solution of the decision problem for a certain additional class of wffs **L**, which is not easily characterized in a simple manner, but which includes the following:

CASE 12. In each elementary part of **L**, at most one variable has occurrences which are bound occurrences in **L**.

Namely if the given wff **L** falls under Case 12, then $\mathbf{L}_*$ necessarily falls under Case 8.

Finally we mention the following subcases of Case 8 and Case 12, because in these subcases certain simplifications of the decision procedure turn out to be possible, as compared to the general solutions of Case 8 and Case 12:

CASE 8a. There are no free individual variables, and every constituent either is a propositional variable or has a prefix consisting of a single quantifier and a matrix containing no propositional variables.

CASE 12a.[16] The only functional variables occurring are singulary functional variables.

To solve Case 8a, we first alter the given wff **L** by alphabetic changes of bound

[16]This case of the decision problem was originally solved by Leopold Löwenheim. However, the solution given here is based on that of Heinrich BEHMANN, *Beiträge zur Algebra der Logik*, ***Mathematische Annalen***, 86 (1922) and that of W. V. QUINE, *On the Logic of Quantification*, ***The Journal of Symbolic Logic***, 10 (1945).

variables in such a way that the only individual variable occurring is the variable x, and then we replace (Ex), wherever it occurs, by $\sim (x) \sim$. The resulting wff $\mathbf{L}'$ is equivalent to $\mathbf{L}$. Let the P-constituents of $\mathbf{L}'$ be numbered in such an order that $\mathbf{P}_1, \mathbf{P}_2, \ldots, \mathbf{P}_\lambda$ are propositional variables, and $\mathbf{P}_{\lambda+1}, \mathbf{P}_{\lambda+2}, \ldots, \mathbf{P}_\mu$ are respectively $(x)\mathbf{M}_{\lambda+1}, (x)\mathbf{M}_{\lambda+2}, \ldots, (x)\mathbf{M}_\mu$. From a particular falsifying system of truth-values $\tau_1^1, \tau_2^i, \ldots, \tau_\mu^i$ of $\mathbf{P}_1, \mathbf{P}_2, \ldots, \mathbf{P}_\mu$ select the last $\mu - \lambda$ truth-values $\tau_{\lambda+1}^i, \tau_{\lambda+2}^i, \ldots, \tau_\mu^i$ and suppose that, among these, $\tau_{t_1}^i, \tau_{t_2}^i, \ldots, \tau_{t_\chi}^i$ are truth and $\tau_{f_1}^i, \tau_{f_2}^i, \ldots, \tau_{f_{\mu-\chi-\lambda}}^i$ are falsehood. Then in order that $\mathbf{P}_1^i \supset . \mathbf{P}_2^i \supset . \ldots \mathbf{P}_\mu^i \supset f$ shall be a theorem it is necessary and sufficient that at least one of the wffs $\mathbf{M}_{t_1} \supset . \mathbf{M}_{t_2} \supset . \ldots \mathbf{M}_{t_\chi} \supset \mathbf{M}_{f_k}$ shall be tautologous $(k = 1, 2, \ldots, \mu - \chi - \lambda)$.

To solve Case 12a, we apply to the given wff $\mathbf{L}$ the inverse of reduction to prenex normal form. The resulting wff $\mathbf{L}_*$ necessarily falls under Case 8a.

b	c	f(b, b)	g(b, b)	f(b, c)	g(b, c)	f(c, b)	g(c, b)	f(c, c)	g(c, c)
0	1	$\varphi(0,0)$	$\psi(0,0)$	$\varphi(0,1)$	$\psi(0,1)$	$\varphi(1,0)$	$\psi(1,0)$	$\varphi(1,1)$	$\psi(1,1)$
1	2	$\varphi(1,1)$	$\psi(1,1)$	$\varphi(1,2)$	$\psi(1,2)$	$\varphi(2,1)$	$\psi(2,1)$	$\varphi(2,2)$	$\psi(2,2)$
2	3	$\varphi(2,2)$	$\psi(2,2)$	$\varphi(2,3)$	$\psi(2,3)$	$\varphi(3,2)$	$\psi(3,2)$	$\varphi(3,3)$	$\psi(3,3)$
3	4	$\varphi(3,3)$	$\psi(3,3)$	$\varphi(3,4)$	$\psi(3,4)$	$\varphi(4,3)$	$\psi(4,3)$	$\varphi(4,4)$	$\psi(4,4)$
4	5	$\varphi(4,4)$	$\psi(4,4)$	$\varphi(4,5)$	$\psi(4,5)$	$\varphi(5,4)$	$\psi(5,4)$	$\varphi(5,5)$	$\psi(5,5)$
...		...	...	...	...	...	...	...	...

TABLE 1. — Prefix $(E\mathbf{b})(\mathbf{c})$, two binary functional variables.

a	c_1	c_2	f(b, b)	f(b, c_1)	f(b, c_2)	f(c_1, b)	f(c_1, c_1)	f(c_1, c_2)	f(c_2, b)	f(c_2, c_1)	f(c_2, c_2)
0	1	2	$\varphi(0,0)$	$\varphi(0,1)$	$\varphi(0,2)$	$\varphi(1,0)$	$\varphi(1,1)$	$\varphi(1,2)$	$\varphi(2,0)$	$\varphi(2,1)$	$\varphi(2,2)$
1	3	4	$\varphi(1,1)$	$\varphi(1,3)$	$\varphi(1,4)$	$\varphi(3,1)$	$\varphi(3,3)$	$\varphi(3,4)$	$\varphi(4,1)$	$\varphi(4,3)$	$\varphi(4,4)$
2	5	6	$\varphi(2,2)$	$\varphi(2,5)$	$\varphi(2,6)$	$\varphi(5,2)$	$\varphi(5,5)$	$\varphi(5,6)$	$\varphi(6,2)$	$\varphi(6,5)$	$\varphi(6,6)$
3	7	8	$\varphi(3,3)$	$\varphi(3,7)$	$\varphi(3,8)$	$\varphi(7,3)$	$\varphi(7,7)$	$\varphi(7,8)$	$\varphi(8,3)$	$\varphi(8,7)$	$\varphi(8,8)$
4	9	10	$\varphi(4,4)$	$\varphi(4,9)$	$\varphi(4,10)$	$\varphi(9,4)$	$\varphi(9,9)$	$\varphi(9,10)$	$\varphi(10,4)$	$\varphi(10,9)$	$\varphi(10,10)$
...			...	...	...	...	...	...	...	...	...

TABLE 2. — Prefix $(E\mathbf{b})(\mathbf{c}_1)(\mathbf{c}_2)$, one binary functional variable.

a	b	c	$\mathbf{f}(\mathbf{a},\mathbf{a})$	$\mathbf{f}(\mathbf{a},\mathbf{b})$	$\mathbf{f}(\mathbf{a},\mathbf{c})$	$\mathbf{f}(\mathbf{b},\mathbf{a})$	$\mathbf{f}(\mathbf{b},\mathbf{b})$	$\mathbf{f}(\mathbf{b},\mathbf{c})$	$\mathbf{f}(\mathbf{c},\mathbf{a})$	$\mathbf{f}(\mathbf{c},\mathbf{b})$	$\mathbf{f}(\mathbf{c},\mathbf{c})$
0	0	1	$\varphi(0,0)$	$\varphi(0,0)$	$\varphi(0,1)$	$\varphi(0,0)$	$\varphi(0,0)$	$\varphi(0,1)$	$\varphi(1,0)$	$\varphi(1,0)$	$\varphi(1,1)$
0	1	2	$\varphi(0,0)$	$\varphi(0,1)$	$\varphi(0,2)$	$\varphi(1,0)$	$\varphi(1,1)$	$\varphi(1,2)$	$\varphi(2,0)$	$\varphi(2,1)$	$\varphi(2,2)$
0	2	3	$\varphi(0,0)$	$\varphi(0,2)$	$\varphi(0,3)$	$\varphi(2,0)$	$\varphi(2,2)$	$\varphi(2,3)$	$\varphi(3,0)$	$\varphi(3,2)$	$\varphi(3,3)$
0	3	4	$\varphi(0,0)$	$\varphi(0,3)$	$\varphi(0,4)$	$\varphi(3,0)$	$\varphi(3,3)$	$\varphi(3,4)$	$\varphi(4,0)$	$\varphi(4,3)$	$\varphi(4,4)$
0	4	5	$\varphi(0,0)$	$\varphi(0,4)$	$\varphi(0,5)$	$\varphi(4,0)$	$\varphi(4,4)$	$\varphi(4,5)$	$\varphi(5,0)$	$\varphi(5,4)$	$\varphi(5,5)$
0	5	6	$\varphi(0,0)$	$\varphi(0,5)$	$\varphi(0,6)$	$\varphi(5,0)$	$\varphi(5,5)$	$\varphi(5,6)$	$\varphi(6,0)$	$\varphi(6,5)$	$\varphi(6,6)$
0	6	7	$\varphi(0,0)$	$\varphi(0,6)$	$\varphi(0,7)$	$\varphi(6,0)$	$\varphi(6,6)$	$\varphi(6,7)$	$\varphi(7,0)$	$\varphi(7,6)$	$\varphi(7,7)$
0	7	8	$\varphi(0,0)$	$\varphi(0,7)$	$\varphi(0,8)$	$\varphi(7,0)$	$\varphi(7,7)$	$\varphi(7,8)$	$\varphi(8,0)$	$\varphi(8,7)$	$\varphi(8,8)$
0	8	9	$\varphi(0,0)$	$\varphi(0,8)$	$\varphi(0,9)$	$\varphi(8,0)$	$\varphi(8,8)$	$\varphi(8,9)$	$\varphi(9,0)$	$\varphi(9,8)$	$\varphi(9,9)$
	...		...	...	...	...	...	...	...	...	...

TABLE 3. — Prefix $(\mathbf{a})(E\mathbf{b})(\mathbf{c})$, one binary functional variable.

$\mathbf{a}_1$	$\mathbf{a}_2$	$\mathbf{a}_3$	$\mathbf{b}_1$	$\mathbf{b}_2$	$\mathbf{f}(\mathbf{a}_1)$	$\mathbf{g}(\mathbf{a}_1)$	$\mathbf{f}(\mathbf{a}_2)$	$\mathbf{g}(\mathbf{a}_2)$	$\mathbf{f}(\mathbf{a}_3)$	$\mathbf{g}(\mathbf{a}_3)$	$\mathbf{f}(\mathbf{b}_1)$	$\mathbf{g}(\mathbf{b}_1)$	$\mathbf{f}(\mathbf{b}_2)$	$\mathbf{g}(\mathbf{b}_2)$
0	1	2	0	0	$\varphi(0)$	$\psi(0)$	$\varphi(1)$	$\psi(1)$	$\varphi(2)$	$\psi(2)$	$\varphi(0)$	$\psi(0)$	$\varphi(0)$	$\psi(0)$
0	1	2	0	1	$\varphi(0)$	$\psi(0)$	$\varphi(1)$	$\psi(1)$	$\varphi(2)$	$\psi(2)$	$\varphi(0)$	$\psi(0)$	$\varphi(1)$	$\psi(1)$
0	1	2	1	0	$\varphi(0)$	$\psi(0)$	$\varphi(1)$	$\psi(1)$	$\varphi(2)$	$\psi(2)$	$\varphi(1)$	$\psi(1)$	$\varphi(0)$	$\psi(0)$
0	1	2	1	1	$\varphi(0)$	$\psi(0)$	$\varphi(1)$	$\psi(1)$	$\varphi(2)$	$\psi(2)$	$\varphi(1)$	$\psi(1)$	$\varphi(1)$	$\psi(1)$
0	1	2	0	2	$\varphi(0)$	$\psi(0)$	$\varphi(1)$	$\psi(1)$	$\varphi(2)$	$\psi(2)$	$\varphi(0)$	$\psi(0)$	$\varphi(2)$	$\psi(2)$
0	1	2	1	2	$\varphi(0)$	$\psi(0)$	$\varphi(1)$	$\psi(1)$	$\varphi(2)$	$\psi(2)$	$\varphi(1)$	$\psi(1)$	$\varphi(2)$	$\psi(2)$
0	1	2	2	0	$\varphi(0)$	$\psi(0)$	$\varphi(1)$	$\psi(1)$	$\varphi(2)$	$\psi(2)$	$\varphi(2)$	$\psi(2)$	$\varphi(0)$	$\psi(0)$
0	1	2	2	1	$\varphi(0)$	$\psi(0)$	$\varphi(1)$	$\psi(1)$	$\varphi(2)$	$\psi(2)$	$\varphi(2)$	$\psi(2)$	$\varphi(1)$	$\psi(1)$
0	1	2	2	2	$\varphi(0)$	$\psi(0)$	$\varphi(1)$	$\psi(1)$	$\varphi(2)$	$\psi(2)$	$\varphi(2)$	$\psi(2)$	$\varphi(2)$	$\psi(2)$
0	1	2	0	3	$\varphi(0)$	$\psi(0)$	$\varphi(1)$	$\psi(1)$	$\varphi(2)$	$\psi(2)$	$\varphi(0)$	$\psi(0)$	$\varphi(3)$	$\psi(3)$
0	1	2	1	3	$\varphi(0)$	$\psi(0)$	$\varphi(1)$	$\psi(1)$	$\varphi(2)$	$\psi(2)$	$\varphi(1)$	$\psi(1)$	$\varphi(3)$	$\psi(3)$
		...			...	...	...	...	...	...	...	...	...	...

TABLE 4. — Prefix $(\mathbf{a}_1)(\mathbf{a}_2)(\mathbf{a}_3)(E\mathbf{b}_1)(E\mathbf{b}_2)$, two singulary functional variables.

$\mathbf{b}_1$	$\mathbf{b}_2$	c	$\mathbf{f}(\mathbf{b}_1,\mathbf{b}_1)$	$\mathbf{f}(\mathbf{b}_1,\mathbf{b}_2)$	$\mathbf{f}(\mathbf{b}_1,\mathbf{c})$	$\mathbf{f}(\mathbf{b}_2,\mathbf{b}_1)$	$\mathbf{f}(\mathbf{b}_2,b_2)$	$\mathbf{f}(\mathbf{b}_2,\mathbf{c})$	$\mathbf{f}(\mathbf{c},\mathbf{b}_1)$	$\mathbf{f}(\mathbf{c},\mathbf{b}_2)$	$\mathbf{f}(\mathbf{c},\mathbf{c})$
0	0	1	$\varphi(0,0)$	$\varphi(0,0)$	$\varphi(0,1)$	$\varphi(0,0)$	$\varphi(0,0)$	$\varphi(0,1)$	$\varphi(1,0)$	$\varphi(1,0)$	$\varphi(1,1)$
0	1	2	$\varphi(0,0)$	$\varphi(0,1)$	$\varphi(0,2)$	$\varphi(1,0)$	$\varphi(1,1)$	$\varphi(1,2)$	$\varphi(2,0)$	$\varphi(2,1)$	$\varphi(2,2)$
1	0	3	$\varphi(1,1)$	$\varphi(1,0)$	$\varphi(1,3)$	$\varphi(0,1)$	$\varphi(0,0)$	$\varphi(0,3)$	$\psi(3,1)$	$\varphi(3,0)$	$\varphi(3,3)$
1	1	4	$\varphi(1,1)$	$\varphi(1,1)$	$\varphi(1,4)$	$\varphi(1,1)$	$\varphi(1,1)$	$\varphi(1,4)$	$\varphi(4,1)$	$\varphi(4,1)$	$\varphi(4,4)$
0	2	5	$\varphi(0,0)$	$\varphi(0,2)$	$\varphi(0,5)$	$\varphi(2,0)$	$\varphi(2,2)$	$\varphi(2,5)$	$\varphi(5,0)$	$\varphi(5,2)$	$\varphi(5,5)$
1	2	6	$\varphi(1,1)$	$\varphi(1,2)$	$\varphi(1,6)$	$\varphi(2,1)$	$\varphi(2,2)$	$\varphi(2,6)$	$\varphi(6,1)$	$\varphi(6,2)$	$\varphi(6,6)$
2	0	7	$\varphi(2,2)$	$\varphi(2,0)$	$\varphi(2,7)$	$\varphi(0,2)$	$\varphi(0,0)$	$\varphi(0,7)$	$\varphi(7,2)$	$\varphi(7,0)$	$\varphi(7,7)$
2	1	8	$\varphi(2,2)$	$\varphi(2,1)$	$\varphi(2,8)$	$\varphi(1,2)$	$\varphi(1,1)$	$\varphi(1,8)$	$\varphi(8,2)$	$\varphi(8,1)$	$\varphi(8,8)$
2	2	9	$\varphi(2,2)$	$\varphi(2,2)$	$\varphi(2,9)$	$\varphi(2,2)$	$\varphi(2,2)$	$\varphi(2,9)$	$\varphi(9,2)$	$\varphi(9,2)$	$\varphi(9,9)$
0	3	10	$\varphi(0,0)$	$\varphi(0,3)$	$\varphi(0,10)$	$\varphi(3,0)$	$\varphi(3,3)$	$\varphi(3,10)$	$\varphi(10,0)$	$\varphi(10,3)$	$\varphi(10,10)$
1	3	11	$\varphi(1,1)$	$\varphi(1,3)$	$\varphi(1,11)$	$\varphi(3,1)$	$\varphi(3,3)$	$\varphi(3,11)$	$\varphi(11,1)$	$\varphi(11,3)$	$\varphi(11,11)$
2	3	12	$\varphi(2,2)$	$\varphi(2,3)$	$\varphi(2,12)$	$\varphi(3,2)$	$\varphi(3,3)$	$\varphi(3,12)$	$\varphi(12,2)$	$\varphi(12,3)$	$\varphi(12,12)$
3	0	13	$\varphi(3,3)$	$\varphi(3,0)$	$\varphi(3,13)$	$\varphi(0,3)$	$\varphi(0,0)$	$\varphi(0,13)$	$\varphi(13,3)$	$\varphi(13,0)$	$\varphi(13,13)$
3	1	14	$\varphi(3,3)$	$\varphi(3,1)$	$\varphi(3,14)$	$\varphi(1,3)$	$\varphi(1,1)$	$\varphi(1,14)$	$\psi(14,3)$	$\varphi(14,1)$	$\varphi(14,14)$
3	2	15	$\varphi(3,3)$	$\varphi(3,2)$	$\varphi(3,15)$	$\varphi(2,3)$	$\varphi(2,2)$	$\varphi(2,15)$	$\varphi(15,3)$	$\varphi(15,2)$	$\varphi(15,15)$
3	3	16	$\varphi(3,3)$	$\varphi(3,3)$	$\varphi(3,16)$	$\varphi(3,3)$	$\varphi(3,3)$	$\varphi(3,16)$	$\varphi(16,3)$	$\varphi(16,3)$	$\varphi(16,16)$
0	4	17	$\varphi(0,0)$	$\varphi(0,4)$	$\varphi(0,17)$	$\varphi(4,0)$	$\varphi(4,4)$	$\varphi(4,17)$	$\varphi(17,0)$	$\varphi(17,4)$	$\varphi(17,17)$
1	4	18	$\varphi(1,1)$	$\varphi(1,4)$	$\varphi(1,18)$	$\varphi(4,1)$	$\varphi(4,4)$	$\varphi(4,18)$	$\varphi(18,1)$	$\varphi(18,4)$	$\varphi(18,18)$
2	4	19	$\varphi(2,2)$	$\varphi(2,4)$	$\varphi(2,19)$	$\varphi(4,2)$	$\varphi(4,4)$	$\varphi(4,19)$	$\varphi(19,2)$	$\varphi(19,4)$	$\varphi(19,19)$
3	4	20	$\varphi(3,3)$	$\varphi(3,4)$	$\varphi(3,20)$	$\varphi(4,3)$	$\varphi(4,4)$	$\varphi(4,20)$	$\varphi(20,3)$	$\varphi(20,4)$	$\varphi(20,20)$
4	0	21	$\varphi(4,4)$	$\varphi(4,0)$	$\varphi(4,21)$	$\varphi(0,4)$	$\varphi(0,0)$	$\varphi(0,21)$	$\varphi(21,4)$	$\varphi(21,0)$	$\varphi(21,21)$
4	1	22	$\varphi(4,4)$	$\varphi(4,1)$	$\varphi(4,22)$	$\varphi(1,4)$	$\varphi(1,1)$	$\varphi(1,22)$	$\varphi(22,4)$	$\varphi(22,1)$	$\varphi(22,22)$
4	2	23	$\varphi(4,4)$	$\varphi(4,2)$	$\varphi(4,23)$	$\varphi(2,4)$	$\varphi(2,2)$	$\varphi(2,23)$	$\varphi(23,4)$	$\varphi(23,2)$	$\varphi(23,23)$
4	3	24	$\varphi(4,4)$	$\varphi(4,3)$	$\varphi(4,24)$	$\varphi(3,4)$	$\varphi(3,3)$	$\varphi(3,24)$	$\varphi(24,4)$	$\varphi(24,3)$	$\varphi(24,24)$
4	4	25	$\varphi(4,4)$	$\varphi(4,4)$	$\varphi(4,25)$	$\varphi(4,4)$	$\varphi(4,4)$	$\varphi(4,25)$	$\varphi(25,4)$	$\varphi(25,4)$	$\varphi(25,25)$
	...		...	...	...	...	...	...	...	...	...

TABLE 5. — Prefix $(E\mathbf{b}_1)(E\mathbf{b}_2)(\mathbf{c})$, one binary functional variable.

b	**c**	**d**	**f**(**b**, **b**)	**f**(**b**, **c**)	**f**(**b**, **d**)	**f**(**c**, **b**)	**f**(**c**, **c**)	**f**(**c**, **d**)	**f**(**d**, **b**)	**f**(**d**, **c**)	**f**(**d**, **d**)
0	1	0	$\varphi(0,0)$	$\varphi(0,1)$	$\varphi(0,0)$	$\varphi(1,0)$	$\varphi(1,1)$	$\varphi(1,0)$	$\varphi(0,0)$	$\varphi(0,1)$	$\varphi(0,0)$
0	1	1	$\varphi(0,0)$	$\varphi(0,1)$	$\varphi(0,1)$	$\varphi(1,0)$	$\varphi(1,1)$	$\varphi(1,1)$	$\varphi(1,0)$	$\varphi(1,1)$	$\varphi(1,1)$
1	2	0	$\varphi(1,1)$	$\varphi(1,2)$	$\varphi(1,0)$	$\varphi(2,1)$	$\varphi(2,2)$	$\varphi(2,0)$	$\varphi(0,1)$	$\varphi(0,2)$	$\varphi(0,0)$
1	2	1	$\varphi(1,1)$	$\varphi(1,2)$	$\varphi(1,1)$	$\varphi(2,1)$	$\varphi(2,2)$	$\varphi(2,1)$	$\varphi(1,1)$	$\varphi(1,2)$	$\varphi(1,1)$
0	1	2	$\varphi(0,0)$	$\varphi(0,1)$	$\varphi(0,2)$	$\varphi(1,0)$	$\varphi(1,1)$	$\varphi(1,2)$	$\varphi(2,0)$	$\varphi(2,1)$	$\varphi(2,2)$
1	2	2	$\varphi(1,1)$	$\varphi(1,2)$	$\varphi(1,2)$	$\varphi(2,1)$	$\varphi(2,2)$	$\varphi(2,2)$	$\varphi(2,1)$	$\varphi(2,2)$	$\varphi(2,2)$
2	3	0	$\varphi(2,2)$	$\varphi(2,3)$	$\varphi(2,0)$	$\varphi(3,2)$	$\varphi(3,3)$	$\varphi(3,0)$	$\varphi(0,2)$	$\varphi(0,3)$	$\varphi(0,0)$
2	3	1	$\varphi(2,2)$	$\varphi(2,3)$	$\varphi(2,1)$	$\varphi(3,2)$	$\varphi(3,3)$	$\varphi(3,1)$	$\varphi(1,2)$	$\varphi(1,3)$	$\varphi(1,1)$
2	3	2	$\varphi(2,2)$	$\varphi(2,3)$	$\varphi(2,2)$	$\varphi(3,2)$	$\varphi(3,3)$	$\varphi(3,2)$	$\varphi(2,2)$	$\varphi(2,3)$	$\varphi(2,2)$
0	1	3	$\varphi(0,0)$	$\varphi(0,1)$	$\varphi(0,3)$	$\varphi(1,0)$	$\varphi(1,1)$	$\varphi(1,3)$	$\varphi(3,0)$	$\varphi(3,1)$	$\varphi(3,3)$
1	2	3	$\varphi(1,1)$	$\varphi(1,2)$	$\varphi(1,3)$	$\varphi(2,1)$	$\varphi(2,2)$	$\varphi(2,3)$	$\varphi(3,1)$	$\varphi(3,2)$	$\varphi(3,3)$
2	3	3	$\varphi(2,2)$	$\varphi(2,3)$	$\varphi(2,3)$	$\varphi(3,2)$	$\varphi(3,3)$	$\varphi(3,3)$	$\varphi(3,2)$	$\varphi(3,3)$	$\varphi(3,3)$
3	4	0	$\varphi(3,3)$	$\varphi(3,4)$	$\varphi(3,0)$	$\varphi(4,3)$	$\varphi(4,4)$	$\varphi(4,0)$	$\varphi(0,3)$	$\varphi(0,4)$	$\varphi(0,0)$
. . .			. . .	. . .	. . .	. . .	. . .	. . .	. . .	. . .	. . .

TABLE 6. — Prefix $(E\mathbf{b})(\mathbf{c})(E\mathbf{d})$, one binary functional variable.

Alonzo CHURCH.

Princeton University.

SPECIAL CASES OF THE DECISION PROBLEM

A CORRECTION

(1952)

In my paper of the above title, which appeared in this *Revue*, volume 49 (1951), pages 203–221, there is an error regarding Case 7 which is called to my attention by a review of the paper by Wilhelm Ackermann in *The Journal of Symbolic Logic*, volume 17 (1952), pages 73–74. Namely Case 7 has not been solved in its full generality, but only the following subcase of Case 7:

CASE 7a. Prefix $(E\mathbf{b}_1)(E\mathbf{b}_2)\ldots(E\mathbf{b}_m)$, where $m \leqq 4$, and a matrix of the form $\mathbf{N} \supset \mathbf{f}(\mathbf{b}_1, \mathbf{c})$, where $\mathbf{N}$ does not contain the variable $\mathbf{c}$ and does not contain any functional variable other than $\mathbf{f}$.

Ackermann's paper in volume 112 of the *Mathematische Annalen* contains a solution of Case 7a (not of Case 7). The paper of Gégalkine in volume 6 of the *Recueil Mathématique* has a simplification of Ackermann's decision procedure, but does not extend the solution beyond Case 7a.

The second sentence of my footnote 12 remains correct, as applied to Case 7a instead of Case 7. But the last sentence of this footnote requires a modification which may best be explained by stating explicitly the reduction to Skolem normal form that is used.

For a wff of the form

$$(E\mathbf{b}_1)(\mathbf{b}_2)(E\mathbf{b}_3)\ldots(E\mathbf{b}_m)\mathbf{M}, \tag{7.1}$$

where there are no free individual variables, and where $\mathbf{M}$ is the matrix, the Skolem normal form is

$$(E\mathbf{b}_1)(E\mathbf{b}_2)(E\mathbf{b}_3)\ldots(E\mathbf{b}_m)(\mathbf{c}) \,.\, \mathbf{M} \supset \mathbf{f}(\mathbf{b}_1, \mathbf{b}_2) \supset \mathbf{f}(\mathbf{b}_1, \mathbf{c}),$$

where $\mathbf{f}$ is a binary functional variable not occurring in $\mathbf{M}$. The Skolem normal form is not, in general, equivalent to the original wff from which it was obtained. But, as is well known, the Skolem normal form is a theorem if and only if the original wff is a theorem — so that the decision problem for the original wff is reduced to the decision problem for its Skolem normal form.[17]

Thus — as appears correctly in Ackermann's paper — the decision problem for wffs of the form (7.1) is reduced to Case 7 if $m \leqq 4$, but not to Case 7a. It is not known whether or not the solution of the decision problem is possible in Case 7.

But even more than Case 7, it seems worth while to call attention to Case 7.1, i.e., the case of wffs of the form (7.1), as a case in which the decision problem is still open. Even in the subcase $m = 3$ of Case 7.1, the case of wffs with prefix $(E\mathbf{b})(\mathbf{c})(E\mathbf{d})$ and with no free individual variables, neither a decision procedure is known nor a proof that the decision procedure does not exist.

I take the opportunity also to explain that the "more general form for the terms of the disjunction $\mathbf{M}_1 \vee \mathbf{M}_2 \vee \ldots \vee \mathbf{M}_k$" which is introduced on page 215, and the method

---

Originally published in ***Revue Philosophique de Louvain***, vol. 50 (1952), pp. 270–272. The footnote numbering here continues the numbering of the original article.

[17]The result is due to Skolem. A convenient source from which to obtain a statement of the reduction to Skolem normal form of an arbitrary wff, and a proof of Skolem's result, is in the writer's ***Introduction to Mathematical Logic*** (see §42 of the forthcoming revised edition of Volume I).

of attack on the decision problem which is associated with it, are (though differently formulated) the same in substance as a method which was used by Herbrand in his dissertation at the University of Paris[18] and in his paper of 1931 which is referred to in footnote 8. The germ of this method, which is used by Gödel in his paper of 1930 and by Herbrand in his dissertation, is to be found already in Skolem's paper of 1928, referred to in footnote 8.

In particular, therefore, the method of solution of Case 6 which is explained on page 214 for the subcase characterized by the prefix $(E\mathbf{b}_1)(E\mathbf{b}_2)(\mathbf{c})$, and which the reader is expected to extend by analogy to other subcases,[19] is a reformulation of Herbrand's method of solving Case 6.

Finally the following typographical errors require correction:[20]

Alonzo CHURCH.

Princeton University.

---

[18] ***Recherches sur la Théorie de la Démonstration***, published in Warsaw in 1930.

[19] E.g., for wffs with prefix $(E\mathbf{b})(\mathbf{c})(E\mathbf{d})$, which fall under Case 6 the solution of the decision problem is given by $K = 6$, as may be seen in Table 6 (or in tables obtained from Table 6 by adding columns corresponding to other functional variables). As solution of the general Case 6, an effective procedure may be stated by which to find the value of K when the particular prefix is given; but this is not easily reduced to the form of an equation for K.

[20] *Editor's note*: This list has been omitted, the errors having been corrected in this volume.

SOME THEOREMS ON DEFINABILITY AND DECIDABILITY

(1952)

Alonzo Church and W. V. Quine

1. Reducibility of numerical relations to symmetric ones. In this paper a theorem about numerical relations will be established and shown to have certain consequences concerning decidability in quantification theory, as well as concerning the foundation of number theory. The theorem is that relations of natural numbers are reducible in elementary fashion to symmetric ones; i.e.:

THEOREM I. *For every dyadic relation R of natural numbers there is a symmetric dyadic relation H of natural numbers such that R is definable in terms of H plus just truth-functions and quantification over natural numbers.*

To state the matter more fully, there is a (well-formed) formula ϕ of pure quantification theory, or first-order functional calculus, which meets these conditions:

(a) ϕ has 'x' and 'y' as sole free individual variables;

(b) ϕ contains just one predicate letter, and it is dyadic;

(c) for every dyadic relation R of natural numbers there is a symmetric dyadic relation H of natural numbers such that, when the predicate letter in ϕ is interpreted as expressing H, ϕ comes to agree in truth-value with 'x bears R to y' for all values of 'x' and 'y'.

For each choice of R, one symmetric relation H which proves to be suited to the above purpose is the relation which relates, backwards and forwards, all and only the following natural numbers, for all choices of z: z with $z + 1$, $z + 2$, and $6z$; $6z$ with itself; and $6z + 1$ with $6w + 1$ for all w and with $6u + 4$ for all u to which z bears R. In other words, writing '$\cong$' to mean 'congruent modulo 6,' we may describe H as the relation of any natural number x to any natural number y such that

$$
\begin{array}{ll}
(1) \quad x = y \cong 0, \text{ or} & (2) \quad x \cong y \cong 1, \text{ or} \\
(3) \quad y = 6x, \text{ or} & (4) \quad x = 6y, \text{ or} \\
(5) \quad y = x + 1, \text{ or} & (6) \quad x = y + 1, \text{ or} \\
(7) \quad y = x + 2, \text{ or} & (8) \quad x = y + 2, \text{ or} \\
(9) \quad x \cong 1, y \cong 4, \text{ and } \tfrac{1}{6}(x - 1) \text{ bears } R \text{ to } \tfrac{1}{6}(y - 4), \text{ or} \\
(10) \quad y \cong 1, x \cong 4, \text{ and } \tfrac{1}{6}(y - 1) \text{ bears } R \text{ to } \tfrac{1}{6}(x - 4).
\end{array}
$$

Interpreting 'Hxy' accordingly as meaning that at least one of the conditions (1)–(10) holds, we proceed now to show R definable. (In effect we proceed to the construction of a ϕ fulfilling (a)–(c).)

To begin with we shall show how the familiar notation '$x = y$', used above in the informal explanation of H, can be formally defined on the basis of H. The arrow '$\rightarrow$' is used as a sign of definition.

Definition: $x = y \rightarrow (z)(Hxz \equiv Hyz)$.

Originally published in *The Journal of Symbolic Logic*, vol. 17 (1952), pp. 179–187. © Association for Symbolic Logic. Reprinted by permission.
Received March 9, 1951.

Justification: Under equality of x and y, obviously $(z)(Hxz \equiv Hyz)$; so it will be sufficient to establish the converse—to *assume* that $(z)(Hxz \equiv Hyz)$ and show that x and y must be equal. By (3), x bears H to $6x$; so, by our assumption, y must bear H to $6x$. Now we see from a survey of (1)–(10) that y, in order to bear H to $6x$ (which $\cong 0$), must be either $6x$ or $36x$ or x or $6x+1$ or $6x-1$ or $6x+2$ or $6x-2$. But, for any natural number x, each of these seven alternatives is greater than or equal to x. (Where x is 0, the alternatives $6x-1$ and $6x-2$ of course drop out as non-existent.) So $y \geqq x$. But the same argument with 'x' and 'y' switched shows that $x \geqq y$. So $x = y$.

We next proceed to show how, now that '$=$' is at hand, the notations '$x \cong 0$', '$x \cong 1$', etc. can be defined in turn.

Definition: $x \cong 0 \rightarrow Hxx \,.\, (\exists y_1)(\exists y_2)(\exists y_3)(\exists y_4)(\exists y_5)(\exists y_6)(z)[Hxz \supset (z = x \vee z = y_1 \vee z = y_2 \vee z = y_3 \vee z = y_4 \vee z = y_5 \vee z = y_6)]$.

Justification: What the definiens says is that x bears H to itself and to at most six other numbers. This is true when $x \cong 0$; for, when $x \cong 0$, we know from (1)–(10) that x bears H to itself and to no other numbers except $6x$, $\frac{1}{6}x$, $x+1$, $x-1$, $x+2$, and $x-2$ (indeed only to itself and 1 and 2 when x is 0). Conversely, when the definiens is true, $x \cong 0$; this is seen as follows. We see from (1)–(10) that if x bears H to itself then $x \cong 1$ or $x \cong 0$; but if $x \cong 1$ then (2) tells us that x bears H not to merely six other numbers but to 1, 7, 13, 19, 25, etc. ad infinitum.

Definition: $x \cong 1 \rightarrow Hxx \,.\, x \ncong 0$.

Justification evident.

Definition: $x \cong 3 \rightarrow \bar{H}xx \,.\, (\exists y)(z)[(z \cong 0 \,.\, Hxz) \supset z = y]$.

Justification: The clause '$\bar{H}xx$' says in effect 'x is congruent to 2 or 3 or 4 or 5', and the rest of the definiens may be read 'x bears H to at most one number congruent to 0'. But we know from (3), (5), (7), and (8) that x bears H to $6x$, $x+1$, $x+2$, and $x-2$, all of which are distinct. Then, since $6x \cong 0$, x will fulfill the definiens only if none of $x+1$, $x+2$, and $x-2$ is congruent to 0; hence only if $x \cong 3$. Conversely, if $x \cong 3$ then we see from (1)–(10) that x will bear H only to one number $\cong 0$ (viz. $6x$), thus fulfilling the definiens.

Definition: $x \cong 4 \rightarrow \bar{H}xx \,.\, x \ncong 3 \,.\, (y)(z)[(\bar{H}yy \,.\, \bar{H}zz \,.\, y \ncong 3 \,.\, z \ncong 3 \,.\, Hxz \,.\, Hyz) \supset x = y]$.

Justification: The first two clauses, '$\bar{H}xx \,.\, x \ncong 3$', say in effect '$x$ is congruent to 2 or 4 or 5', and the ensuing quantification amounts to:

(i) $(y)(z)[(y$ and z are congruent to 2 or 4 or 5 $.\, Hxz \,.\, Hyz) \supset x = y]$.

So it remains to show that (i) is true for all x congruent to 4 and false for all x congruent to 2 or 5.

Then suppose $x \cong 4$; to show that (i) holds. I.e., supposing any y and z such that Hxz and Hyz and

(ii) y and z are congruent to 2 or 4 or 5,

to show that $x = y$. Since $x \cong 4$ and Hxz, survey of (1)–(10) shows that z must be

$6x$ or $x+1$ or $x-1$ or $x+2$ or $x-2$ or $\cong 1$; hence, eliminating by (ii), we conclude that z is $x+1$ or $x-2$. Hence $z \cong 5$ or $z \cong 2$. Then, since Hyz, survey of (1)–(10) shows that y must be $6z$ or $z+1$ or $z-1$ or $z+2$ or $z-2$; hence, eliminating by (ii), we conclude that y is $z-1$ or $z+2$. So, since z was $x+1$ or $x-2$, y is x or $x-3$ or $x+3$. But if y were $x-3$ or $x+3$ then y would be $\cong 1$, contrary to (ii). So $x = y$.

It remains to show (i) false for all x congruent to 2 or 5. I.e., for each x congruent to 2 or 5 we have to produce a y and z such that (ii) holds and Hxz and Hyz and $x \neq y$.

Case where $x \cong 2$: Take y as $x+3$ and z as $x+2$.

Case where $x \cong 5$: Take y as $x-3$ and z as $x-1$.

This completes the justification of the definition of '$x \cong 4$'.

Definition: $x \cong 2 \rightarrow \bar{H}xx \,.\, x \ncong 4 \,.\, (\exists y)(\exists z)(y \cong 3 \,.\, z \cong 1 \,.\, Hxy \,.\, Hyz \,.\, Hxz)$.

Justification: Any value of 'x' congruent to 2 obviously satisfies the definiens; for, we can take y and z as $x+1$ and $x-1$. Moreover, any value of 'x' congruent to 0 or 1 or 4 obviously fails to satisfy the definiens, in view of the clauses '$\bar{H}xx \,.\, x \ncong 4$'. So what we have to show is that the rest of the definiens, viz.:

(iii) $$(\exists y)(\exists z)(y \cong 3 \,.\, z \cong 1 \,.\, Hxy \,.\, Hyz \,.\, Hxz),$$

fails when $x \cong 3$ and when $x \cong 5$.

Case where $x \cong 3$: It is clear from (1)–(10) that x bears H to nothing y congruent to 3; so (iii) fails.

Case where $x \cong 5$: If x is to bear H to y, and $y \cong 3$, then we know from (1)–(10) that y must be $x-2$. If x is to bear H to z, and $z \cong 1$, then we know from (1)–(10) that z must be $x+2$. So $y \cong 3$ and $z = y+4$. But then it is clear from (1)–(10) that $\bar{H}yz$. So (iii) fails.

Definition: $x \cong 5 \rightarrow \bar{H}xx \,.\, x \ncong 2 \,.\, x \ncong 3 \,.\, x \ncong 4$.

Justification evident, since '$\bar{H}xx$' excludes congruence to 0 and 1.

It is now possible to define successor.

Definition: $y = x+1 \rightarrow Hxy \,.\{(x \cong 1 \,.\, y \cong 2) \vee (x \cong 2 \,.\, y \cong 3) \vee (x \cong 3 \,.\, y \cong 4) \vee (x \cong 4 \,.\, y \cong 5) \vee [x \cong 5 \,.\, y \cong 0 \,.\, (\exists z)(z \cong 4 \,.\, Hxz \,.\, Hyz)] \vee [x \cong 0 \,.\, y \cong 1 \,.\, (\exists z)(z \cong 5 \,.\, Hxz \,.\, Hyz)]\}$.

Justification: It is evident from an examination of (1)–(10) that, in the case where $x \cong 1$, y will be $x+1$ if and only if Hxy and $y \cong 2$. Correspondingly for the cases where $x \cong 2, 3$, or 4. So our definition clearly serves its purpose except perhaps where $x \cong 5$ or $x \cong 0$. Where $x \cong 5$, y need not be $x+1$ in order that Hxy and $y \cong 0$; for y could be $6x$. Where $x \cong 0$, again, y need not be $x+1$ in order that Hxy and $y \cong 1$, for y could be $\frac{1}{6}x$. It is in order to eliminate these unwanted alternatives $6x$ and $\frac{1}{6}x$ that the supplementary clauses:

(iv) $$(\exists z)(z \cong 4 \,.\, Hxz \,.\, Hyz),$$

(v) $$(\exists z)(z \cong 5 \,.\, Hxz \,.\, Hyz)$$

have been inserted in the above definiens. It is clear from (6) and (8) that (iv) does hold when $x \cong 5$ and $y = x+1$, and that (v) holds when $x \cong 0$ and $y = x+1$; for we can take z in both cases as $x-1$. So in order to complete the justification of our definition it remains only to show that (iv) fails when $x \cong 5$ and $y = 6x$, and that (v) fails when $x \cong 0$ and $y = \frac{1}{6}x$.

Where $z \cong 4$, we know from (1)–(10) that z bears H only to $6z$, $z+1$, $z-1$, $z+2$, $z-2$, and perhaps various numbers $\cong 1$. Of all these, clearly no two can be x and $6x$ where $x \cong 5$. So (iv) fails when $x \cong 5$ and $y = 6x$.

Where $z \cong 5$, z will bear H only to $6z$, $z+1$, $z-1$, $z+2$, and $z-2$; and clearly no two of these can be x and $\frac{1}{6}x$. So (v) fails when $y = \frac{1}{6}x$.

So the definition of '$y = x+1$' is justified. We next define '$y = x+4$' and '$y = x+6$' in obvious fashion.

Definitions:

$$y = x + 4 \rightarrow (\exists z)(\exists w)(\exists u)(z = x + 1 \,.\, w = z + 1 \,.\, u = w + 1 \,.\, y = u + 1).$$

$$y = x + 6 \rightarrow (\exists z)(\exists w)(z = x + 4 \,.\, w = z + 1 \,.\, y = w + 1).$$

It is now possible to define multiplication by 6.

Definition:

$$y = 6x \rightarrow y \cong 0 \,.\, Hxy \,.(\exists z)(\exists w)(z = x + 1 \,.\, w = y + 6 \,.\, z \neq w \,.\, Hzw).$$

Justification: The definiens says, in brief, that

(vi) $$y \cong 0 \,.\, Hxy \,.\, H(x+1)(y+6) \,.\, x + 1 \neq y + 6.$$

In view of (3), clearly (vi) holds when y is $6x$. So it remains conversely to assume (vi) and show that y must be $6x$. Since $y \simeq 0$ and Hxy, we know from (1)–(10) that

(vii) $$y \text{ is } x \text{ or } 6x \text{ or } \tfrac{1}{6}x \text{ or } x+1 \text{ or } x-1 \text{ or } x+2 \text{ or } x-2.$$

Also, since $y + 6 \cong 0$ and $H(x+1)(y+6)$, we know from (1)–(10) that $y+6$ is $x+1$ or $6x+6$ or $\frac{1}{6}(x+1)$ or $x+2$ or x or $x+3$ or $x-1$; but, by (vi) $y + 6 \neq x + 1$; so

(viii) $$y \text{ is } 6x \text{ or } \tfrac{1}{6}(x+1) - 6 \text{ or } x-4 \text{ or } x-6 \text{ or } x-3 \text{ or } x-7.$$

But it is readily verified that there can be no x such that either x or $\frac{1}{6}x$ or $x+1$ or $x-1$ or $x+2$ or $x-2$ is equal to $\frac{1}{6}(x+1)-6$ or $x-4$ or $x-6$ or $x-3$ or $x-7$; so (vii) and (viii) compel y to be $6x$.

We are at last ready to express 'Rxy'.

Definition:

$$Rxy \rightarrow (\exists z)(\exists w)(\exists u)(\exists v)(z = 6x \,.\, w = z + 1 \,.\, u = 6y \,.\, v = u + 4 \,.\, Hwv).$$

Justification: The definiens says, in brief, that $6x+1$ bears H to $6y+4$. But since $6x + 1 \cong 1$ and $6y + 4 \cong 4$, we see from a survey of (1)–(10) that $6x+1$ will bear H to $6y+4$ if and only if (9) holds (with '$6x+1$' for 'x' therein and '$6y+4$' for 'y'); i.e., if and only if $\frac{1}{6}(6x+1-1)$ bears R to $\frac{1}{6}(6y+4-4)$; i.e., if and only if x bears R to y.

The proof of Theorem I is now complete. 'Rxy' expands through the above dozen definitions into a formula ϕ of pure quantification theory, or first-order functional calculus, having dyadic 'H' as sole predicate letter and 'x' and 'y' as sole free individual variables. Given any relation R of natural numbers, we can give 'H' a symmetric interpretation (viz. as in (1)–(10)) which will cause ϕ to agree in truth-value with 'x

bears R to y' for all values of 'x' and 'y', as has been established by the justification of the twelve definitions.

2. Consequences concerning decidability. Kalmár[1] has presented an effective method whereby, given any formula ψ of quantification theory, or first-order functional calculus, another such formula ψ' can be found which meets the following conditions: ψ' contains just one predicate letter, and that letter is dyadic; and ψ is valid if and only if ψ' is valid.

With help of Theorem I we can strengthen Kalmár's result to read as follows:

THEOREM II. *There is an effective method whereby, given any formula ψ of quantification theory, a formula χ of quantification theory can be found which meets the following conditions: χ contains just one predicate letter, and that letter is dyadic; and ψ is valid if and only if χ comes out true under all symmetric interpretations of its predicate letter in all non-empty universes.*

Proof. We construct χ as follows. Given ψ, we first apply Kalmár's method to obtain a formula ψ' whose sole predicate letter is dyadic 'R'. Then we supplant 'Rxy' throughout ψ' by the ϕ of Theorem I (and correspondingly for $\ulcorner R\alpha\beta \urcorner$ where α and β are any variables). The result, containing dyadic 'H' as sole predicate letter, is χ. It remains to show that ψ is valid if and only if χ comes out true under all symmetric interpretations of 'H' in all non-empty universes. Or, since ψ is valid if and only if ψ' is valid (Kalmár's result), it will be sufficient to show that ψ' is valid (true under all interpretations of 'R' in all non-empty universes) if and only if χ is true under all symmetric interpretations of 'H' in all non-empty universes. But instead of speaking here of all non-empty universes we can without loss of generality limit ourselves to the universe of natural numbers, in view of the Skolem-Löwenheim theorem.[2] So it will be sufficient to show that, when the natural numbers are taken as the range of individual variables (as is to be tacitly understood henceforward), ψ' is true under all interpretations of 'R' if and only if χ is true under all symmetric interpretations of 'H'.

If ψ' is true under all interpretations of 'R', it is true in particular when, choosing any interpretation of 'H', we interpret 'Rxy' as ϕ. So, if ψ' is true under all interpretations of 'R', then χ is true under all interpretations of 'H' (and hence under all symmetric interpretations of 'H'). Conversely, we know from Theorem I that for every interpretation of 'R' there is a symmetric interpretation of 'H' which makes ϕ a necessary and

[1] László Kalmár, *Zurückführung des Entscheidungsproblems auf den Fall von Formeln mit einer einzigen, binären, Funktionsvariablen*, ***Compositio mathematica***, vol. 4 (1936), pp. 137–144. As actually stated by him, Kalmár's method accomplishes the reduction of a formula ψ of quantification theory to a formula ψ' of quantification theory meeting the conditions that ψ' contains just one predicate letter, that that letter is dyadic, and that ψ is satisfiable if and only if ψ' is satisfiable; however, as is well known, this is easily modified to yield the reduction method described in the text by making use of the equivalence between validity of a formula and unsatisfiability of its negation. Moreover, Kalmár's procedure begins with a preliminary reduction of the given formula to a formula in prenex normal form with a prefix of specified form, and in the further reduction to a formula having a single binary predicate letter in the matrix the same form of prefix is preserved. Thus Kalmár's result is stronger than our present purpose requires. A proof of the (weaker) result stated in the text, by a reduction method which is independent of reduction of the prefix to a special form, appears in the forthcoming revised and enlarged edition of Church, ***Introduction to mathematical logic I***, §47.

[2] See Hilbert and Ackerman, ***Grundzüge der theoretischen Logik***, Chapter 3, §10 or Church, op. cit., revised edition, **450.

sufficient condition for 'Rxy'; so, if χ is true for all symmetric interpretations of 'H', ψ' is true for all interpretations of 'R'. This completes the proof of Theorem II.

Church[3] has proved the impossibility of a decision procedure for quantification theory. Combining this result with Kalmár's, we know that there can be no decision procedure even for the quantification theory of a single dyadic predicate letter. But now Theorem II leads, by parallel reasoning, to:

THEOREM III. *There can be no decision procedure for the quantification theory of a dyadic predicate letter 'H' even when no distinction is made between 'Hxy' and 'Hyx'* (nor in general between $\ulcorner H\alpha\beta \urcorner$ and $\ulcorner H\beta\alpha \urcorner$).

The undecidability of quantification theory depends essentially on the presence of polyadic predicate letters; for, since Löwenheim we have known decision procedures for monadic quantification theory.[4] However, Theorem III shows that the undecidability of quantification theory does not depend upon the existence of non-symmetric interpretations of the predicate letters.

As an immediate consequence (little more than a restatement of Theorems II and III in a different phraseology) we have also:

COROLLARY. *There is an effective method whereby the problem of validity of an arbitrary formula of quantification theory (or pure functional calculus of first order) can be reduced to that of validity of a conditional having '$(x)(y)(Hxy \supset Hyx)$' as antecedent and containing no predicate letter but 'H'. And hence the special case of the decision problem which concerns the validity of such conditionals is unsolvable.*

The next theorem on undecidability will have to do with what may be called elementary set theory, or the second-order monadic functional calculus. The notation of this theory may be thought of as identical with that of monadic quantification theory except that the predicate letters are allowed in quantifiers as bound variables. Alternatively, instead of writing 'Fx', 'Fy', etc., we may prefer to write '$x \in f$', $y \in f$', etc. and think of 'f' thus as a class variable.[5]

This theory, however one may care to frame and name it, is known to admit—like monadic quantification theory itself—of a decision procedure.[6] However, we can now show that if the theory is supplemented to the extent merely of allowing one unquantifiable monadic second-level predicate letter, standing for predicates of classes, then the resulting theory admits no decision procedure.

---

[3]Alonzo Church, *A note on the Entscheidungsproblem*, this JOURNAL, vol. 1 (1936), pp. 40–41, 101–102.

[4]Leopold Löwenheim, *Über Möglichkeiten im Relativkalkül*, ***Mathematische Annalen***, vol. 76 (1915), pp. 447–470. Other decision procedures for monadic quantification theory have been presented by Thoralf Skolem, *Untersuchugen über die Axiome des Klassenkalkuls*, ***Skrifter utgit av Videnskapsselskapet i Kristiania***, I, ***Mat. naturvid. Klasse*** 1919, no. 3; Heinrich Behmann, *Beiträge zur Algebra der Logik, insbesondere zum Entscheidungsproblem*, ***Mathematische Annalen***, vol. 86 (1922), pp. 163–229; Paul Bernays and Moses Schönfinkel, *Zum Entscheidungsproblem der mathematischen Logik*, ibid., vol. 99 (1928), pp. 342–372; Jacques Herbrand, ***Recherches sur la théorie de la démonstration*** (Warsaw, 1930), Chapter 2, §9.2. For convenient expositions of these procedures and further variants, see also Hilbert and Ackermann, op. cit., Chapter 3, §12; Hilbert and Bernays, ***Grundlagen der Mathematik***, vol. I, pp. 193–195; G. H. von Wright, ***Form and content in logic*** (Cambridge, England, 1949, 35 pp.); Quine, ***O sentido da nova logica*** (São Paulo, 1944), pp. 126–129; Quine, *On the logic of quantification*, this JOURNAL, vol. 10 (1945), pp. 1–12; ***Methods of logic***, pp. 101–117, 192–194; Church, ***Introduction to mathematical logic I***, revised edition, §46.

[5]In the opinion of one of the authors, it is only the inclusion of predicate letters within quantifiers which makes it preferable to view such letters as class variables. See Quine, ***Methods of logic***, §38.

[6]See Löwenheim, op. cit.; or, for better accounts, Skolem, op. cit., and Behmann, op. cit.

THEOREM IV. *There can be no decision procedure for the theory which consists of elementary set theory plus an unquantifiable monadic second-level predicate letter.* Or, in the terminology of Church,[7] there is no solution of the decision problem for the monadic functional calculus of third order, or even of the decision problem for the subclass of formulas of the monadic functional calculus of third order in which there do not occur more than one functional variable of the highest type.

Proof. One of the formulas falling within the notation of the theory under consideration is

$$\text{(i)} \qquad (\exists f)(Kf \,.\, (z)\{z \in f \equiv [(g)(z \in g \supset x \in g) \vee (g)(z \in g \supset y \in g)]\}),$$

wherein 'K' figures as a second-level predicate letter. Since '$(g)(z \in g \supset x \in g)$' is equivalent to '$z = x$', (i) says in effect merely that 'K' is true of the class $\{x, y\}$ whose members are x and y. So, if we are given any symmetric relation H and then interpret 'K' as true of just the classes $\{x, y\}$ such that x bears H to y, thereupon (i) becomes a necessary and sufficient condition that x bear H to y. Conversely, also, for every interpretation of 'K' there is a symmetric relation H (viz. $\hat{x}\hat{y}(K\{x, y\})$) such that (i) is a necessary and sufficient condition that x bear H to y. Therefore, if throughout a formula of quantification theory we supplant 'Hxy' by (i) (and correspondingly supplant $\ulcorner H\alpha\beta \urcorner$ for all variables α and β), then the resulting formula will be true for all interpretations of 'K' if and only if the original was true for all symmetric interpretations of 'H'. Accordingly Theorem IV follows from Theorem III.

3. A simplest primitive for elementary number theory. Mrs. Robinson has shown[8] that elementary number theory can be expressed in terms of just the predicates of divisibility ('x is a factor of y') and succession ('$x = y + 1$') together with identity, quantification, and the truth functions. However, all of her primitives can be generated from a simple symmetric dyadic predicate together with quantification and the truth-functions, if we follow the constructions in §1 above, interpreting 'Rxy' as 'x is a factor of y'. We have an appropriate predicate for the purpose when we construe 'Hxy' according to (1)–(10) of §1 above, with 'bears R to' changed to 'is a factor of'. Thus

THEOREM V. *Elementary number theory can be expressed in terms of just quantification and the truth-functions and a single symmetric dyadic predicate.*[9]

Note that the above interpretation of 'H' can be described briefly as the relation which relates, backwards and forwards, for all choices of z and w, all and only the following natural numbers: z with $z + 1$, $z + 2$, and $6z$; $6z$ with itself; and $6z + 1$ with $6w + 1$ and $6zw + 4$.

If a symmetric dyadic predicate be thought of as simpler than a non-symmetric one, and if the next higher stage of simplicity be conceived to consist in a multitude of monadic predicates,[10] then we have here achieved the simplest possible primitive for elementary number theory within the framework of quantification theory. For, it will

[7] Church, ***Introduction to mathematical logic I***, edition of 1944, and forthcoming revised edition.

[8] Julia Robinson, *Definability and decision problems in arithmetic*, this JOURNAL, vol. 14 (1949), pp. 98–114. Her construction is geared to just the positive integers, but is easily adjusted to include 0.

[9] A reduction of elementary number theory to a single non-symmetric dyadic predicate was given by John R. Myhill, *A reduction in the number of primitive ideas of arithmetic*, this JOURNAL, vol. 15 (1950), p. 130.

[10] See Nelson Goodman, *The logical simplicity of predicates*, this JOURNAL, vol. 14 (1949), pp. 32–41.

now be shown that no set of monadic predicates can suffice for elementary number theory.

THEOREM VI. *If* $\mathfrak{S}$ *is the system of notation of quantification theory with interpreted predicates* 'F', 'G', $\cdots$ (finite or infinite in number), *and if elementary number theory can be expressed in* $\mathfrak{S}$ (i.e., if '$x = y + z$' and '$x = y \cdot z$' can be translated into $\mathfrak{S}$), *then some of* 'F', 'G', $\cdots$ *are polyadic.*

Proof. Let 'Tx' mean that the formula numbered x in a given Gödel numbering[11] of quantification theory is a theorem of quantification theory. Then, as is clear from Gödel's work, 'Tx' is expressible in elementary number theory, and furthermore a finite set of axioms of number theory can be given from which each instance of 'Tx' (with a numeral for 'x') can, if and only if true, be deduced. Let us abbreviate the conjunction of those axioms as 'A'. Now deduction from 'A' proceeds in turn purely by quantification theory. Hence a formula ϕ of quantification theory is a theorem of quantification theory if and only if the number-theoretic statement '$A \supset Tx$', with the Gödel numeral of ϕ in place of 'x', is itself quantificationally valid (i.e., valid for quantification theory independently of the number-theoretic interpretation of the primitive predicates). Moreover, if we can expand '$A \supset Tx$' (with the aforementioned numeral in place of 'x') into a primitive notation $\mathfrak{S}$ involving none but monadic predicates then we can decide its quantificational validity; for, a decision procedure is available for the quantification theory of monadic predicates.[4] We thus achieve a decision procedure for recognizing theorems of quantification theory in general. But Gödel has shown[12] that all valid formulas of quantification theory are theorems; so this decision procedure amounts to a decision procedure for validity in quantification theory. But Church has shown[3] that no such procedure is possible. Our assumption regarding $\mathfrak{S}$ is therefore wrong.

Theorem III, it will be recalled, was proved with help of Kalmár's theorem. But in conclusion it may be worth noting that by following the line rather of the proof of Theorem VI we could obtain a new proof of Theorem III not presupposing Kalmár's work. The argument is as follows. If, contrary to Theorem III, there were a decision procedure for the quantification theory of a single symmetric dyadic predicate, then the argument of Theorem VI could be repeated with the mere change, twice, of "monadic predicates" to "a single symmetric dyadic predicate." This argument would establish (in place of Theorem VI) the impossibility of basing elementary number theory on a single symmetric dyadic predicate, contrary to Theorem V. So Theorem III follows. Theorem IV, which was proved from Theorem III, then likewise becomes independent of Kalmár's argument.

PRINCETON UNIVERSITY and HARVARD UNIVERSITY

[11]Kurt Gödel, *Über formal unentscheidbare Sätze der Principia Mathematica und verwandter Systeme*, ***Monatshefte für Mathematik und Physik***, vol. 38 (1931), pp. 173–198.

[12]Kurt Gödel, *Die Vollständigkeit der Axiome des logischen Funktionenkalküls*, ibid., vol. 37 (1930), pp. 349–360.

NON-NORMAL TRUTH-TABLES FOR THE PROPOSITIONAL CALCULUS (1953)

Besides the usual two-valued truth-tables for the propositional calculus, it is known that there are many characteristic[1] systems of truth-tables (characteristic matrices[2]) in which there are more than two truth-values.

In particular, since the two-valued truth-tables constitute a two-element Boolean algebra,[3] any system of truth-tables having the two-element Boolean algebra as a homomorphic image will be a characteristic system if the designated truth-values are taken to be those which have the unit of the Boolean algebra as their image. In this way characteristic systems of truth-tables may be obtained with any number of truth-values (not less than two[4]). Characteristic systems of truth-values of this kind we shall call *normal in the sense of Carnap*, and all others will be called *non-normal in the sense of Carnap*,[5] or *weakly non-normal.*

A characteristic system of truth-tables may also be obtained from an arbitrary Boolean algebra by taking the unit, 1, of the algebra as the single designated value.[6] If the Boolean algebra has more than two elements, the resulting characteristic system of truth-tables is weakly non-normal. And we may obtain additional weakly non-normal characteristic systems by means of homomorphisms, as before, taking any system of truth-tables to which the Boolean algebra is homomorphic, and taking the designated truth-values to be those which have the unit of the Boolean algebra as image.

Originally published in ***Boletín de la Sociedad Matemática Mexicana***, vol. 10 (1953), pp. 41–52.

Recibido para el Congreso Científico Mexicano, Septiembre 1951.

[1]In any particular system of propositional calculus (we shall here, however, be concerned with only one such system, the classical propositional calculus), an expression of the calculus is called a *tautology* according to a particular system of truth-tables if, for every system of truth-values of its variables, the truth-values of the expression, as obtained from the truth-tables, belongs to the class of *designated* truth-values. And a system of truth-tables is called *characteristic* of a particular system of propositional calculus if the theorems of the propositional calculus are the same as the tautologies according to the truth-tables.

[2]For a system of truth-tables, the term "matrix" is becoming usual, in spite of the awkward conflict between this use of the word "matrix" and the quite different use of the same word which was introduced in ***Principia Mathematica***. (The long-established use of this word in algebra is of course still a third use.)

[3]In this paper, whenever a Boolean algebra is spoken of as being a system of truth-tables for the propositional calculus or as being homomorphic to such a system of truth-tables, it is to be understood that negation, conjunction, and disjunction are represented respectively by the Boolean complement, the Boolean product, and the Boolean sum. The Boolean representatives of (material) implication and equivalence are then obtained by rewriting, $P \supset Q$ and $P \equiv Q$ as $\sim P \vee Q$ and $PQ \vee \sim P \sim Q$ respectively.

[4]Some examples of this kind are given, e.g., by K. Schröter in ***Zentralblatt für Mathematik und ihre Grenzgebiete***, vol. 37 (1951), p. 4.

[5]This is a minor modification of the terminology of Carnap, who (in his ***Formalization of Logic***, Cambridge, Mass., 1943) speaks rather of normal and non-normal true interpretations where any system of truth-tables will provide an interpretation of the propositional calculus, which will be a true interpretation if only every theorem is a tautology.

[6]This was perhaps first pointed out explicitly by B. A. Bernstein (in a different terminology) in ***Bulletin of the American Mathematical Society***, vol. 38 (1932), pp. 390 and 592.

Characteristic systems of truth-tables obtained in this way are necessarily *regular*,[7] in the sense that $p \supset q$ never has a designated truth-value when p has a designated truth-value and q a non-designated truth-value.

An example of such a characteristic system of truth-tables is provided in Tables I, at the end of this paper. Indeed, Tables I exhibit a four-element Boolean algebra, with the unit, 1, of the Boolean algebra as the designated truth-value. Thus they are the simplest example of a weakly non-normal characteristic system of truth-tables for the propositional calculus (or, as we shall say briefly, of weakly non-normal truth-tables for the propositional calculus).

Tables II and III provide examples of weakly non-normal truth-tables for the propositional calculus which are of a different kind,[8] since they are non-regular.

However, all three examples, Tables I, II, and III, are in a certain sense trivial, since they become normal in the sense of Carnap if (without other change) the designated truth-values are taken to be b and 1, instead of 1 alone. In fact, in each case, a simple proof that the tables are characteristic can be obtained by first showing that the tautologies are the same whether b and 1 or 1 alone are taken as the designated truth-values, and then observing that, when b and 1 are taken as designated, the tables are normal in the sense of Carnap.

Thus every example so far found of a characteristic system of truth-tables for the propositional calculus either is a Boolean algebra or reduces to such under a homomorphism. It is moreover well known that the propositional calculus is *formally* a Boolean algebra, in the sense that every identically true Boolean equation becomes a theorem of the propositional calculus if the variables in the Boolean equation are replaced by (or reconstrued as) propositional variables, the signs for the Boolean complement, the Boolean product, and the Boolean sum are replaced respectively by the signs of negation, conjunction, and disjunction, the signs, 0 and 1, for the Boolean zero and unit are replaced by (e.g.) $p \sim p$ and $p \vee \sim p$ respectively, and the sign of equality, $=$, is replaced by the sign of material equivalence, $\equiv$. Likewise every statement of inclusion which is identically true for Boolean algebra becomes a theorem of the propositional calculus if the same replacements are made as just described and at the same time the sign of inclusion, $\subset$, is replaced by the sign of material implication, $\supset$.

For this reason it might be natural to suppose that every characteristic system of truth-tables for the propositional calculus either is a Boolean algebra (of two or more elements) or reduces to such under a homomorphism.

And indeed this does follow if we assume that the truth-table of $\supset$ is regular (in the sense already explained). For since

$$[p \equiv q] \supset [q \equiv p]$$

and

$$[p \equiv q] \supset [[q \equiv r] \supset [p \equiv r]]$$

[7]We adopt this term from McKinsey and Tarski—see ***The Journal of Symbolic Logic***, vol. 13 (1948), p. 11.

[8]Tables III are a simplified version of the tables given in exercise 19.11 of the forthcoming revised edition of the writer's ***Introduction to Mathematical Logic***, Volume I. (The latter tables, unlike Tables III, are so arranged that symmetry fails in the truth table of $\equiv$.)

are theorems of the propositional calculus and therefore tautologies according to the given system of truth-tables, it follows from the regularity of the truth-table of $\supset$ that the truth-table of $\equiv$ is symmetric and transitive. I.e., if, for particular truth-values of p and q, $p \equiv q$ has a designated value, then $q \equiv p$ must have a designated value; and if, for particular truth-values of p, q, and r, both $p \equiv q$ and $q \equiv r$ have designated values, then $p \equiv r$ must have a designated value. The truth-values are thus divided into equivalence-classes, two different truth-values x and y belonging to the same equivalence-class if and only if $p \equiv q$ has a designated truth-value for the values x, y of p, q. Moreover the equivalence-class to which the value of an expression P belongs will remain unchanged if the value x of one of the variables, say p, is altered in such a way as to leave unchanged the equivalence-class to which x belongs—as follows from the regularity of the truth-table of $\supset$, together with the fact that

$$[p \equiv q] \supset [P \equiv Q]$$

is a theorem of the propositional calculus if Q is obtained from P by substituting q for p. Hence a homomorphism such that two truth-values have the same image if and only if they belong to the same equivalence-class will be a homomorphism of the given system of truth-tables onto a Boolean algebra.[9]

If, however, regularity fails in the truth-table of $\supset$, there is the possibility of obtaining a characteristic system of truth-tables for the propositional calculus of such a sort that no Boolean algebra is homomorphic to it. Such a characteristic system of truth-values we shall call *strongly non-normal.*

Tables IV are a rather obvious example of strongly non-normal truth-tables for the propositional calculus, being so constructed that they follow the usual two-valued truth-tables with regard to the values 0 and 1, and that the value of an expression is always h for any system of values of the propositional variables that includes the value h for one of the variables. A large variety of more elaborate variations on this theme are evidently possible.

Perhaps more interesting as an example of strongly non-normal truth-tables for the propositional calculus are Tables V, due to Z. P. Dienes.[10]

In order to see that Tables V are characteristic of the propositional calculus, notice first that the usual two-valued truth-tables are followed in the case of the values 0 and 1, and hence that no non-theorem can be a tautology. Now for an expression P, consider a system S of values of its variables that includes the value h for one or more of them. At each occurrence of a variable having the value h, replace the h by 0 or 1, taking the various occurrences independently, and abandoning the requirement (as regards these variables) that the same value be assigned to different occurrences of the same variable. If on doing this in all possible ways the value of P is always 0 or always 1, then the value of P is 0 or 1, respectively, for the system S of values of its variables; but otherwise the value of P is h for the system S of values of its variables. I.e., Tables V

[9]Here we make use of the fact that a Boolean algebra can be characterized by conditions which have exclusively the form of identically true equations of the algebra, together with the conditions that the complement, sum, and product exist and that there are at least two elements. (Some of Huntington's system of postulates for Boolean algebra are, for example, substantially, in this form.) Also use is made of the fact that the propositional calculus is formally a Boolean algebra, in the sense explained above.

[10]In ***The Journal of Symbolic Logic***, vol. 14 (1949), pp. 95–97. The remark that these truth-tables are characteristic for the propositional calculus, and that no Boolean algebra is homomorphic to them, was made by Church and Rescher, ibid., vol. 15 (1950), pp. 69–70.

are so constructed that this will be the case, as may readily be verified. Since h is a designated value, it follows that every tautology in the truth-values 0 and 1 remains a tautology when the additional truth-value h is admitted. Hence every theorem of the propositional calculus is a tautology (in the three truth-values, $0, h, 1$).

These examples are brought together here in order to raise the question whether there exist other strongly non-normal truth-tables for the propositional calculus, beyond those cited (together with direct products and other obvious elaborations); or more generally, to raise the question of a survey or characterisation in some sense of the possible strongly non-normal truth-tables for the propositional calculus.

Another motive is the suggestion, which was made to me by Paco Lagerström ten years or more ago, that use may be made of non-normal truth-tables for the propositional calculus in order to extend to the functional calculi of first and higher orders, and other related systems, the method of proving independence of axioms which is familiar in the case of the propositional calculus.

In question at that time were only Boolean algebras, in the role of weakly non-normal truth-tables. And the remark was made in particular by Lagerström[11] that in my *Formulation of the simple theory of types*[12] the independence of the axioms 9^α from axioms 1–8 and $10^{\alpha\beta}$ can be established by means of a complete non-atomic Boolean algebra. For this purpose, axioms 9^α are to be rewritten in the weaker form

$$(\exists t_{\alpha(o\alpha)})(f_{o\alpha})(x_\alpha) \centerdot f_{o\alpha}x_\alpha \supset \centerdot (y_\alpha)[f_{o\alpha}y_\alpha \supset x_\alpha = y_\alpha] \supset f_{o\alpha}(t_{\alpha(o\alpha)}f_{o\alpha}),$$

since the question of independence would otherwise be trivial. The range of the variables of type o is to be the Boolean algebra in question; the range of the variables of type ι is to be the natural numbers; and the range of the variables of the type $\alpha\beta$ is to be the functions from B to A, where A and B are the ranges of the variables of types α and β respectively. The universal quantifier is to correspond to the infinite Boolean product, so that the value of, say, $(x_\iota)M$ for a given system of values of the free variables is

$$\prod_{i=0}^{\infty} \Phi(i),$$

where $\Phi(i)$ is the value of M for the value i of x_ι.

In a similar fashion, the independence of the axioms 6^α can be established by means of a four-element Boolean algebra (Tables I). The value of $(x_\alpha)M$ is to be taken as 1 in case the value of M is 1 for all values of x_α, and as 0 in all other cases.

[11] It has never been published.

[12] ***The Journal of Symbolic Logic***, vol. 5 (1940), pp. 56–68. See errata, ibid., vol. 6, p. iv.

p	q	$p \supset q$	pq	$p \vee q$	$p \equiv q$	$\sim p$
0	0	1	0	0	1	1
0	a	1	0	a	b	
0	b	1	0	b	a	
0	1	1	0	1	0	
a	0	b	0	a	b	b
a	a	1	a	a	1	
a	b	b	0	1	0	
a	1	1	a	1	a	
b	0	a	0	b	a	a
b	a	a	0	1	0	
b	b	1	b	b	1	
b	1	1	b	1	b	
1	0	0	0	1	0	0
1	a	a	a	1	a	
1	b	b	b	1	b	
1	1	1	1	1	1	

Tables I. Designated value 1.

p	q	$p \supset q$	pq	$p \vee q$	$p \equiv q$	$\sim p$
0	0	1	0	0	1	1
0	b	1	0	1	0	
0	1	1	0	1	0	
b	0	0	0	1	0	0
b	b	1	1	1	1	
b	1	1	1	1	1	
1	0	0	0	1	0	0
1	b	1	1	1	1	
1	1	1	1	1	1	

Tables II. Designated value 1.

p	q	$p \supset q$	pq	$p \vee q$	$p \equiv q$	$\sim p$
0	0	1	0	0	1	1
0	a	1	0	0	1	
0	b	1	0	1	0	
0	1	1	0	1	0	
a	0	1	0	0	1	b
a	a	1	a	a	1	
a	b	b	0	1	0	
a	1	1	0	1	0	
b	0	0	0	1	0	a
b	a	a	0	1	0	
b	b	1	b	b	1	
b	1	1	1	1	1	
1	0	0	0	1	0	0
1	a	0	0	1	0	
1	b	1	1	1	1	
1	1	1	1	1	1	

Tables III. Designated value 1.

p	q	$p \supset q$	pq	$p \vee q$	$p \equiv q$	$\sim p$
0	0	1	0	0	1	1
0	h	h	h	h	h	
0	1	1	0	1	0	
h	0	h	h	h	h	h
h	h	h	h	h	h	
h	1	h	h	h	h	
1	0	0	0	1	0	0
1	h	h	h	h	h	
1	1	1	1	1	1	

Tables IV. Designated values h and 1.

p	q	$p \supset q$	pq	$p \vee q$	$p \equiv q$	$\sim p$
0	0	1	0	0	1	1
0	h	1	0	h	h	
0	1	1	0	1	0	
h	0	h	0	h	h	h
h	h	h	h	h	h	
h	1	1	h	1	h	
1	0	0	0	1	0	0
1	h	h	h	1	h	
1	1	1	1	1	1	

Tables V. Designated values h and 1.

SOME EXAMPLES OF STRONGLY NON-NORMAL CHARACTERISTIC SYSTEMS OF TRUTH-TABLES FOR THE PROPOSITIONAL CALCULUS*

(1954)

EXAMPLE 1. The truth-values are the integers. The designated truth-values are the non-negative integers. The disjunction of two integers is the greater of the two. The negation of n is $1 - n$ if n is positive, $-1 - n$ if n is negative, 0 if n is 0.

EXAMPLE 2. The truth-values are the integers between -9 and 9 inclusive. The designated truth-values are those between 0 and 9 inclusive. The disjunction and negation are as in Example 1, except that the negation of 9 is -9 and the negation of -9 is 9. (Example 2, unlike Example 1, has a subsystem which is isomorphic to the usual two-valued truth-tables, namely with -9 corresponding to falsehood and 9 to truth.)

EXAMPLE 3. Given any characteristic system S of truth-tables, normal or non-normal, construct as follows another characteristic system S', which will be strongly non-normal. The truth-values of S' are the non-empty sets C of truth-values of S. The designated truth-values of S' are those in which there is at least one designated truth-value of S. The negation of a truth-value C of S' is the set of negations of elements of C. The disjunction $C \vee D$ of C and D is the set of all elements $c \vee d$, where c is an element of C and d is an element of D. (This construction applied to the usual two-valued truth-tables yields Dienes' three-valued tables.)

EXAMPLE 4. Given any characteristic system S of truth-tables, apply the same construction as in Example 3, except that the empty set of truth-values of S is also to be used, and treated as a designated truth-value of S'.

EXAMPLE 5. After applying the construction of Example 3 to a particular system S, there may appear subsystems of S' which provide new examples of strongly non-normal characteristic systems. This may be illustrated by taking S to be a four-element Boolean algebra and writing out in full the fifteen-valued truth-tables S'.

PROBLEM. In Example 3, let S be an infinite complete Boolean algebra, and consider the subsystem S'' of S' characterized by the two conditions on the truth-values C of S' that the union of all the elements of C shall be 1 and the intersection of all the elements of C shall be 0. Is (or in what cases is) S'' a characteristic system of truth-tables for the propositional calculus?

PROBLEM. To provide in some sense a list or survey of strongly non-normal characteristic systems of truth-tables for the propositional calculus, which is complete to within homomorphisms and direct products (all Boolean algebras being supposed known).

REFERENCES: Z. P. Dienes, ***The Journal of Symbolic Logic***, vol. 14 (1949) pp. 95–97. Alonzo Church and Nicholas Rescher, ***The Journal of Symbolic Logic***, vol. 15 (1950), pp. 69–70. Alonzo Church, ***Boletín de la Sociedad Matemática Mexicana***, vol. 10 (1953), pp. 41–52.

*Previously unpublished typescript, dated March 12, 1954. Printed by permission of the Alonzo Church estate.

INTENSIONAL ISOMORPHISM AND IDENTITY OF BELIEF (1954)

I

The criterion that beliefs expressed by given sentences are identical if and only if the sentences are intensionally isomorphic is contained in Carnap's analysis of belief statements (in *Meaning and Necessity*, §§13–15). And it may be advantageous to separate this criterion from other features of Carnap's analysis, in order to examine it independently.

For our present purpose it will be sufficient to confine attention to a single language, which we may take to be Carnap's S_1 with various individual and predicator constants added to it as required,[1] and to consider L-equivalence and intensional isomorphism, only of designator matrices[2] within this one language and containing the same free variables. It will be recalled that Carnap's definition of 'intensionally isomorphic' depends on a definition of the semantical term 'L-true'.[3] The designator matrices A and B, containing the same free variables, are then said to be L-equivalent if and only if the closure of $A \equiv B$ is L-true.[4] And two designator matrices containing the same free variables are said to be intensionally isomorphic if one can be obtained from the other by a series of steps which consist of (1) alphabetic changes of bound variable, (2) replacements of one individual constant by another which is L-equivalent to it, and (3) replacements of one predicator constant by another which is L-equivalent to it.[5]

Originally published in ***Philosophical Studies*** vol. 5 (1954), pp. 65–73.

[1] We also suppose that Carnap's sign '$\equiv$' of identity of individuals is a predicator constant, and that when A and B are individual expressions, $A \equiv B$ is to be understood merely as an abbreviation or alternate way of writing $\equiv AB$. This modification of S_1 serves to simplify the discussion but is not otherwise essential to the conclusions we reach.

[2] I follow Carnap's terminology, in spite of my own preference for a somewhat different terminology—e.g., 'well-formed formula' instead of 'designator matrix.'

[3] The definition of 'L-true' need not be repeated here. But notice should be taken of two necessary corrections to the definition as it is developed in §§1–2 of Carnap's book.

In 2-2 the correction of Kemeny must be adopted (***Journal of Symbolic Logic***, vol. 16 (1951), p. 206); i.e., in place of "every state-description" the restriction must be made to non-contradictory state-descriptions. Otherwise consequences will follow that are certainly not intended by Carnap, for instance that no two different atomic sentential matrices (and no two different predicator constants) can be L-equivalent.

In the rules of designation 1-1 and 1-2, the way in which the English language and certain phrases of the English language are mentioned, rather than used, is inadmissible—as may be seen by the fact that it forces the tacit use, in 1-3 and 1-4, of certain rules of designation *of the English language*, which, if stated, would have a quite different form from 1-1 and 1-2. For example, Carnap's rule of designation, " 's' is a symbol translation [i.e., from English] of 'Walter Scott'," should be changed to a rule which mentions the man Walter Scott rather than the words 'Walter Scott'; perhaps it should be simply " 's' refers to Walter Scott," in order to justify the inference from 1-3 to 1-4.

These corrections are not directly relevant to the present paper, but our discussion presupposes that suitable corrections have been made.

[4] Carnap uses '$\equiv$' not only between sentential matrices as a sign of material equivalence, but also between other designator matrices as a sign of identity (in place of the usual '$=$').

[5] Because of the restriction to the single language S_1 and to designator matrices containing the same free variables, we have been able to give a simplified form to Carnap's definition of 'L-equivalent' and 'intensionally isomorphic.'

To intensional isomorphism, in this sense, as criterion of identity of belief, there are objections which may be offered on the basis of Carnap's Principle of Tolerance, the principle namely that *every one is at liberty to build his own form of language as he will.*[6]

By the Principle of Tolerance, no one shall forbid us to introduce two completely synonymous predicator constants, or two completely synonymous individual constants, into a language (such as Carnap's S_1), if we choose to do so. Exactly this situation is evidently contemplated in Carnap's definition of 'intensionally isomorphic,' and published informal discussions of the definition have in fact sought out examples of synonymous constants, e.g., in the English language, to be used for purposes of illustration. It is true that formalized languages constructed by logicians rarely contain synonymous primitive constants, as it is clear that the inclusion of such synonyms among the primitive constants would not be consistent with the logician's usual demand for economy of primitives. But to object to a language on the ground of lack of economy is not to say that it is an inadmissible language, but only that it fails to serve a certain purpose. (And the same language which fails to serve one purpose may for that very same reason better serve another.)

However, by the Principle of Tolerance, it is also possible to introduce into a language like S_1 two predicator constants (or two individual constants) which are L-equivalent but not synonymous. For example, let the individuals be the positive integers, and let P and Q be predicator constants, such that Pn expresses that n is less than 3, and Qn expresses that there exist x, y, and z such that $x^n + y^n = z^n$. It is of course permissible to introduce P and Q as primitive constants, together perhaps with axioms containing them, such as may be suggested by their meanings.[7] For the sake of illustration let us suppose that Fermat's claim, to have had a proof of his (now so-called) Last Theorem, was correct. Then P and Q are L-equivalent, and it may even be possible to prove $(n)[Pn \equiv Qn]$ from the axioms. Yet it is evident that one might believe that $(\exists n)[Qn \sim Pn]$ without believing that $(\exists n)[Pn \sim Pn]$, since the proof of Fermat's Last Theorem, though it be possible, is certainly difficult to find (as the history of the matter shows).

Thus if intensional isomorphism is to serve as criterion of identity of belief, Carnap's definition requires the following amendment:

In (2) and (3) as given above, the condition of L-equivalence shall be replaced by that synonymy.

Again by the Principle of Tolerance it is possible to introduce a predicator constant which shall be synonymous with a specified abstraction expression of the form

---

[6]In this form, as applied to the construction of a new language and the determination of what its expressions shall mean, the Principle of Tolerance is hardly open to doubt. The attempt to apply the Principle of Tolerance to the transformation rules of a language after the meaning of the expressions of the language has already been determined (whether by explicit semantical rules or in some looser way) is another matter, and certainly doubtful, but is not at issue here. In fact Carnap (if he ever did) does not now maintain the Principle of Tolerance in this latter and more doubtful form (see §39 of his ***Introduction to Semantics***).

[7]There is no condition to the effect that a predicator constant must express a simple property, rather than such a comparatively complex property as that which is here expressed by Q. In fact some of Carnap's examples of predicator constants express properties which are evidently not especially simple. And it is moreover not clear how the distinction between a simple and a complex property could be made precise in any satisfactory way (except by making it relative to the choice of a particular language).

$(\lambda x)[..x..]$; or to introduce an individual constant synonymous with a specified individual description of the form $(\iota x)[..x..]$. And (unlike the case of synonymous primitive constants) it may be held that something like this actually occurs in formalized languages commonly constructed — namely those in which definitions are treated as introducing new notations into the object language,[8] rather than as metatheoretic abbreviations. But whether or not the process is called definition, it is clear by the Principle of Tolerance that nothing prevents us from introducing (say) a predicator constant R as synonymous with the abstraction expression $(\lambda x)[..x..]$, and taking $R \equiv (\lambda x)[..x..]$ as an axiom.[9] And if this is done, then R must be interchangeable with $(\lambda x)[..x..]$ in all contexts, including belief contexts, being synonymous with $(\lambda x)[..x..]$ by the very construction of the language — by definition, if we choose to call it that.

Thus we are led to a second amendment of Carnap's definition, as follows:

In addition to (1), (2), and (3), as given above, steps of the following kinds shall also be allowed: (4) replacement of an abstraction expression by a synonymous predicator constant; (5) replacement of a predicator constant by a synonymous abstraction expression; (6) replacement of an individual description by a synonymous individual constant; (7) replacement of an individual constant by a synonymous individual description.

For intensional isomorphism as modified by these two amendments of Carnap's definition, let us introduce the name 'synonymous isomorphism.' It is proposed that synonymous isomorphism, as thus defined for the language S_1, and as extended by more or less obvious analogy to many other languages,[10] should replace Carnap's intensional isomorphism as criterion of identity of belief.

In order to make this possible, it is necessary to provide a determination of synonymy as a part of the semantical basis of S_1, or of other language employed. This might be done directly, by means of *rules of synonymy* and *rules of non-synonymy*, or it might be done indirectly by means of *rules of sense*.[11] In either case there are certain obvious limitations upon the Principle of Tolerance which must be taken into account: for example, though we are at liberty in introducing a new constant to fix its meaning in any non-circular fashion that we please, and in particular to make it synonymous with any expression already at hand, we may not by arbitrary convention make the constant synonymous with an expression containing that same constant; and having once fixed the meaning of a constant, we are not then free to make further arbitrary conventions about its meaning (in particular, the same constant may not be made synonymous

[8]This is the account of definition which is given, for example, by Hilbert and Bernays in ***Grundlagen der Mathematik***. In constructing formalized languages, others (including myself) have often preferred to avoid definitions in this sense, which change the object language by adding new notations to it. But such avoidance is on the same ground of economy that underlies the avoidance of synonymous primitive constants, and need not be demanded when economy is not the objective.

[9]Compare Carnap, ***The Logical Syntax of Language***, §22, 1(b).

[10]In particular to any of the languages considered in my paper, *A Formulation of the Logic of Sense and Denotation* (***Structure Method and Meaning***, pp. 3–24), and to languages obtained from these by adding constants of any types, with specified meanings.

It is necessary to explain that the statement on page 5 of that paper, that Alternative (0) "may be described roughly by saying that it makes the notion of sense correspond to Carnap's notion of intensional structure" is an error (unless "roughly" is understood in a very liberal sense). The intention of Alternative (0) is rather that two well-formed formulas shall have the same sense if and only if they are synonymously isomorphic.

[11]See my *The Need for Abstract Entities in Semantic Analysis*, ***Proceedings of the American Academy of Arts and Sciences***, vol. 80 (No. 1), (1951), pp. 110–112.

with two different expressions unless one of these synonymies can be shown to be a consequence of the other).[12]

II

Since our proposal of synonymous isomorphism is almost opposite in tendency to a modification of intensional isomorphism which is proposed in a recent paper by Hilary Putnam,[13] and which seems to be at least partly supported by Carnap,[14] it becomes necessary to consider Putnam's proposal, and in fact to rebut it (in the sense of showing it to be superfluous) if our own is to be maintained. Both Putnam and Carnap rely heavily on a brief remark in a paper of Benson Mates,[15] in such a way that it will be sufficient for our purpose if Mates's remark (as interpreted by Putnam) can be refuted.

Mates introduces two sentences D and D' which shall be particular sentences that are different but intensionally isomorphic. The two sentences D and D' being not otherwise specified by Mates, let us choose them for the purpose of the present discussion as follows:

D. The seventh consulate of Marius lasted less than a fortnight.
D'. The seventh consulate of Marius lasted less than a period of fourteen days.

For the sake of the illustration, we suppose that the word 'fortnight,' in English, means a period of fourteen days and is synonymous with 'a period of fourteen days.'[16] And in order to secure the complete synonymy of D and D', we have used in D' the phrase 'less than a period of fourteen days' rather than the shorter and more natural 'less than fourteen days.'

The sentences D and D', as chosen above, are then not intensionally isomorphic but synonymously isomorphic. They serve our present purpose the better for that very reason. In fact Mates, though directing his remark in the first instance against intensional isomorphism, concludes by saying that it is not affected if 'intensionally isomorphic' is replaced by 'synonymous' throughout. And in reproducing Mates's argument we shall replace his 'intensionally isomorphic' everywhere by 'synonymously isomorphic'—synonymous isomorphism being our proposed explicatum of synonymy.

Consider then, following Mates, the two sentences:

(14) Whoever believes that the seventh consulate of Marius lasted less than a fortnight believes that the seventh consulate of Marius lasted less than a fortnight.

---

[12]Compare the "rules of definition," originally Aristotelian, which are often included in books on traditional logic.

[13]*Synonymity, and the Analysis of Belief Sentences*, **Analysis**, vol. 14 (No. 5), (1954), pp. 114–22.

[14]In a forthcoming paper, *On Belief Sentences: Reply to Alonzo Church*.

[15]*Synonymity*, ***University of California Publications in Philosophy***, vol. 25 (1950), see the lower half of page 215.

[16]To treat the English language as a language for which syntactical and semantical rules have been fully given is of course to make a supposition contrary to fact, but it is one which is very convenient for illustrative purposes and has in fact been adopted in informal discussion by Carnap, Mates, Putnam, and many others. Use of this device has the effect that it may be necessary in the course of the illustration just to invent a rule of English, either to fill a gap in the rules as found in existing grammars and dictionaries or to remove an equivocacy. In the present context, for instance, we have been obliged to decide arbitrarily (or on the basis of mere plausibility) that 'fortnight' is synonymous with 'a period of fourteen days' rather than with 'a period of two weeks'; existing English dictionaries either fail to decide this point or disagree among themselves, probably because universal familiarity with the multiplication table tends to obscure the fact that the two latter (quoted) phrases are not synonymous with each other.

(15) Whoever believes that the seventh consulate of Marius lasted less than a fortnight believes that the seventh consulate of Marius lasted less than a period of fourteen days.

According to Mates, it is true that:

(16) Nobody doubts that whoever believes that the seventh consulate of Marius lasted less than a fortnight believes that the seventh consulate of Marius lasted less than a fortnight.

But it is not true that:

(17) Nobody doubts that whoever believes that the seventh consulate of Marius lasted less than a fortnight believes that the seventh consulate of Marius lasted less than a period of fourteen days.

In fact a counter-example against (17) is evidently provided by philosophers who have considered the question of the criterion of identity of belief, and perhaps in particular by readers of this paper. For by considering this question of philosophical analysis, one is almost inevitably led, at least tentatively, to doubt that (15), or else to entertain an analogous doubt in the case of some other pair of synonymously isomorphic sentences (in place of D and D'). Even if this doubt is afterwards overcome by some counter-argument, the very possibility of entertaining the doubt that (15), without simultaneously doubting that (14), shows (14) and (15) to be non-interchangeable in belief contexts.[17] The historical fact as to who has doubted what or as to the truth of (16) are not really relevant here,[18] but only the *possibility* of doubting that (15) without doubting that (14). Since, according to Mates, (14) and (15) are synonymously isomorphic, the result is to discredit synonymous isomorphism as criterion of identity of belief.

It must be understood that those who are supposed to have doubted that (15) without doubting that (14) are supposed also to have had a sufficient knowledge of the English language so that the doubt was not, for example, a doubt about the meaning of the word 'fortnight' in English.

Nevertheless it is natural to suggest as a means of overcoming Mates's difficulty that it is after all not possible to doubt that (15) without doubting that (14); and that the doubt which has been or may have been sometimes entertained by philosophers in considering the question of the criterion of identity of belief is not the doubt that (15), but a doubt that does have reference to linguistic matters, namely the doubt that

(18) Whoever satisfies in English the sentential matrix '*x* believes that the seventh consulate of Marius lasted less than a fortnight' satisfies in English the sentential matrix '*x* believes that the seventh consulate of Marius lasted less than a period of fourteen days.'[19]

If this suggestion can be supported, the difficulty urged by Mates disappears, as (18) is clearly not synonymously isomorphic either to (14) or to:

---

[17]A context of doubting is of course a belief context, since to doubt is to withhold belief. And a criterion of identity of belief must also be a criterion of identity of doubt.

[18]Doubt being one of the fundamentals of philosophical method, it would be hard indeed to find a proposition that some philosopher might not be found to doubt.

[19]The two occurrences of the phrase 'in English' would usually be omitted, but strictly they are necessary; for the semantical relation of satisfaction (or fulfillment) is a ternary relation among an individual, a sentential matrix, and a language.

(19) Whoever satisfies in English the sentential matrix 'x believes that the seventh consulate of Marius lasted less than a fortnight' satisfies in English the sentential matrix 'x believes that the seventh consulate of Marius lasted less than a fortnight.'[20]

Now the test of translation into another language, originally suggested by C. H. Langford, is often valuable in determining whether a statement under analysis is to be regarded as a statement about some sentence, linguistic expression, or word, or rather as about something which the sentence, expression, or word is being used to mean.[21] I have used this test elsewhere[22] to support the conclusion that the object of a belief shall be taken to be a proposition rather than a sentence, if certain important features of the ordinary usage of indirect discourse are to be preserved. But I say that the same test in the present connection leads to a conclusion of the opposite kind—namely that the doubt whose existence or possibility Mates urges (as a difficulty in the analysis of belief statements) is a doubt about certain sentential matrices, and thus a doubt that (18) rather than a doubt that (15).[23]

Let us therefore translate (14), (15), and (18) into German.

The translation of (18) is:

(18′) Wer auf Englisch die Satzmatrix 'x believes that the seventh consulate of Marius lasted less than a fortnight' erfüllt, erfüllt auf Englisch die Satzmatrix 'x believes that the seventh consulate of Marius lasted less than a period of fourteen days.'

As soon as we set out to translate (14) and (15), our attention is drawn to the fact that the German language has no single word which translates the word 'fortnight,' and that the literal translation of the word 'fortnight' from English into German is 'Zeitraum von vierzehn Tagen.'[24] In consequence the German translations of (14) and (15) are identical, as follows:

(14′) (15′) Wer glaubt dass das siebente Konsulat des Marius weniger als einen Zeitraum von vierzehn Tagen gedauert habe, glaubt dass das siebente Konsulat des Marius weniger als einen Zeitraum von vierzehn Tagen gedauert habe.

Of course we must ask whether the absence of a one-word translation of 'fortnight' is a deficiency of the German language in the sense that there are therefore some things which can be expressed in English but cannot be expressed in German. But it would seem that it can hardly be so regarded—else we should be obligated to call it a deficiency of German also that there is no word to mean a period of fifty-four days and

[20] The point is that names of two different sentences are not synonymous in any sense, and in particular not synonymously isomorphic, even though the sentences themselves be synonymously isomorphic.

[21] (Added August 4, 1954.) The existence of more than one language is not usually to be thought of as a fundamental ground of the conclusions reached by this method. Its role is rather as a useful device to separate those features of a statement which are essential to its meaning from those which are merely accidental to its expression in a particular language, the former but not the latter being invariant under translation. And distinctions (e.g., of use and mention) which are established by this method it should be possible also to see more directly. The point is well illustrated by a paper of Wilfrid Sellars, *Putnam on Synonymity and Belief*, forthcoming in ***Analysis***, in which conclusions the same as or similar to those of Part II of this paper are reached by a more direct analysis. Professor Sellars's paper and mine were written independently, but I saw a copy of it by return mail when my own was submitted to ***Philosophical Studies***.

[22] ***Analysis***, vol. 10 (No. 5), (1950), pp. 97–99.

[23] The object of doubt must still be a proposition, but a proposition *about* certain sentential matrices.

[24] The shorter translation 'vierzehn Tage' would be more usual, but is not quite literal, as may be seen by considering the question of translating the phrase 'three fortnights' into German.

six hours, or that the Latin word 'ero' can be translated only by the three-word phrase 'ich werde sein.' Indeed it should rather be said that the word 'fortnight' in English is not a necessity but a dispensable linguistic luxury.

Granted this, let us translate into German 'Mates doubts that (15) but does not doubt that (14).'[25] As the resulting German sentence is a direct self-contradiction, and as it cannot matter to the soundness of our reasoning whether we carry it out in English or German, we must conclude that Mates (whatever he himself may tell us) does not really so doubt — and that he must have mistaken the doubt that (18) for the doubt that (15).

[25]Of course '(14)' and '(15)' are here used, not as names of the sentences which we have so numbered, but just as convenient abbreviations. The reader must imagine the full sentences written out in place of the '(14)' and '(15).' Indeed throughout the paper such parenthetic numerals are to be understood as *abbreviations* when preceded by the word 'that' — but elsewhere as *names* of their sentences.

PROPOSITIONS AND SENTENCES
(1956)

The meaning of the word *proposition* has an interesting history. In Latin, *propositio* was originally a translation of the Greek *πρότασις*, and seems to have been used at first in the sense of premiss. But already by Boethius the word has come to be used in a sense which it long retained and which I can attempt to express in other words by speaking of a declarative sentence taken together with its meaning. Basically the same sense of the word as Boethius' is intended when Peter of Spain defines, "Propositio est oratio verum vel falsum significans indicando," and when post-scholastic traditional logicians define a proposition as a judgment expressed in words. (The mention of judgment in the definition of proposition is as far as I know post-scholastic, but I think we may ignore it for our present purpose as being a minor change in the definition, not affecting in any essential way the questions with which we shall be concerned tonight.)

Though the terminology is by no means uniform among different writers, it seems fair on the whole to take Peter's definition of *propositio*, just quoted, as representative of the scholastic usage. However, some scholastic logicians use *enuntiatio*, either as an alternative to *propositio*, or in order to reserve the word *propositio* for use in some more special sense. And even Peter of Spain in another passage draws a certain distinction between *propositio* and *enuntiatio*.

Contrasted with this scholastic-traditional use of the word *proposition* is another use of the word which has arisen in more modern times, and which I shall distinguish by speaking of *proposition in the traditional sense* and *proposition in the abstract sense*. It is the latter, abstract, sense of *proposition* which is intended in the title of this lecture.

The difference between the two senses may be explained by supposing that we have before us an English declarative sentence, its translation into Latin, and its translation into German. In the traditional sense there are three different propositions. For though the three sentences have the same meaning (each in its own language), the words used are different in each case, and we must therefore, if we take the traditional definition seriously, speak of three propositions rather than one. In scholastic terminology, *propositio* is *oratio* of a certain kind, and *oratio* in turn is *vox* of a certain kind—to quote Peter of Spain again, *vox significativa ad placitum, cuius partes significant separatae*—and hence if the *vox*, the form of words, is different, the proposition must be said to be different, though all else be the same.

On the other hand, in the abstract sense, the English sentence and its two translations represent just one proposition. A proposition in the abstract sense, unlike the traditional proposition, may not be said to be of any language; it is not a form of words, and is not a linguistic entity of any kind except in the sense that it may be obtained by abstraction from language.

Of some logicians who write of propositions in the traditional sense or of judgments, I think it might reasonably be said that, in some vague way, and in spite of their explicit statement, what is really intended is the more abstract notion. For example in stating a particular syllogism, the minor term, having appeared in full in one of the premisses,

---

Originally published in ***The Problem of Universals: A Symposium***, University of Notre Dame Press, Notre Dame 1956, pp. 1–12.

may be represented in the conclusion only by a pronoun as "it" or "he," or other non-essential changes may be made in the wording of a proposition without any indication that a new proposition has thereby resulted; the fact that no remark or justification is thought to be necessary for this seems to betray, if only by inadvertence, that the writer has in mind the meaning rather than the meaning plus the words. Again, I have heard it argued on behalf of Kant, who makes his logic treat of judgments (*Urtheile*), that what he intends thereby is not a psychological entity, an "act of the mind," but simply a proposition in the abstract sense, as distinguished from the traditional proposition, for which he uses *Satz*.

At any rate I believe that many have found awkward or unsatisfactory the traditional notion of proposition, with its dependence on the particular wording; and that for some purposes at least there is a clear need for the abstract notion — not the declarative sentence, but the content of meaning which is common to the sentence and its translations into other languages — not the particular judgment or thought but, as Frege writes in explaining his term *Gedanke*, the objective content of the thought which is capable of being the common property of many.

An explicit distinction between proposition in the traditional sense and proposition in the abstract sense first appears in Bolzano's *Wissenschaftslehre* of 1837. Bolzano's word is *Satz*, which indeed is the usual German translation of the Latin *propositio*, and the proposition in the abstract sense is distinguished by calling it *Satz an sich*.

In 1892, independently of Bolzano, propositions in the abstract sense were introduced by Frege under the name of *Gedanke*. For Frege, the *Gedanke* is the sense expressed by a declarative sentence (*Behauptungssatz*), as distinguished from the denotation of the sentence, which is its truth-value (i.e., either truth or falsehood).

The abstract notion of proposition appears again in Russell's *The Principles of Mathematics* in 1903. Russell does not mention Bolzano in this connection, but in discussing Frege he explains that Frege's *Gedanke* is approximately the same as his own *unasserted proposition*. Propositions in the abstract sense play an essential role in the *Principia Mathematica* of Whitehead and Russell, as originally written. And though Russell later repudiated the abstract notion — replacing it in *Introduction to Mathematical Philosophy* by a definition of *proposition* which closely follows Peter of Spain, and more recently by a psychological notion of proposition — writers such as Eaton, Cohen and Nagel, Lewis and Langford, Carnap, and many others have followed the early Russell in employing the word *proposition* in the abstract sense.

It should be added that although the use of the particular word *proposition* in this abstract sense is of modern origin, the notion itself is old. In fact the λεκτά of the Stoics are, whenever the λεκτόν of a declarative sense is in question, propositions in the abstract sense.[1] And I am indebted to Professor Bochenski for pointing out to me that the abstract notion appears again in the writings of the later scholastics, beginning with Gregory a Rimini, under the name of *complexe significabile*. Even John of St. Thomas still speaks of the "veritas complexa significata per enuntiationem" in what may be a reference to the *complexe significabile* (though the use of the word "veritas" is very odd in speaking of propositions whose signification may be false as well as true). But these ideas fell into oblivion, and had to be rediscovered in modern times by Bolzano,

[1] It was brought out in the discussion, however, that the Stoics perhaps would not have allowed the existence of a λεκτόν except to correspond to a sentence actually uttered or at least considered by some one.

Frege, and Russell.

The word *sentence* is, of course, originally a term of grammar and linguistics. Its introduction into logic (where it is used to mean declarative sentence) is a recent innovation, and it still seems strange to many to find the word *sentence* where *proposition* might have been expected. I believe that the usage first arose in connection with translation from German into English, as the fact that the German has only one word *Satz* for both sentence and proposition had facilitated a shift in viewpoint on the part of certain logicians which became conspicuous only when the task of translation made it necessary to distinguish the two meanings of the German word. And certainly the use of the word *sentence* has often been the device of nominalistic logicians in order to repudiate propositions. Yet the word cannot be abandoned to the nominalists, as the very decision to use *proposition* in the abstract sense makes it necessary to have another word for the sentence as a purely syntactical entity, taken in abstraction from its meaning. Whatever may be one's philosophical prejudice, nominalistic or Platonistic or other, I believe that this terminology of sentence and proposition will be found superior to the older terminology that emphasizes and gives a special place to the composite entity, sentence plus (abstract) proposition.

The use of the word *propositio* for proposition in the abstract sense is attributed by Bolzano to Leibniz's *Dialogus de Connexione inter Res et Verba*. This would seem to be an exaggeration or a misunderstanding, as it does not appear in this dialogue that Leibniz intends any change in the traditional meaning of *propositio*. Nevertheless the dialogue is of interest in the present connection because it does set forth the essential considerations which tend to show that the duality of sentence and proposition provides a simpler and more satisfactory conceptual scheme than has yet been shown to be possible on a nominalistic basis.

The dialogue begins with A and B agreeing that truth must be supposed to attach, not to thoughts (*cogitationes*) but to things (*res*). For example, if a thread of fixed length is to be laid upon a plane surface so as to enclose a maximum area, the shape must be a circle; and the truth of this, it is agreed, does not depend upon any one's having thought of it, as on the contrary it was true before geometers had proved it or any one had observed it. But, asks A, who seems to speak for Leibniz, can a *thing* be false? B answers that not the thing but some one's thought about it is false. But must it not be the same subject that is capable of truth and of falsehood, as may be seen by considering a case in which one is still in doubt as to whether something is true or false? Thus we are led to say after all that truth must be of thoughts rather than things. How is this to be reconciled with the belief that that can be true which has been thought by no one and perhaps will not be? Leibniz's answer is that truth is neither of thoughts nor of things, but of possible thoughts, or possible propositions.

The remainder of the dialogue is then devoted to arguing that truth cannot be an arbitrary matter, depending upon human conventions about the definition and use of words. For there is but one geometry, the same for the Greeks, the Latins, and the Germans, though expressed in three different languages; and the results of an arithmetic calculation are the same, whether expressed in decimal or duodecimal notation. It is concluded that the basis of truth is not in the notation, not in the symbols or characters themselves, but in something in their use and interconnection which is not arbitrary, a certain relationship (*proportio*) of the characters among themselves and between the characters and things, which under transformation into a different language or notation either remains the same or is transformed into something suitably corresponding.

Leibniz is here very close to the notion of a proposition in the abstract sense, and it remained only for Bolzano to take the final step.

To be sure, instead of transmuting Leibniz's possible thoughts or possible propositions into abstract propositions, we might take them to be the possible sentences of some language, in the sense of all the sentences which the syntactical rules of the language allow as well-formed — whether or not the sentence has actually been written by any one or ever will be, and even if the sentence is so long that there is not space to write it within the confines of our (possibly finite) universe. This might not be wholly in disaccord with Leibniz's own ideas, since he maintains in the same dialogue that no distinct thought or reasoning is possible without words, signs, or characters of some sort. But so far as this device may have a nominalistic motivation, it is defeated by the fact that possible sentences in this sense are not particulars but universals, so that the purposes of nominalism are not served. Moreover if we consider, not the subject (or object) of truth, but rather the object of a belief or an assertion, I believe that the use of possible sentences in place of propositions in the abstract sense, in order to make a purely syntactical analysis, is unsatisfactory. I have tried to show this in a paper in *Analysis* ("On Carnap's analysis of statements of assertion and belief", vol. 10 no. 5, 1950) which I will not discuss now except to express my opinion that, as far as the criticism in this paper concerns the particular analysis of belief statements by Carnap, rebuttals which have been offered are wide of the mark; and that the character of the criticism is such that it seems to leave very little possibility of a successful alternative analysis of belief statements along similar lines.

To return to Leibniz's argument — there are indeed some places in it at which an alternative course to that adopted may be possible or is at least worth consideration. The dialogue form may sometimes conceal this, as when the agreement by A and B that what is thought by no one might still be true tends to deter us from considering the tenability of the contrary opinion. But if we do question this basic assumption, we may then try the possibility of holding that only actual thoughts, or perhaps better, actual concrete sentences, are capable of truth or falsehood. The result is something like the Quine–Goodman finitistic nominalism, since there is possibly only a finite variety of concretely existing sentences — even if we consider the whole extent of past and future time, and even if we count sentences that have not been purposely written by any one but merely happen to exist somewhere and somewhen. On this basis it would seem to become impossible to make certain otherwise ordinary distinctions, such as the distinction between there being in principle no proof of a particular proposed mathematical theorem (for example the Gödel undecidable sentence) and there being no proof of it actually written out because all are too long. But even if we reconcile ourselves to the loss of such distinctions, it is clear from the work of Quine and Goodman that we must at the very best face a very difficult and complicated theory even in the formulation of logical syntax; and the difficulties will certainly be greater in the treatment of even an extensional semantics, and *a fortiori*, of such intensional questions as the logical analysis of belief statements. It is a familiar situation in mathematics generally, and in theoretical physics and other natural sciences, that a theory may be greatly simplified by incorporating into it additional entities beyond those which had originally to be dealt with, and I believe it to be a false economy which would forego simplification of a theory by such means. The notion of a concrete physical object, extended in space and persistent though time, is a case in point, as what had originally to be dealt with by the physical theories in which such objects appear was not these objects themselves but rather certain observations and physical experiences.[2] Indeed the justification would seem to be basically the same for extended physical objects in macrophysical theory and for ideal sentences in logical syntax: both are postulated entities — some may prefer to say inferred entities — without which the theory would be intolerably complex if not impossible.

There is one other point I would like to mention at which the nominalist might attempt to escape the course of Leibniz's argument. This is in the assumption that there is *something* which is the subject of truth — and I would add to this the assumption that there is something which is the object of assertion and of belief. The possibility must indeed be considered that the logical form is to be taken as different. But what has to be said here is that, with one exception, I know of no proposal of a different logical form than this for, say, belief statements which recommends itself as likely to be useful or tenable. There is therefore the chance that some new idea or new direction in the analysis of statements of truth and of assertion, belief, and the like might completely change the present situation.

The one exception I spoke of is the proposal of Israel Scheffler (*Analysis*, vol. 14 no. 4, 1954) according to which names of propositions are to be eliminated in favor of certain predicates of inscriptions, which may be used to assert about an inscription in effect that it expresses such and such a proposition. Scheffler, writing in the context of the Quine–Goodman nominalism, speaks of predicates of inscriptions, i.e., of concrete particular occurrences of sentences, rather than of the ideal sentences of the usual logical syntax, and it seems that this may be essential to his proposal. In fact Scheffler supposes each inscription to have a unique meaning — determined by its context, inclusive of the language of which it forms a part — so that his propositional predicates (as I shall call them) can be taken as predicates of inscriptions only, without introducing the language as a second argument. And because this has the consequence that accidentally occurring inscriptions may not be considered, but only those that have been purposely written by some one in suitable context, the Quine–Goodman undertaking to reconstruct logical syntax on a finitistic basis (see *The Journal of Symbolic Logic*, vol. 12 no. 4, 1947) will certainly be rendered more difficult.

The propositional predicates must of course be either indivisible or at least not analyzable in any way that would reintroduce names of propositions (which would be replaceable by bindable variables). And the propositional predicates themselves must not be allowed to replace or be replaced by bindable variables. In this way Scheffler is able, on Quine's view, to avoid "ontological commitment" to propositions or to abstract entities named by the propositional predicates. But in consequence he is faced with the immediate difficulty that we do often want to make statements which, at least *prima facie*, require in their analysis the use of bound variables taking propositions as values. Examples are, "Church and Goodman have contradicted each other," "Goodman will speak about individuals," "Some assertions of Velikovsky are improbable," and the contention ascribed to Ramus, that all assertions of Aristotle are falsehoods.

[2]Compare a remark made by Kurt Gödel in his paper, *Russell's mathematical logic*, published in the volume ***The Philosophy of Bertrand Russell*** (Library of Living Philosophers).

Of various methods that have occurred to me, to make an attempted nominalistic reproduction of these statements, perhaps the most plausible is to reconstrue them respectively as, "There exists inscriptions i_1 and i_2, such that i_1 has been uttered [spoken or written] by Church, i_2 has been uttered by Goodman, and i_1 and i_2 *are contradictory*." "There exist inscriptions which are *about individuals* and will be uttered by Goodman," "There exist inscriptions which are uttered by Velikovsky and are *improbable*," "All inscriptions uttered by Aristotle *are false*"—where the italicized predicates of inscriptions do not require the language as additional argument, in view of the assumption that the language is uniquely determined by the context. But if one attempts any extensive analysis of these predicates, and many others like them which immediately suggest themselves, it will be difficult to avoid restoring unwanted "ontological commitments." And if on the contrary a large number of these predicates are taken as unanalyzed or primitive, it may be difficult or impossible to provide (axiomatically) for the logical connections among them, between them and the propositional predicates, and between them and the syntactical make-up of the inscriptions they apply to in a specified language.

I conclude that Scheffler has not established his claim to have provided a workable substitute for propositions which is acceptable on the basis of finitistic nominalism. The possibility remains that the claim might be substantiated by a longer and more detailed development, including solutions of the difficulties just discussed and treatment of a compatible finitistic syntax. But objections to finitistic nominalism on the ground that the theory is too complicated in application would in any case not be removed.

It is necessary to consider also the question of an adaptation of Scheffler's device to ordinary non-finitistic syntax, in which sentences are treated rather than inscriptions. And as already suggested, there may be other choices of the logical form to be ascribed to a statement of belief, or the like, that are worth consideration. I want to urge in conclusion that various proposals for the analysis of such statements should be sought, and that those which appear promising should receive a detailed development by the logistic method, as being the only means by which the consequences of the proposal can be satisfactorily brought out and the discussion of it raised above the level of vague and pointless speculation. Sketches, informal suggestions, and general informal surveys such as that I have been making have their place, but in the end are futile, in view of the evident logical difficulty of the problem at hand, unless they issue in a detailed logistic formulation and study of at least one successful solution.

THE EUCLIDEAN PARALLEL POSTULATE (1956)

In order to discuss the parallel postulate it is desirable to have before us a list of postulates (or axioms) for Euclidean plane geometry. It would be appropriate to set down Euclid's own list, except that it is now known that it is incomplete and that some of the postulates in it are inadequately stated. Euclid's list can be amended and completed in various ways, of which for our present purpose we select the following.[1]

1. POSTULATES OF EUCLIDEAN PLANE GEOMETRY

Our undefined terms shall be *point*, *segment* (i.e., of a straight line), *end* point of a segment, *interior* point of a segment, and *congruence* of segments. And our postulates, numbered with Roman numerals, shall be the eighteen following.

I. *Every segment has interior points, and has just two end points, distinct from each other and from the interior points.*

DEFINITION. *If a segment is designated by two letters, as AB, it is meant that A and B are the two end points. The designations AB and BA may be used for the same segment.*

DEFINITION. *One segment is said to contain a second one if all the interior points of the second segment are also interior points of the first one. (In particular, therefore, every segment contains itself.)*

II. *If C is an interior point of the segment AB, there is a unique segment AC contained by AB.*

III. *If the segment AB contains the segment AC and is not the same as AC, then C is an interior point of AB, and the remaining interior points of AB, other than C and the interior points of AC, constitute the interior points (and all the interior points) of a segment CB.*

DEFINITION. *The two segments are said to be linked if the end points of each are end points or interior points of the other. (It will follow from Postulate VI that in Euclidean geometry it is impossible for segments to be linked if they are not the same; but we need the terminology because of the possibility of linked segments in some of the non-Euclidean geometries, as we shall see below, and in order to state Postulates IV and V in a way to make them equally true in Euclidean and non-Euclidean geometry.)*

IV. *If two segments which are not linked have an end point and an interior point in common, one of them contains the other.*

Originally published by the Galois Institute of Mathematics and Art, Brooklyn, 1956.

[1] We have taken these postulates from a set of postulates due to Oswald Veblen (see Veblen and Young, ***Projective Geometry***, volume II); but modifications have been made, with a view to simplifying the explanations and the development of geometry from the postulates, and without regard to the problem of minimizing the assumptions that are made in the postulates. The continuity postulate has been omitted as not properly belonging to elementary geometry; but addition of a continuity postulate to the list would not alter the discussion of non-Euclidean geometry which follows.

V. *If two segments which are not linked have two interior points in common, there exists a segment which contains them both.*

VI. *If two segments have both end points in common they are identical.*

VII. *Given any two distinct points, there exists a segment of which they are the end points.*

VIII. *There exist three points which are not interior points of the same segment.*

DEFINITION. *If the segment AC contains the segment AB, then AC is said to be an extension of AB through B to C.*

IX. *Every segment AB has an extension through B.*

DEFINITION. *The straight line AB consists of all points which are interior points of a given segment AB or of any extension of it through A or through B. These points are said to be points of the straight line, or to lie on it, and the straight line is said to pass through each of the points. Two straight lines are considered to be the same, or to coincide, if their points are the same. And a straight line is said to contain a segment if all interior points of the segment are points of the straight line.*

X. (The parallel postulate.) Through any point not on a straight line AB there is at most one straight line which has no point in common with the straight line AB.

DEFINITION. *A triangle ABC consists of three segments AB, BC, AC (the sides of the triangle) together with the three points A, B, C (the vertices of the triangle), where A, B, C must be points which are not interior points of the same segment. The interior points of the sides of a triangle will be called simply points of the sides.*

XI. *If a straight line does not pass through any of the vertices of a triangle and has a point in common with one of the sides, it must have a point in common with one other side.*

XII. *If one segment is congruent to a second, then the second segment is congruent to the first.*

XIII. *If one segment is congruent to a second, and the second segment is congruent to a third, then the first segment is congruent to the third.*

XIV. *If the segment AB contains the segments AC and CB, and the segment $A'B'$ contains the segments $A'C'$ and $C'B'$, and AC is congruent to $A'C'$ and CB is congruent to $C'B'$, then AB is congruent to $A'B'$.*

XV. *If the three sides of the triangle ABC are congruent respectively to the three sides of the triangle $A'B'C'$, and BD and $B'D'$ are extensions of BC and $B'C'$ respectively, and BD is congruent to $B'D'$, and the segment AD is given, then there is a segment $A'D'$ congruent to AD.*

DEFINITION. *Let AB be a given segment. The end points C of all segments AC congruent to AB constitute a circle whose center is A. The points C are called points of the circle, and the segments AC are called radii of the circle. A segment whose end points are points of the circle and of which the center is an interior point is called a diameter of the circle. The interior points of a circle are the interior points of all its diameters.*

XVI. *If A is the center of a circle, and AD is a segment, then not more than one radius of the circle is contained by AD.*

XVII. *If two circles have distinct centers A and A′, and if at least one point of each circle is an interior point of the other, then the two circles have two points D and E in common, and there is a segment DE one of whose interior points is a point of the straight line AA′.*

XVIII. (The axiom of Archimedes.) *If AB and CD are two segments, there exist a finite number (one or more) of segments* $CD_1, D_1D_2, D_2D_3, \ldots, D_{n-1}D_n$, *of which each is congruent to AB, and all but the last one are contained in CD, the last one has at least some interior points in common with CD, and D is an end point or an interior point of the last one.*

DEFINITION. *An angle BAC is a figure consisting of two non-linked segments AB and AC with a common end point A. The common end point A is called the vertex of the angle, and AB and AC are called the sides of the angle.*

DEFINITION. *The angles BAC and EDF are congruent if there exist extensions AB′, AC′, DE′, DF′, of AB, AC, DE, DF, and also segments B′C′ and E′F′, such that AB′, AC′, and B′C′ are congruent respectively to DE′, DF′, and E′F′.*

DEFINITION. *The point J is interior to the angle BAC if there are segments BC and AJ such that one and only one interior point of BC is an interior point or an end point of some extension of AJ through J.*

DEFINITION. *Angle BAC is the sum of angles EDF and GHI if there is a point J, interior to the angle BAC or interior to one of its sides, such that angles EDF and GHI are congruent to angles BAJ and JAC respectively.*

Having these last definitions, we could go on to obtain the usual theorems about angles (not greater than 180°), including their measurement in degrees.

We carry the development no farther, however. Our present purpose is not to show in detail why it is necessary in the interest of logical accuracy to replace Euclid's original postulates by some such set as these, or how the initial theorems of Euclidean plane geometry follow from them. We intend merely to discuss certain particular questions related to Postulate X, which is equivalent to the postulate of Euclid known as the parallel postulate,[2] and to Postulate VI, whose connection with the parallel postulate will appear. In the course of our discussion we must, for the sake of brevity, assume familiar propositions of geometry as following from the postulates just stated, without having actually exhibited their proofs from these postulates.

The reader may wish to make the attempt to carry out for himself the proofs of, say, the propositions in the first book of Euclid from the above list of postulates. If this is done, it will sometimes be necessary to provide proofs of things which Euclid takes for granted (especially such things as that a certain intersection exists, or that a point falls interior to a certain segment or a certain angle). Again, Euclid's device

[2]Euclid's postulate is, in our terminology: If *AB*, *AC*, and *BD* are segments, and *C* and *D* are on the same side of *AB*, and the sum of the angles *BAC* and *ABD* is less than 180° (less than two right angles), then there are extensions of *AC* and *BD*, through *C* and *D* respectively, which have an interior point in common. Here the condition that *C* and *D* are on the same side of *AB* requires definition, which may be supplied as follows, that there is a segment of which *C* and *D* are interior points and of which none of the interior points lies on the straight line *AB*.

of moving triangles about in order to superpose one on the other may not be used, as we have not included a postulate permitting this, and indeed the accurate statement of such a postulate presents difficulties. And it must also be noticed that the definition of congruence of angles, as given above, does not make it immediately obvious that angles congruent to the same angle are congruent to each other; in the absence of a postulate to this effect it is necessary to make a proof.

2. HYPERBOLIC GEOMETRY

Apparently Euclid himself believed that his parallel postulate did not have the same simple and self-evident character as his other postulates and axioms, since he avoids use of it in the proof of theorems until he reaches a point at which this can no longer conveniently be done. At any rate other geometers after him, in both ancient and modern times, have very generally believed that this proposition ought to be a theorem rather than a postulate, and attempts to prove it as a theorem have been very numerous. All such attempts, however, have either been admitted failures or have contained fallacies, and it is now known with certainty that the postulate which we have numbered X cannot be proved as a consequence of the other postulates. hence there exists besides Euclidean geometry another equally self-consistent geometry, known as hyperbolic geometry, whose postulates (in the plane case) are the same as those which we have listed except that X is replaced by its contrary:

Xa. *Through any point not on a straight line AB there are at least two straight lines which have no point in common with the straight line AB.*

Of fallacious attempts to prove the parallel postulate we may select the following by A. M. Legendre in 1800, which is of special interest as a famous mistake by a great mathematician. If it be admitted that the sum of the angles of a triangle is 180°, then the parallel postulate is readily derived as a consequence (proof of this is left to the reader). Hence in order to prove the parallel postulate it is sufficient to deduce a contradiction from each of the following assumptions: (1) there is a triangle the sum of whose angles is $180° + m$ where m is positive; (2) there is a triangle the sum of whose angles is $180° - m$ where m is positive.

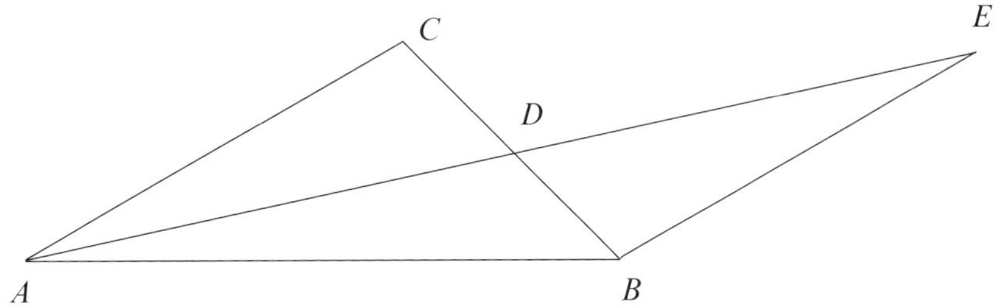

FIGURE 1

To deduce a contradiction from (1) let ABC be a triangle whose angle sum is $180° + m$ and let its smallest angle be at A. Bisect BC at D and extend AD to E so that AD is congruent to ED. Then the triangles ADC and EDB are congruent and their corresponding angles are equal. And from this follows that the angle sum of the triangle ABE is the same as that of the triangle ABC. Hence the process applied to the triangle ABC has yielded another triangle ABE which has again the angle sum $180° + m$ and one of whose angles (at A or at E) is less than or equal to half the smallest angle of the triangle ABC. Let the same process be applied to the triangle ABE, and so on as long as necessary. Ultimately there will be obtained a triangle one of whose angles is less than m but whose angle sum is $180° + m$. This triangle will have two angles whose sum exceeds 180°, contrary to Euclid I 16.[3]

To deduce a contradiction from (2) let ABC be a triangle whose angle sum is $180° - m$ and let its smallest angle be at A. Bisect BC at D and extend AD to E so that AD is congruent to ED. Then the triangles ADC and EDB are congruent, and so are the

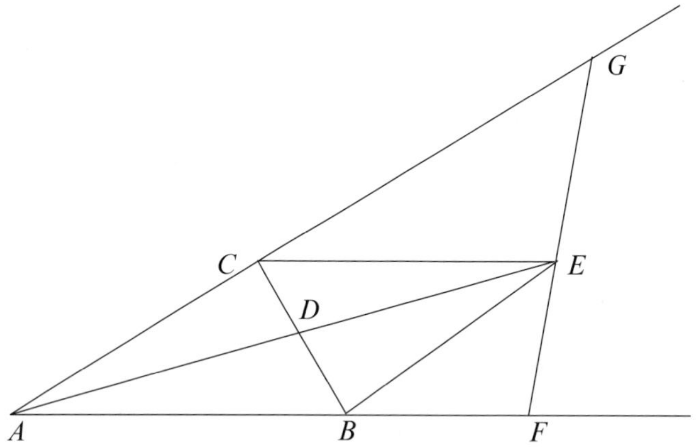

FIGURE 2

triangles ADB and EDC. Hence AC is congruent to EB and AB is congruent to EC. Hence the triangles ABC and ECB are congruent. Extend AB and AC through B and C respectively and through E draw a straight line intersecting AB and AC extended at F and G respectively. Now the sum of the angles of the triangle AFG is equal to the sum of the twelve angles of the four triangles ABC, ECB, CGE, BFE minus three times 180°. Now the sum of the angles of the triangle ABC is $180° - m$ and that of the triangle ECB is $180° - m$, and by the preceding argument (deducing a contradiction from (1)) the triangles CGE and BFE each have angle sums not greater than 180°. Hence the angle sum of the triangle AFG is not greater than

$$180° - m + 180° - m + 180° + 180° - 3 \times 180°,$$

that is, not greater than $180° - 2m$. Repeating on the triangle AFG the same process that we have applied to the triangle ABC, we obtain a triangle whose angle sum is not greater than $180° - 4m$, and so on. Ultimately there will be obtained a triangle whose angle sum is zero or negative. But this is impossible.

---

[3]I 16: In any triangle, if one of the sides is produced, the exterior angle is greater than each of the opposite interior angles.

Indeed in the next proposition, I 17, Euclid uses I 16 to prove explicitly that the sum of two angles of a triangle must be less than two right angles.

Of these two arguments the first is sound (but it should be pointed out that it depends necessarily on Postulate VI, first because without Postulate VI it would be conceivable that the points A and E should coincide and ABE be no triangle at all, or that the segments AD and DE should otherwise be linked, and secondly because the proof of Euclid I 16 depends in an exactly similar way on Postulate VI). The second argument contains a fallacy, which it is left to the reader to discover. It is, as a matter of fact, a theorem of hyperbolic geometry that the angle sum of any triangle is less than 180°.

Examples of fallacious proofs of the parallel postulate could be multiplied almost indefinitely. For the most part they depend, either on a concealed assumption of the parallel postulate itself, or on assumption of some proposition which can be proved only by means of the parallel postulate. Of propositions of the latter class perhaps the most commonly occurring are: (1) through three points not on the same straight line a circle may always be passed; (2) there exist two triangles such that the three angles of one are congruent each to the corresponding angle of the other but whose corresponding sides are not congruent (it is a theorem of the hyperbolic geometry that if the three angles of a triangle are congruent respectively to the corresponding angles of another triangle then the two triangles are congruent).

The first definite development of hyperbolic geometry, based on a postulate asserting the contrary of Euclid's parallel postulate, was made by János Bolyai and N. I. Lobachevsky, in the first half of the nineteenth century, independently of each other and at about the same time.[4] They developed many theorems of hyperbolic geometry without obtaining contradictory results, but of course it by no means follows from their work that hyperbolic geometry is necessarily a self-consistent theory since it would always be possible that some one with more ingenuity or more patience might discover theorems which they had overlooked and that contradictions would arise from these.

The proof that hyperbolic geometry (of either two or three dimensions) is a logically self-consistent theory is contained in results of Arthur Cayley of 1859 and was made explicit by Felix Klein in 1871; and for the two-dimensional case another proof was given by Eugenio Beltrami in 1868. Of the various proofs of this fact which are known, unfortunately all are too long for presentation here.

3. ELLIPTIC GEOMETRY

There is, however, another non-Euclidean geometry, namely elliptic geometry, for which the proof of its self-consistency, while of essentially the same kind as that for hyperbolic geometry, is considerably simpler. This we now proceed to examine.

In our list of postulates, let us replace VI by the following new postulate:

VIa. *There cannot be two straight lines which have no point in common.*

---

[4]In the eighteenth century the attempt of Girolamo Saccheri to demonstrate the parallel postulate by reductio ad absurdum led him to an extensive and accurate development of theorems of hyperbolic geometry; but instead of recognizing the consistency of hyperbolic geometry, Saccheri brings in at the end a fallacious demonstration of inconsistency. (As Halsted remarks, Saccheri's geometry is like a mermaid, somewhat fishy at the end, but oh what an elegant torso!) Many of the ideas involved in hyperbolic geometry were also in the possession of Carl Friedrich Gauss before Bolyai and Lobachevsky, but were not published, and the work of Bolyai and Lobachevsky was independent of Gauss and of Saccheri.

Then the geometry based on the resulting set of postulates is that known as elliptic geometry.

The necessity for omitting Postulate VI in order to introduce V1a can be seen by examining the familiar proof, by reductio ad absurdum, that perpendiculars to the same straight line have no point in common. Let it be granted that two distinct straight lines cannot have more than one point in common (this follows from Postulates IV and V). Let AC and BC be two perpendiculars to the straight line ℓ which have the point C in common. Extend CB to D so that CB is congruent to DB. Angles CBA and DBA

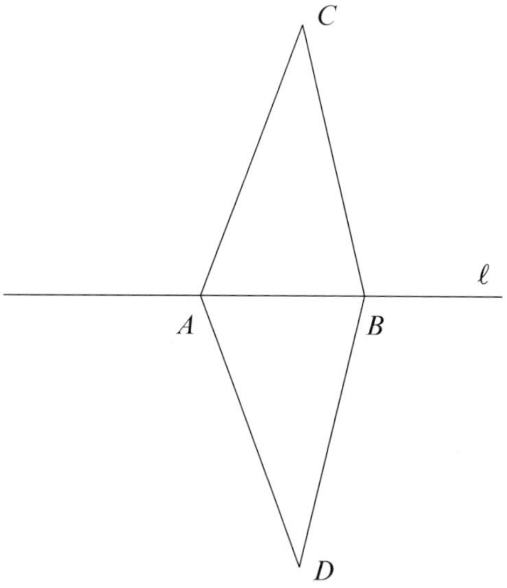

FIGURE 3

are congruent, being both right angles. Hence triangles CBA and DBA are congruent (two sides and included angle). Hence angles CAB and DAB are congruent. Hence both are right angles (since CAB is a right angle). Hence CAD is a straight line. Now CAD and CBD are not in the same straight line, for if they were this straight line would have two points in common with ℓ. Thus we have two straight lines, CAD and CBD, having the two points C and D in common, which is impossible.

This proof would fail without Postulate VI, because it would be conceivable that C and D were the same point. (If C and D were the same point, CB and DB would be non-identical segments having both end points in common, contrary to Postulate VI.)

It follows that Postulate VI is false in elliptic geometry. But it is to be observed that it remains true that two distinct points determine a straight line, and that two distinct straight lines cannot have more than one point in common. Postulate X remains true in elliptic geometry, but has become superfluous, since it makes a weaker statement than Postulate VIa.

In the elliptic geometry it is a theorem that the angle sum of any triangle is greater than 180°.

In order to prove the self-consistency of elliptic geometry, we begin by observing that our axioms contain five undefined terms: point, segment, end point, interior point, congruent. No definitions are given of these five terms, and requirements of rigor compel us in making a proof to be sure that we have not made use of any intuitive conception or pictorial representation of meanings of these terms but have based the proof logically on the statements made in the postulates and those only. Hence if any meanings be assigned to the undefined terms which are such as to make all the postulates true, then all the provable theorems of the geometry will also be true under those same meanings. And this will hold no matter how far-fetched are these meanings assigned to the undefined terms or how widely different from the meanings ordinarily associated with the words in question.

To prove the self-consistency of elliptic geometry it will be sufficient to find meanings for the undefined terms under which all the postulates of elliptic geometry can be shown to be true. For under those meanings of the undefined terms all theorems of elliptic geometry will also be true. And hence there can be no formal contradiction among the theorems of elliptic geometry.

Let the figure composed of two vertical angles (as angles APB and CPD in the figure) be called a *double angle*, and let the two straight lines which contain the sides of the two vertical angles be called the *sides* of the double angle. (Notice in the figure

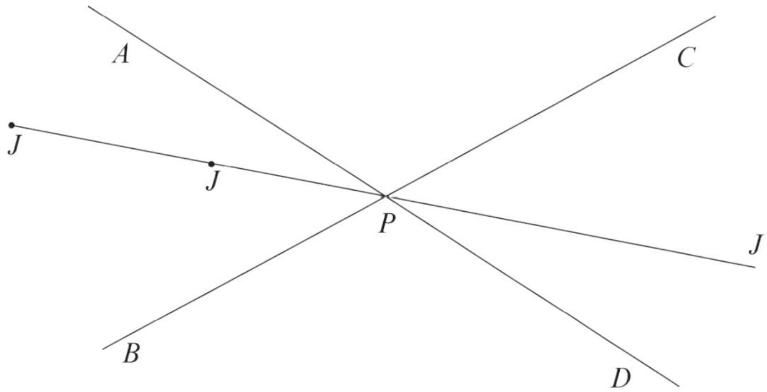

FIGURE 4

that the double angle composed of the two vertical angles APC and BPD is to be considered as different from that composed of the vertical angles APB and CPD.)

Let a straight line be called *interior* to a double angle if it passes through the vertex P and is in the same plane as the two vertical angles, and every point J of the straight line, other than P, is interior to one of the two vertical angles.

Let P be a fixed point in a *Euclidean* three-dimensional space. In the postulates for elliptic geometry let the word "point" be interpreted to mean straight line through P; and let a "segment" be interpreted to consist of all straight lines interior to a given double angle with non-coincident sides and with vertex at P, the "interior points" of the "segment" being the straight lines of which it consists, the "end points" being the sides of the double angle, and two "segments" being "congruent" if the corresponding double angles are congruent. In view of the definitions of "straight line" and "circle" which are adopted, it will be seen that in the postulates for elliptic geometry, a "straight line" must then be interpreted as consisting of all straight lines which pass through P and lie on a certain plane through P, and a "circle" must be interpreted as consisting of the elements of a certain right circular (double) cone with vertex at P, the axis of the cone being the "center" of the circle.[5]

The reader is asked to go through the list of postulates for elliptic geometry which we have given and verify for each one that it is true under the bizarre interpretation of the undefined terms which has just been explained.

It will then follow that elliptic plane geometry is consistent provided Euclidean three-dimensional geometry is consistent, for any inconsistency in elliptic plane geometry would, by means of the interpretation just explained, lead to an inconsistency in Euclidean three-dimensional geometry.

A similar proof of the consistency of elliptic three-dimensional geometry can also be given, but if it be required that the proof be given without making use of a Euclidean space of more than three dimensions then it must be more complicated than that just given for the plane case.

4. GEOMETRY AND COSMOLOGY

Since it can be proved (by methods like those just described) that if Euclidean geometry is consistent with itself, then all three geometries, Euclidean, elliptic, and hyperbolic, are self-consistent, the question arises which of these geometries holds when the undefined terms "point," "segment," "congruent," etc. are given their ordinary meanings,[6] that is, which of them is true of physical space in the real world about us. This question can be answered, if at all, only by the experimental method, not by the logic of the pure geometer.

Here it must be remarked that no geometry can be true of the physical world except in the sense of an approximation, since points and segments exist in the physical world only as approximations. But it is not inconceivable that some experiment might show that Euclidean geometry, say, could not be true of the physical world even in the sense of an approximation. This would be the case, for instance, if accurate measurement of the angles of a large triangle should give an angle sum differing from 180° by an amount too large to be attributed to experimental error.

Until recently (1920 or later) the situation was that no experimental result was known which tended at all to decide the question whether the geometry of the physical world is Euclidean, elliptic, or hyperbolic. For example, if measurement of the angle sum of a triangle gives 180° to within the experimental error, the possibility remains that either more refined methods of measurement or measurement of the angles of a larger triangle would give a result definitely greater or less than 180°.

---

[5]A special cases arises if the "radius" of the "circle" is a 90° double angle, as the cone then reduces to a plane. Correspondingly, in the elliptic geometry proper as distinguished from this interpretation or representation of it, a circle may as a special case reduce to a straight line; such a circle has a center and radii but no diameters and no interior points.

[6]The reader will think of various ways of identifying the ordinary meanings of these terms, by means of physical observations and processes rather than mathematical definitions. For example a rod may be called straight (a segment) if it fits exactly along a plumb line; or if it shows no wobble when the ends are fixed in such a way that the rod is free to rotate; or if it looks straight when one sights along it (in other words if it agrees with the path of a light ray). In astronomy it is most natural to adopt the last of these, to identity a segment with a light ray which, between two points, is not reflected, or refracted, or (for strict accuracy) bent in a gravitational field.

There is now, however, a certain amount of experimental and observational evidence in favor of the General Theory of Relativity, so that it would seem to be indicated that the geometry of the physical world is none of the three which we have discussed but is another and less simple kind of non-Euclidean geometry. This must again be understood in the sense of an approximation. And it must further be remembered that any physical theory, justified by experimental evidence, is subject always to the possibility that it may later be upset by new observations and experiments which are either more accurate or of a different kind than earlier ones.

The question of the geometry of the universe is interesting in particular in connection with the question of its finiteness. For in the elliptic geometry, and in some other more complicated non-Euclidean geometries, a straight line is of finite total length and closes up like a circle–and the total area of a plane and the total volume of three-dimensional space are similarly finite. (Relativity Theory considers a four-dimensional space-time rather than a three-dimensional space, but this also might be either wholly or partly finite.) Of course it may safely be said that no straight line closes up within any distance that a man or any human expedition has actually traveled, or even within the farthest distance which the most powerful existing telescope has seen. But this is not to say what might happen over greater distances. Speculation in this direction is very unsafe, because of the large variety of logically possible geometries; and because of the probability that the same geometry which fits very well, and to within experimental error, not only our local portion of the universe (within the seeing distance of the most powerful telescope) but even perhaps all local portions of the universe (of like size), might still be badly wrong for figures drawn on a grander scale.[7] For this reason cosmology must long remain a far more speculative domain than, say, metaphysics, or theology.

[7]It is true that according to some cosmological theories the local portion of the universe, in the sense here meant, may be as much as some ten thousandths of the whole. But these theories are themselves based on extrapolations of laws of applied geometry and of physics which have to be made *before* there is any estimate of the size of the local portion of the universe in relation to the whole, or even whether the latter is finite or infinite.

INTRODUCTION TO MATHEMATICAL LOGIC, VOLUME I
INTRODUCTION
(1956)

This introduction contains a number of preliminary explanations, which it seems are most suitably placed at the beginning, though many will become clearer in the light of the formal development which follows. The reader to whom the subject is new is advised to read the introduction through once, then return to it later after a study of the first few chapters of the book. Footnotes may in general be omitted on a first reading.

00. Logic. Our subject is *logic*—or, as we may say more fully, in order to distinguish from certain topics and doctrines which have (unfortunately) been called by the same name, it is *formal logic*.

Traditionally, (formal) logic is concerned with the analysis of sentences or of propositions[1] and of proof[2] with attention to the *form* in abstraction from the *matter*. This distinction between form and matter is not easy to make precise immediately, but it may be illustrated by examples.

To take a relatively simple argument for illustrative purposes, consider the following:

I Brothers have the same surname; Richard and Stanley are brothers; Stanley has surname Thompson; therefore Richard has surname Thompson.

Everyday statement of this argument would no doubt leave the first of the three premisses[3] tacit, at least unless the reasoning were challenged; but for purposes of logical analysis all premisses must be set down explicitly. The argument, it may be held, is valid from its form alone, independently of the matter, and independently in particular of the question whether the premisses and the conclusions are in themselves right or wrong. The reasoning may be right though the facts be wrong, and it is just in maintaining this distinction that we separate the form from the matter.

For comparison with the foregoing example consider also:

Originally published as part of ***Introduction to Mathematical Logic***, Volume I, Princeton University Press, Princeton 1956, pp. 1–68.

[1] See §04.

[2] In the light both of recent work and of some aspects of traditional logic we must add here, besides proof, such other relationships among sentences or propositions as can be treated in the same manner, i.e., with regard to form in abstraction from the matter. These include (e.g.) disproof, compatibility; also partial confirmation, which is important in connection with inductive reasoning (cf. C. G. Hempel in ***The Journal of Symbolic Logic***, vol. 8 (1943), pp. 122–143).

But no doubt these relationships both can and should be reduced to that of proof, by making suitable additions to the object language (§07) if necessary. E.g., in reference to an appropriate formalized language as object language, disproof of a proposition or sentence may be identified with proof of its negation. The corresponding reduction of the notions of compatibility and confirmation to that of proof apparently requires modal logic—a subject which, though it belongs to formal logic, is beyond the scope of this book.

[3] Following C. S. Peirce (and others) we adopt the spelling *premiss* for the logical term to distinguish it from *premise* in other senses, in particular to distinguish the plural from the legal term *premises*.

II Complex numbers with real positive ratio have the same amplitude; $i - \sqrt{3}/3$ and ω are complex numbers with real positive ratio; ω has amplitude $2\pi/3$; therefore $i - \sqrt{3}/3$ has amplitude $2\pi/3$.

This may be held to have the same form as I, though the matter is different, and therefore to be, like I, valid from the form alone.

Verbal similarity in the statements of I and II, arranged at some slight cost of naturalness in phraseology, serves to highlight the sameness of form. But, at least in the natural languages, such linguistic parallelism is not in general a safe guide to sameness of logical form. Indeed, the natural languages, including English, have been evolved over a long period of history to serve practical purposes of facility of communication, and these are not always compatible with soundness and precision of logical analysis.

To illustrate this last point, let us take two further examples:

III I have seen a portrait of John Wilkes Booth; John Wilkes Booth assassinated Abraham Lincoln; thus I have seen a portrait of an assassin of Abraham Lincoln.

IV I have seen a portrait of somebody; somebody invented the wheeled vehicle; thus I have seen a portrait of an inventor of the wheeled vehicle.

The argument III will be recognized as valid, and presumably from the logical form alone, but IV as invalid. The superficial linguistic analogy of the two arguments as stated is deceptive. In this case the deception is quickly dispelled upon going beyond the appearance of the language to consider the meaning, but other instances are more subtle, and more likely to generate real misunderstanding. Because of this, it is desirable or practically necessary for purposes of logic to employ a specially devised language, a *formalized language* as we shall call it, which shall reverse the tendency of the natural languages and shall follow or reproduce the logical form—at the expense, where necessary, of brevity and facility of communication. To adopt a particular language thus involves adopting a particular theory or system of logical analysis. (This must be regarded as the essential feature of a formalized language, not the more conspicuous but theoretically less important feature that it is found convenient to replace the spelled words of most (written) natural languages by single letters and various special symbols.)

01. Names. One kind of expression which is familiar in the natural languages, and which we shall carry over also to formalized languages, is the *proper name*. Under this head we include not only proper names which are arbitrarily assigned to denote in a certain way—such names, e.g., as "Rembrandt," "Caracas," "Sirius," "the Mississippi," "The Odyssey," "eight"—but also names having a structure that expresses some analysis of the way in which they denote.[4] As examples of the latter we may cite:

---

[4]We extend the usual meaning of *proper name* in this manner because such alternative terms as *singular name* or *singular term* have traditional associations which we wish to avoid. The single word *name* would serve the purpose except for the necessity of distinguishing from the *common names* (or *general names*) which occur in the natural languages, and hereafter we shall often say simply *name*.

We do use the word *term*, but in its everyday meaning of an item of terminology, and not with any reference to the traditional doctrine of "categorical propositions" or the like.

"five hundred nine," which denotes a certain prime number, and in the way expressed by the linguistic structure, namely as being five times a hundred plus nine; "the author of *Waverley*," which denotes a certain Scottish novelist, namely Sir Walter Scott, and in the particular way expressed by the linguistic structure, namely as having written *Waverley*; "Rembrandt's birthplace"; "the capital of Venezuela"; "the cube of 2."

The distinction is not always clear in the natural languages between the two kinds of proper names, those which are arbitrarily assigned to have a certain meaning (primitive proper names, as we shall say in the case of a formalized language), and those which have a linguistic structure of meaningful parts. E.g., "The Odyssey" has in the Greek a derivation from "Odysseus," and it may be debated whether it is still to be considered in modern English that the name "The Odyssey" has a structure involving the name "Odysseus." This uncertainty is removed in the case of a formalized language by fixing and making explicit the formation rules of the language (§07).

There is not yet a theory of the meaning of proper names upon which general agreement has been reached as the best. Full discussion of the question would take us far beyond the intended scope of this book. But it is necessary to outline briefly the theory which will be adopted here, due in its essentials to Gottlob Frege.[5]

The most conspicuous aspect of its meaning is that a proper name always is, or at least is put forward as, a *name of* something. We shall say that a proper name *denotes*[6] or *names*[7] that of which it is a name. The relation between a proper name and what

---

[5]See his paper, *Ueber Sinn und Bedeutung*, in ***Zeitschrift für Philosophie und philosophische Kritik***, vol. 100 (1892), pp. 25–50. (There are an Italian translation of this by L. Geymonat in ***Gottlob Frege, Aritmetica e Logica*** (1948), pp. 215–252, and English translations by Max Black in ***The Philosophical Review***, vol. 57 (1948), pp. 207–230, and by Herbert Feigl in ***Readings in Philosophical Analysis*** (1949), pp. 85–102. See reviews of these in ***The Journal of Symbolic Logic***, vol. 13 (1948), pp. 152–153, and vol. 14 (1949), pp. 184–185.)

A similar theory, but with some essential differences, is proposed by Rudolf Carnap in his recent book ***Meaning and Necessity*** (1947).

A radically different theory is that of Bertrand Russell, developed in a paper in ***Mind***, vol. 14 (1905), pp. 479–493; in the Introduction to the first volume of ***Principia Mathematica*** (by A. N. Whitehead and Bertrand Russell, 1910); and in a number of more recent publications, among them Russell's book, ***An Inquiry into Meaning & Truth*** (1940). The doctrine of Russell amounts very nearly to a rejection of proper names as irregularities of the natural languages which are to be eliminated in constructing a formalized language. It falls short of this by allowing a narrow category of proper names which must be names of sense qualities that are known by acquaintance, and which, in Fregean terms, have *Bedeutung* but not *Sinn*.

[6]In the usage of J. S. Mill, and of others following him, not only a singular name (proper name in our terminology) but also a common or general name is said to denote, with the difference that the former denotes only one thing, the latter, many things. E.g., the common name "man" is said to denote Rembrandt; also to denote Scott; also to denote Frege; etc.

In the formalized languages which we shall study, the nearest analogues of the common name will be the *variable* and the *form* (see §02). And we prefer to use a different terminology for variables and forms than that of denoting—in particular because we wish to preserve the distinction of a proper name, or constant, from a form which is concurrent to a constant (in the sense of §02), and from a variable which has one thing only in its range. In what follows, therefore, we shall speak of *proper names only* as denoting.

From another point of view common names may be thought of as represented in the formalized languages, not by variables or forms, but by proper names of classes (class constants). Hence the usage has also arisen according to which a proper name of a class is said to denote the various members of the class. We shall not follow this, but shall speak of proper name of a class as denoting the class itself. (Here we agree with Mill, who distinguishes a singular collective name, or proper name of a class, from a common or general name, calling the latter a "name of a class: only in the distributive sense of being a name of each individual.)

[7]We thus translate Frege's *bedeuten* by *denote* or *name*. The verb to *mean* we reserve for general use, in

it denotes will be called the *name relation*,[8] and the thing[9] denoted will be called the *denotation*. For instance, the proper name "Rembrandt" will thus be said to denote or name the Dutch artist Rembrandt, and he will be said to be the denotation of the name "Rembrandt." Similarly, "the author of *Waverley*" denotes or names the Scottish author, and he is the denotation both of this name and of the name "Sir Walter Scott."

That the meaning of a proper name does not consist solely in its denotation may be seen by examples of names which have the same denotation though their meanings are in some sense different. Thus "Sir Walter Scott" and "the author of *Waverley*" have the same denotation; it is contained in the meaning of the first name, but not of the second, that the person named is a knight or baronet and has the given name "Walter" and surname "Scott";[10] and it is contained in the meaning of the second name, but not of the first, that the person named wrote *Waverley* (and indeed as sole author, in view of the definite article and of the fact that the phrase is put forward as a proper name). To bring out more sharply the difference in meaning of the two names let us notice that, if two names are *synonymous* (have the same meaning in all respects), then one may always be substituted for the other without change of meaning. The sentence, "Sir Walter Scott is the author of *Waverley*," has, however, a very different meaning from the sentence, "Sir Walter Scott is Sir Walter Scott": for the former sentence conveys an important fact of literary history of which the latter gives no hint. This difference in meaning may lead to a difference in truth when the substitution of one name for the other occurs within certain contexts.[11] E.g., it is true that "George IV once demanded to know whether Scott was the author of *Waverley*"; but false that "George IV once demanded to know whether Scott was Scott."[12]

Therefore, besides the denotation, we ascribe to every proper name another kind of meaning, the *sense*,[13] saying, e.g., that "Sir Walter Scott" and "the author of *Waverley*" have the same denotation but different senses.[14] Roughly, the sense is what is grasped

---

reference to possible different kinds of meaning.

[8]The name relation is properly a ternary relation, among a language, a word or phrase of the language, and a denotation. But it may be treated as binary by fixing the language in a particular context. Similarly one should speak of the denotation of a name *with respect to a language*, omitting the latter qualifications only when the language has been fixed or when otherwise no misunderstanding can result.

[9]The word *thing* is here used in its widest sense, in short for anything namable.

[10]The term *proper name* is often restricted to names of this, i.e., which have as part of their meaning that the denotation is so called or is or was entitled to be so called. As already explained, we are not making such a restriction.

Though it is, properly speaking, irrelevant to the discussion here, it is of interest to recall that Scott did make use of "the author of ***Waverley***" as a pseudonym during the time that his authorship of the ***Waverley Novels*** was kept secret.

[11]Contexts, namely, which render the occurrences of the names *oblique* in the sense explained below.

[12]The particular example is due to Bertrand Russell; the point which it illustrates, to Frege.

This now famous question, put to Scott himself in the indirect form of a toast "to the author of ***Waverley***" at a dinner at which Scott was present, was met by him with a flat denial, "Sire, I am not the author of ***Waverley***." We may therefore enlarge on the example by remarking that Scott, despite a pardonable departure from the truth, did not mean to go so far as to deny his self-identity (as if he had said "I am not I"). And his hearers surely did not so understand him, though some must have shrewdly guessed the deception as to his authorship of ***Waverley***.

[13]We adopt this as the most appropriate translation of Frege's *Sinn*, especially since the technical meaning given to the word *sense* thus comes to be very close indeed to the ordinary acceptation of the sense of an expression. (Russell and some others following him have used "meaning" as a translation of Frege's *Sinn*.)

[14]A similar distinction is made by J. S. Mill between the denotation and the connotation of a name. And

when one understands a name,[15] and it may be possible thus to grasp the sense of a name without having knowledge of its denotation except as being determined by this sense. If, in particular, the question "Is Sir Walter Scott the author of *Waverley*?" is used in an intelligent demand for new information, it must be that the questioner knows the sense of the names "Sir Walter Scott" and "the author of *Waverley*" without knowing of their denotations enough to identify them certainly with each other.

We shall say that a name *denotes* or *names* its denotation and *expresses*[16] its sense. Or less explicitly we may speak of a name just as *having* a certain denotation and *having* a certain sense. Of the sense we say that it *determines* the denotation, or *is a concept*[17] of the denotation.

Concepts[17] we think of as non-linguistic in character—since synonymous names, in the same or different languages, express the same sense or concept—and since the same name may also express different senses, either in different languages or, by equivocation, in the same language. We are even prepared to suppose the existence of concepts of things which have no name in any language in actual use. But every concept of a thing is a sense of some name of it in some (conceivable) language.

The possibility must be allowed of concepts which are not concepts of any actual thing, and of names which express a sense but have no denotation. Indeed such names, at least on one very plausible interpretation, do occur in the natural languages such as English: e.g., "Pegasus,"[18] "the king of France in A.D. 1905." But, as Frege has observed, it is possible to avoid such names in the construction of formalized

---

in fact we are prepared to accept *connotation* as an alternative translation of *Sinn*, although it seems probable that Frege did not have Mill's distinction in mind in making his own. We do not follow Mill in admitting names which have denotation without connotation, but rather hold that a name must always point to its denotation *in some way*, i.e., through some sense or connotation, though the sense may reduce in special cases just to the denotation's being called so and so (e.g., in the case of personal names), or to its being what appears here and now (as sometimes in the case of the demonstrative "this"). Because of this and other differences, and because of the more substantial content of Frege's treatment, we attribute the distinction between sense and denotation to Frege rather than to Mill. Nevertheless the discussion of names in Mill's ***A System of Logic*** (1843) may profitably be read in this connection.

[15]It is not meant by this to imply any psychological element in the notion of sense. Rather, a sense (or a concept) is a postulated abstract object, with certain postulated properties. These latter are only briefly indicated in the present informal discussion; and in particular we do not discuss the assumptions to be made about equality of senses, since this is unnecessary for our immediate purpose.

[16]This is our translation of Frege's *drückt aus*. Mill's term *connotes* is acceptable here, provided that care is taken not to confuse Mill's meaning of this term with other meanings which it has since acquired in common English usage.

[17]This use of *concept* is a departure from Frege's terminology. Though not identical with Carnap's use of *concept* in recent publications, it is closely related to it, and was suggested to the writer by correspondence with Carnap in 1943. It also agrees well with Russell's use of *class-concept* in ***The Principles of Mathematics*** (1903)—cf. §69 thereof.

[18]While the exact sense of the name "Pegasus" is variable or uncertain, it is, we take it, roughly that of the winged horse who took such and such a part in such and such supposed events—where only such minimum essentials of the story are to be included as it would be necessary to verify in order to justify saying, despite the common opinion, that "Pegasus did after all exist."

We are thus maintaining that, in the present actual state of the English language, "Pegasus" is not just a personal name, having the sense of who or what was called so and so, but has the more complex sense described. However, such questions regarding the natural languages must not be supposed always to have one final answer. On the contrary, the present actual state (at any time) tends to be indeterminate in a way to leave much debatable.

languages.[19] And it is in fact often convenient to do this.

To understand a language fully, we shall hold, requires knowing the senses of all names in the language, but not necessarily knowing which senses determine the same denotation, or even which senses determine denotations at all.

In a well constructed language of course every name should have just one sense, and it is intended in the formalized languages to secure such univocacy. But this is far from being the case in the natural languages. In particular, as Frege has pointed out, the natural languages customarily allow, besides the *ordinary* (*gewöhnlich*) use of a name, also an *oblique* (*ungerade*) use of the name, the sense which the name would express in its ordinary use becoming the denotation when the name is used obliquely.[20]

---

[19]For example, in the case of a formalized language obtained from one of the logistic systems of Chapter X (or of a paper by the writer in ***The Journal of Symbolic Logic***, vol. 5 (1940), pp. 56–68) by an interpretation retaining the principal interpretation of the variables and of the notations λ (abstraction) and ( ) (application of function to argument), it is sufficient to take the following precautions in assigning sense to the primitive constants. For a primitive constant type o or ι the sense must be such as—on the basis of accepted presuppositions—to assure the existence of a denotation in the appropriate domain, $\mathfrak{D}$ (of truth-values) or $\mathfrak{J}$ (of individuals). For a primitive constant of type $\alpha\beta$ the sense must be such as—on the same basis—to assure the existence of a denotation which is in the domain $\mathfrak{A}\mathfrak{B}$, i.e., which is a function from the (entire) domain $\mathfrak{B}$ which is taken as the range of variables of type β, to the domain $\mathfrak{A}$ which is taken as the range of variables of type α.

Then every well-informed formula without free variables will have a denotation, as indeed it must if such interpretation of the logistic system is to accord with formal properties of the system.

As in the case, e.g., of $\iota_{\alpha(o\alpha)}$, it may happen that the most immediate or naturally suggested interpretation of a primitive constant of type $\alpha\beta$ makes it denote a function from a proper part of the domain $\mathfrak{B}$ to the domain $\mathfrak{A}$. In such a case the definition of the function must be extended, by artificial means if necessary, over the remainder of the domain $\mathfrak{B}$, so as to obtain a function having the entire domain $\mathfrak{B}$ as its range. The sense assigned to the primitive constant must then be such as to determine this latter function as denotation, rather than the function which had only a proper part of $\mathfrak{B}$ as its range.

[20]For example, in "Scott is the author of ***Waverley***" the names "Scott," "***Waverley***," "the author of ***Waverley***" have ordinary occurrences. But in "George IV wished to know whether Scott was the author of ***Waverley***" the same three names have oblique occurrences (while "George IV" has an ordinary occurrence). Again, in "Schliemann sought the site of Troy" the names "Troy" and "the site of Troy" occur obliquely. For to seek the site of some other city, determined by a different concept, is not the same as to seek the site of Troy, not even if the two cities should happen as a matter of fact (perhaps unknown to the seeker) to have had the same site.

According to the Fregean theory of meaning which we are advocating, "Schliemann sought the site of Troy" asserts a certain relation as holding, not between Schliemann and the site of Troy (for Schliemann might have sought the site of Troy though Troy had been a purely fabulous city and its site had not existed), but between Schliemann and a certain concept, namely that of the site of Troy. This is, however, not to say that "Schliemann sought the site of Troy" means the same as "Schliemann sought the concept of Troy." On the contrary, the first sentence asserts the holding of a certain relation between Schliemann and the concept of the site of Troy, and is true; but the second sentence asserts the holding of a like relation between Schliemann and the concept of the concept of the site Troy, and is very likely false. The relation holding between Schliemann and the concept of the site of Troy is not quite that of having sought, or at least it is misleading to call it that—in view of the way in which the verb *to seek* is commonly used in English.

(W. V. Quine—in ***The Journal of Philosophy***, vol. 40 (1943), pp. 113–127, and elsewhere—introduces a distinction between the "meaning" of a name and what the name "designates" which parallels Frege's distinction between sense and denotation, also a distinction between "purely designative" occurrences of names and other occurrences which coincides in many cases with Frege's distinction between ordinary and oblique occurrences. For a discussion of Quine's theory and its differences from Frege's see a review by the present writer, in ***The Journal of Symbolic Logic***, vol. 8 (1943), pp. 45–47; also a note by Morton G. White in ***Philosophy and Phenomenological Research***, vol. 9, no. 2 (1948), pp. 305–308.)

Supposing univocacy in the use of names to have been attained (this ultimately requires eliminating the oblique use of names by introducing special names to denote the sense which other names express[21]), we make, with Frege, the following assumptions, about names which have a linguistic structure and contain other names as constituent parts: (1) when a constituent name is replaced by another having the same sense, the sense of the entire name is not changed; (2) when a constituent name is replaced by another having the same denotation, the denotation of the entire name is not changed (though the sense may be).[22]

We make explicit also the following assumption (of Frege), which, like (1) and (2), has been implicit in the foregoing discussion: (3) The denotation of a name (if there is one) *is a function of* the sense of the name, in the sense of §03 below; i.e., given the sense, the existence and identity of the denotation are thereby fixed, though they may not necessarily therefore be known to every one who knows the sense.

02. Constants and variables. We adopt the mathematical usage according to which a proper name of a number is called a *constant*, and in connection with formalized languages we extend this usage by removing the restriction to numbers, so that the term *constant* becomes synonymous with *proper name having a denotation*.

However, the term *constant* will often be applied also in the construction of uninterpreted calculi—logistic systems in the sense of §07—some of the symbols or expressions being distinguished as constants just in order to treat them differently from others in giving the rules of the calculus. Ordinarily the symbols or expressions thus distinguished as constants will in fact become proper names (with denotation) in at least one of the possible interpretations of the calculus.

As already familiar from ordinary mathematical usage, a *variable* is a symbol whose meaning is like that of a proper name or constant except that the single denotation of the constant is replaced by the possibility of various *values* of the variable.

Because it is commonly necessary to restrict the values which a variable may take, we think of a variable as having associated with it a certain non-empty range of possible values, the *range of* the variable as we shall call it. Involved in the meaning of a variable, therefore, are the kinds of meaning which belong to a proper name of the range.[23] But a variable must not be identified with a proper name of its range, since

[21] As an indication of the distinction in question we shall sometimes (as we did in the second paragraph of footnote 20) use such phrases as "the concept of Sir Walter Scott," "the concept of the author of ***Waverley***," "the concept of the site of Troy" to *denote* the same concepts which are *expressed* by the respective names "Sir Walter Scott," "the author of ***Waverley***," "the site of Troy." The definite article "the" sufficiently distinguishes the phrase (e.g.) "the concept of the site of Troy" from the similar phrase "a concept of the site of Troy," the latter phrase being used as a common name to refer to any one of the many different concepts of this same spot.

This device is only a rough expedient to serve the purpose of informal discussion. It does not do away with the oblique use of names because, when the phrase "the concept of the site of Troy" is used in the way described, it contains an oblique occurrence of "the site of Troy."

[22] To avoid serious difficulties, we must also assume when a constituent name has no denotation that the entire name is then likewise without denotation. In the natural languages such apparent examples to the contrary as "the myth of *Pegasus*," "the search by Ponce de Leon for *the fountain of youth*" are to be explained as exhibiting oblique occurrences of the italicized constituent name.

[23] Thus the distinction of sense and denotation comes to have an analogue for variables. Two variables with ranges determined by different concepts have to be considered as variables of different kinds, even if the ranges themselves should be identical. However, because of the restricted variety of ranges of variables

there are also differences of meaning between the two.[24]

The meaning which a variable does possess is best explained by returning to the consideration of complex names, containing other names as constituent parts. In such a complex name, having a denotation, let one of the constituent names be replaced at one or more (not necessarily all) of its occurrences by a variable, say x. To avoid complications, we suppose that x is a variable which does not otherwise occur,[25] and that the denotation of the constituent name which x replaces is in the range of x. The resulting expression (obtained from the complex name by thus replacing one of the constituent names by a variable) we shall call a *form*.[26] Such a form, for each value of x within the range of x, or at least for certain such values of x, has a *value*. Namely, the value of the form, for a given value of x, is the same as the denotation of the expression obtained from the form by substituting everywhere for x a name of the given value of x (or, if the expression so obtained has no denotation, then the form has no value for that value of x).[27]

---

admitted, this question does not arise in connection with any of the formalized languages which are actually considered below.

[24]That such an identification is impossible may be quickly seen from the point of view of the ordinary mathematical use of variables. For two proper names of the range are fully interchangeable if only they have the same sense; but two distinct variables must be kept distinct even if they have the same range determined by the same concept. E.g., if each of the letters x and y is a variable whose range is the real numbers, we are obliged to distinguish the two inequalities $x(x+y) \geqq 0$ and $x(x+x) \geqq 0$ as different—indeed the second inequality is universally true, the first one is not.

[25]This is for momentary convenience of explanation. We shall apply the name *form* also to expressions which are similarly obtained but in which the variable x may otherwise occur, provided the expression has at least one occurrence of x as a free variable (see footnote 28 and the explanation in §06 which is there referred to).

[26]This is a different use of the word *form* from that which appeared in §00 in the discussion of form and matter. We shall distinguish the latter use, when necessary, by speaking more explicitly of *logical form*.

Our present use of the word *form* is similar to that which is familiar in algebra, and in fact may be thought of as obtained from it by removing the restriction to a special kind of expressions (polynomials, or homogeneous polynomials). For the special case of propositional forms (see §04), the word is already usual in logic in this sense, independently of its use by algebraists—see, e.g., J. N. Keynes, ***Formal Logic***, 4th edn., 1906, p. 53; Hugh MacColl in ***Mind***, vol. 19 (1910), p. 193; Susanne K. Langer, ***Introduction to Symbolic Logic***, 1937, p. 91; also Heinrich Scholz, ***Vorlesungen über Grundzüge der Mathematischen Logik***, 1949 (for the use of *Aussageform* in German).

Instead of the word *form*, we might plausibly have used the word *variable* here, by analogy with the way in which we use *constant*. I.e., just as we apply the term *constant* to a complex name containing other names (constants) as constituent parts, so we might apply the term *variable* to an appropriate complex expression containing variables as constituent parts. This usage may indeed be defended as having some sanction in mathematical writing. But we prefer to preserve the better established usage according to which a variable is always a single symbol (usually a letter or letter with subscripts).

The use, by some recent authors, of the word *function* (with or without a qualifying adjective) for what we here call a form is, in our opinion, unfortunate, because it tends to conflict with and obscure the abstract notion of a function which will be explained in §03.

[27]It follows from assumption (2), at the end of §01, that the value thus obtained for the form is independent of the choice of a particular name of the given value of x.

The distinction of sense and denotation is, however, relevant here. For in addition to a *value* of the form in the sense explained in the text (we may call it more explicitly a *denotation value*), a complete account must mention also what we may call a *sense value* of the form. Namely, a sense value of the form is determined by a concept of some value of x, and is the same as the sense of the expression obtained from the form by substituting everywhere for x a name having this concept as its sense.

It should also be noted that a form, in a particular language, may have a value even for a value of x which is without a name in that language: it is sufficient that the given value of x shall have a name in

A variable such as *x*, occurring in the manner just described, is called a *free variable*[28] of the expression (form) in which it occurs.

Likewise suppose a complex name, having a denotation, to contain two constituent names neither of which is a part of the other, and let these two constituent names be replaced by two variables, say *x* and *y* respectively, each at one or more (not necessarily all) of its occurrences. For simplicity suppose that *x* and *y* are variables which do not occur in the original complex name, and that the denotations of the constituent names which *x* and *y* replace are in the ranges of *x* and *y* respectively. The resulting expression (obtained by the substitution described) is a *form*, with two *free variables x* and *y*. For certain pairs of values of *x* and *y*, within the ranges of *x* and *y* respectively, the form has a *value*. Namely, the value of the form, for given values of *x* and *y*, is the same as the denotation of the expression obtained from the form by substituting everywhere for *x* and *y* names of their respective values (or, if the expression so obtained has no denotation, then the form has no value for these particular values of *x* and *y*).

In the same way forms with three, four, and more free variables may be obtained. If a form contains a single free variable, we shall call it a *singulary*[29] form, if just two free variables, *binary*, if three, *ternary*, and so on. A form with exactly *n* different free variables is an *n-ary* form.

Two forms will be called *concurrent* if they agree in value—i.e., either have the same value or both have no value—for each assignment of values to their free variables. (Since the two forms may or may not have the same free variables, all the variables are to be considered together which have free occurrences in either form, and the forms are concurrent if they agree in value for every assignment of values to these variables.) A form will be called *concurrent* to a constant if, for every assignment of values to its free variables, its value is the same as the denotation of the constant. And two constants will be called *concurrent* if they have the same denotation.

---

some suitable extension of the language—say, that obtained by adding to the vocabulary of the language a name of the given value of *x*, and allowing it to be substitutable for *x* wherever *x* occurs as a free variable. Likewise a form may have sense value for a given concept of a value of *x* if some suitable extension of the language contains a name having that concept as its sense.

It is indeed possible, as we shall see later by particular examples, to construct languages of so restricted a vocabulary as to contain no constants, but only variables and forms. But it would seem that the most natural way to arrive at the meaning of forms which occur in these languages is by contemplating languages which are extensions of them and which do contain constants—or else, what is nearly the same thing, by allowing a temporary change in the meaning of the variables ("fixing the values of the variables") so that they become constants.

[28]We adopt this term from Hilbert (1922), Wilhelm Ackermann (1924), J. v. Neumann (1927), Hilbert and Ackermann (1928), Hilbert and Bernays (1934). For what we here call a free variable the term *real variable* is also familiar, having been introduced by Giuseppe Peano in 1897 and afterward adopted by Russell (1908), but is less satisfactory because it conflicts with the common use of "real variable" to mean a variable whose range is the real numbers.

As we shall see later (§06), a free variable must be distinguished from a *bound variable* (in the terminology of the Hilbert school) or *apparent variable* (Peano's terminology). The difference is that an expression containing *x* as a free variable has values for various values of *x*, but an expression, containing *x* as a bound or apparent variable only, has a meaning which is independent of *x*—not in the sense of having the same value for every value of *x*, but in the sense that the assignment of particular values to *x* is not a relevant procedure.

[29]We follow W. V. Quine in adopting this etymologically more correct term, rather than the presently commoner "unary."

Using the notion of concurrence, we may now add a fourth assumption, or principle of meaning, to the assumptions (1)–(3) of the last two paragraphs of §01. This is an extension of (2) to the case of forms, as follows: (4) In any constant or form, when a constituent constant or form is replaced by another concurrent to it, the entire resulting constant or form is concurrent to the original one.[30] The significance of this principle will become clearer in connection with the use of operators and bound variables, explained in §06 below. It is to be taken, like (2), as a part of our explanation of the name relation, and thus a part of our theory of meaning.

As in the case of *constant*, we shall apply the terms *variable* and *form* also in the construction of uninterpreted calculi, introducing them by special definition for each such calculus in connection with which they are to be used. Ordinarily the symbols and expressions so designated will be ones which become variables and forms in our foregoing sense under one of the principal interpretations of the calculus as a language (see §07).

It should be emphasized that a variable, in our usage, is a symbol of a certain kind[31] rather than something (e.g., a number) which is denoted or otherwise meant by such symbol. Mathematical writers do speak of "variable real numbers," or oftener "variable quantities," but it seems best not to interpret these phrases literally. Objections to the idea that real numbers are to be divided into two sorts or classes, "constant real numbers" and "variable real numbers," have been clearly stated by Frege[32] and need not be repeated here at length.[33] The fact is that a satisfactory theory has never been developed on this basis, and it is not easy to see how it might be done.

---

[30]For completeness—using the notion of sense value explained in footnote 27 and extending it in obvious fashion to n-ary forms—we must also extend the assumption (1) to the case of forms, as follows. Let two forms be called *sense-concurrent* if they agree in sense value for each system of concepts of values of their free variables; let a form be called *sense-concurrent* to a constant if, for every system of concepts of values of its free variables, its sense value is the same as the sense of the constant; and let two constants be called *sense-concurrent* if they express the same sense. Then: (5) In any constant or form, when a constituent constant or form is replaced by another which is sense-concurrent to it, the entire resulting constant or form is sense-concurrent to the original one.

[31]Therefore, a variable (or more precisely, particular instances or occurrences of a variable) can be written on paper—just as the figure 7 can be written on paper, though the number 7 cannot be so written except in the indirect sense of writing something which denotes it.

And similarly constants and forms are symbols or expressions of certain kinds. It is indeed usual to speak also of numbers and physical quantities as "constants"—but this usage is not the same as that in which a constant can be contrasted with a variable, and we shall avoid it in this book.

[32]See his contribution to ***Festschrift Ludwig Boltzmann Gewidmet***, 1904. (Frege's theory of functions as "ungesättigt," mentioned at the end of his paper, is another matter, not necessarily connected with his important point about variables. It will not be adopted in this book, but rather we shall take a function—see §03—to be more nearly what Frege would call "Werthverlauf einer Function.")

[33]However, we mention the following parallel to one of Frege's examples. Shall we say that the usual list of seventeen names is a complete list of the Saxon kings of England, or only that it is a complete list of the constant Saxon kings of England, and that account must be taken in addition of an indefinite number of variable Saxon kings? One of these variable Saxon kings would appear to be a human being of a very striking sort, having been, say, a grown man named Alfred in A.D. 876, and a boy named Edward in A.D. 976.

According to the doctrine we would advocate (following Frege), there are just seventeen Saxon kings of England, from Egbert to Harold, and neither a variable Saxon king nor an indeterminate Saxon king is to be admitted to swell the number. And the like holds for the positive integers, for the real numbers, and for all other domains abstract and concrete. Variability or indeterminacy, where such exists, is a matter of language and attaches to symbols or expressions.

The mathematical theory of real numbers provides a convenient source of examples in a system of notation[34] whose general features are well established. Turning to this theory to illustrate the foregoing discussion, we cite as particular examples of constants the ten expressions:

$$0,\ \frac{1}{2},\ e,\ \frac{1}{2\pi},\ \frac{1-4+1}{4\pi},\ 4e^4,\ e^e,\ e-e,\ \frac{\pi}{2\pi},\ \frac{\sin \pi/7}{\pi/7}.$$

Let us say that x and y are variables whose range is the real numbers, and m, n, r are variables whose range is the positive integers.[35] The following are examples of forms:

$$y,\ \frac{1}{y},\ \frac{1}{x},\ \frac{1}{2x},\ \frac{1-4+1}{4x},\ 4e^x,\ xe^x,\ x^x,$$

$$x-x,\ n-n,\ \frac{x}{2x},\ \frac{r}{2r},\ \frac{\sin x}{x},\ \frac{\sin r}{r},$$

$$ye^x,\ \frac{y}{xy},\ \frac{r}{xr},\ \frac{x-m+1}{m\pi}.$$

The forms on the first two lines are singulary, each having one free variable, y, x, n, or r as the case may be. The forms on the third line are binary, the first two having x and y as free variables, the third one x and r, the fourth one x and m.[36]

The constants

$$\frac{1}{2\pi} \text{ and } \frac{1-4+1}{4\pi}$$

are not identical. But they are concurrent, since each denotes the same number.[37] Similarly the constants $e-e$ and 0, though not identical, are concurrent because the numbers $e-e$ and 0 are

[34] We say "system of notation" rather than "language" because only the specifically numerical notations can be regarded as well established in ordinary mathematical writing. They are usually supplemented (for the statement of theorems and proofs) by one or another of the natural languages, according to the choice of the particular writer.

[35] Every positive integer is also a real number. I.e., the terms must be so understood for purposes of these illustrations.

[36] To illustrate the remark of footnote 28, following are some examples of expressions containing bound variables:

$$\int_0^2 x^x\,dx, \qquad \lim_{x\to 0}\frac{\sin x}{x}, \qquad \sum_{n=1}^{\infty}\prod_{m=1}^{m=n}\frac{x-m+1}{m\pi}.$$

The first two of these are constants, containing x as a bound variable. The third is a singulary form, with x as a free variable and m and n as bound variables.

A variable may have both free and bound occurrences in the same expression. An example is $\int_0^x x^x\,dx$, the double use of the letter x constituting no ambiguity. Other examples are the variable Δx in $(D_x \sin x)\Delta x$ and the variable x in $xE(k)$, if the notations $D_x \sin x$ and $E(k)$ are replaced by their equivalents

$$\lim_{\Delta x\to 0}\frac{\sin(x+\Delta x)-\sin x}{\Delta x}$$

and $$\int_0^1 \frac{\sqrt{1-k^2x^2}}{\sqrt{1-x^2}}\,dx$$ respectively.

[37] Whether these two constants have the same sense (as well as the same denotation) is a question which depends for its answer on a general theory of equality of senses, such as we have not undertaken to discuss here—cf. footnote 15. It is clear that Frege, though he formulates no complete theory of equality of senses, would regard these two constants as having different senses. But a plausible case might be made out for supposing that the two constants have the same sense, on some such ground as that the equation between

identical. Similarly $-\pi/2\pi$ and $-1/2$.

The form xe^x, for the value 0 of x, has the value 0. (Of course it is the number 0 that is here in question, not the constant 0, so that it is equally correct to say that the form xe^x, for the value 0 of x, has the value $e - e$; or that, for the value of $e - e$ of x, it has the value 0; etc.) For the value 1 of x the form xe^x has the value e. For the value 4 of x its value is $4e^4$, a real number for which (as it happens) no simpler name is in standard use.

The form ye^x, for the values 0 and 4 of x and y respectively, has the value 4. For the values 1 and 1 of x and y it has the value $1e^1$; or, what is the same thing, it has the value e.

The form $-y/xy$, for the values e and 2 of x and y respectively, has the value $-1/e$. For the values e and e of x and y, it has again the value $-1/e$. For the values e and 0 of x and y it has no value, because of the non-existence of a quotient of 0 by 0.

The form $-r/xr$, for the values e and 2 of x and r respectively, has the value $-1/e$. But there is no value for the values e and e of x and r, because e is not in the range of r (e is not one of the possible values of r).

The forms

$$-\frac{1}{2x} \text{ and } \frac{1-4+1}{4x}$$

are concurrent, since they are both without a value for the value 0 of x, and they have the same value for all other values of x. The forms $-1/x$ and $-y/xy$ fail to be concurrent, since they disagree for the value 0 of y (if the value of x is not 0). But the forms $-1/x$ and $-r/xr$ are concurrent.

The forms $-1/y$ and $-1/x$ are not concurrent, as they disagree, e.g., for the values 1 and 2 of x and y respectively.

The forms $x - x$ and $n - n$ are concurrent to the same constant, namely 0,[38] and are therefore also concurrent to each other.

The forms $-x/2x$ and $-r/2r$ are non-concurrent because of disagreement for the value 0 of x. The latter form, but not the former, is concurrent to a constant, namely to $-1/2$.

03. Functions. By a *function*—or, more explicitly, a *one-valued singulary* function—we shall understand an operation[39] which, when applied to something as *argument*, yields a certain thing as the *value* of the function *for* that argument. It is not required that the function be applicable to every possible thing as argument, but rather it lies in the nature of any given function to be applicable to certain things and, when applied to one of them as argument, to yield a certain value. The things to which the function is applicable constitute the *range of* the function (or the *range of arguments* of the function) and the values constitute the *range of values of* the function. The function itself consists in the yielding or determination[39] of a value from each argument in the range of the function.

---

them expresses a necessary proposition or is true on logical grounds alone or the like. No doubt there is more than one meaning of "sense," according to the criterion adopted for equality of sense, and the decision among them is a matter of convention and expediency.

[38] Or also to any other constant which is concurrent to 0.

[39] Of course the words "operation," "yielding," "determination" as here used are near-synonyms of "function" and therefore our statement, if taken as a definition, would be open to the suspicion of circularity. Throughout this Introduction, however, we are engaged in informal explanation rather than definition, and, for this purpose, must take the notion of function as primitive or undefined, or else some related notion, such as that of a class. (We shall see later how it is possible to think of a class as a special case of a function, and also how classes may be used, in certain connections or for certain purposes, to replace and do the work of functions in general.)

As regards equality or identity of functions we make the decision which is usual in mathematics. Namely, functions are identical if they have the same range and have, for each argument in the range, the same value. In other words, we take the word "function" to mean what may otherwise be called a *function in extension*. If the way in which a function yields or produces its value from its argument is altered without causing any change either in the range of the function or in the value of the function for any argument, then the function remains the same; but the associated *function concept*, or concept determining the function (in the sense of §01), is thereby changed.

We shall speak of a function *from* a certain class *to* a certain class to mean a function which has the first class as its range and has all its values in the second class (though the second class may possibly be more extensive than the range of values of the function).

To denote the value of a function for a given argument, it is usual to write a name of the function, followed by a name of the argument between parentheses. And of course the same notation applies (*mutatis mutandis*) with a variable or a form in place of either one or both of the names. Thus if f is a function and x belongs to the range of f, then $f(x)$ is the value of the function f for the argument x.[40]

This is the usual notation for application of a function to an argument, and we shall often employ it. In some contexts (see Chapter X) we find it convenient to alter the notation by changing the position of the parentheses, so that we may write in the altered notation: if f is a function and x belongs to the range of f, then (fx) is the value of the function f for the argument x.

So far we have discussed only *one-valued singulary functions* (and have used the word "function" in this sense). Indeed no use will be made in this book of many-valued functions,[41] and the reader must always understand "function" to mean a one-valued function. But we go on to explain functions of more than one argument.

A *binary function*, or function *of two arguments*,[42] is characterized by being applicable to two arguments in a certain order and yielding, when so applied, a certain value, the *value of* the function *for* those two arguments in that order. It is not required that the function be applicable to every two things as arguments; but rather, the function

---

[40]This sentence exemplifies the use of variables to make general statements, which we assume is understood from familiar mathematical usage, through it has not yet been explained in this Introduction. (See the end of §06.)

[41]It is the idea of a many-valued (singulary) function that, for a fixed argument, there may be more than one value of the function. If a name of the function is written, followed by a name of an argument between parentheses, the resulting expression is a common name (see footnote 6) denoting the values of the function for that argument.

Though many-valued functions seem to arise naturally in the mathematical theories of real and complex numbers, objections immediately suggest themselves to the idea as just explained and are not easily overcome. Therefore it is usual to replace such many-valued functions in one way or another by one-valued functions. One method is to replace a many-valued singulary function by a corresponding one-valued binary propositional function or relation (§04). Another method is to replace the many-valued function by a one-valued function whose values are classes, namely, the value of the one-valued function for a given argument is the class of the values of the many-valued function for that argument. Still another method is to change the range of the function, an argument for which the function has n values giving way to n different arguments for each of which the function has a different one of those n values (this is the standard role of the Riemann surface in the theory of complex numbers).

[42]Though it is in common use we shall avoid the phrase "function of two variables" (and "function of three variables" etc.) because it tends to make confusion between *arguments* to which a function is applied and *variables* taking such arguments as values.

is applicable in certain cases to an ordered pair of things as arguments, and all such ordered pairs constitute the *range of* the function. The values constitute the *range of values of* the function.

Binary functions are identical (i.e., are the same function) if they have the same range and have, for each ordered pair of arguments which lies in that range, the same value.

To denote the value of a binary function for given arguments, it is usual to write a name of the function and then, between parentheses and separated by a comma, names of the arguments in order. Thus if f is a binary function and the ordered pair of x and y belongs to the range of f, then $f(x, y)$ is the value of the function f for the arguments x and y in that order.

In the same way may be explained the notion of a ternary function, of a quaternary function, and so on. In general, an n-ary function is applied to n arguments in an order, and when so applied yields a value, provided the ordered system of n arguments is in the range of the function. The value of an n-ary function for given arguments is denoted by a name of the function followed, between parentheses and separated by commas, by names of the arguments in order.

Two binary functions ϕ and ψ are called *converses*, each of the other, in case the two following conditions are satisfied: (1) the ordered pair of x and y belongs to the range of ϕ if and only if the ordered pair of y and x belongs to the range of ψ; (2) for all x, y such that the ordered pair of x and y belongs to the range of ϕ,[43]

$$\phi(x, y) = \psi(y, x).$$

A binary function is called *symmetric* if it is identical with its converse. The notions of converse and symmetry may also be extended to n-ary functions, several different converses and several different kinds of symmetry appearing when the number of arguments is three or more (we need not stop over details of this).

We shall speak of a function *of* things of a certain kind to mean a function such that all the arguments to which it is applicable are of that kind. Thus a singulary function of real numbers, for instance, is a function from some class of real numbers to some (arbitrary) class. A binary function of real numbers is a binary function whose range consists of ordered pairs of real numbers (not necessarily all ordered pairs of real numbers).

We shall use the phrase "______ is a function of ______," filling the blanks with forms,[44] to mean what is more fully expressed as follows: "There exists a function f such that

$$\_\_\_\_\_\_ = f(\_\_\_\_\_\_)$$

for all ______," where the first two blanks are filled, in order, with the same forms as before, and the third blank is filled with a complete list of the free variables of those forms. Similarly we shall use "______ is a function of ______ and ______," filling the

[43] The use of the sign = to express that things are identical is assumed familiar to the reader. We do not restrict this notation to the special case of numbers, but use it for identity generally.

[44] Our explanation assumes that neither of these forms has the particular letter f as one of its free variables. In the contrary case, the explanation is to be altered by using in place of the letter f as it appears in the text some variable (with appropriate range) which is not a free variable of either form.

three blanks with forms, to stand for: "There exists a binary function f such that

$$\_\_\_\_ = f(\_\_\_\_, \_\_\_\_)$$

for all ______." where the first three blanks are filled, in order, with the same forms as before, and the last blank is filled with a complete list of the free variables of those forms.[45] And similar phraseology will also be used where the reference is to a function f of more than two arguments.

The phraseology just explained will also be used with the added statement of a condition or restriction. For example, "______ is a function of ______ and ______ if ______," where the first three blanks are filled with forms, and the fourth is filled with the statement of a condition involving some or all of the free variables of those forms,[46] stands for: "There exists a binary function f such that

$$\_\_\_\_ = f(\_\_\_\_, \_\_\_\_)$$

for all ______ for which ______," where the first three blanks are filled, in order, with the same forms as before, the fourth blank is filled with a complete list of the free variables of those forms, and the fifth blank is filled in the same way as the fourth blank was before.[47]

Also the same phraseology, explained in the two preceding paragraphs, will be used with common names[48] in place of forms. In this case the forms which the common names represent have to be supplied from the context. For example, the statement that "*The density of helium* gas is a function of *the temperature* and *the pressure*" is to be understood as meaning the same as "*The density of h* is a function of *the temperature of h* and *the pressure of h*," where the three italicized forms replace the three original italicized common names, and where h is a variable whose values are instantaneous bits of helium gas (and whose range consists of all such). Or to avoid introducing the variable h with so special a range, we may understand instead: "The density of b is a function of the temperature of b and the pressure of b if b is an instantaneous bit of helium gas." Similarly the statement at the end of §01 that the denotation of a name is a function of the sense means more explicitly (the reference being to a fixed language) that there exists a function f such that

$$\text{denotation of } N = f(\text{sense of } N)$$

for all names N for which there is a denotation.

It remains now to discuss the relationship between *functions*, in the abstract sense that we have been explaining, and *forms*, in the sense of the preceding section (§02).

If we suppose the language fixed, every singulary form has corresponding to it a function f (which we shall call the *associated function* of the form) by the rule that the value of f for an argument x is the same as the value of the form for the value x of the

---

[45]The theory of real numbers again serving as a source of examples, it is thus true that $x^3 + y^3$ is a function of $x + y$ and xy. But it is false that $x^3 + x^2y - xy^2 + y^3$ is a function of $x + y$ and xy (as is easily seen on the ground that the form $x^3 + x^2y - xy^2 + y^3$ is not symmetric). Again, $x^4 + y^4 + z^4 + 4x^3y + 4xy^3 + 4x^3z + 4xz^3 + 4y^3z + 4yz^3$ is a function of $x + y + z$ and $xy + xz + yz$. But $x^4 + y^4 + z^4$ is not a function of $x + y + z$ and $xy + xz + yz$.

[46]Thus with a *propositional form* in the sense of §04 below.

[47]Accordingly it is true, for example, that: $x^3 + x^2y - xy^2 + y^3$ is a function of $x + y$ and xy if $x \geqq y$. For the special case that the variables have a range consisting of real or complex numbers, a geometric terminology is often used, thus: $x^3 + x^2y - xy^2 + y^3$ is a function of $x + y$ and xy in the half plane $x \geqq y$.

[48]See footnotes 4, 6.

free variable of the form, the range of f consisting of all x's such that the form has a value for the value x of its free variable.[49] But, still with reference to a fixed language, not every function is necessarily the associated function of some form.[50]

It follows that two concurrent singulary forms with the same free variable have the same associated function. Also two singulary forms have the same associated function if they differ only by alphabetic change of the free variable,[51] i.e., if one is obtained from the other by substituting everywhere for its free variable some other variable with the same range—with, however, the proviso (the need of which will become clearer later) that the substituted variable must remain a free variable at every one of its occurrences resulting from the substitution.

As a notation for (i.e., to denote) the associated function of a singulary form having, say, x as its free variable, we write the form itself with the letters λx prefixed. And of course likewise with any other variable in place of x.[52] Parentheses are to be supplied

---

[49]For example, in the theory of real numbers, the form $\frac{1}{2}(e^x - e^{-x})$ determines the function sinh as its associated function, by the rule that the value of sinh for an argument x is $\frac{1}{2}(e^x - e^{-x})$. The range of sinh then consists of all x's (i.e., all real numbers x) for which $\frac{1}{2}(e^x - e^{-x})$ has a value. In other words, as it happens in this particular case, the range consists of all real numbers.

Of course the free variable of the form need not be the particular letter x, and indeed it may be clearer to take an example in which the free variable is some other letter.

Thus the form $\frac{1}{2}(e^y - e^{-y})$ determines the function sinh as its associated function, by the rule that the value of sinh for an argument x is the same as the value of the form $\frac{1}{2}(e^y - e^{-y})$ for the value x of the variable y. (I.e., in particular, the value of sinh for the argument 2 is the same as the value of the form $\frac{1}{2}(e^y - e^{-y})$ for the value 2 of the variable y; and so on for each different argument x that may be assigned.)

Ordinarily, just the equation

$$\sinh(x) = \frac{1}{2}(e^x - e^{-x})$$

is written as sufficient indication of the foregoing. And this equation may even be called a *definition* of sinh, in the sense of footnote 168, (1) or (3).

[50]According to classical real-number theory, the singulary functions from real numbers to real numbers (or even just the analytic singulary functions) are non-enumerable. Since the forms in a particular language are always enumerable, it follows that there is no language or system of notation in which every singulary function from real numbers to real numbers is the associated function of some form.

Because of the non-enumerability of the real numbers themselves, it is even impossible in any language to provide proper names of all the real numbers. (such a thing as, e.g., an infinite decimal expansion must not be considered a *name* of the corresponding real number, as of course an infinite expansion cannot ever be written out in full, or included as a part of any actually written or spoken sentence.)

[51]E.g., as appears in footnote 49, the forms $\frac{1}{2}(e^x - e^{-x})$ and $\frac{1}{2}(e^y - e^{-y})$ have the same associated function.

[52]Thus the expressions $\lambda x(\frac{1}{2}(e^x - e^{-x}))$, $\lambda y(\frac{1}{2}(e^y - e^{-y}))$, sinh are all three synonymous, having not only the same denotation (namely the function sinh), but also the same sense, even under the severest criterion of sameness of sense.

(In saying this we are supposing a language or system of notation in which the two different expressions sinh and $\lambda x(\frac{1}{2}(e^x - e^{-x}))$ both occur. However, the very fact of synonymy shows that the expression sinh is dispensable in principle: except for considerations of convenience, it could always be replaced by the longer expression $\lambda x(\frac{1}{2}(e^x - e^{-x}))$. In constructing a formalized language, we prefer to avoid such duplications of notation so far as readily possible. See §11.)

The expressions $\lambda x(\frac{1}{2}(e^x - e^{-x}))$, $\lambda y(\frac{1}{2}(e^y - e^{-y}))$ contain the variables x and y respectively, as *bound* variables in the sense of footnotes 28, 36 (and of §06 below). For, according to the meaning just explained for them, these expressions are constants, not singulary forms. But of course the expression $\frac{1}{2}(e^x - e^{-x})$ is a singulary form, with x as a free variable.

The meaning of such an expression as $\lambda x(ye^x)$, formed from the binary form ye^x by prefixing λx, now follows as a consequence of the explanation about variables and forms in §02. In this expression, x is a

as necessary.[53]

As an obvious extension of this notation, we shall also prefix the letters λx (λy, etc.) to any constant as a notation for the function whose value is the same for all arguments and is the denotation of the constant, the range of the function being the same as the range of the variable x.[54] This function will be called an *associated singulary function* of the constant, by analogy with the terminology "associated function of a form," though there is the difference that the same constant may have various associated functions with different ranges. Any function whose value is the same for all arguments will be called a *constant function* (without regard to any question whether it is an associated function of a constant, in some particular language under consideration).[55]

Analogous to the associated function of a singulary form, a binary form has two associated binary functions, one for each of the two orders in which the two free variables may be considered—or better, one for each of the two ways in which a pair of arguments of the function may be assigned as values to the two free variables of the form.

The two associated functions of a binary form are identical, and thus reduce to one function, if and only if they are symmetric. In this case the binary form itself is also called *symmetric*.[56]

Likewise an n-ary form has $n!$ associated n-ary functions, one for each of the permutations of its free variables. Some of these associated functions are identical in certain cases of symmetry.

Likewise a constant has associated m-ary functions, for $m = 1, 2, 3, \ldots$, by an obvious extension of the explanation already made for the special case for $m = 1$. And by a still further extension of this we may speak of the associated m-ary functions of an n-ary form, when $m > n$. In particular a singulary form has not only an associated singulary function but also associated binary functions, associated ternary functions, and so on. (When, however, we speak simply of *the* associated function of a singulary form, we shall mean the associated singulary function.)

The notation by means of λ for the associated functions of a form, as introduced above for singulary functions, is readily extended to the case of m-ary functions,[57] but we shall not have occasion to use such extension in this book. The passage from a form to an associated function (for which the λ-notation provides a symbolism) we

---

bound variable and y is a free variable, and the expression is a singulary form whose values are singulary functions. From it, by prefixing λy, we obtain a constant, denoting a singulary function, and the range of values of this singulary function consists of singulary functions.

[53]In constructing a formalized language, the manner in which parentheses are to be put in has to be specified with more care. As a matter of fact this will be done, as we shall see, not by associating parentheses with the notation λx, but by suitable provision for parentheses (or brackets) in connection with various other notations which may occur in the form to which λx is prefixed.

[54]Thus in connection with real-number theory we use $\lambda x 2$ as a notation for the function whose range consists of all real numbers and whose value is 2 for every argument.

[55]Note should also be taken of expressions in which the variable after λ is not the same as the free variable of the form which follows; thus, for example, $\lambda y(\frac{1}{2}(e^x - e^{-x}))$. As is seen from the explanation in §02, this expression is a singulary form with x as its free variable, the values of the form being constant functions. For the value 0 of x, e.g., the form $\lambda y(\frac{1}{2}(e^x - e^{-x}))$ has as its value the constant function $\lambda y 0$.

In both expressions, $\lambda y(\frac{1}{2}(e^x - e^{-x}))$ and $\lambda y 0$, y is a bound or apparent variable.

[56]We have already used this term, as applied to forms, in footnote 45, assuming the reader's understanding of it as familiar mathematical terminology.

[57]This has been done by Carnap in ***Notes for Symbolic Logic*** (1937) and elsewhere.

shall speak of as *abstraction* or, more explicitly, *m-ary functional abstraction* (if the associated function is *m*-ary).

Historically the notion of a function was of gradual growth in mathematics, and its beginning is difficult to trace. The particular word "function" was first introduced by G. W. v. Leibniz and was adopted from him by Jean Bernoulli. The notation $f(x)$, or fx, with a letter such as f in the role of a function variable, was introduced by A. C. Clairaut and by Leonhard Euler. But early accounts of the notion of *function* do not sufficiently separate it from that of an expression containing free variables (or a *form*). Thus Euler explains a *function of a variable quantity* by identifying it with an analytic expression,[58] i.e., a form in some standard system of mathematical notation. The abstract notion of a function is usually attributed by historians of mathematics to G. Lejeune Dirichlet, who in 1837 was led by his study of Fourier series to a major generalization in freeing the idea of a function from its former dependence on a mathematical expression or law of circumscribed kind.[59] Dirichlet's notion of a function was adopted by Bernhard Riemann (1851),[60] by Hermann Hankel (1870),[61] and indeed by mathematicians generally. But two important steps remained to be taken by Frege (in his ***Begriffsschrift*** of 1879 and later publications): (i) the elimination of the dubious notion of a variable quantity in favor of the variable as a kind of symbol;[62] (ii) the admission of functions of arbitrary range by removing the restriction that the arguments and values of a function be numbers. Closely associated with (ii) is Frege's introduction of the *propositional function* (in 1879), a notion which we go on to explain in the next section.

04. Propositions and propositional functions. According to grammarians, the unit of expression in the natural languages is the *sentence*, an aggregation of words which makes complete sense or expresses a complete thought. When the complete thought expressed is that of an assertion, the sentence is called a *declarative sentence*. In what follows we shall have occasion to refer only to declarative sentences, and the simple

---

[58]"*Functio quantitatis variabilis est expressio analytica quomodocunque composita ex illa quantitate variabili et numeris seu quantitatibus constantibus.* Omnis ergo expressio analytica, in qua praeter quantitatem variabilem z omnes quantitates illam expressionem componentes sunt constantes, erit functio ipsius z ... *Functio ergo quantitatis variabilis ipsa erit quantitas variabilis.*" ***Introductio in Analysin Infinitorum*** (1748), p. 4; ***Opera***, ser. 1, vol. 8, p. 18. See further footnote 62.

[59]See his ***Werke***, vol. 1, p. 135. It is not important that Dirichlet restricts his statement at this particular place to continuous functions, since it is clear from other passages in his writings that the same generality is allowed to discontinuous functions. On page 132 of the same volume is his well-known example of a function from real numbers to real numbers which has exactly two values, one for rational arguments and one for irrational arguments.

Dirichlet's generalization had been partially anticipated by Euler in 1749 (see an account by H. Burkhardt in ***Jahresbericht der Deutschen Mathematiker-Vereinigung***, vol. 10 part 2 (1908), pp. 13–14) and later by J. B. J. Fourier (see his ***Oeuvres***, vol. 1, pp. 207, 209, 230–232).

[60]***Werke***, pp. 3–4.

[61]In a paper reprinted in the ***Mathematische Annalen***, vol. 20 (1882), pp. 63–112.

[62]The passage quoted from Euler in footnote 58 reads as if his *variable quantity* were a kind of symbol or expression. But this is not consistent with statements made elsewhere in the same work which are essential to Euler's use of the notion of function—e.g., "*Si fuerit y functio quaecunque ipsius z, tum vicissim z erit functio ipsius y*" (***Opera***, p. 24), "Sed omnis transformatio consistit in alio modo eandem functionem exprimendi, quemadmodum ex Algebra constat eandem quantitatem per plures diversas formas exprimi posse" (***Opera***, p. 32).

word "sentence" is to be understood always as meaning a declarative sentence.[63]

We shall carry over the term *sentence* from the natural languages also to the formalized languages. For logistic systems in the sense of §07—uninterpreted calculi—the term *sentence* will be introduced by special definition in each case, but always with the intention that the expressions defined to be sentences are those which will become sentences in our foregoing sense under interpretations of the calculus as a formalized language.[64]

In order to give an account of the meaning of sentences, we shall adopt a theory due to Frege according to which sentences are names of a certain kind. This seems unnatural at first sight, because the most conspicuous use of sentences (and indeed the one by which we have just identified or described them) is not barely to name something but to make an assertion. Nevertheless it is possible to regard sentences as names by distinguishing between the assertive use of a sentence on the one hand, and its non-assertive use, on the other hand, as a name and a constituent of a longer sentence (just as other names are used). Even when a sentence is simply asserted, we shall hold that it is still a name, though used in a way not possible for other names.[65]

An important advantage of regarding sentences as names is that all the ideas and explanations of §§01–03 can then be taken over at once and applied to sentences, and related matters, as a special case. Else we should have to develop independently a theory of the meaning of sentences; and in the course of this, it seems, the developments of these three sections would be so closely paralleled that in the end the identification of sentences as a kind of names (though not demonstrated) would be very forcefully suggested as a means of simplifying and unifying the theory. In particular we shall require variables for which sentences may be substituted, forms which become sentences upon replacing their free variables by appropriate constants, and associated functions of such forms—things which on the theory of sentences as names, fit naturally into their proper place in the scheme set forth in §§02–03.

---

[63]The question may be raised whether, say, an interrogative or an imperative logic is possible, in which interrogative or imperative sentences and what they express (questions or commands) have roles analogous to those of declarative sentences and propositions in logic of ordinary kind. And some tentative proposals have in fact been made towards an imperative logic, and also towards an optative logic or logic of wishes. But these matters are beyond the scope of this book.

[64]Cf. the explanation in §02 regarding the use in connection with logistic systems of the terms *constant, variable, form*. An analogous explanation applies to a number of terms of like kind to be introduced below—in particular, *propositional variable, propositional form, operator, quantifier, bound variable, connective*.

[65]To distinguish the non-assertive use of a sentence and the assertive use, especially in a formalized language, Frege wrote a horizontal line, —, before the sentence in the former case, and the character ⊢ before it in the latter case, the addition of the vertical line was thus serving as a sign of assertion. Russell, and Whitehead and Russell in ***Principia Mathematica***, did not follow Frege's use of the horizontal line before non-asserted sentences, but did take over the character ⊢ in the role of an assertion sign.

(Frege also used the horizontal line before names other than sentences, the expression so formed being a false sentence. But this is a feature of his notation which need not concern us here.)

In this book we shall not make use of a special assertion sign, but (in a formalized language) shall employ the mere writing of a sentence displayed on a separate line or lines as sufficient indication of its assertion. This is possible because sentences used non-assertively are always constituent parts of asserted sentences, and because of the availability of a two-dimensional arrangement on a printed page. (In a one-dimensional arrangement the assertion sign would indeed be necessary, if only as punctuation.)

The sign ⊢ which is employed below, in Chapter I and later chapters, is not the Frege-Russell assertion sign, but has a wholly different use.

Granted that sentences are names, we go on, in the light of the discussion in §01, to consider the denotation and the sense of sentences.

As a consequence of the principle (2), stated in the next to last paragraph of §01, examples readily present themselves of sentences which, though in some sense of different meaning, must apparently have the same denotation. Thus the denotation (in English) of "Sir Walter Scott is the author of *Waverley*" must be the same as that of "Sir Walter Scott is Sir Walter Scott," the name "the author of *Waverley*" being replaced by another which has the same denotation. Again the sentence "Sir Walter Scott is the author of *Waverley*" must have the same denotation as the sentence "Sir Walter Scott is the man who wrote twenty-nine Waverley Novels altogether," since the name "the author of *Waverley*" is replaced by another name of the same person; the latter sentence, it is plausible to suppose, if it is not synonymous with "The number, such that Sir Walter Scott is the man who wrote that many Waverley Novels altogether, is twenty-nine," is at least so nearly so as to ensure its having the same denotation; and from this last sentence in turn, replacing the complete subject by another name of the same number, we obtain, as still having the same denotation, the sentence "The number of counties in Utah is twenty-nine."

Now the two sentences, "Sir Walter Scott is the author of *Waverley*" and "The number of counties in Utah is twenty-nine," though they have the same denotation according to the preceding line of reasoning, seem actually to have very little in common. The most striking thing that they do have in common is that both are true. Elaboration of examples of this kind leads us quickly to the conclusion, as at least plausible, that all true sentences have the same denotation. And parallel examples may be used in the same way to suggest that all false sentences have the same denotation (e.g., "Sir Walter Scott is not the author of *Waverley*" must have the same denotation as "Sir Walter Scott is not Sir Walter Scott").

Therefore, with Frege, we postulate[66] two abstract objects called *truth-values*, one of them being *truth* and the other *falsehood*. And we declare all true sentences to denote the truth-value truth, and all false sentences to denote the truth-value falsehood. In alternative phraseology, we shall also speak of a sentence as *having* the truth-value truth (if it is true) or *having* the truth-value falsehood (if it is false).[67]

The sense of a sentence may be described as that which is grasped when one understands the sentence, or as that which two sentences in different languages must have in common in order to be correct translations each of the other. As in the case of names generally, it is possible to grasp the sense of a sentence without therefore necessarily having knowledge of its denotation (truth-value) otherwise than as determined by this sense. In particular, though the sense is grasped, it may sometimes remain unknown whether the denotation is truth.

Any concept of a truth-value, provided that *being a truth-value* is contained in the concept, and whether or not it is the sense of some actually available sentence in a

---

[66]To Frege, as a thoroughgoing Platonic realist, our use of the word "postulate" here would not be acceptable. It would represent his position better to say that the situation indicates that *there are* two such things as truth and falsehood (*das Wahre* and *das Falsche*).

[67]The explicit use of two truth-values appears for the first time in a paper by C. S. Peirce in the ***American Journal of Mathematics***, vol. 7 (1885), pp. 180–202 (or see his ***Collected Papers***, vol. 3, pp. 210–238). Frege's first use of truth-values is in his *Funktion und Begriff* of 1891 and in his paper of 1892 which is cited in footnote 5; it is in these that the account of sentences as names of truth-values is first put forward.

particular language under consideration, we shall call a *proposition*, translating thus Frege's *Gedanke*.

Therefore a proposition, as we use the term, is an abstract object of the same general category as a class, a number, or a function. It has not the psychological character of William of Ockham's *propositio mentalis* or of the traditional *judgment*: in the words of Frege, explaining his term *Gedanke*, it is "nicht das subjective Thun des Denkens, sondern dessen objectiven Inhalt, der fähig ist, gemeinsames Eigenthum von Vielen zu sein."

Traditional (post-Scholastic) logicians were wont to define a proposition as a judgment expressed in words, thus as a linguistic entity, either a sentence or a sentence taken in association with its meaning.[68] But in non-technical English the word has long been used rather for the meaning (in our view the sense) of a sentence,[69] and logicians have latterly come to accept this as the technical meaning of "proposition." This is the happy result of a process which, historically, must have been due in part to sheer confusion between the sentence in itself and the meaning of the sentence. It provides in English a distinction not easily expressed in some other languages, and makes possible a translation of Frege's *Gedanke* which is less misleading than the word "thought."[70]

According to our usage, every proposition determines or is a concept of (or, as we shall also say, has) some truth-value. It is, however, a somewhat arbitrary decision that we deny the name *proposition* to senses of such sentences (of the natural languages) as express a sense but have no truth-value.[71] To this extent our use of *proposition* deviates from Frege's use of *Gedanke*. But the question will not arise in connection with the

[68]E.g., in Isaac Watts's ***Logick***, 1725: "A *Proposition* is a Sentence wherein two or more *Ideas* or *Terms* are joined or disjoined by one Affirmation or Negation . . . . In describing a *Proposition* I use the Word *Terms* as well as *Ideas*, because when mere Ideas are join'd in the Mind without Words, it is rather called a *Judgment*; but when clothed with Words, it is called a *Proposition*, even tho' it be in the Mind only, as well as when it is expressed by speaking or Writing." Again in Richard Whately's *Elements of Logic*, 1826: "The second part of Logic treats of the *proposition*; which is, '*Judgment expressed in words*.' A Proposition is defined logically 'a sentence indicative,' i.e., affirming or denying; (this excludes *commands* and *questions*.)" Here Whately is following in part the Latin of Henry Aldrich (1691). In fact these passages show no important advance over Petrus Hispanus, who wrote a half millennium earlier, but they are quoted here apropos of the history of the word "proposition" in English.

[69]Consider, for example, the incongruous result obtained by substituting the words "declarative sentence" for the word "proposition" in Lincoln's Gettysburg Address.

[70]For a further account of the history of the matter, we refer to Carnap's ***Introduction to Semantics***, 1942, pp. 235–236; and see also R. M. Eaton, ***General Logic***, 1931.

[71]By the remark of footnote 22, such are sentences which contain non-obliquely one or more names that express a sense but lack a denotation—or so, following Frege, we shall take them. Examples are: "The present king of France is bald"; "The present king of France is not bald"; "The author of ***Principia Mathematica*** was born in 1861." (As to the last example, it is true that the phrase "the author of ***Principia Mathematica***" in some appropriate supporting context may be an ellipsis for something like "the author of ***Principia Mathematica*** who was just mentioned" and therefore have a denotation; but we here suppose that there is no such supporting context, so that the phrase can only mean "the one and only author of ***Principia Mathematica***" and therefore have no denotation.)

To sentences as a special case of names, of course the second remark of footnote 22 also applies. Thus we understand as true (and containing oblique occurrences of names) each of the sentences: "Lady Hamilton was like Aphrodite in beauty"; "The fountain of youth is not located in Florida"; "The present king of France does not exist." Cases of doubt whether a sentence has a truth-value or not are also not difficult to find in this connection, the exact meaning of various phraseologies in the natural languages being often insufficiently determinate for a decision.

formalized languages which we shall study, as these languages will be so constructed that every name—and in particular every sentence—has a denotation.

A proposition is then *true* if it determines or has the truth-value truth, *false* if it has the truth-value falsehood. When a sentence expressing a proposition is asserted we shall say that the proposition itself is thereby *asserted*.[72]

A variable whose range is the two truth-values — thus a variable for which sentences (expressing propositions) may appropriately be substituted — is called a *propositional variable*. We shall not have occasion to use variables whose values are propositions, but we would suggest the term *intensional propositional variable* for these.

A form whose values are truth-values (and which therefore becomes a sentence when its free variables are replaced by appropriate constants) is a *propositional form*. Usage sanctions this term[73] rather than "truth-value form," thus naming the form rather by what is expressed, when constants replace the variables, than by what is denoted.

A propositional form is said to be *satisfied by* a value of its free variable, or a system of values of its free variables, if its value for those values of its free variables is truth. (More explicitly, we should speak of a system of values of variables as satisfying a given propositional form *in a given language*, but the references to the particular language may often be omitted as clear from the context.) A propositional form may also be said to be *true* or *false* for a given value of its free variable, or system of values of its free variables, according as its value for those values of its free variables is truth or falsehood.

A function whose range of values consists exclusively of truth-values, and thus in particular any associated function of a propositional form, is a *propositional function*. Here again, established usage sanctions "propositional function"[74] rather than "truth-value function," though the latter term would be the one analogous to, e.g., the term "numerical function" for a function whose values are numbers.

A propositional function is said to be *satisfied by* an argument (or ordered system of arguments) it its value for that argument (or ordered system of arguments) is truth. Or

[72]Notice the following distinction. The statement that a certain proposition was asserted (say on such and such an occasion) need not reveal what language was used nor make any reference to a particular language. But the statement that a certain sentence was asserted does not convey the meaning of the transaction unless it is added what language was used. For not only may the same proposition be expressed by different sentences in different languages, but also the same sentence may be used to assert different propositions according to what language the user intends. It is beside the point that the latter situation is comparatively rare in the principal known natural languages; it is not rare when all possible languages are taken into account.

Thus, if the language is English, the statement, "Seneca said that man is a rational animal," conveys the proposition that Seneca asserted but not the information what language he used. On the other hand the statement, "Seneca wrote, 'Rationale enim animal est homo,' " gives only the information what succession of letters he set down, not what proposition he asserted. (The reader may guess or know from other sources that Seneca used Latin, but this is neither said nor implied in the given statement—for there are many languages besides Latin in which this succession of letters spells a declarative sentence and, for all that thou and I know, one of them may once have been in actual use.)

[73]Cf. footnote 26.

[74]This statement seems to be on the whole just, though the issue is much obscured by divergencies among different writers as to the theory of meaning adopted and in the accounts given of the notions of function and proposition. The idea of the propositional function as an analogue of the numerical function of mathematical analysis originated with Frege, but the term "propositional" function is originally Russell's. Russell's early use of this term is not wholly clear. In his introduction to the second edition of ***Principia Mathematica*** (1925) he decides in favor of the meaning which we are adopting here, or very nearly that.

synonymously we may say that a propositional function *holds for* a particular argument or ordered system of arguments.

From its use in mathematics, we assume that the notion of a *class* is already at least informally familiar to the reader. (The words *set* and *aggregate* are ordinarily used as synonymous with *class*, but we shall not follow this usage, because in connection with the Zermelo axiomatic set theory[75] we shall wish later to give the word *set* a special meaning, somewhat different from that of *class*.) We recall that a class is something which has or may have *members*, and that classes are considered identical if and only if they have exactly the same members. Moreover it is usual mathematical practice to take any given singulary propositional form as having associated with it a class, namely the class whose members are those values of the free variable for which the form is true.

In connection with the functional calculi of Chapters III–VI, or rather, with the formalized languages obtained from them by adopting one of the indicated principal interpretations (§07), it turns out that that we may secure everything necessary about classes by just identifying a class with a singulary propositional function, and membership in the class with satisfaction of the singulary propositional function. We shall consequently make this identification, on the ground that no purpose is served by maintaining a distinction between classes and singulary propositional functions.

We must add at once that the notion of a class obtained by thus identifying classes with singulary propositional functions does not quite coincide with the informal notion of a class which we first described, because it does not fully preserve the principle that classes are identical if they have the same members. Rather, it is necessary to take into account also the *range-members* of a class (constituting, i.e., the range of the singulary propositional function). And only when the range-members are given to be the same is the principle preserved that classes are identical if they have the same members. This or some other departure from the informal notion of a class is in fact necessary, because, as we shall see later,[76] the informal notion—in the presence of some other assumptions difficult to avoid—is self-inconsistent and leads to antinomies. (The *sets* of Zermelo set theory preserve the principle that sets having the same members are identical, but at the sacrifice of the principle that an arbitrary singulary propositional form has an associated set.)

Since, then, a class is a singulary propositional function, we speak of the *range of* the class just as we do of the propositional function (i.e., it is the same thing). We think of the range as being itself a class, having as members the range-members of the given class, and having the same range-members.

(In any particular discussion hereafter in which classes are introduced, and in the absence of any indication to the contrary, it is to be understood that there is a fixed range determined in advance and that all classes have this same range.)

Relations may be similarly accounted for by identifying them with binary propositional functions, the relation being said to *hold between* an ordered pair of things (or the things being said to *stand in* that relation, or to *bear* that relation one to the other) if the binary propositional function is satisfied by the ordered pair. Given that the ranges are the same, this makes two relations identical if and only if they hold between

[75] Chapter XI.

[76] In Chapter VI.

the same ordered pairs, and to indicate this we may speak more explicitly of a *relation in extension*—using this term as synonymous with *relation*.

A *property*, as ordinarily understood, differs from a class only or chiefly in that two properties may be different though the classes determined by them are the same (where the class determined by a property is the class whose members are the things that have that property). Therefore we identify a property with a *class concept*, or concept of a class in the sense of §01. And two properties are said to *coincide in extension* if they determine the same class.

Similarly, a *relation in intension* is a *relation concept*, or concept of a relation in extension.

To turn once more for illustrative purposes to the theory of real numbers and its notations, the following are examples of propositional forms:

$$\sin x = 0, \qquad \sin x = 2,$$
$$e^x > 0, \qquad e^x > 1, \qquad x > 0,$$
$$\varepsilon > 0, \qquad \varepsilon < 0,$$
$$x^3 + y^3 = 3xy, \qquad x \neq y,$$
$$|x - y| < t, \qquad |x - y| < \varepsilon,$$
$$\text{If } |x - y| < \delta \text{ then } |\sin x - \sin y| < \varepsilon.$$

Here we are using x, y, t as variables whose range is the real numbers, and ε and δ as variables whose range is the positive real numbers. The seven forms on the first three lines are examples of singulary propositional forms. Those on the fourth line are binary, on the fifth line ternary, while on the last line is an example of a quaternary propositional form.

Each of the singulary propositional forms has an associated class. Thus with the form $\sin x = 0$ is associated the class of those real numbers whose sine is 0, i.e., the class whose range is the real numbers and whose members are $0, \pi, -\pi, 2\pi, -2\pi, 3\pi$, and so on. As explained, we identify this class with the propositional function $\lambda x(\sin x = 0)$, or in other words the function from real numbers to truth-values which has for any argument x the value $\sin x = 0$.

The two propositional forms $e^x > 1$ and $x > 0$ have the same associated class, namely, the class whose range is the real numbers and whose members are the positive real numbers. This class is identified with either $\lambda x(e^x > 1)$ or $\lambda x(x > 0)$, these two propositional functions being identical with each other by convention about identity of functions adopted in §03.

Since the propositional form $\sin x = 2$ has the value falsehood for every value of x, the associated class $\lambda x(\sin x = 2)$ has no members.

A class which has no members is called a *null class* or *empty class*. From our conventions about identity of propositional functions and of classes, if the range is given, it follows that there is only one null class. But, e.g., the range of the null class associated with the form $\sin x = 2$ and the range of the null class associated with the form $\varepsilon < 0$ are not the same: the former range is the real numbers, and the latter range is the positive real numbers.[77] We shall speak respectively of the "null class of real numbers" and of the "null class of positive real numbers."

A class which coincides with its range is called a *universal class*. For example, the class associated with the form $e^x > 0$ is the universal class of real numbers; and the class associated with the form $\varepsilon > 0$ is the universal class of positive real numbers.

[77]According to the informal notion that classes with the same members are identical, it would be true absolutely that there is only one null class. The distinction of null classes with different ranges was introduced by Russell in 1908 as a part of his theory of types (see Chapter VI). The same thing had previously been done by Ernst Schröder in the first volume of his ***Algebra der Logik*** (1890), though with a very different motivation.

The binary propositional forms $x^3 + y^3 = 3xy$ and $x \neq y$ are both symmetric and therefore each have one associated binary propositional function or relation. In particular, the associated relation of the form $x \neq y$ is the relation of diversity between real numbers; or in other words the relation which has the pairs of real numbers as its range, which any two different real numbers bear to each other, and which no real number bears to itself.

The ternary propositional forms $|x - y| < t$ and $|x - y| < \varepsilon$ have each three associated ternary propositional functions[78] (being symmetric in x and y). All six of these propositional functions are different; but an appropriately chosen pair of them, one associated with each form, will be found to agree in value for all ordered triples of arguments which are in the range of both, differing only in that the first one has the value falsehood for certain ordered triples of arguments which are not in the range of the other.

05. Improper symbols, connectives. When the expressions, especially the sentences, of a language are analyzed into the single symbols of which they consist, symbols which may be regarded as indivisible in the sense that no division of them into parts has relevance to the meaning,[79] we have seen that there are two sorts of symbols which may in particular appear, namely primitive proper names and variables. These we call *proper symbols*, and we regard them as having meaning in isolation, the primitive names as denoting (or at least purporting to denote) something, the variables as having (or at least purporting to have) a non-empty range. But in addition to proper symbols there must also occur symbols which are *improper*—or in traditional (Scholastic and pre-Scholastic) terminology, *syncategorematic*—i.e., which have no meaning in isolation but which combine with proper symbols (one or more) to form longer expressions that do have meaning in isolation.[80]

Conspicuous among improper symbols are parentheses and brackets of various kinds, employed (as familiar in mathematical notation) to show the way in which parts of an expression are associated. These parentheses and brackets occur as constituents in certain combinations of improper symbols such as we now go on to consider—either exclusively to show association and in connection with other improper symbols which carry the burden of showing the particular character of the notation,[81] or

---

[78] We may also occasionally use the term *ternary relation* (and *quaternary relation* etc). But the simple term *relation* will be reserved for the special case of a binary relation or binary propositional function.

[79] The formalized languages are to be so constructed as to make such analysis into single symbols precisely possible. In general it is possible in the natural languages only partially and approximately—or better, our thinking of it as possible involves a certain idealization.

In written English (say), the single symbols obtained are not just the letters with which words are spelled, since the division of a word into letters has or may have no relevance to the meaning. Frequently the single symbols are words. In other cases they are parts of words, since the division, e.g., "of books" into "book" and "s" or of "colder" into "cold" and "er" does have relevance to the meaning. In still other cases the linguistic structure of meaningful parts is an idealization, as when "worse" is taken to have an analysis parallel to that of "colder," or "I went" an analysis parallel to that of "I shall go," or "had I known" parallel to that of "if I should hear." (Less obvious and more complex examples may be expected to appear if analysis is pressed more in detail.)

[80] Apparently the case may be excluded that several improper symbols combine without any proper symbols to form an expression that has meaning in isolation. For the division of that expression into the improper symbols as parts could then hardly be said to have relevance to the meaning.

[81] Thus in the expression $(t - (x - y))$ we may say that the inner parentheses serve exclusively to show the association together of the part $x - y$ of the expression, and that they are used in connection with the sign $-$, which serves to show subtraction.

else sometimes in a way that combines the showing of association with some special meaning-producing character.[82]

Connectives are combinations of improper symbols which may be used together with one or more constants to form or produce a new constant. Then, as follows from the discussion in §02, if we replace one or more of the constants each by a form which has the denotation of that constant among its values, the resulting expression becomes a form (instead of a constant); and the free variables of this resulting form are the free variables of all the forms (one or more) which were united by means of the connective (with each other and possibly also with some constants) to produce the resulting form. In order to give completely the meaning-producing character of a particular connective in a particular language, not only is it necessary to give the denotation[83] of the new constant in every permissible case that the connective is used together with one or more constants to form such a new constant, but also, for every case that the connective may be used with forms or forms and constants to produce a resulting form, it is necessary to give the complete scheme of values of this resulting form for values of its free variables. And this must all be done in a way to conform to the assumptions about sense and denotation at the end of §01, and to the conventions about meaning and values of variables and forms as these were described in §02. Connectives may then be used not only in languages which contain constants but also in languages whose only proper symbols are variables.[84]

The constants or forms, united by means of a connective to produce a new constant or form, are called the *operands*. A connective is called *singulary, binary, ternary*, etc., according to the number of its operands.

A singulary connective may be used with a variable of appropriate range as the operand (this falls under our foregoing explanation since, of course, a variable is a special case of a form). The form so produced is called an *associated form* of the connective if the range of the variable includes the denotations of all constants which may be used as operands of the connective and all the relevant values of all the forms which may be used as operands of the connective (where by a *relevant* value of a form used as operand is meant a value corresponding to which the entire form, consisting of connective and operand, has a value). And the *associated function* of a singulary connective is the associated function of any associated form. The associated function as thus defined is clearly unique.

The notion of the associated function of a singulary connective is possible also in the case of a language containing no variable with a range of the kind required to produce an associated form, namely we may consider an extension of the language obtained by adding such a variable.

---

[82]In real number theory, the usual notation | | for the absolute value is an obvious example of this latter. Again it may be held that the parentheses have such a double use in either of the two notations introduced in §03 for application of a singularly function to its argument.

[83]It is not necessary (or possible) to give the sense of the new constant separately, since the way in which the denotation is given carries with it a sense—the same phrase which is used to name the denotation must also express a sense.

Further questions arise if, besides constants, names having a sense but no denotation are allowed. Such names seem to be used with connectives in the natural languages and in usual systems of mathematical notation, and indeed some illustrations which we have employed depend on this. However, as already explained, we avoid this in the formalized languages which we shall consider.

[84]Cf. footnote 27.

In the same way an n-ary connective may be used together with n different variables as operands to produce a form; and this is called an *associated form* of the connective if, for each variable, the range includes both the denotations of all constants and all relevant values of all forms which may be used as operands at that place. The *associated function* of the connective is that one of the associated n-ary functions of an associated form which is obtained by assigning the arguments of the function, in their order, as values to the free variables of the form in their left-to-right order of occurrence in the form.

In general the meaning-producing character of a connective is most readily given by just giving the associated function, this being sufficient to fix the use of the connective completely.[85]

Indeed there is a close relationship between connectives and *functional constants* or proper names of functions. Differences are that (a) a functional constant *denotes* a function whereas a connective *is associated with* a function, (b) a connective is never replaced by a variable, and (c) the notation for application of a function to its arguments may be paralleled by a different notation when a corresponding connective takes the place of a functional constant. But these differences are from some points of view largely nonessential because (a) notations of course have such meaning as we choose to give them (within limitations imposed by requirements of consistency and adequacy), (b) languages are possible which do not contain variables with functions as values and in which functional constants are never replaced by variables, and (c) the notation for application of a function to its arguments may, like any other, be changed—or even duplicated by introducing several synonymous notations into the same language.[86]

In the case of a language having notations for application of a function to its arguments, it is clear that a connective may often be eliminated or dispensed with altogether by employing instead a name of the associated function—by modifying the language, if necessary, to the extent of adding such a name to its vocabulary. However, the complete elimination of all connectives from a language can never be accomplished in this way. For the notations for applications of a singulary function to its argument,

---

[85]For example, the familiar notation (−) for subtraction of real numbers may be held to be a connective. That is, the combination of symbols which consists of a left parenthesis, a minus sign, and a right parenthesis, in that order, may be considered as a connective—where the understanding is that an appropriate constant or form is to be filled in at each of two places, namely immediately before and immediately after the minus sign. To give completely the meaning-producing character of this connective, it is necessary to give the denotation of the resulting constant when constants are filled in at the two places, and also to give the complete scheme of values of the resulting form when forms are filled in at the two places, or a form at one place and a constant at the other. In order to do this in a way to conform to §§01, 02, it may often be most expeditious first to introduce (by whatever means may be available in the particular context) the binary function of real numbers that is called *subtraction*, and then to declare this to be the associated function of the connective.

[86]Thus, to use once more the example of the preceding footnote, we may hold that the notation (−) is a connective and that the minus sign has no meaning in isolation. Or alternatively we may hold that the minus sign denotes (is a name of) the binary function, *subtraction*, and that in such expressions as, e.g., $(x - y)$ or $(5 - 2)$ we have a special notation for application of a binary function to its arguments, different from the notation for this which was introduced in §03. The choice would seem to be arbitrary between these two accounts of the meaning of the minus sign. But from one standpoint it may be argued that, if we are willing to invent some name for the binary function, then this name might just as well, and would most simply, be the minus sign.

for application of a binary function to its arguments, and so on (e.g., the notations for these which were introduced in §03) are themselves connectives. And though these connectives, like any other, no doubt have their associated functions,[87] nevertheless not all of them can ever be eliminated by the device in question.[88] Connectives other than notations for applications of a function to its arguments are apparently always eliminable in the way described by a sufficient extension of the language in which they occur (including if necessary the addition to the language of notations for application of a function to its arguments). Nevertheless such other connectives are often used—especially in formalized languages of limited vocabulary, where it may be preferred to preserve this limitation of vocabulary, so as to use the language as a means of singling out for separate consideration some special branch of logic (or other subject).

In particular we shall meet with *sentence connectives* in Chapter I. Namely, these are connectives which are used together with one or more sentences to produce a new sentence; or when propositional forms replace some or all of the sentences as operands, then a propositional form is produced rather than a sentence.

The chief *singulary* sentence connective we shall need is one for negation. In this role we shall use, in formalized languages, the single symbol $\sim$, which, when prefixed to a sentence, forms a new sentence that is the negation of the first one. The associated

[87] As explained below, we are for expository purposes temporarily ignoring difficulties or complications which may be caused by the theory of types or by such alternative to the theory of types as may be adopted. On this basis, for the connective which is the notation for application of a singulary function to its argument, we explain the associated function by saying that it is the binary function whose value for an ordered pair of arguments f, x is $f(x)$. But if a name of this associated function is to be used for the purpose of eliminating this connective, then another connective is found to be necessary, the notation, namely, for application of a binary function to its arguments. If the latter connective is to be eliminated by using a name of its associated function, then the notation for application of a ternary function to its arguments becomes necessary. And so on. Obviously no genuine progress is being made in these attempts.

(After studying the theory of types the reader will see that the foregoing statement, and others we have made, remain in some sense essentially true on the basis of that theory. It is only that the connective, e.g., which is the notation for application of a singulary function to its argument must be thought of as replaced by many different connectives, corresponding to different types, and each of these has its own associated function. Or alternatively, if we choose to retain this connective as always the same connective, regardless of considerations of type, then there may well be no variable in the language with a range of the kind required to produce an associated form: an extension of the language by adding such a variable can be made to provide an associated form, but not so easily a name of the associated function. See Carnap, ***The Logical Syntax of Language*** (cited in footnote 131), examples at the end of §53, and references there given; also Bernard Notcutt's proposal of "intertypical variables" in ***Mind***, n.s. vol. 43 (1934), pp. 63–77; and remarks by Tarski in the appendix to his *Wahrheitsbegriff* (cited in footnote 140).)

[88] There is, however, a device which may be used in appropriate context (cf. Chapter X) to eliminate all the connectives except the notation for application of a singulary function to its argument. This is done by reconstruing a binary function as a singulary function whose values are singulary functions; a ternary function as a singular function whose values are binary functions in the foregoing sense; and so on. For it turns out that n-ary functions in the sense thus obtained can be made to serve all the ordinary purposes of n-ary functions (in any sense).

The alternative device of reducing (e.g.) a binary function to a singulary function by reconstruing it as a singulary function whose arguments are ordered pairs is also useful in certain contexts (e.g., in axiomatic set theory). This device does not (at least *prima facie*) serve to reduce the number of connectives to one, as besides the notation for application of a singulary function to its argument there will be required also a connective which unites the names of two things to form a name of their ordered pair (or at least some notation for this latter purpose). Nevertheless it is a device which may sometimes be used to accomplish a reduction, especially where other connectives or operators (§06)—are available.

function of this connective is the function from truth-values to truth-values whose value for the argument *falsehood* is *truth*, and whose value for the argument *truth* is *falsehood*. For convenience in reading orally expressions of a formalized language, the symbol ∼ may be rendered by the word "not" or by the phrase "it is false that."

The principle *binary* sentence connectives are indicated in the table which follows. The notation which we shall use in formalized languages is shown in the first column of the table, with the understanding that each of the two blanks is to be filled by a sentence of the language in question. In the second column of the table a convenient oral reading of the connective is suggested, or sometimes two alternative readings; here the understanding is that the two blanks are to be filled by oral readings of the same two sentences (in the same order) which filled the two corresponding blanks in the first column; and words which appear between parentheses are words which may ordinarily be omitted for brevity, but which are to be supplied whenever necessary to avoid a misunderstanding or to emphasize a distinction. In the third column the associated function of the connective is indicated by means of a code sequence of four letters: in doing this, t is used for truth and f for falsehood, and the first letter of the four gives the value of the function for the arguments t, t, the second letter gives the value for the arguments t, f, the third letter for the arguments f, t, the fourth letter for the arguments f, f. In many cases there is an English name in standard use, which may denote either the connective or its associated function. This is indicated in a fourth column of the table; were alternative names are in use, both are given, and in some cases where none is in use a suggested name is supplied.

[___ ∨ ___]	___ or ___ (or both).	tttf	(Inclusive) disjunction, alternation.
[___ ⊂ ___]	___ if ___.[89]	ttft	Converse implication.
[___ ⊃ ___]	If ___ then ___,[89] ___ (materially) implies ___.[89]	tftt	The (truth-functional) conditional,[90] (material) implication.
[___ ≡ ___]	___ if and only if ___,[89] ___ is (materially) equivalent to ___.[89]	tfft	The (truth-functional) biconditional,[90] (material) equivalence.
[___ ___]	___ and ___.	tfff	Conjunction.
[___ | ___]	Not both ___ and ___.	fttt	Non-conjunction, Sheffer's stroke.
[___ ≢ ___]	___ or ___ but not both, ___ is not (materially) equivalent to ___.[89]	fttf	Exclusive disjunction, (material) non-equivalence.
[___ ⊅ ___]	___ but not ___.	ftff	(Material) non-implication.
[___ ⊄ ___]	Not ___ but ___.	fftf	Converse non-implication.
[___ ⊽ ___]	Neither ___ nor ___.	ffft	Non-disjunction.

[89]The use of the English words "if," "implies," "equivalent" in these oral readings must not be taken as indicating that the meanings of these English words are faithfully rendered by the corresponding connectives in all, or even in most, cases. On the contrary, the meaning-producing character of the connectives is to be learned with accuracy from the third column of the table, where the associated functions are given, and the oral readings supply at best a rough approximation. [*Editor's note*: Footnote continues on p. 401.]

The notations which we use as sentence connectives—and those which we use as quantifiers (see below)—are adaptations of those in Whitehead and Russell's *Principia Mathematica* (some of which in turn were taken from Peano). Various other notations are in use,[91] and the student who would compare the treatments of different authors must learn a certain facility in shifting from one system of notation to another.

The brackets which we indicate as constituents in these notations may in actual use be found unnecessary at certain places, and we may then just omit them at such places (though only as a practically convenient abbreviation).

We shall use the term *truth-function*[92] for a propositional function of truth-values which has as range, if it is n-ary, all ordered systems of n truth-values. Thus every

---

As a matter of fact, the words "if ... then" and "implies" as used in ordinary non-technical English often seem to denote a relation between propositions rather than between truth-values. Their possible meanings when employed in this way are difficult to fix precisely and we shall make no attempt to do so. But we select the one use of the words "if ... then" (or "implies")—their material use, we shall call it—in which they may be construed as denoting a relation between truth-values, and we assign this relation as the associated function for the connective [$\supset$].

As examples of the material use of "if ... then," consider the four following English sentences:

(i) If Joan of Arc was a patriot then Nathan Hale was a patriot.
(ii) If Joan of Arc was a patriot then Vidkun Quisling was a patriot.
(iii) If Vidkun Quisling was a patriot then attar of roses is a perfume.
(iv) If Vidkun Quisling was a patriot then Limburger cheese is a perfume.

For the sake of the illustration let use suppose examination of the historical facts to reveal that Joan of Arc and Nathan Hale were indeed patriots and that Vidkun Quisling was not a patriot. Then (i), (iii), and (iv) are true, and (ii) is false; and to reach these conclusions no examination is necessary of the characteristics of either attar of roses or Limburger cheese. (If the reader is inclined to question the truth of e.g., (iii) on the ground of complete lack of connection between Vidkun Quisling and attar of roses, then this means that he has in mind some other use of "if ... then" than the material use.)

[90]These terms were introduced by Quine, who uses them for "the mode of composition described in" the list of truth-values as given in the third column of the table—i.e., in effect, and in our terminology, for the associated function of the connective rather than for the connective itself. See his ***Mathematical Logic***, 1940, pp. 15, 20.

We prefer the better established terms *material implication* and *material equivalence*, from which the adjective *material* may be omitted whenever there is no danger of confusion with other kinds of implication or equivalence—as, for example, with formal implication and formal equivalence (§06), or with kinds of implication and equivalence (belonging to modal logic) which are relations between propositions rather than between truth-values.

[91]Worthy of special remark is the parenthesis-free notation of Jan Łukasiewicz. In this, the letters N, A, C, E, K are used in the roles of negation, disjunction, implication, equivalence, conjunction respectively. Further letters may be introduced if desired (R has been employed as non-equivalence, D as non-conjunction). In use as a sentence connective, the letter is written first and then in order the sentences or propositional forms together with which it is used. No parentheses or brackets or other notations specially to show association are necessary. E.g., the propositional form

$$[[p \supset [q \vee r]] \supset \sim p]$$

(where p, q, r are propositional variables) becomes, in the Łukasiewicz notation,

$$CCpAqrNp.$$

It is of course possible to apply the same idea to other connectives, in particular to the notation for application of a singulary function to its argument. Hence (see footnote 88) parentheses and brackets may be avoided altogether in a formalized language. The possibility of this is interesting. But the notation so obtained is unfamiliar, and less perspicuous than the usual one.

[92]We adopt this term from ***Principia Mathematica***, giving it substantially the meaning which it acquires through changes in that work that were made (or rather, proposed) by Russell in his introduction to the second edition of it.

associated function of a sentence connective is a truth-function. And likewise every associated function of a form built up from propositional variables solely by iterated use of sentence connectives.[93]

06. Operators, quantifiers. An *operator* is a combination of improper symbols which may be used together with one or more variables—the *operator variables* (which must be fixed in number and all distinct)—and one or more constants or forms or both—the *operands*—to produce a new constant or form. In this new constant or form, however, the operator variables are at certain determinate places not free variables, though they may have been free variables at those places in the operands.

To be more explicit, we remark that, in any application of an operator, the operator variables may (and commonly will) occur as free variables in some of the operands. In the new constant or form produced we distinguish three possible kinds of occurrences of the operator variables, viz.: an occurrence in one of the operands which, when considered as an occurrence in that operand alone, is an occurrence as a free variable; an occurrence in one of the operands, not of this kind; and an occurrence which is an occurrence *as* an operator variable, therefore not in any of the operands. In the new constant or form, an occurrence of one of the two latter kinds is never an occurrence as a free variable, and each occurrence of the first kind is an occurrence as a free variable or not, according to some rule associated with the particular operator.[94] The simplest case is that, in the new constant or form, none of the occurrences of the operator variables are occurrences as free variables. And this is the only case with which we shall meet in the following chapters (though many operators which are familiar as standard mathematical notation fail to fall under this simplest case).

Variables thus having occurrences in a constant or form which are not occurrences as free variables of it are called *bound variables* of the constant or form.[95] The difference is that a form containing a particular variable, say x, *as a free variable* has values for various values of the variable, but a constant or form which contains x *as a bound variable only* has a meaning which is independent of x—not in the sense of having the same value for every value of x, but in the sense that the assignment of particular values to x is not a relevant procedure.[96]

It may happen that a form contains both free and bound occurrences of the same variable. This case will arise, for example, if a form containing a particular variable as a free variable and a form or constant containing that same variable as a bound

[93]For example, the associated function of the propositional form mentioned in footnote 91.

[94]We do require in the case of each operator variable that all occurrences of the first kind shall be occurrences as free variables or else all not, *in any one occurrence of a particular operand* in the new constant or form produced. For operators violating this requirement are not found among existing standard mathematical and logical notations, and it is clear that they would involve certain anomalies of meaning which it is preferable to avoid.

[95]Cf. footnote 28.

[96]Therefore a constant or form which contains a particular variable as a bound variable is unaltered in meaning by alphabetic change of that variable, at all of its bound occurrences, to a new variable (not previously occurring) which has the same range. The condition in parentheses is included only as a precaution against identifying two variables which should be kept distinct, and indeed it may be weakened somewhat—cf. the remark in §03 about alphabetic change of free variables.

E.g., the constant $\int_0^2 x^x dx$ (see footnote 36) is unaltered in meaning by alphabetic change of the variable x to the variable y: it has not only the same denotation but also the same sense as $\int_0^2 y^y dy$.

variable are united by means of a binary connective.[97]

As in the case of connectives, we require that operators be such as to conform to the principles (1)–(3) at the end of §01; also that they conform to the conventions about meaning and values of variables as these were described in §02, and in particular to the principle (4) of §02.[98]

An operator is called *m-ary-n-ary* if it is used with m distinct operator variables and n operands.[99] The most common case is that of a singulary-singulary operator—or, as we shall also call it, a *simple* operator.

In particular, the notation for singulary functional abstraction, which was introduced in §03, is a simple operator (the variable which is placed immediately after the letter λ being the operator variable). We shall call this the *abstraction operator* or more explicitly, the *singulary functional abstraction operator*. In appropriate context, as we shall see in Chapter X, all other operators can in fact be reduced to this one.[100]

Another operator which we shall use—also a simple operator—is the *description operator*, $(\iota\ )$. To illustrate, let the operator variable be x. Then the notation (ιx) is to have as its approximate reading in words, "the x such that"; or more fully the notation is explained as follows. It may happen that a singulary propositional form whose free variable is x has the value truth for one and only one value of x, and in this case a name of that value of x is produced by prefixing (ιx) to the form. In case there is no value of x or more than one for which the form has the value truth, there are

---

[97] See illustrations in the second paragraph of footnote 36.

[98] And also to the principle (5) of footnote 30.

[99] Thus, in the theory of real numbers, the usual notation for definite integration is a singulary-ternary operator. And in, e.g., the form $\int_0^x x^x dx$ (see footnote 36) the operator variable is x and the three operands are the constant 0, the form x, and the form x^x.

Again, the large $\prod$ (product sign), as used in the third example at the beginning of footnote 36, is part of a singulary-ternary operator. The signs $=$ above and below the $\prod$ are not to be taken as equality signs in the ordinary sense (namely that of footnote 43) but as improper symbols, and also part of the operator. In the particular application of the operator, as it appears in this example, the operator variable is m and the operands are 1, n, and

$$\frac{x-m+1}{m\pi}.$$

As another example of application of the same operator, showing both bound and free occurrences of m, we cite

$$\prod_{m=m+1}^{m=m+n+1} \frac{x-m+1}{m\pi}.$$

Examples of operators taking more than one operator variable are found in familiar notations for double and multiple limits, double and multiple integrals.

It should also be noted that n-ary connectives may, if we wish, be regarded as 0-ary-n-ary operators.

[100] In the combinatory logic of H. B. Curry (based on an idea due to M. Schönfinkel) a more drastic reduction is attempted, namely the complete elimination of operators, of variables, and of all connectives, except a notation for application of a singulary function to its argument, so as to obtain a formalized language in which, with the exception of the one connective, all single symbols are constants, and which is nevertheless adequate for some or all of the purposes for which variables are ordinarily used. This is a matter beyond the scope of this book, and the present status of the undertaking is too complex for brief statement. The reader may be referred to a monograph by the present writer, ***The Calculi of Lambda-Conversion*** (1941), which is concerned with a related topic; also to papers by Schönfinkel, Curry, and J. B. Rosser which are there cited, to several papers by Curry and by Rosser in the ***The Journal of Symbolic Logic*** in 1941 and 1942, to an expository paper by Robert Feys in ***Revue Philosophique de Louvain***, vol. 44 (1946), pp. 74–103, 237–270, and to a paper by Curry in ***Synthese***, vol. 7 (1949), pp. 391–399.

various meanings which might be assigned to the name produced by prefixing (ιx) to the form: the analogy of English and other natural languages would suggest giving the name a sense which determines no denotation; but we prefer to select some fixed value of x and to assign this as the denotation of the name in all such cases (this selection is arbitrary, but is to be made once for all for each range of variables which is used).

Of especial importance for our purposes are the *quantifiers*. These are namely operators for which both the operands and the new constant or form produced by application of the operator are sentences or propositional forms.

As the *universal quantifier* (when, e.g., the operator variable is x) we use the notation $(\forall x)$ or (x), prefixing this to the operand. The universal quantifier is thus a simple operator, and we may explain its meaning as follows (still using the particular variable x as an example). (x)______ is true if the value of ______ is truth for all values of x, and (x)______ is false if there is any value of x for which the value of ______ is falsehood. Here the blank is to be filled by a singulary propositional form containing x as a free variable, the same one at all four places. Or if as a special case we fill the blank with a sentence, then (x)______ is true if and only if ______ is true. (The meaning in case the blank is filled by a propositional form containing other variables besides x as free variables now follows by the discussion of variables in §02, and may be supplied by the reader.)

Likewise the existential quantifier is a simple operator for which we shall use the notation $(\exists\ )$, filling the blank space with the operator variable and prefixing the whole to the operand. To take the particular operator variable x as an example, $(\exists x)$______ is true if the value of ______ is truth for at least one value of x, and $(\exists x)$______ is false if the value of ______ is falsehood for all values of x. Here again the blank is to be filled by a singulary propositional form containing x as a free variable. Or if as a special case we fill the blank with a sentence, then $(\exists x)$______ is true if and only if ______ is true.

In words, the notations "(x)" and "$(\exists x)$" may be read respectively as "for all x" (or "for every x") and "there is an x such that."

To illustrate the use of universal and existential quantifiers, and in particular their iterated application, consider the binary propositional form,

$$[xy > 0],$$

where x and y are real variables, i.e., variables whose range is the real numbers. This form expresses about two real numbers x and y that their product is positive, and thus it comes to express a particular proposition as soon as values are given to x and y. If we apply to it the existential quantifier with y as operator variable, we obtain the singulary propositional form,

$$(\exists y)[xy > 0],$$

or as we may also write it, using the device (which we shall find frequently convenient later) of writing a heavy dot to stand for a bracket extending from the place where the dot occurs, forward,

$$(\exists y) \centerdot xy > 0.$$

This singulary form expresses about a real number x that there is some real number with which its product is positive; and it comes to express a particular proposition as soon as a value is given to x. If we apply to it the universal quantifier with x as operator variable, we obtain the sentence,

$$(x)(\exists y) \centerdot xy > 0.$$

This sentence expresses the proposition that for every real number there is some real number such that the product of the two is positive. It must be distinguished from the sentence,

$$(\exists y)(x) \centerdot xy > 0,$$

expressing the proposition that there is a real number whose product with every real number is positive, though it happens that both are false.[101] To bring out more sharply the difference which is made by the different order of the quantifiers, let us replace product by sum and consider the two sentences:

$$(x)(\exists y) \centerdot x + y > 0$$
$$(\exists y)(x) \centerdot x + y > 0$$

Of these sentences, the first one is true and the second one false.[102]

It should be informally clear to the reader that not both the universal and the existential quantifier are actually necessary in a formalized language, if negation is available. For it would be possible, in place of $(\exists x)$______, to write always $\sim(x)\sim$______; or alternatively, in place of (x)______, to write always $\sim(\exists x)\sim$______. And of course likewise with any other variable in place of the particular variable x.

In most treatments the universal and existential quantifiers, one or both, are made fundamental, notations being provided for them directly in setting up a formalized language, and other quantifiers are explained in terms of them (in a way similar to that in which, as we have just seen in the preceding paragraph, the universal and existential quantifiers may be explained, either one in terms of the other). No definite or compelling reason can be given for such a preference of these two quantifiers above others that might equally be made fundamental. But it is often convenient.

---

[101] The single counterexample, of the value 0 for x, is of course sufficient to render the first sentence false. The reader is warned against saying that the sentence $(x)(\exists y) \centerdot xy > 0$ is "nearly always true" or that it is "true with one exception" or the like. These expressions are appropriate rather to the propositional form $(\exists y) \centerdot xy > 0$, and of the sentence it must be said simply that it is false.

[102] A somewhat more complex example of the difference made by the order in which the quantifiers are applied is found in the familiar distinction between continuity and uniform continuity. Using x and y as variables whose range is the real numbers, and ε and δ as variables whose range is the positive real numbers, we may express as follows that the real function f is continuous, on the class F of real numbers (assumed to be an open or closed interval):

$$(y)(\varepsilon)(\exists\delta)(x) \centerdot F(y) \supset \centerdot\ F(x) \supset \centerdot\ |x - y| < \delta \supset \centerdot\ |f(x) - f(y)| < \varepsilon$$

And we may express as follows that f is uniformly continuous on F:

$$(\varepsilon)(\exists\delta)(x)(y) \centerdot F(y) \supset \centerdot\ F(x) \supset \centerdot\ |x - y| < \delta \supset \centerdot\ |f(x) - f(y)| < \varepsilon$$

To avoid complications that are not relevant to the point being illustrated, we have here assumed not only that the class F is an open or closed interval but also that the range of the function f is all real numbers. (A function with more restricted range may always have its range extended by some arbitrary assignment of values; and indeed it is a common simplifying device in the construction of a formalized language to restrict attention to functions having certain standard ranges (cf. footnote 19).)

The application of one or more quantifiers to an operand (especially universal and existential quantifiers) is spoken of as *quantification*.[103]

Another quantifier is a singulary-binary quantifier for which we shall use the notation [——— $\supset$ ———], with the operands in the two blanks, and the operator variable as a subscript after the sign $\supset$. It may be explained by saying that [——— $\supset_x$ ———] is to mean the same as (x)[——— $\supset$ ———], the two blanks being filled with two propositional forms or sentences, the same two in each case (and in the same order); and of course likewise with any other variable in place of the particular variable x. The name *formal implication*[104] is given to this quantifier—or to the associated binary propositional function, i.e., to an appropriate one of the two associated functions of (say) the form $[F(u) \supset_u G(u)]$, where u is a variable with some assigned range, and F and G are variables whose range is all classes (singulary propositional functions) that have a range coinciding with the range of u.

Another quantifier is that which (or its associated propositional function) is called *formal equivalence*.[104] For this we shall use the notation [——— $\equiv$ ———], with the two operands in the two blanks, and the operator variable as a subscript after the sign $\equiv$. It may be explained by saying that [——— $\equiv_x$ ———] is to mean the same as (x)[——— $\equiv$ ———], the two blanks being filled in each case with the two operands in order; and of course likewise with any other variable in place if x.

We shall also make use of quantifiers similar in character to those just explained but having two or more operator variables. These (or their associated propositional functions) we call *binary formal implication, binary formal equivalence, ternary formal implication*, etc. E.g., binary formal implication may be explained by saying that [——— $\supset_{xy}$ ———] is to mean the same as $(x)(y)$[——— $\supset$ ———], the two blanks being filled in each case with the two operands in order; and likewise with any two distinct variables in place of x and y as operator variables. Similarly binary formal equivalence [——— $\equiv_{xy}$ ———], ternary formal implication [——— $\supset_{xyz}$ ———], and so on.[105]

Besides the assertion of a sentence, as contemplated in §04, it is usual also to allow

---

[103]The use of quantifiers originated with Frege in 1879. And independently of Frege the same idea was introduced somewhat later by Mitchell and Peirce. (See the historical account in §49.)

[104]The names *formal implication* and *formal equivalence* are those used by Whitehead and Russell in ***Principia Mathematica***, and have become sufficiently well established that it seems best not to change them—though the adjective *formal* is perhaps not very well chosen, and must not be understood here in the same sense that we shall give it elsewhere.

[105]With the aid of the notations that have now been explained, we may return to §00 and rewrite the examples I–IV of that section as they might appear in some appropriate formalized language.

For this purpose let a and b be variables whose range is human beings. Let v be a variable whose range is words (taking, let us say for definiteness, any finite sequence of letters of the English alphabet as a word). Let B denote the relation of being a brother of. Let S denote the relation of having as surname. Let ϱ and σ denote the human beings Richard and Stanley respectively, and let τ denote the word "Thompson." Then the three premisses and the conclusion of I may be expressed as follows:

$$B(a,b) \supset_{ab} \centerdot\ S(a,v) \equiv_v S(b,v)$$
$$B(\varrho,\sigma)$$
$$S(\sigma,\tau)$$
$$S(\varrho,\tau)$$

Further, let z and w be variables whose range is complex numbers, and x a variable whose range is real numbers, Let R denote the relation of having real positive ratio, and let A denote the relation of having as amplitude. Then the premisses and conclusion if II may be expressed as follows:

assertion of a propositional form, and to treat such an assertion as a particular fixed assertion (in spite of the presence of free variables in the expression asserted). This is common especially in mathematical contexts; where, for instance, the assertion of the equation $\sin(x + 2\pi) = \sin x$ may be used as a means to assert this for all real numbers x; or the assertion of the inequality $x^2 + y^2 \geqq 2xy$ may be used as a means to assert that for any real numbers x and y the sum of the squares is greater than or equal to twice the product.

It is clear that, in a formalized language, if universal quantification is available, it is unnecessary to allow the assertion of expressions containing free variables. E.g., the assertion of the propositional form

$$x^2 + y^2 \geqq 2xy$$

could be replaced by the assertion of the sentence

$$(x)(y) \mathbin{.} x^2 + y^2 \geqq 2xy.$$

But on the other hand it is not possible to dispense with quantifiers in a formalized language merely by allowing the assertion of propositional forms, because, e.g., such assertions as that of

$$\sim(x)(y) \mathbin{.} \sin(x + y) = \sin x + \sin y, \text{[106]}$$

or that of

$$(y)[\,|x| \leqq |y|\,] \supset_x \mathbin{.} x = 0,$$

could not be reproduced.

Consequently it has been urged with some force that the device of asserting propositional forms constitutes an unnecessary duplication of ways of expressing the same

---

$$R(z, w) \supset_{zw} \mathbin{.} A(z, x) \equiv_x A(w, x)$$
$$R(i - \sqrt{3}/3, \omega)$$
$$A(\omega, 2\pi/3)$$
$$A(i - \sqrt{3}/3, 2\pi/3)$$

Here it is obvious that the relation of having real positive ratio is capable of being analyzed, so that instead of $R(z, w)$ we might have written, e.g.:

$$(\exists x)[x > 0][z = xw]$$

Likewise the relation of having as amplitude or (in I) the relation of being a brother of might have received some analysis. But these analyses are not relevant to the validity of the reasoning in these particular examples. And they are, moreover, in no way final or absolute; e.g., instead of analyzing the relation of having real positive ratio, we might with equal right take it as fundamental and analyze instead the relation of being greater than, in such a way that, in place of $x > y$ would be written $R(x - y, 1)$.

In the same way, for III and IV, we make no analysis of the singular propositional functions of having a portrait seen by me, of having assassinated Abraham Lincoln, and of having invented the wheeled vehicle, but let them be denoted just by P, L and W respectively. Then if β denotes John Wilkes Booth, the premisses and conclusion of III may be expressed thus:

$$P(\beta) \qquad L(\beta) \qquad (\exists a)[P(a)L(a)]$$

And the premisses and fallacious conclusion of IV thus:

$$(\exists a)P(a) \qquad (\exists a)W(a) \qquad (\exists a)[P(a)W(a)]$$

When so rewritten, the false appearance of analogy between III and IV disappears. It was due to the logically irregular feature of English grammar by which "somebody" is construed as a substantive.

thing, and ought to be eliminated from a formalized language.[107] Nevertheless it appears that the retention of this device often facilitates the setting up of a formalized language by simplifying certain details; and it also renders more natural and obvious the separation of such restricted systems as propositional calculus (Chapter I) or functional calculus of first order (Chapter III) out from more comprehensive systems of which they are part. In the development which follows we shall therefore make free use of the assertion of propositional forms. However, in the case of such systems as functional calculus of order ω (Chapter VI) or Zermelo set theory (Chapter XI), after a first treatment employing the device in question we shall sketch briefly a reformulation that avoids it.

07. The logistic method. In order to set up a formalized language we must of course make use of a language already known to us, say English or some portion of the English language, stating in that language the vocabulary and rules of the formalized language. This procedure is analogous to that familiar to the reader in language study—as, e.g., in the use of a Latin grammar written in English[108]—but differs in the precision with which the rules are stated, in the avoidance of irregularities and exceptions, and in the leading idea that the rules of the language embody a theory or system of logical analysis (cf. §00).

This device of employing one language in order to talk about another is one for which we shall have frequent occasion not only in setting up formalized languages but also in making theoretical statements as to what can be done in a formalized language, our interest in formalized languages being less often in their actual and practical use as languages than in the general theory of such use and in its possibilities in principle. Whenever we employ a language in order to talk about some language (itself or another[109]), we shall call the latter language the *object language*, and we shall call the former the *meta-language*.[110]

---

[107]The proposal to do this was made by Russell in his introduction to the second edition of ***Principia Mathematica*** (1925). The elimination was actually carried out by Quine in his ***Mathematical Logic*** (1940), and simplifications of Quine's method were effected in papers by F. B. Fitch and by G. D. W. Berry in ***The Journal of Symbolic Logic*** (vol. 6 (1941), pp. 18–22, 23–27).

[108]It is worth remark in passing that this same procedure also enters into the learning of a first language, being a necessary supplement to the method of learning by example and imitation. Some part of the language must first be learned approximately by the method of example and imitation; then this imprecisely known part of the language is applied in order to state rules of the language (and perhaps to correct initial misconceptions); then the known part of the language may be extended by further learning by example and imitation, and so on in alternate steps, until some precision in knowledge of the language is reached.

There is no reason in principle why a first language, learned in this way, should not be one of the formalized languages of this book, instead of one of the natural languages. (But of course there is the practical reason that those formalized languages are ill adapted to purposes of facility of communication.)

[109]The employment of a language to talk about that same language is clearly not appropriate as a method of setting up a formalized language. But once set up, a formalized language with adequate means of expression may be capable of use in order to talk about that language itself; and in particular the very setting up of the language may afterwards be capable of restatement in that language. Thus it may happen that object language and meta-language are the same, a situation which it will be important later to take into account.

[110]The distinction is due to David Hilbert, who, however, speaks of "Mathematik" (mathematics) and "Metamathematik" (metamathematics) rather than "object language" and "meta-language." The latter terms, or analogues of them in Polish or German, are due to Alfred Tarski and Rudolf Carnap, by whom especially (see footnotes 131, 140) the subjects of *syntax* and *semantics* have been developed.

In setting up a formalized language we first employ as meta-language a certain portion of English. We shall not attempt to delimit precisely this portion of the English language, but describe it approximately by saying that it is just sufficient to enable us to give general directions for the manipulation of concrete physical objects (each instance or occurrence of one of the symbols of the language being such a concrete physical object, e.g., a mass of ink adhering to a bit of paper). It is thus a language which deals with matters of everyday human experience, going beyond such matters only in that no finite upper limit is imposed on the number of objects that may be involved in any particular case, or on the time that may be required for their manipulation according to instructions. Those additional portions of English are excluded which would be used in order to treat of infinite classes or of various like abstract objects which are an essential part of the subject matter of mathematics.

Our procedure is not to define the new language merely by means of translations of its expressions (sentences, names, forms) into corresponding English expressions, because in this way it would hardly be possible to avoid carrying over into the new language the logically unsatisfactory features of the English language. Rather, we begin by setting up, in abstraction from all considerations of meaning, the purely formal part of the language, so obtaining an uninterpreted calculus or *logistic system*. In detail, this is done as follows.

The vocabulary of the language is specified by listing the single symbols which are to be used.[111] These are called the *primitive symbols* of the language,[112] and are to be regarded as indivisible in the double sense that (A) in setting up the language no use is made of any division of them into parts and (B) any finite linear sequence of primitive

[111]Notice that we use the term "language" in such a sense that a given language has a given and uniquely determined vocabulary. E.g., the introduction of one additional symbol into the vocabulary is sufficient to produce a new and different language. (Thus the English of 1849 is not the same language as the English of 1949, though it is convenient to call them by the same name, and to distinguish, by specifying the date, only in cases where the distinction is essential.)

[112]The fourfold classification of the primitive notations of a formalized language into constants, variables, connectives, and operators is due in substance to J. v. Neumann in the ***Mathematische Zeitschrift***, vol. 26 (1927), see pp. 4–6. He there adds a fifth category, composed of association-showing symbols such as parentheses and brackets. Our terms "connective" and "operator" correspond to his "Operation" and "Abstraktion" respectively.

Though there is a possibility of notations not falling in any of von Neumann's categories, such have seldom been used, and for nearly all formalized languages that have actually been proposed the von Neumann classification of primitive notations suffices. Many formalized languages have primitive notations of all four (or five) kinds, but it does not appear that this is indispensable, even for a language intended to be adequate for the expressions of mathematical ideas generally.

As an interesting example of a (conceivable) notation not in any of the von Neumann categories, we mention the question of a notation by means of which from a name of a class would be formed an expression playing the role of a variable with that class as its range. Provisions might perhaps be made for the formation from any class name of an infinite number of expressions playing the roles of different variables with the class as their range. But these expressions would have to differ from variables in the sense of §02 not only in being composite expressions rather than single symbols but also in the possibility that the range might be empty. A language containing such a notation has never been set up and studied in detail and it is therefore not certain just what is feasible. (A suggestion which seems to be in this direction was made by Beppo Levi in ***Universidad Nacional de Tucumán, Revista***, ser. A vol. 3 no. 1 (1942), pp. 13–78.)

The use in Chapter X of variables with subscripts indicating the range of the variable (the type) is not an example of a notation of the kind just described. For the variable, letter and subscript together, is always treated as a single primitive symbol.

symbols can be regarded *in only one way* as such a sequence of primitive symbols.[113] A finite linear sequence of primitive symbols is called a *formula*. And among the formulas, rules are given by which certain ones are designated as *well-formed formulas* (with the intention, roughly speaking, that only the well-formed formulas are to be regarded as being genuinely expressions of the language).[114] Then certain among the well-formed formulas are laid down as *axioms*. And finally (primitive) *rules of inference* (or *rules of procedure*) are laid down, rules according to which, from appropriate well-formed formulas as *premisses*, a well-formed formula is *immediately inferred*[115] as *conclusion*. (So long as we are dealing only with a logistic system that remains uninterpreted, the terms *premiss, immediately infer, conclusion* have only such meaning as is conferred upon them by the rules of inference themselves.)

A finite sequence of one or more well-formed formulas is called a *proof* if each of the well-formed formulas in the sequence either is an axiom or is immediately inferred from preceding well-formed formulas in the sequence by means of one of the rules of inference. A proof is called a proof *of* the last well-formed formula in the sequence, and the *theorems* of the logistic system are those well-formed formulas of which proofs exist.[116] As a special case, each axiom of the system is a theorem, that finite sequence being a proof which consists of a single well-formed formula, the axiom alone.

The scheme just described—viz. the primitive symbols of a logistic system, the rules by which certain formulas are determined as well-formed (following Carnap let us call

---

[113]In practice, condition (B) usually makes no difficulty. Though the (written) symbols adopted as primitive symbols may not all consist of a single connected piece, it is ordinarily possible to satisfy (B), if not otherwise, by providing that a sequence of primitive symbols shall be written with spaces between the primitive symbols of fixed width and wider than the space at any place within a primitive symbol.

The necessity for (B), and its possible failure, were brought out by a criticism by Stanislaw Leśniewski against the paper of von Neumann cited in the preceding footnote. See von Neumann's reply in ***Fundamenta Mathematicae***, vol. 17 (1931), pp. 331–334, and Leśniewski's final word in the matter in an offprint published in 1938 as from ***Collectanea Logica***, vol. 1 (cf. ***The Journal of Symbolic Logic***, vol. 5, p. 83).

[114]The restriction to one dimension in combining the primitive symbols into expressions of the language is convenient, and non-essential. Two dimensional arrangements are of course possible, and are familiar especially in mathematical notations, but they may always be reduced to one dimension by a change of notation. In particular the notation in Frege's ***Begriffsschrift*** relies heavily on a two-dimensional arrangement; but because of the difficulty of printing it this notation was never adopted by any one else and has long since been replaced by a one-dimensional equivalent.

[115]No reference to the so-called immediate inferences of traditional logic is intended. We term the inferences *immediate* in the sense of requiring only one application of a rule of inference—not in the traditional sense of (among other things) having only one premiss.

[116]Following Carnap and others, we use the term "language" in such a sense that for any given language there is one fixed notion of a proof in that language. Thus the introduction of one additional axiom or rule of inference, or a change in an axiom or rule of inference, is sufficient to produce a new and different language.

(An alternative, which might be thought to accord better with the every day use of the word "language," would be to define a "language" as consisting of primitive symbols and a definition of well-formed formula, together with an *interpretation* (see below), and to take the axioms and rules of inference as constituting a "logic" for the language. Instead of speaking of an interpretation as *sound* or *unsound* for a logistic system (see below), we would then speak of a logic as being sound or unsound for a language. Indeed this alternative may have some considerations in its favor. But we reject it here, partly because of reluctance to change a terminology already fairly well established, partly because the alternative terminology leads to a twofold division in each of the subjects of syntax and semantics (§§08, 09)—according as they treat of the object language alone or of the object language together with a logic for it—which, especially in the case of semantics, seems unnatural, and of little use so far as can now be seen.)

them the *formulation rules* of the system), the rules of inference, and the axioms of the system—is called the *primitive basis* of the logistic system.[117]

In defining a logistic system by laying down a primitive basis, we employ as meta-language the restricted portion of English described above. In addition to this restriction, or perhaps better as part of it, we impose requirements of *effectiveness* as follows: (I) the specification of the primitive symbols shall be effective in the sense that there is a method by which, whenever a symbol is given, it can always be determined effectively whether or not it is one of the primitive symbols; (II) the definition of a well-formed formula shall be effective in the sense that there is a method by which, whenever a formula is given, it can always be determined effectively whether or not it is well-formed; (III) the specification of the axioms shall be effective in the sense that there is a method by which, whenever a well-formed formula is given, it can always be determined effectively whether or not it is one of the axioms; (IV) the rules of inference, taken together, shall be effective in the strong sense that there is a method by which, whenever a proposed immediate inference is given of one well-formed formula as conclusion from others as premisses, it can always be determined effectively whether or not this proposed immediate inference is in accordance with the rules of inference.

(From these requirements it follows that the notion of a proof is effective in the sense that there is a method by which, whenever a finite sequence of well-formed formulas is given, it can always be determined effectively whether or not it is a proof. But the notion of a theorem is not necessarily effective in the sense of existence of a method by which, whenever a well-formed formula is given, it can always be determined whether or not it is a theorem—for there may be no certain method by which we can always either find a proof or determine that none exists. This last is a point to which we shall return later.)

As to requirement (I), we suppose that we are able always to determine about two given symbol-occurrences whether or not they are occurrences of the same symbol (thus ruling out by assumption such difficulties as that of illegibility). Therefore, if the number of primitive symbols is finite, the requirement may be satisfied just by giving the complete list of primitive symbols, written out in full. Frequently, however, the number of primitive symbols is infinite. In particular, if there are variables, it is desirable that there should be an infinite number of different variables of each kind because, although in any one well-formed formula the number of different variables is always finite, there is hardly a way to determined a finite upper limit of the number

[117]Besides these minimum essentials, the primitive basis may also include other notions introduced in order to use them in defining a well-formed formula or in stating the rules of inference. In particular the primitive symbols may be divided in some way into different categories: e.g., they may be classified as *primitive constants*, *variables*, and *improper symbols*, or various categories may be distinguished of primitive constants, of variables, or of improper symbols. The variables and the primitive constants together are usually called *proper symbols*. Rules may be given for distinguishing an occurrence of a variable in a well-formed formula as being a *free occurrence* or a *bound occurrence*, well-formed formulas being then classified as *forms* or *constants* according as they do or do not contain a free occurrence of a variable. Also rules may be given for distinguishing certain of the forms as *propositional forms*, and certain of the constants as *sentences*. In doing all this, the terminology often is so selected that, when the logistic system becomes a language by adoption of one of the intended principal interpretations (see below), the terms *primitive constant, variable, improper symbol, proper symbol, free, bound, form, constant, propositional form, sentence* come to have meanings in accord with the informal semantical explanations of §§02–06.

The *primitive basis* of a formalized language, or interpreted logistic system, is obtained by adding the semantical rules (see below) to the primitive basis of the logistic system.

of different variables that may be required for some particular purpose in the actual use of the logistic system. When the number of primitive symbols is infinite, the list cannot be written out in full, but the primitive symbols must rather be fixed in some way by a statement of finite length in the meta-language. And this statement must be such as to conform to (I).

A like remark applies to (III). If the number of axioms is finite, the requirement can be satisfied by writing them out in full. Otherwise the axioms must be specified in some less direct way by means of a statement of finite length in the meta-language, and this must be such as to conform to (III). It may be thought more elegant or otherwise more satisfactory that the number of axioms be finite; but we shall see that it is sometimes convenient to make use of an infinite number of axioms, and no conclusive objections appear to doing so if requirements of effectiveness are obeyed.

We have assumed the reader's understanding of the general notion of effectiveness, and indeed it must be considered as an informally familiar mathematical notion, since it is involved in mathematical problems of a frequently occurring kind, namely, problems to find a method of computation, i.e., a method by which to determine a number, or other thing, effectively.[118] We shall not try to give here a rigorous definition of effectiveness, the informal notion being sufficient to enable us, in cases we shall meet, to distinguish given methods as effective or non-effective.[119]

The requirements of effectiveness are (of course) not meant in the sense that a structure which is analogous to a logistic system except that it fails to satisfy these requirements may not be useful for some purposes or that it is forbidden to consider such—but only that a structure of this kind is unsuitable for use or interpretation as a language. For, however indefinite or imprecisely fixed the common idea of a language may be, it is at least fundamental to it that a language shall serve the purpose of communication. And to the extent that requirements or effectiveness fail, the purpose of communication is defeated.

Consider, in particular, the situation which arises if the definition of well-formedness is non-effective. There is then no certain means by which, when an alleged expression of the language is uttered (spoken or written), say as an asserted sentence, the auditor

---

[118] A well-known example from topology is the problem (still unsolved even for elementary manifolds of dimensionalities above 2) to find a method of calculating about any two closed simplicial manifolds, given by means of a set of incidence relations, whether or not they are homeomorphic—or, as it is often phrased, the problem to find a complete classification of such manifolds, or to find a complete set of invariants.

As another example, Euclid's algorithm, in the domain of rational integers, or in certain other integral domains, provides an effective method of calculating for any two elements of the domain their greatest common divisor (or highest common factor).

In general, an effective method of calculating, especially if it consists of a sequence of steps with later steps depending on results of earlier ones, is called on *algorithm*. (This is the long established spelling of this word, and should be preserved in spite of any considerations of etymology.)

[119] For a discussion of the question and proposal of a rigorous definition see a paper by the present writer in the ***American Journal of Mathematics***, vol. 58 (1936), pp. 345–363, especially §7 thereof. The notion of effectiveness may also be described by saying that an effective method of computation, or algorithm, is one for which it would be possible to build a computing machine. This idea is developed into a rigorous definition by A. M. Turing in the ***Proceedings of the London Mathematical Society***, vol. 42 (1936–1937), pp. 230–265 (and vol. 43 (1937), pp. 544–546). See further: S. C. Kleene in the ***Mathematische Annalen***, vol. 112 (1936), pp. 727–742; E. L. Post in the ***The Journal of Symbolic Logic***, vol. 1 (1936), pp. 103–105; A. M. Turing in ***The Journal of Symbolic Logic***, vol. 2 (1937), pp. 153–163; Hilbert and Bernays, ***Grundlagen der Mathematik***, vol. 2 (1939), Supplement II.

(hearer or reader) may determine whether it is well-formed, and thus whether any actual assertion has been made.[120] Therefore the auditor may fairly demand a proof that the utterance is well-formed, and until such proof is provided may refuse to treat is as constituting an assertion. This proof, which must be added to the original utterance in order to establish its status, ought to be regarded, it seems, as part of the utterance, and the definition of well-formedness ought to be modified to provide this, or its equivalent. When such modification is made, no doubt the non-effectiveness of the definition will disappear; otherwise it would be open to the auditor to make further demand for proof of well-formedness.

Again, consider the situation which arises if the notion of a proof is non-effective. There is then no certain means by which, when a sequence of formulas has been put forward as a proof, the auditor may determine whether it is in fact a proof. Therefore he may fairly demand a proof, in any given case, that the sequence of formulas put forward is a proof; and until this supplementary proof is provided, he may refuse to be convinced that the alleged theorem is proved. This supplementary proof ought to be regarded, it seems, as part of the whole proof of the theorem, and the primitive basis of the logistic system ought to be so modified as to provide this, or its equivalent.[121] Indeed it is essential to the idea of a proof that, to any one who admits the presuppositions on which it is based, a proof carries final conviction. And the requirements of effectiveness (I)–(IV) may be thought of as intended just to preserve this essential characteristic of proof.

After setting up the logistic system as described, we still do not have a formalized language until an *interpretation* is provided. This will require a more extensive meta-language than the restricted portion of English used in setting up the logistic system. However, it will proceed not by translations of the well-formed formulas into English phrases but rather by *semantical rules* which, in general, *use* rather than *mention* English phrases (cf. §08), and which shall prescribe for every well-formed formula either how it denotes[122] (so making it a proper name in the sense of §01) or else how it has values[122] (so making it a form in the sense of §02).

---

[120]To say that an assertion has been made if there is a meaning evades the issue unless an effective criterion is provided for the presence of meaning. An understanding of the language, however reached, must include effective ability to recognize meaningfulness (in some appropriate sense), and in the purely formal aspect of the language, the logistic system, this appears as an effective criterion of well-formedness.

[121]Perhaps at first sight it will be thought that the proof as so modified might consist of something more than merely a sequence of well-formed formulas. For instance there might be put in at various places indications in the meta-language as to which rule of inference justifies the inclusion of a particular formula as immediately inferred from preceding formulas, or as to which preceding formulas are the premisses of the immediate inference.

But as a matter of fact we consider this inadmissible. For our program is to express proofs (as well as theorems) in a fully formalized object language, and as long as any part of the proof remains in an unformalized meta-language the logical analysis must be held to be incomplete. A statement in the meta-language, e.g., that a particular formula is immediately inferred from particular preceding formulas—if it is not superfluous and therefore simply omissible—must always be replaced in some way by one or more sentences of the object language.

Though we use a meta-language to set up the object language, we require that, once set up, the object language shall be an independent language capable, without continued support and supplementation from the meta-language, of expressing those things for which it was designed.

[122]Because of the possibility of misunderstanding, we avoid the wordings "what it denotes" and "what

In view of our postulation of two truth-values (§04), we impose the requirement that the semantical rules, if they are to be said to provide an interpretation, must be such that the axioms denote truth-values (if they are names) or have always truth-values as values (if they are forms), and the same must hold of the conclusion of any immediate inference if it holds of the premisses. In using the formalized language, only those well-formed formulas shall be capable of being asserted which denote truth-values (if they are names) or have always truth-values as values (if they are forms); and only those shall be capable of being rightly asserted which denote truth (if they are names) or have always the value truth (if they are forms). Since it is intended that proof of a theorem shall justify its assertion, we call an interpretation of a logistic system *sound* if, under it, all the axioms either denote truth or have always the value truth, and if further the same thing holds of the conclusion of any immediate inference if it holds of the premisses. In the contrary case we call the interpretation *unsound*. A formalized language is called sound or unsound according as the interpretation by which it is obtained from a logistic system is sound or unsound. And an unsound interpretation or an unsound language is to be rejected.

(The requirements, and the definition of soundness, in the foregoing paragraph are based on two truth-values. They are satisfactory for every formalized language which will receive substantial consideration in this book. But they must be modified correspondingly, in case the scheme of two truth-values is modified—cf. the remark in §19.)

The semantical rules must in the first instance be stated in a presupposed and therefore unformalized meta-language, here taken to be ordinary English. Subsequently, for their more exact study, we may formalize the meta-language (using a presupposed meta-meta-language and following the method already described for formalizing the object language) and restate the semantical rules in this formalized language. (This

---

values it has."

For example, in one of the logistic systems of Chapter X we may find a well-formed formula which, under a principal interpretation of the system, is interpreted as denoting: the greatest positive integer n such that $1 + n^r$ is prime, r being chosen as the least even positive integer corresponding to which there is such a greatest positive integer n. Thus the semantical rules do in a sense determine what this formula denotes, but the remoteness of this determination is measured by the difficulty of the mathematical problem which must be solved in order to identify in some more familiar manner the positive integer which the formula denotes, or even to say whether or not the formula denotes 1.

Again in the logistic system F^{1h} of Chapter III (or A^0 of Chapter V) taken with its principal interpretation, there is a well-formed formula which, according to the semantical rules, denotes the truth-value thereof that every even number greater than 2 is the sum of two prime numbers. To say that the semantical rules determine what this formula denotes seems to anticipate the solution of a famous problem, and it may be better to think of the rules as determining indirectly what the formula expresses.

In assigning how (rather than what) a name denotes we are in effect fixing its sense, and in assigning how a form has values we fix the correspondence of sense values of the form (see footnote 27) to concepts of values of its variables. (This statement of the matter will be sufficiently precise for our present purposes, though it remains vague to the extent that we have left the meaning of "sense" uncertain—see footnotes 15, 37.)

It will be seen in particular examples below (such as rules a–g of §10, or rules a–f of §30, or rules α–ζ of §30) that in most of our semantical rules the explicit assertion is that certain well-formed formulas, usually on certain conditions, are to denote certain things or to have certain values. However, as just explained, this explicit assertion is so chosen as to give implicitly also the sense of the sense values. No doubt a fuller treatment of semantics must have additional rules stating the sense or the sense values explicitly, but his would take us into territory still unexplored.

leads to the subject of *semantics* (§09).)

As a condition of rigor, we require that the proof of a theorem (of the object language) shall make no reference to or use of any interpretation, but shall proceed purely by the rules of the logistic system, i.e., shall be a *proof* in the sense defined above for logistic systems. Motivation for this is threefold, three rather different approaches issuing in the same criterion. In the first place this may be considered a more precise formulation of the traditional distinction between form and matter (§00) and of the principle that the validity of an argument depends only on the form—the form of a proof in a logistic system being thought of as something common to its meanings under various interpretations of the logistic system. In the second place this represents the standard mathematical requirement of rigor that a proof must proceed purely from the axioms without use of anything (however supposedly obvious) which is not stated in the axioms; but this requirement is here modified and extended as follows: that a proof must proceed purely from the axioms by the rules of inference, without use of anything not stated in the axioms or any method of inference not validated by the rules. Thirdly there is the motivation that the logistic system is relatively secure and definite, as compared to interpretations which we may wish to adopt, since it is based on a portion of English as meta-language so elementary and restricted that its essential reliability can hardly be doubted if mathematics is to be possible at all.

It is also important that a proof which satisfies our foregoing condition of rigor must then hold under any interpretation of the logistic system, so that there is a resulting economy in proving many things under one process.[123] The extent of the economy is just this, that proofs identical in form but different in matter need not be repeated indefinitely but may be summarized once for all.[124]

Though retaining our freedom to employ any interpretation that may be found useful, we shall indicate, for logistic systems set up in the following chapters, one or more interpretations which we have especially in mind for the system and which shall be called the *principal interpretations*.

The subject of formal logic, when treated by the method of setting up a formalized

---

[123]This remark has now long been familiar in connection with the axiomatic method in mathematics (see below).

[124]The summarizing of a proof according to its form may indeed be represented to a certain extent, by the use of variables, within one particular formalized language. But, because of restricted ranges of the variables, such summarizing is less comprehensive in its scope than is obtained by formalizing in a logistic system whose interpretation is left open.

The procedure of formalizing a proof in a logistic system and then employing the formalized proof under various different interpretations of the system may be thought of as a mere device for brevity and convenience of presentation, since it would be possible instead to repeat the proof in full each time it were used with a new interpretation. From this point of view such use of the meta-language may be allowed as being in principle dispensable and therefore not violating the demand (footnote 121) for an independent object language.

(If on the other hand we wish to deal rigorously with the notion of logical form of proofs, this must be in a particular formalized language, namely a formalized meta-language of the language of proofs. Under the program of §02 each variable of this meta-language will have a fixed range assigned in advance, according, perhaps, with the theory of types. And the notion of form which is dealt with must therefore be correspondingly restricted, it would seem, to proofs of a fixed class, taking no account of sameness of form between proofs of this class and others (in the same or a different language). Presumably our informal references to logical form in the text are to be modified in this way before they can be made rigorous—cf. §09.)

language, is called *symbolic logic*, or *mathematical logic*, or *logistic*.[125] The method itself we shall call the *logistic method*.

Familiar in mathematics is the *axiomatic method*, according to which a branch of mathematics begins with a list of *undefined terms* and a list of assumptions, or *postulates* involving these terms, and the theorems are to be derived from the postulates by the methods of formal logic.[126] If the last phrase is left unanalyzed, formal logic being presupposed as already known, we shall say that the development is by the *informal axiomatic method*.[127] And in the opposite case we shall speak of the *formal axiomatic method*.

The formal axiomatic method thus differs from the logistic method only in the following two ways.

(1) In the logistic system the primitive symbols are given in two categories: the *logical primitive symbols*, thought of as pertaining to the underlying logic, and the *undefined terms*, thought of as pertaining to the particular branch of mathematics. Correspondingly the axioms are divided into two categories: the *logical axioms*, which are well-formed formulas containing only logical primitive symbols, and the *postulates*,[128] which involve also the undefined terms and are thought of as determining the special branch of mathematics. The rules of inference, to accord with the usual conception of the axiomatic method, must all be taken as belonging to the underlying

---

[125]The writer prefers the term "mathematical logic," understood as meaning logic treated by the mathematical method, especially the formal axiomatic or logistic method. But both this term and the term "symbolic logic" are often applied also to logic as treated by a less fully formalized mathematical method, in particular to the "algebra of logic," which had its beginning in the publications of George Boole and Augustus De Morgan in 1847, and received a comprehensive treatment in Ernst Schröder's ***Vorlesungen über die Algebra der Logik*** (1890–1905). The term "logistic" is more definitely restricted to the method described in this section, and has also the advantage that it is more easily made an adjective. (Sometimes "logistic" has been used with special reference to the school of Russell or to the Frege-Russell doctrine that mathematics is a branch of logic—cf. footnote 545. But we shall follow the more common usage which attaches no such special meaning to this word.)

"Logica mathematica" and "logistica" were both used by G. W. v. Leibniz along with "calculus ratiocinator," and many other synonyms, for the calculus of reasoning which he proposed but never developed beyond some brief and inadequate (though significant) fragments. Boole used the expressions "mathematical analysis of logic," "mathematical theory of logic." "Mathematische Logik" was used by Schröder in 1877, "matématičéskaá logika" (Russian) by Platon Poretsky in 1884, "logica matematica" (Italian) by Giuseppe Peano in 1891. "Symbolic logic" seems to have been first used by John Venn (in ***The Princeton Review***, 1880), though Boole speaks of "symbolical reasoning." The word "logistic" and its analogues in other languages originally meant the art of calculation or common arithmetic. Its modern use for mathematical logic dates from the International Congress of Philosophy of 1904, where it was proposed independently by Itelson, Lalande, and Couturat. Other terms found in the literature are "logischer Calcul" (Gottfried Ploucquet 1766), "algorithme logique" (G. F. Castillon 1805), "calculus of logic" (Boole 1847), "calculus of inference" (De Morgan 1847), "logique algorithmique" (J. R. L. Delboeuf 1876), "Logikkalkul" (Schröder 1877), "theoretische Logik" (Hilbert and Ackermann 1928). Also "Boole's logical algebra" (C. S. Peirce 1870), "logique algébrique de Boole" (Louis Liard 1877), "algebra of logic" (Alexander Macfarlane 1879, C. S. Peirce 1880).

[126]Accounts of the axiomatic method may of course be found in many mathematical textbooks and other publications. An especially good exposition is in the Introduction to Veblen and Young's ***Projective Geometry***, vol. 1 (1910).

[127]This is the method of most mathematical treatises, which proceed axiomatically but are not specifically about logic—in particular of Veblen and Young (preceding footnote).

[128]The words "axiom" and "postulate" have been variously used, either as synonymous or with varying distinctions between them, by the present writer among others. In this book, however, the terminology here set forth will be followed closely.

logic. And, though they may make reference to particular undefined terms or to classes of primitive symbols which include undefined terms, they must not involve anything which, subjectively, we are unwilling to assign to the underlying logic rather than to the special branch of mathematics.[129]

(2) In the interpretation the semantical rules are given in two categories. Those of the first category fix those general aspects of the interpretation which may be assigned, or which we are willing to assign, to the underlying logic. And the rules of the second category determine the remainder of the interpretation. The consideration of different representations or interpretations of the system of postulates, in the sense of the informal axiomatic method, corresponds here to varying the semantical rules of the second category while those of the first category remain fixed.

08. Syntax. The study of the purely formal part of a formalized language in abstraction from the interpretation, i.e., of the logistic system, is called *syntax*, or, to distinguish it from the narrower sense of "syntax" as concerned with the formulation rules alone,[130] *logical syntax*.[131] The meta-language used in order to study the logistic system in this way is called the *syntax language*.[131]

We shall distinguish between *elementary syntax* and *theoretical syntax*.

The elementary syntax of a language is concerned with setting up the logistic system and with the verification of particular well-formed formulas, axioms, immediate inferences, and proofs as being such. The syntax language is the restricted portion of English which was described in the foregoing section, or a corresponding restricted formalized meta-language, and the requirements of effectiveness, (I)–(IV), must be observed. The demonstration of derived rules and theorem schemata, in the sense of §§12, 33, and their application in particular cases are also considered to belong to elementary syntax, provided that the requirement of effectiveness holds which is explained in §12.

Theoretical syntax, on the other hand, is the general mathematical theory of a logistic system or systems and is concerned with all the consequences of their formal structure (in abstraction from the interpretation). There is no restriction imposed as to what is available in the syntax language, and requirements of effectiveness are or may

---

[129]Ordinarily, e.g., it would be allowed that the rules of inference should treat differently two undefined terms intended one to denote an individual and one to denote a class of individuals, or two undefined terms intended to denote a class of individuals and a relation between individuals; but not that the rules should treat differently two undefined terms intended both to denote a class of individuals. But no definitive controlling principle can be given.

The subjective and essentially arbitrary character of the distinction between what pertains to the underlying logic and what to the special branch of mathematics is illustrated by the uncertainty which sometimes arises, in treating a branch of mathematics by the informal axiomatic method, as to whether the sign of equality is to be considered as an undefined term (for which it is necessary to state postulates). Again it is illustrated by Zermelo's treatment of axiomatic set theory in his paper of 1908 (cf. Chapter XI) in which, following the informal axiomatic method, he introduces the relation ϵ of membership in a set as an undefined term, though this same relation is usually assigned to the underlying logic when a branch of mathematics is developed by the informal axiomatic method.

[130]Cf. footnote 116.

[131]The terminology is due to Carnap in his ***Logische Syntax der Sprache*** (1934), translated into English (with some additions) as the ***The Logical Syntax of Language*** (1937). In connection with this book see also reviews of it by Saunders MacLane in the ***Bulletin of the American Mathematical Society***, vol. 44 (1938), pp. 171–176, and by S. C. Kleene in ***The Journal of Symbolic Logic***, vol. 4 (1939), pp. 82–87.

be abandoned. Indeed the syntax language may be capable of expressing the whole of extant mathematics. But it may also sometimes be desirable to use a weaker syntax language in order to exhibit results as obtained on this weaker basis.

Like any branch of mathematics, theoretical syntax may, and ultimately must, be studied by the axiomatic method. Here the informal and the formal axiomatic method share the important advantage that the particular character of the symbols and formulas of the object language, as marks upon paper, sounds, or the like, is abstracted from, and the pure theory of the structure of the logistic system is developed. But the formal axiomatic method—the syntax language being itself formalized according to the program of §07, by employing a meta-meta-language—has the additional advantage of exhibiting more definitely the basis on which results are obtained, and of clarifying the way and the extent to which certain results may be obtained on a relatively weaker basis.

In this book we shall be concerned with the task of formalizing an object language, and theoretical syntax will be treated informally, presupposing in any connection such general knowledge of mathematics as is necessary for the work at hand. Thus we do not apply even the informal axiomatic method to our treatment of syntax. But the reader must always understand that syntactical discussions are carried out in a syntax language whose formalization is ultimately contemplated, and distinctions based upon such formalization may be relevant to the discussion.

In such informal development of syntax, we shall think of the syntax language as being a different language from the object language. But the possibility is important that a sufficiently adequate object language may be capable of expressing its own syntax, so that in this case the ultimate formalization of the syntax language may if desired consist in identifying it with the object language.[132]

We shall distinguish between theorems *of* the object language and theorems of the syntax language (which often are theorems *about* the object language) by calling the latter *syntactical theorems*. Though we demonstrate syntactical theorems informally, it is contemplated that the ultimate formalization of the syntax language shall make them theorems in the sense of §07, i.e., theorems of the syntax language in the same sense as that in which we speak of theorems of the object language.

We shall require, as belonging to the syntax language: first, names of the various symbols and formulas of the object language; and secondly, variables which have these symbols and formulas as their values. The former will be called *syntactical constants*, and the latter, *syntactical variables*.[133]

As syntactical variables we shall use the following: as variables whose range is the primitive symbols of the object language, bold Greek small letters ($\boldsymbol{\alpha}$, $\boldsymbol{\beta}$, $\boldsymbol{\gamma}$, etc.); as

[132]Cf. footnote 109. In particular the developments of Chapter VIII show that the logistic system of Chapter VII is capable of expressing its own syntax if given a suitable interpretation different from the principal interpretation of Chapter VII, namely, an interpretation in which the symbols and formulas of the logistic system itself are counted among the individuals, as well as all finite sequences of such formulas, and the functional constant S is given an appropriate (quite complicated) interpretation, details of which may be made out by following the scheme of Gödel numbers that is set forth in Chapter VIII.

[133]Given the apparatus of syntactical variables, we could actually avoid the use of syntactical constants by resorting to appropriate circumlocutions in cases where syntactical constants would otherwise seem to be demanded. Indeed the example of the preceding footnote illustrates this, as will become clear in connection with the cited chapters. But it is more natural and convenient, especially in an informal treatment of syntax, to allow free use of syntactical constants.

variables whose range is the primitive constants and variables of the object language—see footnote 117—bold roman small letters (**a**, **b**, **c**, etc.); as variables whose range is the formulas of the object language, bold Greek capitals (**Γ**, **Δ**, etc.); and as variables whose range is the well-formed formulas of the object language, bold roman capitals (**A**, **B**, **C**, etc.). Wherever these bold letters are used in the following chapters the reader must bear in mind that they are not part of the symbolic apparatus of the object language but that they belong to the syntax language and serve the purpose of talking about the object language. In use of the object language as an independent language, bold letters do not appear (just as English words never appear in the pure text of a Latin author though they do appear in a Latin grammar text written in English).

As a preliminary to explaining the device to which we resort for syntactical constants, it is desirable first to consider the situation in ordinary English, with no formalized object language specially in question. We must take into account the fact that English is not a formalized language and the consequent uncertainty as to what are its formation rules, rules of inference, and semantical rules, the contents of ordinary English grammars and dictionaries providing only some incomplete and rather vague approximations to such rules. But, with such reservations as this remark implies, we go on to consider the use of English in making syntactical statements about the English language itself.

Frequently found in practice is the use of English words *autonymously* (to adopt a terminology due to Carnap), i.e., as names of those same words.[134] Examples are such statements as "The second letter of man is a vowel," "Man is monosyllabic," "Man is a noun with an irregular plural." Of course it is equivocal to use the same word, man, both as a proper name of the English word which is spelled by the thirteenth, first, fourteenth letters of the alphabet in that order, and as a common name (see footnote 6) of featherless plantigrade biped mammals[135]—but an equivocacy which, like many others in the natural languages, is often both convenient and harmless. Whenever there would otherwise be real doubt of the meaning, it may be removed by the use of added words in the sentence, or by use of quotation marks, or of italics, as in: "The word man is monosyllabic"; " 'Man' is monosyllabic"; "*Man* is monosyllabic."

Following the convenient and natural phraseology of Quine, we may distinguish between *use* and *mention* of a word or symbol. In "Man is a rational animal" the word "man" is used but not mentioned. In "The English translation of the French word *homme* has three letters" the word "man" is mentioned but not used. In "Man is a monosyllable" the word "man" is both mentioned and used, though used in an anomalous manner, namely autonymously.

Frege introduced the device of systematically indicating autonymy by quotation marks, and in his later publications (though not in the *Begriffsschrift*) words and symbols used autonymously are enclosed in single quotation marks in all cases. This

---

[134]In the terminology of the Scholastics, use of a word as a name of itself, i.e., to denote itself as a word, was called *suppositio materialis*. Opposed to this as *suppositio formalis* was the use of a noun in its proper or ordinary meaning. This terminology is sometimes still convenient.

The various further distinctions of *suppositiones* are too cumbrous, and too uncertain, to be usable. All of them, like that between *suppositio materialis* and *formalis*, refer to peculiarities and irregularities of meaning which are found in many natural languages but which have to be eliminated in setting up a formalized language.

[135]To follow a definition found in the ***The Century Dictionary***.

has the effect that a word enclosed in single quotation marks is to be treated as a different word from that without the quotation marks—as if the quotation marks were two additional letters in the spelling of the word—and equivocacy is thus removed by providing two different words to correspond to the different meanings. Many recent writers follow Frege in this systematic use of quotation marks, some using double quotation marks in this way, and others following Frege in using single quotation marks for the purpose, in order to reserve double quotation marks for their regular use as punctuation. As the reader has long since observed, Frege's systematic use of quotation marks is not adopted in this book.[136] But we may employ quotation marks or other devices from time to time, especially in cases in which there might otherwise be real doubt of the meaning.

To return to the question of syntactical constants for use in developing the syntax of a formalized object language, we find that there is in this case nothing equivocal in using the symbols and formulas of the object language autonymously in the syntax language, provided that care is taken that no formula of the object language is also a formula of the syntax language in any other wise than as an autonym. Therefore we adopt the following practice:

The primitive symbols of the object language will be used in the syntax language as names of themselves, and juxtaposition will be used for juxtaposition.[137]

---

[136] Besides being rather awkward in practice, such systematic use of quotation marks is open to some unfortunate abuses and misunderstandings. One of these is the misuse of quotation marks as if they denoted a function from things (of some category) to names of such things, or as if such a function might be employed at all without some more definite account of it. Related to this is the temptation to use in the role of a syntactical variable the expression obtained by enclosing a variable of an object language in quotation marks, though such an expression, correctly used, is not a variable of any kind, and not a form but a constant.

Also not uncommon is the false impression that trivial or self-evident propositions are expressed in such statements as the following: ' 'Snow is white' is true if and only if snow is white' (Tarski's example); ' 'Snow is white' means that snow is white'; ' 'Cape Town' is the [or a] name of Cape Town.'

This last misunderstanding may arise also in connection with autonymy. A useful method of combatting it is that of translation into another language (cf. a remark by C. H. Langford in ***The Journal of Symbolic Logic***, vol. 2 (1937), p. 53). For example, the proposition that 'Cape Town' is the name of Cape Town would be conveyed thus to an Italian (whom we may suppose to have no knowledge of English): ' 'Cape Town' è il nome di Città del Capo.' Assuming, as we may, that the Italian words have exactly the same sense as the English words of which we use them as translations—in particular that 'Città del Capo' has the same sense as 'Cape Town' and that ' 'Cape Town' ' has the same sense in Italian as in English—we see that the Italian sentence and its English translation must express the very same proposition, which can no more be a triviality when conveyed in one language than it can in another.

The foregoing example may be clarified by recalling the remark of footnote 8 that the name relation is properly a ternary relation, and may be reduced to a binary relation only by fixing the language in a particular context. Thus we have the more explicit English sentences: ' 'Cape Town' is the English name of Cape Town'; ' 'Città del Capo' is the Italian name of Cape Town.' The Italian translations are: ' 'Cape Town' è il nome inglese di Città del Capo'; ' 'Città del Capo' è il nome italiano di Città del Capo.' Of the two propositions in question, the first one has a false appearance of obviousness when expressed in English, the illusion being dispelled on translation into Italian; the second one contrariwise does not seem obvious or trivial when expressed in English, but on translation into Italian acquires the appearance of being so.

(In the three preceding paragraphs of this footnote, we have followed Frege's systematic use of single quotation marks, and the paragraphs are to be read with that understanding. As explained, we do not follow this usage elsewhere.)

[137] I.e., juxtaposition will be used in the syntax language as a binary connective having the operation of juxtaposition as its associated function. Technically, some added notation is needed to show association, or some convention about the matter, such as that of association to the left (as in §11). But in practice, because

This is the ordinary usage in mathematical writing, and has the advantage of being self-explanatory. Though we employ it only informally, it is also readily adapted to incorporation in a formalized syntax language[138] (and in fact more so than the convention of quotation marks).

As a precaution against equivocation, we shall hereafter avoid the practice—which might otherwise sometimes be convenient—of borrowing formulas of the object language for use in the syntax language (or other meta-language) with the same meaning that they have in the object language. Thus in all cases where a single symbol or a formula of the object language is found as a constituent in an English sentence, it is to be understood in accordance with the italicized rule above, i.e., autonymously.

Since we shall later often introduce conventions for abbreviating well-formed formulas of an object language, some additional explanations will be necessary concerning the use of syntactical variables and syntactical constants (and concerning autonymy) in connection with such abbreviations. These will be indicated in §11, where such abbreviations first appear. But, as explained in that section, the abbreviations themselves and therefore any special usages in connection with them are dispensable in principle, however necessary practically. In theoretical discussions of syntax and in particular in formalizing the syntax language, the matter of abbreviations of well-formed formulas may be ignored.

09. Semantics. Let us imagine the users of a formalized language, say a written language, engaged in writing down well-formed formulas of the language, and in assembling sequences of formulas which constitute chains of immediate inferences or, in particular, proofs. And let us imagine an observer of this activity who not only does not understand the language but refuses to believe that it is a language, i.e., that the formulas have meanings. He recognizes, let as say, the syntactical criteria by which formulas are accepted as well-formed, and those by which sequences of well-formed formulas are accepted as immediate inferences or as proofs; but he supposes that the activity is merely a game—analogous to a game of chess or, better, to a chess problem or a game of solitaire at cards—the point of the game being to discover unexpected theorems or ingenious chains of inferences, and to solve puzzles as to whether and how some given formula can be proved or can be inferred from other given formulas.[139]

To this observer the symbols have only such meaning as is given to them by the rules of the game—only such meaning as belongs, for example, to the various pieces at chess. A formula is for him like a position on a chessboard, significant only as a step in the game, which leads in accordance with the rules to various other steps.

---

of the associativity of juxtaposition, there is no difficulty in this respect.

[138]This is, of course, on the assumption that the syntax language is a different language from the object language.

If on the contrary a formalized language is to contain names of its own formulas, then a name of a formula must ordinarily not be that formula. E.g., a variable of a language must not be, in that same language, also a name of itself; for a proper name of a variable is no variable but a constant (as already remarked, in another connection, in footnote 136).

[139]A comparison of the rules of arithmetic to those of a game of chess was made by J. Thomae (1898) and figures in the controversy between Thomae and Frege (1903–1908). The same comparison was used by Hermann Weyl (1924) in order to describe Hilbert's program of *metamathematics* or syntax of a mathematical object language.

All those things about the language which can be said to and understood by such an observer while he continues to regard the use of the language as merely a game constitute the (theoretical) syntax of the language. But those things which are intelligible only through an understanding that the well-formed formulas have meaning in the proper sense, e.g., that certain of them express propositions or that they denote or have values in certain ways, belong to the semantics of the language.

Thus the study of interpretation of the language as an interpretation is called *semantics*.[140] The name is applied especially when the treatment is in a formalized meta-language. But in this book we shall not go beyond some unformalized semantical discussion, in ordinary English.

Theorems of the semantical meta-language will be called *semantical theorems*, and both semantical and syntactical theorems will be called *metatheorems*, in order to distinguish them from theorems of the object language.

As appears from the work of Tarski, there is a sense in which semantics can be reduced to syntax. Tarski has emphasized especially the possibility of finding, for a given formalized language, a purely syntactical property of the well-formed formulas which coincides in extension with the semantical property of being a true sentence. And in Tarski's *Wahrheitsbegriff*[141] the problem of finding such a syntactical property is solved for various particular formalized languages.[142] But like methods apply to the two semantical concept of denoting and having values, so that syntactical concepts may be found which coincide with them in extension.[143] Therefore, if names expressing these

---

[140]The name (or its analogue in Polish) was introduced by Tarski in a paper in ***Przeglad Filozoficzny***, vol. 39 (1936), pp. 50–57, translated into German as *Grundlegung der wissenschaftlichen Semantik* in ***Actes du Congrès International de Philosophie Scientifique*** (1936). Other important publications in the field of semantics are: Tarski's *Pojecie Prawdy w Jezykach Nauk Dedukcyjnych* (1933), afterwards translated into German (and an important appendix added) as "Der Wahrheitsbegriff in den formalisierten Sprachen" in ***Studia Philosophica***, vol. 1 (1936) pp. 261–405; and Carnap's ***Introduction to Semantics*** (1942). Concerning Carnap's book see a review by the present writer in ***The Philosophical Review***, vol. 52 (1943), pp. 298–304.

The word semantics has various other meanings, most of them older than that in question here. Care must be taken to avoid confusion on this account. But in this book the word will have always the one meaning, intended to be the same (or substantially so) as that in which it is used by Tarski, C. W. Morris (***Foundations of the Theory of Signs***, 1938), Carnap, G. D. W. Berry (***Harvard University, Summaries of Theses 1942***, pp. 330–334).

[141]Cited in the preceding footnote.

[142]Tarski solves also, for various particular formalized languages, the problem of finding a syntactical relation which coincides in extension with the semantical relation of satisfying a propositional form.

In a paper published in ***Monatshefte für Mathematik und Physik***, vol. 42, no. 1 (1935), therefore later than Tarski's *Pojecie Prawdy* but earlier than the German translation and its appendix, Carnap also solves both problems (of finding syntactical equivalents of being a true sentence and of satisfying a propositional form) for a particular formalized language and in fact for a stronger language than any for which this had previously been done by Tarski. Carnap's procedure can be simplified in the light of Tarski's appendix or as suggested by Kleene in his review cited in footnote 131.

On the theory of meaning which we are here adopting, the semantical concepts of being a true sentence and of satisfying a propositional form are reducible to those of denoting and having values, and these results of Tarski and Carnap are therefore implicit in the statement of the following footnote.

[143]More explicitly, this may be done as follows. In §07, in discussing the semantical rules of a formalized language, we thought of the concepts of denoting and of having values as being known in advance, and we used the semantical rules for the purpose of giving meaning to the previously uninterpreted logistic system. But instead of this it would be possible to give no meaning in advance to the words "denote" and "have values" as they occur in the semantical rules, and then to regard the semantical rules, taken together, as constituting definitions of "denote" and "have values" (in the same way that the formation rules of a logistic

two concepts are the only specifically semantical (non-syntactical) primitive symbols of a semantical meta-language, it is possible to transform the semantical meta-language into a syntax language by a change of interpretation which consists only in altering the sense of those names without changing their denotations.

However, a sound syntax language capable of expressing such syntactical equivalents of the semantical concepts of denoting and having values—or even only a syntactical equivalent of the semantical property of truth—must ordinarily be stronger than the object language (assumed sound), in the sense that there will be theorems of the syntax language of which no translation (i.e., sentence expressing the same proposition) as a theorem of the object language. Else there will be simple elementary true propositions about the semantical concepts such that the sentences expressing the corresponding propositions about the syntactical equivalents of the semantical concepts are not theorems of the syntax language.[144]

For various particular formalized languages this was proved (in effect) by Tarski in his *Wahrheitsbegriff*. And Tarski's methods[145] are such that they can be applied to obtain the same result in many other cases—in particular in the case of each of the object languages studied in this book, when a formalized syntax language of it is set up in a straightforward manner. No doubt Tarski's result is capable of precise formulation and proof as a result about a very general class of languages, but we shall not attempt this.

The significance of Tarski's result should be noticed as it affects the question of the use of a formalized language as semantical meta-language of itself. A sound and sufficiently adequate language may indeed be capable of expressing its own syntax

---

system constitute a definition of "well-formed"). The concepts expressed by "denote" and "have values" as thus defined belong to theoretical syntax, nothing semantical having been used in their definition. But they coincide in extension with the semantical concepts of denoting and having values, as applied to the particular formalized language.

The situation may be clarified by recalling that a particular logistic system may be expected to have many sound interpretations, leading to many different assignments of denotations and values to its well-formed formulas. These assignments of denotations and values to the well-formed formulas may be made as abstract correspondences, so that their treatment belongs to theoretical syntax. Semantics begins when we decide the meaning of the well-formed formulas by fixing a particular interpretation of the system. The distinction between semantics and syntax is found in the different significance given to one particular interpretation and to its assignment of denotations and values to the well-formed formulas; but within the domain of formal logic, including pure syntax and pure semantics, nothing can be said about this different significance except to postulate it as different.

Many similar situations are familiar in mathematics. For instance, the distinction between plane Euclidean metric geometry and plane projective geometry may be found in the different significance given to one particular straight line and one particular elliptic involution on it. And it seems not unjustified to say that the sense in which semantics can be reduced to syntax is like that in which Euclidean metric geometry can be reduced to projective geometry.

All this suggests that, in order to maintain the distinction of semantics from syntax, "denote" and "have values" should be introduced as undefined terms and treated by the axiomatic method. Our use of semantical rules is intended as a step towards this. And in fact Tarski's *Wahrheitsbegriff* already contains the proposal of an axiomatic theory of truth as an alternative to that of finding a syntactical equivalent of the concept of truth.

[144]A more precise statement of this will be found in Chapter VIII, as it applies to the special case of the logistic system of Chapter VII when interpreted, in the manner indicated in footnote 132, so as to be capable of expressing its own syntax.

[145]Related to those used by Kurt Gödel in the proof of his incompleteness theorems, set forth in Chapter VIII.

(cf. footnote 132) and its own semantics, in the sense of containing sentences which express at least a very comprehensive class of the propositions of its syntax and its semantics. But among these sentences, if certain very general conditions are satisfied, there will always be true sentences of a very elementary semantical character which are not theorems—sentences to the effect, roughly speaking, that such and such a particular sentence is true if and only if ______, the blank being filled by that particular sentence.[146] Hence, on the assumption that the language satisfies ordinary conditions of adequacy in other respects, not all the semantical rules (in the sense of §07), when written as sentences of the language, are theorems.

On account of this situation, the distinction between object language and meta-language, which first arises in formalizing the object language, remains of importance even after the task of formalization is complete for both the object language and the meta-language.

In concluding this Introduction, let us observe that much of what we have been saying has been concerned with the relation between linguistic expressions and their meaning, and therefore belongs to semantics. However, our interest has been less in the semantics of this or that particular language than in general features common to the semantics of many languages. And very general semantical principles, imposed as a demand upon any language that we wish to consider at all, have been put forward in some cases, notably assumptions (1), (2), (3) of §01 and assumption (4) of §02.[147]

We have not, however, attempted to formalize this semantical discussion, or even to put the material into such preliminary order as would constitute a first step toward formalization. Our purpose has been introductory and explanatory, and it is hoped that ideas to which the reader has thus been informally introduced will be held subject to revision or more precise formulation as the development continues.

From time to time in the following chapters we shall interrupt the rigorous treatment of a logistic system in order to make an informal semantical aside. Though in studying a logistic system we shall wish to hold its interpretation open, such semantical explanations about a system may serve in particular to show a motivation for consideration of it by indicating its principal interpretations (cf. §07). *Except in this Introduction, semantical passages will be distinguished from others by being printed in smaller type, the small type serving as a warning that the material is not part of the formal logistic development and must not be used as such.*

As we have already indicated, it is contemplated that semantics itself should ultimately be studied by the logistic method.

But if semantical passages in this Introduction and in later chapters are to be rewritten in a formalized semantical language, certain refinements become necessary. Thus if the semantical language is to be a functional calculus of order ω in the sense of Chapter VI, or a language like that of Chapter X, then various semantical terms, such as the term "denote" introduced in §01, must give way to a multiplicity of terms

---

[146] A more careful statement is given by Tarski.

By the results of Gödel referred to in the preceding footnote (or alternatively by Tarski's reduction of semantics to syntax), true syntactical sentences which are not theorems must also be expected. But these are of not quite so elementary a character. And the fundamental syntactical rules described in §07 may nevertheless all be theorems when written as sentences of the language.

[147] And assumption (5) of footnote 30.

of different types,[148] and statements which we have made using these terms must be replaced by axiom schemata[149] or theorem schemata[149] with typical ambiguity.[149] Or if the semantical language should conform to some alternative to the theory of types, changes of a different character would be required. In particular, following the Zermelo set theory (Chapter XI), we would have to weaken substantially the assumption made in §03 that every singulary form has an associated function, and explanations regarding the notation λ would have to be modified in some way in consequence.

[148]All the expressions of the language—formulas, or well-formed formulas—may be treated as values of (syntactical) variables of one type. But terms "denote" of different types are nevertheless necessary, because in "______ denotes ______," after filling the first blank with a syntactical variable or syntactical constant, we may still fill the second blank with a variable or constant of any type.

Analogously, various other terms that we have used have to be replaced each by a multiplicity of terms of different types. This applies in particular to "thing," and the consequent weakening is especially striking in the case of footnote 9—which must become a schema with typical ambiguity.

See also the remark in the last paragraph of footnote 87.

[149]The terminology is explained in §§27, 30, 33, and Chapter VI. (The typical ambiguity required here is ambiguity with respect to *type* in the sense described in footnote 578, and is therefore not the same as the typical ambiguity mentioned in footnote 585, which is ambiguity rather with respect to *level*.)

BINARY RECURSIVE ARITHMETIC (1957)

Recursive arithmetic.

The method of treating the axiomatic foundations of the elementary theory of natural numbers which is now usually known as "recursive arithmetic" or "primitive-recursive arithmetic" was introduced by Thoralf Skolem[1] and has been developed and extended by David Hilbert and Paul Bernays,[2] Haskell B. Curry,[3] Skolem himself, and others.

The leading idea of recursive arithmetic is to exclude altogether the use of quantifiers, in favor of the assertion of expressions containing free variables only. The notational apparatus employed consists of numerals (i.e., $0, 1, 2, \ldots$), numerical variables, notations for particular numerical functions, the equality sign, and connectives of the propositional calculus. And the axioms and rules of inference of the system provide: methods of propositional calculus; substitution for free variables; elementary properties of equality; elementary properties of some particular numerical functions which are taken as primitive; the method of proof by mathematical induction; and two methods of introducing new numerical functions, namely *composition*,

$$f(x_1, x_2, \ldots, x_n) = g(h_1(x_1, x_2, \ldots, x_n), h_2(x_1, x_2, \ldots, x_n), \ldots, h_m(x_1, x_2, \ldots, x_n)),$$

and *primitive recursion*,

$$\begin{aligned} f(x_1, x_2, \ldots, x_n, 0) &= g(x_1, x_2, \ldots, x_n), \\ f(x_1, x_2, \ldots, x_n, s(y)) &= h(x_1, x_2, \ldots, x_n, y, f(x_1, x_2, \ldots, x_n, y)), \end{aligned}$$

where in each case f is the new function which is being introduced, and g and $h_1, h_2, \ldots, h_n$, or g and h, are given functions, and $s(y)$ is the successor of y (i.e., $y + 1$).

Recursive arithmetic may be thought of as a branch of intuitionistic arithmetic, since all of the methods just listed are in accord with the mathematical intuitionism of L. E. J. Brouwer — even the law of excluded middle, which is implicit in the use of the classical propositional calculus, is intuitionistically acceptable here, as the lack of quantifiers prevents expression of those propositions in connection with which intuitionistic objections against the law of excluded middle arise. It is even possible to think of recursive arithmetic as representing a tendency more drastically constructivistic than that of intuitionism. Nevertheless, as is now well known, substantially all of elementary number theory which is found informally presented in books on that subject can be reproduced within recursive arithmetic, so receiving a rigorous axiomatic treatment on this very circumscribed basis.

And in fact recursive arithmetic is not of interest exclusively to those who, like Skolem, take an intuitionistic of quasi-intuitionistic view of foundational questions in mathematics. For the most solidly classical mathematician, if he concerns himself with the foundations of his subject, must find important the axiomatic separation of the very elementary part of number theory from the whole — just as, to use a hackneyed

---

Originally published in ***Journal de Mathématiques Pures et Appliquées***, ser. 9 vol. 36 (1957), pp. 39–55. Reprinted by permission from Elsevier.

[1] ***Skrifter utgit av Videnskapsselskapet i Kristiania***, I. Matematisk-Naturvidenskabelig Klasse, 1923, nº 6.

[2] ***Grundlagen der Mathematik***, vol. I, 1934, §7.

[3] ***Amer. J. Math.***, vol. 63, 1941, pp. 263–282.

analogy, he must find important the separation, from the remainder of Euclidean geometry, of that part which is independent of the parallel postulate, and does not thereby commit himself to a preference for non-Euclidean geometry over Euclidean.

In our present formulation we shall add to the usual notations of recursive arithmetic both propositional variables and variables whose values are numerical functions. This has the advantage that certain things which would otherwise have to be embodied in rules of inference or axiom schemata, and thus expressed in the meta-language, may be expressed instead by single axioms in the object language. And the addition of variables of these new kinds does not destroy the intuitionistic character of recursive arithmetic, as they will be used only as free variables, and it will always be possible to suppose the range of the variables restricted to values which are intuitionistically unobjectionable.

We shall also employ explicit notations for the processes of introducing functions by composition and by primitive recursion, so that we shall write, for instance, Cgh_1h_2 to denote the function f introduced by the composition

$$f(x) = g(h_1(x), h_2(x)),$$

and a similar notation to denote functions introduced by recursion. The need for such notations seems to have been first observed by Curry (*loc. cit.*), and in this respect we are following Curry's treatment of recursive arithmetic. It is only in this way that we are able to provide for all primitive recursive functions in a finite list of axioms, instead of requiring new axioms for each new function as it is introduced.

On the other hand we shall not take advantage of the possibility of replacing the simple rule of mathematical induction by any of various special cases of it[4] or of the possibility of eliminating connectives of the propositional calculus from among the primitive notations by using the fact that these connectives can be introduced by contextual definition in terms of the other notations (*see* Hilbert-Bernays, pp. 310-311, and Curry, and Goodstein). For though these matters are of interest in themselves, they seem not to be relevant to our present purpose, and we here prefer the more natural formulation in which propositional calculus and mathematical induction are taken as primitive.

A primitive basis of binary recursive arithmetic.

In this paper we make a formulation of *binary recursive arithmetic*, i.e., of that part of recursive arithmetic which has notations for none but singularly and binary numerical functions. This is not quite an immediate corollary of previous formulations, since the equations of primitive recursion, as given above, make use of a given ternary function (as well as a given singulary function) in order to introduce a binary function by recursion. In fact we shall modify the form of the equations of recursion, so that only singulary and binary functions are used in introducing a binary function g and two given singulary functions h_1 and h_2 are used to introduce a singulary function f by composition.

[4]*See* SKOLEM, ***Det Kongelige Norske Videnskabers Selskab, Forhandlinger***, vol. 22, 1950, pp. 167–170; BERNAYS, ***The Journal of Symbolic Logic***, vol. 16, 1951, pp. 220–221; R. L. GOODSTEIN, ***Mathematica Scandinavica***, vol. 2, 1954, pp. 247–261.

As a step towards showing that the system is adequate, not only for binary recursive arithmetic, but in an indirect sense, also for the whole of recursive arithmetic, we shall show how to obtain functions which provide in the system the essential features of the notion of ordered pair. More explicitly, we shall define — i.e., express in the notation of the system — a singular function τ such that $\tau(x)$ is the xth triangular number, and a binary function ϖ such that $\varpi(x, y) = \tau(x + y) + y$. Then $\varpi(x, y)$ is the number of the ordered pair (x, y) in a certain familiar enumeration of the ordered pairs. And with appropriate definitions of m_1 and m_2 we can then prove the theorems

$$m_1(\varpi(x, y)) = x, \qquad m_2(\varpi(x, y)) = y.$$

Our primitive symbols are parentheses, brackets, comma, the sign =, the numerical constant 0, the four constant functors s, o, u, v, the two functor connectives C, R, the two sentence connectives $\supset$, $\sim$, an infinite list of propositional variables p, q, r, s, $\mathrm{p}_1, \ldots$ (we here use roman letters for these in order to reserve the italic letters for other uses), an infinite list of numerical variables $x, y, z, x_1, y_1, z_1, x_2, y_2, \ldots$, an infinite list of singular numerical-function variables $f^1, g^1, h^1, f^1_1, g^1_1, h^1_1, f^1_2, g^1_2, \ldots$, and an infinite list of binary numerical-function variables $f^2, g^2, h^2, f^2_1, g^2_1, h^2_1, f^2_2, g^2_2, \ldots$.

The well-formed formulas of the system comprise *functors*, *terms*, and *propositional forms*, and are determined by the fourteen following formation rules:

(i) One of the constant functors s, o, u, standing alone is a singular functor.
(ii) The constant functor v standing alone is a binary functor.
(iii) A singular function variable standing alone is a singular functor.
(iv) A binary function variable standing alone is a binary functor.
(v) If **G** is a binary functor and $\mathbf{H}_1$ and $\mathbf{H}_2$ are singular functors, $C\mathbf{G}\mathbf{H}_1\mathbf{H}_2$ is a singular functor.
(vi) If **G** is a singular functor and $\mathbf{H}_1$ and $\mathbf{H}_2$ are binary functors, $R\mathbf{G}\mathbf{H}_1\mathbf{H}_2$ is a binary functor.
(vii) The numerical constant 0 standing alone is a term.
(viii) A numerical variable standing alone is a term.
(ix) If **F** is a singular functor and **A** is a term, **F**(**A**) is a term.
(x) If **F** is a binary functor and **A** and **B** are terms, **F**(**A**, **B**) is a term.
(xi) A propositional variable standing alone is a propositional form.
(xii) If **A** and **B** are terms, **A** = **B** is a propositional form (such a propositional form is called an *equation*).
(xiii) If **P** and **Q** are propositional forms, [**P** $\supset$ **Q**] is a propositional form.
(xiv) If **P** is a propositional form, $\sim$**P** is a propositional form.

The particular terms $0, s(0), s(s(0)), \ldots$ will be called *numerals*.

As already done in stating the formation rules, we shall use bold capital letters in the meta-language as variables for formulas. And hereafter it is to be understood, when the bold capital letters are used, that their values are restricted to *well-formed* formulas.

Similarly we shall use bold small letters, in the meta-language, as variables whose values are single (primitive) symbols of the object language.

As a notation for substitution we shall use an S and a vertical bar — so that, for example, $\mathrm{S}^{\mathbf{a}}_{\mathbf{A}}\mathbf{P}|$ means the result of substituting **A** for all occurrences of **a** throughout **P**, and $\mathrm{S}^{\mathbf{abc}}_{\mathbf{ABC}}\mathbf{P}|$ means the result of substituting simultaneously, **A** for all occurrences of **a**, **B** for all occurrences of **b**, and **C** for all occurrences of **c**, throughout **P**.

In writing well-formed formulas we shall abbreviate by omitting the superscripts which distinguish singulary function variables and binary function variables, whenever this is possible without ambiguity; and also whenever possible by omitting the brackets which according to formation rule (xiii), go with sign $\supset$. To facilitate the omission of brackets we adopt the convention of association to the left, so that, e.g., p $\supset$ **.** q $\supset$ r is to be understood as an abbreviation of [p $\supset$ [q $\supset$ r]].

The primitive rules of inference of the system are the six following:

I. From **P** $\supset$ **Q** and **P** to infer **Q**.

II. From **P** to infer $\mathrm{S}^{\mathbf{a}}_{\mathbf{A}}\mathbf{P}|$, if **a** is a propositional variable and **A** is a propositional form.

III. From **P** to infer $\mathrm{S}^{\mathbf{a}}_{\mathbf{A}}\mathbf{P}|$, if **a** is a numerical variable and **A** is a term.

IV. From **P** to infer $\mathrm{S}^{\mathbf{a}}_{\mathbf{A}}\mathbf{P}|$, if **a** is a singulary function variable and **A** is a singulary functor.

V. From **P** to infer $\mathrm{S}^{\mathbf{a}}_{\mathbf{A}}\mathbf{P}|$, if **a** is a binary function variable and **A** is a binary functor.

VI. From $\mathbf{P} \supset \mathrm{S}^{\mathbf{a}}_{s(\mathbf{a})}\mathbf{P}|$ and $\mathrm{S}^{\mathbf{a}}_{0}\mathbf{P}|$ to infer **P**, if **a** is a numerical variable and **P** is a propositional form.

Of these, rule I is the *rule of modus ponens*. Rules II, III, IV, V are *rules of substitution*. And rule VI is the *simple rule of mathematical induction*, so called to distinguish it from stronger rules of mathematical induction, which are considered by Hilbert and Bernays in connection with recursive arithmetic, but which we do not here take as primitive rules.

To complete the statement of the primitive basis of the system we add finally the twelve following axioms:

1*a*. p $\supset$ **.** q $\supset$ p.

1*b*. s $\supset$ [p $\supset$ q] $\supset$ **.** s $\supset$ p $\supset$ **.** s $\supset$ q.

1*c*. $\sim$p $\supset \sim$q $\supset$ **.** q $\supset$ p.

2. $x = y \supset . \, x = z \supset y = z$.

3*a*. $x = y \supset g(x, z) = g(y, z)$.

4. $\sim s(0) = 0$.

5. $o(x) = 0$.

6. $u(x) = x$.

7. $v(x, y) = y$.

8. $Cgh_1h_2(x) = g(h_1(x), h_2(x))$.

9*a*. $Rgh_1h_2(x, 0) = g(x)$.

9*b*. $Rgh_1h_2(x, s(y)) = h_1(h_2(x, y), Rgh_1h_2(x, y))$.

Standard metatheorems.

From the primitive basis of the system, as just given, there follow a number of metatheorems which are closely analogous to metatheorems familiar in connection with other systems. We state them here without proof, as their proofs do not differ in any important respect from the known proofs of the analogous metatheorems about other systems.

We use $\vdash$ **Q** as a metatheoretic notation to mean that there is a proof of **Q**, or that **Q** is a theorem. And we use $\mathbf{P}_1, \mathbf{P}_2, \ldots, \mathbf{P}_n \vdash \mathbf{Q}$ to mean that there is a proof of $\mathbf{Q}$ from the hypotheses $\mathbf{P}_1, \mathbf{P}_2, \ldots, \mathbf{P}_n$, which must be propositional forms.

[In a proof from hypotheses, the hypotheses $\mathbf{P}_1, \mathbf{P}_2, \ldots, \mathbf{P}_n$ are used as if they were axioms, and in addition the axioms themselves are so used, and any *variant* of an axiom — i.e., any propositional form obtained from an axiom by alphabetic change of the variables, as e.g. $v(y, z) = z$ is a variant of axiom 7. The rules of inference of the system are used with the restriction on the rules of substitution that the variable $\mathbf{a}$ which is substituted for must be a variable that does not occur in $\mathbf{P}_1, \mathbf{P}_2, \ldots, \mathbf{P}_n$, and with the same restriction also on the rule of mathematical induction, i.e., that the variable $\mathbf{a}$ with respect to which the induction is made must be a variable that does not occur in $\mathbf{P}_1, \mathbf{P}_2, \ldots, \mathbf{P}_n$.]

VII. *The rule of simultaneous substitution.* — if $\vdash \mathbf{P}$, then $\vdash \mathrm{S}^{\mathbf{a}_1\mathbf{a}_2\ldots\mathbf{a}_m}_{\mathbf{A}_1\mathbf{A}_2\ldots\mathbf{A}_m}\mathbf{P}|$, if $\mathbf{a}_1, \mathbf{a}_2, \ldots, \mathbf{a}_m$ are any distinct variables and $\mathbf{A}_1, \mathbf{A}_2, \ldots, \mathbf{A}_m$ are well-formed formulas of kinds that correspond to the variables (i.e., if $\mathbf{a}_i$ is a propositional variable, $\mathbf{A}_i$ is a propositional form, and if $\mathbf{a}_i$ is a numerical variable, $\mathbf{A}_i$ is a term, and if $\mathbf{a}_i$ is a singulary function variable, $\mathbf{A}_i$ is a singulary functor, and if $\mathbf{a}_i$ is a binary function variable, $\mathbf{A}_i$ is a binary functor).

VIII. *The deduction theorem.* — If $\mathbf{P}_1, \mathbf{P}_2, \ldots, \mathbf{P}_n \vdash \mathbf{Q}$, then $\mathbf{P}_1, \mathbf{P}_2, \ldots, \mathbf{P}_{n-1} \vdash \mathbf{P}_n \supset \mathbf{Q}$. As a special case, if $\mathbf{P} \vdash \mathbf{Q}$, then $\vdash \mathbf{P} \supset \mathbf{Q}$.

IX. If $\mathbf{A}_1, \mathbf{A}_2, \ldots, \mathbf{A}_m$ are propositional forms, if each of $\mathbf{P}_1, \mathbf{P}_2, \ldots, \mathbf{P}_n$ occurs at least once in the list $\mathbf{A}_1, \mathbf{A}_2, \ldots, \mathbf{A}_m$, and if $\mathbf{P}_1, \mathbf{P}_2, \ldots, \mathbf{P}_n \vdash \mathbf{Q}$, then $\mathbf{A}_1, \mathbf{A}_2, \ldots, \mathbf{A}_m \vdash \mathbf{Q}$.

X. Every tautology of the propositional calculus is a theorem.

Hence by VII, every substitution instance of a tautology of the propositional calculus is a theorem.

XI. If $\vdash \mathrm{S}^{\mathbf{a}}_{0}\mathbf{P}|$ and $\vdash \mathrm{S}^{\mathbf{a}}_{s(\mathbf{a})}\mathbf{P}|$, then $\vdash \mathbf{P}$ (by VI and X).

Properties of equality, other forms of composition.

The axioms allow directly only one form of composition, namely the composition of a binary function and two singulary functions by axiom 8. Some other forms of composition may be obtained quickly from the axioms, however, and with one exception (to be discussed below) all of the elementary properties of equality may also be quickly established. We go on to the proof of theorems of the system which embody these results.

10. $x = x$

Proof by substituting $u(x)$ for x, x for y, and x for z in 2, and then using 6.

11. $x = y \supset y = x.$

Proof by substituting x for z in 2, and then using 10.

12. $x = y \supset . \, y = z \supset x = z.$

Proof by 2 and 11.

13. $Rfvv(x, y) = f(x)$.

Proof as follows (notice that the arrangement used in the first two lines of this proof, in the two lines of the proof of 14, in the first three of the proof of 15, and elsewhere, involves tacit use of 12, as well as the rules of substitution):

$$\begin{aligned} \vdash Rfvv(x, s(y)) &= v(v(x, y), Rfvv(x, y)), && \text{by } 9b; \\ &= Rfvv(x, y), && \text{by } 7. \end{aligned}$$

Hence by 12,

$$\vdash Rfvv(x, y) = f(x) \supset Rfvv(x, s(y)) = f(x).$$

But by 9*a*,

$$\vdash Rfvv(x, 0) = f(x)$$

Hence use mathematical induction (rule VI).

14. $CRfvvgs(x) = f(g(x))$.

Proof:

$$\begin{aligned} \vdash CRfvvgs(x) &= Rfvv(g(x), s(x)), && \text{by } 8; \\ &= f(g(x)), && \text{by } 13 \end{aligned}$$

15. $x = y \supset f(x) = f(y)$.

Proof:

$$\begin{aligned} x = y \vdash f(x) &= Rfvv(x, z), && \text{by } 13; \\ &= Rfvv(y, z), && \text{by } 3a; \\ &= f(y), && \text{by } 13. \end{aligned}$$

Hence use the deduction theorem.

Definition. — $w \rightarrow Ruvv$. (The arrow which is used here and below in stating definitions is to be read as "stands for" or "is an abbreviation for." It indicates that the letter, character, or expression which stands at the left of the arrow will be used as an abbreviation of the well-formed formula which stands at the right of the arrow.)

16. $w(x, y) = x$.

Proof:

$$\begin{aligned} w(x, y) &= u(x), && \text{by } 13; \\ &= x, && \text{by } 6. \end{aligned}$$

17. $CRfwghs(x) = g(h(x), x)$.

Proof:

$$\begin{aligned} CRfwghs(x) &= Rfwg(h(x), s(x)), && \text{by } 8; \\ &= w(g(h(x), x), Rfwg(h(x), x)), && \text{by } 9b; \\ &= g(h(x), x), && \text{by } 16. \end{aligned}$$

18. $y = z \supset g(h(y), y) = g(h(y), z)$.

Proof :

$$\begin{aligned} y = z \vdash g(h(y), y) &= CRswghs(y), && \text{by 17;} \\ &= CRswghs(z), && \text{by 15;} \\ &= g(h(z), z), && \text{by 17.} \\ &= g(h(y), z). \end{aligned}$$

(The last step is by axiom 3a since $y = z \vdash h(z) = h(y)$ by 11 and 15.)

19. $x = f(y) \supset . \, y = z \supset g(x, y) = g(x, z).$

Proof by 18 and 3a.

The ordered pair.

We now proceed as directly as possible to the functions ϖ, m_1, m_2 which were described above, and to the proofs of their essential properties. We give the material in purely formal outline, stating definitions and theorems, and reducing the proof of each theorem usually to a mere indication by number of the principal theorems on which the proof depends.

Definition. — $s_1 \rightarrow Rsvv$.

20. $s_1(x, y) = s(x)$, by 13.

Definition. — $s^2 \rightarrow Cs_1ss$.

21. $s^2(x) = s(s(x))$, by 8, 20 (or by 14).

Definition. — $1 \rightarrow s(0)$.

Definition. — $i \rightarrow Cs_1os$.

22. $i(x) = 1$, by 8, 20, 5, 15.

Definition. — $s_2 \rightarrow RiRs^2vvv$.

23. $s_2(x, y) = s(y)$, by XI, 9, 22, 13, 7, 15, 21.

Definition. — $p_2 \rightarrow Rowv$.

Definition. — $p \rightarrow CRowp_2ss$.

24. $p(0) = 0$, by 17, 9a, 5.
25. $p(s(x)) = x$, by 17, 9b, 16, 7.
26. $p_2(x, y) = p(y)$, by XI, 9, 5, 24, 16, 7, 25.

Definition. — $\mathbf{A} \neq \mathbf{B} \rightarrow \sim \mathbf{A} = \mathbf{B}$.

27. $s(x) \neq 0$, by VI, 15, 24, 25, 4.
28. $s(x) = s(y) \supset x = y$, by 15, 25.

Definition. — $\alpha \rightarrow CRowRoRivvvss$.

Definition. — $\beta \rightarrow CRowRiRovvvss$.

29. $\alpha(0) = 0$, by 17, 5.
30. $\alpha(s(x)) = 1$, by 17, 13, 22.

31. $\beta(0) = 1$, by 17, 22.
32. $\beta(s(x)) = 0$ by 17, 13, 5.

Definition. — $\alpha \to Rus_2v$.

33. $\sigma(x, 0) = x$, by 6.
34. $\sigma(x, s(y)) = s(\sigma(x, y))$, by 23.
35. $\sigma(s(x), y) = s(\sigma(x, y))$, by VI, 15, 34, 33.
36. $\sigma(x, y) = \sigma(y, x)$, by VI, 15, 34, 35, 33.

Definition. — $(\mathbf{A} + \mathbf{B}) \to \sigma(\mathbf{A}, \mathbf{B})$, where the parentheses enclosing $\mathbf{A} + \mathbf{B}$ will be omitted whenever possible without ambiguity.

37. $x = y \supset x + z = y + z$, by 3*a*.
38. $x = y \supset z + x = z + y$, by 3*a*, 36.
39. $x + (y + z) = (x + y) + z$, by VI, 38, 33, 34, 35, 15.
40. $x + z = y + z \supset x = y$, by VI, 34, 28, 33.
41. $1 + x = s(x)$, by 36, 33, 34, 15.
42. $x \neq 0 \supset s(p(x) + y) = x + y$, by XI, 25, 37, 15, 35.

Definition. — $\delta \to Rpp_2v$.

43. $\delta(x, 0) = p(x)$.
44. $\delta(x, s(y)) = p(\delta(x, y))$, by 26.
45. $\delta(x, y) = p(\delta(s(x), y))$, by VI, 44, 15, 43, 25.
46. $\delta(x + y, y) = p(x)$, by VI, 34, 15, 44, 45, 33, 3*a*, 43.

Definition. — $\gamma \to R\beta R\beta vv\delta$.

47. $\gamma(x, 0) = \beta(x)$.
48. $\gamma(x, s(y)) = \beta(\delta(x, y))$, by 13.
49. $\gamma(x, y) = \beta(\delta(s(x), y))$, by XI, 47, 25, 15, 43, 48, 45, 44.
50. $\gamma(s(x), s(y)) = \gamma(x, y)$, by 49, 48.
51. $x \neq 0 \supset \beta(x) = 0$, by XI, 31, 32.

Definition. — $[\mathbf{P} \vee \mathbf{Q}] \to \mathbf{P} \supset \mathbf{Q} \supset \mathbf{Q}$, where the brackets enclosing $\mathbf{P} \vee \mathbf{Q}$ will be omitted whenever possible without ambiguity.

52. $\beta(x) = 0 \vee \beta(x) = 1$, by 51, 31 (or by XI, 31, 32).
53. $\gamma(x, y) = 0 \vee \gamma(x, y) = 1$, by 49, 52.
54. $\gamma(x, s(y)) = 0 \vee \gamma(x, y) = 0$, by 48, 31, 45, 24, 51, 49.

Definition. — $d \to Ro\sigma\gamma$.

55. $d(x, 0) = 0$.
56. $d(x, s(y)) = \sigma(\gamma(x, y), d(x, y))$.
57. $d(s(x), s(y)) = d(x, y)$, by VI, 56, 50, 37, 38, 47, 32, 33, 36, 55.
58. $\gamma(x, y) = 0 \supset d(x, s(y)) = 0$, by VI, 56, 33, 36, 54, 55.
59. $\delta(1, y) = 0$, by VI, 44, 15, 24, 43, 25.
60. $\beta(\delta(1, y)) = 1$, by 59, 15, 31.
61. $d(0, y) = y$, by VI, 56, 49, 60, 38, 41, 55.

62. $d(s(x), y) = p(d(x, y))$, by XI, 55, 24, 53, 50, 54, 58, 24, 57, 25, 41, 37, 56.
63. $d(x, y) = \delta(s(y), x)$, by VI, 62, 44, 61, 25, 43.
64. $\gamma(x, y) = \beta(\delta(y, x))$, by 49, 63.

Definition. — $\mathbf{A} \dot{-} \mathbf{B} \to d(\mathbf{B}, \mathbf{A})$, where parentheses enclosing $\mathbf{A} \dot{-} \mathbf{B}$ will be omitted whenever possible without ambiguity.

65. $x = y \supset z \dot{-} x = z \dot{-} y$, by 3*a*.
66. $x = y \supset x \dot{-} z = y \dot{-} z$, by 63, 3*a*.
67. $x + y \dot{-} y = x$, by 63, 35, 46, 25.
68. $x \dot{-} y \neq 0 \supset x \dot{-} y + y = z$, by VI, 62, 34, 42, 24, 61, 33.
69. $x \dot{-} y = 1 \supset x = s(y)$, by 4, 68, 37, 41.
70. $x \dot{-} y \dot{-} z = x \dot{-} (z + y)$, by VI, 35, 65, 62, 33, 36, 61.

Definition. — $\mathbf{A} \leq \mathbf{B} \to \mathbf{A} \dot{-} \mathbf{B} = 0$.

71. $x = y \supset . x \leq z \supset y \leq z$, by 66.
72. $y = z \supset . x \leq y \supset x \leq z$, by 65.
73. $x \leq x$ by VI, 57, 61.
74. $x \leq y \vee y \leq x$, by VI, 62, 24, 64, 51, 56, 37, 33, 36, 55.
75. $x \leq y \supset . y \leq x \supset x = y$, by 56, 64, 15, 31, 37, 38, 33, 69, 28.
76. $0 \leq x$, by 55.
77. $x \leq 0 \supset x = 0$. by 75, 76, (or by 61).
78. $x \leq y \supset s(x) \leq s(y)$, by 57.
79. $s(x) \leq s(y) \supset x \leq y$, by 57.
80. $s(x) \leq y \supset x \leq y$, by 56, 53, 33, 36, 41, 27.
81. $x \leq y \supset x \leq s(y)$, by 78, 80.
82. $x \leq s(y) \supset . x \leq y \vee x = s(y)$, by 56, 53, 57, 69.
83. $x \leq y \supset . x = y \vee s(x) \leq y$, by XI, 77, 82, 78.
84. $x \leq y \supset . y \leq z \supset x \leq z$, by VI, 82, 72, 81, 77, 76.
85. $x \leq x + y$, by VI, 34, 65, 62, 15, 24, 73, 33.
86. $x \leq y \supset x + z \leq y + z$, by VI, 78, 34, 71, 72, 33.
87. $x \leq y \supset s(y \dot{-} x) = s(y) \dot{-} x$, by 56, 49, 63, 15, 31, 41.
88. $x \leq y \supset x \dot{-} z \leq y \dot{-} z$, by VI, 82, 66, 73, 72, 74, 76, 71, 84, 87, 81, 77, 55.

(The proof of 88, in proceeding by mathematical induction on y, distinguishes three cases: $x = s(y)$; and $\sim z \leq x$; and $z \leq x, x \leq y$.)

Definition. — $\tau_2 \to Ro\sigma s_2$.

Definition. — $\tau \to CRow\tau_2 ss$.

89. $\tau_2(x, y) = \tau(y)$ by VI, 17.
90. $\tau(0) = 0$.
91. $\tau(s(x)) = s(x) + \tau(x)$.
92. $x \leq \tau(x)$, by XI, 76, 85, 91, 72.
93. $\tau(x) \leq \tau(s(x))$, by 91, 85, 36, 72.

94. $s(\tau(x)) \leq \tau(s(x))$, by 85, 78, 35, 72, 91.
95. $x \leq y \supset \tau(x) \leq \tau(y)$, by VI, 82, 93, 84, 15, 73, 71, 77.

Definition. — $\eta \rightarrow Ro R\alpha vv Rsds_2$.

96. $\eta(x, 0) = 0$.
97. $\eta(x, s(y)) = \alpha(s(x) \dot{-} \tau(y))$, by 9*b*, 13, then VI, 70, 91, 61, 90, 15.

Definition. — $\mu \rightarrow Ro\sigma\eta$.

Definition. — $2 \rightarrow s(1)$.

98. $\mu(x, 0) = 0$.
99. $\mu(x, s(y)) = \eta(x, y) + \mu(x, y)$.
100. $\mu(x, 2) = 1$.
101. $\tau(y) \leq x \supset \mu(x, s(s(y))) = s(\mu(x, s(y)))$, by 99, 97, 15, 87, 30, 37, 41.
102. $\tau(y) \leq x \supset \mu(s(x), s(s(y))) = \mu(x, s(s(y)))$, by VI, 81, 101, 93, 84, 15, 100.
103. $s(x) \leq \tau(y) \supset \mu(x, s(s(y))) = \mu(x, s(y))$, by 99, 97, 29, 15, 33, 36.
104. $\tau(y) = s(x) \supset \mu(s(x), s(s(y))) = s(\mu(x, s(s(y))))$, by XI, 90, 27, 73, 71, 103, 101, 94, 84, 79, 102, 15.
105. $y \leq z \supset . \tau(y) \leq x \supset . s(s(x)) \leq \tau(s(y)) \supset \mu(s(x), s(s(z))) = \mu(x, s(s(z)))$, by VI, 82, 78, 95, 84, 103, 80, 15, 71, 102, 77, 90, 91, 72, 79, 27.
106. $y \leq z \supset . \tau(y) = s(x) \supset \mu(s(x), s(s(z))) = s(\mu(x, s(s(z))))$, by VI, 82, 95, 71, 93, 84, 103, 15, 78, 94, 104.

Definition. — $v \rightarrow CRC\mu psvvss$.

107. $v(x) = \mu(x, s(s(x)))$, by 8, then VI, 25.
108. $\tau(y) \leq x \supset . s(s(x)) \leq \tau(s(y)) \supset v(s(x)) = v(x)$, by 105, 92, 84, 81, 103, 107.
109. $\tau(y) = s(x) \supset v(s(x)) = s(v(x))$, by 106, 92, 72, 103, 15, 107.
110. $s(\tau(y) + z) \leq \tau(s(y)) \supset v(\tau(y) + z) = v(\tau(y))$, by VI, 34, 15, 71, 80, 85, 108, 33.
111. $v(\tau(s(y))) = s(v(\tau(y)))$, by 91, 36, 34, 73, 71, 110, 109, 15.
112. $v(\tau(x)) = s(s)$, by VI, 111.

Definition. — $\varpi \rightarrow R\tau\sigma Rs^2s_2v$.

Definition. — $m_2 \rightarrow CRowdC\tau_2sCp_2svs$.

Definition. — $m_1 \rightarrow C\delta vm_2$.

113. $Rs^2s_2v(x, y) = s(s(x + y))$, by VI.
114. $\varpi(x, y) = \tau(x + y) + y$, by VI, 113.
115. $v(\varpi(x, y)) = s(x + y)$, by 114, 110, 112, 85, 34, 36, 39, 91.
116. $m_2(\varpi(x, y)) = y$, by 17, 115, 114, 46.
117. $m_1(\varpi(x, y)) = x$, by 115, 63, 116, 65, 67.

As was pointed out above, theorems 116 and 117 provide the essentials of the notion of ordered pair. It is possible to prove also $\varpi(m_1(z), m_2(z)) = z$, but as the proof is lengthy, and as this theorem is not essential to the use of ϖ as a pair function, we shall not give the details even in outline. On the other hand we add here the following definitions and theorems for the sake of their bearing on the discussion in the next section.

118. $y = z \supset \varpi(x, y) = \varpi(x, z)$, by 114, 15, 38.

Definition. — $\chi \rightarrow Cs_2sC\sigma s\tau$.

Definition. — $\varpi' \rightarrow R\chi\sigma RCs_1s^2ss_2v$.

119. $\chi(x) = s(\tau(s(x)))$.
120. $\varpi(x, 1) = \chi(x)$.
121. $RCs_1s^2ss_2v(x, y) = s(s(s(x + y)))$, by VI.
122. $\varpi'(x, y) = \varpi(x, s(y))$, by VI.

Definition. — $Cm\mathbf{FGH} \rightarrow RC\mathbf{F}C\mathbf{G}m_1m_2C\mathbf{H}m_1m_2vv$.

Definition. — $C_\tau\mathbf{FGH} \rightarrow CC_m\mathbf{FGH}\tau s$.

Definition. — $B\mathbf{FGH} \rightarrow RC_\tau\mathbf{FGH}C_m\mathbf{FGH}\varpi'$.

123. $Bfgh(x, y) = f(g(x, m_2(\varpi(x, y))), Chm_1m_2(\varpi(x, y)))$.

Some open questions.

It is conjectured that the following is not a theorem:

3*b*. $y = z \supset g(x, y) = g(x, z)$.

It would be natural to add this to the system as a further axiom (compare axiom 3*a*), and it is only doubt about the independence of such an added axiom that has deterred us from doing so.

If the conjecture is correct that 3*b* is independent, then the system (without 3*b* as an axiom) provides an interesting illustration of ω-incompleteness. For it is possible by 19 to prove any particular case of 3*b* obtained by substituting a numeral for x.

To show the adequacy of the system for that part of recursive arithmetic in which function variables are not used is would be sufficient to show that we can prove any particular case of 3*b* that is obtained by substituting for g a binary functor **G** containing no variables. This may not be possible. But it becomes easily possible if we add to the primitive bases of the system the following stronger rule of mathematical induction (superseding VI):

VI′. From $\mathbf{P} \supset \mathrm{S}^{\mathbf{a}}_{s(\mathbf{a})}{}^{\mathbf{b}}_{s(\mathbf{b})}\mathbf{P}|$ and $\mathrm{S}^{\mathbf{a}}_{0}\mathbf{P}|$ and $\mathrm{S}^{\mathbf{b}}_{0}\mathbf{P}|$ to infer **P**, if **a** and **b** are different numerical variables and **P** is a propositional form.

The main step in the proof of this metatheorem is namely to show that if we have the particular cases of 3*b* that are obtained by substituting $\mathbf{G}_1$ and $\mathbf{G}_2$, respectively, for g, we can prove the particular case of 3*b* that is obtained by substituting $Rf\mathbf{G}_1\mathbf{G}_2$ for g; and this is done by using 9b and the strengthened rule of induction VI′.

But even if we strengthen the system by adding VI′, it may still be that 3*b* is independent.

By 123, if **F**, **G**, and **H** are three binary functors of which each has the property that the result of substituting it for g in 3*b* is a theorem, we have the composition:

$$B\mathbf{FGH}(x, y) = \mathbf{F}(\mathbf{G}(x, y), \mathbf{H}(x, y)).$$

Reduction of recursions.

As a corollary of our axiomatic treatment of binary recursive arithmetic we have the following (weaker) result in the informal theory of recursive functions, which may be stated without reference to the restricted axiomatic basis:

In order to generate all singularly and binary primitive recursive functions from the four initial functions

$$s(x) = x + 1, \quad o(x) = 0, \quad u(x) = x, \quad v(x, y) = y,$$

it is sufficient to allow the following special form of composition,

$$f(x) = g(h_1(x), h_2(x)),$$

and the following form of recursion.

$$f(x, 0) = g(x),$$
$$f(x, s(y)) = h_1(h_2(x, y), f(x, y)).$$

ONTOLOGICAL COMMITMENT
(1958)

There are familiar philosophical problems which concern what may be called *ontology*, commonly phrased by asking whether something or some category of things is "real," or whether they "exist," or whether they have "real existence." An outstanding example is afforded by the traditional problem of universals, which issues in the nominalist-realist controversy as to the real existence of universals, or of abstract entities such as classes (in the mathematical sense) or propositions (in the abstract sense, i.e., the content of an assertion in abstraction from the particular words used to convey it).

My purpose in this paper is not to attack any of these philosophical problems directly, but to treat a necessary preliminary issue concerning the logic of the matter.

It is the important contribution of W. V. Quine to have pointed out the need for clarifying the logical issue. Quine's criterion of "ontological commitment" is indeed the only existing proposal in this direction. And a critical appraisal of it will be a large part of our task.

Now it may be held that the philosopher who takes the negative side of an ontological dispute thereby debars himself from referring, in certain ways to the entities whose reality he has denied. If he maintains, for example, that numbers do not exist but nevertheless asserts, or is persuaded to admit, '1729 is the sum of two cubes in more than one way,' he is, at least *prima facie*, involved in contradiction, because by naming the number 1729 and attributing a property to it he seems to be committed to the existence of at least this one number. For this reason, if the nominalist is not prepared either to reject mathematics out of hand or to adopt the extreme formalist position which denies all meaning to mathematical formulas, he may be led to the attempt to reformulate the proposition, so that no name of the number 1729 occurs—and if the sequence of digits '1729' is retained at all, it is in some other role than that of a name. Quine remarks in this connection that such ontological commitment to entities, as for example to numbers, may arise not only from using a name of a particular number but also from using variables which have numbers as values. Indeed, from '1729 is the sum of two cubes in more than one way' we can infer by ordinary rules of logic 'There is a number which is the sum of two cubes in more than one way,' or in symbols,

$$(\exists x)(\exists y)(\exists z)(\exists t)(\exists u) \,.\, \mathrm{c}(y)\mathrm{c}(z)\mathrm{c}(t)\mathrm{c}(u) \,.$$
$$t \neq y \,.\, t \neq z \,.\, x = y + z \,.\, x = t + u,$$

where 'c' is to be read as meaning "is a cube [of a positive integer]," and 'x', 'y', 'z', 't', 'u' are variables whose range is the positive integers. This latter clearly still carries ontological commitment to numbers, though not containing a name of any particular number. And if an ontological issue concerns the existence, not of some particular entity, but of entities of a certain category, then the criterion of ontological commitment which has reference to the use of a variable is more direct, and may take precedence over the criterion which has reference to the use of a name.

In the foregoing I have presented what seems to me the most persuasive aspect of

Originally published in ***The Journal of Philosophy***, vol. 55 (1958), pp. 1008–1014.

Quine's proposal, and the one I am most disposed to emphasize. However, in what may be regarded as ordinary standard systems of logic, introduced as principal systems in current works on the subject, the rules of quantifiers compel the variables to have a non-empty range, so that any assertion containing a variable (whether bound by an existential or a universal quantifier) is to be regarded as carrying ontological commitment to entities in the range of the variable. Moreover, it is reasonable to hold that a symbol is genuinely a proper name only if it admits the existential inference typified by the inference from '1729 is the sum of two cubes in more than one way' to '$(\exists x) \,.\, x$ the sum of two cubes in more than one way'—so that ontological commitment through the use of a proper name is reduced to that through the use of a variable. Hence Quine is led to the single very simple criterion that an assertion containing a variable carries ontological commitment to the range of the variable.

Even in this form Quine's proposal seems to me straight-forward and in a sense obvious. Its importance arises from the ease and the frequency with which the proponents of the negative in an ontological debate fall into violation of this simple criterion of logical coherence.

To illustrate the point, let me quote from A. J. Ayer's *Thinking and Meaning* (1947): "... except, perhaps, on certain occasions of reverie, to think of something always involves thinking that something or other is the case. It does not necessarily involve *believing* that anything is the case. It may be that the proposition is only "entertained," as some philosophers have put it, and that the thinker does not actually judge it to be either true or false. He may be imagining that something is so, or wondering whether it is, or raising it as a question. He may be expressing an intention rather than attempting to describe a fact. But, however this may be, what he is thinking may be expressed by a sentence; and what makes his thought a thought of something is simply the fact that this sentence has a meaning.... To be sure, it makes sense to say, in a case where someone is believing or doubting, or whatever it may be, that there is something that he doubts or believes. But it does not follow from this that "doubting" or "believing" are the names of peculiar mental acts, or that something must exist to be doubted or to be believed, in the way that something must exist to be eaten or to be struck."

This is the passage which shows most clearly that Ayer wishes simultaneously to say there is something that some one believes and to deny that there exists something to be believed. But in fact there are many passages throughout the monograph which belie Ayer's nominalistic intentions by requiring for their precise expression the use of bound variables having propositions as their range. To quote one more: "But now it may be asked What happens when I am actually thinking of a proposition which I know or believe or doubt? Allowing that these "cognitive" terms are normally used in a dispositional sense, it will be true that I can, for example, believe something even when I am not thinking of it. I must surely think of it on at least one occasion for the disposition ever to be formed. Now my merely thinking of it may just consist in my using and understanding certain symbols, but my believing it is something more. It is, if you like, a disposition. ... "—In an earlier book, *The Foundation of Empirical Knowledge* (1940), Ayer himself points out that he repeatedly uses the word 'proposition' in spite of his wish to deny propositions, and suggests methods of rephrasing to eliminate the word. But his methods of rephrasing fail of their purpose because they require bound variables that have propositions as their range.

It may well be that Ayer is to be understood as intending consistently to maintain the distinction between 'there is something that' (presumably corresponding to the existential quantifier) and 'something exists to be,' as it is made at the end of the passage I first quoted. But if so it has to be said that neither a logic nor a semantics of this latter notion of existence has even been indicated vaguely, much less received precise formulation. The notion remains mysterious. And Ayer's similes, "something ... to be eaten or to be struck," not only provide no clarification, but momentarily create confusion by suggesting—what must on obvious general grounds be false—that Ayer uses 'something exists' to mean existence as a physical object.

Arthur Pap's *Semantics and Necessary Truth* (1958), in a passage which seems to follow Ayer in some respects, explains "reification" or "Platonic mystification" by saying that one "visualizes the proposition as a *thing*, ready to become object of consciousness" or that "propositions are somehow assimilated to those *temporal* entities which common sense calls "things." But this notion of a thing capable of becoming, or of a temporal entity, is as much in need of clarification as Ayer's distinction between 'something exists' and 'there is something' (which it is, perhaps, intended to explicate). For example, common sense holds that a man who does not love at one time and does love at a later time has become or changed; but if somebody at one time does not love him and later does love him, he has not therefore necessarily changed. Pap similarly holds that a proposition is not to be considered a temporal entity because somebody at one time does not believe it and later does believe it. But there is not presently available a sound and adequate logic which maintains this ordinary-language distinction between the subject and the object of a verb.

As another illustration of inconsistency in regard to ontological commitment, I select for quotation a passage from a paper by Gilbert Ryle which was first published in 1939 and recently reprinted in *Logic and Language (Second Series)*, pp. 65–81: "Now sentences and sentence-factors are English or German, pencilled or whispered or shouted, slangy or pedantic, and so on. What logic is concerned with is something which is indifferent to these differences—namely (it is convenient though often misleading to say), propositions and the parts or factors of propositions. When two sentences of different languages, idioms, authors or dates say the same thing, what they say can be considered in abstraction from the several sayings of it, which does not require us to suppose that it stands to them as a town stands to the several signposts which point to it. And, just as we distinguish propositions from the sentences which propound them, so we must distinguish proposition-factors from the sentence-factors which express them. But again we must not suppose that this means that the world contains cows and earthquakes *and* proposition-factors, any more than we are entitled by the fact that we can distinguish the two faces of a coin to infer that when I have a coin in my hand I have three things in my hand, the coin and its two faces."

The main topic of the paper need not be discussed here except to say that it requires Ryle to speak of propositions, and of what he calls proposition-factors (I take them to be nearly the same as propositional functions in the abstract sense). And statements are not lacking in which the existential quantifier is directly applied—e.g., on page 78, "There are concepts which ... ," "some of which propositions ... ," etc. In the passage quoted and in other passages in the same paper Ryle seeks to avoid commitment to realism, in its original sense of maintaining the real existence of abstract entities. But the supposed philosophical insight of these passages remains unsupported, except by the loose and somewhat dubious analogies which were quoted. And more serious, it appears that this insight is to be kept entirely isolated, without effect on the use which may be made of propositions in logic or the inferences which may be drawn in reasoning that has reference to them. I think it may be argued that Ryle is in this paper a realist, and that in the passages in question his rejection or partial rejection of realism is simply in contradiction with the rest of the paper.

I will not add further quotations to illustrate confusion or inconsistency in regard to ontological commitment, though I believe that many more might be found. In the course of the discussion I have already brought out the main thesis I wish to defend, namely that no discussion of an ontological question, in particular of the issue between nominalism and realism, can be regarded as intelligible unless it obeys a definite criterion of ontological commitment. And though it is open to an author to advocate a different criterion than Quine's, or even to formulate a new system of logic to accommodate his ideas, he must always state his criterion or his logic with sufficient exactness to enable the reader to judge it on its own account.

One important consideration in judging a proposed criterion of ontological commitment is how closely it reproduces the pre-systematically available notion of existence. But it is not profitable to press this test of correspondence with ordinary language beyond a certain degree of approximation. For ordinary language itself is not accurate beyond a certain point, and the obscurity and confusion which may result from seeking in it what is not there seem to me to be illustrated in some of the quotations I have given. It is not a question of reproducing ordinary language but of reforming it. And if in consequence rival criteria of ontological commitment emerge, both agreeing with the pre-systematic notion of existence as far as such agreement can be clearly and reliably recognized, then the two criteria must be considered and compared on their internal merits, without pressing the test of agreement with pre-systematic common-sense notions beyond what it will bear.

Let us return, however, to the particular criterion proposed by Quine. In spite of the importance of Quine's contribution, which I have tried to bring out, I believe that amendments in regard to some points may be desirable.

Especially there are two considerations which combine to suggest that ontological commitment should be associated specifically with the existential quantifier rather than with bound variables generally. One such consideration is that the feature of what were called above "ordinary standard systems of logic," by which the rules of quantifiers compel the variables to have a non-empty range, has by some been regarded as objectionable, and its elimination may be desirable, at least for some purposes.[1] If we are willing to forgo the use of assertions containing free variables, this feature of standard systems of logic is not difficult to eliminate.[2] And in the modified logic which results, it is clear that ontological commitment will attach to an existential statement, but not to the negation of an existential statement or to a universal statement.

[1] Thus in some connection we might wish to use variables with a particular range about which we do not know whether it is empty. And even in the case of variables with such a range as that of physical objects, though the non-emptiness of the range may be regarded as certain, it would seem that the non-emptiness of the range should not be demonstrable on logical grounds alone.

[2] To describe the modification roughly, it consists in abandoning the schema $(\mathbf{x})\mathbf{A} \supset \mathbf{A}$ where $\mathbf{A}$ is an expression which does not contain free variables (and in particular does not contain $\mathbf{x}$ as a free variable). This was first pointed out by myself, in ***Essays in Honor of Henry M. Scheffer***, 1951, see page 18. Later the same idea was used independently by Theodore Hailperin in ***The Journal of Symbolic Logic***, vol. 18 (1953), pp. 197–200.

The second consideration concerns the fact that systems of logic in present use—including "ordinary standard" systems and others—have variables with only a limited variety of ranges. And though it might be possible to allow variables with a much wider variety of ranges, it is not immediately clear that this could be done without introducing excessive complication. In systems in present use, the application of Quine's criterion is indirect in the case of ontological commitment to a proper part of the range of a variable.[3] For example,

$$(\exists x) \,.\, x > 10^{1000},$$

where 'x' is a variable whose range is the positive integers, carries ontological commitment to positive integers, directly by Quine's criterion. But to see that it carries the stronger ontological commitment to positive integers greater than 10^{1000}, it is necessary to supply to a semantical meta-language a demonstration that the truth of the displayed formula requires that the range of the variable 'x' shall include positive integers greater than 10^{1000}.

I therefore propose as an alternative to Quine's criterion the following:[4]

The assertion of $(\exists \boldsymbol{x})\boldsymbol{M}$ carries ontological commitment to entities **x** such that **M**.

This is to be understood as a schema in which '**x**' may be replaced by any variable, '$\boldsymbol{x}$' may be replaced by any name of the same variable, '**M**' may be replaced by any propositional form (open sentence) containing no other free variable but that one, and '$\boldsymbol{M}$' may be replaced by any name of the same propositional form. Moreover, in '$(\exists \boldsymbol{x})\boldsymbol{M}$', for accuracy in regard to distinction of use and mention, '$\exists$' and the parentheses must be understood autonomously, and juxtaposition must be understood as a notation for juxtaposition.

This criterion agrees with Quine's in the case of "ordinary standard systems of logic," but it is applicable to a somewhat wider class of systems, and I believe it to be more direct.

Another, somewhat looser, way of saying the same thing is that those philosophers who speak of "existence," "reality," and the like are to be understood as meaning the existential quantifier, and are to be condemned as inconsistent if on this basis inconsistency appears in their writings. The justification is that no other reasonable meaning of "existence" has ever been provided (which would fit into contexts of the kind that were quoted), and the burden of providing such a second meaning of "existence" rests on those whose writings or whose philosophical views require it.

ALONZO CHURCH

PRINCETON UNIVERSITY

[3] I remark in passing that ontological commitment is an intensional notion, in the sense that ontological commitment must be to a class concept rather than a class. For example, ontological commitment to unicorns is evidently not the same as ontological commitment to purple cows, even if by chance the two classes are both empty and therefore identical.

[4] I have preferred throughout to speak of the ontological commitment of particular assertions rather than of a system of logic or of a language. The connection is of course that a language carries the ontological commitment of every sentence which is analytic in the language, i.e., of every sentence whose truth is a logical consequence of the semantics of the language. In some cases it may be sufficient to say that a language carries the ontological commitments of all its theorems. But to answer particular questions concerning the ontological commitment of a language may in general require reference to analyticity (regardless of whether Quine's criterion of ontological commitment is adopted or my suggested alternative).

MISOGYNY AND ONTOLOGICAL COMMITMENT (1958)

I think I may assume without hesitation that the *second* of the two topics which are named in the title of this lecture is already familiar to members of this audience. Yet a brief review of Quine's criterion of ontological commitment is desirable in order to prepare the ground for two main theses which I should like to offer in regard to it.

There are philosophical problems which concern what may be called ontology, the question whether something or some category of things is real, whether they have real existence. An outstanding example is afforded by the traditional problem of the universals, which issues in the nominalist-realist controversy as to the real existence of universals, or abstract entities such as classes (in the mathematical sense) or propositions (in the abstract sense, i.e., the content of an assertion in abstraction from the particular words used to convey it).

Now it may be held that the philosopher who takes the negative side of an ontological dispute thereby debars himself from referring in certain ways to the entities whose reality he has denied. If he maintains for example that numbers do not exist but nevertheless asserts, or is persuaded to admit, '1729 is the sum of two cubes in more than one way,' he is, at least *prima facie*, involved in contradiction, because by naming the number 1729 and attributing a property to it he seems to be committed to the existence of at least this one number. For this reason the nominalist who is not prepared to reject mathematics out of hand, or to reduce it to the status of meaningless marks on paper, may be led to the attempt to reformulate the proposition, that 1729 is the sum of two cubes in more than one way, so that no name of the number 1729 occurs—and if the sequence of digits '1729' is retained at all it is in some other role than that of a name.

The important contribution of Quine in this connection is to point out that such ontological commitment to entities, as for example to numbers, arises not only from using a name of a particular number but also from using variables which have numbers as values. Indeed from '1729 is the sum of two cubes in more than one way' we can infer by ordinary rules of logic, 'There is a number which is the sum of two cubes in more than one way,' or in symbols,

$$(\exists x)(\exists y)(\exists z)(\exists t)(\exists u) \,.\, x = y^3 + z^3 \,.\, x = t^3 + u^3 \,.\, t \neq y \,.\, t \neq z,$$

where x, y, z, t, u are variables whose range is the positive integers; and this latter clearly still carries ontological commitment to numbers, though not containing a name of any particular number. And if an ontological issue concerns the existence, not of some particular entity, but of entities of a certain category, then the criterion of ontological commitment which has reference to the use of a variable is more direct, and may take precedence over the criterion which has reference to the use of a name.

For the present purpose, and to simplify the discussion, it is desirable to leave aside certain logical systems of unusual kind, which are indeed possible but are not often used as standard. In particular we thus exclude combinatory logic (in which the use of variables is entirely avoided). And we further exclude systems of logic

---

A lecture given at Harvard University, April 18, 1958. Previously unpublished. Substantial parts of this talk overlap *Ontological Commitment*, but the main points of the two pieces are different.

in which the rules regarding quantifiers are such as to allow the possibility that the range of the variables—or of some of the variables if there are variables of more than one kind—may be null. In what we may regard as ordinary standard systems of logic, introduced as principal systems in current works on the subject, the rules of quantifiers compel the variables to have a non-empty range, and hence any assertion containing a variable (whether bound by an existential or a universal quantifier) must be regarded as carrying ontological commitment to entities in the range of the variable. Moreover it is reasonable to hold that a symbol is genuinely a proper name only if it admits the existential inference typified by the inference from '1729 is the sum of two cubes in more than one way' to '$(\exists x)$. x is the sum of two cubes in more than one way' —so that ontological commitment through the use of a proper name is reduced to that through the use of a variable. Altogether then we have the single very simple criterion that an assertion containing a variable carries ontological commitment to the range of the variable.

The foregoing seems to me straightforward and in a sense obvious. Its importance arises from the ease and the frequency with which the proponents of the negative in an ontological debate fall into violation of this simple criterion of logical coherence.

To illustrate the point let me quote from A. J. Ayer's ***Thinking and Meaning*** (1947): " ... it makes sense to say, in a case where someone is believing or doubting, ... that there is something that he doubts or believes. But it does not follow from this that 'doubting' or 'believing' are the names of peculiar mental acts, or that something must exist to be doubted or to be believed, in the way that something must exist to be eaten or to be struck." —This is the passage which shows most clearly that Ayer wishes simultaneously to say there is something that some one believes and to deny that there exists something to be believed. But in fact there are many passages throughout the monograph which belie Ayer's nominalistic intentions by requiring for their precise expression the use of bound variables having propositions as their range. To quote one more: " ... to think of something always involves thinking that something or other is the case. It does not necessarily involve *believing* that anything is the case. It may be that the proposition is only 'entertained,' and that the thinker does not actually judge it to be either true or false. ... But, however this may be, what he is thinking may be expressed by a sentence." —In an earlier book, ***The Foundations of Empirical Knowledge*** (1940), Ayer himself points out that he repeatedly uses the word 'proposition' in spite of his wish to deny propositions, and suggests methods of rephrasing to eliminate the word. But his methods of rephrasing fail of their purpose because they require bound variables that have propositions in their range.

It may well be that Ayer is to be understood as intending consistently to maintain the distinction between 'there is something that' (presumably corresponding to the existential quantifier) and 'something exists to be,' as it is made in the passage I first quoted. But if so it has to be said that neither a logic nor a semantics of this latter notion of existence has ever been indicated vaguely, much less received precise formulation. And Ayer's similes, "something ... to be eaten or to be struck," not only provide no clarification, but momentarily create confusion by suggesting—what must on obvious general grounds be false—that Ayer uses 'something exists' to mean existence as a physical object.

As another illustration of inconsistency in regard to ontological commitment, I select for quotation a passage from a paper by Gilbert Ryle which was first published in 1939 and recently reprinted in ***Logic and Language*** (***Second Series***), pp. 65–81: "Now sentences and sentence-factors are English or German, pencilled or whispered or shouted, slangy or pedantic, and so on. What logic is concerned with is something which is indifferent to these differences—namely (it is convenient though misleading to say), propositions and the parts or factors of propositions. When two sentences of different languages, idioms, authors or dates say the same thing, what they say can be considered in abstraction from the several sayings of it, which does not require us to suppose that it stands to them as a town stands to the several signposts which point to it. And, just as we distinguish propositions from the sentences which propound them, so we must distinguish proposition-factors from the sentence-factors which express them. But again we must not suppose that this means that the world contains cows and earthquakes *and* proposition-factors, any more than we are entitled by the fact that we can distinguish the two faces of a coin to infer that when I have a coin in my hand I have three things in my hand, the coin and its two faces."

The main topic of the paper need not be discussed here except to say that it requires Ryle to speak of propositions, and of what he calls proposition-factors (I take them to be nearly the same as propositional functions in the abstract sense). And statements are not lacking in which the existential quantifier is directly applied—e.g. on page 78, "There are concepts with which ... ," " ... some of which propositions ... ," etc. In the passage quoted and in other passages in the same paper Ryle seeks to avoid commitment to realism, in its original sense of maintaining the real existence of abstract entities. But the supposed philosophical insight of these passages remains unsupported, except by the loose and somewhat dubious analogies which were quoted. And more serious, it appears that this insight is to be kept entirely isolated, without effect on the use which may be made of propositions in logic or the inferences which may be drawn in reasoning that has reference to them. I think it may be argued that Ryle is in this paper a realist, and that in the passages in question his denial of meaning to realism is simply in contradiction with the rest of the paper.

I will not add further quotations to illustrate confusion or inconsistency in regard to ontological commitment, though I believe that many more might be found. In the course of the discussion I have already brought out the first of the two main theses which I wish to defend, namely that no discussion of an ontological question, in particular of the issue between nominalism and realism, can be regarded as intelligible unless it obeys a definite criterion of ontological commitment; and though it may be open to an author to advocate a different criterion than Quine's, or even to formulate a new system of logic (perhaps e.g. with two sorts of existential quantifiers) to accommodate his own special ideas, he must always state his criterion or his logic with sufficient clearness to enable the reader to check whether he adheres to it.

The second thesis is opposite in tendency to the first one. The purely logical criteria for the coherence and intelligibility of an ontological doctrine must not be elevated into an instrument which pretends to select one particular ontology as the true or even the best one. In particular the logical possibility of avoiding ontological commitment to certain entities does not (even in the light of considerations of economy) necessarily establish the ontology which rejects these entities.

An important application of this second thesis concerns the program of Quine and Nelson Goodman for a consistently nominalistic treatment of logic and logical syntax.

The Quine–Goodman proposal involves dealing, not with either propositions or sentences, but with *inscriptions*—where an inscription is a particular occurrence or utterance of a sentence, so that an inscription is, unlike a sentence, a concrete physical object. And for statements which would usually be made with reference to propositions or to sentences, some substitute statement in terms of inscriptions must be found. This is by no means easy; difficulties arise in particular because some sentences are or may be never uttered, some propositions never expressed, while others have various different utterances or different expressions. And the difficulties are increased because Quine and Goodman are unwilling to suppose that the total number of concrete things in the world is necessarily infinite, and they thus superimpose a sort of finitism upon their nominalism. Even the most elementary formalized language, for example, can be expected to have an infinite number of sentences; but the number of inscriptions must not be supposed to be more than finite.

I believe that Quine and Goodman would admit that their program has not yet attained full success in all respects, and that in this connection they would differ with me mainly regarding an estimate of the likelihood that means will be found in the future to clear up all important deficiencies. But rather than go into detail about their methods and results, and the degree of success attained, I conclude by discussing another ontological issue which I believe has an instructive analogy with that of nominalism.

Goodman says somewhere that he finds abstract entities difficult to understand. And from a psychological viewpoint it is certainly his dislike and distrust of abstract entities which leads him to propose an ontology from which they are omitted. Now a misogynist is a man who finds women difficult to understand, and who in fact considers them objectionable incongruities in an otherwise matter-of-fact and hard-headed world. Suppose then that in analogy with nominalism the misogynist is led by his dislike and distrust of women to omit them from his ontology. Women are not real, he tells himself, and derives great comfort from the thought—there *are* no such things. This doctrine let us call *ontological misogyny*.

There are various forms which such a doctrine may take. The misogynist may follow the example of Ryle and say that the world of women has no independent existence, it does not exist in addition to man's world but is an aspect of it; and though it may be convenient to speak of women independently, it is also misleading, and actually one should not ask such questions as whether women exist. But if this doctrine stands in isolation and does not affect the circumstances under which he agrees to my assertion that there is a woman in the room, or admits that some women have made important scientific discoveries, then it is clear that the denial of ontological status to women is only a matter of psychological comfort to the misogynist and has no further significance.

Instead of this the misogynist may take the more profound course which follows Goodman and Quine, attempting to construct a comprehensive theory that is adequate in general for purposes of understanding and communication, but at the same time avoiding ontological commitment to women. It is an interesting logical question how far such a theory is possible (without inconsistency with experimental and observational results). I think it may have at least as much success as has attended the corresponding search for a nominalistic theory, and probably more.

Just as propositions are replaced by inscriptions in order to avoid ontological commitment to the former, so a woman might be replaced by her husband. Instead of saying that a woman is present, we might speak of men as having two kinds of presence, primary presence and secondary presence, the observational criteria for secondary presence of a man being the same which the more usual theory would take as observational criteria for presence of a woman. And similarly in the case of other things that one might think to say about women. Certain difficulties arise over the fact that some women have more than one husband and others none, but these are no greater than the corresponding difficulties in the case of propositions and inscriptions.

Actually the task might be lightened by taking advantage of the fortunate circumstance that every woman has only one father. And for this reason ontological misogyny is a doctrine much easier to put into satisfactory logical order than is the Quine–Goodman finitistic nominalism.

But the question of the logical possibility of such a theory must be separated from the question of the desirability of replacing the ordinary theory by this ontologically more economical variant of it. Quine and Goodman emphasize the economy of nominalism in supposing the existence of fewer entities. But the economy which has commonly been the concern of the logician, and of the mathematician dealing with foundations, has been simply economy of assumption, which might be thought to include (among other things) economy of ontological assumption, but certainly not as its primary or most important element. Surely there are other criteria by which to judge a theory. And though we may be obliged to grant that the ontological misogynist has made a successful application of Ockham's razor, in that he has reduced his ontology without losing the adequacy of his theory, we may still prefer the more usual theory which grants existence to women.

To return to Quine and Goodman, it is possible, even likely, that the failure of their program will demonstrate the untenability of their finitistic nominalism. But the success of their program, like that of the ontological misogynist, would leave us to choose between the rival ontologies on other grounds. It is only in the former case that Quine and Goodman could be said in any sense to have settled the nominalist-realist controversy. But it is in any case a major contribution to have clarified the meaning of the dispute, by putting the opposing doctrines on a sounder basis and showing their relevance to logic.

ARTICLES FROM *DICTIONARY OF PHILOSOPHY*

Edited by

DAGOBERT D. RUNES, PH.D.

(1959)

Absorption: The name *law of absorption* is given to either of the two dually related theorems of the propositional calculus,

$$[p \vee pq] \equiv p, \qquad p[p \vee q] \equiv p,$$

or either of the two corresponding dually related theorems of the algebra of classes,

$$a \cup (a \cap b) = a, \qquad a \cap (a \cup b) = a.$$

Any valid inference of the propositional calculus which amounts to replacing A ∨ AB by A, or A[A ∨ B] by A, or any valid inference of the algebra of classes which amounts to replacing A ∪ (A ∩ B) by A, or A ∩ (A ∪ B) by A, is called *absorption*.

Whitehead and Russell (*Principia Mathematica*) give the name *law of absorption* to the theorem of the propositional calculus,

$$[p \supset q] \equiv [p \equiv pq].$$

—*A.C.*

Abstraction: (Lat. ab, from + trahere, to draw) In logic: Given a relation R which is transitive, symmetric, and reflexive, we may introduce or postulate new elements corresponding to the members of the field of R, in such a way that the same new element corresponds to two members x and y of the field of R if and only if xRy (see the article *Relation*). The new elements are then said to be obtained by *abstraction* with respect to R. Peano calls this a method or kind of definition, and speaks, e.g., of *cardinal numbers* (q.v.) as obtained from classes by abstraction with respect to the relation of equivalence—two classes having the same cardinal number if and only if they are equivalent.

Given a formula A containing a free variable, say x, the process of forming a corresponding monadic *function* (q.v.)—defined by the rule that the value of the function for an argument b is that which A denotes if the variable x is taken as denoting b—is also called *abstraction*, or *functional abstraction*. In this sense, abstraction is an operation upon a *formula* A yielding a function, and is relative to a particular *system of interpretation* for the notations appearing in the formula, and to a particular *variable*, as x. The requirement that A shall contain x as a free variable is not essential: when A does not contain x as a free variable, the function obtained by abstraction relative to x may be taken to be the function whose value, the same for all arguments, is denoted by A.

In articles herein by the present writer, the notation λx[A] will be employed for the function obtained from A by abstraction relative to (or, as we may also say, with respect to) x. Russell, and Whitehead and Russell in *Principia Mathematica*, employ

---

These are the articles signed by Church (or in the case of **Mathematics** unsigned but apparently by Church) that appeared in the ***Dictionary of Philosophy***, edited by Dagobert D. Runes, Littlefield, Adams & Co., Ames, Iowa, 1959. Every effort has been made to contact those who hold the rights. Any rights holders not credited should contact the publisher so that a correction can be made in the next printing.

for this purpose the formula A with a circumflex ˆ placed over each (free) occurrence of x—but only for *propositional* functions. Frege (1893) uses a Greek vowel, say ϵ, as the variable relative to which abstraction is made, and employs the notation $\acute{\epsilon}(A)$ to denote what is essentially the *function in extension* (the "Werthverlauf" in his terminology) obtained from A by abstraction relative to ϵ.

There is also an analogous process of *functional abstraction* relative to two or more variables (taken in a given order), which yields a polyadic function when applied to a formula A.

Closely related to the process of functional abstraction is the process of forming a class by *abstraction* from a suitable formula A relative to a particular variable, say x. The formula A must be such that (under the given system of interpretation for the notations appearing in A) $\lambda x[A]$ denotes a propositional function. Then $x\ni(A)$ (Peano), or $\hat{x}A$ (Russell), denotes the class determined by this propositional function. Frege's $\acute{\epsilon}(A)$ also belongs here, when the function corresponding to A (relative to the variable ϵ) is a propositional function.

Similarly, a relation in extension may be formed by *abstraction* from a suitable formula A relative to two particular variables taken in a given order. —*A. C.*

Scholz and Schweitzer, *Die sogenannten Definitionen durch Abstraktion*, Leipzig, 1935—W. V. Quine, *A System of Logistic*, Cambridge, Mass., 1934. A. Church, review of the preceding, Bulletin of the American Mathematical Society, vol. 41 (1935), pp. 598–603. W. V. Quine, *Mathematical Logic*, New York, 1940.

Affirmation of the consequent: The *fallacy of affirmation of the consequent* is the fallacious inference from B and $A \supset B$ to A. The *law of affirmation of the consequent* is the theorem of the propositional calculus, $q \supset [p \supset q]$. —*A. C.*

Affirmative proposition: In traditional logic, propositions *A*, *I* were called *affirmative*, and *E*, *O*, *negative* (see *Logic, formal*, §4). It is doubtful whether this distinction can be satisfactorily extended to propositions (or even to sentences) generally. —*A. C.*

Aggregate: 1. In a general sense, a collection, a totality, a whole, a class, a group, a sum, an agglomerate, a cluster, a mass, an amount or a quantity of something, with certain definite characteristics in each case.

2. *In Logic and Mathematics*, a collection, a manifold, a multiplicity, a set, an ensemble, an assemblage, a totality of elements (usually numbers or points) satisfying a given condition or subjected to definite operational laws. According to Cantor, an aggregate is any collection of separate objects of thought gathered into a whole; or again, any multiplicity which can be thought as one; or better, any totality of definite elements bound up into a whole by means of a law. Aggregates have several properties: for example, they have the "same power" when their respective elements can be brought into one-to-one correspondence; and they are "enumerable" when they have the same power as the aggregate of natural numbers. Aggregates may be finite or infinite; and the laws applying to each type are different and often incompatible, thus raising difficult philosophical problems. See *One-One; Cardinal Number; Enumerable*. Hence the practice to isolate the mathematical notion of the aggregate from its metaphysical implications and to consider such collections as symbols of a certain kind which are to

facilitate mathematical calculations in much the same way as numbers do. In spite of the controversial nature of infinite sets great progress has been made in mathematics by the introduction of the Theory of Aggregates in arithmetic, geometry and the theory of functions. (German, *Mannigfaltigkeit, Menge*; French, *Ensemble*).

3. *In logic*, an "aggregate meaning" is a form of common or universal opinion or thought held by more than one person.

(*In mathematics*): The concept of an *aggregate* is now usually identified with that of a *class* (q.v.)—although as a historical matter this does not, perhaps, exactly represent Cantor's notion. —*A. C.*

Algebra of logic is the name given to the Nineteenth Century form of the calculi of classes and propositions. It is distinguished from the contemporary forms of these calculi primarily by the absence of formalization as a *logistic system* (q.v.). The propositional calculus was also at first either absent or not clearly distinguished from the class calculus; the distinction between the two was made by Peirce and afterwards more sharply by Schröder (1891) but the identity of notation was retained.

Important names in the history of the subject are those of *Boole* (q.v.), *De Morgan* (q.v.), W. S. Jevons, *Peirce* (q.v.), Robert Grassmann, John Venn, Hugh MacColl, *Schröder* (q.v.), P. S. Poretsky. —*A. C.*

Algorithm (or, less commonly, but etymologically more correctly, *algorism*): In its original usage, this word referred to the Arabic system of notation for numbers and to the elementary operations of arithmetic as performed in this notation. In mathematics, the word is used for a method or process of calculation with symbols (often, but not necessarily, numerical symbols) according to fixed rules which yields effectively the solution of any given problem of some class of problems. —*A. C.*

All: *All* and *every* are usual verbal equivalents of the universal quantifier. See *Quantifier*. —*A. C.*

Amphiboly: Any fallacy arising from ambiguity of grammatical construction (as distinguished from ambiguity of single words), a premiss being accepted, or proved, on the basis of one interpretation of the grammatical construction, and then used in a way which is correct only on the basis of another interpretation of the grammatical construction. —*A. C.*

Analysis (mathematical): The theory of real numbers, of complex numbers, and of functions of real and complex numbers. See *Number; Continuity; Limit*. —*A. C.*

Antecedent: In a sentence of the form A $\supset$ B ("if A then B"), the constituent sentences A and B are called *antecedent* and *consequent* respectively. Or the same terminology may be applied to propositions expressed by these sentences. —*A. C.*

Antilogism: If in the syllogism in Barbara the conclusion is replaced by its contradictory there is obtained the following set of three (formulas representing) propositions,

$$M(x) \supset_x P(x), \quad S(x) \supset_x M(x), \quad S(x) \wedge_x \sim P(x),$$

from any two of which the negation of the third may be inferred. Such an inconsistent triad of propositions is called an *antilogism.*

From the principle of the antilogism, together with obversion, simple conversion of E and I, and the fact that in the pairs, A and O, E and I, each proposition of the pair is equivalent to the negation of the other, all of the traditional valid moods of the syllogism may be derived except those which require a third (existential) premiss (see *Logic, formal*, §§4, 5). With the further aid of subalternation the remaining valid moods may be derived.

This extension of the traditional reductions of the syllogistic moods is due to Christine Ladd Franklin. She, however, stated the matter within the algebra of classes (see *Logic, formal*, §7), taking the three terms of the syllogism as classes. From this point of view the three propositions of an antilogism appear as follows:

$$m \cap -p = \Lambda, \quad s \cap -m = \Lambda, \quad s \cap -p \neq \Lambda.$$

—*A. C.*

Argumentum ad rem: An argument to the point—distinguished from such evasions as *argumentum ad hominem* (q.v.), etc. —*A. C.*

Arithmetic, foundations of: Arithmetic (i.e., the mathematical theory of the non-negative integers, 0, 1, 2, ...) may be based on the five following postulates, which are due to Peano (and Dedekind, from whom Peano's ideas were partly derived):

$$N(0).$$
$$N(x) \supset_x N(S(x)).$$
$$N(x) \supset_x [N(y) \supset_y [[S(x) = S(y)] \supset [x = y]]].$$
$$N(x) \supset_x \sim[S(x) = 0].$$
$$F(0)[N(x)F(x) \supset_x F(S(x))] \supset_F [N(x) \supset_x F(x)].$$

The undefined terms are here O, N, S, which may be interpreted as denoting, respectively, the non-negative integer 0, the propositional function to be a non-negative integer, and the function $+1$ (so that $S(x)$ is $x+1$). The underlying logic may be taken to be the functional calculus of second order (*Logic, formal*, §6), with the addition of notations for descriptions and for functions from individuals to individuals, and the individual constant 0, together with appropriate modifications and additions to the primitive formulas and primitive rules of inference (the axiom of infinity is not needed because the Peano postulates take its place). By adding the five postulates of Peano as primitive formulas to this underlying logic, a logistic system is obtained which is adequate to extant elementary number theory (arithmetic) and to all methods of proof which have found actual employment in elementary number theory (and are normally considered to belong to elementary number theory). But of course, the system, if consistent, is incomplete in the sense of Gödel's theorem (*Logic, formal*, §6).

If the Peano postulates are formulated on the basis of an interpretation according to which the domain of individuals coincides with that of the non-negative integers, the undefined term N may be dropped and the postulates reduced to the three following:

$$(x)(y)[[S(x) = S(y)] \supset [x = y]].$$
$$(x)\sim[S(x) = 0].$$
$$F(0)[F(x) \supset_x F(S(x))] \supset_F (x)F(x).$$

It is possible further to drop the undefined term 0 and to replace the successor function S by a dyadic propositional function S (the contemplated interpretation being that $S(x, y)$ is the proposition $y = x + 1$). The Peano postulates may then be given the following form:

$$(x)(Ey)S(x, y).$$
$$(x)[S(x, y) \supset_y [S(x, z) \supset_z [y = z]]].$$
$$(x)[S(y, x) \supset_y [S(z, x) \supset_z [y = z]]].$$
$$(Ez)[[(x) \sim S(x, y)] \equiv_y [y = z]].$$
$$[(x) \sim S(x, z)] \supset_z [F(z)[F(x) \supset_x [S(x, y) \supset_y F(y)]] \supset_F (x)F(x)].$$

For this form of the Peano postulates the underlying logic may be taken to be simply the functional calculus of second order without additions. In this formulation, numerical functions can be introduced only by contextual definition as incomplete symbols.

In the Frege–Russell derivation of arithmetic from logic (see the article *Mathematics*) necessity for the postulates of Peano is avoided. If based on the theory of types, however, this derivation requires some form of the axiom of infinity—which may be regarded as a residuum of the Peano postulates.

See further the articles *Recursion, definition by*, and *Recursion, proof by*. —*A. C.*

B. Russell, *Introduction to Mathematical Philosophy*, London, 1919.

Assertion: Frege introduced the assertion sign, in 1879, as a means of indicating the difference between asserting a proposition as true and merely naming a proposition (e.g., in order to make an assertion about it, that it has such and such consequences, or the like). Thus, with an appropriate expression A, the notation $\vdash$A would be used to make the *assertion*, "The unlike magnetic poles attract one another," while the notation $-$A would correspond rather to the *noun clause*, "that the unlike magnetic poles attract one another." Later Frege adopted the usage that propositional expressions (as noun clauses) are proper names of truth values and modified his use of the assertion sign accordingly, employing say A (or $-$A) to denote the *truth value thereof that the unlike magnetic poles attract one another* and $\vdash$A to express the assertion that this truth value is truth.

The assertion sign was adopted by Russell, and by Whitehead and Russell in *Principia Mathematica*, in approximately Frege's sense of 1879, and it is from this source that it has come into general use. Some recent writers omit the assertion sign, either as understood, or on the ground that the Frege–Russell distinction between asserted and unasserted propositions is illusory. Others use the assertion sign in a syntactical sense, to express that a formula is a theorem of a *logistic system* (q.v.); this usage differs from that of Frege and Russell in that the latter requires the assertion sign to be followed by a formula denoting a proposition, or a truth value, while the former requires it to be followed by the syntactical name of such a formula.

In the propositional calculus, the name *law of assertion* is given to the theorem:

$$p \supset [[p \supset q] \supset q].$$

(The associated form of inference from A and A $\supset$ B to B is, however, known rather as *modus ponens*.) —*A. C.*

Associative law: Any law of the form,

$$x \circ (y \circ z) = (x \circ y) \circ z,$$

where $\circ$ is a dyadic operation (function) and $x \circ y$ is the result of applying the operation to x and y (the value of the function for the arguments x and y). Instead of the sign of equality, there may also appear the sign of the biconditional (in the propositional calculus), or of other relations having properties similar to equality in the discipline in question.

In arithmetic there are two associative laws, of addition and of multiplication:

$$x + (y + z) = (x + y) + z.$$
$$x \times (y \times z) = (x \times y) \times z.$$

Associative laws of addition and of multiplication hold also in the theory of real numbers, the theory of complex numbers, and various other mathematical disciplines.

In the propositional calculus there are the four following associative laws (two dually related pairs):

$$[p \vee [q \vee r]] \equiv [[p \vee q] \vee r].$$
$$[p\,[qr]] \equiv [[pq]\,r].$$
$$[p + [q + r]] \equiv [[p + q] + r].$$
$$[p \equiv [q \equiv r]] \equiv [[p \equiv q] \equiv r].$$

Also four corresponding laws in the algebra of classes.

As regards exclusive disjunction in the propositional calculus, the caution should be noted that, although $p + q$ is the exclusive disjunction of p and q, and although $+$ obeys an associative law, nevertheless $[p + q] + r$ is not the exclusive disjunction of the three propositions p, q, r—but is rather, "Either all three or one and one only of p, q, r." —*A.C.*

Assumption: A proposition which is taken or posed in order to draw inferences from it; or the act of so taking, posing, or *assuming* a proposition. The motive for an assumption may be (but need not necessarily be) a belief in the truth, or possible truth, of the proposition assumed; or the motive may be an attempt to refute the proposition by *reductio ad absurdum* (q.v.). The word *assumption* has also sometimes been used as a synonym of *axiom* or *postulate* (see the article *Mathematics*). —*A.C.*

Autonymy: In the terminology introduced by Carnap, a word (phrase, symbol, expression) is *autonymous* if it is used as a name for itself—for the geometric shape, sound, etc. which it exemplifies, or for the word as a historical and grammatical unit. Autonymy is thus the same as the Scholastic *suppositio materialis* (q.v.), although the viewpoint is different. —*A.C.*

Biconditional: The sentential connective $\equiv$, "if and only if." See *Logic, formal*, §1. —*A.C.*

Bolzano, Bernard: (1781–1848) Austrian philosopher and mathematician. Professor of the philosophy of religion at Prague, 1805–1820, he was compelled to resign in the latter year because of his rationalistic tendencies in theology, and afterwards held no academic position. His *Wissenschaftslehre* of 1837, while it is to be classed as a work on traditional logic, contains significant anticipations of many ideas which have since become important in symbolic logic and mathematics. In his posthumously published *Paradoxien des Unendlichen* (1851) he appears as a forerunner in some respects of Cantor's theory of transfinite numbers. —*A. C.*

W. Dubislav, *Bolzano als Vorläufer der mathematischen Logik*, Philosophisches Jahrbuch der Görres-Gesellschaft, vol. 44 (1931), pp. 448–456. H. Scholz, *Wissenschaftslehre Bolzanos*, Abhandlungen der Fries'schen Schule, n. s. vol. 6 (1937), pp. 399–472.

Boole, George: (1815–1864) English mathematician. Professor of mathematics at Queen's College, Cork, 1849–1864. While he made contributions to other branches of mathematics, he is now remembered primarily as the founder of the Nineteenth Century algebra of logic and through it of modern symbolic logic. His *Mathematical Analysis of Logic* appeared in 1847 and the fuller *Laws of Thought* in 1854. —*A. C.*

R. Harley, *George Boole*, F.R.S., The British Quarterly Review, vol. 44 (1866), pp. 141–181. Anon., *George Boole*, Proceedings of the Royal Society of London, vol. 15 (1867), Obituary notices of fellows deceased, pp. vi–xi. P. E. B. Jourdain, *George Boole*, The Quarterly Journal of Pure and Applied Mathematics, vol. 41 (1910), pp. 332–352.

Brouwer, Luitzen Egbertus Jan: (1881–*) Dutch mathematician. Professor of mathematics at the University of Amsterdam, 1912–. Besides his work in topology, he is known for important contributions to the philosophy and foundations of mathematics. See *Mathematics* and *Intuitionism* (*mathematical*). —*A. C.*

Calculus: The name *calculus* may be applied to any organized method of solving problems or drawing inferences by manipulations of symbols according to formal rules. Or an exact definition of a *calculus* may be provided by identifying it with a *logistic system* (q.v.) satisfying the requirement of effectiveness.

In mathematics, the word *calculus* has many specific applications, all conforming more or less closely to the above statement. Sometimes, however, the simple phrase "the calculus" is used in referring to those branches of mathematical *analysis* (q.v.) which are known more explicitly as the *differential calculus* and the *integral calculus*. —*A. C.*

Cantor, Georg (Ferdinand Ludwig Philipp), 1845–1918, (Russian born) German mathematician. Professor of mathematics at Halle, 1872–1913. He is known for contributions to the foundations of (mathematical) analysis and as the founder of the theory of transfinite *cardinal numbers* (q.v.) and *ordinal numbers* (q.v.). See *Infinite*. —*A. C.*

Gesammelte Abhandlungen Mathematische und Philosophischen Inhalts, edited by E. Zermelo, and with a life by A. Fraenkel, Berlin, 1932.

Cardinal number: Two classes are *equivalent* if there exists a one-to-one correspondence between them (see *One-one*). *Cardinal numbers* are obtained by *abstraction* (q.v.) with respect to equivalence, so that two classes have the same cardinal number if and only if they are equivalent. This may be formulated more exactly, following Frege, by

---

* *Editor's note*: Brouwer died in 1966.

defining the cardinal number of a class to be the class of classes equivalent to it.

If two classes a and b have no members in common, the cardinal number of the logical sum of a and b is uniquely determined by the cardinal numbers of a and b, and is called the *sum* of the cardinal number of a and the cardinal number of b.

0 is the cardinal number of the null class. 1 is the cardinal number of a unit class (all unit classes have the same cardinal number).

A cardinal number is *inductive* if it is a member of *every* class t of cardinal numbers which has the two properties, (1) $0 \in t$, and (2) for all x, if $x \in t$ and y is the sum of x and 1, then $y \in t$. In other (less exact) words, the inductive cardinal numbers are those which can be reached from 0 by successive additions of 1. A class b is *infinite* if there is a class a, different from b, such that $a \subset b$ and a is equivalent to b. In the contrary case b is *finite*. The cardinal number of an infinite class is said to be infinite, and of a finite class, finite. It can be proved that every inductive cardinal number is finite, and, with the aid of the axiom of choice, that every finite cardinal number is inductive.

The most important infinite cardinal number is the cardinal number of the class of inductive cardinal numbers $(0, 1, 2, \ldots)$; it is called aleph-zero and symbolized by a Hebrew letter aleph followed by an inferior 0.

For brevity and simplicity in the preceding account we have ignored complications introduced by the theory of types, which are considerable and troublesome. Modifications are also required if the account is to be incorporated into the Zermelo set theory.

—*A.C.*

G. Cantor, *Contributions to the Founding of the Theory of Transfinite Numbers*, translated and with an introduction by P. E. B. Jourdain, Chicago and London, 1915. Whitehead and Russell, *Principia Mathematica*, vol. 2.

Categorematic: In traditional logic, denoting or capable of denoting a term, or of standing for a subject or predicate—said of words. Opposite of *syncategorematic* (q.v.). —*A.C.*

Choice, axiom of, or *Zermelo's axiom*, is the name given to an assumption of logical or logico-mathematical character which may be stated as follows: *Given a class K whose members are non-empty classes, there exists a (one-valued) monadic function f whose range is K, such that $f(x) \in x$ for all members x of K*. This had often been employed unconsciously or tacitly by mathematicians—and is apparently necessary for the proofs of certain important mathematical theorems—but was first made explicit by Zermelo in 1904, who used it in a proof that every class can be well-ordered. Once explicitly stated the assumption was attacked by many mathematicians as lacking in validity or as not of legitimately mathematical character, but was defended by others, including Zermelo.

An equivalent assumption, called by Russell the *multiplicative axiom* and afterwards adopted by Zermelo as a statement of his *Auswahlprinzip* is as follows: *Given a class K whose members are non-empty classes no two of which have a member in common, there exists a class A (the Auswahlmenge) all of whose members are members of members of K and which has one and only one member in common with each member of K*. Proof of equivalence of the multiplicative axiom to the axiom of choice is due to Zermelo.

—*A.C.*

E. Zermelo, *Beweis, dass jede Menge wohlgeordnet werden kann*, Mathematische Annalen, vol. 59 (1904), pp. 514–516. B. Russell, *On some difficulties in the theory of transfinite numbers and order types*, Proceedings of the London Mathematical Society, ser. 2, vol. 4 (1906), pp. 29–53. E. Zermelo, *Neuer Beweis für die Möglichkeit einer Wohlordnung*, Mathematische Annalen, vol. 65 (1908), pp. 107–28. K. Gödel, *The Consistency of the Axiom of Choice and of the Generalized Continuum Hypothesis with the Axioms of Set Theory*, Princeton, N.J., 1940.

Class: or *set*, or *aggregate* (in most connections the words are used synonymously) can best be described by saying that classes are associated with monadic propositional functions (in intension—i.e., properties) in such a way that two propositional functions determine the same class if and only if they are formally equivalent. A class thus differs from a propositional function in extension only in that it is not usual to employ the notation of application of function to argument in the case of classes (see the article *Propositional function*). Instead, if a class a is determined by a propositional function A, we say that x *is a member of* a (in symbols, $x \in a$) iff and only if $A(x)$.

Whitehead and Russell, by introducing classes into their system only as incomplete symbols, "avoid the assumption that there are such things as classes." Their method (roughly) is to reinterpret a proposition about a class determined by a propositional function A as being instead an existential proposition, about *some* propositional function formally equivalent to A.

See also *Logic, formal*, §§7, 9. —*A.C.*

Class concept: A monadic propositional function, thought of as determining a *class* (q.v.). —*A.C.*

Combinatory Logic: A branch of mathematical logic, which has been extensively investigated by Curry, and which is concerned with analysis of processes of substitution, of the use of variables generally, and of the notion of a function. The program calls, in particular, for a system of logic in which variables are altogether eliminated, their place being taken by the presence in the system of certain kinds of function symbols. For a more detailed and exact account, reference must be made to the papers cited below. —*A.C.*

M. Schönfinkel, *Über die Bausteine der mathematischen Logik*, Mathematische Annalen, vol. 92 (1924), pp. 305–316. H. B. Curry, *Grundlagen der kombinatorischen Logik*, American Journal of Mathematics, vol. 52 (1930), pp. 509–536, 789–834. H. B. Curry, *The universal quantifier in combinatory logic*, Annals of Mathematics, ser. 2, vol. 32 (1931), pp. 154–180. H. B. Curry, *Apparent variables from the standpoint of combinatory logic*, Annals of Mathematics, ser. 2, vol. 34 (1933), pp. 381–404. H. B. Curry, *Functionality in combinatory logic*, Proceedings of the National Academy of Sciences, vol. 20 (1934) pp. 584–590. J. B. Rosser, *A mathematical logic without variables*, Annals of Mathematics, ser. 2, vol. 36 (1935), pp. 127–150, and Duke Mathematical Journal, vol. 1 (1935), pp. 328–355. H. B. Curry, *A revision of the fundamental rules of combinatory logic*, The Journal of Symbolic Logic, vol. 6 (1941), pp. 41–53. H. B. Curry, *Consistency and completeness of the theory of combinators*, ibid., pp. 54–61.

Commutative law is any law of the form $x \circ y = y \circ x$, or with the biconditional, etc., replacing equality—compare *Associative law*. Commutative laws of addition and multiplication hold in arithmetic, also in the theory of real numbers, etc. In the propositional calculus there are commutative laws of conjunction, both kinds of disjunction, the biconditional, alternative denial and its dual; also corresponding laws in the algebra of classes. —*A. C.*

Completeness: A *logistic system* (q.v.) may be called *complete* if there is no formula of the system which is not a theorem and which can be added to the list of primitive formulas (no other change being made) without rendering the system inconsistent, in one of the senses of *consistency* (q.v.). The pure propositional calculus—as explained under *Logic, formal*, §1—is complete in this sense.

Given the concept of semantical *truth* (q.v.), we may also define a logistic system as *complete* if every true formula of the system is a theorem. This sense of completeness is not, in general, equivalent to the other, and may be the weaker one if formulas containing free variables occur. See *Logic, formal*, §§3, 6. —*A. C.*

Composition is the form of valid inference of the propositional calculus from A ⊃ B and A ⊃ C to A ⊃ BC. The *law of composition* is the theorem of the propositional calculus:

$$[p \supset q][p \supset r] \supset [p \supset qr].$$

—*A. C.*

Concept: In logic syn. either with *propositional function* (q.v.) generally or with *monadic propositional function*. The terminology associated with the word *function* is not, however, usually employed in connection with the word *concept*; and the latter word may serve to avoid ambiguities which have arisen from loose or variant usages of the word *function* (q.v.); or it may reflect a difference in point of view. —*A. C.*

Conditional: The sentential connective ⊃. See *Logic, formal*, §1.

A sentence of the form A ⊃ B (or a proposition expressed by such a sentence)—verbally, "if A then B"—may be called a *conditional* sentence (or proposition). —*A. C.*

Connexity: A dyadic relation R is called *connected* if, for every two different members x, y of its field, at least one of xRy, yRx *holds*.

Consistency: (1) A *logistic system* (q.v.) is *consistent* if there is no theorem whose negation is a theorem. See *Logic, formal*, §§1, 3, 6; also *Proof theory*.

Since this definition of consistency is relative to the choice of a particular notation as representing negation, the following definition is sometimes used instead: (2) A logistic system is *consistent* if not every formula (not every sentence) is a theorem. In the case of many familiar systems, under the usual choice as to which notation represents negation, the equivalence of this sense of consistency to the previous one is immediate.

Closely related to (2), and applicable to logistic systems containing the pure propositional calculus (see *Logic, formal*, §1) or an appropriate part of it, is the notion of consistency in the sense of E. L. Post, according to which a system is consistent if a formula composed of a single propositional variable (say the formula p) is not a theorem. —*A.C.*

Constant: A *constant* is a symbol employed as an unambiguous name—distinguished from a *variable* (q.v.).

Thus in ordinary numerical algebra and in real number theory, the symbols, x, y, z are variables, while 0, 1, 3, $-\frac{1}{2}$, π, e are constants. In such mathematical contexts the term *constant* is often restricted to unambiguous (non-variable) names of *numbers*. But such symbols as $+$, $=$, $<$ may also be called constants, as denoting particular functions and relations.

In various mathematical contexts, the term *constant* will be found applied to letters which should properly be called variables (according to our account here), but which are thought of as constant relatively to other variables appearing. The actual distinction in such cases, as revealed by logistic formalization, either is between free and bound variables, or concerns the order and manner in which the variables are bound by quantifiers, abstraction operators, etc.

In mathematics, the word *constant* may also be employed to mean simply a *number* ("Euler's constant"), or, in the physical sciences, to mean a *physical quantity* ("the gravitational constant," "Planck's constant"). —*A.C.*

Continuity: A class is said to be *compactly* (or *densely*) ordered by a relation R if it is ordered by R (see *Order*) and, whenever xRz and $x \neq z$, there is a y, not the same as either x or z, such that xRy and xRz. (Compact order may thus be described by saying that between any two distinct members of the class there is always a third, or by saying that no member has a *next* following member in the order.)

If a class b is ordered by a relation R, and $a \subset b$, we say that z is an *upper bound* of a if, for all x, $x \in a$ implies xRz; and that z is a *least upper bound* of a if z is an upper bound of a and there is no upper bound y of a, different from z, such that yRz.

A class b ordered by a relation R is said to have *continuous order* (Dedekindian continuity) if it is compactly ordered by R and every non-empty class a, for which $a \subset b$, and which has an upper bound, has a least upper bound.

An important mathematical example of continuous order is afforded by the real numbers, ordered by the relation *not greater than*. According to usual geometric postulates, the points on a straight line also have continuous order, and, indeed, have the same order type as the real numbers.

The term *continuity* is also employed in mathematics in connection with functions of various kinds. We shall state the definition for the case of a monadic function f for which the range of the independent variable and the range of the dependent variable both consist of real numbers (see the article *Function*).

Let us use R for the relation *not greater than* among real numbers. A *neighborhood* of a real number c is determined by two real numbers m and n—both different from c and such that mRc and cRn—and is the class of real numbers x, other than m and n, such that mRx and xRn. The function f is said to be *continuous at* the real number c if the three following conditions are satisfied: (1) c belongs to the range of the independent variable; (2) in every neighborhood of c there are numbers other than c belonging to the range of the independent variable; (3) corresponding to every neighborhood b of

$f(c)$ there is a neighborhood a of c such that, for every real number x belonging to the range of the independent variable, $x \in a$ implies $f(x) \in b$. A function may be called *continuous* if it is continuous at every real number, or at every real number in a certain set determined by the context. —*A.C.*

E. V. Huntington, *The Continuum*, Cambridge, Mass., 1917.

Contradictio in adjecto: A logical inconsistency between a noun and its modifying adjective. A favorite example is the phrase "round square." —*A.C.*

Contradiction, Law of, is given by traditional logicians as "*A* is *B* and *A* is not *B* cannot both be true." It is usually taken to be the theorem of the propositional calculus, $\sim[p \sim p]$. In use, however, the name often seems to refer to the syntactical principle or precept which may be formulated as follows: A logical discipline containing (an applied) propositional calculus, or a set of hypotheses or postulates to be added to such a discipline, shall not lead to two theorems or consequences of the forms A and ~A. The law is explicitly stated in a syntactical form, e.g. by Ledger Wood in his *The Analysis of Knowledge* (1940). —*A.C.*

Contraposition: The recommended use of this word is that according to which the contraposition of $S(x) \supset_x P(x)$ is $\sim P \supset_x \sim S(x)$. This is, however, not quite strictly in accordance with traditional terminology; see *Logic, formal*, §4. —*A.C.*

Conventionalism: Any doctrine according to which *a priori* truth, or the truth of propositions of logic, or the truth of propositions (or of sentences) demonstrable by purely logical means, is a matter of linguistic or postulational convention (and thus not absolute in character). H. Poincaré (q.v.) regarded the choice of axioms as conventional (cf. *Science et hypothèse*, p. 67). —*A.C.*

Coordinates: In mathematics, any system of designating points by means of ordered sets of n numbers may be called an *n-dimensional coordinate system*, and the n numbers so associated with any point are then called its *coordinates*. Coordinates may also be used in like fashion for various other things besides points. —*A.C.*

Copula: The traditional analysis of a proposition into subject and predicate involves a third part, the copula (*is, are, is not, are not*), binding the subject and predicate together into an assertion either of affirmation or of denial. It is now, however, commonly held that several wholly different meanings of the verb to be should be distinguished in this connection, including at least the following: predication of a monadic propositional function of its argument (the sun *is* hot, 7 *is* a prime number, mankind *is* numerous); formal implication (gold *is* heavy, a horse *is* a quadruped, mankind *is* sinful); identity (China *is* Cathay, that *is* the sun, I *am* the State); formal equivalence (lightning *is* an electric discharge between parts of a cloud or a cloud and the earth). —*A.C.*

Dedekind, (Julius Wilhelm) Richard: (1831–1916) German mathematician. Professor of mathematics at Brunswick, 1862–1894. His contributions to the foundations of arithmetic and analysis are contained in his *Stetigkeit und Irrationale Zahlen* (1st edn., 1872, 5th edn., 1927) and *Was Sind und Was Sollen die Zahlen?* (1st edn., 1888, 6th edn., 1930). —*A. C.*

Gesammelte Mathematische Werke, three volumes, Brunswick 1930–1932.

Deduction theorem: In a *logistic system* (q.v.) containing propositional calculus (pure or applied) or a suitable part of the propositional calculus, it is often desirable to have the property that if the inference from A to B is a valid inference then A ⊃ B is a theorem, or, more generally, that if the inference from $A_1, A_2, \ldots, A_n$ to B is valid then the inference from $A_1, A_2 \ldots, A_{n-1}$ to $A_n \supset B$ is valid. The syntactical theorem, asserting of a given logistic system that it has this property, is called the *deduction theorem* for that system. (Certain cautions are necessary in defining the notion of valid inference where free variables are present; cf. *Logic, formal*, §§1, 3.) —*A. C.*

Definition: In the development of a *logistic system* (q.v.) it is usually desirable to introduce new notations, beyond what is afforded by the primitive symbols alone, by means of *syntactical definitions* or *nominal definitions*, i.e., conventions which provide that certain symbols or expressions shall stand (as substitutes or abbreviations) for particular formulas of the system. This may be done either by particular definitions, each introducing a symbol or expression to stand for some one formula, or by *schemata* of definition, providing that any expression of a certain form shall stand for a certain corresponding formula (so condensing many—often infinitely many—particular definitions into a single schema). Such definitions, whether particular definitions or schemata, are indicated, in articles herein by the present writer, by an arrow $\longrightarrow$, the new notation introduced (the *definiendum*) being placed at the left or base of the arrow, and the formula for which it shall stand (the *definiens*) being placed at the right, or head, of the arrow. Another sign commonly employed for the same purpose (instead of the arrow) is the equality sign = with the letters Df, or df, appearing either as a subscript or separately after the definiens.

This use of nominal definition (including contextual definition—see the article *Incomplete symbol*) in connection with a logistic system is extraneous to the system in the sense that it may theoretically be dispensed with, and all formulas written in full. Practically, however, it may be necessary for the sake of brevity or perspicuity, or for facility in formal work.

Such methods of introducing new concepts, functions, etc. as definition by *abstraction* (q.v.), definition by *recursion* (q.v.), definition by *composition* (see *Recursiveness*) may be dealt with by reducing them to nominal definitions; i.e., by finding a nominal definition such that the definiens (and therefore also the definendum) turns out, under an intended interpretation of the logistic system, to mean the concept, function, etc. which is to be introduced.

In addition to syntactical or nominal definition we may distinguish another kind of definition, which is applicable only in connection with *interpreted* logistic systems, and which we shall call *semantical definition*. This consists in introducing a new symbol or notation by assigning a *meaning* to it. In an interpreted logistic system, a nominal definition carries with it implicitly a semantical definition, in that it is intended to give to the definiendum the meaning expressed by the definiens; but two different nominal definitions may correspond to the same semantical definition. Consider, for example, the two following schemata of nominal definition in the propositional calculus (*Logic, formal*, §1):

$$[A] \supset [B] \longrightarrow \sim A \vee B.$$
$$[A] \supset [B] \longrightarrow \sim [A \sim B].$$

As nominal definitions, these are inconsistent, since they represent [A] ⊃ [B] as standing for different formulas: either one, but not both, could be used in a development of the propositional calculus. But the corresponding semantical definitions would be identical if—as would be possible—our interpretation of the propositional calculus were such that the two definientia had the same meaning for any particular A and B.

In the formal development of a logistic system, since no reference may be made to an intended interpretation, semantical definitions are precluded, and must be replaced by corresponding nominal definitions.

Of quite a different kind are so-called *real definitions*, which are not conventions for introducing new symbols or notations—as syntactical and semantical definitions are—but are propositions of equivalence (material, formal, etc.) between two abstract entities (propositions, concepts, etc.) of which one is called the *definiendum* and the other the *definiens*. Not all such propositions of equivalence, however, are real definitions, but only those in which the definiens embodies the "essential nature" (essentia, *ούσία*) of the definiendum. The notion of a real definition thus has all the vagueness of the quoted phrase, but the following may be given as an example. If all the notations appearing, including $\supset_x$, have their usual meanings (regarded as given in advance), the proposition expressed by

$$(F)(G)[[F(x) \supset_x G(x)] \equiv (x)[\sim F(x) \vee G(x)]]$$

is a real definition of formal implication—to be contrasted with the nominal definition of the *notation* for formal implication which is given in the article *Logic, formal*, §3. This formula, expressing a real definition of formal implication, might appear, e.g., as a primitive formula in a logistic system.

(A situation often arising in practice is that a word—or symbol or notation—which already has a vague meaning is to be given a new exact meaning, which is vaguely, or as nearly as possible, the same as the old. This is done by a nominal or semantical definition rather than a real definition; nevertheless it is usual in such a case to speak either of defining the *word* or of defining the associated *notion*.)

Sometimes, however, the distinction between nominal definitions and real definitions is made on the basis that the latter convey an assertion of *existence* of the definiendum, or rather, where the definiendum is a concept, of things falling thereunder (Saccheri, 1697); or the distinction may be made on the basis that real definitions involve the *possibility* of what is defined (Leibniz, 1684). Ockham makes the distinction rather on the basis that real definitions state the whole nature of a thing and nominal definitions state the meaning of a word or phrase, but adds that non-existents (as chimaera) and such parts of speech as verbs, adverbs, and conjunctions may therefore have only nominal definition. —*A.C.*

De Morgan, Augustus: (1806–1871) English mathematician and logician. Professor of mathematics at University College, London, 1828–1831, 1836–1866. His *Formal*

Logic of 1847 contains some points of an algebra of logic essentially similar to that of *Boole* (q.v.) but the notation is less adequate than Boole's and the calculus is less fully worked out and applied. De Morgan, however, had the notion of logical sum for arbitrary classes—whereas Boole contemplated addition only of classes having no members in common. *De Morgan's laws* (q.v.)—as they are now known—were also enunciated in this work. The treatment of the syllogism is original, but has since been superseded and does not constitute the author's real claim to remembrance as a logician. (The famous controversy with Sir William Hamilton over the latter's charge of plagiarism in connection with this treatment of the syllogism may therefore be dismissed as not of present interest.)

Through his paper *On the syllogism, no. IV* in the Transactions of the Cambridge Philosophical Society, vol. 10 (read April 23, 1860), De Morgan is to be regarded as the founder of the logic of relations. —*A. C.*

Sophia Elizabeth De Morgan, *Memoir of Augustus De Morgan*, London, 1882.

De Morgan's laws: Are the two dually related theorems of the propositional calculus,

$$\sim[p \vee q] \equiv [\sim p \sim q],$$
$$\sim[pq] \equiv [\sim p \vee \sim q],$$

or the two corresponding dually related theorems of the algebra of classes,

$$-(a \cup b) = -a \cap -b,$$
$$-(a \cap b) = -a \cup -b.$$

In the propositional calculus these laws (together with the law of double negation) make it possible to define conjunction in terms of negation and (inclusive) disjunction, or, alternatively, disjunction in terms of negation and conjunction. Similarly in the algebra of classes logical product may be defined in terms of logical sum and complementation, or logical sum in terms of logical product and complementation.

As pointed out by Łukasiewicz, these laws of the propositional calculus were known already (in verbal form) to Ockham. The attachment of De Morgan's name to the corresponding laws of the algebra of classes appears to be historically more correct.

Sometimes referred to as generalizations or analogues of De Morgan's laws are the two dually related theorems of the functional calculus of first order,

$$\sim(Ex)F(x) \equiv (x)\sim F(x),$$
$$\sim(x)F(x) \equiv (Ex)\sim F(x),$$

and similar theorems in higher functional calculi. These make possible the definition of the existential quantifier in terms of the universal quantifier (or inversely).

—*A. C.*

Denial of the antecedent: The *fallacy of denial of the antecedent* is the fallacious inference from ~A and A ⊃ B to ~B. The *law of denial of the antecedent* is the theorem of the propositional calculus, $\sim p \supset [p \supset q]$. —*A. C.*

Denotation: In common usage, "denotation" has a less special meaning, *denote* being approximately synonymous with *designate* (q.v.). A proper name may be said to denote that of which it is a name. Or, e.g., in the equation $2 + 2 = 4$, the sign $+$ may be said to denote addition and the sign = to denote equality (even without necessarily intending to construe these signs as proper names).

Concerning Frege's distinction between sense and denotation see the article *Descriptions*. —*A.C.*

Descriptions: Where a formula A containing a free variable—say, for example, x—means a true proposition (is true) for one and only one value of x, the notation (ιx)A is used to mean that value of x. The approximately equivalent English phraseology is "the x such that A"—or simply "the F," where F denotes the concept (monadic propositional function) obtained from A by *abstraction* (q.v.) with respect to x. This notation, or its sense in the sense of Frege, is called a *description*.

In *Principia Mathematica* descriptions (or notations serving the same purpose in context) are introduced as *incomplete symbols* (q.v.). Russell maintains that descriptions not only may but must be thus construed as incomplete symbols—briefly, for the following reasons. The alternative is to construe a description as a proper name, so that, e.g., the description *the author of Waverley* denotes the man Scott and is therefore synonymous with the name *Scott*. But then the sentences "Scott is the author of Waverley" and "Scott is Scott" ought to be synonymous—which they clearly are not (although both are true). Moreover, such a description as *the King of France* cannot be a proper name, since there is no King of France whom it may denote; nevertheless, a sentence such as "The King of France is bald" should be construed to have a meaning, since it may be falsely asserted or believed by one who falsely asserts or believes that there is a King of France.

Frege meets the same difficulties, without construing descriptions as incomplete symbols, by distinguishing two kinds of meaning, the sense (Sinn) and the denotation (Bedeutung) of an expression (formula, phrase, sentence, etc.). *Scott* and *the author of Waverley* have the same denotation, namely the man Scott, but not the same sense. *The King of France* has as sense but no denotation; so likewise the sentence *The King of France is bald*. Two expressions having the same sense must have the same denotation if they have a denotation. When a constituent part of an expression is replaced by another part having the same sense, the sense of the whole is not altered. When a constituent part of an expression is replaced by another having the same denotation, the denotation of the whole (if any) is not altered, but the sense may be. The denotation of an (unasserted) declarative sentence (if any) is a truth-value, whereas the sense is the thought or content of the sentence. But where a sentence is used in indirect discourse (as in saying that so-and-so says that . . . , believes that . . . , is glad that . . . , etc.) the meaning is different: in such a context the denotation of the sentence is that which would be its sense in direct discourse. (In quoting some one in indirect discourse, one reproduces neither the literal wording nor the truth-value, but the sense, of what he said.)

Frege held it to be desirable in a formalized logistic system that every formula should have not only a sense but also a denotation—as can be arranged by arbitrary semantical conventions where necessary. When this is done, Frege's *sense of a sentence* nearly coincides with *proposition* (in sense (b) of the article of that title herein).

Alonzo Church

G. Frege, *Über Sinn und Bedeutung*, Zeitschrift für Philosophie und philosophische Kritik, n.s., vol. 100 (1892), pp. 25–50. B. Russell, *On denoting*, Mind, n.s., vol. 14 (1905), pp. 470–493.

Designate: A word, symbol, or expression may be said to *designate* that object (abstract or concrete) to which it refers, or of which it is a name or sign. See *Name relation*. —*A. C.*

Designatum: The *designatum* of a word, symbol, or expression is that which it *designates* (q.v.). —*A. C.*

Diallelon: A *vicious circle* (q.v.) in definition. —*A. C.*

Diallelus: A *vicious circle* (q.v.) in proof. —*A. C.*

Dialogism: Inference from one premiss of a (categorical) syllogism to the disjunction of the conclusion and the negation of the other premiss is a *dialogism*. Or, more generally, if the inference from A and B to C is a valid inference, that from A to $C \vee \sim B$ may be called a *dialogism*. —*A. C.*

Dictum de omni et nullo: The leading principles of the syllogisms in Barbara and Celarent, variously formulated, and attributed to Aristotle. "Whatever is affirmed (denied) of an entire class or kind may be affirmed (denied) of any part." The four moods of the first figure were held to be directly validated by this dictum, and this was given as the motive for the traditional reductions of the last three syllogistic figures to the first. See also *Aristotle's dictum*. —*A. C.*

Disjunctive: A sentence of either of the forms $A \vee B$, $A + B$ (or a proposition expressed by such a sentence)—see *Logic, formal*, §1—may be called a *disjunctive sentence* (or proposition). —*A. C.*

Distributive law is a name given to a number of laws of the same or similar form appearing in various disciplines—compare *associative law*. A distributive law of multiplication over addition appears in arithmetic:

$$x \times (y + z) = (x \times y) + (x \times z).$$

This distributive law holds also in the theory of real numbers, and in many other mathematical disciplines involving two operations called multiplication and addition. In the propositional calculus there are four distributive laws (two dually related pairs):

$$p[q \vee r] \equiv [pq \vee pr].$$
$$[p \vee qr] \equiv [p \vee q][p \vee r].$$
$$p[q + r] \equiv [pq + pr].$$
$$[p \vee [q \equiv r]] \equiv [[p \vee q] \equiv [p \vee r]].$$

Also four corresponding laws in the algebra of classes. —*A. C.*

Double negation, law of: The theorem of the propositional calculus, $\sim\sim p \equiv p$. —*A. C.*

Eduction: 1. *In logic*, a term proposed by E. E. Constance Jones as a synonym or substitute for the more usual *immediate inference* (see *Logic, formal*, §4). —*A. C.*

Enumerable: A class is *enumerable* if its *cardinal number* (q.v.) is aleph 0. —*A. C.*

Episyllogism: Where the conclusion of one (categorical) syllogism is used as one of the premisses of another, the first syllogism is called a *prosyllogism* and the second one an *episyllogism*. —*A. C.*

Equipollence: Carnap proposes a purely syntactical definition of equipollence by defining two sentences (or two classes of sentences) to be equipollent if they have the same class of non-valid consequences. See the article *Valid*. —*A. C.*

Equivocation is any fallacy arising from ambiguity of a word, or of a phrase playing the rôle of a single word in the reasoning in question, the word or phrase being used at different places with different meanings and an inference drawn which is formally correct if the word or phrase is treated as being the same word or phrase throughout. —*A. C.*

Euler diagram: The elementary operations upon and relations between classes—complementation, logical sum, logical product, class equality, class inclusion—may sometimes advantageously be represented by means of the corresponding operations upon and relations between regions in a plane. (Indeed, if regions are considered as classes of points, the operations and relations for regions become particular cases of those for classes.) By using regions of simple character, such as interiors of circles or ellipses, to stand for given classes, convenient diagrammatic representations are obtained of the possible logical relationships between two or more classes. These are known as *Euler diagrams*, although their employment by Euler in his *Letters to a German Princess* (vol. 2, 1772) was not their first appearance. Or the diagram may be so drawn as to show all possible intersections (2^n intersections in the case of n classes), and then intersections known to be empty may be crossed out, and intersections known not to be empty marked with an asterisk or otherwise (*Venn diagram*). —*A. C.*

Exact: Opposite of *vague* (q.v.). —*A. C.*

Excluded middle, law of, or *tertium non datur*, is given by traditional logicians as "A is B or A is not B." This is usually identified with the theorem of the propositional calculus, $p \vee \sim p$, to which the same name is given. The general validity of the law is denied by the school of mathematical *intuitionism* (q.v.). —*A. C.*

Existential proposition: Traditionally a proposition which directly asserts the existence of its subject, as, e.g., Descartes's "ergo sum" or the Christian's "Good exists." Expressed in symbolic notation, such a proposition has a form like (Ex)M.

By an extension of this, a proposition expressible in the functional calculus of first order may be called *existential* if the prenex normal form has a prefix containing an existential quantifier (see *Logic, formal*, §3).

Brentano (*Psychologie*, 1874) takes an existential proposition (Existentialsatz) to be one that directly affirms *or denies* existence, and shows that each of the four traditional kinds of categorical propositions is reducible (i.e., equivalent) to an existential proposition in this sense; thus, e.g., "all men are mortal" becomes "immortal men do not exist." This definition of an existential proposition and the reduction of categorical propositions to existential appears also in Keynes's *Formal Logic*, 4th edn. (1906).

—*A. C.*

Exportation is the form of valid inference of the propositional calculus from AB ⊃ C to A ⊃ [B ⊃ C]. The *law of exportation* is the theorem of the propositional calculus:

$$[pq \supset r] \supset [p \supset [q \supset r]].$$

—*A. C.*

Fallacy is any unsound step or process of reasoning, especially one which has a deceptive appearance of soundness or is falsely accepted as sound. The unsoundness may consist either in a mistake of formal logic, or in the suppression of a premiss whose unacceptability might have been recognized if it had been stated, or in a lack of genuine adaptation of the reasoning to its purpose. Of the traditional names which purport to describe particular kinds of fallacies, not all have a sufficiently definite or generally accepted meaning to justify notice. See, however, the following: *Affirmation of the consequent; Amphiboly; Denial of the antecedent; Equivocation; Ignoratio elenchi; Illicit process of the major; Illicit process of the minor; Many questions; Non causa pro causa; Non sequitur; Petitio principii; Post hoc ergo propter hoc; Quaternio terminorum; Secundum quid; Undistributed middle; Vicious circle.* —*A. C.*

Figure (syllogistic): The moods of the categorical syllogism (see *Logic, formal*, §5) are divided into four figures, according as the middle term is subject in the major premiss and predicate in the minor premiss (first figure), or predicate in both premisses (second figure), or subject in both premisses (third figure), or predicate in the major premiss and subject in the minor premiss (fourth figure). Aristotle recognized only three figures, including the moods of the fourth figure among those of the first. The separation of the fourth figure from the first (ascribed to Galen) is accompanied by a redefinition of "major" and "minor"—so that the major premiss is that involving the predicate of the conclusion, and the minor premiss is that involving the subject of the conclusion. —*A. C.*

Finite: For the notion of finiteness as applied to classes and cardinal numbers, see the article *Cardinal number*. An ordered class (see *Order*) which is finite is called a *finite sequence* or *finite series*. In mathematical analysis, any fixed real number (or complex number) is called *finite*, in distinction from "infinity" (the latter term usually occurs, however, only as an incomplete symbol, in connection with *limits*, q.v.). Or *finite* may be used to mean *bounded*, i.e., having fixed real numbers as lower bound and upper bound. Various physical and geometrical quantities, measured by real numbers, are called finite if their measure is finite in one of these senses. —*A. C.*

Formalism (mathematical) is a name which has been given to any one of various accounts of the foundations of mathematics which emphasize the formal aspects of mathematics as against content or meaning, or which, in whole or in part, deny content to mathematical formulas. The name is often applied, in particular, to the doctrines of Hilbert (see *Mathematics*), although Hilbert himself calls his method axiomatic, and gives to his syntactical or metamathematical investigations the name Beweistheorie (*proof theory*, q.v.). —*A.C.*

Frege, (Friedrich Ludwig) Gottlob, 1848–1925, German mathematician and logician. Professor of mathematics at the University of Jena, 1879–1918. Largely unknown to, or misunderstood by, his contemporaries, he is now regarded by many as "beyond question the greatest logician of the Nineteenth Century" (quotation from Tarski). He must be regarded—after *Boole* (q.v.)—as the second founder of symbolic logic, the essential steps in the passage from the algebra of logic to the logistic method (see the article *Logistic system*) having been taken in his *Begriffsschrift* of 1879. In this work there appear for the first time the *propositional calculus* in substantially its modern form, the notion of *propositional function*, the use of *quantifiers*, the explicit statement of primitive *rules of inference*, the notion of a *hereditary property* and the logical analysis of proof by mathematical induction or *recursion* (q.v.). This last is perhaps the most important element in the definition of an inductive *cardinal number* (q.v.) and provided the basis for Frege's derivation of arithmetic from logic in his *Grundlagen der Arithmetik* (1884) and *Grundgesetze der Arithmetik*, vol. 1 (1893), and vol. 2 (1903). The first volume of *Grundgesetze der Arithmetik* is the culmination of Frege's work, and we find here many important further ideas. In particular, there is a careful distinction between *using* a formula to express something else and *naming* a formula in order to make a syntactical statement about it, quotation marks being used in order to distinguish the name of a formula from the formula itself. In an appendix to the second volume of *Grundgesetze*, Frege acknowledges the presence of an inconsistency in his system through what is now known as the Russell paradox (see *Paradoxes, logical*), as had been called to his attention by Russell when the book was nearly through the press. —*A.C.*

P. E. B. Jourdain, *Gottlob Frege*, The Quarterly Journal of Pure and Applied Mathematics, vol. 43 (1912), pp. 237–269. H. Scholz, *Was ist ein Kalkül und was hat Frege für eine pünktliche Beantwortung dieser Frage geleistet?*, Semester-Berichte (Münster i.W.), summer 1935, pp. 16–47. Scholz and Bachmann, *Der wissenschaftliche Nachlass von Gottlob Frege*, Actes du Congrès International de Philosophie Scientifique (Paris, 1936), section VIII, pp. 24–30.

Function: In mathematics and logic, an *n-adic function* is a law of correspondence between an ordered set of *n* things (called *arguments* of the function, or *values of the independent variables*) and another thing (the value of the function, or *value of the dependent variable*), of such a sort that, given any ordered set of *n* arguments which belongs to a certain domain (the *range* of the function), the value of the function is uniquely determined. The value of the function is spoken of as obtained by *applying* the function to the arguments. The domain of all possible values of the function is called the *range of the dependent variable*. If F denotes a function and $X_1, X_2, \ldots, X_n$ denote the first argument, second argument, etc., respectively, the notation $F(X_1, X_2, \ldots, X_n)$ is used to denote the corresponding value of the function; or the notation may be

$[F](X_1, X_2, \ldots, X_n)$, to provide against ambiguities which might otherwise arise if F were a long expression rather than a single letter.

In particular, a *monadic function* is a law of correspondence between an *argument* (or *value of the independent variable*) and a *value* of the function (or *value of the dependent variable*), of such a sort that, given any argument belonging to a certain domain (the *range* of the function, or *range of the independent variable*), the value of the function is uniquely determined. If F denotes a monadic function and X denotes an argument, the notation F(X) is used for the corresponding value of the function.

Instead of a *monadic* function, *dyadic* function, etc., one may also speak of a function *of one variable*, a function *of two variables*, etc. The terms *singulary* or *unary* (= monadic), *binary* (= dyadic), etc., are also in use. The phrase, "function from *A* to *B*," is used in the case of a monadic function to indicate that *A* and *B* (or some portion of *B*) are the ranges of the independent and dependent variables respectively—in the case of a polyadic function to indicate that *B* (or some portion of *B*) is the range of the dependent variable while the range of the function consists of ordered sets of *n* things out of *A*.

It is sometimes necessary to distinguish between functions in intension and functions in extension, the distinction being that two *n*-adic functions in extension are considered identical if they have the same range and the same value for every possible ordered set of *n* arguments, whereas some more severe criterion of identity is imposed in the case of functions in intension. In most mathematical contexts the term *function* (also the roughly synonymous terms *operation, transformation*) is used in the sense of function in extension.

(In the case of *propositional functions*, the distinction between intension and extension is usually made somewhat differently, two propositional functions in extension being identical if they have materially equivalent values for every set of arguments.)

Sometimes it is convenient to drop the condition that the value of a function is unique and to require rather that an ordered set of arguments shall determine a set of values of the function. In this case one speaks of a *many-valued function*.

Often the word *function* is found used loosely for what would more correctly be called an ambiguous or undetermined value of a function, an expression containing one or more free variables being said, for example, to denote a function. Sometimes also the word *function* is used in a syntactical sense—e.g., to mean an expression containing free variables.

See the article *Propositional function*. *Alonzo Church*

Functor: In the terminology of Carnap, a *functor* is a sign for a (non-propositional) *function* (q.v.). The word is thus synonymous with *(non-propositional) function symbol*. —*A.C.*

Geometry: Originally abstracted from the measurement of, and the study of relations of position among, material objects, geometry received in Euclid's *Elements* (c. 300 B.C.) a treatment which (despite, of course, certain defects by modern standards) became the historical model for the abstract deductive development of a mathematical discipline. The general nature of the subject of geometry may be illustrated by reference to the *synthetic* geometry of Euclid, and the *analytic* geometry which resulted from the introduction of coordinates into Euclidean geometry by *Descartes* (1637) (q.v.). In the mathematical usage of today the name *geometry* is given to any abstract mathematical discipline of a certain general type, as thus illustrated, without any requirement of applicability to spatial relations among physical objects or the like.

See *Mathematics*, and *Non-Euclidean geometry*. For a very brief outline of the foundations of plane Euclidean geometry, both from the synthetic and the analytic viewpoint, see the Appendix to Eisenhart's book cited below. A more complete account is given by Forder. —*A. C.*

L. P. Eisenhart, *Coordinate Geometry*, 1939. H. G. Forder, *The Foundations of Euclidean Geometry*, Cambridge, England, 1927. T. L. Heath, *The Thirteen Books of Euclid's Elements, translated from the text of Heiberg, with introduction and commentary*, 3 vols., Cambridge, England, 1908.

Gödel, Kurt, 1906–*, Austrian mathematician and logician—educated at Vienna, and now located (1941) at the Institute for Advanced Study in Princeton, N.J.—is best known for his important incompleteness theorem, the closely related theorem on the impossibility (under certain circumstances) of formalizing a consistency proof for a logistic system within that system, and the essentially simple but far-reaching device of *arithmetization of syntax* which is employed in the proof of these theorems (see *Logic, formal*, §6). Also of importance are his proof of the completeness of the functional calculus of first order (see *Logic, formal*, §3), and his recent work on the consistency of the axiom of *choice* (q.v.) and of Cantor's *continuum hypothesis*. —*A. C.*

Hauber's law: Given a set of conditional sentences $A_1 \supset B_1, A_2 \supset B_2, \dots, A_n \supset B_n$, we may infer each of the conditional sentences $B_1 \supset A_1, B_2 \supset A_2, \dots, B_n \supset A_n$, provided we know that $A_1, A_2, \dots, A_n$ are exhaustive and $B_1, B_2, \dots, B_n$ are mutually exclusive—i.e., provided we have also $A_1 \vee A_2 \vee \cdots \vee A_n$ and $\sim[B_1 B_2], \sim[B_1 B_3], \dots, \sim[B_{n-1} B_n]$. This form (or set of forms) of valid inference of the propositional calculus is *Hauber's law*. —*A. C.*

Hilbert, David, 1862–1943, German mathematician. Professor of mathematics at the University of Göttingen, 1895–. A major contributor to many branches of mathematics, he is regarded by many as the greatest mathematician of his generation. His work on the foundations of Euclidean geometry is contained in his *Grundlagen der Geometrie* (1st edn., 1899, 7th edn., 1930). Concerning his contributions to mathematical logic and mathematical philosophy, see the articles *mathematics*, and *proof theory*. —*A. C.*

Gesammelte Abhandlungen, three volumes, with an account of his work in mathematical logic by P. Bernays, and a life by O. Blumenthal, Berlin, 1932–1935.

Hypothetical sentence or proposition is the same as a *conditional* (q.v.) sentence or proposition. —*A. C.*

Identity, law of: Given by traditional logicians as "*A* is *A*." Because of the various possible meanings of the *copula* (q.v.) and the uncertainty as to the range of the variable

---

* *Editor's note*: Gödel died in 1978.

A, this formulation is ambiguous. The traditional law is perhaps best identified with the theorem $x = x$, either of the functional calculus of first order with equality, or in the theory of types (with equality defined), or in the algebra of classes, etc. It has been, or may be, also identified with either of the theorems of the propositional calculus, $p \supset p$, $p \equiv p$, or with the theorem of the functional calculus of first order, $F(x) \supset_x F(x)$. Many writers understand, however, by the law of identity a semantical principle—that a word or other symbol may (or must) have a fixed referent in its various occurrences in a given context (so, e.g., Ledger Wood in his *The Analysis of Knowledge*). Some, it would seem, confuse such a semantical principle with a proposition of formal logic.
—*A. C.*

Ignoratio elenchi: The fallacy of irrelevance, i.e., of proving a conclusion which is other than that required or which does not contradict the thesis which it was undertaken to refute. —*A. C.*

Illative: Having to do with inference. —*A. C.*

Illicit process of the major: In the categorial syllogism (*Logic, formal*, §5), the conclusion cannot be a proposition E or O unless the major term appears in its premiss as *distributed*—i.e., as the subject of a proposition A or E, or the predicate of a proposition E or O. Violation of this rule is the fallacy of *illicit process of the major*.
—*A. C.*

Illicit process of the minor: In the categorial syllogism (*Logic, formal*, §5), the conclusion cannot be a proposition A or E unless the minor term appears in its premiss as *distributed*—i.e., as the subject of a proposition A or E, or the predicate of a proposition E or O. Violation of this rule is the fallacy of *illicit process of the minor*.
—*A. C.*

Importation: The form of valid inference of the propositional calculus from A ⊃ [B ⊃ C] to AB ⊃ C. The *law of importation* is the theorem of the propositional calculus:

$$[p \supset [q \supset r]] \supset [pq \supset r].$$
—*A. C.*

Imposition: In Scholastic logic, grammatical terms such as *noun, pronoun, verb, tense, conjugation* were classed as terms of second imposition, other terms as of first imposition. The latter were subdivided into terms of first and second *intention* (q.v.).
—*A. C.*

Impredicative definition: Poincaré in a proposed resolution (1906) of the paradoxes of Burali-Forti and Richard (see *Paradoxes, logical*), introduced the principle that, in making a definition of a particular member of any class, no reference should be allowed to the totality of members of that class. Definitions in violation of this principle were called *impredicative* (*non prédicatives*) and were held to involve a vicious circle.

The prohibition against impredicative definition was incorporated by Russell into his ramified theory of types (1908) and is now usually identified with the restriction to the ramified theory of types *without the axiom of reducibility*. (Poincaré, however, never made his principle exact and may have intended, vaguely, a less severe restriction than this—as indeed some passages in later writings would indicate.) —*A.C.*

Incomplete symbol: A symbol (or expression) which has no meaning in isolation but which may occur as a constituent part in, and contribute to the meaning of, an expression which does have a meaning. Thus—as ordinarily employed—a terminal parenthesis) is an incomplete symbol; likewise the letter λ which appears in the notation for functional *abstraction* (q.v.); etc.

An expression A introduced by *contextual definition*—i.e., by a definition which construes particular kinds of expressions containing A, as abbreviations or substitutes for certain expressions not containing A, but provides no such construction for A itself—is an incomplete symbol in this sense. In *Principia Mathematica*, notations for classes, and descriptions (more correctly, notations which serve some of the purposes that would be served by notations for classes and by descriptions) are introduced in this way by contextual definition. —*A.C.*

Whitehead and Russell, *Principia Mathematica*, vol. 1.

Inconsistency: As applied to logistic systems, the opposite of *consistency* (q.v.).

A set of propositional functions is *inconsistent* if there is some propositional function such that their conjunction formally implies (see *Logic, formal*, §3) both it and its negation.

A set of sentences is *inconsistent* if there is some sentence A such that there is a valid inference from them to A and also from them to ~A.

If the notion of possibility is admitted, in the sense of a modality (see *Modality*, and *Strict implication*), a set of propositions may be said to be *inconsistent* if their conjunction is impossible. —*A.C.*

Independence: In a set of postulates for a mathematical discipline (see *Mathematics*), a particular postulate is said to be *independent* if it cannot be proved as a consequence of the others. A non-independent postulate is thus superfluous, and should be dropped.

In a *logistic system* (q.v.), a primitive formula or a primitive rule of inference may be said to be *independent* if there are theorems of the system which would cease to be theorems upon omission of the primitive formula or primitive rule of inference. —*A.C.*

Individual: In formal logic, the *individuals* form the first or lowest type of Russell's hierarchy of types. In the *Principia Mathematica* of Whitehead and Russell, individuals are "defined as whatever is neither a proposition nor a function." It is unnecessary, however, to give the word any such special significance, and for many purposes it is better (as is often done) to take the individuals to be an arbitrary—or an arbitrary infinite—domain; or any particular well-defined domain may be taken as the domain of individuals, according to the purpose in hand. When used in this way, the term *domain of individuals* may be taken as synonymous with the term *universe of discourse* (in the sense of Boole) which is employed in connection with the algebra of classes.

See *Logic, formal*, §§3, 6, 7. —*A. C.*

Infinite: Opposite of *finite* (q.v.), as applied to classes, cardinal and ordinal numbers, sequences, etc. See further *Cardinal number; Limit.* —*A. C.*

Infinitesimal: In a phraseology which is logically inexact but nevertheless common, an *infinitesimal* is a quantity, or a variable, whose limit is 0. Thus in considering the limit of $f(x)$ as x approaches c, if this limit is 0 the "quantity" $f(x)$ may be said to be an infinitesimal; or in considering the limit of $f(x)$ as x approaches 0, the "quantity" x may be said to be an infinitesimal. (See the article *Limit.*) —*A. C.*

Intension and extension: The *intension* of a concept consists of the qualities or properties which go to make up the concept. The *extension* of a concept consists of the things which fall under the concept; or, according to another definition, the *extension* of a concept consists of the concepts which are subsumed under it (determine subclasses). This is the old distinction between intension and extension, and coincides approximately with the distinction between a monadic *propositional function* (q.v.) in intension and a *class* (q.v.). The words *intension* and *extension* are also used in connection with a number of distinctions related or analogous to this one, the adjective *extensional* being applied to notions or points of view which in some respect confine attention to truth-values of propositions as opposed to meanings constituting propositions. In the case of (interpreted) calculi of propositions or propositional functions, the adjective *intensional* may mean that account is taken of modality, *extensional* that all functions of propositions which appear are truth-functions. The extreme of the extensional point of view does away with propositions altogether and retains only truth-values in their place. —*A. C.*

The Port-Royal Logic, translated by T. S. Baynes (see *Introduction by the translator*).

Lewis and Langford, *Symbolic Logic*, New York and London, 1932. R. Carnap, *The Logical Syntax of Language*, New York and London, 1937.

Intention: In Scholastic logic, first intentions were properties or classes of, and relations between, concrete things. Second intentions were properties or classes of, and relations between, first intentions.

This suggests the beginning of a simple hierarchy of types (see *Logic, formal*, §6), but actually is not so, because no "third intentions" were separated out or distinguished from second. Thus the general concept of *class* is a second intention, although some particular classes may also be second intentions.

Thomas *Aquinas* (q.v.) defined logic as the science of second intentions applied to first intentions. —*A. C.*

Intuitionism (mathematical): The name given to the school (of mathematics) founded by L. E. J. *Brouwer* (q.v.) and represented also by Hermann Weyl, Hans Freudenthal, Arend Heyting, and others. In some respects a historical forerunner of intuitionism is the mathematician Leopold Kronecker (1823–1891). Views related to intuitionism (but usually not including the rejection of the law of excluded middle) have been expressed by many recent or contemporary mathematicians, among whom are

J. Richard, Th. Skolem, and the French semi-intuitionists—as Heyting calls them—E. Borel, H. Lebesgue, R. Baire, N. Lusin. (Lusin is Russian but has been closely associated with the French school.)

For the account given by Brouwerian intuitionism of the nature of mathematics, and the asserted priority of mathematics to logic and philosophy, see the article *Mathematics*. This account, with its reliance on the intuition of ordinary thinking and on the immediate evidence of mathematical concepts and inferences, and with its insistence on intuitively understandable construction as the only method for mathematical existence proofs, leads to a rejection of certain methods and assumptions of classical mathematics. In consequence, certain parts of classical mathematics have to be abandoned and others have to be reconstructed in different and often more complicated fashion.

Rejected in particular by intuitionism are: (1) the use of *impredicative definition* (q.v.); (2) the assumption that all things satisfying a given condition can be united into a set and this set then treated as an individual thing—or even the weakened form of this assumption which is found in Zermelo's *Aussonderungsaxiom* or axiom of subset formation (see *Logic, formal*, §9); (3) the law of excluded middle as applied to propositions whose expression requires a quantifier for which the variable involved has an infinite range.

As an example of the rejection of the law of excluded middle, consider the proposition, "Either every even number greater than 2 can be expressed as the sum of two prime numbers or else not every even number greater than 2 can be expressed as the sum of two prime numbers." This proposition is intuitionistically unacceptable, because there are infinitely many even numbers greater than 2 and it is impossible to try them all one by one and decide of each whether or not it is the sum of two prime numbers. An intuitionist would accept the disjunction only after a proof had been given of one or other of the two disjoined propositions—and in the present state of mathematical knowledge it is not certain that this can be done (it is not certain that the mathematical problem involved is solvable). If, however, we replace "greater than 2" by "greater than 2 and less than 1,000,000,000," the resulting disjunction becomes intuitionistically acceptable, since the number of numbers involved is then finite.

The intuitionistic rejection of the law of excluded middle is not to be understood as an assertion of the negation of the law of excluded middle; on the contrary, Brouwer asserts the negation of the negation of the law of excluded middle, i.e., $\sim\sim[p \vee \sim p]$. Still less is the intuitionistic rejection of the law of excluded middle to be understood as the assertion of the existence of a third truth-value intermediate between truth and falsehood.

The rejection of the law of excluded middle carries with it the rejection of various other laws of classical propositional calculus and functional calculus of first order, including the law of double negation (and hence the method of indirect proof). In general the double negation of a proposition is weaker than the proposition itself; but the triple negation of a proposition is equivalent to its single negation. Noteworthy also is the rejection of $\sim(x)F(x) \supset (Ex)\sim F(x)$; but the reverse implication is valid. (The sign $\supset$ here does not denote material implication, but is a distinct primitive symbol of implication.) —*A.C.*

L. E. J. Brouwer, *De onbetrouwbaarheid der logische principes*, Tijdschrift voor Wijsbegeerte, vol. 2 (1908), pp. 152–158; reprinted in Brouwer's *Wiskunde, Waarheid, Werkelijkheid*, Groningen, 1919. L. E. J. Brouwer, *Intuitionism and formalism*, English translation by A. Dresden, Bulletin of the American Mathematical Society, vol. 20 (1913), pp. 81–96. H. Weyl, *Consistency in mathematics*, The Rice Institute Pamphlet, vol. 16 (1929), pp. 245–265. A. Heyting, *Mathematische Grundlagenforschung, Intuitionismus, Beweistheorie*, Berlin, 1934.

Lambert, J. H.: (1728–1777) Lambert is known also for important contributions to mathematics, and astronomy; also for his work in logic, in particular his (unsuccessful, but historically significant) attempts at construction of a mathematical or symbolic logic. Cf. C. I. Lewis, *Survey of Symbolic Logic*. —*A.C.*

Leading principle: The general statement of the validity of some particular form of valid inference (see *Logic, formal*) may be called its *leading principle*.

Or the name may be applied to a proposition or sentence of logic corresponding to a certain form of valid inference. E.g., the law of *exportation* (q.v.) may be called the *leading principle* of the form of valid inference known as exportation. —*A.C.*

Lemma: In mathematics, a theorem proved for the sake of its use in proving another theorem. The name is applied especially in cases where the lemma ceases to be of interest in itself after proof of the theorem for the sake of which it was introduced. —*A.C.*

Limit: We give here only some of the most elementary mathematical senses of this word, in connection with real numbers. (Refer to the articles *Number* and *Continuity*.)

The *limit* of an infinite sequence of real numbers $a_1, a_2, a_3, \dots$ is said to be (the real number) b if for *every* positive real number ϵ there is a positive integer N such that the difference between b and a_n is less than ϵ whenever n is greater than N. (By the difference between b and a_n is here meant the non-negative difference, i.e., $b - a_n$ if b is greater than a_n, $a_n - b$ if b is less than a_n, and 0 if b is equal to a_n.)

Let f be a monadic function for which the range of the independent variable and the range of the dependent variable both consist of real numbers; let b and c be real numbers; and let g be the monadic function so determined that $g(c) = b$ and $g(x) = f(x)$ if x is different from c. (The range of the independent variable for g is thus the same as that for f, with the addition of the real number c if not already included.) The *limit* of $f(x)$ as x approaches c is said to be b if g is continuous at c.—More briefly but less accurately, the *limit* of $f(x)$ as x approaches c is the value which must be assigned to f for the argument c in order to make it continuous at c.

The *limit* of $f(x)$ as x approaches *infinity* is said to be b, if the *limit* of $h(x)$ as x approaches 0 is b, where h is the function so determined that $h(x) = f(1/x)$.

In connection with the infinite sequence of real numbers $a_1, a_2, a_3, \dots$, a monadic function a may be introduced for which the range of the independent variable consists of the positive integers $1, 2, 3, \dots$ and $a(1) = a_1$, $a(2) = a_2$, $a(3) = a_3, \dots$. It can then be shown that the limit of the infinite sequence as above defined is the same as the limit of $a(x)$ as x approaches infinity.

(Of course it is not meant to be implied in the preceding that the limit of an infinite sequence or of a function always exists. In particular cases it may happen that there is no limit of an infinite sequence, or no limit of $f(x)$ as x approaches c, etc.)

—*A.C.*

Logic, formal: Investigates the structure of propositions and of deductive reasoning by a method which abstracts from the *content* of propositions which come under consideration and deals only with their logical *form*. The distinction between form and content can be made definite with the aid of a particular language or symbolism in which propositions are expressed, and the formal method can then be characterized by the fact that it deals with the objective form of *sentences* which express propositions and provides in these concrete terms criteria of meaningfulness and validity of inference. This formulation of the matter presupposes the selection of a particular language which is to be regarded as logically exact and free from the ambiguities and irregularities of structure which appear in English (or other languages of everyday use)—i.e., it makes the distinction between form and content relative to the choice of a language. Many logicians prefer to postulate an abstract form for propositions themselves, and to characterize the logical exactness of a language by the uniformity with which the concrete form of its sentences reproduces or parallels the form of the propositions which they express. At all events it is practically necessary to introduce a special logical language, or symbolic notation, more exact than ordinary English usage, if topics beyond the most elementary are to be dealt with (see *Logistic system*, and *Semiotic*).

Concerning the distinction between form and content see further the articles *Formal*, and *Syntax, logical*.

1. The propositional calculus formalizes the use of the sentential connectives *and, or, not, if... then*. Various systems of notation are current of which we here adopt a particular one for purposes of exposition. We use juxtaposition to denote conjunction ("pq" to mean "p and q"), the sign $\vee$ to denote inclusive disjunction ("$p \vee q$" to mean "p or q or both"), the sign $+$ to denote exclusive disjunction ("$p+q$" to mean "p or q but not both"), the sign $\sim$ to denote negation ("$\sim p$" to mean "not p"), the sign $\supset$ to denote the conditional ("$p \supset q$" to mean "if p then q", or "not both p and not-q"), the sign $\equiv$ to denote the biconditional ("$p \equiv q$" to mean "p if and only if q," or "either p and q or not-p and not-q"), and the sign $|$ to denote alternative denial ("$p \mid q$" to mean "not both p and q").—The word *or* is ambiguous in ordinary English usage between inclusive disjunction and exclusive disjunction, and distinct notations are accordingly provided for the two meanings of the word. The notations "$p \supset q$" and "$p \equiv q$" are sometimes read as "p implies q" and "p is equivalent to q" respectively. These readings must, however, be used with caution, since the terms implication and equivalence are often used in a sense which involves some relationship between the logical forms of the propositions (or the sentences) which they connect, whereas the validity of $p \supset q$ and of $p \equiv q$ requires no such relationship. The connective $\supset$ is also said to stand for "material implication," distinguished from *formal implication* (§3 below) and *strict implication* (q.v.). Similarly the connective $\equiv$ is said to stand for "material equivalence."

It is possible in various ways to define some of the sentential connectives names above in terms of others. In particular, if the sign of alternative denial is taken as primitive, all the other connectives can be defined in terms of this one. Also, if the signs of negation and inclusive disjunction are taken as primitive, all the others can be defined in terms of these; likewise if the signs of negation and conjunction are taken as primitive. Here, however, for reasons of naturalness and symmetry, we prefer to take as primitive the three connectives, denoting negation, conjunction, and inclusive disjunction. The remaining ones are then defined as follows:

$$\begin{aligned} \mathrm{A} \mid \mathrm{B} &\longrightarrow \quad {\sim}\mathrm{A} \vee {\sim}\mathrm{B}. \\ \mathrm{A} \supset \mathrm{B} &\longrightarrow \quad {\sim}\mathrm{A} \vee \mathrm{B}. \\ \mathrm{A} \equiv \mathrm{B} &\longrightarrow \quad [\mathrm{B} \supset \mathrm{A}][\mathrm{A} \supset \mathrm{B}]. \\ \mathrm{A} + \mathrm{B} &\longrightarrow \quad [{\sim}\mathrm{B}]\mathrm{A} \vee [{\sim}\mathrm{A}]\mathrm{B}. \end{aligned}$$

The capital roman letters here denote arbitrary formulas of the propositional calculus (in the technical sense defined below) and the arrow is to be read "stands for" or "is an abbreviation for."

Suppose that we have given some specific list of propositional symbols, which may be infinite in number, and to which we shall refer as the *fundamental propositional symbols*. These are not necessarily single letters or characters, but may be expressions taken from any language or system of notation; they may denote particular propositions, or they may contain variables and denote ambiguously any proposition of a certain form or class. Certain restrictions are also necessary upon the way in which the fundamental propositional symbols can contain square brackets [ ]; for the present purpose it will suffice to suppose that they do not contain square brackets at all, although they may contain parentheses or other kinds of brackets. We call *formulas* of the propositional calculus (relative to the given list of fundamental propositional symbols) all the expressions determined by the four following rules: (1) all the fundamental propositional symbols are formulas; (2) if A is a formula, $\sim$[A] is a formula; (3) if A and B are formulas [A][B] is a formula; (4) if A and B are formulas [A] $\vee$ [B] is a formula. The formulas of the propositional calculus as thus defined will in general contain more brackets than are necessary for clarity or freedom from ambiguity; in practice we omit superfluous brackets and regard the shortened expressions as abbreviations for the full formulas. It will be noted also that, if A and B are formulas, we regard [A] | [B], [A] $\supset$ [B], [A] $\equiv$ [B], and [A] + [B], not as formulas, but as abbreviations for certain formulas in accordance with the above given definitions.

In order to complete the setting up of the propositional calculus as a *logistic system* (q.v.) it is necessary to state primitive formulas and primitive rules of inference. Of the many possible ways of doing this we select the following.

If A, B, C are any formulas, each of the seven following formulas is a primitive formula:

$$\begin{array}{ll} [\mathrm{A} \vee \mathrm{A}] \supset \mathrm{A}. & \mathrm{A} \supset [\mathrm{B} \supset \mathrm{AB}]. \\ \mathrm{A} \supset [\mathrm{A} \vee \mathrm{B}]. & \mathrm{AB} \supset \mathrm{A}. \\ [\mathrm{A} \vee \mathrm{B}] \supset [\mathrm{B} \vee \mathrm{A}]. & \mathrm{AB} \supset \mathrm{B}. \\ \multicolumn{2}{c}{[\mathrm{A} \supset \mathrm{B}] \supset [[\mathrm{C} \vee \mathrm{A}] \supset [\mathrm{C} \vee \mathrm{B}]].} \end{array}$$

(The complete list of primitive formulas is thus infinite, but there are just seven possible forms of primitive formulas as above.) There is one primitive rule of inference, as follows: *Given* A *and* A $\supset$ B *to infer* B. This is the inference known as *modus ponens* (see below, §2).

The *theorems* of the propositional calculus are the formulas which can be derived from the primitive formulas by a succession of applications of the primitive rule of inference. In other words, (a) the primitive formulas are theorems, and (b) if A and A $\supset$ B are theorems then B is a theorem. An inference from premisses $A_1, A_2, \ldots, A_n$ to a conclusion B is a *valid inference* of the propositional calculus if B becomes a theorem upon adding $A_1, A_2, \ldots, A_n$ to the list of primitive formulas. In other words, (a) the inference from $A_1, A_2, \ldots, A_n$ to B is a valid inference if B is either a primitive formula or one of the formulas $A_1, A_2, \ldots, A_n$, and (b) if the inference from $A_1, A_2, \ldots, A_n$ to C and the inference from $A_1, A_2, \ldots, A_n$ to $C \supset B$ are both valid inferences then the inference from $A_1, A_2, \ldots, A_n$ to B is a valid inference. It can be proved that the inference from $A_1, A_2, \ldots, A_n$ to B is a valid inference of the propositional calculus if (obviously), and only if (the *deduction theorem*), $[A_1 \supset [A_2 \supset \cdots [A_n \supset B] \cdots]]$ is a theorem of the propositional calculus.

The reader should distinguish between theorems *about* the propositional calculus—the deduction theorem, the principles of duality (below), etc.—and theorems *of* the propositional calculus in the sense just defined. It is convenient to use such words as *theorem, premiss, conclusion* both for propositions (in whatever language expressed) and for formulas representing propositions in some fixed system or calculus.

In the foregoing the list of fundamental propositional symbols has been left unspecified. A case of special importance is the case that the fundamental propositional symbols are an infinite list of variables $p, q, r, \ldots$, which may be taken as representing ambiguously any proposition whatever—or any proposition of a certain class fixed in advance (the class should be closed under the operations of negation, conjunction, and inclusive disjunction). In this case we speak of the *true propositional calculus*, and refer to the other cases as *applied propositional calculus* (although the application may be to something as abstract in character as the pure propositional calculus itself, as, e.g., in the case of the pure functional calculus of first order (§3), which contains an applied propositional calculus).

In formulating the pure propositional calculus the primitive formulas may (if desired) be reduced to a finite number, e.g., to the seven listed above with A, B, C taken to be the particular variables p, q, r. A second primitive rule of inference, the *rule of substitution*, is then required, allowing the inference from a formula A to the formula obtained from A by substituting a formula B for a particular variable in A (the same formula B must be substituted for all occurrences of that variable in A). The definition of a theorem is then given in the same way as before, allowing for the additional primitive rule; the definition of a valid inference must, however, modified.

In what follows (to the end of §1) we shall, for convenience of statement, confine attention to the case of the pure propositional calculus. Similar statements hold, with minor modifications, for the general case.

The formulation which we have given provides a means of *proving* theorems of the propositional calculus, the proof consisting of an explicit finite sequence of formulas, the last of which is the theorem proved, and each of which is either a primitive formula or inferable from preceding formulas by a single application of the rule of inference (or one of the rules of inference, if the alternative formulation of the pure propositional calculus employing the rule of substitution is adopted). The test whether a given finite sequence of formulas is a proof of the last formula of the sequence is *effective*—we have the means of always determining of a given formula whether it is a primitive formula, and the means of always determining of a given formula whether it is inferable form a given finite list of formulas by a single application of *modus ponens* (or substitution). Indeed our formulation would not be satisfactory otherwise. For in the contrary case a proof would not necessarily carry conviction, the proposer of a proof could fairly be asked to give a proof that it was a proof—in short the formal analysis of what constitutes a proof (in the sense of a cogent demonstration) would be incomplete.

However, the test whether a given formula is a theorem, by the criterion that it is a theorem if a proof of it exists, is not effective—since failure to find a proof upon search might mean lack of ingenuity rather than non-existence of a proof. The problem to give an effective test by means of which it can always be determined whether a given formula is a theorem is the *decision problem* of the propositional calculus. This problem can be solved either by the process of reduction of a formula to *disjunctive normal form*, or by the *truth-table decision procedure*. We state the latter in detail.

The three primitive connectives (and consequently all connectives definable from them) denote *truth-functions*—i.e., the *truth-value* (truth or falsehood) of each of the propositions $\sim p$, pq, and $p \vee q$ is uniquely determined by the truth-values of p and q. In fact, $\sim p$ is true if p is false and false if p is true; pq is true if p and q are both true, false otherwise; $p \vee q$ is false if p and q are both false, true otherwise. Thus given a formula of the (pure) propositional calculus and an assignment of a truth-value to each of the variables appearing, we can reckon out by a mechanical process the truth-value to be assigned to the entire formula. If, for all possible assignments of truth-values to the variables appearing, the calculated truth-value corresponding to the entire formula is truth, the formula is said to be a *tautology*. The test whether a formula is a tautology is effective, since in any particular case the total number of different assignments of truth-values to the variables is finite, and the calculation of the truth-value corresponding to the entire formula can be carried out separately for each possible assignment of truth-values to the variables.

Now it is readily verified that all the primitive formulas are tautologies, and that for the rule of *modus ponens* (and the rule of substitution) the property holds that if the premisses of the inference are tautologies the conclusion must be a tautology. It follows that every theorem of the propositional calculus is a tautology. By a more difficult argument it can be shown also that every tautology is a theorem. Hence the test whether a formula is a tautology provides a solution of the decision problem of the propositional calculus.

As corollaries of this we have proofs of the *consistency* of the propositional calculus (if A is any formula, A and $\sim$A cannot both be tautologies and hence cannot both be theorems) and of the *completeness* of the propositional calculus (it can be shown that if any formula not already a theorem, and hence not a tautology, is added to the list of primitive formulas, the calculus becomes inconsistent on the basis of the two rules substitution and *modus ponens*).

As another corollary of this, or otherwise, we obtain also the following theorem about the propositional calculus: If $A \equiv B$ is a theorem, and D is the result of replacing a particular occurrence of A by B in the formula C, then the inference from C to D is a valid inference.

The *dual* of a formula C of the propositional calculus is obtained by interchanging conjunction and disjunction throughout the formula, i.e., by replacing AB everywhere by $A \vee B$, and $A \vee B$ by AB. Thus, e.g., the dual of the formula $\sim[pq \vee \sim r]$ is the formula $\sim[[p \vee q]\sim r]$. In forming the dual of a formula which is expressed with the aid of the defined connectives, $|$, $\supset$, $\equiv$, $+$, it is convenient to remember that the effect of interchanging conjunction and (inclusive) disjunction is to replace $A \mid B$ by $\sim A \sim B$, to replace A ⊃ B by ~A B, and to interchange ≡ and +.

It can be shown that the following *principles of duality* hold in the propositional calculus (where A^* and B^* denote the duals of the formulas A and B respectively): (1) if A is a theorem, then $\sim A^*$ is a theorem; (2) if A ⊃ B is a theorem, then $B^* \supset A^*$ is a theorem; (3) if A ≡ B is a theorem, then $A^* \equiv B^*$ is a theorem.

Special names have been given to certain particular theorems and forms of valid inference of the propositional calculus. Besides §2 following, see: *Absorption; Affirmation of the consequent; Assertion; Associative law; Commutative law; Composition; Contradiction, law of; De Morgan's laws; Denial of the antecedent; Distributive law; Double negation, law of; Excluded middle, law of; Exportation; Hauber's law; Identity, law of; Importation; Peirce's law; Proof by cases; Reductio ad absurdum; Reflexivity; Tautology; Transitivity; Transposition.*

Names given to particular theorems of the propositional calculus are usually thought of as applying to laws embodied in the theorems rather than to the theorems as formulas; hence, in particular, the same name is applied to theorems differing only by alphabetical changes of the variables appearing; and frequently the name used for a theorem is used also for one or more forms of valid inference associated with the theorem. Similar remarks apply to names given to particular theorems of the functional calculus of first order, etc.

Whitehead and Russell, *Principia Mathematica*, 2nd edn., vol. 1, Cambridge, England, 1925. E. L. Post, *Introduction to a general theory of elementary propositions*, American Journal of Mathematics, vol. 43 (1921), pp. 163–185. W. V. Quine, *Elementary Logic*, Boston and New York, 1941. J. Herbrand, *Recherches sur la Théorie de la Démonstration*, Warsaw, 1930. Hilbert and Ackermann, *Grundzüge der theoretischen Logik*, 2nd edn., Berlin, 1938. Hilbert and Bernays, *Grundlagen der Mathematik*, vol. 1, Berlin, 1934; also Supplement III to vol. 2, Berlin, 1939.

2. Hypothetical syllogism, Disjunctive syllogism, Dilemma are names traditionally given to certain forms of inference, which may be identified as follows with certain particular forms of valid inference of the propositional calculus (see §1).

The hypothetical syllogism has two kinds or moods. *Modus ponens* is the inference from a major premiss A ⊃ B and a minor premiss A to the conclusion B. *Modus tollens* is the inference from a major premiss A ⊃ B and a minor premiss ~B to the conclusion ~A.

The disjunctive syllogism has also two moods. *Modus tollendo ponens* is any one of the four following forms of inference:

from A ∨ B and ~B to A;
from A ∨ B and ~A to B;
from A + B and ~B to A;
from A + B and ~A to B.

Modus ponendo tollens is either of the following forms of inference:

from A + B and A to ~B;
from A + B and B to ~A.

In each case the first premiss named is the major premiss and the second one the minor premiss.

Of the dilemma four kinds are distinguished. The simple constructive dilemma has two major premisses A ⊃ C and B ⊃ C, minor premiss A ∨ B, conclusion C. The simple destructive dilemma has two major premisses A ⊃ B and A ⊃ C, minor premiss ~B ∨ ~C, conclusion ~A. The complex constructive dilemma has two major premisses A $\supset$ B and C $\supset$ D, minor premiss A $\vee$ C, conclusion B $\vee$ D. The complex destructive dilemma has two major premisses A $\supset$ B and C $\supset$ D, minor premiss $\sim$B $\vee$ $\sim$D, conclusion $\sim$A $\vee$ $\sim$C. (Since the conclusion of a complex dilemma must involve inclusive disjunction, it seems that the traditional account is best rendered by employing inclusive disjunction throughout.)

The inferences from A $\supset$ B and C $\supset$ A to C $\supset$ B, and from A $\supset$ B and C $\supset$ $\sim$B to C $\supset$ $\sim$A are called pure hypothetical syllogisms, and the above simpler forms of the hypothetical syllogism are then distinguished as mixed hypothetical. Some recent writers apply the names, *modus ponens* and *modus tollens* respectively, also to these two forms of the pure hypothetical syllogism. Other variations of usage or additional forms are also found. Some writers include under these heads forms of inference which belong to the functional calculus of first order rather than to the propositional calculus.

F. Ueberweg, *System der Logik*, 4th edn., Bonn, 1874. H. W. B. Joseph, *An Introduction to Logic*, 2nd edn., Oxford, 1916. R. M. Eaton, *General Logic*, New York, 1931. S. K. Langer, *An Introduction to Symbolic Logic*, 1937, Appendix A.

3. THE FUNCTIONAL CALCULUS OF FIRST ORDER is the next discipline beyond the propositional calculus, according to the usual treatment. It is the first step towards the hierarchy of types (§6) and deals in addition to unanalyzed propositions, with *propositional functions* (q.v.) of the lowest order. It employs the sentential connectives of §1, and in addition the universal *quantifier* (q.v.), written (X) where X is any individual variable, and the existential *quantifier*, written (EX) where X is any individual variable. (The E denoting existential quantification is more often written inverted, as by Peano and Whitehead–Russell, but we here adopt the typographically more convenient usage, which also has sanction.)

For the interpretation of the calculus we must presuppose a certain domain of *individuals*. This may be any well-defined non-empty domain, within very wide limits. Different possible choices of the domain of individuals lead to different interpretations of the calculus.

In order to set the calculus up formally as a logistic system, we suppose that we have given four lists of symbols, as follows: (1) an infinite list of *individual variables* $x, y, z, t, x', y', z', t', x'', \ldots,$ which denote ambiguously any individual; (2) a list of *propositional variables* $p, q, r, s, p', \ldots,$ representing ambiguously any proposition of a certain appropriate class; (3) a list of *functional variables* $F_1, G_1, H_1, \ldots,$ $F_2, G_2, H_2, \ldots,$ $F_3, G_3, H_3, \ldots,$ a variable with subscript n representing ambiguously any n-adic propositional function of individuals; (4) a list of *functional constants*, which denote particular propositional functions of individuals. There shall be an effective notational criterion associating with each functional constant a positive integer n, the functional constant denoting an n-adic propositional function of individuals. One or more of the lists (2), (3), (4) may be empty, but not both (3) and (4) shall be empty. The list (1) is required to be infinite, and the remaining lists may, some or all of them, be infinite. Finally, no symbol shall be duplicated either by appearing twice in the same list or by appearing in two different lists; and no functional constant shall contain braces (or either a left brace or a right brace) as a constituent part of the symbol.

When (2) and (3) are complete—i.e., contain all the variables indicated above (an infinite number of propositional variables and for each positive integer n an infinite number of functional variables with subscript n)—and (4) is empty, we shall speak of the *pure* functional calculus of first order. When (2) and (3) are empty and (4) is not empty, we shall speak of a *simple applied* functional calculus of first order.

Functional variables and functional constants are together called *functional symbols* (the adjective *functional* being here understood to refer to *propositional* functions). Functional symbols are called n-adic if they are either functional variables with subscript n or functional constants denoting n-adic propositional functions of individuals. The *formulas* of the functional calculus of first order (relative to the given lists of symbols (1), (2), (3), (4)) are all the expressions determined by the eight following rules: (1) all the propositional variables are formulas; (2) if F is a monadic functional symbol and X is an individual variable, $\{F\}(X)$ is a formula; (3) if F is an n-adic functional symbol and $X_1, X_2, \ldots, X_n$ are individual variables (which may or may not be all different), $\{F\}(X_1, X_2, \ldots, X_n)$ is a formula; (4) if A is a formula, $\sim[A]$ is a formula; (5) if A and B are formulas, [A][B] is a formula; (6) if A and B are formulas, $[A] \vee [B]$ is a formula; (7) if A is a formula and X is an individual variable, (X)[A] is a formula; (8) if A is a formula and X is an individual variable, $(E\mathrm{X})[A]$ is a formula. In practice, we omit superfluous brackets and braces (but not parentheses) in writing formulas, and we omit subscripts on functional variables in cases where the subscript is sufficiently indicated by the form of the formula in which the functional variable appears. The sentential connectives $|, \supset, \equiv, +$, are introduced as abbreviations in the same way as in §1 for the propositional calculus. We make further the following definitions, which are also to be construed as abbreviations, the arrow being read "stands for":

$$[A] \supset_X [B] \rightarrow (X)[[A] \supset [B]].$$
$$[A] \equiv_X [B] \rightarrow (X)[[A] \equiv [B]].$$
$$[A] \wedge_X [B] \rightarrow (EX)[[A][B]].$$

(Here A and B are any formulas, and X is any individual variable. Brackets may be omitted when superfluous.) If F and G denote monadic propositional functions, we say that $F(X) \supset_X G(X)$ expresses *formal implication of* the function G by the function F, and $F(X) \equiv_X G(X)$ expresses *formal equivalence* of the two functions (the adjective *formal* is perhaps not well chosen here but has become established in use).

A *sub-formula* of a given formula is a consecutive constituent part of the given formula which is itself a formula. An occurrence of an individual variable X in a formula is a *bound* occurrence of X if it is an occurrence in a sub-formula of either of the forms (X)[A] or $(E\mathrm{X})$[A]. Any other occurrence of a variable in a formula is a *free* occurrence. We may thus speak of the *bound variables* and the *free variables* of a formula. (Whitehead and Russell, following Peano, use the terms *apparent variables* and *real variables*, respectively, instead of *bound variables* and *free variables*.)

If A, B, C are any formulas, each of the seven following formulas is a primitive formula:

$$[A \vee A] \supset A. \qquad A \supset [B \supset AB].$$
$$A \supset [A \vee B]. \qquad AB \supset A.$$
$$[A \vee B] \supset [B \vee A]. \qquad AB \supset B.$$
$$[A \supset B] \supset [[C \vee A] \supset [C \vee B]].$$

If X is any individual variable, and A is any formula not containing a free occurrence of X, and B is any formula, each of the two following formulas is a primitive formula:

$$[\mathrm{A} \supset_{\mathrm{X}} \mathrm{B}] \supset [\mathrm{A} \supset (\mathrm{X})\mathrm{B}].$$
$$[\mathrm{B} \supset_{\mathrm{X}} \mathrm{A}] \supset [(E\mathrm{X})\mathrm{B} \supset \mathrm{A}].$$

If X and Y are any individual variables (the same or different), and A is any formula such that no free occurrence of X in A is in a sub-formula of the form (Y)[C], and B is the formula resulting from the substitution of Y for all the free occurrences of X in A, each of the two following formulas is a primitive formula:

$$(\mathrm{X})\mathrm{A} \supset \mathrm{B}. \qquad \mathrm{B} \supset (E\mathrm{X})\mathrm{A}.$$

There are two primitive rules of inference: (1) *Given* A *and* A ⊃ B *to infer* B (the rule of *modus ponens*). (2) *Given* A *to infer* (X)A, *where* X *is any individual variable* (the rule of *generalization*). In applying the rule of generalization, we say that the variable X is *generalized upon*.

The *theorems* of the functional calculus of first order are the formulas which can be derived from the primitive formulas by a succession of applications of the primitive rules of inference. An inference from premisses $A_1, A_2, \ldots, A_n$ to a conclusion B is a *valid inference* of the functional calculus of first order if B becomes a theorem upon adding $A_1, A_2, \ldots, A_n$ to the list of primitive formulas and at the same time restricting the rule of generalization by requiring that the variable generalized upon shall not be any one of the free individual variables of $A_1, A_2, \ldots, A_n$. It can be proved that the inference from $A_1, A_2, \ldots, A_n$ to B is a valid inference of the functional calculus of first order if (obviously), and only if (the *deduction theorem*), $[A_1 \supset [A_2 \supset \cdots [A_n \supset B] \cdots]]$ is a theorem of the functional calculus of first order.

It can be proved that if A ≡ B is a theorem, and D is the result of replacing a particular occurrence of A by B in the formula C, then the inference from C to D is a valid inference.

The *consistency* of the functional calculus of first order can also be proved without great difficulty.

The *dual* of a formula is obtained by interchanging conjunction and (inclusive) disjunction throughout and at the same time interchanging universal quantification and existential quantification throughout. (In doing this the different symbols, e.g., functional constants, although they may consist of several characters in succession rather than a single character, shall be treated as units, and no change shall be made inside a symbol. A similar remark applies at all places where we speak of occurrences of a particular symbol or sequence of symbols in a formula, and the like.) It can be shown that the following *principles of duality* hold (where A^* and B^* denote the duals of the formulas A and B respectively): (1) if A is a theorem, then $\sim A^*$ is a theorem; (2) if A ⊃ B is a theorem, then $B^* \supset A^*$ is a theorem; (3) if A ≡ B is a theorem, then $A^* \equiv B^*$ is a theorem.

A formula is said to be in *prenex normal form* if all the quantifiers which it contains stand together at the beginning, unseparated by negations (or other sentential connectives), and the *scope* of each quantifier (i.e., the extent of the bracket [ ] following the quantifier) is to be the end of the entire formula. In the case of a formula in prenex normal form, the succession of quantifiers at the beginning is called the *prefix*; the remaining portion contains no quantifiers and is the *matrix* of the formula. It can be proved that for every formula A there is a formula B in prenex normal form such that $A \equiv B$ is a theorem; and B is then called a *prenex normal form of* A.

A formula of the pure functional calculus of first order which contains no free individual variables is said to be *satisfiable* if it is possible to determine the underlying non-empty domain of individuals and to give meanings to the propositional and functional variables contained—namely to each propositional variable a meaning as a particular proposition and to each n-adic propositional function of individuals (of the domain in question)—in such a way that (under the accepted meanings of the sentential connectives, the quantifiers, and application of function to argument) the formula becomes *true*. The meaning of the last word, even for abstract, not excluding infinite, domains, must be presupposed—a respect in which this definition differs sharply from most others made in this article.

It is not difficult to find examples of formulas A, containing no free variables, such that both A and ~A are satisfiable. A simple example is the formula $(x)F(x)$. More instructive is the following example,

$$[(x)(y)(z)[[\sim F(x,x)][F(x,y)F(y,z) \supset F(x,z)]]][(x)(Ey)F(x,y)],$$

which is satisfiable in an infinite domain of individuals but not in any finite domain—the negation is satisfiable in any non-empty domain.

It can be shown that all theorems A of the pure functional calculus of first order which contain no free individual variables have the property that ~A is not satisfiable. Hence the pure functional calculus of first order is not complete in the strong sense in which the pure propositional calculus is complete. Gödel has shown that the pure functional calculus of first order is complete in the weaker sense that if a formula A contains no free individual variables and ~A is not satisfiable then A is a theorem.

The *decision problem* of the pure functional calculus of first order has two forms (1) the so-called proof-theoretic decision problem, to find an effective test (decision procedure) by means of which it can always be determined whether a given formula is a theorem; (2) the so-called set-theoretic decision problem, to find an effective test by means of which it can always be determined whether a given formula containing no free individual variables is satisfiable. It follows from Gödel's completeness theorem that these two forms of the decision problem are equivalent: a solution of either would lead immediately to a solution of the other.

Church has proved that the decision problem of the pure functional calculus of first order is unsolvable. Solutions exist, however, for several important special cases. In particular a decision procedure is known for the case of formulas containing only *monadic* function variables (this would seem to cover substantially everything considered in traditional formal logic prior to the introduction of the modern logic of relations).

Finally we mention a variant form of the functional calculus of first order, the *functional calculus of first order with equality*, in which the list of functional constants includes the dyadic functional constant =, denoting equality or identity of individuals. The notation $[X] = [Y]$ is introduced as an abbreviation for $\{=\}(X, Y)$, and primitive formulas are added as follows to the list already given: if X is any individual variable, $X = X$ is a primitive formula; if X and Y are any individual variables, and B results from the substitution of Y for a particular free occurrence of X in A, which is not in a sub-formula of A of the form $(Y)[C]$, then $[X = Y] \supset [A \supset B]$ is a primitive formula.

We speak of the *pure* functional calculus of first order with equality when the lists of propositional variables and functional variables are complete and the only functional constant is =; we speak of a *simple applied* functional calculus of first order with equality when the lists of propositional variables and functional variables are empty.

The addition to the functional calculus of first order of *individual constants* (denoting particular individuals) is not often made—unless symbols for functions from individuals to individuals (so-called "mathematical" or "descriptive" functions) are to be added at the same time. Such an addition is, however, employed in the two following sections as a means of representing certain forms of inference of traditional logic. The addition is really non-essential, and requires only minor changes in the definition of a formula and the list of primitive formulas (allowing the alternative of individual constants at certain places where the above given formulation calls for free individual variables).

Whitehead and Russell, *Principia Mathematica*, 2nd edn., vol. 1, Cambridge, England, 1925. J. Herbrand, *Recherches sur la Théorie de la Démonstration*, Warsaw, 1930. K. Gödel, *Die Vollständigkeit der Axiome der logischen Funktionenkalküls*, Monatshefte für Mathematik und Physik, vol. 37 (1930), pp. 349–360. Hilbert and Ackermann, *Grundzüge der theoretischen Logik*, 2nd edn., Berlin, 1938. Hilbert and Bernays, *Grundlagen der Mathematik*, vol. 1, Berlin, 1934, and vol. 2, Berlin, 1939.

4. OPPOSITION, IMMEDIATE INFERENCE. The four traditional kinds of categorical propositions—all S is P, no S is P, some S is P, some S is not P—customarily designated by the letters *A*, *E*, *I*, *O* respectively—may conveniently be represented in the functional calculus of first order (§3) by the four forms $S(x) \supset_x P(x)$, $S(x) \supset_x \sim P(x)$, $S(x) \wedge_x P(x)$, $S(x) \wedge_x \sim P(x)$, S and P being taken as functional constants. (For brevity, we shall use the notations S, P, $S(x) \supset_x P(x)$, etc., alike for certain formulas and for the propositional functions or propositions expressed by these formulas.)

This representation does not reproduce faithfully all particulars of the traditional account. The fact is that the traditional doctrine, having grown up from various sources and under an inadequate formal analysis, is not altogether coherent or even self-consistent. We here select what seems to be the best representation, and simply note the four following points of divergence:

(1) We have defined the connectives $\supset_x$ and $\wedge_x$ in terms of universal and existential quantification, whereas the traditional account might be thought to be more closely reproduced if they were taken as primitive notations. (It would, however, not be difficult to reformulate the functional calculus of first order so that those connectives would be primitive and the usual quantifiers defined in terms of them.)

(2) The traditional account associates the negation in *E* and *O* with the *copula* (q.v.), whereas the negation symbol is here prefixed to the sub-formula $P(x)$. (Notice that this sub-formula represents ambiguously a proposition and that, in fact, the notation of the functional calculus of first order provides for applying negation only to propositions.)

(3) The traditional account includes under *A* and *E*, respectively, also (propositions denoted by) P(A) and ~P(A), where A is an individual constant. These *singular* propositions are ignored in our account of opposition and immediate inference, but will appear in §5 as giving variant forms of certain syllogisms.

(4) Some aspects of the traditional account require that *A* and *E* be represented as we have here, others that they be represented by $[(Ex)S(x)][S(x) \supset_x P(x)]$ and $[(Ex)S(x)][S(x) \supset_x \sim P(x)]$ respectively. The question concerning the choice between

these two interpretations is known as the problem of *existential import* of propositions. We prefer to introduce $(Ex)S(x)$ as a separate premiss at those places where it is required.

Given a fixed subject S and a fixed predicate P, we have according to the *square of opposition*, that *A* and *O* are *contradictory*, *E* and *I* are *contradictory*, *A* and *E* are *contrary*, *I* and *O* are *subcontrary*, *A* and *I* are *subaltern*, *E* and *O* are *subaltern*. The two propositions in a contradictory pair cannot be both true and cannot be both false (one is the exact negation of the other). The two propositions in a subaltern pair are so related that the first one, together with the premiss $(Ex)S(x)$, implies the second (*subalternation*). Under the premiss $(Ex)S(x)$, the contrary pair, *A*, *E*, cannot be both true, and the subcontrary pair *I*, *O*, cannot be both false.

Simple conversion of a proposition, *A*, *E*, *I* or *O*, consists in interchanging S and P without other change. Thus the converse of $S(x) \supset_x P(x)$ is $P(x) \supset_x S(x)$, and the converse of $S(x) \supset_x \sim P(x)$ is $P(x) \supset_x \sim S(x)$. In mathematics the term *converse* is used primarily for the simple converse of a proposition *A*; loosely also for any one of a number of transformations similar to this (e.g., $F(x)G(x) \supset_x H(x)$ may be said to have the converse $F(x)H(x) \supset_x G(x)$). Simple conversion of a proposition is a valid inference, in general, only in the case of *E* and *I*.

Conversion per accidens of a proposition *A*, i.e., of $S(x) \supset_x P(x)$, yields $P(x) \wedge_x S(x)$.

Obversion of a proposition *A*, *E*, *I*, or *O* consists in replacing P by a functional constant p which denotes the negation of the propositional function (property) denoted by P, and at the same time inserting $\sim$ if not already present or deleting it if present. Thus the obverse of $S(x) \supset_x P(x)$ is $S(x) \supset_x \sim p(x)$ (the obverse of "all men are mortal" is "no men are immortal"). The obverse of $S(x) \supset_x \sim P(x)$ is $S(x) \supset_x p(x)$; the obverse of $S(x) \wedge_x P(x)$ is $S(x) \wedge_x \sim p(x)$; the obverse of $S(x) \wedge_x \sim P(x)$ is $S(x) \wedge_x p(x)$.

The name *immediate inference* is given to certain inferences involving propositions *A*, *E*, *I*, *O*. These include obversion of *A*, *E*, *I*, or *O*, simple conversion of *E* or *I*, conversion *per accidens* of *A*, subalternation of *A*, *E*. The three last require the additional premiss $(Ex)S(x)$. Other immediate inferences (for which the terminology is not wholly uniform among different writers) may be obtained by means of sequences of these: e.g., given that all men are mortal we may take the obverse of the converse of the obverse and so infer that all immortals are non-men (called by some the contrapositive, by others the obverted contrapositive).

The immediate inferences not involving obversion can be represented as valid inferences in the functional calculus of first order, but obversion can be so represented only in an extended calculus embracing functional *abstraction* (q.v.). For the p used above in describing obversion is, in terms of abstraction,

$$\lambda x[\sim P(x)].$$

F. Ueberweg, *System der Logik*, 4th edn., Bonn, 1874. H. W. B. Joseph, *An Introduction to Logic*, 2nd edn., Oxford, 1916. R. M. Eaton, *General Logic*, New York, 1931. Bennett and Baylis, *Formal Logic*, New York, 1939.

5. Categorical syllogism is the name given to certain forms of valid inference (of the functional calculus of first order) which involve as premisses two (formulas representing) categorical propositions, having a term in common—the middle term. Using S, M, P as minor term, middle term, and major term, respectively, we give the

traditional classification into figures and moods. In each case we give the major premiss first, the minor premiss immediately after it, and the conclusion last; in some cases we give a third (existential) premiss which is suppressed in the traditional account. Because of the admission of singular propositions under the heads A, E, two different forms of valid inference appear in some cases under the same figure and mood—these singular forms are separately listed.

First Figure

Barbara: $M(x) \supset_x P(x)$, $S(x) \supset_x M(x)$, $S(x) \supset_x P(x)$.
Celarent: $M(x) \supset_x \sim P(x)$, $S(x) \supset_x M(x)$, $S(x) \supset_x \sim P(x)$.
Darii: $M(x) \supset_x P(x)$, $S(x) \wedge_x M(x)$, $S(x) \wedge_x P(x)$.
Ferio: $M(x) \supset_x \sim P(x)$, $S(x) \wedge_x M(x)$, $S(x) \wedge_x \sim P(x)$.

Second Figure

Cesare: $P(x) \supset_x \sim M(x)$, $S(x) \supset_x M(x)$, $S(x) \supset_x \sim P(x)$.
Camestres: $P(x) \supset_x M(x)$, $S(x) \supset_x \sim M(x)$, $S(x) \supset_x \sim P(x)$.
Festino: $P(x) \supset_x \sim M(x)$, $S(x) \wedge_x M(x)$, $S(x) \wedge_x \sim P(x)$.
Baroco: $P(x) \supset_x M(x)$, $S(x) \wedge_x \sim M(x)$, $S(x) \wedge_x \sim P(x)$.

Third Figure

Darapti: $M(x) \supset_x P(x)$, $M(x) \supset_x S(x)$, $(Ex)M(x)$, $S(x) \wedge_x P(x)$.
Disamis: $M(x) \wedge_x P(x)$, $M(x) \supset_x S(x)$, $S(x) \wedge_x P(x)$.
Datisi: $M(x) \supset_x P(x)$, $M(x) \wedge_x S(x)$, $S(x) \wedge_x P(x)$.
Felapton: $M(x) \supset_x \sim P(x)$, $M(x) \supset_x S(x)$, $(Ex)M(x)$, $S(x) \wedge_x \sim P(x)$.
Bocardo: $M(x) \wedge_x \sim P(x)$, $M(x) \supset_x S(x)$, $S(x) \wedge_x \sim P(x)$.
Feriso or Ferison: $M(x) \supset_x \sim P(x)$, $M(x) \wedge_x S(x)$, $S(x) \wedge_x \sim P(x)$.

Fourth Figure

Bamalip or Bramantip: $P(x) \supset_x M(x)$, $M(x) \supset_x S(x)$, $(Ex)P(x)$, $S(x) \wedge_x P(x)$.
Calemes or Camenes: $P(x) \supset_x M(x)$, $M(x) \supset_x \sim S(x)$, $S(x) \supset_x \sim P(x)$.
Dimatis or Dimaris: $P(x) \wedge_x M(x)$, $M(x) \supset_x S(x)$, $S(x) \wedge_x P(x)$.
Fesapo: $P(x) \supset_x \sim M(x)$, $M(x) \supset_x S(x)$, $(Ex)M(x)$, $S(x) \wedge_x \sim P(x)$.
Fresison: $P(x) \supset_x \sim M(x)$, $M(x) \wedge_x S(x)$, $S(x) \wedge_x \sim P(x)$.

Singular Forms in the First and Second Figures

Barbara: $M(x) \supset_x P(x)$, $M(A)$, $P(A)$.
Celarent: $M(x) \supset_x \sim P(x)$, $M(A)$, $\sim P(A)$.
Cesare: $P(x) \supset_x \sim M(x)$, $M(A)$, $\sim P(A)$.
Camestres: $P(x) \supset_x M(x)$, $\sim M(A)$, $\sim P(A)$.

The five moods of the fourth figure are sometimes characterized instead as indirect moods of the first figure, the two premisses (major and minor) being interchanged, and the names being then given respectively as Baralipton, Celantes, Dabitis, Fapesmo, Frisesomorum. (Some add the five "weakened" moods, Barbari, Celaront, Cesaro, Camestros, Calemos, to be obtained respectively from Barbara, Celarent, Cesare, Camestres, Calemes, by subalternation of the conclusion.) Other variations in the names of the moods are also found. These names have a mnemonic significance, the first three vowels indicating whether the major premiss, minor premiss, and conclusion, in order, are A, E, I, or O; and some of the consonants indicating the traditional reductions of the other moods to the four direct moods of the first figure.

The Port-Royal Logic, translated by T. S. Baynes, 2nd edn., London, 1851. F. Ueberweg, *System der Logik*, 4th edn., Bonn, 1874. H. W. B. Joseph, *An Introduction to Logic*, 2nd edn., Oxford, 1916. R. M. Eaton, *General Logic*, New York, 1931. H. B. Curry, *A mathematical treatment of the rules of the syllogism*, Mind, vol. 45 (1936), pp. 209–216, 416. Hilbert and Ackermann, *Grundzüge der theoretischen Logik*, 2nd edn., Berlin, 1938. Bennett and Baylis, *Formal Logic*, New York, 1939.

6. THEORY OF TYPES. In the functional calculus of first order, variables which appear as arguments of propositional functions or which are bound by quantifiers must be variables which are restricted to a certain limited range, the domain of individuals. Thus there are certain kinds of propositions about propositional functions which cannot be expressed in the calculus. The uncritical attempt to remove this restriction, by introducing variables of unlimited range (the range covering both non-functions and functions of whatever kind) and modifying accordingly the definition of a formula and the lists of primitive formulas and primitive rules of inference, leads to a system which is formally inconsistent through the possibility of deriving in it certain of the logical *paradoxes* (q.v.). The functional calculus of first order may, however, be extended in another way, which involves separating propositional functions into a certain array of categories (the *hierarchy of types*), excluding propositional functions which do not fall into one of these categories, and—besides propositional and individual variables—admitting only variables having a particular one of these categories as range.

For convenience of statement, we confine attention to the pure functional calculus of first order. The first step in the extension consists in introducing quantifiers such as (F_1), (EF_1), (F_2), (EF_3), etc., binding n-adic functional variables. Corresponding changes are made in the definition of a formula and in the lists of primitive formulas and primitive rules of inference, allowing for these new kinds of bound variables. The resulting system is the *functional calculus of second order*. Then the next step consists in introducing new kinds of functional variables; namely for every finite ordered set $k, l, m, \dots, p$ of i non-negative integers ($i = 1, 2, 3, \dots$) an infinite list of functional variables $\mathrm{F}^{klm\cdots p}$, $G^{klm\cdots p}$, …, each of which denotes ambiguously any i-adic propositional function for which the first argument may be any $(k-1)$-adic propositional function of individuals, the second argument any $(l-1)$-adic propositional function of individuals, etc. (if one of the integers $k, l, m, \dots, p$ is 1 the corresponding argument is a proposition—if 0, an individual). Then quantifiers are introduced binding these new kinds of functional variables; and so on. The process of alternately introducing new kinds of functional variables (denoting propositional functions which take as arguments propositional functions of kinds for which variables have already been introduced) and quantifiers binding the new kinds of functional variables, with appropriate extension at each stage of the definition of a formula and the lists of primitive formulas and primitive rules of inference, may be continued to infinity. This leads to what we may call the *functional calculus of order omega*, embodying the (so-called *simple*) *theory of types*.

In the functional calculi of second and higher orders, we may introduce the definitions:

$$\mathrm{X} = \mathrm{Y} \longrightarrow (\mathrm{F})[\mathrm{F}(\mathrm{X}) \supset \mathrm{F}(\mathrm{Y})],$$

where X and Y are any two variables of the same type and F is a monadic functional variable of appropriate type. The notation X = Y may then be interpreted as denoting equality or identity.

The functional calculus of order omega (as just described) can be proved to be *consistent* by a straightforward generalization of the method employed by Hilbert and Ackermann to prove the consistency of the functional calculus of first order.

For many purposes, however, it is necessary to add to the functional calculus of order omega the *axiom of infinity*, requiring the domain of individuals to be infinite.—This is most conveniently done by adding a single additional primitive formula, which may be described by referring to §3 above, taking the formula, which is there given as an example of a formula satisfiable in an infinite domain of individuals but not in any finite domain, and prefixing the quantifier (EF) with scope extending to the end of the formula. This form of the axiom of infinity, however, is considerably stronger (in the absence of the axiom of choice) than the "Infin ax" of Whitehead and Russell.

Other primitive formulas (possibly involving new primitive notations) which may be added correspond to the axiom of *choice* (q.v.) or are designed to introduce *classes* (q.v.) or *descriptions* (q.v.). Functional *abstraction* (q.v.) may also be introduced by means of additional primitive formulas or primitive rules of inference, or it may be defined with the aid of descriptions. Whitehead and Russell employ the axiom of infinity and the axiom of choice but avoid the necessity of special primitive formulas in connection with classes and descriptions by introducing classes and descriptions as incomplete symbols.

With the aid of the axiom of infinity and a method of dealing with classes and descriptions, the non-negative integers may be introduced in any one of various ways (e.g., following Frege and Russell, as finite cardinal numbers), and arithmetic (elementary number theory) derived formally within the system. With the further addition of the axiom of choice, analysis (real number theory) may be likewise derived.

No proof of *consistency* of the functional calculus of order omega (or even of lower order) with the axiom of infinity added is known, except by methods involving assumptions so strong as to destroy any major significance. According to an important theorem of Gödel, the functional calculus of order omega with the axiom of infinity added, if consistent, is *incomplete* in the sense that there are formulas A containing no free variables, such that neither A nor $\sim$A is a theorem. The same thing holds of any logistic system obtained by adding new primitive formulas and primitive rules of inference, provided only that the effective (recursive) character of the formal construction of the system is retained. Thus the system is not only incomplete but, in the indicated sense, incompletable. The same thing holds also of a large variety of logistic systems which could be considered as acceptable substitutes for the functional calculus of order omega with axiom of infinity; in particular the Zermelo set theory (§9 below) is in the same sense incomplete and incompletable.

The formalization as a logistic system of the functional calculus of order omega with axiom of infinity leads, by a method which cannot be given here, to a (definite but quite complicated) proposition of arithmetic which is equivalent to—in a certain sense, expresses—the consistency of the system. This proposition of arithmetic can be represented with the system by a formula A containing no free variables and the following second form of Gödel's incompleteness theorem can then be proved: If the system is consistent, then the formula A, although its meaning is a true proposition of arithmetic, is not a theorem of the system. We might, of course, add A to the system as a new primitive formula—we would then have a new system, whose consistency would correspond to a new proposition of arithmetic, represented by a new formula B (containing no free variables), and we would still have in the new system, if consistent, that B was not a theorem.

Whitehead and Russell, *Principia Mathematica*, 3 vols., Cambridge, England, 1st edn. 1910–13, 2nd edn. 1925–27. R. Carnap, *Abriss der Logistik*, Vienna, 1929. W. V. Quine, *A System of Logistic*, Cambridge, Mass., 1934. Hilbert and Ackermann, *Grundzüge der theoretischen Logic*, 2nd edn., Berlin, 1938. A. Church, *A formulation of the simple theory of types*, The Journal of Symbolic Logic, vol. 5 (1940), pp. 56–68. —A. Tarski, *Einige Betrachtungen über die Begriffe der ω-Widerspruchsfreiheit und der ω-Vollstänndigkeit*, Monatshefte für Mathematik und Physik, vol. 40 (1933), pp. 97–112. G. Gentzen, *Die Widerspruchsfreiheit der Stufenlogik*, Mathematische Zeitschrift, vol. 41 (1936), pp. 357–366. —K. Gödel, *Über formal unentscheidbare Sätze der Principia Mathematica und verwandter Systeme*, Monatshefte für Mathematik und Physik, vol. 38 (1931), pp. 173–198. J. B. Rosser, *Extensions of some theorems of Gödel and Church*, The Journal of Symbolic Logic, vol. 1 (1936), pp. 87–91. J. B. Rosser, *An informal exposition of proofs of Gödel's theorems and Church's theorem*, The Journal of Symbolic Logic, vol. 4 (1939), pp. 53–60. Hilbert and Bernays, *Grundlagen der Mathematik*, vol. 2, Berlin, 1939.

7. Algebra of classes deals with *classes* (q.v.) whose members are from a fixed non-empty class called the *universe of discourse*, and with the operations of complementation, logical sum, and logical product upon such classes. (The classes are to be thought of as determined by propositional functions having the universe of discourse as the range of the independent variable.) The universal class V comprises the entire universe of discourse. The null (or empty) class Λ has no members. The *complement* $-a$ of a class a has as members all those elements of the universe of discourse which are not members of a (and those only). In particular the null class and the universal class are each the complement of the other. The *logical sum* $a \cup b$ of two classes a and b has as members all those elements which are members either of a or of b, not excluding elements which are members of both a and b (and those only). The *logical product* $a \cap b$ of two classes a and b has as members all those elements which are members of both a and b (and those only)—in other words the logical product of two classes is their common part. The *expressions* of the algebra of classes are built up out of *class variables* $a, b, c, \ldots$ and the symbols for the universal class and the null class by means of the notations for complementation, logical sum, and logical product (with parentheses). A *formula* of the algebra of classes consists of two expressions with one of the symbols $=$ or $\neq$ between. ($a = b$ means that a and b are the same class, $a \neq b$ that a and b are not the same class.)

While the algebra of classes can be set up as an independent logistic system, we shall here describe it instead by reference to the functional calculus of first order (§3), using two monadic functional constants, V and Λ, and an infinite list of monadic functional variables $F_1, G_1, H_1, \ldots$ corresponding in order to the class variables $a, b, c, \ldots$ respectively. Given any expression A of the algebra of classes, the corresponding formula A‡ of the functional calculus is obtained by replacing complementation, logical sum, and logical product respectively by negation, inclusive disjunction, and conjunction, and at the same time replacing $\mathrm{V}, \Lambda, a, b, c, \ldots$ respectively by $\mathrm{V}(x), \Lambda(x), F_1(x), G_1(x), H_1(x), \ldots$. Given any formula A = B of the algebra of classes the corresponding formula of the functional calculus is A‡ $\equiv_x$ B‡. Given any formula A $\neq$ B of the algebra of classes, the corresponding formula of the functional calculus is $\sim[\text{A}\ddagger \equiv_x \text{B}\ddagger]$. A formula C is a *theorem* of the algebra of classes if and only if the inference from $(x)\text{V}(x)$ and $(x)\sim\Lambda(x)$ to C‡ (where C‡ corresponds to C) is a valid inference of the functional calculus. The inference from premisses $\text{D}_1, \text{D}_2, \ldots, \text{D}_n$ to a conclusion C is a *valid inference* of the algebra of classes if and only if the inference from $(x)\text{V}(x), (x)\sim\Lambda(x), \text{D}_1\ddagger, \text{D}_2\ddagger, \ldots, \text{D}_n\ddagger$ to C‡ is a valid inference of the functional calculus.

This isomorphism between the algebra of classes and the indicated part of the functional calculus of first order can be taken as representing a parallelism of meaning. In fact, the meanings become identical if we wish to construe the functional calculus *in extension* (see the article *propositional function*); or, inversely, if we wish to construe the algebra of classes *in intension*, instead of the usual construction.

If we deal only with formulas of the algebra of classes which are *equations* (i.e., which have the form A = B), the above description by reference to the functional calculus may be replaced by a simpler description using the applied propositional calculus (§1) whose fundamental propositional symbols are $x \in \text{V}, x \in \Lambda, x \in a, x \in b, x \in c, \ldots$. Given an equation C of the algebra of classes, the corresponding formula C† of the propositional calculus is obtained by replacing equality (=) by the biconditional (≡), replacing complementation, logical sum, and logical product respectively by negation, inclusive disjunction, and conjunction, and at the same time replacing $\text{V}, \Lambda, a, b, c, \ldots$ respectively by $x \in \text{V}, x \in \Lambda, x \in a, x \in b, x \in c, \ldots$. An equation C is a *theorem* of the algebra of classes if and only if the inference from $x \in \text{V}, \sim x \in \Lambda$ to C† is a valid inference of the propositional calculus; analogously for *valid inferences* of the algebra of classes in which the formulas involved are equations.

As a corollary of this, every theorem of the pure propositional calculus (§1) of the form A ≡ B has a corresponding theorem of the algebra of classes obtained by replacing the principal occurrence of ≡ by =, elsewhere replacing negation, inclusive disjunction, and conjunction respectively by complementation, logical sum, and logical product, and at the same time replacing propositional variables by class variables. Likewise, every theorem A of the pure propositional calculus has a corresponding theorem B = V of the algebra of classes, where B is obtained from A by replacing negation, inclusive disjunction, and conjunction respectively by complementation, logical sum, and logical product, and replacing propositional variables by class variables.

The *dual* of a formula or an expression of the algebra of classes is obtained by interchanging logical sum and logical product, and at the same time interchanging V and Λ. The *principle of duality* holds, that the dual of every theorem is also a theorem.

The relation of class inclusion, ⊂, may be introduced by the definition

$$\text{A} \subset \text{B} \longrightarrow \text{A} \cap -\text{B} = \Lambda.$$

Instead of *algebra of classes*, the term *Boolean algebra* is used primarily when it is intended that the formal system shall remain uninterpreted or that interpretations other than that described above shall be admitted. For the related idea of a *Boolean ring* see the paper of Stone cited below.

E. Schröder, *Algebra der Logik*, vol. 1, vol. 2 part 1, and vol. 2 part 2, Leipzig, 1890, 1891, 1905. E. V. Huntington, *Sets of independent postulates for the algebra of logic*, Transactions of the American Mathematical Society, vol. 5 (1904), pp. 288–309. S. K. Langer, *An Introduction to Symbolic Logic*, 1937. M. H. Stone, *The representation of Boolean algebras*, Bulletin of the American Mathematical Society, vol. 44 (1938), pp. 807–816.

8. ALGEBRA OF RELATIONS or *algebra of relatives* deals with *relations* (q.v.) in extension whose domains and converse domains are each contained in a fixed *universe of discourse* (which must be a class having at least two members), in a way similar to that in which the algebra of classes deals with classes. Fundamental ideas involved are those of the universal relation and the null relation; the relations of identity and diversity; the contrary and the converse of a relation; the logical sum, the logical product, the relative sum, and the relative product of two relations.

The *universal relation* V can be described by saying that $x\mathrm{V}y$ holds for every x and y in the universe of discourse. The *null relation* Λ is such that $x\Lambda y$ holds for no x and y in the universe of discourse. The *relation of identity* I is such that xIx holds for every x in the universe of discourse and xIy fails if x is not identical with y. The *relation of diversity* J is such that xJx fails for every x in the universe of discourse and xJy holds if x is not identical with y.

The *contrary* $-R$ of a relation R is the relation such that $x\ -R\ y$ holds if and only if xRy fails (x and y being in the universe of discourse). In particular, the universal relation and the null relation are each the contrary of the other; and the relations of identity and diversity are each the contrary of the other.

The *converse* of a relation R is the relation S such that xSy if and only if yRx. The usual notation for the converse of a relation is obtained by placing a breve ˘ over the letter denoting the relation. A relation is said to be *symmetric* if it is the same as its converse. In particular, the universal relation, the null relation, and the relations of identity and diversity are symmetric.

The *logical sum* $R \cup S$ of two relations R and S is the relation such that $x\ R \cup S\ y$ holds if and only if at least one of xRy and xSy holds. The *logical product* $R \cap S$ of two relations R and S is a relation such that $x\ R \cap S\ y$ if and only if both xRy and xSy. The *relative product* $R \mid S$ of two relations R and S is a relation such that $x\ R \mid S\ y$ if and only if there is a z (in the universe of discourse) such that xRz and zSy. The *relative sum* of two relations R and S is the relation which holds between x and y if and only if for every z (in the universe of discourse) at least one of xRz and zSy holds. The *square* of a relation R is $R \mid R$.

The signs $=$ and $\neq$ are used to form equations and inequations in the same way as in the algebra of classes. An isomorphism between the algebra of relations and an appropriately chosen part of the functional calculus of first order can also be exhibited in the same way as was done in §7 above for the algebra of classes. A *principle of duality* also holds, where duality consists in interchanging logical sum and logical product, relative sum and relative product, V and Λ, I and J.

The portion of the algebra of relations which involves, besides relation variables, only the universal relation and the null relation, and the operations of contrary, logical sum, and logical product, and $=$, $\neq$, is isomorphic with (formally indistinguishable from) the algebra of classes.

Relative inclusion, $\subset$, may be introduced by the definition:

$$R \subset S \longrightarrow R \cap -S = \Lambda.$$

When $R \subset S$, we sat that R *is contained in* S, or that S contains R.

A relation is *transitive* if it contains its square, *reflexive* if it contains I, *irreflexive* if it is contained in J, *asymmetric* if its square is contained in J.

E. Schröder, *Algebra der Logik*, vol. 3, Leipzig, 1895. A. Tarski, *On the calculus of relations*, The Journal of Symbolic Logic, vol. 6 (1941), pp. 73–89.

9. ZERMELO SET THEORY. The attempt to devise a system which deals with the logic of classes in a more comprehensive way than is done by the algebra of classes (§7), and which, in particular, takes account of the relation $\in$ between classes (see the article *class*), must be carried out with caution in order to avoid the Russell paradox and similar logical *paradoxes* (q.v.).

There are two methods of devising such a system which (so at least it is widely held or conjectured) do not lead to any inconsistency. One of these involves the theory of types, which was set forth in §6 above, explicitly for propositional functions, and by implication for classes (classes being divided into types according to the types of the monadic propositional functions which determine them). The other method is the Zermelo *set theory*, which avoids this preliminary division of classes into types, but imposes restrictions in another direction.

Given the relation (dyadic propositional function) $\in$, the relations of equality and class inclusion may be introduced by the following definitions:

$$\mathrm{Z} \in \mathrm{Y} \longrightarrow \in(\mathrm{Z}, \mathrm{Y}).$$
$$\mathrm{Z} = \mathrm{Y} \longrightarrow \mathrm{Z} \in \mathrm{X} \supset_{\mathrm{X}} \mathrm{Y} \in \mathrm{X}.$$
$$\mathrm{Z} \subset \mathrm{Y} \longrightarrow \mathrm{X} \in \mathrm{X} \supset_{\mathrm{X}} \mathrm{X} \in \mathrm{Y}.$$

Here X, Y, and Z are to be taken as individual variables ("individual" in the technical sense of §3), and X is to be determined according to an explicit rule so as to be different from Y and Z.

The Zermelo set theory may be formulated as a simple applied functional calculus of first order (in the sense of §3), for which the domain of individuals is composed of classes, and the only functional constant is $\in$, primitive formulas (additional to those given in §3) being added as follows:

$[x \in z \equiv_x x \in y] \supset z = y$. (Axiom of extensionality)

$(Et)[x \in t \equiv_x [x = y \vee x = z]]$. (Axiom of pairing)

$(Et)[x \in t \equiv_x (Ey)[x \in y][y \in z]]$. (Axiom of summation)

$(Et)[x \in t \equiv_x x \subset z]$. (Axiom of the set of subsets)

$(Et)[x \in t \equiv_x [x \in z]\mathrm{A}]$. (Axiom of subset formation)

$[y \in z \supset_y [y' \in z \supset_{y'} [[x \in y][x \in y'] \supset_x y = y']]]$
$\supset (Et)[y \in z \supset_y [x'' \in y \supset_{x''} (Ex')[[x \in y][x \in t] \equiv_x x = x']]]$.
(Axiom of choice)

$(Et)[z \in t][x' \in t \supset_{x'} (Ey)[y \in t][x \in y \equiv_x x = x']]$. (Axiom of infinity)

$[y \in z \supset_y (Ex')[\mathrm{A} \equiv_x x = x']] \supset (Et)[x \in t \equiv_x (Ey)[y \in z]\mathrm{A}]$.
(Axiom of replacement)

In the axiom of subset formation, A is any formula not containing t as a free variable (in general, A will contain x as a free variable). In the axiom of replacement, A is any formula which contains neither t nor x' as a free variable (in general, A will contain x and y as free variables). These two axioms are thus represented each by an infinite list of primitive formulas—the remaining axioms each by one primitive formula.

The axiom of extensionality as above stated has (incidentally to its principal purpose) the effect of excluding non-classes entirely and assuming that everything is a class. This assumption can be avoided if desired, at the cost of complicating the axioms somewhat—one method would be to introduce an additional functional constant, expressing the property *to be a class* (or *set*), and to modify the axioms accordingly, the domain of individuals being thought of as possibly containing other things besides sets.

The treatment of sets in the Zermelo set theory differs from that of the theory of types in that all sets are "individuals" and the relation $\in$ (of membership in a set) is significant as between any two sets—in particular, $x \in x$ is not forbidden. (We are here using the words *set* and *class* as synonymous.)

The restriction which is imposed in order to avoid paradox can be seen in connection with the axiom of subset formation. Instead of this axiom, an uncritical formulation of axioms for set theory might well have included $(Et)[x \in t \equiv_x \mathrm{A}]$, asserting the existence of a set t whose members are the sets x satisfying an arbitrary condition A expressible in the notation of the system. This, however, would lead at once to the Russell paradox by taking A to be $\sim x \in x$ and then going through a process of inference which can be described briefly by saying that x is put equal to t. As actually proposed, however, the axiom of subset formation allows the use of the condition A only to obtain a set t whose members are the sets x which *are members of a previously given set* z and satisfy A. This is not known to lead to paradox.

The notion of an ordered pair can be introduced into the theory by definition, in a way which amounts to identifying the ordered pair (x, y) with the set z which has two and only two members, x' and y', x' being the set which has x as its only member, and y' being the set which has x and y as its only two members. (This is one of various similar possible methods.) Relations in extension may then be treated as sets of ordered pairs.

The Zermelo set theory has an adequacy to the logical development of mathematics comparable to that of the functional calculus of order omega (§6). Indeed, as here actually formulated, its adequacy for mathematics apparently exceeds that of the functional calculus; however, this should not be taken as an essential difference, since both systems are incomplete, in accordance with Gödel's theorem (§6), but are capable of extension.

Besides the Zermelo set theory and the functional calculus (theory of types), there is a third method of obtaining a system adequate for mathematics and at the same time—it is hoped—consistent, proposed by Quine in his book cited below (1940).—The last word on these matters has almost certainly not yet been said.

Alonzo Church

A. Fraenkel, *Einleitung in die Mengenlehre*, 3rd edn., Berlin, 1928. W. V. Quine, *Set-theoretic foundations for logic*, The Journal of Symbolic Logic, vol. 1 (1936), pp. 45–57. Wilhelm Ackermann, *Mengentheoretische Begründung der Logik*, Mathematische Annalen, vol. 115 (1937), pp. 1–22. Paul Bernays, *A system of axiomatic set theory*, The Journal of Symbolic Logic, vol. 2 (1937), pp. 65–77, and vol. 6 (1941), pp. 1–17. W. V. Quine, *Mathematical Logic*, New York, 1940.

Logic, symbolic, or *mathematical logic*, or *logistic*, is the name given to the treatment of formal logic by means of a formalized logical language or calculus whose purpose is to avoid the ambiguities and logical inadequacy of ordinary language. It is best characterized, not as a separate subject, but as a new and powerful method in formal logic. Foreshadowed by ideas of Leibniz, J. H. Lambert, and others, it had its substantial historical beginning in the Nineteenth Century *algebra of logic* (q.v.), and received its contemporary form at the hands of Frege, Peano, Russell, Hilbert, and others. Advantages of the symbolic method are greater exactness of formulation, and power to deal with formally more complex material. See also *logistic system*.

—*A. C.*

C. I. Lewis, *A Survey of Symbolic Logic*, Berkeley, Cal., 1918. Lewis and Langford, *Symbolic Logic*, New York and London, 1932. S. K. Langer, *An Introduction to Symbolic Logic*, Boston and New York, or London 1937. W. V. Quine, *Mathematical Logic*, New York, 1940. A. Tarski, *Introduction to Logic and to the Methodology of Deductive Sciences*, New York, 1941. W. V. Quine, *Elementary Logic*, Boston and New York, 1941. A. Church, *A bibliography of symbolic logic*, The Journal of Symbolic Logic, vol. 1 (1936), pp. 121–218, and vol. 3 (1938), pp. 178–212. I. M. Bochenski, *Nove Lezioni di Logica Simbolica*, Rome, 1938. R. Carnap, *Abriss der Logistik*, Vienna, 1929. H. Scholz, *Geschichte der Logik*, Berlin, 1931. Hilbert and Ackermann, *Grundzüge der theoretischen Logik*, 2nd edn., Berlin, 1938.

Logic, traditional: the name given to those parts and that method of treatment of formal logic which have come down substantially unchanged from classical and medieval times. Traditional logic emphasizes the analysis of propositions into subject and predicate and the associated classification into the four forms, *A*, *E*, *I*, *O*; and it is concerned chiefly with topics immediately related to these, including opposition, immediate inference, and syllogism (see *Logic, formal*). Associated with traditional logic are also the three so-called laws of thought—the laws of *identity* (q.v.), *contradiction* (q.v.), and *excluded middle* (q.v.)—and the doctrine that these laws are in a special sense fundamental presuppositions of reasoning, or even (by some) that all other principles of logic can be derived from them or are mere elaborations of them. *Induction* (q.v.) has been added in comparatively modern times (dating from Bacon's *Novum Organum*) to the subject matter of traditional logic. —*A. C.*

A. Arnauld and others, *La Logique ou l'Art de Penser*, better known as the Port-Royal Logic, 1st edn., Paris, 1662; reprinted, Paris, 1878; English translation by T. S. Baynes, 2nd edn., London, 1851. F. Ueberweg, *System der Logik und Geschichte der logischen Lehren*, 1st edn., Bonn, 1857; 4th edn., Bonn, 1874. C. Prantl, *Geschichte der Logik im Abendlande*, 4 vols., Leipzig, 1855–1870; reprinted, Leipzig, 1927. H. W. B. Joseph, *An Introduction to Logic*, 2nd edn., Oxford, 1916. F. Enriques, *Per la Storia della Logica*, Bologna, 1922; English translation by J. Rosenthal, New York, 1929. H. Scholz, *Geschichte der Logik*, Berlin, 1931.

Logistic: The old use of the word *logistic* to mean the art of calculation, or common arithmetic, is now nearly obsolete. In Seventeenth Century English the corresponding adjective was also sometimes used to mean simply *logical*. Leibniz occasionally employed *logistica* (as also *logica mathematica*) as one of various alternative names for his *calculus ratiocinator*. The modern use of *logistic* (French *logistique*) as a synonym for *symbolic logic* (q.v.) dates from the International Congress of Philosophy of 1904, where it was proposed independently by Itelson, Lalande, and Couturat. The word *logistic* has been employed by some with special reference to the Frege–Russell doctrine that mathematics is reducible to logic, but it would seem that the better usage makes it simply a synonym of *symbolic logic*. —*A. C.*

L. Couturat, *IIme Congrès de Philosophie*, Revue de Métaphysique et de Morale, vol. 12 (1904), see p. 1042.

Logistic System: The formal construction of a logistic system requires: (1) a list of *primitive symbols* (these are usually taken as marks but may also be sounds or other things—they must be capable of *instances* which are, recognizably, the same or different symbols, and capable of *utterance* in which instances of them are put forth or arranged in an order one after another); (2) a determination of a class of *formulas*, each formula being a finite sequence of primitive symbols, or, more exactly, each formula being capable of instances which are finite sequences of instances of primitive symbols (generalizations allowing two-dimensional arrays of primitive symbols and the like are non-essential); (3) a determination of the circumstances under which a finite sequence of formulas is a *proof* of the last formula in the sequence, this last formula being then called a *theorem* (again we should more exactly speak of proofs as having instances which are finite sequences of instances of formulas); (4) a determination of the circumstances under which a finite sequence of formulas is a *proof* of the last formula of the sequence *as a consequence of* a certain set of formulas (when there is a proof of a formula B as a consequence of the set of formulas $A_1, A_2, \ldots, A_n$, we say that the *inference* from the *premisses* $A_1, A_2, \ldots, A_n$ to the *conclusion* B is a *valid inference* of the logistic system). It is not excluded that the class of proofs in the sense of (3) should be empty. But every proof of a formula B as a consequence of an empty set of formulas, in the sense of (4), must also be a proof of B in the sense of (3), and conversely. Moreover, if to the proof of a formula B as a consequence of $A_1, A_2, \ldots, A_n$ are prefixed in any order proofs of $A_1, A_2, \ldots, A_n$, the entire resulting sequence of formulas must be a proof of B; more generally, if to the proof of a formula B as a consequence of $A_1, A_2, \ldots, A_n$ are prefixed in any order proofs of a subset of $A_1, A_2, \ldots, A_n$ as consequences of the remainder of $A_1, A_2, \ldots, A_n$, the entire resulting sequence must be a proof of B as a consequence of this remainder.

The determination of the circumstances under which a sequence of formulas is a proof, or a proof as a consequence of a set of formulas, is usually made by means of: (5) a list of *primitive formulas*; and (6) a list of *primitive rules of inference* each of which prescribes that under certain circumstances a formula B shall be an *immediate consequence* of a set of formulas $A_1, A_2, \ldots, A_n$. The list of primitive formulas may be empty—this is not excluded. Or the primitive formulas may be included under the head of primitive rules of inference by allowing the case $n = 0$ in (6). A *proof* is then defined as either a primitive formula or an immediate consequence of preceding formulas by one of the primitive rules of inference. A *proof as a consequence* of a set of formulas $A_1, A_2, \ldots, A_n$ is in some cases defined as a finite sequence of formulas each of which is either a primitive formula, or one of $A_1, A_2, \ldots, A_n$, or an immediate consequence of preceding formulas by one of the primitive rules of inference; in other cases it may be desirable to impose certain restrictions upon the application of the primitive rules of inference (e.g., in the case of the functional calculus of first order —*Logic, formal*, §3—that no free variable of $A_1, A_2, \ldots, A_n$ shall be generalized upon).

A logistic system need not be given any meaning or interpretation, but may be put forward merely as a formal discipline of interest for its own sake; and in this case the words *proof, theorem, valid inference*, etc., are to be dissociated from their every-day meanings and taken purely as technical terms. Even when an interpretation of the system is intended, it is a requirement of rigor that no use shall be made of the interpretation (as such) in the determination whether a sequence of symbols is a formula, whether a sequence of formulas is a proof, etc.

The kind of an interpretation, or assignment of meaning, which is normally intended for a logistic system is indicated by the technical terminology employed. This is namely such an interpretation that the formulas, some or all of them, mean or express propositions; the theorems express true propositions; and the proofs and valid inferences represent proofs and valid inferences in the ordinary sense. (Formulas which do not mean propositions may be interpreted as names of things other than propositions, or may be interpreted as containing free variables and having only ambiguous denotation—see *variable*.) A logistic system may thus be regarded as a device for obtaining—or, rather stating—an objective, external criterion for the validity of proofs and inferences (which are expressible in a given notation.)

A logistic system which has an interpretation of the kind in question may be expected, in general, to have more than one such interpretation.

It is usually to be required that a logistic system shall provide an *effective* criterion for recognizing formulas, proofs, and proofs as a consequence of a set of formulas; i.e., it shall be a matter of direct observation, and of following a fixed set of directions for concrete operations with symbols, to determine whether a given finite sequence of primitive symbols is a formula, or whether a given finite sequence of formulas is a proof, or is a proof as a consequence of a given set of formulas. If this requirement is not satisfied, it may be necessary—e.g.—given a particular finite sequence of formulas, to seek by some argument adapted to the special case to prove or disprove that it satisfies the conditions to be a *proof* (in the technical sense); i.e., the criterion for formal recognition of proofs then presupposes, in actual application, that we already know what a valid deduction is (in a sense which is stronger than that merely of the ability to follow concrete directions in a particular case). See further on this point *Logic, formal*, §1.

The requirement of effectiveness does not compel the lists of primitive symbols, primitive formulas, and primitive rules of inference to be finite. It is sufficient if there are effective criteria for recognizing formulas, for recognizing primitive formulas, for recognizing applications of primitive rules of inference, and (if separately needed) for recognizing such restricted applications of the primitive rules of inference as are admitted in *proofs as a consequence* of a given set of formulas.

With the aid of Gödel's device of representing sequences of primitive symbols and sequences of formulas by means of numbers, it is possible to give a more exact definition of the notion of effectiveness by making it correspond to that of *recursiveness* (q.v.) of numerical functions. E.g., a criterion for recognizing primitive formulas is effective if it determines a general recursive monadic function of natural numbers whose value is 0 when the argument is the number of a primitive formula, 1 for any other natural number as argument. The adequacy of this technical definition to represent the intuitive notion of effectiveness as described above is not immediately clear, but is placed beyond any real doubt by developments for details of which the reader is referred to Hilbert–Bernays and Turing (see references below).

The requirement of effectiveness plays an important rôle in connection with logistic systems, but the necessity of the requirement depends on the purpose in hand and it may for some purposes be abandoned. Various writers have proposed non-effective, or non-constructive, logistic systems; in some of these the requirement of finiteness of length of formulas is also abandoned and certain infinite sequences of primitive symbols are admitted as formulas.

For particular examples of logistic systems (all of which satisfy the requirement of effectiveness) see the article *Logic, formal*, especially §§1, 3, 9. *Alonzo Church*

R. Carnap, *The Logical Syntax of Language*, New York and London, 1937. H. Scholz, *Was ist ein Kalkül und was hat Frege für eine pünktliche Beantwortung dieser Frage geleistet?*, Semester-Berichte, Münster i. W., summer 1935, pp. 16–54. Hilbert and Bernays, *Grundlagen der Mathematik*, vol. 2, Berlin, 1939. A. M. Turing, *On computable numbers, with an application to the Entscheidungsproblem*, Proceedings of the London Mathematical Society, ser. 2 vol. 42 (1937), pp. 230–265, and *Correction*, ibid, ser. 2 vol. 43 (1937), pp. 544–546. A. M. Turing, *Computability and λ-definability*, The Journal of Symbolic Logic, vol. 2 (1937), pp. 153–163.

Löwenheim's theorem: The theorem, first proved by Löwenheim, that if a formula of the pure functional calculus of first order (see *Logic, formal*, §3), containing no free individual variables, is *satisfiable* (see ibid.) at all, it is satisfiable in a domain of individuals which is at most enumerable. Other, simpler, proofs of the theorem, were afterwards given by Skolem, who also obtained the generalization that, if an enumerable set of such formulas are simultaneously satisfiable, they are simultaneously satisfiable in a domain of individuals at most enumerable.

There follows the existence of an interpretation of the Zermelo set theory (see *Logic, formal*, §9)—consistency of the theory assumed—according to which the domain of sets is only enumerable; although there are theorems of the Zermelo set theory which, under the *usual interpretation*, assert the existence of the non-enumerable infinite.

A like result may be obtained for the functional calculus of order omega (theory of types) by utilizing a representation of it within the Zermelo set theory.

It is thus in a certain sense impossible to postulate the non-enumerable infinite: any set of postulates designed to do so will have an unintended interpretation within the enumerable. Usual sets of mathematical postulates for the real number system (see *number*) have an appearance to the contrary only because they are incompletely formalized (i.e., the mathematical concepts are formalized, while the underlying logic remains unformalized and indefinite).

The situation described in the preceding paragraph is sometimes called *Skolem's paradox*, although it is not a paradox in the sense of formal self-contradiction, but only in the sense of being unexpected or at variance with preconceived ideas. —*A.C.*

Th. Skolem, *Sur la portée du théorème de Löwenheim–Skolem*, Les Entretiens de Zurich sur les Fondements et la Méthode des Sciences Mathématiques, Zurich 1941, pp. 25–52.

Many questions: The name given to the fallacy—or, rather, misleading device of disputation—which consists in requiring a single answer to a question which either involves several questions that ought to be answered separately or contains an implicit assertion to which any unqualified answer would give assent. —*A.C.*

Mathematics: The traditional definition of mathematics as "the science of quantity" or "the science of discrete and continuous magnitude" is today inadequate, in that modern mathematics, while clearly in some sense a single connected whole, includes many branches which do not come under this head. Contemporary accounts of the nature of mathematics tend to characterize it rather by its method than by its subject matter.

According to a view which is widely held by mathematicians, it is characteristic of a mathematical discipline that it begins with a set of undefined elements, properties, functions, and relations, and a set of unproved propositions (called *axioms* or *postulates*) involving them; and that from these all other propositions (called *theorems*) of the discipline are to be derived by the methods of formal logic. On its face, as thus stated, this view would identify mathematics with *applied logic*. It is usually added, however, that the *undefined terms*, which appear in the rôle of names of undefined elements, etc., are not really names of particulars at all but are variables, and that the theorems are to be regarded as proved for *any* values of these variables which render the postulates true. If then each theorem is replaced by the proposition embodying the implication from the conjunction of the postulates to the theorem in question, we have a reduction of mathematics to *pure logic*. (For a particular example of a set of postulates for a mathematical discipline see the article *Arithmetic, foundations of*.)

There is also another sense in which it has been held that mathematics is reducible to logic, namely that in the expressions for the postulates of a mathematical discipline the undefined terms are to be given definitions which involve logical terms only, in such a way that postulates and theorems of the discipline thereby become propositions of pure logic, demonstrable on the basis of logical principles only. This view was first taken, as regards arithmetic and analysis, by Frege, and was afterwards adopted by Russell, who extended it to all mathematics.

Both views require for their completion an exact account of the nature of the underlying logic, which, it would seem, can only be made by formalizing this logic as a *logistic system* (q.v.). Such a formalization of the underlying logic was employed from the beginning by Frege and by Russell, but has come into use in connection with the other —*postulational* or *axiomatic*—view only comparatively recently (with, perhaps, a partial exception in the case of Peano).

Hilbert has given a formalization of arithmetic which takes the shape of a logistic system having primitive symbols some of a logical and some of an arithmetical character, so that logic and arithmetic are formalized together without taking logic as prior; similarly also for analysis. This would not of itself be opposed to the Frege–Russell view, since it is to be expected that the choice as to which symbols shall be taken as primitive in the formalization can be made in more than one way. Hilbert, however, took the position that many of the theorems of the system are *ideale Aussagen*, mere formulas, which are without meaning in themselves but are added to the *reale Aussagen* or genuinely meaningful formulas in order to avoid formal difficulties otherwise arising. In this respect Hilbert differs sharply from Frege and Russell, who would give a meaning (namely as propositions of logic) to all formulas (sentences) appearing.—Concerning Hilbert's associated program for a consistency proof see the article *Proof theory*.

A view of the nature of mathematics which is widely different from any of the above is held by the school of mathematical *intuitionism* (q.v.). According to this school, mathematics is "identical with the exact part of our thought." "No science, not even philosophy or logic, can be a presupposition for mathematics. It would be circular to apply any philosophical or logical theorem as a means of proof in mathematics, since such theorems already presuppose for their formulation the construction of mathematical concepts. If mathematics is to be in this sense presupposition-free, then there remains for it no other source than an intuition which presents mathematical concepts and inferences to us as immediately clear. . . . [This intuition] is nothing else than the ability to treat separately certain concepts and inferences which regularly occur in ordinary thinking." This is quoted in translation from Heyting, who, in the same connection, characterizes the intuitionistic doctrine as asserting the existence of mathematical objects (Gegenstände), which are immediately grasped by thought, are independent of experience, and give to mathematics more than a mere formal content. But to these mathematical objects no existence is to be ascribed independent of thought. Elsewhere Heyting speaks of a relationship to Kant in the apriority ascribed to the natural numbers, or rather to the underlying ideas of *one* and the process of *adding one* and the *indefinite repetition* of the latter. At least in his earlier writings, Brouwer traces the doctrine of intuitionism directly to Kant. In 1912 he speaks of "abandoning Kant's apriority of space but adhering the more resolutely to the apriority of time" and in the same paper explicitly reaffirms Kant's opinion that mathematical judgments are *synthetic* and *a priori*.

The doctrine that the concepts of mathematics are empirical and the postulates elementary experimental truths has been held in various forms (either for all mathematics, or specially for geometry) by J. S. Mill, H. Helmholtz, M. Pasch, and others. However, the usual contemporary view, especially among mathematicians, is that the propositions of mathematics say nothing about empirical reality. Even in the case of *applied* geometry, it is held, the geometry is used to organize physical measurement, but does not receive an interpretation under which its propositions become unqualifiedly experimental or empirical in character; a particular system of geometry, applied in a particular way, may be wrong (and demonstrably wrong by experiment), but there is not, in significant cases, a *unique* geometry which, when applied in the particular way, is right.

M. Bôcher, *The fundamental conceptions and methods of mathematics*, Bulletin of the American Mathematical Society, vol. 11 (1904), pp. 115–135. J. W. Young, *Lectures on Fundamental Concepts of Algebra and Geometry*, New York, 1911. Veblen and Young, *Projective Geometry*, vol. 1, 1910 (see the Introduction). C. J. Keyser, *Doctrinal functions*, The Journal of Philosophy, vol. 15 (1918), pp. 262–267. — G. Frege, *Die Grundlagen der Arithmetik*, Breslau, 1884; reprinted, Breslau, 1934. G. Frege, *Grundgesetze der Arithmetik*, vol. 1, Jena, 1893, and vol. 2, Jena, 1903. B. Russell, *The Principles of Mathematics*, Cambridge, England, 1903; 2nd edn., London, 1937, and New York, 1938. B. Russell, *Introduction to Mathematical Philosophy*, London, 1919. — R. Carnap, *Die logizistische Grundlegung der Mathematik*, Erkenntnis, vol. 2 (1931), pp. 91–105, 141–144, 145. A. Heyting, *Die intuitionistische Grundlegung der Mathematik*, ibid., pp. 106–115. J. v. Neumann, *Die formalistische Grundlegung der Mathematik*, ibid., pp. 116–121, 144–145, 146, 148. R. Carnap, *The Logical Syntax of Language*, New York and London, 1937. — L. E. J. Brouwer, *Intuitionisme en Formalisme*, Groningen, 1912; reprinted in *Wiskunde, Waarheid, Werkelijkheid*, Groningen, 1919; English translation by A. Dresden, Bulletin of the American Mathematical Society, vol. 20 (1913), pp. 81–96. H. Weyl, *Die heutige Erkenntislage in der Mathematik*, Symposion, vol. 1 (1926), pp. 1–32. D. Hilbert, *Die Grundlagen der Mathematik*, Abhandlungen aus dem Mathematischen Seminar der Hamburgischen Universität, vol. 6 (1928), pp. 65–85; reprinted in Hilbert's *Grundlagen der Geometrie*, 7th edn. A. Heyting, *Mathematische Grundlagenforschung, Intuitionismus, Beweistheorie*, Berlin, 1934. — H. Poincaré, *The Foundations of Science*, English translation by G. B. Halsted, New York, 1913. — E. Nagel, *The formation of modern conceptions of formal logic in the development of geometry*, Osiris, vol. 7 (1939), pp. 142–224. — A. N. Whitehead, *An Introduction to Mathematics*, London, 1911, and New York, 1911. — G. H. Hardy, *A Mathematician's Apology*, London, 1940.

Histories: Moritz Cantor, *Vorlesungen über Geschichte der Mathematik*, 4 vols., Leipzig, 1880–1908; 4th edn., Leipzig, 1921. Florian Cajori, *A History of Mathematics*, 2nd edn., New York and London, 1922. Florian Cajori, *A History of Elementary Mathematics*, revised edn., New York and London, 1917. Florian Cajori, *A History of Mathematical Notations*, 2 vols., Chicago, 1928–1929. D. E. Smith, *A Source Book in Mathematics*, New York and London, 1929. T. L. Heath, *A History of Greek Mathematics*, 2 vols., Oxford, 1921. Felix Klein, *Vorlesungen über die Entwicklung der Mathematik im 19. Jahrhundert*, 2 vols., Berlin, 1926–1927. J. L. Coolidge, *A History of Geometrical Methods*, New York, 1940.

Matrix method: Synonymous with *truth-table method*, q.v. —*A. C.*

Mean: (1) In general, that which in some way mediates or occupies a middle position among various things or between two extremes. Hence (especially in the plural) that through which an end is attained; in mathematics the word is used for any one of various notions of average; in ethics it represents moderation, temperance, prudence, the middle way.

(2) In mathematics:

(A) The *arithmetic mean* of two quantities is half their sum; the *arithmetic mean* of n quantities is the sum of the n quantities, divided by n. In the case of a function $f(x)$ (say from real numbers to real numbers) the *mean value* of the function for the values $x_1, x_2, \ldots, x_n$ of x is the arithmetic mean of $f(x_1), f(x_2), \ldots, f(x_n)$. This notion is extended to the case of infinite sets of values of x by means of integration; thus the *mean value* of $f(x)$ for values of x between a and b is $\int f(x)\,dx$, with a and b as the limits of integration, divided by the difference between a and b.

(B) The *geometric mean* of or between, or the *mean proportional* between, two quantities is the (positive) square root of their product. Thus if b is the geometric mean between a and c, c is as many times greater (or less) than b as b is than a. The geometric mean of n quantities is the nth root of their product.

(C) The *harmonic mean* of two quantities is defined as the reciprocal of the arithmetic mean of their reciprocals. Hence the harmonic mean of a and b is $2ab/(a+b)$.

(D) The *weighted mean* or *weighted average* of a set of n quantities, each of which is associated with a certain number as weight, is obtained by multiplying each quantity by the associated weight, adding these products together, and then dividing by the sum of the weights. As under A, this may be extended to the case of an infinite set of quantities by means of integration. (The weights have the rôle of estimates of relative importance of the various quantities, and if all the weights are equal the weighted mean reduces to the simple arithmetic mean.)

(E) In statistics, given a population (i.e., an aggregate of observed or observable quantities) and a variable x having the population as its range, we have: (a) The *mean value* of x is the weighted mean of the values of x, with the probability (frequency ratio) of each value taken as its weight. In the case of a finite population this is the same as the simple arithmetic mean of the population, provided that, in calculating the arithmetic mean, each value of x is counted as many times over as it occurs in the set of observations constituting the population. (b) In like manner the *mean value* of a function $f(x)$ of x is the weighted mean of the values of $f(x)$, where the probability of each value of x is taken as the weight of the corresponding value of $f(x)$. (c) The *mode* of the population is the most probable (most frequent) value of x, provided there is one such. (d) The *mean* of the population is so chosen that the probability that x be less than the median (or the probability that x be greater than the median) is $\frac{1}{2}$ (or as near $\frac{1}{2}$ as possible). In the case of a finite population, if the values of x are arranged in order of magnitude—repeating any one value of x as many times over as it occurs in the set of observations constituting the population—then the middle term of this series, or the arithmetic mean of the two middle terms, is the median. —*A.C.*

Metalogical: The word is now commonly used as a synonym of *syntactical.* See *Syntax, logical.* —*A.C.*

Modality: Modality is the name given to certain classifications of propositions which are either supplementary to the classification into true and false or intended to provide categories additional to truth and falsehood—namely to classifications of propositions as possible, problematical, and the like. See *Strict implication*, and *Propositional calculus, Many-valued.*

Or, as in traditional logic, modality may refer to a classification of propositions according to the kind of assertion which is contained rather than have the character of a truth-value. From this point of view propositions are classed as *assertoric* (in which something is asserted as true), *problematic* (in which something is asserted as possible), and *apodeictic* (in which something is asserted as necessary). —*A.C.*

Name: A word or symbol which denotes (designates) a particular thing is called a *proper name* of that particular thing.

In English and other natural languages there occur also *common names* (common nouns), such a common name being thought of as if it could serve as a name of anything belonging to a specified class or having specified characteristics. Under usual translations into symbolic notation, common names are replaced by proper names of classes or of class concepts; and this would seem to provide the best logical analysis. In actual English usage, however, a common noun is often more nearly like a *variable* (q.v.) having a specified range. —*A.C.*

Name relation or *meaning relation*: The relation between a symbol (formula, word, phrase) and that which it denotes or of which it is the name.

Where a particular (interpreted) system does not contain symbols for formulas, it may be desirable to employ Gödel's device for associating (positive integral) numbers with formulas, and to consider the relation between a number and that which the associated formula denotes. This we shall call the *numerical name relation* and distinguish it from the relation between a formula and that which it denotes by calling the latter the *semantical name relation*.

In many (interpreted) logistic systems—including such as contain, with their usual interpretations, the Zermelo set theory, or the simple theory of types with axiom of infinity, or the functional calculus of second order with addition of Peano's postulates for arithmetic—it is impossible without contradiction to introduce the numerical name relation with its natural properties, because Grelling's paradox or similar paradoxes would result (see *Paradoxes, logical*). The same can be said of the semantical name relation in cases where symbols for formulas are present.

Such systems may, however, contain partial name relations which function as name relations in the case of some but not all of the formulas of the system (or of their associated Gödel numbers).

In particular, it is normally possible—at least it does not obviously lead to contradiction in the case of such systems as the Zermelo set theory or the simple theory of types (functional calculus of order omega) with axiom of infinity—to extend a system L_1 into a system L_2 (the semantics of L_1 in the sense of Tarski), so that L_2 shall contain symbols for the formulas of L_1, and for the essential syntactical relations between formulas of L_1, and for a relation which functions as a name relation as regards all the formulas of L_1 (or, in the case of the theory of types, one such relation for each type), together with appropriate new primitive formulas. Then L_2 may be similarly extended into L_3, and so on through a hierarchy of systems each including the preceding one as a part.

Or, if L_1 contains symbols for positive integers, we may extend L_1 into L_2 by merely adding a symbol for a relation which functions as a numerical name relation as regards all numbers of formulas of L_1 (or one such relation for each type) together with appropriate new primitive formulas; and so on through a hierarchy of systems $L_1, L_2, L_3, \ldots$.

See further *Semantics*; *Semiotic* 2; *Truth, Semantical.* —*A.C.*

Necessary: According to distinctions of *modality* (q.v.), a proposition is *necessary* if its truth is certifiable on *a priori* grounds, or on purely logical grounds. Necessity is thus, as it were, a stronger kind of truth, to be distinguished from *contingent truth* of a proposition which might have been otherwise. (As thus described, the notion is of course vague, but it may in various ways be given an exact counterpart in one logistic system or another.)

A proposition may also be said to be necessary if it is a consequence of some accepted set of propositions (indicated by the context), even if this accepted set of propositions is not held to be *a priori*. See *Necessity*.

That a propositional function F is necessary may mean simply $(x)F(x)$, or it may mean that $(x)F(x)$ is necessary in one of the preceding senses. —*A.C.*

Necessary condition: F is a *necessary condition* of G if $G(x) \supset_x F(x)$. F is a *necessary and sufficient condition* of G if $G(x) \equiv_x F(x)$. —*A.C.*

Negation: The negation of a proposition p is the proposition $\sim p$ (see *Logic, formal*, §1). The negation of a monadic propositional function F is the monadic propositional function $\lambda x[\sim F(x)]$; similarly for dyadic propositional functions, etc.

Or the word *negation* may be used in a syntactical sense, so that the negation of a sentence (formula) A is the sentence $\sim$A. —*A.C.*

Non causa pro causa, or *false cause*, is the fallacy, incident to the method of proof by *reductio ad absurdum* (q.v.), when a contradiction has been deduced from a number of assumptions of inferring the negation of one of the assumptions, say M, where actually it is one or more of the other assumptions which are false and the contradiction could have been deduced without use of M. This fallacy was committed, e.g., by Burali-

Forti in his paper of 1897 (see *Paradoxes, logical*) when he inferred the existence of ordinal numbers a, b such that a is neither less than, equal to, nor greater than b, upon having deduced what is now known as Burali-Forti's paradox from the contrary assumption: he had used without question the assumption that there is a class of all ordinal numbers. —*A.C.*

Non-contradiction, law of: Same as *Contradiction, law of* (q.v.). —*A.C.*

Non-Euclidean geometry: Euclid's postulates for geometry included one, the *parallel postulate*, which was regarded from earliest times (perhaps even by Euclid himself) as less satisfactory than the others. This may be stated as follows (not Euclid's original form but an equivalent one): *Through a given point P not on a given line l there passes at most one line, in the plane of P and l, which does not intersect l.* Here "line" means a straight line extended infinitely in both directions (not a line segment).

Attempts to prove the parallel postulate from the other postulates of Euclidean geometry were unsuccessful. The undertaking of Saccheri (1733) to make a proof by *reductio ad absurdum* of the parallel postulate by deducing consequences of its negation did, however, lead to his developing many of the theorems of what is now known as hyperbolic geometry. The proposal that this hyperbolic geometry, in which Euclid's parallel postulate is replaced by its negation, is a system equally valid with the Euclidean originated with Bolyai and Lobachevsky (independently, c. 1825). Proof of the self-consistency of hyperbolic geometry, and thus of the impossibility of Saccheri's undertaking, is contained in results of Cayley (1859) and was made explicit by Klein in 1871; for the two dimensional case another proof was given by Beltrami in 1868.

The name *non-Euclidean geometry* is applied to hyperbolic geometry and generally to any system in which one or more postulates of Euclidean geometry are replaced by contrary assumptions. (But geometries of more than three dimensions, if they otherwise follow the postulates of Euclid, are not ordinarily called non-Euclidean.)

Closely related to hyperbolic geometry is elliptic geometry, which was introduced by Klein on the basis of ideas of Riemann. In this geometry lines are of finite total length and closed, and every two coplanar lines intersect in a unique point.

Still other non-Euclidean geometries are given an actual application to physical space—or rather space-time—in the General Theory of Relativity.

Contemporary ideas concerning the abstract nature of *mathematics* (q.v.) and the status of applied geometry have important historical roots in the discovery of non-Euclidean geometries. —*A.C.*

G. Saccheri, *Euclides Vindicatus*, translated into English by G. B. Halsted, Chicago and London, 1920. H. P. Manning, *Non-Euclidean Geometry*, 1901. J. L. Coolidge, *The Elements of Non-Euclidean Geometry*, Oxford, 1909.

Non sequitur is any fallacy which has not even the deceptive appearance of valid reasoning, or in which there is a complete lack of connection between the premisses advanced and the conclusion drawn. By some, however, *non sequitur* is identified with Aristotle's *fallacy of the consequent*, which includes the two fallacies of *denial of the antecedent* (q.v.) and *affirmation of the consequent* (q.v.). —*A.C.*

Notations, logical: There follows a list of some of the logical symbols and notations found in contemporary usage. In each case the notation employed in articles in this dictionary is given first, afterwards alternative notations, if any.

PROPOSITIONAL CALCULUS (see, *Logic, formal*, §1, and *Strict implication*):

pq, the conjunction of p and q, "p and q." Instead of simple juxtaposition of the propositional symbols, a dot is sometimes written between as $p \cdot q$. Or the common abbreviation for *and* may be employed as a logical symbol, $p \,\&\, q$. Or an inverted letter v, usually from a gothic font [$\wedge$], may be used. In the Łukasiewicz notation for the propositional calculus, which avoids necessity for parentheses, conjunction of p and q is Kpq.

$p \vee q$, the inclusive disjunction of p and q, or "p or q." Frequently the letter v is from a gothic font [$\vee$]. In the Łukasiewicz notation, Apq is employed.

$\sim p$, the negation of p, "not p." Instead of $\sim$, a dash – may be used, written either before the propositional symbol or above it. Heyting adds a short downward stroke at the right end of the dash [$\neg$] (a notation which has come to be associated particularly with the intuitionistic propositional calculus and the intuitionistic concept of negation). Also employed is an accent $'$ after the propositional symbol (but this is more usual as a notation for the complement of a class). In the Łukasiewicz notation, the negation of p is Np.

$p \supset q$, the material implication of q by p, "if p then q." Also employed is a horizontal arrow, $p \rightarrow q$. The Łukasiewicz notation is Cpq.

$p \equiv q$, the material equivalence of p and q, "p if and only if q." Another notation which has sometimes been employed is $p \supset\subset q$. Other notations are a double horizontal arrow, with point at both ends [$p \leftrightarrow q$]; and two horizontal arrows, one above the other, one pointing forward and the other back [$\leftrightarrows$]. The Łukasiewicz notation is Epq.

$p + q$, the exclusive disjunction of p and q, "p or q but not both." Also sometimes used is the sign of material equivalence $\equiv$ with a vertical or slanting line [$\not\equiv$] across it (non-equivalence). In connection with the Łukasiewicz notation, Rpq has been employed.

$p \mid q$, the alternative denial of p and q, "not both p and q."—For the dual connective, joint denial ("neither p nor q"), a downward arrow has been used [$p \downarrow q$].

$p \prec q$, the strict implication of q by p, "p strictly implies q."

$p = q$, the strict equivalence of p *and* q, "p strictly implies q and q strictly implies p." Some recent writers employ, for strict equivalence, instead of Lewis's $=$, a sign similar to the sign of material equivalence, $\equiv$, but with four lines instead of three.

Mp, "p is possible." This is Łukasiewicz's notation and has been used especially in connection with his three-valued propositional calculus. For the different notion of possibility which is appropriate to the calculus of strict implication, Lewis employs a diamond [$\Diamond$].

CLASSES (see *Class*, and *Logic, formal*, §§7, 9):

$x \in a$, "x is a member of the class a," or, "x is an a." For the negation of this, sometimes a vertical line across the letter epsilon is employed, or a $\sim$ above it.

$a \subset b$, the inclusion of the class a in the class b, "a is a subclass of b." This notation is usually employed in such a way that $a \subset b$ does not exclude the possibility that $a = b$. Sometimes, however, the usage is that $a \subset b$ ("a is a proper subclass of b") does exclude that $a = b$; and in that case another notation is used when it is not meant that $a = b$ is excluded, the sign $\subset$ being either surcharged upon the sign = or written below it [$\subseteqq$] (or a single horizontal line below the $\subset$ may take the place of = [$\subseteq$]).

$\exists! a$, "the class a is not empty [has at least one member]," or, "a's exist."

ιx, or ι'x, or $\{x\}$—the unit class of x, i.e., the class whose single member is x.

V, the universal class. Where the algebra of classes is treated in isolation, the digit 1 is often used for the universal class.

Λ, the null or empty class. Where the algebra of classes is treated in isolation, the digit 0 is often used.

$-a$, the complement of a, or class of non-members of the class a. An alternative notation is a'.

$a \cup b$, the logical sum, or union of the classes a and b. Alternative notation, $a + b$.

$a \cap b$, the logical product, or intersection, or common part, of the classes a and b. Alternative notation, ab.

Relations (See *Relation*, and *Logic, formal*, §8) (where a notation used in connection with relations is here given as identical with a corresponding notation for classes, the relational notation will also often be found with a dot added to distinguish it from the one for classes):

xRy, "x has [or stands in, or bears] the relation R to y."

$R \subset S$, "the relation R is contained in [implies] the relation S."

$\exists! R$, "the relation R is not null [holds in at least one instance]."

A downward arrow placed between (e.g.) x and y denotes the relation which holds between x and y (in that order) and in no other case.

V, the universal relation. Schröder uses 1.

Λ, the null relation. Schröder uses 0.

$-R$, the contrary, or negation, of the relation R. The dash may also be placed over the letter R (or other symbol denoting a relation) instead of before it [$\overline{R}$].

$R \cup S$, the logical sum of the relations R and S, "R or S." Schröder uses $R + S$.

$R \cap S$, the logical product of the relations R and S, "R and S." Schröder uses $R \cdot S$.

I, the relation of identity—so that xIy is the same as $x = y$. Schröder uses 1'.

J, the relation of diversity—so that xJy is the same as $x \neq y$. Schröder uses 0'.

A breve ˘ is placed over the symbol for a relation to denote the converse relation. An alternative notation for the converse of R is Cnv'R.

$R + S$, the relative sum of R and S. Schröder adds a leftward hook at the bottom of the vertical line in the sign +.

$R \mid S$, the relative product of R and S. Schröder uses a semicolon to symbolize the relative product, but the vertical bar, or sometimes a slanted bar, is now the usual notation.

R^2, the square of the relation R, i.e., $R \mid R$. Similarly for higher powers of a relation, as R^3, etc.

R'y, the (unique) x such that xRy, "the R of y." Frequently the inverted comma is of a bold square (bold gothic) style.

R"b, the class of x's which bear the relation R to at least one member of the class b, "the R's of the b's." Then R"ιy or R"ι'y, is the class of x's such that xRy, "the R's of y."

A forward pointing arrow is placed over (e.g.) R to denote the relation of R"ιy to y. Similarly a backward pointing arrow placed over R denotes the relation of *the class of y's such that xRy* to x.

An upward arrow placed between (e.g.) a and b [$a \uparrow b$] denotes the relation which holds between x and y if and only if $x \in a$ and $y \in b$.

The left half of an upward arrow placed between (e.g.) a and R [$a \upharpoonleft R$] denotes the relation which holds between x and y if and only if $x \in a$ and xRy, in other words, the relation R with its domain limited to the class a.

The right half of an upward arrow placed between (e.g.) R and b [$R \upharpoonright b$] denotes the relation which holds between x and y if and only xRy and $y \in b$; in other words the relation R with its converse domain limited to b.

The right half of a double—upward and downward—arrow placed between (e.g.) R and a denotes the relation which holds between x and y if and only if xRy and both x and y are members of the class a; in other words, the relation R with its field limited to a.

D‘R, the domain of R.

Ɑ‘R, the converse domain of R.

C‘R, the field of R.

R_{po}, the proper ancestral of R—i.e., the relation which holds between x and y if and only if x bears the first or some higher power of the relation R to y (where the first power of R is R).

R_*, the ancestral of R—i.e., the relation which holds between x and y if and only if x bears the zero or some higher power of the relation R to y (where the zero power of R is taken to be, either I, or I with its field limited to the field of R).

QUANTIFIERS (see *Quantifiers*, and *Logic, formal*, §§3, 6):

(x), universal quantification with respect to x—so that (x)M may be read "for every x, M." An alternative notation occasionally met with, instead of (x), is $(\forall x)$, usually with the inverted A from a gothic or other special font [$\forall$]. Another notation is composed of a Greek capital pi with the x placed either after it, or before it, or as a subscript [Π_x].—Negation of the universal quantifier is sometimes expressed by means of a dash, or horizontal line, over it.

(Ex), existential quantification with respect to x—so that (Ex)M may be read "there exists an x such that M." The E which forms part of the notation may also be inverted; and, whether inverted or not, the E is frequently taken from a gothic or other special font [$\exists$]. An alternative notation employs a Greek capital sigma with x either after it or as a subscript [Σ_x].—Negation of the existential quantifier is sometimes expressed by means of a dash over it.

$\supset_x$, formal implication with respect to x. See definition in the article *Logic, formal*, §3.

$\equiv_x$, formal equivalence with respect to x. See definition in *Logic, formal*, §3.

$\bigwedge_x$. See definition in *Logic, formal*, §3.

$\supset_{xy}$ or $\supset_{x,y}$—formal implication with respect to x and y. Similarly for formal implication with respect to three or more variables.

$\equiv_{xy}$ or $\equiv_{x,y}$—formal equivalence with respect to x and y. Similarly for formal equivalence with respect to three or more variables.

ABSTRACTION, DESCRIPTIONS (see articles of those titles):

λx, functional abstraction with respect to x—so that λxM may be read "the (monadic) function whose value for the argument x is M."

$\hat{x}$, class abstraction with respect to x—so that $\hat{x}$M may be read "the class of x's such that M." An alternative notation, instead of $\hat{x}$, is $x\ni$.

$\hat{x}\hat{y}$, relation abstraction with respect to x and y—so that $\hat{x}\hat{y}$M may be read "the relation which holds between x and y if and only if M."

$(\imath x)$, description with respect to x—so that $(\imath x)$M may be read "the x such that M."

E! is employed in connection with descriptions to denote existence, so that E!$(\imath x)$M may be read "there exists a unique x such that M."

OTHER NOTATIONS:

$F(x)$, the result of application of the (monadic, propositional or other) function F to the argument x—the value of the function F for the argument x—"F of x." Sometimes the parentheses are omitted, so that the notation is Fx.—See the articles *Function*, and *Propositional function*.

$F(x, y)$, the result of application of the (dyadic) function F to the arguments x and y. Similarly for larger numbers of arguments.

$x = y$, the identity or equality of x and y, "x equals y." See *Logic, formal*, §§3, 6, 9.

$x \neq y$, negation of $x = y$.

$\vdash$ is the assertion sign. See *Assertion, logical.*

Dots (frequently printed as bold, or bold square, dots) are used in the punctuation of logical formulas, to avoid or replace parentheses. There are varying conventions for this purpose.

$\longrightarrow$ is used to express definitions, the definiendum being placed to the left and the definiens to the right. An alternative notation is the sign $=$ (or, in connection with the propositional calculus, $\equiv$) with the letters Df, or df, written above it, or as a subscript, or separately, after the definiens [$=_{\text{df}}$].

Quotation marks, usually single quotes, are employed as a means of distinguishing the name of a symbol or formula from the symbol or formula itself (see *Syntax, logical*). A symbol or formula between quotation marks is employed as a name of that particular symbol or formula. E.g., 'p' is a name of the sixteenth letter of the English alphabet in small italic type.

The reader will observe that this use of quotation marks has not been followed in the present article, and in fact that there are frequent inaccuracies from the point of view of strict preservation of the distinction between a symbol and its name. These inaccuracies are of too involved a character to be removed by merely supplying quotation marks at appropriate places. But it is thought that there is no point at which real doubt will arise as to the meaning intended. *Alonzo Church*

Nothing: In translation into logical notation, the word *nothing* is usually to be represented by the negation of an existential quantifier. Thus "nothing has the property F" becomes "$\sim(Ex)F(x)$." *—A.C.*

Noun: In English and other natural languages, a word serving as a proper or common *name* (q.v.). *—A.C.*

Number: The number system of mathematical analysis may be described as follows—with reference, not to historical, but to one possible logical order.

First are the *non-negative integers* 0, 1, 2, 3, . . . , for which the operations of addition and multiplication are determined. They are ordered by a relation *not greater than*—

which we shall denote by R—so that, e.g., $0R0, 0R3, 2R3, 3R3, 57R218$, etc.

These are extended by introducing, for every pair of non-negative integers a, b, with b different from 0, the fraction a/b, subject to the following conditions (which can be shown to be consistent): (1) $a/1 = a$; (2) $a/b = c/d$ if and only if $ad = bc$; (3) $a/b\,R\,c/d$ if and only if $ad\,R\,bc$; (4) $a/b + c/d = (ad + bc)/bd$; (5) $(a/b)(c/d) = ac/bd$. The resulting system is that of the *non-negative rational numbers*, which are compactly ordered but not continuously ordered (see *Continuity*) by the relation R (as extended).

Then the next step is to introduce, for every non-negative rational number r, a corresponding negative rational number $-r$, subject to the conditions: (1) $-r = -s$ if and only if $r = s$; (2) $-r = s$ if and only if $r = 0$ and $s = 0$; (3) $-r\,R\,-s$ if and only if sRr; (4) $-r\,R\,s$; (5) $s\,R\,-r$ if and only if $r = 0$ and $s = 0$; (6) $-r + s = s + -r =$ either t, where $r + t = s$, or $-t$ where $s + t = r$; (7) $-r + -s = -(r + s)$; (8) $(-r)s = s(-r) = -(rs)$; (9) $(-r)(-s) = rs$. Here r, s, t are variables whose range is the non-negative rational numbers. The extended system, comprising both non-negative rational numbers and negative rational numbers is the system of *rational numbers*—which are compactly ordered but not continuously ordered by the relation R (as extended).

If we make the *minimum* extension of the system of rational numbers which will render the order continuous, the system of *real numbers* results. Addition and multiplication of real numbers are uniquely determined by the meanings already given to addition and multiplication of rational numbers and the requirement that addition of, or multiplication by, a fixed real number (on right or left) shall be a continuous function (see *Continuity*). Subtraction and division may be introduced as inverses of addition and multiplication respectively.

Finally, the *complex numbers* are introduced as numbers $a + bi$, where a and b are real numbers. There is no ordering relation, but addition and multiplication are determined as follows:

$$(a + bi) + (c + di) = (a + c) + (b + d)i.$$
$$(a + bi)(c + di) = (ac - bd) + (ad + bc)i.$$

In particular i (i.e., $0 + 1i$) multiplied by itself is -1. A number of the form $a + 0i$ may be identified with the real number a; other complex numbers are called *imaginary numbers*, and those of the form $0 + bi$ are called *pure imaginaries*.

(It is, of course, not possible to *define* i as "the square root of -1." The foregoing statement corresponds to taking i as a new, undefined, symbol. But there is an alternative method, of *logical construction*, in which the complex numbers are defined as ordered pairs (a, b) of real numbers, and i is then defined as $(0, 1)$.)

In a mathematical development of the real number system or the complex number system, an appropriate set of postulates may be the starting point. Or the non-negative integers may first be introduced (by postulates or otherwise–see *Arithmetic, foundations of*) and from these the above outlined extensions may be provided for by successive logical constructions, in any one of several alternative ways.

The important matter is not the definition of *number* (or of particular numbers), which may be made in various ways more or less indifferently, but the internal structure of the number *system*.

For the notions of *cardinal number*, *relation-number*, and *ordinal number*, see the articles of these titles. *Alonzo Church*

R. Dedekind, *Essays on the Theory of Numbers*, translated by W. W. Beman, Chicago, 1901. E. V. Huntington, *A set of postulates for real algebra*, Transactions of the American Mathematical Society, vol. 6 (1905), pp. 17–41. E. V. Huntington, *A set of postulates for ordinary complex algebra*, ibid., pp. 209–229. E. Landau, *Grundlagen der Analysis*, Leipzig, 1930.

Object language: A language or logistic system L is called *object language* relatively to another language (metasystem) L' containing notations for formulas of L and for syntactical properties of and relations between formulas of L (possibly also semantical properties and relations). The language L' is called a *syntax language* of L.

See *Name relation; Syntax, Logical; Truth, semantical.* —*A.C.*

One-one: A relation R is *one-many* if for every y in the converse domain there is a unique x such that xRy. A relation R is *many-one* if for every x in the domain there is a unique y such that xRy. (See the article *Relation*). A relation is *one-one*, or *one-to-one*, if it is at the same time one-many and many-one. A one-one relation is said to be, or to determine, a *one-to-one correspondence* between its domain and its converse domain. —*A.C.*

Order: A class is said to be *partially ordered* by a dyadic relation R if it coincides with the field of R, and R is transitive and reflexive, and xRy and yRx never both hold when x and y are different. If in addition R is connected, the class is said to be *ordered* (or simply ordered) by R, and R is called an *ordering relation*.

Whitehead and Russell apply the term *serial relation* to relations which are transitive, irreflexive, and connected (and, in consequence, also asymmetric). However, the use of serial relations in this sense, instead of ordering relations as just defined, is awkward in connection with the notion of order for unit classes.

Examples: The relation *not greater than* among real numbers is an ordering relation. The relation *less than* among real numbers is a serial relation. The real numbers are simply ordered by the former relation. In the algebra of classes (*Logic, formal*, §7), the classes are partially ordered by the relation of class inclusion.

For explanation of the terminology used in making the above definitions, see the articles *Connexity, Reflexivity, Relation, Symmetry, Transitivity.* —*A.C.*

Ordinal number: A class b is *well-ordered* by a dyadic relation R if it is ordered by R (see *Order*) and, for every class a such that $a \subset b$, there is a member x of a, such that xRy holds for *every* member y of a; and R is then called a *well-ordering relation*. The *ordinal number* of a class b well-ordered by a relation R, or of a well-ordering relation R, is defined to be the *relation-number* (q.v.) of R.

The ordinal numbers of finite classes (well-ordered by appropriate relations) are called *finite* ordinal numbers. These are 0, 1, 2, ... (to be distinguished, of course, from the finite cardinal numbers 0, 1, 2, ...).

The first non-finite (transfinite or infinite) ordinal number is the ordinal number of the class of finite ordinal numbers, well-ordered in their natural order, 0, 1, 2, ...; it is usually denoted by the small Greek letter omega. —*A.C.*

G. Cantor, *Contributions to the Founding of the Theory of Transfinite Numbers*, translated and with an introduction by P. E. B. Jourdain, Chicago and London, 1915 (new ed. 1941). Whitehead and Russell, *Principia Mathematica*, vol. 3.

Paradoxes, logical: The ancient paradox of Epimenides the Cretan, who said that all Cretans were liars (i.e., absolutely incapable of telling the truth), was known under numerous variant forms in ancient and medieval times. The medieval name for these was *insolubilia*.

A form of this paradox due to Jourdain (1913) supposes a card upon the front of which are written the words, "On the other side of this card is written a true statement"—and nothing else. It seems to be clear that these words constitute a significant statement, since, upon turning the card over one must either find some statements written or not, and, in the former case, either there will be one of them which is true or there will not. However, on turning the card over there appear the words, "On the other side of this card is written a false statement"—and nothing else. Suppose the statement on the front of the card is true; then the statement on the back must be true; and hence the statement on the front must be false. This is a proof by *reductio ad absurdum* that the statement on the front of the card is false. But if the statement on the front is false, then the statement on the back must be false, and hence the statement on the front must be true. Thus the paradox.

A related but different paradox is Grelling's (1908). Let us distinguish adjectives—i.e., words denoting properties—as *autological* or *heterological* according as they do or do not have the property which they denote (in particular, adjectives denoting properties which cannot belong to words at all will be heterological). Then, e.g., the words *polysyllabic*, *common*, *significant*, *prosaic* are autological, while *new*, *alive*, *useless*, *ambiguous*, *long* are heterological. On their face, these definitions of *autological* and *heterological* are unobjectionable (compare the definition of *onomatopoetic* as *similar in sound to that which it denotes*). But paradox arises when we ask whether the word *heterological* is autological or heterological.

That paradoxes of this kind could be relevant to mathematics first became clear in connection with the paradox of the greatest ordinal number, published by Burali-Forti in 1897, and the paradox of the greatest cardinal number, published by Russell in 1903. The first of these had been discovered by Cantor in 1895, and communicated to Hilbert in 1896, and both are mentioned in Cantor's correspondence with Dedekind of 1899, but were never published by Cantor.

From the paradox of the greatest cardinal number Russell extracted the simpler paradox concerning the class t of all classes x such that $\sim x \in x$. (Is it true or not that $t \in t$?) At first sight this paradox may not seem to be very relevant to mathematics, but it must be remembered that it was obtained by comparing two mathematical proofs, both seemingly valid, one leading to the conclusion that there is no greatest cardinal number, the other to the conclusion that there is a greatest cardinal number.—Russell communicated this simplified form of the paradox of the greatest cardinal number to Frege in 1902 and published it in 1903. The same paradox was discovered independently by Zermelo before 1903 but not published.

Also to be mentioned are König's paradox (1905) concerning the least undefinable ordinal number and Richard's paradox (1905) concerning definable and undefinable real numbers.

Numerous solutions of these paradoxes have been proposed. Many, however, have the fault that, while they purport to find a flaw in the arguments leading to the paradoxes, no effective criterion is given by which to discover in the case of other (e.g., mathematical) proofs whether they have the same flaw.

Russell's solution of the paradoxes is embodied in what is now known as the *ramified theory of types*, published by him in 1908, and afterwards made the basis of *Principia Mathematica*. Because of its complication, and because of the necessity for the much-disputed *axiom of reducibility*, this has now been largely abandoned in favor of other solutions.

Another solution—which has recently been widely adopted—is the *simple theory of types* (see *Logic, formal*, §6). This was proposed as a modification of the ramified theory of types by Chwistek in 1921 and Ramsey in 1926, and adopted by Carnap in 1929.

Another solution is the Zermelo set theory (see *Logic, formal*, §9), proposed by Zermelo in 1908, but since considerably modified and improved.

Unlike the ramified theory of types, the simple theory of types and the Zermelo set theory both require the distinction (first made by Ramsey) between the paradoxes which involve use of the *name relation* (q.v.) or the semantical concept of *truth* (q.v.), and those which do not. The paradoxes of the first kind (Epimenides, Grelling's, König's, Richard's) are solved by the supposition that notations for the name relation and for truth (having the requisite formal properties) do not occur in the logistic system set up—and in principle, it is held, ought not to occur. The paradoxes of the second kind (Burali-Forti's, Russell's) are solved in each case in another way.

Alonzo Church

G. Frege, *Grundgesetze der Arithmetik*, vol. 2, Jena, 1903 (see Appendix). B. Russell, *The Principles of Mathematics*, Cambridge, England 1903; 2nd edn., London, 1937, and New York, 1938. Grelling and Nelson, *Bemerkungen zu den Paradoxieen von Russell und Burali-Forti*, Abhandlungen der Fries'schen Schule, n.s. vol. 2 (1908), pp. 301–334. A. Rüstow, *Der Lügner* (Dissertation Erlangen 1908), Leipzig, 1910. P. E. B. Jourdain, *Tales with philosophical morals*, The Open Court, vol. 27 (1913), pp. 310–315.

Particular proposition: In traditional logic, propositions *A*, *E* (excepting singular forms, according to some) were called *universal* and *I*, *O*, *particular*. See *Logic, formal*, §4. —*A.C.*

Peano, Giuseppe, Professor of mathematics at the University of Turin, 1890–1932. His work in mathematical logic marks a transition stage between the old algebra of logic and the newer methods. It is inferior to Frege's by present standards of rigor, but nevertheless contains important advances, among which may be mentioned the distinction between class inclusion ($\subset$) and class membership ($\in$)—which had previously been confused—and the introduction of a notation for formation of a class by *abstraction* (q.v.). His logical notations are more convenient than Frege's, and many of them are still in common use.

Peano's first publication on mathematical logic was the introduction to his *Calcolo Geometrico*, 1888. His postulates for arithmetic (see *Arithmetic, foundations of*) appeared in his *Arithmetices Principia* (1889) and in revised form in *Sul concetto di numero* (Rivista di Matematica, vol. 1 (1891)) and were repeated in successive volumes (more properly, editions) of his *Formulaire de Mathématiques* (1894–1908). The last-named work, written with the aid of collaborators, was intended to provide a reduction of all mathematics to symbolic notation, and often the encyclopedic aspect was stressed as much as, or more than, that of logical analysis.

Peano is known also for other contributions to mathematics, including the discovery of the area-filling curve which bears his name, and for his advocacy of *Latino sine flexione* as an international language. —*A. C.*

P. E. B. Jourdain, *Giuseppe Peano*, The Quarterly Journal of Pure and Applied Mathematics, vol. 43 (1912), pp. 270–314. *Giuseppe Peano*, supplement to Schola et Vita, Milan, 1928. U. Cassina, *Vita et opera de Giuseppe Peano*, Schola et Vita, vol. 7 (1932), pp. 117–148. E. Stamm, *Józef Peano*, Wiadomości Matematyczne, vol. 36 (1933), pp. 1–56. U. Cassina, *L'opera scientifica di Giuseppe Peano*, Rendiconti del Seminario Matematico e Fisico di Milano, vol. 7 (1933), pp. 323–389. U. Cassina, *L'oeuvre philosophique de G. Peano*, Revue de Métaphysique et de Morale, vol. 40 (1933), pp. 481–491.

Peirce's law: The theorem of the propositional calculus,

$$[[p \supset q] \supset p] \supset p.$$

—*A. C.*

Petitio principii, fallacy involving the assumption as premisses of one or more propositions which are identical with (or in a simple fashion equivalent to) the conclusion to be proved, or which would require the conclusion for their proof, or which are stronger than the conclusion and contain it as a particular case or otherwise as an immediate consequence. There is a fallacy, however, only if the premisses assumed (without proof) are illegitimate for some other reason than merely their relation to the conclusion—e.g., if they are not among the avowed presuppositions of the argument, or if they are not admitted by an opponent in a dispute. —*A. C.*

Planck's constant: In *quantum mechanics* (q.v.), a fundamental physical constant, usually denoted by the letter h, which appears in many physical formulas. It may be defined by the law that the *quantum* (q.v.) of radiant energy of any frequency is equal to the frequency multiplied by h. See further *Uncertainty principle*. —*A. C.*

Polysyllogism: A chain of syllogisms arranged to lead to a single final conclusion, the conclusion of each syllogism except the last serving as premiss of a later syllogism.

In contrast, an argument consisting of a single syllogism is called a *monosyllogism*.

—*A. C.*

Possibility: According to distinctions of *modality* (q.v.), a proposition is *possible* if its negation is not necessary. The word *possible* is also used in reference to a state of knowledge rather than to modality; as a speaker might say, "It is possible that 486763 is a prime number," meaning that he had no information to the contrary (although this proposition is impossible in the sense of modality).

A propositional function F may also be said to be *possible*. In this case the meaning may be either simply $(Ex)F(x)$; or that $(Ex)F(x)$ is possible in one of the senses just described; or that $F(x)$ is permitted under some particular system of conventions or code of laws. As an example of the last we may take: "It is possible for a woman to be President of the United States." Here F is $\lambda x[x$ is a woman and x is a President of the United States], and the code of laws in question is the Constitution of the United States. —*A. C.*

Predicate: The four traditional kinds of categorical propositions (see *Logic, formal,* §4) are: all S is P, no S is P, some S is P, some S is not P. In each of these the concept denoted by S is the *subject* and that denoted by P is the *predicate*.

Hilbert and Ackermann use the word *predicate* for a propositional function of one or more variables; Carnap uses it for the corresponding syntactical entity, the name or designation of such propositional function (i.e., of a property or relation). —*A. C.*

Premiss: A proposition, or one of several propositions, from which an inference is drawn; or the sentence expressing such a proposition. Following C. S. Peirce, we here prefer the spelling *premiss*, to distinguish the word *premise* in other senses (in particular to distinguish the plural from the legal term *premises*). —*A. C.*

Proof by cases: Represented in its simplest form by the valid inference of the propositional calculus, from A $\supset$ C and B $\supset$ C and A $\vee$ B to C. More complex forms involve multiple disjunctions, e.g., the inference from A $\supset$ D and B $\supset$ D and C $\supset$ D and $[A \vee B] \vee C$ to D. The simplest form of proof by cases is thus the same as the simple constructive dilemma (see *Logic, formal,* §2), the former term deriving from mathematical usage and the latter from traditional logic. For the more complex forms of proof by cases, and like generalizations of the other kinds of dilemma to the case of more than two major premisses, logicians have devised the names trilemma, tetralemma, polylemma—but these are not much found in actual use. —*A. C.*

Proof theory: The formalization of mathematical proof by means of a *logistic system* (q.v.) makes possible an objective theory of proofs and provability, in which proofs are treated as concrete manipulations of formulas (and no use is made of meanings of formulas). This is Hilbert's *proof theory*, or *metamathematics*.

A central problem of proof theory, according to Hilbert, is the proof of consistency of logistic systems adequate to mathematics or substantial parts of mathematics.—A logistic system is said to be consistent, relatively to a particular notation in the system called negation, if there is no formula A such that both A and the negation of A are provable (i.e., are theorems). The systems with which Hilbert deals, and the notations in them which he wishes to call negation, are such that, if a formula A and its negation were once proved, every propositional formula could be proved; hence he is able to formulate the consistency by saying that a particular formula (e.g., $\sim[0 = 0]$) is not provable.

A consistency proof evidently loses much of its significance unless the methods employed in the proof are in some sense less than, or less dubitable than, the methods of proof which the logistic system is intended to formalize. Hilbert required that the methods employed in a consistency proof should be *finitary*—a condition more stringent than that of intuitionistic acceptability. See *Intuitionism (mathematical)*.

Gödel's theorems (see *Logic, formal,* §6) are a difficulty for the Hilbert program because they show that the methods employed in a consistency proof must also be in some sense more than those which the logistic system formalizes. Gödel himself remarks that the difficulty may not be insuperable.

Other problems of proof theory are the *decision problem*, and the problem of proving *completeness* (in one of various senses) for a logistic system. Cf. *Logic, formal*, §§1, 3. —*A.C.*

Hilbert and Bernays, *Grundlagen der Mathematik*, vol. 1, Berlin, 1934, and vol. 2, Berlin, 1939. P. Bernays, *Sure le platonisme dans les mathématiques*, and *Quelques points essentiels de la métamathématique*, l'Enseignement Mathématique, vol. 34 (1935), pp. 52–95. W. Ackermann, *Zur Widerspruchsfreiheit der Zahlentheorie*, Mathematische Annalen, vol. 117 (1940), pp. 162–194.

Proposition: This word has been used to mean: (a) a declarative sentence (in some particular language); (b) the content of meaning of a declarative sentence, i.e., a postulated abstract object common not only to different occurrences of the same declarative sentence but also to different sentences (whether of the same language or not) which are synonymous or, as we say, mean the same *thing*; (c) a declarative sentence associated with its content of meaning. Often the word proposition is used ambiguously between two of these meanings, or among all three.

The Port-Royal Logic defines a proposition to be the same as a judgment but elsewhere speaks of propositions as denoting judgments. Traditional logicians generally have defined a proposition as a judgment expressed in words, or as a sentence expressing a judgment, but some say or seem to hold in actual usage that synonymous or intertranslatable sentences represent the same proposition. Recent writers in many cases adopt or tend toward (b).

In articles in this dictionary by the present writer the word proposition is to be understood in sense (b) above. This still leaves an element of ambiguity, since common usage does not always determine of two sentences whether they are strictly synonymous or merely logically equivalent. For a particular language or logistic system, this ambiguity may be resolved in various ways. —*A.C.*

Propositional calculus, many-valued: The truth-table method for the classical (two-valued) propositional calculus is explained in the article *Logic, formal*, §1. It depends on assigning truth-tables to the fundamental connectives, with the result that every formula—of the pure propositional calculus, to which we here restrict ourselves for the sake of simplicity—has one of the two truth-values for each possible assignment of truth-values to the variables appearing. A formula is called a *tautology* if it has the truth-value truth for every possible assignment of truth-values to the variables; and the calculus is so constructed that a formula is a theorem if and only if it is a tautology.

This may be generalized by arbitrarily taking n different truth-values $t_1, t_2, \ldots, t_m$, $f_1, f_2, \ldots, f_{n-m}$, of which the first m are called *designated* values—and then setting up truth-tables (in terms of these n truth-values) for a set of connectives, which usually includes connectives notationally the same as the fundamental connectives of the classical calculus, and may also include others. A formula constructed out of these connectives and variables is then called a *tautology* if it has a designated value for each possible assignment of truth-values to the variables, and the theorems of the n-valued propositional calculus are to coincide with the tautologies.

In 1920, Łukasiewicz introduced a three-valued propositional calculus, with one designated value (interpreted as *true*) and two non-designated values (interpreted as *problematical* and *false* respectively). Later he generalized this to n-valued propositional calculi with one designated value (first published in 1929). Post introduced n-valued propositional calculi with an arbitrary number of designated values in 1921. Also due to Post (1921) is the notion of *symbolic completeness*—an n-valued propositional calculus is symbolically complete if every possible truth-function is expressible by means of the fundamental connectives.

The case of infinitely many truth-values was first considered by Łukasiewicz.

—*A. C.*

J. Łukasiewicz, *O logice trójwartościowej*, Ruch Filozoficzny, vol. 5 (1920), pp. 169–171. E. L. Post, *Introduction to a general theory of elementary propositions*, American Journal of Mathematics, vol. 43 (1921), pp. 163–185. Łukasiewicz and Tarski, *Untersuchungen über den Aussagenkalkül*, Comptes Rendus des Séances de la Sociéte des Sciences et des Lettres de Varsovie, Classe III, vol. 23 (1930), pp. 30–50. J. Łukasiewicz, *Philosophische Bemerkungen zu mehrwertigen Systemen des Aussagenkalküls*, ibid., pp. 51–77. Lewis and Langford, *Symbolic Logic*, New York and London, 1932.

Propositional function is a *function* (q.v.) for which the range of the dependent variable is composed of *propositions* (q.v.). A monadic propositional function is thus in substance a *property* (of things belonging to the range of the independent variable), and a dyadic propositional function a *relation*. If F denotes a propositional function and $X_1, X_2, \ldots, X_n$ denote arguments, the notation $F(X_1, X_2, \ldots, X_n)$—or $[F](X_1, X_2, \ldots, X_n)$—is used for the resulting proposition, which is said to be the *value* of the propositional function for the given arguments, and to be obtained from the propositional function by *applying* it to, or *predicating* it of the given arguments.

Often, however, the assumption is made that two propositional functions are identical if corresponding values are materially equivalent and in this case we speak of propositional functions *in extension* (the definition in the preceding paragraph applying rather to propositional functions *in intension*). The values of a propositional function in extension are *truth-values* (q.v.) rather than propositions. A monadic propositional function in extension is not essentially different from a *class* (q.v.).

Whitehead and Russell use the term *propositional function* in approximately the sense above described, but qualify it by holding, as a corollary of Russell's doctrine of *descriptions* (q.v.), that propositional functions are the fundamental kind from which other kinds of functions are derived—in fact that non-propositional ("descriptive") functions do not exist except as incomplete symbols. For details of their view, which underwent some changes between publication of the first and the second edition of *Principia Mathematica*, the reader is referred to that work.

Historically, the notion of a function was of gradual growth in mathematics. The word *function* is used in approximately its modern sense by John Bernoulli (1698, 1718). The divorce of the notion of a function from that of a particular kind of mathematical expression (analytic or quasi-algebraic) is due to Dirichlet (1837). The general logical notion of a function, and in particular the notion of a propositional function, were introduced by Frege (1879). *Alonzo Church*

Quality: The four traditional kinds of categorical propositions (see *Logic, formal,*

§4) were distinguished according to *quality* as affirmative or negative, and according to *quantity* as particular, singular, or universal. See the articles *Affirmative Proposition* and *Particular Proposition.* —*A. C.*

Quantifier: *Universal quantifier* is the name given to the notation (x) prefixed to a logical formula A (containing the free variable x) to express that A holds for *all* values of x—usually, for all values of x within a certain range or domain of values, which either is implicit in the context, or is indicated by the notation through some convention. The same name is also given to variant or alternative notations employed for the same purpose. And of course the same name is given when the particular variable appearing is some other letter than x.

Similarly, *existential quantifier* is the name given to the notation (Ex) prefixed to a logical formula A (containing the free variable x) to express that A holds for *some* (i.e., at least one) value of x—usually, for some value of x within a certain range or domain. The E which forms part of the notation is often inverted, and various alternative notations also occur.

It may also be allowed to prefix the quantifiers (x) and (Ex) to a formula (sentence) A not containing x as a free variable, (x)A and (Ex)A then having each the same meaning as A.

See *Logic, formal*, §3. —*A. C.*

W. V. Quine, *Elementary Logic*, Boston and New York, 1941.

Quaternio terminorum: In the categorical syllogism (*Logic, formal*, §5), the major and minor premisses must have a term in common, the middle term. Violation of this rule is the fallacy of *quaternio terminorum*, or of *four terms*. It is most apt to arise through *equivocation* (q.v.), an ambiguous word or phrase playing the rôle of the middle term, with one meaning in the major premiss and another meaning in the minor premiss; and in this case the fallacy is called the fallacy of *ambiguous middle*. —*A. C.*

Quantum: An indivisible unit, or atom, of any physical quantity. *Quantum mechanics* (q.v.) is based on the existence of quanta of energy, the magnitude of the quantum of radiant energy (light) of a given frequency—or of the energy of a particle oscillating with a given frequency—being equal to *Planck's constant* (q.v.) multiplied by the frequency. —*A. C.*

Quantum mechanics: An important physical theory, a modification of classical mechanics, which has arisen from the study of atomic structure and phenomena of emission and absorption of light by matter; embracing the matrix mechanics of Heisenberg, the wave mechanics of Schrödinger, and the transformation theory of Jordan and Dirac. The wave mechanics introduces a duality between waves and particles, according to which an electron, or a photon (quantum of light) is to be considered in some of its aspects as a wave, in others as a particle. See further *quantum* and *uncertainty principle*. —*A. C.*

F. A. Lindemann, *The Physical Significance of the Quantum Theory*, Oxford, 1932. J. Frenkel, *Wave Mechanics, Elementary Theory*, Oxford, 1932. Louis de Broglie, *Matter and Light, The*

New Physics, translated by W. H. Johnston, New York, 1939.

Recursion, definition by: A method of introducing, or "defining," functions from non-negative integers to non-negative integers, which, in its simplest form, consists in giving a pair of equations which specify the value of the function when the argument (or a particular one of the arguments) is 0, and supply a method of calculating the value of the function when the argument (that particular one of the arguments) is $x + 1$, from the value of the function when the argument (that particular one of the arguments) is x. Thus a monadic function f is said to be defined by *primitive recursion* in terms of a dyadic function g—the function g being previously known or given—by a pair of equations,

$$f(0) = \mathrm{A},$$
$$f(S(x)) = g(x, f(x)),$$

where A denotes some particular, non-negative integer, and S denotes the *successor* function (so that $S(x)$ is the same as $x + 1$), and x is a variable (the second equation being intended to hold for all non-negative integers x). Similarly the dyadic function f is said to be defined by *primitive recursion* in terms of a triadic function g and a monadic function h by the pair of equations,

$$f(a, 0) = h(a),$$
$$f(a, S(x)) = g(a, x, f(a, x)),$$

the equations being intended to hold for all non-negative integers a and x. Likewise for functions f of more than two variables.—As an example of definition by primitive recursion we may take the "definition" of addition (i.e., of the dyadic function *plus*) employed by Peano in the development of arithmetic from his postulates (see the article *Arithmetic, foundations of*):

$$a + 0 = a,$$
$$a + S(x) = S(a + x),$$

This comes under the general form of definition by primitive recursion just given, with h and g taken to be such functions that $h(a) = a$ and $g(a, x, y) = S(y)$. Another example is Peano's introduction of multiplication by the pair of equations:

$$a \times 0 = 0,$$
$$a \times S(x) = (a \times x) + a,$$

Here addition is taken as previously defined, and $h(a) = 0$, $g(a, x, y) = y + a$.

More general kinds of definition by recursion allow sets of recursion equations of various forms, the essential requirement being that the equations specify the value of the function being introduced (or the values of the functions being introduced), for any given set of arguments, either absolutely, or in terms of the value (values) for preceding sets of arguments. The word *preceding* here may refer to the natural order or order of magnitude of the non-negative integers, or it may refer to some other method of ordering arguments or sets of arguments; but the method of ordering shall be such that *infinite* descending sequences of sets of arguments (in which each set of arguments is *preceded* by the next set) are impossible.

The notion of definition by recursion may be extended to functions whose ranges consist of only a portion of the non-negative integers (in the case of monadic functions) or of only a portion of the ordered sets of n non-negative integers (in the case of n-adic functions); also to functions for which the range of the dependent variable may consist wholly or partly of other things than non-negative integers (in particular, propositional functions—properties, relations—of integers may receive definition by recursion).

The employment of definition by recursion in the development of arithmetic from Peano's postulates, or in the Frege–Russell derivation of arithmetic from logic, requires justification, which most naturally takes the form of finding a method of replacing a definition by recursion by a nominal definition, or a contextual definition, serving the same purpose. In particular it is possible, by a method due to Dedekind or by any one of a number of modifications of it, to prove the existence of a function f satisfying the conditions expressed by an admissible set of recursion equations; and f may then be given a definition employing descriptions, as *the function f such that* the recursion equations, with suitable quantifiers prefixed, hold. See the paper of Kalmár cited below.

See also the article *Recursiveness*. —*A.C.*

L. Kalmár, *On the possibility of definition by recursion*, Acta Scientiarum Mathematicarum (Szeged), vol. 9 (1940), pp. 227–232.

Recursion, proof by, or, as it is more often called, proof by *mathematical induction* or *complete induction*, is in its simplest form a proof that every non-negative integer possesses a certain property by showing (1) that 0 possesses this property, and (2) that, on the hypothesis that the non-negative integer x possesses this property, then $x + 1$ possesses this property. (The condition (2) is often expressed, following Frege and Russell, by saying that the property is *hereditary* in the series of non-negative integers.) The name *proof by recursion*, or *proof by mathematical* or *complete induction*, is also given to various similar but more complex forms.

In Peano's postulates for arithmetic (see *Arithmetic, foundations of*) the possibility of proof by recursion is secured by the last postulate, which, indeed, merely states the leading principle of the simplest form of proof by recursion. In the Frege–Russell derivation of arithmetic from logic, the non-negative integers are identified with the inductive *cardinal numbers* (q.v.), the possibility of proof by recursion being implicit in the definition of *inductive*. —*A.C.*

Recursiveness: The notion of definition by recursion, and in particular of definition by primitive recursion, is explained in the article *Recursion, definition by*. An n-adic function f (from non-negative integers to non-negative integers) is said to be defined by *composition* in terms of the m-adic function g and the n-adic functions $h_1, h_2, \dots, h_m$ by the equation:

$$f(x_1, x_2, \dots, x_n) = g(h_1(x_1, x_2, \dots, x_n), \\ h_2(x_1, x_2, \dots, x_n), \dots, h_m(x_1, x_2, \dots, x_n)).$$

(The case is not excluded that $m = 1$, or $n = 1$, or both.)

A function from non-negative integers to non-negative integers is said to be *primitive recursive* if it can be obtained by a succession of definitions by primitive recursion and composition, from the following list of initial functions: the successor function S, the function C such that $C(x) = 0$ for every non-negative integer x, and the functions U_{in}

$(i \leq n, n = 1, 2, 3, \ldots)$ such that $U_{in}(x_1, x_2, \ldots, x_n) = x_i$. Each successive definition by primitive recursion or composition may employ not only the initial functions but also any of the functions which were introduced by previous definitions.

More general notions of recursiveness result from admitting, in addition to primitive recursion, also more general kinds of definition by recursion, including those in which several functions are introduced simultaneously by a single set of recursion equations. The most general such notion is that of *general recursiveness*—see the first paper of Kleene cited below. Notions of recursiveness may also be introduced for a function whose range consists of only a portion of the non-negative integers (in the case of a monadic function) or of only a portion of the ordered sets of n non-negative integers (in the case of an n-adic function)—see the second paper of Kleene cited.

Concerning the relationship between general recursiveness and the notion of effectiveness, see the article *Logistic system*. —*A. C.*

R. Péter, a series of papers (in German) in the Mathematische Annalen, vol. 110 (1934), pp. 612–632; vol. 111 (1935), pp. 42–60; vol. 113 (1936), pp. 489–527. S. C. Kleene, *General recursive functions of natural numbers*, Mathematische Annalen, vol. 112 (1936), pp. 727–742. S. C. Kleene, *On notation for ordinal numbers*, The Journal of Symbolic Logic, vol. 3 (1938), pp. 150–155.

Reducibility, axiom of: An axiom which (or some substitute) is necessary in connection with the *ramified theory of types* (q.v.) if that theory is to be adequate for classical mathematics, but the admissibility of which has been much disputed (see Paradoxes, logical). An exact statement of the axiom can be made only in the context of a detailed formulation of the ramified theory of types—which will not here be undertaken. As an indication or rough description of the axiom of reducibility, it may be said that it cancels a large part of the restrictive consequences of the prohibition against *impredicative definition* (q.v.) and, in approximate effect, reduces the ramified theory of types to the *simple theory of types* (for the latter see *Logic formal*, §6). —*A. C.*

Reductio ad absurdum: The method of proving a proposition by deducing a contradiction from the negation of the proposition taken together with other propositions which were previously proved or are granted. It may thus be described as the valid inference of the propositional calculus from three premisses, B and B[∼A] ⊃ C and B[∼A] ⊃ ∼C, to the conclusion A (this presupposes the *deduction theorem*, q.v.). Such an argument may be rearranged so that the element of *reductio ad absurdum* appears in the inference from ∼A ⊃ A to A.

The name *reductio ad absurdum* is also given to the method of proving the negation of a proposition by deducing a contradiction from the proposition itself, together with other propositions which were previously proved or are granted.

The first of the two kinds of *reductio ad absurdum*, but not the second, is called *indirect proof*.

Whitehead and Russell give the name *principle of reductio ad absurdum* to the theorem of the propositional calculus:

$$[p \supset \sim p] \supset \sim p.$$

—*A. C.*

Reflexivity: A dyadic relation R is called *reflexive* if xRx holds for all x within a certain previously fixed domain which must include the field of R (cf. *Logic, formal*, §8). In the propositional calculus, the *laws of reflexivity* of material implication and material equivalence (the conditional and biconditional) are the theorems,

$$p \supset p, \qquad\qquad p \equiv p,$$

expressing the reflexivity of these relations. Other examples of reflexive relations are equality; class inclusion, $\subset$ (see *Logic, formal*, §7); formal implication and formal equivalence (see *Logic, formal*, §3); the relation *not greater than* among whole numbers, or among rational numbers, or among real numbers; the relation *not later than* among instants of time; the relation *less than one hour apart* among instants of time.

A dyadic relation R is *irreflexive* if xRx never holds (e.g., the relation *less than* among whole numbers). —*A. C.*

Relation: The same as dyadic *propositional function* (q.v.). The distinction between relations *in intension* and the relations *in extension* is the same as that for propositional functions.—Sometimes the word *relation* is used to mean a propositional function of *two or more* variables, and in this case one distinguishes *binary* (dyadic) relations, *ternary* (triadic) relations, etc.

If R denotes a (binary) relation, and X and Y denote arguments, the notation XRY may be used, instead of R(X, Y), to mean that the two arguments stand in the relation denoted by R. The *domain* of a relation R is the class of things x for which there exists at least one y such that xRy holds. The *converse domain* of a relation R is the class of things y for which there exists at least one x such that xRy. The *field* of a relation is the logical sum of the domain and the converse domain.

See also *Logic, formal*, §8. —*A. C.*

Whitehead and Russell, *Principia Mathematica*, 2nd edn., vol. 1, Cambridge, England, 1925.

Relation-number: Dyadic relations R and R' are said to be *similar* (or *ordinally similar*) if there exists a one-one relation S whose domain is the field of R and whose converse domain is the field of R', such that if aSa' and bSb', then aRb if and only if $a'R'b'$. The *relation-number* of a dyadic relation may then be defined as the class of relations similar to it—cf. *cardinal number*.

The relation-number of an ordering relation (see *order*) is called also an *ordinal type* or *order type*.

The notion of a relation-number may be extended in a straightforward way to polyadic relations. —*A. C.*

Whitehead and Russell, *Principia Mathematica*, vol. 2.

Relative: A concept is *relative* if it is—a word, if it denotes—a polyadic propositional function, or relation, rather than a monadic propositional function. The term *relative* is applied especially to words which have been or might be thought to denote monadic propositional functions, but for some reason must be taken as denoting relations. Thus the word *short* or the notion of shortness may be called relative because as a monadic propositional function it is vague, while as a relation (*shorter than*) it is not vague.

Analogously, the term *relative* may be applied to words erroneously thought of or used as if denoting binary relations, but which actually must be taken as denoting ternary or quaternary relations; etc. E.g., the Special Theory of Relativity is said to make simultaneity relative because, according to it, simultaneity is a function of two events *and a coordinate system or frame of reference*—instead of a function merely of two events, as in the Newtonian or classical theory.

The adjective *relative* is also used in a less special way, to mean simply *relational* or *pertaining to relations*.

In connection with the algebra of relations (see *Logic, formal*, §8), Peirce and Schröder use *relative* as a noun, in place of *relation*. For Schröder, a relative (Relativ) is a relation in extension. Peirce makes a distinction between *relative* and *relation*, not altogether clear; many passages suggest that *relative* is a syntactical term, but others approximate the usage adopted by Schröder. —*A.C.*

Relativity, theory of: A mathematical theory of *space-time* (q.v.), of profound epistemological as well as physical importance, comprising the special theory of relativity (Einstein, 1905) and the general theory of relativity (Einstein, 1914–16). The name arises from the fact that certain things which the classical theory regarded as absolute—e.g., the simultaneity of spatially distant events, the time elapsed between two events (unless coincident in space-time), the length of an extended solid body, the separation of four-dimensional space-time into a three-dimensional space and a one-dimensional time—are regarded by the relativity theory as *relative* (q.v.) to the choice of a coordinate system in space-time, and thus relative to the observer. But on the other hand the relativity theory represents as absolute certain things which are relative in the classical theory—e.g., the velocity of light in empty space. See *Non-Euclidean geometry*. —*A.C.*

Albert Einstein, *Relativity, The Special & The General Theory, A Popular Exposition*, translated by R. W. Lawson, London, 1920. A. S. Eddington, *Space, Time, and Gravitation*, Cambridge, England, 1920. A. V. Vasiliev, *Space, Time, Motion*, translated by H. M. Lucas and C. P. Sanger, with an introduction by Bertrand Russell, London, 1924, and New York, 1924.

Schröder, (Friedrich Wilhelm Karl) Ernst, 1841–1902, German mathematician. Professor of mathematics at Karlsruhe, 1876–1902. His three-volume *Algebra der Logik* (1890–1895, with a posthumous second part of vol. 2 published in 1905) is an able compendium and systematization of the work of his predecessors, with contributions of his own, and may be regarded as giving in nearly all essentials the final form of the Nineteenth Century *algebra of logic* (q.v.), including the algebra of relatives (or relations). —*A.C.*

J. Lüroth, *Ernst Schröder*, Jahresbericht der Deutschen Mathematiker-Vereinigung, vol. 12 (1903), pp. 249–265; reprinted in Schröder's *Algebra der Logik*, vol. 2, part 2.

Secundum quid: Secundum quid, or more fully, *a dicto simpliciter ad dictum secundum quid*, is any fallacy arising from the use of a general proposition without attention to tacit qualifications which would invalidate the use made of it. —*A.C.*

Semantics: The theory of the relation between the formulas of an *interpreted logistic system* (*semantical system* in Carnap's terminology) and their meanings. See *Name relation*; *Semiotic* 2; and *Truth, semantical*. —*A.C.*

R. Carnap, *Foundations of Logic and Mathematics*, International Encyclopedia of Unified Science, vol. 1, no. 3, Chicago, 1939.

Sentence: In connection with logic, and logical syntax, the word *sentence* is used for what might be called more explicitly a *declarative sentence*—thus for a sequence of words or symbols which (in some language or system of notation, as determined by the context) expresses a *proposition* (q.v.), or which can be used to convey an assertion. A sequence of words or symbols which contains free variables and which expresses a proposition when values are given to these variables (see the article *Variable*) may also be called a sentence.

In connection with logistic systems, *sentence* is often used as a technical term in place of *formula* (see the explanation of the latter term in the article *Logistic system*). This may be done when, under the intended interpretation of the system, sentences in this technical or formal sense become sentences in the sense of the preceding paragraph.

—*A. C.*

Sentential calculus: Same as *propositional calculus* (see *Logic, formal*, §1). —*A. C.*

Sentential function has been used by some as a syntactical term, to mean a *sentence* (q.v.) containing free variables. This notion should not be confused with that of a *propositional function* (q.v.); the relationship is that a propositional function may be obtained from a sentential function by *abstraction* (q.v.). —*A. C.*

Signification: *Signify* may be synonymous with *designate* (q.v.), or it may be used rather for the meaning of words which are not or are not thought of as proper names, or it may be used to indicate the intensional rather than the extensional meaning of a word. —*A. C.*

Some: It is now recognized that to construe such a phrase as, e.g., "some men" as a name of an undetermined [non-empty] part of the class of men (thus as a sort of variable) constitutes an inadequate analysis. In translation into an exact logical notation the word "some" is usually to be represented by an existential *quantifier* (q.v.). —*A. C.*

Sorites: A chain of (categorical) syllogisms, the conclusion of each forming a premiss of the next—traditionally restricted to a chain of syllogisms in the first figure (all of which, with the possible exception of the first and last, must then be syllogisms in Barbara).

In the statement of a sorites all conclusions except the last are suppressed, and in fact the sorites may be thought of as a single valid inference independently of analysis into constituent syllogisms. According to the order in which the premisses are arranged, the sorites is called *progressive* (if in the analysis into syllogisms each new premiss after the first is a major premiss, and each intermediate conclusion serves as a minor premiss for the next syllogism) or *regressive or Goclenian* (if each new premiss after the first is a minor premiss, and each intermediate conclusion a major premiss). —*A. C.*

Species: A relatively narrow class—or better, class concept—thought of as included (in the sense of class inclusion, $\subset$) within a wider class—or class concept—the *genus*.
—*A.C.*

Strict implication: As early as 1912, C. I. Lewis projected a kind of implication between propositions, to be called *strict implication*, which should more nearly accord with the usual meaning of "implies" than does *material implication* (see *Logic, formal*, §1), should make "p implies q" synonymous with "q is deducible from p," and should avoid such so-called paradoxes of material implication as the theorem $[p \supset q] \vee [q \supset p]$. The first satisfactory formulation of a calculus of propositions with strict implication appeared in 1920, and this system, and later modified forms of it, have since been extensively investigated. An essential feature is the introduction of modalities through the notation (say) $M[p]$, to mean "p is possible" (Lewis uses a diamond instead of M). The strict implication of q by p is then identified with $\sim M[p \sim q]$, whereas the material implication $p \supset q$ is given by $\sim[p \sim q]$. In 1932 Lewis, along with other modifications, added a primitive formula (involving the binding of propositional variables by existential quantifiers) which renders definitively impossible an interpretation of the system which would make Mp the same as p and strict implication the same as material implication. Consistency of the system, including this additional primitive formula, may be established by means of an appropriate four-valued propositional calculus, the theorems of the system being some among the tautologies of the four-valued propositional calculus. —*A.C.*

Lewis and Langford, *Symbolic Logic*, New York and London, 1932. E. V. Huntington, *Postulates for assertion conjunction, negation, and equality*, Proceedings of the American Academy of Arts and Sciences, vol. 72, no. 1, 1937. W. T. Parry, *Modalities in the Survey system of strict implication*, The Journal of Symbolic Logic, vol. 4 (1939), pp. 137–154.

Subclass: A class a is a *subclass* of a class b if $a \subset b$. See *Logic, formal*, §7.
A class a is a *proper subclass* of a class b if $a \subset b$ and $a \neq b$. —*A.C.*

Sufficient condition: F is a *sufficient condition* of G if $F(x) \supset_x G(x)$. See *Necessary condition*.
—*A.C.*

Suppositio: In medieval logic, the kind of meaning in use which belongs to nouns or substantives: opposed to *copulatio*, belonging to adjectives and verbs. A given noun having a fixed *signification* might nevertheless have different *suppositiones* (stand for different things). Various kinds of *suppositio*, i.e., various ways in which a noun may stand for something, were distinguished. —*A.C.*

Petri Hispani Summulae Logicales cum Versorii Parisiensis clarissima expositione, editions e.g. Venice 1597, 1622. J. Maritain, *Petite Logique*, Paris, 1933.

Suppositio discreta: The kind of *suppositio* belonging to a proper name; opposed to *suppositio communis*.
—*A.C.*

Suppositio materialis: The use of a word autonymously, or as a name for itself (see

Autonymy) —"Home est dissyllabum"; opposed to *suppositio formalis*, the use of a noun in its proper or ordinary signification. —*A. C.*

Suppositio naturalis: The use of a common noun to stand collectively for everything to which the name applies—"Homo est mortalis." It would now usually be held that this involves an inadequate or misleading analysis—see *Copula*. —*A. C.*

Suppositio personalis: The use of a common noun, or class name, to stand for a particular member of the class—"Homo currit." Contemporary logical usage would supply, in such a case, either a description (corresponding in English to the definite article *the*) or an existential quantifier (corresponding to the indefinite article *a*).

Suppositio personalis confusa (opposed to the preceding as *suppositio personalis determinata*) was further ascribed to a common noun used for the subject or predicate of a universal affirmative proposition. The relation of this to *suppositio naturalis* and *suppositio simplex* is not clear, and not uniform among different writers. —*A. C.*

Suppositio simplex: The use of a common noun to stand for the class concept to which it refers—"Homo est species." *Suppositio simplex* was also ascribed to a common noun used for the predicate of an affirmative proposition. —*A. C.*

Symmetry: A dyadic relation R is *symmetric* if, for all x and y in the field of R, $xRy \supset yRx$; it is *asymmetric* if, for all x and y in the field of R, $xRy \supset \sim yRx$; *non-symmetric* if there are x and y in the field of R such that $[xRy][\sim yRx]$. An n-adic propositional function F is *symmetric* if $F(x_1, x_2, \ldots, x_n)$ is materially equivalent to the proposition obtained from it by permuting $x_1, x_2, \ldots, x_n$ among themselves in any fashion—for all sets of n arguments $x_1, x_2, \ldots, x_n$ belonging to the range of F.

A dyadic function f, other than a propositional function, is *symmetric* if, for all pairs of arguments, x, y, belonging to the range of f, $f(x, y) = f(y, x)$. An n-adic function f is symmetric if, for any set of n arguments belonging to the range of f, the same value of the function is obtained no matter how the arguments are permuted among themselves (i.e., if the value of the function is independent of the order of the arguments).

In geometry, a figure is said to be *symmetric* with respect to a point P if the points of the figure can be grouped in pairs in such a way that the straight-line segment joining any pair has P as its mid-point. A figure is *symmetric* with respect to a straight line l if the points can be grouped in pairs in such a way that the straight-line segment joining any pair has l as a perpendicular bisector. These definitions apply in geometry of any number of dimensions. Similar definitions may be given of symmetry with respect to a plane, etc. —*A. C.*

Syncategorematic (word)**:** Approximately a synonym of *incomplete symbol* (q.v.) but usually applied to words of such a language as English rather than to symbols or expressions in a fully formalized logistic system. —*A. C.*

Syntax, logical: "By the *logical syntax* of a language," according to Carnap, "we mean the formal theory of the linguistic forms of that language—the systematic statement of the formal rules which govern it together with the development of the consequences which follow from these rules. A theory, a rule, a definition, or the like is to be called *formal* when no reference is made in it either to the meaning of the symbols or to the sense of the expressions, but simply and solely to the kinds and order of the symbols from which the expressions are constructed."

This definition would make logical syntax coincide with Hilbertian *proof theory* (q.v.), and in fact the adjectives *syntactical, metalogical, metamathematical* are used nearly interchangeably. Carnap, however, introduces many topics not considered by Hilbert, and further treats not only the syntax of particular languages but also *general syntax*, i.e., syntax relating to all languages in general or to all languages of a given kind.

Concerning Carnap's contention that philosophical questions should be replaced by, or reformulated as, syntactical questions, see *scientific empiricism* I C, and Carnap's book cited below. —*A.C.*

R. Carnap, *The Logical Syntax of Language*, New York and London, 1937. Review by S. MacLane, Bulletin of the American Mathematical Society, vol. 44 (1938), pp. 171–176. Review by S. C. Kleene, The Journal of Symbolic Logic, vol. 4 (1939), pp. 82 -87.

Tautology: As a syntactical term of the propositional calculus this is defined in the article on *logic, formal* (q.v.). Wittgenstein and Ramsey proposed to extend the concept of a tautology to disciplines involving quantifiers, by interpreting a quantified expression as a multiple (possibly infinite) conjunction or disjunction; under this extension, however, it no longer remains true that the test of a tautology is effective.

The name *law of tautology* is given to either of the two dually related theorems of the propositional calculus,

$$[p \vee p] \equiv p, \qquad pp \equiv p,$$

or either of the two corresponding dually related theorems of the algebra of classes,

$$a \cup a = a, \qquad a \cap a = a.$$

Whitehead and Russell reserve the name *principle of tautology* for the theorem of the propositional calculus, $[p \vee p] \supset p$, but use *law of tautology* in the above senses. —*A.C.*

L. Wittgenstein, *Tractatus, Logico-Philosophicus*, New York and London, 1922. F. P. Ramsey, *The foundations of mathematics*, Proceedings of the London Mathematical Society, ser. 2, vol. 25 (1926), pp. 338–384; reprinted in his book of the same title, New York and London, 1931.

Term: In common English usage the word "term" is syntactical or semantical in character, and means simply a word (or phrase), or a word associated with its meaning. The phrase "undefined term" as used in mathematical postulate theory (see *mathematics*) is perhaps best referred to this common meaning of "term." In traditional logic, a term is a concept appearing as *subject* or *predicate* (q.v.) of a categorical proposition; also, a word or phrase denoting such a concept. The word "term" has also been employed in a syntactical sense in various special developments of *logistic systems* (q.v.) usually in a way suggested by the traditional usage.

The mathematical use of the word "term" appears in such phrases as "the terms of a sum" (i.e., the separate numbers which are added to form the sum, or the expressions for them), "the terms of a polynomial," "the terms of a proportion," "the terms of an infinite series," etc. Similarly one may speak of "the terms of a logical sum," and the like. —*A. C.*

Transfinite induction: A generalization of the method of proof by mathematical induction or recursion (see *Recursion, proof by*), applicable to a well-ordered class of arbitrary ordinal number—especially one of ordinal number greater than omega (see *Ordinal number*)—in a way similar to that in which mathematical induction is applicable to a well-ordered class of ordinal number omega. —*A. C.*

Transitivity: A dyadic relation R is *transitive* if, whenever xRy and yRz both hold, xRz also holds. Important examples of transitive relations are: the relation of identity or equality; the relation *less than* among whole numbers, or among rational numbers, or among real numbers; the relation *precedes* among instants of time (as usually taken); the relation of class inclusion, $\subset$ (see *Logic, formal*, §7); the relations of material implication and material equivalence among propositions; the relations of formal implication and formal equivalence among monadic propositional functions. In the propositional calculus, the *laws of transitivity* of material implication and material equivalence (the conditional and biconditional) are:

$$[p \supset q][q \supset r] \supset [p \supset r].$$
$$[p \equiv q][q \equiv r] \supset [p \equiv r].$$

Similar laws of transitivity may be formulated for equality (e.g., in the functional calculus of first order with equality), class inclusion (e.g., in the Zermelo set theory), formal implication (e.g., in the pure functional calculus of first order), etc. —*A. C.*

Transposition: The form of valid inference of the propositional calculus from A $\supset$ B to $\sim$B $\supset$ $\sim$A. The *law of transposition* is the theorem of the propositional calculus,

$$[p \supset q] \supset [\sim q \supset \sim p].$$

—*A. C.*

Truth, semantical: Closely connected with the *name relation* (q.v.) is the property of a propositional formula (sentence) that it expresses a true proposition (or if it has free variables, that it expresses a true proposition for all values of these variables). As in the case of the name relation, a notation for the concept of truth in this sense often cannot be added, with its natural properties, to an (interpreted) logistic system without producing contradiction. A particular system may, however, be made the beginning of a hierarchy of systems each containing the truth concept appropriate to the preceding one.

The notion of truth should be kept distinct from that of a theorem, the theorems being in general only some among the true formulas (in view of Gödel's result, *Logic, formal*, §6).

The first paper of Tarski cited below is devoted to the problem of finding a definition of semantical truth for a logistic system L, not in L itself but in another system (metasystem) containing notations for the formulas of L and for syntactical relations between them. This is attractive as an alternative to the method of introducing the concept of truth by arbitrarily adding a notation for it, with appropriate new primitive formulas, to the metasystem; but in many important cases it is possible only if the metasystem is in some essential respect logically stronger than L.

Tarski's concept of truth, obtained thus by a syntactical definition, is closely related to Carnap's concept of *analyticity*. According to Tarski, they are the same in the case that L is a "logical language." See further *Semiotic*. —*A.C.*

A. Tarski, *Der Wahrheitsbegriff in den formalisierten Sprachen*, Studia Philosophica, vol. 1 (1935), pp. 261–405. A. Tarski, *On undecidable statements in enlarged systems of logic and the concept of truth*, The Journal of Symbolic Logic, vol. 4 (1939), pp. 105–112. R. Carnap, *The Logical Syntax of Language*, New York and London, 1937.

Truth-function is either: (1) a *function* (q.v.) from propositions to propositions such that the truth-value of the value of the function is uniquely determined by the *truth-values alone* of the arguments; or (2) simply a function from truth-values to truth-values. —*A.C.*

Truth-value: On the view that every proposition is either true or false, one may speak of a proposition as having one of two *truth-values*, viz. *truth* or *falsehood*. This is the primary meaning of the term *truth-value*, but generalizations have been considered according to which there are more than two truth-values—see *Propositional calculus, many-valued*. —*A.C.*

Uncertainty principle: A principle of *quantum mechanics* (q.v.), according to which complete quantitative measurement of certain states and processes in terms of the usual space-time coordinates is impossible. Macroscopically negligible, the effect becomes of importance on the electronic scale. In particular, if simultaneous measurements of the position and the momentum of an electron are pressed beyond a certain degree of accuracy, it becomes impossible to increase the accuracy of either measurement except at the expense of a decrease in the accuracy of the other; more exactly, if a is the uncertainty of the measurement of one of the coordinates of position of the electron, and b is the uncertainty of the measurement of the corresponding component of momentum, the product ab (on principle) cannot be less than a certain constant h (namely *Planck's constant*, q.v.). On the basis that quantities in principle unobservable are not to be considered physically real, it is therefore held by quantum theorists that simultaneous ascription of an exact position and an exact momentum to an electron is meaningless. This has been thought to have a bearing on, or to limit or modify, the principle of determinism in physics. —*A.C.*

C. G. Darwin, *The uncertainty principle*, Science, vol. 73 (1931), pp. 653–660.

Undistributed middle: In the categorical syllogism (*Logic, formal*, §5), the middle term must appear in at least one of the two premisses (major and minor) as *distributed*—i.e., as denoting the subject of a proposition A or E, or the predicate of a proposition E or O. Violation of this rule is the fallacy of *undistributed middle*. —*A.C.*

Unit class: A class having one and only one member. Or, to give a definition which does not employ the word *one*, a class a is a *unit class* if there is an x such that $x \in a$ and, for all y, $y \in a$ implies $y = x$. —*A. C.*

Vague: A word (or the idea or notion associated with it) is vague if the meaning is so far not fixed that there are cases in which its application is in principle indeterminate—although there may be other cases in which the application is quite definite. Thus *longevity* is vague because, although a man who dies at sixty certainly does not possess the characteristic of longevity, and one who lives to be ninety certainly does, there is doubt about a man who dies at seventy-five. On the other hand, *octogenarian* is not vague, because the precise moment at which a man becomes an octogenarian may (at least in principle) be determined. Of course, the vagueness of *longevity* might be removed by specifying exactly at what age longevity begins, but the meaning of the word would then have been changed. (See further the article *Relative*.)

Similarly a criterion or test, a convention, a rule, a command is vague if there are cases in which it is in principle indeterminate what the result of the test is, or whether the convention has been followed, or whether the rule or command has been obeyed. —*A. C.*

Valid: In the terminology of Carnap, a sentence (or class of sentences) is *valid* if it is a consequence of the null class of sentences, *contravalid* if every sentence is a consequence of it. The notion of consequence here refers to a full set of primitive formulas and rules of inference for the language or *logistic system* (q.v.) in question, known as *c-rules*, and including (in general) non-effective rules. If the notion of consequence is restricted to depend only on the *d-rules*—i.e., the subclass of the c-rules which are effective—it is then called *d-consequence* or *derivability*, and the terms corresponding to *valid* and *contravalid* are *demonstrable* and *refutable* respectively.

The formulas and the c-rules of the language in question may include some which are extra-logical in character—corresponding, e.g., to physical laws or to matters of empirical fact. Carnap makes an attempt (which, however, has been questioned) to define in purely syntactical terms when a relation of consequence is one of *logical consequence*. If the notion of consequence is restricted to that of logical consequence, the terms corresponding to *valid* and *contravalid* are *analytic* and *contradictory* respectively. If the c-rules are purely logical in character, the class of analytic sentences coincides with that of valid sentences, and the class of contradictory sentences with that of contravalid sentences.

The explicit definition of analyticity (etc.) for a particular language of course requires statement of the c-rules. Actually, in the case of his "Language II," Carnap prefers to define *analytic* and *contradictory* first, and *consequence* in terms of these.

Part of the purpose of the definition of analyticity is to secure that every logical sentence is either analytic or contradictory. (The corresponding situation with demonstrability and refutability is impossible in many significant cases in consequence of Gödel's theorem—see *Logic, formal*, §6.)

Refer further to the article *Syntax, logical*, where references to the literature are given. —*A. C.*

Valid inference: In common usage an inference is said to be *valid* if it is permitted by the laws of logic. It is possible to specify this more exactly only in formal terms, with reference to a particular *logistic system* (q.v.).

The question of the validity of an inference from a set of premisses is, of course, independent of the question of the truth of the premisses. —*A.C.*

Variable: A letter occurring in a mathematical or logistic formula and serving, not as a name of a particular, but as an ambiguous name of any one of a class of things—this class being known as the *range* of the variable, and the members of the class as *values* of the variable.

Where a formula contains a variable, say x, as a *free variable*, the meaning of the formula is thought of as depending on the meaning of x. If the formula contains no other free variables than x, then it acquires a particular meaning when x is *given a value*—i.e., when a name of some one value of x is substituted for all free occurrences of x in the formula—or, what comes to the same thing for this purpose, when the free occurrences of x are taken as denoting some one value.

Frequently an (interpreted) *logistic system* (q.v.) is so constructed that the theorems may contain free variables. The interpretation of such a theorem is that, for *any* set of values, of the variables which occur as free variables, the indicated proposition is true. I.e., in the interpretation the free variables are treated as if bound by universal *quantifiers* (q.v.) initially placed.

A *bound variable*, or *apparent variable*, in a given formula, is distinguished from a free variable by the fact that the meaning of the formula does not depend on giving the variable a particular value. (The same variable may be allowed, if desired, to have both bound occurrences and free occurrences in the same formula, and in this case the meaning of the formula depends on giving a value to the variable only at the places where it is free.) For examples, see *Abstraction*, and *Logic, formal*, §3.

For the terminology used in connection with functions, see the article *Function*. Cf. also the articles *Constant* and *Combinatory logic*. —*A.C.*

Verbal: Consisting of or pertaining to words. Having to do (merely) with the use and meaning of words. —*A.C.*

Vicious circle: A vicious circle in proof (*circulus in probando*) occurs if p_1 is used to prove p_2, p_2 to prove $p_3 \ldots p_{n-1}$ to prove p_n, and finally p_n to prove $p_1 - p_1, p_2, \ldots, p_n$ being then taken as all proved. This is a form of the fallacy of *petitio principii* (q.v.).

A vicious circle in definition (*circulus in definiendo*) occurs if A_1 is used in defining A_2, A_2 in defining $A_3, \ldots, A_{n-1}$ in defining A_n, and finally A_n in defining A_1. (The simplest case is that in which $n = 1$, A_1 being defined in terms of itself.) There is, of course, a fallacy if $A_1, A_2, \ldots, A_n$ are then used as defined absolutely. Apparent exceptions, such as definition by *recursion* (q.v.), require special justification, e.g., by finding an equivalent form of definition which is not circular.

The term *vicious circle fallacy* is used by Whitehead and Russell (1910) for arguments violating their *ramified theory of types* (q.v.). Similarly, the name *circulus vitiosus* is applied by Hermann Weyl (1918) to an argument involving *impredicative definition* (q.v.).

—*A. C.*

Zermelo, Ernst (Friedrich Ferdinand), 1871–1953, German mathematician. Professor of mathematics at Zurich, 1910–1916, and at Freiburg, 1926–1953. His important contributions to the foundations of mathematics are the Zermelo axiomatic set theory (see *Logic, formal*, §9), and the explicit enunciation of the axiom of *choice* (q.v.) and proof of its equivalence to the proposition that every class can be well-ordered.

—*A. C.*

LOGIC AND ANALYSIS
(1960)

Problems of philosophical analysis are often in effect problems of logic — not, that is, of logical *inference*, but of formulating some system of ideas in coherent logical form. In such a case the situation usually is that the ideas in question are already known to common sense, as expressed in ordinary language. But the common-sense formulation requires amendment and supplementation, in order to remove uncertainties and perplexities, or even perhaps to resolve paradox. And this process of reformulation is then a matter of logic.

Many analytical philosophers deny so close a tie of their subject to logic, even in special cases. F. Waismann, for example,[1] reports with approval that "philosophers today are getting weaned from casting their ideas into deductive moulds," and holds that philosophical disputes are not settled by merely logical inference from premisses, because "whenever such premisses have been set up ... the discussion at once challenged them and shifted to a deeper level." Gilbert Ryle[2] allows a greater bearing of logic on philosophy but holds — to select the least unfavorable of his similes — that the significance of "formal" logic for the philosopher is "rather like what geometry is to the cartographer." But Waismann seems to overlook an activity of the logician more important than the drawing of particular inferences, and that is the organization of whole theories into logically systematic form. And to Ryle I would say that the position of the philosopher in attacking an open problem may sometimes be like that of the cartographer who has no well settled system of geometry available and must work out his own before drawing a map. Indeed I suggest that the "informal logic" in which Ryle regards the philosopher as engaged is often not different from the informal logic by which the "formal" logician arrives at a proposed logical organization of a previously disorganized or semi-organized body of theory.

The problem of an analysis of believing, which is raised once more in Ayer's report to this Congress, is a case in point. I urge that it is not to be solved by any single statement of analysis, saying in one sentence of ordinary informal language that to believe is so and so and so and so, or that A believes that p if and only if so and so. But rather the solution requires stating a logic of belief in a full and systematic way, removing the uncertainties and variabilities of ordinary language as to what is admissible — first as meaningful, and then as logically compatible — by complete statement of formation and transformation rules.

Such a logic of belief must be distinguished from the question of operational and observational criteria by which belief is to be recognized. Indeed the latter question does not offer the peculiar difficulties of the logical problem, and up to a point the answer to it is straightforward and familiar. We may test a man's belief by putting him in a test situation in which he must reveal his belief by acting on it in his own interest. And in order to isolate the belief of the subject in regard to one particular

Originally published in ***Atti del XII Congresso Internazionale di Filosofia (Venezia, 12–18 settembre 1958)***, Volume 4, ***Logica, linguaggio e communicazione***, Sansoni Editore, Florence 1960, pp. 77–81.

[1] *How I see philosophy*, in ***Contemporary British Philosophy*** (a symposium), London and New York, 1956, p. 488.

[2] *Formal and informal logic*, in ***Dilemmas***, Cambridge, 1954, pp. 111–29.

matter, and eliminate the influence of his belief in other matters, we may use a series of test situations in which the circumstances of the test are appropriately varied. Or we may simply ask him. If we doubt the truthfulness of his answer, we may attempt to trip him on cross-examination, or we may ask others about his reputation for veracity. And if we doubt his understanding, we may ask questions to test understanding, or we may resort to various procedures of explanation, verbal and other, in order to remove misunderstanding. To be sure, this common-sense procedure leaves a considerable fringe of vagueness within which doubt arises whether A believes p or not; and though this fringe may be reduced by devising more efficient testing procedures, it can in principle never be wholly removed. In this respect the notion of belief is not different from other observational notions, and no problem seems to arise that is special to the notion of belief as against others. But the question of a logic of belief is another matter.

The unavoidability of the fringe of vagueness that arises in testing for belief does not show that a precisely formulated logic of belief is impossible, and is not a reason against seeking such a logic. For example there is a fringe of vagueness of the same sort, arising in connection with the observational notions of length and distance. Yet there is a precise theory of these notions available in elementary metric geometry, well known to all of us from our school days (though the usual school treatment of the subject is less fully formal than the logician might ask).

I am aware that I may be accused here of stretching the meaning of the word *logic*.[3] After all, elementary geometry depends not only on logic but on special postulates, which come to have the character of physical postulates when the geometry is applied to the physical notions of length and distance. But I reply that the logic of elementary geometry can be separately given, by stating the formation rules and those of the transformation rules which we regard as logical. And for such an application of the logistic method to a particular theory I know of no other name than logic.

In this sense I speak of a logic of belief. I shall hold that a full analysis must consist of three things, a logic of belief, a theory based on special postulates, and an account of the observational criteria for recognizing or testing belief. But especially the formulation of a satisfactory logic of belief is very much an open problem,[4] and presents peculiar difficulties. As Prof. Ayer has brought out, these difficulties center around what he and Chisholm (following Brentano) call *intentionality* but I would prefer to call *oblique* or indirect use (of sentences, names, variables, etc.). Indeed the principal criterion for intentionality is the same as that for obliqueness (*Ungeradheit*) in the sense of Frege, namely the failure of substitutivity for equality ($=$), and for material equivalence ($\equiv$). And it would seem that intentionality is merely that special case of obliqueness in which the oblique context is introduced by a word (such as *believe*) that has a psychological reference.

[3]Compare the remarks of W. Kneale about "the logic of colour words" in *The province of logic*, in ***Contemporary British Philosophy***, cit., pp. 238, 260–61.

[4]There are some beginnings of a formally stated logic of belief in a paper by A. Pap, *Belief and propositions*, ***Philosophy of Science***, XXIV (1957), pp. 123–36. This paper has a theory involving some special postulates, as well as a rudimentary logic of belief. The treatment, although it is interesting, is not carried far enough to encounter the serious difficulties that have arisen in informal attempts at an analysis of believing. In fact Pap's procedure in expressing belief by a binary predicate 'B' with an individual variable in one argument place and a propositional variable in the other, is already an important step in formulating a logic of belief, and one whose tenability is put into some doubt by difficulties connected with intentionality.

The very special difficulty of the logical problem is indicated by the fact that plausible verbal analyses of believing have repeatedly failed over some point of logic. And the new analysis of believing which Prof. Ayer has offered (though very hesitantly and doubtfully) provides one more illustration of this. Putting his proposed analysis a little more formally, using '$(\exists x)$' as existential quantifier and '$\equiv$' as equivalence sign, I take it that "A believes that p" is analyzed as:

$$(\exists x) \centerdot A \text{ is disposed to behave in way } x \centerdot [x \text{ is appropriate for } A] \equiv p.$$

The characterization of this analysis as non-intentional suggests that the equivalence, $\equiv$, is intended as material equivalence. But if this is intended, the analysis is immediately unacceptable, because it has the consequence that if one thing true is believed by A then all things true are believed by A, and if one thing false is believed by A then all things false are believed by A. Even if something like strict equivalence is understood, there are similar difficulties — for example, that $2^{32} + 1$ is prime is strictly equivalent to $2 + 2 = 5$, yet it is an historical fact that Fermat believed the former without believing the latter. Indeed if we assume transitivity, the equivalence must be of a sort that if A believes that p, and $p \equiv q$, then A believes q. But then the problem of analyzing such a notion of equivalence or providing a logic of it has been left wholly untouched. Moreover intentionality has not been removed in any sense which seems to me significant, as p occurring in the analysis is then still in an oblique if not an intentional context.

In my opinion the failure of the analysis and the fact that the failure concerns a point of logic serve to enforce my contention that a fully formulated logic of belief must be sought, rather than an informal analysis. It would further seem to be unwise to impose restrictions in advance on such a logic, as that it should not be intentional or that it should not involve commitment to objective meanings. It is difficult enough to find any sound logic of belief at all, and the search should be unhampered by extraneous demands. I propose a problem to which I have no solution to offer. But once such a logic were found, it might then be profitable to discuss the question of modification to meet this or that special demand (if it did not already do so). The discovery of some sound logic of belief may well be a necessary first step toward a formulation that meets other demands in addition, or perhaps toward a demonstration that such a formulation is not possible. Meanwhile we should not prescribe in advance, but should be prepared to accept objectively the results of research.

Princeton University
PRINCETON (New Jersey, U.S.A.)

APPLICATION OF RECURSIVE ARITHMETIC TO THE PROBLEM OF CIRCUIT SYNTHESIS (1960)

These notes are a revision of the Cornell notes of 1957 — reference 15 in the bibliography — and are based on lectures given at the University of Michigan in June, 1958 and July 1959. Changes made include corrections of errors, addition of several expository sections, adoption of Wright's modified method for Case 1, and a complete reworking of the final section entitled "An Example Illustrating Case 2."

A Relay Circuit as an Example

The relay circuit shown in the diagram (Figure 1) is not intended to be a familiar circuit, or one which would necessarily be practically useful in any connection, but has merely been drawn for use in illustrating the application of mathematical logic in the analysis of such circuits.

The switches i_1, i_2, i_3, may be hand-operated switches or keys or may be contacts of relays operated from another circuit (it does not matter). Contrary to standard terminology we shall speak of i_1, i_2, i_3, or of their operation, as inputs of the circuit. For they are in fact inputs in the sense of being elements of the circuit whose operation is imposed from outside rather than controlled by the circuit itself.

The large circle marked o in the diagram is a device of unspecified nature which is operated when current passes through it, and which we think of as the output of the circuit. (For example, o may be a relay controlling another circuit, or it may be an output signal of some sort.) The relay r_2 is a simple relay with one make contact, and r_1 is a relay with one make contact and two break contacts.

We begin with the use of propositional calculus, which we assume to be known, but which requires a brief review here.

Many works on circuit theory make use of Boolean algebra rather than propositional calculus. There is indeed a sense in which Boolean algebra and propositional calculus are the same, but also a sense in which they are not the same. For some purposes the more algebraic approach which consists in writing equations between expressions belonging to Boolean algebra is preferable. But in the present connection the viewpoint and notations of propositional calculus will be more useful and more suggestive.

We shall use the small letter $\vee$ as a notation for disjunction (i.e., the inclusive "or"), and juxtaposition of letters or expressions with no symbol between as a notation for conjunction (i.e., "and") For negation ("not") our notation will be a horizontal bar, which is properly to be placed before the letter or expression it affects, but which we may sometimes as a matter of abbreviation or convenience place instead over the letter or expression. The notations for implication and equivalence are $\supset$ and $\equiv$ respectively; thus $p \supset q$ ("if p then q") means the same as $-p \vee q$, and $p \equiv q$ ("p if and only if q") means the same as $-p-q \vee pq$.

Originally published in ***Summaries of talks presented at the Summer Institute for Symbolic Logic, Cornell University, 1957***, 2nd ed., Communications Research Division, Institute for Defense Analyses, Princeton 1960, pp. 3–50.

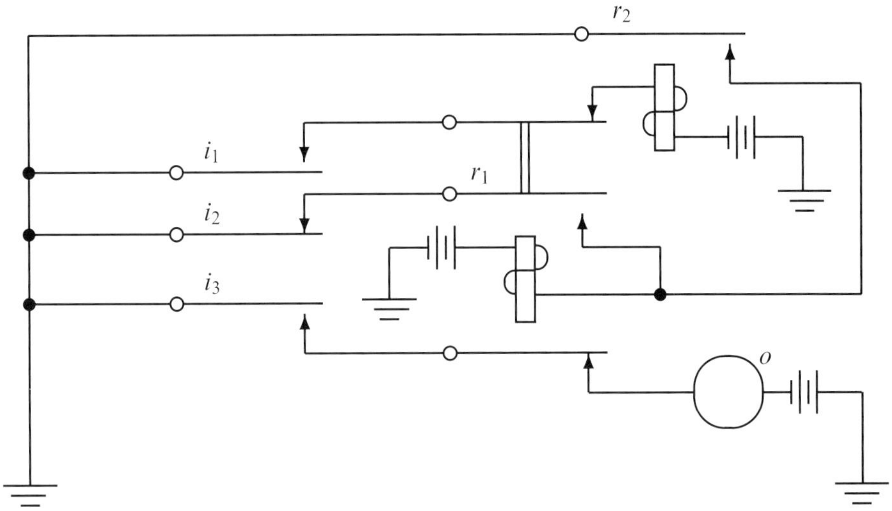

FIGURE 1.

For the two truth-values, truth and falsehood, the digits 0 and 1, or alternatively 1 and 0, are often used. But for our present purpose it is desirable to have distinctive notations. We shall use the capital letter T for truth, and F for falsehood; thus $p \equiv T$ will amount to the same as p, and $p \equiv F$ the same as $-p$.

Any expression built up from a set of letters such as $p, q, r, \ldots$ by means of disjunction, conjunction, negation, implication, and equivalence is said to represent a truth-function of $p, q, r, \ldots$, since for each assignment of truth-values to $p, q, r, \ldots$ the expression has a unique corresponding truth-value. Or in addition to letters $p, q, r, \ldots$ the two truth-value letters T and F may also be used in building up such an expression. Thus, for example, the expression $-p \vee qr \vee ps$ and the expression $p \vee qrs \equiv F$ each of them represent a truth-function of p, q, r, s.

In order to write an analysis of the circuit shown in the diagram (Figure 1), let us use the letter i_1 to mean that the switch i_1 is operated (i.e., closed), i_2 to mean that the switch i_2 is operated (i.e., open), i_3 to mean that switch i_3 is operated (closed), r_2 to mean that relay r_2 is operated (its contact closed), r_1 to mean that relay r_1 is operated, o to mean that the device represented by the large circle is operated. In the familiar application of propositional calculus to analysis of circuits, we then obtain the following equivalences by direct inspection of the diagram:

$$
\begin{aligned}
r_1 &\equiv \bar{i}_2 r_1 \vee r_2 \\
r_2 &\equiv i_1 \bar{r}_1 \qquad (0.1) \\
o &\equiv i_3 \bar{r}_1
\end{aligned}
$$

(Relay r_1 operates if and only if there is current in its coil, i.e., if and only if either i_2 is unoperated and r_1 operated or r_2 is operated; relay r_2 operates if and only if i_1 is operated and r_1 unoperated; o operates if and only if i_3 is operated and r_1 is unoperated.)

This analysis neglects the acting time of the relays, in effect assuming that relays act instantaneously. Such an approximation turns out in many cases, and especially in the absence of feedback, to serve every purpose. But in the present case the inadequacy of the analysis becomes apparent upon taking i_1 and i_2 as both operated, as the first two equivalences, of (4) then reduce to $r_1 \equiv r_2$, $r_2 \equiv -r_1$, which contradict each other.

For a more general analysis of circuits by means of mathematical logic we must go beyond propositional calculus to introduce propositional functions.

Let us therefore amend our previous analysis by taking $i_1(t)$ to mean that the switch i_1 is operated at time t, $i_2(t)$ to mean that i_2 is operated at time t, $i_3(t)$ that i_3 is operated at time t, $r_1(t)$ that r_1 is operated at time t, $r_2(t)$ that r_2 is operated at time t, $o(t)$ that o is operated at time t. If we take $t = 0$ as the instant at which operation of the circuit begins, and choose the acting time of a relay as the unit of time, we then have by inspection of the diagram the following equivalences:

$$\begin{aligned} &r_1(0) \equiv F \\ &r_1(t+1) \equiv \bar{i}_2(t) r_1(t) \vee r_2(t) \\ &r_2(0) \equiv F \\ &r_2(t+1) \equiv i_1(t)\bar{r}_1(t) \\ &o(0) \equiv F \\ &o(t+1) \equiv i_3(t)\bar{r}_1(t) \end{aligned} \tag{0.2}$$

If we abstract from their physical meanings in the particular application we may think of $i_1, i_2, i_3, r_1, r_2, o$ simply as propositional functions of a non-negative integer t, and the above six equivalences (0.2) may then be regarded as *recursion equivalences* by which the functions r_1, r_2, o are defined when the functions i_1, i_2, i_3 are given.

If a circuit is given, we may in the way just illustrated write a set of recursion equivalences to represent it. Or inversely, if recursion equivalences of suitable form are given, we obtain a corresponding circuit. The circuit determined by a given set of recursion equivalences is not unique, but for a particular list of available circuit elements it may often be made unique by the adoption of some convention. For example, if the available circuit elements are relays with operate time 1 and release time 1 (and with single winding), the circuit corresponding to the recursion equivalences (0.2) or to other sets of recursion equivalences of the like form is rendered unique by the convention that conjunction on the right side of an equivalence represents series connection of the indicated contacts, and disjunction similarly represents parallel connection, and each relay r_i is to have one make contact for each occurrence of $r_i(t)$ on the right and one break contact for each occurrence of $\bar{r}_i(t)$ on the right.

Returning once more to the problem of representing the particular circuit in Figure 1 by equivalences which analyze its action, we remark that we assumed, in writing the set of equivalences (0.2) that the operate time of each relay is the same as its release time and the same as the operate time and release time of the other relay and that the time for o to operate or to cease operation is again the same. The equivalences (0.2) accurately represent the action of the circuit only if these assumptions are satisfied. In the contrary case they are only an approximation, at least superior to the assumption of instantaneous action implicit in (4), but perhaps still unsatisfactory.

Generally the operate time and the release time of a relay will not be the same, and relays may also differ among themselves in regard to the operate and release times. In writing recursion equivalences for a circuit we may take these differences into account, to whatever degree of approximation is thought necessary.

As an example suppose that, in the circuit shown in the diagram, r_1 is a relay with slow operate time and r_2 with slow release. Instead of taking all four times as 1, it may be a better approximation to assume, say, 3 and 1 respectively as operate time and release time of r_1, and 1 and 10 as operate time and release time of r_2. And suppose further that the action of o is instantaneous (e.g., $o(t)$ may mean merely a current in the indicated line at time t). Then the recursion equivalences are as follows:

$$
\begin{aligned}
&r_1(0) \equiv F \\
&r_1(1) \equiv F \\
&r_1(2) \equiv F \\
&r_1(t+3) \equiv [\bar{i}_2(t)r_1(t) \vee r_2(t)][\bar{i}_2(t+1)r_1(t+1) \vee r_2(t+1)] \\
&\qquad\qquad [\bar{i}(t+2)r_1(t+2) \vee r_2(t+2)] \\
&r_2(0) \equiv F \\
&r_2(1) \equiv i_1(0) \\
&r_2(2) \equiv i_1(0) \vee i_1(1) \\
&r_2(3) \equiv i_1(0) \vee i_1(1) \vee i_1(2) \\
&r_2(4) \equiv i_1(0) \vee i_1(1) \vee i_1(2) \vee i_1(3) \\
&r_2(5) \equiv i_1(0) \vee i_1(1) \vee i_1(2) \vee i_1(3) \vee i_1(4) \\
&r_2(6) \equiv i_1(0) \vee i_1(1) \vee i_1(2) \vee i_1(3) \vee i_1(4) \vee i_1(5) \\
&r_2(7) \equiv i_1(0) \vee i_1(1) \vee i_1(2) \vee i_1(3) \vee i_1(4) \vee i_1(5) \vee i_1(6) \\
&r_2(8) \equiv i_1(0) \vee i_1(1) \vee i_1(2) \vee i_1(3) \vee i_1(4) \vee i_1(5) \vee i_1(6) \vee i_1(7) \\
&r_2(9) \equiv i_1(0) \vee i_1(1) \vee i_1(2) \vee i_1(3) \vee i_1(4) \vee i_1(5) \vee i_1(6) \vee i_1(7) \vee i_1(8) \\
&r_2(t+10) \equiv i_1(t)\bar{r}_1(t) \vee i_1(t+1)r_1(t+1) \vee \cdots \vee i_1(t+9)\bar{r}_1(t+9) \\
&\ o(t) \equiv i_3(t)\bar{r}_1(t)
\end{aligned}
\tag{0.3}
$$

In this way the behavior of the circuit may be represented by means of recursion equivalences, to as close an approximation as we like, if there is a definite operate time and a definite release time for each relay.

In practice, however, the operate and release times of a relay cannot be standardized closely, and the times must be expected to vary. This is expressed by saying that relay circuits are *asynchronous*.

It may sometimes be sufficient to write the recursion equivalences for such a circuit on the assumption that the action is exactly synchronized, and then afterwards to examine how variations in the acting time of the elements (relays) will affect the behavior of the circuit.

For example, in equivalences (0.2) above we may ask whether there are any circumstances under which the relays r_1 and r_2 change state simultaneously, and if so, how the behavior of the circuit will be affected if one relay acts ahead of the other. Using the sign $\not\equiv$ for the negation of $\equiv$, we have as the condition for r_1 to change state at time t:

$$r_1(t) \not\equiv \bar{\imath}_2(t) r_1(t) \vee r_2(t)$$

Similarly, the condition for r_2 to change state is:

$$r_2(t) \not\equiv i_1(t) \bar{r}_1(t).$$

If we take the conjunction of these two conditions and simplify by elementary laws of propositional calculus, we obtain

$$\bar{\imath}_1(t) \bar{r}_1(t) r_2(t)$$

as the condition for simultaneous change of state. We then ask whether, when this condition is satisfied at t, either the premature operation of relay r_1 or the premature release of relay r_2 might prevent the expected action of the other relay and from (0.2) we obtain the answer no as regards premature operation of r_1, yes as regards premature release of r_2. The behavior of the circuit, including its uncertainties, is therefore more correctly represented if we rewrite (0.2) as follows:

$$\begin{gathered} r_1(0) \equiv F \\ [r_1(t+1) \equiv \bar{\imath}_2(t) r_1(t) \vee r_2(t)] \vee \bar{\imath}_1(t) \bar{r}_1(t) r_2(t) \\ r_2(0) \equiv F \\ r_2(t+1) \equiv i_1(t) \bar{r}_1(t) \\ o(0) \equiv F \\ o(t+1) \equiv i_3(t) \bar{r}_1(t). \end{gathered} \tag{0.4}$$

Finally, if we wish to take into account not only the possibility of variations in the acting times of relays but also the possibility that the different contacts of the relay r_1 may not operate or release quite simultaneously, it becomes necessary first to rewrite (0.2) with three different letters r_{11}, r_{12}, r_{13} to correspond to the three contacts of the relay r_1:

$$\begin{gathered} r_{11}(0) \equiv F \\ r_{12}(0) \equiv F \\ r_{13}(0) \equiv F \\ r_{11}(t+1) \equiv \bar{\imath}_2(t) r_{12}(t) \vee r_2(t) \\ r_{12}(t+1) \equiv \bar{\imath}_2(t) r_{12}(t) \vee r_2(t) \\ r_{13}(t+1) \equiv \bar{\imath}_2(t) r_{12}(t) \vee r_2(t) \\ r_2(0) \equiv F \\ r_2(t+1) \equiv i_1(t) \bar{r}_{11}(t) \\ o(0) \equiv F \\ o(t+1) \equiv i_3(t) \bar{r}_{13}(t). \end{gathered} \tag{0.5}$$

From the form of the equivalences for $r_{12}(t+1)$ and $r_2(t+1)$, we see that the only possibility that failure of simultaneous action of the various contacts of the relay r_1 should make a material difference in the behavior of the circuit is that premature operation or release of r_{11} should cause a premature change of state of the relay r_2, and that moreover the change of state of r_2 takes place rapidly enough so that r_{12} behaves in accordance with the changed state of r_2 rather than the original state of r_2. This gives the condition

$$r_{12}(t) \not\equiv \bar{i}_2(t) r_{12}(t) \vee r_2(t)$$

(that r_{12} would have changed state except for the premature action of r_{11}), and also the condition

$$r_{12}(t) \equiv \bar{i}_2(t) r_{12}(t) \vee i_1(t) r_{11}(t)$$

(that r_{12} in fact did not change state after the premature action of r_{11}). In these two conditions we may replace $r_{11}(t)$ and $r_{12}(t)$ both by $r_1(t)$, since we suppose that at time t (whatever may have happened later) the various contacts of the relay r_1 were all in agreement, i.e., either all of them operated or all of them unoperated. Then if we take the conjunction of the two conditions and simplify by propositional calculus, the result is

$$i_1(t) i_2(t) r_1(t) \bar{r}_2(t) \vee \bar{r}_1(t) r_2(t).$$

From this disjunction, however, the first term may be canceled on the understanding that the diagram means that relay r_1 is so constructed that, on release, r_{12} always breaks before r_{11} breaks. This leaves

$$\bar{r}_1(t) r_2(t)$$

as the condition upon which the action of the relays is uncertain. Hence we are led to write the following as representing the behavior of the circuit:

$$\begin{gathered} r_1(0) \equiv F \\ \left[r_1(t+1) \equiv \bar{i}_2(t) r_1(t) \vee r_2(t)\right] \vee \bar{r}_1(t) r_2(t) \\ r_2(0) \equiv F \\ \left[r_2(t+1) \equiv i_1(t) \bar{r}_1(t)\right] \vee i_1(t) \bar{r}_1(t) r_2(t) \bar{r}_1(t+1) \\ o(0) \equiv F \\ o(1) \equiv i_3(0) \\ \left[o(t+2) \equiv i_3(t+1) \bar{r}_1(t+1)\right] \vee \bar{r}_1(t) r_2(t) i_3(t+1) \bar{r}_1(t+1). \end{gathered} \tag{0.6}$$

Computer Circuits as Examples

For further examples of the same method of applying mathematical logic to circuit analysis, we turn now to circuits of a type commonly used in digital computers. In drawing diagrams we use

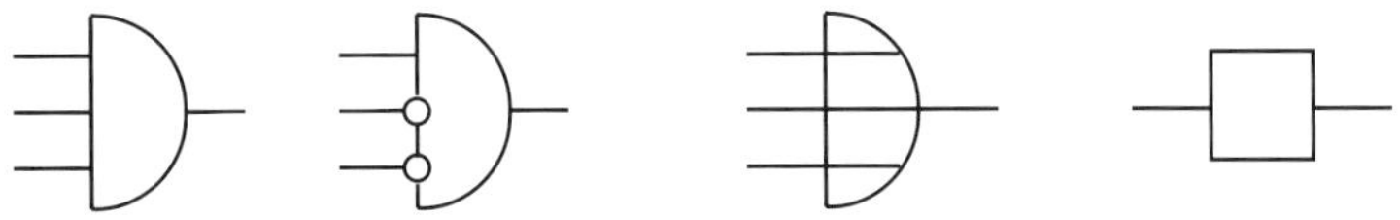

to represent respectively an "and" gate with any number of inputs, an "and" gate with one or more inhibition inputs (marked by the small circle) and any number (0 or more) of other inputs, an "or" gate with any number of inputs, a one-unit delay line. We treat the action of the "and" gates and "or" gates as instantaneous, and we suppose that the circuits are synchronized by a control clock. Diagrams are reduced to the bare essentials necessary for logical analysis, with omission of various details which may sometimes or always be required in the actual construction of the circuit (such as synchronizing connections with the clock, regenerating or amplifying tubes, duplication of lines with pulses of opposite kind in the two lines, extra gates inserted to compensate for the fact that the assumption of instantaneous action of the gates is not quite accurate).

The synchronization of the circuits means that their action is completely deterministic, so that it is always represented by recursion equivalence rather than by expression of such more complicated form as that of (0.6).

In these recursion equivalences, the predicates[1] which occur represent the presence of a pulse in a certain line of the circuit (rather than the operation of a relay or other such element, as in the case of relay circuits). For example, in the circuit of Figure 2 we may let $i_1(t)$ mean that there is a pulse in the input line i_1 at time t, $i_2(t)$ that there is a pulse in the input line i_2 at time t, $i_3(t)$ that there is a pulse in the input line i_3 at time t, and $o(t)$ that there is a pulse in the output line o at time t. And we then have the following recursion equivalences:

$$\begin{aligned} & o(0) \equiv \bar{i}_1(0)i_2(0) \\ (0.7) \qquad & o(t+1) \equiv o(t)\bar{i}_1(t+1)\bar{i}_3(t+1) \vee o(t)i_2(t+1)\bar{i}_3(t+1) \\ & \qquad \vee\ \bar{i}_1(t+1)i_2(t+1). \end{aligned}$$

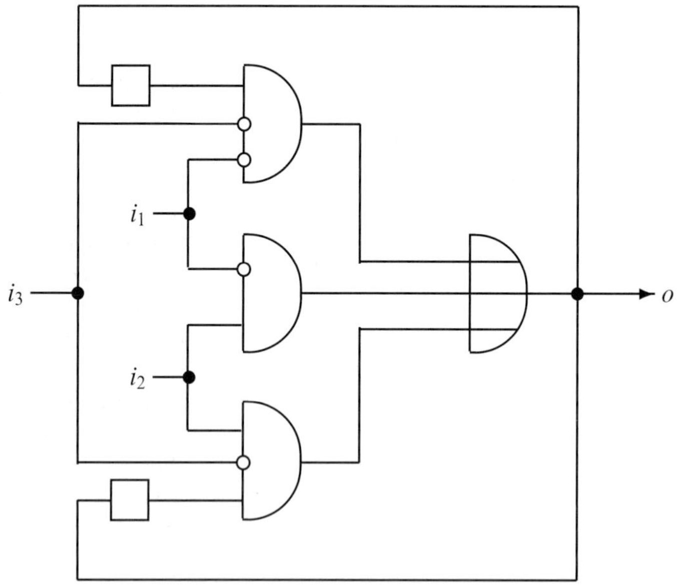

FIGURE 2.

[1]A predicate is a letter which denotes a propositional function, as for example the letters i_1, i_2, i_3, r_1, r_2, o that appear in (0.2), or in (0.6). The reader must bear the terminology in mind, as it is often essential for understanding to distinguish between the predicate and the function it denotes.

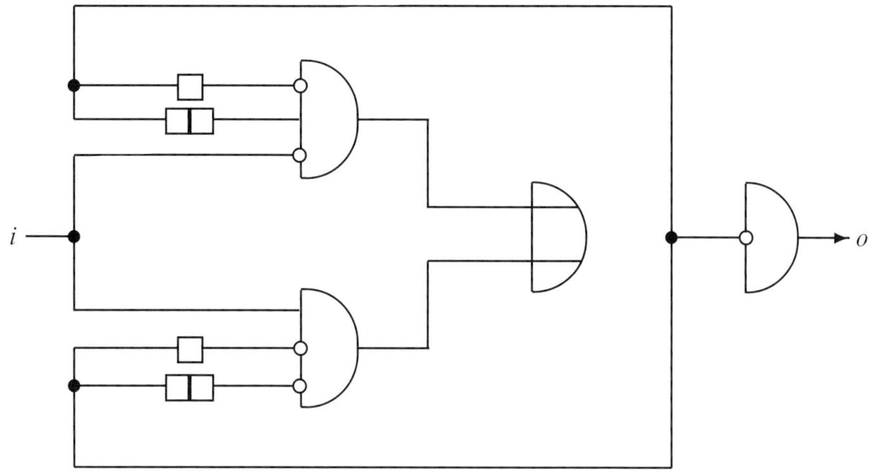

FIGURE 3.

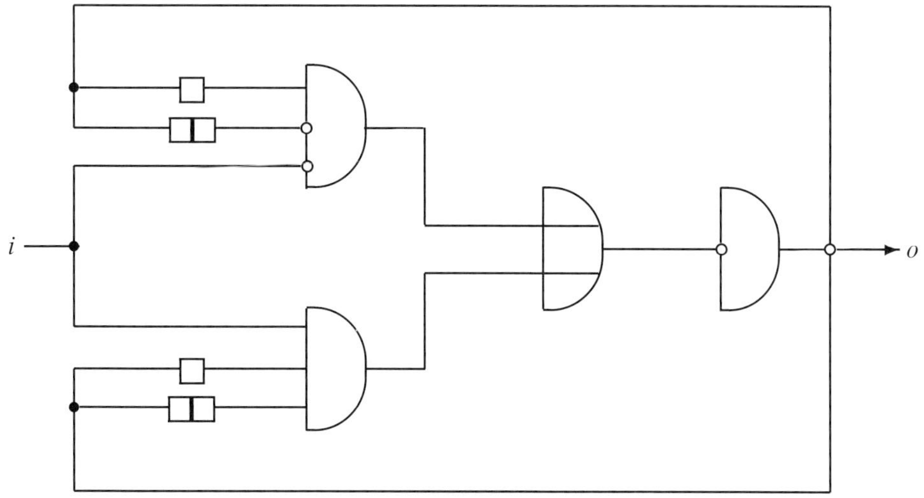

FIGURE 4.

As another example, compare the two closely related circuit diagrams of Figure 3 and Figure 4. For Figure 3 the recursion equivalences are:

$$\begin{aligned} o(0) &\equiv \bar{i}(0) \\ o(1) &\equiv -[\bar{i}(0)i(1)] \\ o(t+2) &\equiv -[o(t)o(t+1)i(t+2) \vee \bar{o}(t)o(t+1)\bar{i}(t+2)]. \end{aligned} \tag{0.8}$$

For the circuit shown in Figure 4, the third equivalence is the same as for Figure 3, but the first two are different:

$$
\begin{aligned}
&o(0) \equiv T \\
(0.9)\qquad &o(1) \equiv i(1) \\
&o(t+2) \equiv -[o(t)o(t+1)i(t+2) \vee \bar{o}(t)o(t+1)\bar{i}(t+2)].
\end{aligned}
$$

The reverse process of finding a circuit which is represented by given recursion equivalences may be illustrated by drawing a circuit diagram to correspond to the equivalences (0.2). This is shown in Figure 5. The reader should similarly find a computer circuit represented by the equivalences (0.3). Also a circuit represented by

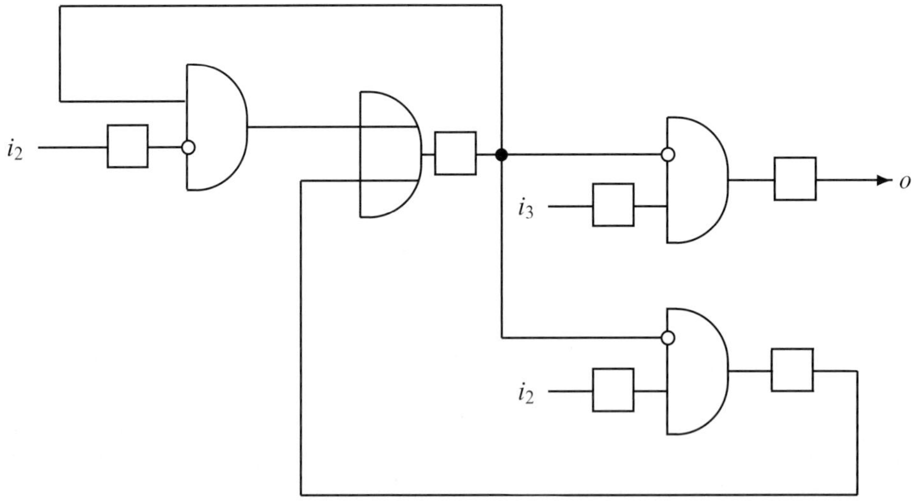

FIGURE 5.

each of the following modifications of the equivalences (0.7):

$$o(0) \equiv F$$
$$o(t+1) \equiv o(t)\bar{i}_1(t+1)\bar{i}_3(t+1) \vee o(t)i_2(t+1)\bar{i}_3(t+1) \vee \bar{i}_1(t+1)i_2(t+1)$$

$$o(0) \equiv \bar{i}_1(0)i_2(0)$$
$$o(t+1) \equiv o(t)\bar{i}_1(t+1)\bar{i}_3(t+1) \vee o(t)i_2(t+1)\bar{i}(t+1) \vee \bar{i}_1(t+1)i_2(t+1) \vee \bar{i}_1(t)\bar{i}_3(t)$$

$$o_1(0) \equiv F$$
$$o_1(t+1) \equiv o_1(t)\bar{i}_1(t+1)\bar{i}_3(t+1) \vee o_2(t)i_2(t+1)\bar{i}_3(t+1) \vee \bar{i}_1(t)i_2(t)$$
$$o_2(0) \equiv \bar{i}_1(0)i_2(0)$$
$$o_2(t+1) \equiv o_1(t)\bar{i}_1(t+1)\bar{i}_3(t+1) \vee o_2(t)i_2(t+1)\bar{i}_3(t+1) \vee \bar{i}_1(t+1)i_2(t+1)$$

$$o_3(0) \equiv i_1(0)i_2(0)$$

$$o_3(t+1) \equiv o_1(t)\bar{i}_1(t+1)\bar{i}_3(t+1) \vee o_2(t)i_2(t+1)\bar{i}(t+1) \vee \bar{i}(t+1)i_2(t+1) \vee \bar{i}(t)i_2(t)$$

Restricted Recursive Arithmetic

In order to make a general formulation of the method of treating circuits which was illustrated in the foregoing examples, we introduce a particular logical system, which is a form of recursive arithmetic (in the sense of Skolem), and which we shall call restricted recursive arithmetic.

The primitive symbols of the system are a list of numerical variables

$$x, y, z, t, \ldots,$$

a list of primitive predicates

$$i_1, i_2, i_3, \ldots,$$

the symbol 0 for the number zero, the accent $'$ as a notation for successor (of a number), the notation ( ) for application of a singulary propositional function to its argument, connectives of the propositional calculus (i.e., the notations for disjunction, conjunction, negation, implication, equivalence, non-equivalence), the letters T and F as notations for the truth-values, and brackets [ ].

An expression which consists of just a numerical variable, or of just the symbol 0, or of a numerical variable followed by any number of accents, or of 0 followed by any number of accents is called a term.

The rules of the system include a rule for so-called definition by recursion, which provides for the introduction of additional predicates, beyond the primitive predicates, in accordance with a specific recursion schema (see below).

The elementary formulas of the system comprise such formulas as

$$i_1(t), i_1(t'), i_2(0), i_2(x'''), i_4(0''),$$

and also such formulas as

$$r_1(t), r_2(y), o(t''''), o_1(0'),$$

where r_1, r_2, o, o_1 are additional predicates that have been introduced by recursion (in accordance with the recursion schema), i.e., any predicate followed by a term between parenthesis is an elementary formula. Also either of the letters T, F standing alone is to be considered an elementary formula.

The well-formed formulas of the system comprise the elementary formulas, and other formulas constructed from them by means of connectives of the propositional calculus, such as, for example:

$$i_1(t) \vee -i_1(t') \equiv F$$

$$[i_2(0) \vee i_2(x''')][i_4(0'')r_1(t) \vee r_2(y)o(t'''')o_1(0')].$$

Notice that each such formula therefore represents a truth-function. For example, the formula $i_1(t) \vee -i_1(t') \equiv F$ represents a truth-function of $i_1(t)$ and $i_1(t')$.

We shall usually write $i_1(t+1)$, $i_2(x+3)$, $i_4(2)$ (and the like) in place of $i_1(t')$, $i_2(x''')$, $i_4(0'')$ respectively. But it must be understood that this way of writing is merely a matter of convenience, and not properly part of the system. In particular there is no notation for addition as part of the system, and, e.g., $i_1(t+x)$ is not a well-formed formula.

Not *as* formulas of the system, but in order to talk *about* the formulas of the system, we shall find it convenient to use such notations as $S(t)$, $R(t)$, $C(t)$—with a capital letter—to mean some unspecified (well-formed) formula of the system, containing the variable t. Similarly we shall use, e.g.,

$$S[i_1(t), i_1(t+1), i_2(0), i_2(x+3), o(t+4)]$$

to mean some unspecified (well-formed) formula of the system, constructed from some or all of the elementary formulas $i_1(t)$, $i_1(t')$, $i_2(0)$, $i_2(x''')$, $o(t'''')$, which are called its elementary parts.

When we do not wish to indicate the variables or the elementary parts which a formula contains, we may also often use just a capital letter with or without subscripts (such as A, S, R, R_1, $P_{1000000}$) to mean an unspecified formula.

The rules of restricted recursive arithmetic allow us:

1. To assert any formula that is tautologous according to laws of propositional calculus.

2. From $A \supset B$ and A to infer B, where A and B are any (well-formed) formulas. (The rule of *modus ponens*.)

3. From any asserted formula to infer the result of substituting any term for any numerical variable throughout—where by a "term" is understood the symbol 0, or 0 followed by any number of accents, or any numerical variable, or a numerical variable followed by any number of accents. (The rule of substitution.)

4. From $S(t) \supset S(t')$ and $S(0)$ to infer $S(t)$. (The rule of mathematical induction.) Here $S(t)$ is any (well-formed) formula, $S(t')$ is the result of substituting t' for t throughout $S(t)$, and $S(0)$ is the result of substituting 0 for t throughout $S(t)$.

5. A rule of definition by recursion, which allows us at any stage to introduce new predicates in accordance with an explicit recursion schema to be given below.

As a matter of fact we shall give, not just one recursion schema, but three alternative schemata. This means that, properly speaking, we consider three different systems under the name of restricted recursive arithmetic, as each different choice of a recursion schema means a different Rule 5 and hence a different logical system (though all else remains the same).

The three recursion schemata are as follows, where in each case $r_1, r_2, \ldots, r_n$ are the new predicates which are being introduced by recursion, and $s_1, s_2, \ldots, s_k$ are given predicates (either primitive or previously introduced by recursion), and where, as already explained, $t+1, t+2, t+3, \ldots$ are used as substitute notations for $t', t'', t''', \ldots$ and $1, 2, 3, \ldots$ are used as substitute notations for $0', 0'', 0''', \ldots$:

A. Restricted Recursion

$$r_1(0) \equiv P_1[s_1(0), s_2(0), \ldots, s_k(0)]$$
$$r_2(0) \equiv P_2[s_1(0), s_2(0), \ldots, s_k(0)]$$

$$\cdots\cdots\cdots\cdots\cdots\cdots\cdots\cdots$$

$$r_n(0) \equiv P_n[s_1(0), s_2(0), \ldots, s_k(0)]$$
$$r_1(t+1) \equiv Q_1[s_1(t), \ldots, s_k(t), s_1(t+1), \ldots, s_k(t+1), r_1(t), \ldots, r_n(t)]$$
$$r_2(t+1) \equiv Q_2[s_1(t), \ldots, s_k(t), s_1(t+1), \ldots, s_k(t+1), r_1(t), \ldots, r_n(t)]$$

$$\cdots\cdots\cdots\cdots\cdots\cdots\cdots\cdots\cdots\cdots\cdots\cdots$$

$$r_n(t+1) \equiv Q_n[s_1(t), \ldots, s_k(t), s_1(t+1), \ldots, s_k(t+1), r_1(t), \ldots, r_n(t)]$$

B. Wider Restricted Recursion

$$r_1(0) \equiv P_{10}[s_1(0), \ldots, s_k(0)]$$
$$r_2(0) \equiv P_{20}[s_1(0), \ldots, s_k(0)]$$

$$\cdots\cdots\cdots\cdots\cdots\cdots\cdots$$

$$r_n(0) \equiv P_{n0}[s_1(0), \ldots, s_k(0)]$$
$$r_1(1) \equiv P_{11}[s_1(0), \ldots, s_k(0), s_1(1), \ldots, s_k(1)]$$

$$\cdots\cdots\cdots\cdots\cdots\cdots\cdots\cdots\cdots\cdots$$
$$\cdots\cdots\cdots\cdots\cdots\cdots\cdots\cdots\cdots\cdots$$

$$r_n(h) \equiv P_{nh}[s_1(0), \ldots, s_k(0), s_1(1), \ldots, s_k(1), \ldots, \ldots, s_k(h)]$$
$$\begin{aligned} r_1(t+h+1) \equiv Q_1[&s_1(t), \ldots, s_k(t), s_1(t+1), \ldots, s_k(t+1), \ldots, \ldots,\\ &s_k(t+h+1), r_1(t), \ldots, r_n(t), r_1(t+1), \ldots, r_n(t+1), \ldots,\\ &\ldots, r_n(t+h)] \end{aligned}$$

$$\cdots\cdots\cdots\cdots\cdots\cdots\cdots\cdots\cdots\cdots$$

$$\begin{aligned} r_n(t+h+1) \equiv Q_n[&s_1(t), \ldots, s_k(t), s_1(t+1), \ldots, s_k(t+1), \ldots, \ldots,\\ &s_k(t+h+1), r_1(t), \ldots, r_n(t), r_1(t+1), \ldots, r_n(t+1), \ldots,\\ &\ldots, r_n(t+h)] \end{aligned}$$

C. Unrestricted Singulary Recursion

$$r_1(0) \equiv P_{10}[s_1(0), \ldots, s_k(0), s_1(1), \ldots, s_k(1), \ldots, \ldots, s_k(a)]$$
$$r_2(0) \equiv P_{20}[s_1(0), \ldots, s_k(0), s_1(1), \ldots, s_k(1), \ldots, \ldots, s_k(a)]$$

$$\cdots\cdots\cdots\cdots\cdots\cdots\cdots\cdots\cdots\cdots$$

$$r_n(0) \equiv P_{n0}[s_1(0), \ldots, s_k(0), s_1(1), \ldots, s_k(1), \ldots, \ldots, s_k(a)]$$
$$r_1(1) \equiv P_{11}[s_1(0), \ldots, s_k(0), s_1(1), \ldots, s_k(1), \ldots, \ldots, s_k(a)]$$

$$\cdots\cdots\cdots\cdots\cdots\cdots\cdots\cdots\cdots\cdots$$

. . . .

$$r_n(h) \equiv P_{nh}[s_1(0), \ldots, s_k(0), s_1(1), \ldots, s_k(1), \ldots, \ldots, s_k(a)]$$

$$r_1(t+h+1) \equiv Q_1[s_1(t), \ldots, s_k(t), s_1(t+1), \ldots, s_k(t+1), \ldots, \ldots, s_k(t+b), r_1(t), \ldots, r_n(t), r_1(t+1), \ldots, r_n(t+1), \ldots, \ldots, r_n(t+h)]$$

. . . .

$$r_n(t+h+1) \equiv Q_n[s_1(t), \ldots, s_k(t), s_1(t+1), \ldots, s_k(t+1), \ldots, \ldots, s_k(t+b), r_1(t), \ldots, r_n(t), r_1(t+1), \ldots, r_n(t+1), \ldots, \ldots, r_n(t+h)]$$

In the schemata B and C the number $h+1$ is called the span of the recursion, and as a special case the value 0 of the span, i.e., $h=-1$, is not excluded.

It will be observed that no provision has been made in the system for the universal quantifier—i.e., the notation (x) ("for all x"), (t) ("for all t")—or for the existential quantifier—the notation $(\exists t)$ ("for some t" or "there is a t such that"). We shall make use of quantifiers later and must assume familiarity with them. But, following Skolem, we reserve the name recursive arithmetic for a logical system which does not have the quantifiers.

Besides the ordinary quantifiers, we shall sometimes have use also for bounded quantifiers, i.e., notations such as $(x)_0^t$ ("for all x from 0 to t"), $(y)_{200}^{t+1}$ ("for all y from 200 to $t+1$"), $(\exists x)_0^t$ ("for some x between 0 and t inclusive"), $(\exists x)_{y+1}^t$ ("for some x between $y+1$ and t inclusive").

Unlike the ordinary quantifiers, the bounded quantifiers with a *constant* lower bound are available in restricted recursive arithmetic, because they can be introduced by recursion. For example, $(x)_0^t S(x)$ can be expressed as $p(t)$, where p is a new predicate introduced by the recursion:

$$p(0) \equiv S(0)$$
$$p(t+1) \equiv p(t)S(t+1).$$

Similarly $(\exists x)_2^t S(x)$ can be expressed as $q(t)$, where q is a new predicate introduced by the recursion:

$$q(0) \equiv F$$
$$q(1) \equiv F$$
$$q(t+2) \equiv q(t+1) \vee S(t+2).$$

(This method of introducing a bounded quantifier, $(x)_a^t$ or $(\exists x)_a^t$, by recursion is directly available only when the scope $S(x)$ of the quantifier is a formula of restricted recursive arithmetic, not itself containing bounded quantifiers, and not containing any numerical variable except x. However, in other cases we may first apply Behmann's procedure of pushing in the quantifiers—see below—in order to reduce to a form in which the scope of every quantifier is a formula of this kind.)

Finite Automata

A *finite automaton* (or finite circuit) consists of a finite number of elements, each of which is capable of only a finite number of different states. Except for the input elements, whose states are imposed from outside, the state of any element at a given instant is completely determined by (1) the states of certain other elements at the given instant and (2) the states of certain elements itself and others, at instants preceding the given instant by no more than a fixed maximum time $h + 1$.

We may assume without loss of generality that each element has exactly two states of which it is capable. (For example, an element having either three or four states is equivalent to two suitably chosen elements which have each just two states.) And in what follows we shall make this assumption, calling the two states "operated" and "unoperated," or "positive" and "negative."

In order to treat finite automata by means of restricted recursive arithmetic, we let the primitive predicates $i_1, i_2, \ldots, i_\mu$ represent the inputs, and we choose other (non-primitive) predicates to represent the remaining elements. If f is a predicate representing a particular element, we use $f(t)$ to express that that element is operated (or positive) at time t — and hence $\overline{f}(t)$ expresses that that element is unoperated (or negative) at time t.

A synchronous circuit in which all elements operate in jumps of exactly one unit of time will then be represented by a series of one or more restricted recursions (of the form A).

The representation of circuits by restricted recursions is not one-to-one. But it is adequate in the sense that the recursion equivalences can be written down when the circuit diagram is given, and inversely a circuit diagram can be drawn to correspond to a given set of recursion equivalences (constituting a restricted recursion) if the admissible kinds of components are known from which the circuit may be constructed.

Restricted recursions correspond to taking $h = 1$ in condition (2) above. If we allow an arbitrary value of h, the restricted recursions are replaced by wider restricted recursions (of the form B). Actually a wider restricted recursion can always be reduced to a series of restricted recursions, by any of various methods. And in this sense we could in principle confine attention to restricted recursions. But we shall find it convenient rather to represent circuits (or finite automata) by wider restricted recursions; and in particular we shall regard the problem of finding a circuit to satisfy a given requirement as having been solved if we find a wider restricted recursion or series of wider restricted recursions to represent the circuit. This may be justified on the ground that it is a substantially straightforward process to find and draw a circuit diagram (with components of ordinary kind) when the circuit has been characterized by recursion equivalences constituting one or more wider restricted recursions — though the relation between the circuit and the corresponding recursion equivalences is not as immediate or direct as in the case of a restricted recursion.

In connection with the examples which were given earlier the reader should now verify that (0.2), (0.7), (0.8), (0.9) are in the form of restricted recursions, with $n = 3$ in the case of (0.2), and $n = 1$ in the other cases. Also (0.3) can be regarded as consisting of two wider restricted recursions, upon rewriting it in the following equivalent form:

$$r_1(0) \equiv F$$

$$r_1(1) \equiv F$$

$$r_1(2) \equiv F$$
$$r_1(3) \equiv F$$
$$r_1(4) \equiv i_1(0)$$
$$r_1(5) \equiv i_1(0) \vee i_1(1)$$
$$r_1(6) \equiv i_1(0) \vee i_1(1) \vee i_1(2)$$
$$r_1(7) \equiv i_1(0) \vee i_1(1) \vee i_1(2) \vee i_1(3)$$
$$r_1(8) \equiv i_1(0) \vee i_1(1) \vee i_1(2) \vee i_1(3) \vee i_1(4)$$
$$r_1(9) \equiv i_1(0) \vee i_1(1) \vee i_1(2) \vee i_1(3) \vee i_1(4) \vee i_1(5)$$
$$r_1(t+10) \equiv [\bar{i}_2(t+7)r_1(t+7) \vee r_2(t+7)][\bar{i}_2(t+8)r_1(t+8) \vee r_2(t+8)] [\bar{i}_2(t+9)r_1(t+9) \vee r_2(t+9)]$$
$$r_2(0) \equiv F$$
$$r_2(1) \equiv i_1(0)$$
$$r_2(2) \equiv i_1(0) \vee i_1(1)$$
$$r_2(3) \equiv i_1(0) \vee i_1(1) \vee i_1(2)$$
$$r_2(4) \equiv i_1(0) \vee i_1(1) \vee i_1(2) \vee i_1(3)$$
$$r_2(5) \equiv i_1(0) \vee i_1(1) \vee i_1(2) \vee i_1(3) \vee i_1(4)$$
$$r_2(6) \equiv i_1(0) \vee i_1(1) \vee i_1(2) \vee i_1(3) \vee i_1(4) \vee i_1(5)$$
$$r_2(7) \equiv i_1(0) \vee i_1(1) \vee i_1(2) \vee i_1(3) \vee i_1(4) \vee i_1(5) \vee i_1(6)$$
$$r_2(8) \equiv i_1(0) \vee i_1(1) \vee i_1(2) \vee i_1(3) \vee i_1(4) \vee i_1(5) \vee i_1(6) \vee i_1(7)$$
$$r_2(9) \equiv i_1(0) \vee i_1(1) \vee i_1(2) \vee i_1(3) \vee i_1(4) \vee i_1(5) \vee i_1(6) \vee i_1(7) \vee i_1(8)$$
$$r_2(t+10) \equiv i_1(t)\bar{r}_1(t) \vee i_1(t+1)\bar{r}_1(t+1) \vee \cdots \vee i_1(t+9)\bar{r}_1(t+9)$$
$$o(t) \equiv i_3(t)\bar{r}_1(t)$$

The two wider restricted recursions may even be reduced to a single wider restricted recursion if we are willing to replace the equivalence $o(t) \equiv i_3(t)r_1(t)$ by the following eleven equivalences:

$$o(0) \equiv i_3(0)$$
$$o(1) \equiv i_3(1)$$
$$o(2) \equiv i_3(2)$$
$$o(3) \equiv i_3(3)$$
$$o(4) \equiv \bar{i}_1(0)i_3(4)$$
$$o(5) \equiv \bar{i}_1(0)\bar{i}_1(1)i_3(5)$$
$$o(6) \equiv \bar{i}_1(0)\bar{i}_1(1)\bar{i}(2)i_3(6)$$
$$o(7) \equiv \bar{i}_1(0)\bar{i}_1(1)\bar{i}(2)\bar{i}_1(3)i_3(7)$$
$$o(8) \equiv \bar{i}_1(0)\bar{i}_1(1)\bar{i}(2)\bar{i}_1(3)\bar{i}_1(4)i_3(8)$$

$$o(9) \equiv \bar{i}_1(0)\bar{i}_1(1)\bar{i}_1(2)\bar{i}_1(3)\bar{i}_1(4)\bar{i}_1(5)i_3(9)$$
$$o(t+10) \equiv i_3(t+10)[i_2(t+7)\bar{r}_2(t+7) \vee \bar{r}_1(t+7)\bar{r}_2(t+7) \vee i_2(t+8)\bar{r}_2(t+8) \vee \bar{r}_1(t+8)\bar{r}_2(t+8) \vee i_2(t+9)\bar{r}_2(t+9) \vee \bar{r}_1(t+9)\bar{r}_2(t+9)]$$

However this last replacement, though it leaves unchanged the functions denoted by the predicates r_1, r_2, o and hence leaves unchanged the action of the corresponding circuit elements, cannot be said to have left the circuit unchanged. For the instantaneously acting element represented by o has been replaced by a more complicated element, which has (with respect to i_2, r_1, r_2) operate time 1 and release time 3, and which is moreover differently connected in the circuit. — It is for this reason that we allow a series of wider restricted recursions, as well as a single wider restricted recursion, in the representation of a circuit.

The representation of a circuit by wider restricted recursions still supposes the circuit to be synchronous, in the sense that the operation of every element is exactly timed (though the time of operation may not always be 1). Extension of the method to asynchronous circuits is a question which we reserve for later discussion.

Recursions of the form C were included above only for comparison with A and B. They do not represent circuits or automata, except in a liberalized sense in which some of the predicates are treated as referring to a future state rather than a present state of an element.

In order to deal with the synthesis problem for circuits, we must extend restricted recursive arithmetic by adding new notations as needed. Such new notations may include quantifiers, or notations for various recursive propositional functions, such as $=, <, \Sigma, \Pi$. [$\Sigma(x, y, z)$ is used to mean $x + y = z$, $\Pi(x, y, z)$ to mean $xy = z$.]

Given a requirement which a circuit is to satisfy we may suppose the requirement expressed in some suitable logical system which is an extension of restricted recursive arithmetic. The *synthesis problem* is then to find recursion equivalences representing a circuit that satisfies the given requirement (or alternatively, to determine that there is no such circuit).

Closely related is the *decision problem*: given a requirement which a circuit is to satisfy, and given a set of recursion equivalences for the circuit, to determine whether or not the circuit satisfies the requirement.

In both cases it is meant that an algorithm shall be supplied, by which to find a required circuit, or to determine whether a given circuit satisfies a given requirement.

Propositional Recursive Arithmetic[2]

In restricted recursive arithmetic, as introduced above, only singulary predicates appear, i.e., predicates denoting propositional functions of one argument.

A generalization of the system obviously suggests itself in which binary predicates, ternary predicates, etc., are allowed and the recursion schema is appropriately generalized so that binary and ternary predicates etc. (as well as singulary), may be introduced by recursion. This generalized system we shall call propositional recursive arithmetic — the adjective "propositional" being in this context an abbreviation of "propositional-functional," intended to distinguish this form of recursive arithmetic, in

[2]It is suggested that, in a first reading of these notes, the reader may omit two sections and proceed at once to the section headed *Some Metatheorems of Propositional Calculus*.

which propositional functions are taken as basic, from the form originally introduced by Skolem, in which numerical functions are basic.[3]

As primitive symbols of propositional recursive arithmetic we shall take all those of restricted recursive arithmetic, as given above, with the modification that the predicates $i_1, i_2, \ldots$ are not necessarily singulary (but may also be binary, ternary, etc.), and with the addition of the notation

$$(\ , \ \ , \ldots , \ )$$

for application of any function (not only singulary but binary, ternary, etc.) to its arguments. The same axioms and rules are retained, except for modification of the recursion schema. And for our present purpose we may leave open the question of possible additional axioms which may be required beyond these.

As a recursion schema the following suggests itself as natural:

D. Unrestricted Propositional Recursion

$$r_1(0, 0, \ldots, 0, 0) \equiv P_{100\ldots00}$$
$$r_2(0, 0, \ldots, 0, 0) \equiv P_{200\ldots00}$$
$$\cdots\cdots\cdots\cdots\cdots\cdots$$
$$r_n(0, 0, \ldots, 0, 0) \equiv P_{n00\ldots00}$$
$$r_1(0, 0, \ldots, 0, 1) \equiv P_{100\ldots01}$$
$$r_2(0, 0, \ldots, 0, 1) \equiv P_{200\ldots01}$$
$$\cdots\cdots\cdots\cdots\cdots\cdots$$
$$r_n(0, 0, \ldots, 0, 1) \equiv P_{n00\ldots01}$$
$$r_1(0, 0, \ldots, 0, 2) \equiv P_{n00\ldots02}$$
$$\cdots\cdots\cdots\cdots\cdots\cdots$$
$$\cdots\cdots\cdots\cdots\cdots\cdots$$
$$r_n(h_1, h_2, \ldots, h_m, h) \equiv P_{nh_1h_2\ldots h_mh}$$
$$r_1(x_1 + h_1 + 1, 0, \ldots, 0, 0) \equiv P_{1(h_1+1)0\ldots00}$$
$$\cdots\cdots\cdots\cdots\cdots\cdots\cdots\cdots\cdots$$
$$\cdots\cdots\cdots\cdots\cdots\cdots\cdots\cdots\cdots$$
$$r_1(x_1 + h_1 + 1, x_2 + h_2 + 1, \ldots, x_m + h_m + 1, 0) \equiv P_{1(h_1+1)(h_2+1)\ldots(h_m+1)0}$$

[3]The two forms of recursive arithmetic, propositional and numerical, are not to be regarded as radically different things, but rather as being in some sense two forms of the same thing. For it has been well known, since ***Principia mathematica*** and earlier, that propositional and numerical functions may replace one another, i.e., either sort of function may be used to do the work of the other. Namely, a propositional function becomes a numerical function if we replace the values T and F by the numerical values 0 and 1. And inversely, in place of a numerical function we may always use a corresponding propositional function with one additional argument. For example the use of the sign $+$, denoting a certain binary numerical function (addition), may be replaced by the use of ternary predicate Σ which was mentioned above, and anything which can be said by use of the sign $+$ can with the aid of some circumlocution also be said by use of Σ instead of $+$. It must be regarded as a historical accident that treatments of quantificational logic, such as that in ***Principia mathematica***, have usually taken propositional functions as basic, while treatments of recursive arithmetic, beginning with Skolem's original paper of 1923, have usually taken numerical functions as basic.

$$r_2(x_1 + h_1 + 1, x_2 + h_2 + 1, \dots, x_m + h_m + 1, 0) \equiv P_{2(h_1+1)(h_2+1)\dots(h_m+1)0}$$
$$\dots\dots\dots\dots\dots\dots\dots\dots\dots\dots\dots\dots\dots\dots$$
$$r_n(x_1 + h_1 + 1, x_2 + h_2 + 1, \dots, x_m + h_m + 1, 0) \equiv P_{n(h_1+1)(h_2+1)\dots(h_m+1)0}$$
$$r_1(x_1 + h_1 + 1, x_2 + h_2 + 1, \dots, x_m + h_m + 1, 1) \equiv P_{1(h_1+1)(h_2+1)\dots(h_m+1)1}$$
$$\dots\dots\dots\dots\dots\dots\dots\dots\dots\dots\dots\dots\dots\dots$$
$$\dots\dots\dots\dots\dots\dots\dots\dots\dots\dots\dots\dots\dots\dots$$
$$r_n(x_1 + h_1 + 1, x_2 + h_2 + 1, \dots, x_m + h_m + 1, h) \equiv P_{n(h_1+1)(h_2+1)\dots(h_m+1)h}$$
$$r_1(0, 0, \dots, 0, t + h + 1) \equiv R_{100\dots0}$$
$$r_2(0, 0, \dots, 0, t + h + 1) \equiv R_{200\dots0}$$
$$\dots\dots\dots\dots\dots\dots\dots$$
$$r_n(0, 0, \dots, 0, t + h + 1) \equiv R_{n00\dots0}$$
$$r_1(0, 0, \dots, 1, t + h + 1) \equiv R_{100\dots1}$$
$$\dots\dots\dots\dots\dots\dots\dots$$
$$\dots\dots\dots\dots\dots\dots\dots$$
$$r_n(h_1, h_2, \dots, h_m, t + h + 1) \equiv R_{nh_1h_2\dots h_m}$$
$$r_1(x_1 + h_1 + 1, 0, \dots, 0, t + h + 1) \equiv R_{1(h_1+1)0\dots0}$$
$$\dots\dots\dots\dots\dots\dots\dots\dots\dots\dots$$
$$\dots\dots\dots\dots\dots\dots\dots\dots\dots\dots$$
$$r_1(x_1 + h_1 + 1, x_2 + h_2 + 1, \dots, x_m + h_m + 1, t + h + 1) \equiv R_{1(h_1+1)\dots(h+1)}$$
$$r_2(x_1 + h_1 + 1, x_2 + h_2 + 1, \dots, x_m + h_m + 1, t + h + 1) \equiv R_{2(h_1+1)\dots(h+1)}$$
$$\dots\dots\dots\dots\dots\dots\dots\dots\dots\dots\dots\dots\dots\dots\dots\dots$$
$$r_n(x_1 + h_1 + 1, x_2 + h_2 + 1, \dots, x_m + h_m + 1, t + h + 1) \equiv R_{n(h_1+1)\dots(h+1)}$$

where the formulas on the right of each equivalence contain no variables other than those (if any) which occur on the left — and though the formulas on the right may contain previously given predicates $s_1, s_2, \dots, s_k$ (which may include predicates previously introduced by recursion, as well as the primitive predicates $i_1, i_2, \dots$), the predicates $r_1, r_2, \dots, r_n$ that are being introduced do not occur on the right except in

$$R_{100\dots0}, \dots\dots\dots, R_{nh_1h_2\dots h_m}, R_{1(h_1+1)0\dots0}, \dots\dots\dots,$$
$$R_{1(h_1+1)(h_2+1)\dots(h+1)}, \dots, R_{n(h_1+1)(h_2+1)\dots(h+1)},$$

and do not occur in these except in elementary parts of the form $r_i(A_1, A_2, \dots, A_m, j)$ and $r_i(A_1, A_2, \dots, A_m, t + j)$, where $0 \leq j \leq h$, and where $A_1, A_2, \dots, A_m$ are any terms.

This schema is meant in the sense that the number $h + 1$ (the span of the recursion) and the numbers $h_1 + 1, h_2 + 1, \dots, h_m + 1$ may any or all of them be 0. In particular if all the numbers $h + 1, h_1 + 1, h_2 + 1, \dots, h_m + 1$ are 0, the schema reduces to the n equivalences

$$r_1(x_1, x_2, \ldots, x_m, t) \equiv R_1$$
$$r_2(x_1, x_2, \ldots, x_m, t) \equiv R_2$$
$$\cdots\cdots\cdots\cdots$$
$$r_n(x_1, x_2, \ldots, x_m, t) \equiv R_n$$

in which the predicates $r_1, r_2, \ldots, r_n$ do not occur in $R_1, R_2, \ldots, R_n$ at all.

In accordance with the schema D, the predicates $=, <\ \Sigma, \Pi$ may be introduced by recursion as follows:

$$0 = 0 \equiv T$$
$$x + 1 = 0 \equiv F$$
$$0 = t + 1 \equiv F$$
$$x + 1 = t + 1 \equiv x = t$$
$$0 < 0 \equiv F$$
$$x + 1 < 0 \equiv F$$
$$0 < t + 1 \equiv T$$
$$x + 1 < t + 1 \equiv x < t$$
$$\Sigma(x, 0, 0) \equiv x = 0$$
$$\Sigma(x, y + 1, 0) \equiv F$$
$$\Sigma(x, 0, t + 1) \equiv x = t + 1$$
$$\Sigma(x, y + 1, t + 1) \equiv \Sigma(x, y, t)$$
$$\Phi(0, 0, z, 0) \equiv z = 0$$
$$\Phi(x + 1, 0, z, 0) \equiv z = 0$$
$$\Phi(0, y + 1, z, 0) \equiv z = 0$$
$$\Phi(x + 1, y + 1, z, 0) \equiv [y = 0][z = x + 1]$$
$$\Phi(0, 0, z, t + 1) \equiv z = 0$$
$$\Phi(x + 1, 0, z, t + 1) \equiv z = 0$$
$$\Phi(0, y + 1, z, t + 1) \equiv z = 0$$
$$\Phi(x + 1, y + 1, z, t + 1) \equiv \Phi(x + 1, y + 1, z, t) \vee \Phi(x + 1, y, t + 1, t)\Sigma(x + 1, t + 1, z)$$
$$\Pi(x, y, t) \equiv \Phi(x, y, t, t)$$

Now it is possible to treat potentially infinite automata by means of propositional recursive arithmetic, in close analogy to the method already explained for treating finite automata by means of restricted recursive arithmetic.

To illustrate the notion of a potentially infinite automaton, consider the case of a computing machine, say a circuit, designed to compute the product of two decimal numbers. As is well known, any particular such machine will always have a limit to its capacity, a certain maximum number of digits such that the machine will not compute a product having more than this number of digits. So the statement that multiplication of decimal numbers is computable does not refer to any particular finite machine or finite automaton, but refers to the existence of a uniform plan for building larger and larger machines, of such a sort that any required product can be computed by a large enough machine built according to this plan. Or instead of a series of separate machines, it will come to the same thing if we have a plan for first building a multiplication machine and then continually adding to it, so that it is converted into a multiplication machine of larger and then still larger capacity, and the size of the product which the machine can compute increases without limit as times goes on. Such a machine of infinitely increasing size — or rather the plan for it, since not all of the infinite series of machines can have physically concrete existence — is an example of what we call a potentially infinite automaton.

Roughly described, then, a potentially infinite automaton consists in a fixed plan for building a finite automaton (or series of finite automata) of infinitely increasing size. The plan must be one which can actually be fixed upon in advance and made definite within a finite time and space, and which can in principle actually be carried out up to an automaton of any arbitrarily large finite size which may then be prescribed.

The recursion schema D in propositional recursive arithmetic is to be thought of as analogous to schema C in restricted recursive arithmetic. And just as for the representation of finite automata we use a suitably restricted special case of schema C, so we must use a special case of D in representing potentially infinite automata.

For the sake of representation of potentially infinite automata we therefore introduce the recursion schema E, which may be regarded as composed of two successive recursions, of which each is a special form of D.

E. Special Propositional Recursion

$\chi_1(0, 0, \ldots, 0, 0) \equiv P_{100\ldots00}$

$\chi_2(0, 0, \ldots, 0, 0) \equiv P_{200\ldots00}$

. . . .

$\chi_N(0, 0, \ldots, 0, 0) \equiv P_{N00\ldots00}$

$\chi_1(0, 0, \ldots, 0, 1) \equiv P_{100\ldots00} \vee P_{100\ldots01}$

$\chi_2(0, 0, \ldots, 0, 1) \equiv P_{200\ldots00} \vee P_{200\ldots01}$

. . . .

$\chi_N(0, 0, \ldots, 0, 1) \equiv P_{N00\ldots00} \vee P_{N00\ldots01}$

$\chi_1(0, 0, \ldots, 0, 2) \equiv P_{100\ldots00} \vee P_{100\ldots01} \vee P_{100\ldots02}$

. . . .

. . . .

$\chi_N(M, M, \ldots, M, g) \equiv P_{NMM\ldots M0} \vee \cdots \vee P_{NMM\ldots Mg}$

$\chi_1(x_1 + M + 1, 0, \ldots, 0, 0) \equiv F$

. . . .

. . . .

$\chi_N(x_1 + M + 1, M, \ldots, M, g) \equiv F$

$\chi_1(x_1 + M + 1, x_2 + M + 1, \ldots, 0, 0) \equiv F$

. . . .

. . . .

$\chi_1(x_1 + M + 1, x_2 + M + 1, \ldots, x_m + M + 1, 0) \equiv F$

$\chi_2(x_1 + M + 1, x_2 + M + 1, \ldots, x_m + M + 1, 0) \equiv F$

. . . .

$\chi_N(x_1 + M + 1, x_2 + M + 1, \ldots, x_m + M + 1, 0) \equiv F$

$$\chi_1(x_1+M+1, x_2+M+1, \ldots, x_m+M+1, 1) \equiv F$$

$$\cdots\cdots\cdots\cdots\cdots\cdots$$

$$\cdots\cdots\cdots\cdots\cdots\cdots$$

$$\chi_N(x_1+M+1, x_2+M+1, \ldots, x_m+M+1, g) \equiv F$$

$$\chi_1(0,0,\ldots,0,t+g+1) \equiv \chi_1(0,0,\ldots,0,t+g) \vee Q_{100\ldots0}\left[\chi_1(0,0,\ldots,0,t),\ldots,\chi_N(0,0,\ldots, 0,t),\chi_1(0,0,\ldots,1,t),\ldots,\ldots,\chi_N(M+1,M+1,\ldots,M+1,t)\right]$$

$$\chi_2(0,0,\ldots,0,t+g+1) \equiv \chi_2(0,0,\ldots,0,t+g) \vee \overline{\chi}_1(0,0,\ldots,0,t+g)Q_{200\ldots0}\left[\chi_1(0,0,\ldots,0,t),\ldots, \chi_N(0,0,\ldots,0,t),\chi_1(0,0,\ldots,1,t),\ldots,\ldots, \chi_N(M+1,M+1,\ldots,M+1,t)\right]$$

$$\cdots\cdots\cdots\cdots\cdots\cdots$$

$$\cdots\cdots\cdots\cdots\cdots\cdots$$

$$\chi_N(M,M,\ldots,M,t+g+1) \equiv \chi_N(M,M,\ldots,M,t+g) \vee \overline{\chi}_1(M,M,\ldots,M,t+g)\overline{\chi}_2(M,M,\ldots,M,t+g)\ldots \overline{\chi}_{N-1}(M,M,\ldots,M,t+g)Q_{NMM\ldots M}\left[\chi_1(0,0,\ldots, 0,t),\ldots,\chi_N(0,0,\ldots,0,t),\chi_1(0,0,\ldots,1,t), \ldots,\ldots,\chi_N(2M+1,2M+1,\ldots,2M+1,t)\right]$$

$$\chi_1(x_1+M+1,0,\ldots,0,t+g+1) \equiv \chi_1(x_1+M+1,0, \ldots,0,t+g) \vee Q_{1(M+1)0\ldots0}\left[\chi_1(x_1,0,\ldots,0,t), \ldots,\chi_N(x_1,0,\ldots,0,t),\chi_1(x_1,0,\ldots,1,t),\ldots,\ldots, \chi_N(x_1+2M+2,M+1,\ldots,M+1,t)\right]$$

$$\chi_2(x_1+M+1,0,\ldots,0,t+g+1) \equiv \chi_2(x_1+M+1,0,\ldots,0, t+g) \vee \overline{\chi}_1(x_1+M+1,0,\ldots,0,t+g)Q_{2(M+1)0\ldots0}\left[\chi_1(x_1, 0,\ldots,0,t),\ldots,\chi_N(x_1,0,\ldots,0,t),\chi_1(x_1,0,\ldots, 1,t),\ldots,\ldots,\chi_N(x_1+2M+2,M+1,\ldots,M+1,t)\right]$$

$$\cdots\cdots\cdots\cdots\cdots\cdots$$

$$\cdots\cdots\cdots\cdots\cdots\cdots$$

$$\chi_1(x_1+M+1,x_2+M+1,\ldots,x_m+M+1,t+g+1) \equiv \chi_1(x_1+M+1,x_2+M+1,\ldots,x_m+M+1,t+g) \vee Q_{1(M+1)(M+1)\ldots(M+1)}\left[\chi_1(x_1,x_2,\ldots,x_m,t),\ldots,\chi_N(x_1,x_2, \ldots,x_m,t),\chi_1(x_1,x_2,\ldots,x_m+1,t),\ldots, \ldots,\chi_N(x_1+2M+2,x_2+2M+2,\ldots, x_m+2M+2,t)\right]$$

$$\chi_2(x_1+M+1,x_2+M+1,\ldots,x_m+M+1,t+g+1) \equiv$$

$$\chi_2(x_1 + M + 1, x_2 + M + 1, \ldots, x_m + M + 1, t + g) \vee$$
$$\overline{\chi}_1(x_1 + M + 1, x_2 + M + 1, \ldots, x_m + M + 1, t + g) Q_{2(M+1)(M+1)\ldots(M+1)}$$
$$\big[\chi_1(x_1, x_2, \ldots, x_m, t), \ldots, \chi_N(x_1, x_2, \ldots, x_m, t),$$
$$\chi_1(x_1, x_2, \ldots, x_m + 1, t), \ldots\ldots\ldots,$$
$$\chi_N(x_1 + 2M + 2, x_2 + 2M + 2, \ldots, x_m + 2M + 2, t)\big]$$

. . . .

$$\chi_N(x_1 + M + 1, x_2 + M + 1, \ldots, x_m + M + 1, t + g + 1) \equiv$$
$$\chi_N(x_1 + M + 1, x_2 + M + 1, \ldots, x_m + M + 1, t + g) \vee$$
$$\overline{\chi}_1(x_1 + M + 1, x_2 + M + 1, \ldots, x_m + M + 1, t + g)\overline{\chi}_2(x_1 + M + 1,$$
$$x_2 + M + 1, \ldots, x_m + M + 1, t + g) \ldots \overline{\chi}_{N-1}(x_1 + M + 1, x_2 + M + 1, \ldots,$$
$$x_m + M + 1, t + g) Q_{N(M+1)(M+1)\ldots(M+1)} \big[\chi_1(x_1, x_2, \ldots, x_m, t), \ldots,$$
$$\chi_N(x_1, x_2, \ldots, x_m, t), \chi_1(x_1, x_2, \ldots, x_m + 1, t), \ldots,$$
$$\ldots, \chi_N(x_1 + 2M + 2, x_2 + 2M + 2, \ldots, x_m + 2M + 2, t)\big]$$

$$r_1(0, 0, \ldots, 0, 0) \equiv \chi_1(0, 0, \ldots, 0, 0) R_{1100\ldots00} \big[i_1'(0, 0, \ldots, 0, 0),$$
$$\ldots, i_\mu'(0, 0, \ldots, 0, 0), i_1'(0, 0, \ldots, 1, 0), \ldots\ldots\ldots,$$
$$i_\mu'(M, M, \ldots, M, 0)\big] \vee \chi_2(0, 0, \ldots, 0, 0) R_{2100\ldots00} \big[i_1'(0,$$
$$0, \ldots, 0, 0,), \ldots\ldots\ldots, i_\mu'(M, M, \ldots, M, 0)\big] \vee \cdots \vee$$
$$\chi_N(0, 0, \ldots, 0, 0) R_{N100\ldots00} \big[i_1'(0, 0, \ldots, 0, 0), \ldots\ldots\ldots,$$
$$i_\mu'(M, M, \ldots, M, 0)\big]$$

. . . .

$$r_n(0, 0, \ldots, 0, 0) \equiv \chi_1(0, 0, \ldots, 0, 0) R_{1n00\ldots00}[i_1'(0, 0, \ldots, 0, 0),$$
$$\ldots\ldots\ldots, i_\mu'(M, M, \ldots, M, 0)] \vee \chi_2(0, 0, \ldots, 0, 0) R_{2n00\ldots00}$$
$$[i_1'(0, 0, \ldots, 0, 0,), \ldots\ldots\ldots, i_\mu'(M, M, \ldots, M, 0)] \vee \cdots \vee$$
$$\chi_N(0, 0, \ldots, 0, 0) R_{Nn00\ldots00}[i_1'(0, 0, \ldots, 0, 0), \ldots\ldots\ldots,$$
$$i_\mu'(M, M, \ldots, M, 0)]$$

$$r_1(0, 0, \ldots, 0, 1) \equiv \chi_1(0, 0, \ldots, 0, 1) R_{1100\ldots01}[i_1'(0, 0, \ldots, 0, 0),$$
$$\ldots\ldots\ldots, i_\mu'(M, M, \ldots, M, 0), i_1'(0, 0, \ldots, 0, 1),$$
$$\ldots\ldots\ldots, i_\mu'(M + 1, M + 1, \ldots, M + 1, 1)] \vee \cdots \vee$$
$$\chi_N(0, 0, \ldots, 0, 1) R_{N100\ldots01}[i_1'(0, 0, \ldots, 0, 0),$$
$$\ldots\ldots\ldots, i_\mu'(M, M, \ldots, M, 0), i_1'(0, 0, \ldots, 0, 1),$$
$$\ldots\ldots\ldots, i_\mu'(M + 1, M + 1, \ldots, M + 1, 1)]$$

. . . .

. . . .

$$r_n(M, M, \ldots, M, h) \equiv \chi_1(M, M, \ldots, M, h) R_{1nMM\ldots Mh}[i_1'(0, 0, \ldots, 0, 0),$$
$$\ldots\ldots\ldots, i_\mu'(M, M, \ldots, M, 0), i_1'(0, 0, \ldots, 0, 1), \ldots\ldots\ldots,$$

$$i'_\mu(2M+1, 2M+1, \ldots, 2M+1, h)] \vee \cdots \vee$$
$$\chi_N(M, M, \ldots, M, h) R_{NnMM\ldots Mh}[i'_1(0,0,\ldots,0,0), \ldots\ldots\ldots,$$
$$i'_\mu(M, M, \ldots, M, 0), i'_1(0,0,\ldots,0,1), \ldots\ldots\ldots,$$
$$i'_\mu(2M+1, 2M+1, \ldots, 2M+1, h)]$$

$$r_1(x_1+M+1, 0, \ldots, 0, 0) \equiv \chi_1(x_1+M+1, 0, \ldots, 0, 0)$$
$$R_{11(M+1)0\ldots 00}[i'_1(x_1, 0, \ldots, 0, 0), \ldots\ldots\ldots, i'_\mu(x_1+2M+2,$$
$$M, \ldots, M, 0)] \vee \chi_2(x_1+M+1, 0, \ldots, 0, 0) R_{21(M+1)0\ldots 00}$$
$$[i'_1(x_1, 0, \ldots, 0, 0), \ldots\ldots\ldots, i'_\mu(x_1+2M+2, M, \ldots,$$
$$M, 0)] \vee \cdots \vee \chi_N(x_1+M+1, 0, \ldots, 0, 0) R_{N1(M+1)0\ldots 00}$$
$$[i'_1(x_1, 0, \ldots, 0, 0), \ldots\ldots\ldots, i'_\mu(x_1+2M+2, M,$$
$$\ldots, M, 0)]$$

. . . .

. . . .

$$r_1(x_1+M+1, x_2+M+1, \ldots, x_m+M+1, 0) \equiv \chi_1(x_1+M+1,$$
$$x_2+M+1, \ldots, x_m+M+1, 0) R_{11(M+1)(M+1)\ldots(M+1)0}[i'_1(x_1, x_2,$$
$$\ldots, x_m, 0), \ldots\ldots\ldots, i'_\mu(x_1+2M+2, x_2+2M+2, \ldots,$$
$$x_m+2M+2, 0)] \vee \cdots \vee \chi_N(x_1+M+1, x_2+M+1, \ldots,$$
$$x_m+M+1, 0) R_{N1(M+1)(M+1)\ldots(M+1)0}[i'_1(x_1, x_2, \ldots, x_m, 0), \ldots,$$
$$\ldots, i'_\mu(x_1+2M+2, x_2+2M+2, \ldots, x_m+2M+2, 0)]$$

. . . .

. . . .

$$r_n(x_1+M+1, x_2+M+1, \ldots, x_m+M+1, h) \equiv$$
$$\chi_1(x_1+M+1, x_2+M+1, \ldots, x_m+M+1, h)$$
$$R_{1n(M+1)(M+1)\ldots(M+1)h} i'_1(x_1, x_2, \ldots,$$
$$x_m, 0), \ldots\ldots\ldots, i'_\mu(x_1+2M+2, x_2+2M+2, \ldots,$$
$$x_m+2M+2, h) \vee \cdots \vee \chi_N(x_1+M+1, x_2+M+1, \ldots,$$
$$x_m+M+1, h) R_{Nn(M+1)(M+1)\ldots(M+1)h}[i'_1(x_1, x_2,$$
$$\ldots, x_m, 0), \ldots\ldots\ldots, i'_\mu(x_1+2M+2, x_2+2M+2,$$
$$\ldots, x_m+2M+2, h)]$$

$$r_1(0, 0, \ldots, 0, t+h+1) \equiv \chi_1(0, 0, \ldots, 0,$$
$$t+h+1) T_{1100\ldots 0}[i'_1(0, 0, \ldots, 0, t), \ldots\ldots\ldots,$$
$$i'_\mu(M+1, M+1, \ldots, M+1, t+h+1), r_1(0, 0, \ldots, 0, t),$$
$$\ldots\ldots\ldots, r_n(M+1, M+1, \ldots, M+1, t+h)] \vee \cdots \vee$$
$$\chi_N(0, 0, \ldots, 0, t+h+1) T_{N100\ldots 0}[i'_1(0, 0, \ldots, 0, t),$$
$$\ldots\ldots\ldots, i'_\mu(M+1, M+1, \ldots, M+1, t+h+1),$$

$$r_1(0,0,\ldots,0,t),\ldots\ldots\ldots,r_n(M+1,M+1,\ldots,M+1,$$
$$t+h)]$$

....

....

$$r_n(M,M,\ldots,M,t+h+1)\equiv\chi_1(M,M,\ldots,M,$$
$$t+h+1)T_{1nMM\ldots M}[i'_1(0,0,\ldots,0,t),\ldots\ldots\ldots,$$
$$i'_\mu(2M+1,2M+1,\ldots,2M+1,t+h+1),$$
$$r_1(0,0,\ldots,0,t),\ldots\ldots\ldots,r_n(2M+1,2M+1,\ldots,$$
$$2M+1,t+h)]\vee\cdots\vee\chi_N(M,M,\ldots,M,$$
$$t+h+1)T_{NnMM\ldots M}[i'_1(0,0,\ldots,0,t),\ldots\ldots\ldots,$$
$$i'_\mu(2M+1,2M+1,\ldots,2M+1,t+h+1),$$
$$r_1(0,0,\ldots,0,t),\ldots\ldots\ldots,r_n(2M+1,2M+1,\ldots,$$
$$2M+1,t+h)]$$

$$r_1(x_1+M+1,0,\ldots,0,t+h+1)\equiv\chi_1(x_1+M+1,0,\ldots,0,$$
$$t+h+1)T_{11(M+1)0\ldots0}[i'_1(x_1,0,\ldots,0,t),\ldots\ldots\ldots,$$
$$i'_\mu(x_1+2M+2,M+1,\ldots,M+1,t+h+1),r_1(x_1,0,$$
$$\ldots,0,t),\ldots,r_n(x_1+2M+2,M+1,\ldots,M+1,$$
$$t+h)]\vee\cdots\vee\chi_N(x_1+M+1,0,\ldots,0,t+h+1)$$
$$T_{N1(M+1)0\ldots0}[i'_1(x_1,0,\ldots,0,t),\ldots\ldots\ldots,i'_\mu(x_1+2M+2,$$
$$M+1,\ldots,M+1,t+h+1),r_1(x_1,0,\ldots,0,t),\ldots,$$
$$r_n(x_1+2M+2,M+1,\ldots,M+1,t+h)]$$

....

....

$$r_1(x_1+M+1,x_2+M+1,\ldots,x_m+M+1,t+h+1)\equiv$$
$$\chi_1(x_1+M+1,x_2+M+1,\ldots,x_m+M+1,t+h+1)T_{11(M+1)(M+1)\ldots(M+1)}$$
$$[i'_1(x_1,x_2,\ldots,x_m,t),\ldots\ldots\ldots,i'_\mu(x_1+2M+2,x_2+2M+2,\ldots,$$
$$x_m+2M+2,t+h+1),r_1(x_1,x_2,\ldots,x_m,t),\ldots,$$
$$\ldots,r_n(x_1+2M+2,x_2+2M+2,\ldots,x_m+2M+2,t+h)$$
$$\vee\cdots\vee\chi_N(x_1+M+1,x_2+M+1,\ldots,x_m+M+1,t+h+1)$$
$$T_{N1(M+1)(M+1)\ldots(M+1)}$$
$$[i'_1(x_1,x_2,\ldots,x_m,t),\ldots\ldots\ldots,i'_\mu(x_1+2M+2,$$
$$x_2+2M+2,\ldots,x_m+2M+2,t+h+1),r_1(x_1,x_2,\ldots,$$
$$x_m,t),\ldots,r_n(x_1+2M+2,x_2+2M+2,\ldots,x_m+2M+2,t+h)]$$

....

$$r_n(x_1+M+1,x_2+M+1,\ldots,x_m+M+1,t+h+1)\equiv$$

$$\chi_1(x_1+M+1, x_2+M+1, \ldots,$$
$$x_m+M+1, t+h+1) T_{1n(M+1)(M+1)\ldots(M+1)}[i_1'(x_1, x_2,$$
$$\ldots, x_m, t), \ldots\ldots, i_\mu'(x_1+2M+2, x_2+2M+2, \ldots,$$
$$x_m+2M+2, t+h+1), r_1(x_1, x_2, \ldots, x_m, t), \ldots\ldots,$$
$$r_n(x_1+2M+2, x_2+2M+2, \ldots, x_m+2M+2, t+h)] \vee$$
$$\cdots \vee \chi_N(x_1+M+1, x_2+M+1, \ldots, x_m+M+1, t+h+1)$$
$$T_{Nn(M+1)(M+1)\ldots(M+1)}[i_1'(x_1, x_2, \ldots, x_m, t), \ldots,$$
$$\ldots, i_\mu'(x_1+2M+2, x_2+2M+2, \ldots, x_m+2M+2, t+h+1), r_1(x_1, x_2, \ldots,$$
$$x_m, t), \ldots\ldots, r_n(x_1+2M+2, x_2+2M+2, \ldots, x_m+2M+2, t+h)]$$

where $P_{100\ldots00}, \ldots\ldots, P_{NMM\ldots Mg}$ are truth-values, so chosen as to make at most one of $\chi_1(j_1, j_2, \ldots, j_m, j), \chi_2(j_1, j_2, \ldots, j_m, j), \ldots, \chi_N(j_1, j_2, \ldots, j_m, j)$ true for each $(m+1)$-tuple $(j_1, j_2, \ldots, j_m, j), 0 \leq j_1 \leq M, 0 \leq j_2, \leq M, \ldots, 0 \leq j_m \leq M, 0 \leq j \leq g$; where the Q's are truth-functions so chosen that the conjunction of any two Q's which differ only in the first subscript is a contradiction; and where $i_j'(A_1, A_2, \ldots, A_m, A)$ is an abbreviation of $i_j(A_1, A_2, \ldots, A_m, A)$ $[\chi_1(A_1, A_2, \ldots, A_m, A) \vee \chi_2(A_1, A_2, \ldots, A_m, A) \vee \cdots \vee \chi_N(A_1, A_2, \ldots, A_m, A)]$, for any terms $A_1, A_2, \ldots, A_m, A$, and for $1 \leq j \leq \mu$.

Potentially Infinite Automata

It can be shown that propositional recursive arithmetic with recursion schema E represents potentially infinite automata in somewhat the same way that restricted recursive arithmetic represents finite automata, where a *potentially infinite automaton* is defined by the following eight conditions:

(1) There are an infinite number of *elements*, each of which is capable of no more than a fixed maximum number of different *states*. There is no loss of generality in assuming that each element is capable of just two states, and in stating the remaining conditions we make this assumption, calling the two states "operated" and "unoperated."

(2) Every element is in a definite *cell*, located by means of an m-tuple of *coordinates*, $x_1, x_2, \ldots, x_m$. Here m is a fixed number, the *dimensionality* of the automaton, and $x_1, x_2, \ldots, x_m$ are integers which may be negative or 0 or positive. Every m-tuple of coordinates determines a cell, and no two cells have the same coordinates. There is moreover a fixed non-negative number M such that two distinct cells are called *neighboring* if each coordinate of one cell differs from the corresponding coordinate of the other by at most $M+1$.

(3) Some cells may be inactive in the sense that no element in them is ever operated. But except inactive cells, every cell has a definite *initial action time* such that no element in it is operated before the initial action time.

(4) A finite number of cells have initial action time 0, and every other cell, unless inactive, has an initial action time that is later by at least a certain fixed length of time $g+1$ than the initial action time of at least one of its neighboring cells. By taking M large enough we may assume that the cells whose initial action time is 0 have coordinates less than $M+1$ in absolute value.

(5) Every cell, unless inactive, has a definite one of a finite list of characters, $1, 2, \ldots, N$. And if the initial action time of the cell is greater than g, its character is uniquely determined by the characters of those of the neighboring cells that have initial action times that are earlier by at least the length of time $g + 1$.

(6) The elements in any cell comprise a fixed number μ of input elements, designated as the 1st, 2nd, ..., μth inputs, and a fixed number n of other elements, designated as the 1st, 2nd, ..., nth.

(7) The *neighboring* elements of an element e are the elements in the neighboring cells and in the same cell, other than e itself. Except for the input elements, whose states are imposed from the outside, the state of a particular element at any instant can depend only on the states of the neighboring input elements at the same instant and on the states of the element itself and its neighboring elements at past instants, for a certain maximum length of time $h + 1$ into the past.

(8) The pattern of action by which the states of the non-input elements of a particular cell at time $t + h + 1$ are determined from the states of these elements and their neighboring elements at times $t, t + 1, \ldots, t + h + 1$ can be summed up in a table analogous to a truth-table. This pattern of action (or the table in which it is summed up) is uniquely determined by the character of the cell and the characters of the neighboring cells.

Conditions (3) and (4) have the consequence that the complete automaton need never be built, but after building a certain finite part, action of the automaton may be started, and thereafter construction must proceed at a sufficient rate to keep ahead of the action of the automaton, It is for this reason that we speak of a *potentially infinite* automaton, rather than infinite.

However, the action of the automaton must not depend on the process of construction, except in the sense that construction has to keep ahead of the action of the automaton. Each new element, as the construction proceeds, is added in unoperated state. And we assume, as evidently we may without loss of generality, that one complete cell is added at a time, and in the action of the automaton the effect of a cell not yet added is exactly the same as the effect of a cell in which all elements are unoperated.

The restriction in (6) to fixed numbers μ and n is justified by the consideration that there must at any rate be a maximum number μ of inputs and a maximum number n of other elements, and these two numbers can be brought to equal μ and n respectively in every cell by adjoining a suitable number of inactive elements of each kind.

Condition (7) is based on the assumption that two elements cannot influence each other sufficiently to control a change of state by either one if they are beyond a certain maximum distance apart; and hence by taking M large enough, that two elements cannot thus influence each other unless they are in the same cell or neighboring cells. The coordinates of the cells need not be Cartesian or the space Euclidean, but it is assumed that, for $j = 1, 2, \ldots, m$, the difference of the jth coordinates of two cells which are of not more than a certain fixed distance apart cannot be unbounded.

If we allow causal influence to pass instantaneously from one element to another, we must impose the restriction that no chain of such instantaneous causal action of one element upon another (and of that element in turn upon another one and so on) can extend beyond a certain maximum distance. And M is then to be taken large enough so that such a chain cannot pass from any element beyond its neighboring elements. As there must not be circles in the causal chain of a kind that would render the action of the automaton nondeterministic, it follows that the instantaneous causal influence upon any element can always be traced back to neighboring *input* elements, as was assumed in condition (7). (I am indebted to John Myhill for calling to my attention the need to make explicit this assumption which was only implicit in my previous treatments of the matter.)

The restrictions in (5) on the characters of cells and the use made of characters in (8) are justified on the ground that the construction of the automaton must follow a preassigned plan, else the automaton would not be a particular potentially infinite automaton, as there would never come a time at which a final choice had been made among all possibilities. Even if we allow the builders in a particular cell to make use of portable computing and recording apparatus of bounded size, the portion of the automaton which they are able to survey in its relation to their own situation, and thus to take into account in carrying out the construction in the particular cell, cannot exceed a certain maximum distance from them — and M may be taken large enough to provide for this. The builders may indeed make preassigned exceptions to the general plan in a finite number of special cells; but this may be allowed for by setting the initial action times of these special cells back to 0, and then assigning to each such cell a special character different from the characters of all other cells. And the builders might also take into account, in determining the character of a cell, not only the characters of the neighboring cells, but also the past history of the neighboring cells (i.e., arrival of the initial action time of the cell, and the succession of states of the elements of the cell) for a certain maximum length of time into the past; but the equivalent of this may evidently be provided instead by the action of elements included in the cells themselves.

(We may not suppose that the builders are guided in the construction by records and computations which either extend indefinitely into the past or increase indefinitely in size as construction proceeds; for if this were the case, then the recording and computing apparatus employed by the builders must be considered as being itself a part of the automaton, and the whole process by which the builders are guided in constructing one part of the automaton by outputs of another part of it must be taken together as a hyper-automaton, in which the action of the builders is mechanized in the form of automatically operating connections from one part of the automaton to the other; and when this is done the hyper-automaton will be constructible without resort to auxiliary recording and computing apparatus of increasing size. On similar grounds we may not suppose that the builders, after constructing a particular element or cell of the automaton, later come back and alter it; for if this is done according to a preassigned plan, then the process can and must be mechanized as part of the automaton.)

The fiction of a crew of builders, engaged to all eternity in constructing an automaton of unbounded size, has been used as a convenient expository device, because it helps in making clear how certain sorts of construction may not be allowed. But as already explained a potentially infinite automaton may be thought of more plausibly just as a plan for building a finite automaton of any required size.

A well-known example of a potentially infinite automaton is of course the Turing machine — which may be treated as a one-dimensional automaton by taking the cells to be the squares on the tape, and the elements of the automaton to be the symbols which may appear in the square, the possible state of the square as being scanned or not scanned, and the possible internal configurations when scanned. It is unimportant that the elements of the automaton are thus taken to be mere potentialities of the square rather than materially solid separate parts. For we might indeed suppose the machine constructed without the moving scanner and with an appropriate collection of relays or other electrically or mechanically operating parts in each square instead.

Modifications of the Turing machine in which more than one tape is provided, or more than one scanner, may also be treated as one-dimensional potentially infinite automata by means of appropriate conventions. The Turing machine with a two-dimensional tape ruled into squares, whether the two-dimensional tape occupies a full plane or only one quadrant, may similarly be regarded as a two-dimensional potentially infinite automaton. And another example of a potentially infinite automaton is a human calculator who has available an unlimited supply of pencils and of sheets of paper of uniform size and proceeds according to a preassigned plan of calculation — a one-dimensional automaton if the sheets of paper are piled in a single pile (or in a finite number of piles) as the calculation proceeds.

In a potentially infinite automaton we may represent the inputs in the cell with coordinates $x_1, x_2, \ldots, x_m$ by $i_1(x_1, x_2, \ldots, x_m, t)$, $\ldots$, $i_\mu(x_1, x_2, \ldots, x_m, t)$, and the remaining elements in the cell by $r_1(x_1, x_2, \ldots, x_m, t)$, $\ldots$, $r_n(x_1, x_2, \ldots, x_m, t)$. And we may use $\chi_c(x_1, x_2, \ldots, x_m, t)$ to mean that the initial action time of the cell is less than or equal to t and the character of the cell is c where $c = 1, 2, \ldots, N$.

If the action of the automaton is confined to a single octant, we may take $x_1, x_2, \ldots, x_m$ to be positive or 0 in the active cells; and the automaton will then be represented within propositional recursive arithmetic by a recursion of the form E — as may be verified directly from the definition of a potentially infinite automaton. And conversely every recursion of the form E represents a potentially infinite automaton.

In the general case, if the action of the automaton is not known to be confined to a single octant, we may replace

$$i_1(x_1, x_2, \ldots, x_m, t), \ldots, i_\mu(x_1, x_2, \ldots, x_m, t),$$
$$r_1(x_1, x_2, \ldots, x_m, t), \ldots, r_n(x_1, x_2, \ldots, x_m, t),$$
$$\chi_1(x_1, x_2, \ldots, x_m, t), \ldots, \chi_N(x_1, x_2, \ldots, x_m, t)$$

by

$$i_1(y_1, y_2, \ldots, y_{2m}, t), \ldots, i_\mu(y_1, y_2, \ldots, y_{2m}, t),$$
$$r_1(y_1, y_2, \ldots, y_{2m}, t), \ldots, r_n(y_1, y_2, \ldots, y_{2m}, t),$$
$$\chi_1(y_1, y_2, \ldots, y_{2m}, t), \ldots, \chi_N(y_1, y_2, \ldots, y_{2m}, t) \text{ respectively;}$$

where y_{2j-1} is 0 if x_j is 0 or positive, and y_{2j-1} is 1 if x; is negative; and where y_{2j} is the same as x_j if x_j is 0 or positive, and y_{2j} is $-x_j - 1$ if x_j is negative. Then in terms of $i_1(y_1, y_2, \ldots, y_{2m}, t)$, etc. — in place of $i_1(x_1, x_2, \ldots, x_m m, t)$, etc. — the recursion equivalences that represent the automaton constitute a special case of a recursion of the form E. (This amounts to replacing an m-dimensional automaton whose action is not necessarily confined to a single octant by a $2m$-dimensional automaton of special kind, whose action is confined to a single octant, and which is equivalent to the given m-dimensional automaton in the sense that the action is cell by cell and element by element the same.)

The definition which we have given of a potentially infinite automaton can certainly be reduced much further in the sense of simplifying the recursion schema (analogous to E) to which the definition leads—even if in the reduction we impose the strong condition that a subset of the elements of the new automaton, obtained by the reduction, shall be in one-to-one correspondence with the full set of elements of the original automaton, in such a way that the action of corresponding elements is at every instant the same. For further treatment of potentially infinite automata, such a reduction must evidently be made. But our purpose here has been to formulate a suitable comprehensive definition from which reduced definitions can be justified by their equivalence to it in some desired sense

The *synthesis problem* and the *decision problem* for potentially infinite automata may now be stated in the same way as was done above for finite automata. But it seems clear that for potentially infinite automata either these problems will be more difficult than for finite automata or unsolvable cases will be reached more quickly or both. And in these notes we shall treat the synthesis problem and the decision problem for the case only of finite automata.

However, before returning to finite automata, we call attention to some open problems in the theory of recursive functions which are suggested by consideration of potentially infinite automata. Taking $\mu = 0$ in propositional recursive arithmetic (so that the system is confined to potentially infinite automata without inputs), consider the class of all recursive propositional functions obtainable by recursions of the form D, and also the class of all those obtainable by recursions of the form E. These two classes suggest themselves as natural, the former class because D is in some sense the most general recursion schema in which the recursion is with respect to one variable only and which can be stated in this notation and in a form of this general kind, and the latter class because it comprises all functions that represent the action of an infinite set of corresponding elements, one in each cell in any potentially infinite automaton. We may ask for simpler characterizations of these classes, and also concerning the relationship of these classes to each other and to the class of primitive recursive functions. Moreover the fact that the second of the two classes is not only narrower than the class of general recursive propositional functions, but also by conjecture narrower than the primitive recursive, suggests a classification of recursive functions, both propositional and numerical, according to the maximum speed of computation by a potentially infinite automaton.

Some Metatheorems of Propositional Calculus

By a solution of a formula X of propositional calculus for a particular propositional variable p we mean an equivalence $p \equiv Y$, where p does not occur in Y, such that X reduces to a tautology upon substituting Y for p throughout.

If we write X in the form $Ap \vee B\overline{p}$, where A and B do not contain p, the condition of possibility of solution for p is:

I. $$A \vee B$$

Subject to this condition, the general solution for p is:

II. $$p \equiv A\theta \vee \overline{B}$$

where θ is an arbitrary parameter. In fact, using the sign $\vDash$ to mean what follows it is tautologous in the sense of propositional calculus, we have:

III. $$\vDash A \vee B \supset .[p \equiv A\theta \vee \overline{B}] \supset Ap \vee B\overline{p}$$

If instead of there being just one variable in question, there are n variables $p_1, p_2, \ldots, p_n$, the simultaneous solution of X for the n variables may be obtained by iterating the above process, i.e., we first solve for p_n in the form II, then the condition I of possibility of this solution is solved for p_{n-1} and so on, until ultimately we solve for p_1. Then in the equivalence which was obtained as solution for p_2, we eliminate p_1 on the right by substituting for it in accordance with the equivalence which was obtained as solution for p_1; in the equivalence which was obtained as solution for p_3, we eliminate p_2 and p_1 on the right by substituting for them in accordance with the equivalences which were obtained as solutions for p_2 and p_1; and so on back, until in the end we find equivalences for $p_1, p_2, \ldots, p_n$ in which none of these variables occur on the right.

Evidently, in the simultaneous solution of X for the n variables $p_1, p_2, \ldots, p_n$ as thus obtained, we may expect in general that n parameters $\theta_1, \theta_2, \ldots, \theta_n$ will occur. Moreover, if the development of X in the n variables $p_1, p_2, \ldots, p_n$ is

$$A_1 p_1 p_2 \ldots p_n \vee A_2 p_1 p_2 \ldots \overline{p}_n \vee \cdots \vee A_{2^n} \overline{p}_1 \overline{p}_2 \ldots \overline{p}_n,$$

the condition of possibility of simultaneous solution for $p_1, p_2, \ldots, p_n$ is:

IV. $$A_1 \vee A_2 \vee \cdots \vee A_{2^n}$$

And the condition that X shall be satisfied identically in $p_1, p_2, \ldots, p_n$ is the conjunction

V. $$A_1 A_2 \ldots A_{2^n}$$

For IV we shall use the notation:

$$\underset{p_1 \cdots p_n}{\mathsf{E}} X$$

and for V we shall use the notation:

$$\underset{p_1 \cdots p_n}{\mathsf{A}} X$$

These notations are intended to suggest a certain analogy of IV and V, respectively, to the existential quantification of X with respect to $p_1, p_2, \ldots, p_n$ and the universal quantification of X with respect to $p_1, p_2, \ldots, p_n$. This analogy extends in fact to the preservation of certain familiar laws of quantification. In particular:

VI. $$\vDash X \supset \underset{p_1 \cdots p_n}{\mathsf{E}} X$$

VII. $$\vDash \underset{p_1 \cdots p_n}{\mathsf{A}} X \supset X$$

VIII. $$\vDash \underset{p_1 \cdots p_n}{\mathsf{E}} [X_1 X_2] \supset \underset{p_1 \cdots p_n}{\mathsf{E}} X_1$$

IX. $$\vDash \underset{p_1 \ldots p_n}{\mathsf{A}} [X_1 X_2] \supset \underset{p_1 \ldots p_n}{\mathsf{A}} X_1$$

Moreover, if X_1 does not contain the variables $p_1, p_2, \ldots, p_n$:

X. $$\vDash \underset{p_1 \ldots p_n}{\mathsf{E}} [X_1 X_2] \equiv X_1 \underset{p_1 \ldots p_n}{\mathsf{E}} X_2$$

XI. $$\vDash \underset{p_1 \ldots p_n}{\mathsf{A}} [X_1 X_2] \equiv X_1 \underset{p_1 \ldots p_n}{\mathsf{A}} X_2$$

Also two or more successive operations E may always be combined into a single operation E in the sense that:

XII. $$\vDash \underset{p_1 \ldots p_m}{\mathsf{E}} \underset{V_1 \ldots V_n}{\mathsf{E}} X \equiv \underset{\substack{p_1 \ldots p_m \\ V_1 \ldots V_n}}{\mathsf{E}} X$$

Similarly two or more successive operations E may always be rearranged in another order. And two or more successive operations A may be either combined into a single operation A or rearranged in another order. But an operation E and an operation A may not in general be interchanged in order.

The Synthesis Problem. Case 1. Synthesis Requirement Expressible in Restricted Recursive Arithmetic with One Free Variable

Subcase 1

Letting $o_1, o_2, \ldots, o_\nu$ represent required outputs, suppose the synthesis requirement

$$\begin{aligned}(1)\ \ & S[i_1(t), i_1(t+1), \ldots, i_1(t+a_1), i_2(t), i_2(t+1), \ldots, i_2(t+a_2), \ldots\ldots\ldots,\\ & i_\mu(t), i_\mu(t+1), \ldots, i_\mu(t+a_\mu), o_1(t), o_1(t+1), \ldots, o_1(t+b), o_2(t),\\ & o_2(t+1), \ldots, o_2(t+b), \ldots\ldots\ldots, o_\nu(t), o_\nu(t+1), \ldots, o_\nu(t+b)],\end{aligned}$$

or as we shall write more briefly, $S(t)$.

If X is any (well-formed) formula, let us use $E_b X$ for the formula obtained as the result of eliminating from X by IV the elementary parts $o_1(t+b), o_2(t+b), \ldots, o_\nu(t+b)$; and let us use $A_{b+1} X$ for the formula obtained as the result of eliminating from X by V all elementary parts $i_k(t+c)$ in which $c > b$, i.e., we let $E_b X$ be the same as

$$\underset{\substack{o_1(t+b), o_2(t+b), \\ \ldots, o_\nu(t+b)}}{\mathsf{E}} X;$$

and we let $A_{b+1} X$ be the same as

$$\underset{\substack{i_1(t+b+1), \ldots, i_1(t+a_1), \\ i_2(t+b+1), \ldots, i_2(t+a_2), \\ \ldots\ldots\ldots, \\ i_\mu(t+b+1), \ldots, i_\mu(t+a_\mu)}}{\mathsf{A}} X.$$

Now let $C_0(t)$ be $A_{b+1} S(t)$. And for $l = 0, 1, 2, \ldots$ let $D_l(t)$ be $E_b C_l(t)$, and let

$C_{l+1}(t)$ be the conjunction $C_l(t)A_{b+1}D_l(t+1)$. We then have:

(2) $$\vDash C_{l+1}(t) \supset C_l(t)$$

(3) $$\vDash C_{l+1}(t) \supset A_{b+1}D_l(t+1)$$

Hence, by VII:

(4) $$\vDash C_{l+1}(t) \supset D_l(t+1)$$

Now the elementary parts of $C_l(t)$ are some or all of $i_1(t)$, $i_1(t+1)$, ..., $i_1(t+b)$, $i_2(t)$, $i_2(t+1)$, ..., $i_2(t+b)$, ..., $i_\mu(t)$, $i_\mu(t+1)$, ..., $i_\mu(t+b)$, $o_1(t)$, $o_1(t+1)$, ..., $o_1(t+b)$, $o_2(t)$, $o_2(t+1)$, ..., $o_2(t+b)$, ..., $o_\nu(t)$, $o_\nu(t+1)$, ..., $o_\nu(t+b)$. The same is true of $C_{l+1}(t)$. And the full disjunctive normal form of $C_{l+1}(t)$ in these elementary parts must be contained in that of $C_l(t)$, in the sense that the terms of the f.d.n.f. of $C_{l+1}(t)$ must be a subset of the terms of $C_l(t)$, as follows from (2). In consequence, there must be some least number n such that the f.d.n.f. of $C_{n+1}(t)$ is in fact identical with the f.d.n.f. of $C_n(t)$, i.e.:

(5) $$\vDash C_{n+1}(t) \equiv C_n(t)$$

Solve $C_n(t)$ for the elementary parts $o_1(t+b), o_2(t+b), \ldots, o_\nu(t+b)$ by iterated applications of II in the manner already described. The solution has the following form, in which $\theta_1, \theta_2, \ldots, \theta_\nu$ are parameters:

(6_1) $$\begin{aligned} o_1(t+b) \equiv R_1[&i_1(t), i_1(t+1), \ldots, i_1(t+b), i_2(t), i_2(t+1), \ldots, \\ &i_2(t+b), \ldots\ldots\ldots, i_\mu(t), i_\mu(t+1), \ldots, i_\mu(t+b), o_1(t), o_1(t+1), \ldots, \\ &o_1(t+b-1), o_2(t), o_2(t+1), \ldots, o_2(t+b-1), \ldots\ldots\ldots, o_\nu(t) \\ &o_\mu(t+1), \ldots, o_\mu(t+b-1), \theta_1(t+b)] \end{aligned}$$

(6_2) $$\begin{aligned} o_2(t+b) \equiv R_2[&i_1(t), i_1(t+1), \ldots, i_1(t+b), i_2(t), i_2(t+1), \ldots, \\ &i_2(t+b), \ldots\ldots\ldots, i_\mu(t), i_\mu(t+1), \ldots, i_\mu(t+b), o_1(t), o_1(t+1), \ldots, \\ &o_1(t+b-1), o_2(t), o_2(t+1), \ldots, o_2(t+b-1), \ldots\ldots\ldots, o_\nu(t) \\ &o_\nu(t+1), \ldots, o_\nu(t+b-1), \theta_1(t+b), \theta_2(t+b)] \end{aligned}$$

....

(6_ν) $$\begin{aligned} o_\nu(t+b) \equiv R_\nu\big[&i_1(t), i_1(t+1), \ldots, i_1(t+b), i_2(t), i_2(t+1), \ldots, i_2(t+b), \\ &\ldots\ldots\ldots, i_\mu(t), i_\mu(t+1), \ldots, i_\mu(t+b), o_1(t), o_1(t+1), \ldots, o_1(t+b-1), \\ &o_2(t), o_2(t+1), \ldots, o_2(t+b-1), \ldots\ldots\ldots, o_\nu(t)o_\nu(t+1), \ldots, o_\nu(t+b-1), \\ &\theta_1(t+b), \theta_2(t+b), \ldots, \theta_\nu(t+b)\big] \end{aligned}$$

And let the conjunction of these equivalences be $E'(t)$. We have

(7) $$\vDash C_{n+1}(t) \supset D_n(t+1)$$

as a special case of (4). Hence, by (5) and (7) we have:

(8) $$\vDash C_n(t) \supset D_n(t+1)$$

Also by IV (and III):

(9) $$\vDash D_n(t) \supset . E'(t) \supset C_n(t)$$

Hence, by (8) and (9):

$$\vDash D_n(t) \supset . E'(t) \supset D_n(t+1) \tag{10}$$

The synthesis requirement $S(t)$ is now reduced to the recursion equivalences (6_1), (6_2), ... , (6_ν) together with the requirement $D_n(0)$.

For if a circuit satisfies the requirement $S(t)$, it must therefore satisfy $A_{b+1}S(t)$, that is, $C_0(t)$, on the ground that outputs at time $t+b$ and earlier cannot be made to depend on inputs at times later than $t+b$; and, hence, by VI it must satisfy $D_0(t)$. Also on the same ground, a circuit which satisfies $C_l(t)$ and $D_l(t)$ for all t must therefore satisfy $A_{b+1}D_l(t+1)$; hence, it must satisfy the conjunction $C_l(t)A_{b+1}D_l(t+1)$, that is, $C_{l+1}(t)$; and hence by VI it must satisfy $D_{l+1}(t)$. By induction with respect to l it follows that a circuit satisfying the requirement $S(t)$ must therefore satisfy $C_l(t)$ and $D_l(t)$ for all l; hence, in particular it must satisfy $C_n(t)$ and $D_n(t)$; hence it must satisfy the equivalences (6_1), (6_2), ... , (6_ν) that were obtained as solution of $C_n(t)$, the condition of possibility of this solution being $D_n(t)$ by IV. That $D_n(0)$ is satisfied, follows from $D_n(t)$ as a special case, i.e., by taking $t = 0$.

Conversely, suppose a circuit satisfies the equivalences (6_1), (6_2), ... , (6_ν), and $D_n(0)$. That is, it satisfies $E'(t)$ for all t, and $D_n(0)$. Then it follows from (10) that the circuit satisfies

$$D_n(t) \supset D_n(t+1) \tag{11}$$

for all t, and hence, by mathematical induction that $D_n(t)$ is satisfied for all t. Hence, by (9) it follows that $C_n(t)$ is satisfied for all t. Hence by (2), which holds for all l and all t, it follows that $C_0(t)$ is satisfied for all t. And since $C_0(t)$ is $A_{b+1}S(t)$, it follows by VII that the circuit satisfies $S(t)$ for all t.

It remains therefore to provide for the satisfaction of $D_n(0)$.

The elementary parts of $D_n(0)$ are some or all of $i_1(0)$, $i_1(1)$, ... , $i_1(b)$, $i_2(0)$, $i_2(1)$, ... , $i_2(b)$, ... , ... , $i_\mu(0)$, $i_\mu(1)$, ... , $i_\mu(b)$, $o_1(0)$, $o_1(1)$, ... , $o_1(b-1)$, $o_2(0)$, $o_2(1)$, ... , $o_2(b-1)$, ... , ... , $o_\nu(0)$, $o_\nu(1)$, ... , $o_\nu(b-1)$. Since the outputs at time $b-1$ and earlier cannot be made to depend on the inputs $i_1(b), i_2(b), \ldots, i_\mu(b)$, it is necessary that $D_n(0)$ shall be satisfied independently of these inputs. Therefore, we solve

$$\mathsf{A}_{i_1(b), i_2(b), \ldots, i_\mu(b)} D_n(0) \tag{12_{b-1}}$$

for $o_1(b-1), o_2(b-1), \ldots, o_\nu(b-1)$ by iterated application of II. The result has the following form, in which $\alpha_{1(b-1)}, \alpha_{2(b-1)}, \ldots, \alpha_{\nu(b-1)}$ are parameters:

$$\begin{aligned} o_1(b-1) \equiv I'_{1(b-1)}[&i_1(0), i_1(1), \ldots, i_1(b-1), i_2(0), i_2(1), \ldots, \\ &i_2(b-1), \ldots, \ldots, i_\mu(0), i_\mu(1), \ldots, i_\mu(b-1), o_1(0), o_1(1), \ldots, \\ &o_1(b-2), o_2(0), o_2(1), \ldots, o_2(b-2), \ldots, \ldots, o_\nu(0), o_\nu(1), \ldots, \\ &o_\nu(b-2), \alpha_{1(b-1)}] \end{aligned} \tag{$13_{1(b-1)}$}$$

$$\begin{aligned} o_2(b-1) \equiv I'_{2(b-1)}[&i_1(0), i_1(1), \ldots, i_1(b-1), i_2(0), i_2(1), \ldots, \\ &i_2(b-1), \ldots, \ldots, i_\mu(0), i_\mu(1), \ldots, i_\mu(b-1), o_1(0), o_1(1), \ldots, \\ &o_1(b-2), o_2(0), o_2(1), \ldots, o_2(b-2), \ldots, \ldots, o_\nu(0), o_\nu(1), \ldots, \\ &o_\nu(b-2), \alpha_{1(b-1)}, \alpha_{2(b-1)}] \end{aligned} \tag{$13_{2(b-1)}$}$$

....

$(13_{v(b-1)})$ $o_v(b-1) \equiv I'_{v(b-1)}[i_1(0), i_1(1), \ldots, i_1(b-1), i_2(0), i_2(1), \ldots, i_2(b-1), \ldots\ldots\ldots, i_\mu(0), i_\mu(1), \ldots, i_\mu(b-1), o_1(0), o_1(1), \ldots, o_1(b-2), o_2(0), o_2(1), \ldots, o_2(b-2), \ldots\ldots\ldots, o_v(0), o_v(1), \ldots, o_v(b-2), \alpha_{1(b-1)}, \alpha_{2(b-1)}, \ldots, \alpha_{v(b-1)}]$

subject to the condition

$$\underset{\substack{o_1(b-1), o_2(b-1),\\ \ldots, o_v(b-1)}}{\mathsf{E}} \quad \underset{\substack{i_1(b), i_2(b),\\ \ldots, i_\mu(b)}}{\mathsf{A}}, \quad D_n(0)$$

which we shall write as

(14_{b-1}) $J_{b-1}[i_1(0), i_1(1), \ldots, i_1(b-1), i_2(0), i_2(1), \ldots, i_2(b-1), \ldots \ldots, i_\mu(0), i_\mu(1), \ldots, i_\mu(b-1), o_1(0), o_1(1), \ldots, o_1(b-2), o_2(0), o_2(1), \ldots, o_2(b-2), \ldots\ldots\ldots, o_v(0), o_v(1), \ldots, o_v(b-2)]$

This condition must be satisfied identically in $i_1(b-1), i_2(b-1), \ldots, i_\mu(b-1)$, as outputs at time $b-2$ and earlier cannot be made to depend on inputs at time $b-1$. Hence, it reduces by V to

$$\underset{\substack{i_1(b-1), i_2(b-1),\\ \ldots, i_v(b-1)}}{\mathsf{A}} \quad J_{b-1}[i_1(0), i_1(1), \ldots\ldots\ldots\ldots, o_v(b-2)],$$

which we shall write as

(12_{b-2}) $J'_{b-1}[i_1(0), i_1(1), \ldots, i_1(b-2), i_2(0), i_2(1), \ldots, i_2(b-2), \ldots\ldots\ldots, i_\mu(0), i_\mu(1), \ldots, i_\mu(b-2), o_1(0), o_1(1), \ldots, o_1(b-2), o_2(0), o_2(1), \ldots, o_2(b-2), \ldots\ldots\ldots, o_v(0), o_v(1), \ldots, o_v(b-2)]$

We then solve (12_{b-2}) for $o_1(b-2), o_2(b-2), \ldots, o_v(b-2)$ by iterated application of II. We thus obtain:

$(13_{1(b-2)})$ $o_1(b-2) \equiv I'_{1(b-2)}[i_1(0), i_1(1), \ldots, i_1(b-2), i_2(0), i_2(1), \ldots, i_2(b-2), \ldots\ldots\ldots, i_\mu(0), i_\mu(1), \ldots, i_\mu(b-2), o_1(0), o_1(1), \ldots, o_1(b-3), o_2(0), o_2(1), \ldots, o_2(b-3), \ldots\ldots\ldots, o_v(0), o_v(1), \ldots, o_v(b-3), \alpha_{1(b-2)}]$

$(13_{2(b-2)})$ $o_2(b-2) \equiv I'_{2(b-2)}[i_1(0), i_1(1), \ldots, i_1(b-2), i_2(0), i_2(1), \ldots, i_2(b-2), \ldots\ldots\ldots, i_\mu(0), i_\mu(1), \ldots, i_\mu(b-2), o_1(0), o_1(1), \ldots, o_1(b-3), o_2(0), o_2(1), \ldots, o_2(b-3), \ldots\ldots\ldots, o_v(0), o_v(1), \ldots, o_v(b-3), \alpha_{1(b-2)}, \alpha_{2(b-2)}]$

....

$(13_{v(b-2)})$ $o_v(b-2) \equiv I'_{v(b-2)}[i_1(0), i_1(1), \ldots, i_1(b-2), i_2(0), i_2(1), \ldots, i_2(b-2), \ldots\ldots\ldots, i_\mu(0), i_\mu(1), \ldots, i_\mu(b-2), o_1(0), o_1(1), \ldots, o_1(b-3), o_2(0), o_2(1), \ldots, o_2(b-3), \ldots\ldots\ldots, o_v(0), o_v(1), \ldots, o_v(b-3), \alpha_{1(b-2)}, \alpha_{2(b-2)}, \ldots, \alpha_{v(b-2)}]$

subject to the condition

(14_{b-2}) $J_{b-2}[i_1(0), i_1(1), \ldots, i_1(b-2), i_2(0), i_2(1), \ldots, i_2(b-2), \ldots \ldots, i_\mu(0), i_\mu(1), \ldots, i_\mu(b-2), o_1(0), o_1(1), \ldots, o_1(b-3),$

$$o_2(0), o_2(1), \ldots, o_2(b-3), \ldots\ldots\ldots, o_\nu(0), o_\nu(1), \ldots, o_\nu(b-3)],$$

which must be satisfied identically in $i_1(b-2), i_2(b-2), \ldots, i_\mu(b-2)$, and therefore, reduces by V to

$$(12_{b-3}) \qquad J'_{b-2}[i_1(0), i_1(1), \ldots, i_1(b-3), i_2(0), i_2(1), \ldots, i_2(b-3), \ldots\ldots\ldots, i_\mu(0), i_\mu(1), \ldots, i_\mu(b-3), o_1(0), o_1(1), \ldots, o_1(b-3), o_2(0), o_2(1), \ldots, o_2(b-3), \ldots\ldots\ldots, o_\nu(0), o_\nu(1), \ldots, o_\nu(b-3)]$$

Then (12_{b-3}) is solved for $o_1(b-3), o_2(b-3), \ldots, o_\nu(b-3)$, and so on. Thus we obtain the succession of equivalences $(13_{1(b-3)}), (13_{2(b-3)}), \ldots, (13_{\nu(b-3)}), (13_{1(b-4)}), \ldots,$ and so on, ending with:

$$(13_{10}) \qquad o_1(0) \equiv I'_{10}[i_1(0), i_2(0), \ldots, i_\mu(0), \alpha_{10}]$$

$$(13_{20}) \qquad o_2(0) \equiv I'_{20}[i_1(0), i_2(0), \ldots, i_\mu(0), \alpha_{10}, \alpha_{20}]$$

$$\ldots\ldots\ldots\ldots\ldots\ldots\ldots\ldots\ldots\ldots$$

$$(13_{\nu 0}) \qquad o_\nu(0) \equiv I'_{\nu 0}[i_1(0), i_2(0), \ldots, i_\mu(0), \alpha_{10}, \alpha_{20}, \ldots, \alpha_{\nu 0}]$$

subject to the condition

$$(14_0) \qquad J_0[i_1(0), i_2(0), \ldots, i_\mu(0)].$$

Then (14_0) must turn out to be tautologous, else the synthesis is impossible. If (14_0) is tautologous, we proceed by substituting in the equivalences $(13_{1(b-1)}), (13_{2(b-1)}), \ldots, (13_{\nu(b-1)})$ in accordance with the equivalences $(13_{1(b-2)}), (13_{2(b-2)}), \ldots, (13_{\nu(b-2)}), (13_{1(b-3)}), (13_{2(b-3)}), \ldots, (13_{\nu(b-3)}), \ldots\ldots\ldots, (13_{10}), (13_{20}), \ldots, (13_{\nu 0})$; substituting in the equivalences $(13_{1(b-2)}), (13_{2(b-2)}), \ldots, (13_{2(b-2)})$ in accordance with the equivalences $(13_{1(b-3)}), (13_{2(b-3)}), \ldots, (13_{\nu(b-3)}), \ldots\ldots\ldots, (13_{10}), (13_{20}), \ldots, (13_{\nu 0})$; and so on. There result the following equivalences—in which $(15_{10}), (15_{20}), \ldots, (15_{\nu 0})$ are the same as $(13_{10}), (13_{20}), \ldots, (13_{\nu 0})$, but the others are new:

$$(15_{10}) \qquad o_1(0) \equiv I'_{10}[i_1(0), i_2(0), \ldots, i_\mu(0), \alpha_{10}]$$

$$(15_{20}) \qquad o_2(0) \equiv I'_{20}[i_1(0), i_2(0), \ldots, i_\mu(0), \alpha_{10}, \alpha_{20}]$$

$$\ldots\ldots\ldots\ldots\ldots\ldots\ldots\ldots\ldots\ldots$$

$$(15_{\nu 0}) \qquad o_\nu(0) \equiv I'_{\nu 0}[i_1(0), i_2(0), \ldots, i_\mu(0), \alpha_{10}, \alpha_{20}, \ldots, \alpha_{\nu 0}]$$

$$(15_{11}) \qquad o_1(1) \equiv I'_{11}[i_1(0), i_1(1), i_2(0), i_2(1), \ldots, i_\mu(0), i_\mu(1), \alpha_{10}, \alpha_{20}, \ldots, \alpha_{\nu 0}, \alpha_{11}]$$

$$(15_{21}) \qquad o_2(1) \equiv I'_{21}[i_1(0), i_1(1), i_2(0), i_2(1), \ldots, i_\mu(0), i_\mu(1), \alpha_{10}, \alpha_{20}, \ldots, \alpha_{\nu 0}, \alpha_{11}, \alpha_{21}]$$

$$\ldots\ldots\ldots\ldots\ldots\ldots\ldots\ldots\ldots\ldots\ldots\ldots$$

$$(15_{\nu 1}) \qquad o_\nu(1) \equiv I'_{\nu 1}[i_1(0), i_1(1), i_2(0), i_2(1), \ldots, i_\mu(0), i_\mu(1), \alpha_{10}, \alpha_{20}, \ldots, \alpha_{\nu 0}, \alpha_{11}, \ldots, \alpha_{\nu 1}]$$

$$\ldots\ldots\ldots\ldots\ldots\ldots\ldots\ldots\ldots\ldots\ldots\ldots$$

$$\ldots\ldots\ldots\ldots\ldots\ldots\ldots\ldots\ldots\ldots\ldots\ldots$$

$$(15_{1(b-1)}) \qquad o_1(b-1) \equiv I'_{1(b-1)}[i_1(0), i_1(1), \ldots, i_1(b-1), i_2(0), i_2(1), \ldots, i_2(b-1), \ldots\ldots\ldots, i_\mu(0), i_\mu(1), \ldots, i_\mu(b-1), \alpha_{10}, \alpha_{20}, \ldots, \alpha_{\nu 0}, \alpha_{11}, \alpha_{21}, \ldots, \alpha_{\nu 1}, \ldots\ldots\ldots, \alpha_{1(b-1)}]$$

$$(15_{2(b-1)}) \qquad o_2(b-1) \equiv I'_{2(b-1)}[i_1(0), i_1(1), \ldots, i_1(b-1), i_2(0), i_2(1), \ldots,$$

$$i_2(b-1), \ldots\ldots\ldots, i_\mu(0), i_\mu(1), \ldots, i_\mu(b-1), \alpha_{10}, \alpha_{20}, \ldots, \alpha_{\nu 0}, \alpha_{11}, \alpha_{21}, \ldots, \alpha_{\nu 1}, \ldots\ldots\ldots, \alpha_{1(b-1)}, \alpha_{2(b-1)}]$$

....

$(15_{\nu(b-1)})$ $$o_\nu(b-1) \equiv I'_{\nu(b-1)}[i_1(0), i_1(1), \ldots, i_1(b-1), i_2(0), i_2(1), \ldots, i_2(b-1), \ldots\ldots\ldots, i_\mu(0), i_\mu(1), \ldots, i_\mu(b-1), \alpha_{10}, \alpha_{20}, \ldots, \alpha_{\nu 0}, \alpha_{11}, \alpha_{21}, \ldots, \alpha_{\nu 1}, \ldots\ldots\ldots, \alpha_{1(b-1)}, \alpha_{2(b-1)}, \ldots, \alpha_{\nu(b-1)}]$$

Then $(6_1), \ldots, (6_\nu)$ and $(15_{10}), \ldots, (15_{\nu(b-1)})$ constitute recursion equivalences for $o_1, o_2, \ldots, o_\nu$; upon taking each of the parameters α_{km} as a truth-function of $i_1(0), i_1(1), \ldots, i_1(m), i_2(0), i_2(1), \ldots, i_2(m), \ldots\ldots\ldots, i_\mu(0), i_\mu(1), \ldots, i_\mu(m)$, and each of the parameters $\theta_k(t+b)$ as a propositional function of the time, not depending on inputs at times later than $t+b$. We may in particular take each $\theta_k(t+b)$ as a truth-function of $i_1(t), i_1(t+1), \ldots, i_1(t+b), i_2(t), i_2(t+1), \ldots, i_2(t+b), \ldots\ldots\ldots, i_\mu(t), i_\mu(t+1), \ldots, i_\mu(t+b), o_1(t), o_1(t+1), \ldots, o_1(t+b-1), o_2(t), o_2(t+1), \ldots, o_2(t+b-1), \ldots\ldots\ldots, o_\nu(t), o_\nu(t+1), \ldots, o_\nu(t+b-1)$, in any arbitrary way. Or some or all of the propositional functions θ_k may be made depending on additional propositional functions, denoted by predicates — say $p_1, p_2, \ldots, p_k$ — which represent *intermediate elements* of the circuit, i.e., elements which are neither inputs nor outputs; this is done by choosing recursion equivalences for $p_1, p_2, \ldots, p_k$, and then taking each $\theta_k(t+b)$ as a truth-function of $i_1(t), i_1(t+1), \ldots, i_1(t+b), i_2(t), \ldots\ldots\ldots, i_\mu(t+b), o_1(t), o_1(t+1), \ldots, o_1(t+b-1), o_2(t), \ldots\ldots\ldots, o_\nu(t+b-1), p_1(t), p_1(t+1), \ldots, p_1(t+b), p_2(t), \ldots\ldots\ldots, p_k(t+b)$.

The general solution of the problem to satisfy the synthesis requirement (1) — or, as we shall say briefly, the general solution of (1) — is given by the recursion equivalences $(6_1), \ldots, (6_\nu)$ and $(15_{10}), \ldots, (15_{\nu(b-1)})$, not in the sense that every circuit satisfying the requirement (1) is directly characterized by these recursion equivalences, but in the sense that if any circuit satisfies (1), its outputs can be expressed in terms of its inputs by the recursion equivalences $(6_1), \ldots, (6_\nu)$ and $(15_{10}), \ldots, (15_{\nu(b-1)})$ with suitable choice of the parameters. In other words, the circuits satisfying the requirement (1) comprise not only the ones directly given by the recursion equivalences $(6_1), \ldots, (6_\nu)$ and $(15_{10}), \ldots, (15_{\nu(b-1)})$ but also all circuits equivalent to these in the usual sense of equivalence of circuits. In order to obtain some particular circuit satisfying (1) it may be necessary not only to make a suitable choice of the parameters, but also to transform the recursion equivalences suitably, by laws of propositional calculus and recursive arithmetic.

There is no reason to suppose that the simplest circuit satisfying the requirement (1), in any sense of simplicity we may wish to consider, is among those which are directly given by the recursion equivalences which we have found as the general solution. But rather the synthesis problem and the simplification problem must be considered as separate problems, and after the problem of synthesis has been solved, that of simplification may well remain.

The simplification problem consists in the following: Having the solution equivalences, we are to make such a choice of the parameters and then to transform the equivalences by laws of propositional calculus and recursive arithmetic in such a way that the simplest circuit is obtained.

The criterion of simplicity in terms of the recursion equivalences will depend upon the circumstances and the purpose, and may well differ from occasion to occasion. Or we may wish to consider the reliability of the circuit as well as the simplicity, and to adopt some compromise between reliability and simplicity if the two cannot be maximized together. Here we may take the reliability to be the probability that the outputs of the circuit will steadily obey the synthesis requirement from the time $t = 0$ to $t = t_0$, as computed in terms of given probabilities associated with the elements of the circuit (namely, for each element the probability in various cases that it will act as it was designed to, i.e., in accordance with the recursion equivalences).

Subcase 2

Consider now the more general synthesis requirement:

(16)
$$S[i_1(0), i_1(1), \ldots, i_1(b), i_1(t), i_1(t+1), \ldots, i_1(t+a_1), i_2(0), i_2(1), \ldots, i_2(b), i_2(t), i_2(t+1), \ldots, i_2(t+a_2), \ldots\ldots\ldots, i_\mu(0), i_\mu(1), \ldots, i_\mu(b), i_\mu(t), i_\mu(t+1), \ldots, i_\mu(t+a_\mu), o_1(0), o_1(1), \ldots, o_1(b-1), o_1(t), o_1(t+1), \ldots, o_1(t+b), o_2(0), o_2(1), \ldots, o_2(b-1), o_2(t), o_2(t+1), \ldots, o_2(t+b), \ldots\ldots\ldots, o_\nu(0), o_\nu(1), \ldots, o_\nu(b-1), o_\nu(t), o_\nu(t+1), \ldots, o_\nu(t+b)]$$

The same procedure may be applied to this which was used in Subcase 1. The formulas $C_n(t)$ and $D_n(t)$, such that $\vDash C_{n+1}(t) \equiv C_n(t)$, are obtained in the same way as before. Upon solving $C_n(t)$ for $o_1(t+b), o_2(t+b), \ldots, o_\nu(t+b)$ we obtained the following analogs of $(6_1), (6_2), \ldots, (6_\nu)$:

(17_1)
$$o_1(t+b) \equiv R'_1[i_1(0), i_1(1), \ldots, i_1(b), i_1(t), i_1(t+1), \ldots, i_1(t+b), i_2(0), i_2(1), \ldots, i_2(b), i_2(t), i_2(t+1), \ldots, i_2(t+b), \ldots\ldots\ldots, i_\mu(0), i_\mu(1), \ldots, i_\mu(b), i_\mu(t), i_\mu(t+1), \ldots, i_\mu(t+b), o_1(0), o_1(1), \ldots, o_1(b-1), o_1(t), o_1(t+1), \ldots, o_1(t+b-1), o_2(0), o_2(1), \ldots, o_2(b-1), o_2(t), o_2(t+1), \ldots, o_2(t+b-1), \ldots\ldots\ldots, o_\nu(0), o_\nu(1), \ldots, o_\nu(b-1), o_\nu(t), o_\nu(t+1), \ldots, o_\nu(t+b-1), \theta_1(t+b)]$$

(17_2)
$$o_2(t+b) \equiv R'_2[i_1(0), i_1(1), \ldots, i_1(b), i_1(t), i_1(t+1), \ldots, i_1(t+b), i_2(0), i_2(1), \ldots, i_2(b), i_2(t), i_2(t+1), \ldots, i_2(t+b), \ldots\ldots\ldots, i_\mu(0), i_\mu(1), \ldots, i_\mu(b), i_\mu(t), i_\mu(t+1), \ldots, i_\mu(t+b), o_1(0), o_1(1), \ldots, o_1(b-1), o_1(t), o_1(t+1), \ldots, o_1(t+b-1), o_2(0), o_2(1), \ldots, o_2(b-1), o_2(t), o_2(t+1), \ldots, o_2(t+b-1), \ldots\ldots\ldots, o_\nu(0), o_\nu(1), \ldots, o_\nu(b-1), o_\nu(t), o_\nu(t+1), \ldots, o_\nu(t+b-1), \theta_1(t+b), \theta_2(t+b)]$$

....

(17_ν)
$$o_\nu(t+b) \equiv R'_\nu[i_1(0), i_1(1), \ldots, i_1(b), i_1(t), i_1(t+1), \ldots, i_1(t+b), i_2(0), i_2(1), \ldots, i_2(b), i_2(t), i_2(t+1), \ldots, i_2(t+b), \ldots\ldots\ldots, i_\mu(0), i_\mu(1), \ldots, i_\mu(b), i_\mu(t), i_\mu(t+1), \ldots, i_\mu(t+b), o_1(0), o_1(1), \ldots, o_1(b-1), o_1(t), o_1(t+1), \ldots, o_1(t+b-1), o_2(0), o_2(1), \ldots, o_2(b-1), o_2(t), o_2(t+1), \ldots, o_2(t+b-1), \ldots\ldots\ldots, o_\nu(0), o_\nu(1), \ldots, o_\nu(b-1), o_\nu(t), o_\nu(t+1), \ldots, o_\nu(t+b-1), \theta_1(t+b), \theta_2(t+b), \ldots, \theta_\nu(t+b)]$$

subject to the condition $D_n(t)$.

Now (2), (9), and (10), or their analogues, will hold, exactly as in Subcase 1, and hence, the condition $D_n(t)$ can be reduced to $D_n(0)$ in the same way as before. The elementary parts of $D_n(0)$ are the same as in Subcase 1. We then proceed in the same way as in Subcase 1 to obtain the following analogues of $(15_{10}), \ldots, (15_{\nu(b-1)})$:

(18_{10}) $$o_1(0) \equiv I'_{10}[i_1(0), i_2(0), \ldots, i_\mu(0), \alpha_{10}]$$

(18_{20}) $$o_2(0) \equiv I'_{20}[i_1(0), i_2(0), \ldots, i_\mu(0), \alpha_{10}, \alpha_{20}]$$

. . . .

$(18_{\nu 0})$ $$o_\nu(0) \equiv I'_{\nu 0}[i_1(0), i_2(0), \ldots, i_\mu(0), \alpha_{10}, \alpha_{20}, \ldots, \alpha_{\nu 0}]$$

(18_{11}) $$o_1(1) \equiv I'_{11}[i_1(0), i_1(1), i_2(0), i_2(1), \ldots, i_\mu(0), i_\mu(1), \alpha_{10}, \alpha_{20}, \ldots, \alpha_{\nu 0}, \alpha_{11}]$$

(18_{21}) $$o_2(1) \equiv I'_{21}[i_1(0), i_1(1), i_2(0), i_2(1), \ldots, i_\mu(0), i_\mu(1), \alpha_{10}, \alpha_{20}, \ldots, \alpha_{\nu 0}, \alpha_{11}, \alpha_{21}]$$

. . . .

$(18_{\nu 1})$ $$o_\nu(1) \equiv I'_{\nu 1}[i_1(0), i_1(1), i_2(0), i_2(1), \ldots, i_\mu(0), i_\mu(1), \alpha_{10}, \alpha_{20}, \ldots, \alpha_{\nu 0}, \alpha_{11}, \alpha_{21}, \ldots, \alpha_{\nu 1}]$$

. . . .

. . . .

$(18_{1(b-1)})$ $$o_1(b-1) \equiv I'_{1(b-1)}[i_1(0), i_1(1), \ldots, i_1(b-1), i_2(0), i_2(1), \ldots, i_2(b-1), \ldots\ldots\ldots, i_\mu(0), i_\mu(1), \ldots, i_\mu(b-1), \alpha_{10}, \alpha_{20}, \ldots, \alpha_{\nu 0}, \alpha_{11}, \alpha_{21}, \ldots, \alpha_{\nu 1}, \ldots\ldots\ldots, \alpha_{1(b-1)}]$$

. . . .

$(18_{\nu(b-1)})$ $$o_\nu(b-1) \equiv I'_{\nu(b-1)}[i_1(0), i_1(1), \ldots, i_1(b-1), i_2(0), i_2(1), \ldots, i_2(b-1), \ldots\ldots\ldots, i_\mu(0), i_\mu(1), \ldots, i_\mu(b-1), \alpha_{10}, \alpha_{20}, \ldots, \alpha_{\nu 0}, \alpha_{11}, \alpha_{21}, \ldots, \alpha_{\nu 1}, \ldots\ldots\ldots, \alpha_{1(b-1)}, \alpha_{2(b-1)} \ldots, \alpha_{\nu(b-1)}]$$

subject to the condition

(19) $$J_0[i_1(0), i_2(0), \ldots, i_\mu(0)],$$

which is the analogue of (14_0).

From the equivalences $(17_1), (17_2), \ldots, (17_\nu)$ it is necessary to eliminate the elementary parts

$$i_j(0), i_j(1), \ldots, i_j(b), o_k(0), o_k(1), \ldots, o_k(b-1)$$

$(j = 1, 2, \ldots, \mu$ and $k = 1, 2, \ldots, \nu)$. In the case of the elementary parts $o_k(0), o_k(1), \ldots, o_k(b-1)$, this is done by substituting in accordance with the equivalences $(18_{10}), \ldots, (18_{\nu(b-1)})$. But in the equivalences which thus result from $(17_1), \ldots, (17_\nu)$ it is then necessary to eliminate $i_j(0), i_j(1), \ldots, i_j(b)$. For this purpose we introduce the set of additional predicates p_{jl} $(j = 1, 2, \ldots, \mu$ and $l = 0, 1, 2, \ldots, b)$, representing intermediate elements of the circuit, and take for them the recursion equivalences:

(20) $$p_{jl}(0) \equiv F, p_{jl}(1) \equiv F, \ldots, p_{jl}(l-1) \equiv F,$$
$$p_{jl}(l) \equiv i_j(l), p_{jl}(t+l+1) \equiv p_{jl}(t+l).$$

Upon substituting $p_{jl}(t+l)$ for $i_j(l)$, for all $j = 1, 2, \ldots, \mu$ and $l = 0, 1, 2, \ldots, b$, the equivalences $(17_1), \ldots, (17_\nu)$ will then have been transformed into:

(21₁) $o_1(t+b) \equiv R_1[i_1(t), i_1(t+1), \ldots, i_1(t+b), i_2(t), i_2(t+1), \ldots, i_2(t+b), \ldots\ldots\ldots, i_\mu(t), i_\mu(t+1), \ldots, i_\mu(t+b), p_{10}(t), p_{11}(t+1), \ldots, p_{1b}(t+b), p_{20}(t), p_{21}(t+1), \ldots, p_{2b}(t+b), \ldots\ldots\ldots, p_{\mu 0}(t), p_{\mu 1}(t+1), \ldots, p_{\mu b}(t+b), o_1(t)o_1(t+1), \ldots, o_1(t+b-1), o_2(t), o_2(t+1), \ldots,, o_2(t+b-1), \ldots\ldots\ldots o_\mu(t), o_\mu(t+1), \ldots, o_\mu(t+b-1), \theta_1(t+b), \alpha_{10}, \ldots,, \alpha_{\nu(b-1)}]$

(21₂) $o_2(t+b) \equiv R_2[i_1(t), i_1(t+1), \ldots, i_1(t+b), i_2(t), i_2(t+1), \ldots, i_2(t+b), \ldots\ldots\ldots, i_\mu(t), i_\mu(t+1), \ldots, i_\mu(t+b), p_{10}(t), p_{11}(t+1), \ldots, p_{1b}(t+b), p_{20}(t), p_{21}(t+1), \ldots, p_{2b}(t+b), \ldots\ldots\ldots, p_{\mu 0}(t), p_{\mu 1}(t+1), \ldots, p_{\mu b}(t+b), o_1(t), o_1(t+1), \ldots, o_1(t+b-1), o_2(t), o_2(t+1), \ldots, o_2(t+b-1), \ldots\ldots\ldots, o_\mu(t), o_\mu(t+1), \ldots, o_\mu(t+b-1), \theta_1(t+b), \theta_2(t+b), \alpha_{10}, \ldots, \alpha_{\nu(b-1)}]$

. . . .

(21$_\nu$) $o_\nu(t+b) \equiv R_\nu[i_1(t), i_1(t+1), \ldots, i_1(t+b), i_2(t), i_2(t+1), \ldots, i_2(t+b), \ldots\ldots\ldots, i_\mu(t), i_\mu(t+1), \ldots, i_\mu(t+b), p_{10}(t), p_{11}(t+1), \ldots, p_{1b}(t+b), p_{20}(t), p_{21}(t+1), \ldots, p_{2b}(t+b), \ldots\ldots\ldots, p_{\mu 0}(t), p_{\mu 1}(t+1), \ldots, p_{\mu b}(t+b), o_1(t), o_1(t+1), \ldots, o_1(t+b-1), o_2(t), o_2(t+1), \ldots, o_2(t+b-1), \ldots\ldots\ldots, o_\mu(t), o_\mu(t+1), \ldots, o_\mu(t+b-1), \theta_1(t+b), \theta_2(t+b), \ldots, \theta_\nu(t+b), \alpha_{10}, \ldots, \alpha_{\nu(b-1)}]$

Our result then is that if (19) is not tautologous, the problem to satisfy the synthesis requirement (16) (independently of any restriction on the inputs) is impossible. If (19) turns out to be tautologous, the general solution is given by the recursion equivalences $(18_{10}), \ldots, (18_{\nu(b-1)})$, (20), (21₁), (21₂), …, (21$_\nu$).

In this form the solution equivalences do not constitute a single wider restricted recursion in the sense of schema B, but rather two successive wider restricted recursions, the first one constituted by the equivalences (20), and the second one by the equivalences $(18_{10}), \ldots, (18_{\nu(b-1)})$ and (21₁), …, (21$_\nu$). This is, however, legitimate in the sense that a definite circuit is represented, namely, the equivalences (20) represent a circuit which inputs $i_j(t)$ and outputs $p_{jl}(t)$; the equivalences $(18_{10}), \ldots, (18_{\nu(b-1)})$, (21₁), …, (21$_\nu$) represent a circuit with inputs $i_j(t)$ and $p_{jl}(t)$, and outputs $o_k(t)$; and the two sets of equivalences taken together represent a circuit obtained by connecting the outputs of the first circuit as inputs of the second circuit [additional to the inputs $i_j(t)$].

Whenever we have two or more successive wider restricted recursions, as we do in this case, it is always possible by an obvious method to replace them by a single wider restricted recursion, which is equivalent in regard to the action of every circuit element represented by a predicate occurring in the recursion equivalences. But from the point of view of the circuit represented it may be undesirable to do this, as the effect on the circuit may be to replace a simpler circuit by a somewhat more complicated one.

As in Subcase 1, we have to remark here that the general solution of the synthesis problem does not mean that we have the solution in simplest form. However, after solution of the synthesis problem, the simplification problem will ordinarily remain.

We also notice, as a special aspect of the simplification problem, that some or all of the additional predicates p_{jl} which were introduced may in particular cases be eliminable. In order to illustrate what may happen, we cite two very simple examples.

Example 1. Suppose as a particular case that the synthesis requirement (16) is:

(22) $$o(t+2) \equiv i(0)i(1)i(t)i(t+1) \vee \overline{o}(t)$$

Here there is just one input and one output. And the equivalences $(17_1), \ldots, (17_\nu)$ become in this case just one equivalence. By following through the prescribed procedure, we find easily that this one equivalence is identical with (22) — or more exactly, after the equivalence is obtained there is an obvious simplification of it by propositional calculus, the parameter $\theta(t+b)$ drops out in the simplification, and the equivalence thus reduces to (22). Moreover, $D_n(0)$ is tautologous, and hence, the equivalences (18) become simply:

(23) $$o(0) \equiv \alpha_0, \qquad o(1) \equiv \alpha_1$$

The equivalences (20) become in this case:

(24) $$\begin{aligned} &p_0(0) \equiv i(0), \qquad p_0(t+1) \equiv p_0(t), \\ &p_1(0) \equiv F, \; p_1(1) \equiv i(1), \; p_1(t+2) \equiv p_1(t+1) \end{aligned}$$

Hence, the equivalences (21) become the single equivalence:

(25) $$o(t+2) \equiv p_0(t)p_1(t+1)i(t)i(t+1) \vee \overline{o}(t)$$

The solution of (22) as found by our general procedure, thus consists in the equivalences (23), (24), and (25). However, since the predicates p_0 and p_1 occur only in the combination $p_0(t)p_1(t+1)$, it is obvious that they may be replaced by a single predicate p, provided that $p(t+1) \equiv p_0(t)p_1(t+1)$ is a consequence of the recursion equivalences. This is accomplished by taking:

(26) $$p(0) \equiv F, \quad p(1) \equiv i(0)i(1), \quad p(t+2) \equiv p(t+1)$$

Then (25) may be replaced by:

(27) $$o(t+2) \equiv p(t+1)i(t)i(t+1) \vee \overline{o}(t)$$

Thus, the solution equivalences (23), (24), (25) are replaced by the simpler equivalences (23), (26), (27), involving (besides i and o) only one additional predicate p, instead of two predicates p_0 and p_1.

It can be shown in this case — we leave the proof to the reader — that the number of additional predicates cannot be reduced further, from one to none at all. That is, the synthesis requirement (22) cannot be satisfied by a set of recursion equivalences, having the form of one or more wider restricted recursions, and containing no other predicates than i and o.

Hence, we turn to another example, in which this final reduction eliminating all the additional predicates is possible.

Example 2. Suppose, as another particular case, that the synthesis requirement (16) is:

(28) $$o(t+2) \equiv i(0)\overline{o}(t) \vee i(1)o(t+1)$$

As before, the equivalences (17) reduce to a single equivalence identical with (28). And since $D_n(0)$ is tautologous, the equivalences (18) become:

(29) $$o(0) \equiv \alpha_0, \qquad o(1) \equiv \alpha_1$$

And the solution of (28), as found by our general procedure, then consists in the equivalences (29) together with the equivalences:

$$(30) \qquad p_0(0) \equiv i(0), \qquad p_0(t+) \equiv p_0(t),$$
$$p_1(0) \equiv F,\ p_1(1) \equiv i(1),\ p_1(t+2) \equiv p_1(t+1)$$
$$(31) \qquad o(t+2) \equiv p_0(t)\bar{o}(t) \vee p_1(t+1)o(t+1)$$

However, since the output is unaffected by inputs at times later than $t = 1$, it is clear that the output must be ultimately periodic. There are sixteen possibilities regarding the truth-values $i(0), i(1), \alpha_0, \alpha_1$: namely, these four truth-values may be T, T, T, T respectively, or they may be T, T, T, F, or ..., or they may be F, F, F, F. If in each of these sixteen cases we follow the course of the outputs, $o(0), o(1), o(2), \ldots$, computing the successive outputs by means of (28), we find in every case that periodicity is established no later than $t = 3$, and the period is either 1 or 4. Hence, as an alternative form of the solution of (28), instead of (29), (30), and (31), we may write the equivalences (29) and the following equivalences:

$$(32) \qquad \begin{aligned} o(2) &\equiv i(0)\bar{\alpha}_0 \vee i(1)\alpha_1 \\ o(3) &\equiv i(0)\bar{\alpha}_1 \vee i(1)\alpha_1 \\ o(4) &\equiv i(0)i(1) \vee i(0)\alpha_0 \vee i(1)\alpha_1 \\ o(5) &\equiv i(0)i(1) \vee i(0)\alpha_1 \vee i(1)\alpha_1 \\ o(6) &\equiv i(0)i(1) \vee i(0)\bar{\alpha}_0 \vee i(1)\alpha_1 \end{aligned}$$
$$(33) \qquad o(t+7) \equiv o(t+3)$$

As a matter of fact (29), (32), and (33) can be derived from (29), (30), and (31) by a transformation process in which only laws of propositional calculus and the rule of substitution are used. But the process described above seems easier to follow.

In (29), (32), (33) the additional predicates p_0 and p_1 have been eliminated. But it may be questioned whether this is a simplification of the solution, since the elimination of p_0 and p_1 has been accomplished at the cost of increasing the span of the recursion from 2 to 7. Whether or not (29), (32), (33) are considered simpler than (29), (30), (31), for some particular determination of α_0 and α_1, may depend perhaps on the choice of a particular definition of simplicity in a particular context.

In still another form of the solution of (28), p_0 and p_1 are eliminated while the span of the recursion is increased only to 6. This form of the solution consists of (29), the first four equivalences of (32), and the equivalence:

$$(34) \qquad o(t+6) \equiv o(t+3)\bar{o}(t+4)\bar{o}(t+5) \vee \bar{o}(t+3)\bar{o}(t+4)o(t+5) \vee o(t+3)o(t+5)$$

We leave it to the reader to verify this, and also to show that no span smaller than 6 is possible if p_0 and p_1 are to be eliminated.

The General Case 1

The most general synthesis requirement belonging to Case 1 has the form:

$$(35) \qquad \begin{aligned} &S[i_1(0), i_1(1), \ldots, i_1(a_1), i_1(t), i_1(t+1), \ldots, i_1(t+a_2), i_2(0), \\ &i_2(1), \ldots, i_2(a_3), i_2(t), i_2(t+1), \ldots, i_2(t+a_4), \ldots\ldots\ldots, i_\mu(0), \\ &i_\mu(1), \ldots, i_\mu(a_{2\mu-1}), i_\mu(t), i_\mu(t+1), \ldots, i_\mu(t+a_{2\mu}), o_1(0), \\ &o_1(1), \ldots, o_1(b_1), o_1(t), o_1(t+1), \ldots, o_1(t+b_2), o_2(0), o_2(1), \ldots, \end{aligned}$$

$o_2(b_3), o_2(t+1), o_2(t+2), \ldots, o_2(t+b_4), \ldots\ldots\ldots, o_\nu(0), o_\nu(1), \ldots,$
$o_\nu(b_{2\nu-1}), o_\nu(t+1), o_\nu(t+2), \ldots, o_\nu(t+b_\nu)]$

This can be brought under Subcase 2 and solved by the method for that subcase if we take b to be the greatest of the numbers $a_1, a_3, \ldots, a_{2\mu-1}, b_1+1, b_3+1, \ldots, b_{2\nu-1}+1, b_2, b_4, \ldots, b_{2\mu}$. For the method of Subcase 2 in no way requires that all of the elementary parts which are indicated in the synthesis requirement (16) shall actually and explicitly occur in a particular case of a formula (16) to which the method is applied.

Subcase 3

Though we have now solved the general Case 1 of the synthesis problem, we wish still to give some special consideration to the subcase in which $i_1, i_2, \ldots, i_\mu$ do not occur in the synthesis requirement, in order to find for this subcase another form of the condition for possibility of the synthesis. In this subcase, which we shall call Subcase 3, the synthesis requirement $S(t)$ may be taken to have the form:

(36) $$S[o_1(0), o_1(1), \ldots, o_1(b-1), o_1(t), o_1(t+1), \ldots, o_1(t+b), o_2(0), o_2(1), \ldots, o_2(b-1), o_2(t), o_2(t+1), \ldots, o_2(t+b), \ldots\ldots\ldots, o_\nu(0), o_\nu(1), \ldots, o_\nu(b-1), o_\nu(t), o_\nu(t+1), \ldots, o_\nu(t+b)]$$

Suppose that we apply to $S(t)$ the synthesis procedure for Subcase 2, as described above. Since no inputs are present in the synthesis requirement $S(t)$, (19) must turn out to be one of the truth-values, viz., T in case the synthesis is possible, and F in the contrary case. We shall, therefore, show that if (19) is F, then the conjunction $S(0)S(1)\ldots S(n)$ is a contradiction (in the sense of propositional calculus), where n is the least subscript for which $\models C_{n+1}(t) \equiv C_n(t)$.

In view of the fact that no inputs are present, the operation A_{b+1} reduces to the identity operation. Hence, $C_0(t)$ is the same as $S(t)$, and $D_0(t)$ is $E_b C_0(t)$, i.e., $E_b S(t)$. And $C_1(t)$ is $C_0(t)D_0(t+1)$, i.e., $S(t)E_{b+1}S(t+1)$.

Thus, we have:

$$\models C_0(t) \equiv S(t)$$
$$\models C_1(t) \equiv E_{b+1}[S(t)S(t+1)], \text{ by } X$$

We shall prove by mathematical induction that, for all l:

(37) $$\models C_l(t) \equiv E_{b+1}E_{b+2}\ldots E_{b+l}[S(t)S(t+1)\ldots S(t+l)]$$

For this purpose, assume (37) as hypothesis of induction. Then:

$$\models C_l(t+1) \equiv E_{b+2}E_{b+3}\ldots E_{b+l+1}[S(t+1)S(t+2)\ldots S(t+l+1)]$$
$$\models E_{b+1}C_l(t+1) \equiv E_{b+1}E_{b+2}\ldots E_{b+l+1}[S(t+1)S(t+2)\ldots S(t+l+1)]$$

Since the operation A_{b+1} reduces to the identity, $C_{l+1}(t)$ is $C_l(t)D_l(t+1)$, i.e., it is $C_l(t)E_{b+1}C_l(t+1)$. So we have that

$$\models C_{l+1}(t) \equiv E_{b+1}E_{b+2}\ldots E_{b+l}[S(t)S(t+1)\ldots S(t+l)]E_{b+1}E_{b+2} \ldots E_{b+l+1}[S(t+1)S(t+2)\ldots S(t+l+1)]$$

From this, by X, we get:

$$\vDash C_{l+1}(t) \equiv S(t)E_{b+1}E_{b+2}\dots E_{b+l}[S(t+1)S(t+2)\dots S(t+l)]E_{b+1}E_{b+2} \\ \dots E_{b+l}[S(t+1)S(t+2)\dots S(t+l)E_{b+l+1}S(t+l+1)]$$

But, by VIII:

$$\vDash E_{b+1}E_{b+2}\dots E_{b+l}[S(t+1)S(t+2)\dots S(t+l)E_{b+l+1}S(t+l+1)] \supset \\ E_{b+1}E_{b+2}\dots E_{b+l}[S(t+1)S(t+2)\dots S(t+l)]$$

Hence:

$$\vDash C_{l+1}(t) \equiv S(t)E_{b+1}E_{b+2}\dots E_{b+l}[S(t+1)S(t+2)\dots S(t+l)E_{b+l+1}S(t+l+1)]$$

Hence, by VIII, we have that

$$\vDash C_{l+1}(t) \equiv E_{b+1}E_{b+2}\dots E_{b+l+1}[S(t)S(t+1)S(t+2)\dots S(t+l)S(t+l+1)],$$

so completing the proof of (37) by mathematical induction.

From (37) we get immediately:

(38) $\vDash D_l(t) \equiv E_bE_{b+1}E_{b+2}\dots E_{b+l}[S(t)S(t+1)\dots S(t+l)]$

And, hence, in particular:

(39) $\vDash D_n(t) \equiv E_bE_{b+1}\dots E_{b+n}[S(t)S(t+1)\dots S(t+n)]$

Let us use E_b^0X to mean

$$\underset{o_1(b),o_2(b),\dots,o_v(b)}{\mathsf{E}} X,$$

and similarly, E_{b+1}^0X, etc. Then:

(40) $\vDash D_n(0) \equiv E_b^0E_{b+1}^0\dots E_{b+n}^0[S(0)S(1)\dots S(n)]$

Upon following through the process by which (19) [or what is the same thing, (14_0)] is obtained from $D_n(0)$, and remembering that all applications of V reduce in the present subcase to the identity operation, we see that (19) is, therefore, equivalent to:

$$E_0^0E_1^0\dots E_{b-1}^0E_b^0E_{b+1}^0\dots E_{b+n}^0[S(0)S(1)\dots S(n)]$$

And this finally, by IV, will be F if and only if $S(0)S(1)\dots S(n)$ is a contradiction in the sense of propositional calculus.

A Further Remark About the General Case 1

The result which was just obtained for Subcase 3 can, as a matter of fact, be generalized to an arbitrary synthesis requirement of the form (16), i.e., in effect, to the general Case 1.

Because in the general case the operations A do not reduce to the identity operation, they therefore, remain in the equivalent which is finally obtained for (19), and this has in consequence a more complicated form. We do not carry this out in detail here, since the result will not be used in these notes. But as a matter of interest we indicate the conclusion obtained.

For the synthesis requirement (16) we have that:

$$\vDash C_n(t) \equiv A_{b+1}E_{b+1}A_{b+2}E_{b+2}\ldots A_{b+n}E_{b+n}[C_0(t)C_0(t+1)\ldots C_0(t+n)]$$

$$\vDash D_n(t) \equiv E_b A_{b+1}E_{b+1}A_{b+2}E_{b+2}\ldots A_{b+n}E_{b+n}[C_0(t)C_0(t+1)\ldots C_0(t+n)]$$

Hence, the condition for existence of a solution can be put in the form:

$$A_0^0E_0^0A_1^0E_1^0\ldots A_{b-1}^0E_{b-1}^0A_b^0E_b^0A_{b+1}^0E_{b+1}^0\ldots A_{b+n}^0E_{b+n}^0[C_0(0)C_0(1)\ldots C_0(n)]$$

Namely this condition will reduce to T if the required synthesis is possible, and to F in the contrary case.

Or in this condition for the existence of a solution of (16), we may replace n by any known upper bound of it. From the argument employed above in connection with Subcase 1, it follows indeed that $2^{(\mu+\nu)(b+1)}$ is an upper bound of n. But better estimates of n are certainly obtainable.

The Decision Problem

Suppose we are given a set of recursion equivalences having the form of a wider restricted recursion, and call the conjunction of the equivalences $E(t)$. In order to determine whether a circuit represented by the given recursion equivalences satisfies a given requirement S, we may proceed by determining whether it is possible to synthesize a circuit satisfying the conjunction of $E(t)$ and S. In this way any case of the decision problem can always be reduced to a corresponding case of the synthesis problem. And in particular we now have, as a corollary of our foregoing solution of Case 1 of the synthesis problem, a solution of the corresponding case of the decision problem, i.e., the case that the requirement S is expressible in restricted recursive arithmetic with one free variable t.

We treat also the case of the decision problem in which the requirement S has the form $(t)H(t) \supset (t)A(t)$, where $H(t)$ and $A(t)$ are formulas of restricted recursive arithmetic having, as indicated, only the one free variable t. If the conjunction of the given recursion equivalences is $E(t)$, we proceed by first synthesizing a circuit which satisfies the requirement $E(t)H(t)$ when the i_j's and o_k's are taken together as outputs. As was shown above (see Subcase 3 of Case 1) this can always be done unless $(t)[E(t)H(t)]$ is logically impossible. Let the conjunction of the recursion equivalences for this circuit be $F(t)$. The answer to the decision problem is in the affirmative (i.e., the originally given circuit satisfies the requirement S) if and only if it is possible, with the parameters $\theta_1, \theta_2, \ldots, \theta_{\mu+\nu}$ that appear in $F(t)$ taken as inputs, and for all values of the parameters $\alpha_{10}, \alpha_{20}, \ldots, \alpha_{(\mu+\nu)(b-1)}$ to synthesize a circuit satisfying $F(t)A(t)$.

For suppose that the circuit represented by $E(t)$ satisfies $(t)H(t) \supset (t)A(t)$. By the way in which $F(t)$ was obtained, we have for every choice of the parameters $\theta_1, \theta_2, \ldots, \theta_{\mu+\nu}, \alpha_{10}, \alpha_{20}, \ldots, \alpha_{(\mu+\nu)(b-1)}$:

(i) $(t)F(t) \supset (t)E(t)$

(ii) $(t)F(t) \supset (t)H(t)$

For any choice for the parameters $\theta_1, \theta_2, \ldots, \theta_{\mu+\nu}, \alpha_{10}, \alpha_{20}, \ldots, \alpha_{(\mu+\nu)(b-1)}$, consider the functions $i_1, i_2, \ldots, i_\mu, o_1, \ldots, o_\nu$ that are determined by the recursion equivalences $F(t)$. By (i) this set of functions satisfies $(t)E(t)$; hence, it satisfies $(t)A(t)$; hence, it satisfies $(t)H(t) \supset (t)A(t)$; but by (ii) it satisfies $(t)H(t)$; hence, it satisfies $(t)[F(t)A(t)]$. This shows that the recursion equivalences $F(t)$ themselves, if we choose $\alpha_{10}, \alpha_{20}, \ldots, \alpha_{(\mu+\nu)(b-1)}$ arbitrarily, and treat $\theta_1, \theta_2, \ldots, \theta_{\mu+\nu}$ as the inputs, represent a circuit satisfying $F(t)A(t)$. Hence, the indicated synthesis is possible.

On the other hand, suppose that for all choices of $\alpha_{10}, \alpha_{20}, \ldots, \alpha_{(\mu+\nu)(b-1)}$ and with $\theta_1, \theta_2, \ldots, \theta_{\mu+\nu}$ treated as inputs, it is possible to synthesize a circuit satisfying $F(t)A(t)$. It then follows that the circuit represented by $F(t)$ satisfies $A(t)$. Take any set of K of functions $i_1, i_2, \ldots, i_\mu$, $o_1, o_2, \ldots, o_\nu$ which is a possible set of inputs and corresponding outputs of the circuit represented by $E(t)$. And suppose that K satisfies $(t)H(t)$. Then K satisfies $(t)[E(t)H(t)]$. Hence K will be given by the recursion equivalences $F(t)$ for some choice of the parameters $\theta_1, \theta_2, \ldots, \theta_{\mu+\nu}$, $\alpha_{10}, \alpha_{20}, \ldots, \alpha_{(\mu+\nu)(b-1)}$. Hence, K satisfies $A(t)$.

The Synthesis Problem. Case 2. Synthesis Requirement Expressible in Restricted Recursive Arithmetic with Two or More Free Variables

For illustration we take the case of two free (numerical) variables, x and t. The procedure is first to reduce the given synthesis requirement to the form

(41) $[A_1(x) \vee B_1(t)][A_2(x) \vee B_2(t)] \ldots [A_n(x) \vee B_n(x)]$

where (as the notation is meant to indicate) $A_j(x)$ has x as its only free variable and $B_j(t)$ has t as its only free variable, $j = 1, 2, \ldots, n$.

Let a_j be the greatest number a, such that either $x + a$, or a, occurs as an argument in any elementary part of $A_j(x)$, and b_j the greatest number b, such that either $t + b$, or b, occurs as an argument in any elementary part of $B_j(t)$. Let $A_j''(x)$ be the result of eliminating from $A_j(x)$ by IV all the elementary parts except $i_k(x), i_k(x+1), \ldots, i_k(x+a_j)$, $k = 1, 2, \ldots, \mu$, and similarly let $B_j''(t)$ be the result of eliminating from $B_j(t)$ by IV all the elementary parts except $i_k(t), i_k(t+1), \ldots, i_k(t+b_j)$. Let n'' be the number of factors in (41) in which either $A_j''(x)$ or $B_j''(t)$ is a nontautology. And let n' be $n - n''$. For the simplicity of the resulting solution of the synthesis problem, it is important to obtain in (41) as small a value of n' as possible. If the reduction to (41) is accomplished by the method of reduction to conjunctive normal form, this raises a question akin to that of the simplest conjunctive normal form. But in some cases, a better reduction to the form (41) is obtainable by methods other than that of reduction to conjunctive normal form, or even by methods other than those of propositional calculus.[4]

The synthesis problem is then reduced to Case 1 by replacing (41) by the following, where $p_1, p_2, \ldots, p_n, q_1, q_2, \ldots, q_n$ are distinct predicates not occurring in (41):

[4]As a simple example of the last, suppose that the synthesis requirement S contains two numerical variables, x and t, but t occurs only in the elementary part $o(t)$. Let S_0 and S_1 be obtained from S by replacing $o(t)$ everywhere by $\overline{o}(0)$ and $o(0)$ respectively. Then the synthesis requirement S can be replaced by the requirement $[S_0 \vee \centerdot\, o(t) \equiv o(0)]S_1$.

(42) $[p_1(t+a_1) \vee q_1(t+b_1)][p_2(t+a_2) \vee q_2(t+b_2)] \ldots [p_n(t+a_n)$
$\vee\, q_n(t+b_n)][p_1(t+1) \supset p_1(t)][p_2(t+1) \supset p_2(t)] \ldots$
$[p_n(t+1) \supset p_n(t)][q_1(t+1) \supset q_1(t)][q_2(t+1) \supset q_2(t)]$
$\ldots [q_n(t+1) \supset q_n(t)][p_1(t+a_1) \supset A_1(t)][p_2(t+a_2)$
$\supset A_2(t)] \ldots [p_n(t+a_n) \supset A_n(t)][q_1(t+b_1) \supset B_1(t)]$
$[q_2(t+b_2) \supset B_2(t)] \ldots [q_n(t+b_n) \supset B_n(t)]$

It is clear that any solution of (42) becomes a solution of (41) upon taking $p_1, p_2, \ldots, p_n, q_1, q_2, \ldots, q_n$ to represent intermediate elements instead of outputs. And on the other hand, any solution of (41) can be made to satisfy (42) by adjoining additional elements (represented by) $p_1, p_2, \ldots, p_n, q_1, q_2, \ldots, q_n$ in accordance with the following recursion equivalences (which have the form of a wider restricted recursion):

$$p_j(0) \equiv T, \ldots, p_j(a_j - 1) \equiv T, p_j(a_j) \equiv A_j(0), p_j(t + a_j + 1) \equiv p_j(t + a_j) A_j(t+1)$$
$$V_j(0) \equiv T, \ldots, V_j(b_j - 1) \equiv T, V_j(b_j) \equiv B_j(0), V_j(t + b_j + 1) \equiv V_j(t + b_j) B_j(t+1)$$

If (41) has factors in which $A_j''(x)$ is a nontautology, then (42) can be simplified by leaving out the corresponding predicates p_j and q_j for each such factor of (41), omitting from (42) the factors containing p_j, q_j, and replacing them by the single factor $B_j(t)$. Similarly if $B_j''(t)$ is a nontautology, we may omit from (42) the factors containing the corresponding predicates p_j, q_j, and replace them by the single factor $A_j(t)$.

We must solve (42) by the method for Case 1. The result is a general solution of (41), with the predicates $p_1, p_2, \ldots, p_n, V_1, V_2, \ldots, V_n$ representing intermediate elements of the circuit.

As before, the simplification problem remains after the synthesis problem has been solved. And one aspect (but only one aspect) of the simplification problem is the problem of eliminating superfluous elements p_j and V_j which may appear, and of finding an effective procedure for recognizing elements p_j and V_j as superfluous.

(As examples of simple special cases in which some of the elements p_j, q_j are superfluous, notice that if $A_1(t) \equiv B_2(t)$ is a tautology, we may replace q_2 by p_1 in (42) and drop redundant factors; if $A_1(t)B_1(t)$ is a contradiction in the sense of propositional calculus, the factors of (42) containing p_1 and q_1 may be replaced by $[A_1(t) \vee B_1(t)][A_1(t) \supset A_1(t+1)][B_1(t) \supset B_1(t+1)]$; if $B_1(t)$ is F, so that the first factor of (41) reduces to $A_1(x)$, and if $A_1(x) \supset . B_2(x+1) \supset B_2(x)$ is a tautology, then q_2 can be eliminated from (42) by replacing $q_2(t+b_2)$ by $B_2(t)$; and many other such cases.)

Behmann's Procedure

In order to deal with the case of the synthesis problem (or of the decision problem) in which quantifiers are added to the notations of restricted recursive arithmetic, we make use of Behmann's procedure of pushing in the quantifiers.

The procedure is applicable primarily to the case in which only singulary predicates occur. And though there is an extension to the case in which the binary predicate = occurs, in addition to the singulary predicates, we shall here confine attention to the case of singulary predicates only. Also we explain the matter with reference to ordinary quantifiers, as distinguished from bounded quantifiers, though the same procedure is in fact applicable to both.

The effect of Behmann's procedure, in the form in which we shall give it here, is to reduce a given formula to an equivalent formula in which each quantifier is isolated—i.e., the scope of the quantifier contains no other quantifier and contains no (numerical) variable except the variable of the quantifier.

Given a formula to which Behmann's procedure is to be applied, there will always be, among the nonisolated quantifiers occurring in it, one at least (usually more) that qualifies as innermost, in the sense that no nonisolated quantifier occurs in its scope. Select one of the innermost nonisolated quantifiers, and distinguish two cases: (i) the quantifier is universal; (ii) the quantifier is existential.

In case (i), say that the quantifier in question, together with its scope, is:

$$(t)R(t)$$

The first step is to reduce $R(t)$ to its conjunctive normal form, i.e., to replace $(t)R(t)$ by

$$(t)\Big\{[A_1 \vee B_1 \vee C_1][A_2 \vee B_2 \vee C_2]\ldots[A_l \vee B_l \vee C_l]\Big\},$$

where $A_1, A_2, \ldots, A_l$ are disjunctions of elementary parts and negations of elementary parts, of which each contains the variable t, and $B_1, B_2, \ldots, B_l$ are disjunctions of elementary parts and negations of elementary parts, of which each contains either some other (numerical) variable than t or no (numerical) variable at all, and $C_1, C_2, \ldots, C_l$ are disjunctions of formulas such as $(t)C_{11}(t)$, $(\exists x)C_{12}(x)$, consisting of a quantifier together with its scope, and negations of such formulas. (As a special case, some of the disjunctions $A_1, B_1, C_1, A_2, \ldots, C_l$ may of course be void.) Since the quantifier (t) at the beginning of $(t)R(t)$ is an innermost nonisolated quantifier, the quantifiers occurring in the disjunctions $C_1, C_2, \ldots, C_l$ must be isolated, and consequently $C_1, C_2, \ldots, C_l$ contain no free (numerical) variables. As a result we may take the further step of replacing $(t)R(t)$ by:

$$\big[(t)A_1 \vee B_1 \vee C_1\big]\ \big[(t)A_2 \vee B_2 \vee C_2\big]\ldots\big[(t)A_l \vee B_l \vee C_l\big]$$

In case (ii), say that the quantifier in question, together with its scope, is:

$$(\exists t)R(t)$$

First reduce $R(t)$ to disjunctive normal form, replacing $(\exists t)R(t)$ by

$$(\exists t)[A_1B_1C_1 \vee A_2B_2C_2 \vee \cdots \vee A_lB_lC_l],$$

where $A_1, A_2, \ldots, A_l$ are conjunctions of elementary parts and negations of elementary parts, of which each contains the variable t, and $B_1, B_2, \ldots, B_l$ are conjunctions of elementary parts and negations of elementary parts not containing the variable t, and $C_1, C_2, \ldots, C_l$ are conjunctions of formulas consisting of a quantifier together with its scope and negations of such formulas. Since the quantifier $(\exists t)$ at the beginning of $(\exists t)R(t)$ is an innermost nonisolated quantifier, the quantifiers occurring in the conjunctions $C_1, C_2, \ldots, C_l$ must be isolated, and consequently $C_1, C_2, \ldots, C_l$ contain no free (numerical) variables. As a result we may take the further step of replacing $(\exists t)R(t)$ by:

$$\big[(\exists t)A_1\big]\,B_1C_1 \vee [(\exists t)\,A_2]\,B_2C_2 \vee \cdots \vee [(\exists t)\,A_l]\,B_lC_l$$

Thus in case (i) and case (ii) alike we have a reduction process that eliminates the selected innermost nonisolated quantifier, replacing it by a number of isolated quantifiers. Repetitions of this reduction will finally eliminate all the nonisolated quantifiers.

The Synthesis Problem. Case 3. Synthesis Requirement Expressible in Restricted Recursive Arithmetic Plus Quantifiers

We may assume without loss of generality that the given synthesis requirement is closed (i.e., it has no free variables, but only bound variables). If we apply Behmann's procedure, just explained, followed by a reduction to conjunctive normal form, and by some further obvious steps, the synthesis requirement may be reduced to the following form, in which t is the only variable occurring:

$$(43)\quad \begin{aligned}&[(t)A_{11}(t) \vee (t)A_{12}(t) \vee \cdots \vee (t)A_{1m_1}(t) \vee \\ &(\exists t)B_1(t)][(t)A_{21}(t) \vee \cdots \vee (t)A_{2m_2}(t) \vee \\ &(\exists t)B_2(t)]\ldots[(t)A_{n1}(t) \vee \cdots \vee (t)A_{nm_n}(t) \vee (\exists t)B_n(t)]\end{aligned}$$

This may be reduced further by the device (akin to that used in Case 2) of replacing each disjunction

$$(t)A_{j1}(t) \vee (t)A_{j2}(t) \vee \ldots (t)A_{jm_j}(t)$$

in which $m_j > 1$ by

$$\begin{aligned}(t)\,.\,&[p_{j1}(t+a_{j1}) \vee p_{j2}(t + a_{j2}) \vee \cdots \vee p_{jm_j}(t+a_{jm_j})]\\ &[p_{j1}(t+1) \supset p_{j1}(t)][p_{j2}(t+1) \supset p_{j2}(t)]\ldots\\ &[p_{jm_j}(t+1) \supset p_{jm_j}(t)][p_{j1}(t+a_{jl}) \supset A_{j1}(t)]\\ &[p_{j2}(t+a_{j2}) \supset A_{j2}(t)]\ldots[p_{jm_j}(t+a_{jm_j}) \supset A_{jm_j}(t)]\end{aligned}$$

where the p_{jk}'s are distinct predicates not occurring in (43), and a_{jk} is the greatest number a, such that either $t + a$, or a, occurs as an argument in any elementary part of $A_{jk}(t)$. (We pass over the question of eliminating superfluous p_{jk}'s, which of course arises here also.) In this way the given synthesis requirement is reduced to the form

$$(44)\quad [(t)A_1(t) \vee (\exists t)B_1(t)][(t)A_2(t) \vee (\exists t)B_2(t)]\ldots[(t)A_n(t) \vee (\exists t)B_n(t)]$$

where again t is the only variable occurring. And this may be rewritten in the form

$$(45)\quad [(t)\overline{B}_1(t) \supset (t)A_1(t)][(t)\overline{B}_2(t) \supset (t)A_2(t)]\ldots[(t)\overline{B}_n(t) \supset (t)A_n(t)]$$

At this point the decision problem for Case 3 is already solved, as it may be solved for each factor of (45) separately by the method that was explained above in the section on the decision problem.

As a means of solving the synthesis problem we attempt to replace each factor of (45) by a factor which has the form of Case 1, i..e, which contains no quantifiers and has only the one free variable t. For if this can be done for every factor of (45), Case 3 of the synthesis problem will have been reduced to Case 1.

Take a particular factor of (45) and for convenience drop the subscripts, so as to write it as:

$$(46)\qquad (t)\overline{B}(t) \supset (t)A(t)$$

We wish to reduce it to the form of Case 1. First we treat two obvious preliminary cases.

Subcase 1: $B(t)$ is F, so that (46) reduces to $(t)A(t)$. In this case replace (46) by $A(t)$.

Subcase 2: (46) holds for any arbitrary circuit. In this case, of course, (46) is to be replaced by T. And as an efficient method to recognize this case, apply the decision procedure already found for requirements of the form (46), but in doing so take the i_j's and o_k's together as inputs, so that the set of recursion equivalences is void.

Then we notice that every solution of

$$(47) \qquad p(b) \equiv \overline{B}(0) \,.\, p(t+b+1) \equiv p(t+b)\overline{B}(t+1) \,.\, p(t+c) \supset A(t),$$

where p is a new predicate not previously used, b is the greatest argument in any elementary part of $B(0)$, and c is arbitrary, becomes a solution of (46) upon taking p to be an intermediate element instead of an output. And every solution of (46) can be made to satisfy (47) by adjoining a new predicate p obeying the recursion equivalences that constitute the first two factors of (47). Hence, we can make the required reduction of (46) to the form of Case 1 if we are able to supply an upper bound of c in (47).

Sometimes it may be possible to supply the required upper bound of c by considering the other factors of (45), in addition to (46). But here we confine attention to two cases in which an upper bound of c can be supplied from (46) alone.

Subcase 3: $B(t)$ does not contain any of the outputs o_k but only the inputs i_k. It may happen that at a certain time $t+b$ the course of the inputs up to that time has been such that, no matter how the inputs are continued from then on, one of the $\overline{B}(t+1), \overline{B}(t+2), \ldots, \overline{B}(t+d)$ must turn out to be false (although, if $d > 1$, the inputs can be so continued as to make all of $\overline{B}(t+1), \overline{B}(t+2), \ldots, \overline{B}(t+d-1)$ true). The maximum D of this number d can be computed if $B(t)$ is given, and is in any case not greater than $2^{\mu(b+1)}$.

If a is the greatest argument in any elementary part of $A(0)$, and if $a \geq b$, the required upper bound of c in (47) is $a+D$.—For suppose $\overline{B}(t)$ has held for all times up to and including $t+b$. At time $t+b$ it is possible that it has already been fated by the past course of the inputs that at least one of $\overline{B}(t+1), \overline{B}(t+2), \ldots, \overline{B}(t+D)$ will be false. If so,

$$B(t+1) \vee B(t+2) \vee \cdots \vee B(t+D)$$

will hold for the particular t. Otherwise there is a course of future inputs that will make $\overline{B}(t)$ always true, and hence, will falsify (46) unless $A(t+b-a)$ holds for the particular t. Thus, if (46) holds,

$$p(t+b) \supset A(t+b-a) \vee B(t+1) \vee B(t+2) \vee \cdots \vee B(t+D)$$

holds at each time $t+b, \geq a$, and hence also $p(t+b+D) \supset A(t+b-a)$. Hence, for all t, $p(t+a+D) \supset A(t)$.

If on the other hand $a < b$, the proof just given shows only that $p(t+a+D) \supset A(t)$ holds for $t \geq b-a$. Hence we also find a number $E \geq 1$, large enough so that if $\overline{B}(0), \overline{B}(1), \ldots, \overline{B}(E-1)$ are all true, there will be a course of future inputs that will make $\overline{B}(t)$ always true. Such a number E can always be found, and will not be greater than $2^{\mu(b+1)}$. For any t less than $b-a$, if (46) is to hold, we must have $A(t) \vee B(0) \vee B(1) \vee \cdots \vee B(E-1)$, which is equivalent to $p(b+E-1) \supset A(t)$. Thus, an upper bound of c in (47) is the greater of $a+D$ and $b+E-1$.

Subcase 4: Let $H(t)$ be the result of eliminating all elementary parts that contain the outputs $o_1, o_2, \ldots, o_\nu$ from $\overline{B}(t)$ by V; then the result of eliminating the elementary parts $i_k(t+b), k = 1, 2, \ldots, \mu$, from $H(t)$ by IV is a tautology. In this case there will always be at time $t+b$ a course of future inputs that will verify all of $H(t+1), H(t+2), \ldots,$ to infinity. Hence, if $\overline{B}(t)$ holds at all times up to and including $t+b$, and $t+b \geq a$, then $A(t+b-a)$ must hold if (46) is not to fail; and for a value of t less than $b-a$, $A(t)$ must always hold if (46) is not to fail. Thus, the required upper bound of c in (47) is the greater of a and b.

It is not difficult to cite examples of synthesis requirements of the form (46) such that there cannot be an upper bound of c in (47). (An obvious but rather trivial example is the synthesis requirement $(t)\overline{o}(t) \supset F$.) In this case the general solution cannot be written in the form we have been using, i.e., by means of recursion equivalences involving a finite number of parameters. We might in this case seek at least a particular solution, and hence, ask for an algorithm by which to find the smallest value of c for which (47) is solvable.

Open problems in regard to Case 3 are then to find algorithms by which to determine whether there is in (47) a maximum value of c (beyond which greater values of c yield no additional solutions), and if so, to find such a value of c, and if not, to find a minimum value of c (for which there are any solutions).

The Synthesis Problem. Case 4. Synthesis Requirement Expressible in Restricted Recursive Arithmetic Plus = and <

We proceed by first reducing the given synthesis requirement to conjunctive normal form, and then seek to replace each factor by a factor of the form of Case 1. It is conjectured that Case 4 is fully solvable by this method. But here we do not go beyond citing a few examples, which will illustrate the method of approach.

Replace the factor $t < 2 \vee A(t)$ by $A(t+2)$.

Replace the factor $t < 2 \vee t > 4 \vee A(t)$ by $A(2)A(3)A(4)$.

Replace the factor $t < 2 \vee t = 4 \vee A(t)$ by $A(2)A(3)A(t+5)$.

Replace the factor $x = t + m \vee A(x) \vee B(t)$ by the conjunction

$[p(t+a+m) \supset p(t+a+m+1) \vee q(t+b)]$

$[q(t+b) \supset p(t+a+m) \vee q(t+b+1)][p(t+1) \supset p(t)]$

$[q(t+1) \supset q(t)][p(t+a) \supset A(t)][q(t+b) \supset B(t)]$

$[p(t+a+m) \vee q(t+b) \vee A(t+m+1)B(t+1)][p(a+m-1) \vee q(b)]$, where a and b have the same meaning as in the previous section, and where the last factor of the conjunction is to be omitted if m is 0.

Replace the factor $x + m = t \vee A(x)$ by $A(t)$.

Replace the factor $x + m \neq t \vee A(x) \vee B(t)$ by $A(t) \vee B(t+m)$.

Replace the factor $x + m < t \vee A(x) \vee B(t)$ by

$[A(t) \vee q(t+b+m)][q(t+1) \supset q(t)][q(t+b) \supset B(t)]$.

Replace the factor $x + m > t \vee B(t)$ by $B(t+m)$.

Replace the factor $x + m_1 < t \vee x > t + m_2 \vee A(x) \vee B(t)$ by

$$[B(0) \vee A(0)A(1)\ldots A(m_2)][B(1) \vee A(0)A(1)\ldots A(m_2+1)]\ldots[B(m_1-1) \vee A(0)A(1)\ldots A(m_1+m_2-1)][B(t+m_1) \vee A(t)A(t+1)\ldots A(t+m_1+m_2)]$$

Replace the factor $x + m_1 < t \vee x = t + m_2 \vee A(x) \vee B(t)$ by

$$[A(0) \vee q(b+m_1)][A(1) \vee q(b+m_1+1)]\ldots[A(m_2-1) \vee q(b+m_1+m_2-1)]$$
$$[A(m_2) \vee q(b+m_1+m_2) \vee \overline{q}(b)B(1)B(2)\ldots B(m_1+m_2)]$$
$$[A(t+m_2+1) \vee q(t+b+m_1+m_2+1) \vee q(t+b)\overline{q}(t+b+1)$$
$$B(t+2)B(t+3)\ldots B(t+m_1+m_2+1)][q(t+1) \supset q(t)][q(t+b) \supset B(t)].$$

Asynchronous Circuits

Let us return to the synthesis requirement (1) which was considered at the beginning of the section on Case 1, and for illustration let us take $\mu = \nu = 2$ and $b = 2$. The recursion equivalences (15_{10}), (15_{20}), (15_{11}), (15_{21}), and

(6_1) $\qquad o_1(t+2) \equiv R_1(t)$
(6_2) $\qquad o_2(t+2) \equiv R_2(t)$

which were obtained in this section constitute the general solution of the synthesis problem in the sense that any circuit satisfying (1) must also satisfy the recursion equivalences (15_{10}), (15_{20}), (15_{11}), (15_{21}), (6_1), and (6_2) for some choice of the parameters. This will hold for the asynchronous circuits as well as the synchronous circuits. And the asynchronous circuits will be distinguished from the others by the fact that the synthesis requirement (1) will still follow as a consequence if (15_{10}), (15_{20}), (15_{11}), (15_{21}), (6_1), and (6_2) are replaced by the weaker conditions:

(15_{10}a) $\quad [o_1(0) \equiv I_{10}] \vee [o_1(0) \equiv F][o_1(1) \equiv I_{11}]$
(15_{20}a) $\quad [o_2(0) \equiv I_{20}] \vee [o_2(0) \equiv F][o_2(1) \equiv I_{21}]$
(15_{11}a) $\quad [o_1(1) \equiv I_{11}] \vee [o_1(1) \equiv o_1(0)][o_1(2) \equiv R_1(0)]$
(15_{21}a) $\quad [o_2(1) \equiv I_{21}] \vee [o_2(1) \equiv o_2(0)][o_2(2) \equiv R_2(0)]$
(6_1a) $\quad [o_1(t+2) \equiv R_1(t)] \vee [o_1(t+2) \equiv o_1(t+1)][o_1(t+3) \equiv R_1(t+1)]$
(6_2a) $\quad [o_2(t+2) \equiv R_2(t)] \vee [o_2(t+2) \equiv o_2(t+1)][o_2(t+3) \equiv R_2(t+1)]$

Let the conjunction of (15_{10}), (15_{20}), (15_{11}), (15_{21}), (6_1), (6_2) be $E(t)$. We must then determine $\alpha_{10}, \alpha_{20}, \alpha_{11}, \alpha_{21}, \theta_1, \theta_2$ so that (1), i.e., $S(t)$, is a consequence of $E(t)$. With any particular choice of $\alpha_{10}, \alpha_{20}, \alpha_{11}, \alpha_{21}$, this is equivalent to the synthesis problem with $(t)E(t) \supset (t)S(t)$ as synthesis requirement and θ_1 and θ_2 as outputs. Or in order to treat $\alpha_{10}, \alpha_{20}, \alpha_{11}, \alpha_{21}$ also as outputs, we may replace them by $\alpha_{10}(0), \alpha_{20}(0), \alpha_{11}(1), \alpha_{21}(1)$, respectively.

If the amount of uncertainty in the time required for the action of an element exceeds one unit of time, this can of course also be expressed, in the same way that an uncertainty of one unit is expressed in (15_{10}a), (15_{20}a), (15_{11}a), (15_{21}a), (6_1a), and (6_2a). Or different amounts of uncertainty in time of action can be ascribed to different elements. But the expressions become more and more complicated as greater amounts of uncertainty are allowed.

Analogous remarks apply in regard to asynchronous circuits in connection with other cases of the synthesis problem.

Notice should be taken, however, of the role of our assumption that a synthesis problem has been solved as soon as a wider restricted recursion has been obtained. In terms of the circuit this can be described as an assumption that circuit elements of arbitrary kind are available, including elements of arbitrarily complicated behavior such as are not likely in practice to be at hand as a single element. Or it may be described by saying that the synthesis of such complicated elements from the types of single elements that are available is left for informal solution as offering no difficulty. But if the circuit is to be asynchronous, there is a stronger assumption, namely, that the synthesis of a complicated element from the simple single elements available can always be accomplished in suitably asynchronous fashion.

Given a particular list of types of single elements, it would be possible to provide a corresponding list of forms of recursion equivalences, and to demand as solution of a synthesis problem a system of recursion equivalences of these particular forms, rather than merely to demand a system constituting a wider restricted recursion. It is conceivable that this might serve some purpose if, after solution of the synthesis problem, some further condition is to be imposed on the circuit, such as that of being asynchronous.

An Example Illustrating Case 2

The following example (which is a modification of one suggested to me by R. J. Nelson) is presumably not of great importance in itself, but will serve to illustrate how conceivably synthesis procedure of the kind we have been discussing might be used in application to actual problems in circuit construction—provided that means can be found to shorten the procedures, and that solutions can be obtained for some of the unsolved cases.

Consider a circuit with four inputs i_1, i_2, i_3, i_4 and three outputs o_1, o_2, o_3. The inputs at $t = 0$ have no significance. Beginning at $t = 1$, the successive truth-values of i_1 represent the successive digits of a number N_1 expressed in binary notation, truth for the digit 1 and falsehood for the digit 0; and the successive truth-values of i_2 represent the successive digits of a binary number N_2, in the same way; and the successive truth-values of i_3 represent the successive digits of a binary number N_3, in the same way. The last digit is signaled by the input i_4, which has the value falsehood until the instant at which the last digit of each of the three binary numbers N_1, N_2, N_3 comes in on the inputs i_1, i_2, i_3, and then at that instant has the value truth. It is required that the successive digits of the smallest of the three binary numbers N_1, N_2, N_3 be reproduced by the input o_1, again with truth for 1 and falsehood for 0; and that the successive digits of the next smallest of the three numbers be reproduced by o_2, in the same way; and the successive digits of the largest of the three numbers, by o_3.—If the requirement is that the indicated action of the circuit shall be instantaneous, it may be written in logistic notation as follows:

$$(48)\qquad \begin{aligned}&(t)_1^u \bar{i}_4(t) \supset . \, i_4(u+1) \supset . \\ &(t)_1^x [\bar{i}_4(t) \, . \, i_1(t) \equiv i_2(t)] \supset .[i_4(x+1) \vee . \, i_1(x+1) \not\equiv i_2(x+1)] \supset . \\ &(t)_1^y [\bar{i}_4(t) \, . \, i_1(t) \equiv i_3(t)] \supset .[i_4(y+1) \vee . \, i_1(y+1) \not\equiv i_3(y+1)] \supset . \\ &(t)_1^z [\bar{i}_4(t) \, . \, i_2(t) \equiv i_3(t)] \supset .[i_4(z+1) \vee . \, i_2(z+1) \not\equiv i_3(z+1)] \supset .\end{aligned}$$

$$[i_1(x+1) \supset i_2(x+1) \supset . \, i_2(z+1) \supset i_3(t+1) \supset (t)_1^{u+1} . \, o_1(t) \equiv i_1(t) . \, o_2(t) \equiv i_2(t) . \, o_3(t) \equiv i_3(t)][i_1(y+1) \supset i_3(y+1) \supset . \, i_3(z+1) \supset i_2(z+1) \supset (t)_1^{u+1} . \, o_1(t) \equiv i_1(t) . \, o_2(t) \equiv i_3(t) . \, o_3(t) \equiv i_2(t)] [i_2(x+1) \supset i_1(x+1) \supset . \, i_1(y+1) \supset i_3(y+1) \supset (t)_1^{u+1} . \, o_1(t) \equiv i_2(t) . \, o_2(t) \equiv i_1(t) . \, o_3(t) \equiv i_3(t)][i_2(z+1) \supset i_3(z+1) \supset . \, i_3(y+1) \supset i_1(y+1) \supset (t)_1^{u+1} . \, o_1(t) \equiv i_2(t) . \, o_2(t) \equiv i_3(t) . \, o_3(t) \equiv i_1(t)][i_3(y+1) \supset i_1(y+1) \supset . \, i_1(x+1) \supset i_2(x+1) \supset (t)_1^{u+1} . \, o_1(t) \equiv i_3(t) . \, o_2(t) \equiv i_1(t) . \, o_3(t) \equiv i_2(t)][i_3(z+1) \supset i_2(z+1) \supset . \, i_2(x+1) \supset i_1(x+1) \supset (t)_1^{u+1} . \, o_1(t) \equiv i_3(t) . \, o_2(t) \equiv i_2(t) . \, o_3(t) \equiv i_1(t)]$$

From this, the bounded quantifiers may be eliminated by standard methods. For example, $(t)_1^x[\bar{i}_4(t) . \, i_1(t) \equiv i_2(t)]$ may be replaced by $r_1(x)$ provided that we adjoin the equivalences $r_1(0) \equiv T, r_1(t+1) \equiv [r_1(t) . \, \bar{i}_4(t+1) . \, i_1(t+1) \equiv i_2(t+1)]$. After thus eliminating the bounded quantifiers, the resulting expression is a formula of restricted recursive arithmetic which contains five variables, t, u, x, y, z, and seventeen predicates, $i_1, i_2, i_3, i_4, o_1, o_2, o_3, r_0, r_1, r_2, r_3, r_4, r_5, r_6, r_7, r_8, r_9$. Omitting the twenty equivalences for $r_0, \ldots, r_9$ (which involve only the variable t), we may write the remainder of this formulas as follows:

$$\bar{r}_0(u) \vee \bar{i}_4(u+1) \vee \bar{r}_1(x) \vee \bar{i}_4(x+1)[i_1(x+1) \equiv i_2(x+1)] \vee \bar{r}_2(y) \vee \bar{i}_4(y+1)[i_1(y+1) \equiv i_3(y+1)] \vee \bar{r}_3(z) \vee \bar{i}_4(z+1)[i_2(z+1) \equiv i_3(z+1)] \vee [i_1(x+1)\bar{i}_2(x+1) \vee i_2(z+1)\bar{i}_3(z+1) \vee r_4(u+1)][i_1(y+1)\bar{i}_3(y+1) \vee i_3(z+1)\bar{i}_2(z+1) \vee r_5(u+1)][i_2(x+1)\bar{i}_1(x+1) \vee i_1(y+1)\bar{i}_3(y+1) \vee r_6(u+1)][i_2(z+1)\bar{i}_3(z+1) \vee i_3(y+1)\bar{i}_1(y+1) \vee r_7(u+1)][i_3(y+1)\bar{i}_1(y+1) \vee i_1(x+1)\bar{i}_2(x+1) \vee r_8(u+1)][i_3(z+1)\bar{i}_2(z+1) \vee i_2(x+1)\bar{i}_1(x+1) \vee r_9(u+1)] \tag{49}$$

To reduce to a single variable t by the method given above for Case 2 of the synthesis problem, therefore, requires the introduction of twenty-four additional predicates $f_1, f_2, f_3, f_4, f_5, f_6, g_1, g_2, g_3, g_4, g_5, g_6, p_1, p_2, p_3, p_4, p_5, p_6, q_1, q_2, q_3, q_4, q_5, q_6$. The resulting formula is too long to be written out in full here, but let us abbreviate it as follows (showing its elementary parts):

$$S[i_1(t+1), i_2(t+1), i_3(t+1), i_4(t+1), o_1(t+1), o_2(t+1), o_3(t+1), r_0(0), r_1(0), r_2(0), r_3(0), r_4(0), r_5(0), r_6(0), r_7(0), r_8(0), r_9(0), r_0(t), r_1(t), r_2(t), r_3(t), r_4(t), r_5(t), r_6(t), r_7(t), r_8(t), r_9(t), r_0(t+1), r_1(t+1), r_2(t+1), r_3(t+1), r_4(t+1), r_5(t+1), r_6(t+1), r_7(t+1), r_8(t+1), r_9(t+1), f_1(t), f_2(t), f_3(t), f_4(t), f_5(t), f_6(t), g_1(t), g_2(t), g_3(t), g_4(t), g_5(t), g_6(t), p_1(t), p_2(t), p_3(t), p_4(t), p_5(t), p_6(t), q_1(t), q_2(t), q_3(t), q_4(t), q_5(t), q_6(t), f_1(t+1), f_2(t+1), f_3(t+1), f_4(t+1), f_5(t+1), f_6(t+1), g_1(t+1), g_2(t+1), g_3(t+1), g_4(t+1), g_5(t+1), g_6(t+1), p_1(t+1), p_2(t+1), p_3(t+1), p_4(t+1), p_5(t+1), p_6(t+1), q_1(t+1), q_2(t+1), q_3(t+1), q_4(t+1), q_5(t+1), q_6(t+1)] \tag{50}$$

In theory we may apply the procedure of Case 1 to this, and so obtain a general solution. In practice the calculation is much too lengthy, even if we suppose the aid of specially constructed computing machinery, and the solution is moreover obtained in an excessively complicated form.

On the other hand, by special methods applicable to the particular case, it is not difficult to obtain a solution of (48) in comparatively simple form. This can be done in various ways. But as one possibility, we introduce six predicates $e_{12}, e_{23}, e_{31}, e_{21}, e_{32}, e_{13}$, letting $e_{jk}(t)$ mean that the course of the inputs has already determined at time t that $N_j < N_k$. The following solution of (48) can then be written by inspection:

(51)
$$
\begin{aligned}
& r(0) \equiv F,\ r(1) \equiv T,\ r(t+2) \equiv r(t+1)\bar{i}_4(t+1) \\
& e_{12}(0) \equiv F,\ e_{23}(0) \equiv F,\ e_{31}(0) \equiv F, \\
& e_{21}(0) \equiv F,\ e_{32}(0) \equiv F,\ e_{13}(0) \equiv F, \\
& e_{12}(t+1) \equiv [e_{12}(t) \vee \bar{i}_1(t+1)i_2(t+1)\bar{e}_{21}(t)]r(t+1) \vee \bar{r}(t+1)\theta_{12}(t+1) \\
& e_{23}(t+1) \equiv [e_{23}(t) \vee \bar{i}_2(t+1)i_3(t+1)\bar{e}_{32}(t)]r(t+1) \vee \bar{r}(t+1)\theta_{23}(t+1) \\
& e_{31}(t+1) \equiv [e_{31}(t) \vee \bar{i}_3(t+1)i_1(t+1)\bar{e}_{13}(t)]r(t+1) \vee \bar{r}(t+1)\theta_{31}(t+1) \\
& e_{21}(t+1) \equiv [e_{21}(t) \vee \bar{i}_2(t+1)i_1(t+1)\bar{e}_{12}(t)]r(t+1) \vee \bar{r}(t+1)\theta_{21}(t+1)^{*} \\
& e_{32}(t+1) \equiv [e_{32}(t) \vee \bar{i}_3(t+1)i_2(t+1)\bar{e}_{23}(t)]r(t+1) \vee \bar{r}(t+1)\theta_{32}(t+1) \\
& e_{13}(t+1) \equiv [e_{13}(t) \vee \bar{i}_1(t+1)i_3(t+1)\bar{e}_{31}(t)]r(t+1) \vee \bar{r}(t+1)\theta_{13}(t+1) \\
& o_1(t) \equiv [\bar{e}_{21}(t)\bar{e}_{31}(t)i_1(t) \vee \bar{e}_{12}(t)\bar{e}_{32}(t)i_2(t) \\
& \qquad \vee\ \bar{e}_{13}(t)\bar{e}_{23}(t)i_3(t)]r(t) \vee \bar{r}(t)\theta_1(t) \\
& o_2(t) \equiv \Big\{[\bar{e}_{12}(t) \vee \bar{e}_{13}(t)][\bar{e}_{21}(t) \vee \bar{e}_{31}(t)]i_1(t) \vee \\
& \qquad [\bar{e}_{21}(t) \vee \bar{e}_{23}(t)][\bar{e}_{12}(t) \vee \bar{e}_{32}(t)]i_2(t) \vee \\
& \qquad [\bar{e}_{31}(t) \vee \bar{e}_{32}(t)][\bar{e}_{13}(t) \vee \bar{e}_{23}(t)]i_3(t)\Big\}r(t) \\
& \qquad \qquad \vee\ \bar{r}(t)\theta_2(t) \\
& o_3(t) \equiv [\bar{e}_{12}(t)\bar{e}_{13}(t)i_1(t) \vee \bar{e}_{21}(t)\bar{e}_{23}(t)i_2(t) \\
& \qquad \vee\ \bar{e}_{31}(t)\bar{e}_{32}(t)i_3(t)]r(t) \vee \bar{r}(t)\theta_3(t)
\end{aligned}
$$

These equivalences do not constitute a single restricted recursion, but by nonessential modifications, which include replacing the first three equivalences by

(51a)
$$
\begin{aligned}
& \xi(0) \equiv T, \quad \xi(t+1) \equiv F, \\
& r(0) \equiv F, \quad r(t+1) \equiv r(t)\bar{i}_4(t) \vee \xi(t),
\end{aligned}
$$

they can be regarded as having the form of three successive restricted recursions. The predicates $r, e_{12}, e_{23}, e_{31}, e_{21}, e_{32}, e_{13}$ represent *intermediate elements* of the circuit in a sense already explained, and we shall, therefore, speak of the functions $r, e_{12}, e_{23}, e_{31}, e_{21}, e_{32}, e_{13}$ (denoted by these predicates) as *intermediate functions*.

Upon choosing values for the parameters $\theta_{12}, \theta_{23}, \theta_{31}, \theta_{21}, \theta_{32}, \theta_{13}$, $\theta_1, \theta_2, \theta_3$ (for example all of them may be taken as being identically F), a circuit diagram may be drawn directly from the equivalences (51). And this is one example of a circuit satisfying the requirement stated at the beginning of this section.

Moreover, (51) fails of being a general solution of (48) only in the rather trivial respect of not showing the complete arbitrariness of the values of $o_1(t)$, $o_2(t)$, and $o_3(t)$ at $t = 0$. A fully general solution of (48) will be obtained from (51) by substituting

---

* *Editor's note*: The third and fifth plus signs in this line are editorial corrections to minus signs in the original.

$t+1$ for t in the last three equivalences and then adding the equivalences

$$o_1(0) \equiv \alpha, \quad o_2(0) \equiv \beta, \quad o_3(0) \equiv \gamma$$

(with α, β, γ as parameters).

In fact, (51) is a general solution of (48) in the sense that any other solution of (48) can be obtained from (51) by some transformation of it, including transformations by which the intermediate functions $r, e_{12}, e_{23}, e_{31}, e_{21}, e_{32}, e_{13}$ are replaced by some other set of intermediate functions.

The contrast between the comparative simplicity of this solution of (48) and the complication of the solution obtained by applying the method of Case 1 to (50) strongly suggests that our general methods for solution of the synthesis problem may be capable of improvement. It is true that when we ask for a method that is effectively applicable in a very general sense, such as Case 1 or Case 2, it may well be that for the generality of the method we have to pay the price of finding it more complicated than particular methods that are available in special cases. But the amount of increase in complication does seem excessive in the present instance. I do not venture to predict what will be possible, but believe that the search for more efficient synthesis procedures is at least worth trying.—Regardless of the issue of such a search, the connection between recursive arithmetic and circuit theory is of theoretical importance for both disciplines. And it may be also be that the notations of recursive arithmetic will be found practically useful in providing a language for circuit theory that is more manageable than that of circuit diagrams and at the same time is not subject to certain limitations that are inherent in the use of propositional calculus (or Boolean algebra) alone.

As exercises to illustrate the way in which other solutions of (48) are obtained by transformations of (51), it is suggested that the reader carry out the following:

1. In the equivalences for $o_1(t+1), o_2(t+1), o_3(t+1)$, substitute $t+1$ for t, and then substitute for $e_{12}(t+1), e_{23}(t+1), e_{31}(t+1), e_{21}(t+1), e_{32}(t+1), e_{13}(t+1)$, in accordance with the equivalences for these latter, and simplify. In the simplification, make use of such properties of the e_{jk}'s as $e_{12}(t) \supset e_{13}(t) \vee e_{32}(t)$, $e_{12}(t)e_{21}(t) \equiv F$, and $e_{12}(t)e_{23}(t)e_{31}(t) \equiv F$, which are evident from the meanings of the e_{jk}'s as given above, and which do in fact follow from the recursion equivalences (51). The result should be:

$$\begin{aligned}
(52) \qquad o_1(t+1) \equiv [&e_{12}(t)e_{13}(t)i_1(t+1) \vee \\
&e_{21}(t)e_{23}(t)i_2(t+1) \vee \\
&e_{31}(t)e_{32}(t)i_3(t+1) \vee \\
&e_{21}(t)e_{31}(t)i_2(t+1)i_3(t+1) \vee \\
&e_{12}(t)e_{32}(t)i_1(t+1)i_3(t+1) \vee \\
&e_{13}(t)e_{23}(t)i_1(t+1)i_2(t+1) \vee \\
&\quad i_1(t+1)i_2(t+1)i_3(t+1)]r(t+1) \vee \bar{r}(t+1)\theta_1(t+1)
\end{aligned}$$

$$\begin{aligned}
o_2(t+1) \equiv \Big\{&[e_{12}(t)e_{32}(t)\bar{e}_{13}(t) \vee e_{13}(t)e_{23}(t)\bar{e}_{12}(t)]i_1(t+1) \vee \\
&[e_{21}(t)e_{31}(t)\bar{e}_{23}(t) \vee e_{13}(t)e_{23}(t)\bar{e}_{21}(t)]i_2(t+1) \vee
\end{aligned}$$

$$[e_{21}(t)e_{31}(t)\overline{e}_{32}(t) \vee e_{12}(t)e_{32}(t)\overline{e}_{31}(t)]i_3(t+1) \vee$$
$$[\overline{e}_{21}(t) \vee \overline{e}_{23}(t)][\overline{e}_{31}(t) \vee \overline{e}_{32}(t)]i_2(t+1)i_3(t+1) \vee$$
$$[\overline{e}_{12}(t) \vee \overline{e}_{13}(t)][\overline{e}_{32}(t) \vee \overline{e}_{31}(t)]i_1(t+1)i_3(t+1) \vee$$
$$[\overline{e}_{13}(t) \vee \overline{e}_{12}(t)][\overline{e}_{23}(t) \vee \overline{e}_{21}(t)]i_1(t+1)i_2(t+1) \vee$$
$$i_1(t+1)i_2(t+1)i_3(t+1)\Big\}r(t+1) \vee \overline{r}(t+1)\theta_2(t+1)$$

$$o_3(t+1) \equiv [\overline{e}_{12}(t)\overline{e}_{13}(t)i_1(t+1) \vee \overline{e}_{21}(t)\overline{e}_{23}(t)i_2(t+1) \vee \overline{e}_{31}(t)\overline{e}_{32}(t)i_3(t+1)]r(t+1) \vee \overline{r}(t+1)\theta_3(t+1)$$

When (51) is modified by replacing its first three equivalences by (51a), replacing $r(t+1)$ by $[r(t)\overline{i}_4(t) \vee \xi(t)]$ in the equivalences for $e_{12}(t+1), \ldots, e_{13}(t+1)$, and replacing the equivalences for $o_1(t+1), o_2(t+1), o_3(t+1)$ by the equivalences (52), the resulting set of recursion equivalences has the form of a single restricted recursion. It may be described either as another solution of (48) or as another form of the solution — according to the point of view. The circuit corresponding to the modified set of recursion equivalences is of course different.

2. Introduce a new set of intermediate functions in accordance with the equivalences:

$$\begin{aligned}
&r_0(t) \equiv r(t+1)\\
&s_1(t) \equiv e_{12}(t)e_{13}(t) \vee e_{31}(t)e_{32}(t)\\
&s_2(t) \equiv e_{21}(t)e_{23}(t) \vee e_{31}(t)e_{32}(t)\\
&s_3(t) \equiv e_{21}(t)e_{31}(t) \vee e_{13}(t)e_{23}(t)\\
&s_4(t) \equiv e_{12}(t)e_{32}(t) \vee e_{13}(t)e_{23}(t)
\end{aligned}$$

Verify the following solutions of these equivalences:

$$\begin{aligned}
&r(0) \equiv T,\\
&r(t+1) \equiv r_0(t)\\
&e_{12}(t) \equiv s_1(t)\overline{s}_2(t) \vee \overline{s}_3(t)s_4(t)\\
&e_{23}(t) \equiv \overline{s}_1(t)s_2(t) \vee s_3(t)s_4(t)\\
&e_{31}(t) \equiv s_1(t)s_2(t) \vee s_3(t)\overline{s}_4(t)\\
&e_{21}(t) \equiv \overline{s}_1(t)s_2(t) \vee s_3(t)\overline{s}_4(t)\\
&e_{32}(t) \equiv s_1(t)s_2(t) \vee \overline{s}_3(t)s_4(t)\\
&e_{13}(t) \equiv s_1(t)\overline{s}_2(t) \vee s_3(t)s_4(t)
\end{aligned}$$

Then transform (51) by eliminating the intermediate functions $r, e_{12}, e_{23}, e_{31}, e_{21}, e_{32}, e_{13}$ in favor of the new intermediate functions r_0, s_1, s_2, s_3, s_4, and simplify as far as possible. (The result will illustrate the point that a reduction in the number of intermediate functions is not necessarily a simplification of the corresponding circuit.)

Such transformations upon the recursion equivalences for a given circuit provide a method of casting about for a circuit that is equivalent to the given one and that may be superior to it in some respect (e.g., in simplicity). But an algorithm by which to find the simplest such equivalent circuit is hardly to be hoped for.

In computations of this sort, it is convenient to abbreviate $s_1(t), e_{12}(t), i_2(t), o_3(t)$, and the like as s_1, e_{12}, i_2, o_3, etc.; then to abbreviate $s_1(t+1), e_{12}(t+1), i_2(t+1), o_3(t+1)$, etc., as $s_1', e_{12}', i_2', o_3'$, etc.; to abbreviate $s_1(t+2)$ as s_1''; and so on. (Whenever the number of accents or "primes" goes above three or four, superscript numerals may be substituted for the series of primes.) These abbreviation shorten the computation by reducing the amount of writing to be done. And whenever it is necessary to deal with $s_1(0), e_{12}(1)$, and the like, or to write formulas which [like (49) above] contain more than one numerical variable, it will always be possible to return to the more explicit notation that is used in these notes.

Another remark to be made about the recursion equivalences (51) is that advantage may be taken of the arbitrariness of the parameters $\theta_{12}, \theta_{23}, \theta_{31}, \theta_{21}, \theta_{32}, \theta_{13}, \theta_1, \theta_2, \theta_3$ in order to impose further restrictions on the circuit.

Our original requirement, which was first given in words and then expressed in logical notation as (48), said nothing about the behavior of the circuit after the last digit of N_1, N_2, N_3 had been signaled by i_4. Therefore, it is now open to us to require (for example) that after this signal on i_4 the inputs i_1, i_2, i_3 shall represent the successive digits of three new binary numbers N_1, N_2, N_3, and that the circuit shall sort these three new numbers into order in such a manner that the digits of the smallest number are reproduced by the output o_1, the digits of the next smallest number are reproduced by o_2, and the digits of the largest number are reproduced by o_3, until the arrival of the last digit of N_1^1, N_2^1, and N_3^1 is indicated by a new signal on i_4. The values of the parameters that will make the circuit satisfy this additional requirement should now be obvious, and it is left to the reader to supply them.

The general solution thus obtained, for a circuit that satisfies both the original requirement and the additional requirement just stated, will still contain parameters, in such fashion as to leave open the behavior of the circuit after the last digit of N_1^1, N_2^1, and N_3^1 has been signaled on i_4. But a specialization is, of course, possible by which the circuit will continue alternately to sort the three numbers coming in on the inputs i_1, i_2, i_3, first in one way and then in the other; and it is left to the reader also to supply recursion equivalences for this specialization. Also recursion equivalences for the simpler specialization by which the circuit continues indefinitely to sort each new triple of binary numbers into order in the same way, namely the smallest on o_1, the next smallest on o_2, and the largest on o_3.

A final remark is that in our original requirement we might relax the demand that the sorting of the numbers N_1, N_2, N_3 be performed instantaneously, and instead require only that the three numbers be reproduced in the right order on the outputs after the lapse of an interval of time, say two units. The effect of this on (48) is just that $o_1(t), o_2(t), o_3(t)$ are replaced everywhere by $o_1(t+2), o_2(t+2), o_3(t+2)$ respectively, with corresponding changes in (49).

From the form of the synthesis algorithms for Case 1 and Case 2, it follows that in one sense no additional solutions are produced by this relaxation of the requirement, and in particular a synthesis problem could not be rendered solvable by such a relaxation of the synthesis requirement if previously it had no solution. For the only difference between the treatment of $o_1(t), o_2(t), o_3(t)$ and that of $o_1(t+2), o_2(t+2), o_3(t+2)$ that is made in the algorithms is that $o_1(t), o_2(t), o_3(t)$ are not allowed to depend on inputs later than $i_1(t), i_2(t), i_3(t), i_4(t)$, whereas $o_1(t+2), o_2(t+2), o_3(t+2)$ may depend on inputs as late as $i_1(t+2), i_2(t+2), i_3(t+2), i_4(t+2)$. But the relaxation of the synthesis requirement (48) which we just described results in a requirement that in effect includes the condition that $o_1(t+2), o_2(t+2), o_3(t+2)$ shall not depend on inputs later than $i_1(t), i_2(t), i_3(t), i_4(t)$. Therefore, it becomes possible in applying the synthesis algorithm to treat $o_1(t+2), o_2(t+2), o_3(t+2)$ in exactly the same way that we previously treated $o_1(t), o_2(t), o_3(t)$.

As a matter of fact, a general solution for the relaxed synthesis requirement (48) is not far to seek, as we have only to replace $o_1(t), o_2(t), o_3(t)$ in (51) by $o_1(t+2)$, $o_2(t+2), o_3(t+2)$ respectively, and add equivalences specifying arbitrary values for $o_1(1), o_2(1), o_3(1), o_1(0), o_2(0), o_3(0)$. It is in this sense that there are no additional solutions. But there will be additional solutions in another sense, namely, that there are additional transformations that can be applied to the recursion equivalences (51b) [obtained from (51) in the way just described].

No very significant transformation of (51b) suggests itself. But the following may be cited for illustration. The intermediate function r in (51b) is to be left unchanged, and $e_{12}, e_{23}, e_{31}, e_{21}, e_{32}, e_{13}$ are to be replaced by three new intermediate functions defined by the following recursions:

$$
\begin{aligned}
\phi_1(0) &\equiv F\\
\phi_1(t+1) &\equiv e_{23}(t+1) \vee e_{32}(t+1)\overline{\phi}(t)\\
\phi_2(0) &\equiv F\\
\phi_2(t+1) &\equiv e_{31}(t+1) \vee e_{13}(t+1)\overline{\phi}_2(t)\\
\phi_3(0) &\equiv F\\
\phi_3(t+1) &\equiv e_{12}(t+1) \vee e_{21}(t+1)\overline{\phi}_3(t)
\end{aligned}
$$

The solutions are for the e_{jk}'s (which have to be verified by mathematical induction) are as follows:

$$
\begin{aligned}
e_{12}(t) &\equiv \phi_3(t)\phi_3(t+1)\\
e_{23}(t) &\equiv \phi_1(t)\phi_1(t+1)\\
e_{31}(t) &\equiv \phi_2(t)\phi_2(t+1)\\
e_{21}(0) &\equiv F\\
e_{21}(t+1) &\equiv \phi_3(t)\overline{\phi}_3(t+1) \vee \phi_3(t+1)\overline{\phi}_3(t+2)\\
e_{32}(0) &\equiv F\\
e_{32}(t+1) &\equiv \phi_1(t)\overline{\phi}_1(t+1) \vee \phi_1(t+1)\overline{\phi}_1(t+2)\\
e_{13}(0) &\equiv F\\
e_{13}(t+1) &\equiv \phi_2(t)\overline{\phi}_2(t+1) \vee \phi_2(t+1)\overline{\phi}_2(t+2)
\end{aligned}
$$

Historical Note

Use of the "algebra of logic" for the analysis of circuits was first suggested by P. Ehrenfest in a review of the Russian translation of Couturat's *L'Algèbre de la Logique* in 1910. Nothing came of this suggestion for many years, and it seems to have remained wholly unknown outside of Russia. In a historical paper of 1948, S. A. Yanovskaya asserts that details of Ehrenfest's proposal were worked out in 1934–35 by V. I. Shestakoff. But Shestakoff's earliest publications in this direction are two papers which appeared in 1941. And meanwhile the idea of using Boolean algebra, or propositional calculus, or class calculus in the treatment of circuits had been reached independently by others, namely by Akira Nakasima and Masao Hanzawa in a series of papers beginning in December, 1936 (the earliest publication in English is in 1938), and by C. E. Shannon in a paper of 1938. For some time the development of this idea proceeded independently in Russia, in Japan, and in the United States, the three lines of development having had at first little or no influence on one another.

In the present context, we are interested primarily in uses of logic which go beyond this. And the bibliography which follows below is intended to be, as far as possible, a complete list of papers and publications in which mathematical logic beyond propositional calculus or Boolean algebra is applied to the theory of circuits and automata. We shall refer to items in this bibliography by number.

The earliest paper listed, McCulloch and Pitts (1), employs first-order arithmetic, i.e., a logical system having numerical variables (taking the values 0,1,2, ..., and here intended to represent time), quantifiers for such variables, numerals 0, 1, 2, etc., predicates, and notations for such elementary arithmetical operations as addition. The application is biological, viz., to analysis of the behavior of a net of neurons and to the question of the existence of (and of finding) a neural net having some specified behavior. But the authors' hypotheses about the nature of a neuron and the way in which it behaves are such that these questions become entirely similar to corresponding questions about computing circuits. And in later papers by Kleene, a neural net is thought of in abstraction from its biological context and essentially as a kind of computing circuit.

Hartree (3) employs a simpler system which we may call a free-variable (singulary) functional calculus, having singulary predicates and numerical variables, but neither quantifiers nor recursions. There is a "delay operator" in the sense that not only a variable t but also $t+1, t+2$, etc., may appear as argument of a predicate. As his sources Hartree refers, not only to McCulloch and Pitts, but also to J. v. Neumann and A. M. Turing (for unpublished suggestions).

The use of such a free-variable functional calculus also developed independently in Japan. The earliest paper which the writer has seen is Gotô (4). But Gotô refers to papers as early as 1941 and 1942, which apparently contained something in this direction.

"Non-Boolean operators" added to Boolean algebra or to free-variable functional calculus appear in such publications as References 6, 7, 8, and 10. And indeed, it can be shown that "Boolean algebra with operators" is capable of reproducing anything available in such standard logical systems as propositional calculus, first-order functional calculus, first-order arithmetic, and recursive arithmetic. But the writer believes the use of these standard systems, of "logistic" rather than algebraic type, to be heuristically preferable for direct applications of mathematical logic such as that to circuits and automata. And the importance of "algebraic logic" is, rather, metatheoretic.

Bounded quantifiers are employed by McCulloch and Pitts, who take the idea from Carnap's *Logical syntax of language*; but the use of bounded quantifiers and their reduction to recursions goes back to Skolem's original paper on recursive arithmetic in 1923.

The use of recursions in the treatment of circuits is found in Berkeley's booklet (6), and more explicitly in Burks and Wright (9). The idea that a particular circuit or finite automaton can be completely characterized by giving a set of recursion equivalences appears first in Reference 9. However, Burks and Wright do not have a recursion schema to which the recursion equivalences are required to conform, and for the distinction between sets of recursion equivalences which are permissible and those which are not, they depend rather on conditions applied directly to the circuit or automaton (to the "logical net" in their terminology).

An abstract definition of finite automaton appears in Kleene (References 5 and 13), and in somewhat more special form, in Burks and Wright (9). An abstract definition of the transformation which can be performed upon a set of inputs by a finite automaton, something which can be accomplished also by means of either of the recursion schemata A or B of these notes, is contained in Kleene (5, 13) and Burks and Wright (9).

The use of recursive arithmetic in the treatment of circuits, and the characterization of a circuit by a set of recursion equivalences governed by an explicit recursion schema, were introduced by the present writer (see References 11, 14, and 15).

Solution of the decision problem of restricted recursive arithmetic in what is called Case 1 in these notes, is due to Joyce Friedman in References 12 and 18. And the solution of Case 1 of the synthesis problem in Reference 15 was, in part, suggested by her methods.

A modification in the method of solving Case 1 of the synthesis problem, by which the procedure is rendered more manageable and, in particular is greatly shortened, was suggested to the writer in 1958 by Jesse B. Wright and has been incorporated in these notes.

BIBLIOGRAPHY

1. McCulloch, W. S., and Pitts, Walter. *A Logical Calculus of the Ideas Immanent in Nervous Activity.* ***Bulletin of Mathematical Biophysics***, **5** (1943), 115–133.
2. Rashevsky, N. *Mathematical Biophysics.* Chicago, 1948.
3. Hartree, D. R. *Calculating Instruments and Machines.* Urbana, 1949.
4. Gotô, Motinori. *Application of Logical Mathematics to the Theory of Relay Networks.* ***The Japan Science Review***, **1**, No. 3 (1950), 35–42.
5. Kleene, S. C. *Representation of Events in Nerve Nets and Finite Automata.* RAND Memorandum, 1951.
6. Berkeley, E. C. *Circuit Algebra, Introduction.* New York, 1952.
7. Robbins, L. C. *An Analysis by Arithmetical Methods of a Calculating Network with Feedback.* ***Proceedings of the Association for Computing Machinery*** (September, 1952), 61–67.
8. Reed, I. S. *Symbolic Synthesis of Digital Computers.* ***Proceedings of the Association for Computing Machinery*** (September, 1952), 90–94.
9. Burks, Arthur W., and Wright, Jesse B. *Theory of Logical Nets.* ***Proceedings of the I.R.E.***, **41** (1953), 1357–1365.
10. Berkeley, E. C. *The Algebra of States and Events.* ***The Scientific Monthly***, **78** (1954), 232–242.
11. Church, Alonzo. Review of the Foregoing. ***The Journal of Symbolic Logic***, **20** (1955), 286–287.
12. Friedman, Joyce. (Abstract.) ***The Journal of Symbolic Logic***, **21** (1956), 219.
13. Kleene, S. C. *Representation of Events in Nerve Nets and Finite Automata.* ***Automata Studies***, Princeton, 1956, 3–41.
14. Church, Alonzo. (Abstract.) ***American Mathematical Monthly***, **64** (1957), 541.
15. Church, Alonzo. *Application of Recursive Arithmetic to the Problem of Circuit Synthesis.* Summaries of Talks Presented at the Summer Institute of Symbolic Logic in 1957, at Cornell University, **I**, 3–50 (Errata, Volume III, 429).
16. Fitch, F. B. *Representation of Sequential Circuits in Combinatory Logic.* Bell Telephone Laboratories Report. September, 1957.
17. Büchi, J. R., Elgot, C. C., and Wright, J. B. (Abstract.) ***Notices of the American Mathematical Society***, **5**, No. 1 (February, 1958), 98.
18. Friedman, Joyce. *Some Results in Church's Restricted Recursive Arithmetic.* ***The Journal of Symbolic Logic***, **22**, No. 4 (1958), 337–342.
19. Fitch, Frederic B. *Representation of Sequential Circuits in Combinatory Logic.* ***Philosophy of Science***, **25**, No. 4 (1958), 263–279.

LOGIC, ARITHMETIC, AND AUTOMATA*
(1962)

This paper is a summary of recent work in the application of mathematical logic to finite automata, and especially of mathematical logic beyond propositional calculus.

To begin with a sketch of the history of the matter, let us recall that application of the "algebra of logic", i.e., elementary Boolean algebra, to the analysis of switching circuits was first suggested by Ehrenfest [A]. Nothing came of Ehrenfest's remark for many years, and it seems to have remained wholly unknown outside of Russia. Yanovskaya [G] says that details of the suggested application were worked out by Shestakoff in 1934–35. However, Shestakoff's candidate's dissertation, embodying the material, was presented to the University of Moscow in 1938, and the earliest publications by Shestakoff are [D] and [E] in 1941. Meanwhile the same idea had occurred independently to Nakasima and Hanzawa [B] and Shannon [C]. For some time the development of the idea proceeded independently in Russia, in Japan, and in the United States, the three lines of development having had at first no influence on one another.

This use of Boolean algebra is now widely familiar, and therefore requires no elaboration here. It is usually taken to be a Boolean algebra of cardinal number 2 that is used, although the character of the application would more naturally suggest propositional calculus. Use of the Boolean algebra and of propositional calculus are equivalent in a way that is well known. The choice of Boolean algebra is advantageous if algebraic methods and results are to be employed. But otherwise there is a certain artificiality in allowing only equations and inequalities to be asserted. And for further application of mathematical logic, the choice of propositional calculus provides a better basis.

Mathematical logic beyond propositional calculus is first applied to automata theory in the paper of McCulloch and Pitts [16], in which the context is biological. The authors are concerned with analyzing the behavior of a net of neurons and with the question of the existence of, and of finding, a neural net having some specified behavior. But their hypotheses about the behavior and the interaction of neurons are such that these questions become entirely similar to corresponding questions about electronic digital computing circuits. The relevance of the ideas of McCulloch and Pitts to the theory of digital computing circuits was noticed by John von Neumann, and it was evidently this that led him to suggest application of mathematical logic to the latter.

The only published reference to von Neumann's contribution to the matter is Hartree [13], where not only McCulloch and Pitts but also unpublished suggestions of both von Neumann and A. M. Turing are referred to. I am indebted to H. H. Goldstine for calling my attention to the privately circulated paper, von Neumann [18], in which there is some detailed discussion of the relationship between the McCulloch–Pitts neural nets and digital computing networks, but without use of mathematical logic.

The logical system employed by McCulloch and Pitts is first-order arithmetic. Both ordinary and bounded quantifiers are used, but not definition by recursion.

Originally published in ***Proceedings of the International Congress of Mathematicians 1962***, pp. 23–35.

*For support of the work represented by this paper the author is indebted to the National Science Foundation of the United States.

Hartree uses a singulary free-variable functional (or "predicate") calculus, having singulary predicates and numerical variables, but neither quantifiers nor recursions. There is a "delay operator" in the sense that not only a variable t but also $t+1$, $t+2$, etc., may appear as argument of a predicate.

The use of (in effect) such a singulary free-variable functional calculus also developed independently in Japan. The earliest paper that I have seen is Gotô [12]. But it is possible that some of the earlier papers to which Gotô refers may constitute at least a partial exception to the statement made above, that the first application of mathematical logic beyond propositional calculus was by McCulloch and Pitts.

The use of recursions in the treatment of circuits is found in Berkeley [1], and more explicitly in Burks and Wright [5]. The idea that a particular circuit or finite automaton can be completely characterized by giving a set of recursion equations— or in this context, since the functions "defined" by the recursions are propositional functions, we shall speak rather of recursion equivalences—appears first in the paper of Burks and Wright, but these authors do not have a recursion schema to which admissible recursions must conform.

The application of restricted recursive arithmetic to automata was introduced in Church [6] and [7]. This system may be described as obtained from singulary free-variable functional calculus by adding a rule of proof by mathematical induction and a rule of definition by recursion, as given below. It is therefore a specialized form of the recursive arithmetic of Skolem [19].

An abstract definition of finite automaton appears in Kleene [14], a paper of restricted circulation, and in Kleene [15], which is the published form of the same paper. The "logical nets" of Burks and Wright [5] may also be regarded as providing such a definition. But Kleene's definition is more satisfactory, and we shall here use it in the following slightly modified and generalized form.

A *finite automaton* consists of a finite number of *elements*, each of which is capable of a finite number, $\leq n$, of different *states*. Time is measured in discrete *instants*, $t = 0, 1, 2, \ldots$, beginning at an *initial instant* and extending indefinitely. The elements are distinguished as *input elements*, *intermediate elements*, and *output elements*. The states of the input elements, at any instant, constitute the *input state*, at that instant; the states of the output elements, at any instant, similarly constitute the *output state*; and the states of the intermediate and output elements constitute the *internal state*. Except the input elements, whose states are imposed from outside, the state of any element at a given instant is completely determined, in some noncircular way, by (1) the states of certain other elements at the given instant, and (2) the states of certain elements, itself and others, at instants which precede the given instant by at least 1 and by not more than a certain fixed maximum *span* $h+1$.

(More fully, the foregoing statement is a definition of "discrete synchronous deterministic finite automaton with outputs." But the longer phrase is for the sake of distinction from a number of other notions with which we are not here concerned, and let us therefore say simply "finite automaton" or "automaton".)

Though there is a certain loss of structure in doing so, it is quite usual to take $n = 2$, $h = 0$. In this lecture we shall take $n = 2$, in order to identify the two states of an element with the two truth-values, T and F. But we preserve a general value of h, ≥ -1.

It is also possible to abstract from the elements of the automaton and to consider only certain numbers called input states, output states, and internal states, and the way in which the successive output states and internal states are determined by the input states. Namely there are to be a finite number each, ≥ 1, of input states, output states, and internal states; the output state at any instant is to be uniquely determined by the internal state at the same instant; and the internal state at any instant is to be uniquely determined by (1) the input state at that instant and (2) the input states and the internal states at instants preceding the given instant by not more than $h + 1$. In this case we have a *finite transition system*.

In either case, automaton or transition system, the sequence of input states from $t = 0$ to $t = t_0$ is called an *input sequence*, and the sequence of output states from $t = 0$ to $t = t_0$ is called an *output sequence*. The function which, for an input sequence from $t = 0$ to $t = t_0$, as argument, has as value the output state at $t = t_0$ is the *behavior function*. The null sequence is also to be counted as an input sequence, but one for which there is no corresponding output state and hence no value of the behavior function.

Two automata, or two transition systems, are said to be *equivalent* if they have the same behavior function. Evidently, two automata are equivalent if and only if their transition systems are equivalent. In particular, two automata with the same transition system are equivalent — but they do not necessarily have the same *structure*, in an obvious sense of that word.

A class of input sequences is an *event*.

An event is *represented* by an automaton if the set of values of the behavior function for input sequences that belong to the event is disjoint from the set of values of the behavior function for other input sequences.

The foregoing (together with the definition of regularity of an event, to be given below) are the essentials of the terminology of automata theory which we shall need. It should be added that the terminology in the field is not in a very settled state, and our terminology here is somewhat eclectic. We have followed Büchi [2] in adopting a terminology that maintains the distinction between automaton and transition system.

We turn now to statement of the primitive basis of restricted recursive arithmetic. And for expository convenience we adopt a formulation which emphasizes brevity of statement rather than economy of primitives.

The primitive symbols of restricted recursive arithmetic are as follows:

i. Numerical variables $t, x, y, z, \ldots$ (having the natural numbers as their range).

ii. The symbol 0 for the number zero.

iii. The accent $'$ as a notation for successor (of a natural number).

iv. Singulary primitive predicates $i_1, i_2, i_3, \ldots$. (The primitive predicates and other singulary predicates introduced by recursion, are to be understood as denoting propositional functions of natural numbers.)

v. Parentheses ( ) as notation for application of a singulary propositional function to its argument.

vi. The letters T and F to denote the truth-values, truth and falsehood respectively.

vii. Connectives of propositional calculus (any sufficient set), and brackets [ ] for use in connection with them.

Then a *term* consists of the symbol 0, or the symbol 0 followed by any number of accents, or a numerical variable, or a numerical variable followed by any number of accents.

An *elementary formula* consists, either of one of the letters T or F, or of a singulary predicate followed by a term between parentheses.

Further *well-formed formulas* are constructed from the elementary formulas by means of connectives of propositional calculus. (We may sometimes abbreviate "well-formed formula" as "wff"; or we may say simply "formula" if well-formedness is clear from the context.)

The primitive rules of restricted recursive arithmetic are the five following:

1. To assert any formula that is tautologous according to the laws of propositional calculus.
2. From $\mathrm{A} \supset \mathrm{B}$ (as major premiss) and A (as minor premiss) to infer B.
3. From any asserted formula to infer the result of substituting any term for any numerical variable throughout.
4. From $\mathrm{S}(t) \supset \mathrm{S}(t')$ (as major premiss) and $\mathrm{S}(0)$ (as minor premiss) to infer $\mathrm{S}(t)$.
5. A rule of "definition" (or better, introduction) by recursion, allowing us at any stage to introduce new predicates in accordance with the schema of wider restricted recursion as given below.

Rules 2, 3, and 4 are, in order, *the rule of modus ponens, the rule of substitution*, and *the rule of mathematical induction.*

In Rule 4, $\mathrm{S}(t)$ is not necessarily an elementary formula, but is any assertable wff; $\mathrm{S}(t')$ is the result of substituting t' for t throughout $\mathrm{S}(t)$; and $\mathrm{S}(0)$ is the result of substituting 0 for t throughout $\mathrm{S}(t)$.

The schema of "wider restricted recursion" referred to in Rule 5 is as follows:

$$\mathrm{r}_i(j) \equiv \mathrm{P}_{ij}[\mathrm{s}_1(0), \mathrm{s}_2(0), \ldots, \mathrm{s}_k(0), \mathrm{s}_1(1), \mathrm{s}_2(1), \ldots, \mathrm{s}_k(1), \ldots, \ldots, \mathrm{s}_1(j), \mathrm{s}_2(j), \ldots, \mathrm{s}_k(j)]$$

$$\mathrm{r}_i(t+h+1) \equiv \mathrm{Q}_i[\mathrm{s}_1(t), \mathrm{s}_2(t), \ldots, \mathrm{s}_k(t), \mathrm{s}_1(t+1), \mathrm{s}_2(t+1), \ldots, \mathrm{s}_k(t+1), \ldots, \ldots, \mathrm{s}_1(t+h+1), \mathrm{s}_2(t+h+1), \ldots, \mathrm{s}_k(t+h+1), \mathrm{r}_1(t), \mathrm{r}_2(t), \ldots, \mathrm{r}_n(t), \mathrm{r}_1(t+1), \mathrm{r}_2(t+1), \ldots, \mathrm{r}_n(t+1), \ldots, \ldots, \mathrm{r}_1(t+h), \mathrm{r}_2(t+h), \ldots, \mathrm{r}_n(t+h)]$$

Here $i = 1, 2, \ldots, n$ and $j = 0, 1, \ldots, h$, so that the schema consists of $(h+1)n$ equivalences altogether. The new predicates introduced by the recursion are $r_1, r_2, \ldots, r_n$, which may be any n new letters not previously used; and $s_1, s_2, \ldots, s_k$ are given predicates, which may be either primitive predicates or predicates which were introduced in previous recursions. For convenience in indicating the form of the equivalences, $t+1$ has been written in place of what should properly be written as t', $t+2$ in place of t'', and so on; thus, e.g., $t+h$ indicates a letter t with h accents after it. Similarly, 1 has been written in place of $0'$, 2 in place of $0''$, j in place of a 0 with j accents after it. Then finally the device has been adopted of indicating the elementary parts of a wff, other than the elementary parts T and F, by writing them in brackets. Thus, e.g.,

$$\mathrm{P}_{12}[\mathrm{s}_1(0), \mathrm{s}_2(0), \ldots, \mathrm{s}_k(0), \mathrm{s}_1(1), \mathrm{s}_2(1), \ldots, \mathrm{s}_k(1), \mathrm{s}_1(2), \mathrm{s}_2(2), \ldots, \mathrm{s}_k(2)]$$

indicates a wff P_{12} that is built up by means of connectives of propositional calculus from some or all of the elementary formulas $\mathrm{s}_1(0), \mathrm{s}_2(0), \ldots, \mathrm{s}_k(0), \mathrm{s}_1(0'), \mathrm{s}_2(0'), \ldots, \mathrm{s}_k(0'), \mathrm{s}_1(0''), \mathrm{s}_2(0''), \ldots, \mathrm{s}_k(0''), T, F$.

Now in order to represent the action of a given automaton by means of restricted recursive arithmetic, we may use a different predicate to correspond to each element of the automaton—primitive predicates $i_1, i_2, \ldots$ for the input elements, and other letters as convenient for the remaining elements. Normally we shall use $o_1, o_2, \ldots$ for the output elements.

If p is the predicate that corresponds in this way to the element e of the automaton, then $\mathrm{p}(t)$ is to denote the state of e at time t. Thus $\mathrm{p}(0)$ denotes T or F according as the state of the element e at time 0 is T or F, $\mathrm{p}(1)$ denotes T or F according as the state of the element e at time 1 is T or F, and so on. And $\mathrm{p}(t)$ may therefore be understood as asserting that the element e is in state T ("is operated," or the like) at time t.

The action of the automaton is then expressed by writing a set of recursion equivalences in the form of one or more wider restricted recursions, constituting definitions by recursion of the intermediate and output predicates (i.e., the predicates that correspond to intermediate and output elements), in terms of the input predicates $i_1, i_2, \ldots, i_\mu$. Given a particular automaton, and the specified rule by which the state of each element at any instant is determined from the states of that element at preceding instants and the states of other elements at the same and preceding instants, the corresponding set of recursion equivalences that characterizes the automaton is then uniquely determined. And conversely, given a set of recursion equivalences of the kind just described, there is then a unique corresponding automaton.

The one-to-oneness of the correspondence between automata and sets of recursion equivalences has been secured by adopting a restricted form of recursion schema. This does not mean any restriction upon the kinds of elements that are available—instead there is an assumption that elements of arbitrary kind are available, within the general definition of automaton as given above, and if we wish a limitation to elements of special kind (e.g., to "and" elements, "or" elements, "not" elements, and one-unit delay elements), it is necessary to impose further restrictions on the recursion schema. But the restrictions which are involved in the schema of wider restricted recursion, as compared to the recursion schema which would be natural if we did not have the application to automata in mind, are dictated rather by two general principles of causality which may be stated briefly as follows: (I) No direct causal action of the future upon the present. (II) No direct causal action of the remote past upon the present. The question, how far back into the past is to be regarded as the immediate rather than the remote past, has an answer in the span $h+1$, which serves as a measure of this.

Two sets of recursion equivalences of the kind here in question are called *equivalent* if the corresponding automata are equivalent. Evidently this notion may have a direct definition, in terms of the recursion equivalences themselves rather than the corresponding automata, provided that some subset of the predicates, not including any of the input predicates $i_1, i_2, \ldots, i_\mu$ is designated as output predicates. Hence we may state:

THE SIMPLIFICATION PROBLEM. Given a set of recursion equivalences of the kind here in question, to find one equivalent to it, which has a specified form, corresponding to a limitation to elements of special kind from which the automaton is to be constructed, and which is simplest in some suitable sense of simplicity.

The definition of simplicity adopted will evidently depend on the kinds of elements which are regarded as available to construct the automaton, and perhaps on some weighting of them. Thus there is properly not one problem but many. But little has been done with the simplification problem for automata, except in the case of combinational circuits, i.e., automata for which the span $h + 1$ is 0, where only propositional calculus is required for the treatment.

In the case of transition systems it is natural to characterize simplicity as minimizing the number of internal states. This has often been followed. But it is clear that such minimization of the number of internal states will be at most a necessary condition of simplicity of a corresponding automaton, and the relevance to the simplification problem for automata is therefore uncertain.

Both the decision problem and the synthesis problem for automata involve a condition which the automaton is required to satisfy, and which we shall call the *synthesis requirement*. The synthesis requirement must be stated in some formalized language L, and we shall in this context suppose that L is either restricted recursive arithmetic itself or an extension of restricted recursive arithmetic obtained by adjoining additional notations—such as, e.g., the binary predicate =, the sign + of addition, the quantifiers—together with rules for them. Then we may state the two problems as follows:

THE SYNTHESIS PROBLEM. Given as synthesis requirement a wff of L, containing the input predicates $i_1, i_2, \ldots, i_\mu$ and the output predicates $o_1, o_2, \ldots, o_\nu$, to find definitions by recursion of $o_1, o_2, \ldots, o_\nu$, and any number of intermediate predicates, from $i_1, i_2, \ldots, i_\mu$, in the form of one or more wider restricted recursions, such that the synthesis requirement is satisfied independently of what particular propositional functions are denoted by $i_1, i_2, \ldots, i_\mu$. Or if the given synthesis requirement is impossible in the sense that there is no automaton, and no set of recursion equivalences that satisfies it, this fact is to be ascertained.

THE DECISION PROBLEM. Given as synthesis requirement a wff of L, as before, and given a proposed solution of the synthesis problem in the form of a set of recursion equivalences (by wider restricted recursion), to determine whether the synthesis requirement is in fact a consequence of the recursion equivalences.

Both the synthesis problem and the decision problem may be divided into cases, according to the form of the synthesis requirement, or the language L in which it is expressed. Solution of any case of the synthesis problem implies, *a fortiori*, solution of the corresponding case of the decision problem, because a decision problem may always be turned into a synthesis problem by including the given recursion equivalences as a part of the synthesis requirement.

Existing results in regard to the synthesis problem and the decision problem are summarized in the following table. But it should be emphasized that the indicated synthesis algorithms and decision algorithms are by no means always of practicable length.

Case 1 of the decision problem was solved by Joyce Friedman [11], the earliest result. The problem of determining the equivalence of two automata which are both given by means of recursion equivalences in the form of wider restricted recursions can be formulated as a subcase of Case 1 of the decision problem, and is therefore covered by Miss Friedman's solution. Her method is also immediately applicable to solution of Case 2 of the decision problem, as pointed out by McNaughton in his review.

Case 1 of the synthesis problem was first solved by me. A modification of the

	Synthesis requirement expressed in restricted recursive arithmetic	Synthesis problem	Decision problem
Case 1	with one numerical variable only	Solved in [7]	Solved in [11]
Case 2	with any number of numerical variables	Solved in [7]	Solvable by the method of [11]
Case 3	plus quantifiers	Partial solution in [7]	Solved in [7]
Case 4	plus = and <	Incomplete sketch of solution in [7], properly an open problem	Solution is a consequence of [23]
Case 5	plus +, =, and quantifiers	Proved unsolvable [4]	Open

synthesis algorithm, by which it is substantially shortened, was then suggested by J. B. Wright, and this is incorporated in the solution as given in [7]. A subcase of this case of the synthesis problem was also solved independently by Wang [22], without recognizing the relationship to recursive arithmetic.

The unsolvability of Case 5 of the synthesis problem is, of course, in the usual sense, that there is no uniform effective procedure which can be given once for all and by which alone every instance of this case can be solved.

A variety of cases of the synthesis problem and of the decision problem, intermediate between Case 2 and Case 5, immediately suggest themselves as being still open. We may, for example, adjoin the sign of addition, $+$, to restricted recursive arithmetic. Or we may adjoin the sign $+$ and the binary predicate $=$ to restricted recursive arithmetic; or, as would be equivalent to this, we may adjoin the ternary predicate Σ, as sign of the ternary relation of addition. Other possibilities are restricted recursive arithmetic plus $+$ and quantifiers; or plus $=$ and quantifiers; or plus $<$ and quantifiers. Or, as suggested by Büchi [2], we may consider adjoining singulary predicate-variables and quantifiers for them.

If to restricted recursive arithmetic we adjoin a sign of addition and a sign of multiplication and the sign $=$ (of equality) and quantifiers, we obtain a formalized language which includes first-order arithmetic as a part. It follows that the decision problem (*Entscheidungsproblem*), and hence *a fortiori* the synthesis problem, is then unsolvable—cf. [4].

Turning now to matters less closely related to recursive arithmetic, we introduce the definition of "regularity of an event" in the sense of Kleene, employing for this purpose the "regular-expression language" of McNaughton and Yamada [17]. We follow the original notation of McNaughton and Yamada, except that we change their $\sim$ to a $-$, out of reluctance to confuse notions of propositional calculus and of class calculus.

We suppose that there are $m + 1$ input states, of an automaton under consideration, and let them be represented by the numbers $0, 1, \ldots, m$. We use (k) as a notation for the unit class of the input sequence which consists of the single input state k. Or, when convenient, and especially if $m \leq 9$, the parentheses enclosing the numeral k may be omitted as an abbreviation.

The notations $\cup$, $\cap$, $-$, and Λ are employed with their usual set-theoretic meanings, i.e., for union of classes, intersection of classes, complement, and null class respectively. The notation ϕ is employed for the unit class of the null sequence.

If a and b are events, then $a \cdot b$ (abbreviated when convenient as ab) is the event whose members are every sequence that is composed of the concatenation of a sequence belonging to a followed by a sequence belonging to b.

If a is an event, then the event a^* is the smallest class of sequences which contains (as members) the null sequence and all sequences belonging to a and is closed under concatenation.

The regular expressions are: (k), for any suitable numeral k; and Λ; and ϕ,and any expression built up out of these by means of $\cup$, $\cap$, $-$, $\cdot$, and * (with, of course, suitable use of brackets).

The regular expressions are intended as names of events, in a way that is implicit in the foregoing explanation. And we may then define an event as being regular if, under some choice of m, there is a regular expression which denotes it. We have the important result:

KLEENE'S THEOREM. *An event can be represented by a finite automaton if and only if it is regular.*

This was first proved by Kleene in [14] and [15], under a definition of regularity equivalent to the one we have just given.

Since the behavior function of an automaton with one binary (i.e., two-state) output is uniquely characterized by giving the event for which the output is T, McNaughton and Yamada propose their regular-expression language for practical use in specifying the behavior of such an automaton, and recommend the redundancy of its notation—as compared, e.g., to the more economical regular-expression languages of [15] and [8]—as facilitating this. For example, if we assume two input states, 0 and 1, the expression

$$-[(0 \cup 1)^*000(0 \cup 1)^*] \cup (0 \cup 1)^*111 - [(0 \cup 1)000(0 \cup 1)^*]$$

specifies an automaton as having output T at time t if and only if either there have never been three consecutive 0's in the input sequence up to and including time t (this accounts for the part of the expression up to the end of the first square bracket) *or* (this accounts for the next symbol, $\cup$, in the expression) there have been three consecutive 1's in the input sequence since the last three consecutive 0's (this accounts for the remainder of the expression).

Kleene proposed the problem of a characterization of regular events directly in terms of their expression in a formalized language of ordinary kind, such as one of the usual formulations of first- or second-order arithmetic. This problem has since been solved by Trachtenbrot [20], and in a different way by Büchi [2] and Elgot [10].

These results of Büchi and Elgot are closely related to one another, and the substance of them was originally announced in the joint abstract [3]. Specifically, representability of an event by a finite automaton is equivalent to expressibility in any one of the three following formalized languages:

(1) *Weak second-order arithmetic* (Büchi). The primitive symbols are: numerical variables; singulary predicate-variables; the notation ( ) for application of a propositional function to its argument; the symbol 0 for the number 0; the accent $'$ as a notation for successor; T, F, and connectives of the propositional calculus; quantifiers on numerical variables; quantifiers on singulary predicate-variables. Terms are the same as in restricted recursive arithmetic. An elementary formula is T or F or a singulary predicate-variable followed by a term between ( ). Other wffs (closed and open sentences) are built up from the elementary formulas by means of connectives of the propositional calculus and both kinds of quantifiers.

(2) *Singulary second-order arithmetic with* $=$ *and* $<$ (Elgot). Differs from (1) by omission of 0 and inclusion of the two binary predicates $=$ and $<$.

(3) *First-order arithmetic with* $+$, *the binary predicate* $=$, *and a singulary predicate meaning "is a power of* 2" (Büchi).

In (1) and (2) the predicate-variables are to be understood as having *finite* sets of natural numbers as their values. And in the application to automata, in order to represent an input sequence, each free predicate-variable of a formula is made to correspond to a finite initial sequence of states of one input element by way of some suitable coding. The simplest convention, following Elgot, is just to take the value of each predicate-variable to be the finite set of instants at which the state of the corresponding input element is T, and then to consider the shortest input sequence which is thus represented. This results in a restriction to a proper subclass of all input sequences, the *admissible* input sequences, but it may be seen that the restriction is not very essential. Büchi uses (in effect) a method of coding that avoids this restriction to admissible input sequences.

In the case of (3), instead of using predicate-variables, each finite set of natural numbers is represented by a natural number, the finite sum $\Sigma 2^t$, where the summation is over all natural numbers t that belong to the set.

That (1) and (2) are equivalent, and that both have a redundant list of primitives, is as a matter of fact quite obvious. Definitions given by Elgot show that we may without loss omit the primitive symbol 0 from (1), or the primitive symbols $=$ and $<$ from (2).

The two notes by Trachtenbrot, [20] and [21], were unknown to me at the time this lecture was delivered, and were called to my attention by several members of the audience. The dates attached to the earlier note [20] place it as approximately simultaneous with [11] and the privately circulated first edition of [7], hence earlier than the abstract [3], later than [4] and the abstract of [11]—all of which it overlaps to some extent, though more in point of view and method than in specific content.

In [20] Trachtenbrot characterizes the behavior of an automaton with one binary output by means of the following formalized language:

(4) *Singulary second-order functional calculus with bounded quantifiers*. The primitive symbols are: numerical variables; singulary predicate variables, the notation ( ) for application of a propositional function to its argument; connectives of the propositional calculus; quantifiers on singulary predicate-variables; and bounded quantifiers on numerical variables. The elementary formulas consist of a singulary predicate-variable followed by a numerical variable between ( ), and other wffs are built up from the elementary formulas by means of connectives of the propositional calculus and quantifiers.

Trachtenbrot introduces four kinds of bounded quantifiers: $(\mathrm{a})_{\mathrm{a}<\mathrm{b}}$, $(\exists \mathrm{a})_{\mathrm{a}<\mathrm{b}}$, $(\mathrm{a})_{\mathrm{a}\leq \mathrm{b}}$,

$(\exists a)_{a \leq b}$, where a and b are any two distinct numerical variables. We shall say in each case that the variable a is bounded by the variable b. Evidently any one of these four kinds of bounded quantifiers would suffice, as the three others would be definable from it. The ordinary, unbounded quantifiers on numerical variables are not used.

In (4) the predicate-variables are to be understood as having arbitrary singulary propositional functions (or arbitrary sets) of natural numbers as their values. And in the application to automata, each free predicate-variable of a formula corresponds to one input element in the same fashion that was explained above for the predicates $i_1, i_2, \ldots$ used in restricted recursive arithmetic. A *t-formula* is a wff of (4) in which the only free numerical variable is the particular variable t and in which—to state it briefly—every bound numerical variable is bounded, either by t, or by a variable that is bounded by t, or by a variable that is bounded by a variable that is bounded by t, or etc. Then we have:

TRACHTENBROT'S THEOREM. The behavior of any finite automaton with one binary output $o(t)$ can be characterized by an equivalence of the form $o(t) \equiv S(t)$, where $S(t)$ is a t-formula whose free predicate-variables correspond to the inputs of the automaton. And conversely every equivalence of this form characterizes the behavior of some finite automaton with one binary output.

As a characterization of the behavior of finite automata with one binary output, Trachtenbrot's theorem is evidently more direct than the results of Kleene, Büchi, and Elgot. As a characterization of events representable by a finite automaton (i.e., of regular events) it is less direct. But it can be made to yield such a characterization by taking each free predicate-variable f in $S(t)$ to stand for a finite initial sequence of input states of length $t + 1$, namely the sequence $f(0), f(1), \ldots, f(t)$. (This amounts to taking the formalized language in such a way that the free predicate-variables have a different range of values from the bound predicate-variables, a device which Trachtenbrot himself does not adopt explicitly in either [20] or [21].)

In order to obtain a language for practical use in specifying the behavior of automata, it may be desirable to modify Trachtenbrot's language (4) by adding the redundant primitive symbols $=$ and $<$, and then to replace the class of t-formulas by the more quickly recognizable class of formulas in which the only free numerical variable is t and all bound numerical variables are bounded by t. The resulting language seems to offer less immediate facility than the regular-expression language of McNaughton and Yamada, but it may be more suitable for use in cases in which some further application of mathematical logic is to be made.

Büchi, Elgot, and Trachtenbrot are all interested, not only in applications of logic to automata theory, but also in the reverse application by which the consideration of automata is made to yield results that belong to the field of mathematical logic. And the main result announced in [21] falls under the latter head. But [21] also announces, in effect, still another characterization of regularity of events by means of expressibility in an appropriate formalized language. Namely in Büchi's characterization, weak second-order arithmetic may be replace by another formalized language, which has the same primitive symbols and the same class of wffs as weak second-order arithmetic, but in which the values of the bound predicate-variables are taken to be arbitrary sets of natural numbers, while the values of the free predicate-variables are taken to be finite sets of natural numbers.

Finally we return to the synthesis problem, to cite some cases which belong in the present context rather than in that of restricted recursive arithmetic.

Case a. Synthesis requirement given by a regular expression, denoting an event which an automaton with one binary output is required to represent. Problem solved originally by Kleene [14], [15]. Revised treatments by Copi, Elgot, and Wright [8]; McNaughton and Yamada [17]. Elgot [10] solves a somewhat generalized form of this synthesis problem.

Case b. Synthesis requirement given by a t-formula that expresses the (one binary) output in terms of the inputs. Solution briefly sketched by Trachtenbrot [20].

Case c. Synthesis requirement given by a wff of the form $(\exists s_1)(\exists s_2)\dots(\exists s_n)(t)M$, belonging to a language which is like weak second-order arithmetic except that the predicate-variables are reinterpreted as ranging over all singulary propositional functions (all sets) of natural numbers. Here $s_1, s_2, \dots, s_n$ are predicate-variables, t is an individual variable, and the quantifier-free matrix M contains no individual variables except t. The free predicate-variables are interpreted, some as corresponding to inputs and some as corresponding to outputs, so that the synthesis requirement expressed is a condition relating the inputs and the outputs. Solved by Elgot [10].

It is clear that, as regards the form of the synthesis requirement, Case c would fit into the table that was given above in connection with restricted recursive arithmetic. But Elgot's method is different, not making use of restricted recursive arithmetic (and in fact leading in the first instance to a transition system rather than an automaton).

Case d. Synthesis requirement given by an open sentence of first-order arithmetic with $'$, $+$, and $=$, and with free singulary predicate-variables, interpreted as outputs, and as ranging over the ultimately periodic propositional functions of natural numbers. Decision problem solved, but synthesis problem unsolvable. See Elgot [10].

Case e. Synthesis requirement given by an arbitrary wff of the same formalized language that is described under Case c. The decision problem is solved by Büchi [23]. The general synthesis problem for this case is still open (as of December 1962), though unpublished results stronger than that of Case c are cited in a letter from Büchi. Evidently Case e would fit into the table that was given above, as intermediate between Cases 3–4 and Case 5.

REFERENCES

Application of propositional calculus

[A] EHRENFEST, P., Review of the Russian translation of Louis Couturat's *L'algèbre de la logique*. Журнал Русскаго Физико-химическаго Общества, section of physics, 42 (1910), part 2, 382–387.

[B] NAKASIMA, AKIRA, & HANZAWA, MASAO, A series of papers published in Japan, some by joint authors, and some by Nakasima alone. The two earliest papers are in Japanese, by Nakasima and Hanzawa, and appeared in *J. Inst. Elect. Comm. Engrs. Japan*, no. 165 (Dec. 1936) and no. 167 (Feb. 1937). I have not seen these papers, but a condensed English translation of them which appeared in *Nippon Elect. Comm. Eng.*, no. 9 (Feb. 1938), 32–39. Later papers appeared in the latter periodical, no. 10 (Apr. 1938), 178–179; no. 12 (Sept. 1938), 310–314; no. 13 (Nov. 1938), 405–412; no. 14 (Dec. 1938), 459–466; no. 24 (Apr. 1941), 203–210. These also are English translations of papers in Japanese which preceded them.

[C] SHANNON, C. E., A symbolic analysis of relay and switching circuits. *Trans. Amer. Inst. Elect. Engrs.*, 57 (1938), 713–723.

[D] SHESTAKOFF, V. I., Алгебра двухполюсных схем, построенных исклычнтелно из двухполысников (Алгебра А-схем). Автоматика и Телемеханика, no. 2 (1941), 15–24.

[E] SHESTAKOFF, V. I., A paper of the same title as [D]. Журнал Технической Физики, 11, 6 (1941), 532–549.

[F] SHESTAKOFF, V. I., Representation of characteristic functions of propositions by expressions realizable by relay-contact circuits. (Russian with English summary.) *Bull. Acad. Sci. URSS* (sér. math.), 10 (1946), 529–554.

[G] YANOVSKAYA, S. A., Основания математики и математическая логика. Математика в СССР за Тридцат' Лет 1917–1947, pp. 9–50. Moscow and Leningrad, 1948.

Application of mathematical logic beyond propositional calculus

[1]. BERKELEY, E. C., *Circuit Algebra—Introduction.* New York, 1952.

[2]. BÜCHI, J. R., Weak second-order arithmetic and finite automata. *Z. Math. Logik Grundlagen Math.*, 6 (1960), 66–92.

[3]. BÜCHI, J. R., & ELGOT, C. C., Decision problems of weak second order arithmetics and finite automata, Part I. Abstract in *Notices Amer. Math. Soc.*, 5,7 (Dec. 1958), 834.

[4]. BÜCHI, J. R., ELGOT, C. C., & WRIGHT, J. B., The non-existence of certain algorithms of finite automata theory. Abstract in *Notices Amer. Math. Soc.*, 5, no. 1 (Feb. 1958), 98.

[5]. BURKS, A. W., & WRIGHT, J. B., Theory of logical nets. *Proc. I.R.E.*, 41 (1953), 1357–1365.

[6]. CHURCH, ALONZO, Review of E. C. Berkeley's "The algebra of states and events". *Journal of Symbolic Logic*, 20 (1956), 286–287.

[7]. CHURCH, ALONZO, Application of recursive arithmetic to the problem of circuit synthesis. *Summaries of Talks Presented at the Summer Institute for Symbolic Logic, Cornell University 1957*, 2nd ed., 3–50. Princeton, 1960.

[8]. COPI, I. M., ELGOT, C. C., & WRIGHT, J. B., Realization of events by logical nets. *J. Ass. Comput. Mach.*, 5 (1958), 181–196.

[9]. Elgot, C. C., Decision problems of weak second order arithmetics and finite automata, Part II. Abstract in *Notices Amer. Math. Soc.*, 6, 1 (Feb. 1958), 48.

[10]. Elgot, C. C., Decision problems of finite automata design and related arithmetics. *Trans. Amer. Math. Soc.*, 98 (1961), 21–51. Errata, ibid., 103 (1962), 558–559.

[11]. Friedman, Joyce, Some results in Church's restricted recursive arithmetic. *Journal of Symbolic Logic*, 22, 4 (for 1957, pub. 1958), 337–342. Reviewed by Robert McNaughton in *Journal of Symbolic Logic*, 24, 3 (for 1959, pub. 1960), 241–242. Abstract in *Journal of Symbolic Logic*, 21 (1956), 219.

[12]. Gotô, Motinori, Application of logical mathematics to the theory of relay networks. *Japan Science Review*, 1, 3 (1950), 35–42.

[13]. Hartree, D. R., Calculating Instruments and Machines. Urbana, 1940.

[14]. Kleene, S. C., Representation of Events in Nerve Nets and Finite Automata. RAND memorandum, Dec. 1951.

[15]. Kleene, S. C., Representation of Events in Nerve Nets and Finite Automata. *Automata Studies*, 3–41. Princeton, 1956.

[16]. McCulloch, W. S., & Pitts, Walter, A logical calculus of the ideas immanent in nervous activity. *Bull. Math. Biophys.*, 5 (1943), 115–133.

[17]. McNaughton, R., & Yamada, H., Regular expressions and state graphs for automata. *I.R.E. Trans.*, EC-9, no. 1 (1960), 39–47.

[18]. von Neumann, John, First Draft of a Report on the EDVAC. Moore School of Electrical Engineering, University of Pennsylvania, June 30, 1945.

[19]. Skolem, Thoralf, *Begründung der elementaren Arithmetik durch die rekurrierende Denkweise ohne Anwendung scheinbarer Veränderlichen mit unendlichem Ausdehnungsbereich.* Skrifter utgit av Videnskapsselskapet i Kristiania, I. matematisk-naturvidenskabelig klasse, no. 6 (1923).

[20]. Trachtenbrot, B. A., Синтез логических сетей, операторы которых описаны средствами исчисления одноместных преднкатов. Доклады Академии Наук СССР, 118, 4 (1958), 646–649. Reviewed by J. C. Shepherdson in *Zbl. Math.*, 84 (1960), 11–12.

[21]. Trachtenbrot, B. A., Некоторые построения в логике одноместных предикатов. Доклады Академии Наук СССР, 138, 2 (1961), pp. 320–321. English translation in *Soviet Math.*, 2,3 (1961), 623–625.

[22]. Wang, Hao, Circuit synthesis by solving sequential Boolean equations. *Z. Math. Logik Grundlagen Math.*, 5 (1959), 291–322. Reviewed by C. C. Elgot in *Journal of Symbolic Logic*, 25, 4 (for 1960, pub. 1962), 373–375.

(Added in proof):

[23]. Büchi, J. R., On a decision method in restricted second-order arithmetic. ***Logic, Methodology and Philosophy of Science, Proceedings of the 1960 International Congress***, Stanford 1962, 1–11.

MATHEMATICS AND LOGIC
(1962)

As the title indicates, this paper concerns, not a contemporary trend in the sense of something that is presently in process, but an old question in regard to which developments have come to a conclusion, or at least a pause. It is not true that opinions now agree. But the cessation of active development means that the matter can be summed up and even that some attempt may be made at adjudication.

There are two senses in which it has been maintained that logic is prior to mathematics. One of these, which I shall call the strong sense, is the doctrine which has come to be known as logicism. And the other, the weak sense, is the sense in which the standard postulational or axiomatic view of the nature of mathematics requires the priority of logic as being the means by which the consequences of a particular system of mathematical postulates are determined.

To take the strong sense first, the logicistic thesis is that logic and mathematics are related, not as two different subjects, but as earlier and later parts of the same subject, and indeed in such a way that mathematics can be obtained entirely from pure logic without the introduction of additional primitives or additional assumptions.

To make this definite, we first require an answer of some sort to the question, What is meant by logic? Certainly, not merely traditional logic, i.e., the logic of Aristotle plus further developments in the same immediate context—else the logicistic thesis is obviously false. If we are to take the logicists seriously, we must concede them a broad sense of the term, logic.

As a descriptive account rather than a definition, and assuming the notion of deductive reasoning as already known, from experience with particular instances of it, let us say that logic consists in a theory of deductive reasoning, plus whatever is required in object language or metalanguage for the adequacy, generality, and simplicity of the theory.

That logic does not therefore consist merely in a metatheory of some object language arises in the following way. It is found that ordinary theories, and perhaps any satisfactory theory, of deductive reasoning in the form of a metatheory will lead to analytic sentences in the object language, i.e., to sentences which, on the theory in question, are consequences of any arbitrary set of hypotheses, or it may be, of any arbitrary non-empty set of hypotheses.[1] These analytic sentences lead in turn to certain generalizations; e.g., the infinitely many analytic sentences $A \vee \sim A$, where A ranges over all sentences of the object language, lead to the generalization $p \vee \sim p$, or more explicitly, $(p) \centerdot p \vee \sim p$; and in similar fashion $(F)(y) \centerdot (x)F(x) \supset F(y)$ may arise by generalization from infinitely many analytic sentences of the appropriate form. These generalizations are common to many object languages on the basis of what is seen to be in some sense the same theory of deductive reasoning for the different languages.

Originally published in ***Logic, Methodology and Philosophy of Science, Proceedings of the 1960 International Congress***, edited by Ernest Nagel, Patrick Suppes, and Alfred Tarski, Stanford University Press, Stanford 1962, pp. 181–186. Reprinted with added bibliography in ***Contemporary Philosophy***, edited by R. Klibansky, La Nuova Italia Editrice, Florence 1968, pp. 232–240.

[1] We may understand consequences either in the sense of provability or in any other, less effective, sense which the particular metatheory may provide, the distinction being not important for the immediate purpose.

Hence they are considered to belong to logic, as not only is natural but has long been the standard terminology.

Against the suggestion, which is sometimes made from a nominalistic motivation, to avoid or omit these generalizations, it must be said that to have, e.g., all of the special cases $A \vee \sim A$ and yet not allow the general law $(p) \centerdot p \vee \sim p$ seems to be contrary to the spirit of generality in mathematics, which I would extend to logic as the most fundamental branch of mathematics. Indeed such a situation would be much as if one had in arithmetic $2 + 3 = 3 + 2$, $4 + 5 = 5 + 4$, and all other particular cases of the commutative law of addition, yet refused to accept or formulate a general law, $(x)(y) \centerdot x + y = y + x$.

For our present purpose it is convenient to regard a language as being given when we have a set of primitive symbols, and formation rules, and in some sense which it is not here necessary to make definite, meanings for the expressions (wffs) of the language. Thus let me speak of the rules of inference, not as constitutive of the language, but rather as belonging to a theory of deductive reasoning for the language—so that there may be different sets of rules of inference for the same language.[2]

There is then another consideration which leads us to count certain generalizations of analytic sentences as belonging to logic. This arises from the idea that logic cannot be taken as exhausted by what is special to particular languages.[3] Abstractly the laws of logic are not dependent on whether, for example, one uses C or $[\quad \supset \quad]$ as a sign of implication, or whether one writes $P(x)$ or xP, the predicate before or after the subject. Thus the metatheoretic principle that from $[B \supset C]$ and $[A \supset B]$ may be inferred $[A \supset C]$ is to be regarded as special to a particular language; and the corresponding general principle of logic is rather something like $(p)(q)(r) \centerdot q \supset r \supset \centerdot p \supset q \supset \centerdot p \supset r$. In this connection the Fregean distinction of two kinds of meaning, the sense and the denotation, has the merit that the abstract general laws of logic may satisfactorily be formulated as extensional, since an intensional aspect of the meaning is still present in the sense. But from the point of view of a Russellian theory of meaning, which denies the sense in favor of the denotation, the variables p, q, r, F, etc., must be taken as intensional variables, having propositions, properties, etc., as their values. From this point of view, if one is inclined to minimize the domain allowed to logic, one might even take the extreme position that such assertions as $(p) \centerdot p \vee \sim p$ and $(F)(y) \centerdot (x)F(x) \supset F(y)$, where p and F are extensional variables, having, respectively, truth-values and classes as their range, though they are analytic, do not belong to logic.[4]

Now from the point of view of such an account of what is meant by logic one may see the logicistic attempt to reduce mathematics to logic in its most favorable light. For it could be said that logic is at any rate one prerequisite for mathematics, since deductive reasoning plays so prominent a role in mathematics. This is the weak sense of the priority of logic, which was referred to above. And if we accept it, we may then

[2]The term, language, is of course often used by logicians in such a sense that a particular language is not determined until inference (or transformation) rules have been determined in addition to the other things named above. There is actually a need for two terms.

[3]Else indeed, since Greek and Latin are different languages, the logic of Aristotle and the logic of Boethius could not be said to have even in part the same content or to treat the same topic.

[4]I do not advocate this extreme position, and Russell did not do so. But it is worth noticing that it provides a very satisfactory motivation for Russell's avoidance of extensionality by means of his contextual definition of class abstracts which is mentioned below.

well say that mathematics would be founded on a minimum basis if it could be reduced to nothing but pure logic.

Historically logicists have not been as explicit as this but have seemed just to assume that the laws of logic have so ultimate and fundamental a character that the reduction to logic is clearly desirable if possible, and that the true nature of mathematics would be revealed by such a reduction.

Now having the case for the logicistic thesis before us, let us proceed immediately to the objections. There are three important ones to be cited, and possibly a fourth.

The first objection is simply that the attempted reduction of mathematics to logic has not been more than half successful. The history is well known. Frege — who held the logicistic thesis only for arithmetic as opposed to geometry — did indeed reduce arithmetic to his own formulation of pure logic. Frege's logic includes the extensional as well as the intensional. This is a feature which I do not regard as being a defect in itself. But it is precisely the extensional part of Frege's logic which, despite the immediate appeal of its principles as intuitively sound, leads to its inconsistency. And the historical fact is that Frege, confronted with the Russell paradox, gave up in despair.

To avoid the paradoxes, Russell, and later Whitehead and Russell in *Principia Mathematica*, introduced the theory of types, and this in turn compelled use of the axiom of infinity. The term "axiom of infinity" is due, not to Russell, but to C. J. Keyser. And in an early paper Russell had argued against Keyser that the axiom of infinity is not a special assumption which is required by mathematics above and beyond the laws of logic, because the axiom (or supposed axiom) can be proved on logical grounds alone. Thus when Russell later reversed himself and adopted an axiom of infinity, it is almost his own admission, it might be said, that he thus went beyond pure logic. Indeed the axiom of infinity might be described as half-logical in character, since it can be stated in the same vocabulary that is used to state the laws of pure logic, but is not analytic according to any known theory of deductive reasoning that I believe can be accepted as adequately and naturally representative of existing standard practice in mathematical reasoning. And though it is known that elementary arithmetic can be obtained without axiom of infinity, it does not appear that mathematics as a whole can dispense with such an axiom if otherwise based only on a standard and naturally acceptable formulation of pure logic.

An incidental point is that Russell showed the derivability of mathematics from intensional logic, plus axiom of infinity.[5] Or more correctly, Russell's logic is neither extensional nor intensional, but neutral. He assumes no principle of extensionality for his propositional and functional variables. Equally he assumes no different and stricter principle of individuation which would require an intensional range for any of these variables. He simply has no need of such an hypothesis, as he says. In accomplishing this, the role of Russell's elimination of classes by the device of contextual definition is perhaps not as widely appreciated as it should be. And though there may be objections to it which I would urge in a different context, they concern such matters as adequacy for the logic of indirect discourse, and certainly not the adequacy for mathematics.

To return, however, to objections against the logistic thesis — the second one concerns the question of founding particular branches of mathematics on a minimum

[5]For a strictly historical account, the axiom of reducibility should also be added. But it is now usual to avoid this by the introduction of simple type theory, and it may therefore be ignored for our present purpose.

basis. If we wish to found the whole of mathematics — or to evade the difficulties connected with the Gödel incompleteness theorem and related questions, let us say rather the whole of presently extant mathematics — it does indeed appear that pure logic in an appropriate formulation,[6] plus axiom of infinity, affords at least an approximation to a minimum basis. But for most particular branches of mathematics, and especially for elementary arithmetic, this is not the case. For the Frege-Russell method requires predicate variables of higher order to obtain even elementary arithmetic. Such higher-order variables may well be regarded as belonging to the vocabulary of pure logic, since they arise naturally from even a first-order language if we first generalize from the analytic sentences in the way that has been described, then extend the language by adding the notations required for these generalizations, then formulate an extended theory of deductive reasoning to apply to the extended language, then generalize again from the analytic sentences, and so on. However, the higher-order predicate (or functional) variables, together with comprehension principles which are required for them, mean in the presence of an axiom of infinity that even the non-denumerably infinite has been admitted. A more satisfactorily economical basis for elementary arithmetic is provided either by one of the standard formulations of first-order arithmetic employing primitive notations special to arithmetic and special arithmetic postulates, or by a weak set theory which is adequate for the treatments of finite sets but omits all the standard set-theoretic axioms that are not needed for this purpose (including the axiom of infinity).

According to a third objection, it is not logic in the sense of a theory of deductive reasoning that is required for mathematics, but only just various concrete instances of deductive reasoning. This third objection might be taken, like the second, as directly against the desirability of deriving mathematics from purely logical primitives, rather than against the success of the logicists in doing so. However, mathematical intuitionists in particular have urged such an objection in a stronger form, maintaining that mathematics is prior to logic in a sense which would make the derivation of mathematics from logic unsound on the ground of circularity. In fact intuitionists go far beyond denying that the strong logic required by Frege and Russell is an acceptable foundation for mathematics, and intend an objection as much against the weak sense of the priority of logic as against the strong sense.

Now it is true that any foundation of either mathematics or logic is in a certain fashion circular. That is, there always remain presuppositions which must be accepted on faith or intuition without themselves being founded. We may seek to minimize the presuppositions, but we cannot do away with them. Whether the minimum of presupposition that remains after reduction is to be called mathematics, or logic, or both, or neither, becomes a question of terminology. But I would remark, in criticism of the intuitionists, that it is at any rate much less than the total content of a mathematics library, or even of a few good mathematics books.

It is not clear to me just how far the intuitionistic denial of priority to logic applies against the use of the logistic method[7] in founding particular theories. For intuitionists, Heyting in particular, have used the logistic method and I think do not dispute its value.

[6]This does not hold if we include in the logic assumptions that require intensional values for the propositional and functional variables. However, as already noted, Russell, and Whitehead and Russell, do not do this, but regard the assumptions of intensionality and of extensionality as alike superfluous.

[7]I.e., the method of metatheoretic statement of the formation and transformation rules of a language.

In fact it is by the logistic method that those who do not share the intuitionistic intuition are nevertheless able to see, at least in the case of particular mathematical and logical theories, what it is to which the intuitionistic intuition would lead. The crucial point in regard to the intuitionists' use of the logistic method is not whether they proceed by direct application of intuition before reaching the logistic formulation of a theory, but whether having once reached the logistic formulation they then regard it as definitive of the theory in the sense that a change in it or addition to it has to be regarded as a change to a different theory. Since it appears that they do not treat logistic formulation as characterizing the theory in this way, or even as adequately representing it, we thus have the situation that no theory is ever fully intersubjectively determinate. If intuitionists are to be so understood, it would seem to me that this is the weakest point of their doctrine, and by no means essential to the rest of it. I believe that there is much more to be said in favor of their objections against particular classical theories, and against conventionalistic and "formalistic" aspects of the axiomatic method as classically understood.

From the point of view of the intuitionistic rejection of certain parts of the logic of classical mathematics, one might add as a fourth objection to the program of logicism that it is simply rendered impossible. And indeed this would seem to be the case, even if an axiom of infinity is conceded. Against the contention on behalf of logicism that logic in the broad sense is a natural extension and an essential completion of even the most rudimentary theory of deductive reasoning, there is the intuitionist contention that the extension has gone too far and in the effort to achieve generality has passed beyond tenability. The objection will carry conviction only for intuitionists. But it agrees with the second objection in emphasizing the strength of the logic required by logicism.

Classically, even if we accept one or more of the first three objections and regard them as cogent, it does not follow that logicism is barren of fruit. Two important things remain. One of these is the reduction of mathematical vocabulary to a surprisingly brief list of primitives, all belonging to the vocabulary of pure logic. The other is the basing of all existing mathematics on one comparatively simple unified system of axioms and rules of inference. Such a reduction of the primitive basis of mathematics might indeed be made differently if one were less exclusively occupied with the logicistic doctrine, but it was nevertheless in the first instance an accomplishment of the logicists.

SELECTED HISTORICAL BIBLIOGRAPHY

FREGE, Gottlob, *Begriffschrift, Eine der arithmetischen nachgebildete Formelsprache des reinen Denkens*, Halle 1879; second edition (edited by I. Angelelli), Hildesheim 1964.

FREGE, Gottlob, *Die Grundlagen der Arithmetik, Eine logisch-mathematische Untersuchung über den Begriff der Zahl*, Breslau 1884; reprinted, Breslau 1934; Italian translation (with notes by L. Geymonat), Torino 1948; reprinted with parallel English translation, Oxford & New York 1950, revised edition 1953; reprinted, Hildesheim 1961.

FREGE, Gottlob, *Grundgesetze der Arithmetik, begriffsschriftlich abgeleitet*, Two volumes, Jena 1893 and 1903; reprinted in: one volume, Hildesheim 1962; English translation (*The Basic Laws of Arithmetic*) of part of volume 1 and the appendix of volume 2 (with an introduction by Montgomery Furth), Berkeley & Los Angeles 1964.

RUSSELL, Bertrand, *The Principles of Mathematics*, Cambridge (England) 1903; second edition, with a new introduction, London 1937 & New York 1938, reprinted London 1950; Spanish translation, Buenos Aires 1948; Italian translation, Milano 1951.

KEYSER, C. J., *The axiom of infinity: A new presupposition of thought*, ***The Hibbert Journal***, vol. 2, (1903/4), pp. 532–552; reprinted in *The Human Worth of Rigorous Thinking, Essays and Addresses*, New York 1916, and second edition, New York 1925.

RUSSELL, Bertrand, *The axiom of infinity*, ***The Hibbert Journal***, vol. 2, (1903/4), pp. 809–812.

KEYSER, C. J., *The axiom of infinity*, ***The Hibbert Journal***, vol. 3 (1904/5), pp. 380–383.

COUTURAT, Louis, *Les Principes des Mathématiques*, Paris 1905.

RUSSELL, Bertrand, *Sur la relation des mathématiques à la logistique*, ***Revue de Métaphysique et de Morale***, vol. 13 (1905), pp. 906–916.

WHITEHEAD, A. N., *Note*, ***Revue de Métaphysique et de Morale***, vol. 13, (1905), pp. 916–917.

POINCARÉ, Henri, *Les mathématiques et la logique*, ***Revue de Métaphysique et de Morale***, vol. 14 (1905), pp. 815–835, and vol. 15 (1906), pp. 17–34, 294–317; reprinted in revised form in Poincaré's *Science et Méthode*, Paris 1908 (and translations of the revised article in Russian, German, and English translations of this book).

COUTURAT, Louis, *Pour la logistique (résponse à M. Poincaré)*, ***Revue de Métaphysique et de Morale*** vol. 14, 208–250 (1906); English translation (with an introductory note by P. E. B. Jourdain), ***The Monist*** vol. 22 (1912), pp. 481–523.

RUSSELL, Bertrand, *Les paradoxes de la logique*, ***Revue de Métaphysique et de Morale***, vol. 14 (1906), pp. 627–650.

POINCARÉ, Henri, *À propos de la logistique*, ***Revue de Métaphysique et de Morale***, vol. 14 (1906), pp. 866–868.

RUSSELL, Bertrand, *Mathematical logic as based on the theory of types*, ***American Journal of Mathematics*** vol. 30 (1908), pp. 222–262; reprinted in: *Logic and Knowledge, Essays 1901–1950* (edited by R. C. Marsh), London 1956, pp. 59–102.

KEYSER, C. J., *The thesis of modern logistic*, ***Science***, vol. 30 (1909), pp. 949–963.

RUSSELL, Bertrand, *La théorie des types logique*, ***Revue de Métaphysique et de Morale***, vol. 18 (1910), pp. 263–301.

WHITEHEAD, A. N., & RUSSELL, Bertrand, *Principia Mathematica*, Three volumes, Cambridge (England) 1910–1913; second edition (with added introduction and appendices by Russell), Cambridge (England) 1925–1927; paperbound reprint of the greater part of volume 1 (second edition), London and New York 1962.

WHITEHEAD, A. N., *Mathematics* in: ***The Encyclopaedia Britannica***, eleventh edition, vol. 17, Cambridge (England) and New York, 1911, pp. 878–883; reprinted in: *Essays in Science and Philosophy*, New York 1947, pp. 269–288, also London and New York 1948, pp. 195–208; German translation in *Philosophie der Mathematik, Vorträge und Essays*, Wien 1949, pp. 125–150.

RUSSELL, Bertrand, *L'importance philosophique de la logistique*, ***Revue de Métaphysique et de Morale***, vol. 19 (1911), 281–291.

POINCARÉ, Henri, *Derniéres Pensées*, Paris 1913; German translation, Leipzig 1913.

PEANO, Giuseppe, Review of *Principia Mathematica*, vols. 1 and 2, ***Bolettino di Bibliografia e Storia delle Scienze Matematiche***, vol. 15 (1913), pp. 47–53.

RUSSELL, Bertrand, *Introduction to Mathematical Philosophy*, London 1919; second edition, London 1920; German translation, München 1923, and second edition München 1930; French translation, Paris 1928; Spanish translation, Buenos Aires 1945; Italian translation, Milano 1947.

RAMSEY, F. P., *The Foundations of Mathematics and Other Logical Essays*, New York and London 1931; reissued, London and New York 1950.

HEYTING, Arend, *Mathematische Grundlagenforschung, Intuitionismus, Beweistheorie*, in: ***Ergebnisse der Mathematik und ihrer Grenzgebiete***, vol. 3, no. 4, Berlin 1934; French translation, with some additions, Paris and Louvain 1955.

QUINE, W. V., *Mathematical Logic*, New York 1940; second printing, Cambridge (Mass.) 1947; revised edition, Cambridge (Mass.) 1951.

QUINE, W. V., *Whitehead and the rise of modern logic*, in: *The Philosophy of Alfred North Whitehead* (edited by P. A. Schilpp), Evanston (Ill.) and Chicago 1941, pp. 127–163.

HEYTING, Arend, *Formal logic and mathematics*, ***Synthese***, vol. 6 (1947/8), pp. 275–282.

RUSSELL, Bertrand, *Whitehead and Principia Mathematica*, ***Mind***, vol. 57 (1948), pp. 137–138.

HEYTING, Arend, *Logique et intuitionnisme*, in: ***Applications Scientifiques de la Logique Mathématique, Actes du Deuxième Colloque International de Logique Mathématique***, Paris and Louvain 1954, pp. 75–83.

HEYTING, Arend, *Intuitionism, An Introduction*, Amsterdam 1956.

AN INDEPENDENCE QUESTION IN RECURSIVE ARITHMETIC*
(1965)

This is a sequel to my paper, "Binary recursive arithmetic," which appeared in *Journal de Mathématiques Pures et Appliquées*, vol. 36 (1957), pp. 39–55. It will be arranged accordingly, and acquaintance with the earlier paper will be presupposed. The theorems and metatheorems here will be numbered consecutively with those of the earlier paper, and all theorems and metatheorems will be referred to simply by number.

The question of the independence of $3b$, which was left open in the earlier paper, is still unanswered. But a proof will here be given of the result (suggested in the earlier paper) that any particular case of $3b$ is a theorem if obtained by substituting for g a binary functor **I** containing no variables.

This result is metatheorem XXV below. It was first proved by Dana Scott (unpublished). The present proof is by a different method than Scott's, and much shorter, but is nevertheless indebted to his in important respects.

For convenience let us call a binary functor **I** *substitutive* if the result of substituting **I** for the variable g in $3b$ is a theorem, i.e., if

$$y = z \supset \mathbf{I}(x, y) = \mathbf{I}(x, z).$$

(This terminology conflicts with other meanings that have been given to the word "substitutive"; but as these other meanings will not be needed here, perhaps it may be used for the purpose of the present paper without confusion.)

Then we have to show that every binary functor **I** containing no variables is substitutive. Or more generally, we have to investigate the question, what binary functors are substitutive.

XII. If **F**, **G**, **H** are substitutive,

$$B\mathbf{FGH}(x, y) = \mathbf{F}\big(\mathbf{G}(x, y), \mathbf{H}(x, y)\big).$$

Proof: This is a consequence of 123, as already noticed.

Definition. — $S\mathbf{FG} \longrightarrow C\,Row\,R\mathbf{FG}vos.$

XIII. If **G** is substitutive, $S\mathbf{FG}(0) = \mathbf{F}(0)$, and

$$\vdash S\mathbf{FG}(s(x)) = \mathbf{G}(x, S\mathbf{FG}(x)).$$

Proof: $\vdash S\mathbf{FG}(x) = R\mathbf{FG}v(o(x), x)$, by 17;
$= R\mathbf{FG}v(0, x)$, by 5.
$\vdash S\mathbf{FG}(x) = R\mathbf{FG}v(0, 0)$, by substitution;
$= \mathbf{F}(0)$, by $9a$.

Originally published in ***Colloquium on the Foundations of Mathematics, Mathematical Machines and Their Applications, Tihany, 11–15 September 1962***, edited by László Kalmár, Akadémiai Kiadó, Budapest 1965, pp. 21–26.

*For support during the preparation of this paper the author is indebted to a grant from the US National Science Foundation.

$$\begin{aligned}\vdash SFG(s(x)) &= RFGv(0, s(x)), \text{ by substitution;}\\ &= G\big(v(0,x), RFGv(0,x)\big), \text{ by } 9b;\\ &= G\big(x, RFGv(0,x)\big), \text{ by } 7;\\ &= G\big(x, SFG(x)\big).\end{aligned}$$

Definition. — $C_1\mathbf{FG} \to CR\mathbf{F}vv\mathbf{G}s$. (Cf. 14.)

Definition. — $A\mathbf{FG} \to RC_1C\mathbf{F}C_1\mathbf{G}m_1m_2svv$.

124. $Afg(x, g) = f\,(g(m_1(s(x))), m_2(s(x)))$.

Proof: $$\begin{aligned}\vdash Afg(x,y) &= C_1CfC_1gm_1m_2s(x), \text{ by } 13;\\ &= CfC_1gm_1m_2\big(s(x)\big), \text{ by } 14;\\ &= f\,\big(C_1gm_1(s(x)), m_2(s(x))\big), \text{ by } 8;\\ &= f\,\big(g(m_1(s(x))), m_2(s(x))\big). \text{ by } 14.\end{aligned}$$

XIV. $A\mathbf{FG}$ is substitutive.

I.e., more fully, if **F** is an binary functor and **G** is any singulary functor,

$$\vdash y = z \supset A\mathbf{FG}(x, y) = A\mathbf{FG}(x, z).$$

The proof is immediate by 124 (and 11, 12, and propositional calculus).

Definition. — $D\mathbf{FG} \to B\mathbf{F}R\mathbf{G}vvv$.

XV. If **F** is substitutive, $\vdash D\mathbf{FG}(x, y) = \mathbf{F}(\mathbf{G}(x), y)$.

Proof: $$\begin{aligned}\vdash D\mathbf{FG}(x,y) &= \mathbf{F}(R\mathbf{G}vv(x,y), v(x,y)), \text{ by XII;}\\ &= \mathbf{F}(\mathbf{G}(x), y), \text{ by } 13, 7.\end{aligned}$$

(The substitutiveness of $R\mathbf{G}vv$ and v, which is assumed in the first step of the proof, is a consequence of 13 and 7.)

XVI. If **F** is substitutive, $D\mathbf{FG}$ is substitutive.

The proof is immediate by XV.

Definition. — $E\mathbf{FGH} \to B\mathbf{F}D\mathbf{GH}RC_1C_1\beta m_2svv$.

XVII. If **F** and **G** are substitutive,

$$\vdash E\mathbf{FGH}(x, y) = \mathbf{F}\big(\mathbf{G}(\mathbf{H}(x), y), \beta(m_2(s(x)))\big).$$

Proof: Since $RC_1C_1\beta m_2svv$ is substitutive by 13,

$$\begin{aligned}\vdash E\mathbf{FGH}(x,y) &= \mathbf{F}\big(D\mathbf{GH}(x,y), RC_1C_1\beta m_2svv(x,y)\big), \text{ by XII;}\\ &= \mathbf{F}\big(\mathbf{G}(\mathbf{H}(x), y), C_1C_1\beta m_2s(x)\big), \text{ by XV, 13;}\\ &= \mathbf{F}\big(\mathbf{G}(\mathbf{H}(x), y), \beta(m_2s(x))\big), \text{ by } 14.\end{aligned}$$

XVIII. If **F** and **G** are substitutive, $E\mathbf{FGH}$ is substitutive.

Definition. — $J\mathbf{FGH}_1\mathbf{H} \to B\sigma A\mathbf{FG}E\mathbf{FH}_1\mathbf{H}$.

XIX. If **F** and $\mathbf{H}_1$ are substitutive, $\vdash J\mathbf{FGH}_1\mathbf{H}(x, y) = \mathbf{F}\big(\mathbf{G}(m_1(s(x)))$ $m_2(s(x))\big) + \mathbf{F}\big(\mathbf{H}_1(\mathbf{H}(x), y), \beta(m_2(s(x)))\big)$.

XX. If $\mathbf{F}$ and $\mathbf{H}_1$ are substitutive, $J\mathbf{FGH}_1\mathbf{H}$ is substitutive.

Definition. — $k \to RuRovvv$.

125. $k(x,0) = x$, by 9*a* and 6.
126. $k(x,s(y)) = 0$, by 9*b*, 13, 5.
127. $y = 0 \supset k(x,y) = x$, by XI, 125, 27.
128. $y = 0 \supset k(x,y) = k(x,0)$, by 127, 125, 11, 12.
129. $y = s(z) \supset k(x,y) = k(x,s(z))$, by XI, 128, 11, 126, 2.
130. $y = z \supset k(x,y) = k(x,z)$, by XI, 128, 129.
131. $y \leqq \varpi(x,y)$, by 114, 85, 36.
132. $y \leqq x \supset x \dot{-} y + y = x$, by 68, 36, 33, 75, 12.
133. $y \leqq x \supset m_2(\tau(x) + y) = y$, by 132, 11, 15, 37, 114, 116.

Definition. — $R_0\mathbf{GH}_1\mathbf{H} \to SGJk\mathbf{GH}_1\mathbf{H}$.

XXI. If $\mathbf{H}_1$ is substitutive, $\vdash R_0\mathbf{GH}_1\mathbf{H}(0) = \mathbf{G}(0)$, and $\vdash R_0\mathbf{GH}_1(s(x)) = k(\mathbf{G}(m_1(s(x))), m_2(s(x))) + k(\mathbf{H}_1(\mathbf{H}(x), R_0\mathbf{GH}_1\mathbf{H}(x)), \beta(m_2(s(x))))$.

Proof by XIII and XIX (since k is substitutive by 130, and hence $Jk\mathbf{GH}_1\mathbf{H}$ is substitutive by XX).

Definition. — $\bar{R}\mathbf{FGH} \to RC_1\mathbf{F}m_1\mathbf{GH}$.

Now associated with any binary functor $\mathbf{I}$ which contains no binary function variables, we determine a binary functor $\bar{\mathbf{I}}$ and a singulary functor $\mathbf{I}_0$ as follows:

$\bar{v}$ is v, and $\overline{R\mathbf{FGH}}$ is $\bar{R}\mathbf{FG}\bar{\mathbf{H}}$.
v_0 is m_2, and if $\mathbf{I}$ is $R\mathbf{FGH}$, then $\mathbf{I}_0$ is $R_0\mathbf{FGH}_0$.

Then $\bar{\mathbf{I}}$ and $\mathbf{I}_0$ will, like $\mathbf{I}$, contain no binary function variables. We may consider the singulary functor $\bar{\mathbf{I}}_0$ which is associated with $\bar{\mathbf{I}}$ in the same way that $\mathbf{I}_0$ is associated with $\mathbf{I}$. And we prove the four following metatheorems simultaneously by induction with respect to the number of occurrences of R in $\mathbf{I}$.

If $\mathbf{I}$ is a binary functor that contains no variables:

XXII. $\vdash \bar{\mathbf{I}}(x,y) = \mathbf{I}(m_1(x),y)$.

XXIII. $\vdash y \leqq x \supset \bar{\mathbf{I}}_0(\tau(x)+y) = \bar{\mathbf{I}}(x,y)$.

XXIV. $\vdash y \leqq x \supset y = z \supset \bar{\mathbf{I}}(x,y) = \bar{\mathbf{I}}(x,z)$.

XXV. $\vdash y = z \supset \mathbf{I}(x,y) = \mathbf{I}(x,z)$.

Proof of XXII: Obvious if the number of occurrences of R in $\mathbf{I}$ is 0, as $\mathbf{I}$ must then be v. Suppose that $\bar{\mathbf{I}}$ is $R\mathbf{FGH}$. Then $\bar{\mathbf{I}}$ is $\bar{R}\mathbf{FG}\bar{\mathbf{H}}$.

$$\begin{aligned}\vdash \bar{\mathbf{I}}(x,0) &= C_1\mathbf{F}m_1(x), \text{ by } 9a;\\ &= \mathbf{F}(m_1(x)), \text{ by } 14;\\ &= \mathbf{I}(m_1(x),0), \text{ by } 9a.\end{aligned}$$

$$\begin{aligned}\vdash \bar{\mathbf{I}}(x,s(y)) &= \mathbf{G}(\bar{\mathbf{H}}(x,y), \bar{\mathbf{I}}(x,y)), \text{ by } 9b;\\ &= \mathbf{G}(\mathbf{H}(m_1(x),y), \mathbf{I}(m_1(x),y)), \text{ by hyp. ind.};\\ &= \mathbf{I}(m_1(x),s(y)), \quad \text{by } 9b.\end{aligned}$$

So the result required follows by induction on y, i.e., by use of VI with **a** taken as y. — In the line for which "by hyp. ind." is given as a reason, there is actually a triple use of hypotheses of induction: namely $\bar{\mathbf{H}}(x, y) = \mathbf{H}(m_1(x), y)$ is by hypothesis of the metatheoretic induction, since **H** contains fewer occurrences of R than **I** does; $\bar{\mathbf{I}}(x, y) = \mathbf{I}(m_1(x), y)$ is by hypothesis of the object-language induction on y; and the substitutiveness of **G** is by hypothesis of the metatheoretic induction, since **G** contains fewer occurrences of R than **I** does.

Proof of XXIII: If the number of occurrences of R in **I** is 0, then **I** is v, and $\bar{\mathbf{I}}$ is v and $\bar{\mathbf{I}}_0$ is m_2. And then:

$$y \leqq x \vdash m_2(\tau(x) + y) = y, \text{ by 133;}$$
$$= v(x, y), \text{ by 7.}$$

If the number of occurrences of R in **I** is greater than 0, suppose that **I** is $R\mathbf{FGH}$. Then $\bar{\mathbf{I}}$ is $RC_1\mathbf{F}m_1\mathbf{G}\bar{\mathbf{H}}$ and $\bar{\mathbf{I}}_0$ is $R_0C_1\mathbf{F}m_1\mathbf{G}\bar{\mathbf{H}}_0$. And then:

$$\vdash \bar{\mathbf{I}}_0(\tau(0) + 0) = \bar{\mathbf{I}}_0(0), \text{ by 90, 33;}$$
$$= C_1\mathbf{F}m_1(0), \text{ by XXI;}$$
$$= \bar{\mathbf{I}}(0, 0), \text{ by } 9a.$$
$$\vdash \bar{\mathbf{I}}_0(\tau(s(x)) + 0) = \bar{\mathbf{I}}_0(s(x + \tau(x))), \text{ by 33, 91, 35;}$$
$$= k(C_1\mathbf{F}m_1(m_1(\tau(s(x)))), m_2(\tau(s(x)))) +$$
$$k(\mathbf{G}(\bar{\mathbf{H}}_0(x + \tau(x)), \mathbf{I}_0(x + \tau(x))), \beta(m_2(\tau(s(x))))),$$
$$\text{by XXI, 91, 35, 130 (and } 3a, 38, 15);$$
$$= k(C_1\mathbf{F}m_1(s(x)), 0) + k(\mathbf{G}(\bar{\mathbf{H}}_0((x + \tau(x)),$$
$$\bar{\mathbf{I}}_0(x + \tau(x)))), 1), \text{ by 33, 114, 116, 117, 31;}$$
$$= C_1\mathbf{F}m_1(s(x)), \text{ by 125, 126, 33;}$$
$$= \bar{\mathbf{I}}(s(x), 0), \text{ by } 9a.$$

$$\vdash \bar{\mathbf{I}}_0(\tau(x) + 0) = \bar{\mathbf{I}}(x, 0), \text{ by XI.}$$
$$\vdash 0 \leqq x \supset \bar{\mathbf{I}}_0(\tau(x) + 0) = \bar{\mathbf{I}}(x, 0), \text{ by propositional calculus.}$$

This completes the proof of XXIII for the case that 0 replaces y. The substitutiveness of **G**, by hypothesis of the metatheoretic induction, has been used in order to apply XXI. — Then we take

$$y \leqq x \supset \bar{\mathbf{I}}_0(\tau(x) + y) = \bar{\mathbf{I}}(x, y)$$

as hypothesis of induction (in the object language) and proceed as follows:

$$s(y) \leqq x \vdash y \leqq x, \text{ by 80.}$$

$$\text{hyp. ind., } s(y) \leqq x \vdash \bar{\mathbf{I}}_0(\tau(x) + y) = \bar{\mathbf{I}}(x, y).$$

Since by hypothesis of the metatheoretic induction, **G** is substitutive,

$$\vdash \bar{\mathbf{I}}_0(\tau(x) + s(y)) = \bar{\mathbf{I}}_0(s(\tau(x) + y)), \text{ by 34;}$$
$$= k(C_1\mathbf{F}m_1(m_1(\tau(x) + s(y))), m_2(\tau(x) + s(y)))$$
$$+ k(\mathbf{G}(\bar{\mathbf{H}}_0(\tau(x) + y), \bar{\mathbf{I}}_0(\tau(x) + y)), \beta(m_2(\tau(x)$$
$$+ s(y)))), \text{ by XXI, 34, 130 (and } 3a, 38, 15).$$

And since by hypothesis of the metatheoretic induction, XXIII holds for **H**,

hyp. ind., $s(y) \leqq x \vdash \bar{\mathbf{I}}_0(\tau(x)+s(y)) = k(C_1\mathbf{F}m_1(m_1(\tau(x)+s(y))), s(y))$
$+ (k\mathbf{G}(\bar{\mathbf{H}}(x,y), \bar{\mathbf{I}}(x,y)), 0)$, by 133, 80, 32, 130;
$= \mathbf{G}(\bar{\mathbf{H}}(x,y), \bar{\mathbf{I}}(x,y))$, by 126, 36, 33, 125;
$= \bar{\mathbf{I}}(x, s(y))$, by 9*b*.

Then XXIII follows by induction on y.

Proof of XXIV: We may now use that XXIII holds for **I**.

$y \leqq x, y = z \vdash z \leqq x$, by 71.
$y \leqq x, y = z \vdash \bar{\mathbf{I}}_0(x,y) = \bar{\mathbf{I}}_0(\tau(x)+y)$;
$= \bar{\mathbf{I}}_0(\tau(x)+z)$, by 38, 15;
$= \bar{\mathbf{I}}(x,z)$.

Proof of XXV: We may now use both that XXII and XXIV hold for **I**.

$y = z \vdash \mathbf{I}(x,y) = \mathbf{I}(m_1(\varpi(x,y)), y)$, by 117, 3*a*;
$= \bar{\mathbf{I}}(\varpi(x,y), y)$;
$= \bar{\mathbf{I}}(\varpi(x,y), z)$, by 131;
$= \bar{\mathbf{I}}(m_1(\varpi(x,y)), z)$;
$= \mathbf{I}(x,z)$, by 117, 3*a*.

And this completes the proof, by simultaneous induction, of the four metatheorems XXII–XXV.

Having completed the proof of XXV, we can now prove a somewhat stronger metatheorem in place of XXIII, namely:

XXVI. If **I** is a binary functor that contains no variables, $\vdash y \leqq x \supset \mathbf{I}_0(\tau(x)+y) = \mathbf{I}(x,y)$.

In fact the proof of XXVI exactly parallels what was done above under the head of proof of XXIII.

Of more interest, however, is the question, to how wide a class of binary functors **I** the proof of substitutiveness can be extended. If we follow the present method for the proof of substitutiveness, it is immediately clear that the condition that **I** shall contain no variables can be weakened to the condition that **I** shall contain no binary function variables. But even this can be extended somewhat. We remark that the condition which was required above in defining $\bar{\mathbf{I}}$ and $\mathbf{I}_0$, that **I** contain no binary function variables, is stronger than necessary, and can itself be weakened. Hence we are able, by the same method which was used in proving XXII–XXV, to obtain the following metatheorem.

XXVII. The class of binary functors which are substitutive includes all those determined recursively by the following statement: v is substitutive; and whenever **G** and **H** are substitutive, *R***FGH** is substitutive.

Also of course XXII and XXVI can be strengthened in the same manner.

REMARKS ON THE ELEMENTARY THEORY OF DIFFERENTIAL EQUATIONS AS AREA OF RESEARCH[1,2] (1965)

As the title indicates, the topic of this chapter is not the (important and difficult) modern theory of differential equations, but something much more modest, the strictly elementary theory, which every mathematician knows, because he has to teach it to elementary students, and every natural scientist, because he has to deal with applications of it. The thesis is that research in this area is not as fully completed as might be supposed from the present lack of activity. And while some ideas and results are sketched which may be new in part, the main purpose is not to present new results but to discuss the significance and the possibilities of research and to indicate some open problems.

I want to begin by reminding you of a problem which is not new but very old, one which mathematicians have not so much forgotten as just become reconciled to supposing unsolvable.

Historians of mathematics say it was about 1800 that mathematicians finally gave up the hope of finding elementary (general) solutions of more than an extremely special subclass of differential equations. But even after the idea of elementary solutions beyond such a special subclass is conceded to be illusory, there remains the question of characterizing the form of the solution for a differential equation of given form.

The paradigm case in which the form of the solution can be given, although an elementary solution is not in general possible, is of course the Riccati equation. And the much greater existing development of the theory of the Riccati equation as compared to that of most others is largely due to this knowledge of the form of the solution.

For the purpose of the present discussion, let us take the Riccati equation in the normal form,

$$\frac{dz}{dx} = z^2 + f(x), \tag{1}$$

since the more general form which is usually given,

$$\frac{dz}{dx} = z^2\phi(x) + z\chi(x) + \psi(x) \tag{2}$$

is reducible to the form 1 by a familiar substitution, so that treatment of 1 is in every relevant sense equivalent to treatment of 2; and it is convenient when possible to deal with a form that involves only one arbitrary function. For the differential equation 1 the form of the solution is

$$z = \frac{\Phi'(x)}{c - \Phi(x)} + \frac{\Phi''(x)}{2\Phi'(x)}, \tag{3}$$

Originally published in ***Information and Prediction in Science***, edited by S. Dockx and P. Bernays, Academic Press, New York and London 1965, pp. 139–177.

[1]Presented at a Colloquium of the Académie Internationale de Philosophie des Sciences in Brussels on September 3, 1962.

[2]The author is indebted for support of the work here represented to the National Science Foundation of the United States.

in the sense that for any given function f there exists a corresponding function θ such that 3 is the general solution of 1—of course under conditions on f that are sufficient for standard existence theorems (a proviso which hereafter will tacitly be taken for granted when relevant).

By a form of a differential equation will be meant simply a differential equation in which one or more arbitrary functions appear, so that (1) and (2) may be regarded as typical example; one might be led to hope or expect that for such a form of a differential equation the corresponding form of the solution would be given by a formula analogous to (3), involving one or more arbitrary functions, derivatives of these function and (conceivably) indefinite integrations of expressions containing these functions and their derivatives. Actually, if attention is confined to differential equations of the first order and first degree, it appears that essentially the only known cases are the Riccati equation and the first-order linear equations; i.e., all other known cases in which the form of the solution can be given in this manner are reducible by a substitution to the Riccati equation or the linear equation. I know of no proof that these are the only cases. But it may plausibly be argued that if any other cases existed in which the form of the solution could be given in a manner even remotely approaching the simplicity and elegance of that for the Riccati equation, they would have been found long ago.

In reviving the problem of the form of the solution of a differential equation I am therefore conscious of venturing on the speculative. I am also unable to make the problem definite by saying exactly what would qualify as a solution of it. But the feeling is persistent that there must be some better way than is at present known to characterize the form of the solution of a differential equation of given form.

In this connection it may be suggestive to ask why there do not exist tables of solutions of differential equations, analogous to the familiar tables of integrals. Such a table would evidently be very useful, even if confined to differential equations of the first order and first degree. The obvious explanation that such a table would require too much space to be practicable is probably not valid. It would certainly be possible to tabulate, on the one hand, reductions of differential equations to certain normal forms by means of substitution, and on the other hand, solutions of differential equations in these normal forms. In this way, at least the table for first order and first degree could be brought within compass and would still retain much of its usefulness.

No doubt the reason for non-existence of such tables is not the space which the table would occupy, but the fact that there would unavoidably be many awkward gaps—places at which for many entries in succession there would be nothing very illuminating to give. Indeed, such a table would be expected to state solutions of differential equations, where appropriate, in terms of specific non-elementary functions which have been studied and tabulated, and even in terms of infinite series representing non-elementary functions of which studies do not exist in literature. But in cases in which the form of the solution is not known except in the case that infinite series can be given with the arbitrary constant or constants appearing in the coefficients of the series, the detailed listing of solutions in series would seem to lose much of its point, and perhaps might be better replaced by general instructions for generating such series.

As distinguished from tables of solutions, there do exist books of the cookbook type, devoted to an extensive collection of recipes for solving differential equations of various given forms. However, when such books come, for example, to Abel's equation, they offer nothing except the advice to expand in series. In the present state of knowledge there is not more that they could profitably do.

As to Abel's equation, it should be said that there is a sense in which it is the principal outstanding case, at least among differential equations of the first order and first degree. The reason is not so much the fact that the form of the equation,

$$\frac{dz}{dx} = z^3\nu(x) + z^2\phi(x) + z\chi(x) + \psi(x), \tag{4}$$

makes it the obvious next case after the Riccati equation, as it is the impressively large variety of differential equations of the first order and first degree which can be reduced to this form by a substitution. It is not too much to say that the satisfactory knowledge of the form of the solution of Abel's equation would make the difference between a stage of development at which it is not profitable to compile a table of solutions of differential equations of the first order and first degree, and a stage at which this would become profitable.

As is well known, Abel's equation can be reduced by an appropriate substitution to either of the normal forms:

$$\frac{dz}{dx} = z^3 + f(x), \tag{5}$$

$$\frac{dz}{dx} = z^{-1} + f(x). \tag{6}$$

The reduction to (6) requires the use of a particular solution of (4). But it is true that if the form were known for the solution of either (5) or (6), the form of the solution of (4) would follow.

Now the problem of finding the form of the solution for an ordinary differential equation of given form is equivalent to the problem of finding the general solution of a certain partial differential equation in two independent variables, which is obtained from the given form of the ordinary differential equation by eliminating the arbitrary functions in usual fashion. We shall here confine attention to the case in which the given form of the ordinary differential equation is of the first order and first degree and contain only one arbitrary function. In this case the associated partial differential equation is of the second order.

As an example, if we eliminate the arbitrary function from (1) we obtain

$$\frac{\partial^2 z}{\partial x \partial y} = 2z\frac{\partial z}{\partial y}. \tag{7}$$

This is the associated partial differential equation. The general solution of the partial differential equation is

$$z = \frac{\Phi'(x)}{\Psi(y) - \Phi(x)} + \frac{\Phi''(x)}{2\Phi'(x)}. \tag{8}$$

And by replacing $\Psi(y)$ in this by c we obtain the form of the solution of (1), i.e. (3).

If we apply the same idea to (5), we obtain, by eliminating the arbitrary function, the associated partial differential equation

$$\frac{\partial^2 z}{\partial x \partial y} = 3z^2\frac{\partial z}{\partial y}. \tag{9}$$

The general solution of this is not known. Perhaps it cannot be expressed by arbitrary functions in the manner that has been described. However, by a known existence theorem for ordinary differential equations, we are assured that (9) has an infinite number of solutions which are partly general in the sense that they contain one arbitrary function $\Psi(y)$. Some suitable way of uniting this infinite totality of solutions of (9) into a single formula or single characterization would constitute the general solution of (9), and upon replacing $\Psi(y)$ by c it would also constitute the general form of the solution of (5).

As found in this way the associated partial differential equation is always an equation which is *mongeable*, i.e., has an intermediate integral in the sense of Monge's method, and it is in effect just this intermediate integral that is the form of ordinary differential equation with which the partial differential equation is associated. However, let us speak more generally of *an associated* partial differential equation for a given form of ordinary differential equation,

$$\frac{dz}{dx} = F(x, z, f(x)), \tag{10}$$

whenever a relationship has been established by which the general solution of the partial differential equation would yield the form of solution of (10). We shall see presently that the notion of intermediate integral can be greatly generalized. In particular, an associated partial differential equation of a given form of ordinary differential equation can often be found by way of the generalized intermediate integral.

Let us remark in passing that the assumption which was made in writing 10, that the arbitrary function is a function of x, involves no loss of generality. This situation can always be attained by a substitution if only one arbitrary function is present. As an example, the homogeneous equation,

$$\frac{dz}{dx} = f\left(\frac{z}{x}\right) \tag{11}$$

can be reduced by the substitution $X = z/x$ to the linear differential equation

$$\frac{dz}{dX} = z\chi(X). \tag{12}$$

This substitution, or a variation of it, is indeed the usual method of solution.

Some writers deprecate concern with general solutions of partial differential equations, on the ground that in cases that are important for applications, use of the general solution, even when it is known, is often not the best method of finding particular solutions to satisfy boundary conditions. But this practical advice to those who work with applications must not be allowed to block any line of research. The connection with the question of the form of solution of ordinary differential equations provides additional motivation, if any is needed, for interest in the general solutions of partial differential equations. Some definite suggestions in regard to the latter will now be discussed. These suggestions are only in the direction of extending as far as possible, known methods of finding and expressing general solutions of partial differential equations, and are, therefore, hardly comparable with foregoing speculation. But in the absence of other ideas, this, too, is necessary.

As already indicated, attention will be confined to partial differential equations of the second order in two independent variables. Among those we shall consider mainly equations which are of the first degree in the three second-order derivatives (or algebraically reducible to the first degree). As is usual, we shall employ the letters p, q, r, s, t as abbreviations of

$$\frac{\partial z}{\partial x}, \frac{\partial z}{\partial y}, \frac{\partial^2 z}{\partial^2 x}, \frac{\partial^2 z}{\partial x \partial y}, \frac{\partial^2 z}{\partial y^2}$$

respectively. When Z is used as the dependent variable we shall similarly employ P, Q, R, S, T as abbreviations of

$$\frac{\partial Z}{\partial x}, \frac{\partial Z}{\partial y}, \frac{\partial^2 Z}{\partial^2 x}, \frac{\partial^2 Z}{\partial x \partial y}, \frac{\partial^2 Z}{\partial y^2}.$$

In a partial differential equation of the second order, say that the independent variables are x, y, and the dependent variable is z. By an *elementary general solution* of the differential equation will be meant an expression of the general solution by means of an equation relating x, y, and z and involving known functions, arbitrary functions (which must be functions of one argument), and possibly arbitrary constants. Derivatives of the arbitrary functions may occur, and indefinite integrations of expressions containing the arbitrary function and their derivatives. It is not meant that the known functions must be elementary functions. However, infinite series and definite integrations are otherwise excluded.

A *parametric general solution* is of the same form as an elementary general solution except that there are $a + 1$ equations in which there are a parameters as well as the variables x, y, z. It is understood that the arguments of the arbitrary function may contain the parameters as well as x, y, z.

We shall further distinguish elementary and parametric general solutions as being either *simple* or *compound*, the adjective *simple* being meant to exclude the occurrence of an arbitrary function within the argument of an arbitrary function. After either of the adjectives, *simple* or *compound*, we may sometimes omit the word *elementary* and speak, e.g., just of a simple general solution.

An example of a simple elementary general solution is the general solution (8) of the partial differential equation (7).

An example of a compound elementary general solution is provided by the partial differential equation,

$$(q - 1)s + (y - p)t - q + 1 = 0 \tag{13}$$

which is mongeable and has the general solution,

$$z = xy + \Phi(x) + (1 - x)\Phi'(x) + \Psi(y - \Phi'(x)). \tag{14}$$

Evidently the compound elementary solution can be changed to a simple parametric solution by taking a parameter $u = y - \Phi'(x)$. (Actually (13) has also the solution $z = y + \Omega(x)$, which must be adjoined to (14) as a part of the general solution.)

Elementary or parametric general solutions involving other than exactly two arbitrary functions or involving the same arbitrary functions with two essentially different arguments appear to be impossible, and we shall assume that they do not occur.

1. Normal Forms of Partial Differential Equations

For heuristic purposes it is often convenient to make use of normal forms to which differential equations may be reduced by substitution, as results found for such a normal form may then be extended to the more general case, from which the normal form was obtained, just by reversing the substitution that led to it.

The most general partial differential equation of the kind we discuss in this section may be written as

$$Ar + Bs + Ct + D = 0 \tag{15}$$

where A, B, C, D are functions of x, y, z, p, q.

In certain cases this general form may be reduced by a substitution on the independent variables to one of the two normal forms:

$$s = f(x, y, z, p, q) \tag{16}$$

$$r = f(x, y, z, p, q) \tag{17}$$

Namely first, if A, B, C are functions of x and y only, the reduction can always be accomplished by a substitution of the form $X = \theta(x, y)$, $Y = \psi(x, y)$, to the normal form (16) if $B^2 - 4AC$ is not identically 0, and to the normal form (17) in the contrary case. Secondly, if A, B, C are not independent of z, p, q, we may try a substitution of the form

$$X = \phi(x, y, z), \; Y = \psi(x, y, z). \tag{18}$$

If the quadratic equation,

$$A\rho^2 + B\rho + C = 0, \tag{19}$$

has distinct roots, $\rho = \rho_{00}$ and $\rho = \rho_0$, we obtain the conditions,

$$\frac{\phi_1(x, y, z) + p\phi_3(x, y, z)}{\phi_2(x, y, z) + q\phi_3(x, y, z)} = \rho_{00}, \tag{20}$$

$$\frac{\psi_1(x, y, z) + p\psi_3(x, y, z)}{\psi_2(x, y, z) + q\psi_3(x, y, z)} = \rho_0, \tag{21}$$

for reduction to (16). Also, if (19) has a double root, $\rho = \rho_0$, we obtain the condition (21) for reduction to (17), $\phi(x, y, z)$ to be chosen arbitrarily, subject to the obvious condition of independence of $\phi(x, y, z)$, $\psi(x, y, z)$, and z. (The subscripts $1, 2, 3$ after the function letters are here used to indicate derivatives with respect to the first, second, third arguments, analogously to the use of the prime, $'$, in the case of functions of one argument.) Because (20) and (21) must hold identically in x, y, z, p, q as five independent variables, they may not always be satisfiable. But the possibility of satisfying (20) and (21), or (21) alone, can be ascertained, and when it exists, the functions θ and ψ can be found, subject to the solution of certain first-order differential equations.

It may sometimes happen when ϕ and ψ are determined from (20) and (21) that $\phi(x, y, z)$, $\psi(x, y, z)$, and z are then not independent. This is not an essential difficulty but may be avoided by making a suitable substitution for the dependent variable before the substitution on the independent variables. Also, it is obvious that the possibility of an infinite root of (19) is not an essential difficulty.

Let us remark that, although the reduction of (15) to one of the forms (16) or (17) is not always possible, it is always possible if (15) has a simple general solution. For $\phi(x, y, z)$ and $\psi(x, y, z)$ will then coincide with the arguments of the two arbitrary functions in the general solution. Or if the two arbitrary functions have the same argument, $\psi(x, y, z)$ will coincide with this argument, and the reduction is to (17).

Similarly, if there is an intermediate integral having a function of x, y, z as argument of the arbitrary function, it will always be possible to satisfy at least one of the conditions (20), (21), and therefore a partial reduction of (15) may be made, reducing one of the coefficients A or C to 0.

Also in some cases, there is a reduction to the form (16) by using a substitution of form more general than (18). For example, in the case of the partial differential equation (13), we have for (19) one infinite root and one root $\rho = (p - y)/(q - 1)$. Hence we take $X = x$, and we find Y from the analog of equation (21), i.e.,

$$\frac{\partial Y}{\partial x} \bigg/ \frac{\partial Y}{\partial y} = \frac{p - y}{q - 1} = \frac{s - 1}{t}.$$

This leads to $Y = q - x$. By this substitution (13) is in fact reduced to the form (16).

Other examples are: $z^3 s + z^4 t + z^2 pq + 2z^3 q^2 - p = 0$, for which the substitution is $X = x, Y = x + \int z^2 q \, dx$; and $s = tf(x, z, p, q)$ for which the substitution is $X = x, Y = q$. Such a reduction exists at least whenever there is a compound general solution obtainable from an intermediate integral, but also in many other cases. A general method of determining whether the reduction exists in a given case is lacking. But in the next section a method is given which finds the reduction whenever the substitution has the simple form

$$X = \phi(x, y, z, p, q), Y = \psi(x, y, z, p, q). \tag{22}$$

The foregoing considerations suggest that there may be heuristic value in beginning with the treatment of differential equations of the forms (16) and (17). In what follows we shall in fact be mainly concerned with (16).

It is familiar, especially in connection with Laplace's transformation, that the reduction of a linear differential equation (15), by a substitution, to one of the normal forms (16) or (17) is always possible, and that when (15) is linear, the reduced equation (16) or (17) is also linear. As a matter of fact, in the special case in which (15) is linear, it is desirable to carry the reduction a step further by adjoining a substitution on the dependent variable. We shall use as normal form of the linear partial differential equation:

$$s = lp + nz + h. \tag{23}$$

Since this is unsymmetric in regard to x and y, it may sometimes be convenient to use the normal form which is analogous to (23) with the roles of x and y interchanged:

$$s = mq + nz + h \tag{24}$$

If $B^2 - 4AC$ is identically 0, in the case of a linear partial differential equation (15), we must replace the normal form (23) by another—we shall use

$$r = mq + nz + h. \tag{25}$$

In Eqs. (23)–(25) it is understood that l, m, n, h are functions of x and y.

2. Intermediate Integral

In the case of a differential equation in one of the normal forms (16) or (17), a necessary and sufficient condition for the existence of an intermediate integral can be given in a very simple form.

In fact (17) is mongeable if and only if $f_5(x, y, z, p, q) = 0$, i.e., q does not occur in the equation.

For (16), a first necessary condition of mongeability is that $f(x, y, z, p, q)$ shall be either of the first degree in p or of the first degree in q. That is, the differential equation must have one of the forms

$$s = \mu + \nu p, \tag{26}$$

where μ and ν are functions of x, y, z, q, or

$$s = \mu + \nu q, \tag{27}$$

where μ and ν are functions of x, y, z, p. Then (26) is mongeable if and only if

$$\frac{\partial \mu}{\partial z} + \nu \frac{\partial \mu}{\partial q} = \mu \frac{\partial \nu}{\partial q} + \frac{\partial \nu}{\partial x}, \tag{28}$$

and (27) is mongeable if and only if

$$\frac{\partial \mu}{\partial z} + \nu \frac{\partial \mu}{\partial p} = \mu \frac{\partial \nu}{\partial p} + \frac{\partial \nu}{\partial y}, \tag{29}$$

When the conditions of mongeability are found to be satisfied, (17) reduces to an ordinary second-order differential equation; (26) must have an intermediate integral of the form

$$\eta(x, y, z, q) = \Psi(y); \tag{30}$$

and (27) must have an intermediate integral of the form

$$\eta(x, y, z, p) = \Phi(x) \tag{31}$$

In the two latter cases we may proceed by eliminating the arbitrary function from (30) or (31) in usual fashion, and then writing the conditions which η must satisfy in order that the resulting equation agree with (26) or (27).[3]

If we wish to find an intermediate integral of a particular differential equation of the more general form (15), we may proceed as follows.

If the quadratic equation (19) has a double root $\rho = \rho_0$, we seek by standard methods to find $\psi(x, y, z)$ to satisfy the conditions (21). If this is determined to be impossible, it follows that there is no intermediate integral having a function of x, y, z as argument of the arbitrary function. If, on the other hand, we find $\psi(x, y, z)$ to satisfy (21), we use the substitution (18) to reduce the given differential equation to form (17), and then we are able to apply the method which was described above.

If the quadratic equation (19) has distinct roots, $\rho = \rho_{00}$ and $\rho = \rho_0$, we seek to find $\phi(x, y, z)$ and $\psi(x, y, z)$ to satisfy the conditions (20) and (21). Three cases arise.

In the first place, if both (20) and (21) are impossible, it follows that there is no intermediate integral having a function of x, y, z as argument of the arbitrary function.

[3]The latter two thirds of the paper, from this point on, have been substantially revised since its presentation at Brussels in September, 1962. The author is indebted to the National Science Foundation of the United States for support in connection with the work represented by this paper.

Secondly, if both (20) and (21) can be satisfied, we use the substitution (18) to reduce the given differential equation to the form (16), in the way that was described in Section 1. We then use the method which was just described above (in the present section).

The third case is that one but not both of (20) and (21) can be satisfied. In this case, as already remarked in Section 1, we may make a partial reduction of the given differential equation by a substitution of the form (18), reducing one of the coefficients A or C in (15) to 0. If after the substitution we make a change of notation, writing either x and y or y and x in place of X and Y, respectively, we may write the reduced differential equation as

$$s = Et + F, \tag{32}$$

where E and F are functions of x, y, z, p, q. Since (32) cannot be reduced to the form (16) by a substitution of the form (18), any intermediate integral that has a function of x, y, z as argument of the arbitrary function must in fact have the form

$$\eta(x, y, z, p, q) = \Phi(x). \tag{33}$$

The condition for this is the conjunction of the two equations,

$$\begin{aligned} \frac{\partial E}{\partial z} + W\frac{\partial E}{\partial p} &= E\frac{\partial W}{\partial p} + \frac{\partial W}{\partial q}, \\ \frac{\partial F}{\partial z} + W\frac{\partial F}{\partial p} &= \frac{\partial W}{\partial y} + q\frac{\partial W}{\partial z} + F\frac{\partial W}{\partial p}, \end{aligned} \tag{34}$$

where

$$W = -\frac{\partial E}{\partial y} - q\frac{\partial E}{\partial z} - F\frac{\partial E}{\partial p} + E\frac{\partial F}{\partial p} + \frac{\partial F}{\partial q}.$$

When this condition is satisfied, we may proceed by eliminating the arbitrary function from (33) in usual fashion, and then writing the conditions which η must satisfy in order that the resulting equation agree with (32).

Finally we must consider the possibility of a Monge intermediate integral of (15) in which the argument of the arbitrary function involves p or q or both. For this we may seek a substitution of the more general form (22) in place of (18). The conditions (20) and (21) are then replaced by the following (in which the arguments of ϕ and ψ are x, y, z, p, q):

$$\frac{\phi_1 + p\phi_3 + r\phi_4 + s\phi_5}{\phi_2 + q\phi_3 + s\phi_4 + t\phi_5} = \rho_{00}$$

$$\frac{\psi_1 + p\psi_3 + r\psi_4 + s\psi_5}{\psi_2 + q\psi_3 + s\psi_4 + t\psi_5} = \rho_0$$

It is clear that the first of these two conditions is a consequence of the given differential equation (15) if and only if we have, independently of (15):

$$\frac{\phi_4}{A} = \frac{\phi_5 - \rho_{00}\phi_4}{B} = \frac{-\rho_{00}\phi_5}{C} = \frac{\phi_1 + p\phi_3 - \rho_{00}(\phi_2 + q\phi_3)}{D} \tag{35}$$

And in the same sense the second condition can be replaced by:

$$\frac{\psi_4}{A} = \frac{\psi_5 - \rho_0\psi_4}{B} = \frac{-\rho_0\psi_5}{C} = \frac{\psi_1 + p\psi_3 - \rho_0(\psi_2 + q\psi_3)}{D} \tag{36}$$

Evidently (35) is equivalent to two (not three) simultaneous partial differential equations to be solved for ϕ, and (36) to two simultaneous partial differential equations to be solved for ψ. By standard methods the existence of solutions of (35) and (36) can be ascertained; and when they exist, ϕ and ψ can be found, subject to the solution of certain first-order differential equations. It can be shown that *a necessary and sufficient condition for the existence of the intermediate integral is that either* (35) *or* (36) *shall have two independent solutions.*[4] E.g. if $\phi_0(x, y, z, p, q)$ and $\phi_{00}(x, y, z, p.q)$ are two independent solutions of (35), the corresponding intermediate integral is $\phi_0(x, y, z, p, q) = \Phi(\phi_{00}(x, y, z, p, q))$. But the reduction of the differential equation (15) to one of the forms (16), (17), (32) is desirable on its own account. If there is no intermediate integral, it may nevertheless be possible to find $\phi(x, y, z, p, q)$ to satisfy (35) or $\psi(x, y, z, p, q)$ to satisfy (36) or both. And the substitution (22) should then be tried, as well as the substitution $X = \phi(x, y, z, p, q)$, $Y = y$, and the substitution $X = \psi(x, y, z, p, q)$, $Y = y$ (as far as the ϕ and ψ exist). The existence of ϕ and ψ does not guarantee the possibility of the corresponding substitution, but this must rather be determined by actually attempting to transform the differential equation (15) by the substitution in question, and seeing whether the requisite eliminations are then possible.

This circumvents altogether the need to consider total differential equations. It is recommended in preference to Monge's method, primarily as leading more quickly to the negative result in cases in which the desired intermediate integral does not exist, but also on the ground that the total differential equations hardly serve any purpose that is not better served by the partial differential equations (35) and (36).

Two possible complications of the present method must, however, be noticed.

First, it may happen that the function $f(x, y, z, p, q)$, or one or both of the two functions $E(x, y, z, p, q)$ and $F(x, y, z, p, q)$ that results by the reduction to (16), (17) or (32) must be expressed in parametric form, because solution of Eqs. (18) for x and y is not elementarily expressible. But this generally makes no difficulty, or none that would not have its parallel if Monge's method were used instead. The alternative of abandoning the reduction by means of a substitution for x and y in this case is possible, as the criterion of two independent solutions of (35) or (36) might then be applied directly; but this is undesirable because it sacrifices other possibilities of reduction or solution of the differential equation that are discussed immediately below.

Secondly, it may happen in exceptional cases, as was already noticed in Section 1, that a preliminary substitution for the dependent variable must be made before the substitution on the independent variables. As a matter of fact, there is a similar complication in connection with Monge's method, though it occurs at a different place (and though textbooks usually fail to point it out). Namely, if either A or C is 0 in (15), it is necessary before applying Monge's method to make a preliminary substitution for the dependent variable in order to obtain an equation in which neither A nor C is 0.

On the other hand, the present method has additional advantages over Monge's method beyond that which is urged above.

One is that when the reduced form (32) or, especially, (16) can be obtained, it is often advantageous as a preliminary to transformations of the kind that are considered

[4]This criterion is also applicable to the case in which the argument of the arbitrary function is independent of p and q. But for this case the criteria given above—i.e., (28), (29), and (34)—are more explicit.

in Section 6.

Another is that the same reduction process which has been described as a means of finding an intermediate integral can also be used to reveal cases in which a given differential equation (15) can be reduced by ordinary substitutions to a linear differential equation. Namely, in any such case, the reduction to (16) by means of a substitution (18) must always be possible, and the reduced equation must take the special form

$$s = \alpha(x, y, z)pq + \beta(x, y, z)p + \gamma(x, y, z)q + \delta(x, y, z). \tag{37}$$

The term in pq can be removed from (37) by the substitution

$$Z = \int e^{-\int \alpha(x,y,z)\,dz}\,dz,$$

and after this final substitution the resulting differential equation must then be linear (if the reduction to linear by ordinary substitution is possible at all).

The case, which is rather overemphasized in most textbooks, that (15) has by Monge's method not one but two intermediate integrals, both having a function of x, y, z as argument of the arbitrary function, is actually a subcase of the case that (15) is reducible by ordinary substitutions to a linear differential equation, and therefore is not of great separate importance.

3. Generalized Intermediate Integral

That the existence of an (ordinary or Monge) intermediate integral of a partial differential equation (15) is not necessary to the existence of an elementary general solution is well known. In the case of linear partial differential equations, this is a consequence of Laplace's transformation. However, a variety of examples which are neither linear nor reducible to linear by substitution are found, for example, in Goursat's "Leçons sur l'Intégration des Équations aux Dérivées Partielles du Second Ordre."

Indeed, independently of whether the intermediate integral leads to an elementary solution, it should be said that the notion of intermediate integral on which Monge's method is based is a very special one, because of the requirement that the variables in the intermediate integral must be the same as the variables in the original second-order partial differential equation. For this reason it is desirable to introduce the notion of *generalized intermediate integral* in which this requirement is abandoned.

It is difficult to say what is the most general possibility that may be significant. But in particular, for a given second-order partial differential equation we consider the possibility of a generalized intermediate integral which consists of two equations, one of which is a first-order differential equation in a dependent variable ζ and the same independent variables x, y as in the given second-order differential equation, and the other an equation expressing the dependent variable z of the given second-order differential equation in terms of x, y, ζ, derivatives of ζ, and indefinite integrations of expressions containing x, y, ζ, and derivatives of ζ. One or both of the two equations is to involve an arbitrary function. Of course the given second-order differential equation must reduce to an identity upon substituting for z and making use of the first-order differential equation.

In the special case of a second-order partial differential equation of the form (16) let us introduce the term *intermediate integral of level n* to mean a generalized intermediate integral of the following special form:

$$\frac{\partial \zeta}{\partial x} = \Gamma(x, y, \zeta, \Phi(x)) \tag{38}$$

$$z = \Theta\left(x, y, \zeta, \frac{\partial \zeta}{\partial y}, \frac{\partial^2 \zeta}{\partial y^2}, \ldots, \frac{\partial^n \zeta}{\partial y^n}\right). \tag{39}$$

Of course the same form is also to be allowed with the roles of x and y interchanged. From this point of view a Monge intermediate integral of a differential equation of the form (16) is regarded as an intermediate integral of level 0.

It is immediately clear that a first necessary condition for the existence of an intermediate integral of level n is the same that we have seen for level 0,i.e., that the differential equation have one of the forms (26) or (27). It can be shown by examples that this is not a necessary condition for a partial differential equation (16) to have a generalized intermediate integral of more general form.

Finally, some mention must be made of intermediate integrals that involve a parameter in a manner analogous to that of a parametric general solution. Such parametric intermediate integrals, not reducible to an ordinary or a generalized intermediate integral without parameter, do occur, at least in the case of second-order partial differential equations which are of higher degree (rather than of the form (15)). As a simple example, consider a differential equation which can be written in the form

$$H\left(x, y, G, \frac{\partial G}{\partial x}, \frac{\partial G}{\partial y}\right) = 0 \tag{40}$$

where G stands for $G(x, y, z, p, q)$ and where $\partial G/\partial x$, $\partial G/\partial y$ are used to mean, respectively

$$G_1(x, y, z, p, q) + pG_3(x, y, z, p, q) + rG_4(x, y, z, p, q) + sG_5(x, y, z, p, q), \tag{41}$$

and

$$G_2(x, y, z, p, q) + qG_3(x, y, z, p, q) + sG_4(x, y, z, p, q) + tG_5(x, y, z, p, q). \tag{42}$$

Such a differential equation (40) is not mongeable unless the function H is of the first degree in its last two arguments. But when H is of higher degree (or even not algebraic) it is clear that the solution of (40) can nevertheless be reduced to the successive solution of two first-order differential equations; as indeed we may expect to obtain a parametric intermediate integral by treating (40) as a first-order differential equation in G and solving for G.

4. Laplace's Transformation

In discussing Laplace's transformation and its analogs we shall for simplicity of statement take h to be 0 in the linear partial differential equation (23). Both Laplace's transformation and its analogs are applicable also in the case in which h is not 0, as will be obvious.

If H is 0, (23) becomes

$$s = lp = nz = 0. \tag{43}$$

From this, by taking partial derivatives with respect to x and y, up to the second order, we obtain the five equations:

$$\frac{\partial r}{\partial y} = lr - (l_x + n)p - n_x z = 0, \tag{44}$$

$$\frac{\partial t}{\partial x} = ls - l_y p - nq - n_y z = 0, \tag{45}$$

$$\frac{\partial^2 r}{\partial x \partial y} - l\frac{\partial r}{\partial x} - (2l_x + n)r - (l_{xx} + 2n_x)p - n_{xx}z = 0, \tag{46}$$

$$\frac{\partial^2 s}{\partial x \partial y} - l\frac{\partial r}{\partial y} - l_y r - (l_x + n)s - (l_{xy} + n_y)p - n_x q - n_{xy}z = 0, \tag{47}$$

$$\frac{\partial^2 t}{\partial x \partial y} - l\frac{\partial t}{\partial x} - 2l_y s - nt - l_{yy}p - 2n_y q - n_{yy}z = 0. \tag{48}$$

If we eliminate z between (43) and (44) we obtain

$$\frac{\partial r}{\partial y} - lr - \frac{n_x}{n}s - (l_x + n - \frac{ln_x}{n})p = 0. \tag{49}$$

Then upon substituting $p = nZ$, (49) becomes

$$S = \left(l - \frac{n_y}{n}\right) P + \left(l_x + n - \frac{nn_{xy} - n_x n_y}{n^2}\right) Z. \tag{50}$$

This is Laplace's transformation of the linear partial differential equation (43).

There is also a second Laplace's transformation of (43) which may be described as being the same with the roles of x and y interchanged. Namely, we substitute in (43)

$$z = we^{\int l\, dy}, \tag{51}$$

with the result

$$\frac{\partial^2 w}{\partial x \partial y} + \frac{\partial w}{\partial y}\int l_x\, dy + (l_x - n)w = 0. \tag{52}$$

We then apply to (52) the same process by which (43) was transformed to (49), but interchanging the roles of x and y. Another substitution,

$$\frac{\partial w}{\partial y} = Ze^{-\int l\, dy}, \tag{53}$$

brings the differential equation back into the same form as (43). The result is

$$S = \left(l + \frac{n_y - l_{xy}}{n - l_x}\right) P + (n - l_x)Z, \tag{54}$$

where $Z = q - lz$.

Now by considering, not the transformation upon z, but just the transformation upon the pair of coefficients l, n, we see that the transformation from (43) to (54) is the inverse of the transformation from (43) to (50).

If we start with a particular partial differential equation (43), and thus a particular pair of coefficients l, n, we obtain by iterations of the two Laplace's transformations an infinite array of pairs of coefficients l, n, extending to infinity in both directions. If any one of this infinite array of pairs of coefficients satisfies the mongeability condition of (43), i.e.

$$n = l_x \text{ or } n = 0, \tag{55}$$

we obtain an elementary general solution of the partial differential equation with which we started.

5. Transformations Analogous to Laplace's

Laplace's transformation has been presented in a form slightly different from the usual one in order to show that there are a number of analogous transformations of linear partial differential equations which immediately suggest themselves.

Taking the linear partial differential equation in the form (43), we have the following possibilities:

(i) To take a linear combination of (43), (44), and (45), with multipliers δ, 1, and ε, then determine δ and ε as functions of x and y in such a way that the resulting differential equation can be rewritten in the form

$$\frac{\partial^2}{\partial x \partial y}(p + \kappa q + \lambda z) + \alpha \frac{\partial}{\partial x}(p + \kappa q + \lambda z) + \beta \frac{\partial}{\partial y}(p + \kappa q + \lambda z) + \gamma (p + \kappa q + \lambda z) = 0; \tag{56}$$

(ii) To take a linear combination of (43), (44), (45), and (46), with multipliers $\delta, \varepsilon, \theta$, and 1, then determine δ, ε, and θ as functions of x and y in such a way that the resulting differential equation can be rewritten in the form

$$\frac{\partial^2}{\partial x \partial y}(r + \kappa p + \iota q + \lambda z) + \alpha \frac{\partial}{\partial x}(r + \kappa p + \iota q + \lambda z) + \beta \frac{\partial}{\partial y}(r + \kappa p + \iota q + \lambda z) + \gamma (r + \kappa p + \iota q + \lambda z) = 0; \tag{57}$$

(iii) To take a linear combination of (43), (44), (45), and (47), with multipliers $\delta, \varepsilon, \theta$, and 1, then determine δ, ε, and θ as functions of x and y in such a way that the resulting differential equation can be rewritten in the form

$$\frac{\partial^2}{\partial x \partial y}(s + \kappa p + \iota q + \lambda z) + \alpha \frac{\partial}{\partial x}(s + \kappa p + \iota q + \lambda z) + \beta \frac{\partial}{\partial y}(s + \kappa p + \iota q + \lambda z) + \gamma (s + \kappa p + \iota q + \lambda z) = 0. \tag{58}$$

Elementary general solutions obtainable from these transformations merely duplicate those obtainable from Laplace's transformation. Nevertheless the transformations are different from Laplace's. And at least the transformations (i) do serve some purposes in connection with linear partial differential equations. But our present interest in them is rather that they may lead to or suggest analogous transformations of non-linear equations.

Under each of (i), (ii), and (iii) an essentially different transformation is obtained by assuming that λ is identically 0 than by the contrary assumption.

The transformations (i) with λ taken as 0 were treated by Darboux, and the transformations (i) with λ different from 0 were treated by Goursat. (See Goursat, *op. cit.*) We may therefore speak of Darboux's and Goursat's transformations, respectively.

All the transformations in question, including Laplace's, can be expressed as an assertion of equivalence between products of linear differential operators, or, what comes to the same thing, as two ways of factoring the same operator.

Thus we may express Laplace's transformation, either by writing

$$\left(\frac{\partial}{\partial x} - \frac{n_x}{n}\right)\left(\frac{\partial^2}{\partial x \partial y} - l\frac{\partial}{\partial x} - n\right) z = \left(\frac{\partial^2}{\partial x \partial y} - l\frac{\partial}{\partial x} - \frac{n_x}{n}\frac{\partial}{\partial y} - l_x - n + \frac{l n_x}{n}\right)\left(\frac{\partial}{\partial x}\right) z, \tag{59}$$

corresponding to the transformation from (43) to (49) above, or by writing

$$\left(\frac{l}{n}\frac{\partial}{\partial x} - \frac{n_x}{n^2}\right)\left(\frac{\partial^2}{\partial x \partial y} - l\frac{\partial}{\partial x} - n\right) z = \left(\frac{\partial^2}{\partial x \partial y} - \left(l - \frac{n_y}{n}\right)\frac{\partial}{\partial x} - l_x - n + \frac{n n_{xy} - n_x n_y}{n^2}\right)\left(\frac{1}{n}\frac{\partial}{\partial x}\right) z, \tag{60}$$

which corresponds to the transformation from (43) to (50). We shall regard (59) and (60) as not essentially different since, in terms of the transformation upon the differential equation, (60) is obtained from (59) by just an ordinary substitution for the dependent variable.

Similarly we may express one of the transformations (i) by writing

$$\left(\frac{\partial}{\partial x} + \varepsilon\frac{\partial}{\partial y} + \delta\right)\left(\frac{\partial^2}{\partial x \partial y} - l\frac{\partial}{\partial x} - n\right) z = \left(\frac{\partial^2}{\partial x \partial y} + \alpha\frac{\partial}{\partial x} + \beta\frac{\partial}{\partial y} + \gamma\right)\left(\frac{\partial}{\partial x} + \kappa\frac{\partial}{\partial y} + \lambda\right) z. \tag{61}$$

For the transformations (ii) we have

$$\left(\frac{\partial^2}{\partial x^2} + \varepsilon\frac{\partial}{\partial x} + \theta\frac{\partial}{\partial y} + \delta\right)\left(\frac{\partial^2}{\partial x \partial y} - l\frac{\partial}{\partial x} - n\right) z = \left(\frac{\partial^2}{\partial x \partial y} + \alpha\frac{\partial}{\partial x} + \beta\frac{\partial}{\partial y} + \gamma\right)\left(\frac{\partial^2}{\partial x^2} + \kappa\frac{\partial}{\partial x} + \iota\frac{\partial}{\partial y} + \lambda\right) z. \tag{62}$$

For the transformation (iii) we have

$$\left(\frac{\partial^2}{\partial x \partial y} + \varepsilon\frac{\partial}{\partial x} + \theta\frac{\partial}{\partial y} + \delta\right)\left(\frac{\partial^2}{\partial x \partial y} - l\frac{\partial}{\partial x} - n\right) z = \left(\frac{\partial^2}{\partial x \partial y} + \alpha\frac{\partial}{\partial x} + \beta\frac{\partial}{\partial y} + \gamma\right)\left(\frac{\partial^2}{\partial x \partial y} + \kappa\frac{\partial}{\partial x} + \iota\frac{\partial}{\partial y} + \lambda\right) z. \tag{63}$$

Transformations analogous to (i)–(iii) exist also in the case of linear partial differential equations that are reducible to the normal form (25) instead of (23). To indicate these it will now be sufficient to write the corresponding equations, analogous to (61)–(63), that express the equivalence of certain products of linear differential operators.

These are

$$\left(\frac{\partial}{\partial x}+\varepsilon\frac{\partial}{\partial y}+\delta\right)\left(\frac{\partial^2}{\partial x^2}-m\frac{\partial}{\partial y}-n\right)z$$
$$=\left(\frac{\partial^2}{\partial x^2}+\alpha\frac{\partial}{\partial x}+\beta\frac{\partial}{\partial y}+\gamma\right)\left(\frac{\partial}{\partial x}+\kappa\frac{\partial}{\partial y}+\lambda\right)z, \tag{64}$$

$$\left(\frac{\partial^2}{\partial x^2}+\varepsilon\frac{\partial}{\partial x}+\theta\frac{\partial}{\partial y}+\delta\right)\left(\frac{\partial^2}{\partial x^2}-m\frac{\partial}{\partial y}-n\right)z$$
$$=\left(\frac{\partial^2}{\partial x^2}+\alpha\frac{\partial}{\partial x}+\beta\frac{\partial}{\partial y}+\gamma\right)\left(\frac{\partial^2}{\partial x^2}+\kappa\frac{\partial}{\partial x}+\iota\frac{\partial}{\partial y}+\lambda\right)z, \tag{65}$$

$$\left(\frac{\partial^2}{\partial x\partial y}+\varepsilon\frac{\partial}{\partial x}+\theta\frac{\partial}{\partial y}+\delta\right)\left(\frac{\partial^2}{\partial x^2}-m\frac{\partial}{\partial y}-n\right)z$$
$$=\left(\frac{\partial^2}{\partial x^2}+\alpha\frac{\partial}{\partial x}+\beta\frac{\partial}{\partial y}+\gamma\right)\left(\frac{\partial^2}{\partial x\partial y}+\kappa\frac{\partial}{\partial x}+\iota\frac{\partial}{\partial y}+\lambda\right)z. \tag{66}$$

Evidently (62) and (66) are essentially the same, differing only by an ordinary substitution for z and interchange of the two sides of the equation.

In each of (61)–(66) the problem of finding $\varepsilon, \theta, \delta, \alpha, \beta, \gamma, \kappa, \iota, \lambda$ when l and n, or m and n, are given reduces in an obvious way to a set of simultaneous partial differential equations which $\varepsilon, \theta, \delta, \alpha, \beta, \gamma, \kappa, \iota, \lambda$ must be made to satisfy. We do not treat here the task of finding particular solutions of these partial differential equations except to say that it must apparently be considered separately in each case. But in each of the two cases of (61), leading to the transformations of Darboux and Goursat, the task can be reduced to that of finding particular solutions of (43), as appears in Goursat, *op. cit.*

Parallel to these transformations of second-order linear partial differential equations, there is a transformation of first-order linear partial differential equations which by using operators can be expressed in the following form:

$$\left(\varepsilon\frac{\partial}{\partial x}+\theta\frac{\partial}{\partial y}+\delta\right)\left(l\frac{\partial}{\partial x}+m\frac{\partial}{\partial y}+n\right)z$$
$$=\left(\alpha\frac{\partial}{\partial x}+\beta\frac{\partial}{\partial y}+\gamma\right)\left(\kappa\frac{\partial}{\partial x}+\iota\frac{\partial}{\partial y}+\lambda\right)z, \tag{67}$$

where $\varepsilon, \theta, \delta, l, m, n, \alpha, \beta, \gamma, \kappa, \iota, \lambda$ are functions of x and y. It may easily be shown that the most general non-trivial identity of the form (67) can be obtained by substitutions for x, y, and z from

$$\left(\frac{\partial}{\partial x}+\omega\frac{\partial}{\partial y}-\frac{\omega_x}{\omega}+\omega\upsilon\right)p=\left(\frac{\partial}{\partial x}-\frac{\omega_x}{\omega}\right)(p+\omega q+\omega\upsilon z), \tag{68}$$

where ω is a function of x and y and υ is a function of y only.

6. Extensions of Laplace's Transformation to Non-linear Equations

The suggestion is immediate that perhaps Laplace's transformation and its analogs might be extended to some non-linear partial differential equations. And indeed, since this lecture was originally delivered, a number of such extensions have been found, and the present section is now revised to include a brief sketch of them.

We begin with differential equations of the form (16), and (26) and (27) as special cases of (16). For in these cases the transformations work out much more simply in their application to particular examples.

(Ia) In the case of a partial differential equation of the form (26) we may, by solving an ordinary differential equation, find a function $G(x, y, z, q)$ such that

$$v(x, y, z, q) = -\frac{G_3(x, y, z, q)}{G_4(x, y, z, q)}.$$

Then since the function G will not be independent of its fourth argument, the given differential equation (26) can be rewritten in the form

$$s = \frac{M(x, y, z, G) - G_1(x, y, z, q) - pG_3(x, y, z, q)}{G_4(x, y, z, q)},$$

and hence in the form

$$\frac{\partial G}{\partial x} = M(x, y, z, G).$$

If the function M is independent of its third argument, this last equation can be solved as an ordinary differential equation for G. This is the case in which the given differential equation (26) is mongeable (and it is worth notice in passing that if we set out to apply the present transformation to a differential equation (26) which is mongeable, the mongeability will automatically be discovered at this point without the need for separate application of the method of Section 2). In the contrary case we may let

$$Z = G(x, y, z, q)$$

and the transformed differential equation is then obtained by eliminating z and q among this and the two equations:

$$P = M(x, y, z, Z),$$
$$S = M_2(x, y, z, Z) + qM_3(x, y, z, Z) + QM_4(x, y, z, Z).$$

If the transformed differential equation is of the form (26), in the new dependent variable Z, it may be mongeable, in which case we are able to obtain an intermediate integral of level 1 for the originally given differential equation; or if it is not mongeable, we can apply the same transformation again to the transformed differential equation. On the other hand if the transformed differential equation is not of the form (26), neither mongeability nor iteration of the transformation is possible.

(Ib) In the case of a partial differential equation of the form (27), of course a transformation applies which is exactly like (Ia) except that the roles of x and y are interchanged.

(IIa) In the case of a partial differential equation of the form

$$s = f(x, y, p, q),$$

the special case of (16) in which z itself (as distinguished from derivatives of z) is not present, we may proceed by taking the partial derivative with respect to x, to obtain

$$\frac{\partial r}{\partial y} = f_1(x, y, p, q) + rf_3(x, y, p, q) + sf_4(x, y, p, q).$$

Then by elimination of q we get a second-order partial differential equation with p as the dependent variable. [The transformation (IIa) is found in Goursat, as well as special cases of some of the other transformations that are listed in the present section.]

(IIb) This case is the same as (IIa) with the roles of x and y interchanged.

We now turn to transformations which are applicable to differential equations of more general form than (16), including transformations (I), (II), (II′) which are generalizations of (Ia), (Ib), (IIa), (IIb).

(I) If a second-order partial differential equation can be written in the form

$$H\left(x, y, G, \frac{\partial G}{\partial x}, \frac{\partial G}{\partial y}\right) = z,$$

where G stands for $G(x, y, z, p, q)$ and where $\partial G/\partial x$ and $\partial G/\partial y$ are used to mean (41) and (42), respectively, we may let

$$Z = G(x, y, z, p, q),$$

and so obtain the transformed differential equation

$$G\left(x, y, H, \frac{\partial H}{\partial x}, \frac{\partial H}{\partial y}\right) = Z,$$

where H stands for $H(x, y, Z, P, Q)$ and where $\partial H/\partial x$, $\partial H/\partial y$ are used to mean, respectively,

$$H_1(x, y, Z, P, Q) + PH_3(x, y, Z, P, Q) + RH_4(x, y, Z, P, Q) + SH_5(x, y, Z, P, Q),$$

$$H_2(x, y, Z, P, Q) + QH_3(x, y, Z, P, Q) + SH_4(x, y, Z, P, Q) + TH_5(x, y, Z, P, Q).$$

(II) If a second-order partial differential equation can be written in the form

$$H\left(x, y, g, \frac{\partial g}{\partial x}, \frac{\partial g}{\partial y}\right) = p,$$

where g stands for $g(x, y, p, q)$ and where $\partial g/\partial x$ and $\partial g/\partial y$ have meanings analogous to those explained above for $\partial G/\partial x$ etc., we may let

$$Z = g(x, y, p, q)$$

and so obtain a transformed differential equation by eliminating q between the two equations:

$$Z = g(x, y, H, q),$$

$$P = g_1(x, y, H, q) + g_3(x, y, H, q)\frac{\partial H}{\partial x} + g_4(x, y, H, q)\frac{\partial H}{\partial y},$$

where H stands for $H(x, y, Z, P, Q)$ and where $\partial H/\partial x$ and $\partial H/\partial y$ have the same meanings as above.

(II′) This case is the same as (II) with the roles of x and y interchanged.

(III) Essentially two different transformations have been described so far in this section. Both of them have obvious generalizations to differential equations of higher order, but we have neglected these to concentrate on the second order. In stating the third transformation it will be convenient to state it first in a way that covers some cases of differential equations of higher order, and then afterwards specialize to second order. If $\mathfrak{J}, \mathfrak{K}, \mathfrak{L}, \mathfrak{N}$ are four linear differential operators such that

$$\mathfrak{J}\mathfrak{K}z = \mathfrak{L}\mathfrak{N}z$$

holds as an identity in x, y, z, and the derivatives of z, and if a partial differential equation can be written in the form

$$H\left(x, y, \mathfrak{K}z, \frac{\partial}{\partial x}(\mathfrak{K}z), \frac{\partial}{\partial y}(\mathfrak{K}z)\right) = \mathfrak{N}z,$$

we may let $Z = \mathfrak{K}z$ and so obtain the transformed differential equation

$$\mathfrak{L}H(x, y, Z, P, Q) = \mathfrak{J}Z.$$

This holds independently of the order of the differential equation. To restrict to second order we must require $\mathfrak{J}, \mathfrak{K}, \mathfrak{L}, \mathfrak{N}$ to be operators of not higher than second order; and if either $\mathfrak{K}$ or $\mathfrak{L}$ is of second order, the function H must be independent of its fourth and fifth arguments.

For examples of linear differential operators $\mathfrak{J}, \mathfrak{K}, \mathfrak{L}, \mathfrak{N}$ such that the identity $\mathfrak{J}\mathfrak{K}z = \mathfrak{L}\mathfrak{N}z$ holds we may refer to Section 5. Indeed Eqs. (59)–(68) provide many such examples. And others may be obtained by composition of these, especially composition of (67) with itself and with other examples.

The reasons for restricting the differential operators $\mathfrak{J}, \mathfrak{K}, \mathfrak{L}, \mathfrak{N}$ to be linear are not very compelling but are just that many examples are known of quadruples of linear operators satisfying $\mathfrak{J}\mathfrak{K}z = \mathfrak{L}\mathfrak{N}z$, whereas non-trivial examples involving operators that are not linear seem to be vanishingly few. Except for the paucity of examples we might well consider generalizing (III) as follows.

If we have five 8-ary functions H, J, K, L, N (not excluding that some of the functions may be independent of, say, the last three or the last five arguments), we may define five corresponding differential operators $\mathfrak{H}, \mathfrak{J}, \mathfrak{K}, \mathfrak{L}, \mathfrak{N}$, not necessarily linear, by the five following equations:

$$\mathfrak{H}z = H(x, y, z, p, q, r, s, t)$$
$$\mathfrak{J}z = J(x, y, z, p, q, r, s, t)$$
$$\mathfrak{K}z = K(x, y, z, p, q, r, s, t)$$
$$\mathfrak{L}z = L(x, y, z, p, q, r, s, t)$$
$$\mathfrak{N}z = N(x, y, z, p, q, r, s, t)$$

Then if $\mathfrak{J}\mathfrak{K}z = \mathfrak{L}\mathfrak{N}z$ holds as an identity, and if a given partial differential equation can be written in the form $\mathfrak{H}\mathfrak{K}z = \mathfrak{N}z$, we may let $Z = \mathfrak{K}z$ and so obtain the transformed differential equation $\mathfrak{L}\mathfrak{H}Z = \mathfrak{J}Z$. The transformation (I) may be brought under this head by taking $\mathfrak{K}$ and $\mathfrak{L}$ to be the same and taking $\mathfrak{J}$ and $\mathfrak{N}$ to be each the identity operator.

It should be noticed finally, for the various transformations described in this section, that the form which is required for the given differential equation is not always one that is invariant under ordinary substitutions for x, y, and z. The most important exception is the form

$$H\left(x, y, G, \frac{\partial G}{\partial x}, \frac{\partial G}{\partial y}\right) = z$$

for the transformation (I), which is invariant under substitutions for z and under substitutions for x and y that do not involve z. But in general the application of one of the transformations to a given differential equation, if not possible directly, may become possible after a preliminary substitution in the differential equation.

7. Examples

We add some examples illustrating the application of some of the transformations just described to particular differential equations. The last example illustrates the reduction of a partial differential equation of the second order and higher degree to one of the form (15), i.e. to the first degree; this is evidently often possible but can be expected to lead to a solution only in special cases.

(Ex. 1)
$$s = f(y, q)e^z.$$

This equation is of the form (26), μ being $f(y, q)e^z$ and ν being 0. We may therefore apply the transformation (Ia). In the notation used in Section 6, $G(x, y, z, q)$ is q. And by letting $Z = q$ we get the transformed differential equation

$$Sf(y, Z) - PQf_2(y, Z) = [f_1(y, Z) + Zf(y, Z)]P.$$

This is mongeable. We find $1/f(y, Z)^2$ as an integrating factor, so that the intermediate integral is

$$\frac{Q}{f(y, Z)} = \int \frac{f_1(y, Z) + Zf(y, Z)}{f(y, Z)^2} dZ + \Psi(y).$$

This enables us to obtain an intermediate integral of level 1 for (Ex. 1). An elementary general solution is obtained only in special cases, in particular if $f(y, q)$ is independent of q, or if $f(y, q)$ has the form $\sqrt{q^2 + f(y)}$.

(Ex. 2)
$$s = ap^2 + (bz + c)p,$$

where a is a function of x, b is a constant, and c is a function of y.

This is of the form (27), and again ν is 0. Applying the transformation (Ib), we let $Z = \log p$ and obtain the transformed differential equation

$$S = (aP + a' + b)e^Z.$$

This is of the form of (Ex. 1) with the roles of x and y interchanged. Hence by letting $\zeta = P$, a second transformed differential equation is obtained, which is mongeable, and hence we find an intermediate integral of level 2 for (Ex. 2).

(Ex. 3)
$$s = (p + c)\frac{q}{z} + azp + bz,$$

where a, b, c are constants.

This is of the form (27), with $\nu = (p + c)/z$. It is mongeable if either a or c is 0. If we apply the transformation (Ib), letting

$$Z = \frac{(ac - b)z}{p + c},$$

we have the transformed differential equation

$$S = (P + ac - b)\frac{Q}{Z} + ZP + bZ,$$

which is of the same form as (Ex. 3). If we apply the transformation (Ib) to (Ex. 3), letting

$$Z = \frac{p+c}{z},$$

we get the transformed differential equation

$$S = (Q + ac - b)\frac{P}{Z} + ZQ - bZ,$$

which is of the same form as (Ex. 3) with the roles of x and y interchanged. By iterations of these transformations, and by the use of the substitution $z = c/\zeta$ in (Ex. 3), we are able to find an elementary general solution of (Ex. 3) whenever the ratio ac/b is an integer.

(Ex. 4) $$r - qs + pt = 0.$$

Goursat solves this example by Darboux's method, a method which we have not discussed here because of a conjecture that it can be replaced altogether by the use of transformations. In order to apply the transformation (II′), we rewrite (Ex. 4) as

$$2\frac{\partial g}{\partial x} + (g - 2q)\frac{\partial g}{\partial y} = 0,$$

where

$$g(x, y, p, q) = q + \sqrt{q^2 - 4p}.$$

Then if we let $Z = g(x, y, p, q)$, we find a transformed differential equation which can be written in the form

$$P\frac{\partial}{\partial y}(ZQ - 2P) - Q\frac{\partial}{\partial x}(ZQ - 2P) = 0.$$

This yields the intermediate integral

$$ZQ - 2P = \Phi(Z).$$

Hence we may go on to obtain an elementary general solution.

(Ex. 5) $$r = \frac{s^2}{t} + f\left(\frac{s}{t}\right) + \alpha(x).$$

We apply the transformation (III), letting the operators $\mathfrak{J}, \mathfrak{K}, \mathfrak{L}, \mathfrak{N}$ be

$$\frac{\partial^2}{\partial x^2}, \frac{\partial}{\partial y}, \frac{\partial}{\partial y}, \frac{\partial^2}{\partial x^2}$$

respectively. By taking $Z = q$ we get the transformed differential equation

$$\frac{\partial}{\partial y}\left[\frac{P^2}{Q} + f\left(\frac{P}{Q}\right)\right] = R.$$

Or, by carrying out the indicated differentiation,

$$\frac{2PQS - P^2T}{Q^2} + \frac{QS - PT}{Q^2}f'\left(\frac{P}{Q}\right) = R.$$

This is mongeable and has the intermediate integral

$$\chi\left(\frac{P}{Q}\right) + \frac{1}{Q} = \Phi'(Z),$$

where the function Φ is arbitrary, and where the function χ is related to the function f by the condition

$$f'\left(\frac{P}{Q}\right)\chi'\left(\frac{P}{Q}\right) = 1.$$

Treating the intermediate integral as a first-order partial differential equation, we find by standard methods the parametric solution:

$$x + uy = u\Phi(Z) - uZ\chi\left(\frac{1}{u}\right) + \Psi(u)$$

$$y = \Phi(Z) - Z\chi\left(\frac{1}{u}\right) + \frac{1}{u}Z\chi'\left(\frac{1}{u}\right) + \Psi'(u).$$

The result of eliminating $\Phi(Z)$ between these equations and then solving for Z is

$$Z = \frac{\Psi(u) - u\Psi'(u) - x}{\chi'\left(\frac{1}{u}\right)}.$$

Hence by eliminating Z we have

$$y = \Phi\left(\frac{\Psi(u) - u\Psi'(u) - x}{\chi'\left(\frac{1}{u}\right)}\right) + \frac{\Psi(u) - x}{u} + [x + u\Psi'(u) - \Psi(u)]\frac{\chi\left(\frac{1}{u}\right)}{\chi'\left(\frac{1}{u}\right)}.$$

Then since $Z = q$, the parametric solution which we find for (Ex. 5) consists of the above equation together with the equation

$$z = \int \frac{\Psi(u) - u\Psi'(u) - x}{\chi'\left(\frac{1}{u}\right)}\, dy + \beta(x).$$

In this we still have to fix the determination of the indicated integration, and find $\beta(x)$ in terms of $\alpha(x)$, by substituting in the given differential equation (Ex. 5). We leave this unfinished. But we notice the following additional solution of (Ex. 5), which is obtained from the "complete integral" of the intermediate integral and is not included in the parametric solution:

$$z = \Omega(y + ax) + \frac{1}{2}x^2 f(a) + cx + \iint \alpha(x)\, dx\, dx,$$

where a and c are constants.

8. Conclusion

In conclusion let me say that the problem of characterizing those partial differential equations of the form (15) which have an elementary or parametric general solution, and of characterizing those which have a generalized intermediate integral in some appropriate sense, would seem to be much less speculative than the problem concerning the form of the solution of ordinary differential equations. To find a satisfactory complete characterization in one or both cases, and a method of finding the solution or the intermediate integral when it exists, may not be too unreasonable a goal to set. And I believe that research in the area should seek general results tending in this direction, rather than isolated solutions of particular equations.

In the special case of linear differential equations it seems at least plausible to suppose that there is neither an elementary general solution nor a generalized intermediate integral unless the equation is either mongeable or reducible to mongeability by iterations of Laplace's transformation. I think it is true that the conjecture has at least this much support, that there is no known counterexample. And Laplace showed in his original memoir that any differential equation (23) that has a solution of the form

$$z = A\Phi(x) + B\Phi'(x) + C\Phi''(x) + \cdots + M\Phi^{(n)}(x),$$

where $A, B, C, \ldots, M$ are particular functions of x and y and $\Phi(x)$ is an arbitrary function of x, must be reducible to mongeability by iterations of his transformation. If the conjecture just mentioned could be proved, this would provide, in the linear case, at least one solution of the problem of characterizing elementary solvability—though perhaps not the best characterization that might be hoped for.

A parallel conjecture for equations of the general form (15) and as regards the transformations listed in Section 6, or some subset of them, would be rash, especially in the absence of knowledge as to whether the list of transformations approaches completeness. However, it is at least a possibility that something of the sort may hold for some suitable list of transformations.

An incidental question is whether there are any cases at all of elementary solvability of differential equations of the form (17), other than the case that $f_5(x, y, z, p, q)$ is identically 0. At least it would be of interest to have a counterexample to the conjecture that this is the only case.

Turning to differential equations of the form (16), we may seek to characterize the case that there exists an intermediate integral of some level n, by a criterion other than, or in some way better than, the existence of a transformation.

Another question concerns the reducibility of a differential equation (15) to one of the normal forms (16) or (17) by at least a substitution of the more general form that was discussed in Section 1 in connection with (13) and (22). It is not immediately clear how closely this is connected with the general problem I am urging. However, the reduction has evident advantages when possible—the transformations (Ia) and (Ib) are, for example, much easier to work with than the general transformation (I) of which they are special cases. A criterion and a general method of finding the reduction would be desirable.

THE HISTORY OF THE QUESTION OF EXISTENTIAL IMPORT OF CATEGORICAL PROPOSITIONS (1965)

This paper is not an original historical investigation in the sense that I have consulted all the original sources at first hand, but is mainly a report, utilizing the work of others in order to select and organize the history of a particular topic. I must acknowledge my indebtedness especially to the publications of I. M. Bocheński, Philotheus Boehner, and Ernest A. Moody.

Since this is a historical report, concerned in large part with traditional logic, I shall use the word "proposition" in its traditional sense. According to explicit definitions that were usual, a proposition is, in effect, a (declarative) sentence taken in conjunction with its meaning. In actual use the word often seems to vacillate between this (explicitly avowed) sense and the more abstract sense according to which different sentences may sometimes express the same proposition. But for the immediate purpose it is not necessary to fix the meaning of the word more precisely than this.

At the beginning of a certain elementary branch of logic (as it is now considered), i.e., the traditional theory of categorical propositions, a decision has to be made as to the meaning to be attached to sentences of categorical form in certain extreme cases in which one of the terms (subject or predicate) is either empty or universal. Especially the case of an empty subject term has proved troublesome or has led to controversy. As the usage of everyday language, out of which the traditional theory arose, is partly vacillating and partly unclear on this point, one simply has to make some convenient decision, subject to obvious conditions of adequacy and internal consistency, and then get on to more important matters. Technically there should be no great difficulty. But the history of the matter has a great deal of human interest, as to how the point could at first be overlooked, then later engender so much heat of controversy, so much plain confusion and stubborn clinging to preconceptions.

Aristotle never considers the question of existential import in connection with categorical inference, and the same is true of the early Scholastics, e.g. Petrus Hispanus.

Perhaps the best rendering of the Aristotelian logic in a way to conform to modern standards of rigor in the statement of it is that by Łukasiewicz.[1] Łukasiewicz's system is internally consistent, as he shows. But in the application it is required that empty terms must not be substituted for the variables. And for this reason the Łukasiewicz version of Aristotelian logic is inadequate for purposes for which that logic was originally intended.[2]

Originally published in ***Logic, Methodology and Philosophy of Science, Proceedings of the 1964 International Congress***, edited by Yehoshua Bar-Hillel, Studies in Logic and the Foundations of Mathematics, North-Holland Publishing Company, Amsterdam 1965, pp. 417–424.

[1] ***Elementy Logiki Matematycznej***, Warsaw 1929, and ***Aristotle's Syllogistic***, Oxford 1951, second edition 1957.

[2] Referring to a suggestion of von Freytag-Löringhoff, and in connection with Łukasiewicz's system, Paul Lorenzen proposes that the particular affirmative and particular negative propositions be defined by $PiQ = (\exists R)(RaP \wedge RaQ)$, $PoQ = (\exists R)(RaP \wedge ReQ)$ (***Archiv für Mathematische Logik und Grundlagenforschung***, vol. 2 (1956), pp. 100–103). Evidently these definitions require the same presupposition of non-emptiness that the Łukasiewicz system does, or else, as indicated in Lorenzen's ***Formale Logik*** (Berlin 1958, second edition 1962), they require "dass nur Pradikate aus einer vorgegebenen Klasse (etwa einer sog. Begriffspyramide) betrachtet werden." In either case they lead to difficulties in the actual use of the

The fact is that ordinary purposes do require reasoning with empty and universal terms, and with terms about which we do not know in advance but what they may be empty or universal. For example, consider the astronomer who infers conditions which life on Mars must satisfy — he does not know whether such life exists but his conclusions may influence our estimate of the probability — or the mathematician who infers one after the other a series of properties that all exponents not obeying Fermat's Last Theorem possess. Or consider the premisses that all men are mortal and that I am a man; it would seem that the usual inference from these premisses might properly and desirably be drawn, even by one who is genuinely in doubt whether immortal beings exist.

However, Łukasiewicz's system is of course not intended as a logic for use, but for a historical purpose, to render the logic of Aristotle into logistic form in such a way as to remain as close as possible to the original. From this point of view, Łukasiewicz may have the best possible formulation (or one such). But there do inevitably remain some questions and uncertainties. In particular, Łukasiewicz takes the position that contraposition and obversion are not Aristotelian; and it is because these are not included in his system that Łukasiewicz needs only the presupposition that terms are not empty, and not the additional presupposition that terms are not universal. Actually this is a little debatable, just because Aristotle is not unmistakably clear and consistent. In one passage Aristotle characterizes a negative name such as "non-man" as being not a name but an *ὄνομα ἀόριστον* (an infinite name, as it is usually translated — but this is quite a different meaning of "infinite" from that familiar in modern mathematics). In spite of this restriction or qualification, examples of contraposition do occur: "If man is an animal, what is not animal is not man," "If the pleasant is good, the non-good is not pleasant." Bocheński points out that these propositions are not quantified — at least not explicitly. Yet the only reasonable understanding of them is as universal affirmative; and the contraposition then seems to be of the sort which changes a universal affirmative into a universal negative. Probably from this source, contraposition appears in Petrus Hispanus as being presumably Aristotelian. But for Petrus Hispanus, contraposition changes a universal affirmative into a universal affirmative, and a particular negative into a particular negative. And I shall hereafter use the term "contraposition" in this sense.

There are also examples of obversion in Aristotle, e.g. the inference of "No man is just" from "Every man is non-just" and the inference of "Not every man is non-just" from "Some man is just." But it is conspicuous that these inferences are sanctioned in only one direction. Obversion of a negative proposition never occurs, and in this respect the Scholastic logic, wittingly or not, definitely goes beyond Aristotle.

P. F. Strawson's proposal[3] for validating the traditional and Aristotelian logic is so close to Łukasiewicz's in spirit that it is convenient to consider it here, out of its chronological order. In this proposal, the "presupposition" of non-emptiness affects only the subject term instead of all terms. It is partly on this account and partly because Strawson, unlike Łukasiewicz, wishes to preserve all the traditional immediate inferences, including contraposition and obversion, that Strawson's system proves to

---

logic as distinguished from its internal consistency. Moreover in the latter work Lorenzen indicates that under these definitions the particular negative PoQ is no longer the contradictory of the universal affirmative PaQ, so that the square of opposition is not preserved.

[3] ***Introduction to Logical Theory***, London and New York 1952.

be internally inadequate. For example, it is held that *No A is B* presupposes rather than asserts that there are *A*'s. There is no presupposition that there are *B*'s, and *No A is B* is allowed to be true even when nothing is *B*. In consequence if we wish to make the traditional inference from *No A is B* to *No B is A*, we must introduce the additional presupposition that there are *B*'s. Strawson is aware of this, but holds that the traditional immediate inferences are nevertheless satisfactorily preserved. Of course this cannot be accepted, and Strawson's position betrays a lamentable lack of concern for the formality of logic. Adequate formality requires that any matter of extralinguistic fact which must be known before an inference can be made shall be stated as a premiss of that inference. And failure to observe this raises difficulties which become more and more serious as one goes on to more advanced parts of logic and more complicated applications of logic. However, let us return to the task of the historical treatment of the question of existential import in more nearly chronological order.

As far as presently known the first logician to consider the question of existential import or to propose a tenable theory of it was William of Ockham, who holds that the affirmative categorical propositions are false and the negative true when the subject term is empty. The same idea occurs also in Buridan, as Moody points out, but I do not know that either Ockham or Buridan ever explicitly considers the particular negative in this regard. Details of the process by which this doctrine became generally accepted among the later Scholastics are a chapter in the history of logic that has still to be written. But it will be convenient to look at the culmination of it in the *Ars Logica* of John of St. Thomas, precisely because John is not an innovator in this area but follows predecessors, as he himself indicates.

John makes it explicit that a particular negative proposition is true if the subject term is empty, and even gives one example of this in which the subject term is self-contradictory.

Prima facie, Ockham's doctrine would invalidate contraposition, and also obversion of negative propositions, but John maintains these inferences by the device of "constantia." The "constantia" is namely the same as Strawson's "presupposition," except that it must be separately stated each time it is required. For example, in order to infer "Everything non-white is a non-man" from "All men are white" we must add the constantia: "And the non-white exists." In modern eyes, the constantia looks much like a second premiss so that contraposition and also obversion of a negative proposition become mediate inferences.

To avoid misleading in regard to the logic of John of St. Thomas it is necessary to add that John has, under the head of ampliation and restriction, a doctrine that rather strongly suggests the "universe of discourse" of the nineteenth century algebraists of logic, although John does not of course use this term or take altogether the point of view that is suggested by it. In particular, in the case of "All white men are men," if the verb is understood as referring to present time, the universe of discourse is that of presently existing things, and if there are no presently existing white men, the proposition is false. If the verb is understood as abstracting from time in order to make a timeless statement, then the universe of discourse is that of things "in intellectu," of essences rather than actualities, and the truth of the proposition requires only the existence of white men "in intellectu." This latter is the case of what John calls natural supposition. But in both these cases, and in the case of other universes of discourse (e.g. the universe of things existing in the past if the verb is in the past tense), the often disputed inference from "All white men are (were) men" to "Some men are (were) white men" is valid, on condition that the universe of discourse remains unchanged.

Of course natural supposition faces all the familiar difficulties in the logic of unactualized possibles, and it would seem that this particular universe of discourse would better be deleted in favor of a variety of others. But the idea of combining the Ockham doctrine of existential import with the admission of various different universes of discourse is entirely sound in itself.

There exists another, much more recent, attempt to save the traditional syllogisms and immediate inferences entire—which, though not all its proponents do this, is most conveniently described in direct reference to John of St. Thomas. Namely the proposal is that, for purposes of logic, all propositions shall be taken in natural supposition. Proponents of this usually add (what is undoubtedly the fact) that in everyday colloquial discourse the existential import of statements that are made varies from case to case and must be judged from the context. Some hold that logic has no business with these extra assertions of existence and should simply ignore them, taking all propositions in natural supposition. Others hold, more reasonably, that these assertions of existence should be taken as separate assertions, which are to be separately stated and treated. But even in this latter form, the restriction which the doctrine imposes upon the application of traditional logic is really quite intolerable—e.g., "Some men (presently existing) are wise" is not adequately represented by "Some men (*in intellectu*) are wise," even if we supply as additional propositions "Men presently exist" and "Wise beings presently exist."

The doctrine of existential import which was just described is referred to by Lewis Carroll in 1897, but he does not quote his source. The earliest proposal of the doctrine that is known to me is in the Cambridge dissertation of Abraham Wolf in 1900. Jacques Maritain attributes a somewhat similar idea to Lachelier (1907).

Maritain's own account[4] of existential import agrees in its main outlines with that of John of St. Thomas, but by comparison with it is seriously deficient in several respects. Namely (1) Maritain's account of the constantia is defective, and in particular he fails to remark that contraposition requires the constantia. (2) Maritain's examples in the discussion of existential import are entirely of affirmative propositions, with one exception; the matter of negative propositions with an empty or non-existent subject is illustrated only by a singular proposition, and it is said only that such a proposition *may* be true, rather than that it *must* be true (as consistency would require, and as John in fact holds). (3) Maritain fails to mention at all the crucial matter of propositions with self-contradictory subjects; it should have been said that, even in natural supposition, these propositions are false if affirmative, true if negative.

Some neo-Scholastics, being perhaps misled by Maritain's discussion of natural supposition, have misunderstood John of St. Thomas more radically, and in such a way that they obtain doctrines of existential import which to a greater or less extent resemble that of Wolf.

In particular F. C. Wade, in the introduction to his translation of the first part of the *Ars Logica*, proposes a doctrine which is quite close to Wolf's. It is to Wade's credit

[4] *Petite Logique* (**Éléments de Philosophie**, IIe fascicule), Paris 1923. In preparing this paper I have used the English translation by Imelda Choquette, which in various printings or editions has sometimes the title *Formal Logic* and sometimes the title *An Introduction to Logic*.

that he observes that there is a difficulty about propositions with self-contradictory subjects. But he seeks to meet the difficulty by excluding such propositions as not being propositions at all, failing to see that the self-contradictory is not always immediately recognizable in a way to make this feasible. Indeed it is often important, especially in mathematics, to be able to reason about a subject which for all one knows may be self-contradictory. And it seems that both Wolf and Wade would do better to have recourse to the Ockham doctrine of existential import in at least the extreme case of a self-contradictory subject.

What is often considered the modern doctrine of existential import, that the universal propositions are true and the particular false when the subject term is empty, appears implicitly in a short paper which was published by the mathematician Arthur Cayley in 1871.[5] In this paper Cayley uses the algebra of logic to work out a scheme of syllogisms in what is actually the obvious way. But he does not point out that the resulting scheme is at variance with the traditional one, and it seems at least possible that he was just not familiar with the traditional scheme. The explicit introduction of the modern doctrine was made independently by Franz Brentano[6] and C. S. Peirce[7] some years later. Brentano maintains some sort of absolute correctness for this account, because he maintains that the categorical propositions are, in an absolute sense, *really* either existential propositions or contradictories of existential propositions. But Peirce, after stating the senses of the categorical propositions "which are traditional" (i.e., Ockham's doctrine) explains the different senses of them which he will adopt—with no justification offered, but presumably just on the ground that he finds them more convenient for his purpose.

It is tempting to conclude that no satisfactory doctrine of existential import can preserve all the traditional immediate inferences and syllogisms, and the square of opposition, since in fact neither that of Ockham nor the modern doctrine does so.[8] The closest approach is the doctrine of Ockham and the late Scholastics, which does preserve all the properly Aristotelian inferences, including even those kinds of obversion and the inference akin to contraposition which Aristotle seems to allow. In this

[5] ***The Quarterly Journal of Pure and Applied Mathematics***, vol. 11 (1871), pp. 282–283. In the first display in the paper, "No *X*'s are *X*'s" is evidently a misprint for "No *X*'s are *Y*'s." The same correction applies also to the reprint of the paper in Volume 8 of Cayley's ***Collected Mathematical Papers***.

[6] Book 11 Chapter VII of ***Psychologie vom Empirischen Standpunkte***, Leipzig 1874. Compare also J. P. N. Land, in ***Mind***, vol. 1 (1876), pp. 289-292, and Franz Hillebrand, ***Die Neuen Theorien der Kategorischen Schlüsse***, Vienna 1891.

[7] ***American Journal of Mathematics***, vol. 3 (1880), pp. 15–57. [*Editor's note*: See Church's review of John Venn's ***Symbolic logic*** for further discussion of the historical priority here.]

[8] Such a conclusion is debatable only because of the vagueness of the adjective "satisfactory." Strawson points out that all the traditional immediate inferences and syllogisms, and the traditional square of opposition can be preserved in the context of a standard (modern) logic of quantifiers if we define PaQ as $\sim(\exists x)(Px \sim Qx) \,.\, (\exists x)Px \,.\, (\exists x) \sim Qx$, and PeQ as $\sim(\exists x)(PxQx) \,.\, (\exists x)Px \,.\, (\exists x)Qx$ and PiQ as $\sim \,.\, PeQ$, and PoQ as $\sim \,.\, PaQ$. This loses the traditional law of identity PaP. But even the law of identity can be maintained too, if we follow a slightly more complicated scheme proposed by H. B. Smith (***The Journal of Philosophy***, vol. 21 (1924), pp. 631–633). In effect, we obtain Smith's scheme as a modification of Strawson's if we allow PaQ to be true also when P and Q are *both* empty, and when P and Q are *both* universal—and modify Strawson's meanings of PeQ, PiQ, and PoQ correspondingly. The same idea was later suggested independently by Stanislaw Jaśkowski (***Studia Societatis Scientiarum Torunensis***, section A, vol. 2 no. 3 (1950) pp. 77–90). But it seems that all those who have proposed or considered schemes of this sort put them forward as mere curiosities and immediately reject them as being too artificial for actual adoption.

latter respect, and quite apart from the matter of application of the logic (which was discussed above), the late Scholastic version is in fact superior to Łukasiewicz's.[9]

If one attaches importance to the purely historical faithfulness of preserving the traditional scheme, either unchanged, or with as little change as possible, then one must decide whether it is the Aristotelian logic proper that is to be preserved, or the Scholastic logic (e.g. of Petrus Hispanus), or one of the post-Scholastic versions of traditional logic. And one must decide how much importance to attach to the linguistic oddity that the late Scholastic doctrine makes it true that "Some chimaeras are not animals" — on the ground alone that there are no chimaeras.[10]

Finally it should be said that if one allows the constantia at appropriate places, in effect a second premiss for certain immediate inferences, then both the late Scholastic version of existential import and the modern version preserve the entire traditional scheme in (what is otherwise) the broadest interpretation.

---

[9]It would indeed be possible to modify the Łukasiewicz version by introducing the additional presupposition that terms are not universal and then to allow all the usual cases of obversion and contraposition. Cf. Ivo Thomas, ***Dominican Studies*** vol. 2 (1949) pp. 145–160, and Anders Wedberg, ***Ajatus***, vol. 15 (1949) pp. 299–314. But the resulting modified Łukasiewicz system is even more deficient than the original from the point of view of its application as a formalized or systematized version of the logic of ordinary discourse.

[10]One might mitigate the oddity by reading the particular negative with the words "not all" — as e.g. "Not all chimaeras are animals" must evidently be counted true if "All chimaeras are animals" is counted false. But the interchangeability of "not all" and "some not" (or more exactly, the analogous interchangeability in Latin) was already laid down by Petrus Hispanus and, as far as I know, was not questioned by later Scholastics. On the other hand, John of St. Thomas finds it necessary to defend the linguistic oddity by remarking that it's as if one had said "The chimaera is not an animal."

A GENERALIZATION OF LAPLACE'S TRANSFORMATION (1966)

In this paper we seek to deal with a question regarding elementary solutions of second-order partial differential equations in two independent variables, namely that of solutions of the form

$$z = A(x, y, \varphi(x), \varphi'(x), \varphi''(x), \ldots, \varphi^{(n)}(x)), \tag{1}$$

where φ is an arbitrary function.

The term "elementary solution" is here used, not with reference to any particular class of functions called elementary functions, but simply to mean a solution involving one or more arbitrary functions and expressed in terms of them by means of particular functions (not necessarily elementary) and the operations of differentiation and indefinite integration.

In place of (1), our results will apply in part also to solutions of the somewhat more general form

$$z = A(x, y, \varphi(x), \varphi_1(x), \varphi_2(x), \ldots, \varphi_n(x)), \tag{2}$$

where φ is an arbitrary function, and $\varphi_1, \varphi_2, \ldots, \varphi_n$ are functions that depend on the function φ in any way at all, subject to the restriction that no relation of the form

$$\begin{aligned} \varphi_n'(x) = B(x, \varphi(x), \varphi_1(x), \varphi_2(x), \\ \ldots, \varphi_n(x), \varphi'(x), \varphi_1'(x), \varphi_2'(x), \ldots, \varphi_{n-1}'(x)) \end{aligned} \tag{3}$$

shall hold for arbitrary choice of the function φ.

The forms of solution (1) and (2) are of course special, both in restricting the argument of the arbitrary function φ to be x and in the restriction that no relation of the form (3) shall hold, if not also otherwise, and in a concluding section we shall give some indications in regard to the possibility of extending our methods so as to remove the two restrictions named.

Following conventions of notation that are standard we shall use z as dependent variable and x and y as independent variables. And then the letters p and q are used to stand for the two first-order partial derivatives (of z with respect to x and y respectively), and the letters r, s, and t are used to stand for the three second-order partial derivatives. Similarly, if new variables Z, X and Y are introduced by a transformation, we shall use P and Q to stand for the two first-order partial derivatives (of Z with respect to X and to Y) and R, S, and T to stand for the three second-order partial derivatives. And if new variables ζ, ξ, and η are introduced, we shall use π, ϙ for the two first-order partial derivatives, and ρ, σ, τ for the three second-order partial derivatives.

Originally published as a monograph in the series ***Annales Academiae Scientiarum Fennicae***, Series A.I, Mathematica, no. 377, Helsinki 1966, 34 pp.

To ROLF NEVANLINNA on his 70th birthday.

For support of the work represented by this paper the author is indebted to the National Science Foundation of the United States.

We shall also employ in connection with n-ary functions a notation, analogous to the standard use of the prime in connection with singulary functions, by which the derived functions are denoted by placing a numerical subscript after the function letter. For example, if the letter f denotes a ternary function, then f_1, f_2, f_3 denote the three derived functions obtained by taking the partial derivative with respect to the first argument, the second argument, and the third argument respectively. And thus, e.g., $f_3(x, y, z)$ expresses what would more usually be expressed by

$$\frac{\delta}{\delta z} f(x, y, z)$$

or by $f_z(x, y, z)$, while $f_3(z, x, y)$ corresponds rather to

$$\frac{\delta}{\delta y} f(z, x, y)$$

or to $f_y(z, x, y)$. Similarly f_{12} denotes the function obtained from f by taking the second partial derivative, with respect to the first argument and then with respect to the second argument; and f_{22} denotes the function obtained from f by taking the second partial derivative, with respect to the second argument twice.

Properly the standard notations such as

$$\frac{\delta}{\delta x}, \frac{\delta}{\delta y}, \frac{\delta^2}{\delta y\, \delta z} \tag{4}$$

are applicable to forms or to letters which stand for forms, while the numerical-subscript notation is applicable rather to letters that denote functions. The distinction is important in principle and must be maintained, although it will be somewhat obscured in the present paper by our practice of omitting the arguments after a function letter, purely as an abbreviation, when it is clear from the context what the arguments are, or when (as frequently happens in the treatment of differential equations) the arguments of a particular function letter remain the same throughout some one context.

This use of, for example, the letter f alone as an abbreviation of $f(x, y, z)$ must of course also be distinguished from the more proper use of the letter f, to denote the function itself as an abstract entity. The use of the abbreviation is justified only so far as it does not engender real confusion.

The numerical subscript notation for the derived functions of an n-ary function has some inconveniences, among them that subscripts used for other purposes must sometimes be enclosed in parentheses to avoid confusing them with subscripts referring to partial derivatives.

However, the numerical-subscript notation avoids the well-known equivocacy of the standard notation (4)—which fails to indicate, when the partial derivative is taken with respect to a particular variable, what the other variables are that are being held constant. This makes the numerical-subscript notation a substantial aid to thought when there are distinctions to be made in this regard, and for this reason we shall tend to use it in preference to the notation (4) whenever it seems to be easily and conveniently possibly to do so.

In the statement of results in the following sections there are certain more or less evident conditions which will generally be left tacit. These include the existence of derivatives which are used, the existence of solutions of certain differential equations (as shown by the context to be required), the existence of certain implicit functions, and the restriction of results to an appropriate neighborhood as may be necessary to secure the foregoing. In general these are conditions which we expect could be secured by imposing appropriate ordinary conditions of regularity on the coefficients of the differential equation for which solutions are sought, and a more thorough account than is attempted in the present paper should spell this out in detail.

1. Preliminary cases

1.1. *If a partial differential equation*

$$f(x, y, z, p, q, r, s, t) = 0 \tag{5}$$

has a solution of the form (1) *or of the form* (2), *then this solution must satisfy the partial differential equation* (5) *identically in* r.

By satisfying (5) "identically in r" it is meant that the solution in fact satisfies

$$f(x, y, z, p, q, u, s, t) = 0 \tag{6}$$

where u is a new independent variable (independent of x and y). And of course it follows as a corollary that any solution of (5) of the form (1) or (2) must be a solution also of the differential equation

$$f_6(x, y, z, p, q, r, s, t) = 0 \tag{7}$$

For a solution of the form (1), theorem 1.1 becomes immediately evident if we substitute the assumed solution in the differential equation (5). In fact substitution of (1) in (5) yields

$$\begin{aligned} f(x, y, A, A_1 + A_3\varphi' + A_4\varphi'' + \cdots + A_{n+3}\varphi^{(n+1)}, A_2, A_{11} + 2A_{13}\varphi' \\ + \cdots + A_{n+3}\varphi^{(n+2)}, A_{12} + A_{23}\varphi' + A_{24}\varphi'' + \cdots \\ + A_{2(n+3)}\varphi^{(n+1)}, A_{22}) = 0 \end{aligned} \tag{8}$$

And since this must hold for an arbitrary function φ, it must hold identically in $x, y, \varphi, \varphi', \varphi'', \ldots, \varphi^{(n+2)}$ as $n + 5$ independent variables. We may of course assume that the function A is not independent of its last argument, and hence that A_{n+3} does not vanish identically. Theorem 1.1 then follows because, in (8), $\varphi^{(n+2)}$ occurs in the sixth argument of f but nowhere else.

On substituting (2) in (5), we see that the condition that no relation of the form (3) shall hold is indispensable as far as 1.1 is concerned. For we may argue that not all of $\varphi'', \varphi_1'', \varphi_2'', \ldots, \varphi_n''$ can be expressed each as a function of $x, \varphi, \varphi_1, \varphi_2, \ldots, \varphi_n, \varphi', \varphi_1', \varphi_2', \ldots, \varphi_n'$, as this would not be compatible with the arbitrariness of the function φ. Then if a solution of the form (2) should satisfy (5) otherwise than identically in r, we would have an equation of the form

$$\begin{aligned} A_3\varphi'' + A_4\varphi_1'' + A_5\varphi_2'' + \cdots + A_{n+3}\varphi_n'' \\ = \theta(x, y, \varphi, \varphi_1, \varphi_2, \ldots, \varphi_n, \varphi', \varphi_1', \varphi_2', \ldots, \varphi_n') \end{aligned} \tag{9}$$

holding identically. From this equation by taking n times in succession the partial derivative with respect to y we get the n equations

$$A_{23}\,\varphi'' + A_{24}\,\varphi_1'' + A_{25}\,\varphi_2'' + \cdots + A_{2(n+3)}\,\varphi_n'' = \theta_2$$
$$A_{223}\,\varphi'' + A_{224}\,\varphi_1'' + A_{225}\,\varphi_2'' + \cdots + A_{22(n+3)}\,\varphi_n'' = \theta_{22}$$
$$\cdots\cdots\cdots\cdots\cdots\cdots\cdots\cdots\cdots\cdots\cdots\cdots\cdots\cdots$$
$$A_{22\ldots23}\,\varphi'' + A_{22\ldots24}\,\varphi_1'' + A_{22\ldots25}\,\varphi_2'' + \cdots + A_{22\ldots2(n+3)}\,\varphi_n'' = \theta_{22\ldots2}$$

These $n+1$ equations, regarded as linear algebraic equations in $\varphi'', \varphi_1'', \varphi_2'', \ldots, \varphi_n''$, must not all be independent, i.e., the determinant

$$\begin{vmatrix} A_3 & A_4 & A_5 & \ldots & A_{n+3} \\ A_{23} & A_{24} & A_{25} & \ldots & A_{2(n+3)} \\ A_{223} & A_{224} & A_{225} & \ldots & A_{22(n+3)} \\ \cdots & \cdots & \cdots & \cdots & \cdots \\ A_{22\ldots23} & A_{22\ldots24} & A_{22\ldots25} & \ldots & A_{22\ldots2(n+3)} \end{vmatrix}$$

must vanish identically. From this it follows that there must be a linear relation

(10) $$a_0 A_3 + a_{(1)} A_4 + a_{(2)} A_5 + \cdots + a_{(n)} A_{n+3} = 0$$

which holds identically and whose coefficients $a_0, a_{(1)}, a_{(2)}, \ldots, a_{(n)}$ are functions of $x, \varphi, \varphi_1, \varphi_2, \ldots, \varphi_n$. By treating (10) as a partial differential equation to be solved for A and considering the form of its solution, we see that it must be possible to rewrite the solution (2) of (5) in such a way that the number n is reduced by 1. This can evidently be iterated until n is reduced to 0, at which point the solution (2) has been reduced to a special case of (1).

This completes the proof of 1.1.

Now for present purposes the problem of finding all the solutions of the form (1) or the form (2) for a given second-order partial differential equation will be dismissed as solved if we have found at least one additional partial differential equation in the same dependent and independent variables that has to be satisfied—provided that the additional partial differential equation is independent of the first one and is of not higher than the second order. This dismissal might be thought too summary, in the absence of a definitive treatment of the question of simultaneous solutions of two second-order partial differential equations in one dependent variable z and the same independent variables x and y. But it will serve to separate this rather different question from the main topic of the present paper. And there is moreover no real difficulty over the matter in the present context, because when simultaneous partial differential equations arise we are concerned, not with all their common solutions, but only with their common solutions of the form (1) or (2), and the theorems of the present section, especially 1.2, may therefore be used to make further reductions.

We shall therefore regard the problem of finding all solutions of the form (1) or (2) for a partial differential equation (5) as solved by 1.1, except in the case in which f_6 vanishes identically, i.e., the case in which (5) is independent of r. In fact we shall think of 1.1 as meaning that, although there do exist cases in which a differential equation (5) not independent of r has a solution of the form (1) or (2), such solutions are in a sense exceptional, and the main case we have to consider is that of a differential equation

$$f(x, y, z, p, q, s, t) = 0 \tag{11}$$

However, we go on immediately to a number of theorems analogous to 1.1, which we regard as showing that (11) is still too general, and that, in a certain sense which we do not attempt to make definite, the main case has to be regarded as consisting only of certain subcases of the case (11).

1.2. *If a partial differential equation*

$$f(x, y, z, p, q, t) = 0 \tag{12}$$

has a solution of the form (1) *or of the form* (2), *then this solution must satisfy the partial differential equation* (12) *identically in* p.

The proof of 1.2 is exactly analogous to that of 1.1.

1.3. *If a partial differential equation*

$$s = f(x, y, z, p, q, t) \tag{13}$$

has a solution of the form (1), *or a solution of the form* (2) *such that no relation* (3) *holds for arbitrary* φ, *then this solution must satisfy*

$$f_{44}(x, y, z, p, q, t) = 0 \tag{14}$$

identically in p.

For if we substitute in (13) an assumed solution of the form (1), we get

$$\begin{aligned} A_{12} + A_{23}\varphi' + A_{24}\varphi'' + \cdots + A_{2(n+3)}\varphi^{(n+1)} \\ = f(x, y, A, A_1 + A_3\varphi' + A_4\varphi'' + \cdots + A_{n+3}\varphi^{(n+1)}, A_2, A_{22}) \end{aligned} \tag{15}$$

This must hold identically in $x, y, \varphi, \varphi', \varphi'', \ldots, \varphi^{(n+1)}$ as $n+4$ independent variables. We may assume that A_{n+3} does not vanish identically, and hence by taking in (15) the second partial derivative with respect to $\varphi^{(n+1)}$,

$$f_{44}(x, y, A, A_1 + A_3\varphi' + A_4\varphi'' + \cdots + A_{n+3}\varphi^{(n+1)}, A_2, A_{22}) = 0 \tag{16}$$

Equation (16) means that (1) satisfies (14), and from 1.2 it then follows that (1) satisfies (14) identically in p.

The argument is analogous in the case of a solution of the form (2), but the condition that there is no relation of the form (3) holding for arbitrary φ is evidently essential.

At this point it remains to consider differential equations of the two forms

$$f(x, y, z, q, t) = 0 \tag{17}$$

$$s = \mu(x, y, z, q, t) + \nu(x, y, z, q, t)p \tag{18}$$

as being the only ones not yet covered. That is, for a second-order differential equation of any form other than (17) or (18), theorems 1.1–1.3 provide the means to find all solutions of the form (1), and at least all those solutions of the form (2) which obey the condition that no relation (3) holds for arbitrary φ.

We shall ignore (17) as being properly an ordinary rather than a partial differential equation, and our main concern will therefore be with the case of differential equations of the form (18). This is the case to which the generalization of Laplace's transformation applies, and we might now turn immediately to consideration of this transformation. But since the method of finding a second differential equation which the required solution must satisfy can as a matter of fact be pressed somewhat further, at least in the case in which the required solution is of the form (1), we proceed to develop this first.

In connection with (18) we shall need to use functions χ and λ determined as follows:

$$(19)\quad \lambda(x, y, z, q, t) = \mu_3 + \mu_4 v - \mu v_4 - v_1 + (v_2 + v_3 q + v_4 t + v^2)\mu_5 - (\mu_5 + \mu_3 q + \mu_4 t + \mu v)v_5$$

$$(20)\quad \chi(x, y, z, q, t) = \lambda_3 + \lambda_4 v - \lambda v_4 + (v_2 + v_3 q + v_4 t + v^2)\lambda_5 - (\lambda_2 + \lambda_3 q + \lambda_4 t + 2\lambda v)v_5$$

If we assume for (18) a solution of the form (1), we get, by substituting the assumed solution in the differential equation (18),

$$(21)\quad A_{12} + A_{23}\varphi' + A_{24}\varphi'' + \cdots + A_{2(n+3)}\varphi^{(n+1)} = \mu(x, y, A, A_2, A_{22}) + v(x, y, A, A_2, A_{22})[A_1 + A_3\varphi' + A_4\varphi'' + \cdots + A_{n+3}\varphi^{(n+1)}]$$

This must hold identically in $x, y, \varphi, \varphi', \varphi'', \ldots, \varphi^{(n+1)}$ as independent variables, and hence we must have separately the two following equations (in which the arguments of μ and v are x, y, A, A_2, A_{22}):

$$(22)\quad A_{12} + A_{23}\varphi' + \cdots + A_{2(n+1)}\varphi^{(n-1)} + A_{2(n+2)}\varphi^{(n)} = \mu + v[A_1 + A_3\varphi' + \cdots + A_{n+1}\varphi^{(n-1)} + A_{n+2}\varphi^{(n)}]$$

$$(23)\quad A_{2(n+3)} = vA_{n+3}$$

From equation (22), by taking the partial derivative with respect to y, and also with respect to $\varphi^{(n)}$, we get the two equations:

$$(24)\quad A_{122} + A_{223}\varphi' + \cdots + A_{22(n+1)}\varphi^{(n-1)} + A_{22(n+2)}\varphi^{(n)} = \mu_2 + \mu_3 A_2 + \mu_4 A_{22} + \mu_5 A_{222} + (v_2 + v_3 A_2 + v_4 A_{22} + v_5 A_{222}) [A_1 + A_3\varphi' + \cdots + A_{n+1}\varphi^{(n-1)} + A_{n+2}\varphi^{(n)}] + v[A_{12} + A_{23}\varphi' + \cdots + A_{2(n+1)}\varphi^{(n-1)} + A_{2(n+2)}\varphi^{(n)}]$$

$$(25)\quad A_{12(n+3)} + A_{23(n+3)}\varphi' + \cdots + A_{2(n+1)(n+3)}\varphi^{(n-1)} + A_{2(n+2)(n+3)}\varphi^{(n)} + A_{2(n+2)} = \mu_3 A_{n+3} + \mu_4 A_{2(n+3)} + \mu_5 A_{22(n+3)} + (v_3 A_{n+3} + v_4 A_{2(n+3)} + v_5 A_{22(n+3)})[A_1 + A_3\varphi' + \cdots + A_{n+1}\varphi^{(n-1)} + A_{n+2}\varphi^{(n)}] + v[A_{1(n+3)} + A_{3(n+3)}\varphi' + \cdots + A_{(n+1)(n+3)}\varphi^{(n-1)} + A_{(n+2)(n+3)}\varphi^{(n)} + A_{n+2}]$$

From equation (23), by applying the operator

$$\frac{\delta}{\delta x} + \varphi'\frac{\delta}{\delta \varphi} + \varphi''\frac{\delta}{\delta \varphi'} + \cdots + \varphi^{(n-1)}\frac{\delta}{\delta \varphi^{(n-2)}}$$

and by taking the partial derivative with respect to y and with respect to $\varphi^{(n-1)}$, we get the three equations:

$$(26)\quad \begin{aligned} &A_{12(n+3)} + A_{23(n+3)}\varphi' + \cdots + A_{2(n+1)(n+3)}\varphi^{(n-1)} \\ &\quad = [\nu_1 + \nu_3(A_1 + A_3\varphi' + \cdots + A_{n+1}\varphi^{(n-1)}) + \nu_4(A_{12} + A_{23}\varphi' + \cdots \\ &\qquad + A_{2(n+1)}\varphi^{(n-1)}) + \nu_5(A_{122} + A_{223}\varphi' + \cdots + A_{22(n+1)}\varphi^{(n-1)}]A_{n+3} \\ &\qquad\qquad + \nu[A_{1(n+3)} + A_{3(n+3)}\varphi' + \cdots + A_{(n+1)(n+3)}\varphi^{(n-1)}] \end{aligned}$$

$$(27)\quad A_{22(n+3)} = (\nu_2 + \nu_3 A_2 + \nu_4 A_{22} + \nu_5 A_{222})A_{n+3} + \nu A_{2(n+3)}$$

$$(28)\quad A_{2(n+2)(n+3)} = (\nu_3 A_{n+2} + \nu_4 A_{2(n+2)} + \nu_5 A_{22(n+2)})A_{n+3} + \nu A_{(n+2)(n+3)}$$

Now we multiply equations (24)–(28) by $-\nu_5 A_{n+3}$, $+1$, -1, $\mu_5 + \nu_5[A_1 + A_3\varphi' + \cdots + A_{n+2}\varphi^{(n)}]$, $\varphi^{(n)}$ respectively, and add. Then use (22) to replace $A_{12} + A_{23}\varphi' + \cdots + A_{2(n+2)}\varphi^{(n)}$ by $\mu + \nu[A_1 + A_3\varphi' + \cdots + A_{n+2}\varphi^{(n)}]$, and (23) to replace $A_{2(n+3)}$ by νA_{n+3}. The result is

$$(29)\quad A_{2(n+2)} = \nu A_{n+2} + \lambda(x, y, A, A_2, A_{22})A_{n+3}$$

We need to look separately at the special case in which n is 0, i.e., the case in which the solution (1) reduces to

$$(30)\quad z = A(x, y, \varphi(x))$$

In this special case equations (22)–(25) become:

$A_{12} = \mu + \nu A_1, \qquad A_{23} = \nu A_3$

$A_{122} = \mu_2 + \mu_3 A_2 + \mu_4 A_{22} + \mu_5 A_{222} + (\nu_2 + \nu_3 A_2 + \nu_4 A_{22} + \nu_5 A_{222})A_1 + \nu A_{12}$

$A_{123} = \mu_3 A_3 + \mu_4 A_{23} + \mu_5 A_{223} + (\nu_3 A_3 + \nu_4 A_{23} + \nu_5 A_{223})A_1 + \nu A_{13}$

For (26) and (27) we use the two equations obtained from $A_{23} = \nu A_3$ by taking the partial derivative with respect to x and with respect to y:

$$A_{123} = (\nu_1 + \nu_3 A_1 + \nu_4 A_{12} + \nu_5 A_{122})A_3 + \nu A_{13}$$
$$A_{223} = (\nu_2 + \nu_3 A_2 + \nu_4 A_{22} + \nu A_{222})A_3 + \nu A_{23}$$

There is no equation (28) in the special case. If we multiply the four last equations, corresponding to (24)–(27), by $-\nu_5 A_3$, $+1$, -1, $\mu_5 + \nu_5 A_1$ respectively and add, and then replace A_{12} by $\mu + \nu A_1$ and A_{23} by νA_3, we get in place of (29):

$$0 = \lambda(x, y, A, A_2, A_{22})A_3$$

As the intention of (30) is of course that A_3 does not vanish identically, i.e., that we actually have a solution depending on an arbitrary function, it follows for any solution of (18) of the form (30) that

$$(31)\quad \lambda(x, y, A, A_2, A_{22}) = 0$$

Now taking in (29) the partial derivative with respect to y and with respect to $\varphi^{(n)}$, we get the two equations:

$$(32)\quad \begin{aligned} A_{22(n+2)} &= (\nu_2 + \nu_3 A_2 + \nu_4 A_{22} + \nu_5 A_{222} + \nu^2)A_{n+2} \\ &\quad + (\lambda_2 + \lambda_3 A_2 + \lambda_4 A_{22} + \lambda_5 A_{222} + 2\lambda\nu)A_{n+3} \end{aligned}$$

(33) $$A_{2(n+2)(n+3)} = (\nu_3 A_{n+3} + \nu_4 A_{2(n+3)} + \nu_5 A_{22(n+3)})A_{n+2} + \nu A_{(n+2)(n+3)} + (\lambda_3 A_{n+3} + \lambda_4 A_{2(n+3)} + \lambda_5 A_{22(n+3)})A_{n+3} + \lambda A_{(n+3)(n+3)}$$

And upon multiplying equations (27), (28), (32), (33) by $\nu_5 A_{n+2} + \lambda_5 A_{n+3}$, -1, $-\nu_5 A_{n+3}$, $+1$ respectively and adding, and then using (23) and (29) to replace $A_{2(n+3)}$ by νA_{n+3} and $A_{2(n+2)}$ by $\nu A_{n+2} + \lambda A_{n+3}$, we get

(34) $$\lambda(x, y, A, A_2, A_{22})A_{(n+3)(n+3)} + \chi(x, y, A, A_2, A_{22})A^2_{n+3} = 0$$

From (34), by taking the partial derivative with respect to y and using (23), there results

(35) $$\lambda A_{2(n+3)(n+3)} + (\lambda_2 + \lambda_3 A_2 + \lambda_4 A_{22} + \lambda_5 A_{222})A_{(n+3)(n+3)} + (\chi_2 + \chi_3 A_2 + \chi_4 A_{22} + \chi_5 A_{222} + 2\chi\nu)A^2_{n+3} = 0$$

Also from (23), by taking the partial derivative with respect to $\varphi^{(n)}$, we get

(36) $$A_{2(n+3)(n+3)} = (\nu_3 A_{n+3} + \nu_4 A_{2(n+3)} + \nu_5 A_{22(n+3)})A_{n+3} + \nu A_{(n+3)(n+3)}$$

Then multiply equations (27), (34)–(36) by $\lambda^2 \nu_5 A_{n+3}$, $\lambda_2 + \lambda_3 A_2 + \lambda_4 A_{22} + \lambda_5 A_{222} + \lambda\nu$, $-\lambda$, λ^2 respectively, and add. Use (23) to replace $A_{2(n+3)}$ by νA_{n+3}, then divide out A^2_{n+3} (since we may assume that A_{n+3} does not vanish identically). The result is

(37) $$\lambda^2\nu_3 + \lambda^2\nu\nu_4 + (\nu_2 + \nu_3 A_2 + \nu_4 A_{22} + \nu_5 A_{222} + \nu^2)\lambda^2\nu_5 + \chi\lambda\nu + \chi_2\lambda - \chi\lambda_2 + (\chi_3\lambda - \chi\lambda_3)A_2 + (\chi_4\lambda - \chi\lambda_4)A_{22} + (\chi_5\lambda - \chi\lambda_5)A_{222} = 0$$

Hence the theorems:

1.4. *If a partial differential equation* (18) *has a solution of the form* (30), *this solution must satisfy also the differential equation*

(38) $$\lambda(x, y, z, q, t) = 0$$

or must satisfy a relation among x, y, z, q, t *for which one of* $\mu, \mu_2, \mu_3, \mu_4, \mu_5, \nu, \nu_1, \nu_2, \nu_3, \nu_4, \nu_5$ *is singular.*

1.5. *If a partial differential equation* (18) *has a solution of the form* (1), *this solution must satisfy also the differential equation*

(39) $$\lambda^2\nu_3 + \lambda^2\nu\nu_4 + (\nu_2 + \nu_3 q + \nu_4 t + \nu^2)\lambda^2\nu_5 + \chi\lambda\nu + \chi_2\lambda - \chi\lambda_2 + (\chi_3\lambda - \chi\lambda_3)q + (\chi_4\lambda - \chi\lambda_4)t + (\lambda^2\nu_5^2 + \chi_5\lambda - \chi\lambda_5)\frac{\delta t}{\delta y} = 0$$

or must satisfy a solution among x, y, z, q, t *for which one of* $\mu, \mu_2, \mu_3, \mu_4, \mu_5, \nu, \nu_1, \nu_2, \nu_3, \nu_4, \nu_5, \lambda_2, \lambda_3, \lambda_4, \lambda_5, \chi_2, \chi_3, \chi_4, \chi_5$ *is singular.*

The alternative which enters in 1.4 and 1.5—that at $z = A, q = A_2, t = A_{22}$ either μ or ν or one (or more) of the partial derivatives which are listed has a singularity—arises because the argument above has tacitly assumed that they are non-singular. The necessity of including this alternative may be shown by examples. For instance the differential equation

$$s = 1 + 3pt^{\frac{1}{3}}$$

has a solution $z = (x + c)y + \varphi(x)$ which, since $\lambda = -3t^{-\frac{1}{3}}$, corresponds to a singularity of v_5 rather than a zero of λ. Again the differential equation

$$(z + yq)s = (z + yq)^2 + 2pq + ypt$$

has a solution $z = y^{-1}\varphi(x)$ which, since $\lambda = -1$, corresponds to a singularity of v rather than a zero of λ.

As a consequence of 1.4 and 1.5, we may expect generally that if a partial differential equation (18) has a two-arbitrary-function solution which involves, besides $\varphi(x)$, one other arbitrary function ψ and which, for all or almost all particular choices of ψ, reduces to the form (30), or the form (1), then $\lambda(x, y, z, q, t)$ will vanish identically, or

$$\lambda^2 v_3 + \lambda^2 v v_4 + (v_2 + v_3 q + v_4 t + v^2)\lambda^2 v_5 + \chi\lambda v + \chi_2\lambda - \chi\lambda_2 + (\chi_3\lambda - \chi\lambda_3)q + (\chi_4\lambda - \chi\lambda_4)t \tag{40}$$

$$\lambda^2 v_5^2 + \chi_5\lambda - \chi\lambda_5 \tag{41}$$

will both vanish identically. But it seems to be difficult to find an exact statement of such a result without imposing some undesirable restriction on the form of the two-arbitrary-function solution (as e.g. that it shall be elementary). And it may be better instead to seek related results about the existence of intermediate integrals of the sort which is illustrated by theorem 2.2 below.

As a special case we notice that (40) and (41) both vanish identically when the differential equation (18) is linear.

2. Generalization of Laplace's transformation, first case

Treating (23) as a partial differential equation to be solved for A, we find that it has an intermediate integral involving an arbitrary function. In fact if we assume an intermediate integral

$$A_2 = \theta(x, y, \varphi, \varphi', \varphi'', \ldots, \varphi^{(n-1)}, A) \tag{42}$$

we get as condition on θ

$$\theta_{n+3} = v(x, y, A, \theta, \theta_2 + \theta\theta_{n+3}) \tag{43}$$

And solution of the first-order differential equation (43) for θ yields the intermediate integral. In this section we deal with the special case in which the function v is of not higher than first degree in its last argument,

$$v(x, y, z, q, t) = \beta(x, y, z, q) + \delta(x, y, z, q)t \tag{44}$$

—the intermediate integral of (23) being in this case of the Monge form.

When v has the form (44), the condition (43) becomes

$$\theta_{n+3} = \beta(x, y, A, \theta) + \delta(x, y, A, \theta)[\theta_2 + \theta\theta_{n+3}] \tag{45}$$

The general solution of the differential equation (45) may be written as

$$G(x, y, A, \theta) = \Phi(x, F(x, y, A, \theta), \varphi, \varphi', \varphi'', \ldots, \varphi^{(n-1)}) \tag{46}$$

where Φ is an arbitrary function and F and G are particular functions satisfying the conditions

$$G_3 = -\beta G_4 + \delta[G_2 + \theta G_3] \tag{47}$$

$$F_3 = -\beta F_4 + \delta[F_2 + \theta F_3] \tag{48}$$

—as (47), (48) are in fact the conditions for $G(x, y, A, \theta) =$ constant, $F(x, y, A, \theta) =$ constant to be solutions of (45), and where of course we must so choose F and G that these are independent solutions of (45), if (46) is to be the general solution of (45).

The partial differential equation (18) has in the present case the special form

$$s = \mu(x, y, z, q, t) + \beta(x, y, z, q)p + \delta(x, y, z, q)pt \tag{49}$$

Generally, for solutions of (49) of the form (1), it follows from (42) and (46) that

$$G(x, y, A, A_2) = \Phi(x, F(x, y, A, A_2), \varphi, \varphi', \varphi'', \dots, \varphi^{(n-1)}) \tag{50}$$

must hold identically in $x, y, \varphi, \varphi', \varphi'', \dots, \varphi^{(n)}$ for some choice of the function Φ. Hence if we make a transformation of the differential equation (49) by letting

$$X = x, \quad Y = F(x, y, z, q), \quad Z = G(x, y, z, q) \tag{51}$$

the solution of the transformed differential equation which corresponds to the solution $z = A$ of (49) may be expected to be

$$Z = \Phi(X, Y, \varphi(X), \varphi'(X), \varphi''(X), \dots, \varphi^{(n-1)}(X)) \tag{52}$$

That is, the effect of the transformation will be to replace any solution of (49) of the form (1) by a solution which, in terms of the new variables X, Y, Z, has the same form with the number n decreased by at least 1.

To find the transformation (51) when the differential equation (49) is given, we have the differential equations (47) and (48) to solve for F and G. As (47) and (48) are conditions on the functions F and G, not on their arguments, we may rewrite (47) and (48) with x, y, z, q as the arguments, in place of x, y, A, θ. The subsidiary equations then are

$$\frac{dx}{0} = \frac{dy}{\delta(x, y, z, q)} = \frac{dz}{q\delta(x, y, z, q) - 1} = -\frac{dq}{\beta(x, y, z, q)} \tag{53}$$

And we must find three independent integrals, $x =$ constant, $F(x, y, z, q) =$ constant, $G(x, y, z, q) =$ constant, of the subsidiary equations (53).

Using (47) and (48) in the forms

$$G_3 + \beta G_4 = \delta(G_2 + G_3 q), F_3 + \beta F_4 = \delta(F_2 + F_3 q) \tag{54}$$

we may work out details of the transformation (51) as follows:

$$\begin{aligned}
P &= \frac{\partial(Z, Y)}{\partial(x, y)} \Big/ \frac{\partial Y}{\partial y} \\
&= G_1 + G_3 p + G_4 s - \frac{(F_1 + F_3 p + F_4 s)(G_2 + G_3 q + G_4 t)}{F_2 + F_3 q + F_4 t} \\
&= G_1 + G_3 p + G_4(\mu + \beta p + \delta pt) \\
&\quad - \frac{[F_1 + F_3 p + F_4(\mu + \beta p + \delta pt)](G_2 + G_3 q + G_4 t)}{F_2 + F_3 q + F_4 t} \\
&= G_1 + G_4 \mu + \delta p(G_2 + G_3 q + G_4 t) \\
&\quad - \frac{[F_1 + F_4 \mu + \delta p(F_2 + F_3 q + F_4 t)](G_2 + G_3 q + G_4 t)}{F_2 + F_3 q + F_4 t}
\end{aligned}$$

(55) $$P = G_1 + G_4\mu - (F_1 + F_4\mu)\frac{G_2 + G_3q + G_4t}{F_2 + F_3q + F_4t}$$

(56) $$Q = \frac{\partial Z}{\partial y}\bigg/\frac{\partial Y}{\partial y} = \frac{G_2 + G_3q + G_4t}{F_2 + F_3q + F_4t}$$

(57) $$t = \frac{G_2 + G_3q - (F_2 + F_3q)Q}{F_4Q - G_4}$$

(58) $$P = G_1 - F_1Q + (G_4 - F_4Q)\mu\left(x, y, z, q, \frac{G_2 + G_3q - (F_2 + F_3q)Q}{F_4Q - G_4}\right)$$

Then letting

(59) $$\Delta(x, y, z, q) = F_4G_2 - F_2G_4 + q(F_4G_3 - F_3G_4) = \frac{F_4G_3 - F_3G_4}{\delta}$$

we get:

(60) $$F_2 + F_3q + F_4t = \frac{\Delta}{F_4Q - G_4}$$

(61) $$\frac{\partial y}{\partial Y} = 1\bigg/\frac{\partial Y}{\partial y} = \frac{F_4Q - G_4}{\Delta}$$

(62) $$\frac{\partial z}{\partial Y} = \frac{\partial z}{\partial y}\bigg/\frac{\partial Y}{\partial y} = \frac{(F_4Q - G_4)q}{\Delta}$$

(63) $$\frac{\partial q}{\partial Y} = \frac{\partial q}{\partial y}\bigg/\frac{\partial Y}{\partial y} = \frac{(F_4Q - G_4)t}{\Delta} = \frac{G_2 + G_3q - (F_2 + F_3q)Q}{\Delta}$$

Unless the Jacobian

(64) $$J = \frac{\partial(Y, Z, P, Q)}{\partial(y, z, q, t)}$$

vanishes identically, the five equations (51), (55), (56) can be solved to express x, y, z, q, t each as a function of X, Y, Z, P, Q. Then the transformed differential equation (that results from (49) by the transformation (51)) can be obtained by taking the partial derivative with respect to Y in (58), using the expressions (61)–(63) for the partial derivatives of y, z, and q with respect to Y, and finally replacing x by X and y, z, q by the expressions just found for them as functions of X, Y, Z, P, Q.

The transformed differential equation will evidently be of the second order and in fact will have the form

(65) $$S = D(X, Y, Z, P, Q) + E(X, Y, Z, P, Q)T$$

It will be possible to iterate the transformation, i.e., to apply the generalized Laplace's transformation again to (65), only if D and E are both of the first degree in P. But if there exist solutions of (49) that are of the form (1), it must be possible to iterate the transformation often enough to find them in accordance with theorem 2.3 below.

By a straightforward computation, expanding the determinant and making use of (54), we find the following expression for the Jacobian, in which λ and Δ are of course the λ and Δ of equations (19) and (59) respectively:

$$J = \frac{\lambda\Delta^3}{(F_2 + F_3 q + F_4 t)^3} \tag{66}$$

That neither Δ nor $F_2 + F_3 q + F_4 t$ can vanish identically follows from the condition that x = constant, F = constant, G = constant are independent integrals of (53). Hence J vanishes identically if and only if λ vanishes identically.

If J, or λ, vanishes identically, a relation of the form

$$f(X, Y, Z, P, Q) = 0 \tag{67}$$

must hold as an identity in x, y, z, q, t. We may regard (67) as a first-order differential equation to be solved for Z; and if in its general solution we replace X by x, Y by $F(x, y, z, q)$, and Z by $G(x, y, z, q)$, in accordance with (51), the result will be an intermediate integral of (49). We rely on the methods of Lagrange and Charpit, not so much as a means of finding an expression of the general solution of (67) in particular cases, but for a proof of the existence of the general solution of (67), in one or other of two forms involving an arbitrary function, according to whether or not (67) is of the first degree in P and Q. And hence we have:

2.1. *If λ vanishes identically, the differential equation* (49) *has a first-order intermediate integral which involves an arbitrary function ψ and which has either the form*

$$H(x, y, z, q, \psi(\iota(x, y, z, q))) = 0 \tag{68}$$

or the parametric form:

$$H(x, y, z, q, u, \psi(u)) = 0 \tag{69}$$

$$H_5(x, y, z, q, u, \psi(u)) + \psi'(u) H_6(x, y, z, q, u, \psi(u)) = 0 \tag{70}$$

(*In the case of the parametric form we must add separately the intermediate integral*

$$H(x, y, z, q, a, b) = 0 \tag{71}$$

obtained from the complete integral of (67), *a and b being arbitrary constants and H the same function as in* (69).)

The converse of 2.1 is:

2.2. *If the differential equation* (49) *has a first-order intermediate integral which involves an arbitrary function ψ and has one of the forms* (68) *or* (69)–(70), *then λ vanishes identically.*

For the proof of 2.2 we notice that, for any particular choice of ψ, the general solution of the intermediate integral (68) or (69)–(70) will ordinarily be of the form (30), so that 1.4 applies. This fails only for those choices of ψ for which (68) or the equation obtained by eliminating u between (69) and (70) is independent of q, and we shall exclude such choices of ψ simply as "exceptional".

If the intermediate integral is (68), the exceptional ψ's are only those, if any, which satisfy $H_4 + \iota_4 H_5 \psi' = 0$ identically in x, y, z, q. And the possibility that H_4 and $\iota_4 H_5$ might be both identically 0 independently of ψ is excluded by the hypothesis that (68) is of first order. If the intermediate integral is (69)–(70), the exceptional ψ's are only those, if any, which satisfy $H_4 = 0$, identically in x, y, z, q when u is treated as a function of x, y, z, q determined by (70). And in this case the possibility that H_4 is identically 0, independently of ψ, is again excluded by the hypothesis that the intermediate integral is of first order.

Thus the exceptional ψ's can be at most only those which satisfy a fixed first-order differential equation.

Now by 1.4, if there are no singularities of μ, v, and their first partial derivatives (other than μ_1), and if λ is not identically 0, the relation $\lambda(x, y, z, q, t) = 0$ holds for all solutions of the intermediate integral for all non-exceptional ψ. That λ is not identically 0 implies that J is not identically 0. Hence the expressions for x, y, z, q, t as functions of X, Y, Z, P, Q exist, and we may use them to express $\lambda(x, y, z, q, t)$ as a function of X, Y, Z, P, Q. Thus we get an equation

$$\lambda(x, y, z, q, t) = \Lambda(X, Y, Z, P, Q) \tag{72}$$

Where λ does not vanish it follows that Λ does not vanish. Hence Λ does not vanish identically. Also J can be expressed as a function of X, Y, Z, P, Q. For any solution of the transformed differential equation (65) that does not correspond to a zero of Λ or a zero or singularity of J we can argue from (72) that λ is non-vanishing for the corresponding solution of (49), and hence that the solution must be one that is obtained from an exceptional ψ. The solutions of (49) arising from exceptional ψ's can evidently be covered by an equation

$$v(x, y, z, c) = 0 \tag{73}$$

obtained from either (68) or (69)–(70) and involving at most one arbitrary constant c. From this by eliminating c we get a first-order differential equation, involving say x, y, z, q, which can be reexpressed in terms of X, Y, Z, P, Q:

$$\Upsilon(X, Y, Z, P, Q) = 0 \tag{74}$$

In a suitably restricted neighborhood within which both Λ and J (as functions of the five variables X, Y, Z, P, Q) are non-zero and non-singular, the solutions of the definitely second-order differential equation (65) are thus included among those of the fixed first-order differential equation (74)—which is impossible.

In the case in which there exist solutions of (49) of the form (30) corresponding to singularities of μ, v, and their first partial derivatives, the above argument may be modified as follows. In addition to the relation $\lambda(x, y, z, q, t) = 0$ we have also a number of relations

$$\lambda_{(i)}(x, y, z, q, t) = 0 \tag{75}$$

which represent the relevant singularities of μ, v, and their first partial derivatives. These may be reexpressed in terms of X, Y, Z, P, Q as

$$\Lambda_{(i)}(X, Y, Z, P, Q) = 0 \tag{76}$$

And we then consider a neighborhood within which $\Lambda, \Lambda_{(i)}, J$ are non-zero and non-singular.

Now returning to the point which was made above in connection with equation (50), that the transformation (51) can be expected to transform a solution of the differential equation (49) of the form (1) into a solution which has the same form with the number n decreased by at least 1, we remark that our proof of this is not yet conclusive, because of the possibility that equation (46) may not include quite all of the solutions of (45). Indeed one class of solutions of (45) which is definitely not included in (46) is given by the equation

$$F(x, y, A, \theta) = \Omega(x, \varphi, \varphi', \varphi'', \dots, \varphi^{(n-1)}) \tag{77}$$

where Ω is an arbitrary function. And correspondingly equation (50) must be supplemented by the equation

$$F(x, y, A, A_2) = \Omega(x, \varphi, \varphi', \varphi'', \dots, \varphi^{(n-1)}) \tag{78}$$

To repair the defect we proceed by finding the following Jacobian, where F stands for $F(x, y, A, A_2)$ and G stands for $G(x, y, A, A_2)$:

$$\begin{aligned}\frac{\partial(F, G)}{\partial(y, \varphi^{(n)})} &= (F_2 + F_3A_2 + F_4A_{22})(G_3A_{n+3} + G_4A_{2(n+3)}) \\ &\quad - (G_2 + G_3A_2 + G_4A_{22})(F_3A_{n+3} + F_4A_{2(n+3)}) \\ &= [F_2G_3 - F_3G_2 + (F_4G_3 - F_3G_4)A_{22}]A_{n+3} \\ &\quad - [F_4G_2 - F_2G_4 + (F_4G_3 - F_3G_4)A_2]A_{2(n+3)} \\ &= [(\beta(x, y, A, A_2) + \delta(x, y, A, A_2)A_{22})A_{n+3} \\ &\quad - A_{2(n+3)}]\Delta(x, y, A, A_2)\end{aligned}$$

In this the factor $(\beta + \delta A_{22})A_{n+3} - A_{2(n+3)}$ vanishes by (23); and the Jacobian $\partial(F, G)/\partial(y, \varphi^{(n)})$ therefore vanishes, with a possible exception if Δ is singular for $z = A, q = A_2$. With this exception we do have, as a consequence of the vanishing of the Jacobian, that either equation (50) must hold for some function Φ or equation (78) must hold for some function Ω.

The exceptional case in which (78) holds is the case in which $F(x, y, A, A_2)$ is independent of y, and hence, taking x, y, z, q as the arguments of F, we may describe it also as the case in which $F_2 + F_3q + F_4t$ vanishes for $z = A$.

The foregoing argument applies as well to the case $n = 0$, i.e., the case of a solution of the form (30), as it does to larger values of n. Thus if the transformation (51) is applied to a differential equation (49) that has a solution $z = A(x, y, \varphi(x))$, and if Δ is not singular and $F_2 + F_3q + F_4t$ does not vanish for $z = A(x, y, \varphi(x))$, the corresponding solution of the transformed differential equation has the form $Z = \Phi(X, Y)$, no longer depending on an arbitrary function $\varphi(x)$ (or $\varphi(X)$). Generally, and within an appropriately restricted region, the transformation (51) effects a one-to-one correspondence between particular solutions of (49) and particular solutions of the transformed differential equation, as is clear from the fact which we found above, that there exist not only the equations (51) expressing X, Y, Z as functions of x, y, z, q but also inverse equations expressing x, y, z as functions of X, Y, Z, P, Q. If this fails in a particular case, as in the reduction of a solution $z = A(x, y, \varphi(x))$ to a solution not involving an arbitrary function, it follows that the Jacobian J has a zero or a singularity.

This suggests as a means of finding solutions of the differential equation (49) of the form (1) that we use iterated application of the transformation (51), examining at each stage the zeros and singularities of the Jacobian. However, we must take into account the exceptional cases that we found, that Δ is singular and that $F_2 + F_3q + F_4t$ vanishes. (For example if we transform $s = zt^3$ by $X = x$, $Y = q$, $Z = y$, we find $J = 1$, and the obvious solution $z = cy + \varphi(x)$ corresponds not to any zero or singularity of J but to a zero of $F_2 + F_3q + F_4t$). Hence in view of the expression (66) for J it will be better not to use J itself at all but to examine at each stage the zeros and singularities of λ, Δ, and $F_2 + F_3q + F_4t$ separately.

To find all the solutions of the form (30) at each stage we might as an alternative, by 1.4, just examine at each stage the zeros of λ and the singularities of μ, ν, and their partial derivatives. But this will not do as a general procedure because there is a possibility that a transformation (51) of a differential equation (49) might transform away not only all the solutions of the form (30) but also a solution of the form (1) with $n = 1$. To illustrate this we may cite the differential equation

$$2zs = -2q + (1 - y)t + 4pq$$

which has a solution

$$z = \frac{x + \varphi(x)}{y + \varphi'(x)}$$

This solution is transformed away—there is in fact no corresponding solution at all of the transformed differential equation—if we apply the transformation $X = x$, $Y = z^2/q$, $Z = y$. To be sure this accident might have been avoided if we had made a different choice of F and G in applying the transformation (51); but this does not destroy the force of the example.

Hence we state our result in the following form:

2.3. *In order to find all solutions of the differential equation* (49) *that are of the form* (1), *for all n not greater than a fixed n_0, it is sufficient to proceed as follows. Apply the transformation* (51) *to the given differential equation* (49), *either n_0 times in succession if this is possible, or else until no further iteration of the transformation is possible* (*either because λ vanishes identically or because the final transformed differential equation is no longer of the form* (49)). *At each stage in the iteration of the transformation examine the zeros and singularities of λ, Δ, and $F_2 + F_3q + F_4t$, in the sense of writing the differential equations which represent these zeros and singularities and determining the solutions which they have in common with the differential equation which, at that stage, has been obtained by transformation of the originally given differential equation* (49). *Then examine the final transformed differential equation which has resulted at the end of the process of iterated application of* (51), *in the following way. If it is of the form* (49) *and λ vanishes identically, solve the corresponding differential equation* (67) *to find an intermediate integral. If it is of the form* (49) *and λ does not vanish identically, make use of* 1.4 *to find its solution of the form* (30). *If it is not of the form* (49), *make use of* 1.3 *to find its solutions of the form* (1).

This theorem solves the problem of finding all solutions of a given differential equation (49) that are of the form (1), for all n not greater than a given n_0, in the sense that the problem is reduced to the problem of solving certain ordinary (as opposed to partial) differential equations. As indicated above our interest is primarily in existence questions rather than in practical solution processes. And the advantage of ordinary differential equations from this point of view is that the elementary solutions (elementary in the sense of the present paper) are known to exist. Nevertheless the method of theorem 2.3 does work out as a practical solution process in some cases.

Parenthetically it should be added that the known existence of elementary solutions of ordinary differential equations applies only to a fixed differential equation, not to an equation-form containing an arbitrary function. For this reason, even if a partial differential equation has an intermediate integral of the form (68) or the form (69)–(70), and hence has almost all its solutions of the elementary form (30), it does not follow that it therefore necessarily has a two-arbitrary-function elementary solution.

In the special case in which the differential equation (49) is linear it is true that if any solution of the form (1) exists, it must be possible to find a two-arbitrary-function elementary solution by iterations of the transformation (51)—which in this case of course reduces to Laplace's transformation.

In connection with 2.3, it is believed that substantial auxiliary theorems can be found by means of 1.5. In particular, in at least the case in which v is independent of t, the equations obtained by setting (40) and (41) equal to 0 can be solved explicitly when regarded as simultaneous differential equations for μ and v; and this results in great simplification in the application of 2.3 to a particular differential equation in this case. But this is a topic which we leave to a possible future paper.

3. Generalization of Laplace's transformation, second case

Now we return to (43), to deal with the case in which $v(x, y, z, q, t)$ does not have the form (44). In this case (43) is a partial differential equation of first order and not of the first degree, and we rely on Charpit's method for the existence of a complete integral. With an appropriate understanding as to what is meant, we may rewrite (43) as

$$\frac{\partial q}{\partial z} = v\left(x, y, z, q, \frac{\partial q}{\partial y} + q\frac{\partial q}{\partial z}\right) \tag{79}$$

Let a complete integral be

$$G(x, y, z, q, c_1) = c \tag{80}$$

The complete integral of (43) as originally written is then

$$G(x, y, A, \theta, c_1) = c \tag{81}$$

And the corresponding general solution is

$$G(x, y, A, \theta, u) = \Phi(x, u, \varphi, \varphi', \varphi'', \dots, \varphi^{(n-1)}) \tag{82}$$

$$G_5(x, y, A, \theta, u) = \Phi_2(x, u, \varphi, \varphi', \varphi'', \dots, \varphi^{(n-1)}) \tag{83}$$

We therefore expect that a solution of the form (1), of the differential equation (18), will satisfy one or other of the two following conditions. Either there exist functions E_0 and Φ_0 such that

$$G(x, y, A, A_2, E_0(x, \varphi, \varphi', \varphi'', \dots, \varphi^{(n-1)})) = \Phi_0(x, \varphi, \varphi', \varphi'', \dots, \varphi^{(n-1)}) \tag{84}$$

or there exist functions E and Φ such that

$$\begin{aligned} &G(x, y, A, A_2, E(x, y, A, A_2, \varphi, \varphi', \varphi'', \dots, \varphi^{(n-1)})) \\ &\quad = \Phi(x, E(x, y, A, A_2, \varphi, \varphi', \varphi'', \dots, \varphi^{(n-1)}), \varphi, \varphi', \varphi'', \dots, \varphi^{(n-1)}) \end{aligned} \tag{85}$$

$$\begin{aligned} &G_5(x, y, A, A_2, E(x, y, A, A_2, \varphi, \varphi', \varphi'', \dots, \varphi^{(n-1)})) \\ &\quad = \Phi_2(x, E(x, y, A, A_2, \varphi, \varphi', \varphi'', \dots, \varphi^{(n-1)}), \varphi, \varphi', \varphi'', \dots, \varphi^{(n-1)}) \end{aligned} \tag{86}$$

In either case, by considering x and y as the independent variables, and taking the partial derivative with respect to y, we get

$$G_2 + G_3 A_2 + G_4 A_{22} = 0 \tag{87}$$

Then by solving (87) for E or E_0 we get

$$E = F(x, y, A, A_2, A_{22}) \text{ or } E_0 = F(x, y, A, A_2, A_{22}) \tag{88}$$

(the same function F in either case). This suggests that if we make a transformation of the differential equation (18) by letting

$$X = x, \quad Y = F(x, y, z, q, t), \quad Z = G(x, y, z, q, F(x, y, z, q, t)) \tag{89}$$

the solution of the transformed differential equation which corresponds to the solution $z = A$ of (18) will be

$$Z = \Phi_0(X, \varphi(X), \varphi'(X), \varphi''(X), \dots, \varphi^{(n-1)}(X)) \tag{90}$$

or

$$Z = \Phi(X, Y, \varphi(X), \varphi'(X), \varphi''(X), \dots, \varphi^{(n-1)}(X)) \tag{91}$$

That is, the effect of the transformation will be to replace any solution of (18) of the form (1) by a solution which, in terms of the new variables X, Y, Z, has the same form with the number n decreased by at least 1.

To find the transformation (89) when the differential equation (18) is given, we must find a complete integral (80) of the differential equation (79). This supplies the function G. And $F(x, y, z, q, t)$ is then found by solving

$$G_2(x, y, z, q, F) + qG_3(x, y, z, q, F) + tG_4(x, y, z, q, F) = 0 \tag{92}$$

The condition that (80) shall be a complete integral of (79) assures that G_4 is not identically 0 and that $(G_2 + qG_3)/G_4$ is not independent of F. Hence (92) can be solved for F, and F so obtained will not be independent of t.

The condition that (80) is an integral of (79) can be expressed by the equation

$$G_3 + G_4 v\left(x, y, z, q, -\frac{G_2 + qG_3}{G_4}\right) = 0 \tag{93}$$

where G stands for $G(x, y, z, q, c_1)$. Hence in consequence of (92) we have the equation

$$G_3(x, y, z, q, F(x, y, z, q, t)) + G_4(x, y, z, q, F(x, y, z, q, t))v(x, y, z, q, t) = 0 \tag{94}$$

By a computation analogous to that by which equations (55) and (56) were found, and using equations (92) and (93), we find

$$P = G_1 + G_4\mu, \quad Q = G_5 \tag{95}$$

where of course the arguments of G are x, y, z, q, F. Then unless the Jacobian

$$J = \frac{\partial(Y, Z, P, Q)}{\partial(y, z, q, t)} \tag{96}$$

vanishes identically, the five equations (89), (95) can be solved to express x, y, z, q, t each as a function of X, Y, Z, P, Q. Then the transformed differential equation (that results from (18) by the transformation (89)) can be obtained by taking the partial derivative with respect to y in equations (89), (95), and eliminating $\partial Y/\partial y$ and $\partial t/\partial y$. The result is

$$(97)\quad S = G_{15} + \mu G_{45} + \mu_5 \frac{G_4}{F_5} + (G_{55} - T)[F_1 + \mu F_4 + \mu v F_5 + F_5(\mu_2 + \mu_3 q + \mu_4 t) - \mu_5(F_2 + F_3 q + F_4 t)]$$

wherein x, y, z, q, t are to be replaced by the expressions for them as functions of X, Y, Z, P, Q.

By expanding the determinant and making use of (92) and (93) we find for J the expression

$$(98)\qquad J = -\lambda G_4^3$$

Hence J vanishes identically if and only if λ vanishes identically.

If J, or λ, vanishes identically, a relation

$$(99)\qquad f(X, Y, Z, P, Q) = 0$$

must hold identically in x, y, z, q, t. If we solve (99) as a first-order differential equation in X, Y, Z as the variables, and if in the result we replace X by x, Y by $F(x, y, z, q, t)$, and Z by $G(x, y, z, q, F(x, y, z, q, t))$, in accordance with (89), the result will be a *second-order intermediate integral* of (18). We understand the italicized phrase as implying that the general solution of (18) is to be found by considering the common solutions of (18) and the intermediate integral. However, if we consider also the equations in x, y, z, q, t which result from any extra solutions (special or singular) that (99) may possess, we can say more strongly that the disjunction of these equations and the second-order intermediate integral is a consequence of the differential equation (18).

3.1. *If v does not have the form* (44), *and if λ vanishes identically, the differential equation* (18) *has a second-order intermediate integral which involves an arbitrary function ψ and which has either the form*

$$(100)\qquad H(x, y, z, q, t, \psi(\iota(x, y, z, q, t))) = 0$$

or the parametric form:

$$(101)\qquad H(x, y, z, q, t, u, \psi(u)) = 0$$

$$(102)\qquad H_6(x, y, z, q, t, u, \psi(u)) + \psi'(u) H_7(x, y, z, q, t, u, \psi(u)) = 0$$

(*In the case of the parametric form we must add separately the intermediate integral*

$$(103)\qquad H(x, y, z, q, t, a, b) = 0$$

obtained from the complete integral of (99), *a and b being arbitrary constants and H the same function as in* (101).)

Now returning to the differential equation (79), we observe that the argument which led us from this equation to the transformation (89), although it was presented heuristically, is as a matter of fact substantially complete. Any solutions of (79) that are not covered by the complete integral (80) and the corresponding general integral can only be one or more singular integrals. These will be equations that may involve x, y, z, q but do not involve any arbitrary constants or arbitrary functions, and therefore they can lead only to solutions of (18) of the form (30). The conclusion above—that solutions of (18) of the form (1) are transformed by (89) to solutions of the same form with the number n decreased by at least 1—can therefore be escaped only at most by some exceptional solutions of the form (30). And in consequence we have:

3.2. *In the case of a differential equation which is of the form* (18), *if* v *is not of the form* (44), *it is sufficient to proceed as follows, in order to find the solutions of the form* (1). *If* λ *vanishes identically, find a second-order intermediate integral by solving the differential equation* (99). *Otherwise use* 1.4 *to find the solutions of the form* (30), *and then apply the transformation* (89). *The zeros and singularities of the Jacobian* J *as given by* (98) *must also be examined for possible additional solutions that might be transformed away. The transformed differential equation is then of a form to which either* 1.3 *or* 2.3 *will be applicable to find the solutions of the form* (1).

4. Quasi-substitutions

We turn now to a class of transformations which are closely associated with the generalized Laplace's transformations and which we shall call quasi-substitutions.

Some of these are obtained as a resultants of a transformation (51) and the inverse of a transformation (51). For example, if the differential equation

$$qs = q^3 - pt$$

is transformed by letting

$$\xi = x, \quad \eta = q, \quad \zeta = z - 2yq$$

the transformed differential equation is

$$\sigma + \eta^2\tau = 0$$

If on the other hand the linear differential equation

$$S = Y^2T - 2YQ + 2Z = 0$$

is transformed by letting

$$\xi = X, \quad \eta = Y, \quad \zeta = Q$$

the result is the same differential equation $\sigma + \eta^2\tau = 0$. Both transformations are instances of (51). The resultant of the first and the inverse of the second is represented by the equations

$$X = x, \quad Y = q, \quad Z = zq - yq^2$$

And this last transformation is an example of a quasi-substitution.

Generally, a quasi-substitution is represented by a triple of equations, expressing the new variables X, Y, Z as functions of x, y, z, p, q. And from these equations alone, independently of any particular differential equation to which the transformation is applied, there follow equations expressing P and Q as functions of x, y, z, p, q, and equations expressing x, y, z, p, q as functions of X, Y, Z, P, Q. Thus the quasi-substitutions share with ordinary substitutions the important property that they may be applied to any arbitrary differential equation of first or higher order, without raising the order of the differential equation. This contrasts with the situation in regard to Laplace's transformation, and its present generalizations, that a particular transformation, represented by a particular triple of equations, is adapted to a particular differential equation and in general cannot be applied to a different differential equation without raising the order, and without destroying the feature that inverse equations exist expressing x, y, z, p, q as functions of X, Y, Z, P, Q.

(In some cases it may happen, as in the example given above, that a quasi-substitution is also a birational transformation in the five-dimensional space whose coordinates are x, y, z, p, q.)

One class of quasi-substitutions is represented by equations of the form

$$X = x, \quad Y = F(x, y, z, q), \quad Z = G(x, y, z, F(x, y, z, q)) \tag{104}$$

where F is to be found from the equation

$$G_2(x, y, z, F) + qG_3(x, y, z, F) = 0 \tag{105}$$

and where G may be chosen arbitrarily, subject to the condition that (105) is a non-trivial equation which has a solution for F, and that F so found is not independent of q. As a consequence of (104) and (105) there follow the equations

$$P = G_1(x, y, z, F) + pG_3(x, y, z, F) \tag{106}$$

$$Q = G_4(x, y, z, F) \tag{107}$$

The resultant of a transformation (104) and an ordinary substitution is of course also a quasi-substitution. These are the only quasi-substitutions that are applicable to ordinary differential equations—in the sense that when they are applied to a partial differential equation that has the property that all derivatives occurring are derivatives with respect to y, the transformed differential equation has the same property.

It can be shown that, besides the foregoing, the only remaining quasi-substitutions are those represented by equations of the form

$$X = E(x, y, z, p, q), \quad Y = F(x, y, z, p, q), \quad Z = G(x, y, z, E, F) \tag{108}$$

where F and F are to be found from the equations

$$G_1(x, y, z, E, F) + pG_3(x, y, z, E, F) = 0 \tag{109}$$

$$G_2(x, y, z, E, F) + qG_3(x, y, z, E, F) = 0 \tag{110}$$

and where G may be chosen arbitrarily, subject to the condition that (109) and (110), regarded as equations to be solved for E and F, are independent, and that E and F as found from (109) and (110) are such that $E_4F_5 - E_5F_4$ does not vanish identically. From (108)–(110) it follows that

$$P = G_4(x, y, z, E, F), \quad Q = G_5(x, y, z, E, F) \tag{111}$$

In order to find a quasi-substitution (108) when the function F is given, not independent of both of its last two arguments, we may treat

$$F\left(x, y, z, -\frac{G_1(x, y, z, c_1, c)}{G_3(x, y, z, c_1, c)}, -\frac{G_2(x, y, z, c_1, c)}{G_3(x, y, z, c_1, c)}\right) = c \tag{112}$$

as a partial differential equation to be solved for G. We must find a solution of (112) such that G_1/G_3 and G_2/G_3 are not both independent of c_1, and we may then expect that (109) and (110) will be compatible and will have a common solution for E.

To find a quasi-substitution (104) when the function F is given, not independent of its last argument, we may express q as a function of x, y, z, F, and then treat (105) as a partial differential equation to be solved for G.

The significance of quasi-substitutions in the present context is that, when the methods of the preceding sections lead to an intermediate integral, it may often happen that the argument of the arbitrary function involves derivatives. In this case the best way to treat the intermediate integral may be to use a quasi-substitution to reduce the argument of the arbitrary function to Y. We may expect to be able to do this whenever the argument of the arbitrary function is a function of x, y, z, p, q. And it may sometimes be possible even when the derivatives of higher order are involved.

Some other applications of quasi-substitutions suggest themselves, which we must leave unexplored in the present paper.

One of these is that it may happen that a given differential equation can be transformed to one that is linear by means of either a generalized Laplace's transformation or a quasi-substitution, even when this is not possible by an ordinary substitution. The example cited at the beginning of this section illustrates this point. In this example the linear differential equation obtained is one that has an elementary general solution. But evidently it may also be that the linear differential equation is not elementarily solvable. And even in this case it seems that the treatment of the given differential equation might be facilitated by the knowledge that it is in a certain sense essentially linear.

Another possible application is to the treatment of the Monge–Ampère equations. It is not immediately clear how substantial this may be. But one example is included in the next section as an indication.

5. Examples

We add some further examples illustrating various points in connection with the preceding sections.

(Ex. 1) $$(x + y)s + p^3t - p + x^2q = 0$$

Theorem 1.3 is applicable, and f_{44} is $-6pt/(x + y)$. To make this 0 identically in p we must have $t = 0$ and hence $z = y\psi(x) + \chi(x)$. Substituting this in the differential equation yields the condition $\chi'(x) = x\psi'(x) + x^2\psi(x)$. Thus we have what may be described as a solution of the form (2), with the condition not satisfied that there is no relation of the form (3). It may, however be reexpressed as a solution of the form (1) by letting

$$\psi(x) = \frac{\varphi'(x)}{x^2 - 1}$$

We then have

$$z = \frac{(x + y)\,\varphi'(x)}{x^2 - 1} + \varphi(x)$$

This is the only solution of the differential equation of the form (1). That it is quite special compared to the general solution is indicated by the fact that we would still get the same solution if we replaced the term p^3t in the differential equation by $tf(p)$, f being chosen as any function not of the first degree or 0 degree.

(Ex. 2) $$s + yqt = 0$$

This has the intermediate integral

$$-xq + \log y = \log\psi(q)$$

By the quasi-substitution

$$X = x, \quad Y = q, \quad Z = z - yq$$

this becomes

$$Q = -e^{XY}\psi(Y)$$

By integration of this with respect to Y

$$Z = -\int e^{XY}\psi(Y)dY + \varphi(X)$$

Hence the general solution of (Ex. 2) may be written in the following parametric form, with q as the parameter:

$$y = e^{xq}\psi(q)$$
$$z = qe^{xq}\psi(q) - \int e^{xq}\psi(q)dq + \varphi(x)$$

To this we must add separately the solution

$$z = cy + \varphi(x)$$

which was lost by the form in which the intermediate integral was written above.

In this case more familiar methods would suffice for the integration of the intermediate integral, but there is an advantage of uniformity in always using a quasi-substitution when the argument of the arbitrary function contains p or q.

(Ex. 3) $$s = (xp + yq - z)t$$

By the same quasi-substitution as was used for (Ex. 2) we get

$$S = XP - Z$$

This can be solved by Laplace's transformation, with the result

$$Z = XI(X, Y) - I_2(X, Y) + X\psi(Y) - \psi'(Y)$$

where $I(X, Y) = \int e^{XY}\varphi(X)dX$. Hence we get for (Ex. 3) the parametric solution

$$y = -xI_2(x, q) + I_{22}(x, q) - x\psi'(q) + \psi''(q)$$
$$z = xI(x, q) - (xq + 1)I_2(x, q) + qI_{22}(x, q)$$
$$+ x\psi(q) - (xq + 1)\psi'(q) + q\psi''(q)$$

To this we must add separately the solution

$$z = cy + \varphi(x)$$

which was lost by the quasi-substitution.

In the former solution we may of course take $\varphi(x)$ to be 0 and so find for (Ex. 3) the more special solution

$$y = -x\psi'(q) + \psi''(q)$$
$$z = x\psi(q) - (xq + 1)\psi'(q) + q\psi''(q)$$

We shall think of this last as a solution of a form that is analogous to the form (1) except with q instead of x as argument of the arbitrary function. For although we must regard q as an unspecified parameter in order to give the pair of equations as solution of the differential equation, it will then follow from these equations that q is in fact $\partial z / \partial y$.

(Ex. 4) $$qs = q^3 - pt$$

Following the method of 2.3 we get from $\lambda = 0$ the solution

$$z = -\frac{y}{x+c} + \varphi(x)$$

Then the remaining solutions are found by the transformation $\xi = x, \eta = q, \zeta = z - 2yq$, which, as we have already seen in section 4, leads to $\sigma + \eta^2 \tau = 0$. For this we get the intermediate integral

$$\text{♀} = \psi\left(\xi + \frac{1}{\eta}\right)$$

and hence the solution

$$\zeta = \int \psi\left(\xi + \frac{1}{\eta}\right) d\eta + \varphi(\xi)$$

Thus as parametric solution for (Ex. 4) we have:

$$\begin{aligned} y &= -\frac{1}{2q^2} I_1(x, q) - \frac{1}{2}\psi\left(x + \frac{1}{q}\right) - \frac{1}{2q^2}\varphi'(x) \\ z &= I(x, q) - \frac{1}{q} I_1(x, q) - q\psi\left(x + \frac{1}{q}\right) + \varphi(x) - \frac{1}{q}\varphi'(x) \end{aligned}$$

where $I(x, q) = \int \psi\left(x + \frac{1}{q}\right) dq$.

(Ex. 5) $$rt - s^2 = \frac{q^2 r - s}{2(z - xp)}$$

If we use the quasi-substitution

$$X = p, \quad Y = q, \quad Z = z - xp - yq$$

we get the transformed differential equation

$$S + Y^2 T - 2YQ + 2Z = 0$$

This is a linear differential equation for which, as we have already seen in section 4, Laplace's transformation is $\xi = X, \eta = Y, \zeta = Q$, reducing it to $\sigma + \eta^2 \tau = 0$. Using for this last the solution which was just obtained above, we find

$$Z = YI(X, Y) - \frac{1}{2} I_1(X, Y) - \frac{1}{2} Y^2 \psi\left(X + \frac{1}{Y}\right) + Y\varphi(X) - \frac{1}{2}\varphi'(X)$$

where $I(X, Y) = \int \psi\left(X + \frac{1}{Y}\right) dY$. Hence for (Ex. 5) we find the parametric solution

$$\begin{aligned} x &= -qI_1(p, q) + \frac{1}{2} I_{11}(p, q) + \frac{1}{2} q^2 \psi'\left(p + \frac{1}{q}\right) - q\varphi'(p) + \frac{1}{2}\varphi''(p) \\ y &= -I(p, q) - \varphi(p) \\ z &= -\left(pq + \frac{1}{2}\right) I_1(p, q) + \frac{1}{2} p I_{11}(p, q) - \frac{1}{2} q^2 \psi\left(p + \frac{1}{q}\right) \\ &\quad + \frac{1}{2} pq^2 \psi'\left(p + \frac{1}{q}\right) - \left(pq + \frac{1}{2}\right)\varphi'(p) + \frac{1}{2} p\varphi''(p) \end{aligned}$$

In addition to this parametric solution we must also look into the question of solutions satisfying either $q = f(p)$ or $p =$ constant, as there is a possibility that such solutions might be lost by the particular quasi-substitution which was used. In fact we find in this way $(c - p)q = 1, q = 0, p = c$. These first-order differential equations are to be solved by familiar methods, and the resulting solutions must be adjoined to the parametric solution above, for the full general solution of (Ex. 5).

(Ex. 6) $$s^2 = a^2 p^2 t$$

To avoid carrying the double sign, let a stand ambiguously for either one of the two square roots of a^2, and so write the differential equation as

$$s = apt^{\frac{1}{2}}$$

Following the method of theorem 3.2 we find

$$G(x, y, z, q, c_1) = \log(q + c_1) - a^2(z + c_1 y)$$
$$F(x, y, z, q, t) = a^{-1}t^{\frac{1}{2}} - q$$

From (95) we find $P = 0$. Hence λ must vanish identically (as is also easily verified directly). The equation $P = 0$ is the equation (99). From it we get $Z = \Phi(Y)$, and hence the second-order intermediate integral is

$$\frac{1}{2}\log t - ayt^{\frac{1}{2}} - a^2(z - yq) = \Phi(a^{-1}t^{\frac{1}{2}} - q) + \log a$$

In order to solve this without first specializing the arbitrary function Φ we proceed by taking the partial derivative with respect to y, obtaining

$$\left[\frac{1}{2}a^{-1}t^{-\frac{1}{2}}\frac{\partial t}{\partial y} - t\right]\left[at^{-\frac{1}{2}} - a^2 y - \Phi'(a^{-1}t^{\frac{1}{2}} - q)\right] = 0$$

From this the differential equation obtained by setting the first factor equal to 0 is easily solved, and its common solutions with (Ex. 6) are found to be:

$$z = -a^{-2}\log(y + \varphi(x)) + by + c$$
$$z = -a^{-2}\log(y + b) + (y + b)\,\varphi(x) + c$$
$$z = by + \varphi(x)$$

Evidently the general solution is to be obtained from the second factor in the equation above. The differential equation obtained by setting it equal to 0 may be rewritten as

$$a^{-1}t^{\frac{1}{2}} - q = \Omega(a^{-1}t^{-\frac{1}{2}} - y)$$

(since no new solutions of (Ex. 6) are obtained by taking Φ' equal to a constant). Then we treat q as the dependent variable and on this basis make the quasi-substitution

$$\xi = x, \quad \eta = a^{-1}t^{-\frac{1}{2}}, \quad \zeta = a^{-1}t^{\frac{1}{2}} + q$$

This yields

$$4\text{♀} = -a^2[\zeta + \Omega(\eta)]^2$$

If we let

$$\Omega(\eta) = a^{-2}\int\left[\frac{\omega''(\eta)}{\omega'(\eta)}\right]^2 d\eta - 2a^{-2}\frac{\omega''(\eta)}{\omega'(\eta)}$$

we may solve this as a Riccati equation, obtaining

$$a^2\zeta + \int \left[\frac{\omega''(\eta)}{\omega'(\eta)}\right]^2 d\eta = \frac{4\omega'(\eta)}{\omega(\eta) + \varphi(\xi)}$$

Hence inverting the quasi-substitution we get the parametric equations

$$y = \frac{\omega'(\eta)[\omega(\eta) + \varphi(x)]}{2\omega'(\eta)^2 - \omega''(\eta)[\omega(\eta) + \varphi(x)]} - \eta$$

$$q = 2a^{-2}\frac{\omega'(\eta)}{\omega(\eta) + \varphi(x)} + a^{-2}\frac{\omega''(\eta)}{\omega'(\eta)} - a^{-2}\int \left[\frac{\omega''(\eta)}{\omega'(\eta)}\right]^2 d\eta$$

We omit the remainder of the work, which is lengthy and serves no further purpose of illustration. It is necessary to find an expression for z in terms of x and η by using

$$z = \int q dy + \psi(x)$$

And as three arbitrary functions are then involved, a relation among them must be found by substituting in the differential equation, i.e., in (Ex. 6).

(Ex. 7) $$s = zq + q^2$$

(Ex. 8) $$(t + y^2)s = t^2 + 2yp - yq$$

We omit details of these last two examples. (Ex. 7) is a very simple case in which a (generalized) intermediate integral, containing an arbitrary function, results by two successive applications of a generalized Laplace's transformation. (Ex. 8) has only a one-arbitrary-function elementary solution, obtained by one application of a generalized Laplace's transformation.

6. Conclusion

The writer has urged elsewhere (*Remarks on the elementary theory of differential equations as area of research*, in ***Information and prediction in science***, New York, 1965) the problem of characterizing the class of partial differential equations that have elementary or elementary-parametric general solutions. It seems quite certain that there are negative results to be obtained, both for this problem and for the problem of elementary and elementary-parametric solutions that involve one arbitrary function, and that the negative results may be expected to begin with differential equations of the second order in two independent variables. But the present paper is directed entirely towards the positive side of these questions.

It is not immediately clear whether the methods of this paper may suffice to treat the positive aspect completely, or how far in this direction they may be expected to reach. But substantial extensions of these methods would seem to be possible, and we conclude by giving some brief indications in this regard.

First, while still keeping to the case in which the argument of the arbitrary function is x, we may seek to use the same or similar methods to treat the question of solutions of the form (2) for which a relation of the form (3) does hold. Or instead of a single relation (3) we may suppose that there are two or more such relations, i.e., relations having the following forms:

$$\varphi_n'(x) = B_{(1)}(x, \varphi(x), \varphi_1(x), \varphi_2(x), \ldots, \varphi_n(x), \varphi'(x), \varphi_1'(x), \varphi_2'(x), \ldots, \varphi_{n-k}'(x))$$
$$\varphi_{n-1}'(x) = B_{(2)}(x, \varphi(x), \varphi_1(x), \varphi_2(x), \ldots, \varphi_n(x), \varphi'(x), \varphi_1'(x), \varphi_2'(x), \ldots, \varphi_{n-k}'(x))$$
$$\cdots\cdots\cdots\cdots\cdots\cdots\cdots\cdots\cdots\cdots\cdots\cdots$$
$$\varphi_{n-k+1}'(x) = B_{(k)}(x, \varphi(x), \varphi_1(x), \varphi_2(x), \ldots, \varphi_n(x), \varphi'(x), \varphi_1'(x), \varphi_2'(x), \ldots, \varphi_{n-k}'(x))$$

A somewhat hasty preliminary survey suggests that such an extension of the methods will indeed be possible, and that it may be expected to lead to a second kind (or kinds) of generalization of Laplace's transformation.

Then a different extension is to allow the argument of the arbitrary function, in the elementary or elementary-parametric solution, to be other than x. Here we must consider not only the case that the argument of the arbitrary function is a function of x, y, z, but also the case that the argument of the arbitrary function involves derivatives of z. (For an illustration of the latter possibility see (Ex. 3) above, where there is an example of an elementary-parametric solution in which the argument of the arbitrary function is q.)

At least if the argument of the arbitrary function involves derivatives of no higher than first order, the expectation is that a substitution or transformation can be used to reduce to the case in which the argument of the arbitrary function is x (or X). Indeed it is clear that if the argument of the arbitrary function is $\iota(x, y, z)$, the required reduction is accomplished by a substitution in which $X = \iota(x, y, z)$; and if the argument of the arbitrary function is $\iota(x, y, z, p, q)$, the required reduction is accomplished by a quasi-substitution in which $X = \iota(x, y, z, p, q)$. The problem which remains is to find the requisite substitution or quasi-substitution when only the differential equation is given, and not the argument of the arbitrary function.

A large part, but not the whole, of the problem of reducing to the case in which the argument of the arbitrary function is x (or X), is included in the problem of reducing a given differential equation, by an appropriate substitution or transformation, to a differential equation in which r (or R) is not present. The latter problem is treated in the paper by the present writer which is referred to above, but the treatment there is inadequate because it fails to take into account the existence of quasi-substitutions.

Princeton University

PAUL J. COHEN AND THE CONTINUUM PROBLEM (1968)

On the occasion of the award of a prize to Paul Cohen, and in spite of significant contributions by him to analysis, to topological groups, and to the theory of differential equations, I believe that the audience will agree that it is appropriate to devote the entire time allowed for exposition to the continuum problem.

For here is another case, of the sort which arises from time to time in the history of mathematics, in which a mathematician who has done important work in other fields turns to a field not properly his own to solve an outstanding problem that has baffled the specialists. As a consequence of the tremendous growth of mathematics the universal mathematician of other days is no longer a possibility — David Hilbert was certainly the last of them. The next best thing is that abler men should not confine themselves too closely to one field or be afraid to turn to an area in which they may not have all the expert knowledge of those who have concentrated their work in it. Certainly Paul Cohen's results have been and will be greatly extended, and the method of his proof greatly improved, by specialists in set theory. But we are concerned today with the initial break-through.

Number one of the Hilbert problems, placed even before the problem of the consistency of arithmetic which occupied so much of Hilbert's own attention in the latter part of his life, is "Cantor's problem of the cardinal number of the continuum." So it is titled in the contemporary English translation of Hilbert's famous paper. Hilbert himself in German uses Cantor's original term "Mächtigkeit," which has no good English translation. Hilbert does not say that the order in which the problems are numbered gauges their relative importance, and it is not meant to suggest that he intended this. But he does mention the arithmetical formulation of the concept of the continuum and the discovery of non-Euclidean geometry as being the outstanding mathematical achievements of the preceding century, and gives this as a reason for putting problems in these areas first.

Already in 1878 Cantor stated the continuum hypothesis as a conjecture. But there is a sense in which the continuum problem dates from Cantor's statement at the end of a paper which appeared in the *Mathematische Annalen* in 1884. Here it is proved that a closed infinite subset of the (linear) continuum must have the cardinal number either of the natural numbers or of the whole continuum. Then it is said that the result can be extended to subsets which are not closed, and a proof will be provided. The paper closes with the worlds "Fortsetzung folgt." But the promised Fortsetzung never did folgen, and it seems clear that the proof Cantor believed he had broke down.

Cantor passes at once from the proposition that there is no cardinal number between that of the natural numbers and that of the continuum to the second form of the continuum hypothesis, that the cardinal number of the continuum is $\aleph_1$. Of course this depends on a tacit assumption of the axiom of choice, in particular as Sierpiński's result of 1947 deriving the axiom of choice from the generalized continuum hypothesis was

---

Originally published in ***Proceedings of the International Congress of Mathematicians***, Izdatél'stvo "MIR", Moscow, 1968.

not then available (or the background that made this result possible).[1] Hilbert is more cautious and states as a separate problem, subsidiary to the continuum problem, the question whether the continuum can be well-ordered. Zermelo's paper which explicitly states the axiom of choice for the first time (in the strong form which Zermelo later called "Prinzip der Auswahl"), and shows as a consequence of it that every set can be well-ordered, followed Hilbert's paper on mathematical problems by only four years.

It was Cantor's original point of view that the transfinite cardinal and ordinal numbers are two different kinds of generalizations of the natural numbers and are to serve the same purpose for transfinite sets which the natural numbers do for finite sets. If this program is to be fulfilled, one evidently must be able to answer at least the simplest and most immediate questions that arise about the cardinal numbers of the most commonly used mathematical sets, among them the continuum. This is clearly the reason why the frustrating difficulties of the continuum problem acquired the importance that they did for the Cantor theory. Surely neither Cantor nor Hilbert could have surmised that the ultimate solution would take the negative form that it has. Yet Hilbert is quite explicit that in general it may happen that the solution of a problem must be in the form of an impossibility proof.

The antinomies of set theory, which first came to the attention of the mathematical public through Burali-Forti's paper of 1897, played an important role in the progress toward the ultimate solution, as it was the antinomies that forced the transition from the older naive and "genetic" use of sets in mathematics to an axiomatic basis for set theory. And it is of course only by the axiomatic method that a proof of the impossibility of a proof becomes possible.

Within axiomatic set theory there was a proof of the independence of the axiom of choice by Fraenkel as early as 1922. But this was unsatisfactory in that it referred to axioms of set theory so formulated as to admit a domain of Urelemente, or non-sets, of unspecified structure, and the possibility remained open that the axiom of choice would lose its independence, upon adding axioms specifying the structure of the domain of Urelemente (or most simply, upon adding as an axiom that there are no Urelemente). Extensions of Fraenkel's result and improvements of his method, by Fraenkel himself, Lindenbaum, Mostowski, and more recently Shoenfield and Mendelson either did not remove this objection or only mitigated it (in the sense that independence from quite the full usual system of axioms for set theory was not yet proved).

A much more important step—which constitutes in fact the first half of the solution of the continuum problem, and on which subsequent work heavily depends—was taken by Kurt Gödel in 1938–40. Abstracts of Gödel's methods and results appeared in 1938 and 1939, and the monograph containing the full proofs, in 1940. Gödel's method is to set up what has since been called an inner model of set theory. I.e., set theory without axiom of choice is used to set up a model of set theory in which both the axiom of choice and the generalized continuum hypothesis hold. (The generalized continuum hypothesis is the proposition that the power set of cardinal number $\aleph_\alpha$ has the cardinal number $\aleph_{\alpha+1}$. Cantor's original continuum hypothesis, in its second form, being the special case of this in which $\alpha = 0$.) The result of Gödel's procedure, setting up an inner model, is a relative consistency proof for the axiom of choice, and for the

[1]Sierpiński points out that this result had been announced by Lindenbaum and Tarski in 1926. Their proof was never published.

generalized continuum hypothesis: If set theory without axiom of choice is consistent, it remains so upon introducing both the axiom of choice and the generalized continuum hypothesis as additional axioms.

After the (relative) consistency proof, the second half of the negative solution of the continuum problem is of course independence. A partial step in this direction was taken by Gödel, who in 1942 found a proof of the independence of the axiom of constructibility in type theory. According to his own statement (in a private communication) he believed that this could be extended to an independence proof of the axiom of choice; but due to a shifting of his interests toward philosophy, he soon afterwards ceased to work in this area, without having settled its main problems. The partial result mentioned was never worked out in full detail or put into form for publication.

These climactic results, the independence in set theory of the axiom of choice (even the weak form of the axiom of choice which concerns a countable set of pairs) and of the continuum hypothesis from the axiom of choice, remained for Paul Cohen in 1963–64. It is no part of our present purpose to describe the details of his method. Let it only be said that, besides the now well-known notion of *forcing*, it depends on an adaptation of Gödel's method of 1940 for setting up models of set theory, on a modification of the earlier methods of Fraenkel, Mostowski, and others in connection with the independence of the axiom of choice, and on the result of Skolem that there exists a countable model of set theory (a model having the cardinal number of the natural numbers).

The feeling that there is an absolute realm of sets, somehow determined in spite of the non-existence of a complete axiomatic characterization, receives more of a blow from the solution (better, the unsolving[2]) of the continuum problem than from the famous Gödel incompleteness theorems. It is not a question of realism (miscalled "Platonism") versus either conceptualism or nominalism, but if one chooses realism, whether there can be a "genetic" realism without axiomatic specification. The Gödel-Cohen results and subsequent extensions of them have the consequence that there is not one set theory but many, with the difference arising in connection with a problem which intuition still seems to tell us must "really" have only one true solution.

I know of mathematicians who hold that the axiom of choice has the same character of intuitive self-evidence that belongs to the most elementary laws of logic on which mathematics depends. It has never seemed so to me. But how shall one argue matters of intuition? The point is, I know of no one who maintains such self-evidence for the continuum hypothesis.

The realist will expect that the reality independent of the human mind which he maintains must have many ramifications, and will take what has now become the classical mathematical view, dating from the nineteenth century discussions of non-Euclidean geometry, that all ramifications equally demand exploration. The same view is possible also for him who takes the intermediate position between radical realism and conceptualism by holding that mathematical and physical objects alike, not excluding such basic logico-mathematical objects as sets, have their reality only relative to and within a certain theory. And if a choice must in some sense be made among the rival set theories, rather than merely and neutrally to develop the mathematical consequences of the alternative theories, it seems that the only basis for it can be the same informal

[2] I borrow this whimsical term from W. W. Boone.

criterion of simplicity that governs the choice among rival physical theories when both or all of them equally explain the experimental facts.

REFERENCES

[1] Cantor G., Ein Beitrag zur Mannigfaltigkeitslehre, *Journal für die reine und angewandte Mathematik*, 84 (1878), 242–258. [See p. 257]

[2] Cantor G., Über unendliche, lineare Punktmannigfaltigkeiten, *Mathematische Annalen*, 23, No. 6 (1884), 453–488.

[3] Burali-Forti Cesare, Una questione sui numeri transfiniti, *Rediconti del Circolo Matemotico di Palermo*, 11 (1897), 154–164.

[4] Hilbert David, Mathematische Probleme, *Nachrichten von der K. Gesellschaft der Wissenschaften zu Göttingen*, Math.-Phys. Kl., 1900, pp. 253–297. Reprinted with additions in *Archiv der Mathematik und Physik*, ser. 3, vol. 1 (1901), 44–63, 213–237. English translation in *Bulletin of the American Mathematical Society*, 8 (1901–2), 437–479.

[5] Zermelo Ernst, Beweis, daß jede Menge wohlgeordnet werden kann, *Mathematische Annalen*, 59 (1904), 514–516.

[6] Fraenkel A., Der Begriff 'definit' und die Unabhängigkeit des Auswahlaxioms, *Sitzungsberichte der Preussischen Akademie der Wissenschaften*, Phys.-Math. Kl., 1922, pp. 253–257.

[7] Skolem Thoralf, Einige Bemerkungen zur axiomatischen Begründung der Mengenlehre, *Wissenschaftliche Vorträge gehalten auf dem Fünften Kongress der Skandinavischen Mathematiker in Helsingfors vom 4, bis 7, Juli 1922*, Helsingfors, 1923, 217–232.

[8] Lindenbaum A., Tarski A., Communication sur les recherches de la théorie des ensembles, *Comptes Rendus des Séances de la Société des Sciences et des Lettres de Varsovie*, Classe III, 19 (1926), see p. 314.

[9] Gödel Kurt, Über formal unentscheidbare Sätze der Principia Mathematica und verwandter Systeme I, *Monatshefte für Mathematik und Physik*, 38 (1931), 173–198.

[10] Fraenkel A., Über eine abgeschwaechte Fassung des Auswahlaxioms, *The Journal of Symbolic Logic*, 2 (1937), 1–25.

[11] Lindenbaum Adolf, Mostowski Andrzej, Über die Unabhängigkeit des Auswahlaxioms und einiger seiner Folgerungen, *Comptes Rendus des Séances de la Société des Sciences et des Lettres de Varsovie*, Classe III, 31 (1938), 27–32.

[12] Gödel Kurt, The consistency of the axiom of choice and of the generalized continuum-hypothesis, *Proceedings of the National Academy of Sciences*, 24 (1938), 556–557.

[13] Gödel Kurt, Consistency-proof for the generalized continuum hypothesis, *Proceedings of the National Academy of Sciences*, 25 (1939), 220–224.

[14] Mostowski Andrzej, Über die Unabhängigkeit des Wohlordnungssatzes vom Ordnungsprinzip, *Fundamenta Mathematicae*, 32 (1939), 201–252.

[15] Gödel Kurt, The consistency of the axiom of choice and of the generalized continuum-hypothesis with the axioms of set theory, Princeton, 1940, 66 pp.

[16] Sierpiński Waclaw, L'hypothèse généralisée du continu et l'axiome du choix, *Fundamenta Mathematicae*, 34 (1947), 1–5.

[17] Shoenfield J. R., The independence of the axiom of choice, Abstract, *The Journal of Symbolic Logic*, 20 (1955), 202.

[18] Mendelson Elliott, The independence of a weak axiom of choice, *The Journal of Symbolic Logic*, 21 (1956), 350–366.

[19] Mendelson Elliott, The axiom of Fundierung and the axiom of choice, *Archiv für mathematische Logik und Grundlagenforschung*, 4 (1958), 65–70.

[20] Cohen Paul J., A minimal model for set theory, *Bulletin of the American Mathematical Society*, 69 (1963), 537–540.

[21] Cohen Paul J., The independence of the continuum hypothesis, *Proceedings of the National Academy of Sciences*, 50 (1963), 1143–1148, and 51 (1964), 105–110.

[22] Cohen Paul J., Independence results in set theory, *The theory of models, Proceedings of the 1963 International Symposium at Berkeley*, Amsterdam (1965), 39–54.

AXIOMS FOR FUNCTIONAL CALCULI OF HIGHER ORDER (1972)

This paper is a contribution to the axiomatic treatment of the simple theory of types, in the sense of functional calculi of third and higher orders, up to and including order ω. This is a topic that seems to have had relatively little attention in the logical literature, at least from the point of view of axiomatic economy and independence of the primitives, although there exist many and detailed axiomatic studies of propositional logic and a lesser number of studies of first-order logic. (By Russell, Łukasiewicz, Tarski, Leśniewski, and others there have also been axiomatic treatments both of propositional calculus as extended by adjunction of quantifiers for propositional variables and of what Leśniewski called protothetic, two calculi which may be regarded as fragments of functional calculi of second and higher orders.)

For references to the literature and for terminology and background material the author's *Introduction to Mathematical Logic* Vol. I (Princeton: Princeton University Press, 1956), is relied on here. Some of the material in the present paper appeared in an abstract by the author in *The Journal of Symbolic Logic*, Vol. 21, page 218.

As regards style of the desired axiomatizations we make the following general remark. The demand for economy in primitive notations, rules of inference, and axioms is not absolute, but may be subject to other demands which for a particular purpose or in a particular connection are thought to be more important.

Especially the demand for economy is always subject to the demand for adequacy of the system formulated. And we regard adequacy as meaning not only that all the expected theorems shall result but also all the expected consequences of any assertion, even of a non-theorem—in some sense of "expected", which is appropriate in the context and which may vary all the way from a truth-definition to merely retaining the consequences of some previously known formulation. This latter requirement of adequacy might be expressed by saying that all the expected consequences of any added postulate shall result; but if it is put this way, it is then necessary to understand that a postulate is not necessarily something that will be firmly maintained, and that a postulate may even be introduced for the purpose of refuting it by showing it to have undesirable consequences.[1]

The foregoing is conclusive, in some cases, against avoiding substitution rules for variables of a particular type by the device of employing axiom schemata in the way that is explained in Sec. 27 and Sec. 30 of *Introduction to Mathematical Logic*. In particular this is clearly inadmissible in the case of the variables of highest type-class in any functional calculus of odd order[2]—with the exception of those functional calculi

Originally published in ***Logic and Art: Essays in Honor of Nelson Goodman***, edited by Richard S. Rudner and Israel Scheffler, Bobbs-Merrill Company, Indianapolis and New York 1972, pp. 197–213.

[1]This point has been made by others, against the idea that an (axiomatically formalized) branch of logic is to be identified with the class of its theorems. The writer is unable to trace the original source. See, however, the first three paragraphs of Sec. 41 of ***Introduction to Mathematical Logic***.

[2]We anticipate the objection that an added postulate might be replaced by a postulate schema. If a single postulate, in the form of an assertion containing a free variable of highest type-class, is unacceptable in the sense of having unacceptable consequences, then the underlying logic ought to yield these consequences.

More loosely put, the objection to the formulation without substitution rules is that it fails to give the variables the status of variables—they could be constants for all that appears in the (syntactical) primitive

that have only constants and not variables in the highest type-class.[3] In this paper we shall employ substitution rules systematically for variables of all types, partly for uniformity, and partly because we prefer to treat first from this simpler point of view the questions of economy that we here raise. We shall leave for later consideration the modifications that result if we seek at the same time to avoid rules of substitution by the device of axiom schemata (due to von Neumann) with or without the additional device of the Gödel-Leśniewski-Henkin[4] "pseudo-definition" (so-called by Leśniewski).

In formulating a primitive basis for one of the functional calculi, it might be natural to demand separation properties analogous to the well-known separation properties of various formulations of propositional calculus which are due to Hilbert, Hilbert-Bernays, and Bernays — among them as an example the System P_H which is described in *Introduction to Mathematical Logic*, Sec. 26. It is further well known that the demand for separation properties may often conflict with the demand for economy; and it is remarkable that in (for example) the system P_H the separation properties are secured without loss of independence of the axioms. But there is of course a deliberate sacrifice of economy in the primitive notations of the system.

For a primitive basis of one of the functional calculi there is no absolute decision as to what separation properties shall be demanded. As one such property, however, it might be asked that a primitive basis for one of the pure functional calculi shall include as a part of it a primitive basis for each of the pure functional calculi of lower order; e.g. from a primitive basis of pure functional calculus of fifth order it shall be possible to obtain a primitive basis for each of the pure functional calculi of lower order; e.g. from a primitive basis for pure functional calculus of fifth order it shall be possible to obtain a primitive basis for pure functional calculus of fourth order (or indeed of any lower order) by merely deleting certain of the primitive notations, rules, and axioms without other change. Or instead of this we may make only the weaker demand, which can perhaps be more simply satisfied, that a primitive basis for a pure functional calculus of odd order shall include as a part of it a primitive basis for each pure functional calculus of lower odd order; and a primitive basis for a pure functional calculus of even order shall include a primitive basis for each pure functional calculus of lower even order.

Where a conflict arises between the demand for economy and such separation demands, the decision between the two may in part depend on purpose and general context and may in part be subjective. It is not the intention of this paper to make such decisions, but rather to call attention to some questions that arise and to make a beginning towards exploration of the possibilities. Such particular primitive bases as we present are mainly in the direction of preferring economy over separation properties when there is a conflict.

The first threatened conflict between economy and separation properties arises from

---

basis. If with each new postulate some of the logic has to be supplied in its application to that new postulate, then the original statement of the underlying logic was defective.

[3]In particular for simple applied functional calculi of first-order, it is well known that the device of axiom schemata which is here in question not only is admissible but, as explained in footnote 250 of ***Introduction to Mathematical Logic***, is hardly avoidable in formulations of standard kind.

[4]Leon Henkin, *Banishing the Rule of Substitution for Functional Variables*, ***The Journal of Symbolic Logic***, Vol. 18 (1953), 201–208.

the remark[5] that in functional calculi of second and higher order it is possible to introduce a notation f, to denote the truth-value falsehood, by the definition

$$f \longrightarrow (s)s$$

If the primitive notations include $\supset$ and the universal quantifier, as it is usual and natural when *modus ponens* and generalization are taken as primitive rules, then this definition of f means that the primitive notations need not include any additional connectives of propositional calculus. In particular, negation may be introduced by the definition $\sim\mathbf{p} \longrightarrow \mathbf{p} \supset f$, or more fully,[5]

$$\sim\mathbf{p} \longrightarrow \mathbf{p} \supset (s)s$$

The theorem $\sim p \supset . \, p \supset q$ is then forthcoming as a consequence of familiar theorems involving $\supset$ and the universal quantifier. In this way introduction of the universal quantifier, at the level of at least second-order functional calculus, results in a reduction in both the primitive connectives and the axioms of propositional calculus that are necessary.[6]

It is only very superficial reflection that could suggest an attempt to avoid the foregoing situation by the device of omitting propositional variables from among the primitive notations of the higher-order functional calculi, or by that of not allowing quantification with respect to propositional variables. For the above definition of f can of course be replaced by

$$f \longrightarrow (F)(x)F(x)$$

Or if we further omit the singulary functional variables from among the primitive notations,[7] we can still make the definition

$$f \longrightarrow (F)(x)(y)F(x, y)\cdot$$

and so on.

A second point of conflict between the demand for economy and that for separation properties arises in connection with the axioms of extensionality. These are the axioms

(1) $$\mathbf{f}(\mathbf{x}, \mathbf{y}, \ldots, \mathbf{u}) \supset_{\mathbf{xy}\ldots\mathbf{u}} \mathbf{g}(\mathbf{x}, \mathbf{y}, \ldots, \mathbf{u}) \supset . \; \mathbf{g}(\mathbf{x}, \mathbf{y}, \ldots, \mathbf{u}) \supset_{\mathbf{xy}\ldots\mathbf{u}} \mathbf{f}(\mathbf{x}, \mathbf{y}, \ldots, \mathbf{u}) \supset . \; \mathbf{h}(\mathbf{f}) \supset \mathbf{h}(\mathbf{g}),$$

[5]Due to Russell, as explained in footnotes 225 and 226 of ***Introduction to Mathematical Logic***.

[6]It is shown by Rasiowa (*O Pewnym Fragmencie Implikacyjnego Rachunku Zdań* in ***Studia Logica***, Vol. 3 (1955), 208–226), crediting the result to Slupecki, that if we begin with a fragment of the implicational propositional calculus characterized by saying that its theorems are all those that remain valid when $\supset$ is replaced everywhere by $\equiv$, the rules of inference being substitution and *modus ponens* as usual, and if we then add quantification with respect to propositional variables, and laws of quantification sufficient to make $f \supset p$ a theorem, the full implicational propositional calculus will follow—and hence of course the full propositional calculus by way of the definition of $\sim\mathbf{p}$ which is given above.

[7]In the same sense in which it is true that we can without loss omit the propositional variables from among the primitive notations, it is true that we can further omit the singulary functional variables without loss. Neither omission would seem to be a genuine economy, as there is always the question: Where shall we stop? If such economies are to be undertaken at all, it would seem to be better to follow Tarski in omitting *all but* the singulary functional variables (although below order ω this interferes with the usual separation of the functional calculi into orders). However, for our present purpose it is not necessary to say more about any of these proposed economies than that they do not interfere in any substantial way with what is said in the text, either about the possibility of defining f or about the effect of the axioms of extensionality.

where $\mathbf{x}, \mathbf{y}, \dots, \mathbf{u}$ are n variables, all different, and of any arbitrary types, and $\mathbf{f}, \mathbf{g}$, and $\mathbf{h}$ are variables of such type as to make the axiom well-formed. (Hence in particular $\mathbf{f}$ and $\mathbf{g}$ are n-ary functional variables of the same type, and $\mathbf{h}$ is a singulary functional variable of such type that $\mathbf{h}(\mathbf{f})$ is well-formed.) Or these axioms may also be written in either of the following forms, which are equivalent to (1) if usual laws of propositional calculus and of quantifiers are granted:

$$(1') \qquad \mathbf{f}(\mathbf{x}, \mathbf{y}, \dots, \mathbf{u}) \equiv_{\mathbf{xy}\dots\mathbf{u}} \mathbf{g}(\mathbf{x}, \mathbf{y}, \dots, \mathbf{u}) \supset . \, \mathbf{h}(\mathbf{f}) \supset \mathbf{h}(\mathbf{g})$$

$$(1'') \qquad \mathbf{f}(\mathbf{x}, \mathbf{y}, \dots, \mathbf{u}) \equiv_{\mathbf{xy}\dots\mathbf{u}} \mathbf{g}(\mathbf{x}, \mathbf{y}, \dots, \mathbf{u}) \supset . \, \mathbf{h}(\mathbf{f}) \equiv \mathbf{h}(\mathbf{g})$$

It is well known that in *Principia Mathematica* axioms of extensionality of the above form (or forms) are not used. In the first edition an intensional interpretation is intended for the propositional and functional variables, and the extensional notions of *class* and *relation* are available only in the sense of Russell's contextual definitions. In the introduction to the second edition axioms of extensionality are proposed as a modification, but because this is done in the context of ramified type theory, the axioms are not entirely the same here and their effect is quite different from that of axioms of extensionality in simple type theory.

In spite of the precedent of *Principia Mathematica*, there are reasons of some force that can be given for including axioms of extensionality in the basis of a functional calculus of third or higher order. One of these is just that the extensional interpretations of the propositional and functional variables are conceptually simpler and are often closer to an intended (especially mathematical) application; if the extensional interpretations are intended, it is better to make this explicit in the axioms and avoid indirection. A second reason is that there is a difficulty in regard to the rule of substitution for functional variables which (it seems) can most simply be resolved by allowing axioms of extensionality.

In connection with the first reason it must be pointed out that adoption of extensional interpretations of the propositional and functional variables forces adoption of a Fregean as against a Russellian theory of meaning. Indeed the intensional interpretations and the resort to contextual definitions in the first edition of *Principia* are Russell's method of dealing with the well-known puzzle about meaning which originated with Frege but is now more famous as illustrated by Russell's example concerning Sir Walter Scott and King George IV.[8] One of the advantages of Frege's theory of meaning is that it leaves us free to formulate a purely extensional object language where the purpose is extensional; then later, if we wish to deal with intensionalities, we may extend the language by introducing appropriate intensional notations.

To explain the second reason let us consider the (metatheoretic) notation

$$\check{S}_{\mathbf{B}}^{\mathbf{f}(\mathbf{x}, \mathbf{y}, \dots, \mathbf{u})} \mathbf{A}|$$

[8]Let us introduce predicates S and W, $S(x)$ to mean that x is Sir Walter Scott (that x scottizes, in Quine's phrase) and $W(x)$ to mean that x wrote ***Waverley***. The predicates S and W are functional constants rather than functional variables, but it is clear that an intensional or extensional interpretation of the functional variables must go together with an intensional or extensional interpretation of any functional constants that are introduced. Since it is a fact that $W(x) \equiv_x S(x)$, the extensional interpretation identifies the denotations of S and W, so that whatever is true of one is true of the other. We may suppose, as seems likely, that King George IV believed that $S(x) \equiv_x S(x)$. On the extensional interpretation it seems to follow that King George IV believed that $W(x) \equiv_x S(x)$. But at the time of the now familiar dinner it is more correct to say that King George suspected that Sir Walter might be the author of ***Waverley*** than that he so believed.

for substitution for any n-ary functional variable **f**, where **x**, **y**, ... , **u** are n different variables of suitable types. The operation on the wff **A** which this notation calls for may conveniently be explained by saying that we first make an ordinary substitution of[9] $\lambda\mathbf{x}\lambda\mathbf{y}\ldots\lambda\mathbf{u}\mathbf{B}$ for all free occurrences of the variable **f** in **A** and then apply a series of contractions by which expressions of the form[10] $\lambda\mathbf{x}\lambda\mathbf{y}\ldots\lambda\mathbf{u}\mathbf{B}(\mathbf{a}_1, \mathbf{a}_2, \ldots, \mathbf{a}_n)$ are replaced by

$$\mathrm{S}^{\mathbf{x}\ \mathbf{y}\ \ldots\mathbf{u}}_{\mathbf{a}_1\mathbf{a}_2\ldots\mathbf{a}_n}\mathbf{B}|$$

These contractions are continued until a wff of the appropriate functional calculus is obtained, and it is this final wff which is denoted by

$$\check{\mathrm{S}}^{\mathbf{f}(\mathbf{x},\mathbf{y},\ldots,\mathbf{u})}_{\mathbf{B}}\mathbf{A}|$$

Restrictions[11] must be imposed to assure that there is no capture of variables either in the original substitution for **f** or in any of the subsequent contractions, and when these restrictions fail, the metatheoretic notation is understood simply as denoting the wff **A**.

The above-mentioned difficulty in regard to the rule of substitution for functional variables arises when the wff **A** contains free occurrences of the functional variable **f** as argument of another functional variable **h**, since in this case the series of contractions that are applied after substitution of $\lambda\mathbf{x}\lambda\mathbf{y}\ldots\lambda\mathbf{u}\mathbf{B}$ for free occurrences of **f** will not result in a wff of functional calculus (not all occurrences of the abstraction operator λ are eliminated by the contractions). We seem to be forced to impose another restriction on the substitution operation S, namely that **f** shall not have in **A** free occurrences as argument of another functional variable, and again to understand the result of the substitution, to be just **A** itself in case the restriction fails. If axioms of extensionality are available, it is easy to see that this new restriction on substitution for functional variables is not in fact an essential restriction; to take for illustration the case in which **h** and **f** are both singulary, we can prove in consequence of the appropriate axiom of extensionality

$$\mathbf{h}(\mathbf{f}) \equiv \centerdot\, \mathbf{f}(\mathbf{x}) \supset_{\mathbf{x}} \mathbf{g}(\mathbf{x}) \supset_{\mathbf{g}} \centerdot\, \mathbf{g}(\mathbf{x}) \supset_{\mathbf{x}} \mathbf{f}(\mathbf{x}) \supset \mathbf{h}(\mathbf{g}),$$

and either side of this equivalence can then be replaced at will by the other (with any choice of the variable **g** as different from but of the same type as **f**). On the other hand, without axioms of extensionality there is a puzzling question as to what can be

[9]The expression $\lambda\mathbf{x}\lambda\mathbf{y}\ldots\lambda\mathbf{u}\mathbf{B}$ is not a wff of any of the functional calculi, at least not in usual or standard formulations of them. But this fact does not prevent the use which we here make of this expression in order to explain an operation by which a wff of a particular functional calculus is transformed into another wff of the same calculus.

The point is conspicuous that if we extend the notation of functional calculus by adjoining the abstraction operator λ as a new primitive, the rule of substitution for functional variables could be replaced by simpler rules of ordinary substitution, alphabetic change of bound variable, and contraction—and some of the complications in connection with the rule of substitution for functional variables could be avoided. In effect a complicated rule of inference is thus broken down into simpler ones. But we here follow the standard usage according to which an abstraction operator is not considered to belong to the notations of functional calculus.

[10]We might write this as $\{\lambda\mathbf{x}\lambda\mathbf{y}\ldots\lambda\mathbf{u}\mathbf{B}\}(\mathbf{a}_1, \mathbf{a}_2, \ldots, \mathbf{a}_n)$. And the braces may indeed be useful for clearness, especially if a complicated particular expression has to be written, but they are not strictly necessary, as in the context of one of the functional calculi the scope of the operators $\lambda\mathbf{x}\lambda\mathbf{y}\ldots\lambda\mathbf{u}$ is determined by the condition that **B** shall be well-formed.

[11]Compare ***Introduction to Mathematical Logic***, Sec. 35.

considered adequate. If intensional interpretations are intended for the propositional variables and the various types of functional variables, perhaps the question can be resolved only by introducing some suitable strict or intensional implication, and some suitable strict or intensional equivalence, and stating analogues of the axioms of extensionality by means of them.

For these reasons, in formulating a primitive basis for one of the higher-order functional calculi, let us proceed on the assumption that axioms of extensionality are to be included. An effect of the axioms of extensionality may be to render non-independent some of the axioms of propositional calculus that would otherwise be used, or at least to open up new possibilities of economy in regard to propositional calculus axioms. For one very important derived rule of propositional calculus, either the principle of substitutivity of mutual implication or the principle of substitutivity of equivalence, follows from the axioms of extensionality by no more than (or little more than) *modus ponens* and substitution for functional variables. This is clearest if we suppose that the case $n = 0$ is admitted in the axiom schema (1) (or (1′) or (1″)), so that we have as one of the axioms that is covered by (e.g.) the schema (1):[12]

$$(1_0) \qquad p \supset q \supset \centerdot q \supset p \supset \centerdot h(p) \supset h(q)$$

For simplicity we shall in fact suppose this in what follows. But the same results can be obtained if we suppose $n \geqq 1$ in schema (1) (or even $n \geqq 2$, etc.), and ultimately the axioms that correspond to $n = 0$ (or $n < 2$, etc.) can then be proved as theorems.

Now therefore we turn to some particular primitive bases for functional calculi as being suggested by the above considerations. We sketch the matter rather than give a fully detailed treatment. We confine attention to pure functional calculi, as for our purpose there is perhaps nothing to be gained in going beyond that. And we further confine attention, in this first treatment, to functional calculi of odd order and of order ω. (For functional calculi of even order the axiomatic basis must in any case be somewhat different; a more extensive use of axiom schemata than in the case of functional calculi of odd order is hardly avoidable, and we may hence prefer to adopt Leśniewski's "pseudo-definitions"[13] as means of excluding the somewhat complicated substitution operation S from at least the primitive rules and axiom schemata.)

For pure functional calculus of order ω we take the following primitive symbols: the sign of material implication $\supset$ and the brackets [ ] that go with it, the universal quantifier, the notation (composed of parentheses and commas) for application of function to argument, an infinite alphabet of individual variables, an infinite alphabet of propositional variables, and for each type of functional variables an infinite alphabet of variables of that type.

This statement presupposes the notion of a type, and must therefore be completed by explaining this notion as follows. There is a type 0, to which the individual variables belong. If $\alpha_1, \alpha_2, \ldots, \alpha_n$ are any given types, there is a corresponding type $(\alpha_1, \alpha_2, \ldots, \alpha_n)$ to which there belong n-ary functional variables. And since

[12]We may without loss specialize to a particular functional variable, say h, so that (1_0) is a single axiom rather than a schema.

[13]They might preferably be called comprehension axioms, as Henkin's explanation tends to suggest, though not quite explicitly. Gödel (***Monatshefte für Mathematik und Physik***, Vol. 38 (1931), 173–198) misleadingly speaks of them also as corresponding to axioms of reducibility.

a pure functional calculus is here in question (i.e., constants are not among the primitive symbols), we have the rule that if **f** is a variable of type $(\alpha_1, \alpha_2, \ldots, \alpha_n)$, then $\mathbf{f}(\mathbf{a}_1, \mathbf{a}_2, \ldots, \mathbf{a}_n)$ is well-formed if and only if $\mathbf{a}_1, \mathbf{a}_2, \ldots, \mathbf{a}_n$ are variables whose types are, in order, $\alpha_1, \alpha_2, \ldots, \alpha_n$.

We think of the propositional variables as included in the foregoing explanation by allowing the case $n = 0$, but for convenience we modify the notation as follows. If **f** is a propositional variable, we call its type 1_0, rather than ( ), and we take **f** standing alone as being well-formed, rather than **f**( ).

Then the types are divided into numbered type-classes as follows. (We may speak either of a type or of a variable of that type as belonging to a certain type-class m.) The type 0 belongs to, or is of, the type-class 0. The type 1_0 belongs to the type-class 1. And beyond that, the type $(\alpha_1, \alpha_2, \ldots, \alpha_n)$ belongs to the type-class $m + 1$, where m is the highest type-class to which any of the types $\alpha_1, \alpha_2, \ldots, \alpha_n$ belong.

This accounts for the primitive symbols of the pure functional calculus of order ω. And for the pure functional calculi of finite order we specify the primitive symbols as follows. The pure functional calculus of order $2m - 1$ and the pure functional calculus of order $2m$ have the same primitive symbols, which are namely the same as those remaining from the primitive symbols of the pure functional calculus of order ω after deleting all variables of type-class higher than m.

The formation rules for all of these functional calculi are now evident, as being implicit in what has already been said. But for the functional calculi of odd order, the formation rule for the universal quantifier, that if **M** is well-formed then (**a**)**M** is well-formed, must have the restriction that the variable **a** is not of the highest type-class.

For the pure functional calculi of finite order and of order ω we then take rules of inference as follows:

The rule of modus ponens, from $\mathbf{A} \supset \mathbf{B}$ and **A** to infer **B**.

The rule of generalization, from **A** to infer (**a**)**A**, provided **a** is a variable of such type that the result of the inference is well-formed.

The rule of substitution for individual variables, from **A** to infer $\underset{\centerdot}{\mathrm{S}}{}^{\mathbf{a}}_{\mathbf{b}}\mathbf{A}|$, where **a** and **b** are individual variables, provided that no bound occurrence of **b** results at any place at which **b** is substituted for a free occurrence of **a** (in other words, provided that there is no capture of the variable **b**).

The rule of substitution for propositional variables, from **A** to infer $\check{\mathrm{S}}^{\mathbf{p}}_{\mathbf{B}}\mathbf{A}|$, where **B** is any wff and **p** is any propositional variable.[14]

The rule of substitution for functional variables, from **A** to infer

$$\check{\mathrm{S}}^{\mathbf{f}(\mathbf{x},\mathbf{y},\ldots,\mathbf{u})}_{\mathbf{B}}\mathbf{A}|,$$

[14] $\check{\mathrm{S}}^{\mathbf{p}}_{\mathbf{B}}\mathbf{A}|$ is the result of substituting **B** for free occurrences of **p** throughout **A**, so that this is the same rule of substitution that is familiar in propositional calculus. Here, however, it is necessary to impose restrictions analogous to those already described in regard to substitution for functional variables, and to understand $\check{\mathrm{S}}^{\mathbf{p}}_{\mathbf{B}}\mathbf{A}|$ as being simply **A** if one of the restrictions fails. There are two such restrictions, the first one being that no capture of variables shall result from the substitution, and the second one that **p** shall not have any free occurrence in **A** as argument of a functional variable.

In connection with the second restriction, observe that we are here following the long-standing restriction on notation in pure functional calculi, that each argument of a functional variable must be a single primitive symbol, hence a variable; as a special case of this, where e.g. h is a variable of such type that $h(p)$ is well-formed, nevertheless $h(f)$ and $h(p \supset q)$ are not well-formed. This is a divergence from the usage of Leśniewski's protothetic, but not an essential one.

where **B** is any wff, **f** is an n-ary functional variable, and $\mathbf{x}, \mathbf{y}, \ldots, \mathbf{u}$ are n distinct variables of such types that $\mathbf{f}(\mathbf{x}, \mathbf{y}, \ldots, \mathbf{u})$ is well-formed.

As axioms there are first of all the axioms of extensionality (1), as given above, including the axiom (1_0) as one of them. Then there are the following axioms of quantifiers:[15]

(2) $$p \supset_{\mathbf{x}} \mathbf{f}(\mathbf{x}) \supset . \, p \supset (\mathbf{x})\mathbf{f}(\mathbf{x})$$

(3) $$(\mathbf{x})\mathbf{f}(\mathbf{x}) \supset \mathbf{f}(\mathbf{y})$$

In (2) and (3), p is the particular propositional variable, and $\mathbf{f}, \mathbf{x}, \mathbf{y}$ are variables of such type as to make the axioms well-formed, with the restriction in the case of (2) that $\mathbf{x}$ is not p.

After that we need only axioms of propositional calculus, and our present purpose is to minimize the number and length of these. Perhaps the simplest choice is to take as the two final axioms:

(4) $$p \supset . \, q \supset p$$

(5) $$p \supset q \supset p \supset p$$

However, we shall also consider replacing axiom (4) by:

(6) $$p \supset [q \supset r] \supset . \, q \supset . \, p \supset r$$

We sketch a proof that the axioms (4) and (5) are sufficient for propositional calculus, in the presence of the other axioms stated.

By substitution for the functional variable h in (1_0) we get:

(7) $$p \supset q \supset . \, q \supset p \supset . \, r \supset r$$

Hence by substitution for q:

$$p \supset [p \supset q \supset p] \supset . \, p \supset q \supset p \supset p \supset . \, r \supset r$$

Hence by two applications of *modus ponens* we get the theorem:

(8) $$r \supset r,$$

the first minor premiss being obtained by substitution in axiom (4) and the second one being axiom (5).[16]

---

[15]For brevity the axiom schemata (1), (2), (3) are here stated in such a way that they include obviously non-independent axioms, even for any particular choice of types of the variables $\mathbf{f}, \mathbf{g}, \mathbf{h}, \mathbf{x}, \mathbf{y}, \ldots, \mathbf{u}$ (or $\mathbf{f}, \mathbf{x}, \mathbf{y}$). In the case of schema (1) it is not difficult to avoid this by restating in such a way that, for any such particular choice of types, particular variables $\mathbf{f}, \mathbf{g}, \mathbf{h}, \mathbf{x}, \mathbf{y}, \ldots, \mathbf{u}$ are prescribed. But in schemata (2) and (3) only $\mathbf{f}$ and $\mathbf{y}$ can be thus particularized; and $\mathbf{x}$ must remain an arbitrary variable of its type — unless a rule of alphabetic change of bound variable is added to the primitive rules of inference.

[16]To get analogous results in a formulation without propositional variables the general plan is to use $F(x)$, $G(x)$, $H(x)$, and so forth in place of the propositional variables p, q, r, and so forth. Thus theorem (8) is replaced by $H(x) \supset H(x)$, and in its proof we use in place of (1_0) the case $n = 1$ in axiom schema (1). Then, e.g. the last two minor premisses in the proof, instead of $p \supset . \, p \supset q \supset p$ and $p \supset q \supset p \supset p$, will be $F(x) \supset_x . \, F(x) \supset G(x) \supset F(x)$ and $F(x) \supset G(x) \supset F(x) \supset_x F(x)$. For this purpose it does not matter whether axioms (4) and (5) are replaced by particular axioms such as $F(x) \supset . \, G(x) \supset F(x)$ and $F(x) \supset G(x) \supset F(x) \supset F(x)$ or whether they are replaced by axiom schemata.

By substituting p for q in (7) and using *modus ponens* (since $p \supset p$ is a theorem by substitution in (8)) we get:

$$(9) \qquad p \supset p \supset . r \supset r$$

Similarly from (7) (or from (9)) we get:

$$(10) \qquad r \supset r \supset . p \supset p$$

By substitution for the functional variable h in (1_0):

$$p \supset q \supset . q \supset p \supset . p \supset s \supset s \supset . q \supset s \supset s$$

Hence by substitution for the propositional variables:

$$p \supset p \supset [r \supset r] \supset . r \supset r \supset [p \supset p] \supset .$$
$$p \supset p \supset p \supset p \supset . r \supset r \supset p \supset p$$

From this by three applications of *modus ponens* we get the theorem:

$$(11) \qquad r \supset r \supset p \supset p$$

(The three successive minor premisses that are required are in order (9), (10), and obtained from (5) by substitution of p for q.)

By substitution in (4) we have:

$$(12) \qquad p \supset . r \supset r \supset p$$

Also from (4) and (8) we can prove:

$$(13) \qquad p \supset . r \supset r$$

In consequence of (11) and (12), and by using the extensionality axiom (1_0), we may replace p by $r \supset r \supset p$ in any context except that of an occurrence of p as argument of a functional variable. We omit details of this, as they are entirely similar to those of the last use of (1_0) which was made above. Equally we may use (11), (12), and (1_0) to replace $p \supset q$ by $r \supset r \supset . p \supset q$ in any context. Hence from the theorem $p \supset q \supset . p \supset q$ (which results from (8) by substitution) we get:

$$(14) \qquad r \supset r \supset [p \supset q] \supset . r \supset r \supset p \supset q$$

Now we prove the following derived rule:

Modified rule of modus ponens. From $\mathbf{C} \supset . \mathbf{A} \supset \mathbf{B}$ and $\mathbf{A}$ we may infer $\mathbf{C} \supset \mathbf{B}$.

For given $\mathbf{A}$ we have, by (12) and (13), $r \supset r \supset \mathbf{A}$ and $\mathbf{A} \supset . r \supset r$. Hence using the extensionality axiom (1_0) we may replace $\mathbf{A}$ by $r \supset r$ in $\mathbf{C} \supset . \mathbf{A} \supset \mathbf{B}$, so obtaining $\mathbf{C} \supset . r \supset r \supset \mathbf{B}$. But (11) and (12) enable us, by another use of (1_0), to replace $r \supset r \supset \mathbf{B}$ by $\mathbf{B}$, so that we get $\mathbf{C} \supset \mathbf{B}$.[17]

Having thus proved the derived rule, we substitute in (1_0) to get the following (first substituting for the functional variable h and then making a succession of substitutions

[17]Obviously we could use a variation of the same argument to prove the much stronger derived rule (not needed here) that, given $\mathbf{A}$ and $\mathbf{C}$ as premisses, we can replace any occurrence of $\mathbf{A} \supset \mathbf{B}$ by $\mathbf{B}$, or of $\mathbf{B}$ by $\mathbf{A} \supset \mathbf{B}$, within $\mathbf{C}$, provided the result is well-formed.

for the propositional variables):

$$r \supset [s \supset r] \supset . s \supset r \supset r \supset .$$
$$r \supset r \supset [p \supset q] \supset [r \supset r \supset p \supset q] \supset .$$
$$s \supset r \supset r \supset [p \supset q] \supset . s \supset r \supset r \supset p \supset q$$

From this by first using *modus ponens*, with $r \supset . s \supset r$ as minor premiss (obtained from (4) by substitution), and then using the modified rule of *modus ponens*, with (14) as minor premiss, we get:

$$s \supset r \supset r \supset . s \supset r \supset r \supset [p \supset q] \supset . s \supset r \supset r \supset p \supset q$$

In this we substitute s for r and then use (11), (12), and (1_0) to replace $s \supset s \supset s$ by s. The result is

(15) $$s \supset . s \supset [p \supset q] \supset . s \supset p \supset q$$

Now (15) is close to the self-distributive law (†103 in *Introduction to Mathematical Logic*); in fact it differs from the self-distributive law only in regard to the order in which the antecedents are arranged. Moreover an examination of the proof of the deduction theorem in Sec. 13 of *Introduction to Mathematical Logic* shows that we may use it here almost unchanged. We have to modify only case 3 in the proof, replacing use of the self-distributive law by use of (15), and hence using the modified rule of *modus ponens* at certain points instead of the rule of *modus ponens*.

By substitution in (3) we have the theorem:

(16) $$f \supset p$$

At this point our proof that axioms (4) and (5) are sufficient for propositional calculus could be considered complete, as it is known that (4), (5), (16), and the transitive law of implication are sufficient—with *modus ponens* and substitution as rules of inference—and the transitive law is of course an easy consequence of the deduction theorem. But to make the connection with *Introduction to Mathematical Logic*, we indicate also the proof of

(17) $$p \supset f \supset f \supset p,$$

so that it can be seen directly that all the axioms of the system P_1 are theorems. Namely we get by substitution in (1_0):

$$p \supset f \supset f \supset . f \supset [p \supset f] \supset . p \supset f \supset p \supset p \supset . f \supset p \supset p$$

But $f \supset . p \supset f$ is a theorem by substitution in (4), $p \supset f \supset p \supset p$ is a theorem by substitution in (5), and $f \supset p$ is (16). Hence we get (17) by three successive applications of the modified rule of *modus ponens*.

Now we turn to demonstration of the sufficiency of axioms (5) and (6) for propositional calculus, in the presence of the axioms of extensionality and the axioms of quantifiers. And for this purpose we follow as closely as we can the above treatment of axioms (4) and (5).

By substitution in (7) we have

$$p \supset [q \supset r] \supset [q \supset . p \supset r] \supset .$$
$$q \supset [p \supset r] \supset [p \supset . q \supset r] \supset . r \supset r,$$

and hence we prove (8) by two applications of *modus ponens*.

By substitution in (8) we have:

$$r \supset r \supset p \supset . r \supset r \supset p$$

Hence by (6):

$$r \supset r \supset . r \supset r \supset p \supset p$$

Hence (11) follows by *modus ponens*.

By substituting in (7) and using *modus ponens*, with (8) as minor premiss, we prove (10). Hence (12) can be proved by use of (6).

By substitution in (7) we have:

$$q \supset [r \supset r] \supset . r \supset r \supset q \supset . p \supset p$$

by (1_0), (11), and (12) we may replace $r \supset r \supset q$ by q, so that we have:

(18) $$q \supset [r \supset r] \supset . q \supset . p \supset p$$

Again by substituting in (1_0) we have:

$$q \supset r \supset . r \supset q \supset . q \supset r$$

Hence by (6):

$$r \supset q \supset . q \supset r \supset . q \supset r$$

Hence by (18):

$$r \supset q \supset . p \supset p$$

Hence by substitution for r:

$$r \supset r \supset q \supset . p \supset p$$

Hence again using (1_0), (11), and (12) to replace $r \supset r \supset q$ by q we have:

$$q \supset . p \supset p$$

From this last we get (13) by substitution and we get (4) by use of (6).

From this point on, the treatment of axioms (5) and (6) may duplicate that already given for axioms (4) and (5). And as Peirce's law, axiom (5), is thus not used until after completion of the proof of the deduction theorem, we have the following separation property of the axioms:

Axioms (1_0) *and* (6) *are sufficient for the positive implicational propositional calculus. The addition of axiom* (5) *then yields the implicational propositional calculus. And the further addition of* (3) *yields the full propositional calculus.*

It would be of interest to determine what is the shortest axiom or the axiom with the smallest number of distinct variables which, in place of (6), preserves this separation property.

Finally, though we do not make a systematic treatment of independence questions in this connection, we cite the two following independence examples, which settle some of the most critical questions.

To show the independence of axiom (5) from axioms (1), (2), (3), (4) we use the truth-table for $\supset$ which is shown in the table below in the column headed (5). The truth value 0 is designated and 1 and 2 are non-designated. And the valuation rule for the universal quantifier is as follows:[18] For a given system of values of the free variables of (**x**)**M**, we give the value 0 to (**x**)**M** if **M** has the value 0 for every value of **x**, the value 1 if **M** has the value 1 for at least one value of **x**, and the value 2 in all remaining cases.

		(4)	(5)
p	q	$p \supset q$	$p \supset q$
0	0	0	0
0	1	0	2
0	2	2	2
1	0	2	0
1	1	0	0
1	2	2	0
2	0	0	0
2	1	0	2
2	2	0	0

This same independence example (or model), since it falsifies axiom (6), shows that the axioms (1), (2), (3), (4), (5) do not have the separation property that was just stated above for the axioms (1), (2), (3), (5), (6).

To show the independence of (4) from axioms (1), (2), (3), (5) we use the truth-table for $\supset$ that is shown in the column headed (4). The truth-values 0 and 1 are designated and 2 is non-designated. And the valuation rule for the universal quantifier is as follows: For a given system of values of the free variables of (**x**)**M**, we give the value 1 to (**x**)**M** if **M** has the value 1 for every value of **x**, the value 2 if **M** has the value 2 for at least one value of **x**, and the value 0 in all remaining cases.

The independence of (6) from axioms (1), (2), (3), (5) is then obvious without further reference to independence examples.

[18]Other details of the intended model are the obvious ones and are omitted for brevity. The model may be varied somewhat, and in particular any non-empty domain of individuals will serve—the suggestion is immediate that the model may be simplified by using the unit domain.

ARTICLES FROM *ENCYCLOPÆDIA BRITANNICA* (1956–1972)

AXIOM. According to Aristotle, every demonstrative science must start from indemonstrable principles. Among these first principles some are peculiar to the particular science but others are common to all sciences and are called axioms. Elsewhere the axioms are characterized as the common opinions from which all demonstration proceeds, and as those things which anyone must hold who is to learn anything at all.

In the *Elements* of Euclid the first principles are listed in two categories, the *postulates* and the *common notions*. The former are the principles peculiar to the particular science of geometry and seem to be thought of just as required assumptions since their statement opens with the word *etestho* ("let there be demanded"). The common notions are evidently the same as Aristotle's axioms, and indeed Proclus (*q.v.*), *On the First Book of Euclid*, tells us explicitly that the two terms are synonymous. However, the principle by which the division between postulates and axioms is to be made seems to have been not very certain; Proclus debates various accounts of it, but among them, that postulates are peculiar to geometric subject matter, while axioms are common, either to all sciences that are concerned with quantity or to all sciences whatever.

In modern times mathematicians have often used the words "postulate" and "axiom" as synonymous. But a distinction which recommends itself as useful, and in part historically justified, is to reserve the name "axiom" for the axioms of logic (*see* LOGIC), and to use "postulate" for those assumptions or first principles (beyond the principles of logic) by which a particular mathematical discipline is defined.

For examples of particular systems of mathematical axioms (or better, postulates) *see* the article POSTULATE.

BIBLIOGRAPHY.—T. L. Heath, *The Thirteen Books of Euclid's Elements* (1908), J. W. Young, *Lectures on Fundamental Concepts of Algebra and Geometry* (1911), Alonzo Church, *Introduction to Mathematical Logic* (1956). (Ao. C.)

CATEGORICAL, in common usage, means "unconditional" or "direct and explicit in statement"; also, "pertaining to a category."

In Traditional Logic.—The word is derived, as Petrus Hispanus says (in *Summulae Logicales*, *c.* 1245), from the Aristotelian *kategorumenon* ("predicate") and thus means "predicative." *See* LOGIC for an account of the traditional categorical propositions and categorical syllogism.

In Mathematics.—A set of postulates is said to be categorical if every two models of the postulates are isomorphic. For example, in the case of Peano's postulates for arithmetic (*see* POSTULATE), a model is a system of meanings for the three primitive terms, "0", "number," "successor," which renders all the postulates true. That the postulates are categorical means that any two such models are isomorphic. That is, given two models, it is always possible to find a correspondence between the numbers of one model and the numbers of the other model, such that every number of either one of the models corresponds to a unique number in the other model, such that the

---

Originally published in ***Encyclopædia Britannica***, from 1956 until 1972, with periodic revisions. The articles **Concrete** and **Mathematical Induction** were added in 1967. The text here follows the 1972 version.

0 of one model corresponds to the 0 of the other model, and such that, if the numbers x, y of one model correspond to the numbers x', y' of the other model respectively, then y' is the successor of x' in the second model if and only if y is the successor of x in the first model. A categorical set of postulates thus determines uniquely the mathematical structure of a model, and in this sense no additional postulates are required (as distinguished from possible additional logical axioms [*see* AXIOM]).

(AO. C.)

CONCRETE (IN PHILOSOPHY). Roughly, such things as numbers, classes, states, qualities and relations are called abstract, while persons, physical things and events are concrete. Logicians have traditionally classified terms or names as being abstract or concrete according as they are intended to denote abstract or concrete things. Many, however, add a third category of collective names, *i.e.*, names of classes or collections of concrete things, distinct from the category of abstract names. *See* ABSTRACT AND ABSTRACTION; and NAME (IN LOGIC).

The distinction between abstract and concrete, though clear enough in general, is not a very sharp one, and borderline cases may be found. For example, consider the series of terms *theory, true proposition, fact, event*, or, to turn to theoretical physics, the series *conductivity, speed, heat, magnetic field, light, electric charge, electron, molecule*. In each case the series begins with an abstract term, and it may be thought that the terms grow successively more concrete. Some, however, may prefer a different order as that of increasing concreteness. In any case, if an absolute separation into abstract and concrete is demanded, it becomes difficult to decide where to draw the line.

See J. S. Mill, *A System of Logic Ratiocinative and Inductive* (1884); W. V. Quine, *Methods of Logic* (1950). (AO. C.)

CONNOTATION. That the meaning of a name may not be identified with its *denotation* (*see* NAME [IN LOGIC]) is readily made clear by means of examples. Thus "the Morning Star" and "the Evening Star" are two names of the same planet. It would be possible, however, to know the meaning of both names and even to have seen and identified the Evening Star on one occasion and the Morning Star on another, without knowing that they are the same. For it is not apparent from casual examination of the heavens that the Morning Star and the Evening Star are the same, but this was rather an early astronomical discovery established by a series of careful observations. Once this discovery is made, it is natural to introduce a third name, "Venus," to mean *the heavenly body which is the Morning Star and is the Evening Star*. But this same planet has also other names, for example, "the second planet from the sun." And to see that this name and the name "Venus" have different meanings, it suffices to remark that if an intra-Mercurial planet were discovered, we would not then say either that Venus does not exist or that Venus is the planet which was previously called Mercury, but only that we were mistaken in supposing Venus to be the second planet from the sun. But if it should be found that, by some unimaginable error, the Morning Star and the Evening Star are after all not the same, we would then indeed be obliged to say, Venus does not exist.

In the light of this, the names "the Morning Star," "the Evening Star," "Venus," "the second planet from the sun" are said to have each a different *connotation* (or *sense*). But if accepted astronomical facts are correct, the four names have the same denotation.

The distinction of connotation and denotation was introduced by J. S. Mill in 1843. And a similar distinction of sense (German *Sinn*) and denotation (German *Bedeutung*) was introduced by Gottlob Frege in 1892, without reference to Mill.

Mill has the credit of having discovered this important distinction of two kinds of meaning. But Mill's treatment is in several respects less satisfactory than Frege's. In particular, Mill applies the distinction primarily to common names and (unlike Frege) denies connotation altogether to a large class of singular names, including all simple abstract singular names such as "courage," "whiteness." And it was Frege who first pointed out the equivocal usage of natural language by which a name, besides its ordinary use, may have in some contexts an *oblique* use, denoting the same which in its ordinary use it connotes.

See further MEANING, and SEMANTICS IN LOGIC. (Ao. C.)

CONVERSION (IN LOGIC AND MATHEMATICS). For the converse of a categorical proposition, and for conversion as an immediate inference, *see* LOGIC, and the section on Aristotle in LOGIC, HISTORY OF.

In mathematics the term *converse* is used for the proposition obtained by the transformation of $\mathbf{AB} \supset \mathbf{C}$ into $\mathbf{AC} \supset \mathbf{B}$, or of $(\mathbf{x}_1)(\mathbf{x}_2)\ldots(\mathbf{x}_n) \,.\, \mathbf{AB} \supset \mathbf{C}$ into $(\mathbf{x}_1)(\mathbf{x}_2)\ldots(\mathbf{x}_n) \,.\, \mathbf{AC} \supset \mathbf{B}$, where $\mathbf{x}_1, \mathbf{x}_2, \ldots, \mathbf{x}_n$ are variables of any kinds; and where **A**, **B**, **C** are propositional forms, of which some may be themselves conjunctions, or **A** may as a special case be absent. (For explanation of the notation and terminology, see LOGIC.) This may in some instances be reduced to the simple converse of an *A* proposition in the sense of traditional logic—for example: *Every equilateral triangle is* [an] *equiangular* [triangle], and conversely, *every equiangular triangle is* [an] *equilateral* [triangle]. But such a reduction becomes either impossible or very artificial in cases like that of Euclid I 25 and its converse (*If two sides of a triangle are equal, respectively, to two sides of another triangle, and the third side of the first is greater than the third side of the second, then the angle opposite the third side of the first triangle is greater than that opposite the third side of the second triangle*, and conversely, *if two sides of a triangle are equal, respectively, to two sides of another triangle, and the angle opposite the third side of the first triangle is greater than that opposite the third side of the second triangle, then the third side of the first triangle is greater than the third side of the second*).

In this sense of conversion, the passage from a proposition to its converse is not in general a valid inference. And though often a mathematical proposition and its converse may both hold, separate proofs must be given for them. (Ao. C.)

DEFINITION, in logic, is qualified as nominal when it deals expressly with meaning (ultimately, that is, with some phenomenon of language) and as real when it deals, or purports to deal, with actual things.

Nominal Definition.—Nominal definition is the act of stating, explaining, or indicating the meaning or use of a *notation* (i.e., word, phrase, symbol, or expression); the act of introducing and fixing the meaning of a notation; or the sentence, in some language, by which such a statement or fixing of meaning is made. Under this broad head are included many procedures, not all closely related to one another which are called or have been called definition. Here will be distinguished (1) abbreviative definition, (2) semantical definition, and (3) ostensive definition.

1. In *abbreviative definition* a notation, the *definiendum*, is introduced as a mere abbreviation of another notation, the definiens, which it may be used to replace (the definiendum being ordinarily a shorter or otherwise more convenient notation than the definiens). The distinguishing feature is that an expression containing the definiendum must be understood as if it had the definiens in place of the definiendum—not only semantically, *i.e.*, in regard to its meaning, but also syntactically. For a fuller explanation and formal examples, *see* LOGIC: *Fundamentals of Modern Logic: The Propositional Calculus.*

2. In *semantical definition* a meaning is directly ascribed or assigned to a notation by a statement that mentions explicitly both the notation and the meaning; *e.g.*, "The word 'pentagon' means a polygon with five sides." In this example the definiendum is "pentagon" and the definiens is "polygon with five sides."

3. In *ostensive definition* a proper name or a common name (*see* NAME [IN LOGIC]) is assigned to a concrete object by physically showing or pointing to the object and naming it. Thus for instance Adam named the beasts of the field as they passed before him (Gen. 2:19–20); and an explorer of unknown territory has to assign names to the geographical features encountered by him.

It is a common remark that without ostensive definition, and related methods of indicating rather than stating meanings, language cannot acquire objective reference at all. Yet, unlike other kinds of definition, the study of ostensive definition is not in the domain of logic; and objective reference, as acquired by ostensive procedures, is rather a presupposition for logic.

A familiar feature of natural language is the use of demonstrative words such as "you," "I," "here," "today," "yesterday," a part of whose meaning is reintroduced by an implicit ostensive definition on each new occasion of their use—and which have as what may be called their secondary meaning the manner in which this implicit ostensive definition is to be understood. For example, "Yesterday was a fine day" may be true when uttered at one time and false when uttered at another, because of the different meaning of the name "yesterday." Or perhaps better, the meaning should be attached, not barely to the word "yesterday," but to the combination consisting of the word plus the occasion and circumstances of its utterance. Such demonstrative words are a linguistic convenience but, unlike ostensive definition generally, are dispensable in principle. For purposes of logic, they must either be eliminated altogether, or treated as used on one fixed occasion in such a way that the meaning remains unchanged.

Real Definition.—Real definition is traditionally distinguished from nominal definition as being the definition of a thing rather than of a notation. Indeed an abbreviative or a semantical definition can always be paralleled by a corresponding statement of identity or equivalence in which the definiendum is used rather than mentioned. For example, the semantical definition of "pentagon" given above can be paralleled by the assertion "A pentagon is a polygon with five sides," which is not about the word "pentagon" but about pentagons (and which therefore does not mention but rather uses the word "pentagon"). While it would seem preferable to speak of this assertion, not as a definition but as being *true by definition*, or as being *analytically true*, some would say that it becomes a real definition if it is supplemented by a proof that pentagons exist or, perhaps, if it is understood as implying that pentagons exist. (By G. G. Saccheri the added demonstration or assumption of existence was even taken as being the only difference of real from nominal definition; *see* LOGIC, HISTORY OF.)

This is not quite the same thing as the Aristotelian real definition, which involves the never very satisfactorily explained notion of essence. For Aristotle a definition is of what a thing is, and of its essential nature or essence.

There is a certain plausibility in the dual contention that (1) "A pentagon is a polygon with five sides" is an Aristotelian real definition because it is of the essence of a pentagon to be a polygon and to have five sides, but that (2) "A pentagon is a polygon the sum of whose interior angles is six right angles" is not an Aristotelian real definition because having the sum of its interior angles equal to six right angles is not of the essence of a pentagon but is merely a property which, though it happens to belong to all and only those polygons which are pentagons, must nevertheless be established by mathematical demonstration. Counterarguments, however, are offered (*e.g.* by Richard Robinson) to show that in such cases the supposed real definition is actually a nominal definition in disguise. And by many logicians the Aristotelian notion of real definition has been abandoned or modified because of the difficulty of a precise account of what is meant by the essence thing, as distinguished from the meaning of some name of thing or an analysis of some concept of it.

BIBLIOGRAPHY.—William Ockham, *Summa Logicae*, i, 26–29; Blaise Pascal, *De l'Esprit géométrique*; J. S. Mill, *A System of Logic*, book i, ch. viii (1884); W. E. Johnson, *Logic*, vol. i, ch. vi–vii (1921); Richard Robinson, *Definition* (1950); Alonzo Church, *Introduction to Mathematical Logic*, vol. i, sec. 11, 55 (1956). (Ao. C.)

DENOTATION. A name is said to *denote* that thing or those things of which it is a name, or to which, in other words, we intend to *refer* when we use the name. The threatened circularity of this definition (the three italicized words are not easily defined except by means of one another) suggests that we are here dealing with a basic concept, for which an axiomatic treatment may be more appropriate than definition. And though, for a fixed formalized language a definition of *denotation* of names in that language is possible, there is no definition available for *denotation* in general except by semantical terms closely related to it (*see* SEMANTICS IN LOGIC). *See* NAME (IN LOGIC), where the denotation of names in natural language is discussed more fully. For the distinction of denotation and connotation, *see* CONNOTATION. (Ao. C.)

DILEMMA. In traditional logic, a dilemma is any one of several forms of inference in which there are two major premisses of hypothetical form and a disjunctive minor premiss; see the full list in the article LOGIC: *Traditional Logic: Hypothetical Syllogism, Disjunctive Syllogism, Dilemma.*

It is not necessary that a dilemma should have an unwelcome conclusion. But this was often the case in illustrations, the dilemma having been originally a device of rhetoric for confuting an opponent. Hence in common usage the word has come to mean a situation in which each of alternative courses of action that are open leads to some unsatisfactory or ill consequence.

Some traditional logicians have used the term *polylemma* for a generalization of the dilemma in which the minor premiss is a disjunction that is more than twofold and the number of major premisses is correspondingly increased. For example, the constructive trilemma: $\mathbf{A} \supset \mathbf{D}$, $\mathbf{B} \supset \mathbf{D}$, $\mathbf{C} \supset \mathbf{D}$, $\mathbf{A} \vee \mathbf{B} \vee \mathbf{C}$, therefore **D**. [For meaning of this notation, *see* LOGIC: *Fundamentals of Modern Logic*.]

As an example of a dilemma taken from number theory, suppose it is required to prove: If an odd integer a is a perfect fourth power, then $a - 1$ is divisible by 16. We argue that, since a is a perfect fourth power and odd, a must be the fourth power of an odd integer b; and b, being odd, must be equal either to $4c + 1$ or to $4c + 3$, for some integer c. This gives us the minor premiss: (1) Either $a = (4c+1)^4$, for some integer c, or $a = (4c + 3)^4$, for some integer c. Then from $a = (4c + 1)^4$ we get, by methods of elementary algebra, $a - 1 = 256c^4 + 256c^3 + 96c^2 + 16c$. Hence: (2) If $a = (4c + 1)^4$, for some integer c, then $a - 1$ is divisible by 16. Again, from $a = (4c + 3)^4$ we get, by methods of elementary algebra, $a - 1 = 256c^4 + 768c^3 + 864c^2 + 432c + 80$. Hence: (3) If $a = (4c+3)^4$, for some integer c, then $a-1$ is divisible by 16. The dilemma which has (2) and (3) as major premisses and (1) as minor premiss yields the conclusion: (4) $a - 1$ is divisible by 16.

Mathematicians call this form of proof *proof by cases*. (Ao. C.)

LOGIC is the systematic study of the structure of propositions and of the general conditions of valid inference by a method which abstracts from the content or *matter* of the propositions and deals only with their logical *form*. This distinction between form and matter is made whenever we distinguish between the logical soundness or validity of a piece of reasoning and the truth of the premisses from which it proceeds, and in this sense is familiar in everyday usage. However, a precise statement of the distinction must be made with reference to a particular language or system of notation, a *formalized language*, which shall avoid the inexactnesses and systematically misleading irregularities of structure and expression that are found in ordinary (colloquial or literary) English and in other natural languages, and shall follow or reproduce the logical form—at the expense, where necessary, of brevity and facility of communication. To adopt a particular formalized language is thus to adopt a particular system or theory of logical analysis. And the formal method may then be characterized by saying that it deals with the objective form of *sentences* which express propositions, and provides in these concrete terms criteria of meaningfulness, of valid inference, and of other notions closely associated with these—among which we mention the notions of logical compatibility, analytic or logical truth, probable inference, and degree of confirmation.

The topics of inductive and probable inference, confirmation, and degree of confirmation belong to *inductive logic*, and are treated in the articles INDUCTION and PROBABILITY. In this article we confine attention to the remaining part of logic, or *deductive logic*. And we also omit treatment of *modal logic*, partly because this branch of deductive logic (though its beginnings go back to Aristotle) is still in a much more unsettled state than the remainder. But works on modal logic and modality are included in the bibliography.

FUNDAMENTALS OF MODERN LOGIC

The Propositional Calculus is, in most developments, the most elementary branch of logic, on which the others are based. It deals with the *sentence connectives*: "and," "or," "if," "not," "if and only if," and others of similar character. And in order to analyze and exhibit the logical properties of the connectives it employs also propositional variables $p, q, r, \cdots$, which are to be thought of as variables replaceable by sentences.

In strictness the sentences by which the propositional variables are replaceable should be sentences of some appropriate formalized language (for instance, one of the functional calculi which are described below, or one of the systems of set theory). But for informal expository purposes we may employ also declarative sentences of the English language. Thus, with $\supset$ as notation for "if ... then," the expression $p \supset q$ of propositional calculus may be thought of as having (by substitution for p and q) such instances as the following: "If you are not satisfied, we will refund your money," "If there are any survivors of the disaster we will be able by prompt action to rescue some," "If Vidkun Quisling was a patriot, then n-butyl mercaptan is a perfume".

Of the various systems of notation which are in use, we here adopt a particular one for purposes of exposition. We shall use the sign $\sim$ to denote *negation* ("$\sim p$" to mean "not p"), the sign $\supset$ to denote the *conditional* ("$p \supset q$" to mean "if p then q" or "not both p and not q"), the sign $\mid$ to denote *non-conjunction* ("$p \mid q$" to mean "not both p and q,"), juxtaposition or a dot to denote *conjunction* ("pq" or "$p \cdot q$" to mean "p and q"), the sign $\equiv$ to denote the *biconditional* ("$p \equiv q$" to mean "p if and only if q"), the sign $\vee$ to denote *inclusive disjunction* ("$p \vee q$" to mean "p or q or both"), the sign $\not\equiv$ to denote *exclusive disjunction* ("$p \not\equiv q$" to mean "p or q but not both"). —As the word "or" is ambiguous in ordinary English usage between inclusive disjunction and exclusive disjunction, the two signs $\vee$ and $\not\equiv$ are provided for the two meanings of the word. The signs $\supset$ and $\equiv$ may often conveniently be read as "implies" (or "implies that") and "is equivalent to" respectively; but these readings must be employed with a certain caution not to misunderstand them, since the terms implication and equivalence in ordinary usage often suggest that there is some relationship between the logical forms of the propositions or the sentences involved, whereas the truth of $p \supset q$ and of $p \equiv q$ (upon substitution of particular sentences for p and q) requires no such relationship. The connective $\supset$ is also said to stand for *material implication*, distinguished from *formal implication* (defined below) and *strict implication* (a notion of modal logic). Similarly, $\equiv$ is said to stand for *material equivalence*.

There are various ways in which some of the sentence connectives named above can be defined in terms of others. If the sign of nonconjunction (*Sheffer's stroke*) is taken as primitive, all the other connectives can be defined from this one. Also, if the signs of negation and inclusive disjunction are taken as primitive, all the others can be defined in terms of these; likewise if the signs of negation and conjunction are taken as primitive. But because of its special role in the rule of *modus ponens* and in the *deduction theorem* (*see* below), we here prefer to include $\supset$ as one of the primitive connectives. Therefore we take $\sim$ and $\supset$ as primitive, and define the remaining connectives as follows:

$$
\begin{array}{ll}
[\mathbf{A} \mid \mathbf{B}] \rightarrow [\mathbf{A} \supset \sim\mathbf{B}] & [\mathbf{A} \equiv \mathbf{B}] \rightarrow [[\mathbf{A} \supset \mathbf{B}][\mathbf{B} \supset \mathbf{A}]] \\
[\mathbf{AB}] \rightarrow \sim[\mathbf{A} \mid \mathbf{B}] & [\mathbf{A} \vee \mathbf{B}] \rightarrow [\sim\mathbf{A} \supset \mathbf{B}] \\
[\mathbf{A} \cdot \mathbf{B}] \rightarrow \sim[\mathbf{A} \mid \mathbf{B}] & [\mathbf{A} \not\equiv \mathbf{B}] \rightarrow \sim[\mathbf{A} \equiv \mathbf{B}]
\end{array}
$$

Here the bold capital letters stand for arbitrary *well-formed formulas* of the propositional calculus (in the technical sense defined below)—or, when appropriate, for arbitrary well-formed formulas of some more extensive formalized language containing the propositional calculus as a part. And the arrow is to be read "is defined as," or "is an abbreviation for."

To formulate the propositional calculus explicitly, we first list the primitive symbols.

These are the two connectives $\sim \supset$, the two brackets [ ], and an infinite list of propositional variables $p, q, r, s, p_1, q_1, r_1, s_1, p_2, \cdots$. Any finite sequence of the primitive symbols is a *formula*. But as some formulas are evidently not meaningful, we define a subclass, the *well-formed formulas*, by the following *formation rules*: i. A propositional variable standing alone is a well-formed formula. ii. If **A** is a well-formed formula, $\sim$**A** is a well-formed formula. iii. If **A** and **B** are well-formed formulas, [**A** $\supset$ **B**] is a well-formed formula.

Hereafter the bold capital letters shall stand for well-formed formulas of the propositional calculus—or later also well-formed formulas of other calculi or formalized languages—and the condition of well-formedness shall be understood without explicit repetition in every case.

In writing well-formed formulas we may abbreviate them by means of the definitions listed above; for instance, $[pq]$ is to be understood as an abbreviation of $\sim[p \supset \sim q]$, and $[p \equiv [p \vee q]]$ is to be understood as an abbreviation of $[[p \supset [\sim p \supset q]][[\sim p \supset q] \supset p]]$, *i.e.*, as an abbreviation of $\sim[[p \supset [\sim p \supset q]] \supset \sim[[\sim p \supset q] \supset p]]$. Also well-formed formulas may be abbreviated by omission of brackets, in accordance with the three following conventions: (1) A bold dot is used to indicate that the scope of an omitted pair of brackets extends from the point where the dot appears, forward to the end of the formula; for example, $p \supset \centerdot\, q \supset \centerdot\, r \supset s$ is an abbreviation of $[p \supset [q \supset [r \supset s]]]$. (2) In restoring omitted brackets, and so far as not otherwise indicated by bold dots, the scope of a pair of brackets belonging to a conditional, a biconditional, or an exclusive disjunction is to be made wider than the scope of a pair of brackets belonging to an inclusive disjunction or a non-conjunction, and the latter in turn wider than the scope of a pair of brackets belonging to a conjunction; for example, $\sim p \vee q \sim r \supset pq \vee rs$ is an abbreviation of $[[\sim p \vee [q \sim r]] \supset [[pq] \vee [rs]]]$, i.e., of $[[\sim\sim p \supset \sim[q \supset \sim\sim r]] \supset [\sim\sim[p \supset \sim q] \supset \sim[r \supset \sim s]]]$. (3) Where neither of the two preceding conventions applies, brackets are restored in accordance with the convention of association to the left; for example, $p \supset [q \supset r] \supset \sim s \supset \centerdot\, p \supset q \supset r$ is an abbreviation of $[[[p \supset [q \supset r]] \supset \sim s] \supset [[p \supset q] \supset r]]$.

It must be understood that these abbreviations are not an actual part of the formalized language—the propositional calculus, or better, the particular formulation of propositional calculus which we are here stating. *E.g.*, the formalized language expresses the inclusive disjunction of p and q by $[\sim p \supset q]$, and it is merely a device for our typographical convenience in writing about the formalized language, serving the purposes of brevity and perspicuity, that we sometimes abbreviate this as $[p \vee q]$, or as $p \vee q$. Especially, the rules of inference stated below are to be applied to the well-formed formulas themselves, and not to their abbreviations. As axioms of the propositional calculus we take the three following:

1. $p \supset q \supset \centerdot\, q \supset r \supset \centerdot\, p \supset r$
2. $\sim p \supset p \supset p$
3. $p \supset \centerdot \sim p \supset q$

The assertion of the axioms is of course meant in the sense that they hold for any p, q, and r whatever. These particular axioms are due to Jan Łukasiewicz; they represent only one of many possible choices of axioms from which the same theorems follow by means of the two rules of inference below. Convenient verbal readings of the axioms, in which advantage is taken of the two readings of $\supset$, are as follows: 1. If p implies q, if q implies r, then p implies r. 2. If not-p implies p, then p. 3. p implies that, if not-p, then q.—Axiom 1 expresses the *transitive law of material implication*, and 2, the *law of Clavius*.

The *rules of inference* are the two following: I. *Rule of substitution*. From **A**, if **B** is any well-formed formula and **b** is a propositional variable, to infer the result of substituting **B** for **b** throughout **A**. II. *Rule of modus ponens*. From [**A** ⊃ **B**], and **A**, to infer **B**.—In an application of I, we call **A** the *premiss*, and the result of the substitution the *conclusion*. In an application of II, [**A** ⊃ **B**] is the *major premiss*, **A** is the *minor premiss*, and **B** is the *conclusion*.

A *theorem* of the propositional calculus is obtained from the axioms by a succession of applications of the rules of inference. Or more explicitly, a *proof* of a theorem is a finite sequence of well-formed formulas, of which the last one is the theorem in question, and each of which either is an axiom or is inferred from earlier formulas in the sequence by one of the two rules of inference; and a *theorem* is then any well-formed formula of which there is a proof.

For example, a proof may be constructed as follows, of the theorem $p \supset p$ (the reflexive law of material implication). Axiom I is first taken as premiss for an application of the rule of substitution. The substitution of $\sim p \supset p$ for q gives $p \supset [\sim p \supset p] \supset . \sim p \supset p \supset r \supset . p \supset r$, and from this by substitution of p for r (another application of the rule of substitution) is inferred $p \supset [\sim p \supset p] \supset . \sim p \supset p \supset p \supset . p \supset p$. This last is to be the major premiss in an application of *modus ponens*. Axiom 3 is the source of the minor premiss $p \supset . \sim p \supset p$ (obtained by substituting p for q). Hence by *modus ponens* is inferred $\sim p \supset p \supset p \supset . p \supset p$. Axiom 2 is then taken as minor premiss, and by another application of *modus ponens* is inferred $p \supset p$.

Other theorems are the following, which may be proved in the order named (though the proofs are generally not as easy as the example just given): $p \supset . q \supset p$ (*law of affirmation of the consequent*), $\sim q \supset \sim p \supset . p \supset q$ (*converse law of contraposition*), $\sim p \supset . p \supset q$ (*law of denial of the antecedent*), $p \supset q \supset p \supset p$ (*Peirce's law*), $p \supset [p \supset q] \supset . p \supset q$, $p \supset . p \supset q \supset q$ (*law of assertion*), $p \supset [q \supset r] \supset . q \supset [p \supset r]$ (*law of commutation*), $q \supset r \supset . p \supset q \supset . p \supset r$, $p \supset [q \supset r] \supset . p \supset q \supset . p \supset r$ (*self-distributive law of material implication*), $\sim\sim p \supset p$ (*law of double negation*), $p \supset \sim\sim p$ (*converse law of double negation*), $p \supset q \supset . \sim q \supset \sim p$ (*law of contraposition*), $p \supset q \supset . p \supset \sim q \supset \sim p$ (*law of reductio ad absurdum*), $\sim . p \sim p$ (*law of contradiction*), $p \vee \sim p$ (*law of excluded middle*), $pq \supset r \supset . p \supset . q \supset r$ (*law of exportation*), $p \supset [q \supset r] \supset . pq \supset r$ (*law of importation*), $[p \supset q][p \supset r] \supset . p \supset qr$ (*law of composition*), $\sim[pq] \equiv p \vee \sim q$ and $\sim[p \vee q] \equiv \sim p \sim q$ (*De Morgan's laws*), $[p \supset q] \vee [q \supset p]$.

The last theorem in the foregoing list is sometimes spoken of as one of the "paradoxes" of material implication. It is not a paradox in the sense of an antinomy and is not inconsistent with important uses of ⊃ as expressing an implication relation. But it shows a divergence of the meaning of ⊃ from some of the various meanings of the words "if … then" in ordinary usage, and especially from their familiar but not very definite meaning in *conditions contrary to fact*. (To illustrate the latter we quote the conditional sentence, "If Vidkun Quisling had been a patriot, then n-butyl mercaptan would be a perfume," which must be distinguished from the corresponding sentence with a simple condition, quoted above, and indeed is clearly false.)

We shall speak of a *valid inference of the propositional calculus* not only within the propositional calculus itself but also within a more extensive formalized language containing the propositional calculus as a part. Namely a *substitution instance* of a theorem of the propositional calculus is any well-formed formula (of the more extensive formalized language) which results from a theorem of the propositional calculus by substituting specific well-formed formulas (of the more extensive language) for one or more of the propositional variables. And an inference from hypotheses $\mathbf{A}_1, \mathbf{A}_2, \cdots, \mathbf{A}_n$ to a conclusion **B** is called a valid inference of the propositional calculus if **B** can be obtained from the hypotheses $\mathbf{A}_1, \mathbf{A}_2, \cdots, \mathbf{A}_n$, together with any number of additional premisses which are either theorems of the propositional calculus or substitution instances of theorems of the propositional calculus, by a succession *of applications of the rule of modus ponens only*.

Within the propositional calculus itself, if the inference from $\mathbf{A}_1, \mathbf{A}_2, \cdots, \mathbf{A}_n$ to **B** is a valid inference of the propositional calculus, it can be shown that $\mathbf{A}_1 \supset . \mathbf{A}_2 \supset . \cdots \mathbf{A}_n \supset \mathbf{B}$ must then be a theorem of the propositional calculus. This general result, which is not a theorem of the propositional calculus but a theorem about the propositional calculus, is known as the *deduction theorem*. We shall see below that it can be extended to other more extensive formalized languages.—It is in the sense that the rule of *modus ponens* and the deduction theorem are both fulfilled that we are able to think of and use $\supset$ as expressing an implication relation (in spite of the so-called paradoxes of material implication).

It is convenient to use such words as "theorem," "proof," "premiss," "conclusion" both for propositions, in whatever language expressed, and for formulas expressing propositions in some fixed formalized language. However, theorems about a particular formalized language are distinguished from theorems of the language by calling the former *metatheorems*. The deduction theorem is, *e.g.*, a metatheorem of the propositional calculus, in contrast with, say, Peirce's law, which is a theorem of the propositional calculus. The metatheorems of a formalized language are said to belong to the *metatheory* of the language; and though we shall here treat the metatheory informally, it may itself be organized as a formalized language, which is then called a *meta-language* of the first language as *object language*. (*See* METATHEORY; SEMANTICS IN LOGIC.)

We go on to explain some additional metatheorems of the propositional calculus.

The two primitive connectives (hence all connectives definable from them) denote *truth-functions: i.e., the truth-value* (truth or falsehood) of $\sim p$ and of $p \supset q$ is uniquely determined by the truth-values of p and q. In fact the truth-value of $\sim p$ is falsehood when the truth-value of p is truth, and truth when the truth-value of p is falsehood. And (corresponding to the four theorems $p \supset . q \supset . p \supset q$, $p \supset . \sim q \supset \sim . p \supset q$, $\sim p \supset . q \supset . p \supset q$, $\sim p \supset . \sim q \supset . p \supset q$) the truth-value of $p \supset q$ is falsehood when the truth-value of p is truth and the truth-value of q is falsehood, and the truth-value of $p \supset q$ is truth in each of the three remaining cases. These determinations of the truth-value of $\sim p$ and of $p \supset q$ in terms of the truth-values of p and q, by using (say) 0 for truth and 1 for falsehood, may conveniently be displayed in the form of tables, which are then called *truth-tables* of negation and material implication.

The truth-tables enable us, given a well-formed formula of the propositional calculus and an assignment of a truth-value to each of its variables, to reckon out by a mechanical process the truth-value of the entire formula. (For example, if the values of p and q are both 1, the value of $p \supset q$ is 0, hence the value of $p \supset q \supset p$ is 1, hence the value of Peirce's law $p \supset q \supset p \supset p$ is 0.) If, for all possible assignments of truth-values to the variables, the calculated truth-value of the entire formula is 0, or truth, the formula is said to be a *tautology*. The test whether a well-formed formula is a tautology is effective, since in any particular case the total number of different assignments of truth-values to the variables is finite, and the calculation of the truth-value of the entire formula can be carried out separately for each assignment of truth-values to the variables.

Now it may be verified that the three axioms, 1–3, given above are tautologies, and that the two rules of inference preserve tautologies in the sense that if the premiss or premisses are tautologies the conclusion must be a tautology. Hence every theorem of the propositional calculus is a tautology. By a more difficult argument it may be shown that every tautology is a theorem. Hence the truth-table test for tautologies (just described) can be used to decide, for any given well-formed formula of the propositional calculus, whether it is a theorem. This is a solution of the *decision problem* of the propositional calculus.

As corollaries of the solution of the decision problem, it follows that the propositional calculus is consistent in the sense that **A** and $\sim$**A** cannot both be theorems, and complete in the sense that if **A** is not a theorem the addition of **A** as an extra axiom would result in inconsistency. By a different method it may be shown that the three axioms and the two rules of inference are *independent* in the sense that completeness is lost if any one of them is omitted.

In one sense the solution of the decision problem renders unnecessary the statement of axioms and rules of inference of the propositional calculus. For we might instead merely stipulate that every tautology shall be a theorem. However, this summary procedure provides no analysis of the logical relationships of the theorems among themselves, and no method by which to trace the consequences of a particular law of the propositional calculus or to formulate the situation which results when particular laws are rejected.

An example of the last is provided by the intuitionistic school of mathematics (*see* MATHEMATICS, FOUNDATIONS OF), which rejects the law of excluded middle and some other related laws of negation. To formulate the propositional calculus of mathematical intuitionism we may modify our above formulation as follows. Axiom 2 is replaced by the weaker axiom $p \supset \sim p \supset \sim p$, and there are then added the two further axioms: $p \supset . q \supset p$, $p \supset [p \supset q] \supset . p \supset q$. These axioms suffice for the intuitionistic laws of implication and negation. But as the definitions of conjunction and inclusive disjunction, given above, no longer serve to yield the characteristic properties of these connectives, it is necessary to introduce these two connectives as additional primitives, and to add also the axioms: $pq \supset p$, $pq \supset q$, $p \supset . q \supset pq$, $p \supset p \vee q$, $q \supset p \vee q$, $p \supset r \supset . q \supset r \supset . p \vee q \supset r$. Then finally, after this last addition, the axiom $p \supset . q \supset p$ may be dropped as not independent: in fact we can prove the theorem $p \supset . p \supset . q \supset p$ using only the axioms $p \supset . q \supset pq$, $pq \supset p$, and the transitive law, and then $p \supset . q \supset p$ follows by the axiom $p \supset [p \supset q] \supset . p \supset q$.

Using this formulation of the intuitionistic propositional calculus we can show that not only the law of excluded middle is not a theorem, but also the law of Clavius, the converse law of contraposition, Peirce's law, and the law of double negation fail; but the introduction of any one of these as an added axiom would restore all the others as theorems. The "paradox" $[p \supset q] \vee [q \supset p]$ is still a consequence in the form $\sim[p \supset q] \supset . q \supset p$, but not in the form containing the disjunction sign.

A Logistic System, or Calculus, is the purely formal part of a formalized language, taken in abstraction from any meaning or interpretation.

A logistic system is determined by giving its *vocabulary*, or list of primitive symbols, defining certain finite sequences of the primitive symbols to be *well-formed formulas*, listing certain well-formed formulas as *axioms*, and finally, stating *rules of inference*, by means of which a well-formed formula may on given conditions be *inferred* (as *conclusion*) from a set of one or more well-formed formulas (as *premisses*). A *theorem* of the logistic system is then defined to be a well-formed formula of which there is a *proof*, i.e., a finite sequence of well-formed formulas each of which is either an axiom or inferred from earlier formulas in the sequence by one of the rules of inference, the last formula in the sequence being the theorem.

It is usually required that the set of rules of inference shall be *effective*, in the sense that there shall be a definite procedure or test by which, whenever a particular proposed conclusion from given premisses is before us, we can always actually decide whether the proposed inference is correctly made in accordance with one of the rules of inference; further that the set of axioms shall be effective, in the sense that there shall be a test by which, whenever a particular well-formed formula is before us, we can always decide whether it is an axiom (this requirement is satisfied when the axioms are a finite list which has been written out in full, but may also be satisfied in some cases of an infinite set of axioms). For these two requirements it will generally be necessary that the definition of well-formedness shall also be effective in the sense that there shall be a test by which, whenever a particular formula is before us, we can always decide whether it is well-formed or not.

The reason for these requirements lies in the nature of the notion of a proof—as it is commonly understood, or as it is needed for the purposes of deductive logic (and, as a special case, in connection with mathematical proof). Namely it is a part of the notion of proof that a proof shall carry final conviction of the theorem proved (*i.e.*, of course, for any one who admits the axioms and rules of inference on which the proof is based). But without the requirements of effectiveness, it might happen that some one confronted with a proof of a theorem (and admitting the axioms and rules of inference) might nevertheless continue to doubt the theorem, because of doubt that the alleged proof actually is a proof in accordance with the axioms and rules. In such a case the proposer of the proof might fairly be asked to give a supplementary proof that it is a proof—and for purposes of deductive logic, this supplementary proof ought then to be treated as part of the whole proof, and included with it when the process of proof is formalized by the logistic method.

Though in the light of the requirements just explained, the notion of being a proof, in a particular logistic system, must be effective, it is clear that in general the notion of being a theorem may not be effective. For a well-formed formula is established as a theorem if a proof of it is found, but the failure of a particular investigator to find a proof may not necessarily mean that there is no proof to be found. It may indeed be possible in particular cases, by special methods, to establish as a metatheorem that a certain well-formed formula is not a theorem. But this is not to say that there is a general test by which it is always possible, whenever a well-formed formula is given, to decide effectively whether or not it is a theorem.

Such a general effective test, by which to recognize any arbitrary given well-formed formula of a particular logistic system as being or not being a theorem, is called a *decision procedure* for the particular system. And the problem to find a decision procedure for a logistic system is called the *decision problem* of the system. As a particular example, we have seen (above) a solution of the decision problem of the full propositional calculus—or, as this particular logistic system is often called because of the character of the decision procedure, the two-valued propositional calculus. There exist also decision procedures for the intuitionistic propositional calculus, but none of them is as simple as that for the two-valued calculus. On the other hand there are logistic systems for which there is no solution of the decision problem, not merely in the sense that none has been found, but in the sense that it has been established as a metatheorem that none can exist; examples of such systems (which will be discussed below) are the pure functional calculus of first order and of all higher orders—but the singulary functional calculi of first and second orders (obtained from the corresponding pure functional calculi by omitting all functional variables which are more than singulary) do have decision procedures.

As already indicated, the two-valued propositional calculus, the intuitionistic propositional calculus, and other systems to be introduced below may be considered as examples of logistic systems. Each is namely a logistic system if regarded as determined by its vocabulary, its *formation rules* (defining the well-formed formulas of the particular system), its axioms, and its rules of inference, in abstraction from any meaning or interpretation. The logistic system becomes a formalized language if suitable meanings are given to its well-formed formulas, as is done informally in this article, and as may be done more accurately by providing definitions of *truth* and of *satisfaction* in the manner outlined in the article SEMANTICS IN LOGIC—or if *sense* as well as *denotation* is to be provided for (*see* the same article), then by *rules of sense*, which are similar in character to the rules that compose the step-by-step definition of satisfaction, but which deal directly with the sense.

In general, if a logistic system has any sound interpretation, *i.e.*, any non-trivial way of giving meanings to its well-formed formulas in conformity with the formation rules, axioms, and rules of inference, it will have many such. Thus the same logistic system is common to many different formalized languages, and the results (theorems and metatheorems) for the logistic system hold equally for all the formalized languages, with appropriate changes of interpretation. —The point may be illustrated by the case of the pure functional calculus of first order (which is treated below), each different domain of individuals providing a different sound interpretation and hence a different formalized language.

The method which makes use of a formalized language, based upon a logistic system with explicitly stated formation rules, axioms, and rules of inference, is known as the *logistic method*. For the treatment of logic by this method, and especially to distinguish it from the less fully formal method of the older logic, the names *symbolic logic, mathematical logic, theoretical logic, logistic* have variously been used.

The Functional Calculus of First Order has, besides the notations of the propositional calculus (propositional variables and sentence connectives), also notations for *propositional functions* and the *quantifiers*.

To explain these new notations and their meanings we return to the notion of a *sentence*, which we have already employed in explaining the use of propositional variables.

A *propositional form* is an expression which is like a sentence except that it contains a number of *variables*, of any kinds, at places at which a corresponding sentence must contain fixed names or terms (*constants*)—the difference being that a sentence may be said simply to be true, or false, but a propositional form must rather be said to be true, or to be false, *for* some *system of values* of its variables. To turn to elementary mathematics for examples, "$49 + 36 = 85$" and "$59 + 36 = 85$" are sentences, the first one true and the second one false. But "$x + 36 = 85$" is a singulary propositional form, the adjective *singulary* indicating that there is one variable of which values are to be considered. It may not be said to be true, or to be false, but rather it is true for the value 49 of x, and false for the value 59 of x; or in a different terminology (*see* SEMANTICS IN LOGIC), "$x + 36 = 85$" is *satisfied* by the value 49 of x, and by no other value of x. Similarly "$x^2 + y^2 = 85$" is a binary propositional form which is true for (satisfied by) the pair of values 7, 6 of x, y; also true for the values 6, 7 of x, y; also true for the values 2, 9 of x, y; but false for the values 5, 8 of x, y. Even "$x + y - x = y$"—being a binary propositional form rather than a sentence—must not be said just to be true, but to be true *for* all values of x, y.

In the foregoing examples the variables are numerical variables, *i.e.*, variables whose values are numbers. As examples of propositional forms containing variables of other kinds we may cite the quaternary form, "The distance from P to Q is less than the distance from R to S," in which the values of the variables P, Q, R, S are points; and the singulary form, "If x is a man, then x is mortal," in which the values of x are, say, concrete material things.

As an extreme case, and as a matter of terminological convenience, it is usual to consider a sentence as being a kind of propositional form, namely a propositional form in which the number of variables is 0 (or more correctly, as we shall see below, in which the number of free variables is 0).

As *universal quantifier* we shall use the notation consisting of a variable between parentheses; *e.g.*, if the variable is x, the notation is (x). The meaning is roughly indicated by saying that the universal quantifier corresponds to such English words as "all," "every"; or better indicated by saying that if a universal quantifier with some particular variable, say x, is prefixed to a singulary propositional form whose variable is x, the resulting expression is a sentence, which is a true sentence if and only if the propositional form is true for all values of x. For example, "(x) [if x is a man, then x is mortal]" is a sentence (of a certain ill-defined language which we here use for temporary illustrative purposes only, a half-formalized version of English); and in fact, on the best available evidence, it is a true sentence. Again, "$(x)[x + 36 = 85]$" is a false sentence. However, a universal quantifier may be prefixed also, with analogous meaning, to a propositional form which is more than singulary—or even, as a quite special but nevertheless allowable case, to a sentence. For example, "$(y)[x^2 + y^2 = 85]$" is a singulary propositional form; and since, as it happens, this singulary form is false for all values of x, the sentence $(x){\sim}(y)[x^2 + y^2 = 85]$" is true. Again "$(y)[x + y - x = y]$" is a singulary propositional form, true for all values of x; and hence "$(x)(y)[x + y - x = y]$" is a true sentence. And again "$(x)[49 + 36 = 85]$" is a true sentence, because "$49 + 36 = 85$" is a true sentence.

As *existential quantifier* we shall use the sign $\exists$ followed by a variable, with parentheses enclosing both. The meaning corresponds roughly to that of the English word "some" (in the sense "at least one") or of the phrase "there is a." And if an existential quantifier with, say, the variable x is prefixed to a singulary propositional form whose variable is x, the resulting expression is a sentence, which is true if and only if the propositional form is true for at least one value of x. The usage is otherwise similar to that of the universal quantifier, as a few examples will suffice to illustrate. If the meaning of the numerical variables x, y is such that they include the negative as well as the positive numbers among their values, "$(\exists y)[x + y = 85]$" is a singulary propositional form which is true for all values of x, and therefore "$(x)(\exists y)[x + y = 85]$" is a true sentence. Also "$(\exists x)(\exists y)[x + y = 85]$" is true. But "$(\exists y)(x)[x + y = 85]$" is a false sentence, as there is no one number y such that $x + y$ is always 85 for every number x. On the other hand, "$(\exists y)(x)[x \times y = 0]$" is true, as there is one number y, namely 0, such that $x \times y$ is always 0.

When a quantifier with a particular variable is prefixed to a propositional form, all occurrences of that variable in the resulting expression, including the occurrence of the variable in the quantifier itself, are said to be *bound* occurrences of the variable. Other occurrences of a variable, not bound, are called *free* occurrences. It is not excluded that a propositional form may contain both bound and free occurrences of the same variable; in fact, if **A** is a propositional form containing free occurrences of a certain variable, and **B** contains bound occurrences of that variable, then $[\mathbf{A} \supset \mathbf{B}]$ and $[\mathbf{B} \supset \mathbf{A}]$ each contain both bound and free occurrences of the variable. The variables which have bound occurrences in a given propositional form are called the *bound variables* of the form, and those which have free occurrences in it are called its *free variables*.

Having these last definitions, we must now go back and make the following amendment to our first account of propositional forms as it was given above. The variables of which values are to be considered in a propositional form are only the free variables of the form. And the form is said to be true or to be false for some system of values of its free variables (or in the other terminology, to be satisfied or not satisfied by a system of values of its free variables). A propositional form is *singulary* if it has just one free variable, *binary* if it has just two different free variables, and so on, regardless of the number of bound variables. A sentence has no free variables; but it may have any number of bound variables, since the presence of bound variables does not prevent the sentence from expressing a particular proposition (making a particular statement or assertion).

For purposes of introductory exposition, we may describe a *propositional function* as obtained by abstraction from a propositional form. A binary propositional form, for instance, determines an association of a truth-value with each system—*i.e.*, each pair—of values of its free variables, since for each such pair of values the propositional form is either true or false. And if this scheme of association of truth-values with pairs of things is taken as an abstract correspondence, independent of its expression in any particular language or of any particular notation for it, we have a binary propositional function. If, say, F is a binary propositional function thus obtained from a binary propositional form, and a, b is a pair of things of appropriate kind, we use the notation $F(a, b)$ to express that truth is associated with the pair a, b in the abstract correspondence; hence we have that $F(a, b)$ if and only if the propositional form is true for the system of values a, b of its free variables.

Similarly, if F is a singulary propositional function obtained by abstraction from a singulary propositional form, we have that $F(a)$—or, as we shall say in words, that

F holds of the argument a—if and only if the propositional form is true for the value a of its free variable. And generally, if F is an n-ary propositional function obtained by abstraction from an n-ary propositional form, we have that $F(a_1, a_2, \cdots, a_n)$—or in words, that F *holds among* the *arguments* $a_1, a_2, \cdots, a_n$—if and only if the propositional form is true for the system of values $a_1, a_2, \cdots, a_n$ of its free variables.

A propositional function obtained by abstraction from a propositional form must not be identified with the latter, since in fact different propositional forms may well sometimes determine the same scheme of association of truth-values with ordered sets of n arguments, that is, the same propositional function. A propositional form must also not be used as a name of the corresponding propositional function, since the propositional function is a fixed particular thing, of which no expression containing free variables may serve as name; but the propositional form with an abstraction operator prefixed (*see* ABSTRACT AND ABSTRACTION) is rather to be used as a name of the propositional function, where the abstraction operator, like the quantifiers, has the effect of changing free variables to bound variables. For example, if **A** is a singulary propositional form with free variable x, then $\hat{x}\mathbf{A}$ or $\lambda x\mathbf{A}$ (we shall here use the latter) is a name of the singulary propositional function obtained from **A** by abstraction.

Moreover, propositional functions must not be limited to those obtained or obtainable from propositional forms by abstraction, especially if only propositional forms belonging to one particular (formalized or other) language are considered. But abstractly, any scheme of the sort described (*i.e.*, by which truth-values are made to correspond to ordered sets of n arguments) is a propositional function, however it may have come to be known, or even if the particular propositional function never comes to be known.

It should be added that the term "propositional function" is used with various meanings and shades of meaning by different writers, the terminology being not yet fixed. Sometimes "propositional function" is used to mean what we here call a propositional form, or in a way that involves confusion between this and one of the more abstract meanings.

But as the term has just been explained, and as it will be used in this article, a singulary propositional function may be identified with a class or set, and a binary propositional function may be identified with a (binary) relation. Thus the notation "$F(a)$" may be read not only as "F holds of the argument a" but also as "a belongs to the class F" and "$F(a, b)$" may be read as "the relation F holds between a and b" or "a bears the relation F to b."

In the functional calculus of first order, the arguments of the propositional functions considered are taken as belonging to a fixed domain, the domain of *individuals*. Any well-defined class of things may be chosen as the domain of individuals, subject to the one restriction that it shall not be empty, *i.e.*, that there shall be some individuals. (And we shall here use the term *domain* of *individuals* to mean a domain that is not empty.) But a definite domain of individuals must be fixed upon in order to have an interpretation of the calculus and thus a particular formalized language.

Individual constants are symbols used as names of particular individuals. *Individual variables* are variables which have individuals as values, and which are therefore under appropriate circumstances replaceable by individual constants in the same sense in which propositional variables (*see* above) are replaceable by sentences. Likewise, *n-ary functional constants* (or *n-ary predicates*, as they are also called) are symbols used to denote particular n-ary propositional functions. And *n-ary functional variables* are variables which have n-ary propositional functions as values, and which are therefore under appropriate circumstances replaceable by n-ary functional constants.

In our present formulation of functional calculus of first order we shall use as individual variables the letters $x, y, z, t, u, v, w, x_1, y_1, z_1, t_1, u_1, v_1, w_1, x_2, \cdots$; we shall use as propositional variables the same which were used in the propositional calculus; as singulary functional variables, $F^1, G^1, H^1, F_1^1, G_1^1, H_1^1, F_2^1, \cdots$; as binary functional variables $F^2, G^2, H^2, F_1^2, \cdots$, and so on for ternary functional variables, quaternary functional variables, $\cdots$. The superscripts $1, 2, 3, \cdots$ upon the functional variables, distinguishing them as singulary, binary, ternary, $\cdots$, are necessary in principle, especially in connection with the semantics of the language, in order to avoid using the very same symbol with two or more different meanings. But in practice and as an abbreviation, the superscript may usually be omitted as being uniquely determined by the context.

The *alphabetic order* of the individual variables is the order in which they were just listed, *i.e.*, x, y, z, t, u, and so on. Likewise in the case of the functional variables of each kind, the *alphabetic order* is that in which they were just listed.

The name *functional calculus of first order* is applied to any one of many (different but closely related) logistic systems, which differ only in regard to the list of primitive symbols. All of these systems include as primitive symbols a sufficient list of sentence connectives and quantifiers, and also the notation which was explained above (consisting of parentheses and commas) for the holding of a propositional function, among certain arguments or of a certain argument. The individual variables are, further always among the primitive symbols. And the remaining primitive symbols must belong to one of the categories of propositional variables, functional variables, individual constants, or functional constants.

The *pure functional calculus of first order* has all the propositional variables and functional variables as primitive symbols and no individual constants or functional constants. An *applied* functional calculus of first order contains individual constants or functional constants or both, among its primitive symbols. A *simple applied* functional calculus of first order is an applied functional calculus of first order in which there are no propositional or functional variables among the primitive symbols.

As primitive connectives and quantifiers for functional calculus of first order we shall here use the signs of negation and material implication, and the universal quantifier. However, as in the case of the propositional calculus, other choices are possible. (And in fact the article SEMANTICS IN LOGIC makes use of negation, inclusive disjunction, and universal quantification as primitive.)

The well-formed formulas are defined by formation rules, which we here state in such a form that they can be used equally for any of the different functional calculi of first order: i_0. A propositional variable standing alone is a well-formed formula. i_1. If $\mathbf{f}$ is a singulary functional variable or a singulary functional constant, and $\mathbf{x}$ is an individual variable or an individual constant, then $\mathbf{f}(\mathbf{x})$ is a well-formed formula. i_n. If f is an n-ary functional variable or an n-ary functional constant, and $x_1, x_2, \cdots, x_n$ are individual variables or individual constants or both, not necessarily all different, then $\mathbf{f}(\mathbf{x}_1, \mathbf{x}_2, \cdots, \mathbf{x}_n)$ is a well-formed formula. ii. If $\mathbf{A}$ is a well-formed formula, $\sim\mathbf{A}$ is a well-formed formula. iii. If $\mathbf{A}$ and $\mathbf{B}$ are well-formed formulas, $[\mathbf{A} \supset \mathbf{B}]$ is a well-formed formula. iv. If $\mathbf{A}$ is a well-formed formula, and $\mathbf{x}$ is an individual variable, then $(\mathbf{x})\mathbf{A}$ is a well-formed formula.

Every well-formed formula in a functional calculus of first order is a propositional form or a sentence, as in the case of the propositional calculus and other formalized languages which we shall consider in this article. However, there are many important cases (not here treated) in which the well-formed formulas include formulas that are not propositional forms or sentences but, *e.g.*, names of propositional functions, or of numbers, or of concrete material things. And indeed it would not be unnatural, in the case of the functional calculi of first order, to take the individual constants, individual variables, functional constants, and functional variables to be well-formed formulas when they stand alone—although the standard convention, in this particular case, is rather that which is given by the formation rules above.

In abbreviating well-formed formulas of functional calculus of first order, we omit superscripts upon functional variables (as already described), we use the same conventions regarding omission of brackets which were explained above in connection with the propositional calculus, we use the same definitions (of $[\mathbf{A}|\mathbf{B}]$ etc.), also the definition of the existential quantifier,

$$(\exists\mathbf{x})\mathbf{A} \rightarrow \sim(\mathbf{x})\sim\mathbf{A},$$

where $\mathbf{x}$ is an individual variable, also further the following definitions:

$$\begin{array}{lcl}
[\mathbf{A} \supset_{\mathbf{x}} \mathbf{B}] & \rightarrow & (\mathbf{x})[\mathbf{A} \supset \mathbf{B}] \\
[\mathbf{A} \equiv_{\mathbf{x}} \mathbf{B}] & \rightarrow & (\mathbf{x})[\mathbf{A} \equiv \mathbf{B}] \\
[\mathbf{A} \wedge_{\mathbf{x}} \mathbf{B}] & \rightarrow & (\exists\mathbf{x})[\mathbf{AB}] \\
[\mathbf{A} \supset_{\mathbf{xy}} \mathbf{B}] & \rightarrow & (\mathbf{x})(\mathbf{y})[\mathbf{A} \supset \mathbf{B}] \\
[\mathbf{A} \equiv_{\mathbf{xy}} \mathbf{B}] & \rightarrow & (x)(y)[\mathbf{A} \equiv \mathbf{B}] \\
[\mathbf{A} \wedge_{\mathbf{xy}} \mathbf{B}] & \rightarrow & (\exists\mathbf{x})(\exists\mathbf{y})[\mathbf{AB}]
\end{array}$$

and likewise with three or more subscripts $\mathbf{x}, \mathbf{y}, \mathbf{z}$, etc., where the subscripts $\mathbf{x}, \mathbf{y}, \mathbf{z}, \ldots$ must be individual variables and all different in each case.

The relation between propositional functions that is expressed, in accordance with the above definitions, by the sign $\supset$ followed by one or more individual variables as subscripts is called *formal implication* (the standard term, though the adjective "formal" has here a rather different meaning from that which we have given to it elsewhere). For example, in the case of singulary propositional functions, $F(x) \supset_x G(x)$ (*i.e.*, the well-formed formula which is thus abbreviated) is said to express that F formally implies G, or in different words, that $F(x)$ formally implies $G(x)$ with respect to x; and in the case of binary propositional functions, $F(x, y) \supset_{xy} G(x, y)$ expresses that F formally implies G, or that $F(x, y)$ formally implies $G(x, y)$ with respect to x and y.—In similar fashion the sign $\equiv$ followed by individual variables as subscripts is said to express the relation of *formal equivalence* between propositional functions.

Now by including rules of substitution among the rules of inference, it is possible to make a formulation of the pure functional calculus of first order in which the number of axioms is finite. But because of rather complicated explanations which would be necessary in connection with one of the substitution rules in particular (that for functional variables) we here employ a formulation without rules of substitution and with an infinite number of axioms. The infinite set of axioms is then given by means of five *axiom schemata*, which we shall state. And this procedure has the further advantage that the same five axiom schemata (though not the same infinite set of axioms) are sufficient for any one of the other first-order functional calculi, as well as for the pure calculus.

The first three of the axiom schemata are as follows:

1. $\mathbf{A} \supset \mathbf{B} \supset . \mathbf{B} \supset \mathbf{C} \supset . \mathbf{A} \supset \mathbf{C}$
2. $\sim\mathbf{A} \supset \mathbf{A} \supset \mathbf{A}$
3. $\mathbf{A} \supset . \sim\mathbf{A} \supset \mathbf{B}$

The first axiom schema means, *e.g.*, that if **A**, **B**, **C** are any well-formed formulas of the first-order functional calculus under consideration (possibly all three different, or possibly some of them the same), then $\mathbf{A} \supset \mathbf{B} \supset . \mathbf{B} \supset \mathbf{C} \supset . \mathbf{A} \supset \mathbf{C}$ is an axiom. In the case of the pure first-order functional calculus, two of the axioms which are instances of the first axiom schema are, for instance, $p \supset q \supset . q \supset r \supset . p \supset r$ and $\sim q \supset [F(x) \supset G(x)] \supset . F(x) \supset G(x) \supset (x)H(x, y) \supset . \sim q \supset (x)H(x, y)$.

The fourth axiom schema is the following:

4. $\mathbf{A} \supset_{\mathbf{x}} \mathbf{B} \supset . \mathbf{A} \supset (\mathbf{x})\mathbf{B}$, provided that **x** is an individual variable which is not a free variable of **A**.

In order to state the fifth axiom schema, it is necessary first to explain another metatheoretic notation. In using bold capital letters to stand for well-formed formulas, we shall sometimes add after the letter an indication of one of the free variables it may contain, providing by this device a convenient notation for the process of substituting one free variable for another. For example, if we use $\mathbf{A}x$ to stand for a well-formed formula which has or may have x as a free variable, then $\mathbf{A}y$ shall stand for the well-formed formula obtained from $\mathbf{A}x$ by substituting y for all free occurrences of x in $\mathbf{A}x$—provided, however, that if any of the free occurrences of x in $\mathbf{A}x$ are in a well-formed part of $\mathbf{A}x$ of the form $(y)\mathbf{C}$ then $\mathbf{A}y$ shall be obtained from $\mathbf{A}x$ by first substituting **z** for all bound occurrences of y in $\mathbf{A}x$ and then substituting y for all free occurrences of x in $\mathbf{A}x$, z being the first individual variable in alphabetic order that does not occur in $\mathbf{A}x$. In the special case that $\mathbf{A}x$ does not actually contain x as a free variable, $\mathbf{A}y$ is the same as $\mathbf{A}x$. And the same notation may be used also for substitution of a constant for a free variable, so that, *e.g.*, if **a** is any individual constant, **Aa** stands for the result of substituting **a** for all free occurrences of x in $\mathbf{A}x$.

Employing this notation, we state the fifth axiom schema:

5. $(\mathbf{x})\mathbf{Ax} \supset \mathbf{Ay}$, where **x** is an individual variable, **Ax** is a well-formed formula which may have **x** as a free variable, and **y** is an individual variable or an individual constant.

To use the case of the pure functional calculus of first order for illustration, following are some examples of instances of axiom schema 5, which are therefore axioms:

$$(x)F(x) \supset F(y)$$
$$(y)F(y) \supset F(y)$$
$$(y)F(x, y) \supset F(x, x)$$
$$(x)(y)F(x, y) \supset (z)F(y, z)$$
$$(z)[F(z) \supset G(z)] \supset . F(x) \supset G(x)$$
$$(z)[F(x) \supset G(x)] \supset . F(x) \supset G(x)$$

To complete the formulation of first-order functional calculus as a logistic system (whether the pure first-order functional calculus or one of the others), it remains only to state the rules of inference. These are the two following: II. *Rule of modus ponens.* From $\mathbf{A} \supset \mathbf{B}$ and **A**, to infer **B**. III. *Rule of generalization.* From **A**, if **x** is any individual variable, to infer $(\mathbf{x})\mathbf{A}$.

In an application of the rule of generalization, the variable $\mathbf{x}$ is said to be *generalized upon*.

Where $\mathbf{A}_1, \mathbf{A}_2, \cdots, \mathbf{A}_n, \mathbf{B}$ are well-formed formulas of one of the functional calculi of first order, we say that the inference from the hypotheses $\mathbf{A}_1, \mathbf{A}_2, \cdots, \mathbf{A}_n$ to the conclusion B is a *valid inference of first-order functional calculus* if $\mathbf{B}$ can be obtained from the hypotheses $\mathbf{A}_1, \mathbf{A}_2, \cdots, \mathbf{A}_n$, together with the axioms of the particular functional calculus of first order, by a succession of applications of the two rules of inference, *modus ponens* and generalization, subject to the restriction that no variable shall be generalized upon which is a free variable of any of the hypotheses $\mathbf{A}_1, \mathbf{A}_2, \cdots, \mathbf{A}_n$. (The distinction which is made here between "hypotheses" and "premisses" may be ignored only if the hypotheses are without free variables.)

For example, from the single hypothesis $F(x)$ there is a valid inference to the conclusion $(y) \mathbin{.} {\sim}F(x) \supset F(y)$, namely by taking as major premiss the axiom $F(x) \supset \mathbin{.} {\sim}F(x) \supset F(y)$, which is an instance of axiom schema 3, and applying *modus ponens*, and then generalizing upon y. On the other hand there is no valid inference from the hypothesis $F(x)$ to the conclusion $(x)F(x)$, and indeed it is informally evident that the definition ought not to allow this as a valid inference—i.e., from the hypothesis that a particular individual x belongs to the class F, it does not in general follow that all individuals belong to the class F.

There is not space to treat in detail particular theorems and valid inferences of first-order functional calculus. But some of the simplest of the latter will be mentioned below in the discussion of traditional logic. And we list here the following important metatheorems (omitting their proofs, with one exception):

The Deduction Theorem.—If there is a valid inference (of first-order functional calculus) from the hypotheses $\mathbf{A}_1, \mathbf{A}_2, \cdots, \mathbf{A}_n$ to the conclusion $\mathbf{B}$, there is a valid inference also from the hypotheses $\mathbf{A}_1, \mathbf{A}_2, \cdots \mathbf{A}_{n-1}$ to the conclusion $\mathbf{A}_n \supset \mathbf{B}$. Hence if there is a valid inference from the hypotheses $\mathbf{A}_1, \mathbf{A}_2, \cdots, \mathbf{A}_n$ to the conclusion $\mathbf{B}$, then $\mathbf{A}_1 \supset \mathbin{.} \mathbf{A}_2 \supset \mathbin{.} \cdots \mathbf{A}_n \supset \mathbf{B}$ is a theorem. And, as a special case, if there is a valid inference from the single hypothesis $\mathbf{A}$ to the conclusion $\mathbf{B}$, then $\mathbf{A} \supset \mathbf{B}$ is a theorem.

Tautologies.—Every *tautologous* well-formed formula is a theorem, *i.e.*, every well-formed formula which is a tautology of the propositional calculus or is obtained from such a tautology by substitutions for the propositional variables.

Substitutivity of Equivalence.—If $\mathbf{M} \equiv \mathbf{N}$ is a theorem, if A is a theorem, and if $\mathbf{B}$ is obtained from $\mathbf{A}$ by replacing $\mathbf{M}$ by $\mathbf{N}$ at one or more places (not necessarily at all occurrences of $\mathbf{M}$ in $\mathbf{A}$), then $\mathbf{B}$ is a theorem.

Reduction to Prenex Normal Form.—There is an effective procedure by which, given a well-formed formula $\mathbf{A}$ of a first-order functional calculus, there may be found a well-formed formula $\mathbf{B}$ of the same first-order functional calculus, such that $\mathbf{A} \equiv \mathbf{B}$ is a theorem, and $\mathbf{B}$ is in *prenex normal form*—*i.e.*, $\mathbf{B}$ consists of a quantifier-free well-formed part, called the *matrix*, and of a preceding part which is called the *prefix* and which consists of a number of universal and existential quantifiers prefixed to the matrix, subject to the condition that the variables occurring in the quantifiers in the prefix shall be all different and shall all occur also in the matrix. (As special cases, the quantifiers constituting the prefix may be all universal, or they may be all existential, or the prefix may even be null so that $\mathbf{B}$ consists entirely of the quantifier-free matrix.)

Consistency.—**A** and $\sim$**A** cannot both be theorems of a first-order functional calculus. The proof of this is by considering, for any well-formed formula of first-order functional calculus, a corresponding well-formed formula of the propositional calculus which is obtained as follows. From the given formula of first-order functional calculus, first delete (universal) quantifiers, so as to obtain a quantifier-free formula, then replace every propositional variable and every well-formed part of the form $\mathbf{f}(\mathbf{x})$ or $\mathbf{f}(\mathbf{x}_1, \mathbf{x}_2, \cdots, \mathbf{x}_n)$—*cf.* the formation rules i_0, i_1, i_n—each by the propositional variable p. (For example, if the given formula is $(x)(y)F(x, y, z) \supset . p \supset q \supset (\exists z)G(z)$, the deletion of quantifiers yields $F(x, y, z) \supset . p \supset q \supset \sim\sim G(z)$, and hence the corresponding formula of the propositional calculus is $p \supset . p \supset p \supset \sim\sim p$.) As the reader may verify, every axiom of first-order functional calculus has the property that the corresponding formula of the propositional calculus is a tautology; moreover, the rules of inference preserve this property, *i.e.*, if in an application of one of the rules of inference the premiss or premisses have this property, the conclusion must have it also; hence every theorem has this property. But **A** and $\sim$**A** cannot both have this property (indeed, of the two corresponding formulas of the propositional calculus, the second one is obtained from the first by prefixing the sign $\sim$, so that not both can be tautologies). Hence not both **A** and $\sim$**A** can be theorems.

The Metatheory of the Pure Functional Calculus of First Order contains many results of greater depth than the elementary fundamentals which have been discussed above. Only a few of these will be briefly indicated here.

A well-formed formula is said to be *satisfiable* in a particular non-empty domain if, when that domain is taken as the domain of individuals, the formula is satisfied by at least one system of values of its free variables which has individuals (of the domain in question) as values of the individual variables, n-ary propositional functions as values of the n-ary functional variables, and truth-values as values of the propositional variables. Similarly a well-formed formula is said to be *valid in* a particular nonempty domain if, when that domain is taken as the domain of individuals, the formula is satisfied by every such system of values of its free variables. A well-formed formula is said to be *satisfiable* if it is satisfiable in some non-empty domain, *valid* if it is valid in every non-empty domain. Evidently, **A** is satisfiable in a particular domain if and only if $\sim$**A** is not valid in that domain, and **A** is satisfiable if and only if $\sim$**A** is not valid.

The validity of a well-formed formula in a domain of individuals depends only on the number of individuals in the domain. And in fact well-formed formulas (*i.e.*, of the pure functional calculus of first order) can be classified as follows: there are those which are valid in no domain of individuals; for every positive integer n, there are those which are valid in domains of not more than n individuals but not valid in larger domains; there are well-formed formulas which are valid in every finite domain of individuals but not valid in an infinite domain; and there are the valid well-formed formulas. It can be shown that every well-formed formula belongs to one of these classes.—As examples we cite $(\exists x)(y) . [F(x, y) \supset F(y, x)] \vee [F(x, x) \equiv F(y, y)]$, valid in domains of not more than three individuals, but not in larger domains, and $(\exists x)(y)(\exists z) . F(z, x) \supset F(z, y) \supset . F(x, y) \supset F(x, x)$, valid in all finite domains of individuals, but not in infinite domains.

The above statement includes the following theorem of Leopold Löwenheim: If a well-formed formula is valid in the domain of positive integers (*i.e.*, when the positive integers $1, 2, 3, \cdots$ are taken as the individuals), it is valid. As a corollary, or as another form of the metatheorem, if a well-formed formula is satisfiable, it is satisfiable in the domain of positive integers.

It can be shown that every theorem of the pure functional calculus of first order is valid (since the axioms are valid and the rules of inference preserve validity). The converse of this is the completeness (meta)theorem of Kurt Gödel: Every valid well-formed formula is a theorem.

A class of well-formed formulas (finite or infinite in number) is said to be *consistent* if it does not contain formulas $\mathbf{A}_1, \mathbf{A}_2, \cdots, \mathbf{A}_n$ from which as hypotheses there is a valid inference to a conclusion $\mathbf{B}$ and also to $\sim\mathbf{B}$. A class of well-formed formulas is said to be *simultaneously satisfiable in* a particular non-empty domain if, when that domain is taken as the domain of individuals, the formulas are *simultaneously satisfied* by at least one system of values of the free variables. And a class of well-formed formulas is said to be simultaneously satisfiable if it is simultaneously satisfiable in some non-empty domain.

According to an extension of Löwenheim's theorem by Thoralf Skolem, if a class of well-formed formulas is simultaneously satisfiable, it is simultaneously satisfiable in the domain of positive integers. And according to a closely related metatheorem due to Gödel, if a class of well-formed formulas is consistent, it is simultaneously satisfiable.

TRADITIONAL LOGIC

The name *traditional logic* is given to that part of the ancient and medieval logic which survived the decline of scholasticism and long remained with little change as a traditionally important part of philosophy. Though historically an independent doctrine, with a viewpoint and method that differ from the logistic method explained above, it can be exhibited as a part of modern logic, and this is the course which we shall follow here. In doing this it is necessary to make certain changes (as noted below), of which some remove uncertainties or correct confusions of the traditional doctrine, and others, though not required from the point of view of traditional logic itself, are desirable in order to incorporate it into the body of modern logic and to give it its proper place in relation to the remainder.

Categorical Propositions are propositions of the traditional subject-predicate form, having a subject S and a predicate P. The four forms—All S is P, No S is P, Some S is P, Some S is not P—are traditionally designated by the letters A, E, I, O respectively. Examples are: All men are mortal (A), All men die (A), No man can serve two masters (E), Some prime numbers are odd (I), A large island is in the bay (I), All that glitters is not gold (O). Propositions of the forms A and I are called *affirmative*; E and O, *negative*; A and E, *universal*; I and O, *particular*.

The subject and predicate of a categorical proposition are together called the *terms* of the proposition Thus in the third example above, the terms are the subject, man, and the predicate, able to serve two masters.

In writing categorical propositions in logistic form we shall use functional constants $\mathbf{s}$ and $\mathbf{p}$ to stand for the subject and predicate, so that the four forms appear as follows: $\mathbf{s}(x) \supset_x \mathbf{p}(x)$, $\mathbf{s}(x) \supset_x \sim\mathbf{p}(x)$, $\mathbf{s}(x) \wedge_x \mathbf{p}(x)$, $\mathbf{s}(x) \wedge_x \sim\mathbf{p}(x)$.

This manner of representing the categorical propositions is not faithful in all particulars to the traditional account. But among various possibilities it seems to be on the whole the best, and we shall employ it here, noting the four following points of divergence:

1. We have defined the notations $\supset_{\mathbf{x}}$ and $\wedge_{\mathbf{x}}$ in terms of the universal quantifier and the two primitive connectives $\sim$ and $\supset$, whereas the traditional account might be thought to be more closely reproduced if $\supset_{\mathbf{x}}$ and $\wedge_{\mathbf{x}}$ were primitive notations.

2. The traditional account associates the negation in E and O with the *copula*, that is, with the words "is" or "is not" that join the subject and predicate, whereas here we prefix the sign $\sim$ to the subformula $\mathbf{p}(x)$.—In regard to 1 and 2 it would be possible to reproduce the traditional account more closely by using four primitive notations A_x, E_x, I_x, O_x, where $[\mathbf{s}(x)A_x\mathbf{p}(x)]$, $[\mathbf{s}(x)E_x\mathbf{p}(x)]$, $[\mathbf{s}(x)I_x\mathbf{P}(x)]$, $[\mathbf{S}(x)O_x\mathbf{p}(x)]$ are to have the meanings of $\mathbf{s}(x) \supset_x \mathbf{p}(x)$, $\mathbf{s}(x) \supset_x {\sim}\mathbf{p}(x)$, $\mathbf{s}(x) \wedge_x \mathbf{p}(x)$, $\mathbf{s}(x) \wedge_x {\sim}\mathbf{p}(x)$ respectively (and the usual quantifiers and sentence connectives could then be defined in terms of these). But it seems preferable not to complicate the formulation of first-order functional calculus in this way.

3. The traditional account includes also, under A and E, propositions expressed in the forms $\mathbf{p}(\mathbf{a})$ and ${\sim}\mathbf{p}(\mathbf{a})$ respectively, where $\mathbf{a}$ is an individual constant. For example, "Socrates is mortal" is considered as expressing an A proposition, and "Socrates is not mortal" an E proposition, the subject being the *singular term*, Socrates. These *singular propositions*, as they are called, will be ignored in our account of opposition and immediate inference, but will appear in connection with the categorical syllogism as giving special forms (called *singular forms*) of certain syllogisms.

4. Some aspects of the traditional account require that A and E be represented as we have here, others that they be represented by the conjunctions $(\exists x)\mathbf{s}(x) \mathbin{.} \mathbf{s}(x) \supset_x \mathbf{p}(x)$ and $(\exists x)\mathbf{s}(x) \mathbin{.} \mathbf{s}(x) \supset_x {\sim}\mathbf{p}(x)$ respectively. The problem of choosing between these two interpretations (or finding a satisfactory third alternative) is known as the problem of *existential import* of categorical propositions. In our account below we shall meet the difficulty by introducing $(\exists x)\mathbf{s}(x)$ as a separate premiss at those places where it is required.

Opposition, Immediate Inference.—According to the square of opposition, if the subject and predicate are fixed, A and O (*i.e.*, $\mathbf{s}(x) \supset_x \mathbf{p}(x)$ and $\mathbf{s}(x) \wedge_x {\sim}\mathbf{p}(x)$) are *contradictory*, E and I are *contradictory*, A and E are *contrary*, I and O are *subcontrary*, A and I are *subaltern*, E and O are *subaltern*. The two propositions of a contradictory pair cannot be both true and cannot be both false. Under the premiss $(\exists x)\mathbf{s}(x)$, the contrary pair, A, E, cannot be both true, the subcontrary pair, I, O, cannot be both false, and each of the propositions A and E has its subaltern proposition as a consequence.

Simple conversion of a categorical proposition consists in interchanging the subject and predicate. Thus the converses of $\mathbf{s}(x) \supset_x \mathbf{p}(x), \mathbf{s}(x) \supset_x {\sim}\mathbf{p}(x), \mathbf{s}(x) \wedge_x \mathbf{p}(x)$, and $\mathbf{s}(x) \wedge_x {\sim}\mathbf{p}(x)$ are respectively $\mathbf{p}(x) \supset_x \mathbf{s}(x), \mathbf{p}(x) \supset_x {\sim}\mathbf{s}(x), \mathbf{p}(x) \wedge_x \mathbf{s}(x)$, and $\mathbf{p}(x) \wedge_x {\sim}\mathbf{s}(x)$. Simple conversion is a generally valid inference only in the case of E and I.

Obversion of a categorical proposition is effected by replacing $\mathbf{p}$ by a functional constant $\mathbf{q}$ which denotes the negation of the propositional function (the complement of the class) that is denoted by $\mathbf{p}$, and at the same time inserting $\sim$ if not already present or deleting it if present. In terms of the abstraction operator, the functional constant $\mathbf{q}$ is $\lambda x{\sim}\mathbf{p}(x)$. Thus the obverse of $\mathbf{s}(x) \supset_x \mathbf{p}(x)$ is $\mathbf{s}(x) \supset_x {\sim}\mathbf{q}(x)$ (the obverse of *All men are mortal is No men are immortal*). Similarly, the obverse of $\mathbf{s}(x) \supset_x {\sim}\mathbf{p}(x)$ is $\mathbf{s}(x) \supset_x \mathbf{q}(x)$, the obverse of $\mathbf{s}(x) \wedge_x \mathbf{p}(x)$ is $\mathbf{s}(x) \wedge_x {\sim}\mathbf{q}(x)$, and that of $\mathbf{s}(x) \wedge_x {\sim}\mathbf{p}(x)$ is $\mathbf{s}(x) \wedge_x \mathbf{q}(x)$.

The name "immediate inference" is given to certain inferences from one categorical proposition as premiss to another as conclusion, all of them being valid inferences either of first-order functional calculus or of an extended calculus embracing the abstraction operator λ. The immediate inferences include obversion of A, E, I, O, simple conversion of E, I, and subalternation of A, E—of which subalternation requires the additional premiss $(\exists x)\mathbf{s}(x)$. Other immediate inferences may be obtained by means of sequences of these; *e.g.*, given that all men are mortal we may take the obverse of the converse of the obverse and so infer that all immortals are non-men (called by some the contrapositive, by others the obverted contrapositive).

Conversion *per accidens*, or *by limitation*, of a proposition A may be described as consisting of subalternation followed by simple conversion. Thus from the premisses $\mathbf{s}(x) \supset_x \mathbf{p}(x)$ and $(\exists x)\mathbf{s}(x)$ it yields the conclusion $\mathbf{p}(x) \wedge_x \mathbf{s}(x)$. Conversion *per accidens* of E is also possible, by a simple conversion followed by subalternation.

Categorical Syllogism.—The name "categorical syllogism" is given to certain valid inferences of first-order functional calculus which involve as premisses two categorical propositions having a term in common—the *middle term*. Using functional constants $\mathbf{s}$, $\mathbf{m}$, $\mathbf{p}$ to stand for the *minor term*, the middle term, and the *major term* respectively, we give the traditional classification into figures and moods. In each case we give the *major premiss* first, the *minor premiss* immediately after it, and the conclusion last; in some cases we give a third (existential) premiss which is suppressed in the traditional account. Where in consequence of the admission of singular propositions (as noted above) two different forms of valid inference appear under the same figure and mood, we give the singular forms in a separate list.

First Figure

Barbara: $\mathbf{m}(x) \supset_x \mathbf{p}(x)$, $\mathbf{s}(x) \supset_x \mathbf{m}(x)$; $\mathbf{s}(x) \supset_x \mathbf{p}(x)$.
Celarent: $\mathbf{m}(x) \supset_x \sim\mathbf{p}(x)$, $\mathbf{s}(x) \supset_x \mathbf{m}(x)$; $\mathbf{s}(x) \supset_x \sim\mathbf{p}(x)$.
Darii: $\mathbf{m}(x) \supset_x \mathbf{p}(x)$, $\mathbf{s}(x) \wedge_x \mathbf{m}(x)$; $\mathbf{s}(x) \wedge_x \mathbf{p}(x)$.
Ferio: $\mathbf{m}(x) \supset_x \sim\mathbf{p}(x)$, $\mathbf{s}(x) \wedge_x \mathbf{m}(x)$; $\mathbf{s}(x) \wedge_x \sim\mathbf{p}(x)$.

Second Figure

Cesare: $\mathbf{p}(x) \supset_x \sim\mathbf{m}(x)$, $\mathbf{s}(x) \supset_x \mathbf{m}(x)$; $\mathbf{s}(x) \supset_x \sim\mathbf{p}(x)$.
Camestres: $\mathbf{p}(x) \supset_x \mathbf{m}(x)$, $\mathbf{s}(x) \supset_x \sim\mathbf{m}(x)$; $\mathbf{s}(x) \supset_x \sim\mathbf{p}(x)$.
Festino: $\mathbf{p}(x) \supset_x \sim\mathbf{m}(x)$, $\mathbf{s}(x) \wedge_x \mathbf{m}(x)$; $\mathbf{s}(x) \wedge_x \sim\mathbf{p}(x)$.
Baroco: $\mathbf{p}(x) \supset_x \mathbf{m}(x)$, $\mathbf{s}(x) \wedge_x \sim\mathbf{m}(x)$; $\mathbf{s}(x) \wedge_x \sim\mathbf{p}(x)$.

Third Figure

Darapti: $\mathbf{m}(x) \supset_x p(x)$, $\mathbf{m}(x) \supset_x \mathbf{s}(x)$, $(\exists x)\mathbf{m}(x)$; $\mathbf{s}(x) \wedge_x \mathbf{p}(x)$.
Disamis: $\mathbf{m}(x) \wedge_x \mathbf{p}(x)$, $\mathbf{m}(x) \supset_x \mathbf{s}(x)$; $\mathbf{s}(x) \wedge_x \mathbf{p}(x)$.
Datisi: $\mathbf{m}(x) \supset_x \mathbf{p}(x)$, $\mathbf{m}(x) \wedge_x \mathbf{s}(x)$; $\mathbf{s}(x) \wedge_x \mathbf{p}(x)$.
Felapton: $\mathbf{m}(x) \supset_x \sim\mathbf{p}(x)$, $\mathbf{m}(x) \supset_x \mathbf{s}(x)$, $(\exists x)\mathbf{m}(x)$; $\mathbf{s}(x) \wedge_x \sim\mathbf{p}(x)$.
Bocardo: $\mathbf{m}(x) \wedge_x \sim\mathbf{p}(x)$, $\mathbf{m}(x) \supset_x \mathbf{s}(x)$; $\mathbf{s}(x) \wedge_x \sim\mathbf{p}(x)$.
Feriso (or Ferison): $\mathbf{m}(x) \supset_x \sim\mathbf{p}(x)$, $\mathbf{m}(x) \wedge_x \mathbf{s}(x)$; $\mathbf{s}(x) \wedge_x \sim\mathbf{p}(x)$.

Fourth Figure

Bamalip (or Bramantip): $\mathbf{p}(x) \supset_x \mathbf{m}(x)$, $\mathbf{m}(x) \supset_x \mathbf{s}(x)$, $(\exists x)\mathbf{p}(x)$; $\mathbf{s}(x) \wedge_x \mathbf{p}(x)$.
Calemes (or Camenes): $\mathbf{p}(x) \supset_x \mathbf{m}(x)$, $\mathbf{m}(x) \supset_x \sim\mathbf{s}(x)$; $\mathbf{s}(x) \supset_x \sim\mathbf{p}(x)$.
Dimatis (or Dimaris): $\mathbf{p}(x) \wedge_x \mathbf{m}(x)$, $\mathbf{m}(x) \supset_x \mathbf{s}(x)$; $\mathbf{s}(x) \wedge_x \mathbf{p}(x)$.

Fesapo: $\mathbf{p}(x) \supset_x \sim\mathbf{m}(x), \mathbf{m}(x) \supset_x \mathbf{s}(x), (\exists x)\mathbf{m}(x); \mathbf{s}(x) \wedge_x \sim\mathbf{p}(x)$.
Fresison: $\mathbf{p}(x) \supset_x \sim\mathbf{m}(x), \mathbf{m}(x) \wedge_x \mathbf{s}(x); \mathbf{s}(x) \wedge_x \sim\mathbf{p}(x)$.

Singular Forms

Barbara: $\mathbf{m}(x) \supset_x \mathbf{p}(x), \mathbf{m}(a); \mathbf{p}(\mathbf{a})$.
Celarent: $\mathbf{m}(x) \supset_x \sim\mathbf{p}(x), \mathbf{m}(a); \sim\mathbf{p}(\mathbf{a})$.
Cesare: $\mathbf{p}(x) \supset_x \sim\mathbf{m}(x), \mathbf{m}(a); \sim\mathbf{p}(\mathbf{a})$.
Camestres: $\mathbf{p}(x) \supset_x \mathbf{m}(x), \sim\mathbf{m}(a); \sim\mathbf{p}(\mathbf{a})$.
Darapti: $\mathbf{p}(\mathbf{a}), \mathbf{s}(\mathbf{a}); \mathbf{s}(x) \wedge_x \mathbf{p}(x)$.
Felapton: $\sim\mathbf{p}(\mathbf{a}), \mathbf{s}(\mathbf{a}); \mathbf{s}(x) \wedge_x \sim\mathbf{p}(x)$.

The last two singular forms, in which the middle term is singular, are classed separately as the *expository* syllogism.

Some add the five so-called weakened moods, Barbari, Celaront, Cesaro, Camestros, Calemos, to be obtained by subalternation of the conclusion from Barbara, Celarent, Cesare, Camestres, Calemes respectively. The five moods of the fourth figure are sometimes classed instead as indirect moods of the first figure, the major and minor premisses being interchanged, and the names being then given as Baralipton, Celantes, Dabitis, Fapesmo, Frisesomorum.

The names of the moods have a mnemonic significance, in which the first three vowels indicate whether the major premiss, minor premiss, and conclusion, in order, are A, E, I, or O. and some of the consonants indicate the traditional reductions of the other moods to the four direct moods of the first figure (*see* LOGIC, HISTORY OF).

Hypothetical Syllogism, Disjunctive Syllogism, Dilemma.—Besides the categorical syllogism, the traditional logic treats also a number of other kinds of *mediate* inference (*i.e.*, inference from two or more premisses), including especially the *hypothetical syllogism* (or conditional syllogism), the *disjunctive syllogism*, and the *dilemma*. All of these can be exhibited as valid inferences of the propositional calculus, and we shall give them here in this form.

In particular, we shall render the words "if" or "if ... then," as used in stating the traditional hypothetical (or conditional) propositions, by the sign $\supset$ of material implication, although this is certainly contrary to the intention of traditional writers. In the Port-Royal Logic, for example, the contradictory negative of *If you eat of the forbidden fruit you will die* is given as *Although you eat of the forbidden fruit you will not die*—whereas if material implication were intended, the contradictory should be rather, *You will eat of the forbidden fruit and you will not die*. Yet in the same work it is explained that a conditional proposition may be true although both parts of it (both the antecedent and the consequent) are false, provided only that the consequence is correctly drawn.

By this it is not meant that the consequent must follow logically from the antecedent, but, as is clear from the discussion and the examples given, that the consequent must follow from the antecedent together with known truths which are taken into consideration. However, this account overlooks that, if two propositions are both false, it is always possible to infer one from the other by means of appropriately chosen truths, which may then be supposed to be known and taken into consideration. It is in this way that the traditional account leads to material implication as the best means of representing it consistently within a more complete theory. An alternative amendment of the traditional account indeed suggests itself, by seeking a satisfactory logistic theory of an implication connective which shall reproduce more closely the ordinary usage of "if ... then" in future conditions and conditions contrary to fact, but this has never been successfully carried out. (Compare the discussion of material implication in the first part of this article.)

There is also a question whether the word "or," as used in stating the traditional disjunctive propositions, shall be understood as denoting inclusive disjunction or exclusive disjunction. Here the traditional logic is more explicitly inconsistent—since, although most of the inferences are valid inferences under either interpretation of the word "or," one of them, the *modus ponendo tollens*, is valid only for exclusive disjunction, and the two complex dilemmas are valid only for inclusive disjunction. We shall use exclusive disjunction in *modus ponendo tollens*, but inclusive disjunction elsewhere.

In each entry in the following table the major premiss is given first, or the two major premisses in the case of the dilemmas, then the minor premiss, then the conclusion.

Hypothetical Syllogism

Modus ponens: $\mathbf{A} \supset \mathbf{B}, \mathbf{A}; \mathbf{B}.$
Modus tollens: $\mathbf{A} \supset \mathbf{B}, \sim\mathbf{B}; \sim\mathbf{A}.$

Disjunctive Syllogism

Modus tollendo ponens: $\mathbf{A} \vee \mathbf{B}, \sim\mathbf{A}; \mathbf{B}.$
Modus tollendo ponens: $\mathbf{A} \vee \mathbf{B}, \sim\mathbf{B}; \mathbf{A}.$
Modus ponendo tollens: $\mathbf{A} \not\equiv \mathbf{B}, \mathbf{A}; \sim\mathbf{B}.$
Modus ponendo tollens: $\mathbf{A} \not\equiv \mathbf{B}, \mathbf{B}; \sim\mathbf{A}.$

Dilemma

Simple constructive dilemma: $\mathbf{A} \supset \mathbf{C}, \mathbf{B} \supset \mathbf{C}, \mathbf{A} \vee \mathbf{B}; \mathbf{C}.$
Simple destructive dilemma: $\mathbf{A} \supset \mathbf{B}, \mathbf{A} \supset \mathbf{C}, \sim\mathbf{B} \vee \sim\mathbf{C}; \sim\mathbf{A}.$
Complex constructive dilemma: $\mathbf{A} \supset \mathbf{B}, \mathbf{C} \supset \mathbf{D}, \mathbf{A} \vee \mathbf{C}; \mathbf{B} \vee \mathbf{D}.$
Complex destructive dilemma: $\mathbf{A} \supset \mathbf{B}, \mathbf{C} \supset \mathbf{D}, \sim\mathbf{B} \vee \sim\mathbf{D}; \sim\mathbf{A} \vee \sim\mathbf{C}.$

The inferences from $\mathbf{A} \supset \mathbf{B}$ and $\mathbf{C} \supset \mathbf{A}$ to $\mathbf{C} \supset \mathbf{B}$, and from $\mathbf{A} \supset \mathbf{B}$ and $\mathbf{C} \supset \sim\mathbf{B}$ to $\mathbf{C} \supset \sim\mathbf{A}$ are sometimes added as *pure hypothetical* syllogisms, and the above simpler forms of the hypothetical syllogism are then distinguished as *mixed hypothetical*. Some older works, including the Port-Royal Logic, add a *copulative* syllogism, with major premiss $\sim . \mathbf{AB}$, minor premiss $\mathbf{A}$ and conclusion $\sim\mathbf{B}$, or minor premiss $\mathbf{B}$ and conclusion $\sim\mathbf{A}$.

TYPE THEORY AND SET THEORY

In the functional calculus of first order there is provision only for individuals as arguments of propositional functions, and it is only individual variables which may be bound by quantifiers. Since the domain of individuals is fixed in advance, and since the classes and relations (and propositional functions generally) which are values of the functional variables are not among the individuals of the fixed domain, this imposes a limitation upon what can be expressed in the notation of the calculus. Indeed it may be thought that the propositions of logic have a character of universal generality which demands for its expression the use of variables of unrestricted range, whose values are not confined to a fixed domain but may be either individuals or classes or relations or anything else whatever. And just this was maintained in particular by Gottlob Frege.

However, the uncritical attempt to introduce such variables of unrestricted range leads to a system which is inconsistent, if besides the variables of unrestricted range, say $x, y, z, x_1, \ldots$, we employ also the usual sentence connectives and quantifiers, and a notation, say $\epsilon(x, y_1, y_2, \ldots, y_n)$, to mean that x *is an n-ary propositional function and holds among the arguments* $y_1, y_2, \ldots, y_n$, and adopt what seem to be the natural and obvious axioms and rules of inference. Namely in such a system the antinomy of Bertrand Russell or that of C. Burali-Forti (*see* ANTINOMY) can be reproduced formally, *i.e.*, in the form of proofs of contradictory theorems of the system. And in fact the significance of these antinomies lies less in their informal statement in words than in their influence upon the construction of logistic systems, through the possibility of their logistic formalization.

In order to obtain a system more comprehensive than functional calculus of first order without falling into antinomy, there are two directions which have been followed. One of these, the direction of type theory, avoids altogether the use of variables of unrestricted range, and introduces instead many different kinds of variables with different restricted ranges. The other, the direction of set theory, has variables of unrestricted range, but imposes restrictions in regard to the existence of sets determined by given conditions. (The word "set" is here used to mean a class which can belong to another class as a member of it, since some forms of set theory admit also classes which are not sets, but outside the context of set theory the words "class" and "sets are generally used synonymously.)

As typical of the two directions we shall describe briefly the simple theory of types, and a form of set theory which is based on the system proposed by Ernst Zermelo in 1908.

The Simple Theory of Types can be described as obtained from the pure functional calculus of first order by successive additions to it.

The first step is to modify the formation rule iv to provide that $(\mathbf{x})\mathbf{A}$ shall be well-formed when $\mathbf{x}$ is a propositional or functional variable as well as when $\mathbf{x}$ is an individual variable. The resulting system is the *pure functional calculus of second order*, if suitable axioms and rules of inference are provided (which will not be stated here, but are similar to those for functional calculus of first order).

Then functional variables of various higher types are adjoined. Namely if $c_1, c_2, \ldots, c_n$ are non-negative integers, then $(c_1, c_2, \ldots, c_n)$ is a type, and an infinite list of variables of this type is provided: $F(c_1, c_2, \ldots, c_n), G(c_1, c_2, \ldots, c_n), \ldots$. A formation rule provides that, if $\mathbf{f}$ is of type $(c_1, c_2, \ldots, c_n)$, then $\mathbf{f}(\mathbf{x}_1, \mathbf{x}_2, \ldots, \mathbf{x}_n)$ is well-formed on condition that each variable $\mathbf{x}_i$ is of the type indicated by the corresponding integer c_i, *i.e.*, that $\mathbf{x}_i$ is an individual variable if c_i is 0, $\mathbf{x}_i$ is a propositional variable if c_i is 1 and $\mathbf{x}_i$ is an m-ary functional variable of lower type (one of the variables $F^m, G^m, H^m, F_1^m, \ldots$) if c_i is $m+1$. The values of the variables of type $(c_1, c_2, \ldots, c_n)$ are propositional functions of the kind which is indicated by this formation rule. And the resulting system, if the new variables of higher types occur only as free variables, and if again suitable axioms and rules of inference are provided, is the *pure functional calculus of third order*. Upon further modifying the formation rule iv, to allow the new variables of higher types to occur also as bound variables, the *pure functional calculus of fourth order* is obtained.

Then the next step is to adjoin functional variables of still higher types; *e.g.*, one of these new types is $(0, (2, 3), (3))$, and if f is a variable of this type, then $\mathbf{f}(\mathbf{x}_1, \mathbf{x}_2, \mathbf{x}_3)$ is well-formed on condition that $\mathbf{x}_1$ is an individual variable, $\mathbf{x}_2$ is a variable of type $(2, 3)$, and $\mathbf{x}_3$ is a variable of type (3).

In this way, by successive adjunctions of functional variables of higher and higher types—with appropriate changes, at each stage, in the formation rules, axioms, and rules of inference—the pure functional calculi of all finite orders are obtained. The logistic system which embraces all of these in a single system is the *pure functional calculus of order* ω.

A notation $=$, to express that two things are the same or identical, may be introduced by definition:

$$\mathbf{x} = \mathbf{y} \rightarrow (F^e) \,.\, F^e(\mathbf{x}) \supset F^e(\mathbf{y}),$$

where **x** and **y** are variables of the same type, and where the particular letter F is used, with type superscript e so chosen as to make $F^e(\mathbf{x})$ and $F^e(\mathbf{y})$ well-formed.

Finally, to the axioms of the pure functional calculus of order ω, there are added one special axiom, *the axiom of infinity*, and two special axiom schemata, which provide the *axioms of extensionality* and the *axioms of choice*.

The axiom of infinity is to the effect that the number of individuals is infinite. There are many ways in which an axiom having this as a consequence may be formulated in the notation of the system, of which the following (based on an idea of Kurt Schütte) is perhaps the briefest:

$$(\exists F)(x)(\exists y)(z) \,.\, {\sim}F(x,x)F(x,y) \,.\, F(z,x) \supset F(z,y)$$

The axioms of extensionality are that two n-ary propositional functions are identical if they hold among exactly the same ordered sets of n arguments:

$$\mathbf{f}(\mathbf{x},\mathbf{y},\cdots,\mathbf{u}) \equiv_{\mathbf{xy}\ldots\mathbf{u}} \mathbf{g}(\mathbf{x},\mathbf{y},\cdots,\mathbf{u}) \supset \mathbf{f} = \mathbf{g},$$

where $\mathbf{x},\mathbf{y},\cdots,\mathbf{u}$ is any ordered set of n different variables (which may be of various types, the same or different, n being any positive integer), and where **f** and **g** are variables of such type as to make $\mathbf{f}(\mathbf{x},\mathbf{y},\cdots,\mathbf{u})$ and $\mathbf{g}(\mathbf{x},\mathbf{y},\cdots,\mathbf{u})$ well-formed.

The axioms of choice, or multiplicative axioms as they are also called are:

$$\mathbf{c}(\mathbf{f})\mathbf{c}(\mathbf{g}) \supset_{\mathbf{fg}} [\mathbf{f}(\mathbf{x})\mathbf{g}(\mathbf{x}) \supset_{\mathbf{x}} \mathbf{f} = \mathbf{g}] \supset (\exists \mathbf{h}) \,.\, \mathbf{c}(\mathbf{f}) \supset_{\mathbf{f}} \,.\, \mathbf{f}(\mathbf{z}) \supset_{\mathbf{z}} (\exists \mathbf{y}) \,.\, \mathbf{f}(\mathbf{x})\mathbf{h}(\mathbf{x}) \equiv_{\mathbf{x}} \mathbf{x} = \mathbf{y},$$

where **c**, **f**, **g**, **h**, **x**, **y**, **z** are variables which must be no two the same, and of such types that the formula is well-formed.

For the simple theory of types without axiom of infinity a proof of consistency is possible which is similar to that given above for the functional calculus of first order, and which, although more complicated, employs only methods of essentially the same elementary character.

On the other hand it is a consequence of Gödel's incompleteness theorem (*see* the explanation below) that any proof of consistency of the system with axiom of infinity cannot be of such elementary character, but must employ methods of proof that in some sense surpass anything available in the simple theory of types itself.

The Set Theory of Zermelo can be described as obtained by adjoining additional axioms to a simple applied functional calculus of first order which has one binary functional constant ϵ and one individual constant Λ. The constant ϵ denotes the relation of membership in a set, so that $\epsilon(x,y)$ may be read as "x belongs to the set y" or "y is a set and x is a member of y." The constant Λ denotes the empty set (the set which has no members). The individual variables may in this theory be regarded as variables of unrestricted range, in the sense that the theory assumes the existence of no entities (sets or others) that cannot be values of the individual variables.

In writing well-formed formulas we shall abbreviate $\epsilon(\mathbf{x}, \mathbf{y})$ as $\mathbf{x} \epsilon \mathbf{y}$; *i.e.*, to state it as a definition: $\mathbf{x} \epsilon \mathbf{y} \rightarrow \epsilon(\mathbf{x}, \mathbf{y})$.

The notation $=$ may be introduced by a definition analogous to that given above in connection with the simple theory of types: $\mathbf{x} = \mathbf{y} \rightarrow \mathbf{x} \epsilon \mathbf{z} \supset_{\mathbf{z}} \mathbf{y} \epsilon \mathbf{z}$. A notation $\subset$, to express that one set is a part of (a subset of) another, may be introduced by a similar definition: $\mathbf{x} \subset \mathbf{y} \rightarrow \mathbf{x} \epsilon \mathbf{z} \supset_{\mathbf{z}} \mathbf{y} \epsilon \mathbf{z}$. In both cases **z** is to be chosen as the first individual variable (in alphabetic order) which is different from both **x** and **y**.

To provide for the notion of an ordered pair, a notation $\{x, y, z\}$ may be introduced to express that z is the ordered pair of x and y: $\{\mathbf{x}, \mathbf{y}, \mathbf{z}\} \rightarrow \mathbf{u} \epsilon \mathbf{z} \equiv_{\mathbf{u}} .[\mathbf{v} \epsilon \mathbf{u} \equiv_{\mathbf{v}} \mathbf{v} = \mathbf{x}] \vee . \mathbf{v} \epsilon \mathbf{u} \equiv_{\mathbf{v}} . \mathbf{v} = \mathbf{x} \vee \mathbf{v} = \mathbf{y}$, where **u** and **v** are to be chosen as the first two individual variables different from **x**, **y**, and **z**. Relations may then be dealt with in the theory by understanding a relation to be a set of ordered pairs.

The axioms of the theory are those of the functional calculus of first order, and in addition the following axioms and axiom schemata:

Axiom of extensionality: $z \epsilon x \supset . z \epsilon x \equiv_z z \epsilon y \supset x = y$.

Axiom of the empty set: $\sim x \epsilon \Lambda$.

Axiom of pairing: $(\exists t) . z \epsilon t \equiv_z . z = x \vee z = y$.

Axiom of summation of sets: $(\exists t) . z \epsilon t \equiv_z (\exists y) . z \epsilon y \cdot y \epsilon x$.

Axiom of the set of subsets: $(\exists t) . z \epsilon t \equiv_z z \subset x$.

Axioms of subset formation: $(\exists t) . z \epsilon t \equiv_z . z \epsilon x \cdot \mathbf{A}z$, where $\mathbf{A}z$ is a well-formed formula which may have z as a free variable but does not have t as a free variable.

Axiom of choice: $y \epsilon x \cdot z \epsilon x \supset_{yz} [u \epsilon y \cdot u \epsilon z \supset_u y = z] \supset (\exists t) . y \epsilon x \supset_y . w \epsilon y \supset_w (\exists v) . u \epsilon y \cdot u \epsilon t \equiv_u u = v$.

Axiom of infinity: $(\exists t) . \Lambda \epsilon t . x \epsilon t \supset_x (\exists y) . y \epsilon t . z \epsilon y \equiv_z . z \epsilon x \vee z = x$.

Axioms of replacement: $y \epsilon x \supset_y (\exists u)[\mathbf{A}yz \equiv_z z = u] \supset (\exists t) . z \epsilon t \equiv_z (\exists y) . y \epsilon x \cdot \mathbf{A}yz$, where $\mathbf{A}yz$ is a well-formed formula which may have y and z as free variables but does not have t or u as a free variable.

Axioms of excluded infinite regress: $(\exists x)\mathbf{A}x \supset (\exists x) . \mathbf{A}x . y \epsilon x \supset_y \sim\mathbf{A}y$, where $\mathbf{A}x$ is a well-formed formula which may have x as a free variable but does not have y as a free variable.

The way in which the theory seeks to avoid antinomy may be seen in particular in connection with the axioms of subset formation. An uncritical formulation might well have included the stronger axiom schema $(\exists t) . z \epsilon t \equiv_z \mathbf{A}z$, providing for the existence of a set t of those things z (sets and others) which satisfy an arbitrary given condition $\mathbf{A}z$. But this axiom schema would lead directly to the Russell antinomy, upon taking $\mathbf{A}z$ to be $\sim z \epsilon z$. The weaker schema actually used (the axioms of subset formation) provides only for the existence of a set t of those things z which belong to a previously given set x and satisfy the condition $\mathbf{A}z$. This is not known to lead to antinomy; but $\sim(\exists t) . z \epsilon t \equiv_z \sim z \epsilon z$ is a theorem.

Concerning the history of these axioms for set theory, *see* the paragraph about Zermelo in the article LOGIC, HISTORY OF. The axioms of replacement and of excluded infinite regress are additions to the original axiom system of Zermelo, and are sometimes omitted, as although they are independent and for some purposes important, there are also many purposes for which they are not needed.

On the other hand, if the axioms of replacement are retained in the form in which we have here stated them, they have the effect of rendering the axioms of subset formation non-independent; *i.e.*, the axioms of subset formation may then be omitted from the list on the ground that they can be proved as theorems by using the axioms of replacement.

An extension of the Zermelo set theory which is due to John von Neumann and Paul Bernays (*see* the paper of Bernays cited in the bibliography) adds, to the sets of the Zermelo theory, also classes which are not sets, *i.e.*, which cannot be members of other classes or sets. And for every condition **A**z expressed in the notation of the Zermelo theory there is a class of those elements z which satisfy **A**z—where an *element* is a set or anything capable of being a member of a set. (In particular there is, according to this extended theory, a class of elements z such that $\sim z \in z$, but not a set of such elements.)

Two other, different, systems of set theory are treated by W. V. Quine, one in his paper cited in the bibliography, and the other in the 1951 edition of his book. The latter, due jointly to Quine and Hao Wang, is related to Quine's system of 1937. These are not forms of the Zermelo theory but seek to exclude the antinomies by different means.

Gödel's Incompleteness Theorem.—In contrast with the completeness theorem for the pure functional calculus of first order, all known systems of type theory or set theory which are of sufficient strength to provide a logical foundation of mathematics are incomplete. We state this incompleteness theorem, due to Gödel, in the slightly stronger form which was given to it by Barkley Rosser. On the hypothesis that the system in question—it may be in particular either the simple theory of types or any of the systems of set theory mentioned above—is consistent, there is a well-formed formula **A** which is a sentence (hence without free variables) such that neither **A** nor $\sim$**A** is a theorem.

The proof of the metatheorem proceeds by constructing a particular well-formed formula **A**, and then showing that, if the system is consistent, **A** cannot be a theorem, and $\sim$**A** cannot be a theorem. However, it can also be shown that, if the system is consistent, the proposition expressed by **A** is true. More accurately, we can show, by means that are formalizable in the system itself, "If the system is consistent, then ____," filling the blank with a statement of the proposition that is expressed by **A**. Hence the further metatheorem follows that the consistency of the system cannot be proved by means that are formalizable in the system itself. (This brief statement of the matter is incomplete, and some essential explanations which are here omitted will be found in the article MATHEMATICS, FOUNDATIONS OF.)

Both the incompleteness theorem and the further theorem about the possibility of a consistency proof have the striking feature that they hold not only for a particular system but for any arbitrary extension of it that may be obtained by adjoining additional axioms or rules of inference or both, provided only that the requirement is satisfied that the axioms and rules of inference shall be effective (as described in the first part of this article, in explaining the notion of a logistic system). Having a particular system, which we suppose consistent, and having constructed the sentence **A** which is then true but not a theorem, we may indeed strengthen the system by adding **A** to it as an axiom. But the resulting stronger system is still incomplete: in it a particular sentence **B** may again be constructed by the same method, and it may be shown on the hypothesis of consistency that neither **B** nor $\sim$**B** is a theorem, and that the proposition expressed by **B** is true.

See also references under "Logic" in the Index.

BIBLIOGRAPHY.—Alfred Tarski *Introduction to Logic*, 2nd ed. (1946); S. C. Kleene, *Introduction to Metamathematics* (1952); Alonzo Church, *Introduction to Mathematical Logic* (1956); D. Hilbert and P. Bernays, *Grundlagen der Mathematik*, 2 vol. (1934, 1939); A. A. Fraenkel, *Abstract Set Theory*, 2nd ed. (1961); Paul Bernays, "A System of Axiomatic Set Theory," *Journal of Symbolic Logic*, vol. 2-19 (1937-54); Hao Wang and Robert McNaughton, *Les Systèmes axiomatiques de la théorie des ensembles* (1953); W. V. Quine, "Element and Number," Journal of Symbolic Logic, vol. 6 (1941), *Mathematical Logic*, rev. ed. (1951); Barkley Rosser, "An Informal Exposition of Proofs of Gödel's Theorems and Church's Theorem," *Journal of Symbolic Logic*, vol. 4 (1939), *Logic for Mathematicians* (1953); C. I. Lewis and C. H. Langford, *Symbolic Logic* (1932; reissued 1951); Ruth C. Barcan, papers on modal logic in *Journal of Symbolic Logic*, vol. 11 (1946) and vol. 12 (1947); Rudolf Carnap, "Modalities and Quantification," *Journal of Symbolic Logic*, vol. 11 (1946), *Introduction to Symbolic Logic and its Applications* (1958); Kurt Schütte, *Beweistheorie* (1960). *See* also LOGIC, HISTORY OF and bibliographies given there. (Ao. C.)

LOGIC, HISTORY OF.

IV. MODERN LOGIC

In this section we outline the development of deductive logic from the 16th to the middle of the 20th century. The history of inductive logic is treated in the articles INDUCTION; PROBABILITY; and SCIENTIFIC METHOD.

1. Ramée.—Pierre de la Ramée, or Petrus Ramus, whose anti-Aristotelian position was widely influential, is perhaps the first of the post-scholastic logicians who deserves mention here. One of the earliest noteworthy works on logic in a modern language is his *Dialectique* of 1555, though other works are in the traditional Latin. His *Dialecticae Libri Duo* of 1556, long used as standard (John Milton's *Artis Logicae Plenior Institutio*, 1672, is, *e.g.*, but a revised edition of it), has a much modified and abbreviated version of the older logic, omitting the immediate inferences altogether and greatly simplifying the rules of the categorical syllogism. In an altered classification into moods or kinds, a special place is provided for syllogisms having two singular premisses, including not only the two moods of the expository syllogism (*see* LOGIC) but inferences involving a singular predicate, such as: Octavius is heir of Caesar, I am Octavius, therefore I am heir of Caesar; Judas that wrote the Epistle was the brother of James, Judas Iscariot was not the brother of James, therefore Judas Iscariot was not the Judas that wrote the Epistle.

Propositions with singular predicates had been treated by Ockham, Pseudo-Scotus, and other scholastics—as well as the expository syllogism, by some as confined to the third figure, so that the singular middle term occurs only as subject, but by others as admitting moods also in other figures. And indeed Pseudo-Scotus already maintained that syllogisms with singular premisses should be brought into the Aristotelian form by rereading in the following way: Everything which is Octavius is heir of Caesar, everything which is I is Octavius, therefore everything which is I is heir of Caesar. The Ramistic treatment of these singular inferences as non-Aristotelian was later countered on similar grounds by Thomas Spencer (*The Art of Logick delivered in the precepts of Aristotle and Ramus*, London, 1628) and, more sharply, by the mathematician John Wallis (*q.v.*) in his Cambridge thesis of 1639, published in 1643, and in his *Institutio Logicae* of 1687. Wallis reached the conclusion that singular propositions are to be assimilated to universal propositions for purposes of the syllogism, on the ground that the predicate of a singular proposition and of a universal proposition are alike predicated of all, of the whole rather than a part of the subject. Following, not necessarily Wallis, but earlier writers generally, the authors of the Port-Royal Logic (*see* below) give a similar account of singular propositions, so that it is said that all propositions can be reduced to the four forms A, E, I, O. And this remained the generally accepted doctrine until the traditional point of view was superseded by the development of modern logic in the 19th and 20th centuries.

2. Jungius.—Joachim Jungius, in his *Logica Hamburgensis* of 1638, brought forward and discussed a number of forms of inference that are not reducible to the traditional immediate inferences and syllogisms. Relations especially are involved in these, and if widespread attention and study had been given to the matter, the logic of relations might have been developed two centuries in advance of the actual event. But Jungius remained without important influence, except upon Leibniz, and his contributions to logic were ignored by his successors or (*e.g.*, by Wallis) dismissed as reducible to the traditional doctrine.

One of these forms of inference which Jungius considers is that of the so-called *oblique inferences*, which had already been treated by Aristotle. An example is: The square of an even number is even, 6 is an even number, therefore the square of 6 is even. The significance of this, as it appeared in the light of much later developments, may be seen by attempting to reduce to traditional syllogistic form as follows: Every even number has an even number as square, 6 is an even number, therefore 6 has an even number as square. In this way the conclusion has 6 as its subject. And if we wish then to infer, in another syllogism, that 36 is the square of 6 and therefore 36 is even, we must first transform the conclusion so that it will have as its subject, the square of 6. The inference which is missing from the traditional catalogue of forms is the immediate inference (if we wish to call it that) from the proposition, 6 has an even number as square, to the proposition, the square of 6 is an even number.

To make the issue clearer we may put the inferences into modern notation, using $e(x)$ to mean that x is even, and $s(x, y)$ to mean that the square of x is y. The first inference is then from the premisses $e(x) \supset_x . s(x, y) \supset_y e(y)$ and $e(6)$ to the conclusion $s(6, y) \supset_y e(y)$, and the second inference then uses the additional premiss $s(6, 36)$ to draw the conclusion $e(36)$. Relational inferences of this sort—here illustrated in a case so simple as to be almost trivial—are as a matter of fact essential to nearly all important mathematical reasoning.

Other examples of Jungian relational inferences are the following. A circle is a figure, therefore who draws a circle draws a figure: $c(x) \supset_x f(x)$, therefore $c(x) \wedge_x d(x, y) \supset_y . f(x) \wedge_x d(x, y)$. A reptile is an animal, therefore Who created every animal created every reptile: $r(x) \supset_x a(x)$, therefore $a(x) \supset_x C(x, y) \supset_y . r(x) \supset_x C(x, y)$. David is the father of Solomon, therefore Solomon is the son of David; *i.e.*, upon supplying a tacit premiss $f(X, y) \supset_{xy} s(y, x)$, $f(S, D)$, therefore $s(D, S)$. (This last had been considered also by Galen.)

3. Geulincx.—Arnold Geulincx published in 1662 *Logica Fundamentis Suis, a quibus hactenus collapse fuerat, Restituta*. Though the title indicates the author's intention to restore neglected fundamentals, this is not just a reproduction of the scholastic logic but an original work. Following are some points of special interest. The doctrine of *suppositio*, largely ignored by other logicians of the period, is here reproduced in altered terminology and with some alteration in content; the possibility of using the same term with different kinds of *suppositio* is treated as being (with a suitable exception to allow for syllogistic inference) a form of equivocacy, and the device is mentioned in particular of underlining in manuscript or using a distinct style of type in print, to distinguish "grammatical" (i.e., material) *suppositio* and "logical" (i.e., simple) *suppositio* from others. It is argued that negation may be applied only to propositions, and not to terms. Detailed rules (not new except in viewpoint) are given for the reduction of negations, including what are now known as the De Morgan laws of the propositional calculus, the law of double negation, and such reductions as that of *not all not* to *some*, *i.e.*, of $\sim(x)\sim$ to $(\exists x)$. Treatment of the categorical syllogism is brief but adequate, and is supplemented by mention of the inference known as the antisyllogism (example: Peter is not an animal, therefore either Peter is not a man or some man is not an animal). What would now be called valid inferences of the propositional calculus are especially well handled, in spite of the handicap of doing it entirely in words, without the aid of special symbols other than the use of capital letters to stand for propositions Disjunction is explained clearly as inclusive, and *modus ponendo tollens* is rejected (in contrast with the uncritical acceptance of this inference by Ramus and by the Port-Royal Logic). The copulative syllogism (*see* LOGIC) is treated by use of the appropriate De Morgan law to reduce it to *modus tollendo ponens*. No reduction of the negation is given in the case of conditional propositions, but negations of such propositions are considered, and the following inferences are allowed: from $\sim . \mathbf{A} \supset \mathbf{B}$ to $\sim . \sim\mathbf{B} \supset \sim\mathbf{A}$; from $\mathbf{B} \supset \mathbf{C}$ and $\sim . \mathbf{A} \supset \mathbf{C}$ to $\sim . \mathbf{A} \supset \mathbf{B}$; from $\mathbf{A} \supset \mathbf{B}$ and $\sim . \mathbf{A} \supset \mathbf{C}$ to $\sim . \mathbf{B} \supset \mathbf{C}$; from $\mathbf{A}\sim\mathbf{B}$ to $\sim . \mathbf{A} \supset \mathbf{B}$.

Geulincx considers at one point the inference: Every white man is white, every white man is a man, therefore some man is white. He notes that the premisses are necessary, but the conclusion contingent. And he rejects the inference on the ground that the word "white" is used equivocally—in the conclusion to refer to the present time, and in the premisses absolutely, *i.e.*, without reference to time—so that the syllogism has four terms. The weakness of this is that there are still only three terms in the premisses and on the traditional doctrine a conclusion should follow in Darapti. But it remained for Leibniz some decades later to give more serious consideration to the problem of existential import.

4. Port-Royal Logic.—Port-Royal Logic or *La Logique ou l'Art de Penser*, the original title, was written by Antoine Arnauld and Pierre Nicole, possibly with the collaboration of some others, and was first published in 1662. It was, like Ramus, long and widely used (both Latin and English translations appeared in England) and though the authors follow Ramus in the fourfold division of logic into idea, judgment, reasoning and method, they restore the Aristotelian-scholastic treatment of opposition, conversion and the categorical syllogism.

An important original contribution of the Port-Royal Logic is the distinction between *comprehension* and *extension*—*of ideas*, as the authors say—or *of concepts*, or *of terms*, as is more usually said by later writers. The comprehension of a concept consists namely of all those attributes which are contained in it and cannot be removed from it without destroying it—as, *e.g.*, the authors say, the idea (concept) of a triangle contains extension, figure, three sides, three angles, the equality of the sum of these three angles to two right angles, etc. And the extension of a concept consists of the subjects to which it extends, or of what were traditionally called the inferiors of a general term—so that, e.g., the extension of the concept of a triangle consists of all the different species of triangles: isosceles, right, obtuse, etc. This distinction was widely adopted by later writers and became a standard part of the traditional logic. Along with Mill's connotation and denotation of names it may be thought of as a forerunner of Frege's distinction of sense and denotation (*see* SEMANTICS IN LOGIC; CONNOTATION; and DENOTATION).

The Port-Royal Logic also contains the first publication of the ideas of Blaise Pascal (*q.v.*) about the nature of definition. Namely it was Pascal who first observed that every demonstrative science must begin, not only with unproved propositions (*see* AXIOM, and POSTULATE) but also with undefined terms, or *termes primitifs* as they are called in the Port-Royal Logic. Pascal also restricts definitions in mathematics to nominal definitions, *définitions de nom*, which consist merely in the giving of a name to something which has been clearly designated in terms completely known, the use and purpose of such definitions is the abbreviation of discourse; they are entirely free, and never subject to being contradicted; and, as Pascal wrote in a letter to Le Pailleur, one may equally well define something impossible as something actual. These ideas were taken up by the authors of the Port-Royal Logic, who lay great emphasis on the need for careful definition as a means of avoiding confusion of thought. However they retain also the notion of a real definition, or *définition de chose*, as a proposition which explains the nature of a thing by means of its essential attributes, and which is therefore capable of being confirmed or disputed. And the idea, which arose much later, that primitive terms and axioms (or postulates) are, like nominal definitions, arbitrary, is not to be found in the Port-Royal Logic; on the contrary the authors demand, following Pascal, that primitive terms shall be completely known, and axioms completely evident.

5. Saccheri.—Giovanni Girolamo Saccheri's *Logica Demonstrativa* of 1697 and 1701 also requires mention, though the author is (deservedly) better known for the partial anticipation of non-Euclidean geometry in his *Euclides Vindicatus* of 1733 (*see* GEOMETRY, NON-EUCLIDEAN). The *Logica* provides a treatment of the traditional logic in the form of a series of demonstrations based on postulates and definitions, in the manner of works on geometry. Two points are emphasized which afterward played an important role in the *Euclides Vindicatus*, namely: (1) the law of Clavius (*see* LOGIC) and the associated method of proving a proposition by showing it to be a consequence of its own negation; and (2) the distinction between nominal definitions and real definitions, on the basis that the former merely state the meaning of a term, whereas the latter carry also the assertion of existence.

(1) is the *consequentia mirabilis* of 17th-century scholastic writers. Geronimo Cardano (*q.v.*) writes in his *De Proportionibus* of the striking character of this method of proof, which he thought to be his own discovery. Knowledge of the method came to Saccheri through Christopher Clavius, by whose name the law is sometimes known, and who, in commentaries on Euclid and on Theodosius, points to its use by Euclid, by Theodosius, and by Cardano. But Saccheri was the first to exploit this method of proof systematically—in his *Logica* and in *Euclides Vindicatus*.

Significant in (2) is that definitions are always of terminology, and real definitions are distinguished from nominal by the added demonstration or assumption of existence. Otherwise Saccheri had been anticipated by Pascal and in the Port-Royal Logic, in regard to the nature of nominal definitions and the fallacy of misusing a nominal definition in the role of a real definition. (But Saccheri, in his Latin terminology and in his notion of nominal definition, is evidently following Ockham and other scholastics rather than Pascal.)

6. Leibniz.—Gottfried Wilhelm von Leibniz (*q. v.*; Leibnitz) brought forward already in a work published in his youth (*Dissertatio de Arte Combinatoria*, 1666) the project of constructing a universal exact system of notation, a symbolic language, in which all concepts would be so analyzed into their ultimate constituents, by the notation itself, that it would be the means for a fundamental knowledge of all things. This project of a *lingua characteristica universalis* was partly based on the ideas of Lull, to whom Leibniz refers. Leibniz continued to advocate it throughout his life, later adding as an important part the project of a *calculus ratiocinator*, or calculus of reasoning. And though the total program of Leibniz, as just described, is no doubt not even theoretically fulfillable, the *calculus ratiocinator* remains an important forerunner of the logistic method in logic.

Leibniz's contributions to logic remained largely unpublished during his lifetime. The most important waited publication much more than a century. But Raspe's *Oeuvres Philosophiques de Leibniz*, published in 1765, contains a purely descriptive paper about the *lingua characteristica universalis*, a paper which treats the difficulty concerning existential import, and the *Nouveaux Essais sur l'Entendement Humain*.

In the paper on existential import, Leibniz is led to hold that all four of the categorical forms *A*, *E*, *I*, *O* are to be understood on the basis of a tacit presupposition that the terms which enter are existent (are not empty). This Leibnizian solution of the difficulty has often been adopted since; and if class variables are used to formulate the traditional logic of categorical propositions, the solution may be put in the form of restricting the range of values of the variables to *non-empty* classes. There is, however, the objection that this seriously reduces the applicability of the traditional logic, as it may be impracticable to ascertain in advance that all terms are existent, and syllogistic reasoning may serve a substantial purpose even in cases in which it is possible or probable that some of the terms are empty.

In a passage in the *Nouveaux Essais*, Leibniz discusses the relational inferences of Jungius, correctly remarking that they cannot be reduced to any traditional syllogism except by a *changement des termes* which renders the total process of inference asyllogistic (*see* above).

The same work also contains a statement of the law of identity in the form, "*A* is *A*," or "All *A* is *A*." This law must not be credited exclusively to Leibniz—since Leibniz refers to its use by Ramus to treat the Aristotelian conversions as special cases of categorical syllogisms—and since indeed this possibility was known in the 13th century, and there is mention already by Boëthius of what John Locke (in 1690) called "identical propositions." But it was Leibniz who first ascribed to the law of identity a special status as a "primitive truth of reason."

In many of his works Leibniz seems to overestimate the importance of one or both of the laws of contradiction and identity. In the *Nouveaux Essais* Leibniz avoids saying that these laws are alone a sufficient basis for the whole of logic, and even may be thought to imply the contrary. But elsewhere he is less cautious, especially in writings that were not intended for publication. Thus in *Réflexions sur l'Essai de l'Entendement Humain de Mr. Locke* (published with Locke's letters in 1708) and in the reply to the first letter of S. Clarke (published in 1717 after Leibniz's death) it is said that the law of contradiction, so stated as to include the law of identity, suffices to demonstrate either all truths independent of experience or all principles of mathematics. It would seem that Leibniz at times really hoped that all necessary truths might be demonstrably reduced to these very simple ones—though indeed with the aid of certain principles of inference, especially the syllogism in Barbara (*cf.* the letter quoted by J. E. Erdmann, vol. 1, p. 81). But some of his successors made the idea into an unsupported and almost meaningless item of doctrine—*see* THOUGHT, LAWS OF.

Leibniz's actual attempts at the construction of a *calculus ratiocinator* were made in the period from 1679 to 1690 and are published, some in Erdmann's *Opera Philosophica* (vol. 1, 1840) and C. I. Gerhardt's *Philosophische Schriften* (vol. 7, 1890), the remainder in L. A. Couturat's *Opuscules et Fragments Inédits* (1903). These show beginnings from which the modern treatment of logic might well have developed. But their content remained unknown for 150 years, and historically only the generalities of Leibniz's program exerted any influence.

In the century following Leibniz there were many attempts at a logical calculus, of which the most widely known and discussed were those of Gottfried Ploucquet (in 1763) and of Johann Heinrich Lambert (*q.v.*) (in 1767, and in a number of fragments published in *Lamberts Logische und Philosophische Abhandlungen*, vol. 1, 1782). Lambert's are noteworthy as containing some beginnings of a logic of relations. But none of the attempts produced a satisfactory calculus, and the main trend of opinion was against this direction.

7. Euler.—Leonhard Euler (*q.v.*), in his *Lettres à une princesse d'Allemagne* (vol. 2, 1770), illustrated his treatment of the categorical syllogism by using the interiors of three circles to represent the minor term, the middle term and the major term. Thus $\mathbf{s}(x) \supset_x \mathbf{m}(x)$ is pictured by showing the circle for **s** entirely within that for **m**, and $\mathbf{s}(x) \supset_x \sim\mathbf{m}(x)$ by showing the two circles as nonoverlapping, etc. This method of visually checking the validity of syllogisms was brought into general use through its adoption by Euler, and came to be known as the *Euler diagram*, though the device did not originate with Euler. Such circle diagrams had been employed in special cases by Johann Christoph Sturm (*Universalia Euclidea*, 1661); and they were used by Leibniz to treat the categorical syllogisms systematically, in a fragment not published until 1903; but their first systematic use for this purpose in a published treatise seems to have been by Johann Christian Lange (*Nucleus Logicae Weisianae*, 1712).

8. Kant.—Immanuel Kant (*q.v.*) contributed little to logic. Indeed it was his opinion that logic had made no important step either forward or backward since Aristotle, and seemed to all appearance to be finished and complete (*Kritik der reinen Vernunft*, preface to 2nd ed., 1787). But his influence was great because of his reputation in other fields. In particular the general acceptance of the term, *analytic*, for propositions that are true on logical grounds alone is traceable to Kant, although Kant's own definition of the term would restrict it to a narrow subclass of such propositions, and although the term, *analytic*, and its opposite, *synthetic*, had been used already by Christian August Crusius in 1747. The now familiar contention that "existence is not a predicate" is due to Kant who used it (*op. cit.*) as an objection against the so-called ontological proof of the existence of God; but a satisfactory positive analysis of the notion of existence had to await the introduction of the quantifiers by Frege and Mitchell.

9. Hegel.—Georg Wilhelm Friedrich Hegel (*q.v.*), in *Wissenschaft der Logik* (1812–16), denounces the Leibnizian project of a universal symbolic language as shallow and senseless, and singles out for special attack Ploucquet's recommendation of his calculus as making possible the mechanical performance of logical inference without danger of error if the rules of the calculus are followed. Hegel is similarly critical of the Euler diagrams, and even of the long-established formal treatment of the syllogism as it appeared in the traditional logic of his day. It is not without some justice that he reproaches the latter with being in an ossified and contemptible state. But Hegel represents, as an extreme example, the tendency which long prevailed to hold logic itself in low esteem and to devote the greater part of a work on logic to other subjects, especially to topics in epistemology and metaphysics that bear upon the traditional logic or are suggested by it.

10. Bolzano.—Bernard Bolzano's *Wissenschaftslehre* of 1837 contains many original contributions of which the importance was long overlooked, and its proper place in the history of logic came to be seen only in the light of much later developments. In this brief account we confine attention to a single point, Bolzano's treatment of the notion of analyticity.

Bolzano introduces notions of analyticity in a *wider* and in a *narrower* sense; but of these only the latter is free from serious objection, and will be described here. It will be convenient to state Bolzano's definition with respect to a formalized language, though it must be remembered that Bolzano did not have this means available, and such restatement of his definition gives it an appearance of greater rigour than was possible for Bolzano himself. We may take this language to be the set theory of Zermelo (*see* LOGIC), with a large number of individual constants added, corresponding to the words of various kinds that appear in an English dictionary, so that the language becomes (theoretically) usable for purposes of ordinary discourse as well as for expressing propositions of pure logic and of mathematics. Also we state the definition as for an *analytically true sentence*, although Bolzano deals rather with the proposition ("Satz an sich") expressed by a sentence; and we employ the terminology explained in the article LOGIC, instead of Bolzano's own. From any sentence let a corresponding propositional form be obtained by replacing every extra-logical constant (Bolzano says extra-logical concept) by a variable, two or more occurrences of the same constant being replaced always by the same variable, and different constants by different variables. A sentence is then analytically true (in the narrower sense) if either: (1) the corresponding propositional form is true for all values of the variables; or (2) the sentence can be reduced to one satisfying condition (1), by a series of steps which consist in replacing an occurrence of an individual constant by a synonymous constant or a well-formed part by a synonymous well-formed part.

This definition is not without its difficulties. In particular no explication is offered of the notions of synonymy and of being true for all values of the variables (or satisfied by all values of the variables—*see* SEMANTICS IN LOGIC), but these notions are taken for granted; indeed the notion of synonymy is not explicitly present at all, but is implicit in the treatment of propositions rather than sentences, and its role is indicated only by some examples. Bolzano himself calls attention to the possibility of dispute as to which constants, or concepts, shall be recognized as logical and which as extra-logical (though in the case of the particular language which we have here selected for the purpose of illustration, it might be possible so to control the vocabulary of added individual constants that all constants but Λ and ϵ would be clearly extra-logical). In spite of such difficulties, Bolzano must be credited with having proposed the first definition of a distinction between analytic and synthetic propositions that deserves serious consideration from the point of view of its logical adequacy.

A minor point is that Bolzano completed Kant's classification of propositions into analytic (analytically true) and synthetic by adding a third category of the analytically false.

11. Mill.—John Stuart Mill's *A System of Logic, Ratiocinative and Inductive* (1843) is remembered for its contributions to inductive logic, which are outside the scope of this article, and for its introduction of the distinction of CONNOTATION and DENOTATION, which is treated in the articles of those titles.

12. Algebra of Logic.—The algebra of logic had its beginning in publications of George Boole (*q.v.*) and Augustus De Morgan (*q.v.*) which appeared simultaneously in 1847. There are two main divisions. —The algebra of classes has three basic operations, the logical sum (or union) $F + G$ of two classes F and G, the logical product (or intersection) FG of F and G, and the complement F' of a class F. In the notation introduced in the article LOGIC, $F + G$ may be explained as meaning $\lambda x[F(x) \vee G(x)]$, FG as meaning $\lambda x[F(x)G(x)]$, and F' as meaning $\lambda x{\sim}F(x)$. The notation 0 was used for the empty class; and 1 for the universal class, *i.e.* the class which coincides with the domain of individuals so that all individuals belong to it. And equations and inequalities were written, such as $G + H = F' + G'$ to mean that the logical sum of G and H is the same as the logical sum of the complements of F and G, or $FG \neq 0$ to mean that the logical product of F and G is not empty, or $F \leqq G$ to mean that F is contained in G in the sense that all individuals belonging to F belong also to G.—The algebra of relations has six basic operations, the logical sum $F + G$ of two relations F and G, the logical product FG, the contrary F' (or $-F$), the relative sum $F \dagger G$, the relative product $F{:}G$, and the converse $\breve{F}$. These may be explained as meaning respectively $\lambda x \lambda y[F(x, y) \vee G(x, y)]$, $\lambda x \lambda y[F(x, y)G(x, y)]$, $\lambda x \lambda y{\sim}F(x, y)$, $\lambda x \lambda y(z)[F(x, z) \vee G(z, y)]$, $\lambda x \lambda y(\exists z)[F(x, z)G(z, y)]$, $\lambda x \lambda y F(y, x)$. The notation 0 was used for the empty relation, 1 for the universal relation, $0'$ for the relation of diversity (which holds between x and y if and only if x and y are different individuals), $1'$ for the relation of identity. And again equations and inequalities may be written.

The three basic operations of the algebra of classes, and the six basic operations of the algebra of relations, obey laws which are of much the same kind as familiar laws of the algebra of numbers (and in part coincide with them). By using these a formal algebra or calculus of classes may be set up, and an algebra of relations. These were the first successful calculi of logic. As treated by 19th-century writers, neither is yet a logistic system in the sense of modern logic, but a calculus in a less rigorous sense.

Various notations were used by different writers. And when the two algebras were incorporated into *Principia Mathematica* (see below) as parts of the logistic system of that work, the authors, partly following Giuseppe Peano, changed the old notations completely. In the algebra of classes the logical sum, logical product, and complement are expressed by $F \cup G$, $F \cap G$, and $-F$ respectively, and the notations 0, 1, $\leqq$ are changed to Λ, V, $\supset$ respectively. In the algebra of relations the same notations are used with a dot added, as $F \dot{\cup} G$, etc.

The method of Boole, in 1847 and in 1854, is not an algebra of classes in the sense described above, but an application of ordinary numerical algebra to the logic of classes—as is possible if the logical sum of F and G is written as $F + G - FG$, and the complement of F is written as $1 - F$. Yet Boole was able to obtain in this way the essential results of the algebra of classes, and indeed worked them out more fully than De Morgan. In De Morgan's *Formal Logic* of 1847 are, however, some beginnings of the algebra of classes in the more proper sense, including in particular the De Morgan laws $(FG)' = F' + G'$, $(F + G)' = F'G'$ (it was only later that De Morgan's name came to be applied also to the analogous laws of the propositional calculus). From these beginnings the algebra of classes developed into its classical form through contributions by William Stanley Jevons (*q.v.*) (*Pure Logic*, 1864, and later works), Charles Sanders Peirce (in a series of papers beginning in 1867), Ernst Schröder (*Der Operationskreis des Logikkalkuls*, 1877), John Venn (*Symbolic Logic*, 1881), and Platon Poretsky (in papers published in the period from 1884 to 1908).

The algebra of relations had its beginnings in publications by De Morgan (*Syllabus of a Proposed System of Logic*, 1860, and a paper in the *Transactions of the Cambridge Philosophical Society*, vol. 10, 1864) and received its major development at the hands of Peirce (in papers beginning in 1870) and Schröder (in *Algebra der Logik*, vol. 3, 1895).

The standard reference work on the algebra of logic is Schröder's three-volume *Algebra der Logik* (1890-1905). However, Schröder's axiomatic basis of the algebra is deficient, and should be replaced, in the case of the algebra of classes by E. V. Huntington's (*Transactions of the American Mathematical Society*, vol. 5, pp. 288–309, 1904), and in the case of the algebra of relations by Tarski's (*The Journal of Symbolic Logic*, vol. 6, pp. 73–89, 1941).

There is a sense in which the algebra of logic has also a third division, the algebra of propositions, though not always clearly distinguished from the algebra of classes. Boole had already considered the alternative of interpreting the variables of his algebra as propositional variables instead of class variables. But the first true calculus of propositions appears in papers of Hugh MacColl, beginning in 1877; and in particular it was MacColl who rediscovered the so-called De Morgan laws of the propositional calculus (known already to the scholastics). The second volume of Schröder's *Algebra* has a combined algebra of classes and propositions.

13. Peirce.—Charles Sanders Peirce (*q.v.*), besides the matters already mentioned, has the credit of having taken the first steps in many things which afterward became important in the development of logic. These include the first definition of the notion of simple order (1881); the first treatment of the propositional calculus as a calculus of two truth-values (1885); the definition of $=$ (1885) which is given above in the article LOGIC (though this had been partly anticipated by Leibniz's informal definition, "Things are identical of which one can be substituted for the other with preservation of truth"); and the definition of finiteness (1885) which would be expressed in modern notation as follows: G is a finite class if $(F) \mathrel{.} F(x,y)F(x,z) \supset_{xyz} y = z \supset \mathrel{.} F(x,z)F(y,z) \supset_{xyz} x = y \supset \mathrel{.} G(x) \supset_x (\exists z)[G(z)F(x,z)] \supset \mathrel{.} G(z) \supset_z (\exists x)[G(x)F(x,z)]$. Peirce also initiated in 1881 the method of treating the foundations of arithmetic which was afterward developed by Julius (Wilhelm Richard) Dedekind (*Was sind and was sollen die Zahlen?*, 1888) and Giuseppe Peano (*Arithmetices Principia*, 1889).

The *Insolubilia* were discussed by Peirce (1869, 1901) with direct reference to the medieval sources. It is quite possibly through Peirce that the Liar first came to the attention of Bertrand Russell, but it required the independent discovery of Richard's antinomy (*see* below) to make clear the importance of what are now known as the *semantical* antinomies or paradoxes.

In a paper of 1880, in treating the categorical syllogism, Peirce says that traditionally the affirmative (categorical) propositions imply that their subjects are existent, while the negative ones do not; but he will assume rather that particular propositions imply the subjects existent, universal propositions not. Indeed this change of the convention about existential import is strongly indicated from the point of view of the algebra of classes; it was implicit in a short note published by Arthur Cayley in 1871, but was first made explicit from this point of view by Peirce, and by Venn in 1881. It is true that under the changed convention, certain traditional inferences (*e.g.*, conversion *per accidens*, syllogism in Darapti) require an added existential premiss to validate them. But the convention which Peirce describes as the older one also renders some of the immediate inferences invalid without an added existential premiss, those namely (including contraposition) that involve obversion of a negative proposition.

14. Brentano.—Franz Brentano, in his *Psychologie vom Empirischen Standpunkte* of 1874, made it a part of his psychology of judgment that every proposition can be reduced to an (affirmative or negative) existential proposition, so that, *e.g.*, "Some man is ill" and "All men are mortal" are said to have the same sense as "An ill man is" and "An immortal man is not" respectively. In modern notation the four categorical forms A, E, I, O thus become $\sim(\exists x) \centerdot \mathbf{s}(x){\sim}\mathbf{p}(x)$, $\sim(\exists x) \centerdot \mathbf{s}(x)\mathbf{p}(x)$, $(\exists x) \centerdot \mathbf{s}(x)\mathbf{p}(x)$, $(\exists x) \centerdot \mathbf{s}(x){\sim}\mathbf{p}(x)$. Hence quite independently of the algebra of logic (to which he was opposed) Brentano arrived at the same doctrine of existential import which was later introduced by Peirce as described above, and at the same modification of the traditional logic, including rejection of the syllogisms in Darapti, etc. But Brentano seems to defend his own revision of the traditional rules as the only right one, whereas Peirce sees more clearly that it is a matter of choosing the convention under which the formal treatment of logic can proceed most efficiently.

An exposition of Brentano's logical innovations was published by J. P. N. Land in the first volume of the British periodical *Mind* in 1876. And a detailed treatment of immediate inference and the syllogism from Brentano's point of view is in Franz Hillebrand's *Die Neuen Theorien der Kategorischen Schluesse* of 1891.

15. Frege.—Gottlob Frege is the founder of modern logic in a sense in which neither Leibniz nor De Morgan and Boole can be so considered, and as such is unquestionably the greatest logician of modern times. The essential steps in the introduction of the logistic method were taken in his *Begriffsschrift* of 1879. In the same work there appear for the first time the propositional calculus in its modern (logistic) form, the notion of a propositional function, the use of quantifiers and the logical analysis of proof by mathematical induction in terms of the notion of a hereditary property—or hereditary class, as we shall here prefer to say.

This last was important in Frege's definition of an inductive cardinal number, and thus provided the basis for the derivation of arithmetic from logic which was described and defended in his *Grundlagen der Arithmetik* (1884) and carried through rigorously in the first volume of his *Grundgesetze der Arithmetik* (1893). To state the definition in the slightly modified form in which it was later used by A. N. Whitehead and Bertrand Russell in *Principia Mathematica*, let us call two classes G and H similar if there is a one-to-one correspondence between them, *i.e.*, if $(\exists F) \centerdot F(x, y)F(x, z) \supset_{xyz} y = z \centerdot F(x, z)F(y, z) \supset_{xyz} x = y \centerdot G(x) \supset_x (\exists z)[H(z)F(x, z)] \centerdot H(z) \supset_z (\exists x)[G(x)F(x, z)]$. As the cardinal number of a class G, we might think to take the concept, or property, of similarity to G. But if a property is understood (as usual) in intension, this has the defect that the cardinal numbers of two different but similar classes will not be actually identical. Hence we are led to take the corresponding extension, *i.e.*, to define the cardinal number of G as the class of classes which are similar to G. The cardinal number of H is successor of the cardinal number of G if, for some class H_1 similar to H, $(\exists y) \,.\, H_1(y) \,.\, G(x) \equiv_x \,.\, H_1(x) \sim x = y$; a class of cardinal numbers is hereditary if all successors of cardinal numbers belonging to it also belong to it, the cardinal number of the empty class Λ is the number 0, and the inductive cardinal numbers (the non-negative integers) are those which belong to every hereditary class to which 0 belongs.

From an antinomy due to Georg Cantor, concerning the question of the greatest (infinite) cardinal number, Russell extracted the simpler antinomy which is now known by Russell's name (*see* ANTINOMY), and communicated it to Frege in 1902. As the antinomy can be put into the form of a demonstration of contradictory theorems from Frege's logical axioms, Frege felt that one of the foundations of his construction had been shattered. Indeed it must have seemed to Frege and his contemporaries that his great work was a failure. For although Frege's axioms can be amended to remove the inconsistency—and Frege himself made some suggestions in this direction in the Appendix to the second volume of his *Grundgesetze der Arithmetik* (1903)—it must then seem that the axioms have been artificially designed for a purpose, and it is less easy to maintain that they have been taken once for all from a realm of eternal truth. It is only in the perspective of time that we are able to see that Frege's positive contributions far outweigh the inconsistency of the particular system in which they were embodied.

Independently of Frege, the use of quantifiers was suggested also by O. H. Mitchell, to whom the idea is credited by Peirce in his paper of 1885.

Other important contributions of Frege are the distinction of *sense* and *denotation* (*see* SEMANTICS IN LOGIC), and the device of systematically employing quotation marks to distinguish the *mention* of a term, expression or symbol from its *use*. The latter distinction is in a sense obvious, and was already recognized among the medieval distinctions of *suppositiones*; yet failure to observe it has been the source of much confusion, in Frege's day and since.

16. Venn.—John Venn, besides contributions already mentioned, introduced in 1880 the *Venn diagram*. This is a modification of the Euler diagram (*see* above) in which the three circles are so drawn as to overlap in all possible ways, thus dividing their plane into eight regions; and to represent the given premisses, some of these regions are shaded as a sign that the classes they stand for are known to be empty, and others are marked with a star as a sign of non-emptiness. If an inference is to be dealt with (not necessarily syllogistic) that involves n basic classes, $n > 3$, the corresponding regions may be drawn as ellipses or other more complicated shape, instead of circles, and so placed as to divide their plane into 2^n regions; and then the same process of shading and starring is followed.

17. Peano.—Giuseppe Peano is important for his influence on Russell, and for his contribution in devising a scheme of logical notation which is more convenient than those of Peirce, Schröder and Frege, and much of which (through its adoption by Russell) is still in use. Peano's postulates for arithmetic (*see* POSTULATE) are due to Dedekind in all but the point of view of taking them as postulates. His *Formulaire de mathématiques*, written with the aid of collaborators and published in five "volumes" (or editions) from 1894 to 1908, is intended as a compendium of mathematics, developed from its postulational beginnings with the aid of Peano's logical notation. His treatment of logic had begun to go beyond the algebra of logic but he did not yet have the logistic method.

18. Burali-Forti.—Cesare Burali-Forti stumbled upon the antinomy which now bears his name (*see* ANTINOMY), in a paper of 1897 in which he seems to have been not aware that he had revealed antinomy at all.

This antinomy and that of the greatest cardinal number were known already to Cantor but had not been published, and it was through Burali-Forti's paper that there first came to general attention the threat to the foundations of mathematics that is constituted by the antinomies.

19. Russell.—Bertrand Russell (*q.v.*) adopted Frege's thesis that arithmetic is a branch of logic, in the sense that all the terms of arithmetic can be defined with the aid of logical terms only (*see* the brief statement above as to how, in part, this is to be done), and all the theorems of arithmetic can be proved from logical axioms only. And the same thesis was extended by Russell to the whole of mathematics—a doctrine which later came to be known as logicism. The project of carrying this out is the subject of Russell's *The Principles of Mathematics* (1903). The theory of types as a means of avoiding the inconsistency of such a system as Frege's, though the idea of it was discussed in the *Principles*, was first put into satisfactory form by Russell in a paper of 1908, and on this basis the detailed logicistic development of a large part of mathematics is in the three volumes of Whitehead and Russell's *Principia Mathematica* (1910–13). Thus Whitehead and Russell succeeded where Frege failed. But the theory of types compels the use of the axiom of infinity—which plays no role in Frege's work—and some have objected that this is not properly an axiom of logic.

The type theory of *Principia* is the so-called *ramified theory of types* (concerning which *see* MATHEMATICS, FOUNDATIONS OF). The reduction of this to the simple theory of types was mentioned by Leon Chwistek in 1921; independently of Chwistek, it was advocated more seriously by F. P. Ramsey in 1926, and thence came into general use through its adoption by Rudolf Carnap (*Abriss der Logistik*, 1929) and Kurt Gödel (1931).

Another major contribution of Russell is the device (1905) by which descriptive phrases such as "the 32nd president of the U.S.A." may be eliminated if desired (*see* SEMANTICS IN LOGIC). A technical improvement of Russell's device, important especially when it is employed in a formalized language, was introduced by W. V. Quine in 1940.

20. Zermelo.—Ernst Zermelo stated the axiom of choice in 1904—the use of which in mathematical reasoning had previously been tacit and unrecognized. The first formulation of axioms for set theory was by Zermelo in 1908. Zermelo's axioms involve, however, an unexplained notion of a "definite property." Different proposals for overcoming this difficulty were made by A. A. Fraenkel and Thoralf Skolem in 1922 and 1923, of which Skolem's has the advantage that it leads more directly to logistic formalization of Zermelo's verbally stated axioms. The axioms of set theory as given in the article LOGIC are obtained from those of Zermelo by Skolem's method, with addition of the axioms of replacement, due to Fraenkel, and the axioms of excluded infinite regress due to John von Neumann (later introduced independently also by Zermelo).

21. Richard.—Jules Richard published in 1905 the antinomy now known by his name. Semantical paradoxes related to Richard's were afterward proposed by various other authors, including the paradox of Kurt Grelling in 1908 (*see* SEMANTICS IN LOGIC).

22. Hilbert.—David Hilbert's contributions to logic arose from his work in the foundations of mathematics (*see* MATHEMATICS, FOUNDATIONS OF). Important among them are the program of proof theory and in particular of a metatheoretic consistency proof, dating from 1905, and in connection with this the sharp distinction between object language and meta-language (in Hilbert's terminology, between mathematics and metamathematics) *Grundlagen der Mathematik* (1934, 1939), a comprehensive treatise of modern logic containing Hilbert's ideas in their final form, was written in collaboration with Paul Bernays, to whom the detailed content of the work is largely due.

23. Brouwer.—Luitzen Egbertus Jan Brouwer is the founder of mathematical intuitionism (*see* MATHEMATICS, FOUNDATIONS OF; and THOUGHT, LAWS OF). His publications in this field began with his dissertation in 1907 and a paper on the law of excluded middle in 1908 and extended through 1954. The logistic formalization of intuitionism is, however, due to Arend Heyting (1930) and others.

24. Lewis.—Clarence Irving Lewis was led by the "paradoxes" of material implication (*see* LOGIC) to seek a notion of implication, *strict implication*, which shall correspond rather to the relation of logical consequence in the sense that, if $<$ is the sign of strict implication, and if **A** and **B** are any sentences, $\mathbf{A} < \mathbf{B}$ shall be true if and only if **B** is a logical consequence of **A**. Lewis's publications about the matter begin in 1912. But the first satisfactory formulation of a propositional calculus with strict implication was in 1920. The book of Lewis and C. H. Langford, *Symbolic Logic* (1932), treating the subject at length, has become a classic in the field of modal logic, and the starting point of many more recent investigations.

25. Löwenheim.—Leopold Löwenheim in a paper of 1915 proved the theorem which is now known as Löwenheim's theorem (*see* LOGIC), and several other important results in the metatheory of the functional calculus of first order.

26. Skolem.—Thoralf Skolem, besides the contribution to set theory already mentioned, gave a new and better proof of Löwenheim's theorem, established the extension of this theorem which is stated in the article LOGIC, contributed results connected with the decision problem of the functional calculus of first order (some of which were later important in the proof of Gödel's completeness theorem), and discovered also the following metatheorem, that no set of postulates, finite in number or enumerably infinite, expressible in the notation of a simple applied functional calculus of first order can be adequate for arithmetic in the sense of characterizing completely the system of non-negative integers.

27. Post.—Emil L. Post's dissertation of 1920, published in 1921, contains the first comprehensive metatheoretic treatment of a logistic formalization of the two-valued propositional calculus, including proofs of consistency and completeness; also the first formulation of a many-valued propositional calculus from a point of view which is abstract in the sense of being concerned with the form of the calculus independently of any particular interpretation.

28. Łukasiewicz.—Jan Łukasiewicz, in a paper of 1920, introduced a three-valued propositional calculus based on Aristotle's doctrine of future contingents (*see*

THOUGHT, LAWS OF). This was later generalized to an analogous n-valued propositional calculus, different from that of Post. Much important work is also due to Łukasiewicz in the two-valued propositional calculus, and in the history of logic.

29. Tarski.—Alfred Tarski contributed extensively to two-valued and many-valued propositional calculus, taking his departure from the work of Łukasiewicz. However, his most noteworthy contributions, beginning in 1930, are to the general metatheory of logistic systems, a domain in which many important new ideas are due to him. Especially semantics, in the sense of the metatheoretic treatment of notions related to those of meaning and truth, is the creation of Tarski (*see* SEMANTICS IN LOGIC). Much of the more recent work of Tarski has been in the boundary region between logic and mathematics, or has applied methods and results of modern logic to special branches of mathematics.

30. Carnap.—Rudolf Carnap, in his *Der logische Aufbau der Welt* (1928), *Testability and Meaning* (1936–37) and many other publications, was a pioneer in the systematic application of the methods of modern logic in epistemology and philosophy of science—making in this a contribution to philosophic method which in the eyes of many exceeds in importance his support of a particular philosophical outlook (that of logical positivism, *q.v.*). Carnap's contributions to the study of the metatheory of logistic systems begin in his *Logische Syntax der Sprache* (1934, published, 1937, in English with some additions as *The Logical Syntax of Language*). In a paper of 1935, somewhat later than Tarski but independently (and in a different terminology), Carnap introduced the idea of syntactical definitions of the semantical notions of truth and satisfaction, and in particular was the first to make such definitions for the full simple theory of types, as distinguished from a functional calculus of finite order. And concerning Carnap's contributions to intentional semantics, *see* SEMANTICS IN LOGIC.

31. Herbrand.—Jacques Herbrand in his short life—he was killed in a mountain-climbing accident in 1931 at the age of 23—made extensive contributions to Hilbertian proof theory and to the metatheory of the functional calculus of first order. The most important of these cannot be stated here. But the deduction theorem of first-order functional calculus should be mentioned as Herbrand's.

32. Gödel.—Kurt Gödel proved the completeness theorem of the pure functional calculus of first order (1930), and the famous incompleteness theorem (1931). For these, *see* LOGIC and (especially for the latter) MATHEMATICS, FOUNDATIONS OF.

The bearing of the incompleteness theorem on Hilbert's program of a metatheoretic consistency proof for mathematics is obvious; but more far-reaching is the consequence that no single logistic system, satisfying certain very general conditions, can tenably claim to embrace only logical truth *and* the whole of logical truth (if indeed the latter phrase has a meaning at all).

Also due to Gödel (1940) is the metatheorem that if the system of set theory with omission of the axiom of choice (*see* LOGIC) is consistent, it remains so upon addition of the axiom of choice or an axiom expressing the generalized continuum hypothesis or both. (For a statement of the continuum hypothesis *see* MATHEMATICS, FOUNDATIONS OF.) As Gödel pointed out, his result is applicable alike to various forms of set theory and to type theory. The solution of the continuum problem, as the problem concerning the continuum hypothesis is known, was completed in 1964 by Paul J. Cohen, who showed on the same assumption of consistency that the axiom of choice is independent of the other axioms of set theory, and the continuum hypothesis is then independent of the axioms of set theory including the axiom of choice.

See also references under "Logic, History of" in the Index. (Ao. C.)

BIBLIOGRAPHY.—*Modern Logic:* Heinrich Scholz, *Geschichte der Logik* (1931), "Pascals Forderungen an die Mathematische Methode," *Festschrift Andreas Speiser* (1945), "Die Wissenschaftslehre Bolzanos," *Ahhandlungen der Fries'schen Schule*, new series, vol. 6 (1937); I. M. Bocheński, "Spitzfindigkeit," *Festgabe an die Schweizerkatholiken* (1954), *Formale Logik* (1956); Karl Duerr, "Die Mathematische Logik des Arnold Geulincx," *The Journal of Unified Science (Erkenntnis)*, vol. 8 (1940), *Girolamo Saccheri's Euclides Vindicatus*, ed. by George Bruce Halsted (1920); Nicholas Rescher, "Leibniz's Interpretation of His Logical Calculi," *Journal of Symbolic Logic*, vol. 19 (1954); Y. Bar-Hillel, "Bolzano's Definition of Analytic Propositions," *Theoria*, vol. 16 (1950), "Bolzano's Propositional Logic," *Archiv für Philosophie*, vol. 4 (1952); Alonzo Church, "Brief Bibliography of Formal Logic," *Proceedings of the American Academy of Arts and Sciences*, vol. 80, no. 2 (1952) and works on the history and bibliography of logic which are there cited, "A Bibliography of Symbolic Logic," *Journal of Symbolic Logic*, vol. 1 and 3 (1937–39); Wilbur S. Howell, *Logic and Rhetoric in England, 1500–1700* (1956).

MATHEMATICAL INDUCTION, one of various methods of proof of mathematical propositions, based on the following principle.

The Principle of Mathematical Induction.—A class of integers is called hereditary if, whenever any integer x belongs to the class, the successor of x (*i.e.*, the integer $x + 1$) also belongs to the class. The principle of mathematical induction is then: If the integer O belongs to the class F and F is hereditary, every non-negative integer belongs to F. Alternatively, if the integer 1 belongs to the class F and F is hereditary, then every positive integer belongs to F. The principle is stated, sometimes in one form, sometimes in the other: and as either form of the principle is easily proved as a consequence of the other, it is not necessary here to distinguish between the two. (*See* also NUMBER.)

The principle is also often stated in intentional form: A property of integers is called hereditary if, whenever any integer x has the property its successor has the property. If the integer 1 has a certain property and this property is hereditary, every positive integer has the property.

Proof by Mathematical Induction.—An example of the application of mathematical induction in the simplest case is the proof that the sum of the first n odd positive integers is n^2; that is, that

$$1 + 3 + 5 + \ldots + (2n - 1) = n^2 \tag{1}$$

for every positive integer n. Let F be the class of integers for which equation (1) holds; then the integer 1 belongs to F, since $1 = 1^2$. If any integer x belongs to F, then

$$1 + 3 + 5 + \ldots + (2x - 1) = x^2 \tag{2}$$

The next odd integer after $2x - 1$ is $2x + 1$, and when this is added to both sides of equation (2), the result is

$$1 + 3 + s + \ldots + (2x + 1) = x^2 + 2x + 1 = (x + 1)^2 \tag{3}$$

Equation (2) is called the hypothesis of induction and states that equation (1) holds when n is x, while equation (3) states that equation (1) holds when n is $x + 1$. Since equation (3) has been proved as a consequence of equation (2), it has been proved that whenever x belongs to F the successor of x belongs to F. Hence by the principle of mathematical induction all positive integers belong to F.

The foregoing is an example of simple induction; an illustration of the many more complex kinds of mathematical induction is the following method of proof by double induction. To prove that a particular binary relation F holds among all positive integers it is sufficient to show, first that the relation F holds between 1 and 1; secondly that whenever F holds between x and y, it holds between x and $y + 1$; and thirdly that whenever F holds between x and a certain positive integer z (which may be fixed or may be made to depend on x), it holds between $x + 1$ and 1.

The logical status of the method of proof by mathematical induction is still a matter of disagreement among mathematicians.

Giuseppe Peano (*q.v.*) included the principle of mathematical induction as one of his five postulates for arithmetic (see POSTULATE). Many mathematicians agree with Peano in regarding this principle just as one of the postulates characterizing a particular mathematical discipline (arithmetic) and as being in no fundamental way different from other postulates of arithmetic or of other branches of mathematics.

Henri Poincaré (*q.v.*) maintained that mathematical induction is synthetic and a priori; that is, it is not reducible to a principle of logic or demonstrable on logical grounds alone, and yet is known independently of experience or observation. Thus mathematical induction has a special place as constituting mathematical reasoning par excellence, and permits mathematics to proceed from its premisses to genuinely new results, something that supposedly is not possible by logic alone. In this doctrine Poincaré has been followed by the school of mathematical intuitionism (*see* MATHEMATICS, FOUNDATIONS OF) which treats mathematical induction as an ultimate foundation of mathematical thought, irreducible to anything prior to it, and synthetic a priori in the sense of Immanuel Kant (*q.v.*).

Directly opposed to this is the undertaking of Gottlob Frege (*q.v.*), later followed by Alfred North Whitehead and Bertrand Russell (*q.v.*) in *Principia Mathematica*, to show that the principle of mathematical induction is analytic, in the sense that it is reduced to a principle of pure logic by suitable definitions of the terms involved. A sketch of the method especially the definitions, by which this is to be accomplished is given under Frege's name in the article LOGIC, HISTORY OF.

Transfinite Induction.—A generalization of mathematical induction applicable to any well-ordered class or domain D, in place of the domain of positive integers, is the method of proof by transfinite induction.

The domain D is said to be well-ordered if the elements (numbers or entities of any other kind) belonging to it are in, or have been put into, an order in such a way that: 1. no element precedes itself in order; 2. if x precedes y in order, and y precedes z, then x precedes z; 3. in every non-empty subclass of D there is a first element (one that precedes all other elements in the subclass). From 3. it follows in particular that the domain D itself, if it is not empty, has a first element.

When an element x precedes an element y in the order just described, it may also be said that y follows x. The successor of an element x of a well-ordered domain D is defined as the first element that follows x (since by 3., if there are any elements that follow x, there must be a first among them). Similarly, the successor of a class E of elements of D is the first element that follows all members of E. A class F of elements of D is called hereditary if, whenever all the members of a class E of elements of D

belong to F, the successor of E, if any, also belongs to F (and hence in particular, whenever an element x of D belongs to F, the successor of x, if any, also belongs to F).

Proof by transfinite induction then depends on the principle that if the first element of a well-ordered domain D belongs to a hereditary class F, all elements of D belong to F.

One way of treating mathematical induction is to take it as a special case of transfinite induction. For example, there is a sense in which simple induction may be regarded as transfinite induction applied to the domain D of positive integers. But the actual reduction of simple induction to this special case of transfinite induction requires the use of principles which themselves are ordinarily proved by mathematical induction especially the ordering of the positive integers, and the principle that the successor of a class of positive integers, if there is one, must be the successor of a particular integer (the last or greatest integer) in the class. And there is therefore also a sense in which mathematical induction is not reducible to transfinite induction.

The point of view of transfinite induction is, however, useful in classifying the more complex kinds of mathematical induction. In particular, double induction may be thought of as transfinite induction applied to the domain D of ordered pairs (x, y) of positive integers, where D is well-ordered by the rule that the pair (x_1, y_1) precedes the pair (x_2, y_2) if $x_1 < x_2$ or if $x_1 = x_2$ and $y_1 < y_2$.

BIBLIOGRAPHY.—G. Frege, *Die Grundlagen der Arithmetik* (1884) trans. as *The Foundations of Arithmetic* (1950), *Grundgesetze der Arithmetik* (1893 and 1903, reprinted 1962); E. Landau, *Grundlagen der Analysis* (1930) trans. as *Foundations of Analysis* (1951), H. Poincaré, *Science et Hypothèse* (1902) trans. as *Science and Hypothesis* (1905); B. Russell, *Introduction to Mathematical Philosophy*, 2nd ed. (1920). (Ao. C.)

NAME (IN LOGIC). A *name* is a word, symbol or expression having the kind of meaning which can be explained only by saying that a sentence in which the name is used is intended to be about something which the name denotes. (*See* DENOTATION.)

By some logicians the term *name* is restricted to the narrow category of what are sometimes called *logically proper names; i.e.*, names which denote uniquely, which are simple in the sense of having no analysis into meaningful parts, and which supposedly have denotation without connotation. But by others the word *name* is used rather in a broader sense than is customary in everyday usage and in linguistics; and it is this broader sense which will be treated in the present article.

Historical changes in the meaning of names (*see* the discussion of particular examples in NAME [IN LINGUISTICS] above), and variabilities and uncertainties of meaning which occur in the natural languages are the concern of the linguist rather than the logician. And many complicating features of the meaning and usage of names in the natural languages, however unavoidable in the actual use and development of a language, are best regarded for purposes of logical analysis as irregularities that have to be removed. This leads ultimately to a formalized language (*see* LOGIC). But we here use for illustration a natural language, English, abstracting from uncertainties of meaning which might be urged as doubts against some of the examples.

Traditionally, names have been classified according to their denotation, as *concrete* names denoting physically concrete things, *collective* names, denoting collections or classes of concrete things, and *abstract* names denoting abstract entities, or according to the connotation (*q.v.*) as *negative* (if they connote the not having of a specific property) or *positive* (in the contrary case), as *relative* (if, like "father," "friend," "cause," they connote the bearing of a specific relation) or *nonrelative*.

More fundamental, however, is the distinction between *singular* names, or *constants*, which have as a part of their meaning that there is one unique thing denoted, and *common* names, or *general* names, which do not have uniqueness as part of their meaning. (*Proper name* is convenient as a synonym of *singular name*, in view of other meanings of the word *singular*; but by many the term *proper name* is used rather for some more restricted class of names.)

Examples of singular names are the place name "Dartmouth," which denotes the town Dartmouth; the personal name "James Mill," which denotes the man James Mill, the phrase, "the father of John Stuart Mill," which (as would be said even by those logicians who do not class such descriptive phrases as names) denotes the same man James Mill; the personal name "Chiron," which purports to denote a certain centaur; the collective name "mankind," which denotes the class of men; the abstract name "courage," which denotes the property or quality of being courageous. The fact that more than one town has the name of "Dartmouth" and more than one man that of "James Mill" constitutes an equivocacy of these names, but does not prevent them from being singular names, as each one purports to denote uniquely. On the other hand, "Englishman" is a common name since, without equivocacy, it denotes James Mill, and also denotes John Stuart Mill, and Guy Fawkes, and Alexander Pope etc. "Son of Henry VIII" is a common name, as it is no part of the meaning of the name that there is only one; but "the son of Henry VIII" is a singular name. "Bachelor" is a concrete common name; "centaur" is a concrete common name, though denoting nothing; "throng" is a collective common name; "virtue" is an abstract common name (as both courage and honesty, *e.g.*, are virtues).

In the construction of formalized languages both common names and names denoting nothing have generally been avoided, in the interest of simplicity except so far as the variables which occur in most formalized languages (*see* LOGIC) may be regarded as partial analogues of the common names. It is an interesting question whether these familiar features of natural language can be reproduced in a formalized language, or whether they have some inherent inconsistency or unsuitability that prevents it. But statements containing either common names or denotationless names can always be reformulated so as to use instead only the constants and variables of the usual formalized languages (an example is the reformulation of "All bachelors are red-haired" as $(x)[B(x) \supset R(x)]$ in the language **L** described in SEMANTICS IN LOGIC). (AO. C.)

POSTULATE, a proposition assumed without proof in founding a mathematical discipline. Though the propositions of a particular mathematical discipline, or branch of mathematics, are generally established by proof, there must be some unproved first principles, if the process of basing one proposition on others from which it is proved is not to be an infinite regress. Similarly, although the terms used are generally introduced by definition, there must be some undefined terms. These unproved first principles are the *postulates*, and the terms not defined are the *primitive terms*. (*See* AXIOM; LOGIC, HISTORY OF.)

For example, in the case of Peano's postulates for arithmetic (so called after Giuseppe Peano, who proposed them as first principles of arithmetic, though the postulates themselves are due rather to C. S. Peirce and Richard Dedekind), the primitive terms are "0," "number" (in the sense of non-negative whole number) and "successor" (in the sense that $x + 1$ is the successor of x). And the postulates themselves are: (1) 0 is a number. (2) Every number has a number as its unique successor. (3) Two numbers having the same successor are identical. (4) 0 is not successor of any number. (5) If 0 belongs to a class F, and if whenever a number x belongs to F the successor of x belongs also to F, then all numbers belong to F.

These may be expressed in logistic notation (*see* LOGIC) by using an individual constant 0, and two functional constants N and S ("$N(x)$" to mean "x is a number," and "$S(x, y)$" to mean "y is successor of x"). However, the implicit assumption that it is only numbers that have successors is best represented by defining $N(x)$ to stand for $(\exists y)S(x, y)$ (and $N(y)$ to stand for $(\exists x)S(y, x)$, etc.), so that the number of primitive terms is reduced to two The statement of some of the postulates may then be simplified and the second postulate is also conveniently divided into two parts: (1) $N(0)$. (2_1) $S(x, y) \supset N(y)$. (2_2) $S(x, y) \supset . S(x, z) \supset y = z$. (3) $S(y, x) \supset . S(z, x) \supset y = z$. (4) $\sim S(x, 0)$. (5) $F(0) \supset . F(x)S(x, y) \supset_{xy} F(y) \supset . N(x) \supset_x F(x)$. The postulates have to be added to an underlying logic, which in this case must be a functional calculus of at least second order (*see* LOGIC) to make definitions possible of addition and multiplication of numbers.

Another familiar example of mathematical postulates is that of the postulates for elementary geometry. As in other cases, the postulates for a particular discipline are not uniquely determined and various alternative systems of postulates are known. The original axioms and postulates of Euclid are the prototype of axiomatic treatments of geometry, and are in a certain sense still standard, though they do not satisfy modern requirements of rigour. For modern systems of postulates for geometry *see* the works of Hilbert and of Veblen, cited in the bibliography below.

According to the view now usual, the postulates constitute a definition of the particular mathematical discipline; and a change of one of the postulates to something different or contrary would not be wrong, but would merely lead to the definition of a different discipline. This contrasts with an older view according to which the postulates of arithmetic or of Euclidean geometry are a priori truths. The abandonment of the older view was brought about largely by the discovery of non-Euclidean geometry (*see* GEOMETRY, NON-EUCLIDEAN; MATHEMATICS, FOUNDATIONS OF), since Euclid's parallel postulate and its contradictory lead to two geometries which not only are equally sound logically but, in the absence of experimental evidence against the Euclidean postulate, would be equally applicable to physical space.

A set of postulates is called *consistent*—if, in the mathematical discipline which the postulates determine, not both A and ~A are ever theorems. A particular postulate is *independent* if it is not a theorem of the discipline determined by the remaining postulates.

In general, a mathematical discipline based on a consistent set of postulates can be strengthened (without adding to the list of primitive terms) by adding new, independent postulates in various alternative ways. But if the set of postulates is *categorical*, there is a sense in which this is not possible. (*See* CATEGORICAL.)

Use of axioms and postulates for a rigorous account of the foundations of a branch of natural science is less usual than in mathematics, and its appropriateness is still the subject of controversy. Some reject the postulational form for theoretical science altogether. And some regard it as an ideal to which the development of a science approaches, not yet attained except in the case of applied geometry and possibly mechanics, and not a useful form in at least the present stage of most sciences. But a noteworthy attempt to apply the axiomatic method in biology is contained in the writings of J. H. Woodger, whose book (cited below) may be consulted both for this and for an expository account of the axiomatic or postulational method in general.

BIBLIOGRAPHY.—D. Hilbert, *Grundlagen der Geometrie* (1899, with later editions) and *The Foundations of Geometry*, Eng. trans. by E. J. Townsend (1902); O. Veblen, "A System of Axioms for Geometry," *Trans. Amer. Math. Soc.*, vol. 5, pp. 343–384 (1904); J. H. Woodger, *The Axiomatic Method in Biology* (1937). (Ao. C.)

PROPOSITION. In scholastic and traditional logic a proposition was understood as an expression in words having the meaning of an assertion. An example is Petrus Hispanus, *c.* 1245, "Propositio est oratio verum vel falsum significans indicando" ("A proposition is a statement indicating a true or false meaning"). Thus a proposition is not simply a declarative sentence, in the grammatical sense, but is such a sentence taken together with its meaning. Consequently, propositions may be different even though the sentences are the same (*e.g., I am hungry*, uttered by two different persons), and, although this consequence is less emphasized by traditional writers, propositions are different when the sentences are different and even if the meaning is the same (*e.g., Tempus fugit* and *Time flies*).

The usual scholastic term was the Latin *propositio*—first found in this meaning in the writings of Lucius Apuleius, *c.* 150, and Manlius Severinus Boethius, *c.* 500. However, *enuntiatio* (enunciation)—a term taken from Cicero—was also employed, and some of the scholastics used this as the general term, reserving *propositio* for some more special meaning.

This scholastic-traditional notion of a proposition is inconvenient or unsatisfactory in many contexts because of its dependence on the particular form of expression in words, or on a particular language. Hence the different notion of a mental proposition (*propositio mentalis*) came to be introduced, and also, chiefly later, that of a judgment (*iudicium*).

According to William Ockham, and later scholastics who followed him, the mental proposition must be formed internally before a corresponding proposition in words is put forward. It is not of any language. Its parts are mental terms or concepts, which are analogous to the spoken or written terms and share with them all properties essential to the meaning, but not such purely grammatical properties as having a particular (grammatical) gender or belonging to a particular declension or conjugation.

The notion of a judgment, as the mental act of assent or dissent has some mention in the writings of various scholastics, and was made an explicit part of the treatment of logic by Petrus Ramus and later by such logicians as Isaac Watts (1725) and Christian Wolff (1728). It was still later that the definition of a proposition as a "judgment expressed in words" became a commonplace.

Immanuel Kant, and many traditional logicians who have followed him, replace the consideration of propositions almost entirely by that of judgments. Thus Kant speaks usually, though not always, of an analytic or synthetic judgment (German *Urtheil*) rather than proposition (German, *Satz*), of a categorical judgment, etc.

However, the mental proposition and the judgment have a psychological reference which may often be as unsatisfactory as the dependence on a particular language which is involved in the traditional notion of a proposition. For some purposes at least there is needed a more abstract notion, independent alike of any particular expression in words and of any particular psychological act of judgment or conception—not the particular declarative sentence, but the content of meaning which is common to the sentence and its translation into another language—not the particular judgment, but the objective content of the judgment, which is capable of being the common property of many. Such an abstract notion may be seen in the Stoic *Lekta* (*see* LOGIC, HISTORY OF: *Stoics and Megarians*) and the *complexe significabilia* of some 14th and 15th century scholastics but these ideas fell into oblivion and were reintroduced in different terminology in modern times. By modern logicians the word "proposition" has come to be used for the abstract notion, and we shall therefore here distinguish between *proposition in the traditional sense* and *proposition in the abstract sense*.

Bernard Bolzano attributes the use of *propositio* for proposition in the abstract sense to G. W. von Leibniz in *Dialogus de Connexione inter Res et Verba*. This may be a misunderstanding. But this dialogue does set forth clearly one important ground of the need for the abstract notion. An explicit distinction between the sentence and the proposition (in the abstract sense) which the sentence expresses is made by Bolzano in his *Wissenschaftslehre* of 1837, where *Satz an sich* is used for the latter. Gottlob Frege in 1892 uses *Gedanke* for the sense of a declarative sentence, giving to this German word (as he explains) an objective, rather than its more natural subjective. meaning. *Proposition* is used in the abstract sense by Bertrand Russell in *The Principles of Mathematics* (1903), where Russell recognizes that Frege's *Gedanke* is approximately "what I have called an unasserted proposition," and in *Principia Mathematica* A. N. Whitehead and Russell speak of "what we call a 'proposition' (in the sense in which this is distinguished from the phrase expressing it)." Russell also uses proposition in the traditional sense, *e.g.*, in *Introduction to Mathematical Philosophy*, defining a proposition as "a form of words which expresses what is either true or false" (very close to the translation of the Latin of Petrus Hispanus as quoted above); but more recent writers have generally followed him in the abstract usage. *See* also LOGIC. (Ao. C.)

BIBLIOGRAPHY.—R. M. Eaton, *General Logic*, i, 3-5 (1931); Alonzo Church, *Introduction to Mathematical Logic*, sec. 04 (1956); Ludwig Wittgenstein, *Philosophical Investigations*, sec. 134 *et seq*. (1953).

SORITES. In traditional logic a chain of successive categorical syllogisms in the first figure (*see* LOGIC: *Traditional Logic*) may be so related that either the conclusion of each except the last is the minor premiss of the next, or the conclusion of each except the last is the major premiss of the next. If then the intermediate conclusions are suppressed (*i.e.*, the conclusions of all the successive syllogisms except the last), and only the remaining premisses and the final conclusion are stated, the resulting argument is a valid inference from the stated premisses, which may be considered independently of its analysis into syllogisms, and which is called a sorites. In the case of the first alternative we have the so-called Aristotelian sorites, and in that of the second, the Goclenian sorites (enunciated by the German philosopher Rudolph Goclenius, 1598). Considered as independent inferences, the Aristotelian and Goclenian sorites are identical, except for the trivial matter of the order in which the premisses are stated. But the Aristotelian and Goclenian analyses of the sorites into syllogisms are different.

An example, taken from Lewis Carroll, is the following. The premisses are: (1) No one takes in [subscribes to] the *Times*, unless he is well-educated; (2) No hedgehogs can read; (3) Those who cannot read are not well-educated. The expected conclusion is: (5) No hedgehog takes in the *Times*. To analyze this into syllogisms, we restate the premisses, with obversion of (2) and (3), as follows: (la) None who are not well-educated take in the *Times*, (2a) All hedgehogs are unable to read; (3a) All who are unable to read are not well-educated. In the Goclenian analysis, the first syllogism has (la) as major premiss, and (3a) as minor premiss and hence the intermediate conclusion: (4) None who are unable to read take in the *Times*. Then the second syllogism has (4) as major premiss and (2a) as minor premiss, hence (5) as final conclusion. In a more complicated example there might be more than three premisses. In general, there may be $n+1$ premisses, and analysis then yields a chain of n successive syllogisms.

The sorites is one case in which a more complex piece of reasoning can be analyzed into syllogisms. But the idea that all deductive reasoning can or must be so analyzed has generally been abandoned.

See Lewis Carroll (Charles Lutwidge Dodgson), *Symbolic Logic*, 4th ed. (1897, reprinted 1958); J. N. Keynes, *Studies and Exercises in Formal Logic* (1884 and later editions to 1906).

(Ao. C.)

TAUTOLOGY, from a Greek word meaning "saying the same thing," is a word used technically in logic in a sense somewhat different from that in which it is used, pejoratively, in grammar. In the latter, a tautology is a needless repetition of a single idea in several words, as if more than one idea were being expressed (*see* FIGURES OF SPEECH).

In logic, the notion of a tautology in the propositional calculus is due to C. S. Peirce (1885). For this notion, and for the symbols used in this article, *see* LOGIC: *The Propositional Calculus*. The name tautology, however, was introduced in 1921 by Ludwig Wittgenstein, whose philosophical use of the notion, outlined in the article LOGICAL POSITIVISM, requires its extension from the propositional calculus to the full theory of types, including the functional calculi of first and higher orders. This extended notion, which was explained somewhat cryptically by Wittgenstein, and more satisfactorily by F. P. Ramsey in 1926, is, in fact, a less precise forerunner of what is now more usually called "validity" (*see* LOGIC: *The Metatheory of the Pure Functional Calculus of First Order*). Later logical positivists, especially Rudolf Carnap, amended Wittgenstein's doctrine in the light of the distinction that there is an effective test of tautology in the propositional calculus but no such test of validity even in first-order functional calculus.

As an example of a tautology in the extended sense of Wittgenstein–Ramsey, consider the formula $p \supset_x F(x) \supset . \ p \supset (x)F(x)$ which is a theorem (in fact an axiom) of the pure functional calculus of first order. Suppose that there is available a list, possibly infinite, of names for all the individuals, say: $i_1, i_2, i_3, \ldots$. Then replace the two universal quantifiers that occur in the formula, each by a corresponding conjunction, so that we have:

$$[p \supset F(i_1)][p \supset F(i_2)][p \supset F(i_3)] \ldots \supset . \ p \supset F(i_1)F(i_2)F(i_3) \ldots$$

This last may not be properly a formula, that is to say, it may be infinite in length, so that it is impossible actually to write it out. Nevertheless, we may abstractly consider such quasi-formulas; and in the same abstract way we may consider truth-table computations (or better, quasi-computations) applied to them, which may be infinite in width and length but are otherwise just like the finite truth-table computations that were explained in the article LOGIC in connection with propositional calculus. Moreover, in the case of the particular quasi-formula in question, it is clear that it is a tautology in the sense that, for every assignment of truth-values to its elementary parts $p, F(i_1), F(i_2), \ldots$, the truth-table computation assigns to the whole quasi-formula the value truth. The analogy to the notion of tautology in the propositional calculus fails in only one respect: because the truth-table computation may be infinite, effectiveness is lost. Wittgenstein and Ramsey characterized logic by the property that its theorems are tautologies in this extended sense. (Ao. C.)

TERM. In traditional logic (*see* LOGIC: *Traditional Logic*) a term is properly the subject or predicate of a categorical proposition. Aristotle so used the Greek word *horos* (limit), apparently by an analogy between the terms of a proportion (*see* below) and those of a syllogism. *Terminus* is the Latin translation of this, used for example by Boëthius. Hence in medieval logic the word came to be used also for common and proper names generally (*see* NAME [IN LOGIC]); and even for what were called syncategorematic terms—words such as *and, if, not, some, only, except*, which are incapable of being used for the subject or predicate of a proposition (*see* MEANING). In everyday English, the meaning of term has been broadened still further, to that of an item of terminology, any word or phrase with a particular meaning.

In mathematics, the terms of a fraction (*see* FRACTION) are the numerator and denominator. The terms of a proportion are the four numbers which enter into the proportion (*see* ALGEBRA) or the corresponding numerical expressions that denote such numbers. Similarly the terms of a sum are the numbers which are added together to constitute the sum, or the numerical expressions denoting them. In this sense an infinite series (*see* SERIES) is thought of as a sum of an infinite number of terms; and a polynomial (*q.v.*) is a sum of a finite number of monomials which are the terms of the polynomial. (Ao. C.)

THOUGHT, LAWS OF. Traditionally, special importance has been attached to three of the simplest laws of logic, the *law of identity*, the *law of contradiction*, and the *law of excluded middle* (or *tertium datur*). These were called the "laws of thought," a name which may conveniently be retained even by those who do not accept the implications which the name suggests. (For "thinking" as studied in psychology, *see* THINKING AND PROBLEM SOLVING.)

As examples of axioms, or indemonstrable principles (*see* AXIOM), Aristotle cities that, of two contradictories, one must be true, and that it is impossible for anything both to be and not to be. These are the principles which were afterward known as the laws of excluded middle and of contradiction respectively. They occur in Aristotle also in other forms, which are regarded as being expressions of the same laws; *e.g.*, that there is no middle ground between contradictories, and that it is impossible for anything both to be predicated and not to be predicated of the same thing in the same sense. Aristotle partly exempted future contingents from the law of excluded middle, holding that it is not (now) either true or false that there will be a naval battle tomorrow, but that it is (now) true that either there will be a naval battle tomorrow or not; against this, Chrysippus (*see* LOGIC, HISTORY OF) maintained that all propositions, even about the future, are either true or false.

The law of identity has usually been stated (for example by Leibniz) as: A is A. Stated in ordinary language, particular cases of the law would then be such things as: gold is gold, men are men, love is love, etc.

To render the three laws in modern logical notation (*see* LOGIC: *The Propositional Calculus*), the best choices would seem to be: $F(x) \supset_x F(x)$ (law of identity), $\sim . p{\sim}p$ (*law of contradiction*), $p \vee {\sim}p$ (*law of excluded middle*). However, the law of identity is also often given as $(x)x = x$, a form which differs from the other (on the basis of the definition of = in the article LOGIC) only by the insertion of a quantifier.

The three laws may also be understood in a semantical sense (*see* SEMANTICS IN LOGIC), the traditional statements of them generally not making a sharp distinction between such metatheoretic principles and principles which are rather to be expressed in the notation of propositional calculus or functional calculus. Thus the law of identity would be that a term must preserve the same denotations in all its occurrences, at least throughout any one context, the law of contradiction, that a sentence and its negation, S and ~S, are not both true; the law of excluded middle, that either S or ~S is true. In this interpretation the "laws of thought" are to be regarded as being about a particular language—or else as general principles of sound notation, demands which any acceptable language or system of notation must satisfy.

The doctrine that the laws of thought—or even only the laws of identity and contradiction—are a sufficient foundation for the whole of logic, or that all other principles of logic are mere elaborations of them, is traceable to Leibniz. Leibniz left a few actual examples of demonstrations of principles of logic and mathematics from only these laws, but all are quite minor in character and moreover involve tacit use of at least syllogistic inference in addition to the laws of thought. It would seem that Leibniz intended originally to express only a hope or an expectation. But later he was more incautious, writing, for instance, in the second letter to Samuel Clarke: "The great foundation of mathematics is *the principle of contradiction, or identity*, that is, that a proposition cannot be true and false at the same time; and that therefore A is A, and cannot be *not* A. This single principle is sufficient to demonstrate every part of arithmetic and geometry, that is, all mathematical principles."

Leibniz's successors applied the doctrine mainly to logic as distinguished from mathematics, but at the same time elevated it into an assertion of established fact. In this form it was common especially among traditional logicians of the 19th century. And though the doctrine appears already in the *Vernunftlehre* (1756) of H. S. Reimarus, its widespread adoption is perhaps due to Kant's characterization of an analytic judgment (*see* LOGIC, HISTORY OF) as one for recognition of whose truth the law of contradiction suffices.

It is difficult to give this doctrine a precise meaning in such a way as to make it tenable. For example it is not true that the three laws, say in the forms $F(x) \supset_x F(x)$, $\sim . p{\sim}p$, $p \vee {\sim}p$, are a sufficient set of axioms for logic, or even for the most elementary branch of logic, the propositional calculus, or for the traditional theory of the categorical syllogism.

Criticisms and Rejections of the Laws of Thought.—The law of excluded middle and some related laws of propositional calculus and functional calculus of first order are rejected by L. E. J. Brouwer and the school of mathematical intuitionism (*see* LOGIC; MATHEMATICS, FOUNDATIONS OF), not in the sense that the negation of the law is asserted, but in the sense that use of the law is not admitted as a valid method of mathematical proof in cases in which all members of an infinite class are involved. For example, Brouwer would not accept the disjunction that either there occur ten consecutive 7's somewhere in the decimal expansion of the number π or else not (since there is no proof known of either alternative), but he would accept that there occur ten consecutive 7's somewhere in the first 10^{100} digits of the decimal expansion of π or else not (since this could in principle be settled by carrying out the computation of the decimal expansion to the required point).

On the basis of Aristotle's doctrine of future contingents, but not without some modification of it, Jan Łukasiewicz was led in 1920 to formulate a propositional calculus which has a third truth-value (for future contingents) in addition to the usual two, and in which the laws of contradiction and excluded middle alike fail. Propositional calculi with n truth-values, $n \geqq 3$, were published independently by E. L. Post in 1921 and by Łukasiewicz in 1930. Corresponding many-valued functional calculi of first order were formulated still later by other writers.

Other criticisms or rejections of one or more of the laws of thought—*e.g.*, by Epicurus (*q.v.*), by Hegel (*q.v.*), by Alfred Korzybski (*see* SEMANTICS, GENERAL)—are in a different category from those of Brouwer and Łukasiewicz, as they offer no precise formulation of a usable logic in which these laws fail or are modified.

BIBLIOGRAPHY.—M. R. Cohen and Ernest Nagel, *An Introduction to Logic and Scientific Method* (1934); Donald C. Williams, "The Sea Fight Tomorrow," *Structure, Method, and Meaning Essays in Honor of Henry M. Sheffer* (1951); S. C. Kleene, *Introduction to Metamathematics* (1952), *see* III, §13, "Intuitionism"; Jan Łukasiewicz "Die Logik und das Grundlagenproblem," *Les Entretiens de Zurich* (1941); A. N. Prior "Three-Valued Logic and Future Contingents," *Philosophical Quarterly*, vol. 3 (1953) and *Time and Modality* (1957); G. W. F. Hegel, *Science of Logic*, tr. by W. H. Johnston and L. G. Struthers, 2 vol. (1929; 1952). (Ao. C.)

ON SCHEFFLER'S APPROACH TO INDIRECT QUOTATION (1973)

In order to comment on the two papers of Scheffler[1] I should like to introduce a terminology which Scheffler himself does not use but which seems to me enlightening as regards the comparison between the Quine-Goodman-Scheffler nominalistic analysis of language and other more standard approaches to linguistic analysis such as that of Carnap[2] or Tarski[3]. Namely, the Goodman predicates which consist of inscriptions framed by quotation marks and which are described in the first paragraph of Scheffler's earlier paper, I shall call *sentence-predicates*. And what Scheffler speaks of as "my peculiar cross-linguistic that-clause predicates" I shall call *proposition predicates*.

It is clear that a terminology is needed for these two kinds of predicates, and the one chosen holds the advantage that it reminds us that the sentence-predicates are intended by the nominalist to take the place of, and at least in part to serve the same purposes as, what appears in more standard treatments of logical syntax as names of sentences; and similarly that the proposition-predicates are intended to serve instead of names of propositions. What the nominalist is doing is thus described in terms that suggest the corresponding realistic notions, but it seems that Scheffler cannot in principle object. It is a matter of history that the Quine-Goodman school of nominalism started from accomplishments in linguistic analysis that had been made on a realistic basis, and undertook to reproduce as much of them as possible nominalistically. Scheffler in the opening paragraphs of his first paper repeatedly speaks of sentences, although his nominalism will not allow him that there are such things. And in a later footnote he explains that various passages are couched for brevity in familiar "Platonistic"[4] (or more correctly, realistic) terms, on the ground that the transition to a nominalistic statement is readily made.

To see the effect of the replacement of names of sentences by sentence-predicates, and of names of propositions by proposition-predicates, we must recall that Goodman and Scheffler confine themselves to a strictly first-order language, which has individual variables, but from which variables replaceable by predicates are excluded, and hence also quantification with respect to such variables. In a sense the Quine-Goodman nominalism had its beginning in Quine's criterion of "ontological commitment," which gave

---

Originally published in Spanish, *Sobre el análisis del discurso indirecto propuesto por Scheffler*, ***Semántica filosófica: Problemas y discusiones***, edited by Thomas Moro Simpson, Siglo XXI Argentina Editores S.A., Buenos Aires 1973, pp. 363–369. First publication in English. Reprinted by permission of the Alonzo Church estate.

[1] *An inscriptional approach to indirect quotation*, ***Analysis***, vol. 14, no. 4 (1954), pp. 83–90, and *Inscriptionalism and indirect quotation*, ***Analysis***, vol. 19 no. 1 (1958), pp. 12–18.

[2] ***Logische Syntax der Sprache***, translated as ***The Logical Syntax of Language***, and later publications.

[3] E.g. ***Logic, Semantics, Mathematics: Papers from 1923 to 1938***, and *The semantic conception of truth and the foundations of semantics*, ***Philosophy and phenomenological research***, vol. 4 no. 3 (1944) pp. 341–376, reprinted in ***Readings in philosophical analysis***, selected and edited by Herbert Feigl and Wilfrid Sellars, Appleton-Century-Crofts, Inc., New York, 1949, pp. 52–84.

[4] Roughly, realism is the doctrine that universals are real, whereas Platonism is the doctrine that they are the only, or at least the ultimate, reality. If the reader finds this statement vague, he is reminded that the traditional philosophical doctrines are themselves vague. Quine's criterion of ontological commitment constitutes a noteworthy attempt or proposal by which precision is to be given to them.

a more precise meaning to nominalism by specifying that there is ontological commitment to an entity if and only if a language is adopted which admits variables having that entity among their values, and quantification with respect to such variables. Thus in a first-order language there is no ontological commitment to anything supposed to be named by the predicates; and Scheffler may safely introduce proposition-predicates without therefore having to call himself a realist.[5]

The essential point is the avoidance of quantification over sentences (i.e., quantification with respect to variables having sentences in their range) and likewise quantification over propositions; and the shift from names to predicates is significant mainly as a device to bring this about. The logical problem may be put by asking how far the results of standard logical syntax can be reproduced without allowing quantification over sentences. And specifically for the Scheffler papers there is the second problem, how far the analysis of statements of assertion and belief, which *prima facie* requires propositions, can be successfully carried through without quantification over propositions. The "problem-sentences" which I presented to Scheffler were selected because they are sentences for which the obvious analysis does require quantification over propositions. In view of the reliance on the Quine criterion they are not, as Scheffler suggests, something new or just a further problem to be solved. On the contrary they are the crux of the matter, and these or similar examples should have been considered in his first paper.

As far as concerns "Church's original arguments[6]", it seems that they are unimpeached as against their main original target—the practice of taking sentences, in place of propositions, as the objects of assertion and belief. The idea that at least statements of assertion (though perhaps not those of belief[7]) might be analyzed in terms of inscriptions as the object was new to me and impressed me as interesting quite independently of the nominalistic motivation. But Scheffler's introduction of proposition-predicates was nevertheless on the edge of admitting propositions, in the sense that it would lead to something not significantly different from this if quantification with respect to variables replaceable by proposition-predicates could be shown to be required for a satisfactory theory.

Let us therefore examine Scheffler's analysis (a') of the problem sentence (a), the only one that he makes explicit. Scheffler himself seems to be in doubt, since he says only "it is conceivable" and "might be interpreted"; but perhaps this weak phraseology is intended only as an admission that he has not actually formulated the proposed base language B, and of course this must (in some sense) be done before his case is complete. Even in advance of the formulation of B, however, difficulties suggest themselves.

In the first place (a') definitely includes an assertion that is not made in (a), namely the existence of inscriptions z and w which are B-rephrasals of the inscriptions x and

---

[5]Hempel objects to this as only "technically nominalistic," and Scheffler's reply in effect defends the Quine criterion without naming it.

[6]In *On Carnap's analysis of statements of assertion and belief*, ***Analysis***, vol. 10 no. 5 (1950), pp. 97–99. This was reprinted, with a reply by Carnap, in ***Philosophy and analysis, A selection of articles published in Analysis between 1933–40 and 1947–53***, edited by Margaret Macdonald; Basil Blackwell, Oxford 1954, and Philosophical Library, New York 1954, pp. 125–128. Carnap's reply is unrelated to the Scheffler papers and is therefore no part of our present topic, but for the record, I do not think the reply sufficient to reinstate an analysis according to which belief is a relation between a man and a sentence.

[7]Since a belief may not have a naturally or immediately associated inscription in the way that an assertion must. Compare Scheffler's reference to Meckler.

y actually used by Church and Goodman. By admitting even accidentally occurring inscriptions as candidates for B-rephrasals of x and y, Scheffler has increased the likelihood that (a') will be true if (a) is. But this leaves unchanged the fact that what is actually asserted in (a) does not include the existence of inscriptions other than x and y.

Moreover the desired inscriptions in the language B will tend to be lengthy because of the many different inscriptions in many different languages for which rephrasals must be provided. For example, how many names of different individuals must B contain, and how long must these names be to secure that each individual has a different name?[8] For this reason it may be difficult to exclude cases in which (a) may be true but (a') false because z and w are too long to occur in the physical world, even accidentally.

In his footnote Scheffler, following Martin and Woodger, suggests it may be possible to "formulate a sufficiently rich base language to incorporate rephrasals of all actual inscriptions (or those we care about)." But if (in spite of my doubts) such a rich base language B should be found, then contrary to Scheffler we must be careful not to provide a nominalized version of Tarski's definition of truth for it. For a nominalized version of Tarski's theorem about truth could hardly be avoided. The very availability of the language together with the definition of truth would lead logicians to make assertions about truth in the language, and so to produce inscriptions which have no rephrasals in the language but which no doubt they do care about.

It is not at all a fantastic supposition that the definition of truth in B might be too long to be written out on the surface of this little globe we inhabit, or even anywhere in the physical world. What if the definition of truth in B can nevertheless be indicated in the sense that we state, in reasonably limited space, instructions for an effective procedure or algorithm which, *given sufficient time and space and physical materials*, we could follow through with the assurance that it would ultimately result in our writing down the explicit truth definition? Does the definition of truth then exist, although not existing as a physical object? The realist would of course answer yes. I fear the nominalist of the Quine-Goodman school must answer no, the sufficient time and space and physical materials being sadly not given. And it is in just such situations as this that his difficulty lies.

Another point is that the use of the notion of rephrasal[9] in such crucial analyses as (a') (if they are accepted at all) makes it more urgent to provide a criterion for rephrasal. And as far as I know, this has never been attempted, whether by Scheffler or others. Scheffler is quite right in saying that there is a parallel problem about propositions, not the problem quoted from Hempel "of specifying the propositional meaning"—which I find unclear—but the problem of an identity criterion for propositions. That is, under what conditions do two sentences[10] of a given language express the same proposition?

---

[8]Or singular predicates such as "pegasizes" if they are preferred to names.

[9]It should be noticed that the relation denoted by "Reph" in (a') is not the same as the relation of rephrasal introduced in Scheffler's first paper. But rather, if I understand the explanation correctly, the relation Reph —or B-rephrasal, as I have called it in the text —holds between x and z if and only if z is an inscription which, taken regardless of context and as in the language B, is a rephrasal of the inscription x taken in context and as in the language which is indicated by its context. It is B-rephrasal rather than the original rephrasal relation which requires analysis in order to support (a'), but the threatened difficulty seems to be about the same.

[10]Except for the nominalist, it is preferable to put this question in regard to sentences rather than inscriptions, and then to treat in the first instance only those languages in which the meaning of a sentence

But this problem about propositions has not been neglected, there have been some attempts and some progress[11]. And the corresponding problem of a criterion for rephrasal threatens greater difficulty because of the in-context feature, which is essential to Scheffler's use of the notion.

Because of these doubtful points I contend that Scheffler has not convincingly solved the central logical problem that is raised by the proposal in his first paper, and especially that he has not established the answer to it which he favors for philosophical reasons.

I have already said that I find the logical problem interesting beyond its philosophical motivation. There are related problems which might be of similar interest—for example obtained by allowing quantification over sentences but not over propositions.

Apropos of the philosophical aspect I can only express my surprise that Scheffler, after admitting that the existence of physical objects raises a serious philosophical problem, nevertheless thinks it *philosophically* important to find a way to eliminate propositions and sentences in favor of inscriptions. He should be asked whether his inscriptions are physical objects, and if not, what. If they are sense-data, or human experiences or observations, the Quine-Goodman use of accidentally occurring inscriptions breaks down. If they are physical events of some kind (strictly instantaneous, or of momentary short duration, or of long duration), the same problem "how to understand the existence of 'external' physical reality" affects them.

Perhaps the explanation is that Scheffler maintains a hierarchy of degrees of intelligibility with physical objects falling somewhat higher in the scale than sentences. This is plausible. But he should then explain the principle of the ranking, and he should avoid such absolute terms as "pseudo-entity," speaking rather of greater and less pseudo-ness.

(once the language is known) is independent of its context.

[11] The criterion that two sentences express the same proposition if and only if they are logically equivalent, proposed by Carnap and others, leads to a notion of proposition which serves some purposes, including that of modal logic, but is not suitable for an analysis of indirect quotation. It is not surprising that there should be more than one notion of proposition, according to what identity criterion is adopted. And Carnap's intensional isomorphism, though not so put forward by Carnap himself, might be taken as another proposed identity criterion for propositions, leading to a notion of proposition which may be used also in analyzing indirect quotation. In my paper, *Intensional isomorphism and identity of belief*, ***Philosophical Studies***, vol. 5 (1954), pp. 65–73, I have suggested some changes in the criterion of intensional isomorphism which seem to me necessary for the purpose.

OUTLINE OF A REVISED FORMULATION OF THE LOGIC OF SENSE AND DENOTATION (1973, 1974)

Part I

In this paper I return to and reexamine my "A Formulation of the Logic of Sense and Denotation". For brevity I shall generally not repeat passages and material from this latter paper [4] or its abstract [2], but shall merely refer to the paper (as *A Formulation*) or to the abstract, often in such a way that the reader must have at least [4] before him in order to follow what is being said.

In fact *A Formulation* is unsound or faulty in many ways, some major and some minor, and though there has never been a detailed published account of the faults, I believe that many have been aware of their existence. Frege's theory of meaning has a very strong intuitive appeal when presented informally as a resolution of the difficulties which Frege considers about the meaning of identity statements, about denotationless names, and especially about the logic of indirect discourse and of other contexts involving what Frege calls the *ungerade* ("oblique" or "indirect") occurrence of names. But in spite of this the conclusion has no doubt often been drawn that a satisfactory treatment of the intensional logic demanded by Frege's theory, if it is possible at all, is at least not possible along the general lines that I proposed. I confess that I was at one time of this opinion myself, and I have on several occasions advised others accordingly (to my present regret). Nevertheless, I now outline an attempt to set matters right.

The principal flaw in the original axioms for Alternative (2) was first called to my attention by A. F. Bausch not long after the publication of *A Formulation*. In effect, the discussion on pages 15–16 [pages 300–301 in present edition] of *A Formulation*, leading up to the assumption that *ϕ' is a concept of a function ϕ if and only if it is a characterizing function of ϕ*, is incompatible with the principle of Alternative (2), that the senses of names **A** and **B** are the same if and only if the equation $\mathbf{A} = \mathbf{B}$ is logically valid. Because *this* assumption is embodied in the axioms as Axioms $15^{\alpha\beta}$ and $16^{\alpha\beta}$, the result is that the axioms lead, not quite to an inconsistency, but to a reduction to extensionality, in the sense that it is a theorem in each type α that two concepts of the same thing are always identical.

As a means of finding a suitable amendment of the axioms for Alternative (2), I tried unsuccessfully in lectures at the University of California, at Berkeley in 1960 and at Los Angeles in 1961, to construct a satisfactory model along the lines of the now familiar "possible worlds" approach. Concerning the origins of this approach to the construction of models in intensional logic, see Carnap [1] and Kaplan [9].[1]

To overcome the difficulties of 1960–1961, one thing that seems to be necessary is to allow the occurrence of vacuous concepts in the model more freely than was done at that time. For though it was known that such concepts are unavoidable (cf. footnotes

---

Originally published in ***Noûs***, vol. 7 (1973), pp. 24–33, and vol. 8 (1974), pp. 135–156. Errata noted in Church's *Revised Formulation of the Logic of Sense and Denotation: Alternative (1)*, ***Noûs***, vol. 27 (1993), pp. 141–157, have been entered. Reprinted by permission.

[1] David Kaplan's unpublished dissertation (University of California at Los Angeles, 1964), should also be mentioned as an early study of the question of formalization of the logic of sense and denotation and of the use, in this connection, of models along the lines which had been suggested by Carnap.

4, 18 of *A Formulation*), there was an attempt to minimize them that is now seen to have been ill judged.

However, a more important amendment is that the single primitive λ is replaced by an infinite list of primitive symbols $\lambda_0, \lambda_1, \lambda_2, \cdots$. Of these, λ_0 is called the *abstraction operator*; and the subscript 0 may be omitted as an abbreviation, so that λ is written to stand for λ_0. Then the formation rule (2) (at the bottom of page 8 [page 295 in present edition] of *A Formulation*) is replaced by the following, where $n = 0, 1, 2, \cdots$:

> (2_n) if $\mathbf{x}_{\beta_n}$ is a variable of type β_n and $\mathbf{M}_{\alpha_n}$ is a well-formed formula of type α_n, then $(\lambda_n \mathbf{x}_{\beta_n} \mathbf{M}_{\alpha_n})$ is a well-formed formula having the type $\alpha_n\beta_n$.

An occurrence of a variable $\mathbf{x}_\delta$ in a well-formed formula is *bound* or *free* according as it is or is not an occurrence in a well-formed part of the formula having the form $(\lambda_n \mathbf{x}_\delta \mathbf{M}_\gamma)$.

That is, this change in the primitive basis of the system, which was originally intended only for Alternative (0), is now made also for Alternative (2). And for Alternative (2) there is no other change in the list of primitive symbols or in the formation rules, as these are given in *A Formulation*.

1. MODELS FOR ALTERNATIVE (2). On this basis let us now go on to describe a class (or better, an approximate class) of miniature models which we propose for Alternative (2), and to illustrate by setting up in detail two examples of such models.

These models are extensional in the sense that the intensional notion of concept is not presupposed but is provided for by an extensional construction that uses the "possible worlds" of the model. Otherwise they follow the general plan of the "intended interpretation" as this is described on pages 10–14, 16–17 of *A Formulation* [pages 297–299, 300–301 in present edition]. But they are miniature in the sense that they use only a small number of possible worlds and have only a small number of individuals (members of the type ι).

Such miniature models may well be suggestive, and they may enable us to control the rules of inference and the axioms in the sense of showing that (or amending them so that) they do not lead to an inconsistency or a reduction to extensionality or other undesired result. But any one of the miniature models or even all of them together may have some property that does not accord with the intended interpretation in general; and hence they must always be used in conjunction with heuristic considerations and informal insights about the latter. Such appeal to intuition cannot be avoided by constructing a more elaborate model or a larger variety of models, as it will still be needed in order to decide which of the various models that suggest themselves are admissible. And so the models are to be regarded as an aid in formulating a satisfactory system, rather than as providing a final criterion.

We name the models by the number of individuals in each domain of individuals (in each possible world). For example, Model 2-2 makes use of two possible worlds, which we shall call the principal world, or simply the world, and the antiworld, each having two individuals. This method of naming the models may be inconvenient if we wish to allow an infinite number of possible worlds, or if we wish to allow two models which have the same domains of the individuals in each possible world but differ in regard to the membership of other types. Nevertheless we shall use it at least temporarily. And to remove equivocacy we specify that in Model l_1-l_2-$\cdots$-l_n there are n possible worlds, of which the first is taken as the principal world and the others as antiworlds; that in the ith world there are exactly l_i individuals as members of the type ι; that the members of the type o are exactly t and f, the two truth-values; that the membership of any type $\alpha\beta$ consists exactly of all functions from the type β into the type α; and that the membership of any simple type γ_1 (i.e., any type γ_1 that is not a type $\alpha\beta$) consists of the maximum number of concepts that is permissible under the guiding principles of Alternative (2), including one and only one concept $(0, 0, \cdots, 0)$ that is vacuous in all n worlds.

Heuristically, we think of the individuals as contingent things, having contingent existence, but of (e.g.) the truth-values as having necessary existence. Admittedly this heuristic notion must not be pressed too far without more precise specification of it. But the axioms and discussion in sections 5 and 7 below may serve to give it at least some of the needed precision. And meanwhile we follow it in the models by taking always t and f as the members of the type o but allowing various different domains of individuals in different worlds.

When, however, two worlds in a particular model have the same number of individuals, we shall by means of some arbitrary one-to-one correspondence identify each individual in one world with a corresponding individual in the other, and so treat the two domains of individuals as the same. Generally this will simplify the description of the model and will not change the class of closed formulas of type o which are *validated* by the model (in the sense that they denote t in every world). Indeed it may be argued that relations of identity or non-identity between individuals in different worlds have no genuine significance, except by associating with each individual in one of the worlds a particular concept of it (in that world) in some arbitrary way;[2] and hence for the particular purpose of setting up a model, these relations of identity and non-identity between individuals in different worlds are a matter of indifference and are to be disposed of as convenient.

2. MODEL 2-2. Nearly the simplest model that has more than minor value for our purpose is Model 2-2, details of which we now go on to give. As already stated, there are two worlds, which we call the (principal) *world* and the *antiworld*. We shall speak of a member a_{α_1} of the type α_1 as being a *concept of* a_α and as an *anticoncept of* b_α, meaning that it is a concept of a_α in the world and a concept of b_α in the antiworld. And we shall speak of the *value* and *antivalue* of any well-formed formula (closed or not) for a system of values of any list of variables that includes all its free variables, meaning its value in the world and in the antiworld. But since in Model 2-2 the membership of any type is the same in world and antiworld, we speak simply of the *members* of the type, not needing to distinguish members or antimembers.

[2]That is, with every individual a_ι in one of the worlds let a concept a_{ι_1} be associated in such a way that a_{ι_1} is in this first world a concept of a_ι. Then an individual d_ι in the second world is identified with the individual a_ι in the first world if and only if a_{ι_1} is in the second world a concept of d_ι. Depending on how the concepts of the individuals in the first world are chosen, this may sometimes result in identifying two different individuals in the first world with the same individual in the second world. We regard this as not excluded. But our point here is a different one, that there is no basis at all for deciding relations of identity and non-identity between individuals in different worlds, except with respect to some choice of concepts of them.

That a_{α_1} is a concept of a_α implies that any closed formula which expresses a_{α_1} as its sense denotes a_α (i.e., in the world). And that a_{α_1} is an anticoncept of b_α implies that any closed formula expressing a_{α_1} as its sense antidenotes b_α. The sense of a closed formula is to be (unlike the denotation) always the same in both worlds.

As we have done in what was just said, we shall sometimes employ lower-case italic letters for elements of the model, usually with a subscript to indicate the type. But as this threatens confusion with object-language variables, we shall (as at least a temporary expedient) use only the letters a, b, c, d and various lower-case Greek letters for elements of the model, and then largely avoid using a, b, c, d (with subscripts) as object-language variables. A few exceptions to this last may be made in cases in which no confusion can arise.

In Model 2-2 the members of the type o are t and f, and the members of the type ι are 1 and 2. If α is any simple type, the members of the type α_1 are all ordered pairs (a_α, b_α), where a_α and b_α are any members of the type α; all ordered pairs $(a_\alpha, 0)$ where a_α is any member of the type α; all ordered pairs $(0, b_\alpha)$ where b_α is any member of the type α; and the ordered pair $(0, 0)$.

For example, the members of the type o_1 are (t,t), (t,f), (f,t), (f,f), (t,0), (f,0), (0,t), (0,f), (0,0); the members of the type ι_1 are (1,1), (1,2), (2,1), (2,2), (1,0), (2,0), (0,1), (0,2), (0,0); and there are one hundred members of the type ι_2, including e.g., ((1,2),(2,2)), ((1,2), (2,0)), ((1,2),(0,0)), ((1,2),0), ((0,0), (0,0)), (0,(0,0)), (0,0).

As already indicated, the members of any type $\alpha\beta$ are all the functions from the members of the type β into members of the type α. (Thus there are four members of the type $o\iota$, sixteen members of the type $o\iota\iota$, five hundred twelve members of the type $o\iota_1$.)

As a recursive definition, let $(0,0)_{\gamma_1}$ be the ordered pair $(0,0)$ if γ_1 is a simple type, and let $(0,0)_{\alpha_1\beta_1}$ be the function $\psi_{\alpha_1\beta_1}$ such that $\psi_{\alpha_1\beta_1} a_{\beta_1} = (0,0)_{\alpha_1}$ for every member a_{β_1} of the type β_1.

The concept relation and the anticoncept relation must now both be determined in the model. This will require another recursive definition; as a first step toward this we specify that if γ is a simple type, then (a_γ, b_γ) is a concept of a_γ and an anticoncept of b_γ; $(a_\gamma, 0)$ is a concept of a_γ and is not an anticoncept of anything; $(0, b_\gamma)$ is an anticoncept of b_γ and not a concept of anything; and $(0,0)$ is neither a concept nor an anticoncept of anything.

When γ is not a simple type, it will be convenient to introduce the notations $(a_\gamma, b_\gamma), (a_\gamma, 0), (0, b_\gamma)$ in such a way that an analogue of the foregoing holds: then (a_γ, b_γ) is a concept of a_γ and an anticoncept of b_γ; $(a_\gamma, 0)$ is a concept of a_γ and is not an anticoncept of anything; $(0, b_\gamma)$ is an anticoncept of b_γ and not a concept of anything; and all remaining members of the type γ_1 (not of one of these three forms) are neither concepts nor anticoncepts of anything. Thus the notations $(a_\gamma, b_\gamma), (a_\gamma, 0), (0, b_\gamma)$ will stand for ordered pairs only in case γ is a simple type symbol; we provide for the contrary case by the following recursive definition:

1. $(\phi_{\alpha\beta}, \psi_{\alpha\beta})$ is the function $\phi_{\alpha_1\beta_1}$ whose value for any argument of the form (a_β, b_β) is $(\phi_{\alpha\beta}a_\beta, \psi_{\alpha\beta}b_\beta)$, whose value for any argument of the form $(a_\beta, 0)$ is $(\phi_{\alpha\beta}a_\beta, 0)$, whose value for any argument of the form $(0, b_\beta)$ is $(0, \psi_{\alpha\beta}b_\beta)$, and whose value for all remaining arguments a_{β_1} of type β_1 is $\phi_{\alpha_1\beta_1}a_{\beta_1} = (0,0)_{\alpha_1}$.
2. $(\phi_{\alpha\beta}, 0)$ is the function $\phi_{\alpha_1\beta_1}$ whose value for an argument of either of the forms

(a_β, b_β) or $(a_\beta, 0)$ is $(\phi_{\alpha\beta} a_\beta, 0)$, and whose value for all remaining arguments of the type β_1 is $(0, 0)_{\alpha_1}$.

3. $(0, \psi_{\alpha\beta})$ is the function $\phi_{\alpha_1\beta_1}$ whose value for an argument of either of the forms (a_β, b_β) or $(0, b_\beta)$ is $(0, \psi_{\alpha\beta} b_\beta)$, and whose value for all remaining arguments of type β_1 is $(0, 0)_{\alpha_1}$.

We say that the function $\phi_{\alpha_1\beta_1}$ *characterizes* the function $\phi_{\alpha\beta}$ if, for every a_β and every concept a_{β_1} of a_β, $\phi_{\alpha_1\beta_1} a_{\beta_1}$ is a concept of $\phi_{\alpha\beta} a_\beta$. And we say that $\phi_{\alpha_1\beta_1}$ *anticharacterizes* $\psi_{\alpha\beta}$ if, for every a_β and every anticoncept a_{β_1} of a_β, $\phi_{\alpha_1\beta_1} a_{\beta_1}$ is an anticoncept of $\psi_{\alpha\beta} a_\beta$. From the foregoing definitions of the concept relation and of the anticoncept relation in the model it is clear that (in the model) every concept of $\phi_{\alpha\beta}$ characterizes $\phi_{\alpha\beta}$ and every anticoncept of $\psi_{\alpha\beta}$ anticharacterizes $\psi_{\alpha\beta}$; but the converses do not hold.

Now to complete the account of Model 2-2 we turn to what we shall call the *model-semantics*. Heuristically this is the semantics of the object language as it would be if Model 2-2 were the true model. But we observe that there is a sense in which the actual intended meaning of the object language is not given by the model-semantics of any one model, even the supposed true model, as the object language is designed for users who do not know the true model or know it only in part.[3]

We state only the semantical rules that determine the value and the antivalue of any well-formed formula for a system of values of its free variables. It is understood that a value of a variable must always be a member of the type to which the variable belongs (as indicated by its subscript) and that the value and the antivalue of a well-formed formula must always be members of the type to which the formula belongs. Moreover, the value (the antivalue) of a well-formed formula, for a given system of values of any list of variables that includes all its free variables, is the same as its value (its antivalue) for the system of values that results by deleting from the list all the variables that are not free variables of the given formula. The denotation and the antidenotation of a closed formula are then the same as its value and its antivalue for any system of values of any list of variables (not excluding the empty list of variables). The sense of a closed formula of type α is (a_α, b_α), where a_α is its denotation and b_α is its antidenotation.[4] And the sense-value of a well-formed formula for a given system of sense-values of its

[3]Thus the users may have a concept, and extend the language by introducing a name that expresses this concept as its sense, and yet in some sense they may not know what the concept is a concept of or what the name denotes; e.g. they may not be able to determine the identity or non-identity of the denotation with other members of the appropriate type which they regard as previously known. On this ground it may be argued that the sense of a name is prior to its denotation, and that the construction of the model misleads by reversing this order, using the members of (say) the type ι as a means of obtaining the concepts of them in the type ι_1. This may be admitted, and yet the models serve their purpose.

The reality to which the artificial models correspond is of course that our ignorance and varying degrees of uncertainty about details of *this* world force us to think of ourselves as being in some unknown one of a large number of partly comprehensible possible worlds.

[4]Because the object language is designed to express the purely logical and nothing else (as far as this is possible) and to have only necessary truths as theorems, we do not admit into it names which lack either a denotation or an antidenotation. If we extend the language in order to be able to express contingent matters, and add postulates that constitute a theory about the real but contingent world, we still (following Frege) confine ourselves to names which, according to the theory, have a denotation. (Indeed, the validity of each of the rules of inference III and IV requires this.) But in this case there will at least arise names that in some models lack an antidenotation in some of the antiworlds. The statement in the text must then be supplemented appropriately; e.g., in a two-world model, if a closed formula (a name) has a_α as its denotation but lacks an antidenotation, its sense is $(a_\alpha, 0)$. [*Editor's note*: This footnote continues.]

free variables may be determined analogously.[5]

The denotation and the antidenotation of the primitive constant C_{ooo} is the function c_{ooo} such that $c_{ooo}a_o b_o$ is f if a_o is t and b_o is f, and $c_{ooo}a_o b_o$ is t if either a_o is f or b_o is t. If the denotation and the antidenotation of $C_{o_n o_n o_n}$ is $c_{o_n o_n o_n}$, the denotation and the antidenotation of $C_{o_{n+1}o_{n+1}o_{n+1}}$ is $(c_{o_n o_n o_n}, c_{o_n o_n o_n})$.

The denotation and the antidenotation of the constant $\Pi_{o(o\alpha)}$ is the function $\pi_{o(o\alpha)}$ whose value $\pi_{o(o\alpha)}\phi_{o\alpha}$, with any member $\phi_{o\alpha}$ of the type $o\alpha$ as argument, is t if $\phi_{o\alpha}$ is the function whose value $\phi_{o\alpha}a_\alpha$ is t for all arguments a_α, and is f in all remaining cases. If the denotation and the antidenotation of $\Pi_{o_n(o_n\alpha_n)}$ is $\pi_{o_n(o_n\alpha_n)}$, the denotation and the antidenotation of $\Pi_{o_{n+1}(o_{n+1}\alpha_{n+1})}$ is $(\pi_{o_n(o_n\alpha_n)}, \pi_{o_n(o_n\alpha_n)})$.

The denotation of the constant $\Delta_{o\alpha_1\alpha}$ is the function $\delta_{o\alpha_1\alpha}$ such that $\delta_{o\alpha_1\alpha}a_\alpha a_{\alpha_1}$ is t if a_{α_1} is a concept of a_α and is f in the contrary case. The antidenotation of $\Delta_{o\alpha_1\alpha}$ is the function $\delta'_{o\alpha_1\alpha}$ such that $\delta'_{o\alpha_1\alpha}a_\alpha a_{\alpha_1}$ is t if a_{α_1} is an anticoncept of a_α and f in the contrary case. The denotation and the antidenotation of $\Delta_{o_1\alpha_2\alpha_1}$ is $(\delta_{o\alpha_1\alpha}, \delta'_{o\alpha_1\alpha})$. If the denotation and antidenotation of $\Delta_{o_n\alpha_{n+1}\alpha_n}$ is $\delta_{o_n\alpha_{n+1}\alpha_n}$ (where n is not 0), the denotation and antidenotation of $\Delta_{o_{n+1}\alpha_{n+2}\alpha_{n+1}}$ is $(\delta_{o_n\alpha_{n+1}\alpha_n}, \delta_{o_n\alpha_{n+1}\alpha_n})$.

For the *principal member* of a type we make a definition by recursion as follows. The principal member of the type o is t. The principal member of the type ι is 1. In the case of any other simple type the principal member is $(0, 0)$. The principal member of the type $\alpha\beta$ is the function $\psi_{\alpha\beta}$ whose value $\psi_{\alpha\beta}a_\beta$ is, for every member a_β of the type β, the principal member of the type α.

For those type symbols β for which $\iota_{\beta(o\beta)}$ is a primitive constant, the denotation and antidenotation of the constant $\iota_{\beta(o\beta)}$ is the function $\theta_{\beta(o\beta)}$ whose value $\theta_{\beta(o\beta)}\phi_{o\beta}$, for any member $\phi_{o\beta}$ of the type $o\beta$, is the member a_β of the type β such that $\phi_{o\beta}a_\beta$ is t, provided there is a unique such member a_β of the type β, and is in the contrary case the principal member of the type β. If the denotation and antidenotation of $\iota_{\beta_n(o_n\beta_n)}$ is $\theta_{\beta_n(o_n\beta_n)}$, the denotation and antidenotation of $\iota_{\beta_{n+1}(o_{n+1}\beta_{n+1})}$ is $(\theta_{\beta_n(o_n\beta_n)}, \theta_{\beta_n(o_n\beta_n)})$.

If a formula consists of a variable $\mathbf{x}_\alpha$ standing alone, its value and its antivalue for the value a_α of $\mathbf{x}_\alpha$ is a_α.

For any system of values of the free variables of $\mathbf{F}_{\alpha\beta}\mathbf{A}_\beta$, if the value of $\mathbf{F}_{\alpha\beta}$ is $\phi_{\alpha\beta}$ and the value of $\mathbf{A}_\beta$ is a_β, the value of $\mathbf{F}_{\alpha\beta}\mathbf{A}_\beta$ is $\phi_{\alpha\beta}a_\beta$; and if the antivalue of $\mathbf{F}_{\alpha\beta}$ is $\psi_{\alpha\beta}$ and the antivalue of $\mathbf{A}_\beta$ is b_β, the antivalue of $\mathbf{F}_{\alpha\beta}\mathbf{A}_\beta$ is $\psi_{\alpha\beta}b_\beta$.

For a given system of values of its free variables, the value of $\lambda\mathbf{x}_\beta\mathbf{M}_\alpha$ is the function $\phi_{\alpha\beta}$ such that, for any member a_β of the type β, the function-value $\phi_{\alpha\beta}a_\beta$ is the same as the value of $\mathbf{M}_\alpha$ for the value a_β of $\mathbf{x}_\beta$ and for the given system of values of the remaining free variables; and the antivalue of $\lambda\mathbf{x}_\beta\mathbf{M}_\alpha$ is the function $\psi_{\alpha\beta}$ such that, for any member a_β of the type β, the function-value $\psi_{\alpha\beta}a_\beta$ is the same as the antivalue of $\mathbf{M}_\alpha$ for the value a_β of $\mathbf{x}_\beta$ and for the given system of values of the remaining free variables.

We now introduce the notion of the *n-fold conceptualization* $\phi^n_{\alpha_n\beta_n}$ of a function $\phi_{\alpha_n\beta_n}$. The definition is by recursion as follows. $\phi^0_{\alpha\beta}$ is the same as $\phi_{\alpha\beta}$. If $\phi_{\alpha_{n+1}\beta_{n+1}}$

(A modified formulation, with modified rules of inference, to allow denotationless as well as antidenotationless names may indeed be worth working out, but it is a complication which we prefer to avoid in the present treatment.)

[5]See the discussion of the notion of sense-value in [5]. We observe that if the sense-value assigned to one of the variables is not a concept of anything (not an anticoncept of anything), then the sense-value of the entire formula must correspondingly not be a concept of anything (an anticoncept of anything).

characterizes $\phi_{\alpha_n\beta_n}$ and anticharacterizes $\psi_{\alpha_n\beta_n}$, then $\phi^{n+1}_{\alpha_{n+1}\beta_{n+1}}$ is $(\phi^n_{\alpha_n\beta_n}, \psi^n_{\alpha_n\beta_n})$. If $\phi_{\alpha_{n+1}\beta n+1}$ characterizes $\phi_{\alpha_n\beta_n}$ and does not anticharacterize any function, then $\phi^{n+1}_{\alpha_{n+1}\beta_{n+1}}$ is $(\phi^n_{\alpha_n\beta_n}, 0)$. If $\phi_{\alpha_{n+1}\beta_{n+1}}$ anticharacterizes $\psi_{\alpha_n\beta_n}$ and does not characterize any function, then $\phi^{n+1}_{\alpha_{n+1}\beta_{n+1}}$ is $(0, \psi_{\alpha_n\beta_n})$. If $\phi_{\alpha_{n+1}\beta_{n+1}}$ neither characterizes nor anticharacterize any function, $\phi^{n+1}_{\alpha_{n+1}\beta_{n+1}}$ is $(0,0)_{\alpha_{n+1}\beta_{n+1}}$.

Then, if $\phi_{\alpha_{n+1}\beta_{n+1}}$ characterizes $\phi_{\alpha_n\beta_n}$, we have that $\phi^1_{\alpha_{n+1}\beta n+1}$ is a concept of $\phi_{\alpha_n\beta_n}$, and that $\phi^{m+1}_{\alpha_{n+1}\beta n+1}$ is a concept of $\phi^m_{\alpha_n\beta_n}$ if $m \leqq n$. Similarly, if $\phi_{\alpha_{n+1}\beta_{n+1}}$ anticharacterizes $\psi_{\alpha_n\beta_n}$, we have that $\phi^1_{\alpha_{n+1}\beta n+1}$ is an anticoncept of $\psi_{\alpha_n\beta_n}$, and that $\phi^{m+1}_{\alpha_{n+1}\beta_{n+1}}$ is an anticoncept of $\psi^m_{\alpha_n\beta_n}$ if $m \leqq n$.

For a given system of values of free variables, if the value and the antivalue of $\lambda \mathbf{x}_{\beta_n}\mathbf{M}_{\alpha_n}$ are $\phi_{\alpha_n\beta_n}$ and $\psi_{\alpha_n\beta_n}$ respectively, the value of $\lambda_n \mathbf{x}_{\beta_n}\mathbf{M}_{\alpha_n}$ is $\phi^n_{\alpha_n\beta_n}$ and the antivalue is $\psi^n_{\alpha_n\beta_n}$.

This now completes the statement of the details of Model 2-2, including the model-semantics.

We shall rely on Model 2-2 as our principal test of proposed axioms, and as a means of establishing about any system of axioms we are thus led to that it is consistent and does not reduce to extensionality. The value of other models is to supplement intuition in enabling us to reject axioms which may be validated by the particular model but seem to be in some degree contrary to the original intention or otherwise doubtful. And as an illustration we go on to outline one additional model.

Part II[6]

3. MODEL 2-0. In Model 2-0, we must distinguish in every type the *members* and the *antimembers*. In the types o, o_1, o_2, ... the members and the antimembers are the same, and are the same as in Model 2-2. In the type ι the members are 1 and 2; and there are no antimembers. In the type ι_{n+1} the members are all the ordered pairs $(a_{\iota_n}, b_{\iota_n})$ where a_{ι_n} may be 0 or any member of the type ι_n and b_{ι_n} may be 0 or any antimember of the type ι_n; and the antimembers are the same as the members. In any type $\alpha\beta$ the members are all functions from the members of the type β to the members of the type α; and the antimembers are all functions from the antimembers of the type β to the antimembers of the type α.

As an example, the type ιo therefore has no antimembers, but the type $o\iota$ has one antimember, the null function from antimembers of the type ι to antimembers of the type o. This latter function is identified, under the original program of *A Formulation*, with the null class of individuals (i.e., of antimembers of the type ι).

Then we make the following recursive definition. Let $(0,0)_{\gamma_1}$ be the ordered pair (0,0) if γ_1 is a simple type; and let $(0,0)_{\alpha_1\beta_1}$ be the function $\psi_{\alpha_1\beta_1}$, which is both a member and an antimember of the type $\alpha_1\beta_1$, such that $\psi_{\alpha_1\beta_1} a_{\beta_1} = (0,0)_{\alpha_1}$ for every member (or antimember) a_{β_1} of the type β_1. Where $\phi_{\alpha\beta}$ is a member of the type $\alpha\beta$ and $\psi_{\alpha\beta}$ is an antimember of the type $\alpha\beta$, let $(\phi_{\alpha\beta}, \psi_{\alpha\beta})$ be the function $\phi_{\alpha_1\beta_1}$, which is both a member and an antimember of the type $\alpha_1\beta_1$, such that, if a_β is any member of the type β and b_β is any antimember of the type β, we have $\phi_{\alpha_1\beta_1}(a_\beta, b_\beta) = (\phi_{\alpha\beta} a_\beta, \psi_{\alpha\beta} b_\beta)$, $\phi_{\alpha_1\beta_1}(a_\beta, 0) = (\phi_{\alpha\beta} a_\beta, 0)$, $\phi_{\alpha_1\beta_1}(0, b_\beta) = (0, \psi_{\alpha\beta} b_\beta)$, and $\phi_{\alpha_1\beta_1} a_{\beta_1} = (0,0)_{\alpha_1}$ in all

[6][Part I was originally published in 1973; Part II was originally published in 1974.]

remaining cases; further let $(\phi_{\alpha\beta}, 0)$ be the function $\phi_{\alpha_1\beta_1}$, which is both a member and an antimember of the type $\alpha_1\beta_1$, such that $\phi_{\alpha_1\beta_1}(a_\beta, b_\beta) = \phi_{\alpha_1\beta_1}(a_\beta, 0) = (\phi_{\alpha\beta}a_\beta, 0)$, and $\phi_{\alpha_1\beta_1}a_{\beta_1} = (0, 0)_{\alpha_1}$ in all remaining cases; and let $(0, \psi_{\alpha\beta})$ be the function $\phi_{\alpha_1\beta_1}$, which is both a member and an antimember of the type $\alpha_1\beta_1$, such that $\phi_{\alpha_1\beta_1}(a_\beta, b_\beta) = \phi_{\alpha_1\beta_1}(0, b_\beta) = (0, \psi_{\alpha\beta}b_\beta)$, and $\phi_{\alpha_1\beta_1}a_{\beta_1} = (0, 0)_{\alpha_1}$ in all remaining cases.

Then for every type γ, (a_γ, b_γ) is a concept of a_γ and an anticoncept of b_γ; $(a_\gamma, 0)$ is a concept of a_γ and is not an anticoncept of anything; $(0, b_\gamma)$ is an anticoncept of b_γ and is not a concept of anything; and all remaining members (and antimembers) of the type γ_1 are neither concepts nor anticoncepts of anything. But observe that (a_γ, b_γ), $(a_\gamma, 0)$, $(0, b_\gamma)$ are ordered pairs only if γ is a simple type.

The denotation and antidenotation of $C_{o_n o_n o_n}$ is the same as in Model 2-2.

The denotation of $\Pi_{o(o\alpha)}$ is the function $\pi_{o(o\alpha)}$ and the antidenotation is the function $\pi'_{o(o\alpha)}$; where if $a_{o\alpha}$ is any member of the type $o\alpha$, the function-value $\pi_{o(o\alpha)}a_{o\alpha}$ is t if and only if $a_{o\alpha}b\alpha$ is t for all members b_α of the type α; and if $a_{o\alpha}$ is any antimember of the type $o\alpha$, the function-value $\pi'_{o(o\alpha)}a_{o\alpha}$ is t if and only if $a_{o\alpha}b_\alpha$ is t for all antimembers b_α of the type α. The denotation and antidenotation of $\Pi_{o_1(o_1\alpha_1)}$ is $(\pi_{o(o\alpha)}, \pi'_{o(o\alpha)})$. If the denotation and antidenotation of $\Pi_{o_n(o_n\alpha_n)}$ is $\pi_{o_n(o_n\alpha_n)}$, the denotation and antidenotation of $\Pi_{o_{n+1}(o_{n+1}\alpha_{n+1})}$ is $(\pi_{o_n(o_n\alpha_n)}, \pi_{o_n(o_n\alpha_n)})$.

The denotations and the antidenotations of the constants $\Delta_{o_n\alpha_{n+1}\alpha_n}$ may be described in the same words as was done in the case of Model 2-2 (though they are not in general the same functions).

The principal member and principal antimember of the type o is t. Of the type ι the principal member is 1 and there is no principal antimember. In the case of any other simple type, the principal member and principal antimember is (0,0). The principal member of the type $\alpha\beta$ is the function $\psi_{\alpha\beta}$ whose value $\psi_{\alpha\beta}a_\beta$ is, for every member a_β of the type β, the principal member of the type α. The principal antimember of the type $\alpha\beta$ is the function $\psi_{\alpha\beta}$ whose value $\psi_{\alpha\beta}a_\beta$ is, for every antimember a_β of the type β, the principal antimember of the type α (or if the type $\alpha\beta$ has no antimembers, there is no principal antimember).

For those type symbols β for which $\iota_{\beta(o\beta)}$ is a primitive constant, the denotation and antidenotation of $\iota_{\beta(o\beta)}$ are the functions $\theta_{\beta(o\beta)}$ and $\theta'_{\beta(o\beta)}$ described as follows. For any member $\phi_{o\beta}$ of the type $o\beta$ the function value $\theta_{\beta(o\beta)}\phi_{o\beta}$ is the member α_β of the type β, such that $\phi_{o\beta}a_\beta$ is t, provided that there is a unique such member a_β of the type β, and is in the contrary case the principal member of the type β. For any antimember $\phi_{o\beta}$ of the type $o\beta$ the function-value $\theta'_{\beta(o\beta)}\phi_{o\beta}$ is the antimember a_β of the type β such that $\phi_{o\beta}a_\beta$ is t, provided there is a unique such antimember a_β of the type β, and is in the contrary case the principal antimember of the type β. The denotation and antidenotation of $\iota_{\beta_1(o_1\beta_1)}$ is $(\theta_{\beta(o\beta)}, \theta'_{\beta(o\beta)})$. If the denotation and antidenotation of $\iota_{\beta_n(o_n\beta_n)}$ is $\theta_{\beta_n(o_n\beta_n)}$, the denotation and antidenotation of $\iota_{\beta_{n+1}(o_{n+1}\beta_{n+1})}$ is $(\theta_{\beta_n(o_n\beta_n)}, \theta_{\beta_n(o_n\beta_n)})$.

In dealing with variables, we must distinguish between a *value* of a variable $\mathbf{x}_\alpha$, which may be any member of the type α, and an *antivalue* of $\mathbf{x}_\alpha$, which may be any antimember of the type α.

Of the formula which consists of the variable $\mathbf{x}_\alpha$ standing alone, the value for the value a_α of $\mathbf{x}_\alpha$ is a_α, and the antivalue for the antivalue b_α of $\mathbf{x}_\alpha$ is b_α.

For any system of values of the free variables of $\mathbf{F}_{\alpha\beta}\mathbf{A}_\beta$, if the value of $\mathbf{F}_{\alpha\beta}$ is $\phi_{\alpha\beta}$

and the value of $\mathbf{A}_\beta$ is a_β, the value of $\mathbf{F}_{\alpha\beta}\mathbf{A}_\beta$ is $\phi_{\alpha\beta}a_\beta$. For any system of antivalues of the free variables of $\mathbf{F}_{\alpha\beta}\mathbf{A}_\beta$, if the antivalue of $\mathbf{F}_{\alpha\beta}$ is $\psi_{\alpha\beta}$ and the antivalue of $\mathbf{A}_\beta$ is b_β, the antivalue of $\mathbf{F}_{\alpha\beta}\mathbf{A}_\beta$ is $\psi_{\alpha\beta}b_\beta$.

For a given system of values of the free variables, the value of $\lambda\mathbf{x}_\beta\mathbf{M}_\alpha$ is the function $\phi_{\alpha\beta}$ such that, for any member a_β of the type β, the function-value $\phi_{\alpha\beta}a_\beta$ is the same as the value of $\mathbf{M}_\alpha$ for the value a_β of $\mathbf{x}_\beta$ and for the given system of values of the remaining free variables. For a given system of antivalues of the free variables, the antivalue of $\lambda\mathbf{x}_\beta\mathbf{M}_\alpha$ is the function $\psi_{\alpha\beta}$ such that, for any antimember b_β of the type β, the function-value $\psi_{\alpha\beta}b_\beta$ is the same as the antivalue of $\mathbf{M}_\alpha$ for the antivalue b_β of $\mathbf{x}_\beta$ and for the given system of antivalues of the remaining free variables.

For a given system of values of the free variables, if the value of $\lambda\mathbf{x}_{\beta_n}\mathbf{M}_{\alpha_n}$ is $\phi_{\alpha_n\beta_n}$, the value of $\lambda_n\mathbf{x}_{\beta_n}\mathbf{M}_{\alpha_n}$ is $\phi^n_{\alpha_n\beta_n}$. For a given system of antivalues of the free variables, if the antivalue of $\lambda\mathbf{x}_{\beta_n}\mathbf{M}_{\alpha_n}$ is $\psi_{\alpha_n\beta_n}$, the antivalue of $\lambda_n\mathbf{x}_{\beta_n}\mathbf{M}_{\alpha_n}$ is $\psi^n_{\alpha_n\beta_n}$.

4. COMPLETENESS OF THE QUANTIFICATION AXIOMS. I am further indebted to A. F. Bausch (1955) for a number of corrections and remarks regarding the primitive basis of the object language as given in *A Formulation*—these are reproduced with some minor changes as i–iii below—and for the suggestion (in early 1956) of a method by which the completeness of the quantification axioms $1^{\alpha\beta}$–7 can be proved.

i. As it is intended that only closed formulas shall be asserted, Rule III on page 17 [page 301 in present edition] is incorrectly stated. It is necessary to add to the rule: "and provided that the formula which results by the inference is closed."

ii. The characterization of preferred type symbols on page 8 [page 295 in present edition] is inaccurate. The intention is that the preferred type symbols correspond to types that are non-empty in all interpretations regarded as admissible, whereas other types may possibly be empty (in models which we presently propose, there may be some worlds in which they are empty). Hence, Bausch's correction is as follows. First, the simple type symbols are divided into preferred and non-preferred in some way that accords with the intended admissible interpretations. Then we reconstrue the type symbols, simple and composite, as expressions of propositional calculus, taking the preferred simple type symbols as denoting t, taking the non-preferred simple type symbols as propositional variables, and taking $(\alpha\beta)$ as a notation for the converse implication $[\alpha \subset \beta]$. The preferred type symbols are those which become tautologies when thus reconstrued. For example, since $p \supset t$ is a tautology (where t denotes t), $o\iota$ and $oo\iota$ are preferred type symbols; and since Peirce's law $p \supset q \supset p \supset p$ is a tautology, all type symbols of the form $\alpha(\alpha(\beta\alpha))$ are preferred type symbols, regardless of the choice of the type symbols α and β.

iii. If we consider the quantification axioms $1^{\alpha\beta}$–7 in isolation, i.e., with the rules I–V (Rule III amended as just described) but without the other axioms, it becomes natural also to reduce the vocabulary of the language to only that which is needed to express the axioms $1^{\alpha\beta}$–7. With the language thus restricted, it is no longer possible to prove the theorems $(p) \,.\, (x_\beta)p \supset p$ in the case of every preferred type symbol β. To remedy this Bausch suggested the axioms:

$$0^{\alpha\beta}. \qquad (p) \,.\, (x_\beta)p \supset \,.\, (f_{\alpha\beta})p \supset p$$

In our present revised formulation, we intend further to take $\iota_1, \iota_2, \iota_3, \ldots$ as preferred type symbols and hence must add also the axioms:[7]

$0^{\iota_{n+1}}$. $\quad (p) \mathbin{.} (x_{\iota_{n+1}})p \supset p$

But these axioms all become superfluous upon restoring to the language the primitive symbols $\iota_{\beta(o\beta)}$ for preferred type symbols β. Indeed, in view of Rule IV, the mere presence in the language of a closed formula of type β is always sufficient to enable us to prove the theorem $(p) \mathbin{.} (x_\beta)p \supset p$ which expresses the non-emptiness of the type β.[8]

Axioms $1^{\alpha\beta}$–7, taken in isolation, have no specifically intensional character and might equally be used for an extensional as an intensional logic—either by restricting the type symbols α, β to those corresponding to extensional types or by reinterpreting the system in such a way that all of the underlying types (even, e.g., such types as o_2 or ι_1 or $o_2\iota_1$) come to be extensional types. The axioms are intended to be adequate for propositional calculus and laws of quantifiers, with the variables of type o in the role of (what are usually called) propositional variables and with the standard logic of quantifiers modified by the restrictions: (a) that only closed formulas shall be theorems, and (b) that all theorems shall be valid even under an interpretation in which certain of the types, especially the type ι of individuals, may be empty.

There is a completeness proof by Hailperin [8] for a formulation of first-order logic which obeys the restrictions (a) and (b). This is the first completeness proof for a formulation of this sort, as Bausch's method not only remained uncompleted and unpublished but was in any case of later date. Nevertheless, Bausch's method is quite different and seems to have been suggested to him, not by Hailperin's paper, but by the well-known completeness proofs of Leon Henkin for functional calculi of first and higher order (cf. [7], §54).

Where $\mathbf{a}, \mathbf{b}, \mathbf{c}, \ldots, \mathbf{u}, \mathbf{v}, \mathbf{w}$ is a finite list of variables of any types, where Γ is a finite set of well-formed formulas of the type o having no other free variables than $\mathbf{a}, \mathbf{b}, \mathbf{c}, \ldots, \mathbf{u}, \mathbf{v}, \mathbf{w}$, and where $\mathbf{A}_o$ is a well-formed formula of type o having no other free variables than $\mathbf{a}, \mathbf{b}, \mathbf{c}, \ldots, \mathbf{u}, \mathbf{v}, \mathbf{w}$, let the notation $\Gamma \vdash_{\mathbf{abc}\ldots\mathbf{uvw}} \mathbf{A}_o$ mean that we have a proof of $\mathbf{A}_o$ from the formulas Γ in which the variables $\mathbf{a}, \mathbf{b}, \mathbf{c}, \ldots, \mathbf{u}, \mathbf{v}, \mathbf{w}$ are treated temporarily as constants—i.e., in place of the axioms (including $1^{\alpha\beta}$–7 and any other axioms admitted), variants of the axioms are used in which $\mathbf{a}, \mathbf{b}, \mathbf{c}, \ldots, \mathbf{u}, \mathbf{v}, \mathbf{w}$ are not bound variables; the variables $\mathbf{a}, \mathbf{b}, \mathbf{c}, \ldots, \mathbf{u}, \mathbf{v}, \mathbf{w}$ are not used as bound variables in the proof; Rules I–V are used with the condition that a formula is closed replaced everywhere by the condition that it is well-formed and has no other free variables than $\mathbf{a}, \mathbf{b}, \mathbf{c}, \ldots, \mathbf{u}, \mathbf{v}, \mathbf{w}$; and the formulas of Γ, after making alphabetic changes of bound variables (if necessary) so that none of the variables $\mathbf{a}, \mathbf{b}, \mathbf{c}, \ldots, \mathbf{u}, \mathbf{v}, \mathbf{w}$ shall have bound

---

[7]The assumption made in the paper of 1951, that the type ι_1 will be empty if the type ι is, has to be corrected. It is an illustration of the unfortunate tendency at that time to avoid allowing vacuous concepts. Frege advised that the primitive basis of a language should be simplified by excluding denotationless names, and indeed it seems that this can be done without loss. But it by no means follows that vacuous concepts can be excluded or even that it is desirable to minimize their occurrence. In retrospect, this seems to be obvious.

[8]To avoid denotationless names, the constants $\iota_{\alpha(o\alpha)}$, where α is a non-preferred type symbol, must not belong to the vocabulary of the purely logical language. But rather the constant $\iota_{\alpha(o\alpha)}$ is to be added to the language (perhaps together with other new primitives) when postulates about contingent matters are introduced which have the non-emptiness of the type α as a consequence. (In our present array of types, if α is a non-preferred type symbol, the emptiness or non-emptiness of the type α is always equivalent to that of the type ι.)

occurrences in them, are used in the proof in the same way as if they were additional axioms. As a special case of this, if Γ is empty, we may write simply $\vdash_{\mathbf{abc...uvw}} \mathbf{A}_o$.[9]

The two following metatheorems were suggested to me by Bausch (I have made non-essential modifications in his statement of them and have verified that they can be proved along lines he indicated):

Deduction Theorem. If $\Gamma, \mathbf{A}_o \vdash_{\mathbf{abc...uvw}} \mathbf{B}_o$, then $\Gamma \vdash_{\mathbf{abc...uvw}} \mathbf{A}_o \supset \mathbf{B}_o$.

Generality Theorem. If $\mathbf{w}$ is not free in any of the formulas of Γ and if $\Gamma \vdash_{\mathbf{abc...uvw}} \mathbf{A}_o$, then $\Gamma \vdash_{\mathbf{abc...uv}} (\mathbf{w})\mathbf{A}_o$.

Proofs of these two metatheorems are by essentially familiar methods and we here give them only in outline. For proof of the Deduction Theorem we must use Axioms $1^{\alpha\beta}$–7 to establish the two following schemata:

VI. If $\mathbf{A}_o$ is obtained from a tautology of the implicational propositional calculus by substituting for each variable of type o a well-formed formula of type o having no other free variables than $\mathbf{a}, \mathbf{b}, \mathbf{c}, \ldots, \mathbf{u}, \mathbf{v}, \mathbf{w}$, then $\vdash_{\mathbf{abc...uvw}} \mathbf{A}_o$.

VII. $\vdash_{\mathbf{abc...uvw}} \mathbf{P}_o \supset \Pi_{o(o\alpha)}\mathbf{F}_{o\alpha} \supset . \mathbf{P}_o \supset \mathbf{F}_{o\alpha}\mathbf{A}_\alpha$.

For the Generality Theorem the critical schemata, again provable from $1^{\alpha\beta}$–7, are:

VIII. $\vdash_{\mathbf{abc...uvw}} (\mathbf{x}_\alpha)[\mathbf{A}_o \supset \mathbf{B}_o] \supset . (\mathbf{x}_\alpha)\mathbf{A}_o \supset (\mathbf{x}_\alpha)\mathbf{B}_o$.

IX. $\vdash_{\mathbf{abc...uvw}} \mathbf{A}_o \supset (\mathbf{x}_\alpha)\mathbf{A}_o$.

X. $\vdash_{\mathbf{abc...uvw}} (\mathbf{x}_\beta)\Pi_{o(o\alpha)}\mathbf{F}_{o\alpha} \supset (\mathbf{x}_\beta)\mathbf{F}_{o\alpha}\mathbf{A}_\alpha$.

In each of the schemata VII–X it is understood that the well-formed formulas represented by the bold capital letters are to be so chosen that the entire formula has no other free variables than $\mathbf{a}, \mathbf{b}, \mathbf{c}, \ldots, \mathbf{u}, \mathbf{v}, \mathbf{w}$. Also, in VI–X the variables $\mathbf{a}, \mathbf{b}, \mathbf{c}, \ldots, \mathbf{u}, \mathbf{v}, \mathbf{w}$ must not occur as bound variables. E.g., in X, $\mathbf{F}_{o\alpha}$ and $\mathbf{A}_\alpha$ may not have $\mathbf{a}, \mathbf{b}, \mathbf{c}, \ldots, \mathbf{u}, \mathbf{v}, \mathbf{w}$ as bound variables and may have no other free variables than $\mathbf{a}, \mathbf{b}, \mathbf{c}, \ldots, \mathbf{u}, \mathbf{v}, \mathbf{w}$, and $\mathbf{x}_\beta$; moreover, $\mathbf{x}_\beta$ may not be the same as any of $\mathbf{a}, \mathbf{b}, \mathbf{c}, \ldots, \mathbf{u}, \mathbf{v}, \mathbf{w}$.

Once the Deduction Theorem and the Generality Theorem have been proved, we may proceed, as Bausch suggested, to imitate Henkin's completeness proof. Or we may just use the two metatheorems to obtain a more direct comparison with some standard formulation of quantifier logic which does not obey the two restrictions (a) and (b) which were stated above.

5. DEFINITIONS. Some use has already been made (in the preceding section) of the definitions which were given on pages 9–10 [page 296 in present edition] of *Formulation*. But since our revised formulation requires that revisions be made in some of these definitions, we restate them here, taking the opportunity also to make some additions to the list:

$[\mathbf{A}_{o_n} \supset \mathbf{B}_{o_n}] \quad \to \quad C_{o_n o_n o_n}\mathbf{A}_{o_n}\mathbf{B}_{o_n}$

$(\mathbf{x}_{\alpha_n})\mathbf{A}_{o_n} \quad \to \quad \Pi_{o_n(o_n\alpha_n)}(\lambda_n \mathbf{x}_{\alpha_n}\mathbf{A}_{o_n})$

$T_{o_n} \quad \to \quad (a_{o_n}) . a_{o_n} \supset a_{o_n}$

$F_{o_n} \quad \to \quad (a_{o_n})a_{o_n}$

$\sim \mathbf{A}_{o_n} \quad \to \quad \mathbf{A}_{o_n} \supset F_{o_n}$

$[\mathbf{A}_{o_n}\mathbf{B}_{o_n}] \quad \to \quad \sim C_{o_n o_n o_n}\mathbf{A}_{o_n} \sim \mathbf{B}_{o_n}$

[9]As here used, the sign $\vdash$, with or without subscripts, is not the Fregean assertion sign but belongs rather to the (syntactical) meta-language.

$$
\begin{array}{lcl}
[\mathbf{A}_{o_n} \equiv \mathbf{B}_{o_n}] & \rightarrow & [[\mathbf{A}_{o_n} \supset \mathbf{B}_{o_n}][\mathbf{B}_{o_n} \supset \mathbf{A}_{o_n}]] \\
(\exists \mathbf{x}_{\alpha_n})\mathbf{A}_{o_n} & \rightarrow & \sim(\mathbf{x}_{\alpha_n}) \sim \mathbf{A}_{o_n} \\
(\iota \mathbf{x}_{\beta_n})\mathbf{A}_{o_n} & \rightarrow & \iota_{\beta_n(o_n\beta_n)}(\lambda_n \mathbf{x}_{\beta_n} \mathbf{A}_{o_n}) \\
Q_{o_n\alpha_n\alpha_n} & \rightarrow & \lambda_n a_{\alpha_n} \lambda_n b_{\alpha_n} (f_{o_n\alpha_n}) \centerdot f_{o_n\alpha_n} b_{\alpha_n} \supset f_{o_n\alpha_n} a_{\alpha_n} \\
[\mathbf{A}_{\alpha_n} =_n \mathbf{B}_{\alpha_n}] & \rightarrow & Q_{o_n\alpha_n\alpha_n} \mathbf{B}_{\alpha_n} \mathbf{A}_{\alpha_n} \\
[\mathbf{A}_{\alpha_n} \neq_n \mathbf{B}_{\alpha_n}] & \rightarrow & \sim \centerdot \mathbf{A}_{\alpha_n} =_n \mathbf{B}_{\alpha_n} \\
N_{o_m o_n} & \rightarrow & Q_{o_m o_n o_n} T_{o_n}, \text{ where } m \leqq n \\
\mathrm{n}_{o_n o_{n+1}} & \rightarrow & \lambda_n p_{o_{n+1}} \centerdot p_{o_{n+1}} =_n \centerdot p_{o_{n+1}} \supset p_{o_{n+1}} \\
\Diamond_{o_n o_{n+1}} & \rightarrow & \lambda_n p_{o_{n+1}} \centerdot \sim \mathrm{n}_{o_n o_{n+1}} \sim p_{o_{n+1}} \\
E_{o_n\alpha_{n+1}} & \rightarrow & \lambda_n x_{\alpha_{n+1}} N_{o_n o_{n+1}} \centerdot x_{\alpha_{n+1}} =_{n+1} x_{\alpha_{n+1}} \\
\mathrm{e}_{o_n\alpha_{n+1}} & \rightarrow & \lambda_n x_{\alpha_{n+1}} (\exists x_{\alpha_n}) \Delta_{o_n\alpha_{n+1}\alpha_n} x_{\alpha_n} x_{\alpha_{n+1}} \\
(\mathbf{F}^n_{\alpha_n\beta_n}) & \rightarrow & \lambda_n a_{\beta_n} (\mathbf{F}_{\alpha_n\beta_n} a_{\beta_n})
\end{array}
$$

The parentheses enclosing $\mathbf{F}^n_{\alpha_n\beta_n}$ in this last definition are often unnecessary and may be omitted.

When T or F is written without type subscript, it is understood that the subscript is o. When n or $\Diamond$ is written without type subscript, it is understood that the subscript is oo_1. When N is written without a type subscript, it is understood that the subscript is oo_n, where n is to be determined from the context (usually by the condition that the formula within which N occurs is to be well-formed). Similarly, when E or e is written without a type subscript, it is understood that the subscript is $o\alpha_1$, with α_1 determined from the context.

For some additional conventions by which type subscripts may be omitted, especially in writing well-formed formulas, see further page 10 [pages 296–297 in present edition] of *A Formulation*.

When either of the signs $=$ or $\neq$ is written without a subscript, it is understood to be 0. This is analogous to the convention already explained for the symbol λ, that when the subscript is omitted, it is understood to be 0.

The notation $(\iota \mathbf{x}_\beta)$, prefixed to a well-formed formula of type o, is a description operator. We use the upright iota to distinguish the Frege-like descriptions $(\iota \mathbf{x}_\beta)\mathbf{A}_o$, whose denotation (or value) lies always in the range of the bound variable $\mathbf{x}_\beta$ even when the description is improper, reserving the inverted iota to be used for Russell's contextually defined descriptions. (Nothing prevents using both kinds of descriptions in the same context.)

The formulas N_{oo_1}, n_{oo_1}, $\Diamond_{oo_1}$ may be regarded as expressing what we shall call strong necessity, weak necessity, and possibility respectively. In the models, strong necessity of a *proposition* (i.e., of a concept of type o_1) corresponds to truth in every world, weak necessity to truth in every world in which the proposition *has* (i.e., is a concept of) a truth-value at all, and possibility to truth in some world.

$E_{o\alpha_1}$ we regard as expressing necessary existence (of something of type α as presented by a concept of it of type α_1), and we regard $\mathrm{e}_{o\alpha_1}$ as similarly expressing actual existence. In the models, necessary existence is existence in every world, actual existence is existence in the principal world; and to these we may add possible existence as existence in some world.

For example, let P_{ι_1} be added to our object language as a name of the Pegasus concept. Then $E_{o\iota_1} P_{\iota_1}$ expresses the (certainly false) proposition that Pegasus has necessary existence; $\mathrm{e}_{o\iota_1} P_{\iota_1}$ expresses the (probably false) proposition that Pegasus has or had actual existence; and to express that Pegasus has possible existence we may write $\Diamond_{oo_1} \cdot P_{\iota_1} =_1 P_{\iota_1}$.

However, it must not be supposed that the relation of identity has some peculiarly appropriate role in expressing existence propositions. We might if we wished even bring in the identity concept (denoted by $Q_{o_1\alpha_1\alpha_1}$) in expressing actual existence, writing, e.g., $\Delta_{oo_1o}T_o[P_{\iota_1} =_1 P_{\iota_1}]$ to express the actual existence of Pegasus. On the other hand, we might express the necessary existence of Pegasus by $N(\exists f_{o_1\iota_1})f_{o_1\iota_1}P_{\iota_1}$, and his possible existence by $\Diamond(\exists f_{o_1\iota_1})f_{o_1\iota_1}P_{\iota_1}$.

Observe that a name P_ι which expresses P_{ι_1} as its sense and therefore is or purports to be a name of Pegasus must not be added to the object language unless we are prepared to assume the actual existence of Pegasus.[10] And even if the name P_ι is added to the language, the existence of Pegasus must still be expressed by such a sentence as $e_{o\iota_1}P_{\iota_1}$ rather than by, e.g., $(\exists f_{o\iota})f_{o\iota}P_\iota$. The fault of the latter sentence as a means of expressing the existence of Pegasus is that the man who disagrees cannot reasonably use the negation of the sentence to assert his disagreement—as he would thereby be using the name P_ι, which he believes to be denotationless. In Fregean terminology, the name "Pegasus" has an *ungerade* occurrence in such English sentences as "Pegasus exists," "Pegasus does not exists [never existed]"; and because the *ungerade* usage is logically anomalous, it is to be eliminated in a formalized language.[11]

6. METATHEOREMS CONCERNING MODEL 2-2. In treating two-world models it will be convenient to use the notations

$$(\mathbf{M}_\alpha)^{\mathbf{x}_\beta}_{\alpha_\beta}, \qquad [\mathbf{M}_\alpha]^{\mathbf{x}_\beta}_{b_\beta}$$

to mean respectively the value of $\mathbf{M}_\alpha$ for the value α_β of $\mathbf{x}_\beta$ and the antivalue of $\mathbf{M}_\alpha$ for the (anti)value b_β of $\mathbf{x}_\beta$. Besides these we use also similar notations for the case of more than one variable; for instance,

$$(\mathbf{M}_\alpha)^{\mathbf{x}_\beta\mathbf{y}_\gamma\mathbf{z}_\delta}_{a_\beta b_\gamma c_\delta}$$

is the value of $\mathbf{M}_\alpha$ for the system of values $a_\beta, b_\gamma, c_\delta$ of the variables $\mathbf{x}_\beta, \mathbf{y}_\gamma, \mathbf{z}_\delta$.

Still in relation to a two-world model, if $\mathbf{M}_{\alpha_1}$ and $\mathbf{M}_\alpha$ are closed formulas, we say that $\mathbf{M}_{\alpha_1}$ is *sense-related* to $\mathbf{M}_\alpha$ if what $\mathbf{M}_{\alpha_1}$ denotes is both a concept of what

---

[10] As already indicated in Note 4, it is not meant to raise objection to the undertaking to devise a formalized language in which denotationless names occur, and in which sentences containing denotationless names are therefore without truth-value. But the simplest formalized language results if the semantics is such that denotationless names are avoided. And it is believed that cases in which a natural-language sentence that contains a denotationless name seems nevertheless to have a truth-value are always better explained as instances of some logical or semantical anomaly, often *ungerade* usage.

[11] A more familiar method of dealing with the existence of Pegasus in the formalized language is to introduce a constant $P_{o\iota}$ which may be read as "is Pegasus" (or "pegasizes"); then to express that Pegasus exists (since $P_{o\iota}$ already has such a sense that the function denoted can have the value t for no more than one argument) we write simply $(\exists x_\iota)P_{o\iota}x_\iota$. This way of making the assertion has the advantage that intensionalities are avoided for the particular purpose, without unjustifiably depriving the unbeliever of the vocabulary by which to assert his dissenting opinion. But lest it be thought that there is a conflict between the two ways of expressing the existence of Pegasus, it should be noticed that $P_{o\iota}$ may have such a sense that it is capable of definition in terms of P_{ι_1} as $\lambda x_\iota(\Delta_{o\iota_1\iota}x_\iota P_{\iota_1})$, and that under this definition $e_{o\iota_1}P_{o\iota}$ and $(\exists x_\iota)P_{\iota_1}x_\iota$ are trivially equivalent.

It is in this sense that existence is not a predicate. That is, it is not a predicate of P_ι, but of either P_{ι_1} or $P_{o\iota}$.

$\mathbf{M}_\alpha$ denotes and an anticoncept of what $\mathbf{M}_\alpha$ antidenotes and what $\mathbf{M}_{\alpha_1}$ antidenotes is both a concept of what $\mathbf{M}_\alpha$ denotes and an anticoncept of what $\mathbf{M}_\alpha$ antidenotes. Under Alternative (2) and because the object language has neither denotationless nor antidenotationless names, this is equivalent to saying that $\mathbf{M}_{\alpha_1}$ both denotes and antidenotes the sense of $\mathbf{M}_\alpha$ (compare Note 4).

If $\mathbf{M}_{\alpha_1}$ has no free variable except $\mathbf{x}_{\beta_1}$ and $\mathbf{M}_\alpha$ has no free variable except $\mathbf{x}_\beta$, we say that $\mathbf{M}_{\alpha_1}$ is *sense-related* to $\mathbf{M}_\alpha$ if (1) for every member a_β of the type β and every concept a_{β_1} of a_β,

$$(\mathbf{M}_{\alpha_1})^{\mathbf{x}_{\beta_1}}_{a_{\beta_1}} \quad \text{and} \quad [\mathbf{M}_{\alpha_1}]^{\mathbf{x}_{\beta_1}}_{a_{\beta_1}}$$

are the same and are a concept of

$$(\mathbf{M}_\alpha)^{\mathbf{x}_\beta}_{a_\beta}$$

and (2) for every antimember b_β of the type β and every anticoncept b_{β_1} of b_β,

$$(\mathbf{M}_{\alpha_1})^{\mathbf{x}_{\beta_1}}_{b_{\beta_1}} \quad \text{and} \quad [\mathbf{M}_{\alpha_1}]^{\mathbf{x}_{\beta_1}}_{b_{\beta_1}}$$

are the same and are an anticoncept of

$$[\mathbf{M}_\alpha]^{\mathbf{x}_\beta}_{b_\beta}.$$

The definition of sense-relatedness is then extended in an obvious way to the case of any finite list $\mathbf{x}_{\beta_1}, \mathbf{y}_{\gamma_1}, \mathbf{z}_{\delta_1}, \ldots$ of free variables of $\mathbf{M}_{\alpha_1}$ and corresponding list $\mathbf{x}_\beta, \mathbf{y}_\gamma, \mathbf{z}_\delta, \ldots$ of free variables of $\mathbf{M}_\alpha$ ($\mathbf{x}_\beta$ corresponding to $\mathbf{x}_{\beta_1}$ and $\mathbf{y}_\gamma$ corresponding to $\mathbf{y}_{\gamma_1}$ and so on).

Now, let us call $\mathbf{M}_{\alpha_n}$ the *nth ascendant* of a well-formed formula $\mathbf{M}_\alpha$ if it is obtained from $\mathbf{M}_\alpha$ by increasing all subscripts of type symbols by n and simultaneously increasing all subscripts of λ by n. For example, the second ascendant of $\lambda x_{\iota_1} x_{\iota_1}$ is $\lambda_2 x_{\iota_3} x_{\iota_3}$, and the first ascendant of $C_{ooo} p_o q_o$ is $C_{o_1 o_1 o_1} p_{o_1} q_{o_1}$.

This definition of the ascendants of $\mathbf{M}_\alpha$ applies primarily to the case of an unabbreviated well-formed formula $\mathbf{M}_\alpha$. But the (object-language) definitions in Section 5 above have been so arranged that the same method of obtaining ascendants by increasing subscripts can be applied directly to well-formed formulas as abbreviated by means of these definitions, provided that the subscripts are increased also after the signs $=$ and $\neq$. For example, the first ascendant of $(\exists \mathbf{x}_\iota) \centerdot \mathbf{x}_\iota = \mathbf{x}_\iota$ is $(\exists \mathbf{x}_{\iota_1}) \centerdot \mathbf{x}_{\iota_1} =_1 \mathbf{x}_{\iota_1}$.

The two following metatheorems, which are to be understood as concerning the particular Model 2-2, will often greatly shorten the process of verifying that a proposed axiom is validated by this model:

Sense-Relationship Theorem. The first ascendant $\mathbf{M}_{\alpha_1}$ of $\mathbf{M}_\alpha$ is sense-related to $\mathbf{M}_\alpha$, with each free variable $\mathbf{x}_{\beta_1}$ corresponding to the variable $\mathbf{x}_\beta$ from which it was obtained by increasing the type-symbol subscripts by 1.

Antivalue Theorem. If $\mathbf{M}_\alpha$ does not contain any of the primitive symbols $\Delta_{o\beta_1\beta}$, its value and antivalue are the same for any given system of values of its free variables.

The Sense-Relationship Theorem presumably generalizes (under appropriate generalization of the definition of *sense-related*) to any model which we would regard as admissible.[12] But the Antivalue Theorem is of course special to models which have the same domain of individuals in all worlds.

7. AXIOMS FOR ALTERNATIVE (2). We find that all the axioms $1^{\alpha\beta}$–17^{α} (from *A Formulation*) are validated by Model 2-2 with the critical exception of the axioms $16^{\alpha\beta}$. Axioms $16^{\alpha\beta}$ are therefore revised as follows:[13]

$16^{\alpha\beta}$. $$(f_{\alpha\beta})(f_{\alpha_1\beta_1}) \centerdot (x_\beta)(x_{\beta_1})[\Delta_{o\beta_1\beta} x_\beta x_{\beta_1} \supset \Delta_{o\alpha_1\alpha}(f_{\alpha\beta}x_\beta)(f_{\alpha_1\beta_1}x_{\beta_1})] \supset \Delta_{o(\alpha_1\beta_1)(\alpha\beta)} f_{\alpha\beta} f^1_{\alpha_1\beta_1}$$

Then our proposal is to retain the axioms $1^{\alpha\beta}$–17^{α} with this one amendment for Alternative (2); and also with some additional amendments for Alternative (0).

However, we should notice at once (as in effect was already pointed out to me by John G. Kemeny in 1952) that we have as a consequence the theorems

$$(x_\beta)(\exists x_{\beta_1})\Delta_{o\beta_1\beta} x_\beta x_{\beta_1}$$

—in words, for each type β that there are no conceptless things. The intention of 1951, to make the non-existence of conceptless things an optional axiom schema, which might or might not be included, is therefore not possible except by a modification of the basic axioms $1^{\alpha\beta}$–17^{α} which threatens to be undesirably or untenably drastic.

Of the remaining axioms for Alternative (2) which were suggested in *A Formulation*, we remark that a few schemata fail to be validated by Model 2-2, notably $33^{*\alpha\beta}$, and a few others have superfluous antecedents (i.e., the axioms were expressed with excessive caution).

---

[12] If $\mathbf{M}_\alpha$ is a closed formula, let us call it a *necessary name* if its sense, denoted by $\mathbf{M}_{\alpha_1}$, satisfies the condition $E_{o\alpha_1}\mathbf{M}_{\alpha_1}$; and in the contrary case (and unless $\sim\Diamond \centerdot \mathbf{M}_{\alpha_1} =_1 \mathbf{M}_{\alpha_1}$ is satisfied), a *contingent name*.

As regards closed formulas, what is essential to the Sense-Relationship Theorem is that for every $\mathbf{M}_\alpha$ there is a corresponding $\mathbf{M}_{\alpha_1}$ which is a necessary name of the sense of $\mathbf{M}_\alpha$. In our present object language we have moreover that every closed formula, of whatever type, is a necessary name (according to the informally intended interpretation of the language).

If we extend the object language in the way described in Note 4, so that contingent names are admitted, the Sense-Relationship Theorem may then be thought to be counterintuitive. This is debatable. But experimentally it may be worth while in this case to look for models in which the Sense-Relationship Theorem fails, and to consider what follows if these are treated as admissible models.

All the models which we have so far considered seem to verify the Sense-Relationship Theorem even upon the addition of contingent names, e.g., of type ι. Or, more accurately, the notation in the extended object language can be so arranged that the Sense-Relationship Theorem is verified. The models for which the Sense-Relationship Theorem fails may be such that, e.g., an individual concept c_{ι_1} does not in general have a corresponding c_{ι_2} which is a concept of it in every world. We do not now pursue this question further, as our first task must be to put into good order the basic object language which excludes contingent names.

However, we anticipate the objection that if $\mathbf{C}_\iota$ is a name expressing c_{ι_1} as its sense, the phrase "the sense of ______", where the blank is to be filled with a name of $\mathbf{C}_\iota$, must express a sense which is a concept of c_{ι_1} in every world. The objection fails because it tacitly assumes that the name of $\mathbf{C}_\iota$ will remain a name of $\mathbf{C}_\iota$ in every world. Indeed the objection is plausible only if we confuse use and mention by filling the blank in the quoted phrase with the formula $\mathbf{C}_\iota$ itself, rather than name of $\mathbf{C}_\iota$.

(Apropos of this last we remark that if the individuals are supposed to be physical things, then names of formulas must no doubt have the type $o_1\iota_1$—the type, not of classes, but of class concepts of physical things.)

[13] In stating $16^{\alpha\beta}$ we retain the subscripts on the Δ's for clearness, though they may be omitted by a convention of abbreviation which was introduced in *A Formulation*.

In this outline, as was done in *A Formulation*, we shall not undertake a deductive organization in the sense of providing a full list of independent axioms, but shall merely list some tentatively proposed axioms for Alternative (2), which are validated by Model 2-2 and by informal intuition, and which might be used as axioms if they do not become theorems in consequence of other axioms chosen.

We also notice that, because of our intention that only truths of logic shall be theorems, it should turn out that the necessitation of every theorem is a theorem—where by the *necessitation* of a well-formed formula of type o we mean the result of applying N_{oo_1} to its first ascendant. And for an analogous reason it should turn out that the necessitation of each of the primitive rules of inference is a derived rule—where the *necessitation* of a rule of inference is obtained by replacing the premisses and the conclusion by their respective necessitations. We must leave open the question, how best to provide axioms or added rules of inference so that these two conditions are fulfilled. But we remark that it is hardly permissible to cut the Gordian knot by adding a primitive *rule of necessitation* to the effect that from $\mathbf{P}_o$ we may infer its necessitation $N\mathbf{P}_{o_1}$.[14]

Axiom 36 of *A Formulation*,

36. $$(p_{o_1})(q_{o_1}) \,\centerdot\, N[p_{o_1} \supset q_{o_1}] \supset \centerdot\, Np_{o_1} \supset Nq_{o_1},$$

is validated by the model, and yields the necessitation of Rule V.

Axioms 35^α of *A Formulation* are preferably modified as follows:

35^α. $$(f_{o_1\alpha_1})(x_{\alpha_1}) \,\centerdot\, Ex_{\alpha_1} \supset \centerdot\, N(\Pi_{o_1(o_1\alpha_1)} f_{o_1\alpha_1}) \supset N(f_{o_1\alpha_1} x_{\alpha_1})$$

These are evidently closely related to the necessitation of Rule IV, but they do not directly yield the necessitation of Rule IV except in cases in which $E\mathbf{A}_{\alpha_1}$ can be proved. Indeed, we might assume as an axiom schema that $E\mathbf{A}_{\alpha_1}$ is an axiom if $\mathbf{A}_{\alpha_1}$ is the first ascendant of a closed formula $\mathbf{A}_\alpha$, but not without first asking whether there is a better method of securing the necessitation of Rule IV.

There are similar difficulties in regard to the necessitation of Rules II and III. But the following axioms are partly sufficient for this if we either assume the axiom schema $E\mathbf{A}_{\alpha_1}$, where $\mathbf{A}_{\alpha_1}$ is a first ascendant and closed, or find the means of proving it as a theorem schema. All of the axioms are validated by Model 2-2, but to secure validation also by Model 2-0 Axioms $56^{\alpha\beta}$ must be restricted to the case that β is a preferred type symbol:

$54^{\alpha\beta}$. $$(f_{\alpha_1\beta_1})(x_{\beta_1}) \,\centerdot\, Ef_{\alpha_1\beta_1} \supset \centerdot\, Ex_{\beta_1} \supset E(f_{\alpha_1\beta_1} x_{\beta_1})$$

$55^{\alpha\beta}$. $$(f_{\alpha_1\beta_1})(x_{\beta_1}) \,\centerdot\, Ef^1_{\alpha_1\beta_1} \supset \centerdot\, Ex_{\beta_1} \supset \centerdot\, f^1_{\alpha_1\beta_1} x_{\beta_1} = f_{\alpha_1\beta_1} x_{\beta_1}$$

$56^{\alpha\beta}$. $$(f_{\alpha_1\beta_1})(g_{\alpha_1\beta_1}) \,\centerdot\, (x_{\beta_1})[Ex_{\beta_1} \supset \centerdot\, f_{\alpha_1\beta_1} x_{\beta_1} = g_{\alpha_1\beta_1} x_{\beta_1}] \supset \centerdot\, Ef^1_{\alpha_1\beta_1} \supset \centerdot\, Eg^1_{\alpha_1\beta_1} \supset \centerdot\, f^1_{\alpha_1\beta_1} = g^1_{\alpha_1\beta_1}$$

[14] Objection to the suggested rule of necessitation is simply that (as elementary textbooks are often at pains to point out) a legitimate principle of inference must not be such that it can lead from true premiss or premisses to false conclusion. Even in our present object language, it is possible to express things which, though they may be true, are not necessary—e.g., that the number of individuals is infinite, or that it is finite, or that it is greater than 0. And one of these may well come to be a theorem when we extend the language to add postulates about contingent matters.

$57^{nm\alpha\beta}$. $(f_{\alpha_n\beta_n}) \centerdot (f^n_{\alpha_n\beta_n})^m = f^n_{\alpha_n\beta_n}$ if $m \leqq n$.

$57^{mn\alpha\beta}$. $(f_{\alpha_n\beta_n}) \centerdot (f^m_{\alpha_n\beta_n})^n = f^n_{\alpha_n\beta_n}$ if $m \leqq n$.

The following proposed axioms, which are validated by both Models 2-2 and 2-0, seem to be also necessary for an adequate theory:

58^{kmn}. $(p_{o_n}) \centerdot N_{o_k o_m}(N_{o_m o_n} p_{o_n}) = N_{o_k o_n} p_{o_n}$ if $k < m < n$.

$59^{n\alpha\beta}$. $(f_{\alpha_n\beta_n})(f_{\alpha_{n+1}\beta_{n+1}}) \centerdot \Delta f_{\alpha_n\beta_n} f_{\alpha_{n+1}\beta_{n+1}} \supset \Delta f^n_{\alpha_n\beta_n} f^{n+1}_{\alpha_{n+1}\beta_{n+1}}$

$60^{\alpha\beta}$. $(f_{\alpha\beta})(f_{\alpha_1\beta_1})(y_\alpha)(x_{\beta_1}) \centerdot \Delta f_{\alpha\beta} f_{\alpha_1\beta_1} \supset \centerdot \Delta y_\alpha(f_{\alpha_1\beta_1} x_{\beta_1}) \supset (\exists x_\beta)\Delta x_\beta x_{\beta_1}$

Of the two proposed axiom schemata that are introduced on page 21 [page 304 in present edition] of *A Formulation* as expressing (in each type α_1) the principle of Alternative (2), only the first one is retained here, and it is modified as follows:

61^{α}. $(x_{\alpha_1})(y_{\alpha_1}) \centerdot N[x_{\alpha_1} =_1 y_{\alpha_1}] \supset \centerdot x_{\alpha_1} = y_{\alpha_1}$

This is validated by Model 2-2; but considered as expressing Alternative (2) it seems undesirably weak, as we may illustrate by taking the particular axiom 61^{ι}.

If we think of the individuals as contingent things, as was suggested above, it is natural to say, given any particular individual, that it might not have existed — or, in reference to the models, that there is a possible world in which it does not exist. Indeed, this assertion may be defended as being a part of what is contained in saying that the individuals are contingent things — but only after it is corrected in form in accordance with a Fregean analysis.

For example, there is a possible world in which *Sir Walter Scott* exists but *the author of Waverley* does not exist (because in that world nobody writes Waverley). Equally, there is a possible world in which *the author of Waverley* exists but *Sir Walter Scott* does not (because in that world the elder Scotts have only one son, who is named James and dies in infancy, and Robert Louis Stevenson writes Waverley).[15] The existence

[15]In these parenthetic descriptions of possible worlds, all the names have *ungerade* occurrences, and for this reason there is no conflict between this method of giving examples and what was said in the last paragraph of Section 1 above.

Nevertheless, the way in which the examples are given calls to attention that such names as "the author of ***Waverley***" or "Robert Louis Stevenson" do not in the natural language have their sense so exactly determined as to enable us always — even in the most remotely possible and unlikely worlds — to pick out the individual named or to decide whether he exists. We may indeed set out to fix the sense of these names more precisely, by deciding successive questions as to what, in pseudo-traditional terminology, is *essential* to being (e.g.) the author of ***Waverley*** and what is merely *accidental* to it. But in a typical case into which contingencies enter, we must not expect that this series of more and more precisely fixed senses will terminate in a completely precised sense.

We mention this essentially familiar point only to remark that theories generally must abstract from such imprecision — e.g., geometry if regarded as a theory of physical measurement — and not excluding logic, if regarded as a theory of natural or possible language.

To see that the formality of logic *presupposes* that ambiguity of sense has been completely removed consider the inference: "Sir Walter Scott is a poet, the author of ***Waverley*** is buried at Dryburgh Abbey, therefore a poet is buried at Dryburgh Abbey." This is fallacious, as there are possible worlds in which the premisses are true but the conclusion false. Hence, it remains fallacious if the names "Sir Walter Scott" and "the author of ***Waverley***" are replaced by a single name W that is ambiguous in sense between these two names — notwithstanding that the name W has no equivocacy of denotation in the real world.

In the foregoing example, the name which is of uncertain sense but determinate denotation is (perhaps) an individual name, but parallel examples can evidently be given in which the ambiguous name is a class

assertions must therefore refer to the concepts expressed by the two italicized names and not to the man Sir Walter Scott known to us in *this* world, who is the common denotation of both names.

Instead of saying that for every individual, there are possible worlds in which it does not exist, we should more properly say that for every individual concept, there are possible worlds in which it is vacuous. And this does seem to be supported by informal intuition.[16]

The bearing of this on Axiom 61^{ι} is the following. The antecedent $N[x_{\iota_1} =_1 y_{\iota_1}]$ has the value t only if the values of x_{ι_1} and y_{ι_1} are such that we have in every world that they are concepts of the same individual. Thus, if every individual concept is vacuous in some world, $N[x_{\iota_1} =_1 y_{\iota_1}]$ is false for all values of x_{ι_1} and y_{ι_1}. Contrary to what holds in Model 2-2, it therefore suggests itself that in a model which better reflects the intention that the individuals are to be contingent things we would have Axiom 61^{ι} vacuously true. We conclude that the axiom says nothing about the conditions under which contingent (i.e., possibly but not necessarily vacuous) individual concepts are identical.

Even if, instead of the type ι, we consider a type α whose members (or some of whose members) are assumed to have necessary existence, we may still see along similar lines that Axiom 61^{α} is too weak because it says nothing about the identity of concepts that are vacuous in some worlds. We need a stronger axiom which shall say roughly: "If about two concepts of type α_1 it is necessary that if either one is a concept of some thing, the other one is a concept of that same thing, then the concepts are identical."[17] And for this the following axiom is suggested as a possibility:

62^{α}. $$(x_{\alpha_1})(y_{\alpha_1}) \mathbin{\centerdot} \Diamond[x_{\alpha_1} =_1 x_{\alpha_1}] \supset \mathbin{\centerdot} [x_{\alpha_1} =_1 x_{\alpha_1}] = [y_{\alpha_1} =_1 y_{\alpha_1}] \supset \mathbin{\centerdot} \mathrm{n}[x_{\alpha_1} =_1 y_{\alpha_1}] \supset \mathbin{\centerdot} x_{\alpha_1} = y_{\alpha_1}.$$

In 62^{α}, if a_{α_1} and b_{α_1} are the values of x_{α_1} and y_{α_1} respectively, the first antecedent has value t if and only if there is some world in which a_{α_1} is a non-vacuous concept; the second antecedent has value t if and only if the worlds in which a_{α_1} is a non-vacuous concept are the same as the worlds in which b_{α_1} is a non-vacuous concept; and the third antecedent corresponds to the antecedent of 61^{α}, with strong necessity replaced by weak necessity.

---

name, etc. Even the traditional "All men are mortal, Socrates is a man, therefore Socrates is mortal" is fallacious if the word "man" (here treated as a class name for the sake of the illustration) is ambiguous in sense.

[16]This can be expressed in the object language as $(x_{\iota_1}) \sim E_{o\iota_1} x_{\iota_1}$. But to take this as an axiom would be contrary to the tendency of other axioms, which is to leave open what the individuals (and individual concepts) are, rather than to make any special assumptions about them (even that they are contingent).

[17]In the axiom itself, as distinguished from the heuristic preliminary, there can be no mention of the possible worlds or of the members of the various types in these worlds and relations which hold among them. Indeed, if reference is intended to a true model, it is clear that it must be supposed that all but one of its possible worlds are unreal. And it is essential to Fregean semantics that nothing can be said truly of what does not exist.

In the miniature models, we have been at pains to secure existence in all the possible worlds by using numbers, functions, and ordered pairs in the construction—excluding such things as, e.g., Halley's Comet, Gondwanaland, and the Great Pyramid, about whose existence we might just possibly be mistaken. But the miniature models are not meant as principal interpretations of the axioms. They serve rather as means by which certain metatheorems regarding consistency and independence can be proved. And their resemblance to a true model can be explained only in indirect discourse: they have certain features which the true model, though never more than partly determinate, is supposed also to have.

If α is a composite type, the first antecedent of 62^α may not be omitted. And indeed an axiom to the effect that if two concepts of type α_1 are both necessarily vacuous, they are identical, cannot be expressed without introducing a new notation to distinguish between concepts of type α_1 (including those which are necessarily vacuous) and members of the type α_1 which are not concepts.

Finally it may be worth noticing that, without changing the effect of the axioms, the first and second antecedents of 62^α can be replaced by $\Diamond(\exists f_{o_1\alpha_1})f_{o_1\alpha_1}x_{\alpha_1}$ and $(\exists f_{o_1\alpha_1})f_{o_1\alpha_1}x_{\alpha_1} = (\exists f_{o_1\alpha_1})f_{o_1\alpha_1}y_{\alpha_1}$ respectively. Compare the discussion at the end of Section 5.

8. ALTERNATIVE (0). As is clear from page 6 [page 294 in present edition] of *A Formulation* (and footnotes 6 and 7), it was seen already at that time that analogues of the semantical antinomies may well arise in connection with either Alternative (0) or Alternative (1), and that caution is therefore necessary in formulating axioms for these alternatives.

Indeed, the stronger are the conditions required in order that two names shall express the same sense, or that concepts shall be identical, the more closely will the abstract theory of concepts resemble the more concrete theory of the names themselves — with the relations symbolized by $\Delta_{o\alpha_1\alpha}$ serving as analogues of the relations of denoting in the semantical theory. Hence, the stronger these conditions are, the greater is the danger of antinomies analogous to Richard's or Grelling's or the Epimenides.

On this basis Alternative (1) might seem to be safer than Alternative (0). However, John Myhill showed me in 1960 that an antinomy of the kind described does in fact follow from the axioms for Alternative (1) as they appear in *A Formulation*. And I shall not now attempt a repair for Alternative (1), but let us turn rather to the intrinsically more important Alternative (0).

It appears that any remedy for the antinomies must parallel one of those that are usual in formulating the (extensional) semantics of a particular language. And with this in mind we abandon the array of primitive symbols

$$\Delta_{o_n\alpha_{n+1}\alpha_n} \qquad n = 0, 1, 2, \ldots$$

in favor of the more complicated array of primitive symbols

$$\Delta^m_{o_n\alpha_{n+1}\alpha_n}$$

where $n = 0, 1, 2, \ldots$ and $m = l, l+1, l+2, \ldots$, where l is the greatest subscript occurring in the type symbol α. No other change is made in the primitive symbols that were used above for Alternative (2). (The superscript m may be omitted as an abbreviation when it is 0.)

The intention is a sort of ramified type theory — but Tarski-like rather than Russell-like, in the sense that the ramification affects only the symbols Δ instead of occurrences of bound variables.

Rules I–V and Axioms $1^{\alpha\beta}$–10^β are retained, the same that were used above for Alternative (2).

Axioms 11^n, $12^{n\alpha}$, $13^{n\beta}$ are unchanged except by supplying a superscript on the Δ, the least possible in each case (i.e., the least superscript such that the resulting primitive symbol Δ is one of those listed above as among our primitive symbols). Axioms $14^{n\alpha}$ are changed to:

$14^{mn\alpha}$. $\quad \Delta^{m+1}\Delta^{m}_{o_n\alpha_{n+1}\alpha_n}\Delta^{m}_{o_{n+1}\alpha_{n+2}\alpha_{n+1}}$.

Here the ramification begins to appear. But to simplify matters we make the ramification affect truth rather than meaningfulness as far as this is possible; if, e.g., in $14^{mn\alpha}$ we replace the superscript $m+1$ on the first Δ by some smaller superscript, leaving the rest unchanged, the resulting formula is still well-formed, provided all the symbols Δ that are called for are in fact among our primitive symbols, but it is counted as false. Also we add the theorems:

$14\frac{1}{2}^{m\alpha}$. $\quad (x_\alpha)(x_{\alpha_1}) \centerdot \Delta^{m}_{o\alpha_1\alpha}x_\alpha x_{\alpha_1} \supset \Delta^{m+1}_{o\alpha_1\alpha}x_\alpha x_{\alpha_1}$

Hence, the result of replacing the superscript $m+1$ on the first Δ in $14^{mn\alpha}$ by some larger superscript will be a theorem.

In Axioms $14^{mn\alpha}$ and $14\frac{1}{2}^{m\alpha}$ and in all the axioms which are introduced below, it is understood that the superscript m must be such that Δ^m is in fact a primitive symbol.

Some form of Axioms $15^{\alpha\beta}$ and $16^{\alpha\beta}$ is evidently essential, to be able to prove theorems of the form $\Delta\mathbf{M}_\alpha\mathbf{M}_{\alpha_1}$ that are called for by the informal semantical background, and Axioms 17^α are basic to the notion of concept. Tentatively, we retain these axioms in the following forms:

$15^{m\alpha\beta}$. $\quad (f_{\alpha\beta})(f_{\alpha_1\beta_1})(x_\beta)(x_{\beta_1}) \centerdot \Delta^m f_{\alpha\beta}f_{\alpha_1\beta_1} \supset \centerdot \Delta^m x_\beta x_{\beta_1} \supset \Delta^m(f_{\alpha\beta}x_\beta)(f_{\alpha_1\beta_1}x_{\beta_1})$

$16^{m\alpha\beta}$. $\quad (f_{\alpha\beta})(f_{\alpha_1\beta_1}) \centerdot (x_\beta)(x_{\beta_1})[\Delta^m x_\beta x_{\beta_1} \supset \Delta^m(f_{\alpha\beta}x_\beta)(f_{\alpha_1\beta_1}x_{\beta_1})] \supset \Delta^m f_{\alpha\beta}f^1_{\alpha_1\beta_1}$

$17^{m\alpha}$. $\quad (x_\alpha)(y_\alpha)(x_{\alpha_1}) \centerdot \Delta^m x_\alpha x_{\alpha_1} \supset \centerdot \Delta^m y_\alpha x_{\alpha_1} \supset \centerdot x_\alpha = y_\alpha$

In Axioms $16^{m\alpha\beta}$, $f^1_{\alpha_1\beta_1}$ is to be read in accordance with the definition in Section 5. In fact, all of the definitions of Section 5 may here be maintained, except that in the definition of $e_{o_n\alpha_{n+1}}$ a superscript m must be supplied on the Δ and hence also on the e. But the definitions that were meant to provide notations for modalities cease to be of interest in connection with Alternative (0) and should preferably be dropped for that reason.

The principle of Alternative (0) is that two names express the same sense if and only if they are synonymously isomorphic in the sense of [6]. Guided by this principle, and considering questions as to when two names of the forms $\mathbf{F}_{\alpha\beta}\mathbf{A}_\beta$ and $\mathbf{G}_{\alpha\gamma}\mathbf{B}_\gamma$ express the same sense, or two names of the form $\lambda_n\mathbf{x}_{\beta_n}\mathbf{M}_{\alpha_n}$ and $\lambda_k\mathbf{x}_{\beta_n}\mathbf{N}_{\alpha_n}$, or two names of the form $\lambda_n\mathbf{x}_{\beta_n}\mathbf{M}_{\alpha_n}$ and $\mathbf{G}_{\alpha_n\beta_n\gamma}\mathbf{B}_\gamma$, we are led to the following axioms:

$63^{m\alpha\beta}$. $\quad (f_{\alpha\beta})(f_{\alpha_1\beta_1})(g_{\alpha\beta})(g_{\alpha_1\beta_1})(x_\beta)(x_{\beta_1})(y_\beta)(y_{\beta_1}) \centerdot$
$\Delta^m f_{\alpha\beta}f_{\alpha_1\beta_1} \supset \centerdot \Delta^m g_{\alpha\beta}g_{\alpha_1\beta_1} \supset \centerdot \Delta^m x_\beta x_{\beta_1} \supset \centerdot$
$\Delta^m y_\beta y_{\beta_1} \supset \centerdot f_{\alpha_1\beta_1}x_{\beta_1} = g_{\alpha_1\beta_1}y_{\beta_1} \supset \centerdot f_{\alpha_1\beta_1} = g_{\alpha_1\beta_1}$

$64^{m\alpha\beta}$. $\quad (f_{\alpha\beta})(f_{\alpha_1\beta_1})(x_\beta)(x_{\beta_1})(y_\beta)(y_{\beta_1}) \centerdot \Delta^m f_{\alpha\beta}f_{\alpha_1\beta_1} \supset \centerdot$
$\Delta^m x_\beta x_{\beta_1} \supset \centerdot \Delta^m y_\beta y_{\beta_1} \supset \centerdot f_{\alpha_1\beta_1}x_{\beta_1} = f_{\alpha_1\beta_1}y_{\beta_1} \supset \centerdot x_{\beta_1} = y_{\beta_1}$

$65^{m\alpha\beta\gamma}$. $\quad (f_{\alpha\beta})(f_{\alpha_1\beta_1})(g_{\alpha\gamma})(g_{\alpha_1\gamma_1})(x_\beta)(x_{\beta_1})(y_\gamma)(y_{\gamma_1}) \centerdot$
$\Delta^m f_{\alpha\beta}f_{\alpha_1\beta_1} \supset \centerdot \Delta^m g_{\alpha\gamma}g_{\alpha_1\gamma_1} \supset \centerdot \Delta^m x_\beta x_{\beta_1} \supset \centerdot$
$\Delta^m y_\gamma y_{\gamma_1} \supset \centerdot f_{\alpha_1\beta_1}x_{\beta_1} \neq g_{\alpha_1\gamma_1}y_{\gamma_1}$
if γ is not the same as β

$66^{mn\alpha\beta}$. $\quad (f_{\alpha_n\beta_n})(f_{\alpha_{n+1}\beta_{n+1}})(g_{\alpha_{n+1}\beta_{n+1}})(x_{\beta_n})(x_{\beta_{n+1}}) \centerdot$

$$\Delta^m f_{\alpha_n\beta_n} f^{n+1}_{\alpha_{n+1}\beta_{n+1}} \supset . \Delta^m x_{\beta_n} x_{\beta_{n+1}} \supset . \\ f^{n+1}_{\alpha_{n+1}\beta_{n+1}} = g^{n+1}_{\alpha_{n+1}\beta_{n+1}} \supset . f_{\alpha_{n+1}\beta_{n+1}} x_{\beta_{n+1}} = g_{\alpha_{n+1}\beta_{n+1}} x_{\beta_{n+1}}$$

$67^{mnk\alpha\beta}$. $(f_{\alpha_n\beta_n})(f_{\alpha_{n+1}\beta_{n+1}})(g_{\alpha_{n+1}\beta_{n+1}}) . \Delta^m f_{\alpha_n\beta_n} f^{n+1}_{\alpha_{n+1}\beta_{n+1}} \supset . f^{n+1}_{\alpha_{n+1}\beta_{n+1}} \neq g^{k}_{\alpha_{n+1}\beta_{n+1}}$ if $1 \leqq k \leqq n$

$68^{mn\alpha\beta\gamma}$. $(f_{\alpha_{n+1}\beta_{n+1}})(g_{\alpha_n\beta_n\gamma})(g_{\alpha_{n+1}\beta_{n+1}\gamma_1})(x_\gamma)(x_{\gamma_1}) .$
$\Delta^m g_{\alpha_n\beta_n\gamma} g_{\alpha_{n+1}\beta_{n+1}\gamma_1} \supset . \Delta^m x_\gamma x_{\gamma_1} \supset . f^{n+1}_{\alpha_{n+1}\beta_{n+1}} \neq g_{\alpha_{n+1}\beta_{n+1}\gamma_1} x_{\gamma_1}$

These axioms may still be insufficient, especially in connection with the question of vacuous concepts. Moreover, in order to be able to prove the theorem $\Delta^m \mathbf{M}_\alpha \mathbf{M}_{\alpha_1}$, with appropriate superscript m, in every case in which $\mathbf{M}_\alpha$ is closed and $\mathbf{M}_{\alpha_1}$ is its first ascendant — and related theorems of this sort in which $\mathbf{M}_\alpha$ and hence its first ascendant $\mathbf{M}_{\alpha_1}$ have free variables — we must add at least the following axioms, generalizing $16^{m\alpha\beta}$:

$16^{mn\alpha\beta}$. $(f_{\alpha_n\beta_n})(f_{\alpha_{n+1}\beta_{n+1}}) . (x_{\beta_n})(x_{\beta_{n+1}})[\Delta^m x_{\beta_n} x_{\beta_{n+1}} \supset$
$\Delta^m (f_{\alpha_n\beta_n} x_{\beta_n})(f_{\alpha_{n+1}\beta_{n+1}} x_{\beta_{n+1}})] \supset \Delta^m f^{n}_{\alpha_n\beta_n} f^{n+1}_{\alpha_{n+1}\beta_{n+1}}$

However, before a definite set of axioms is proposed with assurance we need the guidance of a model, as we had in the case of Alternative (2).[18] The possible-worlds approach to the construction of a model is no longer suitable. But it is believed that a model can be obtained along different lines, by utilizing the analogy that was mentioned above between the theory of concepts in the sense of Alternative (0) and extensional semantics. The fact that the names themselves cannot satisfactorily be used to take the place of concepts need not be a bar to the use of the analogy for the purpose of model construction.

Approximately, the method is that we start with our present object language, supply an extensional semantics for the extensional part of it in the way that is standard for simple type theory (perhaps based on some finite domain of individuals), and then extend the language in such a way that every entity whose existence is contemplated by the semantics shall have a name in the language. A convention is introduced by which every well-formed formula has a *standardization*, obtained from it by only alphabetical changes of bound variable, two formulas having the same standardization if they differ only by alphabetic changes of bound variable.[19] Then, to provide a semantics for the intensional part of the language, we proceed by systematically changing the relation

$\mathbf{A}_{\alpha_1}$ denotes the concept that is expressed by the closed formula $\mathbf{A}_\alpha$,

as this appears in the informally intended semantics, to the relation

$\mathbf{A}_{\alpha_1}$ denotes the standardization of the closed formula $\mathbf{A}_\alpha$.

This description is intended only as a heuristic sketch, and it is not known at this writing whether details can be satisfactorily filled in. But since the membership of some types must be infinite (e.g., the members of the type o_1 are the standardizations

[18]There has been made in proof (February, 1974) a systematic alternation of the axioms which appear in Section 8, correcting an oversight which was called to my attention by C. A. Anderson.

[19]This assumes that, in the particular language, formulas are synonymous (synonymously isomorphic) only when they differ by no more than alphabetic changes of bound variable and that extensions of the language which are made preserve this feature. But we avoid including in the axioms for Alternative (0) any assumption according to which this (necessarily) is the case in any language which the intended system of concept fits.

of all closed formulas of type o), and since the extension of the language to provide names of everything supposed to exist must be repeated at every level in the hierarchy of concepts, it seems likely that the language will in the end be extended to an infinitary language,[20] in which, e.g., disjunctions of infinite sets of formulas (of type o) are allowed as well-formed. It is intended that a model of the extended, possibly infinitary, language shall thus be found, and that the desired model of the original language shall then appear as a submodel.

UNIVERSITY OF CALIFORNIA, LOS ANGELES

REFERENCES

[1] Carnap, Rudolf, "Replies and Systematic Expositions," *The Philosophy of Rudolf Carnap* (La Salle, Illinois: 1963): 859–1013; see 889–900, and also 1045, 1052 of the same book.

[2] Church, Alonzo, "A Formulation of the Logic of Sense and Denotation," abstract, *The Journal of Symbolic Logic* 11 (1946): 31.

[3] ———, "On Carnap's Analysis of Statements of Assertion and Belief," *Analysis* 10 (1950): 97–99.

[4] ———, "A Formulation of the Logic of Sense and Denotation," *Structure, Method, and Meaning, Essays in Honor of Henry M. Sheffer* (New York: 1951): 3–24.

[5] ———, "The Need for Abstract Entities in Semantic Analysis," *Proceedings of the American Academy of Arts and Sciences* 80 (1951): 100–112.

[6] ———, "Intensional Isomorphism and Identity of Belief," *Philosophical Studies* 5 (1954): 65–73.

[7] ———, *Introduction to Mathematical Logic*, Vol. I (Princeton: 1956).

[8] Hailperin, Theodore, "Quantification Theory and Empty Individual-Domains," *The Journal of Symbolic Logic* 18 (1953): 197–200.

[9] Kaplan, David, Review of Saul Kripke's "Semantical Analysis of Modal Logic I," *The Journal of Symbolic Logic* 31 (1966): 120–122.

[20] An infinitary language (so called) is not properly a language, as it could not even theoretically be used for the purpose of human communication. Nevertheless, it is a definite mathematical structure whose existence can be set-theoretically certified, and which is therefore available for use in model building. Our confidence in the resulting model is as great as our confidence in the existence of a model for the requisite portion of axiomatic set theory.

RUSSELLIAN SIMPLE TYPE THEORY (1974)

In this paper I advocate a reexamination of mathematical logic in the tradition of *Principia Mathematica* and the early writings of Russell ([PoM], [OD], [ML]), to ask in what form it should now be studied in order to preserve certain important contributions of Russell and in order to compare it with the somewhat different Fregean tradition. This should not be as ramified type theory with axioms of reducibility, as this involves excessive complication,[1] and Russell himself conceded the objections against the axioms of reducibility (see [Intr2nd]); and not ramified type theory without axioms of reducibility, as this is insufficient for classical mathematics even if one adds the axioms of extensionality which Russell, influenced by Wittgenstein, suggests in [Intr2nd].

The modified version of Russell's mathematical logic which is now usual is simple theory of types ([Chwistek 1, 2], [Ramsey], [Carnap1], [Gödel]), with reliance on Tarski's resolution ([Tarski 2], [Church 6]) of the semantical antinomies.[2] This is often taken as an extensional theory with explicit addition of axioms of extensionality ([Carnap 1], [Gödel], [Tarski 1], [Quine 1], [Henkin], [Church 5]). If one is concerned primarily, or exclusively, with extensional logic, this is indeed the best approach, but it loses some of the characteristically Russellian contributions to logic. Especially the motivation for Russell's contextual definitions of class abstracts and of descriptions largely disappears, and it may seem better just to introduce primitive notations for both of these and to adopt a theory of descriptions which is more like Frege's than Russell's.[3] Having formulated extensional logic in this way, if we then later wish to deal with intensional logic while maintaining the same basis for extensional logic, it is hardly avoidable to follow Frege's distinction of sense and denotation instead of the Russell approach.

I believe that, following [Quine 2, 3], there is now rather wide recognition that Russell's logic must be understood intensionally if some of its significant features are to be preserved. This means in the first place that the values of the propositional-functional (for short "functional") variables are understood to be *properties*, in the case of singulary functional variables, or binary *relations in intension* in the case of binary functional variables, *n-ary relations in intension* in the case of n-ary variables.[4]

Originally published in ***Proceedings and Addresses of the American Philosophical Association***, vol. 47 (1974), pp. 21–33. Presidential address delivered at the Forty-eighth Annual Meeting of the Pacific Division of the American Philosophical Association in San Francisco, March 29, 1974. This paper was afterwards put into form for publication with the support of National Science Foundation Grant No. GP-43517. Reprinted by permission.

[1]Though ramified type theory with axioms of reducibility is now mainly of historical interest, I nevertheless believe it has often been underestimated. See the treatment of this theory in [Church 6].

[2]Not that the resolution of the semantical antinomies by means of ramified type theory is unsound, but that Tarski's is simpler if in conformity with Russell's original purpose a logic adequate for classical mathematics is to be maintained.

[3]Or in this case the notation for class abstracts may be defined by means of that for descriptions.

[4]The italicized terms are to a large extent Russell's own. In particular he makes frequent use of "property" see [ML] sections I–IV and first paragraph of section V, [PM 1] pp. 56–57, 166, 168, and [Intr2nd] pp. xliii-xliv — and this term, as being the usual one for the intensional notion that is in question, is preferable to Quine's "attribute." Moreover Russell's use of "relation in extension" — see [PM 1] p. 81 and [Intr2nd]

To go with this the values of Russell's propositional variables must also be taken as intensional,[5] that is, as propositions in the abstract sense[6] rather than either sentences on the one hand or truth-values on the other. Then sentences, being substitutable for propositional variables, must be taken as names of propositions, notwithstanding Russell's explicit denial in [AppC].[7]

The foregoing refers primarily to the first edition of *Principia Mathematica*, and indeed in the second edition the main text of the work remains unchanged except for the correction of misprints and minor errors. But the second edition adds [Intr2nd] and three appendices, in which various substantial changes are proposed.[8] One of these is that objections which had been made against the axioms of reducibility are conceded and it is suggested to replace these axioms by an axiom of extensionality in each type (without, however, abandoning ramified type theory). And since Russell had previously argued[9] that there are intensional contexts which show that axioms of extensionality fail, he is concerned in [AppC] to rebut this.

However, the arguments in [AppC] are hardly convincing. The strange analysis of

---

p. xxxix — at least suggests the parallel term "relation in intension." But in [Intr2nd] (see pp. xv, xxxii) he introduces "relation in intension" in a specialized sense according to which it is only in atomic propositions $R_2(a,b), R_3(a,b,c), \ldots$ that $R_2, R_3, \ldots$ are relations in intension, and these are said to be "universals" whereas truth-functions of such atomic propositions do not, it seems, give rise in this way to new universals that might be values of Russell's functional variables. (The last point is no doubt connected with Russell's doctrine that propositions are incomplete symbols, described already in [PM 1], pp. 43–44, but this doctrine has to be abandoned in the reconstruction of Russell's logic that is proposed in this paper, because the resulting fragmentation of propositions is incompatible with the use of bound propositional and functional variables.)

[5]From the syntax of Russell's formalized language together with unproblematical aspects of the intended meaning it is clear that the values of the functional variables are functions *from* the members of the appropriate type — e.g. the type of individuals — *to* whatever serve as values of the propositional variables. Such functions can be intensional either by their values' being intensional or by abandonment of the usual principle of extensionality for functions (according to which two functions are identical if they have the same range of arguments and have for the same arguments always the same value). The former alternative must be adopted here in order to accord with Russell's actual use of propositional variables in 1908–1910 (see e.g. sections I and II of [ML], Chapter I of the *Introduction* of [PM 1]), and in view of the serious difficulties of taking sentences as values of the propositional variables.

[6]Abstracting, that is, from the particular words or symbols and the particular language in which the proposition is or might be expressed.

[7]Discussing the axioms of extensionality (for ramified type theory) which were suggested in [Intr2nd], Russell writes in [AppC]: "On grounds not connected with our present question, we cannot regard propositions as names; but that does not decide the question whether [materially] equivalent propositions are identical." Here the word "propositions" must evidently mean *sentences* at its first occurrence, but must have some more abstract meaning at its second occurrence. Thus the quoted sentence not only shows Russell rejecting the idea of treating sentences as names but also illustrates the pervasive use-mention-like confusion in ***Principia Mathematica***. Later in [AppC] Russell distinguishes among (1) "the verbal proposition" (i.e., the sentence), (2) "an instance of the verbal proposition" (i.e., a sentence-token or inscription), and (3) "the fact." But he does not recognize propositions in any sense in which it would be intelligible to ask whether materially equivalent propositions are identical — or in which propositions could be values of the propositional variables in a formalized language such as that of ***Principia Mathematica***. And in the passage in the *Introduction* of [PM 1] that was referred to in footnote 4 it is declared "that what we call a 'proposition' (in the sense in which this is distinguished from the phrase expressing it) is not a single entity at all."

[8]Russell writes: "Some of these are scarcely open to question; others are, as yet, a matter of opinion" ([Intr2nd], p. xiii). And the change regarded as most doubtful ([Intr2nd], footnote * on page xxix) is the same one that concerns us here, the replacement of the axioms of reducibility by axioms of extensionality.

[9]See the beginning of section VII of [ML] and a longer parallel passage on page 73 of [PM 1].

belief statements on the fourth page of the appendix leaves no possibility of quantifying with respect to propositional variables that occur in a belief context,[10] although there are many entirely ordinary assertions whose restatement in a *Principia*-like language would seem to require this;[11] for example "a is sometimes mistaken," which (ignoring for simplicity the possibility that a's beliefs may differ at different times) we may write as[12]

$$(\exists p) \,.\, B^n(a, p) \,.\, {\sim}p,$$

or "a is not omniscient,"

$${\sim}(p) \,.\, B^n(a, p) \equiv p,$$

or "a believes everything asserted by Aristotle,"

$$A^n(\alpha, p) \supset_p B^n(a, p).$$

On the seventh page of the appendix, what seems at first sight to be a bound propositional variable p is found on a more careful reading to be a sentence variable, and this explains the occurrence of "p is true" where, if p were a propositional variable, one would expect simply "p." In fact on this page Russell analyzes statements of the form "a asserts p" and "p is about a" as referring to sentences rather than propositions in any more abstract sense,[13] and his analysis is therefore open to the objection that is raised in [Church 2].

In the present connection a more important consideration than the insufficiency of [AppC] is the point already made, that if Russell is successful in explaining away intensional contexts, he has destroyed the need and the original purpose of his contextual definition of descriptions. For the matter of improper descriptions is in that case quite as well (or better) taken care of by the use of Frege descriptions. And the central point in Russell's original purpose was rather to provide an answer, in his opinion a better answer than Frege's, to Frege's question about the meaning of identity statements:

---

[10]This fragmentation of the object of a belief must be seen as analogous to the fragmentation of the object of a judgment that is maintained in the passage referred to in footnotes 4, 7.

[11]The case in which bound propositional or functional variables are or seem to be involved must always be considered in any proposed analysis of statements of assertion and belief, whether or not the purpose is to explain away intensionalities. In particular the attempt in [Ajdukiewicz] to provide an extensional analysis seems to fail on just this point, as it calls for a fragmentation of the object of knowledge (or belief) which is much like Russell's but without the psychological trappings.

[12]If antinomy is to be avoided, we must not simply introduce two primitive predicates, A ("asserts") and B ("believes"), but rather each must be separated into an infinite hierarchy of predicates of orders 1, 2, 3, . . . ; with, say, the understanding that $A^n(x, p)$ and $B^n(x, p)$ are both false (independently of the value of x) when p has a value that involves either assertions or beliefs of order $\geqq n$. Otherwise antinomy would result if we suppose—as seems to be possible even if unlikely—that a believes that he is sometimes mistaken, while all his remaining beliefs that are not logical consequences of this one are in fact true; and we get an exactly analogous antinomy if we replace a's beliefs by his assertions, or by his assertions on a particular day.

This explains the superscripts that are used after A and B in the text, to show the order.

It is indeed counterintuitive that we are unable to say that a is sometimes mistaken, or that a believes all assertions of Aristotle, without restricting to beliefs and assertions not exceeding some fixed order. But this can be defended on the ground that it is essentially the same restriction that is imposed by ramified type theory. Both this restriction and the awkward necessity of having to reconsider for each new primitive intensional predicate, as it is introduced, whether and how a division into orders is required, I believe can be mitigated to some extent by using a Fregean rather than a Russellian approach to intensional logic—but this is a matter of which further study is still necessary.

[13]The occurrence of "believes" in line 9 of this page seems to be an accidental mistake for "asserts."

How can $\mathbf{A} = \mathbf{B}$, if true, ever differ in meaning from $\mathbf{A} = \mathbf{A}$? If Scott is indeed the same as the author of Waverley, how could King George know that Scott = Scott, and yet not know that Scott = the author of Waverley?

The decision that, in the light of later developments, the logic of *Principia Mathematica* requires modification in the sense of replacing ramified by simple theory of types still leaves open the decision between extensional and intensional values of the variables. Therefore as a means of preserving some of Russell's significant contributions to logic, and for the purpose of comparing Russellian logic with the more extensionally oriented logic of Frege, I propose a simple type theory in which the propositional and functional variables have intensional values but the values of the individual variables remain extensional. Let us call this *Russellian simple type theory* (not of course "Russell's simple type theory," as Russell did not himself propose a simple type theory).

In a formulation R of Russellian simple type theory, let us determine the syntax as follows. There is a type i, the type of individuals. If $\beta_1, \beta_2, \ldots, \beta_m$ are types, there is a type $(\beta_1, \beta_2, \cdots, \beta_m)$. As an abbreviation of $(i, i, \cdots, i)$, with m i's occurring, or as an alternative notation for this type, we use the numeral m; this includes using the numeral 0 for ( ).[14] There is an infinite alphabet of variables of each type, the type being indicated by a superscript.[15] Other primitives are $\sim$, $\vee$, the universal quantifier,[16] brackets, parentheses, comma, and an unspecified list of primitive constants (each of which must have a definite type).[17] Where $\mathbf{f}, \mathbf{a}_1, \mathbf{a}_2, \ldots, \mathbf{a}_m$ are variables or constants, the condition for $\mathbf{f}(\mathbf{a}_1, \mathbf{a}_2, \ldots, \mathbf{a}_m)$ to be well-formed is that $\mathbf{f}$ shall be of type $(\beta_1, \beta_2, \ldots, \beta_m)$ where $\beta_1, \beta_2, \ldots, \beta_m$ are the types of $\mathbf{a}_1, \mathbf{a}_2, \ldots, \mathbf{a}_m$ respectively. Other formation rules are the obvious ones. $[\mathbf{P} \supset \mathbf{Q}]$ is defined as $[\sim\mathbf{P} \vee \mathbf{Q}]$.[18] Axioms of propositional calculus are $*1{\cdot}2$, $*1{\cdot}3$, $*1{\cdot}4$, $*1{\cdot}6$ of [PM 1], omitting $*1{\cdot}5$ which is now known not to be independent ([Bernays]).[19] The other axioms and the rules of inference are as in [Church 5] with omission of the axioms of extensionality. In the rule of substitution for functional variables,

$$\check{S}_{\mathbf{B}}^{\mathbf{f}(\mathbf{x}_1, \mathbf{x}_2, \cdots, \mathbf{x}_m)}\mathbf{A}|$$

[14]This is a revision of the notation used in [Church 5].

[15]This superscript may often be omitted—as an abbreviation if clear from the context—or for the purpose of typical ambiguity.

[16]Following Russell in [ML] and the "alternative method" of $*10$ in [PM 1], we do not take the existential quantifier as primitive but define it as in $*10{\cdot}01$.

[17]Therefore R is not strictly a single language, but many different languages, one for each choice of the list of primitive constants.

[18]The use of dots as brackets is a convention of abbreviation rather than part of the primitive notation. It follows a simplified form of the Peano-Russell convention that is explained in [Church 4]. Briefly a bold dot stands for an omitted pair of brackets with the scope extending from the point at which the dot appears, forward, either to the end of the innermost explicitly written pair of brackets, [ ], within which the dot appears or to the end of the formula if there is no such explicitly written pair of brackets. And if square brackets are omitted without replacement by a dot, the restoration of the brackets is to follow the convention of association to the left (and also, though not needed in this paper, the convention about categories which is explained on pages 74–80, 171).

We also allow that any well-formed part of a formula may first be enclosed in square brackets (if not already so enclosed) and then these square brackets may be eliminated by replacing the first of the pair by a bold dot in accordance with the conventions just described.

[19]The same axioms are also in [ML].

is to be understood as reducing simply to **A** if there is in **A** any occurrence of **f** as argument of another functional variable[20] (as well as in any case in which the substitution would otherwise result in capture of a variable).

For our present purpose we shall think of extensional simple type theory as obtained from R by adding axioms of extensionality in the form given in [Church 5].

In formulating a semantics for R we consider only the case that there are no primitive constants. And since Russell, until he comes to propose axioms of extensionality in [Intr2nd], hardly considers criteria of identity of propositions,[21] we proceed in a way that leaves open all possibilities in this regard, even the most bizarre.

A model for R is based on a domain $\mathfrak{I}$ playing the role of a domain of individuals, a domain $\mathfrak{T}$ playing the domain of true propositions, and a domain $\mathfrak{F}$ playing the role of a domain of false propositions. These three domains may be chosen arbitrarily except that no one of them may be empty, and the domains $\mathfrak{T}$ and $\mathfrak{F}$ must be disjoint.[21½]

For each type there is a corresponding domain whose members are called *the members of the type in the model*; and the values of a variable of any type are restricted to the corresponding domain, i.e., to the members of that type in the model. For the type i the corresponding domain is $\mathfrak{I}$; for the type 0 the corresponding domain is $\mathfrak{T} \cup \mathfrak{F}$; and the members of the type $(\beta_1, \beta_2, \ldots, \beta_m)$ are all m-ary functions for which the appropriate arguments are, in order, the members of the types $\beta_1, \beta_2, \ldots, \beta_m$ and the value is always in the domain $\mathfrak{T} \cup \mathfrak{F}$.[22]

Then truth-tables are chosen for each of $\sim$, $\vee$, and the universal quantifier (though the "tables" may of course be infinite in extent if some of the three domains are infinite). These truth-tables, or (better) tables of values, are arbitrary except that they must agree in extension with the usual extensional truth-tables for $\sim$, $\vee$, and the universal quantifier; i.e., the following conditions must be satisfied (for any system of values of the free variables present):

---

[20] The rule of substitution for propositional variables is meant to be included in this statement as corresponding to the case $m = 0$.

The simple substitution of one propositional or functional variable for another free variable of the same type (cf. footnote 23) may of course be made even in the case that the variable in question occurs as argument of another functional variable. However, this requires no separately stated rule of inference, as it follows from the rule of generalization and the axiom schema $(\mathbf{x})\mathbf{f}(\mathbf{x}) \supset \mathbf{f}(\mathbf{y})$.

[21] An exception is §500 of [PoM], which proposes no specific criterion of identity of propositions but evidently intends a very strong criterion. That Russell does not return to this topic is explained by the fact that he later came to distrust the notion of proposition and sought to explain it away in the manner described in footnotes 4,7.

[21½] *Editor's note*: Church credits John M. Vickers, March 1975, with the addition "and the domains ... must be disjoint". Church adds, in a handwritten note, "A criterion of identity of propositions that allows propositions of opposite truth-value to be identical is of course not a possibility, even a bizarre one".

Church added in his hand: In *Comparison of Russell's Resolution of the Semantical Antinomies with That of Tarski*, the writer notes that instead of this correction it would be possible to correct by changing the words "and otherwise the value of $(\mathbf{a})\mathbf{P}$ is in $\mathfrak{F}$" in line 15 on p. 27 [*Editor's note*: This is line 6 of p. 778 in the present edition.] to "and the value of $(\mathbf{a})\mathbf{P}$ is in $\mathfrak{F}$ if the value of $\mathbf{P}$ is in $\mathfrak{F}$ for some value of $\mathbf{a}$." This alternative correction may not be without interest on its own account. But it was not the original intention of the paper. Moreover, it is not historically accurate; i.e., it is not anything which might be supposed to have been intended by Russell, even implicitly, or even at a time when he still maintained the notion of proposition.

[22] Partial functions are of course excluded; i.e., each member of the type $(\beta_1, \beta_2, \ldots, \beta_m)$ is a function which has a value in the domain $\mathfrak{T} \cup \mathfrak{F}$ for every m-tuple of arguments that are members of the respective types $\beta_1, \beta_2, \ldots, \beta_m$.

If the value of **P** is in $\mathfrak{T}$, the value of ~**P** is in $\mathfrak{F}$.
If the value of **P** is in $\mathfrak{F}$, the value of ~**P** is in $\mathfrak{T}$.
If either the value of **P** is in $\mathfrak{T}$ or the value of **Q** is in $\mathfrak{T}$, the value of [**P** ∨ **Q**] is in $\mathfrak{T}$.
If the value of **P** is in $\mathfrak{F}$ and the value of **Q** is in $\mathfrak{F}$, the value of [**P** ∨ **Q**] is in $\mathfrak{F}$.
Where **a** is a variable of any type, the value of (**a**)**P** is in $\mathfrak{T}$ if the value of **P** is in $\mathfrak{T}$ for all values of **a**, and otherwise the value of (**a**)**P** is in $\mathfrak{F}$.

A particular model is determined when the three non-empty domains $\mathfrak{I}, \mathfrak{T}, \mathfrak{F}$ are chosen, and three truth-tables subject to the above conditions. The definition of the value of a well-formed formula for a given system of values of its free variables, in the particular model, then proceeds in a routine way. A formula is *valid in* a particular model if it is well-formed and, in that model, its value is in the domain $\mathfrak{T}$ for every system of values of its free variables. And a formula is *intensionally valid* if it is valid in every model (of the kind just described).

It can be shown as a consequence of Gödel's incompleteness theorem ([Gödel]) that not all intensionally valid formulas of R can be theorems (if the system obtained by adding an axiom of infinity to R is consistent). But Henkin's proof of a weak completeness theorem for extensional simple type theory (see [Church 4], §54, and [Henkin]) raises the question of an analogous completeness theorem for Russellian simple type theory. Using Henkin's result, we can most easily formulate this by asking whether every theorem of extensional simple type theory which is intensionally valid is also a theorem of R.[23]

Although Russell in [ML] and Whitehead and Russell in *Principia Mathematica* do not (or do not fully and clearly) state formation rules, it is generally understood that the arguments of a functional variable may be only variables and not more complex expressions; or if primitive constants are added to the system, the arguments of a functional variable must be either variables or primitive constants (and not more complex expressions) and the same must hold regarding the arguments of a functional constant. Indeed this seems to be quite clearly implied in [PM 1], pages 50–5 and 161–165, especially when these passages are taken in the light of what is done in ∗9, and when it is remembered that expressions like $\phi\hat{x}$ or $\phi!\hat{z}$ are actually simple variables;[24] and occasional passages which may seem to contradict this (such as primitive proposition (11) in section VI of [ML], or in [PM 1] the paragraph following ∗10·4 or the explanation regarding the use of circumflexed letters at the beginning of ∗21) are probably to be accounted for as loosely worded explanations in which it is not meant that the expressions which literally occur are to be considered as being themselves well-formed formulas.[25]

---

[23]Or if this is not the case, can it be made so by adding some specific list of axioms to R? The restriction on the rule of substitution for functional variables that was noticed above may cause some uneasiness about the completeness of R, as it means that there are some theorems containing free functional variables, such as $H(F) \supset_H H(G) \supset . F(x) \supset_x G(x)$, for which no substitution for one of the free functional variables is even expressible in the language, except the comparatively trivial substitution of one functional variable for another—as e.g., $H(F) \supset_H H(F) \supset . F(x) \supset_x F(x)$. But this reflection is by no means conclusive against the completeness of R (in a sense analogous to Henkin's), and the completeness is rather an open problem.

[24]E.g. that $\phi!\hat{z}$ is a variable is said explicitly on pages 51–52, and again on page 165.

[25]Indeed if such formulas as $F((y)G(x, y))$ or $F((\exists x)G(x, y))$ were admitted as well-formed, the reduction to prenex normal form would be lost, although this reduction is implicit in ∗9 of [PM 1] and is essential to the reduction to matrices which is claimed in the Introduction, pp. 50–51, and in ∗12, last paragraph on

This restriction, that the arguments of a functional variable or functional constant must be either variables or primitive constants, has been followed in formulating the language R above — in order to preserve Russell's doctrine of descriptions and in order to remain as close as possible to *Principia Mathematica* except in those respects in which definite reasons can be given demanding a change. But this formal faithfulness to *Principia Mathematica* has some awkward consequences which Russell perhaps did not foresee. For example if we accept as premiss the dictum of Petrus Ramus that all assertions of Aristotle are false, $A^n(\alpha, p) \supset_p \sim p$, and if we have a particular assertion of Aristotle expressed in the language R (with added primitive constants as necessary) by a well-formed formula $\mathbf{P}$ (not containing any of the functional constants A^m, B^m with $m \geqq n$), we expect to be able to infer $\sim\mathbf{P}$; but this is not possible in the language R, because the formula $A^n(\alpha, \mathbf{P})$ is not well-formed. Or again we can express in R that Scott is the author of Waverley by writing $W(y) \equiv_y y = s$, where s and W are primitive constants, "Scott" and "wrote Waverley";[26] but in spite of the availability of functional constants B^n such that $B^n(x, p)$ expresses that x believes p at order n, we are still unable to express in R that a believes that Scott is the author of Waverley. As still another example, we are unable to express in R that something is true of Scott which a disbelieves (at order n).

To remove the difficulty the best means which suggests itself is to add to the language R a primitive notation $\equiv$ for the relation of identity between propositions. This notation is to be distinguished from the equality sign, $=$, that is introduced in [PM 1] by definition $*13{\cdot}01$ (and in R by the same definition, adapted to simple type theory by replacing the predicate variable $\phi!$ by just a variable of the appropriate type). For the equality sign may stand only between two variables (of the same type) or between a variable and a constant (of the same type) or between two constants (of the same type). But the sign $\equiv$ is intended to have the syntax of a connective of propositional calculus, and the formation rules are to provide that if $\mathbf{\Gamma}$ and $\mathbf{\Delta}$ are well-formed, then $[\mathbf{\Gamma} \equiv \mathbf{\Delta}]$ is well-formed.

To distinguish it from the equality sign, let us call $\equiv$ the sign of strict equivalence. The axioms about it will be very meager if we follow only what is implicitly required by Russell's logic (Russell being hardly an advocate of modal logic in the style of C. I. Lewis, even unwittingly). Perhaps the following are sufficient:

$p \equiv p$

$p \equiv q \equiv . \, p = q$

$p \equiv q \supset . \, p = q$

$(\mathbf{x})\mathbf{f}(\mathbf{x}) \equiv (\mathbf{y})\mathbf{f}(\mathbf{y})$

$\mathbf{f}(\mathbf{x}_1, \mathbf{x}_2, \ldots, \mathbf{x}_m) \equiv_{\mathbf{x}_1, \mathbf{x}_2, \ldots, \mathbf{x}_m} \mathbf{g}(\mathbf{x}_1, \mathbf{x}_2, \ldots, \mathbf{x}_m) \supset . \, \mathbf{f} = \mathbf{g}$

The last two of these are axiom schemata, not only because m is left indeterminate but also because of typical ambiguity: the variables $\mathbf{f}$, $\mathbf{g}$, $\mathbf{x}$, $\mathbf{y}$, $\mathbf{x}_1$, $\mathbf{x}_2$, …, $\mathbf{x}_m$ ($\mathbf{x}_1$, $\mathbf{x}_2$, …, $\mathbf{x}_m$ all different) may be of any types that are compatible with well-formedness. Also the subscripts after the sign $\equiv$ stand for universal quantifiers, in the same manner as in *Principia Mathematica* after the signs $\supset$ and $\equiv$ (definitions $*10{\cdot}02$, $*10{\cdot}03$, $*11{\cdot}05$, $*11{\cdot}06$).[27]

---

page 162.

[26] For the sake of the illustration, we assume that s is a Russellian logical proper name. Otherwise we must use a primitive functional constant S, "scottizes," and write $(\exists x) . \, S(y) \equiv_y y = x . \, W(y) \equiv_y y = x$.

[27] In the present context the notion of synonymous isomorphism of [Church 3] requires an amendment,

Addition of the sign $\equiv$ (with appropriate formation rules and axioms) to the language R enables us to express a comprehension schema

$$(\exists \mathbf{f}) \centerdot \mathbf{P} \equiv_{\mathbf{x}_1\mathbf{x}_2\ldots\mathbf{x}_m} \mathbf{f}(\mathbf{x}_1, \mathbf{x}_2, \ldots, \mathbf{x}_m),$$

where the variables $\mathbf{x}_1, \mathbf{x}_2, \ldots, \mathbf{x}_m$ are all different and $\mathbf{f}$ is of such type as to secure well-formedness. This can be proved as a theorem schema by making use of the rule of substitution for functional variables. And it enables us to use contextual definition to explain $A^n(\alpha, \mathbf{P})$ as standing for $\mathbf{P} \equiv p \supset_p A^n(\alpha, p)$ and (e.g.) $B^n(a, W(y) \equiv_y y = s)$ as standing for $W(y) \equiv_y [y = s] \equiv p \supset_p B^n(a, p)$. This removes the difficulty which we noticed above, that certain things (such as "*a* believes that Scott is the author of Waverley") are not expressible in the language of *Principia Mathematica*, or in the language R, even after supplying appropriate primitive constants. Moreover the missing case in the rule of substitution for functional variables can now be supplied by defining[28]

$$\lambda \mathbf{x}_1 \lambda \mathbf{x}_2 \cdots \lambda \mathbf{x}_m \mathbf{P} \to (\imath \mathbf{f}) \centerdot \mathbf{P} \equiv_{\mathbf{x}_1\mathbf{x}_2\ldots\mathbf{x}_m} \mathbf{f}(\mathbf{x}_1, \mathbf{x}_2, \ldots, \mathbf{x}_m).$$

Or as an alternative we can make a direct contextual definition of the notation $\lambda \mathbf{x}_1 \lambda \mathbf{x}_2 \cdots \lambda \mathbf{x}_m \mathbf{P}$. Namely if **M** is the scope[29] and **N** is obtained from **M** by replacing the particular occurrence of $\lambda \mathbf{x}_1 \lambda \mathbf{x}_2 \cdots \lambda \mathbf{x}_m \mathbf{P}$ by some variable **f** of appropriate type, not occurring in **M**, then

$$\mathbf{M} \to \mathbf{P} \equiv_{\mathbf{x}_1\mathbf{x}_2\ldots\mathbf{x}_m} \mathbf{f}(\mathbf{x}_1, \mathbf{x}_2, \cdots, \mathbf{x}_m) \supset_{\mathbf{f}} \mathbf{N}$$

For the language, call it R*, that is obtained from R by adjoining the sign $\equiv$, as just described, together with suitable axioms about it, the definition of intensional validity may closely parallel that which has been outlined for R. In each model there must be, besides truth-tables for $\sim$, $\vee$, and the universal quantifier, also a truth-table for $\equiv$, and it must obey the conditions that the value of $[\mathbf{P} \equiv \mathbf{Q}]$ is in $\mathfrak{T}$ if the values of **P** and **Q** are the same, and the value of $[\mathbf{P} \equiv \mathbf{Q}]$ is in $\mathfrak{F}$ if the values of **P** and **Q** are different; then the definitions of value of a well-formed formula for a given system of values of its free variables, of validity in a model, and of intensional validity may proceed as before.[30½]

In order still to leave open all possibilities in regard to criteria of identity of proposi-

because it is meant that $\mathbf{p} \equiv \mathbf{q}$ shall be synonymous with $F(\mathbf{p}) \supset_F F(\mathbf{q})$ when **p** and **q** are propositional variables, but there is no parallel synonymy for $\mathbf{M} \equiv \mathbf{N}$ in other cases. The intention of the axioms is that $\mathbf{P} \equiv \mathbf{Q}$ shall be a theorem if and only if **P** and **Q** are synonymously isomorphic in the appropriate modified sense—provided that no two different primitive constants are synonymous, or that $\mathbf{a} = \mathbf{b}$ is taken as an axiom in every case in which **a** and **b** are different but synonymous primitive constants.

[28] Where **f** is some variable of appropriate type that does not occur in **P**, say the first in alphabetic order. We prefer the notation $\lambda \mathbf{x}_1 \lambda \mathbf{x}_2 \cdots \lambda \mathbf{x}_m \mathbf{P}$ to Russell's notation which consists in placing a circumflex accent over all free occurrences of $\mathbf{x}_1, \mathbf{x}_2, \ldots, \mathbf{x}_m$ in **P**, and which requires the rather awkward convention that is explained in section VII of [ML] and on page 200 of [PM 1] (following $*21{\cdot}01$). For example the formula appearing in $*9{\cdot}63$ of [PM 1] would be rewritten in our notation as $\lambda x \centerdot (y) F(x, y) \vee (z) G(x, z)$; and the first displayed formula at the top of page 81 would be rewritten as $\lambda F(\exists G) \centerdot F(x) \equiv_x G(x) \centerdot H(G)$, with Russell's exclamation points omitted in consequence of the change to simple type theory.

[29] The scope must be small enough that none of the free ("real") variables in $\lambda \mathbf{x}_1 \lambda \mathbf{x}_2 \cdots \lambda \mathbf{x}_m \mathbf{P}$ are bound ("apparent") in **M**. And if the function abstract $\lambda \mathbf{x}_1 \lambda \mathbf{x}_2 \cdots \lambda \mathbf{x}_m \mathbf{P}$ occurs within another function abstract $\lambda \mathbf{y}_1 \lambda \mathbf{y}_2 \cdots \lambda \mathbf{y}_m \mathbf{Q}$ (or class abstract $\hat{\mathbf{y}}\mathbf{Q}$ or description $(\imath \mathbf{y})\mathbf{Q}$), then the outer function abstract (or class abstract or description) must be eliminated first. Or more generally, if there are various function abstracts, class abstract, and descriptions occurring in **M**, their elimination must proceed in the left-to-right order of occurrence of their initial symbols.

[30½] *Added by Church*: A correction pointed out to me by John C. Ruttenberg, February 14, 1977: The value of $p \equiv q$ for given values of p and q, must always be the same as the value of $p = q$, as otherwise the axiom $p \equiv q \equiv \centerdot p = q$ would not be intensionally valid. In the printed paper, the lines from "must obey"

tions, the desired (weak) completeness theorem for R^* is that every intensionally valid formula of R^* shall be a theorem if it becomes a theorem of extensional simple type theory upon replacing $\equiv$ by $\equiv$ throughout.[30]

In summary, in order to preserve some of Russell's significant contributions to logic, and especially for the sake of his resolution, by means of the contextual definition of descriptions and of class abstracts,[31] of what Carnap calls the antinomy of the name relation,[32] and in order to compare it with Frege's resolution of the same problem ([Frege 1], [Frege 2]), it is desirable to formulate a modern version of Russell's mathematical logic as it appears in [ML] and [PM 1].[33] This revision of Russell is to make such changes as have come to be considered desirable by a consensus of logicians, while at the same time maintaining certain essential features. A first proposal in this direction, remaining closest to the original, is embodied in the language R as formulated in outline above. But certain lacunae in R, which seem to be undesirable from a Russellian viewpoint, lead to the extended language R^* — obtained by adding a connective $\equiv$ which is a sign of equality or identity but which, unlike Russell's defined equality sign $=$, may stand not only between propositional (or other) variables but also between more complex expressions which are of the same type as propositional variables (i.e., sentences or propositional forms).[34]

to "value of a well-formed" may be corrected to read as follows: "must obey the following condition: For a given system of values of the free variables in $[\mathbf{P} \equiv \mathbf{Q}]$, if the value of $\mathbf{P}$ is v and the value of $\mathbf{Q}$ is w, the value of $[\mathbf{P} \equiv \mathbf{Q}]$ is the same as the value of $[p = q]$ for the values v of p and w of q. Then the definitions of value of a well-formed".

[30]We should notice that there are nevertheless important differences between R and R^*. In particular the reduction to prenex normal form is not generally possible in R^*.

[31]The contextual definition of class abstracts is as much necessary to Russell's resolution of the antinomy of the name relation as his better known (or at least more widely debated) contextual definition of descriptions. For example, if we use the functional constants S and W with the same meaning as we did before, and if we treat class abstracts as genuine names, then since in fact $\hat{x}S(x)$ and $\hat{x}W(x)$ are names of the same class (the class of those who scottize is the same as the class of those who write Waverley) and since King George knew that $\hat{x}S(x) = \hat{x}S(x)$, we seem to be able to infer that King George knew that $\hat{x}S(x) = \hat{x}W(x)$.

In addition the contextual definition of class abstracts serves a purpose of economy, greatly reducing the array of types as compared to what is needed for a Fregean resolution of the antinomy of the name relation. Whether Russell's resolution of the antinomy is fully satisfactory as compared to Frege's remains in doubt, but it is just for this reason that the Russellian language requires a careful formulation.

[32]Not properly an antinomy, since it leads, not to actual contradiction, but to counterintuitive results in a naively formulated logic of such intensionalities as knowledge, assertion, belief, necessity. It should therefore rather be called a paradox, but still a paradox that a satisfactory formulation must take into account. It should also be said that, on Carnap's account of it, the antinomy of the name relation concerns, not quite substitutivity of equality, but the closely related question: How can two names of the same thing ever fail to be interchangeable?

[33]Important features of Russell's logic which appear already in [ML] and later reappear in ***Principia Mathematica*** are of course to be credited to Russell rather than to Whitehead and Russell jointly. The introduction and appendices to the second edition of ***Principia Mathematica*** are also the work of Russell — according to ***Mind***, n.s. vol. 35 (1926), p. 130 — although worded as being by the joint authors.

[34]Curiously, Russell's sign of definition "= Df" (replaced in this paper by the arrow, to emphasize asymmetry) may stand with a sentence or propositional form on the right and an abbreviation of it on the right, thus differing from his defined equality sign.

It may also be worth noticing that if the formation rules of Russell's mathematical logic are liberalized to allow $\mathbf{f}(\mathbf{P})$ to be well-formed, where $\mathbf{P}$ is any sentence or propositional form and $\mathbf{f}$ is a variable of appropriate type, then the connective $\equiv$ can be defined by obvious analogy with Russell's definition of the sign $=$. But to use this approach in extending the language R seems to be a greater departure from the Russellian original than is the extension to R^*.

REFERENCES

[PoM] Bertrand Russell. The Principles of Mathematics. Cambridge, England, 1903.

[OD] Bertrand Russell. *On denoting*. Mind, n.s. vol. 14 (1905), pp. 479–493.

[ML] Bertrand Russell. *Mathematical logic as based on the theory of types*. American Journal of Mathematics, vol. 30 (1908), pp. 222–262.

[PM1] Alfred North Whitehead and Bertrand Russell. Principia Mathematica, Vol. 1. Cambridge, England, 1910.

[Intr2nd] Bertrand Russell. *Introduction to the second edition*. Principia Mathematica, second edition, Vol. 1, Cambridge, England, 1925, pp. xxiii–xlvi.

[AppC] Bertrand Russell. *Appendix C. Truth-functions and others*. Ibid., pp. 659–666.

[Ajdukiewicz] Kazimierz Ajdukiewicz. *A method of eliminating intensional sentences and sentential formulae*. Atti del XII Congresso Internazionale di Filosofia, Vol. 5, Florence 1960, pp. 17–24.

[Bernays] Paul Bernays. *Axiomatische Untersuchung des Aussagen-Kalkuls der "Principia Mathematica"*. Mathematische Zeitschrift, vol. 25 (1926), pp. 305–320.

[Carnap 1] Rudolf Carnap. Abriss der Logistik. Vienna, 1929.

[Carnap 2] Rudolf Carnap. Meaning and Necessity. Chicago, 1947, second edition, 1956.

[Church 1] Alonzo Church. *A formulation of the simple theory of types*. The Journal of Symbolic Logic, vol. 5 (1940), pp. 56–68.

[Church 2] Alonzo Church. *On Carnap's analysis of statements of assertion and belief*. Analysis, vol. 10 no. 5 (1950), pp. 97–99.

[Church 3] Alonzo Church. *Intensional isomorphism and identity of belief*. Philosophical Studies, vol. 5 no. 5 (1954), pp. 65–73.

[Church 4] Alonzo Church. Introduction to Mathematical Logic, Volume I. Princeton 1956.

[Church 5] Alonzo Church. *Axioms for functional calculi of higher order*. Logic & Art, Essays in Honor of Nelson Goodman, edited by Richard Rudner and Israel Scheffler, Indianapolis and New York 1972, pp. 197–213.

[Church 6] Alonzo Church. *Comparison of Russell's resolution of the semantical antinomies with that of Tarski*. Forthcoming.

[Chwistek 1] Leon Chwistek. *Antynomje logiki formalnej*. Przegląd Filozoficzny, vol. 24 (1921), pp. 164–171.

[Chwistek 2] Leon Chwistek. *Über die Antinomien der Prinzipien der Mathematik*. Mathematische Zeitschrift, vol. 14 (1922), pp. 236–243.

[Frege 1] Gottlob Frege. *Über Sinn und Bedeutung*. Zeitschrift für Philosophie und philosophische Kritik, n.s. vol. 100 (1892), pp. 25–50.

[Frege 2] Gottlob Frege. Grundgesetze der Arithmetik, Band I. Jena 1893.

[Gödel] Kurt Gödel. *Über formal unentscheidbare Sätze der Principia Mathematica und verwandter Systeme I*. Monatshefte für Mathematik und Physik, vol. 38 (1931), pp. 173–198.

[Henkin] Leon Henkin. *Completeness in the theory of types*. The Journal of Symbolic Logic, vol. 15 (1950), pp. 81–91.

[Quine 1] Willard Van Orman Quine. A System of Logistic. Cambridge, Massachusetts, 1934.

[Quine 2] Willard Van Orman Quine. *Whitehead and the rise of modern logic*. The Philosophy of Alfred North Whitehead, edited by Paul Arthur Schilpp, Evanston and Chicago 1941, pp. 127–163.

[Quine 3] Willard Van Orman Quine. *Logic and the reification of universals*. From a Logical Point of View, by Willard Van Orman Quine, Cambridge, Massachusetts, 1953, pp. 102–129; reprinted with revisions in the second edition, ibid. 1961, pp. 102–129.

[Ramsey] F. P. Ramsey. *The foundations of mathematics*. Proceedings of the London Mathematical Society, ser. 2, vol. 25 (1926), pp. 338–384.

[Tarski 1] Alfred Tarski. *Einige Betrachtungen über de Begriffe der ω-Widerspruchsfreiheit und der ω-Vollständigkeit*. Monatshefte für Mathematik und Physik, vol. 40 (1933), pp. 97–112.

[Tarski 2] Alfred Tarski. *Der Wahrheitsbegriff in den formalisierten Sprachen*. Studia Philosophica, Vol. 1 (1936), pp. 261–405.

University of California at Los Angeles

SET THEORY WITH A UNIVERSAL SET (1974)

1. Heuristic preliminaries. For what is now usually known as ZF set theory. or often simply as axiomatic set theory, there seem to have been historically two heuristic principles which underlay the selection of the axioms, and it may be argued that these same two principles still control many of the current attempts to extend the theory by finding additional axioms. They are:

1. *The principle of specialized comprehension.* As it is known that the general comprehension axiom for sets leads to antinomy, when taken in conjunction with other things that naive set theory accepts, we seek axioms which are special cases of the general comprehension axiom and which promise to maintain consistency while at the same time being adequate for a large variety of mathematical purposes.
2. *The principle of limitation of size.* We must avoid axioms leading to sets that are too large, especially to sets having a one-to-one or many-one relation with the universal class.

The principle of specialized comprehension may also be described roughly by saying that we accept all of naive set theory except the comprehension axiom, and then assume as many special cases of the comprehension axiom as we dare. This principle indeed cannot be changed without changing the basic character of the theory (e.g. to a type theory). But it is the thesis of this paper that the principle of limitation of size was never very well supported, and that it should now be abandoned to clear the way for consideration of axioms which violate it.

In the early discussion of the antinomies, after they first came to the attention of mathematicians, the suggestion appears more than once that (in effect) such very large sets as the universal set and the set of ordinal numbers are incomprehensible to sound mathematical intuition, so that the antinomies are traceable to a departure from intuition in countenancing these monstrous sets. Cantor's antinomy of the greatest cardinal number and Burali-Forti's antinomy may well suggest this idea, and Zermelo was certainly influenced by it.

Let us define a *low set* as a set which has a one to-one relation with a well-founded set, a *high set* as a set which is the complement of a low set, and an *intermediate set* as a set which is neither a low set nor a high set. To justify the terminology in the sense that the classification of a set as low, intermediate, or high may be thought of as an estimate of the size of the set, we may prove in consequence of the basic axioms of set theory, as these are given below, that: (1) the complement of a low set (whether set or class) has a one-to-one relation with the universal class; (2) a low set has no one-to-one relation with the universal class; and with a strong axiom of choice, (3) if a set u is not a low set, or if a class u does not coincide with a low set, then every ordinal has a one-to-one relation with some subset of u. It is true that not only a high set but also an intermediate set may have a one-to-one relation with the universal class (or with the universal set if it exists), but the intermediate set is nevertheless lower than

Originally published in ***Proceedings of the Tarski Symposium*** in American Mathematical Society: Proceedings of Symposia in Pure Mathematics, vol. 25 (1974).

the high set in the sense of being further removed from the universal class (or set); its complement is higher than that of the high set.

Now Cantor's antinomy concerns the universal set — which, on the basic axioms as given below, is the same as the set of all sets. The reference to cardinal number is not essential, as the antinomy arises just from applying Cantor's theorem, that no set can have a one-to-one relation with its power set, to the case of the universal set. For the power set of the universal set is in fact the same set, so that the relation of identity is a one-to-one relation between the universal set and its power set.

If we apply Cantor's diagonal procedure to this one-to-one relation between the universal set and its power set, we should obtain a member of the power set which is excluded from the one-to-one relation. This is Cantor's argument to show that the relation which was supposed to be one-to-one between the set and its power set could not in fact be so. But in the case before us the set which is obtained by the diagonal procedure is Russell's set of all non-self-membered sets.

Because Russell's set is known to lead to antinomy independently of Cantor's theorem and of considerations concerning one-to-one relations, the argument is persuasive that we ought to think of the Russell set rather than the universal set as being the source of Cantor's antinomy. The bearing of this on the principle of limitation of size is that, although the universal set is of course a high set, the Russell set may not be. If we abandon the principle of limitation of size by allowing the complement of any set to be a set, the Russell set will then be an intermediate set. (This last assertion is a paradoxical way of speaking, as it can be shown by first-order logic, independently of any axioms of set theory, that the Russell set does not exist; but the assertion intended can be made rigorous by extending the distinction of low, intermediate, and high to classes as well as sets.)

The fact is that there is no known case in which we must regard a high set as being the source of an antinomy. Indeed the principal antinomic sets besides the Russell set are the set of ordinals and the set of well-founded sets, which lead to the Burali-Forti antinomy and a variation of it. And these also may be regarded as intermediate sets — or perhaps better, as balanced on the hazardous edge between low sets and intermediate sets, since in both cases the antinomy can be put in the form of showing that the set both is and is not well-founded. (The set of ordinals is easily shown to satisfy the defining conditions of being an ordinal, i.e., it is transitive, connected, and well-founded, but it is therefore a member of itself and so is not well-founded; the set of well-founded sets is well-founded because all its members are, but it is therefore a member of itself and so not well-founded.)

There do appear antinomic sets which are in some sense intermediate between low sets and high sets, as we have just seen. But it by no means follows that all intermediate sets lead to antinomy. And there is room for exploration of the axiomatic possibilities.

2. Notations and terminology. Before stating the basic axioms from which our proposed search for additional axioms is to make its start, we first supply some definitions and explanations of notation and terminology. These are in part familiar, but their explicit statement will serve the purpose of precision, removing possible uncertainties as to just what is meant in the statement and discussion of the axioms.

For sentence connectives and quantifiers we use the standard Peano-Russell notations, with a simplified convention about replacement of brackets by dots — namely a bold dot stands for a bracket which extends from the point where it occurs, forward either to the end of the formula or to the end of any explicitly written bracket $[\cdots]$ within which the bold dot appears–and otherwise, if square brackets, [,], have been omitted without the insertion of a dot, they are to be restored under the convention of association to the left.

As extreme rigor is not our present purpose, we may when convenient mix English words and formulas of first-order logic as grammatically connected parts of a single sentence. Also we shall generally avoid the use of syntactical variables. And in particular if a definition is expressed with object-language variables, there is always a tacit addition to be understood which may be indicated roughly by saying "and likewise with any other variables in place of these."

However, in order to express axiom schemata we make use of a bold capital letter **A** as a syntactical variable. This **A** always represents a first-order formula. And if one or more set variables are written as subscripts after the syntactical variable **A**, the meaning is as follows. If, e.g., $\mathbf{A}_y$ represents a certain first-order formula, $\mathbf{A}_z$ represents the first-order formula which is obtained from this by substituting z for all free occurrences of y, after first making such alphabetic changes of bound variable that no capture of the variable z results by the substitution. Again if $\mathbf{A}_{xy}$ represents a certain first-order formula, $\mathbf{A}_{xz}$ represents the formula obtained by substituting z for the free occurrences of y, and $\mathbf{A}_{zy}$ represents the formula obtained by substituting z for the free occurrences of x, in each case after appropriate alphabetic changes of bound variable. And likewise with any other set variables in place of x, y, z.

Observe that use of the notation (e.g.) $\mathbf{A}_y$ is not to be understood as meaning that the formula $\mathbf{A}_y$ may not contain other free set variables than y. But if (e.g.) the notations $\mathbf{A}_y$ and $\mathbf{A}_z$ occur in the same context, we shall require that z is not a free variable of the formula $\mathbf{A}_y$ and that y is not a free variable of the formula $\mathbf{A}_z$. These conventions are used in the statement of the axiom schemata A, H, I below.

We use the notations $=, \in, \notin, \subseteq$ with their standard meanings, taking $=$ and $\in$ as primitive (or undefined), and then defining $x \notin y$ as $\sim \mathbf{.}\, x \in y$ and $a \subseteq b$ as $x \in a \supset_x x \in b$.

There are many notations that must be understood as descriptions. Because they or the notions they express are essentially familiar, we may often give the definitions only in words or explain in a way that presupposes definitions as already known. But for example we use 0 for the empty set; this means that we define the symbol 0 as standing for $(\imath u)(x) \mathbf{.}\, x \notin u$, and the description symbol $(\imath u)$ which this contains is then to be understood as eliminated by means of Russell's contextual definition of descriptions.

We use the notations $\bar{a}$ for the complement of a, and $\iota`a$ for the unit set of $a$, and $a \cup b$ for the union of $a$ and $b$. For the finite set whose members are $a_1, a_2, \ldots, a_k$ we use the notation $\{a_1, a_2, \ldots, a_k\}$. (Hence $\{a\}$ and $\iota`a$ are synonymous.) And for the ordered k-tuple of $a_1, a_2, \ldots, a_k$ we use $\langle a_1, a_2, \ldots, a_k\rangle$.

We use $\sum`a$ for the sum set of $a$, that is, $(\imath u) \mathbf{.}\, x \in u \equiv_x (\exists y) \mathbf{.}\, y \in a \mathbf{.}\, x \in y$. Similarly the product set $\prod`a$ of a is $(\imath u) \mathbf{.}\, x \in u \equiv_x y \in a \supset_y x \in y$. And the power set $\mathbf{P}`a$ of a is $(\imath u) \mathbf{.}\, x \in u \equiv_x x \subseteq a$.

For iteration of operations whose application we express by means of the inverted comma (as $\iota`$, $\sum`$, etc.) we use an exponent, then dropping the inverted comma. For example $\sum^2 a$ is the sum set of the sum set of a; and $\iota^m a$ is the result of iterating the operation $\iota`$, m times on a.

The members of a at level j are the members of $\sum^{j-1} a$, and a set a is said to be *empty at level j* if $\sum^{j-1} a$ is empty. In particular the members of a at level 1 are the same as the members of a, and to be empty at level 1 is the same as to be empty.

The *successor* of a set a is $a \cup \iota`a$.

A set a is called *transitive* if $x \in y \supset_{xy} . y \in a \supset x \in a$, *connected* if $x \in a \supset_x . y \in a \supset_y . x \in y \vee y \in x \vee x = y$, *well-founded* if $a \in r \supset_r (\exists x) . x \in r . y \in r \supset_y y \notin x$. An *ordinal* is a set which is transitive, connected, and well-founded.[1] And a *finite ordinal* is an ordinal of which no subset other than 0 is closed with respect to successor. In symbols we write: trans(a), conn(a), wf(a), ord(a), finord(a).

As usual we denote the finite ordinals by 0, 1, 2, 3, ... , 0 being the empty set, 1 the successor of 0, and so on. And though we shall not require a notation for addition of ordinals in general, we use $+1$ as notation for the successor of an ordinal, $+2$ as notation for the successor of the successor of an ordinal.

A set c is called a *set of ordered pairs* if $z \in c \supset_x (\exists x)(\exists y) . z = \langle x, y\rangle$, a *one-many correspondence* if it is a set of ordered pairs and $\langle x, y\rangle \in c \supset_{xy} . \langle x_1, y\rangle \in c \supset_{x_1} x = x_1$, a *many-one correspondence* if it is a set of ordered pairs and $\langle x, y\rangle \in c \supset_{xy} . \langle x, y_1\rangle \in c \supset_{y_1} y = y_1$, a *one-to-one correspondence* if it is both a one-many correspondence and a many-one correspondence. A set c which is a one-to-one (one many, many-one) correspondence is said to be a one-to-one (one-many, many-one) correspondence *between the sets a and b* if $x \in a \equiv_x (\exists y) . \langle x, y\rangle \in c$ and $y \in b \equiv_y (\exists x) . \langle x, y\rangle \in c$; in the same case we say also that c is a one-to-one (one-many, many-one) correspondence *of the set a with the set b*.

For the now usual adjunction of classes to set theory perhaps the best approach is to think of the classes as values of singulary functional (alias predicate) variables of the underlying logic. The original Zermelo-Fraenkel-Skolem axiomatic set theory is a first-order theory in the sense that the underlying logic is a first-order logic without functional variables. But it may be argued that if the first-order theory is consistent, it has a model, and it is therefore possible to treat this same model in a logic of higher order. And the advantages of doing this have become increasingly clear since the first step in this direction by von Neumann.[2]

In adjoining, to the original first-order theory, variables whose values are propositional functions of the individuals (i.e., of the sets) we may as a matter of economy allow only singulary functional variables, relying on the availability of ordered k-tuples in the first-order theory to provide the equivalent of k-ary functional variables. But it is very convenient to allow also at least binary functional variables, and we shall do this, speaking not only of classes, taken as values of the singularly functional variables of the underlying logic, but also of [binary] relations, as values of the binary functional variables of the underlying logic.

If in adjoining such functional variables to the first-order theory we allow them

[1]This notion of an ordinal is due to J. von Neumann, *Zür Einführung der transfiniten Zahlen*, ***Acta Sci. Math.*** (Szeged) 1 (1923), 199–208, in the sense that this class of sets is identified with the ordinals. According to Bernays (see ***J. Symbolic Logic*** vol. 6 (1941), p. 6) something similar had been done by Zermelo, about 1915, but was not published. The definition which we use is a simplification of von Neumann's definition that is due to Raphael M. Robinson, *The theory of classes. A modification of von Neumann's system*, ***J. Symbolic Logic***, vol. 2 (1937), pp. 29–36.

[2]J. von Neumann, *Eine Axiomatisierung der Mengenlehre*, ***J. Reine Angew. Math.***, vol. 154 (1925), pp. 219–240; and *Berichtigung*, ibid., vol. 155 (1926), pp. 128.

to occur only as free variables, the result is of course still a first-order theory—and indeed the functional variables might be explained away as being a mere manner of speaking, to replace the explicit use of axiom schemata and theorem schemata.

The next step beyond this supplies the equivalent of the Bernays set theory[3] by taking the underlying logic to be that of a *predicative functional calculus of second order* in the sense of ramified type theory.[4] For our present purpose it is not necessary to go further than a predicative calculus of second order in allowing quantification with respect to functional variables.

A [binary] relation F is said to be a *one-many relation* if $F(x, y) \supset_{xy} . F(x_1, y) \supset_{x_1} x = x_1$, a *many-one relation* if $F(x, y) \supset_{xy} . F(x, y_1) \supset_{y_1} y = y_1$, a *one-to-one relation* if it is both a one-many and a many-one relation. A relation F is said to be a relation *between the classes G and H*, or *of the class G with the class H*, if $G(x) \equiv_x (\exists y)F(x, y)$ and $H(y) \equiv_y (\exists x)F(x, y)$; or we may use a like terminology with sets replacing one or both of the classes G and H. (This terminology has already appeared in the preliminary heuristic discussion above.)

In this paper we omit the word "binary" before "relation." But in another context it might be preferable to restore it in order to be able to speak of ternary relations, etc.

We say that sets a and b are *equivalent*, or that a equiv b, if there is a one-to-one correspondence between them. And the set of sets that are equivalent to a given set b we call the *cardinal number* of b (understanding this term as a description, so as not to decide in advance the question of existence).

We generalize the notions of equivalence and cardinal number to notions of j-equivalence and j-cardinal, for any finite ordinal j except 0. We need the auxiliary notion of *the one-to-one correspondence between a and b that is determined* by a given j-equivalence between them. Then the definition of j-equivalence is by recursion as follows:

A 1-equivalence between sets a and b is a one-to-one correspondence c_1 between a and b, and the one-to-one correspondence between a and b that is determined by the 1-equivalence is the same one-to-one correspondence c_1. An $(i + 1)$-equivalence between sets a and b is an i-equivalence f between $\sum$'a and $\sum$'b such that, if c_i is the one-to-one correspondence between $\sum$'a and $\sum$'b that is determined by f, there exists a one-to-one correspondence c_{i+l} between a and b such that

$$\langle x, y\rangle \in c_{i+1} \equiv_{xy} . x \in a . y \in b . \langle u, v\rangle \in c_i \supset_{uv} . u \in x \equiv v \in y.$$

And then c_{i+1} is the one-to-one correspondence between a and b that is determined by f.

We may evidently generalize the foregoing by allowing one-to-one relations everywhere in place of one-to-one correspondences, but for our present purpose we shall not need this.

We say that sets a and b are *j-equivalent*, or that a equiv^j b, if there exists a j-equivalence between them. And the set of sets that are j-equivalent to a given set b is

[3]Though better known from Gödel's monograph (footnote 6), this form of ZF set theory is due to Paul Bernays, *A system of axiomatic set theory*, ***J. Symbolic Logic***, vol. 2 (1937), pp. 65–77; ibid., vol. 6 (1941), pp. 1–17; ibid., vol. 7 (1942), pp. 65–89, 133–145; ibid., vol. 8 (1943), pp. 89–106; ibid., vol. 13 (1948), pp. 65–79; and ibid., vol. 19 (1954), pp. 81–96.

[4]See the writer's ***Introduction to Mathematical Logic***, Vol. I, §58. The equivalence mentioned in the text (above) is clearer if we follow Bernays in writing $x \in c$ when x and c are sets, but $x\ \eta\ C$ when C is a class. [*Editor's note*: see correction, page 793]

called the *j-cardinal*, or *j-cardinal number*, of b.

Observe that the notion of j-cardinal number includes that of the order type of an ordered set, or more generally, of a relation number in the sense of *Principia Mathematica* (provided that the relation is represented by a low set of ordered pairs), or still more generally, of the isomorphism type of a relational structure (alias relational system, with similar proviso), and so on. Details of this depend on just what definition of ordered pair is adopted. But if for example the ordered pair $\langle x, y\rangle$ is $\{\{x\}, \{x, y\}\}$, relation numbers appear as 3-cardinal numbers of sets of ordered pairs.

3. Basic axioms. As *basic axioms of set theory* we take the following axioms A–J. (The number of axioms is in fact infinite, since A, H, and I are axiom schemata rather than single axioms.)

A. *Substitutivity*. $y = z \supset . \mathbf{A}_y \supset \mathbf{A}_z$.
B. *Extensionality*. $x \in y \equiv_x x \in z \supset y = z$.
C. *Pair set*. $(\exists u) . x \in u \equiv_x . x = y \vee x = z$.
D. *Sum set*. $(\exists u) . x \in u \equiv_x (\exists y) . y \in z . x \in y$.
E. *Product set*. $y \in z \supset (\exists u) . x \in u \equiv_x y \in z \supset_y x \in y$.
F. *Infinity*. $(\exists u) . x \in u \equiv_x \text{finord}(x)$.
G. *Axiom of choice*.
H. *Aussonderung*. $\text{wf}(w) \supset (\exists u) . x \in u \equiv_x . x \in w . \mathbf{A}_x$ where u is not free in $\mathbf{A}_x$.
I. *Replacement*.
$\mathbf{A}_{xy} \supset_{xy} [\mathbf{A}_{xz} \supset_z y = z] \supset . \mathbf{A}_{xy} \supset_{xy} [\mathbf{A}_{zy} \supset_z x = z] \supset . y \in w \equiv_y (\exists x)\mathbf{A}_{xy} \supset . \text{wf}(w) \supset (\exists u) . x \in u \equiv_x (\exists y)\mathbf{A}_{xy}$, where u is not free in $\mathbf{A}_{xy}$.
J. *Power set*. $\text{wf}(w) \supset (\exists u) . x \in u \equiv_x x \subseteq w$.

These axioms yield all the sets of ZF set theory, in the sense that any model of the axioms contains a model of the standard axioms of ZF set theory (which is to be obtained by deleting all the non-well-founded sets). If we assume the consistency of ZF set theory, the consistency of the basic axioms follows. Our proposed program is to seek new axioms which maintain this consistency when added to the basic axioms, though perhaps violating the principle of limitation of size—and as far as may be to survey the possibilities of doing this.

We remark that, although H, I, J have the condition $\text{wf}(w)$, that the set w is well-founded, it follows at once that the three principles of *Aussonderung*, *replacement*, and *power set* can be extended to any low set w, so that the possibility of using one of these three principles to obtain a new set from a given set w depends in the end, not on the well-foundedness of w, but on the size of w.

As the set u of Axiom F can be shown to be well-founded, it is possible to prove Axiom C as a consequence of A, B, F, H, I, J. We retain Axiom C in spite of its not being independent, because we think of the list of basic axioms as tentative and possibly subject to some reconsideration or modifications.

Axiom E might be omitted without losing the feature that any model contains a model of ZF set theory. But it also seems desirable to retain this axiom among the basic axioms because of the importance of the Boolean operations on sets and because of the likelihood that Axiom E will hold (as an axiom or theorem) in any extended system of set theory which may seriously be considered as a replacement for or alternative to the standard ZF set theory. In particular Axiom E provides a weakened principle of *Aussonderung* applicable in cases in which the familiar Zermelo-Skolem *Aussonderungsprinzip* is not available (as indeed the latter almost certainly cannot be extended beyond low sets without restoring antinomy).

This last, parenthetic, remark must be qualified by observing that not only Axiom H (for *Aussonderung*) but also Axioms I and J might be strengthened by replacing the condition wf(w) by the following weaker condition on w: The members of w at level j do not (for any j) include a set a_1 for which there is a *membership cycle*, i.e., a finite number of sets $a_1, a_2, \ldots, a_n$ such that $a_1 \in a_2$, $a_2 \in a_3, \ldots, a_{n-1} \in a_n$, and $a_n \in a_1$. However, the effect of this strengthening of H, I, J is very problematical, because it is not clear whether sets w that obey this weaker condition without being well-founded are reasonably possible—in the sense that they might be introduced by some added axiom of set theory that accords with the principle of specialized comprehension.[5]

Instead of strengthening H, I, J we might weaken them by strengthening the definition of well-foundedness, replacing the set variable r in the definition by a class variable. That is, we might define Wf(w) as $F(w) \supset_F (\exists x) \,.\, F(x) \,.\, F(y) \supset_y y \notin x$, and then modify Axioms H, I, J by replacing wf(w) by Wf(w).

Our reason for not making this otherwise natural modification of Axioms H, I, J is that, tentatively, we prefer axioms of set theory in which class and relation variables are not used. Indeed this is the original approach to axiomatic set theory by its founders, and it is on this basis that we have the clearest justification for extending set theory by the introduction of classes, as explained in §2 above.

This also explains why in the list of basic axioms we have left open the form in which the axiom of choice is to be stated. The preference just indicated suggests adopting one of the usual forms of the axiom of choice in which only set variables appear. However, in this particular case we might also appeal to Gödel's relative consistency proof[6] as justifying the introduction of the axiom of choice in the strong form that there exists a relation by which the universal class is well-ordered, that is,

$$G^*\!\!: \; (\exists F) \,.\, (x) \sim F(x,x) \,.\, F(x,y) \supset_{xy} [F(y,z) \supset_x F(x,z)] \,.\, G(x) \supset_{Gx} \\ (\exists y) \,.\, G(y) \,.\, G(z) \supset_z \,.\, F(y,z) \vee y = z.$$

4. Proposed additional axioms. Though the main purpose of this paper is a program of seeking new axioms and no final set of axioms for set theory is offered, we nevertheless go on to state some of the axioms which most immediately suggest themselves as additions to the basic axioms. We have already mentioned the axiom:

K. *Complement.* $(\exists u) \,.\, x \in u \equiv_x x \notin z$.

The addition of Axiom K to the basic axioms provides for a Boolean algebra of sets. In fact after adding Axiom K we may, if we wish, describe the entire theory as a Boolean algebra with operator, replacing the primitive predicate $\in$ by the Boolean operations and the singulary operator ι‘.

An axiom which it is very tempting to add to the basic axioms is an *axiom of definition by abstraction*, according to which, if F is any equivalence relation, its

[5]On the other hand noncyclic infinite descending membership sequences do of course arise if we accept that every set has a complement. For example the universal set has as member the complement of 1, which has as member the complement of 2, which in turn has as member the complement of 3, and so on.

[6]Kurt Gödel, ***The consistency of the axiom of choice and of the generalized continuum hypothesis***, Ann. of Math. Studies, no. 3, Princeton Univ. Press, Princeton, N.J., 1940; second printing, 1951. MR 2, 66.

equivalence classes coincide with sets. But such an axiom leads quickly to antinomy,[7] and must therefore be restricted in some way. As one way of making this restriction, and pending the introduction of a possibly more general axiom or axiom schema, we adopt the following infinite list of axioms:

L_j. *The j-cardinal number.* $\mathrm{wf}(w) \supset (\exists u) \centerdot x \in u \equiv_x x \text{ equiv}^j\, w$ $(j = 1, 2, 3, \ldots)$.

As already observed in §2 these axioms are sufficient for some of the most needed definitions by abstraction. And though it is known that the definition by abstraction can often be accomplished in the case of a given equivalence relation F by replacing a given equivalence class by the set of well-founded sets of lowest rank belonging to the equivalence class,[8] it is thought that there may be some greater convenience and simplicity in allowing the original Frege-Russell method of definition by abstraction, as well as some interest in the possibility of doing so without inconsistency (at least in some of the important cases).

5. A relative consistency proof. To show that the Axioms A–K, L_j are consistent, it suffices to show that A–K, $L_1, L_2, \ldots, L_m$ are consistent, for arbitrary finite m. Taking a fixed m, we proceed by setting up a model in ZF set theory with a strong axiom of choice G^*. Thus we obtain a consistency proof relative to ZF set theory with the strong axiom of choice. And to reduce this to a consistency proof relative to ZF set theory without the axiom of choice, we may then use the method of Gödel.[9]

In setting up the model we shall need to distinguish the set-membership relation of ZF set theory from the set-membership relation of the model. Hence we express the former by $\in_0$, the latter by $\in$. Similarly we use ι_0‘ for the unit-set operator of ZF set theory, so that e.g. ι_0‘a is the unit set of a in the sense of ZF set theory, as distinguished from ι‘a, which is the unit set of a in the sense of the model. And again we use in ZF set theory the notation $[a_1, a_2, \ldots, a_k]$ for the finite set whose members are $a_1, a_2, \ldots, a_k$, and the notation $(a_1, a_2, \ldots, a_k)$ for the ordered k-tuple of $a_1, a_2, \ldots, a_k$, reserving the notations $\{a_1, a_2, \ldots, a_k\}$ and $\langle a_1, a_2, \ldots, a_k\rangle$ for use in reference to the model.

In ZF set theory we use the axiom of choice to provide, for every set x, a unique j-cardinal representative x^j, such that x^j is j-equivalent to x, and all sets j-equivalent to x have the same j-cardinal representative. And we say that the sets $l_1, l_2, \ldots, l_m$ *constitute a cardinal m-tuple* if they are not all empty and all members[10] of l_j are j-cardinal representatives which are not empty at level j.

Also in ZF set theory we define the 1-analogue of a set by the three following conditions:

(i) If x_1 is the $(m + 2)$-tuple $(c, 0, 0, \ldots, 0, n)$, where c is any set and n is a finite ordinal, the 1-analogue x of x_1 is the $(m + 2)$-tuple $(c, 0, 0, \ldots, 0, n + 1)$.

(ii) If x_1 is $(c, l_1, l_2, \ldots, l_m, n)$, where c is any set, $l_1, l_2, \ldots, l_m$ constitute a cardinal m-tuple, and n is a finite ordinal, the 1-analogue x of x_1 is $(c, l_1, l_2, \ldots, l_m, n + 2)$.

[7] For let us take F to be the relation of equivalence as to well-foundedness, that is, $F(x, y) \equiv_{xy} \centerdot \mathrm{wf}(x) \equiv \mathrm{wf}(y)$. This is an equivalence relation and one of its equivalence classes is the class of well-founded sets. But the set of well-founded sets is known to be antinomic.

[8] Dana Scott, *The notion of rank in set theory*, ***Summaries of Talks Presented at the Summer Institute for Symbolic Logic***, Cornell University 1957, second edition 1960, pp. 267–269.

[9] See footnote 6.

[10] I.e., the $\in_0$-members, since the definition is being made for ZF set theory.

(iii) If x_1 is of neither of the forms (i), (ii), the 1-analogue x of x_1 is x_1 itself, i.e., $x = x_1$.

Then to introduce the notion of an *i-analogue* of a set we make a definition by recursion as follows: Let x_1 be the set whose members[10] are the i-analogues of the members[10] of x_{i+1}; then the $(i+1)$-analogue x of x_{i+1} is the 1-analogue of x_1.

To express that x is an i-analogue of y we write also x anal$_i$ y.

Now to set up the model we take the sets of the model to be the same as the sets of ZF set theory, and the identity relation, =, of the model to be the same as the identity relation of ZF set theory. But a new set-membership relation $\in$ is introduced in the model, defined in terms of ZF set theory by the following equivalences—where the notation (, , ..., ,) stands always for an ordered $(m+2)$-tuple, where c is any set, where $l_1, l_2, \ldots, l_m$ constitute a cardinal m-tuple, and where n is a finite ordinal:

$x \in (c, 0, 0 \ldots, 0, 0) \equiv_x x \notin_0 c$,

$x \in (c, 0, 0, \ldots, 0, n+1) \equiv_x x \in_0 (c, 0, 0, \ldots, 0, n)$,

$x \in (c, l_1, l_2, \ldots, l_m, 0) \equiv_x . x \notin_0 c \not\equiv (\exists x_1)[x \text{ anal}_1\, x_1 . x_1^1 \in_0 l_1] \not\equiv (\exists x_2)[x \text{ anal}_2\, x_2 . x_2^2 \in_0 l_2] \not\equiv \cdots \not\equiv (\exists x_m)[x \text{ anal}_m\, x_m . x_m^m \in_0 l_m]$,

$x \in (c, l_1, l_2, \ldots, l_m, 1) \equiv_x . x \in_0 c \not\equiv (\exists x_1)[x \text{ anal}_1\, x_1 . x_1^1 \in_0 l_1] \not\equiv (\exists x_2)[x \text{ anal}_2\, x_2 . x_2^2 \in_0 l_2] \not\equiv \cdots \not\equiv (\exists x_m)[x \text{ anal}_m\, x_m . x_m^m \in_0 l_m]$,

$x \in (c, l_1, l_2, \ldots, l_m, n+2) \equiv_x x \in_0 (c, l_1, l_2, \ldots, l_m, n)$, and

$x \in c \equiv_x x \in_0 c$ in all remaining cases.

Here x_1^1 is, in the notation which was introduced above, the 1-cardinal representative of x_1, and x_2^2 is the 2-cardinal representative of x_2, and so on. Notice also that the use of the connective $\not\equiv$ (nonequivalence, or exclusive disjunction) corresponds to the symmetric difference of certain sets (or classes). The heuristic considerations by which the model was suggested should then be clear.[11]

So far we have not fixed a definition of the ordered $(m+2)$-tuple. Evidently the structure of the model will not be changed by the choice of the particular definition, as long as the essential properties of the ordered $(m+2)$-tuple are preserved. But the proof that all the required axioms are satisfied in the model may well be simplified by suitable choice of the definition of the ordered $(m+2)$-tuple.

With this in view we now define $(a_1, a_2, \ldots, a_{m+2})$ in ZF set theory as $[\iota_0^m[a_1, 0], \iota_0^{m+1}[a_2, 0], \ldots, \iota_0^{2m+1}[a_{m+2}, 0], \iota_0^{2m+3}0]$. This definition has the consequence that no set is an $(m+2)$-tuple if it is an ordinal or if its members at level m or at any level $i < m$ include at least one ordinal. Also no unit set ι_0'x is an $(m+2)$-tuple, and no pair $[x, y]$ is an $(m+2)$-tuple. And the ordinals are *self-analogous* in the sense that if α is an ordinal, then α anal$_i$ α for all subscripts i.

We omit details of the verification that all of the Axioms A–K, $L_1, L_2, \ldots, L_m$ are satisfied in the model.[12] They are straightforward but (if $m > 0$) laborious.

Generally if a consistency proof is to be supplied for proposed additions to the basic Axioms A–J and if this is done by setting up a model, a lengthy proof may be expected because most of the axioms must be treated separately in verifying that they are satisfied in the model. Nevertheless it is desirable when possible to make the consistency proof by means of a model, or at least to know either that there is or that there is not a model

---

[11]A minor but essential modification has been made in the equivalences characterizing the model, since this paper was presented to the Symposium in June, 1971.

[12]The axiom of choice is obviously satisfied in the model in the form G^*, also in the form that every well-founded set has a one-to-one correspondence with an ordinal, but not in the form that every set is well-ordered by a set of ordered pairs.

in ZF set theory. For it may well be conjectured that the structure needed to provide for sets that violate the principle of limitation of size is to a large extent already available in ZF set theory, so that it may not be necessary to change or extend the structure but only to change the way in which the theory is applied to the preformal notions of set and set membership. In any particular case the model substantiates this conjecture by showing how the change is to be made in the application of the theory to the preformal notions.

In the present case this program is not quite fulfilled, because the model given is such that only a finite subclass of the added Axioms K, L_j is satisfied. But at the cost of some additional labor in the verification of the model it appears that a modification to satisfy all the axioms together can be found without much difficulty.

As already indicated, the purpose in suggesting the Axioms K, L_j and the consistency proof for them is illustrative, as the set theory based on only these added axioms is probably not very significant if they stand alone. And the search for suitable axioms to be added to the basic axioms may very possibly reveal alternative set theories among which there must be a choice — based e.g. on the usefulness of the theory for mathematical purposes, or for the purpose of some special branch of mathematics.

One source of added axioms to be studied for their consistency with the basic axioms is Hailperin's axioms[13] for the Quine set theory,[14] which might be used either as they stand or in modified and perhaps weakened forms. Indeed an interesting possibility which must not at this stage be excluded is a synthesis or partial synthesis of ZF set theory and Quine set theory.

Correction to *Set Theory with a Universal Set* (Correction by Church in typescript, undated; probably mid- or late 1970s):

The statement in footnote 4 and in the paragraph of the text to which the footnote is attached is overhasty, as far as it seems to say (erroneously) that an *axiomatic* equivalent of the von Neumann or the Bernays set theory is obtained by taking the axioms of ZF set theory in usual form, with the axioms of *Aussonderung* and replacement formulated as axiom schemata, and embedding them in a predicative second-order logic. What is essential at the stage of the discussion at which the footnote occurs (prior to stating axioms) is rather the conceptual and notational equivalence.

However, if we take a first-order logic with free functional variables as underlying the logic of the ZF set theory, we may then state the axioms of *Aussonderung* and replacement each as a single axiom by using a free functional variable. And under this reformulation it is true that an axiomatic equivalent of the Bernays set theory is obtained by merely changing the underlying logic to a predicative second-order logic.

From this reformulation of ZF set theory it is of course also possible to obtain what Quine calls "the impredicative extension of the von Neumann-Bernays system" by changing the underlying logic to a standard (impredicative) second-order logic. For the original sources of this see Quine, *Set Theory and its Logic*, 1963, Chapter XIV.

Alonzo Church

[13] Theodore Hailperin, *A set of axioms for logic*, ***J. Symbolic Logic***, vol. 9 (1944), pp. 1–19. MR 5, 197.

[14] W. V. Quine, *New foundations for mathematical logic*, ***Amer. Math. Monthly***, vol. 44 (1937), pp. 70–80. Reprinted in ***From a Logical Point of View***, Harvard Univ. Press, Cambridge, Mass., 1953 and 1961. MR 15, 845

COMPARISON OF RUSSELL'S RESOLUTION OF THE SEMANTICAL ANTINOMIES WITH THAT OF TARSKI (1976)

§1. **Ramified theory of types.** In this paper we treat the ramified type theory of Russell [6], afterwards adopted by Whitehead and Russell in *Principia mathematica* [12], so that we may compare Russell's resolution of the semantical antinomies by ramified type theory with the now widely accepted resolution of them by the method of Tarski in [7], [8], [9].

To avoid impredicativity the essential restriction is that quantification over any domain (type) must not be allowed to add new members to the domain, as it is held that adding new members changes the meaning of quantification over the domain in such a way that a vicious circle results. As Whitehead and Russell point out, there is no one particular form of the doctrine of types that is indispensable to accomplishing this restriction, and they have themselves offered two different versions of the ramified hierarchy in the first edition of *Principia* (see Preface, p. vii).[1] The version in §§58–59 of the writer's [1], which will be followed in this paper, is still slightly different.[2]

To distinguish Russellian types or types in the sense of the ramified hierarchy from types in the sense of the simple theory of types,[3] let us call the former *r-types*.

There is an r-type i to which the individual variables belong. If $\beta_1, \beta_2, \ldots, \beta_m$ are any given r-types, $m \geq 0$, there is an r-type $(\beta_1, \beta_2, \ldots, \beta_m)/n$ to which there belong m-ary functional variables of level n, $n \geq 1$. The r-type $(\alpha_1, \alpha_2, \ldots, \alpha_m)/k$ is said to be *directly lower* than the r-type $(\beta_1, \beta_2, \ldots, \beta_m)/n$ if $\alpha_1 = \beta_1, \alpha_2 = \beta_2, \ldots, \alpha_m = \beta_m, k < n$.

The intention is that the levels shall be cumulative in the sense that the range of a variable of given r-type shall include the range of every variable of directly lower r-type.

Originally published in ***The Journal of Symbolic Logic***, vol. 41 (1976), pp. 747–760. Reprinted with corrections in ***Recent Essays on Truth and the Liar Paradox***, edited by Robert L. Martin, Oxford University Press, 1984.

This research has been supported by the National Science Foundation, grant no. GP-43517.

[1]Russell's earlier version of the ramified type hierarchy is in [6] and in the Introduction to the first edition of [12]. The latter version is in ∗12 and in (Russell's) Introduction to the second edition of [12].

[2]Differences among the three versions of ramified type theory are unimportant for the purpose of resolving the antinomies. The version which is here adopted from [1] is close to Russell's earlier version. But by using "levels" in addition to, and partly in place of, Russell's "orders" and by allowing levels and orders to be cumulative in a sense in which Russell's orders are not, it facilitates comparison both with the simple theory of types and with the hierarchy of languages and meta-languages that enters into Tarski's resolution of the semantical antinomies.

[3]It is types in the sense of the simple type theory that are called simply *types* in [1]. See footnote 578 on page 349.

The writer takes the opportunity to make a correction to his *Russellian simple type theory* (***Proceedings and Addresses of the American Philosophical Association***, vol. 47 (1974), pp. 21–33), the need for which was called to his attention by John M. Vickers. In line 21 on p. 26 [eleventh paragraph of the paper], after the words "may be empty" it is necessary to add "and the domains $\mathfrak{T}$ and $\mathfrak{F}$ must be disjoint."—Instead of this it would be possible to correct by changing the words "and otherwise the value of (**a**)**P** is in $\mathfrak{F}$" in line 15 on p. 27 [end of thirteenth paragraph] to "and the value of (**a**)**P** is in $\mathfrak{F}$ if the value of **P** is in $\mathfrak{F}$ for some value of **a**." This alternative correction may not be without interest on its own account. But it was not the original intention of the paper. Moreover, it is not historically accurate; i.e., it is not anything which might be supposed to have been intended by Russell, even implicitly, or even at a time when he still maintained the notion of proposition.

The *order* of a variable is defined recursively as follows. The order of an individual variable is 0. The order of a variable of r-type $(\beta_1, \beta_2, \dots, \beta_m)/n$ is $N + n$, where N is the greatest of the orders that correspond to the types $\beta_1, \beta_2, \dots, \beta_m$ (and $N = 0$ if $m = 0$). This is Russell's notion of order as modified by the cumulative feature which was just described.

The notations for r-types are abbreviated by writing the numeral m to stand for $(i, i, \dots, i)$, m being the number of i's between the parentheses. For example $(\;)/n$ is abbreviated as $0/n$, $(i, i, i)/n$ is abbreviated as $3/n$, and $((i)/2, (\;)/2)/1$ is abbreviated as $(1/2, 0/2)/1$.

There must be a separate alphabet of variables for each r-type, the r-type being indicated by a superscript on the letter. In writing well-formed formulas (wffs) we may often omit these r-type-superscripts as an abbreviation, if it is clear from the context what the superscript should be or if explained in words accompanying the formula. Or we may write the superscript on only the first occurrence of a particular letter understanding the superscript to be the same on all later occurrences of the same letter — not only in a particular formula but even throughout a particular passage such as a proof.

We take the range of a variable of r-type $0/n$ as propositions of level n, counting propositions as 0-ary propositional functions.[4] And the range of a variable of r-type $(\beta_1, \beta_2, \dots, \beta_m)/n$ where $m > 0$, is to consist of m-ary propositional functions which are of level n and for which the appropriate arguments are of r-types $\beta_1, \beta_2, \dots, \beta_m$ respectively.

The formation rules provide that a propositional variable (i.e., a variable of one of the r-types $0/n$) shall constitute a wff when standing alone. Also a formula $\mathbf{f}(\mathbf{x}_1, \mathbf{x}_2, \dots, \mathbf{x}_m)$ is well-formed (wf) if and only if $\mathbf{f}$ is a variable (or a primitive constant) of some r-type $(\beta_1, \beta_2, \dots, \beta_m)/n$ where $m > 0$, and $\mathbf{x}_1$ is a variable (or a primitive constant) whose r-type is β_1 or directly lower than β_1, and $\mathbf{x}_2$ is a variable (or a primitive constant) whose r-type is β_2 or directly lower than β_2, and ..., and $\mathbf{x}_m$ is a variable (or a primitive constant) whose r-type is β_m or directly lower than β_m.

---

[4]Thus we take propositions as values of the propositional variables, on the ground that this is what is clearly demanded by the background and purpose of Russell's logic, and in spite of what seems to be an explicit denial by Whitehead and Russell in [12], pp. 43–44.

In fact Whitehead and Russell make the claim: "that what we call a 'proposition' (in the sense in which this is distinguished from the phrase expressing it) is not a single entity at all. That is to say, the phrase which expresses a proposition is what we call an 'incomplete' symbol ... ." They seem to be aware that this fragmenting of propositions requires a similar fragmenting of propositional functions. But the contextual definition or definitions that are implicitly promised by the "incomplete symbol" characterization are never fully supplied, and it is in particular not clear how they would explain away the use of bound propositional and functional variables. If some things that are said by Russell in IV and V of *Introduction to the second edition* may be taken as an indication of what is intended, it is probable that the contextual definitions would not stand scrutiny. [*Editor's note*: In the Martin ed. reprint, the preceding three sentences are replaced by the following two sentences: This seems to mark the beginning of Russell's long search for a substitute for propositions, or other way to be rid of them. It is probably a late addition to the *Introduction* of [12], as no trace of it appears in [6] or in the main text of [12].]

Many passages in [6] and [12] may be understood as saying or as having the consequence that the values of the propositional variables are sentences. But a coherent semantics of Russell's formalized language can hardly be provided on this basis (notice in particular that, since sentences are also substituted for propositional variables, it would be necessary to take sentences as names of sentences). And since the passages in question seem to involve confusions of use and mention or kindred confusions that may be merely careless, it is not even certain that they are to be regarded as precise statements of a semantics.

Besides an infinite alphabet of variables in each r-type and the notation for application of a function to its arguments (already used in the preceding paragraph), the primitive symbols comprise an unspecified list of primitive constants,[5] each of definite r-type and the usual notations for negation, disjunction, and the universal quantifier.[6] The remaining formation rules, not already stated, provide that $\sim \mathbf{P}$, $[\mathbf{P} \vee \mathbf{Q}]$, and $(\mathbf{a})\mathbf{P}$ are wf whenever **P** and **Q** are wf and **a** is a variable.

In abbreviating wffs we follow the conventions of [1], as adapted to the present context.[7] The signs of material implication, conjunction,[8] and material equivalence are of course introduced by the definitions ∗1·01, ∗3·01, ∗4·01 of [12]. And as explained in footnote 6, ∗10·01 is used as definition of the existential quantifier.

We do not follow the rules of inference and axioms of either [6] or [12], as these are in some respects insufficient and also involve some oddities[9] (as they would now seem). But rather we suppose that a system of rules and axioms for propositional calculus

---

[5]It is intended that additions to the list of primitive constants may be made from time to time, so that Russell's formalized language is an open language rather than a language of fixed vocabulary.

[6]This means that we use the definition ∗10·01 of the existential quantifier rather than to take it as primitive.

The Frege-Russell assertion sign, ⊢, should also properly be listed as one of the primitives. But we here follow [1] in taking the mere writing of a wff on a separate line or lines as a sign of assertion (unless the context shows otherwise), and in introducing the sign ⊢ in a different (syntactical) sense.

Historically, it must be confessed, this change of notation is unfortunate. For Frege is right that an asserted sentence has a different meaning from a sentence occurring e.g. as antecedent or consequence of an implication. And the assertion *of* a wff (sentence or propositional form) is of course not the same as the assertion *about* the wff that it is a theorem, or that it is a demonstrable consequence of certain listed wffs.

[7]The use of dots as brackets follows a simplified form of the Peano-Russell conventions that is explained in [1]. Briefly a bold dot stands for an omitted pair of brackets with the scope extending from the point at which the dot appears, forward, either to the end of the innermost explicitly written pair of brackets, [ ], within which the dot appears or to the end of the formula if there is no such explicitly written pair of brackets; and if square brackets are omitted without replacement by a dot, the restoration of the brackets is to follow the convention of association to the left and the convention about categories, as these are explained on pp. 74–79, 171.

We also allow that any wf part of a wff may first be enclosed in square brackets (if not already so enclosed) and then these square brackets may be eliminated by replacing the first of the pair by a bold dot in accordance with the same conventions about scope of the omitted brackets that are used in other cases of replacement of brackets by dots. This may often increase the perspicuity of the abbreviated formula when dots are used for brackets.

The foregoing statement of the conventions about the use of dots as brackets is sufficient for our present purpose, and indeed is sufficient in most contexts in which very complicated formulas are not used. For cases of the kind represented by the displayed formulas on p. 80 of [1], the convention which is there intended requires some restatement for accuracy (as was pointed out by Philip Tartaglia in 1963); perhaps the shortest way of putting the required amendment is to provide that the conventions about higher and lower categories that is introduced on page 79 shall be used only when none of the connectives (with which the affected bracket-pairs belong) is written with a bold dot after it.

[8]The conjunction of **P** and **Q** is to be written simply as [**PQ**], or when the brackets are omitted, as **PQ**. If in the expression of a conjunction a bold dot appears between **P** and **Q**, this dot represents an omitted pair of brackets in accordance with the conventions explained in footnote 7—not excluding, however, the case described in the second paragraph of footnote 7, in which the dot, by representing a fictitious pair of brackets, serves only the purpose of perspicuity.

[9]Some of the oddities arise from the fact that Russell does not use a different alphabet of variables for each r-type but in effect has only one alphabet, thus leaving the r-types (or rather the relative r-types) to be determined from the wff itself in which the variables appear. And this seems to be due in turn to Russell's intention (see §II of [6]) that, in an asserted wff, although each bound variable must be restricted to a particular r-type as its range, the free variables may have a wider range.

and laws of quantifiers is adopted from some standard source. And to these we adjoin the two following comprehension axiom schemata:[10]

$$(\exists \mathbf{p}) \,.\, \mathbf{p} \equiv \mathbf{P}, \qquad \mathbf{p} \text{ not free in } \mathbf{P}$$

where **p** is a propositional variable of r-type $0/n$, the bound (in Russell's terminology, "apparent") variables of **P** are all of order less than n, and the free (in Russell's terminology, "real") variables of **P** and the constants of **P** are all of order not greater than n;

$$(\exists \mathbf{f}) \,.\, \mathbf{f}(\mathbf{x}_1, \mathbf{x}_2, \cdots, \mathbf{x}_m) \equiv_{\mathbf{x}_1\mathbf{x}_2\ldots\mathbf{x}_m} \mathbf{P}, \qquad \mathbf{f} \text{ not free in } \mathbf{P}$$

where **f** is a functional variable of r-type $(\beta_1, \beta_2, \cdots, \beta_m)/n$, and $\mathbf{x}_1, \mathbf{x}_2, \cdots, \mathbf{x}_m$ are distinct variables of r-types $\beta_1, \beta_2, \cdots, \beta_m$ and the bound variables of **P** are all of order less than the order of **f**, and the free variables of **P** (among which of course some or all of $\mathbf{x}_1, \mathbf{x}_2, \cdots, \mathbf{x}_m$ may be included) and the constants occurring in **P** are all of order not greater than the order of **f**.

From the comprehension axiom schemata there follow rules of substitution for propositional and functional variables[11] which are like $*510$ in [1], as generalized to higher types, but have the two following restrictions: (i) the wff **P** which is substituted for **p** or the wff **P** which is substituted for $\mathbf{f}(\mathbf{x}_1, \mathbf{x}_2, \cdots, \mathbf{x}_m)$ must obey the same conditions that are attached to the comprehension axiom schemata; (ii) all occurrences of the variable **p** or **f** in the wff **A** into which the substitution is made must be at extensional places.[12]

Using this reconstruction of the logic of *Principia mathematica* (with ramified type theory), as just outlined, we shall present proofs in the manner of [1]—making use in particular of the deduction theorem. The following abbreviations will be used to refer to certain primitive and derived rules of inference, as indicated:

mod. pon.: The rule of *modus ponens*, from $\mathbf{P} \supset \mathbf{Q}$ and **P** to infer **Q**.

P: Laws of propositional calculus.

ded. thm.: The deduction theorem.

univ. inst.: The rule of universal instantiation, from (**a**)**P** to infer the result of substituting **b** for all free occurrences of **a** throughout **P**, if **a** is a variable, if **b** is a variable or a constant and is either of the same r-type as **a** or of r-type directly lower than that of **a**, and (in case **b** is a variable) if there is no capture of **b** that results by the substitution described.

ex. gen.: The rule of existential generalization, from **Q** to infer $(\exists \mathbf{a})\mathbf{P}$, where **Q** is the result of substituting **b** for all free occurrences of **a** throughout **P**, and where **a**, **b**, and **P** obey the same conditions which were just stated in connection with the rule of universal instantiation.

ex. inst.: The rule of existential instantiation,[13] if $\mathbf{P}_1, \mathbf{P}_2, \cdots, \mathbf{P}_n, \mathbf{Q} \vdash \mathbf{S}$ and if the variable **a** is free in none of the wffs except **Q**, then $\mathbf{P}_1, \mathbf{P}_2, \cdots, \mathbf{P}_n, (\exists \mathbf{a})\mathbf{Q} \vdash \mathbf{S}$; also if

[10] Readers not previously familiar with ramified type theory should notice that the significance of the notion of order, which we have not yet explained, first becomes clear in the restrictions that are attached to these schemata.

[11] The proof is similar to that in [5].

[12] I.e., places at which substitutivity of material equivalence holds.

[13] For want of a better we adopt this name from "natural inference" logic, notwithstanding its inappropriateness in the present context.

$\mathbf{P}_1, \mathbf{P}_2, \cdots, \mathbf{P}_n, \mathbf{QR} \vdash \mathbf{S}$ and if the variable **a** is free in none of the wffs except **Q** and **R**, then $\mathbf{P}_1, \mathbf{P}_2, \cdots, \mathbf{P}_n, (\exists \mathbf{a}) \,.\, \mathbf{QR} \vdash \mathbf{S}$.

§2. **Grelling's antinomy.** As an example of one of the semantical antinomies we select Grelling's[14] as being perhaps the simplest to reproduce in a formalized language, although it is not one of the antinomies ("contradictions") that are discussed in [6] and [12]. Applied to one of the familiar natural languages, such as English or German, Grelling's antinomy is concerned with *adjectives* and with *properties* which the adjectives express.[15] In the formalized language which we are here treating it is propositional forms with one free variable that most nearly take the place of adjectives. But the semantics appropriate to such a *propositional form* is rather that it has a *value* for each value of its free variable, and we shall follow this in reproducing Grelling's antinomy.

We assume that symbols and formulas are to be counted among the individuals. This choice is convenient for our purpose and is allowable on the ground that any well-defined domain may be taken as the individuals. It is believed that a different choice will make no important difference in what follows.[16]

It will be sufficient to deal with the case of propositional forms having just one free variable and to assume that this is an individual variable, although there are evident generalizations of the antinomy which concern the case of propositional forms having more than one free variable or free variables of higher r-type or both. Therefore we introduce the infinite list of primitive constants val^2, val^3, val^4, $\cdots$, with the intension that $\mathrm{val}^{n+1}(a^i, v^i, F^{1/n})$ shall mean that a^i is an individual variable and v^i is a wff (propositional form) having no other free variable than a^i and for every value x^i of the variable a^i the value of v^i is $F^{1/n}(x^i)$.[17]

As no reason to the contrary appears, we take the constants val^{n+1} to be of level 1 (i.e., in Russell's terminology, predicative). The r-type of val^{n+1} is therefore $(i, i, 1/n)/1$, and it may indeed be convenient to regard the notation val^{n+1} as an abbreviation for $\mathrm{val}^{(i,i,1/n)/1}$. The order of val^{n+1} is $n+1$.

Based on the intended meaning the following postulates involving the constants val^{n+1} suggest themselves as evident.

First there is the principle of univocacy, which may be taken in the strong form:[17a]

---

[14]This first appears in [4], where it is credited to Grelling.

[15]An adjective is called autological if it has the property which it expresses, and otherwise it is heterological. E.g., if the language is English, the adjective 'polysyllabic' and 'unequivocal' are autological, while the adjectives 'long' and 'unusual' are heterological. Then is the adjective 'heterological' autological or heterological?

[16]Not even a choice that puts primitive symbols and formulas into different types, or different r-types.

[17]In this sentence, those who wish to be very accurate about use-mention distinctions may enclose the constants 'val^1', 'val^2', 'val^3' and the wff '$\mathrm{val}^n(a^i, v^i, F^{1/n})$' in Frege's single quotation marks to show that they are mentioned rather than used. But observe that nothing else in the sentence is to be enclosed in single quotation marks. There should be no confusion over the point that, since individual variables are included among the individuals, the individual variable 'a^i' may have an individual variable as value (which latter individual variable is then spoken of as "the variable a^i") [*Added in proof*: The writer has just noticed that the use of quotation marks that is suggested in the first sentence of this footnote is itself inaccurate; but the footnote may nevertheless serve its purpose of clearing up a possible misunderstanding of what is said in the text.]

[17a]I am indebted to Angela De Paola, a student in philosophy at the University of Florence (in Italy), for pointing out an error in the proof of theorem (6) as it appeared in the original publication of this paper.

(1) $\text{val}^{m+1}(a, v, F^{1/m}) \supset_{avF} . \text{val}^{n+1}(b, v, G^{1/n}) \supset_{bG} . F = G.$

Since $[F = G]$ is defined as $(H) . H(F) \supset H(G)$, where the type of H is $(1/k)/1$ (with k chosen as the greater of m and n), we may infer from (1):

(2) $\text{val}^{m+1}(a, v, F^{1/m}) \supset_{avF} . \text{val}^{n+1}(b, v, G^{1/n}) \supset_{bG} . F(x) \equiv_x G(x).$

We shall need only the weak, or extensional principle of univocacy (2). Then there is the following postulate schema, which (without extending the formalized language) can be stated only in the extensional form shown, and whose truth (for each **P**) may be seen informally by taking v to be the propositional form **P** and a to be the individual variable 'x':

(3) $(\exists a)(\exists v)(\exists F^{1/n}) . \text{val}^{n+1}(a, v, F) . F(x) \equiv_x \mathbf{P},$

where **P** is a wff in which there is no free variable other than 'x', in which all the bound variables are of order less than n, and in which all the constants are of order not greater than n. And finally there are the following postulates which express the cumulative character of the constants val^{n+1} that was implicit in our informal explanation of the meaning:

(4) $\text{val}^{n+1}(a, v, F^{1/n}) \supset \text{val}^{m}(a, v, F^{1/n})$, where $m > n + 1$.

Corresponding to the word 'heterological' that appears in the verbal statement of Grelling's antinomy, we make the definition:[18]

$$\text{het}^{n+1}(v) \rightarrow (\exists a)(\exists F^{1/n}) . \text{val}^{n+1}(a, v, F) . \sim F(v)$$

Then we prove the following theorems:

(5) $\text{het}^{n+1}(v) \supset \text{het}^{m+1}(v)$, if $m \geq n$.

PROOF. Suppose that $m \geq n$.
By (4) and P, $\text{val}^{n+1}(a, v, F^{1/n}) \vdash \text{val}^{m+1}(a, v, F)$.
Hence by P, $\text{val}^{n+1}(a, v, F) . \sim F(v) \vdash \text{val}^{m+1}(a, v, F) . \sim F(v)$.
Hence by ex. gen.,[19] $\text{val}^{n+1}(a, v, F) . \sim F(v) \vdash \text{het}^{m+1}(v)$.
Hence by ex. inst., $\text{het}^{n+1}(v) \vdash \text{het}^{m+1}(v)$.
Hence (5) follows by ded. thm.

(6) $[\text{val}^{m+2}(a, v, G^{1/m+1}) . G(x) \equiv_x \text{het}^{m+1}(x)] \supset \sim \text{het}^{n+1}(v)$, if $m \geq n$.

PROOF. Suppose that $m \geq n$.
By (2), univ. inst., and mod. pon., $\text{val}^{m+2}(a, v, G^{1/m+1})$,

$$\text{val}^{n+1}(b, v, F^{1/n}) \vdash F(x) \equiv_x G(x).$$

Hence by univ. inst. and P, $\text{val}^{m+2}(a, v, G)$,

$$\text{val}^{n+1}(b, v, F), \sim F(v) \vdash \sim G(v).$$

Hence by ex. inst., $\text{val}^{m+2}(b, v, G), \text{het}^{n+1}(v) \vdash \sim G(v)$.
Hence by ded. thm., $\text{val}^{m+2}(b, v, G) \vdash \text{het}^{n+1}(v) \supset \sim G(v)$.

In order to correct the error it has been necessary to strengthen postulate (1) and its consequence (2), but it is thought that the strengthened (1) and (2) are still informally evident.

[18] For strict accuracy the letters v and a should be in bold type—i.e., syntactical variables, as in [1]. If we here follow the common informal practice of using object-language variables, it is to avoid obtruding use-mention distinctions where they are not in fact important for understanding of what is being said.

[19] We here follow the strong form of the rule of existential generalization as this is stated above, taking **b** to be $F^{1/n}$ and **a** to be $F^{1/m}$.

Hence by univ. inst. and P, $\mathrm{val}^{m+2}(b, v, G)$,

$$G(x) \equiv_x \mathrm{het}^{m+1}(x) \vdash \mathrm{het}^{n+1}(v) \supset \sim \mathrm{het}^{m+1}(v).$$

Hence by (5) and P,

$$\mathrm{val}^{m+2}(b, v, G), G(x) \equiv_x \mathrm{het}^{m+1}(x) \vdash \sim \mathrm{het}^{n+1}(v).$$

Hence (6) follows by P, ded. thm and substitution.

(7) $[\mathrm{val}^{m+2}(a, v, G^{1/m+1}) \,.\, G(x) \equiv_x \mathrm{het}^{m+1}(x)] \supset \mathrm{het}^{n+1}(v)$, if $m < n$.

PROOF. Suppose that $m < n$.
By P and ex. gen.,

$$\mathrm{val}^{m+2}(a, v, G^{1/m+1}), \sim G(v) \vdash \mathrm{het}^{m+2}(v).$$

Hence by univ. inst. and P,

$$\mathrm{val}^{m+2}(a, v, G), G(x) \equiv_x \mathrm{het}^{m+1}(x), \sim \mathrm{het}^{m+1}(v) \vdash \mathrm{het}^{m+2}(v).$$

Hence by (5) (used twice) and P,

$$\mathrm{val}^{m+2}(a, v, G), G(x) \equiv_x \mathrm{het}^{m+1}(x), \sim \mathrm{het}^{n+1}(v) \vdash \mathrm{het}^{n+1}(v).$$

Hence by ded. thm. and P,

$$\mathrm{val}^{m+2}(a, v, G), G(x) \equiv_x \mathrm{het}^{m+1}(x) \vdash \mathrm{het}^{n+1}(v).$$

Hence (7) follows by P and ded. thm.

Also as an instance of (3) we have:

(8) $(\exists a)(\exists v)(\exists G^{1/m+1}) \,.\, \mathrm{val}^{m+2}(a, v, G) \,.\, G(x) \equiv_x \mathrm{het}^{m+1}(x)$.

If we reduce to simple type theory by dropping all level indicators, the infinitely many constants val^{n+1} coalesce into a single constant, val, whose type (in the sense of simple type theory) is $(i, i, (i))$. The informal explanation of the meaning of val^{n+1} then becomes an explanation of the meaning of val, and the postulates (1), (3) still seem to be evident from this intended meaning; moreover (2) is still a consequence of (1), and (4) becomes tautologous. The proofs of theorems (5)–(8) still hold, after dropping the level indicators, and the last three theorems then constitute a contradiction. This is Grelling's antinomy, as it arises in simple type theory.

The resolution of the antinomy by ramified type theory consists not merely in the fact that, after restoration of the level indicators, theorems (6)–(8) are no longer a contradiction, but also in that the question "Is the propositional form $\mathrm{het}^{m+1}(x)$ autological or heterological?" can now be answered: namely it is (by (6)) autological at all levels $\leq m+1$, and it is (by (7)) heterological at all levels $> m+1$.

§3. The language L. Now let L be the language of ramified type theory (as here formulated) with addition of all the constants

$$\mathrm{val}^{(i,i,\cdots,i(\beta_1,\beta_2,\cdots,\beta_m)/n)/1},$$

and appropriate postulates involving them. Here m is any nonnegative integer, $\beta_1, \beta_2, \cdots, \beta_m$ are any m r-types, n is any level ≥ 1, and in the superscript

$$(i, i, \cdots, i, (\beta_1, \beta_2, \cdots, \beta_m)/n)/1$$

that indicates the r-type of the constant there are to be exactly $m + 1$ i's preceding the r-type-symbol

$$(\beta_1, \beta_2, \cdots, \beta_m)/n.$$

The constants val^{n+1} that were introduced above are special cases, corresponding to $m = 1, \beta_1 = i$. Generally, the wff

$$\mathrm{val}^{(i,i,\cdots,i,(\beta_1,\beta_2,\cdots,\beta_m)/n)/1}(a_1, a_2, \cdots, a_m, v, F^{(\beta_1,\beta_2,\cdots,\beta_m)/n})$$

shall mean that $a_1, a_2, \cdots, a_m$ are distinct variables of types $\beta_1, \beta_2, \cdots, \beta_m$ respectively, and v is a wff having no other free variables than $a_1, a_2, \cdots, a_m$, and for every system of values

$$x_1^{\beta_1}, x_2^{\beta_2}, \cdots, x_m^{\beta_m}$$

of the variables $a_1, a_2, \cdots, a_m$ the value of v is

$$F^{(\beta_1,\beta_2,\cdots,\beta_m)/n}(x_1^{\beta_1}, x_2^{\beta_2}, \cdots, x_m^{\beta_m}).$$

(Compare footnote 17.)

We put down only the postulates that are generalizations of (1), (3), (4) above.[20] As an abbreviation in stating these we take the r-type of F to be always

$$(\beta_1, \beta_2, \cdots, \beta_m)/n;$$

and the r-type of G is to be

$$(\beta_1, \beta_2, \cdots, \beta_m)/k;$$

and the r-type of val is to be, at each occurrence, the lowest that is compatible with its arguments, unless the contrary is said. The postulates are:

(9) $\mathrm{val}(a_1, a_2, \cdots, a_m, v, F) \supset_{a_1 a_2 \cdots a_m v F} . \mathrm{val}(a_1, a_2, \cdots, a_m, v, G) \supset_G . F = G,$

(10) $(\exists a_1)(\exists a_2) \cdots (\exists a_m)(\exists v)(\exists F) . \mathrm{val}(a_1, a_2, \cdots, a_m, v, F) .$
$F(x_1, x_2, \cdots, x_m) \equiv_{x_1 x_2 \cdots x_m} \mathbf{P},$

where **P** is a wff in which there are no free variables other than 'x_1', 'x_2', $\cdots$, 'x_m' and in which all the bound variables are of order less than the order of 'F' and all the constants of order not greater than the order of 'F'.

(11) $\mathrm{val}(a_1, a_2, \cdots, a_m, v, F) \supset \mathrm{val}(a_1, a_2, \cdots, a_m, v, F),$

where the constant, val, on the right is of the lowest r-type that is compatible with the arguments it has, while that on the left is of any other r-type that is compatible with these arguments.

For $n = 1, 2, 3, \cdots$, let L_n be the sublanguage of L obtained by deleting all variables and constants of order greater than n, and allowing the variables of order n to occur only as free variables.

---

[20]Additional postulates relating the semantical propositional functions, val, to the syntax of L can be expressed only after adding still further primitives enabling us to express the syntax of L. There are indeed some additional things holding that can be expressed without introducing new primitives, for example:

$\mathrm{val}(a, v, F) \supset . \mathrm{val}(b, v, F) \supset . a = b \vee (x)F(x) \vee (x) \sim F(x),$
$\mathrm{val}(a, b, v, F) \supset . \mathrm{val}(b, a, v, G) \supset . F(x, y) \equiv_{xy} G(y, x),$
$a = b \supset \sim \mathrm{val}(a, b, v, F).$

But our present purpose does not make it necessary to explore the question of additional postulates which may therefore be wanted.

Then L_1 is a functional calculus of first order[21] in the presently standard sense, i.e., only individual variables occur as bound variables. None of the semantical constants, val, are in L_1, as the lowest order of these is 2. And the propositional and functional variables in L_1 are of first order, having superscripts of the form $m/1$ $(m = 0, 1, 2, \cdots)$.

In L_2 there are propositional and functional variables with superscripts $m/2$, where $m \geq 0$; and also propositional and functional variables with superscripts $(\beta_1, \beta_2, \cdots, \beta_m)/1$, $m \geq 0$, where each of $\beta_1, \beta_2, \cdots, \beta_m$ is either i or of the form $k/1$, $k \geq 0$. But only individual variables and propositional and functional variables of first order are used as bound variables. And the semantical constants, val, which are present in L_2 are those having the r-types $(i, i, \cdots, i, m/1)/1$, $m = 0, 1, 2, \cdots$, and thus are precisely those needed for the semantical metatheory of L_1.

We may therefore regard L_2 as a semantical meta-language of L_1. However, L_2 is stronger than L_1 not only in having the semantical predicates (semantical functional constants) that are needed for the semantics of L_1 but also in having additional free variables beyond those of L_1, namely the variables of second order, and in admitting as bound variables certain variables which appear only as free variables in L_1, namely the variables of first order.

And so we may continue through the hierarchy of languages L_1, L_2, L_3, $\cdots$, the situation being always that L_{n+1} is a semantical meta-language of L_n, containing the same semantical predicates that are applicable to L_n, and containing also L_n itself plus additional r-types of free variables and additional r-types of bound variables that are not present in L_n. Moreover it is quite indifferent whether we speak of a single language L and a hierarchy of orders of variables and predicates within it or whether we speak of an infinite hierarchy of languages L_1, L_2, L_3, $\cdots$, as it is evident that the distinction is merely terminological.

§4. Comparison with Tarski. It is Tarski's solution of the problem of the semantical antinomies that the semantical predicates for a particular language must be contained, not in the language itself, but always in a meta-language. Indeed the semantical predicates val are intensional, whereas Tarski at the date of [7] and [8] is concerned only with the extensional semantical notions of truth and satisfaction and perhaps would have denied corresponding intensional notions. But it is intensional semantical predicates that are primarily appropriate to the language L (or to L_1, L_2, etc.). The essential point of the resolution of the semantical antinomies by Tarski is unrelated to a distinction of intension and extension, and is simply that the semantical predicates (and propositional forms) appropriate to a language must be put into a meta-language of it.

In the light of this it seems justified to say that Russell's resolution of the semantical antinomies is not a different one than Tarski's but is a special case of it.

[21] The terminology, functional calculus of first order, second order, etc., is appropriate primarily to simple type theory and represents a different meaning of the word 'order' from that which is needed in connection with ramified type theory. The hierarchy of languages L_1, L_2, L_3, $\cdots$ is in fact quite different from the hierarchy of functional calculi of first, second, third, $\cdots$ orders. And except in such phrases as 'functional calculus of second order' we shall always use the word 'order' in the sense (essentially Russell's) which was defined at the beginning of this paper. (Footnote 578 of [1] overlooks that the notion of "level," though useful, cannot wholly supersede Russell's notion of "order" in treating ramified type theory.)

This conclusion may be supported by supplying in L (or in L_1, L_2, etc.) the following definitions to express the extensional semantical notions, that v is a true sentence of L_n,

$$\mathrm{tr}^{n+1}(v) \rightarrow (\exists p^{0/m}) \,.\, \mathrm{val}^{(i,0/n)/1}(v, p) \,.\, p,$$

and that v is a propositional form of L_{N+n} and is satisfied by the values $x_1, x_2, \cdots, x_m$ of the variables $a_1, a_2, \cdots, a_m$.

$$\mathrm{sat}^{N+n+1}(a_1, a_2, \cdots, a_m, x_1, x_2, \cdots, x_m, v) \rightarrow (\exists F) \,.\, \mathrm{val}(a_1, a_2, \cdots, a_m, v, F) \,.\, F(x_1, x_2, \cdots, x_m)$$

where in the latter definition, F is of r-type $(\beta_1, \beta_2, \cdots, \beta_m)/n$ and order $N + n$, and val is of r-type $(i, i, \cdots, i, (\beta_1, \beta_2, \cdots, \beta_m)/n)/1$, and $x_1, x_2, \cdots, x_m$ are of r-types $\beta_1, \beta_2, \cdots, \beta_m$ respectively (where $m \geq 1, n \geq 1$).

As the propositional form $\mathrm{tr}^{n+1}(v)$ contains a bound variable of order n, a constant of order $n + 1$, and as its only free variables, the individual variable 'v', it follows that $\mathrm{tr}^{n+1}(v)$ belongs to the language L_{n+1}, but not to L_n. Also similarly the propositional form $\mathrm{sat}^{N+n+1}(a_1, a_2, \cdots, a_m, x_1, x_2, \cdots, x_m, v)$ belongs to the language L_{N+n+1}, but not to L_{N+n}. And this is just what Tarski's resolution of the semantical antinomies requires.

It should be remarked that if we *begin* with extensional semantics, we may then naturally take as primitive a predicate or infinite list of predicates, tr, which require an individual variable as argument, and an infinite list of predicates, sat, which require $2m + 1$ arguments of r-types $i, i, \cdots, i, \beta_1, \beta_2, \cdots, \beta_m, i$. In this case the levels of the primitive predicates, tr and sat, must be assigned *ad hoc* to avoid antinomy, and the ramified theory of types may seem to play only a secondary role.

However, such priority of extensional semantics is just not appropriate to the language L—and does not accord with the way in which the resolution of the "contradictions" is informally explained in [6] and [12].[22] Moreover, as we have just seen, we are able by taking the intensional predicates val as primitive to supply definitions of $\mathrm{tr}^{n+1}(v)$ and $\mathrm{sat}^{N+n+1}(a_1, a_2, \cdots, a_m, x_1, x_2, \cdots, x_m, v)$. But the reverse does not hold: if the predicates tr and sat of various r-types are primitive, we are unable from them to define $\mathrm{val}(a_1, a_2, \cdots, a_m, v, F)$ suitably, but only an extensional analogue which we may call $\mathrm{valext}(a_1, a_2, \cdots, a_m, v, F)$. For example, if

$$\mathrm{valext}^{n+1}(a, v, F^{1/n}) \rightarrow F(x) \equiv_x \mathrm{sat}^{(i,i,i)/n+1}(a, x, v),$$

we have for $\mathrm{valext}^{n+1}(a, v, F)$, unlike $\mathrm{val}^{n+1}(a, v, F)$, that if it is satisfied by given values of the variable a, v, F, then it is satisfied also by the same values of a and v and any coextensive (or formally equivalent) value of F.

By taking the intensional semantics as prior, and proceeding from it to the extensional semantical notions, ramified type theory resolves the semantical antinomies in

[22] For example, in his explanation regarding the Epimenides antinomy in §I of [6] or p. 62 of [12], when Russell says that the notion of "all propositions" is illegitimate and that a statement about all propositions of some order must be itself of higher order, we may take a "statement" or a "proposition" (the two words seem to be synonymous) to be "in the sense in which this is distinguished from the phrase expressing it," or we may take it to be a declarative sentence *considered together* with its meaning. But for a sentence, either as a finite sequence of sounds (or of printed or written characters) or as a class or class concept of such, organized by a particular syntax but not yet associated with a meaning, there is in the nature of the case nothing illegitimate in the totality of all sentences. And in setting up the language L we have in fact taken the symbols, sentences, and propositional forms of L as being, all of them, members of a single r-type.

a straightforward way, without *ad hoc* additional assumptions, and it is seen only after the event that the resolution is a subcase of Tarski's.

§5. Axioms of reducibility. To secure adequacy for classical mathematics it is necessary to adjoin to the language L an axiom of infinity, axioms of choice in some form, and the axioms of reducibility.[23]

As our concern is with the resolution of semantical antinomies by ramified type theory, it is only the axioms of reducibility that need concern us here. They are:

$$(F^{(\beta_1,\beta_2,\cdots,\beta_m)/n})(\exists G^{(\beta_1,\beta_2,\cdots,\beta_m)/1}) \,.\, F(x_1, x_2, \cdots, x_m) \equiv_{x_1x_2\cdots x_m} G(x_1, x_2, \cdots, x_m),$$

where $m = 1, 2, 3, \cdots,$[24] and the variables $x_1, x_2, \cdots, x_m$ are of r-type $\beta_1, \beta_2, \cdots, \beta_m$ respectively.

The effect of the axioms is that the range of the functional variables is already extensionally complete at level 1, in the sense that it contains a propositional function that is extensionally (or in the terminology of [12], "formally") equivalent to any propositional function which enters as a value of the functional variables at any higher level; and that it is only in intension that we are to think of additional values of the functional variables as arising at each new level. Thus the rejection of impredicative definition is annulled in extensional but not in intensional matters.[25] And this much is enough for classical mathematics, especially mathematical analysis.

The danger may be feared that the axioms of reducibility will restore the semantical antinomies which it was intended to avoid by means of the ramified type hierarchy. But this does not appear to be realized, at least not in any obvious way.

Let us take the case of Grelling's antinomy as an illustration. By using the appropriate one of the axioms of reducibility, we may indeed prove:

(12) $(\exists H^{1/1}) \,.\, H(v) \equiv_v (\exists a)(\exists F^{1/n}) \,.\, \text{val}^{n+1}(a, v, F) \,.\, {\sim} F(v).$

Then if we take as hypothesis

(13) $H^{1/1}(v) \equiv_v (\exists a)(\exists F^{1/n}) \,.\, \text{val}^{n+1}(a, v, F) \,.\, {\sim} F(v),$

---

[23] We treat the system of the first edition of [12], with ramified type theory and axioms of reducibility. The modification which is suggested in Russell's Introduction to the second edition (and is based on ideas of Wittgenstein), to replace the axioms of reducibility by axioms of extensionality, one in each type, is unsatisfactory—because the resulting system is not adequate for classical mathematics (as Russell admits, see pp. xiv, xxix, xliv–xlv, and compare Weyl [10]), and because if Russell's attempt is successful, in Appendix C of the second edition of [12], to be rid of intensional contexts, he thereby abandons some of his own important contributions to logic. As regards this last, we have already seen how a strictly extensional approach prejudices the resolution of the semantical antinomies by ramified type theory; and Russell's theory of descriptions loses its point as a solution of the puzzle about King George IV and the author of ***Waverley*** if there are no intensional contexts.

[24] Only the cases $m = 1, 2$ are used in [6] and [12], but the case of greater m is referred to briefly.

[25] That the restoration of impredicative definition is confined to extensional contexts might be defended on the ground that there are antinomies which are about intensional matters but are not semantical in character. For example, Bouleus believes that he is sometimes mistaken, but (with the possible exception of some that are logically implied by this one together with his true beliefs) all his other beliefs are in fact true. Is it then true that Bouleus is sometimes mistaken?

This is implicitly a correction of the first paragraph of §59 of [1], as it is only by confining attention to extensional logic that it can be said that ramified type theory with axioms of reducibility has no interest as an intermediate position between pure ramified type theory and simple type theory.

we may repeat the proofs of (6) and (7) in modified form, treating them as proofs from the hypothesis (13). In this way we get, as proved from the hypothesis (13), both

$$[\mathrm{val}^2(a, v, G^{1/1}) \,.\, G(x) \equiv_x H(x)] \supset \sim H(v)$$

and

$$[\mathrm{val}^2(a, v, G^{1/1}) \,.\, G(x) \equiv_x H(x)] \supset H(v).$$

No contradiction results, as it does not appear that a similar analogue of (8) can be obtained. But from the last two formulas we get by propositional calculus

$$G(x) \equiv_x H(x) \supset \sim \mathrm{val}^2(a, v, G),$$

still as proved from the hypothesis (13); then deduction theorem and a substitution for the functional variable H enable us to prove the theorem:

(14) $G^{1/1}(x) \equiv_x \mathrm{het}^{n+1}(x) \supset_{avG} \sim \mathrm{val}^2(a, v, G).$

Also by alphabetic changes of bound variable in (12):

(15) $(\exists G^{1/1}) \,.\, G(x) \equiv_x \mathrm{het}^{n+1}(x).$

This is an empiric justification of the axioms of reducibility, based on the failure of the direct attempt to restore Grelling's antinomy by means of them, and on the fact that the resulting situation as expressed in (14) and (15) not only is intelligible but even is to be expected in the light of Tarski's theorem about truth.[26] If the axioms of reducibility are included in the hierarchy of languages L_1, L_2, L_3, $\cdots$, each at its appropriate place, it therefore seems that the resulting hierarchy of languages will still conform to Tarski's resolution of the semantical antinomies.[27] But this again is only an empiric justification because the actual conformity to Tarski's plan of resolution of the antinomies depends on the unprovability of certain theorems,[28] which must here

[26]In fact consider the case that the superscript $n+1$ in (14) and (15) is 2, and let (14′) and (15′) be the sentences obtained from (14) and (15) respectively by replacing '$\mathrm{het}^2(x)$' by '$\mathrm{tr}^2(x)$'. Suppose further that the languages L_1 and L_2 are consistent and that '$\mathrm{tr}^2(x)$', as a propositional form of L_2, is satisfied by those and only those values of 'x' which are true sentences of L_1. Then (15′) must hold if the range of the singular functional variables of L_1 is to be extensionally complete; and (14′) can be taken as an expression of Tarski's theorem.

And that what holds of truth—i.e., expressibility only in a meta-language—must be expected to hold also of other semantical notions, including heterologicality, is already implicit in the description of Tarski's resolution of the semantical antinomies as we gave it above.

[27]Weyl's use [10] of Grelling's antinomy to support what is in effect ramified type theory without axioms of reducibility is therefore not in itself compelling. That is, it is not demonstrated that a system which allows impredicative definition must therefore be inconsistent. But if one agrees with Weyl [10], [11] that impredicative definition is intrinsically unsound, a *circulus vitiosus* whether or not it leads to antinomy, then indeed ramified type theory without axioms of reducibility is what results; and while first-order arithmetic can be obtained by adjoining either Peano's postulates for the natural numbers (under some appropriate choice of r-types for the variables and constants occurring in them) or an axiom of infinity strong enough to yield this, it will still be impossible to obtain more than a weakened form of the classical theory of real numbers.

[28]For example $\mathrm{sat}^2(a, x, v)$, defined as $(\exists F^{1/1}) \,.\, \mathrm{val}^2(a, v, F) \,.\, F(x)$, is supposed to express the semantical satisfaction relation only as it applies to L_1; if theorems could be proved by which it could be regarded as expressing this relation also for L_2, the Tarski plan for avoiding antinomy would be violated. Something may depend on whether postulates (4) and (11) are assumed only in the weak form which was given to them above or whether they are strengthened by putting $\equiv$ in place of $\supset$. It is conjectured that L remains consistent after adjoining both the axioms of reducibility and the strong form of postulates (4) and (11). But this conjecture, if correct, deserves support by a relative consistency proof, relative perhaps to the consistency of the simple theory of types or of standard axiomatic set theory.

remain a conjecture.

The principle significance of theorems (14) and (15) is that there must be, among the values of the variables in L_1 of r-type 1/1, propositional functions such that no coextensive function is expressible by a propositional form in L_1, but only in some language arbitrarily far along in the hierarchy $L_1, L_2, L_3, \cdots$.

REFERENCES

[1] ALONZO CHURCH, ***Introduction to mathematical logic***, Volume 1, Princeton, 1956.

[2] IRVING M. COPI, ***The theory of logical types***, London, 1971.

[3] ABRAHAM A. FRAENKEL and YEHOSHUA BAR-HILLEL, ***Foundations of set theory***, Amsterdam, 1958; second edition 1973, by Fraenkel, Bar-Hillel, and Azriel Levy with collaboration of Dirk van Dalen.

[4] KURT GRELLING and LEONARD NELSON, *Bemerkungen zu den Paradoxieen von Russell und Burali-Forti*, ***Abhandlungen der Fries'schen Schule***, n.s. vol. 2 (1907–08), pp. 301–324.

[5] LEON HENKIN, *Banishing the rule of substitution for functional variables*, ***The journal of symbolic logic***, vol. 18 (1953), pp. 201–208.

[6] BERTRAND RUSSELL, *Mathematical logic as based on the theory of types*, ***American journal of mathematics***, vol. 30 (1908), pp. 222–262.

[7] ALFRED TARSKI, *Pojęcie prawdy w językach nauk dedukcyjnych*, ***Travaux de la Société des Sciences et des Lettres de Varsovie, Classe III***, no. 34, Warsaw, 1933.

[8] ———, *Der Wahrheitsbegriff in den formalisierten Sprachen* (German translation of [7] with added *Nachwort*), ***Studia philosophica***, vol. 1 (1936), pp. 261–405.

[9] ———, *The concept of truth in formalized languages* (English translation of [8]), ***Logic, semantics, metamathematics, Papers from 1923 to 1938, by Alfred Tarski***, London, 1956, pp. 152–278.

[10] HERMANN WEYL, *Das Kontinuum*, ***Kritische Untersuchungen über die Grundlagen der Analysis***, Leipzig, 1918.

[11] ———, *Der circulus vitiosus in der heutigen Begründung der Analysis*, ***Jahresbericht der Deutschen Mathematiker-Vereinigung***, vol. 28 (1919), pp. 85–92.

[12] A. N. WHITEHEAD and BERTRAND RUSSELL, ***Principia mathematica*** (three volumes), Cambridge, 1910–1913; second edition, Cambridge, 1925–1927.

UNIVERSITY OF CALIFORNIA
LOS ANGELES, CALIFORNIA 90024

HOW FAR CAN FREGE'S INTENSIONAL LOGIC BE REPRODUCED WITHIN RUSSELL'S THEORY OF DESCRIPTIONS? (1979)

Footnote 11 of my forthcoming paper *A remark concerning Quine's paradox about modality* shows the context within which the idea arises that Frege's theory might be wholly or partly reproduced within Russell's by making appropriate definitions.

Here it is not Fregean and Russellian semantics that are in question but the corresponding intensional logics. And indeed it is the intensional logic which mainly requires attention, as once the difficulties in this are resolved it should be a comparatively simple matter to devise a language with the desired sort of semantics (Fregean or Russellian). The intensional entities are of course the propositional functions in the case of Russell's logic; and in the case of Frege's logic they are the things which are capable of serving as senses of names, and which I like to call concepts (in order to be able to speak of the sense of a name as being a *concept of* the denotation). It is unfortunate that Frege used the German word "Begriff" (naturally translated into English as "concept") for something quite different — nearly the same as Russell's propositional function, except that the value of the function is a truth-value rather than a proposition. I would propose to distinguish by using the German word "Begriff," even in English, for Frege's Begriffe, so saving the English word "concept" for the other use just described.

Footnote 11 suggests that concepts of things of type α might be identified with singulary Russellian propositional functions over the type α. Then the propositional functions which are unitary are the concepts which succeed in being concepts of something actual, and so may serve as senses of names that succeed in denoting. The propositional functions which are false of all things of type α are vacuous concepts, appropriate as senses of denotationless names. And propositional functions which are true of more than one thing of type α must again be taken as vacuous concepts; the alternative of identifying them as senses of common names (of general names in Mill's sense) is tempting but is for several reasons unacceptable, among them that no satisfactory formalized language is available in which common names appear.

This will of course require that Russell's propositional functions are used in a dual role, once as concepts and once as taking the place of Frege's Begriffe.

An important effect of a successful representation of Frege's theory within Russell's will be to tie together various open or otherwise difficult problems about the two theories. In particular there is a difficulty about the principle of individuation for concepts. My *Outline of a revised formulation of the logic of sense and denotation* attempts to deal with this, but success is only partial. The problem, the partial success in dealing with it, and the remaining difficulty will all be transferred from Frege's theory to Russell's.

For Russell's propositional functions must equally have a principle of individuation if one is to believe in them.

Previously unpublished abstract of a lecture, March 23, 1979. Printed with permission of the Alonzo Church estate.

I have not worked out details of the representation of the one theory within the other. But if we use the formulation of the Fregean theory that is in my paper in *Noûs* (vol. 7, pp. 24–33, and vol. 8, pp. 135–156), it seems that it will be necessary to allow wffs of the Fregean theory to correspond not only to wffs of the Russellian theory but also in some cases (e.g. the Frege-like descriptions of page 142 of my paper just cited) to contextually defined expressions. We may in the first instance take the Russellian theory as a simple type theory. But it will be desirable to add the connective $\equiv$ to it, as in my *Russellian simple type theory*, so that the abstraction operator λ becomes contextually definable. And at least if Alternative (0) is adopted for the Fregean theory, some modification of the simple array of types in the direction of ramification will be needed.

As a proposed representation of Russellian simple type theory within the Fregean theory we first let the Russellian individuals correspond to Fregean individuals or Gegenstände. (It is irrelevant whether Russell's notion of an individual is the same as Frege's notion of a Gegenstand, as we are interested only in a correspondence that is formally accurate and truth-preserving, so that whenever a logical problem has been solved within one theory, its solution may then be transferred to the other.) Russellian propositional functions are then to correspond, not to Begriffe (for which axioms of extensionality may be assumed), but to Begriff concepts. And say for example that φ is a singulary propositional function and x an individual, and let φ' be the corresponding Begriff concept and x' the corresponding Gegenstand; then φx, that φ holds of x, is to correspond to: there exists a Begriff f' such that φ' is a concept of f' and $f'x'$. If a Russellian propositional function requires another propositional function (of lower type) as argument, at one of its argument places, the corresponding concept in the Fregean theory must be a concept of a Begriff which requires as argument another concept of a Begriff (of lower type) at the corresponding argument place.

A REMARK CONCERNING QUINE'S PARADOX ABOUT MODALITY (1982)

In what at first sight may seem to be its simplest and most direct form, Quine's paradox about modality may be explained by reference to the following example:

(1) $\Box 9 = 9$,

(2) The number of major planets = 9.

Here '$\Box$' is the notation for necessity in standard modal logic and may be read as "it is necessary that." Both (1) and (2) are commonly accepted as true, the former on logico-mathematical grounds and the latter on astronomical grounds. The paradox arises if, relying on the principle of substitutivity of equality, we make use of (2) in order to substitute 'the number of major planets' for the first occurrence of '9' in (1). For in this way we infer from two accepted truths what is evidently false:

(3) $\Box$ the number of major planets = 9.

Indeed it was at one time believed that the number of major planets is less than 9; and this belief, though factually false, was hardly a belief in an impossibility.

Quine's original formulation of the paradox, in [11] and [15], does not use exactly this example but a number of others —which however, illustrate what is evidently the same logical difficulty. And one of Quine's examples in [15] is in fact the same as above, except that the premiss (1) is replaced by

(4) $\Box 9 > 7$,

so that, in place of (3), the paradoxical conclusion becomes:

(5) $\Box$ the number of major planets > 7.

The inference of (3) from the premisses (1) and (2) has here been chosen as an example in order to exhibit more clearly the very close parallelism between Quine's paradox and the paradox which Russell [17] sought to eliminate by means of his theory of descriptions.

For the latter paradox we may use a minor variation of Russell's example[1] about King George IV. We may assume it true, as of some appropriate date, that

(6) George IV believes that Sir Walter Scott = Sir Walter Scott,

since George IV, being well acquainted with Sir Walter, is unlikely to doubt this. Moreover, it is factually true, as of this same date, that

(7) The author of Waverley = Sir Walter Scott.

By substitutivity of equality it seems to follow that

(8) George IV believes that the author of Waverley = Sir Walter Scott.

But (8) is known to be false, still as of the same date; and even without the factual information it is clearly unreasonable to suppose that (8) is a logical consequence

---

Originally published in Spanish, *Una Observacion respecto a la Paradoja de Quine sobre la Modalidad*, ***Analisis Filosof.*** vol. 2 (1982), pp. 25–32. Printed by permission of the Alonzo Church estate. Written between 1975 and 1979. The English version is previously unpublished. The editors have inserted "major" in the first two occurrences of "number of planets" on this page.

[1] The example is due to Russell [17], the point which it illustrates, to Frege [5].

of (6) and (7).

Or still better we may use the premiss

(9) George IV believes that $9 = 9$,

which is at least probably true. From (9) and (2) there seems to follow by substitutivity of equality:

(10) George IV believes that the number of major planets $= 9$.

But (10) is certainly false, because the discovery of the last two major planets came only after King George's death.

Given the truth of (2), the paradox based on the inference of (3) from (1) differs from that based on the inference of (10) from (9) only in the replacement of 'George IV believes that' by the modal operator '$\Box$'. It may therefore be argued that there are not two genuinely different paradoxes, but only various examples illustrating what must be regarded as a single paradox. And indeed Quine, already in [11], makes a close association between paradoxes about belief and paradoxes about modality.

This paradox may arise not only in connection with 'believes that' and 'it is necessary that' but also with any of various other phrases which we may speak of as introducing intensional contexts. Carnap [2] calls it the *antinomy of the name relation*. And this terminology is useful as a reminder that, in place of substitutivity of equality, the paradox may be made to depend on the semantical principle that (in Carnap's words) *if two expressions name the same entity, then a true sentence remains true when the one is replaced in it by the other*. But since —contrary to Carnap —there is no actual antinomy or contradiction, but only such results as (3), (8), (10), which are factually false or unacceptably counterintuitive, the writer prefers to speak of the *paradox of the name relation*.

Once it is seen that Quine's paradox about modality and the paradox about King George IV and Sir Walter Scott are to be treated as instances of the same paradox, it is not surprising that Smullyan [19], [20] is able to resolve Quine's paradox by means of Russell's theory of descriptions. Moreover it seems certain in advance that whatever objections may apply to Smullyan's modal logic with descriptions (e.g. Quine's objection in [15] that it requires excessive attention to matters of scope) must equally apply to the logic of belief statements with descriptions which is implicit in Russell's resolution in [17] of the paradox about King George and Sir Walter.[2]

Besides the complications regarding scope which arise when the paradox about modality is resolved by Russell's doctrine of descriptions, Quine [15] raises also a different objection, which depends on reformulating the paradox in a way that refers only to variables and makes no use either of names (or naming expressions) or of

[2]In [17] Russell has an informal treatment of the matter of scope of descriptions in connection with belief statements (and the like), using the particular examples "I thought your yacht was larger than it is" and "George IV wished to know whether Scott was the author of ***Waverley***"; but in his later writings, leading up to and including ***Principia Mathematica***, he does not return to this. His formal language is confined to what is needed for the ***Principia*** account of the foundations of mathematics, and he therefore never considers a formalized logic of belief in connection with the theory of descriptions. Nevertheless the complications about scope are already implicit in Russell's paper of 1905 and are not due to changes by Smullyan.

In regard to substitutivity of equality for descriptions in intensional contexts, a matter closely related to that of scope, Quine makes a mistake which invalidates his inference of (4) from (3) on page 149 of [16]. The effect is that Quine's own error reinforces his complaint about the complications which arise if the theory of descriptions is employed in a logic of belief (or of necessity or other intensionalities).

descriptions. The point is that although Russell largely reconstrues names and naming expressions as descriptions, and then eliminates[3] the descriptions by his device of contextual definition, of course the use of variables is not thereby eliminated.

Citing Ruth Barcan (in [1] and earlier papers) Quine calls attention to the theorem

(11) $x = y \supset_{xy} \Box x = y.$

Indeed this theorem follows by elementary logic alone, independently of the exact meaning or definition of the sign '=' of identity or equality, provided only that we have

(12) $x = y \supset_{xy} . F(x) \supset F(y)$

and

(13) $(x)\Box x = x.$

For by substitution in (12) we get

(14) $x = y \supset_{xy} . \Box[x = x] \supset \Box[x = y].$

And (11) then follows by (13) and (14).

The result is perhaps more striking when put in terms of possibility rather than necessity. For

(15) $\sim F(x) \supset_x . F(y) \supset_y x \neq y$

is equally an elementary property of identity which can hardly be denied. And if we allow further that

(16) $(x) \sim \Diamond x \neq x,$

we obtain by substitution in (15) that

(17) $\sim \Diamond[x \neq x] \supset_x . \Diamond[x \neq y] \supset_y x \neq y.$

Then from (16) and (17) we have the following variant of Murphy's Law:

(18) $\Diamond[x \neq y] \supset_{xy} x \neq y.$

(If two things are possibly different, then they *are* different.)

The theorems (11) and (18) are hardly avoidable if modal logic is formulated in such a way that modal operators are prefixed directly to sentences (as is indeed is now usual[4]). Quine in [15] objects to these theorems as compelling acceptance of "Aristotelian essentialism"[5] —which he regards as philosophically suspect and as being moreover incompatible with the idea which many modal logicians have held, that the modal sentence $\Box S$ is true if and only if the corresponding unmodalized sentence S is analytic.

This second form of Quine's paradox about modality, which refers to variables rather than names,[6] can be paralleled by a paradox about belief statements in place of

[3]As Russell writes in [17]: "The phrase *per se* has no meaning, because in any proposition in which it occurs the proposition, fully expressed, does not contain the phrase, which has been broken up."

[4]The writer believes that a more Fregean version of modal logic might be preferable, in which the modal operators are prefixed not directly to a sentence but to any name of the proposition which the sentence expresses. Quine's misgivings about this in [15] can be dispelled only by a detailed development of such a Fregean modal logic, explicitly exhibiting the "interplay" which he fears may be wanting; but Quine's further objection (in [16], cf. printing of 1973, p. 198) that there is some *ad hoc* restriction on quantifying into modal contexts seems to be based on a misunderstanding. (For historical accuracy it should be added that Frege himself disbelieved in modal logic.)

[5]Cf. also [16], §41.

[6]Or naming expressions. No distinction is intended in this paper, or by Carnap, between names and

modality. By substitution in (15) we get:

(19) For every x and every y, if George IV does not believe that $x \neq x$, if George IV believes that $x \neq y$, then $x \neq y$.

This may be thought of as analogous to (17). The second premiss,

(20) For every x, George IV does not believe that $x \neq x$,

differs from (16) in being no more than very likely rather than certain. But if we accept it, we have as analogous to (18) the conclusion:

(21) For every x and every y, if George IV believes that $x \neq y$, then $x \neq y$.

And this otherwise surprising power of King George's beliefs to control the actual facts about x and y can be explained only on the doubtful assumption that belief properly applies "to the fulfillment of conditions by objects" quite "apart from special ways of specifying" the objects.[7] This assumption (let us call it the principle of transparency of belief) is the same thing to the notion of belief that essentialism is to the notion of necessity. But the consequences of the former for the ordinary notion of belief may be thought to be even more repellent than the consequences of essentialism for modal notions.

To illustrate this last, let us suppose that George IV is convinced (and in fact on good evidence) that there is one and only one who wrote Waverley, that there is one and only who wrote Ivanhoe, and that the two authors are the same. Then let us ask for what objects[8] (or individuals) y we have that

(22) George IV believes that y wrote Waverley.

The ordinary notion of belief seems to require that although (22) holds when y is specified in a special way, namely as having written Ivanhoe, it may yet fail when the same y is specified in some other special way, for example as scottizing.

Our conclusion is that Quine's objections against the Russellian treatment[9] of modal logic according to which modal operators are prefixed to sentences, do have some considerable force. But it is better to present them in a way that exhibits the nearly complete parallelism between the objections against a Russellian modal logic and those against a Russellian logic of belief (or of denying, wishing to know, or the like). For this has the effect of putting the objections in perspective and of clarifying both their strengths and their weaknesses.

In summary the significant objections in [15] are two: the complications about scope

naming expressions. And the idea is Russellian rather than Fregean, that a name must be an unanalyzed primitive and hence normally a single word or a single symbol.

[7]The quoted phrases are from Section III of [15] and are used in order to emphasize that what is here said about belief closely parallels what is said by Quine about modality.

[8]We assume that human beings are included among objects, or among individuals (as it may be better to say in order to allow for type theory and its standard terminology). And we follow Quine in avoiding the semantical formulation that consists in asking what values of the variable 'y' satisfy the propositional form 'George IV believes that y wrote ***Waverley***'—although this alternative formulation might otherwise be helpful, e.g., in bringing out that what is at issue concerns the value of a variable, and not (as in the original paradox of the name relation) the denotation of a name or names.

[9]We call it Russellian in spite of Russell's own rejection of modality, because it is the treatment appropriate to Russell's explicit and implicit semantics, especially propositions as values of the propositional variables, and the sort of transparency in belief contexts, modal contexts, and the like that is required by the theory of descriptions as resolution of the paradox of the name relation.

which arise in connection with the use of descriptions, and the transparency of both belief and necessity which is forced by use of the theory of descriptions to resolve the paradox of the name relation.

But finally, it must be pointed out that Quine's objections, though strong, are no firm refutation of the Russellian resolution of the paradox. There may be those who, in the interest of the resolution of the paradox, are willing to accept both the complications about descriptions and the strange transparent notions of belief and necessity which result. And to them it can only be said that well, it does seem strange. The sort of essentialism to which the transparency of notions such as belief and necessity leads does not rise above the level of variables and primitive constants. And the Russellian may proceed with the more confidence because he is able, besides the transparent notions of belief, necessity, and possibility, to express also the more usual non-transparent notions. Namely if $B(x, p)$ is used to mean that x believes that p,[10] $S(x)$ that x scottizes, $W(x)$ that x is author of Waverley, $G(x)$ that x is George IV (or that x georgivizes), he may write:[11]

(23) $B((\iota x)G(x), (\iota x)S(x) = (\iota x)W(x))$.

(24) $\Diamond(\iota x)S(x) \neq (\iota x)W(x)$.

The convention of minimum scope is of course to be understood in (23) and (24). And notwithstanding (18), it does not follow from (24) that

$$(\iota x)S(x) \neq (\iota x)W(x).$$

BIBLIOGRAPHY

[1] Ruth C. Barcan. *The identity of individuals in a strict functional calculus of second order*. ***The Journal of Symbolic Logic***, vol. 12 (1947), pp. 12–15.

[2] Rudolf Carnap. ***Meaning and Necessity***. Second edition, Chicago 1956.

[3] Irving M. Copi and James A. Gould, editors. ***Contemporary Readings in Logical Theory***. New York and London 1967.

[4] Herbert Feigl and Wilfrid Sellars, editors. ***Readings in Philosophical Analysis***. New York 1949.

[5] Gottlob Frege. *Über Sinn und Bedeutung*. ***Zeitschrift für Philosophie und philosophische Kritik***, vol. 100 (1892), pp. 25–50. Reprinted in Patzig [10], pp. 40–65.

[10]To avoid antinomies such as the Epimenides, it may be necessary either to distinguish different orders of propositional variables (by adopting ramified type theory) or to distinguish different orders of belief by writing B^1, B^2, B^3, etc. We ignore this here as not being immediately relevant to what is being said.

[11]These examples illustrate that Russell's unitary propositional functions, where e.g. 'F is unitary' is defined as $(\exists x) \,.\, F(y) \equiv_y y = x$, may serve to a considerable extent as surrogates for the entities which in Frege's theory appear as senses of names. For example in (23) King George's non-transparent belief appears as a relation between the propositional functions S and W (or between $S(\hat{x})$ and $W(\hat{x})$ as Russell would write), and these propositional functions must be unitary if King George's belief is not mistaken. And in a Fregean theory the same belief by George IV would appear as a relation between the sense, which belong to the names 'Sir Walter Scott' and 'the author of ***Waverley***'. It would be of interest to look into the question how far the Fregean theory can be reproduced within the Russellian by identifying the Fregean senses with propositional functions (in Russell's sense, according to which propositional functions are intensional entities). Indeed the Russellian theory might have to be rather drastically mutilated to obtain the Fregean fragment. But it remains true that any significant partial success in representing one theory within the other would throw light on the relationship of the two theories.

English translation in Feigl-Sellars [4], pp. 85–102, reprinted in Copi-Gould [3], pp. 75–92, and Marras [8], pp. 337–361. Spanish translations in ***Dialogos***, vol. 8 no. 22 (1972), pp. 147–170, and in Simpson [18], pp. 3–27.

[6] Leonard Linsky, editor. ***Semantics and the Philosophy of Language***. Urbana 1952.

[7] Leonard Linsky, editor. ***Reference and Modality***. London 1971.

[8] Ausonio Marras, editor. ***Intentionality, Mind, and Language***. Urbana, Chicago and London 1972.

[9] Robert C. Marsh, editor. ***Logic and Knowledge: Essays 1901–1950***, by Bertrand Russell. London 1956.

[10] Günther Patzig, editor. ***Funktion, Begriff, Bedeutung, Fünf logische Studien***, by Gottlob Frege. Second Edition, Göttingen 1966.

[11] W. V. Quine. *Notes on existence and necessity*, ***The Journal of Philosophy***, vol. 40 (1943), pp. 113–127. Reprinted in Linsky [6], pp. 77–91, Spanish translation in Simpson [18], pp. 121–138.

[12] W. V. Quine. ***O Sentido da Nova Lógica***. São Paulo 1944.

[13] W. V. Quine. *The problem of interpreting modal logic*, ***The Journal of Symbolic Logic***, vol. 12 (1947), pp. 43–48. Reprinted in Copi-Gould [3], pp. 267–273.

[14] W. V. Quine. *Three grades of modal involvement*, ***Actes du XIième Congrès International de Philosophie***, Amsterdam and Louvain, vol. 14 (1953), pp. 65–81.

[15] W. V. Quine, *Reference and modality*, ***From a Logical Point of View***, by W. V. Quine, second edition, Cambridge, Mass., 1961, pp. 139–159. Reprinted in Linsky [7], pp. 17–34. Spanish translation in ***Desde un Punto de Vista Lógico***, by W. V. Quine, Barcelona 1962, pp. 201–227.

[16] W. V. Quine. ***Word and Object***, Cambridge, Mass., 1960.

[17] Bertrand Russell. *On denoting*, ***Mind***, vol. 14 (1905), pp. 479–493. Reprinted in Feigl-Sellars [4], pp. 103–115; Marsh [9], pp. 41–56; Copi-Gould [3], pp. 93–105; Marras [8], pp. 362–379. Spanish translation in Simpson [18], pp. 29–48.

[18] Thomas M. Simpson, editor. ***Semántica Filosófica: Problemas y Discusiones***. Buenos Aires 1973.

[19] Arthur F. Smullyan. Review of Quine [13]. ***The Journal of Symbolic Logic***, vol. 12 (1947), pp. 139–141.

[20] Arthur F. Smullyan. *Modality and description*, ***The Journal of Symbolic Logic***, vol. 13 (1948), pp. 31–37. Reprinted in Linsky [7], pp. 35–43. Spanish translation in Simpson [18], pp. 289–299.

[21] A. N. Whitehead and Bertrand Russell. ***Principia Mathematica***. Three volumes, Cambridge, England, 1910–1913.

RUSSELL'S THEORY OF IDENTITY OF PROPOSITIONS (1984)

In *Principia Mathematica*[1] the sign $=$ is defined by taking $\mathbf{x} = \mathbf{y}$ to stand for $(F) \centerdot F(\mathbf{x}) \supset F(\mathbf{y})$ where $\mathbf{x}$ and $\mathbf{y}$ are any variables of the same type, not excluding that (as a special case) $\mathbf{x}$ and $\mathbf{y}$ may be the same variable, and where the variable F is of such type that $F(\mathbf{x})$ and $F(\mathbf{y})$ are well-formed. The variables $\mathbf{x}$ and $\mathbf{y}$ may in particular be propositional variables, say p and q for example; and in this way $p = q$ qualifies as a wff (well-formed formula) in the formalized language of *Principia*. But the syntax of the language is such that the possibilities of substituting in $p = q$ for either of the variables p or q are extremely limited. For example $p = q \supset \centerdot p \supset r \supset \centerdot q \supset r$ is a theorem, and in it we may (correctly) substitute r for p to infer $r = q \supset \centerdot r \supset r \supset \centerdot q \supset r$, but we may not substitute $q \supset r$ for q to infer $p = [q \supset r] \supset \centerdot p \supset r \supset \centerdot [q \supset r] \supset r$. Indeed this last is not even a wff.

To remove this difficulty I suggested, in a paper of 1974,[2] that a formalized language which is like that of *Principia* in regard to the treatment of propositional variables and the sign $=$ might well be enlarged by adding a primitive connective $\equiv$ which may stand between arbitrary wffs to express identity — so that, for example, $(x) f(x) \equiv (x) g(x)$ expresses that $(x) f(x)$ and $(x) g(x)$ are the same proposition, and $f(x) \equiv_x g(x)$ expresses that, for all x, $f(x)$ and $g(x)$ are the same proposition. The language in question, to which the primitive connective $\equiv$ is added, might be that of *Principia Mathematica* itself.[3] But in this paper we consider instead, primarily, the effect of adding the connective $\equiv$ to a simple type theory — not excluding type theories which allow overlapping types, or types that are of infinite order in the sense of being obtained as the union of an infinite sequence or infinite class of lower orders, provided there are restrictions which suffice for consistency of the type theory.

We presuppose an existing simple type theory without the connective $\equiv$ and supply added rules and axioms for the introduction and use of the new connective. In these it is understood that bold lower-case letters stand for any variables that are of such types as to render the formulas well-formed in which they occur, bold Greek capitals stand for arbitrary formulas, and bold Roman capitals for wffs.

---

[1]We make some non-essential departures from the notation as it actually appears in ***Principia***. In particular Russell's exclamation points are omitted, because we shall be concerned almost entirely with simple type theory. A bold dot is used to replace a left bracket, with a corresponding right bracket at the end of the formula in which the dot appears. In the absence of a bold dot to indicate otherwise, brackets are omitted under the convention of association to the left — e.g., $p \supset q \supset r$ is an abbreviation of $[[p \supset q] \supset r]$, and $p \supset \centerdot q \supset r$ is an abbreviation of $[p \supset [q \supset r]]$.

[2]*Russellian Simple Type Theory* in: ***Proceedings and Addresses of the American Philosophical Association***, vol. 47 (1974), pp. 21–33. I take the opportunity to make three minor corrections to this paper. At the end of the third complete paragraph on page 26 (after the words "may be empty") add "and the domains $\mathfrak{T}$ and $\mathfrak{F}$ must be disjoint"; in footnote 26 insert the word "logical" before "proper name"; and in the first line on page 30 insert after "**f**" the words "has no free occurrence in **P** and".

[3]This holds notwithstanding Russell's declaration, on page 44 of the "Introduction" to the first edition, that "the phrase which expresses a proposition is what we call an 'incomplete' symbol", as the actual syntax of the formalized language remains unchanged. Indeed the passage on pages 43–44, possibly a late addition to the "Introduction", marks the beginning of Russell's long attempt to do away with or modify the notion of proposition, but it has fortunately not affected the main text of ***Principia***.

As added formation rule we need: If $\mathbf{\Gamma}$ and $\mathbf{\Delta}$ are well-formed, then $[\mathbf{\Gamma} \equiv \mathbf{\Delta}]$ is well-formed.

And the added axioms and axiom schemata are 1–15 as follows:

1. $\vdash p \equiv p$
2. $\vdash p \equiv q \equiv . p = q$
3. $\vdash p \equiv q \supset . p = q$
4. $\vdash \mathbf{f}(\mathbf{x}_1, \mathbf{x}_2, \ldots, \mathbf{x}_m) \equiv_{\mathbf{x}_1 \mathbf{x}_2 \ldots \mathbf{x}_m} \mathbf{g}(\mathbf{x}_1, \mathbf{x}_2, \ldots, \mathbf{x}_m) \supset . \mathbf{f} = \mathbf{g}$ where the variables $\mathbf{x}_1$, $\mathbf{x}_2, \ldots, \mathbf{x}_m$ are all different.
5. $\vdash (\exists \mathbf{p}) . \mathbf{p} \equiv \mathbf{P}$, where $\mathbf{p}$ is a propositional variable having no free occurrence in the wff $\mathbf{P}$.
6. $\vdash (\exists \mathbf{f}) . \mathbf{f}(\mathbf{x}_1, \mathbf{x}_2, \ldots, \mathbf{x}_m) \equiv_{\mathbf{x}_1 \mathbf{x}_2 \ldots \mathbf{x}_m} \mathbf{P}$, where the variables $\mathbf{x}_1, \mathbf{x}_2, \ldots, \mathbf{x}_m$ are all different and the variable $\mathbf{f}$ has no free occurrence in the wff $\mathbf{P}$.
7. $\vdash (\mathbf{x})\mathbf{f}(\mathbf{x}) \equiv (\mathbf{y})\mathbf{f}(\mathbf{y})$
8. $\vdash \sim p \equiv \sim q \supset . p \equiv q$
9. $\vdash [p \vee q] \equiv [r \vee s] \supset . p \equiv r$
10. $\vdash [p \vee q] \equiv [r \vee s] \supset . q \equiv s$
11. $\vdash (\mathbf{x})\mathbf{f}(\mathbf{x}) \equiv (\mathbf{x})\mathbf{g}(\mathbf{x}) \supset . \mathbf{f}(\mathbf{x}) \equiv_{\mathbf{x}} \mathbf{g}(\mathbf{x})$
12. $\vdash \sim . \sim p \equiv (\mathbf{x})\mathbf{f}(\mathbf{x})$
13. $\vdash \sim . \sim p \equiv . q \vee r$
14. $\vdash \sim . p \vee q \equiv (\mathbf{x})\mathbf{f}(\mathbf{x})$
15. $\vdash \sim . (\mathbf{x})\mathbf{f}(\mathbf{x}) \equiv (\mathbf{y})\mathbf{g}(\mathbf{y})$, where the variables $\mathbf{x}$ and $\mathbf{y}$ are not of the same type.

Axioms 1–3 are necessary and sufficient to make $\mathbf{p} \equiv \mathbf{q}$ and $\mathbf{p} = \mathbf{q}$ synonymous in the case of propositional variables $\mathbf{p}$ and $\mathbf{q}$. In particular the theorem

3′. $\vdash p \equiv q \supset . p \equiv r \supset . r \equiv q$,

which is needed below, is a consequence of axiom 3 and the rule of substitutivity of equality (see footnote 4 regarding the latter).

Axiom schema 4 represents the familiar principle of identity of functions, that if two m-ary functions have the same range of ordered m-tuples of arguments and have, for each ordered m-tuple of arguments in that range the same value, they are the same function. Notice that if in 4 we replace the sign $\equiv$ by the sign $\equiv$ of material equivalence, we obtain axioms that are equivalent to the axioms of extensionality that are introduced by Russell in his *Introduction to the second edition* (of *Principia*), and these are of course to be avoided for our present purpose.

Axiom schemata 5 and 6 represent comprehension principles for propositional variables and m-ary functional variables. By their inclusion, rules of substitution for propositional and functional variables become unnecessary.[4] In consequence of axiom schema 6 and of the theorem $\vdash p \equiv q \supset . p \equiv q$, which is an easy consequence of axiom 3, we have the theorem schemata:

5′. $\vdash (\exists \mathbf{p}) . \mathbf{p} \equiv \mathbf{P}$, where $\mathbf{p}$ is a propositional variable having no free occurrence in the wff $\mathbf{P}$.

6′. $\vdash (\exists \mathbf{f}) . \mathbf{f}(\mathbf{x}_1, \mathbf{x}_2, \ldots, \mathbf{x}_m) \equiv_{\mathbf{x}_1 \mathbf{x}_2 \ldots \mathbf{x}_m} \mathbf{P}$, where the variables $\mathbf{x}_1, \mathbf{x}_2, \ldots, \mathbf{x}_m$ are all different and the variable $\mathbf{f}$ has no free occurrence in the wff $\mathbf{P}$.

---

[4] In what follows we shall therefore freely employ substitution for propositional and functional variables, omitting details of the essentially well-known method by which these rules of substitution follow from axiom schemata 5 and 6. Indeed for the case in which the substitution takes place in extensional contexts, these

Axiom schema 7 represents the principle that if sentences differ only by alphabetic change of bound variables, the corresponding propositions are the same. It is intended to have the effect that if **A** and **B** are two such sentences, we can prove $\vdash \mathbf{A} \equiv \mathbf{B}$.

Axiom schemata 8–15 presuppose that the primitive connectives of propositional calculus are taken as $\sim$ and $\vee$ (as in *Principia*), and that the universal quantifier is taken as primitive and the existential quantifier is defined from it (as in *Principia* $*10$). It is true that in *The Principles of Mathematics* the primitives are quite different; but in the present paper our intention is to follow *Principia Mathematica* more closely in such respects as this, while at the same time presenting Russell's ideas about propositions as they appear in the *Principles*.

Russell does not proceed by stating axioms or axiom schemata such as 1–15, but rather he assumes a certain informal understanding by the reader as to when propositions are identical and when not. And as a means of discovering implicit axioms we therefore try analyzing his text.

In the important[5] second paragraph of §500 of the *Principles*, which sets forth Russell's antinomy about propositions, we pick out six successive sentences or clauses, quoting them in Russell's own words, and giving them the numbers 1–6. Then after each such quotation from Russell the proposed translation into the formalized language of the present paper is given, accompanied often by explanations and details which Russell omits, and which are given still in or with respect to the same formalized language. For example, 1 is quoted from Russell, 1a is as nearly as possible a literal translation of Russell's English sentences into the formalized language, and 1b is an inference from 1a which will be needed in a later step. In the case of the quotation 5 from Russell, Russell's short sentence has left to the reader an easy but moderately lengthy proof, and in the formalized language the steps of this proof have been given in detail in 5a, 5b, ... , 5g—while the nearest to an actual translation of Russell's English words is at 5e.

It is hoped in this way to make it convincing that what is being done is not a later imposition upon Russell of something he would not have recognized as his own thought, but is in some sense a faithful rendering (with added comments).

In many ways the formalized language used for this has followed that of *Principia*

---

rules of substitution follow already from the weaker axiom schemata (coinciding with the theorem schemata $5'$ and $6'$) that are obtained from axiom schemata 5 and 6 by replacing the connective $\equiv$ by the connective $\equiv$ that expresses material equivalence. For details in regard to this extensional case the reader is referred to Leon Henkin, *Banishing the Rule of Substitution for Functional Variables*, in: ***The Journal of Symbolic Logic*** 18 (1953), pp. 201–208, and Alonzo Church, *Comparison of Russell's Resolution of the Semantical Antinomies with that of Tarski*, in: ***The Journal of Symbolic Logic*** 41 (1976), pp. 747–760. See especially the summary of the matter on page 750 of the latter paper, but notice that the two displayed schemata on that page must be corrected by adding to the first one the condition that **p** is not free in **P** and to the second one the condition that **f** is not free in **P**.

Attention should be called to the point that Russell was aware of the need for the intensional comprehension principle for propositional functions and has a statement of it in words on pages 14–15 of the *Introduction* to the first edition of ***Principia Mathematica***. But for lack of the connective $\equiv$ he is unable to make our axiom schema 6 fully explicit or to state instances of it in the formalized language.

We also notice that the rule of substitutivity of equality follows immediately from Russell's definition of $\mathbf{x} = \mathbf{y}$ as $(F) \centerdot F(\mathbf{x}) \supset F(\mathbf{y})$, together with the rule of universal instantiation and the rule of substitution for functional variables.

[5] Historically important as first indication of the need for something beyond simple type theory to resolve antinomies.

Mathematica. The assertion sign $\vdash$ is used in the same way as in *Principia*. In steps 5a–5f the method of proof from hypotheses has been used, by writing before the assertion sign at each step the hypotheses that are applicable at that step; and there is another use of the method of proof from hypotheses in the case of hypotheses 2a and 3a, which are carried to the end of the proof and then eliminated by ex. inst.; at these places we depart from the method and usage of *Principia* itself, but the reader who wishes to rewrite the entire proof in standard *Principia* form should have no difficulty in doing so.

In presenting this proof (establishing Russell's antinomy about propositions) we make use of the following abbreviations: mod. pon., the rule of *modus ponens*; trans. law, the transitive law of $\supset$; **P**, laws of propositional calculus; ded. thm., the deduction theorem; univ. inst., the rule of universal instantiation; ex. inst., the rule of existential instantiation.

1. If n be different from m, "every n is true" is not the same proposition as "every m is true".

   1a. $\vdash \sim(q)[n(q) \equiv m(q)] \supset \sim . (q)[n(q) \supset q] \equiv (q)[m(q) \supset q]$
   1b. $\vdash (q)[n(q) \supset q] \equiv (q)[m(q) \supset q] \supset (q)[n(q) \equiv m(q)]$

2. Consider now the whole class of propositions of the form "every m is true", and having the property of not being members of their respective m's. Let this class be w.

   2a. $w(r) \equiv_r (\exists m) . r \equiv (q)[m(q) \supset q] . \sim m(r)$

3. And let p be the proposition "every w is true".

   3a. $p \equiv (q)[w(q) \supset q]$
   Notice that 2a and 3a have not been proved but rather have been taken as hypotheses. It is to be understood that these are hypotheses for every step which follows, though not explicitly written; and at the end of the proof they are to be eliminated by ex. inst., with the aid of appropriate instances of theorem schema 6′ and axiom schema 5.

4. If p is a w, it must possess the defining property of w.

   4a. $\vdash w(p) \supset (\exists m) . p \equiv (q)[m(q) \supset q] . \sim m(p)$
   from 2a by univ. inst. and P.

5. But this property demands that p should be not be a w.

   5a. $p \equiv (q)[m(q) \supset q] \vdash (q)[w(q) \supset q] \equiv (q)[m(q) \supset q]$ by 3a and theorem 3′.
   5b. $p \equiv (q)[m(q) \supset q] \vdash (q)[w(q) \equiv m(q)]$ from 5a by substituting w for n in 1b.
   5c. $p \equiv (q)[m(q) \supset q] \vdash w(p) \equiv m(p)$ from 5b by univ. inst.
   5d. $p \equiv (q)[m(q) \supset q] . \sim m(p) \vdash w(p)$ from 5c by P.
   5e. $(\exists m) . p \equiv (q)[m(q) \supset q] . \sim m(p) \vdash \sim w(p)$ from 5d by ex. inst.
   5f. $\vdash w(p) \supset \sim w(p)$ from 5e by ded. thm., 4a, and trans. law.
   5g. $\vdash \sim w(p)$ from 5f by P.

6. On the other hand, if p be not a w, then p does possess the defining property of w, and therefore is a w.

   6a. $\vdash p \equiv (q)[w(q) \supset q] . \sim w(p)$ from 3a and 5g.
   6b. $\vdash (\exists m) . p \equiv (q)[m(q) \supset q] . \sim m(p)$ from 6a by ex. gen.
   6c. $\vdash w(p) \equiv (\exists m) . p \equiv (q)[m(q) \supset q] . \sim m(p)$ from 2a by univ. inst.

6d. $\vdash (\exists m)[p \equiv (q)[m(q) \supset q] \,\boldsymbol{.}\, {\sim} m(p)] \supset w(p)$ from 6c by **P**.
6e. $\vdash w(p)$ from 6b and 6d by mod. pon.

After the final elimination of the hypotheses 2a and 3a (as already described), steps 5g and 6e exhibit the antinomy. It will be seen that all steps except 1a depend on logic alone, and there is no violation of simple type theory.

In 1a and later steps we have replaced Russell's "ranges" or classes n, m, w of propositions by propositional functions n, m, w, taking into account the intended extensionality only by writing e.g. ${\sim}(q)[n(q) \equiv m(q)]$ where Russell speaks of n being different from m, and $(q)[w(q) \equiv m(q)]$ where it has to be said that w and m determine the same class of propositions. This modification serves the purpose of brevity and makes no important difference in the demonstration of the antinomy.

Russell's justification of 1a is informal in character and depends on his claim, that "the relation of ranges of propositions to their logical products is one-one" (in a footnote he even answers yes to the question: "Does the logical product of [the three propositions] p and q and r differ from that of [the two propositions] pq and r?"). This justification can possibly be defended from Russell's point of view at the date, but it is not certain,[6] and it will be better to justify 1a by the following semantical principle: *If* **n** *and* **m** *are names of different things, but are of the same type, and if a (declarative) sentence is altered by replacing one occurrence of* **n** *by* **m**, *the original sentence and the altered sentence represent different propositions.*

This principle is indeed very plausible, and that it already leads to antinomy is a puzzle which remained unresolved until Russell's (later) introduction of ramified type theory (compare the last paragraph of §500 of the *Principles*).

The same principle seems to underlie Russell's finding, in §§498–499 of the *Principles*, that $x = x$ is a different proposition for every choice of x, excluding choices of x from different types.[7] And that Russell immediately sees that this leads to a contradiction is not now an objection, as again there is a resolution by ramified type theory.

On the other hand Russell makes no mention of the italicized principle, and there is no reason to think that cases in which he would declare two propositions different always depend on it. On the contrary we see throughout §500 that Russell tends to declare two propositions different if the corresponding sentences have any difference in form not entirely trivial. In particular, in the first paragraph of the section, an absolute distinction is made between propositions which are logical products and those which are not. And in the fourth paragraph it is said that the proposition represented by

[6]Specifically, Russell can be read as saying in this passage (the first paragraph of §500 and its footnote) that *if two classes of propositions are identical as classes but do not have the same defining property, the corresponding logical products are nevertheless the same proposition.* This might be rendered in the formalized language as the converse of 1b, i.e.,

$\vdash (q)[n(q) \equiv m(q)] \supset \,\boldsymbol{.}\, (q)[n(q) \supset q] \equiv (q)[m(q) \supset q]$

But because Russell does not make this assumption explicit, and because it is not in the spirit of his other assumptions about propositions, I hesitate to attribute it to him.

However, even if we abandon the italicized principle just stated (in this footnote), we shall still have as a consequence of the italicized principle in the text, that the relation of ranges of propositions to their logical products is one-many, and this is already enough to lead to the antinomy by Russell's method, as it then follows that there is a one-many relation of all ranges of propositions to some propositions.

[7]I.e., types in the tentative simple type theory which appears in §104 and in these last sections of the ***Principles***.

$(q) \centerdot q \in m \supset q$ and that represented by $(q) \centerdot q \in m \vee [q \equiv (r) \centerdot r \in m \supset r] \supset q$, where m names a class of propositions, are different.

Axioms and axiom schemata 1–15 are intended to provide in the formalized language a full statement of Russell's theory of identity of propositions as it stood at the date that the text and appendices of *The Principles of Mathematics* had just been completed. Russell himself never provided such a fully formalized statement of the theory, but I believe that he has said enough that there is only one such theory that may reasonably be attributed to him, and that is the theory embodied in 1–15, taken as axioms to be added to some appropriate formalized language (such as Russell might be supposed to have had in mind).

For this purpose, because no formalization of the language of the *Principles* is available, we shift to the language of *Principia Mathematica* as modified to embody simple type theory (rather than ramified type theory). Because it is necessary to carry the accuracy of the formalization somewhat further than is done in *Principia* itself, let us use the version that appears in my paper cited in footnote 4. This has the convenience that the shift to simple type theory is accomplished by merely dropping the level indicators from the type superscripts and abandoning the restrictions in regard to order which affect the two comprehension axiom schemata on page 750; and that the later shift back to ramified type theory then requires just that the level indicators and the restrictions on the two axiom schemata shall be restored.

Then we go on to show that Russell's antinomy about propositions, as it is contained in Russell's own statement of it (quoted as 1–6 above), can be reproduced in the formalized language embodying simple type theory, provided that the axioms and axiom schemata 1–15 are added. In large part this has already been done. But we must still obtain either 1a or 1b from 1–15 rather than from the italicized semantical principle that was introduced above (as this semantical principle might at best be represented in the formalized language by an axiom schema $\mathbf{P}_1 \equiv \mathbf{P}_2 \supset \mathbf{x}_1 = \mathbf{x}_2$, where $\mathbf{P}_2$ is obtained from $\mathbf{P}_1$ by replacing one or more free occurrences of a variable $\mathbf{x}_1$ by a variable $\mathbf{x}_2$ which is of the same type and does not otherwise occur—and it would then immediately be seen that this axiom schemata is better analyzed into simpler axioms and axiom schemata, as is done by 1–15). We therefore supply a proof of 1b as follows:

$\vdash p \equiv q \supset \centerdot F(p) \supset F(q)$ from axiom 3.
$\vdash p \equiv q \supset \centerdot p \equiv p \supset \centerdot p \equiv q$ by substitution for F.
$\vdash p \equiv q \supset \centerdot p \equiv q$ by P.
$\vdash n(q) \equiv m(q) \supset \centerdot n(q) \equiv m(q)$ by substituting, first for q, then for p.
$\vdash \sim n(q) \equiv \sim m(q) \supset \centerdot n(q) \equiv m(q)$ by substitution in axiom 8.
$\vdash \sim n(q) \equiv \sim m(q) \supset \centerdot n(q) \equiv m(q)$ by trans. law.
$\vdash \sim n(q) \vee q \equiv [\sim m(q) \vee q] \supset \centerdot \sim n(q) \equiv \sim m(q)$ by substitution in axiom 9.
$\vdash n(q) \supset q \equiv [m(q) \supset q] \supset \centerdot \sim n(q) \equiv \sim m(q)$ by definition of $\supset$.
$\vdash n(q) \supset q \equiv [m(q) \supset q] \supset \centerdot n(q) \equiv m(q)$ by line 6 of this proof and trans. law.
$\vdash (q) \centerdot n(q) \supset q \equiv [m(q) \supset q] \supset \centerdot n(q) \equiv m(q)$
$\vdash (q)[n(q) \supset q \equiv \centerdot m(q) \supset q] \supset (q) \centerdot n(q) \equiv m(q)$
$\vdash (q)[n(q) \supset q] \equiv (q)[m(q) \supset q] \supset (q) \centerdot n(q) \supset q \equiv \centerdot m(q) \supset q$ by axiom schema 11 and substitution.
$\vdash (q)[n(q) \supset q] \equiv (q)[m(q) \supset q] \supset (q)[n(q) \equiv m(q)]$ by trans. law.

This now completes the demonstration of Russell's antinomy about propositions. We remark at once that there are many variations of the antinomy. In particular $n(q) \supset q$ may be replaced by any truth-function of $n(q)$ and q that is not independent of $n(q)$, provided that $m(q) \supset q$ is replaced throughout by the same truth-function of $m(q)$ and q, and $w(q) \supset q$ by the same truth-function of $w(q)$ and q. Especially striking is the variation of the antinomy that is obtained by replacing $n(q) \supset q$, $m(q) \supset q$, and $w(q) \supset q$ by $n(q)$, $m(q)$, and $w(q)$ respectively, because in this variation none of the axioms and axiom schemata 8–15 is used except axiom schema 11.

On the other hand the antinomy is immediately resolved by ramified type theory. For in the demonstration of the antinomy as given above, ramified type theory would require at step 2a that w must be of higher order than m and at step 3a that p must be of higher order than q, and in consequence both of the later steps 5c and 6b would fail. Moreover there is no reason to suppose that adding axioms of reducibility would restore the antinomy.[8]

If, following the early Russell, we hold that the object of an assertion or a belief is a proposition and then impose on propositions the strong conditions of identity which this requires[9] while at the same time undertaking to formulate a logic that will suffice for classical mathematics, we therefore find no alternative except ramified type theory, with axioms of reducibility, and with axioms and axiom schemata 1–15 appropriately modified.

Outwardly the modification of 1–15 that is required for this purpose is quite minor. The lightface letters p, q, r, s are to be replacing everywhere by bold $\mathbf{p}, \mathbf{q}, \mathbf{r}, \mathbf{s}$. The version of ramified type theory that is chosen may preferably be that of my paper cited in footnote 4, see pages 748–749.[10] Generally the condition on the bold lower-case letters occurring in the schemata is just that they are variables of such types as to secure well-formedness. But in the case of 5 and 6 the conditions are more elaborate, as follows:

In axiom schema 5, $\mathbf{p}$ must be a propositional variable of r-type $0/n$ that is not free in $\mathbf{P}$, the apparent variables in $\mathbf{P}$ must be all of order less than n, and the real (or free) variables in $\mathbf{P}$ and the constants in $\mathbf{P}$ must be of order not greater than n. In axiom schema 6, $\mathbf{f}$ must be a functional variable of r-type $(\beta_1, \beta_2, \ldots, \beta_m)/n$ and $\mathbf{x}_1, \mathbf{x}_2, \ldots, \mathbf{x}_m$ must be distinct variables of r-types $\beta_1, \beta_2, \ldots, \beta_m$, the apparent variables in $\mathbf{P}$ must be of order less than the order of $\mathbf{f}$, and the real variables in $\mathbf{P}$ and the constants in $\mathbf{P}$ must be of order no greater than the order of $\mathbf{f}$.

Alonzo Church
Department of Philosophy
University of California at Los Angeles
Los Angeles, CA 90024
USA

[8]Of course the axioms of reducibility must be expressed with the sign $\equiv$, as Russell in fact does in his paper of 1908 and in ***Principia Mathematica***. If analogues of the axioms of reducibility are introduced with the sign $\equiv$ replaced by the sign $\equiv$, the effect is just that the ramified type theory is replaced by an equivalent of simple type theory. It may well be that much of the early adverse criticism of the axioms of reducibility was based on an implicit confusion between material equivalence, as expressed by the sign $\equiv$, and identity of propositions, as expressed by the sign $\equiv$.

[9]Cf., e.g., page 73 of the *Introduction* to the first edition of ***Principia Mathematica***.

[10]Choosing this particular version of ramified type theory of course presupposes a corresponding particular version of simple type theory within which the antinomy about propositions arose, and therefore specializes to this case. In particular the simple type theory has no types of infinite order.

INTERVIEW WITH WILLIAM ASPRAY (1984)

Alonzo Church is interviewed by William Aspray on 17 May 1984 at the University of California at Los Angeles.

Aspray: Could we begin by your describing how you came to Princeton and what caused your interest in Princeton?

Church: I was an undergraduate at Princeton, and I was pressed by the math department to go on to graduate school. Actually they gave me fellowships that paid my way, otherwise I would not have been able to continue.

Aspray: Who was it on the faculty that was encouraging you to go on to graduate school?

Church: Primarily Oswald Veblen, also to some extent Dean [Henry Burchard] Fine and Luther Eisenhart.

Aspray: What years were you a grad student?

Church: I graduated in 1924, as an undergraduate that is, and then immediately went to graduate school and got my degree in 1927.

Aspray: After finishing graduate school did you immediately become an instructor in the department or did you go off some place?

Church: I had two years on a National Research Fellowship. I spent a year at Harvard and a year in Europe, half the year at Göttingen, because [David] Hilbert was there at the time, and half the year in Amsterdam, because I was interested in [L. E. J.] Brouwer's work, as were some of those advising me.

Aspray: Brouwer was there at the time?

Church: Yes. I think he wasn't teaching. He was quite old. I used to take the train out to his residence, way out in the country.

Aspray: Who of Brouwer's group of disciples, whatever you want to call them, were there while you were there?

Church: [Arend] Heyting was not there, and I remember no one except Brouwer himself. He had a secretary who was also a student, but she was not interested in foundations.

Aspray: How did you get interested in foundations?

Church: Well, mainly through Veblen, who was not himself a contributor to foundations in math except in the old-fashioned sense of postulate theory.

Aspray: Geometry and postulate theory?

Church: Yes. His dissertation was about axioms for Euclidean geometry. He did over again what Hilbert had done, so of course it was not wholly original, but I always thought his axioms for geometry were on the whole somewhat better than Hilbert's. Of course Hilbert had prestige and he didn't.

Originally published in ***The Princeton Mathematics Community in the 1930's: An Oral History Project***; Seeley G. Mudd Manuscript Library, Princeton University Library, Transcript Number 5 (PMC5). http://www.princeton.edu/~mudd/math/

Aspray: At least three other people that I've interviewed have said that. Your interest in logic, did it come as an undergraduate or a graduate student?

Church: I was generally interested in things of a fundamental nature. As an undergraduate I even published a minor paper about the Lorentz transformation, the foundation of (special) relativity theory. It was partly through this general interest and partly through Veblen, who was still interested in the informal study of foundations of mathematics. It was Veblen who urged me to study Hilbert's work on the plea, which may or may not have been fully correct, that he himself did not understand it and he wished me to explain it to him.

At any rate, I tried reading Hilbert. Only his papers published in mathematical periodicals were available at the time. Anybody who has tried those knows they are very hard reading. I did not read as much of them as I should have, but at least I got started that way. Veblen was interested in the independence of the axiom of choice, and my dissertation was about that. It investigated the consequences of studying the second number class under each of two assumptions that contradict the axiom of choice.

Aspray: Was there any opposition on the part of the rest of the department to a graduate student doing a dissertation in logic?

Church: Well it was not exactly a dissertation in logic, at least not the kind of logic you would find in [Whitehead and Russell's] *Principia Mathematica* for instance. It looked more like mathematics; no formalized language was used. The only thing that might have annoyed some mathematicians was the presumption of assuming that maybe the axiom of choice could fail, and that we should look into the contrary assumptions.

Aspray: That suggests that if you wanted to do something along the lines of *Principia Mathematica*, you would have had some trouble in the mathematics department doing it.

Church: Quite possibly. I did later try that. I published a paper with serious errors, and generally got in bad because I was hasty and incautious.

Aspray: Can you tell me something about your graduate education, the kinds of things you studied, the people you studied with.

Church: I had an interest in foundational questions, but there were not many courses in that direction. I took essentially the standard curriculum. I could not name all the courses I took, but there was, of course, a general examination to pass, and there were various required subjects including analysis and real-number theory. I forget exactly what else, but I think I still have something listing the courses I took with the signatures of the instructors.

Aspray: You presumably took courses with Eisenhart, Fine, and Veblen.

Church: Yes, and [Einar] Hille and [J. H. M.] Wedderburn. I can't name the exact courses now, but I remember several courses in analysis. James Alexander had a course in topology. He appointed me to take lecture notes. This is something I have somewhere. He spent about half the course on the solution of the problem of classifying closed two-dimensional manifolds. This was done in a highly geometric way, which has much more appeal than the present topology, which consists mainly of incidence tables and something that looks so much like algebra you can't tell the difference unless you go into detail.

I wrote a very careful set of notes on the first half of the course which was on just this problem of classifying closed two-dimensional manifolds. They are around somewhere. There is nothing original in them, but I think they are a careful job of reproducing Alexander's lectures. Sad to say I never got the second half finished. Somehow or other he forgave me for not doing it, probably because he had to, but by the end of the course I had just finished the notes on the first half.

Aspray: Was it standard for grad students to be asked to take lecture notes at that time? I know it was in the '30s.

Church: I assume it was. I don't know for sure, but I did it for Alexander's lectures, and it may be that is the best record of what he was doing at the time. I have not looked into his publications.

Aspray: What do you remember of the various faculty members as teachers at the time you were a grad student? Does anybody stand out one way or another?

Church: Veblen perhaps. I think it was because of his interest in foundational questions that he impressed me. Fine was excellent for teaching undergraduates, especially for the better sort of undergrads who had some idea of what was going on and were not just grinding away at it. He had not done any research since he got his degree, and he did not try to teach any graduate courses, but I had many courses with him as an undergraduate. I thought well of almost everyone who was teaching there at the time. Who were the others? Eisenhart, Wedderburn, and of course Alexander.

Aspray: [Tracy] Thomas came later, is that right?

Church: Yes. He was essentially a contemporary of mine, I think. He got his degree four years before I did.

Aspray: Didn't Einar Hille come sometime while you were a student?

Church: Yes, I don't remember whether he was there when I entered the graduate school or whether he came later.

Aspray: Who were some of your fellow graduate students?

Church: Paul Smith. I remember him as being a graduate student at the same time I was. There are no doubt a couple of others about whom I would say, "Of course I remember a lot about him" when the name came to mind.

Aspray: How closely did you work with Veblen on your own research?

Church: He was really the only man supervising it. I sort of had to convince him about some aspects of the axiom of choice. To deny what seems intuitively natural is rather difficult. You tend to slip back into what informally seems more reasonable. I remember from time to time having to explain things to him, but I convinced him that my arguments were sound.

Aspray: Do you remember who else was on the committee that read your thesis and examined you?

Church: Certainly Veblen, quite likely Eisenhart and Alexander, but I have forgotten.

Aspray: Several people have suggested that Veblen encouraged grad students and visitors and young faculty members to really push their research and not put as much effort into their teaching. How would you react to that?

Church: Well, I don't remember his being negative toward teaching. Of course he did try to get people interested in research, but that is probably not unusual.

Aspray: Though Princeton was a special place at that time.

Church: It had preeminence specifically in math. There were complaints that the University was overemphasizing this one field to the detriment of others.

Aspray: I see, mainly because the University was thought of primarily as an undergraduate institution.

Church: It had been for a long time. The University was developing the grad school. I wasn't one to complain, but there probably was a one-sided emphasis on math because they happened to be able to get a lot of good people in that particular field. My impression at the time was that for teaching grad students there were abler men in mathematics than there were in other departments. That tended to produce an emphasis on math. You can't be preeminent in all fields, so there is something to be said for being preeminent in one.

Aspray: I know that there are certain external ways of judging which do seem to indicate that math was preeminent. For example, there were certain competitive fellowships that seemed to always go to the math department. One—I can't recall the name now—for people coming from Cambridge each year.

Church: Yes. I remember the fellowships that I had, but I don't know whether they were confined to mathematicians or whether they were general fellowships. I probably did not notice very much.

Aspray: Did you do any teaching while you were a grad student?

Church: I think not. I got an appointment as an assistant professor immediately after my two years on a fellowship. I think that was thc first teaching I did.

Aspray: Did you think of going some place other than Princeton after your two years?

Church: I think nobody made me an offer, and I did not go hunting for offers because I saw no reason to leave Princeton. [Afterthought: Belatedly I remember an offer from Johns Hopkins, but I had already accepted at Princeton.]

Aspray: Do you remember much about your teaching responsibilities in your early years as an assistant professor?

Church: I may well have been teaching things like elementary calculus, more or less according to the routine. There would be a large group of students taking their first or second course in calculus, 100 to 200 I suppose. They were divided up into sections of ten originally—the number kept growing. There was one man in charge who coordinated things. There was a complicated method of judging the examinations so as to try to make the grading uniform and at the same time have input from the instructors. I remember sitting through sessions where the grades given to the students in different sections were compared and adjusted by artificial formulas.

Aspray: Did you get a chance to teach any grad courses?

Church: I can't remember when I started teaching grad courses. Rather early I started teaching grad courses in mathematical logic. There was no one else there to do it.

Aspray: What sort of things would you cover in those courses? What would you use as material?

Church: Yes. I gave first an elementary course in mathematical logic. I forget what textbooks I used at first. I worked as rapidly as possible to get at least something of my own written out. My research was unorthodox and some of it unsound, but I was devoted to it and wanted to get my own ideas down and teach them.

Aspray: I am trying to remember what textbooks were available in the late '20s and early '30s. Do you recall?

Church: There were none that I liked. Lewis and Langford's *Symbolic Logic* was around. No, that may have been later, but certainly the book by C. I. Lewis was available. But there was nothing about the sort of thing I wanted to teach, logic directed towards math rather than the philosophical aspects of logic. Well, I am not sure; there may have been a book of that sort. [Church entry: A later check shows that Lewis's *A Survey of Symbolic Logic* was published in 1918 and Lewis and Langford's *Symbolic Logic* was published in 1932.] Of course [David] Hilbert and Wilhelm Ackermann's *Grundzüge der theoretischen Logik* was in existence at that time, but it was in German. While the grad students were supposed to learn German, as a practical matter I could not have used it as a textbook. So I used written notes of my own and things like that.

Aspray: Was there any relationship between the math department and the philosophy department at this time?

Church: No. Nobody in philosophy was interested in that sort of thing at the time.

Aspray: When did an interest in logic develop among philosophers?

Church: That is hard to say. Of course, C. I. Lewis' *A Survey of Symbolic Logic* was published sometime between 1910–1920, and it is very definitely philosophically oriented. So there were philosophers who were interested in symbolic logic from the point of view of its relevance to philosophy rather than to math, and Lewis was one of the leaders in this. He was at Harvard at the time.

Aspray: While we are on the subject, can we talk more about the logic community in the late '20s and '30s, both in the US and overseas? Where were the active centers? Did you have any contact with these people?

Church: I had very little contact with the people at Harvard, where I suppose the logicians were C. I. Lewis and H. M. Sheffer. Those are the ones I remember.

Aspray: Was anyone at Chicago at that time?

Church: Not that I remember. There must have been some other logicians, but the others who were active at that date or earlier were, I think, mathematicians. E. L. Post, for instance, was a mathematician. He did write papers criticizing some of Lewis' work. In fact, Lewis' first set of axioms for his modal logic had a serious error that Post corrected, and then Lewis tried a second time.

Aspray: Did you have close ties with Post?

Church: No. He was at Princeton just before I was. He may have been there at the time I was an undergraduate, but I did not meet him till much later. He had some sort of mental trouble and was inactive for a long time. He finally recovered from it, and that was really when I first heard of him or heard from him.

Aspray: What about in Europe? Were these people you were in contact with? Did you keep up a contact with Brouwer, for example?

Church: Yes, to some extent. To a greater extent with Bernays, who, because Hilbert was old and ill at the time, was the main logician at Göttingen when I was there.

Aspray: Ackermann?

Church: No, Ackermann was not there at the time I was there. He never had a university position, if my information is correct. He had a degree from Hilbert, but that was before I was in Göttingen. Where he was in Germany at that time, I do not know, but much later he was teaching at a *Gymnasium*. He never did really have a university position, though he finally received an honorary professorship at Munich.

Aspray: What about people like Skolem, did you have contacts with Skolem?

Church: Not till very much later.

Aspray: Do you know anything about the discussion there was to bring you back to Princeton as an assistant professor? Maybe you heard this many years later?

Church: I was not let in on their deliberations. I assume it was Veblen's idea, though it is merely an inference. All I really know is that I got an offer while I was still a fellow at Göttingen, I accepted the offer, and I ceased to look after that.

Aspray: As you progressed up the ranks at Princeton, did Veblen continue to be a strong supporter of your moving up? You obviously had to have your own talent to continue to move up.

Church: I assume he was until he resigned at Princeton and joined the Institute. I forget the date of that. It was probably before 1930, but the dates are on record and you can easily check it.

Aspray: '30–'31.

Church: I see.

Aspray: Why don't we turn to your graduate students for awhile. If I remember correctly you had Alfred Foster, Stephen Kleene, and John Barkley Rosser. Did you have other students in the '30s?

Church: None that I remember now. There may have been some, but none of note. There was a gap there until later when Leon Henkin and John Kemeny were there at the same time. There were also Hartley Rogers, Martin Davis, Norman Shapiro, William W. Boone, and (much later) D. J. Collins. My memory is very poor, both as to the names and as to the chronological order, but most of these were later than the '30's.

Aspray: Can you tell me something about these graduate students? Anecdotes, personal stories, things you remember about their research, how they got involved in logic—anything along these lines?

Church: I remember Kleene was slow getting started. It is possible he was trying other fields, but as far as I knew he did almost nothing for quite a time. Then suddenly he began to come up with things that impressed me greatly.

Aspray: Now did all three of them start by working on the same sorts of things you were working on, such as the lambda calculus?

Church: Kleene and Barkley Rosser were there simultaneously, and both started work in connection with recursive functions and the lambda calculus at about the same time. (The notion of a general recursive function originated with Gödel in lectures at Princeton.)

Aspray: How closely did they work with you on projects? Did you suggest problems to them? Did you talk to them regularly?

Church: I did talk with them in a general way, and they took courses in which I was teaching things such as the lambda calculus. Probably in both cases they worked considerably alone before they came to me. I don't remember details now, especially not the chronology, but I remember being quite surprised when they first brought their results to me.

Aspray: I know Rosser quite well because I studied with him when I was a grad student, but I don't know Foster at all. I'll get to meet him.

Church: I didn't think much of him at the time. He has developed since, and I believe he is very well thought of. His field, though, is not exactly logic. He is at Berkeley now as you probably know.

Aspray: I'll see him tomorrow. Did you have many grad students taking your courses in logic?

Church: At one time I did. There was a time when logic was not very well thought of, and the students tended to follow the trend.

Aspray: Can you elaborate on that? I have always thought that was true, but I was not sure.

Church: There is nothing definite that I can put my finger on. I speak of an impression.

Aspray: What period was this?

Church: Oh, the late '30s. Just before the *Journal of Symbolic Logic* began and for a time after that.

Aspray: Now, as the Institute got started, actually even a little before that in the case of von Neumann, you got other people coming into the community who were interested in logic.

Church: Well, as far as I know, at the time when von Neumann came to Princeton his interest was set theory rather than logic. Even that was in the past as he had already turned to other subjects, either that or he did so very soon after he came.

Aspray: You did not have very much contact with him then?

Church: Not too much. Occasionally there would be a question or a paper in set theory I would consult him about; and occasionally he would consult me. This is how I got the impression that he was no longer active in set theory, but was doing something entirely different.

Aspray: Was it not 1933 or 1934 that Gödel came to Princeton?

Church: Yes, that may be right.

Aspray: Did you have close contacts with Gödel then?

Church: I had a lot of conversations with him and a lot of disagreements. Like most others, I was hard to convince about the incompleteness theorem. There was at the time a tendency, which I shared, to think that it was special to a certain type of formalization of logic and that a radical reformalization might have the effect that the Gödel argument did not apply. I persisted in that longer than I should have, and he was always trying to convince me otherwise.

Aspray: I see. Was the lambda calculus one of those that you would have put into that category of being radical enough that the incompleteness theorem would not apply?

Church: Not the lambda calculus alone. In a way that does escape the Gödel theorem, but it does it by not being powerful enough. I had a scheme that had the lambda calculus as part of it. After publishing a couple of attempts that actually lead to inconsistency, I decided that it couldn't be put through, so the lambda calculus is all that is left of that. The sense in which it escapes the Gödel theorem is not significant from the point of view of logic as a foundation of mathematics, though it might be in other directions.

Aspray: Who else came as a visitor or as a member of the Institute or as a university faculty member in the '30s?

Church: Bernays was there on two successive occasions, each time on one-year appointments. I think it was at Princeton University, rather than the Institute. I had a lot of contact with him at the time.

Aspray: Anyone else?

Church: No, I can think of no one else.

Aspray: You said that Henkin and Kemeny were students at the same time?

Church: Yes.

Aspray: This must have been '39, '40, something like that, is that right?

Church: There was a gap between the students that was important enough for me to remember, that is between Kleene and Rosser and the next two to fall into that category, Henkin and Kemeny.

Aspray: Did they both work with you?

Church: Well, Henkin had a new proof of the Gödel completeness theorem and an extension of it to second-order logic. This was quite substantial. Kemeny's dissertation concerned the relative strength of simple type theory and ZF set theory without replacement axiom. He wrote a dissertation which I thought well of, but he did not accomplish very much in research afterwards.

Aspray: Did you direct Alan Turing's thesis?

Church: Well, he was at Princeton, but not only under my supervision, because, of course, he had worked with M. H. A. Newman in England. It was while he was working with Newman that his truly original ideas came out.

Aspray: On effectively computable functions?

Church: Yes. In fact the definitions of effective calculability and the results on the unsolvable decision problems are essentially the same. These were obtained by me and by Turing almost simultaneously. I think I was the earlier by six months or a year. My paper was delayed in publication, but there is an earlier abstract. Turing did not hear of it until it finally appeared. It was, of course, a great disappointment to him. I don't know the date at which he first had the result.

Aspray: If you don't mind, I would like to ask a few more questions about this topic, because it is one of particular interest to me since I wrote my dissertation on Turing. How did you hear about Turing's work?

Church: Well, Turing heard about mine by seeing the published paper in the *American Journal of Mathematics*. At the time his own work was substantially ready for publication. It may already have been ready for publication. At any rate he arranged with a British periodical to get it published rapidly, and about six months later his paper appeared. At the same time, I think, Newman in England wrote to me about it.

Aspray: Now didn't his papers appear in the *Journal of Symbolic Logic*?

Church: No, I guess there wasn't any such journal at that time. It appeared in a British journal.

Aspray: *Proceedings of the London Mathematical Society*.

Church: It is quite likely, yes.

Aspray: That is where it was. Did you know Newman at the time?

Church: Only by correspondence.

Aspray: How was Turing's visit to Princeton arranged?

Church: At Newman's suggestion he applied for admission as a grad student.

Aspray: I thought that he had come on a one-year fellowship and then was encouraged to stay on by Dean Eisenhart for a second year as a regular grad student.

Church: Yes, I forgot about him when I was speaking about my own graduate students. Truth is, he was not really mine. He came to Princeton as a grad student and wrote his dissertation there. This was his paper about ordinal logics.

Aspray: Right. Did you have much contact with him while he was writing his paper?

Church: I had a lot of contact with him. I discussed his dissertation with him rather carefully.

Aspray: Can you tell me something about his personality?

Church: I did not have enough contact with him to know. He had the reputation of being a loner and rather odd.

Aspray: Could you tell me something about the founding of the *Journal of Symbolic Logic*? For example, what was behind your decision to found a new journal?

Church: It was not my doing. Somewhere there is an historical paper about this in the journal itself.

Aspray: I can find that easily. I was not aware of it.

Church: Yes, it is a historical paper about the founding of the Association for Symbolic Logic and of the journal. I was not in on it from the beginning. I was brought in as editor for the journal later. I think the information in that paper is more accurate than I could give you.

Aspray: Could you tell me something about the time you were editing it? How strongly did the math department support the journal? Did they think it was important?

Church: Well, they yielded finally to the fact that it had a big reputation elsewhere. There were not many others interested in this field, and it was thought of as not a respectable field, with some justice. There was a lot of nonsense published under this heading. I definitely had the idea that one of the things the journal had to do was to suppress this. There were some savage reviews that were written of nonsense papers; I kept them polite, but they were still sharp.

Aspray: And you kept a firm hand on what got published and what did not?

Church: Yes.

Aspray: Did you have pretty much entire editorial control over publication at that time?

Church: There were several editors. I did not try to second guess the other editors when they decided to accept or reject contributed papers.

Aspray: I can't recall who the other editors were now. Can you tell me?

Church: At first there was no one who stayed very long, except myself. I persuaded a man to take a three year term, and there were a number who lasted even longer. A library that has a complete set of the journal will quickly answer that question.

Aspray: I can find that out. To what extent did the department provide you with support, such as secretarial help and money for assistants?

Church: I had a half-time secretary, supplied I think, by the department.

Aspray: That could have been supplied by the Association.

Church: Perhaps, but probably not because they were hard up for funds from the beginning.

Aspray: I know that the *Annals of Mathematics* was more or less reviewed in-house in Princeton. Was that true of the *Journal of Symbolic Logic* also?

Church: You mean that they used only Princetonians to referee the papers? That certainly was not true of the *Journal*. There were no other logicians at Princeton, unless you count the visitors like Gödel and Bernays.

Aspray: One of the things that Professor Tucker is most interested in getting on tape are recollections of the Princeton environment, because many people thought of it as a special place in the '30s, especially after the Institute was established.

Church: Yes, long before that there certainly was an intense interest in mathematical research, and Veblen exhibited that spirit.

Aspray: What kind of decisions were made administratively to allow or to foster this kind of environment?

Church: Well, that I really don't know.

Aspray: For example, the new Fine Hall. Can you comment on it?

Church: Somebody gave money for that specific purpose, but I can't remember who it was.

Aspray: It was the Jones family.

Church: Yes, they were probably friendly with some of the mathematicians.

Aspray: I think they were quite friendly with both Dean Fine and also L. P. Eisenhart.

Church: Yes.

Aspray: But, architecturally speaking did the building fit the requirements of a mathematics research group?

Church: Yes. It was fancier than necessary and not strictly utilitarian. But at the time Princeton was going in for Gothic architecture. All the large offices were paneled in wood up nearly to the ceiling and with elaborate carvings. I assume it is still there. The math department moved, of course, so some other department has it.

Aspray: The East Asian Studies Department is there now and still has all the nice carvings. Do you remember going to tea? Was this a regular part of your day?

Church: Yes. Veblen used to run those himself before there was any Fine Hall. He promoted that as a way for people interested in research to get together. So afternoon tea in Fine Hall was really a continuation of Veblen's teas in Palmer Laboratory. It never worked much for me. I was too much of a loner, but I think in other mathematical fields it was a very useful thing.

Aspray: Did you go anyway? Regularly? Or sometimes?

Church: Yes, I used to go to their teas in the afternoon. I never had any mathematical conversations with anybody, because there was nobody else in my field except a student or two. [Afterthought: An exception must be made in the case of Bernays while he was there; with him there were often conversations at tea time.]

Aspray: How was the library?

Church: That was very good from the beginning. I think a lot of effort and probably a lot of money was put into getting a good mathematical library.

Aspray: That reminds me about another question I have been meaning to ask you about the *Journal of Symbolic Logic*. It seems that the journal had interest in historical and bibliographic information. It kept you up to date in those ways, as well as publishing research.

Church: The intention at the time was to review everything that appeared in the field. A bibliography which was meant to be complete of earlier things was published, and the reviews up until about 1950 were quite complete. The field kept growing, and the reviewing got to be too big a job.

Aspray: I want to say to you that when I was coming through graduate school interested in the history of logic, they provided an invaluable source. They were a real service.

Church: If used right they should be very valuable. I have an idea that few people really use them right. At any rate it could not be continued, because the field got too large and there were not funds around to do it.

Aspray: I have the impression that many at Princeton were rather social people, that people like the von Neumanns, the Eisenharts, and the Robinsons were all people that made their homes available regularly for big social occasions.

Church: Yes, that is true. Veblen was a great advocate of getting together informally. His teas were in the same spirit. He believed in taking long walks through the woods to discuss mathematical research. It never worked for me, but maybe it did for others.

Aspray: Was the relationship between the graduate students and the faculty fairly close?

Church: I think so, yes.

Aspray: Can you tell me something about the role the Depression played in the development of mathematics in the '30s? Were there opportunities for getting students funds for research? Did it make any difference in your own career?

Church: As far as I know, it made no difference to me. Neither do I have much of an impression of the general situation. The university was evidently hard up. For example, they postponed promotions for the faculty.

Aspray: Did it affect your being able to help your grad students find positions, for example Rosser and Kleene and Foster?

Church: There was never any real problem.

Aspray: I seem to recall that Kleene told me that he was ready to go out, that there was not anything for him for a year or two, and that Princeton found money somehow for him to stay on.

Church: I don't remember any agonizing delay about his getting a position, but it could have been he stayed a year or two longer than was absolutely necessary.

Aspray: Do you remember any discussions in the '30s about the hiring of immigrant mathematicians or of bringing in a large number of foreigners as researchers? There were big battles going on, maybe just underneath the surface, about . . .

Church: Yes, I wasn't a party to said battles, I am sure. Many were invited to Princeton, and I did not hear any opposition to it.

Aspray: Princeton seems to have been unusual in opening its arms to immigrant mathematicians, unlike certain other centers at the time.

Church: Yes. I suspect that this was partly Veblen's influence, but I don't know.

INTENSIONALITY AND THE PARADOX OF THE NAME RELATION[1] (1989)

The paradox of the name relation[2] may be illustrated by the now familiar example concerning King George IV and Sir Walter Scott. On a certain occasion, King George wished to know whether Sir Walter was the author of *Waverley*. The *Waverley Novels* (of which *Waverley* was the first) had been published anonymously and at the time there was widespread speculation about the authorship, but it is now known that Scott was indeed the author. Thus in fact Sir Walter Scott = the author of *Waverley*, and by the principle of substitutivity of equality, whatever is true of the author of *Waverley* must be true also of Sir Walter Scott. In particular, one thing true of the author of *Waverley* is that (on the occasion in question) King George wished to know whether Sir Walter Scott was he; and by the principle of substitutivity of equality, it follows that King George wished to know whether Sir Walter Scott was Sir Walter Scott. But this last is clearly false, as King George was well acquainted with Sir Walter and therefore certainly knew of his self-identity; moreover, as Russell says,[3] "an interest in the law of identity can hardly be attributed to the first gentleman of Europe."

The term *paradox of the name relation* is taken from Carnap.[4] Indeed Carnap speaks of the "antinomy" of the name relation, but the word "paradox" seems more appropriate, as (without further assumptions) there is no actual contradiction that arises but highly counterintuitive results. For the name relation itself — i.e., the relation between a name and what it is a name of — we shall use also the verb *to denote* and the noun *denotation*. In this we follow J. S. Mill,[5] and Russell from 1905 on,[6] except that we confine the usage to *singular* names, i.e., to names which have as part of their meaning

Originally published in ***Themes from Kaplan***, edited by J. Almog, J. Perry, and H. Wettstein, Oxford University Press, 1989, pp. 151–165.

[1]The content of this paper was presented as an invited lecture at a joint symposium of the American Philosophical Association and the Association for Symbolic Logic, in Berkeley, California, on 26 March 1983.

[2]Bertrand Russell, *On Denoting*, ***Mind***, vol. 14 (1905), pp. 479–493. Reprinted in ***Readings in Philosophical Analysis***, ed. Herbert Feigl and Wilfrid Sellars (New York: Appleton-Century-Crofts, 1949), 103–115; ***Logic and Knowledge***, ed. Robert Charles Marsh (London: Allen & Unwin, 1956), 41–56; ***Readings in Logical Theory***, ed. Irving M. Copi and James A. Gould (New York: The Macmillan Company, and London: Allen & Unwin, 1967), 93–105; ***Intentionality, Mind, and Language***, ed. Ausonio Marras (Urbana: University of Illinois Press, 1972), 362–379.

[3]Quoted from *On Denoting*.

[4]Rudolf Carnap, ***Meaning and Necessity***, (Chicago: University of Chicago Press, 1st ed. 1947 and 2nd ed. 1956), see §31.

[5]John Stuart Mill, ***A System of Logic, Ratiocinative and Inductive***, vol. I, chap. 2 (London: Longmans, Green, and Co., 1884).

[6]Bertrand Russell, *On Denoting*, footnote 10. But as to Russell the statement in the text must be qualified by remarking that after his paper of 1905 Russell continues to hold that "denoting phrases never have any meaning in isolation, but only enter as constituents into the verbal expression of propositions which contain no constituent corresponding to the denoting phrases in question." (Quotation from Russell, *Mathematical Logic as Based on the Theory of Types*, a paper which appeared originally in the ***American Journal of Mathematics*** in 1908 and has been reprinted in the anthologies of Robert Charles Marsh, op. cit., and of Jean van Heijenoort, ***From Frege to Gödel*** (Cambridge, Mass.: Harvard University Press, 1967) — see therein section III, objection (2).)

that there is just one denotation. The common names of the natural languages are left out of account as being unnecessary in the formalized language, except to the extent that their place is taken by variables. For variables we prefer the terminology that the variable may *have a value*, chosen within a certain *range* of values that is predetermined as a part of the meaning of the variable.

The paradox of the name relation originated with Frege,[7] who explains the difference in cognitive significance between $a = a$ and $a = b$, even in cases in which $a = b$ holds, on the ground that the names a and b, though having the same denotation, may nevertheless differ in another aspect of their meaning, the sense. For a resolution of the paradox along these lines we must add that:

1. In the natural language, or at least in all the familiar natural languages, there are certain contexts,[8] called indirect contexts, within which names change their meaning in such a way that what would be the sense in an ordinary context becomes the denotation in the indirect context.
2. In a formalized language the semantical irregularity by which the same name has different denotations (and different senses) in different contexts should be abolished by introducing two different names.

In spite of the difficulty in attaching any very precise meaning in general to the distinction between the intensional and the extensional, it seems clear that the entities (*concepts* as we shall call them) which serve as senses of names must be considered intensional. We proceed to show also that Russell's resolution of the paradox of the name relation depends on intensionality.

Russell's resolution of the paradox depends on his well-known theory of descriptions,[9] of which we shall here use the version that appeared in *Principia Mathematica*.[10] To show that it depends also on the intensionality of Russell's propositional functions, as these appear in *The Principles of Mathematics*[11] and in the first edition of *Principia Mathematica*, we may proceed by showing that the paradox is restored if we accept the axioms of extensionality for propositional functions that were proposed by Russell in the Introduction to the second edition of *Principia Mathematica*.

According to Russell on page 14 of the Introduction to the first edition of *Principia Mathematica*, if an open sentence, or propositional form, is given in which there occurs

[7] Gottlob Frege, *Über Sinn und Bedeutung*, ***Zeitschrift für Philosophie und philosophische Kritik*** 100 (1892): 25–50; and Frege, ***Grundgesetze der Arithmetik***, vol. I (Jena: Verlag Hermann Pohle, 1893), IX, 7, 50–51. *Über Sinn und Bedeutung* is reprinted in Gottlob Frege, ***Funktion, Begriff, Bedeutung. Fünf logische Studien***, ed. Gunther Patzig (Göttingen: Vandenhoeck & Ruprecht, 1962), and rev. ed. (Gottingen: Vandenhoeck & Ruprecht, 1966). English translations of *Über Sinn und Bedeutung* are available in ***Translations from the Philosophical Writings of Gottlob Frege***, eds. Peter Geach and Max Black; and in the anthologies of Feigl and Sellars, of Copi and Gould, and of Marras that are cited in footnote 2; and in ***The Philosophy of Language***, ed. A. P. Martinich (New York: Oxford University Press, 1985), 200–12.

[8] Because the contexts in question often involve what, in the grammar of a natural language, is called indirect discourse (or in German, *ungerade Rede*), and because Frege's use of *ungerade* is meant as suggested by this, it seems better to use *indirect* as the English translation — in place of my former use of *oblique* for this.

[9] Contextual definition, first introduced by Russell in connection with descriptions in his paper *On Denoting*, came to play an essential role in his theory of knowledge. See his *Knowledge by Acquaintance and Knowledge by Description*, and his ***A History of Western Philosophy***.

[10] ***Principia Mathematica***, 30–32, and ∗14, 173–86.

[11] ***The Principles of Mathematics***, chap. VIII and appendices A and B.

a single free variable such as x, there is a corresponding propositional function whose values are obtained by substituting appropriate constants for the variable x (at all free occurrences of x).[12]

At the top of the next page Russell employs 'x is hurt' as an example of an open sentence to illustrate this principle. In reading this passage bear in mind that Frege's systematic use of quotation marks to distinguish mention from use was not adopted by others until much later, and Russell's use of quotation marks here is probably not meant to serve this purpose.[13]

For singulary propositional functions Russell writes the principle of extensionality as

$$\varphi x \equiv_x \psi x \,.\supset .\, \varphi\hat{x} = \psi\hat{x}$$

This is not assumed in the first edition of *Principia Mathematica*, but it is proposed in the Introduction to the second edition as a possible alternative to the axioms of reducibility, in ramified type theory, and more recently it has been used by others in connection with simple type theory.[14] Using my own notation[15] I prefer to write this principle (or axiom) of extensionality as

$$\varphi x \equiv_x \psi x \supset \varphi = \psi$$

in the lowest type, and analogously in higher types.[16]

Now if we write *Sir Walter Scott is the author of Waverley* as

$$(\iota x)S(x) = (\iota y)A(y, (\iota z)W(z))$$

the result of eliminating the descriptions[17] is:

$$\begin{aligned}&(\exists a) \,.\, S(x) \equiv_x x = a \,.\, (\exists b) \,.\\ &(\exists c)[W(z) \equiv_z z = c \,.\, A(y, c)]\\ &\equiv_y y = b \,.\, a = b\end{aligned}$$

Thus on Russell's theory of descriptions *King George wished to know whether Sir Walter Scott is the author of Waverley* must be rewritten as:

King George wished to know whether

$$\begin{aligned}&(\exists a) \,.\, S(x) \equiv_x x = a \,.\, (\exists b) \,.\\ &(\exists c)[W(z) \equiv_z z = c \,.\, A(y, c)]\\ &\equiv_y y = b \,.\, a = b\end{aligned}$$

[12]More accurately "whose value for any argument c of the same type as x is denoted by the result of substituting for the variable x, at all free occurrences of x in the propositional form, a constant denoting c." Generalization to the case of two or more free variables in the propositional form and a corresponding propositional function of two or more arguments is no doubt intended and is implicit in various later passages.

[13]Indeed in *On Denoting* Russell explains that quotation marks are used to distinguish the meaning (i.e., the sense) of a denoting phrase from its denotation. But he simply has no notation or terminology by which conveniently to distinguish a name of the denoting phrase itself from a name of its meaning or sense—although he agrees (see his footnote 10) that, at least in expounding Frege's theory, this distinction must be made.

[14]E.g., by Rudolf Carnap in ***Abriß der Logistik*** (1929), and in ***Logische Syntax der Sprache*** (1934).

[15]Alonzo Church, ***Introduction to Mathematical Logic*** (1956).

[16]***Principia Mathematica***, $*9$ and $*12$, explain the device of typical ambiguity, by which formulas are asserted that have free variables of ambiguous type.

[17]In accordance with the method of ***Principia Mathematica***, $*14$.

Here it is not meant that S, W, and A are names of certain propositional functions, but (for example) $S(x)$ is used rather as a convenient abbreviation to represent the full statement that x scottizes, i.e., that x satisfies just those conditions, whatever they are taken to be, which uniquely characterize being Sir Walter Scott. Similarly $W(z)$ is the full statement that z waverlizes, and $A(y, c)$ the full statement that y authors c.

Now if we let φ be the propositional function that corresponds to the propositional form $S(x)$, and ψ the propositional function that corresponds to the propositional form $(\exists c)[W(z) \equiv_z z = c \,.\, A(y, c)]$, the last display becomes:

King George wished to know whether
$(\exists a) \,.\, \varphi x \equiv_x x = a \,.\, (\exists b) \,.\, \psi y \equiv_y y = b \,.\, a = b$

Then because we know that as a matter of fact $\varphi x \equiv_x \psi x$, there follows by the principle of extensionality for propositional functions and by substitutivity of equality:

King George wished to know whether
$(\exists a) \,.\, \varphi x \equiv_x x = a \,.\, (\exists b) \,.\, \varphi y \equiv_y y = b \,.\, a = b$

This last represents King George as wishing to know something that is logically equivalent to the existence of a unique individual who scottizes. But the fact is that King George's concern was not at all with this, but rather with the authorship of *Waverley*. Indeed the formula in the last display is just the result of eliminating the descriptions from

$$(\iota x)\varphi x = (\iota y)\varphi y$$

and this certainly misrepresents King George's interest in the matter.

Thus Russell's resolution of the paradox of the name relation fails if extensionality is assumed.[18]

Ajdukiewicz's resolution of the paradox was presented to the Twelfth International Congress of Philosophy, at Venice, in 1958.[19]

Ajdukiewicz uses the example "Julius Caesar knew that Rome is situated on the Tiber." The quoted sentence is no doubt true. But the sentence "Rome is the capital of the Popes" is also true,[20] where "is" is the "is" of identity. Does it follow by substitutivity of identity that "Julius Caesar knew that the capital of the Popes is situated on the Tiber"? On the face of it, and taken purely formally, this would seem to follow. Yet the proposed conclusion is seen to be false, on the ground that Caesar had not heard of the Popes or of their capital. This is another instance of the paradox

[18]This conclusion does not mean that Russell's theory of descriptions is without value. On the contrary, a well-known simplifying feature of Russell's formalized language would be lost by abandoning the theory of descriptions, namely that the argument of a functional variable may be restricted to be a single symbol, either a variable or a primitive constant.

[19]See Kazimierz Ajdukiewicz, *A Method of Eliminating Intensional Sentences and Sentential Formulae*, ***Atti del XII Congresso Internazionale di Filosofia*** (Florence, 1960), 17–24. See also Perry Smith's review of this in ***The Journal of Symbolic Logic*** 37 (1972): 179–180.

[20]The word "true" is here used in Tarski's sense, applying to sentences rather than propositions. Strictly, one should therefore say "true in" a certain language, but the qualification may be omitted when obvious from the context. See Tarski's *Der Wahrheitsbegrif in den formalisierten Sprachen*, ***Studia Philosophica*** 1 (1936): 261–405, and its English translation, with revisions, that appeared in Alfred Tarski, ***Logic, Semantics, Metamathematics. Papers from 1923 to 1938***, trans. J. H. Woodger, 2d ed., edited and introduced by John Corcoran (Indianapolis: Hackett Publishing Company; 1983), 152–278.

of the name relation, and it might be resolved by Frege's method or by Russell's, but Ajdukiewicz seeks a resolution of the paradox that does not depend on meanings or on intensionalities.

According to Ajdukiewicz the sentence "Caesar knew that the capital of the Popes is situated on the Tiber" is ambiguous, and it might mean either

(1) "Caesar knew about the capital of the Popes, about the relation of being situated on, and about the Tiber, that the capital of the Popes is situated on the Tiber"

or

(2) "Caesar knew about the function capital of, about the Popes, about the relation of being situated on, and about the Tiber, that the capital of the Popes is situated on the Tiber"

and of these two sentences, (1) is true, on the ground that Caesar knew that Rome is situated on the Tiber, and Rome is in fact the capital of the Popes; but (2) is false on the ground that Caesar had not heard of the Popes. Thus substitutivity of identity is preserved.

A modification of Ajdukiewicz's resolution of the paradox is desirable because familiar formalized languages do not provide for abstraction with respect to a constant in the way that Ajdukiewicz's formulation requires. The presumed equivalent in a formalized language of more standard sort is to replace the constant by an appropriate variable, to abstract with respect to this variable, and then separately to supply the constant.

Thus in a formalized language we might try rewriting "Caesar knew that Rome is situated on the Tiber" as

$$K(C, \lambda\Phi\lambda x\lambda y \,.\, \Phi xy, R, a, b)$$

rewriting (1) as

$$K(C, \lambda\Phi\lambda x\lambda y \,.\, \Phi xy, R, f\pi, b)$$

and rewriting (2) as

$$K(C, \lambda\Phi\lambda\psi\lambda\xi\lambda y \,.\, \Phi(\psi\xi)y, R, f, \pi, b).$$

However, once this is done it is seen that the propositional function K, "knew that," appears with five arguments at one place and six arguments at another. Because usual formalized languages do not admit such a thing as a propositional function that may take a varying number of arguments, we amend by taking K, "knew that," to be a binary propositional function of which the second argument may be an ordered quadruple, ordered quintuple, etc.[21] Thus "Caesar knew that Rome is situated on the

[21] The ordered n-tuples such as

$$\langle\lambda\Phi\lambda x\lambda y \,.\, \Phi xy, R, f\pi, b\rangle$$

or

$$\langle\lambda\Phi\lambda\psi\lambda\xi y \,.\, \Phi(\psi\xi)y, R, f, \pi, b\rangle$$

which may thus serve as object of knowledge (believe, assertion, etc.) we shall call proposition surrogates, because it is thought that they may in some respects perform the office of propositions, although conceptually very different from propositions. Indeed the task of finding a satisfactory axiomatic theory of propositions has seemed very difficult in the past, especially in regard to the question of identity of propositions, and it

Tiber" becomes

$$K(C, \langle \lambda\Phi\lambda x\lambda y \,.\, \Phi xy, R, a, b\rangle)$$

while (1) becomes

$$K(C, \langle \lambda\Phi\lambda x\lambda y \,.\, \Phi xy, R, f\pi, b\rangle)$$

and (2) becomes

$$K(C, \langle \lambda\Phi\lambda\psi\lambda\xi\lambda y \,.\, \Phi(\psi\xi)y, R, f, \pi, b\rangle).$$

If

(3) "Caesar knew that Rome is the capital of the Popes"

is understood in a sense to make it false (on the ground that Caesar had not heard of the Popes), the corresponding ordered n-tuple, or proposition surrogate (see footnote 21), must be

$$\langle \lambda\Phi\lambda x\lambda\psi\lambda\eta \,.\, \Phi x(\psi\eta), I, a, f, \pi\rangle$$

where I is the propositional function, identity, so that Ixy means the same as $x = y$, and as before a is Rome, f is the function *the capital of*, and π is the class of Popes. On the other hand, if (3) is understood in a sense to make it true, the corresponding proposition surrogate must be

$$\langle \lambda\Phi\lambda x\lambda y \,.\, \Phi xy, I, a, f\pi\rangle.$$

This last ordered quadruple, since we know that $a = f\pi$, is demonstrably the same as

$$\langle \lambda\Phi\lambda x\lambda y \,.\, \Phi xy, I, a, a\rangle$$

and this in turn is therefore the proposition surrogate to be used in analyzing

(4) "Caesar knew that Rome is Rome."

Following these examples, let us consider more generally the question of the proposition surrogate corresponding to a given sentence of the formalized language *that is in λ-normal form*.[22] If an ordered $(n+1)$-tuple is to serve as proposition surrogate for such a sentence, its first member must be in the form

$$\lambda\,\mathbf{x}_1\,\lambda\,\mathbf{x}_2\cdots\lambda\,\mathbf{x}_n\,\mathbf{M}$$

where $\mathbf{M}$ is in λ-normal form and has as its only free variables exactly one free occurrence of each of the variables $\mathbf{x}_1, \mathbf{x}_2, \ldots, \mathbf{x}_n$—and the remaining n members of the

may well be rendered much easier by using propositional surrogates to guide the choice of axioms, and in the end perhaps even to provide a relative consistency proof.

[22]To avoid antinomy, the formalized language that we use is to obey a simple type theory, and this may conveniently be the type theory of my paper *A Formulation of the Simple Theory of Types* in ***The Journal of Symbolic Logic*** 5 (1940): 56–68, or (what is the same thing), the type theory of my paper in ***Noûs*** 7 (1973): 24–33, and ***Noûs*** 8 (1974): 135–156, with all the intensional types omitted.

Notice that in consequence of the type theory to which the formalized language conforms, every wff has a λ-normal form, to which it can be reduced by λ-conversion.

Then as definition of the ordered pair $\langle \mathbf{A}^\alpha, \mathbf{B}^\beta\rangle$, where $\mathbf{A}^\alpha$ and $\mathbf{B}^\beta$ are wffs of the indicated types, not containing $\mathbf{f}^{o\beta\alpha}$ as free variable, we may take $\lambda\mathbf{f}^{o\beta\alpha}(\mathbf{f}^{o\beta\alpha}\mathbf{A}^\alpha\mathbf{B}^\beta)$, which is of type $o(o\beta\alpha)$. Thus $\langle \mathbf{A}^\alpha, \mathbf{B}^\beta\rangle$ is a function whose value with argument $\lambda\,\mathbf{x}^\alpha\,\lambda\,\mathbf{y}^\beta \,.\, \mathbf{C}^\alpha = \mathbf{x}^\alpha$ is the truth-value t if $\mathbf{C}^\alpha = \mathbf{A}^\alpha$ holds, and is otherwise the truth-value f, *and* whose value with argument $\lambda\,\mathbf{x}^\alpha\,\lambda\,\mathbf{y}^\beta \,.\, \mathbf{D}^\beta = \mathbf{y}^\beta$ is the truth-value t if $\mathbf{D}^\beta = \mathbf{B}^\beta$ holds, and is otherwise the truth-value f (this on the assumption that the variables $\mathbf{x}^\alpha$ and $\mathbf{y}^\beta$ are so chosen as to have no free occurrences in either $\mathbf{C}^\alpha$ or $\mathbf{D}^\beta$). [*Editor's note*: Church neglects to mention that the variable $\mathbf{f}^{o\beta\alpha}$ should have no free occurrence in $\mathbf{A}^\alpha$ or $\mathbf{B}^\alpha$.]

$(n + 1)$-tuple must be, in order, constants $\mathbf{c}_1, \mathbf{c}_2, \ldots, \mathbf{c}_n$ that are of the same types (see footnote 22) as the variables $\mathbf{x}_1, \mathbf{x}_2, \ldots, \mathbf{x}_n$ but are not necessarily all different.

This determines the proposition surrogate uniquely in the case of a given sentence in λ-normal form, and two such sentences will have the same proposition surrogate if and only if they differ only by (1) alphabetic changes and (2) replacements of one constant or one notation by another that is completely synonymous with it. If, however, an analogous definition of the proposition surrogate is used in the case of sentences that are not in λ-formal form, it is difficult to avoid the situation that in some cases in which two sentences are λ-convertible each to the other, the two corresponding proposition surrogates are the same and in other such cases the two proposition surrogates are different.

As an example of this consider the three sentences,

$$Iaa, \quad (\lambda\Phi\lambda x\lambda y \,.\, \Phi xy)Iaa, \quad (\lambda\Phi\lambda x \,.\, \Phi xx)Ia$$

of which each one is λ-convertible to each of the others. Because I is taken as a primitive (functional) constant and Iaa is in λ-normal form, the corresponding proposition surrogate is, as was found above,

$$\langle \lambda\Phi\lambda x\lambda y \,.\, \Phi xy, I, a, a\rangle.$$

By analogy we might assume that the proposition surrogates for the other two sentences are

$$\langle \lambda\Psi\lambda u\lambda v \,.\, (\lambda\Phi\lambda x\lambda y \,.\, \Phi xy)\Psi uv, I, a, a\rangle$$

and

$$\langle \lambda\Psi\lambda u \,.\, (\lambda\Phi\lambda x \,.\, \Phi xx)\Psi u, I, a\rangle.$$

But by reducing these last two expressions to λ-normal form we find as proposition surrogates for the second and third sentences

$$\langle \lambda\Psi\lambda u\lambda v \,.\, \Psi uv, I, a, a\rangle$$

and

$$\langle \lambda\Psi\lambda u \,.\, \Psi uu, I, a\rangle$$

respectively. Of the three sentences named, we thus find (as at least plausible) that the first two do indeed have the same proposition surrogate, but not the third.

For a sentence not in λ-normal form it will therefore be better to define the proposition surrogate as being the same as the proposition surrogate corresponding to the λ-normal form of the given sentence. As a consequence we will have the situation that two sentences have the same corresponding proposition surrogate (and hence presumably express the same proposition) if and only if they differ only by λ-conversion and by interchange of synonymous constants or other fully synonymous notations. This is the identity criterion that I have called Alternative (1) in my papers about Frege's intensional logic.[23] But the same three alternatives (0), (1), and (2) may well be considered in connection with Russell's notion of proposition,[24] and in connection with proposition surrogates.

---

[23]See in particular the abstract in ***The Journal of Symbolic Logic*** 11 (1946): 31, and the paper in ***Noûs*** that is referred to in footnote 22.

[24]As it appears in the first edition of ***Principia Mathematica*** and in ***The Principles of Mathematics***, after some corrections in at least the latter case.

Connectives and quantifiers that occur in a sentence must be treated as being or involving constants — names of functions of appropriate kinds in order to obtain from the sentence a proposition surrogate that will serve our present purpose.

Consider, for example, Cicero's remark, *If Julius Caesar is a man, then Julius Caesar is mortal.* Putting this into a formalized language, we may write it as

$$Hj \supset Mj$$

where j is an individual constant denoting Julius Caesar, and H and M are functional constants with such meaning that Hx expresses that x is a man and Mx expresses that x is mortal. Then consider the two sentences

$$Hj \supset Mj$$

and

$$Hj \supset Mj \supset Mj \supset Mj.$$[25]

These sentences are equivalent by propositional calculus, yet someone not well versed in propositional calculus may well believe (what is expressed by) the first one, and yet be in doubt about the second one because it seems to him complicated and not easily analyzed.

Thus the two sentences, though logically equivalent, differ as to possibilities of belief, and the corresponding proposition surrogates must therefore be different. It seems that the best way to deal with this is to use, not the connective $\supset$, but a letter, say, C, as name of the binary propositional function associated with the connective.[26] Then the sentence

$$Hj \supset Mj$$

is rewritten as

$$C(Hj)(Mj)$$

and its proposition surrogate is therefore

$$\langle \lambda f \lambda \Phi \lambda x \lambda \Psi \lambda y \,.\, f(\Phi x)(\Psi y), C, H, j, M, j \rangle.$$

Similarly

$$Hj \supset Mj \supset Mj \supset Mj$$

is rewritten as

$$C(C(C(Hj)(Mj))(Mj))(Mj)$$

and so has as its proposition surrogate

$$\begin{aligned}&\langle \lambda f \lambda g \lambda h \lambda \Phi \lambda w \lambda \Psi_1 \lambda x \lambda \Psi_2 \lambda y \lambda \Psi_3 \lambda z \,.\\ &f(g(h(\Phi w)(\Psi_1 x)(\Psi_2 y)(\Psi_3 z))),\\ &C, C, C, H, j, M, j, M, j, M, j \rangle.\end{aligned}$$

For similar reasons the universal quantifier is best treated as a universality function Π, used together with the abstraction operator λ when and as required. For example, "All men are mortal" becomes in the formalized language $\Pi(\lambda x \,.\, C(Hx)(Mx))$, but

[25] Brackets and parentheses are omitted under the convention of association to the left.

[26] We must therefore take (unasserted) sentences as names of something, possibly (following Frege) of truth-values, or possibly (following the early Russell) of propositions.

"Everything is mortal" may be written simply as ΠM. Something similar applies to the description operator ι if it is taken as primitive—but here it may be better to adopt Russell's contextual definition.[27]

Proposition surrogates under Alternative (0), as criterion of identity, may be obtained by the following modification of the method of the two preceding sections. Let **S** be the sentence for which the proposition surrogate is to be found and let there be n occurrences of constants in **S**, call them $\mathbf{c}_1, \mathbf{c}_2, \ldots, \mathbf{c}_n$ in left-to-right order. Thus $\mathbf{c}_1, \mathbf{c}_2, \ldots, \mathbf{c}_n$ are not necessarily all different when taken as constants rather than as occurrences of constants. Let $\mathbf{x}_1, \mathbf{x}_2, \ldots, \mathbf{x}_n$ be n different variables such that $\mathbf{x}_i$ is of the same type as $\mathbf{c}_i$, $i = 1, 2, \ldots, n$, and the replacement of $\mathbf{c}_i$ by $\mathbf{x}_i$ (at the one place which is the occurrence of the constant that is referred to by $\mathbf{c}_i$) will not result in the capture of the variable $\mathbf{x}_i$.[28] Then let **M** be the result of replacing each $\mathbf{c}_i$ in **S** (at the place which is its occurrence in **S**) by the variable $\mathbf{x}_i$. Then the free variables in **M** are $\mathbf{x}_1, \mathbf{x}_2, \ldots, \mathbf{x}_n$, each with exactly one free occurrence in **M**. After thus finding **M**, we find $\mathbf{M}^*$ by replacing every occurrence in **M** of the notation (**FA**), for application of a function **F** to an argument A, by the notation $\langle \mathbf{F}, \mathbf{A} \rangle$ for the ordered pair of **F** and A as this notation is explained in footnote 22. Then the proposition surrogate is

$$\langle \lambda \mathbf{x}_1 \lambda \mathbf{x}_2 \ldots \lambda \mathbf{x}_n \mathbf{M}^*, \mathbf{c}_1, \mathbf{c}_2, \ldots, \mathbf{c}_n \rangle.^{29}$$

For instance, let $\mathbf{S}_1$ be the sentence

$$\Pi(\lambda x \,.\, C(Hx)(Mx))$$

which was introduced as an example in the last paragraph of the preceding section. Because $\mathbf{S}_1$ is in λ-normal form, the proposition surrogate under Alternative (1) is

$$\langle \lambda F \lambda f \lambda \Phi \lambda \Psi \,.\, F(\lambda x \,.\, f(\Phi x)(\Psi x)), \Pi, C, H, M \rangle$$

and the proposition surrogate under Alternative (0) is

$$\langle \lambda F \lambda f \lambda \Phi \lambda \Psi \langle F, \lambda x \langle f, \langle \Phi, x \rangle, \langle \Psi, x \rangle \rangle \rangle, \Pi, C, H, M \rangle.$$

As a very simple example of a sentence obtained from $\mathbf{S}_1$ by λ-conversion let $\mathbf{S}_2$ be the sentence

$$((\lambda \Phi \lambda \Psi \,.\, \Pi(\lambda x \,.\, C(\Phi x)(\Psi x)))HM).$$

The proposition surrogate of $\mathbf{S}_2$ under Alternative (1) is of course still the same as that of $\mathbf{S}_1$; but the proposition surrogate of $\mathbf{S}_2$ under Alternative (0) is found as follows.

---

[27] That Russell's contextual definition of descriptions does not resolve the paradox of the name relation without resort to intensional propositional functions does not mean that it may not be used for other purposes (such as economy). And even the use of intensionalities is not in the writer's view ultimately objectionable, it is merely that it ought not to be resorted to without a carefully formulated theory of the particular intensionalities.

[28] I.e., $\mathbf{c}_i$ is not in a wf part of **S** of the form $\lambda \mathbf{x}_i \mathbf{W}$.

[29] Another possible determination of the proposition surrogate under Alternative (0) is just to replace every occurrence in **S** of the notation (**FA**) for application of function to argument by the notation $\langle \mathbf{F}, \mathbf{A} \rangle$ for the ordered pair of **F** and **A**. But the definition in the text is preferred because of the closer analogy with the proposition surrogate under Alternative (1), an ordered $(n+1)$-tuple of which the first term shows in a certain sense the form of **S**, and the other n terms show the constants occurring in **S**.

First

$$M \text{ is } (\lambda\Phi\lambda\Psi \,.\, F(\lambda x \,.\, f(\Phi x)(\Psi x)))\Theta\Omega.$$

Hence $\mathbf{M}^*$ is

$$\langle\lambda\varphi\lambda\Psi\langle F, \lambda x\langle f, \langle\Phi, x\rangle, \langle\Psi, x\rangle\rangle\rangle, \Theta, \Omega\rangle.$$

Therefore the proposition surrogate of $\mathbf{S}_2$ is

$$\langle\lambda F\lambda f\lambda\Theta\lambda\Omega\langle\lambda\Phi\lambda\Psi\langle F, \lambda x\langle f, \langle\Phi, x\rangle, \langle\Psi, x\rangle\rangle\rangle, \Theta, \Omega\rangle, \Pi, C, H, M\rangle.$$

Two defects of the proposal which we have just outlined, to replace propositions by proposition surrogates, must finally be pointed out.

The first of these is the complicated character of the array of types into which the proposition surrogates fall, under our present scheme. In contrast, the propositions of *Principia Mathematica* are distinguished merely as being of first order, of second order, and so on, thus falling into a simply infinite array of types which has the order type w of the natural numbers. It may be that this is not a defect of major importance, but it also may be that it can be remedied by some change in type theory on which the formalized language is based or by some change in the definition of proposition surrogate or by a combination of the two.[30]

I am indebted to Yoram Gutgeld for calling the second defect to my attention. The formalized language must not have two different primitive constants, say, c_1 and c_2, which denote the same thing but do not have the same sense.[31] For in this case it may well happen that someone who believes that $c_1 = c_1$ nevertheless fails to believe that $c_1 = c_2$. Yet if we make use of proposition surrogates as objects of belief in the way that we have described in this paper, it will then follow, if x believes that $c_1 = c_1$, that x also believes that $c_1 = c_2$. Or, more generally, if x believes anything about c_1, he must also believe the same about c_2, so restoring the paradox of the name relation.

The most immediate repair of this second defect is as follows. From a given sentence a corresponding ordered n-tuple is first obtained in the way that is described in the three preceding sections, and in the expression of this ordered n-tuple each primitive constant is then replaced by a name of its sense. The resulting expression then denotes

[30]The proposal to avoid the defect by changing to a language based on a set-theoretic approach rather than type-theoretic would seem to be not very significant. For this can be done only at the cost of abandoning Russell's comprehension principle for propositional functions (as it is described above in the second section of this paper). Then if the comprehension principle is restored in the weakened form, that an application of the principle yields an entity of a new sort, a class as distinguished from a set, this is (as well known) the beginning of a new hierarchy of types — as classes cannot, without fear of contradiction, be allowed as members of sets but must rather be members of superclasses, and so on.

[31]Let us call two constants *concurrent* if they denote the same thing, and let us call a class of constants a *concurrent* class if all members of the class denote the same thing.

A now familiar example of two such constants c_1 and c_2 in a natural language is that of the names *Φωσφόρος* and *Ἕσπερος* in ancient Greek. For brevity let us use "c_1" and "c_2" respectively for these two Greek words. Then c_1 is the brilliant white planet which, at times, rises in the east before the sun at an interval which varies from more than three hours down to a fraction of a minute; and c_2 is the brilliant white planet which, at times, is similarly seen in the western sky after sunset. We may suppose that these are primitive constants, although this is never definite in a natural language prior to all formalization. The early Greeks, not knowing that $c_1 = c_2$, used the two different names for the supposedly different planets. After the discovery by Pythagoras that $c_1 = c_2$ became known a third name *ὁ τῆς Ἀφροδίτης ἀστήρ* (Aphrodite's Star) was introduced for the (in a sense) newly discovered planet that is c_1 and also is c_2.

an ordered n-tuple which is taken as the proposition surrogate corresponding to the given sentence. In this way the paradox of the name relation is resolved, but not entirely without use of intensionalities — as notwithstanding lack of a general definition of the distinction between what is intensional and what is extensional, the Fregean sense of a name is clearly an intensional entity.

Languages are of course possible within which no two primitive constants denote the same thing. For given any primitive constant we may delete from the language all but one of the class of primitive constants that are concurrent with it (see footnote 31). *Whether it is always effectively possible to cut down the vocabulary of a given language so that no pair of concurrent but non-synonymous primitive constants remains* is an open question. The difficulty lies in a method by which to determine in regard to each pair of primitive constants whether they are concurrent.

REVISED FORMULATION OF THE LOGIC OF SENSE AND DENOTATION ALTERNATIVE (1)

(1993)

Part I

This paper is a sequel to my *Outline of a revised formulation of the logic of sense and denotation*.[1] Like the earlier paper it is a contribution, not to Fregean semantics,[2] but to the logic of the intensional entities (*concepts* as I shall call them) which are to serve as senses of names. But it differs from the earlier paper in taking Alternative (1) as criterion of identity of concepts, rather than Alternative (2).[3]

Under Alternative (1) we identify *propositions with* Frege's *Gedanken*, i.e., concepts[4] of truth-values, and the proposal is that propositions in this sense shall be taken as objects of assertion and belief. This is thought to be preferable to the use of proposition surrogates as treated in my paper, "Intensionality and the paradox of

---

Originally published in ***Noûs***, vol. 27 (1993), pp. 141–157. Reprinted by permission. Several font changes on formulae have been made from the published version in accord with preferences that Church expressed, too late for inclusion in the original publication. Some inconsistencies of notation and typographical errors have been resolved in ways that accord with Church's practice in other work, and a few typographical errors in formulae have been corrected.

[1] ***Noûs***, vol. 7 (1973), pp. 24–33, and vol. 8 (1974), pp. 135–156. Typographical corrections to this paper are as follows. Insert at the end of line 17 on page 31 [thirty-fourth paragraph of the paper] "and its antivalue"; in line 9 on page 140 [twenty-first paragraph of Part II], for "well-informed" read "well-formed"; in line 7 from the bottom of page 140 [thirtieth paragraph, theorem IX.], "A" should be bold; at the end of line 5 from the bottom of page 145 [fifty-sixth paragraph], read "undesirably"; in line 12 on page 150 [seventy-fifth paragraph], to avoid a possible misreading, change "ℓ being" to "where ℓ is"; in line 13 from the bottom of page 150 [seventy-eighth paragraph], in $15^{m\alpha\beta}$ and $16^{m\alpha\beta}$ close up misleading spaces between letters at various places; on page 151, in $63^{m\alpha\beta}$, for "$(f_{\alpha_1 \beta_1})$" read "$(f_{\alpha_1\beta_1})$"; on page 155, in lines 13–12 from the bottom [in note 15], for "nonwithstanding" read "notwithstanding". [*Editor's note*: All these corrections have been entered in the present edition. Church's instructions are entered here for historical accuracy.]

[2] In the list of references to my own papers that is included on page 32 of the paper in ***Noûs***, it is only [5] [*The Need for Abstract Entities in Semantic Analysis*] that has some treatment of the semantics that arises from Frege's distinction of sense (*Sinn*) and denotation (*Bedeutung*).

At the end of the paper of 1973 and 1974 (see note 1) there is a brief preliminary sketch of a treatment of Alternative (0), under which two names express the same sense (and in particular two sentences express the same proposition) if and only if they differ only by alphabetic changes of bound variable and replacements of one primitive constant by another that is synonymous with it. This is probably the alternative under which it will be most difficult to avoid antinomy while at the same time preserving essentials.

Alternative (1) differs from Alternative (0) by allowing also λ-conversion as a transformation under which names preserve the same sense. This may be thought counterintuitive if propositions in the sense of Alternative (1) are to be taken as objects of assertion and belief. But it has to be said that much less disguise of a sentence is possible by λ-conversion within a language that obeys a type theory than it is by the type-free form of λ-conversion. In any case a satisfactory treatment of Alternative (1) is a desirable preliminary to undertaking the (probably) more difficult task of treating Alternative (0).

[3] It is therefore a revision of Church [4] [*A Formulation of the Logic of Sense and Denotation*].

[4] This use of the word *concept* is not to be confused with Frege's use of the word *Begriff* in German for a notion that is at least rather close to Russell's notion of propositional function (of one argument).

the name relation,"[5] because (as explained in the last section of the paper) this latter requires the use of a language in which no two primitive constants have the same denotation.[6]

The present purpose is to provide a system of axioms characterizing Alternative (1), and a set-theoretic consistency proof for them. The consistency proof falls just short of *reducing* Alternative (1) to set theory because it relies, not on a single set-theoretic model, but on an infinite class of such models.

At the end of the paper of 1973 and 1974 there is a brief preliminary discussion of the problem of formulating the logic of sense and denotation under Alternative (0). But the main part of the paper is devoted to formulation under Alternative (2). Under all three alternatives it appears that internal contradiction is avoided by adopting a simple type theory, but that some form of ramified type theory is needed to avoid paradox (as distinguished from outright contradiction or antinomy, a distinction to be explained below). In Part I of the present paper we provide the axioms for Alternative (1), as based on simple type theory, and the promised consistency proof. Part II will follow, with resolution of the paradoxes by ramified type theory.

In the proposed simple type theory the array of types is as follows. There is first the infinite list of types $\iota_0, \iota_1, \iota_2, \ldots$; then the infinite list of types $o_0, o_1, o_2, \ldots$; and then if α and β are any types, there is a type $(\alpha\beta)$. The type symbols ι_0 and o_0 will usually be abbreviated by omitting the subscript 0. And compound type symbols such as $(o\iota)$, $((o\iota)\iota)$, $(o(o\iota))$, $((o\iota)(\iota\iota))$, $((o_1(\iota_1\iota_1))o)$ will usually be abbreviated by omitting parentheses under the convention of association to the left, so that e.g. $o\iota(o_1\iota_1\iota_2)$ is an abbreviation of $((o\iota)((o_1\iota_1)\iota_2))$. The *first ascendant* of a type symbol is obtained from it by increasing every subscript in it by 1, the *second ascendant* by increasing every subscript by 2, and so on. As notation for the n-th ascendant we use the Arabic numeral n as a subscript, so that e.g. $((o\iota)((o_1\iota_1)\iota_2))_n$ is $((o_n\iota_n)((o_{n+1}\iota_{n+1})\iota_{n+2}))$; and generally $(\alpha\beta)_n$ is $(\alpha_n\beta_n)$. And because each type has a uniquely determined type symbol, we may speak also of the first ascendant of a type, the second ascendant, and so on.

Members of the first ascendant α_1 of a type α are the concepts of things of type α. And we shall so choose our notation that for every primitive constant $\mathbf{c}_\alpha$ of type α there is another primitive constant $\mathbf{c}_{\alpha_1}$ of type α_1 (with the same letter $\mathbf{c}$) which denotes the sense of the constant $\mathbf{c}_\alpha$, and which we call the first ascendant of $\mathbf{c}_\alpha$. It then follows that every wff $\mathbf{A}_\beta$ of type β has a first ascendant $\mathbf{A}_{\beta_1}$ of type β_1 that is obtained from $\mathbf{A}_\beta$ by replacing every type symbol in $\mathbf{A}_\beta$ by its first ascendant. By iteration of this we have, for every positive integer n, the n-th ascendant $\mathbf{K}_{\alpha_n}$ of every primitive constant $\mathbf{K}_\alpha$, and the n-th ascendant $\mathbf{A}_{\beta_n}$ of every wff $\mathbf{A}_\beta$.

The primitive symbols of the formalized language include an infinite alphabet of

[5] *Intensionality and the paradox of the name relation*, in ***Themes from Kaplan***, Almog, Perry, and Wettstein eds. (Oxford, Oxford University Press, 1989), pp. 151–165.

[6] We shall translate Frege's *Bedeutung* as *denotation*, as indeed there is some resemblance between Frege's theory of the meaning of names and J. S. Mill's, notwithstanding the obvious differences.

variables in each type, with a subscript on the variable to show the type.[7] They include also the abstraction operator λ, the connective ( ) for application of function to argument, the primitive constants C_{ooo} and N_{oo} (denoting material implication and negation respectively, each treated as a function of truth-values), $\Pi_{o(o\alpha)}$ (denoting the universal quantifier, treated as a function having the indicated type), $\iota_{\alpha(o\alpha)}$ (denoting the description function of the indicated type), and $\Delta_{o\alpha_1\alpha}$ (denoting the concept function of the indicated type). Then following the preceding paragraph we must include as primitive constants, for every positive integer n, the n-th ascendant of each of the primitive constants C_{ooo}, N_{oo}, $\Pi_{o(o\alpha)}$, $\iota_{\alpha(o\alpha)}$, $\Delta_{o\alpha_1\alpha}$; and we write these as $C_{o_no_no_n}$, $N_{o_no_n}$, $\Pi_{o_n(o_n\alpha_n)}$, $\iota_{\alpha_n(o_n\alpha_n)}$, $\Delta_{o_n\alpha_{n+1}\alpha_n}$.

We must notice as a consequence of what was said in the three preceding paragraphs that a concept of a function of type $\alpha\beta$ is itself a function and is of type $\alpha_1\beta_1$. This assumption is perhaps not unavoidable, but it greatly simplifies the theory and we shall follow it.

The formation rules are then as follows:

i. A variable or a constant standing alone is a wff, and the type is the same as that of the variable or constant.

ii. If $\mathbf{F}_{\alpha\beta}$ and $\mathbf{A}_\beta$ are wffs of the indicated types, $(\mathbf{F}_{\alpha\beta}\mathbf{A}_\beta)$ is a wff of type α.

iii. If $\mathbf{M}_\alpha$ is a wff of type α and $\mathbf{x}_\beta$ is a variable of type β, $(\lambda\mathbf{x}_\beta\mathbf{M}_\alpha)$ is a wff of type $\alpha\beta$.

The rules of inference are as follows, comprising the three rules of λ-conversion, the rule of *modus ponens*, and the rule of generalization:

I. From $\mathbf{A}_o$, if $\mathbf{x}_\beta$ is a variable of type β which has no free occurrence in $\mathbf{N}_\gamma$ and $\mathbf{y}_\beta$ is a variable of type β which does not occur in $\mathbf{N}_\gamma$, if $\mathbf{B}_o$ results from $\mathbf{A}_o$ by replacing a particular occurrence of $\mathbf{N}_\gamma$ by the result of substituting $\mathbf{y}_\beta$ for $\mathbf{x}_\beta$ throughout $\mathbf{N}_\gamma$, to infer $\mathbf{B}_o$. (Rule of alphabetic change of bound variable.)

II. From $\mathbf{A}_o$ to infer the result of replacing any wf part $((\lambda\mathbf{x}_\beta\mathbf{M}_\alpha)\mathbf{N}_\beta)$, at any one of its occurrences in $\mathbf{A}_o$, by the result of substituting $\mathbf{N}_\beta$ for $\mathbf{x}_\beta$ throughout $\mathbf{M}_\alpha$, provided that the bound variables of $\mathbf{M}_\alpha$ are distinct both from $\mathbf{x}_\beta$ and from the free variables of $\mathbf{N}_\beta$.

III. To infer $\mathbf{A}_o$ from the result of replacing any wf part $((\lambda\mathbf{x}_\beta\mathbf{M}_\alpha)\mathbf{N}_\beta)$, at any one of its occurrences in $\mathbf{A}_o$, by the result of substituting $\mathbf{N}_\beta$ for $\mathbf{x}_\beta$ throughout $\mathbf{M}_\alpha$, provided that the bound variables of $\mathbf{M}_\alpha$ are distinct both from $\mathbf{x}_\beta$ and from the free variables of $\mathbf{N}_\beta$, and that $\mathbf{A}_o$ is without free variables.

[7]The use of superscripts to show the type is more usual and I have sometimes followed this in brief references to my papers about the logic of sense and denotation. But in this present paper, and in the papers referred to in notes 1 and 2, I have used type subscripts as being more convenient. Such subscripts to show the type are used not only on variables and primitive constants of the formalized language, but also when a bold letter is used as a syntactical variable to stand for a wff or for a variable or primitive constant of the formalized language.

IV. From $\Pi_{o(o\alpha)}\mathbf{F}_{o\alpha}$ to infer $\mathbf{F}_{o\alpha}\mathbf{A}_\alpha$, if $\mathbf{A}_\alpha$ is without free variables.

V. From $C_{ooo}\mathbf{A}_o\mathbf{B}_o$ as major premiss and $\mathbf{A}_o$ as minor premiss to infer $\mathbf{B}_o$.

The statement of the last two of these rules already illustrates the two conventions by which a pair of parentheses ( ) enclosing an entire wff may be omitted as an abbreviation, and other pairs of parentheses (enclosing a wf part of the entire wff) may be omitted under the convention of association to the left—and that these conventions of abbreviation may be applied not only in writing particular wffs, but also in writing expressions which, by containing bold letters, stand for any wff of a certain form.

We also need axioms sufficient for propositional calculus and laws of quantifiers. For these we may take the following:[8]

$1^{\alpha\beta}$. $(f) \mathbin{\centerdot} (x_\alpha)(y_\beta) f x_\alpha y_\beta \supset (y_\beta)(x_a) f x_\alpha y_\beta$
$2^{\alpha\beta}$. $(f) \mathbin{\centerdot} \Pi f \supset (g_{\alpha\beta})(x_\beta) f(g_{\alpha\beta}x_\beta)$
3^{α}. $(f)(g) \mathbin{\centerdot} (x_\alpha)[f x_\alpha \supset g x_\alpha] \supset \mathbin{\centerdot} \Pi f \supset \Pi g$
4^{α}. $(f) \mathbin{\centerdot} (x_\alpha)(y_\alpha) f x_\alpha y_\alpha \supset (x_\alpha) f x_\alpha x_\alpha$
5^{α}. $(p) \mathbin{\centerdot} p \supset (x_\alpha) p$
6^{α}. $(f)(x_\alpha) \mathbin{\centerdot} \Pi f \supset f x_\alpha$
7. $(p)(q)(r)(s) \mathbin{\centerdot} p \supset q \supset r \supset \mathbin{\centerdot} r \supset p \supset \mathbin{\centerdot} s \supset p$
$7\frac{1}{2}$. $(p)(q) \mathbin{\centerdot} {\sim}p \supset {\sim}q \supset \mathbin{\centerdot} q \supset p$

To these we add the following *axioms of extensionality*:

8. $(p)(q) \mathbin{\centerdot} p \supset q \supset \mathbin{\centerdot} q \supset p \supset \mathbin{\centerdot} p = q$
$9^{\alpha\beta}$. $(f_{\alpha\beta})(g_{\alpha\beta}) \mathbin{\centerdot} (x_\beta)[f_{\alpha\beta}x_\beta = g_{\alpha\beta}x_\beta] \supset \mathbin{\centerdot} f_{\alpha\beta} = g_{\alpha\beta}$

Then also there are the following *axioms of choice*:

$9\frac{1}{2}^{\beta}$. $(f)(x_\beta) \mathbin{\centerdot} f x_\beta \supset f(\iota f)$

These have as a consequence the following principles about descriptions:

10^{β}. $(f)(x_\beta) \mathbin{\centerdot} f x_\beta \supset \mathbin{\centerdot} (y_\beta)[f y_\beta \supset \mathbin{\centerdot} y_\beta = x_\beta] \supset f(\iota f)$

---

[8] At this point various familiar notations are introduced by definition, i.e., as conventions of abbreviation, as follows:

$\sim\mathbf{A}_o$ for $N_{oo}\mathbf{A}_o$, $\sim\mathbf{A}_{o_n}$ for $N_{o_n o_n}\mathbf{A}_{o_n}$, $[\mathbf{A}_o \supset \mathbf{B}_o]$ for $C_{ooo}\mathbf{A}_o\mathbf{B}_o$,
$[\mathbf{A}_{o_n} \supset \mathbf{B}_{o_n}]$ for $C_{o_n o_n o_n}\mathbf{A}_{o_n}\mathbf{B}_{o_n}$, $[\mathbf{A}_\alpha = \mathbf{B}_\alpha]$ for $\Pi_{o(o\alpha)}\lambda\mathbf{f}_{o\alpha}[\mathbf{f}_{o\alpha}\mathbf{A}_\alpha \supset \mathbf{f}_{o\alpha}\mathbf{B}_\alpha]$,
$[\mathbf{A}_{\alpha_n} =_n \mathbf{B}_{\alpha_n}]$ for $\Pi_{o_n(o_n\alpha_n)}\lambda\mathbf{f}_{o_n\alpha_n}[\mathbf{f}_{o_n\alpha_n}\mathbf{A}_{\alpha_n} \supset \mathbf{f}_{o_n\alpha_n}\mathbf{B}_{\alpha_n}]$, $(\mathbf{a}_\alpha)\mathbf{A}_o$ for $\Pi_{o(o\alpha)}\lambda\mathbf{a}_\alpha\mathbf{A}_o$,
$(\mathbf{a}_{\alpha_n})\mathbf{A}_{o_n}$ for $\Pi_{o_n(o_n\alpha_n)}\lambda\mathbf{a}_{\alpha_n}\mathbf{A}_{o_n}$, $(\exists\mathbf{a}_\alpha)\mathbf{A}_o$ for ${\sim}(\mathbf{a}_\alpha){\sim}\mathbf{A}_o$, $(\exists\mathbf{a}_{\alpha_n})\mathbf{A}_{o_n}$ for ${\sim}(\mathbf{a}_{\alpha_n}){\sim}\mathbf{A}_{o_n}$,
T_o for $(\exists p_o)p_o$, F_o for $(p_o)p_o$, T_{o_n} for $(\exists p_{o_n})p_{o_n}$, F_{o_n} for $(p_{o_n})p_{o_n}$.

Other conventions of abbreviation are as follows. A type subscript may be omitted from a variable or from a primitive constant or from a defined constant, provided that there is only one way in which the omitted subscript can be restored, subject to the condition that all occurrences of the same letter (in any one formula or in any one closely connected passage) shall have the same type subscript restored.

A pair of brackets enclosing an entire wff may be omitted as an abbreviation. Also a pair of brackets may be replaced by a bold dot replacing the left bracket [; it is then understood in restoring the pair of brackets that the corresponding right bracket falls at the end of the wff, unless there is an explicitly written pair of brackets within which the dot occurs (as happens e.g. in 10^β in the text); and in this latter case we select the innermost explicitly written pair of brackets within which the dot occurs, and we understand that the omitted right bracket falls just before the right bracket of this pair.

And 10^β may be used in place of $9\frac{1}{2}^\beta$ if it is thought better not to assume axioms of choice for intensionalities.

Then all remaining axioms involve the primitive constants $\Delta_{o\alpha_1\alpha}$ and their ascendants. Among these following (it is thought) are unproblematical.

$11^n.\quad \Delta C_{o_n o_n o_n} C_{o_{n+1} o_{n+1} o_{n+1}}$

$12^{n\alpha}.\quad \Delta \Pi_{o_n(o_n\alpha_n)} \Pi_{o_{n+1}(o_{n+1}\alpha_{n+1})}$

$13^{n\beta}.\quad \Delta \iota_{\beta_n(o_n\beta_n)} \iota_{\beta_{n+1}(o_{n+1}\beta_{n+1})}$

$14^{n\alpha}.\quad \Delta \Delta_{o_n\alpha_{n+1}\alpha_n} \Delta_{o_{n+1}\alpha_{n+2}\alpha_{n+1}}$

$15^{\alpha\beta}.\quad (f_{\alpha\beta})(f_{\alpha_1\beta_1})(x_\beta)(x_{\beta_1}) \mathbin{.} \Delta_{o(\alpha_1\beta_1)(\alpha\beta)} f_{\alpha\beta} f_{\alpha_1\beta_1} \supset \mathbin{.} \Delta_{o\beta_1\beta} x_\beta x_{\beta_1} \supset$
$\Delta_{o\alpha_1\alpha}(f_{\alpha\beta}x_\beta)(f_{\alpha_1\beta_1}x_{\beta_1})$

$16^{\alpha\beta}.\quad (f_{\alpha\beta})(f_{\alpha_1\beta_1}) \mathbin{.} (x_\beta)(x_{\beta_1})[\Delta_{o\beta_1\beta} x_\beta x_{\beta_1} \supset \Delta_{o\alpha_1\alpha}(f_{\alpha\beta}x_\beta)(f_{\alpha_1\beta_1}x_{\beta_1})] \supset$
$\Delta_{o(\alpha_1\beta_1)(\alpha\beta)} f_{\alpha\beta} f_{\alpha_1\beta_1}$

On the other hand the following axioms, which might well be thought of as required, will be found to lead to difficulties:

$17^\alpha.\quad (x_\alpha)(y_\alpha)(x_{\alpha_1}) \mathbin{.} \Delta_{o\alpha_1\alpha} x_\alpha x_{\alpha_1} \supset \mathbin{.} \Delta_{o\alpha_1\alpha} y_\alpha x_{\alpha_1} \supset \mathbin{.} x_\alpha = y_\alpha$

To the axioms 11^n–17^α, because we have taken N_{oo} and $N_{o_n o_n}$ as primitive rather than as introduced by definition, we must add the axioms:

$18^n.\quad \Delta N_{o_n o_n} N_{o_{n+1} o_{n+1}}$

Moreover it now seems desirable (notwithstanding the discussion on page 22 of [4]) [the next to last paragraph of *A Formulation of the Logic of Sense and Denotation*] to take it as an axiom that everything of type α has at least one concept of type α_1:

$19^\alpha.\quad (x_\alpha)(\exists x_{\alpha_1})\Delta_{o\alpha_1\alpha} x_\alpha x_{\alpha_1}$

On the other hand the assumption that everything of type α_1 is a concept of something of type α seems counterintuitive. Indeed if α is a composite type, this is already precluded by the assumption that was introduced above that a concept of a function of type $\beta\gamma$ belongs to the type $\beta_1\gamma_1$.

We proceed therefore to seek a consistency proof, by means of a set-theoretic model, for the theory based on Rules I–V, Axioms $1^{\alpha\beta}$–$16^{\alpha\beta}$, 18^n, 19^α, and as many instances (or if necessary, weakened instances) of Axioms 17^α as may be found possible.

First we provide members for all of the simple types $o, o_1, o_2, \ldots,$ and $\iota, \iota_1, \iota_2, \ldots.$ These members shall be sets from standard set theory, none of them shall be functions, and no two types shall have members in common. The members of the type o shall be the empty set and the unit set of the empty set, which are to play the role of the two truth values, t and f respectively. The membership of the other simple types is left open except for considerations about their cardinality which follow, and except that a specified principal member shall be included in each type.

We shall use the notation *card* α for the cardinality of the type α.

The cardinalities of the composite types (i.e., types other than simple types) are then determined from the cardinalities of the simple types by the rule that $\text{card}(\alpha\beta) = (\text{card}\,\alpha)^{\text{card}\,\beta}$ for any types α and β. Indeed in the model the members of the composite type $\alpha\beta$ shall be all functions from type β into the type α, and from this there follows the foregoing rule about the cardinality of the type $\alpha\beta$.

If α is a simple type, it has a principal member, which we suppose determined as each type is introduced. In particular the principal member of the type o is the truth-value falsehood, and the principal member of a type α_1 must be a concept of the principal member of the type α. Then the principal members of the various composite types are determined by the recursive rule that the principal member of the type $\alpha\beta$ is the function whose value for every argument of type β is the principal member of the type α.

In treating the semantics of the language L (as we shall call it)—or more correctly, the model semantics of L—we use small Greek letters (other than $\alpha, \beta, \gamma, \delta, \iota, o$) for members of the various types, with subscripts to show the type.

If α is a simple type, we provide a many-one correspondence of the members of the type α_1, other than the principal member, with the members of the type α, every such member ξ_{α_1} of the type α_1 having exactly one corresponding member ξ_α of the type α, and every member of the type α, including the principal member, having at least one corresponding member of the type α_1 (other than the principal member). This indeed may always be done, by appeal to the axiom of choice if not otherwise, provided that the type α_1 without its principal member is of no lower cardinality than the type α. Then having provided such a many-one correspondence of the members of the type α_1 (other than the principal member) with the members of the type α, we say that ξ_{α_1} *notes* ξ_α to mean that ξ_{α_1} corresponds to ξ_α in this many-one correspondence, and we say that ξ_{α_1} *represents* a closed formula $\mathbf{X}_\alpha$ iff it is denoted by the first ascendant of $\mathbf{X}_\alpha$ and is thus the sense of $\mathbf{X}_\alpha$.

To complete the definition by recursion of the relation of noting, we must add that $\phi_{\alpha_1\beta_1}$ notes $\phi_{\alpha\beta}$ iff, whenever ξ_{β_1} notes ξ_β, $\phi_{\alpha_1\beta_1}\xi_{\beta_1}$ notes $\phi_{\alpha\beta}\xi_\beta$, and everywhere ξ_{β_1} fails to note anything, $\phi_{\alpha_1\beta_1}\xi_{\beta_1}$ is the principal member of the type α_1. Also $\phi_{\alpha_1\beta_1}$ represents a closed formula $\mathbf{F}_{\alpha\beta}$ iff it is denoted by the first ascendant of $\mathbf{F}_{\alpha\beta}$. The relations of noting and representing are intended to correspond in the model to the relations of being a concept of and being the sense of, respectively; but we must still look into the question of how far they have the properties required for this.

We already have in the case of a simple type α that every member ξ_α of the type α is noted by some member ξ_{α_1} of the type α_1. And in consequence of the recursive definition of noting, as just given, we now have that the analogous thing holds in the case of a composite type $\alpha\beta$; i.e., given any function $\phi_{\alpha\beta}$ of type $\alpha\beta$, there exists a function $\phi_{\alpha_1\beta_1}$ of type $\alpha_1\beta_1$ which notes $\phi_{\alpha\beta}$.

But if axioms $17^{\alpha\beta}$ are to hold in the model, it must be that, for every different $\phi_{\alpha\beta}$, the corresponding function $\phi_{\alpha_1\beta_1}$ which notes it is chosen as different. This is equivalent to the condition that, for every different $\phi_{\alpha\beta}$ there must be at least one ξ_β such that the concept $\phi_{\alpha_1\beta_1}\xi_{\beta_1}$ of $\phi_{\alpha\beta}\xi_\beta$ that is chosen from the type α_1 is different from the other choices that are made from the type α_1 in connection with the other functions $\phi_{\alpha\beta}$ (for various different arguments ξ_β). And the condition for the possibility of this is that:

(1) $\quad \operatorname{card}\alpha_1 \geq \operatorname{card}\alpha\beta.$

And this may be reduced to a more manageable form as follows. First, (1) is equivalent to

(2) $\quad \operatorname{card}\alpha_1 \geq (\operatorname{card}\alpha)^{\operatorname{card}\beta}$

And this is in turn equivalent to the conjunction of the two inequalities:

(3) $\operatorname{card}\alpha_1 \geq \operatorname{card}\alpha$.
(4) $\operatorname{card}\alpha_1 \geq 2^{\operatorname{card}\beta}$.

We have already provided [three paragraphs back in this paper] that (3) shall hold in the model in the case of all simple types α. And from this it follows that (3) holds also for composite types α. We also notice that stronger conditions analogous to (3) might be imposed on the model in a similar way. For example:

(3^1) $\operatorname{card}\alpha_1 \geq 2^{\operatorname{card}\alpha}$
(3^2) $\operatorname{card}\alpha_1 \geq 2^{2^{\operatorname{card}\alpha}}$

—and so on.

Now because we shall not be using the negation of the generalized continuum hypothesis in the treatment of the model, we may replace (4) by the weaker condition:

($4'$) $\operatorname{card}\alpha_1 > \operatorname{card}\beta$

And similarly (3^1), (3^2), and so on may be replaced by the following weaker conditions:

($3'$) $\operatorname{card}\alpha_1 > \operatorname{card}\alpha$.
($3''$) $\operatorname{card}\alpha_1 > 2^{\operatorname{card}\alpha}$.

—and so on. It is important to notice that the sequence of conditions (3), ($3'$), ($3''$), ($3'''$), ... can be extended indefinitely into the transfinite. And for this reason, given any type α and β, there must exist one among this transfinite sequence of conditions that has ($4'$) as a consequence. Returning therefore to the third paragraph of this paper,[9] we claim a consistency proof that falls just short of providing a single set-theoretic model that might be used in place of the domain of intensional entities that Frege's theory calls for.

That axioms $15^{\alpha\beta}$ and $16^{\alpha\beta}$ hold in the model, or more correctly in each model in the transfinite sequence of models, is a consequence of [the discussion of the previous pages], especially the recursion clause [in the paragraph that begins "To complete the definition by recursion ... "].

That axioms 19^{α} hold in each model is shown as follows. In $16^{\alpha\beta}$ take $f_{\alpha\beta}$ to be $\lambda x_\beta M_\alpha$ and $f_{\alpha_1\beta_1}$ to be $\lambda x_{\beta_1} M_{\alpha_1}$, so obtaining:

$$(x_\beta)(x_{\beta_1})[\Delta_{o\beta_1\beta}x_\beta x_{\beta_1} \supset \Delta_{o\alpha_1\alpha}M_\alpha M_{\alpha_1}] \supset \Delta_{o(\alpha_1\beta_1)(\alpha\beta)}(\lambda x_\beta M_\alpha)(\lambda x_{\beta_1} M_{\alpha_1}).$$

Here we take M_α as having no free variables other than x_β, and M_{α_1} as [being the first ascendant of M_α and] having no free variables other than x_{β_1}.

Then to show that axioms 11^n, $12^{n\alpha}$, $13^{n\beta}$, $14^{n\alpha}$, 18^n all hold in each of the models we must first supply denotations for the various primitive constants that occur, and this is done as follows. First we must provide a denotation (in the model) for each primitive constant for which this has not yet been done. Namely N_{oo} denotes the function whose value for the argument *truth* (i.e., the truth-value *truth*) is *falsehood* and whose value for the argument *falsehood* is *truth*. Similarly $C_{o(oo)}$ denotes the function whose value

[9]Because (as a step in the proof by mathematical induction):

$\operatorname{card}\alpha_1\beta_1 = (\operatorname{card}\alpha_1)^{\operatorname{card}\beta_1} \geq (\operatorname{card}\alpha)^{\operatorname{card}\beta} = \operatorname{card}\alpha\beta.$

for the pair of arguments *truth* and *truth* is *truth*, whose value for the arguments *truth* and *falsehood* is *falsehood*, whose value for the arguments *falsehood* and *truth* is *truth*, and for the arguments *falsehood* and *falsehood*, *truth*. $\Pi_{o(o\alpha)}$ denotes the function whose value for any argument $\varphi_{o\alpha}$ is truth if the value of $\varphi_{o\alpha}$ for every argument ξ_α is truth, but falsehood if the value of $\varphi_{o\alpha}$ is falsehood for at least one argument ξ_α. $\iota_{\beta(o\beta)}$ denotes the function whose value for any argument $\varphi_{o\beta}$ is the unique member ξ_β of the type β such that the value of $\varphi_{o\beta}$ for the argument ξ_β is truth, provided that there is such a unique member ξ_β of the type β; and whose value for an argument $\varphi_{o\beta}$ is in all other cases the principal member of the type β. And finally $\Delta_{o\alpha_1\alpha}$ denotes the function $\Phi_{o\alpha_1\alpha}$ whose value for the two arguments ξ_α and ξ_{α_1} (i.e., $\Phi_{o\alpha_1\alpha}\xi_\alpha\xi_{\alpha_1}$) is truth if ξ_{α_1} is a concept of ξ_α but falsehood otherwise. (The statement in this paragraph makes more careful and explicit the assignment of denotations to the primitive constants N_{oo}, C_{ooo}, $\Pi_{o(o\alpha)}$, $\iota_{\alpha(o\alpha)}$, and $\Delta_{o\alpha_1\alpha}$ as these were given already [in the principles about descriptions, displays 10^β to 19^β]; notice, however, that the statement is now made explicitly in reference to the model and as a part of the model semantics, rather than in reference to a supposed true or principal interpretation of the axioms.)

Now having denotations in the model for the primitive constants N_{oo}, C_{ooo}, $\Pi_{o(o\alpha)}$, $\iota_{\alpha(o\alpha)}$, and $\Delta_{o\alpha_1\alpha}$, we use the method of the passage [from displayed principle 10^β to displayed principle 19^β] above—to supply denotations for the primitive constants $N_{o_no_n}$, $C_{o_no_no_n}$, $\Pi_{o_n(o_n\alpha_n)}$, $\iota_{\alpha_n(o_n\alpha_n)}$, and $\Delta_{o_n\alpha_{n+1}\alpha_n}$, $n = 1, 2, 3, \ldots$.

This completes the proof of the internal consistency of the system based on Rules I–V and Axioms $1^{\alpha\beta}$–19^α. And indeed therefore Russell's antinomy or Burali-Forti's[10] cannot arise within the system, unless axiomatic set theory itself betrays our trust by leading to inconsistency. But the difficulty remains that the internally consistent system may nevertheless be in conflict with certain matters of external fact, or with things that informal intuition says at least may be matters of external fact–e.g. (1) that Epimenides, a Cretan, once said that all assertions by Cretans are false, and moreover (2) that indeed all assertions by Cretans other than this one by Epimenides *are* false.

Arguments of this kind, which show no internal inconsistency in the formalized language, but which seem to obtain conclusions by logic alone about matters of external fact (e.g., in this case, the conclusion that either Epimenides never said any such thing or some Cretan at some time once told the truth), let us speak of as paradoxes rather than antinomies. Under our present formulation of the logic of sense and denotation, only a few such are known, all closely related to the Epimenides. They include the liar ("What I am now saying is false"), the paradox about Bouleus (who believes, perhaps falsely, that he is sometimes mistaken),[11] and Myhill's paradox.[12]

It may well be that the logic of sense and denotation—or more accurately the logic of concepts and the concept relation in the sense in which those terms are used in

[10]Whitehead and Russell, ***Principia Mathematica***, vol. I, p. 60, nos. (2) and (4).

[11]See Church, *Comparison of Russell's resolution of the semantical antinomies with that of Tarski*, ***The Journal of Symbolic Logic*** vol. 41 (1976), pp. 747–760, reprinted with corrections in ***Recent Essays on Truth and the Liar Paradox***, ed. Robert L. Martin, Oxford University Press, 1984—therein footnote 25.

[12]***Logique et Analyse*** vol. 1 (1958), pp. 78–83. Myhill's paradox is a clever variant of the Epimenides but has no special significance beyond that. The resolution of these paradoxes by a "Tarski-like" form of ramified type theory is briefly indicated in my paper of 1973–1974, pp. 149–151. The account there is concerned primarily with Alternative (2) and is definitely in error in (at least) suggesting that the same method is not applicable to Alternative (1).

the present paper—already has an adequate formulation in the formalized language that has been presented, comprising the appropriate formation rules, the rules of inference I–V, and Axioms $1^{\alpha\beta}$–17^{α}. This can be determined only by trial in the use of the language for this purpose, and in the end it may depend on somewhat doubtful judgments on whether given (declarative) sentences convey exactly the same item of information, or whether instead it is closely related but different such items.

In particular we may add to the rules I–III of λ-conversion two further rules according to which a wf part $\lambda x_\beta(F_{\alpha\beta}x_\beta)$ is replaceable (at any one occurrence) by $F_{\alpha\beta}$, and *vice versa*. This slightly modified version of Alternative (1) may be called Alternative (1′). It is thought that its treatment makes no special difficulty and may be left to the reader.

On the other hand the formulation of Alternative (0) may be more difficult. As a heuristic program (only) we offer the following.

Given any wff M_α of type α, which may or may not take free variables, and which in particular may or may not have x_β as a free variable, we introduce the notation $\Lambda x_\beta M_\alpha$ by definition as standing for

$$(\iota f_{\alpha\beta})(x_\beta) \centerdot f_{\alpha\beta}x_\beta = M_\alpha,$$

or to write it out more fully, as standing for

$$\iota_{\alpha\beta(o(\alpha\beta))}(\lambda f_{\alpha\beta}(\Pi_{o(o\beta)}(\lambda x_\beta \centerdot f_{\alpha\beta}x_\beta = M_\alpha))).$$

The two wffs $\lambda x_\beta M_\alpha$ and $\lambda x_\beta M_\alpha$ then are not synonyms, but they are concurrent in the sense that they have the same denotation or, if there are free variables, they have the same value for every system of values of the free variables. The abstraction operator λ has been taken as obeying Alternative (1), in the sense that the first ascendant of $\lambda x_\beta M_\alpha$ has been taken as $\lambda x_{\beta_1} M_{\alpha_1}$, where M_{α_1} is the first ascendant of M_α, and that $\lambda x_{\beta_1} M_{\alpha_1}$ therefore denotes the sense of $\lambda x_\beta M_\alpha$ if these wffs are closed. On the other hand the first ascendant of $\Lambda x_\beta M_\alpha$ is not $\Lambda x_{\beta_1} M_{\alpha_1}$ but is rather

$$\iota_{\alpha_1\beta_1(o_1(\alpha_1\beta_1))}(\lambda f_{\alpha_1\beta_1}(\Pi_{o_1(o_1\beta_1)}(\lambda x_{\beta_1} \centerdot f_{\alpha_1\beta_1}x_{\beta_1} =_1 M_{\alpha_1}))),$$

which let us abbreviate as $\Lambda_1 x_{\beta_1} M_{\alpha_1}$. Thus in Λ we have one example of an abstraction operator, introduced by definition, which obeys Alternative (0) rather than Alternative (1), and there are no doubt others. These may serve a heuristic purpose in enabling us to find an appropriate theory, to explore alternative theories, and to treat questions of consistency and independence. But the theory or theories should have in the end an axiomatic treatment.

Part II

Returning to the matter of the Epimenides and related paradoxes, we first supply a formal proof of the following theorem, which we may think of as showing the common form of the paradoxes:

69. $(f_{oo_1})(f_{o_1o_2}) \centerdot \Delta_{o(o_1o_2)(oo_1)}f_{oo_1}f_{o_1o_2} \supset \centerdot (p_{o_1})(p_{o_2})[f_{oo_1}p_{o_1} \supset \centerdot \Delta_{oo_2o_1}p_{o_1}p_{o_2} \supset$
$\centerdot\, p_{o_1} = C_{o_1o_1o_1}(f_{o_1o_2}p_{o_2})(\Delta_{o_1o_2o_1}F_{o_1}p_{o_2})] \supset (p_{o_1}){\sim}f_{oo_1}p_{o_1}.$

Using the method of proof from hypotheses, we take as our two hypotheses the two main antecedents in 69, i.e., $\Delta_{o(o_1o_2)(oo_1)} f_{oo_1} f_{o_1o_2}$ and $(p_{o_1})(p_{o_2})$ $[f_{oo_1} p_{o_1} \supset$ $\centerdot\, \Delta_{oo_2o_1}\, p_{o_1} p_{o_2} \supset \centerdot\, p_{o_1} = C_{o_1o_1o_1}\, (f_{o_1o_2} p_{o_2})\, (\Delta_{o_1o_2o_1}\, F_{o_1} p_{o_2})]$, which let us abbreviate as A and B respectively.

As a preliminary we first exclude the possibility that (on the relevant hypotheses) p_{o_1} is without truth-value. By 19^{o_1}, we may choose p_{o_2} to satisfy $\Delta_{oo_2o_1} p_{o_1} p_{o_2}$. Hence p_{o_1}, being the same as $C_{o_1o_1o_1}(f_{o_1o_2} p_{o_2})(\Delta_{o_1o_2o_1} F_{o_1} p_{o_2})$, is a concept of $C_{ooo}(f_{oo_1} p_{o_1})(\Delta_{oo_1o} F_o p_{o_1})$. Thus after eliminating the hypothesis $\Delta_{oo_2o_1} p_{o_1} p_{o_2}$ by ex. inst., we have proved $(\exists p_o)\Delta_{oo_1o} p_o p_{o_1}$.

Then still on the relevant hypotheses, and because the uniqueness of p_{o_1} is still a consequence of these hypotheses, the two remaining cases are that p_{o_1} is true and that p_{o_1} is false. That is, on the hypotheses A, B, $f_{oo_1} p_{o_1}$, and $\Delta_{oo_2o_1} p_{o_1} p_{o_2}$, we must have either $\Delta_{oo_1o} T_o p_{o_1}$ or $\Delta_{oo_1o} F_o p_{o_1}$. Again the hypothesis $\Delta_{oo_2o_1} p_{o_1} p_{o_2}$ may be eliminated by ex. inst. Thus, we have two cases:

Case 1 $\quad$ $\mathrm{A, B} \vdash (p_{o_1}) \centerdot f_{oo_1} p_{o_1} \supset \Delta_{oo_1o} T_o p_{o_1}$

Case 2 $\quad$ $\mathrm{A, B} \vdash (p_{o_1}) \centerdot f_{oo_1} p_{o_1} \supset \Delta_{oo_1o} F_o p_{o_1}$

Taking Case 1 first, we have by first dropping from B the universal quantifiers on p_{o_1} and p_{o_2}, then using the result as major premiss[13] in two successive applications of *modus ponens*:

$$\mathrm{A, B}, f_{oo_1} p_{o_1}, \Delta_{oo_2o_1} p_{o_1} p_{o_2} \vdash T_{o_1} = C_{o_1o_1o_1}(f_{o_1o_2} p_{o_2})(\Delta_{o_1o_2o_1} F_{o_1} p_{o_2})$$

From this, because identical concepts must be concepts of the same thing, we have:

$$\mathrm{A, B}, f_{oo_1} p_{o_1}, \Delta_{oo_2o_1} p_{o_1} p_{o_2} \vdash T_o = C_{ooo}(f_{oo_1} p_{o_1})(\Delta_{oo_1o} f_o p_{o_1})$$

The inference from the first display to the second one is evident enough informally, but for later reference let us write out some of the details of the formal proof.

By the Sense-Relationship Theorem, we have:

$$\mathrm{A}, \Delta_{oo_2o_1} p_{o_1} p_{o_2} \vdash \Delta_{oo_1o}(C_{ooo}(f_{oo_1} p_{o_1})(\Delta_{oo_1o} F_o p_{o_1}))(C_{o_1o_1o_1}(f_{o_1o_2} p_{o_2})(\Delta_{o_1o_2o_1} F_{o_1} p_{o_2}))$$

Then as a consequence of this and the first display above we have, by substitutivity of =:

$$\mathrm{A, B}, f_{oo_1} p_{o_1}, \Delta_{oo_2o_1} p_{o_1} p_{o_2} \vdash \Delta_{oo_1o}(C_{ooo}(f_{oo_1} p_{o_1})(\Delta_{oo_1o} F_o p_{o_1}))p_{o_1}$$

Because this is Case 1 we have also:

$$\mathrm{A, B}, f_{oo_1} p_{o_1} \vdash \Delta_{oo_1o} T_o p_{o_1}$$

From these last two displays we infer again by Axiom 17^0:

$$\mathrm{A, B}, f_{oo_1} p_{o_1}, \Delta_{oo_2o_1} p_{o_1} p_{o_2} \vdash T_o = C_{ooo}(f_{oo_1} p_{o_1})(\Delta_{oo_1o} F_o p_{o_1})$$

Now continuing the proof in Case 1 we first eliminate the hypothesis $\Delta_{oo_2o_1}\, p_{o_1}\, p_{o_2}$ by ex. inst.[14] and hence infer (since T_o is a theorem):

[13]For the premiss or premisses of an inference the writer prefers this spelling—from the Scholastic Latin *praemissa*—not only as a technical term of logic but also for everyday use. For the premises of a deed or similar legal document, and for the derivative use of *premises* for a piece of real estate, the spelling with an undoubled s is of course correct.

[14]See the paper of note 1, ***Noûs*** 1974, at bottom of page 145 [fifty-sixth paragraph].

$$\mathrm{A}, \mathrm{B}, f_{oo_1} p_{o_1} \vdash \Delta_{oo_1 o} F_o p_{o_1},$$

$$\mathrm{A}, \mathrm{B}, f_{oo_1} p_{o_1} \vdash f_{oo_1} p_{o_1} \supset \Delta_{oo_1 o} F_o p_{o_1},$$

$$\mathrm{A}, \mathrm{B} \vdash (p_{o_1}) \centerdot f_{oo_1} p_{o_1} \supset \Delta_{oo_1 o} F_o p_{o_1}.$$

Thus if we start with Case 1, we prove that Case 2 holds. Let us now reverse this, starting with Case 2 and proving that Case 1 holds.

So going back to the beginning of the proof that we made in Case 1, we see that the first display is obtained again, the same as before. But then when we use Case 2 in arguing again that identical concepts must be concepts of the same thing, we get:

$$\mathrm{A}, \mathrm{B}, f_{oo_1} p_{o_1}, \Delta_{oo_2 o_1} p_{o_1} p_{o_2} \vdash F_o = C_{ooo}(f_{oo_1} p_{o_1})(\Delta_{oo_1 o} F_o p_{o_1})$$

Again we may eliminate the hypothesis $\Delta_{oo_2 o_1} p_{o_1} p_{o_2}$ by ex. inst., and then we get in succession:

$$\mathrm{A}, \mathrm{B}, f_{oo_1} p_{o_1} \vdash F_o = \Delta_{oo_1 o} F_o p_{o_1}$$

$$\mathrm{A}, \mathrm{B}, f_{oo_1} p_{o_1} \vdash {\sim}\Delta_{oo_1 o} F_o p_{o_1}$$

$$\mathrm{A}, \mathrm{B}, f_{oo_1} p_{o_1} \vdash \Delta_{oo_1 o} T_o p_{o_1}$$

$$\mathrm{A}, \mathrm{B} \vdash (p_{o_1}) \centerdot f_{oo_1} p_{o_1} \supset \Delta_{oo_1 o} T_o p_{o_1}.$$

It follows that Case 1 and Case 2 both hold, and hence

$$\mathrm{A}, \mathrm{B} \vdash (p_{o_1}) {\sim} f_{oo_1} p_{o_1}.$$

From this last it follows by two applications of the Deduction Theorem that 69 is a theorem.

The significance of Theorem 69 is that it shows the general form of paradoxes analogous to the Epimenides. In principle we may take f_{oo_1} to be anything whatever of the indicated type. But some of the more plausible choices are that $f_{oo_1} p_{o_1}$ means that p_{o_1} is asserted by Epimenides, denied by Hegel, disbelieved by Anderson, doubted by Russell, favored by Myhill, feared by Church (each with a date attached if desired.)

For the moment we may think of the sign '=', as it appears in 69, as having the standard definition, i.e., that $\mathbf{A}_\alpha = \mathbf{B}_\alpha$ stands for $(f_{o\alpha}) \centerdot f_{o\alpha}\mathbf{A}_\alpha \equiv f_{o\alpha}\mathbf{B}_\alpha$.

The *level* of a type symbol is the greatest natural number corresponding to any subscript appearing in the type symbol. For example the levels of $o(o\iota)$, $o(o_1\iota_1)$, $o(o\iota_2)$, $o_1(o_3\iota)$ are 0,1,2,3 respectively. We may also speak of the *level* of a type, taking it to be the same as the level of the corresponding type symbol.

To resolve the Epimenides and related paradoxes we now introduce a ramified type theory, according to which every variable and every constant has not only a type subscript but also a superscript which shows the level in the sense that the level of the sum of two numbers, that denoted by true superscript and the level of the type subscript. The least possible level of a variable or constant of given type is therefore the same as the level of the type and all levels greater than this are admitted.

In adopting the particular formalized language which we are in process of working out and describing, the intention is that it shall have a certain application to the real world which is at least partly determined in advance. Indeed if this were not the case, the paradoxes, as distinguished from antinomies, could not be considered a fault (cf. §6). At least tentatively we shall therefore speak in a realistic way of the intended membership of the various types and levels.

The level of a type symbol is the greatest natural number that corresponds to any subscript appearing in the type symbol. For example, the levels of $o(o\iota)$, $o(o_1\iota_1)$, $o(o\iota_2)$, $o_1(o_3\iota_1)$ are $0, 1, 2, 3$, respectively.

Every variable and every constant is to have not only a type subscript but also one of the numerals $0, 1, 2, 3, \ldots$ as superscript. If the superscript is 0, it will generally be omitted as an abbreviation. Letters k, l, and possibly others will be used for unspecified or undetermined superscripts, and such superscripts as $k+2, k+l, k+l+1$ will be used with the obvious meaning. Then the level of a variable or constant is the sum of two numbers, the level of the type subscript and the number denoted by the superscript.

To this add the restriction that the superscript on $\Delta_{o\alpha_1\alpha}$ may not be 0, but is always at least 1.

To say (in the formalized language) that y^n is a concept of x^m_α we write $\Delta^{k+1}_{o\alpha_1\alpha}x^m_\alpha y^n_{\alpha_1}$, where $m \leq n \leq k$ are natural number superscripts. Thus if $m \leq n \leq k$, $\Delta^{k+1}_{o\alpha_1\alpha}x^m_\alpha y^n_{\alpha_1}$ denotes the truth-value truth iff $y^n_{\alpha_1}$ is a concept of x^m_α, and denotes the truth-value falsehood if $y^n_{\alpha_1}$ is not a concept of x^m_α. But if either $m > n$ or $n > k$, we let $\Delta^{k+1}_{o\alpha_1\alpha}x^m_\alpha y^n_{\alpha_1}$ denote the truth-value falsehood.

We must add that the range of a variable such as x^m_α comprises all members of the type α that are of level not greater than $k+m$, where m is the level of α.

Then $\Delta^{k+1}_{o_1\alpha_2\alpha_1}$, being a concept of $\Delta^{k+1}_{o\alpha_1\alpha}$, must obey the conditions that (1) if $x^k_{\alpha_1}$ and $y^k_{\alpha_2}$ are concepts of x^k_α and $y^k_{\alpha_1}$ respectively, $\Delta^{k+1}_{o_1\alpha_2\alpha_1}x^k_{\alpha_1}y^k_{\alpha_2}$ is a concept of $\Delta^{k+1}_{o\alpha_1\alpha}x^k_\alpha y^k_{\alpha_1}$, and (2) if either $x^k_{\alpha_1}$ or $y^k_{\alpha_2}$ is not a concept of anything, then $\Delta^{k+1}_{o_1\alpha_2\alpha_1}x^k_{\alpha_1}y^k_{\alpha_2}$ is F_{o_2}, and (3) if $k < m$ or $k < n$, and unless $x^m_{\alpha_1}$ and y^n_α reduces to $x^k_{\alpha_1}$ and $y^k_{\alpha_2}$, we let $\Delta^{k+1}_{o_1\alpha_2\alpha_1}x^m_{\alpha_1}y^n_{\alpha_2}$ be F_{o_1}.

We now introduce a new primitive symbol '=', which is to be understood as a connective rather than a primitive constant, and therefore has no type or level of its own. As a new formation rule, [**A**=**B**] is wf if and only if **A** and **B** have the same type; then the type of [**A**=**B**] is o, and the level of [**A**=**B**] is the level of **A** or the level of **B**, whichever is higher, or it is the common level of **A** and **B** if **A** and **B** have the same level.

As new axioms we then introduce the following:

$70^{k\alpha}$. $(x^k_\alpha) \,\mathbf{.}\, x^k_\alpha = x^k_\alpha$

71^{jk}. $(x^k_\alpha)(y^l_\alpha)(f^l_{o\alpha}) \,\mathbf{.}\, x^k_\alpha y^l_\alpha \supset \mathbf{.}\, f^l_{o\alpha}x^k_\alpha \supset f^l_{o\alpha}y^l_\alpha, j \geq k, j \geq l$

$72^{k\alpha}$. $(x^k_\alpha)(\exists y^{k+1}_\alpha) \,\mathbf{.}\, x^k_\alpha = y^{k+1}_\alpha$

$73^{k\alpha}$. $(x^{k+1}_\alpha)(\exists y^0_\alpha) \,\mathbf{.}\, x^{k+1}_\alpha = y^0_\alpha$, α an extensional type.

To these axioms, if we are to avoid an undesirably weak theory, we must add the following Axioms of Reducibility:[15]

$74^{k\alpha}$. $(f^{k+1}_{o\alpha})(\exists f^k_{o\alpha})(x^k_\alpha) \,\mathbf{.}\, f^{k+1}_{o\alpha}x^k_\alpha = f^k_{o\alpha}x^k_\alpha$

---

[15]Here I am indebted to Mark Stephen Mrotek for pointing out that the value F_{o_1} of $\Delta^{k+1}_{o_1\alpha_2\alpha_1}x^k_{\alpha_1}y^k_{\alpha_2}$ in Case 2 is forced by conventions that were introduced already in Part I of this paper.

We need also the following axioms that are concerned with the matter of level:

$75^{jk\alpha}.\quad (f^{\imath}_{o\alpha}) \centerdot (x^{k+1}_{\alpha}) f^{\imath}_{o\alpha} x^{k+1}_{\alpha}$

$76^{jk\alpha\beta}.\quad (f^{\imath}_{\alpha\beta})(x^{k+1}_{\beta}) \centerdot (x^{k}_{\beta}) \sim [x^{k}_{\beta} = x^{k+1}_{\beta}] \supset \centerdot f^{j}_{\alpha\beta} x^{k+1}_{\beta} = (\iota x_{\alpha}) \sim \centerdot x_{\alpha} = x_{\alpha}$, where $f^{\imath}_{\alpha\beta}$ is of lower level than x^{k+1}_{β}

$77^{kl\alpha}.\quad (x^{k}_{\alpha})(y^{k}_{\alpha_1}) \centerdot \Delta^{l+1}_{o\alpha_1\alpha} x^{k}_{\alpha} y^{k}_{\alpha_1} \equiv \Delta^{k+1}_{o\alpha_1\alpha} x^{k}_{\alpha} y^{k}_{\alpha_1}$, where $l > k$

$78^{k\alpha}.\quad (x^{k}_{\alpha})(y^{k}_{\alpha})(z^{k}_{\alpha_1}) \centerdot \Delta^{k+1}_{o\alpha_1\alpha} x^{k}_{\alpha} z^{k}_{\alpha_1} \supset \centerdot \Delta^{k+1}_{o\alpha_1\alpha} y^{k}_{\alpha} z^{k}_{\alpha_1} \supset x^{k}_{\alpha} = y^{k}_{\alpha}$

$79^{kl\alpha}.\quad (x^{k}_{\alpha})(y^{k}_{\alpha_1}) \centerdot \Delta^{l}_{o\alpha_1\alpha} x^{k}_{\alpha} y^{k}_{\alpha_1} \supset \centerdot (\exists x^{k-1}_{\alpha})[x^{k}_{\alpha} = x^{k-1}_{\alpha}] \supset (\exists y^{k-1}_{\alpha_1})[y^{k}_{\alpha_1} = y^{k-1}_{\alpha_1}], l < k+1$

$80^{kmn\alpha}.\ \Delta^{k+1}_{o\alpha_1\alpha} x^{m}_{\alpha} y^{n}_{\alpha_1} \supset (\exists x^{n}_{\alpha})(x^{n}_{\alpha}) \centerdot x^{m}_{\alpha} = x^{n}_{\alpha}$ if $m > n$

We must add the following principles and remarks about the matter of level (including some repetition of what was said above):

1. In general the superscript on a variable or constant may be any of $0, 1, 2, 3, \ldots,$ but this has the exception that the superscript on $\Delta_{o\alpha_1\alpha}$ or on $\Delta_{o_n\alpha_{n+1}\alpha_n}$ may not be 0, and the exception that if the type subscript is of level 0, the level superscript must be 0.

2. When the level superscript is 0, it may be omitted as an abbreviation.

3. For an unknown or unspecified level superscript, letters such as k, l, m, n may be used.

4. The level of a wff is that of the variable or constant of the highest level that occurs in it.

5. The denotation of a wff, or the value of a wff for a given system of values of its free variables, has the same level as that of the wff itself (or lower level, since the levels are cumulative).

6. Two variables with different type subscripts or with different level superscripts are of course different variables, even if the same main letter is used.

7. In the case of alphabetic change of bound variables (by Rule I), both the type and the level of the alphabetically changed variable must of course be preserved.

8. For any system of values of the free variables occurring in $\lambda x_{\beta} \mathbf{M}_{\alpha}$, the value of $\lambda x_{\beta} \mathbf{M}_{\alpha}$ is the function $f_{\alpha\beta}$ such that $f_{\alpha\beta} x_{\beta}$ has always the same value as $\mathbf{M}_{\alpha}$.

9. In the case of λ-contraction of a wf part $((\lambda x_{\beta} \mathbf{M}_{\alpha}) \mathbf{N}_{\beta})$ by rule II, the level of $\lambda x_{\beta} \mathbf{M}_{\alpha}$ must be no lower than the level of $\mathbf{N}_{\beta}$. (Otherwise, $((\lambda x_{\beta} \mathbf{M}_{\alpha}) \mathbf{N}_{\beta})$ denotes the principal member of the type α or has this principal member as its value for every system of values of the free variables that maintains the condition on the levels.)

10. In the case of a λ-expansion by Rule III which introduces a wf part $((\lambda x_{\beta} \mathbf{M}_{\alpha}) \mathbf{N}_{\beta})$, the level of $\lambda x_{\beta} \mathbf{M}_{\alpha}$ must be no lower than the level of $\mathbf{N}_{\beta}$.

11. To Rule IV we must similarly add that the level of $\Pi_{o(o\alpha)}$ must be no lower than the level of $\mathbf{F}_{o\alpha}$.

12. To Rule V we must similarly add that the level of C_{ooo} must be no lower than the level of $\mathbf{A}_{o}$ and no lower than the level of $\mathbf{B}_{o}$.

13. If $\mathbf{G}_{o\alpha} \mathbf{x}^{k}_{\alpha}$ is false for all values of $\mathbf{x}^{k}_{\alpha}$, then $\iota_{\alpha(o\alpha)} \mathbf{G}_{o\alpha}$ is the principle member of the type α. We make it our convention that the principle member of the type α is to be at the lowest level in the type α, and consequently the principle member of the type α may be named by $(\iota \mathbf{x}^{0}_{\alpha})[\mathbf{x}^{0}_{\alpha} \neq \mathbf{x}^{0}_{\alpha}]$—from which, as already explained, the superscript

0 may be omitted as an abbreviation. Compare Axioms $70^{k\alpha}$.

14. It holds of all the axioms of Part II of this paper, including the added axioms $70^{k\alpha} - 80^{kmn\alpha}$, that if all the level superscripts are dropped, the result is either an axiom or a theorem of Part I. In consequence, the proof of the consistency of the formalized language of Part I carries over in an obvious way to show the consistency of the formalized language of Part II.

Now let us return to consider in 69 the clause

$$81. \qquad p_{o_1} = C_{o_1 o_1 o_1}(f_{o_1 o_2} p_{o_2})(\Delta_{o_1 o_2 o_1} F_{o_1} p_{o_2})$$

If we seek to supply level superscripts in this, in accordance with the ramified type theory that has just been outlined, say that the superscript on $\Delta_{o_1 o_2 o_1}$ is $k + 1$. Then the superscripts on F_{o_1} and p_{o_2} must be k or lower, and of course the superscript on p_{o_2} must be the same at both of its occurrences, as otherwise the two occurrences of p_{o_2} could not be regarded as two occurrences of the same variable. The level on the right-hand of the above equation 81 is thus at least $k + 3$; the level on the left of the equation must be the same, as otherwise the identity of the two sides of the equation could not hold; and thus the superscript on p_{o_2} must be $k + 2$ or higher.

Then in the clause

$$82. \qquad \Delta_{o o_2 o_1} p_{o_1} p_{o_2}$$

the superscripts on p_{o_1} and p_{o_2} must still be the same, i.e., $k + 2$ or higher on p_{o_1} and k or lower on p_{o_2}; and the superscript on $\Delta_{o o_2 o_1}$ must therefore be $k + 3$ or higher if 69 is not to be rendered trivially false. Therefore it follows from $80^{kmn\alpha}$, where $m \geq k + 2$ is the superscript on p_{o_1}, and $n \leq k$ is the superscript on p_{o_2}, that we have $(\exists p^n_{o_1}) \centerdot p^m_{o_1} = p^n_{o_1}$. But this is incompatible with the conclusion reached above that the level on the left side of 81 must be at least $k + 3$. Thus lines 2 and 3 of 69 become either $(p_{o_1}) {\sim} f_{o o_1} p_{o_1}$ or else T_o. In the former case 69 becomes (trivially) true and in the latter case (trivially) false.

A THEORY OF THE MEANING OF NAMES (1995)

For the name relation, i.e., the relationship between a name and what it is a name of, the standard words in ordinary English are the verb *to denote* and the noun *denotation*, going back at least to John Stuart Mill,[1] and perhaps earlier.[2] We shall follow this in the present paper, allowing the verb *to designate* and the noun *designation* as occasional alternative terminology, but certainly not *to refer to* and *reference*, which are so contrary to standard English usage (in the ordinary language) as to make them repellent.[3] For those symbols, words, or terms, if any, which have a denotation without an associated *sense* or *connotation* we shall speak of *direct* denotation.

Kazimierz Ajdukiewicz, in his unfinished and somewhat fragmentary papers,[4] "Intensional expressions" (1967a), and "Proposition as the connotation of a sentence" (1967b), clearly believes in direct denotation of primitive constants, since he holds without remark, in his first named paper, that whoever believes that Dr. Jekyll is a gentleman therefore also believes that Mr. Hyde is a gentleman.[5] He does not add that whoever believes that Dr. Jekyll = Dr. Jekyll therefore also believes that Mr. Hyde = Dr. Jekyll, but it seems clear that he must in consequence either hold this or else have some reason to the contrary that is not stated in the paper itself.[6]

Now we introduce a formalized language which is a simple type theory of ordinary kind and for which a straightforward extensional meaning is provided.

There is a type o whose members are the two truth-values, truth and falsehood, or t_o and f_o as we shall write it in the formalized language.

There is a type ι whose members are those of some infinite domain, which we shall call the domain of individuals.

For every pair of types α, β there is a type $\alpha\beta$ whose members are all and only those functions $f_{\alpha\beta}$ which for every argument x_β of type β have a value $(f_{\alpha\beta}x_\beta)$ of type α, and which have no value for arguments of type other than β.

There is an infinite alphabet of variables of each type. The type of the variable is indicated by a subscript, and the values of the variables are confined to things of this type.

Originally published in ***The Heritage of Kazimierz Ajdukiewicz***, Rodopi, 1995. Written in 1987.

[1]In his ***A System of Logic, Ratiocinative and Inductive***, Volume I, Chapter 2. This book had many editions, with many changes from edition to edition. The first edition appeared in 1843, and the last edition which had Mill's personal attention in 1884.

[2]Russell's notion of denoting, with its variations and uncertainties, is in the writer's opinion now better left to historians.

[3]In support of this consider the following anecdote (strictly fictional). In a crowded room at a party one of the guests is heard to remark, "Somebody in this room has been talking about me behind my back." Then this provokes in response the inquiry, "Who(m) are you referring to?"—The point of this anecdote is of course simply that one may refer to someone without naming him (or her).

[4]These were first published in Polish in the periodical ***Studia Logica***, vol. 20 (1967). Both papers are reprinted (with linquistic corrections made by D. Pearce) in K. Ajdukiewicz, ***The Scientific World Perspective and Other Essays, 1931–1963*** (1978).

[5]The case of Dr. Jekyll and Mr. Hyde provides an interesting alternative to the hackneyed example of Phōsphoros and Hesperos.

[6]It is necessary to remember that Ajdukiewicz died before he could finish the work that was begun in the two papers referred to and in his earlier paper, *A method of eliminating intensional sentences and sentential formulae*, ***Atti del XII Congresso Internazionale di Filosofia*** (Florence 1960), pp.17–24.

Every well-formed formula ("wff") has a type under the rules, and when a single letter is used to represent an unknown or unspecified wff, it may often be convenient to indicate the type of the wff by a subscript on the letter. Thus we may say that $(\lambda \mathbf{x}_\beta \mathbf{M}_\alpha)$ is always of type $\alpha\beta$, and $(\mathbf{F}_{\alpha\beta}\mathbf{A}_\beta)$ is always of type α.[7]

In any wff $\mathbf{W}$ an occurrence of a variable $\mathbf{x}_\beta$ is a *bound* occurrence of $\mathbf{x}_\beta$ if it is within a wf part of $\mathbf{W}$ of the form $(\lambda \mathbf{x}_\beta \mathbf{M}_\alpha)$, and otherwise it is a *free* occurrence. The bound variables of $\mathbf{W}$ are all variables that have bound occurrences in $\mathbf{W}$, and the free variables of $\mathbf{W}$ are all those that have free occurrences in $\mathbf{W}$.

The *value* of a wff for a given system of values of its free variables is determined by the two following recursive rules. If for a given system S of values of the free variables in $(\mathbf{F}_{\alpha\beta}\mathbf{A}_\beta)$ the value of $\mathbf{F}_{\alpha\beta}$ is the function $f_{\alpha\beta}$ and the value of $\mathbf{A}_\beta$ is a_β, then the value of the wff $(\mathbf{F}_{\alpha\beta}\mathbf{A}_b)$ is just $(f_{\alpha\beta}a_\beta)$, i.e., the value of the function $f_{\alpha\beta}$ for the argument a_β. If for a given system S of values of the free variables in $(\lambda \mathbf{x}_\beta \mathbf{M}_\alpha)$ and for every value a_β of $\mathbf{x}_\beta$ the value of $\mathbf{M}_\alpha$ is $(f_{\alpha\beta}a_\beta)$, then the value of $(\lambda \mathbf{x}_\beta \mathbf{M}_\alpha)$ for the given system S is the function $f_{\alpha\beta}$.

If a wff is closed (i.e., without free variables), we may speak simply of its value, defined recursively as follows: if the value of $\mathbf{F}_{\alpha\beta}$ is $f_{\alpha\beta}$ and the value of $\mathbf{A}_\beta$ is a_β, the value of $(\mathbf{F}_{\alpha\beta}\mathbf{A}_b)$ is $f_{\alpha\beta}a_\beta$; if for every value a_β of $\mathbf{x}_\beta$ the value of $\mathbf{M}_\alpha$ is $(f_{\alpha\beta}a_\beta)$, then the value of $\lambda \mathbf{x}_\beta \mathbf{M}_\alpha$ is $f_{\alpha\beta}$. (In the special case in which the closed wff is of type o, i.e., a sentence, this is of course Tarski's definition of truth, as adapted to the present formalized language.)

For the purely logical adequacy of the language it is still necessary to add the following primitive constants: $K_{ooo}, N_{oo}, \iota_{\alpha(o\alpha)}, \Pi_{o(o\alpha)}$. Here α is any arbitrary type, so that the list has infinitely many primitive constants. K_{ooo} denotes conjunction in the sense that $K_{ooo}\mathbf{p}_o\mathbf{q}_o$ has the value truth for the values truth, truth of $\mathbf{p}_o, \mathbf{q}_o$ respectively but has the value falsehood in all other cases. N_{oo} denotes negation in the sense that $N_{oo}\mathbf{p}_o$ has the value falsehood for the value truth of $\mathbf{p}_o$ and has the value truth for the value falsehood of $\mathbf{p}_o$. $\iota_{\alpha(o\alpha)}$ denotes the description function of the indicated type, in the sense that, for any value $\Phi_{o\alpha}$ of $\mathbf{f}_{o\alpha}$, $\iota_{\alpha(o\alpha)}\mathbf{f}_{o\alpha}$ has as value the unique member x_α of the type α such that $\Phi_{o\alpha}x_\alpha$ is truth, provided there is such a unique member x_α of the type α, and otherwise $\iota_{\alpha(o\alpha)}\mathbf{f}_{o\alpha}$ has as value the principal member[8] of the type α. $\Pi_{o(o\alpha)}$ denotes the universality function of the indicated type, in the sense that, for any value $\Phi_{o\alpha}$ of $\mathbf{f}_{o\alpha}$, $\Pi_{o(o\alpha)}\mathbf{f}_{o\alpha}$ has the value truth if and only if $\Phi_{o\alpha}x_\alpha$ is truth in the case of every member x_α of the type α, and otherwise $\Pi_{o(o\alpha)}\mathbf{f}_{o\alpha}$ has the value falsehood.

An additional list of primitive constants may be allowed. All primitive constants are to have direct denotation, and whatever ideas one may have about the possibility of direct denotation, the primitive constants are to be restricted accordingly. But some sufficient list of primitive constants must be allowed, sufficient at least for the purpose at hand.

Now we proceed to introduce concept surrogates in a way that imitates exactly the introduction of proposition surrogates in my paper "Intensionality and the paradox of the name relation".[9] Indeed we may just apply the very same method to names of

---

[7] Here we make use of bold letters as syntactical variables, i.e., as variables whose values (in the meta-language) are wffs of the object language. When a lower-case bold letter is used, it is meant to indicate that the values are wffs of length 1, a variable or constant standing alone.

[8] A convention is needed, unavoidably artificial, by which to designate a principal member of each type.

[9] ***Themes from Kaplan***, Oxford University Press, 1989, pp.151–165. [Please note: Footnote continues.]

arbitrary type which in the cited paper was applied to sentences, i.e., to names which are of type o. And for brevity we treat only Alternative (0), and the simpler of the two methods described for that alternative.[10]

The method is simply that the notation ( ) for application of function to argument is replaced everywhere by the ordered-pair notation, without other changes. For example, the concept surrogates of the wffs

$$\lambda \mathbf{x}_\iota \mathbf{x}_\iota, \qquad \lambda \mathbf{x}_\iota \lambda \mathbf{y}_\iota \mathbf{y}_\iota, \qquad ((K_{ooo}(\Pi_{o(oo)} N_{oo}))(\iota_{o(oo)} N_{oo}))$$

are denoted by

$$\lambda \mathbf{x}_\iota \mathbf{x}_\iota, \qquad \lambda \mathbf{x}_\iota \lambda \mathbf{y}_\iota \mathbf{y}_\iota, \qquad \langle\langle K_{ooo}, \langle \Pi_{o(oo)}, N_{oo}\rangle\rangle, \langle \iota_{o(oo)}, N_{oo}\rangle\rangle$$

respectively. Two closed wffs in the formalized language then have the same concept surrogate if and only if they differ only by alphabetic changes of bound variable and replacements of one occurrence of a primitive constant by another that is synonymous with it (e.g., "Phōsphoros" by "Hesperos," or *vice versa*, or "Dr. Jekyll" by "Mr. Hyde"). We may therefore think of the concept surrogate of a closed wff as being to some extent analogous to Frege's sense or Mill's connotation of the closed wff (considered as a name). The analogy with Mill must, however, depend on modifying his theory of names in such a way that his general names are replaced in the formalized language by names of corresponding singulary propositional functions, and his relative names are replaced by names of corresponding functions of appropriate type, as illustrated below—and among these names we take as primitive those and only those which we are willing to allow as having direct denotation.[11]

We conclude by considering the following examples, of which the first two were suggested by one of Mill's examples. In all the examples we avoid the use of indexicals and demonstratives, in order to treat first the simpler case of non-demonstrative names.

1. Sophroniscus is an Athenian sculptor.

$$((K_{ooo}(A_{o\iota}\Sigma_\iota))(S_{o\iota}\Sigma_\iota))$$

Here K_{ooo} is the truth-function *conjunction*, $A_{o\iota}$ is the propositional function *is an Athenian*, $S_{o\iota}$ is the propositional function *is a sculptor*, and Σ_ι is the individual Sophroniscus. The proposition surrogate (under our present version of Alternative (0), see footnote 10) then is:

$$\langle\langle K_{ooo}, \langle A_{o\iota}, \Sigma_\iota\rangle\rangle, \langle S_{o\iota}, \Sigma_\iota\rangle\rangle$$

And for the two constituent clauses $A_{o\iota}\Sigma_\iota$ and $S_{o\iota}\Sigma_\iota$ the propositional surrogates are $\langle A_{o\iota}, \Sigma_\iota\rangle$ and $\langle S_{o\iota}, \Sigma_\iota\rangle$, which appear as constituents in the proposition surrogate for the full sentence.

In the last display on page 159 of that paper and in the five lines that immediately follow it the letters $x_1, x_2, \cdots, x_n, c_1, c_2, \cdots, c_n, M$ should everywhere be bold. A similar correction is to be made in the last paragraph of footnote 22 on the same page. [*Editor's note*: These corrections have been made in the present volume.]

[10]See footnote 29 of the cited paper. Briefly the notation $(\mathbf{FA})$ is replaced everywhere by $\langle \mathbf{F}, \mathbf{A}\rangle$; hence $(\mathbf{FAB})$ is replaced by $\langle\langle \mathbf{F}, \mathbf{A}\rangle, \mathbf{B}\rangle$, which in turn is abbreviated as $\langle \mathbf{F}, \mathbf{A}, \mathbf{B}\rangle$; and so on. — This means of course that we define the ordered triple $\langle \mathbf{F}, \mathbf{A}, \mathbf{B}\rangle$ as being the ordered pair $\langle\langle \mathbf{F}, \mathbf{A}\rangle, \mathbf{B}\rangle$; then the ordered quadruple $\langle \mathbf{F}, \mathbf{A}, \mathbf{B}, \mathbf{C}\rangle$ as being $\langle\langle \mathbf{F}, \mathbf{A}, \mathbf{B}\rangle \mathbf{C}\rangle$; and so on.

[11]John Stuart Mill, ***A System of Logic, Ratiocinative and Inductive***, many editions, 1843–1884.

2. The father of Socrates is an Athenian sculptor.

$$((K_{ooo}(A_{o\iota}(\pi_{\iota\iota}\sigma_\iota)))(S_{o\iota}(\pi_{\iota\iota}\sigma_\iota)))$$

Here σ_ι is the individual Socrates and $\pi_{\iota\iota}$ is the function *father of*, and the other constants occurring have the same meaning as before. (Still under Alternative (0)) the propositional surrogate is:

$$\langle\langle K_{ooo}, \langle A_{o\iota}, \langle \pi_{\iota\iota}, \sigma_\iota\rangle\rangle\rangle, \langle S_{o\iota}, \langle \pi_{\iota\iota}, \sigma_\iota\rangle\rangle\rangle$$

And in this the proposition surrogates for the two constituent clauses $A_{o\iota}(\pi_{\iota\iota}\sigma_\iota)$ and $A_{o\iota}(\pi_{\iota\iota}\sigma_\iota)$ are $\langle A_{o\iota}, \langle \pi_{\iota\iota}, \sigma_\iota\rangle\rangle$ and $\langle S_{o\iota}, \langle \pi_{\iota\iota}, \sigma_\iota\rangle\rangle$ respectively, and these appear as constituents in the proposition surrogate for the full sentence. Moreover the noun phrase $\pi_{\iota\iota}\sigma_\iota$ ("the father of Socrates") has two occurrences in the sentence as written in the formalized language and hence the corresponding concept surrogate $\langle \pi_{\iota\iota}, \sigma_\iota\rangle$ appears twice in the expression for the proposition surrogate as written out above.

3. Xanthippe's father-in-law is a sculptor.

$$(S_{o\iota}(\pi_{\iota\iota}(\alpha_{\iota\iota}\xi_\iota)))$$

Here $\alpha_{\iota\iota}$ is the function *husband of* and ξ_ι is Xanthippe. The proposition surrogate is:

$$\langle S_{o\iota}, \langle \pi_{\iota\iota}, \langle \alpha_{\iota\iota}, \xi_\iota\rangle\rangle\rangle$$

4. The emperor of Rome in the year 1 A.D. is Augustus.

$$((I_{o\iota\iota}((\epsilon_{\iota\iota\iota}\rho_\iota)1_\iota))\alpha_\iota)$$

Here $I_{o\iota\iota}$ is the equality function, so that $I_{o\iota\iota}\mathbf{x}_\iota\mathbf{y}_\iota$ may be read as "$\mathbf{x}_\iota$ is the same as $\mathbf{y}_\iota$," $\epsilon_{\iota\iota\iota}$ is the function *emperor of* – in –, ρ_ι is Rome, 1_ι is the year 1 A.D., and α_ι is Augustus.

Then the proposition surrogate is:

$$\langle\langle I_{o\iota\iota}, \langle\langle \epsilon_{\iota\iota\iota}, \rho_\iota\rangle, 1_\iota\rangle\rangle, \alpha_\iota\rangle$$

5. There is no emperor of Rome in the year 50 B.C.

$$(x_\iota)(N_{oo}((I_{o\iota\iota}((\epsilon_{\iota\iota\iota}\rho_\iota)L_\iota))x_\iota))$$

Or more fully

$$(\Pi_{o(o\iota)}\lambda x_\iota(N_{oo}((I_{o\iota\iota}((\epsilon_{\iota\iota\iota}\rho_\iota)L_\iota))x_\iota)))$$

where L_ι is the year 50 B.C.

Then the proposition surrogate is:

$$\langle \Pi_{o(o\iota)}, \lambda x_\iota\langle N_{oo}, \langle\langle I_{o\iota\iota}, \langle\langle \epsilon_{\iota\iota\iota}, \rho_\iota\rangle, L_\iota\rangle\rangle, x_\iota\rangle\rangle\rangle$$

In the last two examples we have assumed, for the sake of illustration, that 1_ι for the year 1 A.D. and L_ι for the year 50 B.C. are primitive names. This may well be denied, especially in the latter case, but the samples may still serve their illustrative purpose.

In all examples the present tense is intended in its timeless sense rather than as referring to present time.

Alonzo Church
University of California
at Los Angeles

REFERENCES

Ajdukiewicz, K. (1960). A method of eliminating intensional sentences and sentential formulae. *Atti del XII Congresso Internazionale di Filosofia*. Florence, pp. 17–24.
Ajdukiewicz, K. (1967a). Intensional expressions. *Studia Logica* **20**, 63–86.
Ajdukiewicz, K. (1967b). Proposition as the connotation of a sentence. *Studia Logica* **20**, 87–98.
Ajdukiewicz, K. (1978). *The Scientific World-Perspective and Other Essays 1931–1963*. Edited by J. Giedymin. Dordrecht-Boston: Reidel.
Church, A. (1989). Intensionality and the paradox of the name relation. In: *Themes from Kaplan*. Oxford: Oxford University Press.
Mill, J. S. (1843). *A System of Logic, Ratiocinative and Inductive*, vol. I, London.

CHAPTER VII. THE LOGISTIC SYSTEM A^2 (1968)

A^2 is a formulation of *second order arithmetic*. For some general remarks about the formal axiomatic method and some simple examples the reader is referred to §§07, 55.[700]

70. The primitive basis of A^2. A^2 has as its underlying logic a formulation of F^2 (see §50). The primitive symbols of A^2 are all the primitive symbols of F_2^{2p}, together with the single binary constant S. The formation rules and the rules of inference of A^2 are the same as those of F_2^2.

The abbreviations used in the meta-language to facilitate the presentation of A^2 are those used in presentation of F^2, together with others which will be introduced as needed.

We begin by introducing inductively the following definition schemata, where **a** is any individual variable, and **b** is the next individual variable in alphabetic order after **a**:

D30$_0$. $Z_0(\mathbf{a}) \rightarrow (\mathbf{b}){\sim}S(\mathbf{b}, \mathbf{a})$

D30$_{i+1}$. $Z_{i+1}(\mathbf{a}) \rightarrow (\exists\mathbf{b}) \,.\, Z_i(\mathbf{b})S(\mathbf{b}, \mathbf{a})$[701]

The axioms of A^2 are all of the axioms of F_2^2, together with the five following postulates:

†700. $(\exists y)S(x, y)$

†701. $S(x, y) \supset . S(x, z) \supset y = z$

†702. $S(y, x) \supset . S(z, x) \supset y = z$

†703. $(\exists x)Z_0(x)$

†704. $Z_0(x) \supset . F(x) \supset . F(y) \supset_y [S(y, z) \supset_z F(z)] \supset (y)F(y)$

The semantical rules of the principal interpretation of A^2 are those of F_2^{2p} in the case where the domain of individuals is the natural numbers, together with a rule describing how $S(\mathbf{a}_1, \mathbf{a}_2)$ has values.

This work was left partly in typescript, partly in Church's hand. It was projected to be Chapter VII of the second volume of ***Introduction to Mathematical Logic***. The date is approximate. Printed by permission of the Alonzo Church estate.

[700] If desired Chapter VII may be studied immediately after Chapter V (with omission or postponement of Chapter VI). Or as a minimum prerequisite for the study of Chapter VII are suggested §§10–15, 30, 31, 33–36, 40, 50–52, and the first half of §55 (together with the *Introduction*, sections of which may be read from time to time as the questions which they treat come up in the studying of other material).

[701] We assume that the meta-language contains an infinite list of arabic numerals

$$0, 1, 2, 3, \ldots$$

If i is an arabic numeral in this list, then $i + 1$ stands for the next arabic numeral in order after i in this list. We do not intend that this definition is to reflect any more profound properties of mathematical operation of addition. Indeed, we might equally well use any other infinite list of symbols and method of designating "the next in the list", provided only that the meaning was sufficiently clear.

a. The individual variables are the variables having the natural numbers as their range.

b_0. The propositional variables are variables having the range t and f.

b_n. The n-ary functional variables are variables having as their range the n-ary propositional functions whose range is the ordered n-tuples of natural numbers.

c_0. A wff consisting of a propositional variable **a** standing alone has the value t for the value t of **a**, and the value f for the value f of **a**.

c^2. Let $S(\mathbf{a}_1, \mathbf{a}_2)$ be a wff in which $\mathbf{a}_1$ and $\mathbf{a}_2$ are individual variables, not necessarily different. If $\mathbf{a}_1$ is the same as $\mathbf{a}_2$, then the value of $S(\mathbf{a}_1, \mathbf{a}_2)$ is f for each value which is thus given simultaneously to $\mathbf{a}_1$ and $\mathbf{a}_2$. If $\mathbf{a}_1$ is different from $\mathbf{a}_2$, then for the values a_1 of $\mathbf{a}_1$ and a_2 of $\mathbf{a}_2$ the value of $S(\mathbf{a}_1, \mathbf{a}_2)$ is t if a_2 is the successor of a_1; and the value of $S(\mathbf{a}_1, \mathbf{a}_2)$ is f if a_2 is not the successor of a_1.

c_n. Let $\mathbf{f}(\mathbf{a}_1, \mathbf{a}_2, \ldots, \mathbf{a}_n)$ be a wff in which $\mathbf{f}$ is an n-ary functional variable, and $\mathbf{a}_1, \mathbf{a}_2, \ldots, \mathbf{a}_n$ are individual variables, not necessarily all different. Let $\mathbf{b}_1, \mathbf{b}_2, \ldots, \mathbf{b}_m$ be the complete list of different individual variables among $\mathbf{a}_1, \mathbf{a}_2, \ldots, \mathbf{a}_n$. Consider a system of values b of $\mathbf{f}$, and $b_1, b_2, \ldots, b_m$ of $\mathbf{b}_1, \mathbf{b}_2, \ldots, \mathbf{b}_m$; and let $a_1, a_2, \ldots, a_n$ be the values which are thus given to $\mathbf{a}_1, \mathbf{a}_2, \ldots, \mathbf{a}_n$ in order. Then the value of $\mathbf{f}(\mathbf{a}_1, \mathbf{a}_2, \ldots, \mathbf{a}_n)$ for the system values $b, b_1, b_2, \ldots, b_m$ of $\mathbf{f}, \mathbf{b}_1, \mathbf{b}_2, \ldots, \mathbf{b}_m$ (in that order) is $b(a_1, a_2, \ldots a_n)$.

d. For a given system of values of the free variables of $\sim\mathbf{A}$, the value of $\sim\mathbf{A}$ is f if the value of **A** is t; and the value of $\sim\mathbf{A}$ is t if the value of **A** is f.

e. For a given system of values of the free variables of $[\mathbf{A} \supset \mathbf{B}]$, the value of $[\mathbf{A} \supset \mathbf{B}]$ is t if either the value of **B** is t or the value of **A** is f; and the value of $[\mathbf{A} \supset \mathbf{B}]$ is f if the value of **B** is f and the value of **A** is t.

f^2. Let **a** be a variable and let **A** be any wff. For a given system of values of the free variables of $(\forall\mathbf{a})\mathbf{A}$, the value of $(\forall\mathbf{a})\mathbf{A}$ is t if the value of **A** is t for every value of **a**; and the value of $(\forall\mathbf{a})\mathbf{A}$ is f if the value of **A** is f for at least one value of **a**.

71. Some first theorems and theorem schemata of A^2.

†710. $S(y, z) \supset {\sim}Z_0(z)$.

Proof. This theorem is an instance of *330.[702]

†711. $Z_0(x) \supset . Z_0(y) \supset x = y$.

Proof. By †704 and *510_1, $\vdash \check{\mathrm{S}}^{F(y)}_{(x_1){\sim}S(x_1,y) \supset x=y} Z_0(x) \supset$

$$. F(x) \supset . F(y) \supset_y [S(y, z) \supset_z F(z)] \supset (y)F(y) \Big| .$$

I.e., $\vdash Z_0(x) \supset . (x_1){\sim}S(x_1, x) \supset x = x \supset . (x_1){\sim}S(x_1, y) \supset x = y$

$$\supset_y [S(y, z) \supset_z . (x_1){\sim}S(x_1, z) \supset x = z] \supset . (x_1){\sim}S(x_1, y) \supset_y x = y.$$

Hence by *502,

$$\vdash Z_0(x) \supset . Z_0(x) \supset x = x \supset$$

[702] For the meaning of reference to theorems and theorem schemata of F^1 and the phrase "by P" see the explanation following *511 in §51.

$. Z_0(y) \supset x = y \supset_y [S(y,z) \supset_z . Z_0(z) \supset x = z] \supset . Z_0(y) \supset_y x = y.$[703]
By †520 and P, $\vdash Z_0(x) \supset x = x$.
By †710, P, and *modus ponens*, $S(y,z) \vdash Z_0(z) \supset x = z$.[704]
Hence by the deduction theorem, generalization, and P,
$\vdash Z_0(y) \supset x = y \supset_y [S(y,z) \supset_z . Z_0(z) \supset x = z]$.
Hence by *modus ponens*, $Z_0(x) \vdash Z_0(y) \supset_y x = y$.
Then use *306, *modus ponens*, and the deduction theorem.[702]

*712. $\vdash (\exists x) Z_i(x)$ $(i = 0, 1, 2, \ldots)$.

Proof by mathematical induction in the meta-language with respect to i.
By †703, $\vdash (\exists x) Z_0(x)$.
Take as hyp. ind.,[705] $\vdash (\exists x) Z_i(x)$.
By P, $Z_i(y), S(y,x) \vdash Z_i(y) S(y,x)$.
Hence by *519, $Z_i(y), (\exists x) S(y,x) \vdash (\exists x) Z_i(y) S(y,x)$.
By †700, *502, *503, $\vdash (\exists x) S(y,x)$.[706]
Hence, $Z_i(y) \vdash (\exists x) . Z_i(y) S(y,x)$.
By hyp. ind. and *502, $\vdash (\exists y) Z_i(y)$.[706]
Hence by *519, $\vdash (\exists y)(\exists x) . Z_i(y) S(y,x)$.[707]
Then use *375.

*713. $\vdash Z_i(x) \supset . Z_i(y) \supset x = y$ $(i = 0, 1, 2, \ldots)$.

Proof by mathematical induction in the meta-language with respect to i.
By †711, $\vdash Z_0(x) \supset . Z_0(y) \supset x = y$.
Take as hyp. ind., $\vdash Z_i(x) \supset . Z_i(y) \supset x = y$.
Hence by *527, $Z_i(x_1), S(x_1, x), Z_i(z) \vdash S(z, x)$.
Hence by †701, $Z_i(x_1), S(x_1, x), Z_i(z), S(z, y) \vdash x = y$.
Hence by *518, $Z_i(x_1), S(x_1, x), (\exists z) . Z_i(z) S(z, y) \vdash x = y$.
I.e., $Z_i(x_1), S(x_1, x), Z_{i+1}(y) \vdash x = y$.[707]
Then use *518, the deduction theorem, and *502.

*714. $\vdash Z_i(x) \supset . Z_{i+1}(y) \supset S(x,y)$. $(i = 0, 1, 2, \ldots)$.

Proof. By *713, $Z_i(x), Z_i(z) \vdash z = x$.
Hence by *529, $Z_i(x), Z_i(z), S(z,y) \vdash S(x,y)$.
Then use *518 and the deduction theorem.

---

[703] The Š-substitution of $Z_0(x) \supset . Z_0(y) \supset x = y$ for $F(y)$ in †704 is trivial (see §35), because of the occurrence of the bound variable z in $Z_0(y)$ Hence we first make an alphabetic change of the bound variable z to x_1. Then after the horned substitution we replace $(x_1) \sim S(x_1, x)$, $(x_1) \sim S(x_1, y)$, $(x_1) \sim S(x_1, z)$ by $Z_0(x)$, $Z_0(y)$, $Z_0(z)$ respectively, again using alphabetic change of bound variable. Such application of *510 and *502 will be used frequently in the proofs that follow and will be stated more briefly hereafter.

[704] The tautology which is used here (and at similar places in a number of later proofs) is $\sim p \supset . p \supset q$.

[705] "Hypothesis of induction."

[706] Such obvious uses of *502, *503 will be made in the proofs that follow without specific reference. Note that the change of $(\exists x) Z_i(x)$ to $(\exists y) Z_i(y)$ may be made with $i + 1$ successive applications of *502.

[707] Such application and chains of applications of *518 and *519 will be used frequently in the proofs that follow and will be stated more briefly hereafter.

*715. $\vdash Z_i(x) \supset . Z_j(y) \supset x \neq y$, if i and j are different subscripts.

Proof. By P and $x \neq y \supset y \neq x$ it is sufficient to consider only the case where i precedes j in the alphabetic order of subscripts. The proof will be by mathematical induction in the meta-language with respect to i.

Case 1: i is 0 and j is $k+1$.
By *502, $Z_0(x) \vdash (z){\sim}S(z,x)$.
By *517 and *519, $Z_{k+1}(y) \vdash (\exists z)S(z,y)$.
By †526 and *510$_1$, $\vdash \check{S}^{F(x)}_{(z){\sim}S(z,x)} F(x) \supset . {\sim}F(y) \supset x \neq y \mid$.
I.e., $\vdash (z){\sim}S(z,x) \supset . (\exists z)S(z,y) \supset x \neq y$.
Hence by *modus ponens*, $Z_0(x), Z_{k+1}(y) \vdash x \neq y$.
Then use the deduction theorem.

Case 2: Take as hyp. ind. $\vdash Z_i(x) \supset . Z_j(y) \supset x \neq y$.
By *714, $Z_i(x_1), Z_{i+1}(x) \vdash S(x_1, x)$, and $Z_j(y_1), Z_{j+1}(y) \vdash S(y_1, y)$.
Hence by *529, $Z_i(x_1), Z_{i+1}(x), x = y \vdash S(x_1, y)$.
Hence by †702, $Z_{i+1}(x), Z_{j+1}(y), Z_i(x_1), Z_j(y_1), x = y \vdash x_1 = y_1$.
Hence by the deduction theorem, P, †525, and the hyp. ind.,
$Z_{i+1}(x), Z_{j+1}(y), Z_i(x_1), Z_j(y_1) \vdash x \neq y$.[708]
Hence by *518, the deduction theorem, and *712, $Z_{i+1}(x), Z_{j+1}(y) \vdash x \neq y$.
Then use the deduction theorem.

The two following theorems embody minor variations of the principle of mathematical induction (and some other such variations are given in the exercises which follow). Notice in particular that †717 corresponds more closely than does †704 to the form of the principle of mathematical induction which was introduced in §55 as one of the postulates (A_1).

†716. $Z_0(x) \supset_x F(x) \supset . F(y) \supset_y [S(y,z) \supset_z F(z)] \supset (y)F(y)$.

Proof. By *306, $Z_0(x) \supset_x F(x) \vdash Z_0(x) \supset F(x)$.
Hence by †704 and *modus ponens*,
$Z_0(x) \supset_x F(x), Z_0(x) \vdash F(y) \supset_y [S(y,z) \supset_z F(z)] \supset (y)F(y)$.
Then use *518, the deduction theorem, and †703.

*717. $Z_0(x) \supset . F(x) \supset . F(y) \supset_y (\exists z)[S(y,z)F(z)] \supset (y)F(y)$.

Proof. By P, $S(y,z)F(z) \vdash S(y,z)$.
Hence by †701, $S(y,z)F(z), S(y,z_1) \vdash z = z_1$.
By *P*, $S(y,z)F(z) \vdash F(z)$.
Hence by *529, $S(y,z)F(z), S(y,z_1) \vdash F(z_1)$.
Hence by *518, $(\exists z) . S(y,z)F(z), S(y,z_1) \vdash F(z_1)$.
Hence by the deduction theorem, *503, and generalization,
$(\exists z) . S(y,z)F(z) \vdash S(y,z) \supset_z F(z)$.
Hence by *306 and *modus ponens*,
$F(y) \supset_y (\exists z) . S(y,z)F(z), F(y) \vdash S(y,z) \supset_z F(z)$.
Hence by the deduction theorem and generalization,
$F(y) \supset_y (\exists z) . S(y,z)F(z) \vdash F(y) \supset_y . S(y,z) \supset_z F(z)$.

[708] The tautology used here is $p \supset q \supset . {\sim}q \supset {\sim}p$.

Hence by †704, $Z_0(x), F(x), F(y) \supset_y (\exists z) \mathbin{.} S(y,z)F(z) \vdash (y)F(y)$.
Then use the deduction theorem.

72. Primitive Recursion. A function whose range of arguments is the n-tuples of natural numbers and whose range of values is the natural numbers, is called a *numerical function of n arguments.* Numerical functions, or variables having them as range, do not occur among the primitive symbols of A^2. However, there corresponds to a given numerical function f of n arguments an $n+1$-ary relation (propositional function) described in ordinary mathematical notation by the equation

$$c = f(a_1, a_2, \ldots, a_n).^{709}$$

We shall show that under certain conditions there is an effective procedure whereby we can assign to f a wff $\mathbf{U}_{\mathbf{a}_1\mathbf{a}_2\ldots\mathbf{a}_n\mathbf{c}}$ of A^2 which has no other free variables than the distinct individual variables $\mathbf{a}_1, \mathbf{a}_2, \ldots, \mathbf{a}_n, \mathbf{c}$ (though not necessarily having occurrences of all $\mathbf{a}_1, \mathbf{a}_2, \ldots, \mathbf{a}_n, \mathbf{c}$), and which expresses the relation $c = f(a_1, a_2, \ldots, a_n)$ in a suitable sense.[710] Since we shall frequently discuss wffs of this type, it is necessary at this point to adopt a convention with respect to the syntactical variables having such wffs as their range.

Let $\mathbf{A}_{\mathbf{a}_1\mathbf{a}_2\ldots\mathbf{a}_n}$ be a given wff of A^2 having no other free variables than the distinct individual variables $\mathbf{a}_1, \mathbf{a}_2, \ldots, \mathbf{a}_n$ (though not necessarily having occurrences of all of $\mathbf{a}_1, \mathbf{a}_2, \ldots, \mathbf{a}_n$). Let $\mathbf{b}_1, \mathbf{b}_2, \ldots, \mathbf{b}_n$ be distinct individual variables. We distinguish two cases:

Case 1: $\mathbf{b}_i$ is $\mathbf{a}_i$ $(i = 1, 2, \ldots, n)$. Then $\mathbf{A}_{\mathbf{b}_1\mathbf{b}_2\ldots\mathbf{b}_n}$ is $\mathbf{A}_{\mathbf{a}_1\mathbf{a}_2\ldots\mathbf{a}_n}$.

Case 2: For some i, $\mathbf{b}_i$ is not $\mathbf{a}_i$. Then let $\mathbf{c}_1, \mathbf{c}_2, \ldots, \mathbf{c}_m$ be the distinct individual variables having bound occurrences in $\mathbf{A}_{\mathbf{a}_1\mathbf{a}_2\ldots\mathbf{a}_n}$, in the order of their first occurrences. Let $\mathbf{d}_1, \mathbf{d}_2, \ldots, \mathbf{d}_m$ be the first m individual variables in alphabetic order after $\mathbf{b}_1, \mathbf{b}_2, \ldots, \mathbf{b}_n$. Then $\mathbf{A}_{\mathbf{b}_1\mathbf{b}_2\ldots\mathbf{b}_n}$ shall stand for the result of simultaneously substituting $\mathbf{b}_1, \mathbf{b}_2, \ldots, \mathbf{b}_n$ for the free occurrences of $\mathbf{a}_1, \mathbf{a}_2, \ldots, \mathbf{a}_n$ respectively, and $\mathbf{d}_1, \mathbf{d}_2, \ldots, \mathbf{d}_m$ for the bound occurrences of $\mathbf{c}_1, \mathbf{c}_2, \ldots, \mathbf{c}_m$ respectively.

For example, the reader may verify that if $\mathbf{A}_\mathbf{a}$ is $Z_n(\mathbf{a})$, then $\mathbf{A}_\mathbf{b}$ is $Z_n(\mathbf{b})$.

One basic method of defining a numerical function of one argument is recursively by a pair of equations:

$$f(0) = k,$$
$$f(b+1) = \psi(b, f(b)),$$

---

[709] Strictly speaking, since these equations are actually schemata for all equations of the same forms, $a_1, a_2, \ldots a_n, b, c$ should be replaced by variables in the meta-meta-language having the numerical variables of the meta-language as range. However, the formalization of meta-language is too great complication to introduce at this point; we shall instead adopt the usage of conventional mathematics texts and, whenever it is not clear from the context, explain in writing which letters denote constants and which denote variables.

[710] We use this syntactical notation rather than $\mathbf{U}(\mathbf{a}_1, \mathbf{a}_2, \ldots, \mathbf{a}_n, \mathbf{c})$ in order to distinguish $\mathbf{U}_{\mathbf{a}_1\mathbf{a}_2\ldots\mathbf{a}_n\mathbf{c}}$ from syntactical variables whose range is limited to wffs of the form $\mathbf{f}(\mathbf{a}_1, \mathbf{a}_2, \ldots, \mathbf{a}_n, \mathbf{c})$, where $\mathbf{f}$ is an $n+1$-ary functional variable. However, we do use parentheses in cases like $Z_n(\mathbf{a})$ and $\Upsilon(\mathbf{g}, \mathbf{h}, \mathbf{a}, \mathbf{b})$, the *definientia* of which are given much more specifically.

where k is a fixed natural number, b is a numerical variable,[709] and ψ is a previously given numerical function of two arguments. Such a pair of equations is called a *definition by primitive recursion of the numerical function f of one argument.*[711]

More generally, a pair of equations:

$$f(a_1, a_2, \ldots, a_n, 0) = \phi(a_1, a_2, \ldots, a_n),$$
$$f(a_1, a_2, \ldots, a_n, b+1) = \psi\big(a_1, a_2, \ldots, a_n, b, f(a_1, a_2, \ldots, a_n, b)\big),$$

where $a_1, a_2, \ldots, a_n, b$ are numerical variables[709] and ϕ and ψ are previously given numerical functions of n and $n+2$ arguments respectively, is called a *definition by primitive recursion of the numerical function f of $n+1$ arguments.*[711] $\mathbf{a}_1, \mathbf{a}_2, \ldots, \mathbf{a}_n$ may be thought of as parameters which, for each particular set of values, determine a particular definition by primitive recursion of a numerical function of one argument.

Given a particular numerical function ϕ (or, in case $n = 0$, a particular natural number k) and a particular numerical function ψ in a definition by primitive recursion, suppose we have already assigned to ϕ (or k) and ψ the wffs $\mathbf{Q}_{\mathbf{a}_1\mathbf{a}_2\ldots\mathbf{a}_n\mathbf{c}}$ and $\mathbf{R}_{\mathbf{a}_1\mathbf{a}_2\ldots\mathbf{a}_n\mathbf{b}\mathbf{c}\mathbf{c}_1}$ respectively, of A^2, where $\mathbf{Q}_{\mathbf{a}_1\mathbf{a}_2\ldots\mathbf{a}_n\mathbf{c}}$ has no other free variables than the distinct individual variables $\mathbf{a}_1, \mathbf{a}_2, \ldots, \mathbf{a}_n, \mathbf{c}$ and $\mathbf{R}_{\mathbf{a}_1\mathbf{a}_2\ldots\mathbf{a}_n\mathbf{b}\mathbf{c}\mathbf{c}_1}$ has no other free variables than the distinct individual variables $\mathbf{a}_1, \mathbf{a}_2, \ldots, \mathbf{a}_n, \mathbf{b}, \mathbf{c}, \mathbf{c}_1$. Then let

$\mathbf{U}_{\mathbf{a}_1\mathbf{a}_2\ldots\mathbf{a}_n\mathbf{b}\mathbf{c}}$ be $[Z_0(\mathbf{b}) \supset_{\mathbf{b}} \centerdot\, \mathbf{Q}_{\mathbf{a}_1\mathbf{a}_2\ldots\mathbf{a}_n\mathbf{c}} \supset_{\mathbf{c}} F(\mathbf{b}, \mathbf{c})]$
$[F(\mathbf{b}, \mathbf{c}) \supset_{\mathbf{bc}} \centerdot\, S(\mathbf{b}, \mathbf{b}_1) \supset_{\mathbf{b}_1} \centerdot\, \mathbf{R}_{\mathbf{a}_1\mathbf{a}_2\ldots\mathbf{a}_n\mathbf{b}\mathbf{c}\mathbf{c}_1}$
$\supset_{\mathbf{c}_1} F(\mathbf{b}_1, \mathbf{c}_1)] \supset_F F(\mathbf{b}, \mathbf{c})$,

$\mathbf{U}_1$ be $Z_0(\mathbf{b}) \supset \centerdot\, \mathbf{Q}_{\mathbf{a}_1\mathbf{a}_2\ldots\mathbf{a}_n\mathbf{c}} \supset \mathbf{U}_{\mathbf{a}_1\mathbf{a}_2\ldots\mathbf{a}_n\mathbf{b}\mathbf{c}}$,
$\mathbf{U}_2$ be $\mathbf{U}_{\mathbf{a}_1\mathbf{a}_2\ldots\mathbf{a}_n\mathbf{b}\mathbf{c}} \supset \centerdot\, S(\mathbf{b}, \mathbf{b}_1) \supset \centerdot\, \mathbf{R}_{\mathbf{a}_1\mathbf{a}_2\ldots\mathbf{a}_n\mathbf{b}\mathbf{c}\mathbf{c}_1} \supset \mathbf{U}_{\mathbf{a}_1\mathbf{a}_2\ldots\mathbf{a}_n\mathbf{b}_1\mathbf{c}_1}$,
$\mathbf{Q}_3$ be $(\exists\mathbf{c})\mathbf{Q}_{\mathbf{a}_1\mathbf{a}_2\ldots\mathbf{a}_n\mathbf{c}}$
$\mathbf{R}_3$ be $(\exists\mathbf{c}_1)\mathbf{R}_{\mathbf{a}_1\mathbf{a}_2\ldots\mathbf{a}_n\mathbf{b}\mathbf{c}\mathbf{c}_1}$
$\mathbf{U}_3$ be $(\exists\mathbf{c})\mathbf{U}_{\mathbf{a}_1\mathbf{a}_2\ldots\mathbf{a}_n\mathbf{b}\mathbf{c}}$,
$\mathbf{Q}_4$ be $\mathbf{Q}_{\mathbf{a}_1\mathbf{a}_2\ldots\mathbf{a}_n\mathbf{c}} \supset \centerdot\, \mathbf{Q}_{\mathbf{a}_1\mathbf{a}_2\ldots\mathbf{a}_n\mathbf{c}_1} \supset \mathbf{c} = \mathbf{c}_1$,
$\mathbf{R}_4$ be $\mathbf{R}_{\mathbf{a}_1\mathbf{a}_2\ldots\mathbf{a}_n\mathbf{b}\mathbf{c}\mathbf{c}_1} \supset \centerdot\, \mathbf{R}_{\mathbf{a}_1\mathbf{a}_2\ldots\mathbf{a}_n\mathbf{b}\mathbf{c}\mathbf{c}_2} \supset \mathbf{c}_1 = \mathbf{c}_2$, and
$\mathbf{U}_4$ be $\mathbf{U}_{\mathbf{a}_1\mathbf{a}_2\ldots\mathbf{a}_n\mathbf{b}\mathbf{c}} \supset \centerdot\, \mathbf{U}_{\mathbf{a}_1\mathbf{a}_2\ldots\mathbf{a}_n\mathbf{b}\mathbf{c}_1} \supset \mathbf{c} = \mathbf{c}_1$,

where $\mathbf{b}_1$ and $\mathbf{c}_2$ are the first and second individual variables in alphabetic order after $\mathbf{a}_1, \mathbf{a}_2, \ldots, \mathbf{a}_n, \mathbf{b}, \mathbf{c}, \mathbf{c}_1$.

We assign the wff $\mathbf{U}_{\mathbf{a}_1\mathbf{a}_2\ldots\mathbf{a}_n\mathbf{b}\mathbf{c}}$ to the numerical function f. Most of the remainder of this section will be devoted to proving that the definition by primitive recursion can be *represented* in A^2 in the sense that the following are metatheorems:

*720. $\vdash \mathbf{U}_1$.

*721. $\vdash \mathbf{U}_2$.

*722. If $\vdash \mathbf{Q}_3$ and $\vdash \mathbf{R}_3$, then $\vdash \mathbf{U}_3$.

*727. If $\vdash \mathbf{Q}_4$ and $\vdash \mathbf{R}_4$, then $\vdash \mathbf{U}_4$.

---

[711] What is here called *definition* by primitive recursion (in accord with conventional usage) would more correctly be called a *method of introduction* by primitive recursion.

The idea of definition by primitive recursion as a *general* procedure for the introduction of new functions appears first in *Über das Unendliche*, a paper by David Hilbert in ***Mathematische Annalen***, vol. 95 (1925), pp. 161–190. However, the first explicit setting down of recursion equations of the above form occurs in Kurt Gödel's paper of 1931 (***Monatshefte für Mathematik und Physik***, vol. 38, pp. 173–198).

The reader should note that these metatheorems involve only wffs of A^2 and are independent of any particular relationship between

$$\mathbf{Q}_{\mathbf{a}_1\mathbf{a}_2\ldots\mathbf{a}_n\mathbf{c}} \quad \text{or} \quad \mathbf{R}_{\mathbf{a}_1\mathbf{a}_2\ldots\mathbf{a}_n\mathbf{b}\mathbf{c}\mathbf{c}_1}$$

and any numerical functions.

However, suppose $\mathbf{Q}_{\mathbf{a}_1\mathbf{a}_2\ldots\mathbf{a}_n\mathbf{c}}$ is so chosen that its value for the values $a_1, a_2, \ldots, a_n, c$ of $\mathbf{a}_1, \mathbf{a}_2, \ldots, \mathbf{a}_n, \mathbf{c}$ is t if and only if $c = \phi(a_1, a_2, \ldots, a_n)$ (or, in case $n = 0$, if and only if $c = k$), and that $\mathbf{R}_{\mathbf{a}_1\mathbf{a}_2\ldots\mathbf{a}_n\mathbf{b}\mathbf{c}\mathbf{c}_1}$ is analogously chosen with respect to the relation $c_1 = \psi(a_1, a_2, \ldots, a_n, b, c)$. Then if $\vdash \mathbf{U}_1$, $\vdash \mathbf{U}_2$, and $\vdash \mathbf{U}_4$, the reader may easily verify by mathematical induction with respect to b that the value of $\mathbf{U}_{\mathbf{a}_1\mathbf{a}_2\ldots\mathbf{a}_n\mathbf{b}\mathbf{c}}$ for the values $a_1, a_2, \ldots, a_n, b, c$ of $\mathbf{a}_1, \mathbf{a}_2, \ldots, \mathbf{a}_n, \mathbf{b}, \mathbf{c}$ is t if and only if $c = f(a_1, a_2, \ldots, a_n, b)$. Thus if

$$\mathbf{Q}_{\mathbf{a}_1\mathbf{a}_2\ldots\mathbf{a}_n\mathbf{c}} \quad \text{and} \quad \mathbf{R}_{\mathbf{a}_1\mathbf{a}_2\ldots\mathbf{a}_n\mathbf{b}\mathbf{c}\mathbf{c}_1}$$

express respectively the relations

$$c = \phi(a_1, a_2, \ldots, a_n) \text{ (or } c = k) \quad \text{and} \quad c_1 = \psi(a_1, a_2, \ldots, a_n, b, c),$$

if $\vdash \mathbf{U}_1$, $\vdash \mathbf{U}_2$, and $\vdash \mathbf{U}_4$, then $\mathbf{U}_{\mathbf{a}_1\mathbf{a}_2\ldots\mathbf{a}_n\mathbf{b}\mathbf{c}}$ expresses the relation

$$c = f(a_1, a_2, \ldots, a_n, b).$$

In order to simplify the proofs of the metatheorems of this section we will first prove particular theorems of A^2, from which the metatheorems can be obtained more or less directly, using *502, *503, *510, and sometimes generalization and *306. These latter steps are easy and will be left to the reader.

We begin with a definition schema:

D31. $\Upsilon(\mathbf{g}, \mathbf{h}, \mathbf{b}, \mathbf{c}) \rightarrow [Z_0(y) \supset_y \centerdot\, \mathbf{g}(z) \supset_z F(y, z)]$
$[F(y, z) \supset_{yz} \centerdot\, S(y, y_1) \supset_{y_1} \centerdot\, \mathbf{h}(y, z, z_1)$
$\supset_{z_1} F(y_1, z_1)] \supset_F F(\mathbf{b}, \mathbf{c})$

where $\mathbf{g}$ and $\mathbf{h}$ are singulary and ternary functional variables respectively, and $\mathbf{b}$ and $\mathbf{c}$ are individual variables.

Note how $\mathbf{U}_{\mathbf{a}_1\mathbf{a}_2\ldots\mathbf{a}_n\mathbf{b}\mathbf{c}}$ can be obtained from $\Upsilon(\mathbf{g}, \mathbf{h}, \mathbf{b}, \mathbf{c})$ by first using *502 to replace the bound occurrences y, z, y_1, z_1 by $\mathbf{b}, \mathbf{c}, \mathbf{b}_1, \mathbf{c}_1$ respectively, and then making successively the Š-substitutions of $\mathbf{Q}_{\mathbf{a}_1\mathbf{a}_2\ldots\mathbf{a}_n\mathbf{c}}$ and $\mathbf{R}_{\mathbf{a}_1\mathbf{a}_2\ldots\mathbf{a}_n\mathbf{b}\mathbf{c}\mathbf{c}_1}$ for $\mathbf{g}(\mathbf{c})$ and $\mathbf{h}(\mathbf{b}, \mathbf{c}, \mathbf{c}_1)$ respectively.

†720. $Z_0(y) \supset \centerdot\, G(z) \supset \Upsilon(G, H, y, z)$.

Proof. Let A be the antecedent of $\Upsilon(G, H, y, z)$.
By P, $A \vdash Z_0(y) \supset_y \centerdot\, G(z) \supset_z F(y, z)$.
Hence by *306 and *modus ponens*, $Z_0(y), G(z), A \vdash F(y, z)$.
Then use the deduction theorem and generalization.

*720. $\vdash \mathbf{U}_1$.

†721. $\Upsilon(G, H, y, z) \supset \centerdot\, S(y, y_1) \supset \centerdot\, H(y, z, z_1) \supset \Upsilon(G, H, y_1, z_1)$.

Proof. Let A be the antecedent of $\Upsilon(G, H, y, z)$. A has the form UV.
By P, $A \vdash V$.
By $\ast509_2$ and *modus ponens*, $\Upsilon(G, H, y, z), A \vdash F(y, z)$.

By *306 and *modus ponens*, $S(y, y_1), H(y, z, z_1), F(y, z), V \vdash F(y_1, z_1)$.
Hence, $\Upsilon(G, H, y, z), S(y, y_1), H(y, z, z_1), A \vdash F(y_1, z_1)$.
Then use the deduction theorem and generalization.

*721. $\vdash \mathbf{U}_2$.[712]

†722. $(\exists z)G(z) \supset . (y)(z)(\exists z_1)H(y, z, z_1) \supset (y)(\exists z)\Upsilon(G, H, y, z)$.

Proof. By †720, $Z_0(x), G(z) \vdash \Upsilon(G, H, x, z)$.
Hence by *519,

$$Z_0(x), (\exists z)G(z) \vdash (\exists z)\Upsilon(G, H, x, z). \tag{1}$$

By †721, $\Upsilon(G, H, y, z), S(y, y_1), H(y, z, z_1) \vdash \Upsilon(G, H, y_1, z_1)$.
Hence by *519, $\Upsilon(G, H, y, z), S(y, y_1), (\exists z_1)H(y, z, z_1) \vdash (\exists z_1)\Upsilon(G, H, y_1, z_1)$.
Hence by *306 and *502,

$$(y)(z)(\exists z_1)H(y, z, z_1), \Upsilon(G, H, y, z), S(y, y_1) \vdash (\exists z)\Upsilon(G, H, y_1, z).$$

Hence by the deduction theorem, generalization, and *518,

$$(y)(z)(\exists z_1)H(y, z, z_1) \vdash (\exists z)\Upsilon(G, H, y, z) \supset_y . S(y, y_1) \supset_{y_1} (\exists z)\Upsilon(G, H, y_1, z). \tag{2}$$

By †704 and *510_1, $\vdash \check{S}^{F(y)}_{(\exists z)\Upsilon(G,H,y,z)} Z_0(x) \supset . F(x) \supset . F(y) \supset_y [S(y, y_1) \supset_{y_1} F(y_1)] \supset (y)F(y) \mid$.
I.e., $\vdash Z_0(x) \supset . (\exists z)\Upsilon(G, H, x, z) \supset . (\exists z)\Upsilon(G, H, y, z) \supset_y [S(y, y_1) \supset_{y_1} (\exists z)\Upsilon(G, H, y_1, z)] \supset (y)(\exists z)\Upsilon(G, H, y, z)$.
Hence by (1), (2), and *modus ponens*,

$$(\exists z)G(z), (y)(z)(\exists z_1)H(y, z, z_1), Z_0(x) \vdash (y)(\exists z)\Upsilon(G, H, y, z).$$

Then use *518, the deduction theorem, and †703.

*722. If $\vdash \mathbf{Q}_3$ and $\vdash \mathbf{R}_3$, then $\vdash \mathbf{U}_3$.

†723. $Z_0(y) \supset . \Upsilon(G, H, y, z) \supset G(z)$.

Proof. By P and generalization, $\vdash Z_0(y) \supset_y . G(z) \supset_z . Z_0(y) \supset G(z)$.
By †710 and P, $S(y, y_1) \vdash Z_0(y_1) \supset G(z_1)$.
Hence by P, generalization, and the deduction theorem,

$$\vdash Z_0(y) \supset G(z) \supset_{yz} . S(y, y_1) \supset_{y_1} . H(y, z, z_1) \supset_{z_1} . Z_0(y_1) \supset G(z_1).$$

By *509_2 and *modus ponens*,

$$\Upsilon(G, H, y, z) \vdash \check{S}^{F(y,z)}_{Z_0(y) \supset G(z)}[Z_0(y) \supset_y . G(z) \supset_z F(y, z)][F(y, z) \supset_{yz} . S(y, y_1) \supset_{y_1} . H(y, z, z_1) \supset_{z_1} F(y_1, z_1)] \supset F(y, z) \mid.$$

I.e., $\Upsilon(G, H, y, z) \vdash [Z_0(y) \supset_y . G(z) \supset_z . Z_0(y) \supset G(z)]$ $[Z_0(y) \supset G(z) \supset_{yz} . S(y, y_1) \supset_{y_1} . H(y, z, z_1) \supset_{z_1} . Z_0(y_1) \supset G(z_1)] \supset . Z_0(y) \supset G(z)$.

---

[712] In proving this metatheorem the bound variables of $\mathbf{Q}_{\mathbf{a}_1\mathbf{a}_2\ldots\mathbf{a}_n\mathbf{c}}$ and $\mathbf{R}_{\mathbf{a}_1\mathbf{a}_2\ldots\mathbf{a}_n\mathbf{b}\mathbf{c}\mathbf{c}_1}$ may have to be changed before make the Š-substitution, in order to avoid the situation analogous to that described in footnote 703. The reader should verify that it is always possible to restore these bound variables after the substitution in accordance with the conventions regarding wffs of the form $\mathbf{A}_{\mathbf{a}_1\mathbf{a}_2\ldots\mathbf{a}_n}$ described earlier in this section, and familiarize himself with the general procedure as it will be required frequently in what follows.

Hence by *modus ponens*. $Z_0(y), \Upsilon(G, H, y, z) \vdash G(z)$.
Then use the deduction theorem.

†724. $G(z) \supset_z [G(z_1) \supset_{z_1} z = z_1] \supset \mathbf{.}\ Z_0(x) \supset \mathbf{.}\ \Upsilon(G, H, x, z) \supset_z$
$\mathbf{.}\ \Upsilon(G, H, x, z_1) \supset_{z_1} z = z_1$.

Proof. By †723, $Z_0(x), \Upsilon(G, H, x, z) \vdash G(z)$, and $Z_0(x), \Upsilon(G, H, x, z_1) \vdash G(z_1)$.
Hence by *306 and *modus ponens*,

$$G(z) \supset_z \mathbf{.}\ G(z_1) \supset_{z_1} z = z_1, Z_0(x), \Upsilon(G, H, x, z), \Upsilon(G, H, x, z_1) \vdash z = z_1.$$

Then use the deduction theorem and generalization.

†725. $\sim(\exists z)\Upsilon(G, H, y, z) \supset \mathbf{.}\ S(y, y_1) \supset \sim(\exists z)\Upsilon(G, H, y_1, z)$.

Proof. By †702 and *527, $S(y_1, y_2), \Upsilon(G, H, y_1, z_1), S(y, y_2) \vdash \Upsilon(G, H, y, z_1)$.
Hence by the deduction theorem and P, $S(y_1, y_2), \Upsilon(G, H, y_1, z_1)$
$\vdash \sim\Upsilon(G, H, y, z_1) \supset \sim S(y, y_2)$.
By P, *502, and *306, $\sim(\exists z)\Upsilon(G, H, y, z) \vdash \sim\Upsilon(G, H, y, z_1)$.
Hence by *modus ponens* and the deduction theorem,

$$\sim(\exists z)\Upsilon(G, H, y, z), S(y_1, y_2) \vdash \Upsilon(G, H, y_1, z_1) \supset \sim S(y, y_2).$$

By P, $\Upsilon(G, H, y_1, z_1)\sim S(y, y_1) \vdash \Upsilon(G, H, y_1, z_1)$.
Hence by *modus ponens*, †721, and P,

$$\begin{gathered} \sim(\exists z)\Upsilon(G, H, y, z), \\ \Upsilon(G, H, y_1, z_1)\sim S(y, y_1), \\ S(y_1, y_2), H(y_1, z_1, z_2) \vdash \Upsilon(G, H, y_2, z_2)\sim S(y, y_2). \end{gathered}$$

Hence by the deduction theorem and generalization,

$$\begin{aligned} \sim(\exists z)\Upsilon(G, H, y, z) \vdash \Upsilon(G, H, y_1, z_1)\sim S(y, y_1) \supset_{y_1 z_1} \\ \mathbf{.}\ S(y_1, y_2) \supset_{y_2} \mathbf{.}\ H(y_1, z_1, z_2) \supset_{z_2} \\ \Upsilon(G, H, y_2, z_2)\sim S(y, y_2). \end{aligned} \tag{3}$$

By †720, †710, P, the deduction theorem, and generalization,

$$\vdash Z_0(y_1) \supset_{y_1} \mathbf{.}\ G(z_1) \supset_{z_1} \Upsilon(G, H, y_1, z_1)\sim S(y, y_1). \tag{4}$$

By *509$_2$ and *modus ponens*,

$$\begin{aligned} \Upsilon(G, H, y_1, z_1) \vdash \check{S}^{F(y_1, z_1)}_{\Upsilon(G,H,y_1,z_1)\sim S(y,y_1)}[Z_0(y_1) \supset_{y_1} \mathbf{.}\ G(z_1) \supset_{z_1} F(y_1, z_1)] \\ [F(y_1, z_1) \supset_{y_1 z_1} \mathbf{.}\ S(y_1, y_2) \supset_{y_2} \mathbf{.}\ H(y_1, z_1, z_2) \supset_{z_2} F(y_2, z_2)] \\ \supset F(y_1, z_1) \Big| . \end{aligned}$$

I.e., $\Upsilon(G, H, y_1, z_1) \vdash [Z_0(y_1) \supset_{y_1} \mathbf{.}\ G(z_1) \supset_{z_1} \Upsilon(G, H, y_1, z_1)\sim S(y, y_1)]$
$[\Upsilon(G, H, y_1, z_1)\sim S(y, y_1) \supset_{y_1 z_1} \mathbf{.}\ S(y_1, y_2) \supset_{y_2}$
$\mathbf{.}\ H(y_1, z_1, z_2) \supset_{z_2} \Upsilon(G, H, y_2, z_2)\sim S(y, y_2)]$
$\supset \Upsilon(G, H, y_1, z_1)\sim S(y, y_1)$.
Hence by (3), (4), *modus ponens*, and P,

$$\sim(\exists z)\Upsilon(G, H, y, z), \Upsilon(G, H, y_1, z_1) \vdash \sim S(y, y_1).$$

Hence by the deduction theorem, P, and *modus ponens*,

$$\sim(\exists z)\Upsilon(G, H, y, z), S(y, y_1) \vdash \sim\Upsilon(G, H, y_1, z_1).$$

Then use generalization, *502, P, and the deduction theorem.

†726. $H(y,z,z_1) \supset_{yzz_1} [H(y,z,z_2) \supset_{z_2} z_1 = z_2] \supset$
$\centerdot\, \Upsilon(G,H,y,z) \supset_z [\Upsilon(G,H,y,z_1) \supset_{z_1} z = z_1] \supset_y$
$\centerdot\, S(y,y_1) \supset_{y_1} \centerdot\, \Upsilon(G,H,y_1,z) \supset_z \centerdot\, \Upsilon(G,H,y_1,z_1) \supset_{z_1} z = z_1.$

Proof. By *529 and †702, $S(y,y_1), S(y_2,y_3), y_3 = y_1 \vdash y = y_2$.
By *502, *529, *306, and *modus ponens*,

$$\Upsilon(G,H,y,z) \supset_z \centerdot\, \Upsilon(G,H,y,z_1) \supset_{z_1} z = z_1,$$
$$\Upsilon(G,H,y,z), \Upsilon(G,H,y_2,z_2), y = y_2 \vdash z = z_2.$$

Hence by *527,

$$\Upsilon(G,H,y,z) \supset_z \centerdot\, \Upsilon(G,H,y,z_1) \supset_{z_1} z = z_1, \Upsilon(G,H,y,z),$$
$$S(y,y_1), \Upsilon(G,H,y_2,z_2), S(y_2,y_3), H(y_2,z_2,z_3), y_3 = y_1 \vdash H(y,z,z_3).$$

Hence by the deduction theorem, P, and †721,

$$\Upsilon(G,H,y,z) \supset_z \centerdot\, \Upsilon(G,H,y,z_1) \supset_{z_1} z = z_1, \Upsilon(G,H,y,z), S(y,y_1),$$
$$\Upsilon(G,H,y_2,z_2) \,\centerdot\, y_2 = y_1 \supset H(y,z,z_2), S(y_2,y_3), H(y_2,z_2,z_3)$$
$$\vdash \Upsilon(G,H,y_3,z_3) \,\centerdot\, y_3 = y_1 \supset H(y,z,z_3).$$

Hence by the deduction theorem and generalization,

$$\Upsilon(G,H,y,z) \supset_z \centerdot\, \Upsilon(G,H,y,z_1) \supset_{z_1} z = z_1, \Upsilon(G,H,y,z), S(y,y_1)$$
$$\vdash [\Upsilon(G,H,y_2,z_2) \,\centerdot\, y_2 = y_1 \supset H(y,z,z_2)]$$
$$\supset_{y_2z_2} \centerdot\, S(y_2,y_3) \supset_{y_3} \centerdot\, H(y_2,z_2,z_3)$$
$$\supset_{z_3} \centerdot\, \Upsilon(G,H,y_3,z_3) \,\centerdot\, y_3 = y_1 \supset H(y,z,z_3). \qquad (5)$$

By *529, $S(y,y_1), y_2 = y_1 \vdash S(y,y_2)$.
By †710 and P, $Z_0(y_2) \vdash \sim S(y,y_2)$.
Hence by P, $S(y,y_1), Z_0(y_2), y_2 = y_1 \vdash H(y,z,z_2)$.
Hence by the deduction theorem, $S(y,y_1), Z_0(y_2) \vdash y_2 = y_1 \supset H(y,z,z_2)$.
Hence by †720, P, the deduction theorem, and generalization,

$$S(y,y_1) \vdash Z_0(y_2) \supset_{y_2} \centerdot\, G(z_2) \supset_{z_2} \centerdot\, \Upsilon(G,H,y_2,z_2)$$
$$\centerdot\, y_2 = y_1 \supset H(y,z,z_2). \qquad (6)$$

By *509_2 and *modus ponens*,

$$\Upsilon(G,H,y_1,z_1) \vdash \check{\mathrm{S}}^{F(y_2,z_2)}_{\Upsilon(G,H,y_2,z_2) \,\centerdot\, y_2=y_1 \supset H(y,z,z_2)} [Z_0(y_2) \supset_{y_2}$$
$$\centerdot\, G(z_2) \supset_{z_2} F(y_2,z_2)][F(y_2,z_2) \supset_{y_2z_2} \centerdot\, S(y_2,y_3) \supset_{y_3}$$
$$\centerdot\, H(y_2,z_2,z_3) \supset_{z_3} F(y_3,z_3)] \supset F(y_1,z_1) \;\Big|\,.$$

I.e., $\Upsilon(G,H,y_1,z_1) \vdash [Z_0(y_2) \supset_{y_2} \centerdot\, G(z_2) \supset_{z_2} \centerdot\, \Upsilon(G,H,y_2,z_2)$
$\centerdot\, y_2 = y_1 \supset H(y,z,z_2)][[\Upsilon(G,H,y_2,z_2)$
$\centerdot\, y_2 = y_1 \supset H(y,z,z_2)] \supset_{y_2z_2} \centerdot\, S(y_2,y_3) \supset_{y_3}$
$\centerdot\, H(y_2,z_2,z_3) \supset_{z_3} \centerdot\, \Upsilon(G,H,y_3,z_3) \,\centerdot\, y_3 = y_1 \supset H(y,z,z_3)] \supset$
$\centerdot\, \Upsilon(G,H,y_1,z_1) \,\centerdot\, y_1 = y_1 \supset H(y,z,z_1).$

Hence by (5), (6), *modus ponens*, P, and †520,

$$\Upsilon(G,H,y,z) \supset_z \centerdot\, \Upsilon(G,H,y,z_1) \supset_{z_1} z = z_1,$$
$$\Upsilon(G,H,y,z), S(y,y_1), \Upsilon(G,H,y_1,z_1) \vdash H(y,z,z_1).$$

Hence by the deduction theorem, *503, and *modus ponens*,

$$\Upsilon(G,H,y,z) \supset_z \centerdot\, \Upsilon(G,H,y,z_1) \supset_{z_1} z = z_1,$$
$$\Upsilon(G,H,y,z), S(y,y_1), \Upsilon(G,H,y_1,z_2) \vdash H(y,z,z_2).$$

Hence by *306 and *modus ponens*,

$$H(y,z,z_1) \supset_{yzz_1} \centerdot\, H(y,z,z_2) \supset_{z_2} z_1 = z_2,$$
$$\Upsilon(G,H,y,z) \supset_z \centerdot\, \Upsilon(G,H,y,z_1) \supset_{z_1} z = z_1,$$
$$\Upsilon(G,H,y,z), S(y,y_1), \Upsilon(G,H,y_1,z_1), \Upsilon(G,H,y_1,z_2) \vdash z_1 = z_2. \tag{7}$$

By *330, $\Upsilon(G,H,y_1,z_1) \vdash (\exists z)\Upsilon(G,H,y_1,z)$.
Hence by †725 and P, $S(y,y_1), \Upsilon(G,H,y_1,z_1) \vdash (\exists z)\Upsilon(G,H,y,z)$.
Hence by (7) and *518,

$$H(y,z,z_1) \supset_{yzz_1} \centerdot\, H(y,z,z_2) \supset_{z_2} z_1 = z_2,$$
$$\Upsilon(G,H,y,z) \supset_z \centerdot\, \Upsilon(G,H,y,z_1) \supset_{z_1} z = z_1,$$
$$S(y,y_1), \Upsilon(G,H,y_1,z_1), \Upsilon(G,H,y_1,z_2) \vdash z_1 = z_2.$$

Then use deduction theorem, generalization, and *502.[713]

†727. $G(z) \supset_z [G(z_1) \supset_{z_1} z = z_1] \supset \centerdot\, H(y,z,z_1) \supset_{yzz_1} [H(y,z,z_2) \supset_{z_2} z_1 = z_2]$
$\supset \centerdot\, \Upsilon(G,H,y,z) \supset_z \centerdot\, \Upsilon(G,H,y,z_1) \supset_{z_1} z = z_1$.

Proof. By †704 and *510_1, $\vdash \check{\mathrm{S}}^{F(y)}_{\Upsilon(G,H,y,z) \supset_z \centerdot\, \Upsilon(G,H,y,z_1) \supset_{z_1} z=z_1} Z_0(x) \supset \centerdot\, F(x)$
$\supset \centerdot\, F(y) \supset_y [S(y,y_1) \supset_{y_1} F(y_1)] \supset (y)F(y) \,\Big|$.
Then use *modus ponens*, †724, †726, *518, the deduction theorem, †703, and *306.

*727. If $\vdash \mathbf{Q}_4$ and $\vdash \mathbf{R}_4$, then $\vdash \mathbf{U}_4$.
Let $\mathbf{A}_{\mathbf{a}_1\mathbf{a}_2\ldots\mathbf{a}_n}$ be a wff of A^2 having no other free variables than the distinct individual variables $\mathbf{a}_1, \mathbf{a}_2, \ldots, \mathbf{a}_n$; and let Θ be an n-ary relation (n-ary propositional function) of the individuals.
Then: $\mathbf{A}_{\mathbf{a}_1\mathbf{a}_2\ldots\mathbf{a}_n}$ is said to be *governed positively by* Θ *with respect to* $\mathbf{a}_1, \mathbf{a}_2, \ldots, \mathbf{a}_n$ if $\vdash Z_{i_1}(\mathbf{a}_1) \supset \centerdot\, Z_{i_2}(\mathbf{a}_2) \supset \centerdot \ldots Z_{i_n}(\mathbf{a}_n) \supset \mathbf{A}_{\mathbf{a}_1\mathbf{a}_2\ldots\mathbf{a}_n}$ whenever $\Theta(i_1, i_2, \ldots, i_n)$.
$\mathbf{A}_{\mathbf{a}_1\mathbf{a}_2\ldots\mathbf{a}_n}$ is said to be *governed negatively by* Θ *with respect to* $\mathbf{a}_1, \mathbf{a}_2, \ldots, \mathbf{a}_n$ if $\vdash Z_{i_1}(\mathbf{a}_1) \supset \centerdot\, Z_{i_2}(\mathbf{a}_2) \supset \centerdot \ldots Z_{i_n}(\mathbf{a}_n) \supset {\sim}\mathbf{A}_{\mathbf{a}_1\mathbf{a}_2\ldots\mathbf{a}_n}$ whenever not $\Theta(i_1, i_2, \ldots, i_n)$.
Θ is said to be *calculable in* A^2 *as represented by* $\mathbf{A}_{\mathbf{a}_1\mathbf{a}_2\ldots\mathbf{a}_n}$ *with respect to* $\mathbf{a}_1, \mathbf{a}_2, \ldots, \mathbf{a}_n$ if $\mathbf{A}_{\mathbf{a}_1\mathbf{a}_2\ldots\mathbf{a}_n}$ is governed positively and negatively by Θ with respect to $\mathbf{a}_1, \mathbf{a}_2, \ldots, \mathbf{a}_n$.

*728. If $\mathbf{a}_1, \mathbf{a}_2, \ldots, \mathbf{a}_n, \mathbf{c}, \mathbf{c}_1$ are distinct individual variables, if $\mathbf{A}_{\mathbf{a}_1\mathbf{a}_2\ldots\mathbf{a}_n\mathbf{c}}$ is a wff of A^2 having no other free variables than $\mathbf{a}_1, \mathbf{a}_2, \ldots, \mathbf{a}_n, \mathbf{c}$, if

$$\vdash \mathbf{A}_{\mathbf{a}_1\mathbf{a}_2\ldots\mathbf{a}_n\mathbf{c}} \supset \centerdot\, \mathbf{A}_{\mathbf{a}_1\mathbf{a}_2\ldots\mathbf{a}_n\mathbf{c}_1} \supset c = c_1,$$

if Θ is an n-ary numerical function, and if $\mathbf{A}_{\mathbf{a}_1\mathbf{a}_2\ldots\mathbf{a}_n\mathbf{c}}$ is positively governed by the relation $c = \Theta(a_1, a_2, \ldots, a_n)$, then the relation $c = \Theta(a_1, a_2, \ldots, a_n)$ is calculable in A^2 as represented by $\mathbf{A}_{\mathbf{a}_1\mathbf{a}_2\ldots\mathbf{a}_n\mathbf{c}}$.[714]

Proof. Suppose $\vdash \mathbf{A}_{\mathbf{a}_1\mathbf{a}_2\ldots\mathbf{a}_n\mathbf{c}} \supset \centerdot\, \mathbf{A}_{\mathbf{a}_1\mathbf{a}_2\ldots\mathbf{a}_n\mathbf{c}_1} \supset c = c_1$, and that $\mathbf{A}_{\mathbf{a}_1\mathbf{a}_2\ldots\mathbf{a}_n\mathbf{c}}$ is positively governed by the relation $c = \Theta(a_1, a_2, \ldots, a_n)$.
Let $k = \Theta(i_1, i_2, \ldots, i_n)$, and let $l \neq k$.

---

[713]A rearrangement of the final steps of the proof, by which several superfluous steps are avoided, was suggested to the author by J. R. Guard in May, 1959.

[714]The suggestion to state this metatheorem separately and thus to avoid duplication in the proofs of several other metatheorems was given to the author by Herbert A. Forrester in May, 1951.

Then $\vdash Z_{i_1}(\mathbf{a}_1) \supset . Z_{i_2}(\mathbf{a}_2) \supset . \ldots Z_{i_n}(\mathbf{a}_n) \supset . Z_k(\mathbf{c}) \supset \mathbf{A}_{\mathbf{a}_1\mathbf{a}_2\ldots\mathbf{a}_n\mathbf{c}}$,
and we must show that

$$\vdash Z_{i_1}(\mathbf{a}_1) \supset . Z_{i_2}(\mathbf{a}_2) \supset . \ldots Z_{i_n}(\mathbf{a}_n) \supset . Z_l(\mathbf{c}) \supset \sim\mathbf{A}_{\mathbf{a}_1\mathbf{a}_2\ldots\mathbf{a}_n\mathbf{c}}.$$

By *modus ponens*, $Z_{i_1}(\mathbf{a}_1), Z_{i_2}(\mathbf{a}_2), \ldots, Z_{i_n}(\mathbf{a}_n), Z_k(\mathbf{c}) \vdash \mathbf{A}_{\mathbf{a}_1\mathbf{a}_2\ldots\mathbf{a}_n\mathbf{c}}$.
Hence, since $\vdash \mathbf{A}_{\mathbf{a}_1\mathbf{a}_2\ldots\mathbf{a}_n\mathbf{c}} \supset . \mathbf{A}_{\mathbf{a}_1\mathbf{a}_2\ldots\mathbf{a}_n\mathbf{c}_1} \supset c = c_1$, therefore

$$Z_{i_1}(\mathbf{a}_1), Z_{i_2}(\mathbf{a}_2), \ldots, Z_{i_n}(\mathbf{a}_n), Z_k(\mathbf{c}) \vdash \mathbf{A}_{\mathbf{a}_1\mathbf{a}_2\ldots\mathbf{a}_n\mathbf{c}} \supset c = c_1.$$

By *715 and *modus ponens*, $Z_l(\mathbf{c}_1), Z_k(\mathbf{c}) \vdash c \neq c_1$.
Hence by P, $Z_{i_1}(\mathbf{a}_1), Z_{i_2}(\mathbf{a}_2), \ldots, Z_{i_n}(\mathbf{a}_n), Z_l(\mathbf{c}_1), Z_k(\mathbf{c}) \vdash \sim\mathbf{A}_{\mathbf{a}_1\mathbf{a}_2\ldots\mathbf{a}_n\mathbf{c}_1}$.
Then use *518, the deduction theorem, *712, and *503.

*729. If f is defined in terms of ϕ (or, in case $n = 0$, in terms of the fixed natural number k) and ψ by primitive recursion, if $\mathbf{Q}_{\mathbf{a}_1\mathbf{a}_2\ldots\mathbf{a}_n\mathbf{c}}$ and $\mathbf{R}_{\mathbf{a}_1\mathbf{a}_2\ldots\mathbf{a}_n\mathbf{b}\mathbf{c}\mathbf{c}_1}$ are governed positively by the relations $c = \phi(a_1, a_2, \ldots, a_n)$ (or $c = k$) and $c_1 = \psi(a_1, a_2, \ldots, a_n, b, c)$ respectively, and if $\vdash \mathbf{Q}_4$ and $\vdash \mathbf{R}_4$, then the relation $c = f(a_1, a_2, \ldots, a_n, b)$ is calculable in A^2 as represented by $\mathbf{U}_{\mathbf{a}_1\mathbf{a}_2\ldots\mathbf{a}_n\mathbf{b}\mathbf{c}}$.

Proof. By *727 and *728, it is sufficient to show that $\mathbf{U}_{\mathbf{a}_1\mathbf{a}_2\ldots\mathbf{a}_n\mathbf{b}\mathbf{c}}$ is governed positively by the relation $c = f(a_1, a_2, \ldots, a_n, b)$; i.e., that for any natural numbers $i_1, i_2, \ldots, i_n, j, l$ such that $l = f(i_1, i_2, \ldots, i_n, j)$ we have

$$\vdash Z_{i_1}(\mathbf{a}_1) \supset . Z_{i_2}(\mathbf{a}_2) \supset . \ldots Z_{i_n}(\mathbf{a}_n) \supset . Z_j(\mathbf{b}) \supset . Z_l(\mathbf{c}) \supset \mathbf{U}_{\mathbf{a}_1\mathbf{a}_2\ldots\mathbf{a}_n\mathbf{b}\mathbf{c}}.$$

The proof will be by mathematical induction in the meta-language with respect to j.
Case 1: j is 0. Then $l = \phi(i_1, i_2, \ldots, i_n)$ (or, in case $n = 0$, $l = k$). Since $\mathbf{Q}_{\mathbf{a}_1\mathbf{a}_2\ldots\mathbf{a}_n\mathbf{c}}$ is governed positively by the relation $c = \phi(a_1, a_2, \ldots, a_n)$ (or $c = k$), we have by *modus ponens*, $Z_{i_1}(\mathbf{a}_1), Z_{i_2}(\mathbf{a}_2), \ldots, Z_{i_n}(\mathbf{a}_n), Z_l(\mathbf{c}) \vdash \mathbf{Q}_{\mathbf{a}_1\mathbf{a}_2\ldots\mathbf{a}_n\mathbf{c}}$.
Hence by *720 and *modus ponens*,

$$Z_{i_1}(\mathbf{a}_1), Z_{i_2}(\mathbf{a}_2), \ldots, Z_{i_n}(\mathbf{a}_n), Z_0(\mathbf{b}), Z_l(\mathbf{c}) \vdash \mathbf{U}_{\mathbf{a}_1\mathbf{a}_2\ldots\mathbf{a}_n\mathbf{b}\mathbf{c}}.$$

Then use the deduction theorem.

Case 2: Take as hyp. ind. $l = f(i_1, i_2, \ldots, i_n, j)$ and that

$$Z_{i_1}(\mathbf{a}_1) \supset . Z_{i_2}(\mathbf{a}_2) \supset . \ldots Z_{i_n}(\mathbf{a}_n) \supset . Z_j(\mathbf{b}) \supset . Z_l(\mathbf{c}) \supset \mathbf{U}_{\mathbf{a}_1\mathbf{a}_2\ldots\mathbf{a}_n\mathbf{b}\mathbf{c}}.$$

Let $l_1 = f(i_1, i_2, \ldots, i_n, j + 1)$.
Then $l_1 = \psi(i_1, i_2, \ldots, i_n, j, l)$, and since $\mathbf{R}_{\mathbf{a}_1\mathbf{a}_2\ldots\mathbf{a}_n\mathbf{b}\mathbf{c}\mathbf{c}_1}$ is governed positively by the relation $c_1 = \psi(a_1, a_2, \ldots, a_n, b, c)$, we have by *modus ponens*,

$$Z_{i_1}(\mathbf{a}_1), Z_{i_2}(\mathbf{a}_2), \ldots, Z_{i_n}(\mathbf{a}_n), Z_j(\mathbf{b}), Z_l(\mathbf{c}) \vdash \mathbf{U}_{\mathbf{a}_1\mathbf{a}_2\ldots\mathbf{a}_n\mathbf{b}\mathbf{c}},$$

and

$$Z_{i_1}(\mathbf{a}_1), Z_{i_2}(\mathbf{a}_2), \ldots, Z_{i_n}(\mathbf{a}_n), Z_j(\mathbf{b}), Z_l(\mathbf{c}), Z_{l_1}(\mathbf{c}_1) \vdash \mathbf{R}_{\mathbf{a}_1\mathbf{a}_2\ldots\mathbf{a}_n\mathbf{b}\mathbf{c}\mathbf{c}_1}.$$

Hence by *721 and *modus ponens*,

$$Z_{i_1}(\mathbf{a}_1), Z_{i_2}(\mathbf{a}_2), \ldots, Z_{i_n}(\mathbf{a}_n), Z_j(\mathbf{b}), Z_l(\mathbf{c}), Z_{l_1}(\mathbf{c}_1), S(\mathbf{b}, \mathbf{b}_1) \vdash \mathbf{U}_{\mathbf{a}_1\mathbf{a}_2\ldots\mathbf{a}_n\mathbf{b}_1\mathbf{c}_1}.$$

By *714, $Z_j(\mathbf{b}), Z_{j+1}(\mathbf{b}_1) \vdash S(\mathbf{b}, \mathbf{b}_1)$.
Hence by the deduction theorem and *modus ponens*,

$$Z_{i_1}(\mathbf{a}_1), Z_{i_2}(\mathbf{a}_2), \ldots, Z_{i_n}(\mathbf{a}_n), Z_{j+1}(\mathbf{b}_1), Z_{l_1}(\mathbf{c}_1), Z_j(\mathbf{b}), Z_l(\mathbf{c}) \vdash \mathbf{U}_{\mathbf{a}_1\mathbf{a}_2\ldots\mathbf{a}_n\mathbf{b}_1\mathbf{c}_1}.$$

Then use *518, the deduction theorem, *712, and *modus ponens*.

73. Composition. Primitive recursive functions. Another basic method of defining a numerical function f of n arguments in arithmetic is by a single equation

$$f(a_1, a_2, \ldots a_n) = \psi(\phi_1(a_1, a_2, \ldots a_n), \phi_2(a_1, a_2, \ldots a_n), \ldots, \phi_m(a_1, a_2, \ldots a_n)),$$

where ψ and the ϕ_i's are previously given numerical functions of m and n arguments respectively. Such an equation is called a *definition by composition of the numerical function f of n arguments.*

Given particular numerical functions $\phi_1, \phi_2, \ldots, \phi_n$ and ψ, suppose we have already assigned to the wffs $\mathbf{Q}_{1\,\mathbf{a}_1\mathbf{a}_2\ldots\mathbf{a}_n\mathbf{b}_1}$, $\mathbf{Q}_{2\,\mathbf{a}_1\mathbf{a}_2\ldots\mathbf{a}_n\mathbf{b}_2}$, $\ldots$, $\mathbf{Q}_{m\,\mathbf{a}_1\mathbf{a}_2\ldots\mathbf{a}_n\mathbf{b}_m}$, and $\mathbf{R}_{\mathbf{b}_1\mathbf{b}_2\ldots\mathbf{b}_m\mathbf{c}}$, having no other free variables than the distinct individual variables $\mathbf{a}_1, \mathbf{a}_2, \ldots, \mathbf{a}_n, \mathbf{b}_1, \mathbf{b}_2, \ldots, \mathbf{b}_m, \mathbf{c}$, as displayed by their subscripts. Then let

$\mathbf{L}_{\mathbf{a}_1\mathbf{a}2\ldots\mathbf{a}_n\mathbf{c}}$ be $(\exists\mathbf{b}_1)(\exists\mathbf{b}_2)\ldots(\exists\mathbf{b}_m)\,\centerdot\,\mathbf{Q}_{1\,\mathbf{a}_1\mathbf{a}_2\ldots\mathbf{a}_n\mathbf{b}_1}\mathbf{Q}_{2\,\mathbf{a}_1\mathbf{a}_2\ldots\mathbf{a}_n\mathbf{b}_2}\cdots$
$\mathbf{Q}_{m\,\mathbf{a}_1\mathbf{a}_2\ldots\mathbf{a}_n\mathbf{b}_m}\mathbf{R}_{\mathbf{b}_1\mathbf{b}_2\ldots\mathbf{b}_m\mathbf{c}}$,

$\mathbf{L}_1$ be $\mathbf{Q}_{1\,\mathbf{a}_1\mathbf{a}_2\ldots\mathbf{a}_n\mathbf{b}_1} \supset\centerdot\, \mathbf{Q}_{2\,\mathbf{a}_1\mathbf{a}_2\ldots\mathbf{a}_n\mathbf{b}_2} \supset\centerdot \ldots \mathbf{Q}_{m\,\mathbf{a}_1\mathbf{a}_2\ldots\mathbf{a}_n\mathbf{b}_m} \supset\centerdot\, \mathbf{R}_{\mathbf{b}_1\mathbf{b}_2\ldots\mathbf{b}_m\mathbf{c}} \supset \mathbf{L}_{\mathbf{a}_1\mathbf{a}_2\ldots\mathbf{a}_n\mathbf{c}}$

$\mathbf{Q}_{i3}$ be $(\exists\mathbf{b}_i)\mathbf{Q}_{i\,\mathbf{a}_1\mathbf{a}_2\ldots\mathbf{a}_n\mathbf{b}_i}$ $(i = 1, 2, \ldots, m)$,

$\mathbf{R}_3$ be $(\exists\mathbf{c})\mathbf{R}_{\mathbf{b}_1\mathbf{b}_2\ldots\mathbf{b}_m\mathbf{c}}$,

$\mathbf{L}_3$ be $(\exists\mathbf{c})\mathbf{L}_{\mathbf{a}_1\mathbf{a}_2\ldots\mathbf{a}_n\mathbf{c}}$,

$\mathbf{Q}_{i4}$ be $\mathbf{Q}_{i\,\mathbf{a}_1\mathbf{a}_2\ldots\mathbf{a}_n\mathbf{b}_i} \supset\centerdot\, \mathbf{Q}_{i\,\mathbf{a}_1\mathbf{a}_2\ldots\mathbf{a}_n\mathbf{c}_i} \supset \mathbf{b}_i = \mathbf{c}_i$ $(i = 1, 2, \ldots, m)$,

$\mathbf{R}_4$ be $\mathbf{R}_{\mathbf{b}_1\mathbf{b}_2\ldots\mathbf{b}_m\mathbf{c}} \supset\centerdot\, \mathbf{R}_{\mathbf{b}_1\mathbf{b}_2\ldots\mathbf{b}_m\mathbf{c}_1} \supset \mathbf{c} = \mathbf{c}_1$, and

$\mathbf{L}_4$ be $\mathbf{L}_{\mathbf{a}_1\mathbf{a}_2\ldots\mathbf{a}_n\mathbf{c}} \supset\centerdot\, \mathbf{L}_{\mathbf{a}_1\mathbf{a}_2\ldots\mathbf{a}_n\mathbf{c}_1} \supset \mathbf{c} = \mathbf{c}_1$,

where $\mathbf{c}_1, \mathbf{c}_2, \ldots, \mathbf{c}_m$ are the first m individual variables in the alphabetic order after $\mathbf{a}_1, \mathbf{a}_2, \ldots, \mathbf{a}_n, \mathbf{b}_1, \mathbf{b}_2, \ldots, \mathbf{b}_m, \mathbf{c}$.

We assign the wff $\mathbf{L}_{\mathbf{a}_1\mathbf{a}2\ldots\mathbf{a}_n\mathbf{c}}$ to the numerical function f. Analogous to §72, we shall prove that definition by composition can be *represented* in A^2 in the sense that the following are metatheorems:

*731. $\vdash \mathbf{L}_1$.

*733. If $\vdash \mathbf{Q}_{13}, \vdash \mathbf{Q}_{23}, \ldots, \vdash \mathbf{Q}_{m3}$, and $\vdash \mathbf{R}_3$, then $\vdash \mathbf{L}_3$.

*735. If $\vdash \mathbf{Q}_{14}, \vdash \mathbf{Q}_{24}, \ldots, \vdash \mathbf{Q}_{m4}$, and $\vdash \mathbf{R}_4$, then $\vdash \mathbf{L}_4$.

As in the case of definition by primitive recursion, the reader may verify that if each $\mathbf{Q}_{i\,\mathbf{a}_1\mathbf{a}_2\ldots\mathbf{a}_n\mathbf{b}_i}$ is so chosen that it expresses the relation $b_i = \phi_i(a_1, a_2, \ldots, a_n)$ $(i = 1, 2, \ldots, m)$, and $\mathbf{R}_{\mathbf{b}_1\mathbf{b}_2\ldots\mathbf{b}_m\mathbf{c}}$ is so chosen that it expresses the relation $c = \psi(b_1, b_2, \ldots, b_m)$, then $\mathbf{L}_{\mathbf{a}_1\mathbf{a}2\ldots\mathbf{a}_n\mathbf{c}}$ expresses the relation $c = f(a_1, a_2, \ldots, a_n)$.

Paralleling the development of §72, for each $m = 1, 2, \ldots$ we introduce the definition schema:

D32$_m$. $\Lambda(\mathbf{g}_1, \mathbf{g}_2, \ldots, \mathbf{g}_m, \mathbf{h}, \mathbf{c})$
$\rightarrow (\exists\mathbf{b}_1)(\exists\mathbf{b}_2)\ldots(\exists\mathbf{b}_m)\,\centerdot\,\mathbf{g}_1(\mathbf{b}_1)\mathbf{g}_2(\mathbf{b}_2)\ldots\mathbf{g}_m(\mathbf{b}_m)\mathbf{h}(\mathbf{b}_1, \mathbf{b}_2, \ldots, \mathbf{b}_m, \mathbf{c})$
where $\mathbf{g}_1, \mathbf{g}_2, \ldots, \mathbf{g}_m$ are singular functional variables, $\mathbf{h}$ is an $m+1$-ary functional variable, $\mathbf{c}$ is an individual variable, and $\mathbf{b}_1, \mathbf{b}_2, \ldots, \mathbf{b}_m$ are the first m individual variables in alphabetic order after $\mathbf{c}$.[715]

As in §72, the proofs of *731, *733, and *735 as corollaries of *730, *732, and *734 respectively, will be left to the reader.

*730. $\vdash G_1(y_1) \supset\centerdot\, G_2(y_2) \supset\centerdot \ldots G_m(y_m) \supset\centerdot$

[715]The case $m = 0$ corresponds to a function ψ having no argument whatever, and is not necessary for the development of A^2. Nonetheless, if we take $\Lambda(\mathbf{h}, \mathbf{c})$ as standing for $\mathbf{h}(\mathbf{c})$, then *730, *732, *734 remain trivially true.

$$H(y_1, y_2, \ldots, y_m, z) \supset \Lambda(G_1, G_2, \ldots, G_m, H, z) \qquad (m = 1, 2, \ldots).$$

Proof. By P, $G_1(y_1), G_2(y_2), \ldots, G_m(y_m), H(y_1, y_2, \ldots, y_m, z)$
$\vdash G_1(y_1)G_2(y_2)\ldots G_m(y_m)H(y_1, y_2, \ldots, y_m, z)$.
Then use *330, *502, and the deduction theorem.

*731. $\vdash \mathbf{L}_1$.

*732. $\vdash (\exists y)G_1(y) \supset \centerdot\, (\exists y)G_2(y) \supset \centerdot \ldots (\exists y)G_m(y) \supset$
$\centerdot\, (y_1)(y_2)\ldots(y_m)(\exists z)H(y_1, y_2, \ldots, y_m, z)$
$\supset (\exists z)\Lambda(G_1, G_2, \ldots, G_m, H, z) \qquad (m = 1, 2, \ldots).$

Proof. By *730, *modus ponens*, and *519,
$G_1(y_1), G_2(y_2), \ldots, G_m(y_m), (\exists z)H(y_1, y_2, \ldots, y_m, z)$
$\vdash (\exists z)\Lambda(G_1, G_2, \ldots, G_m, H, z)$.
Hence by the deduction theorem, *306, and *modus ponens*,
$G_1(y_1), G_2(y_2), \ldots, G_m(y_m), (y_1)(y_2)\ldots(y_m)(\exists z)H(y_1, y_2, \ldots y_m, z)$
$\vdash (\exists z)\Lambda(G_1, G_2, \ldots, G_m, H, z)$.
Then use the deduction theorem, *518, and *502.

*733. If $\vdash \mathbf{Q}_{13}, \vdash \mathbf{Q}_{23}, \ldots, \vdash \mathbf{Q}_{m3}$, and $\vdash \mathbf{R}_3$, then $\vdash \mathbf{L}_3$.

*734. $\vdash G_1(y) \supset_y [G_1(y_1) \supset_{y_1} y = y_1] \supset \centerdot\, G_2(y) \supset_y [G_2(y_1) \supset_{y_1} y = y_1] \supset \centerdot \ldots$
$G_m(y) \supset_y [G_m(y_1) \supset_{y_1} y = y_1] \supset \centerdot\, H(y_1, y_2, \ldots, y_m, z) \supset_{y_1 y_2 \ldots y_m z}$
$[H(y_1, y_2, \ldots, y_m, z_1) \supset_{z_1} z = z_1] \supset \centerdot\, \Lambda(G_1, G_2, \ldots, G_m, H, z) \supset_z$
$\centerdot\, \Lambda(G_1, G_2, \ldots, G_m, H, z_1) \supset_{z_1} z = z_1 \qquad (m = 1, 2, \ldots).$

Proof. By *502, *306, and *modus ponens*,
$G_i(y) \supset_y \centerdot\, G_i(y_1) \supset_{y_1} y = y_1, G_i(x_i), G_i(y_i) \vdash x_i = y_i \qquad (i = 1, 2, \ldots, m);$
and
$H(y_1, y_2, \ldots, y_m, z) \supset_{y_1 y_2 \ldots y_m z} \centerdot\, H(y_1, y_2, \ldots, y_m, z_1) \supset_{z_1} z = z_1,$
$H(y_1, y_2, \ldots, y_m, z), H(y_1, y_2, \ldots, y_m, z_1) \vdash z = z_1.$
Hence by P, *527, and *modus ponens*,
$G_1(y) \supset_y \centerdot\, G_1(y_1) \supset_{y_1} y = y_1,$
$G_2(y) \supset_y \centerdot\, G_2(y_1) \supset_{y_1} y = y_1, \ldots,$
$G_m(y) \supset_y \centerdot\, G_m(y_1) \supset_{y_1} y = y_1,$
$G_1(x_1)G_2(x_2)\ldots G_m(x_m)H(x_1, x_2, \ldots, x_m, z),$
$G_1(y_1)G_2(y_2)\ldots G_m(y_m)H(y_1, y_2, \ldots, y_m, z_1) \vdash H(y_1, y_2, \ldots, y_m, z).$
Hence by P, *306, and *modus ponens*,
$G_1(y) \supset_y \centerdot\, G_1(y_1) \supset_{y_1} y = y_1,$
$G_2(y) \supset_y \centerdot\, G_2(y_1) \supset_{y_1} y = y_1, \ldots,$
$G_m(y) \supset_y \centerdot\, G_m(y_1) \supset_{y_1} y = y_1,$
$H(y_1, y_2, \ldots, y_m, z) \supset_{y_1 y_2 \ldots y_m z}$
$\centerdot\, H(y_1, y_2, \ldots, y_m, z_1) \supset_{z_1} z = z_1,$
$G_1(x_1)G_2(x_2)\ldots G_m(x_m)H(x_1, x_2, \ldots, x_m, z),$
$G_1(y_1)G_2(y_2)\ldots G_m(y_m)H(y_1, y_2, \ldots, y_m, z_1) \vdash z = z_1.$
Then use *518, the deduction theorem, *502, and generalization.

*735. If $\vdash \mathbf{Q}_{14}, \vdash \mathbf{Q}_{24}, \ldots, \vdash \mathbf{Q}_{m4}$, and $\vdash \mathbf{R}_4$, then $\vdash \mathbf{L}_4$.

*736. If f is defined in terms of $\phi_1, \phi_2, \ldots, \phi_m$, and ψ by composition; if $\mathbf{Q}1_{\mathbf{a}_1\mathbf{a}_2\ldots\mathbf{a}_n\mathbf{b}_1}, \mathbf{Q}2_{\mathbf{a}_1\mathbf{a}_2\ldots\mathbf{a}_n\mathbf{b}_2}, \ldots, \mathbf{Q}m_{\mathbf{a}_1\mathbf{a}_2\ldots\mathbf{a}_n\mathbf{b}_m}$, and $\mathbf{R}_{\mathbf{b}_1\mathbf{b}_2\ldots\mathbf{b}_m\mathbf{c}}$, are governed positively by the relations $b_1 = \phi_1(a_1, a_2, \ldots a_n), b_2 = \phi_2(a_1, a_2, \ldots a_n), \ldots, b_m = \phi_m(a_1, a_2, \ldots a_n)$ and $c = \psi(b_1, b_2, \ldots b_m)$ respectively; and if $\vdash \mathbf{Q}_{14}, \vdash \mathbf{Q}_{24}, \ldots, \vdash \mathbf{Q}_{m4}$, and $\vdash \mathbf{R}_4$; then the relation $c = f(a_1, a_2, \ldots, a_n)$ is calculable in A^2 as represented by $\mathbf{L}_{\mathbf{a}_1\mathbf{a}_2\ldots\mathbf{a}_n\mathbf{c}}$.

Proof. By *735 and *728, it is sufficient to show that $\mathbf{L}_{\mathbf{a}_1\mathbf{a}_2\ldots\mathbf{a}_n\mathbf{c}}$ is governed positively by the relation $c = f(a_1, a_2, \ldots, a_n)$. Let $k = f(i_1, i_2, \ldots, i_n)$.
Then $k = \psi(j_1, j_2, \ldots, j_m)$, where

$$j_l = \phi_l(i_1, i_2, \ldots, i_n) \qquad (l = 1, 2, \ldots, m).$$

By the hypothesis of *736,

$$\vdash Z_{i_1}(\mathbf{a}_1) \supset . Z_{i_2}(\mathbf{a}_2) \supset . \ldots Z_{i_n}(\mathbf{a}_n) \supset . Z_{j_l}(\mathbf{b}_l) \supset \mathbf{Q}l_{\mathbf{a}_1\mathbf{a}_2\ldots\mathbf{a}_n\mathbf{b}_l} \qquad (l = 1, 2, \ldots, m),$$

and

$$\vdash Z_{j_1}(\mathbf{b}_1) \supset . Z_{j_2}(\mathbf{b}_2) \supset . \ldots Z_{j_m}(\mathbf{b}_m) \supset . Z_k(\mathbf{c}) \supset \mathbf{R}_{\mathbf{b}_1\mathbf{b}_2\ldots\mathbf{b}_m\mathbf{c}}.$$

Hence by *731 and *modus ponens*,

$$Z_{i_1}(\mathbf{a}_1), Z_{i_2}(\mathbf{a}_2), \ldots, Z_{i_n}(\mathbf{a}_n), Z_k(\mathbf{c}),$$
$$Z_{j_1}(\mathbf{b}_1), Z_{j_2}(\mathbf{b}_2), \ldots, Z_{j_m}(\mathbf{b}_m) \vdash \mathbf{L}_{\mathbf{a}_1\mathbf{a}_2\ldots\mathbf{a}_n\mathbf{c}}.$$

Then use *518, the deduction theorem, *712, and *modus ponens*.

The process of defining numerical functions by primitive recursion and composition begins with certain basic numerical functions, which we shall call the *fundamental functions in the definition of primitive recursive functions.* They are infinite in number and are described in ordinary mathematical notation as follows:

$$s(a) = a + 1$$
$$o_n(a) = n \qquad (n = 0, 1, 2, \ldots)$$
$$u_m^l(a_1, a_2, \ldots, a_m) = a_l \qquad (m = 1, 2, \ldots;\ l = 1, 2, \ldots, m)$$

Though all of these functions are fundamental in the sense that they are used frequently in the definitions of more complicated functions (and therefore it is convenient to designate them explicitly), they are not all fundamental in the sense of being independent with respect to definition by primitive recursion and composition. In fact:

**737. There is an effective procedure whereby each of the remaining fundamental functions can be obtained from s and $u_2^2, u_4^4, u_6^6, \ldots$ by a sequence of definitions by primitive recursion and composition.

Proof by two mathematical inductions in the meta-language. We define o_0 in terms of u_2^2 by primitive recursion:

$$o_0(0) = 0,$$
$$o_0(b + 1) = u_2^2(b, o_0(b)).$$

Suppose we have defined o_n. Then o_{n+1} is defined in terms of s and o_0 by composition:

$$o_{n+1}(a) = s(o_n(a)).$$

In order to define u_1^1 we first define by composition:

$$v(a_1, a_2) = s(u_2^2(a_1, a_2)).$$

Then we define u_1^1 by primitive recursion:

$$u_1^1(0) = 0,$$
$$u_1^1(b+1) = v\big(b, u_1^1(b)\big).^{716}$$

An analogous definition of u_{m+1}^{m+1} by composition and primitive recursion may be written in condensed form[717] as follows:

$$u_{m+1}^{m+1}(a_1, a_2, \ldots, a_m, 0) = o_0\big(u_m^m(a_1, a_2, \ldots, a_m)\big),$$
$$u_{m+1}^{m+1}(a_1, a_2, \ldots, a_m, b+1) = s\big(u_{m+2}^{m+2}(a_1, a_2, \ldots, a_m, b, u_{m+1}^{m+1}(a_1, a_2, \ldots, a_m, b))\big).$$

These definitions enable us to dispense with every alternate one in the infinite list of fundamental functions $u_1^1, u_2^2, u_3^3, \ldots$. And the functions u_m^l, where l is not the same as m, may then be defined by primitive recursions as follows:

$$u_{m+1}^l(a_1, a_2, \ldots, a_m, 0) = u_m^l(a_1, a_2, \ldots, a_m),$$
$$u_{m+1}^l(a_1, a_2, \ldots, a_m, b{+}1) = u_{m+2}^{m+2}\big(a_1, a_2, \ldots, a_m, b, u_{m+1}^l(a_1, a_2, \ldots, a_m, b)\big).$$

A *primitive recursive numerical function* is any numerical function which is either a fundamental function, or which can be obtained from the fundamental functions by a finite sequence of definitions by primitive recursion and composition.[718]

For our present purpose the following property of primitive recursive functions is important:

*738. There is an effective procedure by which, given a primitive recursive n-ary numerical function f and a sequence of definitions by primitive recursion and composition whereby f is obtained from the fundamental functions, we may assign to f a wff $\mathbf{A}_{x_1x_2\ldots x_ny}$ of A^2 such that the relation $c = f(a_1, a_2, \ldots, a_n)$ is calculable in A^2 as represented by $\mathbf{A}_{x_1x_2\ldots x_ny}$, and such that further, $\vdash (\exists y)\mathbf{A}_{x_1x_2\ldots x_ny}$ and $\vdash \mathbf{A}_{x_1x_2\ldots x_ny} \supset \centerdot\, \mathbf{A}_{x_1x_2\ldots x_nz} \supset y = z$.

Proof by mathematical induction in the meta-language with respect to the total number of steps in the given sequence of definitions by primitive recursion and

[716]Such a succession of definitions by one or more compositions and primitive recursions will henceforth be condensed by incorporating the compositions directly in the primitive recursion. In this case the condensed definition would be:

$$u_1^1(0) = 0,$$
$$u_1^1(b+1) = s\big(u_2^2(b, u_1^1(b))\big).$$

[717]See footnote 716.

[718]The notion of primitive recursive function first appears in Gödel's paper, ibid. footnote 711, although Gödel called them simply "recursive functions". The present convention of attaching the adjective "primitive" to such recursive functions originates in a paper by Rózsa Péter in ***Mathematische Annalen***, vol. 110 (1934), pp. 612–632, and in ***Grundlagen der Mathematik*** by David Hilbert and Paul Bernays, vol. I, p. 326, Berlin 1934, and serves to distinguish primitive recursive functions from a larger class of functions definable by a more general recursive procedure.

Many of the proofs in the following sections that various particular numerical and propositional functions are primitive recursive are taken more or less directly from Gödel's paper, with modifications to fit the present context.

composition.

Case 1: f is a fundamental function. By **737 we need only consider the fundamental functions s and $u_2^2, u_4^4, u_6^6, \ldots$.

Case 1a: f is s. We assign to s the wff $S(x_1, y)$.
By †700, $\vdash (\exists y)S(x_1, y)$.
By †701, $\vdash S(x_1, y) \supset . S(x_1, z) \supset y = z$.
If $j = s(i)$, then $j = i + 1$.
Hence by *714, $\vdash Z_i(x_1) \supset . Z_j(y) \supset S(x_1, y)$.
Hence $S(x_1, y)$ is positively governed by the relation $c = s(a_1)$ with respect to x_1 and y.
Hence by *728, the relation $c = s(a_1)$ is calculable in A^2 as represented by $S(x_1, y)$.

Case 1b: f is u_n^n. We assign to u_n^n the wff $y = x_n$.[719]
By *330 and †520, $\vdash (\exists y) . y = x_n$.
By †521 and †522, $\vdash y = x_n \supset . z = x_n \supset y = z$.
If $j = u_n^n(i_1, i_2, \ldots, i_n)$ then $j = i_n$.
Hence by *713, *503, and P,
$\vdash Z_{i_1}(x_1) \supset . Z_{i_2}(x_2) \supset . \ldots Z_{i_n}(x_n) \supset Z_j(y) \supset y = x_n$.
Hence by *728, the relation $c = u_n^n(a_1, a_2, \ldots a_n)$ is calculable in A^2 as represented by $y = x_n$.

Case 2a: $n = 1$ and f is defined in terms of the fixed natural number k and the binary numerical function ψ by primitive recursion, where the relation $c = \psi(a_1, a_2)$ is calculable in A^2 as represented by $\mathbf{R}_{x_1 x_2 y}$, and such that further,

$$\vdash (\exists y)\mathbf{R}_{x_1 x_2 y} \quad \text{and} \quad \vdash \mathbf{R}_{x_1 x_2 y} \supset . \mathbf{R}_{x_1 x_2 z} \supset y = z.$$

Then by *502 and *503, the relation $c = \psi(a_1, a_2)$ is calculable in A^2 also as represented by $\mathbf{R}_{x_1 yz}$, $\vdash \mathbf{R}_3$, and $\vdash \mathbf{R}_4$. Also by P and *713, the singulary relation $c = k$ is calculable in A^2 as represented by $Z_k(y)$; by *712, $\vdash (\exists y)Z_k(y)$; and by *713, $\vdash Z_k(y) \supset . Z_k(z) \supset y = z$. This case of *738 then follows from *729, *722, and *727.

Case 2b: $n > 1$ and f is defined in terms of ϕ and ψ by primitive recursion, where the relations

$$c = \phi(a_1, a_2, \ldots a_{n-1}) \quad \text{and} \quad c = \psi(a_1, a_2, \ldots a_{n+1})$$

are calculable in A^2 as represented by $\mathbf{Q}_{x_1 x_2 \ldots x_{n-1} y}$ and $\mathbf{R}_{x_1 x_2 \ldots x_{n+1} y}$ respectively, and such that further,

$$\vdash (\exists y)\mathbf{Q}_{x_1 x_2 \ldots x_{n-1} y},$$
$$\vdash (\exists y)\mathbf{R}_{x_1 x_2 \ldots x_{n+1} y},$$
$$\vdash \mathbf{Q}_{x_1 x_2 \ldots x_{n-1} y} \supset . \mathbf{Q}_{x_1 x_2 \ldots x_{n-1} z} \supset y = z, \quad \text{and}$$
$$\vdash \mathbf{R}_{x_1 x_2 \ldots x_{n+1} y} \supset . \mathbf{R}_{x_1 x_2 \ldots x_{n+1} z} \supset y = z.$$

As in case 2a, this case of *738 follows from *502, *503, *729, *732, and *727.

[719]The reader is reminded that according to the conventions adopted in §72, not all of the variables $x_1, x_2, \ldots, x_n, y$ need necessarily have occurrences in $\mathbf{A}_{x_1 x_2 \ldots x_n y}$.

Case 3: f is defined in terms of $\phi_1, \phi_2, \ldots, \phi_m$ and ψ by composition, where the relations

$$c = \phi_i(a_1, a_2, \ldots, a_n) \quad (i = 1, 2, \ldots, m) \quad \text{and} \quad c = \psi(a_1, a_2, \ldots, a_m)$$

are calculable in A^2 as represented by

$$\mathbf{Q}_{i\,x_1x_2\ldots x_n y} \quad (i = 1, 2, \ldots, m) \quad \text{and} \quad \mathbf{R}_{x_1x_2\ldots x_m y}$$

respectively, and such that further,

$$\vdash (\exists y)\mathbf{Q}_{i\,x_1x_2\ldots x_n y} \quad (i = 1, 2, \ldots, m)$$
$$\vdash (\exists y)\mathbf{R}_{x_1x_2\ldots x_m y},$$
$$\vdash \mathbf{Q}_{i\,x_1x_2\ldots x_n y} \supset . \mathbf{Q}_{i\,x_1x_2\ldots x_n z} \supset y = z \quad (i = 1, 2, \ldots, m), \quad \text{and}$$
$$\vdash \mathbf{R}_{x_1x_2\ldots x_m y} \supset . \mathbf{R}_{x_1x_2\ldots x_m z} \supset y = z.$$

Analogous to case 2b, this case of *738 follows from *502, *503, *736, *733, and *735.

Let Θ be an n-ary propositional function of the natural numbers, and let χ be the n-ary numerical function defined by $\chi(i_1, i_2, \ldots i_n) = 0$ or 1 according as $\Theta(i_1, i_2, \ldots i_n)$ or not $\Theta(i_1, i_2, \ldots i_n)$. Then we call χ the *characteristic function* of Θ, and we say that Θ is *primitive recursive* if and only if χ is primitive recursive.

As a corollary of *738 we have

*739. Given a primitive recursive n-ary relation Θ among the natural numbers, and a sequence of definitions by primitive recursion and composition whereby the characteristic function of Θ is obtained from the fundamental functions, there is an effective procedure by which we may find a wff $\mathbf{B}_{x_1x_2\ldots x_n}$ such that Θ is calculable in A^2 as represented by $\mathbf{B}_{x_1x_2\ldots x_n}$.

Proof. By *738, we can find $\mathbf{A}_{x_1x_2\ldots x_n y}$ such that the relation $c = \chi(a_1, a_2, \ldots, a_n)$ is calculable in A^2 as represented by $\mathbf{A}_{x_1x_2\ldots x_n y}$, and such that further,

$$\vdash \mathbf{A}_{x_1x_2\ldots x_n y} \supset . \mathbf{A}_{x_1x_2\ldots x_n z} \supset y = z. \tag{1}$$

Let $\mathbf{B}_{x_1x_2\ldots x_n}$ be $(\exists y) . Z_0(y)\mathbf{A}_{x_1x_2\ldots x_n y}$.
Suppose $\Theta(i_1, i_2, \ldots, i_n)$.
Then $0 = \chi(i_1, i_2, \ldots, i_n)$, so we have
$$\vdash Z_{i_1}(x_1) \supset . Z_{i_2}(x_2) \supset . \ldots Z_{i_n}(x_n) \supset . Z_0(y) \supset \mathbf{A}_{x_1x_2\ldots x_n y}.$$
Hence by *modus ponens* and P,
$$Z_{i_1}(x_1), Z_{i_2}(x_2), \ldots, Z_{i_n}(x_n), Z_0(y) \vdash Z_0(y)\mathbf{A}_{x_1x_2\ldots x_n y}.$$
Then use *518, the deduction theorem, †703, and *modus ponens*.
On the other hand, suppose not $\Theta(i_1, i_2, \ldots, i_n)$.
Then $1 = \chi(i_1, i_2, \ldots, i_n)$, so we have
$$\vdash Z_{i_1}(x_1) \supset . Z_{i_2}(x_2) \supset . \ldots Z_{i_n}(x_n) \supset . Z_1(y) \supset \mathbf{A}_{x_1x_2\ldots x_n y}.$$
Hence by *modus ponens* and (1),
$$Z_{i_1}(x_1), Z_{i_2}(x_2), \ldots, Z_{i_n}(x_n), Z_1(y) \vdash \mathbf{A}_{x_1x_2\ldots x_n z} \supset y = z.$$
By *715, $Z_1(y) \vdash Z_0(z) \supset y \neq z$.
Hence by †525 and P,
$$Z_{i_1}(x_1), Z_{i_2}(x_2), \ldots, Z_{i_n}(x_n), Z_1(y) \vdash {\sim} Z_0(z)\mathbf{A}_{x_1x_2\ldots x_n z}.$$

Then use generalization, *502, P, *518, the deduction theorem, *712 and *modus ponens*.[720]

OUTLINE OF §§74-75.

We go on to the proofs of the primitive recursiveness of particular numerical and propositional functions, treating such results as metatheorems of A^2 because of their corollaries by *738 and *739.

*740. Where k and l are fixed natural numbers, the n-ary numerical function f is primitive recursive which has the value k for each of the finite number of particular n-tuples of arguments $a_1, a_2, \ldots, a_n$ and the value l for all other n-tuples of arguments.

Consider first the case $n = 1$, and let m be the greatest number such that $f(m) = k$. If $m = 0$, the recursion equations are

$$f(0) = k, \quad f(b+1) = o_l\Big(u_2^2(b, f(b))\Big).$$

Or as we shall write them, in still more abbreviated form than that described in footnote 716, they are:

$$f(0) = k, \quad f(b+1) = l.$$

If we then proceed by mathematical induction, we take as hyp. ind. that *740 holds with $n = 1$ and with a given m, and we seek to show that *740 holds with $n = 1$ and with $m + 1$ in place of m. It suffices to give the two recursion equations as:

Either $f(0) = k$ or $f(0) = l$ as required,
$f(b+1) = f_0(b)$.

For if we so choose the function f_0 as to verify the second recursion equation, the primitive recursiveness of f_0 follows from hyp. ind.

Having thus proved *740 in the case that $n = 1$ we proceed by induction with respect to n. We take as hyp. ind. that *740 holds with a given n. And to show that *740 therefore holds with $n + 1$ in place of n we proceed by induction with respect to m, where m is the greatest number such that $f(a_1, a_2, \ldots, a_n, m) = k$ at least once, i.e., for at least one n-tuple of arguments $a_1, a_2, \ldots, a_n$.

If $m = 0$, and with $n + 1$ in place of n, the two recursion equations for f may be written as

$$f(a_1, a_2, \ldots, a_n, 0) = \phi(a_1, a_2, \ldots, a_n),$$
$$f(a_1, a_2, \ldots, a_n, b+1) = l.$$

For if we so chose the function ϕ as to verify the first recursion equation, the primitive recursiveness of ϕ follows by hyp. ind.

Then we take as a second hyp. ind. that *740 holds with a given m and with $n + 1$ in place of n, and we seek to show that *740 therefore holds with $m + 1$ in place of m and

[720] For the organization (from lecture notes) of the material in §§70–73, and for many details of exposition and proofs, the author is heavily indebted to T. T. Robinson.

$n + 1$ in place of n. For this purpose the recursion equations for f may be written as:

$$f(a_1, a_2, \ldots, a_n, 0) = \phi(a_1, a_2, \ldots, a_n),$$
$$f(a_1, a_2, \ldots, a_n, b+1) = f_0(a_1, a_2, \ldots, a_n, b).$$

Then the primitive recursiveness of the functions ϕ and f_0 is a consequence of the first and second hyp. ind. respectively.

*741. If $\Theta(a_1, a_2, \ldots, a_n) = f(\Theta_1(a_1, a_2, \ldots, a_n), \Theta_2(a_1, a_2, \ldots, a_n), \ldots, \Theta_m(a_1, a_2, \ldots, a_n))$, where f is any truth-function and $\Theta_1, \Theta_2, \ldots, \Theta_m$ are primitive recursive n-ary propositional functions, then the n-ary propositional function Θ is primitive recursive.

Proof. This becomes a definition by composition if we replace Θ, f, $\Theta_1, \Theta_2, \ldots, \Theta_m$ by their characteristic functions, as determined by taking 0 to correspond to truth and 1 to falsehood. That the characteristic function of f is primitive recursive — if we take its value to be 1 for any m-tuple of arguments that includes an argument greater than 1 — follows from *740.

*742. If Θ is a primitive recursive n-ary propositional function and ϕ and ψ are primitive recursive n-ary numerical functions, there is a primitive recursive n-ary numerical function f such that:
$f(a_1, a_2, \ldots, a_n) = \phi(a_1, a_2, \ldots, a_n)$ or
$\psi(a_1, a_2, \ldots, a_n)$ according as
$\Theta(a_1, a_2, \ldots, a_n)$ or not.

Proof. Let θ be the characteristic function of Θ, and introduce the primitive recursive function alt by the recursion equations: $\text{alt}(a_1, a_2, 0) = a_1$, $\text{alt}(a_1, a_2, b+1) = a_2$. Then we have as definition by composition:
$f(a_1, a_2, \ldots, a_n) = \text{alt}(\phi(a_1, a_2, \ldots, a_n),$
$\psi(a_1, a_2, \ldots, a_n), \theta(a_1, a_2, \ldots, a_n))$.

*743. An n-ary numerical (or propositional) function which differs from a primitive recursive n-ary numerical (or propositional) function for only a finite number of n-tuple of arguments is itself primitive recursive.

It is sufficient to consider the case of a numerical function. The proof may be by a method exactly paralleling that used in the proof of *740 (as indeed the theorem *740 is a special case of *743). But at the present stage it is simpler to prove *743 in a way that makes use of *740 and *742 as previous results.

Suppose that $f(a_1, a_2, \ldots, a_n)$ differs from $\phi(a_1, a_2, \ldots, a_n)$ for only one n-tuple of arguments, say $a_1 = k_1$, $a_2 = k_2, \ldots, a_n = k_n$ and that

$$f(k_1, k_2, \ldots, k_n) = k.$$

Let $\psi(a_1, a_2, \ldots, a_n) = o_k(u_n^n(a_1, a_2, \ldots, a_n))$ and let θ be the n-ary numerical function which has the value 1 for the n-tuple of arguments $k_1, k_2, \ldots, k_n$ and the value 0 for all other n-tuples of arguments. Then ψ is primitive recursive by composition and θ is primitive recursive by *740. If we assume that ϕ is primitive recursive, the primitive recursiveness of f follows by *742.

This proves *743 in the case of an n-ary numerical function that differs from a primitive recursive function for only one n-tuple of arguments. The case in which the difference is for a larger finite number of n-tuples of arguments then follows by iteration.

*744. If f is a primitive recursive n-ary numerical function or propositional function and k is a fixed natural number, $f(a_1, a_2, \ldots, a_{i-1}, k, a_{i+1}, a_{i+2}, \ldots, a_n)$ is a primitive recursive function of $a_1, a_2, \ldots, a_{i-1}, a_{i+1}, a_{i+2}, \ldots, a_n$
$(i = 1, 2, \ldots, n, \text{ and } n \geqq 2)$.

For the new primitive recursive function g we have

$$\begin{aligned} g(a_1, a_2, \ldots, a_{i-1}, a_{i+1}, a_{i+2}, \ldots, a_n) \\ = f\big(&u^1_{n-1}(a_1, a_2, \ldots, a_{i-1}, a_{i+1}, a_{i+2}, \ldots, a_n), \\ &u^2_{n-1}(a_1, a_2, \ldots, a_{i-1}, a_{i+1}, a_{i+2}, \ldots, a_n), \ldots, \\ &u^{i-1}_{n-1}(a_1, a_2, \ldots, a_{i-1}, a_{i+1}, a_{i+2}, \ldots, a_n), \\ &o_k\big(u^{n-1}_{n-1}(a_1, a_2, \ldots, a_{i-1}, a_{i+1}, a_{i+2}, \ldots, a_n)\big), \\ &u^i_{n-1}(a_1, a_2, \ldots, a_{i-1}, a_{i+1}, a_{i+2}, \ldots, a_n), \\ &u^{i+1}_{n-1}(a_1, a_2, \ldots, a_{i-1}, a_{i+1}, a_{i+2}, \ldots, a_n), \ldots, \\ &u^{n-1}_{n-1}(a_1, a_2, \ldots, a_{i-1}, a_{i+1}, a_{i+2}, \ldots, a_n)\big). \end{aligned}$$

This is a definition by composition (or more accurately, two successive definitions by composition) of a sort which we shall generally use without explicitly stating the composition as a separate step. The statement of *744 as a separate theorem may also similarly be superfluous. But we shall have frequent occasion to use this special case of composition in which one of the variables is replaced by a constant.

*745. If we take $a \dot{-} b$ to be the same as $a - b$ when $a \geqq b$ and to be 0 when $a < b$, then $a \dot{-} b$ is a primitive recursive function of a and b.

The recursion equations are:

$$p(0) = 0, \qquad p(b+1) = b,$$
$$a \dot{-} 0 = a, \qquad a \dot{-} (b+1) = p(a \dot{-} b)$$

*746. If f and g are primitive recursive n-ary numerical functions, each of the following is a primitive recursive propositional function of $a_1, a_2, \ldots, a_n$:

$$f(a_1, a_2, \ldots, a_n) \leqq g(a_1, a_2, \ldots, a_n)$$
$$f(a_1, a_2, \ldots, a_n) \geqq g(a_1, a_2, \ldots, a_n)$$
$$f(a_1, a_2, \ldots, a_n) < g(a_1, a_2, \ldots, a_n)$$
$$f(a_1, a_2, \ldots, a_n) > g(a_1, a_2, \ldots, a_n)$$
$$f(a_1, a_2, \ldots, a_n) = g(a_1, a_2, \ldots, a_n)$$
$$f(a_1, a_2, \ldots, a_n) \neq g(a_1, a_2, \ldots, a_n)$$

It will suffice to show that each of the six binary relations (or binary propositional functions) $\leqq$, $\geqq$, $<$, $>$, $=$, $\neq$ is primitive recursive, as *746 will then follow by composition. Since $c = 0$ is a primitive recursive propositional function of c, by *740, we may show that $a \leqq b$ and $a \geqq b$ are primitive recursive propositional functions of a and b by expressing them as $a \dot{-} b = 0$ and $b \dot{-} a = 0$ respectively. Similarly, because $c > 0$ is a primitive recursive propositional function of c, by *740, we may show the primitive recursiveness of $a < b$ and $a > b$ by expressing them as $b \dot{-} a > 0$ and $a \dot{-} b > 0$ respectively. Then to establish the primitive recursiveness of $a = b$ and $a \neq b$ we make use of *741, since $a = b$ may be expressed as $a \leqq b \quad \text{and} \quad a \geqq b$, and $a \neq b$ may be expressed as $a < b \quad \text{or} \quad a > b$.

Remark. If (temporarily) we use θ for the characteristic function of the propositional function, $= 0$, and $f(a, b)$ for $a \dot{-} b$, the composition by which we reach the characteristic function of $a \dot{-} b = 0$ is simply $\theta(f(a, b))$, but that by which we reach the characteristic function of $b \dot{-} a = 0$ must be

$$\theta\big(f\left(u_2^2(a, b), u_2^1(a, b)\right)\big)$$

(in order to regard it as a function whose arguments are a, b in that order). Such abbreviated statement of definition by composition, as in the proof above, will be employed hereafter without special remark.

*747. If Θ is a primitive recursive $(n+1)$-ary propositional function, the finite disjunction

$$\sum_{c=a}^{c=b} \Theta(a_1, a_2, \ldots, a_n, c)$$

and the finite conjunction

$$\prod_{c=a}^{c=b} \Theta(a_1, a_2, \ldots, a_n, c)$$

are primitive recursive propositional functions of $a_1, a_2, \ldots, a_n, a, b$. The definitions by primitive recursion and composition are:

$$\sum_{c=0}^{c=0} \Theta(a_1, a_2, \ldots, a_n, c) = \Theta(a_1, a_2, \ldots, a_n, 0),$$

$$\sum_{c=0}^{c=b+1} \Theta(a_1, a_2, \ldots, a_n, c) = \Big[\Theta(a_1, a_2, \ldots, a_n, s(b)) \quad \text{or} \sum_{c=0}^{c=b} \Theta(a_1, a_2, \ldots, a_n, c)\Big],$$

$$\prod_{c=0}^{c=0} \Theta(a_1, a_2, \ldots, a_n, c) = \Theta(a_1, a_2, \ldots, a_n, 0),$$

$$\prod_{c=0}^{c=b+1} \Theta(a_1, a_2, \ldots, a_n, c) = \Big[\Theta(a_1, a_2, \ldots, a_n, s(b)) \quad \text{and} \prod_{c=0}^{c=b} \Theta(a_1, a_2, \ldots, a_n, c)\Big],$$

$$\sum_{c=a}^{c=b} \Theta(a_1, a_2, \ldots, a_n, c) = \sum_{c=0}^{c=b} \left[c \geqq a \text{ and } \Theta(a_1, a_2, \ldots, a_n, c)\right],$$

$$\prod_{c=a}^{c=b} \Theta(a_1, a_2, \ldots, a_n, c) = \prod_{c=0}^{c=b} \left[c \geqq a \text{ implies } \Theta(a_1, a_2, \ldots, a_n, c)\right].$$

Remark. Directly, *747 establishes the primitive recursiveness of a finite continued disjunction or conjunction with respect to the last argument of Θ. The same result of course also holds with respect to any other argument of Θ, but it is unnecessary to state this as a separate theorem, as any permutation of the arguments of Θ can be effected by a definition by composition (compare the Remark following *746).

*748. If Θ is a primitive recursive $(n+1)$-ary propositional function, then

$$\mathop{\mathrm{E}}_{c=a}^{c=b} \Theta(a_1, a_2, \ldots, a_n, c)$$

— the least number c such that $a \leqq c \leqq b$ and $\Theta(a_1, a_2, \ldots, a_n, c)$, or 0 if there is no such number — is a primitive recursive numerical function of $a_1, a_2, \ldots, a_n, a, b$.

Using *742, we may write the recursion equations as:

$$\mathop{\mathrm{E}}_{c=a}^{c=0} \Theta(a_1, a_2, \ldots, a_n, c) = 0$$

$$\mathop{\mathrm{E}}_{c=a}^{c=b+1} \Theta(a_1, a_2, \ldots, a_n, c) = s(b) \text{ or}$$

$$\mathop{\mathrm{E}}_{c=a}^{c=b} \Theta(a_1, a_2, \ldots, a_n, c) \text{ according as}$$

$$\left[a \leqq s(b) \text{ and } \Theta(a_1, a_2, \ldots, a_n, s(b)) \text{ and not } \sum_{c=a}^{c=b} \Theta(a_1, a_2, \ldots, a_n, c)\right]$$

or not.

*749. The quotient and remainder on dividing b by a are primitive recursive functions of a and b.

Denoting the two functions by $\mathrm{quot}(a, b)$ and $\mathrm{rem}(a, b)$ we have as recursion equations for them:

$$\mathrm{rem}(a, 0) = 0, \quad \mathrm{quot}(a, 0) = 0,$$
$$\mathrm{rem}(a, b+1) = 0 \text{ or } s(\mathrm{rem}(a, b))$$
$$\text{according as } s(\mathrm{rem}(a, b)) = a \text{ or not},$$
$$\mathrm{quot}(a, b+1) = s(\mathrm{quot}(a, b)) \text{ or } \mathrm{quot}(a, b)$$
$$\text{according as } s(\mathrm{rem}(a, b)) = a \text{ or not}.$$

Because the special case of *749 in which a is 2 will be used frequently, especially in

Chapter VIII, we put

$$\mathrm{par}(b) = \mathrm{rem}(2, b) \quad \text{and} \quad h(b) = \mathrm{quot}(2, b)$$

reading par(b) as "the parity of b" and $h(b)$ as "the half of b".

*750. Addition and multiplication are primitive recursive.

The recursion equations are:

$$a + 0 = a, \quad a + (b + 1) = s(a + b),$$
$$a0 = 0, \quad a(b + 1) = ab + a.$$

*751. If f is a primitive recursive $(n + 1)$-ary numerical function, the finite sum

$$\sum_{c=a}^{c=b} f(a_1, a_2, \ldots, a_n, c)$$

and the finite product

$$\prod_{c=a}^{c=b} f(a_1, a_2, \ldots, a_n, c)$$

are primitive recursive numerical functions of $a_1, a_2, \ldots, a_n, a, b$.

The recursion equations are:

$$\sum_{c=a}^{c=0} f(a_1, a_2, \ldots, a_n, c) = f(a_1, a_2, \ldots, a_n, 0) \text{ or } 0$$

according as $a = 0$ or not,

$$\sum_{c=a}^{c=b+1} f(a_1, a_2, \ldots, a_n, c) = f\left(a_1, a_2, \ldots, a_n, s(b)\right) + \sum_{c=a}^{c=b} f(a_1, a_2, \ldots, a_n, c) \text{ or } 0$$

according as $a \leqq s(b)$ or not,

$$\prod_{c=a}^{c=0} f(a_1, a_2, \ldots, a_n, c) = f(a_1, a_2, \ldots, a_n, 0) \text{ or } 1$$

according as $a = 0$ or not

$$\prod_{c=a}^{c=b+1} f(a_1, a_2, \ldots, a_n, c) = f\left(a_1, a_2, \ldots, a_n, s(b)\right) \prod_{c=a}^{c=b} f(a_1, a_2, \ldots, a_n, c) \text{ or } 1$$

according as $a \leqq s(b)$ or not.

*752. The factorial is a primitive recursive numerical function.

$$a! = \prod_{c=1}^{c=a} c \qquad \text{We notice that } 0! = 1.$$

*753. Exponentiation is primitive recursive in the sense that a^b is a primitive recursive numerical function of a and b.

$$a^b = \prod_{c=1}^{c=b} a$$

We notice that $a^0 = 1$. In particular $0^0 = 1$; but all other powers of 0 are 0.

*754. If the binary numerical function f is obtained from the singulary numerical function ϕ and the ternary numerical function ψ by the recursion equations

$$f(a,0) = \phi(a), \qquad f(a,n+1) = \psi\big(a,n,f(2a,n)\big),$$

there is also a chain of definitions by composition and by primitive recursion by which f can be obtained from ϕ and ψ; and hence if ϕ and ψ are primitive recursive, it follows that f is primitive recursive.

$$H(a,0) = h(a), \qquad H(a,n+1) = h\big(H(a,n)\big),$$
$$g(a,0) = \phi(a), \qquad g(a,n+1) = \psi\big(H(a,n),n,g(a,n)\big),$$
$$f(a,n) = g(2^n a, n).$$

*755. That a is a divisor of b is a primitive recursive relation between a and b.

For the relation $a \mid b$ may be expressed as $\sum_{c=1}^{c=s(b)} [ac = b]$. (This makes 0 the only divisor of 0.)

*756. The class of prime numbers is primitive recursive.

$$\text{prime}(a) = \left[a > 1 \text{ and } \prod_{b=2}^{b=c}\prod_{c=2}^{c=a}[a \neq bc]\right].$$

(The clause, $a > 1$ excludes 0 and 1, so that the least prime number is 2.)

*757. The n-th prime number, p_n, is a primitive recursive function of n.

$$p_0 = 1, \qquad p_{n+1} = \mathop{\mathrm{E}}_{c=s(p_n)}^{c=s(p_n!)} \text{prime}(c).$$

*758. The exponent of p_n in the prime factorization of a is a primitive recursive function of n and a.

$$n\,\mathrm{Gl}\,a = \mathop{\mathrm{E}}_{c=0}^{c=a}\big[p_n^c \mid a \text{ and not } p_n^{s(c)} \mid a\big].$$

We notice that this makes $n\,\mathrm{Gl}\,0 = 0$, $n\,\mathrm{Gl}\,1 = 0$, $0\,\mathrm{Gl}\,a = 0$.

*759. The greatest number n such that $p_n \mid a$ is a primitive recursive function of a.

$$\mathrm{L}(a) = a \dot{-} \mathop{\mathrm{E}}_{c=0}^{c=a}\big[p_{a \dot{-} c} \mid a\big].$$

We notice that this makes $\mathrm{L}(0) = \mathrm{L}(1) = 0$.

PARTIAL OUTLINE OF CHAPTER VIII (1968, 1972, 1976)

Effectiveness of proofs. As a first approximation to the distinction between effective and non-effective existence proofs, we may say that an existence theorem is effectively proved if the proof puts us in possession of a method by which we can obtain at least one particular example of that which was proved to exist. Proofs that particular functions are primitive recursive are proofs that there exists a sequence of primitive recursive introductions which leads from the fundamental functions to the particular function which is shown to be primitive recursive. More generally for any function f, of a certain kind, and any Φ, Ψ which are known to be primitive recursive, we proceed by proving there exists an introduction of f which uses only primitive recursive equations, introduction by composition, Φ, Ψ, and the fundamental functions. Given that Φ, Ψ are previously known to be primitive recursive we are already in possession of a sequence of primitive recursive introductions which lead to Φ, Ψ, respectively.

Hence the particular sort of effectiveness presently under discussion concerns theorems of e.g., the form $(x)(y)(\exists z)\mathbf{A}_{xyz}$, where $\mathbf{A}_{xyz}$ expresses some relation among the numbers x, y, z. To say the proof of a theorem of this form is effective is to say that the proof puts us in possession of a method, given in advance and stateable in a finite number of words non-equivocally, such that whenever particular numbers x, y are given we can compute a particular number z such that $\mathbf{A}_{xyz}$. The method must be one that can be followed in a purely mechanical way.

An especially important case of non-effectiveness is that in which for infinitely many choices of x and y an essentially different unsolved problem (such as e.g., the Fermat problem) has to be solved in order to find z.

Example (from L. E. J. Brouwer, *Besitzt jede reele Zahl eine Dezimalbruchentwicklung?*, ***Math. Annalen***, vol. 83 (1921), pp. 201–210) of a non-effective existence theorem of analysis: Every real number has a decimal expansion. For purposes of this example we take a real number to be an infinite sequence of open intervals on the real line with rational end points, each interval properly contained in the preceding one and the length of the intervals approaching 0. For a real number to be given effectively there must be given an effective method to calculate the end points for every n. Consider the decimal expansion of π:

$$\pi = 3.14159265\ldots$$

An effective method can be given to generate π's decimal expansion (in fact the nth place of π is a primitive recursive function of n, but this will not be required in the following). Slightly modifying Brouwer's presentation, let us set up the infinite

---

This work was left mostly in Church's hand, and partly in student notes dictated by Church. It was projected to be the better part of Chapter VIII of the second volume of ***Introduction to Mathematical Logic***. The student notes are from lectures given at UCLA in the Spring of 1968, Fall of 1968, Spring of 1972, and Spring of 1976. The lectures were part of a course entitled "Gödel Theory" (Philosophy 261). The notes were taken by R. Smith, Bernard Kobes, Fu-tseng Liu, and Palle Yourgrau. The notes amounted to dictation. So it is believed that they represent Church's presentation very closely. The arrangement and initial editing of the material is due to C. Anthony Anderson. The material is sectioned to correspond, as closely as possible, to the format of the first volume of ***Introduction to Mathematical Logic***. Printed by permission of the Alonzo Church estate.

sequence of numbers:

$$a_1, a_2, \ldots, a_i = \begin{cases} 1, & \text{if } n \text{ is odd and the } n\text{th to the } (n+8)\text{th digits in the decimal expansion of } \pi \text{ are } 314159265, \\ -1, & \text{if } n \text{ is even and the } n\text{th to the } (n+8)\text{th digits in the decimal expansion of } \pi \text{ are } 314159265, \\ 0, & \text{in all other cases} \end{cases}$$

Let: $B = \frac{a_1}{2} + \frac{a_2}{4} + \frac{a_3}{8} + \ldots \frac{a_n}{2^n} \ldots$

I.e., $B = \frac{1}{2} + \frac{0}{4} + \frac{0}{8} + \ldots$

B is between: $\frac{7}{16}, \frac{9}{16}, \ldots, \frac{15}{16}, \frac{17}{32}, \ldots$

Suppose we stop at $\frac{a_3}{8}$, then we have two possible cases according as n is odd or even:

(1) $$B \leq \frac{1}{2} + \frac{1}{32} + \frac{1}{128} \ldots$$

(2) $$\frac{1}{2} - \frac{1}{16} - \frac{1}{64} \ldots \leq B,$$

which we can approximate as:

(3) $$\frac{1}{2} - \frac{2}{16} < B < \frac{1}{2} + \frac{2}{32}$$

(4) $$\frac{1}{2} - \frac{2}{16} < B < \frac{1}{2} + \frac{2}{16}$$

The general case for all terms 0 up to n is:

(5) $$\frac{1}{2} - \frac{1}{2^n} < B < \frac{1}{2} + \frac{1}{2^n}$$

There is an effective process for calculating the infinite number of open intervals in B, but to determine B's decimal expansion requires solving the unsolved mathematical problem — does the sequence 314159265 recur in B's decimal expansion a second time? A third time? And so on. As a matter of fact, since we do not know whether this sequence of nine digits occurs even a second time in the decimal expansion of π, and do not know how to decide this question, we are unable even to write the first digit in the decimal expansion of B. (It is either 4 or 5, but which?)

If we were to examine the proof that every real number has a decimal expansion we would find that the non-effectiveness gets into it through the use of the law of excluded middle. The proof is by cases: either there are integers n and a such that the given real number x can be expressed as $\frac{a}{10^n}$ or else not. The set of instructions for finding the decimal expansion of x differs radically according to which case holds. (Indeed the law of excluded middle should have been mentioned above, along with the method of proof by contradiction, as one of the sources of non-effective existence proofs; as laws of propositional calculus the two are of course closely related.)

The decimal expansion of the Brouwer number B provides an example of a function $f(n)$ such that, in order to compute it for all natural numbers n, an infinite number of non-effective decisions *may* have to be made. If we find the digits 314159265 recurring for a second time in the decimal expansion of π, we will then be able to write down the decimal expansion for B for a considerable number of places, but the point will come at which we cannot continue without knowing whether there is a third occurrence of these same digits in the decimal expansion of π; and so on.

As example of a different and simpler kind, in which there is only one decision to be made regarding an unsolved mathematical question, is obtained if we introduce a function $g(n)$ by stipulating that $g(n) = 2^n$ if Fermat's so-called Last Theorem holds, and $g(n) = 3^n$ in the contrary case. This function $g(n)$ can be shown to be primitive recursive by a use of the law of excluded middle, but it is not certain that the proof can be made effective (this will depend on whether or not a solution of the Fermat problem is possible).

Even the function $f(n)$, the nth digit in the decimal expansion of B, if we believe in its existence at all (as Brouwer does not), can be shown by non-effective means to be general recursive. The proof cannot be given without developing details of the theory of general recursive functions. But it will perhaps appear plausible that a proof can be given by specifying an infinite number of alternative computation procedures for $f(n)$, each one separately being an effective (general recursive) computation.

The examples show that we must distinguish between the question of effectiveness of an existence proof (e.g., of the proof that $g(n)$ is primitive recursive) and the question of the effective computability of a numerical function.

These questions regarding effectiveness will be of interest later, in treating the significance of the Gödel incompleteness theorems; and we shall want to verify as we proceed that certain of our metatheorems have been effectively proved. But we turn aside from this discussion now in order to deal with those theorems.

80. Gödel numbering. An important part of the proof of the Gödel theorems is the use of a method of assigning numbers to the formulas of the object language (in our present case the language A^2), which we shall call *Gödel numbers*. This may be done in many ways, and we must (although arbitrarily) select one particular way — which we now proceed to specify. First we choose a Gödel numbering of the primitive symbols of A^2:

$\supset$	$\sim$	[	]	$\forall$	S	(	)	,
5	7	11	13	17	19	23	29	31

Individual variables:	x	y	z	x_1	y_1	z_1	...
	1	2	2^2	2^3	2^4	2^5	...
Propositional variables:	p	q	r	s	p_1	...	
	3	$2 \cdot 3$	$2^2 3$	$2^3 3$	$2^4 3$	...	

m-ary functional variables:
1st m-ary variable: $3 \cdot 5 \cdot 7 \cdot \ldots \cdot \mathrm{pr}(m+2)$
2nd m-ary variable: $2(3 \cdot 5 \cdot 7 \cdot \ldots \cdot \mathrm{pr}(m+2))$

3rd m-ary variable: $2^2(3 \cdot 5 \cdot 7 \cdot \ldots \cdot \mathrm{pr}(m+2))$

.

.

.

Here $\mathrm{pr}(n)$ is the nth prime number.

A formula is a finite sequence of primitive symbols. We next provide a method for assigning a number to any given formula uniquely. Let the following expression be any formula with n symbols: $a_1a_2a_3\ldots a_n$. Each symbol of this formula has associated with it a unique Gödel number; let these numbers be in order: $x_1, x_2, x_3, \ldots, x_n$. The Gödel number of the formula is then $2^{x_1}3^{x_2}5^{x_3}\ldots\mathrm{pr}(n)^{x_n}$. That each formula has a unique number associated with it follows from the unique prime factorization theorem, for statement and proof of which the reader is referred to any standard elementary number theory text. Thus, for example, the Gödel number of the wff $[p \supset q]$ is given by the product of the numbers below it in the following table:

[	p	$\supset$	q	]
2^{11}	3^3	5^5	7^6	11^{13}

We will make use of the foregoing Gödel numbering to transform syntactical relations among the symbols and formulas of A^2 into relations among natural numbers. Consider the syntactical relation among the three formulas **A**,**B**, **C** and the variable **a**: **C** is well-formed and **B** is $\mathrm{S}^{\mathbf{a}}_{\mathbf{A}}\mathbf{C}|$. We will eventually express this syntactical relation as a primitive recursive relation among natural numbers.

We have so far introduced an individual table of Gödel numbers for each of the following syntactical categories of A^2: the improper symbols, the variables, the binary functional constant S, and the formulas. We now introduce a table of Gödel numbers for finite sequences of formulas. Suppose given a finite sequence of formulas in which each formula has one of the Gödel numbers (taken in order of occurrence in the sequence): $x_1, x_2, x_3, \ldots, x_n$. The Gödel number of the finite sequence of formulas shall then be: $2^{x_1}3^{x_2}5^{x_3}\ldots\mathrm{pr}(n)^{x_n}$. That each finite sequence of formulas has a unique number associated with it again follows from the unique prime factorization theorem.

We will also need Gödel numbering for the system of first-order arithmetic A^0. We assign the numbers 37 to $\sum$ and 41 to $\prod$ respectively, and to all the others we assign the same numbers as used for A^2. And for formulas and sequences of formulas we use the same scheme as that for A^2

In the material that follows we shall be concerned primarily with the system A^2, and the terms "formula," "well-formed formula," etc. are therefore to be understood as referring to A^2 unless the contrary is said. We prove a series of metatheorems in which various numerical and propositional functions, defined by reference to the syntax of A^2 and Gödel numbers of symbols and formulas of A^2, are shown to be primitive recursive. In order to abbreviate the statement we shall often merely name the function in question — the metatheorem is always to the effect that it is primitive recursive — and then write the recursion equations or other expression for the function by means of which its primitive recursiveness is demonstrated. Generally the account or description of the function in terms of Gödel numbers and the syntax of A^2 will not be such as to supply a value of the function for all natural-number arguments, or systems of arguments. The statement that the function is primitive recursive is then to be understood as meaning that there is a way of supplying the values of the function for the remaining systems of natural-number arguments so that the resulting function will be primitive recursive.

We notice that *the Gödel number of the nth symbol of formula number a* is $n \,\mathrm{Gl}\, a$ and *the length of formula number a* is $\mathrm{L}(a)$ (*758/*759). Hence the two italicized expressions correspond to primitive recursive functions — of n and a in the first case, and of a in the second.

This remark illustrates in two simple cases the sort of metatheorems of A^2 which we now go on to prove. In fact we shall more often think of $\mathrm{L}(a)$ as "the length of formula no. a" than with the meaning which is given more directly by *759. For when a is the number of a formula, the length of formula no. a is the same as the greatest number n such that $p_n \mid a$. Similarly we shall more often think of $n \,\mathrm{Gl}\, a$ as "the number of the nth symbol of formula no. a" or else as "the number of the nth formula in sequence of formulas no. a".

81. Primitive recursive functions and relations corresponding to the syntax of A^2.

*810. $a * b$, the number of the concatenation of formula no. a and formula no. b; i.e., the number of the formula which consists of the symbols of formula no. a in order followed by the symbols of formula no. b in order.

$$a * b = \left[\prod_{n=1}^{n=\mathrm{L}(a)} \mathrm{p}_n^{n \,\mathrm{Gl}\, a}\right]\left[\prod_{n=1}^{n=\mathrm{L}(b)} \mathrm{p}_{n+\mathrm{L}(a)}^{n \,\mathrm{Gl}\, b}\right]$$

We notice that $0 * 0 = 0 * 1 = 1 * 0 = 1 * 1$; and otherwise $a * 0 = a * 1 = a$ and $0 * b = 1 * b = b$. Also the function $*$ is associative.

*811. $\mathrm{Neg}(a)$, the number of the negation of formula no. a.

$$\mathrm{Neg}(a) = 2^7 * a$$

*812. $a \,\mathrm{Imp}\, b$, the number of the implication whose antecedent is formula no. a and whose consequent is formula no. b.

$$a \,\mathrm{Imp}\, b = 2^{11} * a * 2^5 * b * 2^{13}$$

*813. $v \,\mathrm{Gen}\, a$, the number of the formula obtained from formula no. a by generalizing upon variable no. v.

$$v \,\mathrm{Gen}\, a = 2^{23} * 2^{17} * 2^v * 2^{29} * a$$

*814. $\mathrm{Var}(n, v)$, that v is the number of a variable of category n (where the individual variables are taken as of category 0, the propositional variables of category 1, and the m-ary functional variables of category $m + 1$).

$$\mathrm{Var}(n, v) = \sum_{i=0}^{i=v}\left[v = 2^i \prod_{k=2}^{k=s(n)} \mathrm{p}_k\right]$$

*815. Var(v), that v is the number of a variable.

$$\mathrm{Var}(v) = \sum_{n=0}^{n=v} \mathrm{Var}(n, v)$$

*816. Elf(a), that a is the number of an elementary formula of A^2.

$$\mathrm{Elf}(a) = \Big[\sum_{k=0}^{k=a}[a = 2^{2^k 3}] \text{ or}$$
$$\sum_{n=2}^{n=a}\Big[2n = \mathrm{L}(a) \text{ and } \big[\mathrm{Var}(n, 1 \,\mathrm{Gl}\, a) \text{ or } [19 = 1 \,\mathrm{Gl}\, a \text{ and } n = 3]\big] \text{ and}$$
$$23 = 2 \,\mathrm{Gl}\, a \text{ and } \prod_{m=3}^{m=p(2n)} [[\mathrm{parx}(m) = 1 \text{ and } \mathrm{Var}(0, m \,\mathrm{Gl}\, a)] \text{ or}$$
$$[\mathrm{parx}(m) = 0 \text{ and } 31 = m \,\mathrm{Gl}\, a]\big] \text{ and } 29 = 2n \,\mathrm{Gl}\, a\Big]\Big]$$

*817. Op(a, b, c), that formula no. a is obtained from formula no. b and formula no. c, or else from formula no. b alone, by one of the formation rules 50iii, 50iv, 50v of A^2.

$$\mathrm{Op}(a, b, c) = \Big[a = \mathrm{Neg}(b) \text{ or } a = b \,\mathrm{Imp}\, c \text{ or } \sum_{v=1}^{v=a}\big[\mathrm{Var}(v) \text{ and } a = v \,\mathrm{Gen}\, b\big]\Big]$$

*818. Form(s), that s is the number of a formation sequence, according to the formation rules of A^2.

$$\mathrm{Form}(s) = \prod_{n=1}^{n=\mathrm{L}(s)} \Big[\mathrm{Elf}(n \,\mathrm{Gl}\, s) \text{ or } \sum_{k=1}^{k=p(n)} \sum_{m=1}^{m=p(n)} \mathrm{Op}(n \,\mathrm{Gl}\, s, k \,\mathrm{Gl}\, s, m \,\mathrm{Gl}\, s)\Big]$$

*819. Wff(a), that a is the number of a well-formed formula of A^2.

$$\mathrm{Wff}(a) = \sum_{s=2}^{s=a^a} \big[\mathrm{Form}(s) \text{ and } \mathrm{L}(s) \,\mathrm{Gl}\, s = a\big]$$

To justify the upper bound a^a for s we observe that the length of the formation sequence of number s need not be greater than $\mathrm{L}(a)$. Hence

$$s \leqq 2^a 3^a 5^a \cdots \mathrm{p}_{\mathrm{L}(a)}^a \leqq a^a$$

82. Primitive recursive functions and relations corresponding to definitions and to free and bound variables in A^2.

*820. a Conj b, the number of the conjunction of formula no. a and formula no. b.

$$a \,\mathrm{Cni}\, b = \mathrm{Neg}(b \,\mathrm{Imp}\, a),$$
$$a \,\mathrm{Conj}\, b = (a \,\mathrm{Cni}\, b) \,\mathrm{Cni}\, b$$

*821. $\mathrm{O}(n, v)$, the number of the wff $Z_n(\mathbf{v})$, where $\mathbf{v}$ is variable no. v.
As recursion equations for the function O we have:

$$\mathrm{O}(0, v) = 2v \;\mathrm{Gen}\; (2^7 3^{19} 5^{23} 7^{2v} 11^{31} 13^v 17^{29}),$$

$$\mathrm{O}(n+1, v) = \mathrm{Neg}\; (2v \;\mathrm{Gen}\; \mathrm{Neg}\; (\mathrm{O}(n, 2v) \;\mathrm{Conj}\; 2^{19} 3^{23} 5^{2v} 7^{31} 11^v 13^{29}))$$

These recursion equations are of the special form of *754, so that the primitive recursiveness of O follows by *754.

*822. $\mathrm{u}(n, a)$, the number of the (well-formed) formula $Z_n(x) \supset_x \mathbf{A}$, with the particular variable x, $\mathbf{A}$ being (well-formed) formula no. a.

$$\mathrm{u}(n, a) = 1 \;\mathrm{Gen}(\mathrm{O}(n, 1) \;\mathrm{Imp}\; a)$$

*823. $a \;\mathrm{in}\; b$, that formula no. a occurs in (i.e., as a consecutive part of) formula no. b.

$$\sum_{m=0}^{m=\mathrm{L}(b) \dot{-} \mathrm{L}(a)} \quad \prod_{n=1}^{n=\mathrm{L}(a)} [(m+n) \;\mathrm{Gl}\; b = n \;\mathrm{Gl}\; a]$$

*824. $\mathrm{Bd}(n, v, a)$, that a is the number of a wff of A^2 in which the nth symbol is a bound occurrence of variable no. v.

$$\mathrm{Wff}(a) \text{ and } v = n \;\mathrm{Gl}\; a \text{ and } \mathrm{Var}(v) \text{ and}$$
$$\sum_{b=1}^{b=a} \sum_{c=1}^{c=a} \sum_{d=1}^{d=a} [a = b * v \;\mathrm{Gen}\; c * d \text{ and } \mathrm{Wff}(c) \text{ and}$$
$$\mathrm{L}(b) < n \text{ and } \mathrm{L}(b * v \;\mathrm{Gen}\; c) \geqq n]$$

We notice that $\mathrm{Bd}(o, v, a)$ is always false.

*825. $\mathrm{Fr}(n, v, a)$, that a is the number of a wff of A^2 in which the nth symbol is a free occurrence of variable no. v.

$$\mathrm{Wff}(a) \text{ and } v = n \;\mathrm{Gl}\; a \text{ and } \mathrm{Var}(v) \text{ and not } \mathrm{Bd}(n, v, a)$$

*826. $\mathrm{Bd}(v, a)$, that variable no. v is a bound variable of wff no. a.

$$\sum_{n=1}^{n=\mathrm{L}(a)} \mathrm{Bd}(n, v, a)$$

*827. $\mathrm{Fr}(v, a)$, that variable no. v is a free variable of wff no. a.

$$\sum_{n=1}^{n=\mathrm{L}(a)} \mathrm{Fr}(n, v, a)$$

*828. $\mathrm{St}(k, v, a)$, the number (counting from the beginning) of the $(k+1)$th place from the end of wff no. a at which variable no. v is free.

*829. $\mathrm{A}(v, a)$, the number of free occurrences of variable no. v in wff no. a.

$$\mathrm{St}(0, v, a) = s(\mathrm{L}(a)) \dot{-} \mathop{\mathrm{E}}_{n=1}^{n=\mathrm{L}(a)} \mathrm{Fr}(s(\mathrm{L}(a)) \dot{-} n, v, a)$$

$$\mathrm{St}(k+1, v, a) = \mathrm{St}(k, v, a) \dot{-} \mathop{\mathrm{E}}_{n=1}^{n=\mathrm{L}(a)} \mathrm{Fr}(\mathrm{St}(k, v, a) \dot{-} n, v, a)$$

This makes $\mathrm{St}(k, v, a) = s(\mathrm{L}(a))$ if there are no free occurrences of variable no. v in wff no. a, or if a is not the number of a wff. And in the case in which there are fewer than $k+1$ free occurrences of variable no. v in wff no. a (since $\mathrm{Fr}(0, v, a)$ is always false) it makes $\mathrm{St}(k, v, a) =$ the number of the first place in wff no. a at which variable no. v is free. Hence we may show the primitive recursiveness of $\mathrm{A}(v, a)$ by writing

$$\mathrm{A}(v, a) = \mathop{\mathrm{E}}_{k=1}^{k=\mathrm{L}(a)} [\mathrm{St}(k, v, a) = \mathrm{St}(p(k), v, a) \text{ and } \mathrm{St}(k, v, a) \leqq \mathrm{L}(a)]$$

83. Primitive recursive functions and relations corresponding to substitution in A^2 and to well-formed formulas in A^0.

*830. $\mathrm{Su}\, a \binom{n}{b}$, the number of the formula obtained by substituting formula no. b for the nth symbol of formula no. a.

$$\mathop{\mathrm{E}}_{c=1}^{c=a*b*a} \sum_{a_1=1}^{a_1=a} \sum_{a_2=1}^{a_2=a} [a = a_1 * 2^{n \,\mathrm{Gl}\, a} * a_2 \text{ and } n = \mathrm{L}(a_1) + 1 \text{ and } c = a_1 * b * a_2]$$

*831. $\underset{\centerdot}{\mathrm{S}}(a^v_b)$, the number of the formula obtained by substituting formula no. b for all free occurrences of variable no. v in wff no. a.

$$\underset{\centerdot}{\mathrm{S}}\left(0, a^v_b\right) = a,$$

$$\underset{\centerdot}{\mathrm{S}}\left(k+1, a^v_b\right) = \mathrm{Su}\, \underset{\centerdot}{\mathrm{S}}\left(k, a^v_b\right) \binom{\mathrm{St}(k,v,a)}{b},$$

$$\underset{\centerdot}{\mathrm{S}}\left(a^v_b\right) = \underset{\centerdot}{\mathrm{S}}\left(\mathrm{A}(v, a), a^v_b\right)$$

*832. $\mathrm{Sbd}(a^v_b)$, the number of the formula obtained by substituting formula no. b for all bound occurrences of variable no. v in wff no. a.

To prove this follow the analogy of the proof of *831, after first going back to *828 and *829 to prove analogues of them in which "free" is replaced by "bound" (and in the proof "Fr" is replaced by "Bd").

*833. $\mathrm{Arg}(m, k, v, a)$, the Gödel number of the variable which is the mth argument of functional variable no. v at its $(k+1)$th free occurrence, counting from the end, in wff no. a.

$$\mathrm{Arg}(m, k, v, a) = (\mathrm{St}(k, v, a) + 2m)\, \mathrm{Gl}\, a$$

*834. Del(n, k, a), the number of the formula obtained by deleting from formula no. a the $k \dot{-} 1$ consecutive symbols which immediately follow the nth symbol.

$$\mathop{\mathrm{E}}_{c=1}^{c=a} \sum_{a_1=1}^{a_1=a} \sum_{a_2=1}^{a_2=a} \sum_{a_3=1}^{a_3=a} \Big[a = a_1 * a_2 * a_3 \text{ and } c = a_1 * a_3 \text{ and } \mathrm{L}(a_1) = n \text{ and } \mathrm{L}(a_2) = k \dot{-} 1 \Big]$$

*835. m(e), the number of occurrences of individual variables in elementary formula no. e.

$$\mathrm{m}(e) = h(\mathrm{L}(e) \dot{-} 2)$$

*836. $\check{\mathrm{S}}\left(a^{e}_{b}\right)$, the number of $\check{\mathrm{S}}_{\mathbf{B}}^{\mathbf{f}(\mathbf{x}_1, \mathbf{x}_2, \ldots, \mathbf{x}_n)} \mathbf{A}|$ where a, b, e are the numbers of $\mathbf{A}, \mathbf{B}$, and $\mathbf{f}(\mathbf{x}_1, \mathbf{x}_2, \ldots, \mathbf{x}_n)$ respectively, or the number of $\check{\mathrm{S}}_{\mathbf{B}}^{\mathbf{p}} \mathbf{A}|$ where a, b, e are the numbers of the wffs $\mathbf{A}, \mathbf{B}, \mathbf{p}$ respectively — $\mathbf{f}$ being an n-ary functional variable, $\mathbf{x}_1, \mathbf{x}_2, \ldots, \mathbf{x}_n$ distinct individual variables, and $\mathbf{p}$ a propositional variable.

$$\mathrm{Sb}(0, a, e, b) = b,$$

$$\mathrm{Sb}(n+1, a, e, b) = \underset{\centerdot}{\mathrm{S}}\left(\mathrm{Sb}(n, a, e, b)^{(2n+3)\,\mathrm{Gl}\,e}_{2^{2^{n+a+e+b}}}\right)$$

$$\mathrm{Sb}(0, k, a, e, b) = \mathrm{Sb}(\mathrm{m}(e), a, e, b),$$

$$\mathrm{Sb}(m+1, k, a, e, b) = \underset{\centerdot}{\mathrm{S}}\left(\mathrm{Sb}(m, k, a, e, b)^{2^{m+a+e+b}}_{2^{\mathrm{Arg}(m+1, k, 1\,\mathrm{Gl}\,e, a)}}\right)$$

$$\mathrm{S}\left(k, a, c^{e}_{b}\right) = \mathrm{Su\,Del}\left(\mathrm{St}(k, 1\,\mathrm{Gl}\,e, a), \mathrm{L}(e), c\right)\binom{\mathrm{St}(k, 1\,\mathrm{Gl}\,e, a)}{\mathrm{Sb}(\mathrm{m}(e), k, a, e, b)}$$

$$\mathrm{S}\left(0, a^{e}_{b}\right) = a,$$

$$\mathrm{S}\left(k+1, a^{e}_{b}\right) = \mathrm{S}\left(k, a, \mathrm{S}\left(k, a^{e}_{b}\right)^{e}_{b}\right)$$

$$\mathrm{S}\left(a^{e}_{b}\right) = \mathrm{S}\left(\mathrm{A}(1\,\mathrm{Gl}\,e, a), a^{e}_{b}\right)$$

$$\check{\mathrm{S}}\left(a^{e}_{b}\right) = a \text{ or } \mathrm{S}\left(a^{e}_{b}\right) \text{ according as}$$

$$\left[\sum_{v=1}^{v=b}\sum_{c=1}^{c=a}\sum_{x=1}^{x=a}\sum_{y=1}^{y=a}\sum_{n=1}^{n=\mathrm{L}(a)}\left[a = x * v \operatorname{Gen} c * y \text{ and } \mathrm{Fr}(n, 1\,\mathrm{Gl}\,e, a) \text{ and}\right.\right.$$

$$\left. n > \mathrm{L}(x) \text{ and } n \leqq \mathrm{L}(x * v \operatorname{Gen} c) \text{ and } \mathrm{Fr}(v, b) \text{ and not } \mathrm{Fr}(v, e)\right] \text{ or}$$

$$\sum_{m=1}^{m=m(e)}\sum_{k=0}^{k=\mathrm{L}(a)}\sum_{c=1}^{c=b}\left[\mathrm{Fr}((2m+1)\,\mathrm{Gl}\,e, \mathrm{Arg}(m, k, 1\,\mathrm{Gl}\,e, a) \operatorname{Gen} c) \text{ and}\right.$$

$$\left.\left.\mathrm{Arg}(m, k, 1\,\mathrm{Gl}\,e, a) \operatorname{Gen} c \text{ in } b\right]\right] \text{ or not}$$

The three following metatheorems *837–*839 have reference to the syntax of A^0 rather than of A^2, but in the present context they are regarded as metatheorems of A^2 because they have as effective consequences the existence of certain wffs and the provability of certain theorems in A^2.

*837. $\mathrm{Elf}^0(a)$, that a is the number of an elementary formula of A^0.

$$\mathrm{Elf}^0(a) = [8 = \mathrm{L}(a) \text{ and } [37 = 1\,\mathrm{Gl}\,a \text{ or } 41 = 1\,\mathrm{Gl}\,a] \text{ and}$$
$$23 = 2\,\mathrm{Gl}\,a \text{ and } \mathrm{Var}(0, 3\,\mathrm{Gl}\,a) \text{ and } 31 = 4\,\mathrm{Gl}\,a \text{ and}$$
$$\mathrm{Var}(0, 5\,\mathrm{Gl}\,a) \text{ and } 31 = 6\,\mathrm{Gl}\,a \text{ and}$$
$$\mathrm{Var}(0, 7\,\mathrm{Gl}\,a) \text{ and } 29 = 8\,\mathrm{Gl}\,a]$$

*838. $\mathrm{Form}^0(s)$, that s is the number of a formation sequence, according to the formation rules of A^0.

$\mathrm{Op}^0(a, b, c)$ may be expressed in the same way as $\mathrm{Op}(a, b, c)$ in the proof of *817, except that $\mathrm{Var}(v)$ is replaced by $\mathrm{Var}(0, v)$. Then $\mathrm{Form}^0(s)$ may be expressed in the same way as $\mathrm{Form}(s)$ in the proof of *818, except that Elf is replaced by Elf^0 and Op by Op^0.

*839. $\mathrm{Wff}^0(a)$, that a is the number of a well-formed formula of A^0.

$\mathrm{Wff}^0(a)$ may be expressed in the same way as $\mathrm{Wff}(a)$ in the proof of *819, except that Form is replaced by Form^0.

84. Primitive recursive functions and relations corresponding to the rules of inference and axioms of A^2, to proof in A^2, and to valuations.

*840. $\mathrm{Imm}(a, b, c)$, that formula no. a is an immediate inference from formula no. b, or from formula no. b and formula no. c, by one of the four rules of inference *500–*503.

$$\mathrm{Imm}(a, b, c) = \left[b = c \operatorname{Imp} a \text{ or } \sum_{v=1}^{v=a}\left[\mathrm{Var}(v) \text{ and } a = v \operatorname{Gen} b\right] \text{ or}\right.$$

$$\sum_{b_1=1}^{b_1=b}\sum_{b_2=1}^{b_2=b}\sum_{b_3=1}^{b_3=b}\sum_{v=1}^{v=b_2}\sum_{w=1}^{w=a}[b = b_1 * b_2 * b_3 \text{ and } \mathrm{Wff}(b_2) \text{ and}$$

$$\mathrm{Var}(0, v) \text{ and } \mathrm{Var}(0, w) \text{ and not } \mathrm{Fr}(v, b_2) \text{ and}$$

$$\text{not } 2^w \text{ in } b_2 \text{ and } a = b_1 * \mathrm{Sbd}(b_{2}{}^{v}_{2^w}) * b_3\big]$$

$$\text{or} \sum_{v=1}^{v=b}\sum_{w=1}^{w=a}\big[\mathrm{Var}(0,v) \text{ and } \mathrm{Var}(0,w) \text{ and}$$

$$\mathrm{Wff}(b) \text{ and } a = \underset{\bullet}{\mathrm{S}}(b^{v}_{2^w}) \text{ and}$$

$$\prod_{n=1}^{n=\mathrm{L}(b)} \big[\mathrm{Fr}(n,v,b) \text{ implies } \mathrm{Fr}(n,w,a)\big]\big]\Big]$$

*841. $\mathrm{Ax}_{508}(a)$, that a is the number of an instance of axiom schema *508.

$$\sum_{b=1}^{b=a}\sum_{c=1}^{c=a}\sum_{v=1}^{v=a}\sum_{n=1}^{n=v}\big[\mathrm{Var}(n,v) \text{ and } \mathrm{Wff}(b) \text{ and } \mathrm{Wff}(c) \text{ and}$$

$$a = \big(v\,\mathrm{Gen}(b\,\mathrm{Imp}\,c)\big)\,\mathrm{Imp}\big(b\,\mathrm{Imp}(v\,\mathrm{Gen}\,c)\big) \text{ and}$$

$$\text{not } \mathrm{Fr}(v,b)\big]$$

*842. $\mathrm{Ax}_{509}(a)$, that a is the number of an instance of axiom schema *509.

$$\sum_{b=1}^{b=a^{a^2}}\sum_{c=1}^{c=a}\sum_{e=1}^{e=a^{a^2}}\Big[\mathrm{Wff}(b) \text{ and } \mathrm{Wff}(c) \text{ and } \mathrm{Elf}(e) \text{ and}$$

$$\prod_{m=1}^{m=\mathrm{m}(e)}\prod_{n=1}^{n=\mathrm{m}(e)}\big[(2m+1)\,\mathrm{Gl}\,e = (2n+1)\,\mathrm{Gl}\,e \text{ implies } m = n\big]$$

$$\text{and } \mathrm{Var}(1\,\mathrm{Gl}\,e) \text{ and } a = \big((1\,\mathrm{Gl}\,e)\,\mathrm{Gen}\,c\big)\,\mathrm{Imp}\,\check{\mathrm{S}}\left(c^{e}_{b}\right)\Big]$$

We supply the following justification for using a^{a^2} as upper bound of b and e in this expression for $\mathrm{Ax}_{509}(a)$. Let a be the number of **A**, b the number of **B**, c the number of **C**, e the number of the elementary formula **E**, and let **f** be the variable which is the initial symbol of **E**. Then **A** is to be $(\mathbf{f})\mathbf{C} \supset \check{\mathrm{S}}^{\mathbf{E}}_{\mathbf{B}}\mathbf{C}|$ and we wish to take the upper bounds of b and e large enough so that all possibilities are covered by which **A** might be an instance of axiom schema *509. This does not mean that we must take the upper bounds large enough to cover all possible wffs **B** and **E**. But rather we may suppose that **A** is first given as $(\mathbf{f})\mathbf{C} \supset \check{\mathrm{S}}^{\mathbf{E}_0}_{\mathbf{B}_0}\mathbf{C}|$ and consider changing $\mathbf{B}_0$ and $\mathbf{E}_0$ to a wff **B** and a suitable elementary formula **E** in such a way that **A** is also $(\mathbf{f})\mathbf{C} \supset \check{\mathrm{S}}^{\mathbf{E}}_{\mathbf{B}}\mathbf{C}|$.

Case 1. Let **f** be an n-ary functional variable, so that $\mathbf{E}_0$ is $\mathbf{f}(\mathbf{x}_{01}, \mathbf{x}_{02}, \cdots, \mathbf{x}_{0n})$ ($\mathbf{x}_{01}, \mathbf{x}_{02}, \cdots, \mathbf{x}_{0n}$ being individual variables that are all different) and **E** is $\mathbf{f}(\mathbf{x}_1, \mathbf{x}_2, \cdots, \mathbf{x}_n)$ ($\mathbf{x}_1, \mathbf{x}_2, \cdots, \mathbf{x}_n$ being again individual variables that are all different); and suppose that **A** is not the same as $(\mathbf{f})\mathbf{C} \supset \mathbf{C}$. We may suppose that **B** is

$$\underset{\bullet}{\mathrm{S}}^{\mathbf{x}_{01}\ \mathbf{x}_{02}\ \cdots\ \mathbf{x}_{0n}}_{\mathbf{x}_1\ \mathbf{x}_2\ \cdots\ \mathbf{x}_n}\ \mathbf{B}_0|$$

and $\mathbf{B}_0$ is

$$\underset{\bullet}{\mathrm{S}}^{\mathbf{x}_1\ \mathbf{x}_2\ \cdots\ \mathbf{x}_n}_{\mathbf{x}_{01}\ \mathbf{x}_{02}\ \cdots\ \mathbf{x}_{0n}}\ \mathbf{B}|.$$

Let the elementary parts of **C** which contain free occurrences of **f** be $\mathbf{f}(\mathbf{y}_{i1}, \mathbf{y}_{i2}, \cdots, \mathbf{y}_{in})$, $i = 1, 2, \cdots, k$. Then $k \geq 1$. The corresponding wf parts $\mathbf{B}_i$ of $\check{\mathrm{S}}^{\mathbf{E}}_{\mathbf{B}}\mathbf{C}|$ are

$$\mathrm{S}^{\mathbf{x}_{01}\ \mathbf{x}_{02}\ \cdots\ \mathbf{x}_{0n}}_{\mathbf{y}_{i1}\ \mathbf{y}_{i2}\ \cdots\ \mathbf{y}_{in}} \mathbf{B}_0|$$

and these are the same as

$$\mathrm{S}^{\mathbf{x}_1\ \mathbf{x}_2\ \cdots\ \mathbf{x}_n}_{\mathbf{y}_{i1}\ \mathbf{y}_{i2}\ \cdots\ \mathbf{y}_{in}} \mathbf{B}|,\ i = 1, 2, \cdots, k.$$

The free individual variables in $\mathbf{B}_0$ other than $\mathbf{x}_{01}, \mathbf{x}_{02}, \cdots, \mathbf{x}_{0n}$ and the bound individual variables in $\mathbf{B}_0$ all occur also in the wffs $\mathbf{B}_i$; hence they occur in **A**; hence they have numbers less than a. In choosing $\mathbf{x}_1, \mathbf{x}_2, \cdots, \mathbf{x}_n$ we have only to choose them as individual variables which are all different among themselves and different from all the individual variables in $\mathbf{B}_0$ except (possibly) $\mathbf{x}_{01}, \mathbf{x}_{02}, \cdots, \mathbf{x}_{0n}$. Let us therefore choose $\mathbf{x}_1, \mathbf{x}_2, \cdots, \mathbf{x}_n$ as the first n individual variables in alphabetic order that have numbers greater than a. This has the effect that the numbers of the variables $\mathbf{x}_1, \mathbf{x}_2, \cdots, \mathbf{x}_n$ are respectively less than $2a, 4a, \cdots, 2^n a$. Since $n < \mathrm{L}(a)$, we have $2^n < 2^{\mathrm{L}(a)}$. Hence all the variables $\mathbf{x}_1, \mathbf{x}_2, \cdots, \mathbf{x}_n$ have numbers less than $2^{\mathrm{L}(a)}a$.

The symbols in **B**, other than the variables $\mathbf{x}_1, \mathbf{x}_2, \cdots, \mathbf{x}_n$, all occur also in $\mathbf{B}_i$; hence they occur in **A**; hence they have numbers less than a. The symbols in **E**, other than the variables $\mathbf{x}_1, \mathbf{x}_2, \cdots, \mathbf{x}_n$, all occur in **C**; hence they occur in **A**; hence they have numbers less than a.

Thus altogether, all symbols in **B** and **E** have numbers less than $2^{\mathrm{L}(a)}a$. Hence:

$$\begin{aligned} b &= 2^{1\,\mathrm{Gl}\,b} 3^{2\,\mathrm{Gl}\,b} 5^{3\,\mathrm{Gl}\,b} \cdots \mathrm{p}_{\mathrm{L}(b)}^{\mathrm{L}(b)\,\mathrm{Gl}\,b} \\ &< 2^{2^{\mathrm{L}(a)}a} 3^{2^{\mathrm{L}(a)}a} 5^{2^{\mathrm{L}(a)}a} \cdots \mathrm{p}_{\mathrm{L}(b)}^{2^{\mathrm{L}(a)}a} \\ &< 2^{2^{\mathrm{L}(a)}a} 3^{2^{\mathrm{L}(a)}a} 5^{2^{\mathrm{L}(a)}a} \cdots \mathrm{p}_{\mathrm{L}(a)}^{2^{\mathrm{L}(a)}a} \\ &= (2 \cdot 3 \cdot 5 \cdots \mathrm{p}_{\mathrm{L}(a)})^{2^{\mathrm{L}(a)}a} \\ &< a^{2^{\mathrm{L}(a)}a}, \end{aligned}$$

$$\begin{aligned} e &= 2^{1\,\mathrm{Gl}\,e} 3^{2\,\mathrm{Gl}\,e} 5^{3\,\mathrm{Gl}\,e} \cdots \mathrm{p}_{2n+2}^{(2n+2)\,\mathrm{Gl}\,e} \\ &< 2^{2^{\mathrm{L}(a)}a} 3^{2^{\mathrm{L}(a)}a} 5^{2^{\mathrm{L}(a)}a} \cdots \mathrm{p}_{2n+2}^{2^{\mathrm{L}(a)}a} \\ &< 2^{2^{\mathrm{L}(a)}a} 3^{2^{\mathrm{L}(a)}a} 5^{2^{\mathrm{L}(a)}a} \cdots \mathrm{p}_{\mathrm{L}(a)}^{2^{\mathrm{L}(a)}a} \\ &= (2 \cdot 3 \cdot 5 \cdots \mathrm{p}_{\mathrm{L}(a)})^{2^{\mathrm{L}(a)}a} \\ &< a^{2^{\mathrm{L}(a)}a}. \end{aligned}$$

Case 2. Let **f** be a propositional variable, and suppose that **A** is not the same as $(\mathbf{f})\mathbf{C} \supset \mathbf{C}$. In this case $\mathbf{E}_0$ is the formula consisting of the propositional variable **f** standing alone. We take **E** to be the same as $\mathbf{E}_0$ and **B** the same as $\mathbf{B}_0$. Since **A** is not the same as $(\mathbf{f})\mathbf{C} \supset \mathbf{C}$, there must be at least one free occurrence of **f** in **C**, and at least one occurrence of **B** as a wf part of $\check{\mathrm{S}}^{\mathbf{E}}_{\mathbf{B}}\mathbf{C}|$. Hence there is at least one occurrence of **B** as a wf part of **A**. Hence $b < a < a^{2^{\mathrm{L}(a)}a}$; and $e < a < a^{2^{\mathrm{L}(a)}a}$.

Before treating the next case we need the four following lemmas.

Lemma 1. $\mathrm{p}_{N+1} \leqq 1 + \prod_{i=1}^{i=N} \mathrm{p}_i$.

For the right-hand side of the inequality cannot be divisible by p_i if $i < N+1$. Hence the prime factors of the right-hand side must all be $\geqq \mathrm{p}_{N+1}$.

Lemma 2. $\mathrm{p}_{N+1} < 2^{N!}$ if $N > 2$.

Proof by mathematical induction with respect to N. The lemma holds when $N = 3$. Suppose it holds for a particular N, and consider replacing N by $N+1$. We have by Lemma 1 that $\mathrm{p}_{N+2} \leqq 1 + \prod_{i=1}^{i=N+1} \mathrm{p}_i$. From the product $\prod_{i=1}^{i=N+1} \mathrm{p}_i$ we obtain p_{N+1}^{N+1} by increasing every factor in the product to p_{N+1}. Since N is at least 3, this has the effect of multiplying the product $\prod_{i=1}^{i=N+1} \mathrm{p}_i$ by at least $\frac{7}{2}\frac{7}{3}\frac{7}{5} = \frac{343}{30}$, and hence of increasing it by at least $\frac{343}{30}210 - 210 = 2191$. Thus we get $\mathrm{p}_{N+2} < \mathrm{p}_{N+1}^{N+1} < (2^{N!})^{N+1} < 2^{(N+1)!}$.

Lemma 3. $(2n+2)(2n+3) \leqq 7^{3^{n+1}2}$.

Proof by mathematical induction with respect to n. The lemma holds when $n = 0$. Then

$$
\begin{aligned}
(2n+4)(2n+5) &= \frac{2n+4}{2n+2}\,\frac{2n+5}{2n+3}(2n+2)(2n+3) \\
&< 2 \cdot 2 \cdot 7^{3^{n+1}2} \\
&< 7^{3^{n+1}4} 7^{3^{n+1}2} \\
&= 7^{3^{n+1}6} \\
&= 7^{3^{n+2}2}.
\end{aligned}
$$

Lemma 4. $(2n+1)!(2^n + 31n + 20) < 3^{n+1}7^{3^{n+1}}$.

Proof by mathematical induction with respect to n. The lemma holds when $n = 0$ and when $n = 1$. Suppose it holds for a particular n greater then 0, and consider replacing n by $n+1$.

$$
\begin{aligned}
(2n+3)!(2^{n+1} + 31n + 51) &= (2n+1)!(2n+2)(2n+3)(2^n + 31n + 20 + 2^n + 31) \\
&< (2n+1)!(2n+2)(2n+3)2(2^n + 31n + 20) \\
&< 2(2n+2)(2n+3)3^{n+1}7^{3^{n+1}} \\
&< (2n+2)(2n+3)3^{n+2}7^{3^{n+1}}.
\end{aligned}
$$

Hence by Lemma 3,

$$
\begin{aligned}
(2n+3)!(2^{n+1} + 31n + 51) &< 7^{3^{n+1}2}3^{n+2}7^{3^{n+1}} \\
&= 3^{n+2}7^{3^{n+2}}.
\end{aligned}
$$

Case 3. Let **f** be an n-ary functional variable and suppose that **A** is (**f**)**C** ⊃ **C**. In this case we take **E** to be $\mathbf{f}(\mathbf{x}_1, \mathbf{x}_2, \cdots, \mathbf{x}_n)$, where $\mathbf{x}_1, \mathbf{x}_2, \cdots, \mathbf{x}_n$ are the first n individual variables in alphabetic order, and we take **B** to be the same as **E**. The smallest possible number a for **A** is obtained by taking **C** to be the formula that consists of the particular propositional variable p standing alone; and if **f** is the mth n-ary functional variable in alphabetic order, we then have:

$$a = 2^{11}3^{23}5^{17}7^{2^{m-1}3\cdot 5\cdot 7\cdots \mathrm{p}_{n+2}}11^{29}13^{3}17^{5}19^{3}23^{13}$$
$$b = 2^{2^{m-1}3\cdot 5\cdot 7\cdots \mathrm{p}_{n+2}}3^{23}5^{17}7^{31}11^{2}13^{31}\cdots \mathrm{p}_{2n}^{31}\mathrm{p}_{2n+1}^{2^{n-1}}\mathrm{p}_{2n+2}^{29}$$

and e the same as b. In the expression for b we may replace all the prime factors from 3 on by p_{2n+2}, so obtaining

$$b < 2^{2^{m-1}3\cdot 5\cdot 7\cdots \mathrm{p}_{n+2}}\mathrm{p}_{2n+2}^{2^n+31n+20},$$

and in the right-hand side of this last inequality we have for the first factor that $2^{2^{m-1}3\cdot 5\cdot 7\cdots \mathrm{p}_{n+2}} < a$ (because of the occurrence of the factor $7^{2^{m-1}3\cdot 5\cdot 7\cdots \mathrm{p}_{n+2}}$ in the expression for a) and for the second factor we seek to prove that

$$\mathrm{p}_{2n+2}^{2^n+31n+20} < a^a.$$

By Lemma 4, and since $m \geqq 1$, we have that

$$(2n+1)!(2^n+31n+20) < 2^{m-1}3^{n+1}7^{2^{m-1}3^{n+1}}.$$

Then by Lemma 2, since $2n+1 > 2$, we have that

$$\begin{aligned}\mathrm{p}_{2n+2}^{2^n+31n+20} &< 2^{(2n+1)!(2^n+31n+20)}\\ &< 2^{2^{m-1}3^{m+1}7^{2^{m-1}3^{n+1}}}\\ &< 7^{2^{m-1}3^{n+1}7^{2^{m-1}3^{n+1}}}\\ &= (7^{2^{m-1}3^{n+1}})^{7^{2^{m-1}3^{n+1}}}.\end{aligned}$$

But from the expression for a above we have

$$7^{2^{m-1}3^{n+1}} < 7^{2^{m-1}3\cdot 5\cdot 7\cdots \mathrm{p}_{n+2}} < a.$$

Hence $\mathrm{p}_{2n+2}^{2^n+31n+20} < a^a$. Hence $b < a \cdot a^a = a^{a+1} < a^{2^{\mathrm{L}(a)}a}$. And since $e = b$ we have also $e < a^{2^{\mathrm{L}(a)}a}$.

Case 4. Let **f** be a propositional variable, and suppose that **A** is (**f**)**C** ⊃ **C**. In this case we take **E** to be the formula which consists of the propositional variable **f** standing alone, and **B** to be the same as **E**. Then the formula **B** (hence also **E**) occurs within the formula **A** as a part of it. Hence $b < a < a^{2^{\mathrm{L}(a)}a}$; and $e < a < a^{2^{\mathrm{L}(a)}a}$.

Thus we have shown in all cases that b and e are less than $a^{2^{\mathrm{L}(a)}a}$. But we have:

$$\begin{aligned}2^{\mathrm{L}(a)} &= 2\cdot 2\cdot 2\cdots 2, \text{ to } \mathrm{L}(a) \text{ factors},\\ &< 2\cdot 3\cdot 5\cdots \mathrm{p}_{\mathrm{L}(a)}\\ &< 2^{1\,\mathrm{Gl}\,a}3^{2\,\mathrm{Gl}\,a}5^{3\,\mathrm{Gl}\,a}\cdots \mathrm{p}_{\mathrm{L}(a)}^{\mathrm{L}(a)\,\mathrm{Gl}\,a}\\ &= a\end{aligned}$$

Thus replacing $2^{\mathrm{L}(a)}$ by a, we have that b and e are less than a^{a^2}.

*843. Ax(a), that a is the number of one of the axioms of A^2.

$$\begin{array}{llllll} a = n_{505} & \text{or} & a = n_{506} & \text{or} & a = n_{507} & \text{or} \\ a = n_{508} & \text{or} & a = n_{509} & \text{or} & \mathrm{Ax}_{508}(a) & \text{or} \\ \mathrm{Ax}_{509}(a) & \text{or} & a = n_{700} & \text{or} & a = n_{701} & \text{or} \\ a = n_{702} & \text{or} & a = n_{703} & \text{or} & a = n_{704} & \end{array}$$

Here n_{505}, n_{506}, etc. are the Gödel numbers of the axioms †505, †506, †507, †508, †509, †700, †701, †702, †703, †704. These numbers can of course be computed and written in, and we suppose this done. For example, n_{505} is $2^{11}3^3 5^5 7^{11} 11^6 13^5 17^3 19^{13} 23^{13}$.

*844. B(s), that s is the number of a proof (in A^2).

$$\mathrm{L}(s) \neq 0 \text{ and } \prod_{n=1}^{n=\mathrm{L}(a)} \left[\mathrm{Ax}(n \,\mathrm{Gl}\, s) \text{ or } \sum_{k=1}^{k=p(n)} \sum_{m=1}^{m=p(n)} \mathrm{Imm}(n \,\mathrm{Gl}\, s, k \,\mathrm{Gl}\, s, m \,\mathrm{Gl}\, s)\right]$$

*845. B(s, a), that s is the number of a proof of wff no. a.

$$\mathrm{B}(s) \text{ and } a = \mathrm{L}(s) \,\mathrm{Gl}\, s$$

*846. P(s, a), that wff no. a has a proof whose number is less than s.

$$\sum_{r=0}^{r=p(s)} \mathrm{B}(r, a)$$

*847. num(v), the place in alphabetic order of individual variable no. v.

$$\mathrm{num}(v) = \mathop{\mathrm{E}}_{n=1}^{n=v}\left[v = 2^{p(n)}\right]$$

We shall speak of a number w as determining a *valuation* of, or an assignment of *values* to, the individual variables in the sense that the nth individual variable (in alphabetic order) has assigned as value the number n Gl w. We remark that a valuation of the individual variables has a number w determining it in this way, if and only if there are no more than a finite number of the individual variables to which a value other than 0 is assigned. And we show the two following functions to be primitive recursive.

*848. val(v, w), the value of individual variable no. v according to valuation no. w.

$$\mathrm{val}(v, w) = \mathrm{num}(v) \,\mathrm{Gl}\, w$$

*849. dif(v, w_1, w_2), the valuation no. w_1, and valuation no. w_2 differ only (at most) in regard to the value assigned to variable no. v.

$$\prod_{n=1}^{n=\mathrm{L}(w_1)+\mathrm{L}(w_2)} \left[n \neq \mathrm{num}(v) \text{ implies } n \,\mathrm{Gl}\, w_1 = n \,\mathrm{Gl}\, w_2\right]$$

85. Gödel's first incompleteness theorem. The theorem falls into the following parts, where J is a certain closed wff of A^2 to be defined below.

**851. If A^2 is consistent, J is not a theorem.

**852. If A^2 is ω-consistent, $\sim J$ is not a theorem.

Before beginning the proofs of these theorems, we introduce some preliminary definitions.

Consistency of A^2. (1) *Simple Consistency* There is no propositional form P, such that both P and $\sim P$ are theorems of A^2.

We remark that this notion of consistency is relative to which symbol in the formal system (e.g., the system A^2) is to express negation. And indeed a formal system need not have a negation sign (that is, no reasonable choice for a negation sign), but one may still meaningfully ask if it is consistent.

(2) *Post Consistency* A wff that consists of a propositional variable standing alone is not a theorem of A^2.

A similar remark may be made about this notion of consistency, viz. that it is relative to the specification of what are propositional variables. Indeed a formal system need not have propositional variables — but the question of its consistency may still arise.

(3) *Absolute Consistency* Not every propositional form or sentence of A^2 is a theorem.

In the case of A^2 the propositional forms and the sentences together make up the entire class of wffs. In other cases they may be wffs that are not propositional forms or sentences. In the case of any particular formal (or logistic) system, the standard program of formalization requires that there shall be a syntactical criterion by which those formulas which are propositional forms and those which are sentences are distinguished from others. But to say in general what a propositional form or a sentence shall be, it is difficult to avoid saying something about meaning, i.e., something that is semantical rather than syntactical.

Thus although all three notions of consistency have a semantical background, the three above definitions of consistency for A^2 are syntactical. And for another particular formal system, the analogous definitions will again be syntactical. Note also that in the particular case of A^2 all three notions of consistency are equivalent.

Proof. If not (1), then not (2). Suppose not (1). Then for some propositional form $\mathbf{P}$, $\vdash \mathbf{P}$ and $\vdash \sim\mathbf{P}$. But $\vdash \sim p \supset . p \supset q$. Hence, by substitution for propositional variables (*510_0) $\vdash \sim\mathbf{P} \supset . \mathbf{P} \supset q$. Hence $\vdash q$.

If not (2), then not (3). Suppose not (2). Then for some propositional variable $\mathbf{p}$, $\vdash \mathbf{p}$. Hence every wff of A^2 is theorem, by substitution for propositional variables. But every propositional form and sentence of A^2 is a wff of A^2. So not (3).

If not (3) then not (1). Suppose not (3). Then every propositional form and sentence of A^2 is a theorem. Hence, there is a propositional form $\mathbf{P}$, such that $\vdash \mathbf{P}$ and $\vdash \sim\mathbf{P}$.

From these three implications it follows that not (1), not (2), not (3) are all equivalent. Hence (1), (2), (3) are all equivalent.

But if we define three analogous notions of consistency for some other formal system in place of A^2, it may happen that not all three notions are equivalent. For example, a formal system might be based on a non-classical propositional calculus in which $\sim p \supset . p \supset q$ is not a theorem.

ω-Consistency of A^2. That A^2 is ω-consistent means that there do not exist an individual variable **a** and wff **M** such that all the following are theorems of A^2:

$$(\exists \mathbf{a}) \sim \mathbf{M}, \quad Z_0(\mathbf{a}) \supset \mathbf{M},$$
$$Z_1(\mathbf{a}) \supset \mathbf{M}, \quad Z_2(\mathbf{a}) \supset \mathbf{M}, \quad Z_3(\mathbf{a}) \supset \mathbf{M}, \ldots$$

Note that if all of the above were theorems of A^2, we could not reasonably hold that the range of the individual variable was exactly the natural numbers. For in the principal interpretation of A^2, the value of $(\exists \mathbf{a})\sim\mathbf{M}$, **a** being the only free variable in **M**, is t only if, for at least one natural number n as value of **a**, the value of **M** is f. While the value of all the other wffs mentioned above is t only if, for each natural number n as value of a, the value of **M** is t.

If A^2 is ω-consistent, it is consistent.

Proof. Suppose A^2 is ω-consistent. Then for any individual variable **a** and wff **M** at least one of the following wffs is not a theorem: $(\exists \mathbf{a})\sim\mathbf{M}$, $Z_0(\mathbf{a}) \supset \mathbf{M}$, $Z_1(\mathbf{a}) \supset \mathbf{M}, \ldots$. Thus A^2 is absolutely consistent, hence also simply consistent and Post consistent.

For every theorem in *740 – *849, its proof was so constructed that there is an effective procedure for finding a particular sequence of definitions by primitive recursion and composition from which the function (or the associated characteristic function) is defined from the fundamental functions. For example in *752 the primitive recursiveness of $a!$ as a function of a was established by means of the equation $a! = \prod_{c=1}^{c=a} c$, or as it might have been written more fully, $a! = \prod_{c=1}^{c=a} \mathrm{u}_1^1(c)$. From this equation the primitive recursiveness of $a!$ followed by means of *747, in view of the fact that u_1^1 is one of the fundamental functions. And the proof of *747 (together with the proofs of the previous theorems used in the proof of *747) is effective in such a way that it implicitly provides a specific sequence of definitions by primitive recursion and composition by which the function $a!$ is obtained from the fundamental functions.

Given the recursion equations (as we shall say) for a function whose primitive recursiveness was proved in *740 – *849, there is an effective procedure by which we can obtain a particular wff of A^2 which expresses that function (*738 and *739). $\mathrm{B}(x_1, x_2)$, the propositional function that x_1 is the Gödel number of a proof in A^2 of wff number x_2, is primitive recursive, by *845. $c = \mathrm{u}(n, a)$, the propositional function that c is the Gödel number of the wff $Z_n(x) \supset_x \mathbf{A}$, where **A** is wff number a, is primitive recursive, by *822. Hence we obtain a particular wff of A^2 that expresses the propositional function $\mathrm{B}(x_1, x_2)$; similarly for $c = \mathrm{u}(n, a)$.

Let $B(x_1, x_2)$ be the wff of A^2 that expresses the propositional function $\mathrm{B}(x_1, x_2)$. Let $U(x, y, z)$ be the wff of A^2 that expresses the propositional function $z = \mathrm{u}(x, y)$. (The fact that $B(x_1, x_2)$ and $U(x, y, z)$ might be enormously long, too long perhaps for any of us to write down, has no theoretical importance. We not only know from the above remarks that there is an effective method for finding these wffs, but more than that, the proofs of *738 and *739 have supplied a particular such method. And this is all we need to justify our adoption of abbreviations for these wffs.)

Definition. $I(\mathbf{a}) \rightarrow U(\mathbf{a}, \mathbf{a}, \mathbf{c}) \supset_{\mathbf{c}} (\mathbf{s}) \sim B(\mathbf{s}, \mathbf{c})$, where **a** is an individual variable and **c** and **s** are the next two individual variables in alphabetic order after **a**.

Let i be the Gödel number of the particular formula $I(x)$.

Definition. $J \rightarrow Z_i(x) \supset_x I(x)$, where i is the number i just introduced, i.e., the Gödel number of $I(x)$ (or for strict accuracy, we should say rather that the subscript i is the numeral corresponding to this number).

Let j be the Gödel number of J.

†850. $Z_i(x) \supset . Z_j(y) \equiv U(x, x, y)$

Proof.

(1) $j = \mathrm{u}(i, i)$, because $\mathrm{u}(i, i)$ is the number of the wff $Z_i(x) \supset_x \mathbf{A}$, where **A** is wff number i. But wff number i is $I(x)$. Hence $\mathrm{u}(i, i)$ is the number of the wff $Z_i(x) \supset_x I(x)$. But j is the number of this wff. Hence $j = \mathrm{u}(i, i)$.

(2) $U(x, y, z)$ is true for the values i, i, j of x, y, z respectively. From (1).

(3) $\vdash Z_i(x) \supset . Z_i(y) \supset . Z_j(z) \supset U(x, y, z)$. From (2), by *739.

(4) $\vdash Z_i(x) \supset . Z_i(x) \supset . Z_j(z) \supset U(x, x, z)$. From (3), by *503.

(5) $\vdash Z_i(x) \supset . Z_i(x) \supset . Z_j(y) \supset U(x, x, y)$. From (4), by *503.

(6) $Z_i(x) \vdash Z_j(y) \supset U(x, x, y)$. From (5), using MP twice.

(7) $\vdash U(x_1, x_2, y) \supset . U(x_1, x_2, z) \supset y = z$
By *822 and *738, since $z = \mathrm{u}(x, y)$ is expressed in A^2 by $U(x, y, z)$.

(8) $\vdash U(x, x, y) \supset . U(x, x, z) \supset y = z$. From (7), using *503 twice.

(9) $Z_i(x), Z_j(y), U(x, x, z) \vdash y = z$. From (7) and (5), using MP three times.

(10) $Z_i(x), Z_j(y), U(x, x, z) \vdash Z_j(z)$. From (7), by *529.

(11) $Z_i(x), (\exists y) Z_j(y) \vdash U(x, x, z) \supset Z_j(z)$. From (10), ded. th., and *518.

(12) $Z_i(x) \vdash U(x, x, z) \supset Z_j(z)$. From (11), by *711.

(13) $\vdash Z_i(x) \supset . U(x, x, z) \supset Z_j(z)$. From (12), by ded. th.

(14) $\vdash Z_i(x) \supset . U(x, x, y) \supset Z_j(y)$. From (13), by *503.

(15) $Z_i(x) \vdash U(x, x, y) \supset Z_j(y)$. From (14), by MP.

(16) $Z_i(x) \vdash Z_j(y) \equiv U(x, x, y)$. From (6) and (15).

(17) $\vdash Z_i(x) \supset . Z_j(y) \equiv U(x, x, y)$. From (16), by ded. th.

**851. If A^2 is consistent, J is not a theorem.

Proof.

(1) Assume J is a theorem of A^2.

(2) Let k be the number of a proof of J.

(3) $\vdash Z_k(z) \supset . Z_j(y) \supset B(z, y)$. From (2), by *739 and *503.

(4) $Z_i(x), Z_j(y) \vdash U(x, x, y)$. By †850, by P.

(5) $J, Z_i(x) \vdash I(x)$. By Def. of J, MP, and *306.

(6) $J, Z_i(x), Z_j(y) \vdash (z) \sim B(z, y)$. From (5) and (4) and Def. of $I(x)$ and *306.

(7) $Z_i(x), Z_j(y) \vdash (z) \sim B(z, y)$. From (6) and (1).

(8) $Z_j(y), (\exists x)Z_i(x) \vdash (z){\sim}B(z, y)$. From (7).
(9) $Z_j(y) \vdash (z){\sim}B(z, y)$. From (8) and *711.
(10) $Z_j(y) \vdash {\sim}B(z, y)$. From (9).
(11) $Z_j(y), Z_k(z) \vdash B(z, y)$. From (3), using MP twice.
(12) $Z_j(y), Z_k(z) \vdash {\sim}B(z, y)$. From (11).
(13) $\vdash {\sim}p \supset . p \supset q$
(14) $\vdash {\sim}B(z, y) \supset . B(z, y) \supset q$. From (13), by subst. for prop. variables (*510_0).
(15) $Z_j(y), Z_k(z) \vdash q$. From (11), (12), and (14), using MP twice.
(16) $Z_j(y), (\exists z)Z_k(z) \vdash q$. From (15), by *518.
(17) $Z_j(y) \vdash q$. From (16), by *711.
(18) $(\exists y)Z_j(y) \vdash q$. From (17), *518.
(19) $\vdash q$. From (18), by *711.

(19) establishes Post inconsistency for A^2. Hence, since Post, simple, and absolute inconsistency are equivalent for A^2, (19) establishes that if J is a theorem, A^2 is not consistent. Hence, by contraposition, if A^2 is consistent, J is not a theorem. [Note that we have not derived a contradiction from J as hypothesis. We have not shown for any **A** that $J \vdash \mathbf{A}{\sim}\mathbf{A}$. Hence we cannot conclude that $\vdash {\sim}J$.]

**852. If A^2 is ω-consistent, ${\sim}J$ is not a theorem.

Proof.
(1) Assume A^2 is ω-consistent.
(2) A^2 is consistent. From (1).
(3) No number is the number of a proof of J. From (2) and *851.
(4) $\vdash Z_n(x_1) \supset . Z_j(x_2) \supset {\sim}B(x_1, x_2)$, where j is the number of J, and n is any natural number at all. From (3) and *739.
(5) $\vdash Z_i(x) \supset . Z_j(x_2) \equiv U(x, x, x_2)$. From †850 and *503.
(6) $Z_i(x), U(x, x, x_2) \vdash Z_j(x_2)$. From (5), using P.
(7) $Z_n(x_1), Z_i(x), U(x, x, x_2) \vdash {\sim}B(x_1, x_2)$. From (4) and (6), using MP twice.
(8) $Z_n(x_1), Z_i(x) \vdash U(x, x, x_2) \supset {\sim}B(x_1, x_2)$. From (7) and ded. th.
(9) $Z_n(x_1), Z_i(x) \vdash U(x, x, y) \supset {\sim}B(x, y)$. From (8) and *503, since x_2 does not occur free in the hypotheses (ded. th., *501, *305, MP).
(10) $Z_n(x_1), Z_i(x) \vdash U(x, x, y) \supset_y {\sim}B(x_1, y)$. From (9), since y is not free in the hypotheses.
(11) $Z_n(x_1) \vdash Z_i(x) \supset . U(x, x, y) \supset_y {\sim}B(x_1, y)$. From (10) and ded. th.
(12) $Z_n(x_1) \vdash Z_i(x) \supset_x . U(x, x, y) \supset_y {\sim}B(x_1, y)$. From (11), since x is not free in the hypothesis.
(13) $\vdash Z_n(x_1) \supset . Z_i(x) \supset_x . U(x, x, y) \supset_y {\sim}B(x_1, y)$. From (12) and ded. th.
(14) $\vdash Z_n(z) \supset . \underbrace{Z_i(x) \supset_x . U(x, x, y) \supset_y {\sim}B(z, y)}_{\text{M}}$. From (13) and *503.
(15) Assume $\vdash {\sim}J$.

(16) $\vdash \sim . Z_i(x) \supset_x . U(x, x, y) \supset_y (z) \sim B(z, y)$.
From (15), Def. of J, and Def. of $I(\mathbf{a})$.

(17) $\vdash \sim (z) . Z_i(x) \supset_x . U(x, x, y) \supset_y \sim B(z, y)$.
From (16), since z is not free in $Z_i(x)$ or $U(x, x, y)$.

(18) $\vdash (\exists z) \sim . \underbrace{Z_i(x) \supset_x . U(x, x, y) \supset_y \sim B(z, y)}_{\mathrm{M}}$.
From (17) and elementary laws of quantifiers.

(19) A^2 is not ω-consistent.
From (14), (18), since the n in line (14) was arbitrarily chosen.

Line (19) contradicts line (1). So $\sim J$ is not a theorem of A^2, by reductio ad absurdum, if A^2 is ω-consistent.

Discussion of the first Gödel incompleteness theorem. A formal system in which there are quantifiers available for every kind of variable that is present is *complete* if, for every sentence, either it or its negation is a theorem. For systems like propositional calculus or pure first order functional calculus, in which there are some variables that may appear only as free variables, a more elaborate definition of completeness is necessary. But a formal system is always *incomplete* if there is a sentence (= closed propositional form) belonging to it, such that neither it nor its negation is a theorem. In particular note that $[J \vee \sim J]$ is a theorem of A^2 (it is but an instance of a tautology), but that (unless A^2 is ω-consistent) neither J nor $\sim J$ is a theorem of A^2, although both J and $\sim J$ are closed wffs.

Can one diminish the force of Gödel's First Incompleteness Theorem by strengthening A^2 sufficiently to enable it to escape incompleteness? The most immediate remedy seems to be to add as a postulate either J or $\sim J$. One clear drawback in so doing, is that the new set of postulates will lack the intuitive plausibility that characterizes the Peano Postulates. Still, from a formal point of view, nothing prevents one from making such an addition to the postulates of A^2. There remains, however the question of which to add, J or $\sim J$? The answer lies in the proof given **852. An analysis of this proof reveals that it was shown that (a) if A^2 is consistent,

$$\vdash Z_n(z) \supset . Z_i(x) \supset_x . U(x, x, y) \supset_y \sim B(z, y),$$

for every natural number n; and (b) from $\sim J$ as hypothesis we can infer

$$(\exists z) \sim . Z_i(x) \supset_x . U(x, x, y) \supset_y \sim B(z, y).$$

(a) and (b) remain theorems when we add $\sim J$ as a postulate to A^2. Hence if A^2, supplemented with $\sim J$ as a postulate, is consistent, it follows that it is ω-inconsistent; and if A^2, supplemented with $\sim J$ as a postulate, is inconsistent, then clearly it is ω-inconsistent. We do not, therefore, want to add $\sim J$ as a postulate to A^2.

Suppose we add J as a postulate to A^2. Then, if we examine the proof given for Gödel's Theorem, it become apparent that by changing the proof of *843, the proof of Gödel's Theorem goes through, as before, except now for some other closed wff, call it $\mathbf{J}'$, neither $\mathbf{J}'$ nor $\sim\mathbf{J}'$ is a theorem. The new proof of *843 (that the propositional function is primitive recursive), $\mathrm{Ax}(a)$, that a is the number of an axiom, is as follows: $a = n_{505}$ or $a = n_{506}$ or $a = n_{507}$ or $a = n_{508}$ or $a = n_{509}$ or $\mathrm{Ax}_{508}(a)$ or $\mathrm{Ax}_{509}(a)$ or $a = n_{700}$ or $a = n_{701}$ or $a = n_{702}$ or $a = n_{703}$ or $a = n_{704}$ or $a = j$ (where j is the Gödel number of J). A similar result occurs when we add $\mathbf{J}'$ as a postulate to A^2 supplemented with J as a postulate, except that now, for some wff, call it $\mathbf{J}''$, neither $\mathbf{J}''$ nor $\sim\mathbf{J}''$ is a theorem. Moreover, since there is an effective procedure for generating the series $\mathbf{J}, \mathbf{J}', \mathbf{J}'', \mathbf{J}''', \ldots$, we can add the series of formulas as postulates to A^2, by adding an appropriate axiom schema to A^2. But once again, by suitably modifying the proof of *843, the proof of Gödel's First Incompleteness Theorem goes through.

Gödel's Theorem has therefore demonstrated, we may say, an essential incompleteness in A^2. There still remains the possibility of adding new, more complicated, rules of inference to A^2, possibly rules that, while effective, are only general recursive. It argues against such additions, however, that they impede the formalization of mathematics, in its role as an explication of mathematical argument based on intuition, by a reduction of it to a series of steps that can be mechanically checked. Nevertheless, even if one were to add such rules of inference to A^2, there is a proof of Gödel's Theorem that applies to formal systems with general recursive axiom schemes and rules of inference.

It might seem, on contemplating the heuristic background of Gödel's Theorem, that there was an avenue of escape from incompleteness. Before entering into this heuristic background, we must first distinguish the syntactical meaning of a wff from its literal meaning in the principal interpretation of A^2, though both, of course, describe the same truth value. The syntactical meaning of a wff is in terms of Gödel numbers and the syntax of A^2. The literal meaning of a wff in the principal interpretation of A^2 is the propositional function from natural numbers to truth values that it expresses. With this distinction in mind, consider the syntactical meanings of the following wffs. $U(x_1, x_2, y)$, for values n, a, c for the variables x_1, x_2, y: that c is the number of $Z_n(x) \supset_x \mathbf{A}$, where $\mathbf{A}$ is formula number a, and x is the first individual variable in alphabetic order (cf. *822). $B(x_1, x_2)$, for values s, c, of the variables x_1, x_2: that s is the number of a proof of wff number c (cf. *845). $I(x) \rightarrow U(x, x, y) \supset_y (z)\sim B(z, y)$, for the value i of x, for any i: for every natural number y, if y is the number of $Z_i(x) \supset_x \mathbf{A}$, where $\mathbf{A}$ is formula number i, and the x occurring in this formula is the first individual variable in alphabetical order, then no natural number z is the number of a proof of formula number y; or, more briefly, $I(x)$: no natural number is the number of a proof of the formula $Z_i(x) \supset_x \mathbf{A}$, where $\mathbf{A}$ is formula number i. $J \rightarrow Z_i(x) \supset_x I(x)$, where i is the number of $I(x)$, and x is the first individual variable in alphabetical order: no natural number is the number of a proof of $Z_i(x) \supset_x \mathbf{A}$, where $\mathbf{A}$ is formula number i; or, more briefly, J: no natural number is the number of a proof of $Z_i(x) \supset_x I(x)$ or, most briefly of all, J: no natural number is the number of a proof of J. The syntactical meaning of J, therefore, is that J is not a theorem.

There is clearly a close similarity between the syntactical meaning of J, as given above, and what is known as the Liar Antinomy (in fact, this similarity renders it useful, from a didactic point of view, to initiate the presentation of Gödel's Theorem with a purely formal exposition of the argument, which puts the validity beyond question). One version of the Liar Antinomy involves considering the sentence: "This sentence is not true". If the demonstrative "this" succeeds in referring to the sentence in which it is contained, it follows that if the sentence is true, it is false, and if it is false, it is true. Whether the demonstrative does succeed in so referring, is, however, problematic. It is preferable, therefore, to reformulate this Liar Antinomy as follows: Epimenides, a Cretan, said: "All assertions of Cretans are false". The sentence asserted by Epimenides cannot be true, for then it would be false. Hence, it seems to follow, on purely logical grounds, that some assertions of Cretans are true (though, as we have seen Epimenides' own assertion is not true). By speaking of itself only indirectly, and having only as a logical consequence that some assertions of Cretans are true, this assertion of Epimenides' issues in what is at least paradox if not outright antinomy.

This antinomy is semantical in character (as distinguished e.g., from Russell's Antinomy, which is purely set-theoretic). There are two standard attempts resolve it (and others like it): Russell's Ramified Type Theory, and Tarski's method. These are of no avail in the case of J in the proof of Gödel's Theorem, however, since *directly* in the principal interpretation of A^2, J only expresses a proposition about natural numbers and certain functions of natural numbers which are defined each by a certain set of recursion equations. It is only *indirectly*, via its syntactical meaning, that it assumes a form reminiscent of the Liar Antinomy. If one were to restrict A^2 so that J could not have its literal meaning in the principal interpretation of A^2, by somehow preventing A^2 from being able to express every primitive recursive propositional function from natural numbers to truth values, the resulting system would be even more patently incomplete than A^2.

The consensus seems to be, therefore, that it is best to accept the incompleteness of A^2, since the more or less radical ways of changing A^2 that might be proposed render it yet more clearly incomplete. It must be observed, nevertheless, that from the practical point of view of studying the theorems of number theory found in the literature, A^2 is an adequate formalization of arithmetic; that is, all proofs of theorems of number theory that actually appear in existing books and research papers, it is safe to say, are indeed formalizable in A^2.

Rosser's theorem. Although in **851 we needed only the condition that A^2 is consistent, it was necessary in **852 to introduce the stronger condition that A^2 is ω-consistent. Gödel's first incompleteness theorem, as it is usually called, consists in the two metatheorems **851 and **852 taken together. Abstracting from the particular choice of the wff J, we may state this as follows:

The first Gödel theorem (1931) There exist closed wffs $\mathbf{X}$ of A^2 such that if A^2 is consistent, $\mathbf{X}$ is not a theorem, and if A^2 is ω-consistent, $\sim\mathbf{X}$ is not a theorem.

But since ω-consistency implies consistency, it is then natural to reduce this to the simpler (but weaker) statement: If A^2 is ω-consistent, there exist closed wffs $\mathbf{X}$ of A^2 such that neither $\vdash \mathbf{X}$ nor $\vdash \sim\mathbf{X}$.

In this way the first Gödel theorem involves the condition of ω-consistency. However, it has been shown by Barkley Rosser (***The Journal of Symbolic Logic***, Volume 1(1936)) that Gödel's original proof can be modified in such a way that the existence of an undecidable formula — a closed wff such that neither it nor its negation is a theorem — is shown to follow merely from the simple consistency of A^2. We shall prove Rosser's theorem as **859 below, using Rosser's original proof with only such modifications as are necessary or convenient to make it fit the present context, and omitting nothing essential except the proofs of ①, ②, ③ [see below] as theorems in A^2.

To avoid misunderstanding we should notice at once that Rosser's proof does not use the same wff J that was constructed by Gödel. Instead a wff J' is used, which indeed has a certain analogy with J but is not the same formula. It remains an open question whether it is possible to prove about the original wff J of Gödel that if A^2 is consistent, neither $\vdash J$ nor $\vdash \sim J$.

We define the symbol $<$, corresponding to the binary relation 'less than', as follows:

$$[\mathbf{a} \leqq \mathbf{b}] \rightarrow (\exists \mathbf{c})\Sigma(\mathbf{a}, \mathbf{c}, \mathbf{b})$$
$$[\mathbf{a} < \mathbf{b}] \rightarrow \mathbf{a} \leqq \mathbf{b} \,.\, \sim\mathbf{a} = \mathbf{b}$$

$\Sigma(\mathbf{a}, \mathbf{c}, \mathbf{b})$ is defined for A^2 as in §55 of *IML*. The primitive recursiveness of the relation $<$ is given by the proof of *746, where implicit instructions are given for finding the recursion equations for the characteristic function of the relation $<$.

For the proof of Rosser's theorem we must know that the following are theorems of A^2, though the proofs are not given here:

① $Z_0(z) \supset \sim . \, x < z$

② $S(z, z_1) \supset . \, x < z_1 \equiv . \, x < z \lor x = z$

③ $z_1 < z \lor z < z_1 \lor z = z_1$

We recall that in *846 we showed the primitive recursiveness of the relation $\mathrm{P}(s, a)$ — that some number less than s is the number of a proof of wff number a. We could therefore make use of *739 to find a wff such that this relation is calculable in A^2 as represented by it. But instead of proceeding in this way it is better for our present purpose to adopt the following definition:

> $\mathrm{P}(\mathbf{s}, \mathbf{a}) \rightarrow (\exists \mathbf{r}) \,.\, \mathrm{B}(\mathbf{r}, \mathbf{a}) \,.\, \mathbf{r} < \mathbf{s}$, where $\mathbf{r}$ is the first individual variable in alphabetic order other than $\mathbf{a}, \mathbf{s}$.

The definition of $\Sigma(\mathbf{a}, \mathbf{c}, \mathbf{b})$ as taken from §55 is indeed the same as the definition which is obtained from the proof of *738 (compare footnote 526). But the definitions we have used for $[\mathbf{a} < \mathbf{b}]$ and for $\mathrm{P}(\mathbf{s}, \mathbf{a})$ are not the same as those which would be obtained from the proof of *739, and for this reason it is necessary to prove as a separate metatheorem that the relation P is calculable in A^2 as expressed by the formula $P(\mathbf{s}, \mathbf{a})$.

†853. $Z_0(z) \supset \sim P(z, y)$

Proof.

$Z_0(z) \vdash \sim . \, B(x, y) \,.\, x < z$, by ①, PC.

$Z_0(z) \vdash (x) \sim . \, B(x, y) \,.\, x < z$,

by generalization, since x is not free in the hypothesis.

$Z_0(z) \vdash \sim(\exists x) \,.\, B(x, y) \,.\, x < z$, by first order logic.

$Z_0(z) \vdash \sim P(z, y)$, by the definition.

Now use the deduction theorem.

†854. $S(z, z_1) \supset . \, P(z_1, y) \equiv P(z, y) \lor B(z, y)$

Proof.

$S(z, z_1) \vdash x < z_1 \equiv . \, x < z \lor x = z$, by ②.

$S(z, z_1) \vdash B(x, y)[x < z_1] \equiv B(x, y)[x < z] \lor B(x, y)[x = z]$,

by PC, using the tautology $p \equiv [q \lor r] \supset . \, sp \equiv . \, sq \lor sr$.

$S(z, z_1) \vdash B(x, y)[x < z_1] \equiv_x B(x, y)[x < z] \lor B(x, y)[x = z]$,

by generalization.

$S(z, z_1) \vdash (\exists x)[B(x, y) \,.\, x < z_1] \equiv (\exists x) \,.\, B(x, y)[x < z] \lor B(x, y)[x = z]$, by quantifier logic.

The existential quantifier can be distributed over a disjunction:

$$\vdash (\exists x)\big[B(x,y)[x<y] \vee B(x,y)[x=z]\big]$$
$$\equiv (\exists x)\big[B(x,y) \,.\, x<z\big] \vee (\exists x)\big[B(x,y) \,.\, x=z\big]$$

$$\mathrm{S}(z,z_1) \vdash \underbrace{(\exists x)\big[B(x,y) \,.\, x<z_1\big]}_{1}$$
$$\equiv \,.\, \underbrace{(\exists x)\big[B(x,y) \,.\, x<z\big]}_{2} \vee \underbrace{(\exists x) \,.\, B(x,y) \,.\, x=z}_{3},$$

by first order logic.

We obtain an equivalence for 3 as follows:

$B(z,y) \vdash B(z,y) \,.\, z=z,$ by PC, †520, substituting z for x.

$B(x,y), x=z \vdash B(z,y),$ *529 (substitutivity of equality).

$(\exists x) \,.\, B(x,y) \,.\, x=z \vdash B(z,y),$ *518 (existential instantiation).

Apply the deduction theorem twice, conjoin and apply the definition of $\equiv$ to get:

$$\vdash (\exists x)\big[B(x,y) \,.\, x=z\big] \equiv B(z,y)$$

Further, 1 and 2 are $\mathrm{P}(z_1,y), \mathrm{P}(z,y)$ respectively, as may be seen by taking **r** to be x, **a** to be y, and **s** to be either z_1 or z in the definition of $\mathrm{P}(\mathbf{s},\mathbf{a})$. Apply *513 (substitutivity of equivalence) to get:

$$\mathrm{S}(z,z_1) \vdash P(z_1,y) \equiv P(z,y) \vee B(z,y),$$

then apply the deduction theorem.

*855. The relation P is calculable in A^2 as represented by the wff $P(z,y)$, with z corresponding to the first argument of the relation P and y to the second argument. That is, for every pair of subscripts k and l (arabic numerals), either

$$\vdash Z_k(z) \supset \,.\, Z_l(y) \supset P(z,y)$$
$$\text{or } \vdash Z_k(z) \supset \,.\, Z_l(y) \supset {\sim}P(z,y),$$

according as the relation P actually holds between k and l (natural numbers) or not.

Proof. The proof proceeds by mathematical induction in the metalanguage with respect to k. In the case that $k=0$ we have

$$Z_l(y) \vdash Z_0(z) \supset {\sim}P(z,y)$$

for arbitrary l, by †853. Suppose as hypothesis of induction that, for a particular k and l, we have

$$\text{either (i) } Z_l(y) \vdash Z_k(z) \supset P(z,y)$$
$$\text{or (ii) } Z_l(y) \vdash Z_k(z) \supset {\sim}P(z,y)$$

according as $\mathrm{P}(k,l)$ or not. We must show that the same thing holds with $k+1$ in place of k, and with the same l.

$Z_{k+1}(z_1), Z_k(z) \vdash \mathrm{S}(z, z_1)$, by *714.
(1) $Z_{k+1}(z_1), Z_k(z) \vdash P(z_1, y) \equiv P(z, y) \vee B(z, y)$, by †854.
We distinguish two cases, according as (i) or (ii) holds. Consider first Case (ii), i.e.,

$$Z_l(y) \vdash Z_k(z) \supset \sim P(z, y).$$

Then not $\mathrm{P}(k, l)$, by hypothesis of induction.
(2) $Z_l(y), Z_{k+1}(z_1), Z_k(z) \vdash P(z_1, y) \equiv B(z, y)$, by (1) and PC.
By *845, the relation B is calculable in A^2 as represented by the wff $B(z, y)$. Hence

$$\begin{array}{ll} \text{either} & Z_l(y) \vdash Z_k(z) \supset B(z, y) \\ \text{or} & Z_l(y) \vdash Z_k(z) \supset \sim B(z, y) \end{array}$$

according as $\mathrm{B}(k, l)$ holds or not. But since we have (in Case (ii)) that not $\mathrm{P}(k, l)$, we may say equivalently, according as $\mathrm{P}(k+1, l)$ or not. Either

$$\begin{array}{ll} & Z_l(y), Z_{k+1}(z_1), Z_k(z) \vdash P(z_1, y) \\ \text{or} & Z_l(y), Z_{k+1}(z_1) Z_k(z) \vdash \sim P(z_1, y) \end{array}$$

according as $\mathrm{P}(k+1, l)$ or not, by the above, (2), and PC. Since z is not free in the conclusion, and since $\vdash (\exists z) Z_k(z)$ by *712, we have by *518 (existential instantiation) that

$$\begin{array}{ll} \text{either} & Z_l(y), Z_{k+1}(z_1) \vdash P(z_1, y) \\ \text{or} & Z_l(y), Z_{k+1}(z_1) \vdash \sim P(z_1, y) \end{array}$$

according as $\mathrm{P}(k+1, l)$ or not. Apply the ded. thm. and change the free variable z_1 to z. This completes the induction in Case (ii). Then consider Case (i), i.e.,

$$Z_l(y) \vdash Z_k(z) \supset P(z, y).$$

Then $\mathrm{P}(k, l)$ holds by hyp. ind., and hence P(k+1,l).

$$Z_l(y), Z_{k+1}(z_1), Z_k(z) \vdash P(z_1, y)$$

By (1) and PC. Since z is not free in the conclusion, apply existential instantiation and $\vdash (\exists z) Z_k(z)$ to get

$$Z_l(y), Z_{k+1}(z_1) \vdash P(z_1, y)$$

Apply the ded. thm. and change the free variable z_1 to z. This completes the induction in Case (i). This has proved, by mathematical induction, that we have for all k and l

$$\begin{array}{ll} \text{either} & Z_l(y) \vdash Z_k(z) \supset P(z, y) \\ \text{or} & Z_l(y) \vdash Z_k(z) \supset \sim P(z, y) \end{array}$$

according as $\mathrm{P}(k, l)$ or not. Hence by ded. thm. and PC,

$$\begin{array}{ll} \text{either} & \vdash Z_k(z) \supset . Z_l(y) \supset P(z, y) \\ \text{or} & \vdash Z_k(z) \supset . Z_l(y) \supset \sim P(z, y) \end{array}$$

according as $\mathrm{P}(k, l)$ or not.

$\text{Neg}(a_1)$, the number of the negation of formula number a_1, was shown to be a primitive recursive function of a_1 by *811. Let $N(x_1, y)$ be the wff of A^2 which represents the relation $c = \text{Neg}(a_1)$ in accordance with *738.

$$I'(\mathbf{a}) \to U(\mathbf{a}, \mathbf{a}, \mathbf{c}) \supset_{\mathbf{c}} . B(\mathbf{s}, \mathbf{c}) \supset_{\mathbf{s}} . N(\mathbf{c}, \mathbf{b}) \supset_{\mathbf{b}} P(\mathbf{s}, \mathbf{b}),$$

where **c**, **s**, **b** are the first three individual variables in alphabetic order other than **a**.

Let i' be the Gödel number of $I'(x)$.

$$J' \to Z_{i'}(x) \supset_x I'(x)$$

Let j' be the Gödel number of J'.

†856. $Z_{i'}(x) \supset . Z_{j'}(y) \equiv U(x, x, y)$

Proof. The formula $U(x, y, z)$ represents the relation $c = \text{u}(n, a)$, where n, a, c correspond to x, y, z respectively. $\text{u}(n, a)$ is the number of $Z_n(x) \supset_x \mathbf{A}$, where $\mathbf{A}$ is formula number a, by *822. $\text{u}(i', i')$ is the number of $Z_{i'}(x) \supset_x \mathbf{A}$, where $\mathbf{A}$ is formula number i'. This means that $\text{u}(i', i')$ is the number of $Z_{i'}(x) \supset_x I'(x)$; i.e., it is the number of J'; so $j' = \text{u}(i', i')$. Hence

$\vdash Z_{i'}(x) \supset . Z_{i'}(y) \supset . Z_{j'}(z) \supset U(x, y, z)$, by calculability of the relation $c = \text{u}(n, a)$ in A^2.

$\vdash Z_{i'}(x) \supset . Z_{i'}(x) \supset . Z_{j'}(z) \supset U(x, x, z)$, by substitution of individual variables.

$\vdash Z_{i'}(x) \supset . Z_{j'}(z) \supset U(x, x, z)$, by PC.

$\vdash Z_{i'}(x) \supset . Z_{j'}(y) \supset U(x, x, y)$, by substitution of y for z.

(1) $Z_{i'}(x), Z_{j'}(y) \vdash U(x, x, y)$, by MP.

It remains to show: $Z_{i'}(x), U(x, x, y) \vdash Z_{j'}(y)$

(2) $\vdash U(x_1, x_2, y) \supset . U(x_1, x_2, z) \supset y = z$, by *822 and *738.

$Z_{i'}(x), Z_{j'}(y_2) \vdash U(x, x, y_2)$, from proof of (1), which did not depend on choice of particular variable y instead of y_2.

$Z_{i'}(x), Z_{j'}(y_2), U(x, x, y) \vdash y_2 = y$, by (2), simultaneously substituting y_2 for y, x for x_1, x for x_2 and y for z.

$Z_{i'}(x), Z_{j'}(y_2), U(x, x, y) \vdash Z_{j'}(y)$, by *529 (substitutivity of equality).

By *518 (existential instantiation), and $\vdash (\exists y_2) Z_{j'}(y_2)$, we obtain

$$Z_{i'}(x), U(x, x, y) \vdash Z_{j'}(y)$$

From this and (1) the theorem †856 follows by ded. thm. and PC.

**857. If A^2 is consistent, J' is not a theorem of A^2.

Proof. Suppose $\vdash J'$. Let k be the number of a proof of J'.

(i) Then $Z_{j'}(y), Z_k(z) \vdash B(z, y)$, by calculability of the relation B in A^2.

(ii) $Z_{j'}(y), Z_{\text{Neg}(j')}(y_1) \vdash N(y, y_1)$, by calculability of the relation $c = \text{Neg}(a_1)$ as represented by the wff $N(x_1, y)$.

We show: $Z_{i'}(x), Z_{j'}(y), Z_k(z), Z_{\text{Neg}(j')}(y_1) \vdash P(z, y_1)$, as follows:

Since $\vdash J'$, i.e., $\vdash Z_{i'}(x) \supset_x I'(x)$,
we have $Z_{i'}(x) \vdash I'(x)$, i.e.,

$$Z_{i'}(x) \vdash U(x, x, y) \supset_y . B(z, y) \supset_z . N(y, x_1) \supset_{x_1} P(z, x_1),$$

from the definition of $I'(x)$, where **c** is y, **s** is z, and **b** is y_1.

$Z_{i'}(x), Z_{j'}(y) \vdash B(z, y) \supset_z . N(y, x_1) \supset_{x_1} P(z, x_1)$,
by †856, drop quantifier, MP.

$Z_{i'}(x), Z_{j'}(y), Z_k(z) \vdash N(y, x_1) \supset_{x_1} P(z, x_1)$,
by (i), drop quantifier, MP.

$Z_{i'}(x), Z_{j'}(y), Z_k(z), Z_{\text{Neg}(j')}(y_1) \vdash P(z, x_1)$,
by (ii), instantiate to y_1, MP.

Since the sequence of formulas no. k is a proof of J', the last formula in the sequence is J'; this last formula in the sequence therefore is not $\sim J'$, and therefore the relation P does not hold between k and $\text{Neg}(j')$. Hence

$Z_k(z), Z_{\text{Neg}(j')}(y_1) \vdash \sim P(z, y_1)$, by *855.
$Z_{i'}(x), Z_{j'}(y), Z_k(z), Z_{\text{Neg}(j')}(y_1) \vdash p$, by the tautology $q \supset . \sim q \supset p$.

The hypotheses can be eliminated by *518 (existential instantiation) and *712. We have a Post-inconsistency, which for A^2 is equivalent to simple inconsistency. **857 follows by contraposition.

**858. If A^2 is consistent, $\sim J'$ is not a theorem of A^2.

Proof. Suppose $\vdash \sim J'$.

$$J' \to Z_{i'}(x) \supset_x . U(x, x, y) \supset_y . B(z, y) \supset_z . N(y, x_1) \supset_{x_1} P(z, x_1)$$

The logic of quantifiers yields an equivalent wff in prenex normal form:

$$J' \equiv (x)(y)(z)(x_1) . Z_{i'}(x) \supset . U(x, x, y) \supset . B(z, y) \supset . N(y, x_1) \supset P(z, x_1)$$
$$\sim J' \equiv (\exists x)(\exists y)(\exists z)(\exists x_1) . Z_{i'}(x) U(x, x, y) B(z, y) N(y, x_1) \sim P(z, x_1)$$

From this, and $Z_{i'}(x), U(x, x, y) \vdash Z_{j'}(y)$ (by †856), it follows by the quantifier calculus that

$\vdash (\exists x)(\exists y)(\exists z)(\exists x_1) . Z_{j'}(y) B(z, y) N(y, x_1) \sim P(z, x_1)$

(1) $(\exists y)(\exists z)(\exists y_1) . Z_{j'}(y) B(z, y) N(y, y_1) \sim P(z, y_1)$,
since x does not occur in the matrix, and by change of bound variables.

(2) $Z_{j'}(y), N(y, y_1) \vdash Z_{\text{Neg}(j')}(y_1)$,
by calculability of the relation $c = \text{Neg}(a_1)$ in A^2 as represented by $N(x_1, y)$, as follows:

$Z_{j'}(y), Z_{\text{Neg}(j')}(z) \vdash N(y, z)$, by *738
$\vdash N(y, y_1) \supset . N(y, z) \supset y_1 = z$, by *738
$Z_{j'}(y), Z_{\text{Neg}(j')}(z), N(y, y_1) \vdash y_1 = z$
$Z_{j'}(y), Z_{\text{Neg}(j')}(z), N(y, y_1) \vdash Z_{\text{Neg}(j')}(y_1)$, by substitutivity of equality.

Then use existential instantiation to eliminate the second hypothesis.

Since the hypotheses of (2) occur in the matrix of (1), we infer by quantifier calculus:

(3) $\vdash (\exists y)(\exists z)(\exists y_1) . Z_{j'}(y) B(z, y) Z_{\text{Neg}(j')}(y_1) \sim P(z, y_1)$

Let k be the number of a proof of $\sim J'$. Then

(4) $Z_{\mathrm{Neg}(j')}(y_1), Z_k(z_1) \vdash B(z_1, y_1)$, by calculability.

Since A^2 is consistent, we have by **857 that no number is the number of a proof of J'; and hence by *855

(5) $Z_{j'}(y), Z_k(z_1) \vdash {\sim}P(z_1, y)$

$P(z_1, y) \to (\exists x) \mathbin{.} B(x, y) \mathbin{.} x < z_1$

$\vdash {\sim}P(z_1, y) \equiv (x) \mathbin{.} x < z_1 \supset {\sim}B(x, y)$

$\vdash {\sim}P(z_1, y) \supset \mathbin{.} z < z_1 \supset {\sim}B(x, y)$

Then use (5) and MP to get

(6) $Z_{j'}(y), Z_k(z_1), z < z_1 \vdash {\sim}B(z, y)$.

$P(z, y_1) \to (\exists x) \mathbin{.} B(x, y_1) \mathbin{.} x < z$

${\sim}P(z, y_1) \equiv (x) \mathbin{.} B(x, y_1) \supset {\sim} \mathbin{.} x < z$

${\sim}P(z, y_1) \supset \mathbin{.} B(z_1, y_1) \supset {\sim} \mathbin{.} z_1 < z$

Then use (4) and MP to get

(7) $Z_{\mathrm{Neg}(j')}(y_1), Z_k(z_1), {\sim}P(z, y_1) \vdash {\sim} \mathbin{.} z_1 < z$.

$\vdash z_1 < z \lor z < z_1 \lor z = z_1$ ③, from page 911.

$Z_{\mathrm{Neg}(j')}(y_1), Z_k(z_1), {\sim}P(z, y_1) \vdash z < z_1 \lor z = z_1$, by (7), ③, and PC.

(8) $Z_{j'}(y), Z_{\mathrm{Neg}(j')}(y_1), Z_k(z_1), {\sim}P(z, y_1), {\sim}z = z_1 \vdash {\sim}B(z, y)$, by (6) and PC.

(9) $Z_{j'}(y), Z_k(z_1) \vdash {\sim}B(z_1, y)$, because k, being the number of a proof of ${\sim}J'$, cannot also be the number of a proof of J', and by calculability in A^2 of the relation B as represented by the wff $B(z_1, y)$.

$Z_{j'}(y), Z_{\mathrm{Neg}(j')}(y_1), Z_k(z_1), {\sim}P(z, y_1), B(z, y) \vdash z = z_1$, by (8), ded. thm., contraposition and MP.

$Z_{j'}(y), Z_{\mathrm{Neg}(j')}(y_1), Z_k(z_1), {\sim}P(z, y_1), B(z, y) \vdash B(z_1, y)$, by substitutivity of equality.

$Z_{j'}(y), Z_{\mathrm{Neg}(j')}(y_1), Z_k(z_1), {\sim}P(z, y_1), B(z, y) \vdash p$, by (9) using the tautology $q \supset \mathbin{.} {\sim}q \supset p$.

$Z_{j'}(y), Z_{\mathrm{Neg}(j')}(y_1), {\sim}P(z, y_1), B(z, y) \vdash p$, by existential instantiation and $\vdash (\exists z_1)Z_k(z_1)$.

$(\exists y)(\exists z)(\exists y_1) \mathbin{.} Z_{j'}(y)B(z, y)Z_{\mathrm{Neg}(j')}(y_1){\sim}P(z, y_1) \vdash p$, by existential instantiation repeated three times.

$\vdash p$, by (3) above.

We have a Post-inconsistency, which for A^2 is equivalent to simple inconsistency. **858 follows by contraposition.

From **857 and **858, we can conclude:

859. *Rosser's theorem.* If A^2 is consistent, there are in A^2 wffs **X such that neither **X** nor ~**X** is a theorem of A^2.

As in the case of the Gödel incompleteness theorem, Rosser's theorem is not peculiar to A^2 but extends to a large variety of formalizations of arithmetic, of which A^2 is here taken as illustrative. In particular, Rosser's theorem evidently holds also for any formalized language obtained from A^2 by adding a finite number of axioms, or a finite number of axiom schemata (if each schema is primitive recursive in terms of Gödel numbers), or a finite number of rules of inference (if each rule is primitive recursive in terms of Gödel numbers).

86. Gödel's second incompleteness theorem. The formula J that occurs in **851, **852 is not the only wff of A^2 such that if A^2 is consistent and ω-consistent neither it nor its negation is a theorem of A^2. First, any formula that is logically equivalent to J has this property with regard to provability. Second, if we alter the system of Gödel numbering (as we may do in many ways within very wide limits, provided that the system allows effective methods of determining the Gödel number of a formula or sequence of formulas, and of constructing a unique formula or sequence of formulas from a given Gödel number), the proof can be repeated for a different J which will thereby be shown to have the same property.

Another formula which has the property in question and which differs more significantly from J is one of the formulas which assert that A^2 is consistent. To determine a particular formula it is necessary to provide 1) a particular system of Gödel numbering, as no wff of A^2 literally asserts anything about the syntax of A^2, but can only do so indirectly by means of Gödel numbering, and 2) a particular definition of consistency. Given this we can write a particular wff that can be taken to assert that A^2 is consistent. Toward this end, we use the system of Gödel numbering provided in **80**; and we introduce a notion of consistency closely akin to, and equivalent to, Post consistency, namely that the wff consisting of the propositional variable p standing alone shall not be a theorem. Then consider the following wff of A^2:

$$Z_8(x) \supset (z) \sim B(z, x)$$

We note that literally this wff is about certain numerical functions that are defined in complicated ways, and it is only relative to a system of Gödel numbering that it can be interpreted as asserting that formula no. 8 has no proof. Since $8 = 2^3$, and 3 is the Gödel number of the propositional variable p, the formula can be said to assert that the wff that consists of p standing alone is not a theorem. Gödel's second theorem (for A^2) can then be stated as follows:

**860. If A^2 is consistent, then $Z_8(x) \supset (z) \sim B(z, x)$ is not a theorem of A^2.

**861. If A^2 is ω-consistent, then $\sim . Z_8(x) \supset_x (z) \sim B(z, x)$ is not a theorem of A^2.

We note that the formula in **861 is the negation of the closure of the formula in **860. This theorem has the somewhat paradoxical result that if we were to discover a proof of the above formula that A^2 is consistent, then A^2 would be inconsistent. This should not be entirely surprising, however, because if A^2 is inconsistent, it's unreliable and theoremhood in A^2 would not imply truth.

Significance of the second Gödel theorem. Before turning to the second incompleteness theorem of Gödel and in order to see its full significance, it is necessary to say something about Hilbert's program of a consistency proof for arithmetic.

Because the antinomies of naive set theory had shown that intuitive plausibility of the primitive basis of a mathematical system is no guarantee of the consistency of that system, Hilbert proposed that, as a condition of acceptability of a mathematical system, one should provide a proof that the system is consistent.

The first appearance of this in Hilbert's writings is in his famous lecture of 1900 setting forth the important unsolved problems of mathematics (for an English translation of this see ***Bulletin of the American Mathematical Society***, vol. 8, pp. 437–479). Here Hilbert points out that the consistency of important branches of mathematics has generally been proved by a reduction process — e.g., in what is historically the oldest case, it is shown that a model of Lobachevskian geometry can be found within Euclidean geometry, so that Lobachevskian geometry must be consistent if the Euclidean is. As the chain of reductions leads back ultimately to arithmetic, so making the consistency of mathematics generally depend on that of arithmetic, it becomes urgent to find a proof of the consistency of arithmetic, not by a reduction process, but by some more direct method.

Generally it is higher-order arithmetic on which the consistency of other branches of mathematics depends (e.g., for the theory of functions of a real variable, arithmetic of at least fourth order). But in the later work of the Hilbert school the emphasis was on finding a consistency proof for first-order arithmetic, as a first step toward the final goal. And various partial results were in fact obtained.

As the work developed the program which emerged is the following. In order to prove the consistency of a particular mathematical theory we must first provide a precise delineation or definition of the theory by formalizing it. For example, first-order arithmetic might be represented by the logistic system A^0 (the formalization of first-order arithmetic employed by the Hilbert school was different but equivalent to this). The logistic system is then to be taken directly as an object of mathematical study, in the sense of considering its syntax in abstraction from the intended meanings of the notations. It then becomes an entirely combinatorial problem whether from the given axioms (certain finite sequences of symbols) by the given rules (certain permitted operations upon finite sequences of symbols) it will be possible, for some $\mathbf{A}$, to obtain both $\mathbf{A}$ and $\sim\mathbf{A}$. We seek a solution by mathematical means.

A Hilbertian consistency proof does not take place in the system itself, but in what Hilbert called metamathematics. This distinction between mathematics and metamathematics is now more familiar in Carnap's terminology as the distinction between object language and metalanguage. (The change in terminology arises because Hilbert was concerned entirely with languages designed to express the propositions of some branch of mathematics, whereas Carnap wished to extend the same idea to formalized languages in general.) Although the metalanguage can be a logistic system it is often an informal language which relies on mathematical intuitions in its proofs. The metamathematical theory of proofs that are possible within a mathematical object language (A^0 for instance), including the consistency question, is Hilbert's *Beweistheorie*, or *proof theory*.

Hilbert's critics have urged that the implicit logic involved could very likely include all of the logic of the system under consideration, resulting in a sort of circle. Hilbert was aware of the point that anyone having real doubt of the consistency of first-order arithmetic would hardly be convinced by a metamathematical proof of the consistency of A^0 in which use is made of substantially all the main methods of proof that are available in first-order arithmetic. Such a proof could not correctly be called circular (since it deals only with the syntax of A^0 and not with the meanings of the wffs of A^0). But there is evidently a sense in which it would accomplish much less than Hilbert's purpose, as seen in his lecture of 1900 and later writings.

For this reason Hilbert required that a metamathematical consistency proof must be restricted to methods he called *finit*, usually translated into English as *finitary*. (Occasionally the word has been mistranslated as "finite," but of course the German word for "finite" is rather "*endlich*.")

Hilbert gave no precise criterion as to when a proof is to be considered finitary, but in particular instances the matter is usually clear. An existence proof must proceed in such a way that, at least implicitly, an example is effectively given of what is said to exist. The method of proof by cases has to be avoided unless there is an effective test to distinguish the cases. (E.g., the usual proof that every real number has a decimal expansion, proceeding by cases according as the number does or does not have the form $\frac{m}{10^n}$ with m and n integers, is not finitary, being subject, for those who demand effective existence proofs, to Brouwer's counterexample.) Proofs of universal propositions must be such that every step is subject to check in the sense that, if a counterexample to the universal proposition were found, it would be possible to trace the proof back step by step, confronting each line of the proof with the counterexample in an effective way until at some step the counterexample reveals an error. This last has the consequence that mathematical induction may be accepted as a finitary method of proof, on condition that the property of natural numbers which is established by mathematical induction is one such that there is an effective test for its application to any particular natural number.

Vaguely the intention is that a finitary proof shall have an openness to verification of a sort that must make it acceptable to anyone who is prepared to accept mathematical proofs at all. There is evidently a kinship to restrictions imposed by mathematical intuitionists, but the restriction to finitary proofs is actually more severe. Sometimes the term *finitism* is used for restriction to proofs which are finitary in the sense of Hilbert, but this terminology may be confusing because of other meanings of the same word.

It is a consequence of the second Gödel theorem, as we shall see, that there is one sense in which the Hilbert program for a consistency proof cannot be carried out. Hilbert himself once spoke of the program as a "*Fiasko*" (a German word which needs no translation). But Hilbertian proof theory remains of major importance for the foundations of mathematics, not only because of other topics than that of proofs of consistency, but also because there are other senses in which consistency proofs are possible, including the Gentzen consistency proof, which is described below, and relative consistency proofs such as Gödel's proof of the consistency of the continuum hypothesis (if axiomatic set theory is consistent, it remains so upon adjoining the continuum hypothesis as additional axiom). Even the sense in which the Hilbert program cannot be carried out is intrinsically important, and could not be understood unless the program were first formulated.

The impact of the second Gödel theorem on the Hilbert program is clear, as it means that if we seek a metamathematical consistency proof of A^2, we cannot hope to succeed with methods of proof that are weaker than those available in A^2 itself, but we must on the contrary be prepared to uses something that is stronger in the sense that the axioms and rules of inference of A^2 are not sufficient to provide it. And the like will be true of other systems for which the second Gödel theorem can be proved (among them A^0).

To be sure, $Z_8(x) \supset (z) {\sim} B(z, x)$ is under the principal interpretation of A^2 an assertion of a proposition of arithmetic, being about natural numbers and a certain binary relation between natural numbers which is defined by recursion equations. The correspondence between this arithmetical proposition and a metamathematical proposition is imposed only from outside, by setting up a method of attaching numbers to formulas which is not and cannot be stated in A^2 itself. But this correspondence, artificial though it may be, is nevertheless sufficient so that from impossibility of a proof of $Z_8(x) \supset (z) \sim B(z,x)$ in A^2 itself (if consistent) we may infer the impossibility of a consistency proof of A^2 in a metalanguage which is not in some essential respect stronger than A^2.

In his paper of 1931, after proving the second incompleteness theorem for a system which may be described as arithmetic of order ω, Gödel remarks that the failure of the Hilbert program for a consistency proof does not strictly follow, as there may yet be finitary methods of proof which are not formalizable in this system. This conjecture of Gödel remains undecided but it did have a remarkable confirmation some years later in a proof by Gerhard Gentzen of the consistency of first-order arithmetic.

Gentzen's metamathematical proof of the consistency of a system equivalent to A^0 employs, with one exception, no method of proof that is not available in A^0 itself. The one exception is that transfinite induction is required over the segment of the ordinal numbers that is determined by the ordinal number ε_0, or briefly "transfinite induction up to ε_0." Contrary to what has sometimes been said, there is no reason that this method of proof should not be considered finitary. For transfinite induction up to ε_0 can be reformulated in such a way that no reference is made to ordinal numbers, and what is used is a certain rather complicated but nevertheless straightforward form of mathematical induction within the natural numbers. In this regard the method of transfinite induction up to ε_0 is quite a different matter from, e.g., transfinite induction through the second number class (of Cantor).

The Gentzen consistency proof, taken in conjunction with the second Gödel theorem, shows the existence of a finitary method of proof which is not formalizable in A^0.

Proof of theorem **860. We follow Gödel in leaving somewhat of a lacuna in the proof by assuming

$$\vdash Z_8(x) \supset_x (z) \sim B(z,x) \supset . \, Z_j(y) \supset (z) \sim B(z,y)$$

where j is the Gödel number of the wff J, with the following informal justification: We have given a proof in the unformalized metalanguage of **851, that if A^2 is consistent in the sense of Post, J is not a theorem of A^2. Hence we conclude that the above wff of A^2 has a proof in A^2, obtained by formalizing in terms of Gödel numbers the metatheoretic argument that was used in the proof of **851. Although strictly speaking this should be done, because of the length involved it never has. Though no one has ever seriously doubted that it could be done, the question is now moot because the theorem has been proved by another method.

To continue with the proof:

$Z_i(x), U(x,x,y) \vdash Z_j(y)$, by †850.

$Z_8(x) \supset_x (z) \sim B(z,x), Z_i(x), U(x,x,y) \vdash (z) \sim B(z,y)$

$Z_8(x) \supset_x (z) \sim B(z,x), Z_i(x) \vdash U(x,x,y) \supset_y (z) \sim B(z,y)$,

by deduction theorem and generalization.

$$Z_8(x) \supset_x (z) \sim B(z, x) \vdash Z_i(x) \supset_x U(x, x, y) \supset_y (z) \sim B(z, y),$$

by deduction theorem and generalization, i.e.

$$Z_8(x) \supset_x (z) \sim B(z, x) \vdash J$$

Hence by supposing $\vdash Z_8(x) \supset_x (z) \sim B(z, x)$, then $\vdash J$, then A^2 is inconsistent. Therefore if A^2 is consistent, $Z_8(x) \supset_x (z) \sim B(z, x)$ is not a theorem of A^2.

Proof of theorem **861. If A^2 is ω-consistent, and therefore consistent, and since B is calculable in A^2 as represented by the wff $B(x, y)$,

$$\vdash Z_n(z) \supset . Z_8(x) \supset_x \sim B(z, x)$$

for any value of n. If we assume that $\vdash \sim . Z_8(x) \supset_x (z) \sim B(z, x)$, then

$$\vdash (\exists z) \sim . Z_8(x) \supset_x \sim B(z, x)$$

This last together with the theorems $Z_n(z) \supset . Z_8(x) \supset_x \sim B(z, x)$ constitutes an ω-inconsistency. Hence if A^2 is ω-consistent, then $\sim . Z_8(x) \supset_x .(z) \sim B(z, x)$ is not a theorem of A^2.

Tarski's theorem on truth. Since 1) the consistency of A^2 is only conjecture and 2) if A^2 is consistent then it is incomplete, truth and provability in A^2 are not equivalent. At best the provable formulas of A^2 are a subset of the true formulas of A^2. As we have seen, the notion of provability can be expressed indirectly in A^2 by means of Gödel numbering. Can we express in A^2 that x is the number of a true formula of A^2?

Let $\mathbf{V}_x$ be a wff of A^2, with no free variables other than x, such that $\mathbf{V}_x$ expresses in A^2 (relative to the choice of Gödel numbering) that x is the number of a wff that is true in A^2. As a minimum criterion of the acceptability of any such wff $\mathbf{V}_x$ as expressing truth in A^2, Tarski proposed the following:

> In all cases in which **A** is a sentence (closed wff) of A^2, and a is a Gödel number of **A**, it must be the case that $\vdash Z_a(x) \supset_x \mathbf{V}_x \equiv \mathbf{A}$.

Though this condition on $\mathbf{V}_x$ has sometimes erroneously been called a definition of truth , it is rather Tarski's criterion of the acceptability of a proposed truth definition.

One common natural-language formulation of Tarski's criterion is:

"p" is true if and only if p.

But according to the standard (or Frege) convention about use of quotation marks for the use-mention distinction, this is sheer nonsense on the ground that the expression which consists of the letter p standing between quotation marks must be understood as a name of the particular letter p and hence as a constant; moreover, if we seek to modify the standard convention in such a way that the expression which consists of the letter p standing between quotation marks can be construed as containing an occurrence of p as a free variable, we are confronted with serious immediate difficulties which are not easily overcome. What is no doubt meant by the above natural-language formulation of Tarski's criterion is better expressed as follows:

> If A is any sentence and if in the expression "______ is true if and only if ______", we replace the first blank by the sentence A enclosed in quotation marks and the second blank by the sentence A, the result holds (either in the sense of being simply true or in the sense of being a consequence of some specified system of axioms).

Here the sentence A enclosed in quotation marks is understood as being a name of the sentence A; and if we are to follow Tarski's program, we must further construe the sentence A enclosed in quotation marks as being an abbreviation of what Tarski calls a structural-descriptive name of A.

Just as Gödel's proof of the first incompleteness theorem involved producing a formula that says of itself (by means of Gödel numbering) that it is not provable, so the proof of Tarski's theorem involves producing a formula that says of itself (by means of Gödel numbering) that it is not true. This strategy traces back to the Greeks with the Antinomy of the Liar. Once the system of Gödel numbering allows us to produce a formula that says of itself that it is not true, assuming that there is a definition of a true sentence of A^2 formulable in A^2, then we need only reproduce the Liar Antinomy in A^2 to prove that A^2 is inconsistent. Hence, if we assume A^2 to be consistent, it follows by contraposition that there is no truth definition in A^2 that satisfies Tarski's criterion.

In our present reformulation of Tarski's treatment of truth, the device of Gödel numbering short-cuts the use of structural-descriptive names, and Tarski's criterion of truth then appears in the form that we have given above.

That is, we are unable, even using the indirect correspondence between formulas and Gödel numbers, to give a definition of truth for A^2 within the system A^2. More precisely we have a metatheorem (which is due to Tarski):

**862. If A^2 is consistent, there is no wff $\mathbf{V}_x$ of A^2 (having x as its only free variable) such that: $\vdash Z_a(x) \supset_x \mathbf{V}_x \,.\!\equiv \mathbf{A}$ in all cases in which $\mathbf{A}$ is a closed wff of A^2 and a is the Gödel number of $\mathbf{A}$.
(Heuristic note: Loosely read $\mathbf{V}_x$ as — the wff of A^2 which has Gödel number x, under our scheme of numbering, is true.)

The proof of this metatheorem is simplified by using the following lemma, which is a theorem of F^2 (functional calculus of second order).

Lemma. $\vdash F(y) \supset_{y,z} [F(z) \supset y = z] \supset .(\exists y)[F(y)G(y)] \supset . F(x) \supset_x G(x)$.

Proof.

$F(y) \supset_{x,z} [F(z) \supset y = z], F(y), G(y), F(z) \vdash y = z$,
by F^I (functional calculus of first order with equality) and *modus ponens*.
$F(y) \supset_{y,z} [F(z) \supset y = z], F(y), G(y), F(z) \vdash G(z)$, by *529.
$F(y) \supset_{y,z} [F(z) \supset y = z], G(y), F(z) \vdash F(z) \supset G(z)$, by Deduction Theorem.
$F(y) \supset_{y,z} [F(z) \supset y = z], F(y), G(y) \vdash F(z) \supset_z G(z)$, by *501.
$F(y) \supset_{y,z} [F(z) \supset y = z], (\exists y) . F(y)G(y) \vdash F(z) \supset_z G(z)$, by *518.
$F(y) \supset_{y,z} [F(z) \supset y = z], (\exists y) . F(y)G(y) \vdash F(x) \supset_x G(x)$, by *502.
$\vdash F(y) \supset_{y,z} [F(z) \supset y = z] \supset .(\exists y)[F(y)G(y)] \supset . F(x) \supset_x G(x)$,
by Deduction Theorem.

Proof of Tarski's theorem (*862). Suppose, for a *reductio ad absurdum*, that there is such a formula as $\mathbf{V}_x$. Let $\mathbf{I}_x$ be the formula $U(x, x, y) \supset_y {\sim}\mathbf{V}_y$. Let the Gödel number of $\mathbf{I}_x$ be i''. Let $\mathbf{J}$ be $Z_{i''}(x) \supset_x \mathbf{I}_x$, i.e., $\mathbf{J}$ is $Z_{i''}(x) \supset_x . U(x, x, y) \supset_y {\sim}\mathbf{V}_y$. Let j'' be the Gödel number of $\mathbf{J}$.

By Tarski's criterion $\vdash Z_{j''}(x) \supset_x \mathbf{V}_x \,.\!\equiv \mathbf{J}$. $j'' = \mathrm{u}(i'', i'')$ because $\mathrm{u}(i'', i'')$ is the number of the formula $Z_{i''}(x) \supset_x \mathbf{A}$, where $\mathbf{A}$ is formula no. i'', i.e., $\mathrm{u}(i'', i'')$ is the number of the formula $Z_{i''}(x) \supset_x \mathbf{I}_x$, i.e., $\mathrm{u}(i'', i'')$ is the number of $\mathbf{J}$.

Hence, $\vdash Z_{i''}(x_1) \supset . Z_{i''}(x_2) \supset . Z_{j''}(y) \supset U(x_1, x_2, y)$ because u is calculable in A^2 as represented by $U(x_1, x_2, y)$. Hence,

(1) $Z_{i''}(x), Z_{j''}(y) \vdash U(x, x, y)$, by ^{*503}, *modus ponens.*

(2) J is $Z_{i''}(x) \supset_x . U(x, x, y) \supset_y \sim \mathbf{V}_y$.

$J, Z_{i''}(x), Z_{j''}(y) \vdash \sim \mathbf{V}_y$,

by (1), (2), F^1 (functional calculus of first-order), *modus ponens.*

$J, Z_{j''}(y) \vdash \sim \mathbf{V}_y$, by ^{*518}.

(3) $Z_{j''}(y) \vdash J \supset \sim \mathbf{V}_y$, by Deduction Theorem

$Z_{j''}(y) \supset_y \mathbf{V}_y, Z_{j''}(y) \vdash \mathbf{V}_y$, by F^1, *modus ponens.*

$\vdash Z_{j''}(y) \supset_y \mathbf{V}_y . \equiv J$, by our *reductio* assumption.

$J, Z_{j''}(y) \vdash \mathbf{V}_y$, by P, F^1, *modus ponens.*

(4) $Z_{j''}(y) \vdash J \supset \mathbf{V}_y$, by Deduction Theorem.

$Z_{j''}(y) \vdash \sim J$, by (3), (4), and P.

(5) $\vdash \sim J$, by ^{*518}, Deduction Theorem, ^{*712}, *modus ponens.*

(6) $\vdash \sim J \equiv .(\exists x)(\exists y) . Z_{i''}(x) U(x, x, y) \mathbf{V}_y$, by definition of J and F^1.

(a) $\vdash (\exists x)(\exists y) . Z_{i''}(x) U(x, x, y) \mathbf{V}_y$ by (5), (6), and P.

(7) $\vdash U(x_1, x_2, y) \supset . U(x_1, x_2, z) \supset y = z$, by ^{*738}.

$Z_{i''}(x), Z_{j''}(z) \vdash U(x, x, z)$, by ^{*738} (calculability), ^{*503}, and *modus ponens.*

(8) $Z_{i''}(x), Z_{j''}(z), U(x, x, y) \vdash U(x, x, z)$, by ^{*517}.

$Z_{i''}(x), Z_{j''}(z), U(x, x, y) \vdash y = z$, by (7), (8), ^{*503}, and *modus ponens.*

$Z_{i''}(x), Z_{j''}(z), U(x, x, y) \vdash Z_{j''}(y)$, by ^{*529}, ^{*502}.

$Z_{i''}(x), U(x, x, y) \vdash Z_{j''}(y)$, by ^{*518}, Deduction Theorem, ^{*712}, *modus ponens.*

$Z_{i''}(x), U(x, x, y), \mathbf{V}_y \vdash Z_{j''}(y)$, by ^{*517}.

$Z_{i''}(x), U(x, x, y), \mathbf{V}_y \vdash Z_{j''}(y) \mathbf{V}_y$, by P.

$(\exists y) . Z_{i''}(x) U(x, x, y), \mathbf{V}_y \vdash (\exists y) . Z_{j''}(y) \mathbf{V}_y$, by ^{*519}.

$(\exists x)(\exists y) . Z_{i''}(x), U(x, x, y), \mathbf{V}_y \vdash (\exists y) . Z_{j''}(y) \mathbf{V}_y$, by ^{*518}.

(b) $\vdash (\exists y) . Z_{j''}(y) \mathbf{V}_y$, by (a).

$\vdash Z_{j''}(x) \supset . Z_{j''}(y) \supset x = y$, by ^{*713}.

(c) $\vdash Z_{j''}(y) \supset_{y,z} . Z_{j''}(z) \supset y = z$ by ^{*503}, generalization twice.

$\vdash Z_{j''}(y) \supset_{y,z} [Z_{j''}(z) \supset y = z] \supset . (\exists y)[Z_{j''}(y) \mathbf{V}_y] \supset . Z_{j''}(x) \supset_x \mathbf{V}_x$,

by the lemma and $^*352_1$ (Rule of substitution for functional variables), since x is the sole free variable of $\mathbf{V}_x$, and we may suppose that y does not occur in $\mathbf{V}_y$ except at those places where x has free occurrences in $\mathbf{V}_x$.

$\vdash Z_{j''}(x) \supset_x \mathbf{V}_x$, by (c), (b) *modus ponens* twice, i.e.

$\vdash J$, by our *reductio* assumption.

But line (5) was $\vdash \sim J$, and this contradicts the assumed consistency of A^2.

$\mathbf{I}_x$ (syntactical meaning) — According to the correspondence of Gödel numbers and formulas we can read $U(x, x, y) \supset_y \mathbf{V}_y$ as: If $y = \mathrm{u}(x, x)$ and if y is the Gödel number of a closed wff, then that wff is not true.

$\mathbf{J}$ (syntactical meaning) — If $y = \mathrm{u}(i'', i'')$ and y is the number of a closed wff, then that wff is not true. Recall that $j'' = \mathrm{u}(i'', i'')$. **J** (analogous to Gödel's J) says about itself that it is not true. More precisely, **J** says (via the indirect correspondence between Gödel numbers and formulas) that the formula which has j'' as Gödel number is not true. It is only later (and as it were accidentally) that we find $j'' = \mathrm{u}(i'', i'')$. If there were a $\mathbf{V}_x$ such that the contradictory of the metatheorem we have just proved were a theorem, the Liar's Antinomy could be formally reproduced in A^2. The Tarski theorem can be generalized as with the Gödel theorems, in particular the Tarski theorem holds for A^0 also.

Metatheorem of A^0: If A^0 is consistent there is no wff $\mathbf{V}_x$ of A^0, having x as its only free variable, and such that $\vdash Z_a \supset_x \mathbf{V}_x \,.\equiv \mathbf{A}$ in all cases in which **A** is a closed wff of A^0 and a is the Gödel number of **A**.

The proof of this metatheorem would be analogous to the one we have done for A^2; that is, if we were to carry through a proof that primitive recursion is representable in A^0 (in the sense of Chapter VII), which would be an even more arduous task than its analogue for A^2, we could use our proof of the Tarski theorem for A^2 as a template to obtain its analogue for A^2.

Hilbert and Bernays in *Grundlagen der Mathematik* sketch a method for obtaining a metatheorem which deals with representing primitive recursion in a system analogous to A^0. This method more or less duplicates the treatment of primitive recursion for A^2. In particular the proof of the analogue of *738 is much more complicated. Our treatment depend upon the use and quantification of functional variables. An insight to the Hilbert-Bernays method can be obtained by recalling that (for example) there is an infinite set of ordered pairs associated with a binary relation on the natural numbers. Instead of considering infinite sets of ordered pairs, Hilbert and Bernays use sets of ordered pairs which are bounded by some upper limit and which can, therefore, be represented by a single natural number. Then, instead of using functional variables as we have done in A^2, they us variables whose range is bounded sets of ordered pairs. Of course, the original proof that primitive recursion can be represented in the first order language analogous to A^0 is due to Gödel.

Definition of Truth for A^0 in A^2. The positive side of Tarski's result is that it is possible to give a definition of truth that satisfies his criterion by giving the definition in a different language from that for which it is to be a definition of truth. Our next project will be to do this for A^0 and A^2. First we will need a method for finding an analogue of each wff of A^0 and A^2:

> If A is an elementary formula of A^0, then the transformation A^* in A^2 of A shall be:
> (i) Propositional variables standing alone transform into themselves,
> (ii) Elementary formulas of the form $\sum(\mathbf{a,b,c})$ and $\prod(\mathbf{a,b,c})$ transform, respectively, into the formulas of A^2 as given by the two definition schemata in §55 (p. 322 of *Introduction to Mathematical Logic*).
> If A is built up of elementary formulas and logical constants, then
> (iii) $(\forall a)M$ — transforms into $(\forall a)M^*$ (where M^* is the transformation of M).
> (iv) $M \supset N$ — transforms into $M^* \supset N^*$ (where M^*, N^* are the transformations of M, N respectively).
> (v) $\sim N$ — transforms into $\sim N^*$ (where N^* is the transformation of N).

We shall now introduce by definition a wff $V(x)$ of A^2, having x as its only free variable, and such that in all cases in which **A** is a closed wff of A^0, and a is the Gödel number of **A**, and $\mathbf{A}^*$ is the transformation of **A** into A^2: $\vdash Z_a(x) \supset_x V(x) \,.\equiv \mathbf{A}^*$, given an explicit Gödel numbering of the syntax of A^0 and the notion of a transformation of every formula **A** of A^0 into a formula $\mathbf{A}^*$ of A^2.

The possibility of giving a definition of truth for A^0 in A^2 is dependent upon certain of the syntactical relations of A^0 being primitive recursive, but we do *not* need to prove that the theorems of A^0 form a primitive recursive set. We have shown that the following functions pertaining to the syntax of A^0 are primitive recursive: $\mathrm{Elf}^0(a)$, that a is the Gödel number of an elementary formula of A^0 (*837) and $\mathrm{Wff}^0(a)$, that a is the Gödel number of a wff a of A^0(*839).

In giving the definition of truth (validity) we have to speak of valuations of the free variables of wffs of A^0. This can most conveniently be done by letting a valuation be an infinite sequence of natural numbers such that the nth term in the sequence is to be the value of the nth term in the alphabetic order of variables. Any given wff can have at most a finite number of free variables; hence among its free variables there must be one which is furthest along in alphabetic order. Suppose that one to be the nth variable in alphabetic order. Then any infinite sequence of natural numbers which takes the value 0 after its nth term can serve as a valuation of the given wff. With each such infinite sequence of natural numbers which is 0 after a certain term d $a, b, c, \ldots d$ we associate a number $2^a 3^b 5^c \ldots \mathrm{pr}(n)^d$. That the number associated with a valuation is unique follows, of course, from the prime factorization theorem. Then if v is the number of an individual variable and w is the number of a valuation, $\mathrm{val}(v, w)$ (*848) has as syntactical meaning (via the indirect correspondence between individual variables and valuations established by the Gödel numbering): the value given to the individual variable number v by the valuation number w. We have shown in *848 that $\mathrm{val}(v, h)$ is a binary primitive recursive function of v and w.

In the following table the expressions on the left are primitive recursive relations and functions; those on the right are to be thought of as abbreviations for propositional forms which represent the respective functions or relations in A^2. By the effective methods of *738 and *739 of Chapter VII, we can, given the set of definitions by primitive recursion and composition leading up to each of the primitive recursive functions and relations in the table, find a propositional form which represents the particular recursive function or relation. In each case we are to think of the expression on the right as abbreviating this propositional form.

$n\,\mathrm{Gl}\,a = c$	represented by	$\mathbf{Gl}(x_1, x_2, y)$
$\mathrm{Neg}(a) = c$		$\mathbf{N}(x_1, y)$
$a\,\mathrm{Imp}\,b = c$		$\mathbf{I}(x_1, x_2, y)$
$v\,\mathrm{Gen}\,a = c$		$\mathbf{V}(x_1, x_2, y)$
$\mathrm{var}(v)$		$\mathbf{v}(x_1)$
$\mathrm{Elf}^0(a)$		$\mathbf{E}^0(x_1)$
$\mathrm{Wff}^0(a_1)$		$\mathbf{W}^0(x_1)$
$\mathrm{val}(v, h) = c$		$\mathbf{v}(x_1, x_2, y)$
$\mathrm{dif}(v, h_1, h_2)$		$\mathbf{d}(x_1, x_2, x_3)$

The only free variables of the propositional forms, which we are to think of as abbreviated by the expressions on the right, are those displayed between the respective parentheses. We can now give the definition of validity for A^0 in A^2:

$$\begin{aligned} V(\mathbf{a},\mathbf{b}) \rightarrow {} & [\mathbf{E}^0(x) \supset_x . Z_1(z_1) \supset_{z_1} . Z_3(z_3) \supset_{z_3} . Z_5(z_5) \supset_{z_5} . Z_7(z_7) \supset_{z_7} \\ & . \mathbf{Gl}(z_1, x, x_1) \supset_{x_1} . \mathbf{Gl}(z_3, x, x_3) \supset_{x_3} . \mathbf{Gl}(z_5, x, x_5) \supset_{x_5} . \mathbf{Gl}(z_7, x, x_7) \supset_{x_7} \\ & . [F(x, y) \equiv_y . \mathbf{v}(x_3, y, y_3) \supset_{y_3} . \mathbf{v}(x_5, y, y_5) \supset_{y_5} . \mathbf{v}(x_7, y, y_7) \supset_{y_7} . Z_{37}(x_1) \\ & \textstyle\sum(y_3, y_5, y_7) \vee Z(x_1) \prod(y_3, y_5, y_7)][\mathbf{W}^0(x_1) \supset_{x_1} . \mathbf{N}(x_1, x) \supset_x . F(x, y) \\ & \equiv_y {\sim} F(x_1, y)][W^0(x_1) \supset_{x_1} . \mathbf{W}^0(x_2) \supset_{x_2} . \mathbf{I}(x_1, x_2, x) \supset_x . F(x, y) \\ & \equiv_y . F(x_1, y) \supset F(x_2, y)][\mathbf{v}(z) \supset_z . \mathbf{W}^0(x_1) \supset_{x_1} . \mathbf{V}(z, x_1, x) \supset_x \\ & . F(x, y) \equiv . \mathbf{d}(z, y, y_1) \supset_{y_1} F(x_1, y_1)] \supset_F F(\mathbf{a}, \mathbf{b}) \end{aligned}$$

This definition very closely parallels in form our definition of $\sum$ in §55 to represent addition in A^2. Each conjunct of its antecedent corresponds to one of the rules in the informal semantical definition of truth for A^0. But here no use is made of the semantical notion of denotation. It should be noted that this is an extensional definition of truth for the formulas of A^0. That this reduction of the semantical notion of truth to syntax is possible is dependent upon our use of the higher order variable "F" in A^2 and quantification on it. Via the indirect correspondence between Gödel numbers and formulas and infinite sequences of natural numbers we are able to use "F" to express a relation between formulas and valuations of their free variables. It should be observed that, given the principal or intended interpretation of A^0, the first clause of our definition (which pertains to elementary formulas of A^0) requires that A^2 be adequate for primitive recursive arithmetic.

The general form of the definition is that each of the conjuncts in its antecedent attributes to the value of "F" one of the properties of valid formulas of A^0. In the second, third, and fourth conjuncts the property attributed to the value of "F" is in terms of the given well formed parts (paralleling the informal semantical rules). There are, of course, an infinite number of binary relations which satisfy the antecedent of our definition. By putting "$F(\mathbf{a}, \mathbf{b})$" in the consequent and quantifying on "F" what we essentially do is take the intersection of these infinitely many binary relations, thus obtaining the common part of all of them.

Our use of the indirect correspondence between Gödel numbers and formulas and valuations insures against mention-use errors. It is evident, however, that in a suitable syntax language for A^0 (which would have to include variables ranging over the expressions of A^0 and another class of variables ranging over valuations, and which would have to be, in addition, adequate for primitive recursive arithmetic) we could introduce essentially the same definition of validity.

We are now in a position to give the definition of truth for A^0:

$V(\mathbf{a}) \rightarrow (\mathbf{h}) V(\mathbf{a}, \mathbf{h})$, where **h** is the first variable in alphabetic order after **a**.

History of the Tarski definition of truth. The history of the Tarski definition of truth is somewhat involved; it might better be called the Carnap-Tarski definition of truth. Tarski's first publication of his definition was the monograph: *O pojęciu prawdy w odniesieniu do sformalizowanych nauk dedukcyjnych* (On the notion of truth in reference to formalized deductive sciences), reprinted in English translation as Chapter VIII of ***Logic, Semantics, Metamathematics***, by Alfred Tarski (translated by J. H. Woodger). Carnap's first publication of his definition was *Ein Gültigkeitskriterium für die Sätze der klassischen Mathematik* (A criterion of validity for the sentences of classical mathematics [The present author prefers the translation "propositions" for "sentences" here.]), which appeared in ***Monatshefte für Mathematik und Physik***, 1935. A German translation of Tarski's monograph appeared in ***Studia Philosophica***, vol. 1 (1936) (reprint dated 1935) under the title "Der Wahrheitsbegriff in den formalisierten Sprachen", pp. 261–405 (with appendix and revisions). Woodger's translation is based on the later German version.

In 1933 Tarski gave the definition of truth only for languages of finite order, the reason being that the type theory of a language must be of transfinite order if it is to be used to give a definition of truth for ω-order arithmetic. The 1936 German translation has an appendix which extends the truth definition to arithmetic of order ω, after a discussion of transfinite type theory. That such an extension to transfinite types is possible was noted by Gödel in 1931.

In 1935 Carnap gave a definition of truth for arithmetic of order ω. He first gives the truth definition informally in ordinary language, which involves a discussion of the valuations of the free variables of formulas. To formalize the definition requires a type that does not occur in the object language. Using a formalized meta-language which includes variables of this higher type is to lift the order by one and quantifying on this class of variables lifts the type one more order. Hence the definition of truth for ω-order arithmetic must be given in a meta-language which has a type theory of order $\omega+2$. Carnap refers to Gödel's results and recognizes that the definition of truth must be given in a meta-language which is essentially stronger than the object language. Having obtained a truth definition in $\omega+2$-order arithmetic for ω-order arithmetic, Carnap goes on to give a consistency proof for ω-order arithmetic. The method of the consistency proof is: first prove that the axioms are valid and then that the rules of inference preserve validity; hence (implicitly by induction on the length of a proof) all the theorems are valid. But then the negation of a theorem cannot be a theorem; hence ω-order arithmetic is consistent. Carnap's result does not conflict with Gödel's second theorem since the metatheorem that ω-order arithmetic is consistent is *not* a theorem of ω-order arithmetic, but rather of $\omega+2$-order arithmetic. Carnap does not mention Tarski's criterion for the adequacy of a definition of truth, nor Tarski's theorem. Carnap defines logical consequence in terms of validity: **B** is a logical consequence of **A** if and only if the closure of $\mathbf{A} \supset \mathbf{B}$ is valid. Tarski gives the same definition in the German translation of 1936. Tarski credits his criterion for truth to Lesniewski, his teacher. It is quite possible that the technically erroneous (and actually meaningless because of the fallacy of misquoted variables) version of Tarski's criterion:

"p" is true if and only if p

can be traced to Lesniewski. The 1936 German translation of Tarski refers to Carnap's article of 1935.

RESOLUTION OF BERRY'S ANTINOMY BY RAMIFIED TYPE THEORY

The least positive integer not nameable in English in less than twenty-five syllables is named by the underscored English phrase, and so is nameable in English in less than twenty-five syllables.

This is Berry's antinomy. For rigorous treatment of it we must reformulate it with respect to a language whose syntax and semantics are less problematical than those of the English language. Let this be the language of *Principia Mathematica* with such primitive constants added as are found to be necessary for the purpose.

We assume a modification of the Principia language according to which both the positive integers and the symbols and wffs of the Principia language itself are included among the individuals. This is convenient. But the explanation which follows can be rewritten without difficulty to put both these things into any desired higher type.

The number 25 which appears in the above statement of Berry's antinomy must no doubt be replaced by a larger number, and to be on the safe side we replace it by 10^{100} (selected as being a quite large number that can be characterized by a comparatively short wff). We must add to the Principia language primitive constants sufficient to enable it to express its own syntax. And in terms of these we define a notation, **short(v)**, to mean that **v** is a wff of length less than 10^{100}.

We must also add primitive constants **val**, of all appropriate types (always predicative), such that **val(a, v, F)** means that **a** is an individual variable and **v** is a wff having no other free variable than **a** and the value of **v** for any value x of **a** is $F(x)$.

In the Principia language the names that are referred to in the statement of Berry's antinomy must be replaced by descriptions $(\iota\mathbf{x})\mathbf{M}$, and hence ultimately by propositional functions that are expressed by the wffs **M**. (The treatment of the antinomy in *Principia Mathematica* is in confusion between the wffs **M** and the propositional functions which they express, but as usual in that work the use-mention confusion is superficial and does not destroy the main point of what is being said.)

The Berry number is then the least positive integer x such that $B(x)$, where

$$B(x) = \sim(\exists x) \mathbin{.} F(y) \equiv_y y = x \mathbin{.} (\exists a)(\exists v) \mathbin{.} val(a, v, F) \mathbin{.} short(v)$$

If we reduce to simple type theory by ignoring the question of the order of the bound variable **F**, Berry's antinomy then reappears. But in ramified type theory the antinomy is resolved as follows: the order of **B** must be greater by one than the order of **F** and hence the propositional function, that x is the least positive integer such that $B(x)$, is not an admissible value of the variable **F**.

This paper was found in typescript among Church's papers after his death. The date of composition is not known. Printed by permission of the Alonzo Church estate.

GRELLING'S ANTINOMY
(ca. 1995)

Let the notation $\mathrm{D}(w, \mathrm{F})$ be used to mean that w is a word and denotes F.

To fix the types, we take words as included among the individuals. (It would make no important difference in what follows if we chose a higher type instead.) Hence the first argument of D must be an individual variable (or individual constant). We shall make use only of the case that the second argument of D is a singulary functional variable or singulary functional constant of lowest type, so fixing the type of D. But in principle any constant may count as a word in the formalized language — e.g. not only a singulary functional constant but also individual constant or a binary functional constant — and hence the second argument of D may be a variable or constant of any type, under suitable choice of the type of D.

As a reasonable postulate about D, in view of the intended meaning, we might take the following principle of univocacy:

$$\mathrm{D}(v, \mathrm{F}) \supset_{v\,\mathrm{F}} . \, \mathrm{D}(v, \mathrm{G}) \supset_{\mathrm{G}} . \, \mathrm{F} = \mathrm{G}$$

But in view of complications about the definition of $\mathrm{F} = \mathrm{G}$ in the case of ramified type theory, we use instead the following weaker postulate:

$*$ $$\mathrm{D}(v, \mathrm{F}) \supset_{v\,\mathrm{F}} . \, \mathrm{D}(v, \mathrm{G}) \supset_{\mathrm{G}} . \, \mathrm{F}(x) \equiv_x \mathrm{G}(x).$$

On the same basis of reasonableness we assume also the principle:

$**$ Whenever an individual or a propositional function has been determined in some unique way, it is permissible to extend the language by introducing a constant (a word) which denotes it.

As an illustration of this last consider the abbreviative definition

$$\mathrm{het}(w) \rightarrow (\exists \mathrm{F}) \, . \, \mathrm{D}(w, \mathrm{F}) {\sim} \mathrm{F}(w).$$

This abbreviative definition does not extend the language and it does not give a meaning, even as an abbreviation, to 'het' in isolation. But in accordance with $**$ we may instead extend the language of simple type theory by adjoining a new singulary functional constant 'het' and taking as an axiom:

$***$ $$\mathrm{het}(w) \equiv_w (\exists \mathrm{F}) \, . \, \mathrm{D}(w, \mathrm{F}) {\sim} \mathrm{F}(w).$$

In the resulting language — if we started with pure functional calculus of third or higher order and adjoined only the one functional constant 'het' — there is a unique word that denotes the propositional function het, namely the word 'het'. Therefore to express that the word 'het' is heterological we may use either of

$$\mathrm{D}(v, \mathrm{het}) \supset_v \mathrm{het}(v), \quad (\exists v) \, . \, \mathrm{D}(v, \mathrm{het}) \, \mathrm{het}(v).$$

But to allow for further extensions of the language in which conceivably synonyms of the functional constant 'het' might be introduced, let us choose the latter. (By this means we avoid the possible but unnecessary step of adjoining to the formalized language not only the functional constant 'het' but *also* an individual constant which denotes this functional constant.)

Previously unpublished. Church left a note expressing a desire that this handwritten piece, which he calls "a minor work," be published in the collection. Printed by permission of the Alonzo Church estate.

The antinomy in simple type theory is obtained as follows. We use the abbreviations P, gen., ded.thm., subst.eq., ex.inst., neg.$\exists$, Š to refer to propositional calculus, rule of generalization, deduction theorem and analogues of *513, *518, *338, *509 in functional calculus of higher order.

By $*$, $\mathrm{D}(v, \mathrm{F}), \mathrm{D}(v, \mathrm{het}) \vdash \mathrm{F}(x) \equiv_x \mathrm{het}(x)$
Hence by subst.eq., $\mathrm{D}(v, \mathrm{F}), \sim\mathrm{F}(v), \mathrm{D}(v, \mathrm{het}) \vdash \sim\mathrm{het}(v)$
Hence by ex.inst. and axiom $***$, $\mathrm{het}(v), \mathrm{D}(v, \mathrm{het}) \vdash \sim\mathrm{het}(v)$
Hence by ded.thm., $\mathrm{D}(v, \mathrm{het}) \vdash \mathrm{het}(v) \supset \sim\mathrm{het}(v)$
Hence by P, $\mathrm{D}(v, \mathrm{het}) \vdash \sim\mathrm{het}(v)$
Hence by the axiom $***$, $\mathrm{D}(v, \mathrm{het}) \vdash \sim(\exists\,\mathrm{F}) \,.\, \mathrm{D}(v, \mathrm{F}) \sim\mathrm{F}(v)$
Hence by the neg.$\exists$ and P, $\mathrm{D}(v, \mathrm{het}) \vdash \mathrm{D}(v, \mathrm{F}) \supset_{\mathrm{F}} \mathrm{F}(v)$
Hence by Š, $\mathrm{D}(v, \mathrm{het}) \vdash \mathrm{D}(v, \mathrm{het}) \supset \mathrm{het}(v)$
Hence by modus ponens, $\mathrm{D}(v, \mathrm{het}) \vdash \mathrm{het}(v)$
Thus from the hypothesis $\mathrm{D}(v, \mathrm{het})$ we have a proof both of $\sim\mathrm{het}(v)$ and of $\mathrm{het}(v)$
Hence by ded.thm. and P, $\vdash \sim\mathrm{D}(v, \mathrm{het})$
Hence by gen., $\vdash (v)\sim\mathrm{D}(v, \mathrm{het})$
Hence by neg.$\exists$, $\vdash \sim(\exists v)\, \mathrm{D}(v, \mathrm{het})$
But this contradicts $**$

In ramified type theory the functional variable F in the axiom $***$ must have a particular level m. Hence instead of the single functional constant 'het' we need an infinite list of functional constants 'het^2', 'het^3', $\cdots$, each of the level indicated by its superscript. Then the axiom $***$ must be replaced by an axiom schema:

$$\mathrm{het}^{m+1}(w) \equiv_w (\exists\,\mathrm{F}^{1/m}) \,.\, \mathrm{D}(w, \mathrm{F}) \sim\mathrm{F}(w)$$

Here as an abbreviation the superscript $1/m$ is written only after the first occurrence of the letter F, and the same superscript is then to be understood after all remaining occurrences of the same letter.

The level of the binary functional constant D will not play a critical role in what follows, but it may be taken to be of minimum level, both in the foregoing axiom schema and in $*$. In $*$ the functional variables F and G are of arbitrary levels, not necessarily the same.

By $*$, $\mathrm{D}(v, \mathrm{F}^{1/n}), \mathrm{D}(v, \mathrm{het}^{m+1}) \vdash \mathrm{F}^{1/n}(x) \equiv_x \mathrm{het}^{m+1}(x)$
Hence by subst.eq., $\mathrm{D}(v, \mathrm{F}^{1/n}), \sim\mathrm{F}^{1/n}(v), \mathrm{D}(v, \mathrm{het}^{m+1}) \vdash \sim\mathrm{het}^{m+1}(v)$
Hence by ex.inst., $\mathrm{het}^{n+1}(v), \mathrm{D}(v, \mathrm{het}^{m+1}) \vdash \sim\mathrm{het}^{m+1}(v)$
Hence by ded.thm., $\mathrm{D}(v, \mathrm{het}^{m+1}) \vdash \mathrm{het}^{n+1}(v) \supset \sim\mathrm{het}^{m+1}(v)$
But $\vdash \mathrm{het}^{n+1}(v) \supset \mathrm{het}^{m+1}(v)$ if $m \geqq n$
Hence by P, $\mathrm{D}(v, \mathrm{het}^{m+1}) \vdash \sim\mathrm{het}^{n+1}(v)$ if $m \geqq n$

Also by neg.$\exists$ and P, $\sim\mathrm{het}^{n+1}(v) \vdash \mathrm{D}(v, \mathrm{F}^{1/n}) \supset_{\mathrm{F}} \mathrm{F}(v)$
Hence by Š, $\sim\mathrm{het}^{n+1}(v) \vdash \mathrm{D}(v, \mathrm{het}^{m+1}) \supset \mathrm{het}^{m+1}(v)$ if $m < n$
Hence by modus ponens, $\mathrm{D}(v, \mathrm{het}^{m+1}), \sim\mathrm{het}^{n+1}(v) \vdash \mathrm{het}^{m+1}(v)$ if $m < n$

But $\vdash \text{het}^{m+1}(v) \supset \text{het}^{n+1}(v)$ if $m \leqq n$
Hence $\text{D}(v, \text{het}^{m+1}), {\sim}\text{het}^{n+1}(v) \vdash \text{het}^{n+1}(v)$ if $m < n$
Hence by deduction theorem, $\text{D}(v, \text{het}^{m+1}) \vdash {\sim}\text{het}^{n+1}(v) \supset \text{het}^{n+1}$ if $m < n$
Hence by P, $\text{D}(v, \text{het}^{m+1}) \vdash \text{het}^{n+1}(v)$ if $m < n$

This is Russell's resolution of the antinomy.

CORRECTIONS AND MINOR COMMENTS AFFECTING HEYTING'S *INTUITIONISM*, EDITION OF 1976 (ca. 1995)

Definitions 2 and 3 on page 35 have the consequence that any "mathematical entity" may be a component of an ips. It is said on the same page that rational numbers (including positive integers) are in any case to be allowed as mathematical entities, and then, in effect, that further mathematical entities may be introduced from time to time at will, subject of course to general intuitionistic conditions of acceptability. However, it must surely be said that a spread-species presupposes the elements of the corresponding spread and that these in turn presuppose their components, in such a sense that the spread-species cannot be defined independently of the definitions of these various components. And when a hierarchy of types of species is introduced on page 38, its success in excluding what are there called circular definitions depends on satisfying the condition that the species of each type are definable independently of the definitions of all species of the same or higher type. Hence the statement on page 38 that spread-species are of type 0 seems to be in contradiction with what was said on page 35. If it was meant that species may not be allowed as mathematical entities in the sense of page 35, it seems that this should have been said. A more likely explanation is perhaps that the assignment of type 0 to spread-species is meant to hold only in the context of the theory of real number-generators, where the components of an ips are always rational numbers.

It is unclear whether the non-circularity condition in 3.2.3 on page 38 (that the members of a species S must be definable independently of the definition of S) is meant to include the requirement that the members of a species S shall be definable without using quantification on a variable whose range includes either the species S itself or something which is not definable independently of the definition of S. This requirement—the exclusion of what is usually called impredicative definition—is not made explicit here or elsewhere in the book. But a reason for thinking that it must be meant is that it seems evidently to be among the things which are, in Heyting's words, obvious from the constructive point of view. And elsewhere Heyting sometimes mentions impredicative definition, by that name, as one of the things which are intuitionistically unacceptable; in particular an early expository paper in the German periodical ***Erkenntnis*** (volume 2, 1931, see pages 110–111) has a passage which closely parallels 3.2.3 in the present book, but which speaks of "imprädikative Definitionen" where the passage in the book has "circular definitions." On the other hand Hermann Weyl's criticism of Brouwer in his *Über die neue Grundlagenkrise der Mathematik* (***Mathematische Zeitschrift***, volume 10, 1921, see the next-to-last paragraph of §3) may mean that Brouwer has used impredicative definition.

If Heyting's non-circularity condition is to be read as excluding impredicative definition, the result of this when taken in conjunction with Definition 1 is a hierarchy of species that corresponds to only a fragment of the ramified type hierarchy of *Principia Mathematica*. Namely the species are restricted to those types which in *Principia*

Previously unpublished. Church left a note expressing a desire that this handwritten piece be published in the collection. Printed by permission of the Alonzo Church estate.

Mathematica are called "predicative." (Note that the authors of *Principia* do not use "impredicative" in our present sense, derived from Poincaré, but they speak instead of a "vicious circle," and they then use "predicative" with the different meaning that is explained in Chapter II of their Introduction.) —In the first edition (1956) of Heyting's book, Section 3.2.3 is the same as in the present edition except that the clause "and at least one of its members has type $n - 1$" is omitted from Definition 1. This difference between the first edition and later editions is not of great importance, as it affects only the terminological question whether a species of type n is regarded as being also of every type higher than n. However, if we alter Definition 1 to read "A species is of type n if all its members have type less than n and it is moreover definable in a way that does not presuppose species of type n or higher," we obtain the full ramified hierarchy in a streamlined version that is essentially the same as the one described by Quine in his essay, *Logic and the reification of universals* (in *From a Logical Point of View*, pp. 102–129).

On page 37 Heyting (following Brouwer) defines a species as a property which mathematical entities can be supposed to possess. This has a resemblance to *Principia Mathematica* as the singulary propositional functions of *Principia* are also properties rather than classes. Then when on page 38 two species are defined to be equal if each is a subspecies of the other, it is evidently meant that the relation between species which is thus defined is to be used in subsequent definitions as if it were a true identity relation. In particular the definition of one-to-one correspondence, which is needed on page 39 for the sake of the definition of equivalent species, must make use of some notion of equality or identity, and it is clear that this must consist in the relations which are defined and called *equal* on pages 36 and 38. However, this method is fallacious because it cannot then be proved that if species S and T are in one-to-one correspondence in such a way that corresponding members are equal, then S and T are equal. As a makeshift we may adjoin this last as an axiom, essentially an extensionality axiom for species, and then proceed with Heyting's treatment. Or equivalently we may adjoin as an axiom: $x \in S \supset_x . x = y \supset_y y \in S$. But since Brouwer is unwilling to assume such extensional entities as classes or sets, but only properties in their place, the ultimate remedy will be to import Russell's contextual definition of class abstracts!

It should be noticed that on page 39 the definition of a finite species is such that the null species is not counted as finite. Thus in lines 11–12 on page 40 Heyting speaks of "the null species, a finite species, or an infinite species." And it seems that line 1 on page 40, to save it from triviality, must be corrected to read: "Thus a species that cannot be the null species and cannot be finite is not necessarily infinite."

On page 42 the remark about Kleene's use of the word "finitary" is misleading, as Kleene uses the word with a different meaning.

Finally the last paragraph on page 114 requires comment because it raises once more the question of impredicative definition. If the non-circularity condition on page 38 means no more than that the species must obey a simple type theory, the conjecture made in the last sentence on page 114 is almost certainly correct. But if non-circularity includes avoidance of impredicative definition, the well-known difficulties in obtaining the full classical analysis without using impredicative definition, as they were experienced by Weyl (in *Das Kontinuum*) and by Whitehead and Russell (see pages xliv–xlv of the introduction to the second edition of *Principia Mathematica*), make it almost certain that Heyting's conjecture is wrong.

REVIEWS FROM *THE JOURNAL OF SYMBOLIC LOGIC* (1937–1975)

A. M. Turing. *On computable numbers, with an application to the Entscheidungsproblem.* ***Proceedings of the London Mathematical Society***, 2 s. vol. 42 (1936–7), pp. 230–265.[1]

The author proposes as a criterion that an infinite sequence of digits 0 and 1 be "computable" that it shall be possible to devise a computing machine, occupying a finite space and with working parts of finite size, which will write down the sequence to any desired number of terms if allowed to run for a sufficiently long time. As a matter of convenience, certain further restrictions are imposed on the character of the machine, but these are of such a nature as obviously to cause no loss of generality—in particular, a human calculator, provided with pencil and paper and explicit instructions, can be regarded as a kind of Turing machine. It is thus immediately clear that computability, so defined, can be identified with (especially, is no less general than) the notion of effectiveness as it appears in certain mathematical problems (various forms of the Entscheidungsproblem, various problems to find complete sets of invariants in topology, group theory, etc., and in general any problem which concerns the discovery of an algorithm).

The principal result is that there exist sequences (well-defined on classical grounds) which are not computable. In particular the *deducibility problem* of the functional calculus of first order (Hilbert and Ackermann's engere Funktionenkalkül) is unsolvable in the sense that, if the formulas of this calculus are enumerated in a straightforward manner, the sequence whose nth term is 0 or 1, according as the nth formula in the enumeration is or is not deducible, is not computable. (The proof here requires some correction in matters of detail.)

In an appendix the author sketches a proof of equivalence of "computability" in his sense and "effective calculability" in the sense of the present reviewer (***American journal of mathematics***, vol. 58 (1936), pp. 345–363, see review in this Journal, vol. 1, pp. 73–74). The author's result concerning the existence of uncomputable sequences was also anticipated, in terms of effective calculability, in the cited paper. His work was, however, done independently, being nearly complete and known in substance to a number of persons at the time that the paper appeared.

As a matter of fact, there is involved here the equivalence of three different notions: computability by a Turing machine, general recursiveness in the sense of Herbrand-Gödel-Kleene, and λ-definability in the sense of Kleene and the present reviewer. Of these, the first has the advantage of making the identification with effectiveness in the ordinary (not explicitly defined) sense evident immediately—i.e. without the necessity of proving preliminary theorems. The second and third have the advantage of suitability for embodiment in a system of symbolic logic. Alonzo Church

---

[1] Review originally published in ***The journal of symbolic logic***, vol. 2 (1937), pp. 42–43.

EMIL L. POST. *Finite combinatory processes—formulation 1.* ***The journal of symbolic logic***, vol. 1 (1936), pp. 103–105.[2]

The author proposes a definition of "finite 1-process" which is similar in formulation, and in fact equivalent, to computation by a Turing machine (see the preceding review). He does not, however, regard his formulation as certainly to be identified with effectiveness in the ordinary sense, but takes this identification as a "working hypothesis" in need of continual verification. To this the reviewer would object that effectiveness in the ordinary sense has not been given an exact definition, and hence the working hypothesis in question has not an exact meaning. To define effectiveness as computability by an arbitrary machine, subject to restrictions of finiteness, would seem to be an adequate representation of the ordinary notion, and if this is done the need for a working hypothesis disappears.

The present paper was written independently of Turing's, which was at the time in press but had not yet appeared. ALONZO CHURCH

LEON CHWISTEK. *Überwindung des Begriffsrealismus.* ***Studia philosophica***, vol. 2 (offprint 1937), pp. 1–18.[3]

In the present review we leave aside some philosophical portions of this article in order to confine attention to the account of the author's own work in symbolic logic (which he regards as contributing to the overthrow of *Begriffsrealismus*).

Chwistek's earliest publication in mathematical logic (220***1***) contains an analysis of the system of ***Principia mathematica*** on the basis of which he concludes that that system "implies contradiction." Stated in this form his conclusion is of course wrong; i.e., the system of ***Principia*** contains no *known self*-contradiction. A closer examination of his argument, however, reveals that the genuine content of it is that the system of ***Principia*** contradicts the well known rule of Poincaré, "Ne jamais envisager que des objets susceptibles d'être définis en un nombre fini de mots" (137***6***). Chwistek's extreme statement of his result is apparently based on an acceptance of Poincaré's dictum as an imperative rule of rational (especially mathematical) thinking—a position which, indeed, is not indefensible, if the evidently intended qualification is made that some particulars must be accepted as undefined or primitive ideas.

The remedy proposed is rejection of the axiom of reducibility and acceptance of the ramified theory of types without that axiom.

Nine years later, in 1921, we find Chwistek publishing the first proposal (220***3***) of the now widely accepted simplified theory of types. The following passage is quoted in translation from this publication:

"For the elimination of this antinomy there suffices the simple theory of types, depending on distinction of individuals, functions of individuals, functions of these functions, and so forth. Distinction of orders of functions of a given argument, and introduction thereby of predicative functions, and in further consequence appealing to the principle of reducibility is from this point of view a superfluous complication of the system. It should be noted that removal of the above elements from the theory of types of Whitehead and Russell would render this theory extraordinarily simple and perspicuous. If therefore Whitehead and Russell could not make up their minds to the

[2]Review originally published in ***The journal of symbolic logic***, vol. 2 (1937), p. 43.

[3]Review originally published in ***The journal of symbolic logic***, vol. 2 (1937), pp. 168–170.

simplification, then they undoubtedly did that as a result of the conviction that a system of logic admitting the antinomy of Richard cannot be regarded as a final expression of that which it is possible to attain in the given sphere. Leaving this matter aside, we restrict ourselves to the assertion that the theory of types together with the principle of reducibility cannot be maintained, because either it is false or else it represents in intricate form that which fundamentally is simple."

Credit for the actual proposal of the simplified theory of types, as acceptable to those logicians who deny "theoretical value" to the Richard paradox, belongs, in fact, to Chwistek; although the first sharply drawn distinction between the logical and semantical paradoxes is Ramsey's (295***5***).

Chwistek's proposal of the simplified theory of types was repeated in 1922 in a publication in German (220***6***), and in 1926 was actually employed by him in provisional fashion as basis for an article dealing with "nicht-Cantorsche Mengelehre" (220***9***). He never himself accepted this theory as other than provisional, however, because it presupposes a *Begriffsrealismus*, as he says in the article under review.

In the publication of 1922 the simplified theory of types is overshadowed by the author's return to his proposal of 1912 for a ramified theory of types without the axiom of reducibility. Indeed he charges that the Whitehead and Russell theory with the axiom of reducibility leads to the Richard paradox equally with the simplified theory of types. His argument as regards the former theory appears to be not altogether satisfactory. But in the case of the simplified theory of types it is well known that the Richard paradox does arise upon incorporation into the theory of symbols for certain semantical concepts. And since there are cogent objections which may be brought against the axiom of reducibility, as Russell has admitted (194***4***), Chwistek's proposal retains much of its force.

In a publication of 1924–25 in English (220*7*) Chwistek carried out in formal detail his project for a ramified type theory without the axiom of reducibility. With the introduction of number theory by means of special axioms, his course is found to be less heroic than might be inferred from the comment of Russell (194***4***), who, it would seem, had not seen Chwistek's paper in its final form.

In 1929 there begins a series of publications which mark a new phase in Chwistek's work. These are concerned with the construction of a formal system which shall contain notations for both the concepts of logic and those of semantics. Such a system may be said in at least a vague way to have been already contemplated in the original attitude of Whitehead and Russell towards the semantical paradoxes. Moreover such a system is no doubt consistently possible, on the basis of the ramified theory of types, and its development should be of considerable interest.

Unfortunately this work of Chwistek's has been marred (as it seems to the reviewer) by the introduction of artificialities which are really irrelevant.

An especially troublesome point is the usage, defended in the article under review, by which the same symbol may denote, either (1) an abstract, e.g. logical, concept in usual fashion, or (2) itself (better, the class of symbols of the same shape as itself). It should be remembered, however, that use of one symbol in two senses does not necessarily result in ambiguity, because there may be a syntactical rule which will determine in every case which sense is intended. Chwistek seems to say (page 10 of the article under review) that his usage is free from ambiguity in this way, and the reviewer, while not having made a detailed examination of Chwistek's formal work, nevertheless has the impression that this is true. If it is true, then presumably the system of Chwistek can be reformulated so as to avoid his dual usage of symbols (without other essential changes) by introducing instead, for each order, a symbol for the relation between a concept of that order and a symbol which denotes it. Such a reformulation, if possible, would be valuable as making clear that Chwistek has been led to his form of the theory of types, not by any ambiguity of notation, but partly by his insistence on including semantics and logic in one language, partly by his acceptance of Poincaré's dictum referred to above. But the reformulation would clearly not be acceptable to Chwistek, who apparently maintains, at least as regards logical and mathematical concepts, that the symbol and the concept symbolized are indeed one.

Parenthetically, it should be added that the current view of advocates of the simplified theory of types, whereby the relation between a concept and the symbol which denotes it must appear, not within the original language, but within a meta-language containing the original language, may itself by regarded as a kind of ramified type theory (the distinction between a hierarchy of languages and a hierarchy of types within one language being here a mere matter of terminology). Such a theory has a practical advantage of simplicity in that it reduces to a simple theory of types on elimination of semantical concepts. But of course it violates Poincaré's dictum.

A descriptive account of this more recent work of Chwistek's, somewhat fuller than that in the article before us, will be found in 220***14***. A detailed account is contained in 220***15***, reviewed by Julian Perkal in this JOURNAL (II 140). ALONZO CHURCH

W. V. QUINE. *A logistical approach to the ontological problem.* Preprinted for the members of the Fifth International Congress for the Unity of Science, Cambridge, Mass., 1939, as from ***The journal of unified science (Erkenntnis)***, vol. 9; 6 pp.[4]

The author is concerned with the distinction between symbols (or expressions) of a formalized language which are *names* and so have a meaning in isolation, and those—like the symbol (in most familiar languages—which are "syncategorematic," i.e., which have no meaning in isolation but which nevertheless occur as essential parts of significant sentences. In particular the question is considered whether abstract nouns—e.g., roundness—have designata, i.e., whether they are names or syncategorematic. As a criterion of the non-syncategorematic character of a symbol n in a given language, it is proposed to take the admissibility of the inference from A_n to $(\exists x)A_x$ (where A_x is an expression containing x as a free variable, and A_n is the result of substituting n for x in A_x)—or, equivalently in most familiar languages, the admissibility of the inference from $(x)A_x$ to A_n.

Now it is clear that a language would be inadequate which did not contain some substitute for, e.g., the sentence $(x)(y) . x + y = y + x$ and the inference from this sentence to $5 + 8 = 8 + 5$, x and y being numerical variables. Hence the author contemplates that in a language in which numerals are syncategorematic it should be possible to introduce numerical variables and quantification of such variables by contextual definition, so that expressions containing numerical variables are explained as mere abbreviations for expressions not containing such variables. Apparently it is hoped that an adequate formalized language may thus be devised in which all abstract

[4]Review originally published in ***The journal of symbolic logic***, vol. 4 (1939), p. 170.

nouns are syncategorematic, and the *tenability* of the nominalistic position thereby demonstrated.

It would seem, however, that such a demonstration of the tenability of the nominalistic position must be at the same time a demonstration of its extreme artificiality. In the opinion of the reviewer, the effect is only to emphasize the illusory character of the question whether abstract nouns *really* have designata. For the matter is relative, on the present showing, not only to the choice of a particular language, but also to the choice as to which particular notation or notations in the language shall be regarded as denoting existential quantification (the syntax of the language will ordinarily not determine the latter choice uniquely). ALONZO CHURCH

BARKLEY ROSSER. *The introduction of quantification into a three-valued logic.* Preprinted for the members of the Fifth International Congress for the Unity of Science, Cambridge, Mass., 1939, as from ***The journal of unified science (Erkenntnis)***, vol. 9; 6 pp.[5]

A system of logic is considered, based on an n-valued propositional calculus with one designated value, and a set of axioms and rules is offered for this system, covering not merely the propositional calculus but also quantification, descriptions, classes. In order to avoid paradoxes, and as a substitute for a device such as a theory of types, Quine's "method of partial stratification" (II 86) is employed.

Certain subsets of the axioms are named as adequate for the propositional calculus and for the functional calculus of first order. The set of axioms for the propositional calculus is complicated, and it would be desirable to replace it by a simpler one—in the three-valued case the six axioms of Wajsberg and Slupecki (437*1*, II 46(2)) might be used. the axioms for the functional calculus of first order may be compared with those of Bocvar (IV 98) for the three-valued functional calculus of first order.

The preprint contains a number of errata. In particular, the definition of $J_i(p)$, $\alpha = i$ should appear below the Σ, and in the statement of Axiom I 13, $j \neq i$ should stand after the semicolon. ALONZO CHURCH

SYLVESTER HARTMAN. *Are there any extra-syllogistic forms of reasoning?* ***Proceedings of The American Catholic Philosophical Association***, vol. 15 (pub. 1940), pp. 235–241.[6]

The author proposes a generalization of the notion of (categorical) syllogism, designed to include under this head certain forms of mediate reasoning usually regarded as non-syllogistic. The generalization depends on allowing, besides the traditional analysis of a proposition into subject–copula–predicate, also, in appropriate cases, an alternative analysis into subject–copula–relation–predicate, e.g., *Homer–is–the author of–the Iliad.* In addition to the traditional syllogisms, there are listed six kinds of syllogisms in the generalized sense.

The formal analyses tend to be inexact; e.g., the rôle of quantifiers is not always clear, and the three meanings of the copula ($=$, $\in$, $\subset$) are not distinguished. In the case of some of the inferences offered it may be thought that the validity rests on empirical grounds rather than purely on formal logic.

[5]Review originally published in ***The journal of symbolic logic***, vol. 4 (1939), p. 170.

[6]Review originally published in ***The journal of symbolic logic***, vol. 5 (1940), p. 81.

The reviewer deplores the excessive preoccupation with the syllogism. For the attempt must fail, to give a complete account of deductive reasoning from this point of view, even within (say) the functional calculus of first order. ALONZO CHURCH

HARRY RUJA. *Good logics and bad.* ***The philosophical review***, vol. 49 (1940), pp. 346–355.[7]

The author denies the existence of "alternative logics," and attacks in particular the position taken by C. I. Lewis on this question. Some of his criticism of Lewis seem to have force. But the argument contains errors at particular points in such a way as to render the whole suspect. It is said that "Brouwer's reform is of importance to mathematics but it has little relevance to logic"—it is not clear how this is to be reconciled with the fact that (even within propositional calculus) there are chains of deductive reasoning, classically accepted as valid, but rejected by Brouwer. Of the various alternative geometries it is said that "no more than one can apply to the actual world," i.e., when the abstract terms *point*, *line*, etc., are identified (or better, associated) with physical constructs in a particular way—the quoted assertion is commonly held to be false, and the author offers no refutation of the accepted view. The "principle of identity" is discussed without indicating which of the various things that have been called by this name is meant, and at one point confusion is unmistakable between a proposition of formal logic and a principle which is either metaphysical or vaguely syntactical in character. ALONZO CHURCH

M. LAZEROWITZ. *Self-contradictory propositions* ***Philosophy of science***, vol. 7 (1940), pp. 229–240.[8]

The author argues that a tautologous proposition (i.e., a proposition expressed by a sentence resulting from a tautology of the propositional calculus by substituting particular sentences for the variables) is one to which no alternative is conceivable, and therefore that it has no negation (i.e., the negated sentence expresses no proposition). This is offered in opposition to the view that there are such things as self-contradictory propositions (meanings of self-contradictory sentences).

In the case of the propositional calculus such an argument can be made very plausible by assuming, as the author does, that the truth-table decision procedure is not merely a decision procedure but also an analysis of meaning of the sentence to which it is applied.—But the point remains that difficulties would result in the actual practice of formal logic. Customary formulations put forward, as true, sentences which contain self-contradictory parts (in places at which, it would seem only parts expressing propositions should enter); and an attempt to change this could be expected to have consequences far beyond the propositional calculus. The author ignores this point, presumably on the—in the reviewer's opinion doubtful—ground that the question at issue has an answer in an absolute sense, independent of all considerations of formal convenience of mathematical adequacy.

In the case of more substantial logical disciplines, involving quantifiers and function variables, the corresponding contention is extremely unplausible. For example, Fermat's conjecture that, for all natural numbers n, $2^{2^n} + 1$ is prime as an intelligible

[7] Review originally published in ***The journal of symbolic logic***, vol. 5 (1940), p. 81.

[8] Review originally published in ***The journal of symbolic logic***, vol. 5 (1940), pp. 81–82.

meaning for any one with a knowledge of elementary arithmetic and algebra; since, by calculation, $2^{2^5} + 1 = 641 \times 6700417$, the conjecture is logically impossible, but it would hardly be said to follow that Fermat never made any conjecture on this point.

In the present paper, however, Lazerowitz confines himself to the case of the propositional calculus, employing the word self-contradictory in the very narrow sense of *self-contradictory according to the rules of the propositional calculus.*

ALONZO CHURCH

A. I. MELDEN. *Thought and its objects.* ***Philosophy of science***, vol. 7 (1940), pp. 434–441.[9]

Following a paper of G. E. Moore (***Philosophical studies***, pp. 215–218) and discussion of it by L. S. Stebbing (394*2* A), the author is concerned with the fact that, in common English usage, "I am thinking of a unicorn" or "I am thinking of the table in the next room" may be true although there be no unicorns or no table in the next room. The first of these sentences, e.g., cannot be analyzed as meaning: $(\exists x)$ *x is a unicorn and I am thinking of x*. And it cannot be analyzed as expressing a relation between the speaker and the class of unicorns, because "I am thinking of a unicorn" and "I am thinking of a griffin" do not mean the same thing, although both unicorns and griffins be the null class.

The choice here of the sentence "I am hunting a lion" to contrast with "I am thinking of a unicorn" is perhaps not very fortunate. The big game hunter may not mean that he is hunting a particular lion, but just any lion, and in that case his sentence likewise is incapable of analysis as: $(\exists x)$*x is a lion and I am hunting x*. He may even truthfully say "I am hunting a unicorn"—if he has heard a legend that there are unicorns in that region and has set out to verify it. In fact there are many verbs which share these peculiarities with thinking and hunting—consider, e.g., the sentences: "I am looking for the table in the next room," "I have need of a substance harder than diamond," "He deserves an everlasting monument," "This rhinoceros resembles a unicorn."

The reviewer is fully in sympathy with the suggestion that it is desirable to postulate an intensional meaning for such words as *lion* or *unicorn* (what it is to be a lion, or a unicorn), distinct from the extensional meaning as denoting classes. The whole matter might well be clarified by attention to Frege's distinction between sense and denotation (49*8*).

On the other hand the reviewer is unable to understand just what is meant by the author's conclusion in the last paragraph that thinking is in some sense not an entity. Would he say the same of hunting, needing, deserving, resembling?

The reviewer's conclusion would be rather that these verbs, when used in the way here under consideration, are not to be regarded as expressing a relation between two individuals—between the thinker and a particular unicorn, the hunter and a particular lion, or the great man and a particular monument—but are best analyzed as expressing a relation between an individual and a propositional function in intension.

ALONZO CHURCH

[9]Review originally published in ***The journal of symbolic logic***, vol. 5 (1940), pp. 162–163.

WILLARD V. QUINE. *Whitehead and the rise of modern logic.* ***The philosophy of Alfred North Whitehead***, edited by Paul Arthur Schilpp, Northwestern University, Evanston and Chicago 1941, pp. 127–163.[10]

This essay provides an able historical account of Whitehead's contributions to logic, beginning with his work in the algebra of logic (99***1, 2, 4***, 1898–1903), and going on to describe his subsequent acquaintance with the methods of Peano and of Frege (which first came to Whitehead's attention in 1900 and 1902 respectively) and the ensuing collaboration with Russell which culminated in the publication of three volumes of ***Principia mathematica*** (194***1, 2, 3***, 1910–1913). An especially valuable feature of Quine's account is the clear picture which is given of the purposes, content, and accomplishments of ***Principia mathematica***. More than two thirds of the essay is devoted to ***Principia*** (and related publications), and explains many matters of historical and technical importance in a way calculated to give the reader an understanding of them which might not otherwise easily be obtainable.

Quine's treatment of his topic is not purely historical, but includes many criticisms of Whitehead and Russell, and sometimes also of others. These critical remarks are on a uniformly high level, and some have the status of minor original contributions.

On some questions, indeed, Quine adopts a position to which there will be by no means universal assent among contemporary logicians. But even in these cases it is clear that those who disagree must take serious account of the arguments presented by Quine (here and elsewhere).

The reviewer would take the opportunity to record his own dissent on one particular point–although admitting that some of the issues which are raised have never received the detailed consideration which they demand.

The statement on page 146 that Russell was the first to define the notation or device of *description* in terms of more basic notions, or indeed that Russell made such a definition at all, seems to involve a misleading emphasis. In fact, Russell's well-known method of dealing with descriptions is not a definition of a notation for descriptions, but a proposal to do without descriptions (to accomplish certain purposes without using them). This aspect of contextual definition is, it seems to the reviewer, too often overlooked, or at least underemphasized. A direct nominal definition–say, e.g., that which, in a system containing class abstraction, $\sim$, and $=$ as primitive, construes the letter Λ as an abbreviation for the expression $\hat{x} \sim (x = x)$ —corresponds to an observation that the primitive notation of the system already provides a name of the null class, namely $\hat{x} \sim (x = x)$. The shorter notation Λ is a matter of convenience and is, in the fullest sense, dispensable in principle. On the other hand, Russell's contextual definition of a description–say, e.g., of the expression $(\imath y) \,.\, y \in \mathrm{Cls} \,.\, (x) \sim (x \in y)$ —does not mean that the primitive notation provides an expression having the same meaning, but only that it provides, for every sentence containing this expression, a *substitute sentence*—i.e., a sentence which in a certain way serves the same purpose without containing anything which could be called a description.

This point is relevant to an objection made by Quine in another passage (pp. 141–142, cf. also p. 148) to "non-truth-functional statement composition," in particular to Lewis's use of $\Diamond$. It is objected that the sentences '$\Diamond$(the number of planets < 7)' and '$\Diamond(9 < 7)$' would be judged to have opposite truth values, whereas the terms 'the

[10]Review originally published in ***The journal of symbolic logic***, vol. 7 (1942), pp. 100–101.

number of planets' and '9' denote the same number, and the two sentences should therefore be interdeducible by substituting one term for the other. On the basis of ***Principia*** or of Quine's own ***Mathematical logic*** (V 163), however, the reply to this is immediate. The translation into symbolic notation of the phrase 'the number of planets' would render it either as a description or as a class abstract, and in either case it would be construed contextually; any formal deduction must refer to the unabbreviated forms of the sentences in question, and the unabbreviated form of the first sentence is found actually to contain no name of the number 9.

The reviewer would prefer a system in which class abstracts and descriptions are construed as names, and hence are not contextually defined. In a system of this kind the situation is different. Sentences, if names at all, are perhaps best taken as names of truth values, and must be taken as *expressing* rather than *denoting* propositions (cf. 49***8***); and Quine's argument (just quoted) then shows that a non-truth-functional operator, such as $\Diamond$, if it is admitted, must be prefixed to names of propositions rather than to sentences.

ALONZO CHURCH

WILLARD V. QUINE. *Notes on existence and necessity.* ***The journal of philosophy***, vol. 40 (1943), pp. 113–127.[11]

The content of this paper is announced as mainly a translation of portions of the author's forthcoming book, ***O sentido da nova lógica*** (São Paulo, Brazil).

The author begins with a distinction between what he calls *purely designative* occurrences of names and occurrences which are not purely designative. The distinction depends on the substitution principle for equals, according to which, if A and B are names and $A = B$ is true ('=' representing the 'is' of identity), then A may be substituted for B or B for A in any sentence without altering its truth. It appears that in ordinary English usage there are certain kinds of occurrences of names for which this principle is not valid, and it is proposed to distinguish those occurrences to which the substitution principle is applicable as purely designative .

Some examples of occurrences of names which are not purely designative are supplied by (1) 'Giorgione was so called because of his size' and (2) 'Philip believes that Tegucigalpa is in Nicaragua.' Both of these sentences (let us suppose) are true. Nevertheless, despite the truth of (3) 'Giorgione = Barbarelli' and (4) 'Tegucigalpa = the capital of Honduras,' the sentences are false (5) 'Barbarelli was so called because of his size' and (6) 'Philip believes that the capital of Honduras is in Nicaragua.' Regarding (1), Quine has the immediately acceptable explanation that the sentence must be construed as a mere abbreviation of (7) 'Giorgione was called 'Giorgione' because of his size'; substitution of 'Barbarelli' for the first word of (7) yields an entirely true sentence, and there is no question of substitution of either 'Barbarelli' or ' 'Barbarelli' ' for the fourth word of (7), since there is no occurrence there of 'Giorgione' (but only of ' 'Giorgione' ') and the equality (3) is therefore irrelevant. He goes on to say that it is unnecessary to explain the instance (2) in an analogous fashion by construing it to represent (8) 'Philip believes 'Tegucigalpa is in Nicaragua'.' But he does not add, as the reviewer would, that there are objections to such a construction of (2) which are not easily superable—e.g., it would seem that (8) has the consequence that Philip understands English, whereas (2) does not.

[11]Review originally published in ***The journal of symbolic logic***, vol. 8 (1943), pp. 45–47.

In the distinction between purely designative occurrences of names and other occurrences, and its criterion, Quine is fully anticipated by Frege (49*8*), who distinguishes in the same way between the ordinary (*gewöhnlich*) and the oblique (*ungerade*) use of a name. In fact the relationship between Quine's present paper and Frege's of 1892 is close throughout, even to the use of similar, and in one instance identical, illustrations. Quine's failure to refer to Frege's paper indicates that he is unacquainted with it, but it is probable that he is indirectly indebted to Frege through Russell's 111***9***.

Quine goes on, however, to give an alternative criterion for the distinction in question, which is original with him and which seems to the reviewer to be of considerable importance. An occurrence, namely, of a name A in a sentence M is purely designative if the inference is valid from M to $(\exists x)N$, where N results from M by replacing the given occurrence of A by the variable 'x'. Thus the ontology to which one's use of language commits him is determined not merely by the names which he employs but also by the inferences which he admits involving these names in relation to quantifiers (cf. Quine IV 170 (1), V 27 (2)).

The reviewer is unable to agree with Quine's further comment here that the principle under which $(\exists x)N$ is inferred from M is a principle only by courtesy, being simply the logical content of the idea that a given occurrence is designative, and that it is therefore anomalous as an adjunct to the purely logical theory of quantification. It is indeed possible to follow Quine in taking existential quantification as given, and then to regard this principle as defining the notion of a purely designative occurrence; but it is equally possible to take the latter notion as given and to regard the principle as characterizing existential quantification. The fundamental thing is that logical systems containing names and quantifiers demand such a principle of inference, and it seems immaterial whether the principle is stated directly or in the form of a syntactical criterion for purely designative occurrences.

In the latter half of the paper the author discusses modality, and related intensional concepts, including that of attributes or properties as distinct from classes. The point is brought out that an occurrence of a name within a clause governed by a modal operator—e.g., 'it is necessary that,' 'it is possible that,' 'it is probable that'—is oblique (not purely designative) in the same way as an occurrence of a name within a clause governed by 'believes that,' 'asserts that,' 'rejoices that,' and the like. Also a similar point is made regarding occurrences of names within a context governed by 'the attribute of.' The importance of this as a limitation upon the application of intensional concepts is clear. In particular the author's conclusion seems sound that serious doubt must affect any attempt to overcome the now well-known difficulty of defining adequately the term 'soluble in water' (cf. II 49 (2), III 157 (2), VII 43 (2, 3)) by the device of employing a modal operator as a means of rendering a contrary-to-fact conditional clause; for it is essential that in such a sentence as 'this crystal is soluble in water' the subject shall have a purely designative occurrence.

But the reviewer would question strongly the conclusion which the author draws that no variable within an intensional context (e.g., within the scope of such a modal operator as '$\Diamond$' for 'it is possible that') can refer back to a quantifier prior to that context (outside the scope of the modal operator). The conclusion should rather be that in order to do this a variable must have an intensional range—a range, for instance, composed of attributes rather than classes. To paraphrase an argument which Quine applies to a somewhat different illustration, let 'b', 'f', and 'm' mean respectively the class of bipeds, the class of naturally featherless creatures, and the class of men. Then the sentence is true (9) '$fb = m \,.\, \Diamond fb \neq m$'—the non-existence of featherless bipeds other than men being a zoological accident. But, where 'α' is a class variable, the inference from (9) of the sentence '$(\exists\alpha) : \alpha = m \,.\, \Diamond\alpha \neq m$' must be in error, since, having '$\alpha = m$', we could substitute 'm' for 'α' and infer further the false sentence '$\Diamond m \neq m$'. There is no similar objection, however, to the inference from (9) of (10) '$(\exists\phi) : (x)(\phi x \equiv x \in m) \,.\, \Diamond \sim(x)(\phi x \equiv x \in m)$', where ϕ is a variable for attributes; and it would seem that in a logical system containing both modal operators and quantifiers such inferences should be retained.

This leads naturally to Frege's conclusion that a name in its oblique use does not lack a denotation (or designatum) but rather has a different denotation, namely it has as denotation that which would be its sense in its ordinary use. For this reason it would seem to be desirable, for the purpose of discussions like the present one, to adopt some notational device to distinguish the oblique use of a name from its ordinary use—just as quotation-marks are now commonly employed to distinguish the autonymous use of a name from its ordinary use, on the ground that the denotation is different. In a formalized logical system, a name would be represented by a distinct symbol in its ordinary and its oblique use. (From this point of view our notation in (9) and (10) is inaccurate—it was intended only to serve a temporary expository purpose.)

The reviewer has emphasized before the importance of Frege's distinction between sense and denotation (cf. V 162, V 163, VII 100 (2), also an abstract in this JOURNAL, vol. 7, p. 47). The distinction may be explained by employing a modification of Quine's own words from the paper before us: to determine that two names or other expressions have the same sense it should be sufficient to understand the expressions, but to determine that two names have the same denotation it is commonly necessary to investigate the world. In fact Quine here introduces a distinction between meaning and designation which closely parallels Frege's between sense and denotation. The one significant difference is that Quine regards meaning as exclusively a syntactical or semantical concept (he suggests, e.g., that the meaning of an expression might be identified with the class of all the expressions synonymous with it) while Frege's sense is an abstract object not having a syntactical make-up. In particular, it is not clear that Quine would be willing to identify the meaning of a sentence, as Frege does the sense, with the proposition which the sentence expresses. (The translation of Frege's "Gedanke" as "proposition" is clearly justified by his explanation, "nicht das subjective Thun des Denkens, sondern dessen objectiven Inhalt, der fähig ist, gemeinsames Eigenthum von Vielen zu sein.")

In the reviewer's opinion, the advantage lies with Frege's concept of sense, especially since Quine himself seems willing, at least provisionally, to countenance such intensional abstract objects as attributes. It would thus be necessary in the case of class names to add only that the sense is the associated attribute. Such multiplication of abstract entities as is demanded in order to provide senses for names of all kinds seems a small price to pay for the gain in simplicity and naturalness. (This is not to say, however, that the identification of senses of class names as attributes is necessarily the only, or the best method.)

There remains the important task, which has never been approached, of constructing a formalized semantical system which shall take account of both kinds of meaning, the relation between a name and its denotation, and the relation between a name and its sense. It would be a desideratum for such a system that the object language should contain for every name in it a name of the associated sense, and should be capable of expressing the relation between a sense and the denotation which it determines.

The reviewer believes that many of the questions which Quine raises without being able to answer them will find their answers in the construction of such a system of semantics. Ultimately it is only on the basis of their inclusion in an adequate system of this kind that such otherwise indefensibly vague ideas as "understanding" of an expression, "attribute," "objectiven Inhalt des Denkens" may be regarded as logically significant.

ALONZO CHURCH

Postscript 1968

To avoid misunderstanding the review requires clarification and correction in regard to the following point.

Quine's discussion of modality, which is treated in the seventh and eighth paragraphs of the review, is based on the manner of formulating modal logic which let me speak of as "the usual formulation." According to this usual formulation, which is still the standard one, modal operators such as '$\Diamond$' are prefixed to sentences — so that, e.g., if '$fb = m$' is a sentence which may correctly be used to assert that the class of featherless bipeds is the same as the class of men, then '$\Diamond fb = m$' is again a sentence and may be used to assert the possibility that the class of featherless bipeds is the same as the class of men.

The latter paragraphs of the review carry implicitly the suggestion of a different manner of formulating modal logic, in which modal operators are to be prefixed to names of propositions, and the result of prefixing a modal operator to a sentence is simply not well-formed. That is, instead of allowing the oblique use of the names 'f', 'b', and 'm' which occur in '$\Diamond fb = m$' as usually understood, we introduce three names, say 'ϕ', 'β', and 'μ', to denote the sense which is expressed (rather than denoted or designated) by the respective names 'f', 'b', and 'm'. Thus 'ϕ', 'β', and 'μ' are names of class concepts, which presumably may be identified with what Quine calls attributes. Then we must introduce signs, say 'ι' and 'o', for the sense of the sign of intersection (or logical product) of classes and the sign of equality. As names of the same proposition which is expressed by the sentence '$fb = m$' we then have '$\phi\iota\beta o\mu$. And as a sentence which expresses (and may be used to assert) the possibility that the class of featherless bipeds is the same as the class of men we have '$\Diamond\phi\iota\beta o\mu$.'

On a modified Fregean theory the sign of intersection and the sign of equality are themselves names. For the present purpose the essential thing is not this, but rather merely that the extensional signs of intersection and equality are to be replaced by corresponding intensional signs before the modal operator is prefixed. However, to reconstrue these signs as names has the convenience that the notion used to express e.g., that "the relation concept o is a concept of the relation of equality" may be exactly

Editor's note: In this Postscript, even in its typescript form, Church does not follow the practice of minimal capitalization.

Postscript 1968 was originally published in ***Semántica filosófica: Problemas y discusiones***, edited by Thomas Moro Simpson, Siglo XXI Argentina Editores S.A., Buenos Aires 1973, pp. 147–152. In the present collection, it is placed here because of its relation to the preceding review.

parallel to that which is used to express that "the class concept μ is a concept of the class of men."*

Now the remark about quantifiers in the eighth paragraph of the review must be understood as referring to the usual formulation of modal logic, as it is indeed this usual formulation which Quine's paper discusses and criticizes. Quine later pointed out** that this restriction on the use of quantification in contexts in which modal operators are involved is at best awkward and artificial, but this artificiality is one of the penalties which must be paid for allowing oblique usage as anything else then an informal makeshift device. What is not made clear in the review is that such *ad hoc* restrictions on quantification are quite unnecessary if a Fregean approach to modal logic is adopted.***

Another method by which, up to a point, such restrictions on quantification in modal logic can be avoided is to adopt a language in which names of extensional entities, and variables having extensional entities as values, are entirely disallowed in favor of corresponding intensional names and intensional variables. In part ***Principia Mathematica*** already provides a language of the kind required for this, since (following Quine****) we must understand the singulary predicate letters in ***Principia*** as variables whose values are attributes, and class names are available only in the improper sense that class abstracts are introduced by contextual definition. All binary (ternary, etc.) predicate letters are similarly to be understood as variables whose values are what, extending Quine's terminology, let me call binary (ternary, etc.) attributes. Likewise the propositional variables are intensional, having propositions as their values. But the individual variables of ***Principia*** seem to be intended as extensional. For purposes of modal logic Quine has suggested***** in effect that the language of ***Principia*** be

*Or to express that "the class concept $\phi\iota\beta$ is a concept of the class of men"—which must of course be equally true, if it is true (as we are supposing) that $fb = m$. The word 'concept' is here being used, not as a translation of Frege's '*Begriff*,' but in a sense which is nearer to that of Russell when he speaks of class concepts.

In fact our proposed modification of Frege's theory consists in abandoning his functions, including *Begriffe*—then employing instead names of, and variables for, functions of a different character, which are like Frege's *Werthverläufe* except that they obey the theory of types.

Reference and modality*, in the first edition of ***From a Logical Point of View, 1953, see especially page 157. In the second edition (1961) of the same book, Quine's comments are greatly revised; see pages 152–159. Here Quine's objection to the Fregean approach, in the paragraph beginning at the bottom of page 153, is based on a misunderstanding: "the interplay . . . between occurrences of expressions outside modal contexts and recurrences of them inside modal contexts," which Quine misses, is in fact provided by the presence in the language of a notation for the relation "is a concept of," except so far as Quine himself has shown this interplay to be undesirable. On the other hand the comment in the paragraph which begins at the bottom of page 152 is well taken, and shows the need for further artificial restrictions on the use of quantification if modal logics are to be retained which allow oblique usage and are of sufficiently high order. And though Quine does not state it as such, his comment in this paragraph is a straightforward demonstration of the need for that much misunderstood feature of the Fregean approach which provides second-level concepts (or concepts of concepts), which may serve as senses of names of names of concepts—compare the next-to-last paragraph of review.

***Frege himself never considered modality, but is concerned with intensional contexts which may in a stricter sense be described by the grammatical term 'indirect discourse' (e.g. contexts following 'believes that'). In fact Frege's '*ungerade*,' though it seems better to translate it into English as 'oblique,' is intended by him to suggest the German term '*ungerade Rede*' for 'indirect discourse.'

****E.g. in *Reference and modality*, either edition, page 122.

******The Journal of Symbolic Logic***, vol. 12 (1947), see pages 47–48. Quine regards his own suggestion unfavorably, but on ontological rather than logical grounds.

modified by rejecting individuals in favor of individual concepts. For example, *The Morning Star*, *The Evening Star*, *Venus* and *the second planet from the sun* are accepted as things among which there is indeed an equivalence relation but no strict relation of identity. The single extensional entity which the Frege scheme would provide to go with these is missing. It is in some sense not needed, as everything required can be said (from the point of view of the Frege scheme, said indirectly) without referring to it.

Still another alternative may be to follow ***Principia*** in letting the individual variables be extensional, although other variables are intensional, and then eliminate names of individuals entirely by means of the contextually defined descriptions of ***Principia***. If careful attention is given to the matter of scope, difficulties largely disappear at the level of individuals.* Quine's paradox of modality will show itself as the theorem '$\Diamond x \neq y \,.\supset.\, x \neq y$', since the theorem '$\sim \Diamond x \neq x$' and '$\sim \phi x \,.\supset:\, \phi y \,.\supset.\, x \neq y$' can hardly be avoided.** But conceivably this might be regarded as acceptable in spite of the intended extensional interpretation of the individual variables.

Both these alternatives to the Fregean approach have difficulty with names of attributes, as Quine's paradox and the clearly related problem of Frege about the meaning of identity statements (How can e.g. '$fb = m$', if it is true, differ in meaning from '$m = m$'?) will tend to reappear. Thus the attribute which you and I were discussing yesterday, the attribute which Russell attributes to Sir Walter Scott, the attribute of being author of Waverley, and the attribute denoted by the English phrase 'authorship of Waverley' are, let us say, all four the same, but in the case of any two of them it is possible that they should be different; and Quine's argument can then be repeated in application to two of these names in place of 'fb' and 'm'. Or to avoid the presumption that it suffices to eliminate only those names of attributes which make mention of such semantic and pragmatic matters as denoting, discussing, and attributing, consider e.g. 'The most admirable attribute of Sir Walter Scott was the attribute of scrupulous honesty'; or consider Quine's example in the passage of his revised *Reference and modality* which is referred to at the end of footnote ** [p. 946, above].

To see that it is not possible to accomplish the elimination of all names by the device of contextual definition, suppose that a language is proposed which has the individual name 'Venus' as a primitive name (thus a name which has no analysis into meaningful parts). In order to eliminate the individual name 'Venus' we may introduce a predicate 'V', *to be [the planet] Venus.* Then e.g. if we wish to assert that Venus is brilliant [as seen from the earth], we write, using also a predicate 'B' *to be brilliant*:

$$`(\exists x) :.\, Bx : x = y \,.\equiv_y.\, Vy'$$

This is Russell's device which appears in ***Principia***. The name 'Venus' has been eliminated in favor of the predicate 'V'. But in this context the predicate 'B' and 'V' have to be regarded as names of attributes (because they may be substituted for attribute variables). We may eliminate the attribute names 'V' by repeating the same device of contextual definition, but for this purpose we require a name of a higher-order

*See reviews in ***The Journal of Symbolic Logic***, by the writer (vol. 7 (1942), pp. 100–101), and by A. F. Smullyan (vol. 12 (1947), pp. 139–141), and a paper by Smullyan, *Modality and description*, ibid. vol. 13 (1948), pp. 31–37; also a paper by F. B. Fitch, *The problem of the Morning Star and the Evening Star*, in ***Philosophy of Science***, vol. 16 (1949), pp. 137–141.

**In his *Reference and modality*, Quine discusses what is essentially the same point, in reference to the modal logic of Ruth C. Barcan.

attribute, the attribute of being the attribute V; there is an infinite regression and no final elimination of all names is accomplished. Of course either or both of the names 'Venus' and 'V', instead of being primitive names, might be taken as having some analysis and so becoming complex names. E.g. we might so analyze 'V' that 'Vx' is to be replaced by '$MxEx$' (to be Venus is to be The Morning Star and The Evening Star), and 'M' and 'E' might then in turn have some analysis. But no matter how far such analysis is carried, it is hardly possible that some primitive names will not remain, whether of individuals (or of individual concepts) or of attributes; and to these primitive names our above remark will apply.

If then Frege's distinction between two kinds of meaning is to be avoided, we must suppose that every individual (every individual concept) and every attribute has apart from mere synonyms, at most one true name. Where a pair of names such as 'The Morning Star' and 'The Evening Star' seems to constitute an exception to this, it is necessary to reject one or both of them.

When a particular formalized language is proposed, this means that there is to be a carefully controlled list of primitive names, which may be all of them names of attributes. Then to avoid danger it is desirable that all other names which might otherwise seem to be needed shall be eliminated by various devices of contextual definition. For names which are overtly descriptions, Russell's device of contextual definition of descriptions will effect the elimination without introducing any new names (not already present in the given description). Names of classes may generally be reconstrued as class abstracts and hence eliminated by another device of contextual definition, which is also due to Russell and which has already been mentioned as appearing in ***Principia***. Then finally there are names which I may call attribute abstracts. For example if W is a well-formed formula of propositional type whose free variables are 'x' and 'y', we may think of W as stating a certain binary attribute (or relation in intension) to hold between x and y, and then denote this binary attribute by* $\lambda x \lambda y W$. Suppose, as is natural in connection with modal logic, that attributes are identical if and only if the equivalence between them is necessary. This will enable us to reconstrue $\lambda x \lambda y W$ as the description $(\imath\phi) \sim \Diamond \sim . \, \phi xy \equiv_{xy} W$, and hence to eliminate it by Russell's device of contextual definition of descriptions.

Thus in summary two main approaches to modal logic promise solution, or at least avoidance, of the Quine paradox. The Fregean approach maximizes the category of names and allows names of, and variables for, a wide variety of entities, both intensional and extensional. The alternative approach severely restricts the category of names and allows variables only (or almost only) for intensional entities, with a possible, though debatable, exception in the case of the individual variables. Quine seems to be successful in showing that his strictly extensionalistic preferences** have the effect of excluding modal logic altogether.

---

*Distinguish between this non-Fregean, or quasi-Russellian, theory of meaning, according to which $\lambda x \lambda y W$ denotes the binary attribute which W expresses as holding between x and y, and the Fregean theory, according to which $\lambda x \lambda y W$ denotes the corresponding relation in extension. (*Editor's note*: Footnotes * and **, on this page, were added in February, 1971.)

**The finitistic nominalism which Quine once espoused in a different context is not in question here, but extensionalism, i.e., the rejection of such intensional entities as attributes, propositions, and individual concepts.

MARVIN FARBER. ***The foundation of phenomenology. Edmund Husserl and the quest for a rigorous science of philosophy.*** Harvard University Press, Cambridge, Mass., 1943, xi + 585 pp.[12]

This book offers a detailed historical account of the development of Husserl's philosophy from his early logical psychologism, through simple descriptive phenomenology, to his final transcendental phenomenology, and includes the main content of the ***Logische Untersuchungen*** "in essential fulfillment of a promise made to Husserl to render that work in English." The author is himself an advocate of "a strict interpretation of phenomenology as a descriptive philosophy" and argues against the acceptability of Husserl's later transcendental idealism.

Of especial interest from the point of view of symbolic logic are Chapter II, which deals with Husserl's ***Philosophie der Arithmetik***, and Chapter III, which deals among other things with Husserl's criticism (75*1*) of Schröder's ***Algebra der Logik*** (42*7*), his intensional reinterpretation of the Schröder algebra (75*2*, *3*), and the subsequent controversy with Voigt (74*2*, 75*4*, 74*3*, 75*5*). But there is a remoter relevance at many places throughout, in that Husserl was often concerned with questions which many would now treat by the methods or with the technical aid of symbolic logic.

In Chapter II there is discussed in particular Husserl's criticism of Frege's treatment of number (49*5*), and Frege's criticism of Husserl in his review of ***Philosophie der Arithmetik***. The account here is clear and objective, and is particularly valuable for its success in reproducing and explaining Husserl's point of view. However, the statement of Frege's criticisms and replies to criticisms is not complete. And at a few places there are obscurities which could be cleared up by slight changes in wording. E.g., the sentence at the top of page 39 should read, "The concept F is said to have the same number as the concept G if it is possible to put the objects coming under one concept into one-to-one correspondence with those coming under the other." And further down on the same page, for "unique coördination" read "one-to-one correspondence." At the middle of page 41 the point to be brought out becomes clearer if we read: "Frege's theory explains why one can with the same truth say 'This is one group-of-trees' and 'These are five trees.' The name alone is changed... "

Husserl's principal objections to Frege's concept of number seem to have been the following: (1) He finds in the details of Frege's development only "hypersubtilitäten," resulting in an extravagant definition of number not in accord with the simple notion as he understands it. (2) Frege defines *the number attaching to the concept* F as the extension of the concept "having the same number as the concept F" (where "having the same number" is defined by means of one-to-one correspondence); to this Husserl objects that it does not define the content ("Inhalt") of the concept of this number but only its extension. (3) Husserl holds that a number attaches, not to a concept ("Begriff") as Frege has it, but to an extension ("Inbegriff"); of a concept one may only say indirectly that it has the property of having a certain number attached to its extension.

Answers to (1) are quite sharply stated by Farber in his summary at the end of the chapter. To the two remaining points the reviewer would reply as follows:

(2) Frege's extensional notion of number is exactly defined and is known to serve all the purposes of pure and applied mathematics, not excluding the every-day use

[12]Review originally published in ***The journal of symbolic logic***, vol. 9 (1944), pp. 63–65.

of elementary arithmetic and counting; by contrast Husserl's intensional notion is not clear (Just what are the intensional distinctions overlooked in Frege's extensional definition?) and is open to the suspicion of being superfluous. Frege's own reply to this point, in his review of Husserl, should also be consulted, especially his comparison with the usual definition of a right angle in elementary geometry. (But Frege's further contention that a definition need only reproduce a denotation, without regard to sense or intension, is not acceptable; the definition in Euclidean geometry, e.g., of an equiangular triangle as a triangle having three equal sides would stultify the theorem that every equilateral triangle is equiangular.)

(3) It would seem that of the two notions, number of a concept and number of an extension, each is immediately definable in terms of the other, and we may therefore in a systematic development choose which one shall be taken as prior to the other. But even if Husserl's point be accepted and the notion of number made to apply to an *Inbegriff*, all the features of Frege's treatment which later development of the subject has shown to be valuable may be retained substantially unaltered, provided that we supplement Husserl's psychological account of *Inbegriff* by a suitable logistic formalization. It is not even necessary to modify the notion of *Inbegriff* so as to admit empty and unit *Inbegriffe*, although it would no doubt result in a more elegant formalization to do so.

Farber points out that Frege's criticisms of Husserl had an important influence on Husserl's later development. In particular Husserl later came to agree with Frege in opposing psychologism, and retracted his original criticism of Frege on this point. But it seems that his other criticisms were still maintained.

Concerning the critique of Schröder, Farber remarks with justice that Husserl's estimate of the value of Schröder's calculus of classes is disappointing. Husserl does not indeed deny value to the calculus, but he thinks of it as being merely an art of substituting symbolic reckoning for "real" inference, and he holds further that the process of deduction is incomplete until we include also the preliminary step of translating the given problem into symbolic language and the final step of interpreting the symbolic result. The reason for this seems to be that Husserl conceives a natural language as "a method for the expression of psychical phenomena," but he denies that the Schröder calculus is capable of being such a method and hence is unwilling to treat it as a language in its own right, independently of considerations of translation into or interpretation by some other language. Consequently Husserl does not see that a symbolic language (even the Schröder class calculus) may serve the additional purpose of expressing briefly and perspicuously many propositions and logical analyses which, on account either of their unfamiliarity or of their complexity, a natural language can handle at best very awkwardly.

Of other criticisms of Schröder (here recounted by Farber), Husserl's difficulties over the use of a null class and a universal class, and his objections to Schröder's definition of the equality of classes by means of the relation $\subseteqq$ of subsumption between classes, seem to the reviewer largely pointless. On the other hand Husserl's insistence on a distinction between the relation $\subseteqq$ and the relation $\in$ of membership in a class would now be accepted by most logicians—although Schröder was able to avoid any formal inconsistencies consequent upon his failure to make such a distinction. And the reviewer would also agree with Husserl's criticism of Schröder's rejection of intensional logic. (Nevertheless the possibility is important of a calculus which is so far extensional that the denotations of all terms are extensions.)

This last criticism of Schröder formed the subject of a separate publication (75**2**, **3**), in which Husserl proposed an intensional reinterpretation of the Schröder calculus, with a view to showing—what Schröder denied—that logic can be developed from the outset as a logic of intension. There followed the acrimonious controversy with Voigt, which Farber regards as having ended in a victory for Husserl (despite some errors).

There are two typographical errors which require correction in order to make certain details clear in Farber's account. On page 79 line 10 a typographical error is reproduced which occurred in Husserl's original article; "$A \subset B$" (or "$A \subseteqq B$" in the original article) should be replaced by "$A \subset 1$" (or "$A \subseteqq 1$"). On page 84 line 11 from the bottom there should be two separate formulas "$x < A$" and "$x < B$" instead of a single formula, and the letter x should be a small letter in both places.

The point on page 84 is that Voigt used the notation $x < A$ to mean that the object x falls under the concept A, and proposed to develop a logic of concepts from a logic of propositions by means of a series of definitions, defining, e.g., $A \subseteqq B$ to mean that for every x for which $x < A$, also $x < B$. Husserl then claimed that he had had the same idea himself but did not publish it.—This point is only incidental to the main controversy, for details of which the reader may be referred to Farber's account.

A reference to Frege's ***Begriffsschrift*** (49**1**) in the course of the controversy—cf. Farber, p. 86—is of interest chiefly as suggesting that Husserl's acquaintance with that work, or at least his insight into it, must have been rather superficial. In fairness to Husserl it must be said, however, that failure to appreciate the significance of the work of Frege was common to all of his great contemporaries, even including such men as Schröder and Georg Cantor, until Bertrand Russell (who was only partly a contemporary).

Farber's bibliography of the papers comprising the controversy brings to light an omission from *A bibliography of symbolic logic*. After 74**3** should have been added: "See also A. H. Voigt, *Berichtigung*, ibid., vol. 18 (1894), p. 135, and Richard Avenarius, *Erklärung des Herausgebers*, ibid., vol. 18 (1894), pp. 134–135." These notes are concerned with an explanation of the circumstances under which Voigt's last paper in the controversy was withdrawn before publication and replaced by a shorter version.

ALONZO CHURCH

ERNEST NAGEL. *Logic without ontology*. ***Naturalism and the human spirit***, edited by Yervant H. Krikorian, Columbia University Press, New York 1944, pp. 210–241.[13]

"No one seriously doubts", the author writes, "that logic and mathematics are used in specific contexts in identifiable ways, however difficult it may be to ascertain those ways in any detail." Therefore, he concludes, it seems "reasonable to attempt to understand the significance of logico-mathematical concepts and principles in terms of the operations associated with them in those contexts and to reject interpretations of their 'ultimate meaning' which appear gratuitous and irrelevant in the light of such an analysis."

In support of this point of view, the author examines the claim for the law of contradiction that it is one of the principles which " 'hold good for everything that is' and therefore belong to the science of being qua being." It is remarked that an objection to the law of contradiction on the ground, e.g., that a penny is both visually circular and

[13]Review originally published in ***The journal of symbolic logic***, vol. 10 (1945), pp. 16–18.

visually non-circular would be met, not by abandoning the law of contradiction, but by demanding the correction of a certain vagueness thus revealed in the specification of the attribute *visually circular* and of the conditions under which we may consider that we are dealing with the same attribute ("in the same respect," as the author adds, following Aristotle's phraseology). "The crucial point is that in specifying both the attribute and the conditions, *the principle* [of contradiction] *is employed as a criterion* for deciding whether the specification of the attribute is suitable and whether those conditions are in fact sufficiently determinate .... Accordingly, the interpretation of the principle as an ontological truth neglects its function as a norm or regulative principle for introducing distinctions and for instituting appropriate linguistic usage."

Regarding the often proposed characterization of logical principles as principles which hold for all possible worlds, the author points out that the only available definition of possible world is that it is one in which the principles of logic hold, whereby the proposed characterization is reduced to a triviality. Less convincingly, it is argued that, because logical consistency and inconsistency are relations not among things and facts but among propositions or statements, "the 'pervasive traits' and 'limiting structures' of all 'possible worlds' which logic is alleged to formulate thus appear to be traits of discourse when it has been ordered in a certain way."

To the view that logical principles are inductive truths, established on empirical evidence, the author objects that no instance can be cited from the history of science supporting the idea that the validity of logical principles is ever established empirically and "that whenever consequences derived from premises believed to be true are in disagreement with the facts of experimental observation, it is not the logical principles in accordance with which those consequences were drawn that are rejected as experimentally unwarranted." Further it is maintained that empirical naturalists adopting this interpretation of logic often have too narrow a criterion of meaningful discourse or an inadequate conception of the rôle of symbolic constructions in the conduct of inquiry.

The author's own view is that logical and mathematical principles are regulative in character and conventional, justified neither as ontological truths nor by experimental evidence, but by their adequacy for achieving some systematization of knowledge. For rules of inference, stated in a syntax language, and for syntactical forms of the three "laws of thought" (so called), this seems to be clear. But there are also statements not syntactical in their formulation which are held to be logical truths, e.g., "Nothing has and also lacks a given property," and "If the sun will rise tomorrow then the sun will rise tomorrow." Of these the author asserts that, although their subject matter is not the language of which they are parts, their character — as statements "such that to deny them is to misuse" the language — is brought about by the fixing of formation rules and rules of inference (primitive and derived) which are to hold for the language; and that their function is still regulative in that they serve as instruments for establishing connections between statements which need not be themselves logically necessary. Moreover he maintains that it is theoretically possible — though the cost in terms of resulting inconveniences and complexities might be prohibitive — so to reconstruct the language that these logically necessary statements are eliminated, namely by modifying the formation rules so that these statements are made out to be meaningless, and at the same time introducing new rules of inference to take the place of the logically necessary statements in their rôle as instruments for establishing connections between other statements.

Finally the author examines the use of the real number system in its application to the process of physical measurement, and draws similar conventionalistic conclusions.

The reviewer has the following criticisms — offered not in support of any theory of his own but merely as difficulties which he believes the paper under review has not met.

The author either overlooks or fails to bring out that the essentials of what he says about principles of logic are applicable likewise to principles of natural science generally. Analogously to the situation he finds for the law of contradiction, it would seem that, within the context of classical mechanics, Newton's second law — that rate of change of momentum proportional to the force acting and is in the same direction — cannot be refuted by experimental evidence, since the law itself is employed as a criterion for deciding what force is acting. Something similar is true even of such particular statements as that the sun revolves about the earth or that the area of Manhattan Island is ten square miles. For these statements could be maintained, if one were resolved to do so, by alterations in the natural laws which are presupposed in the experiments usually taken as refuting them.

The author himself says: "Ever since Duhem wrote on the subject, it has become a commonplace to observe that statements in the sciences are systematically connected and that only systems of beliefs can be put to a definitive test." But he seems to assume tacitly that statements the source of whose validity is logical in character can be distinguished in advance in some way and treated differently. For when he considers experimental verification or refutation of logically necessary statements he writes as if a single statement were question. Should he not rather have considered experimental verification or refutation of systems of statements, including together empirical statements and statements thought to be logically necessary?

The assertion that logical principles are never rejected as experimentally unwarranted seems to have a counterexample in various recent proposals to meet certain difficulties in quantum mechanics by employing a systematization in which familiar principles of logic are rejected or modified. The author mentions one of these (II 44(3)) but apparently does not regard it as a counterexample. Of course these proposals are at present tentative in character and may well in the end prove to be unacceptable. But it would seem to the reviewer that, as intended, they are of the same character as alterations of physical theory which have occurred in the past. True, no experimental evidence, not even that which has led to quantum mechanics, compels the rejection of these laws of logic, since other laws of the system can be modified instead. But it is now commonplace to remark that the like holds, e.g., of Ptolemy's principle that the earth is the center about which the heavenly bodies revolve. This principle could still be maintained if we were willing sufficiently to complicate the laws of mechanics and the descriptions of the orbits of the planets. The universal acceptance of the Copernican theory is due to its greater simplicity and formal convenience.

The author's assertion that logically necessary statements can be eliminated by appropriate reconstruction of the language in which they occur requires support by actually exhibiting such reconstruction in a significant case, and this is not done. It seems to the reviewer at least doubtful that such reconstruction is possible even in the case of the functional calculus of first order, because the attempt to make the reconstruction would be likely to run afoul of the unsolvability of the decision problem.

ALONZO CHURCH

ALEXANDRE KOYRÉ. *The liar.* ***Philosophy and phenomenological research***, vol. 6 no. 3 (1946), pp. 344–362.[14]

The author does not share Russell's "belief in the virtue of formalization and symbolization" and therefore undertakes to analyze the antinomies without benefit of the logistic method. In the reviewer's opinion, the attempt is not successful.

Berry's antinomy and several forms of the antinomy of the liar receive consideration.

Of these, the antinomy of Berry is here resolved only by taking a corrupted form of it. Indeed the distinction which is made of "proper names" of positive integers from other English phrases designating positive integers is irrelevant to the demonstration of the antinomy and should preferably not have been introduced at all. Instead, the author might have considered, say, the *least positive integer which cannot be designated by any English phrase of fewer than forty syllables*. (Surely there is one such, since the number of English phrases of fewer than forty syllables, though large, is finite.)

The antinomy concerning the speaker who says "I am at this very instant lying"—analyzed as meaning that (a) there is an assertion that I am making at this instant and (b) this assertion is false—is dismissed as follows. There is (contrary to (a)) no assertion which the speaker is making at the instant in question; and therefore his judgment is a false one. It is not explained how the speaker, who makes no assertion, is nevertheless found to make a judgment (and in fact a false one), nor why the antinomy would not be reinstated merely by substituting the word "judgment" for the word "assertion" from the beginning.

The statement of Epimenides that "All Cretans are utter liars"—where an utter liar is one who lies, not just sometimes, but always—is accepted as meaningful. In the mouth of an Athenian it would clearly have been meaningful; and the *meaningfulness* of it cannot be changed when it is made instead by Epimenides, who chances to be himself a Cretan. But it is said that Epimenides's statement is self-destructive, in the sense that its truth is disproved by the given fact that Epimenides made it.—Apparently M. Koyré is untroubled by a consequence which, though not outright antinomy, might well be classed as paradox. Namely, without factual information about *other* statements by Cretans, it has been proved by pure logic (so it seems) that *some other* statement by a Cretan, not the famous statement of Epimenides, must once have been true.

ALONZO CHURCH

PAUL FINSLER. *Gibt es unentscheidbare Satze?* ***Commentarii mathematici Helvetici***, vol. 16 no. 4 (1944), pp. 310–320.[15]

Repeating his previously expressed opposition to the logistic method as a means of giving a rigorous account of the notion of mathematical proof, the author recommends that, besides the form of a proof, attention be paid also to the "inhaltliche Bedeutung." And, as against the restricted notion of provability in a particular logistic system, he proposes an *absolute* notion of provability, of which he writes: "Dabei sollen jetzt aber nicht nur formale Beweise in Betracht gezogen werden, sondern auch beliebige ideelle, sofern sie nur inhaltlich einwandfrei sind."—The reviewer would comment that the restricted notion has at least the merit of being precisely communicable from one person to another. But Finsler's notion of proof and provability remains subjective,

[14] Review originally published in ***The journal of symbolic logic***, vol. 11 (1946), pp. 131–132.

[15] Review originally published in ***The journal of symbolic logic***, vol. 11 (1946), pp. 131–132.

since he supplies neither a definition of it (beyond the statement just quoted) nor a set of axioms in which it appears as a primitive term.

In discussing Gödel's *Über formal unentscheidbare Sätze* (4183A) it is pointed out that the formal system P of that paper does not allow the inference from Bew $\mathfrak{A}$ to $\mathfrak{A}$. (More correctly, the author should have spoken of the axiom schema Bew $\mathfrak{a} \supset \mathfrak{A}$, where $\mathfrak{a}$ is the "Zahlzeichen" which represents the number of $\mathfrak{A}$.) Then it is said that if this is added to the system, "so ändert sich der Begriff der Beweisbarkeit," and indeed "so stark, daß entweder der formal unentscheidbare Satz nicht mehr darstellbar oder aber das System widerspruchsvoll ... wird." But (if P is ω-consistent) the second of the two quoted statements is false, not only for the particular "Bew" defined by Gödel, but for any known acceptable notion of provability capable of definition in the notation of P — perhaps even for Finsler's absolute provability *if* that is capable of such definition.

For the antinomy of the liar, in the form presented by the speaker who says "Ich lüge," the author proposes the following as a solution. Because every assertion has the sense that what is asserted is true, the speaker also asserts implicitly that he is telling the truth. His total assertion, including the implicit part is therefore a conjunction, $\mathfrak{A}$ and not-$\mathfrak{A}$, false as being self-contradictory. — This has been proposed also by others (Finsler refers to Geulincx). It may be criticized on the ground that the sentence, "A lies, at t_0," if meaningful at all, should have the same total meaning no matter when or by whom asserted, and even if it happens to be asserted by A himself and at the very time t_0. But more serious is the following criticism.

A solution of the antinomies (that of the liar and others), for the purposes of formal logic, and especially of mathematics, must do more than offer an explanation why certain arguments are unsound, including those which lead to the antinomies. Namely it must effectively certify other arguments, in adequate variety, as sound, free in particular of the unsoundnesses to which the antinomies are ascribed. For a supposed proof is actually no proof at all if subject to refutation at any later time by the discovery of antinomy (as happened in the case of Cantor's apparently valid proof that the set of all ordinal numbers is well-ordered in order of magnitude). — It seems to the reviewer that Finsler's solution of the antinomy of the liar is not of the required kind: it does not provide a test for proposed proofs, but rather explains away the antinomy after the fact.

The point is illustrated by an interesting variant of the antinomy of the liar which Finsler introduces in another section. This depends on the sentence, "Die hier stehende Behauptung kann ich nicht beweisen," where the reference is to personal ability and ingenuity to discover a valid proof. This assertion I cannot prove, Finsler reasons, because if I did I would by that very act refute it; hence the assertion is true (and this is my proof). — Finsler has no general criterion to apply, by which the soundness of the argument may be decided in advance. But he is again obliged, after the observation of antinomy, to seek an explanation. This new explanation (that I cannot prove the assertion myself, yet must believe it because I see how almost any one else could prove it) has little similarity to that of the related paradox concerning "Ich lüge," and moreover seems to the reviewer not quite convincing in itself.

Finally, in the reviewer's opinion there is little value in Finsler's proof that *"jeder beliebige eindeutige Satz ideell entscheidbar* ist." In effect, an "eindeutiger Satz" is *defined* as one such that either it or its negation leads to contradiction (in the sense of the absolute notion of provability); and others are classed by fiat as "sinnlos."

ALONZO CHURCH

MORTON G. WHITE. *A note on the "paradox of analysis."* ***Mind***, n.s. vol. 54 (1945), pp. 71–72.[16]

MAX BLACK. *The "paradox of analysis" again: a reply.* Ibid., pp. 272–273.

MORTON G. WHITE. *Analysis and identity: a rejoinder.* Ibid., pp. 357–361.

MAX BLACK. *How can analysis be informative?* ***Philosophy and phenomenological research***, vol. 6 no. 4 (1946), pp. 628–631.

The paradox of analysis, as propounded by Langford (VIII 149 (3)), may be illustrated by reference to Moore's example (VIII 149 (4)): *a brother is a male sibling* (plausibly rendered in the notation of the Boolean algebra of classes — or better, of class concepts — as $b = ms$). Moore accepts this as a valid, though relatively minor, example of analysis in his sense. And he further makes it clear that he regards the analysis as concerning the concept *brother*, rather than merely the word *brother*. The paradox, or puzzle, arises as follows. If b and ms are the same concept, then it must be possible to introduce either in place of the other without alteration of meaning — indeed this is clearly the intention, or part of the intention, of any analysis in the sense of Moore. Hence in particular the analysis $b = ms$ must be not different from $b = b$ or $ms = ms$. This would seem to reduce analysis to something trivial and uninformative. But it is evident otherwise — especially from cases in which, though a concept is in some sense known, its analysis presents difficulty — that analysis is not always trivial.

Black, in IX 104 (2), proposes as a solution of the puzzle that the analysis in question is not correctly expressed by '$b = ms$' but should rather be expressed by '$B(b, m, s)$' where 'B' is used for a ternary relation which holds among three concepts "whenever the first is the conjunct of the remaining two." He says that there is no justification for regarding '$B(b, m, s)$' as expressing the same proposition as that expressed by '$b = b$'. "For no plausible interpretation of 'being the same statement' would seem to permit the assimilation of a mere identity with a proposition in which a non-identical relation is a component."

White, in the first paper listed above, criticizes this on the ground that Black "appears to be arguing that no two sentences can express the same proposition if one expressly mentions a relation which is distinct from the relation expressly mentioned by the other." As a counter-example to such an assumption he adopts Black's example '$A(21, 3, 7)$' where 'A' is used for the ternary relation of *being the product of*. This sentence, he maintains, expresses the same proposition as '$T(21, 7)$' where 'T' is used for the binary relation of *being thrice*. But the two sentences each expressly mention only one relation (A and T respectively), and these relations are distinct.

In his reply, Black repudiates the assumption attributed to him and points out that his argument needed only "the assumption that no strict *identity* could be properly expressed by a sentence synonymous with sentences in which non-identical relations are named."

In the reviewer's opinion the reply is insufficient, because no positive reason is given why the relation of identity should be an exception to the general rule which both White and Black illustrate with other relations. Indeed if we use 'B' for the relation

[16]Review originally published in ***The journal of symbolic logic***, vol. 11 (1946), pp. 132–133.

$\hat{u}\hat{x}\hat{y}[u = zy]$ where 'u', 'x', 'y' are variables for class concepts—and Black's verbal explanation quoted above even suggests that this may be the same as his use of 'B'—then it is at least very plausible that '$B(b, m, s)$' does express the same proposition as '$b = ms$'; and hence by the argument for the paradox of analysis it would express the same proposition as '$b = b$'. This is, in effect, White's rejoinder.

Moreover the reviewer does not share Black's apparent hope that the puzzle may be solved without a satisfactory *general* theory of meaning, and especially of the circumstances under which two sentences express the same proposition.

The paradox of analysis has an obvious analogy with Frege's puzzle (49*8*), as to how an equation, say '$a = b$', can ever be informative—because, it seems, if the equation is true then 'b' is replaceable by 'a', and hence '$a = b$' is the same in meaning as '$a = a$'. In the reviewer's opinion this is not merely an analogy, but the paradox of analysis is a special case of Frege's puzzle and is to be solved in the same way, on Frege's theory of meaning, by the distinction of sense and denotation ("Sinn und Bedeutung").

In fact the analysis that a brother is a male sibling is not about the classes *brother, male*, and *sibling* but about the corresponding concepts (which are the senses of the class names 'brother' etc.). Hence a correct expression of the analysis must employ names of these concepts. If the expression is '$b = ms$', the names 'b', 'm', 's' must denote the concepts *brother, male, sibling* respectively—not the classes. Now like any other name, a name which denotes a concept must have, besides its denotation, also a sense. From the truth of the equation '$b = ms$' it follows that the names 'b' and 'ms' have the same denotation, hence one may replace the other in a sentence without changing the denotation (truth-value). But, if the analysis expressed by '$b = ms$' is non-trivial, the names 'b' and 'ms' have different senses, hence the replacement of one by the other in a sentence may well change the sense (proposition) expressed.

In his *rejoinder*, White discusses the principle: "Given a true statement of identity of attributes, X, one of its two terms may be substituted for the other in any statement Y, and the result will express the same proposition as Y." Of this principle he writes: "It may very well be that it contains the source of the trouble."—But the rejection of this principle is an immediate corollary of Frege's theory of meaning, in the way just explained.

In the fourth paper listed above, Black defends his rendering '$B(b, m, s)$' of Moore's example, against Langford's criticism that, because 'b' is synonymous with 'ms', only two arguments should appear: '$B(m, s)$'. The reference to three concepts rather than two is confirmed, Black says, if there are other ways of identifying the concept of being a brother than by its analysis as ms.—It would seem to the reviewer that a way of identifying a concept, if understood in a logical rather than a psychological sense, is not far different from the sense of a name of the concept, and that some relationship to the Fregean solution thus appears in Black's remark. Alonzo Church

A. D. Ritchie. *A defence of Aristotle's logic.* ***Mind***, n.s. vol. 55 (1946), pp. 256–262.[17]

The defense of Aristotle is against an attack by Russell in ***A history of Western philosophy***. In the reviewer's opinion the defense is successful of the importance of Aristotelian logic in its own historical setting. Even the well-known difficulty over

[17]Review originally published in ***The journal of symbolic logic***, vol. 11 (1946), p. 134.

existential import, though the author does underestimate its seriousness, must be granted to be a relatively minor blemish on the work of the great founder of logic. On the other hand, in his defense of Aristotelian logic for modern use, the author overlooks the fact that it has now been subsumed as a special case under a more general and adequate theory, in such a way that the detailing of "immediate inferences" and moods and figures of the syllogism (unless for historical reasons) has lost its purpose. In a rhetorical query to Russell—as to why, if he knows about other forms of deductive inference, he does not do for them what Aristotle did for the syllogism, set out the valid and invalid forms—the author reveals on the one hand a sound appreciation of the point that the very idea of setting out valid and invalid *forms* of reasoning originated with Aristotle, and on the other a complete misconception of the accomplishments of modern logic in this direction.

Only loosely connected with the defense of Aristotle is the author's discussion of the question whether logic is "purely linguistic." It would seem to the reviewer that this question has not been given a sufficiently definite meaning; and that the brief critique of the theory of types which the author includes in his discussion of this question involves a number of confusions. ALONZO CHURCH

FRANZ BRENTANO. ***Psychologie du point de vue empirique.*** Translation and preface by Maurice de Gandillac. Aubier, Éditions Montaigne, Paris 1944, 461 pp.[18]

Publication of this translation recalls, in particular, Brentano's critique and proposed reform of the traditional formal logic.

In his VII 103(6)—first published in 1874, translated in the present edition as *Représentation et jugement, deux classes fondamentales différentes* (pp. 207–234)—Brentano makes it a part of his psychology of judgment that the four traditional forms A, E, I, O are reduced to existential propositions and negations of existential propositions. To put it in modern notation, "All S is P" thus becomes $\sim (\exists x) \centerdot S(x) \sim P(x)$, and the three remaining traditional forms become respectively $\sim (\exists x) \centerdot S(x)\, P(x)$, $(\exists x) \centerdot S(x)\, P(x)$, and $(\exists x) \centerdot S(x) \sim P(x)$. We may doubt the claim, which seems to be made, that these renderings of the traditional forms are more fundamental than others, but the *validity* of the reductions, and Brentano's conclusions therefrom, remain.

The question of existential import had already received discussion, by Peirce in 28***3***, and by various earlier writers. But Brentano, as a consequence of his reductions of the traditional forms, seems to have been the first explicitly to question the validity of those moods of the categorical syllogism which purport to infer a particular conclusion from two universal premisses, and to point out that the traditional "immediate inferences" sometimes contradict one another (so that not all can be saved, however the issue of existential import may be decided).

Land's exposition (36***1***) of Brentano's logical innovations should be noted, as well as the later detailed development of them by Hillebrand (77***1***).

Brentano's 200***1*** and 200***2***, first published in 1911, are translated in the present edition as *Des objets vrais et des objets fictifs*, pp. 283–292, and *Des essais de mathématisation de la logique*, pp. 292–297. The first is of interest here only because it includes a restatement of the reductions of the traditional forms to existential propositions and their negations. The second attacks proposals to apply the mathematical method to

[18] Review originally published in ***The journal of symbolic logic***, vol. 12 (1947), pp. 56–57.

logic; but (despite its date) the mathematical logic it refers to seems to consist only of certain earlier forms of the "algebra of logic." Interesting is Brentano's protest against the suggestion that his own treatment of the syllogism should be regarded as a contribution to mathematical logic.

The translator's footnote on page 295 is in error in saying that "la logistique refuse l'inférence de la subalternante à la subalternée . . . pour d'autres motifs que Brentano." Actually the reason is the same which Brentano makes explicit, namely that "All S is P" does not exclude that S be empty. ALONZO CHURCH

WILLARD V. QUINE. ***A short course in logic. Chapters I–VII.*** Mimeographed. Harvard Cooperative Society, Cambridge, Mass., 1946, iv + 130 pp.[19]

This is a first draft of the principal portion of a proposed elementary textbook. Additions which are planned in the completed work are indicated briefly in the Foreword.

In the first chapter a number of forms of inference belonging to propositional calculus are explained, and the second chapter introduces the truth-table method. In the illustrations, ordinary English appears as object language, supplemented early by the use of parentheses to show association, and in Chapter II also by special symbols for some particular connectives. In Chapter III, (categorical) syllogisms and Venn diagrams are explained, and their well-known limitations are pointed out. Variables and quantifiers, with usual notations for the latter, are introduced in Chapter IV, and Chapters IV and V are devoted to quantification theory. Chapter VI discusses the matter of analyzing and rewriting sentences of ordinary discourse, so that the formal technique of inference which has been developed may satisfactorily applied to them. (This is excellently done, and a valuable feature of an introductory textbook.) Chapter VII deals with singular terms (names) and designation, existence and non-existence, identity, and descriptions, in part along the same lines which the author has followed in earlier publications.

Nowhere is an explicit object language constructed, with primitive symbols and formation rules completely given. But the method throughout is rather a progressive formalization of ordinary English, by standardizing and restricting its forms of expression and removing irregularities, and by adding special notations from time to time. Letters p, q, etc. are introduced as *syntactical* variables for sentences; similarly the combinations Fx, Gx, etc. are used as syntactical variables for expressions (matrices) containing the letter x as a free variable, and so on. Propositional and functional variables are avoided—a feature which seems to the reviewer undesirably restrictive, because it precludes the statement of general laws (the transitive law of material implication, for instance) and leaves only the substitute of a syntactical rule for stating any chosen particular case.

As appears from discussion in Chapter VII, the purpose of thus avoiding the use of propositional and functional variables is to do without, or at least postpone, the assumption that there are abstract entities. It would seem to the reviewer that the author has construed the notion of a concrete entity so liberally that existence is often at least as doubtful as of abstract entities. But even so the author is in fact obliged to assume abstract entities from the beginning, at least in the syntax language, because he must speak of sentences and use variables for sentences. It is essential that the sentences

[19]Review originally published in ***The journal of symbolic logic***, vol. 12 (1947), pp. 60–61.

are not physical things, particular utterances, but shapes or patterns of such utterances and therefore abstract. — An attempt might be made to resolve the difficulty by taking a sentence to be a complex physical event, composed of all its actual utterances, and thus a concrete entity within Quine's meaning (if the reviewer has understood him correctly). But consequences of this are such as to give one pause. Thus, sentences (in the ordinary sense) which happen never to be uttered would be exempted from the rules, and such now familiar matters as the identification of syntax and arithmetic would be put in doubt.

An important feature of Quine's treatment not yet mentioned is his use of the *deduction theorem* as a primitive rule of inference. By this device the development is greatly simplified throughout, though at the cost of allowing a primitive rule of different and much more complex character than is customary. Advantages for an elementary treatment are obvious.

The first appearance of the deduction theorem in the literature, though not under that name, seems to have been in Herbrand's 382*5*, where it is demonstrated as a derived rule of inference for quantification theory (or functional calculus of first order). It was also proposed by Tarski in 285*7*, independently and nearly simultaneously, as a general methodological postulate. For expositions of the deduction theorem as a derived rule of the propositional calculus and functional calculus, reference may be made to Hilbert-Bernays 507***1***, pages 155, 157, and the reviewer's V 114, page 62, and X 19, pages 9, 45.

Jaśkowski's method of suppositions (514***1***) may also be described as including use of the deduction theorem as a *primitive* rule, and it is thus closely related to Quine's procedure. The same applies to Gentzen's calculi *N* (442*2*). But the relationship to Gentzen's calculi *L* is more remote, since Quine has in his object language nothing comparable to Gentzen's *Sequenzen*. ALONZO CHURCH

FREDERIC B. FITCH. *On God and immortality*. English with Spanish abstract. ***Philosophy and phenomenological research***, vol. 8 no. 4 (1948), pp. 688–693.[20]

CHARLES A. BAYLIS. *Critical comments on Professor Fitch's article "On God and immortality."* Ibid., pp. 694–697.

FREDERIC B. FITCH. *Reply to Professor Baylis' criticisms*. Ibid., pp. 698–699.

Partly on the basis of logical doctrines which he has previously outlined (XI 95, XII 103(15)), Fitch undertakes to establish "that every fact or class of facts has at least one explanation"; hence as a corollary "that the class of all facts has an explanation"; that "furthermore, there can be only one true explanation of all the facts in the universe, for if there were two different explanations, the fact that one of the explanations was true would be a fact to be accounted for by the other explanation, and so the two explanations would imply each other and hence by equivalent." "This ultimate explanation"—it would seem to be a "theory" in the usual sense of scientific methodology and at the same time, in the author's realistic view, also a "fact"—is then identified with "God."

In the reviewer's opinion, Fitch's argument, besides realism, involves an absolutism according to which one ultimate theory just *is* true. For it would be conceivable that two theories which explain all concrete (or in principle observable) facts, as well as all

[20]Review originally published in ***The journal of symbolic logic***, vol. 13 (1948), p. 148.

things which they respectively imply to be facts, might still fail to imply each other. Moreover the one ultimate theory, though Fitch denies that it is the class of all facts, is at least closely associated therewith—since two true theories are held to be equivalent, and hence the same, when they imply the same facts.

The reviewer also agrees in the main with Baylis's criticisms. The logical doctrines on which Fitch relies are not yet past the controversial stage, and have not yet received sufficient study by the logistic method. According to his above *Reply*, Fitch's paper in the last number of this JOURNAL is intended to provide, at least in part, a logistic formalization of them. But (in view of the claim of an all-comprehensive theory) it must still be asked whether the consistency proof 3.2 can be formalized in the logistic system provided; and especially it must be asked whether the argument of the present paper can be so formalized. In any case Fitch's near-pantheistic god is so different from a personal God that it seems misleading to use the same word.

ALONZO CHURCH

MAX BLACK. *A translation of Frege's Ueber Sinn und Bedeutung. Introductory note.* ***The philosophical review***, vol. 57 (1948), pp. 207–208.[21]

GOTTLOB FREGE. *Sense and reference.* Ibid., pp. 209–230.

The unique importance of Frege's 49***8*** has long been generally overlooked by those interested in philosophy of language and the analysis of meaning. An outstanding exception is Russell, whose *On denoting* (111***9***) seems to have been directly inspired by Frege. Otherwise Frege's theory of meaning has (until quite recently) met often with neglect, sometimes with downright misinformation or misunderstanding. The reason has been in part the lack of easy availability of the original paper. In reprinting it in an accessible place, as well as in providing an English translation, Prof. Black and the editors of ***The philosophical review*** have performed a very valuable service.

Notice should be taken of the point that Black has systematically omitted Frege's quotation marks about *displayed* sentences and phrases. Because in the original not all displayed phrases have quotation marks about them, this has resulted at three places (pp. 225, 228, 229) in what is technically a confusion of use and mention. But perhaps there is hardly room for real misunderstanding.

Black's *Introductory note* is devoted in part to discussion of the translations selected for terms used by Frege in technical senses (often divergent from their usual meanings in German). Perhaps even more than from a particular translation the student may benefit from such a discussion of possible translations. And it therefore seems worth while here to make some additional comments on this matter.

Black uses *refer to* as a translation of *bedeuten, designate* for *bezeichnen, reference* for *Bedeutung* (the process), and *referent* for *Bedeutung* (the object). Russell (111***6, 9***) uses *indicate*, or oftener *denote*, for *bedeuten*. Geymonat—see the following review—uses *indicare* for *bezeichnen* (said of persons), *designare* or *denotare* for *bezeichnen* (said of words or phrases), *denotare* for *bedeuten*, and *significato* for *Bedeutung*.

The reviewer prefers *denote* and *denotation* as translations of *bedeuten* and *Bedeutung*. This is the same sense in which J. S. Mill speaks of the *denotation* of a proper name, and is also one of the senses in which Russell uses this word ("Scott is the denotation of 'the author of *Waverley*' ")—except so far as Russell's usage becomes modified

[21] Review originally published in ***The journal of symbolic logic***, vol. 13 (1948), pp. 152–153.

by his theory of denoting phrases as being introduced by contextual definition and being therefore, like defined notations generally, "mere typographical conveniences." The translation of *bedeuten* as *denote* serves especially as a reminder that Frege's theory of denoting phrases is an important alternative to that of Russell. Otherwise it might be translated as *designate*, or as *be a name of* (since Frege does not make or accept Russell's distinction between descriptions and names). The translation *refer to* seems to the reviewer less natural, but it is selected by Black because he holds that *denote* is misleading, and *designatum* (for *Bedeutung* or *referent*) clumsy.

It is true that Mill speaks also of denotation of common or general names, and Russell speaks, in a related sense, of certain phrases as denoting ambiguously—and that this represents a wholly different meaning or meanings of the word *denote* (more like that of Frege's *andeuten*), for which it is unfortunate that the same word was used. But past confusions in use of the word *denote* can be corrected, and need not lead to its total elimination from semantical vocabulary.

Frege himself explains, in the text of the paper, that he uses *bezeichnen* (of words or phrases) as synonymous with *bedeuten*; but it may be helpful to the student to point this out explicitly in connection with the list of translations given in the *Introductory note*.

Black's translation of *Gedanke* as *thought* and Geymonat's translation of it as *pensiero* reproduce the same defect which may be seen in Frege's German terminology, namely that, in spite of the plainest disavowals, the word persistently suggests something psychological. This may be thought a good translation in faithfully reproducing the original. But Russell translates *Gedanke* as *propositional concept*. The reviewer would suggest simply *proposition*.

Frege's *Begriff* is translated by Black as *concept* or *predicate*, by Geymonat as *concetto*. None of these translations is very good, and it might even be better to follow Russell in leaving the word *Begriff* untranslated. Russell also sometimes translates *Begriff* as *concept*, but explains that this is not the same as his own use of the latter word. (And indeed Russell's class-concept is more nearly what Frege would call *Sinn eines Begriffsumfangsnamens*.) ALONZO CHURCH

R. A. KOCOUREK. *An evaluation of symbolic logic*. ***Proceedings of The American Catholic Philosophical Association***, vol. 22 (pub. 1948), pp. 95–104.[22]

This paper takes its departure from the Preface of Reichenbach's book (see the foregoing review), and discusses a number of theses about symbolic logic there put forward by Reichenbach. The author makes a categorical denial that symbolic logic is important as an initiation to philosophy, but (because of limitations of space) does not explain. He also denies that symbolic or mathematical logic should replace Aristotelian logic. But this is based on emphasizing demonstration, in which the reasoning is without error as well *ex parte materiae* as *ex parte formae*, and on the contention "that the 'Posterior' analysis, concerned with the matter, is an essential and even more principal part of logic than the 'Prior' resolution, concerned with the form." Symbolic logic is described as concerned with the form of reasoning; and the reviewer finds in the paper nothing to gainsay the widely held opinion that it ought to replace or has replaced the Aristotelian formal logic, as its modern successor.

[22]Review originally published in ***The journal of symbolic logic***, vol. 14 (1949), p. 52.

To Reichenbach's statement that an analysis of science demands the use of symbolic logic, the author emphatically agrees. "This new organon"—he includes with mathematical logic modern abstract mathematics generally—"is good, and not only good but, for the experimental sciences, necessary." At least as regards the more elementary parts of mathematical logic, it would seem to the reviewer that the author overestimates the special relevance to natural science as against other fields. ALONZO CHURCH

ALFRED JULES AYER. ***Language, truth and logic.*** Second edition (revised and reset). Victor Gollancz, London 1946 (third impression 1948); 160 pp.[23]

Because of the use of thinner paper and smaller type, and in spite of an added *Introduction* (pp. 5–26), the second edition is a much smaller and thinner book than the first edition (X 134(6)). There is no substantial change in the text, except the addition of a number of footnotes, some of which refer back to the new *Introduction*. In the revision of the index, some minor errors and confusions have unfortunately crept in.

This work is well known from the first edition, and its content need not be outlined here. Its main topic is not within the field of this JOURNAL. But there are many points of contact, of which the most prominent is in the use made of contextual definition (or "definition in use") and of Russell's theory of descriptions.

The *Introduction* to the second edition has additional points of contact with mathematical logic. These include a discussion of how the words "sentence" and "proposition" shall be used, to which the author is led in replying to a point made by M. Lazerowitz, *The principle of verifiability*, ***Mind***, n.s. vol. 46 (1937), pp. 372–378; a clarification and amendment of the author's account of *a priori* propositions; remarks about some minor points arising in connection with the verbal statement of Russell's analysis of "The author of Waverley was Scotch"; a brief account of what is meant by saying that "existence is not a predicate." In the reviewer's opinion, the discussion of these matters could have been greatly clarified by use of a formalized language. And in particular the reply to Lazerowitz—because it involves maintaining that some sentences, though well-formed, nevertheless express nothing either true or false—is unsatisfactory without such formalization, and careful critical study by the logistic method.

The *Introduction* discusses also some other objections that have been raised against the author's "principle of verification" and against his definition of verifiability. A criticism by Berlin (V 130(16)) is accepted as valid, according to which the definition of verifiability as actually given in the first edition would make all statements verifiable. And in order to meet this criticism an amended definition of verifiability is offered.

It would seem, however, that the amended definition of verifiability is open to nearly the same objection as the original definition. For let O_1, O_2, O_3 be three "observation-statements" (or "experiential propositions") such that no one of the three taken alone entails any of the others. Then using these we may show of any statement S whatever that either it or its negation is verifiable, as follows. Let $\bar{O}_1$ and $\bar{S}$ be the negations of O_1 and S respectively. Then (under Ayer's definition) $\bar{O}_1 O_2 \vee O_3 \bar{S}$ is directly verifiable, because with O_1, it entails O_3. Moreover S and $\bar{O}_1 O_2 \vee O_3 \bar{S}$ together entail O_2. Therefore (under Ayer's definition) S is indirectly verifiable—unless it happens that

[23] Review originally published in ***The journal of symbolic logic***, vol. 14 (1949), pp. 52–53.

$\bar{O}_1 O_2 \vee O_3 \bar{S}$ alone entails O_2, in which case $\bar{S}$ and O_3 together entail O_2, so that $\bar{S}$ is directly verifiable.

The difficulty here is similar to difficulties dealt with by Hempel in IX 47(1), X 104(3). Though Hempel characterizes his study as "only a first attempt to arrive at a systematic logical theory of confirmation," it is probable that his results will have important applications in the present connection. At all events, it is again the reviewer's judgment that any satisfactory solution of the difficulty will demand systematic use of the logistic method. Ayer's own attitude towards the logistic method is mixed, being divided between his admission that (footnote 1, page 70) "there is a ground for saying that the philosopher is always concerned with an artificial language" and his declaration on page 23 in favor of "ordinary speech" in connection with the analysis of descriptions.

ALONZO CHURCH

FREDERIC B. FITCH. *The problem of the Morning Star and the Evening Star.* ***Philosophy of science***, vol. 16 (1949), pp. 137–141.[24]

The problem in question is that of Quine in his XII 139 (2). It is pointed out that if the phrases 'the Morning Star' and 'the Evening Star' are construed as descriptions in Russell's contextual sense, then Quine's argument fails to show any difficulty in modal logic (compare the review VII 100(2)). In particular, no objections appear from this point of view against the systems $S2^2$ and $S4^2$ of Ruth Barcan's XII 95(4).

Fitch also discusses briefly the possibility of construing these phrases as proper names, but he holds (with Smullyan) that two proper names of the same individual must be synonymous. It would seem to the reviewer that, as ordinarily used, 'the Morning Star' and 'the Evening Star' cannot be taken to be proper names in this sense; for it is possible to understand the meaning of both phrases without knowing that the Morning Star and the Evening Star are the same planet. Indeed, for like reasons, it is hard to find any clear example of a proper name in this sense.

ALONZO CHURCH

L. JONATHAN COHEN. *Are philosophical theses relative to language?* ***Analysis*** (Oxford), vol. 9 no. 5 (1949), pp. 72–77.[25]

The author contends that "Dictionary definitions are untranslatable because they are relative to natural language." It is meant e.g. (though the author does not use this particular illustration) that, if we find in an English dictionary a definition of 'fortnight' as "a period of fourteen days," then the statement which is contained in this definition is one which it is logically impossible to translate correctly into another language.

It would seem to the reviewer that the conclusion which this involves, that there are certain things (namely certain facts about the English language) which cannot be expressed in *any* language but English, is so immediately unacceptable that the author's contention may be rejected on this ground alone. Indeed the statement made in the dictionary, though elliptical, must no doubt be understood to be: 'The word 'fortnight' means in English a period of fourteen days.' And the translation of this

[24]Review originally published in ***The journal of symbolic logic***, vol. 15 (1950), p. 63.

[25]Review originally published in ***The journal of symbolic logic***, vol. 15 (1950), p. 63.

into, say, German is just 'Das Wort 'fortnight' bedeutet auf Englisch einen Zeitraum von vierzehn Tagen.'

Implicitly the author seems to be rejecting the German translation of ''fortnight'' as ''fortnight'' because the German translation of 'fortnight' is not 'fortnight.' But this is a serious misunderstanding of the Fregean convention about the use of quotation marks, which is now adopted by many logicians, including Carnap, and which must for purposes of the present discussion be supposed to have been incorporated into both the English and the German languages.

The author intends his remarks in criticism of Carnap's doctrine (of 1935) concerning the connection between philosophy and logical syntax. But because of the error indicated, they have in the reviewer's opinion no bearing on this.

Alonzo Church

Philotheus Boehner. *Bemerkungen zur Geshichte der De Morgansche Gesetze in der Scholastik.* ***Archiv für Philosophie***, vol. 4 no. 2 (1951), pp. 113–146.[26]

It was pointed out by Łukasiewicz (1861***2, 13***) that the laws of the propositional calculus which have come to be known as De Morgan's laws were stated in verbal form by Ockham, and later by Johannes Versor (or Versorius). In the present paper, Boehner traces the history of these laws in the writings of the Scholastics from their appearance in Ockham's *Summa logicae* to John of St. Thomas (who died in 1644). Though he does not claim completeness—much of the relevant medieval literature being not easily available—Boehner is able to show clearly that the passages cited by Łukasiewicz are not isolated instances, but that the De Morgan laws have had a continuing place in the Scholastic tradition, and even still appear in Scholastic works of the eighteenth and nineteenth centuries.

As Boehner indicates, he was unable to find any statement of the De Morgan laws before Ockham. Petrus Hispanus does not introduce the idea of the contradictory opposite or negation of a conjunction, or of a disjunction. William of Shyreswood raises the question whether a single negation may be applied to a conjunction taken as a whole, as in *Non (Sortes currit et Plato disputat)*; after some discussion, conceding the existence of reasons for the contrary opinion, he concludes that this may significantly be done, but does not go on to the statement of the corresponding De Morgan law.

Boehner reproduces in especial detail the treatment of the De Morgan laws and related matters by Ockham, by Walter Burleigh (in large part from the manuscripts of two different works, both called *De puritate artis logicae*, which have never been put into print except for a recent edition of one of them edited by Boehner himself), and by Paulus Pergulensis. Burleigh was a contemporary of Ockham, though his works here in question are later than the *Summa logicae*, and Paulus Pergulensis belongs to the end of the fifteenth century. Besides the De Morgan laws, Burleigh has a similar law for the contradictory of a conditional—in symbols, $\sim P \supset Q$ eq. $P \cdot \sim Q$. The material quoted from Paulus Pergulensis includes an interesting treatment of modalities, and the law regarding the negation of a conditional appears, not as an equipollence, but as the inference from $\sim P \supset Q$ to $P \cdot \Diamond \sim Q$, the example given being: *Non (si tu es homo, tu vigilas). Ergo, quamvis tu sis homo, stat vel potest esse quod non vigilas.*

[26]Review originally published in ***The journal of symbolic logic***, vol. 17 (1952), pp. 123–124.

As Boehner remarks, the laws actually stated by De Morgan were not those of the propositional calculus, but the corresponding laws of the class calculus, which do not appear in the works of the Scholastics. Since the author does not enter further into the history of the De Morgan laws outside the Scholastic writings, the reviewer would recall also the following facts, which are already known but must be added to the important historical discoveries of the present paper to form a complete picture.

The De Morgan laws of the propositional calculus were given also by Geulincx in 1663, his statement of them being noteworthy (as pointed out by Dürr, VI 104(3)) because of its use of propositional variables.

The relationship of the De Morgan laws of the class calculus to those of the propositional calculus was not as clear in the time of De Morgan and his immediate successors as it is now, in the light of later developments due to C. S. Peirce, Schröder, and others. The laws of the class calculus, after their statement by De Morgan, appear also in the writings of Jevons and of Peirce. But the first statement of the De Morgan laws within a *formalized propositional* calculus was by Hugh MacColl in 1877.

Since Peirce was well familiar with at least the better known Scholastic works, and in particular often quotes in detail from the *Summa logicae*, it seems strange indeed that he never remarked on the occurrence of the De Morgan laws in the latter treatise.

ALONZO CHURCH

A. G. N. FLEW. *Introduction.* ***Essays on logic and language***, edited and with an introduction by Antony Flew, Basil Blackwell, Oxford 1951, and Philosophical Library, New York 1951, pp. 1–10.[27]

GILBERT RYLE. *Systematically misleading expressions.* Ibid., pp. 11–36.

FRIEDRICH WAISMANN. *Verifiability.* Ibid., pp. 117–144. (Reprint of XII 101(2), with some modifications.)

In a paper reprinted from ***Proceedings of the Aristotelian Society***, n.s. vol. 32 (1931–2), Ryle sets forth as a (perhaps the) task of philosophical analysis the detection of cases of systematic misleadingness of language, their understanding and correction. Namely a linguistic expression is systematically misleading if its grammatical form is not appropriate to the logical form of the fact it records. And evidence of such systematic misleadingness is found when a process of inference generally applicable to statements of a certain grammatical form fails in some particular case.

For example, 'Jones is a suspected murderer' is thus systematically misleading, since—unlike 'Jones is a forgotten murderer' or 'Jones is a clever murderer'—it does not have the consequence, 'Jones is a murderer.' Other examples are 'Mr. Pickwick is a fiction' (as against 'Mr. Baldwin is a statesman'), and 'Poincaré is not the King of France' (which, unlike other statements involving the same descriptive phrase, does not have the existence of the King of France as a consequence).

Many of the examples are in the direction of rejection of universals, and are in the reviewer's opinion debatable. In particular Ryle rejects the notion of *the meaning of* an expression, holding that there is no object so describable, and that the doctrine that there are such objects as meanings has arisen from a misleading similarity of grammatical form between statements like 'I have just met the village policeman' and 'I have just grasped the meaning of [such and such an expression].' Yet, unless Ryle's

[27] Review originally published in ***The journal of symbolic logic***, vol. 17 (1952), pp. 284–285.

own language is systematically misleading in a way not clear to the reviewer, he admits such an entity as a "fact or state of affairs" which "is not a thing" and which may have "logical form" or "real form."

As a consequence of his account, Ryle points out that philosophical analysis must "involve the exercise of systematic restatement." The reviewer would add that (in the interest of accuracy, and lest one kind of systematic misleadingness be obscured by the presence of another) this systematic restatement must be cumulative, and the philosopher's language must thereby diverge more and more from natural English as progress is made in analysis. It would seem that the philosopher's language must thus become progressively formalized (since the process of discovering, and treating, systematically misleading expressions will require explicit statement of the relevant rules of inference); and that the grammatical or syntactical features of the language must in at least some respects approach those familiar in object languages studied by mathematical logicians.

To this extent the reviewer is in disagreement with the criterion of avoidance of symbolism which is laid down by Flew in his *Introduction*. For, as has often been remarked, the essential of logical symbolism is not the use of "hard and unusual" characters but the replacement of the systematically and multiply misleading grammatical forms of a natural language by a rigorously formulated syntax. Even a book which, like the present one, is meant to be "intelligible to the layman" might profit by some symbolism, explained as it is introduced.

The reviewer has noticed minor errors and doubtful points in the present book in connection with which some use of logical symbolism might have been helpful. For example, the reviewer does not believe that Ryle's distinction between a "quality" and a "relational character" (p. 33) would survive a more accurate logistic treatment. Again, Flew cites (in support of Ryle) the well-known conversation of Alice and the White King, but adds the misleading comment: "Such is the mistake of the King, who treats 'Nobody' as if it had the logic of 'Somebody' ... " Actually the logic of 'nobody' and of 'somebody'—though not identical—are very similar, since both words represent quantifiers, and in a similar systematically misleading manner, so that the King's mistake about 'nobody' is readily matched by imaginable parallel mistakes about 'somebody.' (Just as one might mistake 'nobody' for the name of a person who has the peculiarity of being a non-existent person, so one might mistake 'somebody' for the name of a person who has the peculiarity of being an indefinite person.)

As explained in Baylis's review (XII 101), Waismann maintains that complete verification of a statement or of a law is defeated by the unavoidable *open texture* of empirical terms. (The present reviewer would prefer to say *vagueness*, but Waismann reserves the latter term for an equivocacy which is in principle completely removable by further specification.)

Because of this same matter of open texture, Waismann also asserts in one passage (p. 129) that a language used to express empirical knowledge cannot be a formalized language, but must remain at the level of a natural language. (Thus presumably he would hold that the process of cumulative restatement, described above in connection with Ryle's paper, can never terminate in a fully formalized language if empirical terms are present.) The reviewer would hold on the contrary that it is the *application* of the language, rather than its formalization, which must remain in some degree imperfect; and indeed that the success of the mathematical method in science depends precisely on creating a language which is internally more rigorous than is (even in principle) possible for our control over the situations to which it is to be applied.

ALONZO CHURCH

NELSON GOODMAN. *On a pseudo-test of translation.* ***Philosophical studies***, vol. 3 (1952), pp. 81–82.[28]

To the question whether two given sentences are synonymous, "we often find a negative answer defended by the argument that we might know the truth of one without knowing the truth of the other." But this presupposes linguistic understanding, i.e., it must not be that the sentences involve words or constructions whose meanings are unknown or obscure to us.

Thus let the question be whether (1) 'James was John's brother' and (2) 'James was John's sibling and a male' are synonymous. It is not a satisfactory test, to ask whether it is possible for us to know that (1) is true (i.e., in English) without knowing that (2) is true, as it is evident that this would be possible merely through ignorance of the English word "sibling." Goodman (though he does not use this particular example) would therefore amend the test by first supposing that we know the English meaning of the two sentences, and then asking whether it is possible for us to know that (1) is true without knowing that (2) is true. Hence he concludes that the supposed test of synonymy is pointless. For "if we understand two sentences fully, if we 'know the meaning' of each, then we already know, without any further test, whether they have the same meaning."

The reviewer would suggest the simpler amendment of asking whether it is possible [not for us but] for any one [regardless of his knowledge of English] to know that James was John's brother without knowing that James was John's sibling and a male, or *vice versa*. In this modified form the test may be useful especially in supporting a negative answer to a question of synonymy; as in order to be sure of circumstances under which some one would know that ______ without knowing that ______, it may be that a partial understanding of the two sentences that fill the two blanks is enough.

ALONZO CHURCH

MAX BLACK. *Frege on functions.* ***Problems of analysis, Philosophical essays***, by Max Black, Cornell University Press, Ithaca 1954, pp. 229–254, 297–298.[29]

This essay is an examination of Frege's account of functions (including propositional functions) as being *ungesättigt* or *ergänzungsbedürftig*. In the reviewer's opinion the author is successful at least in showing that Frege's account leads to complications and perplexities which it were better to avoid. Even if a form of the Fregean doctrine could somehow be made tenable, there is every advantage of naturalness and simplicity in favor of the alternative doctrine according to which a function is an object capable of being denoted or named (in Black's terminology, "referred to") in the same fashion as any other. Indeed the latter doctrine can be described by saying that we abandon Frege's *ungesättigt* functions, and apply the name *function* instead to what Frege called *Wertverlauf*, thus simplifying the theory by deletion of a superfluous portion.

[28] Review originally published in ***The journal of symbolic logic***, vol. 20 (1955), p. 62.

[29] Review originally published in ***The journal of symbolic logic***, vol. 21 (1956), pp. 201–202.

On the other hand in his discussion of Frege's contention that (to quote Black's paraphrase) "no series of designations constitutes a sentence ... they must be united in order to form a whole," the author seems to have missed an important element of truth in Frege's doctrine. For in every attempt that has ever been made to construct a language with precisely stated syntax, a formalized language, the fact is that no sentence of more than one symbol consists of a mere series of designations or names; but there must occur in addition at least one incomplete symbol or connective—some notation which has no meaning in isolation, or at least no meaning as a name, but is capable of combining with meaningful symbols to form a meaningful whole. The connectives employed may perhaps be reduced to a single one, a notation indicating application of function to argument, and for this, mere juxtaposition between parentheses may be used, as in combinatory logic. But in most formalized languages there are connectives in larger variety. E.g., symbols such as $\supset$ and $\vee$ in the propositional calculus, though they might be regarded as names of functions, are more usually treated as connectives, having no meaning in isolation. It would seem to the reviewer that Frege's introduction of *ungesättigt* function signs may be looked upon as an (unsuccessful) attempt to provide a semantic category intermediate between that of connectives and that of names, notations which are like connectives in that their meaning requires completion by combining them with meaningful symbols to form a meaningful whole, but which are like names in that they may be replaced by variables and such variables may then be bound by quantifiers.

It should be noticed that Black is mistaken in saying, in footnote 6, that Frege's doctrine forbids quantification upon functions. But this does not affect the main part of his argument.

A more serious error, in the reviewer's opinion, is the fallacious attempt to refute Frege's view, that sentences are designations of truth-values, by reference to the grammar of the English language. It is pointed out that if the sentence "Three is prime" is a designation of the True (Black uses a capital letter in translating Frege's *das Wahre*), then the expressions "Three is a prime" and "the True" ought to be interchangeable in non-oblique contexts—as indeed Frege himself maintained in analogous cases. Thus from "If three is a prime then three has no factors" we get "If the True then three has no factors." To the latter expression Black objects by calling it nonsense, and by saying that it "has no more use than" the expression, "If seven then three has no factors." In the absence of supporting reasons for such objections, it must be supposed that Black is rejecting the expression in question on the ground that it violates the rules of English grammar. But surely the right question to ask here is not what, by existing custom, are the rules of English, but rather what it is desirable to take as the rules of a formalized language. In a suitable formalized language the analogue of "If the True then three has no factors" does have a use, namely as a designation of the False. And indeed it is not unusual in formulations of propositional calculus, and of formalized languages containing propositional calculus, to introduce primitive constants denoting one or both of the two truth-values, and to allow substitution of such constants (as well as of longer sentences) for the propositional variables. ALONZO CHURCH

LUITGARD WUNDHEILER and ALEX WUNDHEILER. *Some logical concepts for syntax.* ***Machine translation of languages, Fourteen essays***, edited by William N. Locke and A. Donald Booth, John Wiley & Sons, New York 1955 (co-published with The Technology Press), pp. 194–207.[30]

This paper is an incomplete sketch of a sort of general grammar of natural languages, intended to provide syntactical notions which are invariant under translation, and which may therefore be more useful in the problem of machine translation than application of conventional grammar. As the authors borrow their ideas and terms, or many of them, from modern logic — in particular from Ajdukiewicz 225***14*** and Carnap IV 82 — this is in effect an application of logical syntax to the analysis of natural language. Others have also suggested that modern logic may throw some light on this question (e.g. Siro VII 127, Bar-Hillel XV 220, Chomsky in the ***The journal of symbolic logic***, vol. 18, pp. 242–256), and the suggestions in this direction by the present authors and others certainly deserve consideration, though it is not yet clear how extensive the usefulness of ideas from logic and logical syntax will ultimately prove to be in this connection.

Parts of the sentence, parts of speech, and other grammatical distinctions in the currently conventional grammar of English and other natural languages are often logically irregular or confused. They work as well as they do only because these grammatical notions are long established parts of ordinary education and natural languages have developed around them and conformed to them. And indeed, given the present actual state, these logically irregular concepts are in part practically necessary in order to render and to study the language as it is. For this reason a general grammar such as that proposed in the present paper cannot be expected to supersede established grammatical systems, but besides being useful in connection with translation it may serve purposes of clarification, and it may lead to reforms in some cases in which logically ill-founded terminology is not actually very needful for description of the existing state of the language.

It is not to be supposed that there is a unique scheme which constitutes the one true general grammar. It will often rather be a question of relative expediency among various choices. But many of the particular choices made in the present proposal would seem to the reviewer doubtful. For example, in the system of syntactical categories which the authors propose, proper names and common names are classed together in the same category n. A satisfactory category n/n is provided for attributive adjectives, but the authors are inconsistent as to whether predicate adjectives belong in this same category or another one (compare p. 197 with pp. 200, 204). The distinction between transitive and intransitive verbs is said to be rejected, but it would be more accurate to say that it is regularized rather than rejected, and that it reappears in the authors' distinction between the categories s/n and s/nn. The distinction between active and passive of a verb is also rejected; but as this distinction corresponds well to that between a relation and its converse, it would be better again to regularize the distinction rather than to reject it. (Misled by the feature, common to many natural languages, by which omission of one of the complements of a verb may indicate either existential quantification or a suppressed complement to be supplied from the context — compare "The university offered him a professorship" with "The university

[30]Review originally published in ***The journal of symbolic logic***, vol. 21 (1956), pp. 312–313.

offered a professorship" — the authors as a matter of fact exaggerate the extent to which the established terminology of active and passive voice is irregular.)

ALONZO CHURCH

JOHN MYHILL. *A system which can define its own truth.* ***Fundamenta mathematicae***, vol. 37 (for 1950, pub. 1951), pp. 190–192.[31]

According to a well-known result of Tarski (285***14***), no consistent logistic system satisfying certain conditions can contain a truth predicate, Tr, such that, for every sentence **N** of the system, Tr(**n**) ≡ **N** is a theorem of the system, **n** being the numeral that represents the Gödel number of **N**. The principal conditions are the availability in the system of propositional calculus, numerical variables, quantifiers binding numerical variables, and the usual laws of quantifiers, further a system of numerals (a unique one to represent each natural number), and the provability of equations which, by means of numerals, correctly ascribe a particular value to a primitive recursive numerical function for particular arguments. The conditions can be somewhat weakened, but of course the metatheorem may be expected to fail if they are weakened sufficiently.

In the present paper Myhill makes use of known results concerning recursive functions to construct a system which is consistent and complete, in the sense that its theorems coincide with its true sentences, and which can "define its own truth" in the sense that Tr(**n**) is a theorem if and only if **N** is a theorem. This system has an existential quantifier but no universal quantifier, and has no sentence connectives and hence no propositional calculus at all.

The close relationship should be noticed between Myhill's system and the system of §21 of the reviewer's VI 171. The latter system is consistent and complete in the same sense as Myhill's, and can in the same sense define its own truth (by Tr $\rightarrow B(\delta 2)$form). Moreover Myhill's system is apparently translatable into a fragment of it by means of the known equivalence between partial recursiveness and λ-definability of numerical functions. And this observation can be used to extend Myhill's system to a system which contains propositional calculus in a weak sense and can still define its own truth in the same sense as before. But none of these systems can define its own truth in the strong sense that Tr(**n**) ≡ **N** is a theorem (or is true) in all cases.

ALONZO CHURCH

FRANCIS C. WADE. *Translator's introduction.* ***John of St. Thomas, Outlines of formal logic***, translated by Francis C. Wade, Marquette University Press, Milwaukee 1955, pp. 1–24.[32]

Father Wade's introduction comes within the field of this JOURNAL because of its inclusion of a section about the relation of Symbolic to Traditional Logic (we follow the author's capitalization). The conclusion reached "is that Symbolic and Traditional Logic are not opposed, because they do not treat of the same object. Traditional Logic inquires into ... the various ways things are achieved in thought. Symbolic Logic never asks about the forms of thinking, but about the relations between thoughts considered as things."

[31] Review originally published in ***The journal of symbolic logic***, vol. 21 (1956), p. 319.

[32] Review originally published in ***The journal of symbolic logic***, vol. 24 (1959), pp. 81–83.

To support this, however, the author is obliged to confine Traditional Logic to even a narrower domain than is usually understood.

"Traditional Logic never reads the I proposition as existential . . . *Some men are white* means that the quality *white* modifies some cases of man." It does not mean, he says, "that some white men exist. To get this said requires an existential proposition, *e.g. Some white men are*, where there is no strict logical predicate." Apparently it is meant that *Some men are white* (when the proposition "is brought into a logical framework") would still be true if men did not exist. And the specially existential propositions that are referred to seem to be excluded from Traditional Logic on the general principle that "Formal [Traditional] Logic never deals directly with extra-mental existence."

But even this is not yet enough to save all the traditional immediate inferences. The author finds it necessary also to exclude propositions with self-contradictory subjects, such as *Square circles are oblong*, on the ground that they "cannot be thought" and hence presumably may not (or cannot?) be used in any process of reasoning, even as constituents of compound or hypothetical propositions. The view is rendered plausible by selecting an example in which the self-contradictoriness of the subject is directly obvious; and the author fails to take into account that in some (e.g. mathematical) contexts it is quite important to be able to reason about subjects which may later turn out to be self-contradictory.

In the reviewer's opinion, the author is seriously in error in claiming the support of John of St. Thomas for this account of existential import. John's own account of the matter, come down to him from his predecessors (cf. XX 44), is very clear: a proposition with a non-supposing—i.e., non-existent—subject, if affirmative, is false (see p. 60 of the present translation); but "neither supposition nor existence of the extremes is required for negative propositions" (p. 75). As a result, certain otherwise immediate inferences require for their validation an added premiss of "co-existence" (as Wade translates John's term "constantia"). E.g., from *Any horse belonging to man runs* may be inferred *Any horse belonging to this man runs*, "provided you add or understand co-existence: *And this man possesses a horse*" (pp. 56, 130). Again, in *Every man is rational*, John holds (p. 103), "the matter is necessary"—*rational animal* being an essential definition of man (p. 48). From this, instead of having immediately the truth of the universal, *Every man is rational*, we have that "the universal can be inferred from one particular," i.e., there is a valid (so-called material) consequence from *Some man is rational* to *Every man is rational*. In *Quaestiones disputandae* (not included in Wade's translation) we see that contraposition and the inference which later came to be known as obversion of a negative proposition both require the added "constantia"; thus (see q. 7, a. 3) for the inference from *Some man is not white* to *Some man is non-white* there is required the auxiliary premiss *And man exists* ("et homo est"); and for the inference from *Every man is white* to *Everything not white is a non-man* there is required *And the non-white exists* ("et non-album est"). Also in *Quaestiones disputandae* (q. 7, a. 2) John cites the proposition, *Some irrational man is not an animal*, and declares it to be true, precisely on the ground of the self-contradictoriness of the subject.

In fact "constantia" would be better translated as "confirming premiss," though John says "positā constantiā" without directly calling the constantia a premiss.

This doctrine of existential import is entirely sound, even if not in modern eyes the most natural one, and the reviewer has found only one short passage which might be thought to depart from it. Namely in discussing the proposition that *Every white man is a man* (q. 7, a. 3), John says that the proposition is false in case of the non-existence of white men, provided we take the subject with accidental supposition. But we may also understand the subject as having natural supposition and the verb as abstracting from the time (instead of referring to present time); and the proposition is then necessary, and true independently of the existence of white men. John attempts, in this latter case, to explain away the embarrassing inference from *Every white man is a man* to *Some man is a white man*. The intention of the explanation is not entirely clear. But in the light of John's doctrine of ampliation and restriction, it is probable that he intends to adhere to the same rules of existential import which he has maintained uniformly elsewhere, and only (in a terminology not used by John himself) to extend the universe of discourse, to which the existence refers, from the realm of things actual ("in mundo") to that of things possible. If so, John's account of existential import is not yet in agreement with Wade's even in the special case of subjects having natural supposition. For though John holds that *All white men are men* is true when understood timelessly, he must evidently hold that *All irrational men are men* is false in the timeless sense.

Wade also urges differences between Traditional and Symbolic Logic in regard to the meaning to be attached to A propositions, and to conditional propositions.

In the case of the conditional proposition, "there is a necessary knowledge-connection" between the antecedent and the consequent. Here Wade is in agreement with John of St. Thomas, who writes (pp. 96–97) that, for the truth of the "conditional in its strict sense," "a valid consequence is sufficient. ... it follows that every true conditional is necessary and every false conditional is impossible." But Wade is contrasting the traditional conditional with material implication. He does not discuss the question whether Lewis's strict implication or any of the many variations of it that have been proposed would correspond satisfactorily to his understanding of the traditional conditional.

The reviewer would add that there is a place for material implication in John of St. Thomas under the head of either the disjunctive proposition or the negative copulative, since John does not require any sort of inferential connection between parts of the proposition—or between one part and the negation of the other—in either of these cases. Thus it would seem that John could not object to material implication in itself but only to giving it the name of *implication* or *conditional*. And the moderns might retort, "What's in a name?" ALONZO CHURCH

HANS G. HERZBERGER. *The truth-conditional consistency of natural languages*. ***The journal of philosophy***, vol. 64 (1967), pp. 29–35.[33]

Tarski's condition 𝔚 (285***14***, ***16***) on a definition of truth for a language L is, to quote only the more important clause of it, that there shall follow as consequences of the definition all sentences that are obtained from the expression "x is true if and only if p" by replacing "x" by a structural-descriptive name of some sentence of L and "p" by the translation of that sentence into the meta-language. It is clear that *if this condition is satisfied, the truth of a sentence and of its negation cannot both follow*–i.e., if we make in addition to Tarski's condition the clearly reasonable assumptions that the translation of the negation of a sentence is the negation of its translation and that the meta-language is consistent, it can then be shown in a meta-meta-language that two

[33]Review originally published in ***The journal of symbolic logic***, vol. 33 (1968), pp. 146–147.

sentences obtained from "x is true" by replacing "x" by a structural-descriptive name of some sentence of L and by a structural-descriptive name of the negation of that sentence do not both follow in the meta-language. For the significance of the remark we need also the assumptions that every sentence of L has a structural-descriptive name in the meta-language and that the ordinary inference rules of elementary logic hold in the meta-language, as we will then have further that it can be proved in the meta-language about any specific sentence of L that not both it and its negation are true.

Herzberger's conclusions in the paper under review are based in effect on the foregoing italicized remark. However, he does not quote Tarski's condition in its original form, but rather illustrates it by the example, "The German sentence '[Der] Schnee ist Weiss' is true if snow is white," and then introduces the following condition which he claims to have an evident analogy. Let there be a set W of entities w, and let $T(w, z)$ be read as: the sentence z is true if the condition w obtains. Then "in order that a set W and a relation T should specify truth conditions for some language L, it is requisite, whenever $T(w, z)$ and w obtains, that z in fact be true."—The reviewer is unable to see this as an analogue of Tarski's condition, and recommends that the paper be read in the light of the original Tarski condition.

The author defines a language to be *consistent in its truth conditions* if there is no truth condition "which can possibly obtain and which if it obtained would render true each member of some logically inconsistent set of sentences," and *consistent in its analytic sentences* if the set of its analytic sentences "is a logically consistent set." On the assumptions indicated above, and on the further assumption that analytic sentences are true, it follows in an appropriate meta-language that no language is inconsistent in either of these senses. (The word "appropriate" here, which is the reviewer's, covers an omission that the author's trivialization of the "Tarski condition" has concealed; it might be filled in various ways, but to avoid imposing further conditions on truth or on the object language we may short-cut the difficulty by supposing that a logically inconsistent set of sentences is one that actually contains both x and Neg x for some x, and then supplying an ω-rule for the meta-language.)

From this the author draws somewhat divergent conclusions in regard to formalized languages and in regard to natural languages. For the former it is held that "although one is free to set up formal systems as one chooses, only some of these systems can be interpreted as 'artificial languages' in the usual manner." For the latter the conclusion is that "in at least two very basic senses, natural languages are inherently consistent" (this in opposition to the claim sometimes made that natural languages are inconsistent).

At least in the case of natural languages, not even the requirement that there shall exist a complete truth definition satisfying Tarski's condition is available to take the place of the requirement of consistency, as the author wishes to allow rather that there may be "truth-value gaps" in the sense of sentences that have no truth conditions.

The reviewer sees little value in a definition of consistency that forces every language to be consistent if we have an acceptable meta-language, and prefers return to a notion of consistency that belongs to the syntax rather than the semantics of the object language. ALONZO CHURCH

HANS G. HERZBERGER. *The logical consistency of language.* ***Language and learning***, edited by Janet A. Emig, James T. Fleming, and Helen M. Popp, Harcourt, Brace & World, Inc., New York-Chicago-Burlingame 1966, pp. 250–263.[34]

This is an earlier version of the paper reviewed above. Only one notion of consistency is treated, that of *consistency in the analytic sentences* (in the terminology of the later paper), and there is no explicit reference to Tarski's condition on a definition of truth. The author takes this notion of consistency to be the usual or the correct one (p. 255) and hence is able to say in an early paragraph: "The main result of this paper will be a demonstration of the inherent consistency of natural language."

ALONZO CHURCH

G. E. M. ANSCOMBE. *Before and after.* ***The philosophical review***, vol. 73 (1964), pp. 3–24.[35]

This paper introduces and discusses, but does not fully formalize, a language which has such expressions as '$pTqTrTs$', to be read as "It was the case that p and then it was the case that q and then it was the case that r and then it was the case that s." Here 'p', 'q', 'r', $\cdots$ are to be thought of as analogous to propositional variables, but they differ from (orthodox) propositional variables in that the sentences which may be substituted for them have an indexical reference—so that there is no contradiction in '$\sim pTp$' or even in '$\sim(pTq)T(pTq)$'. The 'T' is a temporal connective, treated in a way that makes it somewhat analogous to modal and other non-truth-functional connectives in propositional calculus; its meaning is sufficiently indicated by the reading just given, provided we add that e.g. 'pTr' is understood in such a sense that it is not inconsistent with, and is even a consequence of, '$pT \sim pTr$'.

'p before q' is defined as '$p \sim qTq$'. The meaning of 'p after q' is regarded as more problematical and is debated with much reference to examples from natural language; the analysis finally offered (p. 23) is not expressed in the 'T' language, but it seems to the reviewer that it may be equivalent to defining 'p after q' as the disjunction '$(\sim qTqTp) \vee (q \sim pTp) \vee (qT \sim qTp)$'.

The author devotes a section to defending her "unorthodox conception of proposition" according to which "propositions can change in truth value." It is not clear whether she regards as also admissible "orthodox" propositions, whose truth or falsehood does not change with time, or whether she holds that all assertions have an unavoidable indexical (*now* or *here-now*) reference.

She is also concerned to defend or justify her method as against the more familiar method of using quantification with respect to a time variable. The latter, it is said, "is often a useful and clarifying device." "But it would be absurd to treat it as capable of giving a fundamental explanation of 'before' and 'after'." A distinction is evidently intended here between a "fundamental explanation" and what the reviewer would like to call simply a "theory" of order in time. And subsequent discussion brings out that at least a part of the distinction lies in the point that quantification with respect to a time variable presupposes measured time and hence the use of clocks, so that it "involves empirical facts which go beyond the facts stated in the ordinary 'before' and 'after' propositions supposedly analyzed."

---

[34]Review originally published in ***The journal of symbolic logic***, vol. 33 (1968), p. 147.

[35]Review originally published in ***The journal of symbolic logic***, vol. 36 (1971), pp. 173–175.

To this the reviewer has the objection that it is never fully clear until *after* analysis just what empirical facts are stated in an ordinary-language assertion, and it is therefore not completely possible to make it a *precondition* of the analysis that it shall not go beyond the facts which are stated.

To illustrate, let us consider Einstein's well-known analysis of simultaneity at a distance. The ordinary-language speaker who applies one of the words 'before,' 'after,' 'simultaneous' to events distant from each other is usually thinking neither of clocks, nor of experiments to determine the speed of radio signals (or of light), nor of a geodetic survey to determine the half-way point. *Prima facie* all of these things go beyond the facts stated. Yet for an analysis of what is said at least one of them is necessary. If we accept Newtonian mechanics, rather than Relativity, the simplest analysis depends on transporting an accurately running clock so as to synchronize clocks at the two places.

For proper time of a particular person there are indeed more basic ways of determining "before" and "after" than the use of clocks or radio signals. But the present paper is concerned with public notions of "before" and "after" in such a way that simultaneity at a distance is required. The author's defense of this (p. 7) is insufficient in view of the fact that she renounces appeal to astronomical observations, clocks, and (presumably) other sophisticated physical apparatus as going "beyond the facts stated."

There is another justification which has sometimes been offered for systems of time logic like the present one, and which is perhaps related to the requirement of a "fundamental explanation" but is not the same. Namely it appears that most commonsense notions about order in time do not depend on such mathematical developments as that of the order type of the continuum, or even perhaps of the rational numbers, and that some uses of them may not demand quantification. To substantiate this, and generally to separate out those notions and propositions which do not depend on this or that part of mathematics or of logic, it is desirable to formulate various weak systems of time logic — a sort of continued experimentation in what is coherently possible and in what can be accomplished on various weak bases. It seems to the reviewer that this is a sounder motivation. But it is not yet clear whether such weak time logics are better formulated by means of connectives analogous to those of propositional calculus (as in the present paper) or more in the manner of Russell's I 72, where order in time appears as a relation between events, and event variables are of a different logical type than propositional variables.

The present author does not refer to I 72, but refers rather to Russell's 111***24***, in particular to criticize his definition of an event as "anything that can be simultaneous with something or other." "At that rate the existence of the Parthenon is a long drawn-out event that is still going on. This is of course not an ordinary sense of 'event.' But, more important, it is not the sense of 'event' intended by anyone maintaining that 'before' and 'after' primarily name relations between events, which relations are genuine converses." It is further held that Russell "was wrong in saying that no instantaneous events occur within our experience" — but it is not held that all events are instantaneous. And it is finally argued that "We shall go wrong if, moved by the feeling that events are the proper terms of these relations ["before" and "after"], we insist on looking always for two events wherever we have p before q or p after q."

The paper may well be used as an illustration of the contrast between two different approaches (or interests): (1) that of always looking for the "ordinary sense" of terms, and (2) that of choosing the senses with a view to logical simplicity, with control by "ordinary language" only to the extent that certain possibilities of communication which exist there shall be preserved in principle. For example in the last quotation above "shall go wrong" means not that a logical error will be made but rather that we shall depart from the ordinary sense of the terms under analysis. The reviewer doubts that the ordinary sense of terms can be determined with certainty to such a degree of definiteness as to support, e.g. the analysis of "after" which is rendered in the second paragraph of this review. And this doubt is quite in spite of the fact that the author does make a persuasive case for some of her contentions if one accepts her point of view.

Correction. On page 13 it is said: "If we use quantification over times it will be a complicated job to reach a formulation that is really equivalent to such a narration as 'He wrote some letters before he drank coffee' as opposed to 'He wrote some letters before ever he drank coffee'." If we use 'h' for "he," '$L(x,t)$' for "x writes letters at time t," '$D(x,t)$' for "x drinks coffee at time t," and 0 for the present instant, it seems to the reviewer that the required formulation is '$(\exists t)(\exists u) \mathbin{.} t < u \mathbin{.} u < 0 \mathbin{.} L(h,t) \mathbin{.} \sim D(h,t) \mathbin{.} D(h,u)$'—unless indeed the words "really equivalent" conceal some point which has not been made clear. ALONZO CHURCH

JUDITH SCHOENBERG. *Belief and intention in the Epimenides*. ***Philosophy and phenomenological research***, vol. 30 no. 2 (1969), pp. 270–278.[36]

When Epimenides of Crete writes that "Cretans are always liars" we may understand him to mean either (1) that Cretans always assert what is false, or (2) that Cretans always assert what they believe to be false, or (3) that Cretans always assert what is both false and believed by them to be false. The Greek word which is translated as "liar" is uncertain among these three meanings; and also among several others which the present author does not mention (e.g. we may understand a liar to be one who asserts what is false and is not believed by him). But it is usual in presenting the Epimenides paradox to explain that we must understand Epimenides in sense (1) if a genuine antinomy is to result.

The author leaves (1) out of account ("In this paper I am not concerned with the truth locution version of the Epimenides"). The paper makes a contribution in distinguishing versions (2) and (3) and in giving reasons for concluding that no antinomy arises in either of these versions.

The reviewer would comment that there is nevertheless a certain paradox to be obtained from version (3). For we may suppose that, other than the assertion by Epimenides that is in question, all assertions by Cretans are indeed lies—however practically unlikely, this must be logically possible. That we can then infer by purely deductive means that Epimenides did not believe his assertion to be false is disturbingly paradoxical, though not outright antinomy. The situation is parallel to that in which, by supposing Epimenides to have said that Cretans always have bare faces and assert what is false, we are able to infer that Epimenides's own face was not bare.

In a final paragraph it is suggested as "speculation" that "the paradoxes which involve self-reference" might generally be resolved without "the need for some such

[36]Review originally published in ***The journal of symbolic logic***, vol. 36 (1971), pp. 671–672.

theory as the logical theory of types or linguistic metalanguage theory." The reviewer believes that the better conclusion from the failure of antinomy in versions (2) and (3) is that matters of belief are, like bareness of the face, extraneous to the paradox in such a way that it is preferable to avoid them in a first presentation. Prior (XXII 90) has versions of the Russell antinomy in which such extraneous matters are deliberately introduced for effect, and similar treatment of the Epimenides is evidently possible, but such versions are intended for those who already know the antinomies in their direct form. ALONZO CHURCH

JOHN VENN. ***Symbolic logic.*** Second edition, revised and rewritten. Chelsea Publishing Company, Bronx, N.Y., 1971, xxxviii + 540 pp.[37]

This is a reprint of the second edition of 37***8***, with pagination unaltered and with no changes except on title page and reverse. The first edition appeared originally in 1881 and the second edition in 1894.

Venn's method is that of the algebra of logic—as it is now usually called, after Macfarlane, Peirce, Schröder, and Couturat. Read today he seems to be excessively occupied with the most elementary parts of logic. The first edition included a section (pp. 399–404) in which a logic of relations is dismissed as a hopeless undertaking, except perhaps special theories to deal with relations having particular properties. And though this section has been deleted in the second edition, Venn's pessimism about a logic of relations is still expressed in the *Introduction* (p. xxxiv). Of several references to Frege's 49***1*** the only significant one expresses approval of Frege's adoption of what is now known as material implication (p. 242); like all his contemporaries, Venn fails to appreciate what now seem to be the important aspects of Frege's logic.

Yet Venn has his own importance. His influence was substantial, in particular upon C. I. Lewis and his school. And those who aspire to knowledge of the field can, and should, read him with profit.

There are two terms now very familiar which originated with Venn (as far as the reviewer knows). These are *symbolic logic* (in place of *mathematical logic*, which Venn seems to have regarded as inappropriate) and *intension* (used as expressing the opposite of extension and distinguished by its spelling from *intention*). Venn uses *extensive* and *intensive* in place of the now more usual adjectives *extensional* and *intensional*; also *extent* and *intent*. It is interesting that he regards *intent*, *intensive* as translations of the German *Inhalt*, *inhaltlich*.

The important original contribution to logic by Venn is of course the Venn diagram. In 37**5** and in the first edition of his book he uses the diagram with only the device of shading to mark a compartment as representing an empty class. But in his second edition there is also a device to mark non-emptiness. This is done by putting a figure 1 in each of any set of compartments of which it is known that at least one must represent a non-empty class; then if there is another set of compartments for which there is such an alternative, a figure 2 is placed in each; and so on (see pp. 130–132, 357–358, 376). This rather elaborate method (which is used with some variations not mentioned here) is now often replaced by the simpler but less powerful method of just using a cross or an asterisk to mark non-emptiness in one compartment.

[37]Review originally published in ***The journal of symbolic logic***, vol. 37 (1972), pp. 614–615.

Some recent writers have so far forgotten the history that the Euler diagram appears under the name of Venn diagram. But Venn presents his diagram as a modification and improvement of Euler's. In the last chapter he brings out that, although the widespread use of the Euler diagram derives from its adoption by Euler in his ***Lettres à une princesse d'Allemagne***, there were anticipations; and he even traces the matter back to the doubtful case of Vives, who has, just once, a diagram, but whose explanation of it is unintelligible as printed (in his ***Opera***, 1555).

Venn believes (see p. xxiv) that he is the originator of the interpretation of the existential import of categoricals, often considered to be the modern doctrine of existential import, according to which the universal propositions are true and the particular propositions false when the subject term is non-existent. Venn's first proposal of this is in 37*5*, July 1880, and it seems that Peirce's 28*5* must have appeared nearly simultaneously (see the footnote in ***Collected papers***, vol. 3, p. 104). Venn includes Peirce's paper in the *Index of bibliographical references* even in his first edition (1881) but in both editions he refers to it in the text only for some minor points and overlooks what Peirce says about existential import (***Collected papers***, vol. 3, pp. 114–115). On the other hand Land's exposition (36*1*) of Brentano's logical innovations appeared in the same volume of ***Mind*** as an earlier paper by Venn. And Venn's footnote on page 184 (p. 165 of the first edition) further suggests influence on Venn by Brentano.

In this footnote Venn makes the interesting historical remark that Leibniz anticipated Brentano's interpretation of the four categorical forms by expressing them as *A non B est non-ens, AB est non-ens, AB est ens, A non B est ens*. "But he seems to have rejected this view to some extent afterwards, owing to the consequent difficulties about the conversion of propositions." Then it is said that Brentano in VII 103(6) (first published in 1874) adopts substantially the same arrangement and "announces it as a novelty which is to spread dismay among orthodox logicians." But the conclusion is not drawn that Brentano's account of existential import therefore is the same as Venn's own.

In his first edition Venn followed Boole in allowing $x+y$ to have logical meaning only when the classes x and y are disjoint. But in the second edition "partly ... because the voting has gone this way" and partly because of "the practical advantages" he changes notation so that $x+y$ is the union of x and y in the now usual sense, without restriction that the classes shall be disjoint. ALONZO CHURCH

LUDOVICO GEYMONAT. *Prefazione*. ***Logica e aritmetica***, by Gottlob Frege, writings collected and edited by Corrado Mangione (translation by Ludovico Geymonat and Corrado Mangione), Paolo Boringhieri, Turin 1965, pp. 9–10.[38]

CORRADO MANGIONE. *Introduzione*. Ibid., pp. 15–81.

CORRADO MANGIONE. *L'ideografia nel sistema freghiano*. Ibid., pp. 82–97.

CORRADO MANGIONE. Premessa. Ibid., pp. 101–102.

GOTTLOB FREGE. *Ideografia. Un linguaggio in formule del pensiero puro, a imitazione di quello aritmetico*. Italian translation of 49*1*. Ibid., pp. 103–206.

CORRADO MANGIONE. Premessa. Ibid., pp. 209–210.

GOTTLOB FREGE. *I fondamenti dell'aritmetica. Una ricerca logico-matematica sul concetto di numero*. Italian translation of 49*5*, a reprint (with minor changes of

[38] Review originally published in ***The journal of symbolic logic***, vol. 38 (1973), pp. 532–534.

wording and punctuation) of pp. 17–187 of XIII 153. Ibid., pp. 211–349.

CORRADO MANGIONE. Premessa. Ibid., pp. 353–354.

GOTTLOB FREGE. *Concetto e rappresentozione*. Italian translation of pp. 157–160 of XIV 208(4), a reprint (with minor changes) of pp. 210–214 of XIII 153. Ibid., pp. 355–358.

GOTTLOB FREGE. *Oggetto e concetto*. Italian translation of 49***9***, a reprint (with minor changes) of pp. 191–209 of XIII 153. Ibid., pp. 359–373.

GOTTLOB FREGE. *Senso e significato*. Italian translation of 49***8***, a reprint (with minor changes) of pp. 215–252 of XIII 153. Ibid., pp. 374–404.

CORRADO MANGIONE. Premessa. Ibid., pp. 405–406.

GOTTLOB FREGE. Review of H. Cohen's ***Das Princip der Infinitesimal-Methode and seine Geschichte***. Italian translation. Ibid., pp. 406–411.

CORRADO MANGIONE. Premessa. Ibid., pp. 412–413.

GOTTLOB FREGE. Review of Cantor's ***Zur Lehre vom Transfiniten***. Italian translation. Ibid., pp. 414–417.

CORRADO MANGIONE. Premessa. Ibid., p. 418.

GOTTLOB FREGE. Review of Husserl's ***Philosophie der Arithmetik***. Italian translation. Ibid., pp. 419–437.

CORRADO MANGIONE. Premessa. Ibid., pp. 441–442.

CORRADO MANGIONE. Premessa. Ibid., pp. 443–444.

GOTTLOB FREGE. *Lettera a Giuseppe Peano*. Italian translation of 49***14***. Ibid., pp. 444–452.

CORRADO MANGIONE. Premessa. Ibid., pp. 453–455.

GOTTLOB FREGE and DAVID HILBERT. *Carteggio Frege-Hilbert*. Italian translation by Corrado Mangione of *Unbekannte Briefe Freges über die Grundlagen der Geometrie and Antwortbrief Hilberts an Frege* [*aus dem Nachlaß von Heinrich Liebmann herausgegeben and mit Anmerkungen versehen von Max Steck*] (cf. XXXVI 155). Ibid., pp. 455–473.

CORRADO MANGIONE. Premessa. Ibid., pp. 477–478.

GOTTLOB FREGE. *Logica matematica psicologia*. Italian translation of parts of the Preface of 49***10***, a reprint of pp. 253–269 of XIII 153. Ibid., pp. 479–493.

CORRADO MANGIONE. Premessa. Ibid., pp. 497–499.

GOTTLOB FREGE. *I numeri reali*. Italian translation of parts of the third section of 49***16***, with interspersed notes by Corrado Mangione (some of which summarize omitted paragraphs). Ibid., pp. 500–570.

CORRADO MANGIONE. Premessa. Ibid., pp. 573–574.

GOTTLOB FREGE. *Nota finale*. Italian translation of the Nachwort of 49***16***. Ibid., pp. 574–594.

CORRADO MANGIONE. Premessa. Ibid., p. 597.

GOTTLOB FREGE. *Abbozzo di una lettera a Giuseppe Peano*. Italian translation. Ibid., pp. 598–603.

GOTTLOB FREGE. *Diciassette proposizioni fondamentali della logica*. Italian translation. Ibid., pp. 604–605.

This collection of Italian translations from Frege may be regarded as a revised and much enlarged edition of ***Aritmetica e logica*** (XIII 153). All the translations that were in the latter book have been included, with revisions; and the various introductory notes and footnotes by Geymonat have been retained, with a very few omissions, and with significant alterations only in the footnotes on pages 297, 374. The most valuable additions for Italian readers are the complete translation of Frege's ***Begriffsschrift*** and of the famous *Nachwort* to the second volume of ***Grundgesetze***. And the two last items in the book are previously unpublished material from Frege's "Nachlaß."

Geymonat's preface justifies the additions to the book as being possible because of the completely changed situation among Italian students since 1948, who have become familiar with modern publications in logic and no longer have difficulty (even the philosophers) with varied technical symbolisms.

Mangione's *Introduzione* is partly a biography of Frege and partly a historical and descriptive account of Frege's contributions to logic. His *L'ideografia nel sistema freghiano* is an account of Frege's symbolism and a sketch of Frege's rules of inference and axioms in ***Grundgesetze*** listing with the axioms the early theorems that Frege derives from the axioms in §§49–52.

In *L'ideografia* the reader must be on guard against typographical errors. There are at least three places at which a negation sign has been lost, no doubt a consequence of the typographical difficulty of Frege's symbolism. And on the last line of page 97, '$\mathfrak{a} = \mathfrak{a}$' should be '$a = \mathfrak{a}$'. Also in making notational explanations and in stating rules of inference, there are some awkward use-mention confusions arising from the replacement of Frege's Greek letters (especially Greek capitals) by the same italic letters that Frege reserves for use as free variables in axioms and theorems.

Mangione writes (*Introduzione*, p. 28): "In the ***Begriffsschrift*** Frege establishes an 'intensional' logic in the sense that he makes reference constantly to concepts (predicates) and not to extensions of concepts (classes); in other words he does not employ the notion (at the ideographic level, the symbol) of class." This is possibly misleading, as Frege's concepts ("Begriffe") are distinguished from their extensions only by being, like all Fregean functions, unsaturated ("ungesättigt") and hence not objects. To say that concepts are intensional would ordinarily be understood to mean that two concepts may have exactly the same objects falling under them without therefore being themselves the same. But in a problematical footnote to ***Die Grundlagen der Arithmetik*** (see pp. 32, 306 of the present volume) Frege considers and seems at least tentatively to reject the intensionality of concepts in this sense; and in the review of Husserl (see p. 425) Frege says flatly that coincidence of the extensions is a necessary and sufficient indication that there shall hold between the concepts the relation which corresponds to equality between objects. Indeed this passage in the review of Husserl seems to commit Frege to the law of extensionality

(1) $$\vdash (\mathfrak{a})(f(\mathfrak{a}) = g(\mathfrak{a})) \supset M_\beta(f(\beta)) = M_\beta(g(\beta)),$$

or at least to the special case of this in which the functions f and g are restricted to be concepts, notwithstanding that (1) nowhere appears in the body of the ***Grundgesetze***.—For typographical simplicity in (1), Frege's notations for conditionality and generality (for implication and the universal quantifier) have been replaced by Russell's, but the notations are otherwise those of Frege in ***Grundgesetze***; this is awkward because the signs '$\supset$' and '( )' must then each be understood as including Frege's "Wagerechter" on both sides of them, but for convenient quotation of Frege it is hardly avoidable. To express the special case of (1) in which f and g are restricted to be concepts, we must put '$—f$' and '$—g$' everywhere in place of 'f' and 'g'. ALONZO CHURCH

W. V. QUINE. Introductory note. ***From Frege to Gödel, A source book in mathematical logic, 1879–1931***, edited by Jean van Heijenoort, Harvard University Press, Cambridge, Mass., 1967, pp. 150–152.[39]

BERTRAND RUSSELL. *Mathematical logic as based on the theory of types.* A reprint of 111***16***. Ibid., pp. 152–182.

This historically important paper of Russell appeared originally in 1908, two years before the first volume of ***Principia mathematica***. It was reprinted in 1956 in a volume edited by Robert C. Marsh (XXV 332) and is here reprinted a second time. There is also a partial reprint by Copi and Gould (reviewed below).

Quine's introductory note has much valuable background information, but it must be read with caution because it tends to exaggerate the faults of Russell's theory. Quine is on sound ground in repeating his well-known remark (VII 100 and elsewhere) that a satisfactory reconstruction of the theory requires that Russell's functional variables shall have intensional values, i.e., "attributes" (or better, "properties," as is Russell's own term) and relations in intension. He might have added that a value of a propositional variable must similarly be a proposition "in the sense in which this is distinguished from the phrase expressing it," notwithstanding that Whitehead and Russell call this a "false abstraction" (194***1***, p. 44). There are indeed serious use-mention confusions in Russell's paper, and any reconstruction of his theory must eliminate these; but the question "Is Russell then assigning types to his objects or to his notations?" has, on a reasonable reconstruction, the straightforward answer that there is a notational criterion for distinctions of type among things that are signified by the notation (and Quine himself has elsewhere said this).

Quine's argument that "the axiom of reducibility is self-effacing" is unconvincing. The correct statement would seem to be that the axioms of reducibility have the effect of cancelling the ramification (and so restoring impredicative definition) in extensional contexts but not in intensional. And indeed the matter is better explained by Russell himself in the last paragraph of section *V*.

Besides its historical importance, Russell's paper may serve as a short summary of some of the main ideas in ***Principia mathematica***. In particular the "primitive ideas" and "primitive propositions" are conveniently collected in one place in section *VI*—except the axiom of infinity and the multiplicative axiom, which are discussed in later sections. Thus the general character and at the same time the faults of Russell's primitive basis can be surveyed, the changes that were made in ***Principia*** being not of great significance.

The idea explained in the eighth paragraph of section *IV* (bottom of page 164 of the present reprint) does not reappear in ***Principia***, where in fact the term *matrix* receives a new and different meaning. This paragraph seems to be one of the few places at which use-mention confusion has resulted in an error that is not easily repaired.

The ramified hierarchy of types that is explained in Russell's paper is the same that appears in the *Introduction* of ***Principia***. In $*12$ of the latter work a modified hierarchy is introduced that is "stricter," as Whitehead and Russell say in the *Preface*. Many readers of $*12$ must have been puzzled by places at which the account of the modified hierarchy contradicts itself; the explanation seems to be that passages from section *V* of the paper of 1908 were imported into $*12$ with the intention of amending them to

---

[39]Review originally published in ***The journal of symbolic logic***, vol. 39 (1974), pp. 355–356.

fit the new context, but the amendments remained incomplete, and the result is that the older version of the type hierarchy shows through at some places.

On page 165 of the present reprint, lines 17–16 from below, i.e., near the middle of section *IV*, there is a confusing typographical error which appears also in the original printing and in both the Marsh reprint and the Copi–Gould reprint. Probably Russell meant to say: "... if it contains a first-order function ψ, it is a predicative function of ψ and will be written $f!(\psi!\hat{z})$."

Minor typographical errors in the present reprint include a redundant left-parenthesis in the fourth displayed formula on page 173 and a left-parenthesis instead of a left-brace in line 3 on page 174. At the last-named place the Marsh reprint has a more serious error, Russell's kappa being thrice misprinted as 'x' (page 90, lines 15–14 f.b.). ALONZO CHURCH

W. V. QUINE. Introductory note. ***From Frege to Gödel, A source book in mathematical logic, 1879–1931***, edited by Jean van Heijenoort, Harvard University Press, Cambridge, Mass., 1967, pp. 216–217.[40]

ALFRED NORTH WHITEHEAD and BERTRAND RUSSELL. *Incomplete symbols: Descriptions* Reprinted from 194*1*, pp. 66–71. Ibid., pp. 217–223.

This is a selection from the *Introduction* of ***Principia mathematica*** in which the authors introduce and discuss Russell's contextual definition of descriptions. Quine's introductory note is valuable, in particular for its account of Jeremy Bentham's anticipation of the general idea of contextual definition. But the last two paragraphs of the note are so worded that they seem to suggest that Whitehead and Russell, in the context of ***Principia mathematica***, could have dispensed with the (separate) contextual definition of class abstracts by treating class abstracts as descriptions. The misleading wording is evidently unintentional but is unfortunate. For it is important that Russell's device eliminates not only class names but also class variables (in favor of what Quine calls "attribute" variables). The note moreover errs in saying that Russell's contextual definition of class abstracts assumes a primitive notation for attribute abstracts—though there are remarks by Russell which might suggest this (e.g. on page 173 of the present volume), and though it is indeed true that ***Principia mathematica*** must be read as having a suppressed rule of inference, a rule of substitution for functional variables which is never explicitly stated (or else as having suppressed comprehension axioms which would take the place of such a rule of substitution). ALONZO CHURCH

STEPHEN F. BARKER. *Realism as a philosophy of mathematics.* ***Foundations of mathematics, Symposium papers commemorating the sixtieth birthday of Kurt Gödel***, edited by Jack J. Bulloff, Thomas C. Holyoke, and S. W. Hahn, Springer-Verlag, Berlin, Heidelberg, and New York, 1969, pp. 1–9.[41]

Taking his departure from Gödel's XI 75, the author outlines various objections against realism as a philosophy of mathematics and ends by selecting two, about which he says, not, that they are finally decisive, but that they "do have considerable force." (More exactly, he confines attention to that "form" of realism according to which sets, for example, are "real, non-spatial, non-mental, timeless objects"; but one may doubt

[40] Review originally published in ***The journal of symbolic logic***, vol. 40 (1975), pp. 472–473.

[41] Review originally published in ***The journal of symbolic logic***, vol. 40 (1975), p. 593.

whether there is in fact another form of realism that ought not preferably to be called either materialism or conceptualism.)

It seems to the reviewer that the two objections, as Barker describes them, have no force against the realist who is prepared to maintain both the following. (1) There may under certain circumstances turn out to be alternative set theories whose sets alike have reality. (2) There are indeed many subdomains of the domain of sets (or of one of the alternative domains of sets) which, after appropriate definition of the successor relation in terms of $\in$, have the structure of the natural numbers; all of these are real; and the decision as to which we shall call *the natural numbers* is purely a decision as to terminology.

Barker allows the realist a recourse to something like (2) (see pp. 6–7) but does not state the position in a way to show its full strength. In particular, if it is an objection against realism that any model of set theory must have many subdomains isomorphic to the natural numbers, is it not equally an objection that any model of set theory must have many subdomains which, after appropriate definition of the new $\in$ in terms of the old, are isomorphic to itself?

On the other hand he seems to contradict or to disallow (1) when he says that "all that one need claim is that talk about objects as real cannot make good sense unless methods can be envisaged by means of which alternative theories about those objects can be tested" (p. 8). Strangely, Gödel's own statement of his realism suggests that he might accept this claim—see e.g. the quotation from XIII 116(2) on page 74 of this same volume. In XXXI 484(8), XIII 116(2), and especially XXXIV 108(14), it is Gödel's purpose to show methods by means of which alternative theories about sets can in fact be tested; some of these are no doubt sound, others have the defect that they yield a decision which is relative to a particular situation and cannot be treated as final, and there is in any case no showing that alternative theories can always be tested by such means.

Granted the realistic outlook, it still seems unlikely that anything presently known or fixed upon has singled out a unique reality to be called the domain of sets, or even has selected a small number of such. As far as the cited publications of Gödel fail to take this into account, the reviewer thinks him vulnerable to Barker's criticism.

ALONZO CHURCH

I. THOMAS. *Logic, history of*. ***New Catholic encyclopedia***, prepared by an editorial staff at the Catholic University of America, McGraw-Hill Book Company, New York etc. 1967, vol. 8, pp. 958–962.[42]

This is an excellent short historical article, valuable especially for its account of the ancient, medieval, and post-Renaissance periods. There are nevertheless some errors, which demand correction all the more strongly because of the value of the article as a whole.

In the reviewer's opinion there is an unfortunate underestimate of Junge (or Jungius as his name appears on the title page of $1\frac{1}{2}$***I***, XXXIII 139) when it is said of him, only that he "showed a deductive interest in the syllogism and some appreciation of Aristotle's logic of relations." Indeed one may not speak of "Aristotle's logic of relations" in the same sense as of (e.g.) "Aristotle's logic of the syllogism." And

[42]Review originally published in ***The journal of symbolic logic***, vol. 40 (1975), pp. 596–597.

though Jungius in no way anticipated the fundamental ideas of modern logic, it is not altogether inappropriate to say that the logic of relations begins with Jungius. "The supposedly Aristotelian idea," spoken of elsewhere in the article, "that valid arguments are always syllogistic" owed much of its impetus to a reaction against Jungius, whose contemporaries and successors in the post-Renaissance period considered him un-Aristotelian.

There is a more sharply objective error when it is said that "Jevons showed that inclusive alternation offered some advantages over the exclusive used by Boole." The logic of Boole and Jevons is primarily a logic of classes, whereas the terms "inclusive alternation" and "exclusive alternation" are appropriate rather to the logic of propositions. But presumably we must understand by "inclusive alternation" the ordinary union of classes, and hence by "exclusive alternation" what is now usually called the symmetric difference. The former originated with De Morgan in 20*1*A—see page 115 and, since De Morgan there treats of [common] names rather than classes, see also page 39 for the relationship between names and classes. Later and independently of De Morgan, Jevons introduced the union of classes as an improvement of Boole's method, denoting it in 24*1* by the plus sign (which became the usual notation of the algebra of logic). The origin of the symmetric difference is with Jevons and later (see e.g. the review XVII 144(3)). As brief indication of the correction which must be made to Thomas's statement we recall that according to Boole, if x is any class, we have $x + x = 2x$; according to Jevons, using the plus sign for the ordinary union, we have $x + x = x$; and if we use the plus sign for the symmetric difference, we have $x + x = 0$. Thomas does say correctly that Jevons's modification has the effect of getting rid of coefficients, but there is some confusion nevertheless.

At the very end of the article it is said that Gödel showed "that the system of ***Principia Mathematica*** is undecidable" and that this method was "adapted to show the same for many other systems," especially by Tarski. The author does not make his reference to Tarski explicit, but possibly XXIV 167 is intended. And when he adds that "Church showed that the predicate calculus has this property," confusion becomes evident between Gödel's incompleteness theorem (for arithmetic) and Church's result concerning the non-existence of a uniform effective decision procedure (for arithmetic, then also for first-order functional, or predicate, calculus). The article mentions Gödel's completeness theorem for first-order predicate calculus, but fails to contrast it with the incompleteness proved for arithmetic (or for Gödel's system P as one formulation of ω-order arithmetic, consistency assumed). ALONZO CHURCH

CORRESPONDENCE BETWEEN ALONZO CHURCH AND COLLEAGUES (1935–1959)

I. *Letter regarding the Formation of* The Association for Symbolic Logic

II. *Letters regarding Church's Thesis (1935–1943)*

III. *Correspondence between Alonzo Church and Rudolf Carnap (1943–1954)*

IV. *Correspondence between Alonzo Church and W. V. Quine (1943, 1959)*

LETTER TO OSWALD VEBLEN
(1935)

June 19, 1935

Dear Professor Veblen:

I have received no bill from the National Academy for excess space in the Proceedings. If I do I will give it to Miss Blake as you suggest. I am also expecting from them a bill for reprints, for which, of course, I should pay. I am sorry about the excess length of the paper; I thought when I gave it to you that I had it down within the limit.

Kleene had a letter from the University of Oregon, a little over a week ago, asking whether he would be interested in an instructorship at $1760, inquiring about his "religious and other affiliations", and asking for letters of recommendation. Kleene replied that he would be interested, and gave them the desired information, and I and others here wrote in his behalf. He is evidently being considered for the position but does not yet have it.

Langford has been selected as the other editor of the journal of the proposed logical association, a choice which seems to me unfortunate, but I am inclined to go ahead hoping for the best. The question has arisen of obtaining a president for the organization (who would serve for three years, according to the present proposal, and be ineligible for reelection). The presidency is at present being offered to C. I. Lewis, but there is strong reason to believe that he will decline; he previously declined the editorship of the journal, making a gift of twenty-five dollars towards the enterprise, but saying definitely that he did not wish to participate in any active way. I proposed Weyl's name as president and I believe that if Lewis declined this idea could be put across. It appears, however, that last December, when a mimeographed notice was sent around concerning the proposed organization and journal, Weyl replied that he did not think that there was a need for such a journal, that he would not care to subscribe for the journal or attend meetings of the organization, and that, "In my opinion symbolic logic has become a part of mathematics" (the last was evidently apropos of the fact that the proposal was coming from a group of philosophers). This would seem to indicate that he would decline the presidency of the organization if offered to him. Do you think that there would be any hope of persuading him to change his mind?

There is the further point that it might be difficult to reach Weyl with a letter this summer. The present proposal is to have the set up of the association and its journal completed this summer, and to send out a definite announcement at the beginning of the fall, asking for members and subscriptions. This seems to be very desirable. But, if it were necessary, and if there were any reasonable hope of getting Weyl to accept the presidency, I should advocate waiting until fall.

Weyl is surely mistaken in supposing that there is no need for the proposed journal. Mathematical journals accept papers in symbolic logic with considerable reluctance, and clearly do not have the space to publish all that deserve publication, as I know

---

This letter is of interest because of its bearing on the founding of the Association for Symbolic Logic and ***The Journal of Symbolic Logic***. Printed by permission of the Alonzo Church estate.

from experience, not only with my own papers and those of Rosser and Kleene, but in refereeing papers of others. In regard to Weyl's apparent objection to the part of philosophers, which is probably based on the admittedly considerable amount of nonsense published by philosophers in this field, it seems to me that the sensible thing is to try to direct the present undertaking into worth while channels, rather than to ignore it or try to suppress it. It is precisely for this reason that I would like to see Weyl, or some one else of genuine ability and understanding, as president.

Of course, when you get right down to it, mathematicians also have been known to publish nonsense in the name of symbolic logic.

Anywhere [*sic*], there is undoubtedly a need for the advent of a younger generation of philosophers, especially philosophical logicians, with a really adequate understanding of mathematics. And a joint enterprise between mathematicians and philosophers like the present one would seem to be a good way of contributing towards such a result.

If Weyl were unavailable as president, could you suggest any one else who would be suitable? I would very much appreciate any advice or help you could give.

With best regards to Mrs. Veblen and yourself,

Sincerely yours,

Alonzo Church

Professor Oswald Veblen, Brooklin, Maine.

LETTERS TO PAUL BERNAYS (1935–1936)

Fine Hall,
Princeton, N. J.
January 23, 1935

Dear Dr. Bernays,

I have your letter of December twenty-fourth.

My remarks, in my second paper in the Annals, about the relation of Gödel's Theorem to the formal system which I was proposing, were ill considered, as you very justly remarked. The point I had in mind was that the argument of Gödel, as it appears in the Monatshefte, depends on the following special property of the system of Principia Mathematica: that if **P** and **Q** are any two formulas, subject to certain descriptive conditions dictated by the theory of types, and if the assumption of **P** leads to a proof of **Q**, then **P** $\supset$ **Q** is a provable formula; in particular, if **P** is any formula, subject to these descriptive conditions, then **P** $\supset$ **P** is a provable formula. Regardless of the question whether or not vacuous implications were to be allowed, it seems to me that no analogous property held of my system, and hence that the Gödel argument did not apply. Gödel has since shown me, however, that his argument can be modified in such a way as to make the use of this special property of the system of Principia unnecessary. In a series of lectures here at Princeton last spring he presented this generalized form of his argument, and was able to set down a very general set of conditions such that his theorem would hold of any system of logic which satisfied them. I understood at the time that this was soon to be published, but have heard nothing of it since.

In regard to my own work, as begun in the two papers in the Annals which you have seen, there are two important developments, and I think I can best answer your questions by describing them.

The first of these developments is a proof by J. B. Rosser and S. C. Kleene that the set of postulates which I proposed in the Annals, or indeed merely the subset of it I–V and 1–16, does in fact lead to contradiction in the sense that every well-formed formula not involving free variables is provable (in particular $1 = 2$ is provable). This proof depends on utilizing Gödel's idea of a representation of a system of formal logic within itself, although they do not use the positive integers for representing formulas, as Gödel does, but use instead certain entities they call metads, devised for this special

---

These previously unpublished letters to Bernays are of interest partly because of their connection to the development of Church's Thesis. The first of the letters to Bernays was typed; the other two were handwritten. The corresponding letters from Bernays to Church are not known still to exist. Printed by permission of the Alonzo Church estate. Discussion of the context in which all the letters in this group (the letters involving Bernays, Kleene, Post, Pepis, and Weyl) were written can be found in Martin Davis, *Why Gödel Didn't Have Church's Thesis*, ***Information and Control***, vol. 54 (1982), pp. 3–24; and in Wilfried Sieg, *Mechanical Procedures and Mathematical Experience* in ***Mathematics and Mind***, A. George ed. (Oxford, Oxford University Press, 1994); and *Step by Recursive Step: Church's Analysis of Effective Calculability*, ***The Bulletin of Symbolic Logic***, vol. 3 (1997), pp. 154–180. Short parts of the January 23, 1935 and July 15, 1935 letters to Bernays, and all of the letter to Pepis are quoted in the papers by Wilfried Sieg. Church's letter to Kleene of November 29, 1935 is quoted in the paper by Martin Davis.

purpose. A correspondence is set up between well-formed formulas and metads. And a formula **G** is found such that, if **a** is a metad which represents a formula **A**, then **G**(**a**) is convertible into a formula equivalent to **A** (one formula is said to be convertible into another if it can be changed into the other by a sequence of applications of my Rules I, II, III). Having done this, Rosser and Kleene are then able to obtain a formal equivalent of the intuitive argument which is set forth in my paper in the Monthly. This results in a paradox, essentially the Richard paradox, obtainable formally within the system.

The second development is a proof by Rosser and myself that my Rules I, II, III, taken alone are free from contradiction, in a certain sense which will appear in a moment. This result, together with certain results obtained in Kleene's thesis, also to be discussed in a moment, serve as a basis for the construction, and proof of freedom from contradiction, of a system in some respects resembling the one which I propose in the Annals.

Specifically, the result obtained by Rosser and myself is the following. Let an application of my Rule II be called a reduction, an application of Rule III an expansion, and any sequence of applications of Rules I, II, III a conversion. Let a formula be said to be in normal form if it has no well-formed part of the form $\{\lambda \mathbf{x} \, . \, \mathbf{M}\}(\mathbf{N})$; that is, if no reduction of the formula is possible, either immediately or after applications of Rule I. And let a formula **B** be called a normal form of a formula **A** if **B** is in normal form and **A** is convertible into **B**. Then if a formula **A** has a normal form it has only one normal form, and every sequence of reductions of **A** (or of reductions interspersed with applications of Rule I) terminates, after a finite number of reductions, in the normal form. That is, of course with the understanding that two normal forms are to be regarded as the same if they differ only by applications of Rule I.

Under the definition of the positive integers proposed in my second paper in the Annals, 1 will be represented by the formula $\lambda f \lambda x \, . \, f(x)$, 2 by $\lambda f \lambda x \, . \, f(f(x))$, 3 by $\lambda f \lambda x \, . \, f(f(f(x)))$, and so on. Suppose that a certain function of positive integers is defined intuitively. Then the formula F will be said to represent this intuitively defined function, if, whenever the intuitively defined function has the value m for a certain positive integer n, it will turn out that the formula $F(n)$ is convertible into the formula m. It follows from the result of Rosser and myself concerning uniqueness of normal form, that formulas which represent functions of positive integers, distinct in the sense of having distinct values for at least one positive integer, are not convertible into each other.

The most important results of Kleene's thesis concern the problem of finding a formula to represent a given intuitively defined function of positive integers (it is required that the formula shall contain no other symbols than λ, variables, and parentheses). The results of Kleene are so general and the possibilities of extending them apparently so unlimited that one is led to conjecture that a formula can be found to represent any particular constructively defined function of positive integers whatever. It is difficult to prove this conjecture, however, or even to state it accurately, because of the difficulty in saying precisely what is meant by "constructively defined". A vague description can be given by saying that a function is constructively defined if a method is given by which its value could be actually calculated for any particular positive integer whatever. Every recursive definition, of no matter how high an order, is constructive, and as far as I know, every constructive definition is recursive.

Kleene proves that a formula can be found to represent any function which is recursive in the somewhat limited sense of Gödel (Monatshefte vol. 38 p. 179). Gödel suggested to us a much more general definition of the notion "recursive" (due to Herbrand). After a certain modification in this definition of Herbrand, a modification which, it can be maintained, is necessary if recursiveness is to imply constructiveness, it can be proved that a formula can be found to represent any function which is recursive in this much more general sense (the proof is due to Rosser).

Of the formulas obtained by Kleene, two are of special interest:

> A formula **P**, the predecessor function, such that **P**(1) is convertible into 1, and, for any positive integer **n**, **P**(**n**+1) is convertible into **n**.
>
> A formula $\mathscr{G}$, the Kleene $\mathscr{G}$-function, such that if **n** is any positive integer and **F** is any function of positive integers all of whose values are either 1 or 2, then $\mathscr{G}$(**F**, **n**) is convertible into the least positive integer **m** greater than or equal to **n** such that **F**(**m**) has the value 2. If there is no positive integer **m** greater than or equal to **n** such that **F**(**m**) has the value 2, then $\mathscr{G}$(**F**, **n**) is a formula which has no normal form.

Suppose now that we construct a system of symbolic logic as follows. Undefined terms: δ, λ, [ ], { }, ( ), and an enumerably infinite set of variables. One formal postulate: $\lambda f \lambda x . f(f(x))$ (this is the formula which we abbreviate by 2). And five rules of procedure, of which the first three are identical with Rules I, II, III of my first paper in Annals. Let two formulas be said to be equivalent under Rule I if one of them can be changed into the other by a sequence of applications of Rule I (given two formulas, it can, of course, always be determined just by inspection of them whether or not they are equivalent under Rule I). Let a formula be said to be in δ-normal form if it contains no part of the form {λ**x**. **M**}(**N**), and also no part of the form δ(**M**, **N**) with **M** and **N** containing no free variables. If **M** and **N** are in δ-normal form and contain no free variables, the fourth rule of procedure allows the replacement, in any part of a proved formula, of δ(**M**, **N**) by either 2 or 1, according to whether **M** and **N** are or not equivalent under Rule I. The fifth rule of procedure is the inverse, exactly, of the fourth rule, allowing the replacement of 2 or 1 by δ(**M**, **N**) under appropriate circumstances.

The idea behind this, of course, is the identification of the truth-value *truth* with the positive integer 2 and the truth-value *falsehood* with the positive integer 1. This identification is artificial but apparently harmless.

Observe that under this identification a propositional function of positive integers reduces simply to a function of positive integers whose values are either 1 or 2, and hence that the Kleene $\mathscr{G}$-function can be used to obtain a formal expression for the least positive integer of which a given propositional function is true.

We identify the symbol δ with the intuitive notion of equality.

With the aid of δ and of certain functions defined by Kleene, we can obtain a formula $\sim$ such that $\sim$(1) is convertible into 2 and $\sim$(2) is convertible into 1, and $\sim$(**X**) is not convertible into 2 or 1 unless **X** is convertible into either 1 or 2. Also the formula & such that &(1, 1), &(1, 2), and &(2, 1) are all convertible into 1, and &(2, 2) is convertible into 2, and &(**X**, **Y**) is not convertible into 1 or 2 unless each of **X** and **Y** is convertible into 1 or 2. The formulas $\sim$ and & are identified with the intuitive notions of negation and logical product respectively.

Let us call any sequence of applications of the five rules of procedure just enumerated, a generalized conversion. The result of Rosser and myself concerning uniqueness of normal form under conversion is readily extended to give an analogous theorem concerning uniqueness of δ-normal form under generalized conversion. Now every provable formula has $\lambda f \lambda x \, . \, f(f(x))$ as its δ-normal form. Also every formula **A**, such that $\sim$(**A**) is provable, has $\lambda f \lambda x \, . \, f(x)$ as its δ-normal form. Hence there can be no formula **A** such that both **A** and $\sim$(**A**) are provable.

Let us use the word δ-convertible to refer to a generalized conversion.

By modifying slightly the definition of metad used by Kleene and Rosser, and by making use of the properties of δ, it is possible to obtain a representation of the logic within itself, so there is a one-to-one correspondence between the formulas of the logic and the metads, and so that the following are definable formally: $\mathfrak{G}$, such that if the metad **a** represents a formula **A**, $\mathfrak{G}$(**a**) is δ-convertible into **A**; met, such that, if the formula **A** has a δ-normal form, and contains no free variables, met(**A**) is δ-convertible into the metad which represents the δ-normal form of **A**; $\mathfrak{M}_1$ and $\mathfrak{M}_2$ such that, if the metad a represents the formula **F**(**X**), $\mathfrak{M}_1$(**a**) is the metad which represents the formula **F**, and $\mathfrak{M}_2$(**a**) is the metad which represents the formula **X**; and then, such that the infinite sequence, thm(1), thm(2), thm(3), ... contains every metad which represents a provable formula, and consists entirely of such metads.

The possibility of defining $\mathfrak{G}$, met, etc., formally, means essentially that any metamathematical statement about the system can be translated into a formula within the system which has this metamathematical statement as its intuitive meaning. For example, it is possible to give a metamathematical description of the form which a formula must have in order to be in δ-normal form and stand for a positive integer. Correspondingly we can define a formula N^* such that, if **a** is a metad, N^*(**a**) is δ-convertible into 2 or 1 according to whether **a** does or does not represent a formula which is in δ-normal form and stands for a positive integer.

A formula N, such that N(**X**) has the intuitive meaning, "**X** is a positive integer", is defined as follows:

$$N \rightarrow \lambda x N^*(\mathrm{met}(x)),$$

The existential quantifier Σ is defined as follows:

$$\varepsilon \rightarrow \lambda f \; \mathfrak{G}(\mathfrak{M}_2(\mathrm{thm}(\mathfrak{G}(\lambda n \, \delta(f, \mathfrak{G}(\mathfrak{M}_1(\mathrm{thm}(n)))), 1)))).$$

$$\Sigma \rightarrow \lambda f \, . \, f(\varepsilon(f)).$$

One is tempted to suppose that the possibility of defining ε means that the Zermelo principle is present in the system. But this fails, because it can happen that two propositional functions, **F** and **G**, which are equivalent in the sense that they define the same class, are nevertheless not δ-convertible into each other, and ε(**F**) and ε(**G**) are different.

It is now necessary to say what is meant by the consequences of assuming a given formula. In order to do this it is necessary to introduce certain operations of the same sort as rules of procedure and to which we shall refer as hypothetical rules of procedure. There is a long list of these. The following may be taken as typical: from **X** and **F**(**X**) to infer **F**(2); from &(**X**, **Y**) to infer **X**.

It can be proved metamathematically about each of the hypothetical rules of procedure that in all cases where the premise or premises are provable the conclusion is also provable. That is, the hypothetical rules of procedure need not be added to our list of postulates as actual rules of procedure, because if so added they would be merely superfluous.

By the consequences of assuming a formula **A** we mean the formulas which can be derived from **A** by means of the rules of procedure and the hypothetical rules of procedure together. In this connection the hypothetical rules of procedure are not superfluous.

Using the Kleene $\mathscr{G}$-function and the notion of metad, it is possible to obtain the formula π_1 such that $\pi_1(\mathbf{F}, \mathbf{G})$ is δ-convertible into 2 if $\mathbf{G}(\mathbf{x})$ is one of the consequences of assuming $\mathbf{F}(\mathbf{x})$, and in the contrary case $\pi_1(\mathbf{F}, \mathbf{G})$ has no δ-normal form. It can be proved metamathematically that whenever $\mathbf{F}(\mathbf{X})$ and $\pi_1(\mathbf{F}, \mathbf{G})$ are provable, $\mathbf{G}(\mathbf{X})$ is also provable.

The definition of π_1, however, introduces an additional hypothetical rule of procedure, namely: from $\mathbf{F}(\mathbf{X})$ and $\pi_1(\mathbf{F}, \mathbf{G})$ to infer $\mathbf{G}(\mathbf{X})$. The definition of π_2 is exactly the same as that of π_1 except account is taken of the additional hypothetical rule of procedure and a number of others like it. Then there arise still other hypothetical rules of procedures involving π_2, for instance: from $\mathbf{F}(\mathbf{X})$ and $\pi_2(\mathbf{F}, \mathbf{G})$ to infer $\mathbf{G}(\mathbf{X})$. Hence we define π_3 and so on.

Then it is possible to obtain a formula π such that, for any positive integer n, $\pi(n)$ is convertible into π_n. And we define

$$\pi_\omega \rightarrow \lambda f \lambda g \Sigma(\lambda n \,.\, N(n) \,.\, \pi(n, f, g)).^*$$

Thus arises an infinite list of formulas $\pi_1, \pi_2, \pi_3, \ldots, \pi_\omega, \pi_{\omega+1}, \cdots$, continuing, apparently, as far into the second number-class as we like.

With the aid of the Kleene $\mathscr{G}$-function we can now define an infinite set of formulas $\pi_1^1, \pi_1^2, \pi_1^3, \cdots, \pi_1^\omega, \pi_1^{\omega+1}, \ldots,$ such that π_1^r is convertible into 2 whenever $\pi_r(N, \mathbf{F})$ is convertible into 2, and $\pi_1^r(\mathbf{F})$ is convertible into 1 whenever $\Sigma(\lambda\, \mathbf{x}\, \&(N(\mathbf{x}), \sim(\mathbf{F}(\mathbf{x}))))$ is convertible into 2. Also an infinite set of formulas $\Sigma_1^1, \Sigma_1^2, \Sigma_1^3, \cdots \Sigma_1^\omega, \Sigma_1^{\omega+1}, \cdots$ such that $\Sigma_1^n(\mathbf{F})$ is convertible into 2 whenever $\Sigma(\lambda\, \mathbf{x}\, \&(N(\mathbf{x}), \mathbf{F}(\mathbf{x})))$ is convertible into 2, and $\Sigma_1^n(\mathbf{F})$ is convertible into 1 whenever $\pi_r(N, \lambda\, \mathbf{x} \sim(\mathbf{F}(\mathbf{x})))$ is convertible into 2.

In order to make an intuitive interpretation of our formal system we must select a particular ordinal number s of the second kind, and identify π_1^s with the intuitive notion of the universal quantifier over the class of positive integers and Σ_1^s with the intuitive notion of the existential quantifier over the class of positive integers.

The amount of number theory to which the system is adequate will depend upon the choice of the ordinal number s, and will tend to increase with s. I believe that, with an appropriate and sufficiently large choice of s, the system will turn out to be adequate to substantially all of extant elementary number theory.

In order to obtain analysis from the system, however, a much more elaborate construction appears to be necessary, and I cannot now say with certainty that it is possible at all.

---

* *Editor's note*: In Church's original letter, the last occurrences of "f" and "g" here are occurrences of "**F**" and "**G**". This appears to be a mistake, since the quantifiers would then bind no variable. See the corresponding definition in *A Proof of Freedom from Contradiction*, 1935, reprinted in this volume.

I am afraid the foregoing explanation of my present program is unduly lengthy and detailed, but I have found it hard to answer your question satisfactorily otherwise.

With best regards,

Very sincerely yours,

Alonzo Church

P.S. Since writing the above I have learned that Dr. Gödel has been ill, and has been obliged on account of illness to cancel his plan to spend the spring term in Princeton.

Princeton, N. J.,
July 15, 1935

Dear Dr. Bernays,

I have delayed answering your letter of May twenty-seventh, in order to complete, and send you copies of, several manuscripts which were in preparation. There have been a number of developments since I last wrote you, and I felt that the most satisfactory way of explaining them to you was to send you copies of the manuscripts rather than to attempt a synopsis in a letter.

I am enclosing herewith a manuscript copy of a joint paper by Rosser and myself, also an abstract (as prepared for publication in the Bulletin of the American Mathematical Society) of a joint paper by Kleene and myself. The manuscript of a paper by myself, entitled, "An unsolvable problem of elementary number theory," is in process of being typewritten, and I will mail you a copy within a week or two. All these papers will eventually be published, but it may be a year or more before they appear.

There are also forthcoming two papers by Kleene, entitled, "General recursive functions of natural numbers," and, "λ-definability and recursiveness." In the first he has taken a number of ideas which first arose in connection with the notions of conversion and λ-definability and has formulated them independently of these notions so as to obtain results which involve instead one definition or another of recursiveness. In the second he establishes the equivalence, for functions of positive integers, of λ-definability (that is, formal definability by means of the λ-operator) and recursiveness in the Herbrand-Gödel sense.

An abstract of my paper, "An unsolvable problem of elementary number theory," appeared in the Bulletin of the American Mathematical Society, May 1935, p. 332, and abstracts of the two papers by Kleene will appear in the Bulletin of the American Mathematical Society, July 1935.

I was, of course, aware of the connection between the Hilbert ϵ and the formula which I called ϵ in my letter to you (ι in my recent note in the Proceedings of the National Academy of Sciences), but I had not known of Hilbert's remark concerning the Zermelo principle, to which you call my attention. It is worth while, perhaps, to remark that my ϵ is not entirely the same as Hilbert's, however, the latter being definitely stronger, in a non-constructive direction, and necessarily so, if I understand the matter rightly, in view of the fact that the Hilbert system contains nothing corresponding to absence of normal form under conversion. Thus, if φ has the property that there is no x for which $\varphi(x)$ is true, then $\varphi(\epsilon(\varphi))$ is false under the Hilbert formulation, but meaningless rather than false under mine.

To the question whether the proof of freedom from contradiction of my new system can be used to obtain a proof of freedom from contradiction for the system which you call classical arithmetic, I am afraid the answer is no. For my new system involves the same sort of a denial of the principle of excluded middle as did the former system, proposed in my first paper in the Annals of Mathematics. For example, the proposition that Fermat's Last Theorem is either true or false is not provable formally unless under circumstances which would lead to a proof that Fermat's Last Theorem was true or to a proof that it was false. The situation is complicated in the new system, as compared with the former one, by the fact that there is not one quantifier but an infinite set of quantifiers of different orders. But I can see no reason why it *might* not be true, no matter how high the order of the quantifier selected to be identified with the intuitive quantifier, that the formula expressing Fermat's Last Theorem should still have no normal form.

The denial of the law of excluded middle which is involved here is of a different sort from that which is associated with Intuitionism and has been embodied in the formalism of Heyting. It is not susceptible, so far as I am able to see at the present time, to anything like the Gödel-Gentzen treatment of the Heyting system.

My idea in speaking of the probable adequacy of my system for elementary number theory was not that the system in any way preserved or was equivalent to any established formal system (such as that of "classical arithmetic"), but rather that it contained provable formulas to correspond to "substantially all" of a certain fairly well defined body of (intuitive) mathematical theorems designated as elementary number theory. This point is not made sufficiently clear in my note in the Proceedings of the National Academy of Sciences (which was out before I received your letter), but it can be made so when I publish a longer paper giving details, as I plan to do.

Your suggestion of modifying the notion of conversion, by omitting the requirement on **M** that **x** occur as a free variable in **M**, had occurred to me at the time that I first proposed the rules I, II, III, and was rejected by me because it raised certain difficulties, which I did not then have as clearly formulated as I do now. The same modification of the notion of conversion was also at one time considered by Rosser as possibly useful in certain connections, and it has in addition recently been urged on me by Curry.

The objections to the modified notion of conversion are: (1) that if E is defined by $E \rightarrow \lambda x \Sigma f \,.\, f(x)$ then $E(\mathbf{A})$ is provable no matter what formula **A** is, whereas it seems to me desirable that $E(\mathbf{A})$ should not be provable unless, at least **A** has normal form; (2) that, of the two principal theorems proved in my joint paper with Rosser, the theorem that the normal form of a formula, if it exists, is unique continues to hold, but the theorem that, if the normal form exists, no sequence of reductions can be infinite fails. It is frequently necessary or convenient to make use of this second theorem (often in the form of the inference that a certain formula must have a normal form because it appears as a part of a formula known to have a normal form). In some cases (for example in the proof of the principal property of $\mathfrak{g}$) methods are known of avoiding the use of this second theorem, often at the cost of some additional complication in the argument. And in other cases where such methods are not known they may nevertheless exist. It seems to me probable that the modified notion of conversion, if its use be feasible at all, would in this way introduce more complications than it avoided (but this is, of course, a mere conjecture, in which I may be mistaken).

On the other hand there are many ways in which the modified notion of conversion is attractive.

Of these, the definition of 0 as $\lambda fx \,.\, x$ (or as $\lambda f \,.\, I$), and, in principle at least, your definition of the ordered pair $\{A, B\}$, were already known to me.

Your definitions of $\mathfrak{N}$ and $\mathfrak{Z}$ definitely add to this attractiveness of the modified notion of conversion.

And another point of attractiveness is the ease with which natural definitions of the transfinite ordinals are obtained by analogy with the definitions of the natural numbers. Thus, ignoring non-constructive ordinals, the second number class can be handled by means of the definitions,

$$0 \rightarrow \lambda lfx \,.\, x,$$
$$S \rightarrow \lambda nlfx \,.\, f(n(l, f, x)),$$
$$L \rightarrow \lambda rlfx \,.\, l(\lambda n \,.\, r(n, l, f, x)),$$

where 0 stands for the ordinal zero, if a stands for any ordinal $S(a)$ stands for the next following ordinal; and if $r(0), r(1), r(2), \cdots$ is an increasing infinite sequence of ordinals, where $0, 1, 2, \cdots$ are the finite ordinals, then $L(r)$ stands for the limit of the sequence $r(0), r(1), r(2), \cdots$.

Under this scheme, if a stands for any ordinal, $a(l, f)$ stands for the ath power of the function f relative to the limiting process l (just as, under the scheme for the natural numbers, if a stands for any natural number, $a(f)$ stands for the a^{th}power of the function f).

The scheme of definition of ordinals proposed in my joint paper with Kleene is a more artificial one.

Thank you for your very suggestive and helpful letter.

Sincerely yours,

Alonzo Church

Fine Hall, Princeton, N. J.,
August 6, 1936

Dear Dr. Bernays,

Thank you for your letter of July twenty-third enclosing reviews. I have made grammatical corrections, as requested. I have also with a great deal of hesitation, made a minor change in content, altering your sentence, "This conversion formalism was shown by J. B. Rosser to be equivalent to the combinatory calculus of H. B. Curry," by inserting the word, "essentially," before "equivalent". The point is that, quite apart from difficulties over the constancy function K, the process of conversion provides no equivalent of Curry's axiom $BI = I$, and it is necessary to delete this axiom in order to obtain the equivalence. This does not, however, affect the correctness of your subsequent remarks in this connection.

In the interest of getting these reviews out promptly, I will not send you proof of them for correction, but will read the proof myself. This will, in fact, be our usual practice, even in the case of reviewers resident in this country. (Authors of articles, as distinguished from reviews, receive one proof).

If you wish, however, I will send you a copy of the proof of these reviews, not for correction, but to keep for your own records.

We are assuming that authors of reviews will ordinarily not want reprints. Our offer of fifty gratis reprints applies only to articles. If, however, on occasion, a reviewer should want reprints (of one review or of several reviews together in one reprint), they could, of course, be had at cost (fifty reprints of not over four printed pages would cost $1.70 plus postage). In the present instance, if you should want reprints, it will be necessary to write at once, in order to reach me before type is destroyed.

In the criticism of my papers which you make in your letter you are, I think, right. I have not, however, published corrections in the same number of the Journal as your reviews, because I want first to have time to think the matter over a little more carefully. The necessary corrections will follow in one of the next two numbers of the Journal, with credit, of course, to you for pointing them out to me. Meanwhile, I will add at the head of your review "see correction thereto, forthcoming in this JOURNAL".

As regards your criticism of my misuse of free variables in "A note on the Entscheidungsproblem", there is, so far as I can see at the moment, nothing better to do than to accept your proposed correction in full.

It is true, as you point out, that, in order to render my proof of the unsolvability of the Entscheidungsproblem completely constructive, it is necessary to replace the condition of ω-consistency by what I may call strong ω-consistency, i. e. the condition that if $(\exists x)P$ is provable then there is some positive integer n such that $\overline{P}_n$ is not provable, where P_n means the result of substituting for x throughout P the symbol for n. It is strong ω-consistency that is directly used in the proof. And the inference from ω-consistency to strong ω-consistency cannot be made constructively (at least if one understands $\sim(x)\cdots$ in the intuitionistic sense).

I realized when I wrote that Gödel's completeness theorem afforded only a non-constructive proof of the equivalence of the two Entscheidungsprobleme, but failed to point this out as I should have. Therefore, as regards the problem of Erfüllbarkeit (in the set-theoretic sense) all that is proved is that a recursive solution of it would lead to a contradiction in "classical" (non-constructive) mathematics.

Finally, there are two more papers which I would like to get you to review, if possible. One of these is a paper by W. V. Quine, entitled *Set-theoretic foundations for logic*, to appear in the Journal of Symbolic Logic, vol. 1, pp. 45–57; I am sending you a copy of the corrected proof of this, in order to enable you to review it. The other is a paper by Józef Pepis, *Beiträge zur Reduktionstheorie des logischen Entscheidungsproblems*, Acta Scientiarum mathematicarum (Szeged), vol. 8 (1936), pp. 7–41. Unfortunately I have no copy of the latter paper to send you, but am hoping that either Pepis will have sent you a reprint or you will have access to the journal (Acta Scient. Math.) at Zürich. If you are unable to get hold of a copy of the paper in order to review it, of course let me know (a postal card would be sufficient), and I will find another reviewer for it.

Sincerely yours,

Alonzo Church

P. S. Rosser claims to have duplicated my result about the unsolvability of the Entscheidungsproblem (in the sense of deducibility), as well as Gödel's result about the existence of formally undecidable propositions, without employing any hypothesis of ω-consistency, but only simple consistency. I have not seen his proof, and do not know whether it is constructive.

LETTERS TO STEPHEN C. KLEENE (1935)

November 15, 1935

Dear Kleene:

I have just heard from the Journal that my paper "An unsolvable problem" is accepted for publication. The referee of the papers objected, however, that my footnotes assigning credit to others than myself on certain points were not always clear as to just who was to be credited with what, and in particular that the introduction seemed to claim for myself credit which in the body of the paper I assigned to others. And I have therefore added, at the end of the introduction, another footnote, a copy of which I enclose. My claim on the definition of λ-definability is, however, really somewhat stronger than represented in that footnote, because the observation, in connection with your thesis, that the notion has an interest independent of any system of symbolic logic, first came from me; that is why I originally included in the footnote the words, "and in the present paper", but subsequently crossed them out for fear they would be misunderstood.

In connection with your further ideas on recursive functions, it occurs to me to suggest that immediate pressure might be relieved by publishing an outline paper in the Bulletin setting forth merely the ideas without any attempt at working out formal details. The possibility of this will, of course depend on the nature of the ideas themselves (you never described them to me except vaguely).

Both Gödel and Bernays are very much interested in your result that all general recursive functions can be obtained from primitive recursive functions by means of the epsilon operator. It is, of course, natural that they should be quicker to see the importance of a result like this than of one concerning lambda-definition or the like.

Is there any possibility that a somewhat stronger result than this one could be proved in which the class of primitive recursive functions was replaced by a smaller class? In particular, could the class of primitive recursive functions be replaced by the class of functions definable by such recursions as $f(0) = a$, $f(n+1) = g(f(n))$, or even by the class of functions which correspond to polynomials with non-negative coefficients?

In connection with our joint paper on the ordinals, Gödel makes the following suggestion. Let an ordinal a be called general recursive if there exists a well-ordered sequence s, of ordinal number a, which contains every natural number once and only once, such that the function $f(m, n)$ equal to 0 if m precedes n in s and equal to 1 otherwise is general recursive. Although I have not worked it out in detail, I believe that our function *enm* leads to a proof that every lambda-definable ordinal of the second number-class, less than [blank], is general recursive. But no proof of the converse suggests itself, and my present conjecture is that the converse is not true.

These previously unpublished letters are of interest partly because of their connection to the development of Church's Thesis. Printed by permission of the Alonzo Church estate. For Kleene and Rosser's perspective on the relevant issues, the reader might consult the following: Stephen C. Kleene, *Origins of Recursive Function Theory*, ***Annals of the History of Computing***, vol. 3 (1981), pp. 52–67; J. Barkley Rosser, *Highlights of the History of the Lambda-calculus*, ***Annals of the History of Computing*** vol. 6 (1984), pp. 337–349.

You may remember my suggesting that, in order to do without the requirement that **M** in λ **x M** actually contain **x** as a free variable, one might, in Rules II and III of conversion, require that if **M** did not contain **x** as a free variable **N** must be in normal form. Bernays now makes what seems to me the neater suggestion that, in Rules II and III, it should be required simply that **N** must be in normal form. This is a point worth looking into, because it may very probably resolve most of the difficulties which arise form the absence of 0 in the lambda notation.

I am sending a copy of this letter to Rosser because of the possibility that he will be interested in some of these suggestions.

Sincerely yours,

Alonzo Church

Dr. S. C. Kleene,
University of Wisconsin,
Madison, Wisconsin.

Princeton University
Princeton New Jersey

November 29, 1935.

Dear Kleene:

I have your letter of November eighteenth.

The notion of lambda-definability in its present form is, of course, the result of a gradual development. The first step was my proposal, within the system of formal logic in which I was working, of definitions of the positive integers and of the function S, and remark as to how, in terms of these, a definition by recursion could (at least in certain cases) be translated into a nominal definition, capable of being formalized in the system. The next steps, taken by you, were the restriction to consideration of a part only of my formal system (so that conversion became the only available way of proving equality), the development in the direction of finding general proofs of definability for various classes of functions, the discovery of such powerful instruments of definition as the perpetual motion function, and the remark that the question of the truth of certain propositions of elementary number theory is reducible to the question whether certain particular formulas have a normal form. After these, however, two steps which I think to be of some importance were taken by myself. The first of these was the proposal of the problem of a complete set of invariants of conversion and the remark that the solution of it would imply a solution of the Entscheidungsproblem of Principia Mathematica; this was made not long after our original discussion of the perpetual motion function and was, of course, immediately suggested by that discussion. The second was the proposal to abstract from any formal system of logic and to consider the class of lambda-definable functions of positive integers as a problem of number theory; this was first made to a number of persons in the fall of 1933 in connection with Rosser's discovery of the contradiction in my system, and afterwards to you in connection with the consequent revision of your thesis.

At any rate, we seem to be agreed that the statement that the notion of lambda-definability is jointly due to you and me is fair, and I am content to let it go at that.

In regard to Gödel and the notions of recursiveness and effective calculability, the history is the following. In discussion with him [of] the notion of lambda-definability, it developed that there was no good definition of effective calculability. My proposal that lambda-definability be taken as a definition of it he regarded as thoroughly unsatisfactory. I replied that if he would propose any definition of effective calculability which seemed even partially satisfactory I would undertake to prove that it was included in lambda-definability. His only idea at the time was that it might be possible, in terms of effective calculability as an undefined notion, to state a set of axioms which would embody the generally accepted properties of this notion, and to do something on that basis. Evidently it occurred to him later that Herbrand's definition of recursiveness, which had no regard to effective calculability, could be modified in the direction of effective calculability, and he made this proposal in his lectures. At that time he did specifically raise the question of the connection between recursiveness in this new sense and effective calculability, but said that he did not think that the two ideas could be satisfactorily identified "except heuristically".

In regard to your functions H, Eval_p, and Val, it seems to me that it might be possible to obtain for them definitions which were not really recursive at all but actual nominal definitions in terms of addition, multiplication, and the epsilon operator. Such definitions would be, in a sense, less simple than primitive recursions, since they would involve the epsilon operator. But they would lead to the interesting theorem (if it be a theorem) that for any general recursive function it is possible to give a nominal definition in terms of addition, multiplication, and the epsilon operator.

Gödel is going back to Europe this week. He has been suffering severely with indigestion ever since his arrival, and concluded finally to resign his appointment with the Institute and go home.

I do not expect to be at the St. Louis meeting. Possibly Bernays will attend it, I do not know.

Sincerely yours,

Alonzo Church

Dr. S. C. Kleene,
University of Wisconsin,
Madison, Wisconsin.

December 18, 1935

Dear Kleene:

I have your letter of December fourteenth.

I cannot remember our first conversation on the perpetual motion function in the way you describe it, that is, my recollection is that your original proposition was simply that this was a method by which many problems (all of a certain type not very clearly described) could be reduced to problems of the form whether or not a certain particular formula had a normal form. Nevertheless I certainly wish to avoid any possibility of injustice resulting from a faulty memory on my part, and will therefore insert a footnote in my paper crediting to you the proposal of the problem of invariants of conversion.

I ask in return, however, that you should insert at an appropriate place in your paper on lambda-definability and recursiveness a footnote to the effect that the notion of the normal form of a formula under conversion was introduced by me in lectures at Princeton in the fall of 1931. This seems to me a matter of some importance. The idea was introduced in your paper on proof by cases as if it were your own, and the point escaped me at the time. That the first proposal of this idea was in fact mine is supported by my own positive recollection, and by the appearance of a very clear exposition of the idea in your notes on the lectures in question (which are on file in the library here). Any possibility that the notion of a normal form did not occur in the lectures but was an addition by yourself to the notes is precluded, not only by my recollection to the contrary, but also by the fact that the notion enters as an essential part of a lengthy discussion which could hardly have been made at all without it.

As to the remark connecting the Entscheidungsproblem with the perpetual motion function, it could hardly have been made before the fall of 1933, because none of us until then fully appreciated the Gödel representation, or realized the fact that in any formal system the set of provable formulas is *effectively* enumerable. I surely did not, indeed could not, have made it in connection with our early conversations on the perpetual motion of function, and so it seems to me now that it must have come at a much later date than the proposal of the problem of a complete set of invariants of conversion. I have a clear recollection of being the first to make the point, and I think it must have been in the fall of 1933 at about the time that the materials for my paper later published in the Monthly were taking shape. It is not, however, a matter of sufficient importance to require a reference.

12/19/35

As to your new theorem on general recursive functions, it seems to me that the form in which it is likely to have the most popular appeal is that every recursive function can be defined in terms of addition, multiplication, the Kronecker delta (or your gamma), and the epsilon operator. A paper in the Bulletin such as you suggest would certainly be very much in order. I can see no occasion for its containing a reference to me, but think that there should be one to Gödel's theorem stated at the bottom of page 27 of the mimeographed notes of his 1934 lectures (of which theorem, by the way, I have never seen a proof). Perhaps before publication it would be wise to consider the more obvious aspects of the question of a connection between this theorem of Gödel's and yours. Is it possible that either theorem might be obtained as a corollary of the other?

I have no intention at the present time of publishing in the near future anything about Bernays' suggested modification of the notion of conversion. The remarks in your letter make me cautious about doing so without a thorough investigation of its consequences. But, of course, if you should ever need this modified form of conversion in connection with a paper which you wish to publish, there is no reason why you should not use it, with a reference to Bernays and myself.

I have a vague idea that the question of the connection between Gödel's definition of recursive ordinal and lambda-definability may lead to something of importance. Unfortunately, like yourself, I have been too busy with routine matters to think about it much.

I have not heard further from the Fundamenta in regard to our joint paper.

With best regards,

Sincerely yours,

Alonzo Church

Dr. S. C. Kleene,
Department of Mathematics,
University of Wisconsin,
Madison, Wisconsin.

CORRESPONDENCE WITH OR CONCERNING EMIL POST (1936–1943)

601 W. 148th St.
New York, N. Y.

May 17, 1936

My dear Prof. Church,

I read your note in the Journal of Symbolic Logic with great interest. Could you favor me with a reprint of the American Journal paper you refer to? I am particularly interested in seeing whether as in the case of Gödel's theorem the problem is really a function of the logic it occurs in and unsolvable in that logic, or, as the title of your paper would seem to indicate, it is a fixed problem unsolvable in any logic in which it could find formulation.

My interest in this matter is more than general. As far back as the Fall of 1921 I was led to formulate a problem which, though specific, I had reason to believe unsolvable by any finite processes. But the attempt to achieve complete generality in the proof led me far astray in more ways than one with the result that I really have nothing that would admit of publication in its present form. With Gödel Church et. alia in the field the need of publication may soon disappear.

Sincerely yours,

Emil L. Post

---

601 W. 148th St.
New York, N. Y.

May 24, 1936

Dear Prof. Church,

On the whole your paper leaves me very much in the condition my last letter prognosticated for the future. With the possible exception of the need of loosening up the definition of recursive function I believe my development to be equivalent to yours. I

Previously unpublished, these letters bear on Church's Thesis. Post's letters were handwritten; Church's were typed. The letter from Church to Pepis of June 8, 1937, is included because of its close relation to the letter from Church to Post written on the same day. The exchange between Church and Post contains further letters and postcards, most of which are dated later than the last letters printed here. These further letters were judged to be of lesser philosophical and mathematical interest. The last three letters in this printed series, between Weyl and Church about Post, are included—out of chronological order—for their poignancy and human interest. Some misspellings in Post's letters have been corrected. Reprinted by permission of the Church, Post, and Weyl estates.

Editor's note: All page references are to the handwritten originals.

suppose there would still be left me the proof of this equivalence; and if the criticisms I make below of the obviousness of the identification of effective calculability with recursiveness have any point my attempt to develop a general theory of finite processes may not be without value in the present development. However in any case I believe this development to have a brilliant future. For if symbolic logic has failed to give wings to mathematicians this study of symbolic logic opens up a new field concerned with the fundamental limitations of mathematics, more precisely the mathematics of Homo Sapiens.

As for loosening up the definition of recursive equations we require the last condition merely to hold say for $\imath = 1$, then the argument for the effective calculability of the values of the functions still holds for the function F^1. Now clearly in this case the other functions need not be even potentially recursive. The primary question however is can a set of recursive equations in the restrictive sense of your paper always be given for such a function F^1? If not the fact that the values of F^1 are effectively calculable would necessitate the modification of the definition of a set of recursive equations for F^1 as suggested. With this modification I feel fairly sure of the equivalence of your development and mine and so I would have no further quarrel with the truth of your criterion of effective calculability.

But as for its obviousness. Consider the following. It is pretty obvious that by means of a Gödel representation for elementary equations one could define "a recursive defined set E" of elementary equations, perhaps infinite in number and with even an infinite number of functional symbols $f_\imath$. [For the functions, not these $f_\imath$'s, entering into the definition of E, I think that last condition could be omitted for all the functions, making, I believe, E merely correspond to a recursively enumerable set {corresponding, I believe, to what I would call a generated set}]. If we now assume this infinite set E of elementary equations with their derived equations to satisfy your further conditions for recursiveness, with again the last condition merely required for $\imath = 1$, the function F^1 would again admit its values to be effectively calculated. Now from my experience with my own development and its equivalence to yours, at least with the above weaker form of recursiveness, I believe that such a F^1 could be recursively defined, again with the modified definition of recursiveness, i.e. by means of a finite basis E. But is this obviously so? And if not does this not mean that the correspondence between recursiveness and effective calculability must be reverified every time some formal procedure not obviously in recursive form does obviously lead to effective calculability?

In other words it seems to me that our intuitive concept of effective calculability is so strong that in any precisely formulated class of cases, as indeed in your instance with recursiveness, it suffices to convince us of its existence. Hence the equivalence between recursiveness and effective calculability seems to me rather in the nature of a natural law which must be constantly reverified. Unless indeed it be possible to give so clear and exhaustive a description of the possible modes of mathematical thinking that the equivalence would obviously be universal—again for Homo Sapiens.

Sincerely yours,

Emil L. Post

P. S. I have not communicated above the fine impression your paper did make on me.

That barrage of successive definitions is most imposing and has already in your paper yielded a part of that brilliant future I wrote of. It would really be unfortunate if the above remarks would have the effect on your development that the classic question put to the centipede had on its further progress. As for the need of loosening up the definition of a recursive set of equations for a function were I forced to hazard a guess I would say that it was necessary. Favoring this guess is that when the last condition is removed from all the functions we get something I believe equivalent to recursive enumeration (generation is my language) which Theorems XV and XVIII show is not equivalent to recursive definition (finite test for generability in my view). On the other hand, the proved equivalence of λ-definability and recursiveness and the fact that the former does not seem to admit of loosening up in obvious fashion prevents me from feeling surer of my guess.

E. L. P.

May 26, 1936.

Dear Dr. Post:

It seems to me that the most essential point raised in your letter, just received, is the question whether a more general notion of recursiveness could be obtained by requiring the last condition (in the definition of a set of recursion equations) to hold only for $u = 1$, or by allowing an infinite (but recursively enumerated) set of recursion equations, or both. To this I think the answer is no. For given a set of recursion equations of this more general sort, the set of derived equations is enumerable, and (in terms of Gödel representations) this enumeration is recursive in the narrower sense. Hence an equivalent set of recursion equations in the narrower sense can be obtained by using the construction of Theorem IV (or the much simpler construction appearing in Kleene's proof of Theorem IV) together with certain simple functions of elementary equations (as arguments) which are readily shown to be recursive (in terms of Gödel representations).

Moreover, and this is a point on which I would lay some emphasis, the formal argument just described to prove that a function recursive in the proposed more general sense is recursive merely parallels the obvious intuitive argument by which we convince ourselves that such a function is effectively calculable. My belief is that this will always be the case.

It is not very clear to me just what is the nature of your own work of which you speak or how far it overlaps mine. If, however, you have (as I think you said in your postal) a problem different from mine suspected of being unsolvable, it would surely be worth while to try to obtain a definite proof of this. The pressing problem in this connection is, as I see it, to determine, in some sense, where the line is between solvable and unsolvable problems of the kind in question. Until more is known, each problem added to the list of those known to be unsolvable is a contribution in the desired direction.

Sincerely yours,

Alonzo Church

Dr. E. L. Post,
601 W. 148th Street,
New York City.

---

601 W. 148th St.
New York, N. Y.

May 30, 1936

Dear Prof. Church,

I think I see your argument in connection with the reducibility of my proposed extension of the concept of a set of recursion equations for a function to a set of recursion equations as such. That no difficulty would arise from having an infinite number of initial equations (recursively defined) I had no doubt; but that the reduction could also be made for that last condition required only for $\imath = 1$ I did not anticipate, and correspondingly found your outline of how it could be done most interesting. It adds to my insight into the power of the concept of recursion equations and leads me to waive the question of the obviousness of its generality till I have had more experience with it.

I am construing a certain part of your letter to mean that you have at least a latent desire to know what my work was like. Since certain other work I must get out first will postpone at least for another year or two my return to this work (a return looked forward to since I left it in the summer of 1924) I will presume on your good nature to outline the development that led me to my own unsolvability conclusion.

Prior to the Fall of 1921 my interest in Symbolic Logic was really in the solution of what I now find termed Entscheidungsproblems. As you know the truth table development of my dissertation solves this problem for the $(\sim, \vee)$ subsystems of Principia Mathematica. My further work took two directions, first extension of the truth table method to more and more complicated subsystems of Principia, second an attack on the entscheidungsproblem for the general class of systems I introduce in the third part of my dissertation (Generalization by Postulation). It is the latter development that concerns us here.

For the special case where the primitive functions are all of one variable I effected this solution in June 1920. Though really simple when all the "productions" in III involve but one "premise" the more general case gave me quite a tussle. That's what the presented but to date unpublished paper entitled "On a simple class of deductive systems" is about.

In attempting to extend this solution to primitive functions of more than one variable I tried to isolate the difficulties and hit upon a problem I had previously come upon in connection with the solution of identities between enunciations in Principia for variable functions as unknowns (a problem offering no difficulty in the $(\sim, \vee)$ systems with variable propositions as unknowns). I spent a major portion of my year as Proctor Fellow [1920–21] trying to solve this problem but only succeeded in certain very special cases. This problem itself in its entscheidungsproblem form is a special case of my unsolvable problem (which I hope to get to at least before the end of this letter if not of your patience) and should it too prove unsolvable I will be supplied with the perfect alibi for a year of frustration.

During that year I went through all the details of reducing that part of Principia Mathematica whose basis is in *9 (I think also *10) (my notes say *10–11) which introduces functional symbols and apparent variables the latter however restricted to the range of "individuals", to a system of the type of that third part of my dissertation. The only difficulty was in explicitly formulating what is implicit in Principia Mathematica—its "sub mathematical" part. Once this was supplied the reduction in question was so simple that I was convinced that the whole of Principia Mathematica could be likewise reduced to that form.

When therefore in the late summer of 1921 I saw my way to reducing the systems of that third part of my dissertation themselves to simpler types formally the whole process assumed added significance. These reductions really were of two kinds. First the infinity of variables were disposed of. As a result of this first reduction the system took on the following form.

There were a finite number of symbols $a_1, a_2, \cdots a_m$. The "enunciations" of the new system then consisted of all finite sequences $a_{\imath_1}, a_{\imath_2}, \cdots a_{\imath_n}$ of these symbols. With the g's and h's referring to fixed sequences of these symbols, the P's to variable sequences the productions of the bases of the system were now in the following form

$$\vdash g^{\nu}_{11} P^{\nu}_{\imath_{11}} g^{\nu}_{12} P^{\nu}_{\imath_{12}} \cdots g^{\nu}_{1k\mu_1} P^{\nu}_{\imath_{1\mu_1}} g^{\nu}_{1(\mu_1+1)}$$

- -

- -

$$\vdash g^{\nu}_{k_\nu 1} P^{\nu}_{\imath_{1k_\nu}} g^{\nu}_{k_\nu 2} P^{\nu}_{\imath_{2k_\nu}} \cdots g^{\nu}_{k_\nu \mu_{k_\nu}} P^{\nu}_{\imath_{k_\nu \mu_{k_\nu}}} g^{\nu}_{k_\nu(\mu_{k_\nu+1})} \qquad \nu = 1, 2, \cdots k$$

produce

$$\vdash g^{\nu}_1 P^{\nu}_{\imath_1} g^{\nu}_2 P^{\nu}_{\imath_2} \cdots g^{\nu}_{\mu_\nu} P^{\nu}_{\imath_{\mu_\nu}} g^{\nu}_{\mu_\nu+1}$$

which applied to a finite number of primitive "assertions"

$$\vdash h_1, \vdash h_2, \cdots \vdash h_\lambda$$

generated the system of assertions. An unavoidable feature of the reduction is that the enunciations and assertions of the original system only *correspond* to a subset of the enunciations and assertions of the next systems. However the correspondence is such that one can always "effectively" pass from an enunciation in the original system to its correspondent in the new so that the solution of the entscheidungsproblem for the new system would entail that for the old. (It is also true that given an enunciation in the new system one can effectively determine whether it corresponds to one in the old and if so which one).

There then ensue a series of reductions (which can be carried somewhat further in different directions) with the final system in the following form. The enunciations are again the finite sequences formed from a finite set of symbols $a_1, a_2, \cdots a_m$ (repetitions of course admitted). The productions are now a finite number of the following at least superficially simple form

$$\begin{array}{l} \vdash g_\nu P \\ \quad \text{produces} \qquad \nu = 1, 2, \cdots k \\ \vdash P g'_\nu \end{array}$$

and the primitive assertions again a finite set of asserted sequences $\vdash h'_1, \vdash h'_2, \cdots \vdash h'_{\lambda'}$. (It seems to me now that λ' should be capable of being taken one. Certainly the solution of the entscheidungsproblem for systems with $\lambda' = 1$ entails the solution of the same for arbitrary λ' as the solutions of the latter are the combined assertions of λ' systems each with but one primitive assertion).

An important feature of these reductions is that the m letters of the original system are a subset of the n letters of the final system and the assertions of the original systems are those and only those assertions of the final system which involve no other letters than $a_1, a_2, \cdots a_m$. This fact led to the following line of thought. With $a_1, a_2 \cdots a_l$ a subset of the letters in the systems of the top of p. 3 of this letter† the assertions in such a system involving only the letter $a_1, a_2, \cdots a_l$ are also the assertions in the system of the top of this page†† involving only these letters. Or we might say any set of sequences of letters $a_1, a_2, \cdots a_l$ that can be generated by systems of the former type can also be generated by systems of the reduced type. It then seemed pretty clear that if we modified that development of the third part of my dissertation to allow for such sequences as enunciations and assertions then the reduction to system of the next type could be made without touching such enunciations so that every set of sequences on letters $a_1, a_2, \cdots a_l$ that could thus be generated would be of the "normal form" given by the simple systems of the top of this page. Coupled with the earlier reduction of a portion of Principia Mathematica to the forms of my dissertation this led to the conviction that were the Principia Mathematica used as a logic for the generation of such sets of sequences any such mode of generation could also be reduced to normal form, and indeed of any logic of the type Principia Mathematica. Finally it seemed pretty clear that if the operations themselves of such a logic were such that any individual process were thus in some sense reducible to normal form such a logic also could only generate sets of sequences on $a_1, a_2, \cdots a_l$ whose mode of generation could be reduced to normal form. And so I was led to formulate the proposition that every set of sequences on letters $a_1, a_2, \cdots a_l$ that could be generated by any finite process (i.e. finite for each generated sequence) would be the sequences on these letters which were asserted by some system of the form given on the top of page 4††† for some extended set of letters $a_1, a_2, \cdots a_l \cdots a_m$.

And then a question terminating in the impossibility formulation. The question was could a set of sequences on a single letter a, for simplicity, [May 31] be defined which could not be given in normal form. The obvious answer was yes. For replacing a_1 by

† *Editor's note*: Page 3 starts with the paragraph "There were a finite number of symbols . . . "

†† *Editor's note*: Post is referring to the top of page 4 of the handwritten original. That page begins with the displayed "assertion", $\vdash g_\nu P$.

††† *Editor's note*: As noted, page 4 begins with the displayed "assertion", $\vdash g_\nu P$.

a in the formulation of the systems of the top of page 4 the bases of this formulation, including $n = 1, 2, 3, \cdots$ could obviously be arranged in a sequence. Now define a set of sequences $aa \cdots a$ where the sequence with n a's does or does not belong to the set according as that sequence is not or is in the system determined by that n^{th} basis. The set of a-sequences thus defined is thus different from each set of a sequences that can be generated by a system in normal form and so cannot be reduced to a set given in normal form.

A sleepless night resulted in which the possible connection with Brouwer's views for a while obscured the real issue. Furthermore just a short while before an impetus had been received in connection with the solution of the entscheidungsproblem for the systems of the top of page 4 which had made me again hopeful of its solution. Yet such a solution would give a finite test as to whether a sequence $aa \cdots a$ was or was not in that bizarre set of a sequences so that the resulting set of a sequences could certainly be said to be generated by finite processes. However the argument for the generality of the normal form for generated sets of sequences seemed just as strong as before. Hence the final conclusion. *The entscheidungsproblem for the class of systems of the top of page 4 is unsolvable.* With that conclusion the difficulty introduced by the above set of a sequences was cleared up. While the set was completely defined no finite process for determining whether an a sequence did or did not belong to the set was given so that its definition did not directly give a mode of generation of the set; and in view of the above no such mode existed.

Perhaps if the unsolvability problem had been stated in terms of the systems of the third part of my dissertation I might have had your assurance of its inevitability for those systems have at least the immediate formal potency possessed by your sets of recursion equations. However I would not even now give up the comparative formal simplicity of the bases of the final systems to which they were reduced. But that very reduction, which was completed, made it seem obligatory to do as much for each more general, if only superficially so, mode of generation that might be conceived. As I stated in my first letter this led me very far astray. However in the excitement engendered by this reversal of all my aims in symbolic logic I did carry through pretty much in complete detail the actual formal proof of what I might state as follows. The entscheidungsproblem for the systems of the top of page 4 is unsolvable by any method having the form of one of those systems. Hence of course any method reducible to such a form. Hence any anxiety with regard to the successive verification of what I could strictly call only an hypothesis, to wit that every finite method could be reduced to such a form. Incidentally the proof actually given showed that the unsolvability problem could be restated for a single one of the systems in question.

It will be noted that in the above development the concept of generated set is primary, something which I think corresponds to recursive enumeration. To give what would correspond to a set recursively defined, as in the formulation of method of solving entscheidungsproblems I mention above, I introduce a new letter b, and, say if the set is on the letters $a_1, a_2, \cdots a_m$, ask that for each sequence g on those letters either g or bg but not both is in the generated set.

My development does give one or two suggestions about the line between solvable and unsolvable problems which however may be very superficial. Of course drawing the line exactly is impossible. E. g. consider the class of entscheidungsproblems for the systems of the top of page 4, one for each basis. The problem of determining which of these problems is solvable and which not is clearly unsolvable for its solution would easily yield the solution of the entscheidungsproblems for the class of systems.

However this little point may be of interest. Because of my solution of the entscheidungsproblem for the systems of the third part of my dissertation I was led to see how closely I could recast the systems in normal form i.e. of the top of page 4 in that form. By continuing the "reduction" I found that the operations on the top of page 4 could be replaced by operations of the following four types (all types being used)

$\vdash gP$	$\vdash g_1P$	$\vdash Pg$	$\vdash Pg_1$
produces	$\vdash g_2P$	produces	$\vdash Pg_2$
$\vdash g'P$	produces	$\vdash Pg'$	produce
	$\vdash g'P$		$\vdash Pg'$

It will be seen that while productions with more than one premise are reintroduced thereby, the modification and qualification in each production occurs on one side only. With a trivial modification systems of this type involving but productions of the first two types come under the type for which I solved the entscheidungsproblem. The impetus toward the solution of the whole problem referred to above was that this solution was extended to include all systems with operations of the last three types only. The line here is pretty well drawn.

Returning to the systems of the top of page 4 I would like to tackle the impossibility problem for the following succession of cases

(a) The productions are such that for each sequence of length greater than the largest g one and only one production could use it as premise.

(b) As in (a) with all g_ν's of equal length (but not g'_ν's).

(c) As in (b) with g'_ν determined by its first letter.

(c) is the problem I tackled in Princeton, in a related form. My solutions did cover the cases where the number of letters were two and the length of each g_ν two. The line here is thus very wide.

All entscheidungsproblems are iterative in character. So I think are your problems concerning normal forms. I think the mathematical world would find such developments more fundamental if they disclosed non-iterative problems as unsolvable. To date I have had but the faintest glimmering as to how such problems could be attached.

Sincerely yours,

Emil L. Post

June 26, 1936.

Dear Dr. Post:

My delay in replying to your long letter of May thirty-first has been due to my desire to give its contents some careful thought and my difficulty (still unresolved) in making up my mind about some of the questions which it raises.

In particular, while it is clear that every generated set in your sense is lambda-enumerable (recursively enumerable), I can see no way of proving the converse of this, and at the moment, therefore, it seems to me possible that the notion of a generated set is less general. This question should, I think, be answered one way or the other, as being now the quickest means of arriving at an evaluation of some of the ideas set forth in your letter.

Since, however, I understand you to say that a portion of Principia which is substantially the same as the "engere Funktionen-kalkül" is reducible to the simple form given at the top of page 4 of your letter, it would seem to follow from my work that your proposed unsolvable problem is indeed unsolvable recursively.

In regard to your work which you describe, taken as a whole, it seems to me that it should have been published in 1922 or thereabouts, although in uncompleted form. Of course I might well think differently if I knew all the details of the work and its present state, but this is very definitely the impression which I get from your letter. I suspect strongly that some of your results have in the mean time been duplicated more or less approximately by others, even though under quite a different name and guise, and that by further delay in publication you run the risk of more such duplications.

Sincerely yours,

Alonzo Church

Dr. E. L. Post,
601 W. 148th Street,
New York City.

---

601 W. 148th St.
New York, N. Y.

July 10, 1936

Dear Prof. Church,

After reading your last letter I was sorry I hurried you into making a response. Your previous replies had come with such speed that I was afraid my last communication because of its bulk might have miscarried in the mails. Also since it was the first time I wrote to anyone of that work I was rather anxious to have an acknowledgement of its receipt. [I talked about it a-plenty in Columbia during 1921–22; also to Prof. B. A. Bernstein at the summer meeting of 1923. But after my breakdown in 1924 I decided to hold my peace till I was ready to return to that work; so that when I was introduced to Klein at a New York meeting about 1930 I was stricken dumb for I could talk to him of nothing else and I was not yet ready to talk.]

Your last paragraph did not quite have the effect it probably aimed at. If only I had had such an audience back in Columbia. But now I feel that personally I have lost so much by not having published that work in its incomplete form before not only your paper but that of Gödel that I feel any other losses I may sustain by a further delay of a year or two would be of little moment. Except perhaps for certain general ideas which it certainly would be foolish to bring out half baked. Fortunately my present position at the City College does not require forced publication. And so I am at liberty to school myself to the thought that the advancement of mathematics is secured no matter who leads the way; and that I may be of greater service in joining the procession with health fully guarded than momentarily strut my stuff and perhaps be forever after incapacitated from further research.

And now I shall take the liberty of answering your difficulty concerning the generally [*sic*] of "generated set" compared with "lambda-enumerable set" — at least in part. I mentioned in my letter that the systems of the third part of my dissertation could be reduced to that outwardly simple form I termed normal. The reduction in question consists of a large number of separate reductions each with its own little trick and is hardly suitable material for a letter. Granting this reduction if we can reduce your system to that dissertation form it could therefore be reduced to "normal form" and so to the entscheidungsproblem problem for the latter systems as a whole would be unsolvable as its solutions for such systems would on your definition of unsolvable contradict my theorem XVIII or better still its corollary. Note that to establish this unsolvability on your basis does not require as much finesse as the independent development I suggested in my letter.

While I could stick to the above plan a neater reduction is effected if instead of the type of system I suggested in the postulational development of my dissertation, I use the following modification. Not only is the postulate of substitution to be omitted [I think I wrote it could be done so in the development as given; if so I was thinking of the case of functions of one variable only. Obviously in general we can at most weaken it to allow substitution of any variable for another. Such systems would be "equipotent" with those of the dissertation, and for recursive as opposed to the λ development might be more suitable than what follows.] but the infinite set of variables is also left to be generated when needed. Such a system then is to have a finite number of constant "functions" $f_j(P_1, P_2, \cdots P_{n_j})$ with at least one a function of no variables ($n_j = 0$) and so a constant. I shall then define a possible enunciation of the system as follows. Each f_j of no variables is a possible enunciation, and if $P_1, P_2, \cdots P_{n_j}$ are possible enunciations $f_j(P_1, P_2, \cdots P_{n_j})$ is a possible enunciation, and for each j. For the operations of the system we do allow a finite number of variables $P_1, P_2, \cdots P_n$, and we define possible forms of the system as follows. Each P_j is a possible form, and if $P_1, P_2, \cdots P_{n_j}$ are possible forms or enunciations $f_j(P_1, P_2, \cdots P_{n_j})$ is a possible form.

The "assertions" of such a system are then generated as follows.

A. There are a finite number of operations of the form

$$g_{j1}(P_1, P_2, \cdots P_{k_j})$$
$$g_{j2}(P_1, P_2, \cdots P_{k_j})$$
$$\vdots$$
$$g_{jl_j}(P_1, P_2, \cdots P_{k_j})$$

produce

$$g_j(P_1, P_2, \cdots P_{k_j})$$

where the g's are possible forms of the system, where not all the P's need enter in each, and where the tacit meaning of the operation is that from l_j assertions which can be identified with the l_j forms by giving the P's enunciations as values can be obtained the assertion given by g_j with the P's also thus replaced.

B. There are a finite number of primitive assertions

$$h_1, h_2, \cdots h_l$$

Note that the phrase possible enunciation refers to possible use in some system. In a given system not all such "possible enunciations" have "meaning" as when certain f's play the part of propositional functions of arguments not propositions. A. however takes care of that (& B. I suppose).

And now for your system. I take it in my stride with only a previous thought to guide me (and the success of essentially the same method with *10–11 of the Principia is it) to show that my development too has its "obviousness".

July 11, 1936.

First the variables. Introduce a function of no variables a, a function of one variable [argument would be better] $b(P)$. Also a function of one argument $\alpha(P)$ whose assertion will mean P is a variable. We then put in

A. $\alpha(P)$
produces
$\alpha(b(P))$

B. $\alpha(a)$

Next the well formed formula. [Formula as such doesn't seem to be necessary.] $\beta(P)$ is to mean P is a well formed formula, i.e. the corresponding postulates are to have that effect. $\gamma(P, Q)$ is to mean P occurs as a free variable in Q, $\delta(P, Q)$ that P occurs as a bound variable in Q. $c(P, Q)$ is to correspond to $\{P\}(Q)$, $d(P, Q)$ to $\lambda P[Q]$. Then we put in

A.

$\alpha(P)$
produces
$\beta(P)$
- - - - - - - - -
$\alpha(P)$
produces
$\gamma(P, P)$

$\beta(P)$
$\beta(Q)$
produce
$\beta(c(P, Q))$
= = = = = = = = =
$\alpha(P)$
$\gamma(P, Q)$
produce
$\gamma(P, c(Q, R))$
- - - - - - - - -
$\alpha(P)$
$\gamma(P, R)$
produce
$\gamma(P, c(Q, R))$
- - - - - - - - -
$\alpha(P)$
$\delta(P, Q)$
produce
$\delta(P, c(Q, R))$
- - - - - - - - -
$\alpha(P)$
$\delta(P, R)$
produce
$\delta(P, c(Q, R))$

$\beta(Q)$
$\gamma(P, Q)$
produce
$\beta(d(P, Q))$
= = = = = = = = =
$\beta(d(P, Q))$
produces
$\delta(P, d(P, Q))$
⋮

B. - - - - - - - - -

I find that to complete this part of the basis I must add $\epsilon(P, Q)$ to mean P and Q are distinct variables. To obtain this effect we also introduce the auxiliary $\epsilon'(P, Q)$ to mean P is one or more b's of variable Q. Hence put in

A.

Here Q could be replaced by just a

$\epsilon'(P, Q)$
produces
$\epsilon'(b(P), Q)$
- - - - - - - - -
$\epsilon'(P, Q)$
produces
$\epsilon'(b(P), b(Q))$

$\epsilon'(P, Q)$
produces
$\epsilon(P, Q)$
- - - - - - - - -
$\epsilon'(Q, P)$
produces
$\epsilon(P, Q)$

B. $\epsilon'(b(a), a)$

We now complete the preceding part of the basis.

A.

$\epsilon(P, Q)$
$\gamma(Q, R)$
$\beta(d(P, R))$
produce
$\gamma(Q, d(P, R))$

- - - - - - - - -

$\epsilon(P, Q)$
$\delta(Q, R)$
$\beta(d(P, R))$
produce
$\delta(Q, d(P, R))$

Here $\epsilon(P, Q)$ could be omitted if redundancy is not objected to.

B. - - - - - - - - -

The definitions need not be considered, i.e. of abbreviation. As for $S_N^x M|$ and I, II, III. First the concept part of a formula, constituent I have been used to call it. By constituent I don't just mean the formula which is part of another but its localized occurrence in the larger formula. I interpret your I, II, III a replacement of a constituent in this sense and not of the formula the constituent assumes irrespective of its position. Should I, II, III mean that each part of the formula which is identical with the formula to be so replaced is to be so replaced a different procedure would be needed, as with $S_n^x M|$. In the latter x a variable only seems to be required.

I guess $S_N^x M|$ is most easily disposed of. $\zeta(P, Q, R, S)$ is to mean that S is the result of replacing each occurrence of variable P in R by Q.

July 12, 1936.

My "stride" is a little broken here. I find I have to rewrite a half page I concluded with yesterday. The following will result in the assertion of all $\zeta(P, G, R, S)$'s where Q and R are any well formed formulas and P any variable present in R or not. I shall try to have such $\zeta(P, Q, R, S)$'s the only ones asserted thus, though that really is not essential. Even so "The" S thus determined need not be well formed [I presume P when present in R need not be just a free variable] but I, II, and III should take care of that. Put then in

A.

$\alpha(P)$
$\beta(Q)$
produce
$\zeta(P, Q, P, Q)$

- - - - - - - -

$\epsilon(P, R)$
$\beta(q)$
produce
$\zeta(P, Q, R, R)$

$\zeta(P, Q, R, S)$
$\zeta(P, Q, U, V)$
produce
$\zeta(P, Q, c(R, U), c(s, V)$

$\zeta(P, Q, U, V)$
$\beta(d(P, U))$
produce
$\zeta(P, Q, d(P, U), d(Q, V))$

- - - - - - - - - - - - - - -

$\zeta(P, Q, U, V)$
$\beta(d(R, U))$
$\epsilon(P, R)$
produce
$\zeta(P, Q, d(R, U), d(R, V))$

B. - - - -

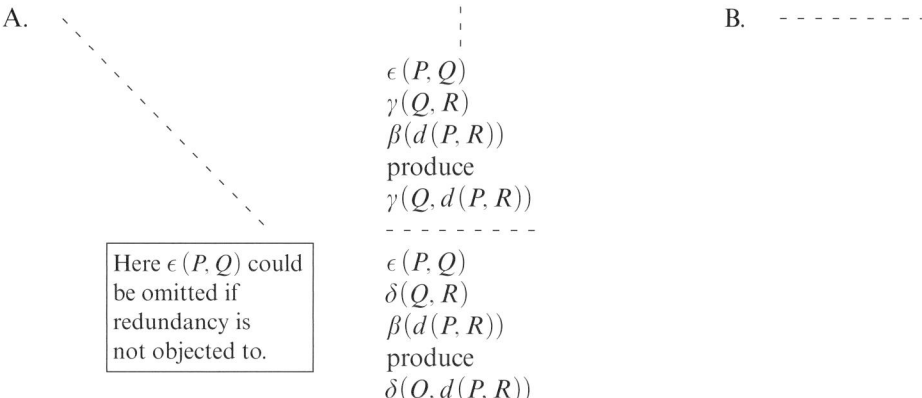

To approach I, II, and III we first introduce $\eta(P, Q, R, S)$ to mean S is a formula obtained from R by replacing a part P thereof by Q. In our notation a part of R is any complete occupant of any of the compartments of R (i.e. all of R between a pair of consecutive commas of (or $\cdots$) corresponding parentheses) [or R itself]. Since in its application the part is never the x following λ in any $\lambda x[M]$ this will be excluded. Otherwise the $\eta(P, Q, R, S)$'s are to be all such with Q and R well formed and P any part of R not of the excluded type. Put then in

A. $\beta(R)$
$\beta(Q)$
produce
$\eta(R, Q, R, Q)$

*material crossed out by Post**

B. - - - - - - - - -

A. $\eta(P, Q, R, S)$
$\beta(c(R, T))$
produce
$\eta(P, Q, c(R, T), c(S, T)$

- - - - - - - - - - - - - - -

$\eta(P, Q, R, S)$
$\beta(c(T, R))$
produce
$\eta(P, Q, c(T, R), c(T, S))$

$\eta(P, Q, R, S)$
$\beta(d(T, R))$
produce
$\eta(P, Q, d(T, R), d(T, S)$

B. - - - - - - - - -

Now for I, II and III. Let $\theta(P, Q)$ mean Q can be obtained from P by an operation of Type I. We shall need the auxiliary $\imath(P, Q)$ to mean P is a variable which does not occur in Q. For $\imath$ put in

A. $\epsilon(P, Q)$
produces
$\imath(P, Q)$

$\imath(P, R)$
$\imath(P, S)$
produce
$\imath(P, c(R, S))$

$\epsilon(P, Q)$
$\imath(P, S)$
$\beta(d(Q, S))$
produce
$\imath(P, d(Q, S))$

B. - - - - - - - - -

---

*The following is the material crossed out by Post.

$\beta(c(P, T))$
$\beta(Q)$
produce
$\eta(P, Q, c(P, T), c(Q, T))$

- - - - - - - - - - - - -

$\beta(c(T, P))$
$\beta(Q)$
produce
$\eta(P, Q, c(T, P), c(T, Q)$

$\beta(d(T, P))$
$\beta(Q)$
produce
$\eta(P, Q, d(T, P), d(T, Q))$

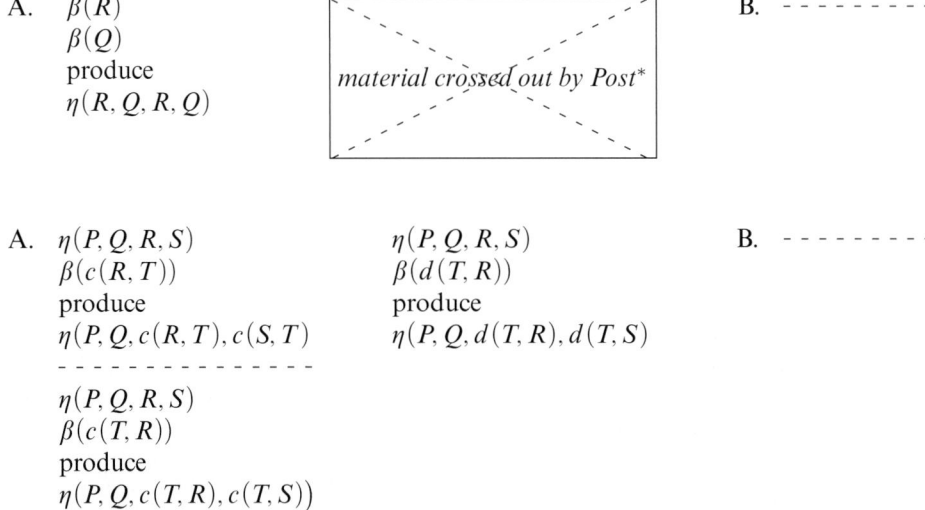

And now for θ, and so for I, put in

A.	$\beta(d(P,Q))$ $\imath(R,Q)$ $\zeta(P,R,Q,S)$ $\eta(d(P,Q),d(R,S),T,U)$ produce $\theta(T,U)$	B. - - - - - - - - -

For II let $k(P,Q)$ be made to mean Q is obtainable from P by an operation of type II. We need the auxiliary $\lambda(P,Q)$ to mean the bound variables in Q are distinct from the free variables in P.

*material crossed out by Post**

λ does not seem to be amenable to the above offhand July 17, 1935 treatment. We therefore draw upon some of the deeper potentialities of the method. We shall take advantage of $c(P,Q)$ to represent a finite class of variables $P_1, P_2, \cdots P_\eta$ in the form $c(P_1, c(P_2, \cdots c(P_{\eta-1}, P_\eta)\cdots))$. For this purpose let $\lambda_1(P)$ be made to mean P thus represents a class of variables. We then put in

A.	$\alpha(P)$ produces $\lambda_1(P)$	$\imath(P,Q)$ $\lambda_1(Q)$ produce $\lambda_1(c(P,Q))$	B. - - - - - - - - -

$\lambda_2(P,Q)$ is to mean P and Q both thus represent the same class of variables, i.e. for each Q with $\lambda_1(Q)$ we shall want asserted all $\lambda_2(P,Q)$'s where P is obtainable from Q by a permutation of its variables. Hence put in

A.	$\lambda_1(P)$ produces $\lambda_2(P,P)$ - - - - - -	$\lambda_1(c(P,c(Q,R)))$ produces $\lambda_2(c(Q,c(P,R)),c(P,c(Q,R)))$ - - - - - - - - - - - -	$\lambda_2(P,Q)$ produces $\lambda_2(Q,P)$ - - - - - -	B. - - - -
	$\epsilon(P,Q)$ produces $\lambda_2(c(P,Q),c(Q,P))$	$\lambda_2(Q,R)$ $\lambda_1(c(P,Q))$ produce $\lambda_2(c(P,Q),c(P,R))$	$\lambda_2(P,Q)$ $\lambda_2(Q,R)$ produce $\lambda_2(P,R)$	

Next $\lambda_3(P,Q,R)$ and $\lambda_4(P,Q,R)$ are to mean R represents the logical sum and logical product respectively of the classes represented by P and Q. We add symbol A to

*The following is the material crossed out by Post.

A.	$\alpha(P)$ $\alpha(Q)$ produce $\lambda(P,Q)$	~~$\beta(P)$~~ ~~$\lambda($~~ $\lambda(P,Q)$ $\lambda(P,R)$ produce $\lambda(P,c(Q,R))$	$\lambda(P,R)$ $\lambda(Q,R)$ produce $\lambda(c(P,Q),R)$	B. - - - - - - - - -

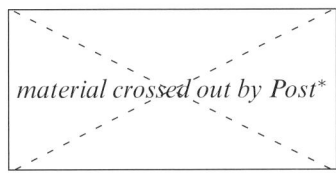

represent the null class and put in

A.	$\lambda_3(P, Q, R)$ produces $\lambda_3(Q, P, R)$ - - - - - - -	$\lambda_4(P, Q, R)$ produces $\lambda_4(Q, P, R)$ - - - - - - -	$\lambda_1(P)$ produces $\lambda_3(P, P, P)$ - - - - - - -	$\lambda_1(P)$ produces $\lambda_4(P, P, P)$ - - - - - - -	$\alpha(P)$ produces $\iota(P, A)$	B.	$\lambda_1(A)$
	$\lambda_2(P, S)$ $\lambda_3(P, Q, R)$ produce $\lambda_3(S, Q, R)$ - - - - - - -	$\lambda_2(P, S)$ $\lambda_4(P, Q, R)$ produce $\lambda_4(S, Q, R)$	$\lambda_1(P)$ produces $\lambda_3(A, P, P)$ - - - - - - -	$\lambda_1(P)$ produces $\lambda_4(A, P, A)$			

*material crossed out by Post**

A.	$\lambda_3(P, Q, R)$ $\lambda_1(c(S, P))$ $\iota(S, Q)$ produce $\lambda_3(c(S, P), Q, c(S, R))$	$\lambda_4(P, Q, R)$ $\lambda_1(c(S, P))$ $\iota(S, Q)$ produce $\lambda_4(c(S, P), Q, R)$	?	B. - - -

I think that takes care of λ_3 and λ_4. Now to get near to what we want. Let $\lambda_5(P, Q)$ mean P represents the class of free variables in well formed Q, $\lambda_6(P, Q)$ P represents the class of bound variables in well formed Q. Put then in

A.	$\lambda_5(P, Q)$ $\lambda_5(R, S)$ $\lambda_3(P, R, T)$ produce $\lambda_5(T, c(Q, S))$	$\lambda_6(P, Q)$ $\lambda_6(R, S)$ $\lambda_3(P, R, T)$ produce $\lambda_6(T, c(Q, S))$	$\beta(d(P, Q))$ $\lambda_5(R, Q)$ $\lambda_4(P, R, T)$ produce $\lambda_5(T, d(P, Q))$	$\beta(d(P, Q))$ $\lambda_6(R, Q)$ $\lambda_3(P, R, T)$ produce $\lambda_6(T, d(P, Q))$	$\alpha(P)$ produces $\lambda_5(P, P)$ - - - - - - $\alpha(P)$ produces $\lambda_6(A, P)$	B. [crossed out]**

*The following is the material crossed out by Post.

$\epsilon(P, Q)$
$\lambda_1(P, R)$
$\lambda_2(Q, R)$
produces
$\lambda_3(c(P, R), c(Q, R), c(P, c(Q, R)))$

**The following is the material crossed out by Post.

$\lambda_5(A, A)$

Finally since we have $\lambda(P, Q)$ add to

A. $\lambda_5(R, P)$
$\lambda_6(S, Q)$
$\lambda_4(R, S, A)$
produce
$\lambda(P, Q)$

B. ---

Let then $\mu(P, Q)$ mean Q can be obtained from well formed P by an operation of type II. We therefore put it

A. $\lambda_6(U, Q)$
$\imath(P, U)$
$\lambda(R, Q)$
$\zeta(P, Q, R, S)$
$\eta(c(d(P, Q), R), S, T, W)$
produce
$\mu(T, W)$

B. ---

Likewise let $\nu(P, Q)$ mean Q can be obtained from well formed P by an operation of type III.

We need the auxiliary $\eta'(P, Q)$ to mean P is a part of well formed Q not immediately following λ. For η' (independently of η) put in

A. $\beta(P)$ produces $\eta'(P, P)$ | $\eta'(P, Q)$ $\beta(c(Q, R))$ produce $\eta'(P, c(Q, R))$ | $\eta'(P, R)$ $\beta(c(Q, R))$ produce $\eta'(P, c(Q, R))$ | $\eta'(P, R)$ $\beta(d(Q, R))$ produce $\eta'(P, d(Q, R))$

B. ---

And so for $\nu(P, Q)$ have in

A. $\lambda_6(U, Q)$
$\imath(P, U)$
$\lambda(R, Q)$
$\zeta(P, Q, R, S)$
$\eta(S, c(d(P, Q), R), T, W)$
$\eta'(S, T)$
produce
$\nu(T, W)$

B. ---

While all of the succeeding definitions could be likewise treated, in answering the questions of your letter I first consider whether your development insures the unsolvability of the problem I proposed. For this purpose your Theorem XIX is sufficient. With this in view let $\xi(P, Q)$ be made to mean P is convertible into Q. We then need merely put in

A. $\theta(P, Q)$ produces $\xi(P, Q)$ | $\mu(P, Q)$ produces $\xi(P, Q)$ | $\nu(P, Q)$ produces $\xi(P, Q)$ | $\beta(P)$ produces $\xi(P, P)$ | $\xi(P, Q)$ $\xi(Q, R)$ produce $\xi(P, R)$

B. ---

And now, speaking informally, my reductions of the late summer of 1921 insure that a system of the type formulated on p 2–3 of this letter is reducible to the 'normal form' of my previous letter. Hence the solution of the entscheidungsproblem for the latter systems would entail its solution for the former. The above system is in the form of p 2–3. Hence its entscheidungsproblem would be solvable. Hence for any two well formed formulas P, Q we would have an effective test as to whether $\xi(P, Q)$ is or is not asserted in the system. But barring careless formulation the above insures that $\xi(P, Q)$ is asserted when and only when P conv B. On the basis of your criterion of effective calculability Theorem XIX would thus be contradicted. Hence your development does imply the unsolvability of my proposed unsolvable problem. Perhaps, however, this need not mean that my own independent development is superfluous. It may have its advantages.

As for my "generated set" being as general as your λ enumerable set I think that while a little of the finesse I mentioned in my last letter might have to be employed the above is sufficient indication that it can be shown to be so. I really had no intention of at this time proving the equivalence of our developments and so I stop now.

Having written so much however I am tempted to go into the question of the obviousness of your criterion of effective calculability and so forth once more. My first argument is a mean one. You state that λ-definability and recursiveness are equally natural. If so, and in the sense that either obviously corresponds to our intuitive notions the proof of their equivalence should be trivial, merely paraphrasing that intuition. This doesn't quite seem to be the case. In both cases obviousness comes only after proof of certain thms. After proof of these theorems, proof of equivalence is trivial.

Again in your answer to one of my questions you state that the affirmative proof merely follows that intuitive concept. Isn't that so only so far as the finiteness of the two processes are concerned, not in their power at each step?

Again you argue that you can always as it were withdraw into your shell and say that that is your definition of effective calculability. But you have a result to the effect that the entscheidungsproblem for the engere funktionenkalkül is unsolvable. Recursively of course makes it O. K. But with your criterion recursively would be omitted. But then you should be able to bar any ambitious young man from attempting its solution by any means just as you would rightfully bar him from attempting the trisection of the general angle by straight edge and compass. But now you see you can't hide behind merely a definition.

In fact I stand greatly in awe of the concept of an absolutely unsolvable problem of the type we have in mind. For I think we can say that the universe as it were does include a solution thereof, i.e. that in the entscheidungsproblem form there is no question that a given enunciation is or is not generated by the postulates, Brouwer possibly to the contrary notwithstanding. [I would not say the same of the truth value of $\aleph = 2^{\aleph_0}$.] And so I am forced again to view the absoluteness of the unsolvability of the problem as absolute so far as human means are concerned. Were the concept of an infinite being philosophically tenable such a being could obviously be conceived who could solve our unsolvable problem. But one need not go so far. We for example are able to reach as it were certain transfinite ordinals of the second type. Analogously an inferior intelligence might be able to count only so far in the scale of finite ordinals. It seems to me not to require the concept of a being with infinite powers to think of one who could reach any transfinite ordinal of the second class as we can reach any finite ordinal. For such a being a problem absolutely unsolvable by us might very well be solvable. And so this suggests the troubling thought have we so fathomed all our own powers as to insure our assertion of absolute unsolvability relative to us.

Sincerely yours,

Emil L. Post

September 18, 1936.

Dear Dr. Post:

In spite of my long delay in answering your letter of July tenth, I fear my answer may have to be somewhat unsatisfactory. It is apparently impossible for me to find time to think with any care about the formal details of your work, and I shall have to be content with such general impressions as I can form without doing so.

When you eventually feel that you are ready to publish, I suggest that your first paper should simply present what you have now, letting the further working out of the matter wait until later papers. You might include a reference to your old abstract in the Bulletin, although, of course, you would not profit by making any specific claims which could not be substantiated. I am still not very clear to just what extent you had anticipated the results of my paper, *An unsolvable problem*. But apart from this I do not believe there is anything very definite in any case — no anticipation for instance, as I understand it, of Gödel's principal result. Nevertheless there seem to be many rather intangible ways in which your ideas overlap the work not only of Gödel, and of myself, but also, I believe, of Curry.

As regards your objections to the identification of recursiveness, and of lamda-definability, with the intuitive notion expressed by "effective", "effectively calculable", or some synonym, you seem to be inclined to put the burden of proof on me, which I do not think justified. The fact is that the intuitive notion is vague and inexact, and that what I proposed was to render it exact by giving a formal definition. No proof can be expected, simply because the intuitive notion is inexact. Those who use this inexact notion are in error to just the extent that the notion is inexact; if they maintain that there is something in their notion which makes it more general than recursiveness or lambda-definability, they can be legitimately asked to produce an example, and failing to do so, stand convicted of making an utterly vague and even pointless assertion. I am thus content to let the matter stand as a challenge.

For such further arguments as can be made, I would refer you to section 7 of my paper.

Of course the relationship of recursiveness and of lambda-definability to the intuitive notion of effective calculability is not immediately clear from the definitions. It becomes clear after proving a number of rather difficult theorems. And after proving these same theorems the equivalence of recursiveness and lambda-definability becomes also clear.

As for the hypothetical ambitious young man, desirous of solving the problem to find a complete set of invariants of conversion, he could not be barred if he were willing to omit the condition "effectively calculable" from the problem. But the burden of proof as regards importance would be on him. An obvious way to get an invariant is to take some particular enumeration of the set of all formulas, and then associate with every formula **A** the number (in this enumeration) of the first formula convertible into **A**. This one invariant alone clearly forms a complete set. The young man could be asked in what way he proposed to make his set of invariants better than this one.

Sincerely yours,

Alonzo Church

Dr. E. L. Post,
601 W. 148th Street,
New York City.

Dr. E. L. Post,
601 W. 148th Street,
New York, N. Y.

May 28, 1937

Dear Prof. Church

May I, in a spirit of complete good fellowship, say that while I may have to eat part of my words in the light of Turing's finite number of mental states formulations you will certainly have to eat yours in thinking that the general concept of a machine is any easier to formulate than that of "effectiveness". Incidentally lack of formulation need not imply non-existence. You can recognize them when you see them.

Apropos of machines one must distinguish between a machine for, say, computing "computable number", *being* a Turing machine and *being equivalent* to a Turing machine. Thus while every rational expression in terms of integers is equivalent to a quotient of two integers even $\frac{\frac{1}{2}}{\frac{2}{3}}$ is not *in* that form. Of course here proving the equivalence is easy. But it has to be proved. So with machines—except for the ease or even possibility of a final complete "proof".

Perhaps the following suggestions will make my position clearer. Except for the final instances there should be no difficulty in making the formulations precise. In each case the machine *is not* a Turing machine. That they are all *equivalent* to Turing machines I am no more doubtful of than are you. But that has to be demonstrated. And not just for my suggestions but for any that may ever or can ever be thought up. ______.

A. Modifications of the "Tape" idea.

1. The tape may be bent to pass through the machine a fixed number of times n; the next event may then depend on the symbols on the tape at n points of it, not relatively fixed or even of bounded distances apart, and changes in those symbols likewise may occur in such places on the tape.

2. There may be a fixed number n of tapes passing through the machine etc.

3. A combination of 1 & 2 plus the following. The machine may admit of *cutting* a tape or tapes at one or more points and *reuniting* with interchange of order of pieces or tapes. Also loops may be formed and opened in the tapes.

4. The tape may be replaced by a two dimensional sheet its passage through the machine being determined by any pre-assigned "triangulation of the plane" — by squares being only the simplest.

5. The sheet may be a cylindrical surface etc.

6. There may be *n* sheets connected in Riemann surface fashion.

7. As in 3 the connections in 6 may themselves be formed and and reformed by the machine.

8. The machine itself may move in a predetermined partitioning of space of three dimensions acting as the symbol space.

9. The machine itself may construct its one partitioning of the symbol space.

10. Conceptually the symbol space need not be completely immersible in ordinary space. Practically that would mean our readjusting the machine when things clog. But the determinateness of the process might not be effected. E. g. in the simplest case imagine a symbol space where from each working box one can proceed to *n* boxes, in this simplest case the resulting symbol space being a "tree".

B. Modification of the "Finite Machine".

Foreword. Since a Turing machine allows an infinite tape there need be no objection to the assumption of an inexhaustible reservoir of molten steel for the construction of machines, or other more undifferentiated "material". Since in industry machines actually are used for the construction of parts of other machines and even assembling the following ensembles are entitled to be called machines, when they are given complete formulation at least.

1. A master machine may keep on constructing Turing machines according to some Turing like law. From a half plane sheet the first Turing machine suitably modified, starts cutting of its strip of tape and operating in that tape; the second does the same from the new edge of the half plane, and so on. As in A2 a unifying tape passes through each separate machine the master machine connecting it with each new formed machine at its completion, the whole to form one (growing) machine.

2. A master machine of the second order may construct master machines which themselves function as in 1, the whole forming a growing machine of type 2. So far any finite type.

3. A master machine of type ω may construct master machines of varying finite type the whole forming one growing machine of transfinite type ω. So for any *constructible* transfinite type. [This idea was largely responsible for my position as stated in my article].

4. Any machine finite at each moment and growing and operating according to any laws *we* can formulate. [For effectiveness the we is essential for the law of growth may be non-constructive without violation to the finite deterministic character of the machine, in which case it could solve what are for us unsolvable problems.]

Now it may be that should the hypothesis of the finite number of mental states of Turing's formulation prove unimpeachable that an equally simple idea may suffice to show in one synthesis that any man-made formulation of a machine of any type or kind must by these very limitations lead to a machine *equivalent* to a Turing machine. Then I would have to give up my whole position. But until then, and may I return to my attitude of complete cordiality, I think your position is untenable.

Sincerely yours,

Emil L. Post

P. S. I forgot to mention other possibilities such as a machine printing a block of raised figures which joined on to the previous printing block forms a new printing block of twice its length and so on. Much of the above applies also to a human computer if you allow him more than a fixed row of squares and only a pencil point and eraser to work with. And if I remember certain suggestions for practical computing writing figures on one card or better strip and moving the whole strip to a new position and so forth, such things actually are done. This then besides generalizations of the symbol space the computer works in. Nevertheless I still have great respect for Turing's finite number of mental states contribution — if sound.

E. L. P.

June 8, 1937

Dear Dr. Post:

In the case of all the methods of mechanical computation which you suggest in your letter, except possibly the last two (B3, B4) I find the equivalence to computation by a Turing machine evident more or less immediately. This is simply because it is clear that, in the case of any particular one of your machines or quasi-mechanical schemes, an effective description of its behavior could be written out in finite space on paper, i.e. practically on a tape, since the paper, by division into lines, is used only one-dimensionally. My doubts in the case of B3 and B4 are due chiefly to uncertainty as to exactly what you mean, but I am quite willing to maintain that these schemes should be regarded as effective calculations only in so far as they are equivalent to computation by a Turing machine (or to general recursiveness, or to general recursiveness[*sic*], or to lambda-definability).

I received today a postal about this same subject from Jozef Pepis. He is of the opinion that numerical functions exist which are effectively calculable but not general recursive, and has an idea which he hopes will enable him to produce particular examples of such functions. He does not describe his idea, and, if I understand him correctly, is not yet actually in possession of such examples.

I have the impression that neither you nor Pepis fully realize what would be the consequences of the discovery of a function which was effectively calculable but not general recursive. Consider for instance the following.

I think we are all agreed that if a numerical function f is effectively calculable then for positive integer a there must exist a positive integer b such that a valid proof can be given of the proposition $f(a) = b$. At least if we are not agreed on this then our ideas of effective calculability are so different as to make further discussion useless.

On the other hand it is proved in my paper in the American Journal that if the numerical function f is not general recursive, and if the system of Principia Mathematica is omega-consistent, then there must exist a positive integer a such that for no positive integer b is the proposition $f(a) = b$ provable within the system of Principia (i.e. this must be true whatever permissible formula we choose as a definition of f within the system of Principia). Moreover this remains true if instead of the system of Principia we substitute any one of the extensions of Principia which have been proposed (e. g. allowing transfinite types), or any one of the forms of the Zermelo set theory, or indeed any system of symbolic logic whatsoever which to my knowledge has ever been proposed.

Therefore to discover a function which was effectively calculable but not general recursive would imply discovery of an utterly new principle of logic, not only never before formulated, but never before actually used in a mathematical proof — since all extant mathematics is formalizable within the system of Principia, or at least within one of its known extensions. Moreover, this new principle of logic must be of so strange, and presumably complicated, a kind that its mathematical expression as a rule of inference was not general recursive (for this reason, if such a proposal of a new principle of logic were ever actually made, I should be inclined to scrutinize the alleged effective applicability of the principle with considerable care).

As regards Pepis I would say that his recent long paper on the Entscheidungsproblem, although its topic may be considered somewhat specialized, is nevertheless undoubtedly an able piece of work. For this reason I am inclined to give his ideas some real consideration, and to be genuinely curious as to what, if anything he has.

Sincerely yours,

Alonzo Church

Dr. E. L. Post,
601 W. 148th Street,
New York City.

June 8, 1937

Dear Mgr. Pepis:

This is to acknowledge receipt of your manuscript, *Ein Verfahren der mathematischen Logik*, offered for publication in the Journal of Symbolic Logic. In accordance with our usual procedure we are submitting this to a referee to determine the question of acceptance for publication, and I will write you further about the matter as soon as I have the referee's report.

In reply to your postal I will say that I am very much interested in your results on general recursiveness, and hope that I may soon be able to see them in detail. In regard to your project to construct an example of a numerical function which is effectively calculable but not general recursive I must confess myself extremely skeptical—although this attitude is of course subject to the reservation that I may be induced to change my opinion after seeing your work.

I would say at the present time, however, that I have the impression that you do not fully appreciate the consequences which would follow from the construction of an effectively calculable non-recursive function.

For instance, I think I may assume that we are agreed that if a numerical function f is effectively calculable then for every positive integer a there must exist a positive integer b such that a valid proof can be given of the proposition $f(a) = b$ (at least if we are not agreed on this then our ideas of effective calculability are so different as to leave no common ground for discussion). But it is proved in my paper in the American Journal of Mathematics that if the system of Principia Mathematica is omega-consistent, and if the numerical function f is not general recursive, then, whatever permissible choice is made of a formal definition of f within the system of Principia, there must exist a positive integer a such that for no positive integer b is the proposition $f(a) = b$ provable within the system of Principia. Moreover this remains true if instead of the system of Principia we substitute any one of the extensions of Principia which have been proposed (e. g., allowing transfinite types), or any one of the forms of the Zermelo set theory, or indeed any system of symbolic logic whatsoever which to my knowledge has ever been proposed.

Therefore to discover a function which was effectively calculable but not general recursive would imply discovery of an utterly new principle of logic, not only never before formulated, but never before actually used in a mathematical proof—since all extant mathematics is formalizable within the system of Principia, or at least within one of its known extensions. Moreover this new principle of logic must be of so strange, and presumably complicated, a kind that its metamathematical expression as a rule of inference was not general recursive (for this reason, if such a proposal of a new principle of logic were ever actually made, I should be inclined to scrutinize the alleged effective applicability of the principle with considerable care).

Sincerely yours,

Alonzo Church

---

601 W. 148th St.
New York, N. Y.

June 24, 1937

Dear Prof. Church,

The letter I meant to write you would, I fear, have been one of those ten page affairs. I will therefore merely go into the one point of your letter where in point of fact as against the evaluation of known facts I cannot see eye to eye with you in that last letter.

I can see no validity whatsoever in your argument for the equivalence of my suggested machines to a Turing machine on the basis that "an effective description of its behaviour could be written out in finite space on paper, i.e. practically on a tape ... ". The role of the tape in a Turing machine is that of the working space. Of what avail is it then to have a static linear description of the behaviour of a machine when the operations thus described, among other things, refer to a space not Turing's.

Before going into this more fully let me recall a point of our conversation of a few months ago. You suggested that an operation carried out in a plane gives rise to just a succession of symbols and so could be carried out on a line. Apart from more fundamental criticisms of this argument observe the following. Suppose that while working in the plane a circuit has been made. The next step will depend on the symbol present in the square we have just returned to. But we may be miles away from that symbol in the straight line. In other words it isn't *that* simple.

To return to your present argument let us consider this relatively simple case of an otherwise Turing like machine whose working space is however an infinite checkerboard instead of an infinite tape. Your linear description of the behaviour of that machine would involve such phrases as "move one square right", "move one square up", "move one square left", "move one square down". How then has your linear description helped you reduce the machine to one which involves only motion right and left?

As far as I can now see the only way to establish this equivalence would be according to the following plan.

A. Set up an effective 1-1 correspondence between the possible symbol complexes in the plane and a class of symbol complexes on the line. Or, more specifically, set up an effective 1–1 correspondence between the squares of the checkerboard and some or all of the squares of the tape, a set of marked squares of the checkerboard then corresponding to the set of corresponding squares of the tape correspondingly marked.

B. In terms of the correspondence translate the set of directions for operating on the checkerboard into a set of directions for operating on the tape.

C. Replace the resulting set of tape directions by an equivalent Turing set.

The carrying out of this plan should not be difficult. But I don't think it is trivial for as I mentioned in our conversation (in a somewhat wider form) there is no way of setting up a 1-1 correspondence between the squares of the infinite checkerboard and those of the infinite tape such that neighboring squares of the checkerboard will be a bounded distance apart on the tape.

Sincerely yours,

Emil L. Post

P. S. From July 1 to about Sept 8 my address will be 107-B, Hampton Ave, Manhattan Beach, New York. E. L. P.

July 10, 1937

Dear Dr. Post:

I am going to try to reply briefly to your letter and postal.

To your position as a whole my reply would be, I think, about as follows. I have been maintaining, "All effectively calculable functions are general recursive." In doing so, however, I have not been trying to establish an empirical proposition, or a mathematical proposition. Instead, I have been proposing an exact definition of a phrase ("effectively calculable") which has hitherto had only a vague meaning. Under such circumstances proofs are not to be expected, but only considerations of convenience and naturalness and facts of historical usage and generally accepted connotations.

Your objections seem to be based on the idea that there is a meaning of the phrase "effectively calculable", which we both understand so well, even without a definition, that, given an utterly new process (method of defining a function) never heard of before, we would both be able to decide whether it was an effective calculation, and would perforce decide alike. But you have not attempted to give reasons why you think this is so.

Finally, to reply to a particular point, it does seem to me that the question whether an effective description can be given in *English* of the behavior of a machine is relevant to the question whether the machine is equivalent to a Turing machine. For the description in English is one-dimensional and so can be represented by a finite series of symbols on the tape of a Turing machine. By saying that the description is effective is meant that a human inquirer could work out from the description the behavior of the machine step by step. And I can see no relevant difference between this human inquirer and a Turing machine which works out the same thing from the equivalent row of symbols on its tape.

This is no less vague than the phrase "could work out" which it contains, but I have explained why I think vagueness must be expected in this argument, which I regard as an argument over the admissibility of a definition.

In practice human calculators always work things out by methods essentially one-dimensional. They use sheets of paper to be sure, but always finite sheets of paper, and they use a one-dimensional succession of these sheets (along the single dimension of time). The sheets are frequently all of the same size; in any case there is a finite upper bound to their size. The complete array of marks on a single sheet may be regarded as one symbol—the total number of available symbols will still be finite. This makes the process clearly one-dimensional. If one of the later sheets of paper is taken back and compared with an early one, this is no more than what a Turing machine can do by scanning a square far along the tape, recording the symbol on the square by a change in its internal configuration, and then reverting a counted number of squares backward along the tape for comparison with an earlier square.

Sincerely yours,

Alonzo Church

Dr. E. L. Post,
601 W. 148th Street,
New York City.

107-B Hampton Ave
Manhattan Beach, N. Y.
(Summer Address)

July 13, 1937

Dear Prof. Church,

I like your "sheets of paper" argument very much. By [*sic*] like all such arguments, with the possible exception of Turing's finite number of mental states formulation, it is of but limited cogency. In particular, neglecting macrocosmic or microcosmic physical considerations, I see no necessity in the assumption of boundedness in the size of the sheets.

You have put my view about our common agreement in the face of a new process as to its being an effective calculation very clearly. Isn't that exactly what happened with these machines and "pseudo-machines" of mine when they were halfway clearly formulated? And I don't think your reaction depended on a possible proof of equivalence to a Turing machine.

As for that one dimensional argument of yours based on a description of the machine I still see nothing in it. It seems to me to be based on a confusion between the static one dimensional description of the laws of the machine (analogous to my set of directions) and the possible reduction of the dynamic working of the machine to a space of one-dimension. In connection with that I cannot be more clear than I was in my last letter, and I can see no rebuttal of that letter in your present argument.

To return to my infinite checker board case may I suggest that for example the process of multiplication in arabic notation or the manipulations of determinants in square notation assume its use. And my position still is that the equivalence of a "Turing machine" with such a working space to a strictly Turing machine needs (and is capable of) proof. Perhaps I shall bring this matter to a test by actually completing the proof which I suggested last time and presenting it as a paper for publication in the Journal of Symbolic Logic.

Sincerely yours,

Emil L. Post

P. S. I am planning to return to Logic research the coming Fall. Lest past mishaps make me wary, so that even with my research stint of two hours a day I plan to give but one week in five to Logic for at least a year. I am not sure therefore when I will issue the above "challenge", especially as I shall be more interested in further objectives than in defending positions already occupied.

E. L. P.

610 W. 173 St
New York, N. Y.

July 1, 1942

Dear Prof. Church,

Except that I have returned to work in Symbolic Logic — with what degree of permanency remains to be seen — the situation has not changed with regard to my undertaking reviewing of papers. I will therefore take the way out you have offered me and decline your invitation with this reservation: that if you think my stand is unjustified then I will capitulate.

When it is a matter of refereeing papers I do not feel justified in refusing to referee the very occasional paper that comes my way if I feel at all competent to do so. As I see it I give others a certain amount of work in refereeing my limited output and I feel duty bound to reciprocate in approximately the same measure. On the other hand reviewing of published work is by comparison a luxury rather than a necessity and so I do not feel the same moral compulsion to reciprocate in the matter of reviews. Moreover while a certain amount of prior knowledge and accompanying literature delving is needed in refereeing to insure the newness of the material, once reasonable certainty is reached the main job is one of internal study, and criticism, of the paper. On the other hand in an adequate review the proper placing of the contribution against the background of existing literature is of primary importance. This requires the scholarly mind and scholarly background that I do not possess and could not acquire without *completely* giving up research for reading.

Whether it is refereeing or reviewing the time for that would have to come out of the ten hours a week I otherwise give to research or the writing up thereof. So long as I am at City college that is all the time my job there will allow for other work. But that puts the cart before the horse; for I went to City College because I realized that two hours a day, five days a week, is all my health would allow me for scientific work. That that stint can't be much increased is evidenced by the fact that while for some weeks of the summer vacations I have been adding a third hour of reading, that, at times, produces a state of excitation and accompanying chronic headache requiring cessation of all work for a couple of weeks. To illustrate this time element, during a refereeing job of some year and a half ago I devoted every other day of my two hours a day, five days a week, to the refereeing job, for about two and a half months. In point of fact that refereeing tended much more to throw me off balance than my own work.

So there is my case. And as I said, if you still feel it my duty say to review at least as much as I force on others then I must submit. But please make it one paper at a time. In the case of the two papers you mentioned, Müller's and Newman's, my own reactions balance to lead to no preference.

Sincerely yours,

Emil L. Post

P. S. I am leaving for the country next Saturday for a week, but expect to be in New York for the six weeks following. — E. L. P.

7 Brighton 10 Terrace
Brighton Beach, N. Y.
(summer address: to labor day)

July 29, 1943

Dear Prof. Church,

I was very much gratified that you thought enough of my "Formal Reductions" to supply that "lacuna".* When I wrote the paper I did not have your recent booklet to consult, and the proof I had in mind was a long letter I once wrote you plus a piece of work in the part of the paper deleted.**

However having the system S′ of your review, its reduction to canonical form can be accomplished much more simply than in your review as follows.

To the primitive letters I, J, (,) of S′ add α. αP will be asserted by the following canonical system when and only when P is a combination involving only I, J and parentheses.

Primitive assertions: αI, αJ

Production: $\alpha P, \alpha Q$ *produce* $\alpha(PQ)$.

Now add to this basis a third primitive assertion, i.e. the combination corresponding to 2, and 37 productions which are those of pp. 50–51 of your booklet, each however having in addition to the premise there given a premise αP for each operational variable P in the production. Thus $(FA) \vdash (F(IA))$ becomes

$$\alpha P,\ \alpha Q,\ (PQ) \text{ produce } (P(IQ))$$

Assuming I have not slipped up on your booklet (and I am nowhere near mastery of it) the validity of the above is obvious. The last thirty seven productions can only yield assertions starting with (. They can therefore never be used as premises of the first production. That is αP when asserted stays what it was by the initial sub-bases, i.e. P is a combination involving only I and J. The added premises of the last 37 productions therefore formally restrict the operational variables as you do informally in system S′.

May I suggest that the tricks employed in my paper, in probably very far from the most economical fashion, were forced on me by the ever more restricted formal means left me by the required even simpler forms of basis. But the canonical form is powerful, and reduction to it should be natural instead of tricky. In this connection may I suggest, from the long version of my paper, that where a system was an infinite number of primitive symbols $p_1, p_2, p_3 \cdots$ (an infinite number of variables say) in addition to a certain finite set, we may introduce one new letter say a and rewrite $p_1, p_2, p_3 \cdots a, aa, aaa, \cdots$. A uniform dot notation can then be introduced, and a new letter b is introduced, ., :, : ., $\cdots$ would be replaced by $b, bb, bbb, \cdots$. Of course if a system allows just p_1, p_2, p_3 say, it should be modified to write the latter, say $p_1 .\ p_2 .\ p_3$ which then becomes $abaabaaa$.

Editor's note*: Post is referring to his paper *Formal Reductions of the General Combinatorial Decision Problem*, ***American Journal of Mathematics* vol. 65 (1943), pp. 197–215; reprinted in Post, E. L., ***Solvability, Provability, Definability: The Collected Works of Emil L. Post***, M. Davis ed., Boston, Birkhäuser, 1994.

Editor's note*: Post is probably referring to Church's ***The Calculi of Lambda-Conversion.

I had hoped you would note the idea of *incompleteness relative* to a given class of propositions suggested in footnote 18. Thus to state Gödel's Theorem in its most general form authors usually indulge in the not very clear condition that a system besides so and so and so and so shall be capable of developing a certain portion of arithmetic. But one can simply state that every finitary symbolic logic is incomplete relative say to the class of arithmetic propositions. Thus the $(\sim, \vee)$ system is so incomplete [even if widened to allow propositional apparent variables so that it is complete in itself in the Gödel sense] either in the sense that none of its assertions represent arithmetical propositions and so of course will not correctly assert p or $\sim p$ for each arithmetical proposition p, or also, and more generally, in the sense that no effective mode of representing arithmetical propositions by $(\sim, \vee)$ enunciations can be introduced such that for each arithmetical proposition p either the representation of p or the representation of $\sim p$ is asserted, it being understood that such a representation if asserted is that of a true arithmetic proposition.

Sincerely yours,

Emil L. Post

---

7 Brighton 10 Terrace
Brighton Beach, N. Y.

August 3, 1943

Dear Prof. Church,

I would not burden you with an immediate reply to your letter* except that I must remove a misunderstanding. It was only the *restricted* functional calculus of Principia Mathematica that I reduced to canonical form; that all of Principia Mathematica could so be reduced was in my own mind as a result obvious, but so far as actual work goes must be described as a pretty good hunch on my part. I should have added in my footnote that with the corrections of the early work I still assumed that as in the propositional calculus all substitutions need only be made in the primitive propositions. Not that it should be very difficult to allow for substitutions at any stage of the game — if the above assumption is wrong.

Perhaps we can discuss the question of whether any other parts of my long "historical" paper other than "Formal Reductions" is worth publishing when we meet at some meeting. But since I am writing to you again there are one or two other things I would like to get off my chest.

Last summer I had an all afternoon gabfest with Tarski in which I let go on some of the things connected with footnote 10; this time, of Formal Reductions. At the end he argued that it was each man having his own system for developing such ideas that made for the neglect of the subject by American mathematicians as a whole. As to what he thought was *the* method, well, general recursive functions, say as presented by

* *Editor's note*: Church's letter has not been found.

Hilbert-Bernays (for my learning the method thereof — I argued that my own scheme was the only one I had any facility in).

Now it seems to me the subject is still too new, and I think important, to freeze its formal development along one line. Far better some confusion for one generation rather than the possibility of following the wrong track for a hundred years. In particular, I doubt if general recursive function is *the* way, just because it draws the inspiration from the algorithms of arithmetic rather than the processes of symbolic logic. Now I have no illusions that the method of canonical systems (the reduction to normal form as part of the *method* may in the long run be a hindrance rather than otherwise) is in any sense definite. (I still think that my Formulation I has better chance for that, though the development essentially of that as carried out by Turing is hardly encouraging.) But it may remain as part of a more natural development. And apart from its ease in use (at least to me) it has the advantage of I think being closer to the spirit of symbolic logic than other formulations.

In particular may I note the following. If as suggested in my last letter, the symbolism of a symbolic logic be changed so that its enunciations are sequences on a fixed finite set of letters, the assertions of that logic, assumed "finitary", would constitute a generated set of such sequences and hence, assuming the generalization referred to in footnote 18, a normal set. The study of normal sets may then be said to be the study of (finitary) symbolic logics. And now a possible advantage of normal set as so developed over recursively enumerable set (as it is developed though the two are identical in fact). It is readily proved that every normal set on one letter a (recursive enumerable set of positive integers therefore) can be given by a normal system on a and one additional letter b. Now *every* normal system basis that we happen to write down on a, b defines a normal set on a. These bases are easily symbolized themselves as sequences on a fixed set of letters and these symbolizations themselves constitute a normal set. By contrast one cannot generate the set of all bases of general recursive functions effectively, since besides the formal set up of recursive equations there is the non-effective condition of one and only one natural number etc. But recursively enumerable as recursively —enumerable is in terms of a recursive function. Hence, as I see it, the treatment of recursively enumerable sets does not *of itself* enable us to generate the set of all bases of such sets, whereas this is possible for the equivalent normal set. Of course the condition on recursive function or λ definable function can be so eased up that when applied to recursively enumerable set each formally written basis without added conditions will define such a set. But that is not in the spirit of the development.

Sincerely yours,

Emil L. Post

August 10, 1943

Dear Professor Post:

I have only a few brief remarks to make in reply to your letter of August third.

The direct reduction to your canonical form of even the functional calculus of first order is still far from clear to me and I suspect it may not be obvious. Of course use may be made of the reduction according to which all substitutions are more directly in the axioms, but I do not see immediately how this helps. On any of the usual formulations of the functional calculus of first order, there is either an infinite number of primitive assertions or else a substitution rule; neither of these things is allowed in your canonical form, as I understand it, and it is moreover not clear how to eliminate them.

As to the question of different methods of defining effectiveness, I am certainly in favor of a reasonable liberality in allowing to different authors their own method of approach. But of course I would not be in favor of the pointless multiplication of different definitions.

It should be noted that Hilbert and Bernays employ two different definitions of effectiveness, introducing the notion of "regelrecht answertbare Funktion" in addition to that of a general recursive function. And on the whole, it seems to me, they tend to emphasize the former notion (which corresponds to the notion of *calculability within* a logic with recursive rules, as introduced in my paper "An unsolvable problem"). Also there are various definitions of general recursiveness, and that of Hilbert and Bernays is not the same as any of Kleene's or as Gödel's. Hilbert and Bernays even use a different term "Quasirekursiv," apparently to distinguish from "allgemein rekursiv," although the two notions are equivalent.

Certainly any enforcement of a rigid restriction to one definition or type of definition of effectiveness would be unfortunate.

Sincerely yours,

Alonzo Church

Professor E. L. Post,
7 Brighton 10 Terrace,
Brighton Beach, New York.

American Journal of Mathematics

December 5, 1941

Dear Professor Church:

The other day I spoke to you about the enclosed manuscript, "Absolutely unsolvable problems and relatively undecidable propositions — Account of an anticipation" by Emil L. Post. You would do me a very great favor indeed by refereeing it for the American Journal. I think it somewhat unfair to request Gödel to do it. Without prejudicing your opinion, I should like to say that under the present crowded conditions it would probably be impossible for the Journal to print a paper of this length, which, if I understand the introduction correctly, contains no essentially new results; and in all likelihood the appendix, consisting of notes from an old diary of Post's, would have to be cut out anyway. On the other hand I should like very much to help Post. In case the manuscript will have to be rejected by the American Journal, could you make any suggestions for another form of publication?

Thanking you in advance for your cooperation,

Sincerely yours,

Hermann Weyl

Professor Alonzo Church
Fine Hall
HW: GB

Jan. 29, 1942

Dear Prof. Weyl,

It is not true that Post's paper contains no essentially new results. The results presented as principal are indeed not now new. But there are new methods and points of view of some interest. And in section 5 there is a substantial result which I believe to be entirely new, and which should certainly not go unpublished; I attach considerable importance to such reductions of the general combinatorial decision problem to special cases of it, for it is precisely in this direction that one might hope ultimately to obtain proofs of unsolvability for various comparatively special mathematical problems — e. g. the word problem of group theory or various well-known unsolved decision problems of topology or number theory (not that Post seems to be near any such result, but I think that the direction in question should be followed up).

It would be possible to present the result of section 5, related results of section 6, and the statement of the "problem of tag" in a shorter paper, relegating all historical remarks to brief footnotes. This would amount to rewriting Part I, with a change in emphasis, partly in the light of my remarks above. I believe that the resulting paper would be much better received by the mathematical public, and in particular would not deceive the reader into thinking that there were no essentially new results. I recommend that the author be asked to do this. If he wishes then afterwards to write up Part II as a separate paper (preferably with omission of the appendix), the question of acceptance of this second paper can be left for later consideration.

I dislike the historical emphasis of Post's paper, and dislike in general such claims of credit for unpublished results. There is always the danger that the claimant, however conscientious, may be subtly deceiving himself as to the extent and definiteness of his anticipation. On the other hand, there is no doubt that Post was very badly treated in the 1920's — not, I hazard the guess, because of any ill will of editors, but simply because of the unavailability of competent referees. Also he deserves sympathy because of his breakdown. (Of course if you decide to quote any part of my report to Post, please omit this last paragraph and quote only from the first two paragraphs.)

Sincerely yours,

Alonzo Church

American Journal of Mathematics

March 2, 1942

Dear Professor Church:

Even though it's more than a month now since I received it, I should like to thank you very much for your constructive report on Post's long paper. I am glad to hear that it contains some substantial new results, and I have acted according to your suggestions.

Sincerely yours,

Hermann Weyl

CORRESPONDENCE WITH RUDOLF CARNAP (1943–1954)

Rudolf Carnap
P.O.B. 1214
Santa Fe, N.M.

May 5, 1943.

Dear Professor Church,

As I mentioned already, your review is of very great interest to me. I regret that it was not published in the JSL. I think that Langford's review there must be rather disappointing for the reader. It gives no information about the content of the book, and I suppose that his discussion will seem to the reader as confusing as it does to me.

I am enclosing a copy of my letter to Quine of January 21st, concerning his paper which in the mean time has appeared in the Journ. of Phil. This letter discusses some of the questions raised in your review and I think you will see that your and my conceptions are not very different (see especially in the table on p. 4 the distinction between designatum and denotatum and the fact that truth values are regarded as denotata of sentences).

A few questions to some special points. You say in your letter p. 8 and in the Dict. of Phil. p. 77 that the denotatum of "that all men are moral" is the proposition. I have Frege's works not here and I wonder whether this view is Frege's or yours. Could you perhaps give me the reasons if that is possible in brief words? To me this distinction between "all men are mortal" and "that all men are mortal" seems an unnecessary complication.

Your statement (p. 12) that on the basis of my definition every language would be extensional is rather puzzling to me because it seems to me obviously erroneous. Is there perhaps a misunderstanding of my terminology? As counter-example, let "A" be L-true and "B" F-true. Hence "A" and "B" are equivalent (D9-4). Let "N" be a sign of logical necessity. Then "$N(A)$" is true but "$N(B)$" is false. Hence these two sentences are non-equivalent. Therefore, "$N(A)$" is non-extensional in relation to "A" (D10-20). By the way, the concept of extensionality and that of L-extensionality are discussed more in detail in Vol. II.

I shall discuss the questions raised in your review and related ones discussed by Quine in a paper which I am preparing now. I believe that the JSL would be the most suitable place of publication for it because it will be somewhat more technical than Quine's paper. Would there be space available in the September or December issues? It might contain between 10 and 15 printed pages. At what time would the MS have to be in your hands for these issues? Before publication, I should like to send the MS to

---

Previously unpublished, these letters are of interest partly because of early formulations of Church's intensional logic, and because of discussion that forms a background for Church and Carnap's disputes over sentences attributing propositional attitudes. There is further correspondence between Church and Carnap dating both before and after the correspondence printed here. It was judged that this further correspondence was of less wide philosophical interest.

you and Quine for comments. You will soon receive a questionnaire on terminology in connection with that paper; if you find time to answer it, it would be very valuable to me.

Very sincerely yours,

Rudolf Carnap

May 10, 1943.

Dear Professor Carnap:

I am replying at once to your letter of May fifth without having been able to give more than a first reading to the copy of your letter to Quine. I am therefore holding the latter temporarily and will return it to you later. I am under heavy pressure of extra work and have to find time for these matters at odd moments as I am able.

I should be very pleased to have you offer your proposed paper to the Journal of Symbolic Logic, especially since the Journal is at the moment rather short of material. We have many things promised and may later be overcrowded if they all materialize, but have little immediately available. For this reason your paper would have a better chance of prompt publication if it came in soon. But of course the first consideration must be to take as much time in the preparation of the paper as is necessary to be sure that you have it in the form that you finally want, and from this point of view I certainly don't want to urge you to haste.

For appearance in the September number of the Journal, manuscripts must be in the hands of the printer by the morning of July fifteenth. This means that they must be mailed from Princeton on the thirteenth, or the fourteenth at the latest. Corresponding dates apply for the December number. The Journal has a fixed rule, however, that all papers offered must be refereed, and I should not want to waive this under any circumstances. This makes it impossible to say what is the latest date of receipt on the basis of which your paper might get into a particular number. I hope that the refereeing might be disposed of quickly, but this would depend on the nature of the paper itself.

The meaning of the word "designatum" proposed on the first page of your letter to Quine seems to me to be a very definite departure from the usage of your Introduction to Semantics. The passage at the top of page 9 of the book seems to make it clear that your *designatum* is the same as Frege's *Bedeutung oder Bezeichnung*, which I have translated "denotation," and this is reinforced by the passage on pages 53–55. Moreover, the very writing (as true) of such things as "Des ('Pferd', horse)" and "Des ('Gross', large)" indicates that your relation of designation is the same as Frege's of denoting—since "Pferd" and "horse" certainly denote the same thing, whether that thing be a class or a property. And "designatum" by its etymology, as well as by your explanation on page 9, means the *designated object*.

Your proposal in your letter to Quine to abandon what Quine calls the principle of substitutivity for equality is a surprise to me, as I had assumed that you would not want to do this. My entire review is on the assumption that this principle holds (as is indicated by my remark about "the inevitable and syntactical rules for '=' "). Hence I will first go on to explain the original intention of my review, as based on this assumption.

On this basis, my demonstration, following Frege, that the denotation of a sentence must be a truth-value seems to me all but inescapable. My criticism of your definition of "extensional" is then an immediate consequence, for if sentences are names of truth-values and synonymous names are interchangeable (cf. page 75 of your book) then extensionality in your sense follows at once. I would make the counter-proposal to define a language as extensional if it contains no two names such that the denotation of one is the sense of the other. This makes the notion of extensionality semantical rather than syntactical, but I believe it must be so. (Or, as an afterthought, perhaps a language should be called extensional if it contains no name whose denotation is a sense.)

In "I believe that all men are mortal" the denotation of the clause "that all men are mortal" must be a proposition rather than a truth-value. For if it were a truth-value, then you could infer that I believe that . . . , replacing the three dots by an *arbitrary* true sentence; and it seems very likely that you would be deceiving yourself if you made such an inference. This observation is Frege's. The word which I translate "proposition" is "Gedanke," explained in a footnote as "objectiver Inhalt des Denkens"; the context leaves little doubt of the justice of the translation.

As briefly indicated in my review, this analysis must be carried further. For example, my young neighbor Mary Jones believes, as I do (truly, let us suppose), that all men are mortal. Moreover it is a fact (let us suppose) that the class of rational animals is identical with the class of men. Nevertheless the inference from

(1) Mary Jones believes that all men are mortal

to

(2) Mary Jones believes that all rational animals are mortal

is grossly in error. For the fact is that Miss Jones (although falsely, let us suppose) believes in the existence of fairies, who (according to her account) are also rational animals but are immortal.

The conclusion must be that (1) is not a statement about the class of men, but rather about the property of being a man. The fact that the English language uses the same word "men" for a statement about the class of men, as e.g., "All men are mortal," and for a statement such as (1) about one of the senses that determines this class, is simply an ambiguity or logical looseness in the language itself. In a careful semantical discussion the distinction should be preserved, say by italicizing the word when it is intended to make the word denote what is ordinarily its sense. The situation is precisely parallel to that in connection with the autonymous use of a word: ordinary English may use the same word sometimes to denote its usual denotation and sometimes to denote itself; in a careful semantical discussion some distinction must be made, say by employing quotation marks. Both of these distinctions, incidentally, are due to Frege—i.e., the distinction between the autonymous use of a word and its ordinary use, and the distinction between the oblique ("ungerade") use of a word and its ordinary use.

The foregoing is on the assumption that the principle of "substitutivity" of equality is required to be preserved unimpaired. (This is the tacit assumption of Frege's paper.)

Now in examining the proposal to modify the principle of "substitutivity" of equality, let us begin with the matter of the autonymous use of a name. I think that the most natural and in the end the simplest course is to hold that a name in its autonymous and in its non-autonymous use has two different denotations and thus is not properly the same name at all, i.e., that autonymy is simply a case of the equivocal use of a name to denote two different things. This would mean that in a formalized language different symbols would be employed to translate the name in its autonymous and in its non-autonymous use. But a conceivable alternative course would be to adopt the Scholastic explanation that a name in its non-autonymous and in its autonymous use has the same "signification" but a different "supposition." This would mean that in a formalized language the same symbol would be employed to translate the name in its two kinds of uses, there would be syntactical criteria to distinguish between "formal" and "material" occurrences of a name (to use the Scholastic terms), and the rule of "substitutivity" of equality would be restricted to "formal" occurrences; and perhaps in addition a second and stronger kind of equality might be introduced which would be substitutive also for "material" occurrences. I believe that something like this was actually proposed by Karl Reach in a paper in the Journal of Symbolic Logic some years ago—I am not familiar with formal details of his proposal, and do not know how successful it was. Moreover it seems to me that Chwistek's "rational semantics" ought to be interpreted along lines of this sort, although it is not clear that Chwistek himself would agree.

Everything I have just said about the non-autonymous and the autonymous use of a name can be paralleled exactly, it seems to me, about the ordinary and the oblique use of a name. There is again the alternative between (a) regarding the two different uses as a case of equivocation and employing two different names in a formalized language, and (b) introducing a syntactical distinction between ordinary and oblique occurrences of a name, restricting the rule of "substitutivity" of equality to ordinary occurrences, and perhaps introducing a stronger kind of equality which would be substitutive also for oblique occurrences. Each of the two courses ought to be followed out in some formal detail, as a check that it does not lead to inconsistency, or (more likely) to some other sort of formal insufficiency for the purpose at hand. The course (a) seems to me the most natural and hence the least likely to lead to formal difficulty. Moreover if both courses should prove to be possible, the course (b) would seem to me the less attractive of the two—unless you can point out some advantage for it which I have not thought of.

In sum, it seems to me that the case for using two different names in place of what appears in common English as the ordinary and the oblique use of the same name is of the same nature and just as strong as the case for using two different names in place of what appears in common English as the non-autonymous and the autonymous [use] of the same name.

My review of your book, by the way, was sent in to the Philosophical Review in September 1942. I had occasion a short time ago to show the review in manuscript to Hempel. It was he who pointed out to me the customary slowness in publication by the Philosophical Review of such material (which I had not at first realized) and suggested my sending you a manuscript copy. I am now promised, however, that the review will appear in their next number.

There is a long review by me of Quine's paper, to appear in the next number of the

Journal of Symbolic Logic. (As a matter of fact the review is disproportionately long, but I felt justified in making an exception because of my conviction of the importance of the comments I wanted to make.) This is immediately relevant to the present discussion, and I will send you a copy, perhaps a spare set of proof, as soon as I have one available. At the moment the printer has my only copy.

Very sincerely yours,

Alonzo Church

Professor Rudolf Carnap
P. O. Box 1214
Sante Fe, New Mexico

May 17, 1943.

Dear Professor Carnap:

I am returning herewith the copy of your letter to Quine, and I enclose also a copy of my review of Quine's paper and answers to your questionnaire about terminology.

There is one further point about the notion of sense which occurred to me only in the course of answering your questions about terminology. For reasons previously discussed, in order to change a sentence into a name of the corresponding proposition it is necessary to change every name appearing in the sentence into a name of the corresponding sense. Whether other changes also should be made was not clear from the previous discussion, but evidently it would be simplest if it could be arranged that a sentence would be changed into a name of the corresponding proposition by merely changing every name in it into a name of the corresponding sense, and generally that a name would be changed into a name of the corresponding sense by changing every constituent name into a name of the sense corresponding to it and making no other changes. I now believe that this can be done, provided that the notation is such, and is so interpreted, that every symbol used is a name, with the exception of the notation for application of function to argument (I include the notation "ϵ" as a special case of this) and the abstraction operator λx with its accompanying parentheses and brackets. It might be necessary—I am uncertain—to analyze the quantifiers in such a way that, e.g., $(\exists x)M$ is construed as an abbreviation for $\Sigma(\lambda x M)$, and "Σ" is interpreted as a name.

In this way, as I said, I think it can be arranged that a sentence is changed into a propositional name by merely changing every name occurring into a name of the corresponding sense. But then it is necessary—as can be seen by taking simple examples—to take the sense of a class name to be, not a property as I previously proposed (i.e., a function from individuals to propositions), but an appropriate special kind of function from individual concepts (sense of individual names) to propositions. In general, the sense of a function name would be a function having as range of independent variable (and as range of dependent variable) the senses of names of elements of the original range of independent variable (and of the original range of dependent variable).

It would indeed be possible (I still think) to follow my original proposal and take the sense of a class name to be a property, but then the change from a sentence to the corresponding propositional name would be more complicated. I now prefer the new proposal and have made out my answer to your questionnaire accordingly.

Very sincerely yours,

Alonzo Church

Professor Rudolf Carnap,
P.O. Box 1214,
Santa Fe, New Mexico.

Rudolf Carnap
P.O.B. 1214
Santa Fe, N.M.

July 12, 1943.

Dear Professor Church,

Thank you very much for your letters of May 10 and 17 with detailed explanations of your view, for your answer to my questionnaire and for the copy of your review of Quine's paper. All of this is very valuable to me in the writing of my article. (Perhaps you overlooked the last question in the questionnaire, whether "confirmancy" seems a suitable term instead of "degree of confirmation"; or perhaps you had no opinion on this point. If, however, you have, I should like very much to know it.)

Today I am writing you in a hurry to let you know that I have not succeeded in finishing my article or part of it in time for the September issue. The article grows very much under my hands. It will certainly be too long for one issue, perhaps even for two. When I can estimate its length I shall write to you and ask for your advice where and how to publish it.

Sincerely yours,

Rudolf Carnap

February 9, 1944.

Dear Professor Carnap:

My delay in commenting on your manuscript "Extension and Intension" is explained by the large amount of extra work which I have had. Even now I am writing without having yet read the manuscript with the care that I want to, but the following remarks occur to me on a first reading.

The entire discussion in your paper must be regarded as preliminary to the construction of a formalized system of logic which shall take account of intension as well as extension, and of a formalized semantical meta-language of this system. To my mind the discussion must be regarded as having failed of its purpose unless it ultimately issues in such a construction, and judgment must be suspended on all the issues which are debated, until we have one or more successful formal constructions of this kind (successful in the sense of a reasonable expectation of consistency together with adequacy for certain purposes). In particular the issue between the method of extension and intension and the method of denotation depends for its decision on the possibility of successful formal constructions embodying these methods and on the relative simplicity, convenience, and elegance of the systems obtained.

Of course I should make the same point in regard to other recent semantical or quasi-semantical publications, including Russell's. It is precisely because your approach seems to me really adapted to the necessary ultimate formalization that I make the point with especial urgency in connection with your paper.

Some preliminary discussion of methods and desiderata is certainly necessary before undertaking the construction of the projected logistic systems. But, granted the point that I am making, I would raise the question whether the actual work of construction might not profitably begin sooner than is the case in your manuscript as it stands. Subsequent discussion could then concern explicit formal systems, and thereby, it would seem to me, the issues would become much clearer and more definite. It may well be that the most efficient progress is to be had by first setting forth one or more specific formal systems intended for the purpose in hand, then subjecting them to debate and criticism and revising them as found necessary, then repeating the process, and so on. This would seem to me preferable to undue prolongation of purely preliminary discussion.

I appreciate that a treatment sufficiently formal to be accurate would simply not be read by many. But I believe that the work of popularization should follow rather than precede the accurate technical treatment.

I have in mind, although only in vague outline, a system based on that of my "Formulation of the Simple Theory of Types" which I would propose as a treatment of extension and intension, or of denotation and sense, along the line of Frege's ideas. The two basic types o and i of that paper would be replaced by two infinite sets of basic types $o_0, o_1, o_2, \ldots$ and $i_0, i_1, i_2, \ldots$, and other types would be provided for (as in the "Formulation") by stipulating that if a and b are type symbols then (ab) is a type symbol. Names of type o_0 (under which head I would include all well-formed formulas of type o_0 not containing free variables) would be construed as names of truth-values, names of type i_0 would be construed as names of individuals, names of type o_{n+1} would be construed as names of senses of names of type o_n, names of type i_{n+1} would be construed as names of senses of names of type i_n, and finally names of type (ab) would be construed as names of functions from the type b to the type a. This is sufficient to provide, corresponding to names of each type, another type for names of senses of such names. The array of types provided is of course more complicated than that of the "Formulation" but it seems to me not excessively so in view of the more complicated material which it is proposed to treat. (The inclusion of senses and their corresponding denotations all together in one type seems to me unattractive, although I have no immediate reason for supposing that it would lead to contradiction. Granted the idea of a theory of types at all, it would seem more natural to think of a sense as belonging to another and higher type than its denotation.)

In the proposed system the axiom of extensionality $p \equiv q \supset p = q$ (which I omitted from the axioms of the "Formulation" because, in effect, I wanted to leave it indefinite whether o was the type o_0 or o_1) would be restored and applied to the type o_0—of course it would not hold in the type o_1. Also for every type a there would be a symbol D_{ab}, b being the type of senses of names of type a, to be read "the denotation of." In particular $D_{o_0o_1} p_{o_1}$ would be read "the truth-value of the proposition p_{o_1}." And appropriate axioms involving the symbols D_{ab} would be necessary.

The following situation in regard to sentence connectives should be noted—I select negation for illustration, but exactly analogous remarks apply to the other connectives. In addition to the strictly extensional negation $N_{o_0o_0}$, construed as a function from truth-values to truth-values, there would be a negation $N_{o_0o_1}$, a function from propositions to truth-values, and a negation $N_{o_1o_1}$, a function from propositions to propositions. The three would be related by the two following equations:

$$N_{o_0o_1} p_{o_1} = N_{o_0o_0}(D_{o_0o_1} p_{o_1})$$
$$N_{o_0o_1} p_{o_1} = D_{o_0o_1}(N_{o_1o_1} p_{o_1}).$$

If desired, modal connectives $M_{o_0o_1}$ and $M_{o_1o_1}$ could be introduced, although of course there is no $M_{o_0o_0}$. (For convenience I replace Lewis's diamond by a letter M.) In the terminology of my reply to your questionnaire, the symbol $M_{o_0o_1}$ would denote a modal function and $M_{o_1o_1}$ would denote a modal function in intension. Lewis's calculus of propositions could be exhibited within the proposed system by identifying his diamond with $M_{o_1o_1}$, his negation with $N_{o_1o_1}$, and likewise for his other connectives (e.g. his conjunction and disjunction would each be construed as of type $((o_1o_1)o_1)$, or, if we use the method of abbreviation employed in my "Formulation," as of type $o_1o_1o_1$). I do not mean that I would necessarily include Lewis's treatment of modality in the proposed system, if I worked it out for publication; but the fact that I could do so with perfect consistency and naturalness (if my hasty conjecture to that effect is correct) is sufficient refutation of your criticism of my "Method III," on pp. 157–158 of your manuscript: In particular there would be no objection to $(p)(Np \supset p)$ with N for "necessary" (your (19-7)); this would appear as

$$[p_{o_1}] \cdot C_{o_1o_1o_1}(N_{o_1o_1}(M_{o_1o_1}(N_{o_1o_1} p_{o_1})))p_{o_1}$$

with C, N, M representing implication, negation, possibility respectively. The square brackets are intended to represent universal quantification of appropriate type, related to the extensional quantification represented by parentheses in the same fashion in which $N_{o_1 o_1}$ is related to $N_{o_0 o_0}$; in fact, if we use A as a syntactical variable,

$$D_{o_0 o_1}([p_{o_1}]A_{o_1}) = (p_{o_1}) \,.\, D_{o_0 o_1} A_{o_1}.$$

I am convinced that Lewis, and indeed practically all authors of modal calculi of propositions, would agree that their propositional variables have propositions, not truth-values, as their range; that their implication, whether strict or material or what not, is a relation between propositions rather than truth-values; and that the formulas which they assert denote propositions rather than truth-values.

Lewis's "diamond diamond *p*," "*p* and diamond not *p*," etc. could be dealt with in the same fashion.

With your ideas about logical necessity in Part IV my impression is that I would largely agree (they can all be appropriately rephrased in terms of the method of denotation). In fact there seems to be here a clear advantage for the semantical approach.

On the other hand I disagree strongly with your explanation on page 137 of the manuscript as to why Frege holds that an expression in an oblique context has not only a different denotation but also a different sense from that which it has in an ordinary context—or, more generally, that we must distinguish, in the case of names of senses just as in the case of all other names, between the sense and the denotation. I object especially to the word "merely" in line 6 from the bottom. There are in fact two much better reasons than the one you give, as follows. (1) Frege holds that the denotation of an expression is uniquely determined by its sense (in my notation as above proposed, that the function D_{ab} is one-valued), indeed it is hard to see how his theory could be made coherent otherwise; but it would be an obvious violation of this if an expression in an oblique and an ordinary context had the same sense but a different denotation. (2) The necessity for distinguishing between sense and denotation in the case of names of senses can be shown directly in the same way that it was first shown for names of extensional objects, namely by considering two names which (like "Scott" and "the author of Waverley") have the same denotation but are not interchangeable for all purposes.

I believe that the same considerations which make it necessary on the basis of the method of denotation to introduce not only senses of names but also senses of names of such senses, and so on to infinity, will compel us on the basis of the method of extension and intension to introduce not only intensional places but also intensionally intensional places, and so on to infinity. Hence I venture to prophesy that the method of extension and intension, when fully worked out and formalized, will be found to be less simple and convenient than the method of denotation. Admittedly this prophecy is premature in advance of the explicit formal development of logistic systems representing the two methods; it is precisely for this reason that I am insisting on the explicit formal development as a prerequisite to deciding the point.

With the aid only of verbal formulations it is difficult to get examples showing the necessity for intensionally intensional places, and I expect this necessity to become clearer after completion of or during the work of formalization. Nevertheless I will make some attempts at illustrations.

According to your D24-20, or (21-18), whenever two properties are the same this can be demonstrated on the basis of linguistic (syntactical and semantical) rules only. Likewise for propositions and other intensions. But there seem to be cases where this leads to difficulties or inadequacies.

(a) Brown and Smith, for a certain special purpose in connection with their manufacturing business are seeking an alloy having the property P (here "P" must be understood as an abbreviation for a specific phrase, say "of melting point between ... and ______ and hardness greater than $* * *$)." Then it would seem that the identity statement is true:

The property which Brown and Smith are seeking = the property P.

Nevertheless a person whose knowledge of syntax and semantics was perfect, but who was misinformed about the motives and business necessities of Brown and Smith, might without logical inconsistency disbelieve this statement. To press the point further, suppose that Brown, speaking of a metallurgist whose competence and information they both respect, says to Smith: "Robinson says that Jones's metal has the property which we are seeking." Brown's sentence is unconsciously ambiguous between two very different meanings. It may be that Brown has consulted Robinson, describing the special needs of the firm of Brown and Smith, and received the reply: "Jones's metal has the property which you are seeking." Or it may be that Brown has not consulted Robinson personally but, purely by accident, has heard a lecture by Robinson containing the sentence: "Jones's metal has the property P." In the latter case it may be that Robinson, without any inconsistency, but purely through misinformation on certain matters of fact, disbelieves and denies "Jones's metal has the property which Brown and Smith are seeking." An exact formal language must be capable of distinguishing the two possible meanings of Brown's sentence spoken to Smith.

(b) Chemical elements have a certain property called their valency. As it happens, the statement is true:

The valency of silver = the valency of sodium.

Nevertheless an expert in semantics (including the semantics of chemical terms) might without logical inconsistency disbelieve this statement, entirely from ignorance of certain matters of non-linguistic fact.

(c) Compare the two statements:

(I) The content of Homer's Iliad is false but not impossible.

(II) It is false but not impossible that ... (where the three dots represent an English translation of all the declarative sentences in the Iliad, arranged in order and connected by "and").

These statements are equivalent but not L-equivalent, for their equivalence depends on the contingent fact that Homer included such and such things in his Iliad. But they differ only by having one name of a certain proposition replaced by another name of that proposition.

From the point of view of Frege's distinction between sense and denotation the explanation of these examples is immediate. In (a) and (b) we are dealing with two names which *denote* the same property but have different *senses*; in (c) we are dealing with two names which denote the same proposition but have different senses. On the basis of your method of extension and intension I see (at this writing) no recourse except to introduce a stricter kind of equivalence than L-equivalence, say LL-equivalence, which has intensionally intensional places; or else at least in these instances to resort to something like Russell's contextual elimination of descriptions (which you were unwilling to admit on the lower level).

For the English translation of Frege's "Sinn," I strongly prefer "sense" to "meaning." Dictionaries give "sense" as the first or primary English translation of the German word, and it seems that this translation well represents Frege's intention. In fact, in places where you do use the word "sense" (e.g. on page 128 of your manuscript, lines 8,9), I should be inclined to say that you are using it in Frege's sense—i.e., in the sense of his "Sinn." The English word "meaning" seems to me too vague, and to have too many meanings, to be satisfactory as a translation of either of Frege's terms. This is borne out by the fact that Gödel, in a shortly forthcoming paper in which he discusses Frege's ideas, uses "meaning" as a translation of "Bedeutung." I objected to this translation also, when Gödel showed me the manuscript of his paper, but he made the point that "mean" is the ordinary translation of the German "bedeuten" in its every-day use, and he preferred to preserve a literal translation. I think, however, that Frege himself may not have been very well satisfied with his choice of the term "bedeuten," since he later suggests "bezeichnen" ("designate") as a synonym. I had forgotten when I talked with Gödel that Russell uses "meaning" as a translation of "Sinn," else I would have reminded him of that.

As to your term "designate," I must say that for a long time I understood you to be using this word in about the sense in which I would use "denote"—as evidence of this, see my definitions of "designate" and "designatum" in Runes's Dictionary. I think that my impression was justified by your explanations. In particular, at the top of page 55 of your "Introduction to Semantics," note that you contemplate taking classes and truth-values as designata in languages having an appropriate structure with respect to extensionality. And cf. also D12-B. At any rate, in my review of your book, I substituted "designate" for "denote" purely out of a desire to conform to your terminology for purposes of this review.

In reply to your footnote on page 60 of the manuscript, I want to say that I still consider my criticism of your definition of "extensional" justified. (1) My criticism was based on the supposition that you intended to preserve the rule of substitutivity for "=" and the principle that expressions having the same denotation or designatum form a true sentence when written with "=" between them; on the whole this supposition seems to be substantiated by your book (all that I had to go on), cf., e.g., page 75, and I believe that if you had intended your definition to require abandonment of these principles there should have been some such indication; and finally, on this supposition it is not true that "any assumption as to what are the designata of sentences is irrelevant." (2) At all events, it is certainly true that your definition characterizes as extensional certain systems (languages, if you wish) which would ordinarily be called intensional. E.g., the system which I outlined at the beginning of this letter could be so set up that all asserted formulas are of type o_0 (formulas intended to represent those of Lewis's propositional calculus would then have to have D_{oo} prefixed before they could be asserted). This system, with its intended semantics (as I have somewhat vaguely described it), would then be intensional according to ordinary use of the word, but extensional according to your definition.

Finally some quite minor points. On page 117 line 7, the word "something" should be replaced by "anything." At several places I have been unable to verify cross-reference and I believe that the entire manuscript should be checked over from this point of view—e.g., I do not understand the reference to section 22 in the footnote on page 60, or the references to (21-25) and (21-26) on page 205. Instead of coining the new word "denotatum," I think my preference would be to preserve existing usage by employing "denotation" for the thing denoted and "denoting" for the relation itself.

By the way, I quite agree that Frege probably did not have either Mill or the Port-Royal Logic in mind when he proposed his distinction between *Sinn* and *Bedeutung*.

Very sincerely yours,

Alonzo Church

Professor Rudolf Carnap,
P. O. Box 1214,
Santa Fe, New Mexico.

Rudolf Carnap
P.O.B. 1214
Santa Fe, N.M.

February 18, 1944.

Dear Professor Church,

Thank you very much for your letter of February 9th. Your detailed discussion is very interesting and valuable to me.

I think you are right that discussions like those in my MS. must ultimately lead to the construction of a system, and that the final judgment on the merits of competitive methods, in this case the method of denotation and the method of extension and intension, must be based on a comparison of completely constructed systems rather than on preliminary discussions alone. On the other hand, I feel that it might save a lot of time if before the final construction of systems the underlying general questions are discussed and clarified as far as possible. I think that this holds especially in the case of systems of a new kind where very little work has been done so far, as in the case of a logic of intensions. I have myself constructed tentative systems of this kind, as I have briefly indicated towards the end of the MS. In connection with this work, and then again in connection with Quine's last paper and your two reviews, I got the impression that it might be advisable to clarify some underlying questions concerning denotation and intensions etc. by preliminary discussions; I am certain that the results of such discussions would help very much when it comes to taking the specific decisions involved in working out the systems. In particular, I think that such discussions can help to avoid waste of energy by showing that certain ways lead to unnecessary complications. The practical question how far one should go with the preliminary discussions is, of course, not easy to answer. At present, I intend to send copies of the MS. to a few interested people and ask for their reactions; I am no yet clear whether I should go further than that, e.g. by making the MS. available to a larger circle by a mimeographed or planographed edition or something like that.

The features of your system which you outline interest me very much, and I hope you will be able to work it out some time in the near future. But my doubts concerning this method are confirmed by the enormous complications which you describe. There is an infinite set of types instead of the one type of sentences in my system, and another infinite set of types instead of the one type of individual expressions. You explain three signs of negation; I suppose there will be an infinite number of signs of negation for the other types. Likewise, an infinite number of modal signs corresponding to Lewis' diamond. Of course, we have to accept these complications if your belief is right that they are inevitable and that corresponding complications would arise in my method.

This leads us to that point in your discussion which interests me most of all (p. 4 middle to p. 6 middle).* I do not understand what you mean by "intensionally intensional places"; but your three examples (a), (b), and (c) are clear and illuminating.

1. Before I explain how these examples would look like in my system, I shall try to apply my method to your formulations or to formulations as closely to yours as possible, imagining a system in accordance with my method but more liberal than my actual system. Since these forms would not occur in my system, my remarks about them are only tentative, without a sufficient investigation of their implications.

Ex. (a). I should write "≡" in place of your "=", and I should say that the two expressions thus connected have the same extension but not the same intension. The intension of the description

(1) "$(\iota \mathrm{F})$(B. and S. are seeking F)"

is not the property P but the property

(2) $(\lambda x)[(\exists \mathrm{F})$(B. and S. are seeking F and no other property, and $\mathrm{F}(x))]$,

which is equivalent with P. The two meanings of Brown's ambiguous statement can now easily be distinguished; one is formulated with "P", the other with (1) or (2).

Ex. (b). Here likewise "=" is replaced by "≡", and the two expressions have the same extension (which is a certain number) but different intensions (number concepts, if we wish to say so, see my MS. p. 245). (I think the two expressions are to be regarded as numerical expressions, not as property expressions).

Ex. (c). I take "the Iliad" as abbreviation for

(3) 'the sentence$_{\mathrm{G}}$ spoken by Homer at such and such a time',

where "G" refers to the Greek language. I understand "the content of the Iliad" as

(4) 'the proposition L-designated$_{\mathrm{G}}$ by the Iliad'.

Then I should say that the Iliad, i.e. the long sentence

(5) 'Agamemnon went to Ilion and ... and ... '

and the propositional description (4) have the same extension (viz. the truth-value falsity) but different intensions. The intension of (5) is obvious. The intension of (4) is another one; it is the same as the intension of the sentence

(6) 'The Iliad is true$_{\mathrm{G}}$',

which is, like (4), in the semantics language. (A propositional description like (4) would here, in distinction to English, be admitted as a sentence.)

2. In my systems, the operand of a description is required not to contain non-extensional signs. Consequently, descriptions with predicate variables are unnecessary because transformable into λ-expressions; likewise those with propositional variables

---

* *Editor's note*: This reference is to the passage in Church's letter of February 9, 1944, that extends from the paragraph that begins "I believe that the same considerations ... " through the paragraph that begins "From the point of view of Frege's distinction ... ".

because transformable into (ordinary) sentences. Thus, only individual descriptions occur. I do not think that this is an essential restriction of the expressive power of the language because it seems to me that all properties and all propositions about which we want to speak can be expressed with the help of λ-expressions and (ordinary) sentences respectively (in both of which non-extensional signs are permitted). It is easily seen that the sentences of your examples can be translated into a language of this kind.

Ex. (a). The description (1) does not occur; but it is translatable into the λ-expression (2).

Ex. (b). No difficulty because no non-extensional signs occur.

Ex. (c). The description (4) does not occur. Your sentence (I), whose reformulation contains (4), is translated first into the sentence

(7) '$(\exists p)[p$ is the only proposition L-designated$_G$ by the Iliad, and not-p, and p is not impossible]'.

If we wish to, we can transform this modal sentence about a proposition into a semantical sentence about a sentence:

(8) 'the Iliad is a sentence$_G$ which is false$_G$ but not L-false$_G$'.

I maintain that, if two expressions have the same intension, this can be shown on the basis of the semantical rules alone. In all three examples, the decisive point of my analysis is that the two expressions in question do not have the same intension.

My analysis of your examples needs only the concepts of extension and intension, not a distinction between L-equivalence and LL-equivalence or between intensional and intensionally intensional places.

Thank you for pointing out wrong cross-references. In the footnote on p. 60, the reference should be to §25 (viz. p. 209 footnote) instead of to §22. On p. 205, instead of "(21-25) and (21-26)" read "(21-17) and (21-18)".

I appreciate it very much that you have given me your detailed comments in spite of your having so much extra work to do. I do not expect a long reply to this letter. But if you could formulate briefly your reaction to what I have said here on your chief problem, I should be very much obliged to you.

When you do not need any longer my MS. or if for the time being you are too much occupied with other work, will you please send it to Ernest Nagel? He wrote me that he would be interested to read it.

With best regards,

Sincerely yours,

Rudolf Carnap

March 10, 1944.

Dear Professor Carnap:

I will make the following comments in reply to your letter, especially in regard to the matter of my examples (a), (b), and (c).

As to (b), I think there are two senses of the word "valency", according to one of which the valency of an element is a certain number (namely, to simplify matters a little for convenience, let us say that number of atoms of the element which combine with one atom of hydrogen), and according to the other of which the valency is a certain property (namely the property of combining with hydrogen in a certain ratio by atoms). Ordinary language does not always distinguish the two senses. Of course I meant valency in the latter sense.

As to (a), your solution of the difficulty here is, if I understand you, simply not adequate. As pointed out in my letter, Brown's statement,

(A) "Robinson says that Jones's metal has the property which we are seeking,"

is ambiguous between two different meanings (senses). But neither of these senses is the same as that of

(B) "Robinson says that Jones's metal has the property P."

This distinction may be obscured when Brown's statement is addressed solely to Smith, since Smith knows quite well that in fact the property P *is* the property which Brown and Smith are seeking. But suppose that there is a third person present who knows that there is a property which Brown and Smith are seeking in an alloy but does not know specifically what it is. The information conveyed to this third person by Brown's assertion (A), no matter in which of the two possible senses he understands it, is quite different from the information which would be conveyed if Brown asserted (B) instead of (A). —Or we might even go a step further and suppose the third person to be *misinformed* about the property which Brown and Smith are seeking; in that case the assertion of (B) might have the effect of leading him to believe the falsehood of (A), in either one or both of the senses of (A).

I would also add that it is essential that the sentence,

"(ιF)(Brown and Smith are seeking F) $= P$",

be true as an equality and not merely as an equivalence. It may happen, in consequence of certain laws of chemistry and metallurgy, that the property P is equivalent to a certain other property Q having to do with the composition of the alloy. Then the property which Brown and Smith are seeking *is* P but is merely *equivalent* to Q. This distinction must be maintained if the situation of Brown and Smith is to be correctly represented.

I am also not satisfied with your treatment of (c). I should like to be able to say that both of the sentences, "Homer asserted the proposition expressed$_G$ [L-designated$_G$, if you wish] by the Iliad," and "Homer asserted that ... ," where the dots are to be replaced by an English translation of the Iliad, are true but have different meanings or senses. The difference in meaning can be seen by considering the different effect of the two sentences on a person who knows the Iliad by description, say as the great epic poem of the Greeks, but does not know the author and is not acquainted in detail with the content. And I should further like to say that the sentence, "Homer asserted that the Iliad is true_G", is probably false, on the ground that Homer was not much concerned with semantics.

Since I wrote before, a further point has occurred to me in support of the hierarchy of senses which results from Frege's analysis.

I think it will be admitted that there are two meanings of the word believe, which are sometimes confused, and which let me distinguish by writing "BELIEVE" and "believe" respectively. According to the first meaning if I BELIEVE a proposition by one name or description I must also BELIEVE it by any other name or description; in particular, if " ... " and "______" are L-equivalent sentences (and hence express the same proposition), and if I BELIEVE that ... , then I must also BELIEVE that ______. According to the second meaning, I may believe a proposition by one name or description without believing it by another; e.g., if I have not seen a proof of the Pythagorean theorem, I may believe that *ABC is a Euclidean right triangle* and at the same time not believe that *ABC is a Euclidean triangle in which the square on one side is equal to the sum of the squares on the other two sides*, and this despite the L-equivalence of the two underscored sentences.

On the sense-denotation basis the explanation here is, I think, that the object of a BELIEF is a proposition, but the object of a belief is rather the sense of a name of a proposition. Since this is an addition to what is in Frege's paper, it requires further thought and examination, but my immediate impression is that it works out satisfactorily. If so, then I should claim it as an advantage for the sense-denotation analysis that it covers this point automatically. You speak of what is in substance the same point, in your manuscript, and outline a proposal to meet it; I do not understand the proposal fully, but I think it is clear that it is an ad hoc addition to your scheme.

I have, as a matter of fact, already sent your manuscript to Prof. Nagel, as you request. My references to it in this letter are from memory, but I hope that this will be accurate enough for the immediate purpose.

Very sincerely yours,

Alonzo Church

Prof. Rudolf Carnap,
Santa Fe, New Mexico.

Rudolf Carnap
P.O.B. 1214
Santa Fe, N.M.

April 10, 1944.

Dear Professor Church,

Thank you for your letter of March 10th. I do not wish to take up your time by further correspondence about these problems. But I think you would like to hear my reaction to your remarks.

To your *example (a)*. I had misunderstood you as to the second meaning of Brown's statement. You do not say now what you take as the second meaning, but I suppose I understand now what you mean. Writing 'b' for "Brown and Smith", 'r' for "Robinson", 'j' for "Jones's metal", I should translate

(9) "Brown and Smith seek P and only P"

into symbols in this way:

(10) '$(G)[\text{Seek}(b, G) \equiv (G \doteq P)]$'.

Let me abbreviate this by

(11) '$S(P)$'.

Then the first meaning of Brown's statement is this:

(12) '$\text{Say}(r, (\exists F)[S(F) \,.\, F(j)])$'.

Now, in the second case, Brown knows this:

(13) '$S(P) \,.\, \text{Say}(r, P(j))$'.

However, what he says (second meaning) is only a weaker consequence of (13):

(14) '$(\exists F)[S(F) \,.\, \text{Say}(r, F(j))]$'.

In this way, my language can clearly show the distinction between the two meanings.

My sentence with '$\equiv$' is not meant as a translation of your sentence with '$=$'; it is clearly weaker than yours. There is in my language no direct translation of your sentence, i.e. one preserving the structure of the sentence, but (10) is a translation in the sense of having the same factual content.

For your new property Q, (15) and (16) hold:

(15) '$Q \equiv P$'.

(16) '$\sim(Q \not\doteq P)$'.

From (16) and (10) we obtain:

(17) '$\sim \text{Seek}(b, Q)$'.

Thus the difference in the attitude of B. and S. towards P and towards Q is clearly expressible in my language.

To your *example (c)*. That your two sentences with "Homer asserted ... " have the same truth-value but different meanings, results likewise from my analysis, since my sentences (4) and (5) have, as I said, the same extension but different intensions.

I admit that the concept of conceptual composition is an addition to my method but, it seems to me, a rather natural one. (The sentence about Homer which you regard as probably false can be accounted for by my method, since my sentences (4) and (6) have, although the same intension, different conceptual compositions.) However, at present, I am not yet clear as to what is the best way of analyzing belief-sentences; I do not yet see clearly enough either for your solution or for mine how it will work out.

With best regards,

Sincerely yours,

Rudolf Carnap

May 11, 1944.

Dear Professor Carnap:

In belated reply to your letter of April tenth I have only the following comment. I think it is clear that in your revised rendering of the two meanings of Brown's statement you have simply adopted Russell's contextual treatment of descriptions in order to eliminate the description "the property which Brown and Smith are seeking." A follower of Russell might well argue: "Since you are willing to adopt Russell's method of eliminating descriptions in the case of intensional entities, why not adopt it also in the case of extensional entities, and so avoid altogether the complication of distinguishing between intensional and extensional places (or between sense and denotation)?" And in fact Russell's method of eliminating descriptions is not different for intensional entities from what it is for extensional. —But for my part, I am inclined to think that your reasons for not adopting Russell's elimination of descriptions are valid, and that they remain valid in the case of descriptions of a property *P*.

My remarks about the connection between senses of names of senses and the rendering of belief propositions were hastily written and I am not sure that they will bear analysis. This is a matter which will require looking into.

I am enclosing a copy of my review of your "Formalization of Logic," which will appear in the Philosophical Review. This was undertaken more than a year ago, at the time that your book first appeared, but because of pressure of other matters has only just been completed.

Very sincerely yours,

Alonzo Church

Professor Rudolf Carnap,
P. O. Box 1214,
Santa Fe, New Mexico.

January 3, 1950

Dear Professor Carnap:

I am enclosing two copies of the paper which I presented at the recent meeting of the Association for Symbolic Logic in Worcester. One is a mimeographed abstract which was distributed to the audience at the meeting, and the other is a longer version which I am offering for publication in "Analysis." A short abstract (three hundred words) will also be published in the Journal of Symbolic Logic.

As stated in the paper itself, it is my present belief that the point which I make may compel you to modify your treatment by admitting propositions in the sense of my first paragraph—i.e., in effect, to admit "intensional structure" as a primitive notion, rather than to introduce it only within the semantical meta-language and by a contextual definition.

Perhaps it is unnecessary to add that my attempted criticism implies no lack of admiration of your book as a whole—which is, I think, the only treatment of its subject sufficiently clear and precise to make criticism profitably possible.

With best wishes for the New Year, I remain,

Very sincerely yours,

Alonzo Church

AC:MMM
Enc.

Professor Rudolf Carnap
Department of Philosophy
University of Chicago
Chicago, Illinois

January 14, 1950.

Professor Alonzo Church
Fine Hall, Box 708,
Princeton, N.J.

Dear Professor Church,

Thank you for sending me a typescript "On Carnap's Analysis of Statements of Assertion and Belief" (four pages), and a mimeographed abstract (three pages). I should like to make some comments on your very interesting discussion. My main point is that, although your final criticism is perhaps correct, I have some doubts concerning your argumentation. My remarks are to the typescript.

In the *first part* your reference to me is misleading because your argument does not hold for the discussion in my book. I have explicitly stated (p. 53) that my analysis does not concern English as a historically given language, but a semantical system *S* containing English words. The sentence "'Man is a rational animal' means in *S* that man is a rational animal" is an analytic sentence of pure semantics, based upon the semantic rules which constitute the definition of *S*, and hence is not an item of factual information. Therefore it seems to me that my name should not occur in the first part but only in the second as example for what you call "the alternative approach". (Since the mimeographed abstract contains only the first part of your discussion, it is, as a criticism of my analysis, erroneous.)

In the *second part* you intend to raise a "modified objection"; but this objection is not clear to me. More specifically, your last argumentation in your last paragraph (with the exception of the last sentence, see below) is not clear to me. You explain here a proposal for German, namely, the analysis of (1′) as (7″). Then you say correctly that "the two proposals stand or fall together". This means that if you can show (7″) as inadequate, then the same holds for the original proposal. But instead of showing the inadequacy of (7″) you merely say that it is neither a translation of (7) nor intensionally isomorphic to it; with this I agree. But where is the "modified objection"?

Thus it seems to me that only the very *last sentence* remains. The point you make here may indeed contain a serious objection against my analysis. But the formulation here is extremely condensed. Both for me and for other readers it will be possible to judge and discuss your argument only if you formulate it in much more extended form.

I am especially interested in the remark in your second paragraph concerning the concept of proposition. You write: "as stated in the paper itself". But the typescript you sent me does not contain this point. Does your remark perhaps refer to an extended version of the paper? You say: "proposition in the sense of my first paragraph". But the first paragraph in the typescript does not indicate the sense except for saying that it is different from my sense. I wonder whether this too indicates that there is a more extended version. Is your sense of "proportion" approximately the same as my sense of "intensional structure", at least as far as the criterion of identity is concerned?

Do you expect to publish soon an exposition of your general theory of meaning? I and many others are waiting for it eagerly. Will it perhaps constitute a part of the book you are writing now or is this restricted to the foundations of mathematics without going into the more general problems of logic and semantics?

If I am not mistaken, the Journal of Symbolic Logic has not yet published a review of my book. In this case and in the case of books by other authors I have sometimes thought that it might be advisable to separate the review of a book from a detailed discussion of its contents. This latter could better be given in an independent paper. This would make it possible to write the review in a shorter form and publish it sooner, mainly for the information of the reader concerning the general content without giving a detailed critical discussion.

With best regards,

Yours,

Rudolf Carnap

January 20, 1950

Dear Prof. Carnap:

Your letter of January fourteenth led me to look up again the passage on page 53 of your book, but even on a careful rereading I find nothing to indicate unmistakably that you intend either of the phrases 'the English language' or 'the semantical system S' in a sense which involves no reference to matters of pragmatics. At least it seems to me that the exact sense which you intend to attach to these phrases is left sufficiently uncertain to justify my considering in my paper various different possibilities, and even to justify my taking the sense which involves a reference to matters of pragmatics as the primary possibility and the other one as secondary.

In defense of the contention, not that you *did* mean these phrases in a sense involving a reference to matters of pragmatics, but that you *can well* be understood as so meaning them, let me make the two following points:

(I) In ordinary every-day usage, of course the sense of the word 'English' in English, as well as that of the word 'Englisch' in German, does certainly involve a reference to matters of pragmatics. This is clear from the fact that it would ordinarily be taken for granted that a German may correctly understand the meaning of the word 'Englisch' in his own language without knowing any English words; and that moreover it would be possible to explain to such a German in his own language the meaning of the word 'English' in English, without telling him the meaning of any other word in English or explaining any of the rules of English grammar. From this point of view it seems to me that to take the word 'English' in English as having the sense of 'the language such that ______,' where the blank is filled by the conjunction of a long list of semantical rules having no reference to matters of pragmatics, is so drastic a departure from the ordinary meaning of this word in English that, if intended, it should have been explained in detail and labeled *explicitly as* a departure from ordinary usage.

(II) Not only a language which has had use in historical fact, but any semantical system may be identified by a reference to matters of pragmatics. And on page 53 the presumption seems to me to be that you intend to identify the semantical system S in this way. For though it is true that, in addition to describing S as a part of the English language, you also speak of semantical rules of S, nevertheless the only examples of semantical rules which have been given thus far in the book are those on pages 4–5, which actually do involve a reference to matters of pragmatics. For example the rules 1-1 when written in full must be: 's' as symbol of S_1 is a translation of 'Walter Scott' as English phrase; 'w' as symbol of S_1 is a translation of 'the book Waverley' as English phrase. And in the absence of any other available meaning for the word 'English' at this point, it must be taken with its ordinary (pragmatic) meaning, something like the language current in such and such a country at such and such a date. And again, the rules 1-3 have an indirect pragmatic reference, since they presuppose the idea of what certain constants refer to, and this in turn has been determined only by rules like 1-1 and 1-2.

To this I would add that my own preference in terminology is to restrict the term 'semantical rules' to rules which have no reference *of the kind here in question* to matters of pragmatical fact, but are more like the semantical rules which occur at a later point in your book. Some examples of such semantical rules would be: (i) In S_1, the symbol 's' denotes Walter Scott; (ii) 'Man is a rational animal' is true in English if and only if man is a rational animal; (iii) 'Man is a rational animal' means in English that man is a rational animal. Here 'S_1' or 'English' is to be taken as a primitive term for which axioms are being given; or alternatively in the description 'the language L such that ______,' where the blank is filled by a conjunction of semantical rules, the term 'S_1' or 'English' (or the like) is to be replaced everywhere just by the variable 'L'. It is to semantical rules of this kind that I mean to refer in §2 of my paper, not to semantical rules like those on pages 4–5 of your book.

Examples (i), (ii), and (iii) will serve well enough here as illustrations of semantical rules, to indicate approximately the kinds of rules that I mean. It is true that (ii) and (iii) would be more likely to appear as consequences of the semantical rules than among the (primitive) semantical rules themselves, but I think this doesn't matter for the immediate purpose.

In order that a description of the form 'the language L such that ______' shall be satisfied by only one language, I strongly suspect that the semantical rules which are filled into the blank must include rules like (iii), and not merely rules like (i) and (ii)—or in other words, as I shall say, that the semantical rules must include rules of sense, and not merely rules of denotation and rules of truth. E.g., it would seem that by rules of denotation and rules of truth alone it is impossible to distinguish between English and the language which is exactly like English except that to the word 'man' is given the sense of featherless biped (instead of its usual English sense).

If we understand the word 'English' (in English) in the sense of the language L such that ______, filling the blank with a conjunction of semantical rules that includes rules of sense, then the proposal to analyze (1) as (7), or to analyze (A) as (C), becomes circular—at least in so far as the analysis is intended to eliminate noun clauses beginning with the conjunction 'that' (and thus to avoid the question whether such noun clauses are to be taken as denoting something like propositions in the sense of the first paragraph of my paper). I discussed this point in the lecture as I delivered it at Worcester; but I said nothing about it in the paper as I finally wrote it for publication, because it seems to me that the question of defining languages by means of semantical rules of various kinds has not yet been sufficiently investigated to support a surmise as to what can or cannot be done in this direction. From this point of view, my criticism of your book at this time would be only that, if you intend 'English' in the sense of the language L such that ______, where the blank is to be filled with a conjunction of semantical rules with no reference to matters of pragmatics, then you have not sufficiently discussed the question what list of semantical rules should be used in order to determine the language *L uniquely*, and whether any of these rules would themselves contain noun clauses beginning with 'that.'

For the limited purpose of showing that there must be some flaw in your analysis of statements of assertion and belief as it now stands, the foregoing criticism seems to me superfluous. I.e., I hold that §2 of my paper is conclusive without it. For as soon as it is seen that (1) and (1′) are translations of each other, whereas the proposed analyses (7) and (7″) are not translations of each other, it would seem to be clear that not both analyses can be maintained. To be sure, as you say, this does not demonstrate directly the simple proposition that the analysis (7″) is inadequate; but it does demonstrate the disjunction, that either (7) or (7″) must be rejected, and this is surely enough to sustain my objection.

The final sentence at the end of §2 of my paper was added as an afterthought, in order to give greater force to my objection. I do not agree with you that it is essential to the objection.

Also I cannot agree that the content of this final sentence requires a more extended formulation. For once it is granted that (7) and (7″) are not intensionally isomorphic, the possibility follows by your own discussion of the notion of belief that 'John believes that (7)' and 'John glaubt dass (7″)' may have opposite truth-values (as English sentence and as German sentence respectively). I might have elaborated by writing out in full the analysis of (α) which is given by your proposal for analyzing statements of assertion and belief, the German translation (α') of (α), and the parallel analysis of (α'); but as these things are automatic and mechanical, isn't it reasonable to ask the reader to supply them for himself?

I would also like to say that the device of translating from English into German, in spite of the prominent place which is given to it in my paper, is not the main point of the paper, but is merely a device for bringing out and enforcing what I regard as the main point. This main point is that (6) or (7) is inadequate as an analysis of (1) because it conveys less information than (1) does. The missing item of information is that 'Man is a rational animal' means in English that man is a rational animal; and it is for the present purpose largely irrelevant whether this is an item of factual information (in case 'English' is defined by reference to matters of pragmatics) or whether it is a logical (but somewhat remote) consequence of an unusual definition of the word 'English.' The consideration of what a German translation of (6) or (7) would mean to a German serves to bring this out more clearly, and it really does not matter whether the information, that 'Man is a rational animal' means in English that man is a rational animal, is an item of factual information which the German would have to look up in a reference work on the English language, or whether it is a logical conclusion which he might with sufficient trouble draw for himself (in the light of the very meaning of the 'Englisch' in German) but which he may nevertheless in practice fail to draw.

Thus it may be thought of as accidental that §1 of my paper assumes a meaning for the word 'English' which has a reference to matters of pragmatics. This feature of §1 can be eliminated by altering the last sentence of the fourth paragraph to read as follows: "For in order to infer (1) from (6) it is necessary to make use of the item of information, which is not stated in (6), that 'Man is a rational animal' means in English that man is a rational animal." As thus modified I believe that §1 is a valid objection against your analysis, independently of the choice between the two suggested meanings of the word 'English' in English.

You are of course right in saying that the first paragraph of my paper makes no positive contribution towards introducing the notion of proposition in a precise or rigorous way. The wording is intended to admit that. And in fact the purpose of the paper is purely negative—not to offer a theory of meaning which makes use of the notion of proposition, but to point out what I believe to be a difficulty for theories of meaning that do not make use of this or some related notion.

As a matter of fact I believe that there are various notions of proposition (according to the criterion of identity adopted) which deserve study. But for the first paragraph of my present paper the relevant notion of proposition is that according to which two sentences express the same proposition if and only if they are intensionally isomorphic. At least I am willing to accept that tentatively, as a first approximation, and subject to later revision if found necessary. However, besides the criterion of identity, one other thing is important if the notion of proposition is to serve the purpose of avoiding the objection that I raise against theories of meaning that do not imply this notion: namely there must be (at least in the meta-language, but preferably also in the object language) some means of naming a proposition without the intermediation of semantical ideas, and in particular without naming or referring to a sentence that expresses the proposition.

Now as to the publication of my own (positive) theory of meaning—so far as I have one.

There is an unformalized descriptive account (not intended to be final) in the "Introduction" of the forthcoming revised edition of my "Introduction to Mathematical Logic." The book itself can hardly appear for at least another year. But the "Introduction" and the first four chapters are available in the form of bound manuscript in the Fine Hall Library at Princeton; and if you would like to see them they could be obtained on inter-library loan (one chapter at a time) by applying through your own library.

The formalization of the Fregean theory of sense and denotation which is sketched in my abstract in the Journal of Symbolic Logic, volume 11, page 31, I have been unable to complete or work out in detail, because of the very heavy pressure of other obligations. I finally decided that it would be desirable to publish what I have, in spite of its chaotic state, in the hope of enabling others to take the matter up and develop a more complete and coherent theory from my beginning. I am doing this with a great deal of hesitation, because I fear that complications of the theory in the form in which I have it may produce an unfavorable reaction to it in some quarters, and because I believe that after working the theory out more fully it may become possible to put it into a more elegant and less complicated form. Nevertheless the paper has now been written and is in process of being printed. I am enclosing a carbon copy of the manuscript ["Logic of Sense and Denotation"].

In addition to this copy of the paper itself, I enclose also copies of two letters about it to my former student, Mr. A. F. Bausch, who is now an instructor at the University of Chicago. All of these enclosures you may keep if you wish. The letters to Bausch contain some additional ideas about the matter of sense and denotation which I felt were not clear and definite enough to put into a printed paper (even one explicitly labeled as a tentative exposition of uncompleted work).

About Bausch himself. He was here for two years (1946–1948) as a graduate student in mathematics, and though he failed to complete a dissertation I formed a favorable opinion of his intelligence and capability. He will never be in a class with such unusual men as Henkin and Kemeny, but I believe that with sufficient initial stimulation and supervision he can develop into a man who will do worth-while research. The pressure which ordinarily forces students to move on if at the end of two years they show no progress towards a dissertation is unfortunate precisely in such cases as that of Bausch. But he may still obtain a degree from Princeton if he submits an acceptable dissertation. And the manuscript of my paper on sense and denotation was sent to him because he expressed a wish to work on this topic.

I have no desire to burden you with the supervision of Mr. Bausch's work, as he is of course my problem. But in case you do come in contact with him, naturally any help you can give him will be appreciated.

The review of your book is in the December number of the Journal of Symbolic Logic, which has been delayed but should appear soon. In spite of the long delay I fear you will find that (except for a few particular passages) it is rather a lengthy abstract of your book than a critical review.

Very sincerely yours,

Alonzo Church

AC:MMM
Enc.

Professor Rudolf Carnap
Department of Philosophy
University of Chicago
Chicago 37, Illinois

Rudolf Carnap
208 W. High St.
Urbana, Illinois

May 8, 1950.

Dear Professor Church,

Thank you very much for your letter of January 20, the copy of your manuscript on sense and denotation, and the copies of two letters to Mr. Bausch. I read all of it with great interest, and I regret that we have no opportunity to talk about these problems orally, because some of them need detailed discussion which cannot easily be carried out by correspondence.

I shall now write some comments. First on the point of your objection against my analysis of belief sentences and then on your manuscript.

1. As I wrote you earlier, I think that the core of your objection is correct, but that the statements in the earlier part of your note do not apply to my conception. Concerning this latter point, I cannot agree with the arguments in your last letter. But since this point is not of great importance, I will be brief here. I wish to stress chiefly that a semantical system in my sense is a system of semantical rules; it does not contain any references to pragmatical features. This is clearly explained and emphasized in "Introduction"; but some of the examples in "Meaning" are formulated in a loose way. Especially the formulation of rules 1-1 and 1-2 on p. 4 is misleading. They ought to be formulated not by reference to a translation but to a designation relation like all the examples of designation rules in the earlier book.

2. My rules of designation are not meant as rules of denotation. They specify not the extension but the intension of the expressions in question.

3. It still seems to me urgently desirable that the extremely condensed formulation of your objection in the second part of your earlier note be considerably expanded. This is, of course, not a logical comment, but a merely psychological one. I believe that the great majority of the readers will not be able to understand your objection in its present formulation.

4. You say on p. 4 of your letter that it is your main point that (6) or (7) conveys less information than (1). This is, of course, correct if the word "English" is defined historically. But now you add that the same holds likewise if "English" refers to a semantical system, in which case the additional information is not of a factual but a logical nature. It is not clear to me how this is meant. In my usage "information" is approximately synonymous with "content"; therefore I would say that two sentences give the same information if they are L-equivalent, and that an L-true sentence gives no information because it has the null content. Presumably you mean "information" in a different sense; perhaps in such a way that two sentences give the same information only if they are intensionally isomorphic?

5. Consequently I cannot agree with what you say in the subsequent paragraph on p. 4. Your §1 may be valid as an objection against other conceptions but not against mine. And the same holds for the formulation of your objection in footnote 5 in the new manuscript because you refer here to English instead of to a semantical system.

6. Perhaps this question hinges on a difference between us with respect to what may be regarded as an adequate analysis. I should accept a proposed analysis or interpretation, perhaps formulated as a general rule, as adequate if in application to any sentence it leads to an L-equivalent sentence. I should not require intensional isomorphism. I believe I am here in agreement with ordinary procedures by scientists and logicians. In this sense I should say, for instance, that material implication may be analysed as "not ... or".

7. I read your ms. on Sense and Denotation with very great interest. Since this article is meant to stimulate investigations by other logicians I think it would be desirable if you were to give, perhaps in a supplementary paper, examples with detailed discussions which would show the necessity or at least desirability of the distinctions you make, and which would make clear your reasons for believing that a simpler distinction, e.g., my distinction of extension and intension, would not be sufficient.

8. Your alternative (0) seems to me very interesting. I hope that you or Bausch will soon make a more detailed investigation of this system, especially with respect to its application to the formulation of statements in modal logic, epistemology and psychology. At the present moment I am not yet able to see clearly to what extent a system of this kind would be not only interesting but necessary for certain purposes. It is presumably not necessary for the purpose of modal logic. Are there any other statements than those of a psychological nature (statements on beliefs, etc.) for which you regard a language of this form as necessary or more convenient?

I regret that I was not able to attend the conference of the Institute for the Unity of Science. I would, of course, have been especially interested to attend and take part in the discussion between you and Quine on abstract entities.

With best regards,

Yours,

Rudolf Carnap

July 11, 1950

Dear Professor Carnap:

I must apologize for having delayed so long before replying to your letter of May 8th. The explanation is as you know that I have been under so much pressure with my work for the Journal and other matters that not only the reply to your letter but also many other urgent things have had to be delayed.

In the meantime, my paper in Analysis has appeared and I am enclosing a copy of it herewith for your files.

With many of the points made in your letter I am quite in agreement and I go on to comment only on those where there seems still to be some difference of opinion.

As regards point 3, of course it is now too late to do anything about your suggestion to expand the paper, unless in connection with some future publication.

In regard to your question as to what "information" means, of course you are right that this requires to be made more precise, especially if a positive contribution is undertaken, as distinguished from the merely negative contribution of objecting to existing analyses. However, I think that you yourself admit that there is something of the sort, as soon as you concede that it is possible to believe one of two L-equivalent sentences and at the same time to disbelieve the other (or, I would prefer here to speak of believing one of two equivalent propositions and disbelieving the other, or else at least to speak of belief with respect to a language). My intention was of course to phrase my criticism in such a way that it would not necessarily presuppose as given the abstract concept of the information conveyed by a sentence or of the proposition expressed by a sentence, or however you want to put it. But then the effect of the criticism, it seems to me, is to bring out the need for admitting such an abstract concept.

The most serious difference of opinion between us seems to concern the question of the meaning of the word "English". Granted that a semantical system in your sense is a system of purely semantical rules, and granted (as I think we both agree) that the word "English" as used in ordinary English, or the corresponding word in another natural language, has a meaning which refers to matters of pragmatics; nevertheless it does not follow that English is not a semantical system. Of course there may be a certain objection against saying that English is a semantical system, namely that the requisite semantical rules have never been formulated with sufficient precision. However, for the purpose of illustration it was my intention to assume that English may be considered to be a semantical system, and it was my impression that you would be willing to agree to that. The point is, that just because the usual sense of the word English refers to matters of pragmatics whereas the sense of the term "semantical system" does not refer to matters of pragmatics, nevertheless this is no reason at all against the proposition that English is a semantical system. In other words, I feel that the statement made in your letter involves a confusion between the sense (in Frege's sense) of the word "English" and the sense of the term "semantical system". It is entirely possible that one of these senses might include a reference to matters of pragmatics and not the other. Indeed any semantical system whatever might have many names with many different senses, and the sense of one of these names might refer to matters of pragmatics and the sense of another name might not.

I must insist that formulation of rules 1-1 and 1-2 on page 4 of your book, instead of being characterized as merely misleading, is one which you are now resolved to change. For as matters stand these rules now refer to English, and the presumption is that the word "English" is used in its ordinary sense since there is no provision for anything to the contrary. That is, the rules as they stand refer to matters of pragmatics, and though I fully agree with you that they ought to be amended so as not to refer to matters of pragmatics, nevertheless my paper is directed toward your book as it stands, not towards amendments of it which you may decide to make. On that basis, the semantical system of which you make use in your discussion and analysis of belief statements, though it remains a semantical system in your sense is certainly named by a name which refers to matters of pragmatics. This is the reason why my paper is written in two sections, the first and longer section referring to what I take to be actually in your book, and the second section referring to an amendment of it which I anticipated that you might wish to make.

This protest that my criticism must be taken as applying to your book as it stands, and not to a later amended version, as a matter of fact I would like to apply also to Footnote 37 on page 50 or your "Meaning and Necessity". Here again I would say that the point in my review which you criticize is correct when taken on the basis of your earlier book as it was actually written, though it is made incorrect by your decision to change the meaning of the word "designate" from that which you give it in your article on semiotic in Runes's Dictionary, and also in various passages in your earlier book, to a meaning which coincides with that of Frege's "express". Of course in one sense it might be said, since there was admitted confusion in your book between "designate" in the sense of denote and "designate" in the sense of express, that you had your choice as to which of the two meanings to adopt in revising the content of the book. However, I think there is a certain objection when this new decision as to the meaning of the word is used to criticize a passage in my review which was intended to apply to the old meaning of the word as I understood it—that is to "designate" in the sense of denote.

I make this protest with a great deal of hesitation, in view of the fact that in your "Meaning and Necessity" you have been very generous in giving me credit at many places and I feel that sometimes you have definitely given me more credit than I deserve. Thus my purpose is certainly not to claim more credit for myself, but rather merely to set the record straight—and at that on what may admittedly be a rather minor point.

With best regards, I remain,

Very sincerely yours,

Alonzo Church

Professor Rudolf Carnap
Post Office Box 1401
Sante Fe, New Mexico.

P.S. To clarify the statement about the definition of extensionality which stands in my review: If sentences designate ("bezeichnen oder bedeuten") truth-values, then—as argued earlier in the review—sentences having the same truth-value, if the sign of equality is placed between them, will make up a true sentence; hence such sentences are always intersubstitutable, and D10–20 is always satisfied. A.C.

P.O. Box 1401
Santa Fe, N. M.
August 25, 1950.

Dear Professor Church:

The University of Chicago Press informs me that a review copy of my latest book "Logical Foundations of Probability" which has just appeared, has been sent to the "Journal of Symbolic Logic". May I use this occasion to make a suggestion concerning not only my book, but in general concerning the reviews of books in the "Journal of Symbolic Logic". To me it would seem advisable to publish at first a short review which merely characterizes the content and nature of the book and which might bring some critical remarks about the book as a whole without, however, going into critical discussions of detail. This detailed critical discussion might then be published separately as an article, no matter whether by the same writer or by another one. It seems to me that this procedure would have the following important advantages. First, it would have the effect that a review would appear earlier, which seems to me a great advantage for the reader. In my opinion the purpose of a review is to inform the readers soon about the existence and general nature of the book. If the reviewers regard it as their task to give detailed critical discussions, the appearance of the review is often unduly delayed. You may remember that the review in the J.S.L. of my "Meaning and Necessity" appeared two years after the book (and it was a relatively small and easy to read book). The second advantage would be that the publication of the critical discussion in the form of an independent article would make it easier for the author to reply. It is, of course, not impossible to reply to criticisms made in a review, and some authors even regularly make such replies; but, on the other hand, many authors, and I among them, hesitate very much to do so because it might give the impression that the author is unwilling to accept critical remarks in reviews. And such replies are sometimes interpreted as attempts to intimidate the reviewers.

Thanks for your letter of July 11, which makes some points in our discussion more clear. I will now comment only on a few points. The paragraph on the word "English" (p. 1–2) is not clear to me. Presumably we use our terms somehow in different ways, which might easily be cleared up in a conversation but makes mutual understanding in correspondence difficult. From the point of view of my conception the essential point is that the word "English" is defined in *descriptive* pragmatics, while the term "semantical System E", as I use the term, is defined in *pure* semantics. Consequently theorems on the system E are analytic, while the statements, at least the more important ones, on English are synthetic and empirical. Thus in my sense of the term "semantical system",

English is certainly not a semantical system. There is not even a semantical system uniquely determined by English; since English is a system of habits there are many semantical systems which in some sense may rightly be regarded as "corresponding" to English. (See "Introduction to Semantics" pp. 13, 14; today I would make a distinction between pure and descriptive pragmatics.) I cannot agree with your remark on footnote 7, p. 81. It seems to me that these remarks, including the handwritten P.S., are not relevant to the reasons stated in the last sentence of my footnote. With respect to the question itself as to the existence of non-extensional sentences, it seems to me that my proof for theorem 11-6 (p. 80) is correct and so obvious that it seems to me rather trivial. I even believe that this result is generally accepted. Perhaps also in this point there exist mutual misunderstandings between us, which could easily be cleared up in a conversation.

Thank you very much for the reprint from "Analysis".

With best regards,

Sincerely yours,

Rudolf Carnap

September 8, 1950

Professor Rudolf Carnap
P. O. Box 1401
Santa Fe, New Mexico

Dear Professor Carnap:

Thank you for the review copy of your book "Logical foundations of probability" and for your letter of August twenty-fifth about it.

I appreciate the fact that you have suffered especially from what seemed to me unjustified slowness of our reviewers in several cases. It appears that the difficulty was that the reviewers found or claimed to find your books hard to follow, and insisted that more time was necessary in order to write an intelligible review of them at all. It was not clear to me that this was justified, and I think that perhaps the fault may have been mine in an unfortunate choice of reviewers. At any rate, I will certainly make every possible effort to see that there is not a similar delay in the review of your present book.

Our first consideration must however be, as far as I am able to arrange it, to get a competent critical judgment of the book. As you know, we have never been satisfied with a mere abstract as a substitute for a review, and we have always felt that to offer such an abstract as a review was a mistake on the part of the reviewer, unless it was actually true that there was nothing that required criticism.

Your suggestion to transform the Reviews section of the Journal into a collection of abstracts, without any attempt on the part of the reviewer to make any sort of judgment of the work for the benefit of the reader, seems to me so completely contrary to the original purpose of the Reviews section of the Journal that I would be very reluctant to consider it at all. The situation in mathematical logic has always been much worse than in other branches of mathematics, in the matter of the frequency of publication of matter containing errors and even absurdities, with the effect that there is serious danger of the field itself falling into disrepute, and of a situation being created in which the careful and competent workers in the field are unable to make themselves heard amid the confusion of incompetent and trivial publications about foundations. I believe that the Journal has made already considerable progress, not only in reducing somewhat the proportion of erroneous or purposeless publications, but more important, in making it possible for the general body of mathematicians and philosophers to have at least some approximate idea as to how the publications in the field are to be sorted out and which ones are worth giving attention to.

The argument is perhaps somewhat weaker for publishing critical remarks about publications which are themselves on a general level of competence, though the reviewer may find mistakes, or matters of opinion with respect to which he wishes to express disagreement. I believe however that it is impossible in practice to draw the line, and that there is also a very useful service to be performed by critical reviewers of publications whose general competence is admitted by every one, including the reviewer.

The suggestion to put critical remarks into an article rather than a review is one which we have used on occasion, when the length of a review threatened to exceed our maximum allowance. However, generally speaking, I believe that this is not practicable, and an attempt to enforce a proposal that all criticism, or even all criticism at any length, should be made into an article rather than a review would simply result in the elimination of criticism in many cases, and the reduction, as I said, of our reviews section to a mere collection of uncritical abstracts.

I realize that probably you have in mind a distinction between two different kinds of criticism. But again it is my belief that this distinction is impossible to enforce in practice, and that our only choice is between allowing critical reviews, and restricting ourselves to a collection of mere abstracts in place of reviews.

I realize that (as you say) a comprehensive program of critical reviews does make many difficulties, and it is even possible that we may be forced to give it up. But, personally, I haven't sufficient interest in a program of publishing mere abstracts to give my own time to it, and I should certainly withdraw from the enterprise at once on this account, if it should appear that we would be forced to reduce our Reviews section to just a collection of abstracts.

As to my letter of July eleventh and your comments on it in your present letter, I think perhaps I ought not to try to write about this further at length, but I will make two brief comments that occur to me immediately on reading your letter.

I would of course admit that to use the word "English" as a name of semantical system is to make use of a very vague term, since it has never been completely fixed exactly what the rules of this semantical system are. However, for the sake of facility in giving illustrations, I understood a tacit agreement between you and myself to ignore the vagueness in the definition of "English" as a semantical system and to use it for purposes of illustration as if English were a well-defined semantical system. The point made in my previous letter as I recall it was just this, that although the class-name "semantical system" involves no reference to descriptive pragmatics, nevertheless particular names of semantical systems may do so. To object to the word "English" as an example of this, on the ground of the vagueness of the word, is beside the point, as it is easy enough to supply other examples to which the same objection would not apply—for example, "the semantical system which is used by Carnap on such and such a page of such and such a book."

As to the matter of your definition of an "extensional language" (or I think you say "extensional system") in your book, "Introduction to Semantics," I would make only the further remark that the languages which are discussed in my paper "A formulation of the theory of sense and denotation," of which I sent you a manuscript copy, seem to me to constitute a clear counter-example against the definition. I mean, of course, not that they show the definition to be false, since truth or falsehood is not to be predicated of definitions, but rather that they show the definition to be so far out of accord with the usual meaning of the word "extensional" that theorems based on this definition are very misleading. For example, in the language which constitutes the main subject of my paper, it is apparently possible to formulate a more or less complete analogue of the Lewis calculus of strict implication, and to introduce modal operators entirely analogous to those of Lewis. Nevertheless this language would be according to your definition extensional.

Very sincerely yours,

Alonzo Church

AC:EW

February 24, 1954

Professor Rudolf Carnap
School of Mathematics
Institute for Advanced Study
Princeton, N. J.

Dear Professor Carnap:

Since our last telephone conversation I have looked again at the passage in Mates's paper which was under discussion (the lower half of page 215 of his paper on "Synonymity") and will try now to offer a definite opinion about it. I am sending a copy of this letter to Professor Putnam, since you told me that he has been interested in the matter.

First, it does seem to me that the device of raising the question as to what might be doubted is not merely incidental to Mates's argument, as he himself intends it, but is actually the central point of the argument. Namely he first puts down the two sentences:

(14) Whoever believes that D, believes that D.
(15) Whoever believes that D, believes that D'.

These, he asserts, are intensionally isomorphic. And he then goes on to say: "But nobody doubts that whoever believes that D believes that D. Therefore nobody doubts that whoever believes that D believes that D'. This seems to suggest that ... if anybody even doubts that whoever believes that D believes that D', then Carnap's explication is incorrect."

For convenience of reference, let me also give numbers to the two sentences about doubt:

(16) Nobody doubts that whoever believes that D believes that D.

(17) Nobody doubts that whoever believes that D believes that D'.

Now Mates's assertion that (14) and (15) are intensionally isomorphic is erroneous, since he is explicitly speaking of your theory in "Meaning and Necessity." For on your theory, (14) and (15) contain, not the sentences D and D', but names of those sentences. And names of intensionally isomorphic sentences are of course not themselves intensionally isomorphic. —Nevertheless it is true that (14) and (15) are L-equivalent, and let us go ahead on that basis.

From Mates's treatment of the matter, it seems clear to me that he does not regard the L-equivalence (or as he says, the intensional isomorphism) of (14) and (15) to be in itself an objection, or at least not unmistakably an objection. But rather he considers that the objectionableness of this requires further support. And this further support he attempts to supply by means of the argument about (16) and (17), which I understand as follows.

Since (according to Mates) the sentences (14) and (15) are intensionally isomorphic, therefore (17) is a logical consequence of (16). But this shows an inadequacy in Carnap's theory, because (16) is almost certainly true and (17) is probably false. Indeed some of those who have considered the question of the analysis of belief statements have probably actually entertained the doubt that is spoken of in (17); or if not, it is clear at least that they might have entertained this doubt without at the same time entertaining the doubt that is spoken of in (16).

Now it seems to me that this argument of Mates is fallacious. For (14) and (15) are not intensionally isomorphic but only L-equivalent. And (17) is therefore not a logical consequence of (16). In fact one of the purposes of your analysis of belief statements was precisely to assure that one who believes a particular sentence need not therefore believe all of its logical consequences, or even all its L-equivalent sentences.

So much for Mates's own argument.

However, I understood you to say that—independently of Mates's argument about (16) and (17)—you felt it to be an objection in itself that (14) and (15) should be L-equivalent (and (15) therefore L-true). Now as long as the notion which we seek to explicate or analyze remains at the vague unanalyzed level, there is no *absolute* criterion of what is an objection in itself, independently of supporting reasons, and what is not. But I point out that it was also one of the purposes of your analysis of belief statements to secure the L-truth of (15), or more generally, to secure that whoever believes a sentence must therefore also believe all intensionally isomorphic sentences. And indeed good reasons can be given why this is necessary, which I needn't now repeat. I only express my surprise that the very thing which you originally sought to secure by your theory should now be regarded by you as being an objection against the theory, not because of any supporting reasons or further argument, but just in itself.

Of course it is true that under some circumstances it is possible that the same person (although fully understanding the language) may make opposite responses to two intensionally isomorphic sentences. But I think it can be argued that it is reasonable in this case just to accept the consequence of the theory, that despite the overt responses he really either believes both sentences or believes neither.

Indeed there does seem to be a difference in principle between the case of the man who admits that he has misappropriated his employer's funds but denies that he has stolen his employer's money (I assume for the sake of the argument that the two sentences are intensionally isomorphic), and the case of Fermat, who believed that $2^{32} + 1$ is prime, but disbelieved the L-equivalent sentence, that $2 + 2 = 5$. For it is reasonable to say of the thief that he knows in his heart that he has stolen, and is engaging in empty verbalism in refusing to use the word. But it would be very unreasonable to say the parallel thing about Fermat, that because he knew that $2 + 2 \neq 5$, therefore he really knew that $2^{32} + 1$ is composite and only refused to admit it. On the contrary, Fermat's belief was genuine, and the later proof by Euler that $2^{32} + 1$ is composite rather than prime has the status of a minor mathematical discovery.

One could even parallel Mates's argument about (16) and (17), applying it to the two (vaguely) analogous sentences:

(16′) Nobody doubts that if D then D.

(17′) Nobody doubts that if D then D'.

Here (unlike (16) and (17) in the original argument) it is true that (16′) has (17′) as a logical consequence. Again I think it is necessary just to accept this—and to bring the charge of not knowing his own mind against the man who claims that he doesn't doubt that if D then D, but does doubt that if D then D'.

It remains to discuss one other tentative suggestion of yours, and that is (as I understood it) that the notion of absolute synonymy might be abandoned in favor of a notion of degrees of synonymy—or in Goodman's phraseology, of a notion of "likeness of meaning" of varying degrees—and that any one who makes a statement about assertion or belief or the like should be required to include as part of his statement the degree of synonymy intended. (The degree of synonymy might not necessarily be a number, but perhaps some other sort of indication of the class of sentences accepted as synonymous for the particular purpose or in the particular context.)

I have no objection in principle to such a theory. And I do not claim to know whether in the end such a theory of likeness of meaning would prove to be superior or inferior to a theory of straight synonymy. But I do say that such a theory must actually be developed in some detail, and in fact treated by the logistic method, so as to have a definite theory available, before it can be judged with any satisfaction, or taken seriously as an alternative to existing theories.

This is my chief criticism of Goodman, that he demands that the notion of synonymy be abandoned in favor of likeness of meaning, but he offers no specific theory of likeness of meaning, and as far as I know has not begun to develop one. Obviously the opposition has no fair chance to criticize the theory or to seek out its weaknesses, as long as the theory is kept so completely vague and unspecified, a mere three words (or at best a short paragraph) at the end of Goodman's paper.

For the analysis of "Columbus believed the world to be round" I suppose your proposal would be that the intended degree d of likeness of meaning should first be specified, and the statement would then be analyzed, roughly as follows:

(18) There is a sentence S in a language L such that Columbus was disposed to a favorable response to S as sentence of L, and the meaning of S as sentence of L resembles the meaning of "The world is round" as English sentence, to within the degree d.

So far I have kept in abeyance entirely the arguments of my paper in "Analysis" which purport to show that, in analyzing statements of assertion and belief, the object of an assertion or belief must be taken to be a proposition rather than a sentence. I now return to these arguments to point out that they apply as well to (18) as to all other analyses which take sentences as objects of belief. The need of propositions (or abstract non-linguistic entities of some kind) for the analysis of belief statements is, as far as I can see, not in the least circumvented by the resort to likeness of meaning in place of synonymy.

Finally, I enclose a copy of a review which may be of some interest, though only rather loosely connected with the discussion in this letter.

Very sincerely yours,

Alonzo Church

AC:vn
Enc.
cc: Dr. Benson Mates
Professor H. Putnam

CORRESPONDENCE WITH W. V. QUINE (1943, 1959)

Tak. Pk.
July 20, 1943

Dear Prof. Church,

Thanks for the excellent review of my "Notes on existence and necessity." Correspondence with Carnap has led me to feel that my point missed him; thus your very understanding review pleases me greatly.

I am much interested in your remarks on anticipation by Frege. Such is my admiration of Frege that I find only pleasure over this discovery, rather than any proprietary regret. Though I should have preferred to keep the history straight.

I think that your point in lines 6–15 of p. 46 is well taken, but would mention that there was really a further consideration in the back of my mind which encourages me to think of the principle of existential generalization as extraneous to the theory of quantification in the strictest sense. It is the eliminability of names altogether, via contextual definition (as in Math. Logic). This is of course no rebuttal of your criticism.

As to your paragraph beginning at middle of p. 46, I agree that variables restricted to an intensional range would conform as you say. I had been thinking in terms of variables of unrestricted range (including intensional values insofar as such entities are countenanced at all in the universe in question, but including also all other values), and I would have done well to make this point explicit in the article. I think that my point, when thus amplified, remains important. Note that intensional contexts of the kind under consideration resist not only quantification involving classes (as your example shows) but also quantification involving individuals, as appears from the example of Evening Star vs. Morning Star.

Note that even your (10) can lead to awkward situations in other connections. Suppose, e. g., we write 'ext ϕ' to designate the extension of ϕ, i. e., the class determined by the attribute ϕ. I take it you would recognize such an operator 'ext', under the program suggested at the end of your review. But then your (10) can be falsified by replacing 'm' therein on the basis of the truth identity '$m = \text{ext}\,\phi$'. We must conclude that the second occurrence of 'm' in (10) is oblique, or not purely designative.

(But there is perhaps no issue here, since you haven't denied that the occurrences of class names in intensional contexts are oblique.)

I have a couple of Bedenken against the last complete paragraph of p. 46. In my sense of "purely designative occurrence", no occurrence of 'Pegasus' is designative, whereas some occurrences of 'Bucephalus' are; yet I shouldn't want to reconstrue the former as designating the sense of 'Pegasus', and introduce a distinguishing notation to record the fact, while leaving the syntactically quite parallel occurrences of 'Bucephalus' unmolested. Whether an occurrence of an expression is purely designative is not in

---

Previously unpublished, these letters bear on a variety of issues in the philosophy of language and philosophy of logic, particularly the nature and prospects of intensional logic, the analysis of belief sentences, nominalism, and a criterion for ontological commitment. Reprinted with permission of the Alonzo Church estate and of the Houghton Library, Harvard University, representing the W. V. Quine estate. Material comes from the Houghton Library shelf mark *91M-68 box 2.

general decided by syntax, you see, but may, as in this example, depend on matters of fact.

This consideration leads me to wonder, moreover, whether my distinction *is* quite the same as Frege's. I don't know Frege's paper, but I wonder whether his "oblique occurrence" may not be narrower than my "not purely designative occurrence", and such that the question of obliquity is indeed syntactically decidable. If so, then I can approve your suggestions (cited paragraph) in relation to oblique occurrence.

Note that my "not purely designative occurrence" is broader than "oblique occurrence" (in the presumed sense) also in other respects than that exemplified above. Another example would be 'cat' in the context 'cataclysm'. A not purely designative occurrence; for I take 'occurrence' in the simple-minded spatial sense, and naturally leave as "not purely designative" all but the very special cases which *are* purely designative. On the other hand presumably 'cat' in 'cataclysm' is neither oblique nor ordinary; for I gather that oblique occurrence is tied up with Sinn.

Obliquity, in a sense tied up with Sinn, would remain a useful and important *subdivision* of those occurrences which aren't purely designative. But I am still a little troubled, having no access to the Frege paper. I gather from the juxtaposition of paragraphs 4 and 3 of your review that the occurrence of 'Giorgione' in (1) is oblique, and that the same would be true of the occurrence of the personal name 'Cicero' in the context ' 'Cicero' ' (or in any broader context of which this latter forms a part). However, the context ' 'Cicero' has 6 letters' clearly relates no more to the Sinn than to the Bedeutung of the personal name which occurs just after the initial quotation mark of that context. Or would you (and Frege) draw a line between (1) and this last example, relegating the latter to the status of 'cat' in 'cataclysm'?

Regarding lines 10–11 of p. 47, and the ensuing paragraph, I am more in agreement than you think. This much I can concede: I am as ready to countenance propositions as I am to countenance attributes. I regard the two as on the same footing. I am as ready to identify the proposition with the sense or meaning of the statement, moreover, as I am to identify the attribute with the sense or meaning of the predicate. I have played with the idea of doing this and still keeping my semiotical notion of meanings as classes of expressions. A proposition thus becomes a class of sense-equivalent statements,* and an attribute becomes a class of sense-equivalent predicates. (For 'sense-equivalent' read 'synonymous'; a better term.)

One further small point in closing: I don't find the last sentence of the third paragraph of your review ("But he does not add ... ") quite conclusive. If we allow 'believes' in (8) to carry the sense of your 'believes-the-proposition-which-is-the-sense-of', then surely (8) does justice to (2) even by your own standards, and does not have the consequence which you impute to it at the end of the paragraph. Of course the hypothetical adherent of the interpretation (8) of (2) could regard the above long hyphenated version of 'believes' as expressing a fundamental relation from his point of view, so as not to presuppose your separate notions of 'proposition' or 'sense' or 'believes' (in your meaning).

Sincerely yours,

W. V. Quine

[In handwriting:] *Such an identification was once suggested by Ayer, I think, independently of the rest of this theory.

July 26, 1943

Dear Professor Quine,

I am planning as soon as I have them to send you reprints of my review of your paper and a related review of Carnap, and at the same time to return the extra reprint of your paper (which, as it happened, I did not need, since I reviewed your paper myself). Meanwhile I am writing in reply to your letter.

When your Brazilian book appears, could you obtain a review copy of it for the Journal? It will evidently be important for us to review it, in spite of any difficulties that may be caused by the language.

From recent correspondence with Carnap, apropos of my review, I have the impression that he may now be more ready to admit at least the force (if not the complete validity) of points made in your paper and in my review of his book, in relation to the distinction between the ordinary and the oblique use of a name or the like. He speaks of publishing a paper on the subject; it remains to be seen just what position he will take.

In your letter, of course you are right in saying that the second occurrence of m in (10) is oblique (if admitted at all — cf. my remark below in the review about the inaccuracy of the notation in (10)).

As to Frege's notion of the oblique use of a name, I think it is true that this is only a special case of a not purely designative use in your sense. Frege introduces the distinction between the ordinary and the oblique use of a name by reference to the substitution rule for equality, but then goes on to say that a name in its oblique use has for denotation that which would be its sense in its ordinary use. Thus he excludes what I think must be recognized in actually existing English (or German) usage as other sorts of not purely designative occurrences of names, namely the autonymous occurrence of a name, occurrences which are mixed between autonymous and ordinary (as in your example which is numbered (1) in my review), and occurrences which are similarly mixed between oblique and ordinary (Frege gives examples).

On the other hand I am unwilling to admit that either 'cat' or 'haven' occurs in (a) 'I have now received your catalogue.' In order for a word to occur in an English sentence, further conditions are necessary besides the physical presence of the letters in consecutive order. These include: (i) the word shall be preceded by a space or by one of a certain list of punctuation marks, or else shall fall at the beginning of the sentence and have its first letter capitalized; (ii) the word shall be followed by a space or by one of the certain list of punctuation marks; (iii) the letters of the word shall not have a space between them at any point; (iv) the word shall be one of the words in a certain complete subdivision of the sentence into words which makes the sentence well-formed according to rules of English syntax. (I add (iv) largely out of caution, lest you propose a modified form of English in which spaces and punctuations are omitted.) In the formation and transformation rules of English, to use Carnap's terminology, the word 'occurs' is needed only in this sense. The other notion of occurrence, according to which 'cat' occurs in (a), is irrelevant to English syntax, and should not be brought into the discussion at all.

Similarly, in that modified form of English in which single quotes are employed, following Frege, to form a name of the word exhibited between them, I should deny that 'Cicero' occurs in (b) ' 'Cicero' has six letters.'

If you define a pure designative occurrence by applicability of the substitution rule for equality, or alternatively by validity of the inference from **M** to $(\exists x)$**N** (notation of my review), I cannot agree that no occurrence of 'Pegasus' is purely designative. On the contrary, the inference from (c) 'Pegasus was a magnificent animal' to (d) $(\exists x)x$ was a magnificent animal' is entirely valid — whether you mean inference in the sense of passage from sentence to sentence or from proposition to proposition. For any one who believed (c) would thereby be justified in believing (d), and in fact would convict himself of inconsistency if he did not also believe (d). Thus the occurrence of 'Pegasus' in (c) is purely designative, in a fashion exactly parallel to the pure designative occurrence of 'Bucephalus' in (e) 'Bucephalus was a magnificent animal.' The fact is that I believe (e), and should draw the inference (d) from it. If through some unexpected historical discovery you should prove to me that Bucephalus never existed, you would thereby convict me of error in my premiss, but not in my inference.

To be sure the occurrence of 'Pegasus' in (f) 'Pegasus never existed' is oblique. But — despite the real existence of Bucephalus — the same must be said of the occurrence of 'Bucephalus' in (g) 'Bucephalus never existed.' The syntactical parallelism is exact throughout.

Thus I disagree with your denial of a syntactical criterion for the distinction between ordinary and oblique occurrences of a name. On the contrary I should say that to just the extent that there failed to be such a syntactical criterion in any language, that language would be unsatisfactory and potentially ambiguous.

I think what you really have in mind is quite a different point, namely that given a name having a certain sense, it may still require an empirical investigation to determine, not only what its denotation is, but even whether it has a denotation. For example, (e) has for denotation the truth-value truth, while (c), although it has a sense, has no denotation at all; but neither of these things could be learned from purely linguistic considerations. However, I consider this to be, not an objection to the last complete paragraph on my page 46, but rather an added point in its favor.

Or you may have in mind the point that — to use again the notation at the top of my page 46 — the inference from (x)**N** to **M** requires as additional premiss some sentence in which **A** has a purely designative occurrence. This is a complication of the rules of inference, but not a serious one. Moreover, it is the same whether **A** is 'Pegasus' or 'Bucephalus', and there is no lack of parallelism.

As a matter of fact, Frege held it to be desirable in a formalized language — not out of necessity, but as a technical device to simplify the rules of inference — to construe all names in such a way that the denotation is the null class in every case where there would otherwise be no denotation. I. e., the sense of every name is to be altered to that extent, and the rules of inference are to be simplified correspondingly.

Now as to your comment on the last sentence of my third paragraph. I think I cannot allow your hypothetical adherent to (8) as an interpretation of (2) to reconstrue 'believes' as 'believes the sense of.' For it so happens that (h) 'Tegucigalpa is in Nicaragua' is a sentence not only in English but also in Gothic, and the sense which it has in Gothic is quite different from that in English. (I trust that you are sufficiently ignorant of Gothic to be unable to dispute me on this point — if otherwise, I shall shift my ground to Old Iranian.) We must therefore distinguish between (j) 'Philip believes the English sense of (h)' or, as I shall put it, (k) 'Philip Englishly believes (h)' and (m) 'Philip Gothicly believes (h),' since these two sentences correspond to entirely different states of mind on the part of Philip. Thus your hypothetical adherent obtains only a series of distinct notions, Englishly believes, Gothicly believes, Germanly believes, etc., and is still unable to free himself of the dependence on a particular language. He may not interpret (2) as (k), because the German translation of (2) is (n) 'Philipp glaubt dass Tegucigalpa in Nicaragua sei,' which by analogy he must then interpret as (p) 'Philipp deutsch glaubt 'Tegucigalpa ist in Nicaragua',' whereas the German translation of (k) is (q) 'Philipp englisch glaubt 'Tegucigalpa is in Nicaragua'.' The fact that (p) and (q) are materially equivalent does not help, as the senses are certainly different (compare the effect of (p) and (q) on a German ignorant of English).

I have a similar objection to identifying a proposition with a class of synonymous sentences. If you identify the proposition expressed by a sentence with the class of synonymous sentences in some language, then you get an English proposition that all men are mortal, a Gaelic proposition that all men are mortal, etc.; there is no single notion of a proposition that all men are mortal, freed of reference to a particular language. If you identify the proposition expressed by a sentence with the class of all synonymous sentences in all languages, you then identify every proposition with the universal class of individuals (or of the appropriate type). If you identify the proposition expressed by a sentence with a class of ordered pairs, each pair consisting of a synonymous sentence and the language in which it is synonymous, you thereby set yourself the perhaps impossible task of defining in general terms the notion of translation between arbitrary languages under the restriction that the notion of proposition shall not be used in the definition.

Sincerely yours,

Alonzo Church

Lt. W. V. Quine
1006 Elm Avenue,
Takoma Park, Maryland

[Quine's handwritten notes in the margins of his copy of Church's letter:
paragraph. 2: easy
paragraph 6: Difficulty of an interlinguistically valid formulation
paragraphs 7–12: erledigt
paragraph. 14: Let it be thus in Gothic; then the sentence is ambiguous, and belongs to 2 propositions; no harm. Inter-linguistic analyticity.
paragraph 15: Do you think it better to assume proposition than to assume translation outright? Surely no clarification.

1006 Elm Av.
Takoma Park, Md.
Aug. 14, 1943

Dear Prof. Church,

I am finding our discussion decidedly valuable. There is a good deal that I must say in answer to your letter. I'll take as my starting point your page 2.

When we take "validity of the inference from **M** to $(\exists x)$**N**" in the sense of logical validity, as you do (mid. paragraph. p. 2), then certainly a notion of purely designative occurrence based on such validity would be syntactically determinate, just as you say (p. 2, last whole paragraph.).

However, consider now my "Designation and existence", pp. 706 f (leaving "Notes on existence and necessity" till later). In the cited pages I intended "validity" (as of the above quote from you) not in the logical sense, but in a certain semantical sense, as follows:

(i) No falsehood $(\exists x)$**N** is such that substitution of the term in question (e. g., 'Pegasus') for the free occurrences of 'x' in **N** turns **N** into a truth.

No such accurate phrasing occurs in "Designation and existence", for I was concerned to avoid excess of technical matter. It was explicit in my classroom lectures, though readers are not expected to have divined this. And it is readily seen that this does draw the required distinction between 'Bucephalus', as naming, and 'Pegasus' as not naming. For take **N** as '$(y)(y \neq x)$', and note that $(\exists x)$**N** is false but **M**, i. e., '$(y)(y \neq$ Pegasus)', is true. This is the example used on p. 706, "Designation and existence".

In "Notes on existence and necessity", on the other hand, I am concerned for the first time with designativity not of terms but of occurrences. In the above semantical sense, "validity of the inference from **M** to $(\exists x)$**N**" is not available as a criterion of designativity in this new connection, since the generality "no falsehood $(\exists x)$**N**" in the above formulation (i) is not available when we are speaking of a specified occurrence in a specified context.

But — and here, now that I've smuggled in the aside on "Designation and existence", comes the direct answer to the mid. paragraph. of p. [illegible] of your letter — I did not intend "validity of the inference from **M** to $(\exists x)$**N**" as a necessary and sufficient condition of designativity of an occurrence, in *any* of the possible senses of "validity". In other words, I deny the initial 'if'-clause of your paragraph.

The structure of "Notes on existence and necessity", pp. [illegible] ff., is rather as follows. First I explain in rough intuitive terms (p. 114) what I mean by "purely designative occurrences", viz. "occurrence in which the name refers simply to the object designated". This is the nearest I come to an outright definition. Then I point out that substitutivity of identity is to be expected and demanded for designative occurrences (p. 114), simply in view of the intended intuitive meanings of the words involved. Next I speak of a basic *connection* between designativity and existential generalization (pp. 116f). In these pages it is implicit of course that existential generalization will *always hold* (i. e., never a true premise and false conclusion) when designative occurrences are concerned. As to the converse, I go only so far as to claim that "clearly the inference [by existential generalization] loses its justification [viz., that "whatever is true of the object designated by a given substantive is true of something"] when the substantive is question does not happen to designate" (p. 116); and again that the inference "is of course equally unwarranted [so far as the above justifying principle is concerned] in the case of an indesignative occurrence" (p. 116). Thus, I might have gone on to say, someone who did not believe in Pegasus, but did believe a certain statement S containing 'Pegasus', would not reasonably attach belief to another statement S′ merely because it followed from S by existential generalization, though indeed S′ might be true and demonstrable in other ways. Your example (c)–(d) does not controvert this, for the man now in question would not believe (c).

I don't deny that there *are* cases of existential generalization which are logically valid despite non-designativity. Your example (c)-(d) is one. But the general case does not hold, failing designativity; a counter-instance is the one earlier in this letter. Your (c)-(d) holds because the context (c) of 'Pegasus' is one which is true for any non-designating substantives (supposing substantives handled as descriptions and descriptions handled as in Russell's theory. If we adopt rather the description theory of my *Math. Logic*, which presupposes class theory, then all substantives designate the null class, so that the problem does not arise.)

There is a reason why I haven't tarried to sharpen my formal criteria of designativity and analyze its relation to quantification etc. more fully. The reason is that when it comes to strict analysis I feel it is simpler and clearer to imagine all terms (and therewith all designation) eliminated entirely, as urged in my *Math. Logic*, p. [no entry]. (The type of contextual definition of terms used need not presuppose class theory as in *Math. Logic*, of course, but alternatively could jump direct to descriptions, essentially a la Russell). Then we can explore the ontological import of a language rather by inquiring into the values of its variables.

Incidentally in the last whole paragraph. of your p. 2 there is probably an inadvertency of phrasing; for I *did* suppose (lines 3–4 of p. 2 of my letter) that the question of obliquity had a syntactical criterion.

The ingenious penultimate paragraph of your letter clinches your point regarding belief, I have no doubt. I fear I'm going to have to leave the poor fellow (my "hypothetical adherent to (8) as an interpretation of (2)") in the lurch, and agree with you that the object of a belief is a proposition. We thus get on to the question what is a proposition, discussed in your last paragraph. Here again I concede the difficulties which you raise, up to but excluding the last sentence. But the definition of proposition mentioned in your last sentence is tenable, it seems to me, if we merely assume an appropriately broad sense for the presupposed (and undefined) notion of synonymy. Viz., let synonymy of statements be a relation of pairs, each the pair of a statement with a language, with no assumption of identity of language from pair to pair. Then a proposition is any largest class of pairs bearing to one another the relation of synonymy of statements.

Thus we merely assume, undefined, the notion which you shy at in your last sentence: synonymy (or translation) between arbitrary languages. No, I don't like this particularly, but then we are assuming intralinguistic synonymy anyway, which is about as bad. (For I am not persuaded by Carnap's talk of "semantic rules", somehow "given" with a language, which define synonymy etc. for the language. This leaves the general idea "synonymy for language x" undefined except in the form "interchangeable according to the semantical rules of language x", and I don't understand the phrase "the semantical rules of language x" in any ordinary pre-Carnap sense of "language".) The essential point is that I don't find the above notion of synonymy (inter- and intralinguistic) as objectionable a starting point as the notion of proposition itself. It is a relation between individually clear and acceptable things; the only question which it leaves unanswered is "just which of those things does it relate?" On the other hand propositions, if assumed unexplained as a starting point, are a fundamentally new and additional kind of entity, neither statement nor class nor etc. etc.

I would hope eventually for an empirical definition or criterion of synonymy as applied to natural languages, and would consider this a solution. For, my attitude toward "formal" languages is very different from Carnap's. Serious artificial notations, e. g. in mathematics or in your logic or mine, I consider supplementary but integral parts of natural language. When an artificial language is set up merely as a semiotic example for study rather than use, I consider it in the spirit merely of a hypothetical argument, as follows: Suppose there were a people (as indeed there theoretically might be, now or in the future) whose natural language had the following characteristics: .... Thus it is that I would consider an empirical criterion, such as alluded to above, a solution of the problem of synonymy in general. And thus it is also that (as noted parenthetically earlier) I am unmoved by constructions by Carnap in terms of so-called "semantical rules of a language" so long as no indication is given of how, empirically, to decide whether or not such and such is a semantical rule of the natural language of the inhabitants of such and such a valley.

Incidentally these latter reflections raise Bedenken against your appeal to the universal class, in the last paragraph of your letter. But I grant that the limits of the class in question would be pretty indeterminate, since we want to allow for all languages past, present, and future. I shouldn't want to build on the non-existence (for all time) of a language and for this reason the argument in your last paragraph carried weight.

Note that the constructions considered in top half this page would be simplified by construing a statement itself as an ordered pair of verbal expression and language, then reverting to statement-synonymy and propositions as relation and classes of statements themselves.

Your restricted sense of 'occurrence', pp. 1–2 of your letter (in the carried over paragraph), is unquestionably an important sense, within the framework of English orthography and syntax. But it was in the absence of any general formulation of this idea, independent of details of a particular language, that I have preferred to get along with "the physical presence of the letters in consecutive order" as my basic notion of occurrence. I am afraid the general formulation of your more limited sense of occurrence might presuppose semantic ideas of the very sort that we have been trying to construct on the basis of "occurrence" and other notions.

Yes, when the Brazilian book appears I will see that the Journal gets a copy. But I have no idea when that will be. It is nearly a year now since I left the finished MS in Sao Paulo. Every time I hear from Brazil it is with the assurance that things are moving nicely and the book is about to be set up and the first proofs should be ready in a couple of weeks; but nothing happens.

Sincerely yours,

W. V. Quine

November 9, 1943

Dear Professor Quine:

I have delayed shamefully in replying to your letter of August fourteenth (it has been on my desk all that time, under my continuing intention to answer it). I must apologize, and explain that I have been under such heavy pressure from routine work that all my correspondence has been neglected.

It is clear from the first part of your letter that I partly misunderstood your intended meaning in distinguishing between purely designative occurrences of names and others. I should still maintain, however, that the simplest and most satisfactory solution of the questions arising in this connection is to follow Frege's distinction between sense and denotation, and to avoid oblique occurrences of names simply by taking as name of the sense of a name **A** always some name other than **A**. A language in which oblique occurrences of names are admitted but in which there is some syntactical criterion for distinguishing oblique occurrences from others is also acceptable. But a language in which there is no syntactical criterion for distinguishing between oblique and ordinary occurrences of names must almost certainly be logically unacceptable; for it is hard to see how cases could be avoided in which it would be indeterminate whether a certain inference (e. g., from **M** to $(\exists x)$**N**) were valid, whereas of course in a logically acceptable language the relation of validity of inference must not be subject to such vagueness.

Your example involving "$(y)(y \neq \text{Pegasus})$" as **M** does not invalidate this. From the Frege point of view the explanation is that you must distinguish not only between 'Pegasus' and '$\underline{\text{Pegasus}}$' (underscored), the latter being a name of the sense of 'Pegasus', but also between '$=$' (denoting the usual relation of identity) and say '$\equiv$' (denoting the relation which holds between two senses if and only if they determine the same denotation). Then '$(y)(y \neq \text{Pegasus})$' is not true and not false; rather it is, like (c) of my letter of July twenty-sixth, an example of a sentence which, although it has a sense, has no denotation at all. As to '$(y)(y \neq \underline{\text{Pegasus}})$', this is false. And as to '$(y)(y \not\equiv \underline{\text{Pegasus}})$', this is true, hence also the consequence '$(\exists x)(y)(y \not\equiv x)$'.

In a language in which the distinction between 'Pegasus' and '$\underline{\text{Pegasus}}$' is dropped in favor of a syntactical distinction between ordinary and oblique occurrences of 'Pegasus', the explanation is similar, since '$=$' and '$\equiv$' must still be distinguished as non-synonymous.

In short I dispute the assertion on page 2 line 24 of your letter, "But the general case does not hold, failing designativity" — at least as applied to any logically acceptable language now known to me (even in outline).

Of course I agree that in the English sentence 'Everything is other than Pegasus' it is vague as to whether 'Pegasus' has a purely designative occurrence or an oblique one, and whether 'is other than' corresponds to '$\neq$' or to '$\not\equiv$'; and hence that the English language requires modification in order to make it logically acceptable.

Now as to your proposal to define 'proposition' in terms of 'synonymy' (of ordered pairs composed of a sentence and a language) rather than vice versa. If this were based on an abstraction notion of language (like Carnap's though perhaps with necessary substantial modifications) my comment would be that your choice of primitive terms, while no doubt admissible, seems to me a strange one, in that you prefer a remote and difficult notion to one which I regard as familiar and ordinary. The economy of categories attained, in that you succeed in this way in subsuming the category of proposition under that of class, does not seem to me of great importance; if it is attained at the price of removing 'proposition' and names of particular propositions from the object language one step up into the hierarchy of meta-languages (this may well prove to be necessary), then the price is too high.

However, as I understand you, your proposal is based not on an abstract notion of language but on an empirical one (a language is something which is actually spoken or written by a particular group of people as a matter of historical fact, past, present, or future). And on this basis I think there are more serious objections.

In the first place there may be some difficulty in producing even one example of an empirical language in this sense. The English language, in the sense of the language which you will find described as such in conventional English grammars and dictionaries, is not such an example, but is as much an abstract language as any of Carnap's most fanciful constructions (the chief difference from Carnap's constructions is that the syntactical rules have been less exactly formulated by the English grammarians and the definition of the language is therefore comparatively vague). There are two reasons for this. (i) The English of the grammarians is based, not on all English sentences which have been spoken or written in historical fact, but only on those spoken or written by good users of English in their more serious moments — the distinction between good users and bad being only in part, or at best very vaguely, an empirical one. (ii) The class of English sentences according to the rules of the grammarians is a much wider one than the class of English sentences which have actually been spoken or written. Thus the rules of the grammarians are not a statement of observed fact but an induction from observed fact; and, like all inductions based on a limited number of observations, this one can be made in more than one way. The grammarians have perhaps chosen the simplest and most natural way (though this may be doubted as regards some special points); certainly they have not chosen the only way.

If we abandon the grammarians and try to go to English as it is actually spoken and written, we are faced again with the double difficulty that existing usage is conflicting and that the class of historically actual English sentences is incomplete. If we will not have the abstract construction of the grammarians, we must make our own.

Of course in addition to the historical fact that certain sentences have been uttered, there is the historical fact that certain rules of grammar have been taught in the schools. But this teaching of rules involving universal quantifiers itself amounts to an abstract construction of the language, analogous to the act of the logician (also a historical fact) in publishing a paper describing the artificial or formal language which he has constructed, and circulating it among his colleagues.

In the second place, even if the existence of empirical languages as a genuinely separate species is granted, their [sic] remain objections of vagueness in the notion of an empirical language which defeat its use for defining 'proposition'.

For example, what shall be said of an artificially invented proposed international language? Some of these have been promoted by an international society, which published a journal in the language in question, and whose members corresponded in that language and spoke in it at meetings (some indeed having no other language in common). Surely this constitutes sufficient sanction of actual use to make this a language in your empirical sense? If your answer is no, then I will press you by asking what if in the distant future such an artificially invented language becomes the universal language of the world, and other languages are ultimately forgotten even by antiquarians — and just where in this direction would you draw the line? If your answer is yes, then I will press you in the other direction by putting cases of progressively smaller and smaller sanction in use, ending with the case of a language which exists only in one book published by its inventor giving the grammar and vocabulary of the language and perhaps a few sample paragraphs of composition in it. Just where in this direction would you draw the line?

Or, to bring out the vagueness in another direction, what of dialects of the same language which are different but sufficiently similar to be inter-intelligible? How different must a dialect be in order to be a different language? If your answer is that any slightest difference makes a different language, then how many people must speak a dialect in order to make it a language? Is one person sufficient? Is my variety of English a different language from ordinary English because it happens that I habitually mispronounce or misspell a particular word in a fashion in which no one else does? Is English spoken with a German accent a language in its own right? (It does have considerable sanction of use.) If the kind of Germanized English spoken by one particular German is different from that spoken by any other, is he the possessor of a one-man language?

Let me cite a case which seems to me analogous. In connection with Euclidean geometry it is tempting to introduce a definition of 'straight line' to the effect that a straight line is a tightly stretched fine wire, or perhaps the shape of a tightly stretched fine wire. But it is generally conceded that this is unacceptable, since the fine wire is different from the straight line in having thickness (further, under sufficient magnification, it is seen that the edges of the wire are not sharp, and the omnipresence of some at least slight gravitational field means that the wire must be bowed slightly in spite of tightest stretching unless aligned exactly with the field). In fact if 'straight line' is defined in this way substantially all accepted theorems of Euclidean geometry become simply false. And the like holds for other attempted empirical definitions of 'straight line.'

The usual opinion is that 'straight line' and other primitive terms of Euclidean geometry have no direct empirical meanings but represent abstract concepts. There are indeed applications of Euclidean geometry to observable reality. And one of these does bring straight lines into a certain relationship with tightly stretched fine wires, but this relationship is not anything as simple as identification of the two.

I think I should hold that the abstract concept of proposition developed as a part of the subject of logic has a relationship to your empirical concept of a proposition (as a class of sentence-language pairs, 'language' being given an empirical meaning) entirely analogous to the relationship between the geometric concept of a straight line and the empirical concept of a tightly stretched fine wire.

I have not taken pains in the foregoing to distinguish between oblique occurrences of names and other sorts of non-designative occurrences, because other non-designative occurrences (such as autonymous occurrences) are easily eliminated by methods which are not in dispute, and I prefer to think of this as having first been done. As to the alleged occurrence of 'cat' in 'catalogue,' this is not an occurrence in my terminology; and I am prepared to defend my terminology, with its more complicated notion of occurrence, against yours.

In general a language may be expected to have its own notion of occurrence which is appropriate to it and is necessary in order to state its syntactical rules. In practice this notion of occurrence appropriate to the language is usually included in your simple notion of occurrence as a special case, but there is no reason why this has to be so (I can devise examples to the contrary). The essential thing is that there should be a set of atomic symbols and a recursive method of constructing well-formed formulas, including sentences, from the atomic symbols; a well-formed formula occurs in another one if it is used at some state in the process of recursive construction of it.

For English the appropriate notion of occurrence is that described roughly in my last letter. Its general formulation does not as you fear presuppose semantical ideas, but is purely syntactical in character. Or rather I should say that it must become purely syntactical in character as a part of the process of removing existing vaguenesses in the rules of English grammar and syntax.

This more complicated meaning of occurrence is required in order to state rules of English grammar and syntax which it is necessary to know in order to speak English correctly. Thus, even if you do not like its complication, you must have this more complicated notion of occurrence in order to discuss and study the English language effectively. And once the more complicated notion of occurrence is given, your simpler notion becomes superfluous — I think it should be abandoned as irrelevant to English syntax.

Very sincerely yours,

Alonzo Church

Lt. W. V. Quine
1006 Elm Avenue
Takoma Park, Md.

1006 Elm Avenue
Takoma Park, Md.

November 13, 1943

Dear Prof. Church,

I was glad to get your letter of November 9. Your remarks fall into three main parts, separated in your letter by ink lines. I shall answer them under heads I, II, and III.

I

Frege, I gather, regarded his oblique occurrences as designating the Bedeutung of the expression, as opposed to the Sinn. Now it appears to me that you, especially in the third paragraph of your letter, are imputing a similar conception to me; whereas in point of fact I have intended no designata at all, neither Bedeutungen nor otherwise, for my not purely designative occurrences. In handling my example '$(y)\neg(y =$ Pegasus)' therefore, I have had no interpretation in mind which could be made explicit by your proposed recourse to 'Pegasus' or '$\equiv$', though I regard these signs, with the meanings which you give them, as useful in their own proper places.

No, what I had in mind in connection with my example did not go so far afield. I was imagining 'Pegasus' in effect as a description, say, '$(\iota x)(x$ is winged. x is a horse. x is captured by Bellerephon)' (see my "Designation and existence", top of p. 703). I was thinking of descriptions, in turn, as handled by contextual definition according to Russell's theory (see my letter of Aug. 14, p. 2, just below middle). Under this theory the example '$(y)\neg(y =$ Pegasus)' becomes true, as I had maintained, rather than neither true nor false as maintained in your last letter.

Nor am I according 'Pegasus' any very special treatment by this descriptional analysis, for, as emphasized in my *Mathematical Logic* pp. 149ff, this treatment is generally available to names.

Actually, I should prefer for this purpose the simplification of Russell's contextual definition of descriptions given in my "New foundations for mathematical logic", coupled with the definition of identity used in *Mathematical Logic*. On this basis equally, '$(y)\neg(y =$ Pegasus)' becomes true. This in particular is my answer to your remark at foot of p. 1 of your letter, in re "logically acceptable language". But I feel incidentally that truth of '$(y)\neg(y =$ Pegasus)' is well suited to the Umgangssprache, vague matter though this is.

A description, according to the above theory, would have a designatum at all, for me (and hence admit of purely designative occurrences among others), if and only if the condition on 'x' in the description is fulfilled by exactly one entity.

If on the other hand we adopt the description theory of *Mathematical Logic* (or Frege's own), which presupposes class theory and endows every description with a designatum trivially if not otherwise, the situation becomes quite different — as remarked in my letter of Aug. 14, p. 2. In particular, '$(y)\neg(y =$ Pegasus)' then becomes false. [2]*

*Numerals in brackets indicate beginnings of pages in the original.

II

Your comparison of language with geometry, p. 4, rather than controverting my point of view toward language, provides an illuminating way of presenting the latter.

In geometry (Euclidean geometry under the standard "interpretation" or quasi-interpretation, as opposed to a completely uninterpreted formal calculus) we understand 'line' by extension or idealization or schematization of the properties of tightness and thinness in a string. Correspondingly for 'surface', 'plane', 'point', and any other terms that we may treat as primitive within the geometry. This, essentially, is your point of view, I gather, and I share it. And it is only by some such process of conceptual extrapolation of physical properties that I understand geometry as a "geometrically interpreted" system, at all. Otherwise the words 'line' etc. are only obscure synonyms for neutral class-parameters 'K' etc.

Now the case is parallel for language. The artificial formal languages, and to a lesser degree even, as you say, the pedantic natural languages such as "standard English", are idealizations or schematizations of language in the more direct empirical sense studied by field linguists.

Language in the direct empirical sense last alluded to is, I should say, *the pattern of actual and potential linguistic behavior of one individual at one state of his development.* (A vague formulation, but right, I think, as far as it goes.) I say 'potential' because at one momentary stage of development the individual has time only to say one syllable at most. Even the discovery of his "language" in the present narrowly empirical sense of the term requires empirical hypotheses as to what he *would* have said at that same stage if other circumstances had been varied thus and so. Such hypotheses rest on observation of behavior of the same individual at other stages believed similar in relevant respects, or of other individuals believed similar in relevant respects; also upon drawing smooth curves through scattered points; but all this is typical of scientific method, and does not belie the formulation underlined at the beginning of this paragraph.

Then comes community dialect, as an extension or idealization already of language in the above narrowest sense; then "standard English" etc. as a more abstract idealization; and so on, even unto (interpreted) symbolic languages never to be proposed for actual use at all.

But I can understand such terms as 'statement', 'analytic', or 'synonymous', e. g. in the context 'What statements are analytic in your new language S?' or 'The analytic statements of any language x are mutually synonymous in x', only insofar as I know some rough empirical criterion for analogues of 'statement', 'analytic', and 'synonymous' applying to languages in the more direct empirical sense of 'language'. The exigencies here are much like those in geometry, where, as noted on 2nd paragraph of this page, I cannot understand 'line' without knowing how to recognize the empirical tightness and thinness which were being idealized.

Such is the nature of my empirical preconceptions in formal semantics. The geometrical analogy illuminates it very well, as seen. On the other hand, failing some general conception of 'statement', 'analytic', etc. such as I have wanted, the situation is this: Granted you can formulate 'statement' and possibly even 'synonymous' and 'analytic', for a *given* language, by recursion in syntactical terms, this provides no explanation of the general relation 'x is a statement of language y', 'x is analytic in y", etc., where 'x' and 'y' are variables. [3] This has long been an issue between Carnap and me. If 'statement', 'analytic', etc. are to be defined *ad hoc* for each new language, with no general principle in the background, there is then no sense to the question "What is a statement (or, what is analytic) for your new language?"; nor should any expectation be raised by such a promise as: 'Now I will explain what are to be statements (or analytic) in the new language,' beyond what would be raised by: 'Now I will define a certain class alpha of which you never heard before.'

III

What I have said in the foregoing paragraph of 'statement' and 'analytic' applies, mutatis mutandis, also to 'occurrence'. So applied, it constitutes an expansion of what I meant in the next to last paragraph of my letter of Aug. 14. Compare also your own remark, foot p. 4 of your last letter: "In general a language may be expected to have its own notion of occurrence which is appropriate to it." True, the syntactical conditions of occurrence will vary from language to language (granted your type of occurrence-concept rather than the more immediate physical sense in which I had used the term), just as the syntactical or other conditions of 'statement', 'analytic', etc. vary from language to language; but there remains the question of the general relation 'x occurs in y for language z', with three variables, analogously to what has been said of 'statement', 'analytic', etc. Similarly one may ask the meaning of your own *general* phrase 'own notion of occurrence'.

Manufacture of my Brazilian book has been progressing, though slowly. Typesetting was finished Sept. 24. First proofs were to be corrected down there, and second proofs were thereupon to be sent to me. I await them now.

Sincerely yours,

Willard V. Quine

1006 Elm Avenue
Takoma Park, Md.

Dec. 1, 1943

Dear Professor Church,

Enclosed is a short paper setting forth an idea which I found in some forgotten notes of seven years ago. The idea is so obvious that I think it must, in its essentials, be old to the literature. Hence I am anxious to have an appraisal. I'd like to see it published in the Journal only in case it is found not merely passable but interesting.

Navy regulations require that I get naval permission on anything I release for publication, but this is presumably a mere rubber-stamp technicality in the case of a paper like this, and I will make the arrangements on the basis of my carbon copy after (if and/or when) publication is found desirable.

Sincerely yours,

Willard V. Quine

December 7, 1943.

Dear Professor Quine:

It is strange how many people (including myself at one time) have known of Tarski's paper of 1933 without realizing that it contains a consistency proof for the form of the system of Principia which is there presented, the leading idea of the proof being somewhat the same as that in your proposed note. I once told Tarski that I thought this might be due to the unnecessarily difficult notation he employs in that paper.

Essentially the same consistency proof for the system of Principia as Tarski's was later published by Gentzen and Beth, independently of one another, and without knowing of Tarski's proof. Gentzen and Beth apply the method directly to less simplified forms of the system of Principia than Tarski's, and they point out that the axiom of choice need not be excluded from the system in order to make the consistency proof possible (but only the axiom of infinity). See my reviews in the Journal, I 199, II 44.

A distinction should be made between (a) consistency proofs which presuppose elementary arithmetic in the sense that they are carried out within a system adequate for elementary arithmetic, and (b) consistency proofs which presuppose elementary arithmetic both in this sense and in the sense that the consistency of elementary arithmetic appears as a hypothesis in the proof. A stronger presupposition is, in general, made in case (b), because by Gödel's theorem a system barely adequate to elementary arithmetic is not adequate to the proposition that elementary arithmetic is consistent.

Now the consistency proofs of Tarski, Gentzen, and Beth for Principia without axiom of infinity are of kind (a), and hence better than yours, which is of kind (b). On the other hand it seems that for set theory without axiom of infinity a consistency proof of kind (b) is the best that can be hoped for. Here, however, your consistency proof is anticipated by Ackermann, who applies the method of finding an arithmetic model to a set-theoretic system somewhat stronger than yours — see Bernays's review III 85.

You have the consistency proof for your system Γ (without axiom of infinity) very clearly and concisely arranged. It might well be included in some future book or longer expository paper, if occasion arises, but I do not think its publication as a separate note in the Journal would be justified.

I go on to reply briefly to your letter of November thirteenth.

In discussing your example '$(y)(y \neq \text{Pegasus})$' from the Frege point of view, of course I did not intend to impute this point of view to you, but rather only to show that it is in no way controverted by your example. I still persist in my opinion that Frege's explanation of the logical [2] difficulties which are in question here, although not the only possible one, is on the whole the most satisfactory one so far proposed. Also my grave doubt remains whether a logically acceptable language is possible in which there are both designative and non-designative occurrences of names but no syntactical criterion of distinction between the two kinds of occurrences.

Your retort that you had in mind a language in which all names are construed as descriptions, which in turn are defined contextually, is beside the point. For contextual definition is merely a device employed in the presentation of a language, and notations and expressions introduced by contextual definition are not themselves part of the language. If the distinction between designative and non-designative occurrences of names is to be discussed, we must begin by assuming a language which actually contains names!

In reply to section II of your letter, I would say that even the field linguist, as soon as he begins to speak of the potential as well as the actual linguistic behavior of an individual is making an abstract construction of a language. This point is not altered by the fact that his procedure is defensible as good scientific method. Thus I still cannot follow you in drawing as sharp a distinction as in your previous letter between natural languages and artificial formal languages.

I will try to restate concisely and clarify my fundamental point here (which I may previously have labored too much). Despite its inexactness, your proposed empirical definition of a proposition, or something like it, no doubt has a necessary place in connection with applied semantics, as indicating how the term "proposition" of abstract semantics is to be applied. Just so, the inexact physical definition of a straight line has a necessary place in connection with applied geometry, as indicating how the term "straight line" of abstract geometry is to be applied. But this does not alter the fact that in abstract semantics "proposition" must either be an undefined term or be given a definition by means of abstract semantical and logical terms; or the fact that in geometry "straight line" must either be an undefined term or be given a definition by means of abstract geometry terms.

A similar point seems to apply to other semantical and syntactical terms, such as "analyticity," "synonymy," "occurrence." In general syntax, "occurrence" or some related term must very likely be undefined, although for any particular language "occurrence" be capable of a purely syntactical definition.

Sincerely yours,

Alonzo Church

Lt. W. V. Quine,
1006 Elm Avenue,
Takoma Park,
Maryland

1006 Elm Avenue
Takoma Park, Md.
December 10, 1943

Dear Professor Church,

Thanks for indicating the many anticipations of my consistency proof. I felt the thing must have been done before, and would have searched a bit if it hadn't been that I knew you'd have the answer pat; but even so I'm shocked at the multiplicity of my oversight. The paper itself represents no lost effort: a casual rough note in 1936 plus a pleasant and instructive evening or two in 1943.

As to the top third of p. 2 of your letter, I think we must distinguish between two distinctions: (a) substantives as names and not names, and (b) designative and non-designative occurrences of names. It is (a) that is relevant to 'Pegasus', which, under the approach I had been urging, is not a name. Now that approach, though advanced in terms of contextual definition of descriptions in my last letter,* does not require the latter. Thus, in conformity with lines 7–14 of p. 2 of your letter, I can let substantives (e. g., 'Pegasus') be primitive, and syntactically recognizable as substantives, and then I can draw the distinction between names and other substantives in this non-syntactical way: *a substantive is a name if and only if the result of inserting it in the blank of* '$(\exists x)(x = \quad)$' *is true*. This is verified by reduction to descriptions (as seen in my last letter*), but does not presuppose it.

Thus the logical manipulation of a substantive as if it named something would be logically dependent on a premise (usually extra-logical) '$(\exists x)(x = \ldots)$'.

Regarding (b), on the other hand, you might still hold to lines 3–6 of p. 2 of your letter, to the following extent: The distinction between those occurrences of a substantive which will be designative if the substantive is a name, and those occurrences which will be non-designative under all circumstances, should be syntactical. This is perhaps necessary; I grant at least its convenience.

As to your remarks on sec. II of my last letter,* I believe we are (contrary, evidently, to your own impression) in substantial agreement. I agree with everything in your next to last paragraph, and find it all consonant with my letters of Nov. 13 and Aug. 14 (see particularly line 7 of p. 3 of the latter). My interest in an empirical definition of an empirical prototype of "proposition" (or of "synonymy") is for precisely the purpose which you formulate in your next to last paragraph: "as indicating how the term .... of abstract semantics is to be applied."

Nor do I see myself drawing a line between "natural" and "artificial" to which you might object (3d from last paragraph of your letter). Thus in sec. II of my letter of Nov. 13 I came out for the absence of a sharp line, allowing for an artificial component even in the field linguist's construction; and in my previous letter I emphasized, as against Carnap, my tendency to view the "artificial" [2] and "natural" from a single point of view: "Serious artificial notations, e. g. in mathematics or in your logic or mine, I consider supplementary but integral parts of natural language." I did make a distinguishing remark about languages set up for semiotic study rather than use, but this is another question, aside, I think, from our discussion.

Sincerely yours,

W. V. Quine

Dec. 11

P. S. - You will be amused to hear that just today I received a refund from Harvard Library together with a note to the effect that Leonard had finally returned the Tseng and Poirier. Three and a half years.

**Editor's note*: Nov. 13. Last except for note accompanying MS on consistency.

January 20, 1959

Professor W. V. Quine
Center for Advanced Study in the Behavioral Sciences
202 Junipero Serra Blvd.
Stanford, California

Dear Professor Quine,

I am sending you separately a copy of the paper which I presented at the recent meeting in Burlington.

My purpose in writing is to call your attention to a point which is not clear in the paper itself but was brought in the discussion. Namely without my having realized that I was doing so, my paper implicitly corrects what I now take to be an error in your statement of your criterion of ontological commitment.

I assume that the italicized sentence on page 103 of "From a Logical Point of View" is meant as the authoritative statement of your criterion, and I select it for illustration of my point. I claim that this criterion, taken literally, has consequences which simply cannot be in accordance with your intentions.

For let A be the statement:

$$(\exists x) \,.\, x \text{ is a dog} \,.\, x \text{ is white}$$

And let T be the theory obtained by adding this statement (as an axiom, or at least as true) to a suitable first-order logic. I say that the theory T has by your criterion on page 103 no ontological commitments whatever.

Fido, it happens, is a white dog. Rover is another white dog, not the same as Fido. Now Fido is not assumed by the theory T, because it is not true that he "must be counted among the values of the variables in order that the statements affirmed in the theory be true" (even if we don't count Fido among the values of the variables, there is still Rover). For a similar reason Rover is not assumed by the theory T (even if we don't count Rover among the values of the variables, there is still Fido). And running through all the white dogs in the universe, we thus find for each one that he is not assumed by the theory T. And as to entities that are not white dogs, it is quite obvious that none of them is assumed by the theory T. Hence T has no ontological commitment to any entity.

So much for the unacceptability of your criterion on page 103 as it stands.

The obvious remedy for the difficulty is to take ontological commitment as being, not to an *entity* but to a *class* of entities. And I see that I have inadvertently, at least twice in the course of the paper, put into your mouth the idea that ontological commitment is to a class — namely in line 8 from the bottom of page 1009, and in the paragraph which occupies the last ten lines of text on page 1013 and the first seven lines on page 1014.

Now the idea of ontological commitment to a class (e.g., of white dogs) can be argued for as at least better than that of ontological commitment to an entity (e.g., Fido). Yet I can understand that the idea of ontological commitment to a class would not be welcome to you — because on this basis it becomes necessary, at least prima facie, to use a class variable in order even to state the criterion of ontological commitment, and we get the result that if there are (as nominalists maintain) no classes, then there is no ontological commitment.

If ontological commitment must not be directly to an entity, and also not to a class, then I think the next try would be to use a predicate (in your sense) in place of a class. This idea leads to a schema, rather than a single statement, to embody the criterion of ontological commitment. In fact it leads exactly to the displayed schema on page 1014 of my paper. And I persist in thinking this schema a good try, in spite of the fact that there are still difficulties.

To see that there are still difficulties, let me first rewrite the schema somewhat more formally as:

(1) The assertion of $(\exists \mathbf{x})\mathbf{M}$ carries ontological commitment to $\hat{\mathbf{x}}\mathbf{M}$.

Suppose we understand $\hat{\mathbf{x}}\mathbf{M}$ as a class name, or otherwise extensionally. And as a particular case let M_1 be "x is Sir Walter Scott" and let M_2 be "x is Sir Walter Scott and x wrote Waverley." By the schema (1) we have:

(2) The assertion of $(\exists \underline{\mathrm{x}})\underline{\mathrm{M}}_1$ carries ontological commitment to xM_1.

As a matter of historical fact we have:

(3) $M_1 \equiv M_2$

Hence by the extensionality principle we have:

(4) The assertion of $(\exists \mathbf{x})M_1$ carries ontological commitment to $\hat{x}M_2$.

Since (4) is clearly false by any reasonable notion of ontological commitment, it follows that $\hat{\mathbf{x}}\mathbf{M}$ cannot be taken extensionally in (1). It must have, in Frege's terminology, an oblique occurrence — or in your terminology it must have an occurrence that is not purely designative, or that is referentially opaque.

This conclusion is in effect pointed out in footnote 3 of my paper. But it should have been pointed out quite specifically in connection with the schema (1) itself; indeed I did do this in the discussion at Burlington, but I regard it as an oversight that I failed to do it in print.

It follows further that in order to deal satisfactorily with the notion of ontological commitment it is necessary not just barely to recognize the phenomenon of oblique (or not purely designative) occurrence, but to have a detailed logic applicable to such occurrences.

Now the way of getting a logic applicable to such oblique occurrences (of names, noun clauses, etc.) that seems to me most likely of success is just to admit the notion of a class concept, in addition perhaps to that of a class, and to try to formulate a logic of class concepts along more or less natural lines. Of course this involves heavy ontological commitments, which it becomes necessary to take upon oneself in order to be able even to deal with the idea of ontological commitment at all. But conceivably it is the only possibility. (And if it proves to be so, the necessity for intensional notions is thus very dramatically brought out.)

The only alternative I see is to try to find a nominalistic analysis of oblique occurrence generally, more or less along the lines that Scheffler followed in proposing a nominalistic analysis of belief statements. I doubt that this could be successful, if one really tried to carry it out in detail. But even granted its success, the conclusion would remain that the notion of ontological commitment belongs to intensional semantics — in your terminology, to the theory of meaning. For Scheffler's that-clause-predicates do of course clearly belong to the theory of meaning.

Your statement at the bottom of page 130 of "From a Logical Point of View," that the notion of ontological commitment (in an explicitly quantificational form of language) belongs to extensional semantics, is in fact unsupported. The only reason that you give is the first sentence on page 131, and this is clearly a mistake, since it has the consequence that if *any* existential quantification presupposes objects of a given kind, then objects of that kind do in fact exist.

In a paper by Scheffler which I have just recently seen, he reaches conclusions about ontological commitment very similar to mine as just set forth, and by a very similar line of argument. Scheffler's solution of the difficulty is, I gather, to reject the notion of ontological commitment altogether: since it cannot be analyzed nominalistically, there is no such thing. But I would suppose that for a nominalist this solution simply will not do, as it leaves nominalism entirely without a general program.

Indeed if there are any ontological commitments which it is actually necessary to make in order to deal with the notion of ontological commitment, then these are in a very peculiar position. The non-philosopher may reject them with impunity. But the philosopher is obliged to accept them in order to be a philosopher.

Very sincerely yours,

Alonzo Church

February 6, 1959

Professor Alonzo Church
Fine Hall
Box 708
Princeton, New Jersey

Dear Professor Church:

Many thanks for your painstaking letter of January 20 and the copy of the Journal of Philosophy. The point that occupies the first pages of your letter is one that I was aware of while writing *From a Logical Point of View*. It is what motivated the formulation in lines 1–4 of page 131; also the less explicit one at the foot of page 8. By no means would I say that your example '$(\exists x)(x$ is a dog, x is white)' carries commitment to the existence of any one dog; it carries commitment to the existence of dogs, and the relevant criterion is not the singular one on page 103 but the plural or general one on page 131. That on page 103 is equally correct, but it defines a different thing, viz., commitment to *an* entity (not entities of a kind). Such commitment occurs when the truth of a theory depends on some irreplaceable entity, as is not uncommon.

What is unfortunate is that the passage on page 103 was italicized in the company of the numerical example and unaccompanied by the complementary criterion of page 131. This is a mistake of editing on my part.

A schematic formulation of the plural case is what I had in mind at the top of page 131. That is, my intention was to explain ontological commitment to dogs, to numbers, etc., in that uniform pattern, the general term being supposed given each time rather than represented by a class variable. Hence my phrase 'a given kind'.

What you advance as a second difficulty at the foot of page 2 of your letter is another point that I had been aware of all along, but I never regarded it as undesirable, and do not now. In talking about the ontological commitments of a theory my concern has been with the objects, however described; not with the meaning of the theory in other respects. I have an example on page 132 that illustrates substantially this point (though introduced to illustrate something else): theory T carries commitment to whole numbers even though 'x is a whole number' is untranslatable into it.

Under my intended usage, a theory that implies '$(\exists x)(x$ is Scott$)$' and '$-(\exists x)(x$ wrote Waverley$)$' carries ontological commitment to the author of Waverley. I would never have hesitated over this conclusion.

That you find the point objectionable suggests that you are construing "is committed to objects of kind K" as "implies a sentence synonymous with '$(\exists x)(x$ is of kind $K)$'". I have never wanted that.

I must concede that the situation changes in the case of ontological commitment to non-existents. Here note 3 of page 1013 of your article has its bearing: the unicorn example. As suggested briefly in lines 4–14 of page 16 of *From a Logical Point of View*, I agree that we have to talk of sentences in other than referential respects when it comes to taking account of ontological commitments that we do not accept.

This bears out your criticism, in the last paragraph of page 3 of your letter, of my statement of page 130 of *From a Logical Point of View* to the effect that ontological commitment belongs to the theory of reference. But my remark is true in this sense: the ontological commitments of a theory can be described in the theory of reference of a metatheory whose universe includes that of the object theory. Similarly truth for a theory can be described in the theory of reference of a metatheory whose notation includes that of the object theory, or an equivalent.

The analogy can be pushed a bit. Take a metatheory with too narrow a universe (as we are bound to do if part of the ontological commitment of the object theory is wrong and our metatheory is to be right), and we can no longer deal with those ontological commitments save in ways other than theory of reference, if at all. But similarly, take a metatheory with too narrow a notation (as we are bound to do if part of the notation of the object theory is meaningless and our metatheory is to be wholly meaningful), and Tarski's truth construction will fail to carry through.

Where the analogy fails is that there are too interesting cases where we would want to brand ontological commitments as wrong. Consequently a qualifying remark would have been desirable; also a more emphatic cleavage between ontological commitments acceptable to the critic and those not acceptable.

The motive behind my remarks on ontological commitment from the beginning has been, not development of a theoretical chapter of semantics, but correction of confusions that kept obstructing discussion: such confusions as you expose so effectively in your references to Ryle and Ayer in your paper. Carnap, in particular, was a sometime offender that I had much in mind.

So far as theory goes, my feeling was that the ontological issues need not detach themselves in any basic ways from issues generally over the truth of assertions, which may in particular be existential assertions. My attitude was very much the one suggested at the end of the paper of Scheffler and Chomsky, and at the end of your paper.

I felt that what I was saying should be too obvious to need saying; but there were those persistent confusions in the literature (and some cases of outright dissent, e. g. Bergmann). You understand this perfectly. Incidentally it is why I didn't fashion my expository remarks on this subject more in the spirit of an explicit theory of ontological commitment; I have been unmotivated to such a theory. The paper by Scheffler and Chomsky, and its predecessor by Cartwright, are aimed at the question of such a theory; I feel none is wanted, though I fear my writing ill reflects the feeling.

Sincerely yours,

W. V. Quine

April 13, 1959

Professor Alonzo Church
Fine Hall
Princeton, New Jersey

Dear Professor Church:

At one point in your review of my Encyclopedia article, namely in paragraph two of p. 208, I am startled to find an abrupt misunderstanding of my intention. You represent me as adopting an attitude of formalism, or disinterpretation, toward the predicate calculus. Such has never been my intention, in this paper or elsewhere. I can't think what created this impression. Certainly I deal formally with those expressions much of the time, as you do, but I intend them to keep their standard interpretations regardless — just as I intend 'Tully' to keep on naming Cicero even when I ponder its spelling or etymology. Or have I misled you with my doctrine of schematic letters as dummy predicates? I suddenly wonder how sweeping a failure of communication between us is symptomatized here.

Sincerely yours,

W. V. Quine

April 16, 1959

Dear Prof. Quine,

This is a belated reply to your letter of February sixth. With regret I have to admit that I am under such pressure that all my correspondence takes that long or longer. And indeed I haven't had time to think about the matter as carefully as I would like to. But there are some points that ought to be made without delaying further.

The situation is that you are threatened with the emergence of a serious incompatibility between your aversion to intensional notions and your wish to enforce certain standards of coherence and responsibility in regard to the matter of ontological commitment. It's true that I think your first-mentioned aversion wrong-headed, whereas I think your wish to enforce standards highly commendable. But let's ignore that and see what can be had on purely logical grounds, sticking to points of logic and metalogic on which I suppose that you and I would agree.

I really think I see a way out of the difficulty for you. But before coming to that, I must strongly protest that the way out which is attempted in your letter just won't do—and I am convinced that on reflection you yourself will not continue to maintain it.

For what I understand you to be saying in the latter part of your letter is that the notion of ontological commitment is so straightforward and informally obvious (except of course to a few knotheads like Bergmann who don't really count) that no *theory* of it is wanted. And by this you do not mean that it would be so trivial and routine a matter to formulate a theory that it is not worth the purely mechanical labor of making such a formulation explicitly (this contention might at least have been plausible before the papers of Cartwright and others). But no, you mean the exact opposite of this; you are so willing to rely on what seems to you the straightforward informal obviousness of the notion of ontological commitment that you will continue to talk about ontological commitment even in the face of the discovery (if discovery it be) that a theory of ontological commitment is on your own principles impossible.

By taking such an attitude you put yourself in my eyes in a class with Ayer and Ryle. Indeed if my paper in the Journal of Philosophy gets any attention from the partisans of the British school, I am fully expecting to hear from one of them the retort that in "ordinary language," the notion of existence which Ayer wishes to use and defend—and which is not the existential quantifier but something else—is so straightforward and informally obvious (except of course to a few knotheads like you and me who don't really count) that no *theory* of it is wanted. Surely if I concede to you and you concede to yourself the use of the notion of ontological commitment without having a theory of it, and without having even an outline plan according to which one might begin to formulate a theory of it, then we must both make a like concession to Ayer in regard to his notion of existence.

That you should be willing to take such a position, even provisionally and in a letter to me, illustrates to my mind the extremes to which philosophers are willing to go when one of their fundamental philosophical convictions is threatened. For such convictions when firmly held come to be like religious convictions, not to be upset by the contingencies of everyday argument. And for the majority of philosophers, who have the misfortune not to have any religion in the proper sense, I suspect that psychologically their metaphysical doctrines (and I include nominalism as one kind of metaphysics) come to be a sort of substitute for a religion. It is one of the reasons that philosophers seldom or never can reach general agreement. I don't claim to be exempt myself from the human failing that is in question—but only that I do sometimes try to be objective. Perhaps you will make the same claim for yourself. But I think I have caught you in a conspicuous lapse.

The ultimate of this sort of thing is the philosopher who, when faced with a logical inconsistency in his basic doctrine, will deny that logic is applicable, and will proceed generally to repudiate and belittle logic. To my mind, the position taken in the last paragraph of your letter of February sixth comes shockingly close to this.

But to come back to a more specific formulation of the criticism—you write in your letter that "a theory that implies that '$(\exists x)(x$ is Scott)' and '$-(\exists x)(x$ wrote Waverley)' carries ontological commitment to the author of Waverley." How can this possibly be taken, if not as a fragment of a *theory* of ontological commitment? Surely you expect that this, and the many statements about ontological commitment that are made at various places in your book, will have logical consequences, that they will be found compatible with some things and incompatible with others. If not, you cannot expect me or any one else to take them seriously, to argue either for or against them, or to apply them to anything whatever, even to the exhibition of alleged confusions in the writings of Carnap or of Ayer. On the other hand if your statements do have logical consequences and logical connections with other statements, it can only be because they belong in a theory; and if on demand you cannot produce the theory, then you must back down. If your theory consists only of first-order logic plus some obvious added primitives, postulates, and definitions, well and good. If your theory is only a special chapter of some general semantical metatheory, obtained from it by some obvious definitions, well and good. But you must have a theory, or more correctly you must at least envisage a theory—in the sense that you know how to go about producing the theory and are not at a loss actually to formulate it on demand.

Now my preference is for an intensional theory of ontological commitment, one according to which ontological commitment to Sir Walter Scott is not the same as (and not equivalent to) ontological commitment to the author of Waverley, and, as I say in the footnote to my paper, ontological commitment to unicorns is not the same as ontological commitment to purple cows. Such a theory would cause me no embarrassment as it would you, because I have never held that an intensional theory must automatically be rejected as bad metaphysics, but rather that each theory is to be examined on its merits. An intensional theory of ontological commitment would seem to me superior, as preserving certain distinctions which seem to me and others to be significant.

But as I said at the beginning, I think I see a way out that will preserve a notion of ontological commitment without either sacrificing your extensionalistic prejudices or getting you into logical inconsistency. For the sentence I quoted above from your letter indicates an extensional notion of ontological commitment, one according to which ontological commitment to Sir Walter Scott is the same as (at least in the sense of being equivalent to) ontological commitment to the author of Waverley. Why not just carry this extensionality of the notion all the way through to its conclusion? Since in fact no cow is purple, I assume that you would further hold that '$(\exists x)(x$ is a cow)' carries ontological commitment to things that are not purple—in spite of the lack of any logical inconsistency between '$(\exists x)(x$ is a cow)' and '$(x)(x$ is purple)'. Indeed you must so hold if you continue to maintain the first sentence on page 131 of "From a Logical Point of View" (as being your only support for the contention that the notion of ontological commitment, as applied to a quantificational form of language, belongs to the "theory of reference").

The question is, why do you begin to have doubts about the appropriateness of this extensionality property of your notion of ontological commitment, just when you come to the special case of the null class? Do you after all, as I suspect, have in your mind some lingering recognition of the significance of intensional notions? If so, wouldn't you do well to join some of your colleagues in the search for a satisfactory intensional logic? And if not, shouldn't you just apply the extensionality of ontological commitment without hesitation to the null class as well as to others?

If Ryle uses bound variables which require propositions as values, why not then denounce him for ontological commitment to purple cows? For the fact is, according to your ideas, that *propositions are a kind of purple cows*. Namely the underscored assertion is, on your ideas, true in extension. And as for the question whether it's true in intension—well, I understand that you don't believe in intension anyway.

Of course this whole-heartedly extensional notion of ontological commitment doesn't accord very well intuitively with your original motivation in introducing the notion—because it has the consequence that a man's ontological commitments cannot be ascertained solely by examining the assertions he makes and their logical consequences, but inquiry into external matters of fact is necessary in addition.

Nevertheless the extensional notion might be made to serve your main purpose. Whether it agrees with everything you've said in your book—including places at which the reference to ontological commitment is implicit rather than explicit—I don't know without a more careful search than I have made. But even if you have to make some changes, this would seem to be preferable from your point of view to either going over to the intensionalists (horrible thought!) or abandoning the notion of ontological commitment as unworkable because no logically coherent theory of it can be formulated.

Now finally in reply to your letter of April thirteenth. I did and do understand you as wanting to replace the free functional variables in the first-order functional (miscalled predicate) calculus by "schematic letters" in order to avoid ontological commitments which you would prefer, not to renounce, but to postpone. Indeed in my "Introduction to Mathematical Logic" the intended interpretation of the pure functional calculus of first order is such that the assertion of, e. g., '$(\exists x)(y)F(x,y) \supset (y)(\exists x)F(x,y)$' means the same as the assertion of '$(F)[(\exists x)(y)F(x,y) \supset (y)(\exists x)F(x,y)]$'. I would suppose that you would construe the assertion of expressions containing free *variables*, as opposed to "schematic letters," in the same way, and hence would understand my calculus as carrying (in its intended interpretation) ontological commitment to classes and relations. And it is one of my objections to your program that you simply have no propositional calculus, and analogously no pure first-order functional calculus, but only a sort of meaningless "schematic" substitute for them, until (at most) belatedly, after you have allowed yourself to get into set theory.

If you mean something different, it's high time you explained yourself better.

Very sincerely yours,

Alonzo Church

Prof. W. V. Quine,
Center for Advanced Study in the Behavioral Sciences,
202 Junipero Serra Boulevard,
Stanford, California

Editor's note: In Quine's handwriting, in red pencil next to the paragraph beginning "If Ryle": "fallacy of subtraction". The letter is in Church's handwriting. It is the only letter in the series that is handwritten. All the others are typewritten.

April 20, 1959

Professor Alonzo Church
Fine Hall
Box 708
Princeton University
Princeton, New Jersey

Dear Professor Church,

Many thanks for your long letter of April 16. On its main topic, ontological commitment, I shall want to take some time and care in order to minimize cross purposes. Meanwhile the present note is on the point treated at the end of your letter.

In a way I am relieved to find, in that paragraph of your letter, so understanding a statement of my position. You are right: the logic of truth functions and quantification as I present them are not theories of propositions and attributes. They are not substantive theories at all, but schemata of theories, as you say, until I get into set theory. (Of course instead of set theory the theory concerned could as well be identity theory, or the mereology of bodies, etc., if one happened to develop them.)

But my puzzlement remains. For my question of April 16* concerned your imputation of *formalism*, in your review; and I am as much in the dark as before on what makes you associate my view with formalism. There is in my approach no disinterpretation of constants, either before or after the advent of set theory. For me set theory departs from the schematic predicate calculus simply in having a single interpreted two-place predicate instead of the dummy predicates. Similarly for identity theory, and similarly for mereology.

Sincerely yours,

W. V. Quine

April 27, 1959

Professor W. V. Quine
Center for Advanced Study in the Behavioral Sciences
202 Junipero Serra Blvd.
Stanford, California

Dear Professor Quine:

In answer to your letter of April twentieth, and in regard to my imputation of *formalism* to your versions of propositional calculus and first-order functional calculus (as you would have them prior to your belated and reluctant introduction of set

* *Editor's note*: Quine must mean April 13.

theory), I persist in thinking my point not only sound in itself but clear enough from my statement of it in the review.

In the standard version of pure functional calculus of first order (to concentrate on this case) there are functional variables which occur in the formulas of the calculus and in particular in theorems. These are in all versions meaningful variables. For those who, like myself, prefer an extensional interpretation, the values of the variables are classes and relations in extension. But an intensional interpretation is also possible, as in Principia Mathematica, the values of the variables being then, as you say, "attributes" (i. e., properties and relations in intension). For the present immediate purpose the choice between the two interpretations does not matter.

Now in your elementary logic you have no pure first-order functional calculus. But you do have something which you contrive to make serve the same purpose, or better, many of the same purposes. This is namely a system of what you call "schemata." These schemata can be matched one for one with the well-formed formulas of the pure functional calculus of first order, indeed they may be identical with the well-formed formulas of the latter calculus if only the functional variables are reconstrued as schematic letters. The schemata obey the same formation rules as the formulas of the pure functional calculus of first order. In an appropriately reconstrued sense, they obey the same transformation rules. In short the system of schemata is syntactically identical with the pure first-order functional calculus. the difference is that the schemata do not have a meaning. They may be written down, but not written down as meaningful. In meaningful statements they occur only between quotation marks.

Now this replacement of a meaningful calculus by a system of expressions which are syntactically identical with those of the calculus but meaningless, and which are made, to a large extent, to serve the same purposes despite their meaninglessness, is *formalism*. I cannot conceive of a clearer case.

My point is, if you are willing to carry formalism this far, why not carry it all the way, and apply it to the whole of mathematics, or the whole of set theory? I can see no reason except the comparative simplicity of the relationship between the schemata and expressions which you do regard as meaningful. And this seems to me a very weak reason. The relationship between the meaningless expressions and ones accepted as meaningful gets more and more complicated as one goes on, but there is no sharp line to be drawn.

Finally, while I am writing, I take the opportunity to add one more explanation to what I said in my last letter about ontological commitment. I think I failed to make clearly enough a point that I wanted to make. In my opinion the idea that there is something to be gained by having a meta-language whose ontological commitments include all those of the object language is at least misleading, if not just simply mistaken. For whatever objection (or lack of objection) there may be to your inability, within the "theory of reference", to distinguish between ontological commitment to unicorns in first century India and ontological commitment to snakes in nineteenth century Ireland, surely there is exactly equal objection (or lack of objection) to your inability, within the "theory of reference", to distinguish between ontological commitment to cows and ontological commitment to cows that are not purple. The nature of the difficulty (or alleged difficulty, if you wish) is the same in both cases: the two classes which are in question are the same, though the corresponding class concepts are different — i. e., there is in each case identity in extension but difference in intension.

A remedy which removes the difficulty in one case but not in the other serves no genuine purpose.

Very sincerely yours,

Alonzo Church

May 5, 1959

Professor Alonzo Church
Fine Hall
Box 708
Princeton, New Jersey

Dear Professor Church,

For readers given, like me, to associating the term "formalism" with views attributed to Hilbert, the word may be expected to connote uninterpreted systems, specified formally and not semantically. My own approach to the logic of quantification is, on the contrary, primarily semantical. The task of that logic is, for me, to identify those true sentences whose truth depends (roughly speaking) only on the meaning of the quantifiers and truth-function signs; i. e., (more exactly) those truths that stay true under all uniform changes of component predicates. Schematic letters are for me a device to this end. Deductive formalization under syntactically stated rules is for me no more essential to the logic of quantification than to arithmetic, set theory, or mereology, though rewarding in all. I regularly begin with intuitive arguments for the laws of quantification, not with formal postulation.

You are right in saying that in the logic of quantification as I present it the formulas are not sentences and the predicate letters are not (in the value-taking sense) variables. The formulas are diagrams for real sentences with arbitrary predicates. You prefer another course; fair enough. What I protest is your terminologically linking this preference with the very different issue between what have been called formalist and logicist philosophies of mathematics. Speaking as a logicist myself, I find something startling in the suggestion that one must either conceive the logic of quantification as a calculus of classes, attributes, or the like, or be a formalist.

I suspect there is more than terminological trouble here — that you do picture me as somehow more "formalistic," even as I understand the word, than I am. This would account for your having a stronger sense of omission than I in that matter of modus ponens; my exposition was more intuitive in intent, more integral to the semantics of truth, than you supposed. It would also account for the feeling on your part, so odd to me, that I might as well go on and disinterpret set theory and arithmetic. Hence I do not hope to clear up the misunderstanding by just questioning the word.

I think our trouble may arise from a difference in focus. In categorizing my view of the logic of quantification you focus on my formulas with their 'F's and 'G's and say they are not sentences, hence meaningless and hence that I am formalistic. My focus is beyond these formulas, on the infinite totality of sentences with (usually) extra-logical predicates, all interpreted, and quantifiers and truth functions, interpreted as always. The study of the logic of quantification is for me a study of the invariance of truth (absolute and relative, or conditional) of *those* sentences under uniform changes of predicates; and my formulas with their 'F's and 'G's are diagrammatic aids to grouping those sentences in appropriate ways.

One more thought: I wonder if you realize that when I say 'valid' I mean it semantically, in the sense of "allgemeingültig"; not formally, in the sense of 'demonstrable'.

Turning now to ontological commitment, I am shocked to discover from your letter of April 16 that my effort of February 6 to clarify my position has only further obscured it. This time I shall try brevity, if only as a means of minimizing the further places where communication can go wrong.

The last paragraph of my letter of February 6 was meant as a modest disclaimer. You have read it as a denunciation of rational inquiry into ontological commitment. The key to the intended sense of my paragraph is the first of my two references in that paragraph to Scheffler and Chomsky, together with the reference to your paper. The relevant passage in their paper, with which I was trying to express agreement, is their final paragraph. The one in yours with which I was trying to express agreement is your final paragraph, p. 1014.

For the rest, what needs to be brought out is that I already have explicitly recognized those main substantive points, about ontological commitment, which you suspect me of suppressing on religious grounds.

To help bring that out, let us distinguish between two senses of "ontological commitment" and call them the *intensional* and *extensional* senses, or briefly OC_i and OC_e. I am willing to consider the term "ontological commitment" ill chosen in one or both senses. The fact remains that theories can without conflict be investigated in respect of OC_i and in respect of OC_e. These are two overlapping topics of inquiry.

Your example about purple cows well illustrates the domain where OC_e is useless and OC_i is required. The example is representative of what I was talking about in the already cited passage of "On what there is" (viz., *From a Logical Point of View*, p. 16) where I spoke in effect of the inadequacy of OC_e for ontological commitments unaccepted in the metatheory, and of the need of switching to discourse of words and their use (which would be my approach to OC_i).

When you express a "preference for an intensional theory of ontological commitment" you do go farther than I; I think OC_e and OC_i are both needed, whether or not by name. OC_e covers matters that cannot, at any rate, be handled in the *pure* theory of meaning. This point is illustrated in the following excerpt from a transcript of my 1953 Harvard lectures:

> The ontological presuppositions of a theory or of a statement are those things which have to be within the range of the values of its variables in order for the statement or the set of statements to be true. Then, take the statement:
>
> $(\exists x)$(x is in Emerson A on April 18, 1953, and x is less than 18 years old).
>
> And the question whether Bernard J. Ortcutt has to be a value of the variables in order that this statement be true, in other words, the question whether Bernard J. Ortcutt *is* one of the ontological presuppositions of this statement, is a question of fact. The question of whether he's presupposed

> by the truth of the statement is a question of fact that we can't decide simply by reflecting on the logic of quantification or on the things that are said in the statement. It depends on whether there's anybody else in the room under 18 years old; it depends on whether Bernard J. Ortcutt is in the room — otherwise he wouldn't help; it depends on whether Bernard J. Ortcutt is less than 18 years old.

Finally, one general remark: I am not, as you repeatedly hint, unaware of the urgency of a tenable theory of intension or meaning. When you say of yourself "I have never held that an intensional theory must automatically be rejected as bad metaphysics," I can echo you (unless 'intensional' is given excessive specificity). I have been known to challenge uncritically accepted intensional notions, such as analyticity, and for known reasons of no nominalistic kind I have consistently opposed intensional values of variables. But to repudiate over-facile answers to problems of meaning, and to be unable to substitute good ones, is not to deny the problems. Even to quit the field is not to deny the problems, but in fact I am working in it.

Sincerely yours,

W. V. Quine

DOCTORAL DISSERTATION STUDENTS OF ALONZO CHURCH (1931–1985)

At Princeton University:

Alfred L. Foster (1931)
Stephen C. Kleene (1934)
John Barkley Rosser (1934)
Alan M. Turing (1938)
Enrique Bustamente-Llaca (1944)
Leon A. Henkin (1947)
John G. Kemeny (1949)
Martin D. Davis (1950)
Maurice L'Abbé (1951)
William W. Boone (1952)
Hartley Rogers, Jr. (1952)
Norman Shapiro (1955)
Michael O. Rabin (1957)
Dana S. Scott (1958)
Aubert Daigneault (1959)
Simon B. Kochen (1959)
Raymond M. Smullyan (1959)
Robert W. Ritchie (1960)
James R. Guard (1961)
James H. Bennett (1962)
Robert O. Winder (1962)
Gustav B. Hensel (1963)
Wayne H. Richter (1963)
T. Thacher Robinson (1963)
Peter B. Andrews (1964)
William B. Easton (1964)
Joel W. Robbin (1965)
Donald J. Collins (1967)

At University of California, Los Angeles:

Richard Jay Malitz (1976)
Curtis Anthony Anderson (1977)
Gary Ronald Mar (1985)

BIBLIOGRAPHY OF CHURCH'S REVIEWS IN *THE JOURNAL OF SYMBOLIC LOGIC*

Ackermann, W., *Konstruktiver Aufbau Eines Abschnitts der Zweiten Cantorschen Zahlenklasse*, reviewed in ***The Journal of Symbolic Logic***, vol. 17, no. 2 (1952), pp. 152–153

Aiken, H. H., Burkhart, W., Kalin, T., & Strong, P. F., ***Synthesis of Electronic Computing and Control Circuits***, reviewed in ***The Journal of Symbolic Logic***, vol. 18, no. 4 (1953), p. 347

Alexandrov, A. D., *L'Idéalisme de la Théorie des Ensembles*, reviewed in ***The Journal of Symbolic Logic***, vol. 23, no. 1 (1958), pp. 30–32

Allendoerfer, C. B., & Oakley, C. O., ***Principles of Mathematics***, reviewed in ***The Journal of Symbolic Logic***, vol. 19, no. 1 (1954), pp. 64–65

Anderson S. B., *Problem Solving in Multiple-Goal Situations*, reviewed in ***The Journal of Symbolic Logic***, vol. 24, no. 1 (1959), p. 86

Angell, R. B., *Note on a Less Restricted Type of Rule of Inference*, reviewed in ***The Journal of Symbolic Logic***, vol. 40, no. 4 (1975), pp. 602–603

Angell, R. B., *The Sentential Calculus using Rule of Inference* R_e, reviewed in ***The Journal of Symbolic Logic***, vol. 40, no. 4 (1975), pp. 602–603

Anonymous, *Inexhaustible*, reviewed in ***The Journal of Symbolic Logic***, vol. 17, no. 4 (1952), p. 289

Anonymous, *Problemas de "Theoria." Problema n.º 1*, reviewed in ***The Journal of Symbolic Logic***, vol. 20, no. 3 (1955), p. 304

Anonymous, *Foreword*, ***Bibliography of Polish Mathematics 1944–1954***, reviewed in ***The Journal of Symbolic Logic***, vol. 31, no. 3 (1966), p. 517

Anscombe, G. E. M., *Before and After*, reviewed in ***The Journal of Symbolic Logic***, vol. 36, no. 1 (1971), pp. 173–175

Archer, A., *A Venn Diagram Analogue Computer*, reviewed in ***The Journal of Symbolic Logic***, vol. 16, no. 1 (1951), p. 62

Ayer, A. J., *Editor's Introduction*, ***Logical Positivism***, reviewed in ***The Journal of Symbolic Logic***, vol. 35, no. 2 (1970), p. 312

Ayer, A. J., ***The Foundations of Empirical Knowledge***, reviewed in ***The Journal of Symbolic Logic***, vol. 6, no. 3 (1941), p. 108

Ayer, A. J., ***Language, Truth and Logic***, reviewed in ***The Journal of Symbolic Logic***, vol. 14, no. 1 (1949), pp. 52–53

Bachhuber, A. H., ***Introduction to Logic***, reviewed in ***The Journal of Symbolic Logic***, vol. 24, no. 1 (1959), pp. 83–84

Bachmann, H., *Die Normalfunktionen und das Problem der ausgezeichneten Folgen von Ordnungszahlen*, reviewed in ***The Journal of Symbolic Logic***, vol. 22 (1957), no. 3, p. 336

Bachmann, H., *Vergleich und Kombination zweier Methoden von Veblen und Finsler zur*

Lösung des Problems der ausgezeichneten Folgen von Ordnungszahlen, reviewed in ***The Journal of Symbolic Logic***, vol. 22 (1957), no. 3, p. 336

Bachmann, H, *Normalfunktionen und Hauptfolgen*, reviewed in ***The Journal of Symbolic Logic***, vol. 22 (1957), no. 3, p. 336

Baker, A., *Effective Methods in the Theory of Numbers*, reviewed in ***The Journal of Symbolic Logic***, vol. 37, no. 3 (1972), p. 606

Bar-Hillel, Y., *Mr. Weiss on the Paradox of Necessary Truth*, reviewed in ***The Journal of Symbolic Logic***, vol. 21, no. 1 (1956), p. 83

Barker, C. C. H., *Some Calculations in Logic*, reviewed in ***The Journal of Symbolic Logic***, vol. 22, no. 4 (1957), p. 379

Barker, S. F., *Realism as a Philosophy of Mathematics*, reviewed in ***The Journal of Symbolic Logic***, vol. 40, no. 4 (1975), p. 593

Barzin, M., *Note sur la Démonstration de M. E. W. Beth*, reviewed in ***The Journal of Symbolic Logic***, vol. 1, no. 3 (1936), p. 118

Bâscă, O. C., *La Synthèse des Automates Finis par la Méthode de A. Church*, reviewed in ***The Journal of Symbolic Logic***, vol. 37, no. 3 (1972), pp. 625–626

Basson, A. H., *Logic and Fact*, reviewed in ***The Journal of Symbolic Logic***, vol. 13, no. 3 (1948), p. 158

Basson, A. H., & O'Connor, D. J., ***Introduction to Symbolic Logic***, reviewed in ***The Journal of Symbolic Logic***, vol. 20, no. 1 (1955), pp. 84–86

Basson, A. H., & O'Connor, D. J., ***Introduction to Symbolic Logic***, 2nd edition, reviewed in ***The Journal of Symbolic Logic***, vol. 23, no. 4 (1958), pp. 434–435

Basson, A. H., & O'Connor, D. J., ***Introduction to Symbolic Logic***, 3rd edition, reviewed in ***The Journal of Symbolic Logic***, vol. 28, no. 2 (1963), p. 169

Bauer, F. L., *Zur Algebraik des Logikkalküls*, reviewed in ***The Journal of Symbolic Logic***, vol. 16, no. 1 (1951), p. 62

Baylis, C. A., *Critical Comments on Professor Fitch's Article "On God and Immortality"*, reviewed in ***The Journal of Symbolic Logic***, vol. 13, no. 3 (1948), p. 148

Bell, E. T., ***Men of Mathematics***, reviewed in ***The Journal of Symbolic Logic***, vol. 2, no. 2 (1937), p. 95

Bell, E. T., ***The Development of Mathematics***, reviewed in ***The Journal of Symbolic Logic***, vol. 5, no. 4 (1940), pp. 152–153

Bell, E. T., ***The Development of Mathematics***, 2nd edition, reviewed in ***The Journal of Symbolic Logic***, vol. 12, no. 2 (1947), pp. 61–62

Bell, E. T., *Mathematics, History of*, in ***Encyclopaedia Britannica***, reviewed in ***The Journal of Symbolic Logic***, vol. 16, no. 3 (1951), p. 223

Bentley, A. F., *On a Certain Vagueness in Logic*, reviewed in ***The Journal of Symbolic Logic***, vol. 10, no. 4 (1945), pp. 132–133

Bergmann, G., *Notes on Identity*, reviewed in ***The Journal of Symbolic Logic***, vol. 8, no. 3 (1943), p. 86

Berkeley, E. C., *Boolean Algebra (The Technique for Manipulating "and," "or," "not," and Conditions) and Applications to Insurance*, reviewed in ***The Journal of Symbolic Logic***, vol. 3, no. 2 (1938), p. 90

Berkeley, E. C., ***Giant Brains. Or Machines That Think***, reviewed in ***The Journal of Symbolic Logic***, vol. 15, no. 3 (1950), pp. 202–203

Berkeley, E. C., ***A Summary of Symbolic Logic and its Practical Applications***, reviewed in ***The Journal of Symbolic Logic***, vol. 18, no. 1 (1953), p. 68

Berkeley, E. C., ***A Summary of Symbolic Logic and its Practical Applications***, 2nd printing, reviewed in ***The Journal of Symbolic Logic***, vol. 18, no. 1 (1953), p. 68

Berkeley, E. C., *The Algebra of States and Events*, reviewed in ***The Journal of Symbolic Logic***, vol. 20, no. 3 (1955), pp. 286–287

Bernays, P., Freudenthal, H., Gonseth, F., Fréchet, M., Ladrière, J., & Segre, B., *Discussion*, reviewed in ***The Journal of Symbolic Logic***, vol. 36, no. 3 (1971), pp. 514–515

Bernstein, B. A., *Postulate-sets for Boolean Rings*, reviewed in ***The Journal of Symbolic Logic***, vol. 9, no. 2 (1944), p. 54

Beth, E. W., *Démonstration d'un Théorème Concernant le Principe du Tiers Exclu*, reviewed in ***The Journal of Symbolic Logic***, vol. 1, no. 3 (1936), p. 118

Beth, E. W., *Une Démonstration de la Non-Contradiction de la Logique des Types au Point de Vue Fini*, reviewed in ***The Journal of Symbolic Logic***, vol. 2, no. 1 (1937), p. 44

Beth, E. W., *Mathematics, Logic, and Philosophy of Science at the Congrès-Descartes*, reviewed in ***The Journal of Symbolic Logic***, vol. 3, no. 2 (1938), pp. 87–88

Beth, E. W., *The Paradoxes*, reviewed in ***The Journal of Symbolic Logic***, vol. 4, no. 3 (1939), p. 125

Beth, E. W., *Logic and Foundations-Research 1940–1945*, reviewed in ***The Journal of Symbolic Logic***, vol. 11, no. 4 (1946), p. 134

Beth, E. W., *For Guidance*, reviewed in ***The Journal of Symbolic Logic***, vol. 21, no. 2 (1956), p. 187

Beth, E. W., *Decision Problems of Logic and Mathematics*, reviewed in ***The Journal of Symbolic Logic***, vol. 22, no. 4 (1957), p. 359

Bhattacharyya, S., *Symmetry, Transitivity and Reflexivity*, reviewed in ***The Journal of Symbolic Logic***, vol. 25, no. 3 (1960), pp. 263–264

Bhattacharyya, S., *Post-Script*, reviewed in ***The Journal of Symbolic Logic***, vol. 25, no. 3 (1960), pp. 263–264

Billing, J., *A Failure of the Bolzano-Weierstrass Lemma*, reviewed in ***The Journal of Symbolic Logic***, vol. 12, no. 3 (1947), p. 94

Birkhoff, G., *An Extended Arithmetic*, reviewed in ***The Journal of Symbolic Logic***, vol. 7, no. 3 (1942), pp. 125–126

Birkhoff, G., *Generalized Arithmetic*, reviewed in ***The Journal of Symbolic Logic***, vol. 7, no. 3 (1942), pp. 125–126

Birkhoff, G., ***Lattice Theory***, reviewed in ***The Journal of Symbolic Logic***, vol. 15, no. 1 (1950), pp. 59–60

Birkhoff, G., *Théorie et Applications des Treillis*, reviewed in ***The Journal of Symbolic Logic***, vol. 15, no. 2 (1950), p. 136

Birkhoff, G., *Mathematics, Foundations of*, in ***Encyclopaedia Britannica***, reviewed in ***The Journal of Symbolic Logic***, vol. 16, no. 3 (1951), p. 223

Birkhoff, G., & Mac Lane, S., *Algebra of Classes*, reviewed in ***The Journal of Symbolic Logic***, vol. 6, no. 4 (1941), p. 165

Birkhoff, G., & Mac Lane, S., *Algebra of Classes*, revised edition, reviewed in ***The Journal of Symbolic Logic***, vol. 19, no. 2 (1954), p. 140

Birkhoff, G., & Neumann, J. v., *The Logic of Quantum Mechanics*, reviewed in ***The Journal of Symbolic Logic***, vol. 2, no. 1 (1937), pp. 44–45

Birkhoff, G. D., *Number*, in ***Encyclopaedia Britannica***, reviewed in ***The Journal of Symbolic Logic***, vol. 16, no. 3 (1951), p. 223

Black, M., *A New Method of Presentation of the Theory of the Syllogism*, reviewed in ***The Journal of Symbolic Logic***, vol. 10, no. 4 (1945), pp. 133–134

Black, M., *The "Paradox of Analysis" Again: A Reply*, reviewed in ***The Journal of Symbolic Logic***, vol. 11, no. 4 (1946), pp. 132–133

Black, M., *How can Analysis be Informative?*, reviewed in ***The Journal of Symbolic Logic***, vol. 11, no. 4 (1946), pp. 132–133

Black, M., *A Translation of Frege's Ueber Sinn und Bedeutung. Introductory Note*, reviewed in ***The Journal of Symbolic Logic***, vol. 13, no. 3 (1948), pp. 152–153

Black, M., *Frege on Functions*, reviewed in ***The Journal of Symbolic Logic***, vol. 21, no. 2 (1956), pp. 201–202

Black, M., ***Linguaggio e Filosofia, Studi Metodologici***, translated by F. Salvoni, reviewed in ***The Journal of Symbolic Logic***, vol. 22, no. 4 (1957), p. 407

Black, M., *Language and Reality*, reviewed in ***The Journal of Symbolic Logic***, vol. 35, no. 2 (1970), p. 313

Blanché, R., *Logique 1900–1950*, reviewed in ***The Journal of Symbolic Logic***, vol. 19, no. 3 (1954), p. 235

Blanshard, B., *The Escape from Philosophic Futility*, reviewed in ***The Journal of Symbolic Logic***, vol. 13, no. 1 (1948), p. 55

Blanshard, B., *Reason and Analysis*, reviewed in ***The Journal of Symbolic Logic***, vol. 35, no. 2 (1970), p. 313

Bocheński, I. M., *De Consequentiis Scholasticorum Earumque Origine*, reviewed in ***The Journal of Symbolic Logic***, vol. 3, no. 1 (1938), pp. 45–46

Bocheński, I. M., *L'État et les Besoins de l'Histoire de la Logique Formelle*, reviewed in ***The Journal of Symbolic Logic***, vol. 14, no. 2 (1949), p. 132

Bocheński, I. M., *The Condition and the Needs of the History of Formal Logic*, reviewed in ***The Journal of Symbolic Logic***, vol. 15, no. 1 (1950), p. 64

Bocheński, I. M., *On the Categorical Syllogism*, reviewed in ***The Journal of Symbolic Logic***, vol. 15, no. 2 (1950), pp. 140–141

Bočvar, D. A., *On a Three-Valued Logical Calculus and its Application to the Analysis of Contradictions*, reviewed in ***The Journal of Symbolic Logic***, vol. 4, no. 2 (1939), pp. 98–99

Bočvar, D. A., *Über Einen Aussagenkalkül mit Abzählbaren Logischen Summen und Produkten*, reviewed in ***The Journal of Symbolic Logic***, vol. 5, no. 3 (1940), p. 119

Boehm, K., ***Axiome der Arithmetik***, reviewed in ***The Journal of Symbolic Logic***, vol. 20, no. 4 (1955), p. 307

Boehner, P., *El Sistema de Lógica Escolástica. Estudio Histórico y Crítico*, reviewed in ***The Journal of Symbolic Logic***, vol. 12, no. 3 (1947), p. 98

Boehner, P., *Bemerkungen zur Geschichte der De Morgansche Gesetze in der Scholastik*, reviewed in ***The Journal of Symbolic Logic***, vol. 17, no. 2 (1952), pp. 123–124

Boehner, P., *Introduction*, ***Ockham, Philosophical Writings***, reviewed in ***The Journal of Symbolic Logic***, vol. 23, no. 3 (1958), p. 351

Boicescu, V., *Sur les Algèbres de Lukasiewicz*, reviewed in ***The Journal of Symbolic Logic***, vol. 39, no. 1 (1974), p. 184

Boll, M., ***Les Étapes des Mathématiques***, reviewed in ***The Journal of Symbolic Logic***, vol. 14, no. 2 (1949), p. 126

Boole, G., ***The Mathematical Analysis of Logic, Being an Essay Towards a Calculus of Deductive Reasoning***, reviewed in ***The Journal of Symbolic Logic***, vol. 13, no. 4 (1948), p. 216

Boole, G., ***An Investigation of the Laws of Thought, on Which are Founded the Mathematical Theories of Logic and Probabilities***, reviewed in ***The Journal of Symbolic Logic***, vol. 16, no. 3 (1951), pp. 224–225

Borel, E., *Sur l'Illusion des Définitions Numériques*, reviewed in ***The Journal of Symbolic Logic***, vol. 12, no. 3 (1947), p. 94

Borel, E., *Analyse et Géométrie Euclidiennes*, reviewed in ***The Journal of Symbolic Logic***, vol. 15, no. 3 (1950), p. 202

Bouligand, G., *Les Crises de l'Unité dans la Mathématique*, reviewed in ***The Journal of Symbolic Logic***, vol. 11, no. 3 (1946), p. 100

Bourbaki, N. (pseudonym), ***Théorie des Ensembles (Fascicule de Résultats)***, reviewed in ***The Journal of Symbolic Logic***, vol. 11, no. 3 (1946), p. 91

Bradley, R. D., *Geometry and Necessary Truth*, reviewed in ***The Journal of Symbolic Logic***, vol. 34, no. 3 (1969), pp. 496–497

Braithwaite, R. B., *Introduction*, Gödel, K., ***On Formally Undecidable Propositions of Principia Mathematica and Related Systems***, reviewed in ***The Journal of Symbolic Logic***, vol. 30, no. 3 (1965), pp. 357–359

Brentano, F., ***Psychologie du Point de Vue Empirique***, reviewed in ***The Journal of Symbolic Logic***, vol. 12, no. 2 (1947), pp. 56–57

Breuer, J., ***Introduction to the Theory of Sets***, translated by H. F. Fehr, reviewed in ***The***

Journal of Symbolic Logic, vol. 23, no. 1 (1958), pp. 32–33

Brouwer, L. E. J., *Richtlijnen der Intuitionistische Wiskunde*, reviewed in ***The Journal of Symbolic Logic***, vol. 13, no. 3 (1948), p. 174

Brouwer, L. E. J., *Consciousness, Philosophy, and Mathematics*, reviewed in ***The Journal of Symbolic Logic***, vol. 14, no. 2 (1949), pp. 132–133

Brown, D. G., *Misconceptions of Inference*, reviewed in ***The Journal of Symbolic Logic***, vol. 20, no. 3 (1955), p. 301

Brumbaugh, R. S., *An Aristotelian Defense of "Non-Aristotelian" Logics*, reviewed in ***The Journal of Symbolic Logic***, vol. 17, no. 3 (1952), p. 217

Brunschvicg, L., ***Les Étapes de la Philosophie Mathématique***, 3rd edition, reviewed in ***The Journal of Symbolic Logic***, vol. 13, no. 4 (1948), p. 216

Burckhardt, J. J., *Zur Neubegründung der Mengenlehre*, reviewed in ***The Journal of Symbolic Logic***, vol. 3, no. 4, (1938), pp. 165–166

Burks, A. W., & Wright, J. B., *Theory of Logical Nets*, reviewed in ***The Journal of Symbolic Logic***, vol. 19, no. 2 (1954), pp. 141–142

Bykhovsky, B., *The Morass of Modern Bourgeois Philosophy*, reviewed in ***The Journal of Symbolic Logic***, vol. 15, no. 3 (1950), pp. 235–236

Cantor, G., ***Contributions to the Founding of the Theory of Transfinite Numbers***, translated by P. E. B. Jourdain, reviewed in ***The Journal of Symbolic Logic***, vol. 17, no. 3 (1952), p. 208

Carmichael, P. A., *The Null Class Nullified*, reviewed in ***The Journal of Symbolic Logic***, vol. 8, no. 1 (1943), p. 35

Carmichael, P. A., *Animadversion on the Null Class*, reviewed in ***The Journal of Symbolic Logic***, vol. 8, no. 2 (1943), p. 48

Carnap, R., ***Notes for Symbolic Logic***, reviewed in ***The Journal of Symbolic Logic***, vol. 4, no. 1 (1939), pp. 29–30

Carnap, R., *Testability and Meaning*, reviewed in ***The Journal of Symbolic Logic***, vol. 21, no. 2 (1956), p. 212

Carnap, R., *Überwindung der Metaphysik durch logische Analyse der Sprache*, reviewed in ***The Journal of Symbolic Logic***, vol. 35, no. 2 (1970), p. 312

Carnap, R., *Empiricism, Semantics, and Ontology*, reviewed in ***The Journal of Symbolic Logic***, vol. 35, no. 2 (1970), p. 313

Carnap, R., ***Logische Syntax der Sprache***, 2nd edition, reviewed in ***The Journal of Symbolic Logic***, vol. 40, no. 3 (1975), p. 472

Carpenter, J. A., Moore, O. K., Snyder, C. R., & Lisansky, E. S., *Alcohol and Higher-Order Problem Solving*, reviewed in ***The Journal of Symbolic Logic***, vol. 30, no. 2 (1965), p. 243

Carroll, L., ***Symbolic Logic and The Game of Logic***, reviewed in ***The Journal of Symbolic Logic***, vol. 25, no. 3 (1960), pp. 264–265

Carruccio, E., *Considerazioni Sulla Compatibilità di un Sistema di Postulati e Sulla*

Dimostrabilità delle Formule Matematiche, reviewed in ***The Journal of Symbolic Logic***, vol. 13, no. 4 (1948), pp. 225–226

Carruccio, E., *Alcune Conseguenze di un Risultato del Gödel e la Razionalità del Reale*, reviewed in ***The Journal of Symbolic Logic***, vol. 14, no. 2 (1949), p. 142

Carruccio, E., *Il Problema della Razionalità del Reale*, reviewed in ***The Journal of Symbolic Logic***, vol. 15, no. 2 (1950), p. 143

Carruccio, E., *Sulla Potenza dell'Insieme delle Proposizioni di un Dato Sistema Ipotetico-Deduttivo*, reviewed in ***The Journal of Symbolic Logic***, vol. 15, no. 2 (1950), p. 143

Cartan, H., *Sur le Fondement Logique des Mathématiques*, reviewed in ***The Journal of Symbolic Logic***, vol. 11, no. 3 (1946), pp. 91–92

Casari, E., *La Logique en Italie*, reviewed in ***The Journal of Symbolic Logic***, vol. 40, no. 3 (1975), p. 472

Cassina, U., *Sulla Critica di Grandjot all'Arimetica di Peano*, reviewed in ***The Journal of Symbolic Logic***, vol. 20, no. 2 (1955), pp. 175–176

Cassina, U., *Prefazione*, reviewed in ***The Journal of Symbolic Logic***, vol. 27, no. 4 (1962), p. 471

Cassina, U., *Introduzione*, reviewed in ***The Journal of Symbolic Logic***, vol. 27, no. 4 (1962), p. 471

Cassina, U., *Note*, reviewed in ***The Journal of Symbolic Logic***, vol. 27, no. 4 (1962), p. 471

Cavaillès, J., *Mathematiques et Formalisme*, reviewed in ***The Journal of Symbolic Logic***, vol. 15, no. 2 (1950), pp. 143–144

Chandrasekharan, K., *The Logic of Intuitionistic Mathematics*, reviewed in ***The Journal of Symbolic Logic***, vol. 7, no. 4 (1942), p. 171

Chandrasekharan, K., *Partially Ordered Sets and Symbolic Logic*, reviewed in ***The Journal of Symbolic Logic***, vol. 11, no. 3 (1946), pp. 100–101

Charlesworth, M. J., *Analytical Philosophy*, in ***New Catholic Encyclopedia***, reviewed in ***The Journal of Symbolic Logic***, vol. 40, no. 4 (1975), p. 595

Cherciu, M., *Filtres de Stone dans les Treillis Distributifs*, reviewed in ***The Journal of Symbolic Logic***, vol. 39, no. 1 (1974), p. 184

Christian, C., *Zwei Theoreme über die Einerklasse*, reviewed in ***The Journal of Symbolic Logic***, vol. 21, no. 3 (1956), p. 322

Christian, C., *A Proof of the Inconsistency of Quine's System "Mathematical Logic (1951)"*, reviewed in ***The Journal of Symbolic Logic***, vol. 21, no. 3 (1956), p. 322

Churchman, C. W., *Towards a General Logic of Propositions*, reviewed in ***The Journal of Symbolic Logic***, vol. 8, no. 2 (1943), pp. 53–54

Churchman, C. W., *Reply to Comments on "Statistics, Pragmatics, Induction"*, reviewed in ***The Journal of Symbolic Logic***, vol. 15, no. 1 (1950), pp. 62–63

Churchman, C. W., & Cowan, T. A., *A Challenge*, reviewed in ***The Journal of Symbolic Logic***, vol. 10, no. 4 (1945), p. 133

Chwistek, L., *Überwindung des Begriffsrealismus*, reviewed in ***The Journal of Symbolic Logic***, vol. 2, no. 4 (1937), pp. 168–170

Clark, J. T., *Contemporary Science and Deductive Methodology*, reviewed in ***The Journal of Symbolic Logic***, vol. 22, no. 4 (1957), p. 359

Cohen, L. J., *Are Philosophical Theses Relative to Language?*, reviewed in ***The Journal of Symbolic Logic***, vol. 15, no. 1 (1950), p. 63

Cohen, M. R., & Nagel, E., ***An Introduction to Logic and Scientific Method***, reviewed in ***The Journal of Symbolic Logic***, vol. 11, no. 3 (1946), p. 100

Copi, I. M., *Modern Logic and the Synthetic A Priori*, reviewed in ***The Journal of Symbolic Logic***, vol. 15, no. 3 (1950), p. 221

Copi, I. M., *The Inconsistency or Redundancy of Principia Mathematica*, reviewed in ***The Journal of Symbolic Logic***, vol. 16, no. 2 (1951), pp. 154–155

Cowan, T. A., *A Note on Churchman's "Statistics, Pragmatics, Induction"*, reviewed in ***The Journal of Symbolic Logic***, vol. 15, no. 1 (1950), pp. 62–63

Crespo Pereira, R., *Sobre el Álgebra de la Lógica de Schröder*, reviewed in ***The Journal of Symbolic Logic***, vol. 17, no. 2 (1952), p. 154

Crystal, D., ***Linguistics***, reviewed in ***The Journal of Symbolic Logic***, vol. 37, no. 2 (1972), p. 420

Curry, H. B., *The Inconsistency of Certain Formal Logics*, reviewed in ***The Journal of Symbolic Logic***, vol. 7, no. 4 (1942), pp. 170–171

Curry, H. B., *Some Advances in the Combinatory Theory of Quantification*, reviewed in ***The Journal of Symbolic Logic***, vol. 8, no. 2 (1943), p. 52

Curry, H. B., *Remarks on the Definition and Nature of Mathematics*, reviewed in ***The Journal of Symbolic Logic***, vol. 22, no. 1 (1957), pp. 85–86

Curvelo, E., ***Introdução à Lógica***, reviewed in ***The Journal of Symbolic Logic***, vol. 13, no. 3 (1948), p. 144

Czeżowski, T., *On Certain Peculiarities of Singular Propositions*, reviewed in ***The Journal of Symbolic Logic***, vol. 21, no. 2 (1956), p. 207

Dassen, C. C., *Sobre una Objeción a la Lógica Brouweriana*, reviewed in ***The Journal of Symbolic Logic***, vol. 6, no. 3 (1941), pp. 106–107

Daya, *Symmetry, Transitivity and Reflexivity*, reviewed in ***The Journal of Symbolic Logic***, vol. 25, no. 3 (1960), pp. 263–264

Daya, *Concluding Note*, reviewed in ***The Journal of Symbolic Logic***, vol. 25, no. 3 (1960), pp. 263–264

De Morgan, A., *On the Syllogism*, reviewed in ***The Journal of Symbolic Logic***, vol. 29, no. 3 (1964), p. 135

De Morgan, A., *On the Syllogism: I. On the Structure of the Syllogism*, reviewed in ***The Journal of Symbolic Logic***, vol. 41, no. 2 (1976), p. 546

De Morgan, A., *On the Syllogism: II. On the Symbols of Logic, the Theory of the Syllogism, and in Particular of the Copula*, reviewed in ***The Journal of Symbolic***

Logic, vol. 41, no. 2 (1976), p. 546

De Morgan, A., *Some Suggestions in Logical Phraseology*, reviewed in ***The Journal of Symbolic Logic***, vol. 41, no. 2 (1976), p. 546

De Morgan, A., *On the Syllogism: III. And on Logic in General*, reviewed in ***The Journal of Symbolic Logic***, vol. 41, no. 2 (1976), p. 546

De Morgan, A., *Syllabus of a Proposed System of Logic*, reviewed in ***The Journal of Symbolic Logic***, vol. 41, no. 2 (1976), p. 546

De Morgan, A., *On the Syllogism: IV. And on the Logic of Relations*, reviewed in ***The Journal of Symbolic Logic***, vol. 41, no. 2 (1976), p. 546

De Morgan, A., *Logic*, reviewed in ***The Journal of Symbolic Logic***, vol. 41, no. 2 (1976), p. 546

De Morgan, A., *On the Syllogism: V. And on Various Points of the Onymatic System*, reviewed in ***The Journal of Symbolic Logic***, vol. 41, no. 2 (1976), p. 546

De Morgan, A., *On the Syllogism: VI. On Quantity and Location; on Identity in Relation to the Syllogism; on the History of the Mnemonic Words, of the Fourth Figure, of Logic at Cambridge, of the Occult Quality; on the Triadic System of Enunciation*, reviewed in ***The Journal of Symbolic Logic***, vol. 41, no. 2 (1976), p. 546

de Raeymaeker, L., ***Introduction to Philosophy***, translated by H. McNeill, reviewed in ***The Journal of Symbolic Logic***, vol. 13, no. 2 (1948), p. 123

de Raeymaeker, L., ***Inleiding tot de Wijsbegeerte***, 2nd edition, reviewed in ***The Journal of Symbolic Logic***, vol. 14, no. 3 (1949), p. 186

de Raeymaeker, L., ***Introduction à la Philosophie***, 3rd edition, reviewed in ***The Journal of Symbolic Logic***, vol. 14, no. 3 (1949), p. 186

De Cesare, E. A., *Evolución de la Lógica*, reviewed in ***The Journal of Symbolic Logic***, vol. 7, no. 4 (1942), p. 174

Denis-Papin, M., Kaufmann, A., & Faure, R., ***Cours de Calcul Booléien Applique (Notions sur les Ensembles et les Treillis, Algèbres Booléiennes, Algèbre Binaire)***, reviewed in ***The Journal of Symbolic Logic***, vol. 33, no. 1 (1968), p. 127

Denjoy, A., ***L'Énumération Transfinie. Livre I. La Notion de Rang***, reviewed in ***The Journal of Symbolic Logic***, vol. 13, no. 3 (1948), p. 144

Destouches, P., *La Logique Symbolique en France et les Récentes Journées de Logique*, reviewed in ***The Journal of Symbolic Logic***, vol. 11, no. 3 (1946), p. 91

Dewey, J., & Bentley, A. F., *A Search for Firm Names*, reviewed in ***The Journal of Symbolic Logic***, vol. 10, no. 4 (1945), pp. 132–133

Dewey, J., & Bentley, A. F., *A Terminology for Knowings and Knowns*, reviewed in ***The Journal of Symbolic Logic***, vol. 10, no. 4 (1945), pp. 132–133

Dewey, J., & Bentley, A. F., *Postulations*, reviewed in ***The Journal of Symbolic Logic***, vol. 10, no. 4 (1945), pp. 132–133

Dienes, Z. P., *On an Implication Function in Many-Valued Systems of Logic*, reviewed by A. Church and N. Rescher in ***The Journal of Symbolic Logic***, vol. 15, no. 1

(1950), pp. 69–70

Dieudonné, J., *Les Méthodes Axiomatiques Modernes et les Fondements des Mathématiques*, reviewed in ***The Journal of Symbolic Logic***, vol. 4, no. 4 (1939), p. 163

Dilworth, R. P., *Lattices with Unique Complements*, reviewed in ***The Journal of Symbolic Logic***, vol. 10, no. 3 (1945), p. 104

Dodd, S. C., ***Dimensions of Society. A Quantitative Systematics for the Social Sciences***, reviewed in ***The Journal of Symbolic Logic***, vol. 7, no. 3 (1942), pp. 128–129

Drobot, S., & Straszewicz, S., *XI. History, Teaching, Popularization and Organization of Mathematics*, reviewed in ***The Journal of Symbolic Logic***, vol. 31, no. 3 (1966), p. 517

Dubreil, P., ***Algèbre. Tome I. Équivalences, Opérations, Groupes, Anneaux, Corps***, reviewed in ***The Journal of Symbolic Logic***, vol. 12, no. 3 (1947), p. 94

Ducasse, C. J., *Some Observations Concerning the Nature of Probability*, reviewed in ***The Journal of Symbolic Logic***, vol. 6, no. 3 (1941), pp. 108–109

Dumitriu, A., *The Antinomy of the Theory of Types*, reviewed in ***The Journal of Symbolic Logic***, vol. 37, no. 1 (1972), p. 194

Duncan-Jones, A., *Fugitive Propositions*, reviewed in ***The Journal of Symbolic Logic***, vol. 15, no. 2 (1950), p. 151

Dürr, K., *Die Mathematische Logik des Arnold Geulincx*, reviewed in ***The Journal of Symbolic Logic***, vol. 6, no. 3 (1941), p. 104

Dürr, K., *Die Logistik Johann Heinrich Lamberts*, reviewed in ***The Journal of Symbolic Logic***, vol. 12, no. 4 (1947), pp. 137–138

Encinas del Pando, J., *La Lógica de Bertrand Russell*, reviewed in ***The Journal of Symbolic Logic***, vol. 8, no. 2 (1943), p. 50

Errera, A., *Sur les Fondements de l'Arithmétique*, reviewed in ***The Journal of Symbolic Logic***, vol. 5, no. 3 (1940), pp. 119–120

Esser, G., ***Logica, in Usum Scholarum***, reviewed in ***The Journal of Symbolic Logic***, vol. 8, no. 2 (1943), pp. 47–48

Evans, E., *On the Language of Converse Relations*, reviewed in ***The Journal of Symbolic Logic***, vol. 21, no. 3 (1956), pp. 318–319

Fairthrone, R. A., *The Mathematics of Classification*, reviewed in ***The Journal of Symbolic Logic***, vol. 13, no. 3 (1948), pp. 158–159

Farber, M., ***The Foundation of Phenomenology. Edmund Husserl and the Quest for a Rigorous Science of Philosophy***, reviewed in ***The Journal of Symbolic Logic***, vol. 9, no. 3 (1944), pp. 63–65

Farinelli, U., & Gamba, A., *Physics and Mathematical Logic*, reviewed in ***The Journal of Symbolic Logic***, vol. 20, no. 3 (1955), p. 285

Farre, G. L., *Boole, George*, in ***New Catholic Encyclopedia***, reviewed in ***The Journal of Symbolic Logic***, vol. 40, no. 4 (1975), p. 596

Farre, G. L., *De Morgan, Augustus*, in ***New Catholic Encyclopedia***, reviewed in ***The***

Journal of Symbolic Logic, vol. 40, no. 4 (1975), p. 596

Farre, G. L., *Frege, Gottlob*, in ***New Catholic Encyclopedia***, reviewed in ***The Journal of Symbolic Logic***, vol. 40, no. 4 (1975), p. 596

Farre, G. L., *Peano, Giuseppe*, in ***New Catholic Encyclopedia***, reviewed in ***The Journal of Symbolic Logic***, vol. 40, no. 4 (1975), p. 598

Fehr, H. F., *Meaning in Algebra*, reviewed in ***The Journal of Symbolic Logic***, vol. 12, no. 3 (1947), p. 96

Feys, R., *Logistique*, reviewed in ***The Journal of Symbolic Logic***, vol. 21, no. 3 (1956), p. 309

Feys, R., *Une Théorie Formalisée Demontrée sans Symboles*, reviewed in ***The Journal of Symbolic Logic***, vol. 23, no. 3 (1958), p. 344

Findlay, J., *Goedelian Sentences: A Non-Numerical Approach*, reviewed in ***The Journal of Symbolic Logic***, vol. 7, no. 3 (1942), pp. 129–130

Finsler, P., *Gibt es Unentscheidbare Sätze?*, reviewed in ***The Journal of Symbolic Logic***, vol. 11, no. 4 (1946), pp. 131–132

Finsler, P., *Das Kontinuumproblem*, reviewed in ***The Journal of Symbolic Logic***, vol. 15, no. 3 (1950), p. 230

Finsler, P., *Eine transfinite Folge arithmetischer Operationen*, reviewed in ***The Journal of Symbolic Logic***, vol. 22, no. 3 (1957), p. 336

Fischer, H., *Zum Problem der Übertragung Mathetischer Prinzipien: Die "Allgemeine Semantik." Eine Nichtaristotelische Wertungslehre Alfred Korzybskis*, reviewed in ***The Journal of Symbolic Logic***, vol. 19, no. 1 (1954), p. 65

Fitch, F. B., *On God and Immortality*, reviewed in ***The Journal of Symbolic Logic***, vol. 13, no. 3 (1948), p. 148

Fitch, F. B., *Reply to Professor Baylis' Criticisms*, reviewed in ***The Journal of Symbolic Logic***, vol. 13, no. 3 (1948), p. 148

Fitch, F. B., *The Problem of the Morning Star and the Evening Star*, reviewed in ***The Journal of Symbolic Logic***, vol. 15, no. 1 (1950), p. 63

Fogarasi, B., ***Logik***, translated by S. Szemere, reviewed in ***The Journal of Symbolic Logic***, vol. 21, no. 3 (1956), p. 314

Fraenkel, A. A., *Diskrete und Kontinuierliche Gebilde*, reviewed in ***The Journal of Symbolic Logic***, vol. 4, no. 4 (1939), p. 163

Fraenkel, A. A., ***Abstract Set Theory***, 2nd edition, reviewed in ***The Journal of Symbolic Logic***, vol. 28, no. 2 (1963), pp. 168–169

Frege, G., *Sense and Reference*, reviewed in ***The Journal of Symbolic Logic***, vol. 13, no. 3 (1948), p. 152

Frege, G., ***Aritmetica e Logica***, L. Geymonat, editor, reviewed in ***The Journal of Symbolic Logic***, vol. 13, no. 3 (1948), p. 153

Frege, G., ***Translations from the Philosophical Writings of Gottlob Frege***, P. Geach and M. Black, editors, reviewed in ***The Journal of Symbolic Logic***, vol. 18, no. 1 (1953),

p. 92

Frege, G., *Definitions*, reviewed in ***The Journal of Symbolic Logic***, vol. 29, no. 3 (1964), p. 135

Frege, G., ***Begriffsschrift, a Formula Language, Modeled upon that of Arithmetic, for Pure Thought***, reviewed in ***The Journal of Symbolic Logic***, vol. 37, no. 2 (1972), p. 405

Frege, G., *Letter to Russell*, reviewed in ***The Journal of Symbolic Logic***, vol. 39, no. 2 (1974), p. 355

Freudenthal, H., *Analyse Mathématique de Certaines Structures Linguistiques*, reviewed in ***The Journal of Symbolic Logic***, vol. 36, no. 3 (1971), pp. 514–515

Freund, J. E., *Statistical vs. Pragmatic Inference*, reviewed in ***The Journal of Symbolic Logic***, vol. 15, no. 1 (1950), pp. 62–63

Freund, J. E., *Chapter 23: Logic*, reviewed in ***The Journal of Symbolic Logic***, vol. 22, no. 4 (1957), p. 407

Friedberg, R. M., *4-Quantifier Completeness: A Banach-Mazur Functional not Uniformly Partial Recursive*, reviewed in ***The Journal of Symbolic Logic***, vol. 24, no. 1 (1959), p. 52

Gardner, M., *Logic Machines*, reviewed in ***The Journal of Symbolic Logic***, vol. 17, no. 3 (1952), p. 217

Geach, P. T., *Designation and Truth*, reviewed in ***The Journal of Symbolic Logic***, vol. 13, no. 3 (1948), pp. 151–152

Geach, P. T., *Mr. Ill-Named*, reviewed in ***The Journal of Symbolic Logic***, vol. 14, no. 2 (1949), p. 136

Geach, P. T., *On Rigour in Semantics*, reviewed in ***The Journal of Symbolic Logic***, vol. 15, no. 2 (1950), p. 151

Geach, P. T., *Russell's Theory of Descriptions*, reviewed in ***The Journal of Symbolic Logic***, vol. 15, no. 3 (1950), p. 217

Geach, P. T., *Designation and Truth—A Reply*, reviewed in ***The Journal of Symbolic Logic***, vol. 17, no. 1 (1952), pp. 70–71

Geach, P. T., *On Insolubilia*, reviewed in ***The Journal of Symbolic Logic***, vol. 20, no. 2 (1955), p. 192

Gentzen, G., *Die Widerspruchsfreiheit der Stufenlogik*, reviewed in ***The Journal of Symbolic Logic***, vol. 1, no. 3 (1936), p. 119

Georgescu, G., *Les Algèbres de Lukasiewicz θ-Valentes*, reviewed in ***The Journal of Symbolic Logic***, vol. 39, no. 1 (1974), p. 184

Germansky, B., *Axiomes des Nombres Naturels*, reviewed in ***The Journal of Symbolic Logic***, vol. 12, no. 2 (1947), p. 58

Germansky, B., *Supplement to my Paper "Axioms of the Natural Numbers"*, reviewed in ***The Journal of Symbolic Logic***, vol. 15, no. 4 (1950), p. 282

Germansky, B., *A New Set of Axioms Sufficient for the Development of the Theory of*

Natural Numbers, reviewed in ***The Journal of Symbolic Logic***, vol. 15, no. 4 (1950), p. 282

Germansky, B., *An Alternative Proof of a Theorem of Equivalence Concerning Axioms of Natural Numbers*, reviewed in ***The Journal of Symbolic Logic***, vol. 15, no. 4 (1950), p. 282

Germansky, B., *Problem*, reviewed in ***The Journal of Symbolic Logic***, vol. 18, no. 3 (1953), p. 263

Gerneth, D. C., *Generalization of Menger's Result on the Structure of Logical Formulas*, reviewed in ***The Journal of Symbolic Logic***, vol. 13, no. 4 (1948), p. 224

Geymonat, L., *Difficoltà del Concetto di "Insieme"*, reviewed in ***The Journal of Symbolic Logic***, vol. 13, no. 2 (1948), pp. 126–127

Geymonat, L, *Le Origini della Metodologia Moderna*, reviewed in ***The Journal of Symbolic Logic***, vol. 13, no. 4 (1948), p. 226

Geymonat, L, *La Crisi Della Logica Formale*, reviewed in ***The Journal of Symbolic Logic***, vol. 13, no. 4 (1948), pp. 226–227

Gilbert, E. N., *N-Terminal Switching Circuits*, reviewed in ***The Journal of Symbolic Logic***, vol. 30, no. 2 (1965), p. 248

Giorgi, G., *A Proposito di Alcune Discussioni Recenti sui Problemi della Logica Dedut-tiva*, reviewed in ***The Journal of Symbolic Logic***, vol. 14, no. 2 (1949), p. 141

Glassen, P., *Some Questions about Relations*, reviewed in ***The Journal of Symbolic Logic***, vol. 32, no. 3 (1967), p. 408

Goddard, L., *'True' and 'Provable'*, reviewed in ***The Journal of Symbolic Logic***, vol. 25, no. 1 (1960), pp. 85–86

Gödel, K., *Some Metamathematical Results on Completeness and Consistency*, reviewed in ***The Journal of Symbolic Logic***, vol. 37, no. 2 (1972), p. 405

Gödel, K., *On Formally Undecidable Propositions of Principia Mathematica and Related Systems I*, reviewed in ***The Journal of Symbolic Logic***, vol. 37, no. 2 (1972), p. 405

Gödel, K., *On Completeness and Consistency*, reviewed in ***The Journal of Symbolic Logic***, vol. 37, no. 2 (1972), p. 405

Gonzalez, M. O., & Mancill, J. D., *On the System of Natural Numbers*, reviewed in ***The Journal of Symbolic Logic***, vol. 15, no. 2 (1950), pp. 138–139

Gonzalez, M. O., & Mancill, J. D., *Remarks on Natural Numbers*, reviewed in ***The Journal of Symbolic Logic***, vol. 16, no. 3 (1951), p. 223

Goodell, J. D., *Notes on Decision Element Systems using Various Practical Techniques*, reviewed in ***The Journal of Symbolic Logic***, vol. 19, no. 2 (1954), p. 143

Goodman, N., *On Likeness of Meaning*, reviewed in ***The Journal of Symbolic Logic***, vol. 15, no. 2 (1950), pp. 150–151

Goodman, N., *On a Pseudo-Test of Translation*, reviewed in ***The Journal of Symbolic Logic***, vol. 20, no. 1 (1955), p. 62

Goodman, N., *On Likeness of Meaning*, reviewed in ***The Journal of Symbolic Logic***, vol. 21, no. 1 (1956), pp. 76–77

Gorman, M., *Semantics*, in ***New Catholic Encyclopedia***, reviewed in ***The Journal of Symbolic Logic***, vol. 40, no. 4 (1975), pp. 598–599

Götlind, E., *A Note on an Article by R. K. P. Singh and R. Shukla*, reviewed in ***The Journal of Symbolic Logic***, vol. 17, no. 4 (1952), p. 277

Götlind, E., *A Note on Chwistek and Hetper's Foundation of Formal Metamathematics*, reviewed in ***The Journal of Symbolic Logic***, vol. 19, no. 2 (1954), p. 140

Gotô, M., *Application of Logical Mathematics to the Theory of Relay Networks*, reviewed in ***The Journal of Symbolic Logic***, vol. 20, no. 3 (1955), pp. 285–286

Graham, E., *Logic and Semiotic. Some Comments Regarding the Treatment of Logical Signs in Charles Morris' Signs, Language, and Behavior*, reviewed in ***The Journal of Symbolic Logic***, vol. 13, no. 4 (1948), p. 218

Graves, L. M., ***The Theory of Functions of Real Variables***, reviewed in ***The Journal of Symbolic Logic***, vol. 12, no. 3 (1947), p. 96

Greenwood, T., *Les Principes de la Logique Mathématique*, reviewed in ***The Journal of Symbolic Logic***, vol. 8, no. 1 (1943), pp. 28–29

Grelling, K., & Oppenheim, P., *Der Gestaltbegriff im Lichte der Neuen Logik*, reviewed in ***The Journal of Symbolic Logic***, vol. 15, no. 1 (1950), p. 61

Grelling, K., & Oppenheim, P., *Supplementary Remarks on the Concept of Gestalt*, reviewed in ***The Journal of Symbolic Logic***, vol. 15, no. 1 (1950), p. 61

Griss, G. F. C., *Negationless Intuitionistic Mathematics*, reviewed in ***The Journal of Symbolic Logic***, vol. 13, no. 3 (1948), p. 174

Gromska, D., *Philosophes Polonais Morts entre 1938 et 1945*, reviewed in ***The Journal of Symbolic Logic***, vol. 18, no. 1 (1953), pp. 93–94

Gutiérrez Novoa, L., *La Ley de Dualidad de los Conjuntos de Puntos*, reviewed in ***The Journal of Symbolic Logic***, vol. 5, no. 1 (1940), pp. 36–37

Guy, W. T., Jr., *On Equivalence Relations*, reviewed in ***The Journal of Symbolic Logic***, vol. 21, no. 2 (1956), p. 207

Hadamard, J., *La Géometrie non Euclidienne et les Définitions Axiomatiques*, reviewed in ***The Journal of Symbolic Logic***, vol. 23, no. 1 (1958), pp. 30–32

Hadamard, J., *Sur l'Impossibilité de Demontrer la Compatibilité des Axiomes de l'Arithmétique*, reviewed in ***The Journal of Symbolic Logic***, vol. 23, no. 1 (1958), pp. 30–32

Hailperin, T., *An Incorrect Theorem*, reviewed in ***The Journal of Symbolic Logic***, vol. 31, no. 1 (1966), p. 128

Hall, E. W., *Some Dangers in the Use of Symbolic Logic in Psychology*, reviewed in ***The Journal of Symbolic Logic***, vol. 7, no. 2 (1942), p. 100

Halldén, S., *A Note Concerning the Paradoxes of Strict Implication and Lewis's System S1*, reviewed in ***The Journal of Symbolic Logic***, vol. 14, no. 1 (1949), p. 69

Halldén, S., *Certain Problems Connected with the Definitions of Identity and of Definite*

Descriptions Given in Principia Mathematica, reviewed in ***The Journal of Symbolic Logic***, vol. 14, no. 2 (1949), pp. 136–137

Halmos, P. R., *The Foundations of Probability*, reviewed in ***The Journal of Symbolic Logic***, vol. 9, no. 4 (1944), p. 106

Hampshire, S., *Multiply General Sentences*, reviewed in ***The Journal of Symbolic Logic***, vol. 15, no. 3 (1950), p. 216

Handy, R., & Kurtz, P., ***A Current Appraisal of the Behavioral Sciences***, reviewed in ***The Journal of Symbolic Logic***, vol. 29, no. 3 (1964), pp. 135–136

Hanson, N. R., *The Gödel Theorem. An Informal Exposition*, reviewed in ***The Journal of Symbolic Logic***, vol. 27, no. 4 (1962), pp. 471–472

Hanson, N. R., *A Note on the Gödel Theorem*, reviewed in ***The Journal of Symbolic Logic***, vol. 28, no. 4 (1963), p. 295

Hartman, S., *Are There any Extra-Syllogistic Forms of Reasoning?*, reviewed in ***The Journal of Symbolic Logic***, vol. 5, no. 2 (1940), p. 81

Hartree, D. R., ***Calculating Instruments and Machines***, reviewed in ***The Journal of Symbolic Logic***, vol. 18, no. 4 (1953), p. 347

Hasenjaeger, G., *Eine Bemerkung zu Henkin's Beweis für die Vollständigkeit des Prädikatenkalküls der ersten Stufe*, reviewed in ***The Journal of Symbolic Logic***, vol. 31, no. 2 (1966), p. 268

Heath, P., *Introduction*, reviewed in ***The Journal of Symbolic Logic***, vol. 41, no. 2 (1976), p. 546

Heath, P., *Select Bibliography*, reviewed in ***The Journal of Symbolic Logic***, vol. 41, no. 2 (1976), p. 546

Hendrix, G., *Developing a Logical Concept in Elementary Mathematics*, reviewed in ***The Journal of Symbolic Logic***, vol. 21, no. 3 (1956), p. 336

Henkin, L., *Banishing the Rule of Substitution for Functional Variables*, reviewed in ***The Journal of Symbolic Logic***, vol. 20, no. 2 (1955), pp. 179–180

Herzberger, H. G., *The Truth-Conditional Consistency of Natural Languages*, reviewed in ***The Journal of Symbolic Logic***, vol. 33, no. 1 (1968), pp. 146–147

Herzberger, H. G., *The Logical Consistency of Language*, reviewed in ***The Journal of Symbolic Logic***, vol. 33, no. 1 (1968), p. 147

Heyting, A., *The Development of Intuitionistic Mathematics*, reviewed in ***The Journal of Symbolic Logic***, vol. 2, no. 2 (1937), p. 89

Heyting, A., *La Conception Intuitionniste de la Logique*, reviewed in ***The Journal of Symbolic Logic***, vol. 23, no. 3 (1958), pp. 344–345

Heyting, A., *Intuitionism in Mathematics*, reviewed in ***The Journal of Symbolic Logic***, vol. 40, no. 3 (1975), p. 472

Hilbert, D., & Ackermann, W., ***Grundzüge der theoretischen Logik***, 3rd edition, reviewed in ***The Journal of Symbolic Logic***, vol. 15, no. 1 (1950), p. 59

Hille, E., *Such Stuff as Dreams are Made on — in Mathematics*, reviewed in ***The Journal***

of Symbolic Logic, vol. 18, no. 2 (1953), p. 183

Hiz, H., *Statement by Dr. Hiz*, reviewed in ***The Journal of Symbolic Logic***, vol. 18, no. 2 (1953), p. 183

Hoberman, S., & McKinsey, J. C. C., *A Set of Postulates for Boolean Algebra*, reviewed in ***The Journal of Symbolic Logic***, vol. 2, no. 4 (1937), pp. 172–173

Howes, D., *On Mr. Korzybski's Vision*, reviewed in ***The Journal of Symbolic Logic***, vol. 18, no. 2 (1953), p. 183

Hu, S.-T., ***Elementary Functions and Coordinate Geometry***, reviewed in ***The Journal of Symbolic Logic***, vol. 34, no. 3 (1969), pp. 520–521

Hu, S.-T., ***Threshold Logic***, reviewed in ***The Journal of Symbolic Logic***, vol. 40, no. 2 (1975), p. 250

Huntington, E. V., & Ladd-Franklin, C., *Logic, Symbolic*, reviewed in ***The Journal of Symbolic Logic***, vol. 18, no. 2 (1953), p. 183

Hurley, R. B., ***Transistor Logic Circuits***, reviewed in ***The Journal of Symbolic Logic***, vol. 33, no. 1 (1968), pp. 126–127

Hutten, E. H., *A Note on Semantics*, reviewed in ***The Journal of Symbolic Logic***, vol. 15, no. 2 (1950), p. 152

Jaśkowski, S., *Trois Contributions au Calcul des Propositions Bivalent*, reviewed in ***The Journal of Symbolic Logic***, vol. 13, no. 3 (1948), pp. 164–165

Johnston, L. S., *Another Form of the Russell Paradox*, reviewed in ***The Journal of Symbolic Logic***, vol. 5, no. 4 (1940), p. 157

Jørgensen, J., ***Outline of the Recent Development of the Theory of Deduction***, reviewed in ***The Journal of Symbolic Logic***, vol. 3, no. 1 (1938), p. 43

Jungius, J., ***Logica Hamburgensis***, R. W. Meyer, editor, reviewed in ***The Journal of Symbolic Logic***, vol. 33, no. 1 (1968), p. 139

Kaczmarz, S., *Axioms for Arithmetic*, reviewed in ***The Journal of Symbolic Logic***, vol. 20, no. 4 (1955), p. 307

Kalin, T. A., *Formal Logic and Switching Circuits*, reviewed in ***The Journal of Symbolic Logic***, vol. 18, no. 4 (1953), pp. 345–346

Kalmár, L., *Zur Reduktion des Entscheidungsproblems*, reviewed in ***The Journal of Symbolic Logic***, vol. 3, no. 1 (1938), p. 46

Kalmár, L, *Contributions to the Reduction Theory of the Decision Problem. First Paper. Prefix* $(x_1)(x_2)(Ex_3)\cdots(Ex_{n-1}(x_n)$*, a Single Binary Predicate*, reviewed in ***The Journal of Symbolic Logic***, vol. 17, no. 1 (1952), p. 73

Kalmár, L., *Contributions to the Reduction Theory of the Decision Problem. Third Paper. Prefix* $(x_1)(Ex_2)\cdots(Ex_{n-2})(x_{n-1})(x_n)$*, a Single Binary Predicate*, reviewed in ***The Journal of Symbolic Logic***, vol. 18, no. 3 (1953), p. 264

Kalmár, L., *Contributions to the Reduction Theory of the Decision Problem. Fourth Paper. Reduction to the Case of a Finite Set of Individuals*, reviewed in ***The Journal of Symbolic Logic***, vol. 18, no. 3 (1953), pp. 264–265

Kamke, E., ***Theory of Sets***, translated by F. Bagemihl, reviewed in ***The Journal of Symbolic Logic***, vol. 15, no. 3 (1950), p. 201

Kattsoff, L. O., *Modality and Probability*, reviewed in ***The Journal of Symbolic Logic***, vol. 2, no. 1 (1937), p. 44

Kattsoff, L. O., *What is Behavior?*, reviewed in ***The Journal of Symbolic Logic***, vol. 13, no. 4 (1948), p. 218

Katz, J. J., ***The Problem of Induction and Its Solution***, reviewed in ***The Journal of Symbolic Logic***, vol. 36, no. 2 (1971), p. 320

Kauppi, R., *Note on Philosophical Trends in Finland*, reviewed in ***The Journal of Symbolic Logic***, vol. 32, no. 1 (1967), p. 106

Keister, W., Ritchie, A. E., & Washburn, S. H., ***The Design of Switching Circuits***, reviewed in ***The Journal of Symbolic Logic***, vol. 18, no. 4 (1953), pp. 347–348

Keyser, C. J., *Charles Sanders Peirce as a Pioneer*, reviewed in ***The Journal of Symbolic Logic***, vol. 6, no. 4 (1941), pp. 161–162

Kiely, E. R., *Mathematics, History of*, in ***New Catholic Encyclopedia***, reviewed in ***The Journal of Symbolic Logic***, vol. 40, no. 4 (1975), pp. 597–598

Kim-Bradley, C., *Symbolic Logic and Metamathematics*, reviewed in ***The Journal of Symbolic Logic***, vol. 17, no. 2 (1952), p. 154

Kim-Bradley, C., Letter, reviewed in ***The Journal of Symbolic Logic***, vol. 17, no. 4 (1952), p. 289

Kim-Bradley, C., *The Paradoxes*, reviewed in ***The Journal of Symbolic Logic***, vol. 19, no. 3 (1954), p. 236

Kleene, S. C., *A Note on Recursive Functions*, reviewed in ***The Journal of Symbolic Logic***, vol. 1, no. 3 (1936), p. 119

Kleene, S. C., *Recursive Predicates and Quantifiers*, reviewed in ***The Journal of Symbolic Logic***, vol. 8, no. 1 (1943), pp. 32–34

Kleene, S. C., *Errata. Arithmetical Predicates and Function Quantifiers*, reviewed in ***The Journal of Symbolic Logic***, vol. 22, no. 4 (1957), p. 375

Kleene, S. C., ***Representation of Events in Nerve Nets and Finite Automata***, reviewed in ***The Journal of Symbolic Logic***, vol. 23, no. 1 (1958), pp. 58–59

Kloyda, T. À. K., *Bolzano, Bernhard*, in ***New Catholic Encyclopedia***, reviewed in ***The Journal of Symbolic Logic***, vol. 40, no. 4 (1975), p. 596

Knox, J., Jr., *Material Implication and "if . . . then"*, reviewed in ***The Journal of Symbolic Logic***, vol. 37, no. 1 (1972), p. 185

Kocourek, R. A., *An Evaluation of Symbolic Logic*, reviewed in ***The Journal of Symbolic Logic***, vol. 14, no. 1 (1949), p. 52

Kokoszyńska, M., *Kazimierz Ajdukiewicz*, reviewed in ***The Journal of Symbolic Logic***, vol. 40, no. 3 (1975), p. 472

Körner, S., *Entailment and the Meaning of Words*, reviewed in ***The Journal of Symbolic Logic***, vol. 15, no. 3 (1950), pp. 217–218

Kossel, C. G., *The Problem of Relation in Some Non-Scholastic Philosophies*, reviewed in ***The Journal of Symbolic Logic***, vol. 11, no. 3 (1946), pp. 82–83

Koyré, A., *The Liar*, reviewed in ***The Journal of Symbolic Logic***, vol. 11, no. 4 (1946), p. 131

Kraft, V., *Die Moderne und die Traditionelle Logik*, reviewed in ***The Journal of Symbolic Logic***, vol. 16, no. 1 (1951), p. 78

Kraft, V., ***The Vienna Circle. The Origin of Neo-Positivism. A Chapter in the History of Recent Philosophy***, reviewed in ***The Journal of Symbolic Logic***, vol. 20, no. 1 (1955), pp. 62–63

Kurtz, P., editor, *Introduction*, ***American Philosophy in the Twentieth Century***, reviewed in ***The Journal of Symbolic Logic***, vol. 35, no. 2 (1970), pp. 312–313

Kurtz, P., *Charles S. Peirce*, reviewed in ***The Journal of Symbolic Logic***, vol. 35, no. 2 (1970), pp. 312–313

Kurtz, P., *Alfred North Whitehead*, reviewed in ***The Journal of Symbolic Logic***, vol. 35, no. 2 (1970), pp. 312–313

Kurtz, P., *Morris R. Cohen*, reviewed in ***The Journal of Symbolic Logic***, vol. 35, no. 2 (1970), pp. 313

Kurtz, P., *Clarence Irving Lewis*, reviewed in ***The Journal of Symbolic Logic***, vol. 35, no. 2 (1970), pp. 313

Kurtz, P., *Rudolf Carnap*, reviewed in ***The Journal of Symbolic Logic***, vol. 35, no. 2 (1970), pp. 313

Kurtz, P., *Willard Van Orman Quine*, reviewed in ***The Journal of Symbolic Logic***, vol. 35, no. 2 (1970), pp. 313

Kurtz, P., *Max Black*, reviewed in ***The Journal of Symbolic Logic***, vol. 35, no. 2 (1970), pp. 313

Kurtz, P., *Ernest Nagel*, reviewed in ***The Journal of Symbolic Logic***, vol. 35, no. 2 (1970), pp. 313

Labérenne, P., *Deux Études sur les Fondements des Mathématiques*, reviewed in ***The Journal of Symbolic Logic***, vol. 23, no. 1 (1958), pp. 30–32

Ladrière, J. A., *Axiomatic System*, in ***New Catholic Encyclopedia***, reviewed in ***The Journal of Symbolic Logic***, vol. 40, no. 4 (1975), p. 596

Lalan, V., *Définition de Deux Structures d'Anneau dans une Algèbre de Boole*, reviewed in ***The Journal of Symbolic Logic***, vol. 12, no. 2 (1947), p. 58

Lambert, K., *Existential Import Revisited*, reviewed in ***The Journal of Symbolic Logic***, vol. 30, no. 1 (1965), pp. 103–104

Landau, E., ***Grundlagen der Analysis. (Das Rechnen mit ganzen, rationalen, irrationalen, komplexen Zahlen.) Ergänzung zu den Lehrbüchern der Differential- und Integralrechnung***, reviewed in ***The Journal of Symbolic Logic***, vol. 11, no. 4 (1946), p. 126

Landau, E., ***Foundations of Analysis***, translated by F. Steinhardt, reviewed in ***The Journal of Symbolic Logic***, vol. 16, no. 3 (1951), p. 223

Langer, S. K., ***An Introduction to Symbolic Logic***, 2nd edition, reviewed in ***The Journal of Symbolic Logic***, vol. 21, no. 2 (1956), p. 187

Lasley, J. W., Jr., *The Revolt Against Aristotle*, reviewed in ***The Journal of Symbolic Logic***, vol. 7, no. 4 (1942), p. 171

Lawrence, N., *Heterology and Hierarchy*, reviewed in ***The Journal of Symbolic Logic***, vol. 15, no. 3 (1950), pp. 216–217

Lazerowitz, M., *Self-Contradictory Propositions*, reviewed in ***The Journal of Symbolic Logic***, vol. 5, no. 2 (1940), pp. 81–82

Leavitt, W. G., *Boolean Algebra and Circuit Analysis*, reviewed in ***The Journal of Symbolic Logic***, vol. 23, no. 1 (1958), p. 62

Leggett, H. W., ***Bertrand Russell, O. M. A Pictorial Biography***, reviewed in ***The Journal of Symbolic Logic***, vol. 16, no. 3 (1951), p. 223

Leonard, H. S., *The Logic of Existence*, reviewed in ***The Journal of Symbolic Logic***, vol. 28, no. 3 (1963), pp. 259–261

Levi, B., *A Propósito de la Nota del Dr. Pi Calleja. Sobre Paradojas Lógicas y Principio del Tertium Non Datur*, reviewed in ***The Journal of Symbolic Logic***, vol. 17, no. 3 (1952), pp. 200–201

Levi, B., *Intorno alla Teoria degli Aggregati*, reviewed in ***The Journal of Symbolic Logic***, vol. 40, no. 3 (1975), p. 527

Lévy, P., *Axiome de Zermelo et Nombres Transfinis*, reviewed in ***The Journal of Symbolic Logic***, vol. 15, no. 3 (1950), pp. 201–202

Lewis, C. I., *The Modes of Meaning*, reviewed in ***The Journal of Symbolic Logic***, vol. 9, no. 1 (1944), pp. 28–29

Lewis, C. I., *An Analysis of Knowledge and Valuation*, reviewed in ***The Journal of Symbolic Logic***, vol. 35, no. 2 (1970), pp. 313

Lewis, C. I., & Langford, C. H., ***Symbolic Logic***, reviewed in ***The Journal of Symbolic Logic***, vol. 16, no. 3 (1951), p. 225

Lieber, L. R., ***Mits, Wits and Logic***, reviewed in ***The Journal of Symbolic Logic***, vol. 13, no. 1 (1948), p. 55

Lindley, T. F., *Moore's Nominal Definitions of 'Culture'*, reviewed in ***The Journal of Symbolic Logic***, vol. 24, no. 1 (1959), pp. 85–86

Linsky, L., *On Using Inverted Commas*, reviewed in ***The Journal of Symbolic Logic***, vol. 16, no. 3 (1951), pp. 208–209

Linsky, L., *Preface*, ***Semantics and the Philosophy of Language***, reviewed in ***The Journal of Symbolic Logic***, vol. 21, no. 1 (1956), p. 76

Linsky, L., *Introduction*, ***Semantics and the Philosophy of Language***, reviewed in ***The Journal of Symbolic Logic***, vol. 21, no. 1 (1956), p. 76

Locher, L., *Die Finsler'schen Arbeiten zur Grundlegung der Mathematik*, reviewed in ***The Journal of Symbolic Logic***, vol. 3, no. 2 (1938), pp. 89–90

Lombardo-Radice, L., *Ordinali Transfiniti e Principio del Terzo Escluso (A Proposito di un Ragionamento del Gödel)*, reviewed in ***The Journal of Symbolic Logic***, vol. 23, no. 2 (1958), pp. 214–215

Lonergan, B., *The Form of Inference*, reviewed in ***The Journal of Symbolic Logic***, vol. 8, no. 2 (1943), p. 48

Lorenzen, P., *Die Allgemeingültigkeit der Logischen Regeln*, reviewed in ***The Journal of Symbolic Logic***, vol. 30, no. 1 (1965), p. 104

Łoś, J., & Suszko, R., *Remarks on Sentential Logics*, reviewed in ***The Journal of Symbolic Logic***, vol. 40, no. 4 (1975), pp. 603–604

Łukasiewicz, J., *Logistic and Philosophy*, reviewed in ***The Journal of Symbolic Logic***, vol. 1, no. 3 (1936), p. 118

Łukasiewicz, J., *The Shortest Axiom of the Implicational Calculus of Propositions*, reviewed in ***The Journal of Symbolic Logic***, vol. 13, no. 3 (1948), p. 164

Łukasiewicz, J., *On Variable Functors of Propositional Arguments*, reviewed in ***The Journal of Symbolic Logic***, vol. 16, no. 3 (1951), pp. 229–230

Łuszczewska-Romahnowa, S., *An Analysis and Generalization of Venn's Diagrammatic Decision Procedure*, reviewed in ***The Journal of Symbolic Logic***, vol. 21, no. 2 (1956), p. 193

MacColl, H., *Symbolic Reasoning*, reviewed in ***The Journal of Symbolic Logic***, vol. 41, no. 3 (1976), pp. 701–702

MacColl, H., *Three Notes from 'Mind'*, reviewed in ***The Journal of Symbolic Logic***, vol. 41, no. 3 (1976), pp. 701–702

Mac Lane, S., *Symbolic Logic*, reviewed in ***The Journal of Symbolic Logic***, vol. 4, no. 3 (1939), pp. 125–126

MacNeille, H. M., *Extensions of Partially Ordered Sets*, reviewed in ***The Journal of Symbolic Logic***, vol. 1, no. 2 (1936), p. 73

Maehara, S., *Logic in Japan*, reviewed in ***The Journal of Symbolic Logic***, vol. 40, no. 3 (1975), p. 472

Marc-Wogau, K., *Remarks Concerning the Latest Discussion on Sense-Data*, reviewed in ***The Journal of Symbolic Logic***, vol. 14, no. 3 (1949), p. 183

Marhenke, P., *The Criterion of Significance*, reviewed in ***The Journal of Symbolic Logic***, vol. 21, no. 1 (1956), pp. 77–78

Markov, A. A., *On the Impossibility of Certain Algorithms in the Theory of Associative Systems*, reviewed in ***The Journal of Symbolic Logic***, vol. 13, no. 3 (1948), pp. 170–171

Markov, A. A., ***On the Representation of Recursive Functions***, reviewed in ***The Journal of Symbolic Logic***, vol. 17, no. 1 (1952), pp. 72–73

Markov, A. A., *Téoriá Algorifmov (Az Algoritmusok Elmélete)*, reviewed in ***The Journal of Symbolic Logic***, vol. 20, no. 1 (1955), p. 73

Martin, G., *Über ein Zweiwertiges Modell einer Vierwertigen Logik*, reviewed in ***The Journal of Symbolic Logic***, vol. 16, no. 2 (1951), p. 150

Martin, N. M., *On Completeness of Decision Element Sets*, reviewed in ***The Journal of Symbolic Logic***, vol. 19, no. 2 (1954), p. 143

Martin, N. M., *Note on the Completeness of Decision Element Sets*, reviewed in ***The Journal of Symbolic Logic***, vol. 32, no. 1 (1967), p. 134

Martin, R. M., *Mr. Geach on Mention and Use*, reviewed in ***The Journal of Symbolic Logic***, vol. 15, no. 2 (1950), p. 151

Martin, R. M., *Some Comments on Truth and Designation*, reviewed in ***The Journal of Symbolic Logic***, vol. 17, no. 1 (1952), pp. 70–71

Martin, R. M., *On Types, Denotation, and Truth*, reviewed in ***The Journal of Symbolic Logic***, vol. 19, no. 2 (1954), pp. 139–140

Mates, B., *Stoic Logic and the Text of Sextus Empiricus*, reviewed in ***The Journal of Symbolic Logic***, vol. 15, no. 1 (1950), pp. 63–64

Mates, B., ***Stoic Logic***, 2nd printing, reviewed in ***The Journal of Symbolic Logic***, vol. 28, no. 4 (1963), p. 295

Mates, B., ***Elementary Logic***, 2nd edition, reviewed in ***The Journal of Symbolic Logic***, vol. 37, no. 2 (1972), pp. 419–420

Mates, B., ***Elementare Logik (Prädikatenlogik der Ersten Stufe)***, translated by A. Oberschelp, reviewed in ***The Journal of Symbolic Logic***, vol. 37, no. 3 (1972), pp. 615–616

Mates, B., ***Elementary Logic***, 2nd edition reviewed in ***The Journal of Symbolic Logic***, vol. 38, no. 4 (1973), p. 647

Matijasevič, Y. V., *Diophantine Representation of Recursively Enumerable Predicates*, reviewed in ***The Journal of Symbolic Logic***, vol. 37, no. 3 (1972), pp. 606–607

Mays, W., Letter, reviewed in ***The Journal of Symbolic Logic***, vol. 17, no. 3 (1952), p. 217

Mays, W., & Henry, D. P., *Exhibition of the Work of W. Stanley Jevons*, reviewed in ***The Journal of Symbolic Logic***, vol. 18, no. 1 (1953), p. 69

Mays, W., & Prinz, D. G., *A Relay Machine for the Demonstration of Symbolic Logic*, reviewed in ***The Journal of Symbolic Logic***, vol. 15, no. 2 (1950), p. 138

McGuinness, B. F., *Russell, Bertrand, Philosophy of*, in ***New Catholic Encyclopedia***, reviewed in ***The Journal of Symbolic Logic***, vol. 40, no. 4 (1975), p. 598

McKinsey, J. C. C., *Reducible Boolean Functions*, reviewed in ***The Journal of Symbolic Logic***, vol. 1, no. 2 (1936), p. 69

McNaughton, R., *Logical and Combinatorial Problems in Computer Design*, reviewed in ***The Journal of Symbolic Logic***, vol. 22, no. 2 (1957), p. 222

Melden, A. I., *Thought and Its Objects*, reviewed in ***The Journal of Symbolic Logic***, vol. 5, no. 4 (1940), pp. 162–163

Meltzer, B., *Preface*, Gödel, K., ***On Formally Undecidable Propositions of Principia Mathematica and Related Systems***, reviewed in ***The Journal of Symbolic Logic***, vol. 30, no. 3 (1965), pp. 357–359

Menger, K., ***Algebra of Analysis***, reviewed in ***The Journal of Symbolic Logic***, vol. 10, no. 3 (1945), p. 103

Menger, K., *Are Variables Necessary in Calculus?*, reviewed in ***The Journal of Symbolic Logic***, vol. 15, no. 1 (1950), p. 61

Menger, K., *On Variables in Mathematics and in Natural Science*, reviewed in ***The Journal of Symbolic Logic***, vol. 22, no. 3 (1957), pp. 300–301

Menger, K., *Variables, de Diverses Natures*, reviewed in ***The Journal of Symbolic Logic***, vol. 22, no. 3 (1957), pp. 300–301

Menger, K., *What are Variables and Constants?*, reviewed in ***The Journal of Symbolic Logic***, vol. 22, no. 3 (1957), pp. 300–301

Meredith, C. A., *On an Extended System of the Propositional Calculus*, reviewed in ***The Journal of Symbolic Logic***, vol. 16, no. 3 (1951), pp. 229–230

Meredith, C. A., *Single Axioms for the Systems* (C, N), $(C, 0)$ *and* (A, N) *of the Two-Valued Propositional Calculus*, reviewed in ***The Journal of Symbolic Logic***, vol. 19, no. 2 (1954), pp. 143–144

Meredith, C. A., *A Single Axiom of Positive Logic*, reviewed in ***The Journal of Symbolic Logic***, vol. 19, no. 2 (1954), p. 144

Meschkowski, H., ***Wandlungen des Mathematischen Denkens. Eine Einführung in die Grundlagenprobleme der Mathematik***, 2nd edition, reviewed in ***The Journal of Symbolic Logic***, vol. 31, no. 1 (1966), p. 111

Mihailescu, E. G., *Recherches sur l'Équivalence et la Réciprocité dans le Calcul des Propositions*, reviewed in ***The Journal of Symbolic Logic***, vol. 3, no. 1 (1938), p. 55

Minogue, G. P., *The Three Fundamental Laws of Thought in Their Metaphysical and Logical Aspects*, reviewed in ***The Journal of Symbolic Logic***, vol. 12, no. 3 (1947), pp. 98–99

Miro Quesada, O., ***El Número y la Realidad***, reviewed in ***The Journal of Symbolic Logic***, vol. 11, no. 3 (1946), p. 96

Moch, F., *On Peut Éviter les Antinomies Classiques Sans Restreindre la Notion d'Ensemble*, reviewed in ***The Journal of Symbolic Logic***, vol. 21, no. 3 (1956), p. 322

Moisil, G. C., *Sur le Syllogisme Hypothétique dans la Logique Intuitioniste*, reviewed in ***The Journal of Symbolic Logic***, vol. 3, no. 3 (1938), p. 118

Moisil, G. C., *Les Logiques à Plusieurs Valeurs et l'Automatique*, reviewed in ***The Journal of Symbolic Logic***, vol. 36, no. 3 (1971), pp. 546–547

Moisil, G. C., ***Teoria Algebrica dei Meccanismi Automatici***, reviewed in ***The Journal of Symbolic Logic***, vol. 36, no. 3 (1971), p. 547

Moisil, G. C., *Les États Transitoires dans les Circuits Séquentiels*, reviewed in ***The Journal of Symbolic Logic***, vol. 37, no. 3 (1972), p. 626

Mokre, J., *Zu den Logischen Paradoxien*, reviewed in ***The Journal of Symbolic Logic***, vol. 28, no. 1 (1963), p. 106

Montague, R., & Tarski, J., *On Bernstein's Self-Dual Set of Postulates for Boolean*

Algebras, reviewed in ***The Journal of Symbolic Logic***, vol. 27, no. 4 (1962), p. 472

Moore, E. F., *Gedanken-Experiments on Sequential Machines*, reviewed in ***The Journal of Symbolic Logic***, vol. 23, no. 1 (1958), p. 60

Moore, O. K., *Nominal Definitions of 'Culture'*, reviewed in ***The Journal of Symbolic Logic***, vol. 24, no. 1 (1959), pp. 85–86

Moore, O. K., *Dr. Lindley and "Nominal Definitions of 'Culture'"*, reviewed in ***The Journal of Symbolic Logic***, vol. 24, no. 1 (1959), pp. 85–86

Moore, O. K., *Problem Solving and the Perception of Persons*, reviewed in ***The Journal of Symbolic Logic***, vol. 24, no. 1 (1959), p. 86

Moore, O. K., & Anderson, S. B., *Modern Logic and Tasks for Experiments on Problem Solving Behavior*, reviewed in ***The Journal of Symbolic Logic***, vol. 24, no. 1 (1959), p. 86

Moore, O. K., & Anderson, S. B., *Search Behavior in Individual and Group Problem Solving*, reviewed in ***The Journal of Symbolic Logic***, vol. 24, no. 1 (1959), p. 86

Moore, O. K., & Lewis, D. J., *Learning Theory and Culture*, reviewed in ***The Journal of Symbolic Logic***, vol. 24, no. 1 (1959), p. 85

More, T., Jr., *On the Construction of Venn Diagrams*, reviewed in ***The Journal of Symbolic Logic***, vol. 27, no. 1 (1962), p. 108

Moritz, R. E., ***On Mathematics and Mathematicians***, reviewed in ***The Journal of Symbolic Logic***, vol. 24, no. 3 (1959), p. 216

Morris, C., *Signs About Signs About Signs*, reviewed in ***The Journal of Symbolic Logic***, vol. 13, no. 4 (1948), p. 218

Morris, C., ***Signs, Language, and Behavior***, reviewed in ***The Journal of Symbolic Logic***, vol. 22, no. 1 (1957), p. 88

Morris, C., ***Signification and Significance. A Study of the Relation of Signs and Values***, reviewed in ***The Journal of Symbolic Logic***, vol. 33, no. 2 (1968), p. 317

Mostowski, A., & Łoś, J., *I. Foundations of Mathematics, Theory of Sets and Mathematical Logic*, reviewed in ***The Journal of Symbolic Logic***, vol. 31, no. 3 (1966), p. 517

Müller, H. R., *Algebraischer Aussagenkalkül*, reviewed in ***The Journal of Symbolic Logic***, vol. 7, no. 3 (1942), p. 126

Muller-Oikonomou, S., *The Three Basic Directions in the Foundations of Mathematics*, reviewed in ***The Journal of Symbolic Logic***, vol. 12, no. 3 (1947), p. 94–95

Murphy, A. E., *American Philosophy in the Twentieth Century*, reviewed in ***The Journal of Symbolic Logic***, vol. 29, no. 1 (1964), pp. 48–49

Myhill, J., *A System Which can Define Its Own Truth*, reviewed in ***The Journal of Symbolic Logic***, vol. 21, no. 3 (1956), p. 319

Nagel, E., *Logic Without Ontology*, reviewed in ***The Journal of Symbolic Logic***, vol. 10, no. 1 (1945), pp. 16–18

Nagel, E., *Sovereign Reason*, reviewed in ***The Journal of Symbolic Logic***, vol. 13, no. 1

(1948), p. 55

Nagel, E., & Newman, J. R., ***Gödel's Proof***, reviewed in ***The Journal of Symbolic Logic***, vol. 21, no. 4 (1956), p. 374

Nakamura, K., *Zum Logischen Funktionsbegriffe des Wiener Kreises*, reviewed in ***The Journal of Symbolic Logic***, vol. 6, no. 1 (1941), p. 36

Nakamura, K., ***The Theory of Logic — A Contribution to Scientific Logic***, reviewed in ***The Journal of Symbolic Logic***, vol. 15, no. 1 (1950), pp. 61–62

Nakamura, K., *Zum Logistischen Gestaltbegriffe*, reviewed in ***The Journal of Symbolic Logic***, vol. 15, no. 1 (1950), pp. 61–62

Nakasima, A., & Hanzawa M., *The Theory of Equivalent Transformation of Simple Partial Paths in the Relay Circuit*, et al., reviewed in ***The Journal of Symbolic Logic***, vol. 18, no. 4 (1953), p. 346

Nakasima, A., *The Theory of Four-Terminal Passive Networks in Relay Circuit*, reviewed in ***The Journal of Symbolic Logic***, vol. 18, no. 4 (1953), p. 346

Nakasima, A., *Algebraic Expressions Relative to Simple Partial Paths in the Relay Circuit*, reviewed in ***The Journal of Symbolic Logic***, vol. 18, no. 4 (1953), p. 346

Nakasima, A., *The Theory of Two-Point Impedance of Passive Networks in the Relay Circuit*, reviewed in ***The Journal of Symbolic Logic***, vol. 18, no. 4 (1953), p. 346

Nakasima, A., *The Transfer Impedance of Four-Terminal Passive Networks in the Relay Circuit*, reviewed in ***The Journal of Symbolic Logic***, vol. 18, no. 4 (1953), p. 346

Nakasima, A., & Hanzawa M., *Expansion Theorem and Design of Two-Terminal Relay Networks (Part I)*, reviewed in ***The Journal of Symbolic Logic***, vol. 18, no. 4 (1953), p. 346

Natucci, A., *2 + 2 Fanno 4*, reviewed in ***The Journal of Symbolic Logic***, vol. 17, no. 1 (1952), p. 77

Neder, L., *Über den Aufbau der Arithmetik*, reviewed in ***The Journal of Symbolic Logic***, vol. 20, no. 4 (1955), p. 307

Nelson, E. J., *A Note on Contradiction: A Protest*, reviewed in ***The Journal of Symbolic Logic***, vol. 1, no. 3 (1936), p. 117

Newman, M. H. A., *The Calculus of Sets*, reviewed in ***The Journal of Symbolic Logic***, vol. 4, no. 3 (1939), p. 126

Newman, M. H. A., *Stratified Systems of Logic*, reviewed in ***The Journal of Symbolic Logic***, vol. 9, no. 2 (1944), pp. 50–52

Nielsen, H. A., *Language as Existent*, reviewed in ***The Journal of Symbolic Logic***, vol. 27, no. 1 (1962), p. 118

Nielsen, H. A., *Antinomy*, in ***New Catholic Encyclopedia***, reviewed in ***The Journal of Symbolic Logic***, vol. 40, no. 4 (1975), p. 595

Nielsen, H. A., *Linguistic Analysis*, in ***New Catholic Encyclopedia***, reviewed in ***The Journal of Symbolic Logic***, vol. 40, no. 4 (1975), p. 596

Northrop, F. S. C., *The Neurological and Behavioristic Psychological Basis of the*

Ordering of Society by Means of Ideas, reviewed in ***The Journal of Symbolic Logic***, vol. 13, no. 3 (1948), pp. 157–158

Novikoff, P. S., *Sur Quelques Théorèmes d'Existence*, reviewed in ***The Journal of Symbolic Logic***, vol. 5, no. 2 (1940), pp. 69–70

Novikoff, P. S., *On the Consistency of Certain Logical Calculus*, reviewed in ***The Journal of Symbolic Logic***, vol. 11, no. 4 (1946), pp. 129–131

Padoa, A., *Essai d'une Théorie Algébrique des Nombres Entiers, Précédé d'une Introduction Logique à une Théorie Déductive Quelconque*, reviewed in ***The Journal of Symbolic Logic***, vol. 40, no. 3 (1975), p. 527

Palmer, F., ***Grammar***, reviewed in ***The Journal of Symbolic Logic***, vol. 37, no. 2 (1972), p. 420

Pankajam, S., *On Euler's ϕ-Function and its Extensions*, reviewed in ***The Journal of Symbolic Logic***, vol. 1, no. 3 (1936), p. 118

Parkinson, G. H. R., editor and translator, *Introduction*, ***Leibniz, Logical Papers***, reviewed in ***The Journal of Symbolic Logic***, vol. 33, no. 1 (1968), pp. 139–140

Parsons, C., & Kohl, H. R., *Self-reference, Truth, and Provability*, reviewed in ***The Journal of Symbolic Logic***, vol. 25, no. 1 (1960), p. 86

Peano, G., ***Formulario Mathematico***, reviewed in ***The Journal of Symbolic Logic***, vol. 27, no. 4 (1962), p. 471

Pedoe, D., ***The Gentle Art of Mathematics***, reviewed in ***The Journal of Symbolic Logic***, vol. 31, no. 4 (1966), p. 675

Peirce, C. S., *Scientific Method*, reviewed in ***The Journal of Symbolic Logic***, vol. 32, no. 3 (1967), p. 421

Peirce, C. S., ***Collected Papers of Charles Sanders Peirce***, C. Hartshorne & P. Weiss, editors, reviewed in ***The Journal of Symbolic Logic***, vol. 34, no. 3 (1969), pp. 494–495

Perelman, C., *Une Solution des Paradoxes de la Logique et ses Conséquences pour la Conception de l'Infini*, reviewed in ***The Journal of Symbolic Logic***, vol. 2, no. 4 (1937), p. 174

Perelman, C., *L'Équivalence, la Définition et la Solution du Paradoxe de Russell*, reviewed in ***The Journal of Symbolic Logic***, vol. 3, no. 2 (1938), p. 88

Perlman, D., *A Milestone in Math–Professor's New Concept*, reviewed in ***The Journal of Symbolic Logic***, vol. 28, no. 4 (1963), p. 295

Péter, R., *Über die Verallgemeinerung der Theorie der Rekursiven Funktionen für Abstrakte Mengen Geeigneter Struktur als Definitionsbereiche*, reviewed in ***The Journal of Symbolic Logic***, vol. 40, no. 4 (1975), pp. 620–621

Péter, R., *Zum Beitrag von F. Schwenkel "Rekursive Wortfunktionen über Unendlichen Alphabeten"*, reviewed in ***The Journal of Symbolic Logic***, vol. 40, no. 4 (1975), p. 622

Petrescu, I., *Algèbres de Morgan Injectives*, reviewed in ***The Journal of Symbolic Logic***, vol. 39, no. 1 (1974), p. 184

Philipov, A., ***Logic and Dialectic in the Soviet Union***, reviewed in ***The Journal of Symbolic Logic***, vol. 18, no. 3 (1953), pp. 272–273

Pi Calleja, P., *La Objeción de Grandjot a la Teoria de Peano del Número Natural*, reviewed in ***The Journal of Symbolic Logic***, vol. 17, no. 3 (1952), pp. 199–200

Pi Calleja, P., *El Tercero Incluído en la Contraparadoja de Russell*, reviewed in ***The Journal of Symbolic Logic***, vol. 17, no. 3 (1952), p. 200

Piesch, H., *Begriff der allgemeinen Schaltungstechnik*, reviewed in ***The Journal of Symbolic Logic***, vol. 30, no. 2 (1965), pp. 247–248

Piesch, H., *Über die Vereinfachung von allgemeinen Schaltungen*, reviewed in ***The Journal of Symbolic Logic***, vol. 30, no. 2 (1965), pp. 247–248

Plochmann, G. K., & Lawson, J. B., ***Terms in Their Propositional Contexts in Wittgenstein's Tractatus***, reviewed in ***The Journal of Symbolic Logic***, vol. 36, no. 3 (1971), p. 551

Poincaré, H., ***Science and Hypothesis***, reviewed in ***The Journal of Symbolic Logic***, vol. 18, no. 4 (1953), p. 327

Poincaré, H., ***Science and Method***, reviewed in ***The Journal of Symbolic Logic***, vol. 18, no. 4 (1953), p. 327

Popper, K. R., ***The Logic of Scientific Discovery***, original and revised editions, reviewed in ***The Journal of Symbolic Logic***, vol. 40, no. 3 (1975), p. 471–472

Popper, K. R., *Addendum, 1964*, reviewed in ***The Journal of Symbolic Logic***, vol. 40, no. 3 (1975), p. 471–472

Popper, K. R., *Addendum, 1967*, reviewed in ***The Journal of Symbolic Logic***, vol. 40, no. 3 (1975), p. 471–472

Popper, K. R., *Addendum, 1968*, reviewed in ***The Journal of Symbolic Logic***, vol. 40, no. 3 (1975), p. 471–472

Porte, J., *La Logique Mathématique et le Calcul Mécanique*, reviewed in ***The Journal of Symbolic Logic***, vol. 24, no. 1 (1959), p. 70

Post, E. L., *Finite Combinatory Processes—Formulation 1*, reviewed in ***The Journal of Symbolic Logic***, vol. 2, no. 1 (1937), p. 43

Post, E. L., *Formal Reductions of the General Combinatorial Decision Problem*, reviewed in ***The Journal of Symbolic Logic***, vol. 8, no. 2 (1943), pp. 50–52

Post, E. L., *A Variant of a Recursively Unsolvable Problem*, reviewed in ***The Journal of Symbolic Logic***, vol. 12, no. 2 (1947), pp. 55–56

Potter, A., *Statement by Dr. Potter*, reviewed in ***The Journal of Symbolic Logic***, vol. 18, no. 2 (1953), p. 183

Prior, A. N., *The Parva Logicalia in Modern Dress*, reviewed in ***The Journal of Symbolic Logic***, vol. 19, no. 1 (1954), pp. 73–74

Prior, A. N., *Symmetry, Transitivity and Reflexivity*, reviewed in ***The Journal of Symbolic Logic***, vol. 25, no. 3 (1960), pp. 263–264

Puig Adam, P., *Métodos Gráfico y Algebraico Para el Proyecto de Circuitos Electrónicos de Cálculo*, reviewed in ***The Journal of Symbolic Logic***, vol. 22, no. 4 (1957), p. 377

Putnam, H., *An Unsolvable Problem in Number Theory*, reviewed in ***The Journal of Symbolic Logic***, vol. 37, no. 3 (1972), pp. 601–602

Quine, W. V., *A Logistical Approach to the Ontological Problem*, reviewed in ***The Journal of Symbolic Logic***, vol. 4, no. 4 (1939), p. 170

Quine, W. V., ***Mathematical Logic***, reviewed in ***The Journal of Symbolic Logic***, vol. 5, no. 4 (1940), pp. 163–164

Quine, W. V., *Reply to Professor Usenko*, reviewed in ***The Journal of Symbolic Logic***, vol. 7, no. 1 (1942), pp. 45

Quine, W. V., *Whitehead and the Rise of Modern Logic*, reviewed in ***The Journal of Symbolic Logic***, vol. 7, no. 2 (1942), pp. 100–101

Quine, W. V., *On Existence Conditions for Elements and Classes*, reviewed in ***The Journal of Symbolic Logic***, vol. 8, no. 1 (1943), pp. 31–32

Quine, W. V., *Notes on Existence and Necessity*, reviewed in ***The Journal of Symbolic Logic***, vol. 8, no. 2 (1943), pp. 45–47

Quine, W. V., ***Mathematical Logic***, 2nd printing, reviewed in ***The Journal of Symbolic Logic***, vol. 12, no. 2 (1947), p. 56

Quine, W. V., ***A Short Course in Logic. Chapters I–VII***, reviewed in ***The Journal of Symbolic Logic***, vol. 12, no. 2 (1947), pp. 60–61

Quine, W. V., *On Natural Deduction*, reviewed in ***The Journal of Symbolic Logic***, vol. 17, no. 1 (1952), pp. 76–77

Quine, W. V., *Two Theorems about Truth-Functions*, reviewed in ***The Journal of Symbolic Logic***, vol. 19, no. 2 (1954), pp. 142–143

Quine, W. V., ***A Theorem on Parametric Boolean Functions***, reviewed in ***The Journal of Symbolic Logic***, vol. 23, no. 1 (1958), pp. 58–59

Quine, W. V., ***Commutative Boolean Functions***, reviewed in ***The Journal of Symbolic Logic***, vol. 23, no. 1 (1958), pp. 58–59

Quine, W. V., ***On Functions of Relations, with Especial Reference to Social Welfare***, reviewed in ***The Journal of Symbolic Logic***, vol. 23, no. 1 (1958), pp. 58–59

Quine, W. V., *Logic, Symbolic*, in ***Encyclopedia Americana***, reviewed in ***The Journal of Symbolic Logic***, vol. 23, no. 2 (1958), pp. 207–209

Quine, W. V., *Speaking of Objects*, reviewed in ***The Journal of Symbolic Logic***, vol. 35, no. 2 (1970), p. 313

Quine, W. V., *Introductory Note*, to Russell, B., *Mathematical Logic as Based on the Theory of Types*, reviewed in ***The Journal of Symbolic Logic***, vol. 39, no. 2 (1974), pp. 355–356

Quine, W. V., *Introductory Note*, to Whitehead, A. N., & Russell, B., *Incomplete Symbols: Descriptions*, reviewed in ***The Journal of Symbolic Logic***, vol. 40, no. 3 (1975), pp. 472–473

Quinton, A., *Philosophy in Great Britain*, reviewed in ***The Journal of Symbolic Logic***, vol. 32, no. 1 (1967), p. 106

Ramsey, F. P., ***The Foundations of Mathematics and Other Logical Essays***, reviewed in ***The Journal of Symbolic Logic***, vol. 15, no. 2 (1950), p. 157

Rapoport, A., *What is Semantics?*, reviewed in ***The Journal of Symbolic Logic***, vol. 17, no. 3 (1952), pp. 216–217

Rapoport, A., *Comments by Dr. Rapoport*, reviewed in ***The Journal of Symbolic Logic***, vol. 18, no. 2 (1953), p. 183

Rapoport, A., *Statement by Dr. Rapoport*, reviewed in ***The Journal of Symbolic Logic***, vol. 18, no. 2 (1953), p. 183

Rapoport, A., & Hayakawa, S. I., *Semantics, General Semantics Again*, reviewed in ***The Journal of Symbolic Logic***, vol. 15, no. 3 (1950), p. 235

Rashevsky, N., ***Mathematical Biophysics***, reviewed in ***The Journal of Symbolic Logic***, vol. 14, no. 2 (1949), p. 128

Rawlins, I., *Natural Philosophy and the Fine Arts*, reviewed in ***The Journal of Symbolic Logic***, vol. 27, no. 1 (1962), pp. 126–127

Rawlins, I., *The Philosophy of Science and Art*, reviewed in ***The Journal of Symbolic Logic***, vol. 27, no. 1 (1962), pp. 126–127

Rawlins, I., *The Functional "A Priori"*, reviewed in ***The Journal of Symbolic Logic***, vol. 27, no. 1 (1962), pp. 126–127

Rawlins, I., *Definition in Philosophy*, reviewed in ***The Journal of Symbolic Logic***, vol. 27, no. 1 (1962), pp. 126–127

Read, A. W., *An Account of the Word 'Semantics'*, reviewed in ***The Journal of Symbolic Logic***, vol. 14, no. 2 (1949), p. 135

Reichenbach, H., ***The Rise of Scientific Philosophy***, reviewed in ***The Journal of Symbolic Logic***, vol. 21, no. 4 (1956), p. 396

Restle, F., ***Psychology of Judgment and Choice: A Theoretical Essay***, reviewed in ***The Journal of Symbolic Logic***, vol. 25, no. 3 (1960), p. 257

Rey Pastor, J., ***Apuntes de Teoría de los Conjuntos Abstractos***, reviewed in ***The Journal of Symbolic Logic***, vol. 28, no. 3 (1963), pp. 250–251

Rey Pastor, J., Pi Calleja, P., & Trejo, C. A., ***Análisis Matemático. Vol. I. Análisis Algebraico–Teoria de Ecuaciones–Cálculo Infinitesimal de una Variable***, reviewed in ***The Journal of Symbolic Logic***, vol. 17, no. 3 (1952), p. 201

Riabouchinsky, D., *Les Diviseurs de Zéro et le Concept de l'Origine d'un Nombre*, reviewed in ***The Journal of Symbolic Logic***, vol. 13, no. 1 (1948), pp. 55–56

Riabouchinsky, D., *Sur le Concept de l'Origine d'un Nombre et le Problème du Continu*, reviewed in ***The Journal of Symbolic Logic***, vol. 13, no. 1 (1948), pp. 55–56

Riabouchinsky, D., *Sur Les Nombres d'Origine Imaginaire et la Notion de Signe d'un Nombre Complexe*, reviewed in ***The Journal of Symbolic Logic***, vol. 13, no. 2 (1948), p. 122

Ricci, G., *Elementi di Teoria dei Numeri*, reviewed in ***The Journal of Symbolic Logic***, vol. 17, no. 3 (1952), p. 201

Ridder, J., *Ueber den Aussagen- und den Engeren Prädikatenkalkül*, reviewed in ***The Journal of Symbolic Logic***, vol. 13, no. 3 (1948), p. 174

Ritchie, A. D., *A Defence of Aristotle's Logic*, reviewed in ***The Journal of Symbolic Logic***, vol. 11, no. 4 (1946), p. 134

Robinson, R., *What is Independence?*, reviewed in ***The Journal of Symbolic Logic***, vol. 5, no. 2 (1940), p. 81

Rose, A., *Remarque sur les Notions d'Indépendance et de Non-Contradiction*, reviewed in ***The Journal of Symbolic Logic***, vol. 16, no. 4 (1951), p. 279

Rose, A., *A Formalization of the C-0 Propositional Calculus*, reviewed in ***The Journal of Symbolic Logic***, vol. 17, no. 1 (1952), p. 66

Rose, A., *Self-Dual Primitives for Modal Logic*, reviewed in ***The Journal of Symbolic Logic***, vol. 18, no. 3 (1953), pp. 282–283

Rose, A., *Sur un Ensemble de Fonctions Primitives pour le Calcul des Prédicats du Premier Ordre lequel constitue son Propre Dual*, reviewed in ***The Journal of Symbolic Logic***, vol. 18, no. 4 (1953), pp. 343–344

Rose, A., *A Formalisation of the 2-Valued Propositional Calculus with Self-Dual Primitives*, reviewed in ***The Journal of Symbolic Logic***, vol. 19, no. 4 (1954), p. 295

Rosenbaum, R. A., *Remark on Equivalence Relations*, reviewed in ***The Journal of Symbolic Logic***, vol. 21, no. 2 (1956), p. 207

Ross, A., *Imperatives and Logic*, reviewed in ***The Journal of Symbolic Logic***, vol. 9, no. 2 (1944), p. 48

Rosser, B., *The Introduction of Quantification into a Three-Valued Logic*, reviewed in ***The Journal of Symbolic Logic***, vol. 4, no. 4 (1939), p. 170

Rosser, B., *On the Many-Valued Logics*, reviewed in ***The Journal of Symbolic Logic***, vol. 6, no. 3 (1941), p. 109

Rosser, B., *Problems on Natural Numbers*, reviewed in ***The Journal of Symbolic Logic***, vol. 15, no. 3 (1950), p. 236

Rougier, L., *La Relativité de la Logique*, reviewed in ***The Journal of Symbolic Logic***, vol. 11, no. 3 (1946), p. 100

Ruja, H., *Good Logics and Bad*, reviewed in ***The Journal of Symbolic Logic***, vol. 5, no. 2 (1940), p. 81

Runes, D. D., ***Pictorial History of Philosophy***, reviewed in ***The Journal of Symbolic Logic***, vol. 24, no. 3 (1959), p. 216

Runes, D. D., editor, ***Classics in Logic, Readings in Epistemology, Theory of Knowledge and Dialectics***, reviewed in ***The Journal of Symbolic Logic***, vol. 29, no. 3 (1964), p. 135

Russell, B., ***Introduction to Mathematical Philosophy***, 6th impression, reviewed in ***The Journal of Symbolic Logic***, vol. 15, no. 2 (1950), p. 157

Russell, B., *Letter to Frege*, reviewed in ***The Journal of Symbolic Logic***, vol. 39, no. 2 (1974), p. 355

Russell, B., *Mathematical Logic as Based on the Theory of Types*, reviewed in ***The Journal of Symbolic Logic***, vol. 39, no. 2 (1974), p. 355–356

Russell, B., *Mathematical Logic as Based on the Theory of Types*, reviewed in ***The Journal of Symbolic Logic***, vol. 39, no. 2 (1974), p. 356

Russell, B., ***Essays in Analysis by Bertrand Russell***, D. Lackey, editor, reviewed in ***The Journal of Symbolic Logic***, vol. 41, no. 3 (1976), pp. 700–702

Russell, B., *The Existential Import of Propositions*, reviewed in ***The Journal of Symbolic Logic***, vol. 41, no. 3 (1976), pp. 701–702

Russell, B., *Is Mathematics Purely Linguistic?*, reviewed in ***The Journal of Symbolic Logic***, vol. 41, no. 3 (1976), pp. 701–702

Russell, B., *On Some Difficulties in the Theory of Transfinite Numbers and Order Types*, reviewed in ***The Journal of Symbolic Logic***, vol. 41, no. 3 (1976), pp. 701–702

Russell, B., *On the Substitutional Theory of Classes and Relations*, reviewed in ***The Journal of Symbolic Logic***, vol. 41, no. 3 (1976), pp. 701–702

Russell, B., *The Theory of Logical Types*, reviewed in ***The Journal of Symbolic Logic***, vol. 41, no. 3 (1976), pp. 701–702

Ryle, G., *Systematically Misleading Expressions*, reviewed in ***The Journal of Symbolic Logic***, vol. 17, no. 4 (1952), pp. 284–285

Rynin, D., *A Critical Essay on Johnson's Philosophy of Language*, reviewed in ***The Journal of Symbolic Logic***, vol. 12, no. 3 (1947), pp. 96–98

Rynin, D., *Introduction*, ***A Treatise on Language***, reviewed in ***The Journal of Symbolic Logic***, vol. 31, no. 4 (1966), pp. 670–671

Saarnio, U., *Über die Konverse der Relation*, reviewed in ***The Journal of Symbolic Logic***, vol. 15, no. 1 (1950), p. 62

Sánchez de Bustamento, T., ***El Infinito***, reviewed in ***The Journal of Symbolic Logic***, vol. 7, no. 4 (1942), p. 174

Sanchis, L. E., *Nueva Demonstración de la Completicidad Funcional del Cálculo Proposicional Bivalente*, reviewed in ***The Journal of Symbolic Logic***, vol. 27, no. 4 (1962), p. 471

Schmidt, A., *Mathematische Grundlagenforschung*, reviewed in ***The Journal of Symbolic Logic***, vol. 17, no. 3 (1952), pp. 198–199

Schock, R., *Some Remarks on Russell's Treatment of Definite Descriptions*, reviewed in ***The Journal of Symbolic Logic***, vol. 28, no. 1 (1963), pp. 105–106

Schoenberg, J., *Belief and Intention in the Epimenides*, reviewed in ***The Journal of Symbolic Logic***, vol. 36, no. 4 (1971), pp. 671–672

Schutz, W. C., ***FIRO, A Three-Dimensional Theory of Interpersonal Behavior***, reviewed in ***The Journal of Symbolic Logic***, vol. 24, no. 3 (1959), pp. 216–217

Schwenkel, F., *Rekursive Wortfunktionen über Unendlichen Alphabeten*, reviewed in ***The Journal of Symbolic Logic***, vol. 40, no. 4 (1975), pp. 621–622

Schwenkel, F., *Entgegnung*, reviewed in ***The Journal of Symbolic Logic***, vol. 40, no. 4

(1975), p. 622

Scott, D., *Existence and Description in Formal Logic*, reviewed in ***The Journal of Symbolic Logic***, vol. 38, no. 1 (1973), pp. 166–169

Scott, J. W., ***A Synoptic Index to the Proceedings of the Aristotelian Society 1900–1949***, reviewed in ***The Journal of Symbolic Logic***, vol. 20, no. 1 (1955), pp. 57–58

Searles, H. L., ***Logic and Scientific Methods. An Introductory Course***, 2nd edition, reviewed in ***The Journal of Symbolic Logic***, vol. 23, no. 4 (1958), p. 436

Sebastião e Silva, J., *A Lógica Matemática e o Ensino Médio*, reviewed in ***The Journal of Symbolic Logic***, vol. 11, no. 3 (1946), p. 101

Sellars, W., *Acquaintance and Description Again*, reviewed in ***The Journal of Symbolic Logic***, vol. 15, no. 3 (1950), p. 222

Serrell, R., *Elements of Boolean Algebra for the Study of Information-Handling Systems*, reviewed in ***The Journal of Symbolic Logic***, vol. 19, no. 2 (1954), p. 142

Sesmat, A., & Lalan, V., *Équations dans le Corps de Boole et Relations Entre Propositions*, reviewed in ***The Journal of Symbolic Logic***, vol. 12, no. 3 (1947), p. 94

Shannon, C. E., ***A Symbolic Analysis of Relay and Switching Circuits***, reviewed in ***The Journal of Symbolic Logic***, vol. 18, no. 4 (1953), p. 347

Shannon, C. E., *Computers and Automata*, reviewed in ***The Journal of Symbolic Logic***, vol. 19, no. 2 (1954), pp. 140–141

Shannon, C. E., & Moore, E. F., *Machine Aid for Switching Circuit Design*, reviewed in ***The Journal of Symbolic Logic***, vol. 19, no. 2 (1954), p. 141

Sheffer, H. M., *Quantifiers*, reviewed in ***The Journal of Symbolic Logic***, vol. 13, no. 1 (1948), pp. 54–55

Shirai, T., *On the Pseudo-Set*, reviewed in ***The Journal of Symbolic Logic***, vol. 19, no. 3 (1954), p. 221

Shwayder, D., *Some Remarks on "Synonymity" and the Language of the Semanticists*, reviewed in ***The Journal of Symbolic Logic***, vol. 19, no. 2 (1954), p. 139

Sierpiński, W., *L'Axiome de M. Zermel et Son Rôle dans la Théorie des Ensembles et l'Analyse*, reviewed in ***The Journal of Symbolic Logic***, vol. 16, no. 3 (1951), p. 234

Sierpiński, W., *Les Exemples Effectifs et l'Axiome du Choix*, reviewed in ***The Journal of Symbolic Logic***, vol. 16, no. 3 (1951), p. 234

Sierpiński, W., *Sur les Ensembles de Points qu'on Sait Definir Effectivement*, reviewed in ***The Journal of Symbolic Logic***, vol. 16, no. 3 (1951), p. 234

Sierpiński, W., ***L'Hypothèse Généralisée du Continu et l'Axiome du Choix***, reviewed in ***The Journal of Symbolic Logic***, vol. 23, no. 2 (1958), p. 215

Sierpiński, W., ***Hypothèse du Continu***, 2nd edition, reviewed in ***The Journal of Symbolic Logic***, vol. 23, no. 2 (1958), p. 215

Skolem, T., *Explanation to the Foregoing Paper of L. Kalmár*, reviewed in ***The Journal of Symbolic Logic***, vol. 3, no. 1 (1938), p. 46

Skolem, T., *On the Proofs of Independence of the Axioms of the Classical Sentential*

Calculus, reviewed by A. Church and N. Rescher in ***The Journal of Symbolic Logic***, vol. 18, no. 1 (1953), p. 67

Skolem, T., *The Logical Nature of Arithmetic*, reviewed in ***The Journal of Symbolic Logic***, vol. 23, no. 2 (1958), p. 237

Słupecki, J., *Logic in Poland*, reviewed in ***The Journal of Symbolic Logic***, vol. 40, no. 3 (1975), p. 472

Smart, H. R., *The Alleged Predicament of Logic*, reviewed in ***The Journal of Symbolic Logic***, vol. 9, no. 4 (1944), p. 103

Smart, H. R., *Frege's Logic*, reviewed in ***The Journal of Symbolic Logic***, vol. 10, no. 3 (1945), pp. 101–103

Smart, J. J. C., *Whitehead and Russell's Theory of Types*, reviewed in ***The Journal of Symbolic Logic***, vol. 15, no. 3 (1950), p. 218

Smart, J. J. C., editor, *Introduction*, ***Problems of Space and Time***, reviewed in ***The Journal of Symbolic Logic***, vol. 38, no. 1 (1973), p. 146

Smart, J. J. C., *Theory Construction*, reviewed in ***The Journal of Symbolic Logic***, vol. 38, no. 4 (1973), pp. 665–668

Smiley, T., *On Łukasiewicz's Ł-Modal System*, reviewed in ***The Journal of Symbolic Logic***, vol. 27, no. 1 (1962), p. 113

Smith, H. B., *The Law of Transitivity*, reviewed in ***The Journal of Symbolic Logic***, vol. 1, no. 1 (1936), p. 43

Smith, H. B., *The Algebra of Propositions*, reviewed in ***The Journal of Symbolic Logic***, vol. 2, no. 1 (1937), pp. 43–44

Smith, H. B., *Modal Logic—A Revision*, reviewed in ***The Journal of Symbolic Logic***, vol. 2, no. 4 (1937), p. 173

Sobociński, B., ***An Investigation of Protothetic***, reviewed in ***The Journal of Symbolic Logic***, vol. 15, no. 1 (1950), p. 64

Sobociński, B., *On a Universal Decision Element*, reviewed in ***The Journal of Symbolic Logic***, vol. 18, no. 3 (1953), pp. 284–285

Stakelum, J. W., *Galen and the Logic of Propositions*, reviewed in ***The Journal of Symbolic Logic***, vol. 7, no. 1 (1942), p. 46

State, L., *Quleques Propriétés des Algèbres de Morgan*, reviewed in ***The Journal of Symbolic Logic***, vol. 39, no. 1 (1974), p. 184

Stebbing, L. S., *Logistic*, in ***Encyclopaedia Britannica***, reviewed in ***The Journal of Symbolic Logic***, vol. 16, no. 3 (1951), p. 223

Stone, M. H., *The Theory of Representations for Boolean Algebras*, reviewed in ***The Journal of Symbolic Logic***, vol. 1, no. 3 (1936), pp. 118–119

Storer, T., *MINIAC: World's Smallest Electronic Brain*, reviewed in ***The Journal of Symbolic Logic***, vol. 34, no. 3 (1969), p. 521

Strauss, M., *Zur Begründung der Statistischen Transformationstheorie der Quantenphysik*, reviewed in ***The Journal of Symbolic Logic***, vol. 2, no. 1 (1937), p. 45

Studley, D., *Algebra of Neural Nets*, reviewed in ***The Journal of Symbolic Logic***, vol. 14, no. 2 (1949), p. 128

Suppes, P., ***Introduction to Logic***, reviewed in ***Science***, vol. 126 (1957), pp. 1250–1251*

Surányi, J., *Contributions to the Reduction Theory of the Decision Problem. Second Paper. Three Universal, One Existential Quantifiers*, reviewed in ***The Journal of Symbolic Logic***, vol. 18, no. 3 (1953), p. 264

Surányi, J., *Contributions to the Reduction Theory of the Decision Problem. Fifth Paper. Ackermann Prefix with Three Universal Quantifiers*, reviewed in ***The Journal of Symbolic Logic***, vol. 18, no. 3 (1953), p. 265

Tarski, A., ***Introduction to Logic and to the Methodology of Deductive Sciences***, translated by O. Helmer, reviewed in ***The Journal of Symbolic Logic***, vol. 6, no. 1 (1941), pp. 30–32

Tarski, A., ***Introduction to Logic and to the Methodology of Deductive Sciences***, 2nd edition, translated by O. Helmer, reviewed in ***The Journal of Symbolic Logic***, vol. 12, no. 2 (1947), p. 61

Tarski, A., ***Introducción a la Lógica y a la Metodologia de las Ciencias Deductivas***, translated by T. R. Bachiller and J. R. Fuentes, reviewed in ***The Journal of Symbolic Logic***, vol. 16, no. 4 (1951), pp. 283–284

Tarski, A., ***Inleiding tot de Logica en tot de Methodeleer der Deductieve Wetenschappen***, reviewed in ***The Journal of Symbolic Logic***, vol. 21, no. 2 (1956), p. 187

Tarski, A., ***The Completeness of Elementary Algebra and Geometry***, reviewed in ***The Journal of Symbolic Logic***, vol. 34, no. 2 (1969), p. 302

Thomas, I., *CS(n): An Extension of CS*, reviewed in ***The Journal of Symbolic Logic***, vol. 15, no. 2 (1950), pp. 141–142

Thomas, I., *Logic and Theology*, reviewed in ***The Journal of Symbolic Logic***, vol. 15, no. 2 (1950), pp. 142–143

Thomas, I., *Some Laws of the Calculus of Quantifiers*, reviewed in ***The Journal of Symbolic Logic***, vol. 15, no. 2 (1950), p. 143

Thomas, I., *Saint Vincent Ferrer's De Suppositionibus*, reviewed in ***The Journal of Symbolic Logic***, vol. 19, no. 1 (1954), p. 74

Thomas, I., *Logic, History of*, in ***New Catholic Encyclopedia***, reviewed in ***The Journal of Symbolic Logic***, vol. 40, no. 4 (1975), pp. 596–597

Thomas, I., *The Rule of Excision in Positive Implication*, reviewed in ***The Journal of Symbolic Logic***, vol. 40, no. 4 (1975), p. 603

Toms, E., *The Law of Excluded Middle*, reviewed in ***The Journal of Symbolic Logic***, vol. 6, no. 1 (1941), p. 35

Trakhtenbrot, B. A., ***Algorithms and Automatic Computing Machines***, reviewed in ***The Journal of Symbolic Logic***, vol. 28, no. 1 (1963), pp. 104–105

Turán, P., *On the Work of Alan Baker*, reviewed in ***The Journal of Symbolic Logic***, vol. 37, no. 3 (1972), p. 606

* *Editor's note*: This review appeared in ***Science***, not ***The Journal of Symbolic Logic***.

Turing, A. M., *On Computable Numbers, with an Application to the Entscheidungsproblem*, reviewed in ***The Journal of Symbolic Logic***, vol. 2, no. 1 (1937), pp. 42–43

Turquette, A. R., *Gödel and the Synthetic A Priori*, reviewed in ***The Journal of Symbolic Logic***, vol. 15, no. 3 (1950), pp. 221–222

Ullian, J., *Mathematical Objects*, reviewed in ***The Journal of Symbolic Logic***, vol. 40, no. 4 (1975), pp. 593–595

Ullian, J., *Is Any Set Theory True?*, reviewed in ***The Journal of Symbolic Logic***, vol. 40, no. 4 (1975), pp. 593–595

Ushenko, A., *Dr. Quine's Theory of Truth-Functions*, reviewed in ***The Journal of Symbolic Logic***, vol. 7, no. 1 (1942), p. 45

Vaccarino, G., *La Scuola Polacca di Logica*, reviewed in ***The Journal of Symbolic Logic***, vol. 14, no. 2 (1949), p. 127

Vaidyanathaswami, R., *Inaugural Address*, reviewed in ***The Journal of Symbolic Logic***, vol. 4, no. 3 (1939), p. 126

Vaidyanathaswami, R., *On Disjunction in Intuitionist Logic*, reviewed in ***The Journal of Symbolic Logic***, vol. 9, no. 2 (1944), p. 48

van der Poel, W. L., *Some Selected Topics on Switching Algebra*, reviewed in ***The Journal of Symbolic Logic***, vol. 24, no. 1 (1959), p. 78

van Heijenoort, J., editor and translator, ***From Frege to Gödel, A Source Book in Mathematical Logic 1879–1931***, reviewed in ***The Journal of Symbolic Logic***, vol. 37, no. 2 (1972), p. 405

van Heijenoort, J., editor and translator, ***Frege and Gödel***, reviewed in ***The Journal of Symbolic Logic***, vol. 37, no. 2 (1972), p. 405

van Heijenoort, J., Introductory Note, ***Frege and Gödel***, reviewed in ***The Journal of Symbolic Logic***, vol. 37, no. 2 (1972), p. 405

van Heijenoort, J., Introductory Note, ***From Frege to Gödel***, reviewed in ***The Journal of Symbolic Logic***, vol. 39, no. 2 (1974), p. 355

Varet, G., & Kurtz, P., editors, ***International Directory of Philosophy and Philosophers***, reviewed in ***The Journal of Symbolic Logic***, vol. 32, no. 1 (1967), p. 106

Vaz Ferreira, C., ***Transcendentalizaciones Matemáticas Ilegitimas y Falacias Correlacionadas. Versión Taquigráfica de las Conferencias Pronunciadas en la Facultad de Filosofia y Letras de Buenos Aires***, reviewed in ***The Journal of Symbolic Logic***, vol. 17, no. 1 (1952), p. 64

Veatch, H., *Basic Confusions in Current Notions of Propositional Calculi*, reviewed in ***The Journal of Symbolic Logic***, vol. 17, no. 1 (1952), pp. 64–66

Venn, J., ***Symbolic Logic***, 2nd edition, reviewed in ***The Journal of Symbolic Logic***, vol. 37, no. 3 (1972), pp. 614–615

von Mises, R., *Scientific Conception of World. On a New Textbook of Positivism*, reviewed in ***The Journal of Symbolic Logic***, vol. 13, no. 2 (1948), p. 127

von Mises, R., *Scientific Conception of World. On a Textbook of Positivism*, reviewed in ***The Journal of Symbolic Logic***, vol. 13, no. 2 (1948), p. 127

von Wright, G. H., ***On the Idea of Logical Truth (I)***, reviewed in ***The Journal of Symbolic Logic***, vol. 15, no. 1 (1950), pp. 58–59, vol. 15, no. 3 (1950), p. 199, and vol. 15, no. 4 (1950), p. 280

von Wright, G. H., ***Form and Content in Logic***, reviewed in ***The Journal of Symbolic Logic***, vol. 15, no. 1 (1950), pp. 58–59, vol. 15, no. 3 (1950), p. 199, and vol. 15, no. 4 (1950), p. 280

Wade, F. C., *Translator's Introduction*, ***John of St. Thomas, Outline of Formal Logic***, reviewed in ***The Journal of Symbolic Logic***, vol. 24, no. 1 (1959), pp. 81–83

Waismann, F., ***Einführung in Das Mathematische Denken. Die Begriffsbildung der Modernen Mathematik***, reviewed in ***The Journal of Symbolic Logic***, vol. 13, no. 2 (1948), p. 117

Waismann, F., ***Introduction to Mathematical Thinking. The Formation of Concepts in Modern Mathematics***, translated by T. J. Benac, reviewed in ***The Journal of Symbolic Logic***, vol. 17, no. 3 (1952), p. 208

Waismann, F., *Verifiability*, reviewed in ***The Journal of Symbolic Logic***, vol. 17, no. 4 (1952), pp. 284–285

Waismann, F., *Language Strata*, reviewed in ***The Journal of Symbolic Logic***, vol. 38, no. 4 (1973), p. 663

Waismann, F., ***How I See Philosophy***, R. Harré, editor, reviewed in ***The Journal of Symbolic Logic***, vol. 38, no. 4 (1973), pp. 663–665

Wallace, W. A., *Logic, Symbolic*, in ***New Catholic Encyclopedia***, reviewed in ***The Journal of Symbolic Logic***, vol. 40, no. 4 (1975), p. 597

Wang, H., *A Proof of Independence*, reviewed in ***The Journal of Symbolic Logic***, vol. 15, no. 2 (1950), p. 138

Wang, H., *Note on Rules of Inference*, reviewed in ***The Journal of Symbolic Logic***, vol. 40, no. 4 (1975), p. 604

Webb, D. L., *Definition of Post's Generalized Negative and Maximum in Terms of One Binary Operation*, reviewed in ***The Journal of Symbolic Logic***, vol. 1, no. 1 (1936), p. 42

Wedberg, A., *The Aristotelian Theory of Classes*, reviewed in ***The Journal of Symbolic Logic***, vol. 15, no. 2 (1950), p. 142

Weinberg, J., *A Possible Solution of the Heterological Paradox*, reviewed in ***The Journal of Symbolic Logic***, vol. 3, no. 1 (1938), p. 46

Weiss, P., *The Paradox of Necessary Truth*, reviewed in ***The Journal of Symbolic Logic***, vol. 21, no. 1 (1956), p. 83

Weiss, P., *The Paradox of Necessary Truth, Once More*, reviewed in ***The Journal of Symbolic Logic***, vol. 23, no. 4 (1958), p. 442

Wellmuth, J. J., *Some Comments on the Nature of Mathematical Logic*, reviewed in ***The Journal of Symbolic Logic***, vol. 7, no. 1 (1942), pp. 39–40

Wendelin, H., *Ein Kriterium für die Erweiterbarkeit einer Implikation zu Einer*

Äquivalenz, reviewed in ***The Journal of Symbolic Logic***, vol. 23, no. 2 (1958), p. 215

Wernick, W., *An Enumeration of Logical Functions*, reviewed in ***The Journal of Symbolic Logic***, vol. 5, no. 1 (1940), p. 31

Weyl, H., *David Hilbert and His Mathematical Work*, reviewed in ***The Journal of Symbolic Logic***, vol. 9, no. 4 (1944), p. 98

White, M. G., *A Note on the "Paradox of Analysis"*, reviewed in ***The Journal of Symbolic Logic***, vol. 11, no. 4 (1946), pp. 132–133

White, M. G., *Analysis and Identity: A Rejoinder*, reviewed in ***The Journal of Symbolic Logic***, vol. 11, no. 4 (1946), pp. 132–133

Whitehead, A. N., & Russell, B., *Incomplete Symbols: Descriptions*, reviewed in ***The Journal of Symbolic Logic***, vol. 40, no. 3 (1975), pp. 472–473

Whiteman, A., *Postulates for Boolean Algebra in Terms of Ternary Rejection*, reviewed in ***The Journal of Symbolic Logic***, vol. 2, no. 2 (1937), p. 91

Whittaker, E. T., *The New Algebras and Their Significance for Physics and Philosophy*, reviewed in ***The Journal of Symbolic Logic***, vol. 9, no. 2 (1944), p. 48

Wiener, N., ***Cybernetics. Or Control and Communication in the Animal and the Machine***, reviewed in ***The Journal of Symbolic Logic***, vol. 14, no. 2 (1949), p. 127

Wilder, R. L., *Introduction to the Foundations of Mathematics*, reviewed in ***The Mathematics Student***, vol. 22 (1954), pp. 109–112*

Winthrop, H., *Metalypsis and Paradox in the Concept of Metalanguage*, reviewed in ***The Journal of Symbolic Logic***, vol. 12, no. 2 (1947), p. 55

Wittgenstein, L., ***Tractatus Logico-Philosophicus***, 4th impression, reviewed in ***The Journal of Symbolic Logic***, vol. 15, no. 2 (1950), p. 157

Wittgenstein, L., ***Tractatus Logico-Philosophicus***, reviewed in ***The Journal of Symbolic Logic***, vol. 23, no. 2 (1958), p. 213

Wolf, A., ***Textbook of Logic***, 2nd edition, reviewed in ***The Journal of Symbolic Logic***, vol. 14, no. 3 (1949), p. 186

Wundheiler, L., & Wundheiler, A., *Some Logical Concepts for Syntax*, reviewed in ***The Journal of Symbolic Logic***, vol. 21, no. 3 (1956), pp. 312–313

Yesenin-Volpin, A. S., *Svobodny Filosofskij Traktat* (*A Free Philosophical Treatise*), reviewed in ***The Journal of Symbolic Logic***, vol. 30, no. 1 (1965), pp. 104–105

Yule, G. U., & Kendall M. G., ***An Introduction to the Theory of Statistics***, reviewed in ***The Journal of Symbolic Logic***, vol. 16, no. 1 (1951), p. 51

Zemanek, H., *Automaten und Denkprozesse*, reviewed in ***The Journal of Symbolic Logic***, vol. 30, no. 3 (1965), p. 382

[Exchange among unspecified authors, reported by editors of ***Synthese***, in ***Synthese***, vol. 7 (1948–1949), pp. 229–232] *Semantics, General Semantics, Semiotic. New Acceptations of Old Terms*, reviewed in ***The Journal of Symbolic Logic***, vol. 15,

* *Editor's note*: This review appeared in ***The Mathematics Student***, not in ***The Journal of Symbolic Logic***.

no. 3 (1950), p. 235

BIBLIOGRAPHY OF ITEMS CITED BY CHURCH

Ackermann, W., *Begründung des "tertium non datur" mittels der Hilbertschen Theorie der Widerspruchsfreiheit*, ***Mathematische Annalen***, vol. 93 (1924–5), pp. 1–136

Ackermann, W., *Zum Hilbertschen Aufbau der reellen Zahlen*, ***Mathematische Annalen***, vol. 99 (1928), pp. 118–133

Ackermann, W., *Über die Erfüllbarkeit gewisser Zählausdrücke*, ***Mathematische Annalen***, vol. 100 (1928), pp. 638–649

Ackermann, W., *Beiträge zum Entscheidungsproblem der Mathematischen Logik*, ***Mathematische Annalen***, vol. 112 (1936), pp. 419–433

Ackermann, W., *Mengentheoretische Begründung der Logik*, ***Mathematische Annalen***, vol. 115 (1937), pp. 1–22

Ackermann, W., *Zur Widerspruchsfreiheit der Zahlentheorie*, ***Mathematische Annalen***, vol. 117 (1940), pp. 162–194

Ackermann, W., Review of Church *Special Cases of the Decision Problem*, ***The Journal of Symbolic Logic***, vol. 17 (1952), pp. 73–74

Ajdukiewicz, K., *A Method of Eliminating Intensional Sentences and Sentential Formulae*, ***Atti del XII Congresso Internazionale di Filosofia***, Florence, 1960, pp. 17–24

Ajdukiewicz, K., *Intensional Expressions*, ***Studia Logica***, vol. 20 (1967), pp. 63–86

Ajdukiewicz, K., *Proposition as the Connotation of a Sentence*, ***Studia Logica***, vol. 20 (1967), pp. 87–98

Ajdukiewicz, K., ***The Scientific World-Perspective and Other Essays 1931–1963***, J. Giedymin, editor, translations by David Pearce, Dordrecht and Boston, 1978

Aldrich, H., ***Artis Logicae Compendium***, 1691

Anonymous, *George Boole*, ***Proceedings of the Royal Society of London***, vol. 15 (1867), pp. vi–xi

Aristotle, ***Prior and Posterior Analytics*** with introduction and commentary by Sir W. D. Ross, 1949

Arnauld, A., & Nicole, P., ***Port-Royal Logic*** or ***La Logique ou l'Art de Penser***, 1662

Ayer, A. J., ***The Foundation of Empirical Knowledge***, London, 1940

Ayer, A. J., ***Language, Truth and Logic***, second edition (revised and re-set) London 1946, third impression 1948

Ayer, A. J., ***Thinking and Meaning***, London, 1947

Barcan, R. C., *A Functional Calculus of First Order Based on Strict Implication*, ***The Journal of Symbolic Logic***, vol. 11 (1946), pp. 1–17

Barcan, R. C., *The Identity of Individuals in a Strict Functional Calculus of Second Order*, ***The Journal of Symbolic Logic***, vol. 12 (1947), pp. 12–15

Bar-Hillel, Y., *Bolzano's Definition of Analytic Propositions*, ***Theoria***, vol. 16 (1950), pp. 91–117

Bar-Hillel, Y., *Bolzano's Propositional Logic*, ***Archiv für Philosophie***, vol. 4 (1952), pp. 305–338

Barzin, M., & Errera, A., *Sur la Logique de M. Brouwer*, Académie Royale de Belgique, ***Bulletins de la Classe des Sciences***, vol. 13 (1927), pp. 56–71

Baylis, C. A., *Critical Comments on Professor Fitch's Article "On God and Immortality"*, ***Philosophy and Phenomenological Research***, vol. 8 no. 4 (1948), pp. 694–697

Behmann, H., *Beiträge zur Algebra der Logik, insbesondere zum Entscheidungsproblem*, ***Mathematische Annalen***, vol. 86 (1922), pp. 163–229

Behmann, H., *Zu den Widersprüchen der Logik und der Mengenlehre*, ***Jahresbericht der deutschen Mathematiker Vereinigung***, vol. 40 (1931), pp. 37–48

Bennett, A. A., & Baylis, C. A., ***Formal Logic***, New York, 1939

Berkeley, E. C., ***Circuit Algebra—Introduction***, New York, 1952

Berkeley, E. C., *The Algebra of States and Events*, ***The Scientific Monthly***, vol. 78 (1954), pp. 232–242

Bernays, P., *Die Bedeutung Hilberts für die Philosophie der Mathematik*, ***Die Naturwissenschaften***, vol. 10 (1922), pp. 97–98

Bernays, P., *Quelques Points Essentiels de la Métamathématique*, ***Mathematische Zeitschrift***, vol. 25 (1926), pp. 70–95

Bernays, P., *Axiomatische Untersuchungen des Aussagen-Kalküls der "Principia Mathematica"*, ***Mathematische Zeitschrift***, vol. 25 (1926), pp. 305–320

Bernays, P., *Sur le Platonisme dans les Mathématiques*, and *Quelques Points Essentiels de la Métamathématique*, ***L'Enseignement Mathématique***, vol. 34 (1935), pp. 52–95

Bernays, P., Review of Church and Rosser, *Some Properties of Conversion*, ***The Journal of Symbolic Logic***, vol. 1 (1936), pp. 74–75

Bernays, P., *A System of Axiomatic Set Theory*, ***The Journal of Symbolic Logic***, vol. 2 (1937), pp. 65–77

Bernays, P., *A System of Axiomatic Set Theory–Part II*, ***The Journal of Symbolic Logic***, vol. 6 (1941), pp. 1–17

Bernays, P., *A System of Axiomatic Set Theory*, ***The Journal of Symbolic Logic***, vol. 7 (1942), pp. 65–89, 133–145

Bernays, P., *A System of Axiomatic Set Theory*, ***The Journal of Symbolic Logic***, vol. 8 (1943), pp. 89–106

Bernays, P., *A System of Axiomatic Set Theory*, ***The Journal of Symbolic Logic***, vol. 13 (1948), pp. 65–79

Bernays, P., Review of Skolem, *A Remark on the Induction Scheme*, ***The Journal of Symbolic Logic***, vol. 16, (1951), pp. 220–221

Bernays, P., *A System of Axiomatic Set Theory*, ***The Journal of Symbolic Logic***, vol. 19 (1954), pp. 81–96

Bernays, P., & Schönfinkel, M., *Zum Entscheidungsproblem der mathematischen Logik*, ***Mathematische Annalen.***, vol. 99 (1928), pp. 342–372

Bernstein, B. A., *On the Serial Relation in Boolean Algebras*, ***Bulletin of the American Mathematical Society***, vol. 32 (1926), pp. 523–524, 712–713

Bernstein, B. A., Review of Whitehead and Russell, *Principia Mathematica*, ***Bulletin of the American Mathematical Society***, vol. 32 (1926), pp. 711–713

Bernstein, B. A., *Whitehead and Russell's Theory of Deduction as a Non-Mathematical Science*, ***Bulletin of the American Mathematical Society***, vol. 37 (1931), pp. 480–488

Bernstein, B. A., *On Proposition *4.78 of Principia Mathematica*, ***Bulletin of the American Mathematical Society***, vol. 38 (1932), pp. 388–391

Bernstein, B. A., *Relation of Whitehead and Russell's Theory of Deduction to the Boolean Logic of Proposition*, ***Bulletin of the American Mathematical Society***, vol. 38 (1932), pp. 589–593

Bernstein, F., *Untersuchungen aus der Mengenlehre*, ***Mathematische Annalen***, vol. 61 (1905), pp. 140–145

Berry, G. D. W., ***Harvard University, Summaries of Theses 1942***, Cambridge, Mass., 1942, pp. 330–334

Berry, G. D. W., *On Quine's Axioms of Quantification*, ***The Journal of Symbolic Logic***, vol. 6 (1941), pp. 23–27

Black, M., *The "Paradox of Analysis"*, ***Mind***, n.s. vol. 53 (1944), pp. 263–268

Black, M., *The "Paradox of Analysis" Again: A Reply*, ***Mind***, n.s. vol. 54 (1945), pp. 272–273

Black, M., *How Can Analysis Be Informative?*, ***Philosophy and Phenomenological Research***, vol. 6 no. 4 (1946), pp. 628–631

Black, M., *A Translation of Frege's Über Sinn und Bedeutung Introductory Note*, ***The Philosophical Review***, vol. 57 (1948), pp. 207–208

Black, M., *Frege on Functions*, in ***Problems of Analysis, Philosophical Essays***, Ithaca, 1954, pp. 229–254, 297–298

Bocheński, I. M., ***Nove Lezioni di Logica Simbolica***, Rome, 1938

Bocheński, I. M., ***La Logique de Théophraste***, Fribourg, 1947

Bocheński, I. M., ***Ancient Formal Logic***, in Studies in Logic and the Foundations of Mathematics, Amsterdam, 1951

Bocheński, I. M., *Spitzfindigkeit*, ***Festgabe an die Schweizerkatholiken***, Freiburg, 1954, pp. 334–352

Bocheński, I. M., ***Formale Logik***, Munich, 1956

Bôcher, M., *The Fundamental Conceptions and Methods of Mathematics*, ***Bulletin of the American Mathematical Society***, vol. 11 (1904), pp. 115–135

Boehner, P., *Bemerkungen zur Geschichte der De Morgansche Gesetze in der Scholastik*, ***Archiv für Philosophie***, vol. 4 no. 2 (1951), pp. 113–146

Bolzano, B., ***Wissenschaftslehre***, 1837

Boole, G., ***Mathematical Analysis of Logic***, 1847

Boole, G., ***Laws of Thought***, 1854

Borel, E., *Les Probabilités dénombrables et Leurs Applications Arithmétiques*, ***Rendiconti del Circolo Matematico di Palermo***, vol. 27 (1909), pp. 247–271, reprinted as Note V to Borel's ***Leçons sur la Théorie des Fonctions***, second edition, 1914, and third edition, 1928

Borel, E., ***Leçons sur la Theorie des Fonctions***, Paris, 1928

Brentano, F., ***Psychologie vom Empirischen Standpunkte***, Leipzig, 1874

Brentano, F., ***Psychologie du Point de Vue Empirique***, translation and preface by Maurice de Gandillac, Paris 1944

Brouwer, L. E. J., *De Onbetrouwbaarheid der logische Principes*, ***Tijdschrift voor Wijsbegeerte***, vol. 2 (1908), pp. 152–158

Brouwer, L. E. J., *Intuitionisme en Formalisme*, Groningen, 1912, reprinted in ***Wiskunde, Waarheid, Werkelijkheid***, Groningen, 1919, English translation by A. Dresden, ***Bulletin of the American Mathematical Society***, vol. 20 (1913), pp. 81–96

Brouwer, L. E. J., *Intuitionistische Mengenlehre*, ***Jahresbericht der Deutschen Mathematiker-Vereinigung***, vol. 28 (1920), pp. 203–208

Brouwer, L. E. J., *Besitzt jede reelle Zahl eine Dezimalbruchentwicklung?*, ***Mathematische Annalen***, vol. 83 (1921), pp. 201–210

Brouwer, L. E. J., *Mathematik, Wissenschaft und Sprache*, ***Monatshefte für Mathematik und Physik***, vol. 36 (1929), pp. 153–164

Büchi, J. R., *Weak Second-order Arithmetic and Finite Automata*, ***Zeitschrift für mathematische Logik und Grundlagen der Mathematik***, vol. 6 (1960), pp. 66–92

Büchi, J. R., *On a Decision Method in Restricted Second-order Arithmetic*, ***Logic, Methodology and Philosophy of Science, Proceedings of the 1960 International Congress***, Stanford 1962, pp. 1–11

Büchi, J. R., & Elgot, C. C., *Decision Problems of Weak Second Order Arithmetics and Finite Automata, Part I*, Abstract in ***Notices of the American Mathematical Society***, vol. 5, 7 (1958), p. 834

Büchi, J. R., Elgot, C. C., & Wright, J. B., *The Non-existence of Certain Algorithms of Finite Automata Theory*, Abstract in ***Notices of the American Mathematical Society***, vol. 5 (1958), p. 98

Burali-Forti, C., *Una Questione sui Numeri Transfiniti*, ***Rediconti del Circolo Matematico di Palermo***, vol. 11 (1897), pp. 154–164

Burkhardt, H., *Entwicklungen nach oszillierenden Funktionen und Interation der Differentialgleichungen der mathematischen Physik*, ***Jahresbericht der Deutschen Mathematiker-Vereinigung***, vol. 10 (1908)

Burks, A. W., & Wright, J. B., *Theory of Logical Nets*, ***Proceedings of the Institute of Radio Engineers***, vol. 41 (1953), pp. 1357–1365

Cajori, F., ***A History of Elementary Mathematics***, revised edition, New York and London, 1917

Cajori, F., ***A History of Mathematics***, second edition, New York and London, 1922

Cajori, F., ***A History of Mathematical Notations***, 2 vols., Chicago, 1928–1929

Cantor, G., *Ein Beitrag zur Mannigfaltigkeitslehre*, ***Journal für die reine und angewandte Mathematik***, vol. 84 (1878), pp. 242–258

Cantor, G., *Über unendliche, lineare Punktmannigfaltigkeiten*, ***Mathematische Annalen***, vol. 23, no. 6 (1884), pp. 453–488

Cantor, G., *Beiträge zur Begründung der transfiniten Mengenlehre*, erster Artikel, ***Mathematische Annalen***, vol. 46 (1895), p. 492

Cantor, G., *Beiträge zur Begründung der transfiniten Mengenlehre*, zweiter Artikel, ***Mathematische Annalen***, vol. 49 (1897), pp. 207–218, and pp. 231–235

Cantor, G., ***Contributions to the Founding of the Theory of Transfinite Numbers***, translated and with an introduction by P. E. B. Jourdain, Chicago and London, 1915 (new edition 1941)

Cantor, G., ***Gesammelte Abhandlungen Mathematische und Philosophischen Inhalts***, edited by E. Zermelo, and with a life by A. Fraenkel, Berlin, 1932

Cantor, M., ***Vorlesungen über Geschichte der Mathematik***, 4 vols., Leipzig, 1880–1908; fourth edition, Leipzig, 1921

Cardano, G., ***De Proportionibus***, 1570

Carnap, R., ***Der logische Aufbau der Welt***, Berlin, 1928

Carnap, R., ***Abriß der Logistik***, Vienna, 1929

Carnap, R., *Die logizistische Grundlegung der Mathematik*, ***Erkenntnis***, vol. 2 (1931), pp. 91–105, 141–144, 145

Carnap, R., ***Logische Syntax der Sprache***, Vienna, 1934

Carnap, R., *Ein Gültigkeitskriterium für die Sätze der klassischen Mathematik*, ***Monatshefte für Mathematik und Physik***, vol. 42 (1935), pp. 163–190

Carnap, R., *Testability and Meaning*, ***Philosophy of Science***, vol. 3 (1936), pp. 419–471; and vol. 4 (1937), pp. 1–40

Carnap, R., ***The Logical Syntax of Language***, London and New York, 1937

Carnap, R., ***Notes for Symbolic Logic***, Chicago, 1937

Carnap, R., *Foundations of Logic and Mathematics*, ***International Encyclopedia of Unified Science***, vol. 1, Chicago, 1939

Carnap, R., ***Introduction to Semantics***, Cambridge, 1942

Carnap, R., ***Formalization of Logic*** in Studies in Semantics, Volume II. Cambridge, Mass., 1943

Carnap, R., *Modalities and Quantification*, ***The Journal of Symbolic Logic***, vol. 11 (1946), pp. 33–64

Carnap, R., ***Meaning and Necessity***, Chicago, 1947, second edition 1956

Carnap, R., *Reply* [to Church's *On Carnap's Analysis of Statements of Assertion and Belief*], ***Philosophy and Analysis, A Selection of Articles Published in Analysis Between 1933–40 and 1947–53***, edited by Margaret Macdonald; Oxford 1954

Carnap, R., ***Introduction to Symbolic Logic and its Applications***, New York, 1958

Carnap, R., *Replies and Systematic Expositions* in ***The Philosophy of Rudolf Carnap***, edited by Paul Arthur Schilpp, La Salle, Illinois, 1963

Carroll, L. (Charles Lutwidge Dodgson), ***Symbolic Logic***, 1897, reprinted 1958

Cassina, U., *Vita et Opera de Giuseppe Peano*, ***Schola et Vita***, vol. 7 (1932), pp. 117–148

Cassina, U., *L'Oeuvre Philosophique de G. Peano*, ***Revue de Métaphysique et de Morale***, vol. 40 (1933), pp. 481–491

Cassina, U., *L'Opera Scientifica di Giuseppe Peano*, ***Rendiconti del Seminario Matematico e Fisico di Milano***, vol. 7 (1933), pp. 323–389

Castillon, G. F., *Memoire sur un Nouvel Algorithme Logique* (read 3 Nov. 1803), ***Memoires de l'Académie Royal des Sciences et Belles Lettres***, vol. 53 (1805), pp. 3–24

Cayley, A., *Note on the Calculus of Logic*, ***The Quarterly Journal of Pure and Applied Mathematics***, vol. 11 (1871), pp. 282–283

Cayley, A., ***Collected Mathematical Papers***, Cambridge, 1895

Champernowne, D. G., *The Construction of Decimals Normal in the Scale of Ten*, ***Journal of the London Mathematical Society***, vol. 8 (1933), pp. 254–260

Chwistek, L., *Antynomje Logiki Formalnej*, ***Przegląd Filozoficzny***, vol. 24 (1921), pp. 164–171

Chwistek, L., *Über die Antinomien der Prinzipien der Mathematik*, ***Mathematische Zeitschrift***, vol. 14 (1922), pp. 236–243

Chwistek, L., *Neue Grundlagen der Logik und Mathematik*, ***Mathematische Zeitschrift***, vol. 30 (1929), pp. 704–724

Chwistek, L., *Überwindung des Begriffsrealismus*, ***Studia Philosophica***, vol. 2 (offprint 1937), pp. 1–18

Chwistek, L., & Hetper, W., *New Foundation of Formal Metamathematics*, ***The Journal of Symbolic Logic***, vol. 3 (1938), pp. 1–37

Chwistek, L., Hetper, W., & Herzberg, J., *Fondements de la Métamathématique Rationnelle*, ***Bulletin International de l'Académie Polonaise des Sciences et des Lettres***, série A (1933), pp. 253–264

Cohen, J., *Are Philosophical Theses Relative to Language?*, ***Analysis***, vol. 9 no. 5 (1949), pp. 72–77

Cohen, M. R., & Nagel, E., ***An Introduction to Logic and Scientific Method***, New York, 1934

Cohen, P. J., *A Minimal Model for Set Theory*, ***Bulletin of the American Mathematical Society***, vol. 69 (1963), pp. 537–540

Cohen, P. J., *The Independence of the Continuum Hypothesis*, ***Proceedings of the National Academy of Sciences of the United States of America***, vol. 50 (1963), 1143–1148, and vol. 51 (1964), pp. 105–110

Cohen, P. J., *Independence Results in Set Theory*, ***The Theory of Models, Proceedings***

of the 1963 International Symposium at Berkeley, Amsterdam, 1965, pp. 39–54

Coolidge, J. L., ***The Elements of Non-Euclidean Geometry***, Oxford, 1909

Coolidge, J. L., ***A History of Geometrical Methods***, New York, 1940

Copeland, A. H., *Admissible Numbers in the Theory of Probability*, ***American Journal of Mathematics***, vol. 50 (1928), pp. 535–552

Copeland, A. H., *Point Set Theory Applied to the Random Selection of the Digits of an Admissible Number*, ***American Journal of Mathematics***, vol. 58 (1936), pp. 181–192

Copi, I. M., ***The Theory of Logical Types***, London, 1971

Copi, I. M., Elgot, C. C., & Wright, J. B., *Realization of Events by Logical Nets*, ***Journal for the Association for Computing Machinery***, vol. 5 (1958), pp. 181–196

Copi, I. M., & Gould, J. A., editors, ***Contemporary Readings in Logical Theory***, New York and London, 1967

Couturat, L. A., ***Opuscules et Fragments Inédits***, 1903

Couturat, L., *IIme Congrès de Philosophie*, ***Revue de Métaphysique et de Morale***, vol. 12 (1904)

Couturat, L., ***Les Principes des Mathématiques***, Paris, 1905

Couturat, L., *Pour la Logistique (Résponse á M. Poincaré)*, ***Revue de Métaphysique et de Morale*** vol. 14, 208–250 (1906); English translation (with an introductory note by P. E. B. Jourdain), ***The Monist***, vol. 22 (1912), pp. 481–523

Couturat, L., ***L'Algèbre de la Logique***, Paris, 1910

Csillig, P., *Eine Bemerkung zur Auflösung der eingeschachtelten Rekursion*, ***Acta Scientiarum Mathematicarum***, vol. 11 (1947), pp. 169–173

Curry, H. B., *An Analysis of Logical Substitution*, ***American Journal of Mathematics***, vol. 51 (1929), pp. 365–384

Curry, H. B., *Grundlagen der kombinatorischen Logik*, ***American Journal of Mathematics***, vol. 52 (1930), pp. 509–536, 789–834

Curry, H. B., *The Universal Quantifier in Combinatory Logic*, ***Annals of Mathematics***, ser. 2, vol. 32 (1931), pp. 154–180

Curry, H. B., *Some Additions to the Theory of Combinators*, ***American Journal of Mathematics***, vol. 54 (1932), pp. 551–558

Curry, H. B., *Apparent Variables from the Standpoint of Combinatory Logic*, ***Annals of Mathematics***, ser. 2, vol. 34 (1933), pp. 381–404

Curry, H. B., *Functionality in Combinatory Logic*, ***Proceedings of the National Academy of Sciences of the United States of America***, vol. 20 (1934), pp. 584–590

Curry, H. B., *Some Properties of Equality and Implication in Combinatory Logic*, ***Annals of Mathematics***, ser. 2, vol. 35 (1934), pp. 849–860

Curry, H. B., *First Properties of Functionality in Combinatory Logic*, ***The Tôhoku Mathematical Journal***, vol. 41 (1936), pp. 371–401

Curry, H. B., *A Mathematical Treatment of the Rules of the Syllogism*, ***Mind***, n.s. vol. 45

(1936), pp. 209–216, 416

Curry, H. B., Review of Church *Mathematical Logic* (mimeographed lecture notes), ***The Journal of Symbolic Logic***, vol. 2 (1937), pp. 39–40

Curry, H. B., *Consistency and Completeness of the Theory of Combinators*, ***The Journal of Symbolic Logic***, vol. 5 (1941), pp. 54–61

Curry, H. B., *A Formalization of Recursive Arithmetic*, ***American Journal of Mathematics***, vol. 63 (1941), pp. 263–282

Curry, H. B., *The Paradox of Kleene and Rosser*, ***Transactions of the American Mathematical Society***, vol. 50 (1941), pp. 454–516

Curry, H. B., *A Revision of the Fundamental Rules of Combinatory Logic*, ***The Journal of Symbolic Logic***, vol. 6 (1941), pp. 41–53

Curry, H. B., *Some Advances in the Combinatory Theory of Quantification*, ***Proceedings of the National Academy of Sciences of the United States of America***, vol. 28 (1942), pp. 564–569

Curry, H. B., *The Combinatory Foundations of Mathematical Logic*, ***The Journal of Symbolic Logic***, vol. 7 (1942), pp. 49–64; see erratum, ibid., vol. 8, p. iv

Curry, H. B., *The Inconsistency of Certain Formal Logics*, ***The Journal of Symbolic Logic***, vol. 7 (1942), pp. 115–117; see erratum, ibid., p. iv

Curry, H. B., *A Simplification of the Theory of Combinators*, ***Synthese***, vol. 7 (1949), pp. 391–399

Darwin, C. G., *The Uncertainty Principle*, ***Science***, vol. 73 (1931), pp. 653–660

Davis, M., ***On the Theory of Recursive Unsolvability***, dissertation, Princeton, 1950

de Broglie, L., ***Matter and Light, The New Physics***, translated by W. H. Johnston, New York, 1939

Dedekind, R., ***Stetigkeit und Irrationale Zahlen***, Braunschweig, first edition, 1872, fifth edition, 1927

Dedekind, R., ***Was Sind und Was Sollen die Zahlen?***, second edition, Braunschweig, 1888, sixth edition, 1930

Dedekind, R., ***Essays on the Theory of Numbers***, translated by W. W. Beman, Chicago, 1901

Dedekind, R., ***Gesammelte Mathematische Werke***, 3 vols., Brunswick, 1930–1932

Delboeuf, J. R. L., *Logique Algorithmique*, ***Revue Philosophique de la France et de l'Etranger***, vol. 2, (1876), pp. 225–252, 335–355, 545–595; reprinted as ***Logique Algorithmique***, Liege and Brussels, 1877

De Morgan, A., ***Formal Logic***, 1847

De Morgan, A., ***Syllabus of a Proposed System of Logic***, 1860

De Morgan, A., *On the Syllogism, No. IV*, ***Transactions of the Cambridge Philosophical Society***, vol. 10 (read April 23, 1860)

De Morgan, S. E., ***Memoir of Augustus De Morgan***, London, 1882

Destouches-Février, Paulette,* ***Comptes Rendus des Séances de l'Académie des Sciences***, Paris, vol. 226 (1948), pp. 38–39

Dienes, Z. P., *On an Implication Function in Many-Valued Systems of Logic*, ***The Journal of Symbolic Logic***, vol. 14 (1949), pp. 95–97

Dirichlet, G. L., ***G. Lejeune Dirichlet's Werke***, vol. 1, Berlin, 1889

Doob, J. L., *Note on Probability*, ***Annals of Mathematics***, vol. 37 (1936), pp. 363–367

Douglas, J., *The Most General Geometry of Paths*, read before the American Mathematical Society May 7, 1927

Dubislav, W., *Bolzano als Vorläufer der mathematischen Logik*, ***Philosophisches Jahrbuch der Görres-Gesellschaft***, vol. 44 (1931), pp. 448–456

Dürr, K., *Die Mathematische Logik des Arnold Geulincx*, ***The Journal of Unified Science*** (***Erkenntnis***), vol. 8 (1940), pp.361–369

Eaton, R. M., ***General Logic***, London, 1931

Eddington, A. S., ***Space, Time, and Gravitation***, Cambridge, England, 1920

Ehrenfest, P., Review of the Russian Translation of Louis Couturat, *L'Algèbre de la Logique*, ***Žurnal Russkago Fiziko-khimicheskago Obshchsstva***, section of physics, vol. 42 (1910), pp. 382–387

Einstein, A., ***Relativity, The Special and The General Theory, A Popular Exposition***, translated by R. W. Lawson, London, 1920

Eisenhart, L. P., ***Coordinate Geometry***, New York, 1939

Elgot, C. C., *Decision Problems of Weak Second Order Arithmetics and Finite Automata, Part II*. Abstract in ***Notices of the American Mathematical Society***, vol. 6 (1958), p. 48

Elgot, C. C., Review of Wang, ***The Journal of Symbolic Logic***, vol. 25 no. 4 (1960), pp. 373–375

Elgot, C. C., *Decision Problems of Finite Automata Design and Related Arithmetics*, ***Transactions of the American Mathematical Society***, vol. 98 (1961), pp. 21–51. Errata, ibid., vol. 103 (1962), pp. 558–559

Enriques, F., ***Per la Storia della Logica***, Bologna, 1922; English translation by J. Rosenthal, New York, 1929

Erdmann, J. E., ***Opera Philosophica***, vol. 1, 1840

Euclid, ***Elements***, ca. 300 B. C.

Euler, L., ***Introductio in Analysin Infinitorum***, 1748

Euler, L., ***Lettres à une Princesse d'Allemagne***, vol. 2, 1770

Euler, L., ***Opera***, ser. 1, vol. 8, 1911

Farber, M., ***The Foundation of Phenomenology. Edmund Husserl and the Quest for a Rigorous Science of Philosophy***, Cambridge, Mass., 1943

Feigl, H., & Sellars, W., editors, ***Readings in Philosophical Analysis***, New York, 1949

* *Editor's note*: The title of this article has not been determined.

Feys, R., *La Technique de la Logique Combinatoire*, ***Revue philosophique de Louvain***, vol. 44 (1946), pp. 74–103, 237–270

Finsler, P., *Gibt es unentscheidbare Sätze?*, ***Commentarii Mathematici Helvetici***, vol. 16 no. 4 (1944), pp. 310–320

Fitch, F. B., *A System of Formal Logic Without an Analogue to the Curry ω operator*, ***The Journal of Symbolic Logic***, vol. 1 (1936), pp. 92–100

Fitch, F. B., *A Basic Logic*, ***The Journal of Symbolic Logic***, vol. 7 (1942), pp. 105–114; see erratum, ibid., p. iv

Fitch, F. B., *Representations of Calculi*, ***The Journal of Symbolic Logic***, vol. 9 (1944), pp. 57–62; see errata, ibid., p. iv

Fitch, F. B., *A Minimum Calculus for Logic*, ***The Journal of Symbolic Logic***, vol. 9 (1944), pp. 89–94; see erratum, ibid., vol. 10, p. iv

Fitch, F. B., *An Extension of Basic Logic*, ***The Journal of Symbolic Logic***, vol. 13 (1948), pp. 95–106

Fitch, F. B., *On God and Immortality*, ***Philosophy and Phenomenological Research***, vol. 8 no. 4 (1948), pp. 688–693

Fitch, F. B., *Reply to Professor Baylis' Criticisms*, ***Philosophy and Phenomenological Research***, vol. 8 no. 4 (1948), pp. 698–699

Fitch, F. B., *Intuitionistic Modal Logic with Quantifiers*, ***Portugaliae Mathematica***, vol. 7 no. 2 (1948)*

Fitch, F. B., *The Heine-Borel Theorem in Extended Basic Logic*, ***The Journal of Symbolic Logic***, vol. 14 (1949), pp. 9–15

Fitch, F. B., *On Natural Numbers, Integers, and Rationals*, ***The Journal of Symbolic Logic***, vol. 14 (1949), pp. 81–84

Fitch, F. B., *A Further Consistent Extension of Basic Logic*, ***The Journal of Symbolic Logic***, vol. 14 (1949), pp. 209–218

Fitch, F. B., *The Problem of the Morning Star and the Evening Star*, ***Philosophy of Science***, vol. 16 (1949), pp. 137–141

Fitch, F. B., *A Demonstrably Consistent Mathematics – Part I*, ***The Journal of Symbolic Logic***, vol. 15 (1950), pp. 17–24

Fitch, F. B., *Representation of Sequential Circuits in Combinatory Logic*, Bell Telephone Laboratories Report, September, 1957, and ***Philosophy of Science***, vol. 25 (1958), pp. 263–279

Flew, A. G. N., ***Introduction***, ***Essays on Logic and Language***, edited and with an introduction by Antony Flew, Oxford and New York, 1951

Forder, H. G., ***The Foundations of Euclidean Geometry***, Cambridge, England, 1927

Fourier, C., ***Oeuvres Complètes de Ch. Fourier***, vol. 1, 1841

Fraenkel, A. A., *Der Begriff 'definit' und die Unabhängigkeit des Auswahlaxioms*,

* *Editor's note*: This is one of Fitch's papers published in vol. 7 of ***Portugaliae Mathematica***. It is not certain that this is the paper Church was citing.

Sitzungsberichte der Preussischen Akademie der Wissenschaften, Phys.-Math. Kl., 1922, pp. 253–257

Fraenkel, A. A., ***Einleitung in die Mengenlehre***, third edition, Berlin, 1928

Fraenkel, A. A., *Über eine abgeschwaechte Fassung des Auswahlaxioms*, ***The Journal of Symbolic Logic***, vol. 2 (1937), pp. 1–25

Fraenkel, A. A., ***Abstract Set Theory***, second edition, Amsterdam, 1961

Fraenkel, A. A., & Bar-Hillel, Y., ***Foundations of Set Theory***, Amsterdam, 1958; second edition 1973, by Fraenkel, Bar-Hillel, and Azriel Levy with collaboration of Dirk van Dalen

Frege, G., ***Begriffsschrift, Eine der arithmetischen nachgebildete Formelsprache des reinen Denkens***, Halle 1879; second edition, edited by I. Angelelli, Hildesheim, 1964

Frege, G., ***Die Grundlagen der Arithmetik, Eine logisch-mathematische Untersuchung über den Begriff der Zahl***, Breslau, 1884; reprinted, Breslau, 1934; Italian translation (with notes by L. Geymonat), Torino, 1948; reprinted with parallel English translation, Oxford and New York, 1950, revised edition 1953; reprinted, Hildesheim, 1961

Frege, G., *Funktion und Begriff*, Address given to Jaenaische Gesellschaft für Medicin und Naturwissenschaft, 1891

Frege, G., *Über Sinn und Bedeutung*, ***Zeitschrift für Philosophie und philosophische Kritik***, vol. 100 (1892), pp. 25–50

Frege, G., ***Grundgesetze der Arithmetik, begriffsschriftlich abgeleitet***, 2 vols., Jena 1893 and 1903; reprinted in one volume, Hildesheim 1962; English translation (***The Basic Laws of Arithmetic***) of part of volume 1 and the appendix of volume 2 (with an introduction by Montgomery Furth), Berkeley and Los Angeles 1964

Frege, G., *Kritische Beleuchtung einiger Punkte in E. Schröder's Vorlesungen über die Algebra der Logik*, ***Archiv für systematische Philosophie***, vol. 1 (1895), pp. 433–456

Frege, G., *Was ist eine Funktion?*, in ***Festschrift Ludwig Boltzmann gewidmet zum sechzigsten Geburtstage, 20 Februar 1904***, 1904

Frege, G., *Sense and Reference*, ***The Philosophical Review***, vol. 57 (1948), pp. 209–230

Frenkel, J., ***Wave Mechanics, Elementary Theory***, Oxford, 1932

Friedman, J., Abstract of *Some Results in Church's Restricted Recursive Arithmetic*, ***The Journal of Symbolic Logic***, vol. 21 (1956), p. 219

Friedman, J., *Some Results in Church's Restricted Recursive Arithmetic*, ***The Journal of Symbolic Logic***, vol. 22 (1958), pp. 337–342; Abstract in ***The Journal of Symbolic Logic***, vol. 21 (1956), p. 219. Review by R. McNaughton, ***The Journal of Symbolic Logic***, vol. 24 (1960), pp. 241–242

Gégalkine, I., *Sur l'Entscheidungsproblem*, ***Recueil Mathématique***, vol. 6 (1939), pp. 185–198

Gentzen, G., *Die Widerspruchsfreiheit der reinen Zahlentheorie*, ***Mathematische Annalen***, vol. 112 (1936), pp. 493–565

Gentzen, G., *Die Widerspruchsfreiheit der Stufenlogik*, ***Mathematische Zeitschrift***, vol. 41 (1936), pp. 357–366

Gerhardt, C. I., ***Philosophische Schriften***, vol. 7, 1890

Geulincx, A., ***Logica Fury damentis Suis, a quibus hactenus collapse fuerat, Restituta***, 1662

Gödel, K., *Die Vollständigkeit der Axiome des Logischen Funktionenkalküls*, ***Monatshefte für Mathematik und Physik***, vol. 37 (1930), pp. 349–360

Gödel, K., *Über formal unentscheidbare Sätze der Principia Mathematica und verwandter Systeme I*, ***Monatshefte für Mathematik und Physik***, vol. 38 (1931), pp. 173–198

Gödel, K., *Ein Spezialfall des Entscheidungsproblem der theoretischen Logik*, ***Ergebnisse eines mathematischen Kolloquiums***, vol. 2 (1932), pp. 27–28

Gödel, K., *Zur intuitionistischen Logik*, ***Ergebnisse eines mathematischen Kolloquiums***, Wien, 1932

Gödel, K., *Zum Entscheidungsproblem des logischen Funktionenkalküls*, ***Monatshefte für Mathematik und Physik***, vol. 40 (1933), pp. 433–443

Gödel, K., *On Undecidable Propositions of Formal Mathematical Systems*, mimeographed lecture notes, The Institute for Advanced Study, Princeton, N. J., 1934

Gödel, K., *Über die Länge von Beweisen*, ***Ergebnisse eines mathematischen Kolloquiums***, no. 7 (1936), pp. 23–24

Gödel, K., *The Consistency of the Axiom of Choice and of the Generalized Continuum-Hypothesis*, ***Proceedings of the National Academy of Sciences of the United States of America***, vol. 24 (1938), pp. 556–557

Gödel, K., *Consistency-Proof for the Generalized Continuum Hypothesis*, ***Proceedings of the National Academy of Sciences of the United States of America***, vol. 25 (1939), pp. 220–224

Gödel, K., ***The Consistency of the Axiom of Choice and of the Generalized Continuum-Hypothesis with the Axioms of Set Theory***, Annals of Mathematics Studies, no. 3, Princeton, N.J., 1940; 2nd printing, 1951

Gödel, K., *Russell's Mathematical Logic*, in ***The Philosophy of Bertrand Russell***, Paul A. Schilpp, editor (Library of Living Philosophers, 111), New York, 1944

Goodman, N., *The Logical Simplicity of Predicates*, ***The Journal of Symbolic Logic***, vol. 14 (1949), pp. 32–41

Goodman, N., *On a Pseudo-test of Translation*, ***Philosophical Studies***, vol. 3 (1952), pp. 81–82

Goodstein, R. L., *Logic-Free Formalizations of Recursive Arithmetic*, ***Mathematica Scandinavica***, vol. 2, (1954), pp. 247–261

Gotô, M., *Application of Logical Mathematics to the Theory of Relay Networks*, ***The Japan Science Review***, vol. 1 (1950), pp. 35–42

Goursat, E., ***Leçons sur l'Intégration des Équations aux Dérivées Partielles du Second Ordre***, Paris, vol. 1 (1896), vol. 2 (1898)

Grelling, K., & Nelson, L., *Bemerkungen zu den Paradoxieen von Russell und Burali-Forti*, ***Abhandlungen der Fries'schen Schule***, n.s. vol. 2 (1907–08), pp. 301–324

Hailperin, T., *A Set of Axioms for Logic*, ***The Journal of Symbolic Logic***, vol. 9 (1944), pp. 1–19

Hailperin, T., *Quantification Theory and Empty Individual-Domains*, ***The Journal of Symbolic Logic***, vol. 18 (1953), pp. 197–200

Halsted, G. B., ***Girolamo Saccheri's Euclides Vindicatus***, 1920

Hankel, H., *Untersuchungen über die unendlich oft oscillirenden und unstetigen Functionen*, ***Mathematische Annalen***, vol. 20 (1882), pp. 63–112

Hardy, G. H., *A Theorem Concerning the Infinite Cardinal Numbers*, ***Quarterly Journal of Pure and Applied Mathematics***, vol. 35 (1903), pp. 87–94

Hardy, G. H., ***A Mathematician's Apology***, London, 1940

Harley, R., *George Boole, F.R.S.*, ***The British Quarterly Review***, vol. 44 (1866), pp. 141–181

Hartman, S., *Are There Any Extra-Syllogistic Forms of Reasoning?*, ***Proceedings of The American Catholic Philosophical Association***, vol. 15 (pub. 1940), pp. 235–241

Hartree, D. R., ***Calculating Instruments and Machines***, Urbana, 1940

Heath, T. L., ***The Thirteen Books of Euclid's Elements***, translated from the text of Heiberg, with introduction and commentary, 3 vols., Cambridge, 1908

Heath, T. L., ***A History of Greek Mathematics***, 2 vols., Oxford, 1921

Hegel, G. W. F., ***Wissenschaft der Logik***, 1812–1816

Hegel, G. W. F., ***Science of Logic***, translated by W. H. Johnston and L. G. Struthers, 2 vols. 1929; 1952

Hempel, C. G., *A Purely Syntactical Definition of Confirmation*, ***The Journal of Symbolic Logic***, vol. 8 (1943), pp. 122–143

Henkin, L., *Completeness in the Theory of Types*, ***The Journal of Symbolic Logic***, vol. 15 (1950), pp. 81–91

Henkin, L., *Banishing the Rule of Substitution for Functional Variables*, ***The Journal of Symbolic Logic***, vol. 18 (1953), pp. 201–208

Herbrand, J., ***Recherches sur la Théorie de la Démonstration***, Warsaw, 1930

Herbrand, J., *Sur la Non-Contradiction de l'Arithmétique*, ***Journal für die Reine und Angewandte Mathematik***, vol. 166 (1931–2), pp. 1–8

Herbrand, J., *Sur le Problème Fondamental de la Logique Mathématique*, ***Comptes Rendus des Séances de la Société des Sciences et des Lettres de Varsovie***, Classe III, vol. 24 (1931), pp. 12–56

Hermes, H., *Definite Begriffe und berechenbare Zahlen*, Semester-Berichte (Münster 1. W.), summer 1937, pp. 110–123

Herzberg, J., *Sur la Notion de Collectif*, ***Annales de la Société Polonaise de Mathématiques*** (1939)

Hetper, W., *Simple Systems and Metasystems and Generalized Metasystems* (à paraître)*

Heyting, A., *Die formalen Regeln der intuitionistischen Logik*, ***Sitzungsberichte der Preussischen Akademie der Wissenschaften zu Berlin***, Phys.-Math. Klasse, (1930), pp. 42–71, 158–169

Heyting, A., *Die intuitionistische Grundlegung der Mathematik*, ***Erkenntnis***, vol. 2 (1931), pp. 106–115

Heyting, A., *Mathematische Grundlagenforschung, Intuitionismus, Beweistheorie*, in ***Ergebnisse der Mathematik und ihrer Grenzgebiete***, vol. 3, no. 4, Berlin, 1934; French translation, with some additions, Paris and Louvain, 1955

Heyting, A., *Formal Logic and Mathematics*, ***Synthese***, vol. 6 (1947/8), pp. 275–282

Heyting, A., *Logique et Intuitionnisme*, in ***Applications Scientifiques de la Logique Mathématique, Actes du Deuxiéme Colloque International de Logique Mathématique***, Paris and Louvain, 1954, pp. 75–83

Heyting, A., ***Intuitionism, An Introduction***, Amsterdam, 1956

Hilbert, D., ***Grundlagen der Geometrie***, first edition, 1899, seventh edition, 1930

Hilbert, D., *Mathematische Probleme*, ***Nachrichten von der K. Gesellschaft der Wissenschaften zu Göttingen***, Math.-Phys. Kl., 1900, pp. 253–297. Reprinted with additions in ***Archiv der Mathematik und Physik***, ser. 3, vol. 1 (1901), pp. 44–63, 213–237. English translation in ***Bulletin of the American Mathematical Society***, vol. 8 (1901–1902), pp. 437–479

Hilbert, D., ***The Foundations of Geometry***, English translation by E. J. Townsend, 1902

Hilbert, D., *Über die Grundlagen der Logik und der Arithmetik* in ***Verhandlungen des dritten Internationalen Mathematiker-Kongresses***, A. Krazer, editor, Heidelberg, 1904, p. 185

Hilbert, D., *Die logischen Grundlagen der Mathematik*, ***Mathematische Annalen***, vol. 88 (1923), pp. 151–165

Hilbert, D., *Über das Unendliche*, ***Mathematische Annalen***, vol. 95 (1926), pp. 161–190

Hilbert, D., *Die Grundlagen der Mathematik*, ***Abhandlungen aus dem Mathematischen Seminar der Hamburgischen Universität***, vol. 6 (1928), pp. 65–85; reprinted in ***Grundlagen der Geometrie***, seventh edition, pp. 289–312

Hilbert, D., ***Gesammelte Abhandlungen***, 3 vols., additions by P. Bernays and O. Blumenthal, Berlin, 1932–1935

Hilbert, D., & Ackermann, W., ***Grundzüge der theoretischen Logik***, Berlin, 1928, second edition, 1938

Hilbert, D., & Bernays, P., ***Grundlagen der Mathematik***, vol. 1, Berlin, 1934; vol. 2, Berlin, 1939

Hillebrand, F., ***Die Neuen Theorien der Kategorischen Schlüsse***, Vienna, 1891

Hispanus, P., ***Petri Hispani Summulae Logicales cum Versorii Parisiensis Clarissima***

---

* *Editor's note*: It is not known when and where this paper was published.

Expositione, various editions (e.g. Venice), 1597, 1622

Howell, W. S., ***Logic and Rhetoric in England, 1500–1700***, Princeton, 1956

Huntington, E. V., *Sets of Independent Postulates for the Algebra of Logic*, ***Transactions of the American Mathematical Society***, vol. 5 (1904), pp. 288–309

Huntington, E. V., *A Set of Postulates for Real Algebra*, ***Transactions of the American Mathematical Society***, vol. 6 (1905), pp. 17–41

Huntington, E. V., *A Set of Postulates for Ordinary Complex Algebra*, ***Transactions of the American Mathematical Society***, vol. 6 (1905), pp. 209–229

Huntington, E. V., ***The Continuum***, Cambridge, Mass., 1917

Huntington, E. V., *A New Set of Postulates for Betweenness, with Proof of Complete Independence*, ***Transactions of the American Mathematical Society***, vol. 26 (1924), pp. 257–282

Huntington, E. V., *A New Set of Independent Postulates for the Algebra of Logic*, ***Transactions of the American Mathematical Society***, vol. 35 (1933), pp. 274–304

Huntington, E. V., *The Relation Between Lewis's Strict Implication and Boolean Algebra*, ***Bulletin of the American Mathematical Society***, vol. 40 (1934), pp. 729–735

Huntington, E. V., *Postulates for Assertion Conjunction, Negation, and Equality*, ***Proceedings of the American Academy of Arts and Sciences***, vol. 72, no. 1, (1937)

Jaśkowski, S., *Un Calcul des Propositions pour les Systèmes déductifs Contradictoires*, ***Studia Societatis Scientiarum Torunensis***, section A, vol. 2 no. 3 (1950) pp. 57–77

Jevons, W. S., ***Pure Logic***, 1864

Johansson, I., *Der Minimalkalkül, ein reduzierter intuitionistischer Formalismus*, ***Compositio Mathematica***, vol. 4 (1936), pp. 119–136

Johnson, W. E., ***Logic***, Cambridge, 1921

Joseph, H. W. B., ***An Introduction to Logic***, second edition, Oxford, 1916

Jourdain, P. E. B., *George Boole*, ***The Quarterly Journal of Pure and Applied Mathematics***, vol. 41 (1910), pp. 332–352

Jourdain, P. E. B., *Gottlob Frege*, ***The Quarterly Journal of Pure and Applied Mathematics***, vol. 43 (1912), pp. 237–269

Jourdain, P. E. B., *Giuseppe Peano*, ***The Quarterly Journal of Pure and Applied Mathematics***, vol. 43 (1912), pp. 270–314

Jourdain, P. E. B., *Tales with Philosophical Morals*, ***The Open Court***, vol. 27 (1913), pp. 310–315

Jourdain, P. E. B., *Giuseppe Peano*, supplement to ***Schola et Vita***, Milan, 1928

Jungius, J., ***Logica Hamburgensis***, 1638

Kalmár, L., *Über die Erfüllbarkeit derjenigen Zählausdrücke, welche in der Normalform zwei Benachbarte Allzeichen Enthalten*, ***Mathematische Annalen***, vol. 108 (1933), pp. 466–484

Kalmár, L., *Zurückführung des Entscheidungsproblems auf den Fall von Formeln mit einer einzigen, binären, Funktionsvariablen*, ***Compositio Mathematica***, vol. 4 (1936), pp. 137–144

Kalmár, L., *On the Possibility of Definition by Recursion*, ***Acta Scientiarum Mathematicarum*** (Szeged), vol. 9 (1940), pp. 227–232

Kamke, E., *Über neuere Begründungen der Wahrscheinlichkeitsrechnung*, ***Jahresbericht der deutschen Mathematiker-Vereinigung***, vol. 42 (1932–1933), pp. 14–27

Kant, I., ***Kritik der reinen Vernunft***, second edition, 1787

Kaplan, D., ***Foundations of Intensional Logic***, University of California at Los Angeles, 1964

Kaplan, D., Review of Kripke, *Semantical Analysis of Modal Logic I*, ***The Journal of Symbolic Logic***, vol. 31 (1966), pp. 120–122

Kemeny, J., Review of Carnap, *Logical Foundations of Probability*, ***The Journal of Symbolic Logic*** vol. 16 (1951), pp. 205–207

Keynes, J. N., ***Studies and Exercises in Formal Logic***, 1884 and later editions to 1906

Keyser, C. J., *The Axiom of Infinity: A New Presupposition of Thought*, ***The Hibbert Journal***, vol. 2, (1903/4), pp. 532–552; reprinted in ***The Human Worth of Rigorous Thinking, Essays and Addresses***, New York, 1916, and second edition, New York, 1925

Keyser, C. J., *The Axiom of Infinity*, ***The Hibbert Journal***, vol. 3 (1904/5), pp. 380–383

Keyser, C. J., *The Thesis of Modern Logistic*, ***Science***, vol. 30 (1909), pp. 949–963

Keyser, C. J., *Doctrinal Functions*, ***The Journal of Philosophy***, vol. 15 (1918), pp. 262–267

Kleene, S. C., *Proof by Cases in Formal Logic*, ***Annals of Mathematics***, ser. 2, vol. 35 (1934), pp. 529–544

Kleene, S. C., *A Theory of Positive Integers in Formal Logic*, ***American Journal of Mathematics***, vol. 57 (1935), pp. 153–173, 219–244

Kleene, S. C., *λ-Definability and Recursiveness* ***Duke Mathematical Journal***, vol. 2 (1936), pp. 340–353 (abstract in ***Bulletin of the American Mathematical Society***, vol. 41 (1935), no. 7)

Kleene, S. C., *General Recursive Functions of Natural Numbers*, ***Mathematische Annalen***, vol. 112 (1936), pp. 727–742

Kleene, S. C., *A Note on Recursive Functions*, ***Bulletin of the American Mathematical Society***, vol. 42 (1936), pp. 544–546

Kleene, S. C., *On Notation for Ordinal Numbers*, ***The Journal of Symbolic Logic***, vol. 3 (1938), pp. 150–155, (abstract in ***Bulletin of the American Mathematical Society***, vol. 43 (1937), p. 41)

Kleene, S. C., Review of Carnap, ***The Logical Syntax of Language***, ***The Journal of Symbolic Logic***, vol. 4 (1939), pp. 82–87

Kleene, S. C., *Recursive Predicates and Quantifiers*, ***Transactions of the American Mathematical Society***, vol. 53 (1943), pp. 41–73

Kleene, S. C., *On the Forms of the Predicates in the Theory of Constructive Ordinals*, ***American Journal of Mathematics***, vol. 66 (1944), pp. 41–58

Kleene, S. C., *On the Interpretation of Intuitionistic Number Theory*, ***The Journal of Symbolic Logic***, vol. 10 (1945), pp. 109–124

Kleene, S. C., *On the Intuitionistic Logic*, ***Proceedings of the Tenth International Congress of Philosophy***, Amsterdam, 1949, pp. 741–743

Kleene, S. C., *Representation of Events in Nerve Nets and Finite Automata*, ***RAND Memorandum***, 1951

Kleene, S. C., ***Introduction to Metamathematics***, Amsterdam and New York, 1952

Kleene, S. C., *Representation of Events in Nerve Nets and Finite Automata*, ***Automata Studies***, Princeton, 1956, pp. 3–41

Kleene, S. C., & Rosser, J. B., *The Inconsistency of Certain Formal Logics*, ***Annals of Mathematics***, vol. 36 (1935), pp. 630–636

Klein, F., ***Vorlesungen über die Entwicklung der Mathematik im 19. Jahrhundert***, 2 vols., Berlin, 1926–1927

Kneale, W., *The Province of Logic*, ***Contemporary British Philosophy***, London and New York, 1924

Kocourek, R. A., *An Evaluation of Symbolic Logic*, ***Proceedings of The American Catholic Philosophical Association***, vol. 22 (pub. 1948), pp. 95–104

Kolmogoroff, A., *O Principé tertium non datur*, ***Recueil Mathématique de la Société Mathématique de Moscou***, vol. 32 (1925), pp. 646–667

König, J., *Sur la Théorie des Ensembles*, ***Comptes Rendus***, vol. 143 (1906), pp. 110–112

Korselt, A., *Über einen Beweis des Äquivalenzsatzes*, ***Mathematische Annalen***, vol. 70 (1911), pp. 294–296

Koyré, A., *The Liar*, ***Philosophy and Phenomenological Research***, vol. 6 no. 3 (1946), pp. 344–362

Lambert, J. H., ***Lamberts Logische und Philosophische Abhandlungen***, vol. I, 1782

Land, J. P. N., *Brentano's Logical Innovations*, ***Mind***, vol. 1 no. 2 (1876), pp. 289–292

Landau, E., ***Grundlagen der Analysis***, Leipzig, 1930, translated as ***Foundations of Analysis***, 1951

Lange, C., ***Nucleus Logicae Weisianae***, 1712

Langer, S. K., ***Introduction to Symbolic Logic***, Boston and New York, 1937

Langford, C. H., Review of Beth, *The Significs of Pasigraphic Systems*, ***The Journal of Symbolic Logic***, vol. 2 (1937), pp. 53–54

Lazerowitz, M., *The Principle of Verifiability*, ***Mind***, n.s. vol. 46 (1937), pp. 372–378

Lazerowitz, M., *Self-contradictory Propositions*, ***Philosophy of Science***, vol. 7 (1940), pp. 229–240

Leibniz, G. W. von, ***Dissertatio de Arte Combinatoria***, 1666

Leibniz, G. W. von, ***Nouveaux Essais sur l'Entendement Humain***, 1705

Leibniz, G. W. von, ***Réflexions sur l'Essai de l'Entendement Humain de Mr. Locke***, published with Locke's letters, 1708

Leibniz, G. W. von, Reply to the first letter of S. Clarke, 1717

Leśniewski, S., ***Collectanea Logica***, vol. 1, 1938

Levi, B.,* ***Universidad Nacional de Tucumán, Revista***, ser. A vol. 3 (1942), pp. 13–78

Lewis, C. I., ***A Survey of Symbolic Logic***, Berkeley, Cal., 1918

Lewis, C. I., & Langford, C. H., ***Symbolic Logic***, New York, 1932; reissued 1951

Liard, L., *Un Nouveau Système de Logique Formelle M. Stanley Jevones*, ***Revue Philosophique de la France et de l'Etranger***, vol. 3 (1877), pp. 277–293

Liard, L., *La Logique Algébrique de Boole*, ***Revue Philosophique de la France et de l'Etranger***, vol. 4 (1877), pp. 285–317

Lindelöf, E., *Sur Quelques Points de la Théorie des Ensembles*, ***Comptes Rendus***, vol. 137 (1903), pp. 697–700

Lindelöf, E., *Remarques sur un Théorème Fondamental de la Théorie des Ensembles*, ***Acta Mathematica***, vol. 29 (1905), pp. 187–189

Lindemann, F. A., ***The Physical Significance of the Quantum Theory***, Oxford, 1932

Lindenbaum, A., & Mostowski, A., *Über die Unabhängigkeit des Auswahlaxioms und einiger seiner Folgerungen*, ***Comptes Rendus des Séances de la Société des Sciences et des Lettres de Varsovie***, Classe III, vol. 31 (1938), pp. 27–32

Lindenbaum, A., & Tarski, A., *Communication sur les Recherches de la Théorie des Ensembles*, ***Comptes Rendus des Séances de la Société des Sciences et des Lettres de Varsovie***, Classe III, vol. 19 (1926), p. 314

Linsky, L., editor, ***Semantics and the Philosophy of Language***, Urbana, 1952

Linsky, L., editor, ***Reference and Modality***, London, 1971

Lorenzen, P., *Zur Interpretation der Syllogistik*, ***Archiv für Mathematische Logik und Grundlagenforschung***, vol. 2 (1956), pp. 100–103

Lorenzen, P., ***Formale Logik***, Berlin, 1958, second edition 1962

Löwenheim, L., *Über Möglichkeiten im Relativkalkül*, ***Mathematische Annalen***, vol. 76 (1915), pp. 447–470

Łukasiewicz, J., *O Logice Trójwartościowej*, ***Ruch Filozoficzny***, vol. 5 (1920), pp. 169–171

Łukasiewicz, J., ***Elementy Logiki Matematycznej***, Warsaw, 1929

---

* *Editor's note*: The title of this article has not been determined.

Łukasiewicz, J., *Philosophische Bemerkungen zu mehrwertigen Systemen des Aussagenkalküls*, ***Comptes Rendus des Séances de la Société des Sciences et des Lettres de Varsovie***, Classe III, vol. 23 (1930), pp. 51–77

Łukasiewicz, J., *Die Logik und das Grundlagenproblem*, ***Les Entretiens de Zurich***, 1941

Łukasiewicz, J., *The Shortest Axiom of the Implicational Calculus of Propositions*, ***Proceeding of the Royal Irish Academy***, vol. 52 (1948), pp. 25–33

Łukasiewicz, J., ***Aristotle's Syllogistic From the Standpoint of Modern Formal Logic***, Oxford, 1951

Łukasiewicz, J., & Tarski, A., *Untersuchungen über den Aussagenkalkül*, ***Comptes Rendus des Séances de la Société des Sciences et des Lettres de Varsovie***, Classe III, vol. 23 (1930), pp. 30–50

Lüroth, J., *Ernst Schröder*, ***Jahresbericht der Deutschen Mathematiker-Vereinigung***, vol. 12 (1903), pp. 249–265; reprinted in Schröder's ***Vorlesungen über der Algebra der Logik***, vol. 2, part 2

MacColl, H., *Linguistic Misunderstandings*, ***Mind***, n.s. vol. 19 (1910), p. 193

Macfarlane, A., ***Principles of the Algebra of Logic With Examples***, Edinburgh, 1879

Mac Lane, S., Review of Carnap ***The Logical Syntax of Language***, ***Bulletin of the American Mathematical Society***, vol. 44 (1938), pp. 171–176

Manning, H. P., ***Non-Euclidean Geometry***, 1901

Maritain, J., *Petite Logique* (***Éléments de Philosophie***, IIe fascicule), Paris, 1923

Markoff, A., *Névozmožnost' Nékotoryh Algorifmov v Téorii Associativnyh Sistém*, ***Doklady Akadémii Nauk SSSR***, vol. 55 (1947), pp. 587–590, vol. 58 (1947), pp. 353–356, and vol. 77 (1951), pp. 19–20

Markoff, A., *On the Impossibility of Certain Algorithms in the Theory of Associative Systems*, ***Comptes Rendus (Doklady) de l'Académie des Sciences de l'URSS***, n.s. vol. 55 no. 7 (1947), pp. 583–586

Markoff, A., *O Nékotoryh Nérazréšimyh Problémah Kasaúščihsá Matric*, ***Doklady Akadémii Nauk SSSR***, vol. 57 (1947), pp. 539–542

Markoff, A., *O Prédstavlénii Rékursivnyh Funkcij*, ***Doklady Akadémii Nauk SSSR***, vol. 58 (1947), pp. 1891–1892

Marras, A., editor, ***Intentionality, Mind, and Language***, Urbana, Chicago, and London, 1972

Marsh, R. C., editor, ***Logic and Knowledge: Essays 1901–1950, by Bertrand Russell***, London, 1956

Martin, R. L., editor, ***Recent Essays on Truth and the Liar Paradox***, Oxford, 1984

Martinich, A. P., editor, ***The Philosophy of Language***, New York, 1985

Mates, B., *Synonymity*, ***University of California Publications in Philosophy***, Berkeley, Calif. (1950), reprinted in Linsky, L., ***Semantics and the Philosophy of Language***, Urbana, 1952

Mates, B., ***Stoic Logic***, Berkeley, 1953

McCulloch, W. S., & Pitts, W., *A Logical Calculus of the Ideas Immanent in Nervous Activity*, ***Bulletin of Mathematical Biophysics***, vol. 5 (1943), pp. 115–133

McKinsey, J. C. C., & Tarski, A., *Some Theorems About the Sentential Calculi of Lewis and Heyting*, ***The Journal of Symbolic Logic***, vol. 13 (1948), pp. 1–15

McNaughton, R., & Yamada, H., *Regular Expressions and State Graphs for Automata*. ***Institute of Radio Engineers Transactions***, EC-9, no. 1 (1960), pp. 39–47

Melden, A. I., *Thought and Its Objects*, ***Philosophy of Science***, vol. 7 (1940), pp. 434–441

Mendelson, E., *The Independence of a Weak Axiom of Choice*, ***The Journal of Symbolic Logic***, vol. 21 (1956), pp. 350–366

Mendelson, E., *The Axiom of Fundierung and the Axiom of Choice*, ***Archiv für mathematische Logik und Grundlagenforschung***, vol. 4 (1958), pp. 65–70

Mill, J. S., ***A System of Logic, Ratiocinative and Inductive***, 1843, 1884

Milton, J., ***Artis Logicae Plenior Institutio***, 1672

Moore, E. H., ***Introduction to a Form of General Analysis***, New Haven Colloquium, 1906

Morris, C. W., ***Foundations of the Theory of Signs***, Chicago, 1938

Mostowski, A., *Über die Unabhängigkeit des Wohlordnungssatzes vom Ordnungsprinzip*, ***Fundamenta Mathematicae***, vol. 32 (1939), pp. 201–252

Mostowski, A., *On Definable Sets of Positive Integers*, ***Fundamenta Mathematicae***, vol. 34 (1946), pp. 81–112

Mostowski, A., *On a Set of Integers Not Definable by Means of One-Quantifier Predicates*, ***Annales de la Société Polonaise de Mathématique***, vol. 21 (1948), pp. 114–119

Mostowski, A., *Sur l'Interprétation Géométrique et Topologique des Notions Logiques*, ***Proceedings of the Tenth International Congress of Philosophy***, Amsterdam, 1949, pp. 767–769

Myhill, J. R., *Note on an Idea of Fitch*, ***The Journal of Symbolic Logic***, vol. 14 (1949), pp. 175–176

Myhill, J. R., *A Reduction in the Number of Primitive Ideas of Arithmetic*, ***The Journal of Symbolic Logic***, vol. 15 (1950), p. 130

Myhill, J. R., *A System Which Can Define Its Own Truth*, ***Fundamenta Mathematicae***, vol. 37 (for 1950, pub. 1951), pp. 190–192

Myhill, J. R., *The Problems Arising in the Formalization of Intensional Logic*, ***Logique et Analyse***, vol. 1 (1958), pp. 78–83

Nagel, E., *The Formation of Modern Conceptions of Formal Logic in the Development of Geometry*, ***Osiris***, vol. 7 (1939), pp. 142–224

Nagel, E., *Logic Without Ontology*, in ***Naturalism and the Human Spirit***, edited by Y. H. Krikorian, New York, 1944

Nakasima, A., & Hanzawa, M., *Theory of Equivalent Transformation of Simple Partial Paths in a Relay Circuit*, ***Nippon Electrical Communication Engineering***, no. 9 (1938), pp. 32–39

Nakasima, A., & Hanzawa, M., *Theory of Four-Terminal Passive Networks in Relay Circuit*, ***Nippon Electrical Communication Engineering***, no. 10 (1938), pp. 178–179

Nakasima, A., & Hanzawa, M., *Algebraic Expressions Relative to Simple Partial Paths in the Relay Circuit*, ***Nippon Electrical Communication Engineering***, no. 12 (1938), pp. 310–314

Nakasima, A., & Hanzawa, M., *The Theory of Two Point Impedance of Passive Networks in the Relay Circuit*, ***Nippon Electrical Communication Engineering***, no. 13 (1938), pp. 405–412

Nakasima, A., & Hanzawa, M.,* ***Nippon Electrical Communication Engineering***, no. 14 (1938), pp. 203–210

Nakasima, A., & Hanzawa, M., *Transfer Impedance of Four-terminal Passive Networks in the Relay Circuit*, ***Nippon Electrical Communication Engineering***, no. 14 (1938), pp. 459–466

Nelson, D., *Recursive Functions and Intuitionistic Number Theory*, ***Transactions of the American Mathematical Society***, vol. 61 (1947), pp. 307–368; see errata, ibid., p. 556

Nelson, D., *Constructible Falsity*, ***The Journal of Symbolic Logic***, vol. 14 (1949), pp. 16–26

Nelson, E. J., *Whitehead and Russell's Theory of Deduction as a Mathematical Science*, ***Bulletin of the American Mathematical Society***, vol. 40 (1934), pp. 478–486

Newman, M. H. A., *On Theories with a Combinatorial Definition of "Equivalence"*, ***Annals of Mathematics***, ser. 2, vol. 43 (1942), pp. 223–243

Newman, M. H. A., *Stratified Systems of Logic*, ***Proceedings of the Cambridge Philosophical Society***, vol. 39 (1943), pp. 69–83

Notcutt, B., *A Set of Axioms for the Theory of Deductions*, ***Mind***, n.s. vol. 43 (1934), pp. 63–77

Ockham, W., ***Summa Logicae***, 1328

Padoa, A., *Un Nouveau Système Irréductible de Postulats pour l'Algèbre*, ***Deuxiéme Congrés International des Mathématiciens***, Paris, 1900, pp. 249–256

Padoa, A., *Numeri Interi Relativi*, ***Rivista di Matematica***, vol. 7 (1901), pp. 73–84

Pap, A., *Belief and Propositions*, ***Philosophy of Science***, vol. 24 (1957), pp. 123–136

Pap, A., ***Semantics and Necessary Truth***, New Haven, 1958

Parry, W. T., *Modalities in the Survey System of Strict Implication*, ***The Journal of Symbolic Logic***, vol. 4 (1939), pp. 137–154

Pascal, B., ***De l'Esprit Géométrique***, 1657

Patzig, G., editor, ***Funktion, Begriff, Bedeutung: Fünf logische Studien, by Gottlob Frege***, second edition, Göttingen, 1966

Peano. G., ***Calcolo geometrico secondo l'Ausdehnungslehre di H. Grassman, preceduto dalle operazioni della logica deduttiva***, Torino, Bocca, 1888

---

* *Editor's note*: The title of this article has not been determined.

Peano, G., ***Arithmetices Principia, nova methodo exposito***, Augustae Taurinorum, Ed. Fratres Bocca -XVI 20 pp., 1889

Peano, G., *Sul Concetto di Numero*, ***Revista di Matematica***, vol. 1 (1891), pp. 87–102, 256–267

Peano, G., *Formulaire de Mathématiques*, ***Revista di Matematica***, supplementary volume (1899), Paris and Turin

Peano, G., Review of *Principia Mathematica*, vols. 1 and 2, ***Bolettino di Bibliografia e Storia delle Scienze Matematiche***, vol. 15 (1913), pp. 47–53

Peirce, C. S., *On the Algebra of Logic*, ***American Journal of Mathematics***, vol. 3 (1880), pp. 15–57

Peirce, C. S., *On the Algebra of Logic: A Contribution to the Philosophy of Notation*, ***American Journal of Mathematics***, vol. 7 (1885), pp. 180–202

Peirce, C. S., ***Collected Papers***, vol. 3, Cambridge, 1934–35

Pepis, J., *Beiträge zur Reduktionstheorie des logischen Entscheidungsproblems*, ***Acta Scientiarum Mathematicarum (Szeged)***, vol. 8 (1936), pp. 7–41

Péter, R., *Über den Zusammenhang der Verschiedenen Begriffe der rekursiven Funktion*, ***Mathematische Annalen***, vol. 110 (1934), pp. 612–632

Péter, R., *Konstruktion nichtrekursiver Funktionen*, ***Mathematische Annalen***, vol. 11 (1935), pp. 42–60

Péter, R., *A rekurzív függvények elméletéhez (Zur Theorie der rekursiven Funktionen)*, ***Matematikai és fizikai lapok***, vol. 42 (1935), pp. 25–44

Péter, R., *Über die mehrfache Rekursion*, ***Mathematische Annalen***, vol. 113 (1936), pp. 489–527

Péter, R., Review of Kleene, *General Recursive Functions of Natural Numbers*, ***The Journal of Symbolic Logic***, vol. 2 (1937), p. 38; see errata, ibid., vol. 4 (1939), p. iv

Péter, R., *Zum Begriff der rekursiven reellen Zahl*, ***Acta Scientiarum Mathematicarum***, vol. 12 part A (1950), pp. 239–245

Péter, R., *Zusammenhang der mehrfachen und transfiniten Rekursionen*, ***The Journal of Symbolic Logic***, vol. 15 (1950), pp. 248–272

Péter, R., ***Rekursive Funktionen***, Budapest, 1951

Ploucquet, G., ***Antwort auf die von Herrn Professor Lambert Gemachten Erinnerungen und diesseitiger Beschluss der logicalischen Rechnungsstreitigkeiten***, 1766

Poincaré, H., ***La Science et l'Hypothèse***, Paris, 1902, translated as ***Science and Hypothesis***, London and New York, 1905

Poincaré, H., *Les Mathématiques et la Logique*, ***Revue de Métaphysique et de Morale***, vol. 14 (1905), pp. 815–835, and vol. 14 (1906), pp. 17–34, 294–317; reprinted in revised form in Poincaré's ***Science et Méthode***, Paris 1908 (and translations of the revised article in Russian, German, and English translations of this book)

Poincaré, H., *Á Propos de la Logistique*, ***Revue de Métaphysique et de Morale***, vol. 14 (1906), pp. 866–868

Poincaré, H., ***Derniéres Pensées***, Paris, 1913; German translation, Leipzig, 1913

Poincaré, H., ***The Foundations of Science***, English translation by G. B. Halsted, New York, 1913

Poretsky, P., *Sobranie protokolov zasedanij sekcii fiziko-matematiceskin nauk Obscestva* (*On Methods of Solution of Logical Equations and on the Inverse Method of Mathematical Logic*), ***Estestvoispytatelej pri Imperatorskom Kazanskom Universitete***, vol. 2 (1884), pp. 161–330

Post, E. L., *Introduction to a General Theory of Elementary Propositions*, ***American Journal of Mathematics***, vol. 43 (1921), pp. 163–185

Post, E. L., *Finite Combinatory Processes—Formulation 1*, ***The Journal of Symbolic Logic***, vol. 1 (1936), pp. 103–105

Post, E. L., ***The Two-Valued Iterative Systems of Mathematical Logic***, Annals of Mathematics Studies, no. 5, Princeton, N.J., 1941

Post, E. L., *Formal Reductions of the General Combinatorial Decision Problem*, ***American Journal of Mathematics***, vol. 65 (1943), pp. 197–215

Post, E. L., *Recursively Enumerable Sets of Positive Integers and Their Decision Problems*, ***Bulletin of the American Mathematical Society***, vol. 50 (1944), pp. 284–316

Post, E. L., *Recursive Unsolvability of a Problem of Thue*, ***The Journal of Symbolic Logic***, vol. 11 (1946), pp. 1–11

Post, E. L., *Note on a Conjecture of Skolem*, ***The Journal of Symbolic Logic***, vol. 11 (1946), pp. 73–74

Post, E. L., *A Variant of a Recursively Unsolvable Problem*, ***Bulletin of the American Mathematical Society***, vol. 52 (1946), pp. 264–269

Prantl, C., ***Geschichte der Logik im Abendlande***, 4 vols., Leipzig, 1855–1870; reprinted, Leipzig, 1927

Prior, A. N., *Three-Valued Logic and Future Contingents*, ***Philosophical Quarterly***, vol. 3 (1953), pp. 317–327

Prior, A. N., ***Time and Modality***, Oxford, 1957

Putnam, H., *Synonymity and the Analysis of Belief Sentences*, ***Analysis***, vol. 14 (1954), pp. 114–122.

Quine, W. V., ***A System of Logistic***, Cambridge, Massachusetts, 1934

Quine, W. V., *Set-Theoretic Foundations for Logic*, ***The Journal of Symbolic Logic***, vol. 1 (1936), pp. 47–57

Quine, W. V., *On Cantor's Theorem*, ***The Journal of Symbolic Logic***, vol. 2 (1937), pp. 120–124

Quine, W. V., *New Foundations for Mathematical Logic*, ***The American Mathematical Monthly***, vol. 44 (1937), pp. 70–80. Reprinted in ***From a Logical Point of View***, Cambridge, Mass., 1953 and 1961

Quine, W. V., *Completeness of the Propositional Calculus*, ***The Journal of Symbolic Logic***, vol. 3 (1938), pp. 37–40

Quine, W. V., *A Logistical Approach to the Ontological Problem*, Preprinted for the members of the Fifth International Congress for the Unity of Science, Cambridge, Mass., 1939; also in W. V. Quine, ***The Ways of Paradox***, New York (1966), pp. 64–70

Quine, W. V., ***Mathematical Logic***, New York 1940; 2nd printing, Cambridge, Mass., 1947; revised edition, Cambridge, Mass., 1951

Quine, W. V., *Element and Number*, ***The Journal of Symbolic Logic***, vol. 6 (1941), pp. 135–149

Quine, W. V., ***Elementary Logic***, Boston and New York, 1941

Quine, W. V., *Whitehead and the Rise of Modern Logic*, in ***The Philosophy of Alfred North Whitehead***, edited by P. A. Schilpp, Evanston, Ill. and Chicago, 1941, pp. 127–163

Quine, W. V., *Notes on Existence and Necessity*, ***The Journal of Philosophy***, vol. 40 (1943), pp. 113–127

Quine, W. V., ***O Sentido da Nova Lógica***, São Paulo, 1944

Quine, W. V., *On the Logic of Quantification*, ***The Journal of Symbolic Logic***, vol. 10 (1945), pp. 1–12

Quine, W. V., ***A Short Course in Logic, Chapters I–VII***, mimeographed, Cambridge, Mass., 1946

Quine, W. V., *The Problem of Interpreting Modal Logic*, ***The Journal of Symbolic Logic***, vol. 12 (1947), pp. 43–48

Quine, W. V., ***Methods of Logic***, New York, 1950

Quine, W. V., *Logic and the Reification of Universals*, in ***From a Logical Point of View***, by W. V. O. Quine, Cambridge, Mass., 1953, pp. 102–129; reprinted with revisions in the second edition, ibid. 1961, pp. 102–129

Quine, W. V., *Three Grades of Modal Involvement*, ***Actes du XIième Congrès International de Philosophie***, Amsterdam and Louvain, vol. 14 (1953), pp. 65–81

Quine, W. V., ***Word and Object***, Cambridge, Mass., 1960

Quine, W. V., *Reference and Modality*, in ***From a Logical Point of View***, second edition, Cambridge, Mass., 1961

Quine, W. V., ***Set Theory and its Logic***, Cambridge, Mass., 1963

Quine, W. V., & Goodman, N., *Steps Toward a Constructive Nominalism*, ***The Journal of Symbolic Logic***, vol. 12 (1947)

Ramsey, F. P., *The Foundations of Mathematics*, ***Proceedings of the London Mathematical Society***, n.s., vol. 25 (1926), pp. 338–384

Ramsey, F. P., ***The Foundations of Mathematics and Other Logical Essays***, New York and London, 1931; reissued, London and New York, 1950

Ramus, P., ***Dialectique***, 1555

Ramus, P., ***Dialecticae Libri Duo***, 1556

Rashevsky, N., ***Mathematical Biophysics***, Chicago, 1948

Rasiowa, H., *O Pewnym Fragmencie Implikacyjnego Rachunku Zdań*, ***Studia Logica***, vol. 3 (1955), 208–226

Raspe, R. E., ***Oeuvres Philosophiques de Leibniz***, 1765

Reed, I. S., *Symbolic Synthesis of Digital Computers*, ***Proceedings of the Association for Computing Machinery*** (September, 1952), pp. 90–94

Reichenbach, H., *Axiomatik der Wahrscheinlichkeitsrechnung*, ***Mathematische Zeitschrift***, vol. 34 (1931–1932), pp. 568–619

Reichenbach, H., *Les Fondements Logiques du Calcul des Probabilités*, ***Annales de l'Institut Henri Poincaré***, vol. 7 (1937), pp. 267–348

Rescher, N., *Leibniz's Interpretation of His Logical Calculi*, ***The Journal of Symbolic Logic***, vol. 19 (1954), pp. 1–14

Riemann, B., ***Bernhard Riemann's gesammelte mathematische Werke und wissenschaftlicher Nachlass***, Leipzig, 1876

Ritchie, A. D., *A Defence of Aristotle's Logic*, ***Mind***, n.s. vol. 55 (1946), pp. 256–262

Robbins, L. C., *An Analysis by Arithmetical Methods of a Calculating Network with Feedback*, ***Proceedings of the Association for Computing Machinery*** (September, 1952), pp. 61–67

Robinson, J., *Definability and Decision Problems in Arithmetic*, ***The Journal of Symbolic Logic***, vol. 14 (1949), pp. 98–114

Robinson, J., *General Recursive Functions*, ***Proceedings of the American Mathematical Society***, vol. 1 (1950), pp. 703–718

Robinson, R., ***Definition***, Oxford, 1950

Robinson, R. M., *The Theory of Classes. A Modification of von Neumann's System*, ***The Journal of Symbolic Logic***, vol. 2 (1937), pp. 29–36

Robinson, R. M., *Primitive Recursive Functions*, ***Bulletin of the American Mathematical Society***, vol. 53 (1947), pp. 925–942

Robinson, R. M., *Recursion and Double Recursion*, ***Bulletin of the American Mathematical Society***, vol. 54 (1948), pp. 987–993

Rosser, J. B., *A Mathematical Logic Without Variables. I*, ***Annals of Mathematics***, ser. 2, vol. 36 (1935), pp. 127–150; *A Mathematical Logic Without Variables. II*, ***Duke Mathematical Journal***, vol. 1 (1935), pp. 328–355

Rosser, J. B., *Extensions of Some Theorems of Gödel and Church*, ***The Journal of Symbolic Logic***, vol. 1 (1936), pp. 87–91

Rosser, J. B., *An Informal Exposition of Proofs of Gödel's Theorems and Church's Theorem*, ***The Journal of Symbolic Logic***, vol. 4 (1939), pp. 53–60

Rosser, J. B., Review of Church, ***The Calculi of λ-Conversion***, ***The Journal of Symbolic Logic***, vol. 6 (1941), p. 171

Rosser, J. B., *New Sets of Postulates for Combinatory Logics*, ***The Journal of Symbolic Logic***, vol. 7 (1942), pp. 18–27; see errata, ibid., vol. 7, p. iv, and vol. 8, p. iv

Rosser, J. B., ***Logic for Mathematicians***, New York, 1953

Rüstow, A., ***Der Lügner***, dissertation, Erlangen, 1908; Leipzig, 1910

Ruja, H., *Good Logics and Bad*, ***The Philosophical Review***, vol. 49 (1940), pp. 346–355

Russell, B., ***The Principles of Mathematics***, Cambridge, England, 1903; second edition, with a new introduction, London, 1937 and New York, 1938, reprinted London, 1950; Spanish translation, Buenos Aires, 1948; Italian translation, Milano, 1951

Russell, B., *The Axiom of Infinity*, ***The Hibbert Journal***, vol. 2, (1903/4), pp. 809–812

Russell, B., *On Denoting*, ***Mind***, n.s. vol. 14 (1905), pp. 479–493

Russell, B., *Sur la Relation des Mathématiques á la Logistique*, ***Revue de Métaphysique et de Morale***, vol. 13 (1905), pp. 906–916

Russell, B., *Les Paradoxes de la Logique*, ***Revue de Métaphysique et de Morale***, vol. 14 (1906), pp. 627–650

Russell, B., *On Some Difficulties in the Theory of Transfinite Numbers and Order Types*, ***Proceedings of the London Mathematical Society***, ser. 2, vol. 4 (1906), pp. 29–53

Russell, B., *The Theory of Implication*, ***American Journal of Mathematics***, vol. 28 (1906), pp. 159–202

Russell, B., *Mathematical Logic as Based on the Theory of Types*, ***American Journal of Mathematics*** vol. 30 (1908), pp. 222–262; reprinted in ***Logic and Knowledge, Essays 1901–1950***, edited by R. C. Marsh, London, 1956, pp. 59–102

Russell, B., *La Théorie des Types Logique*, ***Revue de Métaphysique et de Morale***, vol. 18 (1910), pp. 263–301

Russell, B., *L'importance Philosophique de la Logistique*, ***Revue de Métaphysique et de Morale***, vol. 19 (1911), 281–291

Russell, B., *Knowledge by Acquaintance and Knowledge by Description*, ***The Problems of Philosoply***, London, 1912

Russell, B., ***Introduction to Mathematical Philosophy***, London, 1919; second edition, London, 1920; German translation, München, 1923, and second edition München, 1930; French translation, Paris, 1928; Spanish translation, Buenos Aires, 1945; Italian translation, Milano, 1947

Russell, B., ***An Inquiry into Meaning and Truth***, Hammondsworth, 1940

Russell, B., ***A History of Western Philosophy***, New York, 1945

Russell, B., *Whitehead and Principia Mathematica*, ***Mind***, n.s. vol. 57 (1948), pp. 137–138

Ryle, G., ***Systematically Misleading Expressions, Essays on Logic and Language***, edited and with an introduction by Antony Flew, Oxford and New York 1951, pp. 11–36

Ryle, G., *Categories*, in ***Logic and Language***, edited by Anthony Flew, Oxford, 1953

Ryle, G., *Formal and Informal Logic*, ***Dilemmas***, Cambridge, 1954

Saccheri, G. G., ***Logica Demonstrativa***, 1697 and 1701

Saccheri, G. G., ***Euclides Vindicatus***, 1733, English translation by G. B. Halsted, Chicago and London, 1920

Scheffler, I., *An Inscriptional Approach to Indirect Quotation*, ***Analysis***, vol. 14, no. 4 (1954), pp. 83–90

Scheffler, I., *Inscriptionalism and Indirect Quotation*, ***Analysis***, vol. 19, no. 1 (1958), pp. 12–18

Scholz, H., ***Geschichte der Logik***, Berlin, 1931

Scholz, H., *Was ist ein Kalkül und was hat Frege für eine pünktliche Beantwortung dieser Frage geleistet?*, ***Semester-Berichte Münster i.W.***, summer 1935, pp. 16–47

Scholz, H., *Die Wissenschaftslehre Bolzanos*, ***Ahhandlungen der Fries'schen Schule***, new series, vol. 6 (1937), pp. 399–472

Scholz, H., *Pascals Forderungen an die Mathematische Methode*, in ***Festschrift Andreas Speiser, der Mathematikunterricht*** (1945), pp. 115–127

Scholz, H., ***Vorlesungen über Grundzüge der Mathematischen Logik***, Münster, 1949

Scholz, H., & Bachmann, F., *Der wissenschaftliche Nachlass von Gottlob Frege*, ***Actes du Congrès International de Philosophie Scientifique***, Paris, 1936, section VIII, pp. 24–30

Scholz, H., & Schweitzer, H., ***Die sogenannten Definitionen durch Abstraktion***, Leipzig, 1935

Schönfinkel, M., *Über die Bausteine der Mathematischen Logik*, ***Mathematische Annalen***, vol. 92 (1924), pp. 305–316

Schröder, E., ***Der Operationskreis des Logikkalküls***, 1877

Schröder, E., ***Algebra der Logik***, 1890–1905

Schröder, E., *Über zwei Definitionen der Endlichkeit und G. Cantor'sche Sätze*, ***Nova Acta Academiae Caesareae Leopoldino-Carolinae Germanicae Naturae Curiosorum***, vol. 71 (1898), pp. 336–340

Schröter, K.,* ***Zentralblatt für Mathematik und ihre Grenzgebiete***, vol. 37 (1951), p. 4

Schütte, K., *Untersuchungen zum Entscheidungsproblem der Mathematischen Logik*, ***Mathematische Annalen***, vol. 109 (1934), pp. 572–603

Schütte, K., ***Beweistheorie***, Berlin, 1960

Scott, D., *The notion of rank in set theory*, ***Summaries of Talks Presented at the Summer Institute for Symbolic Logic***, Cornell University 1957, second edition 1960, pp. 267–269

Sellars, W., *Putnam on Synonymity and Belief*, ***Analysis***, vol. 15 (1955), pp. 117–120

Shannon, C. E., *A Symbolic Analysis of Relay and Switching Circuits*, ***Transactions of the American Institute of Electrical Engineers***, vol. 57 (1938), pp. 713–723

Sheffer, H. M., *The General Theory of Notational Relativity*, privately circulated paper

Sheffer, H. M., *Mutually Prime Postulates*, ***Bulletin of the American Mathematical Society***, vol. 22 (1916), p. 287

Shestakoff, V. I., *Алгебра Двухполюсных Цхем, Построенпых исклычнтел но*

* *Editor's note*: The title of this article has not been determined.

из Двухнолысников (Алгебра А-схем), **Журнал Техническои Физики**, no. 11 (1941), pp. 532–549, and **Автоматика и Телемеханика**, no. 2 (1941), pp. 15–24

Shestakoff, V. I., *Representation of Characteristic Functions of Propositions by Expressions Realizable by Relay-Contact Circuits* (Russian with English summary), ***Bulletin of the Academy of Science, URSS*** (sér. math.), vol 10 (1946), pp. 529–554

Shoenfield, J. R., *The Independence of the Axiom of Choice, Abstract*, ***The Journal of Symbolic Logic***, vol. 20 (1955), p. 202

Sierpiński, W., *Un Théoréme sur le Ensembles Fermés*, ***Bulletin International de l'Académie des Sciences de Cracovie***, 1918, p. 110

Sierpiński, W., *L'Hypothèse Généralisée du Continu et l'Axiome du Choix*, ***Fundamenta Mathematicae***, vol. 34 (1947), pp. 1–5

Simpson, T. M., editor, ***Semántica Filosófica: Problemas y Discusiones***, Buenos Aires, 1973

Skolem, T., *Untersuchugen über die Axiome des Klassenkalküls*, ***Skrifter utgit av Videnskapsselskapet i Kristiania***, I, ***Mat. naturvid. Klasse***, no. 3 (1919)

Skolem, T., *Logisch-kombinaritorische Untersuchungen über die Erfüllbarkeit oder Beweisbarkeit mathematischer Sätze nebst ein Theorem über dichte Mengen*, ***Videnskappselskapets Skrifter Oslo, Mat.-Naturv. Klasse***, vol. 4 (1920), pp. 103–136

Skolem, T., *Begründung der elementaren Arithmetik durch die rekurrierende Denkweise ohne Anwendung scheinbarer Veränderlichen mit unendlichem Ausdehnungsbereich*, ***Skrifter utgit av Videnskapsselskapet i Kristiania***, I. Matematisk-Naturvidenskabelig Klasse, no. 6 (1923)

Skolem, T., *Einige Bemerkungen zur axiomatischen Begründung der Mengenlehre*, ***Wissenschaftliche Vorträge gehalten auf dem Fünften Kongress der Skandinavischen Mathematiker in Helsingfors vom 4, bis 7, Juli 1922***, Helsingfors, 1923, pp. 217–232

Skolem, T., *Über die Mathematische Logik*, ***Norsk Matematisk Tidsskrift***, vol. 10 (1928), pp. 125–142

Skolem, T., *Über die Zurückführbarkeit einiger durch Rekursionen definierter Relationen auf "arithmetische"*, ***Acta Scientiarum Mathematicarum***, vol. 8 (1937), pp. 73–88

Skolem, T., *Einfacher Beweis der Unmöglichkeit eines allgemeinen Lösungsverfahrens für arithmetische Probleme*, ***De Kongelige Norske Videnskabers Selskab, Forhandlinger***, vol. 13 (1940), pp. 1–4

Skolem, T., *A Note on Recursive Arithmetic*, ***De Kongelige Norske Videnskabers Selskab, Forhandlinger***, vol. 13 (1940), pp. 107–109

Skolem, T., *Sur la portée du théorème de Löwenheim–Skolem*, ***Les Entretiens de Zurich sur les Fondements et la Méthode des Sciences Mathématiques***, Zurich 1941, pp. 25–52

Skolem, T., *Remarks on Recursive Functions and Relations*, ***De Kongelige Norske Videnskabers Selskab, Forhandlinger***, vol. 17 (1944), pp. 89–92

Skolem, T., *Some Remarks on Recursive Arithmetic*, ***De Kongelige Norske Videnskabers Selskab, Forhandlinger***, vol. 17 (1944), pp. 103–106

Skolem, T., *Some Remarks on the Comparison Between Recursive Functions*, ***De Kongelige Norske Videnskabers Selskab, Forhandlinger***, vol. 17 (1944), pp. 126–129

Skolem, T., *Den rekursive Aritmetik*, ***Norsk Matematisk Tidsskrift***, vol. 28 (1946), pp. 1–12

Skolem, T., *The Development of Recursive Arithmetic*, ***Den 10. Skandinaviske Matematiker, Kongrees***, Jul. Gjellerups Forlag, Copenhagen 1947, pp. 1–16

Skolem, T., *A Remark on the Induction Scheme*, ***De Kongelige Norske Videnskabers Selskab, Forhandlinger***, vol. 22 (1950), pp. 167–170

Smith, D. E., editor, ***A Source Book in Mathematics***, New York and London, 1929

Smith, H. B., *A Further Note on Subalternation and the Disputed Syllogistic Moods*, ***The Journal of Philosophy***, vol. 21 (1924), pp. 631–633

Smith, P., Review of Ajdukiewicz, ***The Journal of Symbolic Logic***, vol. 37 (1972), pp. 179–180

Smullyan, A. F., Review of Quine, *The Problem of Interpreting Modal Logic*, ***The Journal of Symbolic Logic***, vol. 12 (1947), pp. 43–48; review in ***The Journal of Symbolic Logic***, vol. 12 (1947), pp. 139–141

Smullyan, A. F., *Modality and Description*, ***The Journal of Symbolic Logic***, vol. 13 (1948), pp. 31–37

Specker, E., *Nicht konstruktiv beweisbare Sätze der Analysis*, ***The Journal of Symbolic Logic***, vol. 14 (1949), pp. 145–158

Spencer, T., ***The Art of Logick Delivered in the Precepts of Aristotle and Ramus***, London, 1628

Stamm, E., *Józef Peano*, ***Wiadomości Matematyczne***, vol. 36 (1933), pp. 1–56

Stone, M. H., *The Representation of Boolean Algebras*, ***Bulletin of the American Mathematical Society***, vol. 44 (1938), pp. 807–816

Strawson, P. F., ***Introduction to Logical Theory***, London and New York, 1952

Sturm, C., ***Universalia Euclidea***, 1661

Sudan, G., *Sur le Nombre Transfini* ω^{ω}, ***Bulletin Mathématique de la Société Roumaine des Sciences***, vol. 30 (1927), pp. 11–30

Tarski, A., *Einige Betrachtungen über die Begriffe der* ω*-Widerspruchsfreiheit und der* ω*-Vollständigkeit*, ***Monatshefte für Mathematik und Physik***, vol. 40 (1933), pp. 97–112

Tarski, A., *Pojecie Prawdy w Jezykach Nauk Dedukcyjnych*, ***Travaux de la Société des Sciences et des Lettres de Varsovie, Classe III, Sciences mathématiques et physiques***, no. 34, Warsaw, 1933

Tarski, A., *Der Wahrheitsbegriff in den formalisierten Sprachen* (German translation of the preceding with added *Nachwort*), ***Studia Philosophica***, vol. 1 (1936), pp. 261–405

Tarski, A., *Grundlegung der wissenschaftlichen Semantik*, ***Actes du Congrès International de Philosophie Scientifique*** (1936)

Tarski, A., *On Undecidable Statements in Enlarged Systems of Logic and the Concept of Truth*, ***The Journal of Symbolic Logic***, vol. 4 (1939), pp. 105–112

Tarski, A., *On the Calculus of Relations*, ***The Journal of Symbolic Logic***, vol. 6 (1941), pp. 73–89

Tarski, A., *The Semantic Conception of Truth and the Foundations of Semantics*, ***Philosophy and Phenomenological Research***, vol. 4 no. 3 (1944), pp. 341–376, reprinted in ***Readings in Philosophical Analysis***, selected and edited by H. Feigl and W. Sellars, Appleton-Century-Crofts, Inc., New York, 1949, pp. 52–84

Tarski, A., ***Introduction to Logic and to the Methodology of Deductive Sciences***, second edition, Oxford, 1946

Tarski, A., *The Concept of Truth in Formalized Languages* in ***Logic, Semantics, Metamathematics, Papers from 1923 to 1938, by Alfred Tarski***, London, 1956

Tarski, A., Mostowski, A., & Robinson, J., Abstracts in ***The Journal of Symbolic Logic***, vol. 14 (1949), pp. 75–78

Taylor, J. S., *Complete Existential Theory of Bernstein's Set of Four Postulates for Boolean Algebras*, ***Annals of Mathematics***, vol. 19 (1917), pp. 64–69

Thomae, J., ***Elementare Theorie der analytischen Functionen einer complexen Veränderlichen***, 2nd edition, Halle, 1898

Thomas, I., *An Extension of CS*, ***Dominican Studies***, vol. 2 (1949), pp. 145–160

Tornier, E., *Wahrscheinlichkeitsrechnung und Zahlentheorie*, ***Journal für die Reine und Angewandte Mathematik***, vol. 160 (1929), pp. 177–198

Trachtenbrot, B. A., *Синтез логических цетеЮ, операторы которых описаны средствами исчисления одноместхых преднкатов*, **Доклады Академии Наук СССР**, vol. 118 (1958), pp. 646–649. Reviewed by J. C. Shepherdson in ***Zentralblatt für Mathematik und ihre Grenzgebiete***, vol. 84 (1960), 11–12

Trachtenbrot, B. A., *Некоторые построения в логике одноместиых предикатов*, **Доклады Академии Наук СССР**, vol. 138 (1961), pp. 320–321. English translation in ***Journal of Soviet Mathematics***, vol. 2 (1961), pp. 623–625

Turing, A. M., *On Computable Numbers with an Application to the Entscheidungsproblem*, ***Proceedings of the London Mathematical Society***, ser. 2 vol. 42 (1936–1937), pp. 230–265

Turing, A., M., *Computability and λ-definability*, ***The Journal of Symbolic Logic***, vol. 2 (1937), pp. 153–163

Turing, A. M., *On Computable Numbers, with an Application to the Entscheidungsproblem, A Correction*, ***Proceedings of the London Mathematical Society***., ser. 2 vol. 43 (1937), pp. 544–546

Turing, A. M., *Systems of Logic Based on Ordinals*, ***Proceedings of the London Mathematical Society***, ser. 2 vol. 45 (1939), pp. 161–228

Ueberweg, F., ***System der Logik***, fourth edition, Bonn, 1874 van Heijenoort, J., editor, ***From Frege to Gödel***, Cambridge, Mass., 1967

Vasiliev, A. V., ***Space, Time, Motion***, translated by H. M. Lucas and C. P. Sanger, with an introduction by Bertrand Russell, London, 1924, and New York, 1924

Veblen, O., *A System of Axioms for Geometry*, ***Transactions of the American Mathematical Society***, vol. 5 (1904), pp. 343–384

Veblen, O., *Definition in Terms of Order Alone in the Linear Continuum and in Well-Ordered Sets*, ***Transactions of the American Mathematical Society***, vol. 6 (1905), pp. 165–171

Veblen, O., *Continuous Increasing Functions of Finite and Transfinite Ordinals*, ***Transactions of the American Mathematical Society***, vol. 9 (1908), pp. 280–292

Veblen, O., *Remarks of the Foundations of Geometry*, ***Bulletin of the American Mathematical Society***, vol. 31 (1925), p. 128

Veblen, O., and Thomas, J. M., *Projective Invariants of Affine Geometry of Paths*, ***Annals of Mathematics***, ser. 2 vol. 27 (1926), p. 295

Veblen, O., & Young, W. H., ***Projective Geometry***, Boston, vol. I 1910, vol. II 1918

Venn, J., *Symbolic Logic*, ***Princeton Review*** (1880), pp. 247–267

von Mises, R., *Grundlagen der Wahrscheinlichkeitsrechnung*, ***Mathematische Zeitschrift***, vol. 5 (1919), pp. 52–99

von Mises, R., ***Wahrscheinlichkeit, Statistik und Wahrheit***, Vienna, 1928

von Mises, R., ***Wahrscheinlichkeitsrechnung***, Leipzig and Vienna, 1931

von Mises, R., *Über Zahlenfolgen, die ein kollektive-ähnliches varhalten zeigen*, ***Mathematische Annalen***, vol. 108 (1933), pp. 757–772

von Neumann, J., *Zür Einführung der transfiniten Zahlen*, ***Acta Scientiarum Mathematicarum*** (Szeged) vol. 1 (1923), 199–208

von Neumann, J., *Eine Axiomatisierung der Mengenlehre*, ***Journal für die Reine und Angewandte Mathematik***, vol. 154 (1925), pp. 219–240

von Neumann, J., *Berichtigung*, ***Journal für die Reine und Angewandte Mathematik***, vol. 155 (1926), pp. 128

von Neumann, J., *Zur Hilbertschen Beweistheorie*, ***Mathematische Zeitschrift***, vol. 26 (1927), pp. 1–46

von Neumann, J., *Die formalistische Grundlegung der Mathematik*, ***Erkenntnis***, vol. 2 (1931), pp. 116–121, 144–145, 146, 148

von Neumann, J., *Reply to Lesniewski*, ***Fundamenta Mathematicae***, vol. 17 (1931), pp. 331–334

von Neumann, J., *First Draft of a Report on the EDVAC*, Moore School of Electrical Engineering, University of Pennsylvania, June 30, 1945

von Wright, G. H., ***Form and Content in Logic***, Cambridge, 1949

Wade, F. C., translator, ***Ars Logica***, Milwaukee, 1955

Wade, F. C., *Translator's Introduction*, ***John of St. Thomas, Outlines of Formal Logic***, translated by F. C. Wade, Milwaukee, 1955, pp. 1–24

Waismann, F., *Verifiability*, in ***Essays on Logic and Language***, edited and with an introduction by Antony Flew, Oxford and New York, 1951, pp. 117–144

Waismann, F., *How I See Philosophy*, in ***Contemporary British Philosophy*** (a symposium), London and New York, 1956

Wajsberg, M., *Beiträge zum Metaaussgenkalkül*, ***Monatshefte für Mathematik und Physik***, vol. 42 (1935), pp. 221–242

Wajsberg, M., *Untersuchung über Unabhängigkeitsbeweise nach der Matrizenmethode*, ***Wiadomości Matematyczne***, vol. 43 (1936), pp. 131–168

Wald, A., *Sur la Notion de Collectif dans le Calcul des Probabilités*, ***Comptes Rendus des Séances de l'Académie des Sciences***, vol. 202 (1936), pp. 180–183

Wald, A., *Die Widerspruchsfreiheit des Kollektivbegriffes der Wahrscheinlichkeitsrechnung*, ***Ergebnisse eines mathematischen Kolloquiums***, vol. 8 (1937), pp. 38–72

Wallis, J., Cambridge thesis, 1639, published in 1643

Wallis, J., ***Institutio Logicae***, Oxford, 1687

Wang, H., *Circuit Synthesis by Solving Sequential Boolean Equations.* ***Zeitschrift für mathematische Logik und Grundlagen der Mathematik***, vol. 5 (1959), pp. 291–322. Reviewed by C. C. Elgot in ***The Journal of Symbolic Logic***, vol. 25 (1962), pp. 373–375

Wang, H., & McNaughton, R., *Les Systèmes Axiomatiques de la Théorie des Ensembles*, ***Collection of Logique Mathématique***, Serie A, No. 4, Paris, 1963

Watts, I., ***Logick: or, The Right Use of Reason in the Enquiry after Truth. With a Variety of Rules to Guard against Error, in the Affairs of Religion and Human Life, as well as in the Sciences***, 1725

Wedberg, A., *The Aristotelian Theory of Classes*, ***Ajatus***, vol. 15 (1948), pp. 299–314

Wernick, W., *Complete Sets of Logical Functions*, ***The Transactions of the American Mathematical Society***, vol. 51 (1942), pp. 117–132

Weyl, H., ***Das Kontinuum, Kritische Untersuchungen über die Grundlagen der Analysis***, Leipzig, 1918

Weyl, H., *Der circulus vitiosus in der heutigen Begründung der Analysis*, ***Jahresbericht der Deutschen Mathematiker-Vereinigung***, vol. 28 (1919), pp. 85–92

Weyl, H., *Über die neue Grundlagenkrise der Mathematik*, ***Mathematische Zeitschrift***, vol. 10 (1921), pp. 39–37

Weyl, H., *Randbemerkungen zu Hauptproblemen der Mathematik*, ***Mathematische Zeitschrift***, vol. 20 (1924), pp. 131–150

Weyl, H., *Die heutige Erkenntislage in der Mathematik*, ***Symposion***, vol. 1 (1926), pp. 1–32

Weyl, H., *Philosophie der Mathematik und Naturwissenschaften*, ***Handbuch der Philosophie***, vol. 2 (1926), pp. 1–162

Weyl, H., *Consistency in Mathematics*, ***The Rice Institute Pamphlet***, vol. 16 (1929), pp. 245–265

Weyl, H., ***Die Stufen des Unendlichen***, Vortrag, Jena, 1931

Whately, R., *Elements of Logic*, in ***Encyclopaedia Metropolitarta***, 1826

White, M. G., *A Note on the "Paradox of Analysis"*, ***Mind***, n.s. vol. 54 (1945), pp. 71–72

White, M. G., *Analysis and Identity: A Rejoinder*, ***Mind***, n.s. vol. 54 (1945), pp. 357–362

White, M. G., *On the Church–Frege Solution of the Paradox of Analysis*, ***Philosophy and Phenomenological Research***, vol. 9 (1948), pp. 305–308

Whitehead, A. N., *Note*, ***Revue de Métaphysique et de Morale***, vol. 13 (1905), pp. 916–917

Whitehead, A. N., *An Introduction to Mathematics*, in ***The Home University Library of Modern Knowledge XV***, London and New York, 1911

Whitehead, A. N., *Mathematics*, in ***The Encyclopædia Britannica***, eleventh edition, vol. 17, Cambridge and New York, 1911, pp. 878–883; reprinted in ***Essays in Science and Philosophy***, New York, 1947, pp. 269–288, also London and New York, 1948, pp. 195–208; German translation in ***Philosophie der Mathematik, Vorträge und Essays***, Wien, 1949, pp. 125–150

Whitehead, A. N., & Russell, B., ***Principia Mathematica***, 3 vols., Cambridge, 1910–1913; second edition (with added introduction and appendices by Russell), Cambridge, 1925–1927; paperbound reprint of the greater part of volume 1 (second edition), London and New York, 1962

Williams, D. C., *The Sea Fight Tomorrow*, in ***Structure, Method, and Meaning: Essays in Honor of Henry M. Sheffer***, 1951

Wilson, E. B., *Logic and the Continuum*, ***Bulletin of the American Mathematical Society***, vol. 14 (1908), pp. 432–443

Wittgenstein, L., ***Tractatus Logico-Philosophicus***, London, 1922

Wittgenstein, L., ***Philosophical Investigations***, G. E. M. Anscome, translator, Oxford, 1953

Wood, L., ***The Analysis of Knowledge***, Princeton, 1940

Woodger, J. H., ***The Axiomatic Method in Biology***, London, 1937

Wundheiler, L., & Wundheiler, A., *Some Logical Concepts for Syntax*, in ***Machine Translation of Languages, Fourteen Essays***, edited by W. N. Locke and A. D. Booth, New York 1955, pp. 194–207

Yanovskaya, S. A., *Основания Математики и Математическая Логика*, **Математика а СССР за Тридцат Лет 1917–1947**, pp. 9–50, Moscow and Leningrad, 1948

Young, J. W., ***Lectures on Fundamental Concepts of Algebra and Geometry***, New York, 1911

Young, W. H., *Overlapping Intervals*, ***Proceedings of the London Mathematical Society***, vol. 35 (1903), pp. 384–388

Young, W. H., & Young, G. C., ***The Theory of Sets of Points***, Cambridge, 1906

Zermelo, E., *Beweis, daß jede Menge wohlgeordnet werden kann*, ***Mathematische Annalen***, vol. 59 (1904), pp. 514–516

Zermelo, E., *Neuer Beweis für die Möglichkeit einer Wohlordnung*, ***Mathematische Annalen***, vol. 65 (1908), pp. 107–28

Zermelo, E., *Untersuchungen über die Grundlagen der Mengenlehre*, ***Mathematische Annalen***, vol. 65 (1908), pp. 261–281

INDEX OF SUBJECTS

INDEX OF NAMES